MECHANICAL DESIGN HANDBOOK

MECHANICAL DESIGN HANDBOOK

Measurement, Analysis, and Control of Dynamic Systems

Harold A. Rothbart Editor

Dean Emeritus
College of Science and Engineering
Fairleigh Dickinson University
Teaneck, N.J.

Thomas H. Brown, Jr. Editor

Faculty Associate
Institute for Transportation Research and Education
North Carolina State University
Raleigh, N.C.

Second Edition

McGRAW-HILL

New York Chicago San Francisco Lisbon London Madrid
Mexico City Milan New Delhi San Juan Seoul
Singapore Sydney Toronto

The McGraw·Hill Companies

Cataloging-in-Publication Data is on file with the Library of Congress

3 4 5 6 7 8 9 0 IBT/IBT 0 1 2 1 0 9 8 7

ISBN 0-07-146636-3

The sponsoring editor for this book was Kenneth P. McCombs and the production supervisor was Richard C. Ruzycka. It was set in Times Roman by International Typesetting and Composition. The art director for the cover was Anthony Landi.

Printed and bound by IBT Global.

This book is printed on acid-free paper.

CONTENTS

Part 3 Mechanical Subsystem Components

CONTRIBUTORS

William J. Anderson *Vice President, NASTEC Inc., Cleveland, Ohio* (Chap. 15, Rolling-Element Bearings)

William H. Baier *Director of Engineering, The Fitzpatrick Co., Elmhurst, Ill.* (Chap. 19, Belts)

Stephen B. Bennett *Manager of Research and Product Development, Delaval Turbine Division, Imo Industries, Inc., Trenton, N.J.* (Chap. 2, Mechanics of Materials)

Thomas H. Brown, Jr. *Faculty Associate, Institute for Transportation Research and Education, North Carolina State University, Raleigh, N.C.* (Co-Editor)

John J. Coy *Chief of Mechanical Systems Technology Branch, NASA Lewis Research Center, Cleveland, Ohio* (Chap. 21, Gearing)

Thomas A. Dow *Professor of Mechanical and Aerospace Engineering, North Carolina State University, Raleigh, N.C.* (Chap. 17, Friction Clutches, and Chap. 18, Friction Brakes)

Saul K. Fenster *President Emeritus, New Jersey Institute of Technology, Newark, N.J.* (App. A, Analytical Methods for Engineers)

Ferdinand Freudenstein *Stevens Professor of Mechanical Engineering, Columbia University, New York, N.Y.* (Chap. 3, Kinematics of Mechanisms)

Theodore Gela *Professor Emeritus of Metallurgy, Stevens Institute of Technology, Hoboken, N.J.* (Chap. 6, Properties of Engineering Materials)

Herbert H. Gould *Chief, Crashworthiness Division, Transportation Systems Center, U.S. Department of Transportation, Cambridge, Mass.* (App. A, Analytical Methods for Engineers)

Bernard J. Hamrock *Professor of Mechanical Engineering, Ohio State University, Columbus, Ohio* (Chap. 15, Rolling-Element Bearings)

John E. Johnson *Manager, Mechanical Model Shops, TRW Corp., Redondo Beach, Calif.* (Chap. 16, Power Screws)

Sheldon Kaminsky *Consulting Engineer, Weston, Conn.* (Chap. 8, System Dynamics)

Kailash C. Kapur *Professor and Director of Industrial Engineering, University of Washington, Seattle, Wash.* (Chap. 13, System Reliability)

Robert P. Kolb *Manager of Engineering (Retired), Delaval Turbine Division, Imo Industries, Inc., Trenton, N.J.* (Chap. 2, Mechanics of Materials)

Leonard R. Lamberson *Professor and Dean, College of Engineering and Applied Sciences, West Michigan University, Kalamazoo, Mich.* (Chap. 13, System Reliability)

Thomas P. Mitchell *Professor, Department of Mechanical and Environmental Engineering, University of California, Santa Barbara, Calif.* (Chap. 1, Classical Mechanics)

Burton Paul *Asa Whitney Professor of Dynamical Engineering, Department of Mechanical Engineering and Applied Mechanics, University of Pennsylvania, Philadelphia, Pa.* (Chap. 12, Machine Systems)

J. David Powell *Professor of Aeronautics/Astronautics and Mechanical Engineering, Stanford University, Stanford, Calif.* (Chap. 10, Digital Control Systems)

Abillo A. Relvas *Manager—Techical Assistance, Associated Spring, Barnes Group, Inc., Bristol, Conn.* (Chap. 22, Springs)

Harold A. Rothbart *Dean Emeritus, College of Science and Engineering, Fairleigh Dickenson University, Teaneck, N.J.* (Chap. 14, Cam Mechanisms, and Co-Editor)

Andrew R. Sage *Associate Vice President for Academic Affairs, George Mason Universtiy, Fairfax, Va.* (Chap. 9, Continuous Time Control Systems)

Warren J. Smith *Vice President, Research and Development, Santa Barbara Applied Optics, a subsidiary of Infrared Industries, Inc., Santa Barbara, Calif.* (Chap. 11, Optical Systems)

David Tabor *Professor Emeritus, Laboratory for the Physics and Chemistry of Solids, Department of Physics, Cambridge University, Cambridge, England* (Chap. 7, Friction, Lubrication, and Wear)

Steven M. Tipton *Associate Professor of Mechanical Engineering, University of Tulsa, Tulsa, Okla.* (Chap. 5, Static and Fatigue Design)

George V. Tordion *Professor of Mechanical Engineering, Université Laval, Quebec, Canada* (Chap. 20, Chains)

Dennis P. Townsend *Senior Research Engineer, NASA Lewis Research Center, Cleveland, Ohio* (Chap. 21, Gearing)

Eric E. Ungar *Chief Consulting Engineer, Bolt, Beranek, and Newman, Inc., Cambridge, Mass.* (Chap. 4, Mechanical Vibrations)

C. C. Wang *Senior Staff Engineer, Central Engineer Laboratories, FMC Corporation, Santa Clara, Calif.* (App. B, Numerical Methods for Engineers)

Erwin V. Zaretsky *Chief Engineer of Structures, NASA Lewis Research Center, Cleveland, Ohio* (Chap. 21, Gearing)

FOREWORD

Mechanical design is one of the most rewarding activities because of its incredible complexity. It is complex because a successful design involves any number of individual mechanical elements combined appropriately into what is called a *system.* The word *system* came into popular use at the beginning of the space age, but became somewhat overused and seemed to disappear. However, any modern machine is a system and must operate as such. The information in this handbook is limited to the *mechanical* elements of a system, since encompassing all elements (electrical, electronic, etc.) would be too overwhelming.

The purpose of the *Mechanical Design Handbook* has been from its inception to provide the mechanical designer the most comprehensive and up-to-date information on what is available, and how to utilize it effectively and efficiently in a single reference source. Unique to this edition, is the combination of the fundamentals of mechanical design with a systems approach, incorporating the most important mechanical subsystem components. The original editor and a contributing author, Harold A. Rothbart, is one of the most well known and respected individuals in the mechanical engineering community. From the First Edition of the *Mechanical Design and Systems Handbook* published over forty years ago to this Second Edition of the *Mechanical Design Handbook,* he has continued to assemble experts in every field of machine design—mechanisms and linkages, cams, every type of gear and gear train, springs, clutches, brakes, belts, chains, all manner of roller bearings, failure analysis, vibration, engineering materials, and classical mechanics, including stress and deformation analysis. This incredible wealth of information, which would otherwise involve searching through dozens of books and hundreds of scientific and professional papers, is organized into twenty-two distinct chapters and two appendices. This provides direct access for the designer to a specific area of interest or need.

The *Mechanical Design Handbook* is a unique reference, spanning the breadth and depth of design information, incorporating the vital information needed for a mechanical design. It is hoped that this collection will create, through a system perspective, the level of confidence that will ultimately produce a successful and safe design and a proud designer.

Harold A. Rothbart
Thomas H. Brown, Jr.

PREFACE

This Second Edition of the *Mechanical Design Handbook* has been completely reorganized from its previous edition and includes seven chapters from the *Mechanical Design and Systems Handbook*, the precursor to the First Edition. The twenty-two chapters contained in this new edition are divided into three main sections: Mechanical Design Fundamentals, Mechanical System Analysis, and Mechanical Subsystem Components. It is hoped that this new edition will meet the needs of practicing engineers providing the critical resource of information needed in their mechanical designs.

The first section, Part I, *Mechanical Design Fundamentals*, includes seven chapters covering the foundational information in mechanical design. Chapter 1, *Classical Mechanics*, is one of the seven chapters included from the Second Edition of the *Mechanical Design and Systems Handbook*, and covers the basic laws of dynamics and the motion of rigid bodies so important in the analysis of machines in three-dimensional motion. Comprehensive information on topics such as stress, strain, beam theory, and an extensive table of shear and bending moment diagrams, including deflection equations, is provided in Chap. 2. Also in Chap. 2 are the equations for the design of columns, plates, and shells, as well as a complete discussion of the finite-element analysis approach. Chapter 3, *Kinematics of Mechanisms*, contains an endless number of ways to achieve desired mechanical motion. Kinematics, or the geometry of motion, is probably the most important step in the design process, as it sets the stage for many of the other decisions that will be made as a successful design evolves. Whether it's a particular multi-bar linkage, a complex cam shape, or noncircular gear combinations, the information for its proper design is provided. Chapter 4, *Mechanical Vibrations*, provides the basic equations governing mechanical vibrations, including an extensive set of tables compiling critical design information such as, mechanical impedances, mechanical-electrical analogies, natural frequencies of basic systems, torsional systems, beams in flexure, plates, shells, and several tables of spring constants for a wide variety of mechanical configurations. Design information on both static and dynamic failure theories, for ductile and brittle materials, is given in Chap. 5, *Static and Fatigue Design*, while Chap. 6, *Properties of Engineering Materials*, covers the issues and requirements for material selection of machine elements. Extensive tables and charts provide the experimental data on heat treatments, hardening, high-temperature and low-temperature applications, physical and mechanical properties, including properties for ceramics and plastics. Chapter 7, *Friction, Lubrication, and Wear*, gives a basic overview of these three very important areas, primarily directed towards the accuracy requirements of the machining of materials.

The second section, Part II, *Mechanical System Analysis*, contains six chapters, the first four of which are from the Second Edition of the *Mechanical Design and Systems Handbook*. Chapter 8, *Systems Dynamics*, presents the fundamentals of how a complex dynamic system can be modeled mathematically. While the solution of such systems will be accomplished by computer algorithms, it is important to have a solid foundation on how all the components interact—this chapter provides that comprehensive analysis. Chapter 9, *Continuous Time Control Systems*, expands on the material in Chap. 8 by introducing the necessary elements in the analysis when there is a time-dependent input to the mechanical system. Response to feedback loops, particularly for nonlinear damped systems, is also presented. Chapter 10, *Digital Control Systems*, continues with the system analysis presented in Chaps. 8 and 9 of solving the mathematical equations for a complex dynamic system on a computer. Regardless of the hardware used, from personal desktop computers to supercomputers, digitalization of the equations must be carefully considered to avoid errors being introduced by the analog to digital conversion. A comprehensive discussion of the basics of optics and the passage of light through common elements of optical systems is provided in Chap. 11, *Optical Systems*, and Chap. 12, *Machine Systems*, presents the dynamics of mechanical systems primarily from an energy approach, with an extensive discussion of Lagrange's equations for three-dimensional motion. To complete this section, Chap. 13, *System Reliability*, provides a *system* approach rather than addressing single mechanical elements. Reliability testing is discussed along with the Weibull distribution used in the statistical analysis of reliability.

The third and last section, Part III, *Mechanical Subsystem Components*, contains nine chapters covering the most important elements of a mechanical system. Cam layout and geometry, dynamics, loads, and the accuracy of motion are discussed in Chap. 14 while Chap. 15, *Rolling-Element Bearings*, presents ball and roller bearing, materials of construction, static and dynamic loads, friction and lubrication, bearing life, and dynamic analysis. Types of threads available, forces, friction, and efficiency are covered in Chap. 16, *Power Screws*. Chapter 17, *Friction Clutches*, and Chap. 18, *Friction Brakes*, both contain an extensive presentation of these two important mechanical subsystems. Included are the types of clutches and brakes, materials, thermal considerations, and application to various transmission systems. The geometry of belt assemblies, flat and v-belt designs, and belt dynamics is explained in Chap. 19, *Belts*, while chain arrangements, ratings, and noise are dealt with in Chap. 20, *Chains*. Chapter 21, *Gearing*, contains every possible gear type, from basic spur gears and helical gears to complex hypoid bevel gears sets, as well as the intricacies of worm gearing. Included is important design information on processing and manufacture, stresses and deflection, gear life and power-loss predictions, lubrication, and optimal design considerations. Important design considerations for helical compression, extension and torsional springs, conical springs, leaf springs, torsion-bar springs, power springs, constant-force springs, and Belleville washers are presented in Chap. 22, *Springs*.

This second edition of the *Mechanical Design Handbook* contains two new appendices not in the first edition: App. A, *Analytical Methods for Engineers*, and App. B, *Numerical Methods for Engineers*. They have been provided so that the practicing engineer does not have to search elsewhere for important mathematical information needed in mechanical design.

It is hoped that this Second Edition continues in the tradition of the First Edition, providing relevant mechanical design information on the critical topics of interest to the engineer. Suggestions for improvement are welcome and will be appreciated.

Harold A. Rothbart
Thomas H. Brown, Jr.

ACKNOWLEDGMENTS

Our deepest appreciation and love goes to our families, Florence, Ellen, Dan, and Jane (Rothbart), and Miriam, Sianna, Hunter, and Elliott (Brown). Their encouragement, help, suggestions, and patience are a blessing to both of us.

To our Senior Editor Ken McCombs, whose continued confidence and support has guided us throughout this project, we gratefully thank him. To Gita Raman and her wonderful and competent staff at International Typesetting and Composition (ITC) in Noida, India, it has been a pleasure and honor to collaborate with them to bring this Second Edition to reality.

And finally, without the many engineers who found the First Edition of the *Mechanical Design Handbook*, as well as the First and Second Editions of the *Mechanical Design and Systems Handbook*, useful in their work, this newest edition would not have been undertaken. To all of you we wish the best in your career and consider it a privilege to provide this reference for you.

Harold A. Rothbart
Thomas H. Brown, Jr.

MECHANICAL DESIGN HANDBOOK

P • A • R • T • 1

MECHANICAL DESIGN FUNDAMENTALS

CHAPTER 1
CLASSICAL MECHANICS

Thomas P. Mitchell, Ph.D.
Professor
Department of Mechanical and Environmental Engineering
University of California
Santa Barbara, Calif.

The aim of this chapter is to present the concepts and results of newtonian dynamics which are required in a discussion of rigid-body motion. The detailed analysis of particular rigid-body motions is not included. The chapter contains a few topics which, while not directly needed in the discussion, either serve to round out the presentation or are required elsewhere in this handbook.

1.1 INTRODUCTION

The study of classical dynamics is founded on Newton's three laws of motion and on the accompanying assumptions of the existence of absolute space and absolute time. In addition, in problems in which gravitational effects are of importance, Newton's law of gravitation is adopted. The objective of the study is to enable one to predict, given the initial conditions and the forces which act, the evolution in time of a mechanical system or, given the motion, to determine the forces which produce it.

The mathematical formulation and development of the subject can be approached in two ways. The vectorial method, that used by Newton, emphasizes the vector quantities force and acceleration. The analytical method, which is largely due to Lagrange, utilizes the scalar quantities work and energy. The former method is the more physical and generally possesses the advantage in situations in which dissipative forces are present. The latter is more mathematical and accordingly is very useful in developing powerful general results.

1.2 THE BASIC LAWS OF DYNAMICS

The "first law of motion" states that a body which is under the action of no force remains at rest or continues in uniform motion in a straight line. This statement is also known as the "law of inertia," inertia being that property of a body which demands that a force is necessary to change its motion. "Inertial mass" is the numerical measure of inertia. The conditions under which an experimental proof of this law could be carried out are clearly not attainable.

In order to investigate the motion of a system it is necessary to choose a frame of reference, assumed to be rigid, relative to which the displacement, velocity, etc., of the system are to be measured. The law of inertia immediately classifies the possible frames of reference into two types. For, suppose that in a certain frame S the law is found to be true; then it must also be true in any frame which has a constant velocity vector relative to S. However, the law is found not to be true in any frame which is in accelerated motion relative to S. A frame of reference in which the law of inertia is valid is called an "inertial frame," and any frame in accelerated motion relative to it is said to be "noninertial." Any one of the infinity of inertial frames can claim to be at rest while all others are in motion relative to it. Hence it is not possible to distinguish, by observation, between a state of rest and one of uniform motion in a straight line. The transformation rules by which the observations relative to two inertial frames are correlated can be deduced from the second law of motion.

Newton's "second law of motion" states that in an inertial frame the force acting on a mass is equal to the time rate of change of its linear momentum. "Linear momentum," a vector, is defined to be the product of the inertial mass and the velocity. The law can be expressed in the form

$$d/dt(m\mathbf{v}) = \mathbf{F} \tag{1.1}$$

which, in the many cases in which the mass m is constant, reduces to

$$m\mathbf{a} = \mathbf{F} \tag{1.2}$$

where a is the acceleration of the mass.

The "third law of motion," the "law of action and reaction," states that the force with which a mass m_i acts on a mass m_j is equal in magnitude and opposite in direction to the force which m_j exerts on m_i. The additional assumption that these forces are collinear is needed in some applications, e.g., in the development of the equations governing the motion of a rigid body.

The "law of gravitation" asserts that the force of attraction between two point masses is proportional to the product of the masses and inversely proportional to the square of the distance between them. The masses involved in this formula are the gravitational masses. The fact that falling bodies possess identical accelerations leads, in conjunction with Eq. (1.2), to the proportionality of the inertial mass of a body to its gravitational mass. The results of very precise experiments by Eotvös and others show that inertial mass is, in fact, equal to gravitational mass. In the future the word mass will be used without either qualifying adjective.

If a mass in motion possesses the position vectors $\mathbf{r}_1$ and $\mathbf{r}_2$ relative to the origins of two inertial frames S_1 and S_2, respectively, and if further S_1 and S_2 have a relative velocity $\mathbf{V}$, then it follows from Eq. (1.2) that

$$\begin{aligned} \mathbf{r}_1 &= \mathbf{r}_2 + \mathbf{V}t_2 + \text{const} \\ t_1 &= t_2 + \text{const} \end{aligned} \tag{1.3}$$

in which t_1 and t_2 are the times measured in S_1 and S_2. The transformation rules Eq. (1.3), in which the constants depend merely upon the choice of origin, are called "galilean transformations." It is clear that acceleration is an invariant under such transformations.

The rules of transformation between an inertial frame and a noninertial frame are considerably more complicated than Eq. (1.3). Their derivation is facilitated by the application of the following theorem: a frame S_1 possesses relative to a frame S an angular velocity $\boldsymbol{\omega}$ passing through the common origin of the two frames. The time rate of change of any vector **A** as measured in S is related to that measured in S_1 by the formula

$$(d\mathbf{A}/dt)_S = (d\mathbf{A}/dt)_{S_1} + \boldsymbol{\omega} \times \mathbf{A} \tag{1.4}$$

The interpretation of Eq. (1.4) is clear. The first term on the right-hand side accounts for the change in the magnitude of **A**, while the second corresponds to its change in direction.

If S is an inertial frame and S_1 is a frame rotating relative to it, as explained in the statement of the theorem, S_1 being therefore noninertial, the substitution of the position vector **r** for **A** in Eq. (1.4) produces the result

$$\mathbf{v}_{abs} = \mathbf{v}_{rel} + \boldsymbol{\omega} \times \mathbf{r} \tag{1.5}$$

In Eq. (1.5) $\mathbf{v}_{abs}$ represents the velocity measured relative to S, $\mathbf{v}_{rel}$ the velocity relative to S_1, and $\boldsymbol{\omega} \times \mathbf{r}$ is the transport velocity of a point rigidly attached to S_1. The law of transformation of acceleration is found on a second application of Eq. (1.4), in which **A** is replaced by $\mathbf{v}_{abs}$. The result of this substitution leads directly to

$$(d^2\mathbf{r}/dt^2)_S = (d^2\mathbf{r}/dt^2)_{S_1} + \boldsymbol{\omega} \times (\boldsymbol{\omega} \times \mathbf{r}) + \dot{\boldsymbol{\omega}} \times \mathbf{r} + 2\boldsymbol{\omega} \times \mathbf{v}_{rel} \tag{1.6}$$

in which $\dot{\boldsymbol{\omega}}$ is the time derivative, in either frame, of $\boldsymbol{\omega}$. The physical interpretation of Eq. (1.6) can be shown in the form

$$\mathbf{a}_{abs} = \mathbf{a}_{rel} + \mathbf{a}_{trans} + \mathbf{a}_{cor} \tag{1.7}$$

where $\mathbf{a}_{cor}$ represents the Coriolis acceleration $2\boldsymbol{\omega} \times \mathbf{v}_{rel}$. The results, Eqs. (1.5) and (1.7), constitute the rules of transformation between an inertial and a nonintertial frame. Equation (1.7) shows in addition that in a noninertial frame the second law of motion takes the form

$$m\mathbf{a}_{rel} = \mathbf{F}_{abs} - m\mathbf{a}_{cor} - m\mathbf{a}_{trans} \tag{1.8}$$

The modifications required in the above formulas are easily made for the case in which S_1 is translating as well as rotating relative to S. For, if $\mathbf{D}(t)$ is the position vector of the origin of the S_1 frame relative to that of S, Eq. (1.5) is replaced by

$$\mathbf{V}_{abs} = (d\mathbf{D}/dt)_S + \mathbf{v}_{rel} + \boldsymbol{\omega} \times \mathbf{r}$$

and consequently, Eq. (1.7) is replaced by

$$\mathbf{a}_{abs} = (d^2\mathbf{D}/dt^2)_S + \mathbf{a}_{rel} + \mathbf{a}_{trans} + \mathbf{a}_{cor}$$

In practice the decision as to what constitutes an inertial frame of reference depends upon the accuracy sought in the contemplated analysis. In many cases a set of axes rigidly attached to the earth's surface is sufficient, even though such a frame is noninertial to the extent of its taking part in the daily rotation of the earth about its axis and also its yearly rotation about the sun. When more precise results are required, a set of axes fixed at the center of the earth may be used. Such a set of axes is subject only to the orbital motion of the earth. In still more demanding circumstances, an inertial frame is taken to be one whose orientation relative to the fixed stars is constant.

1.3 THE DYNAMICS OF A SYSTEM OF MASSES

The problem of locating a system in space involves the determination of a certain number of variables as functions of time. This basic number, which cannot be reduced without the imposition of constraints, is characteristic of the system and is known as its number of degrees of freedom. A point mass free to move in space has three degrees of freedom. A system of two point masses free to move in space, but subject to the constraint that the distance between them remains constant, possesses five degrees of freedom. It is clear that the presence of constraints reduces the number of degrees of freedom of a system.

Three possibilities arise in the analysis of the motion-of-mass systems. First, the system may consist of a small number of masses and hence its number of degrees of freedom is small. Second, there may be a very large number of masses in the system, but the constraints which are imposed on it reduce the degrees of freedom to a small number; this happens in the case of a rigid body. Finally, it may be that the constraints acting on a system which contains a large number of masses do not provide an appreciable reduction in the number of degrees of freedom. This third case is treated in statistical mechanics, the degrees of freedom being reduced by statistical methods.

In the following paragraphs the fundamental results relating to the dynamics of mass systems are derived. The system is assumed to consist of n constant masses m_i $(i = 1, 2, \ldots, n)$. The position vector of m_i, relative to the origin O of an inertial frame, is denoted by $\mathbf{r}_i$. The force acting on m_i is represented in the form

$$\mathbf{F}_i = \mathbf{F}_i^e + \sum_{j=1}^{n} \mathbf{F}_{ij} \tag{1.9}$$

in which $\mathbf{F}_i^e$ is the external force acting on m_i, $\mathbf{F}_{ij}$ is the force exerted on m_i by m_j, and $\mathbf{F}_{ii}$ is zero.

1.3.1 The Motion of the Center of Mass

The motion of m_i relative to the inertial frame is determined from the equation

$$\mathbf{F}_i^e + \sum_{j=1}^{n} \mathbf{F}_{ij} = m_i \frac{d\mathbf{v}_i}{dt} \tag{1.10}$$

On summing the n equations of this type one finds

$$\mathbf{F}^e + \sum_{i=1}^{n} \sum_{j=1}^{n} \mathbf{F}_{ij} = \sum_{i=1}^{n} m_i \frac{d\mathbf{v}_i}{dt} \tag{1.11}$$

where $\mathbf{F}^e$ is the resultant of all the external forces which act on the system. But Newton's third law states that

$$\mathbf{F}_{ij} = -\mathbf{F}_{ji}$$

and hence the double sum in Eq. (1.11) vanishes. Further, the position vector $\mathbf{r}_c$ of the center of mass of the system relative to O is defined by the relation

$$\mathcal{M}\mathbf{r}_c = \sum_{i=1}^{n} m_i \mathbf{r}_i \tag{1.12}$$

in which $\mathcal{M}$ denotes the total mass of the system. It follows from Eq. (1.12) that

$$\mathcal{M}\mathbf{v}_c = \sum_{i=1}^{n} m_i \mathbf{v}_i \tag{1.13}$$

and therefore from Eq. (1.11) that

$$\mathbf{F}^e = \mathcal{M}\, d^2\mathbf{r}_c/dt^2 \tag{1.14}$$

which proves the theorem: the center of mass moves as if the entire mass of the system were concentrated there and the resultant of the external forces acted there.

Two first integrals of Eq. (1.14) provide useful results [Eqs. (1.15) and (1.16):

$$\int_{t_1}^{t_2} \mathbf{F}^e \, dt = \mathcal{M}\mathbf{v}_c(t_2) - \mathcal{M}\mathbf{v}_c(t_1) \tag{1.15}$$

The integral on the left-hand side is called the "impulse" of the external force. Equation (1.15) shows that the change in linear momentum of the center of mass is equal to the impulse of the external force. This leads to the conservation-of-linear-momentum theorem: the linear momentum of the center of mass is constant if no resultant external force acts on the system or, in view of Eq. (1.13), the total linear momentum of the system is constant if no resultant external force acts:

$$\int_1^2 \mathbf{F}^e \cdot \mathbf{r}_c = \frac{1}{2}\mathcal{M}v_c^2 \Big]_1^2 \tag{1.16}$$

which constitutes the work-energy theorem: the work done by the resultant external force acting at the center of mass is equal to the change in the kinetic energy of the center of mass.

In certain cases the external force $\mathbf{F}_i^e$ may be the gradient of a scalar quantity V which is a function of position only. Then

$$\mathbf{F}^e = -\partial V/\partial \mathbf{r}_c$$

and Eq (1.16) takes the form

$$\left[\frac{1}{2}\mathcal{M}\mathbf{v}_c^2 + V\right]_1^2 = 0 \tag{1.17}$$

If such a function V exists, the force field is said to be conservative and Eq. (1.17) provides the conservation-of-energy theorem.

1.3.2 The Kinetic Energy of a System

The total kinetic energy of a system is the sum of the kinetic energies of the individual masses. However, it is possible to cast this sum into a form which frequently makes the calculation of the kinetic energy less difficult. The total kinetic energy of the masses in their motion relative to O is

$$T = \frac{1}{2}\sum_{i=1}^{n} m_i \mathbf{v}_i^2$$

but

$$\mathbf{r}_i = \mathbf{r}_c + \sigma_i$$

where $\boldsymbol{\sigma}_i$ is the position vector of m_i relative to the system center of mass C (see Fig. 1.1).

Hence

$$T = \frac{1}{2}\sum_{i=1}^{n} m_i \dot{\mathbf{r}}_c^2 + \sum_{i=1}^{n} m_i \dot{\mathbf{r}}_c \cdot \dot{\boldsymbol{\sigma}}_i + \frac{1}{2}\sum_{i=1}^{n} m_i \dot{\boldsymbol{\sigma}}_i^2$$

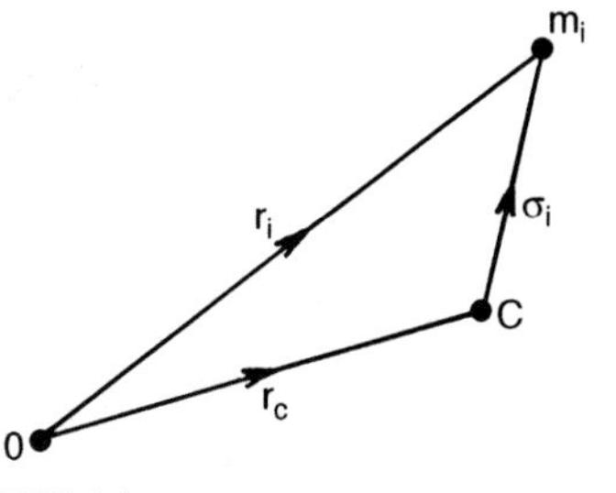

FIG. 1.1

but

$$\sum_{i=1}^{n} m_i \boldsymbol{\sigma}_i = 0$$

by definition, and so

$$T = \frac{1}{2}\mathcal{M}\dot{\mathbf{r}}_c^2 + \frac{1}{2}\sum_{i=1}^{n} m_i \dot{\boldsymbol{\sigma}}_i^2 \tag{1.18}$$

which proves the theorem: the total kinetic energy of a system is equal to the kinetic energy of the center of mass plus the kinetic energy of the motion relative to the center of mass.

1.3.3 Angular Momentum of a System (Moment of Momentum)

Each mass m_i of the system has associated with it a linear momentum vector $m_i\mathbf{v}_i$. The moment of this momentum about the point O is $\mathbf{r}_i \times m_i\mathbf{v}_i$. The moment of momentum of the motion of the system relative to O, about O, is

$$\mathbf{H}(O) = \sum_{i=1}^{n} \mathbf{r}_i \times m_i \mathbf{v}_i$$

It follows that

$$\frac{d}{dt}\mathbf{H}(O) = \sum_{i=1}^{n} \mathbf{r}_i \times m_i \frac{d^2\mathbf{r}_i}{dt^2}$$

which, by Eq. (1.10), is equivalent to

$$\frac{d}{dt}\mathbf{H}(O) = \sum_{i=1}^{n} \mathbf{r}_i \times \mathbf{F}_i^e + \sum_{i=1}^{n} \mathbf{r}_i \times \sum_{j=1}^{n} F_{ij} \tag{1.19}$$

It is now assumed that, in addition to the validity of Newton's third law, the force F_{ij} is collinear with $\mathbf{F}_{ji}$ and acts along the line joining m_i to m_j, i.e., the internal forces are central forces. Consequently, the double sum in Eq. (1.19) vanishes and

$$\frac{d}{dt}\mathbf{H}(O) = \sum_{i=1}^{n} \mathbf{r}_i \times \mathbf{F}_i^e = \mathbf{M}(O) \tag{1.20}$$

where $\mathbf{M}(O)$ represents the moment of the external forces about the point O. The following extension of this result to certain noninertial points is useful.

Let A be an arbitrary point with position vector $\mathbf{a}$ relative to the inertial point O (see Fig. 1.2). If $\boldsymbol{\rho}_i$ is the position vector of m_i relative to A, then in the notation already developed

$$\mathbf{H}(A) = \sum_{i=1}^{n} \boldsymbol{\rho}_i \times m_i \frac{d\mathbf{r}_i}{dt} = \sum_{i=1}^{n} (\mathbf{r}_i - \mathbf{a}) \times m_i \frac{d\mathbf{r}_i}{dt} = \mathbf{H}(O) - \mathbf{a} \times \mathcal{M}\mathbf{v}_c$$

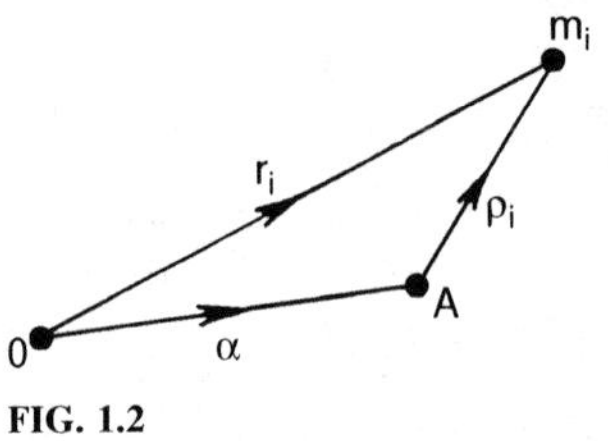

FIG. 1.2

Thus $(d/dt)\,\mathbf{H}(A) = (d/dt)\mathbf{H}(O) - \dot{\mathbf{a}} \times \mathcal{M}\mathbf{v}_c - \mathbf{a} \times \mathcal{M}(d\mathbf{v}_c/dt)$, which reduces on application of Eqs. (1.14) and (1.20) to

$$(d/dt)\mathbf{H}(A) = \mathbf{M}(A) - \dot{\mathbf{a}} \times \mathcal{M}v_c$$

The validity of the result

$$(d/dt)\mathbf{H}(A) = \mathbf{M}(A) \tag{1.21}$$

is assured if the point A satisfies either of the conditions

1. $\dot{\mathbf{a}} = 0$; i.e., the point A is fixed relative to O.
2. $\dot{\mathbf{a}}$ is parallel to $\mathbf{v}_c$; i.e., the point A is moving parallel to the center of mass of the system.

A particular, and very useful case of condition 2 is that in which the point A is the center of mass. The preceding results [Eqs. (1.20) and (1.21)] are contained in the theorem: the time rate of change of the moment of momentum about a point is equal to the moment of the external forces about that point if the point is inertial, is moving parallel to the center of mass, or is the center of mass.

As a corollary to the foregoing, one can state that the moment of momentum of a system about a point satisfying the conditions of the theorem is conserved if the moment of the external forces about that point is zero.

The moment of momentum about an arbitrary point A of the motion relative to A is

$$\mathbf{H}_{\text{rel}}(A) = \sum_{i=1}^{n} \boldsymbol{\rho}_i \times m_i \frac{d\boldsymbol{\rho}_i}{dt} = \sum_{i=1}^{n} \boldsymbol{\rho}_i \times m_i(\dot{\mathbf{r}}_i - \dot{\mathbf{a}}) = H(A) + \dot{\mathbf{a}} \times \sum_{i=1}^{n} m_i\boldsymbol{\rho}_i \quad (1.22)$$

If the point A is the center of mass C of the system, Eq. (1.22) reduces to

$$\mathbf{H}_{\text{rel}}(C) = \mathbf{H}(C) \quad (1.23)$$

which frequently simplifies the calculation of $\mathbf{H}(C)$.

Additional general theorems of the type derived above are available in the literature. The present discussion is limited to the more commonly applicable results.

1.4 THE MOTION OF A RIGID BODY

As mentioned earlier, a rigid body is a dynamic system that, although it can be considered to consist of a very large number of point masses, possesses a small number of degrees of freedom. The rigidity constraint reduces the degrees of freedom to six in the most general case, which is that in which the body is translating and rotating in space. This can be seen as follows: The position of a rigid body in space is determined once the positions of three noncollinear points in it are known. These three points have nine coordinates, among which the rigidity constraint prescribes three relationships. Hence only six of the coordinates are independent. The same result can be obtained otherwise.

Rather than view the body as a system of point masses, it is convenient to consider it to have a mass density per unit volume. In this way the formulas developed in the analysis of the motion of mass systems continue to be applicable if the sums are replaced by integrals.

The six degrees of freedom demand six equations of motion for the determination of six variables. Three of these equations are provided by Eq. (1.14), which describes the motion of the center of mass, and the remaining three are found from moment-of-momentum considerations, e.g., Eq. (1.21). It is assumed, therefore, in what follows that the motion of the center of mass is known, and the discussion is limited to the rotational motion of the rigid body about its center of mass C.*

Let $\boldsymbol{\omega}$ be the angular velocity of the body. Then the moment of momentum about C is, by Eq. (1.3),

$$\mathbf{H}(C) = \int_V \mathbf{r} \times (\boldsymbol{\omega} \times \mathbf{r})\rho \, dV \quad (1.24)$$

*Rotational motion about any *fixed* point of the body is treated in a similar way.

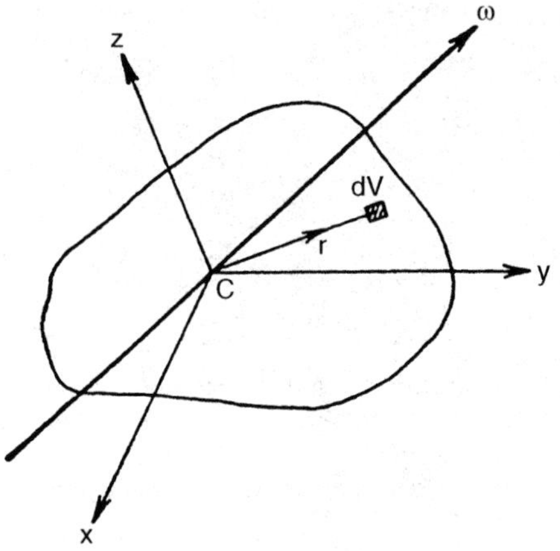

FIG. 1.3

where **r** is now the position vector of the element of volume dV relative to C (see Fig. 1.3), ρ is the density of the body, and the integral is taken over the volume of the body. By a direct expansion one finds

$$\begin{aligned}\mathbf{r} \times (\boldsymbol{\omega} \times \mathbf{r}) &= \mathbf{r}^2\boldsymbol{\omega} - \mathbf{r}(\mathbf{r} \cdot \boldsymbol{\omega}) = \mathbf{r}^2\boldsymbol{\omega} - \mathbf{rr} \cdot \boldsymbol{\omega} \\ &= \mathbf{r}^2\mathbf{I} \cdot \boldsymbol{\omega} - \mathbf{rr} \cdot \boldsymbol{\omega} = (\mathbf{r}^2\mathbf{I} - \mathbf{rr}) \cdot \boldsymbol{\omega}\end{aligned} \tag{1.25}$$

and hence

$$\mathbf{H}(C) = \boldsymbol{I}(C) \cdot \boldsymbol{\omega} \tag{1.25}$$

where

$$\boldsymbol{I}(C) = \int_V \rho(\mathbf{r}^2\mathbf{I} - \mathbf{rr})\, dV \tag{1.26}$$

is the inertia tensor of the body about C.

In Eq. (1.26), **I** denotes the identity tensor. The inertia tensor can be evaluated once the value of ρ and the shape of the body are prescribed. We now make a short digression to discuss the structure and properties of $\boldsymbol{I}(C)$.

For definiteness let x, y, and z be an orthogonal set of cartesian axes with origin at C (see Fig. 1.3). Then in matrix notation

$$\boldsymbol{I}(C) = \begin{pmatrix} I_{xx} & -I_{xy} & -I_{xz} \\ -I_{yx} & I_{yy} & -I_{yz} \\ -I_{zx} & -I_{zy} & I_{zz} \end{pmatrix}$$

where

$$I_{xx} = \int_V \rho(y^2 + z^2)\, dV$$

$$I_{xy} = \int_V \rho xy\, dV$$

. . . .

It is clear that:

1. The tensor is second-order symmetric with real elements.

2. The elements are the usual moments and products of inertia.

3. The moment of inertia about a line through C defined by a unit vector $\mathbf{e}$ is

$$\mathbf{e} \cdot \boldsymbol{I}(C) \cdot \mathbf{e}$$

4. Because of the property expressed in condition 1, it is always possible to determine at C a set of mutually perpendicular axes relative to which $\boldsymbol{I}(C)$ is diagonalized.

Returning to the analysis of the rotational motion, one sees that the inertia tensor $\boldsymbol{I}(C)$ is time-dependent unless it is referred to a set of axes which rotate with the body. For simplicity the set of axes S_1 which rotates with the body is chosen to be the orthogonal set in which $\boldsymbol{I}(C)$ is diagonalized. A space-fixed frame of reference with origin at C is represented by S. Accordingly, from Eqs. (1.4) and (1.21),

$$[(d/dt)\mathbf{H}(C)]_S = [(d/dt)\mathbf{H}(C)]_{S_1} + \boldsymbol{\omega} \times \mathbf{H}(C) = \mathbf{M}(C) \tag{1.27}$$

which, by Eq. (1.25), reduces to

$$\boldsymbol{I}(C) \cdot (d\boldsymbol{\omega}/dt) + \boldsymbol{\omega} \times \boldsymbol{I}(C) \cdot \boldsymbol{\omega} = \mathbf{M}(C) \tag{1.28}$$

where

$$\mathbf{H}(C) = \mathbf{i}I_{xx}\omega_x + \mathbf{j}I_{yy}\omega_y + \mathbf{k}I_{zz}\omega_z \tag{1.29}$$

In Eq. (1.29) the x, y, and z axes are those for which

$$\boldsymbol{I}(C) = \begin{pmatrix} I_{xx} & 0 & 0 \\ 0 & I_{yy} & 0 \\ 0 & 0 & I_{zz} \end{pmatrix}$$

and $\mathbf{i}$, $\mathbf{j}$, $\mathbf{k}$ are the conventional unit vectors. Equation (1.28) in scalar form supplies the three equations needed to determine the rotational motion of the body. These equations, the Euler equations, are

$$\begin{aligned} I_{xx}(d\omega_x/dt) + \omega_y\omega_z(I_{zz} - I_{yy}) &= M_x \\ I_{yy}(d\omega_y/dt) + \omega_z\omega_x(I_{xx} - I_{zz}) &= M_y \\ I_{zz}(d\omega_z/dt) + \omega_x\omega_y(I_{yy} - I_{xx}) &= M_z \end{aligned} \tag{1.30}$$

The analytical integration of the Euler equations in the general case defines a problem of classical difficulty. However, in special cases solutions can be found. The sources of the simplifications in these cases are the symmetry of the body and the absence of some components of the external moment. Since discussion of the various possibilities lies outside the scope of this chapter, reference is made to Refs. 1, 2, 6, and 7 and, for a survey of recent work, to Ref. 3. Of course, in situations in which energy or moment of momentum, or perhaps both, are conserved, first integrals of the motion can be written without employing the Euler equations. To do so it is convenient to have an expression for the kinetic energy T of the rotating body. This expression is readily found in the following manner.

The kinetic energy is

$$\begin{aligned} T &= \frac{1}{2}\int_V \rho(\boldsymbol{\omega} \times \mathbf{r})^2 \, dV \\ &= \frac{1}{2}\int_V \rho\boldsymbol{\omega} \cdot [\mathbf{r} \times (\boldsymbol{\omega} \times \mathbf{r})] \, dV \end{aligned}$$

which, by Eqs. (1.24), (1.25), and (1.26), is

$$T = \frac{1}{2}\boldsymbol{\omega} \cdot \boldsymbol{I}(C) \cdot \boldsymbol{\omega} \tag{1.31}$$

or, in matrix notation,

$$2T = (\omega_x \omega_y \omega_z)\begin{pmatrix} I_{xx} & 0 & 0 \\ 0 & I_{yy} & 0 \\ 0 & 0 & I_{zz} \end{pmatrix}\begin{pmatrix} \omega_x \\ \omega_y \\ \omega_z \end{pmatrix}$$

Equation (1.31) can be put in a simpler form by writing

$$T = \frac{1}{2}\omega^2(\boldsymbol{\omega}/\omega)\cdot \boldsymbol{I}(C)\cdot(\boldsymbol{\omega}/\omega)$$

and hence

$$T = \frac{1}{2} I_{\omega\omega}\omega^2 \tag{1.32}$$

In Eq. (1.32) $I_{\omega\omega}$ is the moment of inertia of the body about the axis of the angular velocity vector $\boldsymbol{\omega}$.

1.5 ANALYTICAL DYNAMICS

The knowledge of the time dependence of the position vectors $\mathbf{r}_i(t)$ which locate an n-mass system relative to a frame of reference can be attained indirectly by determining the dependence upon time of some parameters q_j ($j = 1, \ldots, m$) if the functional relationships

$$\mathbf{r}_i = \mathbf{r}_i(q_j, t) \qquad i = 1, \ldots, n; \quad j = 1, \ldots, m \tag{1.33}$$

are known. The parameters q_j which completely determine the position of the system in space are called "generalized coordinates." Any m quantities can be used as generalized coordinates on condition that they uniquely specify the positions of the masses. Frequently the q_j are the coordinates of an appropriate curvilinear system.

It is convenient to define two types of mechanical systems:

1. A "holonomic system" is one for which the generalized coordinates and the time may be arbitrarily and independently varied without violating the constraints.
2. A "nonholonomic system" is such that the generalized coordinates and the time may not be arbitrarily and independently varied because of some (say s) nonintegrable constraints of the form

$$\sum_{i=1}^{m} A_{ji}\, dq_i + A_j\, dt = 0 \qquad j = 1, 2, \ldots, s \tag{1.34}$$

In the constraint equations [Eq. (1.34)] the A_{ji} and A_j represent functions of the q_k and t. Holonomic and nonholonomic systems are further classified as "rheonomic" or "scleronomic," depending upon whether the time t is explicitly present or absent, respectively, in the constraint equations.

1.5.1 Generalized Forces and d'Alembert's Principle

A virtual displacement of the system is denoted by the set of vectors $\delta\mathbf{r}_i$. The work done by the forces in this displacement is

$$\delta W = \sum_{i=1}^{n} \mathbf{F}_i \cdot \delta\mathbf{r}_i \tag{1.35}$$

If the force $\mathbf{F}_i$, acting on the mass m_i, is separable in the sense that

$$\mathbf{F}_i = \mathbf{F}_i^a + \mathbf{F}_i^c \tag{1.36}$$

in which the first term is the applied force and the second the force of constraint, then

$$\delta W = \sum_{i=1}^{n} (\mathbf{F}_i^a + \mathbf{F}_i^c)\left[\sum_{j=1}^{m} \frac{\partial \mathbf{r}_i}{\partial q_j}\,\delta q_j + \frac{\partial \mathbf{r}_i}{\partial t}\,\delta t\right] \tag{1.37}$$

The generalized applied forces and the generalized forces of constraint are defined by

$$Q_j^a = \sum_{i=1}^{n} \mathbf{F}_i^a \cdot \frac{\partial \mathbf{r}_i}{\partial q_j} \tag{1.38}$$

and

$$Q_j^c = \sum_{i=1}^{n} \mathbf{F}_i^c \cdot \frac{\partial \mathbf{r}_i}{\partial q_j} \tag{1.39}$$

respectively. Hence, Eq. (1.37) assumes the form

$$\delta W = \sum_{j=1}^{m} Q_j^a\,\delta q_j + \sum_{j=1}^{m} Q_j^c\,\delta q_j + \sum_{i=1}^{n} (\mathbf{F}_i^a + \mathbf{F}_i^c)\cdot \frac{\partial \mathbf{r}_i}{\partial t}\,\delta t \tag{1.40}$$

If the virtual displacement is compatible with the instantaneous constraints $\delta t = 0$, and if in such a displacement the forces of constraint do work, e.g., if sliding friction is absent, then

$$\delta W = \sum_{j=1}^{m} Q_j^a\,\delta q_j \tag{1.41}$$

The assumption that a function $V(q_j, t)$ exists such that

$$Q_j^a = -\partial V/\partial q_j$$

leads to the result

$$\delta W = -\delta V \tag{1.42}$$

In Eq. (1.42), $V(q_j, t)$ is called the potential or work function.

The first step in the introduction of the kinetic energy of the system is taken by using d'Alembert's principle. The equations of motion [Eq. (1.10)] can be written as

$$\mathbf{F}_i - m_i\ddot{\mathbf{r}}_i = 0$$

and consequently

$$\sum_{i=1}^{n} (\mathbf{F}_i - m_i\ddot{\mathbf{r}}_i)\cdot \delta\mathbf{r}_i = 0 \tag{1.43}$$

The principle embodied in Eq. (1.43) constitutes the extension of the principle of virtual work to dynamic systems and is named after d'Alembert. When attention is confined to $\delta\mathbf{r}_i$ which represent virtual displacements compatible with the instantaneous constraints and to forces $\mathbf{F}_i$ which satisfy Eqs. (1.36) and (1.41), the principle states that

$$\sum_{j=1}^{m} Q_j^a\,\delta q_j = \sum_{i=1}^{n} m_i\ddot{\mathbf{r}}_i \cdot \delta\mathbf{r}_i \tag{1.44}$$

1.5.2 The Lagrange Equations

The central equations of analytical mechanics can now be derived. These equations, which were developed by Lagrange, are presented here for the general case of a rheonomic nonholonomic system consisting of n masses m_i, m generalized coordinates q_i, and s constraint equations

$$\sum_{j=1}^{m} A_{kj}\,dq_j + A_k\,dt = 0 \qquad k = 1, 2, \ldots, s \tag{1.45}$$

The equations are found by writing the acceleration terms in d'Alembert's principle [Eq. (1.43)] in terms of the kinetic energy T and the generalized coordinates. By definition

$$T = \frac{1}{2}\sum_{1}^{n} m_i \dot{\mathbf{r}}_i^2$$

where
$$\dot{\mathbf{r}}_i = \sum_{j=1}^{m} \frac{\partial \mathbf{r}_i}{\partial q_j}\frac{dq_j}{dt} + \frac{\partial \mathbf{r}_i}{dt} \qquad i = 1, 2, \ldots, n$$

Thus
$$\partial \dot{\mathbf{r}}_i/\partial \dot{q}_j = \partial \mathbf{r}_i/\partial q_j \qquad \partial \dot{\mathbf{r}}_i/\partial q_j = (d/dt)(\partial \mathbf{r}_i/\partial q_j)$$

$$\partial T/\partial q_j = \sum_{i=1}^{n} m_i \dot{\mathbf{r}}_i \cdot \frac{d}{dt}\frac{\partial \mathbf{r}_i}{\partial q_j} \qquad \text{and} \qquad \frac{\partial T}{\partial \dot{q}_j} = \sum_{i=1}^{n} m_i \dot{\mathbf{r}}_i \cdot \frac{\partial \mathbf{r}_i}{\partial q_j}$$

Accordingly,

$$\frac{d}{dt}\frac{\partial T}{\partial \dot{q}_j} - \frac{\partial T}{\partial q_j} = \sum_{i=1}^{n} m_i \ddot{\mathbf{r}}_i \cdot \frac{\partial \mathbf{r}_i}{\partial q_j} \qquad j = 1, 2, \ldots, m \tag{1.46}$$

and by summing over all values of j, one finds

$$\sum_{j=1}^{m}\left(\frac{d}{dt}\frac{\partial T}{\partial \dot{q}_j} - \frac{\partial T}{\partial q_j}\right)\delta q_j = \sum_{i=1}^{n} m_i\,\ddot{\mathbf{r}}_i \cdot \delta \mathbf{r}_i \tag{1.47}$$

because
$$\delta \mathbf{r}_i = \sum_{j=1}^{m} \frac{\partial \mathbf{r}_i}{\partial q_j}\delta q_j$$

for instantaneous displacements. From Eqs. (1.44) and (1.47) it follows that

$$\sum_{j=1}^{m}\left(\frac{d}{dt}\frac{\partial T}{\partial \dot{q}_j} - \frac{\partial T}{\partial q_j} - Q_j^a\right)\delta q_j = 0 \tag{1.48}$$

The δq_j which appear in Eq. (1.48) are not independent but must satisfy the instantaneous constraint equations

$$\sum_{j=1}^{m} A_{kj}\,\delta q_j = 0 \qquad k = 1, 2, \ldots, s \tag{1.49}$$

The "elimination" of s of the δq_j between Eqs. (1.48) and (1.49) is effected, in the usual way, by the introduction of s Lagrange multipliers $\lambda_k (k = 1, 2, \ldots, s)$. This step leads directly to the equations

$$\frac{d}{dt}\frac{\partial T}{\partial \dot{q}_j} - \frac{\partial T}{\partial q_j} = Q_j^a - \sum_{k=1}^{s} \lambda_k A_{kj} \qquad j = 1, 2, \ldots, m \tag{1.50}$$

These m second-order ordinary differential equations are the Lagrange equations of the system. The general solution of the equations is not available.* For a holonomic system with n degrees of freedom, Eq. (1.50) reduces to

$$\frac{d}{dt}\frac{\partial T}{\partial \dot{q}_j} - \frac{\partial T}{\partial q_j} = Q_j^a \qquad j = 1, \ldots, n \tag{1.51}$$

In the presence of a function V such that

$$Q_j^a = -\partial V/\partial q_j$$

and

$$\partial V/\partial \dot{q}_j = 0$$

Eqs. (1.51) can be written in the form

$$\frac{d}{dt}\frac{\partial \mathcal{L}}{\partial \dot{q}_j} - \frac{\partial \mathcal{L}}{\partial q_j} = 0 \qquad j = 1, 2, \ldots, n \tag{1.52}$$

in which

$$\mathcal{L} = T - V$$

The scalar function $\mathcal{L}$—the lagrangian—which is the difference between the kinetic and potential energies is all that need be known to write the Lagrange equations in this case.

The major factor which contributes to the solving of Eq. (1.52) is the presence of ignorable coordinates. In fact, in dynamics problems, generally, the possibility of finding analytical representations of the motion depends on there being ignorable coordinates. A coordinate, say q_k, is said to be ignorable if it does not appear explicitly in the lagrangian, i.e., if

$$\partial \mathcal{L}/\partial q_k = 0 \tag{1.53}$$

If Eq. (1.53) is valid, then Eq. (1.52) leads to

$$\partial \mathcal{L}/\partial \dot{q}_k = \text{const} = c_k$$

and hence a first integral of the motion is available. Clearly the more ignorable coordinates that exist in the lagrangian, the better. This being so, considerable effort has been directed toward developing systematic means of generating ignorable coordinates by transforming from one set of generalized coordinates to another, more suitable, set. This transformation theory of dynamics, while extensively developed, is not generally of practical value in engineering problems.

1.5.3 The Euler Angles

To use lagrangian methods in analyzing the motion of a rigid body one must choose a set of generalized coordinates which uniquely determines the position of the body relative to a frame of reference fixed in space. It suffices to examine the motion of a body rotating about its center of mass.

An inertial set of orthogonal axes ξ, η, and ζ with origin at the center of mass and a noninertial set x, y, and z fixed relative to the body with the same origin are adopted. The required generalized coordinates are those which specify the position of the x, y, and z axes relative to the ξ, η, and ζ axes. More than one set of coordinates which achieves this purpose can be found. The most generally useful one, viz., the Euler angles, is used here.

*Nonholonomic problems are frequently more tractable by vectorial than by lagrangian methods.

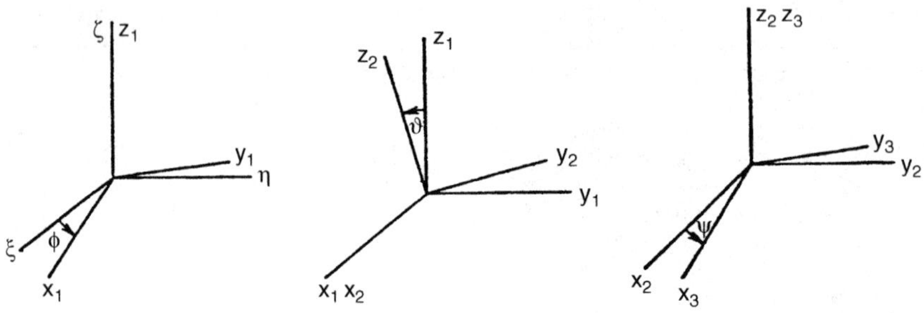

FIG. 1.4

The frame ξ, η, and ζ can be brought into coincidence with the frame x, y, and z by three finite rigid-body rotations through angles ϕ, θ, and ψ,* in that order, defined as follows (see Fig. 1.4):

1. A rotation about the ζ axis through an angle ϕ to produce the frame x_1, y_1, z_1
2. A rotation about the x_1 axis through an angle θ to produce the frame x_2, y_2, z_2
3. A rotation about the z_2 axis through an angle ψ to produce the frame x_3, y_3, z_3, which coincides with the frame x, y, z

Each rotation can be represented by an orthogonal matrix operation so that the process of getting from the inertial to the noninertial frame is

$$\begin{pmatrix} x_1 \\ y_1 \\ z_1 \end{pmatrix} = \begin{pmatrix} \cos\phi & \sin\phi & 0 \\ -\sin\phi & \cos\phi & 0 \\ 0 & 0 & 1 \end{pmatrix} \begin{pmatrix} \xi \\ \eta \\ \zeta \end{pmatrix} = A \begin{pmatrix} \xi \\ \eta \\ \zeta \end{pmatrix} \tag{1.54a}$$

$$\begin{pmatrix} x_2 \\ y_2 \\ z_2 \end{pmatrix} = \begin{pmatrix} 1 & 0 & 0 \\ 0 & \cos\theta & \sin\theta \\ 0 & -\sin\theta & \cos\theta \end{pmatrix} \begin{pmatrix} x_1 \\ y_1 \\ z_1 \end{pmatrix} = B \begin{pmatrix} x_1 \\ y_1 \\ z_1 \end{pmatrix} \tag{1.54b}$$

$$\begin{pmatrix} x_3 \\ y_3 \\ z_3 \end{pmatrix} = \begin{pmatrix} \cos\psi & \sin\psi & 0 \\ -\sin\psi & \cos\psi & 0 \\ 0 & 0 & 1 \end{pmatrix} \begin{pmatrix} x_2 \\ y_2 \\ z_2 \end{pmatrix} = C \begin{pmatrix} x_2 \\ y_2 \\ z_2 \end{pmatrix} \tag{1.54c}$$

Consequently,

$$\begin{pmatrix} x \\ y \\ z \end{pmatrix} = CBA \begin{pmatrix} \xi \\ \eta \\ \zeta \end{pmatrix} = D \begin{pmatrix} \xi \\ \eta \\ \zeta \end{pmatrix} \tag{1.55}$$

where

$$D = CBA = \begin{pmatrix} \cos\psi\cos\phi - \cos\theta\sin\phi\sin\psi & \cos\psi\sin\phi + \cos\theta\cos\phi\sin\psi & \sin\psi\sin\theta \\ -\sin\psi\cos\phi - \cos\theta\sin\phi\cos\psi & -\sin\psi\sin\phi + \cos\theta\cos\phi\cos\psi & \cos\psi\sin\theta \\ \sin\theta\sin\phi & -\sin\theta\cos\phi & \cos\theta \end{pmatrix}$$

*This notation is not universally adopted. See Ref. 5 for discussion.

Since A, B, and C are orthogonal matrices, it follows from Eq. (1.55) that

$$\begin{pmatrix} \xi \\ \eta \\ \zeta \end{pmatrix} = D^{-1} \begin{pmatrix} x \\ y \\ z \end{pmatrix} = D' \begin{pmatrix} x \\ y \\ z \end{pmatrix} \tag{1.56}$$

where the prime denotes the transpose of the matrix. From Eq. (1.55) one sees that, if the time dependence of the three angles ϕ, θ, ψ is known, the orientation of the x, y, z and axes relative to the ξ, η, and ζ axes is determined. This time dependence is sought by attempting to solve the Lagrange equations.

The kinetic energy T of the rotating body is found from Eq. (1.31) to be

$$2T = I_{xx}\omega_x^2 + I_{yy}\omega_x^2 + I_{zz}\omega_z^2 \tag{1.57}$$

in which the components of the angular velocity ω are provided by the matrix equation

$$\begin{pmatrix} \omega_x \\ \omega_y \\ \omega_z \end{pmatrix} = CB \begin{pmatrix} 0 \\ 0 \\ \dot{\phi} \end{pmatrix} = C \begin{pmatrix} \dot{\theta} \\ 0 \\ 0 \end{pmatrix} + \begin{pmatrix} 0 \\ 0 \\ \dot{\psi} \end{pmatrix} \tag{1.58}$$

It is to be noted that if

$$I_{xx} \neq I_{yy} \neq I_{zz} \tag{1.59}$$

none of the angles is ignorable. Hence considerable difficulty is to be expected in attempting to solve the Lagrange equations if this inequality, Eq. (1.59), holds. A similar inference could be made on examining Eq. (1.30). The possibility of there being ignorable coordinates in the problem arises if the body has axial, or so-called kinetic, symmetry about (say) the z axis. Then

$$I_{xx} = I_{yy} = I$$

and, from Eq. (1.57),

$$2T = I(\dot{\phi}^2 \sin^2\theta + \dot{\theta}^2) + I_{zz}(\dot{\phi}\cos\theta + \dot{\psi})^2 \tag{1.60}$$

The angles ϕ and ψ do not occur in Eq. (1.60). Whether or not they are ignorable depends on the potential energy $V(\phi, \theta, \psi)$.

1.5.4 Small Oscillations of a System near Equilibrium

The Lagrange equations are particularly useful in examining the motion of a system near a position of equilibrium. Let the generalized coordinates $q_1, q_2, \ldots, q_n$—the explicit appearance of time being ruled out—represent the configuration of the system. It is not restrictive to assume the equilibrium position at

$$q_1 \text{ and } q_2 = \cdots = q_n = 0$$

and, since motion near this position is being considered, the q_i and $\dot{q}_i$ may be taken to be small.

The potential energy can be expanded in a Taylor series about the equilibrium point in the form

$$V(q_1 \cdots q_n) = V(0) + \sum_{i=1}^{n} \left(\frac{\partial V}{\partial q_i}\right)_0 q_i + \frac{1}{2}\sum_i \sum_j \left(\frac{\partial^2 V}{\partial q_i\, \partial q_j}\right)_0 q_i q_j + \cdots \tag{1.61}$$

In Eq. (1.61) the first term can be neglected because it merely changes the potential energy by a constant and the second term vanishes because $\partial V/\partial q_i$ is zero at the equilibrium point. Thus, retaining only quadratic terms in q_i, one finds

$$V(q_1 \cdots q_n) = \frac{1}{2} \sum_i \sum_j V_{ij} q_i q_j \tag{1.62}$$

in which

$$V_{ij} = (\partial^2 V/\partial q_i \, \partial q_j)_0 = V_{ji} \tag{1.63}$$

are real constants.

The kinetic energy T of the system is representable by an analogous Taylor series

$$T(\dot{q}_i \cdots \dot{q}) = \frac{1}{2} \sum_i \sum_j T_{ij} \dot{q}_i \dot{q}_j \tag{1.64}$$

where

$$T_{ij} = T_{ji} \tag{1.65}$$

are real constants. The quadratic forms, Eqs. (1.62) and (1.64), in matrix notation, a prime denoting transposition are

$$V = \frac{1}{2} q' \mathcal{V} q \tag{1.66}$$

and

$$T = \frac{1}{2} \dot{q}' \mathcal{T} \dot{q} \tag{1.67}$$

In these expressions $\mathcal{V}$ and $\mathcal{T}$ represent the matrices with elements V_{ij} and T_{ij}, respectively, and q represents the column vector $(q_1, \ldots, q_n)$. The form of Eq. (1.67) is necessarily positive definite because of the nature of kinetic energy. Rather than create the Lagrange equations in terms of the coordinates q_i, a new set of generalized coordinages ρ_i is introduced in terms of which the energies are simultaneously expressible as quadratic forms without cross-product terms. That the transformation to such coordinates is possible can be seen by considering the equations

$$\mathcal{V} b_j = \lambda_j \mathcal{T} b_j \qquad j = 1, 2, \ldots, n \tag{1.68}$$

in which λ_j, the roots of the equation

$$|\mathcal{V} - \lambda \mathcal{T}| = 0$$

are the eigenvalues—assumed distinct—and b_j are the corresponding eigenvectors. The matrix of eigenvectors b_j is symbolized by B, and the diagonal matrix of eigenvalues λ_j by Λ. One can write

$$b_k' \mathcal{V} b_j = \lambda_j b_k' \mathcal{T} b_j$$

and

$$b_k' \mathcal{V} b_j = \lambda_k b_k' \mathcal{T} b_j$$

because of the symmetry of $\mathcal{V}$ and $\mathcal{T}$. Thus, if $\lambda_j \neq \lambda_k$, it follows that

$$b_k' \mathcal{T} b_j = 0 \qquad k \neq j$$

and, since the eigenvectors of Eq. (1.68) are each undetermined to within an arbitrary multiplying constant, one can always normalize the vectors so that

$$b_i' \mathcal{T} b_i = 1$$

Hence

$$B' \mathcal{T} B = \mathrm{I} \tag{1.69}$$

where I is the unit matrix. But

$$\mathcal{V}B = \mathcal{T}B\Lambda \tag{1.70}$$

and so

$$B'\mathcal{V}B = B'\mathcal{T}B\Lambda = \Lambda \tag{1.71}$$

Furthermore, denoting the complex conjugate by an overbar, one has

$$\mathcal{V}\bar{b}_j = \bar{\lambda}_j \mathcal{T}\bar{b}_j$$

and

$$b_j'\mathcal{V}\bar{b}_j = \bar{\lambda}_j b_j'\mathcal{T}\bar{b}_j \tag{1.72}$$

since $\mathcal{V}$ and $\mathcal{T}$ are real. However,

$$b_j'\mathcal{V}\bar{b}_j = \lambda_j b_j'\mathcal{T}\bar{b}_j \tag{1.73}$$

because $\mathcal{V}$ and $\mathcal{T}$ are symmetric. From Eqs. (1.72) and (1.73) it follows that

$$(\lambda_j - \bar{\lambda}_j)b_j'\mathcal{T}\bar{b}_j = 0 \tag{1.74}$$

The symmetry and positive definiteness of $\mathcal{T}$ ensure that the form $b_j'\mathcal{T}\bar{b}_j$ is real and positive definite. Consequently the eigenvalues λ_j, and eigenvectors b_j, are real. Finally, one can solve Eq. (1.68) for the eigenvalues in the form

$$\lambda_j = b_j'\mathcal{V}b_j / b_j'\mathcal{T}b_j \tag{1.75}$$

The transformation from the q_i to the ρ_i coordinates can now be made by writing

$$q = B\rho$$

from which

$$V = \frac{1}{2}q'\mathcal{V}q = \frac{1}{2}\rho'B'\mathcal{V}B\rho = \frac{1}{2}\rho'\Lambda\rho \tag{1.76}$$

and

$$T = \frac{1}{2}\dot{q}'\mathcal{T}\dot{q} = \frac{1}{2}\dot{\rho}'B'\mathcal{T}B\dot{\rho} = \frac{1}{2}\dot{\rho}'\mathrm{I}\dot{\rho} \tag{1.77}$$

It is seen from Eqs. (1.76) and (1.77) that V and T have the desired forms and that the corresponding Lagrange equations (1.52) are

$$d^2\rho_i/dt^2 + \omega_i^2\rho_i = 0 \qquad i = 1, \ldots, n \tag{1.78}$$

where $\omega_i^2 = \lambda_i$. If the equilibrium position about which the motion takes place is stable, the ω_i^2 are positive. The eigenvalues λ_i must then be positive, and Eq. (1.75) shows that V is positive definite. In other words, the potential energy is a minimum at a position of stable equilibrium. In this case, the motion of the system can be analyzed in terms of its normal modes—the n harmonic oscillators Eq. (1.78). If the matrix V is not positive definite, Eq. (1.75) indicates that negative eigenvalues may exist, and hence Eqs. (1.78) may have hyperbolic solutions. The equilibrium is then unstable. Regardless of the nature of the equilibrium, the Lagrange equations (1.78) can always be arrived at, because it is possible to diagonalize simultaneously two quadratic forms, one of which (the kinetic-energy matrix) is positive definite.

1.5.5 Hamilton's Principle

In conclusion it is remarked that the Lagrange equations of motion can be arrived at by methods other than that presented above. The point of departure adopted here is Hamilton's principle, the statement of which for holonomic systems is as follows.

Provided the initial (t_1) and final (t_2) configurations are prescribed, the motion of the system from time t_1 to time t_2 occurs in such a way that the line integral

$$\int_{t_1}^{t_2} \mathcal{L}\, dt = \text{extremum}$$

where $\mathcal{L} = T - V$. That the Lagrange equations [Eq. (1.52)] can be derived from this principle is shown here for the case of a single-mass, one-degree-of-freedom system. The generalization of the proof to include an n-degree-of-freedom system is made without difficulty.

The lagrangian is

$$\mathcal{L}(q, \dot{q}, t) = T - V$$

in which q is the generalized coordinate and $q(t)$ describes the motion that actually occurs. Any other motion can be represented by

$$\bar{q}(t) = q(t) + \varepsilon f(t) \tag{1.79}$$

in which $f(t)$ is an arbitrary differentiable function such that $f(t_1)$ and $f(t_2) = 0$ and ε is a parameter defining the family of curves $\bar{q}(t)$. The condition

$$\int_{t_1}^{t_2} \mathcal{L}(q_1, \dot{q}_1, t)\, dt = \text{extremum}$$

is tantamount to

$$\frac{\partial}{\partial \varepsilon} \int_{t_1}^{t_2} (\bar{q}_1, \dot{\bar{q}}_1, t)\, dt = 0 \qquad \varepsilon = 0 \tag{1.80}$$

for all $f(t)$. But

$$\frac{\partial}{\partial \varepsilon} \int_{t_1}^{t_2} \mathcal{L}(\bar{q}_1, \dot{\bar{q}}_1, t)\, dt = \int_{t_1}^{t_2} \left(\frac{\partial \mathcal{L}}{\partial \bar{q}} \frac{\partial \bar{q}}{\partial \varepsilon} + \frac{\partial \mathcal{L}}{\partial \dot{\bar{q}}} \frac{\partial \dot{\bar{q}}}{\partial \varepsilon} \right) dt$$

which, by Eq. (1.79), is

$$\frac{\partial}{\partial \varepsilon} \int_{t_1}^{t_2} (\bar{q}_1, \dot{\bar{q}}_1, t)\, dt = \int_{t_1}^{t_2} \left[f(t) \frac{\partial \mathcal{L}}{\partial \bar{q}} + \dot{f}(t) \frac{\partial \mathcal{L}}{\partial \dot{\bar{q}}} \right] dt \tag{1.81}$$

Its second term having been integrated by parts, Eq. (1.81) reduces to

$$\frac{\partial}{\partial \varepsilon} \int_{t_1}^{t_2} \mathcal{L}(\bar{q}, \dot{\bar{q}}, t)\, dt = \int_{t_1}^{t_2} f(t) \left(\frac{\partial \mathcal{L}}{\partial \bar{q}} - \frac{d}{dt} \frac{\partial \mathcal{L}}{\partial \dot{\bar{q}}} \right) dt$$

because $f(t_1) = f(t_2) = 0$. Hence Eq. (1.80) is equivalent to

$$\int_{t_1}^{t_2} f(t) \left(\frac{\partial \mathcal{L}}{\partial q} - \frac{d}{dt} \frac{\partial \mathcal{L}}{\partial \dot{q}} \right) dt = 0 \tag{1.82}$$

for all $f(t)$. Equation (1.82) can hold for all $f(t)$ only if

$$\frac{d}{dt} \frac{\partial \mathcal{L}}{\partial \dot{q}} - \frac{\partial \mathcal{L}}{\partial q} = 0$$

which is the Lagrange equation of the system.

The extension to an n-degree-of-freedom system is made by employing n arbitrary differentiable functions $f_k(t)$, $k = 1, \ldots, n$ such that $f_k(t_1) = f_k(t_2) = 0$. For the generalizations of Hamilton's principle which are necessary in treating nonholonomic systems, the references should be consulted.

The principle can be extended to include continuous systems, potential energies other than mechanical, and dissipative sources. The analytical development of these and other topics and examples of their applications are presented in Refs. 4 and 8 through 12.

REFERENCES

1. Routh, E. J.: "Advanced Dynamics of a System of Rigid Bodies," 6th ed., Dover Publications, Inc., New York, 1955.
2. Whittaker, E. T.: "A Treatise on Analytical Dynamics," 4th ed., Dover Publications, Inc., New York, 1944.
3. Leimanis, E., and N. Minorsky: "Dynamics and Nonlinear Mechanics," John Wiley & Sons, Inc., New York, 1958.
4. Corben, H. C., and P. Stehle: "Classical Mechanics," 2d ed., John Wiley & Sons, Inc., New York, 1960.
5. Goldstein, H.: "Classical Mechanics," 2d ed., Addison-Wesley Publishing Company, Inc., Reading, Mass, 1980.
6. Milne, E. A.: "Vectorial Mechanics," Methuen & Co., Ltd., London, 1948.
7. Scarborough, J. B.: "The Gyroscope," Interscience Publishers, Inc., New York, 1958.
8. Synge, J. L., and B. A. Griffith: "Principles of Mechanics," 3d ed., McGraw-Hill Book Company, Inc., New York, 1959.
9. Lanczos, C.: "The Variational Principles of Mechanics," 4th ed., University of Toronto Press, Toronto, 1970.
10. Synge, J. L.: "Classical Dynamics," in "Handbuch der Physik," Bd III/I, Springer-Verlag, Berlin, 1960.
11. Crandall, S. H., et al.: "Dynamics of Mechanical and Electromechanical Systems," McGraw-Hill Book Company, Inc., New York, 1968.
12. Woodson, H. H., and J. R. Melcher: "Electromechanical Dynamics," John Wiley & Sons, Inc., New York, 1968.

CHAPTER 2
MECHANICS OF MATERIALS

Stephen B. Bennett, Ph.D.
Manager of Research and Product Development
Delaval Turbine Division
Imo Industries, Inc.
Trenton, N.J.

Robert P. Kolb, P.E.
Manager of Engineering (Retired)
Delaval Turbine Division
Imo Industries, Inc.
Trenton, N.J.

2.1 INTRODUCTION

The fundamental problem of structural analysis is the prediction of the ability of machine components to provide reliable service under its applied loads and temperature. The basis of the solution is the calculation of certain performance indices, such as stress (force per unit area), strain (deformation per unit length), or gross deformation, which can then be compared to allowable values of these parameters. The allowable values of the parameters are determined by the component function (deformation constraints) or by the material limitations (yield strength, ultimate strength, fatigue strength, etc.). Further constraints on the allowable values of the performance indices are often imposed through the application of factors of safety.

This chapter, "Mechanics of Materials," deals with the calculation of performance indices under statically applied loads and temperature distributions. The extension of the theory to dynamically loaded structures, i.e., to the response of structures to shock and vibration loading, is treated elsewhere in this handbook.

The calculations of "Mechanics of Materials" are based on the concepts of force equilibrium (which relates the applied load to the internal reactions, or stress, in the body), material observation (which relates the stress at a point to the internal deformation, or strain, at the point), and kinematics (which relates the strain to the gross deformation of the body). In its simplest form, the solution assumes linear relationships between the components of stress and the components of strain (hookean material models) and that the deformations of the body are sufficiently small that linear relationships exist between the components of strain and the components of deformation. This linear elastic model of structural behavior remains the predominant tool used today for the design analysis of machine components, and is the principal subject of this chapter.

It must be noted that many materials retain considerable load-carrying ability when stressed beyond the level at which stress and strain remain proportional. The modification of the material model to allow for nonlinear relationships between stress and strain is the principal feature of the theory of plasticity. Plastic design allows more effective material utilization at the expense of an acceptable permanent deformation of the structure and smaller (but still controlled) design margins. Plastic design is often used in the design of civil structures, and in the analysis of machine structures under emergency load conditions. Practical introductions to the subject are presented in Refs. 6, 7, and 8.

Another important and practical extension of elastic theory includes a material model in which the stress-strain relationship is a function of time and temperature. This "creep" of components is an important consideration in the design of machines for use in a high-temperature environment. Reference 11 discusses the theory of creep design. The set of equations which comprise the linear elastic structural model do not have a comprehensive, exact solution for a general geometric shape. Two approaches are used to yield solutions:

The geometry of the structure is simplified to a form for which an exact solution is available. Such simplified structures are generally characterized as being a level surface in the solution coordinate system. Examples of such simplified structures include rods, beams, rectangular plates, circular plates, cylindrical shells, and spherical shells. Since these shapes are all level surfaces in different coordinate systems, e.g., a sphere is the surface r = constant in spherical coordinates, it is a great convenience to express the equations of linear elastic theory in a coordinate invariant form. General tensor notation is used to accomplish this task.

The governing equations are solved through numerical analysis on a case-by-case basis. This method is used when the component geometry is such that none of the available beam, rectangular plate, etc., simplifications are appropriate. Although several classes of numerical procedures are widely used, the predominant procedure for the solution of problems in the "Mechanics of Materials" is the finite-element method.

2.2 STRESS

2.2.1 Definition[2]

"Stress" is defined as the force per unit area acting on an "elemental" plane in the body. Engineering units of stress are generally pounds per square inch. If the force is normal to the plane the stress is termed "tensile" or "compressive," depending upon whether the force tends to extend or shorten the element. If the force acts parallel to the elemental plane, the stress is termed "shear." Shear tends to deform by causing neighboring elements to slide relative to one another.

2.2.2 Components of Stress[2]

A complete description of the internal forces (stress distributions) requires that stress be defined on three perpendicular faces of an interior element of a structure. In Fig. 2.1 a small element is shown, and, omitting higher-order effects, the stress resultant on any face can be considered as acting at the center of the area.

The direction and type of stress at a point are described by subscripts to the stress symbol σ or τ. The first subscript defines the plane on which the stress acts and the second indicates the direction in which it acts. The plane on which the stress acts is indicated by the normal axis to that plane; e.g., the x plane is normal to the x axis. Conventional notation omits the second subscript for the normal stress and replaces the σ by a τ for the shear stresses. The "stress components" can thus be represented as follows:

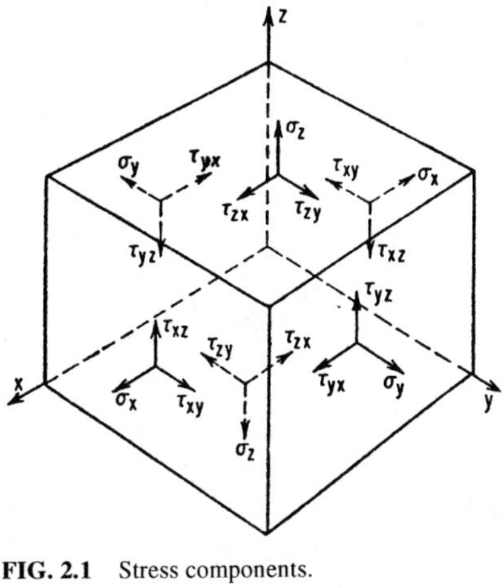

FIG. 2.1 Stress components.

Normal stress:

$$\sigma_{xx} \equiv \sigma_x \qquad \sigma_{yy} \equiv \sigma_y \qquad \sigma_{zz} \equiv \sigma_z \tag{2.1}$$

Shear stress:

$$\sigma_{xy} \equiv \tau_{xy} \qquad \sigma_{yz} \equiv \tau_{yz}$$

$$\sigma_{xz} \equiv \tau_{xz} \qquad \sigma_{zx} \equiv \tau_{zx} \tag{2.2}$$

$$\sigma_{yx} \equiv \tau_{yz} \qquad \sigma_{zy} \equiv \tau_{zy}$$

In tensor notation, the stress components are

$$\sigma_{ij} = \begin{pmatrix} \sigma_x & \tau_{xy} & \tau_{xz} \\ \tau_{yx} & \sigma_y & \tau_{yz} \\ \tau_{zx} & \tau_{zy} & \sigma_z \end{pmatrix} \tag{2.3}$$

Stress is "positive" if it acts in the "positive-coordinate direction" on those element faces farthest from the origin, and in the "negative-coordinate direction" on those faces closest to the origin. Figure 2.1 indicates the direction of all positive stresses, wherein it is seen that tensile stresses are positive and compressive stresses negative.

The total load acting on the element of Fig. 2.1 can be completely defined by the stress components shown, subject only to the restriction that the coordinate axes are mutually orthogonal. Thus the three normal stress symbols σ_x, σ_y, σ_z and six shear-stress symbols τ_{xy}, τ_{xz}, τ_{yx}, τ_{yz}, τ_{zx}, τ_{zy} define the stresses of the element. However, from equilibrium considerations, $\tau_{xy} = \tau_{yx}$, $\tau_{yz} = \tau_{zy}$, $\tau_{xz} = \tau_{zx}$. This reduces the necessary number of symbols required to define the stress state to σ_x, σ_y, σ_z, τ_{xy}, τ_{xz}, τ_{yz}.

2.2.3 Simple Uniaxial States of Stress[1]

Consider a simple bar subjected to axial loads only. The forces acting at a transverse section are all directed normal to the section. The uniaxial normal stress at the section is obtained from

$$\sigma = P/A \tag{2.4}$$

where P = total force and A = cross-sectional area.

"Uniaxial shear" occurs in a circular cylinder, loaded as in Fig. 2.2*a,* with a radius which is large compared to the wall thickness. This member is subjected to a torque distributed about the upper edge:

$$T = \Sigma Pr \tag{2.5}$$

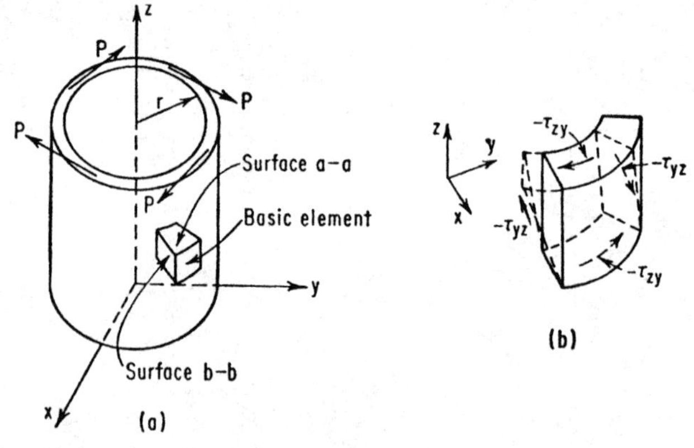

FIG. 2.2 Uniaxial shear basic element.

Now consider a surface element (assumed plane) and examine the stresses acting. The stresses τ which act on surfaces *a-a* and *b-b* in Fig. 2.2*b* tend to distort the original rectangular shape of the element into the parallelogram shown (dotted shape). This type of action of a force along or tangent to a surface produces shear within the element, the intensity of which is the "shear stress."

2.2.4 Nonuniform States of Stress[1]

In considering elements of differential size, it is permissible to assume that the force acts on any side of the element concentrated at the center of the area of that side, and that the stress is the average force divided by the side area. Hence it has been implied thus far that the stress is uniform. In members of finite size, however, a variable stress intensity usually exists across any given surface of the member. An example of a body which develops a distributed stress pattern across a transverse cross section is a simple beam subjected to a bending load as shown in Fig. 2.3*a*. If a section is then taken at *a-a*, F'_1 must be the internal force acting along *a-a* to maintain equilibrium. Forces F_1 and F'_1 constitute a couple which tends to rotate the element in a clockwise direction, and therefore a resisting couple must be developed at *a-a* (see Fig. 2.3*b*). The internal effect at *a-a* is a stress distribution with the upper portion of the beam in tension and the lower portion in compression, as in Fig. 2.3*c*. The line of zero stress on the transverse cross section is the "neutral axis" and passes through the centroid of the area.

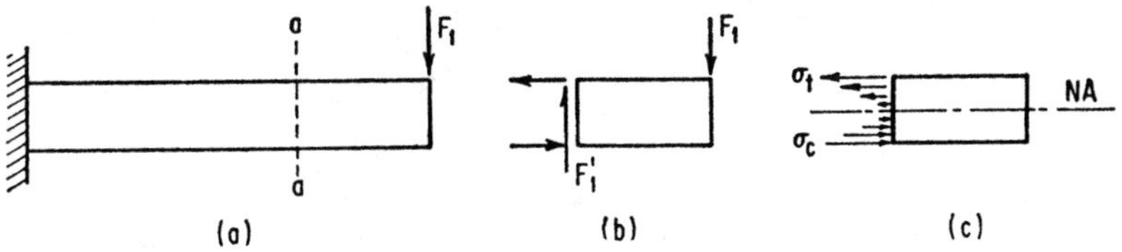

FIG. 2.3 Distributed stress on a simple beam subjected to a bending load.

2.2.5 Combined States of Stress

Tension-Torsion. A body loaded simultaneously in direct tension and torsion, such as a rotating vertical shaft, is subject to a combined state of stress. Figure 2.4*a* depicts such a shaft with end load *W*, and constant torque *T* applied to maintain uniform rotational velocity. With reference to *a-a*, considering each load separately, a force system

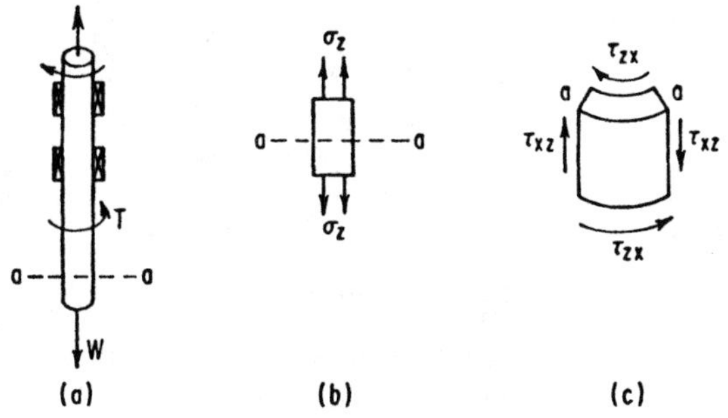

FIG. 2.4 Body loaded in direct tension and torsion.

as shown in Fig. 2.2*b* and *c* is developed at the internal surface *a-a* for the weight load and torque, respectively. These two stress patterns may be superposed to determine the "combined" stress situation for a shaft element.

Flexure-Torsion. If in the above case the load *W* were horizontal instead of vertical, the combined stress picture would be altered. From previous considerations of a simple beam, the stress distribution varies linearly across section *a-a* of the shaft of Fig. 2.5*a*. The stress pattern due to flexure then depends upon the location of the element in question; e.g., if the element is at the outside (element *x*) then it is undergoing maximum tensile stress (Fig. 2.5*b*), and the tensile stress is zero if the element is located on the horizontal center line (element *y*) (Fig. 2.5*c*). The shearing stress is still constant at a given element, as before (Fig. 2.5*d*). Thus the "combined" or "superposed" stress state for this condition of loading varies across the entire transverse cross section.

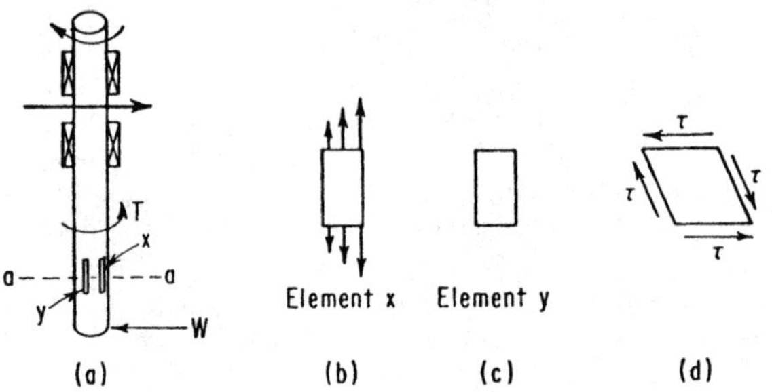

FIG. 2.5 Body loaded in flexure and torsion.

2.2.6 Stress Equilibrium

"Equilibrium" relations must be satisfied by each element in a structure. These are satisfied if the resultant of all forces acting on each element equals zero in each of three mutually orthogonal directions on that element. The above applies to all situations of "static equilibrium." In the event that some elements are in motion an inertia term must be added to the equilibrium equation. The inertia term is the elemental mass multiplied by the absolute acceleration taken along each of the mutually perpendicular axes. The equations which specify this latter case are called "dynamic-equilibrium equations" (see Chap. 4).

Three-Dimensional Case.[5,13] The equilibrium equations can be derived by separately summing all *x, y,* and *z* forces acting on a differential element accounting for the incremental variation of stress (see Fig. 2.6). Thus the normal forces acting on areas $dz\,dy$ are $\sigma_x\,dz\,dy$ and $[\sigma_x + (\partial\sigma_x/\partial x)\,dx]\,dz\,dy$. Writing *x* force-equilibrium equations, and by a similar process *y* and *z* force-equilibrium equations, and canceling *higher-order* terms, the following three "cartesian equilibrium equations" result:

$$\partial\sigma_x/\partial x + \partial\tau_{xy}/\partial y + \partial\tau_{xz}/\partial z = 0 \tag{2.6}$$

$$\partial\sigma_y/\partial y + \partial\tau_{yz}/\partial z + \partial\tau_{yx}/\partial x = 0 \tag{2.7}$$

$$\partial\sigma_z/\partial z + \partial\tau_{zx}/\partial x + \partial\tau_{zy}/\partial y = 0 \tag{2.8}$$

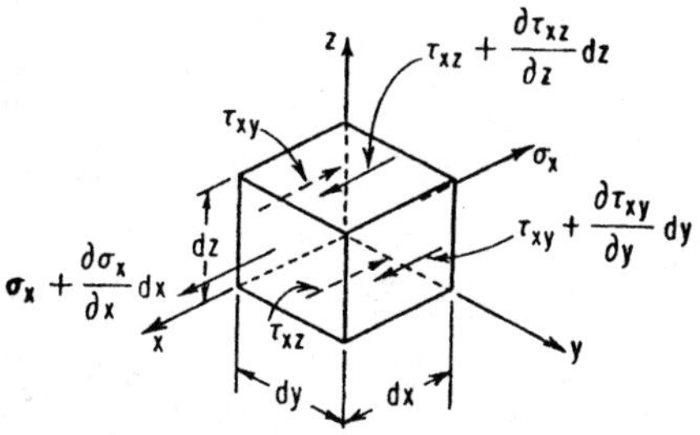

FIG. 2.6 Incremental element (dx, dy, dz) with incremental variation of stress.

or, in cartesian stress-tensor notation,

$$\sigma_{ij,j} = 0 \qquad i,j = x,y,z \tag{2.9}$$

and, in general tensor form,

$$g^{ik}\sigma_{ij,k} = 0 \tag{2.10}$$

where g^{ik} is the contravariant metric tensor.

"Cylindrical-coordinate" equilibrium considerations lead to the following set of equations (Fig. 2.7):

$$\partial\sigma_r/\partial r + (1/r)(\partial\tau_{r\theta}/\partial\theta) + \partial\tau_{rz}/\partial z + (\sigma_r - \sigma_\theta)/r = 0 \tag{2.11}$$

$$\partial\tau_{r\theta}/\partial r + (1/r)(\partial\sigma_\theta/\partial\theta) + \partial\tau_{\theta z}/\partial z + 2\tau_{r\theta}/r = 0 \tag{2.12}$$

$$\partial\tau_{rz}/\partial r + (1/r)(\partial\tau_{\theta z}/\partial\theta) + \partial_{\sigma z}/\partial z + \tau_{rz}/r = 0 \tag{2.13}$$

The corresponding "spherical polar-coordinate" equilibrium equations are (Fig. 2.8)

$$\frac{\partial\sigma_r}{\partial r} + \frac{1}{r}\frac{\partial\tau_{r\theta}}{\partial\theta} + \left(\frac{1}{r\sin\theta}\right)\frac{\partial\tau_{r\phi}}{\partial\phi} + \frac{1}{r}(2\sigma_r - \sigma_\theta - \sigma_\phi + \tau_{r\theta}\cot\theta) = 0 \tag{2.14}$$

$$\frac{\partial\tau_{r\theta}}{\partial r} + \frac{1}{r}\frac{\partial\sigma_\theta}{\partial\theta} + \left(\frac{1}{r\sin\theta}\right)\frac{\partial\tau_{\theta\phi}}{\partial\phi} + \frac{1}{r}[(\sigma_\theta - \sigma_\phi)\cot\theta + 3\tau_{r\theta}] = 0 \tag{2.15}$$

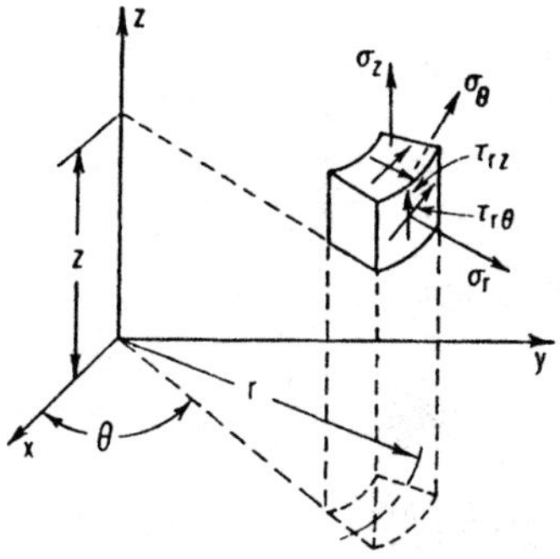

FIG. 2.7 Stresses on a cylindrical element.

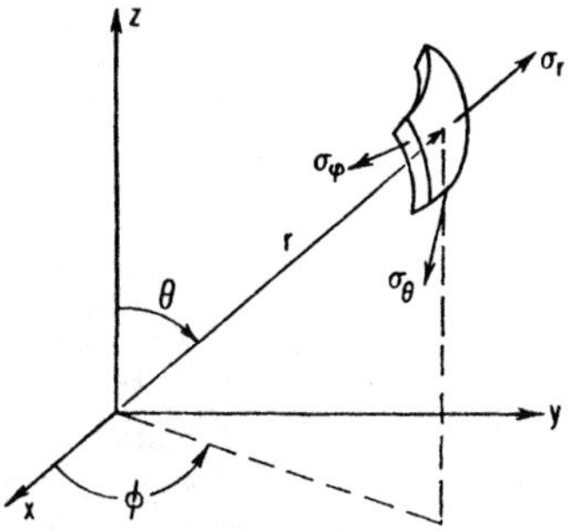

FIG. 2.8 Stresses on a spherical element.

$$\frac{\partial \tau_{r\phi}}{\partial r} + \frac{1}{r}\frac{\partial \tau_{\theta\phi}}{\partial \theta} + \left(\frac{1}{r \sin \theta}\right)\frac{\partial \sigma_{\phi}}{\partial \phi} + \frac{1}{r}(3\tau_{r\phi} + 2\tau_{\theta\phi} \cot \theta) = 0 \tag{2.16}$$

The general orthogonal curvilinear-coordinate equilibrium equations are

$$h_1h_2h_3\left(\frac{\partial}{\partial \alpha}\frac{\sigma_\alpha}{h_2h_3} + \frac{\partial}{\partial \beta}\frac{\tau_{\alpha\beta}}{h_3h_1} + \frac{\partial}{\partial \gamma}\frac{\tau_{\gamma\alpha}}{h_1h_2}\right) + \tau_{\alpha\beta}h_1h_2\frac{\partial}{\partial \beta}\frac{1}{h_1}$$

$$+ \tau_{\gamma\alpha}h_1h_3\frac{\partial}{\partial \gamma}\frac{1}{h_1} - \sigma_\beta h_1h_2\frac{\partial}{\partial \alpha}\frac{1}{h_2} - \sigma_\gamma h_1h_3\frac{\partial}{\partial \alpha}\frac{1}{h_3} = 0 \tag{2.17}$$

$$h_1h_2h_3\left(\frac{\partial}{\partial \beta}\frac{\sigma_\beta}{h_3h_1} + \frac{\partial}{\partial \gamma}\frac{\tau_{\beta\gamma}}{h_1h_2} + \frac{\partial}{\partial \alpha}\frac{\tau_{\alpha\beta}}{h_2h_3}\right) + \tau_{\beta\gamma}h_2h_3\frac{\partial}{\partial \gamma}\frac{1}{h_2}$$

$$+ \tau_{\alpha\beta}h_2h_1\frac{\partial}{\partial \alpha}\frac{1}{h_2} - \sigma_\gamma h_2h_3\frac{\partial}{\partial \beta}\frac{1}{h_3} - \sigma_\alpha h_2h_1\frac{\partial}{\partial \beta}\frac{1}{h_1} = 0 \tag{2.18}$$

$$h_1h_2h_3\left(\frac{\partial}{\partial \gamma}\frac{\sigma_\gamma}{h_1h_2} + \frac{\partial}{\partial \alpha}\frac{\tau_{\gamma\alpha}}{h_2h_3} + \frac{\partial}{\partial \beta}\frac{\tau_{\beta\gamma}}{h_3h_1}\right) + \tau_{\gamma\alpha}h_3h_1\frac{\partial}{\partial \alpha}\frac{1}{h_3}$$

$$+ \tau_{\beta\gamma}h_3h_2\frac{\partial}{\partial \beta}\frac{1}{h_3} - \sigma_\alpha h_3h_1\frac{\partial}{\partial \gamma}\frac{1}{h_1} - \sigma_\beta h_3h_2\frac{\partial}{\partial \gamma}\frac{1}{h_2} = 0 \tag{2.19}$$

where the α, β, γ specify the coordinates of a point and the distance between two coordinate points ds is specified by

$$(ds)^2 = (d\alpha/h_1)^2 + (d\beta/h_2)^2 + (d\gamma/h_3)^2 \tag{2.20}$$

which allows the determination of h_1, h_2, and h_3 in any specific case.

Thus, in cylindrical coordinates,

$$(ds)^2 = (dr)^2 + (r\,d\theta)^2 + (dz)^2 \tag{2.21}$$

so that

$$\alpha = r \qquad h_1 = 1$$

$$\beta = \theta \qquad h_2 = 1/r$$

$$\gamma = z \qquad h_3 = 1$$

In spherical polar coordinates,

$$(ds)^2 = (dr)^2 + (r\,d\theta)^2 + (r \sin \theta\, d\phi)^2 \tag{2.22}$$

so that

$$\alpha = r \qquad h_1 = 1$$

$$\beta = \theta \qquad h_2 = 1/r$$

$$\gamma = \phi \qquad h_3 = 1/(r \sin \theta)$$

All the above equilibrium equations define the conditions which must be satisfied by each interior element of a body. In addition, these stresses must satisfy all surface-stress-boundary conditions. In addition to the cartesian-, cylindrical-, and spherical-coordinate systems, others may be found in the current literature or obtained by reduction from the general curvilinear-coordinate equations given above.

In many applications it is useful to integrate the stresses over a finite thickness and express the resultant in terms of zero or nonzero force or moment resultants as in the beam, plate, or shell theories.

Two-Dimensional Case—Plane Stress.[2] In the special but useful case where the stresses in one of the coordinate directions are negligibly small ($\sigma_z = \tau_{xz} = \tau_{yz} = 0$) the general cartesian-coordinate equilibrium equations reduce to

$$\partial\sigma_x/\partial x + \partial\tau_{xy}/\partial y = 0 \qquad (2.23)$$

$$\partial\sigma_y/\partial y + \partial\tau_{yx}/\partial x = 0 \qquad (2.24)$$

The corresponding cylindrical-coordinate equilibrium equations become

$$\partial\sigma_r/\partial r + (1/r)(\partial\tau_{r\theta}/\partial\theta) + (\sigma_r - \sigma_\theta)/r = 0 \qquad (2.25)$$

$$\partial\tau_{r\theta}/\partial r + (1/r)(\partial\sigma_\theta/\partial\theta) + 2(\tau_{r\theta}/r) = 0 \qquad (2.26)$$

This situation arises in "thin slabs," as indicated in Fig. 2.9, which are essentially two-dimensional problems. Because these equations are used in formulations which allow only stresses in the "plane" of the slab, they are classified as "plane-stress" equations.

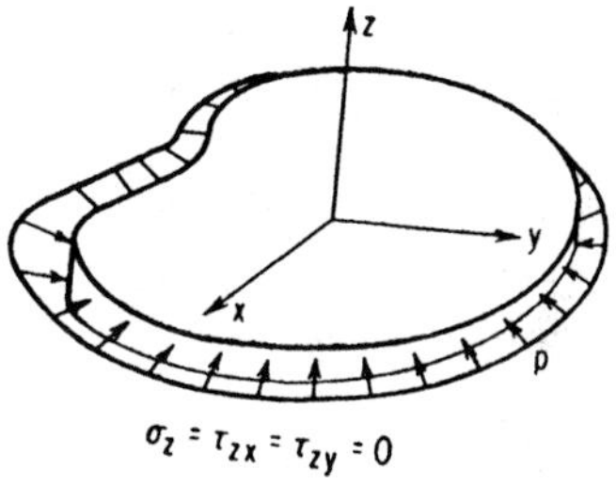

FIG. 2.9 Plane stress on a thin slab.

2.2.7 Stress Transformation: Three-Dimensional Case[4,5]

It is frequently necessary to determine the stresses at a point in an element which is rotated with respect to the *x, y, z* coordinate system, i.e., in an orthogonal *x´, y´, z´* system. Using equilibrium concepts and measuring the angle between any specific original and rotated coordinate by the direction cosines (cosine of the angle between the two axes) the following transformation equations result:

$$\begin{aligned}\sigma_{x'} = {} & [\sigma_x \cos(x'x) + \tau_{xy} \cos(x'y) + \tau_{zx} \cos(x'z)] \cos(x'x) \\ & + [\tau_{xy} \cos(x'x) + \sigma_y \cos(x'y) + \tau_{yz} \cos(x'z)] \cos(x'y) \\ & + [\tau_{zx} \cos(x'x) + \tau_{yz} \cos(x'y) + \sigma_z \cos(x'z)] \cos(x'z) \qquad (2.27)\end{aligned}$$

$$\begin{aligned}\sigma_{y'} = {} & [\sigma_x \cos(y'x) + \tau_{xy} \cos(y'y) + \tau_{zx} \cos(y'z)] \cos(y'x) \\ & + [\tau_{xy} \cos(y'x) + \sigma_y \cos(y'y) + \tau_{yz} \cos(y'z)] \cos(y'y) \\ & + [\tau_{zx} \cos(y'x) + \tau_{yz} \cos(y'y) + \sigma_z \cos(y'z)] \cos(y'z) \qquad (2.28)\end{aligned}$$

$$\begin{aligned}\sigma_{z'} = {} & [\sigma_x \cos(z'x) + \tau_{xy} \cos(z'y) + \tau_{zx} \cos(z'z)] \cos(z'x) \\ & + [\tau_{xy} \cos(z'x) + \sigma_y \cos(z'y) + \tau_{yz} \cos(z'z)] \cos(z'y) \\ & + [\tau_{zx} \cos(z'x) + \tau_{yz} \cos(z'y) + \sigma_z \cos(z'z)] \cos(z'z) \qquad (2.29)\end{aligned}$$

$$\begin{aligned}\tau_{x'y'} = {} & [\sigma_x \cos(y'x) + \tau_{xy} \cos(y'y) + \tau_{zx} \cos(y'z)] \cos(x'x) \\ & + [\tau_{xy} \cos(y'x) + \sigma_y \cos(y'y) + \tau_{yz} \cos(y'z)] \cos(x'y) \\ & + [\tau_{zx} \cos(y'x) + \tau_{yz} \cos(y'y) + \sigma_z \cos(y'z)] \cos(x'z) \qquad (2.30)\end{aligned}$$

$$\tau_{y'z'} = [\sigma_x \cos(z'x) + \tau_{xy} \cos(z'y) + \tau_{zx} \cos(z'z)] \cos(y'x)$$
$$+ [\tau_{xy} \cos(z'x) + \sigma_y \cos(z'y) + \tau_{yz} \cos(z'z)] \cos(y'y)$$
$$+ [\tau_{zx} \cos(z'x) + \tau_{yz} \cos(z'y) + \sigma_z \cos(z'z)] \cos(y'z) \quad (2.31)$$

$$\tau_{z'x'} = [\sigma_x \cos(x'x) + \tau_{xy} \cos(x'y) + \tau_{zx} \cos(x'z)] \cos(z'x)$$
$$+ [\tau_{xy} \cos(x'x) + \sigma_y \cos(x'y) + \tau_{yz} \cos(x'z)] \cos(z'y)$$
$$+ [\tau_{zx} \cos(x'x) + \tau_{yz} \cos(x'y) + \sigma_z \cos(x'z)] \cos(z'z) \quad (2.32)$$

In tensor notation these can be abbreviated as

$$\tau_{k'l'} = A_{l'n} A_{k'm} \tau_{mn} \quad (2.33)$$

where $A_{ij} = \cos(ij) \qquad m,n \rightarrow x,y,z \qquad k',l' \rightarrow x',y',z'$

A special but very useful coordinate rotation occurs when the direction cosines are so selected that all the shear stresses vanish. The remaining mutually perpendicular "normal stresses" are called "principal stresses."

The magnitudes of the principal stresses $\sigma_{x'}$, $\sigma_{y'}$, σ_z are the three roots of the cubic equations associated with the determinant

$$\begin{vmatrix} \sigma_x - \sigma & \tau_{xy} & \tau_{zx} \\ \tau_{xy} & \sigma_y - \sigma & \tau_{yz} \\ \tau_{zx} & \tau_{yz} & \sigma_z - \sigma \end{vmatrix} = 0 \quad (2.34)$$

where $\sigma_x, \ldots, \tau_{xy}, \ldots$ are the general nonprincipal stresses which exist on an element.

The direction cosines of the principal axes x', y' z' with respect to the x, y, z axes are obtained from the simultaneous solution of the following three equations considering separately the cases where $n = x'$, y' z':

$$\tau_{xy} \cos(xn) + (\sigma_y - \sigma_n) \cos(yn) + \tau_{yz} \cos(zn) = 0 \quad (2.35)$$

$$\tau_{zx} \cos(xn) + \tau_{yz} \cos(yn) + (\sigma_z - \sigma_n) \cos(zn) = 0 \quad (2.36)$$

$$\cos^2(xn) + \cos^2(yn) + \cos^2(zn) = 1 \quad (2.37)$$

2.2.8 Stress Transformation: Two-Dimensional Case[2,4]

Selecting an arbitrary coordinate direction in which the stress components vanish, it can be shown, either by equilibrium considerations or by general transformation formulas, that the two-dimensional stress-transformation equations become

$$\sigma_n = [(\sigma_x + \sigma_y)/2] + [(\sigma_x - \sigma_y)/2] \cos 2\alpha + \tau_{xy} \sin 2\alpha \quad (2.38)$$

$$\tau_{nt} = [(\sigma_x - \sigma_y)/2] \sin 2\alpha - \tau_{xy} \cos 2\alpha \quad (2.39)$$

where the directions are defined in Figs. 2.10 and 2.11 ($\tau_{xy} = -\tau_{nt}$, $\alpha = 0$).

The principal directions are obtained from the condition that

$$\tau_{nt} = 0 \qquad \text{or} \qquad \tan 2\alpha = 2\tau_{xy}/(\sigma_x - \sigma_y) \quad (2.40)$$

where the two lowest roots of (first and second quadrants) are taken. It can be easily seen that the first and second principal directions differ by 90°. It can be shown that the *principal* stresses are also the "maximum" or "minimum normal stresses." The "plane of maximum shear" is defined by

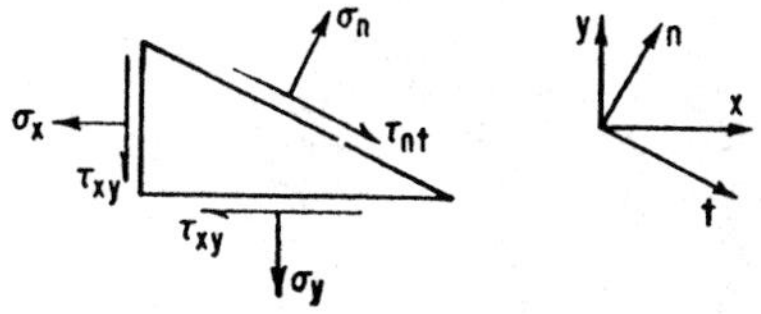

FIG. 2.10 Two-dimensional plane stress.

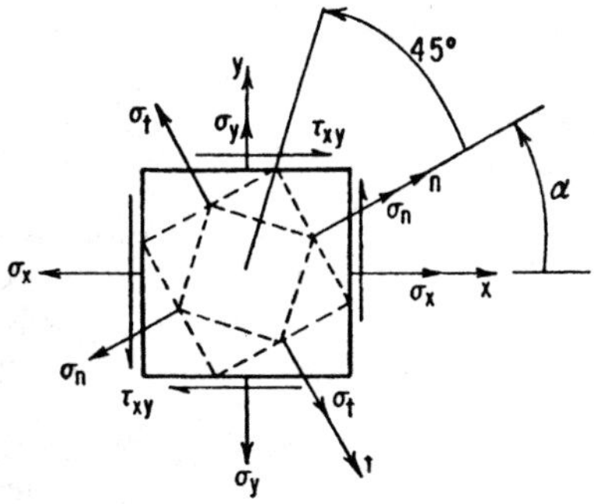

FIG. 2.11 Plane of maximum shear.

$$\tan 2\alpha = -(\sigma_x - \sigma_y)/2\tau_{xy} \tag{2.41}$$

These are also represented by planes which are 90° apart and are displaced from the principal stress planes by 45° (Fig. 2.11).

2.2.9 Mohr's Circle

Mohr's circle is a convenient representation of the previously indicated transformation equations. Considering the x, y directions as positive in Fig. 2.11, the stress condition on any elemental plane can be represented as a point in the "Mohr diagram" (clockwise shear taken positive). The Mohr's circle is constructed by connecting the two stress points and drawing a circle through them with center on the σ axis. The stress state of any basic element can be represented by the stress coordinates at the intersection of the circle with an arbitrarily directed line through the circle center. Note that point x for positive τ_{xy} is below the σ axis and vice versa. The element is taken as rotated counterclockwise by an angle α with respect to the x-y element when the line is rotated counterclockwise an angle 2α with respect to the x-y line, and vice versa (Fig. 2.12).

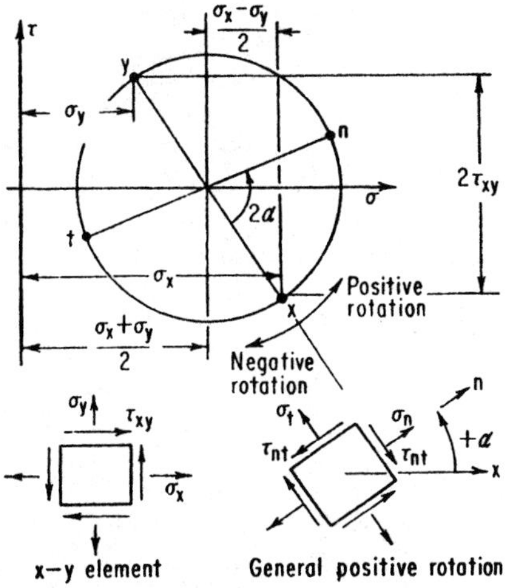

FIG. 2.12 Stress state of basic element.

2.3 STRAIN

2.3.1 Definition[2]

Extensional strain ϵ is defined as the extensional deformation of an element divided by the basic elemental length, $\epsilon = u/l_0$.

In large-strain considerations, l_0 must represent the instantaneous elemental length and the definitions of strain must be given in incremental fashion. In small strain considerations, to which the following discussion is limited, it is only necessary to consider the original elemental length l_0 and its change of length u. Extensional strain is taken positive or negative depending on whether the element increases or decreases in extent. The units of strain are dimensionless (inches/inch).

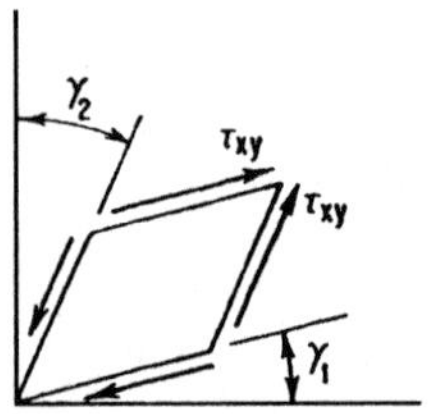

FIG. 2.13 Shear-strain-deformed element.

"Shear strain" γ is defined as the angular distortion of an original right-angle element. The direction of positive shear strain is taken to correspond to that produced by a positive shear stress (and vice versa) (see Fig. 2.13). Shear strain γ is equal to $\gamma_1 + \gamma_2$. The "units" of shear strain are dimensionless (radians).

2.3.2 Components of Strain[2]

A complete description of strain requires the establishment of three orthogonal extensional and shear strains. In cartesian stress nomenclature, the strain components are

Extensional strain:

$$\epsilon_{xx} \equiv \epsilon_x \qquad \epsilon_{yy} \equiv \epsilon_y \qquad \epsilon_{zz} \equiv \epsilon_z \tag{2.42}$$

Shear strain:

$$\epsilon_{xy} = \epsilon_{yx} \equiv \tfrac{1}{2}\gamma_{xy} \qquad \epsilon_{yz} = \epsilon_{zy} \equiv \tfrac{1}{2}\gamma_{yz} \qquad \epsilon_{zx} = \epsilon_{xz} \equiv \tfrac{1}{2}\gamma_{zx} \tag{2.43}$$

where positive ϵ_x, ϵ_y, or ϵ_z corresponds to a positive stretching in the x, y, z directions and positive γ_{xy}, γ_{yz}, γ_{zx} refers to positive shearing displacements in the xy, yz, and zx planes. In tensor notation, the strain components are

$$\epsilon_{ij} = \begin{pmatrix} \epsilon_x & \tfrac{1}{2}\gamma_{xy} & \tfrac{1}{2}\gamma_{zx} \\ \tfrac{1}{2}\gamma_{xy} & \epsilon_y & \tfrac{1}{2}\gamma_{yz} \\ \tfrac{1}{2}\gamma_{zx} & \tfrac{1}{2}\gamma_{yz} & \epsilon_z \end{pmatrix} \tag{2.44}$$

2.3.3 Simple and Nonuniform States of Strain[2]

Corresponding to each of the stress states previously illustrated there exists either a simple or nonuniform strain state.

In addition to these, a state of "uniform dilatation" exists when the shear strain vanishes and all the extensional strains are equal in sign and magnitude. Dilatation is defined as

$$\Delta = \epsilon_x + \epsilon_y + \epsilon_z \tag{2.45}$$

and represents the change of volume per increment volume.

In uniform dilatation,

$$\Delta = 3\epsilon_x = 3\epsilon_y = 3\epsilon_z \tag{2.46}$$

2.3.4 Strain-Displacement Relationships[4,5,13]

Considering only small strain, and the previous definitions, it is possible to express the strain components at a point in terms of the associated displacements and their derivatives in the coordinate directions (e.g., u, v, w are displacements in the x, y, z coordinate system).

Thus, in a "cartesian system" (x, y, z),

$$\begin{aligned} \epsilon_x &= \partial u/\partial x & \gamma_{xy} &= \partial v/\partial x + \partial u/\partial y \\ \epsilon_y &= \partial v/\partial y & \gamma_{yz} &= \partial w/\partial y + \partial v/\partial z \\ \epsilon_z &= \partial w/\partial z & \gamma_{zx} &= \partial u/\partial z + \partial w/\partial x \end{aligned} \tag{2.47}$$

or, in stress-tensor notation,

$$2\epsilon_{ij} = u_{i,j} + u_{j,i} \qquad i,j \rightarrow x,y,z \tag{2.48}$$

In addition the dilatation

$$\Delta = \partial u/\partial x + \partial v/\partial y + \partial w/\partial z \tag{2.49}$$

or, in tensor form,

$$\Delta = u_{i,j} \qquad i \rightarrow x,y,z \tag{2.50}$$

Finally, all incremental displacements can be composed of a "pure strain" involving all the above components, plus "rigid-body" rotational components. That is, in general

$$U = \epsilon_x X + ½\gamma_{xy} Y + ½\gamma_{zx} Z - \bar{\omega}_z Y + \bar{\omega}_y Z \tag{2.51}$$

$$V = ½\gamma_{xy} X + \epsilon_y Y + ½\gamma_{yz} Z - \bar{\omega}_x Z + \bar{\omega}_z X \tag{2.52}$$

$$W = ½\gamma_{zx} X + ½\gamma_{yz} Y + \epsilon_z Z - \bar{\omega}_y X + \bar{\omega}_x Y \tag{2.53}$$

where U, V, W represent the incremental displacement of the point $x + X$, $y + Y$, $z + Z$ in excess of that of the point x, y, z where X, Y, Z are taken as the sides of the incremental element. The rotational components are given by

$$\begin{aligned} 2\bar{\omega}_x &= \partial w/\partial y - \partial v/\partial z \\ 2\bar{\omega}_y &= \partial u/\partial z - \partial w/\partial x \\ 2\bar{\omega}_z &= \partial v/\partial x - \partial u/\partial y \end{aligned} \tag{2.54}$$

or, in tensor notation,

$$2\overline{\omega}_{ij} = u_{i,j} - u_{j,i} \qquad i,j = x,y,z \tag{2.55}$$

$$\overline{\omega}_{zy} \equiv \overline{\omega}_x,\ \overline{\omega}_{xz} \equiv \overline{\omega}_y,\ \omega_{yx} \equiv \overline{\omega}_z$$

In *cylindrical coordinates,*

$$\begin{aligned} \epsilon_r &= \partial u_r/\partial r & \gamma_{\theta z} &= (1/r)(\partial u_z/\partial\theta + \partial u_\theta/\partial z \\ \epsilon_\theta &= (1/r)(\partial u_\theta/\partial\theta) + u_r/r & \gamma_{zr} &= \partial u_r/\partial z + \partial u_z/\partial r \\ \epsilon_z &= \partial u_z/\partial z & \gamma_{r\theta} &= \partial u_\theta/\partial r - u_\theta/r + (1/r)(\partial u_r/\partial\theta) \end{aligned} \tag{2.56}$$

The dilatation is

$$\Delta = (1/r)(\partial/\partial r)(ru_r) + (1/r)(\partial u_\theta/\partial\theta) + \partial u_z/\partial z \tag{2.57}$$

and the rotation components are

$$\begin{aligned} 2\overline{\omega}_r &= (1/r)(\partial u_z/\partial\theta) - \partial u_\theta/\partial z \\ 2\overline{\omega}_\theta &= \partial u_r/\partial z - \partial u_z/\partial r \\ 2\overline{\omega}_z &= (1/r)(\partial/\partial r)(ru_\theta) - (1/r)(\partial u_r/\partial\theta) \end{aligned} \tag{2.58}$$

In *spherical polar coordinates,*

$$\begin{aligned} \epsilon_r &= \frac{\partial u_r}{\partial r} & \gamma_{\theta\phi} &= \frac{1}{r}\left[\frac{\partial u_\phi}{\partial\theta} - u_\phi \cot\theta\right] + \frac{1}{r\sin\theta}\frac{\partial u_\theta}{\partial\phi} \\ \epsilon_\theta &= \frac{1}{r}\frac{\partial u_\theta}{\partial\theta} + \frac{u_r}{r} & \gamma\,\phi_r &= \frac{1}{r\sin\theta}\frac{\partial u_r}{\partial\phi} + \frac{\partial u_\phi}{\partial r} - \frac{u_\phi}{r} \\ \epsilon_\phi &= \frac{1}{r\sin\theta}\frac{\partial u_\phi}{\partial\phi} + \frac{u_\theta}{r}\cot\theta + \frac{u_r}{r} & \gamma_{r\theta} &= \frac{\partial u_\theta}{\partial r} - \frac{u_\theta}{r} + \frac{1}{r}\frac{\partial u_r}{\partial\theta} \end{aligned} \tag{2.59}$$

The dilatation is

$$\Delta = (1/r^2\sin\theta)[(\partial/\partial r)(r^2u_r\sin\theta) + (\partial/\partial\theta)(ru_\theta\sin\theta) + (\partial/\partial\phi)(ru_\phi)] \tag{2.60}$$

The rotation components are

$$\begin{aligned} 2\overline{\omega}_r &= (1/r^2\sin\theta)[(\partial/\partial\theta)(ru_\phi\sin\theta) - (\partial/\partial\phi)(ru_\theta)] \\ 2\overline{\omega}_\theta &= (1/r\sin\theta)[\partial u_r/\partial\phi - (\partial/\partial r)(ru_\phi\sin\theta)] \\ 2\overline{\omega}_\phi &= (1/r)[(\partial/\partial r)(ru_\theta) - \partial u_r/\partial\theta] \end{aligned} \tag{2.61}$$

In general *orthogonal curvilinear coordinates,*

$$\begin{aligned} \epsilon_\alpha &= h_1(\partial u_\alpha/\partial\alpha) + h_1h_2u_\beta(\partial/\partial\beta)(1/h_1) + h_3h_1u_\gamma(\partial/\partial\gamma)(1/h_1) \\ \epsilon_\beta &= h_2(\partial u_\beta/\partial\beta) + h_2h_3u_\gamma(\partial/\partial\gamma)(1/h_2) + h_1h_2u_\alpha(\partial/\partial\alpha)(1/h_2) \\ \epsilon_\gamma &= h_3(\partial u_\gamma/\partial\gamma) + h_3h_1u_\alpha(\partial/\partial\alpha)(1/h_3) + h_2h_3u_\beta(\partial/\partial\beta)(1/h_3) \\ \gamma_{\beta\gamma} &= (h_2/h_3)(\partial/\partial\beta)(h_3u_\gamma) + (h_3/h_2)(\partial/\partial\gamma)(h_2u_\beta) \\ \gamma_{\gamma\alpha} &= (h_3/h_1)(\partial/\partial\gamma)(h_1u_\alpha) + (h_1/h_3)(\partial/\partial\alpha)(h_3u_\gamma) \\ \gamma_{\alpha\beta} &= (h_1/h_2)(\partial/\partial\alpha)(h_2u_\beta) + (h_2/h_1)(\partial/\partial\beta)(h_1u_\alpha) \end{aligned} \tag{2.62}$$

$$\Delta = h_1h_2h_3[(\partial/\partial\alpha)(u_\alpha/h_2h_3) + (\partial/\partial\beta)(u_\beta/h_3h_1) + (\partial/\partial\gamma)(u_\gamma/h_1h_2)] \tag{2.63}$$

$$\begin{aligned} 2\bar{\omega}_\alpha &= h_2h_3[(\partial/\partial\beta)(u_\gamma/h_3) - (\partial/\partial\gamma)(u_\beta/h_2)] \\ 2\bar{\omega}_\beta &= h_3h_1[(\partial/\partial\gamma)(u_\alpha/h_1) - (\partial/\partial\alpha)(u_\gamma/h_3)] \\ 2\bar{\omega}_\gamma &= h_1h_2[(\partial/\partial\alpha)(u_\beta/h_2) - (\partial/\partial\beta)(u_\alpha/h_1)] \end{aligned} \tag{2.64}$$

where the quantities h_1, h_2, h_3 have been discussed with reference to the equilibrium equations.

In the event that one deflection (i.e., w) is constant or zero and the displacements are a function of x, y only, a special and useful class of problems arises termed "plane strain," which are analogous to the "plane-stress" problems. A typical case of plane strain occurs in slabs rigidly clamped on their faces so as to restrict all axial deformation. Although all the stresses may be nonzero, and the general equilibrium equations apply, it can be shown that, after combining all the necessary stress and strain relationships, both classes of *plane* problems yield the same form of equations. From this, one solution suffices for both the related *plane-stress* and *plane-strain* problems, provided that the elasticity constants are suitably modified. In particular the applicable strain-displacement relationships reduce in *cartesian coordinates* to

$$\begin{aligned} \epsilon_x &= \partial u/\partial x \\ \epsilon_y &= \partial v/\partial y \\ \gamma_{xy} &= \partial v/\partial x + \partial u/\partial y \end{aligned} \tag{2.65}$$

and in *cylindrical coordinates* to

$$\begin{aligned} \epsilon_r &= \partial u_r/\partial r \\ \epsilon_\theta &= (1/r)(\partial u_\theta/\partial\theta) + u_r/r \\ \gamma_{r\theta} &= \partial u_\theta/\partial r - u_\theta/r + (1/r)(\partial u_r/\partial\theta) \end{aligned} \tag{2.66}$$

2.3.5 Compatibility Relationships[2,4,5]

In the event that a single-valued continuous-displacement field (u, v, w) is not explicitly specified, it becomes necessary to ensure its existence in solution of the stress, strain, and stress-strain relationships. By writing the strain-displacement relationships and manipulating them to eliminate displacements, it can be shown that the following six equations are both necessary and sufficient to ensure compatibility:

$$\begin{aligned} &\partial^2\epsilon_y/\partial z^2 + \partial^2\epsilon_z/\partial y^2 = \partial^2\gamma_{yz}/\partial y\,\partial z \\ &\quad 2(\partial^2\epsilon_x/\partial y\,\partial z) = (\partial/\partial x)(-\partial\gamma_{yz}/\partial x + \partial\gamma_{zx}/\partial y + \partial\gamma_{xy}/\partial z) \end{aligned} \tag{2.67}$$

$$\begin{aligned} &\partial^2\epsilon_z/\partial x^2 + \partial^2\epsilon_x/\partial z^2 = \partial^2\gamma_{zx}/\partial x\,\partial z \\ &\quad 2(\partial^2\epsilon_y/\partial z\,\partial x) = (\partial/\partial y)(+\partial\gamma_{yz}/\partial x - \partial\gamma_{zx}/\partial y + \partial y_{xy}/\partial z) \end{aligned} \tag{2.68}$$

$$\begin{aligned} &\partial^2\epsilon_x/\partial y^2 + \partial^2\epsilon_y/\partial x^2 = \partial^2\gamma_{xy}/\partial x\,\partial y \\ &\quad 2(\partial^2\epsilon_z/\partial x\,\partial y) = (\partial/\partial z)(+\partial\gamma_{yz}/\partial x + \partial\gamma_{zx}/\partial y - \partial\gamma_{xy}/\partial z) \end{aligned} \tag{2.69}$$

In tensor notation the most general compatibility equations are

$$\epsilon_{ij,kl} + \epsilon_{kl,ij} - \epsilon_{ik,jl} - \epsilon_{jl,ik} = 0 \qquad i,j,k,l = x,y,z \tag{2.70}$$

which represents 81 equations. Only the above six equations are essential.

In addition to satisfying these conditions everywhere in the body under consideration, it is also necessary that all *surface strain* or *displacement boundary conditions* be satisfied.

2.3.6 Strain Transformation[4,5]

As with stress, it is frequently necessary to refer strains to a rotated orthogonal coordinate system (x', y', z'). In this event it can be shown that the stress and strain tensors transform in an identical manner.

$$\sigma_{x'} \rightarrow \epsilon_{x'} \qquad \sigma_x \rightarrow \epsilon_x \qquad \tau_{x'y'} \rightarrow \tfrac{1}{2}\gamma_{x'y'} \qquad \tau_{xy} \rightarrow \tfrac{1}{2}\gamma_{xy}$$

$$\sigma_{y'} \rightarrow \epsilon_{y'} \qquad \sigma_y \rightarrow \epsilon_y \qquad \tau_{y'z'} \rightarrow \tfrac{1}{2}\gamma_{y'z'} \qquad \tau_{yz} \rightarrow \tfrac{1}{2}\gamma_{yz}$$

$$\sigma_{z'} \rightarrow \epsilon_{z'} \qquad \sigma_z \rightarrow \epsilon_z \qquad \tau_{z'x'} \rightarrow \tfrac{1}{2}\gamma_{z'x'} \qquad \tau_{zx} \rightarrow \tfrac{1}{2}\gamma_{zx}$$

In tensor notation the strain transformation can be written as

$$e_{k'l'} = A_{l'n} A_{k'm} \epsilon_{mn} \qquad \begin{matrix} m, n \rightarrow x, y, z \\ l'k' \rightarrow x', y' z' \end{matrix} \tag{2.71}$$

As a result the stress and strain principal directions are coincident, so that all remarks made for the principal stress and maximum shear components and their directions

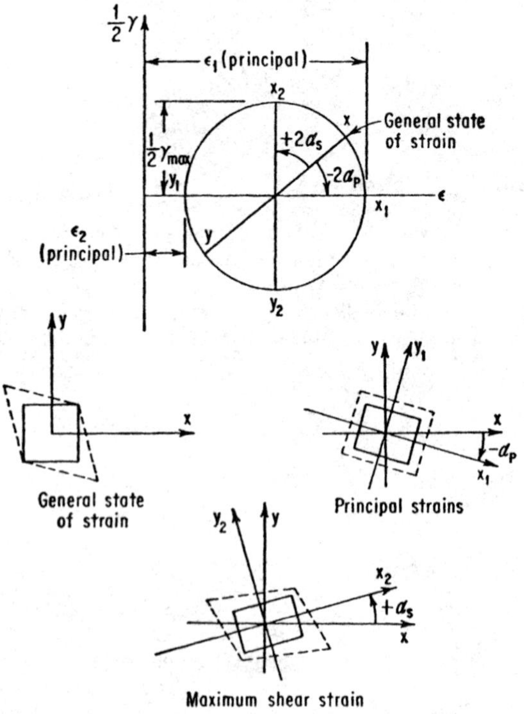

FIG. 2.14 Strain transformation.

apply equally well to strain tensor components. Note that in the use of Mohr's circle in the two-dimensional case one must be careful to substitute $\frac{1}{2}\gamma$ for τ in the *ordinate* and ϵ for σ in the abscissa (Fig. 2.14).

2.4 STRESS-STRAIN RELATIONSHIPS

2.4.1 Introduction[2]

It can be experimentally demonstrated that a one-to-one relationship exists between uniaxial stress and strain during a single loading. Further, if the material is always loaded within its *elastic* or *reversible* range, a one-to-one relationship exists for all loading and unloading cycles.

For stresses below a certain characteristic value termed the "proportional limit," the stress-strain relationship is very nearly linear. The stress beyond which the stress-strain relationship is no longer reversible is called the "elastic limit." In most materials the proportional and elastic limits are identical. Because the departure from linearity is very gradual it is often necessary to prescribe arbitrarily an "apparent" or "offset elastic limit." This is obtained as the intersection of the stress-strain curve with a line parallel to the linear stress-strain curve, but offset by a prescribed amount, e.g., 0.02 percent (see Fig. 2.15*a*). The "yield point" is the value of stress at which continued deformation of the bar takes place with little or no further increase in load, and the "ultimate limit" is the maximum stress that the specimen can withstand.

Note that some materials may show no clear difference between the apparent elastic, inelastic, and proportional limits or may not show clearly defined yield points (Fig. 2.15*b*).

The concept that a useful linear range exists for most materials and that a simple mathematical law can be formulated to describe the relationship between stress and strain in this range is termed "Hooke's law." It is an essential starting point in the "small-strain theory of elasticity" and the associated mechanics of materials. In the above-described tensile specimen, the law is expressed as

$$\sigma = E\epsilon \tag{2.72}$$

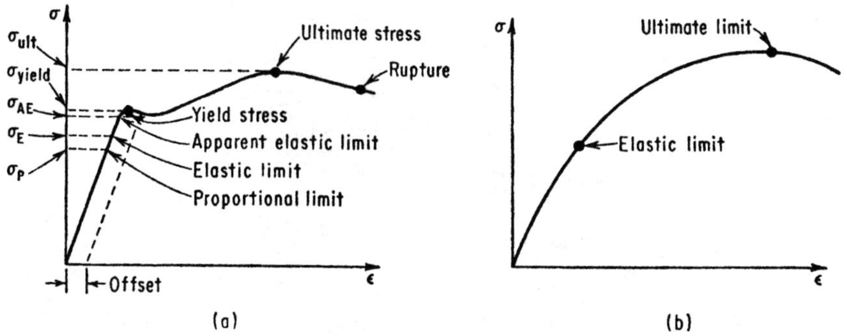

FIG. 2.15 Stress-strain relationship.

as in the analogous torsional specimen

$$\tau = G\gamma \tag{2.73}$$

where E and G are the slope of the appropriate stress-strain diagrams and are called the "Young's modulus" and the "shear modulus" of elasticity, respectively.

2.4.2 General Stress-Strain Relationship[2,4,5]

The one-dimensional concepts discussed above can be generalized for both small and large strain and elastic and nonelastic materials. The following discussion will be limited to small-strain elastic materials consistent with much engineering design. Based upon the above, Hooke's law is expressed as

$$\begin{aligned} \epsilon_x &= (1/E)[\sigma_x - \nu(\sigma_y + \sigma_z)] & \gamma_{xy} &= \tau_{xy}/G \\ \epsilon_y &= (1/E)[\sigma_y - \nu(\sigma_z + \sigma_x)] & \gamma_{yz} &= \tau_{yz}/G \\ \epsilon_z &= (1/E)[\sigma_z - \nu(\sigma_x + \sigma_y)] & \gamma_{zx} &= \tau_{zx}/G \end{aligned} \tag{2.74}$$

where ν is "Poisson's ratio," the ratio between longitudinal strain and lateral contraction in a simple tensile test.

In *cartesian tension form* Eq. (2.74) is expressed as

$$\epsilon_{ij} = [(1 + \nu)/E]\sigma_{ij} - (\nu/E)\delta_{ij}\sigma_{kk} \qquad i,j,k = x,y,z \tag{2.75}$$

where
$$\delta_{ij} = \begin{cases} 0 & i \neq j \\ 1 & i = j \end{cases}$$

The stress-strain laws appear in inverted form as

$$\begin{aligned} \sigma_x &= 2G\epsilon_x + \lambda\,\Delta \\ \sigma_y &= 2G\epsilon_y + \lambda\,\Delta \\ \sigma_z &= 2G\epsilon_z + \lambda\,\Delta \\ \tau_{xy} &= G\gamma_{xy} \\ \tau_{yz} &= G\gamma_{yz} \\ \tau_{zx} &= G\gamma_{zx} \end{aligned} \tag{2.76}$$

where
$$\lambda = \nu/\mathrm{E}/(1 + \nu)(1 - 2\nu)$$
$$\Delta = \epsilon_x + \epsilon_y + \epsilon_z$$
$$G = E/2(1 + \nu)$$

In *cartesian tensor form* Eq. (2.76) is written as

$$\sigma_{ij} = 2G\epsilon_{ij} + \lambda\,\Delta\,\delta_{ij} \qquad i,j = x,y,z \tag{2.77}$$

and in general tensor form as

$$\sigma_{ij} = 2G\epsilon_{ij} + \lambda\,\Delta g_{ij} \tag{2.78}$$

where g_{ij} is the "covariant metric tensor" and these coefficients (stress modulus) are often referred to as "Lamé's constants," and $\Delta = g^{mn}\epsilon_{mn}$.

2.5 STRESS-LEVEL EVALUATION

2.5.1 Introduction[1,6]

The detailed elastic and plastic behavior, yield and failure criterion, etc., are repeatable and simply describable for a simple loading state, as in a tensile or torsional specimen. Under any complex loading state, however, no single stress or strain component can be used to describe the *stress state* uniquely; that is, the *yield, flow,* or *rupture* criterion must be obtained by some combination of all the stress and/or strain components, their derivatives, and loading history. In elastic theory the "yield criterion" is related to an "equivalent stress," or "equivalent strain." It is conventional to treat the *stress* criteria.

An "equivalent stress" is defined in terms of the "stress components" such that plastic flow will commence in the body at any position at which this equivalent stress just exceeds the one-dimensional yield-stress value, for the material under consideration. That is, yielding commences when

$$\sigma_{\text{equivalent}} \geq \sigma_E$$

The "elastic safety factor" at a point is defined as the ratio of the one-dimensional yield stress to the equivalent stress at that position, i.e.,

$$n_i = \sigma_E/\sigma_{\text{equivalent}} \tag{2.79}$$

and the elastic safety factor for the entire structure under any specific loading state is taken as the lowest safety factor of consequence that exists anywhere in the structure.

The "margin of safety," defined as $n - 1$, is another measure of the proximity of any structure to yielding. When $n > 1$, the structure has a *positive* margin of safety and will not yield. When $n = 1$, the margin of safety is zero and the structure just yields. When n <1, the margin of safety is *negative* and the structure is considered unsafe. Note that highly localized yielding is often permitted in ductile materials if it is of such nature as to redistribute stresses without failure, building up a "residual stress state" which allows all subsequent loadings to be accomplished with elastic-stress states. This is the basic concept used in the "autofrettage process" in the strengthening of gun tubes, and it also explains why ductile materials often have low "notch sensitivity" (see Chap. 6).

2.5.2 Effective Stress[1,6]

The concept of effective stress is very closely connected with yield criteria. Geometrically it can be shown that a unique surface can be constructed in stress space in terms of principal stresses (σ_1, σ_2, σ_3) such that all nonyielding states of stress lie within that surface, and yielding states lie on or outside the surface. For *ductile materials* the yield surface may be taken as an infinitely long right cylinder having a center axis defined by $\sigma_1 = \sigma_2 = \sigma_3$. Because the yield criterion is represented by a right cylinder it is adequate to define the yield curve, which is the intersection with a plane normal to the axis of the cylinder. Many yield criteria exist; among these, the "Mises criterion" takes this curve as a circle, and the "Tresca criterion" as a regular hexagon

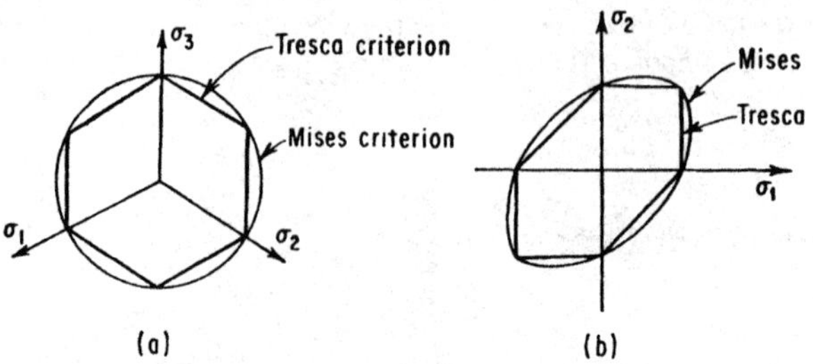

FIG. 2.16 Tresca and Mises criteria. (*a*) General yield criteria. (*b*) One principal stress constant.

(see Fig. 2.16). The former is often referred to by names such as "Hencky," "Hencky-Mises," or "distortion-energy criterion" and the latter by "shear criterion."

Considering any one of the principal stresses as constant, the "yield locus" can be represented by the intersection of the plane σ_i = constant with the cylindrical yield surface, which is represented as an ellipse for the Mises criterion and an elongated hexagon for the Tresca criterion (Fig. 2.16*b*).

In general the yield criteria indicate, upon experimental evidence for a ductile material, that yielding is essentially independent of "hydrostatic compression" or tension for the loadings usually considered in engineering problems. That is, yielding depends only on the "deviatoric" stress component. In general, this can be represented as

$$\sigma'_k = \sigma_k - \tfrac{1}{3}(\sigma_1 + \sigma_2 + \sigma_3) \tag{2.80}$$

where σ' is the principal deviatoric stress component and σ the actual principal stress.

In tensor form

$$\sigma'_{ij} = \sigma_{ij} - \tfrac{1}{3}\sigma_{kk}\delta_{ij} \tag{2.81}$$

The analytical representation of the yield criteria can be shown to be a function of the "deviatoric stress-tensor invariants." In component form these can be expressed as follows, where σ_0 is the yield stress in simple tension:

Mises:

$$\sigma_0 = (1/\sqrt{2})\sqrt{(\sigma_1 - \sigma_2)^2 + (\sigma_2 - \sigma_3)^2 + (\sigma_3 - \alpha_1)^2} \tag{2.82}$$

where σ_1, σ_2, σ_3 are principal stresses.

$$\sigma_0 = (1/\sqrt{2})\sqrt{(\sigma_x - \sigma_y)^2 + (\sigma_y - \sigma_z)^2 + (\sigma_z - \sigma_x)^2 + 6(\tau_{xy}^2 + \tau_{yz}^2 + \tau_{zx}^2)} \tag{2.83}$$

or, in tensor form,

$$\tfrac{2}{3}\sigma_0^2 = \sigma'_{ij}\sigma'_{ij} \tag{2.84}$$

Tresca:

$$\sigma_0 = \sigma_1 - \sigma_3 \tag{2.85}$$

where $\sigma_1 > \sigma_2 > \sigma_3$, or, in general symmetric terms,

$$[(\sigma_1 - \sigma_3)^2 - \sigma_0^2][(\sigma_2 - \sigma_1)^2 - \sigma_0^2][(\sigma_3 - \sigma_2)^2 - \sigma_0^2] = 0 \tag{2.86}$$

2.6 FORMULATION OF GENERAL MECHANICS-OF-MATERIAL PROBLEM

2.6.1 Introduction[2,4,5]

Generally the mechanics-of-material problem is stated as follows: Given a prescribed structural configuration, and surface tractions and/or displacements, find the stresses and/or displacements at any, or all, positions in the body. Additionally it is often desired to use the derived stress information to determine the *maximum load-carrying capacity* of the structure, prior to yielding. This is usually referred to as the problem of *analysis*. Alternatively the problem may be inverted and stated: Given a set of surface tractions and/or displacements, find the geometrical configuration for a constraint such as minimum weight, subject to the yield criterion (or some other general stress or strain limitation). This latter is referred to as the *design* problem.

2.6.2 Classical Formulation[2,4,5]

The classical formulation of the equation for the problem of mechanics of materials is as follows: It is necessary to evaluate the six stress components σ_{ij}, six strain components ϵ_{ij}, and three displacement quantities u_i which satisfy the three equilibrium equations, six strain-displacement relationships, and six stress-strain relationships, all subject to the appropriate stress and/or displacement boundary conditions.

Based on the above discussion and the previous derivations, the most general three-dimensional formulation in cartesian coordinates is

$$\left.\begin{aligned} \partial\sigma_x/\partial x + \partial\tau_{xy}/\partial y + \partial\tau_{xz}/\partial z &= 0 \\ \partial\sigma_y/\partial y + \partial\tau_{yz}/\partial z + \partial\tau_{yx}/\partial x &= 0 \\ \partial\sigma_z/\partial z + \partial\tau_{zx}/\partial x + \partial\tau_{zy}/\partial y &= 0 \end{aligned}\right\} \text{(equilibrium)} \tag{2.87}$$

$$\left.\begin{aligned} \epsilon_x &= \partial u/\partial x & \gamma_{xy} &= \partial v/\partial x + \partial u/\partial y \\ \epsilon_y &= \partial v/\partial y & \gamma_{yz} &= \partial w/\partial y + \partial v/\partial z \\ \epsilon_z &= \partial w/\partial z & \gamma_{zx} &= \partial u/\partial z + \partial w/\partial x \end{aligned}\right\} \text{(strain-displacement)} \tag{2.47}$$

$$\left.\begin{aligned} \epsilon_x &= (1/E)[\sigma_x - \nu(\sigma_y + \sigma_z)] \\ \epsilon_y &= (1/E)[\sigma_y - \nu(\sigma_z + \sigma_x)] \\ \epsilon_z &= (1/E)[\sigma_z - \nu(\sigma_x + \sigma_y)] \\ \gamma_{xy} &= (1/G)\tau_{xy} \\ \gamma_{yz} &= (1/G)\tau_{yz} \\ \gamma_{zx} &= (1/G)\tau_{xz} \end{aligned}\right\} \text{(stress-strain relationships)} \tag{2.88}$$

In cartesian tensor form these appear as

$$\sigma_{ij,j} = 0 \qquad \text{(equilibrium)} \tag{2.89}$$

$$2\epsilon_{ij} = u_{i,j} + u_{j,i} \qquad i,j \rightarrow x,y,z \qquad \text{(strain-displacement)} \tag{2.48}$$

$$\epsilon_{ij} = [(1 + \nu)/E]\sigma_{ij} - (\nu/E)\delta_{ij}\sigma_{kk} \qquad \text{(stress-strain)} \tag{2.90}$$

All are subject to appropriate boundary conditions.

If the boundary conditions are on displacements, then we can define the displacement field, the six components of strain, and the six components of stress uniquely, using the fifteen equations shown above. If the boundary conditions are on stresses, then the solution process yields six strain components from which three unique displacement components must be determined. In order to assure uniqueness, three constraints must be placed on the strain field. These constraints are provided by the compatibility relationships:

$$\left.\begin{aligned}
\partial^2\epsilon_x/\partial y^2 + \partial^2\epsilon_y/\partial x^2 &= \partial^2\gamma_{xy}/\partial x\,\partial y \\
\partial^2\epsilon_y/\partial z^2 + \partial^2\epsilon_z/\partial y^2 &= \partial^2\gamma_{yz}/\partial y\,\partial z \\
\partial^2\epsilon_z/\partial x^2 + \partial^2\epsilon_x/\partial z^2 &= \partial^2\gamma_{zx}/\partial z\,\partial x \\
2(\partial^2\epsilon_x/\partial y\,\partial z) &= (\partial/\partial x)(-\partial\gamma_{yz}/\partial x + \partial\gamma_{zx}/\partial y + \partial\gamma_{xy}/\partial z) \\
2(\partial^2\epsilon_y/\partial z\,\partial x) &= (\partial/\partial y)(\partial\gamma_{yz}/\partial x - \partial\gamma_{zx}/\partial y + \partial\gamma_{xy}/\partial z) \\
2(\partial^2\epsilon_z/\partial x\,\partial y) &= (\partial/\partial z)(\partial\gamma_{yz}/\partial x + \partial\gamma_{zx}/\partial y - \partial\gamma_{xy}/\partial z)
\end{aligned}\right\}\text{(compatibility)} \tag{2.91}$$

In cartesian tensor form,

$$\epsilon_{ij,kl} + \epsilon_{kl,ij} - \epsilon_{ik,jl} - \epsilon_{jl,ik} = 0 \qquad \text{(compatibility)} \tag{2.92}$$

Of the six compatibility equations listed, only three are independent. Therefore, the system can be uniquely solved for the displacement field.

It is possible to simplify the above sets of equations considerably by combining and eliminating many of the unknowns. One such reduction is obtained by eliminating stress and strain:

$$\begin{aligned}
\nabla^2 u + [1/(1 - 2\nu)](\partial\Delta/\partial x) &= 0 \\
\nabla^2 v + [1/(1 - 2\nu)](\partial\Delta/\partial y) &= 0 \\
\nabla^2 w + [1/(1 - 2\nu)](\partial\Delta/\partial z) &= 0
\end{aligned} \tag{2.93}$$

where ∇^2 is the laplacian operator which in cartesian coordinates is $\partial^2/\partial x^2 + \partial^2/\partial y^2 + \partial^2/\partial z^2$; and Δ is the dilatation, which in cartesian coordinates is $\partial u/\partial x + \partial v/\partial y + \partial w/\partial z$.

Using the above general principles, it is possible to formulate completely many of the technical problems of mechanics of materials which appear under special classifications such as "beam theory" and "shell theory." These formulations and their solutions will be treated under "Special Applications."

2.6.3 Energy Formulations[2,4,5]

Alternative useful approaches exist for the problem of mechanics of materials. These are referred to as "energy," "extremum," or "variational" formulations. From a strictly formalistic point of view these could be obtained by establishing the analogous integral equations, subject to various restrictions, such that they reduce to a minimum. This is not the usual approach; instead energy functions U, W are established so that the stress-strain laws are replaced by

$$\sigma_x = \partial U/\partial \epsilon_x \qquad \tau_{xy} = \partial U/\partial \gamma_{xy}$$
$$\sigma_y = \partial U/\partial \epsilon_y \qquad \tau_{yz} = \partial U/\partial \gamma_{yz}$$
$$\sigma_z = \partial U/\partial \epsilon_z \qquad \tau_{zx} = \partial U/\partial \gamma_{zx}$$

or

$$\sigma_{ij} = \partial U/\partial \epsilon_{ij}$$

and

$$\epsilon_x = \partial W/\partial \sigma_x \qquad \gamma_{xy} = \partial W/\partial \tau_{xy}$$
$$\epsilon_y = \partial W/\partial \sigma_y \qquad \gamma_{yz} = \partial W/\partial \tau_{yz}$$
$$\epsilon_z = \partial W/\partial \sigma_z \qquad \gamma_{zx} = \partial W/\partial \tau_{zx}$$

or

$$\epsilon_{ij} = \partial W/\partial \sigma_{ij}$$

The energy functions are given by

$$U = \tfrac{1}{2}[2G(\epsilon_x^2 + \epsilon_y^2 + \epsilon_z^2) + \lambda(\epsilon_x + \epsilon_y + \epsilon_z)^2 + G(\gamma_{xy}^2 + \gamma_{yz}^2 + \gamma_{zx}^2)] \tag{2.94}$$

$$W = \tfrac{1}{2}[(1/E)(\sigma_x^2 + \sigma_y^2 + \sigma_z^2) - (2\nu/E)(\sigma_x\sigma_y + \sigma_y\sigma_z + \sigma_z\sigma_x) + (1/G)(\tau_{xy}^2 + \tau_{yz}^2 + \tau_{zx}^2)] \tag{2.95}$$

The variational principle for strains, or theorem of minimum potential energy, is stated as follows: Among all states of strain which satisfy the strain-displacement relationships and displacement boundary conditions the associated stress state, derivable through the stress-strain relationships, which also satisfies the equilibrium equations, is determined by the minimization of Π where

$$\Pi = \int_{\text{volume}} U\,dV - \int_{\text{surface}} (\bar{p}_x u + \bar{p}_y v + \bar{p}_z w)\,dS \tag{2.96}$$

where $\bar{p}_x, \bar{p}_y, \bar{p}_z$ are the x, y, z components of any prescribed surface stresses.

The analogous variational principle for stresses, or principle of least work, is: Among all the states of stress which satisfy the equilibrium equations and stress boundary conditions, the associated strain state, derivable through the stress-strain relationships, which also satisfies the compatibility equations, is determined by the minimization of I, where

$$I = \int_{\text{volume}} W\,dV - \int_{\text{surface}} (p_x \bar{u} + p_y \bar{v} + p_z \bar{w})\,dS \tag{2.97}$$

where $\bar{u}, \bar{v}, \bar{w}$ are the x, y, z components of any prescribed surface displacements and p_x, p_y, p_z are the surface stresses.

In the above theorems $\Pi_{\min}$ and $I_{\min}$ replace the equilibrium and compatibility relationships, respectively. Their most powerful advantage arises in obtaining approximate solutions to problems which are generally intractable by exact techniques. In this, one usually introduces a limited class of assumed stress or displacement functions for minimization, which in themselves satisfy all other requirements imposed in the statement of the respective theorems. Then with the use of these theorems it is possible to find the best solution in that limited class which provides the best minimum to the associated Π or I function. This in reality does not satisfy the missing equilibrium or compatibility equation, but it does it as well as possible for the class of function assumed to describe the stress or strain in the body, within the framework of the principle established above. It has been shown that most reasonable assumptions, regardless of their simplicity, provide useful solutions to most problems of mechanics of materials.

2.6.4 Example: Energy Techniques[2,4,5]

It can be shown that for beams the variational principle for strains reduces to

$$\Pi_{\min} = \left\{ \int_0^L [\tfrac{1}{2}EI(y'')^2 - qy]\, dx - \Sigma P_i y_i \right\}_{\min} \tag{2.98}$$

where EI is the flexural rigidity of the beam at any position x, I is the moment of inertia of the beam, y is the deflection of the beam, the y' refers to x derivative of y, q is the distributed loading, the P_i's represent concentrated loads, and L is the span length.

If the minimization is carried out, subject to the restrictions of the variational principle for strains, the beam equation results. However, it is both useful and instructive to utilize the above principle to obtain two approximate solutions to a specific problem and then compare these with the exact solutions obtained by other means.

First a centrally loaded, simple-support beam problem will be examined. The function of minimization becomes

$$\Pi = \int_0^{L/2} EI(y'')^2\, dx - Py_{L/2} \tag{2.99}$$

Select the class of displacement functions described by

$$y = Ax(\tfrac{3}{4}L^2 - x^2) \qquad 0 \le x \le L/2 \tag{2.100}$$

This satisfies the boundary conditions

$$y(0) = y''(0) = y'(L/2) = 0$$

In this A is an arbitrary parameter to be determined from the minimization of Π.

Properly introducing the value of y, y'' into the expression for Π and integrating, then minimizing Π with respect to the open parameter by setting

$$\partial\Pi/\partial A = 0$$

yields

$$y = (Px/12EI)\,(\tfrac{3}{4}L^2 - x^2) \qquad 0 \le x \le L/2 \tag{2.101}$$

It is coincidental that this is the exact solution to the above problem.

A second class of deflection function is now selected

$$y = A \sin (\pi x/L) \qquad 0 \le x \le L \tag{2.102}$$

which satisfies the boundary conditions

$$y(0) = y''(0) = y(L) = y''(L) = 0$$

which is intuitively the expected deflection shape. Additionally, $y(L/2) = A$. Introducing the above information into the expression for Π and minimizing as before yields

$$y = (PL^3/EI)[(2/\pi^4) \sin (\pi x/L)] \tag{2.103}$$

The ratio of the approximate to the exact central deflection is 0.9855, which indicates that the approximation is of sufficient accuracy for most applications.

2.7 FORMULATION OF GENERAL THERMOELASTIC PROBLEM[2,9]

A nonuniform temperature distribution or a nonuniform material distribution with uniform temperature change introduces additional stresses and/or strains, even in the absence of external tractions.

Within the confines of the linear theory of elasticity and neglecting small coupling effects between the *temperature-distribution problem* and the *thermoelastic problem* it is possible to solve the general mechanics-of-material problem as the superposition of the previously defined mechanics-of-materials problem and an initially traction-free thermoelastic problem.

Taking the same consistent definition of stress and strain as previously presented it can be shown that the strain-displacement, stress-equilibrium, and compatibility relationships remain unchanged in the thermoelastic problem. However, because a structural material can change its size even in the absence of stress, it is necessary to modify the stress-strain laws to account for the additional strain due to temperature (αT). Thus Hooke's law is modified as follows:

$$\begin{aligned} \epsilon_x &= (1/E)[\sigma_x - \nu(\sigma_y + \sigma_z)] + \alpha T \\ \epsilon_y &= (1/E)[\sigma_y - \nu(\sigma_z + \sigma_x)] + \alpha T \\ \epsilon_z &= (1/E)[\sigma_z - \nu(\sigma_x + \sigma_y)] + \alpha T \end{aligned} \tag{2.104}$$

The shear strain-stress relationships remain unchanged. α is the coefficient of thermal expansion and T the temperature rise above the ambient stress-free state. In uniform, nonconstrained structures this ambient base temperature is arbitrary, but in problems associated with nonuniform material or constraint this base temperature is quite important.

Expressed in cartesian tensor form the stress-strain relationships become

$$\epsilon_{ij} = [(1 + \nu)/E]\sigma_{ij} - (\nu/E)\delta_{ij}\sigma_{kk} + \alpha T\delta_{ij} \tag{2.105}$$

In inverted form the modified stress-strain relationships are

$$\begin{aligned} \sigma_x &= 2G\epsilon_x + \lambda\Delta - (3\lambda + 2G)\alpha T \\ \sigma_y &= 2G\epsilon_y + \lambda\Delta - (3\lambda + 2G)\alpha T \\ \sigma_z &= 2G\epsilon_z + \lambda\Delta - (3\lambda + 2G)\alpha T \end{aligned} \tag{2.106}$$

or, in cartesian tensor form,

$$\sigma_{ij} = 2G\epsilon_{ij} + \lambda\Delta\,\delta_{ij} - (3\lambda + 2G)\alpha T\delta_{ij} \tag{2.107}$$

Considering the equilibrium compatibility formulations, it can be shown that the analogous thermoelastic displacement formulations result in

$$\begin{aligned} (\lambda + G)(\partial\Delta/\partial x) + G\nabla^2 u - (3\lambda + 2G)\alpha(\partial T/\partial x) &= 0 \\ (\lambda + G)(\partial\Delta/\partial y) + G\nabla^2 v - (3\lambda + 2G)\alpha(\partial T/\partial y) &= 0 \\ (\lambda + G)(\partial\Delta/\partial z) + G\nabla^2 w - (3\lambda + 2G)\alpha(\partial T/\partial z) &= 0 \end{aligned} \tag{2.108}$$

A useful alternate stress formulation is

$$(1+\nu)\nabla^2\sigma_x + \frac{\partial^2\Theta}{\partial x^2} + \alpha E\left(\frac{1+\nu}{1-\nu}\nabla^2 T + \frac{\partial^2 T}{\partial x^2}\right) = 0$$

$$(1+\nu)\nabla^2\sigma_y + \frac{\partial^2\Theta}{\partial y^2} + \alpha E\left(\frac{1+\nu}{1-\nu}\nabla^2 T + \frac{\partial^2 T}{\partial y^2}\right) = 0$$

$$(1+\nu)\nabla^2\sigma_z + \frac{\partial^2\Theta}{\partial z^2} + \alpha E\left(\frac{1+\nu}{1-\nu}\nabla^2 T + \frac{\partial^2 T}{\partial z^2}\right) = 0 \qquad (2.109)$$

$$(1+\nu)\nabla^2\tau_{xy} + \partial^2\Theta/\partial x\,\partial y + \alpha E(\partial^2 T/\partial x\,\partial y) = 0$$

$$(1+\nu)\nabla^2\tau_{yz} + \partial^2\Theta/\partial y\,\partial z + \alpha E(\partial^2 T/\partial y\,\partial z) = 0$$

$$(1+\nu)\nabla^2\tau_{zx} + \partial^2\Theta/\partial z\,\partial x + \alpha E(\partial^2 T/\partial z\,\partial z) = 0$$

where $\Theta = \sigma_x + \sigma_y + \sigma_z$.

2.8 CLASSIFICATION OF PROBLEM TYPES

In mechanics of materials it is frequently desirable to classify problems in terms of their geometric configurations and/or assumptions that will permit their codification and ease of solution. As a result there exist problems in plane stress or strain, beam theory, curved-beam theory, plates, shells, etc. Although the defining equations can be obtained directly from the general theory together with the associated assumptions, it is often instructive and convenient to obtain them directly from physical considerations. The difference between these two approaches marks one of the principal distinguishing differences between the *theory of elasticity* and *mechanics of materials.*

2.9 BEAM THEORY

2.9.1 Mechanics of Materials Approach[1]

The following assumptions are basic in the development of elementary beam theory:

1. Beam sections, originally plane, remain plane and normal to the "neutral axis."
2. The beam is originally straight and all bending displacements are small.
3. The beam cross section is symmetrical with respect to the loading plane, an assumption that is usually removed in the general theory.
4. The beam material obeys Hooke's law, and the moduli of elasticity in tension and compression are equal.

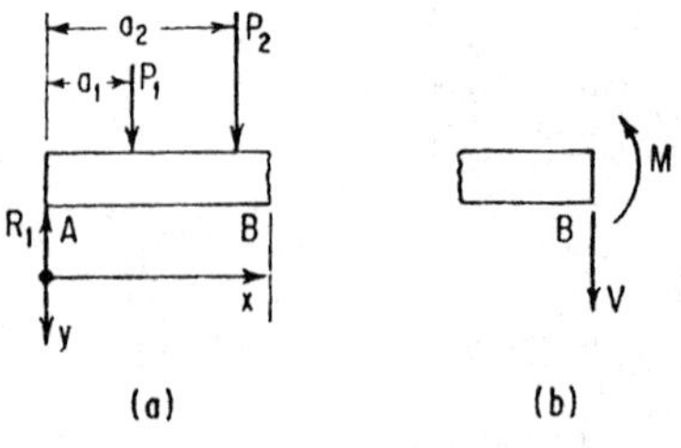

FIG. 2.17 Internal reactions due to externally applied loads. (*a*) External loading of beam segment. (*b*) Internal moment and shear.

Consider the beam portion loaded as shown in Fig. 2.17*a*.

For static equilibrium, the internal actions required at section *B* which are supplied by the immediately adjacent section to the right must consist of a vertical shearing force *V* and an internal moment *M*, as shown in Fig. 2.17*b*.

The evaluation of the shear V is accomplished by noting, from equilibrium $\Sigma F_y = 0$,

$$V = R - P_1 - P_2 \qquad \text{(for this example)} \tag{2.110}$$

The algebraic sum of all the shearing forces at one side of the section is called the shearing force at that section. The moment M is obtained from $\Sigma M = 0$:

$$M = R_1 x - P_1(x - a_1) - P_2(x - a_2) \tag{2.111}$$

The algebraic sum of the moments of all external loads to one side of the section is called the bending moment at the section.

Note the sign conventions employed thus far:

1. Shearing force is positive if the right portion of the beam tends to shear downward with respect to the left.
2. Bending moment is positive if it produces bending of the beam concave upward.
3. Loading w is positive if it acts in the positive direction of the y axis.

In Fig. 2.18*a* a portion of one of the beams previously discussed is shown with the bending moment M applied to the element.

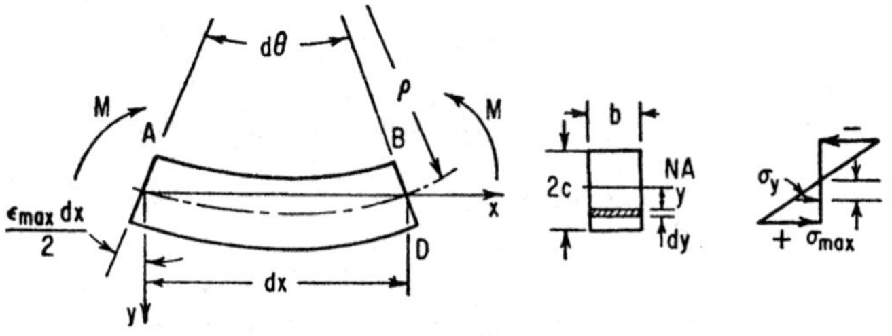

FIG. 2.18 Beam bending with externally applied load. (*a*) Beam element. (*b*) Cross section. (*c*) Bending-stress pattern at section *B–D*.

Equilibrium conditions require that the sum of the normal stresses σ on a cross section must equal zero, a condition satisfied only if the "neutral axis," defined as the plane or axis of zero normal stress, *is also the centroidal axis* of the cross section.

$$\int_{-c}^{+c} \sigma b\, dy = \frac{\sigma}{y}\int_{-c}^{+c} by\, dy = 0 \tag{2.112}$$

where $\sigma/y = (\epsilon/y)E = (\epsilon_{max}/y_{max})E = \text{const.}$

Further, if the moments of the stresses acting on the element dy of the figure are summed over the height of the beam,

$$M = \int_{-c}^{+c} \sigma b y\, dy = \frac{\sigma}{y}\int_{-c}^{+c} by^2\, dy = \frac{\sigma}{y} I \tag{2.113}$$

where y = distance from neutral axis to point on cross section being investigated, and

$$I = \int_{-c}^{+c} by^2\, dy$$

is the area moment of inertia about the centroidal axis of the cross section. Equation (2.113) defines the flexural stress in a beam subject to moment M:

$$\sigma = My/I \tag{2.114}$$

Thus

$$\sigma_{max} = Mc/I \tag{2.115}$$

To develop the equations for shear stress τ, the general case of the element of the beam subjected to a varying bending moment is taken as in Fig. 2.19.

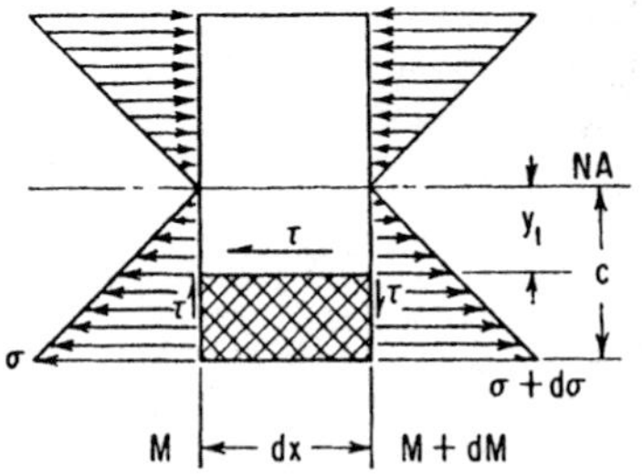

FIG. 2.19 Shear-stress diagram for beam subjected to varying bending moment.

Applying axial-equilibrium conditions to the shaded area of Fig. 2.19 yields the following general expression for the horizontal shear stress at the lower surface of the shaded area:

$$\tau = \frac{dM}{dx}\frac{1}{Ib}\int_{y_1}^{c} y\,dA \tag{2.116}$$

or, in familiar terms,

$$\tau = \frac{V}{Ib}\int_{y_1}^{c} y\,dA = \frac{V}{Ib}Q \tag{2.117}$$

where Q = moment of area of cross section about neutral axis for the shaded area above the surface under investigation
V = net vertical shearing force
b = width of beam at surface under investigation

Equilibrium considerations of a small element at the surface where τ is computed will reveal that this value represents both the vertical and horizontal shear.

For a rectangular beam, the vertical shear-stress distribution across a section of the beam is parabolic. The maximum value of this stress (which occurs at the neutral axis) is 1.5 times the average value of the stress obtained by dividing the shear force V by the cross-sectional area.

For many typical structural shapes the maximum value of the shear stress is approximately 1.2 times the average shear stress.

To develop the governing equation for bending deformations of beams, consider again Fig. 2.18. From geometry,

$$\frac{(\epsilon/2)\,dx}{y} = \frac{dx/2}{\rho} \tag{2.118}$$

Combining Eqs. (2.118), (2.114), and (2.72) yields

$$1/\rho = M/EI \tag{2.119}$$

Since

$$1/\rho \approx -\,d^2y/dx^2 = -y'' \tag{2.120}$$

Therefore

$$y'' = -M/EI \qquad \text{(Bernoulli-Euler equation)} \tag{2.121}$$

In Fig. 2.20 the element of the beam subjected to an arbitrary load $w(x)$ is shown together with the shears and bending moments as applied by the adjacent cross sections of the beam. Neglecting higher-order terms, moment summation leads to the following result for the moments acting on the element:

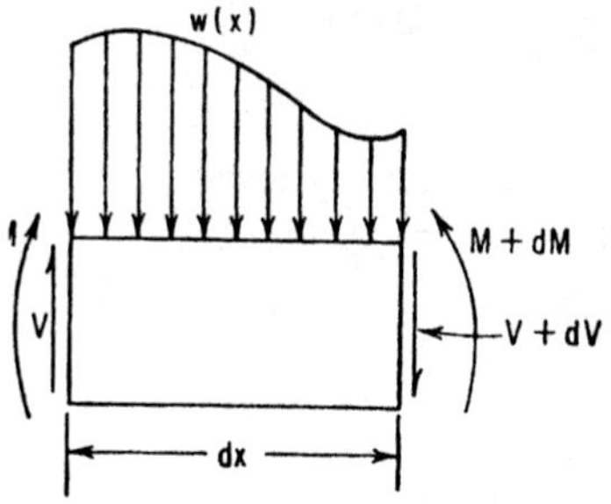

FIG. 2.20 Shear and bending moments for a beam with load $w(x)$ applied.

$$dM/dx = V \tag{2.122}$$

Differentiation of the Bernoulli-Euler equation yields

$$y''' = -V/EI \tag{2.123}$$

In similar manner, the summation of transverse forces in equilibrium yields

$$dV/dx = -w(x) \tag{2.124}$$

or

$$y^{\mathrm{IV}} = \frac{w(x)}{EI} \tag{2.125}$$

where due attention has been given to the proper sign convention.

See Table 2.1 for typical shear, moment, and deflection formulas for beams.

2.9.2 Energy Considerations

The total strain energy of bending is

$$U_b = \int_0^L \frac{M^2}{2EI}\,dx \tag{2.126}$$

The strain energy due to shear is

$$U_s = \int_0^L \frac{V^2}{2GA}\,dx \tag{2.127}$$

In calculating the deflections by the energy techniques, shear-strain contributions need not be included unless the beam is short and deep.

The deflections can then be obtained by the application of Castigliano's theorem, of which a general statement is: The partial derivative of the total strain energy of any structure with respect to any one generalized load is equal to the generalized deflection at the point of application of the load, and is in the direction of the load. The generalized loads can be forces or moments and the associated generalized deflections are displacements or rotations:

$$Y_a = \partial U/\partial P_a \tag{2.128}$$

$$\theta_a = \partial U/\partial M_a \tag{2.129}$$

where U = total strain energy of bending of the beam
P_a = load at point a
M_a = moment at point a
Y_a = deflection of beam at point a
θ_a = rotation of beam at point a

Thus

$$Y_a = \frac{\partial U}{\partial P_a} = \frac{\partial}{\partial P_a}\int_0^L M^2 \frac{dx}{2EI} = \int_0^L \frac{M}{EI}\frac{\partial M}{\partial P_a}\,dx \tag{2.130}$$

TABLE 2.1 Shear, Moment, and Deflection Formulas for Beams[1,12]

Notation: W = load (lb); w = unit load (lb/linear in). M is positive when clockwise; V is positive when upward; y is positive when upward. Constraining moments, applied couples, loads, and reactions are positive when acting as shown. All forces are in pounds, all moments in inch-pounds, all deflections and dimensions in inches. θ is in radians and $\tan\theta = \theta$.

Loading, support, and reference number	Reactions R_1 and R_2, vertical shear V	Deflection y, maximum deflection, and end slope θ
End supports Uniform load 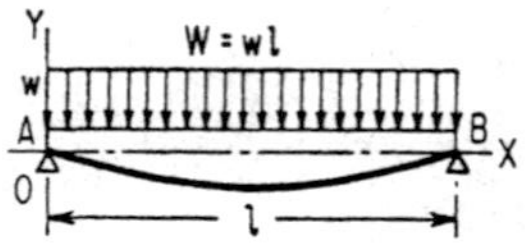 (1)	$R_1 = +\frac{1}{2}W \quad R_2 = +\frac{1}{2}W$ $V = \frac{1}{2}W\left(1 - \frac{2x}{l}\right)$	$y = -\frac{1}{24}\frac{Wx}{EIl}(l^3 - 2lx^2 + x^3)$ Max $y = -\frac{5}{384}\frac{Wl^3}{EI}$ at $x = \frac{1}{2}l$ $\theta = -\frac{1}{24}\frac{Wl^2}{EI}$ at $A \quad \theta = +\frac{1}{24}\frac{Wl^2}{EI}$ at B
End supports Intermediate load 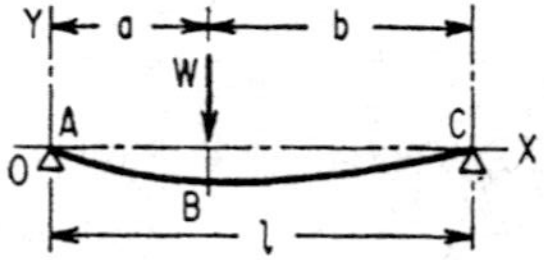 (2)	$R_1 = +W\frac{b}{l} \quad R_2 = +W\frac{a}{l}$ $(A \text{ to } B)\ V = +W\frac{b}{l}$ $(B \text{ to } C)\ V = -W\frac{a}{l}$	$(A \text{ to } B)\ y = -\frac{Wbx}{6EIl}[2l(l - x) - b^2 - (l - x)^2]$ $(B \text{ to } C)\ y = -\frac{Wa(l - x)}{6EIl}[2lb - b^2 - (l - x)^2]$ Max $y = -\frac{Wab}{27EIl}(a + 2b)\sqrt{3a(a + 2b)}$ at $x = \sqrt{\frac{1}{3}a(a + 2b)}$ when $a > b$ $\theta = -\frac{1}{6}\frac{W}{EI}\left(bl - \frac{b^3}{l}\right)$ at A; $\quad \theta = +\frac{1}{6}\frac{W}{EI}\left(2bl + \frac{b^3}{l} - 3b^2\right)$ at C

Supported at both ends Two symmetrical loads 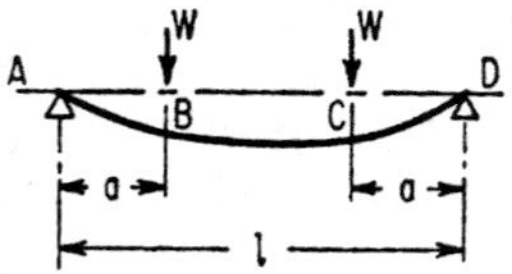 (3)	$R_1 = R_2 = W$ (A to B) and (C to D) $V = W$ (B to C) $V = 0$	(A to B) and (C to D) $y = -\frac{Wx}{6EI}(3\,la - 3a^2 - x^2)$ (B to C) $y = -\frac{Wa}{6EI}(3lx - 3x^2 - a^2)$ $\text{Max } y = -\frac{Wa}{24EI}(3l^2 - 4a^2)$ $\theta_A = -\frac{Pa}{2EI}(l - a)$
Both ends overhanging, supports unsymmetrical Uniform load 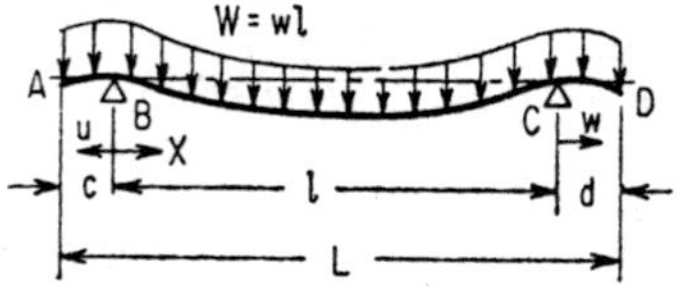 (4)	$R_1 = \frac{W}{2l}(l^2 - d^2 + c^2)$ $R_2 = \frac{W}{2l}(l^2 + d^2 - c^2)$	(A to B) $y = -\frac{Wu}{24EIl}[2l(d^2 + 2c^2) + 6c^2u - u^2(4c - u) - l^3]$ (B to C) $y = -\frac{Wx(l - x)}{24EIl}\left\{x(l - x) + l^2 - 2(d^2 + c^2) - \frac{2}{l}[d^2x + c^2(l - x)]\right\}$ Deflection at end: $y = -\frac{Wc}{24EIl}[2l(d^2 + 2c^2) + 3c^3 - l^3]$

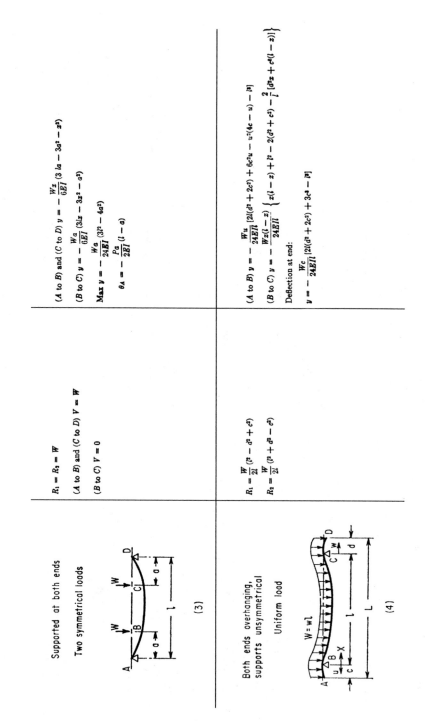

TABLE 2.1 Shear, Moment, and Deflection Formulas for Beams[1,12] (*Continued*)

Loading, support, and reference number	Reactions R_1 and R_2, vertical shear V	Deflection y, maximum deflection, and end slope θ
Both ends overhanging supports, load at any point between 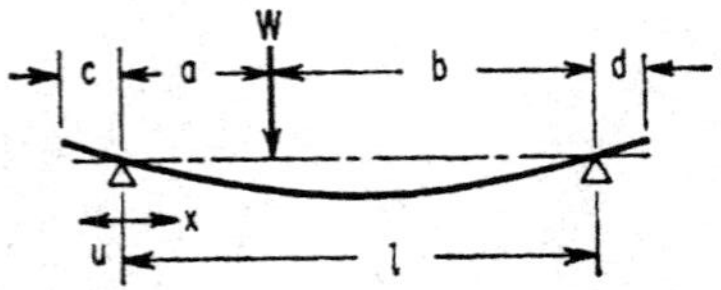 (5)	$R_1 = \frac{Wb}{l}$ $R_2 = \frac{Wa}{l}$	Between supports same as case 1; for overhang $y = \frac{Wabu}{6EIl}(l + b)$ Max y same as case 1 Max y at end Max $y = \frac{Wabc}{6EIl}(l + b)$
Both ends overhanging supports, single overhanging load 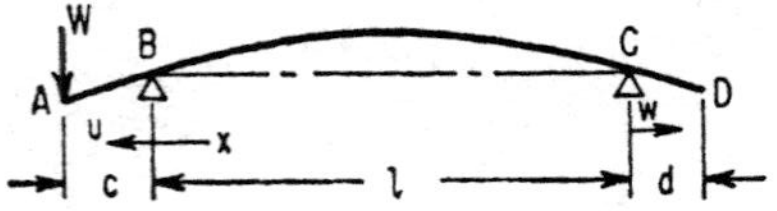 (6)	$R_1 = \frac{W(c + l)}{l}$ $R_2 = \frac{Wc}{l}$	$(A \text{ to } B)\ y = -\frac{Wu}{6EI}(3cu - u^2 + 2cl)$ $(B \text{ to } C)\ y = -\frac{Wcx}{6EIl}(l - x)(2l - x)$ $(C \text{ to } D)\ y = -\frac{Wclw}{6EI}$ $y_A = -\frac{Wc^2}{3EI}(c + l)$ $y_D = -\frac{Wcld}{6EI}$ Max $y = \frac{Wcl^2}{15.55EI}$ at $x = 0.42265L$

Both ends overhanging supports, symmetrical overhanging loads 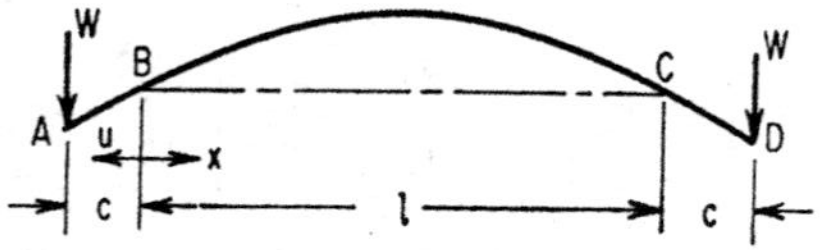 (7)	$R_1 = R_2 = W$	(A to B) and (C to D) $y = -\frac{Wu}{6EI}[3c(l+u) - u^2]$ (B to C) $y = -\frac{Wcx}{2EI}(l - x)$ $y_A = y_D = -\frac{Wc^2}{6EI}(2c + 3l)$ Max $y = \frac{Wcl^2}{8EI}$
Cantilever Uniform load 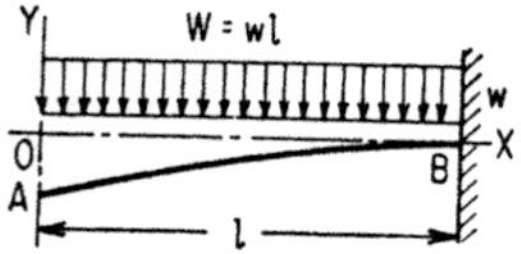 (8)	$R_2 = +W$ $V = -\frac{W}{l}x$	$y = -\frac{1}{24}\frac{W}{EIl}(x^4 - 4l^3x + 3l^4)$ Max $y = -\frac{1}{8}\frac{Wl^3}{EI}$ $\theta = +\frac{1}{6}\frac{Wl^2}{EI}$ at A

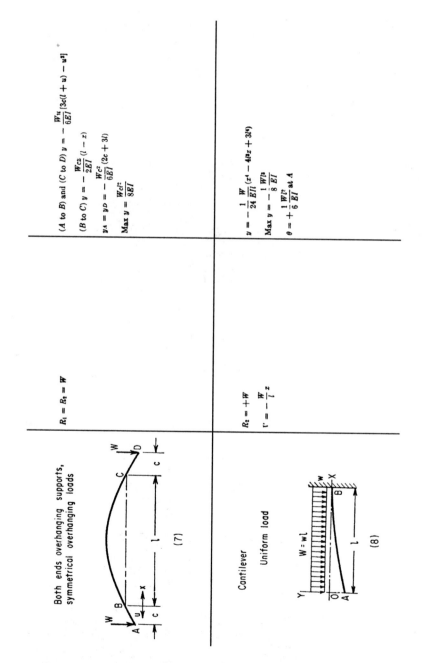

TABLE 2.1 Shear, Moment, and Deflection Formulas for Beams[1,12] (*Continued*)

Loading, support, and reference number	Reactions R_1 and R_2, vertical shear V	Deflection y, maximum deflection, and end slope θ
Cantilever Intermediate load 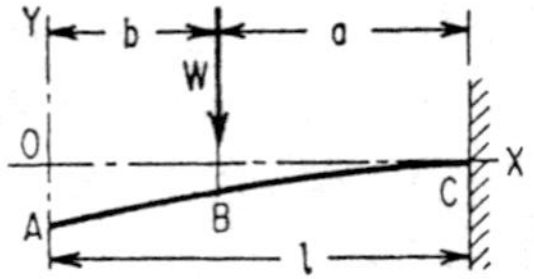 (9)	$R_2 = +W$ (A to B) $V = 0$ (B to C) $V = -W$	(A to B) $y = -\frac{1}{6}\frac{W}{EI}(-a^3 + 3a^2l - 3a^2x)$ (B to C) $y = -\frac{1}{6}\frac{W}{EI}[(x-b)^3 - 3a^2(x-b) + 2a^3]$ Max $y = -\frac{1}{6}\frac{W}{EI}(3a^2l - a^3)$ $\theta = +\frac{1}{2}\frac{Wa^2}{EI}$ (A to B)
One end fixed, one end supported Uniform load 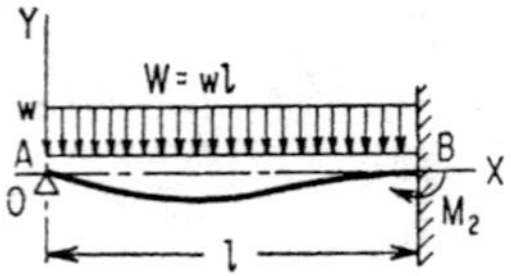 (10)	$R_1 = \frac{3}{8}W$ $R_2 = \frac{5}{8}W$ $M_2 = \frac{1}{8}Wl$ $V = W\left(\frac{3}{8} - \frac{x}{l}\right)$	$y = \frac{1}{48}\frac{W}{EIl}(3lx^3 - 2x^4 - l^3x)$ Max $y = -0.0054\frac{Wl^3}{EI}$ at $x = 0.4215l$ $\theta = -\frac{1}{48}\frac{Wl^2}{EI}$ at A

One end fixed, one end supported Intermediate load 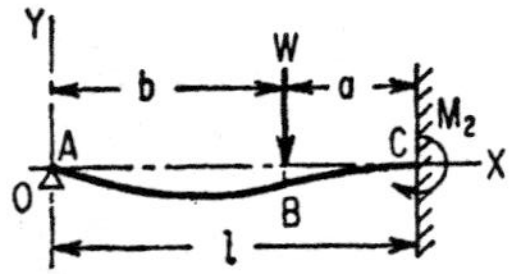 (11)	$R_1 = \frac{1}{2} W \left(\frac{3a^2l - a^3}{l^3}\right) \qquad R_2 = W - R_1$ $M_2 = \frac{1}{2} W \left(\frac{a^3 + 2al^2 - 3a^2l}{l^2}\right)$ $(A \text{ to } B)\ V = +R_1$ $(B \text{ to } C)\ V = R_1 - W$	$(A \text{ to } B)\ y = \frac{1}{6EI}[R_1(x^3 - 3l^2x) + 3Wa^2x]$ $(B \text{ to } C)\ y = \frac{1}{6EI}\{R_1(x^3 - 3l^2x) + W[3a^2x - (x-b)^3]\}$ If $a < 0.586l$, max y is between A and B at: $x = l\sqrt{1 - \frac{2l}{3l - a}}$ If $a > 0.586l$, max y is at: $x = \frac{l(l^2 + b^2)}{3l^2 - b^2}$ If $a = 0.586l$, max y is at B and $= -0.0098\frac{Wl^3}{EI}$, max possible deflection $\theta = \frac{1}{4}\frac{W}{EI}\left(\frac{a^3}{l} - a^2\right)$ at A
Fixed at one end, free but guided at the other Uniform load 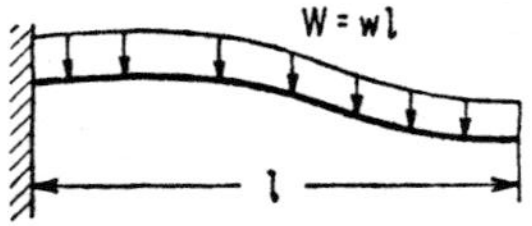 (12)	$R_1 = W$	$y = -\frac{Wx^2}{24EIl}(2l - x)^2$ Max $y = -\frac{Wl^3}{24EI}$

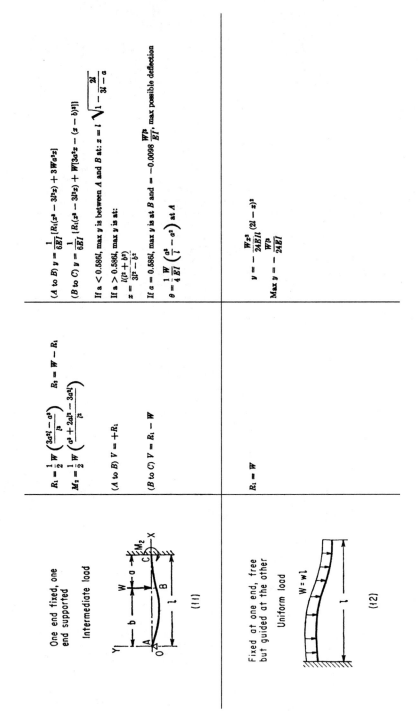

TABLE 2.1 Shear, Moment, and Deflection Formulas for Beams[1,12] (*Continued*)

Loading, support, and reference number	Reactions R_1 and R_2, vertical shear V	Deflection y, maximum deflection, and end slope θ
Fixed at one end, free but guided at the other With load 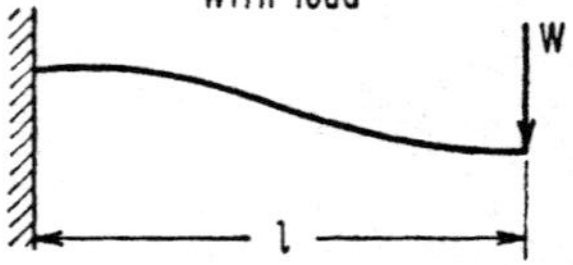 (13)	$R_1 = W$	$y = -\frac{Wx^2}{12EI}(3l - 2x)$ $\text{Max } y = -\frac{Wl^3}{12EI}$
Both ends fixed Uniform load 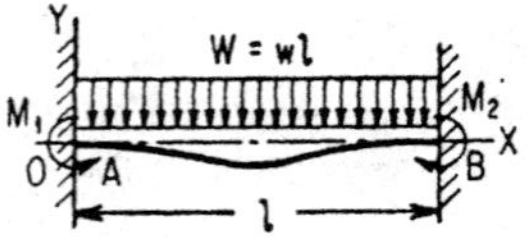 (14)	$R_1 = \frac{1}{2}W \qquad R_2 = \frac{1}{2}W$ $M_1 = \frac{1}{12}Wl \qquad M_2 = \frac{1}{12}Wl$ $V = \frac{1}{2}W\left(\frac{1-2x}{l}\right)$	$y = \frac{1}{24}\frac{Wx^2}{EIl}(2lx - l^2 - x^2)$ $\text{Max } y = -\frac{1}{384}\frac{Wl^3}{EI} \text{ at } x = \frac{1}{2}l$

Both ends fixed Intermediate load 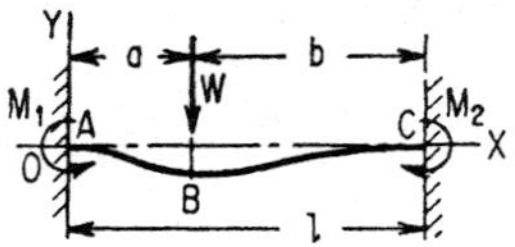 (15)	$R_1 = \frac{Wb^2}{l^3}(3a+b) \qquad R_2 = \frac{Wa^2}{l^3}(3b+a)$ $M_1 = W\frac{ab^2}{l^2} \qquad M_2 = W\frac{a^2b}{l^2}$ $(A \text{ to } B)\ V = R_1$ $(B \text{ to } C)\ V = R_1 - W$	$(A \text{ to } B)\ y = -\frac{1}{6}\frac{Wb^2x^2}{EIl^3}(3ax + bx - 3al)$ $(B \text{ to } C)\ y = -\frac{1}{6}\frac{Wa^2(l-x)^2}{EIl^3}[(3b+a)(l-x) - 3bl]$ $\text{Max } y = \frac{2}{3}\frac{W}{EI}\frac{a^3b^2}{(3a+b)^2} \text{ at } x = \frac{2al}{3a+b} \text{ if } a > b$ $\text{Max } y = \frac{2}{3}\frac{W}{EI}\frac{a^2b^3}{(3b+a)^2} \text{ at } x = l - \frac{2bl}{3b+a} \text{ if } a < b$
Continuous beam, each span uniformly loaded; spans, loads, and sections different 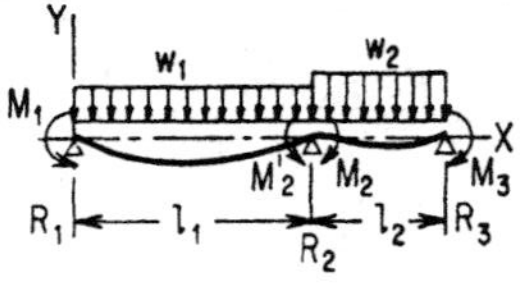 (16)	$\frac{M_1l_1}{I_1} + 2M_2\left[\frac{l_1}{I_1} + \frac{l_2}{I_2}\right] + \frac{M_3l_2}{I_2} = \frac{W_1l_1^3}{4I_1} + \frac{W_2l_2^3}{4I_2}$ (Theorem of Three Moments: I_1 and I_2 refer to 1st and 2d spans. Equation gives M_2 when M_1 and M_3 are known, or can be written for each pair of spans of a continuous beam and resulting equations solved. M_2 acts on span 1, M_2' on span 2.)	Superpose cases 1 and simple beam with end moment
Continuous beam with two unequal spans Unequal loads at any point of each 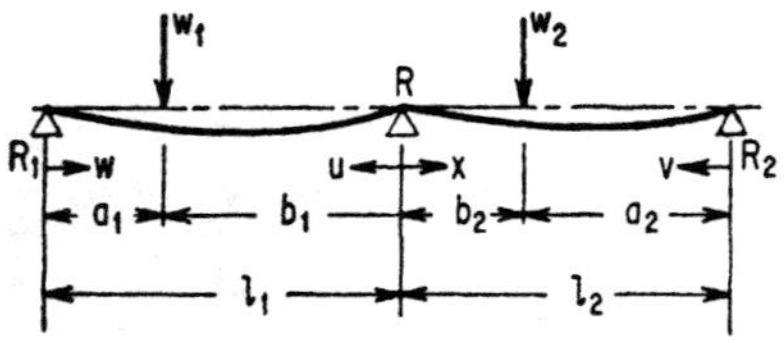 (17)	$R_1 = \frac{W_1b - m}{l_1}$ $R_2 = \frac{W_1a_1 + m}{l_1} + \frac{W_2a_2 + m}{l_2}$ $R^3 = \frac{W_2b_2 - m}{l_2}$ $m = \frac{1}{2(l_1 + l_2)}\left[\frac{W_1a_1b_1}{l_1}(l_1+a_1) + \frac{W_2a_2b_2}{l_2}(l_2+a_2)\right]$	$(R_1 \text{ to } W_1)\quad y = \frac{w}{6EI}\left[(l_1 - w)(l_1 + w)R_1 - \frac{W_1b_1^3}{l_1}\right]$ $(R \text{ to } W_1)\quad y = \frac{u}{6EIl_1}[W_1a_1b_1(l_1 + a_1) - W_1a_1u^2 - m(2l_1 - u)(l_1 - u)]$ $(R \text{ to } W_2)\quad y = \frac{x}{6EIl_2}[W_2a_2b_2(l_2 + a_2) - W_2a_2x^2 - m(2l_2 - x)(l_2 - x)]$ $(R_2 \text{ to } W_2)\quad y = \frac{v}{6EI}\left[(l_2 - v)(l_2 + v)R_2 - \frac{W_2b_2^3}{l_2}\right]$ $y_{W_1} = \frac{a_1b_1}{6EIl_1}[2a_1b_1W_1 - m(l_1 + a_1)]$ $y_{W_2} = \frac{a_2b_2}{6EIl_2}[2a_2b_2W_2 - m(l_2 + a_2)]$

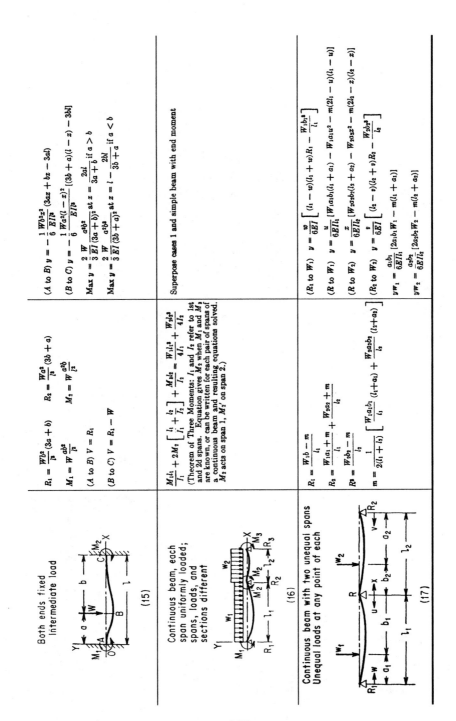

$$\theta_a = \frac{\partial U}{\partial M_a} = \frac{\partial}{\partial M_a}\int_0^L M^2 \frac{dx}{2EI} = \int_0^L \frac{M}{EI}\frac{\partial M}{\partial M_a}\,dx \tag{2.131}$$

An important restriction on the use of this theorem is that the deflection of the beam or structure must be a linear function of the load; i.e., geometrical changes and other nonlinear effects must be neglected.

A second theorem of Castigliano states that

$$P_a = \partial U/\partial Y_a \tag{2.132}$$

$$M_a = \partial U/\partial \theta_a \tag{2.133}$$

and is just the inverse of the first theorem. Because it does not have a "linearity" requirement, it is quite useful in special problems.

To illustrate, the deflection y at the center of wire of length $2L$ due to a central load P will be found.

From geometry, the extension δ of each half of the wire is, for small deflections,

$$\delta \approx y^2/2L \tag{2.134}$$

The strain energy absorbed in the system is

$$U = 2\tfrac{1}{2}(AE/L)\delta^2 = (AE/4L^3)y^4 \tag{2.135}$$

Then, by the second theorem,

$$P = \partial U/\partial y = (AE/L^3)y^3 \tag{2.136}$$

or the deflection is

$$y = L\sqrt[3]{P/AE} \tag{2.137}$$

Among the other useful energy theorems are:

Theorem of Virtual Work. If a beam which is in equilibrium under a system of external loads is given a small deformation ("virtual deformation"), the work done by the load system during this deformation is equal to the increase in internal strain energy.

Principle of Least Work. For beams with statically indeterminate reactions, the partial derivative of the total strain energy with respect to the unknown reactions must be zero.

$$\partial U/\partial P_i = 0 \qquad \partial U/\partial M_i = 0 \tag{2.138}$$

depending on the type of support. (This follows directly from Castigliano's theorems.) The magnitudes of the reactions thus determined are such as to minimize the strain energy of the system.

2.9.3 Elasticity Approach[2]

In developing the conventional equations for beam theory from the basic equations of elastic theory (i.e., stress equilibrium, strain compatibility, and stress-strain relations) the beam problem is considered a plane-stress problem. The equilibrium equations for plane stress are

$$\partial\sigma_x/\partial x + \partial\tau_{xy}/\partial y = 0 \tag{2.139}$$

$$\partial\sigma_y/\partial y + \partial\tau_{xy}/\partial x = 0 \tag{2.140}$$

By using an "Airy stress function" ψ, defined as follows:

$$\sigma_x = \partial^2\psi/\partial y^2 \qquad \sigma_y = \partial^2\psi/\partial x^2 \qquad \tau_{xy} = -\partial^2\psi/\partial x\,\partial y \tag{2.141}$$

and the compatibility equation for strain, as set forth previously, the governing equations for beams can be developed.

The only compatibility equation not identically satisfied in this case is

$$\partial^2\epsilon_x/\partial y^2 + \partial^2\epsilon_y/\partial x^2 = \partial^2\gamma_{xy}/\partial x\,\partial y \tag{2.142}$$

Substituting the stress-strain relationships into the compatibility equations and introducing the Airy stress function yields

$$\partial^4\psi/\partial x^4 + 2\partial^4\psi/\partial x^2\,\partial y^2 + \partial^4\psi/\partial y^4 = \nabla^4\psi = 0 \tag{2.143}$$

which is the "biharmonic" equation where ∇^2 is the Laplace operator.

To illustrate the utility of this equation consider a uniform-thickness cantilever beam (Fig. 2.21) with end load P. The boundary conditions are $\sigma_y = \tau_{xy} = 0$ on the surfaces $y = \pm c$, and the summation of shearing forces must be equal to the external load P at the loaded end,

$$\int_{-c}^{+c} \tau_{xy} b\, dy = P$$

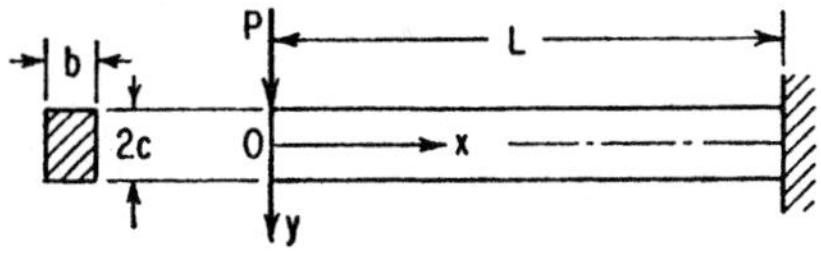

FIG. 2.21 Cantilever beam with end load P.

The solution for σ_x is

$$\sigma_x = \partial^2\psi/\partial y^2 = cxy \tag{2.144}$$

Introducing $b(2c)^3/12 = I$, the final expressions for the stress components are

$$\begin{aligned} \sigma_x &= -Pxy/I = -My/I \\ \sigma_y &= 0 \\ \tau_{xy} &= -P(c^2 - y^2)/2I \end{aligned} \tag{2.145}$$

To extend the theory further to determine the displacements of the beam, the definitions of the strain components are

$$\begin{aligned} \epsilon_x &= \partial u/\partial x = \sigma_x/E = -Pxy/EI \\ \epsilon_y &= \partial v/\partial y = -\nu\sigma_x/E = \nu Pxy/EI \\ \gamma_{xy} &= \partial u/\partial y + \partial v/\partial x = [2(1+\nu)/E]\tau_{xy} = [(1+\nu)P/EI](c^2 - y^2) \end{aligned} \tag{2.146}$$

Solving explicitly for the u and v subject to the boundary conditions

$$u = v = \partial u/\partial x = 0 \qquad \text{at } x = L \text{ and } y = 0$$

there results

$$v = \nu Pxy^2/2EI + Px^3/6EI - PL^2x/2EI + PL^3/3EI \tag{2.147}$$

The equation of the deflection curve at $y = 0$ is

$$(v)_{y=0} = (P/6EI)(x^3 - 3L^2x + 2L^3) \tag{2.148}$$

The curvature of the deflection curve is therefore the Bernoulli-Euler equation

$$1/\rho \approx -(\partial^2 v/\partial^2 x)_{y=0} = -Px/EI = M/EI = -y'' \tag{2.149}$$

EXAMPLE 1 The moment at any point x along a simply supported uniformly loaded beam (w lb/ft) of span L is

$$M = wLx/2 - wx^2/2 \tag{2.150}$$

Integrating Eq. (2.121) and employing the boundary conditions $y(0) = y(L) = 0$, the solution for the elastic or deflection curve becomes

$$y = (wL^4/24EI)(x/L)[1 - 2(x/L)^2 + (x/L)^3] \tag{2.151}$$

EXAMPLE 2 In order to obtain the general deflection curve, a fictitious load P_a is placed at a distance a from the left support of the previously described uniformly loaded beam.

$$\begin{aligned} M &= -wx^2/2 + wLx/2 + [P_a x(L - a)]/L \qquad 0 < x < a \\ M &= -wx^2/2 + wLx/2 + [P_a a(L - x)]/L \qquad a < x < L \end{aligned} \tag{2.152}$$

From Castigliano's theorem,

$$y_a = \frac{\partial U}{\partial P_a} = \int_0^L \frac{M}{EI}\frac{\partial M}{\partial P_a}\,dx \tag{2.153}$$

and therefore

$$y_a = (1/EI)(\tfrac{1}{24}wa^4 - \tfrac{1}{12}wLa^3 + \tfrac{1}{24}waL^3) \tag{2.154}$$

The elastic curve of the beam is obtained by substituting x for a in Eq. (2.154), resulting in the same expression as obtained by the double-integration technique.

EXAMPLE 3 It can be shown, considering stresses away from the beam ends, and essentially considering temperature variations only in the direction perpendicular to the beam axis (or in two or more directions by superposition) that the traction-free thermoelastic stress distribution is given by

$$\sigma_x = -\alpha ET + \frac{1}{2c}\int_{-c}^{+c} \alpha ET\,dy + \frac{3y}{2c^3}\int_{-c}^{+c} \alpha ETy\,dy \tag{2.155}$$

Since this stress distribution results in a net axial force-free, and moment-free, distribution but nonzero bending displacements and slopes, it would be necessary to superpose any additional stresses associated with actual boundary constraints.

Because the stress distribution is independent of any linear temperature gradient, it is always possible to add an arbitrary linear distribution T', such that $T_{tot} = T + T'$ and $\int T_{tot}\,dy = \int T_{tot} y\,dy = 0$.

Thus

$$\sigma_x = -\alpha E T_{tot} \tag{2.156}$$

In general T' can be chosen with sufficient accuracy by *visual examination* of the temperature distribution to make the "total-temperature integral" and its "first moment" equal zero. Thus the simplified formula together with its interpretation presents a useful graphical thermoelastic solution for beam and slab problems.

For the beam with temperature distribution $T = a(c^2 - y^2)$ and a is an arbitrary constant, superimpose a temperature distribution T' such that $A_T = -A_{T'}$, where A_T refers to the area under the temperature distribution curve. Evidently

$$T' = -\tfrac{2}{3}ac^2$$

and

$$T_{tot} = T + T' = a(c^2 - y^2) - \tfrac{2}{3}ac^2$$

and $\int T_{tot}\,dy = \int T_{tot} y\,dy = 0$, evident from the selection of T' and symmetry. The stress is therefore

$$\sigma = -\alpha E T_{tot} = -\alpha E[a(c^2 - y^2) - \tfrac{2}{3}ac^2] \tag{2.157}$$

2.10 CURVED-BEAM THEORY[1]

A "curved beam" (see Fig. 2.22) is defined as a beam in which the line joining the centroid of the cross sections (hereafter referred to as the "center line") is a curve. In the standard developments of the equations for the stresses and deflections for curved beams the following *assumptions* are usually made and represent the *restrictions* on the applicability of curved-beam theory:

> The sections of the beam originally plane and normal to the center line of the beam remain so after bending.

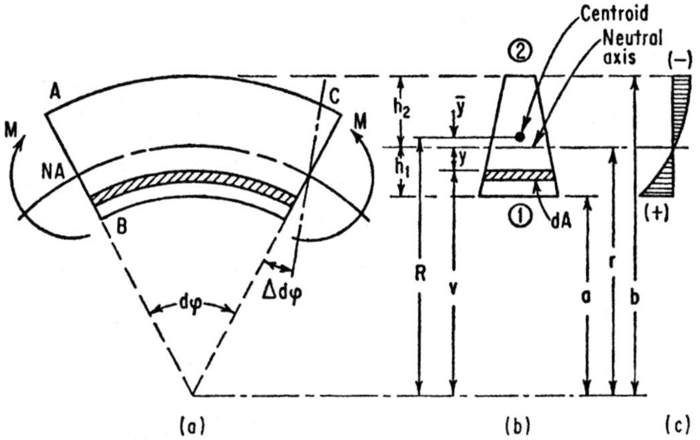

FIG. 2.22 Bending of curved beam element. (*a*) Beam element. (*b*) Cross section. (*c*) Bending-stress pattern at section *C–D*.

All cross sections have an axis of symmetry in the plane of the center line.

The beam is subjected to forces and moments acting in the plane of symmetry.

2.10.1 Equilibrium Approach

In Fig. 2.22 a "positive bending" moment is taken as one which tends to *decrease* the *curvature* of the beam. If R denotes the curvature of the beam at the centroid of a section, then it can be shown that the neutral axis is displaced from R a distance toward the center of curvature. As with straight beams, the "neutral axis" is defined as that axis about which the integrated tangential force is zero, when the external traction is restricted to a bending moment. This distance $\bar{y}$ may be computed from

$$\bar{y} = R - \frac{A}{\int dA/v} \tag{2.158}$$

where A = cross-section area of beam, v the distance from the center of curvature to the incremental area; and the integration $\int dA/v$ is carried out over the entire section of the beam.

The flexural stress at any point a distance y from the neutral axis is

$$\sigma = E\epsilon = Ey(\Delta d\phi)/(r - y)\, d\phi \tag{2.159}$$

which shows that the normal stress distribution over a cross section is not linear, as would be the case in simple beam theory, but hyperbolic in shape. Equating the sum of the moments of each of these segments of normal stress to the applied bending moment on the element,

$$\int \sigma y\, dA = \frac{E(\Delta\, d\phi)}{d\phi} \int \frac{y^2\, dA}{r - y} = M \tag{2.160}$$

yields the resulting expression for the stress

$$\sigma = My/A\bar{y}(r - y) \tag{2.161}$$

with the stresses at the extreme fibers, points 1 and 2, expressed by

$$\sigma_1 = Mh_1/A\bar{y}a \qquad \text{(tensile)} \qquad \sigma_2 = Mh_2/A\bar{y}b \qquad \text{(compressive)} \tag{2.162}$$

The above-described stresses result from pure bending only. If a more general loading condition is given, it must be reduced to the statically equivalent couple and a normal force through the centroid of the section in question.* Then the extensional stresses resulting from the normal force are superposed on the flexural stresses due to the couple.

The development of the governing equation for the displacement of a curved beam, for small deflections with a circular center line, depends on the differential quantities shown in Fig. 2.23, where $1/r$ and $1/r_1$ are the respective curvatures of the undeflected and the deflected beam.

Note that the radial displacement u is taken positive inward in these relations. A comparison of the changes in length $\Delta\, ds$ and central angle $\Delta\, d\phi$ due to deformation leads to

$$1/r_1 - 1/r = u/r^2 + d^2u/ds^2 \tag{2.163}$$

*This also applies in the general case when a shearing force also may be acting on this section.

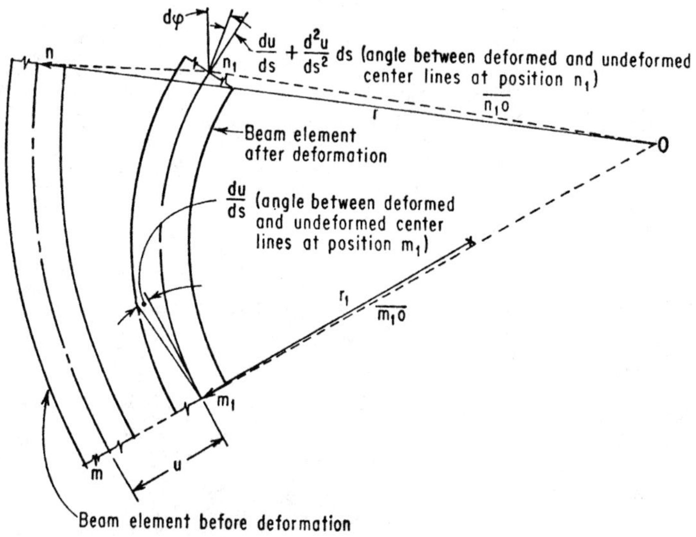

FIG. 2.23 Curved beam with circular center line.

If the thickness of the beam is small compared with the curvature, then

$$1/r_1 - 1/r = -M/EI \tag{2.164}$$

and the differential equation for the deflection curve of a curved beam, which is entirely analogous to that of simple-beam theory, becomes

$$d^2u/ds^2 + u/r^2 = -M/EI \tag{2.165}$$

For an infinitely large r this reduces to the Bernoulli-Euler equation for simple beams.

2.10.2 Energy Approach

A second and more powerful approach, which does not require that the center line of the beam be circular, is essentially the application of Castigliano's theorem. The expression for the strain energy in bending of a curved beam is similar to that of simple-beam theory,

$$U = \int_0^s \frac{M^2\,ds}{2EI} \tag{2.166}$$

where the integration is over the entire length of the beam. The deflection (or rotation) of the beam at a point under a concentrated load (or moment), is

$$\delta_i = \partial U/\partial P_i \tag{2.167}$$

The utility of Eqs. (2.166) and (2.167) in calculating the deflection of curved beams can best be shown by example.

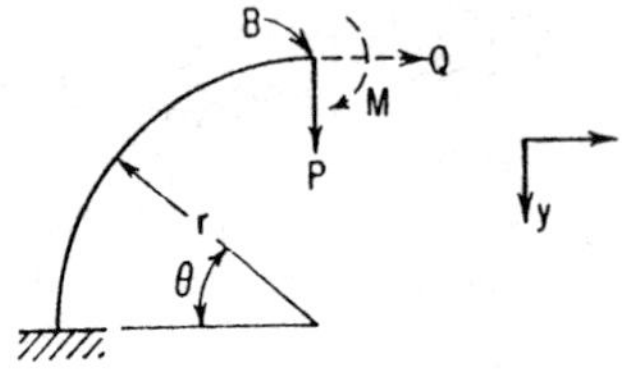

FIG. 2.24 Simple curved beam with vertical load *P*.

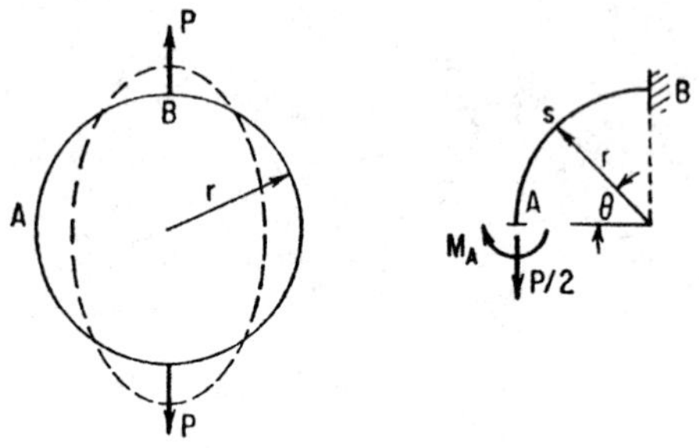

FIG. 2.25 Circular ring.

EXAMPLE 1 The curved beam shown in Fig. 2.24 has a uniform cross section. To determine the horizontal and vertical deflections and rotations, at point *B* fictitious load *Q* and *M* are assumed to act. At any point *s* on the beam,

$$M_s = -[Pr\cos\theta + Qr(1 - \sin\theta) + M] \tag{2.168}$$

$$\partial M_s/\partial P = -r\cos\theta \qquad \partial M_s/\partial Q = -r(1 - \sin\theta) \qquad \partial M_s/\partial M = -1 \tag{2.169}$$

Substituting $Q = 0$, $M = 0$ in the expression M_s we have, for the three deformations with $ds = r\,d\theta$,

$$\delta_y = \left.\frac{\partial U}{\partial P}\right|_{\substack{Q\to 0\\ M\to 0}} = \int_0^s \frac{M_s}{EI}\frac{\partial M_s}{\partial P}\,ds = \int_0^{\pi/2}\frac{Pr^3}{EI}\cos^2\theta\,d\theta = \frac{\pi}{4}\frac{Pr^3}{EI} \tag{2.170}$$

$$\delta_x = \left.\frac{\partial U}{\partial Q}\right|_{\substack{Q\to 0\\ M\to 0}} = \int_0^s \frac{M_s}{EI}\frac{\partial M}{\partial Q}\,ds = \int_0^{\pi/2}\frac{Pr^3}{EI}\cos\theta\,(1 - \sin\theta)\,d\theta = \frac{Pr^3}{2EI} \tag{2.171}$$

$$\theta = \left.\frac{\partial U}{\partial M}\right|_{\substack{Q\to 0\\ M\to 0}} = \int_0^s \frac{M_s}{EI}\frac{\partial M_s}{\partial M}\,ds = \int_0^{\pi/2}\frac{Pr^2}{EI}\cos\theta\,d\theta = \frac{Pr^2}{EI} \tag{2.172}$$

EXAMPLE 2 *Circular Ring-Energy Approach.* The circular ring is subjected to equal and opposite forces *P* as shown in Fig. 2.25. From symmetry considerations an equivalent model may be constructed where the load on the horizontal section is denoted by moment M_A and force *P*/2. From symmetry, there is no rotation of the horizontal section at point *A*. Therefore, by Castigliano's theorem,

$$\theta_A = \partial U/\partial M_A = 0 \tag{2.173}$$

The moment at any point *s* is given by

$$M_s = M_A - (P/2)r(1 - \cos\theta) \qquad \text{and} \qquad \partial M_s/\partial M_A = 1 \tag{2.174}$$

Substituting in Eq. (2.173) and imposing the condition of zero rotation,

$$\theta_A = 0 = \frac{\partial U}{\partial M_A} = \int_0^s \frac{M_s}{EI}\frac{\partial M_s}{\partial M_A}\,ds = \int_0^{\pi/2}\frac{1}{EI}\left[M_A - \frac{Pr}{2}(1 - \cos\theta)\right]r\,d\theta \tag{2.175}$$

which leads to

$$M_A = (Pr/2)(1 - 2/\pi) \tag{2.176}$$

which yields the moment expression

$$M_s = (Pr/2)(\cos\theta - 2/\pi) \tag{2.177}$$

The total strain energy for the entire ring is four times that of the quadrant considered. To obtain the total increase in the vertical diameter, the following steps are taken:

$$U = 4\int_0^{\pi/2} \frac{M_s^2}{2EI}\, r\, d\theta \tag{2.178}$$

$$\delta_y = \frac{\partial U}{\partial P} = \frac{4}{EI}\int_0^{\pi/2} M_s\left(\frac{\partial M_s}{\partial P}\right) r\, d\theta = \frac{Pr^3}{EI}\int_0^{\pi/2}\left(\cos\theta - \frac{2}{\pi}\right)^2 d\theta \tag{2.179}$$

$$\delta_y = \frac{Pr^3}{EI}\left(\frac{\pi}{4} - \frac{2}{\pi}\right) \tag{2.180}$$

Equilibrium Approach: The basic equation for this problem is

$$d^2u/ds^2 + u/r^2 = -M_s/EI \tag{2.181}$$

Substituting Eq. (2.177) yields

$$r^2\, d^2u/ds^2 + u = (Pr^3/2EI)(2/\pi - \cos\theta) \tag{2.182}$$

The boundary conditions, derived from the symmetry of the ring, are

$$u'(\theta = 0) = u'(\theta = \pi/2) = 0$$

The general solution of Eq. (2.182) is

$$u = A\cos\theta + B\sin\theta + Pr^3/EI\pi - (Pr^3/4EI)\,\theta\sin\theta \tag{2.183}$$

$$A = -Pr^3/4EI \qquad \text{and} \qquad B = 0 \tag{2.184}$$

Then $\quad u = Pr^3/EI\pi - (Pr^3/4EI)\cos\theta - (Pr^3/4EI)\theta\sin\theta$

The increase in the vertical radius (point B at $\theta = \pi/2$) becomes

$$u(\theta = \pi/2) = (Pr^3/EI)(1/\pi - \pi/8) \tag{2.185}$$

Thus the total increase in the vertical diameter δ_v is

$$\delta_v = -2u(\theta = \pi/2) = (Pr^3/EI)(\pi/4 - 2/\pi) \tag{2.186}$$

which is in agreement with Eq. (2.180).

2.11 THEORY OF COLUMNS[1]

The equilibrium approach for slender symmetrical columns is based on two sets of assumptions:

1. All the basic assumptions inherent in the derivation of the Bernoulli-Euler equation apply to columns also.
2. The transverse deflection of the column at the point of load application is *not* small when compared with the eccentricity of the applied load.

This theory is best described with the aid of a typical column, taken with a built-in support and subjected to an eccentric load, as illustrated in Fig. 2.26.

The moment at any section a distance x from the base is

$$M = -P(\delta + e - y) \tag{2.187}$$

where the negative sign is in accordance with the sign convention for simple beams. Writing the Bernoulli-Euler equation for the bending deflection of this member with the aid of the substitution $p^2 = P/EI$ gives

$$y'' + p^2 y = p^2(\delta + e) \tag{2.188}$$

The general solution of Eq. (2.188) is

$$y = A \sin px + B \cos px + \delta + e \tag{2.189}$$

From the boundary conditions for a built-in end

$$y(0) = y'(0) = 0$$

the equation for the deflection curve is

$$y = (\delta + e)(1 - \cos px) \tag{2.190}$$

The deflection at the end of the column at $x = L$ is seen to be

$$\delta = e(1 - \cos pL)/(\cos pL) \tag{2.191}$$

The complete description of the deflection curve for any point in the column thus becomes

$$y = e(1 - \cos px)/(\cos pL) \tag{2.192}$$

These results can easily be extended to cover a column hinged at both ends by redefining terms as indicated in Fig. 2.27. Thus, from symmetry, the relation of the deflection at the mid-span can be written directly from the previous results:

$$\delta = e\left[1 - \cos\left(\frac{pL}{2}\right)\right] \Big/ \cos\left(\frac{pL}{2}\right) \tag{2.193}$$

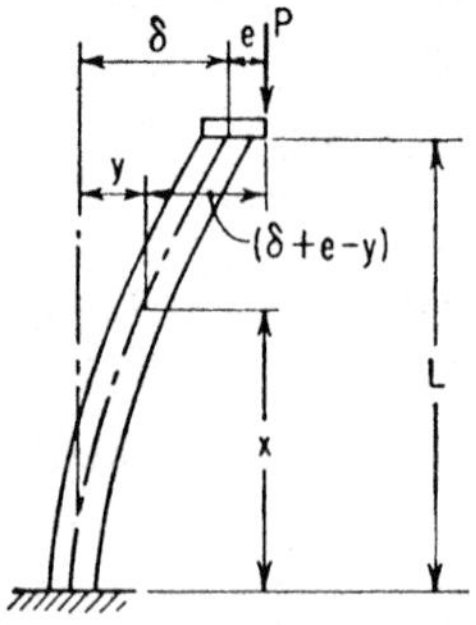

FIG. 2.26 Column subjected to an eccentric load.

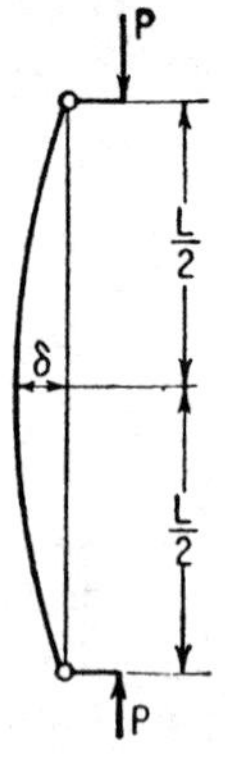

FIG. 2.27 Column hinged at both ends.

In applying these equations note that the deflection is not proportional to the compressive load P. Hence the method of superposing a compressive deflection due to P and a bending deflection due to the couple Pe cannot be used for column action.

In considering column action the concept of "critical load" is of fundamental importance. As the argument of the cosine in the equations for the maximum deflection approaches a value of $\pi/2$, the deflection δ increases without bound and in actual practice the column will fail in a buckling *regardless of the eccentricity e.* Substituting the value $pL = \pi/2$ and $pL/2 = \pi/2$ back into $p^2 = P/EI$ will yield

$$P_{CR} = \pi^2 EI/4L^2 \qquad \text{for the single built-in support} \tag{2.194}$$

$$P_{CR} = \pi^2 EI/L^2 \qquad \text{for the hinged ends} \tag{2.195}$$

$$P_{CR} = 4\pi^2 EI/L^2 \qquad \text{for the both ends built-in} \tag{2.196}$$

The above equations for the critical load (Euler's loads) depend only on the dimensions of the column (I/L^2) and the modulus of the material E. If the moment of inertia is written $I = Ak^2$, where k is the radius of gyration, the critical-load expressions take the form

$$P_{CR} = \frac{C_1 \pi^2 AE}{(L/k)^2} \tag{2.197}$$

where the dependence is now on the material and a slenderness ratio L/k.

The "critical stress" for a column with hinged ends is given by

$$\sigma_{CR} = \frac{P_{CR}}{A} = \frac{\pi^2 E}{(L/k)^2} \tag{2.198}$$

A plot of σ_{CR} vs. L/k is hyperbolic in form, as shown in Fig. 2.28.

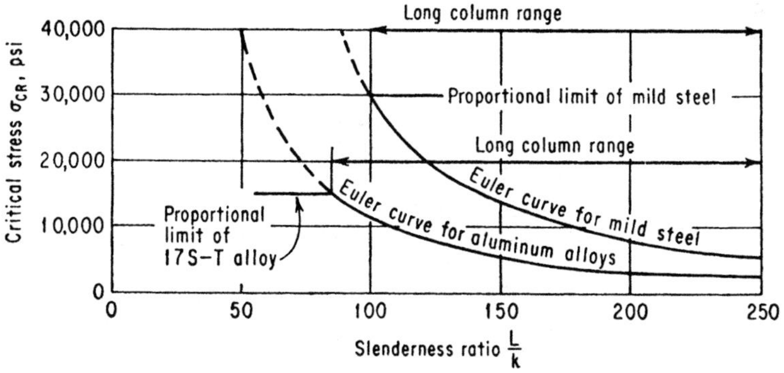

FIG. 2.28 Curve of critical stress vs. slenderness ratio for a column with hinged ends.

The horizontal line in Fig. 2.28 indicates the compressive yield stress of a typical structural steel. For analysis purposes, one uses the compressive yield stress as the design criterion for *small* slenderness ratios and the Euler curve for higher ratios.

Much column design, especially in heavy structural engineering, is accomplished by means of the application of empirical formulas developed as a result of experimental work and practical experience. Several of these formulas are presented below, for hinged bars. Straight-line formulas for structural-steel bars:

$$\sigma_{CR} = P_{CR}/A = 48{,}000 - 210(L/k) \tag{2.199}$$

Parabolic formula for structural steels:

$$\sigma_{CR} = 40{,}000 - 1.35(L/k)^2 \tag{2.200}$$

Gordon-Rankine formula for main members with $120 < L/k < 200$:

$$\sigma_\omega = \frac{18{,}000}{1 + L^2/18{,}000k^2}\left(1.6 - \frac{L}{200k}\right) \tag{2.201}$$

where σ_ω is a working stress.

2.12 SHAFTS, TORSION, AND COMBINED STRESS[1]

2.12.1 Torsion of Solid Circular Shafts

When a solid circular shaft is subjected to a pure torsional load, the reaction of the shaft for small angles of twist is assumed to be subject to the following restraints:

1. Circular cross sections remain circular and their diameters remain unchanged.
2. The axial distances between adjacent cross sections do not change.
3. A lateral surface element of the shaft is in a state of pure shear.

The shearing stresses acting on an element a distance r from the axis are

$$\tau = Gr\theta \tag{2.202}$$

where θ = angle of twist per unit length of shaft
G = modulus of rigidity $= E/2(1 + \nu)$

The shear stress, maximum at the surface is

$$\tau_{max} = \tfrac{1}{2}G\theta D \tag{2.203}$$

where D = diameter of the shaft.

The total torque acting on the shaft can be expressed as

$$T = \int_A r(\tau\, dA) = \int_A G\theta r^2\, dA = G\theta J \tag{2.204}$$

$$T = K\theta \tag{2.205}$$

where J = polar moment of inertia of the circular cross section and K = torsional rigidity = GJ for circular shafts. Therefore, the most useful relations in dealing with circular-shaft torsional problems are

$$\tau_{max} = TD/2J = 16T/\pi D^3 \tag{2.206}$$

$$\phi = TL/GJ = TL/K \tag{2.207}$$

where ϕ = total angle of twist for the shaft of length L.

2.12.2 Shafts of Rectangular Cross Section

For a shaft of rectangular cross section,

$$\tau_{max} = (T/ht^2)[3 + 1.8(t/h)] \qquad (h > t) \tag{2.208}$$

$$\phi = TL/K \tag{2.209}$$

where $K = \beta h t^3 G$ and the constant β is 0.141, 0.249, and 0.333 for various ratios of h/t of 1.0, 2.5, and ∞, respectively.

2.12.3 Single-Cell Tubular-Section Shaft

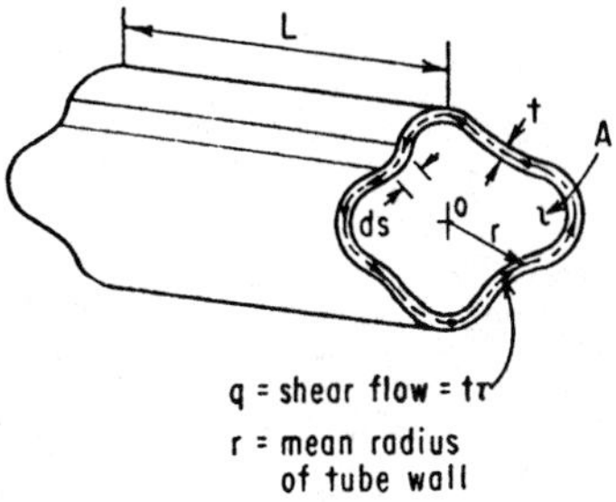

FIG. 2.29 Shaft with general tubular shape.

In Fig. 2.29 a general tubular shape is subjected to a pure torque load; the resultant angular rotations will be about point 0. Considering a slice of the tube of length L, the "shear flow" q can be shown to be constant around the tube. In thin-walled-tube problems the shear is expressed in terms of force per unit length and is thought of as "flowing" from *source* to *sink* in much the same manner as in hydrodynamics problems. Thus it can be shown that

$$q = T/2A \tag{2.210}$$

where A = area enclosed by the median line of the section = πr^2 for a circular tubular section of mean radius r. The total strain energy and total angle of twist (with no warping) for the tube is

$$U = \oint (q^2 L/2tG)\, ds \tag{2.211}$$

$$\phi = \partial U/\partial T = (TL/4A^2G) \oint ds/t$$

$$= (L/2AG) \oint (q/t)\, ds \tag{2.212}$$

The unit angle of twist is

$$\theta = \phi/L = (1/2AG) \oint (q/t)\, ds \tag{2.213}$$

The torsional rigidity of the tubular section may then be defined as

$$K = \frac{4A^2}{\oint (ds/t)} G \tag{2.214}$$

For a circular pipe of mean diameter D_m, the value of the rigidity becomes

$$K = (\pi D_m^3 t/4)G \tag{2.215}$$

2.12.4 Combined Stresses

Frequently problems in shafts involve combined torsion and bending. If the weight of the shaft is neglected relative to load P, there are two major components contributing to the maximum stress:

1. Torsional stress τ, which is a maximum at any point on the surface of the shaft
2. Flexural stress σ, which is a maximum at the built-in end on an element most remote from the shaft axis

The direct stresses due to shear are not significant, since they are a maximum at the axis of the shaft where the flexural stress is zero. The loads on the element at the built-in end are $T = Pr$ and $M = -PL$, and the stresses may be evaluated as follows:

$$\tau = TD/2J = PrD/2J \tag{2.216}$$

$$\sigma = Mc/I = \pm PLD/2I \tag{2.217}$$

where r = load offset distance
D = shaft diameter
L = shaft length

The maximum principle stress on the element in tension is

$$\sigma_{max} = \sigma/2 + \tfrac{1}{2}\sqrt{\sigma^2 + 4\tau^2} \tag{2.218}$$

Noting that for a circular shaft the polar moment is twice the area moment about a diameter ($J = 2I$):

$$\sigma_{max} = D/4I(M + \sqrt{M^2 + T^2}) = (PLD/4I)[1 + \sqrt{1 + (r/L)^2}] \tag{2.219}$$

where $I = \pi D^4/64$

The maximum shear stress on the element is

$$\tau_{max} = D/4I\sqrt{M^2 + T^2} = (PLD/4)\sqrt{1 + (r/L)^2} \tag{2.220}$$

EXAMPLE 1 For a hollow circular shaft the loads and stresses are

$$T = Pr \qquad \tau = PrD/2J$$

$$M = PL \qquad \sigma = PLD/2I$$

$$\sigma_{max} = \frac{16D_0PL}{\pi(D_0^4 - D_i^4)}\left[1 + \sqrt{1 + \left(\frac{r}{L}\right)^2}\right]$$

$$\tau_{max} = \frac{16D_0PL}{\pi(D_0^4 - D_i^4)}\left[\sqrt{1 + \left(\frac{r}{L}\right)^2}\right] \tag{2.221}$$

For the angular deflection,

$$K = \pi D^3tG/4$$

$$\theta = T/K = 4Pr/\pi D^3tG \tag{2.222}$$

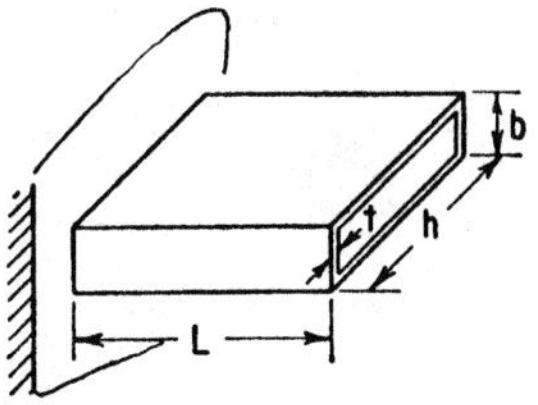

FIG. 2.30 Rectangular thin-walled section.

$$\phi = \theta L = 4PrL/\pi D^3 tG$$

where $D = (D_0 + D_i)/2$.

EXAMPLE 2 The shear stresses and angular deflection of a box as shown in Fig. 2.30 will now be investigated.

The shear flow is given by

$$q = T/2A \tag{2.223}$$

where A = plane area $(b-t)(h-t)$. The shear stress

$$\tau = q/t = T/2t(b-t)(h-t) \tag{2.224}$$

The steps in obtaining the torsional rigidity are

$$\oint ds/t \approx 1/t[2(b-t) + 2(h-t)] \tag{2.225}$$

$$K = \frac{4A^2G}{\oint ds/t} = \frac{2Gt[(b-t)(h-t)]^2}{(b-t)+(h-t)} \tag{2.226}$$

$$\phi = \theta L = \frac{TL}{K} = \frac{TL[(b-t)+(h-t)]}{2Gt[(b-t)(h-t)]^2} \tag{2.227}$$

2.13 PLATE THEORY[3,10]

2.13.1 Fundamental Governing Equation

In deriving the first-order differential equation for a plate under action of a transverse load, the following assumptions are usually made:

1. The plate material is homogeneous, isotropic, and elastic.
2. The least lateral dimension (length or width) of the plate is at least 10 times the thickness h.
3. At the boundary, the edges of the plate are unrestrained in the plane of the plate; thus the reactions at the edges are taken transverse to the plate.
4. The normal to the original middle surface remains normal to the distorted middle surface after bending.
5. Extensional strain in the middle surface is neglected.

The deflected shape of a simply supported plate due to a normal load q is illustrated in Fig. 2.31, which also defines the coordinate system. The positive shears, twists, and moments which act on an element of the plate are depicted in Fig. 2.32.

Application of the equations of equilibrium and Hooke's law to the differential element leads to the following relations:

$$\begin{aligned}
&\partial M_x/\partial x + \partial M_{yx}/\partial y = Q_x && M_{xy} = D(1-\nu)(\partial^2 w/\partial x\,\partial y) = -M_{yx} \\
&\partial M_y/\partial y + \partial M_{yx}/\partial x = Q_y && M_x = -D(\partial^2 w/\partial x^2 + \nu\partial^2 w/\partial y^2) \\
&\partial Q_x/\partial x + \partial Q_y/\partial y = -q && M_y = -D(\partial^2 w/\partial y^2 + \nu\partial^2 w/\partial x^2)
\end{aligned} \tag{2.228}$$

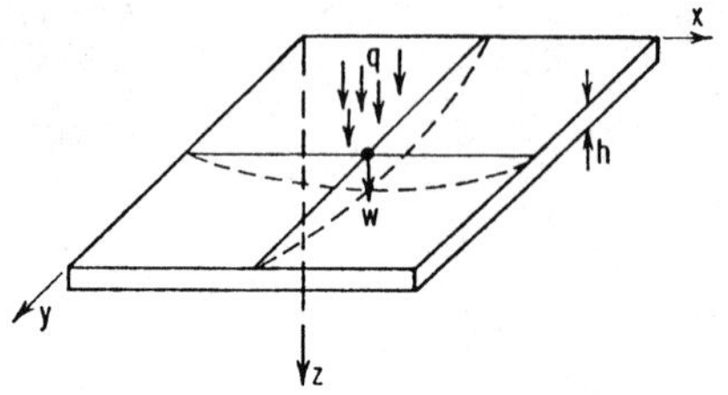

FIG. 2.31 Simply supported plate subjected to normal load q.

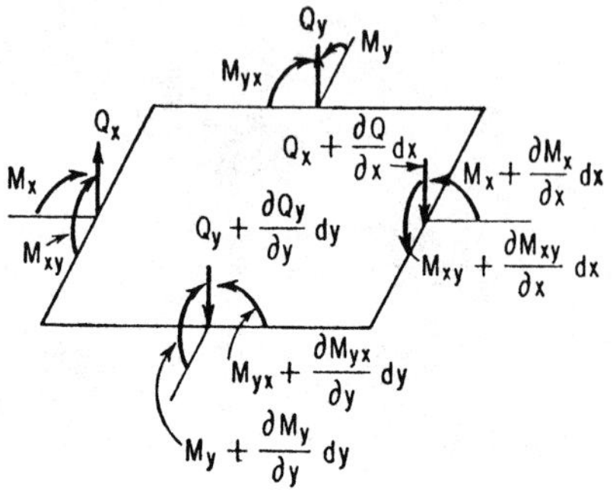

FIG. 2.32 Shears, twists, and moments on a plate element.

where $D = Eh^3/12(1 - \nu^2)$ = plate stiffness and is analogous to the flexural rigidity per unit width (EI) of beam theory.

Properly combining Eqs. (2.228) leads to the basic differential equation of plate theory

$$\nabla^4 w = \partial^4 w/\partial x^4 + 2(\partial^4 w/\partial^2 x\, \partial^2 y) + \partial^4 w/\partial y^4 = q/D \tag{2.229}$$

This may be compared with the similar equation for beams

$$d^4 y/dx^4 = q/EI \tag{2.230}$$

The solution of a specific plate problem involves finding a function w which satisfies Eq. (2.229) and the boundary conditions. With w known, the stresses may be evaluated by employing

$$\begin{aligned} \sigma_{max} &= \pm\, 6M_x/h^2 \\ \sigma_{max} &= \pm\, 6M_y/h^2 \\ \tau_{max} &= 6M_{xy}/h^2 \end{aligned} \tag{2.231}$$

2.13.2 Boundary Conditions

The usual support conditions and the associated boundary conditions are as follows: Simply supported plate:

$w = 0$ (zero deflection)

$M = 0$ (zero moment)

Built-in plate:

$w = 0$ (zero deflection)

$w' = 0$ (zero slope)

Free boundary:

$$M = 0 \qquad \text{(zero moment)}$$

$$V = 0 \qquad \text{(zero reactive shear)}$$

where it is noted that

$$\begin{aligned} V_x &= Q_x - \partial M_{xy}/\partial y = -D[\partial^3 w/\partial x^3 + (2 - \nu)(\partial^3 w/\partial x\, \partial y^2)] \\ V_y &= Q_y - \partial M_{xy}/\partial x = -D[\partial^3 w/\partial y^3 + (2 - \nu)(\partial^3 w/\partial y\, \partial x^2)] \end{aligned} \tag{2.232}$$

where positive V is directed similar to positive Q.

Note that two boundary conditions are necessary and sufficient to solve the problem of bending for plates with transverse loads.

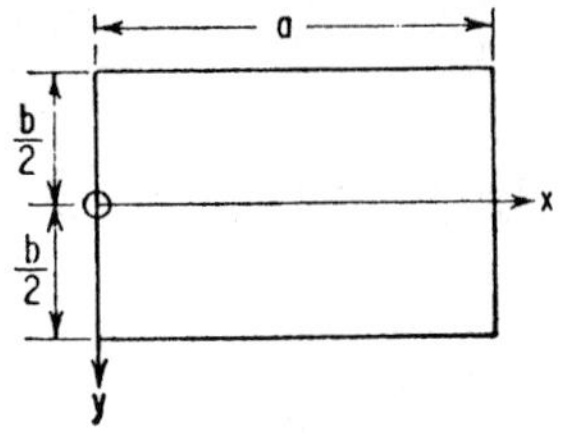

FIG. 2.33 Rectangular plate, simply supported.

EXAMPLE 1 For this problem the loading on a simply supported plate is taken to be distributed uniformly over the surface of that plate, and the coordinate axes are redefined as indicated in Fig. 2.33.

The deflection solution must satisfy Eq. (2.229) and the boundary conditions

$$w = 0 \quad \text{and} \quad \partial^2 w/\partial x^2 = 0 \quad \text{at } x = 0, a$$

$$w = 0 \quad \text{and} \quad \partial^2 w/\partial y^2 = 0 \quad \text{at } y = \pm\, b/2$$

A series solution of the form

$$w = \sum_{m=1}^{\infty} Y_m \sin \frac{m\pi x}{a} \tag{2.233}$$

is assumed, and after manipulation, the following general expression results for the deflection surface:

$$w = \frac{4qa^4}{\pi^5 D} \sum_{m=1,3,\ldots}^{\infty} \frac{1}{m^5}\left(1 - \frac{\alpha_m \tanh \alpha_m + 2}{2 \cosh \alpha_m} \cosh \frac{2\alpha_m y}{b} + \frac{\alpha_m}{2 \cosh \alpha_m} \frac{2y}{b} \sinh \frac{2\alpha_m y}{b}\right) \sin \frac{m\pi x}{a} \tag{2.234}$$

where $\alpha_m = m\pi b/a$. The maximum deflection occurs at $x = a/2$ and $y = 0$:

$$w_{\max} = \frac{4qa^4}{\pi^5 D} \sum_{m=1,3,\ldots}^{\infty} \frac{(-1)^{(m-1)/2}}{m^5}\left(1 - \frac{\alpha_m \tanh \alpha_m + 2}{2 \cosh \alpha_m}\right) = \frac{\alpha q a^4}{D} \tag{2.235}$$

where $\alpha = \alpha(b/a)$. The maximum moments $M_{x\,\max}$ and $M_{y\,\max}$ may be expressed by $\beta q a^2$ and $\beta_1 q a^2$, respectively, where β and β_1 are functions of b/a.

The final group of plate problems deals with edge reactions which are distributed along the edge and the concentrated forces at the corners to keep the corners of the plate from rising, as shown in Fig. 2.34.

Note that the V_x and V_y are negative in sign, whereas R is positive and directed down in accordance with the adopted sign convention for plates.

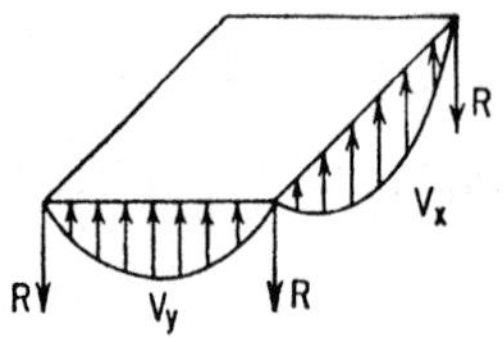

FIG. 2.34 Edge reactions of simply supported plate.

The maximum value of V_x, V_y, and R may be expressed by δqa, $\delta_1 qa$, and ηqa, respectively, where δ, δ_1, and η have the same basic functional relation as the earlier defined coefficients.

Figure 2.35 presents curves of the important plate coefficients.

EXAMPLE 2 The solution for the simple-support uniform-load problem as illustrated above is used as the basis for the solution for a built-in-edge problem. Transposing coordinates as in Fig. 2.36,

$$w = \frac{4qa^4}{\pi^5 D} \sum_{m=1,3,\ldots}^{\infty} \frac{(-1)^{(m-1)/2}}{m^5} \cos \frac{m\pi x}{a} \left(1 - \frac{\alpha_m \tanh \alpha_m + 2}{2 \cosh \alpha_m} \cosh \frac{m\pi y}{a} + \frac{1}{2 \cosh \alpha_m} \frac{m\pi y}{a} \sinh \frac{m\pi y}{a}\right) \tag{2.236}$$

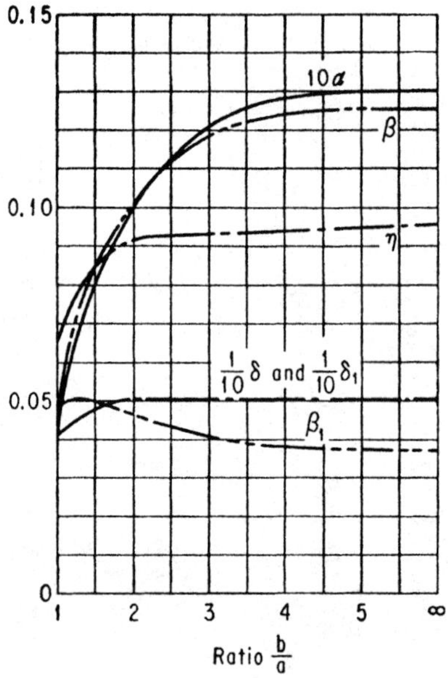

FIG. 2.35 Deflection and moment coefficients for rectangular plate on simple supports.

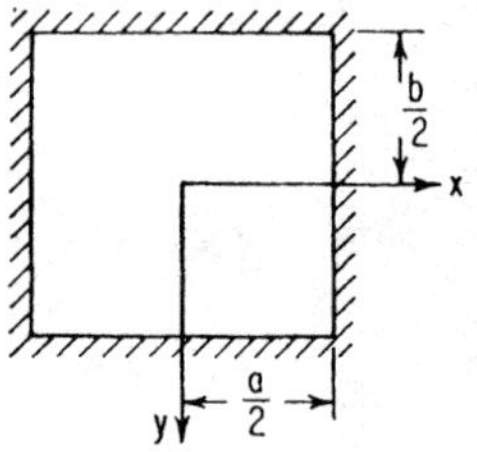

FIG. 2.36 Rectangular plate with built-in edges.

To satisfy the edge restrictions imposed by built-in supports, the deflection of a plate with moments distributed along the edges is superposed on this simple-support solution. These moments are then adjusted to satisfy the built-in boundary condition $\partial w/\partial n = 0$. The procedure finally results in a solution for the maximum deflection at the center of the plate of the form

$$w_{\max} = \alpha' qa^4/D \tag{2.237}$$

The coefficient α' can be evaluated as a function of b/a.

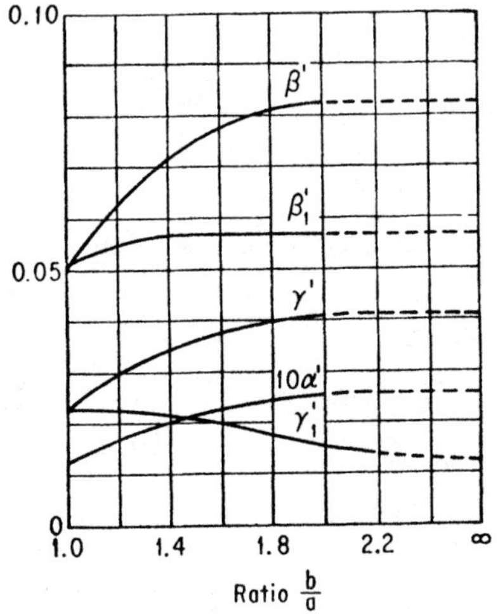

FIG. 2.37 Deflection and moment coefficients for rectangular plate with built-in edges.

Similar expressions for the moments at the center of the plate and at two edges are as follows:

$$\text{Edge:} \quad M_x = -\beta' q a^2 \qquad M_y = -\beta_1' q a^2 \tag{2.238a}$$

$$\text{Center:} \quad M_x = \gamma' q a^2 \qquad M_y = \gamma_1' q a^2 \tag{2.238b}$$

Figure 2.37 presents α', β', β_1', γ', γ_1', in graphical form.

EXAMPLE 3 We shall use a numerical method to solve the simple-support uniform-load problem. The following expressions define the finite-difference form for the second-order partial derivatives in question (see Fig. 2.38).

$$\left.\frac{\partial^2 w}{\partial x^2}\right|_i = \frac{w_{i+1} - 2w_i + w_{i-1}}{h^2} \qquad \left.\frac{\partial^2 w}{\partial y^2}\right|_i = \frac{w_{i'+1} - 2w_i + w_{i'-1}}{k^2} \tag{2.239}$$

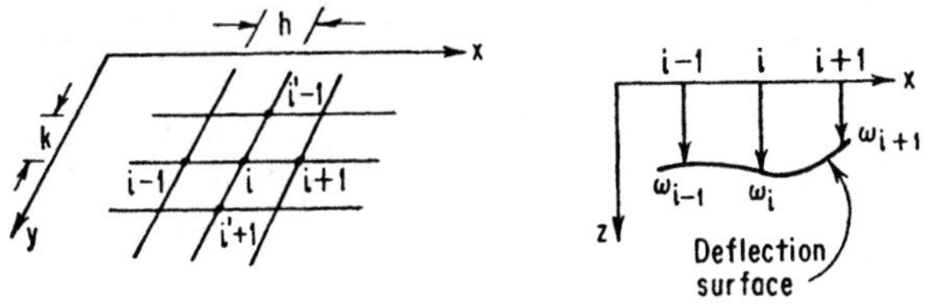

FIG. 2.38 Grid of rectangular plate, simply supported.

For the grid, $h = k$,

$$\nabla^2 w = (1/h^2)(w_{i+1} + w_{i'+1} + w_{i-1} + w_{i'-1} - 4w_i) \tag{2.240}$$

Utilizing

$$\phi = (\nabla^2 w)D \tag{2.241}$$

and

$$\nabla^2 \phi = (1/h^2)(\phi_{i+1} + \phi_{i'+1} + \phi_{i-1} + \phi_{i'-1} - 4\phi_i) \tag{2.242}$$

The governing equation is

$$\nabla^2 \phi = q \tag{2.243}$$

Consider now a square plate coarsely divided into four segments. Then evidently $h = k = a/2$ and $w_{i+1} = w_{i'+1} = w_{i-1} = w_{w'-1} = 0$ since these points are on the boundary. Also $\nabla^2 w_{i+1} = \nabla^2 w_{i'+1} = 0$ and, using the last of the above equations, we find

$$\phi_{i+1} = \phi_{i'+1} = \phi_{i-1} = \phi_{i'-1} = 0$$

Therefore, $$\nabla^2\phi = -4\phi_i/h^2 = q \quad (2.244)$$

Substituting back into the original equation yields

$$\nabla^2 w_i = \phi_i/D = -4w_i/h^2 = -qa^2/16D \quad (2.245)$$

$$w_i = (1/256)(qa^4/D) \quad (2.246)$$

This is approximately 3.5 percent below the maximum deflection obtained by analytical means:

$$w_{max} = 0.00406(qa^4/D) \quad (2.247)$$

Thus, even with this crude grid, the center deflection can be obtained with reasonable accuracy. Since the stress is obtained by combinations of higher derivatives, however, a finer grid is required for a correspondingly accurate stress evaluation.

2.14 SHELL THEORY[3,10]

2.14.1 Membrane Theory: Basic Equation

The general problem of determining stresses in shells may be subdivided into two distinct categories. The first of these is a moment-free "membrane" state of stress, which often predominates over a major portion of the shell. The second, associated with bending effects, is the "discontinuity stress," which affects the shell for a limited segment in the vicinity of a load or profile discontinuity. The final solution often consists of the membrane solution, corrected locally in the regions of the boundaries for the discontinuity effects.

The "membrane solution" for shells in the form of surfaces of revolution, and loaded symmetrically with respect to the axis, often requires the following simplifying assumptions:

1. The thickness h of the shell is small compared with other dimensions of the shell, and the radii of curvature.
2. Linear elements which are normal to the undisturbed middle surface remain straight and normal to the deflected middle surface of the shell.
3. Bending stresses are small and can be neglected; so that only the direct stresses due to strains in the middle surface need be considered.

The basic element for a shell of revolution when subjected to axisymmetric loading is shown in Fig. 2.39. From conditions of static equilibrium the following equations may be derived:

$$(d/d\phi)(N_\varphi r_0) - N_\theta r_1 \cos\phi + Yr_1r_0 = 0 \quad (2.248)$$

$$N_\varphi r_0 + N_\theta r_1 \sin\phi + Zr_1r_0 = 0 \quad (2.249)$$

where Y and Z are the components of the external load parallel to the two-coordinate axis, respectively. Thus, in general, if these components of the load are known and the geometry of the shell is specified, the forces N_ϕ and N_θ can be calculated.

If now the conditions of static equilibrium are applied to a portion of a shell above a parallel circle instead of an element as indicated in Fig. 2.40, then the following equations result:

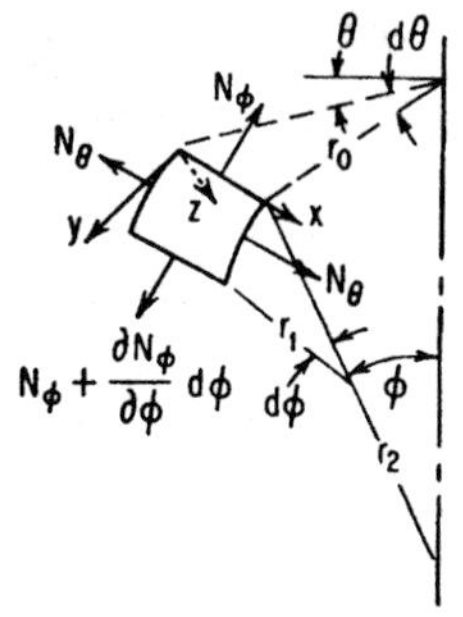

FIG. 2.39 Element of shell of revolution.

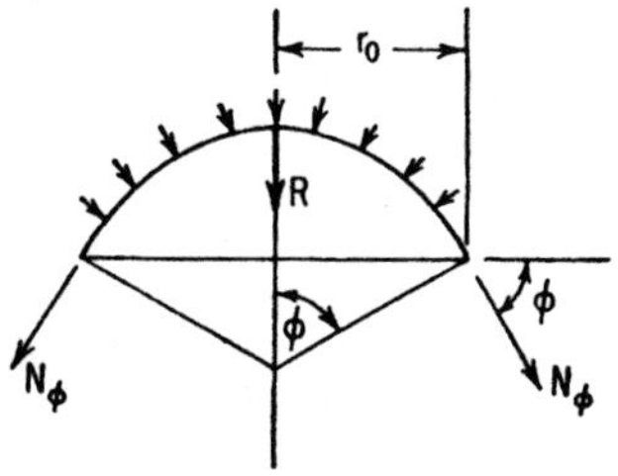

FIG. 2.40 Shell portion above parallel circle.

$$2\pi r_0 N_\phi \sin\phi + R = 0 \tag{2.250}$$

$$N_\phi/r_1 + N_\theta/r_2 = -Z \tag{2.251}$$

where R is the resultant of the total load on that part of the shell in question. Again these two equations suffice for the determination of the two forces N_ϕ and N_θ.

The associated principal stresses for these members are then

$$\sigma_\phi = N_\phi/h \qquad \text{and} \qquad \sigma_\theta = N_\theta/h \tag{2.252}$$

To complete the consideration of membrane action for shells with axisymmetric loading, the concomitant displacements must be computed. The procedure indicated below applies for symmetrical deformations, in which the displacement (along a meridian) is indicated by v and the displacement in the radial direction is w (with an inward displacement taken as positive).

From geometry and Hooke's law,

$$dv/d\phi - v\cot\phi = f(\phi) \tag{2.253}$$

where $$f(\phi) = (1/Eh)[N_\phi(r_1 + \nu r_2) - N_\theta(r_2 + \nu r_1)]$$

The solution of Eq. (2.253) is

$$v = \sin\phi\{\textstyle\int[f(\phi)/\sin\phi]d\phi + C\} \tag{2.254}$$

The constant C is evaluated from the boundary conditions of the problem. The radial displacement is then determined from the following:

$$w = v\cot\phi - (r_2/EH)(N_\theta - \nu N_\phi) \tag{2.255}$$

A similar analysis on a cylindrical shell without the restriction of symmetric loading will yield the basic equations

$$\partial N_x/\partial x + (1/r)(\partial N_{x\phi}/\partial\phi) = -X \tag{2.256}$$

$$\partial N_{x\phi}/\partial x + (1/r)(\partial N_\phi/\partial\phi) = -Y \tag{2.257}$$

$$N_\phi = -Zr \tag{2.258}$$

where the coordinate axes are redefined in Fig. 2.41 to be consistent with usual practice and Eqs. (2.254) and (2.255) are suitably modified.

2.14.2 Example of Spherical Shell Subjected to Internal Pressure

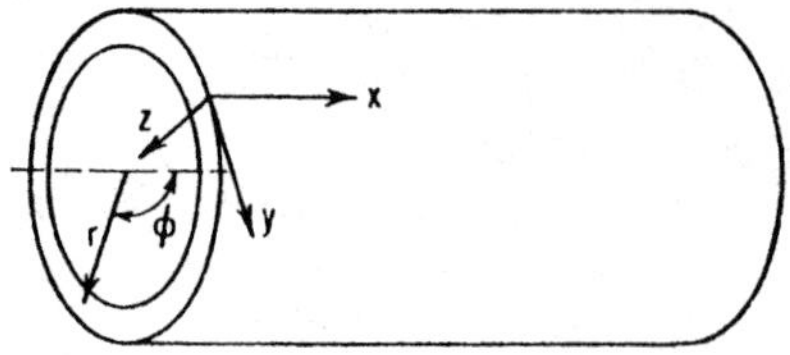

FIG. 2.41 Cylindrical shell.

For a hemispherical section of radius a, the forces N_ϕ distributed along the edges are required to maintain equilibrium. These may be determined from the previous equations. Now

$$2\pi a N_\phi \sin\phi + R = 0 \qquad (2.259)$$

where $\qquad R = -\pi a^2 p \qquad \sin\phi = \sin(\pi/2) = 1$

so that

$$N_\phi = pa/2 \qquad (2.260)$$

From the other equation of equilibrium,

$$N_\phi/a + N_\theta/a = -Z = p \qquad (2.261)$$

where

$$N_\theta = pa/2$$

The stresses may now be written as

$$\sigma_\phi = \sigma_\theta = pa/2h \qquad (2.262)$$

From loading symmetry, it is concluded that $v = 0$ and the increase in the radial direction is given by

$$w = (-a/Eh)(N_\theta - \nu N_\phi) \qquad (2.263)$$

$$w = (-pa^2/2Eh)(1 - \nu) \qquad (2.264)$$

2.14.3 Example of Cylindrical Shell Subjected to Internal Pressure

Using a procedure similar to that outlined above, the following relationships are derived:
For open-ended cylinder:

$$\sigma_\phi = pa/h \qquad (2.265)$$

$$w = -pa^2/Eh \qquad (2.266)$$

For closed-ended cylinder:

$$\sigma_\phi = pa/h \qquad (2.267)$$

$$\sigma_x = pa/2h \qquad (2.268)$$

$$w = -(pa^2/Eh)(1 - \nu/2) \qquad (2.269)$$

2.14.4 Discontinuity Analysis

The membrane solution does not usually satisfy all "edge" conditions, and it is therefore often necessary to superpose a second, "edge-loaded" shell in order to obtain a *complete* solution.

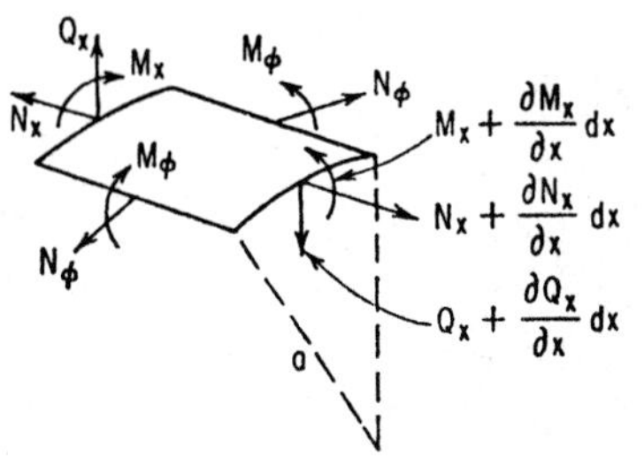

FIG. 2.42 Cylindrical-shell element.

To develop the general equations for the cylindrical shell with axisymmetric load, consider an element as shown in Fig. 2.42, where *bending moments are assumed acting*. The coordinate system is as defined in Fig. 2.41. All forces and moments are shown in their positive directions. Referring to Fig. 2.42, Q_x is the shear force per unit length, M_x is the axial moment per unit length, M_ϕ is the circumferential moment per unit length, and N_x, N_ϕ are the normal forces defined in the discussion of membrane theory.

Applying the equations of equilibrium and Hooke's law and expressing the curvature as a function of moment (as was done in developing the plate equations), the following basic shell differential equation is obtained:

$$d^4w/dx^4 + 4\beta^4 w = Z/D \tag{2.270}$$

where $D = Eh^3/12(1 - \nu^2)$
$\beta^4 = 3(1 - \nu^2)/a^2h^2$
w = radial deflection of shell (positive inward)

The solution of the above equation in any particular case depends on the specific boundary conditions at the ends of the cylinder.

One very useful solution is for the case of a long cylinder, without radial pressure, subjected to uniformly distributed forces and moments along the edge, $x = 0$. The assumed positive directions of these loads are as shown in Fig. 2.43. The resultant expression for the deflection is

$$w = (e^{-\beta x}/2\beta^3 D)[\beta M_0(\sin \beta x - \cos \beta x) - Q_0 \cos \beta x] \tag{2.271}$$

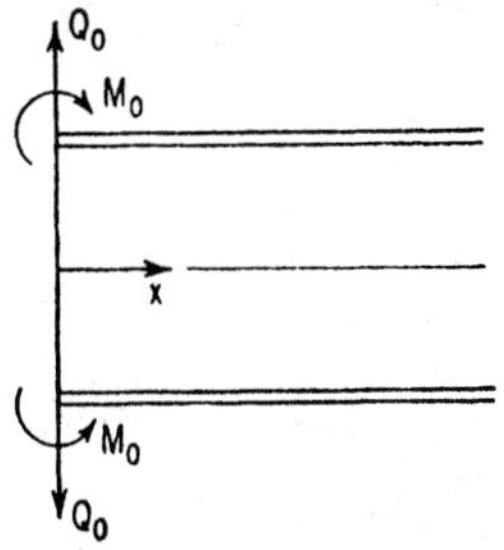

FIG. 2.43 Long-cylinder section.

The maximum deflection occurs at the loaded end and is evaluated as

$$w_{x=0} = -(1/2\beta^3 D)(\beta M_0 + Q_0) \tag{2.272}$$

The accompanying slope at the loaded end is

$$w'_{x=0} = (1/2\beta^2 D)(2\beta M_0 + Q_0) = (dw/dx)_{x=0} \tag{2.273}$$

The successive derivatives of the above expression for deflection can be written in the following simplified form:

$$
\begin{aligned}
w &= -(1/2\beta^3 D)[\beta M_0 \psi(\beta x) + Q_0 \theta(\beta x)] \\
w' &= (1/2\beta^2 D)[2\beta M_0 \theta(\beta x) + Q_0 \phi(\beta x)] \\
w'' &= -(1/2\beta D)[2\beta M_0 \phi(\beta x) + 2Q_0 \xi(\beta x)] \\
w''' &= (1/D)[2\beta M_0 \xi(\beta x) - Q_0 \psi(\beta x)]
\end{aligned} \tag{2.274}
$$

where $\phi(\beta x) = e^{-\beta x}(\cos \beta + \sin \beta x)$

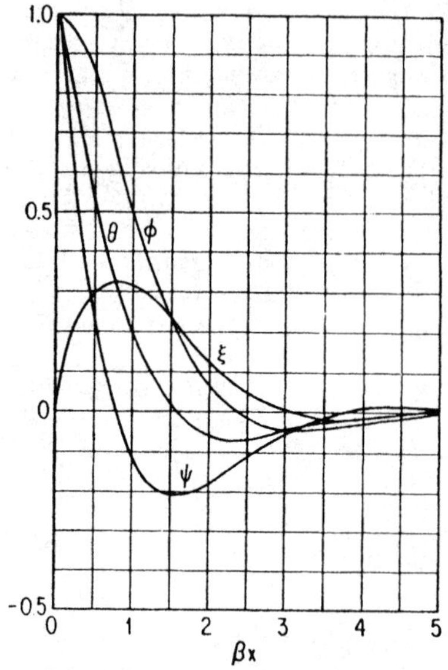

FIG. 2.44 Slope and deflection functions.

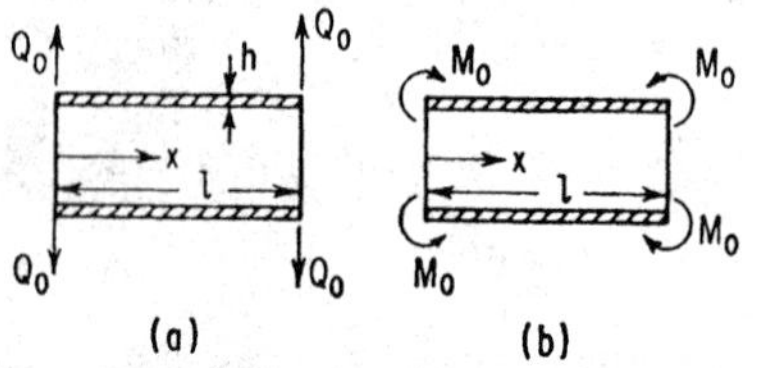

FIG. 2.45 Short shell. (*a*) Bending by shears. (*b*) Bending by moments.

$$\psi(\beta x) = e^{-\beta x}(\cos \beta x - \sin \beta x)$$

$$\theta(\beta x) = e^{-\beta x} \cos \beta x$$

$$\zeta(\beta x) = e^{-\beta x} \sin \beta x$$

Figure 2.44 is a plot of ϕ, ψ, θ, ζ as a function of βx. Because each function decreases in absolute magnitude with increasing βx, in most engineering applications the effect of edge loads may be neglected at locations for which $\beta x > \pi$.

For the short shell, for which opposite end conditions interact, the following results are obtained. For the case of bending by uniformly distributed shearing forces, as shown in Fig. 2.45*a,* the slope and deflection are given by

$$w_{x=0,l} = -(2Q_0\beta a^2/Eh)\chi_1(\beta l) \tag{2.275}$$

$$w'_{x=0,l} = \pm(2Q_0\beta^2 a^2/Eh)\chi_2(\beta l) \tag{2.276}$$

where

$$\chi_1(\beta l) = (\cosh \beta l + \cos \beta l)/(\sinh \beta l + \sin \beta l)$$

$$\chi_2(\beta l) = (\sinh \beta l - \sin \beta l)/(\sinh \beta l + \sin \beta l)$$

For the case of bending by uniformly distributed moments M_0 (as shown in Fig. 2.45*b*), the slope and deflection are given by

$$w_{x=0,l} = -(2M_0\beta^2 a^2/Eh)\chi_2(\beta l) \tag{2.277}$$

$$w'_{x=0,l} = \pm(4M_0\beta^3 a^2/Eh)\chi_3(\beta l) \tag{2.278}$$

where

$$\chi_3(\beta l) = (\cosh \beta l - \cos \beta l)/(\sinh \beta + \sin \beta l)$$

Figure 2.46 is a plot of the functions χ_1, χ_2, and χ_3 as a function of βl. The axial bending moment at any location is given by

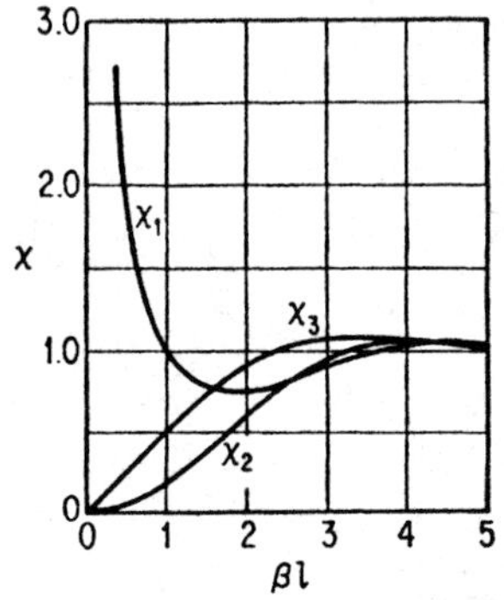

FIG. 2.46 Slope and deflection functions.

$$M_x = -Dw'' \quad (2.279)$$

and the maximum stress which occurs at the inside and outside surfaces of the shell,

$$\sigma_{x,\text{bending}} = \pm(6M_x/h^2) \quad (2.280)$$

Likewise the shear force at any point is given by

$$Q_x = -Dw''' \quad (2.281)$$

and the associated maximum shear stress is $\tau_x = 3Q_x/2h$, which occurs midway between the inner and outer surfaces; the shear stress at the surface is zero.

EXAMPLE Examine the case of a long cylinder subjected to an internal pressure and fixed at the ends as depicted in Fig. 2.47*a*; axial pressure is taken to be zero.

The stress and deformation of this shell can be obtained by the superposition of two distinct problems, the membrane and edge-loaded cylinders. The first presupposes free ends and a membrane action as indicated in Fig. 2.47*b*. The built-in ends resist this membrane deflection at the edges through a system of forces Q_0 and moments M_0 which are required to enforce the boundary conditions of zero deflection and rotation, as shown in Fig. 2.47*c*.

The increase in the radius due to membrane action, as a result of pressure, is then obtained from the membrane solution

$$-w_p = \delta = pa^2/Eh \quad (2.282)$$

The boundary conditions for the edge-loaded problem, based on the actual built-in ends, become

$$w_x = \delta \quad \text{and} \quad w'_{x=0} = 0$$

Hence

$$\delta = (1/2\beta^3 D)(\beta M_0 + Q_0) \quad (2.283)$$

$$0 = (1/2\beta^2 D)(2\beta M_0 + Q_0) \quad (2.284)$$

Solving Eqs. (2.283) and (2.284), using Eq. (2.282)

$$M_0 = p/2\beta^2 \quad \text{and} \quad Q_0 = -p/\beta \quad (2.285)$$

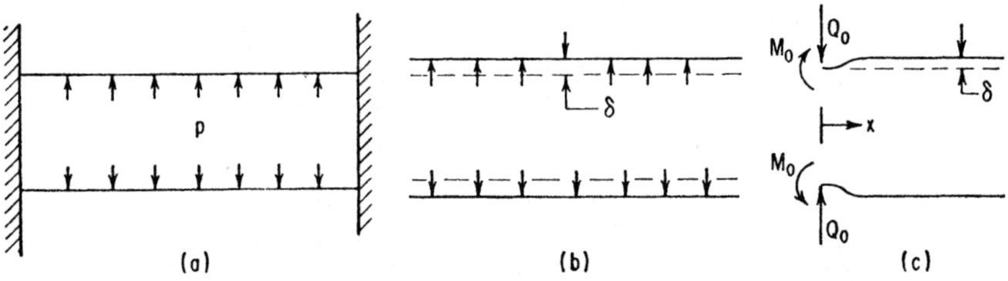

FIG. 2.47 Long cylinder with fixed ends. (*a*) Action of internal pressure. (*b*) Membrane action. (*c*) Discontinuity forces for boundary conditions.

Thus the *complete* solution for the deflection is

$$w = -(1/4\beta^4 D)[p\psi(\beta x) - 2p\theta(\beta x)] - pa^2/Eh \tag{2.286}$$

The axial stresses are given by

$$\sigma_x = \pm(6M_x/h^2) \tag{2.287}$$

where $$M_x = -Dw'' = (p/2\beta^2)[\phi(\beta x) - 2\xi(\beta x)]$$

The mean circumferential stress can be evaluated from

$$\sigma_{\phi,\text{direct}} = -Ew/a \tag{2.288}$$

and the added component of flexural stress due to the Poisson effect is

$$\sigma_{\phi,\text{bending}} = -\nu\sigma_{x,\text{bending}} \tag{2.289}$$

so that $$\sigma_{\phi,\text{total}} = \sigma_{\phi,\text{direct}} + \sigma_{\phi,\text{bending}} \tag{2.290}$$

2.15 CONTACT STRESSES: HERTZIAN THEORY[2]

As discussed in the writings of Hertz (the contact stresses presented here are often termed hertzian stresses), the maximum pressure q due to a compressive force P is given by

$$q = 3P/2\pi a^2 \tag{2.291}$$

and is taken to have a spherical distribution as shown in Fig. 2.48, where

$$a = \sqrt[3]{\frac{3P}{4}\frac{R_1R_2}{R_1 + R_2}\left(\frac{1-\nu_1^2}{E_1} + \frac{1-\nu_2^2}{E_2}\right)} \tag{2.292}$$

These expressions may be simplified, if both spheres are composed of identical materials. For Poisson's ratio ν of approximately 0.3, which is common to steel, iron, aluminum, and most structural materials, there results

$$a = 1.11\sqrt[3]{(P/E)[R_1R_2/(R_1 + R_2)]} \tag{2.292a}$$

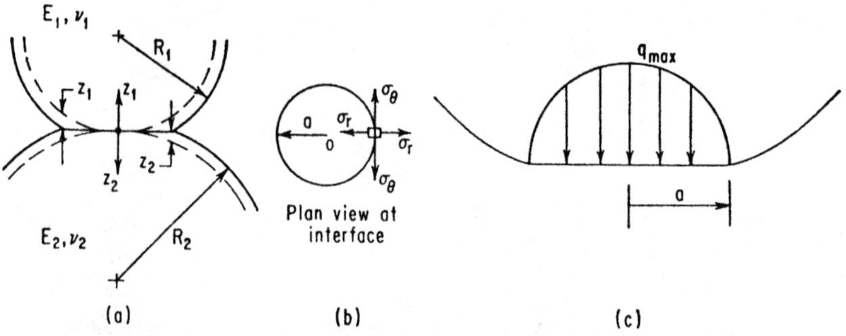

FIG. 2.48 Two spheres in contact. (*a*), (*b*), (*c*) Contact pressure distribution.

and
$$q = 0.388\sqrt[3]{PE^2[(R_1 + R_2)/R_1R_2]^2} \tag{2.292b}$$

The general stress levels in the spheres can now be presented based on the above relations. Maximum compressive stress, which occurs at point O, is

$$\sigma_z = -q \tag{2.293}$$

The maximum tensile stress in the radial direction, which occurs on the periphery of the surface of contact at radius a, is

$$\sigma_r = [(1-2\nu)/3]q \tag{2.294}$$

Maximum shear stress, which occurs under point O of Fig. 2.48*b*, at a depth

$$z_1 = 0.47a$$

is approximately

$$\tau = \tfrac{1}{3}q \tag{2.295}$$

and is in a plane inclined to the z axis. This latter stress is usually the governing criterion in the design for bodies in contact, fabricated from ductile materials. A compilation of important contact-stress cases is given in Table 2.2. Other important cases, associated with rolling-element bearings, are discussed in Chap. 15.

2.16 FINITE-ELEMENT NUMERICAL ANALYSIS[14,15,16]

2.16.1 Introduction

The chapter thus far has dealt with exact solutions to sets of equations which predict the deformation and internal stress distribution of particular bodies such as beams, columns, thin plates, and thin shells. These sets of equations are derived from the same concepts (equilibrium, kinematics, material observation) used to derive the general equations of elastic theory [Eqs. (2.47), (2.76), and (2.87)]. The particular equations for beams, plates, and shells are, in fact, derivable from the general equation by imposing the appropriate limitations with respect to thickness, etc.

In dealing with the more complex structural shapes typical of actual machinery it becomes increasingly difficult to derive appropriate sets of equations for which exact solutions may be found. For such structures predictions of deformation and stress can be effected through numerical solutions of the general equations. Three general numerical techniques are widely used to effect numerical solutions in structural mechanics: transfer-matrix techniques, finite-difference techniques, and finite-element techniques.

Transfer-matrix approaches are typically used to effect solutions in one-dimensional structures. These approaches are widely used in the solution of vibrating shafts and turbine foils. Typical transfer-matrix approaches include, among others, the Holzer method for torsional vibration and the Prohl-Miklestad (Chap. 4) procedure for lateral shaft vibration.

Finite-difference methods[17] are widely used to effect deformation and stress solutions to multidimensional structures. In this technique, the differential operators of the governing equations are replaced with difference operators which relate the values of the unknowns at a gridwork of points in the structure. Example 3 of Sec. 2.13.2 presents a simple illustration of a finite-difference method. The well-known relaxation

TABLE 2.2 Contact Stresses[2,5,12]

Sphere on a sphere, P = total load 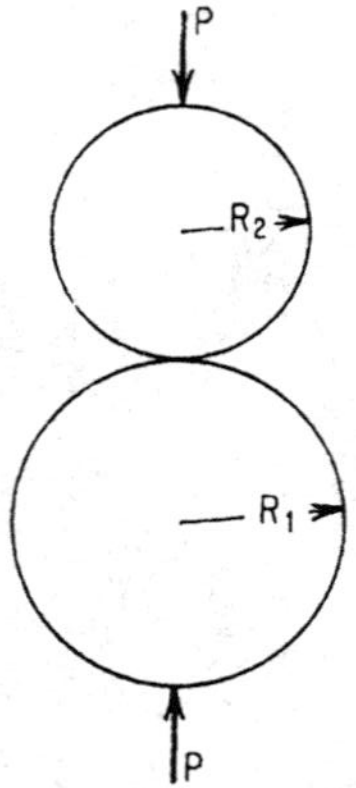	$a = 0.9085 \sqrt[3]{PR_2 \frac{1}{1 + R_2/R_1} \left(\frac{1 - \nu_1^2}{E_1} + \frac{1 - \nu_2^2}{E_2} \right)}$ $\max \sigma_z = 0.580 \sqrt[3]{\frac{P \left(\frac{1 + R_2/R_1}{R_2} \right)^2}{\left(\frac{1 - \nu_1^2}{E_1} + \frac{1 - \nu_2^2}{E_2} \right)^2}}$ $y = 0.8255 \sqrt[3]{\frac{P^2}{R_2} \left(1 + \frac{R_2}{R_1} \right) \left(\frac{1 - \nu_1^2}{E_1} + \frac{1 - \nu_2^2}{E_2} \right)^2}$ For a flat surface $R_1 = \infty$ For a concave surface R_1 is negative

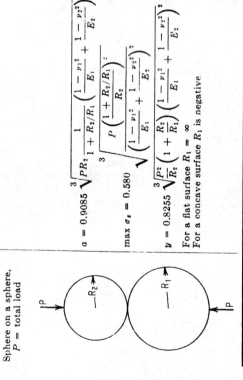

y is the decrease in center-to-center distance between the two spheres.

Cylinder on cylinder axes, parallel, P = load/linear in.

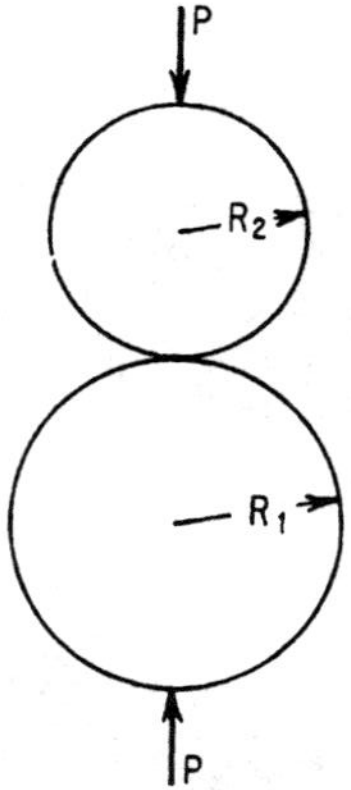

$$b = 1.13\sqrt{PR_2\frac{1}{1+R_2/R_1}\left(\frac{1-\nu_1^2}{E_1}+\frac{1-\nu_2^2}{E_2}\right)}$$

$$\max \sigma_z = 0.564\sqrt{\frac{(P/R_2)(1+R_2/R_1)}{\frac{1-\nu_1^2}{E_1}+\frac{1-\nu_2^2}{E_2}}}$$

General case of two bodies in contact, P = total pressure

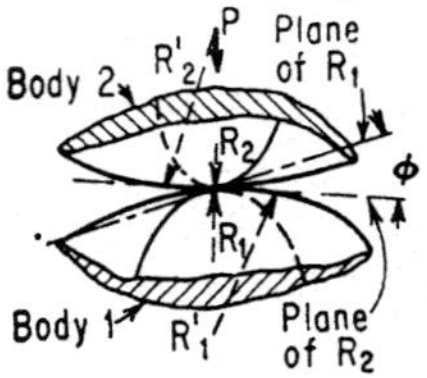

At point of contact minimum and maximum radii of curvature are R_1 and R_1' for body 1, R_2 and R_2' for body 2. Then $1/R_1$ and $1/R_1'$ are *principal curvatures* of body 1, and $1/R_2$ and $1/R_2'$ of body 2, and in each body the principal curvatures are mutually perpendicular. The plane containing curvature $1/R_1$ in body 1 makes with the plane containing curvature $1/R_2$ in body 2 the angle ϕ. Then

$\max \sigma_z = 1.5P/\pi cd$, $c = \alpha\sqrt[3]{P\delta/K}$, $d = \beta\sqrt[3]{P\delta/K}$, and $y = \lambda\sqrt[3]{P^2/K^2\delta}$, where $\delta = \dfrac{4}{1/R_1 + 1/R_2 + 1/R_1' + 1/R_2'}$

and $K = \dfrac{8}{3}\dfrac{E_1E_2}{E_2(1-\nu_1^2)+E_1(1-\nu_2^2)}$

α and β are given by the following table, where

$$\theta = \arccos \tfrac{1}{4}\,\delta\sqrt{(1/R_1 - 1/R_1')^2 + (1/R_2 - 1/R_2')^2 + 2(1/R_1 - 1/R_1')(1/R_2 - 1/R_2')\cos 2\phi}$$

θ	0°	10°	20°	30°	35°	40°	45°	50°	55°	60°	65°	70°	75°	80°	85°	90°
α	∞	6.612	3.778	2.731	2.397	2.136	1.926	1.754	1.611	1.486	1.378	1.284	1.202	1.128	1.061	1.00
β	0	0.319	0.408	0.493	0.530	0.567	0.604	0.641	0.678	0.717	0.759	0.802	0.846	0.893	0.944	1.00
λ	—	0.851	1.220	1.453	1.550	1.637	1.709	1.772	1.828	1.875	1.912	1.944	1.967	1.985	1.996	2.00

Note: y is the decrease in center-to-center distance of the two cylinders. b is the half-width of the contact surface.

method for the solution of the governing equation of multidimensional heat-transfer analysis is an example of a finite-difference solution.

Since the 1960s finite-element methods have become the preeminent tool for the numerical solution of deformation and stress problems in structural mechanics. This popularity arises from the ease with which the most general of structural geometries can be considered. Finite-element analysis replaces the exact structure to be considered with a set of simple structural elements (blocks, plates, shells, etc.) interconnected at a finite set of node points. The set of governing equations for this approximate structure can be solved exactly.

Finite-element analysis deals with the spatial approximation of complex structural shapes. It can be used directly to yield solutions in static elasticity or combined with other numerical techniques to obtain the response of structures with nonlinear material properties (plasticity, creep, relaxation), undergoing finite deformation, or subject to shock and vibration excitation. The finite-element technique has also found a wide application in the analysis of heat transfer and fluid flow in complex multidimensional applications.

Many users of the finite-element method do so through the application of large-scale, general-purpose computer codes.[18–20] These codes are widely available, highly user-oriented, and simple to use. They are also easy to misuse. The consequences of misuse are excess expense and, more important, invalid predictions of the state of stress and deformation of the structure due to the applied loading.

The discussion herein introduces the process of finite-element analysis to enable the prospective user to have a suitable understanding of the calculations being performed. The discussion is limited to the *stiffness* approach, which is the most widely used basis of finite elements. Further, for ease of understanding, the presentation deals with structures of two dimensions. The generalization to three dimensions follows directly.

2.16.2 The Concept of Stiffness

The governing equations of the theory of elasticity relate the loads applied externally to a body to the resulting deformation of that body, using the stress equilibrium equations to relate external forces to internal forces, i.e., stresses. The stress-strain relations relate internal forces to internal strains. The strain-displacement equations relate internal strains to observed deformations. The whole solution process can be stated by the relationship

$$F = k\delta \tag{2.296}$$

where F = externally applied forces
δ = observed deformation
k = stiffness of the structure

Thus, all the material and geometric information for the structure is contained in the stiffness term.

Numerical procedures involve relating the observed deformation at a discrete number of points of the body to the forces applied at these points. The relationship between force and displacement is then expressed most effectively in terms of matrix notation.

$$\{F\} = [k]\{\delta\} \tag{2.297}$$

Thus F becomes the vector of applied forces, δ the vector of displacement response, and k the stiffness matrix of the structure.

Stiffness of Simple Discrete Elements. For simple structures, relationships of the type of Eq. (2.297) can be derived directly.

EXAMPLE 1 The governing equation for the simple spring in Fig. 2.49 is

$$F_1 = k(\delta_1 - \delta_2)$$

$$F_2 = k(\delta_2 - \delta_1)$$

which in matrix notation becomes

$$\begin{Bmatrix} F_1 \\ F_2 \end{Bmatrix} = \begin{bmatrix} k & -k \\ -k & k \end{bmatrix} \begin{Bmatrix} \delta_1 \\ \delta_2 \end{Bmatrix} \tag{2.298}$$

where k is the stiffness of the spring, F_1 and F_2 the applied forces at nodes 1 and 2, and δ_1 and δ_2 the resulting displacements at nodes and 1 and 2.

EXAMPLE 2 The truss element in Fig. 2.50 is limited to stretching-compression response under its applied loads. The general governing equation is

$$F = (AE/L)\,\Delta L$$

where A is the cross-sectional area, E is Young's modulus, and L is the length. ΔL is the change in L under the action of the forces. A series of relationships of the form

$$F_i = k_{ij}\,\delta_j$$

may be developed where F_i is the force at node i and δ_j is the displacement of node j. The resulting set of relationships is

$$\begin{Bmatrix} F_{1x} \\ F_{1y} \\ F_{2x} \\ F_{2y} \end{Bmatrix} = \frac{AE}{L} \begin{bmatrix} \cos^2\alpha & \cos\alpha\sin\alpha & -\cos^2\alpha & -\cos\alpha\sin\alpha \\ \cos\alpha\sin\alpha & \sin^2\alpha & -\cos\alpha\sin\alpha & -\sin^2\alpha \\ -\cos^2\alpha & -\cos\alpha\sin\alpha & \cos^2\alpha & \cos\alpha\sin\alpha \\ -\cos\alpha\sin\alpha & -\sin^2\alpha & \cos\alpha\sin\alpha & \sin^2\alpha \end{bmatrix} \begin{Bmatrix} \delta_{1x} \\ \delta_{1y} \\ \delta_{2x} \\ \delta_{2y} \end{Bmatrix} \tag{2.299}$$

which is the form of Eq. (2.297). In general terms, Eq. (2.299) has the form

$$\begin{Bmatrix} F_{1x} \\ F_{1y} \\ F_{2x} \\ F_{2y} \end{Bmatrix} = \begin{bmatrix} k_{11} & k_{12} & k_{13} & k_{14} \\ k_{21} & k_{22} & k_{23} & k_{24} \\ k_{31} & k_{32} & k_{33} & k_{34} \\ k_{41} & k_{42} & k_{43} & k_{44} \end{bmatrix} \begin{Bmatrix} \delta_{1x} \\ \delta_{1y} \\ \delta_{2x} \\ \delta_{2y} \end{Bmatrix} \tag{2.300}$$

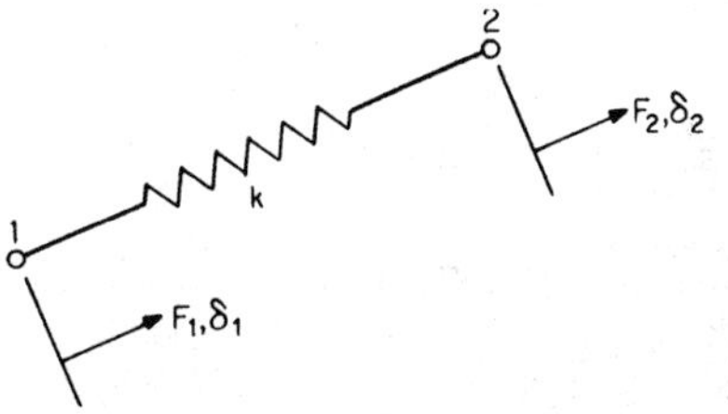

FIG. 2.49 Simple spring finite element.

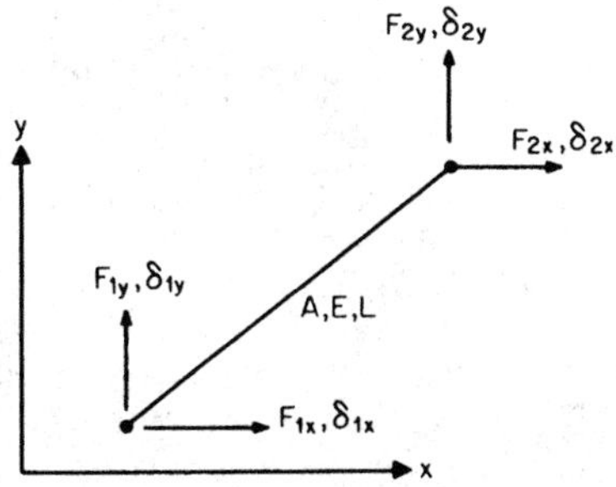

FIG. 2.50 Tension-compression finite element.

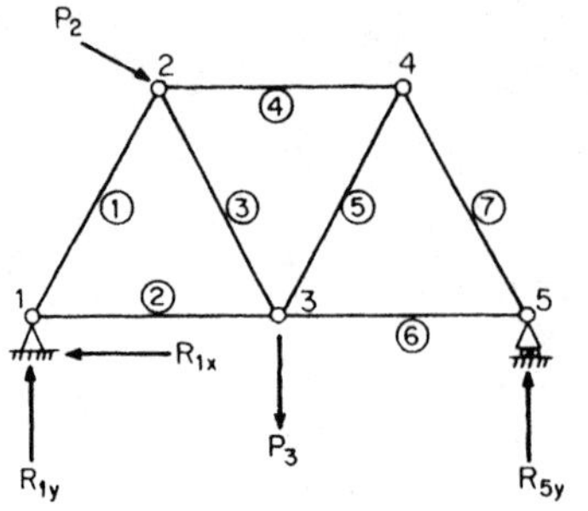

FIG. 2.51 Finite-element model of a simple truss.

Stiffness of a Complex Structure. The simple structures of these examples often comprise the elements of a more complex structure. Thus, the governing equation of the truss of Fig. 2.51 can be developed by combining the relationships of the individual truss elements from Example 2. One simple procedure is to insert the stiffness contribution from each row and column of each truss element to the stiffness of the appropriate row and column of the complex structure. For the truss of Fig. 2.51 the resulting relationship is shown in Eq. (2.301).

$$
\begin{Bmatrix} R_{1x} \\ R_{1y} \\ P_{2x} \\ P_{2y} \\ 0 \\ P_{3y} \\ 0 \\ 0 \\ 0 \\ R_{5y} \end{Bmatrix}
=
\begin{bmatrix}
k^1_{11}+k^2_{11} & k^1_{12}+k^2_{12} & k^1_{13} & k^1_{14} & k^2_{13} & k^2_{14} & 0 & 0 & 0 & 0 \\
k^1_{21}+k^2_{21} & k^1_{22}+k^2_{22} & k^1_{23} & k^1_{24} & k^2_{23} & k^2_{24} & 0 & 0 & 0 & 0 \\
k^1_{31} & k^1_{32} & k^1_{33}+k^3_{11}+k^{14}_{11} & k^1_{34}+k^3_{12}+k^4_{12} & k^3_{13} & k^3_{14} & k^4_{13} & k^4_{14} & 0 & 0 \\
k^1_{41} & k^1_{42} & k^1_{43}+k^3_{21}+k^4_{21} & k^1_{44}+k^3_{22}+k^4_{22} & k^3_{23} & k^3_{24} & k^4_{23} & k^4_{24} & 0 & 0 \\
k^2_{31} & k^2_{32} & k^3_{31} & k^3_{32} & k^2_{33}+k^3_{33}+k^5_{11}+k^6_{11} & k^2_{34}+k^3_{34}+k^5_{12}+k^6_{12} & k^5_{13} & k^5_{14} & k^6_{13} & k^6_{14} \\
k^2_{41} & k^2_{42} & k^3_{41} & k^3_{42} & k^2_{43}+k^3_{43}+k^5_{21}+k^6_{21} & k^2_{44}+k^3_{44}+k^5_{22}+k^6_{22} & k^5_{23} & k^5_{24} & k^6_{23} & k^6_{24} \\
0 & 0 & k^4_{31} & k^4_{32} & k^5_{31} & k^5_{32} & k^4_{33}+k^5_{33}+k^7_{11} & k^4_{34}+k^5_{34}+k^7_{12} & k^7_{13} & k^7_{14} \\
0 & 0 & k^4_{41} & k^4_{42} & k^5_{41} & k^5_{42} & k^4_{43}+k^5_{43}+k^7_{21} & k^4_{44}+k^5_{44}+k^7_{22} & k^7_{23} & k^7_{24} \\
0 & 0 & 0 & 0 & k^6_{31} & k^6_{32} & k^7_{31} & k^7_{32} & k^6_{33}+k^7_{33} & k^6_{34}+k^7_{34} \\
0 & 0 & 0 & 0 & k^6_{11} & k^6_{92} & k^7_{11} & k^7_{42} & k^6_{43}+k^7_{43} & k^6_{44}+k^7_{44}
\end{bmatrix}
\begin{Bmatrix} 0 \\ 0 \\ \delta_{2x} \\ \delta_{2y} \\ \delta_{3x} \\ \delta_{3y} \\ \delta_{4x} \\ \delta_{4y} \\ \delta_{5x} \\ 0 \end{Bmatrix}
\quad (2.301)
$$

2.16.3 Basic Procedure of Finite-Element Analysis

Equation (2.301) comprises 10 linear algebraic equations in 10 unknowns. The unknowns include the forces of reaction R_{1x}, R_{1y}, R_{5y} and the displacements δ_{2x}, δ_{2y}, δ_{3x}, δ_{3y}, δ_{4x}, δ_{4y}, δ_{5x}. Many procedures for the solution of sets of simultaneous, linear algebraic equations are available. One well-known approach is Gauss-Jordan elimination.[17]

Once the nodal displacements are known, the forces acting on each truss element can be determined by solution of the set of equations (2.300) applicable to that element.

The procedure used to determine the displacements and internal forces of the truss of Fig. 2.51 illustrates the procedure used to determine the response to applied load inherent in the stiffness finite-element procedure. These steps include:

1. Divide the structure into an appropriate number of discrete (or finite) elements connected only at a finite set of points in the structure.
2. Develop a load-deflection relationship of the form of Eq. (2.300) for each finite element.
3. Sum up the load-deflection relationships for each element to obtain the load-deflection relationship for the entire structure, as in Eq. (2.301).
4. Obtain the deformation pattern for the entire structure using conventional procedures.
5. Determine the internal force distribution for each element from the known deformations using the element force-deflection relationships.

The key steps in finite-element analysis are the discretization of the structure and the development of load-deflection relationships for the finite element. The subsequent assembly of the structure load-deflection relationship, the solution of the resulting set of simultaneous algebraic equations, and the subsequent determination of internal forces are straightforward mechanical procedures. Thus, it remains to illustrate the approximations associated with developing finite elements to the analysis of complex structures.

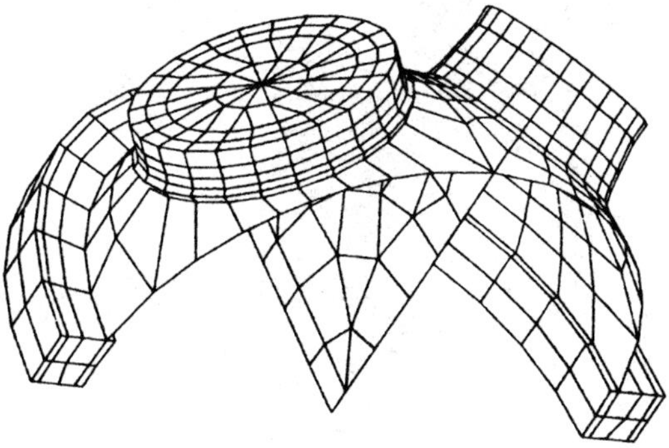

FIG. 2.52 Finite-element model of pressure-vessel head. (*Courtesy of Imo Industries Inc.*)

The truss represents a simplified structure relative to those for which solutions are usually required. A structure such as the pressure-vessel head modeled in Fig. 2.52 is more typical of the component analysis associated with finite-element modeling.

Finite Elements by the Direct Approach. The direct approach to the development of finite elements requires that a complete set of relationships between the internal and externally applied forces be known a priori. For many structural analyses this is not readily available; i.e., the available equilibrium equations are not sufficient. Therefore, the applicability of the direct approach is limited.

The procedure for development of the load-deflection relationships includes:

1. Define the internal displacement field of the element in terms of the nodal displacements. This requires the assumption of a relationship. Usually polynomial expansions are used.

2. Relate the internal displacement field to the internal force field through the strain-displacement and stress-strain equations.
3. Relate the internal force field to the external forces through the force-equilibrium relations.
4. Combine the results of steps 1–4 to obtain a relationship of the form of Eq. (2.300).

EXAMPLE 1 The beam of Fig. 2.53 has length L, Young's modulus E, and area moment of inertia I. At nodes 1 and 2 it is acted upon by external forces and moments F_1, F_2, $\overline{M}_1$, $\overline{M}_2$. As a result, the nodal displacements and rotations are w_1, w_2, θ_1, θ_2.

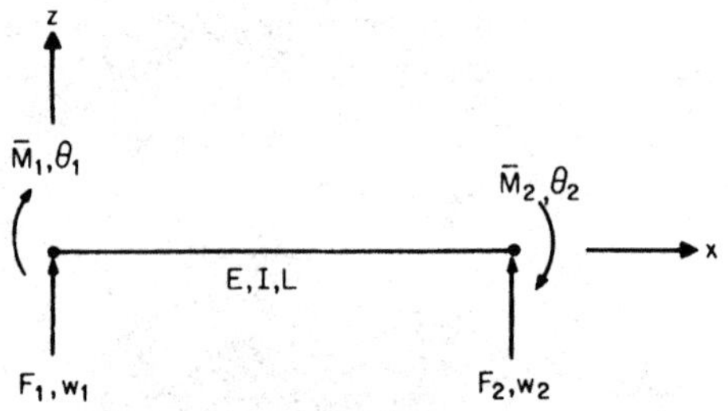

FIG. 2.53 Beam finite element.

Within the beam the deformation pattern is characterized by lateral deflection $w(x)$ and rotation $\theta(x)$, where

$$\theta = -\frac{dw}{dx}$$

Assume that the internal displacement field is governed by the polynomial

$$w(x) = \alpha_1 x^3 + \alpha_2 x^2 + \alpha_3 x + \alpha_4 \tag{2.302}$$

The α's are determined from the boundary conditions on $w(x)$, namely

$$w(0) = w_1$$

$$\theta(0) = \theta_1 = -\left.\frac{dw}{dx}\right|_{x=0}$$

$$w(L) = w_2$$

$$\theta(L) = \theta_2 = -\left.\frac{dw}{dx}\right|_{x=1}$$

The number of terms in the polynomial expansion for $w(x)$ is, in general, limited to the number of nodal degrees of freedom. With the α's known, Eq. (2.300) becomes

$$w(x) = \frac{1}{L^3}[x^3\; x^2\; x\; 1]\begin{bmatrix} 2 & -L & -2 & -L \\ -3L & 2L^2 & 3L & L^2 \\ 0 & -L^3 & 0 & 0 \\ L^3 & 0 & 0 & 0 \end{bmatrix} \tag{2.303}$$

For the special case of a beam, the internal moments are related to the internal displacement field by the Bernoulli-Euler equation

$$M(x) = EI\,(d^2w/dx^2) \tag{2.304}$$

Further, at the nodes

$$M(0) = M_1 = \overline{M}_1 \tag{2.305}$$

$$M(L) = M_2 = -\overline{M}_2 \tag{2.306}$$

For the beam to be in equilibrium under the applied forces and moments it is necessary that

$$F_2 L = \overline{M}_1 + \overline{M}_2 \tag{2.307}$$

$$F_1 L = -\overline{M}_2 - \overline{M}_1 \tag{2.308}$$

Therefore

$$F_2 L = M_1 - M_2 \tag{2.309}$$

$$F_1 L = M_2 - M_1 \tag{2.310}$$

Expressing Eqs. (2.305), (2.306), (2.309), and (2.310), in matrix format

$$\begin{Bmatrix} \mathbf{F}_1 \\ M_1 \\ \mathbf{F}_2 \\ M_2 \end{Bmatrix} = \begin{bmatrix} -1/L & 1/L \\ 1 & 0 \\ 1/L & -1/L \\ 0 & -1 \end{bmatrix} \begin{Bmatrix} M_1 \\ M_2 \end{Bmatrix} \tag{2.311}$$

Combining Eqs. (2.303), (2.304), and (2.311) yields

$$\begin{Bmatrix} \mathbf{F}_r \\ M_1 \\ \mathbf{F}_2 \\ M_1 \end{Bmatrix} = \frac{2EI}{L^3} \begin{bmatrix} 6 & -3L & -6 & -3L \\ -3L & 2L^2 & 3L & L^2 \\ -6 & 3L & 6 & 3L \\ -3L & L^2 & 3L & 2L^2 \end{bmatrix} \begin{Bmatrix} w_1 \\ \theta_1 \\ w_2 \\ \theta_2 \end{Bmatrix} \tag{2.312}$$

which is the required load/deflection relationship.

Finite Elements by Energy Minimization. The principle of stationary potential energy states that, for equilibrium to be ensured, the total potential energy must be stationary with respect to variations of admissible displacement fields. An "admissible displacement field" is one which satisfies the natural boundary conditions of the structure, typically those boundary conditions that constrain displacements and slopes. The exact displacement field will result in the minimum value of potential energy.

This energy principle allows the development of a general load-deflection relationship which, in turn, allows the development of a wide variety of finite elements directly from the assumed displacement field. The total potential energy is, in general, defined by

$$\Pi(u, v, w) = U(u, v, w) - V(u, v, w) \tag{2.313}$$

where Π = total potential energy
U = strain energy of deformation
V = work done by applied loads
u, v, w = components of displacement field within the element

For Π to be stationary it is necessary that

$$\frac{\partial \Pi}{\partial u_i} = 0 \qquad \frac{\partial \Pi}{\partial v_i} = 0 \qquad \frac{\partial \Pi}{\partial w_i} = 0 \qquad i = 1, r \tag{2.314}$$

where the subscript i denotes the ith node of the finite element, and r is the number of nodes. Further, the energy over the volume of the element is

$$U = \int_{\text{vol}} U_0 \, d\bar{v} \tag{2.315}$$

$$V = \{\delta\}^T\{F\} \tag{2.316}$$

where U_0 = strain energy of a unit volume of material
$\{F\}$ = matrix of nodal forces on the element
$\{\delta\}$ = matrix of nodal displacements

If we further express the stress-strain and strain-displacement equations [Eqs. (2.76) and (2.47)] in matrix format:

$$\{\sigma\} = [D]\{\epsilon\} \tag{2.317}$$

$$\{\epsilon\} = [B]\{\delta\} \tag{2.318}$$

where the element $[B]$ are differential operators, then

$$U_0 = \tfrac{1}{2}\{\epsilon\}[D]\{\epsilon\} \tag{2.319}$$

Combining Eqs. (2.313) to (2.319) and performing the indicated operations leads to a relationship of the form

$$\{F\} = \int_{\text{vol}} [B]^T[D][B] \, d\bar{v} \, \{\delta\} \tag{2.320}$$

Equation (2.320) constitutes a general load-deflection relationship which can be particularized to define a wide variety of finite elements.

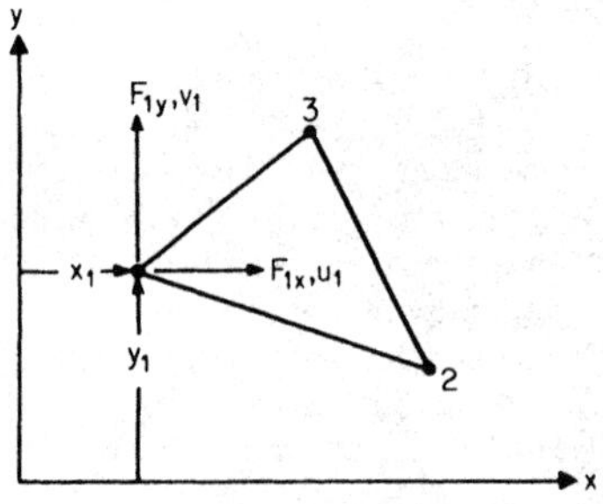

FIG. 2.54 Planar finite element.

EXAMPLE 1 The displacement field within the triangular element in Fig. 2.54 is assumed to be

$$\begin{aligned} u &= \alpha_1 + \alpha_2 x + \alpha_3 y \\ v &= \alpha_4 + \alpha_5 x + \alpha_6 y \end{aligned} \tag{2.321}$$

The six α's may be determined in terms of the six nodal displacement components as was done for the beam element, whence

$$\begin{Bmatrix} u \\ v \end{Bmatrix} = \begin{bmatrix} N_1 & 0 & N_2 & 0 & N_3 & 0 \\ 0 & N_1 & 0 & N_2 & 0 & N_3 \end{bmatrix} \{\delta\} \tag{2.322}$$

$$\text{where } \{\delta\}^T = \{u_1 \; v_1 \; u_1 \; v_2 \; u_3 \; v_3\}^T \tag{2.323}$$

$$N_i = (a_i + b_i x + c_i y)/2\Delta \tag{2.324}$$

The factors a_i, b_i, c_i, and Δ are constants which evolve from the algebraic manipulations. Continuing, for the two-dimensional case

$$\{\epsilon\} = \begin{Bmatrix} \epsilon_x \\ \epsilon_y \\ \epsilon_z \end{Bmatrix} = \begin{Bmatrix} \partial u/\partial x \\ \partial v/\partial y \\ \partial u/\partial y + \partial v/\partial x \end{Bmatrix} \tag{2.325}$$

Substituting Eq. (2.322) into Eq. (2.325) yields

$$\{\epsilon\} = [B]\{\delta\} \tag{2.326}$$

where

$$[B] = \frac{1}{2\Delta}\begin{bmatrix} b_1 & 0 & b_2 & 0 & b_3 & 0 \\ 0 & c_1 & 0 & c_2 & 0 & c_3 \\ c_1 & b_1 & c_2 & b_2 & c_3 & b_3 \end{bmatrix} \tag{2.327}$$

Finally, for the element of Fig. 2.58

$$\{\sigma\} = \begin{Bmatrix} \sigma_x \\ \sigma_y \\ \tau_{xy} \end{Bmatrix} = [D]\begin{Bmatrix} \epsilon_x \\ \epsilon_y \\ \gamma_{xy} \end{Bmatrix} \tag{2.328}$$

where, for plane strain

$$[D] = \frac{E(1-\nu)}{(1+\nu)(1-2\nu)}\begin{bmatrix} 1 & \nu/(1-\nu) & 0 \\ \nu/(1-\nu) & 1 & 0 \\ 0 & 0 & (1-2\nu)/2(1-\nu) \end{bmatrix} \tag{2.329}$$

Therefore, all the terms in Eq. (2.320) have been defined and so the load-deflection relationship for this element is established.

Since all the terms under the integral in Eq. (2.320) are constants, the integral may be evaluated exactly. Note that the resulting matrix equation contains six simultaneous algebraic equations, corresponding to the six degrees of freedom associated with the triangular element of Fig. 2.54.

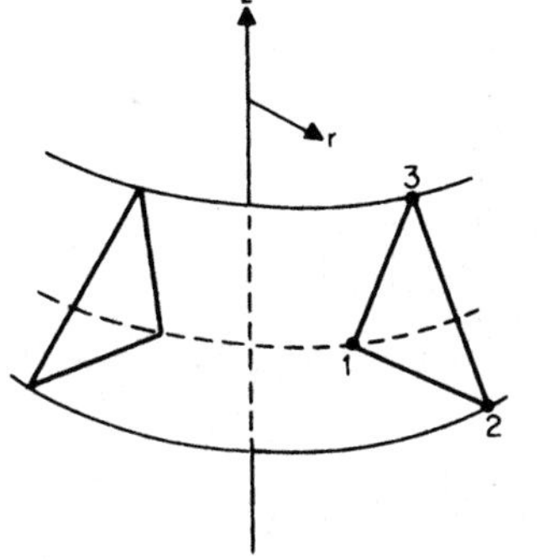

FIG. 2.55 Axisymmetric finite element.

EXAMPLE 2 The displacement function for the axisymmetric element of Fig. 2.55 is

$$\begin{aligned} u &= \alpha_1 + \alpha_2 r + \alpha_3 z \\ v &= \alpha_4 + \alpha_5 r + \alpha_6 z \end{aligned} \tag{2.330}$$

Following the same procedure as in Example 1 we find

$$\{\epsilon\} = \begin{Bmatrix} \epsilon_z \\ \epsilon_r \\ \epsilon_\theta \\ \gamma_{rz} \end{Bmatrix} = \begin{Bmatrix} \partial v/\partial z \\ \partial u/\partial r \\ u/r \\ \partial u/\partial z + \partial v/\partial r \end{Bmatrix} \tag{2.331}$$

whence

$$[B] = \begin{bmatrix} 0 & c_1 & 0 & c_2 & 0 & c_3 \\ b_1 & 0 & b_2 & 0 & b_3 & 0 \\ e_1 & 0 & e_2 & 0 & e_3 & 0 \\ c_1 & b_1 & c_2 & b_2 & c_3 & b_3 \end{bmatrix} \tag{2.332}$$

where $e_i = a_i/r + b_i + c_i(z/r)$. Further,

$$[D] = \frac{E(1-\nu)}{(1+\nu)(1-2\nu)}\begin{bmatrix} 1 & \nu/(1-\nu) & \nu/(1-\nu) & 0 \\ \nu/(1-\nu) & 1 & \nu/(1-\nu) & 0 \\ \nu/(1-\nu) & \nu/(1-\nu) & 1 & 0 \\ 0 & 0 & 0 & (1-2\nu)/2(1-\nu) \end{bmatrix} \quad (2.333)$$

The integral in Eq. (2.320) now has the form

$$2\pi \int_{\text{vol}} [B]^T[D][B]r\,dr\,dz$$

However, $[B]$ is no longer a constant array, i.e., $[B] = [B(r,z)]$ so that integration is a complex process. For many elements, the integrand is sufficiently complex that the integration must be carried out numerically. This numerical integration is a wholly different problem from the numerical analysis that is the finite-element method.

The three-dimensional analog to the triangle element of Example 1 is a four-node tetrahedron. A basic feature of these elements is that the strain field within the element is constant. Thus, to model a structure in which the strains vary considerably throughout the body, a large number of elements are required. Constant-strain elements are most useful for modeling thick-walled bodies in which the main action is stretching. Analysis of more flexible bodies in which bending is significant requires elements in which the strain can vary. These higher-order elements contain higher-order terms in the polynomial displacement expressions, e.g., Eq. (2.321).

Higher-Order Elements. The key ingredient in the development of a finite element is the selection of the shape function, that function which relates the internal-element displacement field to the nodal displacement field, e.g., Eq. (2.322). The remainder of the development is a mechanical process.

The shape function may be selected directly to establish some desired element characteristics or it may evolve from the selection of the displacement function as in the elements developed above. If the displacement function approach is used then the size of the polynomial is limited by the number of nodal degrees of freedom of the element, since the $\mathcal{L}$'s must be uniquely expressed in terms of the nodal degrees of freedom. Thus, in the examples above, the beam element is limited to a cubic polynomial, the triangular plane elements to linear polynomials. Higher-order polynomials require the insertion of additional nodes in the elements or of additional degrees of freedom at the existing nodes. Some typical higher-order elements involving additional nodes are shown in Fig. 2.56. An element involving additional nodal degrees of freedom is shown in Fig. 2.57. This latter type is commonly used to model shell- and plate-type structures.

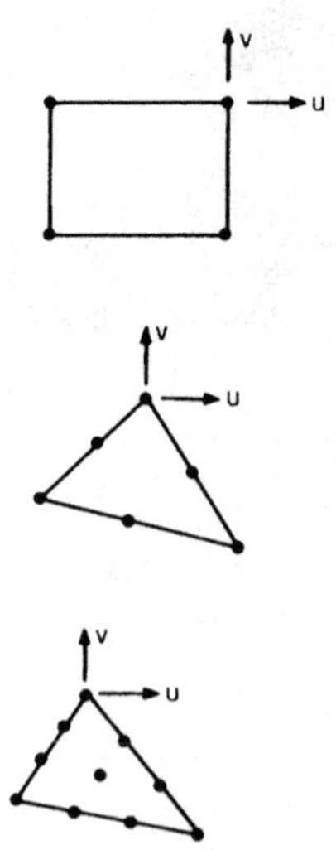

FIG. 2.56 Higher-order finite elements.

A widely used class of elements in which the shape function is chosen directly is the isoparametric elements. The key feature

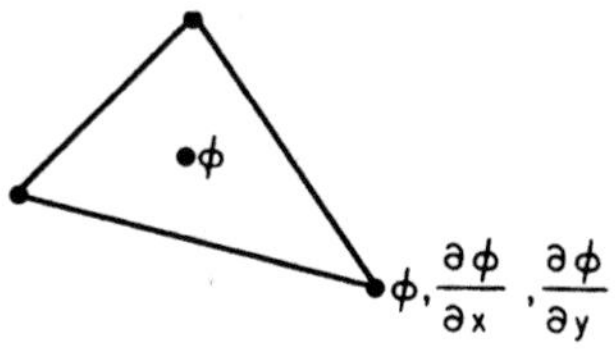

FIG. 2.57 Shell-type finite element.

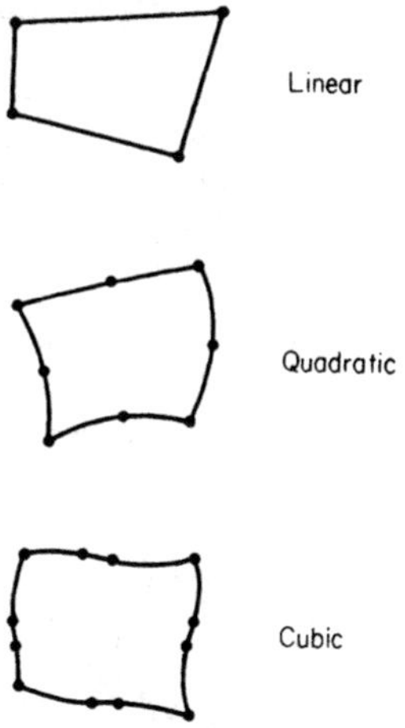

FIG. 2.58 Isoparametric finite elements.

of isoparametric elements is that the elements can have curved sides (Fig. 2.58). This feature allows the element to follow the flow of the structure more readily so that significantly fewer elements are needed to achieve a successful model.

2.16.4 Nature of the Solution

Unless the displacement function used constitutes the exact solution, the equilibrium equation applied within the finite element, or to the total structure, will not be satisfied, i.e., only the exact solution satisfies the equilibrium equations. Further, equilibrium is not satisfied across element boundaries. For example, two adjacent constant-strain (and hence constant-stress) elements cannot correctly represent a continuously varying strain field. Given the approximate nature of the solution, it is appropriate to question whether the response to applied load is at least approximately correct. It can, in fact, be shown that, subject to certain conditions on the finite elements, that the solution will converge to the exact solution with increasing grid refinement. Thus, if questions of accuracy in the analysis of a structure exist, one need only subdivide the critical areas into successively finer element grids. The solutions from these refined analyses will converge toward the correct answer.

The conditions on the elements to assure convergence can be satisfied if the displacement functions used are continuous polynomials of at least the first order within the element and if the elements are compatible. Compatibility requires that at least the nodal variables vary continuously along the boundary between adjacent elements, e.g., the displacement along edge 1-2 of the triangle element of Fig. 2.54 must be the same as along the edge of any other similar element attached to nodes 1 and 2. For the triangular element, since the displacements along edges are straight lines, compatibility is assured.

The above discussion does not preclude the successful use of nonpolynomial displacement functions or nonconverging elements or incompatible elements. However, such elements must be used with great care.

2.16.5 Finite-Element Modeling Guidelines

General rules for finite-element modeling do not exist. However, some reasonable guidelines have evolved to aid the analyst in developing a model which will yield accurate results with a reasonable effort. The more important of these guidelines include:

1. If at all possible, use converging, compatible elements.
2. Grids can be relatively coarse in regions where the state of strain varies slowly. In regions where strains change rapidly, e.g., strain concentrations and structural discontinuities, the grid should be refined.
3. Quadrilateral elements should be used wherever possible in place of triangular elements.
4. Accurate determination of forces and displacements can be accomplished with a more coarse grid than needed for accurate determination of strains and stresses.
5. Prediction of modes of vibration requires a more refined grid than that needed for prediction of natural frequencies.
6. Higher-order elements are generally preferable to constant strain elements.
7. Aspect ratios of multisided two- or three-dimensional elements should be kept below 5.
8. When the accuracy of the solution from a grid is in doubt, the grid should be refined in the critical regions and the analysis rerun.

2.16.6 Generalizations of the Applications

The finite-element method has applications in mechanics of materials beyond the static, linear elastic, isothermal, small-strain class of analyses discussed herein. The generalizations can be classified as related to generalizations of the stress-strain equations and generalizations of the equilibrium equations.

Generalizations of the Stress-Strain Relations. A more general statement of the stress-strain relations of linear elasticity (Eq. 2.74) is

$$
\begin{aligned}
\epsilon_x - \epsilon_x^0 - \alpha(T - T_0) &= (1/E)\{(\sigma_x - \sigma_x^0) - \nu[(\sigma_y - \sigma_y^0) + (\sigma_z - \sigma_z^0)]\} \\
\epsilon_y - \epsilon_y^0 - \alpha(T - T_0) &= (1/E)\{(\sigma_y - \sigma_y^0) - \nu[(\sigma_z - \sigma_z^0) + (\sigma_x - \sigma_x^0)]\} \\
\epsilon_z - \epsilon_z^0 - \alpha(T - T_0) &= (1/E)\{(\sigma_z - \sigma_z^0) - \nu[(\sigma_x - \sigma_x^0) + (\sigma_y - \sigma_y^0)]\} \\
\gamma_{xy} - \gamma_{xy}^0 &= (1/G)(\tau_{xy} - \tau_{xy}^0) \\
\gamma_{yz} - \gamma_{yz}^0 &= (1/G)(\tau_{yz} - \tau_{yz}^0) \\
\gamma_{zx} - \gamma_{zx}^0 &= (1/G)(\tau_{zx} - \tau_{zx}^0)
\end{aligned}
\tag{2.334}
$$

The strain terms with a superscript 0 represent a possible general state of initial strain. The stress terms with a superscript 0 represent a possible general state of initial stress. The strain terms $\alpha(T - T_0)$ represent a possible state of temperature-induced strain. Inclusion of these terms in the development of the finite-element results in a set of additional terms in the load-deflection relationship, which takes on the form

$$\{F\}_{\sigma 0} + \{F\}_{\epsilon 0} + \{F\}_{\alpha T} + \{F\} = \int_{\text{vol}} [B]^T[D][B]\, d\bar{v}\, \{\delta\} \tag{2.335}$$

where
$$\{F\}_{\sigma 0} = \int_{\text{vol}} [B]^T\{\sigma^0\}\, dv$$
$$\{F\}_{\epsilon 0} = -\int_{\text{vol}} [B]^T[D]\{\epsilon^0\}\, d\bar{v}$$

and similarly for $\{F\}_{\alpha T}$.

Nonlinear stress-strain relationships, i.e.,

$$[\sigma] = F[\epsilon] \tag{2.336}$$

are generally incorporated into finite-element analysis in terms of the incremental plasticity formulation (see Refs. 6 to 8). Solutions are effected by applying load to the structure in additive increments. For each load increment a modified linear analysis is performed. Thus the numerical analysis in the space defined by the finite-element model is supplemented by a numerical analysis in the load dimension to yield an analysis of the total problem.

Similarly, creep problems, for which the stress-strain relation is of the form

$$[\sigma] = f([\epsilon], [\partial\epsilon/\partial t]) \tag{2.337}$$

are solved using a numerical analysis in the time domain to supplement the finite-element models in space.

Generalizations of the Equilibrium Equations. The equilibrium equations, with the addition of body force terms, such as gravitational or inertia, have the form

$$\begin{aligned} \partial\sigma_x/\partial x + \partial\tau_{xy}/\partial y + \partial\tau_{xz}/\partial z + \bar{F}_x &= 0 \\ \partial\sigma_y/\partial y + \partial\tau_{yz}/\partial z + \partial\tau_{yx}/\partial x + \bar{F}_y &= 0 \\ \partial\sigma_z/\partial z + \partial\tau_{zx}/\partial x + \partial\tau_{zy}/\partial y + \bar{F}_z &= 0 \end{aligned} \tag{2.338}$$

With these terms, the load-deflection relationship now has the form

$$\{\bar{F}\}_{BF} + \{F\} = \int_{\text{vol}} [B]^T[D][B]\, d\bar{v}\, \{\delta\} \tag{2.339}$$

where
$$\{\bar{F}\}_{BF} = -\int_{\text{vol}} [N]^T\{\bar{F}\}\, d\bar{v}$$

$[N]^T$ = shape-function matrix, analogous to Eq. (2.322)

If $\{\bar{F}\}_{BF}$ represents an acceleration force per unit volume, then

$$\{\bar{F}\}_{BF} = \rho[N](\partial^2/\partial t^2)\{\delta\} \tag{2.340}$$

where ρ is the mass per unit volume.

If we further define

$$\begin{aligned} [M] &= \rho\int_{\text{vol}} [N]^T[N]\, d\bar{v} \\ [k] &= \int_{\text{vol}} [B]^T[D][B]\, d\bar{v} \end{aligned} \tag{2.341}$$

then Eq. (2.297) takes the form

$$[M]\{\ddot{\delta}\} + [k]\{\delta\} = \{F\} \tag{2.342}$$

which is the matrix statement of the general vibration problem discussed in Chap. 4. Therefore, all the solution techniques noted therein are applicable to the spatial finite-element model. A damping force vector can also be developed for Eq. (2.342).

2.16.7 Finite-Element Codes

Structural analysis by the finite-element method contains two major engineering steps: the design of the grid and the use of an appropriate finite element. Finite elements have been developed to represent a broad range of structural configurations, including constant strain and higher-order two- and three-dimensional solids, shells, plates, beams, bars, springs, masses, damping elements, contact elements, fracture mechanics elements, and many others. Elements have been designed for static and dynamic analysis, linear and nonlinear material models, linear and nonlinear deformations.

The finite element depends upon the selection of an appropriate shape or displacement function. The remainder of the analysis is a mechanical process. The element stiffness matrix calculations, including any numerical integrations required, the assembly of the structural load-deflection relationship, the solution of the structure equations for loads and deflections, and the back substitution into the individual element relationships to obtain stress and strain fields require a huge number of calculations but no engineering judgment. The calculation procedure is clearly suited to the "number crunching" digital computer. Effective use of the computer for models of any substantial size requires that efficient computer-oriented numerical integration and simultaneous equation solvers be incorporated into the solution process.

To this end many large, general-purpose, finite-element-based computer codes have been developed[18,19,20] and are available in the marketplace. These codes feature large element libraries, extremely efficient solution algorithms, and a broad range of applications. The code developers strive to make these codes "user friendly" to minimize the effort required to assemble the computer input once the engineering decisions of grid design and element selection from the element library have been made.

Many special-purpose codes with unique finite elements are available to solve problems beyond the range of the general-purpose codes. Beyond the contents of the marketplace, the creation of a finite-element program for any particular application is a relatively simple process once the required finite element has been designed.

REFERENCES

1. Timoshenko, S.: "Strength of Materials," 3d ed., Parts I and II, D. Van Nostrand Company, Princeton, NJ, 1955.
2. Timoshenko, S., and J. N. Goodier: "The Theory of Elasticity," 2d ed., McGraw-Hill Book Company, Inc., New York, 1951.
3. Timoshenko, S., and S. Woinowsky-Krieger: "Theory of Plates and Shells," 2d ed., McGraw-Hill Book Company, Inc., New York, 1959.
4. Sokolnikoff, I. S.: "Mathematical Theory of Elasticity," 2d ed., McGraw-Hill Book Company, Inc., New York, 1956.

5. Love, A. E. H.: "A Treatise on the Mathematical Theory of Elasticity," 4th ed., Dover Publications, Inc., New York, 1944.

6. Prager, W.: "An Introduction to Plasticity," Addison-Wesley Publishing Company, Inc., Reading, MA, 1959.

7. Mendelson, A.: "Plasticity: Theory and Application," The Macmillan Company, Inc., New York, 1968.

8. Hodge, P. G.: "Plastic Analysis of Structures," McGraw-Hill Book Company, Inc., New York, 1959.

9. Boley, B. A., and J. H. Weiner: "Theory of Thermal Stresses," John Wiley & Sons, Inc., New York, 1960.

10. Flugge, W.: "Stresses in Shells," Springer-Verlag OHG, Berlin, 1960.

11. Hult, J. A. H.: "Creep in Engineering Structures," Blaisdell Publishing Company, Waltham, MA, 1966.

12. Roark, R. J.: "Formulas for Stress and Strain," 3d ed., McGraw-Hill Book Company, Inc., New York, 1954.

13. McConnell, A. J.: "Applications of Tensor Analysis," Dover Publications, Inc., New York, 1957.

14. Cook, R. D.: "Concepts and Applications of Finite Element Analysis," 2d ed., John Wiley & Sons, Inc., New York, 1981.

15. Zienkiewicz, O. C.: "The Finite Element Method," 3d ed., McGraw-Hill Book Company, (U.K.) Ltd., London, 1977.

16. Gallagher, R. H.: "Finite Element Analysis: Fundamentals," Prentice-Hall, Inc., Englewood Cliffs, NJ, 1975.

17. Hildebrand, F. B.: "Methods of Applied Mathematics," Prentice-Hall, Inc., Englewood Cliffs, NJ, 1952.

18. "ANSYS Engineering Analysis System," Swanson Analysis Systems, Inc., Houston, Pa.

19. "The NASTRAN Theoretical Manual," NASA-SP-221(03), National Aeronautics and Space Administration, Washington, D.C.

20. "The MARC Finite Element Code," MARC Analysis Research Corporation, Palo Alto, CA.

CHAPTER 3
KINEMATICS OF MECHANISMS

Ferdinand Freudenstein, Ph.D.
Stevens Professor of Mechanical Engineering
Columbia University
New York, N.Y.

George N. Sandor, Eng.Sc.D., P.E.
Research Professor Emeritus of Mechanical Engineering
Center for Intelligent Machines
University of Florida
Gainesville, Fla.

3.1 DESIGN USE OF THE MECHANISMS SECTION 3.2
3.2 BASIC CONCEPTS 3.2
3.2.1 Kinematic Elements 3.2
3.2.2 Degrees of Freedom 3.4
3.2.3 Creation of Mechanisms According to the Separation of Kinematic Structure and Function 3.5
3.2.4 Kinematic Inversion 3.6
3.2.5 Pin Enlargement 3.6
3.2.6 Mechanical Advantage 3.6
3.2.7 Velocity Ratio 3.6
3.2.8 Conservation of Energy 3.7
3.2.9 Toggle 3.7
3.2.10 Transmission Angle 3.7
3.2.11 Pressure Angle 3.8
3.2.12 Kinematic Equivalence 3.8
3.2.13 The Instant Center 3.9
3.2.14 Centrodes, Polodes, Pole Curves 3.9
3.2.15 The Theorem of Three Centers 3.10
3.2.16 Function, Path, and Motion Generation 3.11
3.3 PRELIMINARY DESIGN ANALYSIS: DISPLACEMENTS, VELOCITIES, AND ACCELERATIONS 3.11
3.3.1 Velocity Analysis: Vector-Polygon Method 3.11
3.3.2 Velocity Analysis: Complex-Number Method 3.12
3.3.3 Acceleration Analysis: Vector-Polygon Method 3.13
3.3.4 Acceleration Analysis: Complex-Number Method 3.14
3.3.5 Higher Accelerations 3.14
3.3.6 Accelerations in Complex Mechanisms 3.15
3.3.7 Finite Differences in Velocity and Acceleration Analysis 3.15
3.4 PRELIMINARY DESIGN ANALYSIS: PATH CURVATURE 3.16
3.4.1 Polar-Coordinate Convention 3.16
3.4.2 The Euler-Savary Equation 3.16
3.4.3 Generating Curves and Envelopes 3.19
3.4.4 Bobillier's Theorem 3.20
3.4.5 The Cubic of Stationary Curvature (the k_u Curve) 3.21
3.4.6 Five and Six Infinitesimally Separated Positions of a Plane 3.22
3.4.7 Application of Curvature Theory to Accelerations 3.22
3.4.8 Examples of Mechanism Design and Analysis Based on Path Curvature 3.23
3.5 DIMENSIONAL SYNTHESIS: PATH, FUNCTION, AND MOTION GENERATION 3.24
3.5.1 Two Positions of a Plane 3.25
3.5.2 Three Positions of a Plane 3.26
3.5.3 Four Positions of a Plane 3.26
3.5.4 The Center-Point Curve or Pole Curve 3.27
3.5.5 The Circle-Point Curve 3.28
3.5.6 Five Positions of a Plane 3.29
3.5.7 Point-Position Reduction 3.30
3.5.8 Complex-Number Methods 3.30
3.6 DESIGN REFINEMENT 3.31
3.6.1 Optimization of Proportions for Generating Prescribed Motions with Minimum Error 3.32
3.6.2 Tolerances and Precision 3.34
3.6.3 Harmonic Analysis 3.35
3.6.4 Transmission Angles 3.35
3.6.5 Design Charts 3.35
3.6.6 Equivalent and "Substitute" Mechanisms 3.36
3.6.7 Computer-Aided Mechanism Design and Optimization 3.37
3.6.8 Balancing of Linkages 3.38
3.6.9 Kinetoelastodynamics of Linkage Mechanisms 3.38
3.7 THREE-DIMENSIONAL MECHANISMS 3.30
3.8 CLASSIFICATION AND SELECTION OF MECHANISMS 3.40

3.9 KINEMATIC PROPERTIES OF MECHANISMS 3.46
3.9.1 The General Slider-Crank Chain 3.46
3.9.2 The Offset Slider-Crank Mechanism 3.46
3.9.3 The In-Line Slider-Crank Mechanism 3.48
3.9.4 Miscellaneous Mechanisms Based on the Slider-Crank Chain 3.49
3.9.5 Four-Bar Linkages (Plane) 3.51
3.9.6 Three-Dimensional Mechanisms 3.59
3.9.7 Intermittent-Motion Mechanisms 3.62
3.9.8 Noncircular Cylindrical Gearing and Rolling-Contact Mechanisms 3.64
3.9.9 Gear-Link-Cam Combinations and Miscellaneous Mechanisms 3.68
3.9.10 Robots and Manipulators 3.69
3.9.11 Hard Automation Mechanisms 3.69

3.1 DESIGN USE OF THE MECHANISMS SECTION

The design process involves intuition, invention, synthesis, and analysis. Although no arbitrary rules can be given, the following design procedure is suggested:

1. Define the problem in terms of inputs, outputs, their time-displacement curves, sequencing, and interlocks.
2. Select a suitable mechanism, either from experience or with the help of the several available compilations of mechanisms, mechanical movements, and components (Sec. 3.8).
3. To aid systematic selection consider the creation of mechanisms by the separation of structure and function and, if necessary, modify the initial selection (Secs. 3.2 and 3.6).
4. Develop a first approximation to the mechanism proportions from known design requirements, layouts, geometry, velocity and acceleration analysis, and path-curvature considerations (Secs. 3.3 and 3.4).
5. Obtain a more precise dimensional synthesis, such as outlined in Sec. 3.5, possibly with the aid of computer programs, charts, diagrams, tables, and atlases (Secs. 3.5, 3.6, 3.7, and 3.9).
6. Complete the design by the methods outlined in Sec. 3.6 and check end results. Note that cams, power screws, and precision gearing are treated in Chaps. 14, 16, and 21, respectively.

3.2 BASIC CONCEPTS

3.2.1 Kinematic Elements

Mechanisms are often studied as though made up of rigid-body members, or "links," connected to each other by rigid "kinematic elements" or "element pairs." The nature and arrangement of the kinematic links and elements determine the kinematic properties of the mechanism.

If two mating elements are in surface contact, they are said to form a "lower pair"; element pairs with line or point contact form "higher pairs." Three types of lower pairs permit relative motion of one degree of freedom ($f = 1$), turning pairs, sliding pairs, and screw pairs. These and examples of higher pairs are shown in Fig. 3.1. Examples of element pairs whose relative motion possesses up to five degrees of freedom are shown in Fig. 3.2.

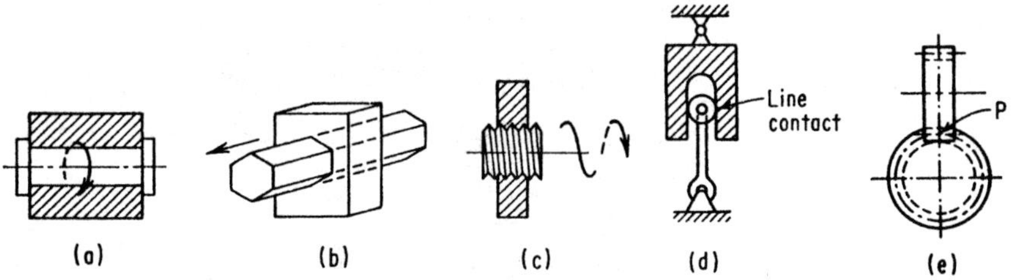

FIG. 3.1 Examples of kinematic-element pairs: lower pairs *a, b, c,* and higher pairs *d* and *e.* (*a*) Turning or revolute pair. (*b*) Sliding or prismatic pair. (*c*) Screw pair. (*d*) Roller in slot. (*e*) Helical gears at right angles.

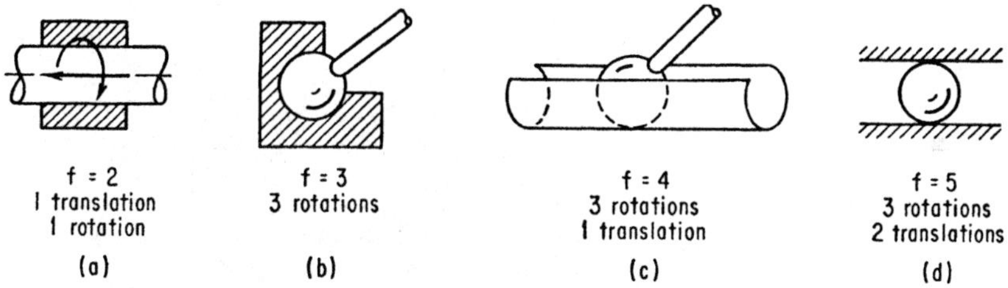

FIG. 3.2 Examples of elements pairs with $f > 1$. (*a*) Turn slide or cylindrical pair. (*b*) Ball joint or spherical pair. (*c*) Ball joint in cylindrical slide. (*d*) Ball between two planes. (Translational freedoms are in mutually perpendicular directions. Rotational freedoms are about mutually perpendicular axes.)

A link is called "binary," "ternary," or "*n*-nary" according to the number of element pairs connected to it, i.e., 2, 3, or *n*. A ternary link, pivoted as in Fig. 3.3*a* and *b,* is often called a "rocker" or a "bell crank," according to whether α is obtuse or acute.

A ternary link having three parallel turning-pair connections with coplanar axes, one of which is fixed, is called a "lever" when used to overcome a weight or resistance (Fig. 3.3*c, d,* and *e*). A link without fixed elements is called a "floating link."

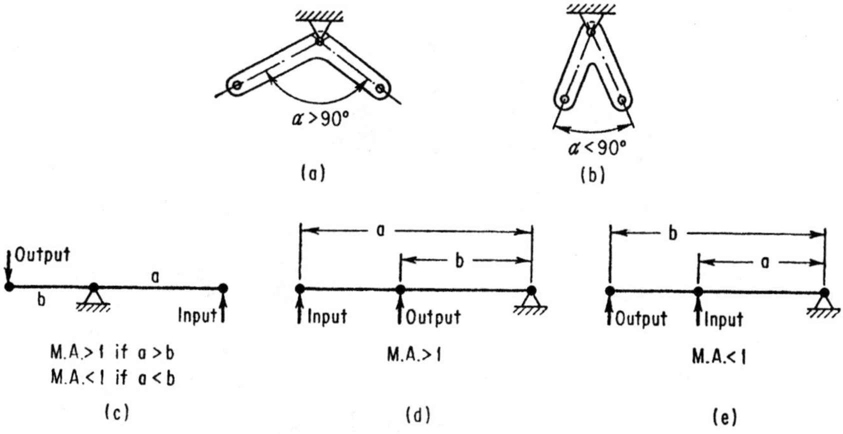

FIG. 3.3 Links and levers. (*a*) Rocker (ternary link). (*b*) Bell crank (ternary link). (*c*) First-class lever. (*d*) Second-class lever. (*e*) Third-class lever.

Mechanisms consisting of a chain of rigid links (one of which, the "frame," is considered fixed) are said to be closed by "pair closure" if all element pairs are constrained by material boundaries. All others, such as may involve springs or body forces for chain closure, are said to be closed by means of "force closure." In the latter, nonrigid elements may be included in the chain.

3.2.2 Degrees of Freedom[6,9,10,13,94,111,154,242,368]

Let F = degree of freedom of mechanism
l = total number of links, including fixed link
j = total number of joints
f_i = degree of freedom of relative motion between element pairs of ith joint

Then, *in general,*

$$F = \lambda(l - j - 1) + \sum_{i=1}^{j} f_i \tag{3.1}$$

where λ is an integer whose value is determined as follows:

$\lambda = 3$: Plane mechanisms with turning pairs, or turning and sliding pairs; spatial mechanisms with turning pairs only (motion on sphere); spatial mechanisms with rectilinear sliding pairs only.

$\lambda = 6$: Spatial mechanisms with lower pairs, the axes of which are nonparallel and nonintersecting; note exceptions such as listed under $\lambda = 2$ and $\lambda = 3$. (See also Ref. 10.)

$\lambda = 2$: Plane mechanisms with sliding pairs only; spatial mechanisms with "curved" sliding pairs only (motion on a sphere); three-link coaxial screw mechanisms.

Although included under Eq. (3.1), the motions on a sphere are usually referred to as special cases. For a comprehensive discussion and formulas including screw chains and other combinations of elements, see Ref. 13. The freedom of a mechanism with higher pairs should be determined from an equivalent lower-pair mechanism whenever feasible (see Sec. 3.2).

Mechanism Characteristics Depending on Degree of Freedom Only. For plane mechanisms with turning pairs only and one degree of freedom,

$$2j - 3l + 4 = 0 \tag{3.2}$$

except in special cases. Furthermore, if this equation is valid, then the following are true:

1. The number of links is even.
2. The minimum number of binary links is four.
3. The maximum number of joints in a single link cannot exceed one-half the number of links.
4. If one joint connects m links, the joint is counted as $(m - 1)$-fold.

In addition, for nondegenerate plane mechanisms with turning and sliding pairs and one degree of freedom, the following are true:

1. If a link has only sliding elements, they cannot all be parallel.
2. Except for the three-link chain, binary links having sliding pairs only cannot, in general, be directly connected.

3. No closed nonrigid loop can contain less than two turning pairs.

For plane mechanisms, having any combination of higher and/or lower pairs, and with one degree of freedom, the following hold:

1. The number of links may be odd.
2. The maximum number of elements in a link may exceed one-half the number of links, but an upper bound can be determined.[154,368]
3. If a link has only higher-pair connections, it must possess at least three elements.

For constrained spatial mechanisms in which Eq. (3.1) applies with $\lambda = 6$, the sum of the degrees of freedom of all joints must add up to 7 whenever the number of links is equal to the number of joints.

Special Cases. F can exceed the value predicted by Eq. (3.1) in certain special cases. These occur, generally, when a sufficient number of links are parallel in plane motion (Fig. 3.4*a*) or, in spatial motions, when the axes of the joints intersect (Fig. 3.4*b*—motion on a sphere, considered special in the sense that $\lambda \neq 6$).

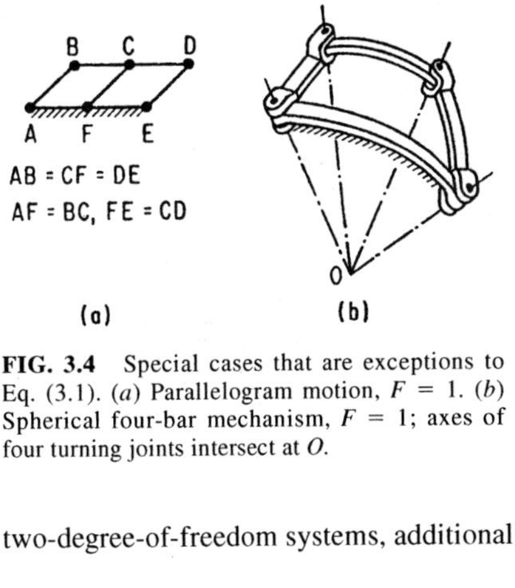

FIG. 3.4 Special cases that are exceptions to Eq. (3.1). (*a*) Parallelogram motion, $F = 1$. (*b*) Spherical four-bar mechanism, $F = 1$; axes of four turning joints intersect at O.

The existence of these special cases or "critical forms" can sometimes also be detected by multigeneration effects involving pantographs, inversors, or mechanisms derived from these (see Sec. 3.6 and Ref. 154). In the general case, the critical form is associated with the singularity of the functional matrix of the differential displacement equations of the coordinates;[130] this singularity is usually difficult to ascertain, however, especially when higher pairs are involved. Known cases are summarized in Ref. 154. For two-degree-of-freedom systems, additional results are listed in Refs. 111 and 242.

3.2.3 Creation of Mechanisms According to the Separation of Kinematic Structure and Function[54,74,110,132,133]

Basically this is an unbiased procedure for creating mechanisms according to the following sequence of steps:

1. Determine the basic characteristics of the desired motion (degree of freedom, plane or spatial) and of the mechanism (number of moving links, number of independent loops).
2. Find the corresponding kinematic chains from tables, such as in Ref. 133.
3. Find corresponding mechanisms by selecting joint types and fixed link in as many inequivalent ways as possible and sketch each mechanism.
4. Determine functional requirements and, if possible, their relationship to kinematic structure.
5. Eliminate mechanisms which do not meet functional requirements. Consider remaining mechanisms in greater detail and evaluate for potential use.

The method is described in greater detail in Refs. 110 and 133, which show applications to casement window linkages, constant-velocity shaft couplings, other mechanisms, and patent evaluation.

3.2.4 Kinematic Inversion

Kinematic inversion refers to the process of considering different links as the frame in a given kinematic chain. Thereby different and possibly useful mechanisms can be obtained. The slider crank, the turning-block and the swinging-block mechanisms are mutual inversions, as are also drag-link and "crank-and-rocker" mechanisms.

3.2.5 Pin Enlargement

Another method for developing different mechanisms from a base configuration involves enlarging the joints, illustrated in Fig. 3.5.

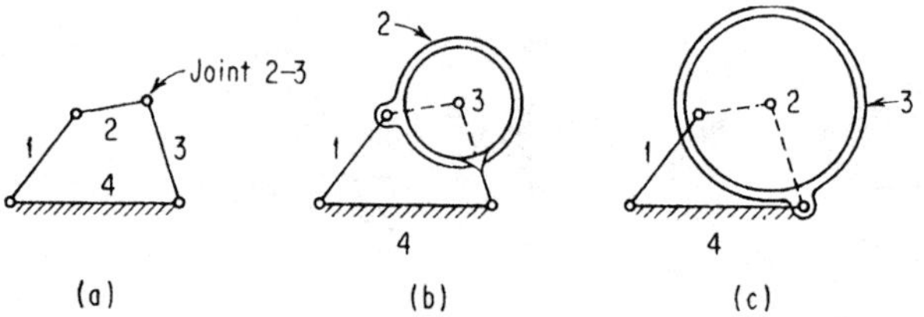

FIG. 3.5 Pin enlargement. (*a*) Base configuration. (*b*) Enlarged pin at joint 2–3; pin part of link 3. (*c*) Enlarged pin at joint 2–3; pin takes place of link 2.

3.2.6 Mechanical Advantage

Neglecting friction and dynamic effects, the instantaneous power input and output of a mechanism must be equal and, in the absence of branching (one input, one output, connected by a single "path"), equal to the "power flow" through any other point of the mechanism.

In a single-degree-of-freedom mechanism without branches, the power flow at any point J is the product of the force F_j at J, and the velocity V_j at J in the direction of the force. Hence, for any point in such a mechanism,

$$F_j V_j = \text{constant} \tag{3.3}$$

neglecting friction and dynamic effects. For the point of input P and the point of output Q of such a mechanism, the mechanical advantage is defined as

$$MA = F_Q / F_P \tag{3.4}$$

3.2.7 Velocity Ratio

The "linear velocity ratio" for the motion of two points P and Q representing the input and output members or "terminals" of a mechanism is defined as V_Q/V_P. If input and output terminals or links P and Q rotate, the "angular velocity ratio" is defined as ω_Q/ω_P, where ω designates the angular velocity of the link. If T_Q and T_P refer to torque output and input in single-branch rotary mechanisms, the power-flow equation, in the absence of friction, becomes

$$T_P\omega_P = T_Q\omega_Q \tag{3.5}$$

3.2.8 Conservation of Energy

Neglecting friction and dynamic effects, the product of the mechanical advantage and the linear velocity ratio is unity for all points in a single-degree-of-freedom mechanism without branch points, since $F_QV_Q/F_PV_P = 1$.

3.2.9 Toggle

Toggle mechanisms are characterized by sudden snap or overcenter action, such as in Fig. 3.6*a* and *b*, schematics of a crushing mechanism and a light switch. The mechanical advantage, as in Fig. 3.6*a*, can become very high. Hence toggles are often used in such operations as clamping, crushing, and coining.

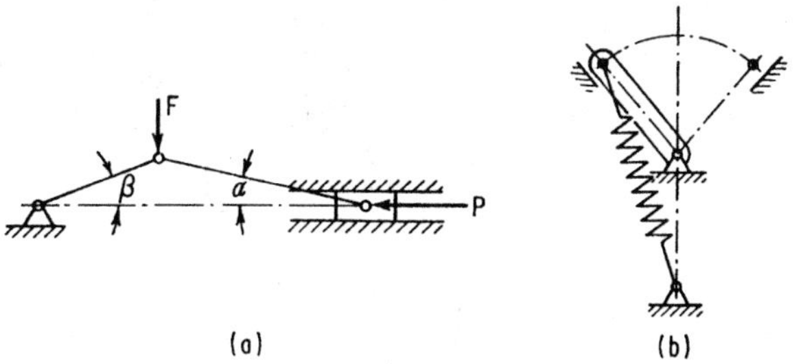

FIG. 3.6 Toggle actions. (*a*) $P/F = (\tan\alpha + \tan\beta)^{-1}$ (neglecting friction). (*b*) Schematic of a light switch.

3.2.10 Transmission Angle[15,159,160,166–168,176,205] (see Secs. 3.6 and 3.9)

The transmission angle μ is used as a geometrical indication of the ease of motion of a mechanism under static conditions, excluding friction. It is defined by the ratio

$$\tan\mu = \frac{\text{force component tending to move driven link}}{\text{force component tending to apply pressure on driven-link bearing or guide}} \tag{3.6}$$

where μ is the transmission angle.

In four-link mechanisms, μ is the angle between the coupler and the driven link (or the supplement of this angle) (Fig. 3.7) and has been used in optimizing linkage proportions (Secs. 3.6 and 3.9). Its ideal value is 90°; in practice it may deviate from this value by 30° and possibly more.

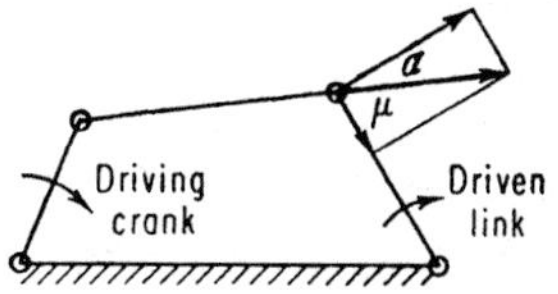

FIG. 3.7 Transmissional angle μ and pressure angle α (also called the deviation angle) in a four-link mechanism.

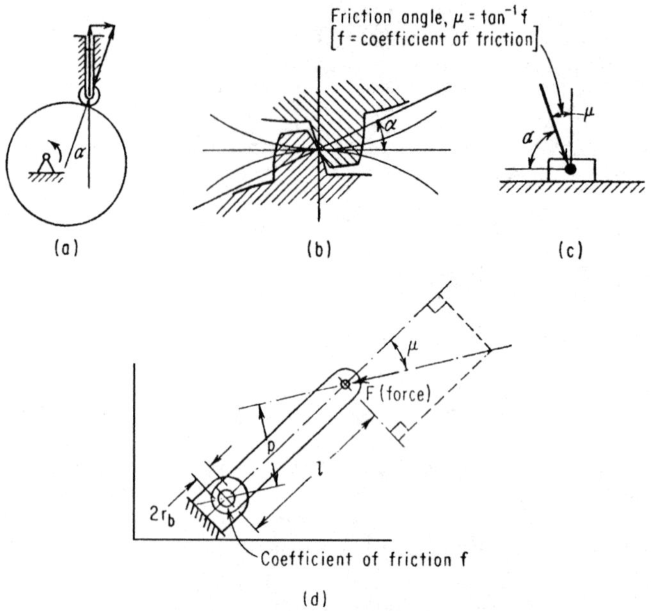

FIG. 3.8 Pressure angle (*a*) Cam and follower. (*b*) Gear teeth in mesh. (*c*) Link in sliding motion; condition of locking by friction $(\alpha + \mu) \geq 90°$. (*d*) Conditions for locking by friction of a rotating link: $\sin \mu \leq fr_b/l$.

3.2.11 Pressure Angle

In cam and gear systems, it is customary to refer to the complement of the transmission angle, called the pressure angle α, defined by the ratio

$$\tan \alpha = \frac{\text{force component tending to put pressure on follower bearing or guide}}{\text{force component tending to move follower}} \tag{3.7}$$

The ideal value of the pressure angle is zero; in practice it is frequently held to within 30° (Fig. 3.8). To ensure movability of the output member the ultimate criterion is to preserve a sufficiently large value of the ratio of driving force (or torque) to friction force (or torque) on the driven link. For a link in pure sliding (Fig. 3.8*c*), the motion will lock if the pressure angle and the friction angle add up to or exceed 90°.

A mechanism, the output link of which is shown in Fig. 3.8*d*, will lock if the ratio of p, the distance of the line of action of the force F from the fixed pivot axis, to the bearing radius r_b is less than or equal to the coefficient of friction, f, i.e., if the line of action of the force F cuts the "friction circle" of radius fr_b, concentric with the bearing.[171]

3.2.12 Kinematic Equivalence[159,182,288,290,347,376] (see Sec. 3.6)

"Kinematic equivalence," when applied to two mechanisms, refers to equivalence in motion, the precise nature of which must be defined in each case.

The motion of joint C in Fig. 3.9*a* and *b* is entirely equivalent if the quadrilaterals $ABCD$ are identical; the motion of C as a function of the rotation of link AB is also

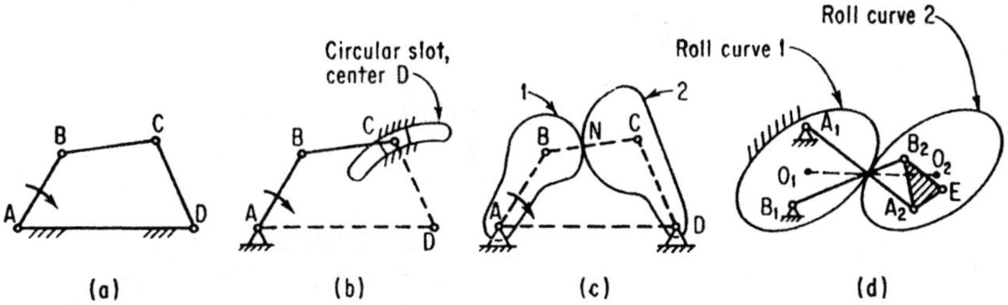

FIG. 3.9 Kinematic equivalence: (*a*), (*b*), (*c*) for four-bar motion; (*d*) illustrates rolling motion and an equivalent mechanism. When O_1 and O_2 are fixed, curves are in rolling contact; when roll curve 1 is fixed and rolling contact is maintained, O_2 generates circle with center O_1.

equivalent throughout the range allowed by the slot. In Fig. 3.9*c*, *B* and *C* are the centers of curvature of the contacting surfaces at *N*; *ABCD* is one equivalent four-bar mechanism in the sense that, if *AB* is integral with body 1, the angular velocity and angular acceleration of link *CD* and body 2 are the same in the position shown, but not necessarily elsewhere.

Equivalence is used in design to obtain alternate mechanisms, which may be mechanically more desirable than the original. If, as in Fig. 3.9*d*, A_1A_2 and B_1B_2 are conjugate point pairs (see Sec. 3.4), with A_1B_1 fixed on roll curve 1, which is in rolling contact with roll curve 2 (A_2B_2 are fixed on roll curve 2), then the path of *E* on link A_2B_2 and of the coincident point on the body of roll curve 2 will have the same path tangent and path curvature in the position shown, but not generally elsewhere.

3.2.13 The Instant Center

At any instant in the plane motion of a link, the velocities of all points on the link are proportional to their distance from a particular point *P*, called the instant center. The velocity of each point is perpendicular to the line joining that point to *P* (Fig. 3.10).

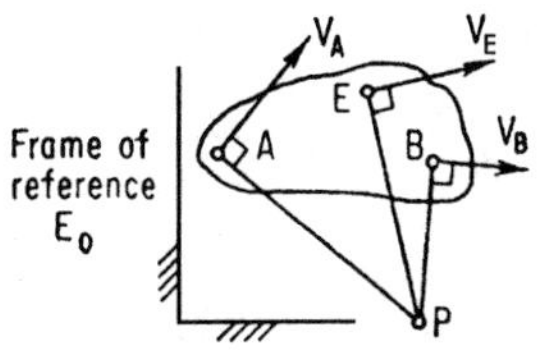

FIG. 3.10 Instant center, *P*. $V_E/V_B = EP/BP$, $V_E \perp EP$, etc.

Regarded as a point on the link, *P* has an instantaneous velocity of zero. In pure rectilinear translation, *P* is at infinity.

The instant center is defined in terms of velocities and is not the center of path curvature for the points on the moving link in the instant shown, except in special cases, e.g., points on common tangent between centrodes (see Sec. 3.4).

An extension of this concept to the "instantaneous screw axis" in spatial motions has been described.[38]

3.2.14 Centrodes, Polodes, Pole Curves

Relative plane motion of two links can be obtained from the pure rolling of two curves, the "fixed" and "movable centrodes" ("polodes" and "pole curves," respectively),

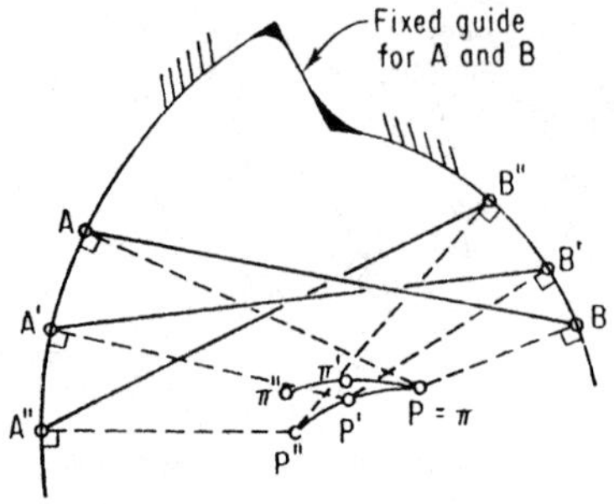

FIG. 3.11 Construction of fixed and movable centroides. Link AB in plane motion, guided at both ends; $PP' = P\pi'$; $\pi'\pi'' = P'P''$, etc.

which can be constructed as illustrated in the following example.

As shown in Fig. 3.11, the intersections of path normals locate successive instant centers P, P', P'', ..., whose locus constitutes the fixed centrode. The movable centrode can be obtained either by inversion (i.e., keeping AB fixed, moving the guide, and constructing the centrode as before) or by "direct construction": superposing triangles $A'B'P'$, $A''B''P''$, ..., on AB so that A' covers A and B' covers B, etc. The new locations thus found for P', P'', ..., marked π', π'', ..., then constitute points on the movable centrode, which rolls without slip on the fixed centrode and carries AB with it, duplicating the original motion. Thus, for the motion of AB, the centrode-rolling motion is kinematically equivalent to the original guided motion.

In the antiparallel equal-crank linkage, with the shortest link fixed, the centrodes for the coupler motion are identical ellipses with foci at the link pivots (Fig. 3.12); if the longer link AB were held fixed, the centrodes for the coupler motion of CD would be identical hyperbolas with foci at A, B, and C, D, respectively.

In the elliptic trammel motion (Fig. 3.13) the centrodes are two circles, the smaller rolling inside the larger, twice its size. Known as "cardanic motion," it is used in press drives, resolvers, and straight-line guidance.

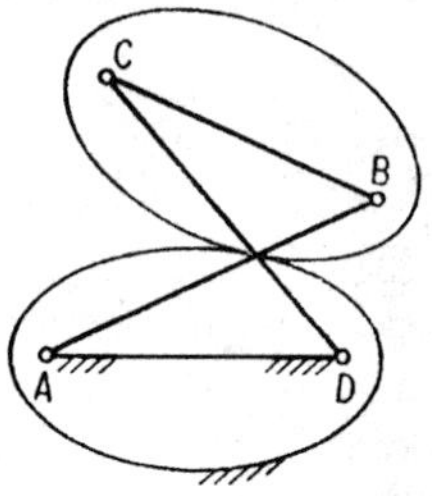

FIG. 3.12 Antiparallel equal-crank linkage; rolling ellipses, foci at A, D, B, C; $AD < AB$.

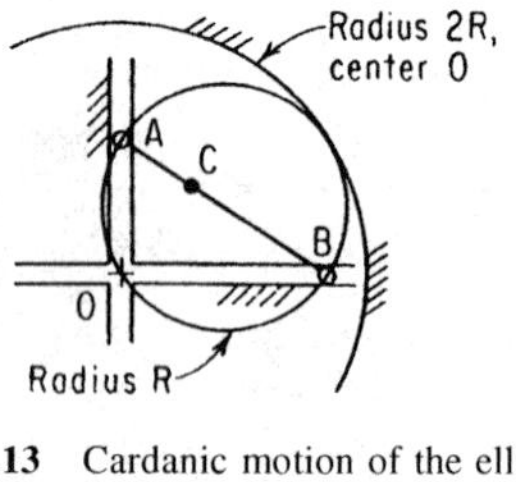

FIG. 3.13 Cardanic motion of the elliptic trammel, so called because any point C of AB describes an ellipse; midpoint of AB describes circle, center O (point C need not be collinear with AB).

Apart from their use in kinematic analysis, the centrodes are used to obtain alternate, kinematically equivalent mechanisms, and sometimes to guide the original mechanism past the "in-line" or "dead-center" positions.[207]

3.2.15 The Theorem of Three Centers

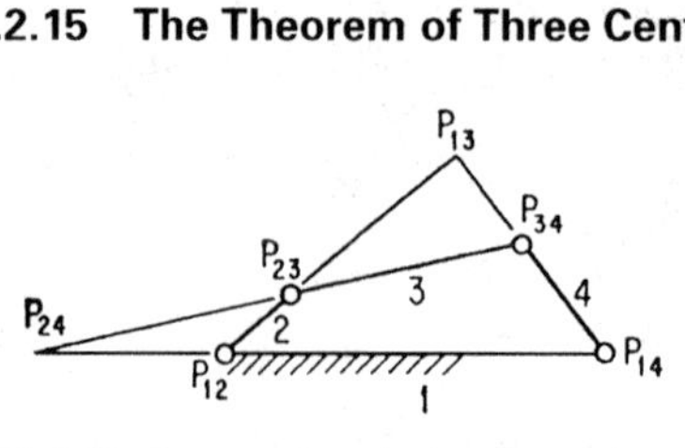

FIG. 3.14 Instant centers in four-bar motion.

Also known as Kennedy's or the Aronhold-Kennedy theorem, this theorem states that, for any three bodies i, j, k in plane motion, the relative instant centers P_{ij}, P_{jk}, P_{ki} are collinear; here P_{ij}, for instance, refers to the instant center of the motion of link i relative to link j, or vice versa. Figure 3.14 illustrates the theorem with respect to four-bar motion. It is used in determining the location of instant centers and in planar path curvature investigations.

3.2.16 Function, Path, and Motion Generation

In "function generation" the input and output motions of a mechanism are linear analogs of the variables of a function $F(x,y, \ldots) = 0$. The number of degrees of freedom of the mechanism is equal to the number of independent variables.

For example, let ϕ and ψ, the linear or rotary motions of the input and output links or "terminals," be linear analogs of x and y, where $y = f(x)$ within the range $x_0 \leq x \leq x_{n+1}$, $y_0 \leq y \leq y_{n+1}$. Let the *input* values ϕ_0, ϕ_j, ϕ_{n+1} and the *output* values ψ_0, ψ_j, ψ_{n+1} correspond to the values x_0, x_j, x_{n+1} and y_0, y_j, y_{n+1}, of x and y, respectively, where the subscripts 0, j, and $(n + 1)$ designate starting, jth intermediate, and terminal values. Scale factors r_ϕ, r_ψ are defined by

$$r_\phi = (x_{n+1} - x_0)/(\phi_{n+1} - \phi_0) \qquad r_\psi = (y_{n+1} - y_0)/(\psi_{n+1} - \psi_0)$$

(it is assumed that $y_0 \neq y_{n+1}$), such that $y - y_i = r_\psi(\psi - \psi_j)$, $x - x_j = r_\phi(\phi - \phi_j)$, whence

$$d\psi/d\phi = (r_\phi/r_\psi)(dy/dx),\ d^2\psi/d\phi^2 = (r_\phi^2/r_\psi)(d^2y/dx^2), \text{ and generally,}$$

$$d^n\psi/d\phi^n = (r_\phi^n/r_\psi)(d^ny/dx^n)$$

In "path generation" a point of a floating link traces a prescribed path with reference to the frame. In "motion generation" a mechanism is designed to conduct a floating link through a prescribed sequence of positions (Ref. 382). Positions along the path or specification of the prescribed motion may or may not be coordinated with input displacements.

3.3 PRELIMINARY DESIGN ANALYSIS: DISPLACEMENTS, VELOCITIES, AND ACCELERATIONS (Refs. 41, 58, 61, 62, 96, 116, 117, 129, 145, 172, 181, 194, 212, 263, 278, 298, 302, 309, 361, 384, 428, 487; see also Sec. 3.9)

Displacements in mechanisms are obtained graphically (from scale drawings) or analytically or both. Velocities and accelerations can be conveniently analyzed graphically by the "vector-polygon" method or analytically (in case of plane motion) via complex numbers. In all cases, the "vector equation of closure" is utilized, expressing the fact that the mechanism forms a closed kinematic chain.

3.3.1 Velocity Analysis: Vector-Polygon Method

The method is illustrated using a point D on the connecting rod of a slider-crank mechanism (Fig. 3.15). The vector-velocity equation for C is

$$\mathbf{V}_C = \mathbf{V}_B + \mathbf{V}_{C/B}^n + \mathbf{V}_{C/B}^t = \text{a vector parallel to line AX}$$

where $\mathbf{V}_C$ = velocity of C (Fig. 3.15)
$\mathbf{V}_B$ = velocity of B

$\mathbf{V}^n_{C/B}$ = normal component of velocity of C relative to B = component of relative velocity along BC = zero (owing to the rigidity of the connecting rod)

$\mathbf{V}^t_{C/B}$ = tangential component of velocity of C relative to B, value $(\omega_{BC})\, BC$, perpendicular to BC

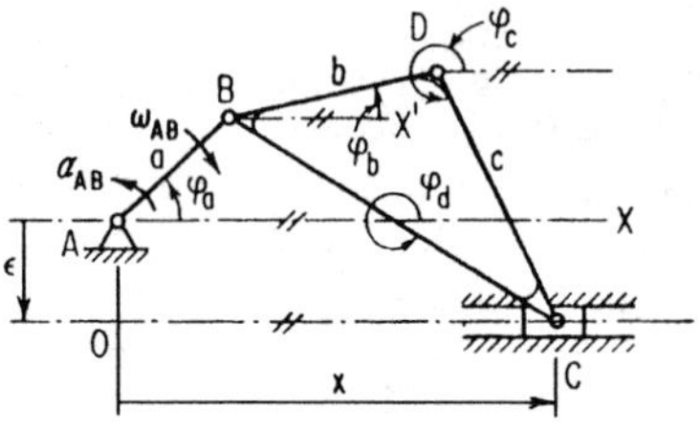

FIG. 3.15 Offset slider-crank mechanism.

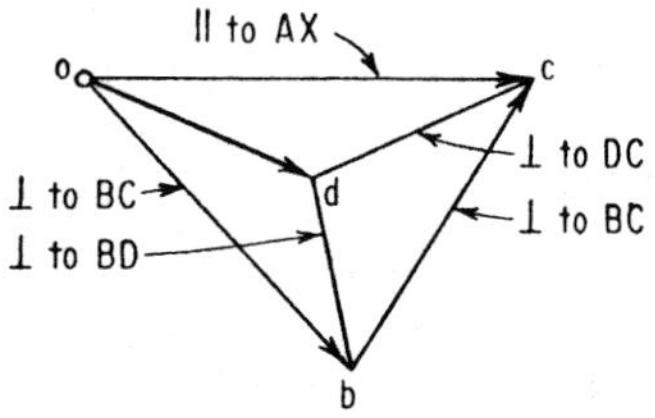

FIG. 3.16 Velocity polygon for slider-crank mechanism of Fig. 3.15.

The velocity equation is now "drawn" by means of a vector polygon as follows:

1. Choose an arbitrary origin o (Fig. 3.16).
2. Label terminals of velocity vectors with lowercase letters, such that absolute velocities start at o and terminate with the letter corresponding to the point whose velocity is designated. Thus $\mathbf{V}_B = \mathbf{ob}$, $\mathbf{V}_c = \mathbf{oc}$, to a certain scale.
3. Draw $\mathbf{ob} = (\omega_{AB})\, AB/k_v$, where k_v is the velocity scale factor, say, inches per inch per second.
4. Draw $\mathbf{bc} \perp BC$ and $\mathbf{oc} \| AX$ to determine intersection c.
5. Then $\mathbf{V}_C = (\mathbf{oc})/k_v$; absolute velocities always start at o.
6. Relative velocities $\mathbf{V}_{C/B}$, etc., connect the terminals of absolute velocities. Thus $\mathbf{V}_{C/B} = (\mathbf{bc})/k_v$. Note the reversal of order in C/B and **bc**.
7. To determine the velocity of D, one way is to write the appropriate velocity-vector equation and draw it on the polygon: $\mathbf{V}_D = \mathbf{V}_C + \mathbf{V}^n_{D/C} + \mathbf{V}^t_{D/C}$; the second is to utilize the "principle of the velocity image." This principle states that Δbcd in the velocity polygon is similar to ΔBCD in the mechanism, and the sense $b \to c \to d$ is the same as that of $B \to C \to D$. This "image construction" applies to any three points on a rigid link in plane motion. It has been used in Fig. 3.16 to locate d, whence $\mathbf{V}_D = (\mathbf{od})/k_v$.
8. The angular velocity ω_{BC} of the coupler can now be determined from

$$|\omega_{BC}| = \frac{|\mathbf{V}_{B/C}|}{BC} = \frac{|(\boldsymbol{cb})/k_v|}{BC}$$

The sense of ω_{BC} is determined by imagining B fixed and observing the sense of $\mathbf{V}_{C/B}$. Here ω_{BC} is counterclockwise.

9. Note that to determine the velocity of D it is easier to proceed in steps, to determine the velocity of C first and thereafter to use the image-construction method.

3.3.2 Velocity Analysis: Complex-Number Method

Using the slider crank of Fig. 3.15 once more as an illustration with x axis along the center line of the guide, and recalling that $i^2 = -1$, we write the complex-number equations as follows, with the equivalent vector equation below each:

Displacement: $$ae^{i\varphi_a} + be^{i\varphi_b} + ce^{i\varphi_c} = x \tag{3.8}$$

$$\mathbf{AB} + \mathbf{BD} + \mathbf{DC} = \mathbf{AC}$$

Velocity: $$iae^{i\varphi_a}\omega_{AB} + (ibe^{i\varphi_b} + ice^{i\varphi_c})\omega_{BC} = dx/dt \qquad (t = \text{time}) \tag{3.9}$$

$$\mathbf{V}_B + \mathbf{V}_{D/B} + \mathbf{V}_{C/D} = \mathbf{V}_C$$

Note that $\omega_{AB} = d\varphi_a/dt$ is positive when counterclockwise and negative when clockwise; in this problem ω_{AB} is negative.

The complex conjugate of Eq. (3.9)

$$-iae^{-i\varphi_a}\omega_{AB} - (ibe^{-i\varphi_b} + ice^{-i\varphi_c})\omega_{BC} = dx/dt \tag{3.10}$$

From Eqs. (3.9) and (3.10), regarded as simultaneous equations:

$$\frac{\omega_{BC}}{\omega_{AB}} = \frac{ia(e^{i\varphi_a} + e^{-i\varphi_a})}{-ib(e^{i\varphi_b} + e^{-i\varphi_b}) - ic(e^{i\varphi_c} + e^{-i\varphi_c})} = \frac{a \cos \varphi_a}{-(b \cos \varphi_b + c \cos \varphi_c)}$$

$$\mathbf{V}_D = \mathbf{V}_B + \mathbf{V}_{D/B} = iae^{i\varphi_a}\omega_{AB} + ibe^{i\varphi_b}\omega_{BC}$$

The quantities φ_a, φ_b, φ_c are obtained from a scale drawing or by trigonometry.

Both the vector-polygon and the complex-number methods can be readily extended to accelerations, and the latter also to the higher accelerations.

3.3.3 Acceleration Analysis: Vector-Polygon Method

We continue with the slider crank of Fig. 3.15. After solving for the velocities via the velocity polygon, write out and "draw" the acceleration equations. Again proceed in order of increasing difficulty: from *B* to *C* to *D*, and determine first the acceleration of point *C:*

$$\mathbf{A}_C = \mathbf{A}_C^n = \mathbf{A}_C^t = \mathbf{A}_B^n + \mathbf{A}_B^t + \mathbf{A}_{C/B}^n + \mathbf{A}_{C/B}^t$$

where $\mathbf{A}_C^n$ = acceleration normal to path of *C* (equal to zero in this case)
$\mathbf{A}_C^t$ = acceleration parallel to path of *C*
$\mathbf{A}_B^n$ = acceleration normal to path of *B*, value $\omega_{AB}^2(\overline{AB})$, direction *B* to *A*
$\mathbf{A}_B^t$ = acceleration parallel to path of *B*, value $\alpha_{AB}(AB)$, $\perp AB$, sense determined by that of α_{AB} (where $\alpha_{AB} = d\omega_{AB}/dt$)
$\mathbf{A}_{C/B}^n$ = acceleration component of *C* relative to *B*, in the direction *C* to *B*, value $(BC)\omega_{BC}^2$
$\mathbf{A}_{C/B}^t$ = acceleration component of *C* relative to *B*, $\perp \overline{BC}$, value α_{BC} $(\overline{BC})$. Since α_{BC} is unknown, so is the magnitude and sense of $\mathbf{A}_{C/B}^t$

The acceleration polygon is now drawn as follows (Fig. 3.17):

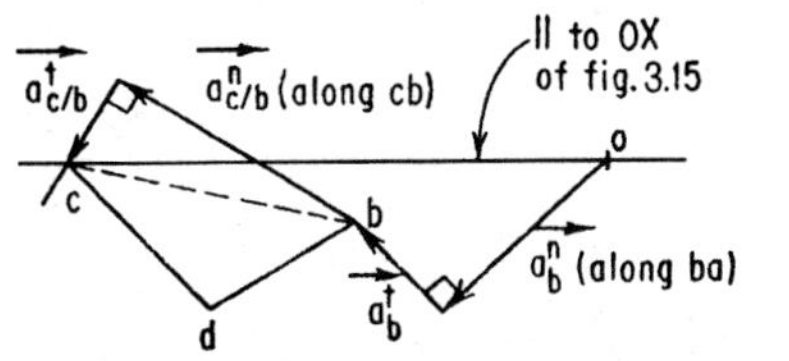

FIG. 3.17 Acceleration polygon for slider crank of Fig. 3.15. $\Delta bcd \approx \Delta BCD$ of Fig. 3.15.

1. Choose an arbitrary origin *o*, as before.
2. Draw each acceleration of scale k_a (inch per inch per second squared), and label the appropriate vector terminals with the lowercase letter corresponding to the point whose acceleration is designated, e.g., $\mathbf{A}_B = (\mathbf{ob})/k_a$. Draw $\mathbf{A}_B^n$, $\mathbf{A}_B^t$, and $\mathbf{A}_{C/B}^n$.

3. Knowing the direction of $\mathbf{A}^t_{C/B}(\perp \overline{BC})$, and also of $\mathbf{A}_C$ (along the slide), locate c at the intersection of a line through o, parallel to AX, and the line representing $\mathbf{A}^t_{C/B}$. $\mathbf{A}_C = (\mathbf{oc})/k_a$.
4. The acceleration of D is obtained using the "principle of the acceleration image," which states that, for any three points on a rigid body, such as link BCD, in plane motion, Δbcd and ΔBCD are similar, and the sense $b \rightarrow c \rightarrow d$ is the same as that of $B \rightarrow C \rightarrow D$. $\mathbf{A}_D = (\mathbf{od})/k_a$.
5. Relative accelerations can also be found from the polygon. For instance, $\mathbf{A}_{C/D} = (\mathbf{dc})/k_a$; note reversal of order of the letters C and D.
6. The angular acceleration α_{BC} of the connecting rod can now be determined from $|\alpha_{BC}| = |\mathbf{A}^t_{C/B}|/BC$. Its sense is determined by that of $\mathbf{A}^t_{C/B}$.
7. The acceleration of D can also be obtained by direct drawing of the equation $\mathbf{A}_D = \mathbf{A}_C + \mathbf{A}_{D/C}$.

3.3.4 Acceleration Analysis: Complex-Number Method (see Fig. 3.15)

Differentiating Eq. (3.9), obtain the acceleration equation of the slider-crank mechanism:

$$ae^{i\varphi_a}(i\alpha_{AB} - \omega^2_{AB}) + (be^{i\varphi_b} + ce^{i\varphi_c})(i\alpha_{BC} - \omega^2_{BC}) = d^2x/dt^2 \quad (3.11)$$

This is equivalent to the vector equation

$$\mathbf{A}^t_B + \mathbf{A}^n_B + \mathbf{A}^t_{D/B} + \mathbf{A}^n_{D/B} + \mathbf{A}^t_{C/D} + \mathbf{A}^n_{C/D} = \mathbf{A}^n_C + \mathbf{A}^t_C$$

Combining Eq. (3.11) and its complex conjugate, eliminate d^2x/dt^2 and solve for α_{BC}. Substitute the value of α_{BC} in the following equation for $\mathbf{A}_D$:

$$\mathbf{A}_D + \mathbf{A}_B + \mathbf{A}_{D/B} = ae^{i\varphi_a}(i\alpha_{AB} - \omega^2_{AB}) + be^{i\varphi_b}(i\alpha_{BC} - \omega^2_{BC})$$

The above complex-number approach also lends itself to the analysis of motions involving Coriolis acceleration. The latter is encountered in the determination of the relative acceleration of two instantaneously coincident points on different links.[106,171,384] The general complex-number method is discussed more fully in Ref. 381. An alternate approach, using the acceleration center, is described in Sec. 3.4. The accelerations in certain specific mechanisms are discussed in Sec. 3.9.

3.3.5 Higher Accelerations (see also Sec. 3.4)

The second acceleration (time derivative of acceleration), also known as "shock," "jerk," or "pulse," is significant in the design of high-speed mechanisms and has been investigated in several ways.[41,61,62,106,298,381,384,487] It can be determined by direct differentiation of the complex-number acceleration equation.[381] The following are the basic equations:

Shock of B Relative to A (where A and B represent two points on one link whose angular velocity is ω_p; $\alpha_p = d\omega_p/dt$).[298] Component along AB:

$$-3\alpha_p\omega_p AB$$

Component perpendicular to AB:

$$AB(d\alpha_p/dt - \omega_p^3)$$

in direction of $\boldsymbol{\omega}_p \times \mathbf{AB}$.

Absolute Shock.[298] Component along path tangent (in direction of $\boldsymbol{\omega}_p \times \mathbf{AB}$):

$$d^2v/dt^2 - v^3/\rho^2$$

where v = velocity of B and ρ = radius of curvature of path of B.

Component directed toward the center of curvature:

$$\frac{v}{\rho}\left[3\frac{dv}{dt} - \frac{v}{\rho}\frac{d\rho}{dt}\right]$$

Absolute Shock with Reference to Rolling Centrodes (Fig. 3.18, Sec. 3.4) [l, m as in Eq. (3.22)]. Component along AP:

$$-3\omega_p^3\delta\left[\delta \sin\psi\left(\frac{1}{m} + \frac{1}{g}\right) + \delta\cos\psi\left(\frac{1}{l} - \frac{1}{\delta}\right) - \frac{r}{g}\right] \qquad g = \frac{\omega_p^2\delta}{\alpha_p}$$

Component perpendicular to AP in direction of $\boldsymbol{\omega}_p \times \mathbf{PA}$:

$$r\left(\frac{d\alpha_p}{dt} - \omega_p^3\right) + 3\delta^2\omega_p^3\left[\cos\psi\left(\frac{1}{m} + \frac{1}{g}\right) - \sin\psi\left(\frac{1}{l} - \frac{1}{g}\right)\right]$$

3.3.6 Accelerations in Complex Mechanisms

When the number of real unknowns in the complex-number or vector equations is greater than two, several methods can be used.[106,145,309] These are applicable to mechanisms with more than four links.

3.3.7 Finite Differences in Velocity and Acceleration Analysis[212,375,419,428]

When the time-displacement curve of a point in a mechanism is known, the calculus of finite differences can be used for the calculation of velocities and accelerations. The data can be numerical or analytical. The method is useful also in ascertaining the existence of local fluctuations in velocities and accelerations, such as occur in cam-follower systems, for instance.

Let a time-displacement curve be subdivided into equal time intervals Δt and define the ith, the general interval, as $t_i \le t \le t_{i+1}$, such that $\Delta t = t_{i+1} - t_i$. The "central-difference" formulas then give the following approximate values for velocities dy/dt, accelerations d^2y/dt^2, and shock d^3y/dt^3, where y_i denotes the displacement y at the time $t = t_i$:

Velocity at $t = t_i + ½\Delta t$:

$$\frac{dy}{dt} = \frac{y_{i+1} - y_i}{\Delta t} \tag{3.12}$$

Acceleration at $t = t_i$:

$$\frac{d^2y}{dt^2} = \frac{y_{i+1} - 2y_i + y_{i-1}}{(\Delta t)^2} \tag{3.13}$$

Shock at $t = t_i + ½\Delta t$:

$$\frac{d^3y}{dt^3} = \frac{y_{i+2} - 3y_{i+1} + 3y_i - y_{i-1}}{(\Delta t)^3} \tag{3.14}$$

If the values of the displacements y_i are known with absolute precision (no error), the values for velocities, accelerations, and shock in the above equations become increasingly accurate as Δt approaches zero, provided the curve is smooth. If, however, the displacements y_i are known only within a given tolerance, say $\pm \epsilon y$, then the accuracy of the computations will be high only if the interval Δt is sufficiently small and, in addition, if

$$2\epsilon y/\Delta t << dy/dt \qquad \text{for velocities}$$

$$4\epsilon y/(\Delta t)^2 << d^2y/dt^2 \qquad \text{for accelerations}$$

$$8\epsilon y/(\Delta t)^3 << d^3y/dt^3 \qquad \text{for shock}$$

and provided also that these requirements are mutually compatible.

Further estimates of errors resulting from the use of Eqs. (3.12), (3.13), and (3.14), as well as alternate formulations involving "forward" and "backward" differences, are found in texts on numerical mathematics (e.g., Ref. 193, pp. 94–97 and 110–112, with a discussion of truncation and round-off errors).

The above equations are particularly useful when the displacement-time curve is given in the form of a numerical table, as frequently happens in checking an existing design and in redesigning.

Some current computer programs in displacement, velocity, and acceleration analysis are listed in Ref. 129; the kinematic properties of specific mechanisms, including spatial mechanisms, are summarized in Sec. 3.9.[129]

3.4 PRELIMINARY DESIGN ANALYSIS: PATH CURVATURE

The following principles apply to the analysis of a mechanism in a given position, as well as to synthesis when motion characteristics are prescribed in the vicinity of a particular position. The technique can be used to obtain a quick "first approximation" to mechanism proportions which can be refined at a later stage.

3.4.1 Polar-Coordinate Convention

Angles are measured counterclockwise from a directed line segment, the "pole tangent" PT, origin at P (see Fig. 3.18); the polar coordinates (r, ψ) of a point A are either $r = |PA|$, $\psi = \angle TPA$ or $r = -|PA|$, $\psi = \angle TPA \pm 180°$. For example, in Fig. 3.18 r is positive, but r_c is negative.

3.4.2 The Euler-Savary Equation (Fig. 3.18)

PT = common tangent of fixed and moving centrodes at point of contact P (the instant center).

PN = principal normal at P; $\angle TPN = 90°$.

PA = line or ray through P.

$C_A(r_c, \psi)$ = center of curvature of path of $A(r, \psi)$ in position shown. A and C_A are called "conjugate points."

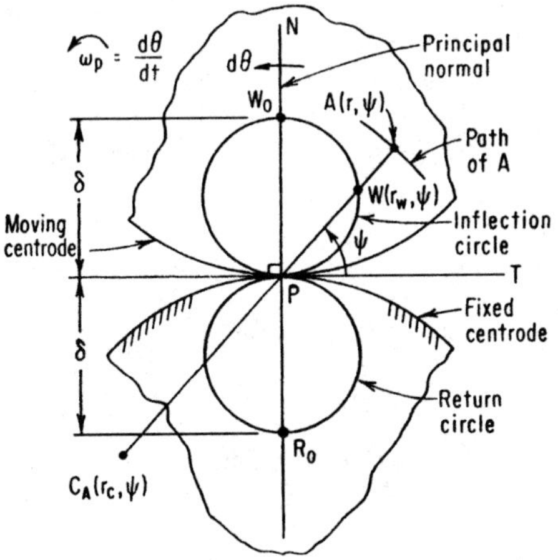

FIG. 3.18 Notation for the Euler-Savary equation.

θ = angle of rotation of moving centrode, positive counterclockwise.

s = arc length along fixed centrode, measured from P, positive toward T.

The Euler-Savary equation is valid under the following assumptions:

1. During an infinitesimal displacement from the position shown, $d\theta/ds$ is finite and different from zero.
2. Point A does not coincide with P.
3. AP is finite.

Under these conditions, the curvature of the path of A in the position shown can be determined from the following "Euler-Savary" equations:

$$[(1/r) - (1/r_c)] \sin \psi = -d\theta/ds = -\omega_p/v_p \tag{3.15}$$

where ω_p = angular velocity of moving centrode
= $d\theta/dt$, t = time
v_p = corresponding velocity of point of contact between centrodes along the fixed centrode
= ds/dt

Let r_w = polar coordinate of point W on ray PA, such that radius of curvature of path of W is infinite in the position shown; then W is called the "inflection point" on ray PA, and

$$1/r - 1/r_c = 1/r_w \tag{3.16}$$

The locus of all inflection points W in the moving centrode is the "inflection circle," tangent to PT at P, of diameter $PW_0 = \delta = -ds/d\theta$, where W_0, the "inflection pole," is the inflection point on the principal normal ray. Hence,

$$[(1/r) - (1/r_c)] \sin \psi = 1/\delta \tag{3.17}$$

The centers of path curvature of all points at infinity in the moving centrode are on the "return circle," also of diameter δ, and obtained as the reflection of the inflection circle about line PT. The reflection of W_0 is known as the "return pole" R_0. For the pole velocity (the time rate change of the position of P along the fixed centrode as the motion progresses, also called the "pole transfer velocity"[421f]) we have

$$v_p = ds/dt = -\,\omega_p \delta \tag{3.18}$$

The curvatures of the paths of all points on a given ray are concave toward the inflection point on that ray.

For the diameter of the inflection and return circles we have

$$\delta = r_p r_\pi / (r_\pi - r_p) \tag{3.19}$$

where r_p and r_π are the polar coordinates of the centers of curvature of the moving and fixed centrodes, respectively, at P. Let $\rho = r_c - r$ be the instantaneous value of the radius of curvature of the path of A, and $w = AW$, then

$$r^2 = \rho w \tag{3.20}$$

which is known as the "quadratic form" of the Euler-Savary equation.

Conjugate points in the planes of the moving and fixed centrodes are related by a "quadratic transformation."[32] When the above assumptions 1, 2, and 3, establishing the validity of the Euler-Savary equations, are not satisfied, see Ref. 281; for a further curvature theorem, useful in relative motions, see Ref. 23. For a computer-compatible complex-number treatment of path curvature theory, see Ref. 421*f*, Chap. 4.

EXAMPLE Cylinder of radius 2 in, rolling inside a fixed cylinder of radius 3 in, common tangent horizontal, both cylinders above the tangent, $\delta = 6$ in, $W_0(6, 90°)$. For point $A_1(\sqrt{2}, 45°)$, $r_{c1} = 1.5\sqrt{2}$, $C_{A1}(1.5\sqrt{2}, 45°)$, $\rho_{A1} = 0.5\sqrt{2}$, $r_{w1} = 3\sqrt{2}$, $v_p = -6\omega_p$. For point $A_2(-\sqrt{2}, 135°)$, $r_{c2} = -0.75\sqrt{2}$, $C_{A2}(-0.75\sqrt{2}, 135°)$, $\rho_{A2} = 0.25\sqrt{2}$, $r_{w2} = 3\sqrt{2}$.

Complex-number forms of the Euler-Savary equation[393,421f] and related expressions are independent of the choice of the x, iy coordinate system. They correlate the following complex vectors on any one ray (see Fig. 3.18): $\mathbf{a} = \mathbf{PA}$, $\mathbf{w} = \mathbf{PW}$, $\mathbf{c} = \mathbf{PC}_A$ and $\boldsymbol{\rho} = \mathbf{C}_A\mathbf{A}$, each expressed explicitly in terms of the others:

1. If points P, A, and W are known, find C_A by

$$\boldsymbol{\rho} = (a^2/|\mathbf{a} - \mathbf{w}|)\, e^{i \arg(\mathbf{a} - \mathbf{w})}$$

where $a = |\mathbf{a}|$.

2. If points P, A, and C_A are known, find W by

$$\mathbf{w} = \mathbf{a} - (a/\rho)^2 \boldsymbol{\rho}$$

where $\rho = |\boldsymbol{\rho}|$.

3. If points P, W, and C_A are known, find A by

$$\mathbf{a} = \mathbf{wc}/(\mathbf{w} + \mathbf{c})$$

4. If points A, C_A, and W are known, find P by

$$\mathbf{a} = |(|\mathbf{WA}|\rho)^{1/2}|(\pm\, e^{i \arg \boldsymbol{\rho}})$$

Note that the last equation yields two possible locations for P, symmetric about A. This is borne out also by Bobillier's construction (see Ref. 421f, Fig. 4.29, p. 329).

5. The vector diameter of the inflection circle, $\boldsymbol{\delta} = \mathbf{PW}_0$, in complex notation:

$$\boldsymbol{\delta} = -\mathbf{r}_p\mathbf{r}_\pi/(\mathbf{r}_p - \mathbf{r}_\pi) \tag{3.19a}$$

where $\mathbf{r}_p = \mathbf{O}_p\mathbf{P}$, $\mathbf{r}_\pi = \mathbf{O}_\pi\mathbf{P}$ and O_p and O_π are the centers of curvature of the fixed and moving centrodes, respectively.

6. The pole velocity in complex vector form is

$$\mathbf{v}_p = i\omega_\pi\boldsymbol{\delta} \tag{3.18a}$$

where ω_π is the angular velocity of the moving centrode.

7. If points P, A, and W_0 are known:

$$\mathbf{w} = \cos(\arg\mathbf{a} - \arg\boldsymbol{\delta})\boldsymbol{\delta}e^{i(\arg\mathbf{a} - \arg\boldsymbol{\delta})}$$

With the data of the above example, letting PT be the positive x axis and PN the positive iy axis, we have $\mathbf{r}_p = -i2$, $\mathbf{r}_\pi = -i3$; $\boldsymbol{\delta} = -(-i2)(-i3)/(-i3 + i2) = i6$, which is the same as the vector locating the inflection pole W_0, $\mathbf{w}_0 = \mathbf{PW}_0 = i6$. For point A_1,

$$\mathbf{a}_1 = \sqrt{2}e^{i45°} \qquad \mathbf{w}_1 = \cos(45° - 90°)i6e^{i(45°-90°)} = 3\sqrt{2}e^{i45°}$$

$$\boldsymbol{\rho}_{A1} = (2/|\sqrt{2}e^{i45°} - 3\sqrt{2}e^{i45°}|)\exp[i\arg(\sqrt{2}e^{i45°} - 3\sqrt{2}^{i45°})] = (\sqrt{2}/2)e^{i(-135°)}$$

$$\mathbf{c}_{A1} = \mathbf{a}_1 - \boldsymbol{\rho}_{A1} = \sqrt{2}e^{i45°} - (\sqrt{2}/2)e^{i(-135°)} = (3\sqrt{2}/2)e^{i45°}$$

$$\mathbf{v}_p = i\omega_\pi i6 = -\omega 6$$

For point A_2,

$$\mathbf{a}_2 = \sqrt{2}e^{i(-45°)} \qquad \mathbf{w}_2 = \cos(-45° - 90°)i6e^{i(-45°-90°)} = 3\sqrt{2}e^{i135°}$$

$$\boldsymbol{\rho}_{A2} = (2/|\sqrt{2}e^{i(-45°)} - 3\sqrt{2}e^{i135°}|)\exp[i\arg(\sqrt{2}e^{i(-45°)} - 3\sqrt{2}e^{i135°})] = (\sqrt{2}/4)e^{i(-45°)}$$

and $\mathbf{C}_{A2} = \mathbf{a}_2 - \boldsymbol{\rho}_{A2} = (\sqrt{2} - \sqrt{2}/4)e^{i(-45°)} = (3\sqrt{2}/4)e^{i(-45°)}$

Note that these are equal to the previous results and are readily programmed in a digital computer.

Graphical constructions paralleling the four forms of the Euler-Savary equation are given in Refs. 394 and 421f, p. 3.27.

3.4.3 Generating Curves and Envelopes[368]

Let g-g be a smooth curve attached to the moving centrode and e-e be the curve in the fixed centrode enveloping the successive positions of g-g during the rolling of the centrodes. Then g-g is called a "generating curve" and e-e its "envelope" (Fig. 3.19).

If C_g is the center of curvature of g-g and C_e that of e-e (at M):

1. C_e, P, M, and C_g are collinear (M being the point of contact between g-g and e-e).
2. C_e and C_g are conjugate points, i.e., if C_g is considered a point of the moving centrode,

the center of curvature of its path lies at C_e; interchanging the fixed and moving centrodes will invert this relationship.

3. Aronhold's first theorem: The return circle is the locus of the centers of curvature of all envelopes whose generating curves are straight lines.
4. If a straight line in the moving plane always passes through a fixed point by sliding through it and rotating about it, that point is on the return circle.
5. Aronhold's second theorem: The inflection circle is the locus of the centers of curvature for all generating curves whose envelopes are straight lines.

EXAMPLE (utilizing 4 above): In the swinging-block mechanism of Fig. 3.20, point C is on the return circle, and the center of curvature of the path of C as a point of link BD is therefore at C_c, halfway between C and P. Thus $ABCC_c$ constitutes a four-bar mechanism, with C_c as a fixed pivot, equivalent to the original mechanism in the position shown with reference to path tangents and path curvatures of points in the plane of link BD.

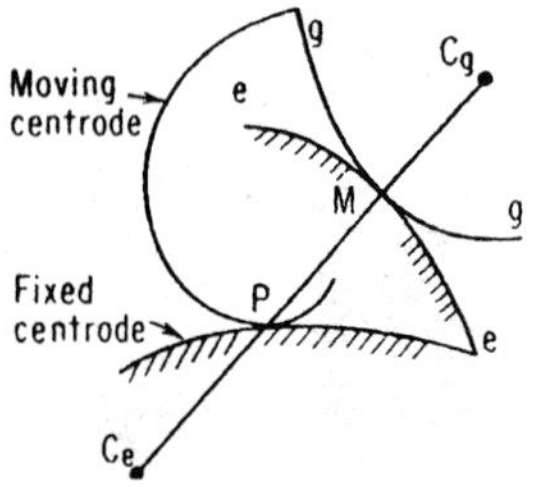

FIG. 3.19 Generating curve and envelope.

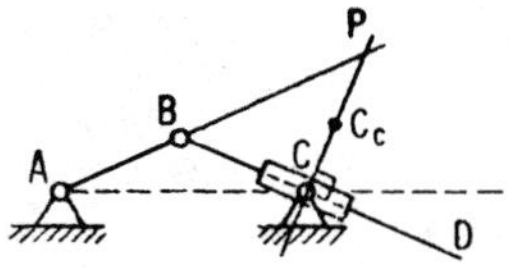

FIG. 3.20 Swinging-block mechanism: $CC_c = C_cP$.

3.4.4 Bobillier's Theorem

Consider two separate rays, 1 and 2 (Fig. 3.21), with a pair of distinct conjugate points on each, A_1, C_1, and A_2, C_2. Let Q_{A1A2} be the intersection of A_1A_2 and C_1C_2. Then the line through PQ_{A1A2} is called the "collineation axis," unique for the pair of rays 1 and 2, regardless of the choice of conjugate point pairs on these rays. Bobillier's theorem states that the angle between the common tangent of the centrodes and one ray is equal to the angle between the other ray and the collineation axis, both angles being described in the same sense.[368] Also see Ref. 421*f*, p. 3.31.

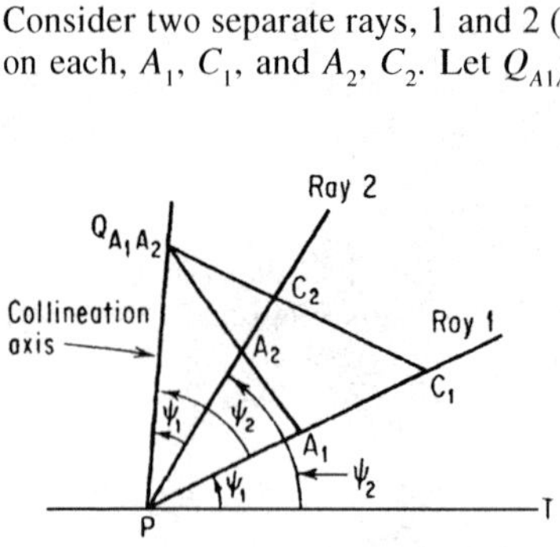

FIG. 3.21 Bobillier's construction.

The collineation axis is parallel to the line joining the inflection points on the two rays.

Bobillier's construction for determining the curvature of point-path trajectories is illustrated for two types of mechanisms in Figs. 3.22 and 3.23.

Another method for finding centers of path curvature is Hartmann's construction, described in Refs. 83 and 421*f*, pp. 332–336.

Occasionally, especially in the design of linkages with a dwell (temporary rest of output link), one may also use the "sextic of constant curvature," known also as the ρ curve,[32,421f] the locus of all points in the moving centrode whose paths at a given instant have the same numerical value of the radius of curvature.

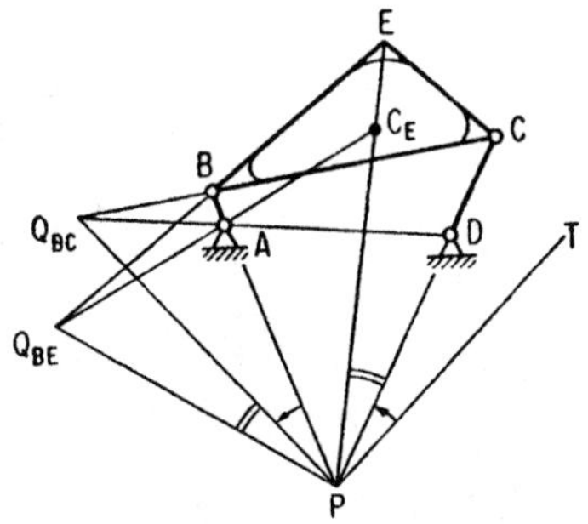

FIG. 3.22 Bobillier's construction for the center of curvature C_E of path of E on coupler of four-bar mechanism in position shown.

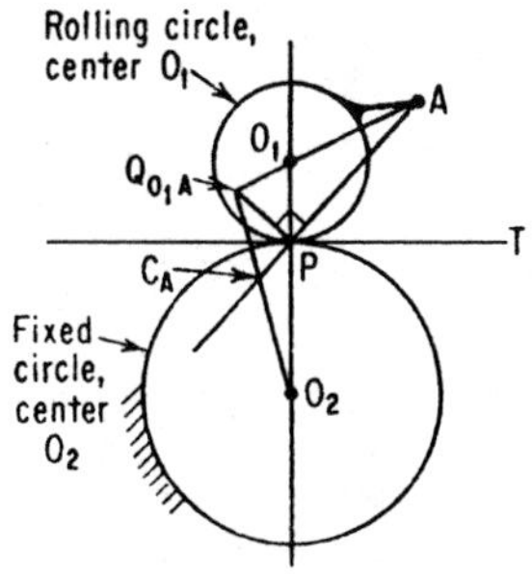

FIG. 3.23 Bobillier's construction for cycloidal motion. Determination of C_A, the center of curvature of the path of A, attached to the rolling circle (in position shown).

The equation of the ρ curve in the cartesian coordinate system in which PT is the positive x axis and PN the positive y axis is

$$(x^2 + y^2)^3 - \rho^2(x^2 + y^2 - \delta y)^2 = 0 \tag{3.20a}$$

where ρ is the magnitude of the radius of path curvature and δ is that of the inflection circle diameter.

3.4.5 The Cubic of Stationary Curvature (the k_u Curve)[421f]

The "k_u curve" is defined as the locus of all points in the moving centrode whose rate of change of path curvature in a given position is zero: $d\rho/ds = 0$. Paths of points on this curve possess "four-point contact" with their osculating circles. Under the same assumptions as in Sec. 3.4.1, the following is the equation of the k_u curve:

$$(\sin \psi \cos \psi)/r = (\sin \psi)/m + (\cos \psi)/l \tag{3.21}$$

where (r, ψ) = polar coordinates of a point on the k_u curve

$$m = -3\delta/(d\delta/ds) \tag{3.22}$$

$$l = 3r_p r_\pi/(2r_\pi - r_p)$$

In cartesian coordinates (x and y axes PT and PN),

$$(x^2 + y^2)(mx + yl) - lmxy = 0 \tag{3.23}$$

The locus of the centers of curvature of all points on the k_u curve is known as the "cubic of centers of stationary curvature,"[421f] or the "k_a curve." Its equation is

$$(x^2 + y^2)(mx + l^*y) - l^*mxy = 0 \tag{3.24}$$

where

$$1/l - 1/l^* = 1/\delta \tag{3.25}$$

The construction and properties of these curves are discussed in Refs. 26, 256, and 421*f*.

The intersection of the cubic of stationary curvature and the inflection circle yields the "Ball point" $U(r_u, \psi_u)$, which describes an approximate straight line, i.e., its path possesses four-point contact with its tangent (Ref. 421*f*, pp. 354–356). The coordinates of the Ball point are

$$\psi_u = \tan^{-1} \frac{2r_p - r_\pi}{(r_\pi - r_p)(d\delta/ds)} \tag{3.26}$$

$$r_u = \delta \sin \psi_u \tag{3.27}$$

In the case of a circle rolling inside or outside a fixed circle, the Ball point coincides with the inflection pole.

Technical applications of the cubic of stationary curvature, other than design analysis in general, include the generation of *n*-sided polygons,[32] the design of intermittent-motion mechanisms such as the type described in Ref. 426, and approximate straight-line generation. In many of these cases the curves degenerate into circles and straight lines.[32] Special analyses include the "Cardan positions of a plane" (osculating circle of moving centrode inside that of the fixed centrode, one-half its size; stationary inflection-circle diameter)[49,126] and dwell mechanisms. The latter utilize the "q_1 curve" (locus of points having equal radii of path curvature in two distinct positions of the moving centrode) and its conjugate, the "q_m curve." See also Ref. 395*a*.

3.4.6 Five and Six Infinitesimally Separated Positions of a Plane (Ref. 421*f*, pp. 241–245)

In the case of five infinitesimal positions, there are in general four points in the moving plane, called the "Burmester points," whose paths have "five-point contact" with their osculating circles. These points may be all real or pairwise imaginary. Their application to four-bar motion is outlined in Refs. 32, 411, 469, and 489, and related computer programs are listed in Ref. 129, the last also summarizing the applicable results of six-position theory, insofar as they pertain to four-bar motion. Burmester points and points on the cubic of stationary curvature have been used in a variety of six-link dwell mechanisms.[32,159]

3.4.7 Application of Curvature Theory to Accelerations (Ref. 421*f*, p. 313)

1. The acceleration $\mathbf{A}_p$ of the instant center (as a point of the moving centrode) is given by $\mathbf{A}_p = \omega_p^2(\mathbf{PW}_0)$; it is the only point of the moving centrode whose acceleration is independent of the angular acceleration α_p.
2. The inflection circle (also called the "de la Hire circle" in this connection) is the locus of points having zero acceleration normal to their paths.
3. The locus of all points on the moving centrode, whose tangential acceleration (i.e., acceleration along path) is zero, is another circle, the "Bresse circle," tangent to the principal normal at *P*, with diameter equal to $-\ \omega_p^2\delta/\alpha_p$ where α_p is the angular acceleration of the moving centrode, the positive sense of which is the same as that of θ. In complex vector form the diameter of the Bresse circle is $i\omega_p^2\delta/\alpha_p$ (Ref. 421*f*, pp. 336–338).
4. The intersection of these circles, other than *P*, determines the point *F*, with zero total acceleration, known as the "acceleration center." It is located at the intersection of the inflection circle and a ray of angle γ, where

$$\gamma = \angle\, W_0PF = \tan^{-1}(\alpha_p/\omega_p^2) \qquad 0 \le |\gamma| \le 90^\circ$$

measured in the direction of the angular acceleration (Ref. 421*f*, p. 337).

5. The acceleration $\mathbf{A}_B$ of any point B in the moving system is proportional to its distance from the acceleration center:

$$\mathbf{A}_B = (\mathbf{B\Gamma})(d^{-i\gamma})|(\omega_p^4 + \alpha_p^2)^{1/2}| \tag{3.28}$$

6. The acceleration vector $\mathbf{A}_B$ of any point B makes an angle γ with the line joining it to the acceleration center [see Eq. (3.28)], where γ is measured from $\mathbf{A}_B$ in the direction of angular acceleration (Ref. 421*f*, p. 340).
7. When the acceleration vectors of two points (V, U) on one link, other than the pole, are known, the location of the acceleration center can be determined from item 6 and the equation

$$|\tan \gamma| = \frac{|A^t_{U/V}|}{A^n_{U/V}}$$

8. The concept of acceleration centers and images can be extended also to the higher accelerations[41] (see also Sec. 3.3).

3.4.8 Examples of Mechanism Design and Analysis Based on Path Curvature

1. Mechanism used in guiding the grinding tool in large gear generators (Fig. 3.24): The radius of path curvature ρ_m of M at the instant shown: $\rho_m = (W_1W_2)/(2 \tan^3 \theta)$, at which instant M is on the cubic of stationary curvature belonging to link W_1W_2; ρ_m is arbitrarily large if θ is sufficiently small.

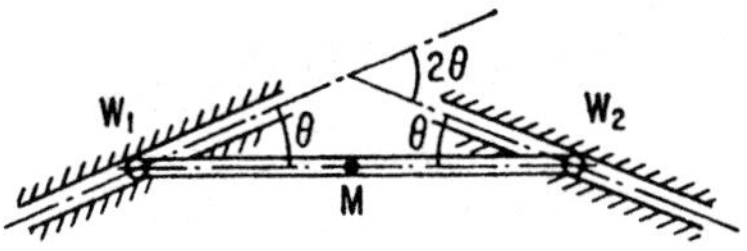

FIG. 3.24 Mechanism used in guiding the grinding tool in large gear generators. (Due to A. H. Candee, Rochester, N.Y.) $MW_1 = MW_2$; link W_1W_2 constrained by straight-line guides for W_1 and W_2.

2. Machining of radii on tensile test specimens[175,488] (Fig. 3.25): C lies on cubic of stationary curvature; AB is the diameter of the inflection circle for the motion of link ABC; radius of curvature of path of C in the position shown:

$$\rho_c = (AC)^2/(BC)$$

3. Pendulum with large period of oscillation, yet limited size[283,434] (Fig. 3.26), as used

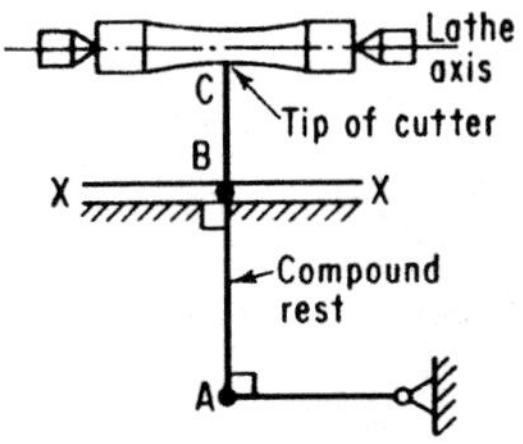

FIG. 3.25 Machining of radii on tensile test specimens. B guided along $X - X$.

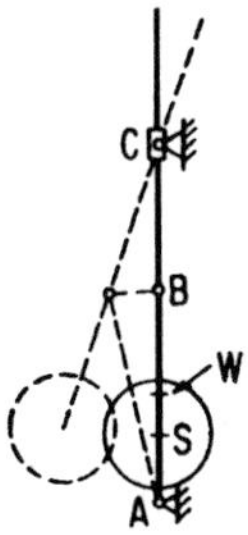

FIG. 3.26 Pendulum with large period of oscillation.

in recording ship's vibrations: $AB = a$, $AC = b$, $CS = s$, r_t = radius of gyration of the heavy mass S about its center of gravity. If the mass other than S and friction are negligible, the length l of the equivalent simple pendulum is given by

$$l = \frac{r_t^2 + s^2}{(b/a)(b - a)} - s$$

where the distance CW is equal to $(b/a)(b - a)$. The location of S is slightly below the inflection point W, in order for the oscillation to be stable and slow.

4. Modified geneva drive in high-speed bread wrapper[377] (Fig. 3.27): The driving pin of the geneva motion can be located at or near the Ball point of the pinion motion; the path of the Ball point, approximately square, can be used to give better kinematic characteristics to a four-station geneva than the regular crankpin design, by reducing peak velocities and accelerations.

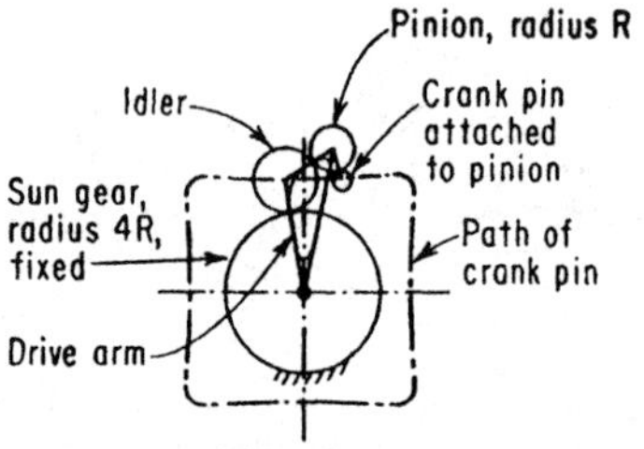

FIG. 3.27 Modified geneva drive in high-speed bread wrapper.

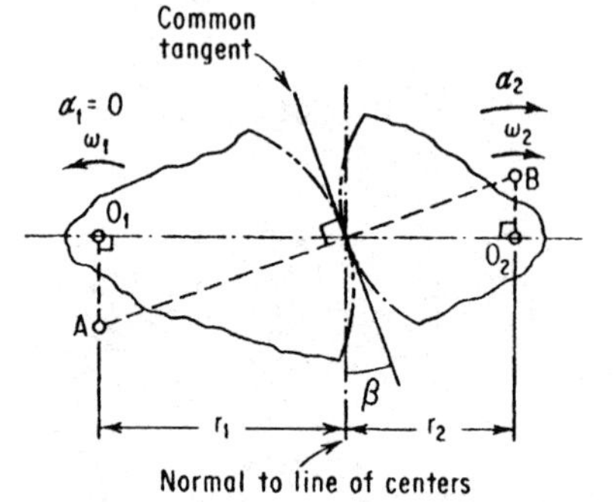

FIG. 3.28 Angular acceleration diagram for noncircular gears.

5. Angular acceleration of noncircular gears (obtainable from equivalent linkage O_1ABO_2) (Ref. 116, discussion by A. H. Candee; Fig. 3.28):

Let ω_1 = angular velocity of left gear, assumed constant, counterclockwise
ω_2 = angular velocity of right gear, clockwise
α_2 = clockwise angular acceleration of right gear

Then
$$\alpha_2 = [r_1(r_1 + r_2)/r_2^2] (\tan \beta)\omega_1^2$$

3.5 DIMENSIONAL SYNTHESIS: PATH, FUNCTION, AND MOTION GENERATION[106,421f]

In the design of automatic machinery, it is often required to guide a part through a sequence of prescribed positions. Such motions can be mechanized by dimensional synthesis based on the kinematic geometry of distinct positions of a plane. In plane motion, a "kinematic plane," hereafter called a "plane," refers to a rigid body, arbitrary in extent. The position of a plane is determined by the location of two of its points, A and B, designated as A_i, B_i in the ith position.

3.5.1 Two Positions of a Plane

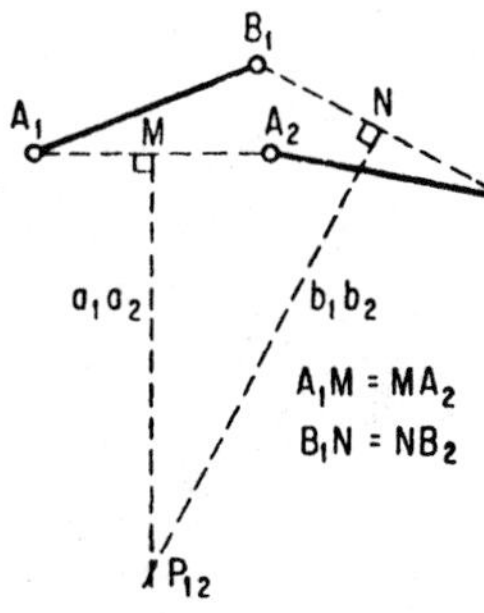

FIG. 3.29 Two positions of a plane. Pole $P_{12} = a_1a_2 \times b_1b_2$.

According to "Chasles's theorem," the motion from A_1B_1 to A_2B_2 (Fig. 3.29) can be considered *as though* it were a rotation about a point P_{12}, called the pole, which is the intersection of the perpendicular bisectors a_1a_2, b_1b_2 of A_1A_2 and B_1B_2, respectively. A_1, A_2, ..., are called "corresponding positions" of point A; B_1, B_2, ..., those of point B; A_1B_1, A_2B_2, ..., those of the plane AB.

A similar construction applies to the "relative motion of two planes" (Fig. 3.30) AB and CD (positions A_iB_i and C_iD_i, i = 1, 2). The "relative pole" Q_{12} is constructed by transferring the figure $A_2B_2C_2D_2$ as a rigid body to bring A_2 and B_2 into coincidence with A_1 and B_1, respectively, and denoting the new positions of C_2, D_2, by C_2^1, D_2^1, respectively. Then Q_{12} is obtained from C_1D_1 and $C_2^1D_2^1$ as in Fig. 3.29.

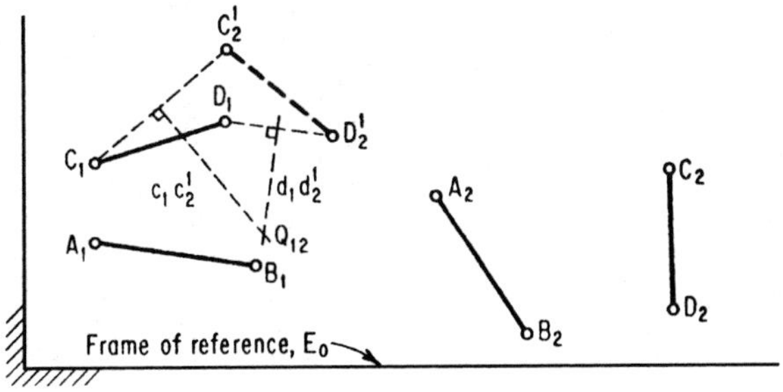

FIG. 3.30 Relative motion of two planes, AB and CD. Relative pole, $Q_{12} = c_1c_2^1 \times d_1d_2^1$.

1. The motion of A_1B_1 to A_2B_2 in Fig. 3.29 can be carried out by four-link mechanisms in which A and B are coupler-hinge pivots and the fixed-link pivots A_0, B_0 are located on the perpendicular bisectors a_1a_2, b_1b_2, respectively.
2. To construct a four-bar mechanism A_0ABB_0 when the corresponding angles of rotation of the two cranks are prescribed (in Fig. 3.31 the construction is illustrated with ϕ_{12} clockwise for A_0A and ψ_{12} clockwise for B_0B):
   a. From line A_0B_0X, lay off angles ½ϕ_{12} and ½ψ_{12} opposite to desired direction of rotation of the cranks, locating Q_{12} as shown.
   b. Draw any two straight lines L_1 and L_2 through Q_{12}, such that

$$\angle L_1Q_{12}L_2 = \angle A_0Q_{12}B_0$$

   in magnitude and sense.
   c. A_1 can be located on L_1, B_1, and L_2, and when A_0A_1 rotates clockwise by ϕ_{12}, B_0B_1 will rotate clockwise by ψ_{12}. Care must be taken, however, to ensure that the mechanism will not lock in an intermediate position.

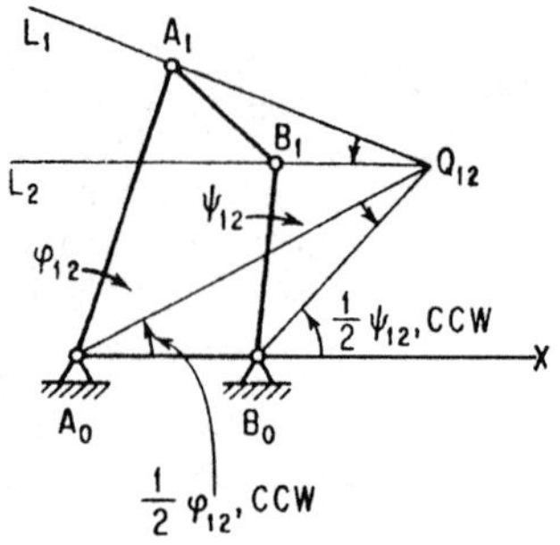

FIG. 3.31 Construction of four-bar mechanism $A_0A_1B_1B_0$ in position 1, for prescribed rotations ϕ_{12} vs. ψ_{12}, both clockwise in this case.

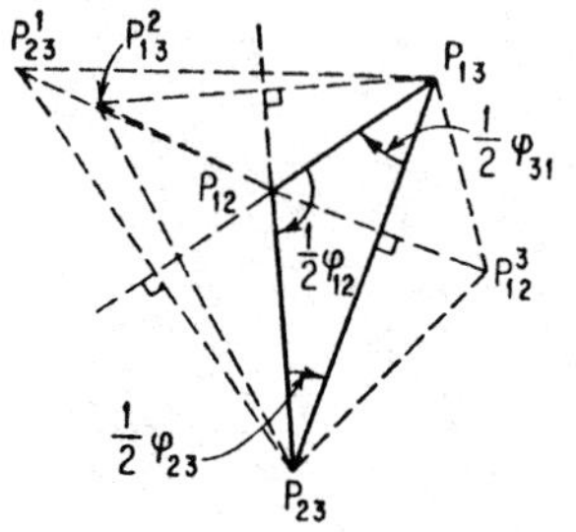

FIG. 3.32 Pole triangle for three positions of a plane. Pole triangle $P_{12}P_{23}P_{13}$ for three positions of a plane; image poles P^3_{12}, P^1_{23}, P^2_{31}; subtended angles $\frac{1}{2}\phi_{23}$, $\frac{1}{2}\phi_{31}$.

3.5.2 Three Positions of a Plane (A_iB_i, $i = 1, 2, 3$)[420]

In this case there are three poles P_{12}, P_{23}, P_{31} and three associated rotations ϕ_{12}, ϕ_{23}, ϕ_{31}, where $\phi_{ij} = \angle A_iP_{ij}A_j = \angle B_iB_{ij}B_j$. The three poles form the vertices of the *pole triangle* (Fig. 3.32). Note that $P_{ij} = P_{ji}$, and $\phi_{ij} = -\phi_{ji}$.

Theorem of the Pole Triangle. The internal angles of the pole triangle, corresponding to three distinct positions of a plane, are equal to the corresponding halves of the associated angles of rotation ϕ_{ij} which are connected by the equation

$$\tfrac{1}{2}\phi_{12} + \tfrac{1}{2}\phi_{23} + \tfrac{1}{2}\phi_{31} = 180^\circ \qquad \angle \tfrac{1}{2}\phi_{ij} = \angle P_{ik}P_{ij}P_{jk}$$

Further developments, especially those involving subtention of equal angles, are found in the literature.[32]

For any three corresponding points A_1, A_2, A_3, the center M of the circle passing through these points is called a "center point." If P_{ij} is considered as though fixed to link A_iB_i (or A_jB_j) and A_iB_i (or A_jB_j) is transferred to position k (A_kB_k), then P_{ij} moves to a new position P^k_{ij}, known as the "image pole," because it is the image of P_{ij} reflected about the line joining $P_{ik}P_{jk}$. $\Delta P_{ik}P_{jk}P^k_{ij}$ is called an "image-pole triangle" (Fig. 3.32). For "circle-point" and "center-point circles" for three finite positions of a moving plane, see Ref. 106, pp. 436–446 and Ref. 421*f*, pp. 114–122.

3.5.3 Four Positions of a Plane (A_iB_i, $i = 1, 2, 3, 4$)

With four distinct positions, there are six poles P_{12}, P_{13}, P_{14}, P_{23}, P_{24}, P_{34} and four pole triangles ($P_{12}P_{23}P_{13}$), ($P_{12}P_{24}P_{14}$), ($P_{13}P_{34}P_{14}$), ($P_{23}P_{34}P_{24}$).

Any two poles whose subscripts are all different are called "complementary poles." For example, $P_{23}P_{14}$, or generally $P_{ij}P_{kl}$, where *i, j, k, l* represents any permutation of the numbers 1, 2, 3, 4. Two complementary-pole pairs constitute the two diagonals of a "complementary-pole quadrilateral," of which there are three: ($P_{12}P_{24}P_{33}P_{13}$), ($P_{13}P_{32}P_{24}P_{14}$), and ($P_{14}P_{43}P_{32}P_{12}$).

Also associated with four positions are six further points Π_{ik} found by intersections of opposite sides of complementary-pole quadrilaterals, or their extensions, as follows: $\Pi_{ik} = P_{il}P_{kl} \times P_{ij}P_{kj}$.

3.5.4 The Center-Point Curve or Pole Curve[32,67,127,421f]

For three positions, a center point corresponds to *any* set of corresponding points; for four corresponding points to have a common center point, point A_1 can no longer be located arbitrarily in plane *AB*. However, a curve exists in the frame of reference called the "center-point curve" or "pole curve," which is the locus of centers of circles, each of which passes through *four* corresponding points of the plane *AB*. The center-point curve may be obtained from any complementary-pole quadrilateral; if associated with positions *i, j, k, l,* the center-point curve will be denoted by m_{ijkl}. Using complex numbers, let $\mathbf{OP}_{13} = \mathbf{a}$, $\mathbf{OP}_{23} = \mathbf{b}$, $\mathbf{OP}_{14} = \mathbf{c}$, $\mathbf{OP}_{24} = \mathbf{d}$, and $\mathbf{OM} = \mathbf{z} = x + iy$, where $\mathbf{OM}$ represents the vector from an arbitrary origin O to a point M on the center-point curve. The equation of the center-point[127] curve is given by

$$\frac{(\bar{\mathbf{z}} - \bar{\mathbf{a}})(\mathbf{z} - \mathbf{b})}{(\mathbf{z} - \mathbf{a})(\bar{\mathbf{z}} - \bar{\mathbf{b}})} = \frac{(\bar{\mathbf{z}} - \bar{\mathbf{c}})(\mathbf{z} - \mathbf{d})}{(\mathbf{z} - \mathbf{c})(\bar{\mathbf{z}} - \bar{\mathbf{d}})} = e^{2i\varphi} \tag{3.29}$$

where $\varphi = \angle P_{16}MP_{23} = \angle P_{14}MP_{24}$. In cartesian coordinates with origin at P_{12}, this curve is given in Ref. 16 by the following equation:

$$(x^2 + y^2)(j_2x - j_1y) + (j_1k_2 - j_2k_1 - j_3)x^2 + (j_1k_2 - j_2k_1 + j_3)y^2 + 2j_4xy$$
$$+ (-j_1k_3 + j_2k_4 + j_3k_1 - j_4k_2)x + (j_1k_4 + j_2k_3 - j_3k_2 - j_4k_1)y = 0 \tag{3.30}$$

where

$$\begin{aligned} k_1 &= x_{13} + x_{24} \\ k_2 &= y_{13} + y_{24} \\ k_3 &= x_{13}y_{24} + y_{13}x_{24} \\ k_4 &= x_{13}x_{24} - y_{13}y_{24} \\ j_1 &= x_{23} + x_{14} - k_1 \\ j_2 &= y_{23} + y_{14} - k_2 \\ j_3 &= x_{23}y_{14} + x_{14}y_{23} - k_3 \\ j_4 &= x_{23}x_{14} - y_{23}y_{14} - k_4 \end{aligned} \tag{3.31}$$

and (x_{ij}, y_{ij}) are the cartesian coordinates of pole P_{ij}. Equation (3.30) represents a third-degree algebraic curve, passing through the six poles P_{ij} and the six points Π_{ij}. Furthermore, any point M on the center-point curve subtends equal angles, or angles differing by two right angles, at opposite sides $(P_{ij}P_{jl})$ and $(P_{ik}P_{kl})$ of a complementary-pole quadrilateral, provided the sense of rotation of subtended angles is preserved:

$$\angle P_{ij}MP_{jl} = \angle P_{ik}MP_{kl} \cdots \tag{3.32}$$

Construction of the Center-Point Curve m_{ijkr}[32] When the four positions of a plane are known ($A_i B_i$, $i = 1, 2, 3, 4$), the poles P_{ij} are constructed first; thereafter, the center-point curve is found as follows:

A chord $P_{ij}P_{jk}$ of a circle, center O, radius

$$R = \overline{P_{ij}P_{jk}}/2 \sin \theta$$

(Fig. 3.33) subtends the angle θ (mod π) at any point on its circumference. For any value of θ, $-180° \le \theta \le 180°$, two corresponding circles can be drawn following Fig. 3.33,

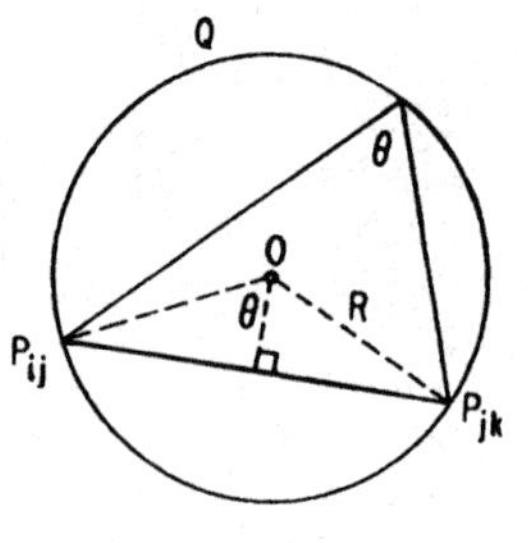

FIG. 3.33 Subtention of equal angles.

using as chords the opposite sides $P_{ij}P_{jk}$ and $P_{il}P_{kl}$ of a complementary-pole quadrilateral; intersections of such corresponding circles are points (M) on the center-point curve, provided Eq. (3.32) is satisfied.

As a check, it is useful to keep in mind the following angular equalities:

$$\tfrac{1}{2}\angle A_iMA_l = \angle P_{ij}MP_{jl} = \angle P_{ik}MP_{kl}$$

Also see Ref. 421*f*, p. 189.

Use of the Center-Point Curve. Given four positions of a plane A_iB_i ($i = 1, 2, 3, 4$) in a coplanar motion-transfer process, we can mechanize the motion by selecting points on the center-point curve as fixed pivots.

EXAMPLE[91] A stacker conveyor for corrugated boxes is based on the design shown schematically in Fig. 3.34. The path of C should be as nearly vertical as possible; if A_0, A_1, AC, $C_1C_2C_3C_4$ are chosen to suit the specifications, B_0 should be chosen on the center-point curve determined from A_iC_i, $i = 1, 2, 3, 4$; B_1 is then readily determined by inversion, i.e., by drawing the motion of B_0 relative to A_1C_1 and locating B_1 at the center of the circle thus described by B_0 (also see next paragraphs).

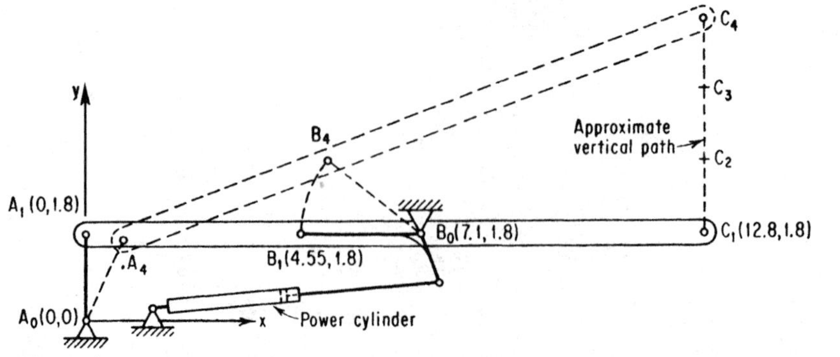

FIG. 3.34 Stacker conveyor drive.

3.5.5 The Circle-Point Curve

The circle-point curve is the kinematical inverse of the center-point curve. It is the locus of all points K in the moving plane whose four corresponding positions lie on one circle. If the circle-point curve is to be determined for positions i of the plane AB, Eqs. (3.29), (3.30), and (3.31) would remain unchanged, except that P_{jk}, P_{kl}, and P_{jl} would be replaced by the image poles P^i_{jk}, P^i_{kl}, and P^i_{jl}, respectively.

The center-point curve lies in the frame or reference plane; the circle-point curve lies in the moving plane. In the above example, point B_1 is on the circle-point curve for plane AC in position 1. The example can be solved also by selecting B_1 on the circle-point curve in A_1C_1; B_0 is then the center of the circle through $B_1B_2B_3B_4$. A computer program for the center-point and circle-point curves (also called "Burmester curves") is outlined in Refs. 383 and 421*f*, p. 184.

SPECIAL CASE If the corresponding points $A_1A_2A_3$ lie on a straight line, A_1 must lie on the circle through $P_{12}P_{13}P_{23}{}^1$; for four corresponding points $A_1A_2A_3A_4$ on one straight line, A_1 is located at intersection, other than P_{12}, of circles through $P_{12}P_{13}P^1_{23}$ and $P_{12}P_{14}P^1_{24}$, respectively. Applied to straight-line guidance in slider-crank and four-bar drives in Ref. 251; see also Refs. 32 and 421*f*, pp. 491–494.

3.5.6 Five Positions of a Plane (A_iB_i, $i = 1, 2, 3, 4, 5$)

In order to obtain accurate motions, it is desirable to specify as many positions as possible; at the same time the design process becomes more involved, and the number of "solutions" becomes more restricted. Frequently four or five positions are the most that can be economically prescribed.

Associated with five positions of a plane are four sets of points $K_u^{(i)}$ ($u = 1, 2, 3, 4$ and i is the position index as before) whose corresponding five positions lie on one circle; to each of these circles, moreover, corresponds a center point M_u. These circle points $K_u^{(1)}$ and corresponding center points M_u are called "Burmester point pairs." These four point pairs may be all real or pairwise imaginary (all real, two-point pairs real and two point pairs imaginary, or all point pairs imaginary).[127,421f] Note the difference, for historical reasons, between the above definition and that given in Sec. 3.4.6 for infinitesimal motion. The location of the center points, M_u, can be obtained as the intersections of two center-point curves, such as m_{1234} and m_{1235}.

A complex-number derivation of their location,[127,421f] as well as a computer program for simultaneous determination of the coordinates of both M_u and $K_u^{(i)}$, is available.[108,127,380,421f]

An algebraic equation for the coordinates (x_u, y_u) of M_u is given in Ref. 16 as follows. Origin at P_{12}, coordinates of P_{ij} are x_{ij}, y_{ij}:

$$x_u = \frac{(u - \tan \tfrac{1}{2}\theta_{12})[l_1(k_2 - k_3 u) - l_2(e_2 - e_3 u)]}{p_1 u^2 + p_2 u + p_3}$$

$$y_u = \frac{(u - \tan \tfrac{1}{2}\theta_{12})[l_1(k_4 + k_1 u) - l_2(e_4 + e_1 u)]}{p_1 u^2 + p_2 u + p_3} \tag{3.33}$$

where

$$\tan \tfrac{1}{2}\theta_{12} = (x_{13}y_{23} - x_{23}y_{13})(x_{13}x_{23} + y_{13}y_{23}) \tag{3.34}$$

and u is a root of

$$m_4u^4 + m_3u^3 + m_2u^2 + m_1u + m_0 = 0$$

wherein

$$\begin{aligned} m_0 &= p_3(q_1 + l_3p_3) \\ m_1 &= p_2(q_1 + 2l_3p_3)p_2 + q_2p_3 - q_3 \tan \tfrac{1}{2}\theta_{12} \\ m_2 &= q_0p_3 + q_2p_2 + q_1p_1 + l_3(p_2^2 + 2p_1p_3) - q_5 \tan \tfrac{1}{2}\theta_{12} + q_3 \\ m_3 &= q_0p_2 + p_1(q_2 + 2l_3p_2) - q_4 \tan \tfrac{1}{2}\theta_{12} + q_5 \\ m_4 &= p_1(q_0 + l_3p_1) + q_4 \end{aligned} \tag{3.35}$$

$$\begin{aligned} q_0 &= d_1h_3 - d_3h_1 & h_1 &= k_1l_1 - e_1l_2 \\ q_1 &= d_2h_4 - d_4h_2 & h_2 &= k_2l_1 - e_2l_2 \\ q_2 &= -d_1h_2 + d_2h_1 - d_3h_4 + d_4h_3 & h_3 &= k_3l_1 - e_3l_2 \\ q_3 &= h_2^2 + h_4^2 & h_4 &= k_4l_1 - e_4l_2 \end{aligned} \tag{3.36}$$

$$
\begin{aligned}
q_4 &= h_1^2 + h_3^2 & p_1 &= k_3e_1 - k_1e_3 \\
q_5 &= 2(h_1h_4 - h_2h_3) & p_2 &= k_3e_4 + k_1e_2 - k_2e_1 - k_4e_3 \\
d_1 &= x_{15} + x_{25} & p_3 &= k_4e_2 - k_2e_4 \\
d_2 &= x_{15} - x_{25} & e_1 &= d_1 - x_{13} - x_{23} \\
d_3 &= y_{15} + y_{25} & e_2 &= d_2 - x_{13} + x_{23} \\
d_4 &= y_{15} - y_{25} & e_3 &= d_3 - y_{13} - y_{23} \\
k_1 &= d_1 - x_{14} - x_{24} & e_4 &= d_4 - y_{13} + y_{23} \\
k_2 &= d_2 - x_{14} + x_{24} & l_1 &= x_{13}x_{23} + y_{13}y_{23} - l_3 \\
k_3 &= d_3 - y_{14} - y_{24} & l_2 &= x_{14}x_{24} + y_{14}y_{24} - l_3 \\
k_4 &= d_4 - y_{14} + y_{24} & l_3 &= x_{15}x_{25} + y_{15}y_{25}
\end{aligned}
\tag{3.36}
$$

The Burmester point pairs are discussed in Refs. 16, 67, and 127 and extensions of the theory in Refs. 382, 400, and 421*f*, pp. 211–230. It is suggested that, except in special cases, their determination warrants programmed computation.[108,421f]

Use of the Burmester Point Pairs. As in the example of Sec. 3.5.4, the Burmester point pairs frequently serve as convenient pivot points in the design of linked mechanisms. Thus, in the stacker of Sec. 3.5.4, five positions of C_i could have been specified in order to obtain a more accurately vertical path for C; the choice of locations of B_0 and B_1 would then have been limited to at most two Burmester point pairs (since A_0A_1 and $C_1C_0^\infty$, prescribed, are also Burmester point pairs).

3.5.7 Point-Position Reduction[2,159,194,421f]

"Point-position reduction" refers to a construction for simplifying design procedures involving several positions of a plane. For five positions, graphical methods would involve the construction of two center-point curves or their equivalent. In point-position reduction, a fixed-pivot location, for instance, would be chosen so that one or more poles coincide with it. In the relative motion of the fixed pivot with reference to the moving plane, therefore, one or more of the corresponding positions coincide, thereby reducing the problem to four or fewer positions of the pivot point; the center-point curves, therefore, may not have to be drawn. The reduction in complexity of construction is accompanied, however, by increased restrictions in the choice of mechanism proportions. An exhaustive discussion of this useful tool is found in Ref. 159.

3.5.8 Complex-Number Methods[106,123,371,372,380,381,421f,435]

Burmester-point theory has been applied to function generation as well as to path generation and combined path and function generation.[106,127,380,421f] The most general approach to path and function generation in plane motion utilizes complex numbers. The vector closure equations are used for each independent loop of the mechanism for every prescribed position and are differentiated once or several times if velocities, accelerations, and higher rates of change are prescribed. The equations are then solved for the

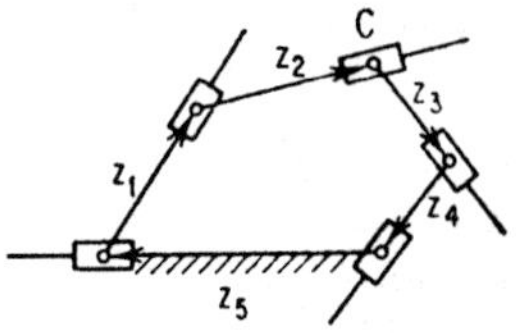

FIG. 3.35 Mechanism derived from a bar-slider chain.

unknown mechanism proportions. This method has been applied to four-bar path and function generators[106,123,127,371,380,384,421f] (the former with prescribed crank rotations), as well as to a variety of other mechanisms. The so-called "path-increment" and "path-increment-ratio" techniques (see below) simplify the mathematics insofar as this is possible. In addition to path and function specification, these methods can take into account prescribed transmission angles, mechanical advantages, velocity ratios, accelerations, etc., and combinations of these.

Consider, for instance, a chain of links connected by turning-sliding joints (Fig. 3.35). Each bar slider is represented by the vector $\mathbf{z}_j = r_j e^{i\theta_j}$. In this case the closure equation for the position shown, and its derivatives are as follows:

Closure: $$\sum_{j=1}^{5} \mathbf{z}_j = 0$$

Velocity: $$\frac{d}{dt}\sum_{j=1}^{5} \mathbf{z}_j = 0 \qquad \text{or} \qquad \sum_{j=1}^{5} \lambda_j \mathbf{z}_j = 0$$

where $$\lambda_j = (1/r_j)(dr_j/dt) + i(d\theta_j/dt) \qquad (t = \text{time})$$

Acceleration: $$\frac{d}{dt}\sum_{j=1}^{5} \lambda_j \mathbf{z}_j = 0 \qquad \text{or} \qquad \sum_{j=1}^{5} \lambda_j \mu_j \mathbf{z}_j = 0$$

where $$\mu_j = \lambda_j + (1/\lambda_j)(d\lambda_j/dt)$$

Similar equations hold for other positions. After suitable constraints are applied on the bar-slider chain (i.e., on r_j, θ_j) in accordance with the properties of the particular type of mechanism under consideration, the equations are solved for the $\mathbf{z}_j$ vectors, i.e., for the "initial" mechanism configuration.

If the path of a point such as C in Fig. 3.35 (although not necessarily a joint in the actual mechanism represented by the schematic or "general" chain) is specified for a number of positions by means of vectors $\boldsymbol{\delta}_1$, $\boldsymbol{\delta}_2$, ..., $\boldsymbol{\delta}_k$, the "path increments" measured from the initial position are $(\boldsymbol{\delta}_j - \boldsymbol{\delta}_1)$, $j = 2, 3, \ldots, k$. Similarly, the "path increment ratios" are $(\boldsymbol{\delta}_j - \boldsymbol{\delta}_1)/(\boldsymbol{\delta}_2 - \boldsymbol{\delta}_1)$, $j = 3, 4, \ldots, k$. By working with these quantities, only moving links or their ratios are involved in the computations. The solution of these equations of synthesis usually involves the prior solution of nonlinear "compatibility equations," obtained from matrix considerations. Additional details are covered in the above-mentioned references. A number of related computer programs for the synthesis of linked mechanisms are described in Refs. 129 and 421*f*. Numerical methods suitable for such syntheses are described in Ref. 372.

3.6 DESIGN REFINEMENT

After the mechanism is selected and its approximate dimensions determined, it may be necessary to refine the design by means of relatively small changes in the proportions, based on more precise design considerations. Equivalent mechanisms and cognates (see Sec. 3.6.6) may also present improvements.

3.6.1 Optimization of Proportions for Generating Prescribed Motions with Minimum Error

Whenever mechanisms possess a limited number of independent dimensions, only a finite number of independent conditions can be imposed on their motion. Thus, if a path is to be generated by a point on a linkage (rather than, say, a cam follower), it is not possible—except in special cases—to generate the curve exactly. A desired path (or function) and the actual, or generated, path (or function) may coincide at several points, called "precision points"; between these, the curves differ.

The minimum distance from a point on the ideal path to the actual path is called the "structural error in path generation." The "structural error in function generation" is defined as the error in the ordinate (dependent variable y) for a given value of the abscissa (independent variable x). Structural errors exist independent of manufacturing tolerances and elastic deformations and are thus inherent in the design. The combined effect of these errors should not exceed the maximum tolerable error.

The structural error can be minimized by the application of the fundamental theorem of P. L. Chebyshev[16,42] phrased nonrigorously for mechanisms as follows:

If n independent, adjustable proportions (parameters) are involved in the design of a mechanism, which is to generate a prescribed path or function, then the largest absolute value of the structural error is minimized when there are n precision points so spaced that the $n + 1$ maximum values of the structural error between each pair of adjacent precision points—as well as between terminals and the nearest precision points—are numerically equal with successive alterations in sign.

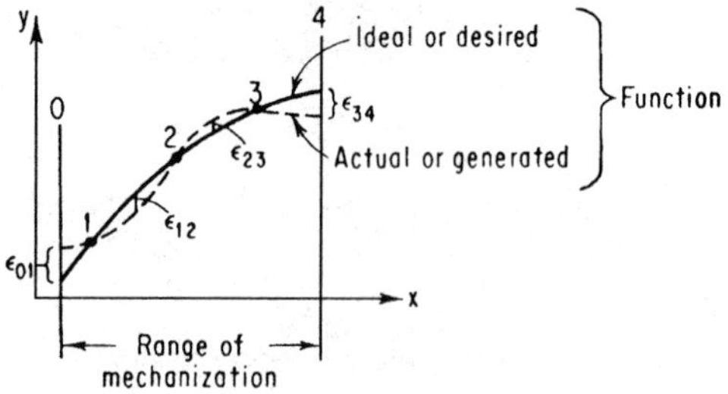

FIG. 3.36 Precision points 1, 2, and 3 and "regions" 01, 12, 23, and 34, in function generation.

In Fig. 3.36 (applied to function generation) the maximum structural error in each "region," such as 01, 12, 23, and 34, is shown as ϵ_{01}, ϵ_{12}, ϵ_{23}, and ϵ_{34}, respectively, which represent vertical distances between ideal and generated functions having three precision points. In general, the mechanism proportions and the structural error will vary with the choice of precision points. The spacing of precision points which yields least maximum structural error is called "optimal spacing." Other definitions and concepts, useful in this connection, are the following:

n-point approximation: Generated path (or function) has n precision points.

nth-order approximation: Limiting case of n-point approximation, as the spacing between precision points approaches zero. In the limit, one precision point is retained, at which point, however, the first $n - 1$ derivatives, or rates of change of the generated path (or function), have the same values as those of the ideal path (or function).

The following paragraphs apply both to function generation and to planar path generation, provided (in the latter case) that x is interpreted as the arc length along the ideal curve and the structural error ϵ_{ij} refers to the distance between generated and ideal curves.

Chebyshev Spacing.[122] For an n-point approximation to $y = f(x)$, within the range $x_0 \le x \le x_{n+1}$, Chebyshev spacing of the n precision points x_j is given by

$$x_j = \tfrac{1}{2}(x_0 + x_{n+1}) - \tfrac{1}{2}(x_{n+1} - x_0)\cos\{[2j - 1)\pi]/2n\} \qquad j = 1, 2, \ldots, n$$

Though not generally optimum for finite ranges, Chebyshev spacing often represents a good first approximation to optimal spacing.

The process of respacing the precision points, so as to minimize the maximum structural error, is carried out numerically[122] unless an algebraic solution is feasible.[42,404]

Respacing of Precision Points to Reduce Structural Error via Successive Approximations. Let $x_{ij}^{(1)} = x_j^{(1)} - x_i^{(1)}$, where $j = i + 1$, and let $x_i^{(1)}(i = 1, 2, \ldots, n)$ represent precision-point locations in a first approximation as indicated by the superscript (1). Let $\epsilon_{ij}^{(1)}$ represent the maximum structural error between points $x_i^{(1)}$, $x_j^{(1)}$, in the first approximation with terminal values x_0, x_{n+1}. Then a second spacing $x_{ij}^{(2)} = x_j^{(2)} - x_i^{(2)}$ is sought for which $\epsilon_{ij}^{(2)}$ values are intended to be closer to optimum (i.e., more nearly equal); it is obtained from

$$x_{ij}^{(2)} = \frac{x_{ij}^{(1)}(x_{n+1} - x_0)}{[\epsilon_{ij}^{(1)}]^m \sum_{i=0}^{n} \{x_{ij}^{(1)}/[\epsilon_{ij}^{(1)}]^m\}} \tag{3.37}$$

The value of the exponent m generally lies between 1 and 3. Errors can be minimized also according to other criteria, for instance, according to least squares.[249] Also see Ref. 404.

Estimate of Least Possible Maximum Structural Error. In the case of an n-point Chebyshev spacing in the range $x_0 \le x \le x_n + 1$ with maximum structural errors ϵ_{ij} ($j = i + 1$; $i = 0, 1, \ldots, n$),

$$\epsilon^2_{\text{opt(estimate)}} = (1/2n)[\epsilon^2_{01} + \epsilon^2_{n(n+1)}] + (1/n)[\epsilon^2_{12} + \epsilon^2_{23} + \cdots + \epsilon^2_{(n-1)n}] \tag{3.38}$$

In other spacings different estimates should be used; in the absence of more refined evaluations, the root-mean-square value of the prevailing errors can be used in the general case. These estimates may show whether a refinement of precision-point spacing is worthwhile.

Chebyshev Polynomials. Concerning the effects of increasing the number of precision points or changing the range, some degree of information may be gained from an examination of the "Chebyshev polynomials." The Chebyshev polynomial $T_n(t)$ is that nth-degree polynomial in t (with leading coefficient unity) which deviates least from zero within the interval $\alpha \le t \le \beta$. It can be obtained from the following differential-equation identity by equating to zero coefficients of like powers of t:

$$2[t^2 - (\alpha + \beta)t + \alpha\beta]\, T_n''(t) + [2t - (\alpha + \beta)]T_n'(t) - 2n^2 T_n(t) \equiv 0$$

where the primes refer to differentiation with respect to t. The maximum deviation from zero, L_n, is given by

$$L_n = (\alpha - \beta)^n / 2^{2n-1}$$

For the interval $-1 \le t \le 1$, for instance,

$$T_n(t) = (1/2^{n-1}) \cos (n \cos^{-1} t) \qquad L_n = 1/2^{n-1}$$

$$T_1(t) = t$$

$$T_2(t) = t^2 - \tfrac{1}{2}$$

$$T_3(t) = t^3 - (\tfrac{3}{4})t$$

$$T_4(t) = t^4 - t^2 + \tfrac{1}{8}$$

....

Chebyshev polynomials can be used directly in algebraic synthesis, provided the motion and proportions of the mechanism can be suitably expressed in terms of such polynomials.[42]

Adjusting the Dimensions of a Mechanism for Given Respacing of Precision Points. Once the respacing of the precision points is known, it is possible to recompute the mechanism dimensions by a linear computation[122,174,249,446] provided the changes in the dimensions are sufficiently small.

Let $f(x)$ = ideal or desired functional relationship.

$g(x) = g(x, p_0^{(1)}, p_1^{(1)}, \ldots, p_{n-1}^{(1)})$
= generated functional relationship in terms of mechanism parameters or proportions $p_j^{(1)}$, where $p_j^{(k)}$ refers to the jth parameter in the kth approximation.

$\epsilon^{(1)}(x_i^{(2)})$ = value of structural error at $x_i^{(2)}$ in the first approximation, where $x_i^{(2)}$ is a new or respaced location of a precision point, such that ideally $\epsilon^{(2)}(x_i^{(2)}) = 0$ (where $\epsilon = f - g$).

Then the new values of the parameters $p_j^{(2)}$ can be computed from the equations

$$\epsilon^{(1)}(x_i^{(2)}) = \sum_{j=0}^{n-1} \frac{\partial g(x_i^{(2)})}{\partial p_j^{(1)}} (p_j^{(2)} - p_j^{(1)}) \qquad i = 1, 2, \ldots, n \tag{3.39}$$

These are n linear equations, one each at the n "precision" points $x_i^{(2)}$ in the n unknowns $p_j^{(2)}$. The convergence of this procedure depends on the appropriateness of neglecting higher-order terms in Eq. (3.39); this, in turn, depends on the functional relationship and the mechanism and cannot in general be predicted. For related investigations, see Refs. 131 and 184; for respacing via automatic computation and for accuracy obtainable in four-bar function generators, see Ref. 122, and in geared five-bar function generators, see Ref. 397.

3.6.2 Tolerances and Precision[17,147,158,174,228,243,482]

After the structural error is minimized, the effects of manufacturing errors still remain.

The accuracy of a motion is frequently expressed as a percentage defined as the maximum output error divided by total output travel (range).

For a general discussion of the various types of errors, see Ref. 482.

Machining errors may cause changes in link dimensions, as well as clearances and backlash. Correct tolerancing requires the investigation of both. If the errors in link dimensions are small compared with the link lengths, their effect on displacements, velocities, and accelerations can be determined by a linear computation, using only first-order terms.

The effects of clearances in the joints and of backlash are more complicated and, in addition to kinematic effects, are likely to affect adversely the dynamic behavior of the mechanism.[147] The kinematic effect manifests itself as an uncertainty in displacements, velocities, accelerations, etc., which, in the absence of load reversal, can be computed as though due to a change in link length, equivalent to the clearance or backlash involved. The dynamic effects of clearances in machinery have been investigated in Ref. 98 to 100.

Since the effect of tolerances will depend on the mechanism and on the "location" of the tolerance in the mechanism, each tolerance should be specified in accordance with the magnitude of its effect on the pertinent kinematic behavior.

3.6.3 Harmonic Analysis (see also Sec. 3.9 and bibliography in Ref. 493)

It is sometimes desirable to express the motion of a machine part as a Fourier series in terms of driving motion, in order to analyze dynamic characteristics and to ensure satisfactory performance at high speeds. Harmonic analysis, for example, is used in computing the inertia forces in slider-crank mechanisms in internal-combustion engines[39,367] and also in other mechanisms.[128,286,289,493]

Generally, two types of investigations arise:

1. Determination of the "harmonics" in the motion of a given mechanism as a check on inertial loads and critical speeds
2. Proportioning to minimize higher harmonics[128]

3.6.4 Transmission Angles (see also Sec. 3.2.10)[134–136,155,467]

In mechanisms with varying transmission angles μ, the optimum design involves the minimization of the deviation ν of the transmission angle from its ideal value. Such a design maximizes the force tending to turn the driven link while minimizing frictional resistance, assuming quasi-static operation.

In plane crank-and-rocker linkages, the minimization of the maximum deviation of the transmission angle from 90° has been worked out for given rocker swing angle ψ and corresponding crank rotation ϕ. In the special case of centric crank-and-rocker linkages ($\phi = 180°$) the solution is relatively simple: $a^2 + b^2 = c^2 + d^2$ (where a, b, c, and d denote the lengths of crank, coupler, rocker, and fixed link, respectively). This yields $\sin \nu = (ab/cd)$, $\mu_{max} = 90° + \nu$, $\mu_{min} = 90° - \nu$. The solution for the general case (ϕ arbitrary), including additional size constraints, can be found in Refs. 134 to 136, 155, and 371, and depends on the solution of a cubic equation.

3.6.5 Design Charts

To save labor in the design process, charts and atlases are useful when available. Among these are Refs. 199 and 210 in four-link motion; the VDI-Richtlinien Duesseldorf (obtainable through Beuth-Vertrieb Gmbh, Berlin), such as 2131, 2132 on the offset turning block and the offset slider crank, and 2125, 2126, 2130, 2136 on the offset slider-crank and crank-and-rocker mechanisms; 2123, 2124 on four-bar mechanisms; 2137 on the in-line swinging block; and data sheets in the technical press.

3.6.6 Equivalent and "Substitute" Mechanisms[106,112,421g]

Kinematic equivalence is explained in Sec. 3.2.12. Ways of obtaining equivalent mechanisms include (1) pin enlargement, (2) kinematic inversion, (3) use of centrodes, (4) use of curvature constructions, (5) use of pantograph devices, (6) use of multigeneration properties, (7) substitution of tapes, racks, and chains for rigid links[84,103,159,189,285] and other ways depending on the inventiveness of the designer.* Of these, (5) and (6) require additional explanation.

The "pantograph" can be used to reproduce a given motion, unchanged, enlarged, reduced, or rotated. It is based on "Sylvester's plagiograph," shown in Fig. 3.37.

AODC is a parallelogram linkage with point *O* fixed with two similar triangles ACC_1, *DBC*, attached as shown. Points *B* and C_1 will trace similar curves, altered in the ratio $OC_1/OB = AC_1/AC$ and rotated relative to each other by an amount equal to the angle α. The ordinary pantograph is the special case obtained when *B, D, C,* and *C, A,* C_1 are collinear. It is used in engraving machines and other motion-copying devices.

Roberts' theorem[32,182,288,347,421f] states that there are three different but related four-bar mechanisms generating the same coupler curve (Fig. 3.38): the "original" *ABCDE*, the "right cognate" *LKGDE*, and the "left cognate" *LHFAE*. Similarly, slider-crank mechanisms have one cognate each.[182]

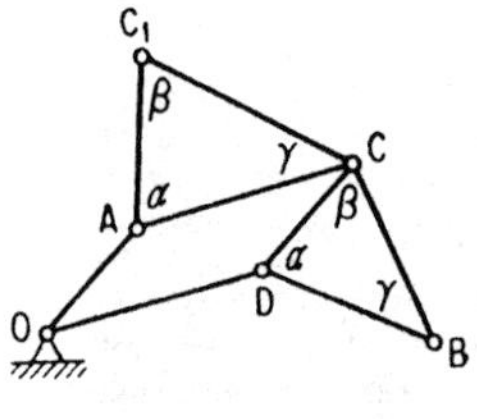

FIG. 3.37 Sylvester's plagiograph or skew pantograph.

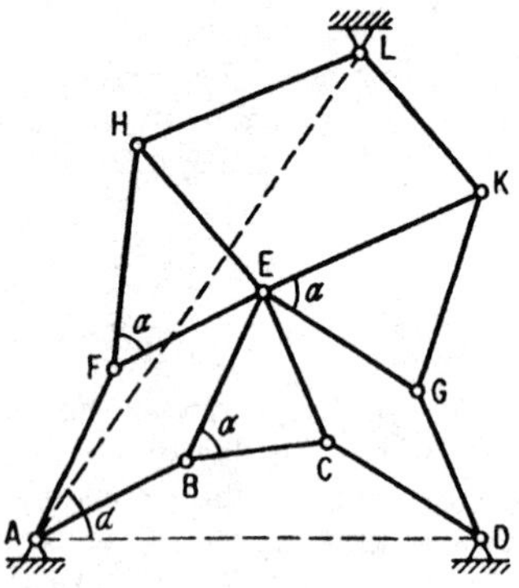

FIG. 3.38 Roberts' theorem. $\Delta BEC \approx \Delta FHE \approx \Delta EKG \approx \Delta ALD \approx \Delta AHC \approx \Delta BKD \approx FLG$; *AFEB, EGDC, HLKE* are parallelograms.

If the "original" linkage has poor proportions, a cognate may be preferable. When Grashof's inequality is obeyed (Sec. 3.9) and the original is a double rocker, the cognates are crank-and-rocker mechanisms; if the original is a drag link, so are the cognates; if the original does not obey Grashof's inequality, neither do the cognates, and all three are either double rockers or folding linkages. Several well-known straight-line guidance devices (Watt and Evans mechanisms) are cognates.

Geared five-bar mechanisms (Refs. 95, 119, 120, 347, 372, 391, 397, 421*f*) may also be used to generate the coupler curve of a four-bar mechanism, possibly with better transmission angles and proportions, as, for instance, in the drive of a deep-draw press. The gear ratio in this case is 1:1 (Fig. 3.39), where *ABCDE* is the four-bar linkage and *AFEGD* is the five-bar mechanism with links *AF* and *GD* geared to each other by 1:1 gearing. The path of *E* is identical in both mechanisms.

*Investigation of enumeration of mechanisms based on degree-of-freedom requirements are found in Refs. 106, 159, and 162 to 165 with application to clamping devices, tools, jigs, fixtures, and vise jaws.

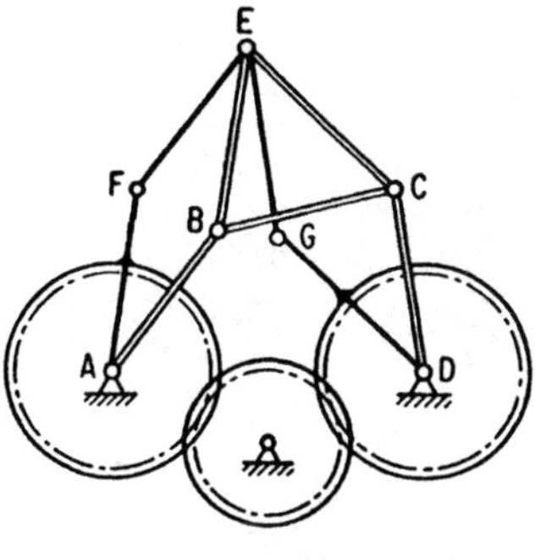

FIG. 3.39 Four-bar linkage *ABCDE* and equivalent 1:1 geared five-bar mechanism *AFEGD*; *AFEB* and *DGEC* are parallelograms.

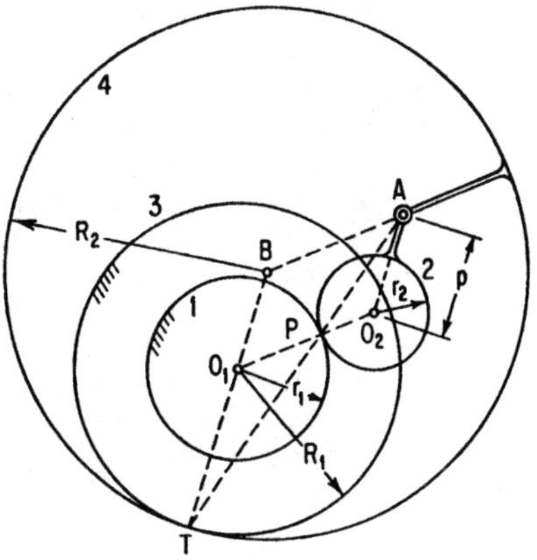

FIG. 3.40 Double generation of a cycloidal path. For the case shown O_1O_2 and AO_2 rotate in the same direction. $R_2/R_1 = r_2/r_1 + 1$; $R_1 = p(r_1/r_2)$; $R_2 = p[1 + (r_1/r_2)]$. Radius ratios are considered positive or negative depending on whether gearing is internal or external.

In Fig. 3.38 each cognate has one such derived geared five-bar mechanism (as in Fig. 3.39), thus giving a choice of six different mechanisms for the generation of any one coupler curve.

Double Generation of Cycloidal Curves.[315,385,386] A given cycloidal motion can be obtained by two different pairs of rolling circles (Fig. 3.40). Circle 2 rolls on fixed circle 1 and point *A*, attached to circle 2, describes a cycloidal curve. If O_1, O_2 are centers of circles 1 and 2, *P* their point of contact, and O_1O_2AB a parallelogram, circle 3, which is also fixed, has center O_1 and radius O_1T, where *T* is the intersection of extensions of O_1B and *AP*; circle 4 has center *B*, radius *BT*, and rolls on circle 3. If point *A* is now rigidly attached to circle 4, its path will be the same as before. Dimensional relationships are given in the caption of Fig. 3.40. For analysis of cycloidal motions, see Refs. 385, 386, and 492.

Equivalent mechanisms obtained by multigeneration theory may yield patentable devices by producing "unexpected" results, which constitutes one criterion of patentability. In one application, cycloidal path generation has been used in a speed reducer.[48,426,495] Another form of "cycloidal equivalence" involves adding an idler gear to convert from, say, internal to external gearing; applied to resolver mechanism in Ref. 357.

3.6.7 Computer-Aided Mechanisms Design and Optimization (Refs. 64–66, 78, 79, 82, 106, 108, 109, 185, 186, 200, 217, 218, 246, 247, 317, 322–324, 366, 371, 380, 383, 412–414, 421*a*, 430, 431, 445, 457, 465, 484, 485)

General mechanisms texts with emphasis on computer-aided design include Refs. 106, 186, 323, 421*f*, 431, 445. Computer codes having both kinematic analysis and synthesis capability in linkage design include KINSYN[217,218] and LINCAGES.[108] Both codes also include interactive computer graphics features. Codes which can perform both kinematic and dynamic analysis for a large class of mechanisms include DRAM and

ADAMS,[64–66,457] DYMAC,[322,324] IMP,[430,465] kinetoelastodynamic codes,[109,421f] codes for the sensitivity analysis and optimization of mechanisms with intermittent-motion elements,[185,186,200] heuristic codes,[78,79,246,247] and many others.[82,317,322,366,484,485]

The variety of computational techniques is as large as the variety of mechanisms. For specific mechanisms, such as cams and gears, specialized codes are available.

In general, computer codes are capable of analyzing both simple and complex mechanisms. As far as synthesis is concerned the situation is complicated by the nonlinearity of the motion parameters in many mechanisms and by the impossibility of limiting most motions to small displacements. For the simpler mechanisms synthesis codes are available. For more complex mechanisms parameter variation of analysis codes or heuristic methods are probably the most powerful currently available tools. The subject remains under intensive development, especially with regard to interactive computer graphics [for example, CADSPAM, *c*omputer-*a*ided *d*esign of *spa*tial *m*echanisms (Ref. 421*a*)].

3.6.8 Balancing of Linkages

At high speeds the inertia forces associated with the moving links cause shaking forces and moments to be transmitted to the frame. Balancing can reduce or eliminate these. An introduction is found in Ref. 421*f.* (See also Refs. to BAL 25–30, 191, 214, 219, 258–260, 379*a,* 452, 452*a,* 452*b,* 459, 460.)

3.6.9 Kinetoelastodynamics of Linkage Mechanisms

Load and inertia forces may cause cyclic link deformations at high speeds, which change the motion of the mechanism and cannot be neglected. An introduction and copious list of references are found in Ref. 421*f.* (See also Refs. 53, 107, 202, 416, 417.)

3.7 *THREE-DIMENSIONAL MECHANISMS*[5,21,35,421f] (Sec. 3.9)

Three-dimensional mechanisms are also called "spatial mechanisms." Points on these mechanisms move on three-dimensional curves. The basic three-dimensional mechanisms are the "spherical four-bar mechanisms" (Fig. 3.41) and the "offset" or "spatial four-bar mechanism" (Fig. 3.42).

The spherical four-bar mechanism of Fig. 3.41 consists of links *AB, BC, CD,* and *DA,* each on a great circle of the sphere with center *O*; turning joints at *A, B, C,* and *D,* whose axes intersect at *O*; lengths of links measured by great-circle arcs or angles α_i subtended at *O*. Input θ_2, output θ_1; single degree of freedom, although $\Sigma f_i = 4$ (see Sec. 3.2.2).

Figure 3.42 shows a spatial four-bar mechanism; turning joint at *D*, turn-slide (also called cylindrical) joints at *B, C,* and *D*; a_{ij} denote minimum distances between axes of joints; input θ_2 at *D*; output at *A* consists of translation *s* and rotation θ_1; $\Sigma f_i = 7$; freedom, $F = 1$.

Three-dimensional mechanisms used in practice are usually special cases of the above two mechanisms. Among these are Hooke's joint (a spherical four-bar, with $\alpha_2 = \alpha_3 = \alpha_4 = 90°$, $90° < \alpha_1 \leq 180°$), the wobble plates ($\alpha_4 < 90°$, $\alpha_2 = \alpha_3 = \alpha_1 = 90°$), the space crank,[332] the spherical slider crank,[304] and other mechanisms, whose analysis is outlined in Sec. 3.9.

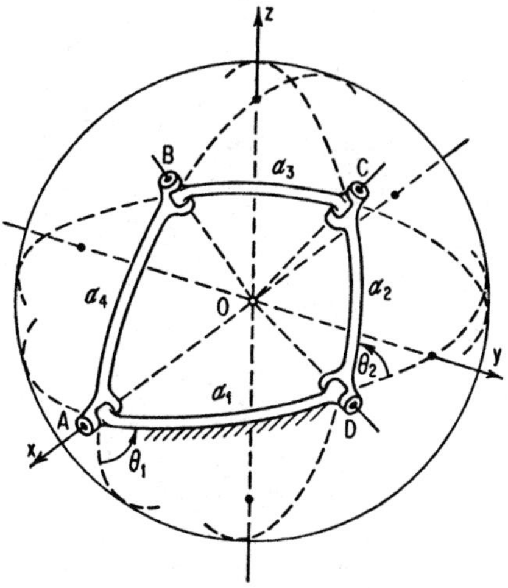

FIG. 3.41 Spherical four-bar mechanism.

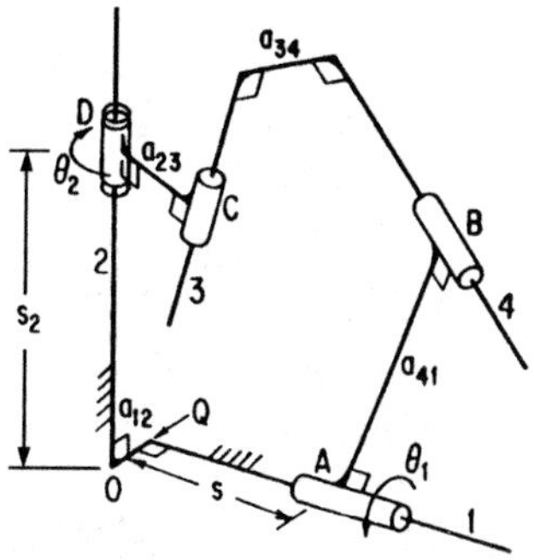

FIG. 3.42 Offset or spatial four-bar mechanism.

The analysis and synthesis of spatial mechanisms require special mathematical tools to reduce their complexity. The analysis of displacements, velocities, and accelerations of the general spatial chain (Fig. 3.42) is conveniently accomplished with the aid of dual vectors,[421c] numbers, matrices, quaternions, tensors, and Cayley-Klein parameters.[85,87,494] The spherical four-bar (Fig. 3.41) can be analyzed the same way, or by spherical trigonometry.[34] A computer program by J. Denavit and R. S. Hartenberg[129] is available for the analysis and synthesis of a spatial four-link mechanism whose terminal axes are nonparallel and nonintersecting, and whose two moving pivots are ball joints. See also Ref. 474 for additional spatial computer programs. For the simpler problems, for verification of computations and for visualization, graphical layouts are useful.[21,33,36,38,462]

Applications of three-dimensional mechanisms involve these motions:

1. Combined translation and rotation (e.g., door openers to lift and slide simultaneously[3])
2. Compound motions, such as in paint shakers, mixers, dough-kneading machines and filing[8,35,36,304]
3. Motions in shaft couplings, such as universal and constant-velocity joints[4,6,21,35,262] (see Sec. 3.9)
4. Motions around corners and in limited space, such as in aircraft, certain wobble-plate engines, and lawn mowers[60,310,332]
5. Complex motions, such as in aircraft landing gear, remote-control handling devices,[71,270] and pick-and-place devices in automatic assembly machines

When the motion is constrained ($F = 1$), but $\sum f_i < 7$ (such as in the mechanism shown in Fig. 3.41), any elastic deformation will tend to cause binding. This is not the case when $\sum f_i = 7$, as in Fig. 3.42, for instance. Under light-load, low-speed conditions, however, the former may represent no handicap.[10] The "degenerate" cases, usually associated with parallel or intersecting axes, are discussed more fully in Refs. 3, 10, 143, and 490.

In the analysis of displacements and velocities, extensions of the ideas used in plane kinematic analysis have led to the notions of the "instantaneous screw axis,"[38] valid for displacements and velocities; to spatial Euler-Savary equations; and to concepts involving line geometry.[224]

Care must be taken in designing spatial mechanisms to avoid binding and low mechanical advantages.

3.8 CLASSIFICATION AND SELECTION OF MECHANISMS

In this section, mechanisms and their components are grouped into three categories:

A. Basic mechanism components, such as those adapted for latching, fastening, etc.

B. Basic mechanisms: the building blocks in most mechanism complexes.

C. Groups or assemblies of mechanisms, characterized by one or more displacement-time schedules, sequencing, interlocks, etc.; these consist of combinations from categories A and B and constitute important mechanism units or independent portions of entire machines.

Among the major collections of mechanisms and mechanical movements are the following:

1M. Barber, T. W.: "The Engineer's Sketch-Book," Chemical Publishing Company, Inc., New York, 1940.

2M. Beggs, J. S.: "Mechanism," McGraw-Hill Book Company, Inc., New York, 1955.

3M. Hain, K.: "Die Feinwerktechnik," Fachbuch-Verlag, Dr. Pfanneberg & Co., Giessen, Germany, 1953.

4M. "Ingenious Mechanisms for Designers and Inventors," vols. 1–2, F. D. Jones, ed.: vol. 3, H. L. Horton, ed.; The Industrial Press, New York, 1930–1951.

5M. Rauh, K.: Praktische Getriebelehre," Springer-Verlag OHG, Berlin, vol. I, 1951; vol. II, 1954.

There are, in addition, numerous others, as well as more special compilations, the vast amount of information in the technical press, the AWF publications,[468] and (as a useful reference in depth), the Engineering Index. For some mechanisms, especially the more elementary types involving fewer than six links, a systematic enumeration of kinematic chains based on degrees of freedom may be worthwhile,[162–165,421c] particularly if questions of patentability are involved. Mechanisms are derived from the kinematic chains by holding one link fixed and possibly by using equivalent and substitute mechanisms (Sec. 3.6.6). The present state of the art is summarized in Ref. 159.

In the following list of mechanisms and components, each item is classified according to category (A, B, or C) and is accompanied by references, denoting one or more of the above five sources, or those at the end of this chapter. In using this listing, it is to be remembered that a mechanism used in one application may frequently be employed in a completely different one, and sometimes combinations of several mechanisms may be useful.

The categories A, B, C, or their combinations are approximate in some cases, since it is often difficult to determine a precise classification.

Adjustments, fine (A, 1M)(A, 2M)(A, 3M)

Adjustments, to a moving mechanism (A, 2M)(AB, 1M); see also Transfer, power

Airplane instruments and linkages (C, 3M)[342]
Analog computing mechanisms;[306] see also Computing mechanisms
Anchoring devices (A, 1M)
Automatic machinery, special-purpose;[161] automatic handling[363]
Ball bearings, guides and slides (A, 3M)
Ball-and-socket joints (A, 1M); see also Joints
Band drives (B, 3M); see also Tapes
Bearings (A, 3M)(A, 1M); jewel;[245] for oscillating motion[56]
Belt gearing (B, 1M)
Bolts (A, 1M)
Brakes (B, 1M)
Business machines, bookkeeping and records (C, 3M)
Calculating devices (C, 3M);[277] see also Mathematical instruments
Cameras (C, 3M); see also Photographic devices
Cam-link mechanisms (BC, 1M)(BC, 5M)[421]
Cams and cam drives (BC, 4M)(BC, 5M)(BC, 1M)[375]
Carriages and cars (BC, 1M)
Centrifugal devices (BC, 1M)
Chain drives (B, 1M)(B, 5M)[265]
Chucks, clamps, grips, holders (A, 1M)
Circular-motion devices (B, 1M)
Clock mechanisms;[18] see also Escapements (Ref. 106 and 421*f*, pp. 37–39)
Clutches, overrunning (BC, 1M)(C, 4M); see also Couplings and clutches
Computing mechanisms (BC, 5M)[80,271,338,446]
Couplings and clutches (B, 1M)(B, 3M)(B, 5M);[139,140,225,261,336] see also Joints
Covers and doors (A, 1M)
Cranes (AC, 1M)[77]
Crank and eccentric gear devices (BC, 1M)
Crushing and grinding devices (BC, 1M)
Curve-drawing devices (BC, 1M); see also Writing instruments and Mathematical instruments
Cushioning devices (AC, 1M)
Cutting devices (A, 1M)[329,354]
Derailleurs or deraillers (see Speed-changing mechanisms) (Refs. 106 and 421*f*, pp. 27)
Detents (A, 3M)
Differential motions (C, 1M)(C, 4M)[188]
Differentials (B, 5M)[4]
Dovetail slides (A, 3M)
Drilling and boring devices (AC, 1M)
Driving mechanisms for reciprocating parts (C, 3M)
Duplicating and copying devices (C, 3M)

Dwell linkages (C, 4M)
Ejecting mechanisms for power presses (C, 4M)
Elliptic motions (B, 1M)
Energy storage, instruments and mechanisms involving[434]
Energy transfer mechanism, special-purpose[267]
Engines, rotary (BC, 1M)
Engines, types of (C, 1M)
Escapements (B, 2M); see also Ratchets and Clock mechanisms
Expansion and contraction devices (AC, 1M)
Fasteners[244,423]
Feed gears (BC, 1M)
Feeding, magazine and attachments (C, 4M)[170]
Feeding mechanisms, automatic (C, 4M)(C, 5M)
Filtering devices (AB, 1M)[230]
Flexure pivots[103,141,269]
Flight-control linkages[365]
Four-bar chains, mechanisms and devices (B, 5M)[106,112,421f]
Frames, machine (A, 1M)
Friction gearing (BC, 1M)
Fuses (see Escapements)
Gears (B, 3M)(B, 1M)
Gear mechanisms (BC, 1M)[106]
Genevas; see Intermittent motions
Geodetic instruments (C, 3M)
Governing and speed-regulating devices (BC, 1M)[461]
Guidance, devices for (BC, 5M)
Guides (A, 1M)(A, 3M)
Handles (A, 1M)
Harmonic drives[101]
High-speed design;[47] special application[339]
Hinges (A, 1M); see also Joints
Hooks (A, 1M)
Hoppers, for automatic machinery (C, 4M), and hopper-feeding devices[234,235,236,421f]
Hydraulic converters (BC, 1M)
Hydraulic and link devices[89,90,92,168,337]
Hydraulic transmissions (C, 4M)
Impact devices (BC, 1M)
Indexing mechanisms (B, 2M); see also Sec. 3.9 and Intermittent motions
Indicating devices (AC, 1M); speed (C, 1M)
Injectors, jets, nozzles (A, 1M)
Integrators, mechanical[358]

Interlocks (C, 4M)
Intermittent motions, general;[44,357,468] see also Sec. 3.9
Intermittent motions from gears and cams (C, 4M)[269]
Intermittent motions, geneva types (BC, 4M)(BC, 5M);[211,357] see also Indexing mechanisms
Intermittent motions from ratchet gearing (C, 1M)(C, 4M)
Joints, all types (A, 1M);[239] see also Couplings and clutches and Hinges
Joints, ball-and-socket[69]
Joints, to couple two sliding members (B, 2M)
Joints, intersecting shafts (B, 2M)
Joints, parallel shafts (B, 2M)
Joints, screwed or bolted (A, 3M)
Joints, skew shafts (B, 2M)
Joints, soldered, welded, riveted (A, 3M)
Joints, special-purpose, three-dimensional[272]
Keys (A, 1M)
Knife edges (A, 3M)[141]
Landing gear, aircraft[71]
Levers (A, 1M)
Limit switches[498]
Link mechanisms (BC, 5M)
Links and connecting rods (A, 1M)
Locking devices (A, 1M)(A, 3M)(A, 5M)
Lubrication devices (A, 1M)
Machine shop, measuring devices (C, 3M)
Mathematical instruments (C, 3M); see also Curve-drawing devices and Calculating devices
Measuring devices (AC, 1M)
Mechanical advantage, mechanisms with high value of (BC, 1M)
Mechanisms, accurate;[482] general[21,96,153,159,176,180,193,263,278]
Medical instruments (C, 3M)
Meteorological instruments (C, 3M)
Miscellaneous mechanical movements (BC, 5M)(C, 4M)
Mixing devices (A, 1M)
Models, kinematic, construction of[51,183]
Noncircular gearing (Sec. 3.9)
Optical instruments (C, 3M)
Oscillating motions (B, 2M)
Overload-relief mechanisms (C, 4M)
Packaging techniques, special-purpose[197]
Packings (A, 1M)

Photographic devices (C, 3M);[478,479,486] see also Cameras
Piping (A, 1M)
Pivots (A, 1M)
Pneumatic devices[57]
Press fits (A, 3M)
Pressure-applying devices (AB, 1M)
Prosthetic devices[311,344,345]
Pulleys (AB, 1M)
Pumping devices (BC, 1M)
Pyrotechnic devices (C, 3M)
Quick-return motions (BC, 1M)(C, 4M)
Raising and lowering, including hydraulics (BC, 1M)[88,340]
Ratchets, detents, latches (AB, 2M)(B, 5M);[18,362,468] see also Escapements
Ratchet motions (BC, 1M)[468]
Reciprocating mechanisms (BC, 1M)(B, 2M)(BC, 4M)
Recording mechanisms, illustrations of;[55,203] recording systems[206]
Reducers, speed; cycloidal;[48,331,495] general[308]
Releasing devices and circuit breakers[84,325,458]
Remote-handling robots;[270] qualitative description[364]
Reversing mechanisms, general (BC, 1M)(C, 4M)
Reversing mechanisms for rotating parts (BC, 1M)(C, 4M)
Robots and manipulators (See Sec. 5.9.10)[421f]
Rope drives (BC, 1M)
Safety devices, automatic (A, 1M)(C, 4M)[20,481]
Screening and sifting (A, 1M)
Screw mechanisms (BC, 1M) (See Ref. 468, no. 6071)
Screws (B, 5M)
Seals, hermetic;[59] O-ring;[209] with gaskets;[46,113,346] multistage[422]
Self-adjusting links and slides (C, 4M)
Separating and concentrating devices (BC, 1M)
Sewing machines (C, 3M)
Shafts (A, 1M)(A, 3M); flexible [198,241]
Ship instruments (C, 3M)
Slider-crank mechanisms (B, 5M)
Slides (A, 1M)(A, 3M)
Snap actions (A, 2M)
Sound, devices using (B, 1M)
Spacecraft, mechanical design of[497]
Spanners (A, 1M)
Spatial body guidance (Refs. 421*c*, 421*d*, 421*e*, 421*f*)
Spatial function generators with higher pairs (Ref. 421*b*)

Speed-changing mechanisms (C, 4M); also see Transmissions
Spindles and centers (A, 1M)
Springs (A, 1M)(A, 3M); devices;[81,434] fastening of[343]
Springs and mechanisms (BC, 5M)
Steering mechanisms (BC, 5M)[254,447,495]
Stop mechanisms (C, 4M)
Stops (A, 2M)[343]
Straight-line motions, guides, parallel motions and devices (B, 1M)(BC, 4M),[72,150,157,476] Sec. 3.9
Struts and ties (A, 1M)
Substitute mechanisms[112,421g]
Swivels (A, 1M)
Tape drives and devices (B, 3M)(B, 5M)[21,189]
Threads[343]
Three-dimensional drives;[5,8,304] Sec. 3.9
Time-measuring devices (C, 3M); timers[274]
Toggles[138,144,275,427,498]
Torsion devices[141]
Toys, mechanisms used in[312]
Tracks and rails (A, 1M)
Transducers (AB, 2M)(C, 3M)[1,52,70,105,477]
Transfer, of parts, or station advance (B, 2M)
Transfer, power to moving mechanisms (AB, 2M)
Transmissions and speed changers (BC, 5M);[4,7,437] see also Variable mechanical advantage and Speed-changing mechanisms
Transmissions, special (C, 4M)[179]
Tripping mechanisms (C, 4M)
Typewriting devices (C, 3M)[192,266]
Universal joints[21,262,264,379,443]
Valve gear (BC, 1M)
Valves (A, 1M); design of nonlinear[335,374]
Variable mechanical advantage and power devices (A, 1M);[268,441,499] see also Transmissions and speed changers
Washing devices (A, 1M)
Wedge devices (A, 3M)(B, 5M)
Weighing devices (AB, 1M)
Weights, for compensation and balancing (A, 1M)[421f]
Wheels (A, 1M)
Wheels, elastic (A, 1M)
Windmill and feathering devices (A, 1M)
Window-regulating mechanisms[106,142,421f]
Woodworking machines[273]

Writing instruments (C, 3M); see also Curve-drawing devices and Mathematical instruments

3.9 KINEMATIC PROPERTIES OF MECHANISMS

For a more complete literature survey, see Refs. 1 to 6 cited in Ref. 124, and the Engineering Index, currently available computer programs are listed in Ref. 129.

3.9.1 The General Slider-Crank Chain (Fig. 3.43)

Nomenclature

A = crankshaft axis
B = crankpin axis
C = wrist-pin axis
FD = guide
AF = e = offset, $\perp FD$
AB = r = crank
BC = l = connecting rod
x = displacement of C in direction of guide, measured from F
t = time

Block at C = slider
s = stroke
θ = crank angle
ϕ = angle between connecting rod and slide, pressure angle
τ = $\angle ABC$
η = $\angle BGP$ = auxiliary angle
PG = collineation axis; $CP \perp FD$

The following mechanisms are derivable from the general slider-crank chain:

1. The *slider-crank mechanism*; guide fixed; if $e \neq 0$, called "offset," if $e = 0$, called "in-line"; $\lambda = r/l$; in case of the in-line slider crank, if $\lambda < 1$, AB rotates; if $\lambda > 1$, AB oscillates.
2. *Swinging-block mechanism*; connecting rod fixed; "offset" or "in-line" as in 1.
3. *Turning-block mechanism*; crank fixed; exact kinematic equivalent of 2; see Fig. 3.44.
4. The *standard geneva mechanism* is derivable from the special case, $e = 0$ (see Fig. 3.45*b*).
5. Several *variations* of the *geneva mechanism* and other pin-and-slot or block-and-slot drives.

3.9.2 The Offset Slider-Crank Mechanism (see Fig. 3.43 with *AFD* stationary)

Let

$$\lambda = r/l \qquad \epsilon = e/l \tag{3.40}$$

where l is the length of the connecting rod, then

$$s = l[(1 + \lambda)^2 - \epsilon^2]^{1/2} - l[(1 - \lambda)^2 - \epsilon^2]^{1/2} \tag{3.41}$$

$$\sin \phi = \epsilon + \lambda \sin \theta \tag{3.42}$$

$$\tau = \pi - \phi - \theta \tag{3.43}$$

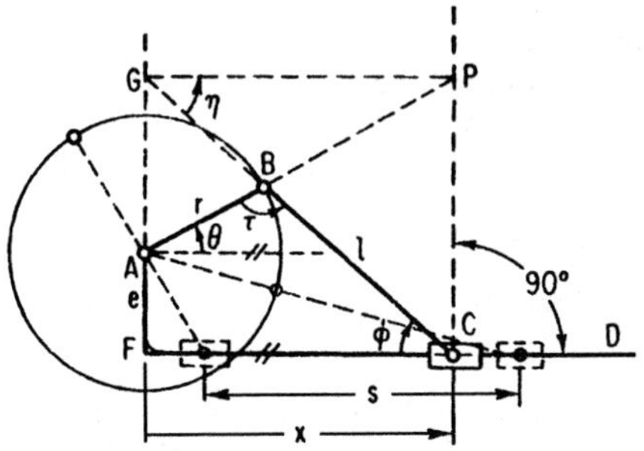

FIG. 3.43 General slider-crank chain.

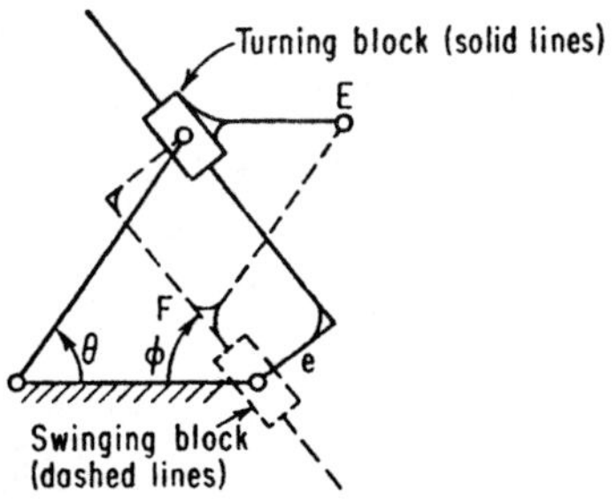

FIG. 3.44 Kinematic equivalence of the swinging-block and turning-block mechanisms, shown by redundant connection *EF*.

and $$x = r\cos\theta + l\cos\phi \tag{3.44}$$

Let the angular velocity of the crank be $d\theta/dt = \omega$; then the slider velocity is given by

$$dx/dt = r\omega\,[-\sin(\theta+\phi)/\cos\phi] \tag{3.45}$$

Extreme value of dx/dt occurs when the auxiliary angle $\eta = 90°$.[116]

Slider acceleration (ω = constant):

$$\frac{d^2x}{dt^2} = r\omega^2\left[\frac{-\cos(\theta+\phi)}{\cos\phi} - \lambda\frac{\cos^2\theta}{\cos^3\phi}\right] \tag{3.46}$$

Slider shock (ω = constant):

$$\frac{d^3x}{dt^3} = r\omega^3\left[\frac{\sin(\theta+\phi)}{\cos\phi} + \frac{3\lambda\cos\theta}{\cos^5\phi}(\sin\theta\cos^2\phi - \lambda\sin\phi\cos^2\theta)\right] \tag{3.47}$$

For the angular motion of the connecting rod, let the angular velocity ratio,

$$m_1 = d\phi/d\theta = \lambda(\cos\theta/\cos\phi) \tag{3.48}$$

Then the angular velocity of the connecting rod

$$d\phi/dt = m_1\omega \tag{3.49}$$

Let $$m_2 = d^2\phi/d\theta^2 = m_1(m_1\tan\phi - \tan\theta) \tag{3.50}$$

Then the angular acceleration of the connecting rod, at constant ω, is given by

$$d^2\phi/dt^2 = m_2\omega^2 \tag{3.51}$$

In addition, let

$$m_3 = d^3\phi/d\theta^3 = 2m_1m_2\tan\phi - m_2\tan\theta + m_1^3\sec^2\phi - m_1\sec^2\theta \tag{3.52}$$

Then the angular shock of the connecting rod, at constant ω, becomes

$$d^3\phi/dt^3 = m_3\omega^3$$

In general, the $(n-1)$th angular acceleration of the connecting rod, at constant, ω, is given by

$$d^n\phi/dt^n = m_n\omega^n \tag{3.53}$$

where $m_n = dm_{n-1}/d\theta$. In a similar manner, the general expression for the $(n-1)$th linear acceleration of the slider, at constant ω, takes the form

$$d^nx/dt^n = r\omega^n M_n \tag{3.54}$$

where $M_n = dM_{n-1}/d\theta$,

$$M_1 = -\sin(\theta + \phi)/\cos\phi \tag{3.55}$$

and M_2 and M_3 are the bracketed expressions in Eqs. (3.46) and (3.47), respectively.

Kinematic characteristics are governed by Eqs. (3.40) to (3.55). Examples for path and function generation, Ref. 432. Harmonic analyses, Refs. 39 and 296. Coupler curves, Ref. 104. Cognates, Ref. 182. Offset slider-crank mechanism can be used to reduce the friction of the slider in the guide during the "working" stroke; transmission-angle charts, Ref. 467.

Amplitudes of the harmonics are slightly higher than for the in-line slider-crank with the same λ value.

For a nearly constant slider velocity $(1/\omega)(dx/dt) = k$ over a portion of the motion cycle, the proportions[42]

$$12k = 3e \pm \sqrt{9e^2 - 8(l^2 - 9r^2)}$$

may be useful.

3.9.3 The In-Line Slider-Crank Mechanism

$(e = 0)$[32,39,73,104,129,182,296,472,467]

If $\omega \neq$ constant, see Ref. 21. In general, see Eqs. (3.44) to (3.55).

Equations (3.56) and (3.57) give approximate values when $\lambda < 1$, and with $\omega =$ constant. (For nomenclature refer to Fig. 3.43, with $e = 0$, and guide fixed.)

Slider velocity: $$dx/dt = r\omega(-\sin\theta - \tfrac{1}{2}\lambda\sin 2\theta) \tag{3.56}$$

Slider acceleration: $$d^2x/dt^2 = r\omega^2(-\cos\theta - \lambda\cos 2\theta) \tag{3.57}$$

Extreme Values. $(dx/dt)_{max}$ occurs when the auxiliary angle $\eta = 90°$. For a prescribed extreme value, $(1/r\omega)(dx/dt)_{max}$, λ is obtainable from Eq. (22) of Ref. 116.

At extended dead center: $$d^2x/dt^2 = -r\omega^2(1 + \lambda) \tag{3.58}$$

At folded dead center: $$d^2x/dt^2 = r\omega^2(1 - \lambda) \tag{3.59}$$

Equations (3.58) and (3.59) yield exact extreme values whenever $0.264 < \lambda < 0.88$.[472]

Computations. See computer programs in Ref. 129, and also Kent's "Mechanical Engineers Handbook," 1956 ed., Sec. II, Power, Sec. 14, pp. 14-61 to 14-63, for displacements, velocities, and accelerations vs. λ and θ; similar tables, including also kinematics of connecting rod, are found in Ref. 73 for $0.2 \leq \lambda \leq 0.7$ in increments of 0.1.

Harmonic Analysis[39]

$$x/r = A_0 + \cos\theta + \tfrac{1}{4}A_2\cos 2\theta - \tfrac{1}{16}A_4\cos 4\theta + \tfrac{1}{36}A_6\cos 6\theta - \cdots \tag{3.60}$$

If ω = constant,

$$-(1/r\omega^2)(d^2x/dt^2) = \cos\theta + A_2\cos 2\theta - A_4\cos 4\theta + A_6\cos 6\theta - \cdots \qquad (3.61)$$

where A_j are given in Table 3.1[39]

For harmonic analysis of $\phi(\theta)$, and for inclusion of terms for $\omega \neq$ constant, see Ref. 39;

TABLE 3.1 Values of A^*_j

l/r	A_2	A_4	A_6
2 5	0.4173	0.0182	0.0009
3.0	0.3431	0.0101	0.0003
3.5	0.2918	0.0062	0.0001
4.0	0.2540	0.0041	0.0001
4.5	0.2250	0.0028	0.0000
5.0	0.2020	0.0021	
5.5	0.1833	0.0015	
6.0	0.1678	0.0012	

*Biezeno, C. B., and R. Grammel, "Engineering Dynamics," (Trans. M. P. White), vol. 4, Blackie & Son, Ltd., London and Glasgow, 1954.

coupler curves (described by a point in the plane of the connecting rod) in Refs. 32 and 104; "cognate" slider-crank mechanism (i.e., one, a point of which describes the same coupler curve as the original slider-crank mechanism), Ref. 182; straight-line coupler-curve guidance, see VDI—Richtlinien No. 2136.

3.9.4 Miscellaneous Mechanisms Based on the Slider-Crank Chain[19,39,42,104,289,291,292,297,299,301,349,367,432,436]

1. In-line swinging-block mechanism
2. In-line turning-block mechanism
3. External geneva motion
4. Shaper drive
5. Offset swinging-block mechanism
6. Offset turning-block mechanism
7. Elliptic slider-crank drive

For "in-line swinging-block" and "in-line turning block" mechanisms, see Fig. 3.45*a* and *b*. The following applies to both mechanisms. $\lambda = r/a$; θ is considered as input, with ω_{AB} = constant.

Displacement:

$$\phi = \tan^{-1}\frac{\lambda\sin\theta}{1-\lambda\cos\theta} \qquad (3.62)$$

Angular velocities (positive clockwise):

$$\frac{d\phi}{dt} = \omega_{AB}\frac{\lambda\cos\theta - \lambda^2}{1+\lambda^2-2\lambda\cos\theta} \qquad (3.63)$$

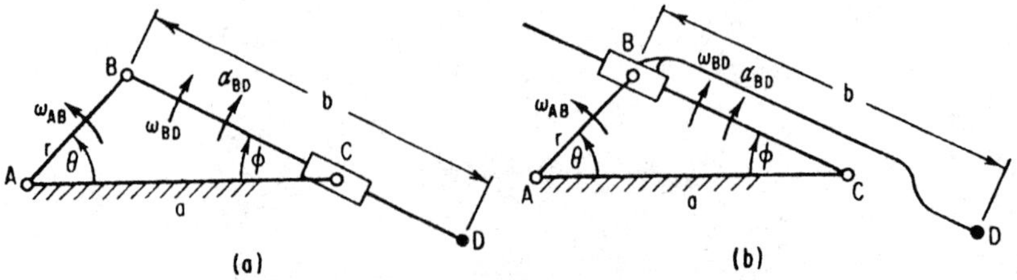

FIG. 3.45 (*a*) In-line swinging-block mechanism. (*b*) In-line turning-block mechanism.

$$\frac{1}{\omega_{AB}}\left|\frac{d\phi}{dt}\right|_{\min,\theta=180°} = \frac{-\lambda}{\lambda+1} \tag{3.64}$$

$$\frac{1}{\omega_{AB}}\left(\frac{d\phi}{dt}\right)_{\max,\theta=0°} = \frac{\lambda}{1-\lambda} \tag{3.65}$$

Angular acceleration:
$$\alpha_{BD} = \frac{d^2\phi}{dt^2} = \frac{(\lambda^3-\lambda)\sin\theta}{(1+\lambda^2-2\lambda\cos\theta)^2}\,\omega_{AB}^2 \tag{3.66}$$

Extreme value of α_{BD} occurs when $\theta = \theta_{\max}$, where

$$\cos\theta_{\max} = -G + (G^2+2)^{1/2} \tag{3.67}$$

and
$$G = \tfrac{1}{4}(\lambda + 1/\lambda) \tag{3.68}$$

Angular velocity ratio ω_{BD}/ω_{AB} and the ratio $\alpha_{BD}/\omega_{AB}^2$ are found from Eqs. (3.63) and (3.66), respectively, where $\omega_{BD} = d\phi/dt$. See also Sec. 3.9.7.

Straight-Line Guidance.[42,467] Point *D* (see Fig. 3.45*a* and *b*) will generate a close point-approximation to a straight line for a portion of its (bread-shaped) path, when

$$b = 3a - r + \sqrt{8a(a-r)} \tag{3.69}$$

Approximate Circular Arc (for a portion of motion cycle).[42] Point *D* (Fig. 3.45*a* and *b*) will generate an approximately circular arc whose center is at a distance *c* to the right of *A* (along *AC*) when

$$[b(a+c) - c(a-r)]^2 = 4bc(c-a)(a+r)$$

with $b > 0$ and $|c| > a > r$.

Proportions can be used in intermittent drive by attachment of two additional links (*VDI-Berichte,* vol. 29, 1958, p. 28).[42,473]

Harmonic Analysis[39,286,289,301] (see Fig. 3.45*a* and *b*). Case 1, $\lambda<1$:

$$\phi = \sum_{n=1}^{\infty}\frac{\lambda^n \sin n\theta}{n} \tag{3.70}$$

$$\frac{\omega_{BD}}{\omega_{AB}} = \frac{d\phi}{d\theta} = \sum_{n=1}^{\infty}\lambda^n\cos n\theta \tag{3.71}$$

Case 2, $\lambda > 1$:

$$\phi = \pi - \theta - \sum_{n=1}^{\infty} \frac{\sin n\theta}{n\lambda^n} \tag{3.72}$$

$$\frac{d\phi}{d\theta} = -1 - \sum_{n=1}^{\infty} \frac{\cos n\theta}{\lambda^n} \tag{3.73}$$

Note that in case 1 *AB* rotates and *BC* oscillates, while in case 2 both links perform full rotations.

External Geneva Motion. Equations (3.62) to (3.68) apply. For more extensive data, including tables of third derivatives and various numerical values, see Intermittent-Motion Mechanisms, Sec. 3.9.7, and Refs. 252, 253, and 352.

An analysis of the "shaper drive" involving the turning-block mechanism is described in Ref. 289, part 2; see also Ref. 42.

Offset Swinging-Block and Offset Turning-Block Mechanisms.[301] (See Fig. 3.44.) Synthesis of offset turning-block mechanisms for path and function generation described in Ref. 432; see also Ref. 436 for velocities and accelerations; extreme values of angular velocity ratio $d\phi/d\theta = q$ are related by the equation $q_{\max}^{-1} + q_{\min}^{-1} = -2$; these occur for the same position of the driving link or for those whose crank angles add up to two right angles, depending on whether the driving link swings (oscillates) or rotates, respectively[436]; for graphical analysis of accelerations involving relative motion between two (instantaneously) coincident points on two moving links, use Coriolis's acceleration, or complex numbers in analytical approach.[106]

For the "elliptic slider-crank drive" see Refs. 293, 295, and 297.

3.9.5 Four-Bar Linkages (Plane) (Refs. 2, 12, 16, 32, 42, 58, 61, 62, 104, 106, 115, 118, 122, 123, 125, 127–129, 159, 160, 166, 170, 173, 176, 180, 194, 199, 201, 205, 209, 210, 223, 249, 279, 284, 298, 300, 303, 307, 315, 380, 421*f*, 432, 435, 448, 453, 467, 471, 473, 476, 483, 489, 491)

See Fig. 3.46. Four-bar mechanism, *ABCD, E* on coupler; *AB* = crank *b*; *BC* = coupler *c*; *CD* = crank or link *d*; *AD* = fixed link *a*; *AB* is assumed to be the driving link.

Grashof's Inequality. Length of longest link + length of shortest link < sum of lengths of two intermediate links.

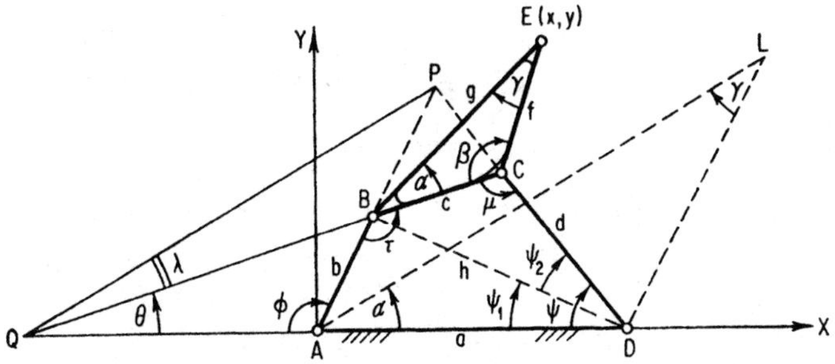

FIG. 3.46 Four-bar mechanism.

Types of Mechanisms

1. If Grashof's inequality is satisfied and b or d is the shortest link, the linkage is a "crank and rocker"; the shortest link is the "crank," and the opposite link is the "rocker."
2. If Grashof's inequality is satisfied and the fixed link is the shortest link, the linkage is a "drag linkage"; both cranks can make complete rotations.
3. All other cases except 4: the linkage is a "double-rocker" mechanism (cranks can only oscillate); this will be the case, for instance, whenever the coupler is the smallest link.
4. Special cases: where the equal sign applies in Grashof's inequality. These involve "folding" linkages and "branch positions," at which the motion is not positive. Example: parallelogram linkage; antiparallel equal-crank linkage ($AB = CD$, $BC = AD$, but AB is not parallel to CD).[207]

Angular Displacement. In Fig. 3.46, $\psi = \psi_1 + \psi_2$ (a minus sign would occur in front of ψ_2 when a mechanism lies entirely on one side of diagonal BD).

$$\psi = \cos^{-1}\frac{h^2 + a^2 - b^2}{2ah} + \cos^{-1}\frac{h^2 + d^2 - c^2}{2hd} \tag{3.74}$$

$$h^2 = a^2 + b^2 + 2ab\cos\phi \tag{3.75}$$

For alternative equation between tan ½ψ and tan ½φ (useful for automatic computation) see Ref. 87.

The general closure equation:[115]

$$R_1\cos\phi - R_2\cos\psi + R_3 = \cos(\phi - \psi) \tag{3.76}$$

where $$R_1 = a/d \qquad R_2 = a/b \qquad R_3 = (a^2 + b^2 - c^2 + d^2)/2bd \tag{3.77}$$

The ϕ, τ equation:

$$p_1\cos\phi - p_2\cos\tau + p_3 = \cos(\phi - \tau) \tag{3.78}$$

where $$p_1 = b/c \qquad p_2 = b/a \qquad p_3 = (a^2 + b^2 + c^2 - d^2)/2ac \tag{3.79}$$

$$\theta = \tau - \phi = \angle AQB \tag{3.80}$$

Extreme rocker-angle values in a crank and rocker:

$$\psi_{max} = \cos^{-1}\{[a^2 + d^2 - (b + c)^2]/2ad\} \tag{3.81}$$

$$\psi_{min} = \cos^{-1}\{[a^2 + d^2 - (c - b)^2]/2ad\} \tag{3.82}$$

Total range to rocker: $\Delta\psi = \psi_{max} - \psi_{min}$

To determine inclination of the coupler $\angle AQB = \theta$, determine length of AC:

$$\overline{AC}^2 = a^2 + d^2 - 2ad\cos\psi = k^2 \quad \text{(say)} \tag{3.83}$$

Then compute $\angle ABC = \tau$ from

$$\cos\tau = (b^2 + c^2 - k^2)/2bc \tag{3.84}$$

and use Eq. (3.80). See also Refs. 307 and 448 for angular displacements; for extreme positions see Ref. 284; for geometrical construction of proportions for given ranges and extreme positions, see Ref. 173. For analysis of complex numbers see Ref. 106.

Velocities.

Angular velocity ratio:

$$\omega_{CD}/\omega_{AB} = QA/QD \tag{3.85}$$

Velocity ratio:

$$V_C/V_B = PC/PB \qquad P = AB \times CD \qquad V_p\ (P \text{ on coupler}) = 0 \tag{3.86}$$

Velocity ratio of tracer point E:

$$V_E/V_B = PE/PB \tag{3.87}$$

Angular velocity ratio of coupler to input link:

$$\omega_{BC}/\omega_{AB} = BA/BP \tag{3.88}$$

When cranks are parallel, B and C have the same linear velocity, and $\omega_{BC} = 0$.

When coupler and fixed link are parallel, $\omega_{CD}/\omega_{AB} = 1$.

At an extreme value of angular velocity ratio, $\lambda = 90°$.[116,433] When ω_{BC}/ω_{AB} is at a maximum or minimum, $QP \perp CD$.

Angular velocity ratio of output to input link is also obtainable by differentiation of Eq. (3.76):

$$m_1 = \frac{\omega_{CD}}{\omega_{AB}} = \frac{d\psi}{d\phi} = \frac{\sin(\phi - \psi) - R_1 \sin\phi}{\sin(\phi - \psi) - R_2 \sin\psi} \tag{3.89}$$

Accelerations (ω_{AB} = constant, t = time)

$$m_2 = \frac{d^2\psi}{d\phi^2} = \frac{1}{\omega_{AB}^2}\frac{d^2\psi}{dt^2} = \frac{(1 - m_1)^2 \cos(\phi - \psi) - R_1 \cos\phi + m_1^2 R_2 \cos\psi}{\sin(\phi - \psi) - R_2 \sin\psi} \tag{3.90}$$

Alternate formulation:

$$\frac{1}{\omega_{AB}^2}\frac{d^2\psi}{dt^2} = m_1(1 - m_1)\cot\lambda \tag{3.91}$$

(useful when $m_1 \neq 1, 0$, and $\lambda \neq 0°, 180°$).

On extreme values, see Ref. 116; velocities, accelerations, and point-path curvature are discussed via complex numbers in Refs. 58, 106, and 427*f*; computer programs are in Ref. 129.

Second Acceleration or Shock (ω_{AB} = *const*)

$$\begin{aligned}\frac{1}{\omega_{AB}^3}\frac{d^3\psi}{dt^3} &= \frac{d^3\psi}{d\phi^3} = m_3 \\ &= [R_1 \sin\phi - (m_1^3 \sin\psi - 3m_1m_2\cos\psi)R_2 - 3m_2(1 - m_1)\cos(\phi - \psi) \\ &\quad -(1 - m_1)^3 \sin(\phi - \psi)]/[\sin(\phi - \psi) - R_2 \sin\psi]\end{aligned} \tag{3.92}$$

Coupler Motion. Angular velocity of coupler:

$$d\theta/dt = (n_1 - 1)\omega_{AB}$$

where

$$n_1 = 1 + \frac{d\theta}{d\phi} = \frac{\sin(\phi - \tau) - p_1 \sin \phi}{\sin(\phi - \tau) - p_2 \sin \tau} \tag{3.93}$$

where p_1 and p_2 are as before [Eq. (3.79)]. Let

$$n_2 = \frac{d^2\theta}{d\phi^2} = \frac{(1 - n_1)^2 \cos(\phi - \tau) - p_1 \cos \phi + n_1^2 p_2 \cos \tau}{\sin(\phi - \tau) - p_2 \sin \tau} \tag{3.94}$$

Then the angular acceleration of the coupler,

$$d^2\theta/dt^2 = n_2 \omega_{AB}^2 \qquad \omega_{AB} = \text{const} \tag{3.95}$$

If

$$\begin{aligned} n_3 &= \frac{d^3\theta}{d\phi^3} \\ &= [p_1 \sin \phi - (n_1^3 \sin \tau - 3n_1 n_2 \cos \tau) p_2 - 3n_2(1 - n_1) \cos(\phi - \tau) \\ &\quad - (1 - n_1)^3 \sin(\phi - \tau)]/[\sin(\phi - \tau) - p_2 \sin \tau] \end{aligned} \tag{3.96}$$

The angular shock of the coupler $d^3\theta/dt^3$ at ω_{AB} = const is given by

$$d^3\theta/dt^3 = n_3 \omega_{AB}^3 \tag{3.97}$$

See also Refs. 61 and 62 for angular acceleration and shock of coupler; for shock of points on the coupler see Ref. 298.

Harmonic Analysis (ψ vs. ϕ). Literature survey in Ref. 493. General equations for crank and rocker in Ref. 125. Formulas for special crank-and-rocker mechanisms designed to minimize higher harmonics:[128] Choose $0° << \nu << 90°$, and let

$$AB = \tan \tfrac{1}{2}\nu \qquad BC = (1/\sqrt{2}) \sec \tfrac{1}{2}\nu = CD \qquad AD = 1$$

$$\mu_{max} = 90° + \nu \qquad \mu_{min} = 90° - \nu$$

$$\psi = \text{const} + \sum_{m=1}^{\infty} \frac{(-\tan \tfrac{1}{2}\nu)^m}{-m} \sin m\phi - \frac{C_0}{4} \sin \nu \cos \phi - \sum_{m=1}^{\infty} \frac{\sin \nu}{4m} (C_{m-1} - C_{m+1}) \cos m\phi$$

where letting

$$\sin \nu = p \qquad a_2 = \frac{1}{4}p^2 \qquad a_4 = \frac{3}{64}p^4 \qquad a_6 = \frac{5}{512}p^6 \qquad a_8 = \frac{35}{128^2}p^8$$

C_m (m odd) = 0

and $\quad C_0 = 1 + C_2 - C_4 + C_6 - C_8 + \cdots \qquad C_2 = a_2 + 4C_4 - 9C_6 + 16C_8 + \cdots$

$C_4 = a_4 + 6C_6 - 20C_8 + \cdots \qquad C_6 = a_6 + 8C_8 + \cdots$

$C_8 = a_s + \cdots$

For numerical tables see Ref. 128. For four-bar linkages with adjacent equal links (driven crank = coupler), as in Ref. 128; see also Refs. 45, 300, 303.

Three-Point Function Synthesis. To find mechanism proportions when (ϕ_i, ψ_i) are prescribed for i = 1, 2, 3 (see Fig. 3.46).

$$a = 1 \qquad b = \frac{w_2w_3 - w_1w_4}{w_1w_6 - w_2w_5} \qquad d = \frac{w_2w_3 - w_1w_4}{w_3w_6 - w_4w_5}$$

$$c^2 = 1 + b^2 + d^2 - 2bd\cos(\phi_i - \psi_i) - 2d\cos\psi_i + 2b\cos\phi_i \qquad i = 1, 2, 3$$

where

$$w_1 = \cos\phi_1 - \cos\phi_2 \qquad w_2 = \cos\phi_1 - \cos\phi_3$$

$$w_3 = \cos\psi_1 - \cos\psi_2 \qquad w_4 = \cos\psi_1 - \cos\psi_3$$

$$w_5 = \cos(\phi_1 - \psi_1) - \cos(\phi_2 - \psi_2) \qquad w_6 = \cos(\phi_1 - \psi_1) - \cos(\phi_3 - \psi_3)$$

Four- and Five-Point Synthesis. For maximum accuracy, use five points; for greater flexibility in choice of proportions and transmission-angle control, choose four points.

Four-Point Path and Function Generation. Path generation together with prescribed crank rotations in Refs. 106, 123, 371, and 421*f.* Function generation in Ref. 106, 381, and 421*f.*

Five-Point Path and Function Generation. See Refs. 123, 380, and 421*f*; the latter reference usable for five-point path vs. prescribed crank rotations, for Burmester point-pair determinations pertaining to five distinct positions of a plane, and for function generation with the aid of Ref. 127; additional references include 381, 435, and others at beginning of section; minimization of structural error in Refs. 16, 122, 249, and 421*f,* the latter with least squares; see Refs. 118, 122, and 194 for minimum-error function generators such as log x, sin x, tan x, e^x, x^n, tanh x; infinitesimal motions, Burmester points in Refs. 421*f,* 469, and 489.

General. Atlases for path generation (Ref. 199) and for function generation via "trace deviation" (Refs. 210 and 471); point-position-reduction discussed in Refs. 2, 106, 159, 194, 421*f,* and Sec. 3.5.7; nine-point path generation in Ref. 372.

Coupler Curve.[32,104,315] Traced by point E, in cartesian system with origin at A, and x and y axes as in Fig. 3.46:

$$U = f[(x - a)\cos\gamma + y\sin\gamma](x^2 + y^2 + g^2 - b^2) - gx[(x - a)^2 + y^2 + f^2 - d^2]$$

$$V = f[(x - a)\sin\gamma - y\cos\gamma](x^2 + y^2 + g^2 - b^2) + gy[(x - a)^2 + y^2 + f^2 - d^2]$$

$$W = 2gf\sin\gamma[x(x-a)+y^2-ay\cot\gamma]$$

With these

$$U^2 + V^2 = W^2 \tag{3.98}$$

Equation (3.98) is a tricircular, trinodal, sextic, algebraic curve. Any intersection of this curve with circle through ADL (Fig. 3.46) is a double point, in special cases a cusp; coupler curves may possess up to three real double points or cusps (excluding curves traced by points on folding linkages); construction of coupler curves with cusps and application to instrument design (dwells, noiseless motion reversal, etc.) described in Refs. 32, 159, 279, theory in Ref. 63; detailed discussion of curves, including Watt straight-line motion and equality of two adjacent links, in Ref. 104; instant center (at intersection of cranks, produced if necessary) describes a cusp.

Radius of Path Curvature R for Point E (Fig. 3.46). In this case, as in other linkages, analytical determination of R is readily performed *parametrically.* Parametric equations of the coupler curve:

$$x = x(\phi) = -b \cos \phi + g \cos (\theta + \alpha) \tag{3.99}$$

$$y = y(\phi) = b \sin \phi + g \sin (\theta + \alpha) \tag{3.100}$$

where $\theta = \tau - \phi$, $\tau = \tau(\phi)$ is obtainable from Eqs. (3.74), (3.75), (3.83), and (3.84) and $\cos \alpha = (g^2 - f^2 + c^2)/2gc$.

$$x' = dx/d\phi = b \sin \phi - g(n_1 - 1) \sin (\theta + \alpha) \tag{3.101}$$

$$y' = dy/d\phi = b \cos \phi + g(n_1 - 1) \cos (\theta + \alpha) \tag{3.102}$$

$$x'' = d^2x/d\phi^2 = b \cos \phi - gn_2 \sin (\theta + \alpha) - g(n_1 - 1)^2 \cos (\theta + \alpha) \tag{3.103}$$

$$y'' = d^2y/d\phi^2 = -b \sin \phi + gn_2 \cos (\theta + \alpha) - g(n_1 - 1)^2 \sin (\theta + \alpha) \tag{3.104}$$

where n_1 and n_2 are given in Eqs. (3.93) and (3.94).

$$R = \frac{(x'^2 + y'^2)^{3/2}}{x'y'' - y'x''} \tag{3.105}$$

Equivalent or "Cognate" Four-Bar Linkages. For Roberts' theorem, see Sec. 3.6, Fig. 3.38. Proportions of the cognates are as follows (Figs. 3.38 and 3.46).

Left cognate:

$$\mathbf{AF} = \mathbf{BC}z \qquad \mathbf{HF} = \mathbf{AB}z \qquad \mathbf{HL} = \mathbf{CD}z \qquad \mathbf{AL} = \mathbf{AD}z$$

where $z = (g/c)e^{i\alpha}$ $\quad \alpha = \angle\, CBE$

and where **AF,** etc., represent the complex-number form of the vector $\overrightarrow{AF}$, etc.

Right cognate:

$$\mathbf{GD} = \mathbf{BC}u \qquad \mathbf{GK} = \mathbf{CD}u \qquad \mathbf{LK} = \mathbf{AB}u \qquad \mathbf{LD} = \mathbf{AD}u$$

where $u = (f/c)e^{-i\beta}$ $\quad \beta = \angle\, ECB$

The same construction can be studied systematically with the "Cayley diagram."

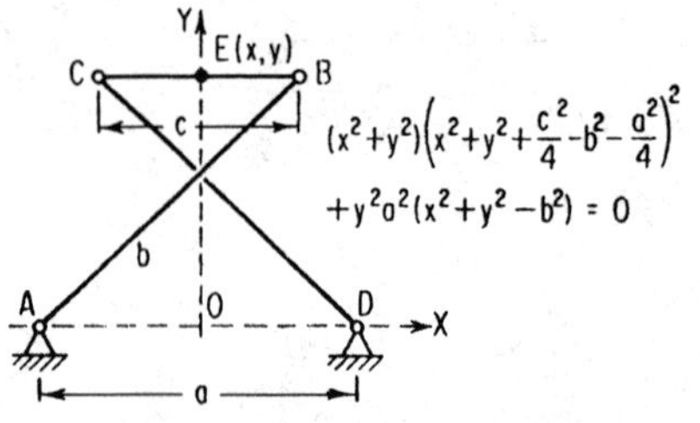

FIG. 3.47 Equal-crank linkage showing equation of symmetric coupler curve generated by point E, midway between C and B.

Symmetrical Coupler Curves. Coupler curves with an axis of symmetry are obtained when $BC = CD = EC$ (Fig. 3.46); also by cognates of such linkages; used by K. Hunt for path of driving pin in geneva motions;[201] also for dwells and straight-line guidance (see Sec. 3.8, 5M). Symmetrical coupler-curve equation[42] for equal-crank linkage, traced by midpoint E of coupler in Fig. 3.47.

Transmission Angles. Angle μ (Fig. 3.46) should be as close to 90° as possible; nontrivial extreme values occur when AB and AD are parallel or antiparallel ($\phi = 0°, 180°$). Generally

$$\cos \mu = [(c^2 + d^2 - a^2 - b^2)/2cd] - (2ab/2cd) \cos \phi \tag{3.106}$$

$\cos \mu_{max}$ occurs when $\phi = 180°$, $\cos \phi = -1$; $\cos \mu_{min}$ when $\phi = 0°$; $\cos \phi = 1$. Good crank-and-rocker proportions are given in Sec. 3.6:

$$b^2 + a^2 = c^2 + d^2 \qquad |\mu_{min} - 90°| = |\mu_{max} - 90°| = \nu \qquad \sin \nu = ab/cd$$

A computer program for path generation with optimum transmission angles and proportions is described in Ref. 371. Charts for optimum transmission-angle designs are as follows: drag links in Ref. 160; double rockers in Ref. 166; general, in Refs. 170 and 205. See also VDI charts in Ref. 467.

Approximately Constant Angular-Velocity Ratio of Cranks over a Portion of Crank Rotation (see also Ref. 42). In Fig. 3.46, if $d = 1$, a three-point approximation is obtained when

$$a^2 = c^2 \frac{(1 - 2m_1)(m_1 - 2)}{9m_1} + b^2 \frac{(1 - m_1)(2 - m_1)}{m_1(m_1 + 1)} + \frac{(1 - m_1)(1 - 2m_1)}{(m_1 + 1)}$$

where the angular-velocity ratio m_1 is given by Eq. (3.89). Useful only for limited crank rotations, possibly involving connection of distant shafts, high loads.

Straight-Line Mechanisms. Survey in Refs. 72, 208; modern and special applications in Refs. 223, 473, 476, theory and classical straight-line mechanisms in Ref. 42; see also below; order-approximation theory in Refs. 421*f* and 453.

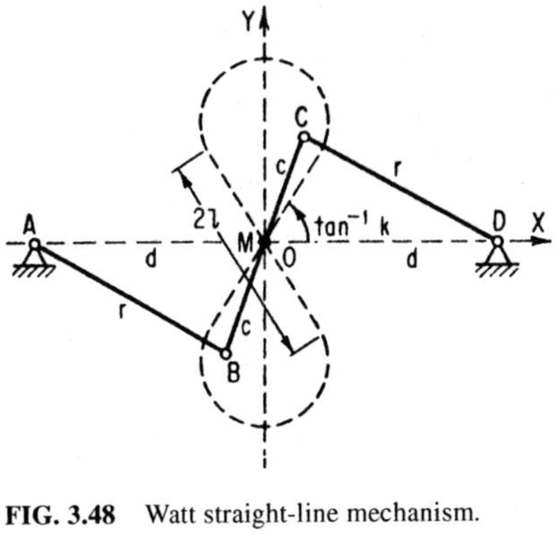

FIG. 3.48 Watt straight-line mechanism.

Fifth-Order Approximate Straight Line via a Watt Mechanism.[42] "Straight" line of length $2l$, generated by M on coupler, such that $y \cong kx$ (Fig. 3.48).

Choose k, l, r; let $\beta = l(1 + k^2)^{-1/2}$; then maximum error from straight-line path $\cong 0.038(1 + k^2)^3\beta^6$. To compute d and c:

$$(d^2 - c^2) = [r^4 + 6(7 - 4\sqrt{3})l^4 + 3(3 - 2\sqrt{3})l^2r^2]^{1/2}$$

$$p_2 = 3(3 - 2\sqrt{3})\beta$$

$$4k^2d^2 = 2(1 + k^2)(d^2 - c^2 + r^2) + p_2(1 + k^2)^2$$

For less than fifth-order approximation, proportions can be simpler: $AB = CD$, $BM = MC$.

Sixth-Order Straight Line via a Chebyshev Mechanism.[42,491] M will describe an approximate horizontal straight line in the position shown in Fig. 3.49, when

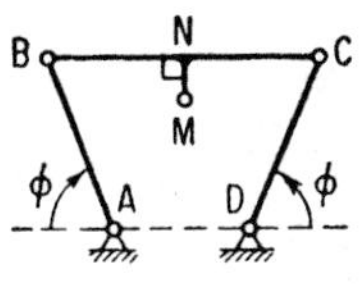

FIG. 3.49 Chebyshev straight-line mechanism. $BN = NC = \frac{1}{2}AD = b$; $AB = CD = r$; $NM = C$ (+ downward). In general, $\phi = 90° - \theta$, where θ is the angle between axis of symmetry and crank in symmetry position.

$$a/r = (2\cos^2\phi \cos 2\phi)/\cos 3\phi$$

$$b/r = -\sin^2 2\phi/\cos 3\phi$$

$$c/r = (\cos^2\phi \cos 2\phi \tan 3\phi)/\cos 3\phi$$

when $\phi = 60°$, $NM = 0$.

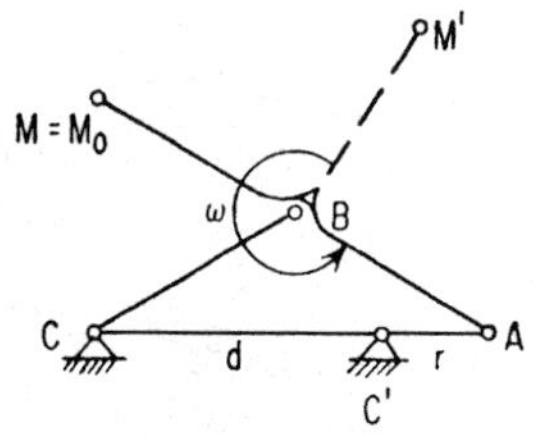

FIG. 3.50 Lambda mechanism.

Lambda Mechanism*[12,42] *and a Related Motion. The four-bar lambda mechanism of Fig. 3.50 consists of crank $AC' = r$, fixed link $CC' = d$, coupler AB, driven link BC, with generating point M at the straight-line extension of the coupler, where $BC = MB = BA = 1$. M generates a symmetrical curve. In a related mechanism, $M'B = BA$, $\omega = \angle M'BA$ as shown, and M' generates another symmetrical coupler curve.

Case 1. Either coupler curve of M contained between two concentric circles, center O_1, O_1M_0C collinear. $M = M_0$ when $AC'C$ are collinear as shown.

Let ψ'' be a parameter, $0 \le \psi'' \le 45°$. Then a six-point approximate circle is generated by M with least maximum structural error when

$$r = 2\sin\psi'' \sin 2\psi'' \sqrt{2\cos 2\psi''}/\sin 3\psi''$$

$$d = \sin 2\psi''/\sin 3\psi''$$

$$O_1C = 2\cos^2\psi''/\sin 3\psi''$$

Radius R of generated circle (at precision points):

$$R = r\cot\psi''$$

Maximum radial (structural) error:

$$2\cos 2\psi''/\sin 3\psi''$$

For table of numerical values see Ref. 12.

Case 2. Entire coupler curve contained between two straight lines (six-point approximation of straight line with least maximum structural error). In the equations above, M' generates this curve when $\angle M'BA = \omega = \pi + 2\psi''$. Maximum deviation from straight line:[12]

$$2\sin 2\psi'' \sqrt{2\cos^3 2\psi''}/\sin 3\psi''$$

Case 3. Six-point straight line for a portion of the coupler curve of M':

$$r = 1/4 \qquad d = 3/4$$

Case 4. Approximate circle for a *portion* of the coupler curve of M. Any proportions for r and d give reasonably good approximation to *some* circle because of symmetry. Exact proportions are shown in Refs. 12 and 42.

Balancing of Four-Bar Linkages for High-Speed Operation.[421f,448] Make links as light as possible; if necessary, counterbalance cranks, including appropriate fraction of coupler on each.

Multilink Planar Mechanisms. Geared five-bar mechanisms;[95,119,120,421f] six-link mechanisms.[348,421f]

Design Charts.[467] See also section on transmission angles.

3.9.6 Three-Dimensional Mechanisms (Refs. 33–37, 60, 75, 85, 86, 204, 224, 226, 250, 262, 294, 295, 302, 304, 305, 310, 327, 332, 361, 421*f*, 462–464, 466, 493, 495)

Spherical Four-Bar Mechanisms[85] (θ_1, output vs. θ_2, input) (Fig. 3.51)

$$A \sin \theta_1 + B \cos \theta_1 = C$$

where $A = \sin \alpha_2 \sin \alpha_4 \sin \theta_2$ (3.107)

$B = -\sin \alpha_4(\sin \alpha_1 \cos \alpha_2 + \cos \alpha_1 \sin \alpha_2 \cos \theta_2)$ (3.108)

$C = \cos \alpha_3 - \cos \alpha_4(\cos \alpha_1 \cos \alpha_2 - \sin \alpha_1 \sin \alpha_2 \cos \theta_2)$ (3.109)

Other relations given in Ref. 85. Convenient equations between tan $\frac{1}{2}\theta_1$ and tan $\frac{1}{2}\theta_2$ given in Ref. 87.

Maximum angular velocity ratio $d\theta_1/d\theta_2$ occurs when $\lambda = 90°$.

Types of mechanisms: Assume α_i ($i = 1, 2, 3, 4$) < 180° and apply Grashof's rule (p. 3.51) to equivalent mechanism with identical axes of turning joints, such that all links except possibly the coupler, <90°.

Harmonic analysis: see Ref. 493.

Special Cases of the Spherical Four-Bar Mechanism

Hooke's Joint. $\alpha_2 = \alpha_3 = \alpha_4 = 90°$, $90° < \alpha_1 \le 180°$; in practice, if $\alpha_1 = 180° - \beta$, then $0 \le \beta \le 37\frac{1}{2}°$. If angles θ, $\varphi(\theta = \theta_1 - 90°$, $\varphi = 180° - \theta_2)$ are measured from a starting position (shown in Fig. 3.52) in which the planes *ABO* and *OCD* are perpendicular, then

$$\tan \varphi / \tan \theta = \cos \beta \qquad \beta = 180° - \alpha_1$$

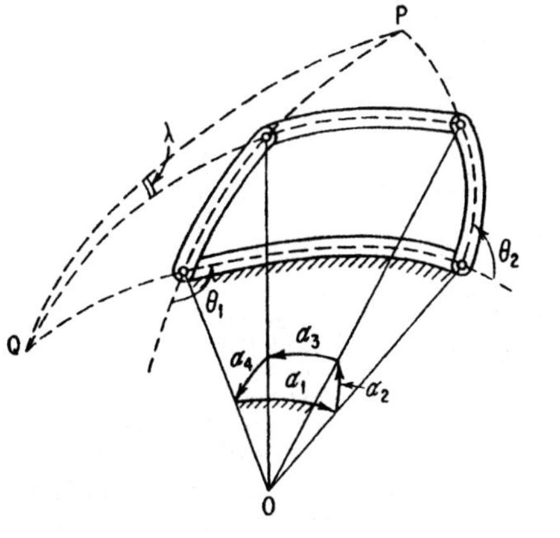

FIG. 3.51 Spherical four-bar mechanism.

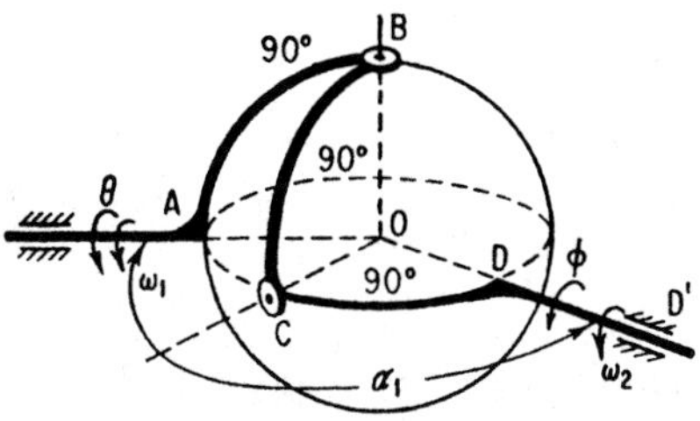

FIG. 3.52 Hooke's joint.

Angular velocity ratio:

$$\omega_2/\omega_1 = \cos\beta/(1 - \sin^2\theta \sin^2\beta)$$

Maximum value, $(\cos\beta)^{-1}$, occurs at $\theta = 90°, 270°$; minimum value, $\cos\beta$, occurs at $\theta = 0°, 180°$.

A graph of θ vs. φ will show two "waves" per revolution.

If ω_1 = constant, the angular-acceleration ratio of shaft DD' is given by

$$\frac{1}{\omega_1^2}\frac{d^2\varphi}{dt^2} = \frac{\cos\beta \sin^2\beta \sin 2\theta}{(1 - \sin^2\beta \sin^2\theta)^2} \qquad \text{maximum at } \theta = \theta_{max}$$

where

$$\cos 2\theta_{max} = G - (G^2 + 2)^{1/2} \tag{3.110}$$

and

$$G = (2 - \sin^2\beta)/2\sin^2\beta \tag{3.111}$$

For $d^2\varphi/dt^2$ as a function of β, see Ref. 264.

Harmonic Analysis of Hooke's Joint.[264,493,495] If $\alpha_1 = 180° - \beta$, the amplitude ϵ_m of the mth harmonic, in the expression $\varphi(\theta)$, is given by $\epsilon_m = 0$, m odd; and $\epsilon_m = (2/m)(\tan \frac{1}{2}\beta)^m$, m even. Two Hooke's joints in series[96,327] can be used to transmit constant (1:1) angular velocity ratio between two intersecting or nonintersecting shafts 1 and 3, provided that the angles between shafts 1 and 3 and the intermediate shaft are the same (Fig. 3.53) and that when fork 1 lies in the plane of shafts 1 and 2, fork 2 lies in the plane of shafts 2 and 3; thus in case shafts 1 and 3 intersect, forks 1 and 2 are coplanar; see also Refs. 85 and 327 when $\alpha \neq \alpha'$, which may arise due to misalignment or the effect of manufacturing tolerances, or may be intentional for use as a vibration-excitation drive.

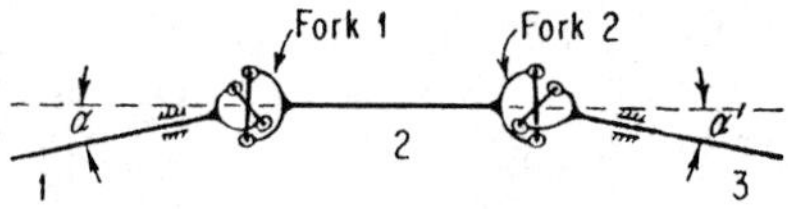

FIG. 3.53 Two Hooke's joints in series.

Other Special Cases of the Spherical Four-Bar Mechanism

Wobble-plate mechanism: $\alpha_4 < 90°$, $\alpha_2 = \alpha_3 = \alpha_1 = 90°$; Ref. 35 gives displacements, velocities, and equation of "coupler curve." See also Refs. 36, 294, and 493, the latter giving harmonic analysis.

Two angles 90°, two angles arbitrary[36,295,305]

Fixed link = 90°: Reference 332 includes applications to universal joints, gives velocities and accelerations ("space crank").

Spherical Four-Bar Mechanisms with Dwell. See Ref. 35.

Spherical Slider-Crank Mechanism. Three turning pairs, one moving joint a turn slide, input pair and adjacent pair at right angles: see Ref. 304 for displacements; if input and output axes intersect at right angles, obtain "skewed Hooke's joint."[294,304]

Spatial Four-Bar Mechanisms (see Fig. 3.42). To any spatial four-bar mechanism a corresponding spherical four-bar can be assigned as follows: Through O (Fig. 3.41) draw four radii, parallel to the axes of joints A, B, C, and D in Fig. 3.42 to intersect the surface of a sphere in four points corresponding to the joints of the spherical four-bar. The rotations of the spatial four-bar (Fig. 3.42) are the same as those of the corresponding spherical four-bar and are independent of the offsets a_{ij}, the minimum distances between the (noninteresting) axes i and j (ij = 12, 23, 34, and 41) of the spatial four-bar (Fig. 3.42). The input and output angles θ_2, θ_1, of the spatial four-bar, can be measured as in Fig. 3.41 for the corresponding spherical four-bar; for s, the sliding at the output joint, measured from A to Q in Fig. 3.42, the general displacement equations[85,86] are (for rotations see Fig. 3.51 and accompanying equations)

$$s = \frac{A_1 \sin\theta_1 + B_1 \cos\theta_1 - C_1}{-A\cos\theta_1 + B\sin\theta_1} \tag{3.112}$$

where A and B are given in Eqs. (3.107) and (3.108) and

$$A_1 = (a_2 \cos\alpha_2 \sin\alpha_4 + a_4 \sin\alpha_2 \cos\alpha_4)\sin\theta_2 + s_2 \sin\alpha_2 \sin\alpha_4 \cos\theta_2 \tag{3.113}$$

$$\begin{aligned} B_1 = {} & -a_4 \cos\alpha_4(\sin\alpha_1 \cos\alpha_2 + \cos\alpha_1 \sin\alpha_2 \cos\theta_2) \\ & -a_1 \sin\alpha_4(\cos\alpha_1 \cos\alpha_2 - \sin\alpha_1 \sin\alpha_2 \cos\theta_2) \\ & + a_2 \sin\alpha_4(\sin\alpha_1 \sin\alpha_2 - \cos\alpha_1 \cos\alpha_2 \cos\theta_2) \\ & + s_2 \sin\alpha_4 \cos\alpha_1 \sin\alpha_2 \sin\theta_2 \end{aligned} \tag{3.114}$$

$$\begin{aligned} C_1 = {} & -a_3 \sin\alpha_3 + a_4 \sin\alpha_4(\cos\alpha_1 \cos\alpha_2 - \sin\alpha_1 \sin\alpha_2 \cos\theta_2) \\ & + a_1 \cos\alpha_4(\sin\alpha_1 \cos\alpha_2 + \cos\alpha_1 \sin\alpha_2 \cos\theta_2) \\ & + a_2 \cos\alpha_4(\cos\alpha_1 \sin\alpha_2 + \sin\alpha_1 \cos\alpha_2 \cos\theta_2) \\ & - s_2 \cos\alpha_4 \sin\alpha_1 \sin\alpha_2 \sin\theta_2 \end{aligned} \tag{3.115}$$

where $a_1 = a_{12}$ of Fig. 3.42, and similarly $a_2 = a_{23}$, $a_3 = a_{34}$, $a_4 = a_{41}$, α_1 = angle between axes 1 and 2, α_2 = angle between axes 2 and 3, α_3 = angle between axes 3 and 4, and α_4 = angle between axes 4 and 1 (Figs. 3.41 and 3.42). For complete nomenclature and sign convention, see Ref. 494. These equations are used principally in special cases, in which they simplify.

Special Cases of the Spatial Four-Bar and Related Three-Dimensional Four-Bar Mechanisms[8]

1. Spatial four-bar with two ball joints on coupler and two turning joints: displacements and velocities,[60,201,310,361] synthesis for function generation,[86,250,361] computer programs for displacements and synthesis according to Ref. 86 are listed in Ref. 129; forces and torques are listed in Ref. 60.
2. Spatial four-bar: three angles 90°.[34,35]
3. Spatial four-bar with one ball joint and two turn slides (three links).[35]
4. Spatial four-bar with one ball joint, one turn slide, two turning joints.[462]
5. The "3-D crank slide," one ball joint, one turn slide, intersecting axes; used for agitators.[304]
6. "Degenerate" mechanisms, wherein $F = 1$, $\Sigma f_i < 7$; conditions for,[490] practical constructions in;[3] see also Refs. 143, 194, and 464.
7. Spherical geneva.[37,352]
8. Spatial five-link mechanisms.[6,93]
9. Spatial six-link mechanisms.[9,93]

Harmonic Analysis. Rotations θ_1 vs. θ_2 in Ref. 493, which also includes special cases, such as Hooke's joint, wobble plates, and spherical-crank drive.

Special applications.[35,36,294,304,332]

Miscellaneous special shaft couplings;[35,294] Cayley-Klein parameters and dual vectors;[85] screw axes and graphical methods.[36,151,196,462]

3.9.7 Intermittent-Motion Mechanisms (Refs. 11, 16, 21, 22, 44, 148, 149, 195, 201, 204, 213, 214, 227, 252, 253, 307, 351, 352, 355, 377, 424, 425, 426, 429, 442, 450, 451, 466, 468)

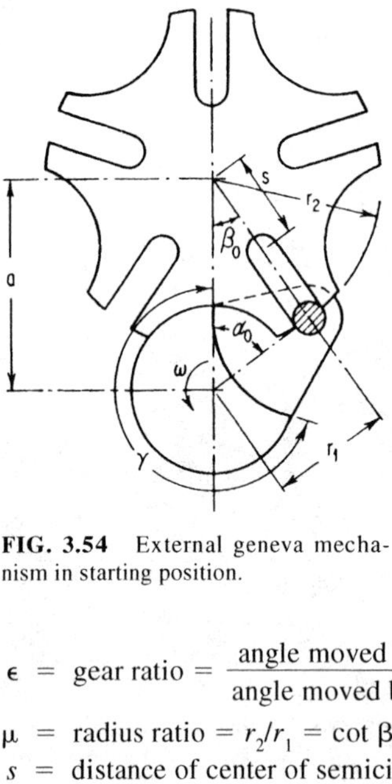

FIG. 3.54 External geneva mechanism in starting position.

The external geneva[252,253,352] is an intermittent-motion mechanism. In Fig. 3.54,

a = center distance

α = angle of driver, radians ($|\alpha| = \alpha_0$)

β = angle of driven or geneva wheel, radians [α and β measured from center-line in the direction of motions ($|\beta| = \beta_0$)]

r_1 = radius to center of driving pin

γ = locking angle of driver, radians

n = number of equally spaced slots in geneva (≥ 3) at start: $|\alpha| = \alpha_0$, $|\beta| = \beta_0 = \pi/2 - \alpha_0$ so that driving pin can enter slot tangentially to reduce shock

r_2 = radius of geneva = $a \cos \beta_0$

r_2' = outside radius of geneva wheel, with correction for finite pin diameter[351] $= r_2 \sqrt{1 + r_p^2/r_2^2}$; r_p = pin radius

$$\epsilon = \text{gear ratio} = \frac{\text{angle moved by driver during motion}}{\text{angle moved by geneva during motion}} = \frac{n-2}{2}$$

μ = radius ratio = $r_2/r_1 = \cot \beta_0$

s = distance of center of semicircular end of slot from center of geneva $\leq a(1 - \sin \beta_0)$

γ = angle of locking action = $(\pi/n)(n + 2)$ rad; note that classical locking action shown is subject to play in practice; better constructions in 4M, Sec. 3.8

ν = ratio, time of motion of geneva wheel to time for one revolution of driver = $(n - 2)/2n$ ($< \frac{1}{2}$)

$r_1 = a \sin \beta_0$, $\beta_0 = \pi/n$

Let ω = angular velocity of driving wheel, assumed constant, t = time.

Displacement (β vs. α). Let $r_1/a = \lambda$; then

$$\beta = \tan^{-1} [\lambda \sin \alpha/(1 - \lambda \cos \alpha)] \tag{3.116}$$

Velocities

$$\frac{1}{\omega}\frac{d\beta}{dt} = \frac{\lambda \cos \alpha - \lambda^2}{1 - 2\lambda \cos \alpha + \lambda^2} \tag{3.117}$$

$$\frac{1}{\omega}\left(\frac{d\beta}{dt}\right)_{max} = \frac{\lambda}{1 - \lambda} \quad \text{at } \alpha = 0 \tag{3.118}$$

Accelerations

$$\frac{1}{\omega^2}\frac{d^2\beta}{dt^2} = \frac{(\lambda^3 - \lambda)\sin\alpha}{(1 + \lambda^2 - 2\lambda\cos\alpha)^2} \tag{3.119}$$

$$(1/\omega^2)(d^2\beta/dt^2)_{\text{initial}\alpha = -\alpha 0} = \tan\beta_0 = r_1/r_2 \tag{3.120}$$

Maximum acceleration occurs at $\alpha = \alpha_{max}$, where

$$\cos\alpha_{max} = -G + (G^2 + 2)^{1/2} \qquad G = \tfrac{1}{4}(\lambda + 1/\lambda) \tag{3.121}$$

Second Acceleration or Shock

$$\frac{1}{\omega^3}\frac{d^3\beta}{dt^3} = \frac{\lambda(\lambda^2 - 1)[2\lambda\cos^2\alpha + (1 + \lambda^2)\cos\alpha - 4\lambda]}{(1 + \lambda^2 - 2\lambda\cos\alpha)^3} \tag{3.122}$$

$$\frac{1}{\omega^3}\frac{d^3\beta}{dt^3}_{(\alpha = 0)} = \frac{\lambda(\lambda + 1)}{(\lambda - 1)^3} \tag{3.123}$$

Starting Shock ($|\beta| = \beta_0$)

$$\frac{1}{\omega^3}\frac{d^3\beta}{dt^3} = \frac{3\lambda^2}{1 - \lambda^2}$$

Design Procedure (ω = const)

1. Select number of stations ($n \geq 3$).
2. Select center distance a.
3. Compute: $r_1 = a\sin\beta_0$; $r_2'^2 = (a\cos\beta_0)^2 + r_p^2$; $s \leq a(1 - \sin\beta_0)$;

$$|\alpha_0| = (\pi/2)[(n - 2)/n] \qquad \rho|\beta_0| = \pi/2 - \alpha_0 \qquad \text{rad}$$

$$\gamma = (\pi/n)(n + 2) \qquad \text{rad}$$

4. Determine kinematic characteristics from tables, including maximum velocity, acceleration, and shock: $d^n\beta/dt^n = \omega^n\, d^n\beta/d\alpha^n$, taking the last fraction from the tables.

Check for resulting forces, stresses, and vibrations. See Tables 3.2 and 3.3.

TABLE 3.2 External Geneva Characteristics[253]*

n	β_0	α_0	ϵ	(r_1/a)	(r_2/a)	μ	s_{min}/a	γ	ν	$(d\beta/d\alpha)_{max}$
3	60°	30°	0.5	0.8660	0.5000	0.5774	0.13397	300°	0.1667	6.46
4	45°	45°	1	0.7071	0.7071	1.0000	0.2929	270°	0.2500	2.41
5	36°	54°	1.5	0.5878	0.8090	1.3764	0.4122	252°	0.3000	1.43
6	30°	60°	2	0.5000	0.8660	1.7320	0.5000	240°	0.3333	1.00
7	25°43′	64°17′	2.5	0.4339	0.9009	2.0765	0.5661	231°26′	0.3571	0.766
8	22°30′	67°30′	3	0.3827	0.9239	2.4142	0.6173	225°	0.3750	0.620
9	20°	70°	3.5	0.3420	0.9397	2.7475	0.6580	220°	0.3889	0.520
10	18°	72°	4	0.3090	0.9511	3.0777	0.6910	216°	0.4000	0.447
∞	0°	90°	∞	0	1	∞	1	180°	0.5000	0

*O. Lichtwitz, Getriebe fuer Aussetzende Bewegung, Springer-Verlag OHG, Berlin, 1953.

TABLE 3.3 External Geneva Characteristics[253*]

n	$(d^2\beta/d\alpha^2)_{\text{initial}}$ $\alpha=-\alpha_0$	$\alpha_{\max}$	$(d^2\beta/d\alpha^2)_{\max}$	$(d^3\beta/d\alpha^3)_{\alpha=0}$
3	1.732	4°46′	31.44	−672
4	1.000	11°24′	5.409	− 48.04
5	0.7265	17°34′	2.299	− 13.32
6	0.5774	22°54′	1.350	− 6.000
7	0.4816	27°33′	0.9284	− 3.429
8	0.4142	31°38′	0.6998	− 2.249
9	0.3640	35°16′	0.5591	− 1.611
10	0.3249	38°30′	0.4648	− 1.236
∞	0	90°	0	0

*O. Lichtwitz, Getriebe fuer Aussetzende Bewegung, Springer-Verlag OHG, Berlin, 1953.

Modifications of Standard External Geneva. More than one driving pin;[252,253] pins not equally spaced;[16,252,253] designs for all small indexing mechanisms;[355] for high-speed indexing;[213] mounting driving pin on a planet pinion to reduce peak loading;[11,253,377] pin guided on four-bar coupler[201] (see also Sec. 3.8, Ref. 5M); double rollers and different entrance and exit slots, especially for starwheels;[227,466] eccentric gear drive for pin.[204]

Internal Genevas.[252,253,352] Used when $\nu > ½$; better kinematic characteristics, but more expensive.

Star Wheels.[227,252,253,466] Both internal and external are used; permits considerable freedom in choice of ν, which can equal unity, in contrast to genevas. Kinematic properties of external star wheels are better or worse than of external geneva with same n, according as the number of stations (or shoes), n, is less than six or greater than five, respectively.

Special Intermittent and/or Dwell Linkages. The three-gear drive[21,114,195,215,424,442] cardioid drive (slotted link driven by pin on planetary pinion);[352,425,426] link-gear (and/or) -cam mechanisms to produce dwell, reversal, or intermittent mo-tions[22,449,450,451] include link-dwell mechanisms;[148] eccentric-gear mechanisms.[149] These special motions may be required when control of rest, reversal, and kinematic characteristics exceeds that possible with the standard genevas.

3.9.8 Noncircular Cylindrical Gearing and Rolling-Contact Mechanisms
(Refs. 16, 43, 76, 117, 255, 256, 292, 314, 318, 319, 326, 341, 350, 356, 429, 438, 483)

Most of the data for this article are based on Refs. 43 and 318. Noncircular gears can be used for producing positive unidirectional motion; if the pitch curves are closed curves, unlimited rotations may be possible; only externally meshed, plane spur-type gearing will be considered; point of contact between pitch surfaces must lie on the line of centers.

A pair of roll curves may serve as pitch curves for noncircular gears (see Table 3.4): C = center distance; β = angle between the common normal to roll-curves at contact and the line of centers. Angular velocities ω_1 and ω_2 measured in opposite directions; polar-coordinate equations of curves; $R_1 = R_1(\theta_1)$, $R_2 = R_2(\theta_2)$, such that the points $R_1(\theta_1 = 0)$, $R_2(\theta_2 = 0)$ are in mutual contact, where θ_1 and θ_2 are respective oppositely directed rotations from a starting position. Centers O_1, O_2 and contact point Q are collinear.

Twin Rolling Curves. Mating or pure rolling of two identical curves, e.g., two ellipses when pinned at foci.

Mirror Rolling Curves. Curve mates with mirror image.

Theorems

1. Every mating curve to a mirror (twin) rolling curve is itself a mirror (twin) rolling curve.
2. All mating curves to a given mirror (twin) rolling curve will mate with each other.
3. A closed roll curve can generally mate with an entire set of different closed roll curves at varying center distances, depending upon the value of the "average gear ratio."

Average Gear Ratio. For mating closed roll curves: ratio of the total number of teeth on each gear.

Rolling Ellipses and Derived Forms. If an ellipse, pivoted at the focus, mates with a roll curve so that the average gear ratio n (ratio of number of teeth on mating curve to number of teeth on ellipse) is integral, the mating curve is called an "nth-order ellipse." The case $n = 1$ represents an identical (twin) ellipse; second-order ellipses are oval-shaped and appear similar to ordinary ellipses; third-order ellipses appear pear-shaped with three lobes; fourth-order ellipses appear nearly square; nth-order ellipses appear approximately like n-sided polygons. Equations for several of these are found in Ref. 43. Characteristics for five noncircular gear systems are given in Table 3.4.[43]

Design Data.[43] Data are usually given in one of three ways:

1. Given $R_1 = R_1(\theta_1)$, C. Find R_2 in parametric form: $R_2 = R_2(\theta_1)$; $\theta_2 = \theta_2(\theta_1)$.

$$\theta_2 = -\theta_1 + C\int_0^{\theta_1} \frac{d\theta_1}{C - R_1(\theta_1)} \qquad R_2 = C - R_1(\theta_1)$$

2. Given $\theta_2 = f(\theta_1)$, C. Find $R_1 = R_1(\theta_1)$, $R_2 = R_2(\theta_2)$.

$$R_1 = \frac{(df/d\theta_1)C}{1 + df/d\theta_1} \qquad R_2 = C - R_1$$

3. Given $\omega_2/\omega_1 = g(\theta_1)$, C. Find $R_1 = R_1(\theta_1)$, $R_2 = R_2(\theta_2)$.

$$R_1 = \frac{Cg(\theta_1)}{1 + g(\theta_1)} \qquad R_2 = C - R_1 \qquad \theta_2 = \int_0^{\theta_1} g(\theta_1)\, d\theta_1$$

4. *Checking for closed curves:* Let $R_1 = R_1(\theta_1)$ be a single-turn closed curve; then $R_2 = R_2(\theta_2)$ will be a single-turn closed curve also, if and only if C is determined from

$$4\pi = C\int_0^{2\pi} \frac{d\theta_1}{C - R_1(\theta_1)}$$

TABLE 3.4 Characteristics of Five Noncircular Gear Systems[43]

Type	Comments	Basic equations	Velocity equation
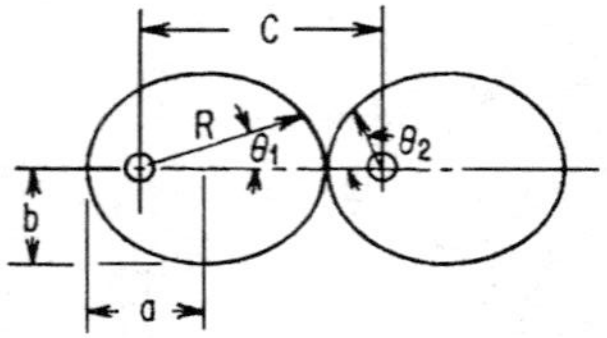 Two ellipses pivoted at foci	Gears are identical, easy to manufacture. Used for quick-return mechanisms, printing presses, automatic machinery.	$R = \dfrac{b^2}{a(1 + \epsilon \cos \theta)}$ assuming $\theta = 0$ at $R = R_{\min}$ ϵ = eccentricity $= \sqrt{1 - (b/a)^2}$ $a = \frac{1}{2}$ major axis $b = \frac{1}{2}$ minor axis	$\omega_2 = \omega_1 \left[\dfrac{r^2 + 1 + (r^2 - 1)\cos\theta_2}{2r} \right]$ where $r = \dfrac{R_{max}}{R_{min}}$ 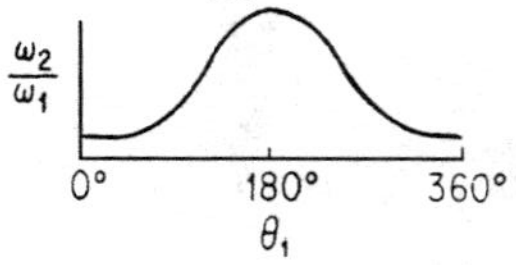
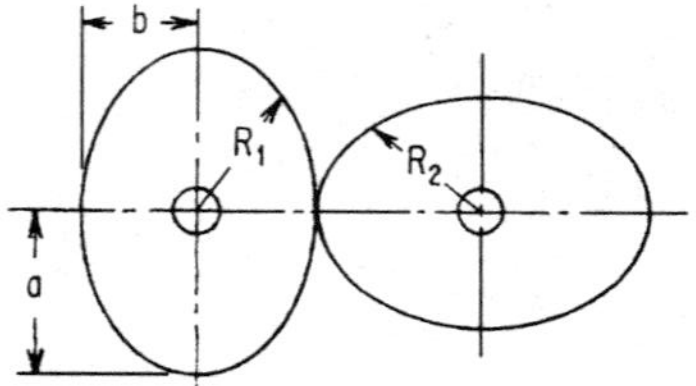 Second-order ellipses rotated about their geometric centers	Gears are identical. Well-known geometry. Better balance than true elliptical. Two complete speed cycles in one revolution.	$R = \dfrac{2ab}{(a + b) - (a - b)\cos 2\theta}$ assuming $\theta = 0$ at $R = R_{\min}$ $C = a + b$ a = max radius b = min radius	$\omega_2 = \omega_1 \left[\dfrac{r + 1 + (r^2 - 1)\cos 2\theta_2}{2r} \right]$ where $r = \dfrac{a}{b}$ 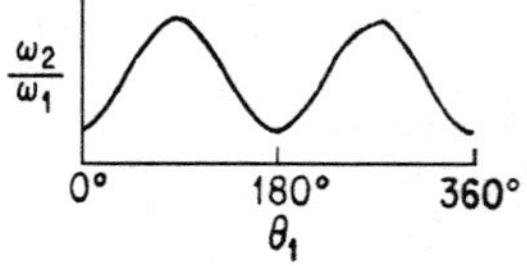

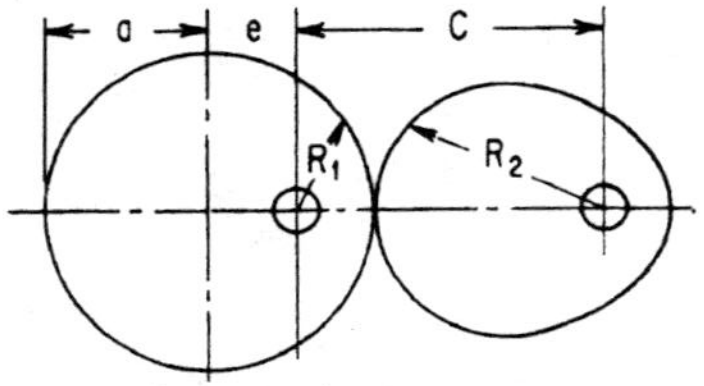 Eccentric circular gear meshing with its conjugate	Standard circular spur gear can be employed as the eccentric. Mating gear has special shape.	$R_1^2 - 2eR_1 \cos \theta_1 = a^2 - e^2$ $\theta_2 = \theta_1 + C \int_0^{\theta_1} \frac{d\theta_1}{C - R_1}$ $C = R_1 + R_2$	$\frac{\omega_2}{\omega_1} = \frac{R_1}{C - R_1}$ 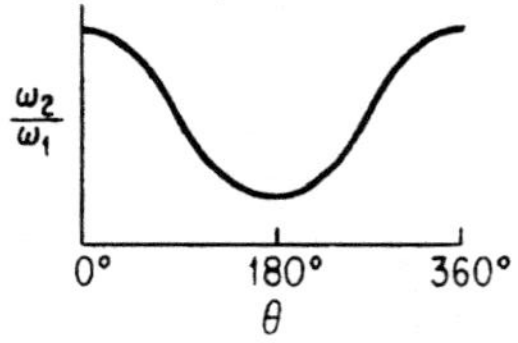
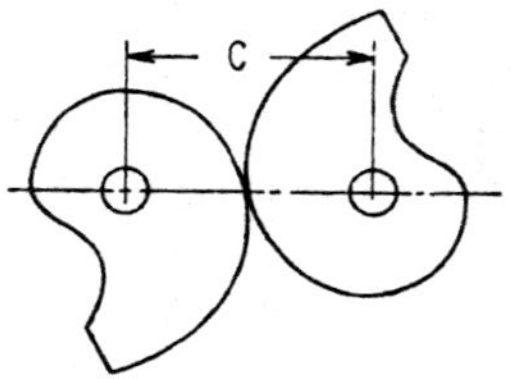 Logarithmic spirals	Gears can be identical or in combinations to give variety of functions. Must be open gears.	$R_1 = Ae^{k\theta_1}$ $R_2 = C - R_1 = Ae^{k\theta_2}$ $\theta_2 = \frac{1}{k}[\log (C - Ae^{k\theta_1}) - \log A]$ e = base of natural logarithm A = radius at $\theta = 0$	$\frac{\omega_2}{\omega_1} = \frac{R_1}{C - R_1}$ 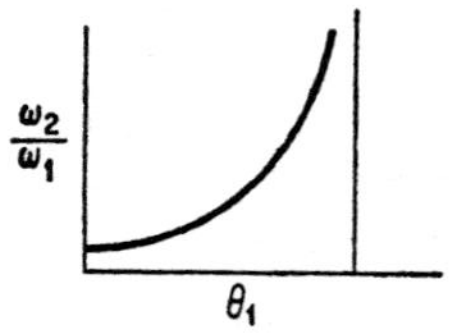
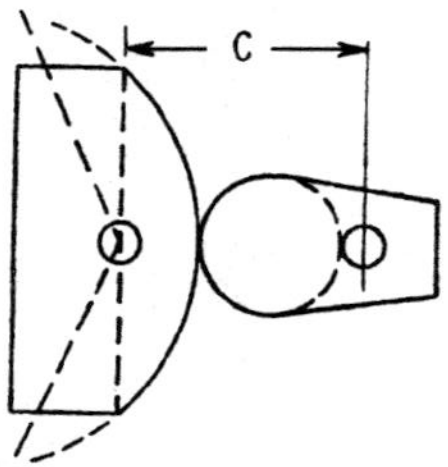 Sine-function gears	For producing angular displacement proportional to sine of input angle. Must be open gears.	$\theta_2 = k \sin \theta_1$ $R_2 = \frac{C}{1 + k \cos \theta_1}$ $R_1 = C - R_2$	$\frac{\omega_2}{\omega_1} = k \cos \theta_1$ 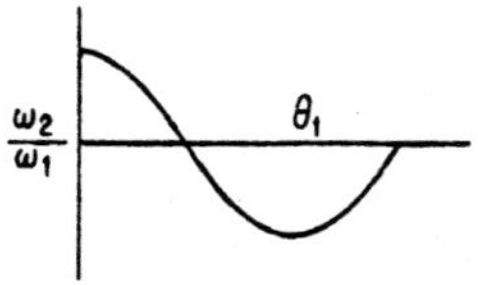

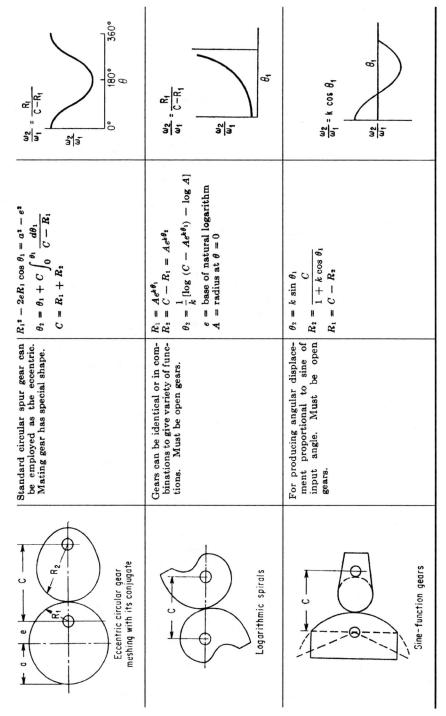

When the average gear ratio is not unity, see Ref. 43.

5. *Checking angle* β, also called the angle of obliquity:

$$\beta = \tan^{-1}\frac{1}{R_i}\frac{dR_i}{d\theta_i} \qquad i = 1 \text{ or } 2.$$

Values of β between 0 and 45° are generally considered reasonable.

6. *Checking for tooth undercut:* Let ρ = radius of curvature of pitch curve (roll curve)

$$\rho = \frac{[R_i^2 + (dR_i/d\theta_i)^2]^{3/2}}{R_i^2 - R_i(d^2R_i/d\theta_i^2) + 2(dR_i/d\theta_i)^2} \qquad i = 1 \text{ or } 2$$

Number of teeth:

$$T_{\min} = \begin{cases} 32 & \text{for } 14\tfrac{1}{2}^\circ \text{ pressure angle cutting tool} \\ 18 & \text{for } 20^\circ \text{ pressure angle cutting tool} \end{cases}$$

Condition to avoid undercut in noncircular gears:

$$\rho > \frac{T_{\min}}{2 \times \text{diametral pitch}}$$

7. *Determining length S of roll curves:*

$$S = \int_0^{2\pi}\left[R^2 + \left(\frac{dR}{d\theta}\right)^2\right]^{1/2} d\theta$$

This is best computed automatically by numerical integration or determined graphically by large-scale layout.

8. *Check on number of teeth:* For closed single-turn curves,

$$\text{Number of teeth} = (S \times \text{diametral pitch})/\pi$$

Diametral pitch should be integral but may vary by a few percentage points. For symmetrical twin curves, use odd number of teeth for proper meshing following identical machining.

Manufacturing information in Ref. 43.

Special Topics in Noncircular Gearing. Survey;[341] elliptic gears;[292,350,356] noncircular cams and rolling-contact mechanisms, such as in shears and recording instruments;[117,256,314] noncircular bevel gears;[319] algebraic properties of roll curves;[483] miscellaneous.[255,326,438]

3.9.9 Gear-Linked-Cam Combinations and Miscellaneous Mechanisms[396]

Two-gear drives;[287,391,421f,474,475] straight-type mechanisms in which rack on slide drives output gear (Refs. 285, 330, 390); mechanical analog computing mechanisms;[21,40,137,216,306,328,338,446] three-link screw mechanisms;[359] ratchets;[18,362,370] function generators with two four-bars in series;[232,248] two-degree-of-freedom computing mechanisms;[328] gear-train calculations;[21,24,178,280,282,360] the harmonic drive;[68,126,316] design of variable-speed drives;[31] rubber-covered rollers;[382] eccentric-gear drives.[152]

3.9.10 Robots and Manipulators[97,102,156,187,240,320,373,421f,444,454,470]

Robots are used for production, assembly, materials handling, and other purposes. Mechanically most robots consist of computer-controlled, joint-actuated, open kinematic chains terminating in an "end effector," such as a gripper, hand, or a tool adaptor, which is used for motion transfer. The gripper may have many degrees of freedom, just as does the human hand. The mechanical design of robot mechanisms can include both rigid and elastic elements and involves the determination of kinematic structure, ranges of motion, useful work space, dexterity, kinematics, joint actuation, mechanical advantage, dynamics, power requirements, optimization, and integration with the electronic and computer portions of the robot. A general survey can be found in Heer[187] and Roth,[373] while more specialized investigation can be found in Refs. 97, 102, 156, 240, 320, 421*f*, 444, 454, and 470. The subject is extensive and continuously expanding.

3.9.11 Hard Automation Mechanisms[421c,421f]

For highly repetitive spatial automation tasks, robotic devices with their multiple programmed inputs are greatly "over-qualified." For these tasks, single-input, purely mechanical spatial mechanisms can be more economical and efficient. For design-synthesis of these see Refs. 421*c* and 421*f*.

REFERENCES

1. Alban, C. F.: "Thermostatic Bimetals," *Machine Design,* vol. 18, pp. 124–128, Dec. 1946.
2. Allen, C. W.: "Points Position Reduction," *Trans. Fifth Conf. Mechanisms,* Purdue University, West Lafayette, Ind., pp. 181–193, October 1958.
3. Altmann, F. G.: "Ausgewaehlte Raumgetriebe," *VDI-Berichte,* vol. 12, pp. 69–76, 1956.
4. Altmann, F. G.: "Koppelgetriebe fuer gleichfoermige Uebersetzung," *Z. VDI,* vol. 92, no. 33, pp. 909–916, 1950.
5. Altmann, F. G.: "Raumgetriebe," *Feinwerktechnik,* vol. 60, no. 3, pp. 1–10, 1956.
6. Altmann, F. G.: "Raümliche fuenfgliedrige Koppelgetriebe," *Konstruktion,* vol. 6, no. 7, pp. 254–259, 1954.
7. Altmann, F. G.: "Reibgetriebe mit stufenlos einstellbarer Übersetzung," *Maschinenbautechnik,* supplement *Getriebetechnik,* vol. 10, no. 6, pp. 313–322, 1961.
8. Altmann, F. G.: "Sonderformen raümlicher Koppelgetriebe und Grenzen ihrer Verwendbarkeit," *VDI-Berichte,* Tagungsheft no. 1, *VDI,* Duesseldorf, pp. 51–68, 1953; also in *Konstruktion,* vol. 4, no. 4, pp. 97–106, 1952.
9. Altmann, F. G.: "Ueber raümliche sechsgliedrige Koppelgetriebe," *Z. VDI,* vol. 96, no. 8, pp. 245–249, 1954.
10. Altmann, F. G.: "Zur Zahlsynthese der raümlichen Koppelgetriebe," *Z. VDI,* vol. 93, no. 9, pp. 205–208, March 1951.
11. Arnesen, L.: "Planetary Gears," *Machine Design,* vol. 31, pp. 135–139, Sept. 3, 1959.
12. Artobolevskii, I. I., S. Sh. Blokh, V. V. Dobrovolskii, and N. I. Levitskii: "The Scientific Works of P. L. Chebichev II—Theory of Mechanisms" (Russian), Akad. Nauk, Moscow, p. 192, 1945.
13. Artobolevskii, I. I.: "Theory of Mechanisms and Machines" (Russian). State Publishing House of Technical-Theoretical Literature, Moscow-Leningrad, 1940 ed., 762 pp.; 1953 ed., 712 pp.

14. Artobolevskii, I. I. (ed.): "Collected Works (of L. V. Assur) on the Study of Plane, Pivoted Mechanisms with Lower Pairs," Akad. Nauk, Moscow, 1952.

15. Artobolevskii, I. I.: "Das dynamische Laufkriterium bei Getrieben," *Maschinenbautechnik (Getriebetechnik),* vol. 7, no. 12, pp. 663–667, 1958.

16. Artobolevskii, I. I., N. I. Levitskii, and S. A. Cherkudinov: "Synthesis of Plane Mechanisms" (Russian), Publishing House of Physical-Mathematical Literature, Moscow. 1959.

17. "Transactions of Second U.S.S.R. Conference on General Problems in the Theory of Mechanisms and Machines" (Russian), Board of Editors: I. I. Artobolevskii, S. I. Artobolevskii, G. G. Baranov, A. P. Bessonov, V. A. Gavrilenko, A. E. Kobrinskii, N. I. Levitskii, and L. N. Reshetov, Publishing House of Scientific Technical Studies on Machine Design, Moscow, 1960. 4 Volumes of 1958 Conference: A. Analysis and Synthesis of Mechanisms. B. Theory of Machine Transmissions. C. Dynamics of Machines. D. Theory of Machines in Automatic Operations and Theory of Precision in Machinery and Instrumentation.

18. Assmus, F.: "Technische Laufwerke," Springer-Verlag OHG, Berlin, 1958.

19. Bammert, K., and A. Schmidt: "Die Kinematik der mittelbaren Pleuelanlenkung in Fourier-Reihen," *Ing.-Arch.,* vol. 15, pp. 27–51, 1944.

20. Bareiss, R. A., and P. A. Brand, "Which Type of Torque-limiting Device," *Prod. Eng.,* vol. 29, pp. 50–53, Aug. 4, 1958.

21. Beggs, J. S.: "Mechanism," McGraw-Hill Book Company, Inc., New York, 1955.

22. Beggs, J. S., and R. S. Beggs: "Cams and Gears Join to Stop Shock Loads," *Prod. Eng.,* vol. 28, pp. 84–85, Sept, 16, 1957.

23. Beggs, J. S.: "A Theorem in Plane Kinematics," *J. Appl. Mechanics,* vol. 25, pp. 145–146, March 1958.

24. Benson, A.: "Gear-train Ratios," *Machine Design,* vol. 30, pp. 167–172, Sept. 18, 1958.

25. Berkof, R. S.: "Complete Force and Moment Balancing of in-line Four-Bar Linkages," *Mechanism and Machine Theory,* vol. 8, pp. 397–410, 1973.

26. Berkof, R. S.: "Force Balancing of a Six-Bar Linkage," *Proceedings,* Fifth World Congress on Theory of Machines and Mechanisms, ASME, pp. 1082–1085, 1979.

27. Berkof, R. S.: "The Input Torque in Linkages," *Mechanism and Machine Theory,* vol. 14, pp. 61–73, 1979.

28. Berkof, R. S., and G. G. Lowen: "A New Method for Completely Force Balancing Simple Linkages," *J. Eng. Ind., Trans. ASME,* vol. 91B, pp. 21–26, 1969.

29. Berkof, R. S.: "On the Optimization of Mass Distribution in Mechanisms," Ph.D. thesis, City University of New York, 1969.

30. Berkof, R. S., and G. G. Lowen: "Theory of Shaking Moment Optimization of Force-Balanced Four-Bar Linkages," *J. Eng. Ind., Trans. ASME,* vol. 93B, pp. 53–60, 1971.

31. Berthold, H.: "Stufenlos verstellbare mechanische Getriebe," VEB Verlag Technik, Berlin, 1954.

32. Beyer, R.: "Kinematische Getriebesynthese," Springer-Verlag OHG, Berlin, 1953.

33. Beyer, R., and E. Schoerner: "Raumkinematische Grundlagen," Johann Ambrosius Barth, Munich, 1953.

34. Beyer, R.: "Zur Geometrie und Synthese eigentlicher Raumkurbelgetriebe," *VDI-Berichte,* vol. 5, pp. 5–10, 1955.

35. Beyer, R.: "Zur Synthese und Analyse von Raumkurbelgetrieben," *VDI-Berichte,* vol. 12, pp. 5–20, 1956.

36. Beyer, R.: "Wissenschaftliche Hilfsmittel und Verfahren zur Untersuchung räumlicher Gelenkgetriebe," *VDI Zeitschrift,* vol. 99, pt. I, no. 6, pp. 224–231, February, 1957; pt. 2, no. 7, pp. 285–290, March 1957.

37. Beyer, R.: "Räumliche Malteserkreutzgetriebe," *VDI-Forschungsheft,* no. 461, pp. 32–36, 1957 (see Ref. 120).

38. Beyer, R.: Space Mechanisms, *Trans. Fifth Conf. Mechanisms,* Purdue University, West Lafayette, Ind., pp. 141–163, Oct. 13–14, 1958.

39. Biezeno, C. B., and R. Grammel: "Engineering Dynamics" (Trans. M. P. White), vol. 4, Blackie & Son, Ltd., London and Glasgow, 1954.

40. Billings, J. H.: "Review of Fundamental Computer Mechanisms," *Machine Design,* vol. 27, pp. 213–216, March 1955.

41. Blaschke, W., and H. R. Mueller: "Ebene Kinematik," R. Oldenbourg KG, Munich, 1956.

42. Blokh, S. Sh.: "Angenaeherte Synthese von Mechanismen," VEB Verlag Technik, Berlin, 1951.

43. Bloomfield, B.: "Non-circular Gears," *Prod. Eng.,* vol. 31, pp. 59–66, Mar. 14, 1960.

44. Bogardus, F. J.: "A Survey of Intermittent-motion Mechanisms," *Machine Design,* vol. 28, pp. 124–131, Sept. 20, 1956.

45. Bogdan, R. C., and C. Pelecudi: "Contributions to the Kinematics of Chebichev's Dyad," *Rev. Mecan. Appliquee,* vol. V, no. 2, pp. 229–240, 1960.

46. Bolz, W.: "Gaskets in Design," *Machine Design,* vol. 17, pp. 151–156, March 1945.

47. Bolz, R. W., J. W. Greve, K. R. Harnar, and R. E. Denega: "High Speeds in Design," *Machine Design,* vol. 22, pp. 147–194, April 1950.

48. Botstiber, D. W., and L. Kingston: "Cycloid Speed Reducer," *Machine Design,* vol. 28, pp. 65–69, June 28, 1956.

49. Bottema, O.: "On Cardan Positions for the Plane Motion of a Rigid Body," *Koninkl. Ned. Akad. Watenschap. Proc.,* pp. 643–651, June 1949.

50. Bottema, O.: "On Gruebler's Formulae for Mechanisms," *Appl. Sci. Research,* vol. A2, pp. 162–164, 1950.

51. Brandenberger, H.: "Kinematische Getriebemodelle," Schweizer, Druck & Verlagshaus, AG, Zurich, 1955.

52. Breunich, T. R.: "Sensing Small Motions," *Prod. Eng.,* vol. 21, pp. 113–116, July 1950.

53. Broniarek, C. A., and G. N. Sandor: "Dynamic Stability of an Elastic Parallelogram Linkage," *J. Nonlinear Vibration Problems,* Polish Academy of Science, Institute of Basic Technical Problems, Warsaw, vol. 12, pp. 315–325, 1971.

54. Buchsbaum, F., and F. Freudenstein: "Synthesis of Kinematic Structure of Geared Kinematic Chains and Other Mechanisms," *J. Mechanisms,* vol. 5, pp. 357–392, 1970.

55. Buffenmyer, W. L.: "Pressure Gages," *Machine Design,* vol. 31, p. 119, July 23, 1959.

56. Bus, R. R.: "Bearings for Intermittent Oscillatory Motions," *Machine Design,* vol. 22, pp. 117–119, January 1950.

57. Campbell, J. A.: "Pneumatic Power—3," *Machine Design,* vol. 23, pp. 149–154, July 1951.

58. Capitaine, D.: "Zur Analyse and Synthese ebener Vier und Siebengelenkgetriebe in komplexer Behandlungsweise," doctoral dissertation, Technische Hochschule, Munich, 1955–1960.

59. Carnetti, B., and A. J. Mei: "Hermetic Motor Pumps for Sealed Systems," *Mech. Eng.,* vol. 77, pp. 488–494, June 1955.

60. Carter, B. G.: "Analytical Treatment of Linked Levers and Allied Mechanisms," *J. Royal Aeronautical Society,* vol. 54, pp. 247–252, October 1950.

61. Carter, W. J.: "Kinematic Analysis and Synthesis Using Collineation-Axis Equations," *Trans. ASME,* vol. 79, pp. 1305–1312, 1957.

62. Carter, W. J.: "Second Acceleration in Four-Bar Mechanisms as Related to Rotopole Motion," *J. Appl. Mechanics,* vol. 25, pp. 293–294, June 1958.

63. Cayley, A.: "On Three-Bar Motion," *Proc. London Math. Soc.,* vol. 7, pp. 135–166, 1875–1876.

64. "DRAM and ADAMS" (computer codes), Chace, M. A.: Mechanical Dynamics Inc., 55 South Forest Avenue, Ann Arbor, Mich. 48104.

65. Chace, M. A., and Y. O. Bayazitoglu: "Development and Application of a Generalized d'Alembert Force for Multifreedome Mechanical Systems," *J. Eng. Ind., Trans ASME,* vol. 93B, pp. 317–327, 1971.

66. Chace, M. A., and P. N. Sheth: "Adaptation of Computer Techniques to the Design of Mechanical Dynamic Machinery," ASME Paper 73-DET-58, 1973.

67. Cherkudinov, S. A.: "Synthesis of Plane, Hinged, Link Mechanisms" (Russian), Akad, Nauk, Moscow, 1959.

68. Chironis, N.: "Harmonic Drive," *Prod. Eng.,* vol. 31, pp. 47–51, Feb. 8, 1960.

69. Chow, H.: "Linkage Joints," *Prod. Eng.,* vol. 29, pp. 87–91, Dec. 8, 1958.

70. Conklin, R. M., and H. M. Morgan: "Transducers," *Prod. Eng.,* vol. 25, pp. 158–162, December 1954.

71. Conway, H. G.: "Landing-Gear Design," Chapman & Hall, Ltd., London, 1958.

72. Conway, H. G.: "Straight-Line Linkages," *Machine Design,* vol. 22, p. 90, 1950.

73.Creech, M. D.: "Dynamic Analysis of Slider-Crank Mechanisms," *Prod. Eng.,* vol. 33, pp. 58–65, Oct. 29, 1962.

74. Crossley, F. R. E.: "The Permutations or Kinematic Chains of Eight Members or Less from the Graph-Theoretic Viewpoint," *Developments in Theoretical and Applied Mechanics,* vol. 2, Pergamon Press, Oxford, pp. 467–486, 1965.

75. Crossley, F. R. E., and F. W. Keator: "Three Dimensional Mechanisms," *Machine Design,* vol. 27, no. 8, pp. 175–179, 1955; also, no. 9, pp. 204–209, 1955.

76. Cunningham, F. W.: "Non-circular Gears," *Machine Design,* vol. 31, pp. 161–164, Feb. 19, 1959.

77. "Skoda Harbor Cranes for the Port of Madras, India," *Czechoslovak Heavy Industry*; pp. 21–28, February 1961.

78. Datseris, P.: "Principles of a Heuristic Optimization Technique as an Aid in Design: An Overview," ASME Paper 80-DET-44, 1980.

79. Datseris, P., and F. Freudenstein: "Optimum Synthesis of Mechanisms Using Heuristics for Decomposition and Search," *J. Mechanical Design, ASME Trans.,* vol. 101, pp. 380–385, 1979.

80. Deimel, R. F., and W. A. Black: "Ball Integrator Averages Sextant Readings," *Machine Design,* vol. 18, pp. 116–120, March 1946.

81. Deist, D. H.: "Air Springs," *Machine Design,* vol. 30, pp. 141–146, Jan. 6, 1958.

82. Dix, R. C., and T. J. Lehman: "Simulation of the Dynamics of Machinery," *J. Eng. Ind., Trans. ASME,* vol. 94B, pp. 433–438, 1972.

83.de Jonge, A. E. R.: "A Brief Account of Modern Kinematics," *Trans. ASME,* vol. 65, pp. 663–683, 1943.

84. de Jonge, A. E. R.: "Kinematic Synthesis of Mechanisms," *Mech. Eng.,* vol. 62, pp. 537–542, 1940 (see Refs. 167 and 168).

85. Denavit, J.: "Displacement Analysis of Mechanisms Based on 2×2 Matrices of Dual Numbers" (English), *VDI-Berichte,* vol. 29, pp. 81–88, 1958.

86. Denavit, J., and R. S. Hartenberg: "Approximate Synthesis of Spatial Linkages," *J. Appl. Mechanics,* vol. 27, pp. 201–206, Mar. 1960.

87. Dimentberg, F. M.: "Determination of the Motions of Spatial Mechanisms," Akad. Nauk, Moscow, p. 142, 1950 (see Ref. 78).

88. *Design News:* "Double-Throw Crankshaft Imparts Vertical Motion to Worktable," p. 26, Apr. 14, 1958.

89. *Design News:* "Sky-Lift"-Tractor and Loader with Vertical Lift Attachment," pp. 24–25, Aug. 3, 1959.

90. *Design News:* "Hydraulic Link Senses Changes in Force to Control Movement in Lower Limb," pp. 26–27, Oct. 26, 1959.

91. *Design News:* "Four-Bar Linkage Moves Stacker Tip in Vertical Straight Line," pp. 32–33, Oct. 26, 1959.

92. *Design News:* "Linkage Balances Forces in Thrust Reverser on Jet Star," pp. 30–31, Feb. 15, 1960.

93. Dimentberg, F. M.: "A General Method for the Investigation of Finite Displacements of Spatial Mechanisms and Certain Cases of Passive Constraints" (Russian), Akad. Nauk, Moscow, *Trudi Sem. Teor. Mash. Mekh.,* vol. V, no. 17, pp. 5–39, 1948 (see Ref. 72).

94. Dobrovolskii, V. V.: "Theory of Mechanisms" (Russian), State Publishing House of Technical Scientific Literature, Machine Construction Division, Moscow, p. 464, 1951.

95. Dobrovolskii, V. V.: "Trajectories of Five-Link Mechanisms," *Trans. Moscow Machine-Tool Construction Inst.,* vol. 1, 1937.

96. Doughtie, V. L., and W. H. James: "Elements of Mechanism," John Wiley & Sons, Inc., New York, 1954.

97. Dubowsky, D., and T. D. DesForges: "Robotic Manipulator Control Systems with Invariant Dynamic Characteristics," *Proc. Fifth World Congress on the Theory of Machines and Mechanisms,* Montreal, pp. 101–111, 1979.

98. Dubowsky, S.: "On Predicting the Dynamic Effects of Clearances in Planar Mechanisms," *J. Eng. Ind., Trans. ASME,* vol. 96B, pp. 317–323, 1974.

99. Dubowsky, S., and F. Freudenstein: "Dynamic Analysis of Mechanical Systems and Clearances I, II," *J. Eng. Ind., Trans. ASME,* vol. 93B, pp. 305–316, 1971.

100. Dubowsky, S., and T. N. Gardner: "Dynamic Interactions of Link Elasticity and Clearance Connections in Planar Mechanical Systems," *J. Eng. Ind., Trans. ASME,* vol. 97B, pp. 652–661, 1975.

101. Dudley, D. W.: "Gear Handbook," McGraw-Hill Book Company, Inc., New York, 1962.

102. Duffy, J.: "Mechanisms and Robot Manipulators," Edward Arnold, London, 1981.

103. Eastman, F. S.: "Flexure Pivots to Replace Knife-Edges and Ball Bearings," *Univ. Wash. Eng. Expt. Sta. Bull. 86,* November 1935.

104 Ebner, F.: "Leitfaden der Technischen Wichtigen Kurven," B. G. Teubner Verlagsgesellschaft, GmbH, Leipzig, 1906.

105. Ellis, A. H., and J. H. Howard: "What to Consider When Selecting a Metallic Bellows," *Prod. Eng.,* vol. 21, pp. 86–89, July 1950.

106. Erdman, A. G., and G. N. Sandor: "Mechanism Design, Analysis and Synthesis," vol. 1, Prentice-Hall, Inc., Englewood Cliffs, N.J., 1984.

107. Erdman, A. G., and G. N. Sandor: "Kineto-Elastodynamics—A Frontier in Mechanism Design," *Mechanical Engineering News,* ASEE, vol. 7, pp. 27–28, November 1970.

108. Erdman, A. G.: "LINCAGES," Department of Mechanical Engineering, University of Minnesota, 111 Church Street S.E., Minneapolis, Minn. 55455.

109. Erdman, A. G., G. N. Sandor, and R. G. Oakberg: "A General Method for Kineto-Elastodynamic Analysis and Synthesis of Mechanisms," *J. Eng. Ind., Trans. ASME,* vol. 94B, pp. 1193–1205, 1972.

110. Erdman, A. G., and J. Bowen: "Type and Dimensional Synthesis of Casement Window Mechanisms," *Mech. Eng.,* vol. 103, no. 12, pp. 46–55, 1981.

111. Federhofer, K.: "Graphische Kinematik und Kinetostatik," Springer-Verlag OHG, Berlin, 1932.

112. Franke, R.: "Vom Aufbau der Getriebe," vol. I, Beuth-Vertrieb, GmbH, Berlin, 1948; vol. II, VDI Verlag, Duesseldorf, Germany, 1951.

113. Nonmetallic Gaskets, based on studies by E. C. Frazier, *Machine Design,* vol. 26, pp. 157–188, November 1954.

114. Freudenstein, F.: "Design of Four-Link Mechanisms," doctoral dissertation, Columbia University, University Microfilms, Ann Arbor, Mich., 1954.

115. Freudenstein, F.: "Approximate Synthesis of Four-Bar Linkages," *Trans. ASME,* vol. 77, pp. 853–861, August 1955.

116. Freudenstein, F.: "On the Maximum and Minimum Velocities and the Accelerations in Four-Link Mechanisms," *Trans. ASME,* vol. 78, pp. 779–787, 1956.

117. Freudenstein, F.: "Ungleichfoermigkeitsanalyse der Grundtypen ebener Getriebe," *VDI-Forschungsheft,* vol. 23, no. 461, ser. B, pp. 6–10, 1957.

118. Freudenstein, F.: "Four-Bar Function Generators," *Machine Design,* vol. 30, pp. 119–123, Nov. 27, 1958.

119. Freudenstein, F., and E. J. F. Primrose: "Geared Five-Bar Motion I—Gear Ratio Minus One," *J. Appl. Mech.,* vol. 30: *Trans. ASME,* vol. 85, ser. E, pp. 161–169, June 1963.

120. Freudenstein, F., and E. J. F. Primrose: "Geared Five-Bar Motion II—Arbitrary Commensurate Gear Ratio," *ibid.,* pp. 170–175 (see Ref. 268a).

121. Freudenstein, F., and B. Roth: "Numerical Solution of Systems of Nonlinear Equations," *J. Ass. Computing Machinery,* October 1963.

122. Freudenstein, F.: "Structural Error Analysis in Plane Kinematic Synthesis," *J. Eng. Ind., Trans. ASME,* vol. 81B, pp. 15–22, February 1959.

123. Freudenstein, F., and G. N. Sandor: "Synthesis of Path-Generating Mechanisms by Means of a Programmed Digital Computer," *J. Eng. Ind., Trans. ASME,* vol. 81B, pp. 159–168, 1959.

124. Freudenstein, F.: "Trends in the Kinematics of Mechanisms," *Appl. Mechanics Revs.,* vol. 12, no. 9, September 1959, survey article.

125. Freudenstein, F.: "Harmonic Analysis of Crank-and-Rocker Mechanisms with Application," *J. Appl. Mech.,* vol. 26, pp. 673–675, December 1959.

126. Freudenstein, F.: "The Cardan Positions of a Plane," *Trans. Sixth Conf. Mechanisms,* Purdue University, West Lafayette, Ind., pp. 129–133, October 1960.

127. Freudenstein, F., and G. N. Sandor: "On the Burmester Points of a Plane," *J. Appl. Mechanics,* vol. 28, pp. 41–49, March 1961; discussion, September 1961, pp. 473–475.

128. Freudenstein, F., and K. Mohan: "Harmonic Analysis," *Prod. Eng.,* vol. 32, pp. 47–50, Mar. 6, 1961.

129. Freudenstein, F.: "Automatic Computation in Mechanisms and Mechanical Networks and a Note on Curvature Theory," presented at the International Conference on Mechanisms at Yale University, March, 1961; Shoestring Press, Inc., New Haven, Conn., 1961, pp. 43–62.

130. Freudenstein, F.: "On the Variety of Motions Generated by Mechanisms," *J. Eng. Ind., Trans. ASME,* vol. 81B, pp. 156–160, February 1962.

131. Freudenstein, F.: "Bi-variate, Rectangular, Optimum-Interval Interpolation," *Mathematics of Computation,* vol. 15, no. 75, pp. 288–291, July 1961.

132. Freudenstein, F., and L. Dobrjanskyj: "On a Theory for the Type Synthesis of Mechanisms," *Proc. Eleventh International Congress of Applied Mechanics,* Springer-Verlag, Berlin, pp. 420–428, 1965.

133. Freudenstein, F., and E. R. Maki: "The Creation of Mechanisms According to Kinematic Structure and Function," *J. Environment and Planning B,* vol. 6, pp. 375–391, 1979.

134. Freudenstein, F., and E. J. F. Primrose: "The Classical Transmission Angle Problem," *Proc. Conf. Mechanisms,* Institution of Mechanical Engineers (London), pp. 105–110, 1973.

135. Freudenstein, F.: "Optimum Force Transmission from a Four-Bar Linkage," *Prod. Eng.,* vol. 49, no. 1, pp. 45–47, 1978.

136. Freudenstein, F., and M. S. Chew: "Optimization of Crank-and-Rocker Linkages with Size and Transmission Constraints," *J. Mech. Design, Trans. ASME,* vol. 101, no. 1, pp. 51–57, 1979.

137. Fry, M.: "Designing Computing Mechanisms," *Machine Design,* vol. 17–18, 1945–1946; I, August, pp. 103–108; II, September, pp. 113–120; III, October, pp. 123–128; IV, November, pp. 141–145; V, December, pp. 123–126; VI, January, pp. 115–118; VII, February, pp. 137–140.

138. Fry, M.: "When Will a Toggle Snap Open," *Machine Design,* vol. 21, p. 126, August 1949.

139. Gagne, A. F., Jr.: "One-Way Clutches," *Machine Design,* vol. 22, pp. 120–128, Apr. 1950.

140. Gagne, A. F., Jr.: "Clutches," *Machine Design,* vol. 24, pp. 123–158, August 1952.

141. Geary, P. J.: "Torsion Devices," Part 3, Survey of Instrument Parts, British Scientific Instrument Research Association, Research Rep. R. 249, 1960, South Hill, Chislehurst, Kent, England; see also Rep. 1, on Flexture Devices; 2 on Knife-Edge Bearings, all by the same author.

142. *General Motors Eng. J.:* "Auto Window Regulator Mechanism," pp. 40–42, January, February, March, 1961.

143. Goldberg, M.: "New Five-Bar and Six-Bar Linkages in Three Dimensions," *Trans. ASME,* vol. 65, pp. 649–661, 1943.

144. Goodman, T. P.: "Toggle Linkage Applications in Different Mechanisms," *Prod. Eng.,* vol. 22, no. 11, pp. 172–173, 1951.

145. Goodman, T. P.: "An Indirect Method for Determining Accelerations in Complex Mechanisms," *Trans. ASME,* vol. 80, pp. 1676–1682, November 1958.

146. Goodman, T. P.: "Four Cornerstones of Kinematic Design," *Trans. Sixth Conf. Mechanisms,* Purdue University, West Lafayette, Ind., pp. 4–30, Oct. 10–11, 1961; also published in *Machine Design,* 1960–1961.

147. Goodman, T. P.: "Dynamic Effects of Backlash," *Trans. Seventh Conf. Mechanisms,* Purdue University, West Lafayette, Ind., pp. 128–138, 1962.

148. Grodzenskaya, L. S.: "Computational Methods of Designing Linked Mechanisms with Dwell" (Russian), *Trudi Inst. Mashinoved.,* Akad. Nauk, Moscow, vol. 19, no. 76, pp. 34–45, 1959; see also vol. 71, pp. 69–90, 1958.

149. Grodzinski, P.: "Eccentric Gear Mechanisms," *Machine Design,* vol. 25, pp. 141–150, May 1953.

150. Grodzinski, P.: "Straight-Line Motion," *Machine Design,* vol. 23, pp. 125–127, June 1951.

151. Grodzinski, P., and E. M. Ewen: "Link Mechanisms in Modern Kinematics," *Proc. Inst. Mech. Eng. (London),* vol. 168, no. 37, pp. 877–896, 1954.

152. Grodzinski, P.: "Applying Eccentric Gearing," *Machine Design,* vol. 26, pp. 147–151, July 1954.

153. Grodzinski, P.: "A Practical Theory of Mechanisms," Emmett & Co., Ltd., Manchester, England, 1947.

154. Gruebler, M.: "Getriebelehre," Springer-Verlag OHG, Berlin, 1917/21.

155. Gupta, K. C.: "Design of Four-Bar Function Generators with Minimax Transmission Angle," *J. Eng. Ind., Trans. ASME,* vol. 99B, pp. 360–366, 1977.

156. Gupta, K. C., and B. Roth: "Design Consideration for Manipulator Workspace," ASME paper 81-DET-79, 1981.

157. Hagedorn, L.: "Getriebetechnik und Ihre praktische Anwendung," *Konstruktion,* vol. 10, no. 1, pp. 1–10, 1958.

158. Hain, K.: "Der Einfluss der Toleranzen bei Gelenkrechengetrieben," *Die Messtechnik,* vol. 20, pp. 1–6, 1944.

159. Hain, K.: "Angewandte Getriebelehre," 2d ed., VDI Verlag, Duesseldorf, 1961, English translation "Applied Kinematics," McGraw-Hill Book Company, Inc., New York, 1967.

160. Hain, K.: "Uebertragungsguenstige unsymmetrische Doppelkurbelgetriebe," *VDI-Forschungsheft,* no. 461, supplement to *Forsch. Gebiete Ingenieurw.,* series B, vol. 23, pp. 23–25, 1957.

161. Hain, K.: "Selbsttaetige Getriebegruppen zur Automatisierung von Arbeitsvorgaengen," *Feinwerktechnik,* vol. 61, no. 9, pp. 327–329, September 1957.

162. Hain, K.: "Achtgliedrige kinematische Ketten mit dem Freiheitsgrad F = 1 fuer gegebene Kraeftever-haeltnisse," *Das Industrieblatt,* vol. 62, no. 6, pp. 331–337, June 1962.

163. Hain, K.: Beispiele zur Systematik von Spannvorrichtungen aus sechsgliedrigen kinematischen Ketten mit dem Freiheitsgrad F = 1," *Das Industrieblatt,* vol. 61, no. 12, pp. 779–784, December 1961.

164. Hain, K.: "Die Entwicklung von Spannvorrichtungen mit mehrenren Spannstellen aus kinematischen Ketten," *Das Industrieblatt,* vol. 59, no. 11, pp. 559–564, November 1959.

165. Hain, K.: "Entwurf viergliedriger kraftverstaerkender Zangen fuer gegebene Kraefteverhaeltnisse," *Das Industrieblatt,* pp. 70–73, February 1962.

166. Hain, K.: "Der Entwurf Uebertragungsguenstiger Kurbelgetriebe mit Hilfe von Kurventafeln," *VDI-Bericht,* Getriebetechnik und Ihre praktische Anwendung, vol. 29, pp. 121–128, 1958.

167. Hain, K.: "Drag Link Mechanisms," *Machine Design,* vol. 30, pp. 104–113, June 26, 1958.

168. Hain, K.: "Hydraulische Schubkolbenantriebe fuer schwierige Bewegungen," *Oelhydraulik & Pneumatik,* vol. 2, no. 6, pp. 193–199, September 1958.

169. Hain, K., and G. Marx: "How to Replace Gears by Mechanisms," *Trans. ASME,* vol. 81, pp. 126–130, May 1959.

170. Hain, K.: "Mechanisms, a 9-Step Refresher Course," (trans. F. R. E. Crossley), *Prod. Eng.,* vol. 32, 1961 (Jan. 2, 1961–Feb. 27, 1961, in 9 parts).

171. Ham, C. W., E. J. Crane, and W. L. Rogers: "Mechanics of Machinery," 4th ed., McGraw-Hill Book Company, Inc., New York, 1958.

172. Hall, A. S., and E. S. Ault: "How Acceleration Analysis Can Be Improved," *Machine Design,* vol. 15, part I, pp. 100–102, February, 1943; part II, pp. 90–92, March 1943.

173. Hall, A. S.: "Mechanism Properties," *Machine Design,* vol. 20, pp. 111–115, February 1948.

174. Hall, A. S., and D. C. Tao: "Linkage Design—A Note on One Method," *Trans. ASME,* vol. 76, no. 4, pp. 633–637, 1954.

175. Hall, A. S.: "A Novel Linkage Design Technique," *Machine Design,* vol. 31, pp. 144–151, July 9, 1959.

176. Hall, A. S.: "Kinematics and Linkage Design," Prentice-Hall, Inc., Englewood Cliffs, N.J., 1961.

177. Hannula, F. W.: "Designing Non-circular Surfaces," *Machine Design,* vol. 23, pp. 111–114, 190, 192, July 1951.

178. Handy, H. W.: "Compound Change Gear and Indexing Problems," The Machinery Publishing Co., Ltd., London.

179. Harnar, R. R.: "Automatic Drives," *Machine Design,* vol. 22, pp. 136–141, April 1950.

180. Harrisberger, L.: "Mechanization of Motion," John Wiley & Sons, Inc., New York, 1961.

181. Hartenberg, R. S.: "Complex Numbers and Four-Bar Linkages," *Machine Design,* vol. 30, pp. 156–163, Mar. 20, 1958.

182. Hartenberg, R. S., and J. Denavit: "Cognate Linkages," *Machine Design,* vol. 31, pp. 149–152, Apr. 16, 1959.

183. Hartenberg, R. S.: "Die Modellsprache in der Getriebetechnik," *VDI-Berichte,* vol. 29, pp. 109–113, 1958.

184. Hastings, C., Jr., J. T. Hayward, and J. P. Wong, Jr.: "Approximations for Digital Computers," Princeton University Press, Princeton, N.J., 1955.

185. Haug, E. J., R. Wehage, and N. C. Barman: "Design Sensitivity Analysis of Planar Mechanisms and Machine Dynamics," *J. Mechanical Design, Trans. ASME,* vol. 103, pp. 560–570, 1981.

186. Haug, E. J., and J. S. Arora: "Applied Optimal Design," John Wiley & Sons, Inc., New York, 1979.

187. Heer, E.: "Robots and Manipulators," *Mechanical Engineering,* vol. 103, no. 11, pp. 42–49, 1981.

188. Heidler, G. R.: "Spring-Loaded Differential Drive (for tensioning)," *Machine Design,* vol. 30, p. 140, Apr. 3, 1958.

189. Hekeler, C. B.: "Flexible Metal Tapes," *Prod. Eng.,* vol. 32, pp. 65–69, Feb. 20, 1961.

190. Herst, R.: "Servomechanisms (types of), *Electrical Manufacturing,* pp. 90–95, May 1950.

191. Hertrich, F. R.: "How to Balance High-Speed Mechanisms with Minimum-Inertia Counterweights," *Machine Design,* pp. 160–164, Mar. 14, 1963.

192. Hildebrand, S.: "Moderne Schreibmaschinenantriebe und Ihre Bewegungsvorgänge," *Getriebetechnik, VDI-Berichte,* vol. 5, pp. 21–29, 1955.

193. Hildebrand, F. B.: "Introduction to Numerical Analysis," McGraw-Hill Book Company, Inc., New York, 1956.

194. Hinkle, R. T.: "Kinematics of Machines," 2d ed., Prentice-Hall, Inc., Englewood Cliffs, N.J., 1960.

195. Hirschhorn, J.: "New Equations Locate Dwell Position of Three-Gear Drive," *Prod. Eng.,* vol. 30, pp. 80–81, June 8, 1959.

196. Hohenberg, F.: "Konstruktive Geometrie in der Technik," 2d ed., Springer-Verlag OHG, Vienna, 1961.

197. Horsteiner, M.: "Getriebetechnische Fragen bei der Faltschachtel-Fertigung," *Getriebetechnik, VDI-Berichte,* vol. 5, pp. 69–74, 1955.

198. Hotchkiss, C., Jr.: "Flexible Shafts," *Prod. Eng.,* vol. 26, pp. 168–177, February 1955.

199. Hrones, J. A., and G. L. Nelson: "Analysis of the Four-Bar Linkage," The Technology Press of the Massachusetts Institute of Technology, Cambridge, Mass., and John Wiley & Sons, Inc., New York, 1951.

200. Huang, R. C., E. J. Haug, and J. G. Andrews: "Sensitivity Analysis and Optimal Design of Mechanical Systems with Intermittent Motion," *J. Mechanical Design, Trans. ASME,* vol. 100, pp. 492–499, 1978.

201. Hunt, K. H.: "Mechanisms and Motion," John Wiley & Sons, Inc., New York, pp. 108, 1959; see also *Proc. Inst. Mech. Eng.* (London), vol. 174, no. 21, pp. 643–668, 1960.

202. Imam, I., G. N. Sandor, W. T. McKie, and C. W. Bobbitt: "Dynamic Analysis of a Spring Mechanism," *Proc. Second OSU Applied Mechanisms Conference,* Stillwater, Okla., pp. 31-1–31-10, Oct. 7–9, 1971.

203. *Instruments and Control Systems:* "Compensation Practice," p. 1185, August 1959.

204. Jahr, W., and P. Knechtel: "Grundzuege der Getriebelehre," Fachbuch Verlag, Leipzig, vol. 1, 1955; vol. II, 1956.

205. Jensen, P. W.: "Four-Bar Mechanisms," *Machine Design,* vol. 33, pp. 173–176, June 22, 1961.

206. Jones, H. B., Jr.: "Recording Systems," *Prod. Eng.,* vol. 26, pp. 180–185, March 1955.

207. de Jonge, A. E. R.: "Analytical Determination of Poles in the Coincidence Position of Links in Four-Bar Mechanisms Required for Valves Correctly Apportioning Three Fluids in a Chemical Apparatus," *J. Eng. Ind., Trans. ASME,* vol. 84B, pp. 359–372, August 1962.

208. de Jonge, A. E. R.: "The Correlation of Hinged Four-Bar Straight-Line Motion Devices by Means of the Roberts Theorem and a New Proof of the Latter," *Ann. N.Y. Acad. Sci.,* vol. 84, pp. 75–145, 1960 (see Refs. 68 and 69).

209. Johnson, C.: "Dynamic Sealing with O-rings," *Machine Design,* vol. 27, pp. 183–188, August 1955.

210. Johnson, H. L.: "Synthesis of the Four-Bar Linkage," M.S. dissertation, Georgia Institute of Technology, Atlanta, June 1958.

211. Johnson, R. C.: "Geneva Mechanisms," *Machine Design,* vol. 28, pp. 107–111, Mar. 22, 1956.

212. Johnson, R. C.: "Method of Finite Differences in Cam Design—Accuracy—Applications," *Machine Design,* vol. 29, pp. 159–161, Nov. 14, 1957.

213. Johnson, R. C.: "Development of a High-Speed Indexing Mechanism," *Machine Design,* vol. 30, pp. 134–138, Sept. 4, 1958.

214. Kamenskii, V. A., "On the Question of the Balancing of Plane Linkages," *Mechanisms,* vol. 3, pp. 303–322, 1968.

215. Kaplan, J., and H. North: "Cyclic Three-Gear Drives," *Machine Design,* vol. 31, pp. 185–188, Mar. 19, 1959.

216. Karplus, W. J., and W. J. Soroka: "Analog Methods in Computation and Simulation," 2d ed., McGraw-Hill Book Company, Inc., New York, 1959.

217. Kaufman, R. E.: "KINSYN: An Interactive Kinematic Design System," *Trans. Third World Congress on the Theory of Machines and Mechanisms,* Dubrovnik, Yugoslavia; September 1971.

218. Kaufman, R. E.: "KINSYN II: A Human Engineered Computer System for Kinematic Design and a New Least Squares Synthesis Operator," *J. Mechanisms and Machine Theory,* vol. 8, no. 4, pp. 469–478, 1973.

219. Kaufman, R. E., and G. N. Sandor: "Complete Force Balancing of Spatial Linkages," *J. Eng. Ind., Trans. ASME,* vol. 93B, pp. 620–626, 1971.

220. Kaufman, R. E., and G. N. Sandor: "The Bicycloidal Crank: A New Four-Link Mechanism," *J. Eng. Ind., Trans. ASME,* vol. 91B, pp. 91–96, 1969.

221. Kaufman, R. E., and G. N. Sandor: "Complete Force Balancing of Spatial Linkages," *J. Eng. Ind., Trans. ASME,* vol. 93B, pp.620–626, 1971.

222. Kaufman, R. E., and G. N. Sandor: "Operators for Kinematic Synthesis of Mechanisms by Stretch-Rotation Techniques," ASME Paper No. 70-Mech-79, *Proc. Eleventh ASME Conf. on Mechanisms,* Columbus, Ohio, Nov. 3–5, 1970.

223. Kearny, W. R., and M. G. Wright: "Straight-Line Mechanisms," *Trans. Second Conf. Mechanisms,* Purdue University, West Lafayette, Ind., pp. 209–216, Oct. 1954.

224. Keler, M. K.: "Analyse und Synthese der Raumkurbelgetriebe mittels Raumliniengeometrie und dualer Groessen," *Forsch. Gebiele Ingenieurw.,* vol. 75, pp. 26–63, 1959.

225. Kinsman, F. W.: "Controlled-Acceleration Single-Revolution Drives," *Trans. Seventh Conf. Mechanisms,* Purdue University, West Lafayette, Ind., pp. 229–233, 1962.

226. Kislitsin, S. G.: "General Tensor Methods in the Theory of Space Mechanisms" (Russian), Akad. Nauk, Moscow, *Trudi Sem. Teor. Mash. Mekh.,* vol. 14, no. 54, 1954.

227. Kist, K. E.: "Modified Starwheels," *Trans. Third Conf. Mechanisms,* Purdue University, West Lafayette, Ind., pp. 16–20, May 1956.

228. Kobrinskii, A. E.: "On the Kinetostatic Calculation of Mechanisms with Passive Constraints and with Play," Akad. Nauk, Moscow, *Trudi Sem. Teor. Mash. Mekh.,* vol. V, no. 20, pp. 5–53, 1948.

229. Kobrinskii, A. E., M. G. Breido, V. S. Gurfinkel, E. P. Polyan, Y. L. Slavitskii, A. Y. Sysin, M. L. Zetlin, and Y. S. Yacobson: "On the Investigation of Creating a Bioelectric Control System," Akad. Nauk, Moscow, *Trudi Inst. Mashinoved,* vol. 20, no. 77, pp. 39–50, 1959.

230. Kovacs, J. P., and R. Wolk: "Filters," *Machine Design,* vol. 27, pp. 167–178, Jan. 1955.

231. Kramer, S. N., and G. N. Sandor: "Finite Kinematic Synthesis of a Cycloidal-Crank Mechanism for Function Generation," *J. Eng. Ind., Trans. ASME,* vol. 92B, pp. 531–536, 1970.

232. Kramer, S. N., and G. N. Sandor: "Kinematic Synthesis of Watt's Mechanism," ASME Paper No. 70-Mech-50, *Proc. Eleventh Conf. Mechanisms,* Columbus, Ohio, Nov. 3–5, 1970.

233. Kramer, S. N., and G. N. Sandor: "Selective Precision Synthesis—A General Method of Optimization for Planar Mechanisms," *J. Eng. Ind., Trans. ASME,* vol. 97B, pp. 689–701, 1975.

234. Kraus, C. E.: "Chuting and Orientation in Automatic Handling," *Machine Design,* vol. 21, pp. 95–98, Sept. 1949.

235. Kraus, C. E., "Automatic Positioning and Inserting Devices," *Machine Design,* vol. 21, pp. 125–128, Oct. 1949.

236. Kraus, C. E., "Elements of Automatic Handling," *Machine Design,* vol. 23, pp. 142–145, Nov. 1951.

237. Kraus, R.: "Getriebelehre," vol. I, VEB Verlag Technik, Berlin, 1954.

238. Kuhlenkamp, A.: "Linkage Layouts," *Prod. Eng.,* vol. 26, no. 8, pp. 165–170, Aug. 1955.

239. Kuhn, H. S.: "Rotating Joints," *Prod. Eng.,* vol. 27, pp. 200–204, Aug. 1956.

240. Kumar, A., and K. J. Waldron: "The Workspace of a Mechanical Manipulator," *J. Mechanical Design, Trans. ASME,* vol. 103, pp. 665–672, 1981.

241. Kupfrian, W. J.: "Flexible Shafts, *Machine Design,* vol. 26, pp. 164–173, October 1954.

242. Kutzbach, K.: "Mechanische Leitungsyerzweigung, ihre Gesetze und Anwendungen, Maschinenbau," vol. 8, pp. 710–716, 1929; see also *Z. VDI,* vol. 77, p. 1168, 1933.

243. Langosch, O.: "Der Einfluss der Toleranzen auf die Genauigkeit von periodischen Getrieben," *Konstruktion,* vol. 12, no. 1, p. 35, 1960.

244. Laughner, V. H., and A. D. Hargan: "Handbook of Fastening and Joining Metal Parts," McGraw-Hill Book Company, Inc., New York, 1956.

245. Lawson, A. C.: "Jewel Bearing Systems," *Machine Design,* vol. 26, pp. 132–137, April 1954.

246. Lee, T. W., and F. Freudenstein: "Heuristic Combinatorial Optimization in the Kinematic Design of Mechanisms I, II," *J. Eng. Ind., Trans. ASME,* vol. 98B, pp. 1277–1284, 1976.

247. Lee, T. W.: "Optimization of High-Speed Geneva Mechanisms," *J. Mechanical Design, Trans. ASME,* vol. 103, pp. 621–630, 1981.

248. Levitskii, N. I.: "On the Synthesis of Plane, Hinged, Six-Link Mechanisms," pp. 98–104. See item *A,* Ref. 17, in this section of book.

249. Levitskii, N. I.: "Design of Plane Mechanisms with Lower Pairs" (Russian), Akad. Nauk, Moscow-Leningrad, 1950.

250. Levitskii, N. I., and Sh. Shakvasian: "Synthesis of Spatial Four-Link Mechanisms with Lower Paris," Akad. Nauk, Moscow, *Trudi Sem. Teor. Mash. Mekh.,* vol. 14, no. 54, pp. 5–24, 1954.

251. Lichtenheldt, W.: "Konstruktionslehre der Getriebe," Akademie Verlag, Berlin, 1964.

252. Lichtwitz, O.: "Mechanisms for Intermittent Motion," *Machine Design,* vol. 23–24, 1951–1952; I, December, pp. 134–148; II, January, pp. 127–141; III, February, pp. 146–155; IV, March, pp. 147–155.

253. Lichtwitz, O.: "Getriebe fuer Aussetzende Bewegung," Springer-Verlag OHG, Berlin, 1953.

254. Lincoln, C. W.: "A Summary of Major Developments in the Steering Mechanisms of American Automobiles," *General Motors Eng. J.,* pp. 2–7, March–April 1955.

255. Litvin, F. L.: "Design of Non-circular Gears and Their Application to Machine Design," Akad. Nauk, Moscow, *Trudi Inst. Machinoved,* vol. 14, no. 55, pp. 20–48, 1954.

256. Lockenvitz, A. E., J. B. Oliphint, W. C. Wilde, and J. M. Young: "Non-circular Cams and Gears," *Machine Design,* vol. 24, pp. 141–145, May 1952.

257. Loerch, R., A. G. Erdman, G. N. Sandor, and A. Midha: "Synthesis of Four-Bar Linkages with Specified Ground Pivots," *Proceedings, Fourth OSU Applied Mechanisms Conference,* pp. 10-1–10-6, Nov. 2–5, 1975.

258. Lowen, G. G., and R. S. Berkof: "Determination of Force-Balanced Four-Bar Linkages with Optimum Shaking Moment Characteristics," *J. Eng. Ind., Trans. ASME,* vol. 93B, pp. 39–46, 1971.

259. Lowen, G. G., and R. S. Berkof: "Survey of Investigations into the Balancing of Linkages," *J. Mechanisms,* vol. 3, pp. 221–231, 1968.

260. Lowen, G. G., F. R. Tepper, and R. S. Berkof: "The Qualitative Influence of Complete Force Balancing on the Forces and Moments of Certain Families of Four-Bar Linkages," *Mechanism and Machine Theory,* vol. 9, pp. 299–323, 1974.

261. Lundquist, I.: "Miniature Mechanical Clutches," *Machine Design,* vol. 28, pp. 124–133, Oct. 18, 1956.

262. Mabie, H. H.: "Constant Velocity Universal Joints," *Machine Design,* vol. 20, pp. 101–105, May 1948.

263. Mabie, H. H., and F. W. Ocvirk: "Mechanisms and Dynamics of Machinery," John Wiley & Sons, Inc., New York, 1957.

264. *Machine Design,* "Velocity and Acceleration Analysis of Universal Joints," vol. 14 (data sheet), pp. 93–94, November 1942.

265. *Machine Design,* "Chain Drive," vol. 23, pp. 137–138, June 1951.

266. *Machine Design,* "Constant Force Action" (manual typewriters), vol. 28, p. 98, Sept. 20, 1956.

267. *Machine Design*, vol. 30, p. 134, Feb. 20, 1958.

268. *Machine Design*, "Constant-Force Output," vol. 30, p. 115, Apr. 17, 1958.

269. *Machine Design*, "Taut-Band Suspension," vol. 30, p. 152, July 24, 1958, "Intermittent-Motion Gear Drive," vol. 30, p. 151, July 24, 1958.

270. *Machine Design*, "Almost Human Engineering," vol. 31, pp. 22–26, Apr. 30, 1959.

271. *Machine Design*, "Aircraft All-Mechanical Computer, vol. 31, pp. 186–187, May 14, 1959.

272. *Machine Design*, "Virtual Hinge Pivot," vol. 33, p. 119, July 6, 1961.

273. Mansfield, J. H.: "Woodworking Machinery," *Mech. Eng.*, vol. 74, pp. 983–995, December 1952.

274. Marich, F.: "Mechanical Timers," *Prod. Eng.*, vol. 32, pp. 54–56, July 17, 1961.

275. Marker, R. C.: "Determining Toggle-Jaw Force," *Machine Design*, vol. 22, pp. 104–107, March 1950.

276. Martin, G. H., and M. F. Spotts: "An Application of Complex Geometry to Relative Velocities and Accelerations in Mechanisms," *Trans. ASME*, vol. 79, pp. 687–693, April 1957.

277. Mathi, W. E., and C. K. Studley, Jr.: "Developing a Counting Mechanism," *Machine Design*, vol. 22, pp. 117–122, June 1950.

278. Maxwell, R. L.: "Kinematics and Dynamics of Machinery," Prentice-Hall, Inc., Englewood Cliffs, N.J., 1960.

279. Mayer, A. E.: "Koppelkurven mit drei Spitzen und Spezielle Koppelkurvenbueschel," *Z. Math. Phys.*, vol. 43, p. 389, 1937.

280. McComb, G. T., and W. N. Matson: "Four Ways to Select Change Gears—And the Faster Fifth Way," *Prod. Eng.*, vol. 31, pp. 64–67, Feb. 15, 1960.

281. Mehmke, R.: "Ueber die Bewegung eines Starren ebenen Systems in seiner Ebene," *Z. Math. Phys.*, vol. 35, pp. 1–23, 65–81, 1890.

282. Merritt, H. E.: "Gear Trains," Sir Isaac Pitman & Sons, Ltd., London, 1947.

283. Meyer zur Capellen, W.: "Getriebependel," *Z. Instrumentenk.*, vol. 55, 1935; pt. I, October, pp. 393–408; pt. II, November, pp. 437–447.

284. Meyer zur Capellen, W.: "Die Totlagen des ebenen, Gelenkvierecks in analytischer Darstellung," *Forsch. Ing. Wes.*, vol. 22, no. 2, pp. 42–50, 1956.

285. Meyer zur Capellen, W.: "Der einfache Zahnstangen-Kurbeltrieb und das entsprechende Bandgetriebe," *Werkstatt u. Betrieb*, vol. 89, no. 2, pp. 67–74, 1956.

286. Meyer zur Capellen, W.: "Harmonische Analyse bei der Kurbelschleife," *Z. angew., Math. u. Mechanik*, vol. 36, pp. 151–152, March–April 1956.

287. Meyer zur Capellen, W.: "Kinematik des Einfachen Koppelraedertriebes," *Werkstatt u. Betrieb*, vol. 89, no. 5, pp. 263–266, 1956.

288. Meyer zur Capellen, W.: "Bemerkung zum Satz von Roberts über die Dreifache Erzeugung der Koppelkurve," *Konstruktion*, vol. 8, no. 7, pp. 268–270, 1956.

289. Meyer zur Capellen, W.: "Kinematik und Dynamik der Kurbelschleife," *Werkstatt u. Betrieb*, pt. I, vol. 89, no. 10, pp. 581–584, 1956; pt. II, vol. 89, no. 12, pp. 677–683, 1956.

290. Meyer zur Capellen, W.: "Ueber gleichwertige periodische Getriebe," *Fette, Seifen, Anstrichmittel*, vol. 59, no. 4, pp. 257–266, 1957.

291. Meyer zur Capellen, W.: "Die Kurbelschleife Zweiter Art," *Werkstatt u. Betrieb*, vol. 90, no. 5, pp. 306–308, 1957.

292. Meyer zur Capellen, W.: "Die elliptische Zahnraeder und die Kurbelschleife," *Werkstatt u. Betrieb*, vol. 91, no. 1, pp. 41–45, 1958.

293. Meyer zur Capellen, W.: "Die harmonische Analyse bei elliptischen Kurbelschleifen," *Z. angew, Math. Mechanik*, vol. 38, no. 1/2, pp. 43–55, 1958.

294. Meyer zur Capellen, W.: "Das Kreuzgelenk als periodisches Getriebe," *Werkstatt u. Betrieb*, vol. 91, no. 7, pp. 435–444, 1958.

295. Meyer zur Capellen, W.: "Ueber elliptische Kurbelschleifen," *Werkstatt u. Betrieb*, vol. 91, no. 12, pp. 723–729, 1958.

296. Meyer zur Capellen, W.: "Bewegungsverhaeltnisse an der geschraenkten Schubkurbel," *Forchungs berichte des Landes Nordrhein-Westfalen, West-Deutscher Verlag,* Cologne, no. 449, 1958.

297. Meyer zur Capellen, W.: "Eine Getriebergruppe mit stationaerem Geschwindigkeitsver-lauf" (Elliptic Slidercrank Drive) (in German) *Forschungsberichte des Landes Nordrhein-Westfalen,* West-Deutscher Verlag, Cologne, no. 606, 1958.

298. Meyer zur Capellen, W.: "Die Beschleunigungsaenderung," *Ing-Arch.,* vol. 27, pt. I, no. 1, pp. 53–65; pt. II, no. 2, pp. 73–87, 1959.

299. Meyer zur Capellen, W.: "Die geschraenkte Kurbelschleife zweiter Art," *Werkstatt u. Betrieb,* vol. 92, no. 10, pp. 773–777, 1959.

300. Meyer zur Capellen, W.: "Harmonische Analyse bei Kurbeltrieben," *Forschungsberichte des Landes Nordrhein-Westfalen,* West-Deutscher Verlag, Cologne, pt. I, no. 676, 1959; pt. II, no. 803, 1960.

301. Meyer zur Capellen, W.: "Die geschraenkte Kurbelschleife," *Forschungsberichte des Landes Nordrhein-Westfalen,* West-Deutscher Verlag, Cologne, pt. I, no. 718, 1959; pt. II, no. 804, 1960.

302. Meyer zur Capellen, W.: "Die Extrema der Uebersetzungen in ebenen und sphaerischen Kurbeltrieben," *Ing. Arch.,* vol. 27, no. 5, pp. 352–364, 1960.

303. Meyer zur Capellen, W.: "Die gleichschenklige zentrische Kurbelschwinge," *Z. Prakt. Metallbearbeitung,* vol. 54, no. 7, pp. 305–310, 1960.

304. Meyer zur Capellen, W.: "Three-Dimensional Drives," *Prod. Eng.,* vol. 31, pp. 76–80, June 30, 1960.

305. Meyer zur Capellen, W.: "Kinematik der sphaerischen Schubkurbel," *Forschungsberichte des Landes Nordrhein-Westfalen,* West-Deutscher Verlag, Cologne, no. 873, 1960.

306. Michalec, G. W.: "Analog Computing Mechanisms," *Machine Design,* vol. 31, pp. 157–179, Mar. 19, 1959.

307. Miller, H.: "Analysis of Quadric-Chain Mechanisms," *Prod. Eng.,* vol. 22, pp. 109–113, February 1951.

308. Miller, W. S.: "Packaged Speed Reduces and Gearmotors," *Machine Design,* vol. 29, pp. 121–149, Mar. 21, 1957.

309. Modrey, J.: "Analysis of Complex Kinematic Chains with Influence Coefficients," *J. Appl. Mech.,* vol. 26; *Trans. ASME,* vol. 81, E., pp. 184–188, June 1959; discussion, *J. Appl. Mech.,* vol. 27, pp. 215–216, March, 1960.

310. Moore, J. W., and M. V. Braunagel: "Space Linkages," *Trans. Seventh Conf. Mech.,* Purdue University, West Lafayette, Ind., pp. 114–122, 1962.

311. Moreinis, I. Sh.: "Biomechanical Studies of Some Aspects of Walking on a Prosthetic Device" (Russian), Akad. Nauk, Moscow, *Trudi Inst. Machinoved.,* vol. 21, no. 81–82, pp. 119–131, 1960.

312. Morgan, P.: "Mechanisms for Moppets," *Machine Design,* vol. 34, pp. 105–109, Dec. 20, 1962.

313. Moroshkin, Y. F.: "General Analytical Theory of Mechanisms" (Russian), Akad. Nauk, Moscow, *Trudi Sem. Teor. Mash. Mekh.,* vol. 14, no. 54, pp. 25–50, 1954.

314. Morrison, R. A.: "Rolling-Surface Mechanisms," *Machine Design,* vol. 30, pp. 119–123, Dec. 11, 1958.

315. Mueller, R.: "Einfuehrung in die theoretische Kinematik," Springer-Verlag OHG, Berlin, 1932.

316. Musser, C. W.: "Mechanics Is Not a Closed Book," *Trans. Sixth Conf. Mechanisms,* Purdue University, West Lafayette, Ind., October, 1960, pp. 31–43.

317. Oldham, K., and J. N. Fawcett: "Computer-Aided Synthesis of Linkage—A Motorcycle Design Study," *Proc. Institution of Mechanical Engineers* (London), vol. 190, no. 63/76, pp. 713–720, 1976.

318. Olsson, V.: "Non-circular Cylindrical Gears," *Acta Polytech., Mech. Eng.* Ser., vol. 2, no. 10, Stockholm, 1959.

319. Olsson, V.: "Non-circular Bevel Gears," *Acta Polytech., Mech. Eng. Ser., 5,* Stockholm, 1953.

320. Orlandea, N., and T. Berenyi: "Dynamic Continuous Path Synthesis of Industrial Robots Using ADAMS Computer Program," *J. Mechanical Design, Trans. ASME,* vol. 103, pp. 602–607, 1981.

321. Paul, B.: "A Unified Criterion for the Degree of Constraint of Plane Kinematic Chains," *J. Appl. Mech.,* vol. 27; *Trans. ASME,* ser. E., vol. 82, pp. 196–200, March 1960.

322. Paul, B.: "Analytical Dynamics of Mechanisms—A Computer Oriented Overview," *J. Mechanisms and Machine Theory,* vol. 10, no. 6, pp. 481–508, 1975.

323. Paul, B.: "Kinematics and Dynamics of Planar Machinery," Prentice-Hall, Inc., Englewood Cliffs, N.J., 1979.

324. Paul, B., and D. Krajcinovic: "Computer Analysis of Machines with Planar Motion, I, II," *J. Appl. Mech.,* vol. 37, pp. 697–712, 1979.

325. Peek, H. L.: "Trip-Free Mechanisms," *Mech. Eng.,* vol. 81, pp. 193–199, March 1959.

326. Peyrebrune, H. E.: "Application and Design of Non-circular Gears," *Trans. First Conf. Mechanisms,* Purdue University, West Lafayette, Ind., pp. 13–21, 1953.

327. Philipp, R. E.: "Kinematics of a General Arrangement of Two Hooke's Joints," ASME paper 60-WA-37, 1960.

328. Pike, E. W., and T. R. Silverberg: "Designing Mechanical Computers," *Machine Design,* vol. 24, pt. I, pp. 131–137, July 1952; pt. II, pp. 159–163, August 1952.

329. Pollitt, E. P.: "High-Speed Web-Cutting," *Machine Design,* vol. 27, pp. 155–160, December 1955.

330. Pollitt, E. P.: "Motion Characteristics of Slider-Crank Linkages," *Machine Design,* vol. 30, pp. 136–142, May 15, 1958.

331. Pollitt, E. P.: "Some Applications of the Cycloid in Machine Design," Trans. ASME, *J. Eng. Ind.,* ser. B, vol. 82, no. 4, pp. 407–414, November 1960.

332. Predale, J. O., and A. B. Hulse, Jr.: "The Space Crank," *Prod. Eng.,* vol. 30, pp. 50–53, Mar. 2, 1959.

333. Primrose, E. J. F., and F. Freudenstein: "Geared Five-Bar Motion II—Arbitrary Commensurate Gear Ratio," *J. Appl. Mech.,* vol. 30; *Trans. ASME,* ser. E., vol. 85, pp. 170–175, June 1963.

334. Primrose, E. J. F., F. Freudenstein, and G. N. Sandor: "Finite Burmester Theory in Plane Kinematics," *J. Appl. Mech., Trans. ASME,* vol. 31E, pp. 683–693, 1964.

335. Procopi, J.: "Control Valves," *Machine Design,* vol. 22, pp. 153–155, September 1950.

336. Proctor, J.: "Selecting Clutches for Mechanical Drives," *Prod. Eng.,* vol. 32, pp. 43–58, June 19, 1961.

337. *Prod. Eng.:* "Mechanisms Actuated by Air or Hydraulic Cylinders," vol. 20, pp. 128–129, December 1949.

338. *Prod. Eng.:* "Computing Mechanisms," vol. 27, I, p. 200, Mar.; II, pp. 180–181, April 1956.

339. *Prod. Eng.:* "High-Speed Electrostatic Clutch," vol. 28, pp. 189–191, February 1957.

340. *Prod. Eng.:* "Linkage Keeps Table Flat," vol. 29, p. 63, Feb. 3, 1958.

341. *Prod. Eng.:* vol. 30, pp. 64–65, Mar. 30, 1959.

342. *Prod. Eng.:* "Down to Earth with a Four-Bar Linkage," vol. 31, p. 71, June 22, 1959.

343. *Prod. Eng.:* "Design Work Sheets," no. 14.

344. Radcliffe, C. W.: "Prosthetic Mechanisms for Leg Amputees," *Trans. Sixth Conf. Mechanisms,* Purdue University, West Lafayette, Ind., pp. 143–151, October 1960.

345. Radcliffe, C. W.: "Biomechanical Design of a Lower-Extremity Prosthesis," ASME Paper 60-WA-305, 1960.

346. Rainey, R. S.: "Which Shaft Seal," *Prod. Eng.,* vol. 21, pp. 142–147, May 1950.

347. Rankers, H.: "Vier genau gleichwertige Gelenkgetriebe für die gleiche Koppelkurve," *Des Industrieblatt,* pp. 17–21, January 1959.

348. Rankers, H.: "Anwendungen von sechsgliedrigen Kurbelgetrieben mit Antrieb an einem Koppelpunkt," *Das Industrieblatt,* pp. 78–83, February 1962.

349. Rankers, H.: "Bewegungsverhaeltnisse an der Schubkurbel mit angeschlossenem Kreuzschieber," *Das Industrieblatt,* pp. 790–796, December 1961.

350. Rantsch, E. J.: "Elliptic Gears Depend on Accurate Layout," *Machine Design,* vol. 9, pp. 43–44, March 1937.

351. Rappaport, S.: "A Neglected Design Detail," *Machine Design,* vol. 20, p. 140, September 1948.

352. Rappaport, S.: "Kinematics of Intermittent Mechanisms," *Prod. Eng.,* vols. 20–21, 1949–1951, I, "The External Geneva Wheel," July, pp. 110–112; II, "The Internal Geneva Wheel," August, pp. 109–112; III, "The Spherical Geneva Wheel," October, pp. 137–139; IV, "The Three-Gear Drive," January, 1950, pp. 120–123; V, "The Cardioid Drive," 1950, pp. 133–134.

353. Rappaport, S.: "Crank-and-Slot Drive," *Prod. Eng.,* vol. 21, pp. 136–138, July 1950.

354. Rappaport, S.: "Shearing Moving Webs," *Machine Design,* vol. 28, pp. 101–104, May 3, 1956.

355. Rappaport, S.: "Small Indexing Mechanisms," *Machine Design,* vol. 29, pp. 161–163, Apr. 18, 1957.

356. Rappaport, S.: "Elliptical Gears for Cyclic Speed Variation," *Prod. Eng.,* vol. 31, pp. 68–70, Mar. 28, 1960.

357. Rappaport, S.: "Intermittent Motions and Special Mechanisms," *Trans. Conf. Mechanisms,* Yale University, Shoestring Press, New Haven, Conn., pp. 91–122, 1961.

358. Rappaport, S.: "Review of Mechanical Integrators," *Trans. Seventh Conf. Mechanics,* Purdue University, West Lafayette, Ind., pp. 234–240, 1962.

359. Rasche, W. H.: "Design Formulas for Three-Link Screw Mechanisms," *Machine Design,* vol. 17, pp. 147–149, August 1945.

360. Rasche, W. H.: "Gear Train Design," *Virginia Polytechnic Inst. Eng. Experiment Station Bull.,* 14, 1933.

361. Raven, F. H.: "Velocity and Acceleration Analysis of Plane and Space Mechanisms by Means of Independent-Position Equations," *J. Appl. Mech.,* vol. 25, March 1958; *Trans. ASME,* vol. 80, pp. 1–6.

362. Reuleaux, F.: "The Constructor" (trans. H. H. Suplee), D. Van Nostrand Company, Inc., Princeton, N.J., 1983.

363. Richardson, I. H.: "Trend Toward Automation in Automatic Weighing and Bulk Materials Handling," *Mech. Eng.,* vol. 75, pp. 865–870, November 1953.

364. Ring, F.: "Remote Control Handling Devices," *Mech. Eng.,* vol. 78, pp. 828–831, September 1956.

365. Roemer, R. L.: Flight-Control Linkages, *Mech. Eng.,* vol. 80, pp. 56–60, June 1958.

366. Roger, R. J., and G. C. Andrews: "Simulating Planar Systems Using a Simplified Vector-Network Method," *J. Mechanisms and Machine Theory,* vol. 16, no. 6, pp. 509–519, 1975.

367. Root, R. E., Jr., "Dynamics of Engine and Shaft," John Wiley & Sons, Inc., New York, 1932.

368. Rosenauer, N., and A. H. Willis: "Kinematics of Mechanisms," Associated General Publications, Pty. Ltd., Sydney, Australia, 1953.

369. Rosenauer, N.: "Some Fundamentals of Space Mechanisms," *Mathematical Gazette,* vol. 40, no. 334, pp. 256–259, December 1956.

370. Rossner, E. E.: "Ratchet Layout," *Prod. Eng.,* vol. 29, pp. 89–91, Jan. 20, 1958.

371. Roth, B., F. Freudenstein, and G. N. Sandor: "Synthesis of Four-Link Path Generating Mechanisms with Optimum Transmission Characteristics," *Trans. Seventh Conf. Mechanics,* Purdue University, West Lafayette, Ind., pp. 44–48, October 1962 (available from *Machine Design,* Penton Bldg., Cleveland).

372. Roth, B., and F. Freudenstein: "Synthesis of Path Generating Mechanisms by Numerical Methods," *J. Eng. Ind., Trans. ASME,* vol. 85B, pp. 298–306, August 1963.

373. Roth, B.: "Robots," *Appl. Mechanics Rev.*, vol. 31, no. 11, pp. 1511–1519, 1978.

374. Roth, G. L.: "Modifying Valve Characteristics," *Prod. Eng.*, Annual Handbook, pp. 16–18, 1958.

375. Rothbart, H. A.: "Cams," John Wiley & Sons, Inc., New York, 1956.

376. Rothbart, H. A.: "Equivalent Mechanisms for Cams," *Machine Design*, vol. 30, pp. 175–180, Mar. 20, 1958.

377. Rumsey, R. D.: "Redesigned for Higher Speed," *Machine Design*, vol. 23, pp. 123–129, April 1951.

378. Rzeppa, A. H.: "Universal Joint Drives," *Machine Design*, vol. 25, pp. 162–170, April 1953.

379. Saari, O.: "Universal Joints," *Machine Design*, vol. 26, pp. 175–178, October 1954.

379*a*. Sadler J. P., and R. W. Mayne: "Balancing of Mechanisms by Nonlinear Programming," *Proc. Third Applied Mechanisms Conf.*, 1973.

380. Sandor, G. N., and F. Freudenstein: "Kinematic Synthesis of Path-Generating Mechanisms by Means of the IBM 650 Computer," Program 9.5.003, IBM Library, Applied Programming Publications, IBM, 590 Madison Avenue, New York, N.Y. 10022, 1958.

381. Sandor, G. N.: "A General Complex-Number Method for Plane Kinematic Synthesis with Applications," doctoral dissertation, Columbia University, University Microfilms, Ann Arbor, Mich., 1959, Library of Congress Card No. Mic. 59-2596.

382. Sandor, G. N.: "On the Kinematics of Rubber-Covered Cylinders Rolling on a Hard Surface," ASME Paper 61-SA-67, Abstr., *Mech. Engrg.*, vol. 83, no. 10, p. 84, October 1961.

383. Sandor, G. N.: "On Computer-Aided Graphical Kinematics Synthesis," Technical Seminar Series, Rep. 4, Princeton University, Dept. of Graphics and Engineering Drawing, Princeton, N.J., 1962.

384. Sandor, G. N.: "On the Loop Equations in Kinematics," *Trans. Seventh Conf. Mechanisms*, Purdue University, West Lafayette, Ind., pp. 49–56, 1962.

385. Sandor, G. N.: "On the Existence of a Cycloidal Burmester Theory in Planar Kinematics," *J. Appl. Mech.*, vol. 31, *Trans. ASME*, vol. 86E, 1964.

386. Sandor, G. N.: "On Infinitesimal Cycloidal Kinematic Theory of Planar Motion," *J. Appl. Mech., Trans. ASME*, vol. 33E, pp. 927–933, 1966.

387. Sandor, G. N., and F. Freudenstein: "Higher-Order Plane Motion Theories in Kinematic Synthesis," *J. Eng. Ind., Trans. ASME*, vol. 89B, pp. 223–230, 1967.

388. Sandor, G. N.: "Principles of a General Quaternion-Operator Method of Spatial Kinematic Synthesis," *J. Appl. Mech., Trans. ASME*, vol. 35E, pp. 40–46, 1968.

389. Sandor, G. N., and K. E. Bisshopp: "On a General Method of Spatial Kinematic Synthesis by Means of a Stretch-Rotation Tensor," *J. Eng. Ind., Trans. ASME*, vol. 91B, pp. 115–122, 1969.

390. Sandor, G. N., with D. R. Wilt: "Synthesis of a Geared Four-Link Mechanism," *Proc. Second International Congress Theory of Machines and Mechanics*, vol. 2, Zakopane, Poland, 1969, pp. 222–232, Sept. 24–27; *J. Mechanisms*, vol. 4, pp. 291–302, 1969.

391. Sandor, G. N., et al.: "Kinematic Synthesis of Geared Linkages," *J. Mechanisms*, vol. 5, pp. 58–87, 1970, *Rev. Rumanian Sci. Tech.-Mech. Appl.*, vol. 15, Bucharest, pp. 841–869, 1970.

392. Sandor, G. N., with A. V. M. Rao: "Extension of Freudenstein's Equation to Geared Linkages," *J. Eng. Ind., Trans. ASME*, vol. 93B, pp. 201–210, 1971.

393. Sandor, G. N., A. G. Erdman, L. Hunt, and E. Raghavacharyulu: "New Complex-Number Forms of the Euler-Savary Equation in a Computer-Oriented Treatment of Planar Path-Curvature Theory for Higher-Pair Rolling contact, *J. Mech. Design, Trans. ASME*, vol. 104, pp. 227–232, 1982.

394. Sandor, G. N., A. G. Erdman, and E. Raghavacharyulu: "A Note on Bobillier Constructions," in preparation.

395. Sandor, G. N., with A. V. Mohan Rao and Steven N. Kramer: "Geared Six-Bar Design," *Proc. Second OSU Applied Mechanisms Conference*, Stillwater, Okla, pp. 25-1–25-13, Oct. 7–9, 1971.

395*a*. Sandor, G. N., A. G. Erdman, L. Hunt, and E. Raghavacharyulu: "New Complex-Number Form of the Cubic of Stationary Curvature in a Computer-Oriented Treatment of Planar Path-Curvature Theory for Higher-Pair Rolling Contact," *J. Mech. Design, Trans. ASME,* vol. 104, pp. 233–238, 1982.

396. Sandor, G. N., with A. V. M. Rao: "Extension of Freudenstein's Equation to Geared Linkages," *J. Eng. Ind., Trans. ASME,* vol. 93B, pp. 201–210, 1971.

397. Sandor, G. N., with A. G. Erdman: "Kinematic Synthesis of a Geared Five-Bar Function Generator," *J. Eng. Ind., Trans. ASME,* vol. 93B, pp. 11–16, 1971.

398. Sandor, G. N., with A. D. Dimarogonas and A. G. Erdman: "Synthesis of a Geared *N*-bar Linkage," *J. Eng. Ind., Trans. ASME,* vol. 93B, pp. 157–164, 1971.

399. Sandor, G. N., with A. V. M. Rao, and G. A. Erdman: "A General Complex-Number Method of Synthesis and Analysis of Mechanisms Containing Prismatic and Revolute Pairs," *Proc. Third World Congress on the Theory of Machines and Mechanisms,* vol. D, Dubrovnik, Yugoslavia, pp. 237–249, Sept. 13–19, 1971.

400. Sandor, G. N. et al.: "Synthesis of Multi-Loop, Dual-Purpose Planar Mechanisms Utilizing Burmester Theory," *Proc. Second OSU Applied Mechanisms Conf.,* Stillwater, Okla., pp. 7-1–7-23, Oct. 7–9, 1971.

401. Sandor, G. N., with A. D. Dimarogonas: "A General Method for Analysis of Mechanical Systems," *Proc. Third World Congress for the Theory of Machines and Mechanisms,* Dubrovnik, Yugoslavia, pp. 121–132, Sept. 13–19, 1971.

402. Sandor, G. N., with A. V. M. Rao: "Closed Form Synthesis of Four-Bar Path Generators by Linear Superposition," *Proc. Third World Congress on the Theory of Machines and Mechanisms,* Dubrovnik, Yugoslavia, pp. 383–394, Sept. 13–19, 1971.

403. Sandor, G. N., with A. V. M. Rao: "Closed Form Synthesis of Four-Bar Function Generators by Linear Superposition," *Proc. Third World Congress on the Theory of Machines and Mechanisms,* Dubrovnik, Yugoslavia, pp. 395–405, Sept. 13–19, 1971.

404. Sandor, G. N., with R. S. Rose: "Direct Analytic Synthesis of Four-Bar Function Generators with Optimal Structural Error," *J. Eng. Ind., Trans. ASME,* vol. 95B, pp. 563–571, 1973.

405. Sandor, G. N., with A. V. M. Rao, and J. C. Kopanias: "Closed-form Synthesis of Planar Single-Loop Mechanisms for Coordination of Diagonally Opposite Angles," *Proc. Mechanisms 1972 Conf.,* London, Sept. 5–7, 1972.

406. Sandor, G. N., A. V. M. Rao, D. Kohli, and A. H. Soni: "Closed Form Synthesis of Spatial Function Generating Mechanisms for the Maximum Number of Precision Points," *J. Eng. Ind., Trans. ASME,* vol. 95B, pp. 725–736, 1973.

407. Sandor, G. N., with J. F. McGovern: "Kinematic Synthesis of Adjustable Mechanisms," Part 1, "Function Generation," *J. Eng. Ind., Trans. ASME,* vol. 95B, pp. 417–422, 1973.

408. Sandor, G. N., with J. F. McGovern: "Kinematic Synthesis of Adjustable Mechanisms," Part 2, "Path Generation," *J. Eng. Ind., Trans. ASME,* vol. 95B, pp. 423–429, 1973.

409. Sandor, G. N., with Dan Perju: "Contributions to the Kinematic Synthesis of Adjustable Mechanisms," *Trans. International Symposium on Linkages and Computer Design Methods,* vol. A-46, Bucharest, pp. 636–650, June 7–13, 1973.

410. Sandor, G. N., with A. V. M. Rao: "Synthesis of Function Generating Mechanisms with Scale Factors as Unknown Design Parameters," *Trans. International Symposium on Linkages and Computer Design Methods,* vol. A-44, Bucharest, Romania, pp. 602–623, 1973.

411. Sandor, G. N., J. F. McGovern, and C. Z. Smith: "The Design of Four-Bar Path Generating Linkages by Fifth-Order Path Approximation in the Vicinity of a Single Point," *Proc. Mechanisms 73 Conf.,* University of Newcastle upon Tyne, England, Sept. 11, 1973: *Proc. Institution of Mechanical Engineers* (London), pp. 65–77.

412. Sandor, G. N., with R. Alizade and I. G. Novrusbekov: "Optimization of Four-Bar Function Generating Mechanisms Using Penalty Functions with Inequality and Equality Constraints," *Mechanism and Machine Theory,* vol. 10, no. 4, pp. 327–336, 1975.

413. Sandor, G. N., with R. I. Alizade and A. V. M. Rao: "Optimum Synthesis of Four-Bar and Offset Slider-Crank Planar and Spatial Mechanisms Using the Penalty Function Approach with Inequality and Equality Constraints," *J. Eng. Ind., Trans. ASME,* vol. 97B, pp. 785–790, 1975.

414. Sandor, G. N., with R. I. Alizade and A. V. M. Rao: "Optimum Synthesis of Two-Degree-of-Freedom Planar and Spatial Function Generating Mechanisms Using the Penalty Function Approach," *J. Eng. Ind., Trans. ASME,* vol. 97B, pp. 629–634, 1975.

415. Sandor, G. N., with I. Imam: "High-Speed Mechanism Design—A General Analytical Approach," *J. Eng. Ind., Trans. ASME,* vol. 97B, pp. 609–629, 1975.

416. Sandor, G. N., with D. Kohli: "Elastodynamics of Planar Linkages Including Torsional Vibrations of Input and Output Shafts and Elastic Deflections at Supports," *Proc. Fourth World Congress on the Theory of Machines and Mechanisms,* vol. 2, University of Newcastle upon Tyne, England, pp. 247–252, Sept. 8–13, 1975.

417. Sandor, G. N., with D. Kohli: "Lumped-Parameter Approach for Kineto-Elastodynamic Analysis of Elastic Spatial Mechanisms," *Proc. Fourth World Congress on the Theory of Machines and Mechanisms,* vol. 2, University of Newcastle upon Tyne, England, pp. 253–258, Sept. 8–13, 1975.

418. Sandor, G. N., with S. Dhande: "Analytical Design of Cam-Type Angular-Motion Compensators," *J. Eng., Ind., Trans. ASME,* vol. 99B, pp. 381–387, 1977.

419. Sandor, G. N., with C. F. Reinholtz, and S. G. Dhande: "Kinematic Analysis of Planar Higher-Pair Mechanisms," *Mechanism and Machine Theory,* vol. 13, pp. 619–629, 1978.

420. Sandor, G. N., with R. J. Loerch and A. G. Erdman: "On the Existence of Circle-Point and Center-Point Circles for Three-Precision-Point Dyad Synthesis," *J. Mechanical Design, Trans. ASME,* Oct. 1979, pp. 554–562.

421. Sandor, G. N., with R. Pryor: "On the Classification and Enumeration of Six-Link and Eight-Link Cam-Modulated Linkages," Paper No. USA-66, *Proc. Fifth World Congress on the Theory of Machines and Mechanisms,* Montreal, July, 1979.

421*a.* Sandor, G. N., et al.: "Computer Aided Design of Spatial Mechanisms," (CADSPAM), Interactive Computer Package for Dimensional Synthesis of Function, Path and Motion Generator Spatial Mechanisms, in preparation.

421*b.* Sandor, G. N., D. Kohli, and M. Hernandez: "Closed Form Analytic Synthesis of R-Sp-R Three-Link Function Generator for Multiply Separated Positions," ASME Paper No. 82-DET-75, May 1982.

421*c.* Sandor, G. N., D. Kohli, C. F. Reinholtz, and A. Ghosal: "Closed-Form Analytic Synthesis of a Five-Link Spatial Motion Generator," *Proc. Seventh Applied Mechanisms Conf.,* Kansas City, Mo., pp. XXVI-1–XXVI-7, 1981, "Mechanism and Machine Theory," vol. 19, no. 1, pp. 97–105, 1984.

421*d.* Sandor, G. N., D. Kohli, and Zhuang Xirong: "Synthesis of a Five-Link Spatial Motion Generator for Four Prescribed Finite Positions," submitted to the ASME Design Engineering Division, 1982.

421*e.* Sandor, G. N., D. Kohli, and Zhuang Xirong: "Synthesis of RSSR-SRR Spatial Motion Generator Mechanism with Prescribed Crank Rotations for Three and Four Finite Positions," submitted to the ASME Design Engineering Division, 1982.

421*f.* Sandor, G. N., and A. G. Erdman: "Advanced Mechanism Design Analysis and Synthesis," vol. 2, Prentice-Hall, Inc., Englewood Cliffs, N.J., 1984.

421*g.* Sandor, G. N., A. G. Erdman, E. Raghavacharyulu, and C. F. Reinholtz: "On the Equivalence of Higher and Lower Pair Planar Mechanisms," *Proc. Sixth World Congress on the Theory of Machines and Mechanisms,* Delhi, India, Dec. 15–20, 1983.

422. Saxon, A. F.: "Multistage Sealing," *Machine Design,* vol. 25, pp. 170–172, March, 1953.

423. Soled, J.: "Industrial Fasteners," *Machine Design,* vol. 28, pp. 105–136, Aug. 23, 1956.

424. Schashkin, A. S.: "Study of an Epicyclic Mechanism with Dwell," Ref. 17*B,* pp. 117–132.

425. Schmidt, E. H.: "Cyclic Variations in Speed," *Machine Design,* vol. 19, pp. 108–111, March 1947.

426. Schmidt, E. H.: "Cycloidal-Crank Mechanisms," *Machine Design,* vol. 31, pp. 111–114, Apr. 2, 1959.

427. Schulze, E. F. C.: "Designing Snap-Action Toggles," *Prod. Eng.,* vol. 26, pp. 168–170, November 1955.

428. Shaffer, B. W., and I. Krause: "Refinement of Finite Difference Calculations in Kinematic Analysis," *Trans. ASME,* vol. 82B, no. 4, pp. 377–381, November 1960.

429. Sheppard, W. H.: "Rolling Curves and Non-circular Gears," *Mech. World,* pp. 5–11, January 1960.

430. Sheth, P. N., and J. J. Uicker: "IMP—A Computer-Aided Design Analysis System for Mechanisms and Linkages," *J. Eng. Ind., Trans. ASME,* vol. 94B, pp. 454–464, 1972.

431. Shigley, J., and J. J. Uicker: "Theory of Machines and Mechanisms," McGraw-Hill Book Company, Inc., New York, 1980.

432. Sieker, K. H.: "Kurbelgetriebe-Rechnerische Verfahren," *VDI-Bildungswerk,* no. 077 (probably 1960–1961).

433. Sieker, K. H.: "Extremwerte der Winkelgeschwindigkeiten in Symmetrischen Doppelkurbeln," *Konstruktion,* vol. 13, no. 9, pp. 351–353, 1961.

434. Sieker, K. H.: "Getriebe mit Energiespeichern," C. F. Winterische Verlagshandlung, Fussen, 1954.

435. Sieker, K. H.: "Zur algebraischen Mass-Synthese ebener Kurbelgetriebe," *Ing. Arch.,* vol. 24, pt. I, no. 3, pp. 188–215, pt. II, no. 4, pp. 233–257, 1956.

436. Sieker, K. H.: "Winkelgeschwindigkeiten und Winkelbeschleunigungen in Kurbelschleifen," *Feinverktechnik,* vol. 64, no. 6, pp. 1–9, 1960.

437. Simonis, F. W.: "Stufenlos verstellbare Getriebe," Werkstattbuecher no. 96, Springer-Verlag OHG, Berlin, 1949.

438. Sloan, W. W.: "Utilizing Irregular Gears for Inertia Control," *Trans. First Conf. Mechanisms,* Purdue University, West Lafayette, Ind., pp. 21–24, 1953.

439. Soni, A. H., et al.: "Linkage Design Handbook," ASME, New York, 1977.

440. Spector, L. F.: "Flexible Couplings," *Machine Design,* vol. 30, pp. 101–128, Oct. 30, 1958.

441. Spector, L. F.: "Mechanical Adjustable-Speed Drives," *Machine Design,* vol. 27, I, April, pp. 163–196; II, June, pp. 178–189, 1955.

442. Spotts, M. F.: "Kinematic Properties of the Three-Gear Drive," *J. Franklin Inst.,* vol. 268, no. 6, pp. 464–473, December 1959.

443. Strasser, F.: "Ten Universal Shaft-Couplings," *Prod. Eng.,* vol. 29, pp. 80–81, Aug. 18, 1958.

444. Sugimoto, K., and J. Duffy: "Determination of Extreme Distances of a Robot Hand—I: A General Theory," *J. Mechanical Design, Trans. ASME,* vol. 103, pp. 631–636, 1981.

445. Suh, C. H., and C. W. Radcliffe: "Kinematics and Mechanisms Design," John Wiley & Sons, Inc., New York, 1978.

446. Svoboda, A.: "Computing Mechanisms and Linkages," MIT Radiation Laboratory Series, vol. 27, McGraw-Hill Book Company, Inc., New York, 1948.

447. Taborek, J. J.: "Mechanics of Vehicles 3, Steering Forces and Stability," *Machine Design,* vol. 29, pp. 92–100, June 27, 1957.

448. Talbourdet, G. J.: "Mathematical Solution of Four-Bar Linkages," *Machine Design,* vol. 13, I, II, no. 5, pp. 65–68; III, no. 6, pp. 81–82; IV, no. 7, pp. 73–77, 1941.

449. Talbourdet, G. J.: "Intermittent Mechanisms (data sheets)," *Machine Design,* vol. 20, pt. I, September, pp. 159–162; pt. II, October, pp. 135–138, 1948.

450. Talbourdet, G. J.: "Motion Analysis of Several Intermittent Variable-Speed Drives," *Trans. ASME,* vol. 71, pp. 83–96, 1949.

451. Talbourdet, G. J.: "Intermittent Mechanisms," *Machine Design,* vol. 22, pt. I, September, pp. 141–146, pt. II, October, pp. 121–125, 1950.

452. Tepper, F. R., and G. G. Lowen: "On the Distribution of the RMS Shaking Moment of Unbalanced Planar Mechanisms: Theory of Isomomental Ellipses," ASME technical paper 72-Mech-4, 1972.

452*a.* Tepper, F. R., and G. G. Lowen: "General Theorems Concerning Full Force Balancing of Planar Linkages by Internal Mass Redistribution," *J. Eng. Ind., Trans. ASME,* vol. 94B, pp. 789–796, 1972.

452*b*. Tepper, F. R., and G. G. Lowen: "A New Criterion for Evaluating the RMS Shaking Moment in Unbalanced Planar Mechanisms," *Proc. Third Applied Mechanisms Conf.*, 1973.

453. Tesar, D.: "Translations of Papers (by R. Mueller) on Geometrical Theory of Motion Applied to Approximate Straight-Line Motion," Kansas State Univ. Eng. Exp. Sta., Spec. Rept. 21, 1962.

454. Tesar, D., M. J. Ohanian, and E. T. Duga: "Summary Report of the Nuclear Reactor Maintenance Technology Assessment," *Proc. Workshop on Machines, Mechanisms and Robotics,* sponsored by the National Science Foundation and the Army Research Office, University of Florida, Gainesville, 1980.

455. Tesar, D., and G. Matthews: "The Dynamic Synthesis, Analysis and Design of Modeled Cam Systems," Lexington Books, D. C. Heath and Company, Lexington, Mass., 1976.

456. Thearle, E. L.: "A Non-reversing Coupling," *Machine Design,* vol. 23, pp. 181–184, April 1951.

457. Threlfall, D. C.: "The Inclusion of Coulomb Friction in Mechanisms Programs with Particular Reference to DRAM," *J. Mechanisms and Machine Theory,* vol. 13, no. 4, pp. 475–483, 1978.

458. Thumin, C.: "Designing Quick-Acting Latch Releases," *Machine Design,* vol. 19, pp. 110–115, September 1947.

459. Timoshenko, S., and D. G. Young: "Advanced Dynamics," McGraw-Hill Book Company, Inc., New York, 1948.

460. Timoshenko, S., and D. G. Young: "Engineering Mechanics," McGraw-Hill Book Company, Inc., New York, pp. 400–403, 1940.

461. Tolle, M.: "Regelung der Kraftmaschinen," 3d ed., Springer-Verlag OHG, Berlin, 1921.

462. Trinkl, F.: "Analytische und zeichnerische Verfahren zur Untersuchung eigentlicher Raumkurbelgetriebe," *Konstruktion,* vol. 11, no. 9, pp. 349–359, 1959.

463. Uhing, J.: "Einfache Raumgetriebe, für ungleichfoermige Dreh-und Schwingbewegung," *Konstruktion,* vol. 9, no. 1, pp. 18–21, 1957.

464. Uicker, J. J., Jr.: "Displacement Analysis of Spatial Mechanisms by an Iterative Method Based on 4×4 Matrices," M.S. dissertation, Northwestern University, Evanston, Ill., 1963.

465. Uicker, J. J.: "IMP" (computer code), Department of Mechanical Engineering, University of Wisconsin, Madison.

466. Vandeman, J. E., and J. R. Wood: "Modifying Starwheel Mechanisms," *Machine Design,* vol. 25, pp. 255–261, April 1953.

467. VDI Richtlinien. VDI Duesseldorf; for transmission-angle charts, refer to (*a*) four-bars, *VDI* 2123, 2124, Aug. 1959; (*b*) slider cranks, *VDI* 2125, Aug. 1959. For straight-line generation, refer to (*a*) in-line swinging-blocks, *VDI* 2137, Aug. 1959; (*b*) in-line slider-cranks, *VDI* 2136, Aug. 1959. (*c*) Planar four-bar, *VDI* 2130-2135, August, 1959.

468. "Sperrgetriebe," AWF-VDMA-VDI Getriebehefte, Ausschuss f. Wirtschaftliche Fertigung, Berlin, no. 6061 pub. 1955, nos. 6062, 6071 pub. 1956, no. 6063 pub. 1957.

469. Veldkamp, G. R.: "Curvature Theory in Plane Kinematics," J. B. Wolters, Groningen, 1963.

470. Vertut, J.: "Contributions to Analyze Manipulator Morphology, Coverage and Dexterity," vol. 1, "On the Theory and Practice of Manipulators," Springer-Verlag, New York, pp. 227–289, 1974.

471. Vidosic, J. P., and H. L. Johnson: "Synthesis of Four-Bar Function Generators," *Trans. Sixth Conf. Mechanisms,* Purdue University, West Lafayette, Ind., pp. 82–86, October 1960.

472. Vogel, W. F.: "Crank Mechanism Motions," *Prod. Eng.,* vol. 12, pt. I, June, pp. 301–305; II, July, pp. 374–379; III, pp. 423–428; August, IV, September, pp. 490–493, 1941.

473. Volmer, J.: "Konstruktion eines Gelenkgetriebes fuer eine Geradfuehrung," *VDI-Berichte,* vol. 12, pp. 175–183, 1956.

474. Volmer, J.: "Systematik," Kinematik des Zweiradgetriebes," *Maschinenbautechnik,* vol. 5, no. 11, pp. 583–589, 1956.

475. Volmer, J.: "Raederkurbelgetriebe," *VDI-Forschungsheft,* no. 461, pp. 52–55, 1957.

476. Volmer, J.: "Gelenkgetriebe zur Geradfuehrung einer Ebene" *Z. Prakt. Metallbearbeitung,* vol. 53, no. 5, pp. 169–174, 1959.

477. Wallace, W. B., Jr.: "Pressure Switches," *Machine Design,* vol. 29, pp. 106–114, Aug. 22, 1957.

478. Weise, H.: "Bewegungsverhaeltnisse an Filmschaltgetriebe, Getriebetechnik," *VDI-Berichte,* vol. 5, pp. 99–106, 1955.

479. Weise, H.: "Getriebe in Photographischen und kinematographischen Geraeten," *VDI-Berichte,* vol. 12, pp. 131–137, 1956.

480. Westinghouse Electric Corp., Pittsburgh, Pa., "Stability of a Lifting Rig," "Engineering Problems," vol. II, approx. 1960.

481. Weyth, N. C., and A. F. Gagne: "Mechanical Torque-Limiting Devices," *Machine Design,* vol. 18, pp. 127–130, May 1946.

482. Whitehead, T. N.: "Instruments and Accurate Mechanisms," Dover Publications, Inc., New York, 1954.

483. Wieleitner, H.: "Spezielle ebene Kurven," G. J. Goeschen Verlag, Leipzig, 1908.

484. Williams, R. J., and A. Seireg: "Interactive Modeling and Analysis of Open or Closed Loop Dynamic Systems with Redundant Actuators," *J. Mechanical Design, Trans. ASME,* vol. 101, pp. 407–416, 1979.

485. Winfrey, R. C.: "Dynamic Analysis of Elastic Link Mechanisms by Reduction of Coordinates," *Trans. ASME,* 94B, *J. Eng. Ind.,* 1972, pp. 577–582.

486. Wittel, O., and D. C. Haefele: "A Non-intermittent Film Projector," *Mech. Eng.,* vol. 79, pp. 345–347, April 1957.

487. Wolford, J. C., and A. S. Hall: "Second-Acceleration Analysis of Plane Mechanisms," ASME paper 57-A-52, 1957.

488. Wolford, J. C., and D. C. Haack: "Applying the Inflection Circle Concept," *Trans. Fifth Conf. Mechanisms,* Purdue University, West Lafayette, Ind., pp. 232–239, 1958.

489. Wolford, J. C.: "An Analytical Method for Locating the Burmester Points for Five Infinitesimally Separated Positions of the Coupler Plane of a Four-Bar Mechanism," *J. Appl. Mech.,* vol. 27, *Trans. ASME,* vol. 82E, pp. 182–186, March 1960.

490. Woerle, H.: "Sonderformen zwangläufiger viergelenkiger Raumkurbelgetriebe," *VDI- Berichte,* vol. 12, pp. 21–28, 1956.

491. Wunderlich, W.: "Zur angenaeherten Geradfuehrung durch symmetrische Gelenkvierecke," *Z. angew Math. Mechanik,* vol. 36, no. 3/4, pp. 103–110, 1956.

492. Wunderlich, W.: "Hoehere Radlinien,: *Osterr. Ing.-Arch.,* vol. 1, pp. 277–296, 1947.

493. Yang, A. T.: "Harmonic Analysis of Spherical Four-Bar Mechanisms," *J. Appl. Mech.,* vol. 29, no. 4, *Trans. ASME,* vol. 84E, pp. 683–688, December 1962.

494. Yang, A. T.: "Application of Quaternion Algebra and Dual Numbers to the Analysis of Spatial Mechanisms," doctoral dissertation, Columbia University, New York, 1963. See also *J. Appl. Mech.,* 1964.

495. Yudin, V. A.: "General Theory of Planetary-Cycloidal Speed Reducer," Akad. Nauk, Moscow, *Trudi Sem. Teor. Mash. Mekh.,* vol. 4, no. 13, pp. 42–77, 1948.

496. "Computer Explosion in Truck Engineering," a collection of seven papers, SAE Publ. SP240, December 1962.

497. "Mechanical Design of Spacecraft," Jet Propulsion Laboratory, Seminar Proceedings, Pasadena, Calif., August 1962.

498. "Guide to Limit Switches," *Prod. Eng.,* vol. 33, pp. 84–101, Nov. 12, 1962.

499. "Mechanical Power Amplifier," *Machine Design,* vol. 32, p. 104, Dec. 22, 1960.

CHAPTER 4
MECHANICAL VIBRATIONS

Eric E. Ungar, Eng.Sc.D.
Chief Consulting Engineer
Bolt, Beranek and Newman, Inc.
Cambridge, Mass.

4.1 INTRODUCTION 4.1
4.2 SYSTEMS WITH A SINGLE DEGREE OF FREEDOM 4.2
4.2.1 Linear Single-Degree-of-Freedom Systems 4.2
4.2.2 Nonlinear Single-Degree-of-Freedom Systems 4.14
4.3 SYSTEMS WITH A FINITE NUMBER OF DEGREES OF FREEDOM 4.24
4.3.1 Systematic Determination of Equations of Motion 4.24
4.3.2 Matrix Methods for Linear Systems—Formalism 4.25
4.3.3 Matrix Iteration Solution of Positive-Definite Undamped Systems 4.28
4.3.4 Approximate Natural Frequencies of Conservative Systems 4.31
4.3.5 Chain Systems 4.32
4.3.6 Mechanical Circuits 4.33
4.4 CONTINUOUS LINEAR SYSTEMS 4.37
4.4.1 Free Vibrations 4.37
4.4.2 Forced Vibrations 4.44
4.4.3 Approximation Methods 4.45
4.4.4 Systems of Infinite Extent 4.47
4.5 MECHANICAL SHOCKS 4.47
4.5.1 Idealized Forcing Functions 4.47
4.5.2 Shock Spectra 4.49
4.6 DESIGN CONSIDERATIONS 4.52
4.6.1 Design Approach 4.52
4.6.2 Source-Path-Receiver Concept 4.54
4.6.3 Rotating Machinery 4.55
4.6.4 Damping Devices 4.57
4.6.5 Charts and Tables 4.62

4.1 INTRODUCTION

The field of dynamics deals essentially with the interrelation between the motions of objects and the forces causing them. The words "shock" and "vibration" imply particular forces and motions: hence, this chapter concerns itself essentially with a subfield of dynamics. However, oscillatory phenomena occur also in nonmechanical systems, e.g., electric circuits, and many of the methods and some of the nomenclature used for mechanical systems are derived from nonmechanical systems.

Mechanical vibrations may be caused by forces whose magnitudes and/or directions and/or points of application vary with time. Typical forces may be due to rotating unbalanced masses, to impacts, to sinusoidal pressures (as in a sound field), or to random pressures (as in a turbulent boundary layer). In some cases the resulting vibrations may be of no consequence; in others they may be disastrous. Vibrations may be undesirable because they can result in deflections of sufficient magnitude to lead to malfunction, in high stresses which may lead to decreased life by increasing material fatigue, in unwanted noise, or in human discomfort.

Section 4.2 serves to delineate the concepts, phenomena, and analytical methods associated with the motions of systems having a single degree of freedom and to introduce the nomenclature and ideas discussed in the subsequent sections. Section 4.3 deals similarly with systems having a finite number of degrees of freedom, and Sec. 4.4 with continuous systems (having an infinite number of degrees of freedom). Mechanical shocks are discussed in Sec. 4.5, and in Sec. 4.6 appears additional information concerning design considerations, vibration-control techniques, and rotating machinery, as well as charts and tables of natural frequencies, spring constants, and material properties. The appended references substantiate and amplify the presented material.

Complete coverage of mechanical vibrations and associated fields is clearly impossible within the allotted space. However, it was attempted here to present enough information so that an engineer who is not a specialist in this field can solve the most prevalent problems with a minimum amount of reference to other publications.

4.2 SYSTEMS WITH A SINGLE DEGREE OF FREEDOM

A system with a single degree of freedom is one whose configuration at any instant can be described by a single number. A mass constrained to move without rotation along a given path is an example of such a system; its position is completely specified when one specifies its distance from a reference point, as measured along the path. Single-degree-of-freedom systems can be analyzed more readily than more complicated ones; therefore, actual systems are often approximated by systems with a single degree of freedom, and many concepts are derived from such simple systems and then enlarged to apply also to systems with many degrees of freedom.

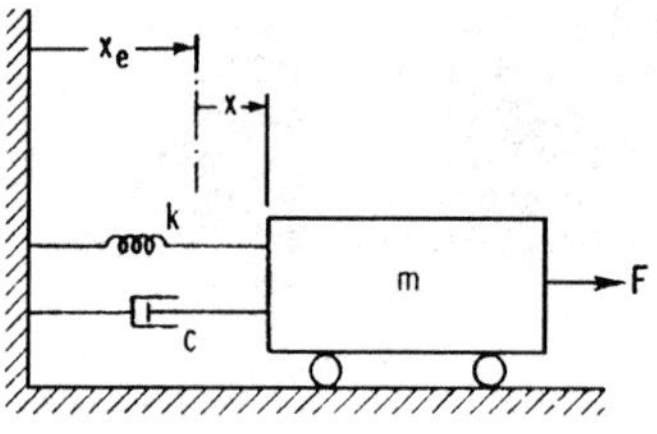

FIG. 4.1 A system with a single degree of freedom.

Figure 4.1 may serve as a model for all single-degree-of-freedom systems. This model consists of a pure inertia component (mass m supported on rollers which are devoid of friction and inertia), a pure restoring component (massless spring k), a pure energy-dissipation component (massless dashpot c), and a driving component (external force F). The inertia component limits acceleration. The restoring component opposes system deformation from equilibrium and tends to return the system to its equilibrium configuration in absence of other forces.

4.2.1 Linear Single-Degree-of-Freedom Systems[8,17,32,40,61,63]

If the spring supplies a restoring force proportional to its elongation and the dashpot provides a force which opposes motion of the mass proportionally to its velocity, then the system response is proportional to the excitation, and the system is said to be linear. If the position x_e indicated in Fig. 4.1 corresponds to the equilibrium position of the mass and if x denotes displacement from equilibrium, then the spring force may be written as $-kx$ and the dashpot force as $-c\ dx/dt$ (where the displacement x and all forces are taken as positive in the same coordinate direction). The equation of motion of the system then is

$$m\ d^2x/dt^2 + c\ dx/dt + kx = F \tag{4.1}$$

Free Vibrations. In absence of a driving force F and of damping c, i.e., with $F = c = 0$, Eq. (4.1) has a general solution which may be expressed in any of the following ways:

$$x = A\cos(\omega_n t - \phi) = (A\cos\phi)\cos\omega_n t + (A\sin\phi)\sin\omega_n t$$

$$= A\sin(\omega_n t - \phi + \pi/2) = A\,\mathrm{Re}\{e^{i(\omega_n t - \phi)}\} \qquad (4.2)$$

A and ϕ are constants which may in general be evaluated from initial conditions. A is the maximum displacement of the mass from its equilibrium position and is called the "displacement amplitude"; ϕ is called the "phase angle." The quantities ω_n and f_n, given by

$$\omega_n = \sqrt{k/m} \qquad f_n = \omega_n/2\pi$$

are known as the "undamped natural frequencies"; the first is in terms of "circular frequency" and is expressed in radians per unit time, the second is in terms of cyclic frequency and is expressed in cycles per unit time.

If damping is present, $c \neq 0$, one may recognize three separate cases depending on the value of the damping factor $\zeta = c/c_c$, where c_c denotes the critical damping coefficient (the smallest value of c for which the motion of the system will not be oscillatory). The critical damping coefficient and the damping factor are given by

$$c_c = 2\sqrt{km} = 2m\omega_n \qquad \zeta = \frac{c}{c_c} = \frac{c}{2\sqrt{km}} = \frac{c}{2m\omega_n}$$

The following general solutions of Eq. (4.1) apply when $F = 0$:

$$c > c_c(\zeta > 1): \qquad x = Be^{(-\zeta + \sqrt{\zeta^2 - 1})\omega_n t} + Ce^{(-\zeta - \sqrt{\zeta^2 - 1})\omega_n t} \qquad (4.3a)$$

$$c = c_c(\zeta = 1): \qquad x = (B + Ct)e^{-\omega_n t} \qquad (4.3b)$$

$$c < c_c(\zeta < 1): \qquad x = Be^{-\zeta\omega_n t}\cos(\omega_d t + \phi) = B\,\mathrm{Re}\,\{e^{i(\omega_N t + \phi)}\} \qquad (4.3c)$$

where $\qquad \omega_d \equiv \sqrt{\omega_n 1 - \zeta^2} \qquad i\omega_N = -\zeta\omega_n + i\omega_d$

denote, respectively, the "undamped natural frequency" and the "complex natural frequency," and the B, C, ϕ are constants that must be evaluated from initial conditions in each case.

Equation (4.3*a*) represents an extremely highly damped system; it contains two decaying exponential terms. Equation (4.3*c*) applies to a lightly damped system and is essentially a sinusoid with exponentially decaying amplitude. Equation (4.3*b*) pertains to a critically damped system and may be considered as the dividing line between highly and lightly damped systems.

Figure 4.2 compares the motions of systems (initially displaced from equilibrium by an amount x_0 and released with zero velocity) having several values of the damping factor ζ.

In all cases where $c > 0$ the displacement x approaches zero with increasing time. The damped natural frequency ω_d is generally only slightly lower than the undamped natural frequency ω_n; for $\zeta \leq 0.5$, $\omega_d \geq 0.87\ \omega n$.

The static deflection x_{st} of the spring k due to the weight mg of the mass m (where g denotes the acceleration of gravity) is related to natural frequency as

$$x_{st} = mg/k = g/\omega_n^2 = g/(2\pi f_n)^2$$

This relation provides a quick means for computing the undamped natural frequency (or for approximating the damped natural frequency) of a system from its static deflection. For x in inches or centimeters and f_n in hertz (cycles per second) it becomes

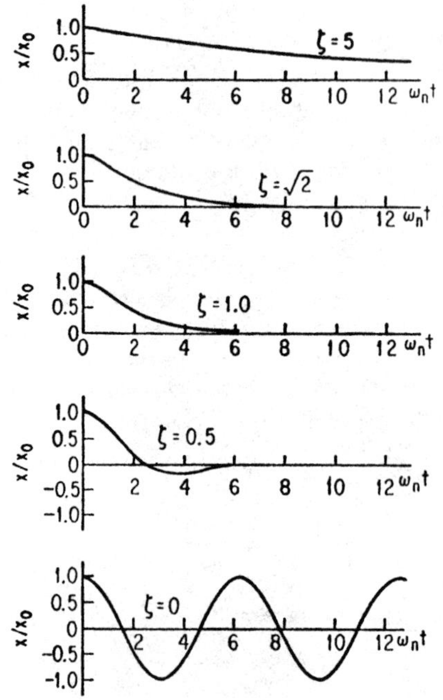

FIG. 4.2 Free motions of linear single-degree-of-freedom systems with various amounts of damping.

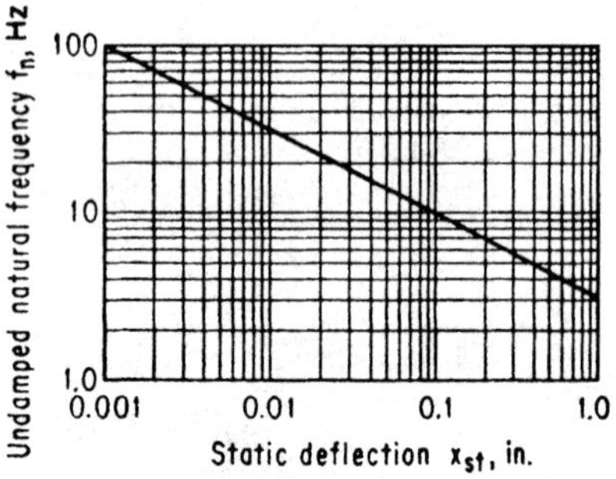

FIG. 4.3 Relation between natural frequency and static deflection of linear undamped single-degree-of-freedom system.

$$f_n^2 \text{ (Hz)} = 9.80/x_{st} \text{ (in)} = 24.9/x_{st} \text{ (cm)}$$

which is plotted in Fig. 4.3.

Forced Vibrations. The previous section dealt with cases where the forcing function F of Eq. (4.1) was zero. The solutions obtained were the so-called "general solution of the homogeneous equation" corresponding to Eq. (4.1). Since these solutions vanish with increasing time (for $c > 0$), they are sometimes also called the "transient solutions." For $F \neq 0$ the solutions of Eq. (4.1) are made up of the aforementioned general solution (which incorporates constants of integration that depend on the initial conditions) plus a "particular integral" of Eq. (4.1). The particular integrals contain no constants of integration and do not depend on initial conditions, but do depend on the excitation. They do not tend to zero with increasing time unless the excitation tends to zero and hence are often called the "steady-state" portion of the solution.

The complete solution of Eq. (4.1) may be expressed as the sum of the general (transient) solution of the homogeneous equation and a particular (steady-state) solution of the nonhomogeneous equation. The associated general solutions have already been discussed; hence the present discussion will be concerned primarily with the steady-state solutions.

The steady-state solutions corresponding to a given excitation $F(t)$ may be obtained from the differential equation (4.1) by use of various standard mathematical techniques[12,20] without a great deal of difficulty. Table 4.1 gives the steady-state responses x_{ss} to some common forcing functions $F(t)$.

TABLE 4.1 Steady-State Responses of Linear Single-Degree-of-Freedom Systems to Several Forcing Functions

$F(t)$	x_{ss}
1	$1/k$
t	$t/k - c/k^2$
t^2	$t^2/k - 2ct/k^2 + 2c^2/k^3 - 2m/k^2$
$e^{\pm\alpha t}$*	$h(\pm\alpha)$
$te^{\pm\alpha t}$*	$h(\pm\alpha)e^{\pm\alpha t} - h^2(\pm\alpha)(c \pm 2m\alpha)te^{\pm\alpha t}$
$e^{i\omega t}$	$(k - m\omega^2 + ic\omega)^{-1}$
$\sin \beta t$†	$g(\beta)[(k - m\beta^2)\sin \beta t - c\beta \cos \beta t]$
$\cos \beta t$†	$g(\beta)[c\beta \sin \beta t + (k - m\beta^2)\cos \beta t]$
$e^{\pm\alpha t} \sin \beta t$‡	$(A \cos \beta t + B \sin \beta t)e^{-\alpha t}/D$
$e^{\pm\alpha t} \cos \beta t$‡	$(A \sin \beta t - B \cos \beta t)e^{-\alpha t}/D$

*$h(\pm\alpha) = (m\alpha^2 \pm c\alpha + k)^{-1}$
†$g(\beta) = [(k - m\beta^2)^2 + (c\beta)^2]^{-1}$
‡$A = k + m(\alpha^2 - \beta^2) \pm \alpha c$
$B = \beta(c \pm 2\alpha m)$
$D = A^2 + B^2$

Superposition. Since the governing differential equation is linear, the response corresponding to a sum of excitations is equal to the sum of the individual responses; or, if

$$F(t) = A_1F_1(t) + A_2F_2(t) + A_3F_3(t) + \cdots$$

where $A_1, A_2, \ldots$ are constants, and if $x_{ss1}, x_{ss2}, \ldots$ are solutions corresponding, respectively, to $F_1(t), F_2(t), \ldots,$ then the steady-state response to $F(t)$ is

$$x_{ss} = A_1x_{ss1} + A_2x_{ss2} + A_3x_{ss3} + \cdots$$

Superposition permits one to determine the response of a linear system to any time-dependent force $F(t)$ if one knows the system's impulse response $h(t)$. This impulse response is the response of the system to a Dirac function $\delta(t)$ of force; also $h(t) = \dot{u}(t)$, where $u(t)$ is the system response to a unit step function of force [$F(t) = 0$ for $t < 0$, $F(t) = 1$ for $t > 0$]. In the determination of $h(t)$ and $u(t)$ the system is taken as at rest and at equilibrium at $t = 0$.

The motion of the system may be found from

$$x_{ss}(t) = \int_0^t F(\tau)h(t - \tau)\,d\tau \tag{4.4}$$

in conjunction with the proper "transient" solution expression, the constants in which must be adjusted to agree with specified initial conditions. For single-degree-of-freedom systems,

$$h(t) = \frac{\omega_n}{2k\sqrt{\zeta^2-1}}\left[e^{[-\zeta+(\zeta^2-1)^{1/2}]\omega_n t} + e^{[-\zeta-(\zeta^2-1)^{1/2}]\omega_n t}\right] \qquad \text{for } \zeta > 1$$

$$h(t) = \frac{\omega_n^2 t}{k} e_0^{-\omega_n t} \qquad \text{for } \zeta = 1$$

$$h(t) = \frac{\omega_n}{k\sqrt{1-\zeta^2}} e^{-\zeta\omega_n t} \sin \omega_\alpha t \qquad \text{for } \zeta < 1$$

$$h(t) = \frac{\omega_n}{k} \sin \omega_n t \qquad \text{for } \zeta = 0$$

Sinusoidal (Harmonic) Excitation. With an excitation

$$F(t) = F_0 \sin \omega t$$

one obtains a response which may be expressed as

$$x_{ss} = X_0 \sin (\omega t - \phi)$$

where $$X_0 = \frac{F_0}{\sqrt{(k - m\omega^2)^2 + (c\omega)^2}} \qquad \tan \phi = \frac{c\omega}{k - m\omega^2} \tag{4.5}$$

The ratio

$$H_s(\omega) = \frac{X_0}{F_0/k} = \left\{\left[1 - \left(\frac{\omega}{\omega_n}\right)^2\right]^2 + \left(2\zeta \frac{\omega}{\omega_n}\right)^2\right\}^{-1/2} \tag{4.6}$$

is called the frequency response or the magnification factor. As the latter name implies, this ratio compares the displacement amplitude X_0 with the displacement F_0/k that a force F_0 would produce if it were applied statically. $H_s(\omega)$ is plotted in Fig. 4.4.

Complex notation is convenient for representing general sinusoids.* Corresponding to a sinusoidal force

$$F(t) = F_0 e^{i\omega t}$$

one obtains a displacement

$$x_{ss} = X_0 e^{i\omega t}$$

where $$\frac{X_0}{F_0/k} = H(\omega) = \left[1 - \left(\frac{\omega}{\omega_n}\right)^2 + 2i\zeta\left(\frac{\omega}{\omega_n}\right)\right]^{-1} \tag{4.7}$$

$H(\omega)$ is called the complex frequency response, or the complex magnification factor† and is related to that of Eq. (4.6) as

$$H_s(\omega) = |H(\omega)|$$

From the model of Fig. 4.1 one may determine that the force F_{TR} exerted on the wall at any instant is given by

$$F_{TR} = kx + c\dot{x}$$

The ratio of the amplitude of this transmitted force to the amplitude of the sinusoidal applied force is called the transmissibility TR_s and obeys

$$\mathrm{TR}_s = \frac{F_{\mathrm{TR}}}{F_0} = \sqrt{\frac{1 + [2\zeta(\omega/\omega_n)]^2}{[1 - (\omega/\omega_n)^2]^2 + [2\zeta(\omega/\omega_n)]^2}} \tag{4.8}$$

---

*In complex notation[40] it is usually implied, though it may not be explicitly stated, that only the *real parts* of excitations and responses represent the physical situation. Thus the complex form $Ae^{i\omega t}$ (where the coefficient $A = a + ib$ is also complex in general) implies the oscillation given by

$$\mathrm{Re}\{Ae^{i\omega t}\} = \mathrm{Re}\{(a + ib)(\cos \omega t + i \sin \omega t)\} = a \cos \omega t - b \sin \omega t$$

†An alternate formulation in terms of mechanical impedance is discussed in Sec. 4.3.6.

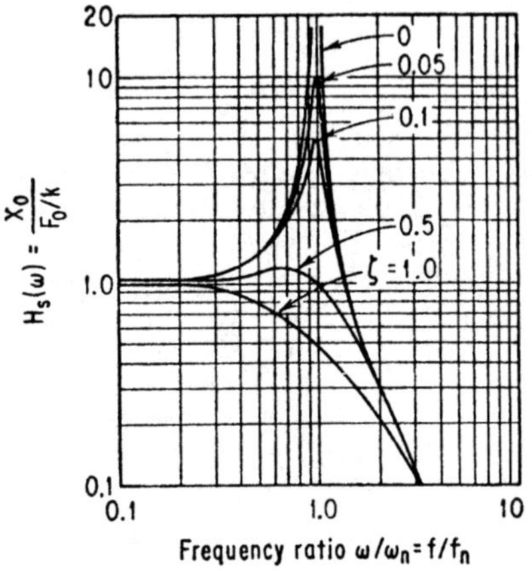

FIG. 4.4 Frequency response (magnification factor) of linear single-degree-of-freedom system.

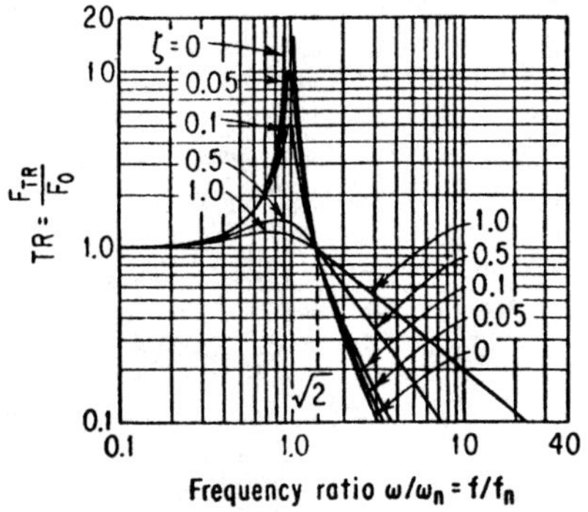

FIG. 4.5 Transmissibility of linear single-degree-of-freedom system.

Transmissibility $TR_s(\omega)$ is plotted in Fig. 4.5.

In complex notation

$$TR = \frac{1 + 2i\zeta(\omega/\omega_n)}{1 - (\omega/\omega_n)^2 + 2i\zeta(\omega/\omega_n)} \qquad TR_s(\omega) = |TR(\omega)|$$

It is evident that

$$|H(\omega)| \approx |TR| \approx \begin{cases} 1 & \text{for } \omega << \omega_n \\ (\omega_n/\omega)^2 & \text{for } \omega >> \omega_n \end{cases}$$

Increased damping ζ always reduces the frequency response H. For $\omega/\omega_n < \sqrt{2}$ increased damping also decreases TR, but for $\omega/\omega_n > \sqrt{2}$ increased damping increases TR.*

The frequencies at which the maximum transmissibility and amplification factor occur for a given damping ratio are shown in Fig. 4.6; the magnitudes of these maxima are shown in Fig. 4.7. For small damping ($\zeta < 0.3$, which applies to many practical

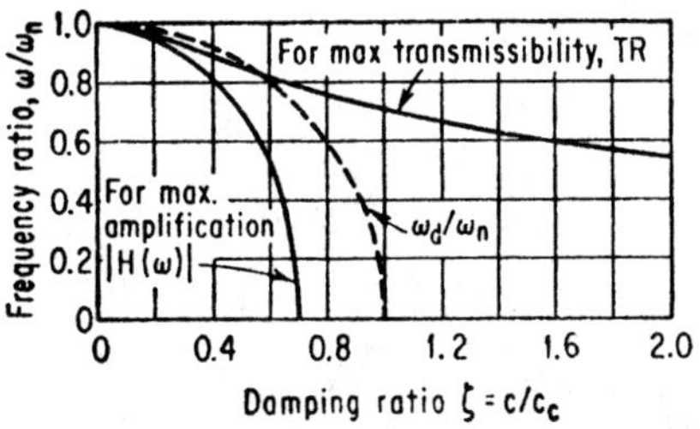

FIG. 4.6 Frequencies at which magnification and transmissibility maxima occur for given damping ratio.

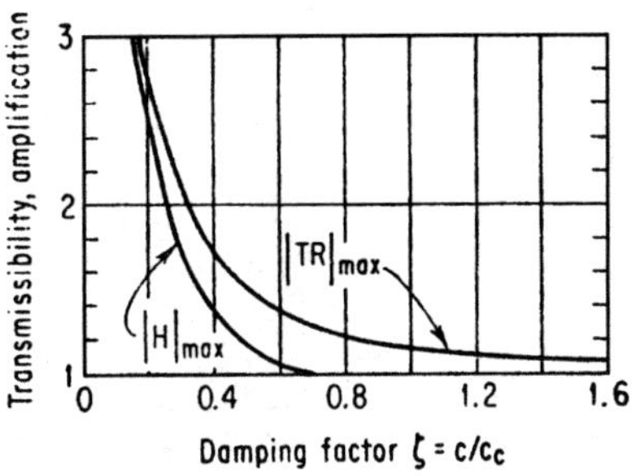

FIG. 4.7 Maximum values of magnification and transmissibility.

*It is important to note that these remarks apply only for the type of damping represented by a viscous dashpot model; different relations generally apply for other damping mechanisms.[49]

problems), the maximum transmissibility $|\mathrm{TR}|_{max}$ and maximum amplification factor $|H|_{max}$ both occur at $\omega_d \approx \omega_n$, and

$$|\mathrm{TR}|_{max} \approx |H|_{max} \approx (2\zeta)^{-1}$$

The quantity $(2\zeta)^{-1}$ is often given the symbol Q, termed the "quality factor" of the system. The frequency at which the greatest amplification occurs is called the resonance frequency; the system is then said to be in resonance. For lightly damped systems the resonance frequency is practically equal to the natural frequency, and often no distinction is made between the two. Thus, for lightly damped systems, resonance (i.e., maximum amplification) occurs essentially when the exciting frequency ω is equal to the natural frequency ω_n.

Equation (4.6) shows that

$$\begin{aligned} X_0/F_0 &\approx 1/k \text{ (system is stiffness-controlled) for } \omega << \omega_n \\ &\approx 1/2k\zeta \text{ (system is damping-controlled) for } \omega \approx \omega_n \ (\zeta << 1) \\ &\approx 1/m\omega^2 \text{ (system is mass-controlled) for } \omega >> \omega_n \end{aligned}$$

General Periodic Excitation. Any periodic excitation may be expressed in terms of a Fourier series (i.e., a series of sinusoids) and any aperiodic excitation may be expressed in terms of a Fourier integral, which is an extension of the Fourier-series concept. In view of the superposition principle applicable to linear systems the response can then be obtained in terms of a corresponding series or integral.

A periodic excitation with period T may be expanded in a Fourier series as

$$F(t) = \frac{A_0}{2} + \sum_{r=1}^{\infty} (A_r \cos r\omega_0 t + B_r \sin r\omega_0 t) = \sum_{r=-\infty}^{\infty} C_r e^{ir\omega_0 t} \tag{4.9}$$

where the period T and fundamental frequency ω_0 are related by

$$\omega_0 T = 2\pi$$

The Fourier coefficients A_r, B_r, C_r may be computed from

$$A_r = \frac{2}{T}\int_t^{t+T} F(t) \cos (r\omega_0 t)\, dt \qquad B_r = \frac{2}{T}\int_t^{t+T} F(t) \sin (r\omega_0 t)\, dt$$

$$C_r = \tfrac{1}{2}(A_r - iB_r) = \frac{1}{T}\int_t^{t+T} F(t) e^{-ir\omega_0 t}\, dt \tag{4.10}$$

Superposition permits the steady-state response to the excitation given by Eq. (4.9) to be expressed as

$$x_{ss} = \frac{1}{k}\sum_{r=-\infty}^{\infty} H_r C_r e^{ir\omega_0 t} \tag{4.11}$$

where H_r is obtained by setting $\omega = r\omega_0$ in Eq. (4.7).

If a periodic excitation contains a large number of harmonic components with $C_r \neq 0$, it is likely that one of the frequencies $r\omega_0$ will come very close to the natural frequency ω_n of the system. If $r_0\omega_0 \approx \omega_n$, $C_{r0} \neq 0$, then $H_{r0}C_{r0}$ will be much greater than the other components of the response (particularly in a very lightly damped system), and

$$\begin{aligned} x_{ss}k &\approx H_{r0}C_{r0}e^{i\omega_n t} + H_{-r0}C_{-r0}e^{-i\omega_n t} + A_0/2 \\ &\approx (1/2\zeta)(A_{r0} \sin \omega_n t - B_{r0} \cos \omega_n t) + A_0/2 \end{aligned}$$

General Nonperiodic Excitation.[2,3,11,16] The response of linear systems to any well-behaved* forcing function may be determined from the impulse response as discussed in conjunction with Eq. (4.4) or by application of Fourier integrals. The latter may be visualized as generalizations of Fourier series applicable for functions with infinite period.

A "well-behaved"* forcing function $F(t)$ may be expressed as†

$$F(t) = \frac{1}{2\pi}\int_{-\infty}^{\infty} \Phi(\omega)e^{i\omega t}\,d\omega \tag{4.12}$$

where

$$\Phi(\omega) = \int_{-\infty}^{\infty} F(t)e^{-i\omega t}\,dt \tag{4.13}$$

[These are analogous to Eqs. (4.9) and (4.10)]. With the ratio $H(\omega)$ of displacement to force as given by Eq. (4.7), the displacement-response transform then is

$$X(\omega) = (1/k)H(\omega)\Phi(\omega)$$

and, analogously to Eq. (4.11), one finds the displacement given by

$$x_{ss}(t) = (1/2\pi)\int_{-\infty}^{\infty} X(\omega)e^{i\omega t}\,d\omega = (1/2\pi k)\int_{-\infty}^{\infty} H(\omega)\Phi(\omega)e^{i\omega t}\,d\omega$$

$$= (1/2\pi k)\int_{-\infty}^{\infty} H(\omega)\left[\int_{-\infty}^{\infty} F(t)e^{-i\omega t}\,dt\right]e^{i\omega t}\,d\omega$$

One may expect the components of the excitation with frequencies nearest the natural frequency of a system to make the most significant contributions to the response. For lightly damped systems one may assume that these most significant components are contained in a small frequency band containing the natural frequency. Usually one uses a "resonance bandwidth" $\Delta\omega = 2\zeta\omega_n$, thus effectively assuming that the most significant components are those with frequencies between $\omega_n(1 - \zeta)$ and $\omega_n(1 + \zeta)$. (At these two limiting frequencies, commonly called the half-power points, the rate of energy dissipation is one-half of that at resonance. The amplitude of the response at these frequencies is $1/\sqrt{2} \approx 0.707$ times the amplitude at resonance.) Noting that the largest values of the complex amplification factor H(ω) occur for $\omega \approx \pm\,\omega_d \approx \pm\,\omega_n$, one may write

$$2\pi x_{ss}k \approx -\,i\omega_n e^{-\zeta\omega_n t}[\Phi(\omega_n)e^{i\omega_n t} + \Phi(-\omega_n)e^{-i\omega_n t}]$$

Random Vibrations: Mean Values, Spectra, Spectral Densities.[2,3,11,16] In many cases one is interested only in some mean value as a characterization of response. The time average of a variable $y(t)$ may be defined as

$$\bar{y} = \lim_{\tau\to\infty} (1/\tau)\int_0^{\tau} y(t)\,dt \tag{4.14}$$

where it is assumed that the limit exists. For periodic $y(t)$ one may take τ equal to a period and omit the limiting process.

*"Well-behaved" means that $|F(t)|$ is integrable and $F(t)$ has bounded variation.

†Other commonly used forms of the integral transforms can be obtained by substituting $j = -i$. Since $j^2 = i^2 = -1$, all the developments still hold. Fourier transforms are also variously defined as regards the coefficients. For example, instead of $1/2\pi$ in Eq. (4.12), there often appears a $1/\sqrt{2\pi}$; then a $1/\sqrt{2\pi}$ factor is added in Eq. (4.13) also. In all cases the product of the coefficients for a complete cycle of transformations is $1/2\pi$.

The mean-square value of $y(t)$ thus is given by

$$\overline{y^2} = \lim_{t\to\infty} (1/\tau) \int_0^\tau y^2(t)\,dt$$

and the root-mean-square value by $y_{rms} = (\overline{y^2})^{1/2}$. For a sinusoid $x = \text{Re}\,\{Ae^{i\omega t}\}$ one finds $\overline{x^2} = ½|A|^2 = ½AA^*$ where A^* is the complex conjugate of A.

The mean-square response $\overline{x^2}$ of a single-degree-of-freedom system with frequency response $H(\omega)$ [Eq. (4.7)] to a sinusoidal excitation of the form $F(t) = \text{Re}\,\{F_0 e^{i\omega t}\}$ is given by

$$k^2\overline{x^2} = H(\omega)F_0 H^*(\omega)F_0^*/2 = |H(\omega)|^2\overline{F^2}$$

Similarly, the mean-square value of a general periodic function $F(t)$, expressed in Fourier-series form as

$$F(t) = \sum_{r=-\infty}^{\infty} C_r e^{ir\omega_0 t}$$

is

$$\overline{F^2} = \sum_{r=-\infty}^{\infty} \frac{C_r C_r^*}{2} = ½ \sum_{r=-\infty}^{\infty} |C_r|^2 \qquad (4.15)$$

The mean-square displacement of a single-degree-of-freedom system in response to the aforementioned periodic excitation is given by

$$k^2\overline{x^2} = ½ \sum_{r=-\infty}^{\infty} |C_r|^2 |H(r\omega_0)|^2$$

where convergence of all the foregoing infinite series is assumed.

If one were to plot the cumulative value of (the sum representing) the mean-square value of a periodic variable as a function of frequency, starting from zero, one would obtain a diagram somewhat like Fig. 4.8. This graph shows how much each frequency (or "spectral component") adds to the total mean-square value. Such a graph* is called the spectrum (or possibly more properly the integrated spectrum) of $F(t)$. It is generally of relatively little interest for periodic functions, but is extremely useful for aperiodic (including random) functions.

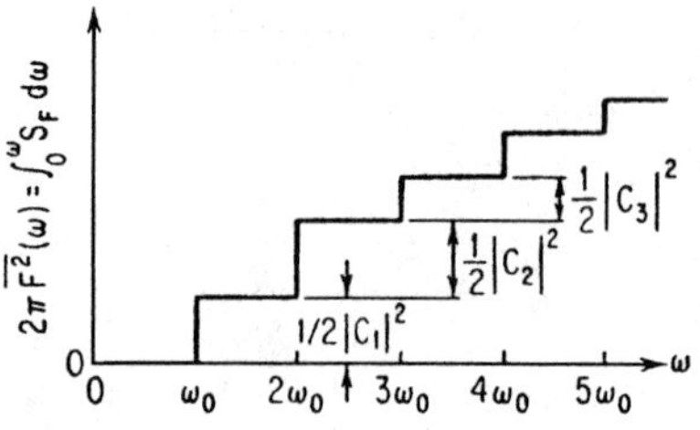

FIG. 4.8 (Integrated) spectrum of periodic function $F(t)$.

The derivative of the (integrated) spectrum with respect to ω is called the "mean-square spectral density" (or power spectral density) of F. Thus the power spectral density S_F of F is defined as*

$$S_F(\omega) = 2\pi d(\overline{F^2})/d\omega \qquad (4.16)$$

where $\overline{F^2}$ is interpreted as a function of ω as in Fig. 4.8. The mean-square value of F is related to power spectral density as

$$\overline{F^2} = (1/2\pi) \int_0^\infty S_F(\omega)\,d\omega \qquad (4.17)$$

*The factor 2π appearing in Fig. 4.8 and Eqs. (4.16) and (4.17) is a matter of definition. Different constants are sometimes used in the literature, and one must use care in comparing results from different sources.

This integral over all frequencies is analogous to the infinite sum of Eq. (4.15). From Fig. 4.8 and Eq. (4.17) one may visualize that power spectral density is a convenient means for expressing the contributions to the mean-square value in any frequency range.

For nonperiodic functions one obtains contributions to the mean-square value over a continuum of frequencies instead of at discrete frequencies, as in Fig. 4.8. The (integrated) spectrum and the power spectral density then are continuous curves. The relations governing the mean responses to nonperiodic excitation can be obtained by the same limiting processes which permit one to proceed from the Fourier series to Fourier integrals. However, the results are presented here in a slightly more general form so that they can be applied also to systems with random excitation.*

Response to Random Excitation: Autocorrelation Functions. For stationary ergodic random processes† whose sample functions are $F(t)$ or for completely specified functions $F(t)$ one may define an autocorrelation function $R_F(\tau)$ as

$$R_F(\tau) = \lim_{T\to\infty} (1/2T) \int_{-T}^{T} F(t)F(t + \tau)\, dt \tag{4.18}$$

This function has the properties

$$R_F(0) = \overline{F^2} \geq R_F(\tau) \qquad R_F(-\tau) = R_F(\tau)$$

For many physical random processes the values of F observed at widely separated intervals are uncorrelated, that is

$$\lim_{\tau\to\infty} R_F(\tau) = (\overline{F})^2$$

or R_F approaches the square of the mean value (not the mean-square value!) of F for large time separation τ. (Many authors define variables measured from a mean value; if such variables are uncorrelated, their $R_F \to 0$ for large τ.) One may generally find some value of τ beyond which R_F does not differ "significantly" from $(\overline{F})^2$. This value of τ is known as the "scale" of the correlation.

The power spectral density of F is given by‡,§

*In the previous discussion the excitation was described as some known function of time and the responses were computed as other completely defined time functions; in each case the values at each instant were specified or could be found. Often the stimuli cannot be defined so precisely; only some statistical information about them may be available. Then, of course, one may only obtain some similar statistical information about the responses.

†A "random process" is a mathematical model useful for representing randomly varying physical quantities. Such a process is determined not by its values at various instants but by certain average and spectral properties. One sacrifices precision in the description of the variable for the sake of tractability.

One may envision a large number of sample functions (such as force vs. time records obtained on aircraft landing gears, with time datum at the instant of landing). One may compute an average value of these functions at any given time instant; such an average is called a "statistical average" and generally varies with the instant selected. On the other hand, one may also compute the time average of any given sample function over a long interval. The statistical average will be equal to the time average of almost every sample function, provided that the sample process is both stationary and ergodic.[3]

A random process is stationary essentially if the statistical average of the sample functions is independent of time, i.e., if the ensemble appears unchanged if the time origin is changed. Ergodicity essentially requires that almost every sample function be "typical" of the entire group. General mathematical results are to a large degree available only for stationary ergodic random processes; hence the following discussion is limited to such processes.

‡Definitions involving different numerical coefficients are also in general use.

§For real $F(t)$ one may multiply the coefficient shown here by 2 and replace the lower limit of integration by zero.

$$S_F(\omega) = 2\int_{-\infty}^{\infty} R_F(\tau)e^{-i\omega\tau}\,d\tau \tag{4.19}$$

and is equal to twice* the Fourier transform of the autocorrelation function R_F. Inversion of this transform gives*,†

$$R_F(\tau) = \frac{1}{4\pi}\int_{-\infty}^{\infty} S_F(\omega)e^{i\omega\tau}\,d\tau$$

whence*,†

$$R_F(0) = \overline{F^2} = \frac{1}{4\pi}\int_{-\infty}^{\infty} S_F(\omega)\,d\omega \tag{4.20}$$

The last of these relations agrees with Eq. (4.17).

System Response to Random Excitation. For a system with a complex frequency response $H(\omega)$ as given by Eq. (4.7) one finds that the power spectral density S_x of the response is related to the power spectral density S_F of the exciting force according to

$$k^2S_x(\omega) = |H(\omega)|^2 S_F(\omega) \tag{4.21}$$

In order to compute the mean-square response of a system to aperiodic (or stationary ergodic random) excitation one may proceed as follows:

1. Calculate $R_F(\tau)$ from Eq. (4.18).
2. Find $S_F(\omega)$ from Eq. (4.19).
3. Determine $S_x(\omega)$ from Eq. (4.21).
4. Find $\overline{x^2}$ from Eq. (4.20) (with F subscripts replaced by x).

"White noise" is a term commonly applied to functions whose power spectral density is constant for all frequencies. Although such functions are not realizable physically, it is possible to obtain power spectra that remain virtually constant over a frequency region of interest in a particular problem (particularly in the neighborhood of the resonance of the system considered, where the response contributes most to the total).

The mean-square displacement of a single-degree-of-freedom system to a (real) white-noise excitation F, having the power spectral density $S_F(\omega) = S_0$, is given by

$$\overline{x^2} = \omega_n S_0/8\zeta k^2 = S_0/4ck$$

Probability Distributions of Excitation and Response.[2,3,11,16] For most practical purposes it is sufficient to define the probability of an event as the fraction of the number of "trials" in which the event occurs, provided that a large number of trials are made. (In throwing an unbiased die a large number of times one expects to obtain a given number, say 2, one-sixth of the time. The probability of the number 2 here is $\frac{1}{6}$.)

If a given variable can assume a continuum of values (unlike the die for which the variable, i.e., the number of spots, can assume only a finite number of discrete values) it makes generally little sense to speak of the probability of any given value. Instead, one may profitably apply the concepts of probability distribution and probability density functions. Consider a continuous random variable x and a certain value x_0 of that

*Definitions involving different numerical coefficients are also in general use.

†For real $F(t)$ one may multiply the coefficient shown here by 2 and replace the lower limit of integration by zero.

variable. The probability distribution function P_{dis} then is defined as a function expressing the probability P that the variable $x \le x_0$. Symbolically,

$$P_{dis}(x_0) = P(x \le x_0)$$

The probability density function P_{dens} is defined by

$$P_{dens}(x_0) = dP_{dis}(x_0)/dx_0$$

so that the probability of x occurring between x_0 and $x_0 + dx_0$ is

$$P(x_0 < x \le x_0 + dx_0) = P_{dens}(x_0)dx_0$$

Among the most widely studied distributions are the gaussian (or normal) distribution, for which

$$P_{dens}(x) = [1/\sigma\sqrt{2\pi}]e^{-(x-M)2/2\sigma 2} \tag{4.22}$$

and the Rayleigh distribution, for which

$$P_{dens}(x) = (x/\sigma^2)e^{-x2/2\sigma 2} \tag{4.23}$$

In the foregoing, M denotes the statistical mean value, defined by

$$M = \int_{-\infty}^{\infty} xP_{dens}(x)\, dx$$

and σ^2 denotes the variance of x and is defined by

$$\sigma^2 = \int_{-\infty}^{\infty} (x-M)^2 P_{dens}(x)\, dx$$

σ is called the "standard deviation" of the distribution. For a stationary ergodic random process

$$\sigma^2 + M^2 = R_x(0) = \overline{x^2}$$

The gaussian distribution is by far the most important, since it represents many physical conditions relatively well and permits mathematical analysis to be carried out relatively simply. With some qualifications, the "central limit theorem" states that any random process, each of whose sample functions is constructed from the sum of a large number of sample functions selected independently from some other random process, will tend to become gaussian as the number of sample functions added tends to infinity. A stationary gaussian random process with zero mean is completely characterized by either its autocorrelation function or its power spectral density. If the excitation of a linear system is a gaussian random process, then so is the system response.

For a gaussian random process $x(t)$ with autocorrelation function $R_x(\tau)$ and power spectral density $S_x(\omega)$ one may find the *average* number of times N_0 that $x(t)$ passes through zero in unit time from

$$(\pi N_0)^2 = -\left.\frac{d^2R_x}{d\tau^2}\right|_{\tau=0} [R_x(0)]^{-1} = \left[\int_0^{\infty} \omega^2 S_x(\omega)\, d\omega\right]\left[\int_0^{\infty} S_x(\omega)\, d\omega\right]^{-1}$$

The quantity N_0 gives an indication of the "apparent frequency" of $x(t)$. A sinusoid crosses zero twice per cycle and has $f = N_0/2$. This relation may be taken as the definition of apparent frequency for a random process.

The average number of times N_α that the aforementioned gaussian $x(t)$ crosses the value $x = \alpha$ per unit time is given by

$$N_\alpha = N_0 e^{-\alpha^2/2R_x(0)}$$

The average number of times per unit time that $x(t)$ passes through α with positive slope is half the foregoing value. The average number of peaks* of $x(t)$ occurring per unit time between $x = \alpha$ and $x = \alpha + d\alpha$ is

$$N_{\alpha,\alpha+d\alpha} = \frac{N_0 \alpha \, d\alpha}{2R_x(0)} e^{-\alpha^2/2R_x(0)}$$

For a linear single-degree-of-freedom system with natural circular frequency ω_n subject to white noise of power spectral density $S_F(\omega) = S_0$, one finds

$$N_0 = \omega_n/\pi \qquad f_{\text{apparent}} = \omega_n/2\pi = f_n$$

The displacement vs. time curve representing the response of a lightly damped system to broadband excitation has the appearance of a sinusoid with the system natural frequency, but with randomly varying amplitude and phase. The average number of peaks per unit time occurring between α and $\alpha + d\alpha$ in such an oscillation is given by

$$(2\pi/\omega_n)N_{\alpha,\alpha+d\alpha} = [\alpha \, d\alpha/R_x(0)]e^{-\alpha^2/2R_x(0)}$$

The term on the right-hand side is, except for the $d\alpha$, the Rayleigh probability density of Eq. (4.23).

4.2.2 Nonlinear Single-Degree-of-Freedom Systems[21,34,60]

The previous discussion dealt with systems whose equations of motion can be expressed as linear differential equations (with constant coefficients), for which solutions can always be found. The present section deals with systems having equations of motion for which solutions cannot be found so readily. Approximate analytical solutions can occasionally be found, but these generally require insight and/or a considerable amount of algebraic manipulation. Numerical or analog computations or graphical methods appear to be the only ones of general applicability.

Practical Solution of General Equations of Motion. After one sets up the equations of motion of a system one wishes to analyze, one should determine whether solutions of these are available by referring to texts on differential equations and compendia such as Ref. 27. (The latter reference also describes methods of general utility for obtaining approximate solutions, such as that involving series expansion of the variables.) If these approaches fail, one is generally reduced to the use of numerical or graphical methods. A wide range of computer-based methods is available.[44,45] In the following pages two generally useful methods are outlined. Methods and results applicable to some special cases are discussed in subsequent sections.

A Numerical Method. The equation of motion of a single-degree-of-freedom system can generally be expressed in the form

$$m\ddot{x} + G(x, \dot{x}, t) = 0 \qquad \text{or} \qquad \ddot{x} + f(x, \dot{x}, t) = 0 \tag{4.24}$$

*Actually average excess of peaks over troughs, but for $\alpha >> x_{\text{rms}}$ the probability of troughs in the interval becomes very small.

where f includes all nonlinear and nonconstant coefficient effects. (f may occasionally also depend on higher time derivatives. These are not considered here, but the method discussed here may be readily extended to account for them.) It is assumed that f is a known function, given in graphical, tabular, or analytic form.

In order to integrate Eq. (4.24) numerically as simply as possible, one assumes that f remains virtually constant in a small time interval Δt. Then one may proceed by the following steps:

1. *a.* Determine $f_0 = f(x_0, \dot{x}_0, 0)$, the initial value of f, from the specified initial displacement x_0 and initial velocity $\dot{x}_0$. Then the initial acceleration is $\ddot{x}_0 = -f_0$.
   b. Calculate the velocity $\dot{x}_1$ at the end of a conveniently chosen small time interval Δt_{0-1}, and the average velocity $\bar{\dot{x}}_{0-1}$ during the interval from

$$\dot{x}_1 = \dot{x}_0 + \ddot{x}_0\,\Delta t_{0-1} \qquad \bar{\dot{x}}_{0-1} = \tfrac{1}{2}(\dot{x}_0 + \dot{x}_1)$$

   c. Calculate the displacement x_1 at the end of the interval Δt_{0-1} and the average displacement $\bar{x}_{0-1}$ during the interval from

$$x_1 = x_0 + \bar{\dot{x}}_{0-1}\,\Delta t_{0-1} \qquad \bar{x}_{0-1} = \tfrac{1}{2}(x_0 + x_1)$$

   d. Compute a better approximation* to the average f and $\ddot{x}$ during the interval Δt_{0-1} by using $x = \bar{x}_{0-1}$, $\dot{x} = \bar{\dot{x}}_{0-1}$, $t = \tfrac{1}{2}\Delta t_{0-1}$ in the determination of f.
2. Repeat steps 1*b* to 1*d*, beginning with the new approximation of f, until no further changes in f occur (to the desired accuracy).
3. Select a second time interval Δt_{1-2} (not necessarily of the same magnitude as Δt_{0-1}) and continue to
   a. Find $\ddot{x}_1 = -f_1 = -f(x_1, \dot{x}_1, t_1)$.
   b. Calculate the velocity $\dot{x}_2$ at the end of the interval Δt_{1-2}, and the average velocity $\bar{\dot{x}}_{1-2}$ during the interval, from

$$\dot{x}_2 = \dot{x}_1 + \ddot{x}_1\,\Delta t_{1-2} \qquad \bar{\dot{x}}_{1-2} = \tfrac{1}{2}(\dot{x}_1 + \dot{x}_2)$$

   c. Similarly, find the final and average displacements for the Δt_{1-2} interval from

$$x_2 = x_1 + \bar{\dot{x}}_{1-2}\,\Delta t_{1-2} \qquad \bar{x}_{1-2} = \tfrac{1}{2}(x_1 + x_2)$$

   d. Compute a better approximation to the average f and $\ddot{x}$ during the interval Δt_{1-2} by using $x = \bar{x}_{1-2}$, $\dot{x} = \bar{\dot{x}}_{1-2}$, $t = \Delta t_{0-1} + \tfrac{1}{2}\Delta t_{1-2}$ in the determination of f.
4. Repeat steps 3*b* to 3*d*, starting with the better value of f, until no changes in f occur to within the desired accuracy.
5. One may then continue by essentially repeating steps 3 and 4 for additional time intervals until one has determined the motion for the desired total time of interest.

Generally, the smaller the time intervals selected, the greater will be the accuracy of the results (regardless of the f-averaging method used). Use of smaller time intervals naturally leads to a considerable increase in computational effort. If high accuracy is required, one may generally benefit by employing one of the many available more sophisticated numerical-integration schemes.[12,44,45] In many practical instances the labor of carrying out the required calculations by "hand" becomes prohibitive, however, and use of a digital computer is indicated.

*It should be noted that evaluation of f at the average values of the variables involved is only one of many possible ways of obtaining an average f for the interval considered. Other averages, for example, can be obtained from $\sqrt{f_0 f_1}$, $\tfrac{1}{2}(f_0 + f_1)$. One can rarely predict which average will produce the most accurate results in a given case.

A Semigraphical Method. One may avoid some of the tedium of the foregoing numerical-solution method and gain some insight into a problem by using the "phase-plane delta" method[25,32] discussed here. This method, like the foregoing numerical one, is essentially a stepwise integration for small time increments. It is based on rewriting the equation of motion (4.24) as

$$\ddot{x} + \omega_0^2(x + \delta) = 0 \qquad \delta = -x + f(x, \dot{x}, t)/\omega_0^2 \tag{4.25}$$

where ω_0 is any convenient constant circular frequency. (Any value may be chosen for ω_0, but it is usually useful to select one with some physical meaning, e.g., $\omega_0 = \sqrt{k_0/m_0}$, where k_0 and m_0 are values of stiffness and mass for small, x, $\dot{x}$, and t.) If one introduces into Eq. (4.25) a reduced velocity v given by

$$v = \dot{x}/\omega_0$$

and assumes that $\delta(x, \dot{x}, t)$ remains essentially constant in a short time interval one may integrate the resulting equation to obtain

$$v^2 + (x + \delta)^2 = R^2 = \text{const}$$

Thus for small time increments the solutions of (4.25) are represented in the xv plane (Fig. 4.9) by short arcs of circles whose centers are at $x = -\delta$, $v = 0$.

The angle $\Delta\theta$ subtended by the aforementioned circular arc is related to the time interval Δt according to

$$\Delta t \approx \Delta x/v\omega_0 \approx \Delta\theta/\omega_0 \tag{4.26}$$

On the basis of the foregoing discussion one may thus proceed as follows:

1. Calculate $\delta(x_0, \dot{x}_0, t_0)$ from Eq. (4.25) using the given initial conditions.
2. Locate the circle center $(-\delta, 0)$ and the initial point (x_0, v_0) on the xv plane; draw a small clockwise arc.
3. At the end of this arc is the point x_1, v_1 corresponding to the end of the first time interval.
4. Measure or calculate (in radians) the angle $\Delta\theta$ subtended by the arc; calculate the length of the time increment from Eq. (4.26).
5. Calculate $\delta(x_1, \dot{x}_1, t_1)$, and continue as before.
6. Repeat this process until the desired information is obtained.
7. Plots, such as those of x, $\dot{x}$, or $\ddot{x}$, against time, may then be readily obtained from the xv curve and the computed time information.

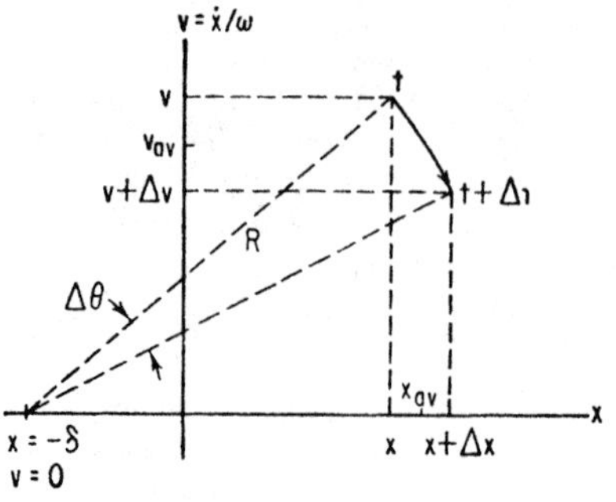

FIG. 4.9 Phase-plane delta method.

If increased accuracy is desired, particularly where δ changes rapidly, δ should be evaluated from average conditions (x_{av}, $\dot{x}_{av}$, t_{av}) during the time increment instead of conditions at the beginning of this increment (see Fig. 4.9). If δ depends on only one variable, a plot of δ against this variable may generally be used to advantage, particularly if it is superposed onto the vx plane.

Mathematical-Approximation Methods. An analytical expression is usually preferable to a series of numerical solutions, since it generally permits greater insight into a given problem. If exact analytical solutions cannot be found, approximate ones may be the next best approach.

Series expansion of the dependent in terms of the independent variable is often a useful expedient. Power series and Fourier series are most commonly used, but occasionally series of other functions may be employed. The approach consists essentially of writing the dependent variable in terms of a series with unknown coefficients, substituting this into the differential equation, and then solving for the coefficients. However, in many cases these solutions may be difficult, or the series may converge slowly or not at all.

Other methods attempt to obtain solutions by separating the governing equations into a linear part (for which a simple solution can be found) and a nonlinear part. The solution of the linear part is then applied to the nonlinear part in some way so as to give a first correction to the solution. The correction process is then repeated until a second better approximation is obtained, and the process is continued. Such methods include:

1. Perturbation,[21,40] which is particularly useful where the nonlinearities (deviations from linearity) are small
2. Reversion,[21] which is a special treatment of the perturbation method
3. Variation of parameters,[21,60] useful where nonlinearities do not result in additive terms
4. Averaging methods, based on error minimization
   a. Galerkin's method[21]
   b. Ritz method[21]

Conservative Systems: The Phase Plane.[21,26,60] A conservative system is one whose equation of motion can be written

$$m\ddot{x} + f(x) = 0 \tag{4.27}$$

In such systems (which may be visualized as masses attached to springs of variable stiffness) the total energy E remains constant; that is,

$$E = V(\dot{x}) + U(x)$$

where $$V(\dot{x}) = \tfrac{1}{2}m\dot{x}^2 \qquad U(x) = \int_{x_0}^{x} f(x)\,dx$$

where $V(\dot{x})$ and $U(x)$ are, respectively, the kinetic and the potential energies and x_0 is a convenient reference value.

The velocity-displacement ($\dot{x}$ vs. x) plane is called the "phase plane"; a curve in it is called a "phase trajectory." The equation of a phase trajectory of a conservative system with a given total energy E is

$$\dot{x}^2 = (2/m)[E - U(x)] \tag{4.28}$$

The time interval $(t-t_1)$ in which a change of displacement from x_1 to x occurs, is given by

$$t - t_1 = \int_{x_1}^{x} \frac{dx}{\dot{x}} = \int_{x_1}^{x} \frac{dx}{\sqrt{(2/m)[E - U(x)]}} \tag{4.29}$$

Some understanding of the geometry of phase trajectories may be obtained with the aid of Fig. 4.10, which shows the dependence of phase trajectories on total energy for a hypothetical potential energy function $U(x)$. For $E = E_1$ the motion is periodic; zero

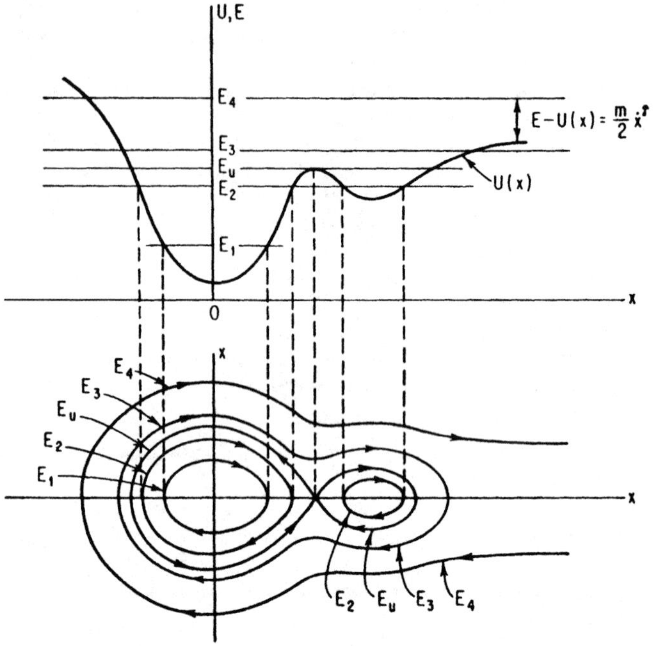

FIG. 4.10 Dependence of phase trajectories on energy.

velocity and velocity reversal occur where $U(x) = E_1$. With $E = E_2$ periodic oscillations are possible about two points; the initial conditions applicable in a given case dictate which type of oscillation occurs in that case. For $E = E_3$ only a single periodic motion is possible; for $E = E_4$ the motion is aperiodic. For $E = E_u$ there exists an instability at x_u; there the mass may move either in the increasing or decreasing x direction. (The arrows on the phase trajectories point in the direction of increasing time.)

For a linear undamped system $f(x) = kx$, $U(x) = \frac{1}{2}kx^2$, and phase trajectories are ellipses with semiaxes $(2E/k)^{1/2}$, $(2E/m)^{1/2}$.

The following facts may be summarized for conservative systems:

1. Oscillatory motions occur about minima in $U(x)$.
2. Phase trajectories are symmetric about the x axis and cross the x axis perpendicularly.
3. If $f(x)$ is single-valued, the phase trajectories for different energies E do not intersect.
4. All finite motions are periodic.

For nonconservative systems the phase trajectories tend to cross the constant-energy trajectories for the corresponding conservative systems. For damped systems the trajectories tend toward lower energy, i.e., they spiral into a point of stability. For excited systems the trajectories spiral outward, either toward a "limit-cycle" trajectory or indefinitely.[21,60]

The period T of an oscillation of a conservative system occurring with maximum displacement (amplitude) x_{max} may be computed from

$$T = \sqrt{2m} \int_{x_{min}}^{x_{max}} \left[\int_{x}^{x_{max}} f(x)\,dx \right]^{-1/2} dx \tag{4.30}$$

where x_{max} and x_{min} are the largest and smallest values of x (algebraically) for which zero velocity $\dot{x}$ occurs. The frequency f may then be obtained from $f = 1/T$.

For a linear spring-mass system with clearance, as shown in Fig. 4.11, the frequency is given by[17,32]

$$f = \sqrt{\frac{k}{m}} \left\{ 2\left[\pi + \frac{2}{(x_{max}/a) - 1} \right] \right\}^{-1}$$

where x_{max} is the maximum excursion of the mass from its middle position.

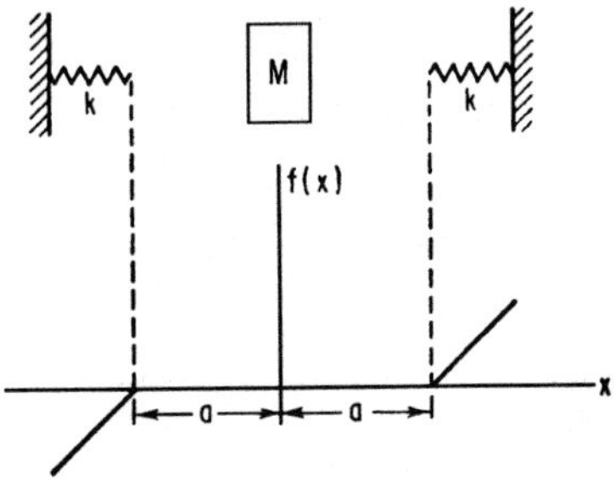

FIG. 4.11 Spring-mass system with clearance.

For a system governed by Eq. (4.27) with $f(x)=kx|x|^{b-1}$ [or $f(x) = kx^b$, if b is odd], the frequency is given by[35]

$$f = \sqrt{\frac{(b+1)k}{8\pi m} x_{max}^{b-1}} \frac{\Gamma[1/(b+1) + \frac{1}{2}]}{\Gamma[1/(b+1)]}$$

in terms of the gamma function Γ, values of which are available in many tables.

Steady-State Periodic Responses. In many cases, particularly in steady-state analyses of periodically forced systems, periodic oscillatory solutions are of primary interest. A number of mathematical approaches are available to deal with these problems. Most of these, including the well-known methods of Stoker[60] and Schwesinger,[21,54] are based on the idea of "harmonic balance."[21] They essentially assume a Fourier expansion of the solution and then require the coefficients to be adjusted so that relevant conditions on the lowest few harmonic components are satisfied.

For example, in order to find a steady-state periodic solution of

$$m\ddot{x} + g(\dot{x}) + f(x) = F \sin(\omega t + \phi)$$

one may substitute an assumed displacement

$$x = A_1 \sin \omega t + A_2 \sin 2\omega t + \cdots + A_n \sin n\omega t$$

and impose certain restrictions on the error ϵ,

$$\epsilon(t) = m\ddot{x} + g(\dot{x}) + f(x) - F \sin(\omega t + \phi)$$

In Schwesinger's method the mean-square value of the error, $\overline{\epsilon^2} = \int_0^{2\pi} \epsilon^2(t)d(\omega t)$, is minimized, and values of F and ϕ are calculated from this minimization corresponding to an assumed A_1.

Systems and Nonlinear Springs. The restoring forces of many systems (particularly with small amounts of nonlinearity) may be approximated so that the equation of motion may be written as

$$\ddot{x} + 2\zeta\omega_0\dot{x} + \omega_0^2 x + (a/m)x^3 = (F/m) \cos \omega t \tag{4.31}$$

in the presence of viscous damping and a sinusoidal force. ζ is the damping factor and ω_0 the natural frequency of a corresponding undamped linear system (i.e., for $a = 0$).

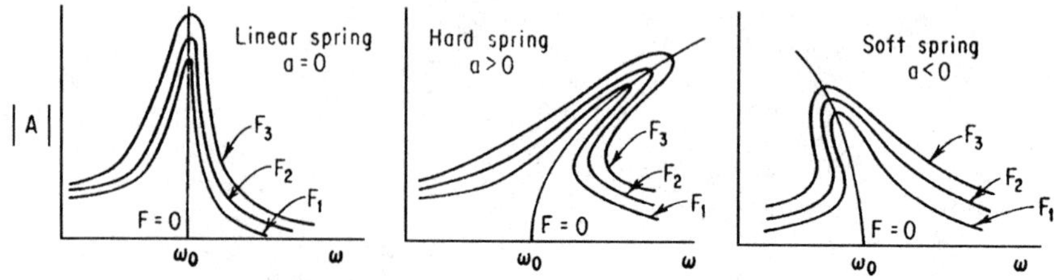

FIG. 4.12 Comparison of frequency responses of linear and nonlinear systems.

For $a > 0$ the spring becomes stiffer with increasing deflection and is called "hard"; for $a < 0$ the spring becomes less stiff and is called "soft."

Figure 4.12 compares the responses of linear and nonlinear lightly damped spring systems. The responses are essentially of the form $x = A \cos \omega t$; curves of response amplitude A vs. forcing frequency ω are sketched for several values of forcing amplitude F, for constant damping ζ. For a linear system the frequency of free oscillations ($F = 0$) is independent of amplitude; for a hard system it increases; for a soft system it decreases with increasing amplitude. The response curves of the nonlinear spring systems may be visualized as "bent-over" forms of the corresponding curves for the linear systems.

As apparent from Fig. 4.12, the response curves of the nonlinear spring systems are triple-valued for some frequencies. This fact leads to "jump" phenomena, as sketched in Figs. 4.13 and 4.14. If a given force amplitude is maintained as forcing frequency is changed slowly, then the response amplitude follows the usual response curve until point 1 of Fig. 4.13 is reached. The hatched regions between points 1 and 3 correspond to unstable conditions; an increase in ω above point 1 causes the amplitude to jump to that corresponding to point 2. A similar condition occurs when frequency is slowly decreased; the jump then occurs between points 3 and 4.

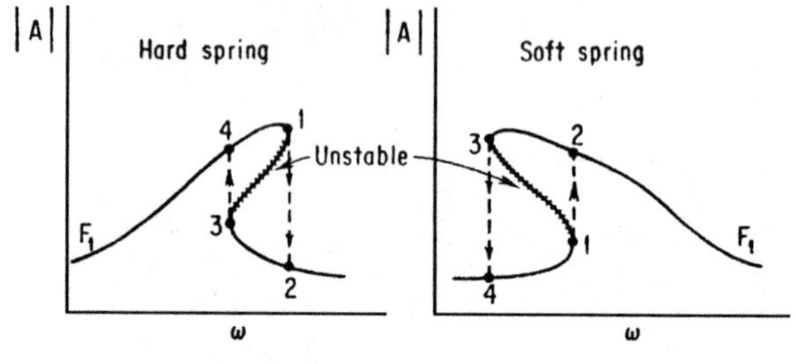

FIG. 4.13 Jump phenomena with variable frequency and constant force.

As also evident from Fig. 4.12, a curve of response amplitude vs. force amplitude at constant frequency is also triple-valued in some regions of frequency. Thus amplitude jumps occur also when one changes the forcing amplitude slowly at constant frequency ω_c. This condition is sketched in Fig. 4.14. For a hard spring this can occur only at frequencies above ω_0, for soft springs below ω_0. The equations characterizing a lightly damped nonlinear spring system and its jumps are summarized in Fig. 4.15.

The previous discussion deals with the system response as if it were a pure sinusoid $x = A \cos \omega t$. However, in nonlinear systems there occur also harmonic components (at frequencies $n\omega$, where n is an integer) and subharmonic components (frequencies

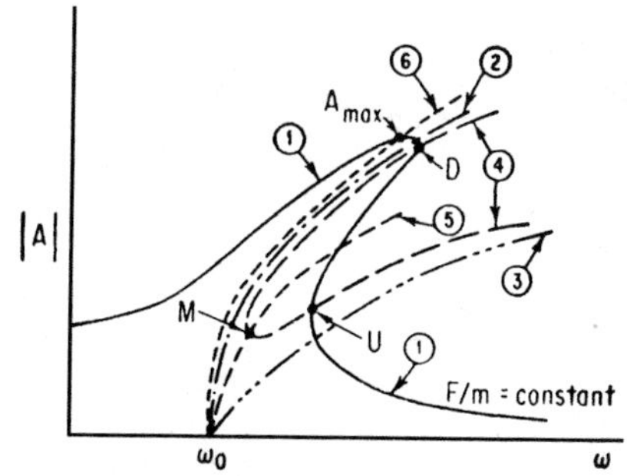

FIG. 4.15 Characteristics of responses of nonlinear springs. Equation of motion: $m\ddot{x} + c\dot{x} + kx + ax^3 = F \cos \omega t$, where ω_0 = undamped natural frequency for linear system (with $a = 0$); ζ = damping ratio for linear system = $c/2m\omega_0$. (1) Response curve for forced vibrations, F/m constant: $[A(\omega_0^2 - \omega^2) + \frac{3}{4}(a/m)A^3]^2 + [2\zeta\omega_0\omega A]^2 = (F/m)^2$. (2) Response curve for undamped free vibrations (approximate locus of downward jump points D, and of A_{max}): $\omega^2 = \omega_0^2 + \frac{3}{4}(a/m)A^2$. (3) Locus of upward jump points U with zero damping; approximate locus of same with finite damping: $\omega^2 = \omega_0^2 + \frac{9}{4}(a/m)A^2$. (4) Locus of upward and downward jump points U and D with finite damping: $[\omega_0^2 - \omega^2 + \frac{3}{4}(a/m)A^2][\omega_0^2 - \omega^2 + \frac{9}{4}(a/m)A^2] + (2\zeta\omega_0\omega)^2 = 0$. (5) Locus of points M below which no jumps occur: $\omega^2 = \omega_0^2 + \frac{9}{8}(a/m)A^2$. (6) Locus of maximum amplitudes A_{max}: $\omega^2 - \frac{3}{4}(a/m)A^2 = \omega_0^2(1 - 2\zeta^2)$.

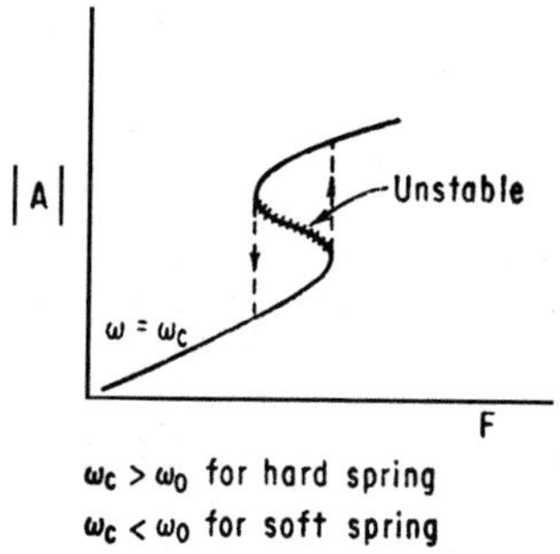

FIG. 4.14 Jump phenomena with variable force and constant frequency.

ω/n). In addition, components occur at frequencies which are integral multiples of the subharmonic frequencies. The various harmonic and subharmonic components tend to be small for small amounts of nonlinearity, and damping tends to limit the occurrence of subharmonics. The amplitude of the component with frequency 3ω (the lowest harmonic above the fundamental with finite amplitude for an undamped system) is given by $aA^3/36m\omega^2$, where A is the amplitude of the fundamental response, in view of Eq. (4.31). For more complete discussions see Ref. 21.

Graphical Determination of Response Amplitudes. A relatively easily applied method for approximating response amplitudes was developed by Martienssen[37] and improved by Mahalingam.[34] It is based on the often observed fact that the response to sinusoidal excitation is essentially sinusoidal. The method is here first explained for a linear system, then illustrated for nonlinear ones.

In order to obtain the steady-state response of a linear system one substitutes an assumed trial solution

$$x = A \cos (\omega t - \phi)$$

into the equation of motion

$$m\ddot{x} + c\dot{x} + kx = P \cos \omega t$$

By equating coefficients of corresponding terms on the two sides of the resulting equation one obtains a pair of equations which may be solved to yield

$$\omega_n^2 A = \omega^2 A + \frac{P}{m}\cos\phi \qquad \tan\phi = \frac{c\omega/m}{(\omega_n/\omega)^2 - 1} \tag{4.32}$$

where ω_n is the undamped natural frequency. One may plot the functions

$$y_1(A) = A\omega_n{}^2(A) \qquad y_2(A) = A\omega^2 + (P/m)\cos\theta$$

and determine for which value of A these two functions intersect. This value of A then is the desired amplitude. Since P and m are given constants, y_2 plots as a straight line with slope ω^2 and y intercept $(P/m)\cos\phi$. For a linear system one may compute $\tan\phi$ directly from Eq. (4.32), but for a nonlinear system ω_n^2 depends on the amplitude A and direct computation of $\tan\phi$ is not generally possible. Use of a method of successive approximations is then indicated.

Figure 4.16*a* shows application of this method to a linear system. After calculating ϕ from Eq. (4.32) one may find the y intercept $(P/m)\cos\phi$. For a given frequency ω one may then draw a line of slope ω^2 through that intercept to represent the function y_2. For a linear system y_1 is a straight line with slope ω_n^2 and passing through the origin. The amplitude of the steady-state oscillation may then be determined as the value A_0 of A where the two lines intersect.

Figure 4.16*b* shows a diagram analogous to Fig. 4.16*a*, but for an arbitrary nonlinear system. The function $y_1 = A\omega_n^2$ is not a straight line in general since ω_n generally is a function of A. This function may be determined from the restoring function $f(x)$ by use of Eq. (4.30). The possible amplitudes corresponding to a given driving frequency ω and force amplitude P are determined here, as before, by the intersection of the y_1 and y_2 curves. As shown in the figure, more than one amplitude may correspond to a given frequency—a condition often encountered in nonlinear systems.

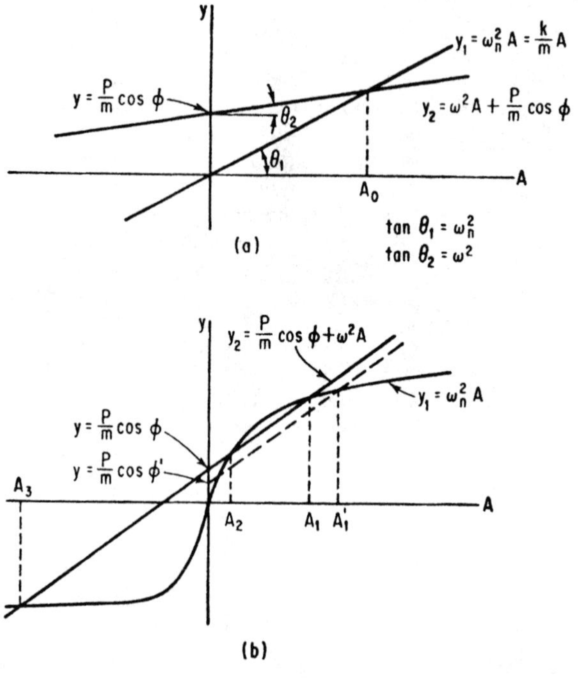

FIG. 4.16 Graphical determination of amplitude. (*a*) Linear system. (*b*) Nonlinear system.

In applying the previously outlined method to nonlinear systems one generally cannot find the correct value of ϕ at once from Eq. (4.32), since ω_n depends on the amplitude A, as has been pointed out. Instead one may assume any value of ϕ, such as ϕ', and determine a first approximation A_1' to A_1. Using the approximate amplitude one may then determine better values of $\omega_n(A)$ and ϕ from Eq. (4.32) and then use these better values to obtain a better approximation to A_1. This process may be repeated until ϕ and A_1 have been found to the desired degree of accuracy. A separate iteration process of this sort is generally required for each of the possible amplitudes. (Different values of ϕ correspond to the different amplitudes A_1, A_2, A_3.)

Systems with Nonlinear Damping. The governing equation in this case may be written as

$$m\ddot{x} + C(x, \dot{x}) + kx = F \sin \omega t \tag{4.33}$$

where $C(x, \dot{x})$ represents the effect of damping. Exact or reasonably good approximate solutions are available only for relatively few cases. However, for many practical cases where the damping is not too great, the system response is essentially sinusoidal. One may then use an "equivalent" viscous-damping term $c_e\dot{x}$ instead of $C(x, \dot{x})$, so that the equivalent damping results in the same amount of energy dissipation per cycle as does the original nonlinear damping. For a response of the form $x = A \sin \omega t$ the equivalent viscous damping may be computed from

$$c_e = \frac{1}{\pi\omega A}\int_0^{2\pi} C(A \sin \omega t, A\omega \cos \omega t)\, d(\omega t) \tag{4.34}$$

In contrast to the usual linear case, c_e is generally a function of frequency and amplitude. Once c_e has been found, the response amplitude may be computed from Eq. (4.6); successive approximations must be used if c_e is amplitude-dependent.

For "dry" or Coulomb friction, where the friction force is $\pm\,\mu$ (constant in magnitude but always directed opposite to the velocity) the equivalent viscous damping c_e and response amplitude A are given by

$$c_e = \frac{4\mu}{\pi A\omega} \qquad A = \frac{F}{k(1-\omega^2/\omega_n^2)}\sqrt{1-\left(\frac{4}{\pi}\frac{\mu}{F}\right)^2}$$

Further details on Coulomb damped systems appear in Refs. 17 and 63.

If complex notation is used, the damping effect may be expressed in terms of an imaginary stiffness term, and Eq. (4.33) may alternatively be written

$$m\ddot{x} + k(1 + i\gamma)x = Fe^{i\omega t}$$

where γ is known as the "structural damping factor."[18,51,68] For a response given by $x = Ae^{i\omega t}$ one finds

$$|A| = F/|-m\omega^2 + k(1 + i\gamma)| \tag{4.35}$$

This reduces identically to the linear case with viscous damping c, if γ is defined so that $k\gamma = \omega c$. The steady-state behavior of a system with any reasonable type of damping may be represented by this complex stiffness concept, provided that γ is prescribed with the proper frequency and amplitude dependence. The foregoing relation and Eq. (4.34) may be used to find the aforementioned proper dependences for a given resisting force $C(x, \dot{x})$. The case of constant γ corresponds to "structural damping" (widely used in aircraft flutter calculations) and represents a damping force proportional to displacement but in phase with velocity. The damping factor γ is particularly useful for describing the damping action of rubberlike materials, for which the damping is virtually independent of amplitude (but not of frequency),[55] since then the response is explicitly given by Eq. (4.35).

Response to Random Excitation. For quantitative results the reader is referred to Refs. 3, 11, and 16.

Qualitatively, the response of a linear system to random excitation is essentially a sinusoid at the system's natural frequency. The amplitudes (i.e., the envelope of this sinusoid) vary slowly and have a Rayleigh distribution. Compared with a linear system a system with a "hard" spring has a higher natural frequency, a lower probability of large excursions, and waves with flattened peaks. ("Soft" spring systems exhibit opposite characteristics.) The effects of the nonlinearities on frequency and on the wave shape are generally very small.

Self-Excited Systems. If the damping coefficient c of a linear system is negative, the system tends to oscillate with ever-increasing amplitude. Positive damping extracts energy from the system; negative damping contributes energy to it. A system (such as one with negative damping) for which the energy-contributing forces are controlled by the system motion is called self-excited. A source of energy must be available if a system is to be self-excited. The steady-state amplitude of a self-excited oscillation may generally be determined from energy considerations, i.e., by requiring the total energy dissipated per cycle to equal the total energy supplied per cycle.

The chatter of cutting tools, screeching of hinges or locomotive wheels, and chatter of clutches are due to self-excited oscillations associated with friction forces which decrease with increasing relative velocity. The larger friction forces at lower relative velocities add energy to the system; smaller friction forces at higher velocities (more slippage) remove energy. While the oscillations build up, the energy added is greater than that removed; at steady state in each cycle the added energy is equal to the extracted energy.

More detailed discussions of self-excited systems may be found in Refs. 21 and 60.

4.3 SYSTEMS WITH A FINITE NUMBER OF DEGREES OF FREEDOM

The instantaneous configurations of many physical systems can be specified by means of a finite number of coordinates. Continuous systems, which have an infinite number of degrees of freedom, can be approximated for many purposes by systems with only a finite number of degrees of freedom, by "lumping" of stiffnesses, masses, and distributed forces. This concept, which is extremely useful for the analysis of practical problems, has been the basis for numerous computer-based "finite-element" and "modal-analysis" methods.[24,44,45]

4.3.1 Systematic Determination of Equations of Motion

Generalized Coordinates: Constraints.[53,64] A set of n quantities q_i ($i = 1, 2, \ldots, n$) which at any time t completely specify the configuration of a system are called "generalized coordinates" of the system. The quantities may or may not be usual space coordinates.

If one selects more generalized coordinates than the minimum number necessary to describe a given system fully, then one finds some interdependence of the selected coordinates dictated by the geometry of the system. This interdependence may be expressed as

$$G(q_1, q_2, \ldots, q_n; \dot{q}_1, \dot{q}_2, \ldots, \dot{q}_n; t) = 0 \tag{4.36}$$

Relations like those of Eq. (4.36) are known as "equations of constraint,"; if no such equations can be formulated for a given set of generalized coordinates, the set is known as "kinematically independent."

The constraints of a set of generalized coordinates are said to be integrable if all equations like (4.36) either contain no derivatives $\dot{q}_i$ or if such $\dot{q}_i$ that do appear can be eliminated by integration. If to the set of n generalized coordinates there correspond m constraints, all of which are integrable, then one may find a new set of $(n - m)$ generalized coordinates which are subject to no constraints. This new system is called "holonomic," and $(n - m)$ is the number of degrees of freedom of the system (that is, the smallest number of quantities necessary to describe the system configuration at any time). In practice one may often be able to select a holonomic system of generalized coordinates by inspection. Henceforth the discussion will be limited to holonomic systems.

Lagrangian Equations of Motion. The equations governing the motion of any holonomic system may be obtained by application of Lagrange's equation

$$(d/dt)(\partial T/\partial \dot{q}_i) - \partial T/\partial \dot{q}_i + \partial U/\partial q_i + \partial F/\partial \dot{q}_i = Q_i \qquad i = 1, 2, \ldots, n \tag{4.37}$$

where T denotes the kinetic energy, U the potential energy of the entire dynamic system, F is a dissipation function, and Q_i the generalized force associated with the generalized coordinate q_i.

The generalized force Q_i may be obtained from

$$\delta W = Q_i \delta q_i \tag{4.38}$$

where δW is the total work done on the system by all external forces not contributing to U when the single coordinate q_i is changed to $q_i + \delta q_i$. The potential energy U accounts only for forces which are "conservative" (that is, for those forces for which the work done in a displacement of the system is a function of only the initial and final configurations). The dissipation function F represents half the rate at which energy is lost from the system; it accounts for the dissipative forces that appear in the equations of motion.

Lagrange's equations can be applied to nonlinear as well as linear systems, but little can be said about solving the resulting equations of motion if they are nonlinear, except that small oscillations of nonlinear systems about equilibrium can always be approximated by linear equations. Methods of solving linear sets of equations of motion are available and are discussed subsequently.

4.3.2 Matrix Methods for Linear Systems—Formalism

The kinetic energy T, potential energy U, and dissipation function F of any *linear* holonomic system with n degrees of freedom may be written as

$$\begin{aligned} T &= \tfrac{1}{2}\sum_{i=1}^{n}\sum_{j=1}^{n} a_{ij}\dot{q}_i\dot{q}_j \qquad a_{ij} = a_{ji} \\ U &= \tfrac{1}{2}\sum_{i=1}^{n}\sum_{j=1}^{n} c_{ij}q_iq_j \qquad c_{ij} = c_{ji} \\ F &= \tfrac{1}{2}\sum_{i=1}^{n}\sum_{j=1}^{n} b_{ij}\dot{q}_i\dot{q}_j \qquad b_{ij} = b_{ji} \end{aligned} \tag{4.39}$$

The equations of motion may then readily be determined by use of Eq. (4.37). They may be expressed in matrix form as

$$A\{\ddot{q}\}+B\{\dot{q}\}+C\{q\}=\{Q(t)\} \tag{4.40}$$

where A, B, C are symmetric square matrices with n rows and n columns whose elements are the coefficients appearing in Eqs. (4.39) and $\{q\}$, $\{\dot{q}\}$, $\{\ddot{q}\}$, $\{Q(t)\}$ are n-dimensional column vectors.

(*Note*: A is called the inertia matrix, B the damping matrix, C the elastic or the stiffness matrix. For example,

$$A = \begin{bmatrix} a_{11} & a_{12} & \cdots & a_{1n} \\ a_{21} & a_{22} & \cdots & a_{2n} \\ \cdot & \cdot & & \cdot \\ \cdot & \cdot & & \cdot \\ \cdot & \cdot & & \cdot \\ a_{n1} & a_{n2} & \cdots & a_{nn} \end{bmatrix} \quad \{q\} = \begin{bmatrix} q_1 \\ q_2 \\ \cdot \\ \cdot \\ \cdot \\ q_n \end{bmatrix} \quad \{\dot{q}\} = \begin{bmatrix} \dot{q}_1 \\ \dot{q}_2 \\ \cdot \\ \cdot \\ \cdot \\ \dot{q}_n \end{bmatrix} \quad \{Q\} = \begin{bmatrix} Q_1 \\ Q_2 \\ \cdot \\ \cdot \\ \cdot \\ Q_n \end{bmatrix}$$

The elements of $\{q\}$ are the coordinates of q_i, the elements of $\{\dot{q}\}$ are the first time derivatives of q_i (i.e., the generalized velocities $\dot{q}_i$); those of $\{\ddot{q}\}$ are the generalized accelerations $\ddot{q}_i$, those of $\{Q\}$ are the generalized forces Q_i.)

Free Vibrations: General System. If all the generalized forces Q_i are zero, then Eq. (4.40) reduces to a set of homogeneous linear differential equations. To solve it one may postulate a time dependence given by

$$\{q\}=\{r\}e^{st} \qquad s = \sigma + i\omega$$

which, when introduced into Eq. (4.40), results in

$$(s^2A + sB + C)\{r\}= 0 \tag{4.41}$$

One may generally find n nontrivial solutions $\{r^{(j)}\}$, $j = 1, 2, \ldots, n$, each corresponding to a specific value $s_{(j)}$ of s. The general solution of the homogeneous equation may then be expressed as

$$\{q\}=\sum_{j=1}^{n} \alpha_j\{r^{(j)}\}e^{s(j)t} = R\{p\} \tag{4.42}$$

in terms of complex constants α_j and the newly defined

$$R = \begin{bmatrix} r_1^{(1)} & r_1^{(2)} & \cdots & r_1^{(n)} \\ r_2^{(1)} & r_2^{(2)} & \cdots & r_2^{(n)} \\ \cdot & \cdot & & \cdot \\ \cdot & \cdot & & \cdot \\ \cdot & \cdot & & \cdot \\ r_n^{(1)} & r_n^{(2)} & \cdots & r_n^{(n)} \end{bmatrix} \qquad \{p(t)\} = \begin{bmatrix} \alpha_1 & e^{s(1)t} \\ \alpha_2 & e^{s(2)t} \\ \cdot & \cdot \\ \cdot & \cdot \\ \cdot & \cdot \\ \alpha_n & e^{s(n)t} \end{bmatrix} \tag{4.43}$$

Initial conditions, e.g., $\{q(0)\}$, $\{\dot{q}(0)\}$ may be introduced into Eq. (4.42) to evaluate the constants α_j.

Forced Vibrations: General System. The forced motion may be described in terms of the sum of two motions, one satisfying the homogeneous equation (with all $Q_i = 0$) and including all the constants of integration, the other satisfying the complete Eq. (4.40) and containing no integration constants. (The constants must be evaluated so that the total solution satisfies the prescribed initial conditions.) The latter constant-free solution is often called the "steady-state solution."

Steady-State Solution for Periodic Generalized Forces. If the generalized forces are harmonic with frequency ω_0 one may set

$$\{Q\}=\{\overline{Q}\}e^{i\omega_0 t} \qquad \{q\}=\{\overline{q}\}e^{i\omega_0 t} \tag{4.44}$$

in Eq. (4.40) and obtain

$$(-\omega_0^2 A + i\omega_0 B + C)\{\overline{q}\}=\{\overline{Q}\}$$

from which $\{\overline{q}\}$ may be determined. Equation (4.44) then gives the steady-state solutions.

If the Q_i are periodic with period T, but not harmonic, one may expand them and the components q_i of the steady-state solution in Fourier series:

$$\{Q\}=\sum_{N=-\infty}^{\infty}\{\overline{Q}^{(N)}\}e^{iN\omega_0 t} \qquad \{q\}=\sum_{N=-\infty}^{\infty}\{\overline{q}^{(N)}\}e^{iN\omega_0 t} \qquad \omega_0=\frac{T}{2\pi}$$

The Fourier components $\{\overline{q}^{(N)}\}$ may be evaluated from

$$(-N^2\omega_0^2 A + iN\omega_0 B + C)\{\overline{q}^{(N)}\} = \{\overline{Q}^{(N)}\}$$

Steady-State Solution for Aperiodic Generalized Forces. For general $\{Q(t)\}$ one may write the solutions of Eq. (4.40) as

$$q_i(t) = \sum_{j=1}^{n}\int_0^t h_i^{(j)}(t-\tau)Q_j(\tau)\,d\tau \tag{4.45}$$

where $h_i^{(j)}(t)$ is the response of coordinate q_i to a unit impulse acting in place of Q_j. The "impulse response function" $h_i^{(j)}$ may be found from

$$h_i^{(j)}(t) = (d/dt)u_i^{(j)}(t)$$

where $u_i^{(j)}$ is the response of q_i to a unit step function acting in place of Q_j. (A unit step function is zero for $t < 0$, unity for $t > 0$.)

If one defines a square matrix $[H(t)]$ whose elements are $h_i^{(j)}$, one may write Eq. (4.45) alternatively as

$$\{q(t)\}=\int_0^t [H(t-\tau)]\{Q(\tau)\}\,d\tau$$

where $$[H(t)] = \frac{1}{2\pi i}\int_{c+i\infty}^{c-i\infty}[T(s)]e^{st}\,ds \qquad [T(s)] = (s^2A + sB + C)^{-1}$$

Undamped Systems. For undamped systems all elements of the B matrix of Eq. (4.40) are zero, and Eq. (4.41) may be rewritten in the classical eigenvalue form

$$E\{r\} = \omega^2\{r\} \qquad E = A^{-1}C \tag{4.46}$$

The eigenvalues $\omega_{(j)}$, i.e., the values for which nonzero solutions $r^{(j)}$ exist, are real. They are the natural frequencies; the corresponding solution vectors $r^{(j)}$ describe the mode shapes.

One may then find a set of principal coordinates ψ_i, in terms of which the equations of motion are uncoupled and may be written in the following forms:

$$\{\ddot{\psi}\} + \Omega\{\psi\} = R^{-1}A^{-1}\{Q\} \qquad \text{or} \qquad M\{\ddot{\psi}\} + K\{\psi\} = \Phi(t)$$

Here $$M = \overline{R}AR \qquad K = \overline{R}CR \qquad \{\Phi(t)\} = \overline{R}\{Q(t)\}$$

Ω is the diagonal matrix of natural frequencies

$$\Omega = \begin{bmatrix} \omega_{(1)}^2 & 0 & \cdots & 0 \\ 0 & \omega_{(2)}^2 & & 0 \\ \cdot & \cdot & & \cdot \\ \cdot & \cdot & & \cdot \\ \cdot & \cdot & & \cdot \\ 0 & 0 & \cdots & \omega_{(n)}^2 \end{bmatrix}$$

R is given by Eq. (4.43), and $\overline{R}$ denotes the transpose of R.

The principal coordinates ψ_i are related to the original coordinates q_i by

$$\{q\} = R\{\psi\} \qquad \{\psi\} = R^{-1}\{q\} \tag{4.47}$$

The response of the system to any forcing function $\{Q(t)\}$ may be determined in terms of the principal coordinates from

$$\psi_j(t) = m_{jj}k_{jj}\int_0^t \Phi_j(\tau)\sin[\omega_{(j)}(t-\tau)]\,d\tau + \psi_j(0)\cos[\omega_{(j)}t] + [1/\omega_{(j)}]\dot{\psi}_j(0)\sin[\omega_{(j)}t] \tag{4.48}$$

The response in terms of the original coordinates q_i may then be obtained by substitution of the results of Eq. (4.48) into Eq. (4.47).

4.3.3 Matrix Iteration Solution of Positive-Definite Undamped Systems

Positive-Definite Systems: Influence Coefficient and Dynamic Matrixes. A system is "positive-definite" if its potential energy U, as given by Eq. (4.39), is greater than zero for any $\{q\} \neq \{0\}$. Systems connected to a fixed frame are positive-definite; systems capable of motion (changes in the coordinates q_i) without increasing U are called "semidefinite."[64] The latter motions occur without energy storage in the elastic elements and are called "rigid-body" motions or "zero modes." (They imply zero natural frequency.)

Rigid-body motions are generally of no interest in vibration study. They may be eliminated by proper choice of the generalized coordinates or by introducing additional relations (constraints) among an arbitrarily chosen system of generalized coordinates by applying conservation-of-momentum concepts. Thus any system of generalized coordinates can be reduced to a positive-definite one.

For positive-definite linear systems C^{-1}, the inverse of the elastic matrix, is known as the influence coefficient matrix D. The elements of D are the influence coefficients; the typical element d_{ij} is the change in coordinate q_i due to a unit generalized force Q_j (applied statically), with all other Q's equal to zero. Since these influence coefficients can be determined from statics, one generally need not find C at all. It should be noted that for systems that are not positive-definite one cannot compute the influence coefficients from statics alone.

For iteration purposes it is useful to rewrite Eq. (4.46), the system equation of free sinusoidal motion, as

$$G\{r\} = (1/\omega^2)\{r\} \tag{4.49}$$

where $$\{q\}=\{r\}e^{i\omega t} \qquad G = C^{-1}A = DA = E^{-1} \tag{4.50}$$

The matrix G is called the "dynamic matrix" and is defined, as above, as the product of the influence coefficient matrix D and the inertia matrix A.

Iteration for Lower Modes. In order to solve Eq. (4.49), which is a standard eigenvalue matrix equation, numerically for the lowest mode one may proceed as follows: Assume any vector $\{r_{(1)}\}$; then compute $G\{r_{(1)}\} = \alpha_{(1)}\{r_{(2)}\}$, where $\alpha_{(1)}$ is a constant chosen so that one element (say, the first) of $\{r_{(2)}\}$ is equal to the corresponding element of $\{r_{(1)}\}$. Then find $G\{r_{(2)}\} = \alpha_{(2)}\{r_{(3)}\}$, with $\alpha_{(2)}$ chosen like $\alpha_{(1)}$ before. Repeat this process until $\{r_{(n+1)}\} \approx \{r_{(n)}\}$ to the desired degree of accuracy. The corresponding constant $\alpha_{(n)}$ which satisfies $G\{r_{(n)}\} = \alpha(n)\{r_{(n+1)}\}$ then yields to lowest natural frequency ω_1 of the system and $\{r_{(n)}\}$ describes the shape of the corresponding (first) mode $\{r^{(1)}\}$. In view of Eq. (4.49)

$$\omega_1^2 = 1/\alpha_{(n)}$$

The second mode $\{r^{(2)}\}$ must satisfy the orthogonality relation

$$\overline{\{r^{(2)}\}}A\{r^{(1)}\} = 0 \qquad \text{or} \qquad \sum_{i=1}^{n}\sum_{j=1}^{n} r_i^{(2)}a_{ij}r_j^{(1)} = 0 \tag{4.51}$$

In order to obtain a vector that satisfies Eq. (4.51) from an arbitrary vector $\{r\}$ one may select $(n - 1)$ components of $\{r_{(2)}\}$ as equal to the corresponding components of $\{r\}$ and then compute the nth from Eq. (4.51). This process may be expressed as

$$\{r^{(2)}\} = S_1\{r\}$$

where S_1 is called the "first sweeping matrix." S_1 is equal to the identity matrix in n dimensions, except for one row which describes the interrelation Eq. (4.51). If A is diagonal one may take, for example,

$$S_1 = \begin{bmatrix} 0 & -\dfrac{a_{22}r_2^{(1)}}{a_{11}r_1^{(1)}} & -\dfrac{a_{33}r_3^{(1)}}{a_{11}r_1^{(1)}} & \cdots & \dfrac{a_{nn}r_n^{(1)}}{a_{11}r_1^{(1)}} \\ 0 & 1 & 0 & \cdots & 0 \\ 0 & 0 & 1 & \cdots & 0 \\ \cdot & \cdot & \cdot & & \cdot \\ \cdot & \cdot & \cdot & & \cdot \\ \cdot & \cdot & \cdot & & \cdot \\ 0 & 0 & 0 & \cdots & 1 \end{bmatrix}$$

To obtain the second lowest mode shape $\{r^{(2)}\}$ and the second lowest natural frequency ω_2 one may form $H_1 = GS_1$, and solve

$$H_1\{r\} = (1/\omega^2)\{r\} \tag{4.52}$$

by iteration. Since Eq. (4.52) is of the same form as Eq. (4.49), one may proceed here as previously discussed, i.e., by assuming a trial vector $\{r_{(1)}\}$, forming $H_1\{r_{(1)}\} = \alpha_{(1)}\{r_{(2)}\}$, so that one element of $\{r_{(2)}\}$ is equal to the corresponding element of $\{r_{(1)}\}$, then forming $H_1\{r_{(2)}\} = \alpha_{(2)}\{r_{(3)}\}$, etc. This process converges to $\{r^{(2)}\}$ and $\alpha = 1/\omega_2^2$.

The third mode $\{r^{(3)}\}$ similarly must satisfy

$$\overline{\{r^{(3)}\}}A\{r^{(1)}\} = 0 \qquad \overline{\{r^{(3)}\}}A\{r^{(2)}\} = 0$$

or

$$\sum_{i=1}^{n}\sum_{j=1}^{n} r_i^{(3)} a_{ij} r_j^{(1)} = 0 \qquad \sum_{i=1}^{n}\sum_{j=1}^{n} r_i^{(3)} a_{ij} r_j^{(2)} = 0 \tag{4.53}$$

One may thus select $n - 2$ components of $\{r^{(3)}\}$ as equal to the corresponding components of an arbitrary vector $\{r\}$ and adjust the remaining two components to satisfy Eq. (4.53). The matrix S_2 expressing this operation, or

$$\{r^{(3)}\} = S_2\{r\}$$

is called the "second sweeping matrix." For diagonal A one possible form of S_2 is

$$S_2 = \begin{bmatrix} 0 & -\dfrac{a_{22}r_2^{(1)}}{a_{11}r_1^{(1)}} & -\dfrac{a_{33}r_3^{(1)}}{a_{11}r_1^{(1)}} & -\dfrac{a_{nn}r_n^{(1)}}{a_{11}r_1^{(1)}} \\ 0 & -\dfrac{a_{22}r_2^{(2)}}{a_{11}r_1^{(2)}} & -\dfrac{a_{33}r_3^{(2)}}{a_{11}r_1^{(2)}} & -\dfrac{a_{nn}r_n^{(2)}}{a_{11}r_1^{(2)}} \\ 0 & 0 & 1 & 0 \\ \vdots & \vdots & \vdots & \vdots \\ 0 & 0 & \cdots & 1 \end{bmatrix}$$

Then one may form $H_2 = GS_2$ and solve

$$H_2\{r\} = (1/\omega^2)\{r\}$$

by iteration. The process here converges to $\{r^{(3)}\}$ and $\alpha = 1/\omega_3^2$.

Higher modes may be treated similarly; each mode must be orthogonal to all the lower ones, so that $p - 1$ relations like Eq. (4.51) must be utilized to find the $(p - 1)$st sweeping matrix. Iteration on $H_{(p-1)} = GS_{(p-1)}$ then converges to the pth mode.

Iteration for the Higher Modes. The previously outlined process begins with the lowest natural frequency and works toward the highest. It is not very useful for the highest few modes because of the tedium and of the accumulation of rounding off errors. Results for the higher modes can be obtained more simply and accurately by starting with the highest frequency and working toward lower ones.

The highest mode may be obtained by solving Eq. (4.46) directly by iteration. This is accomplished by assuming any trial vector $\{r_{(1)}\}$, forming $E\{r_{(1)}\} = \beta_{(1)}\{r_{(2)}\}$ with $\beta_{(1)}$ chosen so that one element of the result $\{r_{(2)}\}$ is equal to the corresponding element of $\{r_{(1)}\}$. Then one may form $E\{r_{(2)}\} = \beta_{(2)}\{r_{(3)}\}$ similarly, and continue until $\{r_{(p+1)}\} \approx \{r_{(p)}\}$ to within the required accuracy. Then $\{r_{(p)}\} = \{r^{(n)}\}$ and $\beta_{(p)} = \omega_n^2$.

The next-to-highest $[(n - 1)$st$]$ mode may be found by writing

$$\{r^{(n-1)}\} = T_1\{r\}$$

where T_1 is a sweeping matrix that "sweeps out" the nth mode. T_1 is equal to the identity matrix, except for one row, which expresses the orthogonality relation

$$\overline{\{r^{(n-1)}\}}A\{r^{(n)}\} = 0 \qquad \text{or} \qquad \sum_{i=1}^{n}\sum_{j=1}^{n} r_i^{(n-1)} a_{ij} r_j^{(n)} = 0$$

Iterative solution of

$$J_1\{r\} = \omega^2\{r\} \qquad \text{where} \qquad J_1 = ET_1$$

then converges to $\{r^{(n-1)}\}$ and $\omega_{(n-1)}^2$.

The next lower modes may be obtained similarly, using other sweeping matrixes embodying additional orthogonality relations in complete analogy to the iteration for lower modes described above.

4.3.4 Approximate Natural Frequencies of Conservative Systems

A conservative system is one which executes free oscillations without dissipating energy. The potential energy $\hat{U}$ that the system has at an instant when its velocity (and hence its kinetic energy) is zero must therefore be exactly equal to the kinetic energy $\hat{T}$ of the system when it occupies its equilibrium position (zero potential energy) during its oscillation.

For sinusoidal oscillations the (generalized) coordinates obey $q_j = \bar{q}_j \sin \omega t$ where $\bar{q}_j$ is a constant (i.e., the amplitude of q_j). For linear holonomic systems, in view of Eq. (4.39),

$$\hat{T} = \tfrac{1}{2}\omega^2 \sum_{i=1}^{n} \sum_{j=1}^{n} a_{ij}\bar{q}_i\bar{q}_j \qquad \hat{U} = \tfrac{1}{2}\sum_{i=1}^{n} \sum_{j=1}^{n} c_{ij}\bar{q}_i\bar{q}_j$$

Rayleigh's Quotient. Rayleigh's quotient RQ, defined as[64]

$$RQ = \frac{\sum_{i=1}^{n} \sum_{j=1}^{n} c_{ij}\bar{q}_i\bar{q}_j}{\sum_{i=1}^{n} \sum_{j=1}^{n} a_{ij}\bar{q}_i\bar{q}_j} \tag{4.54}$$

is a function of $\{\bar{q}_1, \bar{q}_2, \ldots, \bar{q}_n\}$. However, multiplication of each $\bar{q}_i$ by the same number does not change the value of RQ. Rayleigh's quotient has the following properties:

1. The value of RQ one obtains with any $\{\bar{q}_1, \bar{q}_2, \ldots, \bar{q}_n\}$ always equals or exceeds the square of the lowest natural frequency of the system; $RQ \geq \omega_1^2$.
2. $RQ=\omega_n^2$ if $\{\bar{q}_1, \bar{q}_2, \ldots, \bar{q}_n\}$ corresponds to the nth mode shape (eigenvector) of the system, but even fairly rough approximations to the eigenvector generally result in good approximations to ω_n^2.

For systems whose influence coefficients d_{ij} are known, one may substitute an arbitrary vector $\{\bar{q}\} = \{\bar{q}_1, \bar{q}_2, \ldots, \bar{q}_n\}$ into the right-hand side of

$$\{\bar{q}\}_1 = \alpha G\{\bar{q}\}$$

where $G = DA$ is the dynamic matrix of Eq. (4.50) and α is an arbitrary constant. [This relation follows directly from Eq. (4.49).] The resulting vector $\{\bar{q}\}_1$, when substituted into Eq. (4.54), results in a value of RQ which is nearer to ω_1^2 than the value obtained by direct substitution of the arbitrary vector $\{\bar{q}\}$. Often one obtains good results rapidly if one assumes $\{\bar{q}\}$ initially so as to correspond to the deflection of the system due to gravity (i.e., the "static" deflection).

Rayleigh-Ritz Procedure. An alternative method useful for obtaining improved approximations to ω_1^2 from Rayleigh's quotient is the so-called Rayleigh-Ritz procedure. It consists of computing RQ from Eq. (4.54) for a trial vector $\{\bar{q}\}$ made up of a linear combination of arbitrarily selected vectors

$$\{\bar{q}\} = \alpha_1\{\bar{q}\}_1 + \alpha_2\{\bar{q}\}_2 + \cdots$$

then minimizing RQ with respect to the coefficients α of the selected vectors. The resulting minimum value of RQ is approximately equal to ω_1^2. That is, RQ evaluated so that $\partial RQ/\partial\alpha_1 = \partial RQ/\partial\alpha_2 = \cdots = 0$ is approximately equal to ω_1^2:

$$RQ \approx \omega_1^2$$

Dunkerley's Equation.[61] This equation states that

$$1/\omega_1^2 + 1/\omega_2^2 + \cdots + 1/\omega_n^2 = 1/\Omega_1^2 + 1/\Omega_2^2 + \cdots + 1/\Omega_n^2 \tag{4.55}$$

where ω_i denotes the ith natural frequency of an n-degree-of-freedom system, and Ω_i denotes the natural frequency that the ith inertia element would have if all others were removed from the system. Usually $\omega_n^2 >> \cdots >> \omega_2^2 >> \omega_1^2$, so that the left-hand side of Eq. (4.55) is approximately equal to $1/\omega_1^2$ and Eq. (4.55) may be used directly for estimation of the fundamental frequency ω_1. In many cases the Ω_1 are obtainable almost by inspection, or by use of Table 4.8.

4.3.5 Chain Systems

A chain system is one in which the inertia elements are arranged in series, so that each is directly connected only to the one preceding and the one following it. Shafts carrying a number of disks (or other rotational inertia elements) are the most common example and are discussed in more detail subsequently. Translational chain systems, as sketched in Fig. 4.17, may be treated completely analogously and hence will not be discussed separately.

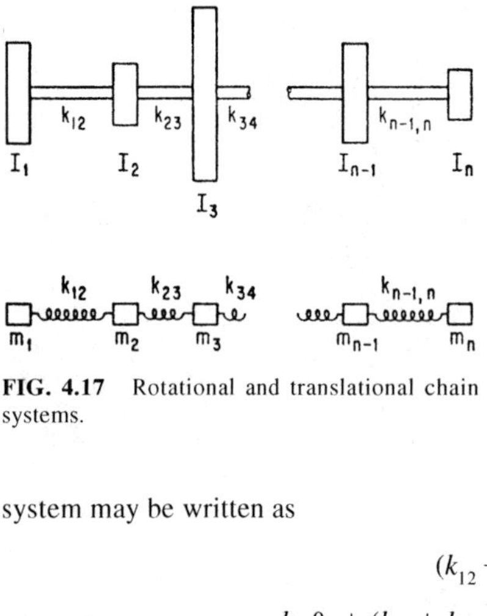

FIG. 4.17 Rotational and translational chain systems.

Sinusoidal Steady-State Forced Motion. If the torque acting on the sth disk is $T_s e^{i\omega t}$, where T_s is a known complex number, then the equations of motion of the system may be written as

$$\begin{aligned}
(k_{12} - I_1\omega^2)\theta_1 - k_{12}\theta_2 &= T_1 \\
-k_{12}\theta_1 + (k_{12} + k_{23} - I_2\omega^2)\theta_2 - k_{23}\theta_3 &= T_2 \\
-k_{23}\theta_2 + (k_{23} + k_{34} - I_3\omega^2)\theta_3 - k_{34}\theta_4 &= T_3 \\
\cdots\cdots\cdots\cdots\cdots\cdots\cdots\cdots\cdots & \\
-k_{n-1,n}\theta_{n-1} + (k_{n-1,n} - I_n\omega^2)\theta_n &= T_n
\end{aligned} \tag{4.56}$$

where $\theta_s e^{i\omega t}$ describes the angular motion of the sth disk, as measured from equilibrium. (Damping in the system may be taken into account by assigning complex values to the k's, as in the last portion of Sec. 4.2.2.)

One may solve this set of equations simply by using each equation in turn to eliminate one of the θ's, so that one may finally solve for the last remaining θ, then obtain the others by substitution of the determined value into the given equations. This procedure becomes prohibitively tedious if more than a few disks are involved.

If all T's and k's are real (i.e., if the driving torques are in phase and if damping is neglected), one may assume a real value for θ_1, then calculate the corresponding value of θ_2 from the first of Eqs. (4.56). Then one may find θ_3 from the second equation, θ_4 from the third, and so on. Finally, one may compute T_n from the last (nth) equation and compare it with the given value. This process may be repeated with different initially assumed θ_1 values until the computed value of T_n comes out sufficiently close to the specified one. After a few computations one may often make good use of a plot of computed T_n vs. assumed θ_1 for determining by interpolation or extrapolation a good approximation to the correct value of θ_1.

If all T's are zero, except T_n, one may proceed as before. But, since θ_1 is proportional to T_n in this case, the correct value of θ_1 may be computed directly after a single complete calculation by use of the proportionality

$$\theta_{1,\text{correct}} = \theta_{1,\text{assumed}}\,(T_{n,\text{specified}}/T_{n,\text{calculated}})$$

The latter approach may be used also for damped systems (i.e., with complex k's).

Natural Frequencies: Holzer's Method. Free oscillations of the system considered obey Eqs. (4.56), but with all $T_s = 0$. To obtain the natural frequencies one may proceed by assuming a value of ω and setting $\theta_1 = 1$, then calculating θ_2 from the first equation, thereafter θ_3 from the second, etc. Finally one may compute T_n from the last equation. If this T_n comes out zero, as required, the assumed frequency is a natural frequency. By repeating this calculation for a number of assumed values of ω one may arrive at a plot of T_n vs. ω, which will aid in the estimation of subsequent trial values of ω. (Natural frequencies are obtained where this curve crosses the ω axis.) One should keep in mind that a system composed of n disks has n natural frequencies. The mode shape (i.e., a set of values of θ's that satisfy the equation of motion) is also obtained in the course of the calculations.

Convenient tabular calculation methods (Holzer tables) may be set up on the basis of the equations of motion Eq. (4.56) rewritten in the following form:

$$I_1\theta_1\omega^2 = k_{12}(\theta_1 - \theta_2)$$

$$(I_1\theta_1 + I_2\theta_2)\omega^2 = k_{23}(\theta_2 - \theta_3)$$

$$\cdots\cdots\cdots\cdots\cdots\cdots\cdots\cdots$$

$$(I_1\theta_1 + I_2\theta_2 + \cdots + I_s\theta_s)\omega^2 = k_{s,s+1}(\theta_s - \theta_{s+1}) \tag{4.57}$$

$$\cdots\cdots\cdots\cdots\cdots\cdots\cdots\cdots\cdots\cdots\cdots\cdots$$

$$\sum_{s=1}^{n} (I_s\theta_s)\omega^2 = 0$$

Tabular formats are given in a number of texts.[17,61,63,64] Methods for obtaining good first trial values for the lowest natural frequency are also discussed in Refs. 17 and 61. Branched systems, e.g., where several shafts are interconnected by gears, may also be treated by this method.[9,61,63] Damped systems (complex values of k) may also be treated with no added difficulty in principle.[61]

4.3.6 Mechanical Circuits[20,30,50]

Mechanical-circuit theory is developed in direct analogy to electric-circuit theory in order to permit the highly developed electrical-network-analysis methods to be applied to mechanical systems. A mechanical system is considered as made up of a number of mechanical-circuit elements (e.g., masses, springs, force generators) connected in series or parallel, in much the same way that an electrical network is considered to be made up of a number of interconnected electrical elements (e.g., resistances, capacitances, voltage sources). From known behavior of the elements one may then, by proper combination according to established rules, determine the system responses to given excitations.

Mechanical-circuit concepts are useful for determination of the equations of motion (which may then be solved by classical or transform techniques[20,30]) for analyzing and visualizing the effects of system interconnections, for dealing with electromechanical systems, and for the construction of electrical analogs by means of which one may evaluate the responses by measurement.

Mechanical Impedance and Mobility. As in electric-circuit theory, the sinusoidal steady state is assumed, and complex notation is used in basic mechanical-impedance analysis. That is, forces F and relative velocities V are expressed as

$$F = F_0 e^{i\omega t} \qquad V = V_0 e^{i\omega t}$$

where F_0 and V_0 are complex in the most general case.

Mechanical impedance Z and mobility Y are complex quantities defined by

$$Z = F_0/V_0 \qquad Y = 1/Z = V_0/F_0 \tag{4.58}$$

If the force and velocity refer to an element or system, the corresponding impedance is called the "impedance of the element" or system; if F and V refer to quantities at the same point of a mechanical network, then Z is called the "driving-point impedance" at that point. If F and V refer to different points, the corresponding Z is called the "transfer impedance" between those points.

Mechanical impedances (and mobilities) of elements can be combined exactly like electrical impedances (and admittances), and the driving-point impedances of composites can easily be obtained. From a knowledge of the elemental impedances and of how they combine, one may calculate (and often estimate quickly) the behavior of composite systems.

Basic Impedances, Combination Laws, Analogies. Table 4.2 shows the basic mechanical-circuit elements and their impedances and summarizes the impedances of some simple systems.

In Table 4.3 are indicated the combination laws for mechanical impedances and mobilities. Table 4.4 is a summary of analogies between translational and rotational mechanical systems and electrical networks. The impedances of some distributed mechanical systems are discussed in Sec. 4.4.3 and summarized in Tables 4.6 and 4.7.

Systems which are a combination of rotational and translational elements or electromechanical systems may be treated as either all-mechanical systems of a single type or as all-electrical systems by suitable substitution of analogous elements and variables.[20] In systems where both rotation and translation of a single mass occur, this single mass may have to be represented by two or more mass elements, and the concept of mutual mass (analogous to mutual inductance) may have to be introduced.[20]

Some Results from Electrical-Network Theory:[20,58] Resonances. Resonances occur at those frequencies for which the system impedances are minimum; antiresonances occur when the impedances are maximum.

Force and Velocity Sources. An ideal velocity generator supplies a prescribed relative velocity amplitude regardless of the force amplitude. A force generator supplies a prescribed force amplitude regardless of the velocity. When a velocity source is "turned off," $V = 0$, it acts like a rigid connection between its terminals. When a force generator is turned off, $F = 0$, it acts like no connection between the terminals.

TABLE 4.2 Mechanical Impedances of Simple Systems

SYSTEM	DIAGRAM	IMPEDANCE Z	FREQUENCY DEPENDENCE
MASS	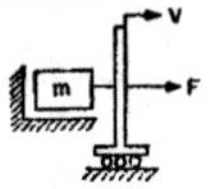	$i\omega m$	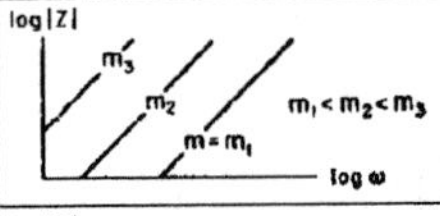
VISCOUS DASHPOT	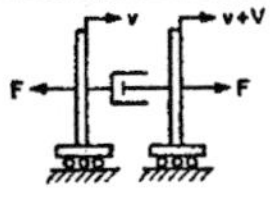	c	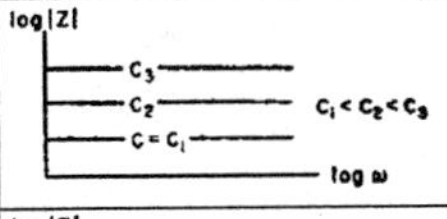
SPRING	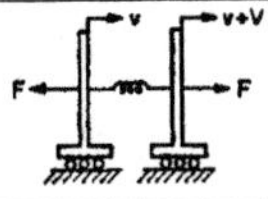	$k/i\omega$	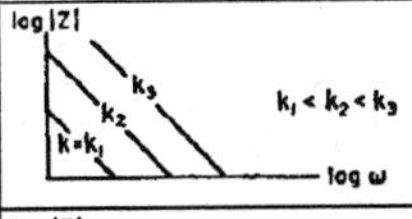
DRIVEN MASS ON SPRING 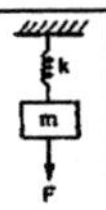	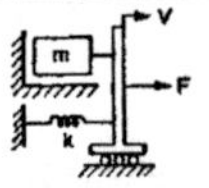	$\frac{k}{i\omega} + i\omega m$	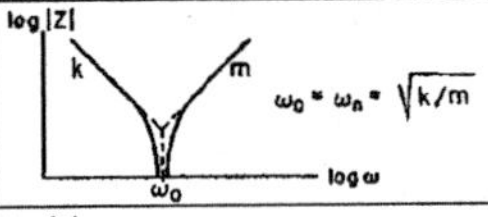
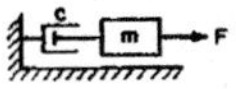 DAMPED MASS	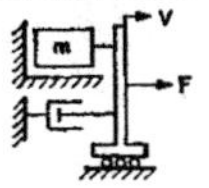	$c + i\omega m$	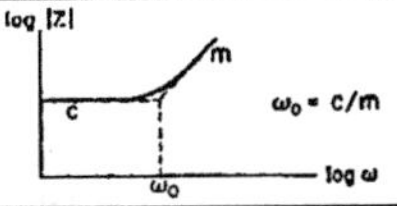
SPRING AND DASHPOT IN PARALLEL	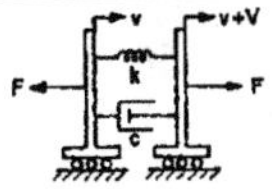	$c + k/i\omega$	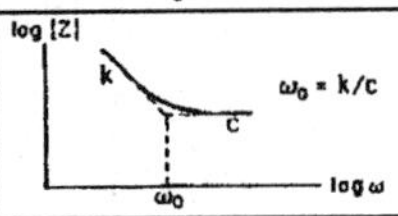
DAMPED SINGLE DEGREE OF FREEDOM SYSTEM	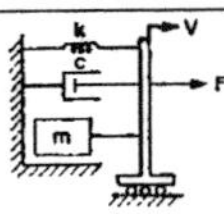	$i\omega m + c + \frac{k}{i\omega}$	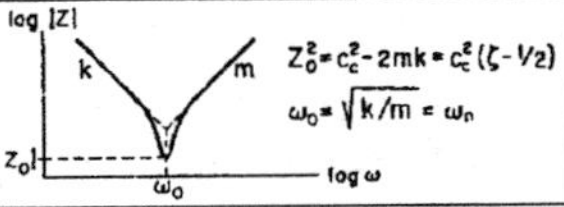
MASS ON DRIVEN SPRING 	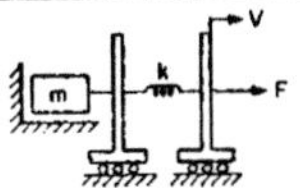	$\dfrac{1}{\frac{i\omega}{k} + \frac{1}{i\omega m}}$	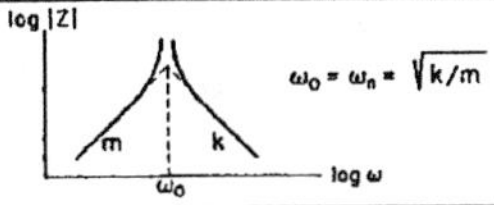
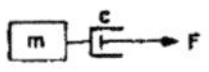 MASS DRIVEN THROUGH DASHPOT	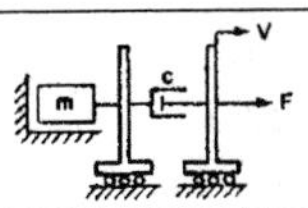	$\dfrac{1}{\frac{1}{c} + \frac{1}{i\omega m}}$	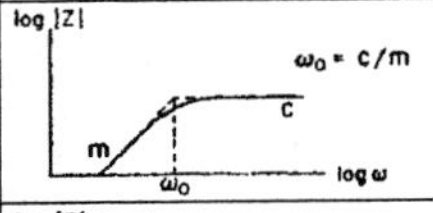
SPRING AND DASHPOT IN SERIES	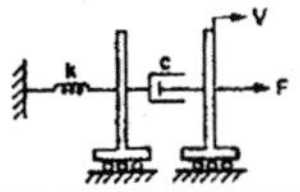	$\dfrac{1}{\frac{1}{c} + \frac{i\omega}{k}}$	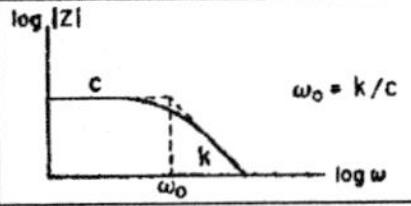
 DAMPED MASS DRIVEN THROUGH SPRING	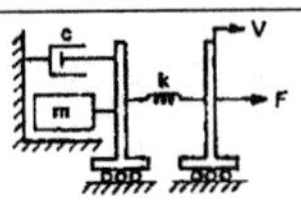	$\dfrac{1}{\frac{1}{c+i\omega m} + \frac{i\omega}{k}} = \dfrac{c_c\left[\zeta + \frac{i}{2}\frac{\omega}{\omega_n}\right]}{1-\left(\frac{\omega}{\omega_n}\right)^2 + 2i\zeta\frac{\omega}{\omega_n}}$	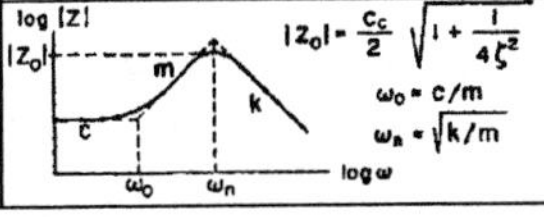

Reciprocity. The transfer impedance Z_{ij} (the force in the jth branch divided by the relative velocity of a generator in the ith branch) is equal to the transfer impedance Z_{ji} (force in the ith branch divided by relative velocity of generator in jth branch).

Foster's Reactance Theorem. For a general undamped system the driving-point impedance can be written $Z = if(\omega)$, where $f(\omega)$ is a real function of (real) ω. The function $f(\omega)$ always has $df/d\omega > 0$ and has a pole or zero at $\omega = 0$ and at $\omega = \infty$. All

TABLE 4.3 Combination of Impedances and Mobilities

Elements or subsystems:

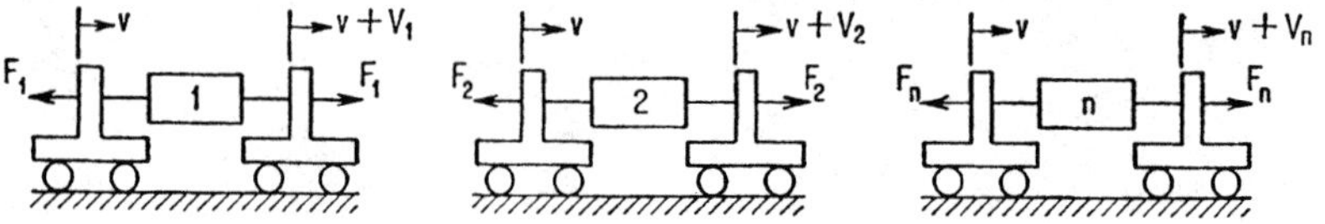

$Z_1 = 1/Y_1 = F_1/V_1 \qquad Z_2 = 1/Y_2 = F_2/V_2 \qquad \cdots \; Z_n = 1/Y_n = F_n/V_n$

All F, V, Z, Y are complex quantities

Combination in	Parallel	Series
Diagram	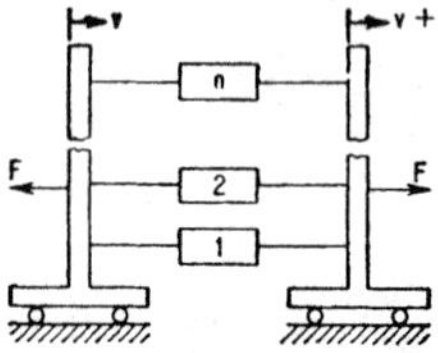	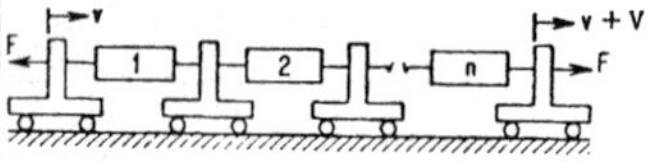
Connection identified by	All components having same V	Same F acting on all components
$Z = F/V =$	$Z_1 + Z_2 + \cdots + Z_n$	$(1/Z_1 + 1/Z_2 + \cdots + 1/Z_n)^{-1}$
$Y = V/F =$	$(1/Y_1 + 1/Y_2 + \cdots + 1/Y_n)^{-1}$	$Y_1 + Y_2 + \cdots + Y_n$
Z's combine like electrical resistances in	Series	Parallel
Y's combine like electrical resistances in	Parallel	Series

poles and zeros are simple (not repeated), poles and zeros alternate (i.e., there is always a zero between two poles), and $f(\omega)$ is determined within a multiplicative factor by its poles and zeros.

Thévenin's and Norton's Theorems. Consider any two terminals of a linear system. Then, as far as the effects of the system at these terminals are concerned, the system may be replaced by (see Fig. 4.18)

1. (Thévenin's equivalent) A series combination of an impedance Z_i and a velocity source V_{oc}
2. (Norton's equivalent) A parallel combination of an impedance Z_i and a force source F_b

Z_i is called the "internal impedance" of the system and is the driving-point impedance obtained at the terminals considered when all sources in the system are turned off (velocity generators replaced by rigid links, force generators replaced by disconnections). V_{oc} is the "open-circuit" velocity, i.e., the velocity occurring between the terminals considered (with all generators active). F_b is the "blocked force," i.e., the force transmitted through a rigid link inserted between the terminals considered, with all generators active.

TABLE 4.4 Mechanical-Electrical Analogies (Lumped Systems)

Mechanical System		Electrical Systems	
Translational	Rotational	Force-voltage (classical) analog	Force-current (mobility) analog
F Force	T Torque	V Voltage	I Current
u Velocity	Ω Angular velocity	I Current	V Voltage
x Displacement	θ Angular displacement	q Charge	ϕ Magnetic flux
m Mass	J Moment of inertia	L Inductance	C Capacitance
c Damping	c_r Rotational damping	R Resistance	$G = \frac{1}{R}$ Conductance
k Spring rate	k_r Torsional spring rate	$S = \frac{1}{C}$ Elastance	$\Gamma = \frac{1}{L}$ Inverse inductance
$Z = \frac{F}{u}$ Mechanical impedance	$Z_r = \frac{T}{\Omega}$ Rotational impedance	$Z = \frac{V}{I}$ Impedance	$Y = \frac{I}{V}$ Admittance
$Y = \frac{u}{F}$ Mobility	$Y_r = \frac{\Omega}{T}$ Rotational mobility	$Y = \frac{I}{V}$ Admittance	$Z = \frac{V}{I}$ Impedance

4.4 CONTINUOUS LINEAR SYSTEMS

4.4.1 Free Vibrations

The equations that govern the deflection $u(x, y, t)$ of many continuous linear systems (e.g., bars, shafts, strings, membranes, plates) in absence of external forces may be expressed as

$$\mathfrak{M}\ddot{u} + \mathfrak{K}u = 0 \tag{4.59}$$

where $\mathfrak{M}$ and $\mathfrak{K}$ are linear differential operators involving the coordinate variables only. Table 4.5 lists $\mathfrak{M}$ and $\mathfrak{K}$ for a number of common systems. Solutions of Eq. (4.59) may be expressed in terms of series composed of terms of the form $\phi(x,y)e^{i\omega t}$, where ϕ satisfies

$$(\mathfrak{K} - \mathfrak{M}\omega^2)\phi = 0 \tag{4.60}$$

TABLE 4.4 Mechanical-Electrical Analogies (Lumped Systems) (*Continued*)

Mechanical Systems		Electrical Systems	
Translational	Rotational	Force-voltage (classical) analog	Force-current (mobility) analog
Kinetic energy $= \frac{1}{2}mu^2$ Potential energy $= \frac{1}{2}kx^2$ Power loss $= u^2c$	Kinetic energy $= \frac{1}{2}I\Omega^2$ Potential energy $= \frac{1}{2}k_r\theta^2$ Power loss $= \Omega^2c_r$	Magnetic energy $= \frac{1}{2}LI^2$ Electrical energy $= \frac{1}{2}\frac{1}{C}q^2$ Power loss $= I^2R$	Electrical energy $= \frac{1}{2}CV^2$ Magnetic energy $= \frac{1}{2}\frac{1}{L}\phi^2$ Power loss $= V^2G$
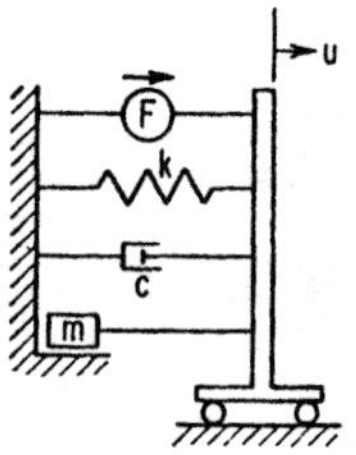	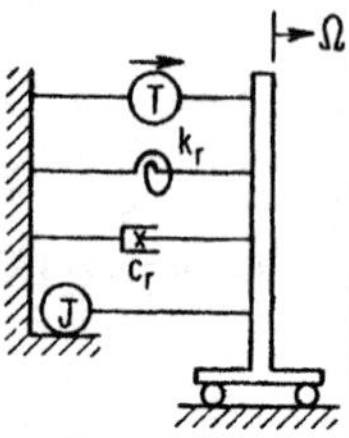	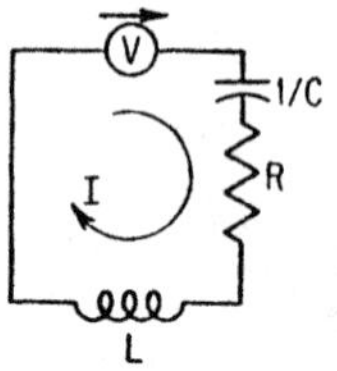	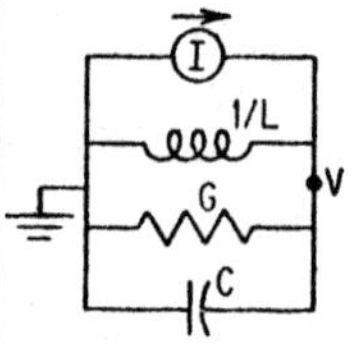
$m\dot{u} + cu + k\int u\,dt = F(t)$	$J\dot{\Omega} + c_r\Omega + k_r\int \Omega\,dt = T(t)$	$L\dot{I} + RI + \frac{1}{C}\int I\,dt = V(t)$	$C\dot{V} + RV + \frac{1}{L}\int V\,dt = I(t)$
Elements connected to same node have same velocity u	Elements connected to same node have same angular velocity Ω	Elements in same loop have same current I	Elements connected to same node have same voltage v
Element connected to ground has one end at zero velocity	Element connected to ground has one end at zero velocity	Element in one loop has only one loop current through it	Element connected to ground has one end at zero (reference) voltage
Elements between two nodes	Elements between two nodes	Elements in two loops	Elements between two nodes
Elements in parallel	Elements in parallel	Elements in series	Elements in parallel

in addition to the boundary conditions of a given problem. In solving the differential equation (4.60) by use of standard methods and introducing the boundary conditions applicable in a given case one finds that solutions ϕ that are not identically zero exist only for certain frequencies. These frequencies are called the "natural frequencies of the system"; the equation that the natural frequencies must satisfy for a given system is called the "frequency equation of the system"; the functions ϕ that satisfy Eq. (4.60) in conjunction with the natural frequencies are called the "eigenfunctions" or "mode shapes" of the system. One-dimensional systems (strings, bars) have an infinite number of natural frequencies ω_n and eigenfunctions ω_n; $n = 1, 2, \ldots$. Two-dimensional systems (membranes, plates) have a doubly infinite set of natural frequencies ω_{mn}, and eigenfunctions ϕ_{mn}; $m, n = 1, 2, \ldots$. Table 4.6 lists eigenfunctions and frequency equations for some common systems.

The eigenfunctions of flexural systems where all edges are either free, built-in, or pinned are "orthogonal," that is,

$$\int_0^L \mu(x)\phi_n(x)\phi_{n'}(x)\,dx = \begin{cases} \overline{\mu}L\Phi_n & \text{for } n = n' \\ 0 & \text{for } n \neq n' \end{cases} \tag{4.61}$$

TABLE 4.4 Mechanical-Electrical Analogies (Lumped Systems) (*Continued*)

Mechanical Systems		Electrical Systems	
Translational	Rotational	Force-voltage (classical) analog	Force-current (mobility) analog
Newton's law: sum of forces on node = 0	Newton's law: sum of torques on node = 0	Kirchhoff's voltage law: sum of voltages around loop = 0	Kirchhoff's current law: sum of currents into node = 0
Ideal lever 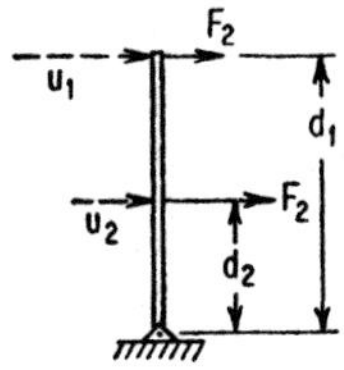	Ideal gear train 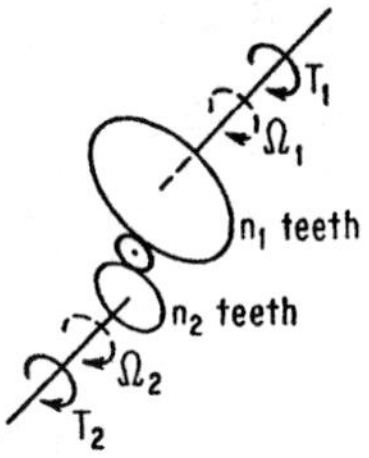	Ideal voltage transformer 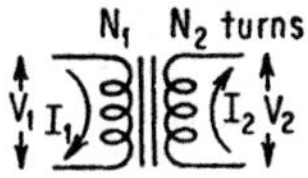	Ideal current transformer 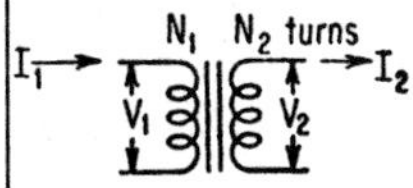
$\frac{u_1}{u_2} = \frac{F_2}{F_1} = \frac{d_1}{d_2}$	$\frac{\Omega_1}{\Omega_2} = \frac{T_2}{T_1} = \frac{n_2}{n_1}$	$\frac{I_1}{I_2} = \frac{V_2}{V_1} = \frac{N_2}{N_1}$	$\frac{V_1}{V_2} = \frac{I_2}{I_1} = \frac{N_1}{N_2}$

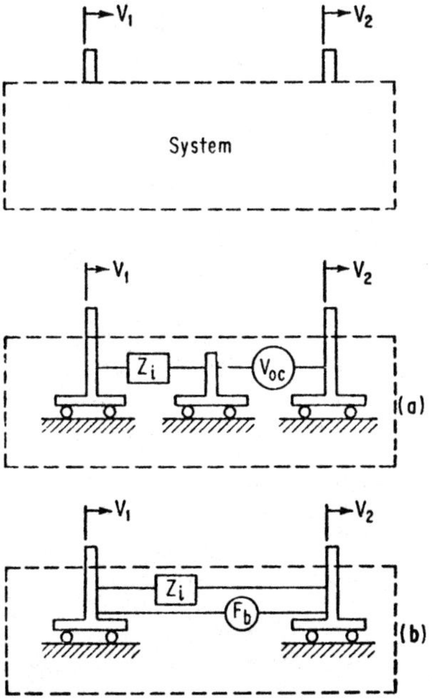

FIG. 4.18 (*a*) Thévenin's and (*b*) Norton's equivalent networks.

TABLE 4.5 Operators [See Eq. (4.59)] for Some Elastic Systems[64]

System	Deflection	$\mathfrak{M} = \mu$	$\mathcal{K}$
String	Lateral	$\rho A(x)$	$T \frac{\partial^2}{\partial x^2}$
Bar	Longitudinal	$\rho A(x)$	$-\frac{\partial}{\partial x}\left(EA \frac{\partial}{\partial x}\right)$
Bar	Torsional (angular)	$\rho J(x)$	$-\frac{\partial}{\partial x}\left(KG \frac{\partial}{\partial x}\right)$
Bar	Lateral (flexure)	$\rho A(x)$	$\frac{\partial^2}{\partial x^2}\left(EI \frac{\partial^2}{\partial x^2}\right)$
Membrane	Lateral	$\rho h(x,y)$	$-S\nabla^2$
Plate	Lateral	$\rho h(x,y)$	$\nabla^2(D\nabla^2)$

Symbols:
- ρ = density
- A = cross-sectional area
- I = centroidal moment of inertia of A
- J = polar moment of inertia of A
- h = plate thickness
- T = tensile force
- S = tension/unit length
- E = Young's modulus
- G = shear modulus
- K = torsional constant of A (= J for circular sections)
- $D = Eh^3/12(1 - \nu^2)$
- ν = Poisson's ratio

TABLE 4.6a Modal Properties for Some One-Dimensional Systems

System	Wave velocity c	Boundary conditions $x = 0$	Boundary conditions $x = L$	Mode shape† $\phi_n(x)$	Natural frequency $\omega_n = 2\pi f_n$	$\frac{\lambda}{L} = \frac{2\pi}{Lk_n}$
String	$\sqrt{T/\rho A}$	Fixed	Fixed	$\sin \frac{n\pi x}{L}$	$\frac{n\pi}{L} c$	$\frac{2}{n}$
Uniform shaft* in torsion	$\sqrt{\frac{GK}{\rho J}}$	Clamped	Free	$\sin \frac{(2n-1)\pi x}{2L}$	$\frac{(2n-1)\pi}{2L} c$	$\frac{4}{2n-1}$
		Free	Free	$\cos \frac{n\pi x}{L}$	$\frac{n\pi}{L} c$	$\frac{2}{n}$
Uniform bar* in longitudinal vibration	$\sqrt{E/\rho}$	Clamped	Clamped	$\sin \frac{n\pi x}{L}$	$\frac{n\pi}{L} c$	$\frac{2}{n}$

Symbols:
- A = cross-sectional area
- ρ = material density
- T = tensile force
- E = Young's modulus
- G = shear modulus
- J = polar moment of inertia of A
- K = torsional constant of A (= J for circular sections)
- GK = torsional rigidity
- λ_n = wavelength
- k_n = wave number
- n = mode number

*The same modal properties, but with different values of c, apply for uniform shafts vibrating torsionally and uniform bars vibrating longitudinally.

†$\Phi_n = (1/L)\int_0^L \phi_n^2\, dx = \frac{1}{2}$.

TABLE 4.6b Modal Properties for Flexural Vibrations of Uniform Beams[6,7]

Natural frequency $\omega_n = 2\pi f_n = k_n^2 c_L r = k_n^2 \sqrt{EI/\rho A}$

Wavelength $\lambda_n = 2\pi / k_n$

Wave velocity $c_n = \omega_n / k_n = k_n c_L r = k_n \sqrt{EI/\rho A} = \sqrt{\omega_n c_L r}$

For mode shapes $\phi_n(x)$ listed below, $\Phi_n = (1/L) \int_0^L \phi_n^2 \, dx = 1$

Boundary conditions		Mode shape $\phi_n(x)$	σ_n	Frequency equation	Roots of frequency equation					
$x = 0$	$x = L$				k_1L	k_2L	k_3L	k_4L	k_5L	$k_nL,\ n > 5$
Pinned	Pinned	$\sqrt{2}\sin(k_nx)$		$\sin(k_nL) = 0$	π	2π	3π	4π	5π	$n\pi$
Clamped	Clamped	$\cosh(k_nx) - \cos(k_nx) - \sigma_n[\sinh(k_nx) - \sin(k_nx)]$	$\dfrac{\cosh(k_nL) - \cos(k_nL)}{\sinh(k_nL) - \sin(k_nL)}$	$\cos(k_nL)\cosh(k_nL) = 1$	4.73	7.85	11.00	14.14	17.29	$\approx \dfrac{2n+1}{2}\pi$
Free	Free	$\cosh(k_nx) + \cos(k_nx) - \sigma_n[\sinh(k_nx) + \sin(k_nx)]$								
Clamped	Free	$\cosh(k_nx) - \cos(k_nx) - \sigma_n[\sinh(k_nx) - \sin(k_nx)]$	$\dfrac{\sinh(k_nL) - \sin(k_nL)}{\cosh(k_nL) + \cos(k_nL)}$	$\cos(k_nL)\cosh(k_nL) = -1$	1.875	4.69	7.85	11.00	14.14	$\approx \dfrac{2n-1}{2}\pi$
Clamped	Pinned	$\cosh(k_nx) - \cos(k_nx) - \sigma_n[\sinh(k_nx) - \sin(k_nx)]$	$\cot(k_nL)$	$\tan(k_nL) = \tanh(k_nL)$	3.93	7.07	10.21	13.35	16.49	$\approx \dfrac{4n+1}{4}\pi$
Free	Pinned	$\cosh(k_nx) + \cos(k_nx) - \sigma_n[\sinh(k_nx) - \sin(k_nx)]$								

Symbols:

k_n = wave number
E = Young's modulus
ρ = material density
$c_L = \sqrt{E/\rho}$ = longitudinal wave velocity
A = cross-section area
I = moment of inertia of A
$r = \sqrt{I/A}$ = radius of gyration
L = beam length

TABLE 4.6c Modal Properties for Some Plates[40]

Frequency	$\omega_{mn} = 2\pi f_{mn} = k_{mn}^2 \sqrt{D/\rho h} \approx k_{mn}^2 c_L r$	$D = Eh^3/12(1-\nu^2)$
Wavelength	$\lambda_{mn} = 2\pi/k_{mn}$	$r = h/\sqrt{12}$
Wave velocity	$c_{mn} = \omega_{mn}/k_{mn} = k_{mn}\sqrt{D/\rho h} \approx k_{mn} c_L r$	$c_L = \sqrt{E/\rho}$

Rectangular, on simple supports; $0 < x < a$, $0 < y < b$:

$$\phi_{mn}(x,y) = \sin\frac{m\pi x}{a}\sin\frac{n\pi y}{b} \qquad \Phi_{mn} = 1/4$$

$$k_{mn} = \pi\sqrt{m^2/a^2 + n^2/b^2}$$

Circular,* clamped at circumference; $0 < r < a$, $0 < \theta < 2\pi$:

$$\phi_{mn_e}(r,\theta) = \cos(m\theta)F_{mn}(r)$$
$$\phi_{mn_o}(r,\theta) = \sin(m\theta)F_{mn}(r)$$

$$F_{mn}(r) = J_m(k_{mn}r) - \frac{J_m(k_{mn}a)}{I_m(k_{mn}a)} I_m(k_{mn}r)$$

Same orthogonality holds as for membranes (see Table 4.6d)
Frequency equation:

$$I_m(k_{mn}a)\left[\frac{d}{dr}J_m(k_{mn}r)\right]_{r=a} = J_m(k_{mn}a)\left[\frac{d}{dr}I_m(k_{mn}r)\right]_{r=a}$$

Table of $k_{mn}a/\pi$ for Clamped Circular Plate

		n			
		1	2	3	>3
m	0	1.015	2.007	3.000	$\approx n + \frac{m}{2}$
	1	1.468	2.483	3.490	
	2	1.879	2.992	4.000	

*For circular plates one finds two sets of ϕ_{mn}, one even (subscript e), one odd (subscript o).

$$\iint_{A_S} \mu(x,y)\phi_{mn}(x,y)\phi_{m'n'}(x,y)\,dA_s = \begin{cases} \bar{\mu}A_s\Phi_{mn} & \text{for } m = m',\ n = n' \\ 0 & \text{otherwise} \end{cases} \tag{4.61}$$

where L is the total length of a (one-dimensional) system, A_s is the total surface area of a (two-dimensional) system, μ is a weighting function, and $\bar{\mu}$ is the mean of value of μ for the system. (The weighting functions of the common systems tabulated in Table 4.5 are equal to the mass operator $\mathfrak{M}$.)

$$\bar{\mu}L = \int_0^L \mu(x)\,dx \qquad \text{for one-dimensional systems}$$

$$\bar{\mu}A_s = \iint_{A_s} \mu(x,y)\,dA_s \qquad \text{for two-dimensional systems}$$

For uniform systems $\mu = \bar{\mu}$ is a constant and may be canceled from Eqs. (4.61).*

*Note that for the systems of Table 4.6 $\mathfrak{M} = \mu$ is a multiplicative factor, not an operator in the more general sense. However, for other systems $\mathfrak{M}$ may be a more general operator (e.g., for lateral vibrations of beams when rotatory inertia is not neglected) and may differ from μ. The distinction between $\mathfrak{M}$ and μ is maintained throughout the subsequent discussion to permit application of the results to systems that are more complicated than those of Table 4.6.

TABLE 4.6*d* Modal Properties for Some Membranes[40]

Wave velocity $c = \sqrt{S/\rho h}$

Rectangular; $0 < x < a$, $0 < y < b$:

$$\phi_{mn}(x,y) = \sin\frac{m\pi x}{a}\sin\frac{n\pi y}{b} \qquad \phi_{mn} = \tfrac{1}{4}$$

$$\omega_{mn} = 2\pi f_{mn} = \pi c\sqrt{(m/a)^2 + (n/b)^2} \qquad k_{mn} = 2\pi/\lambda_{mn} = \omega_{mn}/c$$

Circular,* $0 < r < a$, $0 < \theta < 2\pi$:

$$\phi_{mn_e}(r,\theta) = \cos(m\theta)J_n(\omega_{mn}r/c)$$
$$\phi_{mn_o}(r,\theta) = \sin(m\theta)J_n(\omega_{mn}r/c)$$

$$\iint_{A_s}\phi_{mn_j}\phi_{m'n'_{j'}}\,dA_s = \begin{cases}\pi a^2\Phi_{mn} & \text{for } m' = m,\ n' = n,\ j' = j\\ 0 & \text{otherwise}\end{cases}$$

$$\Phi_{mn} = \begin{cases}[J_1(\omega_{mn}a/c)]^2 & \text{for } m = 0\\ \tfrac{1}{2}[J_{m-1}(\omega_{mn}a/c)]^2 & \text{for } m > 0\end{cases}$$

Frequency equation:

$$J_m(\omega a/c) = 0$$

Table of $\omega_{mn}a/\pi c$ for Circular Membrane

		n: 1	2	3	>3
m	0	0.7655	1.7571	2.7546	$\approx n + \frac{m}{2} - \frac{1}{4}$
	1	1.2197	2.2330	3.2383	
	2	1.6347	2.6793	3.6987	

Table of Φ_{mn} for Circular Membrane

		n: 1	2	3	4
m	0	0.2695	0.1241	0.07371	0.05404
	1	0.08112	0.04503	0.02534	0.02385
	2	0.05770	0.03682	0.02701	0.02134

Symbols for Table 4.6*c* and *d*:

S = tension force/unit length
μ = mass/unit area
D = flexural rigidity = $Eh^3/12(1 - \nu^2)$
k_{mn} = wave number
E = Young's modulus
ν = Poisson's ratio
c_L = longitudinal wave velocity = $\sqrt{E/\rho}$
r = radius of gyration of section
h = plate thickness
ρ = material density
J_m = mth-order Bessel function
I_m = mth-order hyperbolic Bessel function

*For circular membranes one finds two set of ϕ_{mn}, one even (subscript e), one odd (subscript o).

The free motions (or "transient" responses) of two-dimensional* systems with initial displacement $u(x, y, 0)$ from equilibrium and initial velocity $\dot{u}(x, y, 0)$ are given by

$$u_t(x,y,t) = \sum_{m=1}^{\infty} \sum_{n=1}^{\infty} \phi_{mn}(B_{mn} \cos \omega_{mn} t + C_{mn} \sin \omega_{mn} t)$$

$$B_{mn} = \frac{1}{\mu A_s \Phi_{mn}} \iint_{A_s} u(x,y,0)\mu(x,y)\phi_{mn}(x,y)\, dA_s \qquad (4.62)$$

$$C_{mn} = \frac{1}{\mu A_s \Phi_{mn} \omega_{mn}} \iint_{A_s} \dot{u}(x,y,0)\mu(x,y)\phi_{mn}(x,y)\, dA_s$$

4.4.2 Forced Vibrations

Forced vibrations of distributed systems are governed by

$$\mathfrak{M}\ddot{u} + \mathfrak{K}u = F(x,y,t)$$

where the right-hand side represents the applied load distribution in space and time.

General Response: Modal Displacements, Forcing Functions, Masses. The displacement $u(x, y, t)$ of a system† may, in general, be expressed in a modal series as

$$u(x,y,t) = \sum_{m=1}^{\infty} \sum_{n=1}^{\infty} U_{mn}(t)\phi_{mn}(x,y) \qquad (4.63)$$

where the modal displacement $U_{mn}(t)$ is given by

$$U_{mn}(t) = \frac{1}{\omega_{mn} M_{mn}} \int_0^t G_{mn}(\tau) \sin \omega_{mn}(t - \tau)\, d\tau \qquad (4.64)$$

The modal forcing function $G_{mn}(t)$ and the modal mass M_{mn} are given by

$$G_{mn}(t) = \iint_{A_s} F(x,y,t)\phi_{mn}(x,y)\, dA_s \qquad M_{mn} = \iint_{A_s} \phi_{mn}(x,y)\mathfrak{M}\phi_{mn}(x,y)\, dA_s \qquad (4.65)$$

For the special case of a uniform system with $\mathfrak{M} = \mu = \bar{\mu} =$ constant,

$$M_{mn} = \mu A_s \Phi_{mn}$$

where Φ_{mn} is defined as in Eq. (4.61).

The "steady-state" response given by (4.63) and (4.64) must be combined with the "transient" response given by (4.62) if one desires the complete solution.

Sinusoidal Response: Input Impedance. For a sinusoidal forcing function $F(x, y, t) = e^{i\omega t}F_0(x, y)$ one finds a steady-state response given by

$$u_{ss}(x,y,t) = e^{i\omega t} \sum_{m=1}^{\infty} \sum_{n=1}^{\infty} U_{mn}\phi_{mn}(x,y)$$

---

*These equations apply also for one-dimensional systems if all m subscripts and y dependences are deleted, if A_s is replaced by L, and if the double integrations over A_s are replaced by a single integration for Q to L.

†These equations apply also for one-dimensional systems if all m subscripts and y dependences are deleted, if A_s is replaced by L, and if the double integrations over A_s are replaced by a single integration from 0 to L.

with $$U_{mn} = \frac{G_{mn}}{(\omega_{mn}{}^2 - \omega^2)M_{mn}} \qquad G_{mn} = \iint_{A_s} F_0(x,y)\phi_{mn}(x,y)\, dA_s$$

and M_{mn} given by (4.65).

For a point force $F_1 e^{i\omega t}$ applied at $x = x_0$, $y = y_0$, one finds $G_{mn} = F_1\phi_{mn}(x_0, y_0)$. The input impedance $Z(x_0, y_0, \omega)$ of the system at $x = x_0$, $y = y_0$ for frequency ω then may be found from

$$\frac{1}{i\omega Z(x_0,y_0,\omega)} = \sum_{m=1}^{\infty}\sum_{n=1}^{\infty} \frac{\phi_{mn}^2(x_0,y_0)}{(\omega_{mn}^2 - \omega^2)M_{mn}} \tag{4.66}$$

Point-Impulse Response: Alternate Formulation of General Response. The response of a system to a point force applied at $x = x_0$, $y = y_0$ and varying like a Dirac impulse function (of unit magnitude) with time is given by

$$u_\delta(x,y;\, x_0,y_0;\, t) = \sum_{m=1}^{\infty}\sum_{n=1}^{\infty} \frac{\phi_{mn}(x_0,y_0)\phi_{mn}(x,y)}{\omega_{mn}M_{mn}} \sin \omega_{mn} t$$

An alternate expression for the steady-state response of a system to a general distributed force $F(x, y, t)$ may then be written as

$$u_{ss}(x,y,t) = \iint_{A_s}\left[\int_0^t F(x_0,y_0,\tau)u_\delta(x,y;\, x_0,y_0;\, t-\tau)\, d\tau\right] dx_0\, dy_0$$

This expression is entirely analogous to the result one obtains by combining Eqs. (4.63) and (4.64).

4.4.3 Approximation Methods

Finite-Difference Equations, Finite-Element Approximations. One of the most widely applicable numerical methods, particularly if digital-computation equipment is available, consists of replacing the applicable differential equations by finite-difference equations[12,64] which may then be solved numerically.

A second method consists of replacing the continuous (infinite-degrees-of-freedom) system by one made up of a finite number of suitably interconnected elements (masses, springs, dashpots), then applying methods developed for systems with a finite number of degrees of freedom, as outlined in Sec. 4.3. These "finite-element" approximations may be obtained, for example, by dividing a beam or plate to be analyzed into arbitrary segments and assuming the mass of each segment concentrated at its center of gravity or "lumping point." If one then establishes the influence coefficients between the various lumping points, one has enough information to apply directly the methods outlined in Sec. 4.3.[24] In concept, one replaces the structure between lumping points by equivalent springs to obtain a new system analogous to the continuous one; a beam is replaced by a linear array, a plate by a two-dimensional network of masses interconnected by springs. One may then proceed by determining the equations of motion of the new systems and by solving these as discussed in Sec. 4.3. A variety of corresponding computer codes has become available; e.g., see Refs. 44 and 45.

Fundamental Frequencies:* Rayleigh's Quotient. *RQ* for an arbitrary deflection function $u(x, y)$ of a two-dimensional continuous system is given by[64]

---

*These equations apply also for one-dimensional systems if all m subscripts and y dependences are deleted, if A_s is replaced by L, and if the double integrations over A_s are replaced by a single integration from 0 to L.

$$RQ = \frac{\iint_{A_s} u\mathfrak{K}(u)\, dA_s}{\iint_{A_s} u\mathfrak{M}(u)\, dA_s}$$

For $u = \phi_{mn}$ (the mn mode shape) RQ takes on the value ω_{mn}^2. For any function u that satisfies the boundary conditions of the given system

$$RQ \geq \omega_{11}^2$$

Thus RQ produces an estimate of the fundamental frequency ω_{11} which is always too high.

To obtain a better estimate one may use the Rayleigh-Ritz procedure. In this procedure one forms RQ for a linear combination of any convenient number of functions $u_i(x, y)$ that satisfy the boundary conditions of the problem; that is, one forms RQ from

$$u = \alpha_1 u_1 + \alpha_2 u_2 + \cdots$$

where the α are constants. A good approximation to ω_{11}^2 is then obtained, in general, by minimizing RQ with respect to the various α's.

$$RQ \text{ (evaluated so that } \partial RQ/\partial\alpha_1 = \partial RQ/\partial\alpha_2 = \cdots = 0) \approx \omega_{11}^2$$

Special Methods for Lateral Vibrations of Nonuniform Beams. In addition to the foregoing methods a number of others are available that have been developed specially for dealing with the vibrations of beams and shafts. In all these methods the beam mass is replaced by a number of masses concentrated at lumping points.

The Stodola method[53,59,61,63] is related to both the Rayleigh and matrix-iteration methods. In it one may proceed as follows:

1. Assume a deflection curve that satisfies the boundary conditions. (Usually the static-deflection curve gives good results.)
2. Determine a first approximation to ω_1 using Rayleigh's quotient, from

$$\omega_1^2 \approx RQ = \frac{g\Sigma m_i u_i}{\Sigma m_i u_i^2}$$

where g denotes the acceleration of gravity and u_i the assumed deflection of the mass m_i.

3. Calculate the deflection of the beam as if inertia forces $(-m_i u_i \omega_1^2)$ were applied statically. (This is usually done best by graphical or numerical means.) Use these new deflections instead of the original u_i in the foregoing equation. Repeat this process until no further changes in RQ result to the degree of accuracy desired. Then $RQ = \omega_1^2$ to within the desired accuracy.

The Myklestad method[41] is essentially the same as Holzer's method, but considerably simpler to use for flexural vibrations. Extensions of Myklestad's original method also apply to coupled bending-torsion vibrations and to vibrations in centrifugal fields.[41] Some simplifications of Myklestad's method have been developed by Thomson.[61] Because of the details necessary for a sufficient discussion of these procedures the reader is referred to the original sources.

4.4.4 Systems of Infinite Extent

Truly infinite or semi-infinite systems do not occur in reality. However, as far as the local response to a local excitation is concerned, finite systems behave like infinite ones if the ends are far (many wavelengths) removed from the excitation and if there is enough dissipation in the system or at the ends so that little effect of reflected waves is felt near the driving point.

The velocities with which waves travel in infinite systems are listed under the heading of "wave velocity" in Table 4.6, but in infinite systems the frequencies and wave numbers are not restricted, as they are in finite ones. In all cases the wave velocity c is related to wavelength λ, wave number k, frequency f (cycles/time), and circular frequency ω (radians/time) as

$$c = f\lambda = \omega\lambda/2\pi = \omega/k$$

Input impedances of infinite structures are useful for estimation of the responses of mechanical systems that are composed of or connected to one or more structures, if the responses of the latter may be approximated by those of corresponding infinite structures in the light of the first paragraph of this article. These impedances may be used precisely like previously discussed impedances of systems with only a few degrees of freedom.

If a force $F_0e^{i\omega t}$ gives rise to a velocity $V_0e^{i\omega t}$ at its point of application (where F_0 and V_0 may be complex in general), then the driving-point impedance is defined as $Z = F_0/V_0$. Similarly, the driving-point moment impedance is defined by $Z_M = M_0/\Omega_0$, where M_0 denotes the amplitude of a driving moment $M_0e^{i\omega t}$ and where $\Omega_0e^{i\omega t}$ is the angular velocity at the driving point. Table 4.7 lists the driving-point impedances of some infinite and semi-infinite systems.

4.5 MECHANICAL SHOCKS

By a mechanical shock one generally means a relatively suddenly applied transient force or support acceleration. The responses of mechanical systems to shocks may be computed by direct application of the previously discussed methods for determination of transient responses, provided that the forcing functions are known. If the forcing functions are not known precisely, one may approximate them by some idealized functions or else describe them in some rough way, for example, in terms of the subsequently discussed shock spectra.

4.5.1 Idealized Forcing Functions

Among the most widely studied idealized shocks are those associated with sudden support displacements or velocity changes, or with suddenly applied forces. The responses of systems with one or two degrees of freedom to idealized shocks have been studied in considerable detail, since for such systems solutions may be obtained relatively simply by analytical or analog means.

Sudden Support Displacement. Sudden (vertical) support displacements occur, for example, when an automobile hits a sudden change in level of the roadbed. The responses of simple systems to such shocks have been studied in considerable detail. Families of

TABLE 4.7 Driving-Point Impedances of Some Infinite Uniform Systems*

Beams

Loading	Extent: Infinite	Extent: Semi-infinite	
Axial force..........	$Z = 2A\rho c_L$	$Z = A\rho c_L$	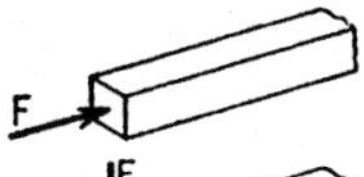
Lateral force.........	$Z = 2mc_B(1 + i)$	$Z = \frac{mc_B}{2}(1 + i)$	
Moment............	$Z_M = \frac{2mc_B}{k^2}(1 + i)$	$Z_M = \frac{mc_B}{2k^2}(1 + i)$	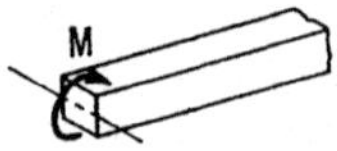
Torsion.............	$Z_M = 2J\rho c_T$	$Z_M = J\rho c_T$	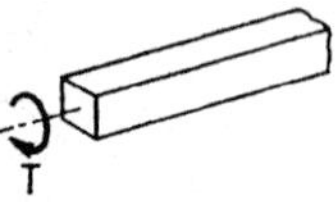

Symbols:

A = cross-section area
ρ = material density
E = Young's modulus
J = polar moment of inertia of A
K = torsional constant
G = shear modulus
$r = \sqrt{I/A}$

$c_L = \sqrt{E/\rho}$ = longitudinal wave velocity
$c_T = \sqrt{GK/\rho J}$ = torsional wave velocity
$c_B = \sqrt{\omega r c_L}$ = flexural wave velocity
$k = \sqrt{\omega / r c_L}$ = flexural wave number
$\frac{c_B}{k^2} = \sqrt{\frac{(rc_L)^3}{\omega}}$

Plates

Infinite isotropic: $Z = 8\sqrt{D\rho h} \approx 2.3h^2\sqrt{\frac{E\rho}{1 - \nu^2}}$

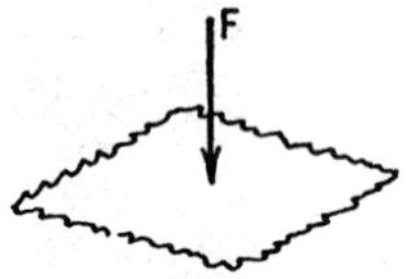

Infinite orthotropic: $Z \approx 8\sqrt[4]{D_x D_y}\sqrt{\rho h}$

Isotropic, infinite in x direction, on simple supports at $y = 0, L$, forced at $(0, y_0)$:

$$\frac{1}{Z} = \frac{i}{2\sqrt{D\rho h}} \sum_{n=1}^{\infty} [(n^2\pi^2 - K)^{-1/2} - (n^2\pi^2 + K)^{-1/2}] \sin^2\left(\frac{n\pi y_0}{L}\right)$$

$$K = (kL)^2 = L^2\omega\sqrt{\frac{\rho h}{D}}$$

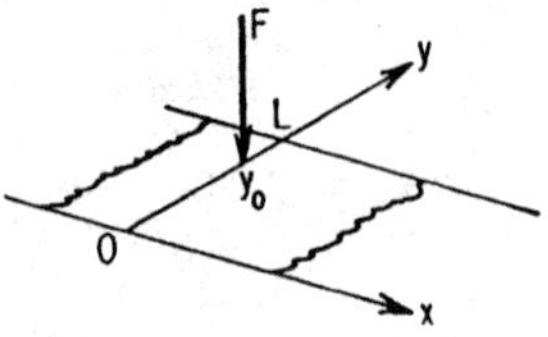

Symbols:

ρ = density of plate material
h = plate thickness
E = Young's modulus of plate material
E_x, E_y = Young's modulus in principal directions
ν = Poisson's ratio of plate material
ν_x, ν_y = Poisson's ratio in principal directions

$k = \sqrt[4]{\omega^2 \rho h / D}$
$D = Eh^3/12(1 - \nu^2)$
$D_x = E_x h^3/12(1 - \nu_x^2)$
$D_y = E_y h^3/12(1 - \nu_y^2)$

*Systems are assumed undamped. For finite systems see Eq. (4.66) and Table 4.6.

curves describing system responses to a certain type of rounded step-function displacement are given in Refs. 19, 48, and 52. It is found among other things that the responses are generally of an oscillatory nature, except for peaks in acceleration that occur during the "rise" time of the displacement.

Sudden Support Velocity Change. Velocity shocks, i.e., those associated with instantaneous velocity changes, have been studied considerably. They approximate a number of physical situations where impact occurs; for example, when a piece of equipment is dropped the velocity of its outer parts changes from some finite value to zero at the instant these parts hit the ground. Reference 39 discusses in detail how one may compute the velocity shock responses of single-degree-of-freedom systems with linear and various nonlinear springs. It also presents charts for the computation of the important parameters of responses of simple systems attached to the aforementioned velocity-shock-excited systems. That is, if one system is mounted inside another, one may first compute the motion of the outer system in response to a velocity shock, then the response of the inner system to this motion (assuming the motion of the inner system has little effect on the shock response of the outer). This permits one to estimate, for example, what happens to an electronic component (inner system) when a chassis containing it (outer system) is dropped. Damping effects are generally neglected in Ref. 39, but detailed curves for lightly damped linear two-degree-of-freedom systems appear in Refs. 43 and 46. Some discussion appears also in Ref. 14.

Suddenly Applied Forces. If a force is suddenly applied to a linear system with a single degree of freedom, then this force causes at most twice the displacement and twice the stress that this same force would cause if it were applied statically. This rule of thumb can lead to considerable error for systems with more than one degree of freedom.[46] One generally does well to carry out the necessary calculations in detail for such systems.

4.5.2 Shock Spectra

The shock-spectrum concept is useful for describing shocks and the responses of simple systems exposed to them, particularly where the shocks cannot be described precisely and where they cannot be reasonably approximated by one of the idealized shocks of Sec. 4.5.1.

Physical Interpretation. Assume that a given (support acceleration) shock is applied to an undamped linear single-degree-of-freedom system whose natural frequency is f_1. The maximum displacement of the system (relative to its supports) in response to this shock then gives one point on a plot of displacement vs. frequency. Repetition with the same shock applied to systems with different natural frequencies gives more points which, when joined in a curve, make up the "displacement shock spectrum" corresponding to the given shock. If instead of the maximum displacement one had noted the maximum velocity or acceleration for each test system, one would have obtained the velocity or acceleration shock spectrum.

Definitions.[1,19] If a force $F(t)$ is applied to an undamped single-degree-of-freedom system* with natural frequency ω and mass m, then the displacement, velocity, and

*The shock spectra defined here are essentially descriptions of the shock in the frequency domain. Shock spectra are related to the Fourier transforms, the latter being lower bounds to the former.[66] Inclusion of damping appears unnecessary for purposes of describing the shock, but most authors[1,19] include damping in their definitions since they tend to be more concerned with descriptions of system responses to shocks than with descriptions of shocks. The term "shock spectra of a system" is often applied to descriptions of responses of a specified system to specified shocks.[62]

acceleration of the system (assumed to be initially at rest and at equilibrium) are given by

$$x(t, \omega) = \frac{1}{m\omega} \int_0^t F(\tau) \sin \omega(t - \tau)\, d\tau$$

$$\dot{x}(t, \omega) = \frac{1}{m} \int_0^t F(\tau) \cos \omega(t - \tau)\, d\tau$$

$$\ddot{x}(t, \omega) = F(t)/m - \omega^2 x(t, \omega)$$

The various shock spectra of the force shock $F(t)$ are then defined as follows:

$$\begin{aligned}
D'_+(\omega) &= \text{timewise maximum of } x(t, \omega) \\
&= \text{positive-displacement spectrum} \\
D'_-(\omega) &= \text{timewise minimum of } x(t, \omega) \\
&= \text{negative-displacement spectrum} \\
D'(\omega) &= \text{timewise maximum of } |x(t, \omega)| \\
&= \text{displacement spectrum}
\end{aligned}$$

Similarly, for example,

$$\begin{aligned}
V'(\omega) &= \text{timewise maximum of } |\dot{x}(t, \omega)| \\
&= \text{velocity spectrum} \\
A'(\omega) &= \text{timewise maximum of } |\ddot{x}(t, \omega)| \\
&= \text{acceleration spectrum}
\end{aligned}$$

Similarly, if the previously discussed test system is exposed to a support acceleration $\ddot{s}(t)$, then the displacement, velocity, and acceleration of the system relative to its supports are given by

$$y(t, \omega) = -\frac{1}{\omega} \int_0^t \ddot{s}(\tau) \sin \omega(t - \tau)\, d\tau = x - s$$

$$\dot{y}(t, \omega) = -\int_0^t \ddot{s}(\tau) \cos \omega(t - \tau)\, d\tau = \dot{x} - \dot{s}$$

$$\ddot{y}(t, \omega) = -\ddot{s}(t) = \omega^2 y(t, \omega) = \ddot{x} - \ddot{s}$$

where $y = x - s$ denotes the relative displacement, x the absolute displacement of the mass, and s the displacement of the support.

The various shock spectra of the acceleration shock $\ddot{s}(t)$ are defined as follows:

$$\begin{aligned}
D(\omega) &= \text{timewise maximum of } |y(t, \omega)| \\
&= \text{relative-displacement spectrum} \\
V(\omega) &= \text{timewise maximum of } |\dot{y}(t, \omega)| \\
&= \text{relative-velocity spectrum} \\
A(\omega) &= \text{timewise maximum of } |\ddot{y}(t, \omega)| \\
&= \text{relative-acceleration spectrum} \\
D_a(\omega) &= \text{timewise maximum of } |x(t, \omega)| \\
&= \text{absolute-displacement spectrum}
\end{aligned}$$

$$V_a(\omega) = \text{timewise maximum of } |\dot{x}(t, \omega)|$$
$$= \text{absolute-velocity spectrum}$$
$$A_a(\omega) = \text{timewise maximum of } |\ddot{x}(t, \omega)|$$
$$= \text{absolute-acceleration spectrum}$$
$$\tilde{V}(\omega) = \omega\, D(\omega) = \text{pseudo-velocity spectrpum}$$

For any $s(t)$ it is true that

$$A_a(\omega) = \omega^2 D(\omega)$$

but $\tilde{V}(\omega) \neq V(\omega)$, $\tilde{V}(\omega) \neq V_a(\omega)$ in general.

One often distinguishes also between system responses during the action of a shock and responses after the shock action has ceased. The spectra associated with system motion during the shock action are called "primary spectra"; those associated with system motion after the shock are called "residual spectra."

Shock spectra are generally reduced to dimensionless "amplification spectra," e.g., by dividing $D'(\omega)$ by the static displacement due to the maximum value of $F(t)$, or $A(\omega)$ by the maximum value of $\ddot{s}(t)$. Frequency is also usually reduced to dimensionless form by division by some suitable frequencylike parameter.

Simple Shocks. A simple shock is one like that sketched in Fig. 4.19; it is generally nonoscillatory, has a unique absolute maximum reached within a finite "rise time" t_m, and a finite duration t_0. Spectra of simple shocks are usually presented in terms of ft_m or ωt_m (and/or ft_0 or ωt_0) or simple multiples of these dimensionless quantities.

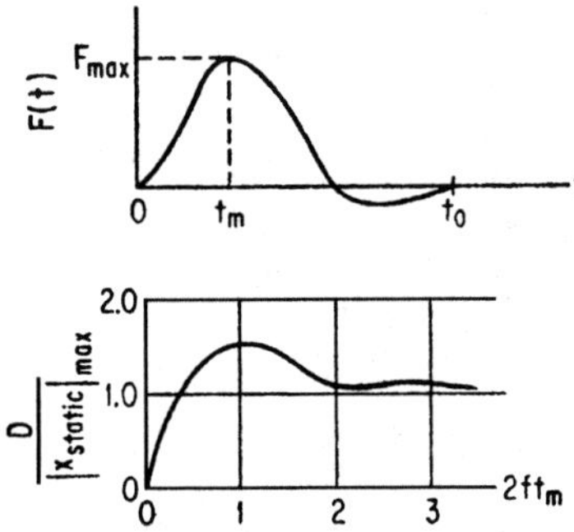

FIG. 4.19 Typical simple shock and dimensionless amplification spectrum.

Shock spectra may, by virtue of their definitions, be used directly to determine the maximum responses of single-degree-of-freedom systems to the shocks to which the spectra pertain.

The main utility of the shock-spectrum concept, however, is due to the fact that the amplification spectra of roughly similar shocks, when presented in terms of the foregoing dimensionless-frequency parameters, tend to coincide very nearly; i.e., amplification spectra are relatively insensitive to details of the pulse form. (Good coincidence of spectra of similar shocks is obtained for small values of f if the spectra are plotted against ft_0, for large values of f if spectra are plotted against ft_m.)

The effect of damping is to reduce system responses in general. The greatest reduction usually occurs near the peaks of the amplification spectra; near relative minima of these spectra the reduction due to damping is generally small.

The amplification spectra of simple shocks have the following properties:

They pass through the origin.

They do not exceed 2.0.

They approach 1.0 for large ft_m.

They generally reach their maxima between $2\,ft_m = 0.7$ and 2.0, but most often near $2\,ft_m = 1$.

Simple shocks occur in drop tests, aircraft landing impact, hammer impact, gun recoil, explosion blasts, and ground shock due to explosive detonations. Details of many spectra appear in Refs. 1, 19, 64, and 72.

4.6 DESIGN CONSIDERATIONS

4.6.1 Design Approach

Since vibration is generally considered an undesirable side effect, it seldom controls the primary design of a machine or structure. Items usually are designed first to fulfill their main function, then analyzed from a vibration viewpoint in regard to possible equipment damage or malfunction, structural fatigue failure, noise, or human discomfort or annoyance.

The most severe effects of vibration generally occur at resonance; therefore, one usually is concerned first with determination of the resonance frequencies of the preliminary design. (Damping is usually neglected in the pertinent calculations for all but the simplest systems, unless a prominent damping effect is anticipated.) If resonance frequencies are found to lie within the intended range of driving frequencies, one should attempt a redesign to shift the resonances out of the driving-frequency range.

If resonances cannot be avoided reasonably, the designer must determine the severity of these resonances. Damping must then be considered in the pertinent calculations, since it is primarily damping that limits response at resonance. If resonant responses are too severe, one must reduce the excitation and/or incorporate increased damping in the system or structure.[69]

Shifting of Resonances. If resonances are found to occur within the range of excitation frequencies, one should try to redesign the system or structure to change its resonance frequencies. Added stiffness with little addition of mass results in shifting of the resonances to higher frequencies. Added mass with little addition of stiffness results in lowering of the resonance frequencies. Addition of damping generally has little effect on the resonance frequencies.

In cases where the vibrating system cannot be modified satisfactorily, one may avoid resonance effects by not operating at excitation frequencies where resonances are excited. This may be accomplished by automatic controls (e.g., speed controls on a machine) or by prescribing limitations on use of the system (e.g., "red lines" on engine tachometers to show operating speeds to be avoided).

Evaluation of Severity of Resonances. The displacement amplitude X_0 of a linear single-degree-of-freedom system (whose natural frequency is ω_n), excited at resonance by a sinusoidal force $F_0 \sin \omega_n t$, is given by

$$kX_0/F_0 = 1/2\zeta = Q = 1/\eta$$

where ζ is the ratio of damping to critical damping, Q the "quality factor," and η the loss factor of the system. The velocity amplitude V_0 and acceleration amplitude A_0 are given by

$$A_0 = \omega_n V_0 = \omega_n^2 X_0$$

The maximum force exerted by the system's spring (of stiffness k) is

$$F_s = kX_0$$

The maximum spring stress may readily be calculated from the foregoing spring force. Multiple-degree-of-freedom systems generally require detailed analysis in accordance with methods outlined in Sec. 4.3. The previous expressions pertaining to single-degree-of-freedom systems hold also for systems with a number of degrees of freedom if X_0 is taken as the amplitude of a generalized (principal) coordinate that is independent of the other coordinates, and if F_0 is taken as the corresponding generalized force. Then $\eta = 2\zeta = Q^{-1}$ must describe the effective damping for that coordinate.

Stresses in the various members may be determined from the mode shapes (vectors) corresponding to the resonant mode; i.e., from the maximum displacements or forces to which the elements are subjected.

For uniform distributed systems* the modal displacement U_{mn} at resonance of the m, n mode due to a modal excitation $G_{mn} \sin \omega_{mn} t$ is given by

$$|U_{mn}| = |G_{mn}|/M_{mn}\omega_{mn}^2\eta(\omega_{mn})$$

and is related to modal velocity and acceleration amplitudes $\dot{U}_{mn}$, $\ddot{U}_{mn}$, according to

$$|\ddot{U}_{m,n}| = \omega_{mn}^2|U_{mn}| = \omega_{mn}|\dot{U}_{mn}|$$

$\eta(\omega)$ denotes the system loss factor, which varies with frequency depending on the damping mechanism present. For beams and plates in flexure[64] η is related to the usual viscous-damping coefficient c and flexural rigidity D as†

$$\eta(\omega) = c\omega/D$$

The maximum stress $\sigma_{\max}$ and maximum strain $\epsilon_{\max}$ that occur at resonance in bending of beams and plates may be approximated by

$$\sigma_{\max} \approx (C/r)|U_{mn}|\omega_{mn}\rho c_{Ln} g_{mn} \approx \epsilon_{\max}\rho c_L^2$$

where $c_L = \sqrt{E/\rho}$ denotes the velocity of sound in the material (E is Young's modulus, ρ the material density) and C denotes the distance from the neutral to the outermost fiber, r the radius of gyration of the cross section. For beams with rectangular cross sections and for plates, $C/r = \sqrt{3}$. For plates $M_{mn} = \rho h$; for beams $M_m = \rho A$, where h denotes plate thickness, A beam cross-section area.

The factor g_{mn} accounts for boundary conditions and takes on the following approximate values:

Boundary conditions on pair of opposite edges	g_{mn}
Pinned-pinned or clamped-free	1.00
Clamped-clamped or clamped-pinned	1.33
Free-free or pinned-free	0.80

For conservative design if the boundary conditions are now known one should use $g_{mn} = 1.33$. Otherwise, one should use the largest value of g_{mn} pertinent to the existing boundary conditions.

Reduction of Severity of Resonances. If redesign to avoid resonances is not feasible, one has only two means for reducing resonant amplitudes: (1) reduction of modal excitation and (2) increase of damping.

*The summary here is presented for two-dimensional systems (plates), and double subscripts are used for the various modal parameters. However, the identical expressions apply also for one-dimensional systems (beams, strings) if the n subscripts are deleted. See also Sec. 4.4.

†For plates $D = Eh^3/12(1 - \nu^2)$; for beams $D = EI$.

Excitation reduction may take the form of running a machine at reduced power, isolating the resonating system from the source of excitation, or shielding the system from exciting pressures. Increased damping may be obtained by addition of energy-dissipating devices or structures. For example, one might use metals with high internal damping for the primary structure, or else attach coatings or sandwich media with large energy-dissipation capacities to a primary structure of common materials.[51,65] Alternately, one might rely on structural joints or shaft bearings to absorb energy by friction, or else attempt to extract energy by means of viscous friction or acoustic radiation in fluids in contact with the resonant system. Simple dashpots, localized friction pads, or magnetically actuated eddy-current-damping devices may also be used, particularly for systems with a few degrees of freedom; however, such localized devices may not be effective for higher modes of distributed systems.

4.6.2 Source-Path-Receiver Concept

If one is called upon to analyze or modify an existing or projected system from the vibration viewpoint, one may find it useful to examine the system in regard to

1. Sources of vibration (e.g., reciprocating engines, unbalanced rotating masses, fluctuating air pressures)
2. Paths connecting sources to critical items (e.g., substructures, vibration mounts)
3. Receivers of vibratory energy, i.e., critical items that malfunction if exposed to too much vibration (e.g., electronic components)

The problem of limiting the effects of vibration may then be attacked by any or all of the following means:

1. Vibration elimination at the source (e.g., designing machines in opposed pairs so that inertia forces cancel, balancing all rotating items, smoothing or deflecting unsteady air flows)
2. Modification of paths (e.g., changing substructures so as to transmit less vibration, introducing vibration mounts)
3. Decreasing vibration sensitivity of critical items (e.g., changing orientation of components with respect to excitation, using more rugged components, using more fatigue-resistant materials, designing more damping into components in order to decrease the effects of internal resonances)

Vibration Reduction at the Source: Vibration Absorbers. The strength of vibration sources may often be reduced by proper design. Such design may include balancing, use of the lightest possible reciprocating parts, arranging components so that inertia forces cancel, or attaching vibration absorbers.

A vibration absorber* is essentially a mass m attached by means of a spring k to a primary mass M in order to reduce the response of M to a sinusoidal force acting directly on M (or to a sinusoidal support acceleration when the M is attached to the support by a spring K). If an absorber for which $k/m = \omega_0^2$ is attached to M, then M will experience no excursion if the excitation occurs at a frequency ω_0. (The displacement amplitude of m will be F/k, where F is the force amplitude.)

For driving frequencies very near ω_0, attachment of the vibration absorber results in small amplitudes for M, but for other frequencies it generally results in amplitudes

*The discussion is presented here only for translational systems. It may be extended by analogy to apply also to rotational systems.

that are greater than those obtained without an absorber. Vibration absorbers are very frequency-sensitive and hence should be used only where there exists essentially a single relatively accurately known driving frequency. The addition of damping to an absorber[17,63] extends to some extent the frequency range over which the absorber results in vibration reduction of the primary mass, but increased damping also results in more motion of M at the optimum frequency ω_0.

Path Modification: Vibration Mounts. The prime means for modifying paths traversed by vibratory energy consists of the addition of vibration mounts. If a mass M is excited by a vibratory force of frequency ω, one may reduce the vibratory force that the mass exerts on a rigid support to which it is attached by inserting a spring of stiffness k between mass and support such that $k/M < \omega^2/2$. Further reduction in k/M reduces the transmitted force further; added damping with k/M held constant increases the transmitted force.

If a mass M is attached by means of a spring k to a support that is vibrating at frequency ω, then M can be made to vibrate less than the support if k is chosen so that $k/M < \omega^2/2$. Further reduction of k/M further reduces the motion of the mass; addition of damping increases this motion.

Discussions of vibration and shock mounting are to be found in Refs. 14, 32, 55, and 56. Data on commercial mounts may be found in manufacturers' literature.

Modification of Critical Items. Critical items may often be made less sensitive to vibration by redesigning them so that all their internal-component resonance frequencies fall outside the excitation frequency range. Occasionally this can be done merely by reorienting a critical item with respect to the direction of excitation; more often it requires stiffening the components (or possibly adding mass). If fatigue rather than malfunction is a problem, one may obtain improved parts by careful redesign to eliminate stress concentrations and/or by using materials with greater fatigue resistance. If internal resonances cannot be avoided, one may reduce their effect by designing damping into the resonant components.

4.6.3 Rotating Machinery

A considerable amount of information has been amassed in relation to reciprocating and turbine machines. Comprehensive treatments of the associated torsional vibrations may be found in Refs. 42 and 72; less detailed discussions of these appear in standard texts.[17,32,61,63] A few of the most important items pertaining to rotating machinery are outlined subsequently.

Vibrations of rotating machines are caused by the following factors; reduction of these generally serves to reduce vibrations:

1. Unbalance of rotating components
2. Reciprocating components
3. Whirling of shafts
4. Gas forces
5. Instabilities, such as those due to slip-stick phenomena

Balancing. A rotor mounted on frictionless bearings so that its axis of rotation is horizontal remains motionless (if subject only to gravity) in any angular position, if the rotor is *statically* balanced. Static balance implies an even mass distribution around the rotational axis. However, even statically balanced rotors may be dynamically

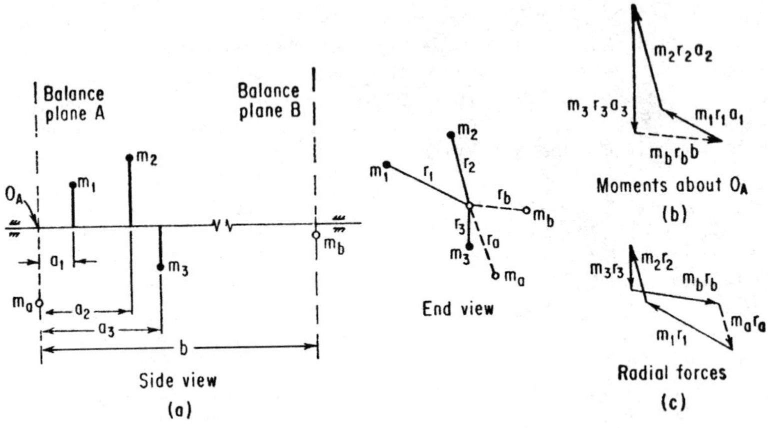

FIG. 4.20 Balancing of rigid rotors.

unbalanced. Dynamic unbalance occurs if the centrifugal forces set up during rotation result in a nonzero couple. (Static balance assures only that the centrifugal forces result in zero net radial force.)

Rigid rotors can always be balanced statically by addition of a single weight in any arbitrarily chosen plane, or dynamically (and statically) by addition of two weights in two arbitrarily chosen planes. The procedure, illustrated in Fig. 4.20, is as follows:[9]

1. Divide the rotor into a convenient number of sections by passing planes perpendicular to the rotational axis.
2. Determine the mass and center-of-gravity position for each section.
3. Draw a diagram like Fig. 4.20*a* where each section is represented by its mass located at the center-of-gravity position.
4. Select balance planes, i.e., planes in which weights are to be attached for balance.
5. Draw a diagram (Fig. 4.20*b*) of centrifugal force moments* about point Q_A (where rotational axis intersects balance plane *A*). Each moment is represented by a vector of length $m_i r_i a_i$ (to some suitable scale) parallel to r_i in the end view. The vector required to close the diagram then is $m_b r_b b$; its direction gives the direction of r_b in the end view; $m_b r_b$ may be calculated and either m_b or r_b may be selected arbitrarily.
6. Draw a diagram (Fig. 4.20*c*) of centrifugal forces* $m_i r_i$, including $m_b r_b$. Each force vector is drawn parallel to r_i in the end view; the vector required to close the diagram is $m_a r_a$ and defines the mass (and its location) to be added in plane *A*. Again, either m_a or r_a may be selected arbitrarily; r_a in the end view must be parallel to the $m_a r_a$ vector, however.

Balancing of flexible rotors or of rotors on flexible shafts (particularly when operating above critical speeds) can generally be accomplished for only one particular speed or for none at all.[17,61]

Balancing machines, their principles and use, and field balancing procedures are discussed in Refs. 17, 41, 61, and 63, among others.

*The common factor ω^2 is omitted from the diagrams.

Balancing of reciprocating engines is treated in Ref. 17 and in a number of texts on dynamics of machines, such as Ref. 23.

Whirling of Shafts: Critical Speeds. Rotating shafts become unstable at certain speeds, and large vibrations are likely to develop. These speeds are known as "critical speeds." At a critical speed the number of revolutions per second is generally very nearly equal to a natural frequency (in cycles per second) of the shaft considered as a nonrotating beam vibrating laterally. Thus critical speeds of shafts may be found by any of the means for calculating the natural frequencies of lateral vibrations of beams (see Sec. 4.4.3).

"Whirling," i.e., violent vibration at critical speeds, occurs in vertical as well as horizontal shafts. In nonvertical shafts gravity effects may introduce additional "critical speeds of second order," as discussed in Ref. 63. Complications may also occur where disks of large inertia are mounted on flexible shafts. Gyroscopic effects due to thin disks generally tend to stiffen the system and thus to increase the critical speeds above those calculated from static flexural vibrations. Thick disks, however, may result in lowering of the critical speeds.[17] Similarly, flexibility in the bearings results in softening of the system and in lowering of the critical speeds.

Turbine Disks. Turbine disks and blades vibrate essentially like disks and beams, and may be treated by previously outlined standard procedures if the rotational speeds are low. However, centrifugal forces result in considerable stiffening effects at high rotational speeds, so that the natural frequencies of rotating disks and blades tend to be considerably higher (and in different ratio to each other) than those of nonrotating assemblies. A brief discussion of these effects and analytical methods may be found in Ref. 63, a more comprehensive one in Ref. 59.

4.6.4 Damping Devices

In this subsection will be presented various devices for the damping or dissipation of mechanical energy. For applications of elastomers employed as damping attachments see Sec. 40, "Dampers and Elastomers," of the Second Edition of the *Mechanical Design and Systems Handbook* (H. Rothbart, ed., McGraw-Hill, New York, 1985).

Electrodynamic Damping. Electrodynamic damping is obtained when a short-circuited electrical conductor is made to move in a magnetic field. An induced current appears in the conductor. The conductor experiences a force proportional to but opposite to the velocity (see Fig. 4.21). The damping force is expressed by

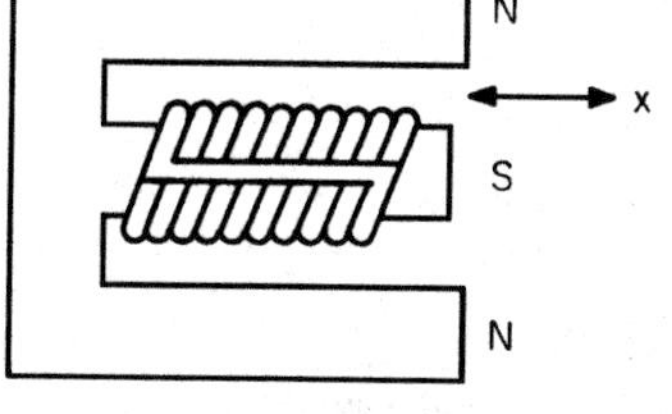

FIG. 4.21 Electrodynamic damper with induced current.

$$F = H^2 A \ell \times 10^{-9}/\rho$$

$$= H^2 V \times 10^{-9}/\rho \qquad (4.67)$$

where F = force
H = average magnetic field strength
A = cross-sectional area of coil wire
ℓ = length of wire
V = volume of wire
ρ = specific resistance

The damping constant ($f = c\dot{x}$) is therefore

$$c = H^2 V \times 10^{-9}/\rho\dot{x} \qquad (4.68)$$

Hydromechanical Damping

Incompressible Fluids (Fig. 4.22). A general form of the damping force may be written as

$$F = k\dot{x}^n \tag{4.69}$$

where $1 < n < 2$. In terms of the pressure drop the equation becomes

$$\Delta p = f(\ell/d)(\dot{x}^2/2g) \tag{4.70}$$

where f = a dimensionless friction coefficient which is a function of Reynolds number
ℓ = length of duct
d = diameter of duct

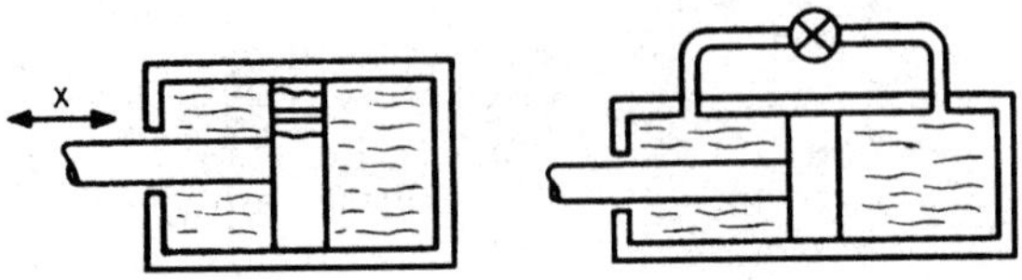

FIG. 4.22 Hydromechanical damping with incompressible fluid.

At low speeds, in which fully developed laminar flow is achieved, Eq. (4.70) assumes the Poiseuille form

$$\Delta p = (32\eta\ell/d^2)\dot{x} \tag{4.71}$$

where η is the viscosity of the liquid.

Compressible Fluids. The principle of damping in this case is similar to that for incompressible fluids. However, the fluid also acts as a spring. The total behavior can be imagined as a spring in parallel with a damper. Assuming adiabatic compression and ideal gas behavior, the effective spring constant of the gas can be evaluated from

$$k = p\gamma A^2/V \tag{4.72}$$

where p = gas pressure
$\gamma = C_p/C_v$
C_p = specific heat at constant pressure
C_v = specific heat at constant volume
V = volume
A = cross-sectional area of duct

Untuned Viscous Shear Damper (Fig. 4.23). This damper (also viscous-fluid damper, untuned damped vibration absorber) consists of an annular mass enclosed by an annular ring. A viscous fluid fills the volume between mass and casing. Since there is no elastic connection between the casing, which is attached to the torsionally vibrating system, and the mass, the damper is untuned. Reduction in amplitudes and a lowering of the natural frequency are obtained without introduction of additional resonances.

Reduction of a complex system to an equivalent two-mass system yields a useful design approximation. With $J_F = \infty$, the system and its equivalent are shown in Fig. 4.24. The impedance equations for steady state can be written as

$$(-\omega^2 J_D + j\omega c)\theta_1 - (j\omega c)\theta_2 = 0 \tag{4.73}$$

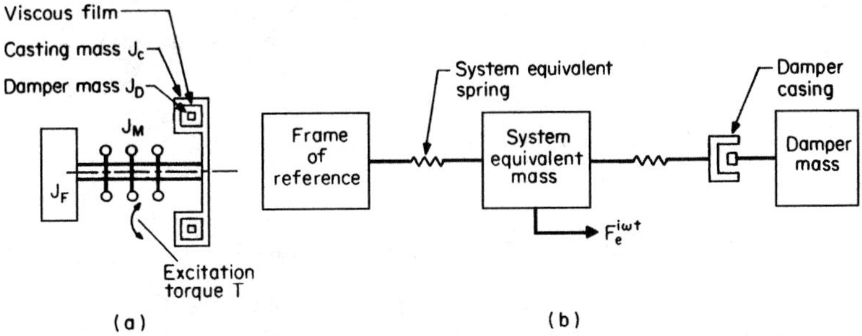

FIG. 4.23 Damping devices. (*a*) Torsional system. (*b*) Translational system.

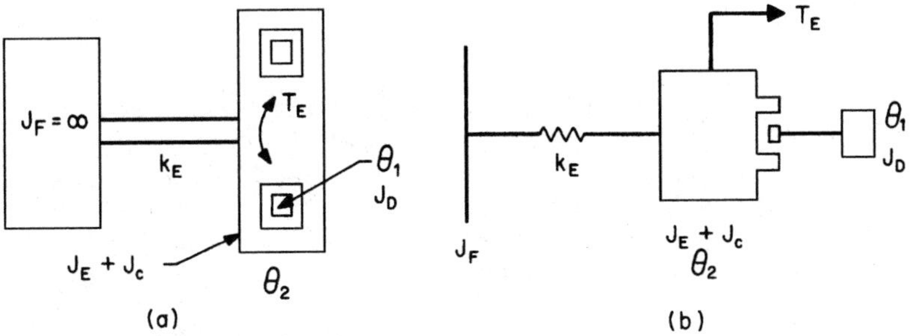

FIG. 4.24 (*a*) Complex system. (*b*) Equivalent two-mass system.

$$-(j\omega c)\theta_1 + [K_2 - \omega^2(J_E + J_c) + j\omega c]\theta_2 = T_E \tag{4.74}$$

where J_D = mass moment of inertia of damper
J_E = equivalent mass moment of inertia of torsionally vibrating systems
J_c = mass moment of inertia of damper casing
ω = circular natural frequency of vibrating system with damper
θ_1 = steady-state vibration amplitude at damper mass
θ_2 = steady-state vibration amplitude at damper casing

Figure 4.25 is a plot of θ_2 as a function of ω for $c = 0$ and $c = \infty$. For optimum

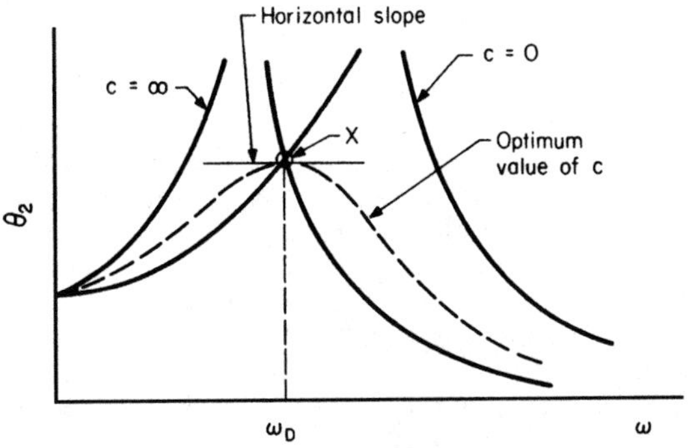

FIG. 4.25 Plot of θ_2 vs. ω for $c = 0$ and $c = \infty$.

damping, the curve of θ versus ω has a horizontal slope at point X. Setting $\mu = J_D/J_E$ and solving for conditions at point X, the following useful formulas result:

$$\omega_D/\omega_2 = \sqrt{2/(2 + \mu)} \tag{4.75}$$

$$\theta_2/(T_0/K_E) = M = (2 + \mu)/\mu \tag{4.76}$$

$$c/(2J_D\omega_2) = \gamma = 1/\sqrt{2(1 + \mu)(2 + \mu)} \tag{4.77}$$

$$|\theta_{SH}/\theta_2| = |\theta_{SH}/(T_0/K_E)|/M = \sqrt{(1 + \mu)/(2 + \mu)} \tag{4.78}$$

where ω_2 = circular natural frequency of torsionally vibrating system without damper
ω_D = circular frequency corresponding to point X in Fig. 4.25
c = optimum value of coefficient of viscous damping
T_0 = excitation torque
$\theta_{SH} = |\theta_1 - \theta_2|$
M = dynamic magnifier

Knowledge of such items as θ_2, T_0, K_E, and ω_2 permits the determination of μ, from which J_D and c may be calculated.

In reciprocating engines μ tends to fall in the range 0.4 to 1.0. Also, J_c/J_D usually lies between 0.35 and 0.8. The dimensions are so proportioned that sufficient surface is provided for proper heat dissipation. Determination of the necessary fluid viscosity is based upon a modified value of c which accounts for varying shear rates on the lateral and peripheral surfaces of the damper.

Slipping-Torque-Type Dampers. In this type of damper, relative motion between a shaft-fixed hub and a damping mass occurs only when the relative acceleration of the two exceeds a predetermined value. An effective change of natural frequency occurs as the damping mass "locks" and "unlocks" from the hub during each oscillation. Dissipation of energy by damping occurs during the intervals of relative motion.

An analysis of the input and dissipated energy can be made by assuming continual slip, sinusoidal motion of the hub, and a linear time variation of the damper mass. The maximum energy dissipated is

$$U_{D,\max} = (4/\pi)\omega^2\theta_1^2 J_R \tag{4.79}$$

where ω = circular natural frequency of vibrating system without damper
θ_1 = permissible oscillation amplitude at damper hub
J_R = moment of inertia of damper

Energy input $U_{in} = T_0\theta_1$, where T_0 is the peak value excitation torque at the hub. Equating input and dissipated energy yields the useful design equation

$$J_R = \pi^2 T_0/4\omega^2\theta_1 \tag{4.80}$$

The Sandner Damper (Pumping-Chamber Type). A rim and side plates, which act at the damper mass, and a hub are arranged to form internal cavities at certain points of the interfaces. The cavities act as pumping chambers. Oil passes through radial passages starting at the hub, moves to the pumping chambers, returns to the hub to pass through spring-loaded relief valves, and then is discharged into some convenient space. The slipping torque is accurately set by adjustment of the relief valves.

The damper moment of inertia is determined from Eq. (4.80). Maximum torque at which slipping occurs is calculated from

$$T_{R,\max} = (\sqrt{2}/\pi)\omega^2\theta_1 J_R \tag{4.81}$$

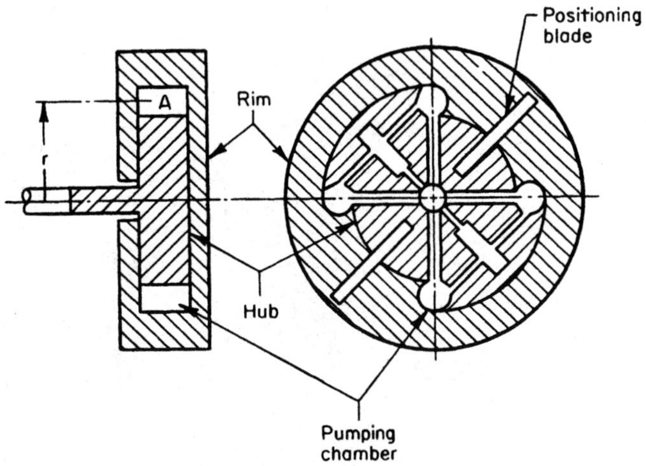

FIG. 4.26 The Sandner damper (pumping-chamber type).

Determination of the relief-valve spring pressure can be calculated from

$$p = T_{R,\max}/2rA$$

where A is the cross-sectional area of one of the pumping chambers and r is the mean radius to the area A, as shown in Fig. 4.26.

The Sandner Damper (Gear-Wheel Type). In this type of damper a rim is cut with gear teeth on its inner surface. Pinions mesh with this gear and are enclosed in special recesses in the hub. Passages connect opposite sides of each recess to centrally located relief valves.

The rim tends to rotate the pinions because of its inertia torque. Actual rotation takes place beyond a certain critical value as predetermined by the relief-valve adjusted pressure. Oil is passed from one gear chamber to the next as the pinions rotate, as shown in Fig. 4.27. Calculations for this damper are the same as for the pumping-chamber type. However, the pressure is computed as $p = T_{R,\max}/8rA$, where A is the effect area of the gear-pump recess.

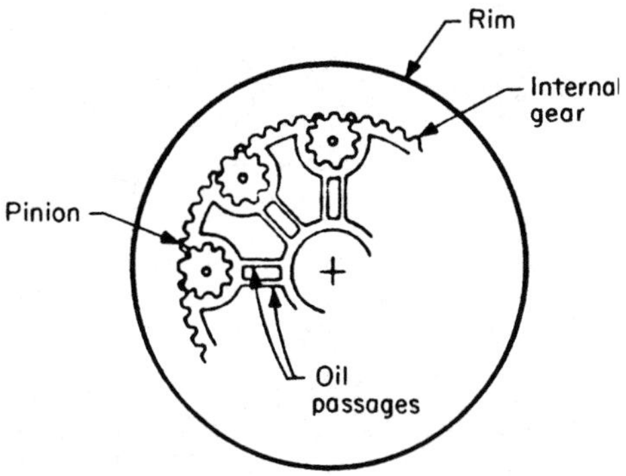

FIG. 4.27 The Sandner damper (gear-wheel type).

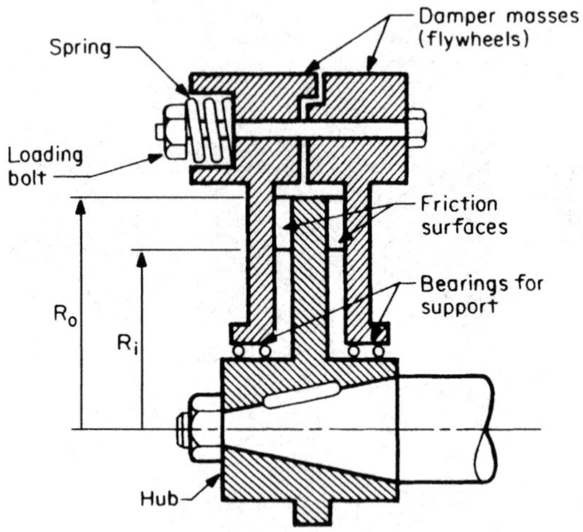

FIG. 4.28 The Lanchester damper (semi-dry-friction type).

The Lanchester Damper (Semi-Dry-Friction Type). In this type of damper, a hub fixed coaxially to a vibrating shaft, carries friction plates on an annulus near its rim. The damper mass, consisting of two flywheels and loading bolts, presses against the friction surfaces. The damper mass lies coaxial with the shaft but is coupled to it only through the friction surfaces, as shown in Fig. 4.28.

The moment of inertia J_{2R} of the two flywheels is calculated by Eq. (4.80). Maximum torque is computed from $T_{R,\max} = (\sqrt{2}/\pi)\omega^2\theta_1 J_{2R}$ and is equated to the friction torque T_f. Spring load can then be computed from

$$T_f = 4/3\mu P(R_o^3 - R_i^3)/(R_o^2 - R_i^2) \tag{4.82}$$

where μ is the coefficient of friction, R_i and R_o are the inner and outer radii of the friction surfaces, respectively, and P is the total spring load.

4.6.5 Charts and Tables

Information on natural frequencies, spring constants, and material properties appears in the tables listed below.

Characteristic	Table
Natural frequencies	
Simple translational systems	4.8*a*
Simple torsional systems	4.8*b*
Beams, bars, shafts, (uniform section unless noted otherwise):	
Uniform and variable section, in flexure	4.8*c*
Flexure; mode shapes	4.6*b*
Torsion; mode shapes	4.6*a*
Longitudinal vibration; mode shapes	4.6*a*
On multiple evenly spaced supports	4.8*d*
Free-free, on elastic foundation	4.8*h*

(Continued)

Characteristic	Table
Natural frequencies (*continued*)	
Strings, uniform; mode shapes	4.6*a*
Membranes, uniform:	
Circular	4.8*g*
Rectangular and circular; mode shapes	4.6*d*
Plates, uniform:	
Cantilever, various shapes	4.8*e*
Circular, various boundary conditions	4.8*f*
Rectangular, on simple supports; mode shapes	4.6*c*
Circular, clamped; mode shapes	4.6*c*
Cylindrical shells	4.8*h*
Rings	4.8*i*
Mass free to rotate and translate in plane	4.8*i*
Mass on spring of finite mass	4.8*i*
Spring constants	
Combinations of springs	4.9*a*
Round-wire helical springs	4.9*b*
Beams with force or moment inputs	4.9*c*
Torsion springs	4.9*d*
Shafts in torsion	4.9*d,e*
Plates loaded at centers	4.9*f*
Miscellaneous	
Longitudinal wavespeed and K_m (for Table 4.8) for engineering materials	4.10

Tables 4.8, 4.9, and 4.10 begin on pages 4.64, 4.74, and 4.79, respectively.

TABLE 4.8*a* Natural Frequencies of Simple Translational Systems[7,17]

System	$\omega^2 =$	Remarks
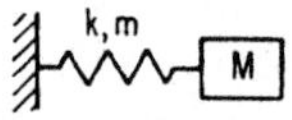	$\frac{k}{M+0\cdot 33m}$	
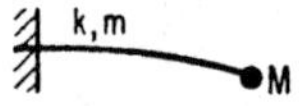	$\frac{k}{M+0\cdot 23m}$	Point mass on cantilever beam
	$\frac{k}{M+0\cdot 375m}$	Point mass at center of clamped beam; See Table 4.9c for k
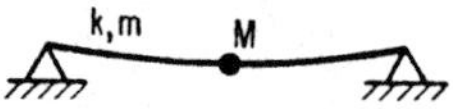	$\frac{k}{M+0\cdot 50m}$	Point mass at center of simply supported beam
	$k\left[\frac{1}{M_1}+\frac{1}{M_2}\right]$	
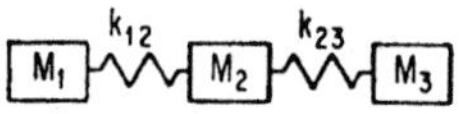	$B \pm \sqrt{B^2-C}$	$2B = \frac{k_{12}}{M_1} + \frac{k_{23}}{M_3} + \frac{k_{12}+k_{23}}{M_2}$ $C = \frac{k_{12}+k_{23}}{M_1\,M_2\,M_3}(M_1+M_2+M_3)$
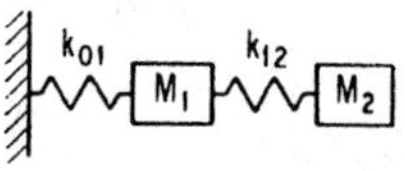	$B \pm \sqrt{B^2-C}$	$2B = \frac{k_{01}+k_{12}}{M_1} + \frac{k_{12}}{M_2}$ $C = \frac{k_{01}\,k_{12}}{M_1\,M_2}$
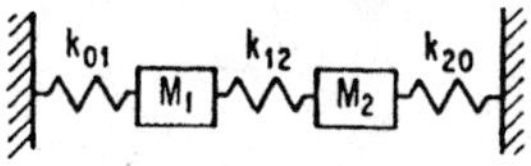	$B \pm \sqrt{B^2-C}$	$2B = \frac{k_{01}+k_{12}}{M_1} + \frac{k_{12}+k_{20}}{M_2}$ $C = \frac{1}{M_1 M_2}(k_{01}k_{12}+k_{12}k_{20}+k_{20}k_{01})$
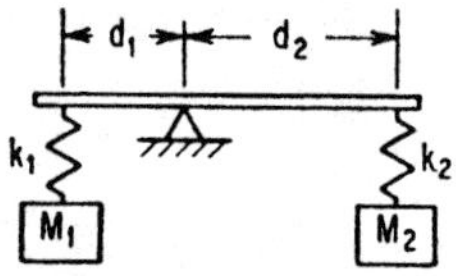	$\dfrac{\frac{1}{M_1}+\frac{1}{R^2M_2}}{\frac{1}{k_1}+\frac{1}{R^2k_2}}$	Inertia of lever negligible $R = \frac{d_2}{d_1}$
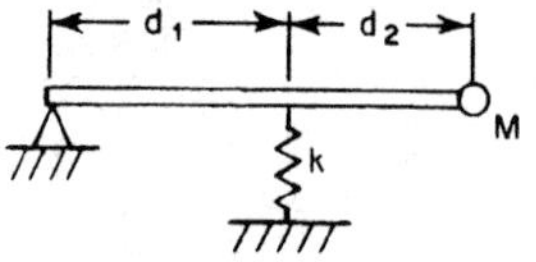	$\frac{k}{M}\left(\frac{d_1}{d_1+d_2}\right)^2$	Inertia of lever negligible.

Symbols:
ω = circular natural frequency = $2\pi f$
M = mass
m = total mass of spring element
k = spring constant

TABLE 4.8*b* Natural Frequencies of Simple Torsional Systems[7,17]

System	$\omega^2 =$	Remarks
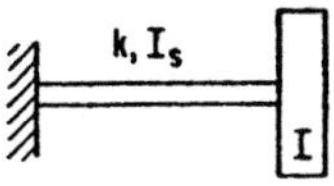	$\dfrac{k}{I + I_s/3}$	
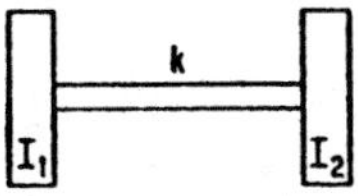	$k\left(\dfrac{1}{I_1} + \dfrac{1}{I_2}\right)$	
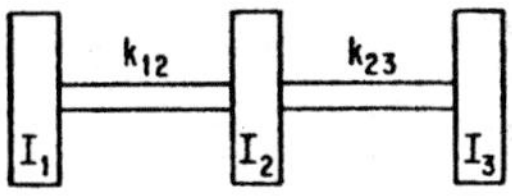		$2B = \dfrac{k_{12}}{I_1} + \dfrac{k_{23}}{I_3} + \dfrac{k_{12}+k_{23}}{I_2}$ $C = \dfrac{k_{12}\,k_{23}}{I_1\,I_2\,I_3}(I_1 + I_2 + I_3)$
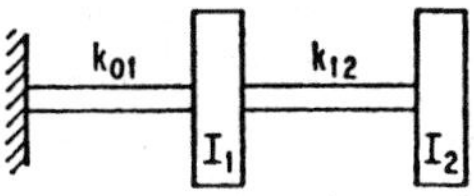	$B \pm \sqrt{B^2 - C}$	$2B = \dfrac{k_{01}+k_{12}}{I_1} + \dfrac{k_{12}}{I_2}$ $C = \dfrac{k_{01}\,k_{12}}{I_1\,I_2}$
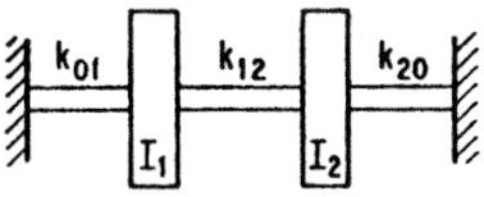		$2B = \dfrac{k_{01}+k_{12}}{I_1} + \dfrac{k_{12}+k_{20}}{I_2}$ $C = \dfrac{1}{I_1\,I_2}(k_{01}k_{12} + k_{12}k_{20} + k_{20}k_{01})$
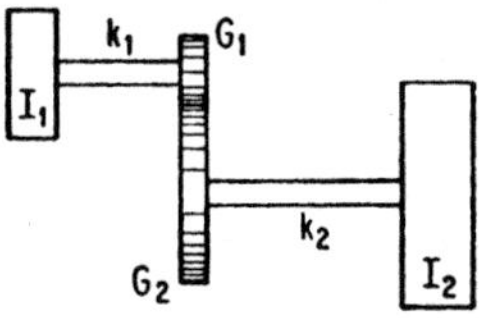	$\dfrac{\dfrac{1}{I_1} + \dfrac{1}{R^2 I_2}}{\dfrac{1}{k_1} + \dfrac{1}{R^2 k_2}}$	Inertia of gears G_1, G_2 assumed negligible $R = \dfrac{\text{number of teeth on gear 1}}{\text{number of teeth on gear 2}}$ $= \dfrac{\text{rpm of shaft 2}}{\text{rpm of shaft 1}}$

Symbols:
ω = circular natural frequency = $2\pi f$
k = torsional spring constant; see Table 4.9*d*
I = polar mass moment of inertia
I_s = polar mass moment of inertia of entire shaft

TABLE 4.8c Natural Frequencies of Beams in Flexure[7,33]

$$f_n = C_n \frac{r}{L^2} \times 10^4 \times K_m$$

f_n = nth natural frequency, hertz
C_n = frequency constant listed in these tables
r = radius of gyration of cross section = $\sqrt{I/A}$, inches*
L = beam length, inches*
K_m = material constant (Table 4.10) = 1.00 for steel

Uniform-section beams		n				
		1	2	3	4	5
 Clamped–Clamped	 Free–Free	71.95	198.29	388.73	642.60	959.94
 Clamped–Free		11.30	70.85	198.30	388.73	642.60
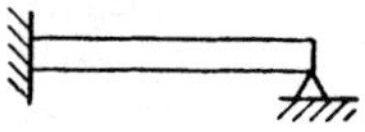 Clamped–Hinged	 Free–Hinged	49.57	160.65	335.17	573.20	874.65
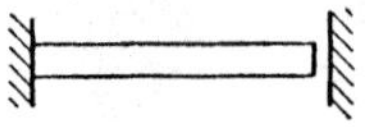 Clamped–Guided	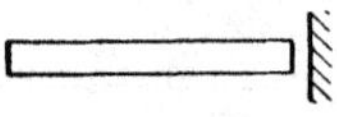 Free–Guided	17.98	97.18	239.98	446.25	715.98
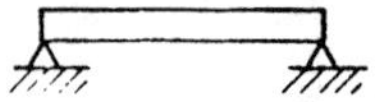 Hinged–Hinged	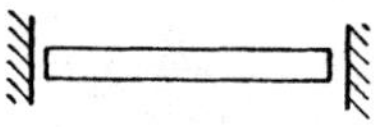 Guided–Guided	31.73	126.93	285.60	507.73	793.33
 Hinged–Guided		7.93	71.40	198.33	388.73	642.60

*For r and L in centimeters, use C_n values listed in table multiplied by 2.54.

TABLE 4.8c Natural Frequencies of Beams in Flexure (*Continued*)

Variable-section beams (Use maximum r to calculate f_n)	Shape		n		
	b/b_0	h/h_0	1	2	3
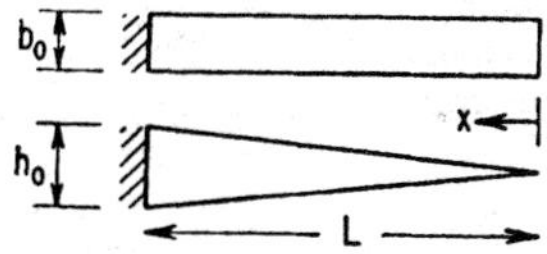	1	x/L	17.09	48.89	96.57
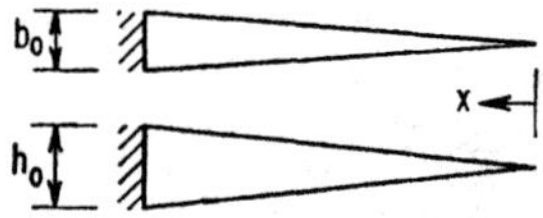	x/L	x/L	26.08	68.08	123.64
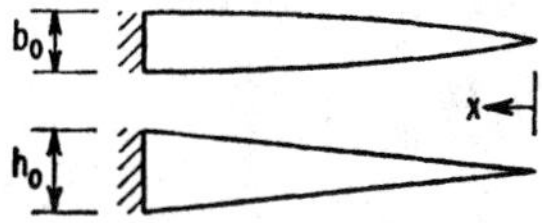	$\sqrt{x/L}$	x/L	22.30	58.18	109.90
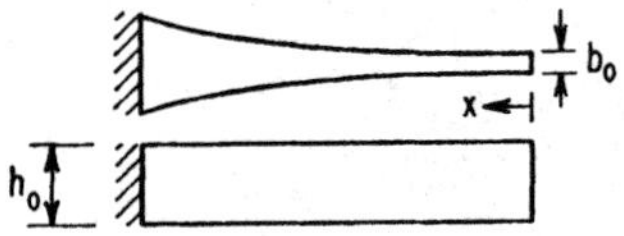	exp x/L	1	15.23	77.78	206.07
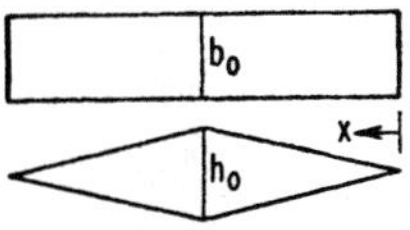	1	x/L	21.21* 35.05†	56.97*	
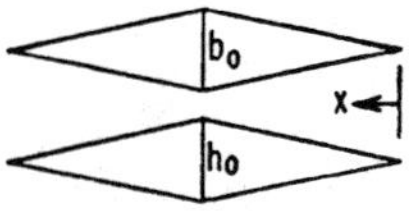	x/L	x/L	32.73* 49.50†	76.57*	
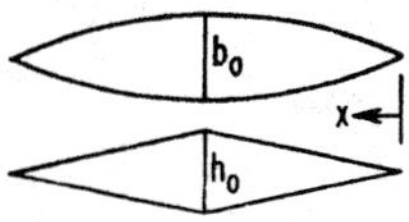	$\sqrt{x/L}$	x/L	25.66* 42.02†	66.06*	

*Symmetric mode.
†Antisymmetric mode.

TABLE 4.8*d* Natural Frequencies of Uniform Beams on Multiple Equally Spaced Supports[7,33]

$$f_n = C_n \frac{r}{L^2} \times 10^4 \times K_m$$

f_n = nth natural frequency, hertz
C_n = frequency constant listed in these tables
r = radius of gyration of cross section = $\sqrt{I/A}$, inches*
L = span length, inches*
K_m = material constant (Table 4.10) = 1.00 for steel

	Number of spans	*n*				
		1	2	3	4	5
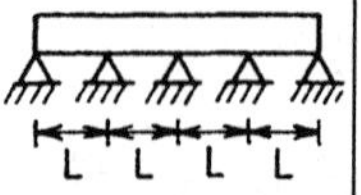 Ends simply supported	1	31.73	126.94	285.61	507.76	793.37
	2	31.73	49.59	126.94	160.66	285.61
	3	31.73	40.52	59.56	126.94	143.98
	4	31.73	37.02	49.59	63.99	126.94
	5	31.73	34.99	44.19	55.29	66.72
	6	31.73	34.32	40.52	49.59	59.56
	7	31.73	33.67	38.40	45.70	53.63
	8	31.73	33.02	37.02	42.70	49.59
	9	31.73	33.02	35.66	40.52	46.46
	10	31.73	33.02	34.99	39.10	44.19
	11	31.73	32.37	34.32	37.70	41.97
	12	31.73	32.37	34.32	37.02	40.52
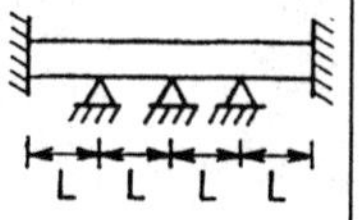 Ends clamped	1	72.36	198.34	388.75	642.63	959.98
	2	49.59	72.36	160.66	198.34	335.20
	3	40.52	59.56	72.36	143.98	178.25
	4	37.02	49.59	63.99	72.36	137.30
	5	34.99	44.19	55.29	66.72	72.36
	6	34.32	40.52	49.59	59.56	67.65
	7	33.67	38.40	45.70	53.63	62.20
	8	33.02	37.02	42.70	49.59	56.98
	9	33.02	35.66	40.52	46.46	52.81
	10	33.02	34.99	39.10	44.19	49.59
	11	32.37	34.32	37.70	41.97	47.23
	12	32.37	34.32	37.02	40.52	44.94
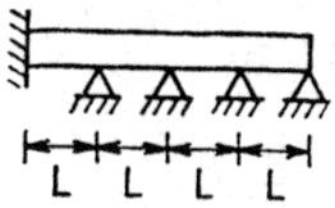 Ends clamped-supported	1	49.59	160.66	335.2	573.21	874.69
	2	37.02	63.99	137.30	185.85	301.05
	3	34.32	49.59	67.65	132.07	160.66
	4	33.02	42.70	56.98	69.51	129 49
	5	33.02	39.10	49.59	61.31	70.45
	6	32.37	37.02	44.94	54.46	63.99
	7	32.37	35.66	41.97	49.59	57.84
	8	32.37	34.99	39.81	45.70	53.63
	9	31.73	34.32	38.40	43.44	49.59
	10	31.73	33.67	37.02	41.24	46.46
	11	31.73	33.67	36.33	39.81	44.19
	12	31.73	33.02	35.66	39.10	42.70

*For r and L in centimeters, use C_n values multiplied by 2.54.

TABLE 4.8e Natural Frequencies of Cantilever Plates[7,33]

$$f_n = C_n \frac{h}{a^2} \times 10^4 \times K_m$$

f_n = nth natural frequency, hertz
C_n = frequency constant listed in table
h = plate thickness, inches*
a = plate dimension, as shown inches*
K_m = material constant (Table 4.10) = 1.00 for steel

Boundary conditions F = free C = clamped S = simply supported		n = 1	2	3	4	5
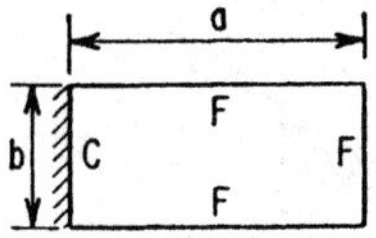	$a/b = 1/2$	3.41	5.23	9.98	21.36	24.18
	$a/b = 1$	3.40	8.32	26.71	20.86	30.32
	$a/b = 2$	3.38	14.52	91.92	21.02	47.39
	$a/b = 5$	3.36	33.79	548.60	20.94	103.03
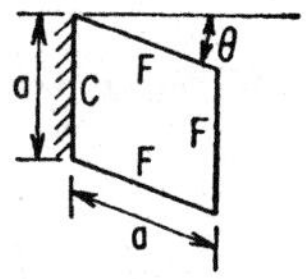	$\theta = 15°$	3.50	8.63			
	$\theta = 30°$	3.85	9.91			
	$\theta = 45°$	4.69	13.38			
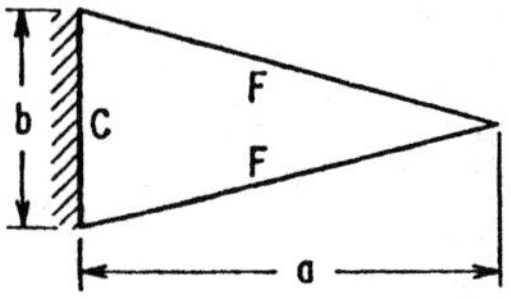	$a/b = 2$	6.7	28.6	56.6	137	
	$a/b = 4$	6.6	28.5	83.5	240	
	$a/b = 8$	6.6	28.4	146.1	457	
	$a/b = 14$	6.6	28.4	246	790	
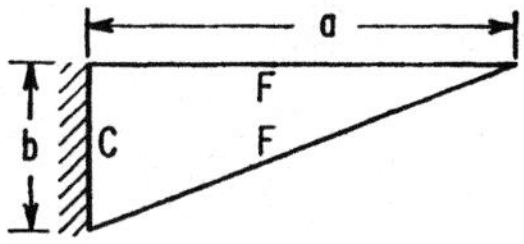	$a/b = 2$	5.5	23.6			
	$a/b = 4$	6.2	26.7			
	$a/b = 7$	6.4	28.1			

*For h and a in centimeters, use given C_n values multiplied by 2.54.

TABLE 4.8f Natural Frequencies of Circular Plates[7,33]

$$f = C\frac{h}{r^2} \times 10^4 \times K_m$$

f = natural frequency, hertz
C = frequency constant listed in table
h = plate thickness, inches*
r = plate radius, inches*
K_m = material constant (Table 4.10) = 1.00 for steel

Boundary conditions	Number of nodal circles	Number of nodal diameters 0	1	2	3
Clamped at circumference 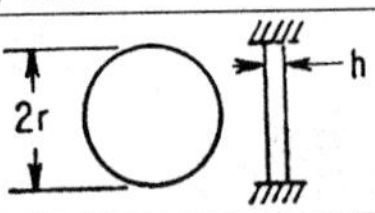	0	9.94	20.67	33.91	49.61
	1	38.66	59.12	82.22	107.9
	2	86.61	116.8	149.5	185.0
	3	153.8	193.5	235.9	281.1
Free 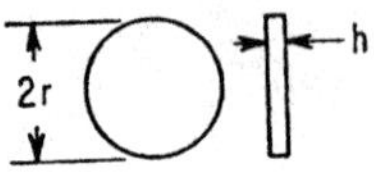	0			5.11	11.89
	1	8.83	19.94	34.27	51.43
	2	37.47	58.19	81.6	108.2
	3	85.35	115.68	149.7	186.1
Clamped at center; free at circumference 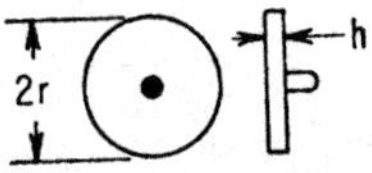	0	3.65			
	1	20.33			
	2	59.49			
	3	117.2			
Simply supported at circumference 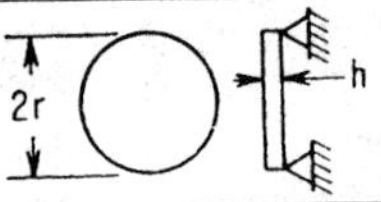	0	4.84	13.55	24.93	
	1	28.93	47.16	68.18	
	2	72.13	99.93	130.6	
	3	134.4	171.9	212.1	
Clamped at center; simply supported at circumference 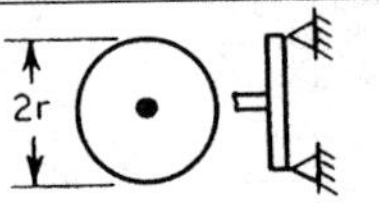	0	14.4			
	1	48.0			
Clamped at center and at circumference 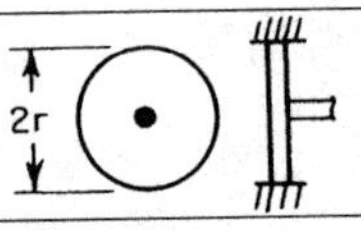	0	22.1			
	1	60.2			

Simply supported at radius a; free at circumference 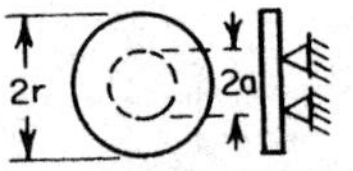

Fundamental mode						
a/r	0	0.2	0.4	0.6	0.8	1.0
C	3.64	4.4	6.5	8.6	7.2	4.84

*For h and r in centimeters, use given C_n values multiplied by 2.54.

TABLE 4.8*g* Natural Frequencies of Circular Membranes[33]

$$f = C\sqrt{T/h\rho r^2}$$

f = natural frequency, hertz
C = frequency constant listed in table
r = membrane radius, inches*
h = membrane thickness, inches*
T = tension at circumference, lb_f/in*
ρ = material density, lb_m/in^3 or $lb_f \cdot s^2/in^4$*

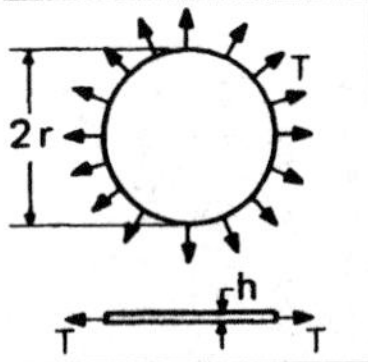

Number of nodal circles	Number of nodal diameters			
	1	2	3	4
1	0.383	0.610	0.817	1.015
2	0.887	1.116	1.340	1.553
3	1.377	1.619	1.849	2.071
4	1.877	2.120	2.355	2.582

*Any set of consistant units may be used that causes all units except *s* to cancel in the square root.

TABLE 4.8*h* Natural Frequencies of Cylindrical Shells[7]

f_{ij} = natural frequency for mode shape ij, hertz
h = shell thickness, inches*
L = length, inches*
i = number of circumferential waves in mode shape
j = number of axial half-waves in mode shape
c_L = Longitudinal wavespeed in shell material, in/s* (See Table 4.10)
ν = Poisson's ratio, dimensionless

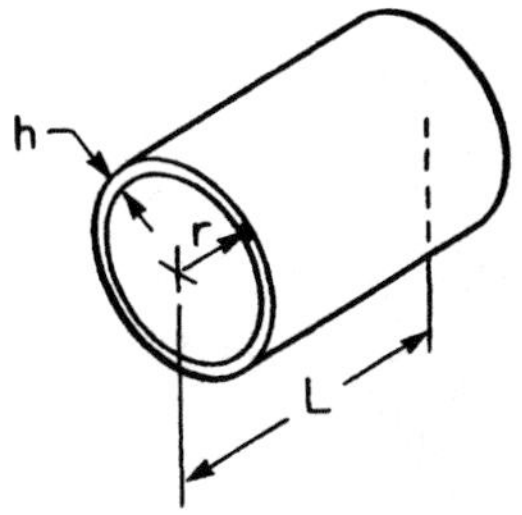

$$f_{ij} = \frac{\lambda_{ij}}{2\pi r}\frac{c_L}{\sqrt{1 - \nu^2}}$$

*Any other consistent set of units may be used. For example, h, r, L in centimeters and c_L in centimeters per second.

TABLE 4.8*h* Natural Frequencies of Cylindrical Shells (*Continued*)

Infinitely long		
Axial modes:	$\lambda_{ij} = i\sqrt{\dfrac{1-\nu}{2}},$	$i = 1, 2, 3, \ldots$
Radial (extensional) modes:	$\lambda_{ij} = \sqrt{1+i^2},$ *Note:* $i = 0$ corresponds to breathing mode	$i = 0, 1, 2, \ldots$
Radial-circumferential flexural modes:	$\lambda_{ij} = \dfrac{h/r}{\sqrt{12}}\dfrac{i(i^2-1)}{\sqrt{i^2+1}},$	$i = 2, 3, 4, \ldots$
Finite length, simply supported edges without axial constraint		
Torsional modes:	$\lambda_{ij} = \dfrac{j\pi r}{L}\sqrt{\dfrac{1-\nu}{2}}$	$i = 0$; $j = 1, 2, 3$
Axial modes†:	$\lambda_{ij} = \dfrac{j\pi r}{L}\sqrt{1-\nu^2}$	$i = 0$; $j = 1, 2, 3$
Radial modes†:	$\lambda_{ij} = 1$	$i = 0$; $j = 1, 2, 3$
Bending modes†:	$\lambda_{ij} = \left(\dfrac{j\pi r}{L}\right)^2\sqrt{\dfrac{1-\nu}{2}}$	$i = 1$; $j = 1, 2, 3$
Radial-axial modes:	$\lambda_{ij} = \dfrac{\sqrt{(1-\nu^2)\beta_j^4 + \dfrac{h^2}{12r^2}(i^2+\beta_j^2)^4}}{i^2+\beta_j^2}$	$\beta_j = j\pi r/L$; $i = 2, 3, 4, \ldots$; $j = 1, 2, 3, \ldots$

†Values given here apply for long shells, for which $L > 8jr$.

TABLE 4.8*i* Natural Frequencies of Miscellaneous Systems

Simple Point-Mass Pendulum[7]

$$\omega = 2\pi f = \sqrt{g/L}$$

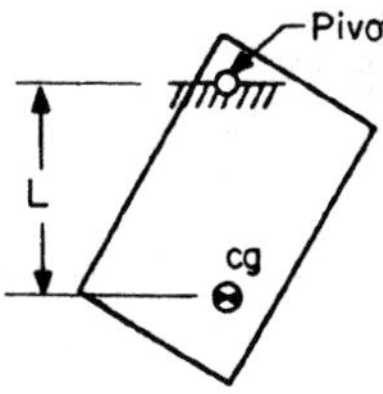

Pendulum with Distributed Mass[7]

$$\omega = 2\pi f = \sqrt{MLg/I_0}$$

M = mass
I_0 = mass moment of inertia about pivot
g = acceleration of gravity
L = distance from pivot to center of gravity

Mass Free to Rotate and Translate in Plane[63]

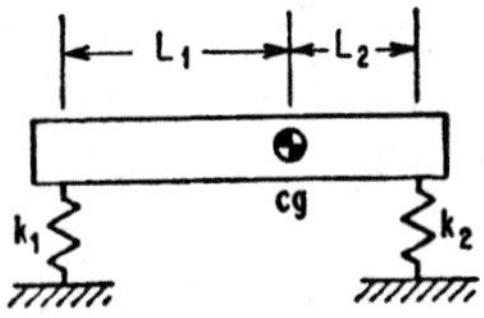

Mass = M, moment of inertia about c.g. = r^2M

$$2\omega^2 = a + c \pm \sqrt{(a-c)^2 + (2b/r)^2}$$

$$a = \frac{k_1 + k_2}{M} \qquad b = \frac{L_2k_2 - L_1k_1}{M}$$

$$c = \frac{L_1^2k_1 + L_2^2k_2}{r^2M}$$

Thin Rings[17]

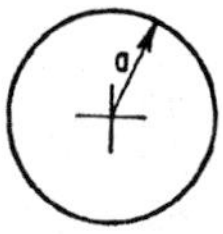

Extension (pure radial mode): $\omega = \frac{1}{a} c_L$

Bending in its own plane: $\omega_n = \frac{n(n^2-1)}{\sqrt{n^2+1}} \frac{r}{a^2} c_L$

$n = 2, 3, \ldots$

$r = \sqrt{I/A}$

I = moment of inertia of ring section area A for bending in its own plane
For out-of-plane bending and partial rings, see Den Hartog[17]

Free-free Beam on Elastic Foundation[63]

Same mode shapes as without support, but frequencies increased:

$$\omega^2_{n,\text{ on support}} = \omega^2_{n,\text{ unsupported}} + \frac{k'}{\rho A}$$

$$\frac{k'}{\rho A} = \frac{\text{support spring constant/unit length}}{\text{beam mass/unit length}} = \omega^2_{\text{rigid beam on elastic foundation}}$$

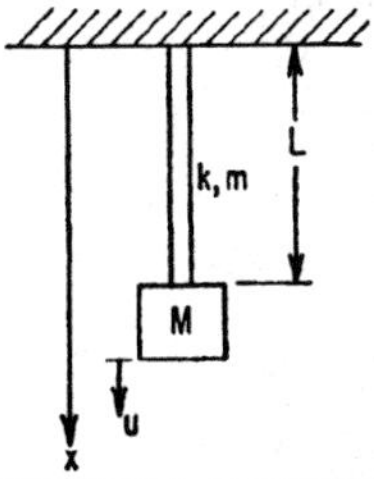

Mass on Rod or Spring of Static Stiffness k, Spring Total Mass m Uniformly Distributed[63]

Frequency equation:

$$(M/m)(\omega\sqrt{m/k}) = \cot(\omega\sqrt{m/k})$$

$$\phi_n(x) = \sin(\omega_n x/L \sqrt{m/k})$$

Weighting function $\mu(x) = m/L + M\delta(x - L)$
[see Sec. 4.4]

M/m	$\omega_1\sqrt{m/k}$	$\omega_2\sqrt{m/k}$	$\omega_3\sqrt{m/k}$	$\omega_4\sqrt{m/k}$	$\omega_5\sqrt{m/k}$
0*	$\pi/2$	$3\pi/2$	$5\pi/2$	$7\pi/2$	$9\pi/2$
$\frac{1}{2}$	1.077	3.644	6.579	9.630	12.722
1	0.960	3.435	6.437	9.426	12.645
∞ †	0	π	2π	3π	4π

*Corresponds to spring free at $x = L$.
†Corresponds to spring clamped at $x = L$.

TABLE 4.9a Combination of Spring Constants

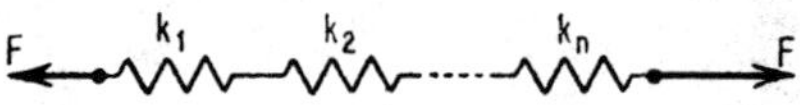	Series combination	$1/k_{total} = 1/k_1 + 1/k_2 + \cdots + 1/k_n$
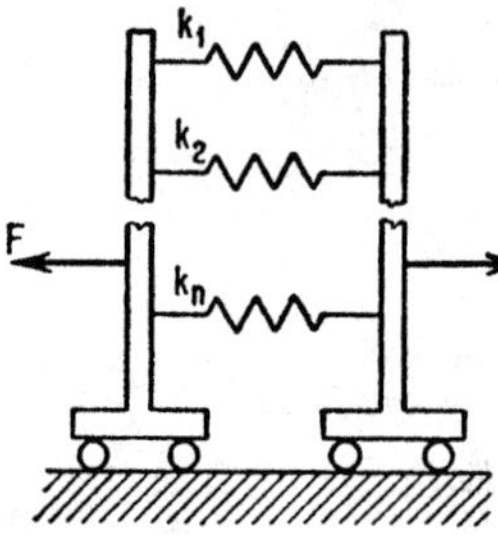	Parallel combination	$k_{total} = k_1 + k_2 + \cdots + k_n$

TABLE 4.9b Spring Constants of Round-Wire Helical Springs[17]

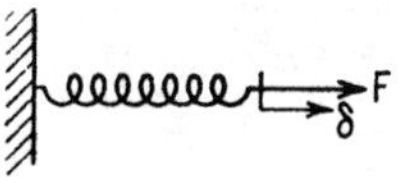	Axial loading	$\frac{F}{\delta} = k = \frac{Gd^4}{8nD^3}$
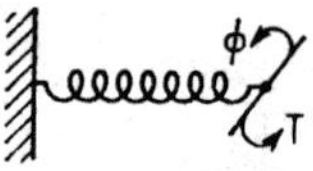	Torsion	$\frac{T}{\phi} = k_r = \frac{Ed^4}{64nD}$
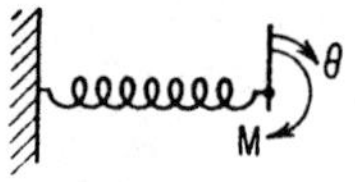	Bending	$\frac{M}{\theta} = k_b = \frac{2}{2+\nu} k_r$

Symbols:

F = axial force
δ = axial deflection
T = torque
ϕ = torsion angle
M = bending moment
θ = flexure angle
G = shear modulus
E = elastic modulus
ν = Poisson's ratio
d = wire diameter
D = mean coil diameter
n = number of coils

TABLE 4.9c Spring Constants of Beams[69]

	Translational (force F)	Rotational (moment M)
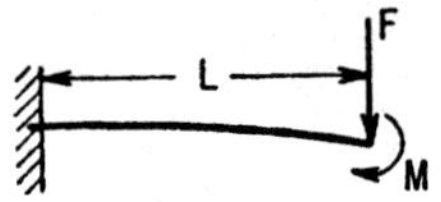	$k = \frac{3EI}{L^3}$	$k_r = \frac{EI}{L}$
	$k = \frac{12EI}{L^3}$	
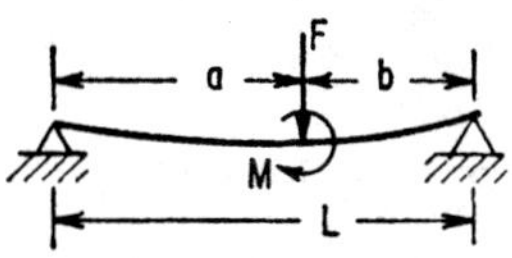	$k = \frac{3EIL}{(ab)^2}$ For $a = b$ $k = \frac{48EI}{L^3}$	$k_r = \frac{3EIL}{L^2 - 3ab}$ For $a = b$ $k_r = \frac{12EI}{L}$
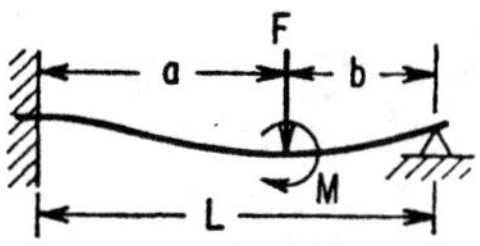	$k = \frac{12EIL^3}{a^3b^2(3L + b)}$ For $a = b$ $k = \frac{768EI}{7L^3}$	$k_r = \frac{4EIL^3/b}{4L^3 - 3b(L + a)^2}$ For $a = b$ $k_r = \frac{64EI}{5L}$
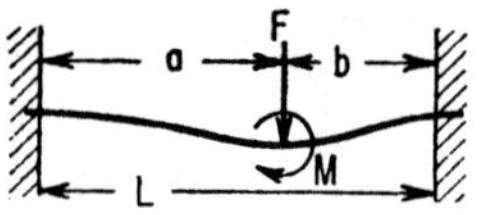	$k = \frac{3EIL^3}{(ab)^3}$ For $a = b$ $k = \frac{192EI}{L^3}$	$k_r = \frac{EIL^3}{ab(L^2 - 3ab)}$ For $a = b$ $k_r = \frac{16EI}{L}$
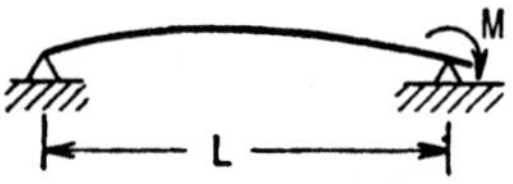		$k_r = \frac{3EI}{L}$
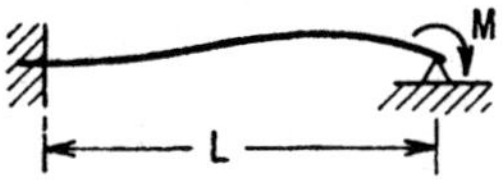		$k_r = \frac{4EI}{L}$

Symbols:

E = modulus of elasticity
I = centroidal moment of inertia of cross section
F = force
M = bending moment
$k = F/\delta$, translational spring constant
$k_r = M/\theta$, rotational spring constant
δ = lateral deflection at F
θ = flexural angle at M

TABLE 4.9*d* Torsion Springs

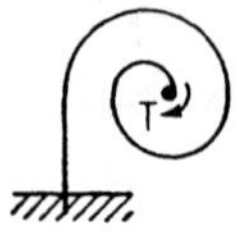	Spiral spring[17] $k_r = T/\phi = EI/L$ E = modulus of elasticity L = total spring length I = moment of inertia of cross section
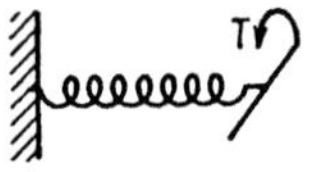	**Helical spring** **See Table 4.9*b***
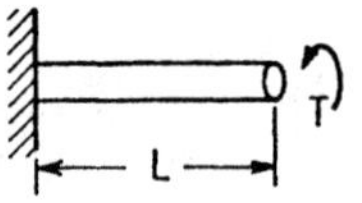	Uniform shaft $k_r = GJ/L$ G = shear modulus L = length J = torsional constant (Table 4.9*e*)
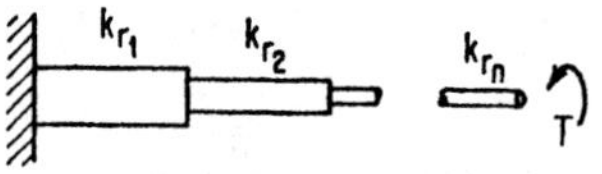	Stepped shaft $1/k_r = 1/k_{r_1} + 1/k_{r_2} + \cdots 1/k_{r_n}$ $k_{r_j} = k_r$ of jth uniform part by itself

TABLE 4.9e Torsional Constants J of Common Sections[47]

Circle	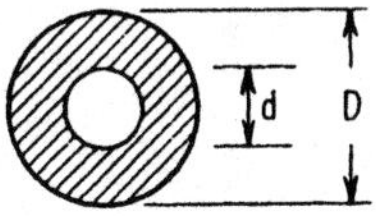	$J = \frac{\pi}{32}(D^4 - d^4)$	
Ellipse	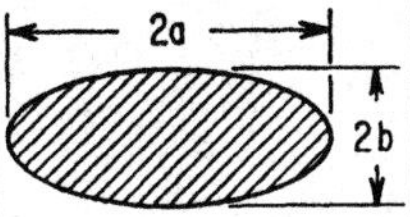	$J = \frac{\pi a^3 b^3}{a^2 + b^2}$	
Square	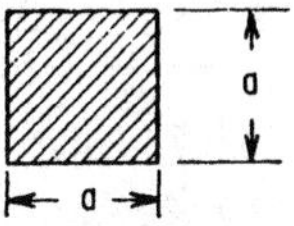	$J = 0.1406\, a^4$	
Rectangle	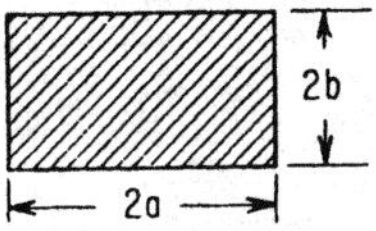	$J = ab^3 \left[\frac{16}{3} - 3.36\,\frac{b}{a}\left(1 - \frac{b^4}{12a^4}\right)\right]$	
Equilateral triangle		$J = \frac{\sqrt{3}\, a^4}{80}$	
Any solid compact section without reentrant angles	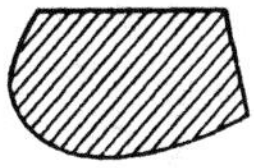	$J \approx \frac{A^4}{40I}$	
Any thin closed tube of uniform thickness t	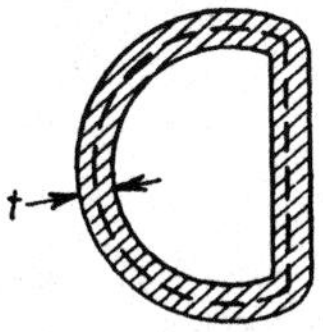	$J \approx \frac{4A^2 t}{U}$	A = cross-section area U = mean circumferential length (length of dotted lines shown) I = polar moment of inertia of section about its centroid
Any thin open tube of uniform thickness t	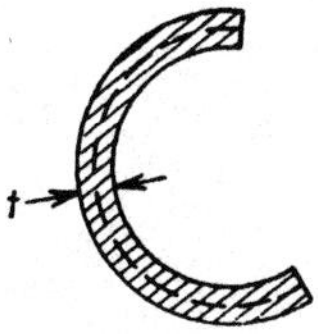	$J \approx \frac{Ut^3}{3}$	

TABLE 4.9f Spring Constants of Centrally Loaded Plates[47]

Circular	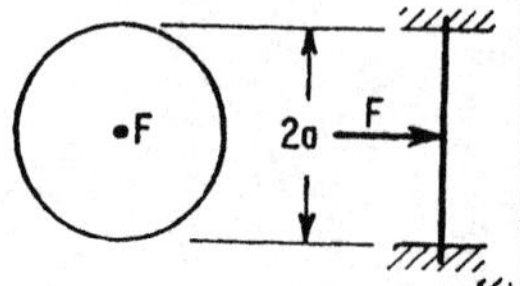	Clamped	$k = \frac{16\pi D}{a^2}$
	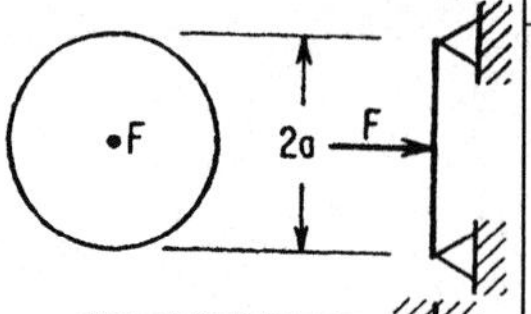	Simply supported	$k = \frac{16\pi D}{a^2}\left(\frac{1+\nu}{3+\nu}\right)$
	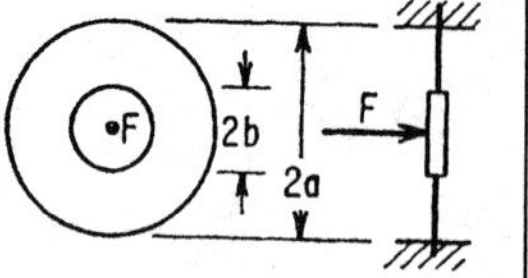	Clamped, with concentric rigid insert	$k = \frac{(16\pi D/a^2)(1-c^2)}{(1-c^2)^2 - 4c^2(\ln c)^2}$ $c = b/a$
Square	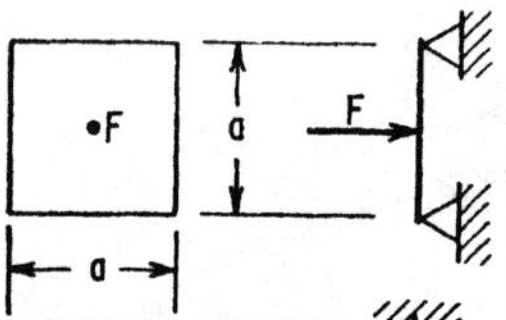	All edges simply supported	$k = 86.1D/a^2$
	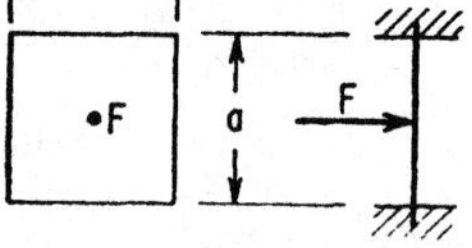	All edges clamped	$k = 192.2D/a^2$
Rectangular	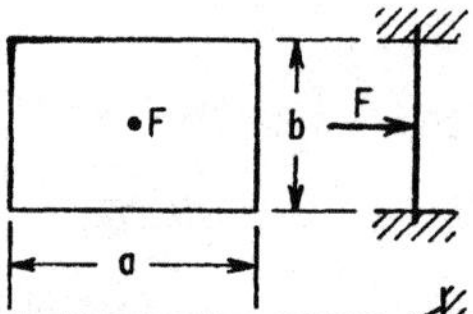	All edges clamped	$k = D\alpha/b^2$ a/b: 4 | 2 | 1 α: 167 | 147 | 192
	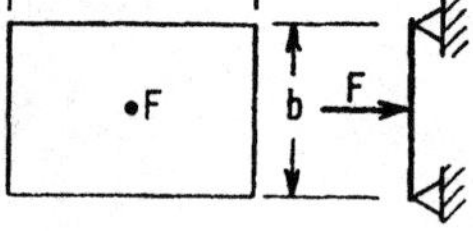	All edges simply supported	$k = 59.2(1 + 0.462c^4)D/b^2$ $c = b/a < 1$
Equilateral triangular	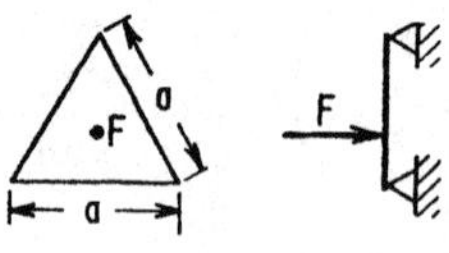	All edges simply supported	$k = 175D/a^2$

Symbols:

k = force/deflection, spring constant

$D = \frac{Eh^3}{12(1-\nu^2)}$

E = modulus of elasticity

ν = Poisson's ratio

h = plate thickness

TABLE 4.10 Longitudinal Wavespeed and K_m for Engineering Materials

Material	Temperature, °C	Longitudinal wavespeed $c_L = \sqrt{E/\rho}$ 10^5 in/s*	$K_m = \frac{(c_L)_{\text{material}}}{(c_L)_{\text{structural steel}}}$
Metals			
Aluminum	20	2.00	1.04
Beryllium	20	4.96	2.57
Brass, bronze	20	1.38–1.57	0.715–0.813
Copper	20	1.34	0.694
Copper	100	1.21	0.627
Copper	200	1.16	0.601
Cupro-nickel	20	1.47–1.61	0.762–0.834
Iron, cast	20	1.04–1.64	0.539–0.850
Lead	20	0.49	0.25
Magnesium	20	2.00	1.04
Monel metal	20	1.76	0.912
Nickel	20	1.24–1.90	0.642–0.984
Silver	20–100	1.03	0.534
Tin	20	0.98	0.51
Titanium	20	1.06–2.00	1.02–1.04
Titanium	90	1.90–1.96	0.984–1.02
Titanium	200	1.84–1.88	0.953–0.974
Titanium	325	1.68–1.75	0.870–0.907
Titanium	400	1.57–1.68	0.813–0.870
Zinc	20	1.46	0.756
Steel, typical	20	2.00	1.04
Steel, structural	−200	2.00	1.04
	−100	1.97	1.02
	25	1.93	1.00
	150	1.92	0.994
	300	1.86	0.963
	400	1.77	0.917
Steel, stainless	−200	1.98–2.04	1.03–1.06
	−100	1.94–2.02	1.005–1.047
	25	1.92–2.06	0.995–1.07
	150	1.89–2.02	0.979–1.05
	300	1.84–1.98	0.953–0.979
	400	1.76–1.94	0.912–1.005
	600	1.70–1.87	0.881–0.969
	800	1.57–1.76	0.813–0.912
Aluminum Alloys	−200	2.08–2.13	1.078–1.104
	−100	2.01–2.06	1.04–1.07
	25	1.97–2.02	1.02–1.05
	200	1.79–1.87	0.927–0.969
Plastics†			
Cellulose acetate	25	0.40	0.21
Methyl methacrylate	25	0.60	0.31
Nylon 6, 6	25	0.54	0.78
Vinyl chloride	25	0.26	0.13
Polyethylene	25	0.41	0.21
Polypropylene	25	0.42	0.22
Polystyrene	25	0.68	0.35

TABLE 4.10 Longitudinal Wavespeed and K_m for Engineering Materials (*Continued*)

Material	Temperature, °C	Longitudinal wavespeed $c_L = \sqrt{E/\rho}$ 10^5 in/s*	$K_m = \frac{(c_L)_{\text{material}}}{(c_L)_{\text{structural steel}}}$
Teflon	25	0.17	0.088
Epoxy resin	25	0.61	0.32
Phenolic resin	25	0.57	0.30
Polyester resin	25	0.61	0.32
Glass-fiber–epoxy laminate	25	1.72	0.89
Glass-fiber–phenolic laminate	25	1.44	0.75
Glass-fiber–polyester laminate	25	1.25	0.65
Other			
Concrete	25	1.44–1.80	0.75–0.93
Cork	25	0.20	0.104
Glass	25	1.97–2.36	1.02–1.22
Granite	25	2.36	1.22
Marble	25	1.50	0.78
Plywood	25	0.84	0.44
Woods, along fibers		1.32–1.92	0.68–0.99
Woods, across fibers		0.48–0.54	0.25–0.28

*To convert to centimeters per second, multiply by 2.54
†Values given here are typical. Wide variations may occur with composition, temperature, and frequency.

REFERENCES

1. Barton, M. V. (ed.): "Shock and Structural Response," American Society of Mechanical Engineers, New York, 1960.
2. Bendat, J. S.: "Principles and Applications of Random Noise Theory," John Wiley & Sons, Inc., New York, 1958.
3. Bendat, J. S., and A. G. Piersol: "Engineering Applications of Correlation and Spectral Analysis," John Wiley & Sons, Inc., New York, 1980.
4. Beranek, L. L. (ed.): "Noise and Vibration Control," McGraw-Hill Book Company, Inc., New York, 1971.
5. Biezeno, C. B., and R. Grammel: "Engineering Dynamics," vol. III, "Steam Turbines," Blackie & Son, Ltd., London, 1954.
6. Bishop, R. E. D., and D. C. Johnson: "Vibration Analysis Tables," Cambridge University Press, Cambridge, England, 1956.
7. Blevins, R. D.: "Formulas for Natural Frequency and Mode Shape," Van Nostrand Reinhold Company, Inc., New York, 1979.
8. Burton, R.: "Vibration and Impact," Addison-Wesley Publishing Company, Inc., Reading, Mass., 1958.
9. Church, A. H., and R. Plunkett: "Balancing Flexible Rotors," *Trans. ASME* (*Ser. B*), vol. 83, pp. 383–389, November, 1961.
10. Craig, R. R., Jr.: "Structural Dynamics: An Introduction to Computer Methods," John Wiley & Sons, Inc., New York, 1981.
11. Crandall, S. H., and W. D. Mark: "Random Vibration in Mechanical Systems," Academic Press, Inc., New York, 1963.
12. Crandall, S. H.: "Engineering Analysis," McGraw-Hill Book Company, Inc., New York, 1956.
13. Crede, C. E.: "Theory of Vibration Isolation," chap. 30, "Shock and Vibration Handbook," C. M. Harris, and C. E. Crede, eds., McGraw-Hill Book Company, Inc., New York, 1976.

14. Crede, C. E.: "Vibration and Shock Isolation," John Wiley & Sons, Inc., New York, 1951.
15. Cremer, L., M. Heckl, and E. E. Ungar: "Structure-Borne Sound," Springer-Verlag, New York, 1973.
16. Davenport, W. B., Jr.: "Probability and Random Processes," McGraw-Hill Book Company, Inc., New York, 1970.
17. Den Hartog, J. P.: "Mechanical Vibrations," 4th ed., McGraw-Hill Book Company, Inc., New York, 1956.
18. Fung, Y. C.: "An Introduction to the Theory of Aeroelasticity," John Wiley & Sons, Inc., New York, 1955.
19. Fung, Y. C., and M. V. Barton: "Some Shock Spectra Characteristics and Uses," *J. Appl. Mech.*, pp. 365–372, September, 1958.
20. Gardner, M. F., and J. L. Barnes: "Transients in Linear Systems," vol. I, John Wiley & Sons, Inc., New York, 1942.
21. Hagedorn, P.: "Non-Linear Oscillations," Clarendon Press, Oxford, England, 1981.
22. Himelblau, H., Jr., and S. Rubin: "Vibration of a Resiliently Supported Rigid Body," chap. 3, "Shock and Vibration Handbook," C. M. Harris and C. E. Crede, eds., McGraw-Hill Book Company, Inc., New York, 1976.
23. Holowenko, R.: "Dynamics of Machinery," John Wiley & Sons, Inc., New York, 1955.
24. Hueber, K. H.: "The Finite Element Method for Engineers," John Wiley & Sons, Inc., New York, 1975.
25. Jacobsen, L. S., and R. S. Ayre, "Engineering Vibrations," McGraw-Hill Book Company, Inc., New York, 1958.
26. Junger, M. C., and D. Feit: "Sound, Structures, and their Interaction," The MIT Press, Cambridge, Mass., 1972.
27. Kamke, E.: "Differentialgleichungen, Lösungsmethoden and Lösungen," 3d ed., Chelsea Publishing Company, New York, 1948.
28. Leissa, A. W.: "Vibration of Plates," NASA SP-160, U.S. Government Printing Office, Washington, D.C., 1969.
29. Leissa, A. W.: "Vibration of Shells," NASA SP-288, U.S. Government Printing Office, Washington, D.C., 1973.
30. LePage, W. R., and S. Seely: "General Network Analysis," McGraw-Hill Book Company, Inc., New York, 1952.
31. Lowe, R., and R. D. Cavanaugh: "Correlation of Shock Spectra and Pulse Shape with Shock Environment," *Environ. Eng.*, February 1959.
32. Macduff, J. N., and J. R. Curreri: "Vibration Control," McGraw-Hill Book Company, Inc., New York, 1958.
33. Macduff, J. N., and R. P. Felgar: "Vibration Design Charts," *Trans. ASME*, vol. 79, pp. 1459–1475, 1957.
34. Mahalingam, S.: "Forced Vibration of Systems with Non-linear Non-symmetrical Characteristics," *J. Appl. Mech.*, vol. 24, pp. 435–439, September 1957.
35. Major, A.: "Dynamics in Civil Engineering," Akadémiai Kiadó, Budapest, 1980.
36. Marguerre, K., and H. Wölfel: "Mechanics of Vibration," Sijthoff & Noordhoff, Alphen aan den Rijn, The Netherlands, 1979.
37. Martienssen, O.: "Über neue Resonanzerscheinungen in Wechselstromkreisen," *Physik. Z.*, vol. 11, pp. 448–460, 1910.
38. Meirovitch, L.: "Analytical Methods in Vibration," The Macmillan Company, Inc., New York, 1967.
39. Mindlin, R. D.: "Dynamics of Package Cushioning," *Bell System Tech. J.*, vol. 24, nos. 3, 4, pp. 353–461, July–October 1945.
40. Morse, P. M.: "Vibration and Sound," 2d ed., McGraw-Hill Book Company, Inc., New York, 1948.

41. Myklestad, N. O.: "Fundamentals of Vibration Analysis," McGraw-Hill Book Company, Inc., New York, 1956.

42. Nestroides, E. J. (ed.): "Bicera: Handbook on Torsional Vibrations," Cambridge University Press, New York, 1958 (British Internal Combustion Engine Research Association).

43. Ostergren, S. M.: "Shock Response of a Two-Degree-of-Freedom System," Rome Air Development Center Rep. RADC TN 58-251.

44. Perrone, N., and W. Pilkey, and B. Pilkey, (eds.): "Structural Mechanics Software Series," University of Virginia Press, Charlottesville, vols. I–III, 1977–1980.

45. Pilkey, W., and B. Pilkey, (eds.): "Shock and Vibration Computer Programs," The Shock and Vibration Information Center, Naval Research Laboratory, Washington, D.C., 1975.

46. Pistiner, J. S., and H. Reisman: "Dynamic Amplification Factor of a Two-Degree-of-Freedom System," *J. Environ. Sci.*, pp. 4–8, October 1960.

47. Plunkett, R. (ed.): "Mechanical Impedance Methods for Mechanical Vibrations," American Society of Mechanical Engineers, New York, 1958.

48. Richart, F. E., Jr., J. R. Hall, Jr., and R. D. Woods: "Vibration of Soils and Foundations," Prentice-Hall, Inc., Englewood Cliffs, N.J., 1970.

49. Roark, R. J.: "Formulas for Stress and Strain," 3d ed., McGraw-Hill Book Company, Inc., New York, 1954.

50. Rubin, S.: "Concepts in Shock Data Analysis," chap. 23, "Shock and Vibration Handbook," C. M. Harris and C. E. Crede, eds., McGraw-Hill Book Company, Inc., New York, 1976.

51. Ruzicka, J. (ed.): "Structural Damping," American Society of Mechanical Engineers, New York, 1959.

52. Ruzicka, J. E., and T. F. Derby: "Influence of Damping in Vibration Isolation," SVM-7, Shock and Vibration Information Center, Washington, D.C., 1971.

53. Scanlan, R. H., and R. Rosenbaum: "Introduction to the Study of Aircraft Vibration and Flutter," The Macmillan Company, Inc., New York, 1951.

54. Schwesinger, G.: "On One-term Approximations of Forced Nonharmonic Vibrations," *J. Appl. Mech.*, vol. 17, no. 2, pp. 202–208, June 1950.

55. Snowdon, J. C.: "Vibration and Shock in Damped Mechanical Systems," John Wiley & Sons, Inc., New York, 1968.

56. Snowdon, J. C.: "Vibration Isolation: Use and Characterization," NBS Handbook 128, U.S. Department of Commerce, National Bureau of Standards, Washington, D.C., May 1979.

57. Snowdon, J. C., and E. E. Ungar (eds.): "Isolation of Mechanical Vibration, Impact, and Noise," AMD-vol. 1, American Society of Mechanical Engineers, New York, 1973.

58. Skudrzyk, E.: "Simple and Complex Vibratory Systems," The Pennsylvania State University Press, University Park, Pa., 1968.

59. Stodola, A. (Transl. L. C. Lowenstein): "Steam and Gas Turbines," McGraw-Hill Book Company, Inc., New York, 1927.

60. Stoker, J. J.: "Nonlinear Vibrations," Interscience Publishers, Inc., New York, 1950.

61. Thomson, W. T.: "Theory of Vibration with Applications," 2d ed., Prentice-Hall, Inc., Englewood Cliffs, N.J., 1981.

62. Thomson, W. T.: "Shock Spectra of a Nonlinear System," *J. Appl. Mech.*, vol. 27, pp. 528–534, September 1960.

63. Timoshenko, S.: "Vibration Problems in Engineering," 2d ed., Van Nostrand Company, Inc., Princeton, N.J., 1937.

64. Tong, K. N.: "Theory of Mechanical Vibration," John Wiley & Sons, Inc., New York, 1960.

65. Torvik, P. J. (ed.): "Damping Applications for Vibration Control," AMD-vol. 38, American Society of Mechanical Engineers, New York, 1980.

66. Trent, H. M.: "Physical Equivalents of Spectral Notions," *J. Acoust. Soc. Am.*, vol. 32, pp. 348–351, March 1960.

67. Ungar, E. E.: "Maximum Stresses in Beams and Plates Vibrating at Resonance," *Trans. ASME, Ser. B,* vol. 84, pp. 149–155, February 1962.

68. Ungar, E. E.: "Damping of Panels," chap. 14, "Noise and Vibration Control," L. L. Beranek, ed., McGraw-Hill Book Company, Inc., New York, 1971.

69. Ungar, E. E., and R. Cohen: "Vibration Control Techniques," chap. 20, "Handbook of Noise Control," C. M. Harris, ed., McGraw-Hill Book Company, Inc., New York, 1979.

70. Vigness, I.: "Fundamental Nature of Shock and Vibrations," *Elec. Mfg.,* vol. 63, pp. 89–108, June 1959.

71. Warburton, G. B.: "The Dynamical Behavior of Structures," Pergamon Press, Oxford, 1964.

72. Wilson, W. K.: "Practical Solution of Torsional Vibration Problems," John Wiley & Sons, Inc., New York, 1956.

CHAPTER 5
STATIC AND FATIGUE DESIGN

Steven M. Tipton, Ph.D., P.E.
Associate Professor of Mechanical Engineering
University of Tulsa
Tulsa, Okla.

SYMBOLS 5.1
NOTATIONS 5.2
5.1 INTRODUCTION 5.3
5.2 ESTIMATION OF STRESSES AND STRAINS IN ENGINEERING COMPONENTS 5.4
5.2.1 Definition of Stress and Strain 5.4
5.2.2 Experimental 5.9
5.2.3 Strength of Materials 5.11
5.2.4 Elastic Stress-Concentration Factors 5.11
5.2.5 Finite-Element Analysis 5.13
5.3 STRUCTURAL INTEGRITY DESIGN PHILOSOPHIES 5.14
5.3.1 Static Loading 5.15
5.3.2 Fatigue Loading 5.16
5.4 STATIC STRENGTH ANALYSIS 5.18
5.4.1 Monotonic Tensile Data 5.19
5.4.2 Multiaxial Yielding Theories (Ductile Materials) 5.20
5.4.3 Multiaxial Failure Theories (Brittle Materials) 5.21
5.4.4 Summary Design Algorithm 5.23
5.5 FATIGUE STRENGTH ANALYSIS 5.24
5.5.1 Stress-Life Approaches (Constant-Amplitude Loading) 5.25
5.5.2 Strain-Life Approaches (Constant-Amplitude Loading) 5.37
5.5.3 Variable-Amplitude Loading 5.52
5.6 DAMAGE-TOLERANT DESIGN 5.58
5.6.1 Stress-Intensity Factor 5.58
5.6.2 Static Loading 5.59
5.6.3 Fatigue Loading 5.60
5.7 MULTIAXIAL FATIGUE LOADING 5.62
5.7.1 Proportional Loading 5.62
5.7.2 Nonproportional Loading 5.65

SYMBOLS

α = "characteristic length" (empirical curve-fit parameter)
a = crack length
a_f = final crack length
a_i = initial crack length
A = cross-sectional area
A_f = Forman coefficient
A_p = Paris coefficient
A_w = Walker coefficient
b = fatigue strength exponent
b' = baseline fatigue exponent
c = fatigue ductility exponent
C = $2\tau_{xy,a}/\sigma_{x,a}$ (during axial-torsional fatigue loading)
C' = baseline fatigue coefficient
CM = Coulomb Mohr Theory
d = diameter of tensile test specimen gauge section
DAM_i = cumulative fatigue damage for a particular (ith) cycle
Δ = range of (e.g., stress, strain, etc.) = maximum − minimum
e = nominal axial strain
e_{offset} = offset plastic strain at yield
e_u = engineering strain at ultimate tensile strength
E = modulus of elasticity
E_c = Mohr's circle center
ε = normal strain
ε_a = normal strain amplitude

ε_f = true fracture ductility
ε_f' = fatigue ductility coefficient
ε_u = true strain at ultimate tensile strength
FS = factor of safety
G = elastic shear modulus
γ = shear strain
K = monotonic strength coefficient
K = stress intensity factor
K' = cyclic strength coefficient
K_c = fracture toughness
K_f = fatigue notch factor
K_{Ic} = plane-strain fracture toughness
K_t = elastic stress-concentration factor
MM = modified Mohr Theory
MN = maximum normal stress theory
n = monotonic strain-hardening exponent
n' = cyclic strain-hardening exponent
n_f = Forman exponent
n_p = Paris exponent
n_w and m_w = Walker exponents
N = number of cycles to failure in a fatigue test
N_T = transition fatigue life
P = axial load
ϕ = phase angle between σ_x and τ_{xy} stresses (during axial-torsional fatigue loading)
r = notch root radius
R = cyclic load ratio (minimum load over maximum load)
R = Mohr's circle radius
RA = reduction in area
S = stress amplitude during a fatigue test
S_e = endurance limit
$S_{eq,a}$ = equivalent stress amplitude (multiaxial to uniaxial)
$S_{eq,m}$ = equivalent mean stress (multiaxial to uniaxial)
S_{nom} = nominal stress
$S_u = S_{ut}$ = ultimate tensile strength
S_{uc} = ultimate compressive strength
S_y = yield strength
SALT = equivalent stress amplitude (based on Tresca for axial-torsional loading)
SEQA = equivalent stress amplitude (based on von Mises for axial-torsional loading)
σ = normal stress
σ_a = normal stress amplitude
σ_{eq} = equivalent axial stress
σ_f = true fracture strength
σ_f' = fatigue strength coefficient
σ_m = mean stress during a fatigue cycle
σ_{norm} = normal stress acting on plane of maximum shear stress
σ_{notch} = elastically calculated notch stress
σ_A = maximum principal stress
σ_B = minimum principal stress
σ_u = true ultimate tensile strength
t = thickness of fracture mechanics specimen
τ = shear stress
τ_{max} = maximum shear stress
θ_σ = orientation of maximum principal stress
θ_τ = orientation of maximum shear stress
ν = Poisson's ratio

NOTATIONS

1, 2, 3 = subscripts designating principal stress
I, II, III = subscripts denoting crack loading mode

a,eff = subscript denoting effective stress amplitude (mean stress to fully reversed)

f = subscript referring to final dimensions of a tension test specimen

bend = subscript denoting bending loading

max = subscript denotes maximum or peak during fatigue cycle

min = subscript denoting minimum or valley during fatigue cycle

o = subscript referring to original dimensions of a tension test specimen

Tr = subscript referring to Tresca criterion

vM = subscript referring to von Mises criterion

x, y, z = orthogonal coordinate axes labels

5.1 INTRODUCTION

The design of a component implies a design framework and a design process. A typical design framework requires consideration of the following factors: component function and performance, producibility and cost, safety, reliability, packaging, and operability and maintainability.

The designer should assess the consequences of failure and the normal and abnormal conditions, loads, and environments to which the component may be subjected during its operating life. On the basis of the requirements specified in the design framework, a design process is established which may include the following elements: conceptual design and synthesis, analysis and gathering of relevant data, optimization, design and test of prototypes, further optimization and revision, final design, and monitoring of component performance in the field.

Requirements for a successful design include consideration of data on the past performance of similar components, a good definition of the mechanical and thermal loads (monotonic and cyclic), a definition of the behavior of candidate materials as a function of temperature (with and without stress raisers), load and corrosive environments, a definition of the residual stresses and imperfections owing to processing, and an appreciation of the data which may be missing in the trade-offs among parameters such as cost, safety, and reliability. Designs are typically analyzed to examine the potential for fracture, excessive deformation (under load, creep), wear, corrosion, buckling, and jamming (due to deformation, thermal expansion, and wear). These may be caused by steady, cyclic, or shock loads, and temperatures under a number of environmental conditions and as a function of time. Reference 92 lists the following failures: ductile and brittle fractures, fatigue failures, distortion failures, wear failures, fretting failures, liquid-erosion failures, corrosion failures, stress-corrosion cracking, liquid-metal embrittlement, hydrogen-damage failures, corrosion-fatigue failures, and elevated-temperature failures.

In addition, property changes owing to other considerations, such as radiation, should be considered, as appropriate. The designer needs to decide early in the design process whether a component or system will be designed for infinite life, finite specified life, a fail-safe or damage-tolerant criterion, a required code, or a combination of the above.[3]

In the performance of design trade-offs, in addition to the standard computerized tools of stress analysis, such as the finite-element method, depending upon the complexity of the mathematical formulation of the design constraints and the function to be optimized, the mathematical programming tools of operations research may apply. Mathematical programming can be used to define the most desirable (optimum) behavior of a component as a function of other constraints. In addition, on a systems level, by assigning relative weights to requirements, such as safety, cost, and life, design parameters can be optimized. Techniques such as linear programming, nonlinear programming, and dynamic programming may find greater application in the future in the area of mechanical design.[93]

Numerous factors dictate the overall engineering specifications for mechanical design. This chapter concentrates on philosophies and methodologies for the design of components that must satisfy quantitative strength and endurance specifications. Only deterministic approaches are presented for statically and dynamically loaded components.

Although mechanical components can be susceptible to many modes of failure, approaches in this chapter concentrate on the comparison of the state of stress and/or strain in a component with the strength of candidate materials. For instance, buckling, vibration, wear, impact, corrosion, and other environmental factors are not considered. Means of calculating stress-strain states for complex geometries associated with real mechanical components are vast and wide-ranging in complexity. This topic will be addressed in a general sense only. Although some of the methodologies are presented in terms of general three-dimensional states of stress, the majority of the examples and approaches will be presented in terms of two-dimensional surface stress states. Stresses are generally maximum on the surface, constituting the vast majority of situations of concern to mechanical designers. [Notable exceptions are contact problems,[1–6] components which are surfaced processed (e.g., induction hardened or nitrided[7]), or components with substantial internal defects, such as pores or inclusions.]

In general, the approaches in this chapter are focused on isotropic metallic components, although they can also apply to homogeneous nonmetallics (such as glass, ceramics, or polymers). Complex failure mechanisms and material anisotropy associated with composite materials warrant the separate treatment of these topics.

Typically, prototype testing is relied upon as the ultimate measure of the structural integrity of an engineering component. However, costs associated with expensive and time consuming prototype testing iterations are becoming more and more intolerable. This increases the importance of modeling durability in everyday design situations. In this way, data from prototype tests can provide valuable feedback to enhance the reliability of analytical models for the next iteration and for future designs.

5.2 ESTIMATION OF STRESSES AND STRAINS IN ENGINEERING COMPONENTS

When loads are imposed on an engineering component, stresses and strains develop throughout. Many analytical techniques are available for estimating the state of stress and strain in a component. A comprehensive treatment of this subject is beyond the scope of this chapter. However, the topic is overviewed for engineering design situations.

5.2.1 Definition of Stress and Strain

An engineering definition of "stress" is the force acting over an infinitesimal area. "Strain" refers to the localized deformation associated with stress. There are several important practical aspects of stress in an engineering component:

1. A state of stress-strain must be associated with a particular location on a component.

2. A state of stress-strain is described by stress-strain components, acting over planes.
3. A well-defined coordinate system must be established to properly analyze stress-strain.
4. Stress components are either normal (pulling planes of atoms apart) or shear (sliding planes of atoms across each other).
5. A stress state can be uniaxial, but strains are usually multiaxial (due to the effect described by Poisson's ratio).

The most general three-dimensional state of stress can be represented by Fig. 5.1*a*. For most engineering analyses, designers are interested in a two-dimensional state of stress, as depicted in Fig. 5.1*b*. Each side of the square two-dimensional element in Fig. 5.1*b* represents an infinitesimal area that intersects the surface at 90°.

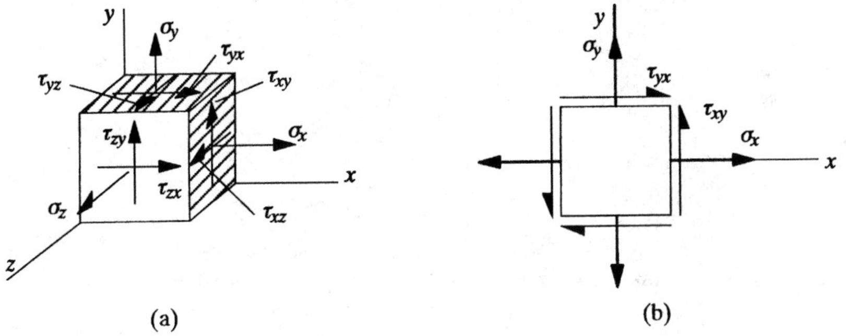

FIG. 5.1 The most general (*a*) three-dimensional and (*b*) two-dimensional stress states.

By slicing a section of the element in Fig. 5.1*b,* as shown in Fig. 5.2, and analytically establishing static equilibrium, an expression for the normal stress σ and the shear stress τ acting on any plane of orientation θ can be derived. This expression forms a circle when plotted on axes of *shear stress* versus *normal stress*. This circle is referred to as "Mohr's circle."

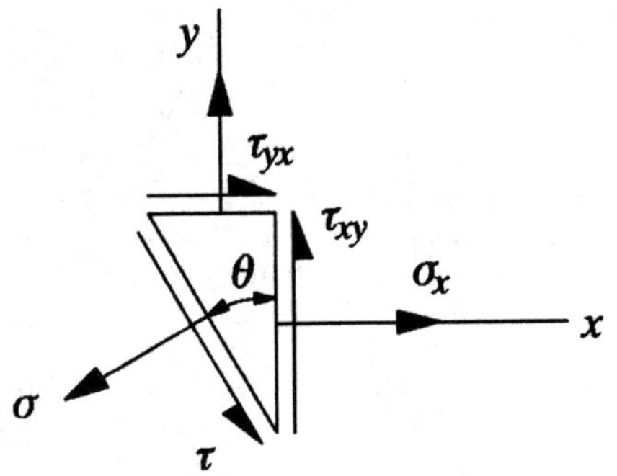

FIG. 5.2 Shear and normal stresses on a plane rotated θ from its original orientation.

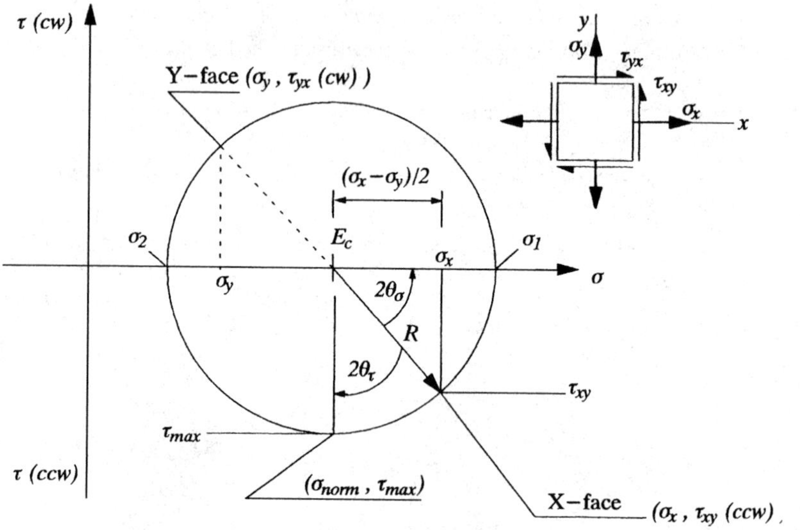

FIG. 5.3 Mohr's circle for a generic state of surface stress.

Mohr's circle is one of the most powerful analytical tools available to a design analyst. Here, the application of Mohr's circle is emphasized for two-dimensional stress states. From this understanding, it is a relatively simple step to extend the analysis to most three-dimensional engineering situations.

Consider the stress state depicted in Fig. 5.1*b* to lie in the surface of an engineering component. To draw the Mohr's circle for this situation (Fig. 5.3), three simple steps are required:

1. Draw the shear-normal axes [τ(cw) positive vertical axis, σ tensile along horizontal axis].
2. Define the center of the circle E_c (which always lies on the σ axis):

$$E_c = (\sigma_x + \sigma_y)/2 \tag{5.1}$$

3. Use the point represented by the "*X*-face" of the stress element to define a point on the circle (σ_x, τ_{xy}). The *X*-face on the Mohr's circle refers to the plane whose normal lies in the *X* direction (or the plane with a normal and shear stress of σ_x and τ_{xy}, respectively).

That's all there is to it. The sense of the shear stress [clockwise (cw) or counterclockwise (ccw)] refers to the direction that the shear stress attempts to rotate the element under consideration. For instance, in Figs. 5.1 and 5.2, τ_{xy} is ccw and τ_{yx} is cw. This is apparent in Fig. 5.3, a schematic Mohr's circle for this generic surface element.

The interpretation and use of Mohr's circle is as simple as its construction. Referring to Fig. 5.3, the radius of the circle *R* is given by Eq. (5.2).

$$R = \sqrt{\left(\frac{\sigma_x - \sigma_y}{2}\right)^2 + \tau_{xy}^2} \tag{5.2}$$

This could suggest an alternate step 3: that is, define the radius and draw the circle with the center and radius. The two approaches are equivalent. From the circle, the following important items can be composed: (1) the principal stresses, (2) the maximum shear stress, (3) the orientation of the principal stress planes, (4) the orientation of the maximum shear planes, and (5) the stress normal to and shear stress acting over a plane of *any* orientation.

1. Principal Stresses. It is apparent that

$$\sigma_1 = E_c + R \tag{5.3}$$

and

$$\sigma_2 = E_c - R \tag{5.4}$$

2. Maximum Shear Stress. The maximum in-plane shear stress at this location,

$$\tau_{max} = R \tag{5.5}$$

3. Orientation of Principal Stress Planes. Remember only one rule: A rotation of 2θ around the Mohr's circle corresponds to a rotation of θ for the actual stress element. This means that the principal stresses are acting on faces of an element oriented as shown in Fig. 5.4. In this figure, a counterclockwise rotation from the *X*-face to σ_1 of $2\theta_\sigma$, means a ccw rotation of θ_σ on the surface of the component, where θ_σ is given by Eq. (5.6):

$$\theta_\sigma = 0.5 \tan^{-1}\left[\frac{2\tau_{xy}}{(\sigma_x - \sigma_y)}\right] \tag{5.6}$$

In Figs. 5.3 and 5.4, since the "*X*-face" refers to the plane whose normal lies in the *x* direction, it is associated with the *x* axis and serves as a reference point on the Mohr's circle for considering normal and shear stresses on any other plane.

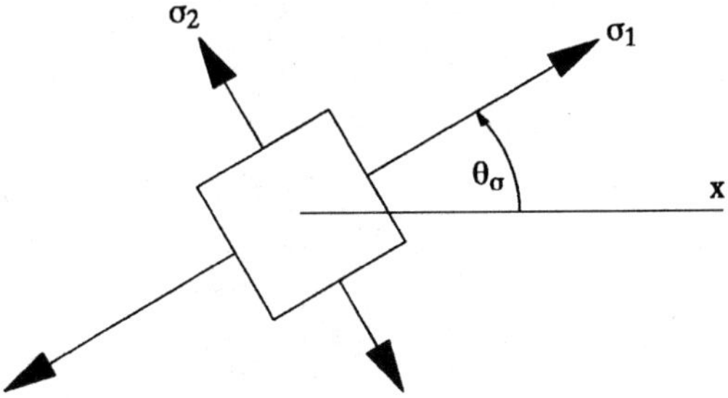

FIG. 5.4 Orientation of the maximum principal stress plane.

4. Orientation of the Maximum Shear Planes. Notice from Fig. 5.3 that the maximum shear stress is the radius of the circle $\tau_{max} = R$. The orientation of the plane of maximum shear is thus defined by rotating through an angle $2\theta_\tau$ around the Mohr's circle, clockwise from the *X*-face reference point. This means that the plane oriented at an angle θ_τ (cw) from the *x* axis will feel the maximum shear stress, as shown in Fig. 5.5. Notice that the sum of θ_τ and θ_σ on the Mohr's circle is 90°; this will always be the case. Therefore, the planes feeling the maximum principal (normal) stress and maximum shear stress always lie 45° apart, or

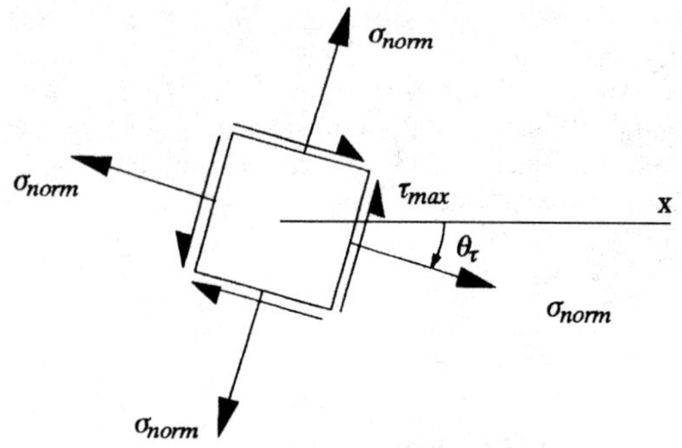

FIG. 5.5 Orientation of the planes feeling the maximum shear stress.

$$\theta_\tau = 45° - \theta_\sigma \tag{5.7}$$

EXAMPLE 1 Suppose a state of stress is given by $\sigma_x = 30$ ksi, $\tau_{xy} = 14$ ksi (ccw) and $\sigma_y = -12$ ksi. If a seam runs through the material 30° from the vertical, as shown, compute the stress normal to the seam and the shear stress acting on the seam.

solution Construct the Mohr's circle by computing the center and radius:

$$E_c = \frac{[30 + (-12)]}{2} = 9 \text{ ksi} \qquad R = \sqrt{\left(\frac{[30 - (-12)]}{2}\right)^2 + 14^2} = 25.24 \text{ ksi}$$

The normal stress σ and shear stress τ acting on the seam are obtained from inspection of the Mohr's circle and shown below:

$$\sigma = E_c + R\cos(33.69° + 60°) = 7.38 \text{ ksi}$$

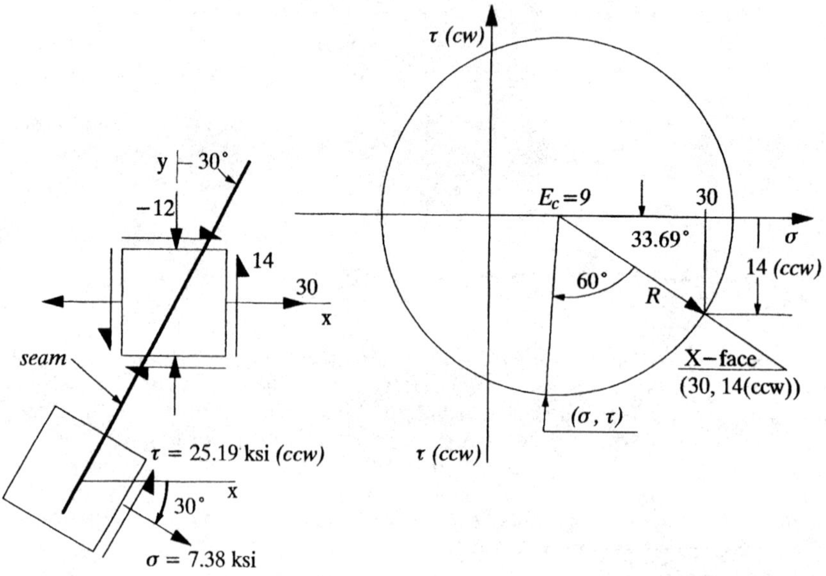

$$\tau = R \sin(33.69° + 60°) = 25.19 \text{ ksi (ccw)}$$

The normal stress σ_{norm} is equal on each face of the maximum shear stress element and $\sigma_{norm} = E_c$, the Mohr's circle center. (This is always the case since the Mohr's circle is always centered on the normal stress axis.)

5. Stress Normal to and Shear Stress on a Plane of Any Orientation. Remember that *the Mohr's circle is a collection of* (σ,τ) *points that represent the normal stress* σ *and the shear stress* τ *acting on a plane at any orientation in the material.* The *X*-face reference point on the Mohr's circle is the point representing a plane whose stresses are (σ_x, τ_{xy}). Moving an angle 2θ in either sense from the *X*-face around the Mohr's circle corresponds to a plane whose normal is oriented an angle θ in the same sense from the x axis. (See Example 1.)

More formal definitions for three-dimensional tensoral stress and strain are available.[5,6,8–13] In the majority of engineering design situations, bulk plasticity is avoided. Therefore, the relation between stress and strain components is predominantly elastic, as given by the generalized Hooke's law (with ε and γ referring to normal and shear strain, respectively) in Eqs. (5.8) to (5.13):

$$\varepsilon_x = \frac{1}{E}[\sigma_x - \nu(\sigma_y + \sigma_z)] \tag{5.8}$$

$$\varepsilon_y = \frac{1}{E}[\sigma_y - \nu(\sigma_z + \sigma_x)] \tag{5.9}$$

$$\varepsilon_z = \frac{1}{E}[\sigma_z - \nu(\sigma_x + \sigma_y)] \tag{5.10}$$

$$\gamma_{xy} = \frac{\tau_{xy}}{G} \tag{5.11}$$

$$\gamma_{yz} = \frac{\tau_{yz}}{G} \tag{5.12}$$

$$\gamma_{zx} = \frac{\tau_{zx}}{G} \tag{5.13}$$

where E is the modulus of elasticity, ν is Poisson's ratio, and G is the shear modulus, expressed as Eq. (5.14):

$$G = \frac{E}{2(1 + \nu)} \tag{5.14}$$

5.2.2 Experimental

Experimental stress analysis should probably be referred to as experimental strain analysis. Nearly all commercially available techniques are based on the detection of local states of strain, from which stresses are computed. For elastic situations, stress components are related to strain components by the generalized Hooke's law as shown in Eqs. (5.15) to (5.20):

$$\sigma_x = \frac{E}{1 + \nu}\varepsilon_x + \frac{\nu E}{(1 + \nu)(1 - 2\nu)}(\varepsilon_x + \varepsilon_y + \varepsilon_z) \tag{5.15}$$

$$\sigma_y = \frac{E}{1 + \nu}\varepsilon_y + \frac{\nu E}{(1 + \nu)(1 - 2\nu)}(\varepsilon_x + \varepsilon_y + \varepsilon_z) \tag{5.16}$$

$$\sigma_z = \frac{E}{1 + \nu}\varepsilon_z + \frac{\nu E}{(1 + \nu)(1 - 2\nu)}(\varepsilon_x + \varepsilon_y + \varepsilon_z) \tag{5.17}$$

$$\tau_{xy} = G\gamma_{xy} \tag{5.18}$$

$$\tau_{yz} = G\gamma_{yz} \tag{5.19}$$

$$\tau_{zx} = G\gamma_{zx} \tag{5.20}$$

For strains measured on a stress-free surface where $\sigma_z = 0$; the in-plane normal stress relations simplify to Eqs. (5.21) and (5.22).

$$\sigma_x = \frac{E}{1 - \nu^2}[\varepsilon_x + \nu\varepsilon_y] \tag{5.21}$$

$$\sigma_y = \frac{E}{1 - \nu^2}[\varepsilon_y + \nu\varepsilon_x] \tag{5.22}$$

Several techniques exist for measuring local states of strain, including electromechanical extensometers, photoelasticity, brittle coatings, moiré methods, and holography.[14,15] Other, more sophisticated approaches such as X-ray and neutron diffraction, can provide measurements of stress distributions below the surface. However, the vast majority of experimental strain data are recorded with electrical resistance strain gauges. Strain gauges are mounted directly to a carefully prepared surface using an adhesive. Instrumentation measures the change in resistance of the gauge as it deforms with the material adhered to its gauge section, and a strain is computed from the resistance change. Gauges are readily available in sizes from 0.015 to 0.5 inches in gauge length and can be applied in the roots of notches and other stress concentrations to measure severe strains that can be highly localized.

As implied by Eqs. (5.15) to (5.22), it can be important to measure strains in more than one direction. This is particularly true when the direction of principal stress is unknown. In these situations it is necessary to utilize three-axis rosettes (a pattern of three gauges in one, each oriented along a different direction). If the principal stress directions are known but not the magnitudes, two-axis (biaxial) rosettes can be oriented along principal stress directions and stresses computed with Eqs. (5.21) and (5.22) replacing σ_x and σ_y with σ_1 and σ_2, respectively. These equations can be used to show that severe errors can result in calculated stresses if a biaxial stress state is assumed to be uniaxial. (See Example 2.)

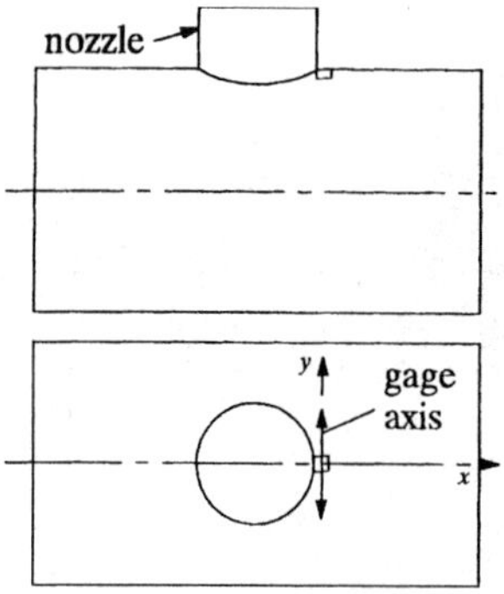

EXAMPLE 2 This example demonstrates how stresses can be underestimated if strain is measured only along a single direction in a biaxial stress field. Compute the hoop stress at the base of the nozzle shown if (1) a hoop strain of 0.0023 is the only measurement taken and (2) an axial strain measurement of +0.0018 is also taken.

solution For a steel vessel (E = 30,000 ksi and ν = 0.3), if the axial stress is neglected, the hoop stress is calculated to be

$$\sigma_y = E\varepsilon_y = 69 \text{ ksi}$$

However, if the axial strain measurement of +0.0018 is used with Eq. (5.22), then the hoop stress is given by

$$\sigma_x = \frac{E}{1 - \nu^2}[0.0023 + 0.3(0.0018)] = 93.63 \text{ ksi}$$

In this example, measuring only the hoop strain caused the hoop stress to be underestimated by over 26 percent.

Obviously, in order to measure strains, prototype parts must be available, which is generally not the case in the early design stages. However, rapid prototyping techniques, such as computer numerically controlled machining equipment and stereolithography, can greatly facilitate prototype development. Data from strain-gauge testing of components in the final developmental stages should be compared to preliminary design estimates in order to provide feedback to the analysis.

5.2.3 Strength of Materials

Concise solutions have been developed for pressure vessels; beams in bending, tension and torsion; curved beams; etc.[1–6] These are usually based on considering a section through the point of interest, establishing static equilibrium with externally applied forces, and making assumptions about the distribution of stress or strain throughout the cross section.

Example 3 illustrates the use of traditional bending- and torsional-stress relations, showing how they can be used to improve the efficiency of an experimental strain measurement.

EXAMPLE 3 A steel component is welded to a solid base and loaded as shown. Identify the region of maximum stress. Show where to mount and how to orient a single-axis strain gauge to pick up the maximum signal. Compute the maximum principal stress and strain in the structure for a value of $P = 400{,}000$ lb.

solution The critical section is the cross section defined by $x = 0$, and the maximum stress can be expected at the origin of the coordinate system shown on page 5.14. The cross section feels bending about the centroidal y axis M_y and a torque T. (Transverse shear is neglected since it is zero at the point of maximum stress.)

$$\sigma_x = \frac{M_y c}{I_y} = \frac{(1{,}800{,}000 \text{ in}\cdot\text{lb})(2.5 \text{ in})}{(11 \text{ in})(5 \text{ in})^3/12} = 39.27 \text{ ksi}$$

$$\tau_{xy} = \frac{4{,}760{,}000 \text{ in}\cdot\text{lb}}{(11 \text{ in})(5 \text{ in})^2}[3 + 1.8(5/11)] = 66.09 \text{ ksi (from Ref. 1)}$$

Constructing a Mohr's circle (as in Fig. 5.3) the orientations of the principal stresses and their magnitudes are given by

$$\theta = 36.7^\circ \text{ cw}$$

$$\sigma_1 = 88.58 \text{ ksi}$$

$$\sigma_2 = -49.3 \text{ ksi}$$

and Eqs. (5.21) and (5.22) yield

$$\varepsilon_1 = 0.003446$$

$$\varepsilon_2 = -0.002530$$

5.2.4 Elastic Stress-Concentration Factors

Most mechanical components are not smooth. Practical components typically include holes, keyways, notches, bends, fillets, steps, or other structural discontinuities. Stresses tend to become "concentrated" in such regions such that these stresses are

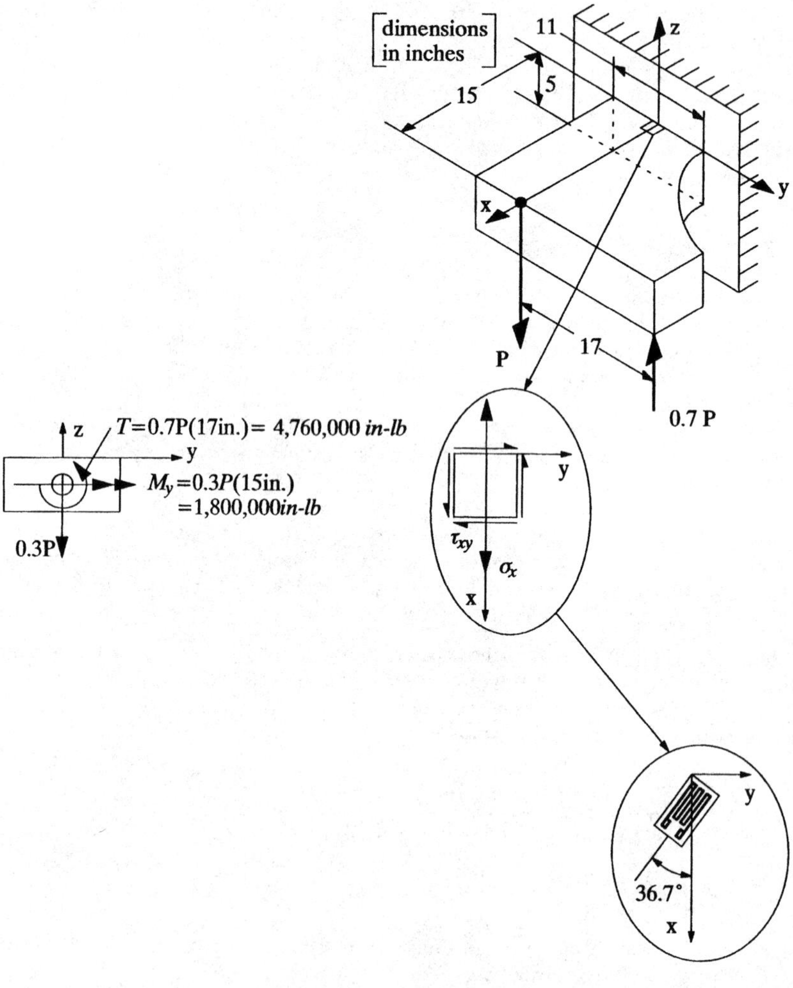

significantly greater than the nominally calculated stresses, based on the external forces and cross-sectional area. For instance, stress distributions are shown in Fig. 5.6 for a uniform rectangular plate in tension and through the identical cross section of a plate with a filleted step. Notice that the maximum elastically calculated stress at the root of the fillet, σ_{notch}, is greater than the nominal stress, S_{nom}. From this, the *stress-concentration factor* K_t is defined as the maximum stress divided by the nominal stress:

$$K_t = \frac{\sigma_{notch}}{S_{nom}} \tag{5.23}$$

Stress-concentration factors are found by a number of techniques including experimental, finite-element analysis, boundary-element analysis, closed-form elasticity solutions, and others. Fortunately, researchers have tabulated K_t values for many generalized geometries.[16,17]

Reference 16, from Peterson, is a compendium of design charts. Reference 17, from Roark, provides useful empirical formulas that can be programmed into spreadsheets

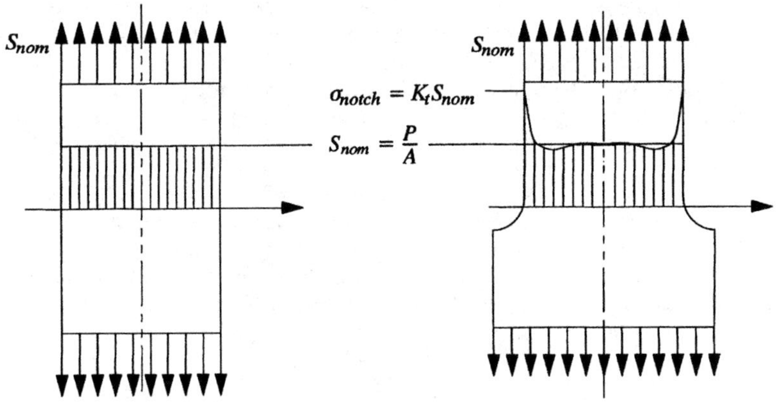

FIG. 5.6 Definition of elastic stress-concentration factor.

or computer routines for design optimization. There are several important points to remember about stress concentration factors:

1. They only apply to elastic states of stress-strain.
2. They are tabulated for a particular mode of loading (axial, bending, torsional, etc.).
3. Since they are elastic, they can be superimposed (i.e., computed separately, then added).

It has been shown that using the full value of stress-concentration factors for evaluating strength can be overly conservative, especially for static design situations. Shigley and Mischke[1] state that one can usually neglect the stress-concentration factor due to the fact that localized yielding can work-harden the material in the notch vicinity and relieve the stresses. On the other hand, neglecting stress-concentration factors can be nonconservative, especially if very brittle material or fatigue loading is involved. As safe design practice, stress-concentration factors should be considered for preliminary analysis. If the resulting solution is unacceptable from a weight, size, or cost standpoint, then reasonable reductions in K_t can be considered based on the potential for the material to deform and locally work-harden. Such decisions can be based on experimental data generated with the material of interest and notches with similar values of K_t. This type of testing can supplement analysis prior to the availability of fully designed prototype parts. Data from testing such as this should be carefully documented, since it can be used as a basis for future design decisions.

5.2.5 Finite-Element Analysis

The use of computers is increasing as rapidly in engineering design as in any other profession. Finite-element analysis (FEA),[18–21] coupled with increasing computational capabilities, is providing increasing analytical power for use on everyday design situations. Commercially available software packages are enabling designers to evaluate states of stress in situations involving complex geometry and loading combinations. However, three important points should be considered when stresses are computed from FEA: (1) Elastic analysis can be straightforward, but the potential for error is great if the analyst has not assured mesh convergence, especially for sharp geometric discontinuities or contact problems. (2) The specification of boundary conditions is critical to obtaining valid results that correlate with the physical stress-strain state in the component being modeled. (3) Elastic-plastic analysis is not yet simplified for everyday design use, particularly for cyclic loading conditions.[22]

These three points are intended to remind the designer not to accept FEA results without an adequate awareness of the assumptions used to implement the analysis (most importantly, mesh density, boundary conditions, and material modeling). Most robust commercial FEA codes provide error estimates associated with their solutions. In high-stress-concentration regions, these errors can be substantial and a locally refined mesh could be called for (often not a simple task). A designer must be careful to avoid the tendency to simply marvel at appealing and colorful FEA output without fully understanding that the results are only as valid as the assumptions used to build the analytical model.

5.3 STRUCTURAL INTEGRITY DESIGN PHILOSOPHIES

An engineer must routinely assure that designs will endure anticipated loading histories with no significant change in geometry or loss in load-carrying capability. Anticipating service-load histories can require experience and/or testing. Techniques for load estimation are as diverse as any other aspect of the design process.[23]

The *design* or *allowable* stress is generally defined as the tension or compressive stress (yield point or ultimate) depending on the type of loading divided by the safety factor. In fatigue the appropriate safety factor is used based on the number of cycles. Also when wear, creep, or deflections are to be limited to a prescribed value during the life of the machine element, the design stress can be based upon values different from above.

The magnitude of the design factor of safety, a number greater than unity, depends upon the application and the uncertainties associated with a particular design. In the determination of the factor of safety, the following should be considered:

1. The possibility that failure of the machine element may cause injury or loss of human life
2. The possibility that failure may result in costly repairs
3. The uncertainty of the loads encountered in service
4. The uncertainty of material properties
5. The assumptions made in the analysis and the uncertainties in the determination of the stress-concentration factors and stresses induced by sudden impact and repeated loads
6. The knowledge of the environmental conditions to which the part will be subjected
7. The knowledge of stresses which will be introduced during fabrication (e.g., residual stresses), assembly, and shipping of the part
8. The extent to which the part can be weakened by corrosion

Many other factors obviously exist. Typical values of design safety factors range from 1.0 (against yield) in the case of aircraft, to 3 in typical machine-design applications, to approximately 10 in the case of some pressure vessels. It is to be noted that these safety factors allow us to compute the allowable stresses given and are not in lieu of the stress-concentration factors which are used to compute stresses in service.

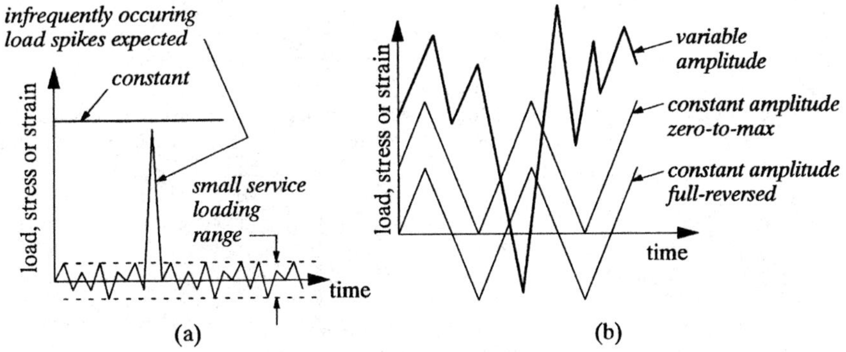

FIG. 5.7 Examples of (*a*) static and (*b*) dynamic, or fatigue, loading.

If the uncertainties are great enough to cause severe weight, volume, or economic penalties, testing and/or more thorough analyses should be performed rather than relying upon very large factors of safety. Factors of safety to be used with standard, commercially available design elements should be those recommended for them by reliable manufacturers and/or by established codes for design of machines.

In probabilistic approaches to design, in terms of stress (or actual load) and strength (or load capability) the safety factor is related to reliability. When a failure may cause injury or otherwise be disastrous, the probability density curves representing the strength of the part and the stress to be sustained should not overlap, and the factor of safety equals the ratio of the mean strength to the mean stress. If the tails of the two curves overlap, a possibility for failure exists.

The "true factor of safety," which may be defined in terms of load, stress, deflection, creep, wear, etc., is the ratio of the magnitude of any of the above parameters resulting in damage to its actual value in service. For example:

$$\text{True factor of safety} = \frac{\text{maximum load part can sustain without damage}}{\text{maximum load part sustains in service}}$$

The true factor of safety is determined after a part is built and tested under service conditions.

In this chapter, loading is classified as "static" or "dynamic." Static loading could be formally defined as loading that remains constant over the life of the component, as depicted in Fig. 5.7*a*. Under this type of loading, the primary concern is avoiding failure by yielding or fracture. Also shown in Fig. 5.7*a* is another type of loading that can be considered "static," from a structural integrity viewpoint. This refers to situations where only a few, *infrequently occurring* load spikes can be expected in service. Dynamic loading fluctuates significantly during the life of the component, as shown in Fig. 5.7*b*. Although peak stresses can remain well below levels associated with yielding, this type of loading can lead to failure by fatigue.

5.3.1 Static Loading

Loading on a mechanical component is rarely steady. However, in many cases, safety factors and service load ratings are used in order to keep in-service load fluctuations small relative to the maximum load the component can sustain. Often this is assured by proof loading a component as the final step in its manufacturing process. Proof loads usually exceed the rated service load by a factor of 2 to 3.5. Examples of this include chains, other lifting hardware, and pressure vessels. Proof loading not only ensures the structural integrity of the part, but can also serve to impart residual stresses that increase the functional elastic limit and increase fatigue life.[24–26]

When a component is designed for static strength, it must be assured that service loads indeed remain well below the strength of the component. Unfortunately, the user, more than the designer, often dictates the maximum load level a component will experience. Users tend to push designs over the limit at every available opportunity. For safety-critical components, designers should consider mechanisms to ensure that loadings do not exceed safe operating levels. For instance, rupture disks are effective "weak links" in the design of pressure systems. Should the operating pressure be exceeded, the rupture disk fails by design, into a discharge tank. Another example would be the use of redundant, or backup, elements. When a specimen begins to deform under too great a load, it gains support when it encounters a backup element, thus avoiding complete fracture (and possibly alerting the end user).

Care must be taken when the loading on a component is classified as "static" for design purposes. The approach is only safe when the static limit is rarely seen in service, as depicted by the load spike in Fig. 5.7*a*. For example, suppose a nozzle discharges under a constant internal pressure. There is a tendency to utilize that pressure for static design (the constant loading line in Fig. 5.7*a*). However, if the nozzle discharges for 30 minutes, drains, then repeats, on a regular basis, then fatigue could be important.

5.3.2 Fatigue Loading

Under fatigue loading, cracks develop in high-stress regions which were initially free of any macroscopic defect. A component can endure numerous cycles of loading before the crack is detectable. Once this occurs, a dominant crack usually propagates progressively to fracture. The relative life spent in developing a crack of "engineering size" (usually defined as 1–2 mm in surface length) and then propagating the crack to fracture can define the fatigue design philosophy, as overviewed below.

Infinite-Life Design. This philosophy is based on the concept of the *fatigue limit,* or the stress amplitude below which fatigue will not occur. For high-cycle components like valve springs, turbo machinery, and other high-speed rotating equipment, this is still a very widely utilized concept. However, the approach is going out of style due to cost- and weight-reduction requirements. There are also problems pertaining to the definition of a fatigue limit for a particular material, since numerous factors have proven influential. These factors include heat treatment, surface condition, residual stresses, temperature, environment, etc. (Furthermore, aluminum and other nonferrous alloys do not exhibit a fatigue limit.) One final cautionary note: Intermittent overloads can reduce or eliminate the fatigue limit.[13]

Finite-Life (Safe-Life) Design. Instead of designing a component to never fail, parts are designed for a specified life deemed "safe," or unlikely to occur during the rated life of the machine, except in cases of abusive loading. For instance, even a safety-critical automobile suspension component might be designed to sustain only 1000 of the most severe impact loads corresponding to the worst high-speed curb strike on the proving ground. However, the vast majority of automobiles will experience nowhere near this many occurrences. Pressure vessels are sometimes designed to lives on the order of a few hundred cycles, corresponding to cleanout cycles that will occur only a few times annually. Ball joints in automobiles and landing-gear parts in aircraft are other examples of finite-life design situations.

Fail-Safe Design. This strategy, developed primarily in the aircraft industry, should be implemented whenever possible. The approach invokes measures to ensure that, if cracks initiate, they will grow in a controlled manner. Then, measures are taken to ensure that catastrophic failure is avoided, including the use of redundant elements, backup elements, crack-arrest holes positioned at strategic locations, and the use of multiple load paths. This approach is referred to as "leak before burst" in the pressure-vessel industry.

Damage-Tolerant Design. Also developed in the aircraft industry, this philosophy assumes that cracks exist *before* a component is put into service. For instance, cracks are assumed to exist underneath rivet heads or behind a seam, anywhere that they might be concealed during routine inspection. Then, the behavior of the crack is predicted from flight-loading spectra anticipated for the aircraft. Analyses of many key locations on an aircraft are used to schedule maintenance and inspections.

The four methods discussed above cover most design situations, but which of these is utilized depends on the design criteria. It is typical for more than one (sometimes all) of the strategies to be utilized in a single design. For instance, in the design of an aircraft, the fail-safe approach is routinely applied to wings, fuselages, and control surfaces. However, a landing gear and a rotor in the jet engine are designed for finite life.

To provide some size scale to the issue of fatigue crack development, refer to Fig. 5.8.

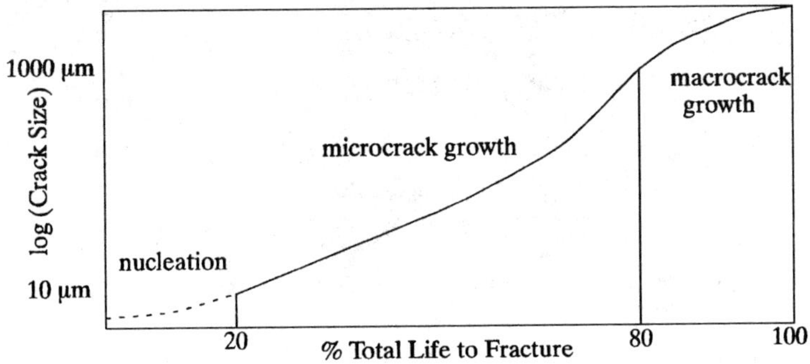

FIG. 5.8 Size scale associated with fatigue crack development.

Crack development can be divided into three separate regimes: nucleation, microcrack growth, and macrocrack growth. The first two regimes are often referred to together as Region I, or *crack initiation.* This is still a very unclear area. Although numerous theories exist, experimental verification is difficult. Obtaining repeatable data in this region has proven difficult and active research is underway. The macrocrack growth regime is referred to as Region II. This region is associated with linear elastic fracture mechanics (LEFM).

Region I: This region involves the formation of surface cracks of the order of 1 mm in length. It is considered to be controlled by the maximum shear stress fluctuation, $\Delta\tau_{max}$, since cracks on this small scale tend to originate on planes experiencing maximum shear stress. Two kinds of maximum shear planes are shown in Fig. 5.9. One intersects the surface at 45° (Sec. A-A) and the other at 90° (Sec. B-B). Note that both are inclined 45° to the applied stress axis. The enlarged views in Fig. 5.9 represent the intersection of slip bands with the free surface. "Slip bands" refer to multiple parallel planes, each accommodating massive dislocation movement, and associated plastic slip. The cross sections depict discontinuities (called intrusions and extrusions) created on the free surface that eventually lead to a macroscopic crack.

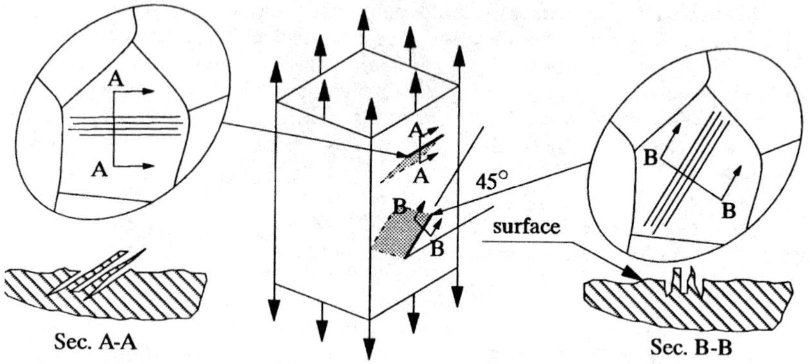

FIG. 5.9 Schematic microscopic shear cracks intersecting the surface at 45° (Sec. A-A) and 90° (Sec. B-B).

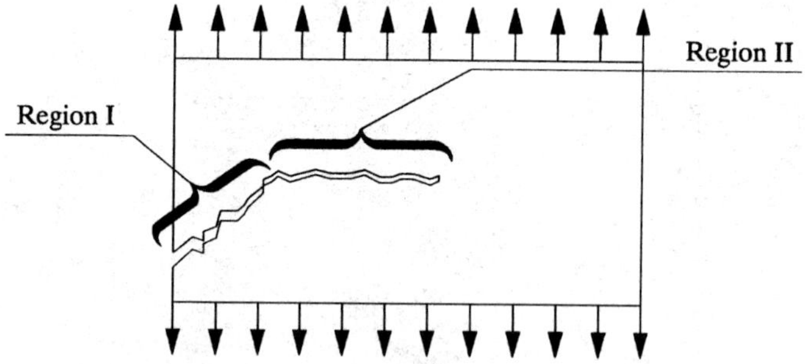

FIG. 5.10 Macroscopic cracks typically propagate perpendicular to maximum principal stress.

Region II: The surface steps created by the slip-band intersection are forced open and decohesion of slip planes forms small cracks oriented 45° to the loading axis. Upon growth, cracks typically turn to become oriented 90° to the principal stress direction. This is illustrated in Fig. 5.10. The crack is forced open and the crack tip blunts. This causes *striations* to form, which result in the *beach marks* that characterize fatigue fracture surfaces. This region is controlled by the range of principal stress ($\Delta\sigma_1$) acting normal to the plane of the crack. Generally, once this stage is reached, fracture mechanics are used to describe subsequent behavior.

5.4 STATIC STRENGTH ANALYSIS

In this section, a state of stress such as that depicted in Fig. 5.1*b*, is considered to be known. Based on this state of stress, the structural integrity of a component is assessed by comparing the stress state to the strength of the material. Although the methodology shown here is applicable to any three-dimensional stress state, surface stress states are emphasized since these comprise the vast majority of engineering design situations. Approaches are described in only a cursory manner, with more detailed references given. Emphasis is given to the application of the approaches to design situations.

5.4.1 Monotonic Tensile Data

The tensile test provides the input data for conducting static strength analysis. A sample of material (usually round or rectangular in cross section) is pulled apart under a monotonically increasing tensile load until failure occurs. Guidelines for conducting tensile tests are found in the American Society for Testing and Materials (ASTM) Specification E-8, "Standard Test Methods of Tension Testing of Metallic Materials."[27] A stress-strain curve from a tensile test is illustrated in Fig. 5.11, with a list of important points corresponding to "properties" measured by the test.

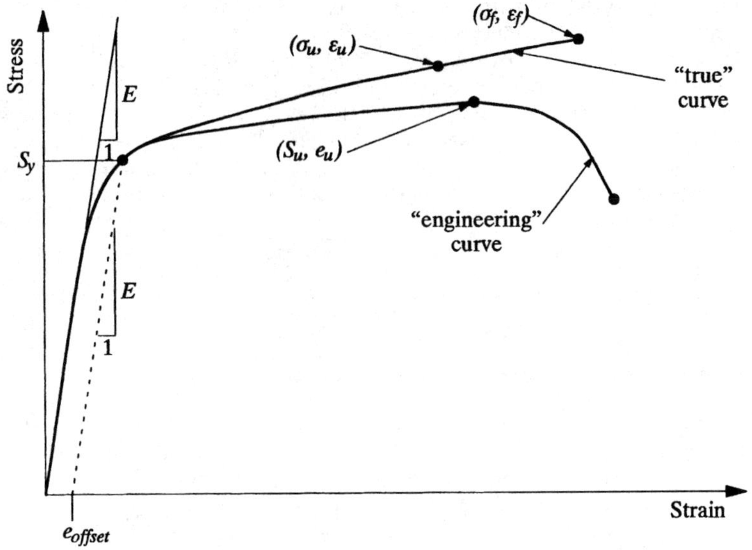

FIG. 5.11 Schematic engineering and true stress-strain curves, with list of properties.

Another important parameter is the *reduction in area.* It is determined from a measurement of the minimum diameter of the broken specimen, and the relation

$$\text{RA} = \frac{A_o - A_f}{A_o} = \frac{d_o^2 - d_f^2}{d_o^2} \tag{5.24}$$

where A_o and d_o are the initial specimen cross-sectional area and diameter, respectively, and the f subscript refers to those dimensions at fracture.

The engineering stress is computed by dividing the applied load by the original gauge section cross-sectional area A_o. Engineering strain is computed by dividing the change in gauge-section length by the initial gauge-section length. The calculation of "true" stress and strain quantities accounts for the fact that, as the loading increases, the cross-sectional area decreases and the gauge length increases. However, the need to distinguish between the two is rare, for everyday engineering design. The two curves are virtually identical up to plastic strains on the order of 10 percent.

A relation was developed by Ramberg and Osgood[28] to describe the stress-strain curve for many metallic engineering alloys. This relation is usually expressed as

$$\varepsilon = \frac{\sigma}{E} + \left(\frac{\sigma}{K}\right)^{1/n} \tag{5.25}$$

where K = strength coefficient
n = strain-hardening exponent

For situations involving large plasticity (such as forming operations) the approximate log-log linear (power log) relation between stress and plastic strain can sometimes be quite inaccurate over a wide range of plastic strains. Therefore, it can be important to specify the plastic strain range over which n is defined. A designer interested in moderate plastic strain in a notch might be concerned with the range 0.002 to 0.02. However, a manufacturing engineer interested in a forming operation might need more accurate stress-strain information over a range from 0.05 to 0.15. The ASTM Specification E-646, "Tensile Strain-Hardening Exponents (n-Values) of Metallic Sheet Materials,"[29] deals with this issue specifically.

Data from compression tests for engineering materials can be equally important for conducting a static strength assessment. ASTM Specification E-9, "Standard Test Methods of Compression Testing of Metallic Materials at Room Temperature,"[30] describes this type of testing. Data from such a test can be important when attempting to classify a material as ductile or brittle. Failure of "brittle" materials in tension is usually associated with internal stress risers, such as voids or inclusions. Under compression such stress concentrations are less influential and the strength of a brittle material can considerably exceed its own tensile strength (for instance, by a factor of over 4 for some cast irons).

5.4.2 Multiaxial Yielding Theories (Ductile Materials)

Ductile materials are considered to be able to exhibit notable plasticity in a tensile test prior to fracture. No rigorous definition of "ductile" exists. Generally, however, a material is considered ductile if the percent reduction in area is greater than 15 to 20 percent, and the ultimate tensile strength exceeds the yield strength by a notable amount. Another important indicator used to classify a material as ductile is the relation between magnitudes of the tensile and compressive yield strengths. Ductile materials tend to yield in compression at nearly the same stress level as they do in tension, whereas brittle materials are typically quite a bit stronger in compression.

For the design of ductile machine components, two theories are typically utilized: (1) the Tresca criterion (maximum shear stress) and (2) von Mises' criterion (equivalently, the octahedral shear-stress or distortion-energy theory). These approaches can be depicted as safe operating envelopes on axes of minimum versus maximum principal stress (Fig. 5.12). Notice that the Tresca approach is smaller and therefore more conservative than the von Mises.

The Tresca (Maximum Shear-Stress Theory) Criterion. This approach is based on the premise that yielding will occur when the maximum shear stress under multiaxial loading, τ_{max}, is equal to the maximum shear stress imposed during a tensile test at yield. In other words, yielding occurs when

$$\tau_{max} = \frac{\sigma_A - \sigma_B}{2} = \frac{S_y}{2} \tag{5.26}$$

where σ_A and σ_B are the maximum and minimum principal stresses, respectively. This approach can be restated in terms of an "equivalent stress,"

$$\sigma_{eq,Tr} = \sigma_A - \sigma_B \tag{5.27}$$

which is directly comparable to the axial yield strength of the material. In this way,

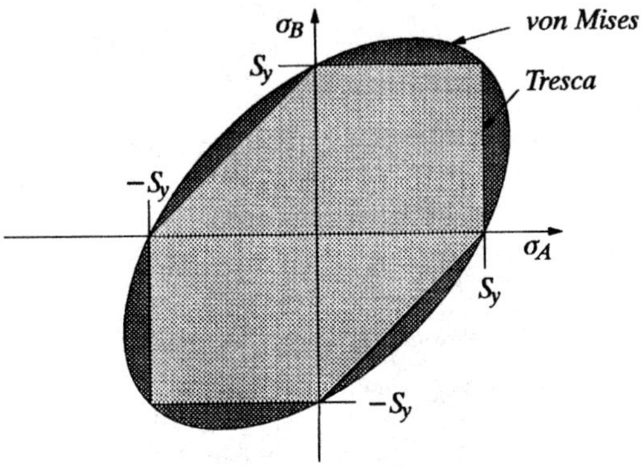

FIG. 5.12 Safe operating regions for von Mises (octahedral shear stress) and Tresca (maximum shear stress) criteria.

the "factor of safety" for the stress state is straightforwardly defined from the Tresca criterion by

$$FS_{Tr} = \frac{S_y}{\sigma_{eq,Tr}} \tag{5.28}$$

The von Mises Criterion. The von Mises criterion refers to any of several approaches shown to be essentially identical. These include the distortion energy, octahedral shear stress, and the Mises-Henkey theories. In terms of an equivalent stress, the von Mises approach is given by

$$\sigma_{eq,vM} = \frac{1}{\sqrt{2}}\sqrt{(\sigma_1 - \sigma_2)^2 + (\sigma_2 - \sigma_3)^2 + (\sigma_3 - \sigma_1)^2} \tag{5.29}$$

Conceptually, the approach can be considered a root-mean-square average of the principal shear stresses, with a scaling factor to assure that the equivalent stress is equal to σ_1 for a uniaxial stress state.

The factor of safety for the von Mises approach is thus given by

$$FS_{vM} = \frac{S_y}{\sigma_{eq,vM}} \tag{5.30}$$

Experiments have shown that the von Mises criterion is more accurate in terms of describing data trends, but the Tresca approach is a more conservative design option.[13]

5.4.3 Multiaxial Failure Theories (Brittle Materials)

In this section, the use of three design criteria is demonstrated. These approaches are referred to (in order of decreasing conservatism) as the Coulomb-Mohr, modified Mohr, and the maximum normal fracture criteria.[13] Each can be considered to define safe operating envelopes on axes of minimum versus maximum principal stress (Fig. 5.13). The most notable difference between Figs. 5.12 and 5.13 is the typically greater compressive strength S_{uc} exhibited by a brittle material relative to its tensile strength S_{ut}. Also,

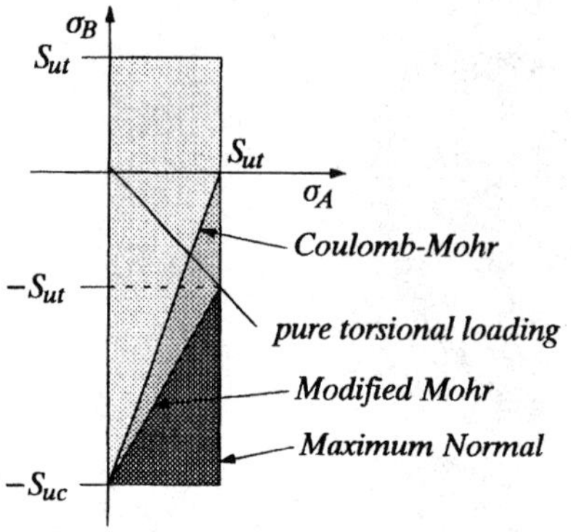

FIG. 5.13 Safe operating regions for the Coulomb-Mohr, modified Mohr, and maximum normal failure theories for brittle materials.

notice how only the first and fourth quadrants of principal stress space are depicted in Fig. 5.13. This is because the vast majority of all engineering stress states of concern to mechanical designers lie in these quadrants, with the vast majority located in the fourth. (With the exception of the deepest points in the ocean, it is difficult to imagine practical engineering states of surface stress that do not reside in or along the fourth quadrant.)

The three theories are described below, followed by the presentation of a static strength design algorithm. For all three theories, the factor of safety for a state of stress is defined as the ratio of the radial distance to the boundary (through the state of stress) to the radial distance defined by the state of stress. This is depicted in Fig. 5.14.

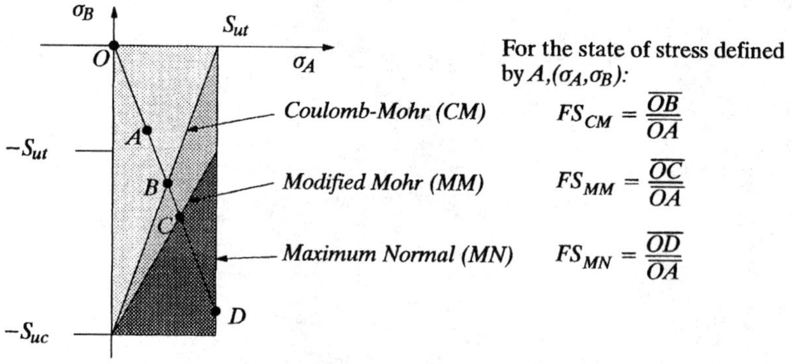

FIG. 5.14 Definition of factor of safety based on the Coulomb-Mohr, modified Mohr, and maximum normal failure theories for brittle materials.

The Coulomb-Mohr Fracture Theory. The Coulomb-Mohr theory is based on the concept that certain combinations of shear stress and stress normal to the plane of maximum shear are responsible for failure. This is manifested in the fourth quadrant of principal stress space by the line from S_{ut} on the tensile stress axis to $-S_{uc}$ on the compressive strength axis.

As is apparent from Figs. 5.13 and 5.14, the Coulomb-Mohr theory is the most conservative design approach. Experimental results have indicated that the approach is typically conservative for design applications.

The Modified Mohr Fracture Theory. This theory is based on empirical observations that the maximum principal stress tends to define failure under torsional loading (or along a line 45° through the fourth quadrant of principal stress space). However, when significant compression accompanies torsion (stress states below the 45° torsion line), the maximum normal stress theory becomes nonconservative. Therefore, the line in the fourth quadrant defined by $(S_{ut}, -S_{ut})$ and $(0, -S_{uc})$ is used as the boundary for the modified Mohr theory. Since the approach is formulated from empirical observations, it tends to correlate well with data.

The Maximum Normal Fracture Theory. Conceptually, this is the simplest of the theories in this section. If the magnitude of the maximum (or minimum) principal stress in the material exceeds the material's tensile (or compressive) strength, failure is predicted. Unfortunately, experiments have shown it to be nonconservative for situations involving substantial compression (states of stress in the fourth quadrant, below the line of pure torsion).

5.4.4 Summary Design Algorithm

In practice, the designation of a material as ductile or brittle and the selection of an appropriate failure criterion can be subjective. Major factors include whether or not compressive strength data are available, and whether or not compression constitutes a major portion of the loading. In situations where the choices are not clear, it is advisable to conduct analyses based on limiting assumptions, implementing all potential approaches to bound a solution. To assist in this, an algorithm is presented in the form of a flowchart in Table 5.1 that can be easily coded into a computer program or applied using a computer spreadsheet. The design engineer is responsible for supplying the correct input information (including the classification of the material as ductile or brittle) and for interpreting the output. Several techniques are used to evaluate strength and the designer must decide which is the most appropriate. Output from such a routine is presented in Example 4.

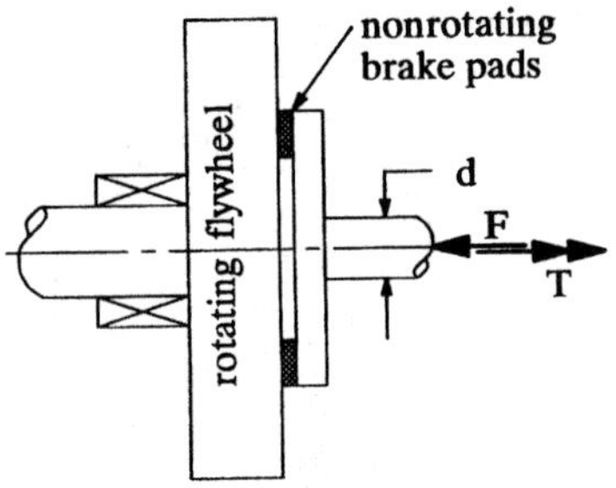

EXAMPLE 4 A round shaft is to be used to apply brake pads to the side of a large flywheel. The shaft is to experience a compressive load of $F = 22{,}000$ lb, and corresponding torsional load of $T = 23{,}100$ in·lb. Specify the diameter of the shaft d (to the nearest one-eighth inch) for a safety factor of at least 2.0 using the following materials:

1. ASTM #40 cast iron ($S_{ut} = 42.5$ ksi, $S_{uc} = 140$ ksi)
2. 1020 steel ($S_y = 65$ ksi)
3. Q&T 4340 steel ($S_y = 240$ ksi)

solution For each material, the three failure theories were used from Table 5.1. Diameters (in inches) and safety factors (in parentheses) estimated for each material are presented in tabular form, below.

	1. #40 iron	2. 1020	3. 4340
Tr	N/A	2.0 (2.15)	1.375 (2.62)
vM	N/A	1.875 (2.04)	1.25 (2.27)
CM	1.875 (2.02)	N/A	N/A
MM	1.75 (2.06)	N/A	N/A
MN	1.75 (2.38)	N/A	N/A

TABLE 5.1 Static Strength Analysis

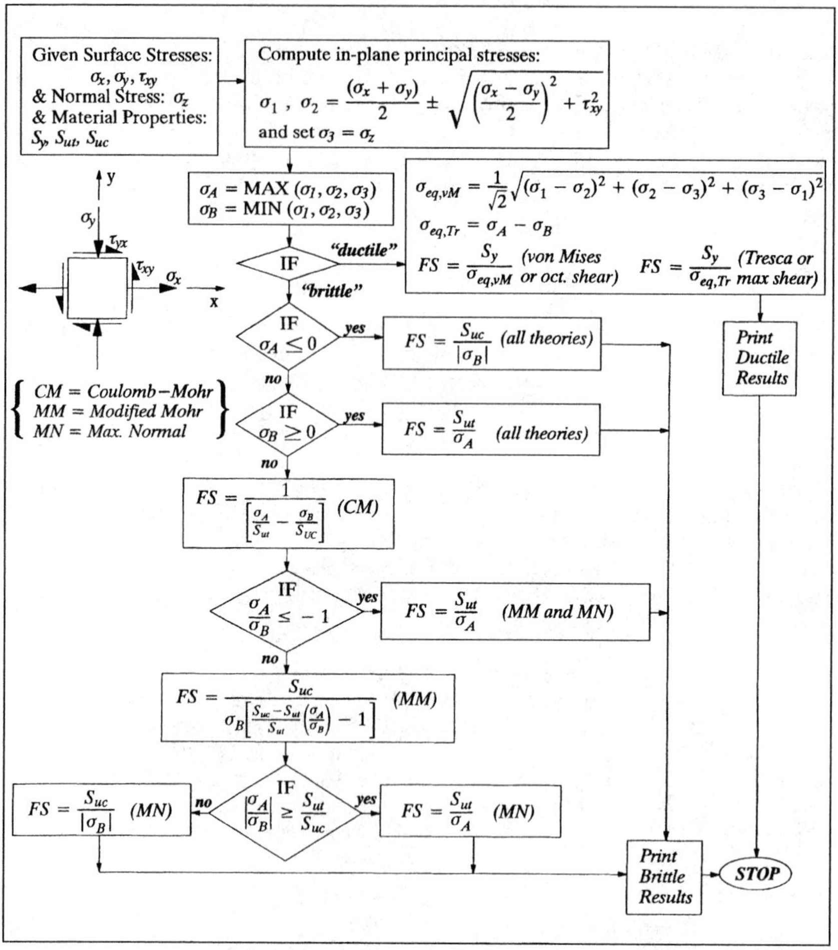

5.5 FATIGUE STRENGTH ANALYSIS

The subject of fatigue analysis is considered in this section from the point of view of an engineering designer. Although this subject is still actively researched, a great deal of solid engineering methodology has been developed. The fatigue design strategy to be described in this section is outlined below.

Crack initiation is defined as the occurrence of a crack of engineering size, usually 1 to 2 mm in surface length. The basis of this definition is illustrated in Fig. 5.15. To obtain baseline fatigue data (stress-life or strain-life), tests are usually conducted on small specimens, 0.25 in (6 mm) in diameter. Usually, "failure" in these tests can be associated with complete fracture of the specimen. *It is assumed that a component experiencing a localized stress-strain history equivalent to the axial specimen will develop a crack of approximately the same size in approximately the same number of cycles.* This concept is often referred to as the local strain approach.[13,23,31,32]

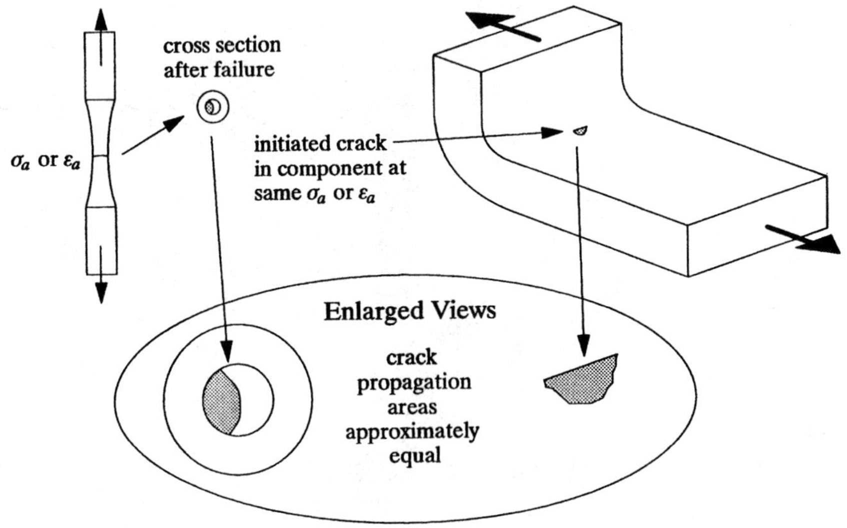

FIG. 5.15 Similitude between failure in a baseline test specimen and crack initiation in an actual engineering component.

In the remainder of this chapter, "failure" will refer to the occurrence of an engineering-sized crack, roughly the same size as that found in an axial specimen upon its failure in a standard fatigue test. The propagation of the crack due to subsequent fatigue loading is considered separately using fracture-mechanics techniques for damage-tolerant design.

5.5.1 Stress-Life Approaches (Constant-Amplitude Loading)

In this section, the *stress-life* (*S-N*) *approach* to fatigue design is overviewed. This is one of the earliest fatigue design approaches to be developed and can still be a useful tool. Its success is based on the fact that, for predominantly elastic loading, the state of stress in a component can often be characterized quite accurately. *As long as the state of fluctuating stress can be accurately estimated, the S-N approach can do a good job of predicting fatigue.* However, fatigue cracks usually develop at structural discontinuities, or notches. In these regions, localized cyclic plastic strains can develop and the task of estimating the state of stress becomes far more difficult. Without a reliable knowledge of the stress state, the utility of the *S-N* approach becomes limited and a strain-based approach (described later) becomes more useful.

Stress-Life Curve. Baseline data are generated by imposing fully reversed fluctuating stress in a standard specimen, as shown below in Fig. 5.16. This can be done via axial loading or rotating-bending.

Fully reversed loading refers to the fact that $\sigma_{max} = -\sigma_{min}$ (or, the alternating stress, $\sigma_a = \sigma_{max}$). Tests are conducted by applying loading as shown in Fig. 5.16 until the specimen "fails," usually by fracturing into two separate pieces. Typically, the gauge section ranges in size from 0.25 to 0.5 inch in diameter (6 to 12 mm). To generate *S-N* data for fatigue design purposes, a number of specimens must be tested at varying stress levels. Applicable ASTM guidelines are listed below.

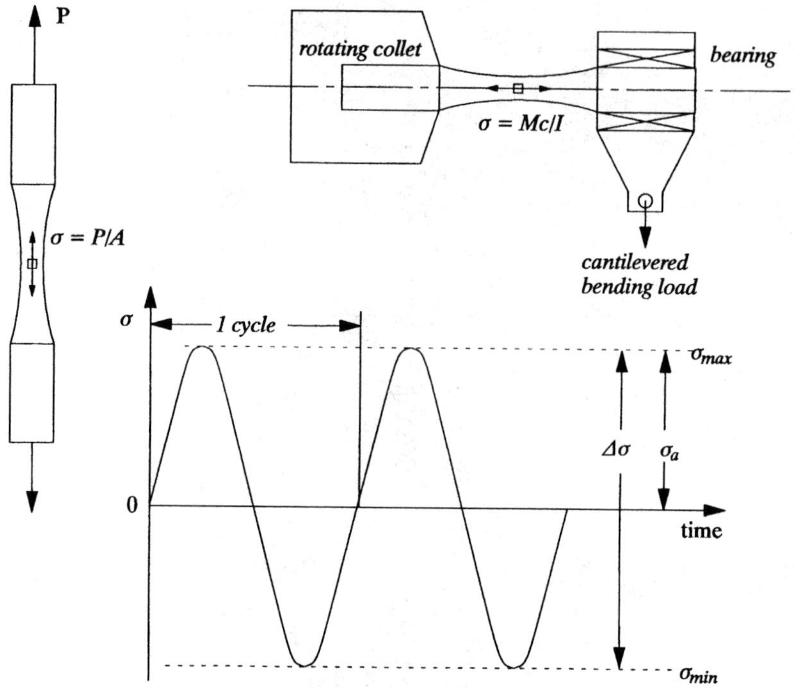

FIG. 5.16 Baseline *S-N* fatigue testing.

- Details for conducting S-N tests are presented in the ASTM E-466-82, "Standard Practice for Conducting Constant Amplitude Axial Stress-Life Tests of Metallic Materials."[33]
- Data from fatigue tests are analyzed according to the ASTM E-739, "Standard Practice for Statistical Analysis of Linear or Linearized Stress-Life (*S-N*) and Strain Life (ε-*N*) Fatigue Data."[34]
- Data are typically presented according to ASTM E-468-82, "Standard Practice for Presentation of Constant Amplitude Fatigue Test Results for Metallic Materials."[35]

There are some fundamental differences between baseline data obtained from rotating-bending and axial testing. The stress amplitude for rotating-bending is computed elastically, even though severe plastic deformation occurs at higher load levels. Therefore, the quantity

$$S = \frac{Mc}{I} \tag{5.31}$$

is actually only a parameter with units of stress, indicating the severity of bending. A plasticity analysis would be required to estimate the actual stress at the specimen surface. And even for high-cycle tests (lower load levels), there is a bending-stress gradient as depicted in Fig. 5.17. For this reason, bending tests are less severe than axial tests and can make the material *appear* stronger. This is due to two factors, both related to the bending versus axial stress distribution: (1) Physically, more of the gauge section is subjected to the maximum stress in an axial test than in a bending test. This

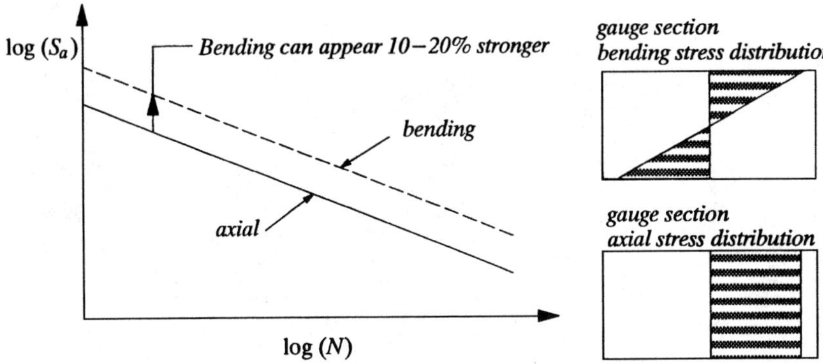

FIG. 5.17 Comparison of stress-life data from axial and rotating-bending test.

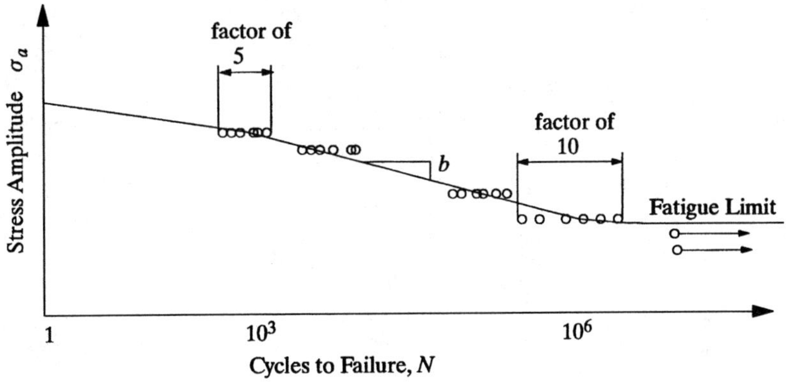

FIG. 5.18 Stress-life fatigue data.

increases the likelihood that a critically sized material defect or properly oriented slip system will experience the most severe stress fluctuation. (2) The bending stress distribution is less severe from the standpoint of crack propagation during microcrack development.

A major benefit to the use of a rotating-bending machine is speed. Motors are used to drive the specimen at a very high rpm, generating data very quickly (e.g., 10,000 rpm). A schematic set of *S-N* data from such a machine is shown in Fig. 5.18 to illustrate some more fatigue data trends. In Fig. 5.18 and all subsequent *S-N* plots, axes are logarithmic.

Scatter can plague fatigue data. Factors of 10 or more are not unusual in the high-cycle regime. Scatter is very dependent on cleanliness of material (pores, inclusions, and other microstructural defects). Statistical guidelines from ASTM E-739[34] can be very useful in understanding and utilizing fatigue data.

One of the most utilized features of the *S-N* curve limit is the fatigue limit. It is important to remember that aluminum and other nonferrous metals do not exhibit a fatigue limit. (Fatigue limits are quoted in the literature for aluminum as the stress amplitude corresponding to a very large number of cycles, such as $5 \cdot 10^7$ to $5 \cdot 10^8$.) For ferrous alloys, fatigue limits can be affected by many factors, as outlined below.

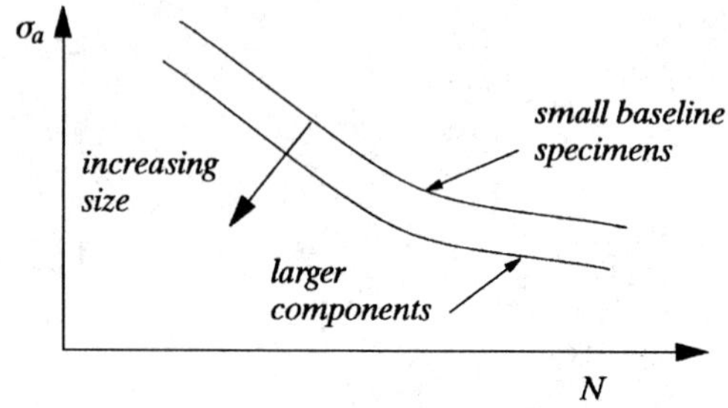

FIG. 5.19 Increasing component size decreases fatigue strength, relative to data generated with small specimens.

- *Size effects.* When a component is considerably larger than the specimen used to generate the baseline fatigue data, a greater volume of material is subjected to a particular stress amplitude. This increases the statistical probability that a microscopic flaw, defect, or slip system will exist that is susceptible to fatigue-crack development. For this reason larger components often fail sooner than smaller specimens, as depicted in Fig. 5.19. This discrepancy is affected by other factors, such as inhomogeneity of microstructure.
- *Type of loading.* Differences between bending and axial loading have already been discussed (Fig. 5.17).
- *Surface processing.* Besides surface roughness in general, plating, nitriding, induction hardening, rolling, shot peening, or any other surface modification can drastically affect the fatigue behavior of a part. Generally, processes improve fatigue resistance if they increase hardness, impose residual compressive surface stresses, and/or reduce surface roughness.
- *Grain size.* This is particularly important for high-cycle fatigue. Typically, smaller grain size means longer fatigue lives. (This is not surprising, since smaller grain size usually means higher yield strength.)
- *Material processing.* The "cleaner" the material, the better its fatigue resistance. For instance, vacuum-melt steel exhibits fatigue lives longer by 50 percent relative to furnace-melt steels. Wrought metals show better fatigue resistance than cast metals (Fig. 5.20). Crack nucleation and microcrack propagation time is avoided since microscopic defects such as inclusions or pores act as instant crack growth sites. (The same can be true for powdered-metal parts.)

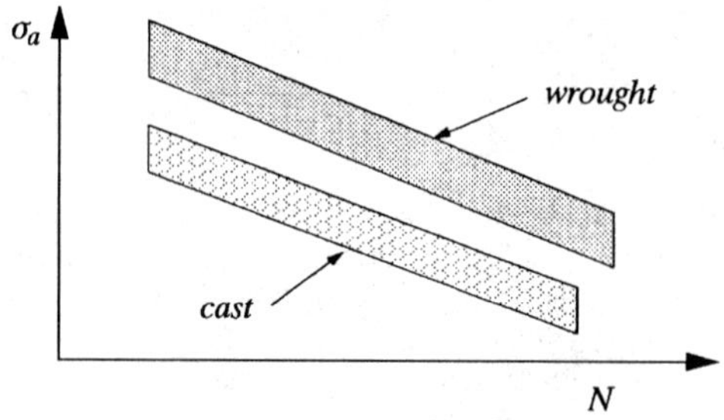

FIG. 5.20 Wrought material is generally more fatigue resistant than cast material.

- *Temperature and environment.* Both of these factors can exhibit profound negative synergism with fatigue mechanisms. (The sum of the two effects can be more than a simple superposition.) When these variables are important, loading frequency and waveform must be considered influential. This is not the case ordinarily (when environmental concerns are not considered influential).
- *Intermittent overloads.* Suppose a component operates in service at a stress level below the fatigue limit, but experiences occasional overloads. Even though the overloads are infrequent and cause no macroscopic plasticity, they can serve to reduce or eliminate the endurance limit.[13]

Numerous attempts have been made to quantify the effects just described.[1–3,31,32] These are generally presented as empirical factors used to reduce the endurance limit. These factors tend to reduce the high-cycle, finite portion of the *S-N* curve as well.

Finally, designers are frequently forced to evaluate the endurance of a part for which *S-N* data are not available. Therefore, several textbooks have suggested empirical approaches to estimate the *S-N* curve from monotonic tensile data.[1,3,12] A comprehensive overview of many of these can be found in Dowling.[13] For example, data have suggested the following relation between the endurance limit and ultimate tensile strength:[1] For wrought steels,

$$\begin{aligned} S_e &= 0.5\, S_u \qquad &&\text{for } S_u \le 200 \text{ ksi} \\ S_e &= 100 \text{ ksi} \qquad &&\text{for } S_u > 200 \text{ ksi} \end{aligned} \tag{5.32}$$

For cast iron,

$$\begin{aligned} S_e &= 0.45\, S_u \qquad &&\text{for } S_u \le 88 \text{ ksi} \\ S_e &= 40 \text{ ksi} \qquad &&\text{for } S_u > 88 \text{ ksi} \end{aligned} \tag{5.33}$$

***S-N* Finite-Life Prediction.** Many factors have caused infinite-life design to become impractical, weight and cost being the primary motivators. It has become more common for designers to anticipate typical service-load histories and design for adequate service lives, building in a reasonable allowance for occasional abusive loading. This can result in components without unreasonably high safety factors that are therefore lighter and less expensive. The methodology to be presented here is intended primarily for use in high-cycle fatigue situations ($N > 10^5$ cycles), although it can be useful in other situations so long as stresses can be accurately determined.

For fatigue design based on finite life, the sloping portion of the curve from $10^3 \le N \le 10^6$ in Fig. 5.18 must be known from testing or estimated. If data are available, the log-log linear portion of the curve can be characterized by a power law relation,

$$S_a = C'(N)^{b'} \tag{5.34}$$

where C' and b' are curve-fit parameters used to relate the stress amplitude S_a and number of cycles to failure N. In the absence of fatigue data, the following procedure can be used to estimate these parameters:

- Assume the fatigue limit occurs at a life of 10^6 cycles. [If no fatigue limit data are available, estimate S_e from Eq. (5.32) or (5.33).]
- Assume a stress amplitude of $0.9S_u$ corresponding to a life of 1000 cycles, S_{1000}.

This results in a curve as shown in Fig. 5.21. The coefficient and exponent in Eq. (5.34) are therefore given by

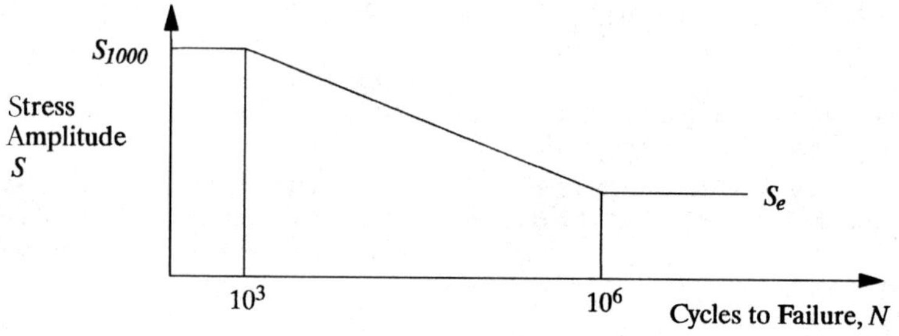

FIG. 5.21 Curve used to approximate *S-N* data.

$$C' = \frac{(S_{1000})^2}{S_e} \tag{5.35}$$

$$b' = -\tfrac{1}{3}\log\left(\frac{S_{1000}}{S_e}\right) \tag{5.36}$$

If S_{1000} is assumed to be $0.9S_u$ and S_e to be $0.5S_u$, then $C' = 1.62S_u$ and $b' = -0.0851$.

An equivalent way to express an *S-N* relation is through the use of the following axial fatigue parameters:

$$S = \sigma_f'(2N)^b \tag{5.37}$$

where σ_f' = fatigue strength coefficient
b = fatigue strength exponent

This relation is referred to in the literature as the Basquin relation, and its parameters will be discussed in Sec. 5.5.2.

Notice the factor of two that appears in Eq. (5.37). The quantity $2N$ is considered the number of *stress reversals* to failure, since there are two reversals for every cycle (see Fig. 5.22). This is a consequence of some early work on variable-amplitude loading that was taking place while the concept of a "fatigue strength coefficient and exponent" was being developed to characterize fatigue data. At the time, it was felt that considering stress reversals instead of cycles could expedite cumulative fatigue damage analysis. This later proved not to be the case, and consequently, the factor of two must now be accounted for somewhat meaninglessly. This situation is discussed further in Bannantine.[32]

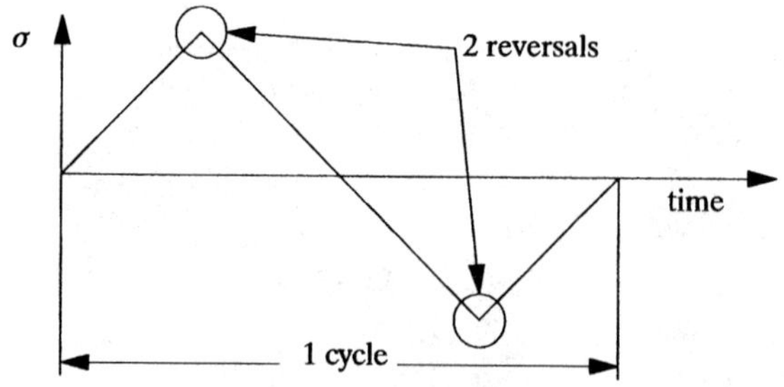

FIG. 5.22 Number of reversals = 2 (number of cycles).

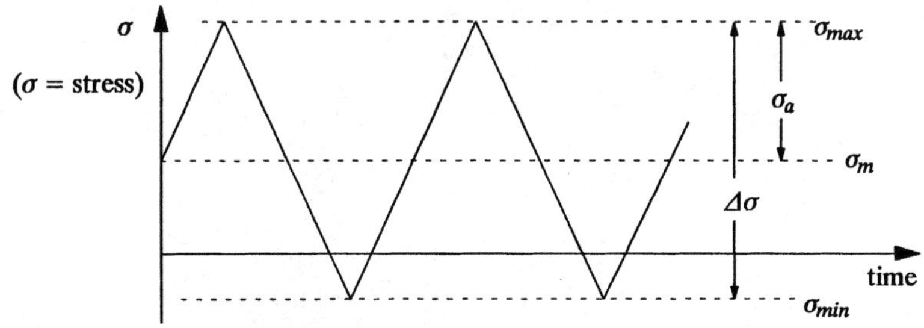

FIG. 5.23 Constant-amplitude loading with a mean stress.

The remainder of this section will be devoted to specifying how to use experimental or estimated baseline *S-N* data (from constant-amplitude, fully reversed specimen loading) on more complex uniaxial stress histories.

Mean Stress Effects. Baseline data are fully reversed ($R = -1$) but actual engineering components are often subjected to loading with *nonzero* mean stress as depicted in Fig. 5.23. From this figure, several parameters are defined, including the stress ratio,

$$R = \frac{\sigma_{\min}}{\sigma_{\max}} \tag{5.38}$$

stress range,

$$\Delta\sigma = \sigma_{\max} - \sigma_{\min} \tag{5.39}$$

stress amplitude,

$$\sigma_a = \frac{\sigma_{\max} - \sigma_{\min}}{2} \tag{5.40}$$

and mean stress,

$$\sigma_m = \frac{\sigma_{\max} + \sigma_{\min}}{2} \tag{5.41}$$

Mean stresses can act to shorten or lengthen fatigue life, depending on (1) whether the mean stress is positive or negative and (2) whether the loading is predominantly elastic or plastic. This is depicted schematically in Fig. 5.24.

Tensile mean stresses superimpose with applied loading to decrease fatigue life while compressive mean stress decreases the applied loading to increase fatigue life.

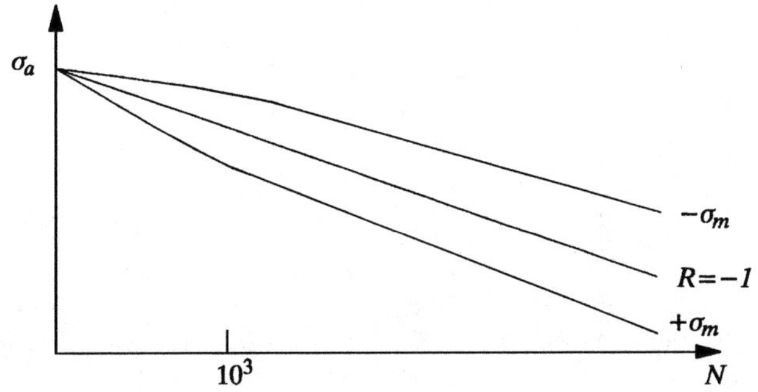

FIG. 5.24 Mean stress effect on *S-N* curve.

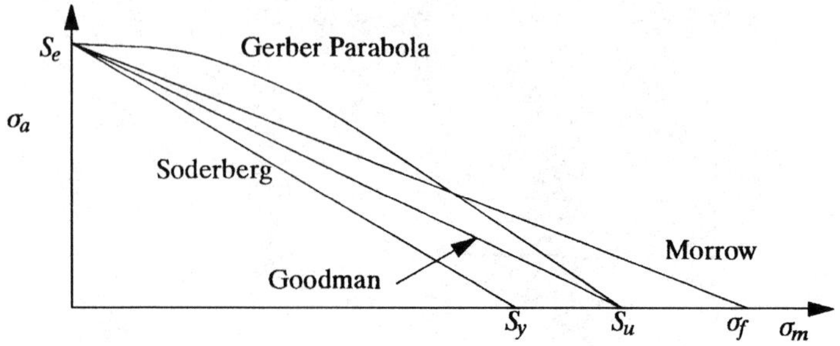

FIG. 5.25 Mean stress constant-life plots (endurance limit).

Mean stress relaxation can occur at higher load levels, diminishing the effect of mean stress at lower lives. This is particularly true when a component has a notch, as discussed later.

Mean stress data are often presented as plots of stress amplitude versus mean stress, corresponding to a particular life. For instance, Fig. 5.25 shows plots of several empirical relations to account for mean stress that have been suggested from testing at endurance-limit load levels. The curves represent combinations of mean stress and stress amplitude (σ_m and σ_a) that correspond to the fatigue limit S_e. Data sets have indeed been shown to lie in the vicinity of these lines and occasionally suggest that particular relations do a better job than others. However, in practice, none of these has been universally agreed upon as superior. In general, the Soderberg line has been determined to be too conservative for practical design use. The Goodman line and Gerber parabola are often more accurate than the Morrow relation. Another popular parameter was proposed by Smith, Watson, and Topper (SWT).[36] This relation is shown schematically with a Goodman line in Fig. 5.26.

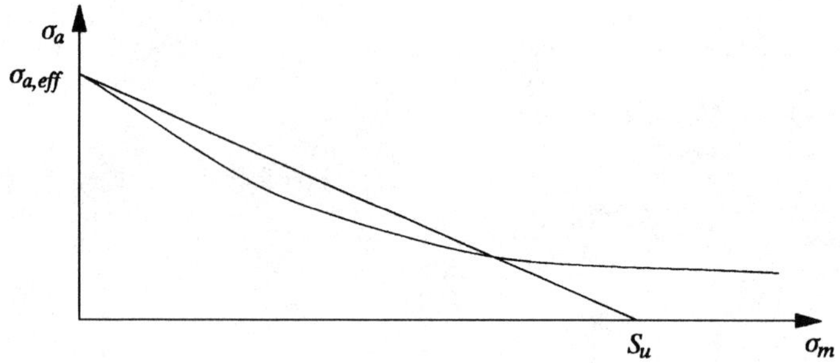

FIG. 5.26 Constant-life plots for Goodman and Smith-Watson-Topper relations.

The curves in Fig. 7.25, considered to describe the *fatigue limit,* can be extended to the finite-life regime by considering combinations of stress amplitude and mean stress that result in a particular life corresponding to a fully reversed test conducted at a stress amplitude of $\sigma_{a,\text{eff}}$. Schematically, this *effective, fully reversed stress amplitude* concept is depicted in Fig. 5.27. The effective stress amplitude, $\sigma_{a,\text{eff}}$, provides a conceptually straightforward approach to account for mean stress effect based on fully reversed baseline data. Relations for $\sigma_{a,\text{eff}}$ are given in Eqs. (5.42) to (5.46) for the criteria illustrated in Figs. 5.25 and 5.26:

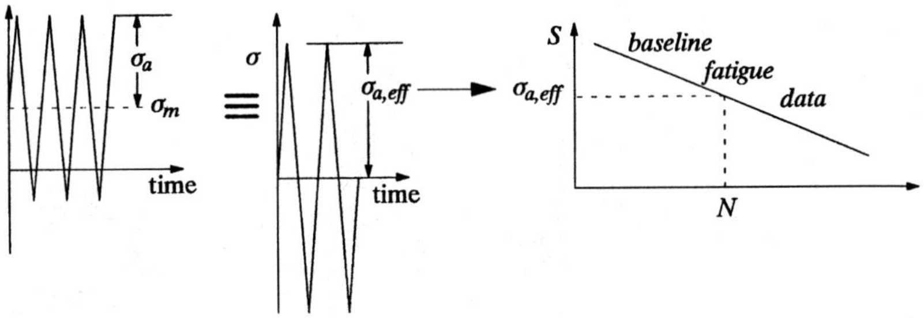

FIG. 5.27 Effective stress-amplitude concept.

Goodman:
$$\sigma_{a,\text{eff}} = \sigma_a\left(\frac{S_u}{S_u - \sigma_m}\right) \tag{5.42}$$

Soderberg:
$$\sigma_{a,\text{eff}} = \sigma_a\left(\frac{S_y}{S_y - \sigma_m}\right) \tag{5.43}$$

Morrow:
$$\sigma_{a,\text{eff}} = \sigma_a\left(\frac{\sigma_f}{\sigma_f - \sigma_m}\right) \tag{5.44}$$

Gerber:
$$\sigma_{a,\text{eff}} = \sigma_a\left(\frac{S_u^2}{S_u^2 - \sigma_m^2}\right) \tag{5.45}$$

Smith-Watson-Topper:
$$\sigma_{a,\text{eff}} = \sqrt{\sigma_a(\sigma_a + \sigma_m)} \tag{5.46}$$

Equations (5.42) to (5.45) are illustrated again in Fig. 5.28. These curves differ from those in Fig. 5.25. Each curve is based on the same input point, that is, the state of stress in the engineering component defined by σ_a and σ_m. But, each implies a different $\sigma_{a,\text{eff}}$ corresponding to the applied stress state. These curves make it apparent that Soderberg is the most conservative from a designer's perspective, since it specifies the highest effective stress, and Gerber is the least conservative.

One final note should be made before leaving mean stress effects. The relations illustrated so far have been discussed primarily in the context of positive mean stress.

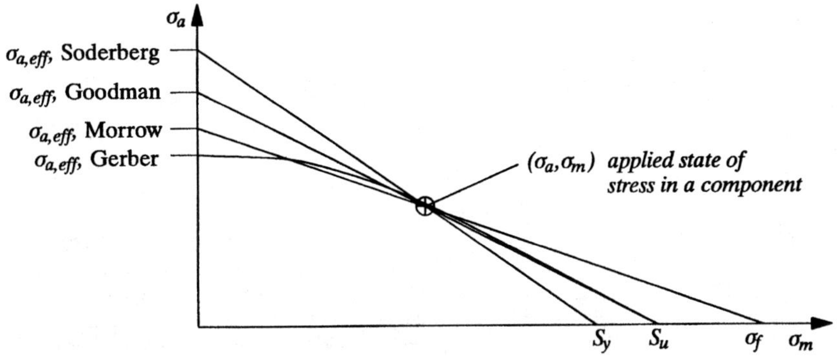

FIG. 5.28 Definition of $\sigma_{a,\text{eff}}$ from relations in Fig. 5.27.

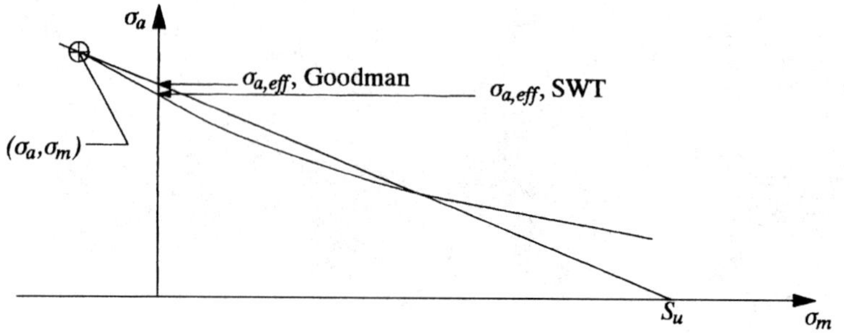

FIG. 5.29 Extension of Goodman and SWT relations into compressive mean stress regime.

The fact that tensile mean stresses have a deleterious effect on fatigue is modeled by the $\sigma_{a,\text{eff}}$ concept. (For example, increasing σ_m increases $\sigma_{a,\text{eff}}$ and decreases estimated life.) However, there are valuable data[37] that demonstrate the beneficial effect of compressive mean stress on fatigue. Therefore, a compressive mean stress should decrease $\sigma_{a,\text{eff}}$. (This is not the case for the Gerber parabola.) The Goodman and SWT relations have been shown to do a good job for small compressive stresses, as illustrated in Fig. 5.29. ("Small" is defined as having a magnitude of less than about $0.5S_y$. A more comprehensive treatment of compressive mean stress effects can be found in Ref. 31.)

The fact that compression can enhance fatigue life can be taken advantage of through material processes such as shot peening, proof loading, carburizing, nitriding, and induction hardening. All of these processes impose large compressive residual stresses at the surface of the material, reducing effective stress amplitudes and increasing fatigue life. (The latter three also considerably harden the surface layer.) Thread rolling and hole stretching are other processes that enhance fatigue resistance by inducing residual surface compression.

Notches. Figure 5.6 illustrated the concept of an elastic stress-concentration factor K_t defined as the maximum elastic stress at the notch root, divided by the nominal stress (based on net section area). Since the notch stresses increase according to K_t, it would be convenient, analytically, if fatigue strengths were reduced proportionally. However, the effect that a notch has on fatigue is dependent on

- Notch severity (magnitude of K_t)
- Material strength and ductility
- The applied nominal stress magnitude

Figure 5.30 illustrates how a notch can affect a set of fatigue data, relative to smooth-specimen data. Stress-concentration factor effects tend to diminish at lower lives since localized plastic flow can reduce the stress amplitude at the notch root, as shown in Fig. 5.31. At longer lives, K_t does a better job describing notch fatigue strength, but tends to overestimate the effect. Several factors can explain the reduced effect of K_t on fatigue. These include (1) the fact that localized stresses are reduced by yielding, (2) the effect of subsurface stress gradient (microcracks growing into a decreasing stress field), and (3) the fact that only a small volume of material experiences the extreme localized concentrated stresses. From a design point of view, using the full value of a stress-concentration factor to compute notch stresses ($\sigma_{\text{notch}} = K_t S_{\text{nom}}$) is a very safe way to operate, since notch effects are overestimated.

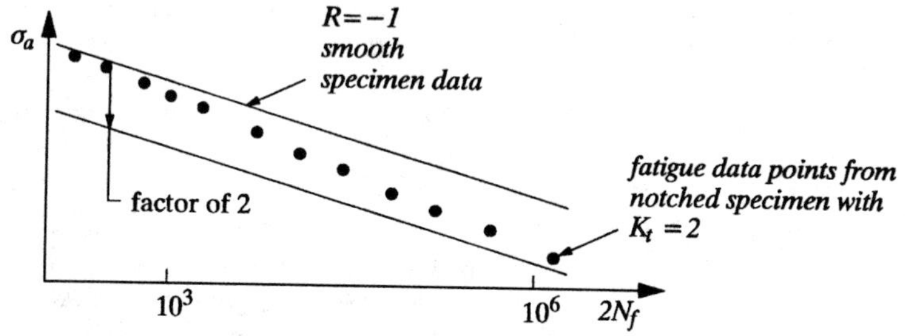

FIG. 5.30 Notch effect on *S-N* behavior.

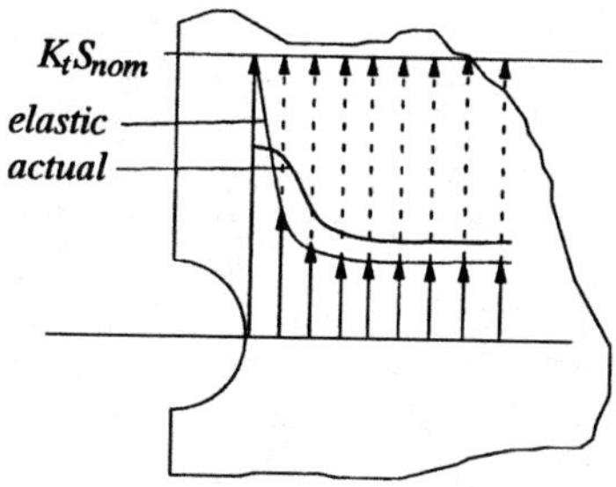

FIG. 5.31 Illustration explaining how elastically calculated K_t can overestimate the effect of a notch on fatigue behavior. Localized yielding and subsurface gradients are apparent.

Fatigue Notch Factors. Recognizing that K_t overestimates fatigue-strength reduction, the concept of an empirical fatigue notch factor (also called a fatigue-strength reduction factor) was developed. The fatigue notch factor K_f is defined from a comparison of fatigue data generated with smooth and notched specimens, as shown in Fig. 5.32.

Unfortunately, the use of fatigue notch factors in design is not straightforward. In many instances, K_f has been shown to vary with life, as is apparent in Fig. 5.32. Furthermore, it can only be reliably determined empirically (by experiment) for the material, geometry, and surface processing of interest.

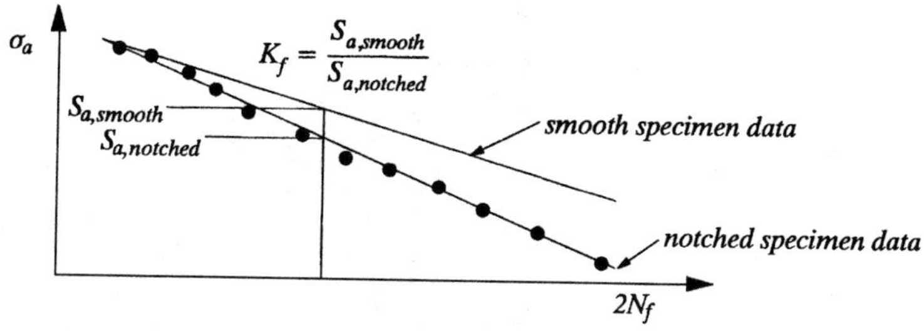

FIG. 5.32 Definition of fatigue notch factor.

To quantify the fatigue-strength reduction associated with a notch, a notch-sensitivity factor was developed as defined in Eq. (5.47):

$$q = \frac{K_f - 1}{K_t - 1} \tag{5.47}$$

where q varies from 0 to 1: $q = 0$ no notch effect ($K_f = 1$)

$q = 1$ full elastic effect ($K_f = K_t$)

Some researchers have attempted to formulate empirical relations for K_f, based on K_t, for fatigue-limit load levels. One approach, proposed by Peterson, is given by

$$K_f = 1 + \frac{K_t - 1}{1 + (a/r)} \tag{5.48}$$

where r = the notch root radius
a = a "characteristic length" (empirical curve-fit parameter)

$$\alpha \cong \left(\frac{300}{S_{ut}\,(\text{ksi})}\right)^{1.8} \times 10^{-3}\ \text{in} \tag{5.49}$$

For steels, a rule of thumb assessment of the parameter α is often cited:

Annealed steel	Quenched and tempered	Highly hardened
$\alpha \approx 0.010$ in	$\alpha \approx 0.0025$ in	$\alpha \approx 0.001$ in

Consistent with this is the general assessment that harder materials are more notch sensitive than softer materials. Example 5 uses Eq. (5.48) to illustrate this point. It should be remembered that no such empirical relations have been proposed for aluminum or other nonferrous materials.

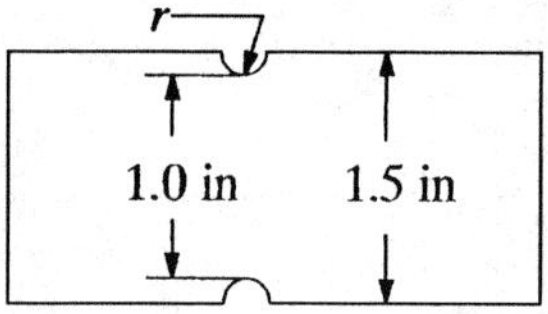

EXAMPLE 5 This example illustrates the effects of tensile strength and notch severity on the estimated values of the fatigue notch factor. Use Eqs. (5.48) and (5.49) to compute K_f for the two different steels and three different notch root radii.

solution The values of K_f are tabulated below for two steels and three values of r:

Material A	Material B
$S_u = 68$ ksi $\alpha = 0.015$ in	$S_u = 180$ ksi $\alpha = 0.0025$ in

		Material A	Material B
r (in)	K_t	$K_f (S_u = 68)$	$K_f (S_u = 180)$
0.2	2.05	1.98	2.03
0.05	3.5	2.92	3.38
0.01	6.0	3.0	5.0

The K_f relations and the example shown above are valid only for fully reversed loading ($R = -1$). Mean stresses can affect notched components differently than smooth ones. To accurately analyze a particular situation, empirical data are usually necessary.

A great deal of data have been generated in the aerospace industry and are published in the form of plots[38,39] such as the one shown in Fig. 5.33. These plots provide direct information on the combined effects of the mean stress and the stress-concentration factor.

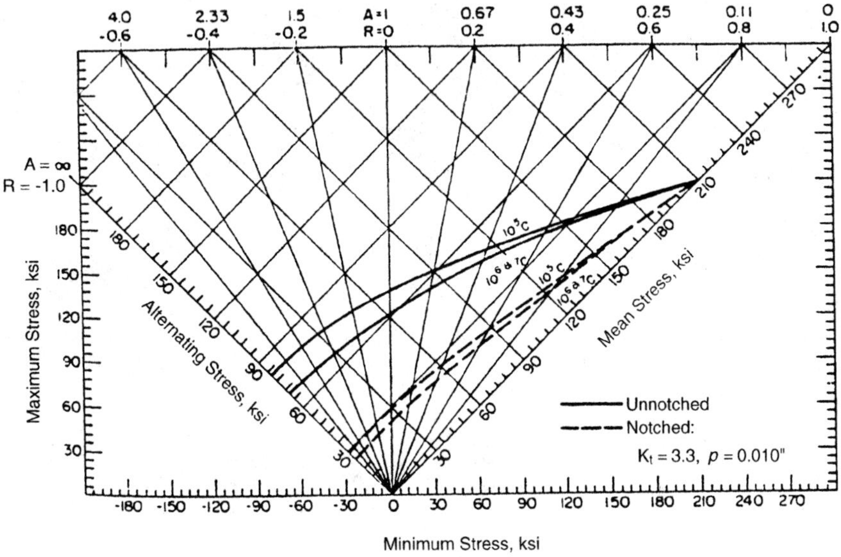

FIG. 5.33 Constant-life plots from MIL Handbook 5 (AISI 4340); $S_u = 208$ ksi.

A final important trend is noted by those who have studied fatigue notch factors: there appears to be an upper limit to K_f of about 5 or 6 for very sharp notches.[32] Two possible explanations for this are: (1) the notch tip blunts, reducing K_f, or (2) the notch constitutes a crack and removes the initiation life of the component.

A safe, recommended approach suitable for design is outlined in Table 5.2 for uniaxial loading situations. If more detailed data are available, they can be incorporated into the approach as outlined below:

- Use a measured rather than estimated K_f over the entire range of life.
- Estimate the variation of K_f with life experimentally, or from a source such as that found in "MIL Handbook 5," Fig. 5.33.
- If estimates are unduly conservative, use only the nominal mean stresses ($\sigma_{m,\text{notch}} = S_{m,\text{ax}} + S_{m,\text{bend}}$) to compute the notch mean stress.
- Use only nominal stress amplitudes, and modify baseline *S-N* data using approaches detailed in Refs. 13 and 32.

5.5.2 Strain-Life Approaches (Constant-Amplitude Loading)

Cyclic Stress-Strain Relation. A standard, low-cycle fatigue specimen is fabricated and tested according to ASTM E-606, "Standard Recommended Practice for Constant-Amplitude Low-Cycle Fatigue Testing,"[40] in strain control. A typical specimen is depicted in Fig. 5.34, along with a stabilized cyclic stress-strain loop.

When fatigue testing is conducted, several specimens (ideally, at least 20) are tested at varying strain amplitudes. At each strain amplitude, a different stabilized loop forms, as depicted in Fig. 5.35. From these loops, the cyclic stress-strain curve may be

TABLE 5.2 Elastic Uniaxial Stress-Life Design Approach

(1) Obtain or otherwise estimate fully−reversed S−N data for material of interest using

$$S_a = C'(N)^{b'} \quad \text{OR} \quad S_a = \sigma_f'(2N_f)^b$$

(2) Use K_f to estimate stresses at notch from in−phase bending and axial loading.

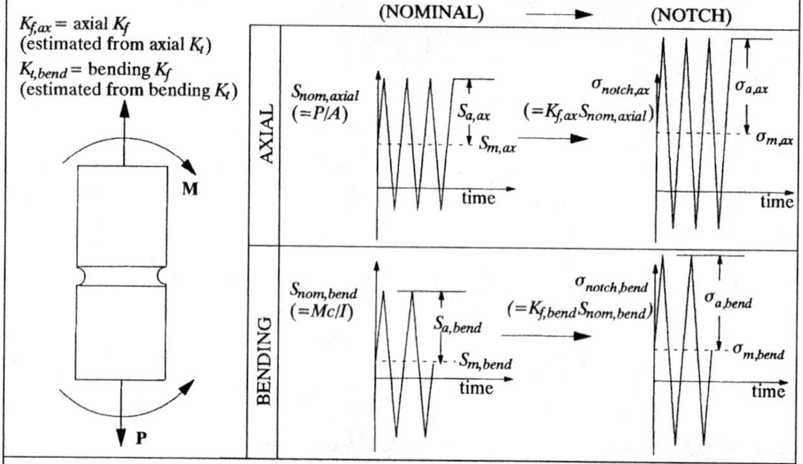

(3) Combine axial and bending notch stresses into uniaxial notch stress history.

NOTCH STRESS AMPLITUDE: $\sigma_{a,notch} = \sigma_{a,ax} + \sigma_{a,bend}$

NOTCH MEAN STRESS: $\sigma_{m,notch} = \sigma_{m,ax} + \sigma_{m,bend}$

(4) Compute ***effective fully−reversed notch stress amplitude***, using appropriate mean stress relation.

(Goodman) (SWT)

$$\sigma_{a,eff} = \sigma_{a,notch}\left(\frac{S_u}{S_u - \sigma_{m,notch}}\right) \quad \text{OR} \quad \sigma_{a,eff} = \sqrt{\sigma_{a,notch}(\sigma_{a,notch} + \sigma_{m,notch})}$$

(5) Use effective notch stress amplitude to enter baseline $S-N$ data to predict life

$$N_f = \left(\frac{\sigma_{a,eff}}{C'}\right)^{1/b'} \quad \text{OR} \quad N_f = \frac{1}{2}\left(\frac{\sigma_{a,eff}}{\sigma_f'}\right)^{1/b}$$

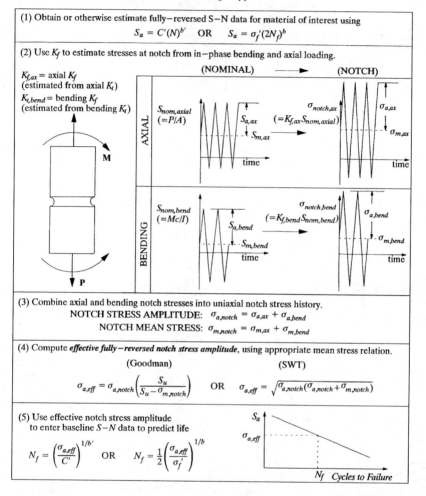

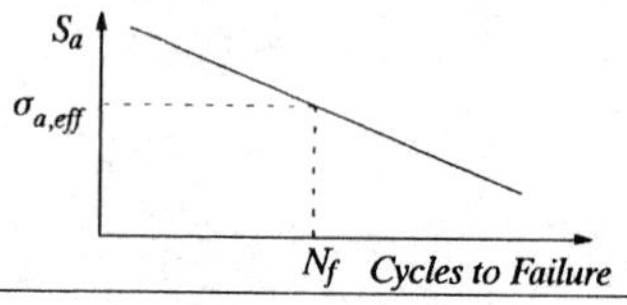

defined from the locus of the tips of the stabilized hysteresis loops, and expressed using a Ramberg-Osgood[28] relation:

$$\varepsilon_a = \frac{\sigma_a}{E} + \left(\frac{\sigma_a}{K'}\right)^{1/n'} \tag{5.50}$$

where K' = cyclic strength coefficient
n' = cyclic strain-hardening exponent

In this relation, ε_a and σ_a represent strain and stress *amplitudes*, respectively. The curve therefore represents a relation between stress and elastic-plastic strain amplitudes that form during fully reversed strain-controlled testing.

The formation of the stabilized hysteresis loops depicted in Figs. 5.34 and 5.35 usually requires a substantial number of cycles, during which transient softening or hardening may occur. Such behavior is depicted in Fig. 5.36. This can cause the cyclic stress-strain relation to lie below or above the monotonic curve. If the cyclic curve is

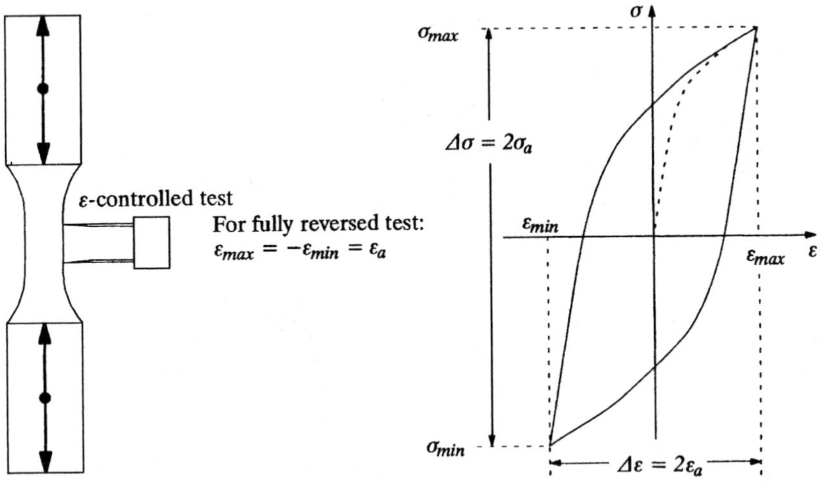

FIG. 5.34 Stabilized stress-strain hysteresis loop from typical ASTM E-606 fatigue test.

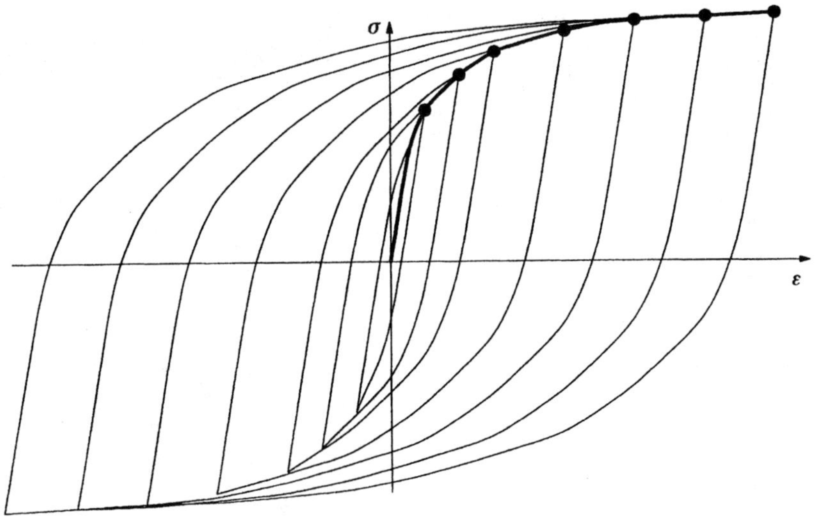

FIG. 5.35 Cyclic stress-strain curve from stable hysteresis loops.

below the monotonic curve, the material can be called a "cyclic softening material." If the cyclic curve is above the monotonic curve, the material "cyclically hardens." Mixed behavior is also observed, depending on the strain amplitude. Examples of each situation are shown in Fig. 5.37.

The transient stress behavior during a typical strain-controlled test is depicted differently in Fig. 5.38 for a cyclically softening material. This figure is a plot of peak and valley stress components at each reversal point throughout the life of the material. There are several noteworthy features to this plot. First, cyclic stabilization is shown to occur within about 10–20 percent of the total life. This depends on the material

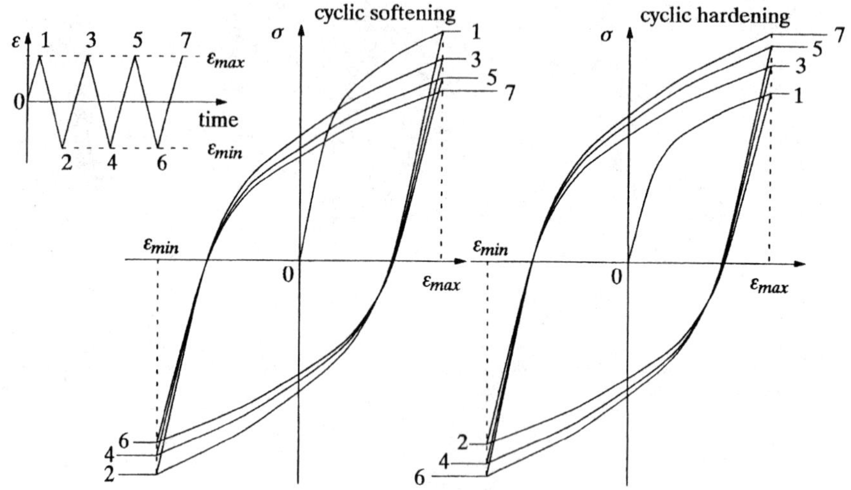

FIG. 5.36 Transient softening and hardening occurring on the first few cycles.

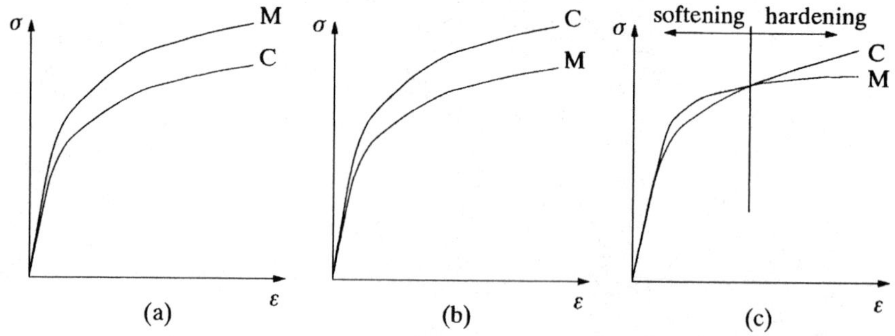

FIG. 5.37 Monotonic (*M*) and cyclic (*C*) stress-strain curves for (*a*) cyclic softening material, (*b*) cyclic hardening material, and (*c*) mixed transient behavior.

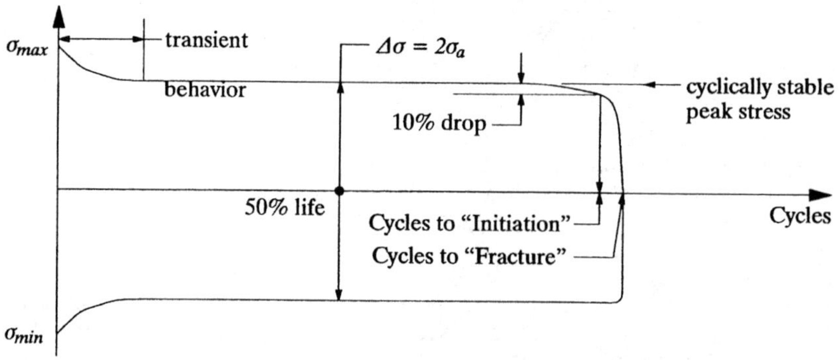

FIG. 5.38 Typical peak and valley stresses versus cycle for a strain-controlled test.

being tested, and the strain amplitude. At larger applied strains (lower lives), stabilization can be less pronounced (that is, the maximum stress can change gradually over the entire test). For this reason, the stabilized stress amplitude is usually defined as the stress amplitude at the "half-life" of the specimen (at near 50 percent of its total life). Near the end of the life of the specimen, notice in Fig. 5.38 how the peak tensile stress drops just prior to final fracture. This results from the decrease in specimen stiffness associated with crack formation. Therefore, this drop in the maximum stress is typically used to define the "crack initiation life" of the specimen, as opposed to the total number of cycles to fracture. (Notice how the compressive valley stress is maintained throughout the test, since the crack faces can sustain the compressive loading.) Some testing laboratories use a peak load drop of 10 percent from the half-life value to define initiation, others use a larger value, such as 50 percent, while others simply use the life to fracture. The discrepancy this causes is usually considered negligible, since the life of a specimen after a discernable crack ("engineering-sized," on the order of 1 to 2 mm in surface length) has formed is generally a small percentage of the life to fracture. However, the subjectivity associated with reducing low-cycle fatigue data is apparent, especially in the low-cycle regime. It is advised that stress-versus-time data be obtained and reviewed by the engineer when low-cycle fatigue testing is conducted.

The definition of the cyclic stress-strain curve requires the testing of several specimens. This is referred to as "companion specimen" testing. Attempts have been made to define the curve from a single test called the "incremental step test."[41] It should be noted that this technique can only approximate the curve and not enough data exist to assess its general reliability.

Refer again to Fig. 5.35, and recall that the dark cyclic stress-strain curve [Eq. (5.50)] is defined by the tips of the hysteresis loops. The light curves (referred to as hysteresis curves) can be approximated well by scaling the dark curve, geometrically, by a factor of two. This is referred to as Massing's hypothesis.[42] To demonstrate this, refer to Fig. 5.39.

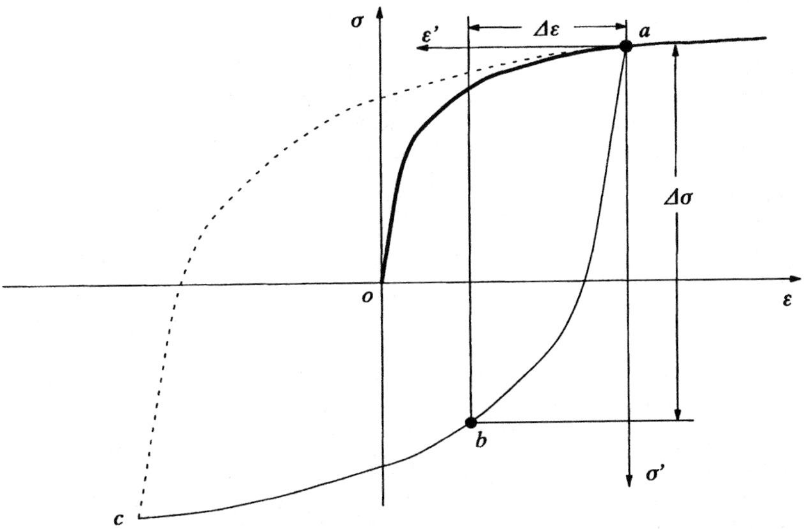

FIG. 5.39 Demonstration of Massing's hypothesis: Solid curve *a-b-c* is equal to dark curve *o-a*, geometrically doubled by a factor of two. The dashed curve *c-a* is obtained by rotating the solid curve, *a-b-c*, by 180°.

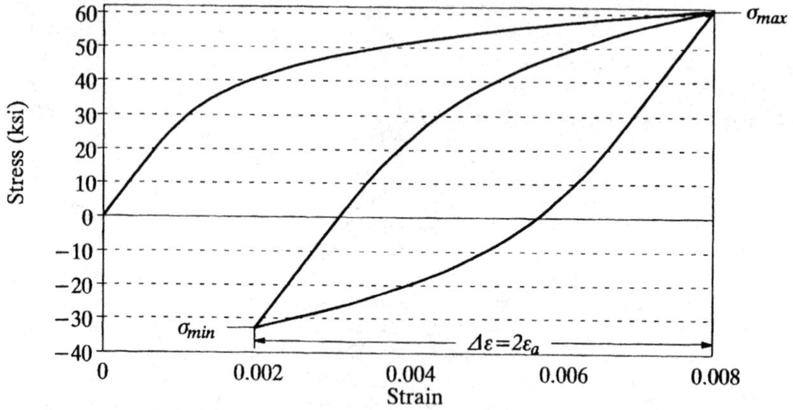

FIG. 5.40 Cyclically stable hysteresis loop computed for Example 6.

In Fig. 5.39, the dark curve from *o* to *a* is the cyclic stress-strain curve. The light curve from *a* to *b* to *c* is the reverse-loading hysteresis curve. The expression for the hysteresis curve on σ′-ε′ axes is given by

$$\Delta\varepsilon = \frac{\Delta\sigma}{E} + 2\left(\frac{\Delta\sigma}{2K'}\right)^{1/n'} \tag{5.51}$$

Notice that Eq. (5.51) is given in terms of stress and strain *ranges* (denoted by "Δ"). This is illustrated for point *b* along the curve *a*-*b*-*c* in Fig. 5.40. For fully reversed loading, the reversal at *c* would be followed by a stress-strain path along the dashed line from *c* back to *a*.

This is important, since it can be used to reveal the location (on σ-ε axes) where a stable hysteresis loop will form during constant-amplitude strain-controlled loading that is *not* fully reversed. The ability to estimate the form of the hysteresis curve provides a mechanism to estimate the path-dependent plasticity behavior of the material under axial loading. The use of this approach is demonstrated in Example 6 for constant-amplitude loading. Life prediction for this example will be discussed later (as will the use of this approach with variable-amplitude loading).

EXAMPLE 6 Consider an axial specimen with the following properties:

$$E = 30{,}000 \text{ ksi}$$

$$K' = 156.88 \text{ ksi}$$

$$n' = 0.184$$

The specimen is to be subjected to strain-controlled cyclic loading between maximum and minimum values of 0.008 and 0.002. Compute the corresponding maximum and minimum stress and plot the cyclically stable hysteresis loop.

solution Using Eq. (5.50), the stress amplitude is computed that would correspond to a strain amplitude of 0.008, if the loading were fully reversed. By trial and error, this value is found to be 61.13 ksi:

$$0.008 = \frac{61.13 \text{ ksi}}{E} + \left(\frac{61.13 \text{ ksi}}{K'}\right)^{1/n'}$$

Now, the stress range $\Delta\sigma$ corresponding to a strain range, $\Delta\varepsilon$, of 0.006 is computed. This corresponds to $\varepsilon_{max} - \varepsilon_{min} = 0.008 - 0.002 = 0.006$. Equation (5.51) is used to define a value of $\Delta\sigma = 94.05$ ksi:

$$0.006 = \frac{94.05 \text{ ksi}}{E} + 2\left(\frac{94.05 \text{ ksi}}{2K'}\right)^{1/n'}$$

From these values, the stress is estimated to fluctuate from a maximum of 61.13 ksi to a minimum of (61.13 − 94.05 =) −32.92 ksi. The corresponding stable hysteresis loop is shown in Fig. 5.40.

Strain-Life Relation. As discussed in the preceding section, strain-controlled companion specimen fatigue testing per ASTM E-606[40] results in strain-amplitude versus cycles-to-failure data (defined as complete specimen fracture or the formation of detectable cracks). As shown in Fig. 5.41, the cyclically stable total strain amplitude can be divided into elastic and plastic components. This can be expressed as

$$\varepsilon_a = \varepsilon_a^e + \varepsilon_a^p \tag{5.52}$$

where the superscripts e and p represent elastic and plastic components, respectively. In 1910 Basquin[43] is credited with the observation that log-log plots of stress amplitude (and, therefore, elastic strain amplitude) versus life data behaved linearly. Manson[44] and Coffin,[45] working independently, later observed that log-log plots of plastic strain versus life were also linear. These two observations were combined into the now familiar form

$$\varepsilon_a = \frac{\sigma_f'}{E}(2N_f)^b + \varepsilon_f'(2N_f)^c \tag{5.53}$$

$$\varepsilon_a = \quad \varepsilon_a^e \quad + \quad \varepsilon_a^p$$

where σ_f' = fatigue strength coefficient
b = fatigue strength exponent

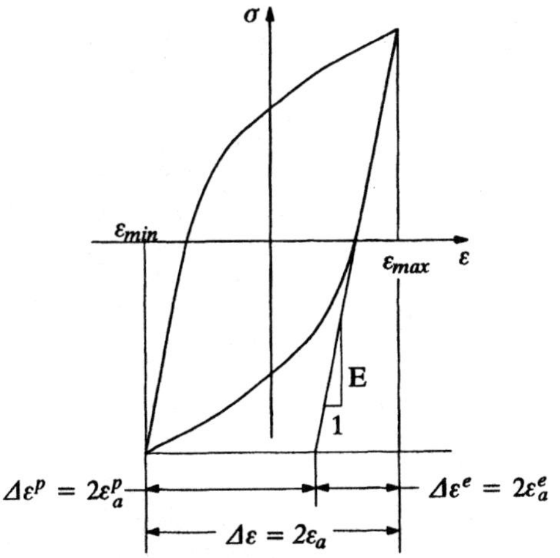

FIG. 5.41 Definition of elastic and plastic strain amplitude from total strain amplitude.

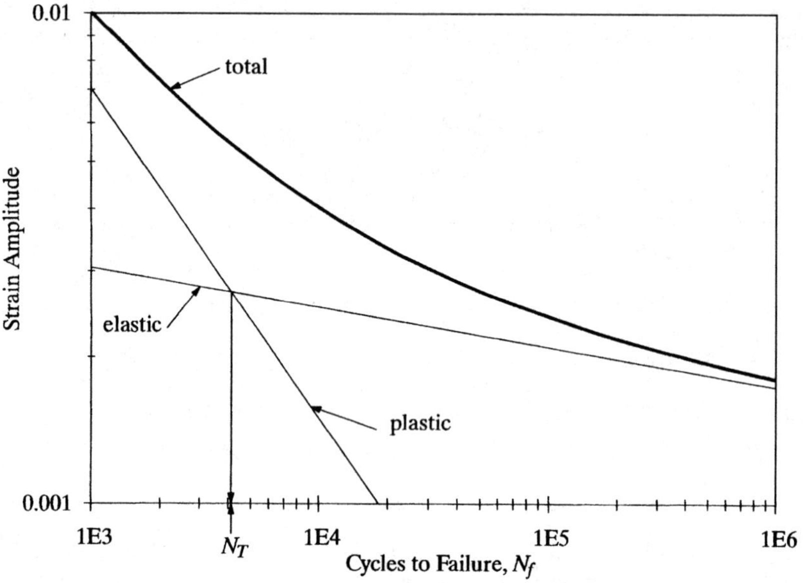

FIG. 5.42 Strain-life relation for a medium-strength steel.[49]

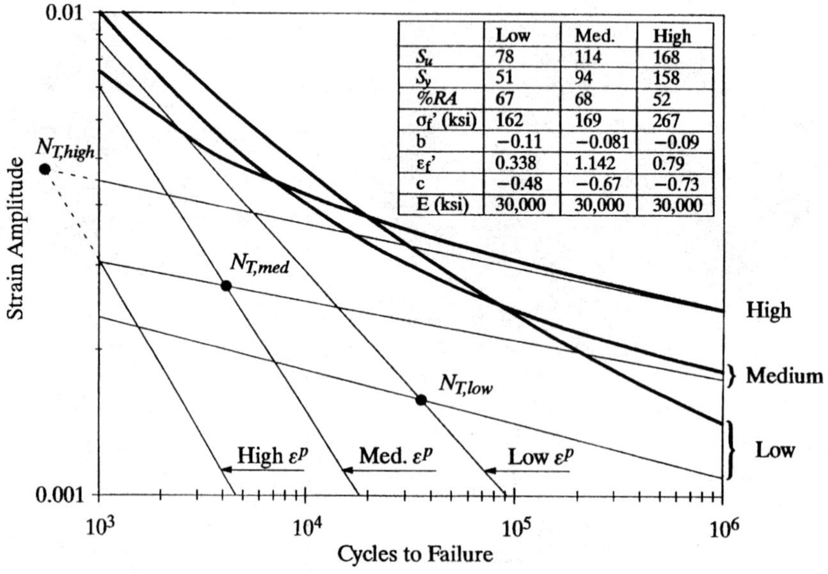

	Low	Med.	High
S_u	78	114	168
S_y	51	94	158
%RA	67	68	52
σ_f' (ksi)	162	169	267
b	−0.11	−0.081	−0.09
ε_f'	0.338	1.142	0.79
c	−0.48	−0.67	−0.73
E (ksi)	30,000	30,000	30,000

FIG. 5.43 Strain-life relation for a low-strength,[50] medium-strength,[49] and high-strength[51] steel.

ε'_f = fatigue ductility coefficient
c = fatigue ductility exponent
E = modulus of elasticity
N_f = number of cycles to failure

Equation (5.53) serves as the foundation of local strain-based fatigue analysis.[13,23,31,32,46–48] It is plotted in Fig. 5.42 for a medium-strength steel.[49] The "transition life," N_t, is also depicted in Fig. 5.42, as the life at which the elastic and plastic strain amplitudes are equal. The transition life can provide an indication of whether straining over a life regime of interest is more elastic or plastic. In general, the higher the tensile strength of the materials, the lower the transition life, and elastic strains tend to dominate for a greater portion of the overall life regime. To demonstrate this, Fig. 5.43 shows the same strain-life curve from Fig. 5.42, replotted with curves for a low-strength steel[50] and a high-strength steel.[51]

Inspection of Fig. 5.43 is interesting from the standpoint of selecting a material for maximum fatigue resistance. For example, in the higher-cycle life regime ($>10^5$ cycles) the higher-strength material provides the greatest fatigue strength. But for a component that must operate in the lower life regime ($<10^4$ cycles) the selection of the lowest-strength steel might be warranted. The medium-strength material could represent a more suitable selection if a combination of high strength and fatigue resistance is required.

As a consequence of Eqs. (5.52) and (5.53), and the Ramberg-Osgood formulation of the cyclic stress-strain curve [Eq. (5.50)], it is apparent that the cyclic strength coefficient and cyclic strain-hardening exponent may be expressed in terms of the fatigue strength and ductility coefficients and exponents as follows:

$$K' = \frac{\sigma'_f}{(\varepsilon'_f)^{n'}} \tag{5.54}$$

$$n' = \frac{b}{c} \tag{5.55}$$

However, experience has shown that parameters formed this way can result in a stress-strain curve that *does not* correlate well with actual stress versus strain-amplitude data points. For this reason, use of Eqs. (5.54) and (5.55) is not recommended if K' and n' may be fit directly to data.

Values for cyclic stress-strain and fatigue parameters can be found in Refs. 52 through 54. Unfortunately, tabulated data are available for only a fraction of available engineering alloys and heat treatments. Although such data are becoming more and more available, situations routinely arise where a designer must estimate fatigue properties without the availability of fatigue testing data.

Several attempts have been made to correlate fatigue parameters with monotonic tensile properties,[46,55–57] but no clearly superior approach has been identified. Table 5.3 summarizes several approaches.

These relations should be used *only with a great deal of caution.* The universal slopes[55] approach, the Socie et al.[57] approach, and a more complicated four-point correlation approach were compared to data in Ref. 58. The four-point correlation method provided the best approximation but requires the true fracture stress, a quantity not generally reported during tensile testing. The potential for any of the approaches to provide poor life estimates is great. Considering the differences in deformation and failure mechanisms (on a microscopic level) between monotonic and cyclic loading, limited correlation is not surprising.

Some final notes about the limitations of strain-life data are in order. Equation (5.53) is based on data generated with polished specimens. (ASTM E-606 recommends

TABLE 5.3 Approximate Relations between Tensile Properties and Fatigue Parameters

	Universal slopes[55]	Modified universal slopes[56]	Socie et al.[57]
σ_f'	$1.9018\, S_u$	$0.6227E\left[\frac{S_u}{E}\right]^{0.832}$	$S_u + 345$ MPa
b	-0.12	-0.09	$-\frac{1}{6}\log\left(\frac{2(S_u + 345\text{ MPa})}{S_u}\right)$
ε_f'	$0.7579\left[\ln\left(\frac{1}{1-\text{RA}}\right)\right]^{0.6}$	$0.01961\left[\ln\left(\frac{1}{1-\text{RA}}\right)\right]^{0.155}\left[\frac{S_u}{E}\right]^{-0.53}$	$\ln\left(\frac{1}{1-\text{RA}}\right)$
c	-0.6	-0.56	-0.6

that specimens be ground and longitudinally lapped to a surface finish of 8μin or better, with no circumferential grinding marks observable at 20 × magnification.) In the higher life regime, strain-controlled fatigue data are affected by the same factors that can affect stress-life data (surface finish, inclusions, defects, size, etc.). Although it is recognized in the literature that adjustments are necessary to account for surface effects, no accepted methodology to accomplish this exists. It is suggested[23] that the elastic portion of Eq. (5.53) can be modified, by adjusting the fatigue strength exponent *b*. However, this often affects the total strain-life curve into the lower-cycle regime as well, as is apparent in Fig. 5.44.

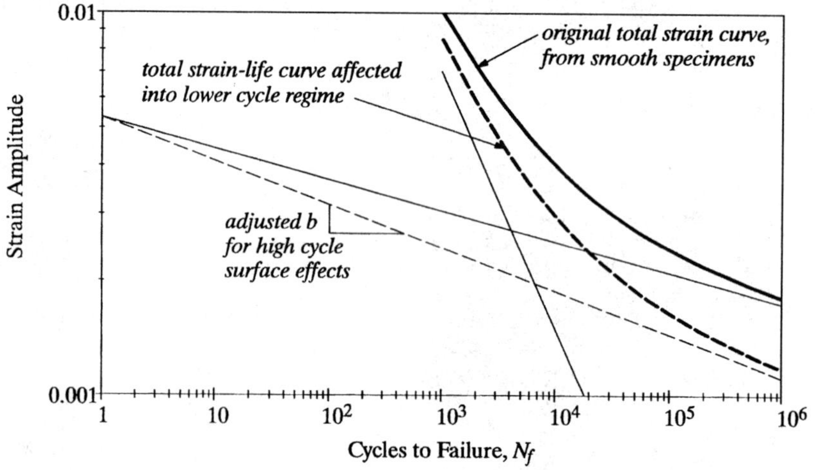

FIG. 5.44 Modifying the strain-life relation by adjusting b to account for surface roughness effects can influence the curve over the entire life regime.

Along these same lines, the strain-life equation itself has no mechanism to incorporate a fatigue limit. When lives in the vicinity of the endurance limit are being considered ($\geq 10^6$–10^7 cycles), the use of stress-based approaches should be considered.

Finally, it can be useful to obtain as much information about a particular data set as possible. Recall that "failure" in strain-controlled tests can be defined as complete fracture, or a drop in the cyclically stable maximum stress (typically 10 to 50 percent). This can be significant, especially if there is a substantial difference between the size of the component being analyzed and the size of the specimens used to generate the data. Published data sets are often based on a very small number of data, generated over a fairly narrow range. There is scatter associated with all fatigue data, and generating more data could change associated strain-life curves drastically. Furthermore, additional uncertainty is associated with utilizing the strain-life relation outside the range over which data were collected. Knowing the range of experimental lives could be very worthwhile toward assessing the reliability of a fatigue life estimate.

Mean Stress Effects. The standard strain-life relation, Eq. (5.53), is formulated from fully reversed ($R = -1$) strain-controlled test data. However, engineering components are often subjected to loading that induces mean stresses (strains). Although the effect of mean stress on fatigue has been discussed, mean strain, in general, has little effect. For example, consider the stress-strain response during a mean strain axial fatigue test as depicted in Fig. 5.45. The mean stress in the specimen is shown to relax to a cyclically stable response by the seventh reversal. (This response is somewhat exaggerated, as many more cycles are usually required.) This should not be confused with cyclic softening, since cyclically stable materials can exhibit relaxation.

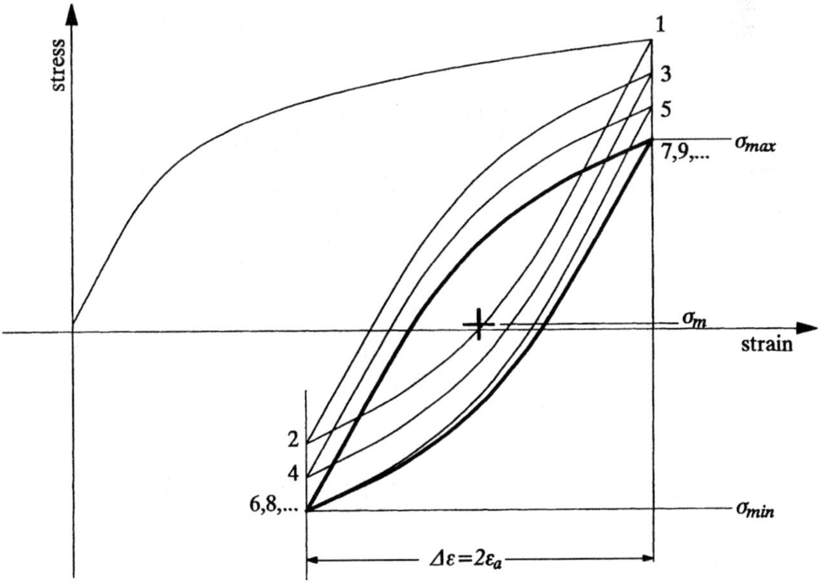

FIG. 5.45 Stress relaxation during a mean strain axial fatigue test.

Mean stress relaxation is the main reason that mean stresses tend to have more of an effect on high-cycle fatigue than on low-cycle fatigue. For high-cycle situations, loading is typically elastic and mean stresses relax little (if any).

Several approaches have been proposed to account for mean stresses with strain-based analysis.[13,31,32] The most effective of these is the Smith-Watson-Topper[36] relation, already discussed in terms of stresses. This approach is based on empirical observations that the product of the stabilized maximum stress and the strain range (or amplitude) during a cycle is proportional to life [Eq. (5.56)]. From fully reversed data, the maximum stress can be approximated in terms of life as Eq. (5.57).

$$\sigma_{max}\varepsilon_a \propto N_f \tag{5.56}$$

$$\sigma_{max} = \sigma_f'(2N_f)^b \tag{5.57}$$

Equations (5.53), (5.56), and (5.57) are combined to form the local-strain-based expression for the SWT relation,

$$\sigma_{max}\varepsilon_a = \frac{(\sigma_f')^2}{E}(2N_f)^{2b} + \sigma_f'\varepsilon_f'(2N_f)^{b+c} \tag{5.58}$$

It should be kept in mind that this expression is strictly empirically based. For extreme cases (e.g., high mean stress) outside the range of data for which the expression was established, the validity of this approach has not been assessed. For purposes of bounding a life estimate, an approximation may be made based on the strain amplitude only, neglecting the mean stress effect [i.e., using Eq. (5.53)]. This is demonstrated in Example 7.

EXAMPLE 7 Compute the life corresponding to the stable hysteresis loop from Example 6 (Fig. 5.40). The low-cycle fatigue parameters for the specimen are given as

$$\sigma_f' = 110 \text{ ksi} \qquad b = -0.105$$
$$\varepsilon_f' = 0.55 \qquad c = -0.625$$

solution The cyclically stable hysteresis loop fluctuated between maximum and minimum strain values of 0.008 and 0.002, with stresses ranging between $\sigma_{max} = 61.13$ ksi and $\sigma_{min} = -32.92$ ksi.

For the strain amplitude of $\varepsilon_a = 0.003$ [= (0.008−0.002)/2], and maximum stress of $\sigma_{max} = 61.13$ ksi, Eq. (5.58) appears as

$$(61.13)(0.003) = \frac{(\sigma_f')^2}{E}(2N_f)^{2b} + \sigma_f'\varepsilon_f'(2N_f)^{b+c}$$

This equation is solved iteratively for a life of $N_f = 2618$ cycles. If Eq. (5.53) is solved with no mean stress effect, an upper-bound life estimate of $N_f = 5590$ results.

Notches. Many mechanical components operate with elastically calculated stresses that exceed the yield strength of the material at a notch. However, once yielding occurs the elastic stress concentration factor K_t cannot be expected to accurately predict localized stresses. Since the material has yielded, local stress values are less than the elastically computed quantities. Similarly, strains computed elastically are exceeded in notches. This situation is depicted in Fig. 5.46. Equations (5.59) and (5.60) define notch stress- and strain-concentration factors, where σ_n and ε_n are the actual notch stress and strain, respectively, and S and e are elastically calculated nominal notch stress and strain, respectively.

$$K_\sigma = \frac{\sigma_n}{S} \tag{5.59}$$

$$K_\varepsilon = \frac{\varepsilon_n}{e} \tag{5.60}$$

To conduct a strain-based fatigue life estimate of a notch, the state of stress and strain *at the notch root* must be estimated. To achieve this, three approaches could be considered:

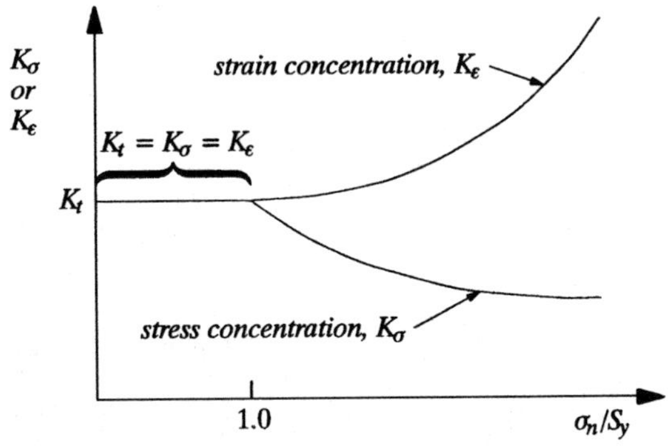

FIG. 5.46 Stress and strain concentration at a notch.

1. Place a strain gauge in the notch root.
2. Conduct elastic-plastic finite-element analysis.
3. Estimate notch root stress-strain from nominal, elastically calculated values.

The first two approaches sound straightforward enough, but both are extremely time consuming and complex. Notch geometries do not usually facilitate the placement of a strain gauge. Furthermore, a part must exist. This is not usually the case early in the design process. The application of finite-element analysis is getting easier for elastic situations, but is still a formidable task for elastic-plastic situations, especially under cyclic loading.[22] Mesh density requirements are great and codes are simply not set up to easily accept cyclic input loading.

For this reason, item 3 in the above list is emphasized here. Specifically, the application of Neuber's rule[59] is discussed. Neuber originally noticed that the geometric mean of the stress- and strain-concentration factors remains approximately constant as plasticity occurs in a notch. This means that the product of notch root stress and strain can be determined from elastically calculated quantities. From this observation, Neuber's rule is stated as

$$K_t^2 Se = \sigma_n \varepsilon_n \tag{5.61}$$

If nominal stress is elastic, the left-hand side of the equation above can be restated as

$$\frac{(K_t S)^2}{E} = \sigma_n \varepsilon_n \tag{5.62}$$

Notice that the left-hand side of this equation represents the applied loading (S) and notch geometry (K_t). The right-hand side is simply the product of the actual stress and strain in the notch root. If the left side is considered as known input, one more relation is needed to solve for the two unknowns. This relation is the cyclic stress-strain curve:

$$\varepsilon_n = \frac{\sigma_n}{E} + \left(\frac{\sigma_n}{K'}\right)^{1/n'} \tag{5.63}$$

Figure 5.47 illustrates the approach for elastic nominal stress-strain. Neuber's rule is represented by the hyperbola. The product of stress and strain is a constant, defined by the geometry and applied loading. The cyclic stress-strain curve provides the second

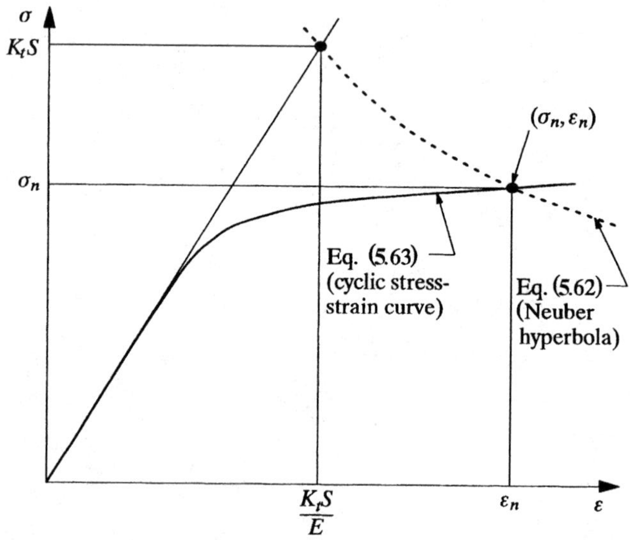

FIG. 5.47 Neuber's rule specifies the point on the cyclic stress-strain curve corresponding to the notch root.

relation necessary to obtain the two unknowns, the stress and strain in the notch root, represented by the intersection of the two curves in Fig. 5.47.

In some instances, a component can be loaded such that the nominal stress S or nominal stress range ΔS is itself large enough to cause plastic straining. In this instance, the material's cyclic stress-strain curve should be used to compute the nominal strain, e, according to

$$e = \frac{S}{E} + \left(\frac{S}{K'}\right)^{1/n'} \tag{5.64}$$

This is demonstrated in Example 8 for applied loading with a mean stress.

EXAMPLE 8 A notched steel bar has an elastic stress-concentration factor of $K_t = 2.42$. The material properties for the steel are given by

$$E = 30{,}000 \text{ ksi} \qquad K' = 154 \text{ ksi} \qquad n' = 0.123$$

$$\sigma_f' = 169 \text{ ksi} \qquad \varepsilon_f' = 1.14$$

$$b = -0.081 \qquad c = -0.67$$

For applied loading represented by the nominal stress history depicted on page 5.53, compute the cycles to crack initiation for constant-amplitude loading from 50 ksi to −30 ksi.

solution To conduct the analysis, consider loading from (o) to (a), (a) to (b), and finally (b) to (c). Subsequent loadings are repeated (a)-(b)-(c) cycles. First, consider loading from (o) to (a): The nominal stress changes from 0 to 50 ksi, or $\Delta S = 50 - 0$ ksi. For this initial loading segment, Neuber's rule is expressed as

$$K_t^2 \Delta S \Delta e = \Delta\sigma\Delta\varepsilon$$

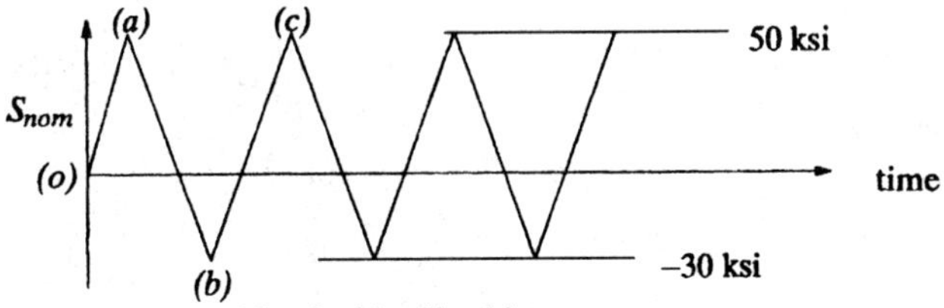

Nominal loading history

On the left side of the Neuber equation, the nominal stress and strain ranges are related by

$$\Delta e = \frac{\Delta S}{E} + \left(\frac{\Delta S}{K'}\right)^{1/n'}$$

On the right side of the Neuber equation, the notch strain and stress ranges are related by

$$\Delta \varepsilon = \frac{\Delta \sigma}{E} + \left(\frac{\Delta \sigma}{K'}\right)^{1/n'}$$

Combining these relations,

$$K_t^2(\Delta S)\left[\frac{\Delta S}{E} + \left(\frac{\Delta S}{K'}\right)^{1/n'}\right] = \Delta\sigma\left[\frac{\Delta \sigma}{E} + \left(\frac{\Delta \sigma}{K'}\right)^{(1/n')}\right]$$

$$(2.42)^2(50)(0.001773) = \Delta\sigma\left[\frac{\Delta \sigma}{E} + \left(\frac{\Delta \sigma}{K'}\right)^{(1/n')}\right]$$

Solving iteratively, $\Delta\sigma = 78.19$ ksi. This corresponds to a strain of

$$\Delta \varepsilon = \frac{78.19}{E} + \left(\frac{78.19}{K'}\right)^{1/n'} = 0.00665$$

Therefore, at point (a), the notch root stress and strain are defined (78.19 ksi, 0.00665).

Next, to get from (a) to (b) the approach is similar, but the stress and strain changes occur along the hysteresis curve, Eq. (5.51), rather than along the cyclic stress-strain curve, Eq. (5.50). The nominal stress changes from 50 to −30 ksi: $\Delta S = 50-(-30) = 80$ ksi. For this segment, Neuber's rule is expressed as

$$K_t^2 \Delta S \Delta e = \Delta\sigma\Delta\varepsilon$$

On the left side of the Neuber equation, the nominal strain change is related to the nominal stress change by

$$\Delta e = \frac{\Delta S}{E} + 2\left(\frac{\Delta S}{2K'}\right)^{1/n'}$$

On the right side of the Neuber equation, the notch strain range is related to the notch stress by

$$\Delta \varepsilon = \frac{\Delta \sigma}{E} + 2\left(\frac{\Delta \sigma}{2K'}\right)^{1/n'}$$

Combining these relations,

$$K_t^2(\Delta S)\left[\frac{\Delta S}{E} + 2\left(\frac{\Delta S}{2K'}\right)^{1/n'}\right] = \Delta\sigma\left[\frac{\Delta \sigma}{E} + 2\left(\frac{\Delta \sigma}{2K'}\right)^{(1/n')}\right]$$

$$(2.42)^2(80)(0.002701) = \Delta\sigma\left[\frac{\Delta\sigma}{E} + \left(\frac{\Delta\sigma}{2K'}\right)^{(1/n')}\right]$$

Solving iteratively, $\Delta\sigma = 143.55$ ksi.

The strain change corresponding to this stress change is given by

$$\Delta\varepsilon = \frac{143.55}{E} + 2\left(\frac{143.55}{2K'}\right)^{1/n'} = 0.008817$$

Therefore, at point (*b*), the stress is given by 78.19 − 143.55 = −65.36 ksi, and the strain is given by 0.00665 − 0.008817 = −0.00217.

Finally, to get from (*b*) to (*c*), notice that the nominal stress change is the same as in the proceeding step. The hysteresis loop for the cycle is shown below.

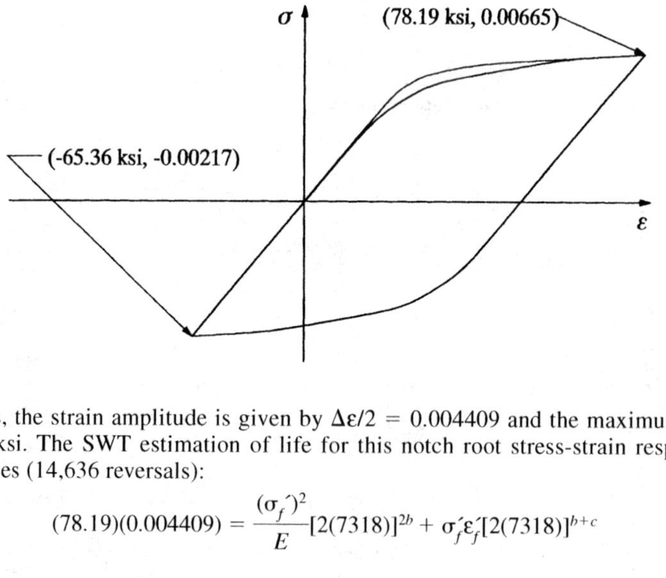

From this, the strain amplitude is given by $\Delta\varepsilon/2 = 0.004409$ and the maximum stress is 78.19 ksi. The SWT estimation of life for this notch root stress-strain response is 7318 cycles (14,636 reversals):

$$(78.19)(0.004409) = \frac{(\sigma_f')^2}{E}[2(7318)]^{2b} + \sigma_f'\varepsilon_f'[2(7318)]^{b+c}$$

5.5.3 Variable-Amplitude Loading

Thus far, the fatigue-life prediction approaches presented in this section have only been discussed in the context of constant-amplitude loading. However, engineering components are seldom subjected to constant-amplitude loading for their entire life. More often, load fluctuations occur with means and amplitudes that vary. In this section, a methodology is presented for applying the approaches presented earlier to variable-amplitude fatigue-life prediction.

Cumulative Damage. In order to compute the life of a component subjected to irregular loading, the concept of "fatigue damage" was developed, whereby each cycle is considered to expend a finite fraction of the overall life. The most widely utilized damage concept is that of linear damage (also called Miner's rule or the Palmgren-Miner rule).

Linear damage is best explained by considering a single "cycle" of loading (from a minimum to maximum and back to a minimum). If this cycle were applied under constant amplitude to a new component, the techniques described previously in this chapter could be used to compute the number of times N that cycle could be applied before a crack could be expected to develop. With the linear damage concept, each cycle of loading is considered to impose an amount of fatigue "damage" equal to $1/N$. (Equivalently, each cycle is considered to expend $1/N$th of the component life.) If this damage is counted for each cycle of loading, then failure would occur when the cumulative damage reached a value of unity. At this point, 100 percent of the fatigue life has been expended. In other words, damage imposed by the ith cycle of loading DAM_i is expressed by

$$\text{DAM}_i = \frac{1}{N_i} \tag{5.65}$$

where N_i is the number of cycles a new component would be computed to sustain under constant-amplitude loading. Failure is considered to occur when

$$\sum^{i=N_f} \text{DAM}_i = 1 \tag{5.66}$$

Other, more sophisticated nonlinear cumulative-damage approaches have been proposed.[12] Also, several effects, such as sequence of loading or overloads, have been shown to bias linear damage summation in either the conservative or nonconservative directions. However, given all the uncertainties associated with fatigue-life estimation, linear damage theory has been successfully applied to a wide range of engineering design situations, especially when loading is highly irregular and in the absence of major overloads.

Cycle Counting. In order to sum damage for variable-amplitude load histories, it is necessary to define "cycles" from those histories. For certain situations, this can be a fairly straightforward problem. For instance, consider that n_1 cycles of constant-amplitude loading at an effective stress amplitude of $(\sigma_a)_1$ are applied to a component (Fig. 5.48). Following this, the peak and valley (maximum and minimum) values of stress change, such that an effective stress amplitude of $(\sigma_a)_2$ is now being applied. If this loading continues for n_2 cycles, how many cycles (n_3) could a third range of loading

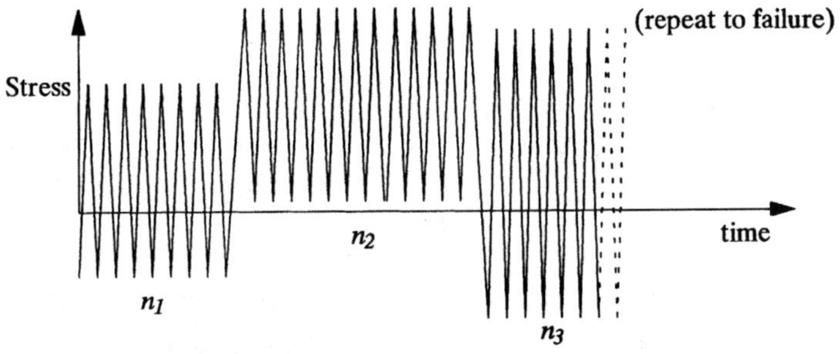

FIG. 5.48 Block loading problem.

be applied until failure is expected? For each effective stress amplitude, a corresponding fatigue life of N_1, N_2, and N_3 can be calculated. Therefore, n_3 can be computed from Eq. (5.66), written for this problem as

$$\sum \text{DAM}_i = \frac{n_1}{N_1} + \frac{n_2}{N_2} + \frac{n_3}{N_3} = 1 \tag{5.67}$$

For situations where loading is more irregular, numerous cycle-counting procedures have been proposed. Several approaches are presented in ASTM E-1049, "Recommended Practices for Cycle Counting in Fatigue Analysis."[60] Of these, rainflow cycle counting is the most popular and widely used approach.

Rainflow Cycle Counting. The concept of rainflow cycle counting has been described by two completely different approaches (referred to as the "falling rain" and "ASTM" approaches) that yield the same results. In each of these, it is necessary to define a "block" of irregular loading. This block could correspond to measurements or estimates over a representative period of service operation (one day, one job, one proving-ground lap, etc.). Furthermore, the block must be reorganized such that the largest peak in the history occurs as the first reversal. Cycles occurring prior to this are simply moved to the end such that the overall peak-valley content of the block remains the same. The "falling rain" approach is described first by considering the stress history, shown in Fig. 5.49. It must be visualized that "gravity" acts along the time axis, as shown in the figure. With this basis, the approach is described below.

1. Define a block (or history) with the highest peak occurring first.
2. Assuming "gravity" acts along the time axis, start a "flow of water droplets" at each peak and valley, *in succession*. The droplets flow along the profile and over the edge.
3. Flow originating at a peak (or valley) stops (⊣) if it "sees" a higher peak (or a deeper valley) than the one from which it started.
4. Flow also stops (>) to avoid collision with rain from a previous flow.

Stress "ranges" (valley-to-peak) are defined by the amount of vertical travel incurred by a particular flow. For instance, the rainflows in Fig. 5.49 define the eight ranges shown at the top of page 5.57.

To predict life, damage imposed by each of the four rainflow counted ranges is computed appropriately. This is demonstrated in Example 9.

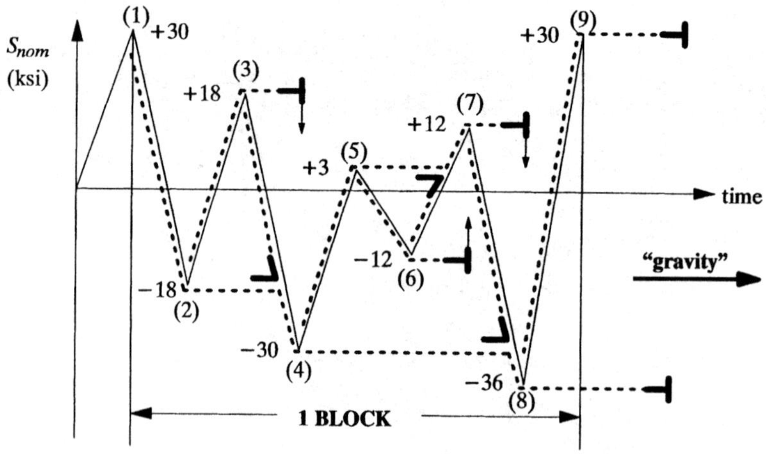

FIG. 5.49 Rainflow cycle counting ("falling rain" analogy).

Each of the ranges corresponds to one-half cycle, or one-half of a range pair

Ranges

1. (1) to (8)
2. (2) to (3)
3. (3) to (2)
4. (4) to (7)
5. (5) to (6)
6. (6) to (5)
7. (7) to (4)
8. (8) to (9)

Therefore, the four "cycles" or range pairs counted by the routine are

1. (1) to (8) & (8) to (9)
2. (2) to (3) & (3) to (2)
3. (4) to (7) & (7) to (4)
4. (5) to (6) & (6) to (5)

(as indicated by the dashed lines connecting the ranges)

EXAMPLE 9 For a forged 1040 steel component with a stress concentration factor of 1.8, conduct an elastic analysis and compute the number of times the nominal stress-loading block shown in Fig. 5.49 can be sustained by the component. The material properties for the material are given by

$$\sigma'_f = 223 \text{ ksi}$$

$$b = -0.14$$

solution The rainflow-cycle-counted nominal and notch stresses are tabulated below. The effective stress amplitude is based on the SWT parameter [Eq. (5.46)].

Range pair	S_{max} (ksi)	S_{min} (ksi)	σ_{max} (ksi)	σ_{min} (ksi)	σ_a (ksi)	$\sigma_{eff,a}$ (ksi)	N (cycles)	D (cyc^{-1})
(1–8)(8–9)	30	−36	54	−64.8	59.4	56.64	8922	1.12E−4
(2–3)(3–2)	18	−18	32.4	−32.4	32.4	32.4	4.82E5	2.08E−6
(4–7)(7–4)	12	−30	21.6	−54	37.8	28.57	1.18E6	8.46E−7
(5–6)(6–5)	3	−12	5.4	−21.6	13.5	8.54	6.61E9	1.5E−10

$\Sigma D = 0.000115$

The life of the component, in number of blocks to failure, is computed as shown below:

$$\text{Life} = \frac{1}{\Sigma D} = 8696 \text{ blocks}$$

The ASTM rainflow approach is conceptually quite different from the falling rain approach, but both give identical answers. The ASTM procedure is extremely beneficial analytically, since it can be implemented using a few lines of computer code. A BASIC listing to implement rainflow cycle counting is presented below, as Table 5.4.

Notch Strain Analysis. The estimated life in Example 9 is somewhat low and likely not in the elastic regime, indicating that cyclic plasticity may be occurring in the notch root. To analyze such a situation, a computer program can be written to implement a Neuber analysis (similar to Example 8) for every nominal stress range in the block (RANGE in Table 5.4). This is demonstrated in Example 10.

EXAMPLE 10 The nominal stress history from Fig. 5.49 is to be applied to a forged 1045 steel shaft with a K_t value of 1.8 and again for a K_t value of 3.0. Compute the notch stress-strain response. Rainflow-cycle-count the strain history and estimate the life of each component. The properties are given below.

TABLE 5.4 BASIC Computer Listing for ASTM Rainflow Approach

Nominal stress reversals are stored in an array, Stress(NPTS), where NPTS=number of reversals in the history. Since the history must begin and end with the maximum peak, NPTS will always be an odd integer.

```
      Event(1) = Stress(1)
      N = 1 : 'Counter N initialized
      J = 1 : 'Counter J initialized
1     N = N + 1
      J = J + 1
      IF (J = NPTS + 1) THEN GOTO 3
      Event(N) = Stress(J)
2     IF (N < 3) GOTO 1
      X = ABS(Event(N) - Event(N-1))
      Y = ABS(Event(N-1) - Event(N-2)
      IF (X < Y) THEN GOTO 1
      RANGE = Y
      Xmean = (Event(N-1) + Event(N-2)/2
      Xamplitude = RANGE/2
      N=N-2
      Event(N)=Event(N+2)
      GOTO 2
3     'Finished
      END
```

$$K' = 171.4 \text{ ksi} \qquad \sigma'_f = 223 \text{ ksi}$$

$$n' = 0.18 \qquad b = -0.14$$

$$E = 30{,}000 \text{ ksi} \qquad \varepsilon'_f = 0.61$$

$$c = -0.57$$

solution The rainflow-cycle-counted strain amplitude and maximum stress for each range pair in the stress history are tabulated below for each shaft. The estimated number of cycles corresponding to each range pair are computed using the SWT parameter, Eq. (5.58).

	For $K_t = 1.8$			For $K_t = 3.0$		
Range pair	ε_a	σ_{max} (ksi)	N (cycles)	ε_a	σ_{max} (ksi)	N (cycles)
(1–8)(8–9)	0.002619	46.25	35,760	0.005737	59.60	4,018
(2–3)(3–2)	0.001352	25.08	9.31E5	0.002149	43.87	6.75E4
(4–7)(7–4)	0.001124	29.80	9.66E5	0.002693	39.88	3.19E4
(5–6)(6–5)	0.00045	11.74	4.39E8	0.000756	26.59	9.40E6

The life in number of blocks for each shaft is computed as shown below. For $K_t = 1.8$,

$$\text{Life} = \frac{1}{\sum 1/N} = 33{,}248 \text{ blocks}$$

For $K_t = 3.0$,

$$\text{Life} = \frac{1}{\sum 1/N} = 3842 \text{ blocks}$$

The mean stress effect on life predictions can be assessed by assuming the strain range to be fully reversed (neglecting mean stress). To do this, the N for each range pair is computed using Eq. (5.53). This results in life predictions of 31,230 blocks for K_t = 1.8 and 3387 blocks for K_t = 3.0.

In Example 9, life predictions are based on notch root stresses and strains estimated from the applied nominal stress ranges and Neuber's rule. The estimated notch root stress-strain history can be plotted as a set of repeating hysteresis loops, such as presented in Fig. 5.50. This illustrates an important physical feature associated with rainflow

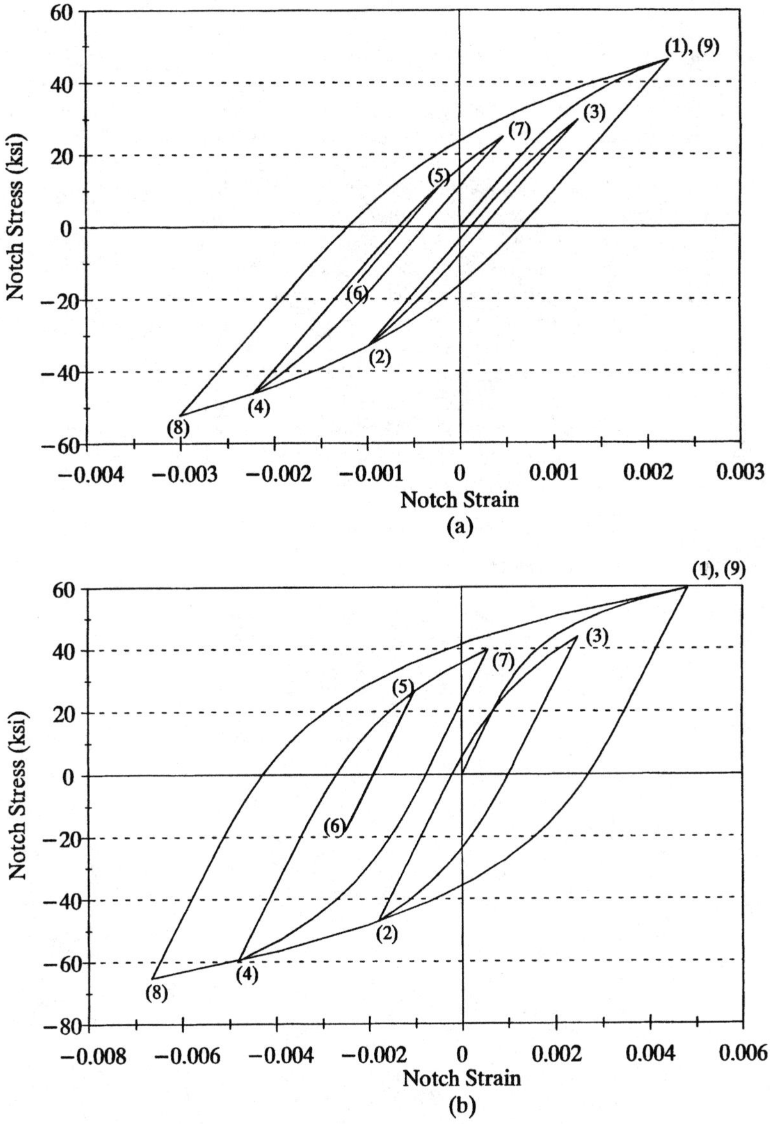

FIG. 5.50 Closed hysteresis loops from Example 10 for (*a*) K_t = 1.8 and (*b*) K_t = 3.0.

cycle counting, in that each event defined by the routine corresponds to a closed hysteresis loop shown in the figure. Comparing Fig. 5.50 and the nominal stress history, Fig. 5.49, notice that each closed hysteresis loop can be directly identified with each rainflow-cycle-counted range pair. Also note that, for this example load history, the largest strain range imposes the vast majority of damage. In general, this may not be the case.

5.6 DAMAGE-TOLERANT DESIGN

The structural integrity assessment methodologies presented in preceding sections of this chapter assume the material to be free of substantial flaws or defects. Stresses and strains are computed and compared to the strength of homogeneous engineering material. Damage-tolerant design recognizes that flaws, specifically cracks, can and do exist, even before a component is placed into service. Therefore, the focus of this design philosophy is on estimating behavior of a crack in an engineering material under service loading, and *fracture mechanics* provide the analytical tools.

Macroscopic cracks are assumed to exist in regions where detection may be difficult or impossible (e.g., under a flange or rivet head), and the behavior of the crack is predicted under anticipated service loading. The estimated behavior is used to schedule inspection and maintenance in order to assure that defects do not propagate to a catastrophic size.

Only a brief overview is presented in this section. This important method, used extensively in the aerospace industry, is covered comprehensively with example applications in several references.[9,13,31,32,61–65] The discussion here is limited to linear elastic approaches, although research on elastic-plastic fracture mechanics is very active, particularly regarding the behavior of very small cracks in locally plastic strain fields.[66–68]

5.6.1 Stress-Intensity Factor

An "ideal" crack in an engineering structure has been modeled analytically as a notch with a root radius of zero (Fig. 5.51). When a stress is applied perpendicular to the

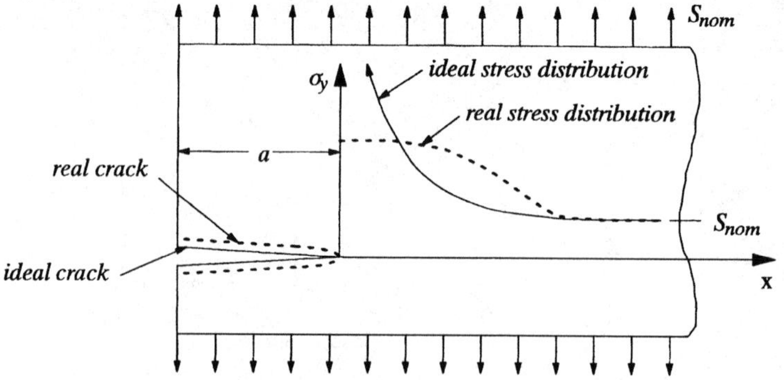

FIG. 5.51 Schematic crack opening and stress distribution from the tip of an edge crack of length *a*, in a theoretical and real engineering material.

ideal crack, stresses approach infinity at the crack tip. This is evident in Fig. 5.51 and Eq. (5.68) for the stress along the x axis:

$$\sigma_y = \frac{K_I}{\sqrt{2\pi x}} \tag{5.68}$$

The parameter K_I is referred to as the "stress-intensity factor" and it lies at the center of all fracture mechanics analyses.

The stress-intensity factor can be considered a quantitative measure of the severity of a crack in material attempting to sustain a particular state of stress. It also can be thought of as the rate at which the stresses approach infinity at the tip of a theoretical crack. Since real material cannot sustain infinite stress, intense localized deformation occurs, causing the crack tip to blunt and stresses to redistribute, as depicted in Fig. 5.51. Even though crack-tip deformation is plastic, as long as the size of the plastic zone at the crack tip is small relative to crack dimensions, the elastically calculated K_I has been successfully used to describe the strength of an engineering component. Therefore, the term "linear elastic fracture mechanics" (LEFM) is typically applied to analyses based on K_I.

The significance of the subscript I is shown in Fig. 5.52. Since cracks are considered analytically as planar defects, the subscript refers to the mode in which the two crack faces are displaced. Mode I is the normal loading mode (crack faces are pulled apart) while II and III are shear loading modes (crack faces slide relative to each other). The discussion in the remainder of this section is directly in terms of mode I analyses. However, the extension of the approaches to the shear modes is straightforward.

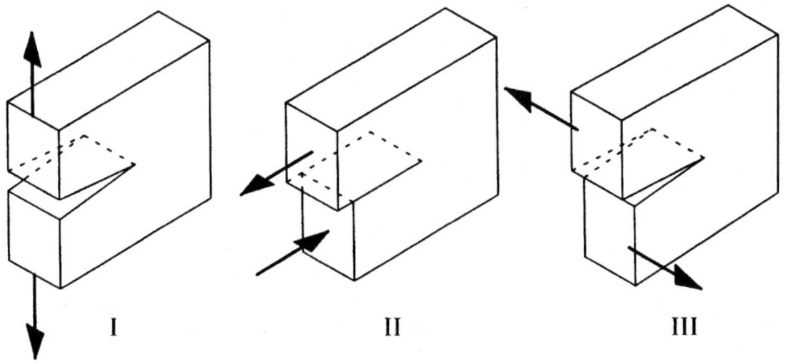

FIG. 5.52 Crack loading modes: (*left*) I, normal; (*center*) II, sliding; (*right*) III, tearing.

Stress-intensity factors can be computed for any geometry and loading combination using finite-element analysis, but not without considerable analytical effort. More typically in design, stress-intensity factors are found from a tabulated reference.[69–72] They are usually expressed as in Eq. (5.69), in terms of the gross nominal stress S_{nom}, crack length, and geometry for a particular type of loading (e.g., bending or tension). Since they are elastically calculated, they can be superimposed.

$$K_I = f(a,\text{geometry})S_{\text{nom}}\sqrt{a} \tag{5.69}$$

5.6.2 Static Loading

The premise of damage-tolerant design is that engineering materials with macroscopic cracks can sustain stresses without failure, up to a point. Obviously, the larger a crack in a component, the less load it can sustain. One of the important features of K_I is its ability to characterize combinations of loading and crack size that correspond to unstable crack propagation (or the strength of the cracked component). Experiments have shown that cracks in engineering materials propagate catastrophically at a certain critical value of the stress-intensity factor, K_c. The critical value is referred to as the *fracture toughness* (not directly associated with the impact strength or area under a stress-strain curve) and is considered a property of the material. In other words, failure is predicted when

$$K_I = f(a,\text{geometry})S_{\text{nom}}\sqrt{a} = K_c \tag{5.70}$$

Like other "material properties," K_c is influenced by numerous factors including environment, rate of loading, etc. But an important distinction for K_c arises from its observed dependence on specimen size. As the thickness t of the specimen increases along the dimension of the crack front, increasing constraint develops, making plastic deformation more difficult and causing the material to behave in a more brittle manner. Therefore, as shown in Fig. 5.53, as the thickness increases, K_c decreases asymptotically to a value referred to as K_{IC}, the *plane-strain fracture toughness*. Specifications for the experimental determination of K_{IC} are found in ASTM E-399, "Standard Test Method for Plane-Strain Fracture Toughness of Metallic Materials,"[71] and tabulated values can be found in Refs. 73 through 80 for many engineering alloys.

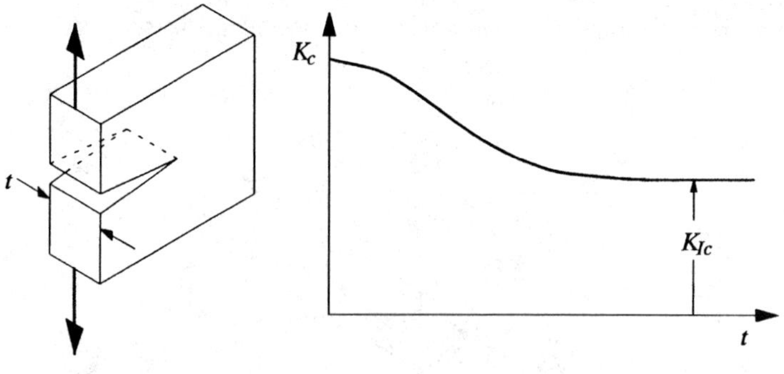

FIG. 5.53 As specimen thickness increases, fracture toughness K_c decreases asymptotically to the value K_{Ic}, the plane-strain fracture toughness.

5.6.3 Fatigue Loading

Cracks in engineering components tend to propagate progressively under fatigue loading. The damage-tolerant design philosophy takes advantage of this phenomenon to schedule periodic inspection and maintenance of components to assure against the growth of a crack to the catastrophic size associated with Eq. (5.70).

Paris and Erdogan[81] are credited with first making the observation that the stress-intensity factor range can be used effectively to correlate crack propagation rates for a particular material under a wide variety of crack geometry and loading combinations. As shown schematically in Fig. 5.54, suppose three separate specimens are subjected to the $R = 0$ loading at different applied loading levels. A stress-intensity factor range is computed from each nominal stress range. Cracks grow through each specimen at different rates, as shown in Fig. 5.54*b*. However, if the crack propagation rate, *da/dN*, is plotted versus stress-intensity factor range ΔK_I (on log-log coordinates), then the data from all three tests collapse to the single curve, Fig. 5.54*c*. This curve is considered to

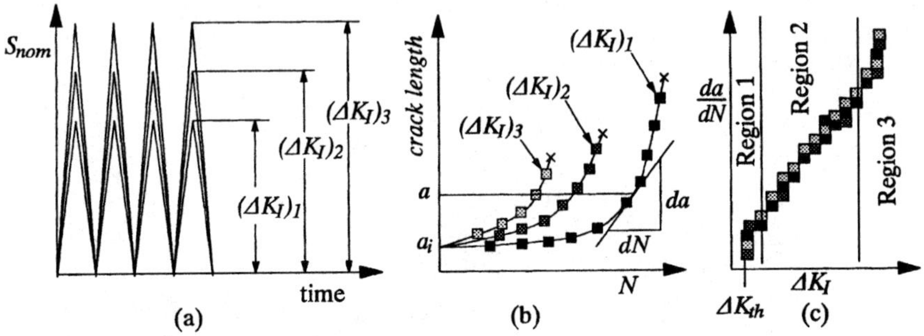

FIG. 5.54 Schematic data from three separate specimens: (*a*) Different nominal stress ranges lead to three different ΔK ranges for a given crack length; (*b*) crack length versus cycles of applied loading; (*c*) crack growth rate versus ΔK for all three tests.

be a characteristic of the material. Guidelines for conducting this type of testing are specified in ASTM E-647, "Standard Test Method for Measurement of Fatigue Crack Growth Rates."[82]

The shape associated with the crack growth rate curve (Fig. 5.54*c*) has been referred to as *sigmoidal,* having three distinct regions. Region 1 is associated with another property called the threshold stress-intensity factor range, ΔK_{th}. This is analogous to the endurance limit for local stress-based fatigue design. Applied loading cycles below ΔK_{th} are considered not to result in any crack advancement. Values of ΔK_{th} for various materials can be found in Refs. 73 through 80. At the upper region of the curve (called Region 3), peak values of K_I are approaching K_c. At this point crack propagation is rapid and growing beyond the limits of LEFM validity. The middle portion of the curve, Region 2, is important since it encompasses a substantial portion of the propagation life. Data in this regime usually appear somewhat linear on logarithmic coordinates, leading to the Paris relation,

$$\frac{da}{dN} = A_p(\Delta K)^{n_p} \tag{5.71}$$

Life prediction based on LEFM is based upon the integration of Eq. (5.71), or any other relation describing the data in Fig. 5.54*c*. This is illustrated in Eq. (5.72) for constant-amplitude loading:

$$N = \int_0^N dN = \int_{a_i}^{a_f} \frac{da}{\Delta K} = \int_{a_i}^{a_f} \frac{da}{f(\text{geometry},a)\sqrt{a}} \tag{5.72}$$

As implied in Eq. (5.72) and Fig. 5.54*b,* a fracture mechanics analysis of crack propagation must begin with the assumption of an initial crack of length a_i. [The final crack length, a_f in Eq. (5.72), can be directly calculated from the maximum anticipated loading and K_c.] The definition of the initial crack size for testing purposes is well defined (ASTM E-647, Ref. 82) but for design purposes can be less objective. Sometimes, cracks are considered to exist where they may be difficult to detect, such as underneath a seam. However, for macroscopically smooth surfaces, the assumption of very small initial crack lengths (cracks that would be difficult to detect without the aid of a microscope, on the order of 0.001 in) can significantly affect the analysis, since small cracks can result in extremely low estimated propagation rates. Very small crack lengths, on the order of typical surface roughness values or surface scratches, are not considered to lie within the valid domain of LEFM.[66–68] The threshold stress intensity factor can be used to define valid initial crack lengths.

Mean stress effects on crack propagation are accounted for with empirical relations usually defined in terms of the stress ratio R [Eq. (5.38)]. Two widely referenced equations were developed by Foreman,[83]

$$\frac{da}{dN} = \frac{A_f(\Delta K)^{n_f}}{(1 - R)K_c - \Delta K} \tag{5.73}$$

and Walker,[84]

$$\frac{da}{dN} = A_w\left(\frac{\Delta K}{(1 - R)^{1-m_w}}\right)^{n_w} \tag{5.74}$$

Constants in Eqs. (5.73) and (5.74) are empirically defined by comparison to constant-amplitude data generated over a range of load ratios. The relations can then be used to estimate crack growth during variable-amplitude load histories, usually by assuming that only certain segments in the history will result in incremental crack advancements. For example, damaging segments can be defined as only those that are both *tensile* and *increasing*. Maximum and minimum stresses for a particular segment are used to compute ΔK and R for use in Eq. (5.73) or (5.74), which is thus considered to compute the amount of crack growth resulting from that segment. References 9, 13, 31, 32, and 61 through 65 provide more detailed explanations of the implementation and use of fracture mechanics, including coverage of more advanced topics such as crack closure and sequence effects.

5.7 MULTIAXIAL LOADING

Fatigue under multiaxial loading is an extremely complex phenomenon. Evidence of this is the fact that even though the topic has been actively researched for more than a century, new theories on multiaxial fatigue are still emerging.[85] However, recent work has led to advances in understanding of the mechanisms of multiaxial fatigue and addressed some of the problems associated with implementing those advances for practical design problems.

In this section, only multiaxial fatigue-life prediction approaches are presented that are considered somewhat established. Although such approaches only exist for fairly simplistic multiaxial loading (e.g., constant amplitude, proportional loading, high-cycle regime), they still cover a substantial number of design situations.

5.7.1 Proportional Loading

Loading is defined as proportional when the ratios of principal stresses remains fixed with time. A consequence of this is that principal stress directions do not rotate. The simplest example would be a situation where the three components of surface stress are in phase and fully reversed, as shown in Fig. 5.55*a*. Figure 5.55*b* depicts a situation where stresses fluctuate in phase about mean values. In this case, whether or not the loading is purely proportional depends on the ratios of the mean stresses. The mean stresses can cause a slight oscillation of the principal stress axes over a cycle. In either case, a modified Sines approach has been successfully applied to this type of loading in the high-cycle regime.

Sines[37] observed that *mean torsional stresses do not influence fatigue behavior*, while mean tensile stresses decrease life and mean compression improves life. Sines' approach can be summarized as follows:

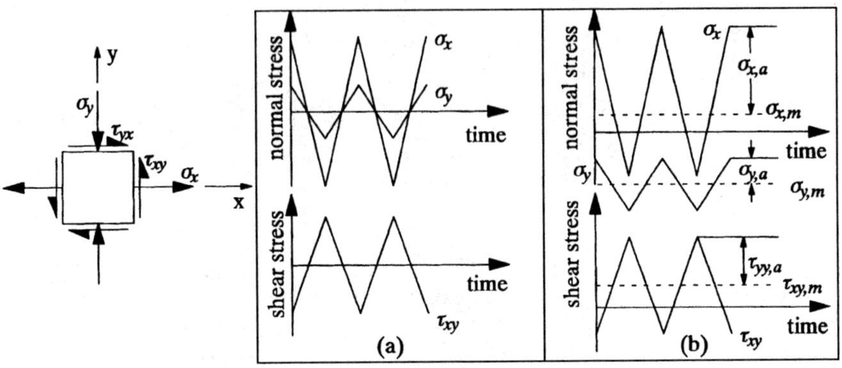

FIG. 5.55 (*a*) Pure proportional loading, fully reversed stress components. (*b*) Proportional stress amplitudes, with mean stresses.

- Compute the amplitude and mean of each normal stress:

$$\sigma_{x,a} = (\sigma_{x,\max} - \sigma_{x,\min})/2$$

$$\sigma_{x,m} = (\sigma_{x,\max} + \sigma_{x,\min})/2$$

$$\sigma_{y,a} = (\sigma_{y,\max} - \sigma_{y,\min})/2$$

$$\sigma_{y,m} = (\sigma_{y,\max} + \sigma_{y,\min})/2$$

$$\sigma_{z,a} = (\sigma_{z,\max} - \sigma_{z,\min})/2$$

$$\sigma_{z,m} = (\sigma_{z,\max} + \sigma_{z,\min})/2$$

- Compute the amplitude of the shear stress:

$$\tau_{xy,a} = (\tau_{xy,\max} - \tau_{xy,\min})/2$$

- Compute the principal stress amplitude from the amplitudes of the normal and shear applied stresses:

$$\sigma_{1,a} = \left(\frac{\sigma_{x,a} + \sigma_{y,a}}{2}\right) + \sqrt{\left(\frac{\sigma_{x,a} - \sigma_{y,a}}{2}\right)^2 + \tau_{xy,a}^2} \tag{5.75}$$

$$\sigma_{2,a} = \left(\frac{\sigma_{x,a} + \sigma_{y,a}}{2}\right) - \sqrt{\left(\frac{\sigma_{x,a} - \sigma_{y,a}}{2}\right)^2 + \tau_{xy,a}^2} \tag{5.76}$$

$$\sigma_{3,a} = \sigma_{z,a} \tag{5.77}$$

- Calculate the equivalent stress amplitude according to a von Mises and an equivalent mean stress as given by

$$S_{\text{eq},a} = \frac{1}{\sqrt{2}}\sqrt{(\sigma_{1,a} - \sigma_{2,a})^2 + (\sigma_{2,a} - \sigma_{3,a})^2 + (\sigma_{3,a} - \sigma_{1,a})^2} \tag{5.78}$$

$$S_{\text{eq},m} = \sigma_{x,m} + \sigma_{y,m} + \sigma_{z,m} \tag{5.79}$$

- Use $S_{eq,a}$ and $S_{eq,m}$ as an amplitude and mean stress (in place of σ_a and σ_{mean}) to form an effective, fully reversed stress amplitude, such as described in Fig. 5.27 or given in Eqs. (5.42) to (5.46).

Note that $\tau_{xy,a}$ influences neither $S_{eq,a}$ nor $S_{eq,m}$, and thus does not affect the life estimate. If a pressure-free surface element is being considered, then $\sigma_{3,a} = 0$, and Eq. (5.77) can be simplified to

$$S_{eq,a} = \sqrt{\sigma_{x,a}^2 - (\sigma_{x,a})(\sigma_{y,a}) + \sigma_{y,a}^2 + 3\tau_{xy,a}^2} \tag{5.80}$$

eliminating the intermediate principal stress-amplitude calculations.

An important point must be kept in mind when implementing this approach: *it can be crucial to keep track of the algebraic sign of the amplitudes of the normal stress components* $\sigma_{x,a}$, $\sigma_{y,a}$, *and* $\sigma_{z,a}$, *as well as the principal stresses* $S_{1,a}$ *and* $S_{2,a}$. Generally, "amplitudes" are considered positive. However, considering amplitude as always positive here can lead to nonconservative life estimates! When two normal stresses *peak* at the same time, then both amplitudes should be considered positive (or the two should have the same algebraic sense). When one is at a *valley* while the other is at a *peak,* then the amplitude of the valley signal is negative while amplitude of the peak signal is positive (or the two should have the opposite algebraic sense). Example 11 serves to illustrate that point.

EXAMPLE 11 Estimate the fatigue life for two materials experiencing the Case A and B stress histories below. The only difference between the two stress histories is the phase relation of the normal stress in the y direction. Properties for the two materials (1045 steels) are as follows: (1) $S_u = 220$ ksi, $\sigma_f' = 843$ ksi, $b = -0.1538$; (2) $S_u = 137$ ksi, $\sigma_f' = 421$ ksi, $b = -0.1607$.

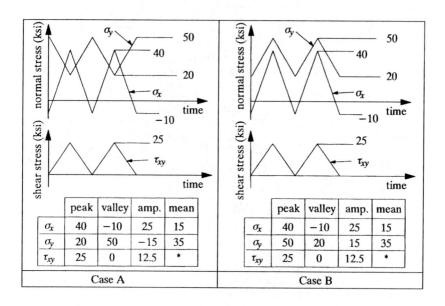

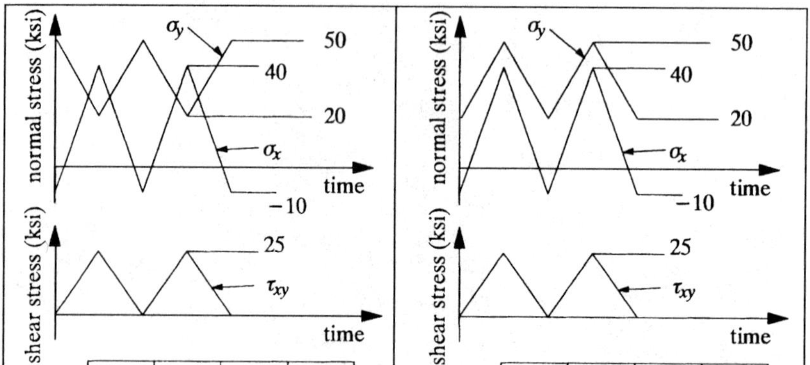

Case A

	peak	valley	amp.	mean
σ_x	40	−10	25	15
σ_y	20	50	−15	35
τ_{xy}	25	0	12.5	*

Case B

	peak	valley	amp.	mean
σ_x	40	−10	25	15
σ_y	50	20	15	35
τ_{xy}	25	0	12.5	*

solution The peak, valley, amplitude, and mean stresses are tabulated above for each load history. For Case A, notice the negative amplitude for $\sigma_{y,a}$. For Case A, $S_{eq,a} = 41.16$ ksi. For Case B, $S_{eq,a} = 30.7$ ksi. For both cases, $S_{eq,m} = 50$ ksi.

Goodman:
$$\sigma_{a,\mathrm{eff}} = S_{\mathrm{eq},a}\left(\frac{S_u}{S_u - S_{\mathrm{eq},m}}\right)$$

SWT:
$$\sigma_{a,\mathrm{eff}} = \sqrt{S_{\mathrm{eq},a}(S_{\mathrm{eq},a} + S_{\mathrm{eq},m})}$$

Using the equivalent amplitude and mean stresses, the Goodman and SWT parameters were used to compute effective stress amplitudes and corresponding lives. Results are tabulated below.

	Case A		Case B	
	$\sigma_{a,\mathrm{eff}}$ (ksi)	N (cycles)	$\sigma_{a,\mathrm{eff}}$ (ksi)	N (cycles)
Material 1				
Goodman	53.26	31.4×10^6	39.76	211×10^6
SWT	61.25	12.7×10^6	49.80	48×10^6
Material 2				
Goodman	64.81	5.7×10^4	48.38	3.52×10^5
SWT	61.25	8.1×10^4	49.80	2.94×10^5

Several important observations can be made from Example 11. Estimated lives are significantly shorter for Case A relative to Case B. This is because the von Mises equivalent stress amplitude reflects the fact that principal shear stresses fluctuate more in Case A than they do in Case B. Also, whether or not the Goodman approach is more or less conservative than the SWT approach depends on the material and load history. For Case B, the SWT prediction is more conservative than Goodman for either material. But, for Case A, SWT is more conservative for the harder material (1), but less so for the softer material (2).

5.7.2 Nonproportional Loading

For nonproportional loading no consensus exists on the most suitable design approach. The ASME Boiler and Pressure Vessel Code[86] presents a generalized procedure, given in Table 5.5. Based on this procedure, an equivalent stress parameter, SEQA, has been derived[87] for combined, constant-amplitude, out-of-phase bending (or axial) and torsional stress:

$$\mathrm{SALT} = \frac{\sigma_{x,a}}{\sqrt{2}}\sqrt{1 + C^2 + \sqrt{1 + 2C^2\cos(2\phi) + C^4}} \tag{5.81}$$

where $\sigma_{x,a}$ and $\tau_{xy,a}$ = elastically calculated notch bending (or axial) and torsional shear-stress amplitudes, respectively, $C = 2\tau_{xy,a}/\sigma_{x,a}$, and ϕ = the phase angle between bending and torsion. This parameter reduces to the Tresca (maximum shear) equivalent stress amplitude for in-phase loading. A similar relation can be defined based on the von Mises criterion:[87]

$$\mathrm{SEQA} = \frac{\sigma}{\sqrt{2}}\sqrt{1 + \tfrac{3}{4}C^2 + \sqrt{1 + \tfrac{3}{2}C^2\cos(2\phi) + \tfrac{9}{16}C^4}} \tag{5.82}$$

For bending-torsion cases, these relations can simplify analysis. For a more random load history, the applications of the ASME approach is demonstrated in Example 12.

EXAMPLE 12 Use the procedure outlined in ASME Code Case NB-3216.2 to compute the Tresca-based S_{alt} for the bending-torsion operating cycle shown on page 5.69. Also,

TABLE 5.5 Excerpt from ASME Boiler and Pressure Vessel Code, Sec. III: Multiaxial Fatigue Evaluation

NB-3216.1 Constant Principal Stress Direction.

For any case in which the directions of the principal stresses at the point being considered do not change during the cycle, the steps stipulated in the following subparagraphs shall be taken to determine the alternating stress intensity.

(a) Principal Stresses—Consider the values of the three principal stresses at the point versus time for the complete stress cycle taking into account both the gross and local structural discontinuities and the thermal effects which vary during the cycle. These are designated as σ_1, σ_2 and σ_3 for later identification.

(b) Stress Differences—Determine the stress differences $S_{12} = \sigma_1 - \sigma_2$, $S_{23} = \sigma_2 - \sigma_3$ and $S_{31} = \sigma_3 - \sigma_1$ versus time for the complete cycle. In what follows, the symbol S_{ij} is used to represent any one of these stress differences.

(c) Alternating Stress Intensity—Determine the extremes of the range through which each stress difference (S_{ij}) fluctuates and find the absolute magnitude of this range for each S_{ij}. Call this magnitude $S_{r\,ij}$ and let $S_{\text{alt}\,ij} = 0.5\,S_{r\,ij}$. The alternating stress intensity S_{alt}, is the largest of the $S_{\text{alt}\,ij}$'s.

NB-3216.2 Varying Principal Stress Direction.

For any case in which the directions of the principal stresses at the point being considered do change during the stress cycle, it is necessary to use the more general procedure of the following subparagraphs.

(a) Consider the values of the six stress components, σ_t, σ_l, σ_r, τ_{tl}, τ_{lr}, τ_{rt}, versus time for the complete stress cycle, taking into account both the gross and local structural discontinuities and the thermal effects which vary during the cycle. (*The subscripts t, l and r represent the tangential, longitudinal and radial directions, respectively.*)

(b) Choose a point in time when the conditions are one of the extremes for the cycle (either maximum or minimum, algebraically) and identify the stress components at this time by the subscript i. In most cases, it will be possible to choose at least one time during the cycle when the conditions are known to be extreme. In some cases it may be necessary to try different points in time to find the one which results in the largest value of alternating stress intensity.

(c) Subtract each of the six stress components σ_{ti}, σ_{li}, σ_{ri}, τ_{tli}, τ_{lri}, τ_{rti}, from the corresponding stress components σ_t, σ_l, σ_r, τ_{tl}, τ_{lr}, τ_{rt}, at each point in time during the cycle and call the resulting components σ'_t, σ'_l, σ'_r, τ'_{tl}, τ'_{lr}, τ'_{rt}.

(d) At each point in time during the cycle, calculate the principal stresses, σ'_1, σ'_2 and σ'_3, derived from the six stress components, σ'_t, σ'_l, σ'_r, τ'_{tl}, τ'_{lr}, τ'_{rt}. Note that the directions of the principal stresses may change during the cycle but the principal stress retains its identity as it rotates.

(e) Determine the stress differences $S'_{12} = \sigma'_1 - \sigma'_2$, $S'_{23} = \sigma'_2 - \sigma'_3$ and $S'_{31} = \sigma'_3 - \sigma'_1$ versus time for the complete cycle and find the largest absolute magnitude of any stress difference at any tie. The alternating stress intensity, S_{alt}, is one-half of this magnitude.

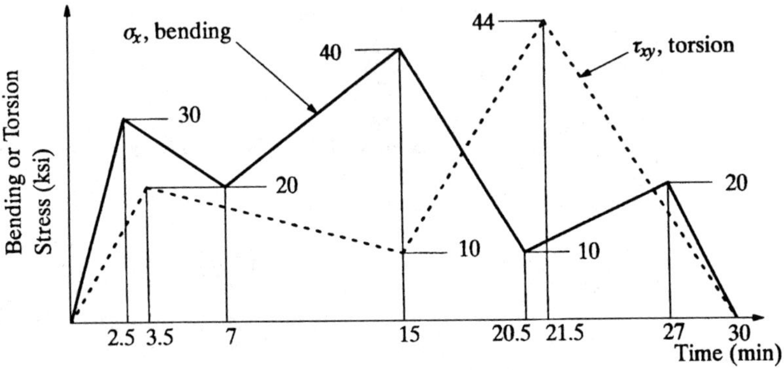

modify the final step (e) of the approach to compute an equivalent stress amplitude based on the von Mises criterion and compute that quantity for the same load history.

solution The difficult step for implementing the procedure shown in Table 5.5 is step b, selecting the critical time in the load history. In this example, either $t = 0$ ($\sigma_{xi} = 0$ and $\tau_{xyi} = 0$) or $t = 21.5$ min ($\sigma_{xi} = 11.54$ ksi and $\tau_{xyi} = 44$ ksi) give identical results. The Tresca-based equivalent stress amplitude is given by

$$\text{SALT} = 44.38 \text{ ksi}$$

Step e in the procedure can be modified to define a von Mises equivalent stress amplitude:

$$\text{SEQA} = \frac{1}{\sqrt{2}}\sqrt{(\sigma_1' - \sigma_2')^2 + (\sigma_2' - \sigma_3')^2 + (\sigma_3' - \sigma_1')^2}$$

This results in

$$\text{SEQA} = 38.54 \text{ ksi}$$

Although the ASME approach demonstrated in Example 12 is straightforward in its implementation, it has the potential to make very nonconservative life estimates for long loading histories. In fact, predictions made by the approach have been shown to be nonconservative, even for relatively simple constant-amplitude, out-of-phase bending, and torsional loading.[87,88] Active research in multiaxial fatigue-life prediction is still underway to address practical design concerns such as notches[89] and variable-amplitude loading.[90,91]

REFERENCES

1. Shigley, J. E. and C. R. Mischke: "Mechanical Engineering Design," 5th ed., McGraw-Hill Book Company, New York, 1989.
2. Spotts, M. F.: "Design of Machine Elements," 6th ed., Prentice-Hall, Englewood Cliffs, N.J., 1985.
3. Juvinall, R. C.: "Fundamentals of Machine Component Design," John Wiley and Sons, New York, 1983.
4. Slocum, A. H.: "Precision Machine Design," Prentice-Hall, Englewood Cliffs, N.J., 1992.

5. Timoshenko, S. P., and J. N. Goodier: "Theory of Elasticity," 3d ed., McGraw-Hill Book Company, New York, 1970.
6. Boresi, A. P., and O. M. Sidebottom: "Advanced Mechanics of Materials," 4th ed., John Wiley and Sons, New York, 1985.
7. Landgraf, R. W., and R. A. Chernenkoff: "Residual Stress Effects on Fatigue of Surface Processed Steels," "Analytical and Experimental Methods for Residual Stress Effects in Fatigue," ASTM STP 1004, ASTM, Philadelphia, 1986.
8. Fung, Y. C.: "A First Course in Continuum Mechanics," 2d ed., Prentice-Hall, Englewood Cliffs, N.J., 1977.
9. Hertzberg, R. W.: "Deformation and Fracture Mechanics of Engineering Materials," 3d ed., John Wiley and Sons, New York, 1989.
10. Dieter, G. E.: "Mechanical Metallurgy," 3d ed., McGraw-Hill Book Company, New York, 1986.
11. Beer, F. P., and E. R. Johnson: "Mechanics of Materials," 2d ed., McGraw-Hill Book Company, New York, 1992.
12. Collins, J. A.: "Failure of Materials in Mechanical Design," John Wiley and Sons, New York, 1981.
13. Dowling, N. E.: "Mechanical Behavior of Materials," Prentice-Hall, Englewood Cliffs, N.J., 1993.
14. A. S. Cobayashi (ed.): "Manual on Experimental Stress Analysis," Prentice-Hall, Englewood Cliffs, N.J., 1987.
15. Dally, J. W., and W. F. Riley: "Experimental Stress Analysis," 3d ed., McGraw-Hill Book Company, New York, 1991.
16. Peterson, R. E.: "Stress Concentration Factors," John Wiley and Sons, New York, 1974.
17. Young, Y. C.: "Roark's Formulas for Stress and Strain," 6th ed., McGraw-Hill Book Company, New York, 1989.
18. Zienkiewicz, O. C.: "The Finite Element Method," 3d ed., McGraw-Hill Book Company, New York, 1987.
19. Cook, R. D., D. S. Malkus, and M. E. Plesha: "Concepts and Applications of Finite Element Analysis," 3d ed., John Wiley and Sons, New York, 1989.
20. Reddy, J. W.: "An Introduction to the Finite Element Method," McGraw-Hill Book Company, New York, 1984.
21. Bickford, W. B.: "A First Course in the Finite Element Analysis," Richard D. Irwin, Homewood, Ill., 1988.
22. Sorem, J. R., Jr., and S. M. Tipton: "The Use of Finite Element Codes for Cyclic Stress-Strain Analysis," "Fatigue Design," European Structural Integrity Society, vol. 16, J. Solin, G. Marquis, A. Siljander and S. Sipilä, eds., Mechanical Engineering Publications, London, 1993, pp. 187–200.
23. "Fatigue Design Handbook," Society of Automotive Engineers, Warrendale, Pa., 1990.
24. Tipton, S. M., and G. J. Shoup: "The Effect of Proof Loading on the Fatigue Behavior of Open Link Chain," *Trans. ASME, J. Fatigue and Fracture Eng. Mat. Technol.*, vol. 114, no. 1, pp. 27–33, January 1992.
25. Shoup, G. J., S. M. Tipton, and J. R. Sorem: "The Effect of Proof Loading on the Fatigue Behavior of Studded Chain," *Int. J. Fatigue*, vol. 14, no. 1, pp. 35–40, January 1992.
26. Tipton, S. M., and J. R. Sorem: "A Reversed Plasticity Criterion for Specifying Optimal Proof Load Levels," "Advances in Fatigue Lifetime Predictive Techniques," vol. 2, ASTM STP 1211, M. R. Mitchell and R. W. Landgraf, eds., American Society for Testing and Materials STP, Philadelphia, 1993, pp. 186–202.
27. "Standard Test Methods of Tension Testing of Metallic Materials," Specification E-8, American Society for Testing and Materials, Philadelphia.
28. Ramberg, W., and W. Osgood: "Description of Stress-Strain Curves by Three Parameters," NACA TN 902, 1943.

29. "Tensile Strain-Hardening Exponents (n-Values) of Metallic Sheet Materials," Specification E-646, American Society for Testing and Materials, Philadelphia.

30. "Standard Test Methods of Compression Testing of Metallic Materials at Room Temperature," Specification E-9, American Society for Testing and Materials, Philadelphia.

31. Fuchs, H. O., and R. I. Stephens: "Metal Fatigue in Engineering," John Wiley and Sons, New York, 1980.

32. Bannantine, J. A., J. J. Comer, and J. H. Handrock: "Fundamentals of Metal Fatigue Analysis," Prentice-Hall, Englewood Cliffs, N.J., 1990.

33. "Standard Practice for Conducting Constant Amplitude Axial Stress-Life Tests of Metallic Materials," Specification E-466, American Society for Testing and Materials, Philadelphia.

34. "Standard Practice for Statistical Analysis of Linear or Linearized Stress-Life (S-N) and Strain Life (ε-N) Fatigue Data," Specification E-739, American Society for Testing and Materials, Philadelphia.

35. "Standard Practice for Presentation of Constant Amplitude Fatigue Test Results for Metallic Materials," Specification E-468, American Society for Testing and Materials, Philadelphia.

36. Smith, K. N., P. Watson, and T. H. Topper: "A Stress-Strain Function for the Fatigue of Metals," *J. Materials,* vol. 5, no. 4, pp. 767–778, December 1970.

37. Sines, G.: "Failure of Metals under Combined Repeated Stresses with Superimposed Static Stresses," NACA Tech. Note 3495, 1955.

38. "MIL Handbook 5," Department of Defense, Washington, D.C.

39. Grover, H. J.: "Fatigue of Aircraft Structures," NAVAIR 01-1A-13, 1966.

40. "Standard Recommended Practice for Constant-Amplitude Low-Cycle Fatigue Testing," Specification E-606, American Society for Testing and Materials, Philadelphia.

41. Martin, J. F.: "Cyclic Stress-Strain Behavior and Fatigue Resistance of Two Structural Steels," Fracture Control Program Report No. 9, University of Illinois at Urbana-Champaign, 1973.

42. Massing, G.: "Eigenspannungen und Verfestigung beim Messing (Residual Stress and Strain Hardening of Brass)," *Proc. Second International Congress for Applied Mechanics,* Zurich, September 1926.

43. Basquin, O. H.: "The Exponential Law of Endurance Tests," *American Society for Testing and Materials Proceedings,* vol. 10, pp. 625–630, 1910.

44. Manson, S. S.: "Behavior of Materials under Conditions of Thermal Stress," *Heat Transfer Symposium,* University of Michigan Engineering Research Institute, pp. 9–75, 1953 (also published as NACA TN 2933, 1953).

45. Coffin, L. F.: "A Study of the Effects of Thermal Stresses on a Ductile Metal," *Trans. ASME,* vol. 76, pp. 931–950, 1954.

46. Mitchell, M. R.: "Fundamentals of Modern Fatigue Analysis for Design," "Fatigue and Microstructure," Society of Automotive Engineers, pp. 385–437, 1977.

47. Socie, D. F.: "Fatigue Life Prediction Using Local Stress-Strain Concepts," *Experimental Mechanics,* vol. 17, no. 2, 1977.

48. Socie, D. F.: "Fatigue Life Estimation Techniques," Electro General Corporation Technical Report 145A-579, Minnetonka, Minn.

49. Socie, D. F., N. E. Dowling, and P. Kurath: "Fatigue Life Estimation of a Notched Member," ASTM STP 833, *Fracture Mechanics; Fifteenth Symposium,* R. J. Sanford, ed., American Society for Testing and Materials, Philadelphia, 1984, pp. 274–289.

50. Sehitoglu, H.: "Fatigue of Low Carbon Steels as Influenced by Repeated Strain Aging," Fracture Control Program Report No. 40, University of Illinois at Urbana-Champaign, June 1981.

51. Kurath, P.: "Investigation into a Non-arbitrary Fatigue Crack Size Concept," Theoretical and Applied Mechanics, Report No. 429, University of Illinois at Urbana-Champaign, October 1978.

52. "Technical Report on Fatigue Properties—SAE J1099," SAE Informational Report, Society of Automotive Engineers, Warrendale, Pa., 1975.

53. Boller, C., and T. Seeger: "Materials Data for Cyclic Loading," Parts A to E, Elsevier Science Publishers B. V., 1987.

54. Boardman, B. E.: "Crack Initiation Fatigue—Data, Analysis, Trends, and Estimations," SAE Technical Paper 820682, 1982.

55. Manson, S. S.: "Fatigue: A Complex Subject—Some Simple Approximations," *Proc. Soc. Experimental Stress Analysis,* vol. 12, no. 2, 1965.

56. Muralidharan, U., and S. S. Manson: "A Modified Universal Slopes Equation for Estimation of Fatigue Characteristics of Metals," *Trans. ASME, J. Eng. Mat. Technol.,* vol. 110, pp. 55–58, January 1988.

57. Socie, D. F., M. R. Mitchell, and E. M. Caulfield: "Fundamentals of Modern Fatigue Analysis," Fracture Control Program Report No. 26, University of Illinois at Urbana-Champaign, 1977.

58. Ong, J. H.: "An Evaluation of Existing Methods for the Prediction of Axial Fatigue Life from Tensile Data," *Int. J. Fatigue,* vol. 15, no. 1, pp. 13–19, 1993.

59. Neuber, H.: "Theory of Stress Concentration for Shear Strained Prismatical Bodies with Arbitrary Stress-Strain Law," *Trans. ASME, J. Appl. Mech.,* vol. 12, pp. 544–550, 1961.

60. "Recommended Practices for Cycle Counting in Fatigue Analysis," Specification E-1049, American Society for Testing and Materials, Philadelphia.

61. Broek, D.: "Elementary Engineering Fracture Mechanics," 4th ed., Kluwer Academic Publishers, Dordrecht, The Netherlands, 1986.

62. Broek, D.: "The Practical Use of Fracture Mechanics," Kluwer Academic Publishers, Dordrecht, The Netherlands, 1988.

63. Barsom, J. M., and S. T. Rolfe: "Fracture and Fatigue Control in Structures," 2d ed., Prentice-Hall, Englewood Cliffs, N.J., 1987.

64. Anderson, T. L.: "Fracture Mechanics: Fundamentals and Applications," CRC Press, Boca Raton, Fla., 1991.

65. Ewalds, H. L., and R. J. H. Wanhill: "Fracture Mechanics," Edward Arnold Publishers, London, 1984.

66. Newman, J. C., Jr.: "A Review of Modelling Small-Crack Behavior and Fatigue Life Predictions for Aluminum Alloys," *Fatigue and Fracture Eng. Mat. Struct.,* vol. 17, no. 4, pp. 429–439, 1994.

67. Miller, K. J.: "Materials Science Perspective of Metal Fatigue Resistance," *Mat. Sci. Technol.,* vol. 9, no. 6, pp. 453–462, 1993.

68. Ahmad, H. Y., and J. R. Yates: "An Elastic-Plastic Model for Fatigue Crack Growth at Notches," *Fatigue and Fracture Eng. Mat. Struct.,* vol. 17, no. 4, pp. 439–447, 1994.

69. Y. Murakami (ed.): "Stress Intensity Factors Handbook," Pergamon Press, Oxford, 1987.

70. Rooke, D. P., and D. J. Cartwright: "Compendium of Stress Intensity Factors," Her Majesty's Stationery Office, London, 1976.

71. "Standard Test Method for Plane-Strain Fracture Toughness of Metallic Materials," Specification E-399, American Society for Testing and Materials, Philadelphia.

72. Tada, H., P. C. Paris, and G. R. Irwin: "The Stress Analysis of Cracks Handbook," 2d ed., Paris Productions, Inc., 226 Woodbourne Dr., St. Louis, Mo., 1985.

73. J. P. Gallagher (ed.): "Damage Tolerant Design Handbook," Metals and Ceramics Information Center, Battelle Columbus Laboratories, Columbus, Ohio, 1983.

74. "Aerospace Structural Metals Handbook," Metals and Ceramics Information Center, Battelle Columbus Laboratories, Columbus, Ohio, 1991.

75. Hudson, C. M., and S. K. Seward: "A Compendium of Sources of Fracture Toughness and Fatigue Crack Growth Rate Data for Metallic Alloys," *Int. J. Fracture,* vol. 14, pp. R151–R184, 1978.

76. Hudson, C. M., and S. K. Seward: "A Compendium of Sources of Fracture Toughness and Fatigue Crack Growth Data for Metallic Alloys—Part II," *Int. J. Fracture,* vol. 20, pp. R59–R117, 1982.

77. Hudson, C. M., and S. K. Seward: "A Compendium of Sources of Fracture Toughness and Fatigue Crack Growth Data for Metallic Alloys—Part III," *Int. J. Fracture,* vol. 39, pp. R43–R63, 1989.

78. Hudson, C. M., and J. J. Ferraino: "A Compendium of Sources of Fracture Toughness and Fatigue Crack Growth Data for Metallic Alloys—Part IV," *Int. J. Fracture,* vol. 48, no. 2, pp. R19–R43, March 1991.

79. Taylor, D.: "A Compendium of Fatigue Threshold and Growth Rates," EMAS, Great Britain, 1985.

80. "Data Book on Fatigue Growth Rates of Metallic Materials," JSMS, Japan, 1983.

81. Paris, P. C., and R. Erdogan: "A Critical Analysis of Crack Propagation Laws," *Trans. ASME, J. Basic Eng.,* vol. 85, p. 528, 1963.

82. "Standard Test Method for Measurement of Fatigue Crack Growth Rates," Specification E-647, American Society for Testing and Materials, Philadelphia.

83. Forman, R. G., V. E. Kearney, and R. M. Engle: "Numerical Analysis of Crack Propagation in Cyclic-Loaded Structures," *Trans. ASME, J. Basic Eng.,* ser. D., vol. 89, pp. 459–464, 1967.

84. Walker, K.: "The Effect of Stress Ratio during Crack Propagation and Fatigue for 2024-T3 and 7075-T6 Aluminum Alloy Specimens," NASA TN-D 5390, National Aeronautics and Space Administration, 1969.

85. Miller, K. J.: "Metal Fatigue—Past, Current and Future," *Proc. Institution of Mechanical Engineers,* vol. 205, 1991.

86. "ASME Boiler and Pressure Vessel Code," Sec. III, Div. I, Subsection NA, American Society of Mechanical Engineers, New York, 1974.

87. Tipton, S. M.: "Fatigue Behavior under Multiaxial Loading in the Presence of a Notch: Methodologies for the Prediction of Life to Crack Initiation and Life Spent in Crack Propagation," Ph.D. dissertation, Stanford University, Stanford, Calif., 1985.

88. Tipton, S. M., and D. V. Nelson: "Fatigue Life Predictions for a Notched Shaft in Combined Bending and Torsion," "Multiaxial Fatigue," ASTM STP 853, K. J. Miller and M. W. Brown, eds., American Society for Testing and Materials, Philadelphia, 1985, pp. 514–550.

89. Tipton, S. M., and D. V. Nelson: "Developments in Life Prediction of Notched Components Experiencing Multiaxial Fatigue," "Material Durability/Life Prediction Modelling," PVP-vol. 290, S. Y. Zamirk and G. R. Halford, eds., ASME, 1994, pp. 35–48.

90. Bannantine, J. A.: "A Variable Amplitude Multiaxial Fatigue Life Prediction Method," Ph.D. dissertation, The University of Illinois, Champaign, 1989.

91. Chu, C. C.: "Programming of a Multiaxial Stress-Strain Model for Fatigue Analysis," SAE Paper 920662, Society of Automotive Engineers, Warrendale, Pa., 1992.

92. "Failure Analysis and Prevention," "Metals Handbook," 8th ed., vol. 10, American Society for Metals, Metals Park, Ohio, 1978.

CHAPTER 6
PROPERTIES OF ENGINEERING MATERIALS

Theodore Gela, D.Eng.Sc.
Professor Emeritus of Metallurgy
Stevens Institute of Technology
Hoboken, N.J.

6.1 MATERIAL-SELECTION CRITERIA IN ENGINEERING DESIGN 6.1
6.2 STRENGTH PROPERTIES: TENSILE TEST AT ROOM TEMPERATURE 6.2
6.3 ATOMIC ARRANGEMENTS IN PURE METALS: CRYSTALLINITY 6.5
6.4 PLASTIC DEFORMATION OF METALS 6.6
6.5 PROPERTY CHANGES RESULTING FROM COLD-WORKING METALS 6.9
6.6 THE ANNEALING PROCESS 6.11
6.7 THE PHASE DIAGRAM AS AN AID TO ALLOY SELECTION 6.12
6.8 HEAT-TREATMENT CONSIDERATIONS FOR STEEL PARTS 6.15
6.9 SURFACE-HARDENING TREATMENTS 9.21
6.10 PRESTRESSING 6.23
6.11 SOME PRACTICAL CONSIDERATIONS OF INDUCED RESIDUAL STRESSES IN ALLOYS 6.23
6.12 NOTCHED IMPACT PROPERTIES: CRITERIA FOR MATERIAL SELECTION 6.25
6.13 FATIGUE CHARACTERISTICS FOR MATERIALS SPECIFICATIONS 6.28
6.14 SHEAR AND TORSIONAL PROPERTIES 6.29
6.15 MATERIALS FOR HIGH-TEMPERATURE APPLICATIONS 6.30
6.15.1 Introduction 6.30
6.15.2 Creep and Stress: Rupture Properties 6.30
6.15.3 Heat-Resistant Superalloys: Thermal Fatigue 6.31
6.16 MATERIALS FOR LOW-TEMPERATURE APPLICATIONS 6.33
6.17 RADIATION DAMAGE 6.35
6.18 PRACTICAL REFERENCE DATA 6.37

6.1 MATERIAL-SELECTION CRITERIA IN ENGINEERING DESIGN

The selection of materials for engineering components and devices depends upon knowledge of material properties and behavior in particular environmental states. Although a criterion for the choice of material in critically designed parts relates to the performance in a field test, it is usual in preliminary design to use appropriate data obtained from standardized tests. The following considerations are important in material selection:

1. Elastic properties: stiffness and rigidity
2. Plastic properties: yield conditions, stress-strain relations, and hysteresis
3. Time-dependent properties: elastic phenomenon (damping capacity), creep, relaxation, and strain-rate effect
4. Fracture phenomena: crack propagation, fatigue, and ductile-to-brittle transition
5. Thermal properties: thermal expansion, thermal conductivity, and specific heat
6. Chemical interactions with environment: oxidation, corrosion, and diffusion

It is good design practice to analyze the conditions under which test data were obtained and to use the data most pertinent to anticipated service conditions.

The challenge that an advancing technology imposes on the engineer, in specifying treatments to meet stringent material requirements, implies a need for a basic approach which relates properties to structure in metals. As a consequence of the mechanical, thermal, and metallurgical treatments of metals, it is advantageous to explore, for example, the nature of induced internal stresses as well as the processes of stress relief. Better material performance may ensue when particular treatments can be specified to alter the structure in metals so that the likelihood of premature failure in service is lessened. Some of the following concepts are both basic and important:

1. Lattice structure of metals: imperfections, anisotropy, and deformation mechanisms
2. Phase relations in alloys: equilibrium diagrams
3. Kinetic reactions in the solid state: heat treatment by nucleation and by diffusionless processes, precipitation hardening, diffusion, and oxidation
4. Surface treatments: chemical and structural changes in carburizing, nitriding, and localized heating
5. Metallurgical bonds: welded and brazed joints

6.2 STRENGTH PROPERTIES: TENSILE TEST AT ROOM TEMPERATURE

The yield strength determined by a specified offset, 0.2 percent strain, from a stress-strain diagram is an important and widely used property for the design of statically loaded members exhibiting elastic behavior. This property is derived from a test in which the following conditions are normally controlled: surface condition of standard specimen is specified; load is axial; the strain rate is low, i.e., about 10^{-3} in/(in·s); and grain size is known. Appropriate safety factors are applied to the yield strength to allow for uncertainties in the calculated stress and stress-concentration factors and for possible overloads in service. Since relatively small safety factors are used in critically stressed aircraft materials, a proof stress at 0.01 percent strain offset is used because this more nearly approaches the proportional limit for elastic behavior in the material. A typical stress-strain plot from a tensile test is shown in Fig. 6.1, indicating the elastic and plastic behaviors. In order to effect more meaningful comparisons in design strength properties among materials having different specific gravities, the strength property can be divided by the specific gravity, giving units of psi per pound per cubic inch.

The modulus of elasticity is a measure of the stiffness or rigidity in a material. Values of the modulus normally are not exactly determined quantities, and typical values are commonly reported for a given material. When a material is selected on the basis of a high modulus, the tendency toward whip and vibration in shaft or rod applications is reduced. These effects can lead to uneven wear. Furthermore the modulus assumes particular importance in the design of springs and diaphragms, which necessitate a definite degree of motion for a definite load. In this connection, selection of a high-modulus material can lead to a thinner cross section.

The ultimate tensile strength and the ductility, percent elongation in inches per inch or percent reduction in area at fracture are other properties frequently reported from tensile tests. These serve as qualitative measures reflecting the ability of a material in deforming plastically after being stressed beyond the elastic region. The strength properties and ductility of a material subjected to different treatments can vary widely. This is illustrated in Fig. 6.2. When the yield strength is raised by treatment to a high

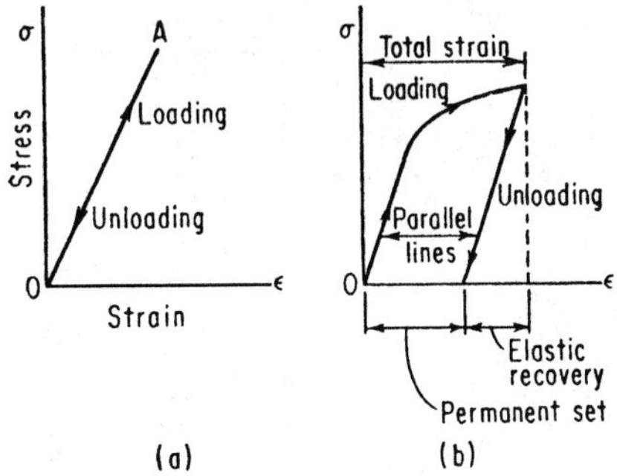

FIG. 6.1 Portions of tensile stress σ-strain ε curves in metals.[1] (*a*) Elastic behavior. (*b*) Elastic and plastic behaviors.

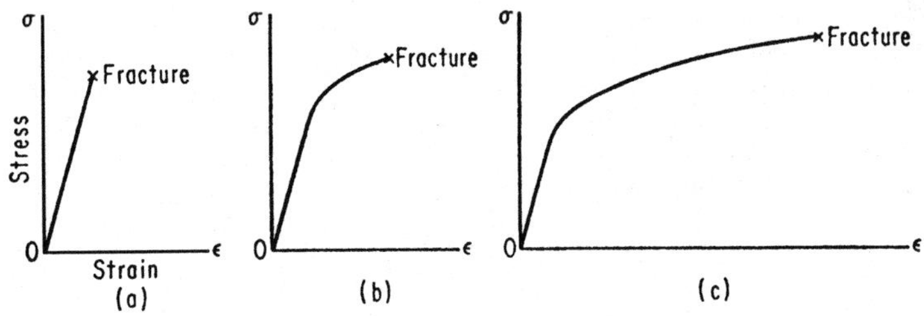

FIG. 6.2 The effects of treatments on tensile characteristics of a metal.[1] (*a*) Perfectly brittle (embrittled)—all elastic behavior. (*b*) Low ductility (hardened)—elastic plus plastic behaviors. (*c*) Ductile (softened)—elastic plus much plastic behaviors.

value, i.e., greater than two-thirds of the tensile strength, special concern should be given to the likelihood of tensile failures by small overloads in service. Members subjected solely to compressive stress may be made from high-yield-strength materials which result in weight reduction.

When failures are examined in statically loaded tensile specimens of circular section, they can exhibit a cup-and-cone fracture characteristic of a ductile material or on the other extreme a brittle fracture in which little or no necking down is apparent. Upon loading the specimen to the plastic region, axial, tangential, and radial stresses are induced. In a ductile material the initial crack forms in the center where the triaxial stresses become equally large, while at the surface the radial component is small and the deformation is principally by biaxial shear. On the other hand, an embrittled material exhibits no such tendency for shear and the fracture is normal to the loading axis. Some types of failures in round tensile specimens are shown in Fig. 6.3.

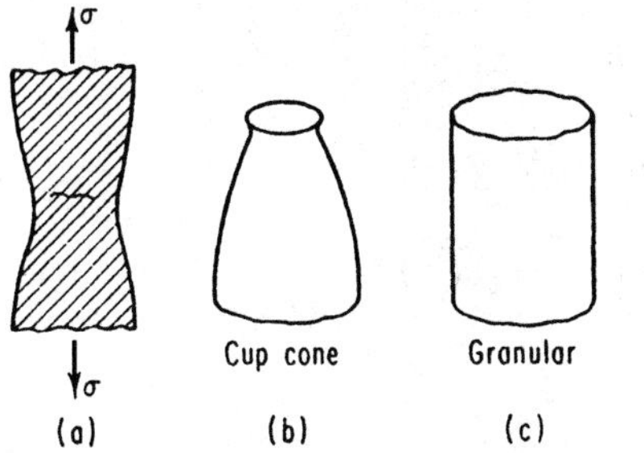

FIG. 6.3 Typical tensile-test fractures.[1] (*a*) Initial crack formation. (*b*) Ductile material. (*c*) Brittle material.

The properties of some wrought metals presented in Table 6.1 serve to show the significant differences relating to alloy content and treatment. Section 6.17 gives more information.

TABLE 6.1 Room-Temperature Tensile Properties for Some Wrought Metals

Metal	Condition	Ultimate tensile σ_{ult}, psi	Yield strength σ_y, psi	% elongation	Modulus of elasticity, psi	Density, lb/cu in.
Aluminum (pure)..	Annealed	13,000	5,000	35	10,000,000	0.098
7075T6...........	Heat-treated	76,000	67,000	11	10,400,000	0.101
Copper (pure).....	Annealed	32,000	10,000	45	17,000,000	0.321
Cu-Be(2 %).......	Heat-treated	200,000	150,000	2	19,000,000	0.297
Magnesium	Annealed	33,000	18,000	17	6,500,000	0.064
Mg-Al(8.5)A780A.	Strain relieved	44,000	31,000	18	6,500,000	0.065
Nickel (pure).....	Annealed	65,000	20,000	40	30,000,000	0.321
K Monel.........	Heat-treated	190,000	140,000	5	....	0.306
Titanium (pure)...	Annealed	59,000	40,000	28	15,000,000	0.163
Ti-Al(6)-V(14)α-β.	Heat-treated	170,000	150,000	7	15,000,000	0.163

The tensile properties of metals are dependent upon the rate of straining, as shown for aluminum and copper in Fig. 6.4, and are significantly affected by the temperature, as shown in Fig. 6.5. For high-temperature applications it is important to base design on different criteria, notably the stress-rupture and creep characteristics in metals, both of which are also time-dependent phenomena. The use of metals at low temperatures requires a consideration of the possibility of brittleness, which can be measured in the impact test.

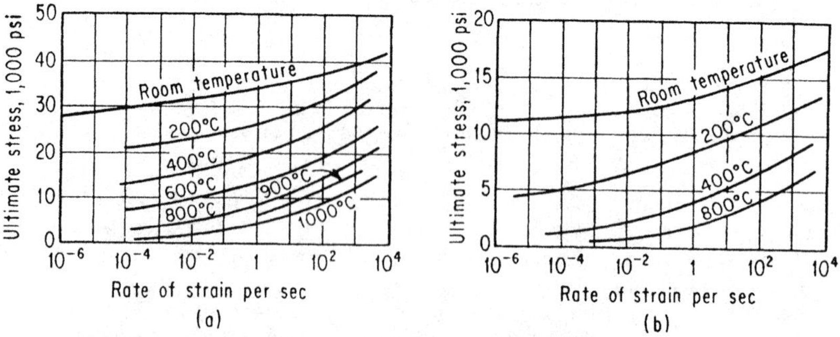

FIG. 6.4 Effects of strain rates and temperatures on tensile-strength properties of copper and aluminum.[1] (*a*) Copper. (*b*) Aluminum.

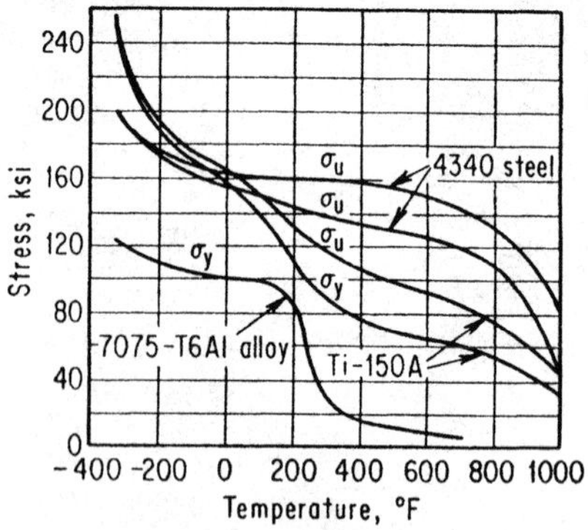

FIG. 6.5 Effects of temperatures on tensile properties. σ_u = ultimate tensile strength; σ_y = yield strength.

6.3 ATOMIC ARRANGEMENTS IN PURE METALS: CRYSTALLINITY

The basic structure of materials provides information upon which properties and behavior of metals may be generalized so that selection can be based on fundamental considerations. A regular and periodic array of atoms (in common metals whose atomic diameters are about one hundred-millionth of an inch) in space, in which a unit cell is the basic structure, is a fundamental characteristic of crystalline solids. Studies of these structures in metals lead to some important considerations of the behaviors in response to externally applied forces, temperature changes, as well as applied electrical and magnetic fields.

The body-centered cubic (bcc) cell shown in Fig. 6.6*a* is the atomic arrangement characteristic of αFe, W, Mo, Ta, ßTi, V, and Nb. It is among this class of metals that transitions from ductile to brittle behavior as a function of temperature are significant to investigate. This structure represents an atomic packing density where about 66 percent of the volume is populated by atoms while the remainder is free space. The elements Al, Cu, γFe, Ni, Pb, Ag, Au, and Pt have a closer packing of atoms in space constituting a face-centered cubic (fcc) cell shown in Fig. 6.6*b*. Characteristic of these are ductility properties which in many cases extend to very low temperatures. Another structure, common to Mg, Cd, Zn, αTi, and Be, is the hexagonal close-packed (hcp) cell in Fig. 6.6*c*. These metals are somewhat more difficult to deform plastically than the materials in the two other structures cited above.

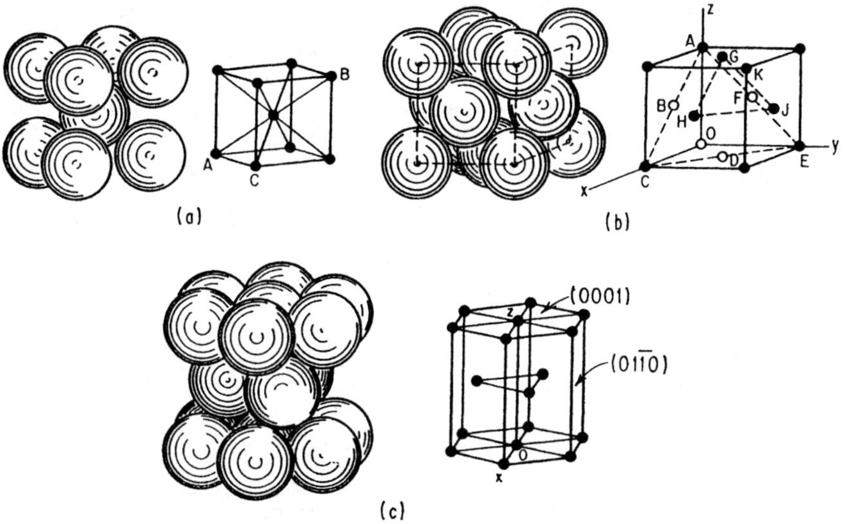

FIG. 6.6 Cell structure. (*a*) Body-centered cubic (bcc) unit cell structure. (*b*) Face-centered cubic (fcc) unit cell structure. (*c*) Hexagonal close-packed (hcp) unit cell structure.

It is apparent, from the atomic arrays represented in these structures, that the closest approach of atoms can vary markedly in different crystallographic directions. Properties in materials are anisotropic when they show significant variations in different directions. Such tendencies are dependent on the particular structure and can be especially pronounced in single crystals (one orientation of the lattices in space). Some examples of these are given in Table 6.2. When materials are processed so that

TABLE 6.2 Examples of Anisotropic Properties in Single Crystals

Property	Material and structure	Properties relation
Elastic module E in tension	αFe (bcc)	$E_{\{AB\}} \sim 2.2E_{\{AC\}}$
Elastic module G in shear	Ag (fcc)	$G_{\{OC\}} \sim 2.3G_{\{OK\}}$
Magnetization	αFe (bcc)	Ease of magnetization
Thermal expansion coefficient—α	Zn (hcp)	$\alpha_{\{OZ\}} \sim 4\alpha_{\{OA\}}$

their final grain size is large (each grain represents one orientation of the lattices) or that the grains are preferentially oriented, as in extrusions, drawn wire, rolled sheet, sometimes in forgings and castings, special evaluation of anisotropy should be made. In the event that directional properties influence design considerations, particular attention must be given to metallurgical treatments which may control the degree of anisotropy. The magnetic anisotropy in a single crystal of iron is shown in Fig. 6.7.

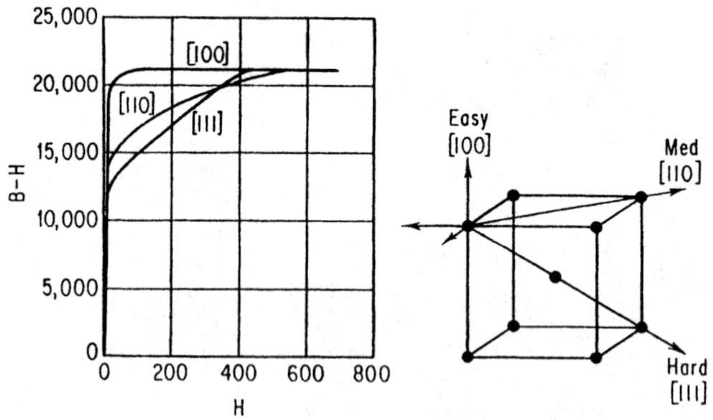

FIG. 6.7 Magnetic anisotropy in a single crystal of iron[2]: $I = (B - H)/4\pi$, where I = intensity of magnetization; B = magnetic induction, gauss; H = field strength, oersteds.

6.4 PLASTIC DEFORMATION OF METALS

When metals are externally loaded past the elastic limit, so that permanent changes in shape occur, it is important to consider the induced internal stresses, property changes, and the mechanisms of plastic deformation. These are matters of practical consideration in the following: materials that are to be strengthened by cold work, machining of cold-worked metals, flow of metals in deep-drawing and impact extrusion operations, forgings where the grain flow patterns may affect the internal soundness, localized surface deformation to enhance fatigue properties, and cold working of some magnetic materials. Experimental studies provide the key by which important phenomena are revealed as a result of the plastic-deformation process. These studies indicate some treatments that may be employed to minimize unfavorable internal-stress distributions and undesirable grain-orientation distributions.

Plastic deformation in metals occurs by a glide or slip process along densely packed planes fixed by the particular lattice structure in a metal. Therefore, an applied

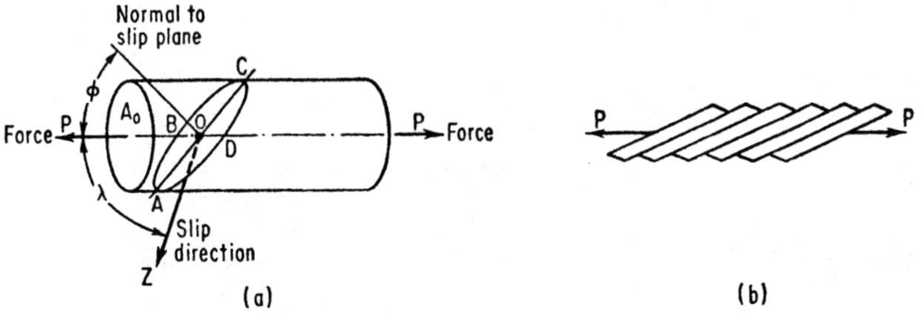

FIG. 6.8 Slip deformation in single crystals. (*a*) Resolved shear stress = $P/A_0 \cos \phi \cos \lambda$. *ABCD* is plane of slip. *OZ* is slip direction. (*b*) Sketch of single crystal after yielding.

load is resolved as a shear stress, on those particular glide elements (planes and directions) requiring the least amount of deformation work on the system. An example of this deformation process is shown in Fig. 6.8. Face-centered cubic (fcc) structured metals, such as Cu, Al, and Ni, are more ductile than the hexagonal structured metals, such as Mg, Cd, and Zn, at room temperature because in the fcc structure there are four times as many possible slip systems as in a hexagonal structure. Slip is initiated at much lower stresses in metals than theoretical calculations based on a perfect array of atoms would indicate. In real crystals there are inherent structural imperfections termed dislocations (atomic misfits) as shown in Fig. 6.9, which account for the observed yielding phenomenon in metals. In addition, dislocations are made mobile by mechanical and thermal excitations and they can interact to result in strain hardening of metals by cold work. Strength properties can be increased while the ductility is

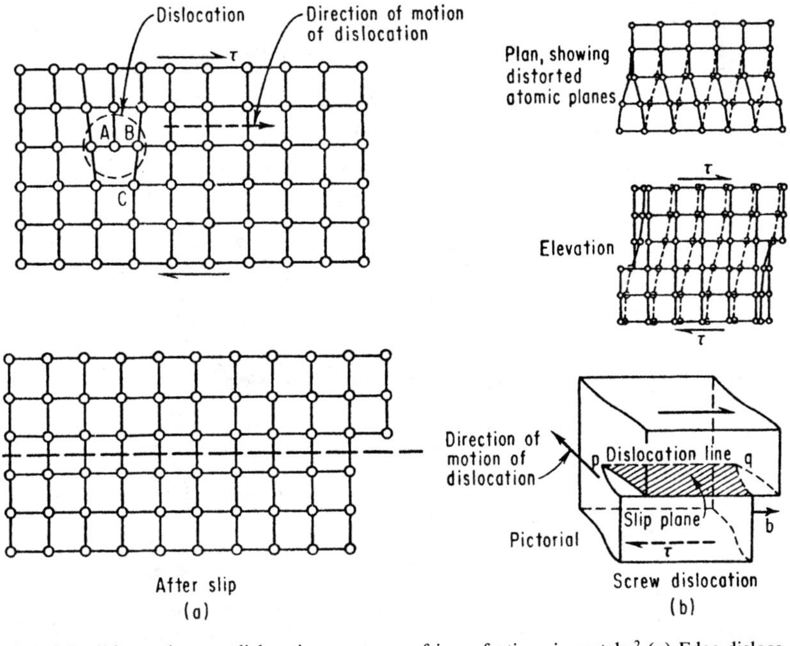

FIG. 6.9 Edge and screw dislocations as types of imperfections in metals.[2] (*a*) Edge dislocation. (*b*) Screw dislocation.

decreased in those metals which are amenable to plastic deformation. Cold working of pure metals and single-phase alloys provides the principal mechanism by which these may be hardened.

The yielding phenomenon is more nonhomogeneous in polycrystalline metals than in single crystals. Plastic deformation in polycrystalline metals initially occurs only in those grains in which the lattice axes are suitably oriented relative to the applied load axis, so that the critically resolved shear stress is exceeded. Other grains rotate and are dependent on the orientation relations of the slip systems and load application; these may deform by differing amounts. As matters of practical considerations the following effects result from plastic deformation:

1. Materials become strain-hardened and the resistance to further strain hardening increases.
2. The tensile and yield strengths increase with increasing deformation, while the ductility properties decrease.
3. Macroscopic internal stresses are induced in which parts of the cross section are in tension while other regions have compressive elastic stresses.
4. Microscopic internal stresses are induced along slip bands and grain boundaries.
5. The grain orientations change with cold work so that some materials may exhibit different mechanical and physical properties in different directions.

The Bauschinger effect in metals is related to the differences in the tensile and compressive yield-strength values, as shown at σ_T and σ_C in Fig. 6.10 when a ductile metal undergoes stress reversal. This change in polycrystalline metals is the result of the nonuniform character of deformation and the different pattern of induced macrostresses. These grains, in which the induced macrostresses are compressive, will yield at lower values upon the application of a reversed compressive stress because they are already part way toward yielding. This effect is encountered in cold-rolled metals where there is lateral contraction together with longitudinal elongation; this accounts for the decreased yield strength in the lateral direction compared with the increased longitudinal yield strength.

The control of metal flow is important in deep-drawing operations performed on sheet metal. It is desirable to achieve a uniform flow in all directions. Cold-rolling

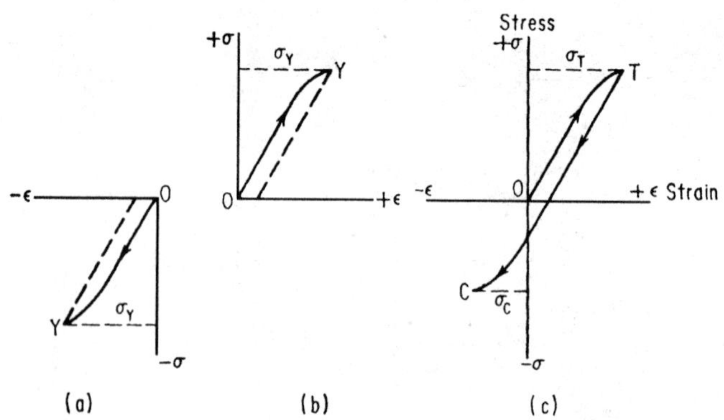

FIG. 6.10 The Bauschinger effect. (*a*) Compression. (*b*) Tension. The application of a compressive stress (*a*) or a tensile stress (*b*) results in the same value of yield strength *y*. (*c*) Stress reversal. A reversal of stress $O \rightarrow T \rightarrow C$ results in different values of tensile and compressive yield strengths; $\sigma_T \neq \sigma_C$.

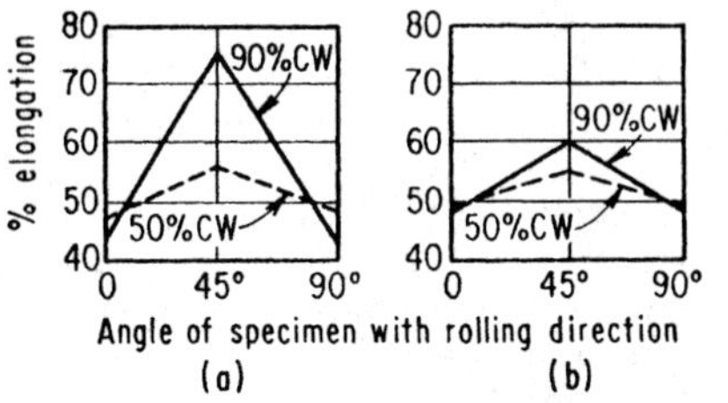

FIG. 6.11 Directionality in ductility in cold-worked and annealed copper sheet.[1] (*a*) Annealed at 1470°F. (*b*) Annealed at 750°F. The variation in ductility with direction for copper sheet is dependent on both the annealing temperature and the amount of cold work (percent CW) prior to annealing.

sheet metal produces a structure in which the grains have a preferred orientation. This characteristic can persist, even though the metal is annealed (recrystallized), resulting in directional properties as shown in Fig. 6.11. A further consequence of this directionality, associated with the deep-drawing operation, is illustrated in Fig. 6.12. The important factors, involved with the control of earing tendencies, are the fabrication practices of the amount of cold work in rolling and duration and temperatures of annealing. When grain textural problems of this kind are encountered, they can be studied by x-ray diffraction techniques and reasonably controlled by the use of optimum cold-working and annealing schedules.

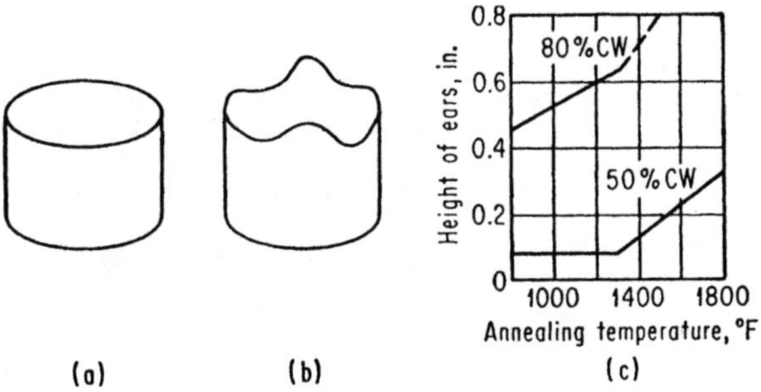

FIG. 6.12 The earing tendencies in cup deep drawn from sheet. (*a*) Uniform flow, nonearing. (*b*) Eared cup, the result of nonuniform flow. (*c*) Height of ears in deep-drawn copper cups related to annealing temperatures and amount of cold work.

6.5 PROPERTY CHANGES RESULTING FROM COLD-WORKING METALS

Cold-working metals by rolling, drawing, swaging, and extrusion is employed to strengthen them and/or to change their shape by plastic deformation. It is used principally on ductile metals which are pure, single-phase alloys and for other alloys which will not crack upon deformation. The increase in tensile strength accompanied by the decrease in ductility characteristic of this process is shown in Fig. 6.13. It is to be noted, especially from the yield-strength curve, that the largest rates of change occur during the initial amounts of cold reduction.

The variations in the macrostresses induced in a cold-drawn bar, illustrated in Fig. 6.14*a*, show that tensile stresses predominate at the surface. The equilibrium state of macrostresses throughout the cross section is altered by removing the surface layers in machining, the result of which may be warping in the machined part. It may be possible, however, to stress-relieve cold-worked metals, which generally have better machinability than softened (annealed) metals, by heating below the recrystallization

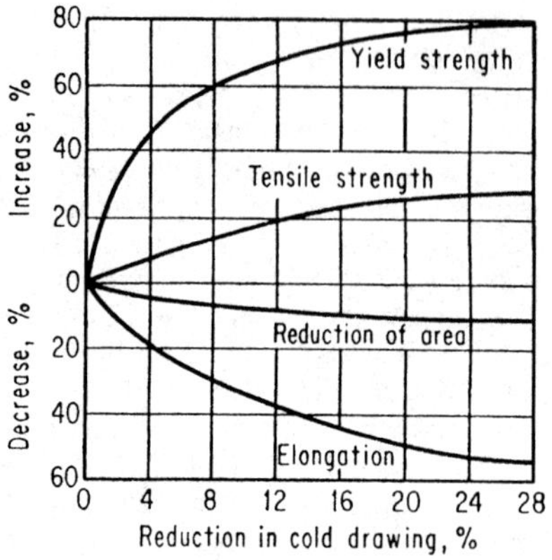

FIG. 6.13 Effect of cold drawing on the tensile properties of steel bars of up to 1-in cross section having tensile strength of 110,000 lb/in^2 or less before cold drawing.[3]

temperature. A typical alteration in the stress distribution, shown in Fig. 6.14*b*, is achieved so that the warping tendencies on machining are reduced, without decreasing the cold-worked strength properties. This stress-relieving treatment may also inhibit season cracking in cold-worked brasses subjected to corrosive environments containing amines. Since stressed regions in a metal are more anodic (i.e., go into solution more readily) than unstressed regions, it is often important to consider the relieving of stresses so that the designed member is not so likely to be subjected to localized corrosive attack.

Changes in electrical resistivity, elastic springback, and thermoelectric force resulting from cold work can be altered

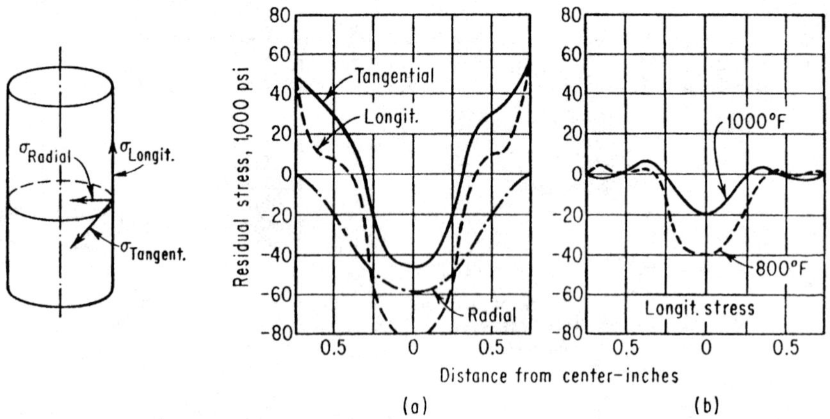

FIG. 6.14 Residual stress.[3] (*a*) In a cold-drawn steel bar 1½ in in diameter 20 percent cold-drawn, 0.45 percent C steel. (*b*) After stress-relieving bar.

by a stress-relieval treatment, in a temperature range from *A* to *B*, as shown in Fig. 6.15. However, the grain flow pattern (preferred orientation) produced by cold working can be changed only by heating the metal to a temperature at which recrystallized stress-free grains will form.

Residual tensile stresses at the surface of a metal promote crack nucleation in the fatigue of metal parts. The use of a localized surface deformation treatment by shot peening, which induces compressive stresses in the surface fibers, offers the likelihood of improvement in fatigue and corrosion properties in alloys. Shot-peening a forging flash line in high-strength aluminum alloys used in aircraft may also lessen the tendency toward stress-corrosion cracking. The effectiveness of this localized surface-hardening treatment is dependent on both the nature of surface discontinuities formed by shot-peening and the magnitude of compressive stresses induced at the surface.

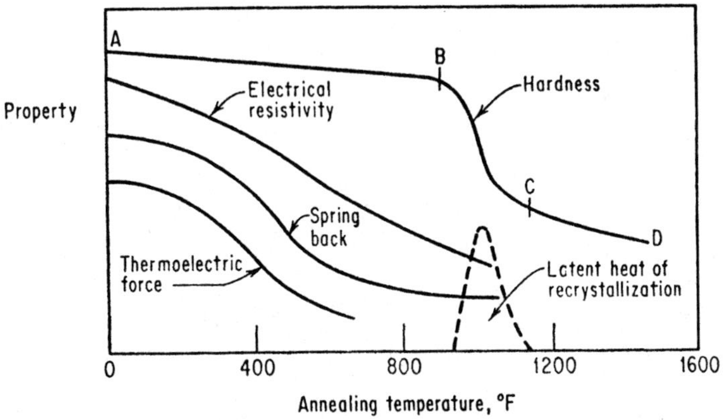

FIG. 6.15 The property changes in 95 percent cold-worked iron with heating temperatures (1 h). The temperature intervals[4]—$A \rightarrow B$, stress relieval; $B \rightarrow C$, recrystallization; and $C \rightarrow D$, grain growth—signify the important phenomena occurring.

6.6 THE ANNEALING PROCESS

Metals are annealed in order to induce softening for further deformation, to relieve residual stresses, to alter the microstructure, and, in some cases after electroplating, to expel by diffusion gases entrapped in the lattice. The process of annealing, the attaining of a strain-free recrystallized grain structure, is dependent mainly on the temperature, time, and the amount of prior cold work. The temperature indicated at C in Fig. 6.15 results in the complete annealing after 1 h of the 95 percent cold-worked iron. Heating beyond this temperature causes grains to grow by coalescence, so that the surface-to-volume ratio of the grains decreases together with decreasing the internal energy of the system. As the amount of cold work (from the originally annealed state) decreases, the recrystallization temperature increases and the recrystallized grain size increases. When a metal is cold-worked slightly (less than 10 percent) and subsequently annealed, an undesirable roughened surface forms because of the abnormally large grain size (orange-peel effect) produced. These aspects of grain-size control in the annealing process enter in material specifications.

The annealing of iron-carbon-base alloys (steels) is accomplished by heating alloys of eutectoid and hypoeutectoid compositions (0.8 percent C and less in plain carbon steels) to the single-phase region; austenite, as shown in Fig. 6.16, above the transition line *GS*; and for hypereutectoid alloys (0.8 to 2.0 percent C) between the transition lines *SK* and *SE* in Fig. 6.16; followed by a furnace cool at a rate of about 25°F per hour to below the eutectoid temperature *SK*. In the annealed condition, a desirable distribution of the equilibrium phases is thereby produced. A control of the microstructure is manifested by this process in steels. Grain-size effects are principally controlled by the high-temperature treatment, grain sizes increasing with increasing temperatures, and in some cases minor impurity additions such as vanadium inhibit grain coarsening to higher temperatures. These factors of grain-size control enter into the considerations of hardening steels by heat treatment.

The control of the atmosphere in the annealing furnace is desirable in order to prevent gas-metal attack. Moisture-free neutral atmospheres are used for steels which oxidize readily. When copper and its alloys contain oxygen, as oxide, it is necessary to keep the hydrogen content in the atmosphere to a minimum. At temperatures lower

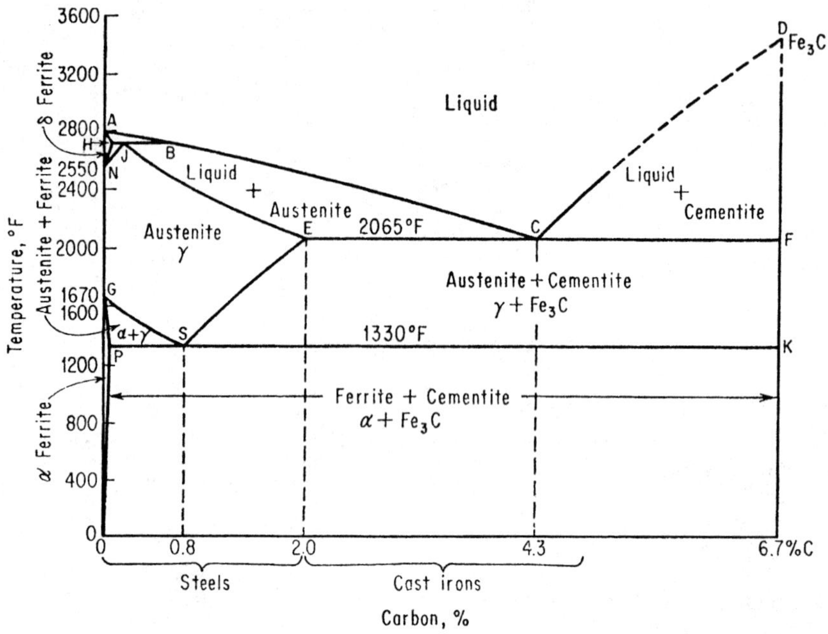

FIG. 6.16 The iron-carbon phase diagram.[4]

than 900°F, the hydrogen should not exceed 1 percent, and as the temperature is increased the hydrogen content should be reduced in order to prevent hydrogen embrittlement. In nickel and its alloys the atmosphere must be free from sulfur and slightly reducing by containing 2 percent or more of CO. Some aluminum alloys containing magnesium are affected by high-temperature oxidation in annealing (and heat treatment) and therefore require atmosphere control.

It is a characteristic property that strengths in all metals decrease with increasing temperatures. The coalescence of precipitate particles is one factor involved, so that material specifications for high-temperature use are concerned with alloy compositions that form particles having lower solubility and lower mobility. A second factor is concerned with the mobility of dislocations which increases at higher temperatures. Since strain hardening is reasoned to be due to the interaction of dislocations, then by the proper additions of solid-solution alloying elements that impede dislocations, resistance to softening will increase at the high temperature. The recrystallization temperature of iron is raised by the addition of 1 atomic percent of Mn, Cr, V, W, Cb, Ta in the same order in which the atomic size of the alloying elements differs from that of iron. The practical implications of these basic atomic considerations are important in selecting metals for high-temperature service.

6.7 THE PHASE DIAGRAM AS AN AID TO ALLOY SELECTION

Phase diagrams, which are determined experimentally and are based upon thermodynamic principles, are temperature-composition representations of slowly cooled alloys (annealed state). They are useful for predicting property changes with composition and selecting feasible fabrication processes. Phase diagrams also indicate the possible response of alloys to hardening by heat treatment. Shown on these diagrams are first-order phase transitions and the phases present. In two-phase regions, the compositions of each phase are shown on the phase boundary lines and the relative amounts of each phase present can be determined by a simple lever relation at a given temperature.

The particular phases that are formed in a system are governed principally by the physical interactions of valence electrons in the atoms and secondarily by atomic-size factors. When two different atoms in the solid state exist on, or where one is in, an atomic lattice, the phase is a solid solution (e.g., γ austenite phase in Fig. 6.16) analogous to a miscible liquid solution. When the atoms are strongly electropositive and strongly electronegative to one another, an intermetallic compound is formed (e.g., Fe_3C, cementite). The two atoms are electronically indifferent to one another and a phase mixture issues (e.g., $\alpha + Fe_3C$) analogous to the immiscibility of water and oil.

The thermodynamic criteria for a first-order phase change, indicated by the solid lines on the phase diagram, are that, at the transition temperature, (1) the change in Gibbs's free energy for the system is zero, (2) there is a discontinuity in entropy (a latent heat of transformation and a discontinuous change in specific heat), and (3) there is a discontinuous change in volume (a dilational effect).

In the selection of alloys for sound castings, particular attention is given that part of the system where the liquidus line (*ABCD*) goes through a minimum. For alloys between the composition limits of $E \rightarrow F$, a eutectic reaction occurs at 2065°F such that

$$\text{liq } C = \gamma + Fe_3C$$

It is for this reason in the iron-carbon system that cast irons are classified as having carbon contents greater than 2 percent. For purposes of controlling grain size, obtaining sound castings free from internal porosities (blowholes) and internal shrinkage cavities, and possessing good mold-filling characteristics, alloys and low-melting solutions are chosen near the eutectic composition (i.e., at *C*). Aluminum-silicon die-casting alloys have a composition of about 11 percent silicon near the eutectic composition. Special considerations need be given to the properties and structures in cast irons because the Fe_3C phase is thermodynamically unstable and decomposition to graphite (in gray cast irons) may result.

The predominant phase-diagram characteristic in steels is the eutectoid reaction, in the solid state, along *GSE* where $\gamma_s = \alpha + Fe_3C$ (pearlite) at 1330°F. Steels are therefore classified as alloys in the Fe-C system having a carbon content less than 2.0 percent C; and furthermore, according to their applications, compositions are designated as hypoeutectoid (C < 0.8 percent), eutectoid (C = 0.8 percent), and hypereutectoid (C < 2 percent > 0.8 percent). Since the slowly cooled room-temperature structures of steels contain a mechanical aggregate of the ferrite and Fe_3C-cementite phases, the property relations vary linearly as shown in Fig. 6.17. The ductility decreases with increasing carbon contents.

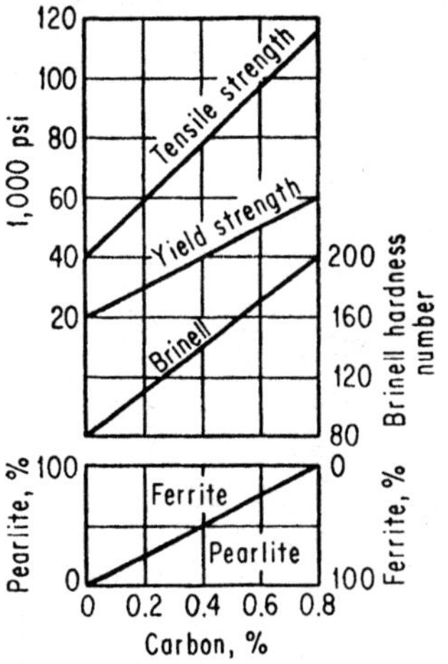

FIG. 6.17 Relation of mechanical properties and structure to carbon content of slowly cooled carbon steels.[4]

Some important characteristics of the equilibrium phases in steels are listed below:

Phase	Characteristics
α ferrite	Low C solubility (less than 0.03%) bcc, ductile, and ferromagnetic below 1440°F
Fe_3C, cementite	Intermetallic compound, orthorhombic, hard, brittle, and fixed composition at 6.7% C
γ austenite	Can dissolve up to 2% C in solid solution, fcc, nonmagnetic, and in this region annealing, hardening, forging, normalizing, and carburizing processes take place

Low-carbon alloys can be readily worked by rolling, drawing, and stamping because of the predominant ductility of the ferrite. Wires for suspension cables having a carbon content of about 0.7 percent are drawn at about 1100°F (patenting) because of the greater difficulty, in room-temperature deformation, caused by the presence of a relatively large amount of the brittle Fe_3C phase.

Extensive substitutional solid-solution alloys form in binary systems when they have similar chemical characteristics and atomic diameters in addition to having the same lattice structure. Such alloys include copper-nickel (monel metal being a commercially useful one), chromium-molybdenum, copper-gold, and silver-gold (jewelry alloys). The phase diagram and the equilibrium-property changes for this system are shown in Fig. 6.18*a*. Each pure element is strengthened by the addition of the other, whereby the strongest alloy is at an equal atom concentration. There are no first-order phase changes up to the start of melting (the solidus line *EHG*), so that these are not hardened by heat treatment but only by cold work. The electrical conductivity decreases

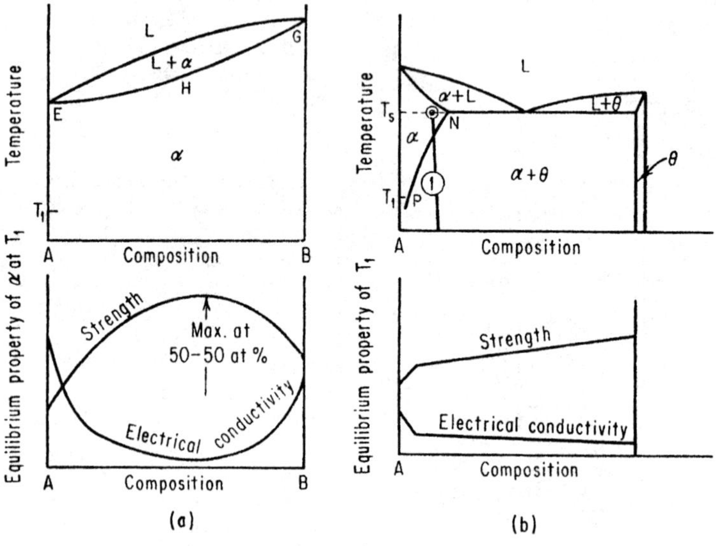

FIG. 6.18 Binary systems.[4] (*a*) Complete solid-solubility phase diagram. (*b*) Partial solid-solubility part of phase diagram. α is a substitutional solid solution, a phase with two different atoms on the same lattice. In the AlCu system θ is an intermediate phase (precipitant) having a composition nominally of $CuAl_2$.

from each end of the composition axis. Because of the presence of but one phase, these alloys are selected for their resistance to electrochemical corrosion. High-temperature-service metals are alloys which have essentially a single-phase solid solution with minor additions of other elements to achieve specific effects.

Another important system is one in which there are present regions of partial solid solubility as shown in Fig. 6.18*b* together with equilibrium-property changes. An important consideration in the selection of alloys containing two or more phases is that galvanic-corrosion attack may occur when there exists a difference in the electromotive potential between the phases in the environmental electrolyte. Sacrificial galvanic protection of the base metal in which the coating is more anodic than the base metal is used in zinc-plating iron-base alloys (galvanizing alloys). The intimate mechanical mixture of phases which are electrochemically different may result in pitting corrosion, or even more seriously, intergranular corrosion may result if the alloy is improperly treated by causing localized precipitation at grain boundaries.

Heat treatment by a precipitation-hardening process is indeed an important strengthening mechanism in particular alloys such as the aircraft aluminum-base, copper-beryllium, magnesium-aluminum, and alpha-beta titanium alloys (Ti, Al, and V). In these alloys a distinctive feature is that the solvus line *NP* in Fig. 6.18*b* shows decreasing solid solubility with decreasing temperature. This in general is a necessary, but not necessarily sufficient, condition for hardening by precipitation since other thermodynamic conditions as well as coherency relations between the precipitated phases must prevail. The sequence of steps for this process is as follows: An alloy is solution heat-treated to a temperature T_s, rapidly quenched so that a metastable supersaturated solid solution is attained, and then aged at experimentally determined temperature-time aging treatments to achieve desired mechanical properties. This is the principal hardening process for those particular nonferrous alloys (including Inconels) which can respond to a precipitation-hardening process.

The engineer is frequently concerned with the strength-to-density ratio of materials and its variation with temperature. A number of materials are compared on this basis in Fig. 6.19 in which the alloys designated by curves 1, 3, 4, 5, and 8 are heat-treatable nonferrous alloys.

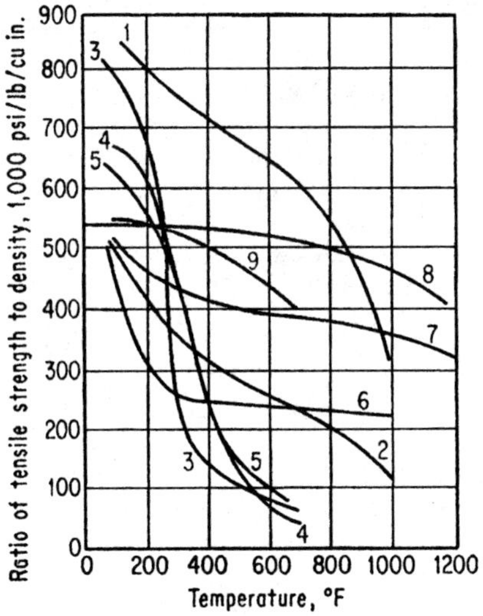

FIG. 6.19 Approximate comparison of materials on a strength-weight basis from room temperature to 1000°F.[6] (1) T_1, 8Mn; (2) 9990 T_1; (3) 75S, T6Al; (4) 24S, T4Al; (5) AZ31A Mg; (6) annealed stainless steel; (7) half-hard stainless steel; (8) Inconel X; (9) glass-cloth laminate.

6.8 HEAT-TREATMENT CONSIDERATIONS FOR STEEL PARTS

The heat-treating process for steel involves heating to the austenite region where the carbon is soluble, cooling at specific rates, and tempering to relieve some of the stress which results from the transformation. Some important considerations involved in specifying heat-treated parts are strength properties, warping tendencies, mass effects (hardenability), fatigue and impact properties, induced transformation stresses, and the use of surface-hardening processes for enhanced wear resistance. Temperature and time factors affect the structures issuing from the decomposition of austenite; for a eutectoid steel (0.8 percent C) they are as follows:

Decomposition product from γ	Structure	Mechanism	Temperature range, °F
Pearlite	Equilibrium ferrite + Fe_3C	Nucleation; growth	1300–1000
Bainite	Nonequilibrium ferrite + carbide	Nucleation; growth	1000–450
Martensite	Supersaturated tetragonal lattice	Diffusionless	M_s (≤450)

The tensile strength of a slowly cooled (annealed) eutectoid steel containing a coarse pearlite structure is about 120,000 lb/in^2. To form bainite, the steel must be cooled rapidly enough to escape pearlite transformation and must be kept at an intermediate temperature range to completion, from which a product having a tensile strength of about 250,000 lb/in^2 can be formed. Martensite, the hardest and most brittle product, forms independently of time by quenching rapidly enough to escape higher-temperature transformation products. The carbon atoms are trapped in the martensite, causing its lattice to be highly strained internally; its tensile strength is in excess of 300,000 lb/in^2. Isothermal transformation characteristics of all steels show the temperature-time and transformation products as in Fig. 6.20, where the lines indicate the start and end of transformation. On the temperature-time coordinates, involved cooling curves can be superimposed which show that, for a 1-in round water-quenched specimen, mixed products will be present. The outside will be martensite and the middle sections will contain pearlite. Alloying elements are added to steels principally to retard pearlite transformation either so that less drastic quenching media can be used or to ensure more uniform hardness throughout. This retardation is shown in Fig. 6.20*b* for an SAE 4340 steel containing alloying additions of Ni, Cr, or Mo and 0.4 percent C.

The carbon content in steels is the most significant element upon which selection for the maximum attainable hardness of the martensite is based. This relation is shown in Fig. 6.21. Since the atomic rearrangements involved in the transformation from the fcc austenite to the body-centered-tetragonal martensite result in a volumetric expansion, on cooling, of about 1 percent (for a eutectoid steel) nonuniform stress patterns can be induced on transformation. As cooling starts at the surface, by the normal process of heat transfer, parts of a member can be expanding, because of transformation, while further inward normal contraction occurs on the cooling austenite. The danger of cracking and distortion (warping) as a consequence of the steep thermal gradients and the transformation involved in hardening steels can be eliminated by using good design and the selection of the proper alloys. Where section size, time factors, and alloy content (as it affects transformation curves) permit, improved practices by martempering shown by *EFGH* in Fig. 6.20, followed by tempering or austempering shown by *EFK*, may be feasible and are worthy of investigation for the particular alloy used.

Uniform mass distribution and the elimination of sharp corners (potential stress raisers) by the use of generous fillets are recommended. Some design features pertinent to the elimination of quench cracks and the minimization of distortion by warping are illustrated by Fig. 6.22 in which *a*, *c*, and *e* represent poor designs in comparison with the suggested improvements apparent in *b, d,* and *f.*

Steels are tempered to relieve stresses, to impart ductility, and to produce a desirable microstructure by a reheating process of the quenched member. The tempering

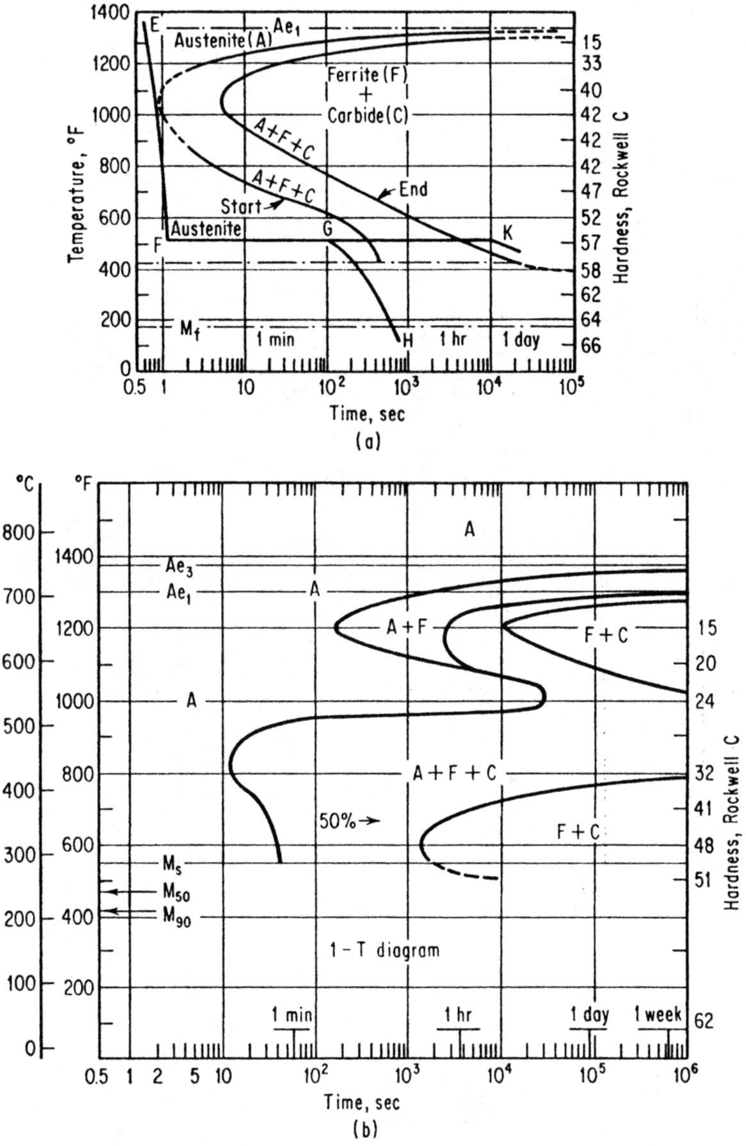

FIG. 6.20 Isothermal transformation diagram.[5] (*a*) Eutectoid carbon steel. (*b*) SAE 4340 steel. M_s = start of martensite temperature; M_f = finish of martensite temperature; *EFGH* = martempering (follow by tempering); *EFK* = austempering. *A*, austenite; F, ferrite; *C*, carbide; M_s and M_f, temperatures for start and finish of martensite transformation; M_{50} and M_{90}, temperatures for 50% and 90% of martensite transformation.

process is dependent on the temperature, time, and alloy content of the steel. Different alloys soften at different rates according to the constitutionally dependent diffusional structure. The response to tempering for 1 h for three different steels of the same carbon content is shown in Fig. 6.23. In addition, the tempering characteristics of a high-speed

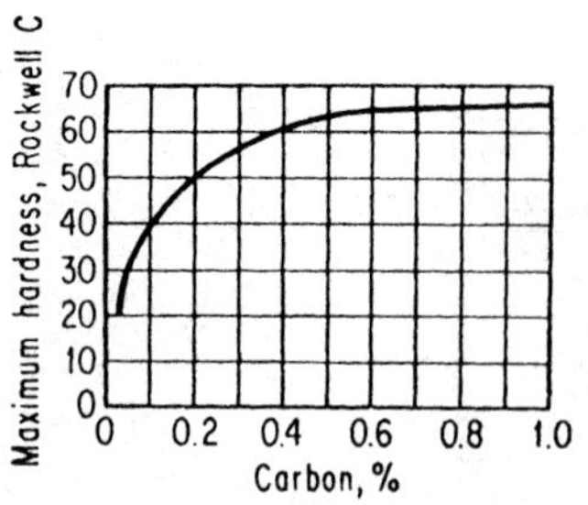

FIG. 6.21 Relation of maximum attainable hardness of quenched steels to carbon content.[7]

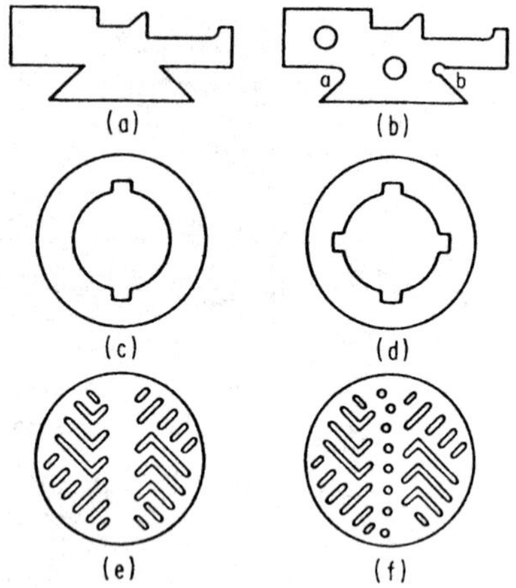

FIG. 6.22 Examples of good (*b*, *d*, and *f*) and bad (*a*, *c*, and *e*) designs for heat-treated parts.[8] (1) *b* is better than *a* because of fillets and more uniform mass distribution. (2) In *c* cracks may form at keyways. (3) Warping may be more pronounced in *e* than in *f*, which are blanking dies.[8]

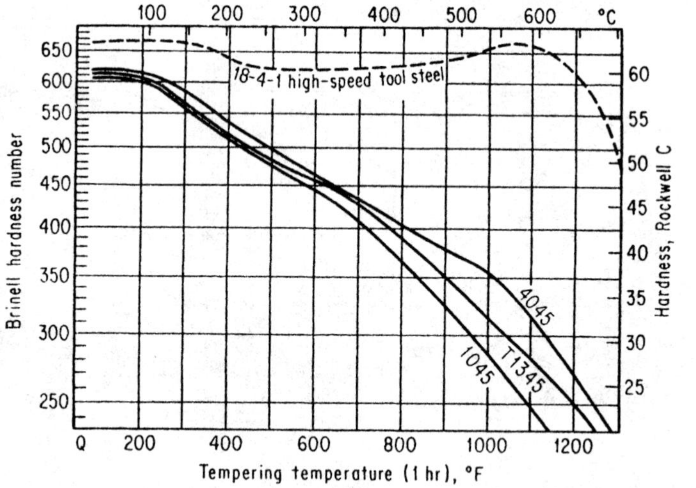

FIG. 6.23 Effect of tempering temperature on the hardnesses of SAE 1045, T1345, and 4045 steels. In the high-speed tool steel 18-4-1, secondary hardening occurs at about 1050°F.[9]

tool steel, 18 percent W, 4 percent Cr, 1 percent V, and 0.9 percent C, are shown to illustrate the secondary hardening at about 1050°F. The pronounced tendency for high-carbon steels to retain austenite on transformation normally has deleterious effects on dimensional stability and fatigue performance. In high-speed tool steel, the

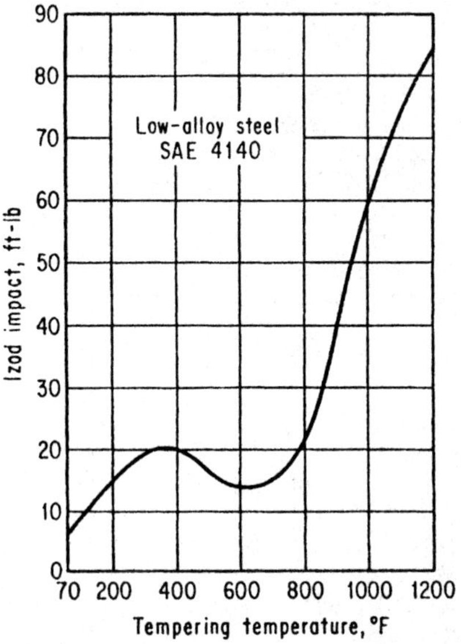

FIG. 6.24 In the tempering of this 4140 steel the notched-bar impact properties decrease in the range of 450 to 650°F.[9]

secondary hardening is due to the transformation of part of the retained austenite to newly transformed martensite. The structure contains tempered and untempered martensite with perhaps some retained austenite. Multiple tempering treatments on this type of steel produce a more uniform product.

In low-alloy steels where the carbon content is above 0.25 percent, there may be a tempering-temperature interval at about 450 to 650°F, during which the notch impact strength goes through a minimum. This is shown in Fig. 6.24 and is associated with the formation of an embrittling carbide network (ϵ carbide) about the martensite subgrain boundaries. Tempering is therefore carried out up to 400°F where the parts are to be used principally for wear resistance, or in the range of 800 to 1100°F where greater toughness is required. In the nomenclature of structural steels adopted by the Society of Automotive Engineers and the American Iron and Steel Institute, the first two numbers designate the type of steel according to the principal alloying elements and the last two numbers designate the carbon content:

Series designation	Types
10xx	Nonsulfurized carbon steels
11xx	Resulfurized carbon steels (free-machining)
12xx	Rephosphorized and resulfurized carbon steels (free-machining)
13xx	Manganese 1.75%
23xx*	Nickel 3.50%
25xx*	Nickel 5.00%
31xx	Nickel 1.25%, chromium 0.65%
33xx	Nickel 3.50%, chromium 1.55%
40xx	Molybdenum 0.20 or 0.25%
41xx	Chromium 0.50 or 0.95%, molybdenum 0.12 or 0.20%
43xx	Nickel 1.80%, chromium 0.50 or 0.80%, molybdenum 0.25%
44xx	Molybdenum 0.40%
45xx	Molybdenum 0.52%
46xx	Nickel 1.80%, molybdenum 0.25%
47xx	Nickel 1.05%, chromium 0.45%, molybdenum 0.20 or 0.35%
48xx	Nickel 3.50%, molybdenum 0.25%
50xx	Chromium 0.25, 0.40, or 0.50%
50xxx	Carbon 1.00%, chromium 0.50%
51xx	Chromium 0.80, 0.90, 0.95, or 1.00%
51xxx	Carbon 1.00%, chromium 1.05%
52xxx	Carbon 1.00%, chromium 1.45%
61xx	Chromium 0.60, 0.80, or 0.95%, vanadium 0.12%, 0.10% min, or 0.15% min

(*Continued*)

Series designation	Types
81xx	Nickel 0.30%, chromium 0.40%, molybdenum 0.12%
86xx	Nickel 0.55%, chromium 0.50%, molybdenum 0.20%
87xx	Nickel 0.55%, chromium 0.50%, molybdenum 0.25%
88xx	Nickel 0.55%, chromium 0.50%, molybdenum 0.35%
92xx	Manganese 0.85%, silicon 2.00%, chromium 0 or 0.35%
93xx	Nickel 3.25%, chromium 1.20%, molybdenum 0.12%
94xx	Nickel 0.45%, chromium 0.40%, molybdenum 0.12%
98xx	Nickel 1.00%, chromium 0.80%, molybdenum 0.25%

*Not included in the current list of standard steels.

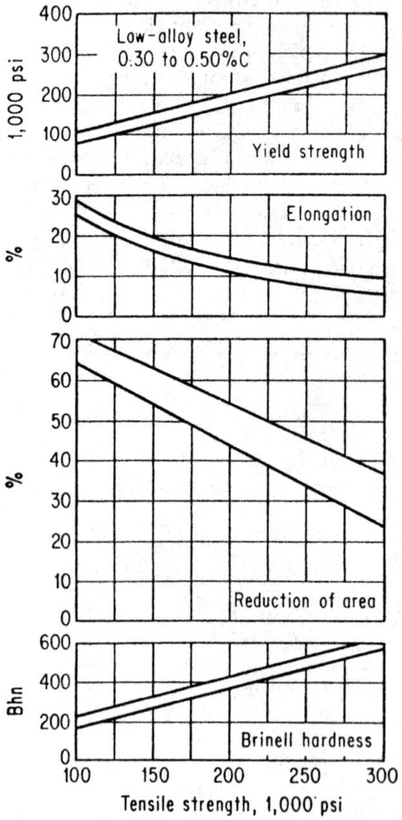

FIG. 6.25 The most probable properties of tempered martensite for a variety of low-alloy steels.[4]

The most probable properties of tempered martensite for low-alloy steels fall within narrow bands even though there are differences in sources and treatments. The relations for these shown in Fig. 6.25 are useful in predicting properties to within approximately 10 percent.

Structural steels may be specified by hardenability requirements, the H designation, rather than stringent specification of the chemistry. Hardenability, determined by the standardized Jominy end-quench test, is a measurement related to the variation in hardness with mass, in quenched steels. Since different structures are formed as a function of the cooling rate and the transformation is affected by the nature of the alloying elements, it is necessary to know whether the particular steel is shallow (A) or deeply hardenable (C), as in Fig. 6.26. The hardenability of a particular steel is a useful criterion in selection because it is related to the mechanical properties pertinent to the section size.

The selection of through-hardened steel based upon carbon content is indicated on the next page for some typical applications.

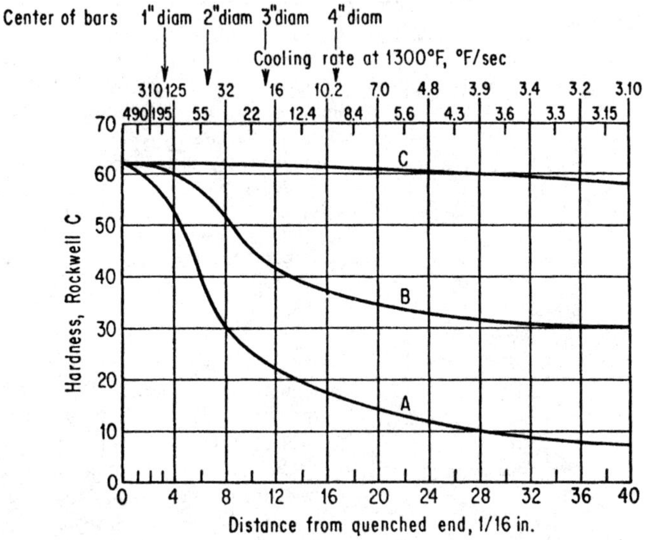

FIG. 6.26 Hardenability curves for different steels with the same carbon content.[7]

Carbon range	Requirement	Approx. tensile strength level, lb/in^2	Applications
Medium, 0.3 to 0.5%	Strength and toughness	150,000	Shafts, bolts, forgings, nuts
Intermediate, 0.5 to 0.7%	Strength	225,000	Springs
High, 0.8 to 1.0%	Wear resistance	300,000	Bearings, rollers, bushings

6.9 SURFACE-HARDENING TREATMENTS

The combination of high surface wear resistance and a tough-ductile core is particularly desirable in gears, shafts, and bearings. Various types of surface-hardening treatments and processes can achieve these characteristics in steels; the most important of these are the following:

Base metal	Process
Low C, to 0.3%	Carburization: A carbon diffusion in the γ-phase region, with controlled hydrocarbon atmosphere or in a box filled with carbon. The case depth is dependent on temperature-time factors. Heat treatment follows process.
Medium C, 0.4 to 0.5%	Localized surface heating by induction or a controlled flame to above the Ac_3 temperature; quenched and tempered.

(Continued)

Base metal	Process
Nitriding (nitralloys, stainless steels)	Formation of nitrides (in heat-treated parts) in ammonia atmosphere at 950 to 1000°F, held for long times. A thin and very hard surface forms and there may be dimensional changes.
Low C, 0.2%	Cyaniding: Parts placed in molten salt baths at heat-treating temperatures; some limited carburization and nitriding occur for cases not exceeding 0.020 in. Parts are quenched and tempered.

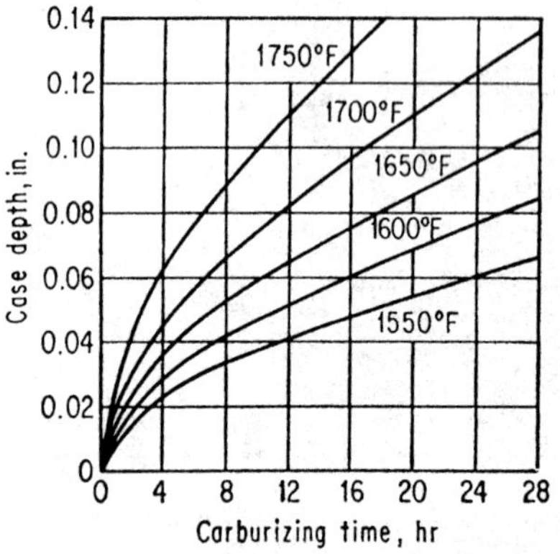

FIG. 6.27 Relation of time and temperature to carbon penetration in gas carburizing.[10]

The carbon penetration in carburization is determined by temperature-time-distance relations issuing from the solution of diffusional equations where *D*, the diffusion coefficient, is independent of concentrations. These relations, shown in Fig. 6.27, permit the selection of a treatment to provide specific case depths. Typical applications are as follows:

Case depth, in	Applications (automotive)
0.020 or more	Push rods, light-load gears, water pump shafts
0.020–0.040	Valve rocker arms, steering-arm bushings, brake and clutch pedal shifts
0.040–0.060	Ring gears, transmission gears, piston pins, roller bearings
0.060 or more	Camshafts

The heat treatments used on carburized parts depend upon grain-size requirements, minimization of retained austenite in the microstructure, amount of undissolved carbide network, and core-strength requirements. As a result of carburization, the surface fiber stresses are compressive. This leads to better fatigue properties. This treatment, which alters the surface chemistry by diffusion of up to 1 percent carbon in a low-carbon steel, gives better wear resistance because the surface hardness is treated for values above Rockwell C 60, while the low-carbon-content core has ductile properties to be capable of the transmission of torsional or bending loads.

Selection of the nitriding process requires careful consideration of cost because of the long times involved in the case formation. A very hard case having a hardness of about Rockwell C 70 ensures excellent wear resistance. Nitrided parts have good corrosion resistance and improved fatigue properties. Nitriding follows the finish-machining and -grinding operations, and many parts can be nitrided without great likelihood of distortion. Long service (several hundred hours) at 500°F has been attained in nitrided gears made from a chromium-base hot-work steel H11. Some typical nitrided steels and their applications are as follows:

Steel	Nitriding treatment h	Nitriding treatment °F	Case hardness, in	Case depth, Rockwell C	Applications
4140	48	975	0.025–0.035	53–58	Gears, shafts, splines
4340	48	975	0.025–0.035	50–55	Gears, drive shafts
Nitralloy 135M	48	975	0.020–0.025	65–70	Valve stems, seals, dynamic faceplates
H11	70	960 + 980	0.015–0.020	67–72	High-temperature power gears, shafts, pistons

6.10 PRESTRESSING

Concrete has a tensile strength which is a small fraction of its compressive strength and it can be prestressed (pretensioned) with high carbon, 0.7 to 0.85 percent C, steel wire, or strands. These have a tensile strength of about 260,000 lb/in^2, a yield strength of at least 80 percent of the tensile strength, and an adequate ductility. This wire or strand is pretensioned and when the concrete is cured the tensioning is relaxed. As a result the concrete is in compression, having more favorable load-carrying characteristics. When the steel is put in tension after concrete is cured, by different techniques, this is called "posttensioning."

When SAE 1010 steel was prestressed by roll-induced tension, different effects were observed. By prestraining this steel to 0.9 of its true fracture strain the fatigue life was increased. Prestraining to 0.95 of the true fracture strain, on the other hand, resulted in a decreased fatigue life. Generalizations on the effects of prestressing can create misleading results. Involved are the nature and amount of prestressing, type of materials, and effects on property changes. Plain carbon steels, when prestressed within narrow limits, may have their torsional fracture strains increased. On the other hand, prestressing 304 stainless steel resulted in stress corrosion cracking in a corrosive environment. Transition temperatures on impact can be increased (an undesirable effect) in prestressed low-carbon steels. Shot-peening gear teeth, which can result in improved fatigue properties because of the induced surface compressive stresses, can be a beneficial effect of prestressing.

6.11 SOME PRACTICAL CONSIDERATIONS OF INDUCED RESIDUAL STRESSES IN ALLOYS

The presence of residual stresses, especially at surfaces, may affect the fatigue and bending properties significantly. These stresses can be induced in varying degrees by processes involving thermal changes as in heat treatments, welding, and exposures to steep temperature gradients, as well as mechanical effects such as plastic deformation, machining, grinding, and pretensioning. Effects of these when superimposed on externally applied stresses may reduce fatigue properties, or these properties may be improved by specific surface treatments. Comprehensive studies have been made to quantitatively measure the type—tensile or compressive—and the magnitude of residual surface elastic stresses so that the design engineer and user of structural members can optimize such material behavior.

When surface elastic stresses are present in crystalline material, they may be measured nondestructively by x-ray diffraction techniques, a review of which is given by D. P. Koistenen et al., in the SAE publication "TR-182." A two-exposure method for locating the shifts in x-ray diffraction peaks at high Bragg angles θ can result in stress measurements accurate to within ±5000 lb/in², using elasticity relations and the fact that elastic atom displacements enable x-ray measurements to act as sensitive nondestructive strain gauges. The relation which is used is

$$\sigma_\phi = \frac{E(\cot\theta)(2\theta_n - 2\theta_i)}{2(1+\nu)\sin^2\psi}\,\frac{1}{57.3}\ \text{lb/in}^2$$

where σ_ϕ = surface elastic stress, lb/in²
E = Young's modulus, lb/in²
θ_n, θ_i = measured Bragg angle at normal (n) and inclined (i) incidence
ν = Poisson's ratio
ψ = angle between normal and inclined incident x-ray beam

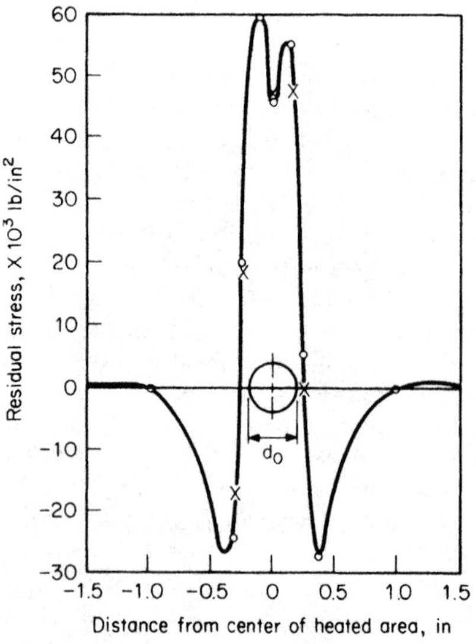

FIG. 6.28 Induced residual stress pattern by localized spot welding.[33]

Typically the back-reflection x-rays used will penetrate the surface by about 0.0002 in. Subsurface stresses may be measured in thick specimens by removing surface layers by electropolishing. These measurements are used for design of gears, bearings, welded plates, aircraft structures, and generally where fatigue and bending properties in use are important. Furthermore, improvements in surface properties can be enhanced by surface treatments which will induce residual compressive stresses.

The effect of localized heating (to about 1300°F) in spot welding a constrained steel specimen had a residual stress pattern as shown in Fig. 6.28.

Surface finishing in heat-treated steel gears, bearings, and dies often entails grinding in the hardened condition because of the greater dimensional tolerances that are desired. Metal removal by the abrasive grinding wheels can cause different degrees of surface residual stresses in the metals because of the frictional heat generated. Very abusive grinding may cause cracks. Examples of the stress patterns and fatigue properties in grinding a hardened aircraft alloy AISI 4340 are shown in Figs. 6.29 and 6.30. The significant decrease in fatigue strength of over 40 percent focuses special attention on the importance of controlling and specifying grinding and metal removal practices in hardened alloys. In addition to this, residual stress patterns may be induced due to the presence of retained austenite in hardened high-carbon steels.

Surface treatments which will induce compressive stresses in surface layers can improve fatigue properties substantially. These include mechanical deformation of surfaces by shot peening and roller burnishing, as well as microstructural effects in carburizing or nitriding steels. The beneficial effects of surface treatments by hardened 4340 steel are shown in Fig. 6.31.

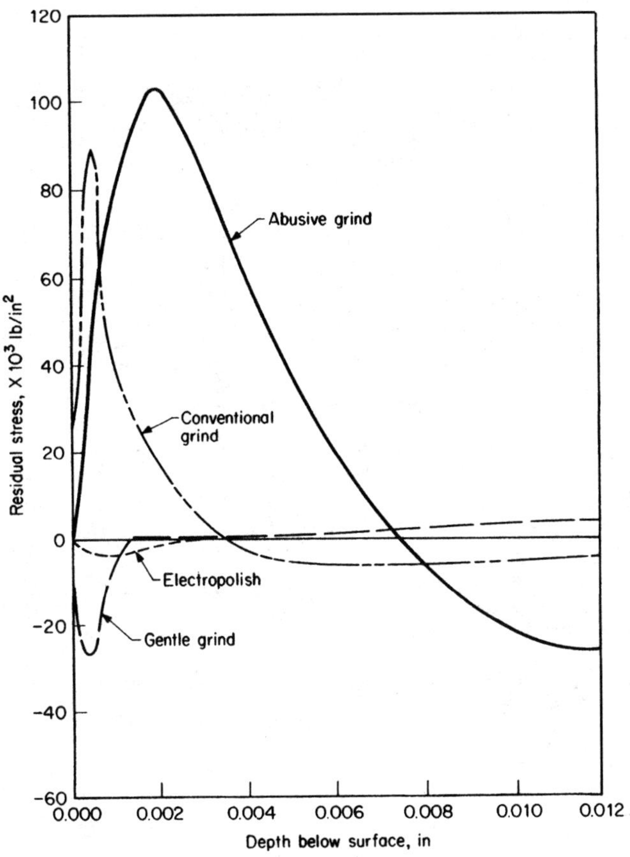

FIG. 6.29 Induced stresses in grinding ASAI 4340 steel heat-treated to Rockwell C 50.

To recover some fatigue properties in decarburized steel surfaces it may be helpful to consider shot peening these. Skin rolling of sheet or tubular products may also induce compressive surface stresses.

6.12 NOTCHED IMPACT PROPERTIES: CRITERIA FOR MATERIAL SELECTION

When materials are subject to high deformation rates and are particularly sensitive to stress concentrations at sharp notches, criteria must be established to indicate safe operating-temperature ranges. The impact test (Izod or Charpy V-notch) performed on notched specimens conducted over a prescribed temperature range indicates the likelihood of ductile (shear-type) or brittle (cleavage-type) failure. In this test the velocity of the striking head at the instant of impact is about 18 fps, so the strain rates are several orders of magnitude greater than in a tensile test. The energy absorbed in fracturing a standard notched specimen is measured by the differences in potential energy from free fall of the hammer to the elevation after fracturing it. The typical effect of temperature upon impact energy for a metal which shows ductile and brittle characteristics is shown

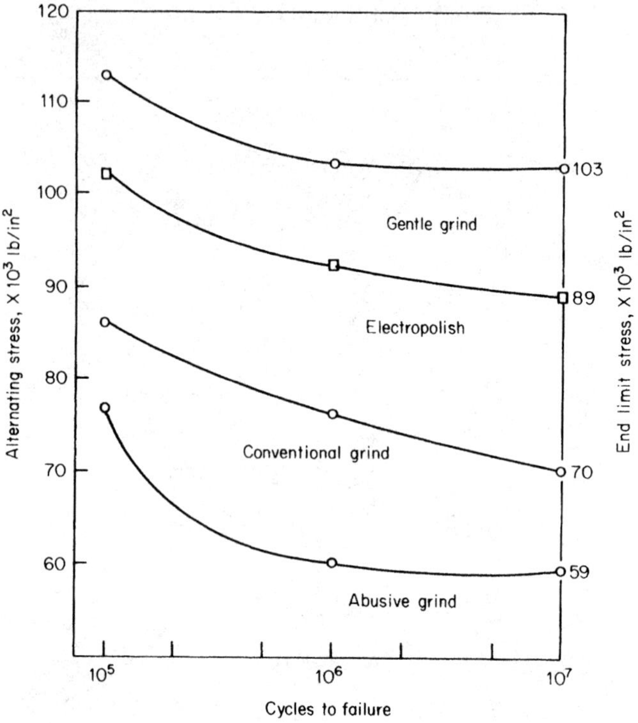

FIG. 6.30 Effect of severity of surface grinding of hardened 4340 Rockwell C (50) steel.[34,35]

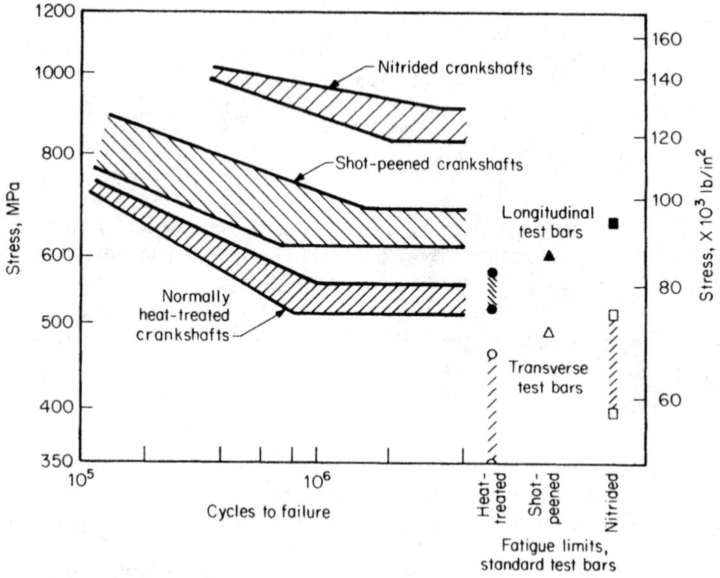

FIG. 6.31 Improved fatigue strength by shot-peening and nitriding 4340 hardened crankshafts.[36]

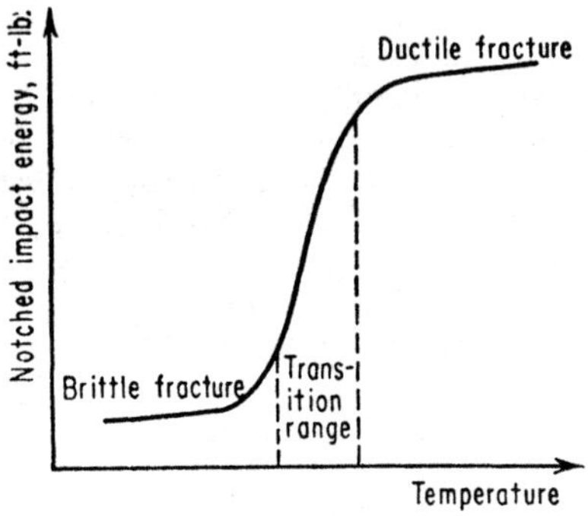

FIG. 6.32 The ductile-to-brittle transition in impact.

in Fig. 6.32. Interest centers on the transition temperature range and the material-sensitive factors, such as composition, microstructure, and embrittling treatments. ASTM specifications for structural steel for ship plates specify a minimum impact energy at a given temperature, as, for example, 15 ft·lb at 40°F. It is desirable to use materials at impact energy levels and at service temperatures where crack propagation does not proceed. Some impact characteristics for construction steels are shown in Fig. 6.33.

It is generally characteristic of pure metals which are fcc in lattice structure to possess toughness (have no brittle fracture tendencies) at very low temperatures. Body-centered-cubic pure metals, as well as hexagonal metals, do show ductile-to-brittle behavior. Tantalum, a bcc metal, is a possible exception and is ductile in impact even at cryogenic temperatures. Alloyed metals do not follow any general pattern of behavior; some specific impact values for these are as follows:

Material	Charpy V notch at 80°F	Impact strength at −100°F, ft·lb
Cu-Be (2%) HT	5.4	5.5
Phosphor bronze (5% Sn):		
Annealed	167	193
Spring temper	46	44
Nilvar Fe-Ni (36%):		
Annealed	218	162
Hard	97	77
2024-T6 aluminum aircraft alloy, HT aged	12	12
7079-T6 aluminum aircraft alloy, HT aged	4.5	3.5
Mg-Al-Zn extruded	7.0	3.0

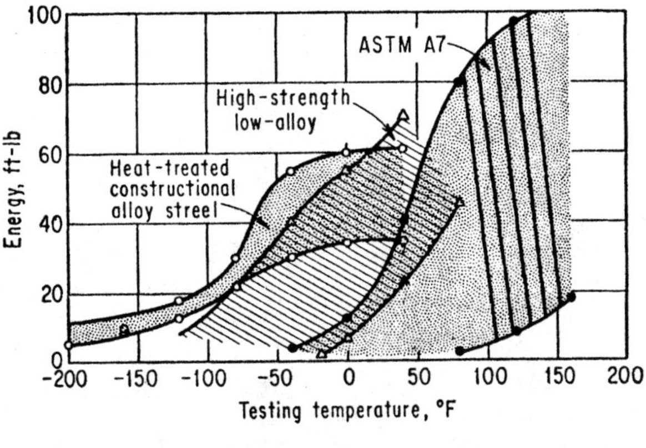

FIG. 6.33 Influence of testing temperature on notch toughness, comparing carbon steel of structural quality (ASTM A7) with high-strength, low-alloy and heat-treated constructional alloy steel.[4] Charpy V notch.

Austenitic stainless steels are ductile and do not exhibit transition in impact down to very low temperatures.

Some brittle service failures in steel structures have occurred in welded ships, gas-transmission pipes, pressure vessels, bridges, turbine generator rotors, and storage tanks. Serious consideration must then be given the effect of stress raisers, service temperatures, tempering embrittling structures in steels, grain-size effects, as well as the effects of minor impurity elements in materials.

6.13 FATIGUE CHARACTERISTICS FOR MATERIALS SPECIFICATIONS

Most fatigue failures observed in service as well as under controlled laboratory tests are principally the result of poor design and machining practice. The introduction of potential stress raisers by inadequate fillets, sharp undercuts, and toolmarks at the surfaces of critically cyclically stressed parts may give rise to crack nucleation and propagation so that ultimate failure occurs. Particular attention should be given to material fatigue properties where rotating and vibrating members experience surface fibers under reversals of stress.

In a fatigue test, a highly polished round standard specimen is subjected to cyclic loading; the number of stress reversals to failure is recorded. For sheets, the standard specimen is cantilever supported. Failure due to tensile stresses usually starts at the surface. Typical of a fatigue fracture is its conchoidal appearance, where there is a smooth region in which the severed sections rubbed against each other and where the crack progressed to a depth where the load could no longer be sustained. From the fatigue data a curve of stress vs. number of cycles to failure is plotted. Note that there can be considerable statistical fluctuation in the results (about ±15 percent variations in stress).

Two characteristics can be observed in fatigue curves with respect to the endurance limit shown by E_a and E_b in Fig. 6.34:

E_a, curve *a*, the asymptotic stress value, typical in most materials

E_b, curve *b*, a stress value taken at an arbitrary number of cycles; e.g., 500,000,000, typical in Al and Mg alloys

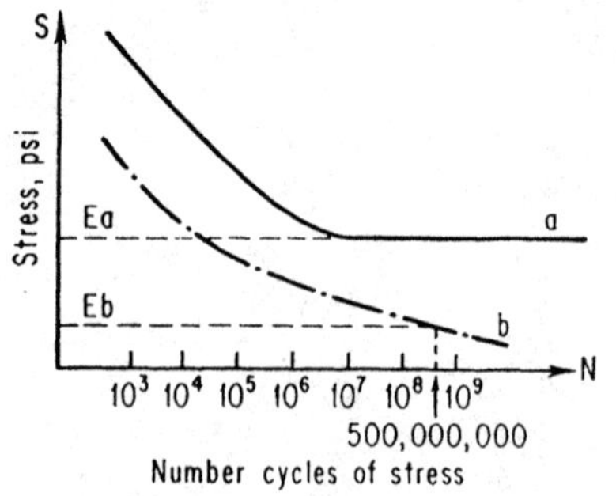

FIG. 6.34 Fatigue curves. (*a*) Most materials have an endurance limit E_a (asymptotic stress). (*b*) Endurance limit E_b (nonasymptotic set at arbitrary value of N).

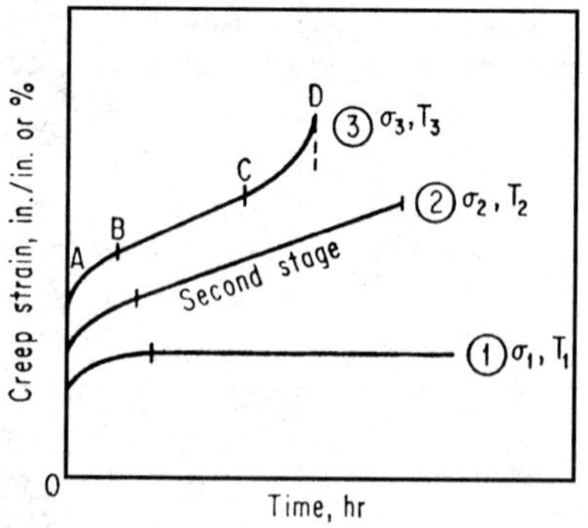

FIG. 6.35 Typical creep curves. At constant temperature, $\sigma_3 > \sigma_2 > \sigma_1$. At constant stress $T_3 > T_2 > T_1$.

For design specifications, the endurance limit represents a safe working stress for fatigue. The endurance ratio is defined as the ratio of the endurance limit to the ultimate tensile strength. These values are strongly dependent upon the presence of notches on the surface and a corrosive environment, and on surface-hardening treatments. In corrosion, the pits formed act as stress raisers leading generally to greatly reduced endurance ratios. References 15 through 20 provide useful information on corrosion. A poorly machined surface or a rolled sheet with surface scratches evidences low endurance ratios, as do parts in service with sharp undercuts and insufficiently filleted changes in section.

Improvements in fatigue properties are brought about by those surface-hardening treatments which produce induced compressive stresses as in steels, nitrided (about 160,000 lb/in^2 compressive stress) or carburized (about 35,000 lb/in^2 compressive stress).

Metallurgical factors related to poorer fatigue properties are the presence of retained austenite in hardened steels, the presence of flakes or sharp inclusions in the microstructure, and treatments which induce preferential corrosive grain-boundary attack. When parts are quenched or formed so that surface tensile stresses are present, stress-relieval treatments are advisable.

6.14 SHEAR AND TORSIONAL PROPERTIES

When rivets, bolts, and fastening pins are used in structural components, their shear and torsional properties are evaluated mainly for ductile materials. The stress which produces shear fracture is reported as the ultimate shear strength or sometimes as shear strength. Some typical and averaged ratios of strength are given below:

Material	Brinell hardness	Ultimate shear strength/ ultimate tensile strength
Ductile cast irons	170–300	0.9–1.0
Aluminum aircraft alloys	120	0.65
Brasses	100	0.7

In the elastic region the shear or torsional properties are related by

$$G = \tau/\gamma$$

where G = modulus of rigidity in shear
τ = shear stress
γ = shear strain

Furthermore, one can calculate with reasonably good accuracy the relation of elastic moduli by

$$G = E/2(1 + \nu)$$

where E = Young's modulus of rigidity in tension
ν = Poisson's ratio

This relation presumes isotropic elastic behavior.

The elastic shear strength, determined at a fixed offset on the shear stress-strain diagram (i.e., 0.04 percent offset), has been related to the tensile yield strength

(0.2 percent offset) such that its ratio is about 0.6 for ductile cast irons and 0.55 for aluminum alloys.

6.15 MATERIALS FOR HIGH-TEMPERATURE APPLICATIONS

6.15.1 Introduction

Selection of materials to withstand stress at high temperatures is based upon experimentally determined temperature stress-time properties. Some useful engineering design criteria follow.

1. Dimensional change, occurring by plastic flow, when metals are stressed at high temperatures for prolonged periods of time, as measured by creep tests
2. Stresses that lead to fracture, after certain set time periods, as determined by stress-rupture tests, where the stresses and deformation rates are higher than in a creep test
3. The effect of environmental exposure on the oxidation or scaling tendencies
4. Considerations of such properties as density, melting point, emissivity, ability to be coated and laminated, elastic modulus, and the temperature dependence on thermal conductivity and thermal expansion

Furthermore, the microstructural changes occurring in alloys used at high temperatures are correlated with property changes in order to account for the significant discontinuities which occur with exposure time. As a result of these evaluations, special alloys that have been (or are being) developed are recommended for use in different temperature ranges extending to about 2800°F (refractory range). Vacuum or electron-beam melting and special welding techniques are of special interest here in fabricating parts.

6.15.2 Creep and Stress: Rupture Properties

In a creep test, the specimen is heated in a temperature-controlled furnace, an axial load is applied, and the deformation is recorded as a function of time, for periods of 1000 to 3000 h. Typical changes in creep strain with time, for different conditions of stress and temperature, are shown in Fig. 6.35. Plastic flow creep, associated with the movement of dislocations by climb sliding of grain boundaries and the diffusion of vacancies, is characterized by:

1. *OA*, elastic extension on application of load
2. *AB*, first stage of creep with changing rate of creep strain
3. *BC*, second stage of creep, in which strain rate is linear and essentially constant
4. *CD*, third stage of increasing creep rate leading to fracture

Increasing stress at a constant temperature or increasing temperature at constant stress results in the transfers from curve 1 to curves 2 and 3 in Fig. 6.35.

The engineering design considerations for dimensional stability are based upon

1. Stresses resulting in a second-stage creep rate of 0.0001 percent per hour (1 percent per 10,000 h or 1 percent per 1.1 years)

2. A second-stage creep rate of 0.00001 percent per hour (1 percent per 100,000 h or 1 percent per 11 years), where weight is of secondary importance relative to long service life, as in stationary turbines

The time at which a stress can be sustained to failure is measured in a stress-rupture test and is normally reported as rupture values for 10, 100, 1000, and 10,000 h or more. Because of the higher stresses applied in stress-rupture tests, shown in Fig. 6.36, some extrapolation of data may be possible and some degree of uncertainty may ensue. Discontinuous changes at points *a* in the stress-rupture data shown in Fig. 6.37 are associated with a change from transgranular to intergranular fracture, and further microstructural changes can occur at increasing times. A composite picture of various high-temperature test results is given in Fig. 6.38 for a type 316 austenitic stainless steel.

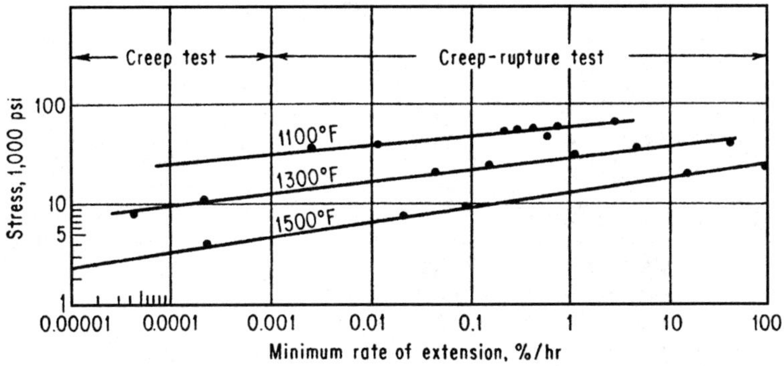

FIG. 6.36 Correlation of creep and rupture test data for type 316 stainless steel (18 Cr, 8 Ni, and Mo).[12]

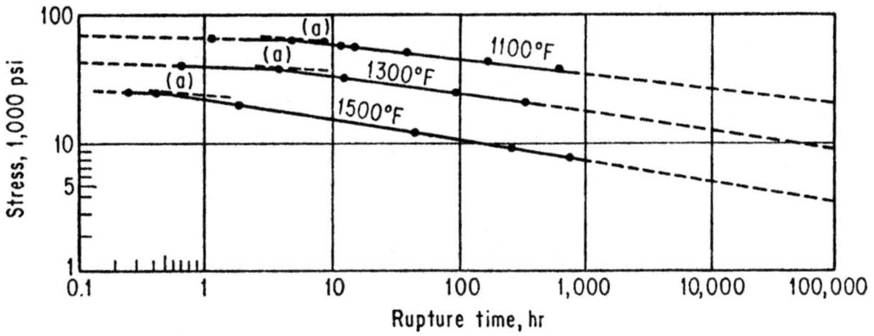

FIG. 6.37 Stress vs. rupture time for type 316 stainless steel.[12] The structural character associated with point (*a*), on each of the three relations, is that the mode of fracture changes from transgranular to intergranular.

6.15.3 Heat-Resistant Superalloys: Thermal Fatigue

The materials especially developed to function at high temperatures for sustained stress application in aircraft (jet engines) are referred to as "superalloys." These are nickel-base alloys designated commercially as Nimonic Hastelloy, Inconel Waspaloy, and René or cobalt-base alloys designated as S-816, HS25, and L605. Significant

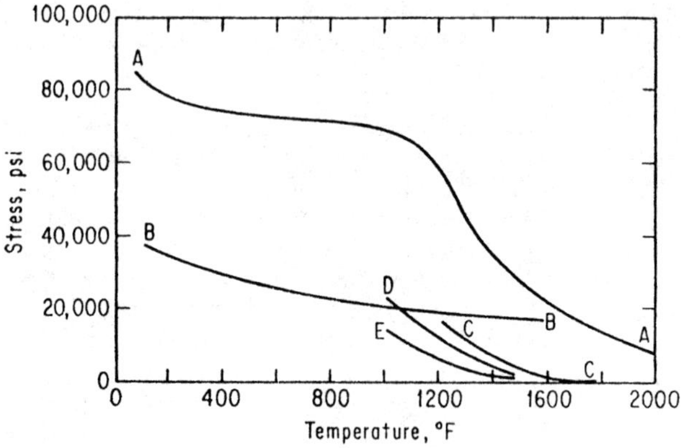

FIG. 6.38 Properties of type 316 stainless steel (18 Cr, 8 Ni, and Mo).[12] A = short-time tensile strength; B = short-time yield strength, 0.2 percent offset; C = stress of rupture, 10,000 h; D = stress for creep rate, 0.0001 percent per hour; E = stress for creep rate, 0.00001 percent per hour.

improvements have been made in engineering design of gas turbine blades (coolant vents) and material processing techniques (directional solidification, single crystal, and protective coating). The developed progress in the past several decades[40] led to increased engine efficiencies and increased time between overhauls from 100 h to more than 10,000 h.

Thermal fatigue is an important high-temperature design property which is related to fracture occurring with cyclical stress applications. This type of fatigue failure, resulting from thermal stresses (constrained expansion and metal structural changes), is also termed low-cycle fatigue. At sustained temperatures above 800°F the *S–N* curves generally do not have an endurance limit (asymptotic stress) and the fatigue stress continually decreases with cycles to failure.

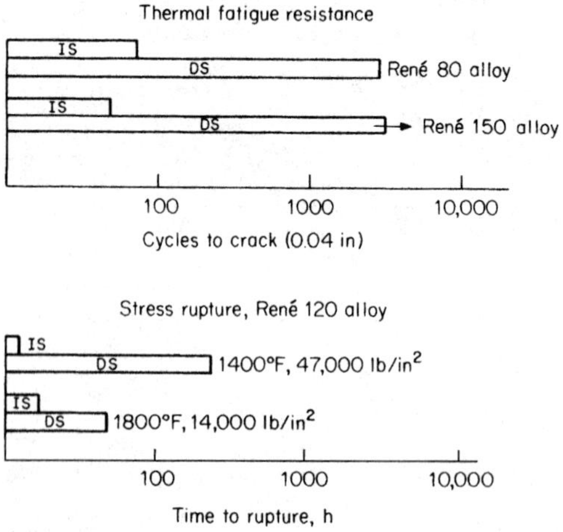

FIG. 6.39 Improved thermal fatigue resistance and stress-rupture properties of directionally solidified (DS) and isotropic superalloy (IS).[37,38]

Processing some superalloy turbine blades by directional solidification (DS) or by eliminating grain-boundary effects in single crystals results in significant improvements in thermal fatigue resistance and stress-rupture lives, as shown in Fig. 6.39. DS is controlled by casting with stationary and movable heat sinks so that all grain boundaries are made to be parallel to the applied stress direction. Crystal growth directions are controlled to take advantage of the favorable anisotropic high-strength property values as well as minimizing deleterious grain-boundary reactions.

Materials selections for aircraft gas turbine systems are typically as follows:

Component	Materials used
Turbofin compressor blades	Titanium alloys: Ti-6Al-4V, Ti-6Al-2Sn-4Mo
Combustor (burner)	Sheet alloys, shaped and joined Hastelloy X, Inconel 617
Gas turbine blades	Cast Waspaloy, René, PNA 1422 (special processes—directional or single-crystal solidification)

Thermal shock failures, as in engine valves, can occur as a result of steep temperature gradients, leading to constrained thermal dilational changes when encountering high stress. These restrained stresses can exceed the breaking point of the material at the operating temperatures. Some superalloy properties useful for design are given in Table 6.3.

TABLE 6.3 Superalloy Properties[39]

	Rupture strength, after 1000 h, lb/in^2			
Alloy	1200°F	1500°F	1800°F	Characteristics and applications
Hastelloy	34,000	10,000	2,100	Jet engine sheet parts, good oxidation resistance
Inconel 617	47,000	14,000	3,600	Gas turbine aircraft engine parts
René 41	10,200	29,000	—	Jet engine blade, parts
Waspaloy	88,000	25,500	—	Jet engine blades
In 102	52,000	7,400	—	Superheater, jet engine part
TAZ 8A	12,000	61,000	14,000	Good thermal fatigue and oxidation resistance
René	—	5,700	10,000	Compressor disk alloy
S816	46,000	18,000	—	Gas turbine blades, bolts, springs
Haynes 150	—	5,800	1,700	Resists thermal shock, high-temperature corrosion

6.16 MATERIALS FOR LOW-TEMPERATURE APPLICATION

Materials for low-temperature application are of increasing importance because of the technological advances in cryogenics. The most important mechanical properties are usually strength and stiffness, which generally increase as the temperature is decreased. The temperature dependence on ductility is a particularly important criterion in design, because some materials exhibit a transition from ductile to brittle behavior with decreasing temperature. Factors related to this transition are microstructure, stress concentrations present in notches, and the effects of rapidly applied strain rates in materials. Mechanical design can also influence the tendency for brittle failure at low temperature, and for this reason, it is essential that sharp notches (which can result from surface-finishing operations) be eliminated and that corners at changes of section be adequately filleted.

Low-temperature tests on metals are made by measuring the tensile and fatigue properties on unnotched and notched specimens and the notched impact strength. Metals exhibiting brittle characteristics at room temperature, by having low values of percent elongation and percent reduction in area in a tensile test as well as low impact strength, can be expected to be brittle at low temperatures also. Magnesium alloys, some high-strength aluminum alloys in the heat-treated condition, copper-beryllium heat-treated alloys, and tungsten and its alloys all exhibit this behavior. At best, applications of these at low temperatures can be made only provided that they adequately fulfill design requirements at room temperature.

When metals exhibit transitions in ductile-to-brittle behavior, low-temperature applications should be limited to the ductile region, or where experience based on field tests is reliable, a minimum value of impact strength should be specified. The failure, by breaking in two, of 19 out of 250 welded transport ships in World War II, caused by the brittleness of ship plates at ambient temperatures, focused considerable attention on this property. It was further revealed in tests that these materials had Charpy V-notch impact strengths of about 11 ft·lb at this temperature. Design specifications for applications of these materials are now based on higher impact values. For temperatures extending from subatmospheric temperatures to liquid-nitrogen temperatures (−320°F), transitions are reported for ferritic and martensitic steels, cast steels, some titanium alloys, and some copper alloys.

Design for low-temperature applications of metals need not be particularly concerned with the Charpy V-notch impact values provided they can sustain some shear deformation and that tensile or torsion loads are slowly applied. Many parts are used successfully in polar regions, being based on material design considerations within the elastic limit. When severe service requirements are expected in use, relative to rapid rates of applied strain on notch-sensitive metals, particular attention is placed on selecting materials which have transition temperatures below that of the environment. Some important factors related to the ductile-to-brittle transition in impact are the composition, microstructure, and changes occurring by heat treatment, preferred directions of grain orientation, grain size, and surface condition.

The transition temperatures in steels are generally raised by increasing carbon content, by the presence of more than 0.05 percent sulfur, and significantly by phosphorus at a rate of 13°F per 0.01 percent P. Manganese up to 1.5 percent decreases the transition temperature and high nickel additions are effective, so that in the austenitic stainless steels the behavior is ductile down to liquid-nitrogen temperatures. In high-strength medium-alloy steels it is desirable, from the standpoint of lowering the transition range, that the structure be composed of a uniformly tempered martensite, rather than containing mixed products of martensite and bainite or martensite and pearlite. This can be controlled by heat treatment. The preferred orientation that can be induced in rolled and forged metals can affect notched impact properties, so that specimens made from the longitudinal or rolling direction have higher impact strengths than those taken from the transverse direction. Transgranular fracture is normally characteristic of low-temperature behavior of metals. The metallurgical factors leading to intergranular fracture, due to the segregation of embrittling constituents at grain boundaries, cause concern in design for low-temperature applications. In addition to the control of these factors for enhanced low-temperature use, it is important to minimize or eliminate notch-producing effects and stress concentration, by specifying proper fabrication methods and providing adequate controls on these, as by surface inspection.

Examples of some low-temperature properties of the refractory metals, all of which have bcc structures, are shown in Fig. 6.40. The high ductility of tantalum at very low temperatures is a distinctive feature in this class that makes it attractive for use as a cryogenic (as well as a high-temperature) material. Based on the increase of yield-strength-to-density ratio with decreasing temperatures shown in Fig. 6.41, the

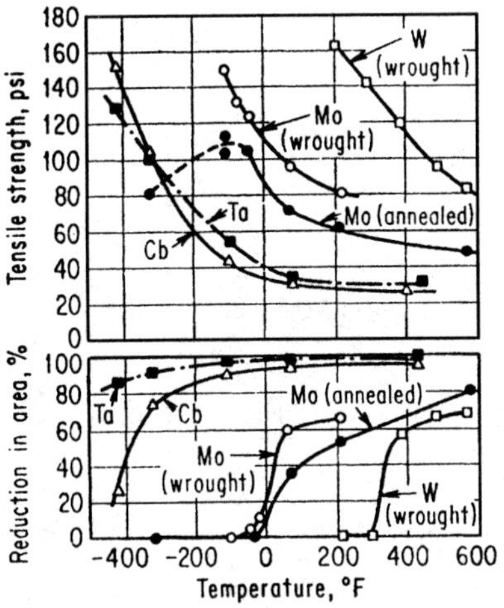

FIG. 6.40 Strength and ductility of refractory metals at low temperatures.[13]

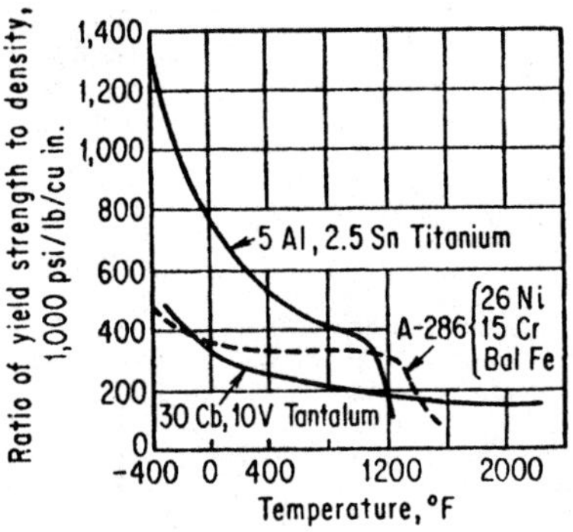

FIG. 6.41 Yield-strength-to-density ratios related to temperature for some alloys of interest in cryogenic applications.[13]

three alloys of a titanium-base Al (5) Sn (2.5), an austenitic iron-base Ni (26), Cr (15) alloy A286, as well as the tantalum-base Cb (30), 10 V alloy, are also useful for cryogenic use.

Comparisons of the magnitude of property changes obtained by testing at room temperature and −100°F for some materials of commercial interest are shown in Table 6.4.

6.17 RADIATION DAMAGE[31]

A close relationship exists between the structure and the properties of materials. Modification and control of these properties are available through the use of various metallurgical processes, among them nuclear radiation. Nuclear radiation is a process whereby an atomic nucleus undergoes a change in its properties brought about by interatomic collisions.

The energy transfer which occurs when neutrons enter a metal may be estimated by simple mechanics, the quantity of energy transferred being dependent upon the atomic mass. The initial atomic collision, or primary "knock-on" as it is called, has enough energy to displace approximately 1000 further atoms, or so-called secondary knock-ons. Each primary or secondary knock-on must leave behind it a resulting vacancy in the lattice. The primaries make very frequent collisions because of their slower movement, and the faster neutrons produce clusters of "damage," in the order of 100 to 1000 Å in size, which are well separated from one another.

Several uncertainties exist about these clusters of damage, and because of this it is more logical to speak of radiation damage than of point defects, although much of the damage in metals consists of point defects. Aside from displacement collisions, replacement collisions are also possible in which moving atoms replace lattice atoms. The latter type of collision consumes less energy than the former.

Another effect, important to the life of the material, is that of transmutation, or the conversion of one element into another. Due to the behavior of complex alloys, the cumulative effect of transmutation over long periods of time will quite often be of

TABLE 6.4 Low-Temperature Test Properties

Tensile Test Properties

Material	Tensile strength, 1,000 psi		Yield strength, 1,000 psi		Elongation, %		Modulus of elasticity, 100 psi	
	80°F	−100°F	80°F	−100°F	80°F	−100°F	80°F	−100°F
Beryllium copper 2% Be—heat-treated sheet...	189	194	154	170	3	3	19.1	19.1
Phosphor bronze 5% Sn; spring temp...........	98	107	90	97	7	11	16.5	17.1
Molybdenum sheet partly recrystallized..........	97	141	75	130	19	3.5	48.7	47.5
Tungsten wire, drawn....	214	...	196	...	2.6	....	56	
Tantalum sheet, annealed	55	71	36	66	27	25	28.2	28.3
Nilvar sheet 36 Ni: 74 Fe, rolled..................	76	101	62	76	27	30	21.8	19.7

Sheet Fatigue Properties

Material	Fatigue strength at 2 × 10^7 cycles 1,000 psi		Endurance ratios	
	80°F	−100°F	80°F	−100°F
Beryllium copper..................	31	45	0.16	0.23
Molybdenum....................	46	65	0.59	0.47
Tantalum......................	35	41	0.64	0.58
Nilvar........................	26	30	0.33	0.30

Charpy V-notch Impact Strength

Material	Impact strength, ft-lb	
	80°F	−100°F
Beryllium copper................	5.4	5.4
Phosphor bronze...............	46	44
Nilvar.........................	97	77

importance. U^{235}, the outstanding example of this phenomenon, has enough energy after the capture of a slow neutron to displace one or more atoms.

Moving charged particles may also donate energy to the valence electrons. In metals this energy degenerates into heat, while in nonconductors the electrons remain in excited states and will sometimes produce changes in properties.

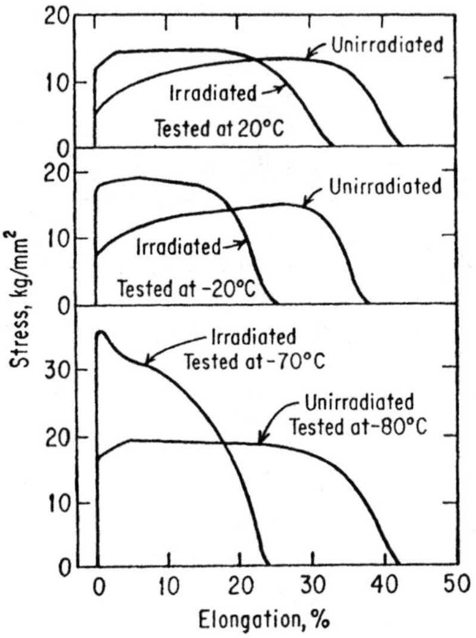

FIG. 6.42 Effect of irradiation on stress-strain curves of Fe single crystals tested at different temperatures. Irradiation dose 8 × 10^{17} thermal n/cm^2. (*Courtesy of D. McLean.*[31])

Figure 6.42 illustrates the effect or irradiation on the stress-strain curve of iron crystals at various temperatures.[32] In metals other than iron irradiation tends to produce a ferrous-type yield point and has the effect of hardening a metal. This hardening may be classified as a friction force and a locking force on the dislocations. Some factors of irradiation hardening are:

1. It differs from the usual alloy hardening in that it is less marked in cold-worked than annealed metals.
2. Annealing at intermediate temperatures may increase the hardening.
3. Alloys may exhibit additional effects, due, for example, to accelerated phase changes and aging.

The most noticeable effect of irradiation is the rise in transition temperatures of metals which are susceptible to cold brittleness. Yet another consequence of irradiation is the development of internal cracks produced by growth stresses. At high enough temperatures gas atoms can be diffused and may set up large pressures within the cracks.

Some other effects of irradiation are swelling, phase changes which may result in greater stability, radiation growth, and creep. The reader is referred to Ref. 31 for an analysis of these phenomena.

6.18 PRACTICAL REFERENCE DATA

Tables 6.5 through 6.9 give various properties of commonly used materials. Figure 6.43 provides a hardness conversion graph for steel. References 21 through 30 yield more information.

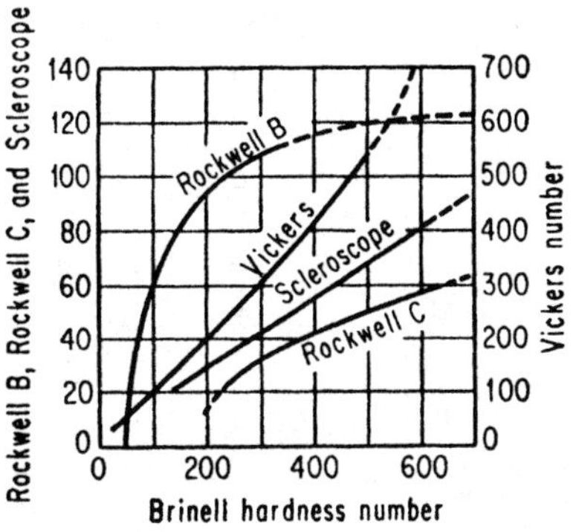

FIG. 6.43 Hardness conversion curves for steel.

TABLE 6.5 Physical Properties of Metallic Elements*

Element	Symbol	Melting point, °F	Boiling point, °F	Specific heat,[a] cal/g/°C	Thermal conductivity,[a] Btu/hr/sq ft/°F/ft	Density,[a] g/cm³	Modulus of elasticity in tension, million psi	Coefficient of linear thermal expansion,[a] μ in./in./°F	Electrical resistivity, microhm-cm	Crystal structure
Aluminum	Al	1220	4442	0.215	128.	2.70	9	13.1	2.65	f.c.c.
Antimony	Sb	1167	2516	0.049	10.8	6.62	11.3	4.7	39	Rhomb.
Arsenic	As	1503 (28 atm)	1135[b]	0.082	.....	5.72	.....	2.6	33.3	Rhomb.
Barium	Ba	1317	2980	0.068	.....	3.5	.....	....	....	b.c.c.
Beryllium	Be	2332	5020	0.45	84.4	1.85	42	6.4	4	h.c.p.
Bismuth	Bi	520	2840	0.029	4.8	9.80	4.6	7.4	107	Rhomb.
Boron	B	3690	......	0.309	.....	2.34	.....	4.6	10^{12}	Orthorhomb.
Cadmium	Cd	610	1409	0.055	53.	8.65	8	16.55	6.83	h.c.p.
Calcium	Ca	1540	2625	0.149	72.3	1.55	3.5	12.4	3.91	f.c.c.
Carbon (graphite)	C	6740[b]	8730	0.165	13.8	2.25	0.7	0.3 to 2.4	1375	Hexag.
Cerium	Ce	1479	6280	0.045	6.6	6.77	6	4.4	75	f.c.c.
Cesium	Cs	84	1273	0.048	.....	1.90	.....	54	20	b.c.c.
Chromium	Cr	3407	4829	0.11	40.3	7.19	36	3.4	12.9	b.c.c.
Cobalt	Co	2723	5250	0.099	41.5	8.85	30	7.66	6.24	h.c.p.
Columbium	Cb	4474	8901	0.065	31.5	8.57	.....	4.06	12.5	b.c.c.
Copper	Cu	1981	4703	0.092	226.	8.96	16	9.2	1.67	f.c.c.
Gallium	Ga	86	4059	0.079	19.4	5.91	.....	10	17.4	Orthorhomb.
Germanium[c]	Ge	1719	5125	0.073	33.7	5.32	.....	3.19	46×10^6	Diam. cubic
Gold	Au	1954	5380	0.031	171.	19.32	11.6	7.9	2.35	f.c.c.
Indium	In	313	3632	0.057	13.8	7.31	1.57	18	8.37	f.c.tetr.
Iridium	Ir	4449	9570	0.031	33.7	22.50	76	3.8	5.3	f.c.c.
Iron	Fe	2798	5430	0.11	43.3	7.87	28.5	6.53	9.71	b.c.c.
Lanthanum	La	1688	6280	0.048	8.	6.19	10.5	2.77	57	Hexag.
Lead	Pb	621	3137	0.031	20.	11.36	2	16.3	20.6	f.c.c.
Lithium	Li	357	2426	0.79	41.	0.534	.....	31	8.55	b.c.c.
Magnesium	Mg	1202	2025	0.245	88.5	1.74	6.35	15.05	4.45	h.c.p.
Manganese	Mn	2273	3900	0.115	.....	7.43	23	12.22	185	Complex cubic
Mercury	Hg	−37	675	0.033	4.7	13.55	.....	....	98.4	Rhomb.

Element	Symbol	Melting point, °F	Boiling point, °F	Specific heat,[a] cal/g/°C	Thermal conductivity,[a] Btu/hr/sq ft/°F/ft	Density,[a] g/cm³	Modulus of elasticity in tension, million psi	Coefficient of linear thermal expansion,[a] μ in./in./°F	Electrical resistivity, microhm-cm	Crystal structure
Molybdenum	Mo	4730	10040	0.066	82.	10.22	47	2.7	5.2	b.c.c.
Nickel	Ni	2647	4950	0.105	53.	8.90	30	7.39	6.84	f.c.c.
Osmium	Os	4900	9950	0.031	.....	22.57	81	2.6	9.5	h.c.p.
Palladium	Pd	2826	7200	0.058	40.5	12.02	16.3	6.53	10.8	f.c.c.
Phosphorus (white)	P	112	536	0.177	.....	1.83	.....	70	10^{17}	Cubic
Platinum	Pt	3217	8185	0.0314	39.8	21.45	21.3	4.9	10.6	f.c.c.
Plutonium	Pu	1184	6000	0.033	4.8	19.00	14	30.55	141.4	Monoclinic
Potassium	K	147	1400	0.177	58.	0.86	.....	46	6.15	b.c.c.
Rhenium	Re	5755	10650	0.033	41.	21.04	66.7	3.7	19.3	h.c.p.
Rhodium	Rh	3571	8130	0.059	50.6	12.44	42.5	4.6	4.51	f.c.c.
Rubidium	Rb	102	1270	0.080	.....	1.53	.....	50	12.5	b.c.c.
Ruthenium	Ru	4530	8850	0.057	.....	12.20	60	5.1	7.6	h.c.p.
Selenium	Se	423	1265	0.084	.....	4.79	8.4	21	12	Hexag.
Silicon[c]	Si	2570	4860	0.162	48.2	2.33	16.35	1.6 to 4.1	10^5	Diam. cubic
Silver	Ag	1761	4010	0.056	242.	10.49	11	10.9	1.59	f.c.c.
Sodium	Na	208	1638	0.295	77.2	0.971	.....	39	4.2	b.c.c.
Strontium	Sr	1414	2520	0.176	.....	2.60	.....	.....	23	f.c.c.
Tantalum	Ta	5425	9800	0.034	31.3	16.60	27	3.6	12.45	b.c.c.
Tellurium	Te	841	1814	0.047	3.3	6.24	6	9.3	4.4×10^5	Hexag.
Thallium	Tl	577	2655	0.031	22.5	11.85	.....	16	18	h.c.p.
Thorium	Th	3182	7000	0.034	21.7	11.66	.....	6.9	13	f.c.c.
Tin	Sn	449	4120	0.054	36.2	7.30	6.3	13	11	Tetrag
Titanium	Ti	3035	5900	0.124	9.8	4.51	16.8	4.67	42	h.c.p.
Tungsten	W	6170	10706	0.033	96.	19.30	50	2.55	5.65	b.c.c.
Uranium	U	2070	6904	0.028	17.1	19.07	24	3.8 to 7.8	30	Orthorhomb.
Vanadium	V	3450	6150	0.119	16.9	6.1	19	4.6	26	b.c.c.
Yttrium	Y	2748	5490	0.071	8.5	4.47	17	....	57	h.c.p.
Zinc	Zn	787	1663	0.092	65.	7.13	.....	22	5.92	h.c.p.
Zirconium	Zr	3366	6470	0.067	9.6	6.49	13.7	3.2	40	h.c.p.

*Courtesy of "Metals Handbook," vol. 1, 8th ed., American Society of Metals, Cleveland, 1961.
[a]Near 68°F (20°C).
[b]Sublimes; triple point at 2028 atm.
[c]Semiconductor.

TABLE 6.6 Typical Mechanical Properties of Cast Iron, Cast Steel, and Other Metals (Room Temperature)*

Metal	Modulus of elasticity E, million psi	Ultimate tensile strength σ_u, thousand psi	Yield strength σ_y, thousand psi	Endurance limit σ_{end}, thousand psi	Hardness, Brinell
Gray cast iron, ASTM 20, med. sec.	12	22	....	10	180
Gray cast iron, ASTM 50, med. sec.	19	53	....	25	240
Nodular ductile cast iron:					
Type 60-45-10	22–25	60–80	45–60	35	140–190
Type 120-90-02	22–25	120–150	90–125	52	240–325
Austenitic	18.5	58–68	32–38	32	140–200
Malleable cast iron, ferritic 32510	25	50	32.5	28	110–156
Malleable cast iron, pearlitic 60003	28	80–100	60–80	39–40	197–269
Ingot iron, hot rolled	29.8	44	23	28	83
Ingot iron, cold drawn	29.8	73	69	33	142
Wrought iron, hot rolled longit	29.5	48	27	23	97–105
Cast carbon steel, normalized 70000	30	70	38	31	140
Cast steel, low alloy, 100,000 norm. and temp	29–30	100	68	45	209
Cast steel, low alloy, 200,000 quench. and temp	29–30	200	170	85	400
Wrought plain C steel:					
C1020 hot rolled	29–30	66	44	32	143
C1045 hard. and temp. 1000°F	29–30	118	88	......	277
C1095 hard. and temp. 700°F	29–30	180	118	......	375
Low-alloy steels:					
Wrought 1330, HT and temp. 1000°F	29–30	122	100	......	248
Wrought 2317, HT and temp. 1000°F	29–30	107	72	......	222
Wrought 4340, HT and temp. 800°F	29–30	220	200	......	445
Wrought 6150, HT and temp. 1000°F	29–30	187	179	......	444
Wrought 8750, HT and temp. 800°F	29–30	214	194	......	423
Ultra high strength steel H11, HT	30	295–311	241–247	132	
300M HT and temp. 500°F	....	289	242	116	
4340 HT and temp. 400°F	30	287	270	107	
25 Ni Maraging	24	319	284		
Austenitic stainless steel 302, cold worked	28	110	75	34	240
Ferritic Stainless Steel 430, cold worked	29	75–90	45–80		
Martensitic Stainless Steel 410, HT	29	90–190	60–145	40	180–390
Martensitic Stainless Steel 440A, HT	29	260	240	......	510
Nitriding Steels, 135 Mod., hard. and temp. (core properties)	29–30	145–159	125–141	45–90	285–320
Nitiding Steels 5Ni-2A, hard. and temp	29–30	206	202	90	
Structural Steel	30	50–65	30–40	......	120
Aluminum Alloys, cast:					
195 SHT and aged	10.1	36	24	8	75
220 SHT	9.5	48	26	8	75
142 SHT and aged	10.3	28–47	25–42	9.5	75–110
355 SHT and aged	10.2	35–42	25–27	9–10	80–90
A13 as cast	10.3	39	21	19	

TABLE 6.6 Typical Mechanical Properties of Cast Iron, Cast Steel, and Other Metals (Room Temperature)* (*Continued*)

Metal	Modulus of elasticity E, million psi	Ultimate tensile strength σ_u, thousand psi	Yield strength σ_y, thousand psi	Endurance limit σ_{end}, thousand psi	Hardness, Brinell
Aluminum Alloys, wrought:					
EC ann	10	12	4		
ECH 19, hard	10	27	24	7	
3003 H 18, hard	10	29	27	10	55
2024 H T (T3)	10.6	70	50	20	120
5052 H 38, hard	10.2	42	37	20	77
7075 HT (T6)	10.4	83	73	23	150
7079 HT (T6)	10.3	78	68	23	145
Copper alloys, cast:					
Leaded red brass BB11-4A	9–14.8	33–46	17–24	......	55
Leaded tin bronze BB11-2A	12–16	36–48	16–21	......	60–72
Yellow brass BB11-7A	12–14	60–78	25–40	......	80–95
Aluminum bronze BB11-9BHT	15	90	40	......	180
Copper alloys, wrought:					
Oxygen-free 102 ann	17	32–35	10	11	
Hard	....	50–55	45	13	
Beryllium copper, 172 HT	19	165–183	150–170	35–40	
Cartridge Brass, 260 hard	16	76	63	21	
Muntz metal, 280 ann	15	54	21		
Admiralty, 442 ann	16	53	22		
Manganese bronze, 675 hard	15	84	60		
Phosphor bronze, 521 spring	16	112	70		
Silicon bronze, 647 HT	18	100	88		
Cupro-Nickel, 715 hard	22	80	73		
Magnesium Alloys, cast:					
AZ63A, aged	6.5	30–40	14–19	11–15	55–73
AZ92A, aged	6.5	26–40	16–21	11–15	80–84
HK31A, T6	6.5	31	16	9–11	55
Magnesium Alloys, wrought:					
AZ61AF, forged	6.5	43	26	17–22	55
ZE-10A-H24	6.5	34–38	19–28	20–24	
HM31A-T5	6.5	42	33	12–14	
Nickel-alloy castings 210	21.5	45–60	20–30	......	80–125
Monel 411 cast	19	65–90	32–45	......	125–150
Inconel 610 cast	23	70–95	30–45	......	190
Nickel alloys, wrought:					
200 Spring	30	90–130	70–115		
Duranickel, 301 Spring	30	155–190			
K Monel, K500 Spring	26	145–165	130–180		
Titanium alloys, wrought, unalloyed	15–16	60–110	40–95	60–70	
5Al-2.5 Sn	16–17	115–140	110–135	95	
13V-11 Cr-3Al HT	14.5–16	190–240	170–220	50–55	
Zinc, wrought, comm. rolled	....	25–31	....	4.1	
Zirconium, wrought:	14	64	53		
Reactor grade	14	49	29		
Zircaloy 2	13.8	68	61		
Pure metals, wrought:					
Beryllium, ann	44	60–90	45–55		
Hafnium, ann	20	77	32		
Thorium, ann	10	34	26		
Vanadium, ann	20	72	64		
Uranium, ann	30	90	25		

TABLE 6.6 Typical Mechanical Properties of Cast Iron, Cast Steel, and Other Metals (Room Temperature)* (*Continued*)

Metal	Modulus of elasticity E, million psi	Ultimate tensile strength σ_u, thousand psi	Yield strength σ_y, thousand psi	Endurance limit σ_{end}, thousand psi	Hardness, Brinell
Precious metals:					
Gold, ann.	12	19	....	46	25
Silver, ann.	11	22	8	....	25–35
Platinum, ann.	21	17–26	2–5.5	....	38–52
Palladium, ann.	17	30	5	....	46
Rhodium, ann.	42	73	....	....	55–156
Osmium, cast.	80	....	....	....	350
Iridium, ann.	74	....	....	....	170

Babbitt has a compressive elastic limit of 1.3 to 2.5 × 10^3 lb/in^2 and a Brinell hardness of 20.
Compressive yield strength of all metals, except those cold-worked = tensile yield strength.
Poisson's ratio is in the range 0.25 to 0.35 for metals.
Yield strength is determined at 0.2 percent permanent deformation.
Modulus of elasticity in shear for metals is approximately 0.4 of the modulus of elasticity in tension E.
Compressive yield strength of cast iron 80,000 to 150,000 × 10^3 lb/in^2.

*From *Materials in Design Engineering*, Materials Selector Issue, vol. 56, no. 5, 1962; courtesy of Reinhold Publishing Corporation, New York.

TABLE 6.7 Mechanical Properties and Applications of Steels

AISI	Condition	Tensile strength, lb/in²	Yield strength, lb/in²	Elongation in 2 in, %	Brinell hardness	Machinability	Comments and uses
Carbon steels*							
1020	Annealed	55,000	30,000	25	111	65	Weldable, structural beams, column: case hardener, gears, camshafts, kingpins (hard surface and soft core)
1040	Annealed	76,000	42,000	18	149	60	Not readily weldable. Hardened bolts (shallow), case-fused axles, shafts, chain pins, hobbing splints
1070	Annealed	102,000	56,000	12	212	55	Hardened springs, hand tools, saw material
1095	Annealed	120,000	66,000	10	248	45	Hardened springs, knives, ore screens

Alloy steels (quenched and tempered)†

AISI	Tempering temperature, °F	Tensile strength, lb/in²	Yield strength, lb/in²	Elongation, % in 2 in	Brinell hardness	Machinability (soft)	Comments and uses
1141	400	237,000	176,000	6	461	70	Free-machining (sulfur); gears, transmission, and spline shafts works
	1000	130,000	111,000	18	262		
1345	400	270,000	240,000	10	530	45	Steering knuckle forgings
	1000	140,000	124,000	16	310		
4140	400	257,000	238,000	8	510	65	Can be nitrided, transmission shafts, tractor pins, pinions
	1000	138,000	121,000	18	285		
4340	400	272,000	243,000	10	520	50	Drums for helicoptor transmission (nitrided)
	1000	170,000	156,000	13	360		
6150	400	280,000	245,000	8	538	55	High-stress, high-temperature springs, die-casting dies, plastic molds
	1000	168,000	155,000	13	345		
8650	400	281,000	243,000	10	525	60	Springs, gears
	1000	170,000	153,000	15	340		
9310	400	174,000	150,000	15	390	50	Carburized aircraft gears, high shock resistance, surface C not to exceed 0.9 percent to avoid retained austenite (poorer wear)

TABLE 6.7 Mechanical Properties and Applications of Steels (*Continued*)

Stainless and heat-resistant steels‡

AISI	Typical composition: Cr	Ni	C	Mechanical properties at 70°F, annealed: Tensile strength, lb/in²	Yield strength, lb/in²	Elongation, %	Maximum continuous service temperature, °F (air)	Characteristics and applications
205	17	1.8	0.20	120,000	69,000	58	—	Spinning and drawing operations, nonmagnetic cryogenic parts
304	18	8	0.08	85,000	35,000	60	1650	Chemical and food processing, cryogenic hypodermic needles
316	17	12	0.08	84,000	32,000	50	1650	Higher corrosion resistance, high creep strength, chemical and pulp equipment
317	18§	10§	0.08§	90,000	35,000	45	1650	Stabilized for weldments, exhaust manifold aircraft, fire walls, pressure vessels
403	12	—	0.15	70,000	45,000	25	1300	Martensitic, turbine quality, jet engine rings
440C	17	—	1.2	110,000	65,000	14	1400	Martensitic, highest hardness, balls, bearings, cases, and nozzles
S15500	17	3.5	0.07	160,000	145,000	15	—	Martensitic, maraging high strength, cams, gears, shafting

Tool Steels¶

AISI	Type	Typical composition: C	W	Nb	Cr	V	Applications
W1	Water hardening	0.6 1.4	— —	— —	— —	— —	Low C, blanking tools, forging dies; high C, lathe tools, reamers, taps
S1	Shock resisting	0.5	2.5	—	1.5	—	Pneumatic tools, swaging dies, shear blades, master hobs
O7	Oil hardening	1.2	1.75	—	0.75	—	Mandrels, slitters, drills, taps, blanking dies
A2	Age hardening	1.0	—	1.0	5.0	—	Thread rolling dies, extrusion dies, coining dies
D2	High C–high Cr	1.5	—	1.0	12.0	—	Blanking dies, roll thread dies, shear blades, slitters

AISI	Type	Typical composition					Applications
		C	W	Nb	Cr	V	
D3	High C–high Cr	2.25	—	—	12.0	—	More wear-resistant than D2
H12	Hot work	0.35	1.5	1.5	5.0	0.4	Extrusion dies, forging die inserts, punches
T1	High speed	0.75	18	—	4	1	Drills, taps, tool cutters, high hot hardness
M4	High speed	1.3	5.5	4.5	4	4	Lathe and planer tools, brooches, swaging dies
P2	Mold dies	0.07	—	0.2	2	0.5 Ni	Plastic molding dies (hobbed and carburized)

*From "Metals Handbook," vol. 1, 8th ed., American Society for Metals, 1961. Machinability values from "SAE Handbook", 1979; refer to a rating of 100 for AISI 1212 (a resulfurized and rephosphorized) free-machining steel in cold-drawn condition. These plain carbon steels have a low or shallow hardenability when heat-treated.

†From "Metal Progress Databook," 1981; "ASM Metals Handbook," vol. 1, 8th ed., American Society for Metals, 1961. These alloyed steels have greater hardenability than the plain carbon steels when heat-treated.

‡"Metal Progress Databook," 1981.

§Plus Ti (5 × C).

¶From "Metal Progress Databook," 1981.

TABLE 6.8 Typical Properties of Refractory Ceramics and Cermets and Other Materials

Type	Specific gravity	Melting point, °F	Mean specific heat, Btu/lb/°F	Mean coefficient of thermal expansion, μ in./in./°F	Mean thermal conductivity, Btu/hr/sq ft/°F/ft	Hardness, Mohs scale	Maximum service temperature (oxidiz.), °F
Alumina (99+) (Al_2O_3)	3.85	3725	0.23	4.3	10.7	9	3540
Beryllia (BeO)	3.0	4620	0.29	5.3	9.5	9	4350
Magnesia (MgO)	3.6	5070	0.26	7.8	1.47	6	4350
Thoria (ThO_2)	....	6000	....	5.28	0.0	7	4890
Zirconia (ZrO_2)	5.5–6	4710	0.16	3.06	0.53	7–8	4530
Quartzite (SiO_2)	2.65	2552	0.26	0.28	0.8		
Silicon carbide (Dens)(SiC)	3.2	......	0.33	2.17	25	9–10	3000
Boron carbide	2.5	......	....	1.73	16	....	1000
Titanium carbide (TiC)	6.5	......	....	4.3–7.5			
Tungsten carbide (WC)	14.3	......	....	2.5–3.9	26–50		
Boron nitride	2.05–2.15	4930	....	5.5	10–20		
Graphite	2.25	......	0.18	1.0–1.3	70–120	1.2	

TABLE 6.9 Typical Properties of Plastics at Room Temperature

Type	Specific gravity	Coefficient of thermal expansion, 10^{-5}/°F	Thermal conductivity, Btu/hr/sq ft/°F/ft	Volume resistivity, ohm-cm	Dialectric strength[a], volts/mil	Modulus of elasticity in tension, 10^5 psi	Tensile strength, 10^3 psi
Acrylic, general purpose, type I	1.17–1.19	4.5	0.12	$>10^{15}$	450–530	3.5–4.5	6–9
Cellulose acetate, type I (med.)	1.24–1.34	4.4–9.0	0.1–0.19	10^{12}	250–600	....	2.7–6.5
Epoxy, general purpose	1.12–2.4	1.7–5.0	0.1–0.8	10^{13}	350–550	....	2–12
Nylon 6	1.13–1.14	4.6–5.4	0.1–0.14	10^{14}	420–485	2.5–3.4	10.2–12
Phenolic, type I (mech.)	1.31	3.3–4.4	....	1.7×10^{12}	350–400	4–5	6–9
Polyester, Allyl type	1.30–1.45	2.8–5.6	0.12	$>10^{13}$	330–500	2–3	4.5–7
Silicone, general (mineral)	1.80–2.0	2.8–3.2	0.09	$>10^{13}$	350–400	....	4.2
Polystyrene, general purpose	1.04–1.07	3.3–4.8	0.06–0.09	10^{18}	>500	4–5	5–8
Polyethylene, low density	0.92	8.9–11	0.19	10^{18}	480	0.22	1.4–2
Polyethylene, medium density	0.93	8.3–16.7	0.19	$>10^{15}$	480	....	2
Polyethylene, high density	0.96	8.3–16.7	0.19	$>10^{15}$	480	....	4.4
Polypropylene	0.89–0.91	6.2	0.08	10^{16}	769–820	1.4–1.7	5

[a] Short time.

REFERENCES

1. Richards, C. W.: "Engineering Materials Science," Wadsworth Publishing Co., San Francisco, 1961.
2. Barrett, C. S.: "Structure of Metals," 2d ed., McGraw-Hill Book Company, Inc., New York, 1952.
3. Sachs and Van Horn: "Practical Metallurgy," American Society for Metallurgy, 1940.

4. "Metals Handbook," vol. 1, 8th ed., American Society for Metals, Cleveland, 1961.
5. "Heat Treatment and Properties of Iron and Steel," *Natl. Bur. Stand. (U.S.) Monograph* 18, 1960.
6. "Metals Handbook," 1954 Supplement, American Society for Metals, Cleveland.
7. "Heat Treatment and Properties of Iron and Steel," *Natl. Bur. Stands (U.S.) Monograph* 18, 1960.
8. Palmer, F. R., and G. V. Luersson: "Tool Steel Simplified," Carpenter Steel Co., 1948.
9. "Suiting the Heat Treatment to the Job," United States Steel Co.
10. "Metals Handbook," American Society for Metals, Cleveland, 1939 ed.
11. "Three Keys to Satisfaction," Climax Molybdenum Co., New York.
12. "Steels for Elevated Temperature Service," United States Steel Co.
13. *Metal Prog.*, vol. 80, nos. 4 and 5, October and November, 1961.
14. Norton, J. T., and D. Rosenthal: *Welding J.*, vol. 2, pp. 295–307, 1945.
15. "ASME Handbook, Metals Engineering—Design," McGraw-Hill Book Company, Inc., New York, 1953.
16. "Symposium on Corrosion Fundamentals," A series of lectures presented at the University of Tennessee Corrosion Conference at Knoxville, The University of Tennessee Press, Knoxville, 1956.
17. Evans, Ulich R.: "The Corrosion and Oxidation of Metals," St. Martin's Press, Inc., New York, 1960.
18. Burns, R. M., and W. W. Bradley: "Protective Coatings for Metals," Reinhold Publishing Corporation, New York, 1955.
19. Bresle, Ake: "Recent Advances in Stress Corrosion," Royal Swedish Academy of Engineering Sciences, Stockholm, Sweden, 1961.
20. "ASME Handbook, Metals Engineering—Design," McGraw-Hill Book Company, Inc., New York, 1953.

Some suggested references recommended for the selections and properties of engineering materials are the following:

21. "Metals Handbook," vol. 1, 8th ed., American Society for Metals, Cleveland, 1961.
22. *Metals Prog.*, vol. 66, no. 1-A, July 15, 1954.
23. *Metals Prog.*, vol. 68, no. 2-A, Aug. 15, 1955.
24. Dumond, T. C.: "Engineering Materials Manual," Reinhold Publishing Corporation, New York, 1951.
25. "Steels for Elevated Temperature Service," United States Steel Co.
26. "Three Keys to Satisfaction," Climax Molybdenum Co., New York.
27. Zwikker, C.: "Physical Properties of Solid Materials," Interscience Publishers, Inc., New York, 1954.
28. Teed, P. L.: "The Properties of Metallic Materials at Low Temperatures," John Wiley & Sons, Inc., New York, 1950.
29. Hoyt, S. L.: "Metals and Alloys Data Book," Reinhold Publishing Corporation, New York, 1943.
30. *Materials in Design Engineering*, Materials Selector Issue, vol. 56, no. 5, Reinhold Publishing Corporation, New York, 1962.
31. McLean, D.: "Mechanical Properties of Metals," John Wiley & Sons, Inc., New York, 1962, pp. 363–382.
32. Edmonson, B.: *Proc. Roy. Soc. (London), Ser. A*, vol. 264, p. 176, 1961.
33. Norton, J. T., and D. Rosenthal: "X-ray Diffraction Measurements," *Welding J.*, vol. 24, pp. 295–307, 1945.

34. Field, Metal: "Machining of High Strength Steels with Emphasis on Surface Integrity," Air Force Machine Data Center, Cincinnati, 1970.

35. Suh, N. P., and A. P. L. Turner: "Elements of Mechanical Behavior of Solids," McGraw-Hill Book Co., Inc., New York, pp. 489–490, 1975.

36. "Metals Handbook," vol. 1, 9th ed. *American Society for Metals*, Cleveland, p. 674, 1978.

37. Woodford, D. A., and D. F. Mawbray: *Mater. Sci. Eng.*, vol. 16, pp. 5–43, 1974.

38. Wright, P. K., and A. F. Anderson: *Met. Tech. G.E.* pp. 31–35, Spring 1981.

39. "Metal Progress Databook," American Society for Metals, Metals Park, Ohio, 1980.

CHAPTER 7
FRICTION, LUBRICATION, AND WEAR

David Tabor, Sc.D.
Professor Emeritus
Laboratory for the Physics and Chemistry of Solids
Department of Physics
Cambridge University
Cambridge, England

7.1 INTRODUCTION 7.1
7.2 DEFINITIONS AND LAWS OF FRICTION 7.2
7.2.1 Definition 7.2
7.2.2 Static and Kinetic Friction 7.2
7.2.3 Basic Laws of Friction 7.2
7.3 SURFACE TOPOGRAPHY AND AREA OF REAL CONTACT 7.2
7.3.1 Profilometry and Asperity Slopes 7.2
7.3.2 Elastic and Plastic Deformation of Conical Indenters 7.3
7.3.3 Elastic and Plastic Deformation of Real Surfaces 7.4
7.4 FRICTION OF CLEAN METALS 7.7
7.4.1 Theory of Metallic Friction 7.7
7.4.2 Microdisplacements before Sliding 7.9
7.4.3 Breakdown of Oxide Films 7.10
7.4.4 Friction of Metals after Repeated Sliding 7.11
7.4.5 Friction of Hard Solids 7.11
7.4.6 Friction of Thin Metallic Films 7.11
7.4.7 Friction of Polymers 7.12
7.4.8 Shear Properties of Thin Polymer Films 7.13
7.4.9 Kinetic Friction 7.14
7.4.10 New Tribological Materials: Composites, Ceramics 7.15
7.5 LUBRICATION 7.16
7.5.1 Hydrodynamic or Fluid Lubrication 7.16
7.5.2 Elastohydrodynamic Lubrication 7.18
7.5.3 Boundary Lubrication 7.18
7.6 WEAR 7.21
7.6.1 Laws of Wear 7.21
7.6.2 Mild and Severe Wear 7.21
7.6.3 Effect of Environment 7.22
7.6.4 Effect of Speed 7.22
7.6.5 Wear by Abrasives 7.23
7.6.6 Wear Behavior of Specific Materials 7.23
7.6.7 Identification of Wear Mechanisms 7.23

7.1 INTRODUCTION

Sliding friction is primarily a surface phenomenon. Consequently it depends very markedly on surface conditions, such as roughness, degree of work hardening, type of oxide film, and surface cleanliness.[4,6,11] In general, in unlubricated sliding the roughness has only a secondary effect, but surface contamination can have a profound influence on friction (and wear), particularly with surfaces that are nominally clean. Because of this the account given here concentrates mainly on the mechanisms involved in friction.[4,11,20a] In this way the reader may be better able to assess the main factors involved in any particular situation. Tables of friction values are given, but they must be used with caution. Very wide differences in friction may be obtained under apparently similar conditions, especially with unlubricated surfaces.

7.2 DEFINITIONS AND LAWS OF FRICTION

7.2.1 Definition

The friction between two bodies is generally defined as the force at their surface of contact which resists their sliding on one another. The friction force F is the force required to initiate or maintain motion. If W is the normal reaction of one body on the other, the coefficient of friction

$$\mu = F/W \tag{7.1}$$

7.2.2 Static and Kinetic Friction

If the force to initiate motion of one of the bodies is F_s and the force to maintain its motion at a given speed is F_k, there is a corresponding coefficient of static friction $\mu_s = F_s/W$ and a coefficient of kinetic friction $\mu_k = F_k/W$. In some cases, these coefficients are approximately equal; in most cases $\mu_s > \mu_k$.

7.2.3 Basic Laws of Friction

The two basic laws of friction, which are valid over a wide range of experimental conditions, state that:[4]

1. The frictional force F between solid bodies is proportional to the normal force between the surfaces, i.e., μ is independent of W.
2. The frictional force F is independent of the apparent area of contact.

These two laws of friction are reasonably well obeyed for sliding metals whether clean or lubricated. With polymeric solids (plastics) the laws are not so well obeyed: in particular, the coefficient of friction usually decreases with increasing load as a result of the detailed way in which polymers deform.

7.3 SURFACE TOPOGRAPHY AND AREA OF REAL CONTACT

7.3.1 Profilometry and Asperity Slopes

When metal surfaces are placed in contact they do not usually touch over the whole of their apparent area of contact.[4,11] In general, they are supported by the surface irregularities which are present even on the most carefully prepared surfaces. Such roughnesses are usually characterized by means of a profilometer in which a fine stylus runs over the surface and moves up and down with the surface contour. The movement is measured electrically and may be recorded digitally for future detailed analysis by appropriate interfaced display units.[28] These units can provide information (see below) concerning the mean asperity heights, the distribution of peaks, valleys, slopes, asperity-tip curvatures, correlation lengths, and other features. Some commercial units display not only the essential parameters but also some which are redundant or even pointless. For visualization of the surface topography it is convenient to display the stylus movements on a chart. Since changes in height are generally very small compared with the horizontal distance traveled by the stylus, it is usual to compress the horizontal movement on the chart by a factor of 100 or more. As a result the chart record appears to suggest that the surface is covered with sharp jagged peaks.[30] In fact, when allowance is made for the difference in vertical and horizontal scales, the average slopes are rarely more than a few degrees (see Fig. 7.1).[30]

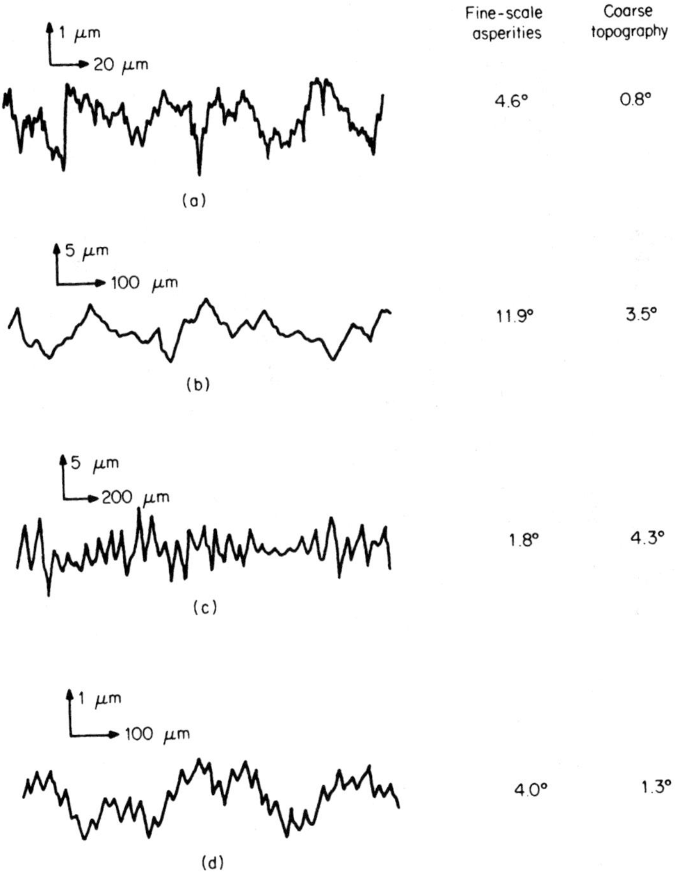

FIG. 7.1 Profilometry traces of surfaces[16] showing the average surface slopes of the fine-scale asperities and of the coarser topography. Surface treatments are (*a*) ground; (*b*) shot peened; (*c*) turned; (*d*) diamond turned.

7.3.2 Elastic and Plastic Deformation of Conical Indenters

The characterization of surface topographies and the detailed way in which the asperities deform under contact have become the subject of a number of specialized studies of varying degrees of sophistication.[28,30] We consider here the simplest case, in which the individual asperity is represented by a right circular cone of a slope θ (semiapical angle $90° - \theta$). If the cone is pressed against a smooth, flat, nondeformable surface and if it deforms elastically, the mean contact pressure p is independent of the load and is given by

$$p = (E \tan \theta)/2(1 - \nu^2) \qquad (7.2)$$

where E is Young's modulus and ν is Poisson's ratio of the cone material.[29] Ignoring the problem of infinite stresses at the cone tip we may postulate that plastic deformation of the cone will occur when p equals the indentation hardness H of the cone,* that is, when

$$(E \tan \theta)/H = 2(1 - \nu^2) \approx 2 \qquad (7.3)$$

Thus the factors favoring elastic deformation are (1) smooth surfaces, that is, low θ, and (2) high hardness compared with modulus, that is, a low value of E/H.* We may at once apply this to various materials to show the conditions of surface roughness under which the asperities will deform elastically (Table 7.1). The results show that with pure metals the surfaces must be extremely smooth if plastic deformation is to be avoided. By contrast, ceramics and polymers can tolerate far greater roughnesses and still remain in the elastic regime. From the point of view of low friction and wear, elastic deformation is generally desirable, particularly if interfacial adhesion is weak (see below). Further, on this model the contact pressure is constant, either $(E \tan \theta)/2(1 - \nu^2)$ for elastic deformation or H for plastic deformation. Thus the area of contact will be directly proportional to the applied load.

TABLE 7.1 Mean Asperity Slope for Conical Asperity Marking Transition from Elastic to Plastic Deformation

Material	Ratio E/H	Critical asperity slope for plastic deformation
Pure metals (Annealed)	200–400	$> \frac{1}{2}°$
Pure metals (Work-hardened)	70–100	1°
Alloys in hardened state	30–50	2°
Ceramics	20–30	5°
Diamond	8–10	10°
Thermoplastics (Below T_g)	5	> 20°
Cross-linked plastics	3–5	> 20°

7.3.3 Elastic and Plastic Deformation of Real Surfaces

Real surfaces are conveniently described by two main classes of representation, each of which requires two parameters.

Random or Stochastic Process. Here the profile is treated as a two-dimensional random process and is described by the root-mean roughness (see below) and the distance

*The indentation hardness for a conical asperity depends on the cone angle, but for shallow asperities ($80° < \theta < 90°$) the variation in H is small compared with $E \tan \theta$ in Eq. (7.3).

over which the autocorrelation function decays to a certain fraction of its initial value. This is sometimes referred to as the "correlation distance," though it is not quite the same thing. This representation is extremely powerful but not so convenient conceptually for the purposes of this section.

Statistical Height Description. Here the two most important parameters are:

1. The radius of curvature ß of the tip of the asperities (sharp-pointed conical asperities are unrealistic). This quantity may be treated as approximately constant for a fixed surface or it may be given a distribution of values.
2. The center-line average R_a, or alternatively the root-mean-square σ of the asperity heights (see Fig. 7.2). If in any length L of the surface the distance of any point from the mean is y,

$$R_a = \frac{1}{L}\int_0^L y\,dL \tag{7.4}$$

$$\text{and } \sigma = \left(\frac{1}{L}\int_0^L y^2\,dL\right)^{1/21} \tag{7.5}$$

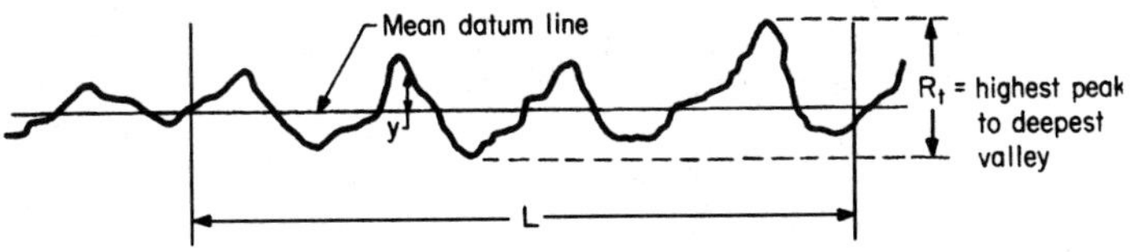

FIG. 7.2 Typical surface profile indicating the main parameters used to describe the heights of the surface roughnesses.[28]

For roughnesses resembling sine functions $\sigma \simeq 1.1R_a$ and for gaussian roughnesses $\sigma \approx 1.25R_a$. In engineering practice it is usual to use the center-line average R_a to specify surface roughness. However, in topographical theories it is more useful to use the rms value σ. (Note that in some conventions the symbol for the rms value is R_q.) However, it is clear that a measure of the mean surface height does not include such features as the spacing between significant peaks. This is particularly important for surfaces which have been milled or planed or turned, for here the topography is very different along or across the direction of machining. The random-process analyses specifically include a correlation distance between asperity peaks as well as a distribution of radii of curvature of the asperity tips. The sampling length is also of great importance.[28]

If an individual asperity is pressed against a hard smooth surface under a load ω and if it deforms elastically, the area of contact is[29]

$$A = \pi\{\tfrac{3}{4}\,\omega\beta[(1 - v^2)/E]\}^{2/3} \tag{7.6}$$

Thus for each asperity A is proportional to $\omega^{2/3}$ and the contact pressure increases as $\omega^{1/3}$. The overall behavior of the real surface then depends on the way in which the asperities deform as the total load is increased. Clearly, existing asperity contacts will grow in size while new asperities will come into contact. Some of the initial asperities may reach a contact pressure exceeding the elastic limit and plastic flow will occur.

Greenwood and Williamson[8] showed that, with surfaces for which the asperity heights followed an exponential distribution, the total area of contact, even if some asperities undergo elastic and others plastic deformation, will be directly proportional to the applied load.[26,28] Physically this means that the distribution of asperity-contact areas remains almost constant so that increasing the load merely increases the number of contacts proportionately. However, real surfaces do not show an exponential distribution. The distribution is more nearly gaussian.* The detailed behavior now depends in the Greenwood-Williamson model on (1) the *surface topography* which can be described by the square root of α/β; (2) the deformation properties of the material as represented by the ratio E'/H, where $E' = E/(1 - \nu^2)$ and H is the contact pressure at which plastic deformation occurs. For a spherical asperity, plastic deformation is initiated when the contact pressure is about $H_v/3$, where H_v is the Vickers indentation hardness, and gradually increases with further deformation.[23] Thus H is not a crisp constant. However, to a good approximation the situation is described analytically in terms of the plasticity index ψ where

$$\psi = \frac{E'H}{(\sigma/\beta)^{1/2}} \tag{7.7}$$

The analysis shows that if $\psi < 0.6$, the deformation will be elastic over an enormous load range. The mean asperity contact pressure increases somewhat as the load is increased, but the change is not large: it is of order $0.3E(\sigma/\beta)^{1/2}$ over a large load range, so that the area of contact is very nearly proportional to the load. For extremely smooth surfaces the true contact pressure turns out to be between 0.1 and $0.3H$.[25]

For most engineering surfaces $\psi > 1$: the deformation is now plastic over an enormous range of loads and the true contact pressure is close to H. The area of contact is again proportional to the load.

The elastic and plastic regimes are shown in Fig. 7.3 for a gaussian distribution of roughnesses on a series of solids of different hardnesses. For a given surface finish the deformation is elastic if the nominal pressure is below each line. If it passes across the line, a fraction of the asperities will begin to deform plastically. This fraction will increase with increasing load until the major part of the contact becomes plastic. It will be noted that with aluminum ($H = 40$ kg/mm^2), a nominal pressure of only 4 kg/mm^2 will give predominantly plastic deformation for even the smoothest surface. Only with ball-bearing steel ($H = 900$ kg/mm^2) is the contact predominantly elastic even for relatively rough surfaces.[25] Similar results have been obtained with a stochastic treatment of surface asperities.

We conclude that for plastic deformation the true contact pressure will be equal to H. For elastic contact, over an extremely wide range it will lie between $0.1H$ and $0.3H$; indeed, it is difficult to envisage any type of asperity distribution involving elastic deformation for which contact pressure is less than about $0.1H$. This provides limiting values to the true area of contact A for the most diverse situations; for metals it will lie between W/H (plastic) and $10W/H$ (elastic), where W is the applied load.

Finally, we may note that $(\sigma/\beta)^{1/2}$ is a direct measure of the average slope of the asperity. In fact, for sine-wave asperities, it is roughly equal to θ. Thus the results obtained in the topographical model assuming spherical asperities [Eq. (7.7)] merge with the results deduced for conical asperities [Eq. (7.3)]. We may also note that if the surface roughness is characterized by a correlation length ℓ and rms asperity height σ (stochastic treatment), the average asperity slope is $2.3\sigma/\ell$.

---

*Note that the tail of a gaussian is approximately exponential, so that the highest asperities would first deform in the way described below.

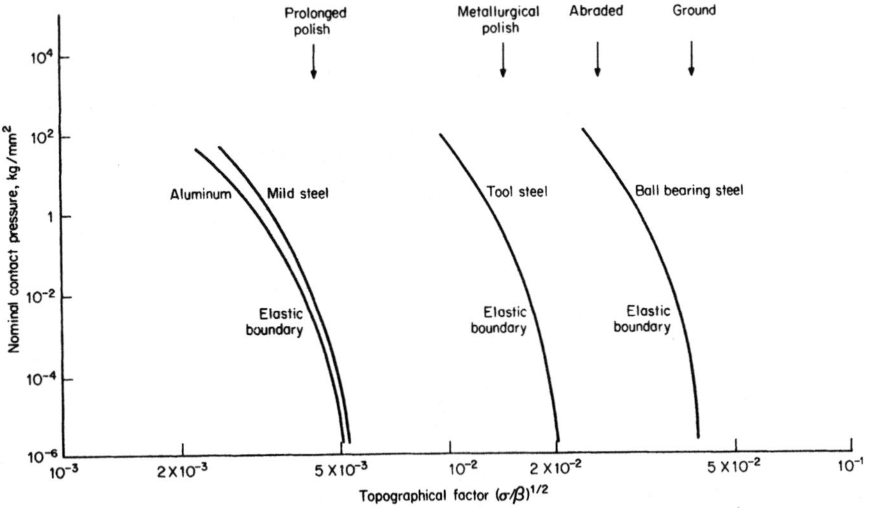

FIG. 7.3 Graph showing the nominal pressure at which the transition from elastic to the onset of plastic deformation occurs for a flat metal of specified roughness pressed on to a flat ideally smooth hard surface.[12] The materials are aluminium (H = 40 kg·mm^{-2}); mild steel (H = 120 kg·mm^{-2}); tool steel (H = 400 kg·mm^{-2}); ball-bearing steel (H = 900 kg·mm^{-2}).

7.4 FRICTION OF CLEAN METALS

7.4.1 Theory of Metallic Friction

Friction involves three major factors: (1) the area of true contact A between the surfaces, (2) the nature of the adhesion or bonding at the regions of real contact, and (3) the way in which the junctions so formed are sheared during sliding.

We have already seen that over a wide range of experimental conditions, A is proportional to the applied load and independent of the size of the bodies. For clean surfaces the adhesion that occurs at these regions is a process resembling the cold welding of metals. Consequently strong junctions are formed at the interface. Then if s is the specific shear strength of the interface, the force to produce sliding is

$$F = As + P$$

where P is a deformation or ploughing term which arises if a harder surface slides over a softer one. In general, for unlubricated surfaces, the adhesion term As is very much larger than the deformation term P, so that $F = As$. Thus the friction is proportional to the load and independent of the size of the bodies. This mechanism also explains the type of adhesive wear and surface damage which occurs between unlubricated surfaces. If the surfaces are contaminated, the adhesion is weaker and the amount of surface plucking and transfer—that is, the wear—is much less. The friction will also be smaller, but the two laws of friction still apply. However, the deformation term P may become more important relative to the adhesion term.

If metal surfaces are thoroughly cleaned in a vacuum, it is almost impossible to slide them over one another.[4,6] An attempt to do so causes further deformation at the regions of contact. The surfaces, being clean, adhere strongly wherever they touch, so that marked junction growth occurs (see Sec. 7.4.2 on microdisplacements before sliding).

TABLE 7.2 Static Friction of Metals (Spectroscopically Pure) in Vacuum (Outgassed) and in Air (Unlubricated)

	Metals											
Conditions	Ag	Al	Co	Cr	Cu	Fe	In	Mg	Mo	Ni	Pb	Pt
μ, metal on itself in vacuo	S*	S	0.6	1.5	S	1.5	S	0.8	1.1	2.4	S	4
μ, metal on itself in air	1.4	1.3	0.3	0.4	1.3	1.0	2	0.5	1.9	1.7	1.5	1.3

*S signifies gross seizure ($\mu = 10$).

The resistance to motion increases the harder the surfaces are pulled, and complete seizure may readily occur (top row, Table 7.2). Hydrogen and nitrogen generally have little effect, but the smallest trace of oxygen or water vapor produces a profound reduction in friction by inhibiting the formation and growth of strong metallic junctions. With most metals in air the surface oxide film serves a similar role, and the friction μ is in the range 0.5 to 1.3 (second row, Table 7.2).

The results in Table 7.2 are for spectroscopically pure metals.[4] The friction values depend crucially on the state of surface cleanliness. Thus, measurements carried out in even better vacuum will give higher values than those quoted in the table.[6]

Small amounts of impurities do not have a marked effect on the friction if (1) they do not produce a second phase, (2) they do not diffuse to the surface and dominate the

TABLE 7.3 Static Friction of Unlubricated Metals and Alloys Sliding on Steel in Air*

Metal or alloy	μ_s
Aluminum (pure)	0.6
Aluminum bronze	0.45
Brass (Cu 70%, Zn 30%)	0.5
Cast iron	0.4
Chromium (pure)	0.5
Constantan	0.4
Copper (pure)	0.8
Copper-lead (dendritic: Pb 20%)	0.2
Copper-lead (nondendritic: Pb 27%)	0.28
Indium (pure)	2
Lead (pure)	1.5
Molybdenum (pure)	0.5
Nickel (pure)	0.5
Phosphor-bronze	0.35
Silver	0.5
Steel (C 0.13%, Ni 3.42%)	0.8
Tin (pure)	0.9
White metal (tin-base): (Sb 6.4%, Cu 4.2%, Ni 0.1%, Sn 89.2%)	0.8
White metal (lead base); (Sb 15%, Cu 0.5%, Sn 6%, Pb 78.5%)	0.5
Wood's alloy	0.7

*The values are for sliders of pure metals and alloys sliding over 0.13 percent C, 3.42 percent Ni normalized steel. The results on mild steel are essentially the same.

surface properties, and (3) they do not appreciably modify the nature of the oxide film. The friction values obtained in air will depend on the extent to which the surface oxide is ruptured by the sliding process itself. Indeed, with metals in air, especially in the absence of lubricant films, the removal of the surface oxide either by an adhesion or ploughing (abrasion) mechanism—and its reformation—often play an important part in both the friction and wear mechanisms.

For softer metals, pickup occurs onto the harder surface and with repeated traversals of the same track, the sliding becomes characteristic of the softer metal sliding on itself.

The data of Tables 7.3–7.5 show that the friction does not vary monotonically with the hardness and is not greatly dependent on it. The reason is that with softer metals the area of contact A is large for a given load, but the interface is weak and therefore s is small. Conversely with hard metals A is small but s is large. Consequently the product $F = As$ is scarcely affected by the hardness. However, the friction of hard metals is, on the whole, somewhat less than that of softer metals. This is partly because of the reduced ductility of the metal junctions, which restricts junction growth, but mainly because the harder substrate provides greater support to the surface oxide film (see results for copper-beryllium alloy in Table 7.5). The friction also depends on the nature and strength of the oxide film itself. With ferrous alloys the homogeneity of the alloy is at least as important a factor, since heterogeneous materials will give weakened junctions.

TABLE 7.4 Static Friction of Unlubricated Ferrous Alloys Sliding on Themselves in Air*

Alloy	VPN kg/mm^2	μ_s
Pure iron	150	1–1.2
Normalized steel (C 0.13%, Ni 3.42%)	170	0.7–0.8
Austenitic steel (Cr 18%, Ni 8%)	200	1
Cast iron (pearlitic)	200	0.3–0.4
Ball race steel (Hoffman)	900	0.7–0.8
Tool steel (C 0.8%, containing carbides)	900	0.3–0.4
Chromium plate (hard bright)	1000	0.6

*The values indicate the effect of structure and the lack of correlation with hardness.[4]

TABLE 7.5 Static Friction of Hard Steel on Beryllium Alloy in Air*

Condition	Approximate VPN, kg/mm^2	μ_s
As quenched	120	1
Fully aged	410	0.4
Overaged	200	0.9–1

*The values are for a hard-steel slider (VPN 464 kg/mm^2) on beryllium alloy (Be 2%, Co 0.25%). They show the effect of heat treatment.

7.4.2 Microdisplacements before Sliding[4,11]

When surfaces are placed in contact under a normal load and a tangential force F is applied, the combined stresses produce further flow in the junctions long before gross sliding occurs. On a microscopic scale the surfaces sink together, increasing the area of contact; at the same time a minute tangential displacement occurs. As the tangential force is increased this process continues until a stage is reached at which the applied shear stress is greater than the strength of the interface. Junction growth comes to an end and gross sliding takes place. The tangential force has its critical value F_s. The tangential displacements before sliding are always very small. The values given below correspond to the stage where F has reached 90 percent of the value necessary to produce gross sliding ($F = 0.9F_s$). The results are for a hemispherically tipped conical slider on a flat, finely abraded surface of the same metal. As a crude approximation these tangential displacements are proportional to the square root of the normal load (Table 7.6).

TABLE 7.6 Microdisplacements before Gross Sliding[4]

Surfaces	Vicker's hardness kg/mm^2	Tangential displacements at $F = 0.9F_s$, 10^{-4} cm		
		1-g load	100-g load	10,000-g load
Indium	1	30	100	
Tin	7	1	20	200
Gold	19	0.5	5	
Platinum	117		1	60
Mild steel	280		1	8

7.4.3 Breakdown of Oxide Films

If the surface deformation produced during sliding is sufficiently small, the surface oxide may not be ruptured so that all the sliding may occur within the oxide film itself. The junctions formed in the oxide film are often weaker than purely metallic junctions, so that the friction may be appreciably less than when the oxide is ruptured. Since the shearing process occurs within the oxide film, the surface damage and wear are always considerably reduced. The criterion for "survival" of the oxide film is that it should be sufficiently soft or ductile compared with the substrate metal itself, so that it deforms with it and is not easily ruptured or fractured. Thus the oxide normally present on copper is not easily penetrated, whereas aluminum oxide, being a hard oxide on a soft substrate, is readily shattered during sliding, and even at the smallest loads there is some metallic interaction (Table 7.7). Thicker oxide films often provide more effective protection to the surfaces. Thus with anodically oxidized surfaces of aluminum or aluminum alloys,

TABLE 7.7 Breakdown of Oxide Films Produced during Sliding

Metal	Vickers hardness, kg/mm		Load, g, at which appreciable metallic contact occurs, g
	Metal	Oxide	
Gold	20		0
Silver	26		0.003
Tin	5	1650	0.02
Aluminum	15	1800	0.2
Zinc	35	200	0.5
Copper	40	130	1
Iron	120	150	10
Chromium plate	800	2000	> 1000

the sliding may be entirely restricted to the oxide layer. Similarly, with very hard metal substrates, such as chromium, the surface deformation may be so small that the oxide is never ruptured.[4]

In Table 7.7 the breakdown is detected by electrical conductance measurements. The results are for a spherical slider on a flat, electrolytically polished surface. The actual breakdown loads will depend on the geometry of the surfaces and on the thickness of the oxide film. The values given in this table provide a relative measure of the protective properties of the oxide film normally present on metals prepared by electrolytic polishing.

7.4.4 Friction of Metals after Repeated Sliding

Much of the earlier basic work on the friction of metals dealt with single traversals of one body over the other. This emphasized the initial deformation and shearing at the interface and tended to give prominence to plastic deformation even if the surfaces were covered with a thin protective film of oxide (see Sec. 7.4.3). In practical systems where repeated traversals take place, work hardening and gradual surface conformity may occur; the asperities may achieve a state of plastic-elastic shakedown and the deformation may be quasi-elastic. Such a condition can never be achieved with thoroughly clean surfaces but with engineering surfaces, in air, where oxide films form this may be the more common mode.

Friction is then primarily due to shearing of oxide layers which can reform if worn away. A small part of the friction may also arise from deformation losses. In the presence of effective lubrication the friction is due to the viscous shearing of the lubricant (see Sec. 7.5). In both cases the asperities are subjected to repeated loading-unloading cycles and, even in the state of plastic-elastic shakedown, they will gradually fail by fatigue. At this stage the surfaces are worn out (see Sec. 7.6).

7.4.5 Friction of Hard Solids

The friction of hard solids in air is generally small, partly because they lack ductility and partly because of the presence of surface films. Further, there is some evidence that with covalent solids the adhesion at the interface will generally be weaker than the cohesion within the solid itself. Higher frictions are observed if the surface films are removed and/or if the sliding occurs at elevated temperatures; with covalent solids, this will favor the formation of strong interfacial bonds (see Table 7.8).

7.4.6 Friction of Thin Metallic Films

If soft metal films of suitable thickness are plated onto a hard metal, the substrate supports the load, while sliding occurs within the soft film. This can give very low coefficients of friction which persist up to the melting point of the surface film. Copper-lead–bearing alloys function in this way. Note, however, that the shear properties of the metallic film may depend on the contact pressure.

Table 7.9 gives typical friction values for thin films of indium, lead, and copper (10^{-3} to 10^{-4} cm thick) deposited on various metal substrates. The other sliding member is a steel sphere 6 mm in diameter. In air, lead films have been found to show remarkable viability.

A very effective low-friction, low-wear combination for sliding electrical contacts consists of a thin flash of gold on plated rhodium.

TABLE 7.8 Friction of Very Hard Solids

Unbonded "hard metals"*			
	Coefficient of friction, μ_s		
	In air	Outgassed and measured in vacuo	
Material	at 20°C	20–1000°	Comments
Boron carbide	0.2	0.9	Rises rapidly above 1800°C
Silicon carbide	0.2	0.6	
Silicon nitride	0.2		
Titanium carbide	0.15	1.0	Rises rapidly above 1200°C
Titanium monoxide	0.2	0.6	
Titanium sesquioxide	0.3		
Tungsten carbide	0.15	0.6	Rises rapidly above 1000°C

Diamond and sapphire†	
Surfaces and conditions	μ_s
Diamond on self:	
In air at 20°C, clean and lubricated	0.05–0.15
Outgassed at 1000°C and measured in vacuo at 20°C	0.5
Surface films worn away be repeated sliding in high vacuum[6]	0.9
Diamond on steel:	
In air at 20°C, clean and lubricated	0.1–0.15
Sapphire on self:	
In air at 20°C, clean	0.2
Outgassed at 1000°C and measured in vacuo at 20°C	0.9
Sapphire on steel	
In air at 20°C, clean and lubricated	0.15–0.2

*Carbides, oxides, and nitrides sliding on themselves in air.

†Diamond shows marked frictional anisotropy. On the cube face {100}, μ_s = 0.05 along the edge direction <100>. As with sapphire, lubrication has little effect in air.[4]

TABLE 7.9 Static Friction of Thin Metallic Films, Unlubricated Room Temperature, in Air, Spherical Steel Slider Diameter 6 mm[4]

	Coefficient of static friction, μ_s			
Load, g	Indium film on steel	Indium film on silver	Lead film on copper	Copper film on steel
4000	0.07	0.1	0.18	0.3
8000	0.04	0.07	0.12	0.2

7.4.7 Friction of Polymers

The friction of polymers is fairly adequately explained in terms of the adhesion theory of friction. There are, however, three main differences from the behavior of metals. First, Amontons' laws are not accurately obeyed; the coefficient of friction tends to decrease with increasing load; it also tends to decrease if the geometric contact area is decreased. Second, if the surfaces are left in contact under load, the area of true contact may increase with time because of creep and the starting friction may be correspondingly larger. Third, the friction may show changes with speed which reflect the viscoelastic properties of the polymer, but the most marked changes occur as a result of frictional heating.[27] Even at speeds of only a few meters per second, the friction of unlubricated polymers can, as a result of thermal softening, rise to very high values.

TABLE 7.10 Friction of Steel on Polymers, Room Temperature, Low Sliding Speeds*

Material	Condition	μ
Nylon	Dry	0.4
Nylon	Wet (water)	0.15
Perspex (Plexiglas)	Dry	0.5
PVC	Dry	0.5
Polystyrene	Dry	0.5
Low-density polythene (No plasticizer)	Dry or wet (water)	0.4
Low-density polythene (With plasticizer)	Dry or wet (water)	0.1
High-density polythene (No plasticizer)	Dry or wet (water)	0.15
Soft wood	Natural	0.25
Lignum vitae	Natural	0.1
PTFE† (low speeds)	Dry or wet (water)	0.06
PTFE (high speeds)	Dry or wet (water)	0.3
Filled PTFE (15% glass fiber)	Dry	0.12
Filled PTFE (15% graphite)	Dry	0.09
Filled PTFE (60% bronze)	Dry	0.09
Rubber (polyurethane)	Dry	1.6
Rubber (isoprene)	Dry	3–10
Rubber (isoprene)	Wet (water-alcohol solution)	2–4

*The slider is 0.13% carbon, 3.42 percent Ni normalized steel. Mild steels give essentially the same result.
†Polytetrafluoroethylene.

On the other hand, at extremely high speeds the friction may fall again because of the formation of a molten lubricating film (see Table 7.10). With very soft rubbers, sliding may occur by a type of ruck moving through the interface so that motion resembles the movement of a caterpillar.

7.4.8 Shear Properties of Thin Polymer Films

The shear strengths of thin films of polymer trapped between hard surfaces have been studied experimentally.[5] (See Table 7.11.) It is found that s depends to some extent on speed and temperature, but most markedly on contact pressure p. To a first approximation

$$s = s_0 + \alpha p \tag{7.8}$$

where s_0 is the shear strength at negligibly small pressure and α is a coefficient which is approximately constant for a given material. This has an interesting relation to the friction coefficient of polymers sliding on themselves or on harder solids.[1]

Under a load W, the true area of contact A is given by $A = W/p$, where p is the true contact pressure acting on the polymer. The frictional force, ignoring the deformation term, is $F = As = (W/p)s = (W/p)(s_0 + \alpha p)$. Consequently,

TABLE 7.11 Shear Properties of Solid Polymer Films at Room Temperature (Low Speeds)[5]

Polymer	s_0, 10^7 N/m_2	α	μ_s*
Polytetrafluoroethylene (PTFE)	0.1	0.08	0.06
High-density polyethylene (HDPE)	0.25	0.10	0.12
Low-density polyethylene (LDPE)	0.6	0.14	0.4[†]
Polystyrene (PS)	1.4	0.17	0.5[†]
Polyvinyl choride (PVC)	0.45	0.18	0.5[†]
Polymethyl methacrylate	2.00	0.77	0.5
Plexiglas, Perspex			
Stearic acid	0.15	0.07	0.07
Calcium stearate	0.10	0.08	0.08

*See Table 7.9.
[†]s_0 is too large to be neglected in Eq. (7.9).

$$\mu_s = F/W = s_{0/p} + \alpha \tag{7.9}$$

The first term is usually small compared with the second, so that

$$\mu_s \approx \alpha \tag{7.10}$$

7.4.9 Kinetic Friction

Kinetic friction is usually smaller than static. The behavior is complicated by frictional heating which may produce structural changes near the surface or influence oxide formation. At speeds of a few meters per second, these effects are not as marked as at very high speeds (compare Tables 7.12 and 7.13), but they may be significant. The results in Table 7.12 are for stationary sliders rubbing on a mild steel disk in air. The materials are grouped in descending order of friction.

At very high sliding speeds the friction generally falls off because of the formation of a very thin molten surface layer which acts as a lubricant film. Other factors may

TABLE 7.12 Kinetic Friction of Unlubricated Metals at Speeds of a Few Meters per Second

Slider	μ_k
Nickel, mild steel	0.55–0.65
Aluminum, brass (70:30), cadmium, magnesium	0.4–0.5
Chromium (hard plate), steel (hard)	0.4
Copper, copper-cadmium alloy	0.3–0.35
Bearing alloys:	
Tin-base	0.46
Lead-base	0.34
Phosphor-bronze	0.34
Copper-lead (Pb 20)	0.18
Nonmetals:	
Brake materials (resin-bonded asbestos)	0.4
Garnet	0.4
Carbon	0.2
Bakelite	0.13
Diamond	0.08

TABLE 7.13 Kinetic Friction of Unlubricated Metals at Very High Sliding Speeds

Surface	Duration of experiment, s	Coefficient of friction, μ_k			
		9 m/s	45 m/s	225 m/s	450 m/s
Bismuth	1–10	0.25	0.1	0.05	—
Lead	1–10	0.8	0.6	0.2	0.12
Cadmium	1–10	0.3	0.25	0.15	0.1
Copper	1–10	>1.5	1.5	0.7	0.25
Molybdenum	1–10	1	0.8	0.3	0.2
Tungsten	1–10	0.5	0.4	0.2	0.2
Diamond	1–10	0.06	0.05	0.1	~ 0.1
Bismuth	10^{-3}	0.25	0.1	0.05	0.3
Lead	10^{-3}	0.7	0.5	0.2	0.3
Copper	10^{-3}	—	0.25	0.15	0.1
Steel (mild)	10^{-3}	—	0.2	0.1	0.08
Nylon	10^{-3}	0.4	0.25	0.1	0.08

*Up to 600 m/s.

also be involved. For example, with steel sliding on diamond, the friction first diminishes and then increases, because at higher speeds steel is transferred to the diamond so that the sliding resembles that of steel on steel.[4,6] In some cases the solids may fragment at these very high speeds, particularly if they are of limited ductility. Again, if appreciable melting occurs, the friction may increase at high speeds because of the viscous resistance of the liquid interface: this occurs with bismuth.

The results in Table 7.13 are for a rapidly rotating sphere of ball-bearing steel rubbing against another surface in a moderate vacuum.[4] The friction is roughly independent of load over the load range examined (10 to 500 *g*). However, it may depend critically on the duration of sliding, since cumulative frictional heating may greatly change the sliding conditions. The duration in the upper part of the table is about 1 to 10 s; in the lower part where a special rebound technique was used[5] the duration was about 10^{-3} s.

7.4.10 New Tribological Materials: Composites, Ceramics

Because most lubricating oils oxidize and form gums at temperatures above 250°C, considerable effort has been expended in developing high-temperature materials which are self-lubricating over a wide temperature range. One very promising approach is the formation of composite surfaces by plasma spraying, electrodeposition, or by "ion-plating." It is now possible to deposit almost any required material onto any substrate and to achieve strong adhesion to the substrate. Ceramics are particularly effective as low-friction, low-wear surfaces for operation at elevated temperatures. They may be made less brittle and tougher by incorporating other constituents during deposition. In some cases the additives act as binders, as in cermets, or as solid lubricants or structural modifiers. Ceramics are often covalent solids. As a result, when ceramics slide on one another, the interfacial adhesion is weaker than the cohesion in the bulk, so that sliding should occur at the interface itself. If they are extremely brittle, interfacial sliding may be regarded as a Type II fracture. The friction will be small compared with metals. The wear will also be smaller unless shear produces interfacial fragmentation. In general some ductility is desirable. However, at very high temperatures interfacial covalent bonding may be activated, the solids may become more ductile, and the behavior will begin to resemble that of metals.[21]

7.5 LUBRICATION

7.5.1 Hydrodynamic or Fluid Lubrication

If a convergent wedge of fluid can be established between surfaces in relative motion it will, because of its viscosity Z, generate a hydrodynamic pressure in the fluid film.[7.18] If there is no solid-solid contact, the whole of the resistance to motion is due to the viscous shear of the fluid. In journal bearings the bearing has a radius of curvature a little bigger than that of the journal (by about one part in 1000). The journal acquires a position which is slightly eccentric relative to the bearing so that lubricant is squeezed through the converging gap between the surfaces. Under properly designed conditions, the hydrodynamic pressure built up in the lubricant film is sufficient to support the normal load W. The friction is very low ($\mu_k \approx 0.001$) and there is, in principle, no wear of the solid surfaces.

Journal bearings operate under average pressures P of order 10^6 to 10^7 N/m^2 and speeds of revolution N of the order of 100 r/min. For a journal of radius R, diameter D, length L, radial clearance c, the torque G to overcome the viscous resistance of the lubricant in a full bearing may be calculated fairly reliably, simply by assuming that the journal and bearing run concentrically:

$$G = (4\pi^2 R^3 L/60c)ZN \tag{7.11}$$

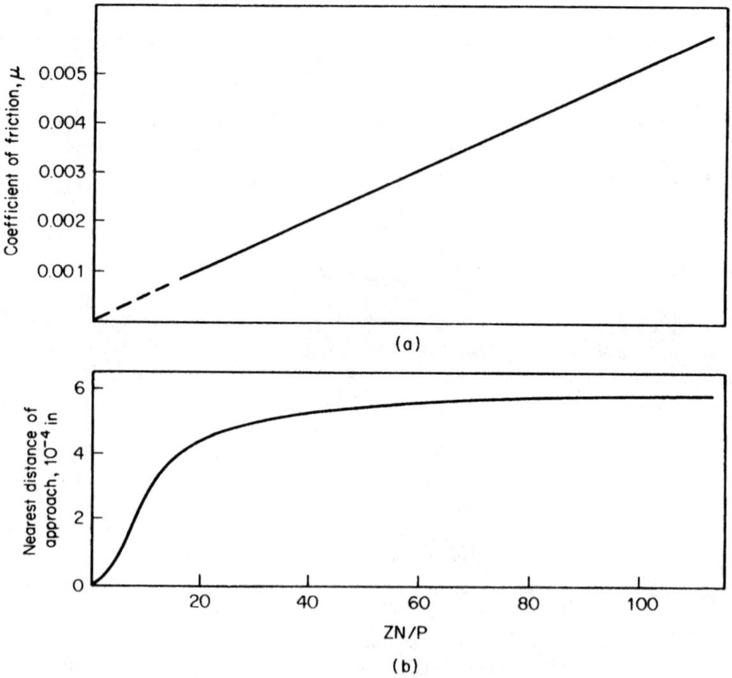

FIG. 7.4 Hydrodynamic lubrication between journal and bearing[18]: (*a*) friction and (*b*) distance of nearest approach. The dimensionless parameter ZN/P is in mixed units, with Z in centipoise, N in r/min, and P in lb/in^2.

Since the nominal pressure $P = W/2RL$, and the couple G may be written $G = \mu WR$, we obtain

$$\mu = (2\pi^2/60)(R/c)(ZN/P) \tag{7.12}$$

where N is in revolutions per minute and all the other parameters are in consistent units.

Results for a typical full oil bearing are shown in Fig. 7.4*a* for $R = 15$ mm and $R/c = 1000$. Evidently μ is very small and can be reduced by working at very low values of ZN/P. However, there is a limit to this. As ZN/P diminishes, the distance h_{min} of nearest approach (Fig. 7.4*b*) diminishes in order to maintain adequate convergence of the lubricant film. The film may then become smaller than the surface roughness and penetration of the film may occur. In engineering practice in oil bearings, the average film thickness is of order 10^{-3} cm, and in air bearings perhaps 10 times smaller. Figure 7.5 shows the distance of nearest approach (h_{min}) for a full bearing (of infinite length) and for a full bearing of length $L = 2R$. The quantities are dimensionless and may therefore be used for both oil and air bearings if self-consistent units are used. This implies that in English units, all lengths should be in inches, forces (and loads) in pound-force, viscosity in reyns, where 1 reyn = 6.9×10^6 cP. In SI units, all forces should be in newtons, viscosity in pascals per second, where 1 Pa/s = 10^3 cP. In both systems N should be in revolutions per second. For further design charts, see Ref. 18.

In hydrodynamic lubrication (HL) it is essential to maintain an adequate value of h_{min} relative to surface roughness. The most important properties of the oil are its viscosity,

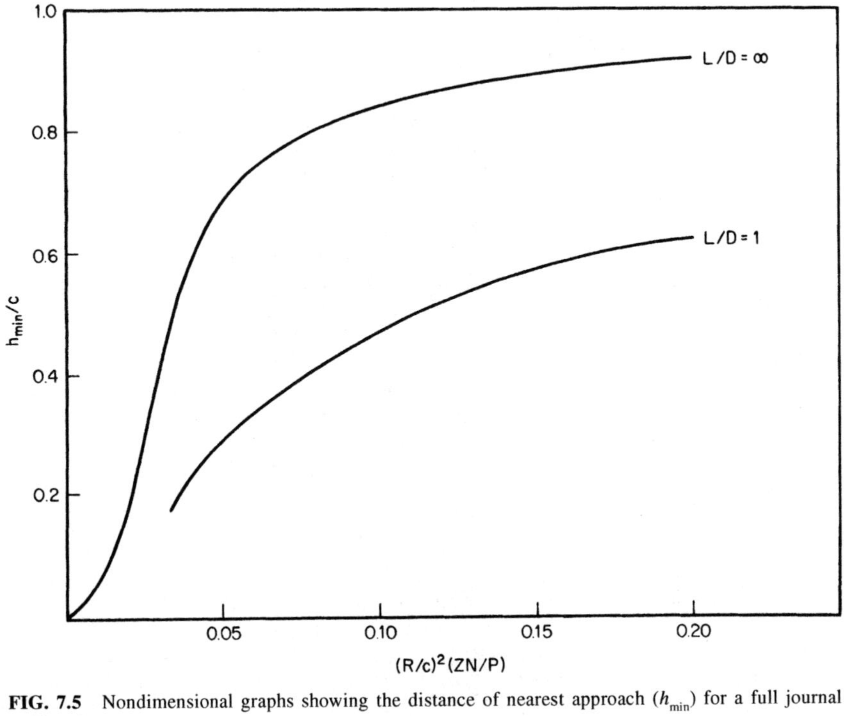

FIG. 7.5 Nondimensional graphs showing the distance of nearest approach (h_{min}) for a full journal and bearing.

its viscosity temperature dependence, and its chemical stability, especially stability against oxidation.[17]

7.5.2 Elastohydrodynamic Lubrication

In normal hydrodynamic lubrication, the hydrodynamic pressures developed in the oil film are too small to produce appreciable elastic deformation of the bearing. However, if rubber bearings are used, appreciable elastic deformation may occur and there may be a significant change in the geometry of the convergent film.[7] The hydrodynamic equations must then be combined with the equations for elastic deformation. If both surfaces are metallic and the contact pressures are high (as in rolling-element bearings or in the contact between gear teeth), there may again be sufficient elastic deformation to produce a significant change in the geometry of the contacting surfaces. A new feature is that with most lubricating oils the high pressures produce a prodigious increase in the viscosity of the oil. Thus at contact pressures of 30, 60, and 100 kg/mm^2 (such as may occur between gear teeth), the viscosity of a simple mineral oil is increased 200-, 400-, and 1000-fold, respectively. The harder the surfaces are pressed together, the harder it is to extrude the lubricant. As a result, effective lubrication may be achieved under conditions where it would normally be expected to break down. The film thickness in elastohydrodynamic lubrication (EHL) is of order 10^{-5} cm (0.1 μm), so that for safe operation, surface finish and alignment are of great importance. A full and detailed account of EHL is given in Chap. 15. It will be observed that the EHL film can show elastic, viscous, and viscoelastic properties. At sufficiently high contact pressures where the lubricant solidifies (that is, below its glass transition temperature), the oil behaves as a solid wax and its shear behavior is essentially that of a plastic solid.

7.5.3 Boundary Lubrication

Under severe conditions the EHL film may prove inadequate. Metallic contact and surface damage may occur, particularly if the oil-film thickness is too small relative to surface roughness. It is then found that the addition of a few percent of a fatty acid, alcohol, or ester may significantly improve the lubrication even if the thickness of the lubricant film is no larger than 100 Å (10^{-2} μm) or so. This is the regime of boundary lubrication (BL).[9] Radioactive tracer experiments show that while a good boundary lubricant may reduce the friction by a factor of about 20 (from $\mu \approx 1$ to $\mu \approx 0.05$), it may reduce the metallic transfer by a factor of 20,000 or more. Under these conditions the metallic junctions contribute very little to the frictional resistance. The friction is due almost entirely to the force required to shear the lubricant film itself. For this reason two good boundary lubricants may give indistinguishable coefficients of friction, but one may easily give 50 times as much metallic transfer (i.e., wear) as the other. Thus with good boundary lubricants the friction may be an inadequate indication of the effectiveness of the lubricant.[4]

In boundary lubrication the film behaves in a manner resembling EHL. There is, however, one marked difference. Because most boundary additives are adsorbed at the surface to form a condensed film or react with the surface to form a metallic soap, they are virtually solid; they do not depend on high contact pressures to achieve the load-bearing capacity of an EHL film. They are able to resist penetration by surface asperities (and here their protective properties may be enhanced by the high contact pressures), and thus they provide protection which cannot be achieved with ordinary EHL films. If, however, the temperature is raised, the boundary film may melt or it may dissolve in the superincumbent bulk fluid, and lubrication may then become far less effective (see Fig. 7.6).

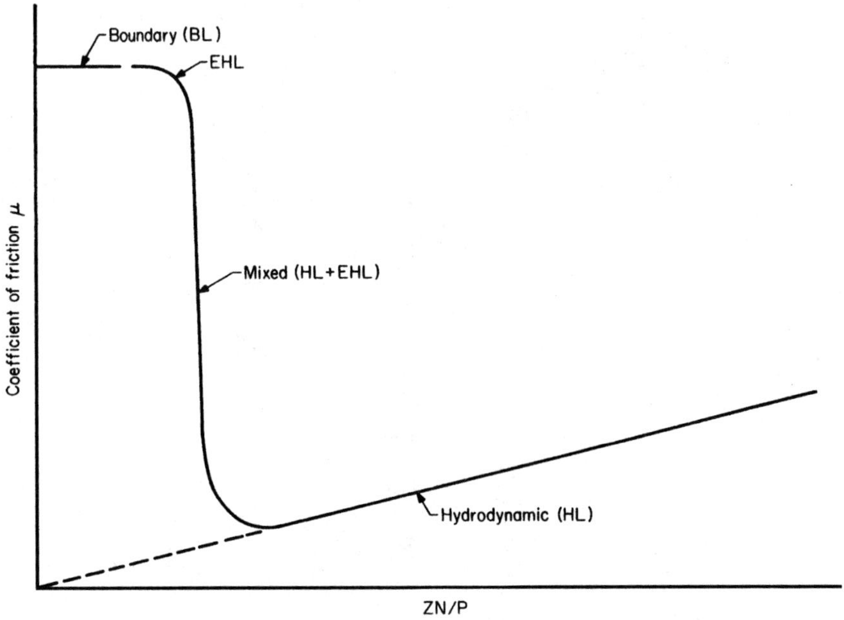

FIG. 7.6 Lubrication of a journal bearing showing the regimes of hydrodynamic (HL), elastohydrodynamic (EHL), and boundary lubrication (BL).

In the older literature the transition from HL to BL was referred to as the regime of "mixed" lubrication. We now recognize that HL gradually merges into EHL and that this then merges into BL. The coefficient of friction with good boundary lubricants ($\mu = 0.05$ to $\mu = 0.1$) is indeed similar to that observed in EHL.

In view of the nature of boundary films it is not surprising to find that their shear behavior resembles that of thin polymeric films.[5] The shear strength s per unit area of film again depends to some extent on speed and temperature, but so long as it is solid it is affected most by the contact pressure p. We find

$$s = s_0 + \alpha p$$

where α for long-chain fatty acids or esters is of order 0.05 to 0.08 and s_0 is very small (see Tables 7.14 to 7.16).

Another approach is to form a protective film by chemical attack, a small quantity of a suitable reactive compound being added to the lubricating oil. The most common materials are additives containing sulfur or chlorine or both. Phosphates are also used. The additive must not be too reactive, otherwise excessive corrosion will occur; only when there is danger of incipient seizure should chemical reaction take place. The earlier work suggested that metal sulfides and chlorides were formed and the results in Table 7.17 are based on idealized laboratory experiments in which metal surfaces were exposed to H_2S or HCl vapor and the frictional properties of the surface examined. The results show that the films formed by H_2S give a higher friction than those formed by HCl. However, in the latter case the films decompose in the presence of water to liberate HCl, and for this reason chlorine additives are less commonly used than sulfur additives.

The detailed behavior of commercial additives depends not only on the reactivity of the metal and the chemical nature of the additive but also on the type of carrier

TABLE 7.14 Static Friction of Pure Metals Sliding on Themselves in Air and When Lubricated with 1 Percent Fatty Acid in Mineral Oil (Room Temperature)[4]

	Coefficient of friction, μ_s	
Metal	Unlubricated	Lubricated
Aluminum	1.3	0.3
Cadmium	0.5	0.05
Chromium	0.4	0.34
Copper	1.4	0.10
Iron	1.0	0.12
Magnesium	0.5	0.10
Nickel	0.7	0.3
Platinum	1.3	0.25
Silver	1.4	0.55

TABLE 7.15 Lubrication of Mild Steel Surfaces by Various Lubricants (Room Temperature)[4]

Lubricant	μ_s
None	0.6
Vegetable oils	0.08–0.10
Animal oils	0.09–0.10
Mineral oils:	
Light machine	0.16
Heavy motor	0.2
Paraffin	0.18
Extreme pressure	0.10
Oleic acid	0.08
Trichloroethylene	0.3
Ethyl alcohol	0.4
Benzene	0.5
Glycerine	0.2

TABLE 7.16 Friction of Metals Lubricated with Certain Protective Films[4]

Protective film	Coefficient of friction μ_s	Temperature up to which lubrication is effective
PTFE (Teflon)	0.05	320°C
Graphite	0.07–0.13	600°C
Molybdenum disulfide	0.07–0.1	800°C

TABLE 7.17 Effect of Sulfide and Chloride Films on Friction of Metals[4]

	Coefficient of friction, μ_s				
		Sulfide films		Chloride films	
Metal	Clean	Dry	Covered with lubricating oil	Dry	Covered with lubricating oil
Cadmium on cadmium	0.5	—	—	0.3	0.15
Copper on copper	1.4	0.3	0.2	0.3	0.25
Silver on silver	1.4	0.4	0.2	—	—
Steel on steel (0.13% C, 3.42% Ni)	0.8	0.2	0.05	0.15	0.05

fluid used (e.g., aromatic, naphthenic, paraffinic). Further, the chemical reactions which occur are far more complicated than originally supposed. With sulfurized additives, oxide formation appears to be at least as important as sulfide formation. With phosphates the surface reaction is still the subject of dispute.[2] The most widespread phosphate is zinc dialkyldithio phosphate (ZDP), and in most applications it provides a low coefficient of friction ($\mu < 0.1$) up to elevated temperatures.

7.6 WEAR

7.6.1 Laws of Wear

Although the laws of friction are fairly well substantiated, there are no satisfactory laws of wear. In general, it is safe to say that wear increases with time of running and that with hard surfaces the wear is less than with softer surfaces, but there are many exceptions, and the dependence of wear on load, nominal area of contact, speed, etc., is even less generally agreed upon. This is because there are many factors involved in wear and a slight change in conditions may completely alter the importance of individual factors or change their mode of interaction.[26,31]

7.6.2 Mild and Severe Wear[10]

One of the most general characteristics of metallic wear, both for clean and for lubricated surfaces, is that below a certain load the wear is small (mild wear); above this load it rises catastrophically to values that may be 1000 or 10,000 times greater ("severe wear"). In severe wear, which occurs most readily with unlubricated surfaces, the wear is mainly due to adhesion and the shearing of the intermetallic junctions so formed.[4,26] If the junctions are very strong, shearing takes place a short distance from the interface; lumps of metal are torn out of one or both of the surfaces and these later appear as wear fragments. This often occurs in the sliding of similar metals since the junctions at the interface are highly work-hardened. If the junctions are weaker than one surface but stronger than the other, fragments will be torn out of the softer metal and the wear will generally be lower. This often occurs in the sliding of dissimilar metals. In this regime of severe wear the wear of various metallic pairs may vary by a factor of say 100 to 1, although the friction may be substantially the same.

Mild wear occurs with metals in the presence of suitable oxide films.[4,10] If the surface deformation is below a critical value, the oxide retains its integrity and the shearing occurs in the oxide film itself. The wear rate is very small and the oxide is able to reform. Mild wear also occurs with lubricated surfaces.

If wear is due to interfacial adhesion and the shearing of junctions, it may be shown on a simple model that wear is proportional to the load, is not greatly dependent on the nominal area, and is little affected by the sliding speed if frictional heating is not excessive.[4] The wear volume Z per unit distance of sliding may be written as

$$Z = K(W/3p) \tag{7.14}$$

where Z is in cubic millimeters per millimeter (or cubic centimeters per meter), the load W is in kilogram-force, and p, the yield pressure or indentation hardness of the softer of the two bodies, is in kilograms per square millimeter. In the earlier work the quantity K was regarded as the fraction of friction junctions which produce a wear fragment. More recent work suggests that it may be more meaningful to regard it as a measure of the rate at which subsurface fatigue causes cracking and the release of a wear fragment, often in the form of a flake (delamination).[22] Table 7.18 shows wear rates of different materials in combination.

There are some combinations of friction pairs in which the interfacial adhesion is weak and sliding appears to occur truly at the interface (e.g., polyethylene on steel). Minute wear rates may then be regarded as long-term fatigue of surface asperities. A similar situation appears to occur in the wear of lubricated metals.

TABLE 7.18 Wear Rates of Various Combinations of Materials

Surfaces	Hardness p, kg/mm^2	K (calculated)
Similar metals, 0.01 cm/s:		
Cadmium	20	10^{-2}
Zinc	38	10^{-1}
Silver	43	10^{-2}
Mild Steel	160	10^{-2}
Dissimilar metals, 0.01 cm/s:		
Cadmium on mild steel	—	10^{-4}
Copper on mild steel	—	10^{-3}
Platinum on mild steel	—	10^{-3}
Mild steel on copper	—	4×10^{-4}
Similar metals, 180 cm/s:		
Mild steel	190	7×10^{-3}
Hardened tool steel	850	10^{-4}
Sintered tungsten carbide	1300	10^{-6}
On hard tool steel, 180 cm/s:		
Brass 60 : 40	90	6×10^{-4}
Silver steel	320	6×10^{-5}
Beryllium copper	210	4×10^{-5}
Stellite 1	690	5×10^{-5}
Nonmetals on hard steel, 180 cm/s:		
PTFE (Teflon, Fluon)	5	2×10^{-6}
Perspex (Plexiglas)	20	7×10^{-6}
Bakelite (poor)	5	7×10^{-6}
Bakelite (good)	30	7×10^{-7}
Polythene	2	1×10^{-7}

7.6.3 Effect of Environment

The surrounding atmosphere can have a marked effect on friction, and in many cases air or oxygen or water vapor reduce the wear rate. However, this is not always the case. If, for example, the metal oxide is hard and the conditions favor abrasive wear, the continuous formation of oxidized wear fragments may lead to a large increase in wear rate. With ferrous materials in air, the atmospheric nitrogen may play an important part. Frictional heating and rapid cooling can produce martensite, but with low-carbon steels, surface hardening can still occur by reaction with nitrogen. When these hard surface films are formed the wear generally decreases.[10]

7.6.4 Effect of Speed

The main effect of speed arises from increased surface temperatures.[4] Four of the most important consequences are:

1. High hot-spot temperatures increase reactivity of the surfaces and the wear fragments with the environment.
2. Rapid heating and cooling of asperity contacts can lead to metallurgical changes which can change the wear process.
3. High temperatures may greatly increase interdiffusion and alloy formation.
4. Surface melting may occur. In some cases, if melting is restricted to the outermost surface layers, the friction and wear may become very low.

7.6.5 Wear by Abrasives

Wear by hard abrasive particles is very common in running machinery. Measurements of the wear rate Z on abrasive papers show that the abrasion resistance $1/Z$ increases almost proportionally with the hardness of the metal.[12] This is shown in Fig. 7.7. The surfaces are grooved by the abrasive particles, but the wear is mainly in the form of fine shavings. It has been estimated that the amount of wear corresponds to about 10 percent of the volume of the material displaced in the grooves.[3]

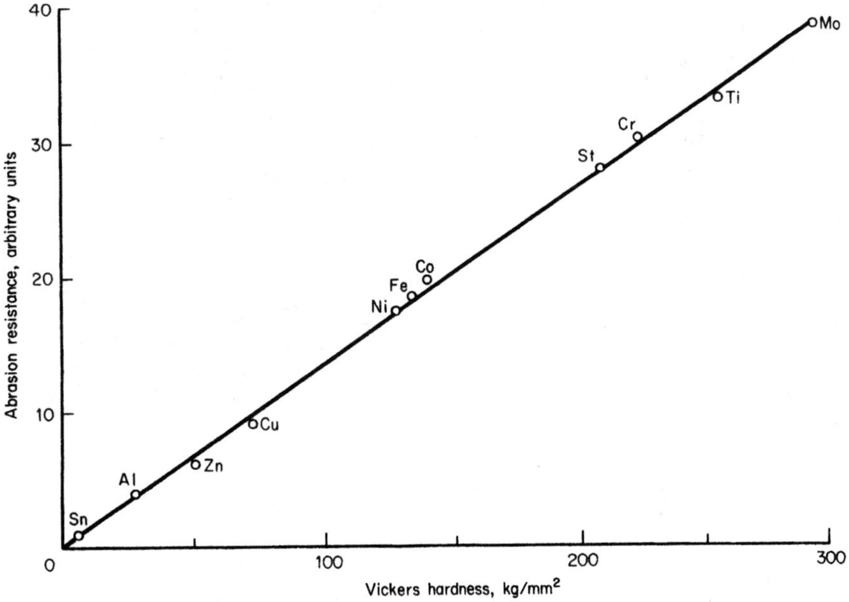

FIG. 7.7 Variation of abrasion resistance (reciprocal of wear rate) as a function of hardness for metals rubbed under standard conditions on dry abrasive paper. (The symbol St refers to 1.2 percent carbon steel.)

7.6.6 Wear Behavior of Specific Materials

There are many current publications, proceedings of conferences, etc., which deal with the wear of specific materials.[13a,13b,21,26,27,31] In addition, there are individual papers which deal, for example, with the wear of aluminum, carbon brushes,[14] bronze bearings,[15] steel,[20] and polymers.[13]

7.6.7 Identification of Wear Mechanisms

The following are possible ways of identifying wear mechanisms in a particular piece of machinery:

1. Examination of the wear debris (collected, for example, from the lubricating oil): large lumps imply adhesive wear; fine particles, oxidative wear; chiplike particles, abrasive wear; flakelike particles, delamination wear.

2. Examination of the worn surfaces: heavy tearing implies adhesive wear; scratches imply abrasive wear; burnishing indicates nonadhesive wear.
3. Metallographic examination of the surface and subsurface structure. This may reveal the type of deformation produced by the sliding process, the generation of subsurface cracks, incipient delamination, etc.

REFERENCES

1. Amuzu, J. K. A., B. J. Briscoe, and M. M. Chaudhri: "Frictional Properties of Explosives," *J. Phys. D. Appl. Phys.*, vol. 9, pp. 133–143, 1976.
2. Barcroft, F. T., K. J. Bird, J. F. Hutton, and D. Parks: "The Mechanism of Action of Zinc Thiophosphates as Extreme Pressure Additives," *Wear*, vol. 77, pp. 355–382, 1982.
3. Battacharya, S.: "Wear and Friction of Aluminum and Magnesium Alloys and Brasses," "Wear of Materials," ASME, New York, p. 40, 1981.
4. Bowden, F. P., and D. Tabor: "Friction and Lubrication of Solids," Clarendon Press, Oxford, England, part 1, 1954; part 2, 1964.
5. Briscoe, B. J., and D. Tabor: "Shear Properties of Polymeric Films," *J. Adhesion*, vol. 9, pp. 145–155, 1978.
6. Buckley, D. H.: "Surface Effects in Adhesion, Friction, Wear and Lubrication," Elsevier Publishing Co., Amsterdam, The Netherlands, 1981.
7. Cameron, A. (ed.): "Principles of Lubrication," Longman, Inc., New York, 1966.
8. Greenwood, J. A., and J. B. P. Williamson: "The contact of nominally flat surfaces," *Proc. Roy. Soc. London*, vol. A295, pp. 300–319, 1966.
9. Hardy, Sir W. B.: "Collected Scientific Papers," Cambridge University Press, Cambridge, England, 1936.
10. Hirst, W.: "Basic Mechanisms of Wear," *Proc. Lubrication and Wear, Proc. Inst. Mech. Eng.*, vol. 182 (3A), pp. 281–292, 1968.
11. Kragelsky, I. V., M. N. Dobychin, and V. S. Kombalov: "Friction and Wear-Calculation Methods" (in English), Pergamon Press, Oxford, England, 1982.
12. Kruschov, M. M.: "Principles of Abrasive Wear," *Wear*, vol. 28, pp. 69–88, 1974.
13. Lancaster, J. K.: "Friction and Wear," in "Polymer Science, a Materials Science Handbook," A. D. Jenkins ed., North Holland Publishing Co., Amsterdam, The Netherlands, pp. 960–1046, 1972.

13*a*. "Selecting Materials for Wear Resistance," Third International Conference on Wear of Materials, ASME, San Francisco, 1981.

13*b*. Ludema, K. C.: The Biennial Wear Conferences organized by ASME.

14. McNab, I. R., and J. L. Johnson: "Brush Wear," in "Wear Control Handbook," ASME, New York, p. 1053, 1980.
15. Murray, S. F., M. B., Peterson, and F. Kennedy: "Wear of Cast Bronze Bearings," INCRA Project 210, International Copper Research Association, New York, 1975.
16. Neale, M. J. (ed.): "Tribology Handbook," Butterworth & Company, London, 1973.
17. Norton, A. E.: "Lubrication," McGraw-Hill Book Company, Inc., New York, p. 24, 1942.
18. O'Connor, J. J., and J. Boyd: "Standard Handbook of Lubrication Engineering," McGraw-Hill Book Company, Inc., New York, 1978.
19. Rabinowicz, E.: "Friction and Wear of Materials," John Wiley & Sons, Inc., New York, 1965.
20. Salesky, W. J.: "Design of Medium Carbon Steels for Wear Applications," "Wear of Materials," ASME, New York, p. 298, 1981.

20*a*. Singer, I. L., and H. M. Pollock: "Fundamentals of Friction: Mocroscopic and Microscopic Processes," Proceedings of a NATO Conference, 1991, Kluwer Academic Publishers, Dordrecht, The Netherlands, 1992.

21. Suh, N. P., and N. Saka (eds.): "International Conference on Fundamentals of Tribology," MIT Press, Cambridge, Mass., 1978.

22. Suh, N. P.: "The Delamination Theory of Wear," *Wear,* vol. 25, pp. 111–124, 1973.

23. Tabor, D.: "Hardness of Metals," Cambridge University Press, Cambridge, England, 1951.

24. Tabor, D.: "A Simplified Account of Surface Topography and the Contact between Solids," *Wear,* vol. 32, pp. 269–271, 1975.

25. Tabor, D.: "Interaction between Surfaces: Adhesion and Friction," chap. 10 in "Surface Physics of Materials," vol. II, J. M. Blakely (ed.), Academic Press, New York, 1975.

26. Tabor, D.: "Wear—A Critical Synoptic View," *J. Lub. Technol.,* vol. 99, pp. 387–395, 1977.

27. Tanaka, K., and Y. Uchiyama: "Friction and Surface Melting of Crystalline Polymers," in "Advances in Polymer Friction and Wear, Polymer Science and Technology," 5A and 5B, pp. 499–532, H. H. Lee (ed.), Plenum Press, Inc., New York, 1974.

28. Thomas, T. R. (ed.): "Rough Surfaces," Longman, Inc., New York, 1982.

29. Timoshenko, S., and J. N. Goodier: "Theory of Elasticity," McGraw-Hill Book Company, Inc., New York, 1951.

30. Whitehouse, D. J., and M. J. Phillips: "Discrete Properties of Random Surfaces," *Phil. Trans. Roy. Soc.* (London), vol. 290, pp. 267–298, 1976.

31. Winer, W. O. (ed.): "Wear Control Handbook," ASME, New York, 1980.

P • A • R • T • 2

MECHANICAL SYSTEM ANALYSIS

CHAPTER 8
SYSTEM DYNAMICS

Sheldon Kaminsky, M.M.E., M.S.E.E.

Consulting Engineer
Weston, Conn.

8.1 INTRODUCTION: PRELIMINARY CONCEPTS 8.3
8.1.1 Degrees of Freedom 8.5
8.1.2 Coupled and Uncoupled Systems 8.6
8.1.3 General System Considerations 8.7
8.2 SYSTEMS OF LINEAR PARTIAL DIFFERENTIAL EQUATIONS 8.8
8.2.1 Elastic Systems 8.8
8.2.2 Inelastic Systems 8.15
8.3 SYSTEMS OF ORDINARY DIFFERENTIAL EQUATIONS 8.17
8.3.1 Fundamentals 8.17
8.3.2 Introduction to Systems of Nonlinear Differential Equations 8.18
8.4 SYSTEMS OF ORDINARY LINEAR DIFFERENTIAL EQUATIONS 8.26
8.4.1 Introduction to Matrix Analysis of Differential Equations 8.29
8.4.2 Fourier-Series Analysis 8.36
8.4.3 Complex Frequency-Domain Analysis 8.38
8.4.4 Time-Domain Analysis 8.44
8.5 BLOCK DIAGRAMS AND THE TRANSFER FUNCTION 8.50
8.5.1 General 8.50
8.5.2 Linear Time-Invariant Systems 8.50
8.5.3 Feedback Control-System Dynamics 8.52
8.5.4 Linear Time-Invariant Control System 8.53
8.5.5 Analysis of Control System 8.54
8.5.6 The Problem of Synthesis 8.62
8.5.7 Linear Discontinuous Control: Sampled Data 8.62
8.5.8 Nonlinear Control Systems 8.70
8.6 SYSTEMS VIEWED FROM STATE SPACE 8.79
8.6.1 State-Space Characterization 8.79
8.6.2 Transfer Function from State-Space Representation 8.82
8.6.3 Phase-State Variable-Form Transfer Function: Canonical (Normal) Form 8.82
8.6.4 Transformation to Normal Form 8.85
8.6.5 System Response from State-Space Representation 8.86
8.6.6 State Transition matrix for Sampled Data Systems 8.87
8.6.7 Time-Varying Linear Systems 8.88
8.7 CONTROL THEORY 8.89
8.7.1 Controllability 8.89
8.7.2 Observability 8.89
8.7.3 Introduction to Optimal Control 8.89
8.7.4 Euler-Lagrange Equation 8.90
8.7.5 Multivariable with Constraints and Independent Variable *t* 8.92
8.7.6 Pontryagin's Principle 8.95

8.1 INTRODUCTION—PRELIMINARY CONCEPTS

A physical system undergoing a time-varying interchange or dissipation of energy among or within its elementary storage or dissipative devices is said to be in a "dynamic state." The elements are in general inductive, capacitative, or resistive—the first two being capable of storing energy while the last is dissipative. All are called "passive," i.e., they are incapable of generating net energy. A system composed of a finite number or a denumerable infinity of storage elements is said to be "lumped" or "discrete," while a system containing elements which are dense in physical space is called "continuous." The mathematical description of the dynamics for the discrete case is a set of ordinary differential equations, while for the continuous case it is a set of partial differential equations.

The mathematical formulation depends upon the constraints (e.g., kinematic or geometric) and the physical laws governing the behavior of the system. For example, the motion of a single point mass obeys $\mathbf{F} = m(d\mathbf{v}/dt)$ in accordance with Newton's second law of motion. Analogously, the voltage drop across a perfect coil of self-inductance L is $V = L(di/dt)$, a consequence of Faraday's law. In the first case the energy-storage element is the mass, which stores $mv^2/2$ units of kinetic energy while the inductance L stores $Li^2/2$ units of energy in the second case. A spring-mass system and its electrical analog, an inductive-capacitive series circuit, represent higher-order discrete systems. The unbalanced force acting on the mass is $F - kx$. Thus

$$F = kx + m\ddot{x} \qquad m, k > 0 \tag{8.1}$$

Analogously for the electrical case,

$$V = L\ddot{q} + q/c \qquad L, c > 0$$

following Kirchhoff's voltage-drop law (i.e., the sum of voltage drop around a closed loop is zero). To show that Eq. (8.1) expresses the dynamic exchange of energy, multiply Eq. (8.1) by $\dot{x}\,dt$ (which is equal to dx) and integrate:

$$\int_0^t F\dot{x}\,dt = \int_{x=x_0}^{x} F\,dx = \int_0^t m\dot{x}\ddot{x}\,dt + \int_0^t k\dot{x}x\,dt$$

$$\text{Work input} = \underbrace{\frac{m\dot{x}^2}{2}\bigg]_0^t}_{\Delta\text{KE}} + \underbrace{\frac{kx^2}{2}\bigg]_0^t}_{\Delta\text{PE}} = \frac{m\dot{x}^2}{2} - \frac{m\dot{x}_0^2}{2} + \frac{kx^2}{2} - \frac{kx_0^2}{2}$$

which is a statement of the law of conservation of energy. This illustrates that work input is divided into two parts, one part increasing the kinetic energy, the remainder increasing the potential energy. The actual partition between the two energy sources at any instant is time-varying, depending on the solution to Eq. (8.1).

If a viscous damping element is added to the system the force equation becomes (see Fig. 8.1*a*)

$$m\ddot{x} + c\dot{x} + kx = F \qquad c > 0$$

and performing the same operation of multiplying by $\dot{x}\,dt$, (dx) and integrating we obtain

$$\int_0^t m\dot{x}\ddot{x}\,dt + \int_0^t c\dot{x}^2\,dt + \int_0^t kx\dot{x}\,dt = \int_{x_0}^{x} F\,dx \tag{8.2a}$$

$$\frac{m\dot{x}^2}{2}\bigg]_0^t + \int_0^t c\dot{x}^2\,dt + \frac{kx^2}{2}\bigg]_0^t = \int_{x_0}^{x} F\,dx \tag{8.2b}$$

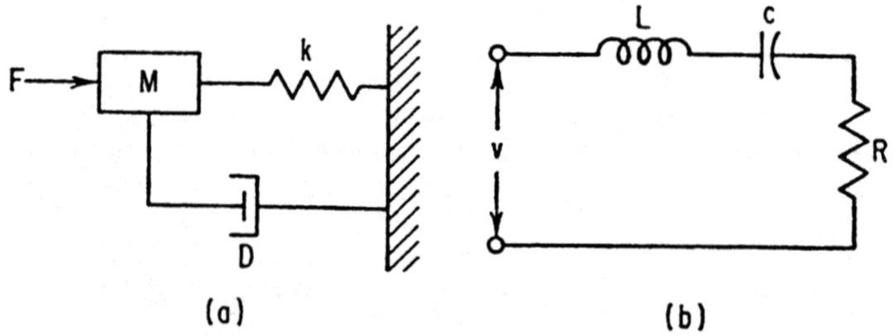

FIG. 8.1 Second-order systems. (*a*) Mechanical system. (*b*) Electrical analog.

again expressing the energy-conservation law. Note that the integrand $c\dot{x}^2 \geq 0$ and that the integral in Eq. (8.2*b*) is thus a monotonically increasing function of time. This condition assures that, for $F = 0$, the free (homogeneous) system must eventually come to rest since under this condition Eq. (8.2*b*) becomes

$$\frac{m\dot{x}^2}{2} + \frac{kx^2}{2} + \int_0^t c\dot{x}^2\,dt = \text{const} = \frac{m\dot{x}_0^2}{2} + \frac{kx_0^2}{2} \tag{8.3}$$

which again is an expression of the law of energy conservation. The first two terms are positive since they contain the squared factors $\dot{x}^2$ and x^2, while the third term, as noted above, increases with time. It follows that the sum of the first two must decrease monotonically in order to satisfy Eq. (8.3); moreover, neither term can be greater than the sum. It follows that, as $t \to \infty$, $x \to 0$ and $\dot{x} \to 0$.

Formulation of the foregoing simple problems was based upon fundamental physical laws. The derivation by Lagrange equations, which in this simple case offers little advantage, is (Chap. 1)

$$L = T - V \quad T = \tfrac{1}{2}m\dot{x}^2 \quad V = kx^2/2$$

For conservative systems (e.g., spring-mass),

$$\frac{d}{dt}\frac{\partial L}{\partial \dot{x}} - \frac{\partial L}{\partial x} = F = m\ddot{x} + kx$$

For nonconservative systems with dissipation function $\mathcal{F}$,

$$\mathcal{F} = c\dot{x}^2/2$$

$$\frac{d}{dt}\frac{\partial L}{\partial \dot{x}} - \frac{\partial L}{\partial x} = -\frac{\partial \mathcal{F}}{\partial \dot{x}} + F$$

$$m\ddot{x} + kx = -c\dot{x} + F$$

$$m\ddot{x} + c\dot{x} + kx = F$$

Precisely the same form is deducible from a Lagrange statement of the electrical equivalent (Fig. 8.1*b*).

8.1.1 Degrees of Freedom

Thus far it has been observed that one independent variable x was employed to describe the system dynamics. In general, however, several variables $x_1, x_2, \ldots, x_n$ are necessary to describe the motion of a complex system. The minimum number of coordinates that are so required is defined as the number of degrees of freedom of the system. Simple examples of two-degree-of-freedom systems are shown in Fig. 8.2. The respective equations of motion are

Mechanical:

$$m_1\ddot{x}_1 + k_1(x_1 - x_2) = F \tag{8.4a}$$

$$m_2\ddot{x}_2 + k_2x_2 + k_1(x_2 - x_1) = 0$$

Electrical:

$$L_1\ddot{q}_1 + (q_1 - q_2)/c_1 = V \tag{8.4b}$$

$$L_2\ddot{q}_2 + q_2/c_2 + (q_2 - q_1)/c_1 = 0$$

derivable from force and loop voltage-drop considerations.

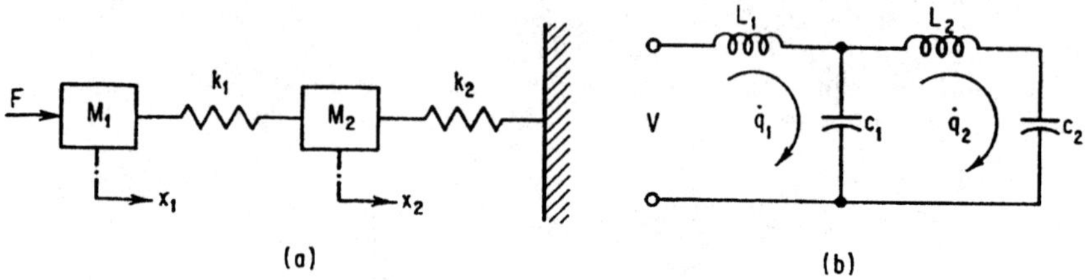

FIG. 8.2 Two-degree-of-freedom systems. (*a*) Mechanical. (*b*) Electrical analog.

Another example of a two-degree-of-freedom system is shown in Fig. 8.3, a compound pendulum constrained to move in a plane. While the system may at first appear to have four degrees of freedom with the positions of m_1 and m_2 given by r_1, θ_1, and r_1, θ_2, r_2, θ_2, respectively, two seemingly trivial expressions of constraint, r_1 = constant and r_2 = constant, show that the motion is describable in terms of θ_1 and θ_2 only. If a spring were interposed between m_1 and the pivot r_1, then r_1 would no longer be a constant and the motion would involve r_1, θ_1, θ_2, or three independent variables, resulting in a three-degree-of-freedom system.

8.1.2 Coupled and Uncoupled Systems

Equations (8.4*a*) or (8.4*b*) also illustrate a coupled system. The term "coupled" is a consequence of having more than one independent variable present in each equation of a set. In Eq. (8.4*a*), x_1 and x_2 and/or their derivatives appear in each of the two dynamic equations, implying that motion of one mass excites motion in the other mass. Only in conservative linear systems is it always possible to uncouple the system by a linear transformation.

An n-degree-of-freedom system requires for description n independent equations, usually of second order or lower. It is sometimes convenient to make changes in variables to facilitate the analysis of complex systems, or indeed to express the motion in terms of parameters, which are more accessible. In any case this amounts to having

$$x_i = x_i(q_1, q_2, \ldots, q_m) \qquad i = 1, 2, \ldots, n$$

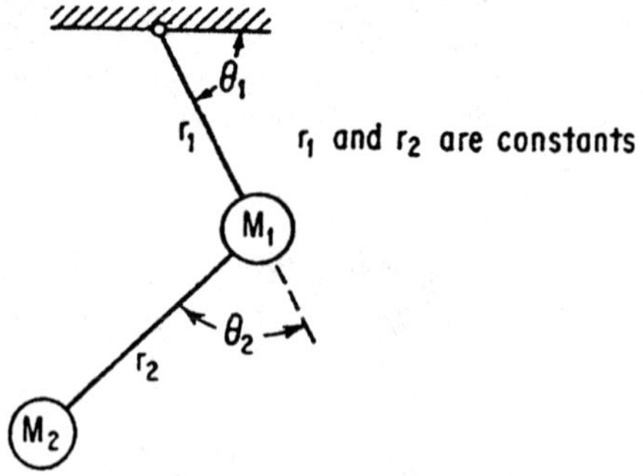

FIG. 8.3 Two-degree-of-freedom system (compound pendulum).

The q's, called "generalized coordinates," when judiciously chosen play a useful role in the analysis of complex systems. The q's need not be independent. This implies $m > n$ and the existence of $m - n$ equations that connect the q's, since the motion must involve only n independent equations. The case of Fig. 8.3 is an example in which $m = 4$ and $n = 2$ with $m - n = 2$ or two constraint equations, namely,

$$r_1 = \text{const} \qquad r_2 = \text{const}$$

8.1.3 General System Considerations

Discrete Systems. The equations for a system of n degrees of freedom can be written as

$$(b_{11}p^2 + c_{11}p + d_{11})x_1 + (b_{12}p^2 + c_{12}p + d_{12})x_2 + \cdots + (b_{1n}p^2 + c_{1n}p + d_{1n})x_n = f_1(t)$$

$$\cdots\cdots\cdots\cdots\cdots\cdots\cdots\cdots\cdots\cdots\cdots\cdots\cdots\cdots\cdots\cdots$$

$$(b_{n1}p^2 + c_{n1}p + d_{n1})x_1 + \cdots + (b_{nn}p^2 + c_{nn}p + d_{nn})x_n = f_n(t) \tag{8.5a}$$

where $p = d/dt$, $p^2 = d^2/dt^2$. Or, more concisely,

$$\sum_{j=1}^{n} (b_{ij}p^2 + c_{ij}p + d_{ij})x_j = f_i(t) \qquad i = 1, \ldots, n \tag{8.5b}$$

where b_{ij}, c_{ij}, d_{ij} are in general functions of x_k, $\dot{x}_k$, $\ddot{x}_k$, $k = 1, \ldots, n$, and time. In terms of generalized coordinates,

$$\sum_{j=1}^{m} (b'_{ij}p^2 + c'_{ij}p + d'_{ij})q_j = Q_i(t) \qquad i = 1, \ldots, m \qquad m \geq n \tag{8.5c}$$

where the Q_i's are the generalized forces (see Chap. 3). The number of degrees of freedom appears to have increased in Eq. (8.5c) for $m > n$, but this really is not the case, because of the existence of $m - n$ constraint equations connecting the q's.

The general form depicted by Eq. (8.5) is nonlinear in view of the b_{ij}, c_{ij}, and d_{ij} dependence on x_k and its time derivatives. Removal of this dependence yields the linear form of Eq. (8.5). Elimination of the time dependence in these coefficients yields the linear constant-coefficient form, which is of greatest engineering interest because it is the only one yielding completely to analysis and because a large class of systems can be approximated by this form. This is in contradistinction to the nonlinear and linear time-variable cases for which analytic solutions are in general not obtainable and not obtainable in closed form, respectively.

The initial state of each coordinate of Eq. (8.5) must be known [i.e., $x_i(0)$, $\dot{x}_i(0)$, $i = 1, 2, \ldots, n$], before the general solution is possible; hence $2n$ initial conditions are available which coincide with the maximum order of the differential equation obtained by eliminating $n - 1$ variables in Eq. (8.5). If the order is less than $2n$, then some of the initial conditions are not independent.

Continuous Systems. In passing from the description of discrete to that of continuous systems, the ordinary differential equation of n degrees of freedom becomes the set of partial differential equations as $n \to \infty$, i.e., the storage and dissipative elements become densely packed. The initial conditions are similar to those of the ordinary differential equation case which required initial velocity and position coordinates of each elementary mass particle; for now, in the limit of continuous systems, the initial displacement from equilibrium $\mathbf{u}(x, y, z, 0)$ and the displacement velocity $(d\mathbf{u}/dt)(x, y, z, 0)$ as well as the conditions $\mathbf{u}(x_b, y_b, z_b, t)$ that bound the system (where x_b, y_b, z_b are the continuous coordinates that bound the unperturbed system) are essential. As an example, consider the propagation of a pressure wave moving longitudinally in an infinite elastic dissipationless medium of small cross section. The equation of motion is derived by considering the elemental width dx having a stress $\sigma_{(x)}$ at the position x. Newton's law of motion applied to the element of mass of unit cross section is written as

$$\begin{aligned} \sigma - [\sigma + (\partial\sigma/\partial x)\,dx] &= m\,dx(\partial^2 u/\partial t^2) \\ -\partial\sigma/\partial x &= m(\partial^2 u/\partial t^2) \end{aligned} \qquad (8.6)$$

where m = mass density and σ = compressive stress.

The displacement from equilibrium u results in strain (compressive)

$$\epsilon = -\partial u/\partial x$$

and by Hooke's law,

$$\sigma = -Y(\partial u/\partial x) \qquad \text{(positive } \sigma \text{ is compressive)} \qquad (8.7)$$

where Y is Young's modulus for solids and a proportionality constant for other elastic media. Substitution of Eq. (8.7) into Eq. (8.6) yields

$$(Y/m)(\partial^2 u/\partial x^2) = \partial^2 u/\partial t^2 \qquad (8.8)$$

which is the simple one-dimensional wave equation.

8.2 SYSTEMS OF LINEAR PARTIAL DIFFERENTIAL EQUATIONS[1–5]

8.2.1 Elastic Systems

That class of systems characterized by interchange of kinetic and elastic energy is termed "elastic." The formulation of the nondissipative (conservative) type leads to the simple wave equation

$$c^2\nabla^2 u = \partial^2 u/\partial t^2 \qquad (8.9)$$

where ∇^2 is the Laplace operator (three-dimensional in general).

Examples of such systems follow.

Hydrodynamics and Acoustics. Applying Newton's second law to an elementary particle yields

$$\rho(dx\,dy\,dz)(d\mathbf{v}/dt) = (dx\,dy\,dz)\mathbf{F} + \sum \mathbf{f} \qquad (8.10)$$

where **F** represents the external forces (body forces) acting per unit volume on the element (e.g., gravity or inertia, using d'Alembert's principle), $\Sigma\mathbf{f}$ the sum of forces acting on the surfaces, and ρ the mass density. We can write

$$\sum \mathbf{f} = -(\partial p/\partial x)\,dx\,dy\,dz\mathbf{i} - (\partial p/\partial y)\,dy\,dx\,dz\mathbf{j} - (\partial p/\partial z)\,dz\,dx\,dy\mathbf{k} \qquad (8.11)$$

$$= -\nabla p\,dx\,dy\,dz \qquad (8.12)$$

where ∇ is the "del" or gradient operator and p is the pressure.

Substituting Eq. (8.12) into (8.10) yields

$$\rho(d\mathbf{v}/dt) = \mathbf{F} - \nabla p \qquad (8.13)$$

Expanding the left-hand side of Eq. (8.13), we obtain

$$\rho(\partial \mathbf{v}/\partial t + \underbrace{V_x \partial \mathbf{v}/\partial x + V_y \partial \mathbf{v}/\partial y + V_z \partial \mathbf{v}/\partial z)}_{(\mathbf{v}\cdot\nabla)\mathbf{v}} = \mathbf{F} - \nabla p \tag{8.14}$$

where $\mathbf{v} = V_x\mathbf{i} + V_y\mathbf{j} + V_z\mathbf{k}$.

From continuity (conservation of mass),

$$\partial\rho/\partial t + \mathrm{div}(\rho\mathbf{v}) = 0 \tag{8.15}$$

$$\partial\rho/\partial t + \mathbf{v}\cdot\nabla\rho + \rho\,\mathrm{div}\,\mathbf{v} = 0 \tag{8.16}$$

Equation (8.15) states that the rate of mass increase in elementary volume $dx\,dy\,dz$ $(\partial\rho/\partial t)$ equals the rate of flow into the same volume, $-dx\,dy\,dz\,\mathrm{div}\,(\rho\mathbf{v})$. If $|\partial\mathbf{v}/\partial t| \gg |(\mathbf{v}\cdot\nabla)\mathbf{v}|$, then to a good approximation of Eq. (8.14)

$$\rho\,\partial\mathbf{v}/\partial t \approx \mathbf{F} - \nabla p \tag{8.17}$$

Let the density be given by

$$\rho = \rho_0(1 + \epsilon) \qquad \rho_0 = \mathrm{const} \tag{8.18}$$

Taking a first differential of Eq. (8.18), we obtain

$$d\rho/\rho_0 = d\epsilon \tag{8.19}$$

Elimination of ρ in Eq. (8.16) yields

$$\partial\epsilon/\partial t + \mathrm{div}\,\mathbf{v} + \mathbf{v}\cdot\nabla\epsilon \approx 0 \tag{8.20}$$

where it is assumed that the variation of density about ρ_0 is small (i.e., $|\epsilon| \ll 1$). As a further consequence the third term of Eq. (8.20), involving space derivatives of ϵ which are of higher order, is accordingly dropped, leaving to a good approximation

$$\partial\epsilon/\partial t + \mathrm{div}\,\mathbf{v} \approx 0 \tag{8.21}$$

which, together with Eq. (8.17) in rearranged form,

$$\partial\mathbf{v}/\partial t - \mathbf{F}/\rho_0 + \nabla p/\rho_0 \approx 0 \tag{8.22}$$

provides two of the three essential relationships for small perturbation analysis; the remaining expression is the equation of state (e.g., $dp = k\,d\epsilon$).

Substituting $\nabla p = k\nabla\epsilon$ (where k is a constant) into Eq. (8.22), the following is obtained:

$$\partial\mathbf{v}/\partial t - \mathbf{F}/\rho_0 + k(\nabla\epsilon/\rho_0) \approx 0 \tag{8.23}$$

which together with Eq. (8.21) forms a fundamental set.

Calling u the particle displacement, $\mathbf{v} = \partial\mathbf{u}/\partial t$. Substituting for $\mathbf{v}$ in Eqs. (8.23) and (8.21) yields

$$\partial^2\mathbf{u}/\partial t^2 - \mathbf{F}/\rho_0 + k\,\nabla\epsilon/\rho_0 = 0 \tag{8.24}$$

$$\partial\epsilon/\partial t + (\partial/\partial t)\,\mathrm{div}\,\mathbf{u} = 0 \tag{8.25}$$

and

$$\epsilon + \mathrm{div}\,\mathbf{u} = 0 \tag{8.26}$$

Taking the divergence of Eq. (8.24) yields

$$-(\partial^2/\partial t^2)\,\mathrm{div}\,\mathbf{u} - \mathrm{div}(\mathbf{F}/\rho_0) + (k/\rho_0)\,\mathrm{div}\,\mathrm{grad}\,\epsilon = 0 \tag{8.27}$$

Next, substituting ϵ for $-\text{div}\,\mathbf{u}$ results in

$$-(\partial^2\epsilon/\partial t^2) - \text{div}(\mathbf{F}/\rho_0) + (k/\rho_0)\nabla^2\epsilon = 0 \tag{8.28}$$

For the case div $\mathbf{F}/\rho_0 = 0$, Eq. (8.28) becomes

$$\partial^2\epsilon/\partial t^2 = (k/\rho_0)\nabla^2\epsilon$$

which is the three-dimensional wave equation in ϵ.

If, in addition, the velocity is derivable from a scalar potential ϕ, i.e.,

$$\mathbf{v} = \text{grad}\,\phi = \partial\mathbf{u}/\partial t \tag{8.29}$$

then substitution in Eq. (8.24) gives

$$(\partial/\partial t)\,\text{grad}\,\phi = -(k/\rho_0)\,\text{grad}\,\epsilon + \mathbf{F}/\rho_0 \tag{8.30}$$

Differentiating with respect to time and assuming $\mathbf{F}$ time-independent,

$$\text{grad}\,[(\partial^2/\partial t^2)\phi - (k/\rho_0)\nabla^2\phi] = 0$$

From Eqs. (8.26) and (8.29),

$$\partial\epsilon/\partial t = -\text{div grad}\,\phi = -\nabla^2\phi$$

whence, by a suitable choice of ϕ,

$$(\partial^2/\partial t^2)\phi - (k/\rho_0)\nabla^2\phi = 0$$

i.e., the velocity-potential function is also of the wave type. For the special case of one-dimensional propagation with $\mathbf{F} = 0$,

$$V_x\mathbf{i} = (\partial u_x/\partial t)\mathbf{i} = \text{grad}\,\phi$$

Differentiating with respect to time and substituting from Eqs. (8.30) and (8.26),

$$\partial^2u_x/\partial t^2 = (k/\rho_0)(\partial^2u_x/\partial x^2) \tag{8.31}$$

Transverse Motion of an Elastic String Due to a Slight Perturbation. Consider a string under uniform tension T_0 initially stretched in a horizontal (x) direction (Fig. 8.4). If the weight of the string is negligible compared with inertia forces and the elongation is

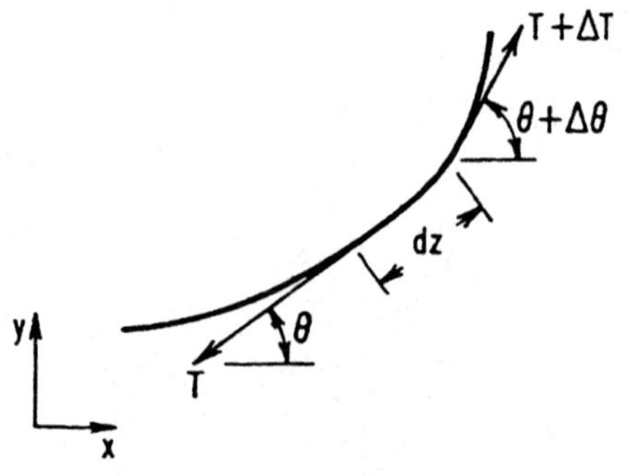

FIG. 8.4 String under tension.

negligible, the force balance in the y direction for an elementary section of length dz is

$$[\partial(T \sin\theta)/\partial z]dz = m\,dz\,\partial^2 y/\partial t^2 \tag{8.32a}$$

where $\sin\theta = \partial y/\partial z$. Considering only small displacements from the unperturbed position, i.e., $|\partial y/\partial z| \ll 1$ and

$$dz = du[1 + (\partial y/\partial x)^2]^{1/2} \qquad \partial x/\partial z \approx 1$$

$$\frac{\partial[T(\partial y/\partial z)]}{\partial z} = \frac{\partial\{T[(\partial y/\partial x)(\partial x/\partial z)]\}}{\partial x}\frac{\partial x}{\partial z} \approx \frac{\partial[T(\partial y/\partial x)]}{\partial x} \tag{8.32b}$$

If tension T is essentially constant and additive elongation is negligible,

$$T = T_0(1 + \epsilon) \qquad |\epsilon| \ll 1$$

then to a first approximation,

$$\frac{\partial[T(\partial y/\partial x)]}{\partial x} \approx T_0\frac{\partial^2 y}{\partial x^2}$$

and Eq. (8.32a) becomes, after dz is canceled, the one-dimensional wave equation

$$(T_0/m)(\partial^2 y/\partial x^2) = \partial^2 y/\partial t^2$$

Transverse Vibration of Stretched Membrane. Consider the stretched membrane of circular cross section (see Fig. 8.5). The transverse motion under a pressure p is found by forming the equation of motion on an elementary annulus of width dr and again ignoring the membrane weight,

$$dr(\partial/\partial r)(T2\pi r \sin\theta) = 2\pi r\,dr\,p + 2\pi\rho r\,dr\,(\partial^2 y/\partial t^2)$$

where $\sin\theta = \partial y/\partial r$
T = tension
ρ = density
whence, for T essentially constant [$T = T_0(1 + \epsilon)$, $|\epsilon| << 1$]

$$\frac{T_0}{\rho}\frac{1}{r}\frac{\partial}{\partial r}\left(r\frac{\partial y}{\partial r}\right) \approx \frac{\partial^2 y}{\partial t^2} + \frac{p}{\rho} = \frac{T_0}{\rho}\nabla^2 y = \frac{\partial^2 y}{\partial t^2} + \frac{p}{\rho} \qquad \text{(inhomogeneous wave equation)}$$

where ∇^2 = Laplacian operator

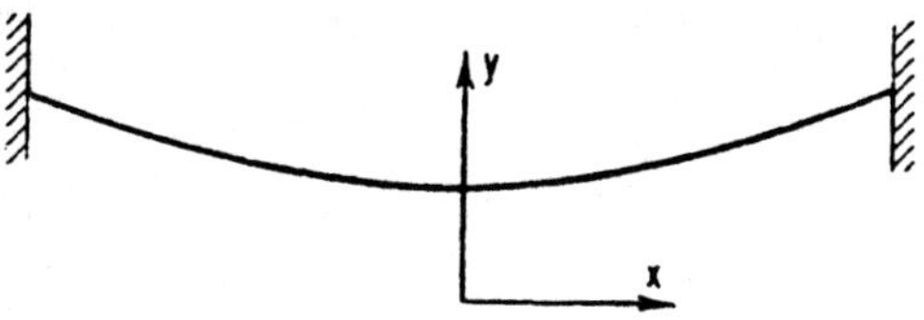

FIG. 8.5 Stretched membrane.

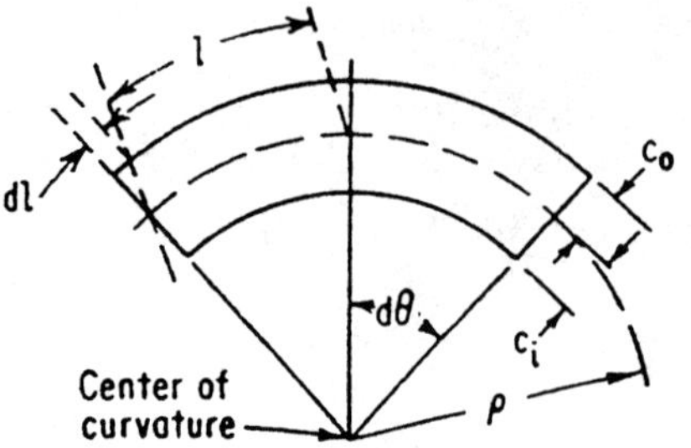

FIG. 8.6 Bending of a bar.

Transverse Vibrations of a Rod. The dynamics of motion of a uniform bar shown in Fig. 8.6 are derived by satisfying $\Sigma M = 0$ and $\Sigma F_y = 0$.

V and M are shear and bending moment shown acting on the elemental section of length dx. F_y are the vertical forces including the d'Alembert inertia force in the y direction $-(m\,dx)(\partial^2 y/\partial t^2)$. Satisfying $\Sigma F_y = 0$ in the positive y direction yields

$$\begin{aligned} -(\partial V/\partial x)\,dx - mg\,dx - m\,dx\,(\partial^2 y/\partial t^2) &= 0 \\ \partial V/\partial x + mg + m\,\partial^2 y/\partial t^2 &= 0 \end{aligned} \tag{8.33a}$$

and satisfying the moment equation about the center of mass of the elementary section results in

$$\partial M/\partial x = V \tag{8.33b}$$

Now a physical relation exists between M and y which is derivable by considering the bent section which is compressed on the inner fiber and stretched on the upper fiber with a "neutral axis," unstressed at the initial length (Fig. 8.6).

From geometric considerations,

$$\begin{aligned} dl &= c_0\,d\theta \qquad \rho \gg c_0 \\ d\theta &= l/\rho \qquad \rho \gg c_i \end{aligned} \tag{8.34a}$$

where ρ is the radius of curvature; c_0 and c_i distances from the neutral axis to the outer and inner fiber, respectively; l the half width of the elementary section; and $d\theta$ the half angle subtended by the section under stressed conditions.

Density ρ is further expressed by (from elementary calculus)

$$\frac{1}{\rho} = \frac{\partial^2 y/\partial x^2}{[1 + (\partial y/\partial x)^2]^{3/2}} \approx \frac{\partial^2 y}{\partial x^2} \qquad \text{for } \frac{\partial y}{\partial x} \ll 1 \tag{8.34b}$$

The strain at the outer fiber is $\epsilon_0 \approx dl/l$. From the geometry, the strain at any other point is $\epsilon_0(y/c_0)$ where y is the position measured from the neutral axis. From Eq. (8.34a), $dl/l = c_0/\rho$ and therefore the strain at y is

$$\epsilon = \epsilon_0(y/c_0) = (dl/l)(y/c_0) = (c_0/\rho)(y/c_0) = y/\rho$$

The stress, following Hooke's law, is

$$\sigma = E\epsilon = Ey/\rho$$

where E is the modulus of elasticity.

The bending moment about the neutral axis is expressed by

$$M = \int_{-c_i}^{c_0} y\sigma b\,dy$$

where b is the depth and $b\,dy$ is the elementary cross-sectional area. Substituting for σ the above expression becomes

$$\int_{-c_i}^{c_0} y\frac{Ey}{\rho} b\,dy = \frac{E}{\rho}\int_{-c_i}^{c_0} y^2 b\,dy$$

The integral on the right is I, the area moment of inertia about the neutral axis, a geometric property.

Thus

$$M = EI/\rho$$

and from Eq. (8.34*b*)

$$M/EI = \partial^2 y/\partial x^2 \quad (8.35)$$

Taking two derivatives of Eq. (8.35) with respect to x, one derivative of Eq. (8.33*b*), and substituting for $\partial y/\partial x$ in Eq. (8.33*a*) yields

$$EI\,\partial^4 y/\partial x^4 + m(g + \partial^2 y/\partial t^2) = 0$$

Torsional Motion of a Rod. Consider an elemental cylindrical section of length dl and a twist angle $d\theta$ (see Fig. 8.7). The strain on an elemental area da is $r\,d\theta/dl$, and the associated stress is

$$\sigma = Gr(d\theta/dl) \quad (8.36)$$

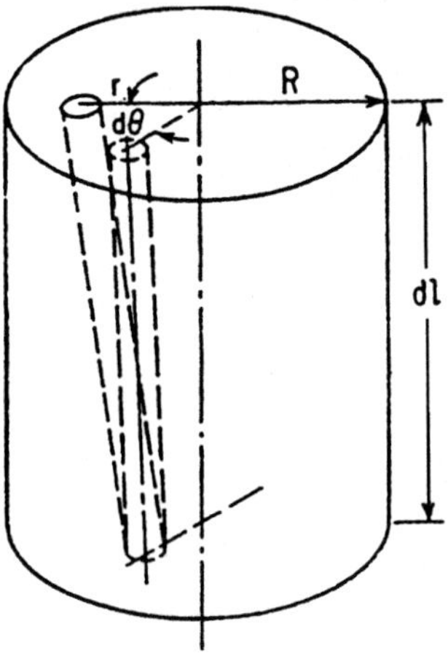

FIG. 8.7 Torsion in a rod.

where G is the shear modulus and r is the radius to the point in question. The total torque

$$T = \int_A r\sigma\,da = \int_A Gr\frac{d\theta}{dl}r\,da = G\frac{d\theta}{dl}\int_A r^2\,da = G\frac{d\theta}{dl}J$$

where J = polar moment of inertia $= \int_A r^2\,da = \int_0^R r^2 2\pi r\,dr$.

The expression for torsional oscillations is obtained from Newton's second law:

$$\begin{aligned} dl(\partial T/\partial l) &= I(\partial^2\theta/\partial t^2)\,dl \\ JG(\partial^2\theta/\partial t^2) &= I(\partial^2\theta/\partial t^2) \end{aligned} \quad (8.37)$$

where I = mass moment of inertia per unit length. For homogeneous media $I = \rho J$ and Eq. (8.37) becomes

$$(G/\rho)(\partial^2\theta/\partial t^2) = \partial^2\theta/\partial t^2 \quad (8.38)$$

Electric-Transmission-Line Equation for Low-Frequency Operation. Consider a section of length dx as shown in Fig. 8.8. From Ohm's law the current density is

$$\mathbf{i} = -k\,\text{grad}\,V$$

If the wire has cross section $\mathbf{A}$ where $\mathbf{A}$ is a vector in the direction normal to the cross section, the total current I is

$$I = \mathbf{i}\cdot\mathbf{A} = -k\mathbf{A}\cdot\text{grad}\,V = -\frac{1}{R}\text{grad}\,V \quad (8.39)$$

In the x direction this becomes

$$-\frac{\partial v}{\partial x} = IR$$

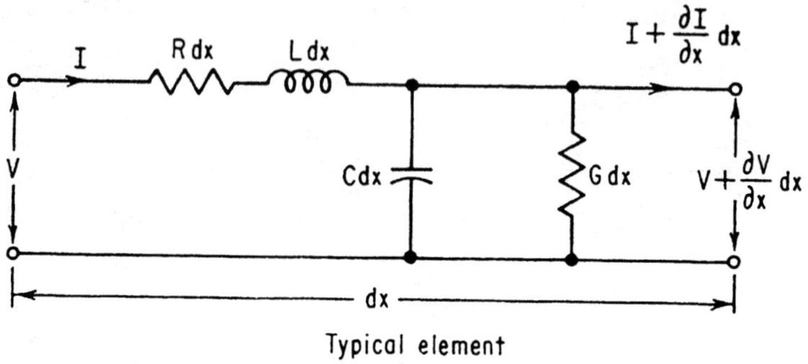

FIG. 8.8 Electric transmission line.

where R is the resistance per unit length. The wire also acts as a distributed capacitance C per unit of length; following Faraday's law,

$$\frac{dQ}{C\,dx} = dV \tag{8.40}$$

Taking the partial differential of Eq. (8.40) with respect to time for the elemental section yields

$$\frac{1}{C\,dx}\frac{\partial Q}{\partial t} = \frac{\partial V}{\partial t} \tag{8.41}$$

The charge Q which collects within the dx section is

$$Q = \int_0^t I\,dt - \int_0^t \left(I + \frac{\partial I}{\partial x}dx\right)dt \tag{8.42}$$

and therefore
$$\frac{\partial Q}{\partial t} = -\frac{\partial I}{\partial x}dx$$

Substitution in Eq. (8.41) yields

$$\frac{1}{C}\frac{\partial I}{\partial x} = -\frac{\partial V}{\partial t} \tag{8.43}$$

If in addition some current leaks off and is proportional to V, then Eq. (8.42) should be modified as follows:

$$Q = \int_0^t I\,dt - \int \left(I + \frac{\partial I}{\partial x}dx\right)dt - \int (GV\,dx)\,dt \tag{8.44}$$

where G is the leakage conductance per unit length.

Taking the partial derivative of Eq. (8.44) with respect to time,

$$\frac{\partial Q}{\partial t} = -\frac{\partial I}{\partial x}dx - GV\,dx$$

and replacing $\partial Q/\partial t$ in Eq. (8.41), leads to the modified equation

$$-\frac{\partial I}{\partial x} = GV + C\frac{\partial V}{\partial t} \tag{8.45}$$

Also, the inductance along the wire owing to Faraday's law introduces an additional voltage drop to modify Eq. (8.39) to read

$$-\frac{\partial V}{\partial x}dx = RI\,dx + L\frac{\partial I}{\partial t}dx \tag{8.46}$$

where L is the inductance per unit length. Then

$$-\frac{\partial V}{\partial x} = RI + L\frac{\partial I}{\partial t}$$

Combining Eqs. (8.45) and (8.46) results in

$$CL\frac{\partial^2 I}{\partial t^2} + (RC + GL)\frac{\partial I}{\partial t} + RGI = \frac{\partial^2 I}{\partial x^2} \tag{8.47}$$

and the identical form in V, i.e.,

$$CL\frac{\partial^2 V}{\partial t^2} + (RC + GL)\frac{\partial V}{\partial t} + RGV = \frac{\partial^2 V}{\partial x^2} \tag{8.48}$$

Equations (8.47) and (8.48) are the telegrapher's equation which was first reported by Kirchhoff. Note that, if R and G are zero, they reduce to the simple wave equation.

8.2.2 Inelastic Systems

Flow of Heat, Electricity, and Fluid. The flow of heat across a boundary as given by Fourier is

$$-k \operatorname{grad} T = \mathbf{Q} \tag{8.49}$$

where $\mathbf{Q}$ = heat flux and T = temperature.

For electricity, Ohm's law is analogous to Fourier's law; thus

$$-k \operatorname{grad} V = \mathbf{i} \tag{8.50}$$

where V = voltage and $\mathbf{i}$ = current flow density.

Following Fick's law[6] for flow of incompressible fluids through finely divided porous media,

$$-k \operatorname{grad} p = \mathbf{v} \tag{8.51}$$

where p = pressure and $\mathbf{v}$ = flow rate per unit area.

Conservation laws applied to Eqs. (8.49) to (8.51) yield the following expressions. For Eq. (8.49), conservation of thermal energy implies

$$\rho c(\partial T/\partial t) = -\text{div}\,\mathbf{Q} \tag{8.52}$$

where c = specific heat per unit mass and ρ = mass density. Similarly, for Eq. (8.50) and Eq. (8.51), conservation of charge and conservation of mass, respectively, imply

$$\partial q/\partial t = -\text{div}\,\mathbf{i} \tag{8.53}$$

where q = charge density, and

$$\partial \rho/\partial t = -\text{div}\,(\rho\mathbf{v}) \tag{8.54}$$

where ρ = mass density.

Q is eliminated between Eqs. (8.49) and (8.52) by taking the divergence of Eq. (8.49):

$$\rho c(\partial T/\partial t) = -(-\text{div}\,k\,\text{grad}\,T) = k\,\nabla^2 T \tag{8.55}$$

Similarly, for Eqs. (8.50) and (8.53),

$$\partial q/\partial t = k\,\nabla^2 V \tag{8.56}$$

And, for Eqs. (8.51) and (8.54),

$$\partial \rho/\partial t = k\,\text{div}\,(\rho\,\text{grad}\,p) \tag{8.57}$$

If ρ = const, Eq. (8.57) reduces to Laplace's equation,

$$\nabla^2 p = 0 \tag{8.58}$$

If Eq. (8.52), (8.53), or (8.54) had volume sources at the points of investigation, for example,

$$\rho c(\partial T/\partial t) = -\text{div}\,\mathbf{Q} + S$$

Then Eqs. (8.55), (8.56), and (8.58) would respectively read

$$k\,\nabla^2 T = \rho c(\partial T/\partial t) - S \tag{8.59}$$

$$k\,\nabla^2 V = \partial q/\partial t - S \tag{8.60}$$

$$k\rho\,\nabla^2 p = -S \tag{8.61}$$

In the absence of time-varying potentials, Eqs. (8.59) and (8.60) reduce to the Poisson form of Eq. (8.61), and where no source is present all reduce to the form of Laplace's equation (8.58). Electrostatic phenomena are closely related to the above developments. The electrostatic field **E** is given by

$$\mathbf{E} = -\text{grad}\,V \tag{8.62}$$

and the flux **D** is linearly related to **E** by

$$\mathbf{D} = \epsilon\mathbf{E} \tag{8.63}$$

where ϵ = dielectric constant. By Gauss's law, which follows from Coulomb's law of forces,

$$\text{div}\,\mathbf{D} = \rho \tag{8.64}$$

where ρ = charge density. Eliminating **D** and **E** among Eqs. (8.62) to (8.64) yields

$$\rho = \text{div } \mathbf{D} = -\epsilon \text{ div grad } V = -\epsilon \nabla^2 V \tag{8.65}$$

$$\nabla^2 V = -\rho/\epsilon$$

which is Poisson's equation, degenerating to Laplace's equation in the absence of sources (i.e., $\rho = 0$).

8.3 SYSTEMS OF ORDINARY DIFFERENTIAL EQUATIONS

8.3.1 Fundamentals

All systems which occur in nature are nonlinear and distributed. To an excellent approximation, many systems can be "lumped," permitting vast simplifications of the mathematical model. For example, the lumped spring-mass-damping system is, strictly speaking, a distributed system with the "mass" composed of an infinity of densely packed elementary springs and masses and damping elements arranged in some uncertain order. Because of the theoretical difficulties encountered in formulating an accurate mathematical model which fits the actual system and the analytical difficulties in attacking the complex problem, the engineer (with experimental justification) makes the "mass" a point mass which cannot be deformed, and the spring a massless spring without damping. If damping is present an element called the *damper* is isolated so that the "lumped" system is composed of discrete elements. Having settled on an equivalent lumped physical model, the equations describing system behavior are next formulated on the basis of known physical laws.

The equations thus derived constitute a set of ordinary differential equations, generally nonlinear, implying the existence of one or more lumped elements which do not behave in a "linear" fashion, e.g., nonlinearity of load vs. deflection of a spring.

In mathematical terms it is easier to define a nonlinear set by first defining what constitutes a linear set and then using the exclusion principle as follows.

A set of ordinary differential equations is linear if terms containing the dependent variable(s) or their time derivatives appear to the first degree only. The physical system it characterizes is termed *linear*. All other systems are nonlinear, and the physical systems they define are nonlinear.

An example of a linear system is the set

$$t^2 \frac{d^2x_1}{dt^2} + t\frac{d^2x_2}{dt^2} + \frac{d^2x_1}{dt^2} + (\sin t)^2 x_2 + G = 0$$

$$\frac{dx_1}{dt} + \frac{d^2x_2}{dt^2} = 0$$

where it is noted that the factors containing functions of t, the independent variable, are ignored in determining linearity, and each term containing one of the dependent variables x_1, x_2, or their derivatives, is of the first degree.

Two examples of nonlinear systems are

$$(dx_1/dt)^2 + 2x_1 = 0 \tag{8.66a}$$

$$x_1(d^3x_2/dt^2) + x_1x_2 = 0 \qquad x_2 + d^2x_1/dt^2 + d^3x_2/dt^3 = 0 \tag{8.66b}$$

In the first system, the square of the first derivative immediately rules it as nonlinear. The first equation of the second system is nonlinear on two counts: first, by virtue of the product of x_1, and a time derivative d^3x_2/dt^2, and second, because of the term containing the product of two dependent variables, x_1, x_2. Despite the linearity of the second equation in the system, the overall system [Eq. (8.66*b*)] is nonlinear.

In general, and with few exceptions, the nonlinear equation does not yield to analysis, so that machine, numerical, or graphical methods must be employed. Wherever possible and under very special circumstances approximations are made to "linearize" a nonlinear system in order to make the problem amenable to analysis.

8.3.2 Introduction to Systems of Nonlinear Differential Equations[7–9]

Perhaps the simplest classic example of a nonlinear system is the undamped free pendulum, the equation of motion of which is

$$\ddot{\theta} + (g/l)\sin\theta = 0 \tag{8.67}$$

This belongs to a class of elastic systems containing nonlinear restoring forces. Here $\sin\theta$ is clearly the nonlinear term. For small displacements of θ, $\sin\theta \approx \theta$ and Eq. (8.67) becomes

$$\ddot{\theta} + (g/l)\theta = 0 \tag{8.68}$$

The general solution of Eq. (8.68) is

$$\theta = A\sin(\sqrt{g/lt}) + B\cos(\sqrt{g/lt}) \tag{8.69}$$

A and B are constants of integration depending upon initial conditions. As θ gets large, Eq. (8.68) no longer holds, and therefore Eq. (8.69) is an invalid approximation to Eq. (8.67). Under this condition Eq. (8.67) cannot be "linearized." Other nonlinear restoring forces are characterized as hard and soft springs whose force F vs. deflection x characteristics are given by $F = ax + bx^2$, $a > 0$ where $b < 0$ for soft springs, $b > 0$ for hard springs, and $b = 0$ for linear springs (see Chap. 2). The degree of nonlinearity is measured by the relative magnitudes of bx^3 and ax and implies some knowledge of x. Linearization of the spring-mass system given by

$$\ddot{x} + ax + bx^3 = 0 \tag{8.70}$$

is possible if

$$|bx^3| \ll |ax|$$

for all x experienced, yielding the approximation

$$\ddot{x} + ax \approx 0$$

An electric analog of this system of Eq. (8.70) exists for an LC circuit where C depends upon q in accordance with $1/C = \alpha + \beta q^2$.

From $q/C + L\,d^2q/dt^2 = 0$ the following is derived after substitution for $1/C$:

$$L\ddot{q} + \alpha q + \beta q^3 = 0$$

Another analog derives from the nonlinear dependence of flux ϕ on current i in an LC circuit given by $i = \alpha\phi + \beta\phi^3$, which when substituted in the first time derivative of the loop-drop equation

$$L\frac{d\phi}{dt} + \frac{q}{C} = 0 \tag{8.71}$$

yields
$$L\frac{d^2\phi}{dt^2} + \frac{i}{c} = L\frac{d^2\phi}{dt^2} + \frac{\alpha\phi + \beta\phi^3}{C}$$

Expressed in generalized form, the foregoing nonlinear spring-mass (capacitance-inductance) systems are given by

$$\ddot{x} + f(x) = 0 \tag{8.72}$$

Multiplying Eq. (8.72) by $\dot{x}\,dt$ and integrating, we have

$$\begin{aligned} \int_0^t \dot{x}\ddot{x}\,dt + \int_0^t f(x)\dot{x}\,dt &= \frac{\dot{x}^2}{2}\bigg]_0^t + \int_{x(0)}^{x(t)} f(x)\,dx \\ \frac{\dot{x}^2}{2} + \int_{x(0)}^{x(t)} f(x)\,dx &= \frac{\dot{x}^2_{(0)}}{2} \end{aligned} \tag{8.73}$$

which is a statement expressing energy conservation. If $V(x)$ is the indefinite integral,

$$\int f(x)\,dx = +V(x)$$

Then Eq. (8.73) becomes

$$\dot{x}^2/2 - x^2(0)/2 + V(x) - V[x(0)] = 0$$
$$\dot{x}^2/2 + V(x) = \dot{x}^2(0)/2 + V[x(0)] \overset{\Delta}{=} E \tag{8.74}$$

$V(x)$ is the potential-energy function which represents stored energy from some arbitrary reference level, and E, a constant, is defined to be the "total energy" at any time. Solving Eq. (8.74), we have

$$\dot{x} = \sqrt{2[E - V(x)]} \tag{8.75}$$

Qualitative Behavior of the Conservative Free System. From Eq. (8.75) it is evident that physically realizable motion demands that $E \geq V(x)$ for all possible x. Consider a possible graph of $V(x)$ (Fig. 8.9) with E_0 drawn intersecting at points 1, 2, 3, 4, which points correspond to $E_0 = V(x)$, and from Eq. (8.75), $\dot{x} = 0$. Since $f(x) = (dV/dx)$, the slopes of the curve at these points give the spring force $f(x)$. From Eq. (8.72)

$$\ddot{x} = f(x) = -dV/dx \tag{8.76}$$

Consequently acceleration corresponds to the direction of arrows shown for the two possible states of motion in Fig. 8.9, implying periodic motion between x_1 and x_2 in one case, and between x_3 and x_4 in the order. To find the period for case 1, for example,

$$\tau = \int_{x_1}^{x_2} \frac{dx}{\dot{x}} + \int_{x_2}^{x_1} \frac{dx}{\dot{x}} = \oint \frac{dx}{\dot{x}} \tag{8.77}$$

where integration is around a cycle loop in a phase-plane plot shown in Fig. 8.9, where $\dot{x}$ is plotted as a function of x, and the sign of $\dot{x}$ equals the sign of dx.

For E_1 as in initial energy level shown in Fig. 8.9, motion is possible when $E_1 \geq V(x)$; it is seen that, for an initial negative velocity, the system will come to rest at point 5 and then, from Eq. (8.76), since the acceleration at that point is positive,

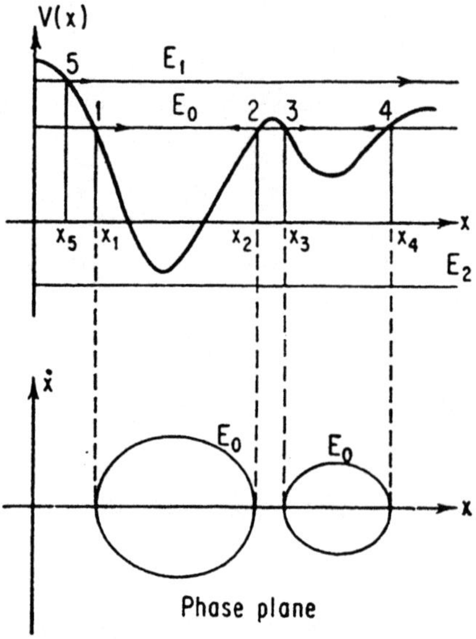

FIG. 8.9 Qualitative behavior of second-order free system.

motion would start to the right. Since $E_1 > V$ for $x > x_5$, it is impossible for x to reach zero again, and hence motion would continue in the positive direction without bound.

If $E = E_2$ as shown in Fig. 8.9, $E_2 < V(x)$ for all x; this cannot correspond to a physical system, a consequence of Eq. (8.75).

As an example of the above, the energy of the simple undamped pendulum is found from Eqs. (8.67) and (8.74):

$$\theta^2/2 - (g/l)\cos\theta = E$$

where motion is indicated between θ_1 and θ_2 for $E = E_0$ (Fig. 8.10). For $E = E_1$ motion continues in a single direction, which physically amounts to putting in more energy than that required to bring the pendulum into the position where it is vertically above its support.

Graphical Analysis of Second-Order Nonlinear Autonomous Differential Equations. Consider the following form of a free second-order equation with time-invariant coefficients:

$$\ddot{x} + f(x, \dot{x}) = 0 \tag{8.78}$$

It is possible to analyze this very restrictive equation by a graphical method called "phase-plane analysis." Equation (8.78) is first rewritten as

$$\begin{aligned} \dot{x}\, d\dot{x}/dx &= -f(x, \dot{x}) \\ d\dot{x}/dx &= -f(x, \dot{x})/\dot{x} \end{aligned} \tag{8.79}$$

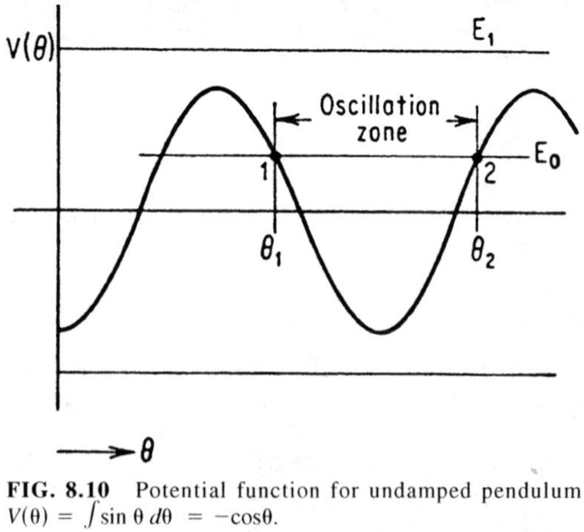

FIG. 8.10 Potential function for undamped pendulum $V(\theta) = \int \sin\theta\, d\theta = -\cos\theta$.

A plot is next made of $\dot{x}$ as a function of x (phase-plane plot). At every point Eq. (8.79) states that the slope is $-f(x, \dot{x})/\dot{x}$.

The initial conditions $\dot{x}(0)$, $x(0)$ place the origin of the system in the phase plane. An arc with slope equal to $-f(x(0), \dot{x}(0))/\dot{x}(0)$ is laid off over a small length terminating at $x(1)$, $\dot{x}(1)$. The process is continued until either a stable point is reached, or a limit cycle is manifest, or indications show the growth without bound of the system parameters x or $\dot{x}$. To find x as a function of time, $t = \int dx/\dot{x}$.

Several convenient techniques are available to facilitate procedures (e.g., the isocline method), the essentials of which were described above. Special cases of phase-plane analyses are given below.

Special Case 1: Linear Spring-Mass System

$$\ddot{x} + kx = 0$$
$$\frac{\dot{x}^2}{2} + \frac{kx^2}{2} = \frac{\dot{x}(0)^2}{2} + \frac{kx^2(0)}{2} = E$$

The equation is of an ellipse in the phase plane $\dot{x}$ versus x, or if we make the following changes of variable:

$$y = x/\sqrt{k} \qquad \tau = \sqrt{k}t$$

and substitute in the above, we obtain

$$\begin{aligned} (dy/d\tau)^2 + y^2 &= 2E/k^2 \\ \dot{y}^2 + y^2 &= 2E/k^2 \end{aligned} \tag{8.80}$$

which represents a circle of radius $\sqrt{2E/k^2}$ about the origin in the phase plane of $\dot{y}$ versus y.

Special Case 2: Spring-Mass-Damper System

$$\ddot{x} + c\dot{x} + kx = 0$$
$$\frac{\dot{x}^2}{2} + \int_0^t c\dot{x}^2\,dt + \frac{kx^2}{2} = E \qquad c > 0$$

Writing the energy form, where the damping integral is greater than zero as shown earlier,

$$\frac{\dot{x}^2}{2} + \frac{kx^2}{2} = E - \int_0^t c\dot{x}^2\,dt$$
$$\frac{k^2\dot{y}^2}{2} + \frac{k^2y^2}{2} = E - \frac{k^2}{k^{1/2}}\int_0^\tau c\dot{y}^2\,d\tau$$

The right side decreases in time, so that in the phase-plane plot the locus must lie on a continuously decreasing radius from the origin as time increases, until the origin is reached. The actual path is a logarithmic spiral. Other systems of the form $\ddot{x} + f(\dot{x}) + kx = 0$ are, by suitable changes of variable, shown equivalent to

$$\ddot{y} + \phi(\dot{y}) + y = 0$$

whence

$$\frac{d\dot{y}}{dy} = \frac{-[\phi(\dot{y}) + y]}{\dot{y}} \tag{8.81}$$

The phase-plane plot of Eq. (8.81) is obtainable by a neat method due to Liénard, described as follows: In Fig. 8.11, first $-\phi(\dot{y})$ is drawn. Then for any point of state, say P_1, the locus has a center of curvature in the phase plane located on the y axis shown by dotted construction. The slope must be that given by Eq. (8.81), tan ω. From geometry,

$$\tan \omega = -\frac{1}{\tan \theta} = \frac{-[\phi(\dot{y}) + y]}{\dot{y}}$$

Special Case 3: Coulomb Damping (Dry Friction), Second-Order System

$$\ddot{x} + c\,\text{sgn}\,\dot{x} + x = 0 \tag{8.82}$$

where sgn = sign of. The phase-plane plot in Fig. 8.12 is accomplished, following Liénard's method, by first plotting $-c$ sgn x and then following in accordance with the above description. The plot consists of arcs of two circles centered at 1 for $\dot{x} > 0$ and 2 for $\dot{x} < 0$. This is shown for two different initial conditions corresponding to p_1 and p_1' in Fig. 8.12. Note that motion stops at a position corresponding to 4 since $\dot{u} = 0$ and the spring force is less than the impending damping force, thus preventing motion. This can also be shown analytically by considering two regions $\dot{x} < 0$ and $\dot{x} > 0$. Rewriting Eq. (8.82), we obtain

$$\dot{x}\,d\dot{x}/dx + (x + c\,\text{sgn}\,\dot{x}) = 0$$

Multiplying Eq. (8.82) by dx and integrating for the two regions,

$$\dot{x}^2/2 + x^2/2 + cx = E_1 \rightarrow \dot{x}^2 + (x + c)^2 = 2E_1 + c^2 \qquad \dot{x} > 0$$

Similarly,
$$\dot{x}^2 + (x - c)^2 = 2E_2 + c^2 \qquad \dot{x} < 0$$

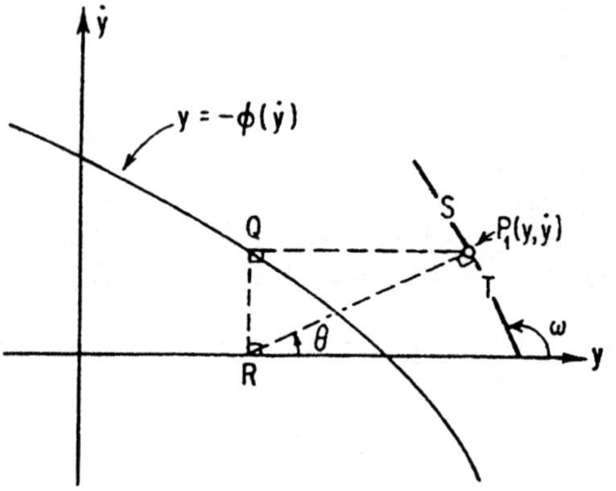

FIG. 8.11 Phase-plane plot of $d\dot{y}/dy = \{-[\phi(\dot{y}) + y]\}/\dot{y}$ (Liénard's construction).

Method: $QR = \dot{y}$

$$QP = y - [-\phi(\dot{y})] = y + \phi(\dot{y})$$

$$\tan \theta = QR/QP = \dot{y}/[y + \phi(\dot{y})]$$

$$= -1/(d\dot{y}/dy)$$

Hence it follows that the slope of line ST, perpendicular to line RP_1 at P_1, is $d\dot{y}/dy$.

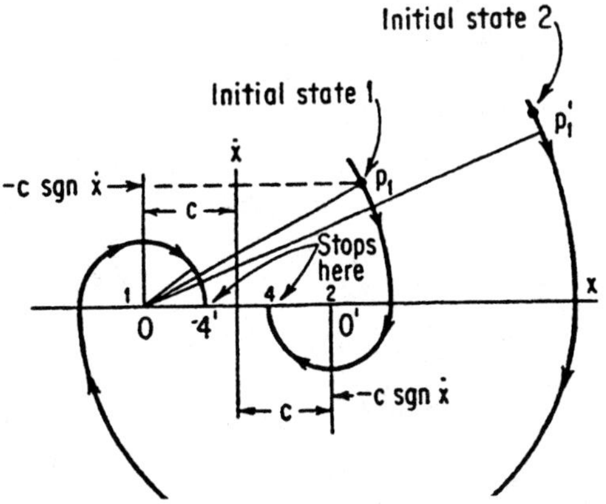

FIG. 8.12 Phase-plane plot for Coulomb damping of spring-mass system.
$\ddot{x} + c \operatorname{sgn} \dot{x} + x = 0$

Limit Cycles and Sustained Oscillations. Consider the system governed by $\ddot{x} + f(x, \dot{x}) + x = 0$. If

$$\dot{x}f(x, \dot{x}) < 0 \quad |x| < \delta \tag{8.83a}$$

$$\dot{x}f(x, \dot{x}) > 0 \quad |x| > \delta \tag{8.83b}$$

where δ is some positive constant, the system will exhibit a limit cycle which corresponds to a closed curve in the phase plane. When Eq. (8.83*a*) holds, there is a net increase in the system energy $\mathscr{E}$:

$$\frac{\dot{x}^2}{2} + \frac{x^2}{2} + \int_0^t f(x, \dot{x})\dot{x}\, dt = E = \frac{\dot{x}^2(0)}{2} + \frac{x^2(0)}{2} = \mathscr{E}(0)$$

$$\mathscr{E} = \frac{\dot{x}^2}{2} + \frac{x^2}{2} = E - \int_0^t f(\dot{x}, x)\dot{x}\, dt \tag{8.84}$$

given by the initial state E minus the integral. Since $2\mathscr{E}$ is the radius squared from the center to the point of state in the phase plane, there is a time rate of increase of radius every time the motion falls within the shaded zone (Fig. 8.13) and a decrease for motion corresponding to points outside the shaded zone. The type of oscillation is self-sustained and will start of its own accord for any initial condition. The van der Pol equation is an example of this type:

$$\begin{aligned} \ddot{x} - \epsilon\dot{x} + \beta x^2\dot{x} + x = 0 \qquad \epsilon > 0 \\ \dot{x}f(x, \dot{x}) = \dot{x}^2(-\epsilon + \beta x^2) \end{aligned} \tag{8.85}$$

The term $f(x, \dot{x})$ changes sign when

$$-\epsilon + \beta x^2 = 0$$

$$x = \pm\sqrt{\epsilon/\beta} = \pm\delta$$

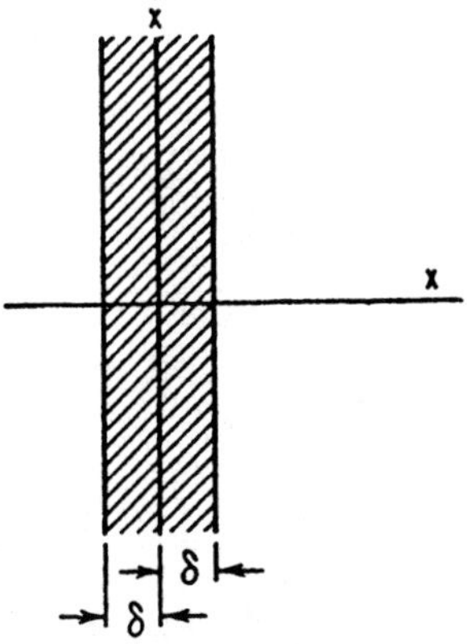

FIG. 8.13 Limit cycles and sustained oscillations.

Limit cycles for higher-order systems are conceptually depicted by closed curves in multidimensional space, which is the generalization of single-degree-of-freedom systems.

Singular Points and Stability. It can be shown[7] that any autonomous set of nonlinear differential equations can be represented by

$$
\begin{aligned}
\dot{x}_1 &= f_1(x_1, \ldots, x_n) \\
\dot{x}_2 &= f_2(x_1, \ldots, x_n) \\
&\cdots\cdots\cdots\cdots \\
\dot{x}_n &= f_n(x_1, \ldots, x_n)
\end{aligned}
\tag{8.86}
$$

The equilibrium positions are given by the roots of

$$
\begin{aligned}
f_1(x_1, \ldots, x_n) &= 0 \\
f_2(x_1, \ldots, x_n) &= 0 \\
\cdots\cdots\cdots&\cdots \\
f_n(x_1, \ldots, x_n) &= 0
\end{aligned}
\tag{8.87}
$$

The roots $x_1, \ldots, x_n$ of this set are called *singular equilibrium points* where all the time derivatives $\dot{x}_1, \dot{x}_2, \ldots, \dot{x}_n$ are equal to zero.

The algebraic solution to Eq. (8.87) gives in general one or more sets of singular points, e.g.,

$$
\begin{array}{llll}
x_1^{(0)} \; x_2^{(0)} & \cdots & x_n^{(0)} & \text{first set} \\
x_1^{(1)} \; x_2^{(1)} & \cdots & x_n^{(1)} & \text{second set} \\
\cdots\cdots & \cdots & \cdots & \cdots\cdots
\end{array}
$$

If the motion at any time corresponds to one of these points, the system is at rest. If left undisturbed, from Eq. (8.86), $\dot{x}_1 = \dot{x}_2 = \cdots = \dot{x}_n = 0$, the point (in phase space), and therefore the corresponding motion, does not change in time; i.e., the system remains at rest. If, however, the point is disturbed from its equilibrium position (perturbation), it is of interest from the stability point of view as to whether or not it will return to the point. If for a small perturbation from equilibrium the system tends to return to the same equilibrium position as $t \to \infty$, the system is said to be "asymptotically stable." If, however, the system diverges from the equilibrium point, it is said to be in a state of "unstable equilibrium," or the point is unstable. Special points in which neither of these events occurs are said to display neutral stability and are exceptional. To test stability, the nonlinear system is "linearized" in the neighborhood of the equilibrium point $\bar{x}_1, \bar{x}_2, \ldots, \bar{x}_n$ by performing a Taylor's-series expansion about the point and ignoring terms higher than the first power of x_i. The typical expansion is

$$
f_i = f_i(\bar{x}_1, \bar{x}_2, \ldots, \bar{x}_n) + \sum_i \left(\frac{\partial f_i}{\partial x_i}\right)_{x_k = \bar{x}_k} (x_i - \bar{x}_i) + \text{higher-order terms} \quad j = 1, 2, \ldots, n
$$

We define

$$
\left(\frac{\partial f_i}{\partial x_i}\right)_{x_k = \bar{x}_k} = a_{ji} \qquad f_i \approx \sum_i a_{ji}(x_1 - \bar{x}_i)
$$

Since the constant term vanishes; i.e.,

$$
f_i(\bar{x}_1, \bar{x}_2, \ldots, \bar{x}_n) = 0
$$

which is a consequence of the definition of the equilibrium point. For convenience, the substitution

$$\begin{aligned} x_1 - \bar{x}_1 &= y_1 \\ x_2 - \bar{x}_2 &= y_2 \\ &\cdots\cdots\cdots \\ x_n = \bar{x}_n &= y_n \end{aligned}$$

placed in Eq. (8.86) yields

$$\begin{aligned} \dot{y}_1 &= a_{11}y_1 + a_{12}y_2 + \cdots + a_{1n}y_n \\ &\cdots\cdots\cdots\cdots\cdots\cdots \\ \dot{y}_n &= a_{n1}y_1 + a_{n2}y_2 + \cdots + a_{nn}y_n \end{aligned} \tag{8.88}$$

This is the well-known linear set whose solution is of the exponential type. Assuming a solution,

$$y_j = A_j e^{\lambda t}$$

and making this substitution in Eq. (8.88) yields

$$\begin{aligned} 0 &= (a_{11} - \lambda)A_1 + a_{12}A_2 + \cdots + a_{1n}A_n \\ 0 &= a_{21}A_1 + (A_{22} - \lambda)A_2 + \cdots + a_{2n}A_n \\ &\cdots\cdots\cdots\cdots\cdots\cdots \\ 0 &= a_{n1}A_1 + a_{n2}A_2 + d \cdots + (a_{nn} - \lambda)A_n \end{aligned} \tag{8.89}$$

From linear theory, the necessary condition for Eq. (8.89) to have nontrivial solutions is

$$\begin{vmatrix} a_{11} - \lambda & a_{12} & \cdots & a_{1n} \\ a_{21} & a_{22} - \lambda & \cdots & a_{2n} \\ \cdots & \cdots & \cdots & \cdots \\ a_{n1} & a_{n2} & \cdots & a_{nn} - \lambda \end{vmatrix} = 0$$

which when expanded leads to an nth-order algebraic equation,

$$b_n\lambda^n + b_{n-1}\lambda^{n-1} + \cdots + b_1 x + b_0 = 0 \tag{8.90}$$

which has n roots for λ, the characteristic roots of the matrix

$$\begin{bmatrix} a_{11} & a_{12} & \cdots & a_{1n} \\ \cdots & \cdots & \cdots & \cdots \\ a_{n1} & a_{n2} & \cdots & a_{nn} \end{bmatrix}$$

Each root corresponds to a solution $y_j = A_j e^{\lambda_i t}$. If λ_i has a real part greater than zero, y_j will grow without bound. Hence the necessary and sufficient condition for stability at the equilibrium point is that Re $\lambda_i < 0$ for all roots, $i = 1, \ldots, n$. As an example, consider the second-order van der Pol equation (8.85) in the form

$$\begin{aligned} \dot{x} &= v \\ \dot{v} &= \epsilon v - \beta x^2 v - x \end{aligned}$$

The only equilibrium point is $x = v = 0$, obtained after invoking Eq. (8.87).

Expanding about $x = 0$, $v = 0$, carrying linear terms

$$\dot{x} = v$$
$$\dot{v} \approx -x + \epsilon v$$

The characteristic roots are found from

$$\begin{vmatrix} -\lambda & 1 \\ -1 & \epsilon - \lambda \end{vmatrix} = 0$$

$$-\lambda(\epsilon - \lambda) + 1 = 0 \qquad \lambda = (\epsilon \pm \sqrt{\epsilon^2 - 4})/2$$

It is evident that one or more roots must satisfy Re $\lambda > 0$; therefore, the system is unstable about the equilibrium point as observed previously. Note that, if ϵ is negative, the system is stable.

8.4 SYSTEMS OF ORDINARY LINEAR DIFFERENTIAL EQUATIONS[11–14]

Formulation of the linearized form of a lumped dynamic system leads to a set of linear differential equations. The question of validity in assuming linearity is, in general, complicated. One method (though not conclusive) is to assume a linear form, solve the set, cast the solution into the original form to measure deviations from linearity, and finally, on this basis, render a decision on validity. An example is the simple pendulum undergoing a forced vibration where the steady-state solution is of interest:

$$\ddot{\theta} + \omega_0^2 \sin\theta = A \sin \omega t \qquad \omega_0^2 = g/l$$

The linearized form and its characteristic solution are

$$\ddot{\theta} + \omega_0^2\,\theta = A \sin \omega t$$
$$\theta = [A/(\omega_0^2 - \omega^2)]\sin \omega t$$

And therefore θ is bounded by

$$|\theta| \leq |A|/(\omega_0^2 - \omega^2)$$

The linear form is valid if $(\sin\theta - \theta)/\theta$ remains small for all motion, i.e.,

$$|(\sin\theta - \theta)/\theta| \ll 1$$

But

$$\sin\theta = \theta - \theta^3/3! + \theta^5/5! - \cdots$$

Therefore

$$\sin\theta/\theta = 1 - \theta^2/3! + \theta^4/5! - \cdots \qquad |(\sin\theta - \theta)/\theta| \leq \theta^2/3!$$

and the necessary condition for validity becomes

$$|\theta^2/3!| \ll 1$$

Linear systems are classified as time-variant or time-invariant and in the former derive from a system containing time-variable parameters leading to product terms in independent and dependent variables, e.g., $\sin[\alpha t(d^3x/dt^3)]$ and t^2x, whereas the time-invariant case shows no parametric dependence on time.

An example of a time-variable linear system is that of a rocket propelled in free unidirectional flight by a jet-exhaust thrust $C(t)$. Fuel expenditure results in a rocket mass loss. The equation of motion is

$$m(t)\, dv/dt = C(t) \tag{8.91}$$

with $m(t)$ and $C(t)$, the mass and thrust, connected under some broad assumptions by the linear differential equation

$$C(t) = -k[dm(t)/dt] \tag{8.92}$$

The solution is

$$v - v(0) = k \ln [m(0)/m]$$

This is an exceptional case since a closed solution is obtainable. More generally, however, time-variable systems are practically invulnerable to analytic attack in contrast to their time-invariant counterparts whose solutions are completely known.

Properties of Linear Differential Equations. The general properties of linear differential equations are stated as follows:

1. The general homogeneous linear equation is expressed in operator form

$$D(Y) = f(t)$$

with initial conditions

$$Y(0) = a_0 \qquad \dot{Y}(0) = a_1 \qquad (d^{n-1}/dt^{n-1})Y(0) = a_{n-1} \tag{8.93}$$

D is the linear operator of nth order:

$$D = \sum_{m=0}^{n} P_m(t) \frac{d^m}{dt^m}$$

where $P_m(t)$ is the coefficient (function of time in general) of the nth derivative term.

2. The linear operator has the properties

$$D(\alpha y) = \alpha\, Dy \qquad (\alpha = \text{const})$$

$$D(y_1 + y_2) = Dy_1 + Dy_2$$

Therefore,

$$D(\alpha y_1 + \beta y_2) = \alpha\, Dy_1 + \beta\, Dy_2$$

3. The complete solution to the homogeneous equation $D(Z) = 0$ of the nth degree is the sum of n linearly independent solutions, viz.,

$$Z = b_1 z_1 + b_2 z_2 + \cdots + b_n z_n$$

where $b_1, \ldots, b_n$ are constants which can be adjusted to satisfy n initial conditions of the problem.

4. The general solution to Eq. (8.93) is the sum of two solutions $Y = Z + X$, where Z is the total solution to the homogeneous equation

$$D(Z) = 0 \tag{8.94}$$

and X is *any* solution to $D(X) = f(t)$ regardless of initial conditions. It should be emphasized, and is implied, that X is not unique, containing any number of solutions to the homogeneous equation, e.g., $c_1 z_1 + \cdots + c_m z_m$.

5. If the solution to $D(X) = f(t)$ is confined to be the asymptotic solution (i.e., the solution as $t \to \infty$, then this solution $X \triangleq X_p$ is called the "particular solution" and the remainder $Z = b_1 z_1 + b_2 z_2 + \cdots + b_n z_n \triangleq Z_c$ is called the "complementary solution" and is called the "transient solution" if it vanishes with time. The terms $b_1, \ldots, b_n$ are chosen such that all initial conditions of Eq. (8.93) are satisfied. Then since the solution to Eq. (8.93) is unique,[11]

$$Y = X_p + Z_c = X + Z$$

6. Another important specialized partitioning of X and Z restricts X to satisfy zero initial conditions, i.e., the solution to $D(x) = f(t)$ for

$$X(0) = \dot{X}(0) = \cdots = \frac{d^{n-1}}{dt^{n-1}} X(0) = 0$$

which is

$$X_{(1)} = \int_0^t W(t, T) f(T)\, dT$$

where $W(t, T)$ is the solution to

$$D(W) = \delta(t - T)$$

$W(t, T) \equiv 0$ for $t \le T$, and $\delta(t - T)$ is defined as the Dirac delta function. The remaining part of the solution is

$$Z_{(1)} = c_1 z_1 + c_2 z_2 + \cdots + c_n z_n$$

where the c's are chosen to satisfy the initial conditions of the problem, which are absorbed in the z's alone.

$$Y(0) = Z_{(1)}(0) = \sum_{i=1}^{n} c_i z_i(0)$$

$$Y'(0) = Z'_{(1)}(0) = \sum_{i=1}^{n} c_i z'_i(0)$$

$$Y''(0) = Z''_{(1)}(0) = \sum_{i=1}^{n} c_i z''_i(0)$$

$$\cdots\cdots\cdots\cdots\cdots\cdots\cdots\cdots$$

$$\frac{d^{n-1}}{dt^{n-1}} Y(0) = \frac{d^{n-1} Z_{(1)}(0)}{dt^{n-1}} = \sum_{i=1}^{n} c_i \frac{d^{n-1}}{dt^{n-1}} z_{(i)}(0)$$

and

$$Y = X_{(1)} + Z_{(1)} = X + Z$$

7. As a direct consequence of property 6, the principle of superposition follows. For a system initially inert, i.e., for zero initial conditions,

y_1 is the response to a forcing function $f_1(t)$

y_2 is the response to a forcing function $f_2(t)$

The response y to forcing function $f(t) = \alpha f_1(t) + \beta f_2(t)$ is $y = \alpha y_1 + \beta y_2$.

8. The necessary and sufficient condition that the solution y be bounded in Eq. (8.93) for any bounded input $f(t)$ is that

$$\int_0^\infty |W(t, T)|\, dT < M < \infty$$

where M is some arbitrarily large positive number. Systems satisfying this criterion are said to be stable.

9. The fact that n solutions to Eq. (8.94) are linearly independent requires that the wronskian be different from zero at any point in the interval $t_1 < t < t_2$ where Eq. (8.94) is valid.

$$W_r(z_1, z_2, \ldots, z_n) = \det \begin{bmatrix} z_1 & z_1 & \cdots & z_n \\ dz_1/dt & dz_2/dt & \cdots & dz_n/dt \\ \cdots & \cdots & \cdots & \cdots \\ -d^{n-1}z_1/dt^{n-1} & d^{n-1}z_2/dt^{n-1} & \cdots & d^{n-1}z_n/dt^{n-1} \end{bmatrix}$$

The wronskian of the solutions to

$$d^n z/dt^n + Q_{(n-1)}(t)(d^{n-1}z)/dt^{n-1}) + \cdots + Q_1(t)\, dz/dt + Q_0(t)z = 0$$

is given by

$$W_r(z_1, \ldots, z_n) = W_r[z_1(\tau), z_2(\tau), \ldots, z_n(\tau)] \exp \int_\tau^t Q_{n-1}(x)\, dx \tag{8.95}$$

As a direct consequence of Eq. (8.95) the wronskian of a set satisfying Eq. (8.94) either does not vanish at all or vanishes identically since the exponential term cannot vanish.

8.4.1 Introduction to Matrix Analysis of Differential Equations[15]

The nth-order differential equation

$$d^n y_1/dt^n + Q_{n-1}(d^{n-1}y_1/dt^{n-1}) + \cdots + Q_0 y_1 = F \tag{8.96}$$

can be written as n first-order differential equations

$$\begin{aligned} dy_1/dt &= y_2 \\ dy_2/dt &= y_3 \\ &\cdots\cdots\cdots\cdots \\ dy_{n-1}/dt^{n-1} &= y_n \\ dy_n/dt &= -Q_{n-1}y_n - Q_{n-2}y_{n-1} - \cdots - Q_0 y_1 + F \end{aligned} \tag{8.97}$$

which in matrix form is written as

$$\frac{d\mathbf{y}}{dt} = \mathbf{A}\mathbf{y} + \mathbf{f} \qquad \mathbf{y} = \begin{vmatrix} y_1 \\ y_2 \\ y_3 \\ \cdot \\ \cdot \\ \cdot \\ y_n \end{vmatrix} \tag{8.98}$$

In Eq. (8.97),

$$\mathbf{A} = \begin{bmatrix} 0 & 1 & 0 & 0 & \cdots & 0 & 0 \\ 0 & 0 & 1 & 0 & \cdots & 0 & 0 \\ 0 & 0 & 0 & 1 & \cdots & 0 & 0 \\ \cdots & \cdots & \cdots & \cdots & \cdots & \cdots & \cdots \\ 0 & 0 & 0 & 0 & \cdots & 1 & 0 \\ 0 & 0 & 0 & 0 & \cdots & 0 & 1 \\ -Q_0 & -Q_1 & -Q_2 & -Q_3 & \cdots & -Q_{n-2} & -Q_{n-1} \end{bmatrix} \qquad \mathbf{f} = \begin{bmatrix} 0 \\ 0 \\ \cdot \\ \cdot \\ \cdot \\ F \end{bmatrix}$$

Equation (8.98) is linear if matrix $\mathbf{A} = \mathbf{A}(t)$ and $\mathbf{f} = \mathbf{f}(t)$ and linear time-invariant if A does not depend on t.

Following classical methods, the total solution to

$$d\mathbf{y}/dt = \mathbf{A}(t)\mathbf{y} + \mathbf{f}(t)$$

is the sum of two solutions, one to the homogeneous equation

$$d\mathbf{z}/dt = \mathbf{A}(t)\mathbf{z} \tag{8.99}$$

plus any solution to

$$d\mathbf{x}/dt = \mathbf{A}(t)\mathbf{x} + \mathbf{f}(t) \tag{8.100}$$

where

$$\mathbf{y} = \mathbf{x} + \mathbf{z}$$

The general solution chosen here will let the homogeneous solution satisfy the initial conditions of the problem, i.e.,

$$\mathbf{y}(0) = \mathbf{z}(0) = \mathbf{c}$$

and the remaining solution satisfy the null-vector condition at $t = 0$,

$$\mathbf{x}(0) = \mathbf{0}$$

so that

$$\mathbf{y}(0) = \mathbf{z}(0) + \mathbf{x}(0) = \mathbf{y}(0) + \mathbf{0} = \mathbf{y}(0) = \mathbf{c}$$

as required.

If the solution to Eq. (8.99) is known, the solution to Eq. (8.100) can be obtained by Lagrange's method of variation of parameters as follows: Consider the matrix differential equation

$$d\mathbf{Z}(t)/dt = \mathbf{A}(t)\mathbf{Z}(t) \qquad \mathbf{Z}(0) = \mathbf{I}(\text{initial conditions}) \tag{8.100a}$$

Postmultiplication by $\mathbf{c}$ yields

$$d(\mathbf{Zc})/dt = \mathbf{A}(\mathbf{Zc})$$

From this and Eq. (8.99) it follows that

$$\mathbf{z} = \mathbf{Z}(t)\mathbf{c} = \mathbf{Z}(t)\mathbf{z}(\mathbf{0})$$

Now let

$$\mathbf{y} \triangleq \mathbf{Z}(t)\mathbf{u} \tag{8.100b}$$

Substituting for $\mathbf{y}$ in Eq. (8.98),

$$\frac{d\mathbf{y}}{dt} = \mathbf{A}(t)[\mathbf{Z}(t)\mathbf{u}] + \mathbf{f} = \mathbf{Z}(t)\frac{d\mathbf{u}}{dt} + \frac{d\mathbf{Z}(t)}{dt}\mathbf{u} = \mathbf{Z}(t)\frac{d\mathbf{u}}{dt} + \mathbf{A}(t)\mathbf{Z}(t)\mathbf{u}$$

whence $\mathbf{f} = \mathbf{Z}\, d\mathbf{u}/dt$. Premultiplying by $\mathbf{Z}^{-1}$ yields $\mathbf{Z}^{-1}\mathbf{f} = d\mathbf{u}/dt$, and integrating after separation of variables gives

$$\mathbf{u} = \mathbf{u}(0) + \int_0^t \mathbf{Z}^{-1}(\tau)\mathbf{f}(\tau)\, d\tau$$

Premultiply by $\mathbf{Z}(t)$ to give

$$\mathbf{y} = \mathbf{Z}(t)\mathbf{u} = \mathbf{Z}(t)\mathbf{c} + \int_0^t \mathbf{Z}(t)\mathbf{Z}^{-1}(\tau)\mathbf{f}(\tau)\, d\tau \tag{8.101}$$

where $\mathbf{u}(0) = \mathbf{y}(0) = \mathbf{c}$ from Eq. (8.100*b*) and $\mathbf{Z}(0) = \mathbf{I}$.

For the time-invariant case,

$$\mathbf{Z}(t)\mathbf{Z}^{-1}(\tau) = \mathrm{Z}(t - \tau)$$

and Eq. (8.101) takes the simpler form

$$\mathbf{y} = \mathbf{Z}(t)\mathbf{c} + \int_0^t \mathbf{Z}(t - \tau)\mathbf{f}(\tau)\, d\tau \tag{8.102}$$

Eqs. (8.101) and (8.102) hinge on the solution to the homogeneous matrix equation

$$d\mathbf{Z}/dt = \mathbf{A}(t)\mathbf{Z} \tag{8.103}$$

However, for $\mathbf{A}(t)$, the time-variable case, a solution is rarely possible. For the time-invariant case,

$$d\mathbf{Z}(t)/dt = \mathbf{A}\mathbf{Z}(t) \qquad \mathbf{Z}(0) = \mathbf{I} \tag{8.104}$$

The formal solution is

$$\mathbf{Z}(t) = e^{\mathbf{A}t}$$

which must be defined as

$$e^{\mathbf{A}t} \triangleq \sum_{n=0}^{\infty} \frac{\mathbf{A}^n t^n}{n!}$$

and

$$\mathbf{Z}(t + \tau) = e^{\mathbf{A}(t+\tau)}$$

Equation (8.102) then becomes

$$y = e^{\mathbf{A}t}\mathbf{c} + \int_0^t e^{\mathbf{A}(t-\tau)}\mathbf{f}(\tau)\, d\tau \tag{8.105}$$

Equation (8.105) is in a form that is not useful for quantitative analysis.

It is desirable to obtain **Z** in closed form. If **Z** is represented by column vectors

$$\mathbf{Z} = \begin{bmatrix} -z_1^{(1)} & \cdots & z_1^{(n)} \\ z_2^{(1)} & \cdots & z_2^{(n)} \\ \cdots & \cdots & \cdots \\ -z_n^{(1)} & \cdots & z_n^{(n)} \end{bmatrix} \triangleq [\mathbf{z}^{(1)} \quad \mathbf{z}^{(2)} \quad \mathbf{z}^{(3)} \quad \ldots \quad \mathbf{z}^{(n)}]$$

then $d\mathbf{Z}/dt = \mathbf{AZ}$

$$\mathbf{Z}(0) = \mathbf{I} = \begin{bmatrix} 1 & 0 & 0 & \cdots & 0 & 0 \\ 0 & 1 & 0 & \cdots & 0 & 0 \\ 0 & 0 & 1 & \cdots & 0 & 0 \\ \cdots & \cdots & \cdots & \cdots & \cdots & \cdots \\ 0 & 0 & 0 & \cdots & 1 & 0 \\ 0 & 0 & 0 & \cdots & 0 & 1 \end{bmatrix}$$

is equivalent to n equations

$$\frac{d\mathbf{z}^{(1)}}{dt} = \mathbf{Az}^{(1)} \qquad \mathbf{z}^{(1)}(0) = \begin{bmatrix} 1 \\ 0 \\ \cdot \\ \cdot \\ \cdot \\ 0 \end{bmatrix}$$

$$\cdots\cdots\cdots\cdots\cdots\cdots\cdots\cdots \tag{8.106}$$

$$\frac{d\mathbf{z}^{(n)}}{dt} = \mathbf{Az}^{(n)} \qquad \mathbf{z}^{(n)}(0) = \begin{bmatrix} 0 \\ 0 \\ \cdot \\ \cdot \\ \cdot \\ 1 \end{bmatrix}$$

The homogeneous matrix equation possesses n possible solution vectors. Assuming one such vector solution

$$\mathbf{z}^{(k)} = \mathbf{c}^{(k)} e^{\lambda_j t} \qquad j = 1, 2, \ldots, n$$

and entering this into Eq. (8.106) yields

$$\lambda_j \mathbf{c}^k = \mathbf{Ac}^{(k)} \tag{8.107}$$

$$(\mathbf{A} - \lambda_j \mathbf{Ic}^k) = 0$$

which is the eigenvector equation that must satisfy

$$|\mathbf{A} - \lambda_j \mathbf{I}| = 0$$

an nth-order equation yielding n roots for λ_j. The discussion here will be limited to distinct roots.

For each λ_j so found there exists a column vector formed from the cofactors of any row of $(A - \lambda_j I)$; let the matrix formed by the n column vectors thus formed be called B:

$$\mathbf{B} = [\mathbf{c}^{k(1)} \quad \mathbf{c}^{k(2)} \quad \cdots \quad \mathbf{c}^{k(n)}]$$

Equation (8.107) can be represented as

$$\mathbf{AB} = \mathbf{B\Lambda} \qquad \mathbf{\Lambda} = \begin{bmatrix} \lambda_1 & 0 & \cdots & 0 \\ 0 & \lambda_2 & \cdots & 0 \\ \cdots & \cdots & \cdots & \cdots \\ 0 & 0 & \cdots & \lambda_n \end{bmatrix}$$

Premultiplying by $\mathbf{B}^{-1}$ yields

$$\mathbf{B}^{-1}\mathbf{AB} = \mathbf{\Lambda}$$

which shows that $\mathbf{A}$ is diagonalized by a linear transformation.

From Eq. (8.100*a*),

$$d\mathbf{Z}/dt = \mathbf{AZ} \qquad \mathbf{Z}(0) = \mathbf{I}$$

Let $\mathbf{W}$ be introduced by defining the transformation

$$\mathbf{Z} = \mathbf{BW}$$

Substituting in Eq. (8.100*a*), we have

$$d(\mathbf{BW})/dt = \mathbf{A}(\mathbf{BW})$$

$$\mathbf{B}(d\mathbf{W}/dt) = \mathbf{ABW}$$

Premultiplying by $\mathbf{B}^{-1}$, we have

$$\frac{d\mathbf{W}}{dt} = \mathbf{B}^{-1}\mathbf{ABW} = \mathbf{\Lambda W} = \begin{bmatrix} \lambda_1 & 0 & \cdots & 0 \\ 0 & \lambda_2 & \cdots & 0 \\ \cdots & \cdots & \cdots & \cdots \\ 0 & 0 & \cdots & \lambda_n \end{bmatrix} \mathbf{W}$$

The solution for $\mathbf{W}$ is easily verified to be

$$\mathbf{W} = \begin{bmatrix} e^{\lambda_1 t} & 0 & \cdots & 0 \\ 0 & e^{\lambda_2 t} & \cdots & 0 \\ \cdots & \cdots & \cdots & \cdots \\ 0 & 0 & \cdots & e^{\lambda_n t} \end{bmatrix} \mathbf{B}^{-1} \qquad \mathbf{W}(0) = \mathbf{B}^{-1}$$

which satisfies its differential equation and the initial conditions, viz.,

$$\begin{bmatrix} \lambda_1 e^{\lambda_1 t} & 0 & \cdots & 0 \\ 0 & \lambda_2 e^{\lambda_2 t} & \cdots & 0 \\ \cdots & \cdots & \cdots & \cdots \\ 0 & 0 & \cdots & \lambda_n e^{\lambda_n t} \end{bmatrix} \mathbf{B}^{-1} \equiv \begin{bmatrix} \lambda_1 & 0 & \cdots & 0 \\ 0 & \lambda_2 & \cdots & 0 \\ \cdots & \cdots & \cdots & \cdots \\ 0 & 0 & \cdots & \lambda_n \end{bmatrix} \begin{bmatrix} e^{\lambda_1 t} & 0 & \cdots & 0 \\ 0 & e^{\lambda_2 t} & \cdots & 0 \\ \cdots & \cdots & \cdots & \cdots \\ 0 & 0 & \cdots & e^{\lambda_n t} \end{bmatrix} \mathbf{B}^{-1}$$

$$d\mathbf{W}/dt = \mathbf{\Lambda W}$$

$$\mathbf{Z}(0) = \mathbf{BW}(0) = \mathbf{BB}^{-1} = I$$

as required.

Linear Time-Invariant Systems.[16–24] The important property that distinguishes these systems from the time-variable systems is the following:

1. If the input $f(t)$ yields the response $y(t)$, the input $f(t+T)$ yields the response $y(t+T)$ (where all initial conditions are zero). As a consequence of property 1 and the superposition property we have the following property.
2. The response to the derivative of an arbitrary input is equal to the derivative of the response. If $f(t) \to y(t)$, then

$$\frac{f(t+\epsilon) - f(t)}{\epsilon} \to \frac{y(t+\epsilon) - y(t)}{\epsilon} \qquad \begin{array}{c} \epsilon \to 0 \\ f'(t) \to y'(t) \end{array}$$

or, vectorially, $\mathbf{f}(t) \to \phi[\mathbf{y}(t)]$ where ϕ is a linear time-invariant operator. Then

$$\frac{\mathbf{f}(t+\epsilon) - \mathbf{f}(t)}{\epsilon} \to \frac{\phi[\mathbf{y}(t+\epsilon) - \mathbf{y}(t)]}{\epsilon} = \phi[\mathbf{y}'(t)] \qquad \begin{array}{c} \epsilon \to 0 \\ \mathbf{f}'(t) \to \phi[\mathbf{y}'(t)] \end{array}$$

3. The solution to a free (homogeneous) time-invariant system of equations is composed of exponential terms, there being as many terms as the highest degree of the differential equation obtained in one dependent variable. These terms must be linearly independent, satisfying $W_r \neq 0$.

The general form of the coupled time-invariant system of n degrees of freedom is

$$\begin{aligned} a_{11}(p)y_1 + a_{12}(p)y_2 + \cdots + a_{1n}(p)y_n &= f_1(t) \\ \cdots\cdots\cdots\cdots\cdots\cdots\cdots\cdots\cdots\cdots \\ a_{n1}(p)y_1 + a_{n2}(p)y_2 + \cdots + a_{nn}(p)y_n &= f_n(t) \end{aligned} \tag{8.108}$$

In matrix-operator form,

$$\mathbf{A}(p)\mathbf{y} = \mathbf{f}(t) \qquad \mathbf{A}(p) = \begin{bmatrix} a_{11}(p) & a_{12}(p) & \cdots & a_{1n}(p) \\ \cdots & \cdots & \cdots & \cdots \\ a_{n1}(p) & a_{n2}(p) & \cdots & a_{nn}(p) \end{bmatrix} \qquad \mathbf{f} = \begin{bmatrix} f_1 \\ f_2 \\ \cdot \\ \cdot \\ \cdot \\ f_n \end{bmatrix} \tag{8.108a}$$

where p is the differential-integral operator defined by

$$p \triangleq \frac{d(\ )}{dt} \qquad \frac{1}{p} \triangleq \int_0^t (\ )dt$$

For linear passive systems (i.e., involving inductance, capacitance, and resistance or inertia, spring, and damping), each coefficient takes the form

$$a_{ij}(p) = L_{ij}p + 1/C_{ij}p + R_{ij}$$

where L_{ij}, C_{ij}, and R_{ij} are constants.

Recall that the general solution to Eq. (8.108) is composed of two solutions, one to the homogeneous system,

$$\mathbf{A}(p)\mathbf{z} = 0 \quad \text{or} \quad \begin{aligned} a_{11}(p)z_1 + a_{12}(p)z_2 + \cdots + a_{1n}(p)z_n &= 0 \\ \cdots\cdots\cdots\cdots\cdots\cdots\cdots\cdots\cdots\cdots \\ a_{n1}(p)z_1 + a_{n2}(p)z_2 + \cdots + a_{nn}(p)z_n &= 0 \end{aligned} \tag{8.109}$$

and one to the inhomogeneous equation,

$$\mathbf{A}(p)\mathbf{x} = \mathbf{f}(t)$$

Thus the total solution is

$$\mathbf{y} = \mathbf{x} + \mathbf{z} \tag{8.110}$$

To find **z** assume, as before, the exponential form [similar to Eq. (8.106*a*)]

$$\mathbf{z} = \begin{vmatrix} z_1 = c_1^{(i)} e^{\lambda t} \\ z_2 = c_2^{(i)} e^{\lambda t} \\ \cdots\cdots\cdots \\ z_n = c_n^{(i)} e^{\lambda t} \end{vmatrix}$$

Substitution in Eq. (8.109) yields

$$\mathbf{A}(\lambda)\mathbf{c} = 0 \quad \text{or} \quad \begin{array}{l} a_{11}(\lambda)c_1 + a_{12}(\lambda)c_2 + \cdots + a_{1n}(\lambda)c_n = 0 \\ \cdots\cdots\cdots\cdots\cdots\cdots\cdots\cdots\cdots \\ a_{n1}(\lambda)c_1 + a_{n2}(\lambda)c_2 + \cdots + a_{nn}(\lambda)c_n = 0 \end{array} \tag{8.111}$$

From linear theory a solution $c_1, c_2, \ldots, c_n$ different from zero (the trivial case) can exist if and only if the determinant vanishes; thus

$$\det[\mathbf{A}(\lambda)] = 0 \qquad \begin{bmatrix} a_{11}(\lambda) & a_{12}(\lambda) & \cdots & a_{1n}(\lambda) \\ a_{21}(\lambda) & a_{22}(\lambda) & \cdots & a_{2n}(\lambda) \\ \cdots & \cdots & \cdots & \cdots \\ a_{n1}(\lambda) & a_{n2}(\lambda) & \cdots & a_{nn}(\lambda) \end{bmatrix} = 0$$

which leads to an *m*th-degree algebraic equation in λ called the characteristic equation of the matrix **A**. In general, $m \neq n$ and $m \leqq 2n$ for passive systems.

The *m*th-degree equation yields *m* roots $\lambda_1, \lambda_2, \ldots, \lambda_m$, each of which satisfies Eq. (8.111). If the roots are distinct, the corresponding **c** column vector

$$\mathbf{c} = \begin{bmatrix} c_1 \\ c_2 \\ \cdot \\ \cdot \\ \cdot \\ c_n \end{bmatrix}$$

can be found for each λ_i, being the cofactors of any row of $\mathbf{A}(\lambda i)$ in order from left to right: $c_1 = \text{cof}(a_{j1})$; $c_2 = \text{cof}(a_{j2})$; $c_n = \text{cof}(a_{jn})$ for any *j*. The solution to Eq. (8.109) is then

$$\mathbf{z} = \sum_{i=1}^{n} \mathbf{c}^{(i)} e^{\lambda_i t} \quad \text{or} \quad \begin{array}{l} z_1 = c_1^{(1)} e^{\lambda_1 t} + c_1^{(2)} e^{\lambda_2 t} + \cdots + c_1^{(m)} e^{\lambda m t} \\ \cdots\cdots\cdots\cdots\cdots\cdots\cdots\cdots \\ z_n = c_n^{(1)} e^{\lambda_1 t} + c_n^{(2)} e^{\lambda_2 t} + \cdots + c_n^{(m)} e^{\lambda m t} \end{array}$$

If some of the roots are repeated, then the above fails and these roots of multiplicity $v_j (v_j > 1)$ yield the solution

$$\sum [\mathbf{c}^{(j)} + \mathbf{d}^{(j)} t + \cdots + \mathbf{h}^{(j)} t^{\lambda_j - 1}] e^{\lambda_j t}$$

The total solution is

$$\mathbf{z} = \sum_{i=l+1}^{m-q+1} \mathbf{c}^{(i)} e^{\lambda_i t} + \sum_{j=1}^{l} [\mathbf{c}^{(j)} + \mathbf{d}^{(j)} t + \cdots + \mathbf{h}^{j} t^{\lambda_j - 1}] e^{\lambda_j t} \qquad q = \sum v_j$$

Alternatively, and if interest is focused on one of the dependent variables in Eq. (8.109), then all other variables can be eliminated to yield

$$\det \begin{bmatrix} a_{11}(p) & \cdots & a_{1n}(p) \\ \cdots & \cdots & \cdots \\ a_{n1}(p) & \cdots & a_{nn}(p) \end{bmatrix} z_i = 0 \quad \text{or} \quad |\mathbf{A}(p)|\, z_i = 0 \qquad i = 1, 2, \ldots, n$$

giving the identical homogeneous equation for each of the dependent variables. If, as before, the exponential form $z_1 = c_1 e^{\lambda t}$ is assumed, substitution gives $|\mathbf{A}(\lambda)| c_1 e^{\lambda t} = 0$, where for $c_1 \neq 0$, $|\mathbf{A}(\lambda)| = 0$, giving the same characteristic equation for the exponential constants as before. The inhomogeneous reduced equations from Eq. (8.108) are formally obtained by purely algebraic considerations as

$$|\mathbf{A}(p)| y_i = \sum_{i=1}^{n} M_{ij}(p) f_i \qquad j = 1, 2, \ldots, n$$

where M_{ij} is the cofactor of the element in the ith row and jth column of A.

Let

$$|\mathbf{A}(p)| \triangleq D(p)$$

$$D(p) y_j = \sum_{i=1}^{n} M_{ij}(p) f_i(t) \qquad j = 1, 2, \ldots, n$$

From the previous considerations,

$$D(p) z_j = 0 \tag{8.112a}$$

$$D(p) x_j = \sum_{i=1}^{n} M_{ij}(p) f_i(t) \qquad j = 1, 2, \ldots, n \tag{8.112b}$$

$$y_j = x_j + z_j$$

To conclude, in general the solution x_j may be determined without regard for the initial conditions. It can be obtained in many ways depending on the character of the f_i's. If the f_i's are known as a finite power series, the method of undetermined coefficients will be expeditious; if it has more general behavior, it may be convenient to use the method of variation of parameters; if f_i's are exponential (including sin, cos, sinh, cosh) then an assumed exponential solution for each exponent will yield the answer; if the f_i's are periodic, by Fourier analysis these can be reconstructed as exponential functions and solved as outlined above; if f_i's are not periodic, having certain restrictive integral-convergence behavior, Fourier integral methods can be utilized. Last and most powerful is the Laplace-transform method, which not only has the widest range of applicability but can be utilized to obtain y (the total solution) directly.

8.4.2 Fourier-Series Analysis

If $f(t)$ is real and periodic of period T, with few restrictions, it can be approximately expressed as a linear sum of sine and cosine terms (Fourier series) or exponential terms. That is, if

$$f(t) = f(t + T)$$

then

$$f(t) = \frac{a_0}{T} + \sum_{n=1}^{\infty} a_n \cos \omega_0 nt + b_n \sin \omega_0 nt$$

where $\omega_0 = 2\pi/T$ and a_n, b_n are real constants. Alternatively,

$$f(t) = \sum_{n=-\infty}^{+\infty} c_n e^{j\omega_0 nt} \tag{8.113}$$

$$\text{where } a_0 = \int_t^{t+T} f(t)\,dt$$

$$a_n = \frac{2}{T}\int_t^{t+T} f(t)\cos n\omega_0 t\,dt$$

$$b_n = \frac{2}{T}\int_t^{t+T} f(t)\sin n\omega_0 t\,dt$$

$$c_n = \frac{1}{T}\int_t^{t+T} f(t)\,e^{-jn\omega_0 t}\,dt$$

$$2c_n = a_n - jb_n$$

$$2c_{-n} = a_n + jb_n$$

Let $e^{j\omega t}$ be the input f_i in Eq. (8.112). For a response $X_{ij}^{(1)} = H_{ij}(j\omega)e^{j\omega t}$ it is required to find $H_{ij}(j\omega)$. From Eq. (8.112) it is clear that all operations on $e^{j\omega t}$ are equivalent to replacing p by $j\omega$ and Eq. (8.112) becomes, for $f_i = e^{j\omega t}$,

$$D(j\omega)H_{ij}(j\omega)e^{j\omega t} = M_{ij}(j\omega)e^{j\omega t}$$

yielding
$$\begin{aligned} X_{ij}^{(1)} &= H_{ij}(j\omega)e^{j\omega t} = \frac{M_{ij}(j\omega)}{D(j\omega)}e^{j\omega t} \\ &= [P(j\omega) + Q(j\omega)]e^{j\omega t} \\ &= R(j\omega)e^{j\phi}e^{j\omega t} \qquad \phi = \tan^{-1}\frac{-jQ(j\omega)}{P(j\omega)} \\ &= R(j\omega)e^{j(\omega t+\phi)} \qquad R(j\omega) = [P^2(j\omega) + Q^2(j\omega)]^{1/2} \end{aligned} \tag{8.114}$$

where $P(j\omega)$ and $R(j\omega)$ are real and therefore even functions of $j\omega$, and $Q(j\omega)$ is imaginary and an odd function of $j\omega$ and the following properties apply:

$$Q(j\omega) = -Q(-j\omega) \qquad \text{odd function}$$

$$\left.\begin{aligned} p(j\omega) &= p(-j\omega) \\ R(j\omega) &= R(-j\omega) \end{aligned}\right\} \qquad \text{even functions}$$

Similarly for an input $e^{-j\omega t}$, the output is

$$X_{ij}^{(2)} = H_{ij}(-j\omega)e^{-j\omega t} = [P(-j\omega) + Q(-j\omega)]e^{-j\omega t} = Re^{j\phi'}e^{-j\omega t}$$

Since
$$\phi' = \tan^{-1}\frac{-jQ(-j\omega)}{P(j\omega)} = \tan^{-1}\frac{jQ(j\omega)}{P(j\omega)} = -\phi$$

the response $X_{ij}^{(2)}$ is

$$X_{ij}^{(2)} = H_{ij}(-j\omega)e^{-j\omega t} = Re^{-j(\omega t+\phi)} \tag{8.115}$$

The sum of the responses is

$$X_{ij}^{(1)} + X_{ij}^{(2)} = H_{ij}(-j\omega)e^{-j\omega t} + H_{ij}(j\omega)e^{j\omega t} = Re^{j(\omega t+\phi)} + Re^{-j(\omega t+\phi)}$$
$$= 2R \cos (\omega t + \phi)$$

which is just twice the real part of either response or

$$X_{ij}^{(1)} + X_{ij}^{(2)} = 2 \operatorname{Re} X_{ij}^{(1)} = 2 \operatorname{Re} X_{ij}^{(2)}$$

showing that the total input, $e^{j\omega t} + e^{-j\omega t} = 2 \cos \omega t$, results in an output of different phase and amplitude.

$$H_{ij}(j\omega) = M_{ij}(j\omega)/D(j\omega) = R(j\omega)e^{j\phi(j\omega)}$$

is called the "transfer function" for sinusoidal inputs (real frequency) containing both amplitude and phase information.

It follows readily from superposition that the periodic responses to forcing functions having Fourier-series representations are available as a sum of responses of the form

$$X = \sum_{n=0}^{\infty} Ra_n \sin (n\omega_0 t + \phi_n) + Rb_n \cos (n\omega_0 t + \phi_n)$$

where $\phi_n = \phi(jn\omega_0)$ and $\phi_{-n} = \phi(-jn\omega_0)$

or

$$X = \sum_{n=-\infty}^{+\infty} Rc_n e^{j(\omega_0 nt+\phi_n)}$$

8.4.3 Complex Frequency-Domain Analysis[16]

It is often convenient to cast the linear system from its time-domain representation into a frequency-domain form in order to simplify analysis or exhibit more clearly certain of its important properties (e.g., spectrum, stability). The Fourier- and Laplace-transform methods are most prominent in this regard.

Fourier-Transform Method. If the input forcing function(s) are not periodic functions of time, the Fourier-transform method may be employed to solve Eq. (8.112) for each input f_i whenever

$$\int_{-\infty}^{+\infty} |f_i(t)|\,dt < \infty \tag{8.116}$$

That is, the absolute convergence of the infinite integral is a sufficient condition for Fourier transformability. Examples of functions not satisfying Eq. (8.116) are the step function,* sinusoid, rising exponentials, and functions containing t to positive exponents, e.g., ramp function (αt). Examples of functions satisfying Eq. (8.116) are pulses of finite duration. The Fourier-integral theorem asserts

$$f(t) = \frac{1}{2\pi}\int_{-\infty}^{+\infty} d\omega \int_{-\infty}^{+\infty} f(T)e^{j\omega(t-T)}dT \tag{8.117}$$

*These have Fourier representations despite violation of Eq. (8.116). Note that Eq. (8.116) is only a sufficient condition.

If $f(t)$ has a finite discontinuity at any point, then this integration will yield the average value of $f(t)$ at the discontinuity. From Eq. (8.117) the Fourier-transform pair is obtained:

$$f(t) = \frac{1}{2\pi}\int_{-\infty}^{+\infty} F(j\omega)e^{j\omega t}d\omega = \int_{-\infty}^{+\infty} F(j2\pi f)e^{j2\pi ft}\,df \qquad (8.118a)$$

$$F(j\omega) = \int_{-\infty}^{+\infty} f(t)e^{-j\omega t}\,dt = \int_{-\infty}^{+\infty} f(t)e^{-j2\pi f}t\,dt \qquad (8.118b)$$

where $F(j\omega)$ is in general complex and is denoted as the complex spectrum of $f(t)$. Equation (8.118a) can be imagined to express $f(t)$ as the infinite sum of Fourier components $F(j\omega)e^{j\omega t}\,d\omega/2\pi$. From the superposition principle, the total response is made up of the sum of each of the responses $H_{ij}(j\omega)e^{j\omega t}\,d\omega/2\pi$, where $H_{ij}(j\omega)$ was defined as the real frequency-transfer function, and the sum is expressed (since it is continuous in ω) as

$$x(t) = \frac{1}{2\pi}\int_{-\infty}^{+\infty} H_{ij}(j\omega)F(j\omega)e^{j\omega t}\,d\omega \qquad (8.119)$$

But $x(t)$ has a transform representation from Eq. (8.118):

$$x(t) = \frac{1}{2\pi}\int_{-\infty}^{+\infty} X(j\omega)e^{j\omega t}\,d\omega \qquad (8.120)$$

The integrals Eqs. (8.119) and (8.120) are evidently identical. Hence

$$X(j\omega) = H_{ij}(j\omega)F(j\omega) \qquad (8.121)$$

which gives the *important property* that the product of the transfer function (at real frequency) and the Fourier transform of the driving function yields the Fourier transform of the response.

Consider the linear time-invariant differential equation

$$\frac{d^n x}{dt^n} + Q_{n-1}\frac{d^{n-1}x}{dt^{n-1}} + \cdots + Q_0 x = f(t) \qquad f(t) = \frac{1}{2\pi}\int_{-\infty}^{+\infty} F(j\omega)e^{j\omega t}\,d\omega$$

In terms of the Fourier transforms of x and $f(t)$, this equation becomes

$$\frac{1}{2\pi}\int_{-\infty}^{+\infty} [(j\omega)^n + Q_{n-1}(j\omega)^{n-1} + \cdots + Q_0]X(j\omega)e^{j\omega t}\,dt = \frac{1}{2\pi}\int_{-\infty}^{+\infty} F(j\omega)e^{j\omega t}\,d\omega$$

where $\qquad x(t) = \frac{1}{2\pi}\int_{-\infty}^{+\infty} X(j\omega)e^{j\omega t}\,d\omega \qquad \frac{d^k}{dt^k}x = \frac{1}{2\pi}\int_{-\infty}^{+\infty} (j\omega)^k X(j\omega)\,d\omega$

whence $\qquad X(j\omega) = \frac{F(j\omega)}{(j\omega)^n + A_{n-1}(j\omega)^{n-1} + \cdots + Q_1(j\omega) + Q_0}$

The Fourier transform, in terms of its real and imaginary parts, is

$$X(j\omega) = M'(j\omega) + N'(j\omega) = M(\omega) + jN(\omega)$$

where M is an even function of ω and N an odd function of ω.

Then

$$x(t) = \frac{1}{2\pi}\int_{-\infty}^{+\infty} [M(\omega) + jN(\omega)]e^{j\omega t}\, d\omega$$

$$= \frac{1}{2\pi}\int_{-\infty}^{+\infty} [M(\omega)\cos \omega t - N(\omega)\sin \omega t]\, d\omega$$

$$= \frac{1}{\pi}\int_{-0}^{+\infty} (M\cos \omega t - N\sin \omega t)\, d\omega \qquad (8.122)$$

where only the even parts of the integrands can contribute because integration of the odd terms vanishes over the infinite limits. Now for the system (causal)

$$x(t) = 0 \qquad \text{for } t < 0$$

which is mathematically equivalent to

$$x(-t) = 0 \qquad \text{for } t > 0$$

Substitution in Eq. (8.122) yields

$$x(-t) = 0 = \frac{1}{\pi}\int_{0}^{\infty} (M\cos \omega t + N\sin \omega t)\, d\omega \qquad t > 0 \qquad (8.123)$$

Adding Eqs. (8.122) and (8.123), we obtain

$$x(t) = \frac{2}{\pi}\int_{0}^{\infty} M\cos \omega t\, d\omega = \frac{1}{\pi}\int_{-\infty}^{+\infty} M\cos \omega t\, d\omega \qquad t > 0 \qquad (8.124a)$$

Subtracting Eq. (8.123) from Eq. (8.122),

$$x(t) = -\frac{2}{\pi}\int_{0}^{\infty} N\sin \omega t\, d\omega = -\frac{1}{\pi}\int_{-\infty}^{+\infty} N\sin \omega t\, d\omega \qquad t > 0 \qquad (8.124b)$$

Since $x(t)$ has two integral representations, they are equal:

$$\frac{2}{\pi}\int_{0}^{\infty} M\cos \omega t\, d\omega = -\frac{2}{\pi}\int_{0}^{\infty} N\sin \omega t\, d\omega$$

$$\int_{0}^{\infty} M(\omega)\cos \omega t\, d\omega = -\int_{0}^{\infty} N(\omega)\sin \omega t\, d\omega$$

$$M(\omega) = \int_{-\infty}^{+\infty} x(t)\cos \omega t\, dt \qquad (8.125a)$$

$$N(\omega) = -\int_{-\infty}^{\infty} x(t)\sin \omega t\, dt \qquad (8.125b)$$

which are the real Fourier-transform coefficients and together with Eq. (8.124) constitute the real Fourier-transform pair. Properties of Fourier transforms are given in Table 8.1.

TABLE 8.1 Properties of Fourier-Transform Pairs

Property	Fourier transform	Time function
Basic pairs	$F(\omega)$ $F(\omega) = \int_{-\infty}^{+\infty} f(t)e^{-j\omega t}\,dt$	$f(t)$ $f(t) = 1/2\pi \int_{-\infty}^{+\infty} F(\omega)e^{j\omega t}\,d\omega$
Linearity, a_1 and a_2 constants	$a_1F_1(\omega) + a_2F_2(\omega)$	$a_1f_1(t) + a_2f_2(t)$
Time multiplication, a real constant	$\frac{1}{\|a\|}F(\omega/a)$	$f(at)$
Time shift	$F(\omega)e^{-j\omega t_1}$	$f(t - t_1)$
Frequency multiplication, a real constant	$F(a\omega)$	$\frac{f(t/a)}{\|a\|}$
Frequency shift	$F(\omega - \omega_1)$	$f(t)e^{j\omega_1 t}$
Time differentiation	$(j\omega)^nF(\omega)$	$(d^n/dt^n)f(t)$
Integration	$(1/j\omega)F(\omega)$	$\int_{-\infty}^{t} f(x)\,dx$
Frequency differentiation	$j^n(d^n/d\omega^n)F(\omega)$	$t^nf(t)$
Convolution time domain	$F_1(\omega)F_2(\omega)$	$\int_{-\infty}^{+\infty} f_1(t - T)f_2(T)\,dT \triangleq f_1(t) * f_2(t)$
Convolution frequency domain	$F_1(\omega) * F_2(\omega) =$ $\frac{1}{2\pi}\int_{-\infty}^{+\infty} F_1(\omega - x)F_2(\omega)\,dx$	$f_1(t)f_2(t)$
Forms for real $f(t)$	$F(\omega) = R(\omega) + jX(\omega)$ $R(\omega) = \int_{-\infty}^{+\infty} f(t)\cos\omega t\,dt$ $X(\omega) = -\int_{-\infty}^{+\infty} f(t)\sin\omega t\,dt$	$f(t) = 1/\pi \int_0^{\infty} [R(\omega)\cos\omega t\,d\omega$ $- X(\omega)\sin\omega t]\,d\omega$ $= 1/\pi \int_0^{\infty} \|F(\omega)\|\cos(\omega t + \phi)\,d\omega$ $\phi = \tan^{-1} X(\omega)/R(\omega)$
Forms for real even $f(t)$	$X(\omega) = 0$ $R(\omega) = -2\int_0^{\infty} f(t)\sin\omega t\,dt$	$f(t) = -1/\pi \int_0^{\infty} R(\omega)\cos\omega t\,d\omega$
Forms for real odd $f(t)$	$R(\omega) = 0$ $X(\omega) = -2\int_0^{\infty} f(t)\sin\omega t\,dt$	$f(t) = -1/\pi \int_0^{\infty} X(\omega)\sin\omega t\,d\omega$
$f(t)$ causal, i.e., $f(t) = 0$ for $t < 0$	$F(\omega) = \int_0^{\infty} f(t)e^{-j\omega t}\,dt$ $= R(\omega) + jX(\omega)$	$f(t) = 2/\pi \int_0^{\infty} R(\omega)\cos\omega t\,d\omega$ $= -2/\pi \int_0^{\infty} X(\omega)\sin\omega t\,d\omega \quad t > 0$
Forms for periodic function $f(t) = f(t + T)$	$2\pi \sum_{n=-\infty}^{+\infty} a_n\delta(\omega - \omega_n)$ $a_n = 1/T \int_0^{T} f(t)e^{-j\omega_n t}\,dt$	$f(t) = f(t + T)$ $= \sum_{n=-\infty}^{+\infty} a_ne^{j\omega_n t}$ $\omega_n = 2\pi n/T$

Laplace-Transform Method. When Eq. (8.116) does not hold for a forcing function $f(t)$, recourse may be taken to the unilateral Laplace-transform method provided that $f(t) \equiv 0$ for $t < 0$ and there exists a positive number c (in most physical problems of interest one exists) such that

$$\int_{-\infty}^{+\infty} |f(t)| e^{-ct}\, dt < \infty \tag{8.126}$$

The minimum value of c for which Eq. (8.126) holds is designated as the abscissa of convergence, equal to c_1. Equation (8.126) ensures that the Laplace transform of $f(t)$, written $\mathcal{L}\{f(t)\}$ and defined by

$$\mathcal{L}\{f(t)\} = \int_0^{\infty} f(t)\, e^{-st}\, dt = F(s) \tag{8.127}$$

will converge to a function of s. s is a complex variable given by

$$s = c + j\omega \qquad c \geq c_1$$

Examples of c_1 are for $f(t) = \sin \omega t$, $c_1 > 0$; for unit step, $c_1 > 0$; for $t^n e^{dt}$, $c_1 > d$, n finite, d real. An example of a function where c cannot be found to satisfy Eq. (8.126) is $f(t) = \exp t^n$ for $n > 1$, and Laplace methods will accordingly fail. Equation (8.127) looks like the Fourier transform of $f(t)e^{-ct}$ if $f(t) = 0$ for $t < 0$:

$$\mathcal{L}\{f(t)\} = \mathcal{F}\{f(t)e^{-ct}\}$$

where $\mathcal{F} \triangleq$ Fourier transform. Invoking the Fourier-transform theorem [Eq. (8.117)] and manipulating, we find

$$f(t)e^{-ct} = \frac{1}{2\pi}\int_{-\infty}^{+\infty} d\omega \int_{-\infty}^{+\infty} f(T)e^{-cT}e^{j\omega(t-T)}\, dT$$

$$= \frac{1}{2\pi}\int_{-\infty}^{+\infty} e^{j\omega t}\, d\omega \int_{-\infty}^{+\infty} f(T)e^{-(c+j\omega)T}\, dT$$

$$f(t) = \frac{1}{2\pi}\int_{-j\infty}^{+j\infty} e^{(c+j\omega)t}\, \frac{d(j\omega)}{j} \int_0^{\infty} f(T)e^{-(c+j\omega)T}\, dT \qquad f(T) = 0 \text{ for } T < 0$$

Since $s = c + j\omega$ and c is a constant, $ds = d(j\omega)$ and the above becomes

$$f(t) = \frac{1}{2\pi j}\int_{c-j\infty}^{c+j\infty} e^{st}\, ds \int_0^{\infty} f(T)e^{-sT}\, dT = \frac{1}{2\pi j}\int_{c-j\infty}^{c+j\infty} \mathcal{L}\{f(t)\}e^{st}\, ds$$

$$= \frac{1}{2\pi j}\int_{c-j\infty}^{c+j\infty} F(s)e^{st}\, ds \tag{8.128}$$

Equation (8.128) is the inversion form for going from $\mathcal{L}\{f(t)\}$ to $f(t)$. Equations (8.127) and (8.128) constitute the Laplace-transform pair.

From Eq. (8.114), the real-frequency transfer function

$$H_{ij}(j\omega) = M_{ij}(j\omega)/D(j\omega)$$

was deduced for sinusoidal inputs. Similarly, consider the response [Eq. (8.112)] x_{ij} due to f_i:

$$D(p)x_{ij}(t) = M_{ij}(p)f_i(t)$$

where $D(p)$ and $M_{ij}(p)$ are linear differential operators. For zero initial conditions the Laplace transformation of both sides yields

$$D(s)X_{ij}(s) = M_{ij}(s)F_i(s) \tag{8.129}$$

Transfer Function. From Eq. (8.129), dropping all subscripts for clarity, $X(s)/F(s) = M(s)/D(s) = H(s)$, which is, by definition, the transfer function where s replaces $j\omega$ in the argument $H(j\omega)$, the real-frequency transfer function for sinusoidal input. $H(s)$ in itself has no physical significance; it contains however, the complete characterization of the system. This is in contrast with $H(j\omega)$ which gives the steady-state response to a sinusoidal input, its amplitude being the gain and its argument the phase difference between output and input.

It follows that, if $H(j\omega)$ is a known analytic function of $j\omega$, then $H(s)$ is immediately available (by analytic continuation) for a complete system description.

The total response $y(t)$ satisfying the equation

$$D(p)y(t) = M(p)f(t) \qquad f(t) = 0 \text{ for } t < 0 \tag{8.130}$$

can be obtained directly by taking the Laplace transform [including initial conditions which result in the polynominal $L(s)$ of lower order than $D(s)$] as follows:

$$D(s)Y(s) - L(s) = M(s)F(s) \tag{8.130a}$$

$$Y(s) = M(s)F(s)/D(s) + L(s)/D(s) = H(s)F(s) + L(s)/D(s)$$

By inversion,

$$y(t) = \underbrace{\frac{1}{2\pi j}\int_{c-j\infty}^{c+j\infty} H(s)F(s)e^{st}ds}_{x(t) = \int_0^t W(t-T)f(T)\,dT} + \underbrace{\frac{1}{2\pi j}\int_{c-j\infty}^{c+j\infty} \frac{L(s)}{D(s)}e^{st}ds}_{z(t)} \tag{8.131}$$

The first integral in Eq. (8.131), $x(t)$, is the solution to Eq. (8.112*b*) for zero initial conditions; the second, $z(t)$, is the solution to the homogeneous form Eq. (8.112*a*) which satisfies the initial conditions of Eq. (8.130). $W(t - T)$ is the response $x(t)$ at time t to a unit impulse input $f = \delta(t - T)$, the system being initially at rest.

Inversion. The transformation (inversion) from the complex frequency representation to the time domain is given by

$$f(t) = \frac{1}{2\pi j}\int_{c-j\infty}^{c+j\infty} F(s)\, e^{st}dt$$

which is a line integral along the line Re $s = c$ in the complex s plane where $c > c_1$, and c_1 is defined as the abscissa of convergence. For the case $|F(s)| \to 0$ as $|s| \to \infty$ the line integral is most readily evaluated by forming a contour including this line and an infinite semicircle connected on the left and considering the contour integral

$$\oint F(s)e^{st}\, ds$$

Since $|F(s)| \to 0$ as $|s| \to \infty$, the line integral around the semicircular portion of this contour vanishes as a consequence of Jordan's lemma.[16] This leads to the equality of the contour integral with the inversion integral, viz.,

$$f(t) = \frac{1}{2\pi j}\int_{c-j\infty}^{c+j\infty} F(s)e^{st}\, ds = \frac{1}{2\pi j}\oint F(s)e^{st}\, ds \qquad t > 0$$

From Cauchy's residue theorem, the right-hand side equals the sum of the residues of $F(s)e^{st}$ enclosed. The residues are evaluated at each simple pole s_k by

$$(s - s_k)F(s)e^{st} = R_{sk} \qquad s \to s_k$$

where R_{sk} = residue at s_k. If $F(s)$ is a fraction, $F(s) = A(s)/B(s)$ where $A(s)$ and $B(s)$ are analytic functions of s inside the contour (excluding poles at infinity), then the poles of $F(s)$ are clearly the zeros of $B(s)$. A pole of multiplicity m is equal to the excess of zeros of $B(s)$ over $A(s)$ at the pole. For simple poles of $F(s)$, i.e., where $m = 1$, the residue at $s = s_k$ is simply

$$(s - s_k)F(s)e^{st} = A(s_k)e^{s_k t}/B'(s_k) \qquad s \to s_k$$

The residue at pole s_j of multiplicity m is

$$\frac{1}{(m-1)!}\frac{d^{m-1}}{ds^{m-1}}\left[(s - s_j)^m F(s)e^{st}\right]_{s=sj}$$

If there are no zeros of $A(s)$ at the point $s = s_j$, then an alternative form of this is

$$\frac{md^{m-1}}{ds_j^{m-1}}\frac{A(s_j)e^{s_j t}}{d^m B(s_j)/ds_j^m}$$

The following are some important properties of $H(s)$ for passive systems of differential equations:

1. $H(s)$ is a rational function with real coefficients.
2. The degree of $D(s)$ is equal to or greater than $M_{ij}(s)$.
3. As a consequence of property 1, the complex zeros and poles occur in conjugate pairs.
4. All poles of $H(s)$ lie in the closed left half plane, a consequence of passivity. Table 8.2 has properties of Laplace-transform pairs.

8.4.4 Time-Domain Analysis

The general solution to Eq. (8.130) given by Eq. (8.131) is

$$y(t) = \int_0^t W(t - T)f(T)\, dT + z(t) \qquad (8.132)$$

where the integral expression alone satisfies the inhomogeneous equation with zero initial conditions, and $z(t)$ contains the linearly independent solutions to the homogeneous

TABLE 8.2 Properties of Laplace-Transform Pairs for Causal Time Function

$f(t) = 0 \quad t < 0$

Property	Laplace transform	Time function
Basic pairs	$F(s)$ $F(s) = \int_0^{\infty} f(t)e^{-st}\,dt$ $= \int_{-\infty}^{+\infty} f(t)e^{-st}\,dt$	$f(t)$ $f(t) = 1/2\pi j \int_{c-j\infty}^{c+j\infty} F(s)e^{st}\,ds$
Linearity a_1 and a_2 constants	$a_1F_1(s)t + a_2F_2(s)$	$a_1f_1(t) + a_2f_2(t)$
Time scale multiplication a positive real constant	$\frac{F(s/a)}{a}$	$f(at)$
Time shift	$F(s)e^{-st_1}$	$f(t - t_1)$
Complex frequency multiplication a positive real constant	$F(as)$	$\frac{f(t/a)}{a}$
Complex frequency shift	$F(s - s_1)$	$f(t)e^{s_1t}$
Time differentiation	$s^nF(s) - \sum_{k=1}^{n} f^{k-1}(0)s^{n-k}$	$\frac{d^n}{dt^n} f(t)$
Integration	$\frac{F(s)}{s} + \frac{f^{-1}(0)}{s}$	$\int_0^t f(t)\,dt$
Complex frequency differentiation	$(-1)^n \frac{d^n}{ds^n} F(s)$	$t^nf(t)$
Convolution time domain	$F_1(s)F_2(s)$	$f_1(t) * f_2(t) = \int_0^{\infty} f_1(t-T)f_2(T)\,dT$ $= \int_0^t f_1(t-T)f_2(T)\,dT$
Convolution frequency domain	$1/2\pi j \int_{c-j\infty}^{c+j\infty} F_1(s-x)F_2(x)\,dx$ $= (1/2\pi j)F_1(s) * F_2(s)$	$f_1(t)f_2(t)$
Forms for periodic function $f_k(t) = f_k(t + T)$	$\frac{1/2\pi j \int_0^T f_k(t)e^{-st}\,dt}{1 - e^{-sT}}$	$f_k(t) = f_k(t + T)$
Initial-value theorem [$f(t)$ and $f'(t)$ are Laplace transformable]	$\lim_{s\to\infty} sF(s)$	$\lim_{t\to 0} f(t)$
Final-value theorem $f(t)$, $f'(t)$ Laplace transformable and $sF(s)$ analytic $Re\ s \geq 0$	$\lim_{s\to 0} sF(s)$	$\lim_{t\to\infty} f(t)$

form fulfilling the initial conditions on $y(t)$, i.e.,

$$y(0) = z(0)$$

$W(t)$ is the inverse transform of the transfer function $H(s)$ which physically is the response to the delta function $\delta(t)$. The Dirac delta function $\delta(t - T)$ (defined as a pulse of infinite height at $t = T$, with unit area) has the following properties:

$$\int_{-\infty}^{+\infty} \delta(t - T)\, dt = 1 \qquad \text{and} \qquad \int_{-\infty}^{+\infty} \delta(t - T)A(t)\, dt = A(T)$$

Examples of the delta function are shown in Fig. 8.14. To find the response $y(t)$ for input $\delta(t - T)$ and zero initial conditions, we first evaluate

$$F(s) = \mathcal{L}\{\delta(t - T)\} = e^{-sT}$$

and substitute in Eq. (8.130*a*)

$$Y(s) = H(s)F(s) = H(s)e^{-sT}$$

where $L(s) = 0$ is a consequence of zero initial conditions. Transforming to the time domain, we have

$$y(t) = \frac{1}{2\pi j}\int_{c-j\infty}^{c+j\infty} H(s)\, e^{-sT} e^{st}\, ds = \frac{1}{2\pi j}\int_{c-j\infty}^{c+j\infty} H(s)\, e^{s(t-T)}\, ds = W(t - T) \qquad t < T$$

Any forcing function can be approximated by an infinite number of delta functions of strength $f(T)\,\Delta T$ with the responses at time t the sum of the responses $W(t - T)$ per unit impulse for each of these pulses which occurred $t - T$ seconds previous to the

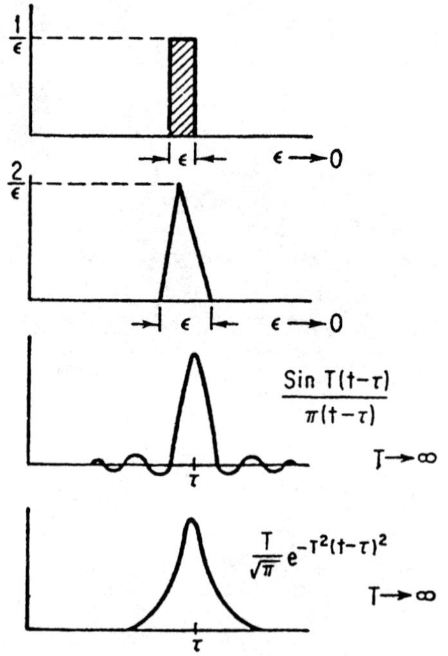

FIG. 8.14 Examples of delta functions, $\delta(t - T)$.

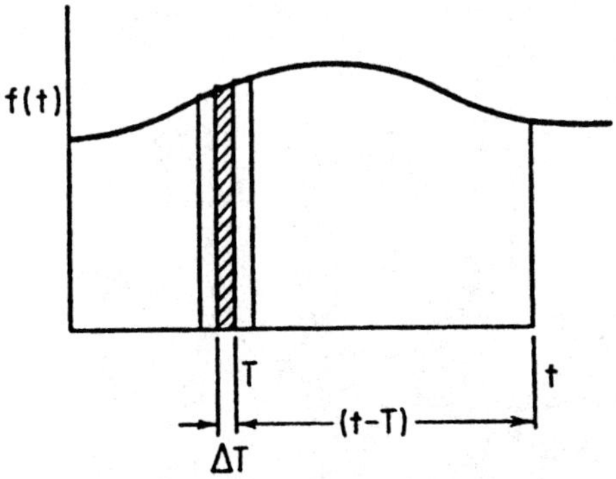

FIG. 8.15 Pulse synthesis for time-convolution theorem.

time of inspection, t. Figure 8.15 shows a graphical construction with equal-duration rectangular pulses, ΔT wide and $f(T)$ high. Each pulse has an area $f(T)\,\Delta T$ so that the function can be approximated by

$$\sum_{i=0}^{i=t/\Delta T} f(T)\,\Delta T\,\delta(t - i\,\Delta T)$$

The transient produced at time t by the pulse $f(T)\,\Delta T\,\delta(t - T)$ is

$$W(t - T)\,f(T)\,\Delta T$$

The total effect of all pulses is by superposition:

$$y(t) = \sum_{T=0}^{t} f(T)\,\Delta T W(t - T) = \sum_{i=0}^{i=t/\Delta T} f(i\,\Delta T)\,\Delta T\,W(t - i\,\Delta t)$$

which in the limit $\Delta T \to 0$ is

$$\int_0^t f(T) W(t - T)\,dT \tag{8.133}$$

$$W(t - T) = 0 \qquad t \leq T \quad t \leq 0$$

$$f(T) = 0 \qquad T < 0$$

consistent with Eq. (8.132).

In view of the restrictions on $W(t - T)$ and $f(t)$ the limits on Eq. (8.133) can be changed to any of the following:

$$y(t) = \int_0^t = \int_{-\infty}^{+\infty} = \int_{-\infty}^{t} = \int_0^{\infty}$$

By a change of variable, y can be represented as

$$y(t) = \int_0^t f(t-T) W(T)\,dT$$

and also is obtainable directly as an alternative form from the convolution integral in going from complex to the real time domain.

Staircase Development. Another manner of depicting the forcing function $f(t)$ is shown in Fig. 8.16 as the synthesis of step functions.

Since the unit step function is the integral of the delta function,

$$U(t - T) = \int_{-\infty}^{+\infty} \delta(t - T)\, dt \qquad \text{and} \qquad \frac{dU(t - T)}{dt} = \delta(t - T)$$

From property 2 of time-invariant systems previously discussed, if the response to the unit step is

$$\text{Input} = U(t - T) \rightarrow Q(t - T) = \text{response}$$

then the response to the derivative $[\partial U(t - T)]/\partial t = \delta(t - T)$ is $[\partial Q(t - T)]/\partial t$. But the response to $\delta(t - T)$ has already been shown to be $W(t - T)$. Therefore,

$$W(t - T) = (\partial/\partial t)Q(t - T)$$

The elementary step functions are of varying amplitudes Δh_i for fixed ΔT. From geometrical considerations

$$\Delta h_i \approx f'(T)\,\Delta T \qquad T = i\,\Delta T$$

and

$$\Delta h_i\, U(t - T) = f'(T)\,\Delta T\, U(t - T)$$

The response to the elementary step $\Delta h_i\, U(t - T)$ is $f'(T)\,\Delta T\, Q(t - T)$. The total response is therefore

$$f(0)Q(t) + \sum_{i=0}^{i=T/\Delta t} f'(T)\,\Delta T\, Q(t - T) = \sum_{i=0}^{i=t/\Delta T} f'(i\,\Delta T)\,\Delta T\, Q(t - i\,\Delta T) + f(0)Q(t)$$

As $\Delta T \rightarrow 0$ this becomes in the limit

$$y(t) = f(0)Q(t) + \int_0^t f'(T)Q(t - T)\, dT \tag{8.134}$$

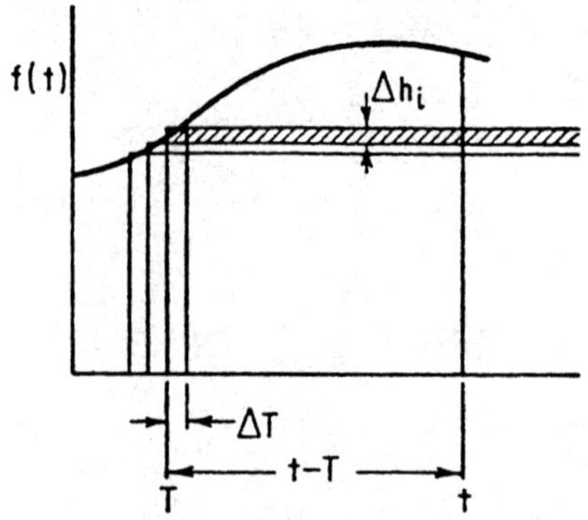

FIG. 8.16 Staircase synthesis.

which is identical to Eq. (8.133). This is shown by integrating Eq. (8.134) by parts.

$$y(t) = f(0)Q(t) + f(T)Q(t - T)\Big|_{T=0}^{T=t} - \int_0^t f(T)\frac{\partial Q(t - T)}{\partial T}\,dT$$

$$= f(0)Q(t) + f(t)Q(0) - f(0)Q(t) + \int_0^t f(T)\frac{\partial Q}{\partial t}(t - T)\,dT$$

since $Q(0) = 0$ and

$$\frac{\partial Q}{\partial T}(t - T) = \frac{-\partial Q}{\partial t}(t - T) = -W(t - T)$$

$$y(t) = \int_0^t f(T)W(t - T)\,dT$$

Stability of Time-Invariant System. From general property 8 (Sec. 8.4) of linear systems the necessary and sufficient condition for $y(t)$ to have a bounded output for any bounded input $f(t)$ is

$$\int_0^\infty |W(t, T)|\,dT < M < \infty$$

which becomes

$$\int_0^\infty W(t - T)\,dT < M < \infty \tag{8.135}$$

for the time-invariant case.

By definition, $W(t - T)$, the weighting function, is the response to the unit delta function $\delta(t - T)$, and it has been shown that $W(t)$ is the inverse of the system transfer function

$$W(t) = H^{-1}(s) = [M(s)/D(s)]^{-1}$$

From the inversion theorem,

$$W(t) = \sum_{k=1}^{m}\sum_{n=1}^{vk} a_{kn}t^{vk-n}e^{s_kt}$$

where vk = multiplicity of roots s_k
s_k = zeros of $D(s)$
m = number of different poles of $H(s)$

Invoking Eq. (8.135) for a typical term $at^qe^{s_kt}$, q = integer,

$$\int_0^\infty |W(t - T)|\,dt = \int_0^\infty at^qe^{\text{Re}s_kt}\,dt < M < \infty$$

which can hold if and only if Re s_k is negative. Hence the criterion for stability for the time-invariant case is simply that all the zeros of $D(s)$ lie in the left half complex plane excluding the imaginary axis. Translated to the time domain, this is equivalent to stating that the real parts of the exponent of each solution to the homogeneous equation $D(p)z = 0$ must be negative so that as $t \to \infty$ they all tend to vanish. Real systems composed only of passive elements are necessarily stable since it can be shown that the poles of their transfer function are restricted to the left-hand plane. The systematic investigation of locating the zeros of $D(s)$ has been motivated by control theory and is discussed subsequently.

8.5 BLOCK DIAGRAMS AND THE TRANSFER FUNCTION

8.5.1 General

A convenient and descriptive way of viewing a system is by use of a block diagram. While a block diagram has little practical value for a simple system, in a complex array of coupled systems it suggests the flow of signals and facilitates analysis. The basic "block" essentially defines the system by giving a description of the physical processes which occur. Specifically for an input i the block gives information on the output o. An example of a block expressing an algebraic relationship is shown in Fig. 8.17*a*. A block of a more general operator relationship which includes the nonlinear differential-integral operator Φ is shown in Fig. 8.17*b*.

8.5.2 Linear Time-Invariant Systems

Single-Degree-of-Freedom Case. The case of the linear time-invariant system (Fig. 8.18) is of special interest because the block-diagram characterization can be very simply shown in terms of the transfer function $H(s)$ in the frequency domain.

The input and output, $I(s)$ and $O(s)$, are Laplace transforms of the input and output signals. From Fig. 8.18*a* we have

$$O(s)/I(s) = H(s) \qquad i(t) = o(t) = 0 \qquad \text{for t} < 0$$

The multiplication property $O(s) = H(s)I(s)$ is of fundamental importance when dealing with linear cascaded systems such as occur in control theory. Moreover, the complete analysis of such systems can be made in this domain without translating to real time. An alternative representation in the time domain is shown in Fig. 8.18*b* where $W(t)$ is the weighting function implying the convolution relation

$$o(t) = \int_0^t W(t - T)i(T)\, d(T)$$

For the frequency-domain representation, the response to cascaded systems is simply

$$O(s) = H_1(s)H_2(s)H_3(s) \cdots H_n(s)I(s)$$

Some care must be exercised in implementing this formula since the $H_1(s)$ (e.g., network systems) sometimes displays a loading effect; i.e., the individual free transfer functions differ from the transfer functions in cascade.

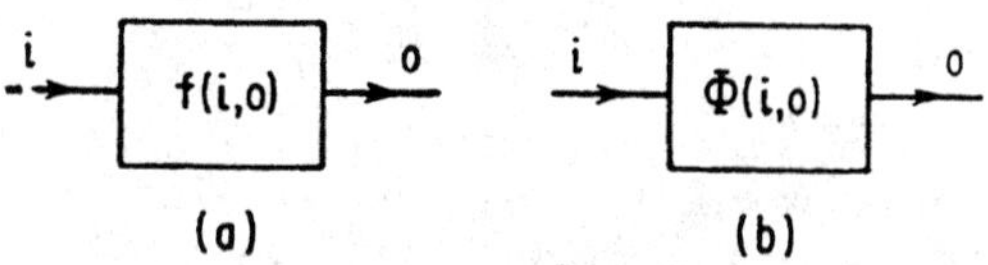

FIG. 8.17 Basic block diagrams. (*a*) $f(i, o)$, algebraic relationship. (*b*) $\delta(i, o)$, general operator relationship.

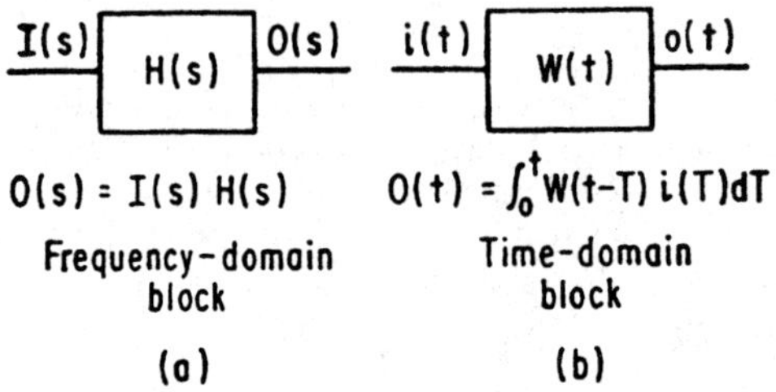

FIG. 8.18 Transfer blocks for linear time-invariant system.

Consider the second-order system

$$d^2x/dt^2 + c\,dx/dt + kx = f(t) \tag{8.136}$$

$$A(p)x = f(t) \qquad A(p) = p^2 + cp + k$$

The system transfer function is obtained by taking the Laplace transform of Eq. (8.136) with zero initial conditions, thus:

$$H(s) = 1/A(s) = 1/(s^2 + cs + k) = X(s)/F(s)$$

Multiple-Degree-of-Freedom Case. For the more general system of n forcing functions f_i, $i = 1, \ldots, n$, e.g., Eq. (8.108) with n outputs $y_1, \ldots, y_n$, and assuming A^{-1} exists, i.e., A is a nonsingular matrix operator, consider

Laplace form

$$\mathbf{A}(p)\mathbf{y} = \mathbf{f}(t) \qquad \mathbf{A}(s)\mathbf{Y}(s) = \mathbf{F}(s)$$

$$\mathbf{y}_j = \sum_{i=1}^{n} \frac{M_{ij}(p) f_i(t)}{D(p)} \qquad Y_j(s) = \sum_{i=1}^{n} \frac{M_{ij}(s)}{D(s)} F_i(s)$$

$$y_{ij} = \frac{M_{ij}(p)\, f_i(t)}{D(p)} \qquad Y_{ij}(s) = \frac{M_{ij}(s)}{D(s)} F_i(s)$$

where y_j is the jth output for all f's and y_{ij} is a component of y_j produced by f_i

The natural extension of the one-dimensional case to this n-dimensional case is made by representing the matrix as follows:

$$\mathbf{Y}(s) = \mathbf{H}(s)\mathbf{F}(s)$$

where

$$\mathbf{Y}(s) = \begin{bmatrix} Y_1(s) \\ Y_2(s) \\ \cdot \\ \cdot \\ \cdot \\ Y_n(s) \end{bmatrix} \qquad \mathbf{F}(s) = \begin{bmatrix} F_1(s) \\ F_2(s) \\ \cdot \\ \cdot \\ \cdot \\ F_n(s) \end{bmatrix}$$

and

$$\mathbf{H}(s) = A^{-1}(s) = \begin{bmatrix} h_{11}(s) & \cdots & h_{1n}(s) \\ \cdots & \cdots & \cdots \\ h_{n1}(s) & \cdots & h_{nn}(s) \end{bmatrix}$$

which formally is identical to Fig. 8.18, the one-dimensional case.

8.5.3 Feedback Control-System Dynamics[10,25–30]

In the feedback control system one or more dependent variables (output) of a dynamic process is controlled. To this end, the difference(s) (error) between the desired value (input) and output is measured and functionally operated on to obtain a correcting signal (e.g., force) which is imparted to the basic system for the purpose of driving the output to correspondence with the input. The following descriptions imply the basic blocks which distinguish a control system:

1. **The plant:** The uncontrolled system
2. **Sensor:** A device to detect the output
3. **Transmitter:** A device to transmit the output or input signals to the comparator
4. **Comparator:** A device to detect differences between output and input
5. **Controller:** A device which takes some useful function of the input and output to correct the "error"

An overall description of the controlled process will yield forms outlined in the foregoing sections which dealt with single systems, where emphasis was placed upon passive types. Stability of these types of systems was assured without recourse to mathematical analysis. On the other hand, the control-system equations relating input to output (overall transfer) are not in general "passive" because the control system contains energy sources such as power amplifiers. The basic problem of control is the synthesis of an optimum control system which exhibits absolute as well as relative stability.

The general equations for control and the corresponding block-diagram representations are as follows:

1. The plant:

$$\omega = \phi(v, d, t) \qquad \begin{matrix} d \rightarrow \\ v \rightarrow \end{matrix} \boxed{\sigma(v, d, t)} \rightarrow \omega$$

where ϕ = general operator
v = input
d = disturbance
t = time

2. Sensor:

$$x = \Omega(\omega) \qquad \xrightarrow{\omega} \boxed{\Omega(\omega)} \xrightarrow{x}$$

3. Transmitter:

$$y = \theta(x) \qquad \xrightarrow{x} \boxed{\theta(x)} \xrightarrow{y}$$

4. Comparator (error device):

$$\epsilon = z - y \qquad z \rightarrow \otimes \rightarrow \epsilon = (z - y), \quad y \uparrow$$

5. Controller:

$$v = \psi(\epsilon) \qquad \xrightarrow{\epsilon} \boxed{\psi(\epsilon)} \xrightarrow{v}$$

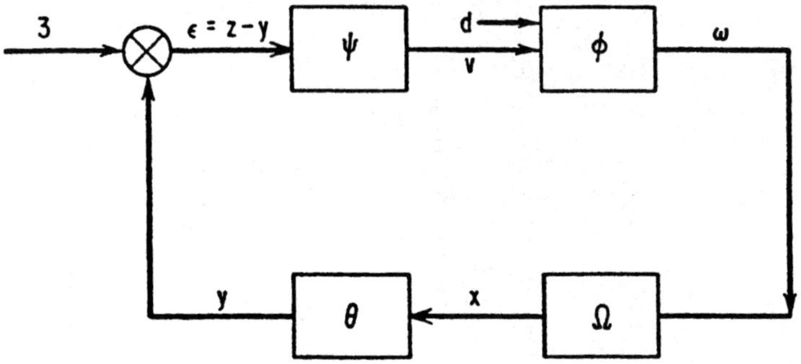

FIG. 8.19 Closed-loop feedback control system.

The total feedback control system for the control of output w is shown in Fig. 8.19. Here ϕ, ψ, Ω, θ are in general nonlinear differential operators. Because of the extraordinary complexity of systems involving nonlinear operators, only the linear time-invariant case in which the operators in the set of control equations are linear with constant coefficients will be presented.

8.5.4 Linear Time-Invariant Control System

Linearity and time invariance admit to simplifying techniques of frequency-domain methods of analyses. In block-diagram form the system transfer functions and the Laplace transforms of the signals entering and leaving are given for each block. The linear time-invariant control-system representation of Eq (8.137) is depicted in Fig. 8.20. The disturbance and prescribed input to the plant are, in the linear case, connected by the differential equation

$$A\,(p)w(t) = B(p)v(t) + K(p)\,d(t)$$

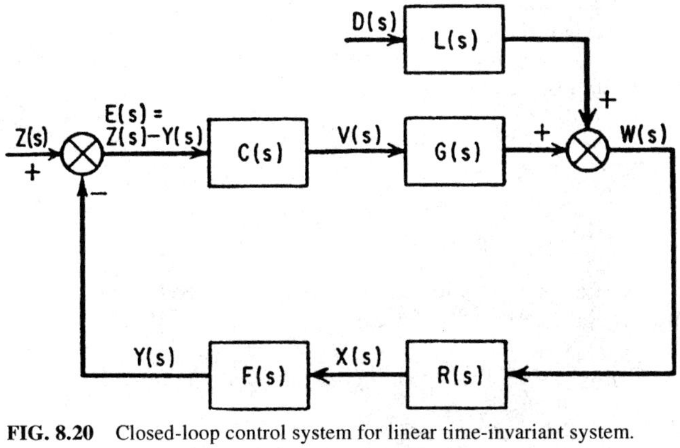

FIG. 8.20 Closed-loop control system for linear time-invariant system.

whence the transfer function derives

$$A(s)W(s) = B(s)V(s) + K(s)D(s)$$
$$W(s) = [B(s)/A(s)]V(s) + [K(s)/A(s)]D(s) = G(s)V(s) + L(s)D(s) \tag{8.138}$$

where $G(s) = B(s)/A(s)$ and $L(s) = K(s)/A(s)$

The equivalent frequency-domain forms of Eq. (8.137) are

$$W(s) = G(s)V(s) + L(s)D(s) \tag{8.138a}$$

$$E(s) = Z(s) - Y(s) \tag{8.138b}$$

$$V(s) = C(s)E(s) \tag{8.138c}$$

$$X(s) = R(s)W(s) \tag{8.138d}$$

$$Y(s) = F(s)X(s) \tag{8.138e}$$

The functional expression relating output to input is obtained by algebraic manipulation of Eq. (8.138).

$$W(s) = \frac{CGZ}{1 + CGFR} + \frac{LD}{1 + CGFR} \tag{8.139}$$

The transfer function between input $z(t)$ and output $w(t)$ in the absence of disturbance $d(t)$ is therefore

$$\frac{W(s)}{Z(s)} = \frac{CG}{1 + CGFR} \tag{8.140}$$

and the transfer function between output and disturbance $d(t)$ in the absence of signal $z(t)$ is

$$\frac{W(s)}{D(s)} = \frac{L}{1 + CGFR} \tag{8.141}$$

The matrix generalization for the simultaneous control of many variables is obtained by considering the transfer functions to be transfer matrices from which

$$\mathbf{W}(s) = (1 + CGF\,R)^{-1}\,CG\mathbf{Z}(s) + (1 + CGF\,R)^{-1}L\mathbf{D}(s)$$

where the column $\mathbf{W}(s)$ implies the separate controlled variables, and $\mathbf{Z}(s)$ and $\mathbf{D}(s)$, the input and disturbance vectors.[28]

Multiloop Systems. In general, control systems are more complicated than the one shown in Fig. 8.20, being composed of many loops, as shown, for example, in Fig. 8.21.

If the inner loop has a transfer function H_2 (dashed box), then the overall transfer function is accordingly

$$\frac{W(s)}{Z(s)} = \frac{H_1H_2}{1 + H_1H_2H_3} \tag{8.142}$$

8.5.5 Analysis of Control System

The transfer-function representations, e.g., Eqs. (8.140) and (8.142), permit the evaluation of three important properties of the control system without transformation to the time domain.

Equation (8.140) or (8.142) can be represented conveniently as

$$\frac{W(s)}{Z(s)} = \frac{Y_1}{1 + Y_1Y_2} \tag{8.143}$$

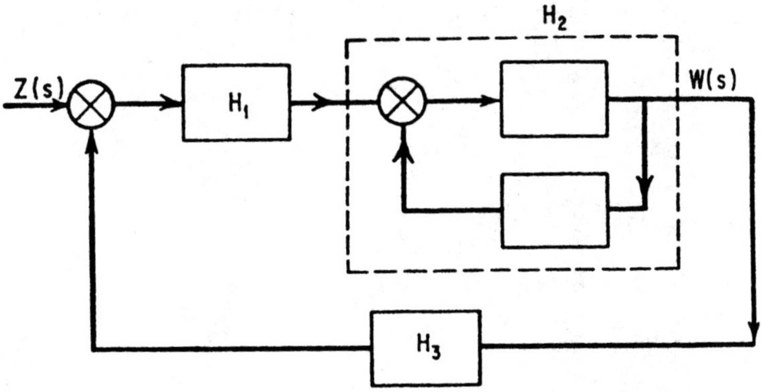

FIG. 8.21 Multiloop control system.

The error-to-input transfer function is obtained by subtracting both sides of the above from unity to give

$$\frac{E(s)}{Z(s)} = \frac{1 + Y_1Y_2 - Y_1}{1 + Y_1Y_2} \tag{8.144}$$

Properties Deducible from Frequency-Domain Representation

Property 1. The steady-state error is

$$\epsilon(t) = sE(s) = s\frac{1 + Y_1(s)Y_2(s) - Y_1(s)}{1 + Y_1(s)Y_2(s)}Z(s) \qquad \begin{matrix} t \to \infty \\ s \to 0 \end{matrix} \tag{8.145}$$

If the input $z(t)$ is the unit step $Z(s) = 1/s$, substitution in Eq. (8.145) yields

$$\epsilon(t) = s\frac{1 + Y_1(0)Y_2(0) - Y_1(0)}{1 + Y_1(0)Y_2(0)}\frac{1}{s} \triangleq \frac{1}{1 + K_p} \qquad \begin{matrix} t \to \infty \\ s \to 0 \end{matrix} \tag{8.146}$$

If the input $z(t)$ is the ramp t, $Z(s) = 1/s^2$ and Eq. (8.145) becomes

$$\epsilon(t) = s\frac{1 + Y_1(s)Y_2(s) - Y_1(s)}{1 + Y_1(s)Y_2(s)}\frac{1}{s^2} = \frac{1 + Y_1(s)Y_2(s) - Y_1(s)}{sY_1(s)Y_2(s)} \triangleq \frac{1}{K_t} \qquad t \to \infty \quad s \to 0 \tag{8.147}$$

If the input $z(t)$ is t^2, $Z(s) = 2/s^3$ and Eq. (8.145) becomes

$$\epsilon(t) = 2\frac{1 + Y_1(s)Y_2(s) - Y_1(s)}{s^2Y_1(s)Y_2(s)} \triangleq \frac{1}{K_a} \qquad \begin{matrix} t \to \infty \\ s \to 0 \end{matrix} \tag{8.148}$$

where the steady-state errors due to a step, ramp, and acceleration of the input are developed in Eqs. (8.146) through (8.148) assuming that they exist in each case.

If $E(s)/Z(s)$ can be represented by a Maclaurin series and is stable.

$$E(s)/Z(s) = a_0 + a_1s + a_2s^2 + \cdots$$

$$E(s) = a_0Z(s) + a_1sZ(s) + \cdots$$

the steady state $e(t)$ for an input $z(t)$ is then

$$e(t) \to a_0z(t) + a_1z'(t) + a_2z''(t) + \cdots \qquad t \to \infty$$

which implicitly ignores all initial conditions, i.e., assumes the initial disturbances vanish as $t \to \infty$.

Property 2. The rms error, a quantitative measure of the effectiveness of control,

$$\overline{\mathscr{E}}^2 = \frac{1}{T}\int_0^T |e(t)|^2 dt \qquad T \to \infty$$

can be obtained from[10,16,27]

$$\overline{\mathscr{E}}^2 = \frac{1}{2}\int_{-\infty}^{+\infty} G_e(f)\,df = \int_0^{\infty} G_e(f)\,df$$

where $G_e(f)$ is the spectral density of the error $e(t)$. G_z for the input $z(t)$ is defined by

$$G_z(j\omega) = \frac{1}{2T}\left|\int_{-T}^{+T} z(t)e^{-j\omega t}dt\right|^2$$

For the input forms which are bounded, G_z exists and the overall G_e is given by

$$G_e = \left|\frac{E(j\omega)}{Z(j\omega)}\right|^2 G_z(j\omega) \qquad T \to \infty$$

The spectral density G_z for random-type inputs and disturbances is also obtainable by statistical methods; then the property becomes the "expected rms error" owing to the nonspecific character of input.

Property 3. The stability of the system, however complex, is completely ascertained by the locations of the poles of the right-hand member of Eq. (8.143), namely,

$$\frac{Y_1(s)}{1 + Y_1(s)Y_2(s)} \tag{8.149}$$

The system is stable if no poles of this function lie in the right half s plane, the imaginary axis included. Otherwise it is unstable.

There are three prominent methods for determining whether or not poles of the function lie in the right-hand plane: (1) Routh-Hurwitz, (2) root locus, and (3) Nyquist criterion.

Routh-Hurwitz Method. The method is applicable to rational fractions

$$Y_1(s) = A(s)/B(s) \qquad Y_2(s) = C(s)/D(s)$$

First the fractions are cleared after substituting in Eq. (8.149), leaving

$$\frac{A/B}{1 + (A/B)(C/D)} = \frac{AD}{AC + BD}$$

assuming the fraction in the lowest form $AC + BD$ cannot coincide with zeros of its numerator. The poles of Eq. (8.149) correspond to the zeros of the denominator. The denominator polynomial can be written as

$$AC + BD = a_n s^n + a_{n-1}s^{n-1} + \cdots + a_1 s + a_0$$

System stability is therefore governed by the location of zeros of this polynomial.

Stability: Routh-Hurwitz Criterion. Given the general nth-order equation,

$$a_n s^n + a_{n-1}s^{n-1} + \cdots + a_1 s + a_0 = 0 \tag{8.150}$$

with real coefficients.

The following statements apply to the roots:

1. The roots occur in conjugate complex pairs.
2. A necessary condition for the real parts of all roots to be negative is that all coefficients have the same sign, and hence is a necessary condition for stability.
3. A necessary condition for all real parts to be nonpositive is that all coefficients a_0, $a_1, \ldots, a_n$ be different from zero, i.e.,

$$a_0, a_1, a_2, \ldots, a_n \neq 0$$

and is a necessary condition for stability.

If conditions 2 or 3 fail, the system is unstable. If, on the other hand, Eq. (8.150) meets conditions 2 and 3, further tests (Routh-Hurwitz) must be made to determine stability.

The procedure is first to arrange the coefficients in two rows as shown followed by a third row developed from the first two rows, viz.,

Row 1:	a_n	a_{n-2}	a_{n-4}	$\cdots$
Row 2:	a_{n-1}	a_{n-3}	a_{n-5}	$\cdots$
Row 3:	b_{n-1}	b_{n-3}	b_{n-5}	$\cdots$

where

$$b_{n-1} \triangleq \frac{-1}{a_{n-1}} \begin{vmatrix} a_n & a_{n-2} \\ a_{n-1} & a_{n-3} \end{vmatrix}$$

$$b_{n-3} \triangleq \frac{-1}{a_{n-1}} \begin{vmatrix} a_n & a_{n-4} \\ a_{n-1} & a_{n-5} \end{vmatrix}$$

$$\cdots\cdots\cdots\cdots\cdots\cdots\cdots\cdots$$

where the bars indicate the determinant of the enclosed array. In a like fashion form a fourth row developed from rows 2 and 3:

Row 4:	c_{n-1}	c_{n-3}	c_{n-5}	$\cdots$

where

$$c_{n-1} = \frac{-1}{b_{n-1}} \begin{vmatrix} a_{n-1} & a_{n-3} \\ b_{n-1} & b_{n-3} \end{vmatrix}$$

$$c_{n-3} = \frac{-1}{b_{n-1}} \begin{vmatrix} a_{n-1} & a_{n-5} \\ b_{n-1} & b_{n-5} \end{vmatrix}$$

$$\cdots\cdots\cdots\cdots\cdots\cdots\cdots\cdots$$

Continue this procedure of forming a new row from the two preceding rows until zeros are obtained; $n + 1$ rows will result.

The Routh-Hurwitz criterion states that the number of roots with positive real parts equals the number of changes of sign in the first column. Since only one root with positive real part is sufficient to cause instability, the following stability criterion may be stated. A system whose characteristic equation is Eq. (8.150) is stable if and only if the elements formed in the first column (a_n, a_{n-1}, b_{n-1}, …) are all of the same algebraic sign.

Root-Locus Method. This method, attributable to Evans, takes the denominator of Eq. (8.149) and factors Y_1Y_2 into zeros and poles.

$$1 + Y_1(s)Y_2(s) = 1 + \frac{K\prod_i (s - z_i)}{\prod_j (s - p_j)}$$

where Π denotes product.

One first explores the zeros of $1 + Y_1Y_2$, which must satisfy two conditions—the amplitude and the phase.

$$1 + \frac{K\prod_i (s - z_i)}{\prod_j (s - p_j)} = 0$$

Amplitude condition: $K \left| \frac{\prod_i (s - z_i)}{\prod_j (s - p_j)} \right| = 1$

Phase condition: $\arg \frac{\prod_i (s - z_i)}{\prod_j (s - p_j)} = \pi(2n + 1)$ n an integer (8.151)

where arg = argument.

The location of all possible s that satisfy the phase condition (8.151) is drawn in the s plane; each corresponds to a K satisfying the amplitude condition. Figure 8.22 shows a typical example where the locus is drawn in solid lines.

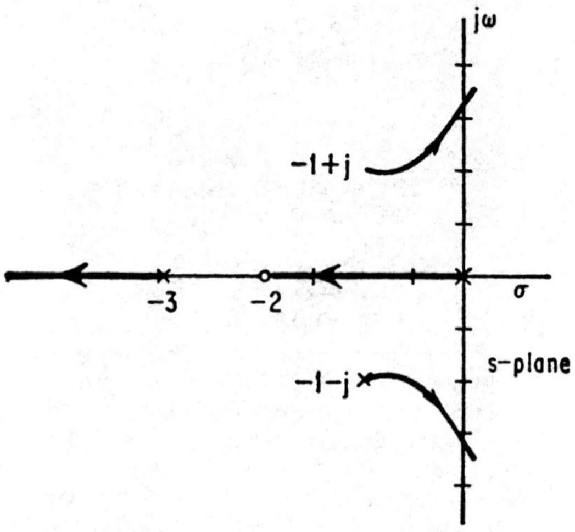

FIG. 8.22 Typical root-locus plot for $H(s)$.

$$H(s) = \frac{(s + 2)K}{s(s + 3)(s^2 + 2s + 2)}$$

Stability for the system is ascertained by the points on the locus (of roots) that apply for the specific K in question. If the locus in question lies in the left half plane the system is stable; otherwise it is unstable. Relative stability is judged by the proximity to the imaginary axis.

Complex-Function Theory for Nyquist Criterion. The Nyquist criterion utilizes complex-function theory and in particular the so-called "argument principle." It is stated and proved as follows: Given a function $F(s)$ regular in a closed region R (except for a finite number of poles) bounded by the closed curve C, then the curve C maps into the curve C' in the plane, $w = F(s)$. The theorem asserts that the number of times the curve C' encircles the origin in the w plane is equal to the difference between the number of zeros and poles of $F(s)$ included within the region R, $N = Z - P$, including the multiplicity of the zeros and poles. In proof:

Near a zero inside R, say near $s = z_i$,

$$F(s) = (s - z_i)^{\gamma_i}[\psi(s)]$$

Near a pole inside R at p_i,

$$F(s) = \frac{1}{(s - p_i)^{\beta_i}}[\Omega(s)]$$

In general, therefore,

$$F(s) = \frac{(s - z_1)^{\gamma_1}(s - z_2)^{\gamma_2} \cdots (s - z_n)^{\gamma_n}}{(s - p_1)^{\beta_1}(s - p_2)^{\beta_2} \cdots (s - p_m)^{\beta_m}}[\phi(s)]$$

where $\phi(s)$ has no zeros or poles inside the region R. Taking the logarithmic derivative of $F(s)$,

$$\frac{d}{ds}\ln F(s) = \frac{F'(s)}{F(s)} = \frac{\gamma_1}{s - z_1} + \frac{\gamma_2}{s - z_1} + \cdots - \frac{\beta_1}{s - p_1} - \frac{\beta_2}{s - p_2} - \cdots + \frac{\phi'(s)}{\phi(s)}$$

Integrating around the closed contour C in the s plane and employing Cauchy's residue theorem we get

$$\oint_C d[\ln F(s)] = \ln F(s)\Big|_C = 2\pi j\left(\sum_i \gamma_i - \sum_k \beta_k\right) + 0 = \ln w\Big|_{\text{contour } c'} = 2\pi jN$$

$$N = \sum_i \gamma_i - \sum_k \beta_k$$

where it is noted that the evaluation of $\ln w$ over the closed curve c' in the w plane yields the change of argument times j or $2\pi jN$ where N is the number of encirclements of the origin with due regard for sign. Also, since ϕ has no zeros or poles in region R, ϕ'/ϕ is regular inside R and its contour integral vanishes over the closed path, viz.,

$$\oint \frac{\phi'}{\phi}ds = 0$$

Nyquist Criterion. This principle is now applied to the transfer function $Y_1(s)/[1 + Y_1(s)Y_2(s)]$ to determine the zeros of $1 + Y_1(s)Y_2(s)$ in the right half s plane. The scanning contour in the s plane is the region bounded by the imaginary axis and the right-hand infinite semicircle. Instead of examining the number of encirclements of $1 + Y_1(s)Y_2(s)$ around the origin of w, by a shift of axis one unit, it is exactly equivalent to examining the contour of $Y_1(s)Y_2(s)$ mapped into w with reference to the -1 point. Now the number of encirclements of $Y_1(s)Y_2(s)$ around -1 is given by

$$N = Z - P \qquad Z = N + P \tag{8.152}$$

If the number of right-half-plane poles P of $Y_1(s)Y_2(s)$ are known (none for passive blocks, but for multiloop systems they may be present) then the number of right half plane zeros Z can be determined from Eq. (8.152) since N, the number of turns, can be counted. If Z has a value different from zero it implies that there are zeros of $1 + Y_1(s)Y_2(s)$ in the right-hand s plane establishing the case of instability. Otherwise the system is stable. The necessary and sufficient condition for stability is $Z \equiv 0$.

The actual plotting of the contour $Y_1(s)Y_2(s)$ (for rational fractions) in the w plane requires only one half imaginary axis, say $s = j\omega$, since rational functions of s display real axis symmetry as follows:

$$Y_1(s)Y_2(s) \rightarrow \begin{cases} Y_1(j\omega)Y_2(j\omega) = M(j\omega) + jN(j\omega) & s = j\omega \\ Y_1(-j\omega)Y_2(-j\omega) = M(j\omega) - jN(j\omega) & s = -j\omega \end{cases}$$

where M and N are real even and odd functions, respectively, of ω. Hence the contour mapping $Y_1(j\omega)Y_2(j\omega)$ is symmetric with $Y_1(-j\omega)Y_2(-j\omega)$ with respect to the real w axis. In plotting the infinite semicircles

$$s = Re^{j\phi} \qquad R \rightarrow \infty$$

the terms in the highest power of the numerator and denominator are retained for evaluation.

Consider, for example,

$$Y_1(s)Y_2(s) = \frac{as^3 + bs^2 + cs + d}{es^5 + fs^4 + gs^3 + hs^2 + ns + m} \rightarrow \frac{as^3}{es^5}$$

$$= \frac{a}{es^2} \rightarrow \frac{a}{eR^2e^{2\phi j}} \qquad \begin{matrix} s = Re^{j\phi} \\ R \rightarrow \infty \end{matrix}$$

A Nyquist plot is shown for a third-order system with a pole at the origin in Fig. 8.23.

Relative Stability. In addition to providing information on stability a measure of relative stability is provided by noting the proximity of the graphical plot to the -1 point in the w plane.

In quantitative terms, relative stability is determined by the relative gain of the open-loop transfer function $Y_1(j\omega)Y_2(j\omega)$ (Nyquist plot) 180° out of phase with the input, shown with amplitude e. The ratio $1/e$ is called the *gain margin*, implying that an increase of gain by this factor would make the system unstable. If $e > 1$, then the system is unstable. Similarly, the angle ϕ at unit distance is called the *phase margin* and indicates the additional amount of phase lag necessary to destabilize a stable system (see Fig. 8.24).

Systems with Feedback Time Lag. Interest is often centered on the destabilizing effect of time delays in the feedback path of a system which is otherwise stable. In some applications they are intentionally introduced for such an effect (oscillator).

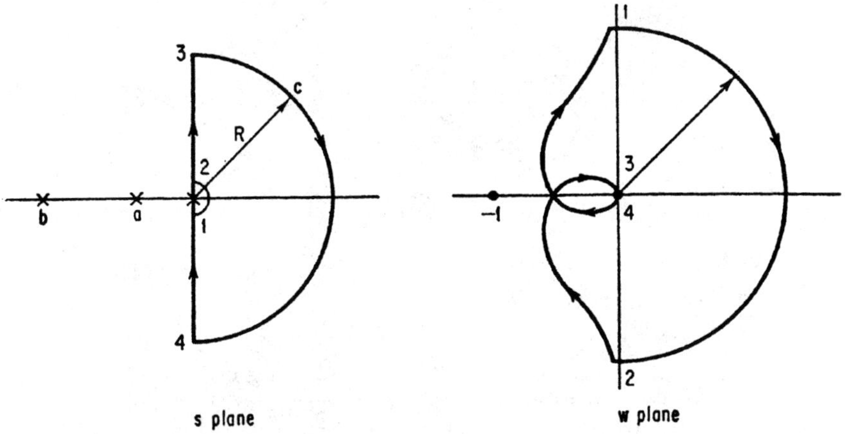

FIG. 8.23 Nyquist plot $Y_1(s)Y_2(s) = \frac{K}{s(s+a)(s+b)}$.

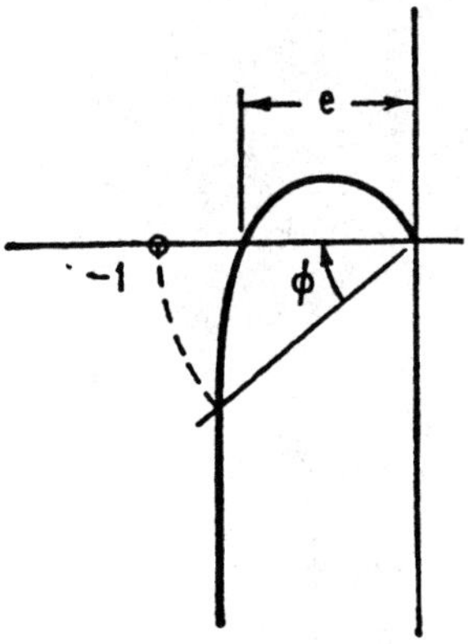

FIG. 8.24 Gain and phase margin. Phase margin = ϕ. Gain margin = 1/*e*.

The basic system is shown in Fig. 8.25 with an overall transfer function $Y_1/(1 + Y_1Y_2)$. Introduction of the time-delay block $\boxed{D}$ which is e^{-sT} changes the transfer function to $Y_1/(1 + Y_1Y_2e^{-sT})$. The system stability is readily evaluated by making the basic system Nyquist plot of Y_1Y_2 which is stable (by hypothesis) and incorporating the e^{-sT} by increasing the phase angle by $-\omega T$ at each point along the existing $Y_1(j\omega)Y_2(j\omega)$ basic plot. The new plot provides the required stability picture.

A method due to Satch, applied to a first-order differential equation with a time lag T, is shown below for the first-order system.

$$dx/dt + \alpha x + \beta x(t - T) = 0$$

The transfer function is $1/(s + \alpha + \beta e^{-sT})$, which in order to exhibit stability must be free of right-hand-plane poles. The equation for poles is

$$-(s + \alpha)/\beta = e^{-sT} \tag{8.153}$$

Let each side map the right-hand s plane, shown superimposed in Fig. 8.26*b* and *c* for two different cases. The right side maps into the unit circle, the left side the half plane

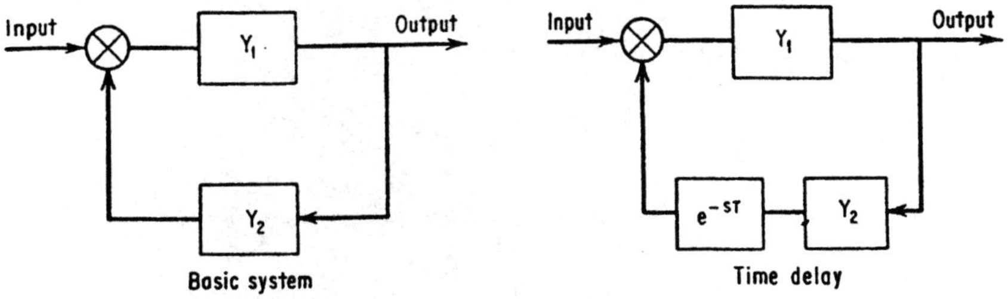

FIG. 8.25 System with feedback time delay.

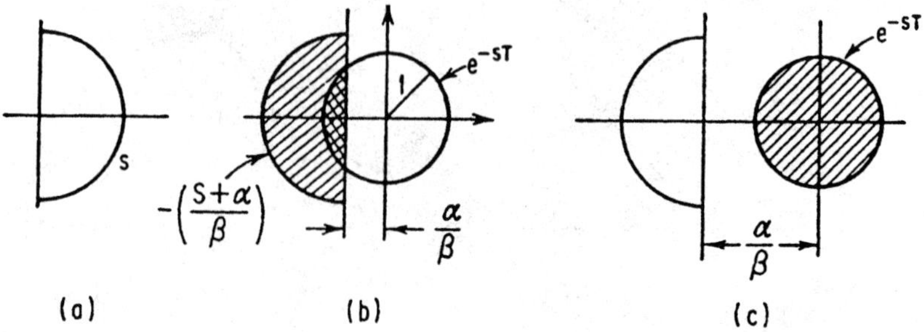

FIG. 8.26 Satch diagram for $dx/dt + \alpha x + \beta x\,(t - T) = 0$. Intersection $\alpha/\beta < 1$. No intersection (unconditionally stable), $\alpha/\beta > 1$.

displaced by $-\alpha/\beta$, shown crosshatched. The two closed curves are shown intersected in Fig. 8.26*b* for $\alpha/\beta < 1$, and therefore the included region corresponds to points in the *s* plane which satisfy Eq. (8.153) and consequently the system may be unstable. Further investigation would be required to ascertain stability or instability. If the regions intersected contain in each case one or more coincident points of the right half *s* plane, the system is unstable; otherwise it is stable. The situation depicted in Fig. 8.26*c*, $\alpha/\beta > 1$, reveals no intersection and illustrates the unconditionally stable system.

8.5.6 The Problem of Synthesis[31–33]

The problem of synthesis is to realize an overall transfer function within a set of specifications which often requires optimization of several conflicting requirements. Consider Fig. 8.20. The designer usually has little control over any block except $C(s)$, the "controller." Hence synthesis involves realization of transfer functions in cascade with fixed elements to produce the desired overall transfer function.

8.5.7 Linear Discontinuous Control: Sampled Data[25,34]

If at once or more points in a linear control system the signal is interrupted intermittently at a prescribed rate, the resultant system is discontinuous and linear. If the rate is constant, the system is called a linear sampled-data control system. Insofar as analysis is concerned, this merely introduced another building block called the "sampler" at each sampling point.

The sampler shown schematically in Fig. 8.27 has the property of taking the input $e(t)$ and periodically sampling it for time durations such that the area under each pulse (strength) is proportional to the instantaneous input. If the proportionality constant is made unity and the sampling pulse duration is small compared with the sampling period T, then to an excellent approximation which offers considerable analytic advantages, the output is assumed to be a train of impulses of strength $e(t)$ at each sampling "instant." The actual and ideal outputs are shown in Fig. 8.27.

The ideal output $e^*(t)$ considered below is given by

$$e^*(t) = e(t) \sum_{n=0}^{\infty} \delta(t - nT) = \sum_{n=0}^{\infty} e(nT)\,\delta(t - nT) \tag{8.154}$$

where $\delta(x)$ = Dirac delta function. According to continuous theory, it is desirable to obtain the frequency-domain behavior of the "sampling" block. To this end consider the Laplace transform of the sampled signal:

$$\mathcal{L}\{e^*(t)\} \overset{\Delta}{=} E^*(s) = \mathcal{L}\left\{e(t) \sum_{n=0}^{\infty} \delta(t - nT)\right\} \tag{8.155}$$

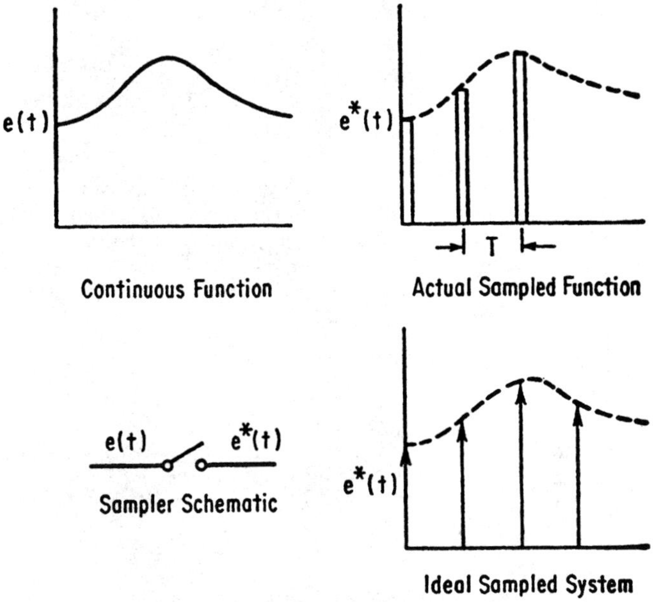

FIG. 8.27 The sampler and a sampled function.

From the convolution theorem,

$$\mathcal{L}\{(x(t)y(t)\} = X(s) * Y(s)$$

Eq. (8.155) becomes

$$E^*(s) = \mathcal{L}\{e^*(t)\} = E(s) * \mathcal{L}\left\{\sum_{n=0}^{\infty} \delta(t - nT)\right\}$$

$$= E(s) * \sum_{n=0}^{\infty} e^{-nTs} = E(s) * I(s) \qquad (8.156)$$

$$I(s) = \sum_{n=0}^{\infty} e^{-nTs}$$

where $I(s)$ is defined as the Laplace transform of the impulse train

$$\sum_{n=0}^{\infty} \delta(t - nT)$$

An alternative and less useful form is derived from the right-side representation of Eq. (8.154).

$$E^*(s) = \mathcal{L}\{e^*(t)\} = \mathcal{L}\left\{\sum_{n=0}^{\infty} e(nT)\,\delta\,(t - nT)\right\} = \sum_{n=0}^{\infty} e(nT)e^{-nTs}$$

Using the closed-form representation of $I(s)$, we find

$$I(s) = \sum_{n=0}^{\infty} e^{-nTs} = \frac{1}{1 - e^{-sT}}$$

Eq. (8.156) is evaluated by a closed-contour integration as follows:

$$E^*(s) = \frac{1}{2\pi j}\int_{c-j\infty}^{c+j\infty} E(w)I(s-w)\,dw = \frac{1}{2\pi j}\oint \frac{E(w)\,dw}{1-e^{-(s-w)T}} \tag{8.157}$$

The abscissa c is chosen so that the poles of $E(w)$ have real parts $<c$ and s is defined for Re $s > c$ where the contour integral shown in Fig. 8.28 is employed since the integral over the infinite right-hand semicircle vanishes. The infinity of poles of the integrand inside this contour are then the zeros of $1 - e^{-(s-w)T}$, the latter corresponding to $s - w = \pm 2n\pi j/T$.

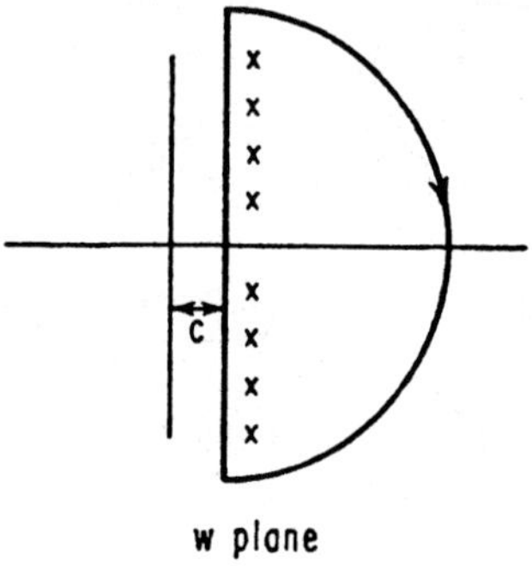

FIG. 8.28 *W*-plane contour for complex convolution.

$$w = s \pm 2n\pi j/T \qquad \text{(poles)}$$

$$2\pi/T = (\omega_s) \qquad \text{(sampling frequency)}$$

which are an infinity of simple poles. Equation (8.157) is evaluated by Cauchy's residue theorem, taking the residues of the infinity of poles yielding

$$E^*(s) = \sum_{n=-\infty}^{+\infty} \frac{E(s + jn\omega_s)}{T} \tag{8.158}$$

If $E(s)$ has no right-half-plane poles, then Eq. (8.158) is defined for the entire right half s plane. Equation (8.158) is clearly a periodic function of s having the complex period $j\omega_s$ as shown:

$$E^*(s + j\omega_s) = \sum_{n=-\infty}^{+\infty} \frac{E[s + j(n+1)\omega_s]}{T} = \sum_{-\infty}^{+\infty} \frac{E(s + jn\omega_s)}{T} = E^*(s) \tag{8.159}$$

From the periodic character of $E(s)$ it follows that if $E^*(s)$ is known in any strip in the complex s plane bounded by

$$jx < \text{Im } s < j(x + \omega_s) \tag{8.160}$$

it is known everywhere in the s plane. The transfer function at real frequency $s = j\omega$ is found directly from Eq. (8.158). Its amplitude spectrum is sketched (Fig. 8.29) for $\omega_s/2 > \omega_0$ where ω_0 is the cutoff frequency of $E(j\omega)$, i.e., for the sampling frequency greater than twice the highest frequency component of $e(t)$. Note that $|E^*(j\omega)|$ yields the infinitely repeated spectrum of $|E(j\omega)|$ attenuated by $1/T$. If $\omega_s/2 < \omega_0$, there is

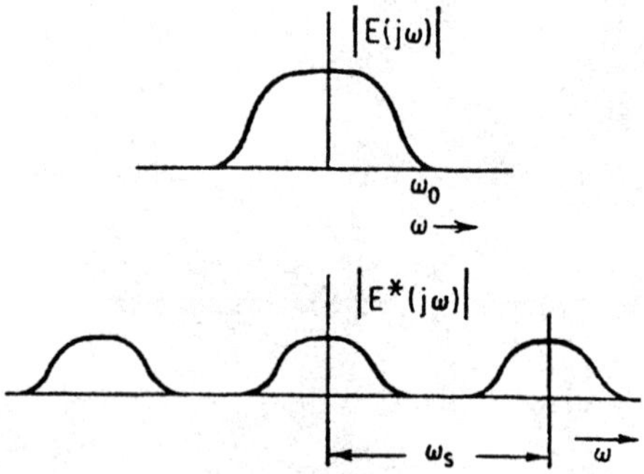

FIG. 8.29 Frequency spectrum for function and for sampled function.

overlapping and resultant distortion of the input signal. Returning to the case $\omega_s/2 > \omega_0$, practically all the input $e(t)$ information is stored in $E^*(j\omega)$ over the frequency range $0 < \omega < \omega_0$. By ideal low-pass filtering of the signal $E^*(j\omega)$, spectral components greater than ω_0 can be eliminated, leaving the fundamental signal shape. The resultant system would then be the equivalent of the continuous system with an attenuator $1/T$ placed after the input signal. In practice, however, deviations from ideally of this filter introduce severe stability problems. A smoothing device is utilized as a compromise between high degree of filtering and its concomitant stability problems.

One such smoothing device is the holding circuit whose transfer function is

$$(1 - e^{-Ts})/s \tag{8.161}$$

which for any impulse input $\delta(t - nT)$ yields the output $u(t - nT) - u[t - (n + 1)T]$, a pulse of unit height starting $t = nT$ and of duration T.

Stability Investigation of Sampled-Data Control Systems. Consider the sampled-data control system shown in Fig. 8.30. The overall transfer function is derived from the basic properties of transfer functions.

$$E = Z - Y_2W \qquad W = Y_1E^* \tag{8.162}$$

Elimination of W in Eq. (8.162) gives

$$E = Z - Y_1Y_2E^* \tag{8.163}$$

Now a fundamental property of sampling is stated and proved as follows:

Given $$A = BC^*$$

then $$A^* = B^*C^*$$

Proof [utilizing Eq. (8.158)]:

$$X^*(s) = \frac{1}{T}\sum_{n=-\infty}^{+\infty} X(s + jn\omega_s)$$

$$A^*(s) = \frac{1}{T}\sum_{n=-\infty}^{+\infty} A(s + jn\omega_s) = \frac{1}{T}\sum_{n=-\infty}^{+\infty} B(s + jn\omega_s)C^*(s + jn\omega_s)$$

But from Eq. (8.159)

$$C^*(s + jn\omega_s) = C^*(s)$$

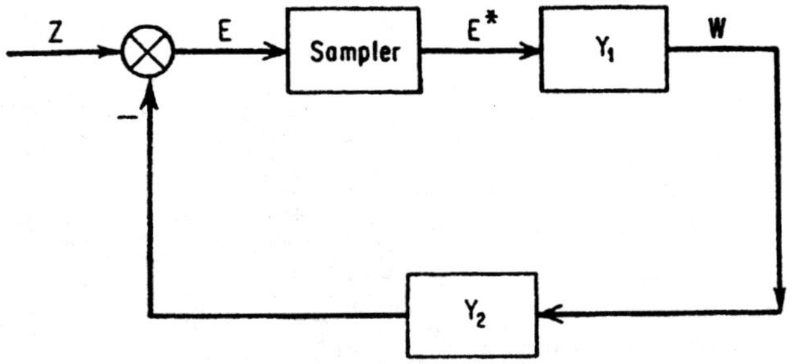

FIG. 8.30 Sampled-data system.

and therefore

$$A^*(s) = \frac{C^*(s)1}{T} \sum_{n=-\infty}^{+\infty} B(s + jn\omega_s) = C^*(s)B^*(s) = B^*(s)C^*(s)$$

Sampling Eq. (8.163) and utilizing the results of this theorem yields

$$E^* = Z^* - (Y_1Y_2)^*E^* \tag{8.163a}$$

or

$$E^* = Z^*/[1 + (Y_1Y_2)^*]$$

and

$$W = Y_1E^* = Z^*Y_1/[1 + (Y_1Y_2)^*]$$

Following previous work, the stability of the system rests with the location of the poles of $1/[1 + (Y_1Y_2)^*]$ or more specifically the zeros of $1 + (Y_1Y_2)^*$.

The transcendental form of $(Y_1Y_2)^*$ makes this evaluation using methods cited earlier extremely difficult to apply, per se. The task is simplified, however, owing to the periodic property of sampled functions embodied in Eq. (8.159) which implies that if one investigates the zeros of $(Y_1Y_2)^* + 1$ bounded by a strip given in Eq. (8.160) in region Re $s > 0$, then this effectively gives the zero configuration in the entire right half plane. In practice the strip chosen is $0 < \text{Im } s < j\omega_s = j2\pi/T$, and Re $s > 0$ shown crosshatched in Fig. 8.31. Now a function z is defined by

$$z = e^{sT} \tag{8.164}$$

$$s = \ln z/T \tag{8.165}$$

with s defined only in the whole strip

$$0 < \text{Im } s < j2\pi/T$$

so that s and z are single-valued analytic functions of each other (except for a branch cut on the real z axis). Equation (8.164) implies a mapping of the whole strip in the s plane onto the entire z plane with the shaded portion mapping outside the unit circle as shown in Fig. 8.31. The location of zeros of $1 + (Y_1Y_2)^*$ in the right half strip of s corresponds to the location of zeros outside the unit circle of the z plane, and their presence or absence is translated as unstable or stable conditions, respectively.

***z* Transform.** If the transformation Eq. (8.164) is applied to the Laplace transformation of a sampled function $x^*(t)$ as follows:

$$X(z) \triangleq X^*(s) \qquad z = e^{sT}$$

then $X(z)$ is defined as the z transform of $x(t)$. $X(z)$ is valid only at the sampling instants despite its general continuity properties. Note that the z transform of $x^*(t)$ is

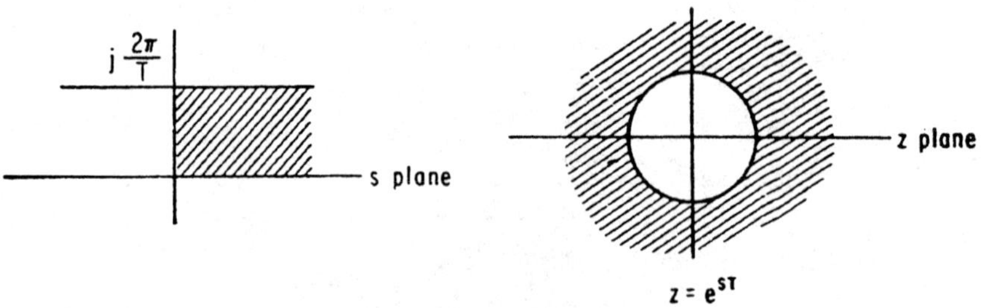

FIG. 8.31 Mapping the strip onto the z plane.

the same as that of $x(t)$ since the sampled $x^*(t)$ is indeed $x^*(t)$. The z transformation of any function is obtained by first sampling the function, then taking its Laplace transform, and finally making the substitution [Eq. (8.165)] to eliminate s in favor of z. From the definition,

$$E(z) = E^*(s) = \sum_{n=0}^{\infty} e(nT)e^{-nTs} = \sum_{n=0}^{\infty} e(nT)z^{-n} \tag{8.166}$$

or from the convolution form [Eq. (8.157)],

$$\begin{aligned} E^*(s) &= \frac{1}{2\pi j}\int_{c-j\infty}^{c+j\infty} E(w)\frac{dw}{1 - e^{-st}e^{wT}} \\ &= \frac{1}{2\pi j}\int_{c-j\infty}^{c+j\infty} E(w)\frac{dw}{1 - e^{wT}z^{-1}} = E(z) \qquad \text{Re } s > c \end{aligned}$$

If this integration is performed over the contour enclosing the left-hand infinite semicircle in contradistinction to the contour used previously so that the zeros of $1 - e^{wt}z^{-1}$ are not contained inside the contour, then application of the residue theorem yields

$$E(z) = \sum \text{residues of}\left[E(w)\frac{1}{1 - e^{wT}z^{-1}}\right]$$

for poles of $E(w)$ only.

Inversion of E(z) to Time Domain. To go from the z domain to the time domain, it is only necessary to consider the coefficients of a Laurent series about $z = 0$ which yield the sampling values, i.e., if

$$F(z) = a_0 + a_1/z + a_2/z^2 + \cdots$$

in accordance with Eq. (8.166)

$$\begin{aligned} a_0 &= e(0) \\ a_1 &= e(T) \\ &\cdots\cdots\cdots\cdot \\ a_n &= e(nT) \end{aligned}$$

Formally using any closed contour around $z = 0$, this is equivalent to the contour integral

$$e(nT) = (1/2\pi j)\oint E(z)z^{n-1}\,dz = \text{residues of } E(z)z^{n-1}$$

Table 8.3 shows some z transforms and their properties.

Example of Stability Investigation. Consider the sampled-proportional-level control system shown schematically and in block form in Fig. 8.32 with the constant input w_R (desired level). The system is described by the following:

1. $q_0(t)$ is an arbitrary-rate flow of effluent.
2. The replenishment rate is proportional to the error existing one sample time prior to

$$q_{in} = -K(w[t/T] - w_R)$$

where $[x]$ is defined as the smallest integral value of x.

TABLE 8.3 Properties of z Transforms of Causal Time Function

$f(t) = 0 \qquad t < 0$

Property	z transform	Time function
Basic pairs.	$F(z) = \sum_{n=0}^{\infty} f(nT)z^{-n}$	$f(t)$ $f(nT) \qquad n = 0,1, \ldots, \infty$ $f(nT) = 1/2\pi j \oint F(z)z^{n-1}\,dz$
Linearity.	$a_1F_1(z) + a_2F_2(z)$	$a_1f_1(t) + a_2f_2(t)$
Time shift.	$z[F(z) - f(0)]$	$f(t + T)$
Initial-value theorem.	$\lim_{z \to \infty} F(z)$	$\lim_{t \to 0} f(t)$
Final-value theorem	$\lim_{z \to 1} (z - 1)F(z)$	$f(t)$ $t \to \infty$

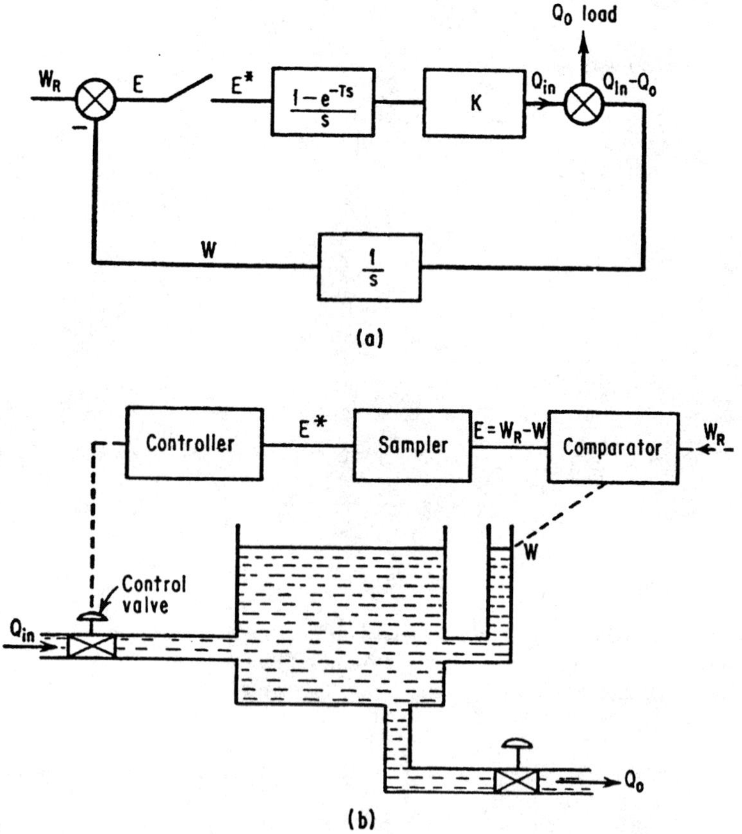

FIG. 8.32 Sampled-level control system. (*a*) Control block diagram. (*b*) System.

Writing the equation conserving mass yields

$$dw/dt = -K(w[t/T] - w_R) - q_0(t)$$

Integrating over sampling times nT and $(n + 1)T$ yields the equation

$$w\{(n + 1)T\} - w(nT) = TK\{w_R - w(nT)\} - \int_{nT}^{(n+1)T} q_0\, dt$$

$$w\{(n + 1)T\} - (1 - TK)w(nT) = TKw_R - \int_{nT}^{(n+1)T} q_0\, dt$$

which is the inhomogeneous difference equation whose theory parallels that of ordinary differential equations. The stability is a function of the solutions to the homogeneous equation

$$w_1\{(n + 1)T\} - (1 - TK)w_1\{nT\} = 0 \tag{8.167}$$

A solution is found by assuming the exponential form

$$w_1(t) = e^{\lambda t}$$

whence

$$e^{\lambda(n+1)T} - (1 - TK)e^{n\lambda T} = 0 \tag{8.168}$$

from which

$$e^{\lambda T} - (1 - TK) = 0$$

$$\lambda = \frac{\log(1 - TK)}{T} = \frac{\log|1 - TK|}{T} + j\arg(1 - TK)$$

The condition for stability is that as $t \to \infty$, $w_1 \to 0$. From Eq. (8.168), this is met by requiring

$$\text{Re}\,\lambda < 0$$
$$\log|1 - TK| < 0$$
$$|1 - TK| < 1$$

Alternatively the z-transform method can be applied directly to Eq. (8.167), which is just written as

$$w_1^*(t + T) - (1 - TK)w_1^*(t) = 0 \tag{8.167a}$$

and is a valid representation of Eq. (8.167) only at the sampling instants.

Taking the z transform of Eq. (8.167a) yields

$$z\{W(z) - w(0)\} - (1 - TK)W(z) = 0$$

$$W(z) = \frac{zw(0)}{z - (1 - TK)} \tag{8.169}$$

Applying the stability condition to Eq. (8.169), namely, that it have no poles outside the unit circle (i.e., no zeros of its denominator outside the unit circle) yields

$$|1 - TK| < 1$$

which has been obtained by classical methods above. Since T and K are real, the condition is

$$-1 < (1 - TK) < 1$$

Finally, and more generally, direct consideration of the control block (Fig. 8.32) and Eq. (8.163*a*) yields

$$E^* = \frac{W_R^*}{1 + K\{(1 - e^{-Ts})/s^2\}^*} + \frac{\{Q_0(s)/s\}^*}{1 - K\{(1 - e^{-Ts})/s^2\}^*} \tag{8.170}$$

From tables in Ref. 10

$$\left(\frac{1 - e^{-Ts}}{s^2}\right) = (1 - z^{-1})\frac{Tz}{(z - 1)^2} = \frac{T}{z - 1}$$

The transform of Eq. (8.170) is

$$E(z) = \frac{W_R(z)}{1 + KT/(z - 1)} + \frac{\{Q_0(s)/s\}^*}{1 + KT/(z - 1)} \qquad s = \ln\frac{z}{T}$$

No poles of the numerator are envisioned for practical systems, so that the zeros of the denominator give all $E(z)$ poles

$$z = 1 - TK$$

which for stability demands $|z| < 1$ or once again

$$-1 < 1 - TK < 1$$

8.5.8 Nonlinear Control Systems[36–39]

The treatment of control systems containing nonlinearities (as defined under general nonlinear systems) is for the most part so formidable that all known analytic methods fail. The very special case of the second-order autonomous (time-invariant coefficients) can be handled most conveniently by graphical methods in the phase plane. Also under very special conditions it is possible for a higher-degree nonlinear system to be analyzed by a "describing-function" technique. Systems which contain switching-function nonlinearities which are otherwise linear can be analyzed (with great difficulty) by linear methods over each (linear) regime of operation satisfying boundaries between regimes. However, as the order of the equation goes beyond three, the difficulty in matching boundaries becomes prohibitive. Often the second-order case is handled most conveniently by phase-plane graphical methods.

Linear Systems with Discontinuous Switching. These systems are characterized by switching operations. If switching occurs at a constant rate, then the system is the linear sampled-data system described above. If on the other hand the switching operation occurs whenever the signal (e.g., error) reaches a prescribed level of some function of the output, then the system is of the relay type and is in general nonlinear. Examples of this type are shown in Fig. 8.33. More generally, the switching points may be mixed functions of the input variable and its derivative; ϵ and $\dot{\epsilon}$ have a phase-plane representation as shown for the example in Fig. 8.33*e*. In addition to the behavior as a function of input, hysteresis, and dead zone, there are time lags inherent in operation because of inertia and inductance.

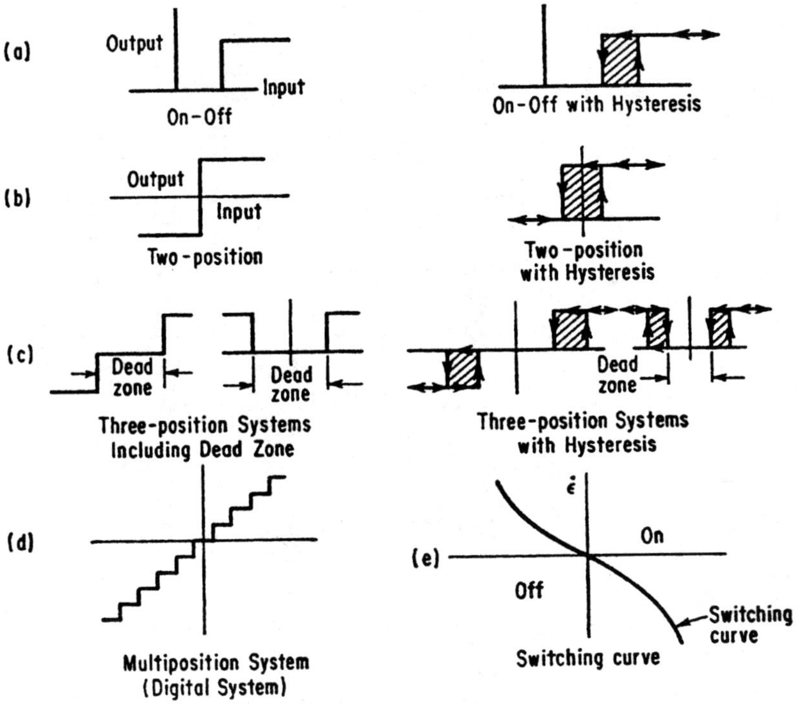

FIG. 8.33 Switching functions.

An illustration of a linear system with switching is a room heater whose block diagram is drawn in Fig. 8.34 with relay characteristics shown in Fig. 8.34*c* indicating that the furnace goes on whenever $T < T_1$ and off whenever $T > T_2$. The furnace and room are two first-order systems connected in tandem. Analytically, this result is expressed by two equations where $p = d/dt$:

$$(p + b)T = By + bT_0 \tag{8.171}$$

$$(p + a)y = Af(T, T_1, \dot{T})$$

where y = furnace output
T_o = outside temperature
$f(T, T_1, \dot{T})$ = switching function

Elimination of y in Eq. (8.171) yields

$$(p + a)(p + b)T = ABf(T, T_1, \dot{T}) + abT_o \tag{8.171a}$$

Note that the function f not only depends upon T and T_1 but also on the sign of $\dot{T}$ as follows:

$$f(T, T_1, \dot{T}) = \begin{cases} Q & T < T_1 \\ Q & T_1 < T < T_2, \dot{T} > 0 \\ 0 & T_1 < T < T_2, \dot{T} < 0 \\ 0 & T > T_2 \end{cases} \tag{8.172}$$

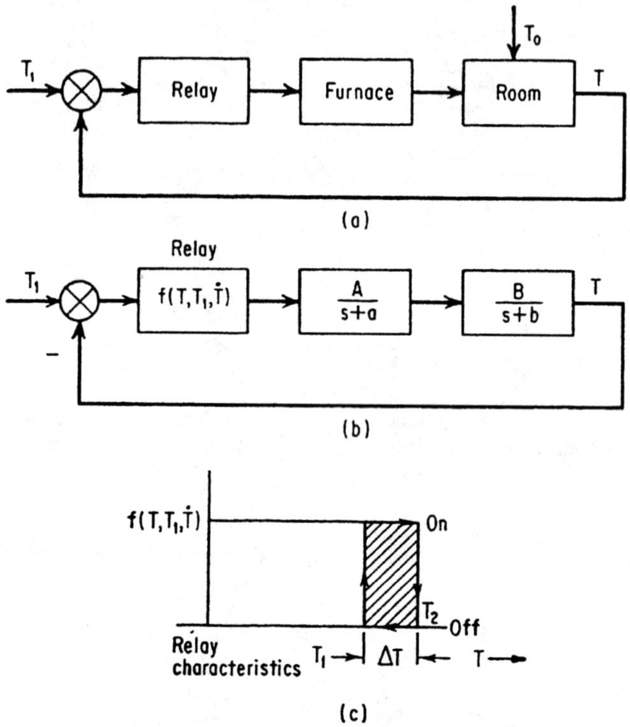

FIG. 8.34 Room-heater control system.

which constitutes four possible regimes of operation. Accordingly, Eq. (8.172) must be solved for the heating cycle $f(T, T_1, \dot{T}) = Q$ and the cooling cycle $f(T, T_1, \dot{T}) = 0$. For heating,

$$T_{\text{heating}} = \alpha e^{-at} + \beta e^{-bt} + T_\infty$$

where T_∞ is the asymptotic temperature for uncontrolled continuous heating. For cooling,

$$T_{\text{cooling}} = \gamma e^{-at} + \delta e^{-bt} + T_o$$

The four constants of integration, α, β, γ, and δ, must be determined by matching conditions at the transitions of any two regimes where continuity of T and its derivative $\dot{T}$ must be preserved. This becomes a most laborious procedure, since these constants change repeatedly, regime after regime, cycle after cycle. Only under constant load T_o will a cycle that is repetitive be eventually reached (limit cycle).

The prohibitive analytic method is seldom justified for second-order systems whose complete graphical solution in the phase plane can be easily generated. As an introduction to the method, consider the first-order temperature-control system

$$\dot{T} + \alpha T = f(T, T_1, \dot{T}) + \alpha T_o \tag{8.173}$$

with its characteristic switching function f. The two modes of operation for Eq. (8.173) are

Heating: $$\dot{T} + \alpha T = Q + \alpha T_o \stackrel{\Delta}{=} \alpha T_\infty$$

Cooling: $$\dot{T} + \alpha T = \alpha T_o$$

In the phase plane for both cases (taking one time derivative), assuming T_o constant,

$$\dot{T}(d\dot{T}/dT)\dot{T} = \ddot{T} = -\alpha\dot{T}$$
$$d\dot{T}/dT = -\alpha$$

The corresponding locus of phase-plane operation therefore consists of two parallel lines of operation of slope $-\alpha$ as drawn in Fig. 8.35 identified by their T intercepts. The limit cycle 1-2-3-4 is shown crosshatched. Jump action occurs at points 2 and 4, the switching points. For any initial $\dot{T}$, T to the left of $T = T_1$ shown for point Q, the point will jump vertically to the heating line and proceed to 3-4-1-2-3, closing the cycle. Similarly for point P at $T > T_2$, the point inside the zone $T_1 < T < T_2$ goes to the cooling curve first. The closed-cycle time is

$$\oint \frac{dT}{\dot{T}} = \int_{T_1}^{T_2} \frac{dT}{\dot{T}} + \int_{T_2}^{T_1} \frac{dT}{\dot{T}}$$

Returning to the second-order temperature system defined by Eq. (8.171a),

$$\ddot{T} + (a + b)\dot{T} + abT = ABf(T, T_1, \dot{T}) + abT_o$$

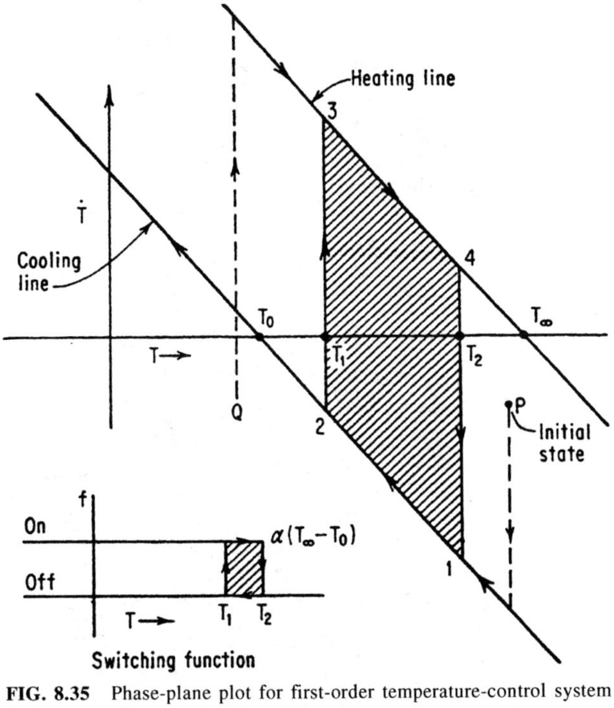

FIG. 8.35 Phase-plane plot for first-order temperature-control system $\dot{T} + \alpha T = f + \alpha T_0$; f = switching function.

By a suitable change of variable, Eq. (8.171) can always be represented as

$$\ddot{T} + a'\dot{T}' + T' = f'(T', T_1', \dot{T}) + T_o'$$

where primes have been appended to imply the transformation. Dropping these primes for convenience leaves

$$\ddot{T} + \alpha\dot{T} + T = f(T, T_1, \dot{T}) + T_o$$

whose representation for the phase-plane plot has been shown to be

$$\begin{aligned} d\dot{T}/dT &= (-\alpha\dot{T} - T + f + T_o)/\dot{T} \\ &= (-\alpha\dot{T} - T + T_\infty)/\dot{T} \qquad \text{(heating cycle)} \\ &= (-\alpha\dot{T} - T + T_o)/\dot{T} \qquad \text{(cooling cycle)} \end{aligned}$$

The phase-plane plot of this system is determined by first drawing the switching lines $T = T_1$ and $T = T_2$, and executing the Liénard construction for each of the two operating regimes. A limit cycle is reached as shown in Fig. 8.36.

Positioning Systems with Dead Zone. The phase-plane constructions for the following second-order positioning systems are shown in Fig. 8.37, each with dead zone:

Spring-mass damping: $\ddot{\theta} + c\dot{\theta} + \theta = -F \operatorname{sgn} \theta$ (8.174*a*)

$$\frac{d\dot{\theta}}{d\theta} = \frac{-c\dot{\theta} - F \operatorname{sgn} \theta - \theta}{\dot{\theta}} \qquad F = \begin{cases} 0 & |\theta| < \delta \\ 1 & |\theta| > \delta \end{cases} \tag{8.174b}$$

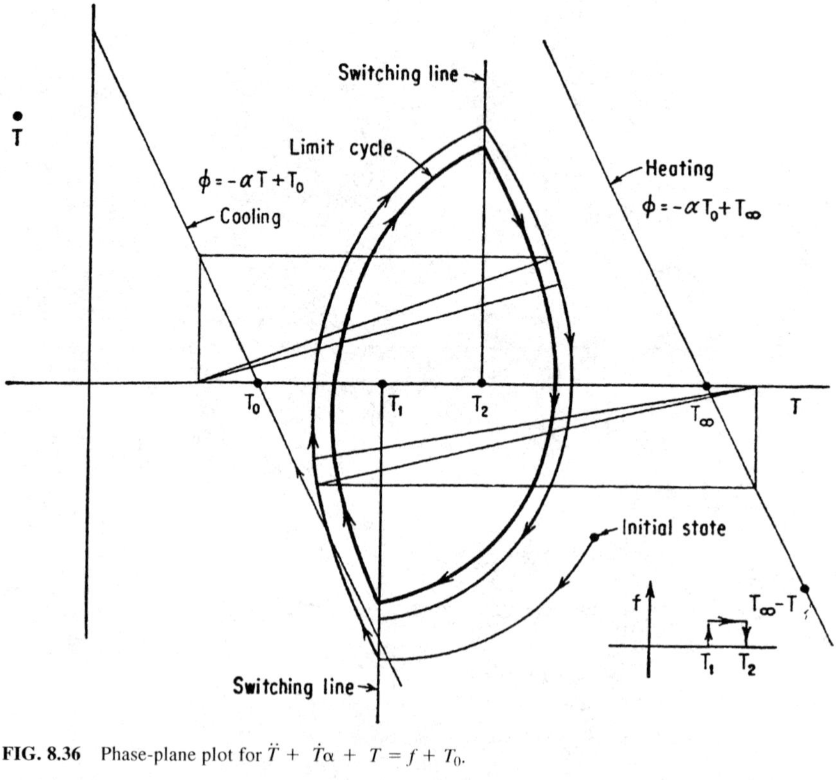

FIG. 8.36 Phase-plane plot for $\ddot{T} + \dot{T}\alpha + T = f + T_0$.

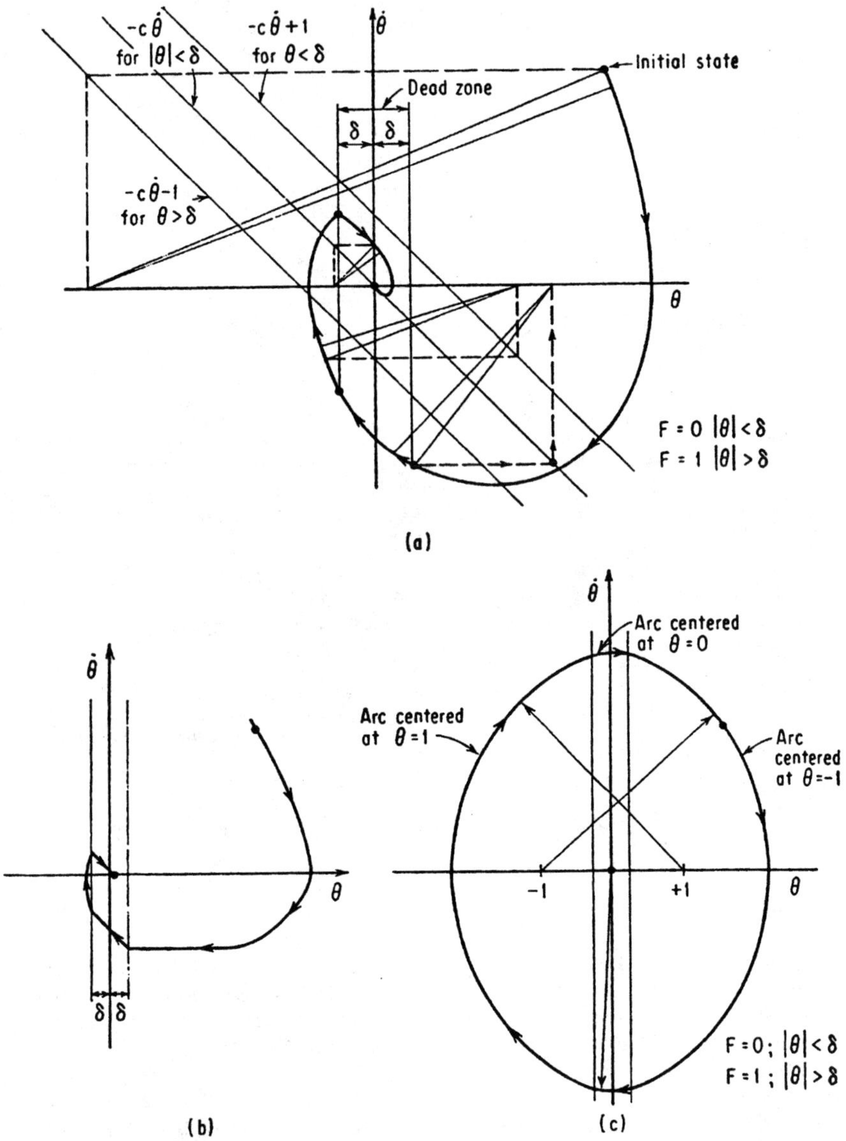

FIG. 8.37 Phase-plane plot for second-order systems.

(a) Spring-mass damping $\ddot{\theta} + c\dot{\theta} + \theta = -F \operatorname{sgn} \theta$

(b) Mass damping $\ddot{\theta} + c\dot{\theta} = -F \operatorname{sgn} \theta$

(c) Mass spring $\ddot{\theta} + \theta = -F \operatorname{sgn} \theta$

Mass damping:

$$\ddot{\theta} + c\dot{\theta} = -F \operatorname{sgn} \theta$$

$$\frac{d\dot{\theta}}{d\theta} = \frac{-c\dot{\theta} - F \operatorname{sgn} \theta}{\dot{\theta}} = -c - \frac{F \operatorname{sgn} \theta}{\dot{\theta}}$$

Mass spring:

$$\ddot{\theta} + \theta = -F \operatorname{sgn} \theta \tag{8.174c}$$

$$\frac{d\dot{\theta}}{d\theta} = \frac{-\theta - F \operatorname{sgn} \theta}{\dot{\theta}}$$

A first integral of Eq. (8.174*c*) gives

$$\dot{\theta}^2/2 + (\theta + F \operatorname{sgn} \theta)^2/2 = k_0^2$$

Indicating that the phase-plane plot consists entirely of arcs of circles centered at 0, −1, +1, as shown in Fig. 8.37.

Systems with Nonlinear Elements. Previously considered were linear systems made nonlinear by switching operations. These could be analyzed by classical analytical techniques. The situation with nonlinear systems is more difficult.

Systems of the second degree and lower can be treated by graphical analysis as shown for linear switching systems. Systems of higher order cannot be studied in general by analytic or graphical methods.

Examples of some usual types of nonlinear frequency-insensitive elements frequently occurring in "linear" systems are shown in Fig. 8.38.

An example of a nonlinear system controlled by a two-position force with dead zone is the second-order Coulomb damped system

$$\ddot{\theta} + c \operatorname{sgn} \dot{\theta} + \theta = F \operatorname{sgn} \theta$$

$$\dot{\theta}(d\dot{\theta}/d\theta) + \theta = F \operatorname{sgn} \theta - c \operatorname{sgn} \dot{\theta}$$

$$\dot{\theta}\, d\dot{\theta} + (\theta - F \operatorname{sgn} \theta) + c \operatorname{sgn} \dot{\theta})\, d\theta = 0$$

$$\dot{\theta}^2/2 + (\theta - F \operatorname{sgn} \theta) + c \operatorname{sgn} \dot{\theta})^2/2 = k^2$$

which describes circular arcs in the phase plane with six centers depending on the signs of θ and $\dot{\theta}$ and the amplitude of θ. Motion in the phase plane is shown in Fig. 8.39.

Describing-Function Analysis.[10] As pointed out above, higher-order nonlinear systems are not amenable to graphical analysis. A method of analysis has evolved which is valid under very restrictive conditions. It employs linear concepts in an attempt to simplify the complex nonlinear problem and bring it within the realm of analysis.

The analysis is limited to systems containing one nonlinearity or where many can be grouped to yield effectively one nonlinear and time-invariant block. For a sinusoidal input, the resultant output will be composed of the fundamental plus higher harmonics. The essence of the analysis is to ignore all harmonics other than the fundamental. This is the most restrictive assumption and can often be justified for slight nonlinearities where the higher harmonics are small to begin with; these are further attenuated since most systems are usually natural low-pass filters.

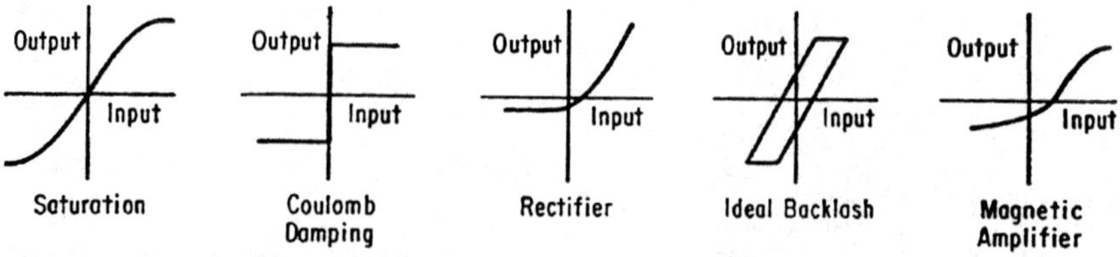

FIG. 8.38 Characteristic nonlinear elements.

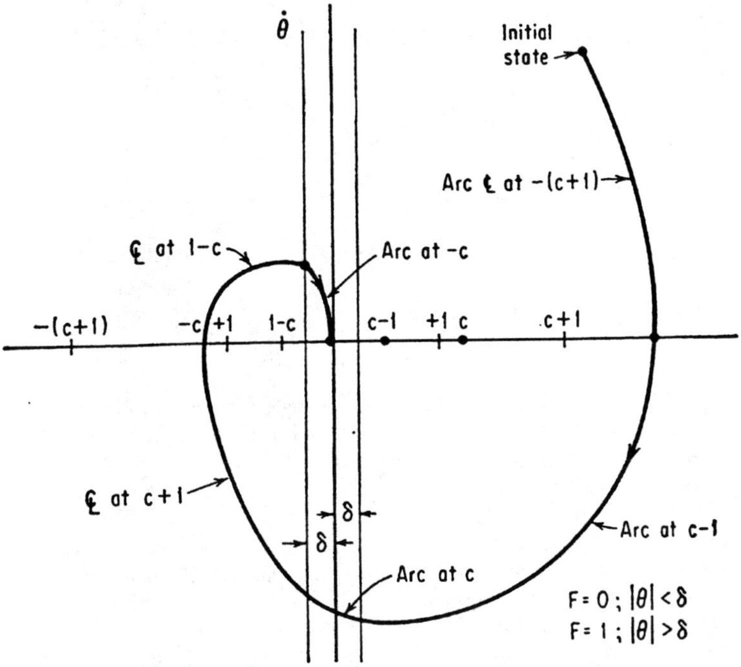

FIG. 8.39 Phase-plane plot for $\ddot{\theta} + c \operatorname{sgn} \dot{\theta} + \theta = -F \operatorname{sgn} \theta$.

Implementing the foregoing description, the object is to obtain for a fundamental input of amplitude A and frequency ω a Fourier series whose fundamental amplitude is $B(A, \omega)$ and phase $\phi(A, \omega)$. If a functional relation connects the input with the output, say

$$x_2 = f(x_1)$$

then for an input

$$x_1 = A \sin \omega t$$

the output is

$$x_2(A \sin \omega t) = f(A \sin \omega t)$$

which shows the x_2 is a periodic function of time, with period $2\pi/\omega$. It therefore has a Fourier series development ($\phi = \omega t$)

$$x_2(\mathrm{A} \sin \phi) = f(A \sin \phi)$$

$$x_2(A \sin \phi) = a_1 \sin \phi + a_2 \sin 2\phi + \cdots + a_n \sin n\phi$$
$$+ b_1 \cos \phi + b_2 \cos 2\phi + \cdots + b_n \cos n\phi$$

For an input $A \cos \phi$, the output would be obtained by adding $\pi/2$ to ϕ:

$$x_2(A \cos \phi) = x_2[A \sin (\phi + \pi/2)] = a_1 \cos \phi + \cdots - b_1 \sin \phi + \cdots$$

The amplitude ratio of the fundamental output to input is

$$|N| = \frac{(a_1^2 + b_1^2)^{1/2}}{A} \qquad \begin{aligned} a_1 &= a_1(A) \\ b_1 &= b_1(A) \end{aligned}$$

and the phase

$$\theta = \tan^{-1} (b_1/a_1)$$

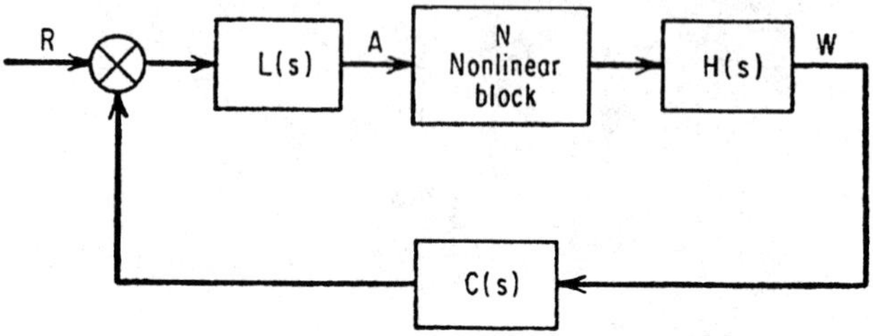

FIG. 8.40 Nonlinear control system.

By definition, the describing function

$$N = (a_1^2 + b_1^2)^{1/2}/A < \theta$$

is the complex ratio of the fundamental component of output to input. For nonlinear systems containing energy storage or dissipative elements the describing-function amplitude is a function not only of amplitude but of frequency as well, i.e.,

$$N = N(A, \omega) < (A, \omega)$$

Stability Analysis. A stability investigation of a system with one nonlinear block characterized by a describing function utilizes graphical techniques.

Consider, for example, the control system shown in Fig. 8.40. Assume a signal whose fundamental amplitude A impinges on the input to the nonlinear block. The transfer functions of the linear blocks have their arguments s replaced by $j\omega$ to obtain overall characteristics for real frequency. It should be noted here that right-half-plane poles no longer have meaning in the usual sense and interest must be necessarily restricted to sinusoidal signals owing to the describing-function definition. The loop gain at frequency ω is obtained by starting a signal amplitude A at zero phase at the input to the nonlinear device, and going completely around the loop giving

$$LHCN \times A$$

If this signal is 180° out of phase and greater than A, the amplitude grows; if it is less than A in amplitude, the signal decays. Grouping the linear blocks $LHC = G$ the condition for a net increase of signal is

Amplitude: $|G(j\omega)N(A, \omega)| > 1$

Argument: $G(j\omega)N(A, \omega) = (2n + 1)\pi \qquad n = \text{integer}$

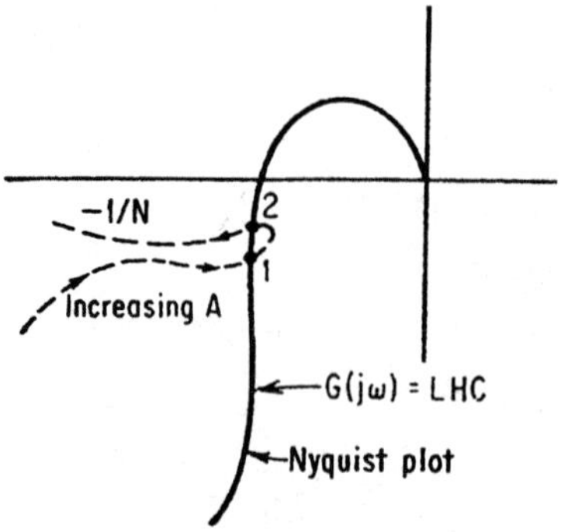

FIG. 8.41 Nyquist plot for nonlinear system.

A convenient way of illustrating this is indicated in Fig. 8.41, where first $G(j\omega)$ is plotted in a usual Nyquist plot. On the same set $-1/N$ is plotted (frequency-independent case) with the arrow in the direction of increasing amplitude A. Intersection for this case corresponds to sustained oscillations. At point 1 the frequency corresponding to the plot $G(j\omega)$ is an unstable point, since if perturbed in the direction where $-1/N > G(j\omega)$, A will decay. If perturbed in the opposite direction $-1/N < G(j\omega)$, the condition for increased growth of A, it will proceed to point 2, the stable oscillation point. If the $-1/N$ curve lies entirely within the $G(j\omega)$ plot, the amplitude

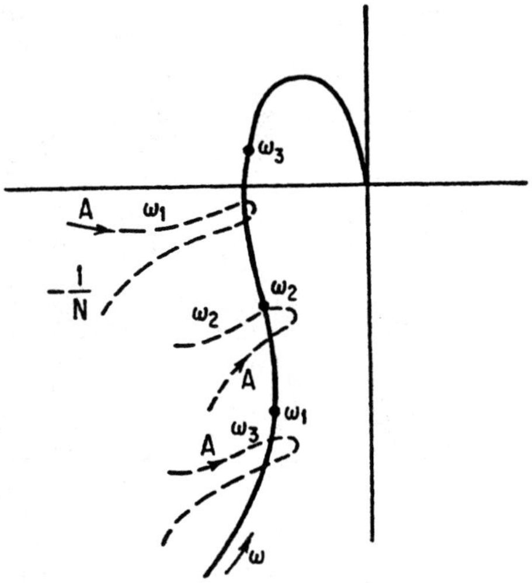

FIG. 8.42 Nyquist plot for nonlinear system where N is a function of frequency and amplitude.

will grow without bound, since $-1/N < G(j\omega)$ for all A; if outside, the system is unconditionally stable.

If N is a function of ω as well, then in the plot of $G(j\omega)$ and N as shown in Fig. 8.42 N is drawn for constant ω, e.g., ω_1, ω_2, ω_3. A qualitative analysis similar to the foregoing can be inferred. Stable or unstable points of intersection take on meaning only where there is correspondence of ω as well as amplitude as shown at ω_2.

8.6 SYSTEMS VIEWED FROM STATE SPACE

8.6.1 State-Space Characterization

The state-space description is a general time-domain representation of discrete systems that yield differential (or difference) equations, both linear and nonlinear. It is the basis of modern control theory and an outgrowth of modern computer technology with its well-known capabilities to solve systems, however formidable, often in real time. Moreover, its concise form is utilized in applications of methods of the calculus of variations for the optimal control of systems.

In its most general form the system is given by

$$\dot{\mathbf{x}}(t) = \mathbf{f}(\mathbf{x}(t), \mathbf{u}(t), t) \qquad \text{(plant equation)} \qquad (8.175a)$$

$$\mathbf{y}(t) = \mathbf{h}(\mathbf{x}(t), \mathbf{u}(t), t) \qquad \text{(output equation)} \qquad (8.175b)$$

where $\mathbf{x}(t)$ = state vector with n elements $x_i(t)$, $i = 1, 2, \ldots, n$
$\mathbf{u}(t)$ = input vector with m elements $u_i(t)$, $i = 1, 2, \ldots, m$
$\mathbf{y}(t)$ = output vector with r elements $y_i(t)$, $i = 1, 2, \ldots, r$
t = independent variable, time
$\mathbf{f}$ = general plant vector of n functions $f_i(\mathbf{x}, \mathbf{u}, t)$, $i = 1, 2, \ldots, n$
$\mathbf{h}$ = output vector of r functions $h_i(\mathbf{x}, \mathbf{u}, t)$, $i = 1, 2, \ldots, r$

The linear form of the system is written as

$$\dot{\mathbf{x}}(t) = \mathbf{A}(t)\,\mathbf{x}(t) + \mathbf{B}(t)\,\mathbf{u}(t) \tag{8.176a}$$

$$\mathbf{y}(t) = \mathbf{C}(t)\,\mathbf{x}(t) + \mathbf{D}(t)\,\mathbf{u}(t) \tag{8.176b}$$

where $\mathbf{A}(t) = n \times n$ system or plant matrix
$\mathbf{B}(t) = n \times m$ input matrix
$\mathbf{C}(t) = r \times n$ output matrix
$\mathbf{D}(t) = r \times n$ input-output coupling matrix

Equation (8.176*a*) is the set of n linear first-order differential equations and is referred to as the "plant equation." Equation (8.176*b*) is a set of r algebraic equations called the "output equation." We will confine our attention mostly to the special case of a linear time-invariant system, owing to its sufficiently useful representation for most systems and its analytic tractability; we thus focus on Eqs. (8.176*a*) and (8.176*b*) with constant matrices **A**, **B**, **C**, and **D**, or

$$\dot{\mathbf{x}}(t) = \mathbf{A}\mathbf{x}(t) + \mathbf{B}\mathbf{u}(t) \tag{8.176c}$$

$$\mathbf{y}(t) = \mathbf{C}\mathbf{x}(t) + \mathbf{D}\mathbf{u}(t) \tag{8.176d}$$

If an nth-order differential equation of a system is known, then an equivalent phase state set of equations can be determined. As an example, consider the equation of the special class of single-input–single-output (SISO) systems,

$$\frac{d^n y}{dt^n} + a_n \frac{d^{n-1} y}{dt^{n-1}} + a_{n-1} \frac{d^{n-2} y}{dt^{n-1}} + \cdots + a_1 y = Ku(t) \tag{8.177}$$

which is equivalent to a transfer function,

$$y(s)/u(s) = K/(s^n + a_n s^{n-1} + a_{n-1} s^{n-2} + \cdots + a_, s + a_1)$$

If it is assumed that

$$\begin{aligned} x_1 &= y \\ \dot{x}_1 &= x_2 \\ \dot{x}_2 &= x_3 \\ &\cdots\cdots \\ \dot{x}_{n-1} &= x_n \end{aligned}$$

then by repeated differentiation and substitution,

$$\dot{x}_n = d^n y/dt^n = -a_n x_n - a_{n-1} x_{n-1} - \cdots - a_1 x_1 + Ku(t)$$

In matrix form this is equivalent to

$$\dot{\mathbf{x}} = \begin{bmatrix} 0 & 1 & 0 & \cdots & 0 \\ 0 & 0 & 1 & \cdots & 0 \\ \cdots & \cdots & \cdots & \cdots & \cdots \\ 0 & 0 & 0 & \cdots & 1 \\ -a_1 & -a_2 & -a_3 & \cdots & -a_n \end{bmatrix} \mathbf{x} + \begin{bmatrix} 0 \\ 0 \\ 0 \\ \cdot \\ \cdot \\ \cdot \\ K \end{bmatrix} \mathbf{u} = \mathbf{A}\mathbf{x} + \mathbf{B}\mathbf{u} \tag{8.178a}$$

$$\mathbf{y} = [1 \quad 0 \quad 0 \quad \cdots \quad 0]\mathbf{x} = \mathbf{C}^T\mathbf{x} \tag{8.178b}$$

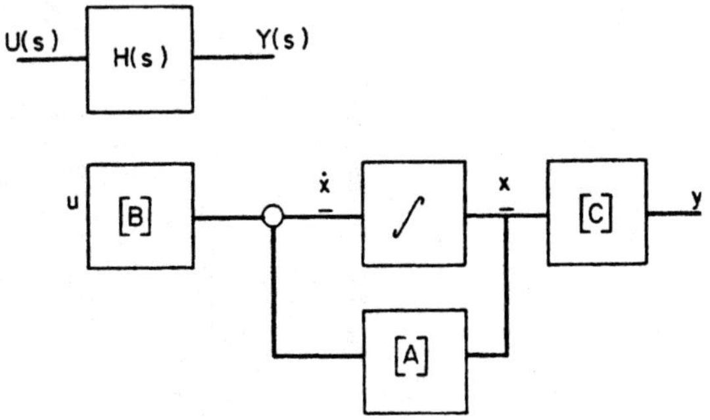

FIG. 8.43 Classical transfer function state-space diagram.

This equation constitutes the state-space set of the ordinary differential equations, Eqs. (8.176*c*) and (8.176*d*). The state variables so chosen, $x_i(t)$, $t = 1, 2, \ldots, n$, lead to a special matrix **A**, called the "companion matrix." It should be pointed out that the state-space set so chosen is not unique for the system, and indeed if one set is determined, then an infinite number of admissible sets exist. Figure 8.43 compares the classical transfer-function block diagram for this type system with the state-variable form. If derivatives of $\mathbf{u}(t)$ appear in the system equation, then the transfer function is given by

$$y(s)/u(s) = Kf_1(s)/f_2(s)$$

where $f_1(s)$ is now a polynominal in s. A selection of state variables as shown for the system of Eq. (8.177) would lead to derivative terms in Eqs. (8.178*a*) and (8.178*b*), thus violating its assumed form. In order to deal with the problem, another choice of state variables must be made. A convenient approach is to factor the transfer function into two parts, viz.,

$$H(s) = \frac{y(s)}{u(s)} = \frac{K_1 f_1(s)}{F_2(s)}$$

$$= \frac{x_1(s)}{u(s)}\frac{y(s)}{x_1(s)} = \frac{K_1}{f_2(s)} f_1(s)$$

And assign each of the factors as follows:

$$x_1(s)/u(s) = K_1/f_2(s) \qquad \text{and} \qquad y(s)/x_1(s) = f_1(s)$$

The transfer function between x_1 and u is now without numerator zero, and the choice of state variables is the same as before. The second transfer function is expanded from transfer-function to state-variable form, yielding

$$y(t) = c_m \frac{d^{m-1}x_1}{dt^{m-1}} + c_{m-1}\frac{d^{m-2}x_1}{dt^{m-2}} + \cdots + c_2\frac{dx_1}{dt} + c_1x_1 = c_1x_1 + c_2x_2 + \cdots + c_mx_m$$

where

$$f_1(s) = c_m s^{m-1} + c_{m-1}s^{m-2} + \cdots + c_2 s + c_1$$

$$f_2(s) = s^n + a_n s^{n-1} + \cdots + a_2 s + a_1 \qquad m \le n$$

In phase-state form

$$\dot{x}_1 = x_2$$
$$\dot{x}_2 = x_3$$
$$\cdots\cdots$$
$$\dot{x}_{n-1} = x_n$$

or

$$\dot{\mathbf{x}} = \mathbf{A}\mathbf{x} + \mathbf{B}u \qquad y = \mathbf{C}^T\mathbf{x}$$

where

$$\mathbf{A} = \begin{bmatrix} 0 & 1 & 0 & \cdots & 0 \\ 0 & 0 & 1 & \cdots & 0 \\ \cdots & \cdots & \cdots & \cdots & \cdots \\ 0 & 0 & 0 & \cdots & 1 \\ -a_1 & -a_2 & -a_3 & \cdots & -a_n \end{bmatrix} \qquad \mathbf{B} = \begin{bmatrix} 0 \\ 0 \\ 0 \\ \cdot \\ \cdot \\ \cdot \\ K \end{bmatrix}$$

$$\mathbf{C}^T = [c_1 \quad c_2 \quad c_3 \quad \cdots \quad c_m \quad 0 \quad \cdots \quad 0]$$

8.6.2 Transfer Function from State-Space Representation

Given the state variable representation

$$\dot{\mathbf{x}}(t) = \mathbf{A}\mathbf{x}(t) + \mathbf{B}\mathbf{u}(t)$$
$$\mathbf{y}(t) = \mathbf{C}\mathbf{x}(t) + \mathbf{D}\mathbf{u}(t)$$

and taking Laplace transforms of this set assuming zero initial conditions to determine the transfer function, we have

$$s\mathbf{x}(s) = \mathbf{A}\mathbf{x}(s) + \mathbf{B}\mathbf{u}(s)$$
$$\mathbf{y}(s) = \mathbf{C}\mathbf{x}(s) + \mathbf{D}\mathbf{u}(s)$$

The solution for $\mathbf{y}(s)$ is

$$\mathbf{y}(s) = [\mathbf{C}(s\mathbf{I} - \mathbf{A})^{-1}\,\mathbf{B} + \mathbf{D}]\mathbf{u}(s)$$

It is clear that the transfer function by its definitions is

$$\mathbf{H}(s) = \mathbf{C}(\mathbf{sI} - \mathbf{A})^{-1}\,\mathbf{B} + \mathbf{D}$$

For the single-input–single-output case, $\mathbf{H}(s)$ reduces to a scalar function, $H(s)$. The matrix $(s\mathbf{I} - \mathbf{A})^{-1}$ is referred to as the "resolvent matrix" and is designated as $\mathbf{\Phi}(s)$, i.e.,

$$\mathbf{\Phi}(s) = \text{adj}\,(\mathbf{sI} - \mathbf{A})/\det\,(\mathbf{sI} - \mathbf{A})$$

Using this notation the transfer function becomes

$$\mathbf{H}(s) = \mathbf{C}\mathbf{\Phi}(s)\mathbf{B} + \mathbf{D}$$

8.6.3 Phase-State Variable-Form Transfer Function: Canonical (Normal) Form

One of the most important representations of the state space is in the decoupled or normal form.

Given the transfer function

$$H(s) = \frac{y(s)}{u(s)} = K\frac{b_m s^m + b_{m-1}s^{m-1} + \cdots + b_0}{(s - \lambda_1)(s - \lambda_2)\cdots(-s - \lambda_n)} \qquad m \le n$$

for a siso system, where the denominator is shown in factored form; a partial fraction expansion yields.

$$\frac{y(s)}{u(s)} = \frac{c_1}{s - \lambda_1} + \frac{c_2}{s - \lambda_2} + \cdots + \frac{c_n}{s - \lambda_n}$$

From the definition

$$z_i(s) \stackrel{\Delta}{=} \frac{u(s)}{s - \lambda_i} \qquad i = 1, 2, \ldots, n$$

$y(s)$ becomes

$$y(s) = c_1 z_1(s) + c_2 z_2(s) + \cdots + c_n z_n(s)$$

In the time domain, the foregoing two expressions transform to

$$\dot{z}_i(t) = \lambda_i Z_i(t) + u(t) \qquad i = 1, 2, 3, \ldots, n$$
$$y(t) = c_1 z_1(t) + c_2 z_2(t) + \cdots + c_n z_n(t)$$

In normal form, the system takes the form

$$\dot{\mathbf{z}}(t) = \mathbf{\Lambda z}(t) + \mathbf{bu}(t)$$
$$y(t) = \mathbf{C}^T\mathbf{z}(t)$$

where

$$\mathbf{\Lambda} = \begin{bmatrix} \lambda & 0 & \cdots & 0 \\ 0 & \lambda_2 & \cdots & 0 \\ \cdots & \cdots & \cdots & \cdots \\ 0 & 0 & \cdots & \lambda_n \end{bmatrix} \qquad \mathbf{b} = \begin{bmatrix} 1 \\ 1 \\ \cdot \\ \cdot \\ \cdot \\ 1 \end{bmatrix} \qquad \mathbf{C} = \begin{bmatrix} c_1 \\ c_2 \\ \cdot \\ \cdot \\ \cdot \\ c_n \end{bmatrix}$$

where it is noted that $[\mathbf{\Lambda}]$ is in diagonal form, the matrix **b** elements are unity, and the matrix **C** elements are residues of the poles of $H(s)$.

As an example consider

$$y(s)/u(s) = H(s) = (s + 1)/(s^2 + 5s + 6)$$

After factoring we have

$$y(s)/u(s) = (s + 1)/(s + 3)(s + 2)$$

and the resulting partial-fraction expansion becomes

$$y(s)/u(s) = 2/(s + 3) - 1/(s + 2)$$

Choosing the Laplace transform of state variables and expanding, we find

$$z_1(s) = u(s)/(s + 3) \qquad sz_1(s) = -3z_1(s) + u(s)$$
$$z_2(s) = u(s)/(s + 2) \qquad sz_2(s) = -2z_2(s) + u(s)$$

Then $$y(s) = 2z_1(s) - z_2(s)$$

and the time-domain transformation is

$$\dot{z}_1(t) = -3z_1(t) + u(t)$$

$$\dot{z}_2(t) = -2z_2(t) + u(t) \qquad \dot{\mathbf{z}}(t) = \begin{bmatrix} -3 & 0 \\ 0 & -2 \end{bmatrix} \mathbf{z}(t) + \begin{bmatrix} 1 \\ 1 \end{bmatrix} u(t)$$

$$y(t) = 2z_1(t) - z_2(t)$$

where the plant matrix

$$\begin{bmatrix} -3 & 0 \\ 0 & -2 \end{bmatrix}$$

is seen to be diagonal.

As an example of a multiple-input–multiple-output system consider a system with two inputs and two outputs having a transfer function

$$\mathbf{H}(s) = \begin{bmatrix} \dfrac{1}{s+1} & \dfrac{2}{(s+1)(s+2)} \\ \dfrac{1}{(s+1)(s+3)} & \dfrac{1}{s+3} \end{bmatrix}$$

Expanding the elements into partial fractions, we rewrite $\mathbf{H}(s)$ as

$$\mathbf{H}(s) = \begin{bmatrix} \dfrac{1}{s+1} & \dfrac{2}{s+1} - \dfrac{2}{s+2} \\ \dfrac{1}{2(s+1)} - \dfrac{1}{2(s+3)} & \dfrac{1}{s+3} \end{bmatrix}$$

and the output transform is

$$\begin{aligned} y_1(s) &= \frac{u_1(s)}{s+1} + \frac{2}{s+1}u_2(s) - \frac{2}{s+2}u_2(s) \\ y_2(s) &= \frac{1}{2(s+1)}u_1(s) - \frac{2}{2(s+3)}u_1(s) + \frac{1}{s+3}u_2(s) \end{aligned} \tag{8.179}$$

Allowing the definitions for x_i and their transformations to the time domain,

$$x_1(s) = \frac{u_1(s)}{s+1} \qquad \dot{x}_1(t) = -x_1(t) + u_1(t)$$

$$x_2(s) = \frac{u_2(s)}{s+1} \qquad \dot{x}_2(t) = -x_2(t) + u_2(t)$$

$$x_3(s) = \frac{u_2(s)}{s+2} \qquad \dot{x}_3(t) = -2x_3(t) + u_2(t)$$

$$x_4(s) = \frac{u_1(s)}{s+3} \qquad \dot{x}_4(t) = -3x_4(t) + u_1(t)$$

$$x_5(s) = \frac{u_2(s)}{s+3} \qquad \dot{x}_5 = -3x_5(t) + u_2(t)$$

Now the output can be represented in both domains as

$$y_1(s) = x_1(s) + 2x_2(s) - 2x_3(s) \qquad y_1(t) = x_1(t) + 2x_2(t) - 2x_3(t)$$
$$y_2(s) = \tfrac{1}{2}x_1(s) - \tfrac{1}{2}x_4(s) + x_5(s) \qquad y_2(t) = \tfrac{1}{2}x_1(t) - \tfrac{1}{2}x_4(t) + x_5(t)$$

By direct substitution, the phase-state canonical form is

$$\dot{\mathbf{x}}(t) = \mathbf{A}\mathbf{x}(t) + \mathbf{B}\mathbf{u}(t)$$
$$\mathbf{y}(t) = \mathbf{C}\mathbf{x}(t)$$

$$\mathbf{A} = \begin{bmatrix} -1 & 0 & 0 & 0 & 0 \\ 0 & -1 & 0 & 0 & 0 \\ 0 & 0 & -2 & 0 & 0 \\ 0 & 0 & 0 & -3 & 0 \\ 0 & 0 & 0 & 0 & -3 \end{bmatrix}$$

$$\mathbf{B} = [1 \quad 1 \quad 1 \quad 1 \quad 1]$$

$$\mathbf{C} = \begin{bmatrix} 1 & 2 & -2 & 0 & 0 \\ -\frac{1}{2} & 0 & 0 & -\frac{1}{2} & 1 \end{bmatrix}$$

8.6.4 Transformation to Normal Form

We begin with the phase-state representation

$$\dot{\mathbf{x}}(t) = \mathbf{A}\mathbf{x}(t) + \mathbf{B}\mathbf{u}(t)$$
$$\mathbf{y}(t) = \mathbf{C}\mathbf{x}(t) \qquad (8.180)$$

which ignores direct coupling between input $\mathbf{U}(t)$ and output $\mathbf{y}(t)$. We next consider a matrix transformation $[\mathbf{P}]$:

$$\mathbf{x}(t) = \mathbf{P}\mathbf{z}(t)$$

After substitution and premultiplication by $\mathbf{P}^{-1}$ in Eq. (8.180),

$$\dot{\mathbf{z}}(t) = \mathbf{P}^{-1}\mathbf{A}\mathbf{P}\mathbf{z}(t) + \mathbf{P}^{-1}\mathbf{B}\mathbf{u}(\mathrm{t})$$
$$\mathbf{y}(t) = \mathbf{C}\mathbf{P}\mathbf{z}(t)$$

If $\mathbf{P}$ is chosen to render $\mathbf{z}(t)$ in normal form, then

$$\mathbf{P}^{-1}\,\mathbf{A}\mathbf{P} = \mathbf{\Lambda}$$

and from Eq. (8.107) $\mathbf{P}$ is determined following the solution to the eigenvector equation

$$|\mathbf{A} - \lambda_i\mathbf{I}| = 0$$
$$\mathbf{P} \triangleq \mathbf{c}_1 \quad \mathbf{c}_2 \quad \cdots \quad \mathbf{c}_n$$

where each $\mathbf{c}_i$ is an eigenvector corresponding to each eigenvalue λ_i. The eigenvector is given by any nonzero column of the matrix:

$$\operatorname{adj}(\mathbf{A} - \lambda_i\mathbf{I})$$

Also the matrix form of the eigenvalue equation is

$$\mathbf{A}\mathbf{P} = \mathbf{P}\mathbf{\Lambda}$$

as previously shown in the development following Eq. (8.107).

The special case of transforming **A** from companion form yields the **P** transformation:

$$\mathbf{P} = \begin{bmatrix} 1 & 1 & \cdots & 1 \\ \lambda_1 & \lambda_2 & \cdots & \lambda_n \\ \lambda_1^2 & \lambda_2^2 & \cdots & \lambda_n^2 \\ \cdots & \cdots & \cdots & \cdots \\ \lambda_1^n & \lambda_2^n & \cdots & \lambda_n^n \end{bmatrix}$$

8.6.5 System Response from State-Space Representation

The response to the homogeneous plant equation is

$$\dot{\mathbf{x}}(t) = \mathbf{A}\mathbf{x}(t) + \mathbf{B}\mathbf{u}(t) \tag{8.181}$$

which is akin to the solution of the first-order differential equation

$$\dot{x} = ax + bu(t)$$

whose general solution is, by Eq. (8.100)

$$x = e^{at}x(0) + b\int_0^t e^{a(b-\tau)}\,u(\tau)\,d\tau$$

Analogous to e^{at}, $e^{\mathbf{A}t}$ is defined by the infinite series

$$e^{\mathbf{A}t} = \mathbf{I} + \mathbf{A}t + \frac{\mathbf{A}^2t^2}{2!} + \cdots = \sum_{n=0}^{\infty}\frac{\mathbf{A}^n t^n}{n!}$$

which converges absolutely for $t < \infty$ and uniformly in any finite interval. Then the homogeneous equation

$$\dot{\mathbf{x}}(t) = \mathbf{A}\mathbf{x}(t) \tag{8.182}$$

has the solution

$$\mathbf{x}(t) = e^{\mathbf{A}t}\mathbf{x}(0) \tag{8.183}$$

$e^{\mathbf{A}t}$ has the property in this case to transform the state of $\mathbf{x}$ at $t = 0$ to the state at any future time $\mathbf{x}(t)$. Accordingly, it is denoted as the state-transition matrix $\mathbf{\Phi}(t)$. Under this designation Eq. (8.183) becomes

$$\mathbf{x}(t) = \mathbf{\Phi}(t)\mathbf{x}(0) \tag{8.183a}$$

If the state is known at $t = t_0$ [i.e., $\mathbf{x} = \mathbf{x}(t_0)$], then at time t

$$\mathbf{x}(t) = \mathbf{\Phi}(t - t_0)\mathbf{x}(t_0)$$

which is essentially the same as the statement of Eq. (8.183) and is the complete solution to Eq. (8.182).

A Laplace transformation of Eq. (8.182) yields

$$s\mathbf{x}(s) - \mathbf{x}(0) = \mathbf{A}\mathbf{x}(s)$$

After rearranging the premultiplying by $(s\mathbf{I} - \mathbf{A})^{-1}$ we have

$$s\mathbf{x}(s) + \mathbf{A}\mathbf{x}(s) = \mathbf{x}(0)$$

$$(s\mathbf{I} + \mathbf{A})\mathbf{x}(s) = \mathbf{x}(0)$$

$$\mathbf{x}(s) = (s\mathbf{I} + \mathbf{A})^{-1}\,\mathbf{x}(0) \tag{8.184}$$

where a Laplace transformation of Eq. (8.183*a*) establishes the identity

$$\mathscr{L}[\mathbf{\Phi}(t)] = \mathbf{\Phi}(s) = (sI + \mathbf{A})^{-1}$$

The total solution of the plant equation, Eq. (8.181), is

$$\mathbf{x}(t) = e^{\mathbf{A}t}\mathbf{x}(0) + \int_0^t e^{\mathbf{A}(t-\tau)}\mathbf{B}\mathbf{u}(\tau)\,d\tau$$

$$= \mathbf{\Phi}(t)\,\mathbf{x}(0) + \int_0^t \mathbf{\Phi}(t-\tau)\mathbf{B}\mathbf{u}(\tau)\,d\tau$$

a result previously obtained in Eq. (8.102).

8.6.6 State-Transition Matrix for Sampled Data Systems

The response to the system equation, Eq. (8.181), where $\mathbf{u}(t)$ is governed by a sampling process, is

$$\mathbf{x}(t) = \mathbf{\Phi}(t)\mathbf{x}(0) + \int_0^t \mathbf{\Phi}(t-\tau)\mathbf{B}\mathbf{u}_s(\tau)\,d\tau \tag{8.185}$$

Since $\mathbf{u}_s(t) = \mathbf{u}(nT)$ = the sampled input for $nT \le t \le (n+1)T$, where $n = 0, 1, 2, \ldots =$ sample number and T = sampling interval, Eq. (8.185) becomes

$$\mathbf{x}(t) = \mathbf{\Phi}(t)\mathbf{x}(0) + \sum_{j=0}^{n-1}\left[\int_{jT}^{(j+1)T} \mathbf{\Phi}(t-\tau)\mathbf{B}\mathbf{u}(jT)\,d\tau\right] + \int_{nt}^{t} \mathbf{\Phi}(t-\tau)\mathbf{B}\mathbf{u}(nT)\,d\tau \tag{8.186}$$

and since $\mathbf{\Phi}(t-\tau) = \mathbf{\Phi}(t)\,\mathbf{\Phi}(-\tau)$, Eq. (8.186) can also be written as

$$\mathbf{x}(t) = \mathbf{\Phi}(t)\mathbf{x}(0) + \mathbf{\Phi}(t)\left[\sum_{j=0}^{n-1}\int_{jT}^{(j+1)T} \mathbf{\Phi}(-\tau)\mathbf{B}\mathbf{u}(jT)\,d\tau\right] + \int_{nt}^{t} \mathbf{\Phi}(t-\tau)\mathbf{B}\mathbf{u}(nT)\,d\tau \tag{8.186a}$$

For notational convenience we define

$$\gamma \triangleq \sum_{j=0}^{n-1}\left[\int_{jT}^{(j+1)} \mathbf{\Phi}(-\tau)\mathbf{B}\mathbf{u}(jT)\,d\tau\right]$$

Then Eq. (8.186*a*) becomes

$$\mathbf{x}(t) = \mathbf{\Phi}(t)\mathbf{x}(0) + \mathbf{\Phi}(t)\gamma + \int_{nt}^{t} \mathbf{\Phi}(t-\tau)\mathbf{B}\mathbf{u}(nT)\,d\tau \tag{8.187}$$

If interest is focused on state values at sampling instants only (i.e., $t = nT$), then Eq. (8.185) becomes

$$\mathbf{x}(nT) = \mathbf{\Phi}(nT)\mathbf{x}(0) + \mathbf{\Phi}(nT)\boldsymbol{\gamma} \tag{8.188}$$

at $t = (n + 1)T$

$$\mathbf{x}[(n+1)T] = \mathbf{\Phi}[(n+1)T]\mathbf{x}(0) + \mathbf{\Phi}[(n+1)T]\boldsymbol{\gamma} + \int_{nT}^{(n+1)T} \mathbf{\Phi}[(n+1)T - \tau]\mathbf{Bu}(nT)\, d\tau$$

Factoring $\mathbf{\Phi}(T)$, we have

$$\mathbf{x}[(n+1)T] = \mathbf{\Phi}(T)[\mathbf{\Phi}(nt)\mathbf{x}(0) + \mathbf{\Phi}(nT)\gamma] + \int_{nT}^{(n+1)T} \mathbf{\Phi}[(n+1)T - \tau]\mathbf{Bu}(nT)\, d\tau$$

and replacing $\mathbf{\Phi}(nt)\mathbf{x}(0) + \mathbf{\Phi}(nt)\gamma$ with its equivalent in Eq. (8.188), $\mathbf{x}(nt)$, yields

$$\mathbf{x}[(n+1)T] = \mathbf{\Phi}(T)\mathbf{x}(nT) + \int_{nT}^{(n+1)T} \mathbf{\Phi}[(n+1)T - \tau]\mathbf{Bu}(nT)\, d\tau$$

The integral term can be simplified by a change of variable:

$$\gamma \triangleq (n+1)T - \tau$$

resulting in

$$\mathbf{x}[(n+1)T] = \mathbf{\Phi}(T)\mathbf{x}(nT) + \left[\int_0^T \mathbf{\Phi}(\lambda)\mathbf{B}\, d\lambda\right]\mathbf{u}(nT) \tag{8.188a}$$

where $\mathbf{u}(nT)$, being a constant, is moved outside the integral. The integral term is recognized as a constant matrix $\mathbf{F}$, defined as

$$\mathbf{F} \triangleq \int_0^T \mathbf{\Phi}(\lambda)\mathbf{B}\, d\lambda$$

and therefore Eq. (8.188a) becomes

$$\mathbf{x}[(n+1)T] = \mathbf{\Phi}(T)\mathbf{x}(nT) + \mathbf{Fu}(nT)$$

8.6.7 Time-Varying Linear Systems

The plant equation with time-dependent coefficients is written as

$$\dot{\mathbf{x}}(t) = \mathbf{A}(t)\,\mathbf{x}(t) + \mathbf{B}(t)\,\mathbf{u}(t) \tag{8.189}$$

Recalling Eq. (8.100a), the homogeneous matrix differential equation, we have

$$d\mathbf{Z}(t)/dt = \mathbf{A}(t)\mathbf{Z}(t) \tag{8.190}$$

with $\mathbf{Z}(t_0) = \mathbf{I}$ (the identity matrix).

The solution to the homogenous equation, Eq. (8.99) or Eq. (8.181), is

$$\dot{\mathbf{x}}(t) = \mathbf{A}(t)\,\mathbf{x}(t) \qquad \mathbf{x}(t) = \mathbf{Z}(t)\,\mathbf{x}(0) \tag{8.191}$$

and the general solution to Eq. (8.189) is given by

$$\mathbf{x}(t) = \mathbf{Z}(t)\mathbf{x}(t_0) + \int_0^t \mathbf{Z}(t)\mathbf{Z}^{-1}(\tau)\,\mathbf{B}(\tau)\mathbf{u}(\tau)\,\mathbf{d}\tau \tag{8.192}$$

Finally, using the notation for the transition matrix, we see that

$$\mathbf{\Phi}(t, t_1) = \mathbf{Z}(t)\,\mathbf{Z}^{-1}(t_1)$$

and Eq. (8.192) becomes

$$\mathbf{x}(t) = \mathbf{\Phi}(t, t_0)\mathbf{x}(t_0) + \int_0^t \mathbf{\Phi}(t, \tau)\,\mathbf{B}(\tau)\mathbf{u}(\tau)\,\mathbf{d}\tau \tag{8.193}$$

after $\mathbf{x}(t_0)$ is replaced in Eq. (8.192) with its identity $\mathbf{Z}^{-1}(t_0)\mathbf{x}(t_0)$, where $\mathbf{Z}(t_0) = \mathbf{Z}^{-1}(t_0) = \mathbf{I}$.

Unfortunately there is no general method to find $\mathbf{\Phi}(t, t_0)$ despite the mathematical compactness of form and similarity with the time-invariant case.

8.7 CONTROL THEORY[40,41,43–58,60,61]

8.7.1 Controllability

In order to control a plant, an input vector $\mathbf{u}(t)$ or sequence $u_i(t)$ must be determined to drive the system from its initial state $\mathbf{x}_0$ to a prescribed final state $\mathbf{x}_f$. If $\mathbf{u}(t)$ can thus be found, then the system is deemed "controllable."

The following is a more general definition of controllability: In a region $\mathcal{R}$ of state space, if the state of a system can be transformed from arbitrary state $\mathbf{x}_0$ at time t_0 to another arbitrary state $\mathbf{x}_f$ in finite time, then the system is completely controllable in that region. Clearly, systems in which elements of the state cannot be independently influenced by the input sector $\mathbf{u}$ are not controllable.

8.7.2 Observability

A closely allied concept is that if the system output is known over some finite time $\mathbf{y}(t_0, t)$, one can determine the state of the system $\mathbf{x}(t_0)$. If so, then the system is called "observable." Stated more precisely, a system is observable if its state at $t = t_0$ can be uniquely determined by observing its output over a finite interval of time $t_0 < t < \infty$.

The mathematical statements governing controllability and observability of the familiar linear time-invariant system, Eq. (8.176), are as follows. The system is controllable if the $n \times rn$ matrix formed by

$$\mathbf{B} \quad \mathbf{AB} \quad \mathbf{A}^2\mathbf{B} \quad \cdots \quad \mathbf{A}^{n-1}\mathbf{B}$$

has rank n, where $\mathbf{A} = n \times n$ matrix and $\mathbf{B} = r \times n$ matrix. The system is observable if the $n \times mn$ matrix

$$\overline{\mathbf{C}}^T \qquad \overline{\mathbf{A}}^T\overline{\mathbf{C}}^T \qquad (\overline{A}^2)^T\overline{\mathbf{C}}^T \cdots (\overline{A}^{n-1})^T\overline{\mathbf{C}} \tag{8.194}$$

has rank n, where $\mathbf{C} = n \times m$ matrix.

8.7.3 Introduction to Optimal Control

The primary objective of optimal control is to choose a function (policy) of one or more parameters such that some meaningful function of these parameters is rendered either a maximum or a minimum (i.e., an extremum).

An elementary example in rocket control would be to choose a thrust attitude and a fuel firing rate that will send a satellite into a prescribed orbit with minimum fuel expenditure; another control policy would allow the achievement of orbit in minimum time. The simplified equations governing rocket motion are given by

$$m\,d\mathbf{v}/dt = \mathbf{f} + \mathbf{f}_{\text{ext}}$$
$$\mathbf{f} = (dm/dt)\mathbf{c} = -\beta\mathbf{c} \tag{8.195}$$

where m = rocket mass
$\beta = -dm/dt$ = fuel burning rate
$\mathbf{c} = c_x, c_y$ = exhaust vector
$\mathbf{v} = v_x, v_y$ = velocity vector
$\mathbf{f}_{\text{ext}}$ = external forces (assumed for simplicity to be $(a - mg)$)

Figure 8.44 indicates the rocket trajectory. The equations in each coordinate are

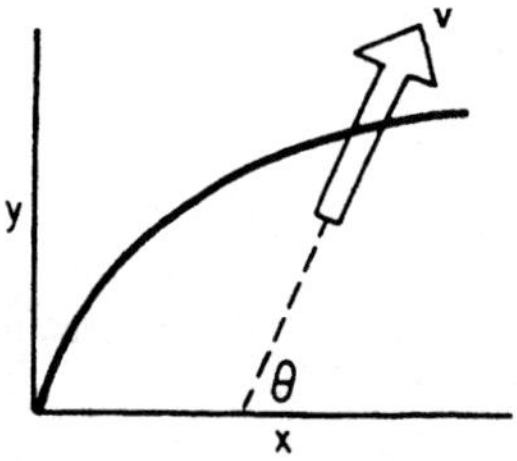

FIG. 8.44 Rocket trajectory.

$$m\,dv_x/dt = c_x\,dm/dt$$
$$m\,dv_y/dt = c_y\,dm/dt - mg$$

where θ = instantaneous rocket attitude angle. The problem here is to choose a firing rate β such that the fuel consumed is a minimum, i.e.,

$$\int_0^T \beta\,dt \text{ min}$$

where T = time for rocket to achieve its final height y.

Many cases of optimal control involve finding extremum values of integrals subject to constraints. The integrals are usually of the type

$$J = \int_A^B f\left(y, \frac{dy}{dx}, x\right) dx \tag{8.196a}$$

$$= \int_{t_0}^{t_1} f\left(x, \frac{dx}{dt}, t\right) dt \tag{8.196b}$$

Of all possible paths $y(x)$ in Eq. (8.196*a*) or $x(t)$ in Eq. (8.196*b*), some (usually one) make J (functional) an extremum.

If we consider the optimal control of a system whose plant equation is

$$\dot{\mathbf{x}} = \mathbf{f}(\mathbf{x}, \mathbf{u}, t)$$

then, in general, the control function $\mathbf{u}$ is chosen (assuming controllability) to transform the state from $\mathbf{x}(t_0)$ to a prescribed final state $\mathbf{x}_f = \mathbf{x}(T)$ such that the functional

$$J = \int_0^T f_0(\mathbf{x}, \mathbf{u}, t)\,dt$$

is rendered an extremum.

8.7.4 Euler-Lagrange Equation

The extremum of J for fixed end points a and b,

$$J = \int_a^b f(y, y', x)\,dx$$

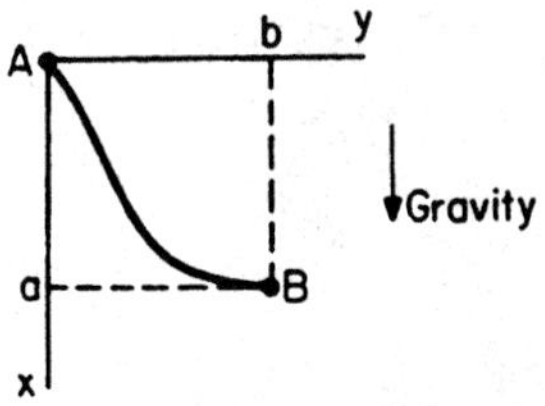

FIG. 8.45 Problem of mass from fixed end A to fixed end B.

corresponds to that path $y(x)$ which conforms to

$$\partial f/\partial y - (d/dx)/(\partial f'/\partial y') = \phi$$

which is the Euler-Lagrange equation taken from the "calculus of variations," a branch of mathematics.

One of the earliest problems posed by Bernoulli was to determine a curve joining two fixed end points A and B in a vertical plane (Fig. 8.45), on which a mass starting from rest at A sliding along that path under the force of gravity alone with reach B in minimum time. Since the velocity of a conservative system is

$$|\mathbf{v}| = (2gx)^{1/2}$$

the transit time is given by

$$J = \int_A^B \frac{ds}{(2gx)^{1/2}} = \int_A^B \left[\frac{(dy/dx)^2 + 1}{2gx}\right]^{1/2} dx$$

Moreover, the integrand does not contain terms in y, so the Euler-Lagrange equation reduces to $\partial f/\partial y' = c$, where c is a constant, or

$$y'/(1 + y'^2)^{1/2} = c$$

The solution in parametric form is

$$x = (1/2c^2)(1 - \cos\theta)$$
$$y = (1/2c^2)(\theta - \sin\theta)$$

which is a cycloid through points a and b, with c chosen to satisfy the end points.

Free-End Conditions. In the absence of end constraints, i.e., for free-end conditions, the extremizing problem is to find $y(x)$ when $a \leq x \leq b$, without any constraints on $y(a)$ and $y(b)$. The results are

$$\partial F/\partial y - (d/dx)(\partial F/\partial y') = 0$$

and

$$\partial F/\partial y' = 0 \qquad \text{at } x = a, b$$

Variable-End-Point Condition. The more general case of variability of one or both end points of the functional integral,

$$J = \int_{t_0}^{t_1} F(\mathbf{x}, \dot{\mathbf{x}}, t)\, dt$$

leads to the following generalized boundary condition at $t = t_1$:

$$\sum \frac{\partial F}{\partial \dot{x}_i} \delta x_i \bigg|_{t_1} + \left[F - \sum \dot{x}_i \frac{\partial F}{\partial \dot{x}_i}\right]\bigg|_{t_1} \partial t_1$$

Similarly at $t = t_0$

$$\sum \frac{\partial F}{\partial \dot{x}_i} \delta x_i \bigg|_{t_0} + \left[F - \sum \dot{x}_i \frac{\partial F}{\partial \dot{x}_i}\right]\bigg|_{t_0}^{\delta t_0}$$

If the end point is free, the δt_0 and $\delta x_i|_{t_0}$ are independent; then each of the coefficients of δx_i and δt_0 must be zero, or

$$\left.\frac{\partial F}{\partial X_i}\right|_{t_0} = 0$$

$$\left.\left[F - \sum \dot{x}_i \frac{\partial F}{\partial \dot{x}_i}\right]\right|_{t_0} = 0$$

If x_j, for example, is constrained to some curve at the right end point, then

$$x_j\Big|_{t_1} = y_j(t) \qquad j = 1, 2, \ldots, n$$

and

$$\delta X\Big|_{t_1} = \dot{y}_j \delta t_1 \qquad j = 1, 2, \ldots, n$$

$$F + \sum_{i=1}^{n} (\dot{y}_i - \dot{x}_i)\frac{\partial F}{\partial x} = 0$$

8.7.5 Multivariable with Constraints and Independent Variable *t*

Consider the functional

$$J = \int_a^b F(x_1, x_2, \ldots, x_n, \dot{x}_1, \dot{x}_2, \ldots, \dot{x}_n, t)\, dt \tag{8.197}$$

where $x_1, x_2, \ldots, x_n$ are independent functions of t and

$$\dot{x}_i = dx_i/dt$$

The extremal path now conforms to n Euler equations

$$\partial F/\partial x_i - (d/dt)(\partial F/\partial \dot{x}_i) = 0 \qquad i = 1, 2, \ldots, n$$

At any end point where x_i is free,

$$\partial F/\partial \dot{x}_i = 0$$

Now if one or more of the variables x_i are functionally related, say in m independent equations

$$g_i(\dot{\mathbf{x}}, \mathbf{x}, t) = 0 \qquad i = 1, 2, \ldots, m \tag{8.198}$$

we first form an augmented function

$$\hat{F} = F + \sum_{i=1}^{n} p_i g_i \tag{8.199}$$

where p_i is a Lagrange multiplier for the function g_i.

The augmented functional is next defined by

$$\hat{J} = \int_a^b \hat{F}\, dt \tag{8.200}$$

The phase form of Eq. (8.197) is

$$J = \int_a^b F(\mathbf{x}, \dot{\mathbf{x}}, t)\, dt \tag{8.201}$$

which is subject to the constraint equations

$$\mathbf{g}(\mathbf{x}, \dot{\mathbf{x}}, t) = 0$$

By substitution for F in Eq. (8.200), the augmented functional becomes

$$\hat{J} = \int_a^b [F(\mathbf{x}, \dot{\mathbf{x}}, t)\, dt - \mathbf{P}\mathbf{g}(\mathbf{x}, \dot{\mathbf{x}}, t)]\, dt$$

where

$$\mathbf{P} = \begin{bmatrix} p_1 & 0 & \cdots & 0 \\ 0 & p_2 & \cdots & 0 \\ \cdots & \cdots & \cdots & \cdots \\ 0 & 0 & \cdots & p_m \end{bmatrix}$$

The extremum for $\hat{J}$ is found by using the Euler-Lagrange equation and treating p_i, $i = 1, \ldots, m$, as m additional variables.

More generally, consider the functional J written as

$$J = \int_0^t F(\mathbf{x}, \mathbf{u}, t)\, dt$$

which is subject to differential constraints

$$\dot{x}_i = f_i(\mathbf{x}, \mathbf{u}, t) \qquad i = 1, 2, \ldots, n \tag{8.202}$$

Rearranging Eq. (8.202) results in

$$g_i = f_i - \dot{x}_i = 0 \qquad i = 1, 2, \ldots, m$$

Now replacing $f_i - \dot{x}_i$ for g_i in Eq. (8.199),

$$\hat{F} = F + \sum p_i\, (f_i - \dot{x}_i) \tag{8.203}$$

and

$$\hat{J} = \int_0^T \hat{F}\, dt = \int_0^T F + \sum p_i\, (f_i - \dot{x}_i)\, dt \tag{8.204}$$

For convenience the hamiltonian H is introduced, defined by

$$H \triangleq F + \sum p_i f_i$$

with this definition, Eq. (8.204) can be rewritten as

$$\hat{J} = \int_0^T H - \sum p_i\, \dot{x}_i\, dt$$

Invoking the Euler-Lagrange equation, we have

$$\partial\hat{F}/\partial x_i - (d/dt)(\partial\hat{F}/\partial\dot{x}_i) = 0 \qquad i = 1, 2 \ldots, n$$

or

$$\dot{p}_i = -\partial H/\partial x_i$$

and $$\partial H/\partial u_j = 0 \qquad j = 1, 2, \ldots, n$$

$$\partial H/\partial p_i = \dot{x}_i \qquad i = 1, 2, \ldots, n$$

yielding $2n + m$ equations in $2n + m$ unknowns.

The boundary conditions for $x_i(0)$, $i = 1, \ldots, n$ are the state at $t = 0$. If the control problem requires that, say, only r final state values be met, then values of $x_l(T)$, $l > r$, are free, yielding the free-end-point conditions

$$\frac{\partial F}{\partial \dot{x}_k} = 0 \qquad \begin{matrix} k = r + 1, \ldots, n \\ t = T \end{matrix}$$

which coincides with

$$p_k(T) = 0 \qquad k = r + 1, \ldots, n$$

As an example, consider the control which minimizes $\int_0^T u^2\,dt$ for the system given by

$$\begin{matrix} \dot{x}_1 = x_2 & x_1(0) = 0, x_1(T) = 1 \\ \dot{x}_2 = u & x_2(0) = 0, x_2(T) = 0 \end{matrix}$$

which describes a control force u accelerating a unit inertial mass starting at rest moving to a unit position at zero velocity at $t = T$.

We first form the hamiltonian

$$H = u^2 + p_1x_2 + p_2u$$

Now for $\partial H/\partial u_j = 0$

$$2u + p_2 = 0 \quad \text{or} \quad u = -\tfrac{1}{2}p_2$$

and for $-\partial H/\partial x_j = \dot{p}_j$

$$\dot{p}_1 = 0 \quad \text{and} \quad \dot{p}_2 = -p_1$$

From the foregoing three equations we can establish that

$$p_1 = \text{const} = -C \qquad \dot{p}_2 = \text{const} = -p_1 = C \tag{8.205}$$

A first integral of Eq. (8.205) yields

$$p_2 = Ct + D$$

Since $u = -\frac{1}{2}p_2$,

$$u = -\tfrac{1}{2}Ct - \tfrac{1}{2}D$$

Since $\dot{x}_2 = u$,

$$x_2 = -\tfrac{1}{4}Ct^2 - \tfrac{1}{2}Dt + E$$

From the boundaries $x_2(T) = 0$ and $x_2(0) = 0$,

$$x_2 = \tfrac{1}{4}C(-t^2 + Tt)$$

$$D = -\tfrac{1}{2}CT$$

$$E = 0$$

and
$$\dot{x}_1 = x_2 \qquad x_1(0) = 0$$
$$x_1 = \tfrac{1}{4}C(-\tfrac{1}{3}t^3 + \tfrac{1}{2}Tt^2)$$

since
$$x_1(T) = 1$$

The required u is

$$u = (12/T^3)(-t + \tfrac{1}{2}T)$$

and elements of the state vector are

$$x_2 = (6/T^3)(-t^2 + Tt) \qquad x_1 = (6/T^3)(-\tfrac{1}{3}t^3 + \tfrac{1}{2}Tt^2)$$

8.7.6 Pontryagin's Principle

Again we consider

$$J = \int_0^t F(\mathbf{x}, \mathbf{u}, t)\, dt$$

Subject to

$$\dot{x}_i = f_i(\mathbf{x}, \mathbf{u}, t) \qquad i = 1, 2, \ldots, n$$

with initial conditions

$$\mathbf{x}(0) = \mathbf{x}_0$$

and end conditions

$$\mathbf{x}(T) = \mathbf{x}_f$$

where $\mathbf{u} \in U$ is the admissible control region.

Now if $\mathbf{u}$ is discontinuous (the general case), $\partial H/\partial u_j$ does not exist at the discontinuities and $\partial H/\partial u_j = 0$, one of the criteria for optimal control of ∂u_j continuous inputs, is no longer valid. In lieu of this condition, u_j, $j = 1, \ldots, n$, are determined to minimize H. This is the fundamental contribution of Pontryagin; the theory admits not only continuous control but bounded and discontinuous controls as well.

As an example consider the system of the previous problem with the functional

$$J = \int_0^T dt$$

being the time it takes to move the mass from $x_1(0) = 0$ to $x_1(T) = 1$ in minimum time subject to the bounds on u:

$$-\alpha < u < \beta$$

We first form the hamiltonian

$$H = 1 + p_1 x_2 + p_2 u$$

and as determined previously,

$$\dot{p}_1 = 0$$
$$-p_1 = \dot{p}_2$$
$$C = \dot{p}_2$$
$$p_2 = Ct + D$$

In accordance with the Pontryagin principle, if

$$p_2 < 0 \qquad u = +\beta$$

and if

$$p_2 > 0 \qquad u = -\alpha$$

Thus p_2u and H are made as small as possible.

Since

$$\dot{x}_2 = u \tag{8.206}$$

$$x_2(T) = \int_0^T u\,dt = 0$$

It appears from Eq. (8.206) that u has at least one positive region and at least one negative region since its total integration over the interval 0, T, is 0. Further, for every sign change of u there must be a sign change of p_2. Now, because it is a linear function of time, p_2 can have at most one change of sign. It follows that there is a single crossover. This fact, considered in relation to Eq. (8.206), leads to a crossover time t_1 derived from

$$\beta t_1 = (T - t_1)\alpha \tag{8.207}$$

Equating the integral from 0 to T with $x_1(T)$ yields

$$x_1(T) = \int_0^T x_2\,dt = 1 \tag{8.208}$$

There are two regimes for x_2, namely,

$$x_2 = \begin{cases} \displaystyle\int_0^t \beta\,dt = \beta t & t \le t_1 \\ \displaystyle\int_{t_1}^t -\alpha\,dt = \beta t_1 - \alpha(t - t_1) & t \ge t_1 \end{cases}$$

Direct substitution in Eq. (8.208) over the whole range 0, T yields

$$\int_0^T x_2 dt = \beta t_1^2/2 + \beta^2 t_1^2/2\alpha = 1 \tag{8.209}$$

Figure 8.46 shows the x_2 and u functions.

From Eqs. (8.207) and (8.209) we obtain the crossover time

$$t_1 = \left[\frac{2}{\beta(1 + \beta/\alpha)}\right]^{1/2}$$

and the excursion time

$$T = [(2/\beta)(1 + \beta/\alpha)]^{1/2}$$

which is a minimum.

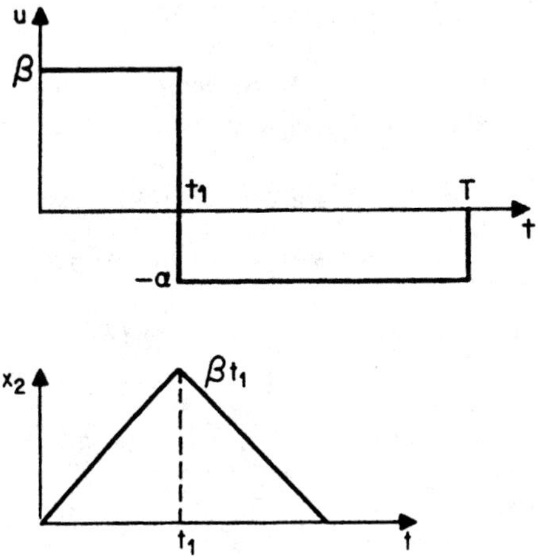

FIG. 8.46 Relationship between functions x_2 and u.

REFERENCES

1. Miller, F. H.: "Partial Differential Equations," John Wiley & Sons, Inc., New York, 1941.
2. Courant, R., and D. Hilberg: "Methods of Mathematical Physics," vol. I, Interscience Publishers, Inc., New York, 1953.
3. Frank, P. L., and R. von Mises: "Differential and Integral Gleichungen der Mechanik und Physik," Friedr. Vieweg & Sohn, Germany, 1935.
4. Sagan, H.: "Boundary and Eigenvalue Problems in Mathematical Physics," John Wiley & Sons, Inc., New York, 1961.
5. Webster, A. G.: "Partial Differential Equations," S. B. Teubner, Leipzig, 1927.
6. Lamb, H.: "Hydrodynamics," Cambridge University Press, New York, 1932.
7. Struble, R. A.: "Nonlinear Differential Equations," McGraw-Hill Book Company, Inc., New York, 1962.
8. Cunningham, W. J.: "Introduction to Nonlinear Analysis," McGraw-Hill Book Company, Inc., New York, 1958.
9. Stoker, J. J.: "Nonlinear Vibrations," Interscience Publishers, Inc., New York, 1950.
10. Truxal, J. G.: "Automatic Feedback Control System Synthesis," McGraw-Hill Book Company, Inc., New York, 1955.
11. Coddington, E. A., and N. Levinson: "Theory of Ordinary Differential Equations," McGraw-Hill Book Company, Inc., New York, 1955.
12. Kumke, E.: "Differential Gleichungen reeler Funktionen," Akademische Verlagsgesellschaft Geest & Portig KG, Leipzig, 1930.
13. Lepschetz, S.: "Lectures on Differential Equations," Princeton University Press, Princeton, N.J., 1946.
14. Kaplan, W.: "Ordinary Differential Equations," Addison-Wesley Publishing Company, Inc., Reading, Mass., 1958.
15. Bellman, R. E.: "Stability Theory of Differential Equations," McGraw-Hill Book Company, Inc., New York, 1953.
16. Papoulis, A.: "The Fourier Integral and Its Applications to Linear Systems," McGraw-Hill Book Company, Inc., New York, 1962.
17. Carslaw, H. S.: "Introduction to the Theory of Fourier Series and Integrals," 3d ed., Dover Publications, Inc., New York, 1930.

18. Churchill, R. V.: "Operational Mathematics," 2d ed., McGraw-Hill Book Company, Inc., New York, 1944.

19. Weber, E.: "Linear Transient Analysis," vol. II, John Wiley & Sons, Inc., New York, 1956.

20. Gardner, M. F., and J. S. Barnes: "Transients in Linear Systems," vol. I, John Wiley & Sons, Inc., New York, 1942.

21. Goldman, S.: "Transformation Calculus and Electrical Transients," Prentice-Hall, Inc., Englewood Cliffs, N.J., 1949.

22. Cheng, D. L.: "Analysis of Linear Systems," Addison-Wesley Publishing Company, Inc., Reading, Mass., 1959.

23. Pipes, L. A.: "Applied Mathematics for Engineers and Physicists," 2d ed., McGraw-Hill Book Company, Inc., New York, 1958.

24. Pfeiffer, P. E.: "Linear Systems Analysis," McGraw-Hill Book Company, Inc., New York, 1961.

25. Scott, E. J.: "Transform Calculus with an Introduction to Complex Variables," Harper & Row, Publishers, Incorporated, New York, 1955.

26. Evans, W. R.: "Control-system Dynamics," McGraw-Hill Book Company, Inc., New York, 1954.

27. Solodovnickoff, V. V.: "Introduction to the Statistical Dynamics of Automatic Control Systems," Dover Publications, Inc., New York, 1960.

28. Tsien, H. S.: "Engineering Cybernetics," McGraw-Hill Book Company, Inc., New York, 1954.

29. Seifert, W. W., and C. W. Steeg: "Control Systems Engineering," McGraw-Hill Book Company, Inc., New York, 1960.

30. Smith, O. J. M.: "Feedback Control Systems," McGraw-Hill Book Company, Inc., New York, 1958.

31. Chang, S. S.: "Synthesis of Optimum Control Systems," McGraw-Hill Book Company, Inc., New York, 1961.

32. Kipiniak, W.: "Dynamic Optimization Use Control," John Wiley & Sons, Inc., New York, 1961.

33. Gibson, J. E. (ed.): "Proceedings of Dynamic Programming Workshop," Purdue University, 1961.

34. Ragazzini, J. R., and Gene F. Franklin: "Sampled-data Control Systems," McGraw-Hill Book Company, Inc., New York, 1958.

35. Jury, E. J.: "Sampled Data Control Systems," John Wiley & Sons, Inc., New York, 1958.

36. McRuer, Graham D.: "Analysis of Nonlinear Control Systems," John Wiley & Sons, Inc., New York, 1961.

37. Cosgriff, R. L.: "Nonlinear Control Systems," McGraw-Hill Book Company, Inc., New York, 1958.

38. Ku, Y. H.: "Analysis and Control of Nonlinear Systems," The Ronald Press Company, New York, 1958.

39. Setov, J.: "Stability in Nonlinear Control Systems," Princeton University Press, Princeton, N.J., 1961.

40. Anand, D. K.: "Introduction to Control Systems," Pergamon Press, New York, 1974.

41. Berkowitz, L. D.: "Optimal Control Theory," Springer-Verlag, Berlin, 1974.

42. Blackman, P. F.: "State Variable Analysis," the Macmillan Company, New York, 1977.

43. Brewer, J. W., "Control System Analyses, Design and Simulation," Prentice-Hall, Inc., Englewood Cliffs, N.J., 1974.

44. Brogan, W. L.: "Modern Control Theory," Quantum Publishers, New York, 1974.

45. Bergess, T., and H. Graham: "Introduction to Control Theory Including Optimal Control," John Wiley & Sons, Inc., New York, 1980.

46. Clark, R. N.: "Introduction to Automatic Control Systems," John Wiley & Sons, Inc., New York, 1966.

47. Fortmann, T. E., and Hitz, K. L.: "An Introduction to Linear Control Systems," Marcel Dekker, New York, 1977.

48. Jacobs, O. L.: "Introduction to Control Theory," Clarendon Press, New York, 1974.

49. Kirk, D. E.: "Optimal Control Theory," Prentice-Hall, Inc., Englewood Cliffs, New Jersey, 1970.

50. Layton, J. M.: "Multivariable Control Theory," Peter Perengrenus, New York, 1976.

51. Marshall, S. A.: "Introduction to Control Theory," The Macmillan Company, New York, 1978.

52. Schultz, D. G., and J. L. Melsa: "State Functions and Linear Control Systems," McGraw-Hill Book Company, Inc., New York, 1967.

53. Ogata, K.: "Modern Control Engineering," Prentice-Hill, Inc., Englewood Cliffs, N.J. 1970.

54. Owens, D. H.: "Feedback and Multivariable Systems," Peter Perengrenus, New York, 1978.

55. Pontryagin, L. S., et al.: "The Mathematical Theory of Optimal Processes," John Wiley & Sons, Inc., New York, 1982.

56. Power, H. M. and R. J. Simpson: "Introduction to Dynamics and Control," McGraw-Hill Book Company, Inc., New York, 1978.

57. Prime, H. A., "Modern Concepts in Control Theory," McGraw-Hill Book Company., Inc., New York, 1967.

58. Sage, A. P.: "Optimum System Control," Prentice-Hall, Inc., Englewood Cliffs, N.J., 1968.

59. Timothy, L. K. and B. E. Bona: "State Space Analysis," McGraw-Hill Book Company, Inc., New York, 1968.

60. Truxal, J. G.: "Introductory Systems Engineering," McGraw-Hill Book Company, Inc., New York, 1972.

61. Uolovich, W. A.: "Linear Multivariable Systems," Springer-Verlag, Berlin, 1974.

62. Zadeh, L. A., and C. A. Desoer: "Linear System Theory: The Static Space Approach," McGraw-Hill Book Company, Inc., New York, 1963.

CHAPTER 9
CONTINUOUS TIME CONTROL SYSTEMS

Andrew P. Sage, Ph.D.
Associate Vice President for Academic Affairs
George Mason University
Fairfax, Va.

9.1 INTRODUCTION 9.1
9.2 MATHEMATICAL BACKGROUND FOR LINEAR CONTROL SYSTEMS 9.2
9.3 SYSTEM CONCEPTS 9.4
9.4 LINEAR CONTROL SYSTEM DESIGN BY BODE DIAGRAMS 9.15
9.4.1 Gain Reduction 9.17
9.4.2 Phase-Lead Compensation 9.19
9.4.3 Phase-Lag Compensation 9.20
9.4.4 Compensation Using Lag-Lead Networks 9.23
9.5 THE ROOT LOCUS METHOD 9.24

9.1 INTRODUCTION

The developments of this chapter are based on the previous work of the author in Ref. 12. Reference 16 is an encyclopedia of control systems. Both provide more detail concerning the topics discussed. References 3 through 17 give more on the subject.

We encounter many feedback processes in our daily lives. A very simple example of a feedback control system is that of a person attempting to adjust the speed of an automobile. First, it is necessary to determine an *objective function*. This is a function of environmental conditions and our estimate of the time that we wish to take in going somewhere. There are constraints as well, with the most significant constraint in this case being the speed limit that has been legally imposed for safety reasons. There exists a *control input*. In this case, it is the position of the driver's foot on the accelerator. The motion of the automobile is some rather complex dynamic function of the mechanics of the automobile. It is important to note that there is some *error detection* device involved. In this particular case, we sense the difference between the *desired speed* and the *actual speed* and generate an *error signal*, which represents the difference between the desired velocity of the automobile and the actual velocity. We adjust the position of the accelerator in accordance with this error signal. If the velocity is too low, we sense a positive error in that the desired velocity is greater than the actual velocity. The control input to the velocity of the automobile is such as to increase the actual velocity. Just the opposite is true if the actual velocity of the automobile is greater than the desired velocity. Figure 9.1 presents a block diagram that is representative of the system that we are describing.

We note the presence of *feedback* in this system. This occurs because of the use of an error detection mechanism and the fact that the control is driven by the signal at the output of the error detector. If we did not have this error detector, we would have an *open-loop* system.

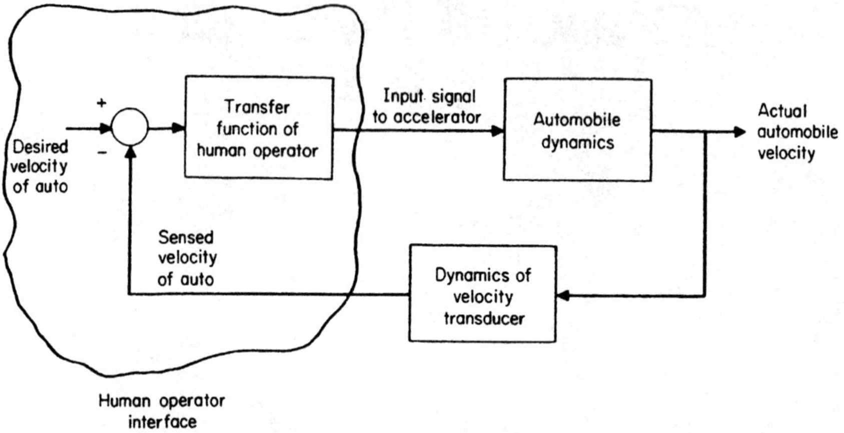

FIG. 9.1 Conceptual block diagram of human-machine interface in control of car speed.

Another common characteristic of a physical control system is the presence of amplification of power. In this case, we sense the error between desired and actual velocity and need only use a very small amount of energy or power in order to adjust the velocity. The advantages of having accuracy dependent upon accuracy of the error detection mechanism, rather than accuracy of knowledge of the dynamics of the entire process, and the fact that the effects of noise, disturbances, and modeling errors are ameliorated through closed-loop control, are compelling reasons to use it for many applications.

9.2 MATHEMATICAL BACKGROUND FOR LINEAR CONTROL SYSTEMS

Consider the linear constant-coefficient differential equation

$$\sum_{i=0}^{n} a_i \frac{d^i z(t)}{dt^i} = \sum_{i=0}^{n} \frac{d^i u(t)}{dt^i} \tag{9.1}$$

Its Laplace transform for the system initially unexcited is

$$\sum_{i=0}^{n} a_i s^i Z(s) = \sum_{i=0}^{m} b_i s^i U(s)$$

Thus, we get

$$Z(s) = H(s)U(s)$$

where

$$H(s) = \sum_{i=0}^{m} b_i s^i \Big/ \sum_{i=0}^{n} a_i s^i \tag{9.2}$$

is called the system transfer function.

Systems are often constructed by interconnecting subsystems. It will be of considerable value, therefore, to be able to have a convenient method to cope mathematically with two or more coupled subsystems. We must be very careful to describe precisely what we mean by the interconnection of systems; the easiest way to accomplish this is to use a differential equation representation of the two subsystems. We assume that

$$\sum_{i=0}^{n_1} a_{1i} \frac{d^i z_1}{dt^i} = \sum_{i=0}^{m_1} b_{1i} \frac{d^i u_1}{dt^i} \tag{9.3}$$

describes the first system. The second system is assumed to be described by

$$\sum_{i=0}^{n_1} a_{2i} \frac{d^i z_2}{dt^i} = \sum_{i=0}^{m_2} b_{2i} \frac{d^i u_2}{dt^i} \tag{9.4}$$

The systems are assumed to be interconnected such that the output from the first system serves as the input to the second system:

$$u_2 = z_1 \tag{9.5}$$

There are some physical requirements, involving such things as nonloading considerations, which must be imposed to ensure that the output (z_1) of the first differential equation can be input to the second differential equation (u_2) without altering the physical characteristics and hence the differential equation of the first system.

We determine the transfer function of the individual subsystems. We have from Eqs. (9.3) through (9.5)

$$\begin{aligned}
Z_1(s)/U_1(s) &= Q_1(s)/P_1(s) = H_1(s) \\
Z_2(s)/U_2(s) &= Q_2(s)/P_2(s) = H_2(s) \\
U_2(s) &= Z_2(s) \\
\frac{Z_2(s)}{U_1(s)} &= \frac{Z_2(s)/U_2(s)}{U_1(s)/Z_1(s)} = \frac{Z_2(s)}{U_2(s)}\frac{Z_1(s)}{U_1(s)} = H_2(s)H_1(s)
\end{aligned}$$

and thus we see that the overall transfer function is just the product of the transfer function of the individual parts of the system.

Two other elements are needed in order to complete our description of fundamental linear block diagram elements, the summer and the pickoff point. We usually place signs alongside a summer to denote whether inputs to the summer are added or subtracted. If no signs are shown, all inputs are generally summed.

A linear system block diagram is a representation of the way in which signals flow in the system. It consists of oriented lines and transfer function boxes, each of which we label with a variable and interconnect using summers and pickoff points. More often than not there will be one or more feedback loops in a typical linear-system block diagram. One very simple such diagram, representative of a linear positioning servomechanism, is shown in Fig. 9.2*a*. The presence of a closed loop, in which we can start at point $A(s)$ and move through the system in the feedforward position and return to point $A(s)$, is noted. The name "closed-loop system" is used to denote systems of this type in which there are feedback elements. Here, we may go through a complete path in the forward transfer functions G_3 and G_4 and redraw the block diagram as in Fig. 9.2*b*. We desire to obtain a single-block diagram which represents the transfer function of this system. Such a single block is shown in Fig. 9.2*c*. We have the transfer function relations, Fig. 9.2*b*,

$$\begin{aligned}
Z(s) &= G_2(s)A(s) \\
A(s) &= U(s) - G_1(s)Z(s)
\end{aligned}$$

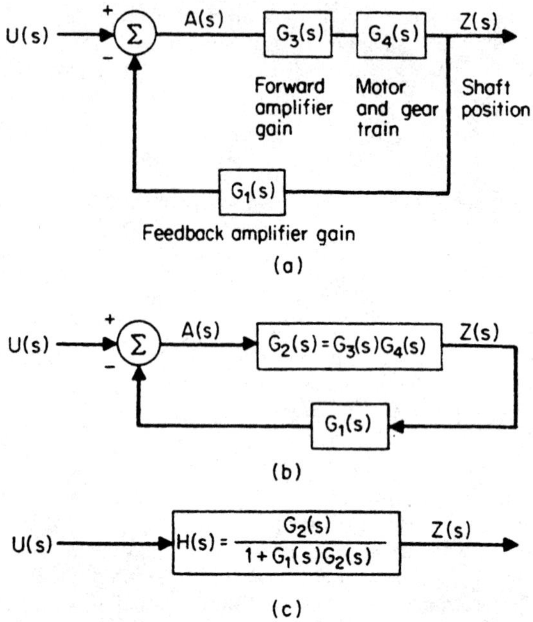

FIG. 9.2 Simple feedback system and equivalents. (*a*) Original block diagram. (*b*) Reduced block diagram. (*c*) Single-block equivalent system.

which we may solve for the desired input-output transfer relation as

$$Z(s)/U(s) = H(s) = G_2(s)/1 + G_1(s)G_2(s) \tag{9.6}$$

Three other rules for block diagram manipulation are helpful. These concern a rule for combining transfer functions in parallel and two rules for moving summing junctions and takeoff points around transfer functions. Figure 9.3 illustrates these three rules as well as rules for cascading transfer functions and feedback loop reduction.

9.3 SYSTEM CONCEPTS

We may characterize a first-order linear system by the differential equation

$$Tdz(t)/dt + z(t) = u(t)$$

The unit impulse response for this system is, for $t \geq 0$,

$$z(t) = (1/\mathrm{T})e^{-t/T}$$

and the unit step response is

$$z(t) = 1 - e^{-t/T}$$

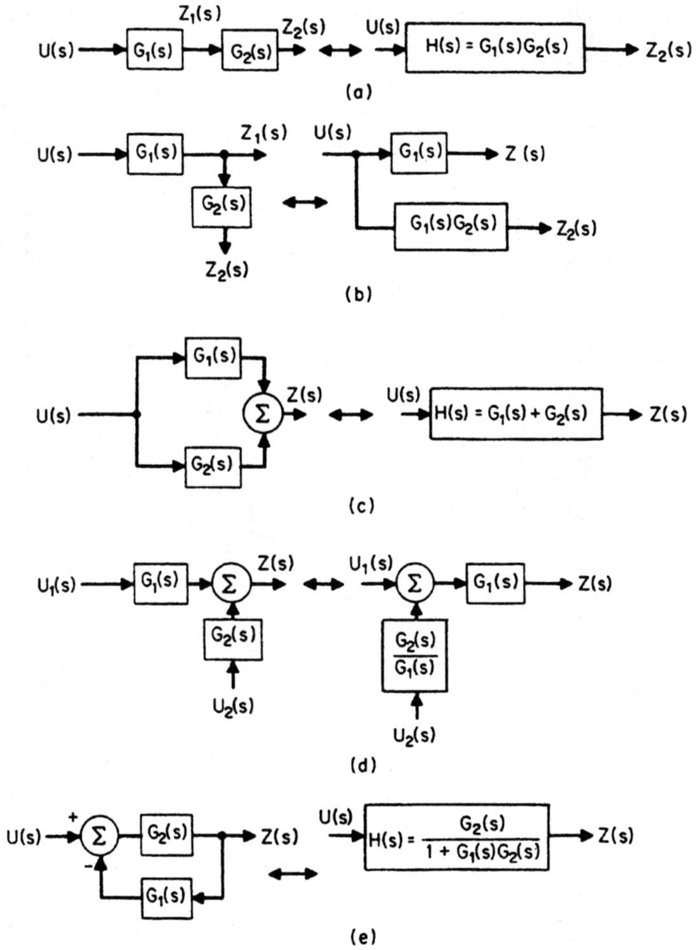

FIG. 9.3 Basic rules for block diagram reduction. (*a*) Combining cascade transfer functions. (*b*) Moving a takeoff point. (*c*) Parallel transfer function. (*d*) Moving a summing junction. (*e*) Single-loop feedback system.

Figure 9.4 illustrates a time sketch of the unit impulse and step response and enables us to define the rise time of this system. The time for the response to rise to 0.632, that is, to $1 - e^{-1}$, of the final value is a very convenient rise time definition here since, for a first-order system, the rise time is just the system time constant.

The ramp $[u(t) = at]$ response of this simple first-order system is

$$z(t) = \alpha(t - T + Te^{-t/T})$$

The error in the response, defined by

$$e(t) = u(t) - z(t)$$

is

$$e(t) = \alpha(T - Te^{-t/T})$$

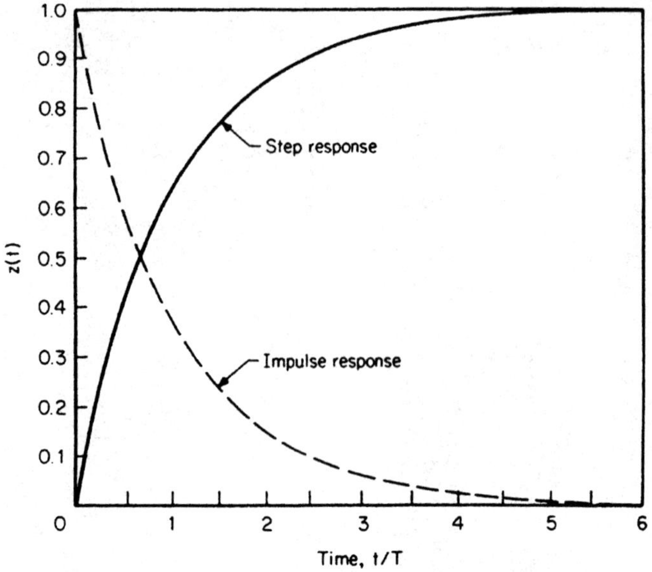

FIG. 9.4 Unit impulse and step response of standard first-order system.

The steady-state error defined by

$$e_{ss} = \lim_{t\to\infty} e(t)$$

is just $e(t) = \alpha T$ for this simple system. We see that this first-order system will have a nonzero steady-state error for a ramp input. This steady-state error, sometimes called the velocity error, can be decreased by decreasing the system time constant or rise time T, which will also increase the speed of response of the system for a step input. Steady-state error is an important concern for many applications, especially precise instrument servos. Many control system design specifications contain minimum requirements on the maximum allowable steady-state error.

The complete response of the first-order system to a sinusoidal input $u(t) = U \sin \omega t$ is

$$z(t) = [UT\omega e^{-t/T}/(1 + T^2\omega^2) + U/(1 + T^2\omega^2)^{1/2}] \sin(\omega t - \phi)$$

where $\phi = \arctan \omega T$. The steady-state response occurs when the exponential term decays to zero. This is given by

$$z(t) = |H(j\omega)|\, U \sin(\omega t + \beta)$$

where

$$|H(j\omega)| = 1/(1 + T^2\omega^2)^{1/2}$$

is the magnitude of the system transfer function $H(s)$ evaluated at $s = j\omega$ and $\beta = -\arctan \omega T$ is the phase angle associated with the transfer function at frequency ω. It is very convenient to plot the magnitude of the transfer function $|H(j\omega)|$, or system frequency response, on log-log coordinate paper and the phase angle curve, or phase response, on semilog paper as illustrated in Fig. 9.5. Also shown in this figure are

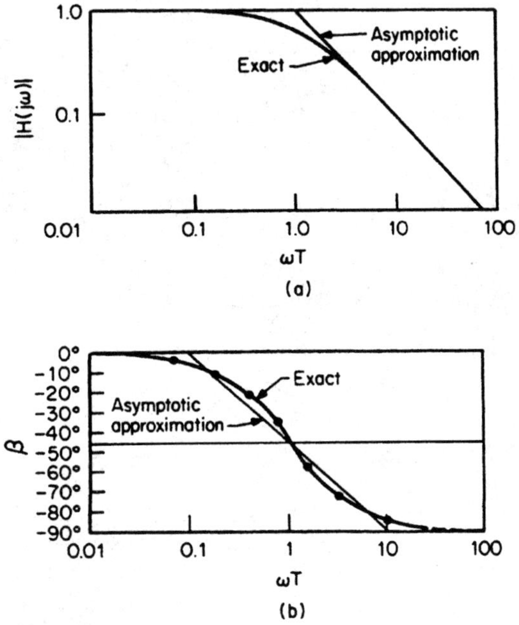

FIG. 9.5 A first-order system. (*a*) Magnitude of frequency response. (*b*) Phase shift curve.

straight-line approximations to the frequency and phase response. The frequency response straight-line approximation has two asymptotes which intersect at the frequency $\omega = 1/T$, often called the "break point" or "break frequency." At the break frequency the actual frequency response magnitude is 0.707 of the asymptotic value. The approximate curve is a better fit to the actual curve at all other frequencies. It will often be very convenient for us to replace an actual frequency response or gain magnitude curve

$$|H(s)| = |H(j\omega)| = 1/|1 + sT| = 1/(1 + \omega^2T^2)^{1/2}$$

by either of its asymptotic values

$$|H(j\omega)| = \begin{cases} 1 & \omega T \le 1 \\ 1/\omega T & \omega T \ge 1 \end{cases}$$

and this will form the basis for a very important design approach. In a similar way we will often find it convenient to use for the exact phase response, which is $\beta = -\arctan \omega T$, the analytical approximation (where β is expressed in radians) given by

$$\beta = \begin{cases} -\omega T & \omega \le 1/T \\ -\pi/2 - 1/\omega T & \omega \ge 1/T \end{cases}$$

which may be obtained by a Taylor series approximation of arctan ωT.

For some applications, a polar plot of the frequency response locus or transfer locus of the sinusoidal steady-state response of this system is desirable. Figure 9.6 illustrates this for the first-order system studied here.

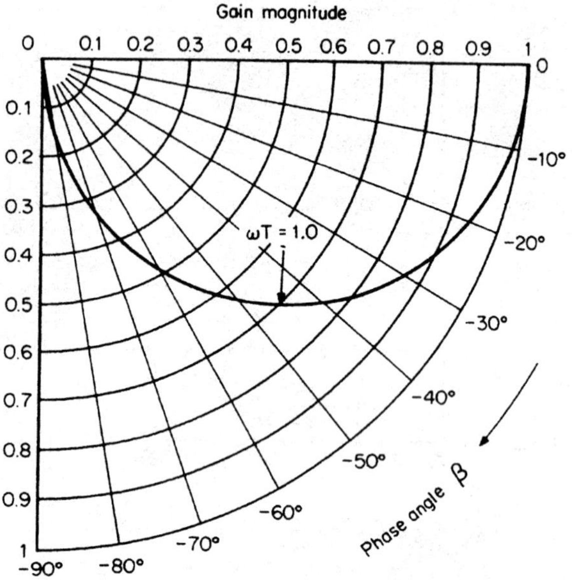

FIG. 9.6 First-order system polar plot of gain vs. frequency.

In a similar way, we may obtain the time and frequency response of second-order systems. Many performance criteria for high-order systems are based upon concepts desired for second-order systems. We consider the input-output differential equation

$$d^2z/dt^2 + 2\zeta\omega_n\,dz/dt + \omega_n^2 z(t) = \omega_n^2 u(t)$$

which can be represented by the transfer function

$$H(s) = Z(s)/U(s) = \omega_n^2/(s^2 + 2\zeta\omega_n s + \omega_n^2)$$

There are no finite zeros for this transfer function. There are two finite poles located at

$$s = -\zeta\omega_n \pm \omega_n\sqrt{\zeta^2 - 1}$$

As the damping ratio is varied, the poles move about a circle in the s plane. For $\zeta > 1$, the system has two negative real roots. The unit impulse response is of the form

$$z(t) = (\omega_n/2\sqrt{\zeta^2 - 1}\,)\exp[-(\zeta + \sqrt{\zeta^2 - 1}\,)\omega_n t] - \exp[-(\zeta - \sqrt{\zeta^2 - 1}\,)\omega_n t]$$

When $\zeta > 1$, the system is said to be overdamped. This relation also holds when $\zeta < -1$.

For the very important underdamped case where $0 < \zeta < 1$, the poles are complex conjugates. The unit impulse response is of the form

$$z(t) = (\omega_n/\sqrt{1 - \zeta^2}\,)e^{-\zeta\omega_n t}\sin(\omega_n\sqrt{1 - \zeta^2}\,t)$$

which represents a sinusoidally damped or decaying exponential. The frequency of oscillation $\omega_n\sqrt{1 - \zeta^2}$ is known as the damped frequency. For $-1 < \zeta < 0$, the foregoing relation holds and the response grows in time since the coefficient of the exponential in this equation is positive for positive time. For $\zeta < -1$, the response also grows as time increases. For each of these cases, that is, whenever the damping ratio is less than zero, we say that the system is unstable.

The unit step response of this second-order system is

$$z(t) = 1 - (1/\sqrt{1 - \zeta^2})e^{-\zeta\omega_n t} \sin(\omega_n\sqrt{1 - \zeta^2}t - \phi - \pi/2)$$

where $\phi = \sin^{-1}\zeta$. Figure 9.7 illustrates the unit step response of this second-order system for various values of the damping ratio ζ. The time to peak value of the output can easily be obtained by setting $dz(t)/dt = 0$, where $z(t)$ is the step response. Alternatively, we can just find the time where the response goes to zero. We obtain for the time peak t_p the value of the peak output z_p and the peak overshoot os, which is $z_p - 1$,

$$t_p = \pi/\omega_n\sqrt{1 - \zeta^2}$$
$$z_p = 1 + \exp(-\zeta\pi/\sqrt{1 - \zeta^2}) = 1 + \text{os}$$

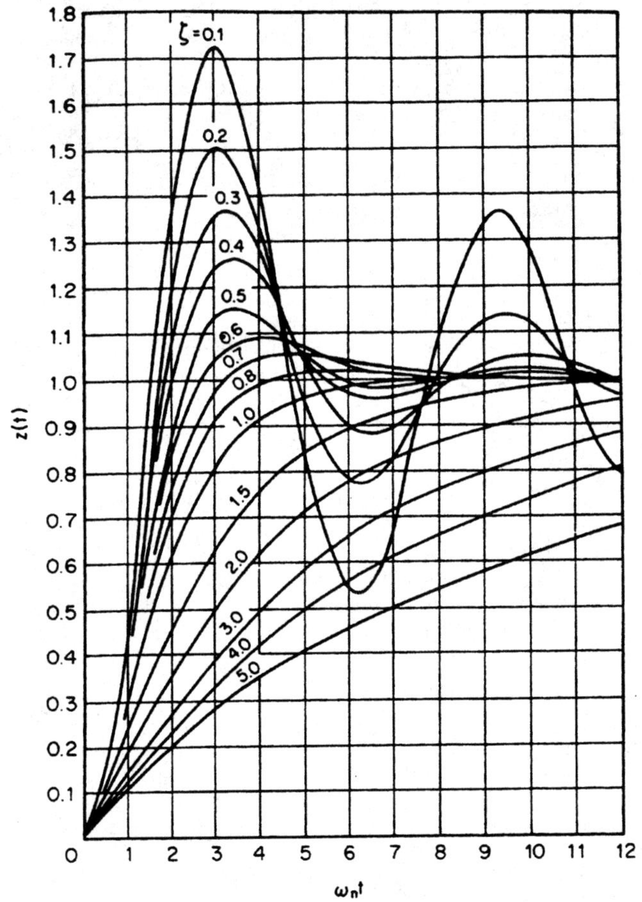

FIG. 9.7 Second-order system unit step response for various damping

The settling time t_s is often defined as the time when the response settles down to 5 percent of its steady-state value. This is, for all practical purposes, the time when $e^{-\omega_n t s} = 0.05$ and this occurs when $\zeta\omega_n t_s = 3$, so that we have

$$t_s = 3/\zeta\omega_n$$

It is of interest to determine the response of this second-order system to a ramp input $u(t) = \alpha t$, for $t > 0$. We easily obtain this response as

$$z(t) = \alpha t - 2\zeta\alpha/\omega_n + \alpha/(\omega_n\sqrt{1 - \zeta^2})e^{-\zeta\omega_n t}\sin(\omega_n\sqrt{1 - \zeta^2}t - \pi - 2\phi)$$

where $\phi = \sin^{-1}\zeta$. This is the most convenient form of the response for the underdamped case where $0 < \zeta < 1$. It can be easily converted to a form convenient for the overdamped case. The last term decays to zero (for the stable case where $\zeta > 0$) and the steady-state error becomes

$$e_{ss} = \lim_{t\to\infty} e(t) = \lim_{t\to\infty}[u(t) - z(t)] = 2\zeta\alpha/\omega_n$$

The steady-state error for this system is a constant. It can be reduced by decreasing ζ. This makes the system less damped, and the system transient response is poor for too small a value of ζ.

We obtain the frequency response by evaluating the magnitude of the transfer function with $s = j\omega$. This yields

$$|H(j\omega)| = \omega_n^2/\sqrt{\omega_n^2 - \omega^2)2 + (2\zeta\omega_n\omega)^2}$$

The corresponding phase-shift characteristic is

$$\beta(\omega) = -\arctan - \left[2\zeta\omega_n\omega/\left(\omega_n^2 - \omega^2\right)\right]$$

Frequency and phase-shift curves for this second-order system are illustrated in Figs. 9.8 and 9.9. These are based on the frequency transformation $v = \omega/\omega_n$. In order to get the asymptotic gain approximation, we assume that for $\zeta < 1$, $v < 1$, we have $|H(jv)| = 1$ and for $v > 1$ we have $|H(jv)| = 1/v^2$ such that we have the normalized frequency response of Fig. 9.8. We note that this is precisely the same asymptotic gain as if we had the transfer function

$$H'(jv) = 1/(jv + 1)^2$$

In a similar way, the asymptotic phase curve is obtained as if the transfer function were given by the foregoing transfer function with two real negative poles, regardless of the value of the damping ratio ζ. Alternatively, we might use the analytical approximation for the phase characteristic

$$\beta(v) = \begin{cases} -2v & v < 1 \\ -\pi - 2/v & v > 1 \end{cases}$$

Although these three approximations for the asymptotic gain, the asymptotic phase, and the analytical phase approximation may appear poor, they are really quite good except at $v = 1$ for very small values of ζ. For values of $\zeta > 1$ the roots are real, and we would use these real distinct roots rather than assume that both roots are at $jv = -1$. Thus we would use

$$|H(j\omega)| \approx \begin{cases} w_1^1/\omega & \omega < \omega_1 \\ & \omega_1 < \omega < \omega_2 \\ \omega_1\,\omega_2/\omega^2 & \omega_2 < \omega \end{cases}$$

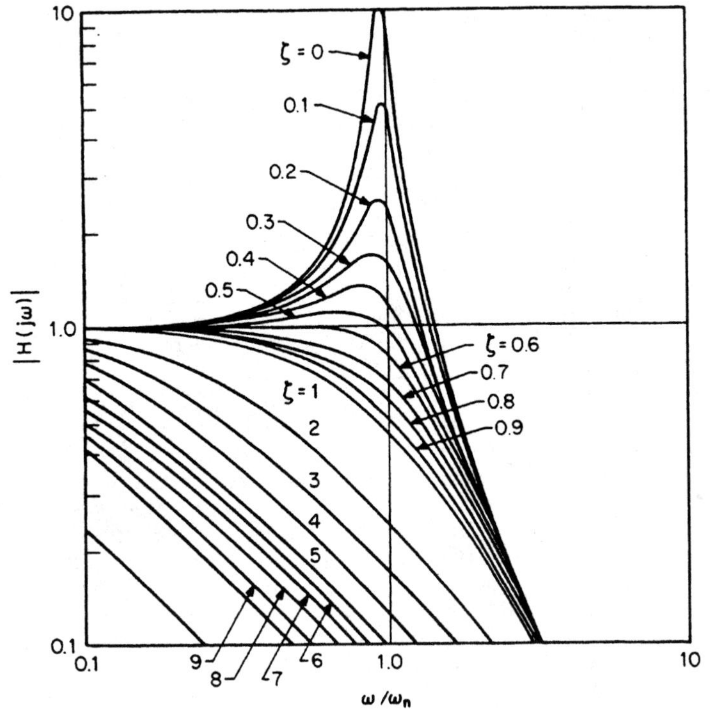

FIG. 9.8 Second-order system normalized frequency-response curve for various values of damping ratio.

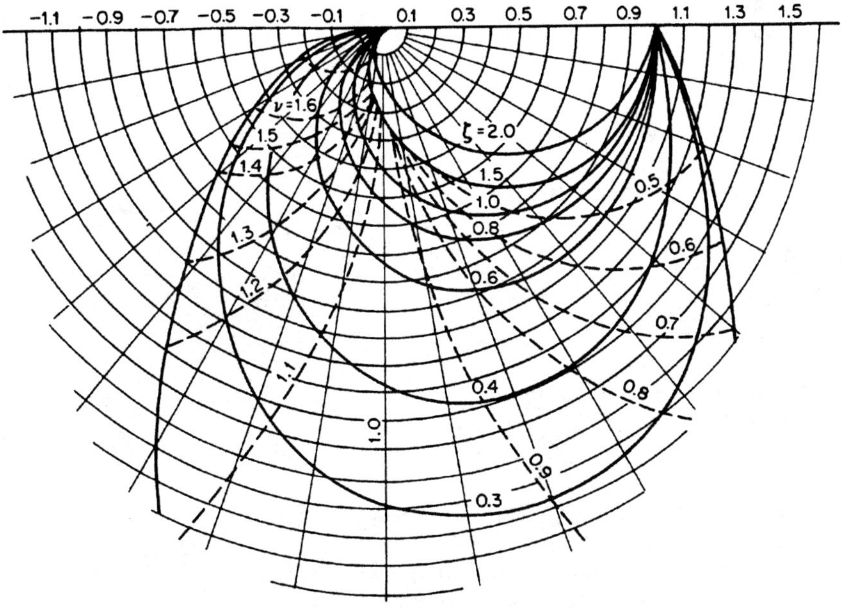

FIG. 9.9 Polar or Nyquist plot of normalized second-order system transfer function $H(j\nu) = 1/[(1 - \nu^2) + 2j\zeta\nu]$.

This asymptotic approximation will match quite well the actual frequency characteristics shown in Fig. 9.8 for the overdamped case ($\zeta > 1$). A similar approximation would be used for the phase-shift characteristics.

We can combine the frequency and the phase response on a polar plot which is called a Nyquist diagram. This is just a graph of the normalized locus:

$$H(jv) = 1/[(1 - v^2) + 2j\zeta v]$$

Figure 9.9 illustrates the loci corresponding to this equation. As we note from the figure, the phase is always negative and decreases from a zero frequency value of 0 to $-180°$. The solid lines in this figure are for constant normalized frequency v. It is interesting and significant to note that at unity normalized frequency ($v = 1$) the phase shift is always $-90°$. For any positive value of ζ, the loci never reach the point where the gain is infinite. This ensures stability of this second-order system.

For linear systems, a growing transient response would be defined as unstable. This condition coincides with one or more right-half-plane poles of the transfer function. Thus if there are no right-half-plane poles, then the system is stable; otherwise the system is unstable. See Chap. 13 for a more complete discussion including Routh-Hurwitz, and Nyquist criteria.

Some very important frequency-domain specifications for linear control systems design are best defined by reference to a relative stability and system performance Nyquist plot. One of these is the phase margin concept. Phase margin is the angle between the phase shift at the crossover frequency (that is, the frequency at which the magnitude of the open-loop gain in a unity ratio feedback system is 1) and the negative real axis. Phase margin is a very important concept, primarily because it is a single figure of merit that is easily calculated and typically gives a very good indication of the peaking that occurs in the closed-loop system frequency response and is the measure of relative stability.

The closed-loop transfer function for a unity-ratio feedback system is related to the open-loop transfer function $G(s)$ by

$$H(s) = G(s)/[1 + G(s)]$$

The magnitude of the closed-loop transfer function $M(\omega) = |H(j\omega)|$ is a very important performance criterion. Below crossover we have $|G(j\omega)| > 1$ and so we have for the approximate $M(\omega)$ curve $M(\omega) \approx 1$ for $\omega < \omega_c$. For $\omega > \omega_c$ we have $|G(j\omega)| < 1$ so that $M(\omega) \approx G(j\omega)$ for $\omega > \omega_c$. Near the crossover frequency neither of these two approximations will generally yield good results.

We may determine a set of loci of constant $M(\omega)$ in the s plane used to represent a Nyquist plot such that the values of $M(\omega)$ for a given system may be determined directly from the Nyquist plot. The loci of constant $M(\omega)$ are easily determined. We let $G(j\omega) = x + jy$ such that from $M^2(\omega) = |H(j\omega)|^2 = H(j\omega)H(-j\omega)$ we obtain

$$M^2 = (x^2 + y^2)/[(1 + x)^2 + y^2]$$

which can be rewritten as the equation for a circle with a center at $x = -M^2/(M^2 - 1)$, $y = 0$ and radius $= |M/(M^2 - 1)|$. This is

$$[x + M^2/(M^2 - 1)]^2 + y^2 = [M/(M^2 - 1)]^2$$

Figure 9.10 illustrates loci of constant M for this example. Typical performance specifications for control systems require peak values of M, denoted M_p, in the range 1.2 to 1.4. In order to ensure that $M_p < 1.4$, for example, we must restrict the Nyquist plot of $G(s)$ from entering the interior of the $M = 1.4$ circle. Normally trial-and-error approaches will be required to accomplish this.

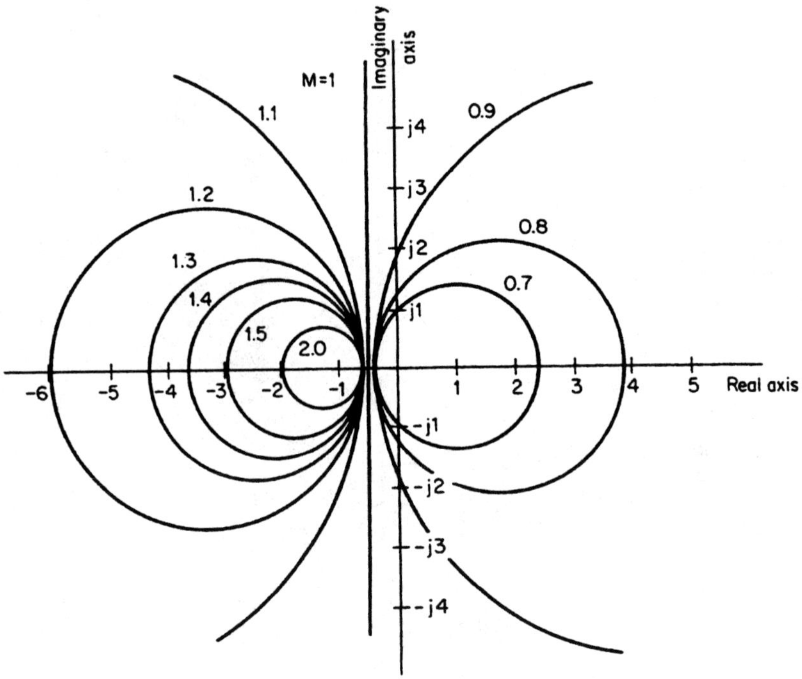

FIG. 9.10 Loci for constant-M circles.

We may determine an approximate value of phase shift at crossover which corresponds to a given M_p as follows. Figure 9.11 illustrates the $M = 1.4$ circle as well as the unit circle about the origin. The phase margin PM will, for well-behaved systems, be approximately the radius of the M_p circle divided by the distance from the origin to the center of the M_p circle. Thus we have the approximation $\sin \text{PM} = 1/M_p$. For the vast majority of control system design efforts, we may rely upon this approximation; and we can then convert specifications in terms of M_p into phase margin specifications.

There are a number of commonly stated performance specifications that are very useful for design purposes.

Damping Factor. The damping factor, or decrement factor as it is sometimes known, is the product of the damping ζ and the natural frequency ω_n for a second-order system. For a second-order system there is an exponential decay term $e^{-\zeta\omega_n t}$ associated with the transient response, and specification of the damping factor $\zeta\omega_n$ controls the envelope of the system transient response. For a second-order system $\omega_0 = \sqrt{1 - \zeta^2}\omega_n$, and we have about $\sqrt{1 - \zeta^2}/\pi\zeta$ discernible oscillations of the system transient response.

Time to Peak. For a second-order system, we can calculate the time to peak and obtain $t_p = \pi/\omega_n\sqrt{1 - \zeta^2}$. Since an equivalent ω_n is not normally calculated for high-order systems, it is convenient to express this in terms of crossover frequency $\omega_c \approx \omega_0 = \omega_n\sqrt{1 - \zeta^2}$ such that we have $t_p \approx \pi/\omega_c$ as a rule-of-thumb approximation for the system time to peak.

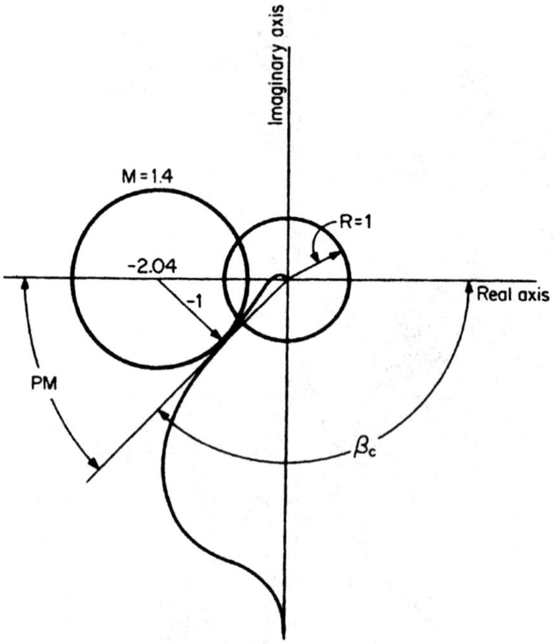

FIG. 9.11 Typical plot of Nyquist diagram with $M_p = 1.4$.

Percentage Overshoot. The peak response z_p for a two-pole-zero system with one open-loop integration can be shown to be given by

$$z_p = 1 + \exp(-\zeta\pi/\sqrt{1 - \zeta^2})$$

Damping Ratio. The term ζ in the expression for the transfer function of the second-order system

$$H(s) = \omega_n^2/(s^2 + 2\zeta\omega_n s + \omega_n^2)$$

is known as the damping ratio. Most systems are more complicated than a single second-order system, and so it is desirable to develop a rule-of-thumb approximation based on the damping ratio concept.

Final Value of Error. The final value theorem,

$$\lim_{t\to\infty} f(t) = \lim_{s\to 0} sF(s)$$

may easily be used to determine the final value of the system error for polynomial inputs $u(t) = t^n$. We can use this criterion to determine minimum gain values and the type of system required to yield specified maximum final steady-state errors. Especially important error coefficients result from the steady-state errors to inputs of the form $u(t) = at^n/n!$. For a unity-ratio system with m open-loop integrations, such that the open-loop transfer function is $G(s) = (K/s^m)F(s)$, where $F(0) = 1$, we obtain the results shown in Table 9.1. The coefficients in this table are called steady-state, or static, error coefficients.

TABLE 9.1 Steady-State Errors for Various System Types and Polynomial Input $u(t) = \alpha t^n/n!$ Open-Loop $G(s) = (K/S^m)F(s)$; $F(0) = 1$

	m			
Input	0	1	2	3
Step, $n = 0$, $u(t) = \alpha$	$\alpha/(1+K)$	0	0	0
Ramp, $n = 1$, $u(t) = \alpha t$	∞	α/K	0	0
Acceleration, $n = 2$, $u(t) = \alpha t^2/2$	∞	∞	α/K	0
Jerk, $n = 3$, $u(t) = \alpha t^3/6$	∞	∞	∞	α/K

Rise Time. A number of different rise times have been defined to enable response-speed specification. The simplest definition for rise time is the reciprocal of the crossover frequency, but this is perhaps best called the equivalent system time constant τ_{eq}. If a high-order system is reasonably well behaved, $1/\omega_c$ is also a good rule-of-thumb indicator of the system equivalent time constant and rise time.

It is perhaps best to define rise time as the time required for the step response to rise from 0 to 90 percent of its final value. For the second-order system response curve for, say, $\zeta = 0.7$, we see that the response reaches 0.9 at time $\omega_n t_r = 3$. The natural frequency ω_n is not very different from the crossover frequency for small ζ, and so we have the rule-of-thumb approximation $t_r \approx 3/\omega_c$.

Settling Time. The settling time for a second-order system is the time required for the transient response to settle down within 5 percent of its final value. When we use the approximate relations $\omega_0 = \omega_n\sqrt{1 - \zeta^2}$, we obtain

$$t_s = 3(\sqrt{1 - \zeta^2})/\omega_0\zeta \approx 4(\sqrt{1 - \zeta^2})/\zeta\omega_c$$

which we may use as a rule-of-thumb approximation for more complex systems.

Bandwidth. The bandwidth of a system is generally defined as the frequency at which the system frequency response is 0.707 of its value at zero frequency. For a first-order system with open-loop transfer function $G(s) = K/s$, the system bandwidth ω_b is just the crossover frequency $\omega_b = \omega_c = K$. For a second- or higher-order system the bandwidth is generally greater than the crossover frequency. A rule-of-thumb approximation for system bandwidth is

$$\omega_b \approx 1.55\omega_n \approx 1.165\omega_c/\sqrt{1 - 2\zeta^2} \approx 1.165\omega_c$$

9.4 LINEAR CONTROL SYSTEM DESIGN BY BODE DIAGRAMS

We now turn our attention to system design concepts. The design of systems often involves trial-and-error repetition of analysis until a set of design specifications has been met; analysis methods are most useful in the design process. The Bode diagram plays a central role in our efforts.

A transfer function can often be written as a ratio of zeros (numerator root locations in the s plane) and poles (denominator root locations in the s plane) as

$$H(s) = \frac{N(s)}{D(s)} = \prod_{i=1}^{m}\left(1 + \frac{s}{i_i}\right)K \Big/ \prod_{i=1}^{n}\left(1 + \frac{s}{d_i}\right)$$

where some of the e_i and d_i may appear in complex conjugate pairs. It is easier to obtain asymptotic approximations for $s = j\omega$ than it is to obtain exact expressions. First we will replace all poles and zeros d_i and e_i with $|d_i|$ and $|e_i|$ such that the resulting expression has poles and zeros on the negative real axis. This is equivalent to regarding all d_i and e_i terms in the foregoing as being positive and real. Next we take the logarithm of the magnitude to obtain

$$\log|H(j\omega)| = \log K + \sum_{i=1}^{m}\log\left|1 + \frac{j\omega}{e_i}\right| - \sum_{i=1}^{n}\log\left|1 + \frac{j\omega}{d_i}\right|$$

Now we consider a typical term in the transfer function and use either of two asymptotic approximations

$$\log|1 + j\omega/p_i| \cong \begin{cases} 0 & \omega < p_i \\ \log \omega/p_i & \omega > p_i \end{cases}$$

and we see that this typical term will graph as a straight line (on logarithmic coordinates). The transfer function will also graph as a straight line if we use logarithmic coordinates and the asymptotic approximation. This result is a valid approximation at frequencies slightly distant from the break frequencies and is usually not a bad approximation even at the break frequencies.

We approximate the phase shift equation given by

$$\beta = \sum_{i=1}^{m}\arctan\frac{\omega}{e_i} - \sum_{i=1}^{n}\arctan\frac{\omega}{d_i}$$

by analytical arctangent approximations. We have for a typical term

$$\arctan\frac{\omega}{\alpha} \cong \begin{cases} \omega/\alpha & \omega < \alpha \\ \pi/2 - \alpha/\omega & \omega > \alpha \end{cases}$$

Errors that result from using this approximation will generally be small. It is this approximation that we will use to develop our Bode diagram design procedure.

There are four types of series equalization approaches:

1. Gain adjustment, normally attenuation by a constant at all frequencies.
2. Increasing the phase lead, or reducing the phase lag, at the crossover frequency by use of a phase-lead network (Fig. 9.12).
3. Attenuation of the gain at middle and high frequencies such that the crossover frequency will be decreased to a lower value where the phase lag is less, by use of a lag network (Fig. 9.13).
4. Composite (lag-lead) equalization (Fig. 9.14).

We will use Bode diagram techniques to develop a design procedure for each of these types of series equalization.

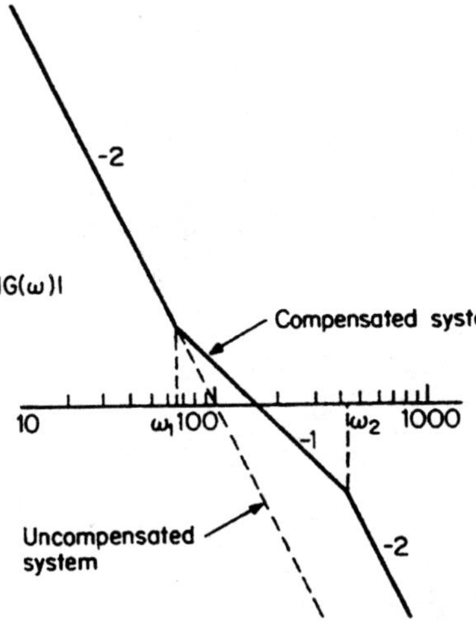

FIG. 9.12 Bode diagram of uncompensated and compensated system with lead network compensation.

9.4.1 Gain Reduction

The majority of linear control systems can be made sufficiently stable merely by reduction of the open-loop system gain to a sufficiently low value. This approach ignores all performance specifications, however, except that of phase margin and is, therefore, usually not a satisfactory approach. It is a very simple one, however, and serves to illustrate the approach to be taken in more complex cases.

We consider a unity-ratio closed-loop system which has an open-loop transfer function

$$G(s) = \frac{10{,}000}{s(1 + s/100)^2}$$

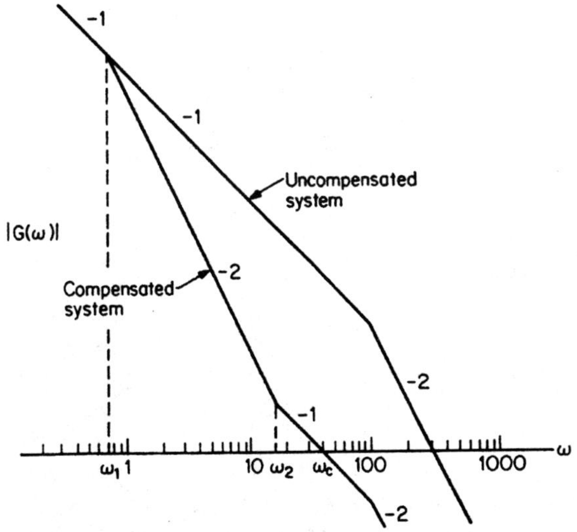

FIG. 9.13 $G_f(s)$ and $G_f(s)G_c(s)$ for lag network compensation.

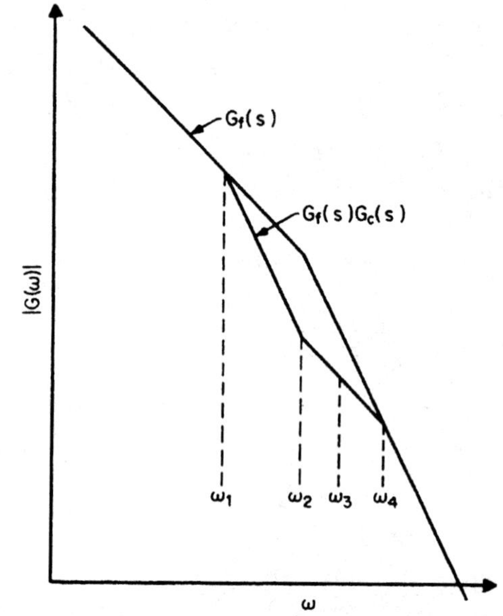

FIG. 9.14 Bode amplitude lag-lead network compensation.

Use of the asymptotic gain approximation for $\omega_c > 100$,

$$|G(\omega)| \cong 10^8/\omega^3$$

results in the crossover frequency $\omega_c \cong 10^{8/3} = 215$. The phase shift at the crossover frequency, obtained as

$$\beta_c = -\pi/2 - 2 \arctan(\omega_c/100) \cong -\pi/2 - 2(\pi/2 - 100/\omega_c)$$

becomes -3.78 rad or $-216.70°$ at the crossover frequency $\omega_c = 215$. This indicates that the closed-loop system is unstable. The phase shift at $\omega = 100$ is $-\pi$ rad, or $-180°$, so that any crossover frequency greater than 100 rad/s is bound to result in an unstable system.

Suppose that we desire a phase margin at $45° = \pi/4$ rad. The crossover frequency must be less than 100 rad/s such that the use of the appropriate arctangent approximation yields

$$\beta = -\pi/2 - 2 \arctan(\omega/100) \cong -\pi/2 - 2\omega/100$$

For a 45° phase margin we have

$$\beta_c = -3\pi/4 = -\pi/2 - 2\omega_c/100$$

Thus we find that the crossover frequency is $\omega_c = 39.27$. The asymptotic gain expression for $G(s)$, which is valid for $\omega < 100$, is given by

$$|G(j\omega)| = 10{,}000\alpha/\omega$$

where α represents the attenuation to be inserted to give a crossover frequency of 39.27 rad/s. We have $1 = 10{,}000\alpha/39.27$, and so we see that we must introduce the rather large attenuation of 254.7 such that $\alpha = 0.00392$. In this example, we have no real control over the closed-loop system bandwidth or the crossover frequency, and the system response will be very sluggish.

The following steps constitute on appropriate Bode diagram design procedure for compensation by gain adjustment:

1. Determine the required phase margin PM and the corresponding phase shift $\beta_c = -\pi + \text{PM}$.
2. Determine the frequency ω_c at which the phase shift is such as to yield the phase shift at crossover required to give the desired PM.
3. Adjust the gain such that the actual crossover frequency occurs at the value computed in step 2.

9.4.2 Phase-Lead Compensation

In compensation using a phase-lead network, we increase the phase lead at the crossover frequency such that we meet a performance specification concerning phase shift. A single-stage phase-lead compensating network transfer function is

$$G_c(s) = \frac{1 + s/\omega_1}{1 + s/\omega_2} \qquad \omega_1 < \omega_2$$

We indicate the suggested lead network design procedure by means of an example. We assume that we have an open-loop system with transfer function

$$G_f(s) = 10^4/s^2$$

such that there will be zero steady-state error for an acceleration input, with a crossover frequency of 100 rad/s. The phase margin is zero for the uncompensated system, and we attempt compensation by means of a lead network.

The asymptotic gain diagram for this example is as shown in Fig. 9.12. We have an open-loop transfer function

$$G(s) = \frac{10^4}{s^2}\,\frac{1 + s/\omega_1}{1 + s/\omega_2}$$

and wish to select the break frequencies ω_1 and ω_2 such that the phase shift at crossover is maximum and, further, we want this maximum phase shift to be $-3\pi/4$ rad, such that we have a 45° phase margin.

Since the crossover frequency is such that $\omega_1 < \omega_c < \omega_2$, we have for the arctangent approximation to the phase shift in the vicinity of crossover

$$\beta(\omega) \approx -\pi + (\pi/2 - \omega_1/\omega) - \omega/\omega_2$$

To maximize the phase shift at crossover we set $d\beta(\omega)/d\omega\,| = 0$ at $\omega = \omega_c$ and obtain $\omega_c = \sqrt{\omega_1\omega_2}$. The crossover frequency is thus halfway between the two break frequencies ω_1 and ω_2 on a logarithmic frequency coordinate. The phase shift at this optimum value of crossover frequency becomes

$$\beta_c = \beta(\omega_c) = -\pi/2 - 2\sqrt{\omega_1/\omega_2}$$

For a phase margin of $-3\pi/4$, we have $-3\pi/4 = -\pi/2 - 2\sqrt{\omega_1/\omega_2}$ and $\omega_1/\omega_2 = 0.1542$. We see that we have need for a lead network with a gain of $\omega_2/\omega_1 = 6.485$. The gain at the crossover frequency is 1, and for $\omega_1 < \omega < \omega_2$ we have $|G(j\omega)| = 10^4/\omega\omega_1$. Solving for the lead network equalizer parameters, we obtain $\omega_c = 159.58$, $\omega_1 = 62.66$, and $\omega_2 = 406.37$.

We note that we have increased the crossover frequency to 159.58 by the design procedure we have employed here. This will result in a system with the required degree of stability but with a faster speed of response because the crossover frequency is higher than that specified. If we wish to retain the $\omega_c = 100$ specification, there are several modified approaches which could be taken. We could note that a gain of $(6.485)^{1/2}$ occurs at the crossover frequency of a lead network compensating network with $\omega_2/\omega_1 = 6.485$. This is the break-point frequency ratio to yield a phase lead of $\omega_c = \pi/4$, the amount required here to obtain a phase margin of $\pi/4$. Thus we should reduce the gain of the open-loop system by $(6.485)^{1/2}$, so that we have

$$G(s) = G_f(s)G_c(s) = \frac{10^4}{2.547s^2}\,\frac{1 + s/\omega_1}{1 + s/\omega_2}$$

To get the $\omega_c = 100$ crossover frequency we use $|G(j\omega_c)| = 1$ and obtain $\omega_1 = 39.27$ from the foregoing. This completes the equalizer design as we obtain for the open-loop transfer function

$$G(s) = \frac{3927}{s^2}\,\frac{1 + s/39.27}{1 + s/254.67}$$

In the direct approach to design for a specific phase margin we assume a single lead network equalizer. We design the equalizer using the following steps:

1. We find an equation for the gain at the crossover frequency in terms of the compensated open-loop-system break frequencies.
2. We find an equation for the phase shift at crossover.
3. We find the relationship between equalizer parameters and crossover frequency such that the phase shift at crossover is the maximum possible and a minimum of additional gain is needed.
4. We determine all parameter specifications to meet the phase margin specifications.
5. We check to see that all design specifications have been met. If they have not, we iterate the design process.

9.4.3 Phase-Lag Compensation

In phase-lag compensation we reduce the gain at low frequencies such that crossover occurs before the phase lag has had a chance to become intolerably large. A simple single-stage phase-lag compensation network transfer function is

$$G_c(s) = \frac{1 + s/\omega_2}{1 + s/\omega_1} \qquad \omega_1 < \omega_2$$

The maximum phase lag obtainable from a phase-lag network depends upon the ratio ω_2/ω_1 used in designing the network. From the expression for the phase shift of the transfer function, $\beta = \arctan(\omega/\omega_2) - \arctan(\omega/\omega_1)$, we see that maximum phase lag occurs at the frequency ω where $\partial\beta/\partial\omega = 0$. We have $\omega_m = \sqrt{\omega_1\omega_2}$, which is at the center of the two break frequencies for the lag network on a Bode diagram log-log asymptotic gain plot.

The maximum value of the phase lag obtained at $\omega = \omega_m$ is $\beta_m(\omega_m) = -\pi/2 - 2 \arctan \sqrt{\omega_1/\omega_2}$. This can be approximated in a more usable form, with the arctangent approximation, as $\beta_m(\omega_m) \cong -(\pi/2) - 2\sqrt{\omega_1/\omega_2}$. The attenuation of the lag network at the frequency of minimum phase shift is obtained from the asymptotic approximation as $|G_c(\omega_m)| = \sqrt{\omega_1/\omega_2}$.

It appears best, just as in our discussion of series equalization by means of lead network compensation, to present the approach for lag network design using a specific example first and then state some general rules. We consider the fixed plant

$$G_f(s) = \frac{100}{s(1 + s/100)}$$

As design specifications we assume

1. The velocity error constant K_i is 1000, equivalent to a steady-state error of 0.001 in for an input $u(t) = t$.
2. The phase margin is 45°.
3. Sinusoidal inputs for any frequency greater than 1000 rad/s must be attenuated by a factor of at least 10.

The third specification can be converted into an open-loop gain approximation from an approximation

$$Z(s)/U(s) = G_f(s)G_c(s)/[1 + G_f(s)G_c(s)] \cong G_f(s)G_c(s)$$

which is valid for $|G_f(\omega)G_c(\omega)| \ll 1$. Thus we see that specification 3 can be met by restricting $|G_f(\omega)G_c(\omega)| < 0.1$ for $\omega > 1000$. From the Bode diagram of the fixed plant, we see that a lead network cannot be used. Thus we attempt design by means of a lag network.

We illustrate a method here that does not employ intuitive guesses concerning the attenuation ratio of the lag network. The approach we use in order to determine the frequencies ω_1, ω_2, and ω_c is based upon satisfying three conditions:

1. The magnitude of the open-loop gain at the crossover frequency is unity.
2. The phase shift at the crossover frequency is that specified.
3. The phase shift at crossover (phase margin) is the minimum (maximum) possible for a minimum-attenuation lag network.

In addition, we set the constant gain such that the static error coefficient specification on the velocity error coefficient is satisfied. This approach leads to an optimum result in terms of a minimum attenuation lag network and is the approach we will develop here.

We use three simple relations to determine the ω_1, ω_2, and ω_c, parameters to satisfy design specifications. We assume that crossover occurs on the -1 slope as shown in Fig. 9.13. The open-loop compensated system transfer function is

$$G_f(s)G_c(s) = \frac{1000(1 + s/\omega_2)}{s(1 + s/100)(1 + s/\omega_1)} \qquad \omega_2 < \omega_1 < \omega_c$$

The asymptotic gain at crossover is

$$1 = 10^3\omega_1/\omega_c\omega_2$$

The phase shift in the vicinity of crossover is

$$\beta(\omega) = -\pi/2 = (\pi/2 - \omega_1/\omega_c) - \omega_c/100 + (\pi/2 - \omega_2/\omega_c)$$

In order to obtain the minimum attenuation lag network which meets specifications, we set

$$\left.\frac{\partial\beta(\omega)}{\partial\omega}\right|_{\omega=\omega_c} = 0 = -\frac{1}{100} + \frac{\omega_2 - \omega_1}{\omega_c^2}$$

to obtain the expression

$$\omega_c = \sqrt{100(\omega_2 - \omega_1)}$$

We satisfy the second specification by setting the phase shift $\beta(\omega)$ at $\omega = \omega_c$ equal to $-\pi/4$. We obtain an equation responsive to the second requirement:

$$-3\pi/4 = -\pi/2 - \omega_c/100 + (\omega_1 - \omega_2)/\omega_c$$

Simultaneous solution of these leads to $\omega_1 = 0.63$, $\omega_2 = 16.04$, and $\omega_c = 39.27$ as the parameters for the lag network and the crossover frequency. Figure 9.13 illustrates the Bode diagram of the compensated open-loop system.

The steps involved in the design procedure for lag network compensation that we have presented here are:

1. Sketch $|G_f(w)|$, adjust gain to meet specifications, and compute performance of the uncompensated system.
2. Determine that compensation is needed and possible.
3. Use the lag network design procedure, and sketch the approximate Bode diagram for a possible lag network compensated system.
4. Obtain the arctangent approximation $\beta(\omega)$ for the compensated system.
5. Set $\partial\beta(\omega)/\partial\beta = 0$ at $\omega = \omega_c$ to determine the optimum crossover frequency.
6. Determine ω_1, ω_2, and ω_3 from the above two steps and the relation $|G_f(\omega)G_c(\omega)| = 1$.
7. Determine the actual performance specifications achieved by the design.
8. Accept the obtained parameters as meeting system design specifications, or iterate to a new solution if they do not satisfy these specifications.

The object of lag network compensation design is to reduce the gain at frequencies lower than the original crossover frequency in order to reduce the open-loop gain to unity before the phase shift becomes so excessive that the system phase margin is too small. A disadvantage of lag network compensation is that the attenuation introduced reduces the crossover frequency and makes the system slower in terms of its transient response. Of course, this would be advantageous if high-frequency noise is present and we wish to reduce its effect. The lag network is an entirely passive device and thus is more economical to instrument than the lead network.

In lead network compensation we actually insert phase lead in the vicinity of the crossover frequency in order to increase the phase margin. Thus we realize a specified phase margin without lowering the medium-frequency system gain. We see that the disadvantages of the lag network are the advantages of the lead network and the advantages of the lag network are the disadvantages of the lead network.

Combination of the lag network with the lead network into an all-passive structure called a lag-lead network is often advantageous. Generally we obtain better results than we can achieve using either a lead or a lag network. In Sec. 9.4.4, we will consider design using lag-lead networks and also outline a procedure for more complex composite equalization efforts.

9.4.4 Compensation Using Lag-Lead Networks

Examination of the characteristics of lead network and lag network compensation suggests that it might be possible to combine the two approaches in order to achieve the desirable features of each approach. Thus we may attempt to provide attenuation below the crossover frequency to decrease the phase lag at crossover and phase lead closer to the crossover frequency in order to increase the phase lead of the uncompensated system at the crossover frequency.

The transfer function of the basic lag-lead network is

$$G_c(s) = \frac{(1 + s/\omega_2)(1 + s/\omega_3)}{(1 + s/\omega_1)(1 + s/\omega_4)}$$

where $\omega_4 > \omega_3 > \omega_2 > \omega_1$. Often it is desirable that $\omega_2\omega_3 = \omega_1\omega_4$ such that the high-frequency gain of the equalizer is unity. It is generally not desirable that $\omega_1\omega_4 > \omega_2\omega_3$ as this indicates a high-frequency gain greater than 1 and this will require an active network, or gain, and a passive equalizer. It is a fact that we should always be able to realize a linear minimum phase network using only passive components if the network has a rational transfer function with a gain magnitude that is no greater than 1 at any (real) frequency.

As a simple example to illustrate lag-lead series equalizer design, let us consider the fixed plant

$$G_f(s) = \frac{1000}{s(1 + s/100)}$$

with design specifications:

1. $K_v = 1000$.
2. Phase margin = 45°.
3. Sinusoidal inputs at frequencies up to 10 rad/s must be passed with less than 2 percent error.
4. Sinusoidal inputs at frequencies greater than 1000 rad/s must be attenuated at the output by a factor of at least 10.

The first specification is satisfied by the open-loop system without compensation. The second specification is not satisfied by the uncompensated system, and so we attempt series equalization. To determine the type of compensation needed we examine specifications 3 and 4. The error transfer function for $|G_f(\omega)G_c(\omega)| > 1$ is

$$E(s)/U(s) = 1/[1 + G_f(s)G_c(s)] \cong 1/G_f(s)G_c(s)$$

so we see that we must require $|G_f(\omega)G_c(\omega)| > 50$ for all $\omega < 10$ in order to satisfy specification 3. Simple lag network compensation will not allow us to meet this specification. To meet specification 4, we use the approximation for the closed-loop transfer function

$$Z(s)/U(s) = G_f(s)G_c(S)/[1 + G_f(s)G_c(s)] \cong G_f(s)G_c(s)$$

which is valid for $|G_f(\omega)G_c(\omega)| < 1$. Thus we see that we must require $|G_f(\omega)G_c(\omega)|$

< 0.10 for all $\omega > 1000$ in order to satisfy specification 4. A simple lead network design will not allow us to satisfy this specification, as we saw in Sec. 9.4.3. Thus we attempt design with a lag-lead network.

Figure 9.14 illustrates the Bode amplitude diagram for the uncompensated system and the suggested equalized system. We will now determine the parameters for this equalized system and will then state a suggested design procedure for lag-lead network design.

The compensated system transfer function is

$$G_c(s)G_f(s) = \frac{1000(1 + s/\omega_2)}{s(1 + s/\omega_1)(1 + s/\omega_4)}$$

which requires the lag-lead network compensation

$$G_c(s) = \frac{(1 + s/\omega_2)(1 + s/100)}{(1 + s/\omega_1)(1 + s/\omega_4)}$$

We need to determine, ω_1, ω_2, ω_4, and ω_c, and so we need four equations. The four equations we use represent the requirements that

1. The magnitude of the gain at the crossover frequency is 1.
2. The phase shift at the crossover frequency is $-3\pi/4$ rad.
3. The phase shift-vs.-frequency curve is flat at crossover so that we get maximum phase shift at crossover.
4. The high-frequency gain of the lag-lead network equalizer is 1.

Requirements 1 and 2 are necessary to satisfy system specifications. Requirement 3 ensures optimum placement of compensating network break frequencies such that we put the maximum phase-lead frequency of the compensated system at the crossover frequency. This results in the smallest reduction in crossover frequency possible. Requirement 4 also keeps the closed-loop system bandwidth as large as possible. This procedure is very general and not at all restricted to the specific example we consider here.

Use of requirement 1, provided that $\omega_2 < \omega_c < \omega_4$, leads to $1 = 10^3\omega_1/\omega_c\omega_2$. The arctangent approximation for the phase shift for $\omega_2 < \omega < \omega_4$ is

$$\beta(\omega) \cong \pi/2 - (\pi/2 - \omega_1/\omega) + (\pi/2 - \omega_2/\omega_1) - \omega/\omega_4$$

To satisfy requirement 2 we have

$$\beta(\omega_c) = -3\pi/4 = -\pi/2 + (\omega_1 - \omega_2)/\omega_c - \omega_c/\omega_4$$

To satisfy requirement 3 we set

$$\left.\frac{\partial\beta(\omega)}{\partial\beta}\right|_{\omega=\omega_c} = 0$$

and obtain $\omega_c^2 = \omega_4(\omega_2 - \omega_1)$. Finally we satisfy requirement 4 by $\omega_1\omega_4 = 100\omega_2$. Simultaneous solution of these last five equations leads to the parameters that satisfy system specifications. We obtain $\omega_1 = 19.23$, $\omega_2 = 97.05$, $\omega_3 = 97.05$, and $\omega_4 = 504.63$.

9.5 THE ROOT LOCUS METHOD

The root locus method provides a useful tool for obtaining roots of the numerator and denominator polynomial of a closed-loop system. These roots, the zero and pole locations for a closed-loop system, provide an indication of stability as well as transient and steady-state system behavior. We consider a unity-feedback-ratio system with open-loop transfer function $G(s)$. The closed-loop transfer function is $Z(s)/U(s) = G(s)/[1 + G(s)]$. We see that the closed-loop zeros are the zeros of $G(s)$ and that the closed-loop poles are the zeros of $1 + G(s)$.

Prior to establishing rules we may use in constructing root locus diagrams, we examine some fundamental relations which determine the locus of roots of a linear control system. We have the closed-loop transfer function of a unity-ratio system

$$Z(s)/U(s) = H(s) = G(s)/[1 + G(s)] = N(s)/[N(s) + D(s)]$$

where $N(s)$ is the numerator polynomial and $D(s)$ the denominator polynomial of $G(s)$.

We obtain the system characteristic equation by setting the numerator of $1 + G(s)$ equal to zero. We assume that the open-loop transfer function is a ratio of rational polynominals in s of the form

$$G(s) = \frac{N(s)}{D(s)} = \frac{K(s^n + a_1 s^{n-1} + a_2 s^{n-2} + \cdots + a_n)}{s^d + b_1 s^{d-1} + b_2 s^{d-2} + \cdots + b_{d-1}s + b_d}$$

Thus we obtain

$$N(s) + D(s) = s^d + b_1 s^{d-1} + \cdots + b_d + K(s^n + a_1 s^{n-1} + \cdots + a_n) = 0$$

as the system characteristic equation. Setting the characteristic equation equal to zero is clearly the same as setting $1 + G(s) = 0$. This will occur only if $G(s) = -1$. The basic root locus will be determined by changing the negative loop gain, or K. Thus it is convenient to define a normalized transfer function with K removed,

$$G_n(s) = \frac{G(s)}{K} = \frac{s^n + a_1 s^{n-1} + \cdots + a_n}{s^d + b_1 s^{d-1} + \cdots + b_d}$$

such that the condition for existence of roots becomes $G_n(s) = 1/K$. This equality may be satisfied, in general, in two ways. First of all we must have

$$|G_n(s)| = 1/|K| \qquad s = \sigma + j\omega$$

and then we must also have for positive K

$$\angle G_n(s) = (2K + 1)\pi \qquad K > 0$$

where $\angle$ denotes "angle of." For negative K we have

$$\angle G_n(s) = 2K\pi \qquad K < 0$$

We assume that the open-loop transfer function $G(s)$ is available in, or can be put into, standard factored form:

$$G(s) = \frac{K(s + n_1)(s + n_2)\cdots(s + n_n)}{(s + d_1)(s + d_2)\cdots(s + d_d)} = KG_n(s)$$

The angle criterion becomes

$$\angle G_n(s) = \sum_{i=1}^{n} \angle s + n_i - \sum_{i=1}^{d} \angle s + d_i = (2K + 1)\pi$$

for $0 < K < \infty$. For negative K we obtain

$$\angle G_n(s) = \sum_{i=1}^{n} \angle s + n_i - \sum_{i=1}^{d} \angle s + d_i = 2K\pi$$

Each of these equations may be used to construct the complete root locus. Often only positive gains are of significance, and, in this case, we need only find the values of $s = \sigma + j\omega$ which satisfy the angle criterion.

There are a number of ways of stating formal rules or theorems useful for root locus construction and, as we might suspect, all are very closely related. We will present 14 aids to root locus construction here:

1. The total numbers of branches of the root locus will be equal to the number of poles in the open-loop transfer function $G(s)$
2. Closed-loop zeros are the same as open-loop zeros.
3. The starting points for the root locus at $K = 0$ are at the poles of the open-loop transfer function $G(s)$.
4. The branches of pole movement on the root locus will terminate at open-loop zeros at infinite gain.
5. The root loci must be symmetrical with respect to the s-plane real axis.
6. A point on the real axis will be a point on the root locus (for $K > 0$) if there is an odd number of open-loop zeros and poles on the real axis to the right of this point.
7. For large values of the gain K, the loci of the system closed-loop poles are asymptotic to the straight lines given by the angle of the asymptotes

   $$\alpha_{as} = (1 + 2K)180°/(n - d)$$

   where $K = 0, 1, 2, \ldots, |d - n| - 1$; d is the number of open-loop poles; and n is the number of open-loop zeros.
8. The origin of the asymptotes, which gives the centroid for the asymptotic behavior of the closed-loop poles, is on the real axis at a point given by

   $$\text{OA} = \frac{a_1 - b_1}{d - n} = \left[\sum_{i=1}^{n} n_1 - \sum_{i=1}^{n} d_i\right] \Big/ (d - n)$$

9. Candidate breakaway points, from the real axis or reentry points at which the locus rejoins the real axis, may be determined by finding the roots of $dG^{-1}(s)/ds = 0$.
10. The angles of departure and angles of arrival on the real axis are separated by an angle of $180°/m$ where m represents the number of root locus branches intersecting at the point in question.
11. The angle of departure of an open-loop pole or the arrival angle of a pole at an open-loop zero location can be determined by use of the angle criterion

    $$\sum_{i=1}^{n} \underline{|s + n_i} - \sum_{i=1}^{d} \underline{|s + d_i} = \pi$$

    where s is a given pole or zero location. This rule is a little more difficult to apply for real-axis breakaway and reentry rule 10. For complex conjugate poles, however, rule 10 is inapplicable and rule 11 becomes the appropriate rule. This rule is easily stated for negative gains as

    $$\sum_{i=1}^{n} \underline{|s + n_i} - \sum_{i=1}^{d} \underline{|s + d_i} = 0$$

12. The root locus diagram crosses the imaginary axis at points where the characteristic equation, $1 + G(s) = 0$, is satisfied. The Routh-Hurwitz criterion or the Bode diagram may be used to obtain the gain and frequency of the sinusoidal oscillation that would result.

13. If the open-loop pole-zero excess is at least 2, then the sum of the closed-loop poles is a constant. In these cases the center of gravity of the poles is preserved.
14. At any complex frequency s that satisfies the angle criterion, the gain K can be determined such that the magnitude criterion is satisfied and $K = |1/G_n(s)|$.

A simple example will illustrate the procedure. We consider a simple lead network design where the fixed plant-transfer function is

$$G_f(s) = K/s^2$$

and we assume a single-state lead network

$$G_c(s) = \frac{1 + s/\omega_1}{1 + s/\omega_2} \qquad \omega_1 < \omega_2$$

such that the open-loop transfer function becomes

$$G(s) = G_f(s)G_c(s) = (K\omega_2/\omega_1)(s + \omega_1)/s^2(s + \omega_2)$$

We normalize this open-loop transfer function in order to obtain the most general results possible. One particularly useful and physically significant way to do this is to let $\omega_2 = \alpha\omega_1$ for $\alpha > 1$ such that α represents the ratio of break frequencies in the lead network and the active gain required. The open-loop transfer function becomes

$$G(s) = aK(s + \omega_1)/s^2(s + \alpha\omega_1)$$

We normalize the effort by change of time scale $t = \tau/\omega_1$ or $s = p\omega_1$, where τ and p are the normalized time and frequency variables. The normalized open-loop transfer function becomes

$$G(p) = (\alpha K/\omega_1^2)(p + 1)/p^2(p + \alpha) = K_n(p + 1)/p^2(p + \alpha)$$

where the normalized gain is

$$K_n = \alpha K/\omega_1^2$$

Before turning to an analysis of the closed-loop transfer function that corresponds to the normalized open-loop system, let us reexamine some of the Bode diagram performance characteristics of the system. First we assume that we wish to find the gain K_n to yield maximum phase margin for a fixed α. The crossover frequency will be assumed to occur on the -1 slope such that we can obtain the crossover frequency from $1 = K_n/\alpha\omega_c$ as $\omega_c = K_n/\alpha$. The phase shift on the -1 slope becomes, from α the arctangent approximation, $\beta(\omega) = -\pi + \pi/2 - 1/\omega - \omega/\alpha$. To maximize the phase margin we set

$$\left.\frac{\partial\beta(\omega)}{\partial\beta}\right|_{\omega=\omega_c} = 0 = +\frac{1}{\omega_c^2} - \frac{1}{\alpha}$$

and we obtain $\omega_c = \alpha^{1/2}$. From these expressions, we obtain the optimum value of gain for maximum phase margin $\hat{K}_n = \alpha^{3/2}$ and the corresponding maximum value of the phase margin $\hat{\text{PM}} = -\pi/2 - 2/\alpha^{1/2}$. As we should expect, the phase margin continues to increase for increases in α.

The root locus analysis proceeds in the following manner.

1. The zeros and poles of the open-loop system are given.
2. The angle of the asymptotes is determined from rule 7 as $\alpha_{as} = (1 + 2K)180°/(1 - 3) = -90°, -270°$.

3. The origin of the asymptotes is determined from rule 8 as OA $= -(1 - \alpha)/(1 - 3) = -(\alpha - 1)/2$.
4. Those portions of the real axis which may form part of the equal root locus are determined as part of the real axis from $-\alpha < \alpha < 1$.
5. Possible locations for multiple poles on the real axis are those points in the interval $-\alpha < \alpha < 1$. Reentry and breakaway points are not guaranteed to exist here. We must determine whether they do. Our first thought is that there are no real-axis breakaways, but it is very dangerous to not examine this further, as we will soon see.
6. Breakaway points and/or reentry points are determined using rule 9 and $dG^{-1}(s)/ds = 0$ as the roots of $2_p^3 + p^2(3 + \alpha) + 2\alpha p = 0$. One loop is $p = 0$; this is a trivial solution since the open-loop transfer function contains the two poles at $p = 0$. They do, of course, break away from the origin as K_n increases from zero. The other roots are

$$p = -3/4 - \alpha/4 \pm \{[(3 + \alpha)^2 - 16\alpha]/16\}^{1/2}$$

Inspection of the root loci of Fig. 9.15 indicates that the system response will not be very good unless α is at least 4 and preferably as large as 5. There is essentially no point in having a value of α greater than $\alpha = 9$ from the point of view of relative stability since the phase margin becomes quite large at $\alpha = 9$, and $K_n = 27$. For the case where $\alpha = 9$ all the poles are on the real axis. It is not at all true that the system response will not overshoot unless there are poles off the real axis. As a case in point, let us examine the unit step response of this system for $K_n = 27$ and $\alpha = 9$. We have

$$\frac{Z(p)}{U(p)} = \frac{27(p + 1)}{(p + 3)^3} = \frac{(1 + p)}{(1 + p/3)^3}$$

and for $U(p) = 1/p$ we have

$$Z(p) = \frac{27(p + 1)}{p(p + 3)^3} = \frac{1}{p} + \frac{18}{(p + 3)^3} - \frac{3}{(p + 3)^3} - \frac{1}{p + 3}$$

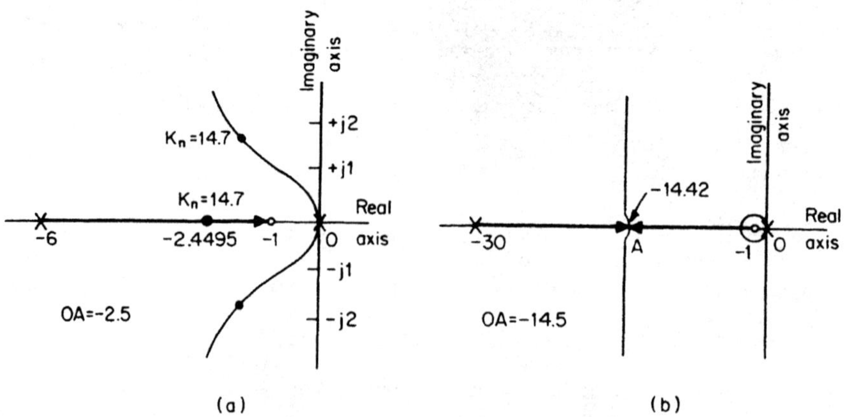

FIG. 9.15 Root locus diagrams for two values of α. (*a*) $\alpha = 3$. (*b*) $\alpha = 30$.

The normalized time response of the system is

$$z(t) = 1 + (18t^2 - 3t - 1)e^{-3t}$$

This response shows that there is a 74 percent overshoot of the step response even though there are poles only on the negative real axis. The closed-loop frequency response also indicates that there is peaking of the system frequency response since the closed-loop gain magnitude is 1.3 at 1.73 rad/s even though there are no poles off the real axis.

Values of α between 5 and 9 are quite reasonable for this system. Either the Bode diagram approach or the root locus approach leads to this conclusion. The basic approach to root locus design suggested here implies, for this example, that we would pick a value of α and then plot the root loci. The best value of K to yield maximum damping would be found for the performance of the closed-loop system calculated. If that is not acceptable, as would be the case if we had selected $\alpha = 2$ initially, the value of α would be changed and the root locus for varying K_n would again be determined. It would be desirable to be able to vary α and plot a root locus for varying α, and perhaps also K. It is possible to modify the root locus approach slightly to that this is possible.

This subsection has concerned the root locus approach and has provided detailed rules for construction of the locus. A design approach was given and illustrated with an example. It is possible to sketch and visualize the behavior characteristics of a closed-loop system in a relatively straightforward manner which allows us to obtain a quick estimate of the transient response modes of a system and to determine the sensitivity of closed-loop pole locations to gain and other parameter variations. Development of practical experience using the technique and use of digital computer routines available for plotting root loci, once preliminary engineering decisions have been made relative to system structure, will equip the user with a very powerful design tool. The combination of the root locus approach and the Bode diagram approach leads to a very potent and useful set of linear systems control design tools as well.

REFERENCES

1. Sage, A.P.: "Linear Systems Control," Matrix Press, Portland, Oreg., 1978.
2. Singh, M.: "Encyclopedia of Systems and Control," Pergammon Press, New York, 1984.
3. Andreson, B. D. O., and J. B. Moore: "Linear Optimal Control," Prentice Hall, Inc., Englewood Cliffs, N.J., 1971.
4. Astrom, K. J.: "Introduction to Stochastic Control," Academic Press, New York, 1970.
5. Athans, M., and P. L. Falb: "Optimal Control," McGraw-Hill Book Company, Inc., New York, 1966.
6. Bryson, A. E. and Y. C. Ho: "Applied Optimal Control," Blaisdell Press, Waltham, Mass., 1969.
7. D'Azzo, J. D., and C. H. Houpis: "Linear Control Systems," McGraw-Hill Book Company, Inc., New York, 1975.
8. Gelb, A. (ed.): "Applied Optimal Estimation," M.I.T. Press, Cambridge, Mass., 1974.
9. Jazwinski, A. H.: "Stochastic Processes and Filtering Theory," Academic Press, Inc., New York, 1970.
10. Kuo, B. C.: "Automatic Control Systems," Prentice-Hall, Inc., Englewood Cliffs, N.J., 1982.
11. Kwakernaak, H., and R. Sivan: "Linear Optimal Control Theory," John Wiley & Sons, Inc., New York, 1972.
12. Luenberger, D. G.: "Dynamic Systems: Control and Modeling," John Wiley & Sons, Inc., New York, 1982.
13. Meditch, J.S.: "Stochastic Optimal Linear Estimation and Control," McGraw-Hill Book Company, Inc., New York, 1969.

14. Sage, A. P.: "Linear Systems Control," Matrix Press, Portland, Oreg., 1978.

15. Sage, A. P., and C. C. White: "Optimum Systems Control," 2d ed., Prentice-Hall, Inc., Englewood Cliffs, N.J., 1977.

16. Sage, A. P., and J. L. Melsa: "Estimation Theory: with Applications to Communications and Control," McGraw-Hill Book Company, Inc., New York, 1971.

17. Sage, A. P., and J. L. Melsa: "System Identification," Academic Press, Inc., New York, 1971.

18. Singh, M., editor-in-chief: "Encyclopedia of Systems and Control," Pergammon Press, Inc., New York, 1984.

19. Truxal, J. G.: "Control System Synthesis," McGraw-Hill Book Company, Inc., New York, 1955.

CHAPTER 10
DIGITAL CONTROL SYSTEMS

J. David Powell, Ph.D.
Professor of Aeronautics/Astronautics and Mechanical Engineering
Stanford University
Stanford, Calif.

10.1 INTRODUCTION 10.1
10.2 THEORETICAL BACKGROUND 10.2
10.2.1 *z* Transform 10.2
10.2.2 *z* Transform Inversion 10.3
10.2.3 Relationship between *s* and *z* 10.6
10.2.4 Final-Value Theorem 10.7
10.3 CONTINUOUS DESIGN AND DIGITIZATION 10.7
10.3.1 Digitization Procedures 10.7
10.3.2 Design Example 10.11
10.3.3 Applicability Limits of Method 10.12
10.4 DISCRETE DESIGN 10.12
10.4.1 Analysis Tools 10.12
10.4.2 Feedback Properties 10.14
10.4.3 Design Example 10.14
10.4.4 Design Comparison 10.15
10.5 HARDWARE CHARACTERISTICS 10.16
10.5.1 Analog-to-Digital Converters 10.16
10.5.2 Digital-to-Analog Converters 10.16
10.5.3 Analog Prefilters 10.17
10.6 WORD SIZE EFFECTS 10.18
10.6.1 Random Effects 10.18
10.6.2 Systematic Effects 10.19
10.7 SAMPLE RATE SELECTION 10.20
10.7.1 Tracking Effectiveness 10.20
10.7.2 Disturbance Rejection 10.21
10.7.3 Parameter Sensitivity 10.21
10.7.4 Control System Modularization 10.22

10.1 INTRODUCTION

The intent of this chapter is to provide the theoretical background and practical tools for the design of a control system which is to be implemented with a computer or microprocessor. The methods to be studied are primarily for closed-loop (feedback) systems in which the dynamic response of the process being controlled is a major consideration in the design. The design methods are applicable to any type of computer (from microprocessors to large-scale computers); however, the effects of small word size and slow sample rates take on a more important role when microprocessors are used.

It will be assumed in the chapter that the reader has some knowledge of control-system design methods for continuous (or analog) systems such as those covered in Chap. 13 and Refs. 1 and 2. Furthermore, a more complete reference for the subject material of this chapter can be found in a digital control textbook like Ref. 3.

A typical topology of the type of system to be considered is shown in Fig. 10.1. There are two fundamentally different methods for the design of digital algorithms:

1. Continuous design and digitization: Perform a continuous design, then digitize the resulting compensation.
2. Direct digital design: Digitize the plant model, then perform a design using discrete analysis methods.

Both methods will be covered and their advantages and disadvantages discussed.

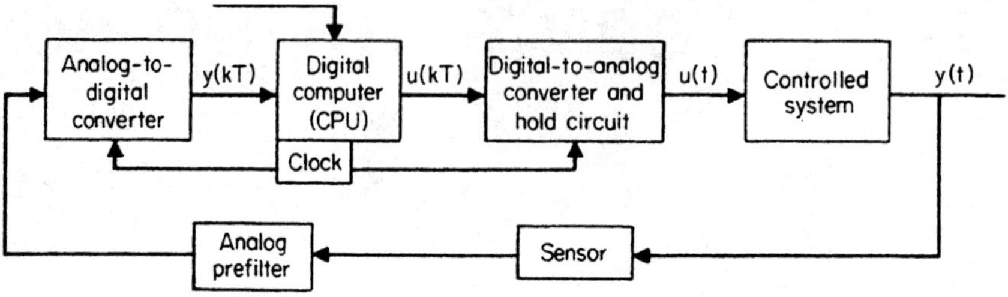

FIG. 10.1 Basic control system block diagram; r = reference or command input, y = output quantities, and u = actuator input signals.

10.2 THEORETICAL BACKGROUND

10.2.1 z Transform

In the analysis of continuous systems, we use the Laplace transform which is defined by

$$\mathcal{L}\{f(t)\} = F(s) = \int_0^{\infty} f(t)e^{-st}\,dt \tag{10.1}$$

which leads directly to the important property that

$$\mathcal{L}\{\dot{f}(t)\} = sF(s) \tag{10.2}$$

This relation enables us to easily find the transfer function of a linear continuous system given the differential equations of that system.

For discrete systems, a very similar procedure is available. The "z transform" is defined by

$$\mathfrak{Z}\{f(n)\} = F(z) = \sum_{n=0}^{\infty} f(n)z^{-n} \tag{10.3}$$

which also leads directly to a property analogous to Eq. (10.2), specifically, that

$$\mathfrak{Z}\{f(n-1)\} = z^{-1}F(z) \tag{10.4}$$

This relation allows us to easily find the transfer function of a discrete system given the difference equations of that system. For example, the general second-order difference equation,

$$y(n) = -a_1y(n-1) - a_2y(n-2) + b_0u(n) + b_1u(n-1) + b_2u(n-2) \tag{10.5}$$

can be converted from $y(n)$, $u(n)$, etc., to the z transform of those variables by invoking Eq. (10.4) once or twice to arrive at

$$Y(z) = (-a_1z^{-1} - a_2z^{-2})Y(z) + (b_0 + b_1z^{-1} + b_2z^{-2})U(z) \tag{10.6}$$

which results in the transfer function

$$Y(z)/U(z) = (b_0 + b_1z^{-1} + b_2z^{-2})/(1 + a_1z^{-1} + a_2z^{-2}) \tag{10.7}$$

10.2.2 z Transform Inversion

A table relating simple discrete time functions to their z transform is contained in Table 10.1 along with the Laplace transform for the same time function. (See also Tables 1.17 and 1.18.)

Given a general z transform, one can break it up into a sum of elementary terms using partial fraction expansion and find the resulting time series from the table. Again, these procedures are exactly the same as those used for continuous systems.

A z transform inversion technique which has no continuous counterpart is called "long division." Given a z transform

$$Y(z) = N(z)/D(z) \tag{10.8}$$

one simply divides the denominator into the numerator using long division. The result is a polynomial (perhaps infinite) in z, from which the time series can be found by using Eq. (10.3).

For example, a first-order system described by the difference equation

$$y(n) = ky(n-1) + u(n) \tag{10.9}$$

yields

$$Y(z)/U(z) = 1/(1 - kz^{-1}) \tag{10.10}$$

for an impulsive input:

$$u(0) = 1$$
$$u(n) = 0 \quad n \neq 0$$

implying

$$U(z) = 1 \quad \text{and} \quad Y(z) = 1/(1 - kz^{-1}) \tag{10.11}$$

Therefore, to find the time series, use long division:

$$\begin{array}{r|l} & 1 + kz^{-1} + k^2z^{-2} + k^3z^{-3} + \cdots \\ \hline 1 - kz^{-1} & 1 \\ & 1 - kz^{-1} \\ \hline & \quad kz^{-1} \\ & \quad kz^{-1} - k^2z^{-2} \\ \hline & \qquad k^2z^{-2} \\ & \qquad k^2z^{-2} - k^3z^{-3} \\ \hline & \qquad\quad k^3z^{-3} \\ & \qquad\quad \cdots\cdots \end{array}$$

The quotient, $1 + kz^{-1} + k^2z^{-2} + k^3z^{-3} + \cdots$, is $Y(z)$, which means that

$$y(0) = 1$$
$$y(1) = k$$
$$y(2) = k^2$$
$$\cdots\cdots\cdots$$
$$y(n) = k^n$$

TABLE 10.1 Transforms[3]

$F(s)$ is the Laplace Transform of $f(t)$ and $F(z)$ Is the z Transform of $f(nT)$.
Unless Otherwise Noted, $f(t) = 0$, $t < 0$.

Number	$F(s)$	$f(nT)$	$F(z)$
1	—	$1, n = 0; 0, n \neq 0$	1
2	—	$1, n = k; 0, n \neq k$	z^{-k}
3	$\frac{1}{s}$	$1(nT)$	$\frac{z}{z-1}$
4	$\frac{1}{s^2}$	nT	$\frac{T_z}{(z-1)^2}$
5	$\frac{1}{s^3}$	$\frac{1}{2!}(nT)^2$	$\frac{T^2 z(z+1)}{2(z-1)^3}$
6	$\frac{1}{s^4}$	$\frac{1}{3!}(nT)^3$	$\frac{T^3}{6}\frac{z(z^2+4z+1)}{(z-1)^4}$
7	$\frac{1}{s^m}$	$\lim_{a \to 0} \frac{(-1)^{m-1}}{(m-1)!}\frac{\partial^{m-1}}{\partial a^{m-1}} e^{-anT}$	$\lim_{a \to 0} \frac{(-1)^{m-1}}{(m-1)!}\frac{\partial^{m-1}}{\partial a^{m-1}} \frac{z}{z-e^{-aT}}$
8	$\frac{1}{s+a}$	e^{-anT}	$\frac{z}{z-e^{-aT}}$
9	$\frac{1}{(s+a)^2}$	nTe^{-anT}	$\frac{Tze^{-aT}}{(z-e^{-aT})^2}$
10	$\frac{1}{(s+a)^3}$	$\frac{1}{2}(nT)^2 e^{-anT}$	$\frac{T^2}{2} e^{-aT} \frac{z(z+e^{-aT})}{(z-e^{-aT})^3}$
11	$\frac{1}{(s+a)^m}$	$\frac{(-1)^{m-1}}{(m-1)!}\frac{\partial^{m-1}}{\partial a^{m-1}}(e^{-anT})$	$\frac{(-1)^{m-1}}{(m-1)!}\frac{\partial^{m-1}}{\partial a^{m-1}}\frac{z}{z-e^{-aT}}$
12	$\frac{a}{s(s+a)}$	$1 - e^{-anT}$	$\frac{z(1-e^{-aT})}{(z-1)(z-e^{-aT})}$

13	$\dfrac{a}{s^2(s+a)}$	$\dfrac{1}{a}(anT - 1 + e^{-anT})$	$\dfrac{z[(aT - 1 + e^{-aT})z + (1 - e^{-aT} - aTe^{-aT})]}{a(z-1)^2(z - e^{-aT})}$
14	$\dfrac{b-a}{(s+a)(s+b)}$	$(e^{-anT} - e^{-bnT})$	$\dfrac{(e^{-aT} - e^{-bT})z}{(z - e^{-aT})(z - e^{-bT})}$
15	$\dfrac{s}{(s+a)^2}$	$(1 - anT)e^{-anT}$	$\dfrac{z[z - e^{-aT}(1 + aT)]}{(z - e^{-aT})^2}$
16	$\dfrac{a^2}{s(s+a)^2}$	$1 - e^{-anT}(1 + anT)$	$\dfrac{z[z(1 - e^{-aT} - aTe^{-aT}) + e^{-2aT} - e^{-aT} + aTe^{-aT}]}{(z-1)(z - e^{-aT})^2}$
17	$\dfrac{(b-a)s}{(s+a)(s+b)}$	$be^{-bnT} - ae^{-anT}$	$\dfrac{z[z(b-a) - (be^{-aT} - ae^{-bT})]}{(z - e^{-aT})(z - e^{-bT})}$
18	$\dfrac{a}{s^2 + a^2}$	$\sin anT$	$\dfrac{z \sin aT}{z^2 - (2\cos aT)z + 1}$
19	$\dfrac{s}{s^2 + a^2}$	$\cos anT$	$\dfrac{z(z - \cos aT)}{z^2 - (2\cos aT)z + 1}$
20	$\dfrac{s+a}{(s+a)^2 + b^2}$	$e^{-anT}\cos bnT$	$\dfrac{z(z - e^{-aT}\cos bT)}{z^2 - 2e^{-aT}(\cos bT)z + e^{-2aT}}$
21	$\dfrac{b}{(s+a)^2 + b^2}$	$e^{-anT}\sin bnT$	$\dfrac{z\, e^{-aT}\sin bT}{z^2 - 2e^{-aT}(\cos bT)z + e^{-2aT}}$
22	$\dfrac{a^2 + b^2}{s((s+a)^2 + b^2)}$	$1 - e^{-anT}\left(\cos bnT + \dfrac{a}{b}\sin bnT\right)$	$\dfrac{z(Az + B)}{(z-1)(z^2 - 2e^{-aT}(\cos bT)z + e^{-2aT})}$ $A = 1 - e^{-aT}\cos bT - \dfrac{a}{b}e^{-aT}\sin bT$ $B = e^{-2aT} + \dfrac{a}{b}e^{-aT}\sin bT - e^{-aT}\cos bT$

10.2.3 Relationship between *s* and *z*

For continuous systems, one often associates certain behavior for different pole locations in the s plane: oscillatory behavior for poles near the imaginary axis, exponential decay for poles on the negative real axis, and unstable behavior for poles with a positive real part. The same kind of association is also useful to designers for discrete systems. The equivalent characteristics in the z plane are related to those in the s plane by the expression

$$z = e^{sT} \tag{10.12}$$

where T = sample period. This is obtained by comparing the z transform of the sampled version of a signal with the Laplace transform of the signal itself. The z transform (Table 10.1) also includes the Laplace transforms, which demonstrate the $z = e^{sT}$ relationship in the denominators of all the table entries.

Figure 10.2 shows the mapping of lines of constant damping ζ and natural frequency ω from the s plane to the upper half of the z plane according to Eq. (10.12). The mapping has several important features:

1. The stability boundary is the unit circle, $|z| = 1$.
2. The small vicinity around $z = 1$ is essentially identical to the vicinity around $s = 0$.
3. z-Plane locations give response information normalized to the sample rate, rather than with respect to time as in the s plane.
4. The negative real z axis always represents a frequency of $\omega_s/2$, where ω_s = sample rate.
5. Vertical lines in the left-hand s plane (constant real part or time constant) map into circles within the unit circle.
6. Horizontal lines in the s plane (constant imaginary part or frequency) map into radial lines in the z plane.

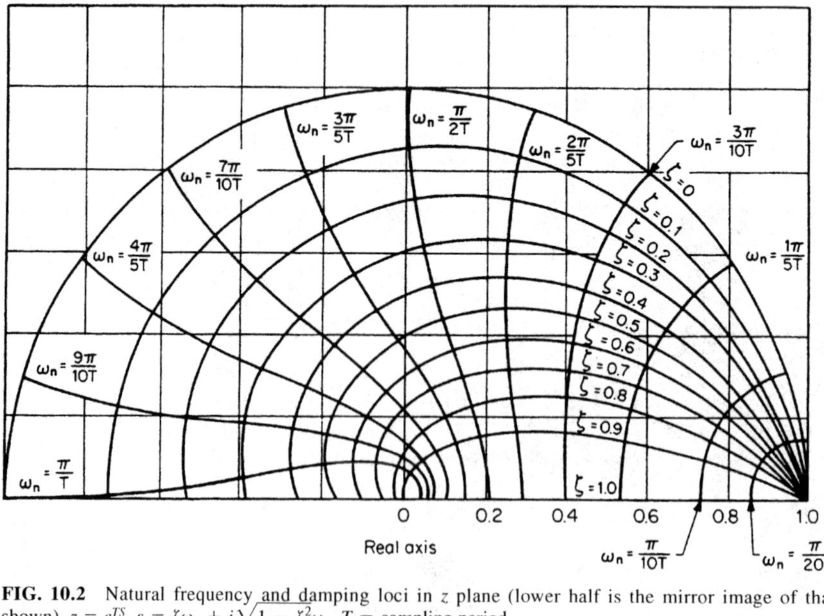

FIG. 10.2 Natural frequency and damping loci in z plane (lower half is the mirror image of that shown). $z = e^{TS}$, $s = \zeta\omega_n + i\sqrt{1 - \zeta^2}\omega_n$, T = sampling period.

7. There is no location in the z plane that represents frequencies greater than $\omega_s/2$. Physically, this is because you must sample at least twice as fast as a signal's frequency to represent it digitally and mathematically because of the nature of the trigonometric functions imbedded in Eq. (10.12).

10.2.4 Final Value Theorem

The final-value theorem for continuous systems

$$x(t) = \lim_{s \to 0} sX(s) \qquad \text{as } t \to \infty \tag{10.13}$$

is often used to find steady-state system errors and/or steady-state gains of portions of a control system. The analog for discrete systems is obtained by noting that a continuous steady response is denoted by $X(s) = A/s$ and leads to the multiplication by s in Eq. (10.13). Therefore, since the steady response for discrete systems is $X(z) = A/(1 - z^{-1})$, the discrete final-value theorem is

$$x(n) = \lim_{z \to 1} (1 - z^{-1})X(z) \qquad \text{as } n \to \infty \tag{10.14}$$

For example, to find the dc gain of the transfer function,

$$G(z) = X(z)/U(z) = 0.58(1 + z)/(z + 0.16),$$

let $u(n) = 1$ for $n \geq 0$, so that

$$U(z) = 1/(1 - z^{-1}) \qquad \text{and} \qquad X(z) = 0.58(1 + z)/(1 - z^{-1})(z + 0.16).$$

Applying the final-value theorem yields

$$x(\infty) = \lim_{z \to 1} 0.58(1 + z)/(z + 0.16) = 1$$

and therefore the dc gain of $G(z)$ is unity. In general, we see that to find the dc gain of any transfer function, we simply substitute $z = 1$ and compute the resulting gain.

Since the gain of a system does not change whether represented continuously or discretely, this calculation is an excellent check on the calculations associated with determining the discrete model of a system.

10.3 CONTINUOUS DESIGN AND DIGITIZATION

The first part of this design procedure is the design of feedback control compensation for a continuous system. This design is carried out as if the system were continuous and no changes are required to represent the fact that the control will eventually be implemented digitally.

10.3.1 Digitization Procedures

The second part of the procedure is to digitize the resulting compensation. Therefore, the problem to be addressed is: Given a $D(s)$, find the best equivalent $D(z)$. Or more exactly, given a $D(s)$ from the control system shown in Fig. 10.3, find the best digital implementation of that compensation. A digital implementation requires that y is sampled at some sample rate and that the computer output samples are smoothed in some manner so as to provide a continuous u. For ease of hardware design, the smoothing operation is almost always a simple hold (or zero-order hold, or ZOH) which is shown in Fig. 10.4.

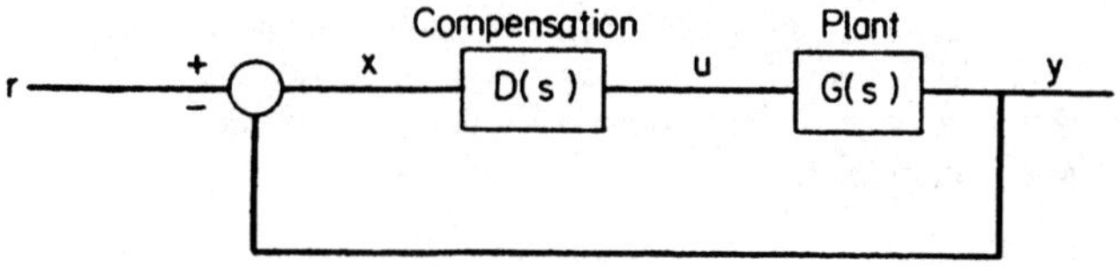

FIG. 10.3 Continuous control system.

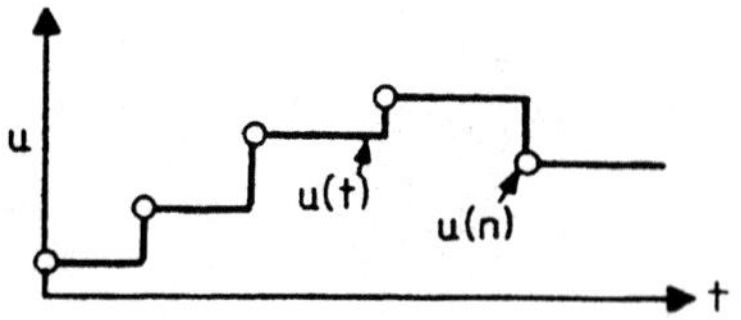

FIG. 10.4 Zero-order hold.

Therefore, we can restate the problem as: Find the best $D(z)$ in the digital implementation shown in Fig. 10.5 to match a desired $D(s)$. It is important to note at the outset that there is no exact solution to this problem because $D(s)$ responds to the complete time history of $x(t)$ whereas $D(z)$ only has access to the samples $x(n)$. In a sense, the various digitization approximations (and approximations they are) simply make different assumptions about what happens to $x(t)$ between the sample points.

Tustin's Method. One digitization method is to approach the problem as one of numerical integration. Suppose

$$U(s)/X(s) = D(s) = 1/s$$

i.e., pure integration. Therefore, as Fig. 10.6 shows,

$$\begin{aligned} u(nT) &= \int_0^{nT-T} x(t)\,dt + \int_{nT-T}^{T} x(t)\,dt \\ &= u(nT - T) + [\text{area under } x(t) \text{ over last } T] \end{aligned} \qquad (10.15)$$

where T = sample period. The function $u(nT)$ is often written $u(n)$ for short. The task at each step is to use trapezoidal integration, i.e., to approximate $x(t)$ by a straight line between the two samples. Therefore Eq. (10.5) becomes

$$u(nT) = u(nT - T) + 1/2T[x(nT - T) + x(nT)] \qquad (10.16)$$

or taking the z transform,

$$\frac{U(z)}{X(z)} = \frac{T}{2}\frac{1 + z^{-1}}{1 - z^{-1}} = \frac{1}{\dfrac{2}{T}\dfrac{1 - z^{-1}}{1 + z^{-1}}} \qquad (10.17)$$

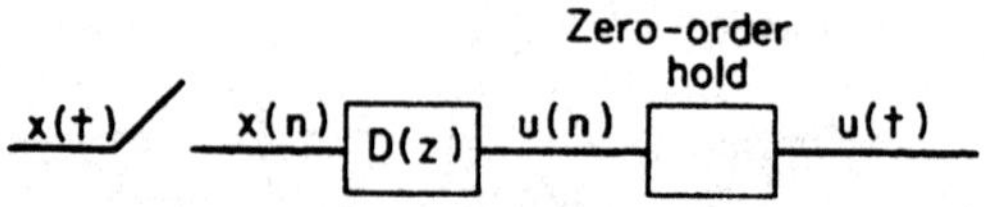

FIG. 10.5 Digital compensation implementation.

For $D(s) = a/(s + a)$, application of the same integration approximation yields

$$D(z) = \frac{a}{\dfrac{2}{T}\dfrac{1 - z^{-1}}{1 + z^{-1}} + a}$$

and, in fact, the substitution

$$s = \frac{2}{T}\frac{1 - z^{-1}}{1 + z^{-1}} \qquad (10.18)$$

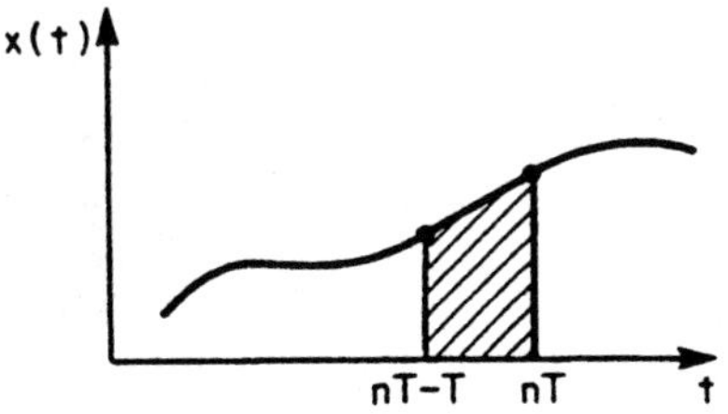

FIG. 10.6 Trapezoidal integration.

in any $D(s)$ yields a $D(z)$ based on the trapezoidal integration formula. This is called "Tustin's" or the "bilinear" approximation.

Matched Pole-Zero Method (MPZ). Another digitization method, called the "matched pole-zero" method, is found by extrapolation of the relation between the s and z planes stated in Eq. (10.12). If we take the z transform of a sampled $x(t)$, then the poles of $X(z)$ are related to the poles of $X(s)$ according to $z = e^{sT}$. However, we must go through the z transform process to locate the zeros of $X(z)$. The idea of the matched pole-zero technique is to apply $z = e^{sT}$ to the poles and zeros of a transfer function. Since physical systems often have more poles than zeros, it is also useful to arbitrarily add zeros of $D(z)$ at $z = -1$ [i.e., a $(1 + z^{-1})$ term] which causes an averaging of the current and past input values as in the trapezoidal integration (Tustin's) method. The gain is selected so that the low-frequency gain of $D(s)$ and $D(z)$ match one another. The method summarized is:

1. Map poles and zeros according to $z = e^{sT}$.
2. Add $(1 + z^{-1})$ or $(1 + z^{-1})^2$, etc., if numerator is lower order than the denominator.
3. Match dc or low-frequency gain.

For example, the matched pole-zero approximation of $D(s) = (s + a)/(s + b)$ is

$$D(z) = k(z - e^{-aT})/(z - e^{bT}) \qquad (10.19)$$

where $k = (a/b)/(1 - e^{-bT})/(1 - e^{-aT})$, and for $D(s) = (s + a)/s(s + b)$,

$$D(z) = k(z + 1)(z - e^{-aT})/(z - 1)(z - e^{-bT}) \qquad (10.20)$$

where $k = (a/2b)/(1 - e^{-bT})/(1 - e^{-aT})$.

In both digitization methods, the fact that an equal power of z appears in numerator and denominator of $D(z)$ implies that the difference equation output at time n will require a sample of the input at time n. For example, the $D(z)$ in Eq. (10.19) can be written

$$U(z)/X(z) = D(z) = k(1 - \alpha z^{-1})/(1 - \beta z^{-1})$$

which results in the difference equation

$$u(n) = \beta u(n - 1) + k[x(n) - \alpha x(n - 1)] \qquad (10.21)$$

Modified Matched Pole-Zero Method (MMPZ). The $D(z)$ in Eq. (10.20) would also result in $u(n)$ being dependent on $x(n)$, the input at the same time point. If the structure of the computer hardware prohibits this relation or if the computations are particularly

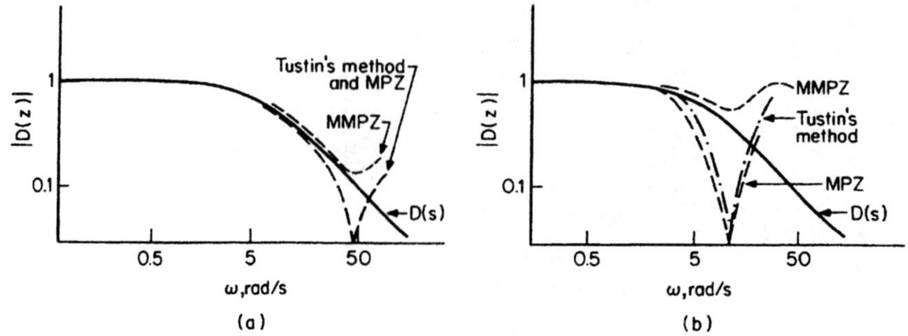

FIG. 10.7 Comparison of discrete approximations. (*a*) ω_s = 15 Hz ≈ 100 rad/s. (*b*) ω_s = 3 Hz ≈ 20*s* rad/s.

lengthy thus rendering Eq. (10.21) impossible to implement, it may be desirable to arrive at a $D(z)$ which has one less power of z in the numerator than denominator and hence the computer output, $u(n)$, requires only input from the previous time, i.e., $x(n - 1)$. To do this, we simply omit step 2 in the matched pole-zero procedure. The second example,

$$D(s) = (s + a)/s(s + b)$$

would then become

$$D(z) = k(z - e^{-aT})/(z - 1)(z - e^{-bT})$$

where $k = (a/b)/(1 - e^{-bT})/(1 - e^{-aT})$, which results in

$$u(n) = (1 + e^{-bT})u(n - 1) - e^{-bT}u(n - 2) + k[x(n - 1) - e^{-aT}x(n - 2)]$$

Method Comparison. A numerical comparison of the magnitude of the frequency response is made in Fig. 10.7 for the three approximation techniques at two sample rates. The results of the $D(z)$ computations used in arriving at Fig. 10.7 are shown in Table 10.2.

The figure shows that all the approximations are quite good at frequencies below about one-fourth the sample rate, $\omega_s/4$. If $\omega_s/4$ is sufficiently larger than the filter break frequency, i.e., if the sampling is fast enough, the break characteristics are accurately reproduced. Tustin's method and the MPZ method show a notch at $\omega_s/2$ due to their zero term $z + 1$. Other than the large difference at $\omega_s/2$ which is typically outside the range of interest, the three methods have similar accuracies. Since the MPZ techniques require much simpler algebra than Tustin's, they are typically preferred.

TABLE 10.2 Digital Approximations

	$D(z)$ for $D(s) = 5/(s+5)$	
	ω_s = 15 Hz	ω_s = 3 Hz
Matched pole-zero (MPZ)	$0.143\dfrac{z+1}{z-0.715}$	$0.405\dfrac{z+1}{z-0.189}$
Modified MPZ (MMPZ)	$0.285\dfrac{1}{z-0.715}$	$0.811\dfrac{1}{z-0.189}$
Tustin's	$0.143\dfrac{z+1}{z-0.713}$	$0.454\dfrac{z+1}{z-0.0914}$

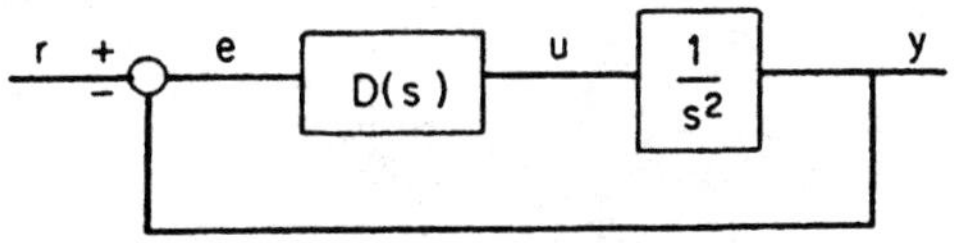

FIG. 10.8 Continuous design statement.

10.3.2 Design Example

For a $1/s^2$ plant, we wish to design a digital controller to have a closed-loop natural frequency ω_n, of ~0.3 rad/s and $\zeta = 0.7$. The first step is to find the proper $D(s)$ defined in Fig. 10.8. The specifications can be met with

$$D(s) = k(s + a)/(s + b) \quad (10.22)$$

where
$$a = 0.2$$
$$b = 2.0$$
$$k = 0.81$$

as can be verified by the root locus in Fig. 10.9. To digitize this $D(s)$, we first need to select a sample rate. For a system with $\omega_n = 0.3$ rad/s a very "safe" sample rate would be a factor of 20 faster than ω_n, yielding

$$\omega_s = 0.3 \times 20 = 6 \text{ rad/s}$$

Thus, let us pick $T = 1$ s. The matched pole-zero digitization of Eq. (10.22) is given by Eq. (10.19) and yields

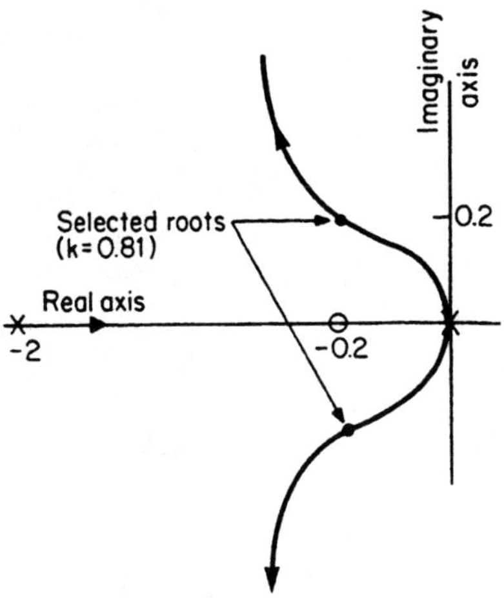

FIG. 10.9 s-Plane locus versus k.

$$D(z) = 0.389(z - 0.82)/(z - 0.135)$$

or

$$D(z) = (0.389 - 0.319z^{-1})/(1 - 0.135z^{-1}) \quad (10.23)$$

which leads to

$$u(n) = 0.135u(n - 1) + 0.389e(n) - 0.319e(n - 1)$$

where

$$e(n) = r(n) - y(n) \quad (10.24)$$

This completes the digital algorithm design. The complete digital system is shown in Fig. 10.10.

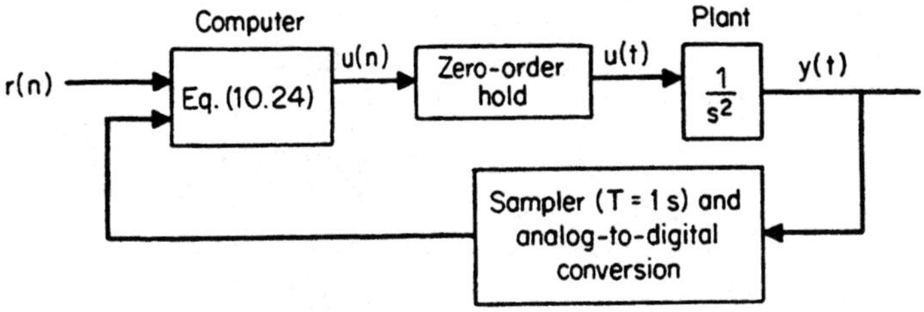

FIG. 10.10 Digital control system.

10.3.3 Applicability Limits of Method

If an exact discrete analysis or a simulation of the system was performed and the digitization was determined for a wide range of rates, the system would be unstable for sample rates slower than approximately $5\omega_n$ and the damping would be substantially degraded for sample rates slower than $10\omega_n$. At sample rates on the order of $20\omega_n$ (or 20 times bandwidth for more complex systems), this design method can be used with confidence.

Basically, the errors come about because the technique ignores the lagging effect of the zero-order hold (ZOH). An approximate method to account for this is to assume that the transfer function of the ZOH is

$$G_{\text{ZOH}}(s) = \frac{2/T}{s + 2/T} \tag{10.25}$$

This is based on the idea that, on the average, the hold delays by $T/2$ and the above is a first-order lag with a time constant of $T/2$, dc gain = 1. We could therefore patch the original $D(s)$ design by inserting this $G_{\text{ZOH}}(s)$ in the original plant model and finding the $D(s)$ that yields satisfactory response.

One of the advantages of using this design method, however, is that the sample rate need not be selected until after the basic feedback design is completed. Therefore the patching eliminates this advantage, although it does partially alleviate the approximate nature of the method, which is the primary disadvantage.

10.4 DISCRETE DESIGN

10.4.1 Analysis Tools

The first step in performing a control design or analysis of a system with some discrete elements in it is to find the discrete transfer function of the continuous portion. For a system similar to that shown in Fig. 10.1, we wish to find the transfer function between $u(kt)$ and $y(kt)$. Unlike the previous section, there is an exact discrete equivalent for this system because the ZOH precisely describes what happens between samples, and the output $y(kt)$ is only dependent on the input at the sample times $u(kt)$.

For a plant described by a $G(s)$ and preceded by a ZOH, the discrete transfer function is

$$G(z) = (1 - z^{-1})\mathscr{Z}\{G(s)/s\} \tag{10.26}$$

Generally, $\mathscr{Z}\{F(s)\}$ means the z transform of the time series whose Laplace transform is $F(s)$, i.e., the same line in Table 10.1. The formula has the term $G(s)/s$ because the control comes in as a step input during each sample period. The term $1 - z^{-1}$ is there because a one-sample-duration step can be thought of as an infinite-duration step followed by a negative step one cycle delayed. This formula [Eq. (10.26)] allows us to replace the mixed (continuous and discrete) system shown in Fig. 10.11*a* with the pure discrete equivalent system shown in Fig. 10.11*b*.

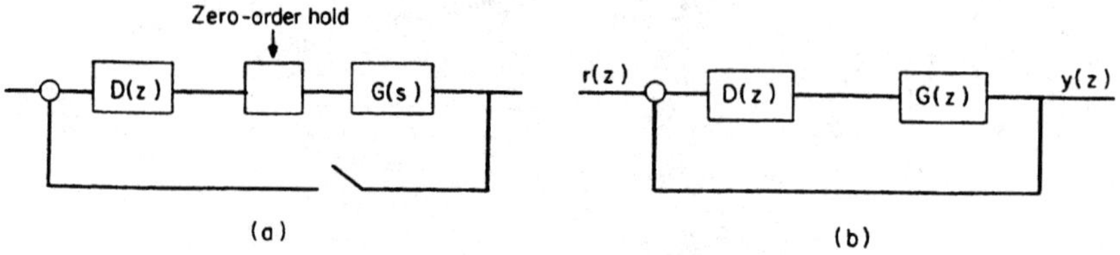

FIG. 10.11 Control system. (*a*) Mixed control system. (*b*) Pure discrete equivalent.

The analysis and design of discrete systems is very similar to continuous ones; in fact, the same rules apply. The closed-loop transfer function of Fig. 10.11*b* is obtained using the same rules of block diagram reduction, i.e.,

$$Y(z)/R(z) = DG/(1 + DG) \tag{10.27}$$

Since we would like to find the characteristic behavior of the closed-loop system, we wish to find the factors of the denominator of Eq. (10.27), i.e., find the roots of the characteristic equation

$$1 + D(z)G(z) = 0 \tag{10.28}$$

The root locus techniques used in continuous systems to find roots of a polynomial in s apply equally well here for the polynomial in z. The rules apply directly without modification; however, the interpretation of the results is quite different, as we saw in Fig. 10.2. A major difference is that the stability boundary is now the unit circle instead of the imaginary axis.

A simple example of the discrete design tools discussed so far follows: Suppose $G(s)$ in Fig. 10.11*a* is

$$G(s) = a/(s + a)$$

It follows from Eq. (10.26) that

$$\begin{aligned} G(z) &= (1 - z^{-1})\mathcal{L}\left\{\frac{a}{s(s + a)}\right\} \\ &= (1 - z^{-1})\left[\frac{(1 - e^{-aT})z^{-1}}{(1 - z^{-1})(1 - e^{-aT}z^{-1})}\right] \\ &= \frac{1 - \alpha}{z - \alpha} \end{aligned} \tag{10.29}$$

where $\alpha = e^{-aT}$.

To analyze the performance of a closed-loop proportional control law, i.e., $D(z) = k$, we use standard root locus rules. The result is shown in Fig. 10.12*a* and for comparison, the root locus for a continuous controller is shown in Fig. 10.12*b*. In contrast to the

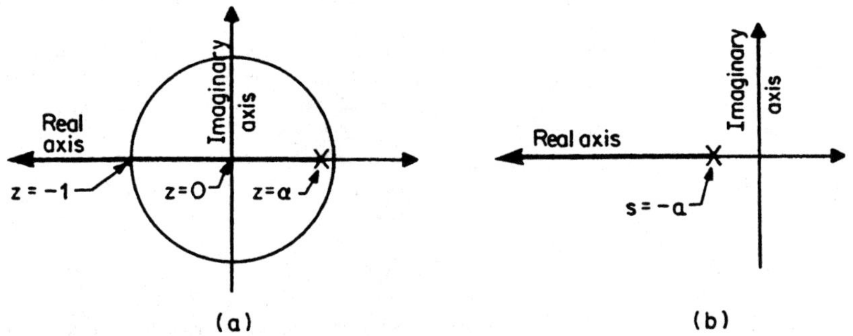

FIG. 10.12 Root locus. (*a*) *z*-Plane root locus. (*b*) *s*-Plane root locus.

continuous case which remains stable for all values of k, the discrete case becomes oscillatory with a decreasing damping ratio as z goes from 0 to -1 and eventually becomes unstable. This instability is caused by the lagging effect of the ZOH which is properly accounted for in the discrete analysis.

10.4.2 Feedback Properties

In continuous systems, we typically start the design process by using proportional, derivative, or integral control laws or combinations of these, sometimes with a lag included. The same ideas are used in discrete designs directly or perhaps the $D(z)$ that results from the digitization of a continuously designed $D(s)$ is used as a starting point.

The discrete control laws follow:

Proportional

$$u(n) = k_p e(n)$$

implies

$$D(z) = k_p \tag{10.30}$$

Derivative

$$u(n) = k_D[e(n) - e(n - 1)]$$

implies

$$D(n) = k_D(1 - z^{-1}) = k_D(z - 1)/z \tag{10.31}$$

Integral

$$u(n) = u(n - 1) + k_I e(n)$$

implies

$$D(z) = k_I/(1 - z^{-1}) = k_I z/(z - 1) \tag{10.32}$$

10.4.3 Design Example

For an example, let us use the same problem as we used for the continuous design: the $1/s^2$ plant. Using Eq. (10.26), we have

$$G(z) = \tfrac{1}{2}T^2(z + 1)/(z - 1)^2 \tag{10.33}$$

which becomes with $T = 1$ s

$$G(Z) = \tfrac{1}{2}(z + 1)/(z - 1)^2 \tag{10.34}$$

Proportional feedback in the continuous case yields pure oscillatory motion, and in the discrete case we should expect even worse results. The root locus in Fig. 10.13 verifies this. For very low values of k (very low frequencies compared with the sample rate) the locus is tangent to the unit circle ($\zeta \approx 0$ and pure oscillatory motion), thus matching the proportional continuous design. For higher values of k, the locus diverges into the unstable region because of the effect of the ZOH and sampling.

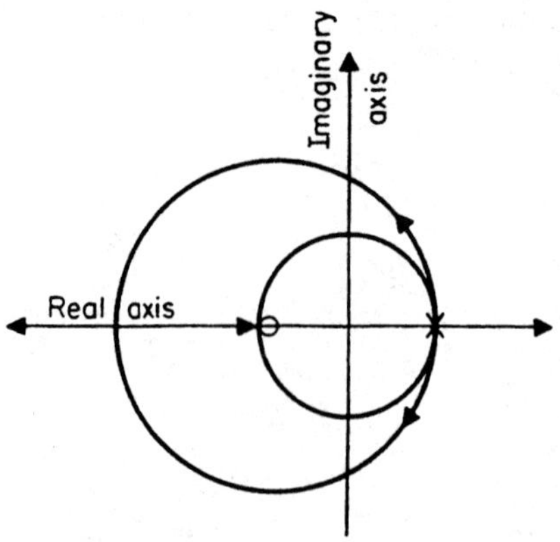

FIG. 10.13 z Plane for $1/s^2$ plant.

To compensate for this, add a velocity term to the control law, or

$$u(z) = k[1 + \gamma(1 - z^{-1})]e(z) \tag{10.35}$$

which yields

$$D(z) = k(1 + \gamma)\frac{z - \gamma/(1 + \gamma)}{z} \tag{10.36}$$

Now the task is to find values of γ and k that yield good performance. When we did this design previously, we wanted $\omega_n = 0.3$ rad/s and $\zeta = 0.7$. Figure 10.2 indicates that this s-plane root location maps into a z-plane location of

$$z = 0.8 \pm 0.2j$$

Figure 10.14 shows that for $\gamma = 4$ and $k = 0.08$, or

$$D(z) = 0.4(z - 0.8)/z \tag{10.37}$$

the roots are at the desired location. Normally, it is not particularly advantageous to match specific z-plane root locations; rather it is only necessary to pick k and γ to obtain acceptable z-plane roots, a much easier task. In this example, we wanted to match a specific location only so we could compare the result with the previous design.

The control law that results is

$$u(z) = 0.08[1 + 4(1 - z^{-1})]$$

or

$$u(n) = 0.4e(n) - 0.32e(n - 1) \tag{10.38}$$

10.4.4 Design Comparison

The controller designed using pure discrete methods, Eq. (10.38), basically only differs from the continuously designed controller, Eq. (10.24), by the absence of the $u(n - 1)$ term. The $u(n - 1)$ term in Eq. (10.24) resulted from the lag term $s + b$ in the compensation Eq. (10.22), which is typically included in analog controllers because of the difficulty in building pure analog differentiators and for noise attenuation. Some equivalent lag in discrete design naturally appears as a pole at $z = 0$ (see Fig. 10.14) and represents the one

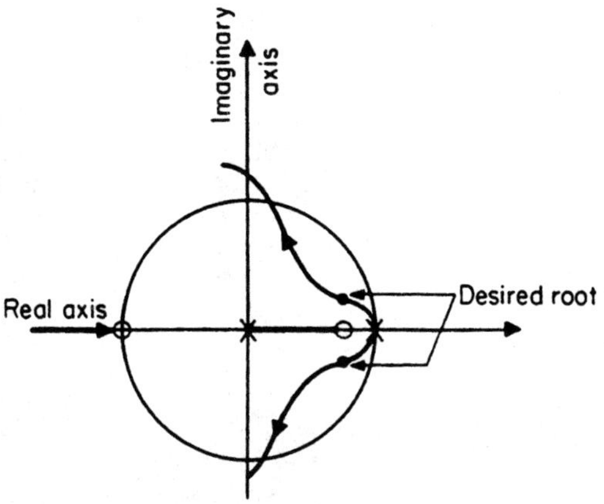

FIG. 10.14 Compensated z plane locus for $1/s^2$ example.

sample delay in computing the derivative by a first difference. For more noise attenuation, the pole could be moved to the right of $z = 0$, thus resulting in less derivative action and more smoothing, the same trade-off that exists in continuous control design.

Other than the $u(n - 1)$ term, the two controllers are very similar [Eqs. (10.24) and (10.38)]. This similarity resulted because the sample rate is fairly fast compared to ω_n, that is, $\omega_s \approx 20\omega_n$. For designs at slower sample rates, the numerical values in the compensations would become increasingly different as the sample rate decreased. For the discrete design, the actual system response would follow that indicated by the z-plane root locations, while the continuously designed system response would diverge from that indicated by the s-plane root locations.

As a general rule, discrete design should be used if sampling is slower than $10\omega_n$. At the very least, a continuous design with slow sampling ($\omega_s < 10\omega_n$) should be verified by a discrete analysis or simulation and the compensation adjusted if needed. A simulation of a digital control system is a good idea in any case. If it properly accounts for all delays and possibly asynchronous behavior of different digital modules, it may expose instabilities that are impossible to detect using continuous or discrete linear analysis.

10.5 HARDWARE CHARACTERISTICS

10.5.1 Analog-to-Digital Converters

These are the devices that convert a voltage level from a sensor to a digital word usable by the computer. At the most basic level, all digital words are binary numbers consisting of many "bits" that are set to either 1 or 0. Therefore, the task of the analog-to-digital (A/D) converter at each sample time is to convert a voltage level to the correct bit pattern and often to hold that pattern until the next sample time.

Of the many A/D conversion techniques that exist, the most common are based on counting schemes or successive approximation schemes. In counting methods, the input voltage may be converted to a train of pulses whose frequency is proportional to the voltage level. The pulses are then counted over a fixed period by a binary counter, which produces a binary representation of the voltage level. A variation on this scheme is to start the count simultaneously with a linear (in time) voltage and to stop the count when the voltage reaches the magnitude of the input voltage to be converted.

The successive approximation technique tends to be much faster than the counting methods. It is based on successively comparing the input voltage to reference levels representing the various bits in the digital word. The input voltage is first compared with a reference which is half the maximum. If greater, the most significant bit is set and the signal is then compared with a reference which is three-fourths of the maximum to determine the next bit, and so on. These converters require one clock cycle to set each bit, thus they would need 2^n cycles for an n-bit converter. At the same clock rate, a counter-based converter might require as many as two cycles, which would usually make the process much slower.

For either technique, it will take longer to perform the conversion for a higher number of bits, i.e., a more accurate converter. Not surprisingly, the price of A/D converters goes up with both speed and bit size.

If more than one channel of data need to be sampled and converted to digital words, it is usually accomplished by use of a "Multiplexer" rather than multiple A/D converters. This device sequentially connects the A/D converter into the channel being sampled.

10.5.2 Digital-to-Analog Converters

These devices are used to convert the digital words from the computer to a voltage level for driving actuators or perhaps a recording device such as an oscilloscope or strip-chart recorder. The basic idea is that the binary bits are used to cause switches (electronic gates) to open or close, thus routing the electric current through an appropriate network of resistors so that the correct voltage level is generated. Since no counting or iteration is required for digital-to-analog (D/A) converters, they tend to be much faster than A/D converters. In fact, D/A converters are a component in the A/D converters based on successive approximation.

10.5.3 Analog Prefilters

This device is often placed between the sensor and the A/D converter. Its function is to reduce the higher-frequency noise components in the analog signal so as to prevent the frequency of the noise from being switched to a lower frequency by the sampling process (called "aliasing" or "folding").

An example of aliasing is shown in Fig. 10.15, where a 60-Hz oscillatory signal is being sampled at 50 Hz. The figure shows that the result from the samples is a 10-Hz signal and the mechanism by which the frequency of the signal was aliased from 60 to 10 Hz. Aliasing will occur any time the sample rate is not at least twice as fast as any of the frequencies in the signal being sampled. Therefore, to prevent aliasing of a 60-Hz signal, the sample rate would have to be faster than 120 Hz rather than the 50 Hz in the figure.

This phenomenon is one of the consequences of the sampling theorem of Nyquist and Shannon. The theorem basically states that a signal must have no frequency components greater than half the sample rate ($\omega_s/2$) for the signal to be accurately reconstructed from the samples. Another consequence is that the highest frequency that can be represented by discrete samples is $\omega_s/2$, an idea that has already been discussed in Sec. 10.2.3.

The consequence of aliasing on a digital control system could be substantial. Noise components with a frequency much higher than the control system bandwidth normally have a small effect because the system will not respond at the high frequency. However, if the frequency of that noise was aliased down to the vicinity of the system bandwidth, the system would respond, thus causing the noise to appear on the system's output.

The solution is to place an analog prefilter before the sampler. In most cases a simple first-order low-pass filter will do, i.e.,

$$H_p(s) = a/(s + a)$$

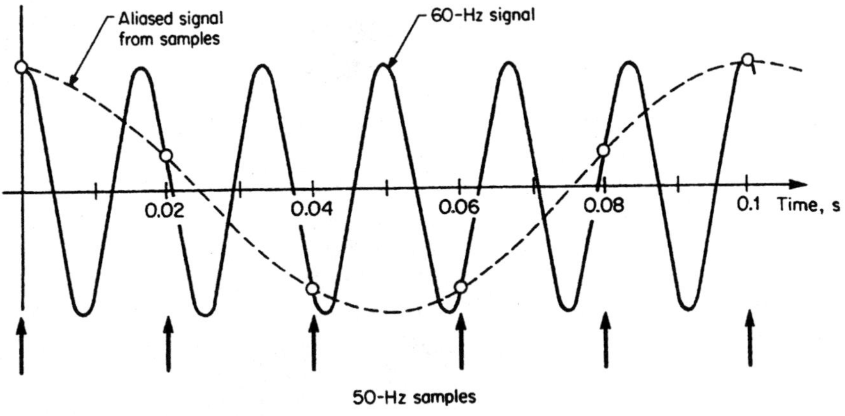

FIG. 10.15 Example of aliasing.

where the break point a is selected lower than $\omega_s/2$ so that any noise present with frequencies greater than $\omega_s/2$ would be attenuated by the prefilter. The lower the break-point frequency selected, the more the noise above $\omega_s/2$ is attenuated. However, too low a break point may reduce the control-system bandwidth. The prefilter does not eliminate the aliasing, but by judicious choice of the prefilter break point and the sample rate, the designer has the ability to reduce the magnitude of the aliased noise to some acceptable level.

10.6 WORD SIZE EFFECTS

A numerical value can be represented only with a limited precision in a digital computer. For fixed-point arithmetic, the resolution is 0.4 percent of full range for 8 bits and 0.1 percent for 10 bits; it drops by a factor of 2 for each additional bit. The effect of this limited precision shows up in the A/D conversion which often has a smaller word size than the computer, multiplication truncation, and parameter storage errors. If the computer uses floating point arithmetic, the resolution of the multiplication and parameter storage changes with the magnitude of the number being stored, the resolution affecting only the mantissa while the exponent essentially continually adjusts the full scale.

10.6.1 Random Effects

As long as a system has varying inputs or disturbances, A/D errors and multiplication errors act in a random manner on the system and essentially produce noise at the output of the system. The output noise due to a particular noise source (A/D or multiplication) has a mean value of

$$\bar{n}_0 = H_{\text{dc}}\bar{n}_I \tag{10.39}$$

where H_{dc} = dc gain of transfer function between noise source and output and $\bar{n}_i$ = mean value of noise source. The mean value of the noise source will be zero for a rounding-off process but has a value

$$\bar{n}_I = q/2$$

where q = resolution level for a truncation process. Although most A/D converters round off without mean error, some truncate, producing an error. The total noise effect is the sum of all noise sources.

The variance of the output noise is always nonzero, irrespective of whether the process truncates or rounds. It is[3]

$$\sigma_0^2 = \sigma_I^2 \frac{1}{2\pi j}\oint H(z)H(z^{-1})\frac{dz}{z} \tag{10.40}$$

where σ_0 = output noise variance
σ_I = input noise variance
$H(z)$ = transfer function between noise input and system output

The input noise variance has magnitude

$$\sigma_I^2 = q^2/12 \tag{10.41}$$

for either rounding off or truncation.

In general, evaluation of noise response using Eq. (10.40) would show that the discrete portion of the controller becomes more sensitive to noise as the sampling rate increases. However, this trend with sampling rate is partly counterbalanced by the decreasing total system response cause by the increasing noise frequency.

The sensitivity of a system of A/D errors can be partly alleviated by adding lag in the digital controller or by adding more bits to the A/D converter. Different structures of the digital controller have no effect.

On the other hand, multiplication errors can be reduced substantially for high-order (greater than second order) controllers by proper structuring of a given control transfer function. For example, a second-order transfer function with real roots,

$$D(z) = U(z)/E(z) = (z - 0.8)/(z - 0.2)(z - 0.3) \tag{10.42}$$

can be implemented in a *direct* manner yielding

$$u(n) = 0.5u(n - 1) - 0.06u(n - 2) + e(n) - 0.8e(n - 1) \tag{10.43}$$

or could be implemented in a *parallel* manner that results from a partial fraction expansion of Eq. (10.42). The result is shown in Fig. 10.16 and yields

$$\begin{aligned} x_1(n) &= 0.2x_1(n - 1) + 6e(n) \\ x_2(n) &= 0.3x_2(n - 1) + 5e(n) \\ u(n) &= x_1(n) - x_2(n) \end{aligned} \tag{10.44}$$

Note that the transfer functions to the output from the multiplications in Eq. (10.43) are substantially different from those in Eq. (10.44). It is also possible to implement the $D(z)$ with a *cascade* factorization which would be two first-order blocks arranged serially for this example.

Either cascade or parallel implementations are preferred to the direct implementation, and for transfer functions higher than third order they are almost mandatory.

10.6.2 Systematic Effects

Parameters such as the numerical values in Eqs. (10.43) and (10.44), if in error, will change dynamic behavior of a system. In a high-order controller with a direct implementation, a very small percentage error in a stored parameter can result in substantial root location changes and sometimes cause instability. In the Apollo command module, a 14-bit word size for parameter storage would have resulted in instability if a direct implementation had been used in the sixth-order compensator. These effects are amplified when there are two compensator poles close together (or repeated) and at fast sample rates where all poles tend to clump around $z = 1$ and are close to one another. These effects can be reduced by using larger word sizes, parallel or cascade implementations, double-precision parameter storage, and slower sample rates.

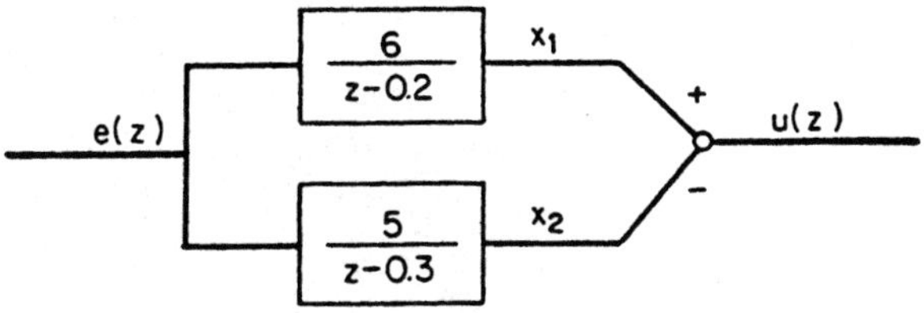

FIG. 10.16 Parallel implementation.

Under conditions of constant disturbances and input commands, multiplication errors can also cause systematic errors. Typically, the result is a steady-state error or possibly a stable limit cycle. The steady-state error results from a dead band that exists in any digital controller. The magnitude of the dead band is proportional to the dc gain from the multiplication error to the output and to the resolution level. Stable limit cycles occur only for controllers with lightly damped poles. These effects are also reduced by larger word sizes, parallel or cascade implementations, and slower sample rates.

10.7 SAMPLE RATE SELECTION

The selection of the best sample rate for a digital control system is a compromise among many factors. The basic motivation to lower the sample rate ω_s is cost. A decrease in sample rate means more time is available for the control calculations, hence slower computers are possible for a given control function or more control capability is available for a given computer. Either result lowers the cost per function. For systems with A/D converters, less demand on conversion speed will also lower cost. These economic arguments indicate that the best engineering choice is the slowest possible sample rate which meets all performance specifications.

Factors which could provide a lower limit to the acceptable sample rate are:

1. Tracking effectiveness as measured by closed-loop bandwidth or by time response requirements, such as rise time and settling time
2. Regulation effectiveness as measured by the error response to random plant disturbances
3. Sensitivity to plant parameter variations
4. Error caused by measurement noise and the associated prefilter design methods

A fictitious limit occurs when continuous design techniques are used. The inherent approximation in the method may give rise to system instabilities as the sample rate is lowered, and this can lead the designer to conclude that a lower limit on ω_s has been reached when in fact the proper conclusion is that the approximations are invalid; the solution is not to sample faster but to switch to the direct digital design method.

10.7.1 Tracking Effectiveness

An absolute lower bound to the sample rate is set by a specification to track a command input with a certain frequency (the system bandwidth). The *sampling theorem*[3] states that in order to reconstruct an unknown band-limited continuous signal from samples of that signal, one must sample at least twice as fast as the highest frequency contained in the signal. Therefore, in order for a closed-loop system to track an input at a certain frequency, it must have a sample rate twice as fast; that is, ω_s must be at least twice the system bandwidth ($\omega_s \geq 2\omega_{BW}$). We also saw from the z-plane mapping, $z = e^{sT}$, that the highest frequency that can be represented by a discrete system is $\omega_s/2$, supporting the conclusion above.

It is important to note the distinction between the closed-loop bandwidth ω_{BW} and the highest frequencies in the open-loop plant dynamics since these two frequencies can be quite different. For example, closed-loop bandwidths could be an order of magnitude *less* than open-loop modes of resonances for some vehicle control problems. Information concerning the state of the plant resonances for purposes of control can be extracted from sampling the output without satisfying the sampling theorem because some a priori knowledge is available (albeit imprecise) concerning these dynamics and the system is *not* required to track these frequencies. Thus a priori knowledge of the dynamic model of the plant can be included in the compensation in the form of a notch filter.

The closed-loop bandwidth limitation provides the fundamental lower bound on the sample rate. In practice, however, the theoretical lower bound of sampling at twice the bandwidth of the reference input signal would not be judged sufficient in terms of the quality of the desired time responses. For a system with a rise time on the order of 1 s and a required closed-loop bandwidth on the order of 0.5 Hz, it is not unreasonable to insist on a sample rate of 2 to 10 Hz, which is a factor of 4 to 20 times ω_{BW}. This will reduce the delay between a command and the system response to the command and will smooth the system output response to the control steps coming out of the ZOH.

10.7.2 Disturbance Rejection

Disturbance rejection is an important aspect of any control system, if not the most important one. Disturbances enter a system with various frequency characteristics ranging from steps to white noise. For purposes of sample rate determination, the higher frequency random disturbances are the most influential.

The ability of the control system to reject disturbances with a good continuous controller represents a lower bound on the error response that can be hoped for when implementing the controller digitally. In fact, some degradation over the continuous design must occur because the sampled values are slightly out of date at all times except precisely at the sampling instants. However, if the sample rate is very fast compared with the frequencies contained in the noisy disturbance, no appreciable loss should be expected from the digital system compared with the continuous controller. At the other extreme, if the sample time is very long compared with the characteristic frequencies of the noise, the response of the system due to noise is essentially the same as would result if there were no control at all. The selection of a sample rate will place the response somewhere in between these two extremes, and thus the impact of sample rate on the disturbance rejection of the system may be very influential to the designer in selecting the sample rate.

Although the best choice of sample rate in terms of the ω_{BW} multiple is dependent on the frequency characteristics of the noise and the degree to which random disturbance rejection is important to the quality of the controller, sample rate requirements of 10 or 20 times ω_{BW} are not uncommon.

10.7.3 Parameter Sensitivity

Any control design relies to some extent on knowledge of the parameters representing plant dynamics. Discrete systems exhibit an increasing sensitivity to parameter errors for a decreasing ω_s when the sample interval becomes comparable to the period of any of the open-loop vehicle dynamics. For systems with all plant dynamics in the vicinity of the closed-loop bandwidth or slower, root location changes due to parameter errors would probably not be a constraining factor unless the parameter error was quite large. However, for systems with a structural (or other) resonance which is stabilized by a notch filter, imperfect knowledge of the plant resonance characteristics will lead to changes in the system roots and possibly instabilities. This sensitivity to plant parameters increases as the sample rate decreases and could limit how slow the sample rate is in some cases. Typically, however, some other factor limits the sample rate and, at worst, some effort needs to be made in the controller design to minimize its sensitivity.

10.7.4 Control System Modularization

Prefilter Design. Digital control systems with analog sensors typically include an analog prefilter between the sensor and the sampler or A/D converter as an antialiasing device. The prefilters are low-pass, and the simplest transfer function is

$$H_p(s) = a(s + a) \tag{10.45}$$

so that the noise above the prefilter break point a is attenuated. The design goal is to provide enough attenuation at half the sample rate ($\omega_s/2$) so that the noise above $\omega_s/2$, when aliased into lower frequencies by the sampler, will not be detrimental to the control system performance.

A conservative design procedure is to select the break point and ω_s sufficiently higher than the system bandwidth so that the phase lag from the prefilter does not significantly alter the system stability, and thus the prefilter can be ignored in the basic control-system design. Furthermore, for a good reduction in the high-frequency noise at $\omega_s/2$, the sample rate is selected about 5 or 10 times higher than the prefilter break point. The implication of this prefilter design procedure is that sample rates need to be on the order of 20 to 100 times faster than the system bandwidth. If done this way, the prefilter design procedure is likely to provide the lower bound on the selection of the sample rate.

An alternative design procedure is to allow significant phase lag from the prefilter at the system bandwidth and thus to require that the control design be carried out with the analog prefilter characteristics included. This procedure allows the designer to use very low sample rates, but at the expense of increased complexity in the original design since the prefilter must be included in the plant transfer function. Using this procedure and allowing low prefilter break points, the effect of sample rate on sensor noise is small and essentially places no limits on the sample rate.

General Comments. Divorcing the prefilter design from the control law design may result in the necessity for a faster sample rate than otherwise. This same result may exhibit itself for other types of modularization. For example, a smart sensor with its own computer running asynchronously from the primary control computer will not be amenable to direct digital design because the overall system transfer function depends on the phasing. Therefore, sample rates on the order of $10\omega_{BW}$ or slower will not be possible.

In fact, any type of modularization which does not provide for one clock synchronizing the entire system will typically necessitate sample rates on the order of $20\omega_{BW}$ or higher.

REFERENCES

1. Dorf, R. C.: "Modern Control Systems," Addison-Wesley Publishing Co., Inc., Reading, Mass., 1980.
2. Ogata, K.: "Modern Control Engineering," Prentice-Hall, Inc., Englewood Cliffs, N.J., 1970.
3. Franklin G. F., and J. D. Powell: "Digital Control of Dynamic Systems," Addison-Wesley Publishing Co., Inc., Reading, Mass., 1980.

CHAPTER 11
OPTICAL SYSTEMS

Warren J. Smith, B.S.
Vice President, Research and Development
Santa Barbara Applied Optics
A Subsidiary of Infrared Industries Inc.
Santa Barbara, Calif.

11.1 BASIC DEFINITIONS AND CONVENTION 11.2
11.2 IMAGE SIZE AND LOCATION (FIRST-ORDER OPTICS) 11.6
11.2.1 Image Position 11.6
11.2.2 Magnification 11.6
11.2.3 Paraxial Ray Tracing 11.7
11.2.4 Element Cardinal Points 11.8
11.2.5 Multielement or Multicomponent Systems 11.8
11.2.6 Two-Component Systems 11.9
11.3 EXACT RAY TRACING 11.10
11.3.1 Spherical Surfaces 11.11
11.3.2 Aspheric Surfaces 11.12
11.3.3 Meridional Ray Tracing and Coddington's Equations 11.13
11.3.4 Graphical Ray Tracing 11.13
11.4 ABERRATION DESCRIPTIONS 11.14
11.4.1 Lateral, Longitudinal, and Angular Aberrations 11.14
11.4.2 Spherical Aberration 11.14
11.4.3 Coma 11.15
11.4.4 Field Curvature and Astigmatism 11.15
11.4.5 Distortion 11.16
11.4.6 Chromatic Aberrations 11.16
11.4.7 Longitudinal Aberration Representation 11.17
11.4.8 Lateral Aberration Representation 11.18
11.4.9 Spot Diagrams, Point- and Line-Spread Functions 11.18
11.4.10 Aplanatic Surfaces 11.19
11.4.11 Symmetrical Systems 11.21
11.5 DIFFRACTION 11.21
11.5.1 Diffraction Image 11.21
11.5.2 Gaussian (Laser) Beams 11.21
11.5.3 Point Resolution: The Rayleigh and Sparrow Criteria 11.23
11.6 IMAGE QUALITY CRITERIA 11.24
11.6.1 Rayleigh Quarter-Wave Limit 11.24
11.6.2 Strehl Definition 11.24
11.6.3 Optical Transfer Function 11.24
11.6.4 Specific Modulation Transfer Functions 11.25
11.6.5 Sine Waves and Square Waves 11.27
11.6.6 Aerial Image Modulation Curve 11.28
11.6.7 Depth of Focus 11.28
11.7 AFOCAL SYSTEMS AND TELESCOPES 11.29
11.7.1 Magnification, Apertures, and Fields of View 11.29
11.7.2 Astronomical Telescope 11.30
11.7.3 Galilean Telescope 11.31
11.7.4 Terrestrial Telescope 11.31
11.7.5 Relay Lenses 11.31
11.7.6 Field Lenses 11.31
11.8 MICROSCOPES 11.32
11.8.1 Simple Microscopes (Magnifying Glass) 11.32
11.8.2 Compound Microscopes 11.32
11.9 DETECTOR OPTICS 11.33
11.9.1 Detector Size Limitations, and Field Lenses 11.33
11.9.2 Immersion Lenses 11.34
11.10 FIBER OPTICS 11.34
11.10.1 Total Internal Reflection 11.34
11.10.2 Optical Fibers 11.34
11.10.3 Numerical Aperture and Cladding 11.34
11.10.4 Resolution of Fiber Optics Images 11.35

11.1 BASIC DEFINITIONS AND CONVENTIONS[47]

This chapter deals with optical systems, that is, lenses, mirrors, and combinations of these elements.* The media of propagation of light are assumed to be isotropic, and optical systems are assumed to be axially symmetrical, that is, to be composed of surfaces which are figures of rotation and whose axes of symmetry coincide with the optical axis.

Anamorphic System: An optical system with a different power (or magnification) in one meridian from that in the other.[15]

Angle of Incidence: The angle between an incident ray and the normal to a surface at the point of incidence of the ray (Fig. 11.1).

Angle of Refraction (Reflection): The angle between an emergent refracted (reflected) ray and the normal to the surface at the point of emergence of the ray (Fig. 11.1).

Axis, Optical: The common axis of symmetry of an optical system; in an element, the line between the centers of curvature of the two (axially symmetric) surfaces.

Back Focal Length (BFL), or Back Focus: The distance from the last vertex of the optical system to the back focal point (Fig. 11.2).

Back (or Second) Focal Point: The focal point to which paraxial rays parallel to the axis and incident on the optical system from the left are converged (Fig. 11.2).

Baffles: Opaque diaphragms which prevent the propagation of light through the system by reflection or scattering from the mechanical (nonoptical) elements.

Brewster's Angle: The light reflected from a surface is completely polarized if the angle of incidence is Brewster's angle, $I = \arctan (n'/n)$.

Cardinal Points: The focal points, the principal points, and the nodal points (Fig. 11.2).

Chief Ray: A ray directed toward the center of the entrance pupil of the optical system.

Component: One or more elements of an optical system which are treated as a unit, e.g., a cemented doublet.

Diopter: See "power."

Effective Focal Length (EFL): The distance from the second principal point to the back (or second) focal point. The distance from the front (or first) focal point to the first principal point (Fig. 11.2).

Element: A single indivisible entity of an optical system, usually a lens composed of refractive material bounded by two surfaces or revolution.

Eye Relief: In a visual instrument (e.g., a telescope or microscope), the distance from the last optical surface to the (usually external) exit pupil. Thus the clearance or "relief" between the instrument and the eye.[24,47]

Focal Length: See "effective focal length."

Focal Point: The point to which (paraxial) rays parallel to the optical axis converge or appear to converge after passing through the optical system (Fig. 11.2).

Front Focal Length (FFL): The distance from the front vertex of the optical system to the front focal point (Fig. 11.2).

---

* See Refs. 8, 25, 26, 30, 36, 39, 47, 51, and 55 for basic information on optics. Other sources of information on optical systems include Refs. 1 to 3, 9, 12, 17, 21, 24, 28, 32, 34, 35, 44, 45, 48 to 50, 56, and 57.

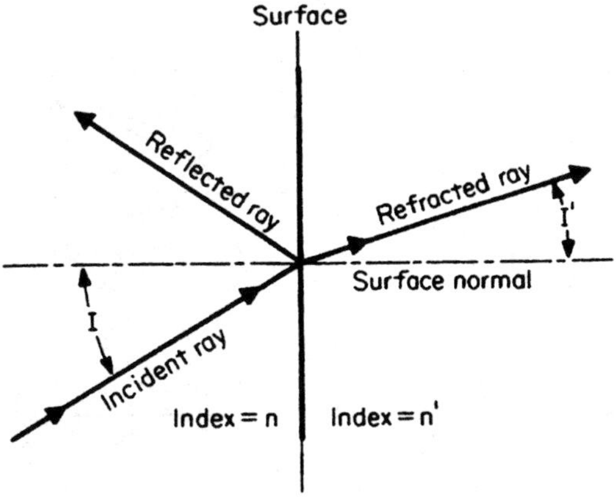

FIG. 11.1 Refraction at an optical surface. The plane of incidence (and refraction) is the plane of the paper.

Front (or First) Focal Point: The focal point to which (paraxial) rays parallel to the axis and incident on the optical system from the right are converged (Fig. 11.2).

Glare Stop: An opaque diaphragm located at the image of the aperture stop, i.e., at a (usually internal) pupil, to intercept light scattered from the walls and other parts of the instrument.

Index: See "Refractive index."

Invariant, Optical: When two unrelated paraxial rays (i.e., with different axial intercepts) are traced through an optical system, their data are sufficient to completely define the system. If the ray height and slope data of the marginal and oblique (chief, or principal) rays are identified by y, u and y_p, u_p, the expression $I = n(y_p u - y u_p)$ (where n is the index of refraction of the medium) is invariant across any surface or space, or series of surfaces and spaces, of the optical system. At an object or image plane (where $y = 0$ and $y_p = h$), the invariant reduces to the *Lagrange invariant* $I = hnu = h'n'u'$, where h and h' are the object and image height, respectively.

Magnification, Angular: The ratio of the angular size of an image (produced by an afocal optical system) to the angular size of the corresponding object.

Magnification, Microscopic: The ratio of the angular size of an image to the angular size of the object viewed at a conventional distance. For visual work, the conventional distance is 250 mm (10 in).

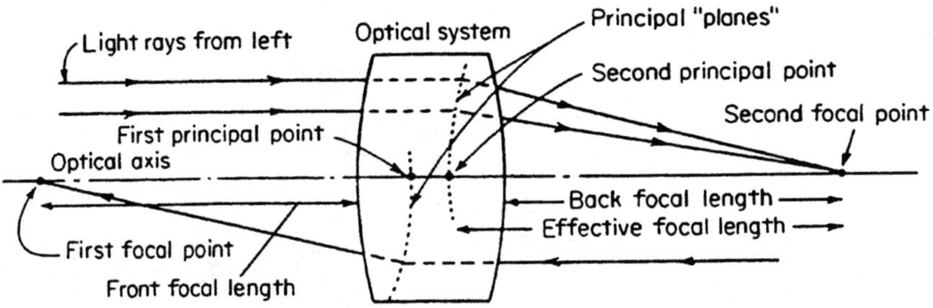

FIG. 11.2 The focal points and principal points (cardinal points) of a generalized optical system.

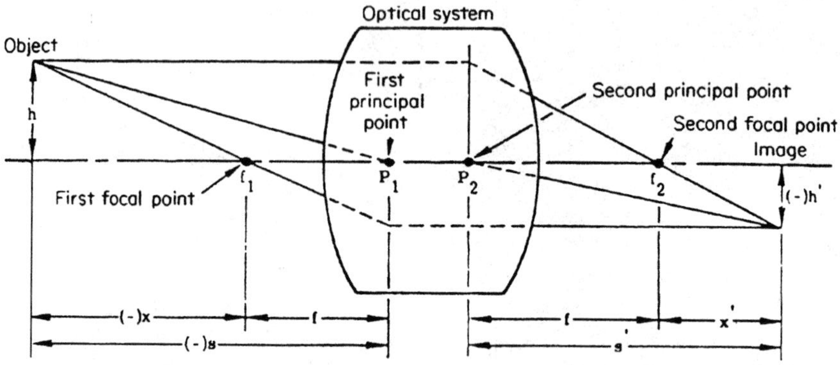

FIG. 11.3 Object and image relationships.

Magnification, Lateral or Linear: The ratio of the size of an image (measured perpendicular to the optical axis) to the size of the object (Fig. 11.3).

Magnification, Longitudinal: The ratio of the length or depth (measured along the optical axis) of an image to the length of the object (Fig. 11.4).

Marginal Ray: The ray from the axial point on the object which intersects the rim of the aperture stop.

Member: All the components either ahead of the aperture stop (front member) or behind it (rear member).

Meridional: The "meridional plane" is any plane which includes the optical axis. A "meriodional ray" is one which lies in the meridional plane and is thus coplanar with the axis.

Nodal Points: Two axial points so located that an oblique ray directed toward one appears to emerge from the other, parallel to its original direction. For optical systems in air, the nodal points coincide with the *principal points*.

Numerical Aperture (NA): NA $= n' \sin U'$, where n' is the final index in an optical system and U' is the slope angle of the marginal ray at the image. If the object is at infinity, NA $= 0.5/(f/\text{number})$.

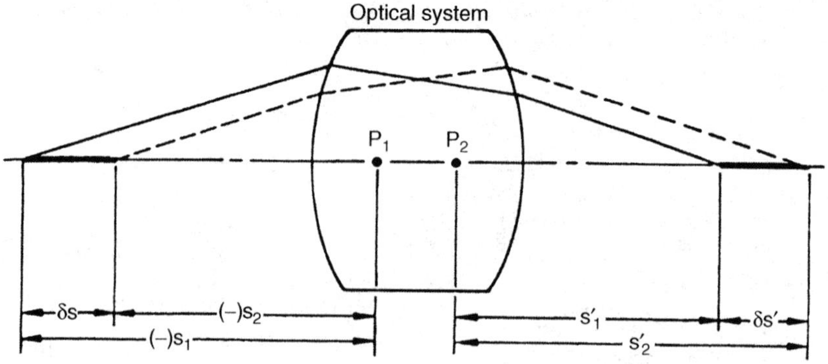

FIG. 11.4 Longitudinal magnification.

Optical Axis: See "axis, optical."

Paraxial: Pertaining to an infinitesimal threadlike region about the optical axis.

Plane of Incidence: The plane in which the incident ray, the normal to the surface at the point of incidence, and the refracted (or reflected) ray all lie in the same plane (Fig. 11.1).

Power: The reciprocal of the effective focal length of an optical element or system. The power is in diopters when the focal length is in meters.

Principal Planes: Planes only in the paraxial region; at a finite distance from the axis they are figures of rotation, frequently approximating spheres. If each ray of a bundle incident on an optical system parallel to the axis is extended to meet the backward extension of the ray after passing through the system, the locus of the intersections of all the rays is called a "principal surface." The second principal plane is defined by rays from the left, the first by rays from the right (Fig. 11.2).

Principal Points: The intersection of the principal planes with the optical axis (Fig. 11.2).

Principal Ray: Strictly, a ray directed toward the first principal point; however the term is frequently used to refer to the *chief ray*.

Pupil: An image of the aperture stop.

Pupil, Entrance: The image of the aperture stop formed by the optical elements between it and the object. It is the image of the stop as "seen" from the object.

Pupil, Exit: The image of the aperture stop formed by the elements behind the stop.

Refractive Index: The ratio of the velocity of light in vacuum to that in the medium.

Relative Aperture, or f/Number: The ratio of effective focal length to entrance pupil diameter. If the object is at infinity, *f*/number = 0.5/NA.

Sagittal: Pertaining to, or rays lying in, a plane normal to the meridional plane.

Sign Conventions: Light rays are assumed to progress from left to right. Radii and curvatures are positive if the center of curvature is to the right of the surface. Surfaces or elements have positive power if they converge light. Distances upward (or to the right) are positive; that is, points which lie above the axis (or to the right of an element, surface, or another point) are considered to be a positive distance away. Slope angles are positive if the ray is rotated counterclockwise to reach the axis. (This is the reverse of the usual geometrical convention.) Angles of incidence, refraction, and reflection are positive if the ray is rotated clockwise to reach the normal to the surface. The index of refraction is positive when light travels in the normal left-to-right direction. When light travels from right to left, as after a reflection, the index is taken as negative, as is the distance to the "next" surface, since it is to the left.

Snell's Law: The angles of incidence and refraction and the refractive indices on either side of an optical surface are related by $n_I \sin I = n_R \sin R$ (Fig. 11.1).

Speed: Relative aperture.

Stop, Aperture: The physical diameter which limits the size of the cone of radiation which an optical system will accept from an axial object point. For off-axis points, the limiting aperture may be defined by more than one physical feature of the system.

Stop, Field: The physical diameter which limits the angular field of view of an optical system. Although the apertures and field stops in a well-designed system are often well-defined, this is not a universal condition. For the field stop especially, there may be more than one physical diameter which limits the field of view; if these are not in an image plane, they produce "vignetting," a gradual reduction or obscuration of the size of the exit pupil as the field angle is increased. In some systems, especially anamorphic systems, the stops in one meridian may be completely different from those in the other.

Total Internal Reflection (TIR): When light is incident on a surface separating two media of indices n and n', the light is totally reflected if the angle of incidence exceeds the critical angle $I_c = \arcsin(n'/n)$. TIR can occur only when n exceeds n', typically when light passes from glass to air.

Varifocal or Zoom Lens: A lens whose effective focal length can be varied by moving one or more elements along the axis; the lens is designed so that the image position is relatively constant as the power is varied.[7,31]

Vignetting: The partial blocking of an oblique bundle of light rays by the limiting diameters of the optical system.

Window, Entrance: The image of the field stop formed by the optical elements between the stop and the object. If the field stop is located in an image plane, the entrance window will lie in the object plane.

Window, Exit: The image of the field stop formed by the elements behind the stop.

11.2 IMAGE SIZE AND LOCATION (FIRST-ORDER OPTICS)

The following equations apply rigorously and exactly to the paraxial characteristics of any optical system, simple or complex. Although these paraxial relationships are strictly valid for only a thin threadlike, infinitesimal region near the optical axis, most well-corrected systems closely approximate the performance given by these relationships.

11.2.1 Image Position (See Fig. 11.3)

$$1/s' = 1/s + 1/f \tag{11.1}$$

$$x' = -f^2/x \tag{11.2}$$

11.2.2 Magnification

Image Size: *Lateral Magnification* (See Fig. 11.3)

$$m = h'/h = s'/s = f/x = -x'/f \tag{11.3}$$

Longitudinal Magnification (See Fig. 11.4)

$$m_L = (s'_2 - s'_1)/(s_2 - s_1) = (s'_1/s_1)(s'_2/s_2) = m_1 m_2 \approx m^2 \tag{11.4}$$

Equations (11.1) to (11.3) can be combined to give

$$s = f(1/m - 1) \qquad s' = f(1 - m) \tag{11.5}$$

Thin-lens overall track length

$$= -s + s' = -(f/m)(m - 1)^2 \tag{11.6}$$

11.2.3 Paraxial Ray Tracing[47]

The paraxial equations can be derived by reducing the exact ray-tracing equations (Sec. 11.3) to the limit (about the axis) where the angles are infinitesimal and are equal to their sines and tangents.

Paraxial Image Location, Single Surface (See Fig. 11.5)

$$n'/l' = n/l + (n' - n)/r = n/l + (n' - n)C \tag{11.7}$$

$$m = h'/h = nl'/n'l \tag{11.8}$$

Paraxial Ray Tracing[47](See Fig. 11.6). Note that the angles u and heights y in the following equations are fictitious and may be canceled to yield Eqs. (11.7) and (11.8) above. The following equations are more convenient for tracing ray paths:

Opening Equations (relating the object to the first surface)

$$n_1u_1 = n_1y_1/l_1 \tag{11.9}$$

$$= n_1h_1/(l_1 - s_1) \tag{11.10}$$

Iterative Equations (applied to each surface in turn, $j = 1, 2, \ldots, k$)

$$n'_ju'_j = n_ju_j + (n'_j - n_j)y_jC_j \tag{11.11}$$

$$y_{j+1} = y_j - t'_jn'_ju'_j/n'_j \tag{11.12}$$

Alternative Iterative Equations

$$i_j = y_jC_j - u_j \tag{11.13}$$

$$i'_j = n_ji_j/n'_j \tag{11.14}$$

$$u'_j = u_j + i_j - i'_j \tag{11.15}$$

$$y_{j+1} = y_j - t'_ju'_j \tag{11.16}$$

Closing Equations (relating last surface to image)

$$l'_k = n'_ky_k/n'_ku'_k = y_k/u'_k \tag{11.17}$$

$$h'_k = y_k - u'_ks'_k = u'_k(l'_k - s'_k) \tag{11.18}$$

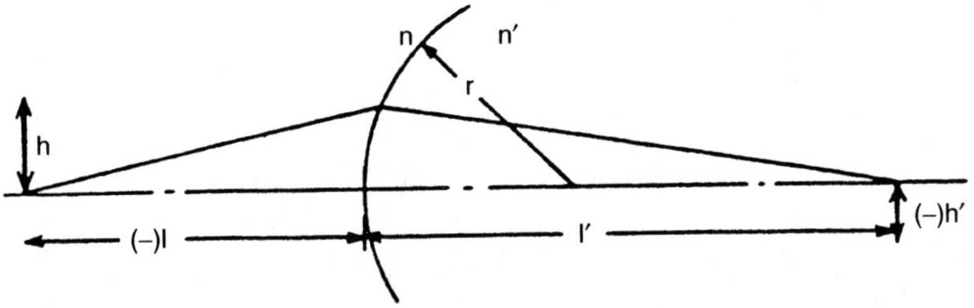

FIG. 11.5 Image formation by a single surface.

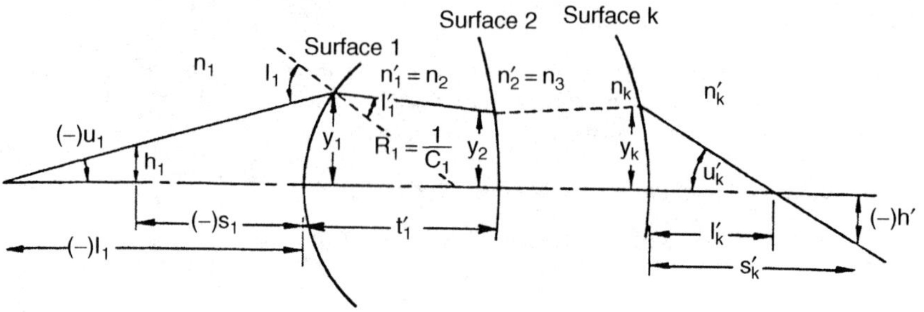

FIG. 11.6 Illustrating the nomenclature of the paraxial ray-tracing equations, Eqs. (11.9) to (11.18).

11.2.4 Element Cardinal Points[47]

To determine the cardinal points, following the definition of the focal points, trace a ray parallel to the axis ($u_1 = 0$) through the system. Then the effective focal length EFL $= y_1/u'_k$ and the back focal length BFL $= y_k/u'_k$. The other focal point is found by repeating the process with the system reversed.

Thick Elements. (See Fig. 11.7) Applied to a single element this procedure yields

$$\phi = 1/f = (n-1)[C_1 - C_2 + tC_1C_2(n-1)/n]$$

$$= (n-1)[1/R_1 - 1/R_2 + t(n-1)/nR_1R_2] \qquad (11.19)$$

$$\text{BFL} = f[1 - tC_1(n-1)/n] = f[1 - t(n-1)/nR_1] \qquad (11.20)$$

Thin Lenses. When the thickness of the element is negligible, Eq. (11.19) reduces to

$$\phi = 1/f = (n-1)(C_1 - C_2) = (n-1)(1/R_1 - 1/R_2) \qquad (11.21)$$

11.2.5 Multielement or Multicomponent Systems[47] (See Fig. 11.8)

Although paraxial rays may be traced through complete systems, one surface at a time, by using Eqs. (11.9) to (11.18), it is frequently more convenient to treat a system as a set of components separated by air. The object and image for each component in turn

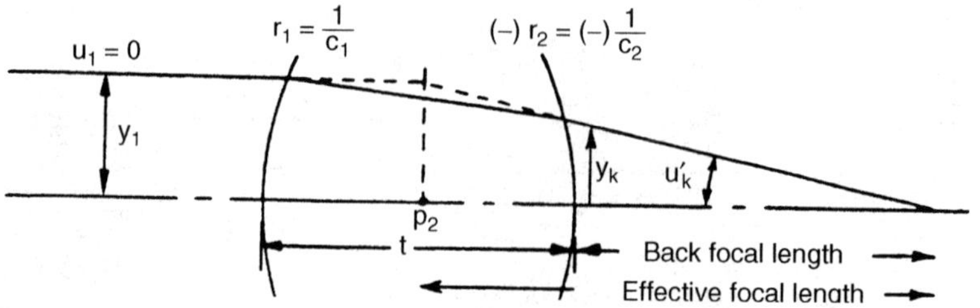

FIG. 11.7 The second set of cardinal points of a thick element.

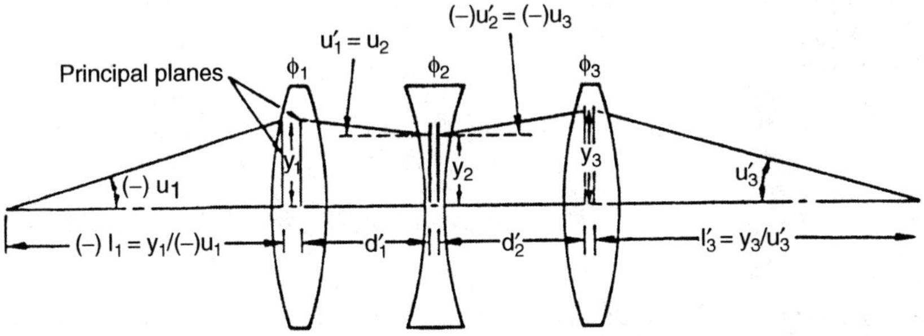

FIG. 11.8 Illustrating the ray-tracing nomenclature for use with component-by-component ray-tracing equations, Eqs. (11.22) and (11.23).

may be determined by using Eq. (11.2). It is usually even more convenient to trace rays through the system by using the following: ϕ_j, the power of the jth component; y_j, the height at which the ray strikes the principal planes of the jth component; and d'_j, the distance from the second principal plane of the jth component to the first principal plane of the $(j + 1)$th component. The ray slope after refraction by the jth component is given by

$$u'_j = u_j + y_j\phi_j \tag{11.22}$$

The ray height at the next component is

$$y_{j+1} = y_j - d'_j u'_j \tag{11.23}$$

If the elements or components are thin, d is simply the space between them. If a ray from the axial intercept of the object has been traced, the magnification can be determined from the Lagrange invariant, $hnu = h'n'u'$.

11.2.6 Two-Component Systems[47] (See Fig. 11.9)

When a system consists of two components a and b, the following explicit expressions may be applied to determine the cardinal points. The components may be simple elements,

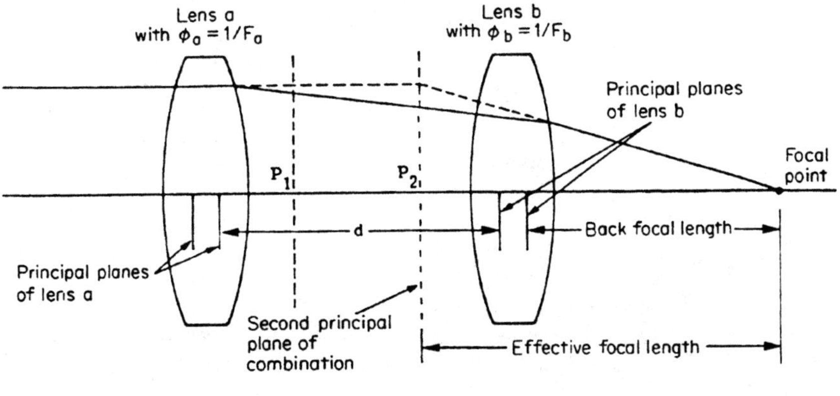

FIG. 11.9 Two-component system.

mirrors, compound lenses, or complex systems in their own right. In Eqs. (11.24) through (11.30), the distances d, BFL, and FFL (front focal length) are measured from the principal points of the components a and b.

$$\phi_{ab} = 1/F_{ab} = \phi_a + \phi_b - d\phi_a\phi_b = 1/F_a + 1/F_b - d/F_aF_b \tag{11.24}$$

$$F_{ab} = \mathrm{EFL}_{ab} = F_aF_b/(F_a + F_b - d) \tag{11.25}$$

$$\mathrm{BFL}_{ab} = F_b[(F_a - d)/(F_a + F_b - d)] = F_{ab}[(F_a - d)/F_a] \tag{11.26}$$

$$-\mathrm{FFL}_{ab} = F_{ab}[(F_b - d)/F_b] \tag{11.27}$$

The component powers which will yield a desired set of system characteristics are given by

$$F_a = dF_{ab}/(F_{ab} - \mathrm{BFL}) \tag{11.28}$$

$$F_b = -d\mathrm{BFL}/(F_{ab} - \mathrm{BFL} - d) \tag{11.29}$$

$$d = F_b\mathrm{BFL}/(F_b - \mathrm{BFL}) = F_a + F_b - F_aF_b/F_{ab} \tag{11.30}$$

11.3 EXACT RAY TRACING[18,39,40,47–50,53]

(See Fig. 11.10)

A general ray, or skew ray, is defined by its direction cosines (X, Y, Z) and by the coordinates (x, y, z) of its intersection with a surface of the optical system. The subscript notation is shown in Fig. 11.11

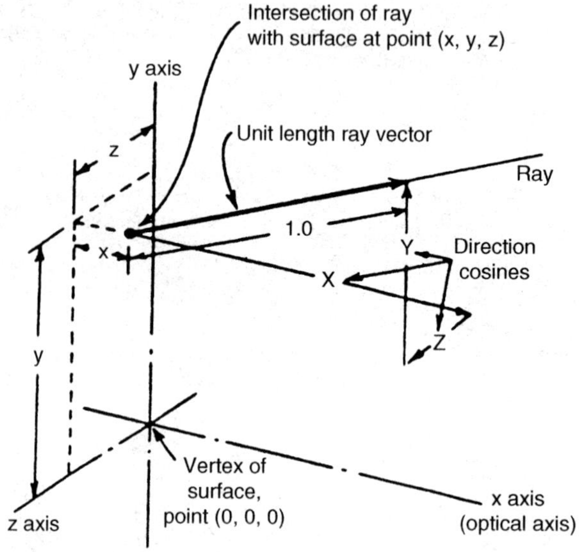

FIG. 11.10 Illustrating the symbols used in the general ray-tracing equations of Sec. 11.3. The spatial coordinates of the intersection of the ray with the surface are x, y, and z. The ray direction cosines are X, Y, and Z.

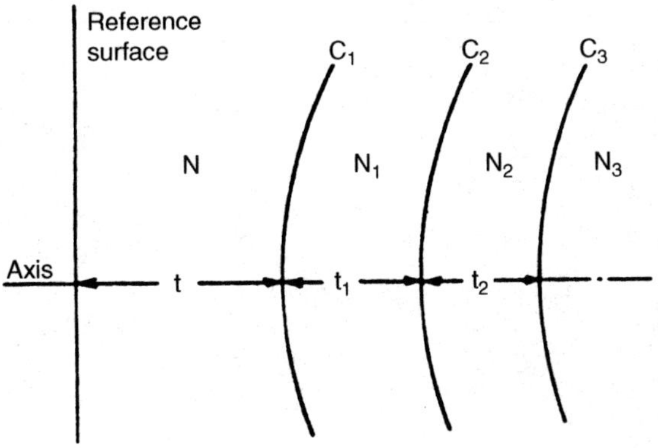

FIG. 11.11 The subscript notation used for the constructional parameters in Sec. 11.3.

11.3.1 Spherical Surfaces

Opening at the initial reference surface:

$$C(x^2 + y^2 + z^2) - 2x = 0 \tag{11.31}$$

$$X^2 + Y^2 + Z^2 = 1 \tag{11.32}$$

Intersection of the ray with next surface:

$$e = tX - (xX + yY + zZ) \tag{11.33}$$

$$M_{1x} = x + eX - t \tag{11.34}$$

$$M_1^2 = x^2 + y^2 + z^2 - e^2 + t^2 - 2tx \tag{11.35}$$

$$\cos I_1 = E_1 = [X^2 - C_1(C_1 M_1^2 - 2M_{1x})]^{0.5} \tag{11.36}$$

$$D_{0,1} = e + [(C_1 M_1^2 - 2M_{1x})/(X + E_1)] \tag{11.37}$$

$$x_1 = x + D_{0,1}X - t \tag{11.38}$$

$$y_1 = y + D_{0,1}Y \tag{11.39}$$

$$z_1 = z + D_{0,1}Z \tag{11.40}$$

Direction cosines of the ray after refraction:

$$\cos I'_1 = E'_1 = [1 - (1 - E^2)(n/n_1)^2]^{0.5} \tag{11.41}$$

$$g_1 = E'_1 - (n/n_1)E_1 \tag{11.42}$$

$$X_1 = (n/n_1)X - g_1 C_1 x_1 + g_1 \tag{11.43}$$

$$Y_1 = (n/n_1)Y - g_1 C_1 y_1 \tag{11.44}$$

$$Z_1 = (n/n_1)Z - g_1 C_1 z_1 \tag{11.45}$$

Equations (11.33) to (11.45) are repeated, incrementing the subscripts by 1, for the next surface. The iteration is continued until the last (image) surface is reached and the ray intersection coordinates (x, y, z) have been determined.

11.3.2 Aspheric Surfaces

An aspheric surface of revolution can be represented as a sphere of curvature C deformed by a series of terms in even powers of the semidiameter r, where $r^2 = y^2 + z^2$ and j is an even integer.

$$x = Cr^2/(1 + \sqrt{1 - C^2r^2} + A_2r^2 + A_4r^4 + \cdots + A_jr^j \tag{11.46}$$

Intersection of Ray and Aspheric. The sphere of curvature C is presumed to be a fair approximation to the aspheric; the intersection of the ray with the sphere, (x_0, y_0, z_0), is found with Eqs. (11.33) to (11.40). The actual x coordinate of the aspheric at this distance from the axis is determined by substituting the y and z coordinates of the ray intersection with the sphere into Eq. (11.46):

$$r_0^2 = y_0^2 + z_0^2 \tag{11.47}$$

$$\bar{x}_0 = Cr_0^2/(1 + \sqrt{1 - C^2r_0^2}) + A_2r_0^2 + A_4r_0^4 + \cdots \tag{11.48}$$

A measure of the approximation error is the difference $\bar{x} - x$ between the true sag of the aspheric and the approximation. The normals to the tangent plane are

$$l_0 = (1 - C^2r_0^2)^{0.5} \tag{11.49}$$

$$m_0 = -y_0[C + l_0(2A_2 + 4A_4r_0^2 + \cdots + jA_jr_0^{(j-2)})] \tag{11.50}$$

$$n_0 = -z_0[C + l_0(2A_2 + 4A_4r_0^2 + \cdots + jAjr_0^{(j-2)})] \tag{11.51}$$

An improved approximation to the intersection of the ray with the aspheric (see Fig. 11.12) can be obtained from the intersection with the tangent plane:

$$G_0 = l_0(\bar{x}_0 - x_0)/(Xl_0 + Ym_0 + Zn_0) \tag{11.52}$$

$$x_1 = G_0X + x_0 \tag{11.53}$$

$$y_1 = G_0Y + y_0 \tag{11.54}$$

$$z_1 = G_0Z + z_0 \tag{11.55}$$

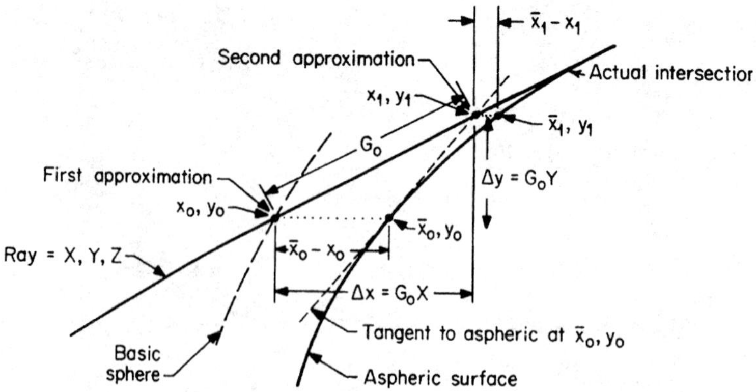

FIG. 11.12 Determination of the ray intersection with an aspheric surface. The intersection of the ray with the surface is found by a convergent series of approximations. Shown here are the relationships used in finding the first approximation after the intersection with the base sphere has been found.

and the new error of the approximation is $\tilde{x}_1 - x_1$. This process is repeated, from Eqs. (11.47) to (11.55), advancing the subscripts at each iteration, until the error $\tilde{x}_k - x_k$ after the kth iteration is negligible. Then the refraction at the aspheric is

$$P^2 = l_k^2 + m_k^2 + n_k^2 \tag{11.56}$$

$$P \cos I = F = Xl_k + Ym_k + Zn_k \tag{11.57}$$

$$P \cos I' = F' = [P^2(1 - (n/n_1)^2) + (n/n_1)^2F^2]^{0.5} \tag{11.58}$$

$$g = \frac{(F' - nF/n_1)}{P^2} \tag{11.59}$$

$$X_1 = X(n/n_1) + gl_k \tag{11.60}$$

$$Y_1 = Y(n/n_1) + gm_k \tag{11.61}$$

$$Z_1 = Z(n/n_1) + gn_k \tag{11.62}$$

11.3.3 Meridional Ray Tracing and Coddington's Equations[7,14,47]

Meridional rays are those which are coplanar with the optical axis. Thus one can always set $z = 0$ and $Z = 0$ in the ray-tracing equations of the preceding section by a simple rotation of the coordinate system about the axis. The equations can be significantly simplified. There are many specialized forms of ray-tracing equations for meridional rays in the literature.

Differential equations[29] can be applied to a meridional ray to determine the image-forming properties of an infinitesimal region about the ray. This is analogous to the paraxial region about the optical axis. A significant simplification of the equations is possible. The classical forms are known as "Coddington's equations."

For differential rays in the sagittal plane,

$$n'/s' = n/s + \phi \tag{11.63}$$

For differential rays in the tangential plane,

$$(n' \cos^2 I')/t' = (n \cos^2 I)/t + \phi \tag{11.64}$$

where s, t = object distances along ray, from the ray-surface intersection
s', t' = corresponding image distances
I, I' = angles of incidence and refraction of the meridional ray
n, n' = indices of refraction
= oblique surface power
$= C(n' \cos I' - n \cos I)$ (11.65)

Note that if the meridional ray is the optical axis, these equations reduce to Eq. (11.7).

11.3.4 Graphical Ray Tracing (see Fig. 11.13)

Meridional rays can be traced using only a scale, straightedge, and compass. The ray is drawn to the surface, and the normal to the surface is erected at the ray-surface intersection. Two circles with radii proportional to the refractive indices n and n' are drawn about the intersection point. A line is drawn parallel to the normal, passing through the intersection of the ray with the n index circle, point A in Fig. 11.13. From the intersection of this parallel, at point B, with the n' index circle, a line through the

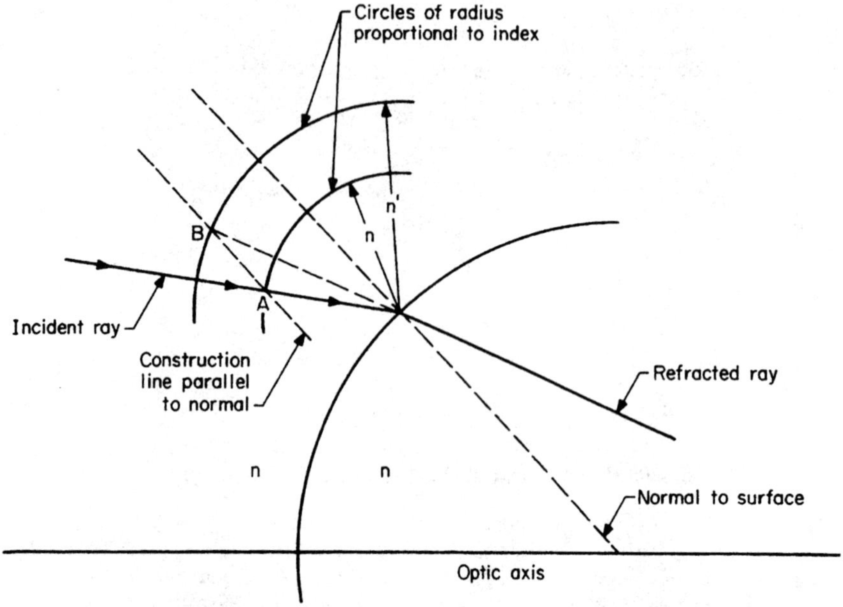

FIG. 11.13 Graphical ray tracing.

ray-surface intersection is the refracted ray. For reflection, $n' = -n$, and a single, full circle is drawn; point B is then the intersection of the parallel with the index circle on the opposite side of the surface.

11.4 ABERRATION DESCRIPTIONS[10,27,47]

Optical aberrations are defects of the image. They are described in terms of the amount by which a geometrically traced ray misses a desired location in the image. The desired location is usually that indicated by the first-order (paraxial) imagery as determined from Eqs. (11.1) to (11.23)

11.4.1 Lateral, Longitudinal, and Angular Aberrations

As indicated in Fig. 11.14 for the case of spherical aberration, an aberration may be measured in a lateral, longitudinal, or angular dimension. Aberrations may also be described as departures from an ideal spherical wave front. Section 11.6.1 indicates the relationships between several aberrations and the departure from a perfect wave front.

11.4.2 Spherical Aberration

Spherical aberration can be defined as the longitudinal variation of focus with aperture. Figure 11.15 shows the undercorrected spherical aberration typical of a simple positive element, in which the rays nearer the edge of the lens, being more strongly refracted than ideal, come to a focus nearer the lens than rays closer to the axis.

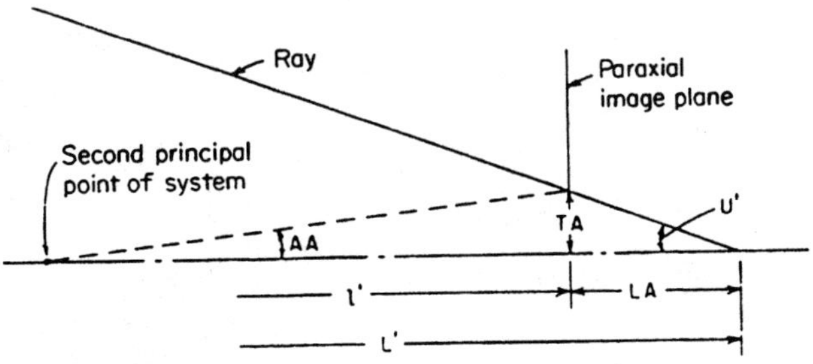

FIG. 11.14 The relationship between the longitudinal (LA), the transverse (TA), and the angular (AA) measures of spherical aberration. TA = LA tan $U' = AAl'$.

Longitudinal spherical aberration is the distance from the paraxial focus to the axial intersection of the ray. Lateral or transverse spherical aberration is the vertical distance from the axis to the intersection of the ray with the paraxial focal plane.

11.4.3 Coma

Coma is the variation of magnification or image size with aperture. Figure 11.16 shows the rays through the outer portions of the aperture intersecting to form an image larger than that formed by rays through the lens center. (The appearance of a comatic point image is simulated in Fig. 11.23.)

11.4.4 Field Curvature and Astigmatism

Field curvature is the amount by which an off-axis image departs longitudinally from the surface in which it should ideally be located. The ideal is usually a flat plane. Field curvature usually differs for fans of rays in different meridians. The tangential focus is that of a fan of rays in the meridional plane; the sagittal focus is that of a fan in a plane normal to the meridional plane. The distance between the tangential and sagittal foci is the astigmatism. Figure 11.17 shows the field curvature of a simple lens. The Petzval surface is 3 times as far from the tangential field as from the sagittal.

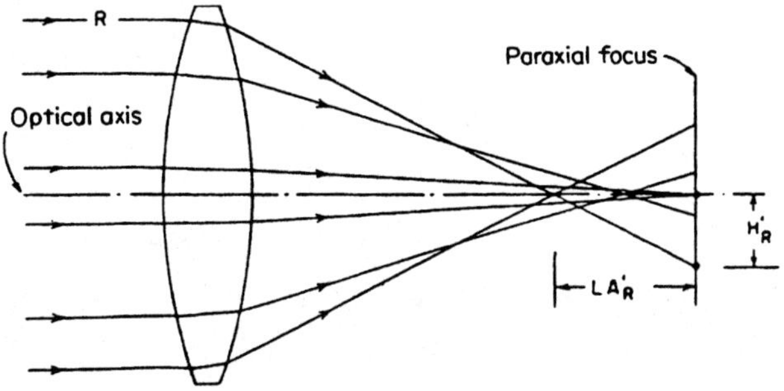

FIG. 11.15 A simple converging lens with undercorrected spherical aberration.

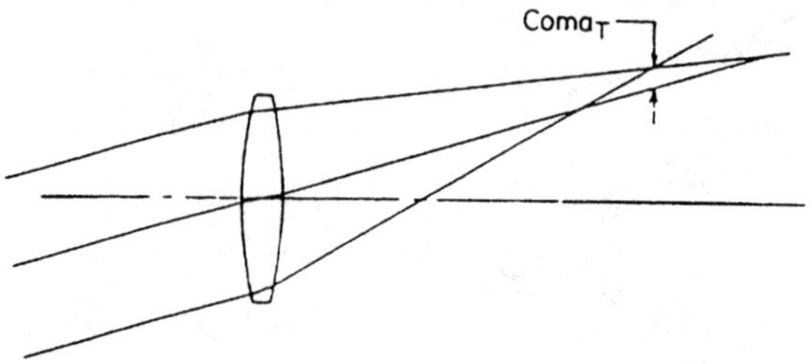

FIG. 11.16 Coma.

11.4.5 Distortion

Distortion is the amount by which an image point is closer to, or further from, its ideal position, usually that defined by first-order optics. The linear amount of simple distortion varies as the cube of the image height; therefore nonradial straight lines are imaged as curved lines, as in Fig. 11.18.

11.4.6 Chromatic Aberrations

The preceding are aberrations which exist in monochromatic light. Although they do vary with wavelength, the change is usually small. In addition, there are aberrations that depend on wavelength:

Axial Chromatic Aberration: The longitudinal variation of focal position with wavelength; it is the distance from the long-wavelength focus to the short (Fig. 11.19).

Lateral Chromatic Aberration: The variation of image size with wavelength; it is the vertical distance from the image of a point in long-wavelength light to the image of the same point in short-wavelength light (Fig. 11.20).

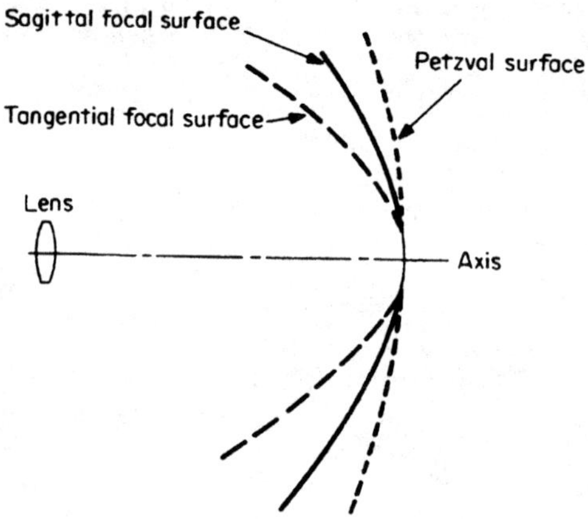

FIG. 11.17 The undercorrected astigmatism and field curvature of a simple lens.

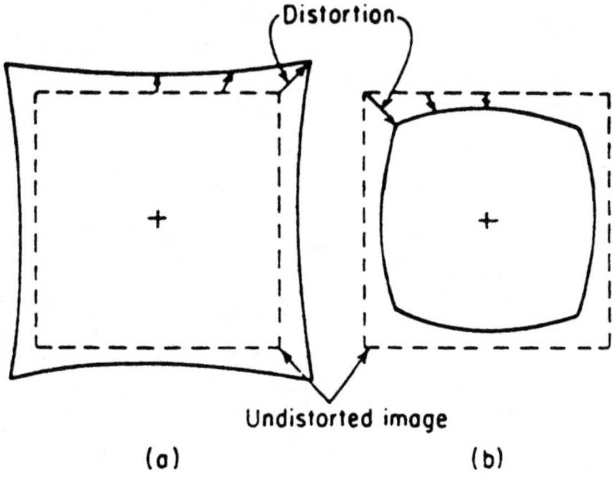

FIG. 11.18 Distortion. (*a*) Pincushion. (*b*) Barrel.

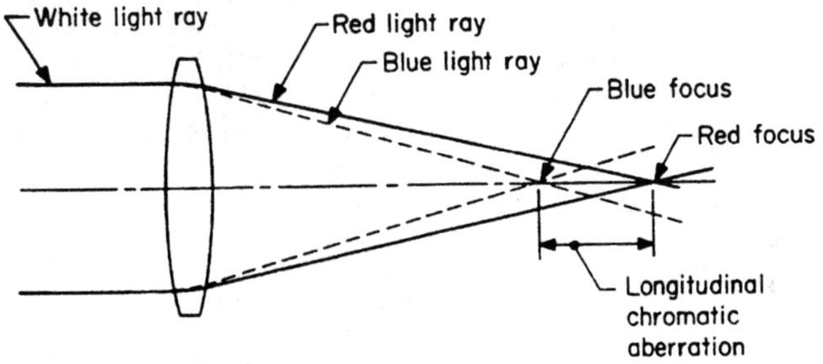

FIG. 11.19 A simple positive lens has undercorrected axial chromatic aberration because the short-wavelength (blue) rays undergo a greater refraction than the long-wavelength (red) rays.

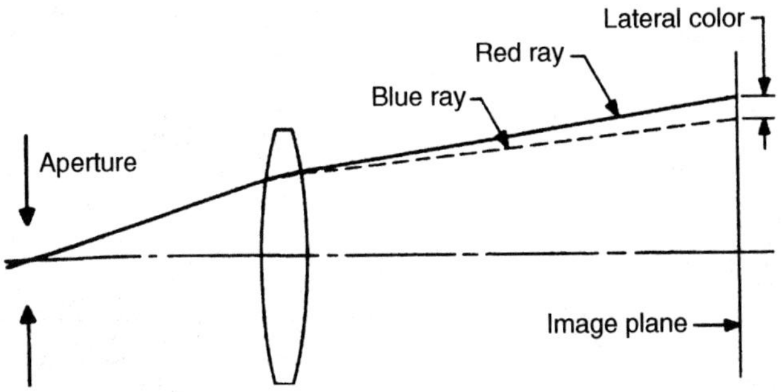

FIG. 11.20 Lateral color, or chromatic difference of magnification.

As indicated above, the monochromatic aberrations do vary slightly with wavelength. The chromatic variation of spherical aberration, or spherochromatism, is the most common; ordinarily the spherical aberration will be overcorrected in short-wavelength light and undercorrected in long. Chromatic variation of distortion or field curvature is occasionally encountered.

11.4.7 Longitudinal Aberration Representation[30,47]

Because aberrations vary with aperture and image size, a simple, single numerical value is seldom an adequate description of the image quality. Several standard methods of representing aberration are in common use.

Spherical aberration is frequently shown by a plot of the longitudinal position of the ray intercept with the optical axis as a function of the height at which the ray passes through the entrance pupil. Figure 11.21 shows such a plot for rays of three wavelengths.

Field curvature is often represented as a longitudinal aberration plotted against image height or obliquity. In Fig. 11.21 the sagittal and tangential field curvatures, as well as the Petzval curvature, are represented.

11.4.8 Lateral Aberration Representation[47]

The lateral representation of aberrations permits a more universal and complete description of the correction of the optical system. A fan of rays from an object point is traced through the system and the coordinates of the ray intersections with the image surface are plotted as a function of the ray position in the aperture, which is usually represented by the slope (tan U) of the ray at the image. In this format the effect of the aberrations can be estimated directly in terms of the size of the blur they produce in the image. Also the effects of refocusing, that is, a longitudinal shift of the reference (image) surface, can be determined by simply rotating the tan U axis. Figure 11.22 illustrates the appearance of this type of plot for several common aberrations. These curves are called "ray-intercept plots."

11.4.9 Spot Diagrams, Point- and Line-Spread Functions

If the aperture of the optional system is divided into a large number of equal small areas and a ray from a selected object point is traced through the center of each small area, a plot of the intersection points of the rays (spots) with the image surface is an approximate

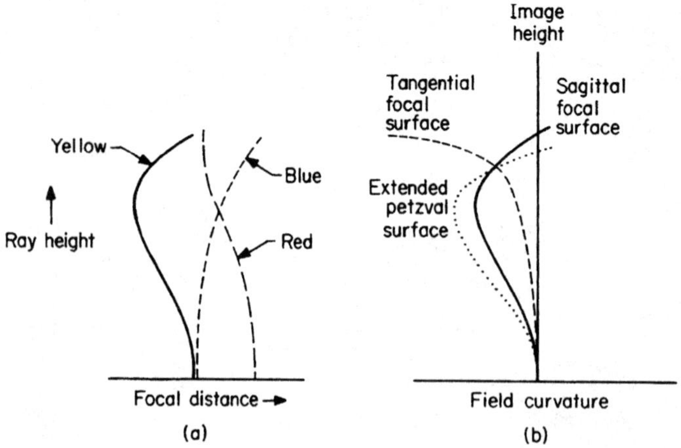

FIG. 11.21 Longitudinal aberration representations of (*a*) zonal spherical aberration and spherochromatism and (*b*) residual astigmatism and curvature of field.

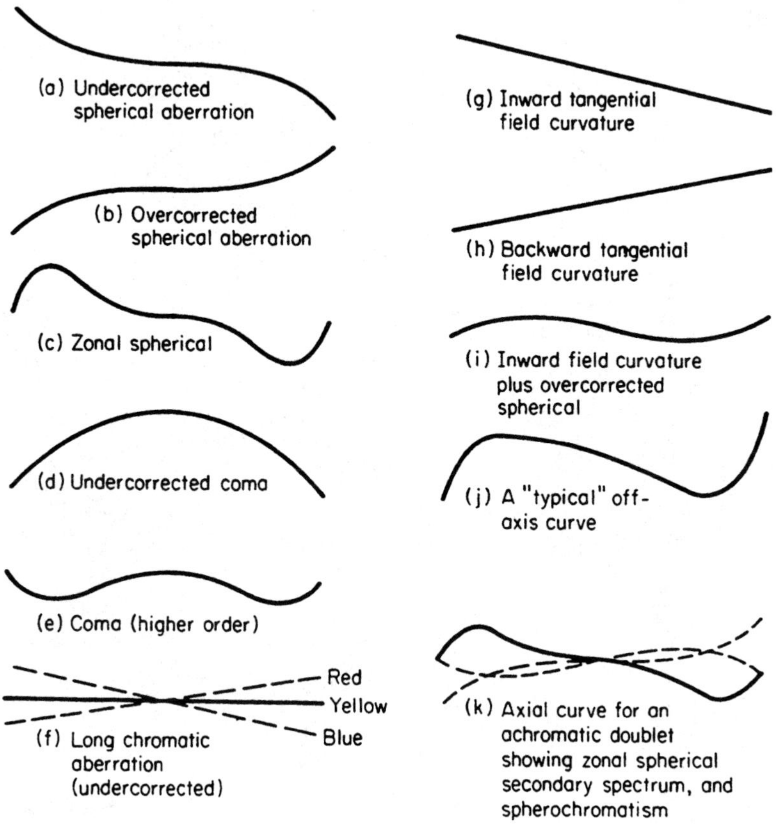

FIG. 11.22 Ray-intercept curves for various aberrations. The ordinate of each curve is H, the height at which the ray intersects the (paraxial) image plane, and the abscissa is tan U', the final slope of the ray with respect to the optical axis.

representation of the (geometric) illuminance distribution in the image. The more rays traced, the better the approximation. Such a plot is called a "spot diagram."

It is also possible to calculate the distribution of illuminance exactly, taking into account the effects of diffraction. The distribution of energy in the image of a point object, regardless of how derived, is the "point-spread function" of the optical system. The "radial energy distribution" is obtained by arbitrarily selecting a center point and plotting the percentage of energy (or the number of spots) encircled within a radius R as a function of R.

The "line-spread function" is the cross section of the illuminance distribution in the image of a line; it is derived from the point-spread function by summing it in one direction. Thus $L(y) = \int P(y, z)\,dz$ and $L(z) = \int P(y, z)\,dy$, where $L(\ )$ is the line-spread function and $P(y, z)$ is the point-spread function. See Fig. 11.23.

11.4.10 Aplanatic Surfaces

There are three object positions for which the spherical aberration and coma of a spherical surface are zero.

1. Object and image at the surface:

$$L = L' = O$$

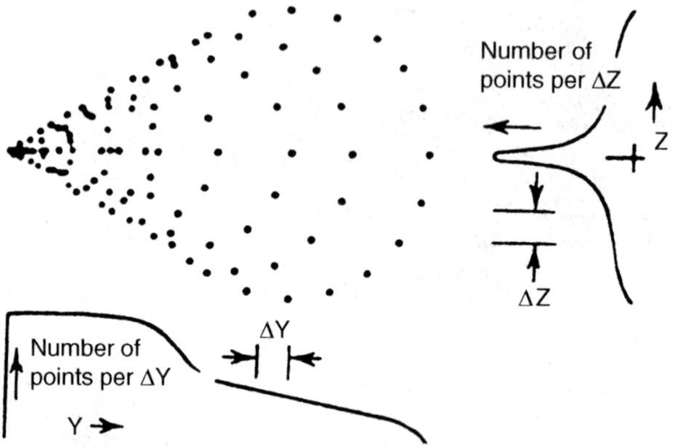

FIG. 11.23 Spot diagram for a system with pure coma, and the two line-spread functions (below and to the right) which can be obtained by counting the spots between parallel lines separated by a small distance, ΔY and ΔZ.

2. Object and image at the center of curvature of the surface:

$$L = L' = R$$

3. The "aplanatic case," where

$$L = R(n' + n)/n = n'L'/n$$

$$n \sin U' = n' \sin U$$

$$U = I' \quad \text{and} \quad U' = I$$

The third-order aberrations of a single air-glass surface are plotted as a function of object position in Fig. 11.24.

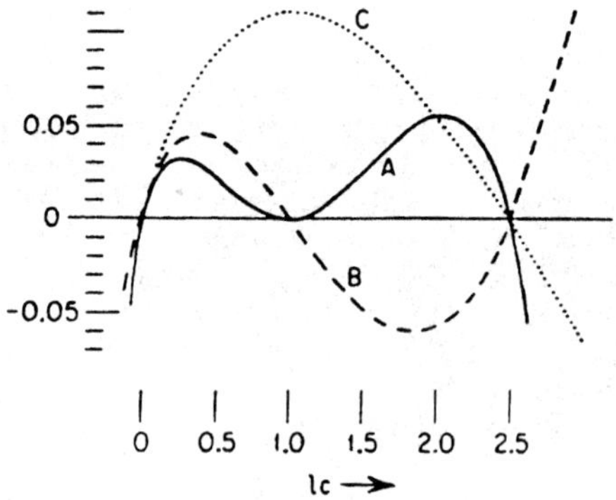

FIG. 11.24 The aberrations of a single refracting surface, with $n = 1.0$, $n' = 1.5$, as a function of the object position. Curve A is the transverse spherical times (c/u^3). Curve B is sagittal coma times (c/u^2u_p). Curve C is transverse astigmatism times (c/cuu_p^2). The symbols are: l is the object position, c is the surface curvature, u is the marginal ray slope, and u_p is the principal ray slope.

11.4.11 Symmetrical Systems[14]

In an optical system which is completely symmetrical about the aperture stop, coma, distortion, and lateral color are identically zero. For complete symmetry the system must work at unit magnification, and the optical system on the image side of the stop must be a mirror image of that on the object side. Symmetrical (or nearly so) construction is often utilized to reduce these aberrations in other than unit power systems, e.g., photographic objectives.

11.5 DIFFRACTION

11.5.1 Diffraction Image[8,54]

If the transmission of an optical system is uniform over its circular aperture and the system is aberration-free, the illuminance distribution in the image of a point object is

$$E(y, z) = (\mathrm{NA}/\lambda)^2\, \phi\, [2J_1(m)/m]^2 = E_0[2J_1(m)/m]^2 \tag{11.66}$$

where ϕ = total power in the point image
$J_1(m)$ = first-order Bessel function
E_0 = peak illuminance in the image
NA = $n' \sin U'$ = numerical aperture
m = normalized radial coordinate

$$m = (2\pi/\lambda)\, \mathrm{NA}(y^2 + z^2)^{0.5} = (2\pi/\lambda)(\mathrm{NA})r \tag{11.67}$$

The fraction of the total power falling within a radial distance r_0 of the center of the pattern is given by $1 - J_0^2(m_0) - J_1^2(m_0)$, where $J_0(m)$ is the zero-order Bessel function.

Equation (11.66) is plotted in Fig. 11.25, which also shows the appearance of the diffraction pattern. The pattern consists of a circular patch of light, called the "Airy disk," surrounded by concentric rings of rapidly decreasing intensity. Table 11.1 indicates the size and distribution of energy in the pattern for both a circular and a slit aperture.

11.5.2 Gaussian (Laser) Beams

Many lasers emit a beam which has a cross-sectional flux density distribution described by the gaussian

$$E(r) = E_0 e^{-2(r/w)^2} \tag{11.68}$$

This nonuniform distribution modifies the diffraction pattern from the "normal" one for a system with a uniformly illuminated transmitting aperture described in Sec. 11.5.1. In the absence of any aberration the diffraction pattern of a gaussian beam is also a gaussian. The size of a gaussian beam is usually described in terms of the semidiameter w at which the flux density falls to $E_0 e^{-2}$, or about 13.5 percent of its central peak value. At large distances the angular spread of the beam is $4\lambda/\pi D$ between the e^{-2} points, where D is the beam diameter. The size of the diffraction image diameter at the

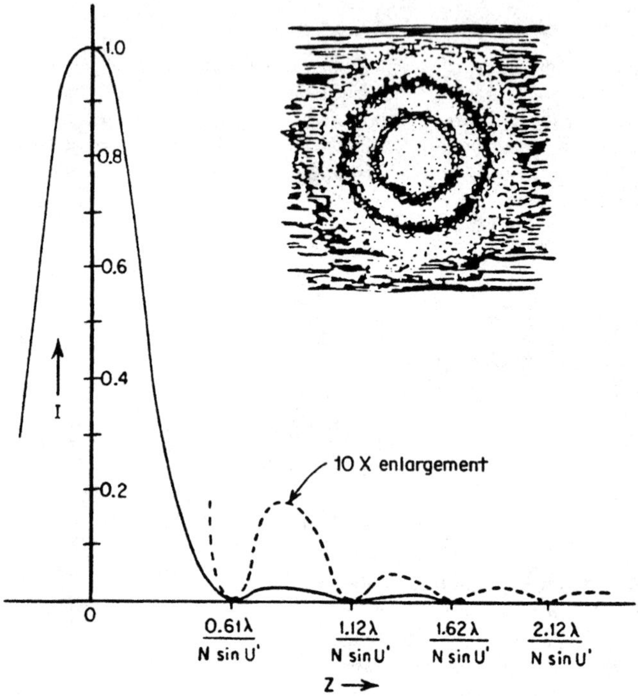

FIG. 11.25 The distribution of illumination in the diffraction pattern formed in the image plane of a perfect lens with a uniformly illuminated aperture.

focus of a lens of effective focal length f is $4\lambda f/\pi D$. If the clear aperture of the optical system is less than twice the e^{-2} diameter, the diffraction-pattern flux distribution will depart from a gaussian toward that described in the preceding section for a uniformly illuminated aperture.

TABLE 11.1 Distribution of Energy in the Diffraction Pattern at the Focus of a Perfect Lens, as a Function of the Distance z from the Center of the Pattern

	Circular aperture			Slit aperture	
Ring (or band)	z	Peak illumination	Energy in ring	z	Peak illumination
Central maximum	0	1.0	83.9%	0	1.0
1st dark ring	$0.61\lambda/(n' \sin U')$	0.0	—	$0.5\lambda/(n' \sin U')$	0.0
1st bright ring	$0.82\lambda/(n' \sin U')$	0.017	7.1%	$0.72\lambda/(n' \sin U')$	0.047
2d dark ring	$1.12\lambda/(n' \sin U')$	0.0	—	$1.0\lambda/(n' \sin U')$	0.0
2d bright ring	$1.33\lambda/(n' \sin U')$	0.0041	2.8%	$1.23\lambda/(n' \sin U')$	0.017
3d dark ring	$1.62\lambda/(n' \sin U')$	0.0	—	$1.5\lambda/(n' \sin U')$	0.0
3d bright ring	$1.85\lambda/(n' \sin U')$	0.0016	1.5%	$1.74\lambda/(n' \sin U')$	0.0083
4th dark ring	$2.12\lambda/(n' \sin U')$	0.0	—	$2.0\lambda/(n' \sin U')$	0.0
4th bright ring	$2.36\lambda/(n' \sin U')$	0.00078	1.0%	$2.24\lambda/(n' \sin U')$	0.0050
5th dark ring	$2.62\lambda/(n' \sin U')$	0.0	—	$2.5\lambda/(n' \sin U')$	0.0

11.5.3 Point Resolution: The Rayleigh and Sparrow Criteria

The images of two point sources are said to be resolved when the combined diffraction pattern can be determined to be due to two objects rather than one. The effect of various source separations on the combined pattern is sketched in Fig. 11.26.

The Rayleigh criterion is that two adjacent, equal-intensity point sources can be considered resolved if the first dark ring of the diffraction pattern of one point image coincides with the center of the other pattern. This is an arbitrary, but useful, resolution limit for an optical system. Thus, in an aberration-free system with a uniformly illuminated pupil, the separation of the two point images is equal to the radius of the first dark ring. From Table 11.1 we get

$$\text{Separation} = 0.61\lambda/(n' \sin U') = 0.61\lambda/\text{NA} \tag{11.69}$$

The Sparrow criterion is that two adjacent point sources be considered resolved if the combined diffraction pattern has no minimum between the two images. This gives

$$\text{Separation} = 0.5\lambda/(n' \sin U') = 0.5\lambda/\text{NA} \tag{11.70}$$

The separations can be converted to object separations either by using the object space NA in the equations, or by dividing the image separations by the system magnification. For distant objects the angular separation is

$$\text{Rayleigh angular resolution} = 1.22\lambda/D \quad \text{rad} \tag{11.71}$$

$$\text{Sparrow angular resolution} = \lambda/D \quad \text{rad} \tag{11.72}$$

where D is the diameter of the system entrance pupil. For visual use, the resolutions become $5.5/D$ and $4.5/D$ seconds, respectively, if D is expressed in inches.

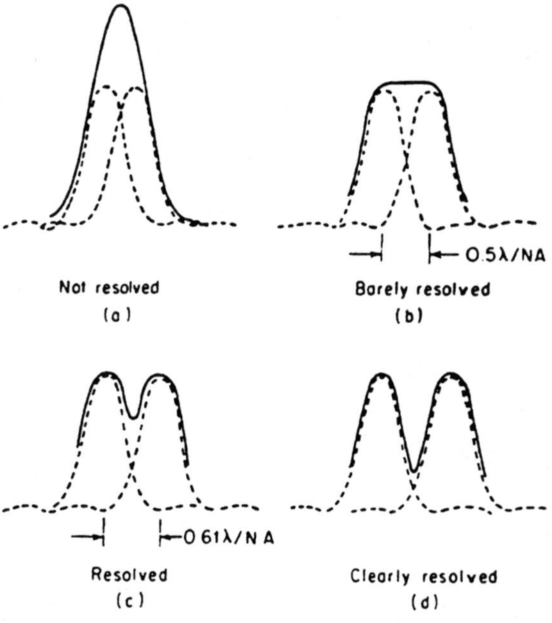

FIG. 11.26 Point resolution: the dashed lines represent the diffraction patterns of two point images at various separations, and the solid line is the combined illumination. (*b*) The Sparrow criterion for resolution. (*c*) The Rayleigh criterion.

Since these resolution criteria are based on aberration- and defect-free systems, they are often used as a standard for excellence of design and construction, as well as an indication of the diffraction limit on the performance of the system.

11.6 IMAGE QUALITY CRITERIA

11.6.1 Rayleigh Quarter-Wave Limit[14, 47]

The Rayleigh limit for image quality indicates that the image will be sensibly perfect if the emerging wavefront departs from a perfect sphere by no more than one-quarter of the wavelength of the radiation forming the image. A system which meets this limit is sometimes descried as "diffraction-limited."

The amounts of certain aberrations which correspond to a maximum wavefront deformation of one-quarter wave are as follows:

Out of focus:	$n\lambda/2(\mathrm{NA})^2$	(11.73)
Spherical aberration:	$4n\lambda/(\mathrm{NA})^2$	(11.74)
Zonal spherical aberration:	$6n\lambda/(\mathrm{NA})^2$	(11.75)
Axial chromatic aberration:	$n\lambda/(\mathrm{NA})^2$	(11.76)
Tangential coma:	$3\lambda/2(\mathrm{NA})$	(11.77)

when the reference point is chosen to minimize the departure of the wavefront from the reference sphere.

The effect of a quarter-wave of aberration on the diffraction pattern is to shift some light from the central patch to the rings. A perfect system has 84 percent of the light in the central patch, 16 percent in the rings. A quarter-wave system has 68 percent in the central disk, 32 percent in the rings, but the diameter of the disk and rings is essentially the same. Such a change is detectable, but with difficulty. For most applications a system with less than a quarter-wave of aberration is a excellent one.

The quarter-wave limit tacitly assumes that the wavefront is relatively smooth. When this is not the case, the root-mean-squared (rms) wavefront deformation is a better measure of system quality than the peak-to-valley deformation. An rms deformation of between a fourteenth ($\lambda/14$) and a twentieth ($\lambda/20$) wave is approximately the equivalent of the classical quarter-wave limit.

11.6.2 Strehl Definition[8]

The Strehl definition is the ratio of the illuminance at the peak of the diffraction pattern of an aberrated image of a point to that at the peak of an aberration-free image. A Strehl ratio of 80 percent is the equivalent of the Rayleigh quarter-wave limit. The Strehl ratio is also equal to the normalized volume under the (three-dimensional) modulation transfer function (MTF). See Secs. 11.6.3 and 11.6.4.

11.6.3 Optical Transfer Function[8,16,42]

The optical transfer function OTF(ν) is a complex function describing the performance of an optical system in terms of its imagery of a linear object pattern whose luminance varies sinusoidally according to the spatial frequency ν (in cycles per unit length). The modulus of the OTF is the modulation transfer MTF(ν), which describes the transfer of contrast (modulation) from object to image; the argument of the OTF is the spatial phase shift $\phi(\nu)$ of the sinusoidal image pattern from its nominal position.

For the one-dimensional case the luminance of the sinusoidal object is

$$G(y) = b_0 + b_1 \cos(2\pi v y) \tag{11.78}$$

where v is the spatial frequency and y is the coordinate in which the luminance varies. Modulation is defined as the peak-to-valley variation of luminance divided by the sum of peak and valley. Thus Eq. (11.78) gives for the object modulation

$$M_o = [(b_0 + b_1) - (b_0 - b_1)]/[b_0 + b_1) + (b_0 - b_1)] = b_1/b_0 \tag{11.79}$$

Each line element making up the object is imaged as the line-spread function $L(y)$ of the optical system (Sec. 11.4.9). Assuming unit magnification and 100 percent transmission to simplify matters, we can express the illuminance at position y in the image as the summation of the product of $G(y)$ and $L(y)$:

$$F(y) = \int L(\delta)G(y - \delta)d\delta = b_0\int L(\delta)d\delta + b_1\int L(\delta)\cos[2\pi v(y - \delta)]\,d\delta \tag{11.80}$$

When normalized by dividing by $\int L(\delta)\,d\delta$, Eq. (11.80) can be transformed into

$$F(y) = b_0 + b_1\,|A(v)|\cos(2\pi v y - \phi) \tag{11.81}$$

$$\text{where } |A(v)| = [A_c^2(v) - A_s^2(v)]^{0.5} \tag{11.82}$$

$$A_c(v) = \int L(\delta)\cos(2\pi v\delta)\,d\delta/\int L(\delta)\,d\delta \tag{11.83}$$

$$A_s(v) = \int L(\delta)\sin(2\pi v\delta)\,d\delta/\int L(\delta)d\delta \tag{11.84}$$

$$\phi = \arctan[A_s(v)/A_c(v)] = \arccos[A_c(v)/|A(v)|] \tag{11.85}$$

$F(y)$ in Eq. (11.81) is the illuminance in the image. Note that it is sinusoidally modulated at the same frequency as the object, Eq. (11.78), and the pattern is shifted a distance represented by the phase angle ϕ [which is zero if $L(y)$ is a symmetrical function]. The modulation in the image is thus

$$M_i = (b_1/b_0)|A(v)| = M_o|A(v)| \tag{11.86}$$

The modulation transfer function (MTF) is, by definition, the ratio of the modulation in the image to that in the object

$$\text{MTF}(v) = M_i/M_o = |A(v)| \tag{11.87}$$

11.6.4 Specific Modulation Transfer Functions[4–26]

The MTF graphs of this section are plotted as functions of v_o, the limiting cutoff frequency. An optical system is a low-pass filter and will not transmit spatial frequencies above v_o, which is given by

$$v_o = \frac{2\text{NA}}{\lambda} = \frac{1}{\lambda(f/\text{number})} \tag{11.88}$$

where NA $= n \sin U$, the numerical aperture, λ is the wavelength of the light forming the image, and *f*/number is the effective speed or relative aperture at the image.

The limiting frequency can also be expressed as an angular frequency of the object, subtended from the entrance pupil: $v_o = D/\lambda$ cycles per radian, where D is the diameter of the entrance pupil.

Perfect System. The MTF at a spatial frequency ν for an optical system with no aberrations and a uniformly illuminated circular aperture is given by

$$\mathrm{MTF}(\nu) = (2/\pi)\{\arccos(\lambda\nu/2\mathrm{NA}) - (\lambda\nu/2\mathrm{NA}) \sin[\arccos(\lambda\nu/2\mathrm{NA})]\} \quad (11.89)$$

Equation (11.89) is plotted in Fig. 11.27.

For a slit or a rectangular aperture the MTF is

$$\mathrm{MTF}(\nu) = 1 - \nu/\nu_o = 1 - \nu\lambda/2\mathrm{NA} \quad (11.90)$$

Annular Aperture. In the presence of a central obscuration of the pupil, as in a Cassegrain mirror system, the MTF of an aberration-free system is reduced greatly at low frequencies and increased slightly at high frequencies, as shown in Fig. 11.28.

Defocused System. The effects of various amounts of defocus are shown in Fig. 11.29. The defocus is expressed as a function of sin U so that the graph may be applied to any optical system. If the defocusing is large, on the order of $4\lambda/(n \sin^2 U)$ or greater, then diffraction effects can be neglected and

$$\mathrm{MTF}(\nu) = 2J_1(\pi B\nu)/\pi B\nu = J_1(2\pi\delta\mathrm{NA}\nu)/\pi\delta\mathrm{NA}\nu \quad (11.91)$$

where $J_1(\,)$ = first-order Bessel function
B = diameter of blur produced by defocusing
δ = longitudinal shift from focus
NA = n sin
U = numerical aperture
ν = spatial frequency in cycles per unit length

Equation (11.91) is the MTF of a uniformly illuminated circular disk image. A system whose image is a uniformly illuminated slit or band of light, as when the image is blurred by motion, has an MTF given by

$$\mathrm{MTF}(\nu) = (\sin \pi W\nu)/\pi W\nu \quad (11.92)$$

where W is the width of the slit or band.

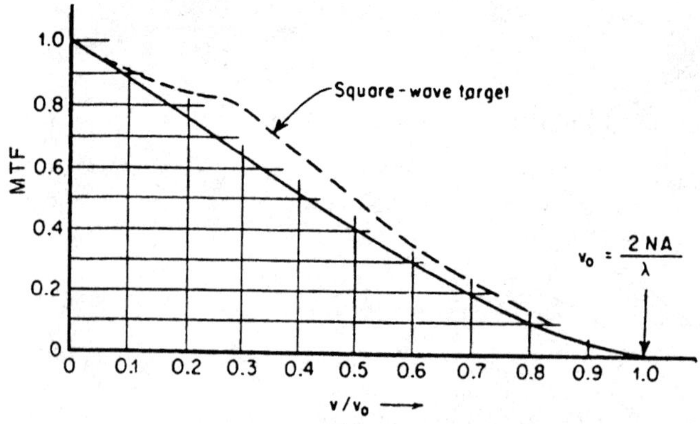

FIG. 11.27 The modulation transfer function (MTF) of an aberration-free optical system with a uniformly illuminated circular aperture.

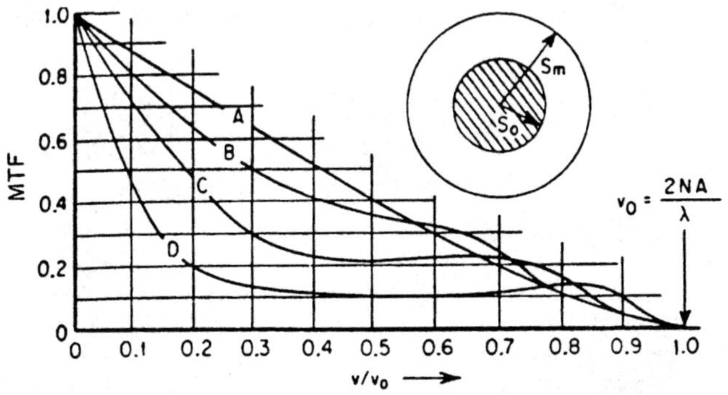

FIG. 11.28 The modulation transfer function of an aberration-free optical system with an annular aperture. Curve A, $S_0/S_m = 0$; curve B, $S_0/S_m = 0.25$; curve C, $S_0/S_m = 0.50$; curve D, $S_0/S_m = 0.75$.

Aberrated Systems. There is a strong similarity between the MTF curves for systems afflicated with one-quarter wave of aberration, regardless of the type of aberration. See Sec. 11.6.1. In general one is fairly safe in assuming that a given amount of wavefront deformation (from a perfect spherical wave front) caused by any aberration will produce an MTF characteristic similar to that resulting from the same amount (measured in wavelengths of deformation) of another aberration; thus Fig. 11.29 is typical of most aberrations.

11.6.5 Sine Waves and Square Waves[13,28]

By definition the OTF and MTF apply to the imagery of a sinusoidally modulated target object. A square-wave target, i.e., a series of alternating light and dark bars, is a convenient and widely used target for testing the performance of optical systems. When the MTF (i.e., the sine-wave response) is known, the modulation transfer of a

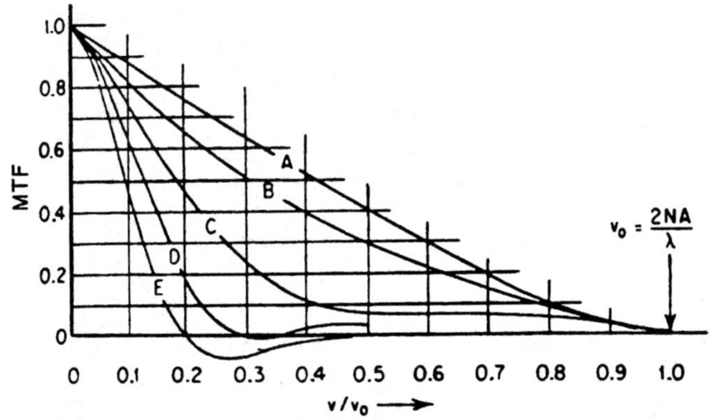

FIG. 11.29 The effect of defocusing on the MTF of an aberration-free optical system. Curve A, in focus, OPD = 0; curve B, defocus = $\lambda/(2n \sin^2 U)$, OPD = $\lambda/4$; curve C, OPD = $\lambda/2$; curve D, OPD = $3\lambda/4$; curve E, defocus = $2\lambda/(n \sin^2 U)$, OPD = λ.

square-wave target can be calculated by summing the response to the Fourier components of the square wave:

$$S(v) = (4/\pi)[M(v) - M(3v)/3 + M(5v)/5 - M(7v)/7 + \cdots] \tag{11.93}$$

where $S(v)$ = square-wave target transfer factor and $M(v)$ = sine wave MTF.

If the square-wave factor is known, the MTF can be found from

$$M(v) = (\pi/4)[S(v) + S(3v)/3 - S(5v)/5 + S(7v)/7 - \cdots] \tag{11.94}$$

In general, the modulation transfer factor is higher for a square-wave target than for a sine-wave target. See Fig. 11.17, for example. The most common form of bar target for lens testing is the USAF/1951 target which consists of only three dark bars on an extended white background (or the reverse) for each frequency. If the target frequency is taken as the reciprocal of the centerline spacing of the bars, the modulation transfer is much higher for the three-bar target than for an extended sine-wave target. This is because the frequency content is heavily concentrated in the subharmonics; i.e., a spectral breakdown (Fourier analysis) of a three-bar target shows much power at frequencies less than v = 1/(bar spacing).

11.6.6 Aerial Image Modulation Curve

The aerial image modulation (AIM) curve is a plot of the minimum image modulation required to produce a response in a sensing element, plotted as a function of spatial frequency. AIM curves are commonly used to describe such detectors as photographic film, image tubes, and the human eye. A typical AIM curve rises with increasing frequency, indicating that a higher image modulation is required to produce a response at higher frequencies. If the AIM curve for a film and the MTF curve of a lens are plotted on the same graph, the intersection of the two curves indicates the limiting resolution of the combination.

11.6.7 Depth of Focus

Photographic Depth of Focus. Assuming that an arbitrarily selected level of blur (of diameter B) caused by a defocusing of the optical system can be accepted, the tolerable depth of focus is then $\pm B/2\text{NA}$, if diffraction effects are totally neglected and if the optical system is free of aberrations. The corresponding depth of field at the object ranges from S_{near} to S_{far}:

$$S_{\text{near}} = fS(D + B)/(fD - SB) \tag{11.95}$$

$$S_{\text{far}} = fS(D - B)/(fD + SB) \tag{11.96}$$

where S = nominal distance of focus
D = diameter of entrance pupil
B = acceptable blur diameter in image

The "hyperfocal distance" is the distance at which the optical system is focused in order to make S_{far} equal to infinity and S_{near} equal one-half the hyperfocal distance.

$$S_{\text{hyperfocal}} = -fD/B \tag{11.97}$$

Physical Depth of Focus. There is actually no sharp demarcation between being in focus and out of focus. The image simply deteriorates gradually as the amount of defocus is increased. The wave-front aberration caused by defocusing is given by

$$\text{OPD} = 0.5(\delta S)n \sin^2 U_m \tag{11.98}$$

where δS = distance from point of best focus
n = index of image medium
U_m = slope angle of marginal ray at image

Thus a depth of focus "tolerance" corresponding to the Rayleigh quarter-wave criterion is

$$\delta S = \pm\lambda/(2n \sin^2 U_m) \tag{11.99}$$

Note that Eq. (11.99) may be used for both depth of focus (at the image) and depth of field (at the object) if U_m is taken as the slope of the marginal ray at the image or object, respectively.

11.7 AFOCAL SYSTEMS AND TELESCOPES

Afocal systems are without focal length. An ordinary telescope is an afocal system. It forms an image of an infinitely distant object at infinity. Thus, the usual definition for effective focal length is meaningless.

11.7.1 Magnification, Apertures, and Fields of View

The angular magnification of an afocal system is the ratio of the angular size of the image to the angular size of the object. An afocal device also forms a finite image of any object at a finite distance. For an afocal system, linear magnification, which is the ratio of the *height* of the image to the *height* of the object, is the reciprocal of the angular magnification; as in ordinary systems the longitudinal magnification is approximately the square of the linear magnification. Since the pupils of an optical system are conjugates of each other, the magnifications, the pupil diameters, and the fields of view for any afocal system are related by

$$m = 1/m_a = D_i/D_o = \tan A_o/\tan A_i \tag{11.100}$$

where m, m_a = linear and angular magnifications
D_i, and D_o = exit and entrance pupil diameters
A_o, A_i = half-field angles in object and image space (see Fig. 11.30)

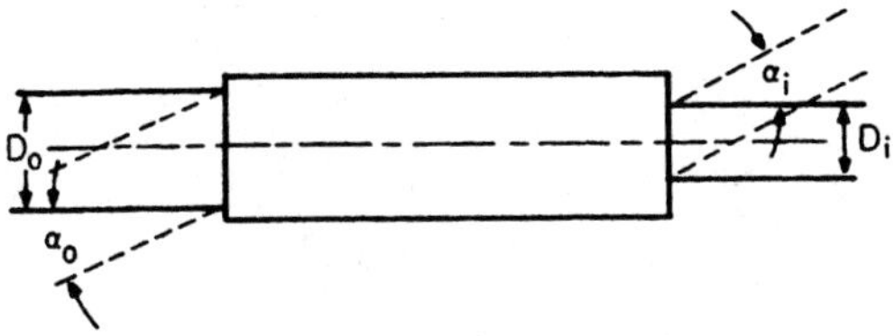

FIG. 11.30 Generalized afocal system.

In any afocal system, such as a telescope, the angular magnification is given by

$$m_a = -F_o/F_i \qquad (11.101)$$

where F_o = effective focal length of the objective system, i.e., that member near the object

F_i = focal length of the member near the image, i.e., the eyepiece

A positive magnification indicates an erect image, a negative an inverted image. If there are three major components in a telescope—the objective, the erector, and the eyepiece—the above can be applied with the erector taken either as a part of the objective or as part of the eyepiece. Either will yield

$$m_a = -(F_o/F_i)/(S_2/S_1) \qquad (11.102)$$

where S_2/S_1 is the linear magnification of the erector component. See Fig. 11.31.

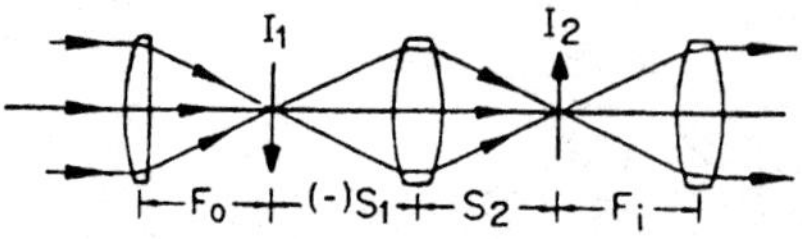

FIG. 11.31 The lens-erecting telescope.

The objective member of an afocal system forms the image of an infinitely distant object at its second focal point. In order that the final image be at infinity, this internal image must lie at the first focal point of the eyepiece member. Thus for a simple telescope as shown in Figs. 11.32 and 11.33, the spacing d between the (adjacent principal points of the) objective and eyepiece equals the algebraic sum of their focal lengths:

$$d = F_o + F_i \qquad (11.103)$$

11.7.2 Astronomical Telescope

The astronomical telescope is an afocal system in which both members have positive focal lengths. A real internal image is formed within the device, and the final image is inverted (and reversed). Figure 11.32 shows schematic diagrams of refracting and

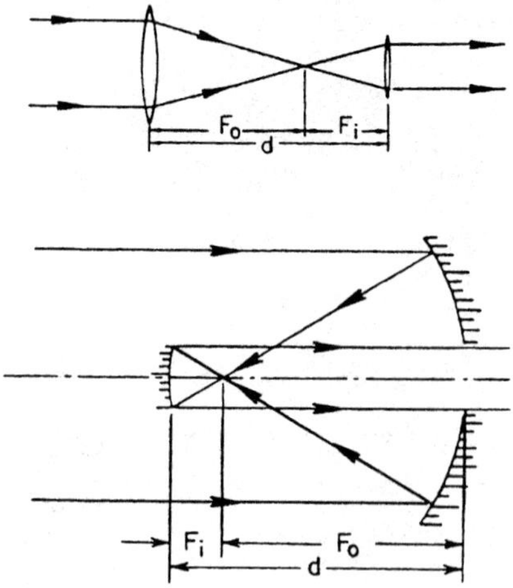

FIG. 11.32 The astronomical telescope, refracting and reflecting versions.

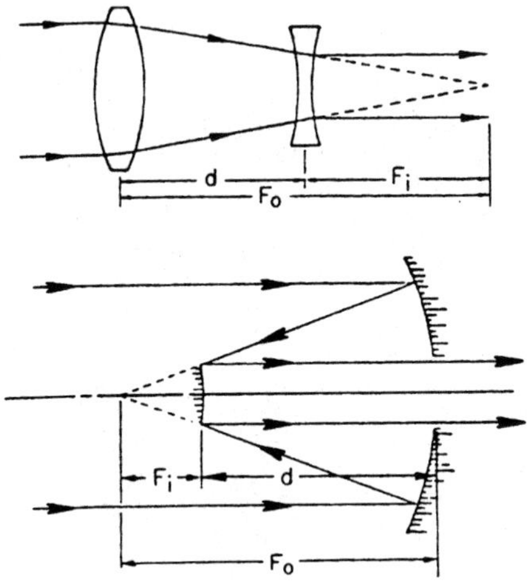

FIG. 11.33 The galilean telescope.

reflecting versions of astronomical telescopes. The normal telescope for visual astronomy has a reflecting objective and a refracting eyepiece. Refracting components are usually of compound construction to correct the aberrations. The aperture stop and the entrance pupil are almost always located at the objective in order to minimize its size and cost; the exit pupil is then the image of the objective formed by the eyepiece.

11.7.3 Galilean Telescope

A galilean telescope has a positive objective and a simple negative eyepiece. Since there is no real internal image, a reticle or cross hair cannot be used as an aiming point. The final image is erect. The galilean telescope is compact, and the field is usually small and the power low. The aperture stop is usually the pupil of the user's eye. Sketches of refracting and reflecting galileans are shown in Fig. 11.33.

11.7.4 Terrestrial Telescope

When an erecting relay component is inserted between the objective and eye lens of the astronomical telescope, the result is the terrestrial telescope, a configuration used for surveying instruments and sailor's spyglasses. A schematic is shown in Fig. 11.31. An erect-image telescope can also be made by erecting the image with an appropriate prism system.

11.7.5 Relay Lenses

A relay lens is used to transfer an image from one location to another and to erect the image in the process. The center element in Fig. 11.31 relays the image from I_1 to I_2. A series of several relay lenses may be used to carry an image through a long hole of limited diameter.

11.7.6 Field Lenses

A field lens is a positive element placed at or near an image for the purpose of shifting the location of the pupil without significantly altering the power of the system. A field

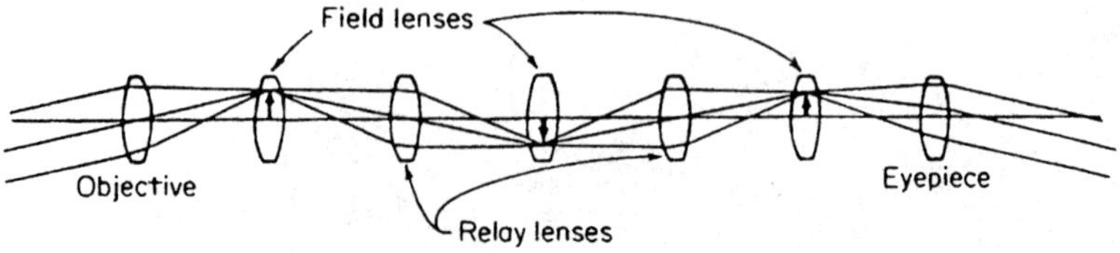

FIG. 11.34 A series of relay and field lenses—used to carry an image through a long, narrow space, as in a periscope.

lens is often designed to image the pupil into the following components of the system so that the diverging beams which form the images in the outer parts of the field will not escape from the system.

Figure 11.34 illustrates the basic principle of the periscope. A system of alternating field and relay lenses transfers both the primary image and the pupil from one end to the other without allowing the edge-of-the-field rays to escape.

11.8 MICROSCOPES

11.8.1 Simple Microscope (Magnifying Glass)

A simple microscope is a single positive component used to magnify an object, when the object is located at or within the focal point of the component. The magnification of microscopes, both simple and compound, is somewhat arbitrarily defined as the ratio of the apparent angular size of the image to the angular size that the object would subtend if viewed from a distance of 10 in, which is assumed to be the distance of most distinct vision (on the average).

Assuming that the eye is located close to the lens, the simple microscope of Fig. 11.35 has a magnification of

$$M = (10 \text{ in})/f - (10 \text{ in})/s' \qquad (11.104)$$

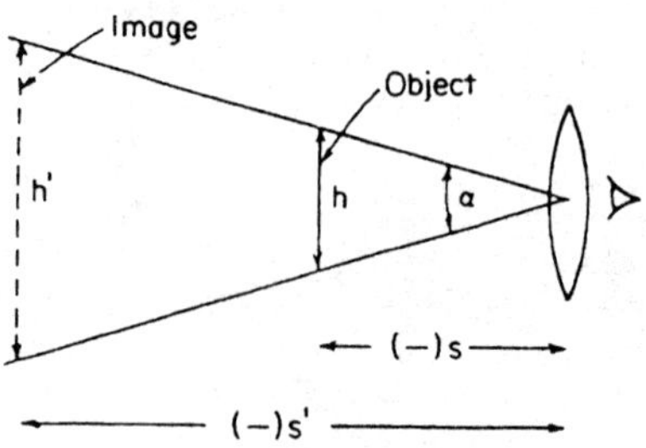

FIG. 11.35 A magnifier, or simple microscope.

When the object is placed at the focal point, s' is infinite, and $M = (10 \text{ in})/f$. If the object is within the focus by a distance such that $s' = -10$ in, then $M = 1 + (10 \text{ in})/f$. These relationships are conventionally used to define the power of magnifiers and eyepieces for telescopes and compound microscopes.

11.8.2 Compound Microscopes

A compound microscope consists of an objective forming an intermediate, usually magnified, image and an eyepiece to view, and further magnify, the image. From the definition of magnification above, and with reference to Fig. 11.36, the magnification of a compound microscope is

$$M = (10 \text{ in})/f_{eo} = (f_e + f_o - d)(10 \text{ in})/f_e f_o \qquad (11.105)$$

or

$$M = M_o M_e = (s_2/s_1)/[(10 \text{ in})/f_e] \qquad (11.106)$$

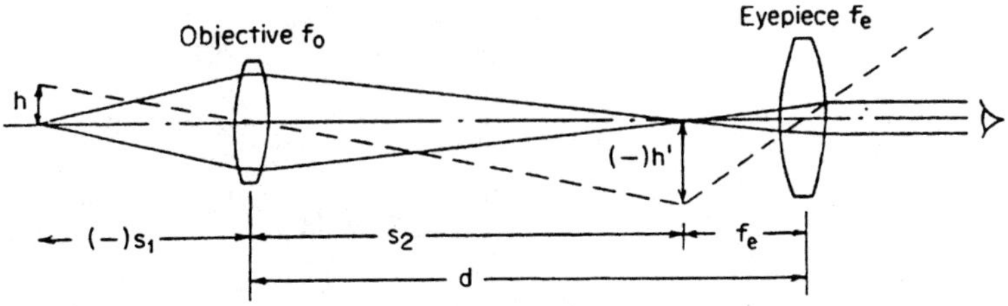

FIG. 11.36 The compound microscope.

Note that there are instances where the magnification may not be clearly defined by either the telescope or microscope conventions. If the object is neither at infinity (or at a great distance) nor close enough that it could be viewed at the distance of most distinct vision (10 in), quite different results will be obtained if the conventional magnification formulas are applied. The proper magnification under such circumstances is still the ratio of angular size of the image to the angular size of the object at a distance determined by the application.

11.9 DETECTOR OPTICS

Optics are used near radiation detectors to spread the incident energy as uniformly as possible over the surface of the detector, to collect a larger amount of energy on a small detector, or to limit or control the field of view of radiation detection.

A field lens can be used to increase the field of view of an instrument without increasing the size of the detector. Figure 11.37 shows a simple radiometer system with a field lens arranged to image the exit pupil of the objective system (which may be refracting as shown, or reflecting) onto the surface of the detector. The field lens, used this way, allows smaller detectors, uniformly illuminates the detector surface, and makes the detector irradiation independent of the position of an image point within the field of view.

11.9.1 Detector Size Limitations and Field Lenses

The detector field lens of Fig. 11.37 is exactly analogous to a projection condenser; both are subject to similar limitations.

For *any* optical system associated with a detector, there is a limit on the minimum size (i.e., smallness) of the detector, given by

$$S_d \geq D\theta/n \tag{11.107}$$

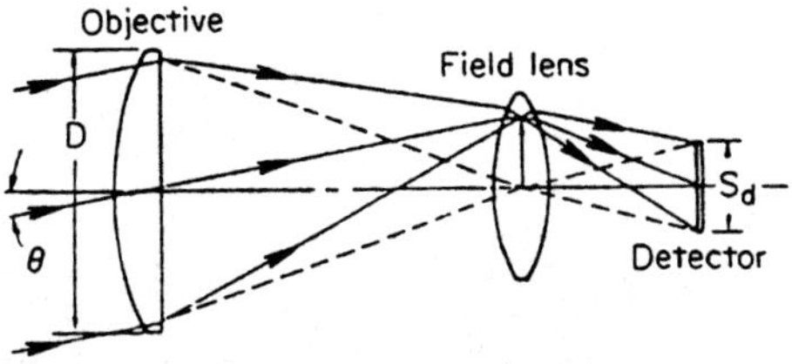

FIG. 11.37 A radiometer with a field lens used to increase the field of view with a small detector.

where S_d = diameter or linear dimension of detector
D = corresponding dimension of entrance pupil of optical system
θ = half-angle of field of view
n = index of refraction of medium at detector

This limit is difficult to achieve in practice. A detector whose size is twice that indicated by Eq. (11.107) is a reasonable practical limit, corresponding to the use of optics with a speed of f/1.0. In projection or illumination systems the same equation can be applied if S_d is taken as the lamp or source dimension.

11.9.2 Immersion Lenses

If a detector is immersed on the rear surface of a hemispherical lens, a small detector may be used in place of a larger. The hemispherical lens, used this way, is aplanatic and introduces no coma or spherical aberration. It increases the apparent size of the detector by a factor of the index, as shown in Fig. 11.38.

The immersion technique can be used in conjunction with a field lens of the type described above, or the immersion lens may be made hyperhemispherical to increase the magnification.

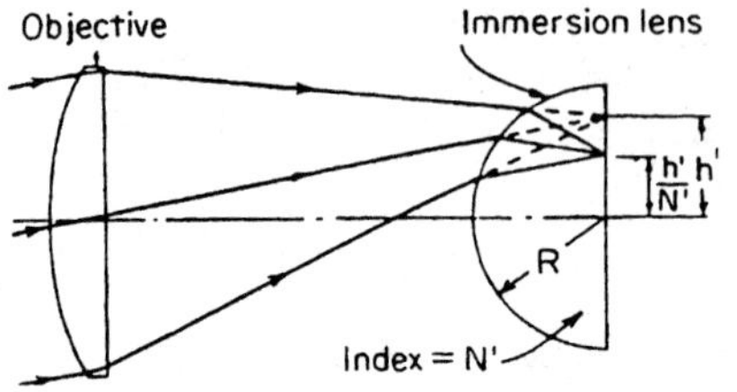

FIG. 11.38 A hemispherical immersion lens, concentric with the focus of an optical system reduces the linear size of the image by a factor equal to the index of the immersion lens.

11.10 FIBER OPTICS

11.10.1 Total Internal Reflection

When light is incident on a surface separating media of indices n and n', the light will be totally reflected at the surface if the angle of incidence equals or exceeds the critical angle (and if $n > n'$).

$$I_c = \arcsin (n'/n) \tag{11.108}$$

11.10.2 Optical Fibers

A long, polished cylinder of glass will transmit light without loss by leakage through the cylinder walls if all surface reflections are at angles larger than I_c, as shown in Fig. 11.39. If the cylinder has a length l, the total path of a ray in the plane of the figure is $l/\cos U'$ and the number of reflections undergone by the ray is $(l/d_f) \tan U'$, where d_f is the cylinder diameter.

11.10.3 Numerical Aperture and Cladding

With reference to Fig. 11.39, the numerical aperture of an optical fiber is

$$\mathrm{NA} = n_0 \sin U = (n_1^2 - n_2^2)^{0.5} \tag{11.109}$$

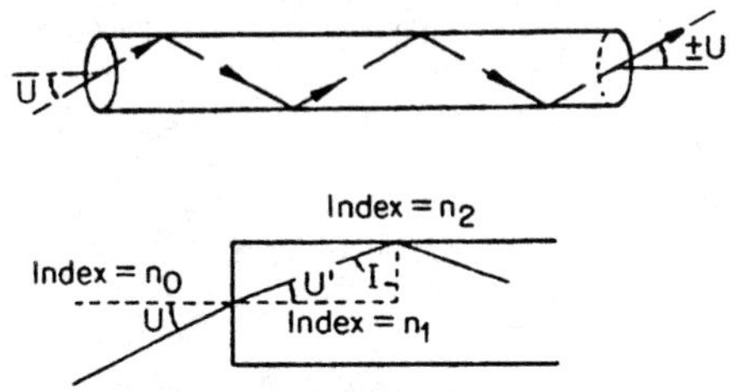

FIG. 11.39 Optical fibers: light can be transmitted through a long polished cylinder by means of total internal reflection (TIR) at the surfaces.

Skew rays (nonmeridional rays which do not lie in the plane of the sketch) have a somewhat greater no-loss transmission angle since they make greater angles of incidence with the surface. They also travel a greater distance and are reflected more often than meridional rays.

Because of the large number of reflections necessary to transmit a ray a useful distance through a small-diameter fiber, fibers are often "clad" with a material of lower index. The total internal reflection occurs at the interface between the core of the fiber and the cladding material. The cladding protects against dirt, oil, or contact with adjacent fibers, any of which will allow the light to leak out of the fiber.

Optical fibers are ordinarily a few thousandths of an inch in diameter and made either of glass or plastic. They are quite flexible and can be used singly or in bundles, either to transmit light or to transmit an image. If the fibers are clad, the cladding material can be fused to make a bundle of fibers into a solid structure, which can be used as a field lens or as a faceplate for a cathode-ray tube or an image intensifier. In this application it is possible for light which has escaped from one fiber to enter an adjacent fiber, producing cross talk and reducing the contrast in the image. The use of an absorbing material between the fibers can prevent this problem.

11.10.4 Resolution of Fiber Optics Images

An image transmitted by a coherent fiber bundle, that is, a bundle whose fibers have the same spatial relationships at both ends, is transmitted in bits, each bit the size of a fiber; within the fiber details smaller than the fiber diameter are lost. The resolution, determined by the diameters of the fibers is approximately $(2d_f)^{-1}$ lines per millimeter, where d_f is the fiber diameter in millimeters.

Figure 11.39 and Eq. (11.109) describe straight fibers. In practice fibers are often curved, and the numerical aperture is reduced because the angle of incidence is smaller than that assumed in Eq. (11.109). The ultimate limit on the bending of an optical fiber is set by the tensile strength of the fibers.

REFERENCES

1. Amon, M., and S. Rosin: "Mangin Mirror Systems," *Applied Optics*, vol. 2, p. 214, 1968.
2. Amon, M., and S. Rosin: "Color Corrected Mangin Mirror," *Applied Optics*, vol. 6, p. 963, 1967.
3. Baker, J.: "On Improving the Effectiveness of Large Telescopes," *IEEE Trans. on Aerospace and Electronic Systems*, March 1969, pp. 261–272.

4. Barakat, R.: "Numerical Results Concerning the Transfer Functions and Total Illuminance for Optimum Balanced Fifth-Order Spherical Aberration," *J. Opt. Soc. Am.*, vol. 54, pp. 38–44, 1964.
5. Barakat, R., and J. Houston: "Diffraction Effects of Coma," *J. Opt. Soc. Am.*, vol. 54, pp. 1084–1088, 1964.
6. Barakat, R., and J. Houston: "The Effect of a Sinusoidal Wavefront on the Transfer Function of a Circular Aperture," *Applied Optics*, vol. 5, pp. 1850–1852, 1966.
7. Bergstein and Motz: "Optically Compensated Varifocal Systems," *J. Opt. Soc. Am.*, vol. 52, pp. 353–388, 1962.
8. Born, M., and E. Wolf: "Principles of Optics," The Macmillan Company, New York, 1964.
9. Bouwers, A.: "Achievements in Optics," Elsevier Publishing Company, New York, 1946.
10. Buchdahl, H.: "Optical Aberration Coefficients," Dover Publications, Inc., New York, 1968.
11. Buchroeder, R.: "Applications of Aspherics for Weight Reduction in Selected Data," Design Examples of Tilted Component Telescopes, Optical Sciences Center, University of Arizona, Tucson, Technical Report 68, 1971.
12. Chretien, H.: "Revue d'Optique," Paris, vol. 1, pp. 13–22 and 49–64.
13. Coltman, J.: "The Specification of Imaging Properties by Response to a Sine Wave Input," *J. Opt. Soc. Am.*, vol. 44, p. 234, 1954.
14. Conrady, A.: "Applied Optics and Optical Design," Dover Publications, Inc., New York, 2 vol., 1957 and 1960.
15. Cooke, G.: "Recent Developments in Anamorphic Systems," *J. Soc. Motion Pict. Telev. Engr.*, vol. 65, pp. 151–154, 1956.
16. Cooley and Tukey: "An Algorithm for the Machine Calculation of Complex Fourier Series," *Math. Comput.*, vol. 19, pp. 297–301, 1965.
17. Cox, A.: "A System of Optical Design," Focal, New York, 1964.
18. Feder, D.: "Optical Calculations with Automatic Computing Machinery," *J. Opt. Soc. Am.*, vol. 41, pp. 630–635, 1951.
19. Feder, D.: "Automatic Optical Design," *Applied Optics*, vol. 2, pp. 1209–1226, 1963.
20. Feder, D.: "Differentiation of Ray-Tracing Equations with Respect to Constructional Parameters of Rotationally Symmetrical Optics," *J. Opt. Soc. Am.*, vol. 58, pp. 1494–1505, 1968.
21. Gascoigne, S.: "Some Recent Advances in the Optics of Larger Telescopes," *Q. J. Royal Astron. Soc.*, vol. 9, 1968.
22. Glatzel, E., and R. Wilson: "Adaptive Automatic Correction in Optical Design," *Applied Optics*, vol. 7, pp. 265–276, 1968.
23. Goodman, J.: "Introduction to Fourier Optics," McGraw-Hill Book Company, Inc., New York, 1968.
24. "Handbook of Optical Design," MIL-HDBK-141, U.S. Government Printing Office, 1962.
25. Hardy, A. and F. Perrin: "The Principles of Optics," McGraw-Hill Book Company, Inc., New York, 1932.
26. Herzberger, M.: "Modern Geometrical Optics," Interscience Publishers, New York, 1958.
27. Hopkins, H.: "Wave Theory of Aberrations," Oxford University Press, London, 1950.
28. Houston, J., et al.: "A Laser Unequal Path Interferometer for the Optical Shop," *Applied Optics*, vol. 6, pp. 1237–1242, 1967.
29. Kelly, D.: "Spatial Frequency, Bandwidth and Resolution," *Applied Optics*, vol. 4, pp. 435–438, 1965.
30. Kingslake, R.: "Applied Optics and Optical Engineering," Academic Press, Inc., 5 vols., 1965 to 1969.
31. Kingslake, R.: "The Development of the Zoom Lens," *J. Soc. Motion Pict. Telev. Eng.*, vol. 69, pp. 534–544, 1960.
32. Lauroesch, T., and C. Wing: "Bouwers Concentric Systems for Materials of High Refractive Index," *J. Opt. Soc. Am.*, vol. 49, pp. 410–411, 1959.

33. Linfoot, E.: "Fourier Methods in Optical Design," Focal Press, New York, 1964.
34. Linfoot, E.: "Recent Advances in Optics," Clarendon Press, Oxford, England, 1955.
35. Maksutov, D.: "New Catadioptric Meniscus Systems," *J. Opt. Soc. Am.*, vol. 34, pp. 270–284, 1944.
36. Martin, L.: "Technical Optics," Sir Isaac Pitman & Sons, Ltd., London, 1960.
37. McNeil, G.: "Optical Fundamentals of Underwater Photography," Photogrammetry, Rockville, Md., 1968.
38. Meiron, J.: "The Use of Merit Functions Based on Wavefront Aberrations in Automatic Lens Design," *Applied Optics*, vol. 7, pp. 667–672, 1968.
39. MIL-HDBK-141, "Handbook of Optical Design," U.S. Government Printing Office, 1962.
40. Montagino, L.: "Ray Tracing in Inhomogenous Media," *J. Opt. Soc. Am.*, vol. 58, pp. 1667–1668, 1968.
41. O'Neill, E.: "Transfer Function for an Annular Aperture," *J. Opt. Soc. Am.*, vol. 46, pp. 285–288, 1956.
42. Perrin, F.: "Methods of Appraising Photographic Systems," *J. Soc. Motion Pict. Telev. Eng.*, vol. 69, pp. 151–156, 239–248, 1960.
43. Potter, R.: "Transmission Properties of Optical Fibers," *J. Opt. Soc. Am.*, vol. 51, pp. 1079–1089, 1961.
44. Ritchey, G. and H. Chretien: *Compes Rendus*, Bucharest, Romania, vol. 185, p. 266, 1927.
45. Ross, F.: "Lens Systems for Correcting Coma of Mirrors," *Mt. Wilson Obs. Yearbook*, vol. 35, p. 191, 1936, and *Astrophysical J., Am. Stron. Soc.*, vol. 81, p. 156. 1935.
46. Schmidt, B.: "Mithandlungen der Hamburger Sternwarde in Bergedorf," vol. 7, p. 15, 1932.
47. Smith, W.: "Modern Optical Engineering: The Design of Optical Systems," McGraw-Hill Book Company, Inc., New York, 1966.
48. Smith, W.: "Handbook of Military Infrared Technology," W. Wolfe, ed., chaps. 9 and 10, "Optics" and "Optical Systems," U.S. Office of Naval Research, Washington, D.C., 1965.
49. Smith, W.: "Handbook of Optics," W. Driscoll, ed., chap. 2, "Image Formation: Geometrical and Physical Optics," McGraw-Hill Book Company, Inc., New York, 1978.
50. Smith, W.: "Infrared Handbook," Wolfe and Zissis, eds., chaps. 8 and 9, "Optical Design" and "Optical Elements—Lenses and Mirrors," U.S. Office of Naval Research, Washington, D.C., 1978.
51. Southall, J.: "Mirrors, Prisms and Lenses," Dover Publications, Inc., New York, 1964.
52. Spencer, G.: "A Flexible Automatic Lens Correction Procedure," *Applied Optics*, vol. 2, pp. 1257–1264, 1963.
53. Spencer, G., and M.V.R.K. Murty: "Generalized Ray-Tracing Procedure," *J. Opt. Soc. Am.*, vol. 52, pp. 672–678, 1962.
54. Stephens, R. and C. Sutton: "Diffraction Images of a Point in the Focal Plane and Several Out-of-Focus Planes," *J. Opt. Soc. Am.*, vol. 58, pp. 1001–1002, 1968.
55. Strong, J.: "Concepts of Classical Optics," W. H. Freeman and Company, San Francisco, 1958.
56. Steel, W.: "Interferometry," Cambridge University Press, New York, 1967.
57. Walles, S. and R. Hopkins: "The Orientation of the Image Formed by a Series of Plane Mirrors," *Applied Optics*, vol. 3, pp. 1447–1452, 1964.
58. Williamson, D.: "Cone Channel Condenser Optics," *J. Opt. Soc. Am.*, vol. 42, pp. 712–715, 1952.

CHAPTER 12
MACHINE SYSTEMS

Burton Paul, Ph.D.
Asa Whitney Professor of Dynamical Engineering
Department of Mechanical Engineering and Applied Mechanics
University of Pennsylvania
Philadelphia, Pa.

12.1 INTRODUCTION 12.1
12.2 ANALYTICAL KINEMATICS 12.2
12.2.1 Lagrangian Coordinates 12.2
12.2.2 Degree of Freedom 12.5
12.2.3 Generalized Coordinates 12.6
12.2.4 Points of Interest 12.8
12.2.5 Angular Position, Velocity, and Acceleration 12.9
12.2.6 Kinematics of Systems with One Degree of Freedom 12.10
12.2.7 Computer Programs for Kinematics Analysis 12.11
12.3 STATICS OF MACHINE SYSTEMS 12.11
12.3.1 Direct Use of Newton's Laws (Vectorial Mechanics) 12.11
12.3.2 Terminology of Analytical Mechanics 12.12
12.3.3 The Principle of Virtual Work 12.14
12.3.4 Lagrange Multipliers 12.15
12.3.5 Conservative Systems 12.16
12.3.6 Reactions at Joints 12.17
12.3.7 Effects of Friction in Joints 12.18
12.4 KINETICS OF MACHINE SYSTEMS 12.18
12.4.1 Lagrange's Form of d'Alembert's Principle 12.18
12.4.2 Equations of Motion from Lagrange's Equations 12.22
12.4.3 Kinematics of Single-Freedom Systems 12.25
12.4.4 Kinetostatics and Internal Reactions 12.27
12.4.5 Balancing of Machinery 12.28
12.4.6 Effects of Flexibility and Joint Clearance 12.30
12.5 NUMERICAL METHODS FOR SOLVING THE DIFFERENTIAL EQUATIONS OF MOTION 12.30
12.5.1 Determining Initial Values of Lagrangian Coordinates 12.31
12.5.2 Initial Values of Lagrangian Velocities 12.31
12.5.3 Numerical Integration 12.31
12.6 SURVEY OF MACHINE DYNAMICS LITERATURE AND COMPUTER PROGRAMS 12.32
12.6.1 Computer Programs 12.32
12.6.2 Survey Papers 12.32
12.6.3 Books on Multibody Dynamics 12.33
12.6.4 Books on Dynamics of Robots 12.33

12.1 INTRODUCTION

The following material is a survey of the theory of machine systems based on analytical mechanics and digital computation. Much of the material to be presented is described in greater detail in Refs. 1, 36, and 64 to 66, but some of it is an extension of previously published works. For a discussion of the subject based on the more traditional approach of vector dynamics and graphical constructions see Refs. 51, 52, and 83.

The systems which we shall consider comprise resistant bodies interconnected in such a way that specified input forces and motions are transformed in a predictable way to produce desired output forces and motions. The term "resistant" body includes both rigid bodies and components such as cables or fluid columns which momentarily serve the same function as rigid bodies. This description of a mechanical system is essentially the definition of the term "machine" given by Reuleaux.[75] When discussing the purely kinematic aspects of such a system, the assemblage of "links" (i.e., resistant bodies) is frequently referred to as a "kinematic chain"; when one of the links of a kinematic chain is fixed, the chain is said to form a "mechanism." Various types of "kinematic pairs" or interconnections (e.g., revolutes, sliders, cams, rolling pairs) between pairs of links are described in Chap. 3. Two or more such kinematic pairs may be superposed at an idealized point, or "joint," of the system.

The organization of this chapter follows that classification[93] of mechanics which divides the subject into "kinematics" (the study of motion irrespective of its cause) and "dynamics" (the study of motion due to forces); dynamics is then subdivided into the categories "statics" (forces in equilibrium) and "kinetics" (forces not in equilibrium).

A Notation. Boldface type is used to represent ordinary vectors, matrices, and column vectors, with their respective scalar components indicated as in the following examples:

$$\mathbf{r} = x\mathbf{i} + y\mathbf{i} + z\mathbf{k} \equiv (x, y, z)$$

$$\mathbf{C} = [c_{ij}] = \begin{bmatrix} c_{11} & \cdots & c_{1N} \\ \cdots & \cdots & \cdots \\ c_{M1} & \cdots & c_{MN} \end{bmatrix}$$

$$\mathbf{x} \equiv \{x_1, x_2, \ldots, x_N\} \equiv [x_1, x_2, \ldots, x_N]^T$$

We shall occasionally refer to a matrix such as **C** as "an $M \times N$ matrix." Note the use of a superscript T to indicate the transpose of a matrix. In what follows, we shall make frequent use of the rule for transposition of a matrix product, i.e., $(\mathbf{AB})^T = \mathbf{B}^T\mathbf{A}^T$.

If $F(\psi) = F(\psi_1, \psi_2, \ldots, \psi_M)$, we will use subscript commas to denote the partial derivatives of F with respect to its arguments, as follows:

$$F_{,i} = \partial F(\psi)/\partial\psi_i \qquad F_{,ij} = \partial^2 F(\psi)/\partial\psi_i\partial\psi_j$$

Differentiation with respect to time t will be denoted by an overdot:

$$\dot{x} = dx/dt$$

12.2 ANALYTICAL KINEMATICS

12.2.1 Lagrangian Coordinates

The cartesian coordinates (x^i, y^i, z^i) of a particle P^i with respect to a fixed right-handed reference frame are called the "global" coordinates of the particle. Occasionally, we shall represent the global coordinates by the alternative notation (x_1^i, x_2^i, x_3^i), abbreviated as

$$\mathbf{x}^i \equiv (x_1^i, x_2^i, x_3^i) = (x^i, y^i, z^i) \tag{12.1}$$

When all particles move parallel to the global xy plane, the motion is said to be "planar."

Since the number of particles in a mechanical system is usually infinite, it is convenient to represent the global position vector $\mathbf{x}^k \equiv (x^k, y^k, z^k)$ of an arbitrary particle P^k in terms of a set of M variables $\psi_1 < \psi_2, \ldots, \psi_M$, called "lagrangian coordinates." By

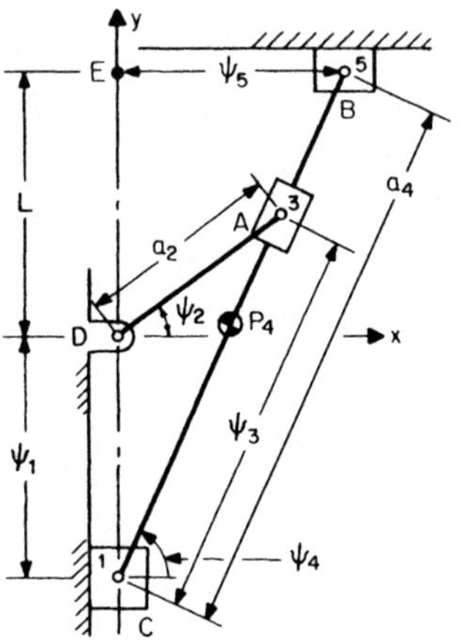

FIG. 12.1 Example of multiloop planar mechanism.

introducing a suitable number of lagrangian coordinates, it is always possible to express the global coordinates of every particle in the system via "transformation equations" of the form

$$\mathbf{x}^k = \mathbf{x}^k(\psi_1, \ldots, \psi_M) \tag{12.2a}$$

or

$$x_i^k = x_i^k(\psi_1, \ldots, \psi_M) \qquad i = 1, 2, 3 \tag{12.2b}$$

where the time t does not appear explicitly. For example, in Fig. 6.1 the global (x,y) coordinates of the midpoint P^4 of bar BC are given by

$$\begin{aligned} x_1^4 &= x^4 = \tfrac{1}{2}a_4 \cos \psi_4 \\ x_2^4 &= y^4 = -\psi_1 + \tfrac{1}{2}a_4 \sin \psi_4 \end{aligned} \tag{12.3}$$

In general, the lagrangian coordinates ψ_i need not be independent, but may be related by "equations of constraint" of the form

$$f_i(\psi_1, \psi_2, \ldots, \psi_M) = 0 \qquad i = 1, 2, \ldots, N_s \tag{12.4}$$

or

$$g_k(\psi_1, \psi_2, \ldots, \psi_M, t) = 0 \qquad k = 1, 2, \ldots, N_t \tag{12.5}$$

Equation (12.4) represents "spatial" (or "scleronomic") constraints because only the space variables ψ_i appear as arguments; there are N_s spatial constraints. On the other hand, time t does appear explicitly in Eq. (12.5), which is therefore said to represent "temporal" (or "rheonomic") constraints; there are N_t such constraints. For example, closure of the two *independent** loops ECBE and DCAD requires that the following *spatial* constraints be satisfied:

$$\begin{aligned} f_1(\psi) &= a_4 \cos \psi_4 - \psi_5 = 0 \\ f_2(\psi) &= a_4 \sin \psi_4 - L - \psi_1 = 0 \\ f_3(\psi) &= \psi_3 \cos \psi_4 - a_2 \cos \psi_2 = 0 \\ f_4(\psi) &= \psi_3 \sin \psi_4 - a_2 \sin \psi_2 - \psi_1 = 0 \end{aligned} \tag{12.6}$$

Other equations of spatial constraint described by Eq. (12.4) might arise from the action of gears (see Sec. 7.30 of Ref. 66), from cams (see Sec. 7.40 of Ref. 66), or from kinematic pairs in three-dimensional mechanisms (see p. 752 of Ref. 62).

As an example of a temporal constraint consider the function

*The equations of constraint corresponding to loop closure will be independent if each loop used is a "simple closed circuit"; i.e., it encloses exactly one polygon with nonintersecting sides. See Chap. 8 of Ref. 66 for criteria of loop independence.

$$g(\psi, t) = \psi_1 - \beta - \omega t - \alpha t^2/2 = 0 \tag{12.7}$$

where α, β, and ω are constants. This constraint specifies that the variable ψ_1 increases with constant acceleration ($\ddot{\psi}_1 = \alpha$), starting from a reference state (at $t = 0$) where its initial value is β and its initial "velocity" is $\dot{\psi}_1 = \omega$. The special case where $\alpha = 0$ corresponds to a link being driven by a constant-speed motor.

A second example is given by the constraint

$$g(\psi, t) = \psi_1 + \psi_2 + \psi_3 - 3\beta - 3\omega t = 0 \tag{12.8}$$

which requires that the average velocity $\frac{1}{3}(\dot{\psi}_1 + \dot{\psi}_2 + \dot{\psi}_3)$ remain fixed at a constant value ω. Such a constraint could conceivably be imposed by a feedback control system.

If the total number of constraints, given by

$$N_c = N_s + N_t \tag{12.9}$$

just equals the number M of lagrangian variables, the M equations, Eqs. (12.4) and (12.5), which can be written in the form

$$F_i(\psi_1, \psi_2, \ldots, \psi_M, t) = 0 \qquad i = 1, 2, \ldots, M \tag{12.10}$$

can be solved for the M unknown values of ψ_j (except for certain *singular* states; see Sec. 8.2 of Ref. 66). Because Eqs. (12.10) are most often highly nonlinear, it is usually necessary to solve them by a numerical procedure, such as the Newton-Raphson algorithm. Details of the technique are given in Chap. 9 of Ref. 66.

To find the lagrangian velocities, it is necessary to differentiate Eqs. (12.10) with respect to time t. The differentiation yields

$$\sum_{j=1}^{M} \frac{\partial F_i}{\partial \psi_j} \dot{\psi}_j + \frac{\partial F_i}{\partial t} = 0 \qquad i = 1, \ldots, M \tag{12.11}$$

In matrix terms this may be expressed as

$$\mathbf{G}\,\dot{\psi} = \mu \tag{12.12}$$

where

$$\mathbf{G} = [\partial F_i/\partial \psi_j] \qquad \mu = -\{\partial F_i/\partial t\} \tag{12.13}$$

In the nonsingular case, Eq. (12.12) can be solved for the $\dot{\psi}_j$.

To find lagrangian accelerations, we may differentiate Eq. (12.11) to obtain

$$\sum_{j=1}^{M} \left(F_{i,j}\ddot{\psi}_j + \sum_{k=1}^{M} F_{i,jk}\dot{\psi}_k\dot{\psi}_j\right) + \sum_{j=1}^{M} \left(\frac{\partial F_i}{\partial t}\right)_{,j}\dot{\psi}_j + \frac{\partial^2 F_i}{\partial t^2} \tag{12.14}$$

where use has been made of the comma notation for denoting partial derivatives.

In matrix notation Eq. (12.14) can be expressed as

$$\mathbf{G}\ddot{\boldsymbol{\psi}} = \mathbf{b} \tag{12.15}$$

where $\mathbf{G}$ has been previously defined and

$$\mathbf{b} = \dot{\boldsymbol{\mu}} - \dot{\mathbf{G}}\dot{\boldsymbol{\psi}} \tag{12.16}$$

or
$$b_i = -\frac{\partial^2 F_i}{\partial t^2} - 2\sum_{j=1}^{M}\left(\frac{\partial F_i}{\partial t}\right)_{,j}\dot{\psi}_j - \sum_{j=1}^{M}\sum_{k=1}^{M} F_{i,jk}\dot{\psi}_k\dot{\psi}_j \tag{12.17}$$

To find velocities of the point of interest $\mathbf{x}^k$, we differentiate Eq. (12.2*b*) to obtain

$$\dot{x}_i^k = \sum_{j=1}^{M} x_{i,j}^k \dot{\psi}_j \tag{12.18}$$

or
$$\dot{\mathbf{x}}^k = \mathbf{P}^k \dot{\boldsymbol{\psi}} \tag{12.19}$$

where the $3 \times M$ matrix $\mathbf{P}^k$ is defined by

$$\mathbf{P}^k = [x_{i,j}^k] \tag{12.20}$$

The corresponding acceleration is given by

$$\ddot{x}_i^k = \sum_{j=1}^{M} x_{i,j}^k \ddot{\psi}_j + \sum_{j=1}^{M}\sum_{k=1}^{M} x_{i,jk}^k \dot{\psi}_k \dot{\psi}_j \tag{12.21}$$

or
$$\ddot{\mathbf{x}}^k = \mathbf{P}^k \ddot{\boldsymbol{\psi}} + \mathbf{p}^k \tag{12.22}$$

where the components of the vector $\mathbf{p}^k$ are given by

$$p_i^k = \sum_{j=1}^{M}\sum_{k=1}^{M} x_{i,jk}^k \dot{\psi}_k \dot{\psi}_j \tag{12.23}$$

12.2.2 Degree of Freedom

For a system described by M lagrangian coordinates, the most general form for the equations of spatial constraint is

$$f_i(\psi_1, \ldots, \psi_M) = 0 \qquad i = 1, \ldots, N_s \tag{12.24}$$

The "rate form" of these equations is

$$\sum_{j=1}^{M} \frac{\partial f_i}{\partial \psi_j}\dot{\psi}_j = 0 \qquad \text{or} \qquad \mathbf{D}\dot{\boldsymbol{\psi}} = \mathbf{0} \tag{12.25}$$

where
$$\mathbf{D} = [D_{ij}] = [\partial f_i / \partial \psi_j] \tag{12.26}$$

is an $N_s \times M$ matrix. If the rank of matrix $\mathbf{D}$ is r, we know from a theorem of linear algebra[9] that we may arbitrarily assign any values to $M - r$ of the $\dot{\psi}_j$ and the others will be uniquely determined by Eq. (12.25). In kinematics terms, we say that the degree of freedom (DOF) or mobility* of the system is

$$F = M - r \tag{12.27}$$

*See Chap. 8 of Ref. 66 for a discussion of mobility criteria.

This result is due to Freudenstein.[24]

In the nonsingular case, $r = N_s$ and the DOF is

$$F = M - N_s \tag{12.28}$$

For example, consider the mechanism of Fig. 12.1, where $M = 5$ and $N_s = 4$ [see Eq. (12.6)]. From Eq. (12.28) we see that this mechanism has one DOF.

If any temporal constraints are also present, we should refer to F as the "unreduced DOF," because the imposition of N_t temporal constraints will lower the degree of freedom still further to the "reduced DOF";

$$F_R = F - N_t = M - N_s - N_t \tag{12.29}$$

For example, if we drive AD of Fig. 12.1 with uniform angular velocity ω, we are in essence imposing the single temporal constraint

$$g_1(\psi,t) = \psi_2 - \omega t = 0 \tag{12.30}$$

Hence the reduced degree of freedom is

$$F_R = 1 - 1 = 0$$

In short, the system is "kinematically determinate." No matter what external forces are applied, the *motion* is always the same. Of course, the internal forces at the joints of the mechanism are dependent on the applied forces, as will be discussed in the following sections on statics and kinetics of machine systems.

12.2.3 Generalized Coordinates

For purposes of kinematics, it is sufficient to work directly with the lagrangian coordinates ψ_i. However, in problems of dynamics, it is useful to express the terms ψ_i in terms of a smaller number of variables q_r, called "generalized coordinates" (also called "primary coordinates"). In problems of kinematics, we may think of the q_r as "driving variables." For example, if a motor drives link DA of Fig. 12.1, $\psi_2 = q_1$ is a good choice for this problem.

In a system with F DOF, we are free to choose any subset of F lagrangian coordinates to serve as generalized coordinates. This choice may be expressed by the matrix relationship

$$\mathbf{T}\boldsymbol{\psi} = \mathbf{q} \tag{12.31}$$

where $\mathbf{T}$ is an $F \times M$ matrix whose elements are defined as

$$T_{ij} = \begin{cases} 1 & \text{if } q_i = \psi_j \quad (12.32a) \\ 0 & \text{otherwise} \quad (12.32b) \end{cases}$$

For example, if $M = 5$ and $F = 2$, we might select ψ_2 and ψ_5 as the primary variables q_1 and q_2. Then Eq. (12.31) becomes

$$\begin{bmatrix} 0\,1\,0\,0\,0 \\ 0\,0\,0\,0\,1 \end{bmatrix} \begin{bmatrix} \psi_1 \\ \psi_2 \\ \psi_3 \\ \psi_4 \\ \psi_5 \end{bmatrix} = \begin{bmatrix} q_1 \\ q_2 \end{bmatrix}$$

If desired, the matrix **T** may be chosen so that any q_i is some desired linear combination of the ψ_i, but the simpler choice embodied in Eqs. (12.32*a*) and (12.32*b*) should suffice for most purposes.

The F equations represented by Eq. (12.31), together with the $M - F$ spatial constraint equations represented by

$$f_i(\psi_1, \ldots, \psi_M) = 0 \qquad i = 1, \ldots, N_s \tag{12.33}$$

constitute a set of M equations in the M variables ψ_k. We can represent the differentiated form of these equations by

$$\sum_{j=1}^{M} f_{i,j}\dot{\psi}_j = 0 \qquad i = 1, \ldots, N_s \tag{12.34}$$

$$\sum_{j=1}^{M} T_{kj}\dot{\psi}_j = \dot{q}_k \qquad k = 1, \ldots, F \tag{12.35}$$

or

$$\mathbf{A}\dot{\boldsymbol{\psi}} = \mathbf{B}\dot{\mathbf{q}} \tag{12.36}$$

where

$$A_{ij} = f_{i,j}, \qquad B_{ij} = 0 \qquad i = 1, \ldots, N_s \tag{12.37a}$$

$$A_{ij} = T_{ij}, \qquad \left.\begin{matrix} B_{ij} = 1 \text{ if } i = j + N_s \\ B_{ij} = 0 \text{ if } i \neq j + N_s \end{matrix}\right\} n \qquad i = N_s + 1, \ldots, M \tag{12.37b}$$

In general, the square matrix **A** will be nonsingular for some choice of driving variables, i.e., for some arrangement of the ones and zeros in the matrix **T.** It may be necessary to redefine the initial choice of the **T** matrix during the course of time in order for **A** to remain nonsingular at certain critical configurations. For example, in a slider-crank mechanism, we could not use the displacement of the slider as a generalized coordinate when the slider is at either of its extreme positions (i.e., in the so-called dead-center positions).

When **A** is nonsingular, we may assign independent numerical values to all the q_i, and solve the $M \times M$ system of nonlinear algebraic equations represented by Eqs. (12.31) and (12.33) for $\psi_1, \ldots, \psi_M$ by means of a numerical technique such as the Newton-Raphson method. In order for real solutions to exist, it is necessary that the q_i lie within geometrically meaningful ranges. Then one may find the lagrangian velocities $\dot{\psi}_k$ from Eq. (12.36) in the form

$$\dot{\boldsymbol{\psi}} = \mathbf{A}^{-1}\mathbf{B}\dot{\mathbf{q}} \equiv \mathbf{C}\dot{\mathbf{q}} \qquad \text{or} \qquad \dot{\psi}_j = \sum_{r=1}^{F} C_{jr}\dot{q}_r \tag{12.38}$$

where

$$\mathbf{C} \equiv \mathbf{A}^{-1}\mathbf{B} \tag{12.39}$$

To find the accelerations associated with given values of $\ddot{\mathbf{q}}_r$, it is only necessary to solve the differentiated form of Eq. (12.36), i.e.,

$$\mathbf{A}\ddot{\boldsymbol{\psi}} = \mathbf{B}\ddot{\mathbf{q}} + \mathbf{v} \tag{12.40}$$

where

$$\mathbf{v} \equiv -\dot{\mathbf{A}}\,\dot{\boldsymbol{\psi}} \tag{12.41}$$

and

$$A_{ij} \equiv \begin{cases} \sum\limits_{k=1} f_{i,jk}\dot{\psi}_k & i = 1, \ldots, N_s \quad (12.42) \\ 0 & i = N_s + 1, \ldots, M \quad (12.43) \end{cases}$$

Note that $\dot{\mathbf{B}} \equiv \mathbf{0}$, since all B_{ij} are constants by definition.

From Eq. (12.40), it follows that

$$\ddot{\boldsymbol{\psi}} = \mathbf{C}\ddot{\mathbf{q}} + \mathbf{e} \tag{12.44}$$

where

$$\mathbf{e} = \mathbf{A}^{-1}\mathbf{v} \tag{12.45}$$

From previous definitions, it may be verified that each component of the vector **e** is a quadratic form in the generalized velocities $\dot{q}_r$. Hence, we see from Eq. (12.44) that any component of lagrangian acceleration is expressible as a linear combination of the generalized accelerations $\ddot{q}_r$ plus a quadratic form in the generalized velocities $\dot{q}_r$. The quadratic velocity terms are manifestations of centripetal and Coriolis accelerations (also called compound centripetal accelerations). We will accordingly refer to them as "centripetal" terms for brevity.

12.2.4 Points of Interest

To express the velocity of point $\mathbf{x}^k$ in terms of $\dot{\mathbf{q}}$ we merely substitute Eq. (12.38) into Eq. (12.19) and find

$$\dot{\mathbf{x}}_k = \mathbf{U}^k\dot{\mathbf{q}} \quad \text{or} \quad \dot{x}_i^k = \sum_{j=1}^{M} U_{ij}^k \dot{q}_j \tag{12.46}$$

where

$$\mathbf{U}^k = \mathbf{P}^k\mathbf{C} \quad \text{or} \quad U_{ij}^k = \sum_{n=1}^{M} P_{in}^k C_{nj} \tag{12.47}$$

Similarly, the acceleration for the same point of interest is found by substituting Eq. (12.44) into Eq. (12.22) to yield

$$\ddot{\mathbf{x}}^k = \mathbf{U}^k\ddot{\mathbf{q}} + \mathbf{v}^k \tag{12.48}$$

where

$$\mathbf{v}^k \equiv \mathbf{P}^k\mathbf{e} + \mathbf{p}^k \tag{12.49}$$

From previous definitions, it is readily verified that each component of the vector $\mathbf{v}^k$ is a quadratic form in $\dot{q}_i$.

Upon multiplying each term in Eq. (12.46) by an infinitesimal time increment δt we find that any set of displacements which satisfies the spatial constraint equations represented by Eq. (12.33) must satisfy the differential relation

$$\delta x_i^k = \sum_{j=1}^{F} U_{ij}^k \delta q_j \tag{12.50}$$

This result will be useful in our later discussion of virtual displacements.

12.2.5 Angular Position, Velocity, and Acceleration

Let (ξ, η, ζ) be a set of right-handed orthogonal axes ("body axes") fixed in a rigid body, and let (x, y, z) be a right-handed orthogonal set of axes parallel to directions fixed in inertial space ("space axes"). Both sets of axes have a common origin. The angular orientation of the body relative to fixed space is defined by the three Euler angles Θ, Ψ, Φ shown in Fig. 12.2.

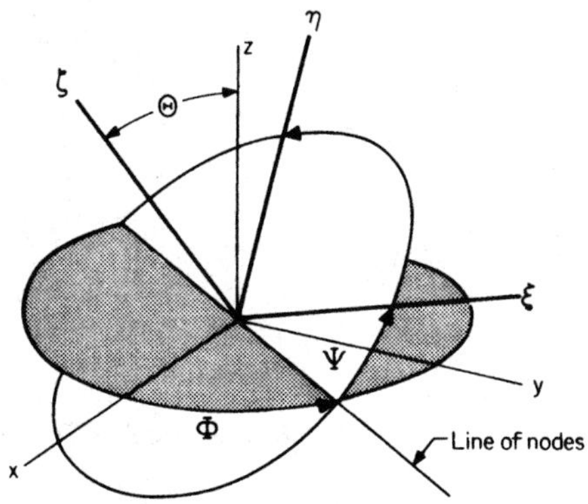

FIG. 12.2 Euler angles.

With this definition of the Euler angles,[15,29,53,61] the body is brought from its initial position (where ξ, η, ζ coincide with x, y, z) to its final position by the following sequence:

1. Rotation by Φ (precession angle) about the z axis, thereby carrying the ξ axis into the position marked "line of nodes"
2. Rotation by Θ (nutation angle) about the line of nodes, thereby carrying the ζ axis into its final position shown
3. Rotation by Ψ (spin angle) about the ζ axis, thereby carrying the ξ and η axes into their final position as shown

It may be shown[15] that the angular velocity components ω_x, ω_y, ω_z relative to the fixed axes x, y, z are given by

$$\begin{bmatrix} \omega_x \\ \omega_y \\ \omega_z \end{bmatrix} = \begin{bmatrix} \cos\Phi & \sin\Theta\sin\Phi & 0 \\ \sin\Phi & -\sin\Theta\cos\Phi & 0 \\ 0 & \cos\Theta & 1 \end{bmatrix} \begin{bmatrix} \dot{\Theta} \\ \dot{\Psi} \\ \dot{\Phi} \end{bmatrix} \tag{12.51}$$

$$\boldsymbol{\omega} = \mathbf{R}\{\dot{\Theta}, \dot{\psi}, \dot{\Phi}\} \tag{12.52}$$

where $\mathbf{R}$ is the square matrix shown in Eq. (12.51).

Now the Euler angles for body k can be defined as a subset of the lagrangian coordinates as follows:

$$\{\Theta, \Psi, \Phi\}^k = \mathbf{E}^k \boldsymbol{\psi} \tag{12.53}$$

where the components of the $3 \times M$ matrices $\mathbf{E}^k$ are all zeros or ones.

Upon noting from Eq. (12.38) that $\dot{\boldsymbol{\psi}} = \mathbf{C}\dot{\mathbf{q}}$, we can write Eq. (12.52) in the form

$$\boldsymbol{\omega}^k = \mathbf{R}^k \mathbf{E}^k \mathbf{C} \dot{\mathbf{q}} = \mathbf{W}^k \dot{\mathbf{q}} \tag{12.54}$$

where

$$\mathbf{W}^k = \mathbf{R}^k \mathbf{E}^k \mathbf{C} \tag{12.55}$$

Later we will be interested in studying infinitesimal rotations $(\delta\theta_1, \delta\theta_2, \delta\theta_3)^k$ of body k about the x, y, z axes. From Eq. (12.54) it follows that

$$\delta\boldsymbol{\theta}^k = \mathbf{W}^k \delta\mathbf{q} \qquad \text{or} \qquad \delta\theta_i^k = \sum_{r=1}^{F} W_{ir}^k \delta q_r \tag{12.56}$$

If desired, the angular accelerations are readily found by differentiation of Eq. (12.51) or (12.54).

It is worth noting that use of Euler angles can lead to singular states when Θ is zero or π, because in such cases there is no clear distinction between the angles Φ and Ψ (see Fig 12.2). For this reason, it is sometimes convenient to work with other measures of angular position, such as the four (dependent) "Euler parameters."[29,41,57,63,78]

12.2.6 Kinematics of Systems with One Degree of Freedom

The special case of $F = 1$ is sufficiently important to introduce a notational simplification which is possible because the vector of generalized coordinates $\mathbf{q}$ has only the single component $q_1 \equiv q$. Similarly, we may drop the second subscript in all the matrices which operate on q. For example, Eq. (12.36) becomes

$$\mathbf{A}\dot{\boldsymbol{\psi}} = \mathbf{b}\dot{q} \tag{12.57}$$

where

$$b_i = B_{i1} \tag{12.58}$$

In like manner we may rewrite Eqs. (12.38) and (12.44) as

$$\dot{\psi}_i = c_i \dot{q} \tag{12.59}$$

and

$$\ddot{\psi}_i = c_i \ddot{q} + e_i = c_i \ddot{q} + c_i' \dot{q}^2 \tag{12.60}$$

where

$$c_i' \equiv - \sum_{j=1}^{M} \frac{\partial c_i}{\partial \psi_j} c_j \tag{12.61}$$

Similarly, the velocities and accelerations, for point of interest k, can be written in the compact forms

$$\dot{\mathbf{x}}^k = \mathbf{u}^k \dot{q} \tag{12.62}$$

and $$\ddot{\mathbf{x}}^k = \mathbf{u}^k\ddot{q} + \mathbf{v}^k = \mathbf{u}^k\ddot{q} + \mathbf{u}'^k\dot{q}^2 \tag{12.63}$$

where $$\mathbf{v}_k \equiv \dot{q}\sum_{j=1}^{M}\frac{\partial \mathbf{u}^k}{\partial \psi_j}\dot{\psi}_j = \dot{q}^2\sum_{j=1}^{M}\frac{\partial \mathbf{u}^k}{\partial \psi_j}c_j$$

and $$\mathbf{u}'^k \equiv \sum_{j=1}^{M}\frac{\partial \mathbf{u}^k}{\partial \psi_j}c_j$$

Finally, we note that the components of the angular velocity of any rigid-body member k of the system are given by Eq. (12.54) in the simplified form

$$\omega_i^k = w_i^k\dot{q} \tag{12.64}$$

and the corresponding angular accelerations are given by

$$\dot{\omega}_i^k = w_i^k\ddot{q} + \dot{w}_i^k\dot{q} \tag{12.65}$$

where $$w_i^k \equiv W_{i1}^k \tag{12.66}$$

12.2.7 Computer Programs for Kinematics Analysis

The foregoing analysis forms the basis of a digital computer program called KINMAC (kinematics of machinery) developed by the author and his colleagues.[1,36,69] This FORTRAN program calculates displacements, velocities, and accelerations for up to 30 points of interest in planar mechanisms with up to 10 DOF, up to 10 loops, and up to 10 user-supplied auxiliary constraint relations (in addition to the automatically formulated loop-closure constraint relations). The program also contains built-in subroutines for use in the modeling of cams and temporal constraint relations. A less sophisticated student-oriented program for kinematics analysis, called KINAL (kinematics analysis), is described and listed in full in Ref. 66.

In Ref. 89, Suh and Radcliffe give listings for a package of FORTRAN subroutines called LINCPAC which can be used to solve a variety of fundamental problems in planar kinematics. They also give listings for a package called SPAPAC to be similarly used for problems of spatial kinematics.

Other programs have been developed primarily for dynamics analysis, but they can function in a so-called kinematic mode; among these are IMP and DRAM which will be discussed further in connection with programs for dynamics analysis. Computer programs which have been developed primarily for the problem of kinematic synthesis, but have some application to analysis and simulation, include R. Kaufman's KINSYN[80] and A. Erdman's LINCAGES[22]; these programs are described in the survey article by Kaufman.[42]

12.3 STATICS OF MACHINE SYSTEMS

12.3.1 Direct Use of Newton's Laws (Vectorial Mechanics)

The most direct way of formulating the equations of statics for a machine is to write equations of equilibrium for each body of the system. For example, Fig. 12.3 shows

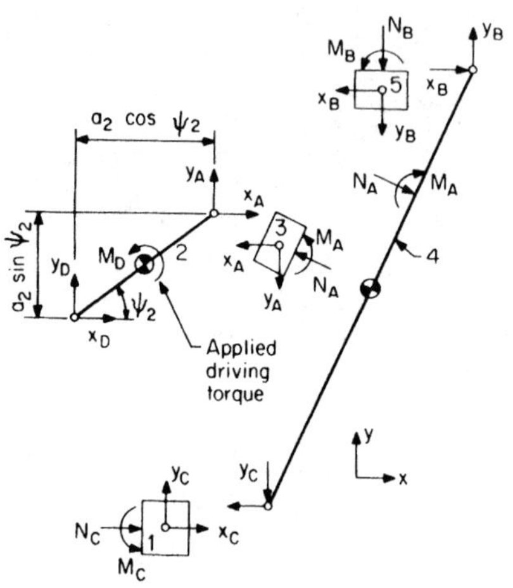

FIG. 12.3 Free-body diagrams for the mechanism of Fig. 12.1.

the set of five free-body diagrams associated with the mechanism of Fig. 12.1. The internal reaction forces at pin joints (e.g., X_A, Y_A) and the normal forces and moments at smooth slider contacts (e.g., N_A, M_A) appear as external loadings on the various bodies of the system. A driving torque M_D, acting on link 2, is assumed to be a known function of position. Thus, there exists a total of 14 unknown internal force components, namely, X_A, Y_A, N_A, M_A, X_B, Y_B, N_B M_B, X_C, Y_C, N_C, M_C, X_D, and Y_D. In addition, there exist five unknown position variables ψ_1, ..., ψ_5. In order to find the 19 unknown quantities, we can write three equations of equilibrium for each of the five "free bodies." A typical set of such equations, for link 2, is

$$X_D + X_A = 0 \tag{12.67a}$$

$$Y_D + Y_A = 0 \tag{12.67b}$$

$$M_D + (X_D - X_A)(a_2/2) \sin \psi_2 + (Y_A - Y_D)(a_2/2) \cos \psi_2 = 0 \tag{12.67c}$$

In all, we can write a total of 15 such equations of equilibrium for the five links. We can eliminate the 14 internal reactions from these equations to yield one master equation involving the five position variables ψ_1, ..., ψ_5. In addition, we have the four equations of constraint in the form of Eqs. (12.6). Thus all the equations mentioned above may be boiled down to five equations for the five unknown lagrangian coordinates ψ_i. These equations are nonlinear in the ψ_i, but a numerical solution is always possible (see Sec. 9.22 of Ref. 66). Having found the ψ_i, one may solve for the 14 unknown reactions from the equilibrium equations which are linear in the reactions.

If one wishes to find only the configuration of a mechanism under the influence of a set of static external forces, the above procedure is unnecessarily cumbersome compared with the method of virtual work (discussed below).

However, if the configuration of the system is known (all ψ_i given), then the 15 equations of equilibrium are convenient for finding the 14 internal reaction components plus the required applied torque M_D. The details of such a solution are given in Sec. 11.52 of Ref. 66 for arbitrary loads on all the links of the example mechanism.

12.3.2 Terminology of Analytical Mechanics

We now define some key terms and concepts from the subject of analytical mechanics. For a more detailed discussion of the subject see any of the standard works, e.g., Refs. 29, 30, 46, 61, and 102 or Refs. 53, 64, and 66.

1. *Virtual displacements:* Any set of infinitesimal displacements which satisfy all *instantaneous* constraints

This means that if there exist F degrees of freedom, the displacements of a typical point P^k must satisfy the differential relationship

$$\delta x_i^k = \sum_{j=1}^{F} U_{ij}^k \delta q_j \tag{12.68}$$

given by Eq. (12.50).

If, in addition, there exist N_t temporal constraints of the form

$$g_k(\psi_1, \ldots, \psi_M, t) = 0 \qquad k = 1, 2, \ldots, N_t \tag{12.69}$$

they can be expressed in the "rate" form

$$\sum_{j=1}^{M} g_{k,j}\dot{\psi}_j + \frac{\partial g_k}{\partial t} = 0 \tag{12.70}$$

However, from Eq. (12.38),

$$\dot{\psi}_j = \sum_{r=1}^{F} C_{jr}\dot{q}_r$$

Therefore,

$$\sum_{r=1}^{F} \alpha_{kr}\dot{q}_r + \beta_k = 0 \qquad k = 1, \ldots, N_t \tag{12.71}$$

where

$$\alpha_{kr} = \sum_{j=1}^{M} g_{k,j}C_{jr} \tag{12.72a}$$

$$\beta_k = \partial g_k/\partial t \tag{12.72b}$$

Since $\dot{q}_r = dq_r/dt$, Eq. (12.71) can be expressed in the form

$$\sum_{r=1}^{F} \alpha_{kr}dq_r + \beta_k dT = 0 \tag{12.73}$$

The virtual displacements δq_r, however, must, *by definition,* satisfy

$$\sum_{r=1}^{F} \alpha_{kr}\delta q_r = 0 \qquad k = 1, \ldots, N_t \tag{12.74}$$

We say that Eq. (12.73) represents the *actual* constraints and Eq. (12.74) represents the *instantaneous* constraints. Any infinitesimals dq_r which satisfy Eq. (12.73) are called "possible displacements," in distinction to the "virtual displacements" which satisfy Eq. (12.74).

In problems of statics, such temporal constraints are not normally present, but they could easily arise in problems of dynamics. Equations of constraint of the form in Eq. (12.77) sometimes arise which cannot be integrated to give a function of the form $g_k(\psi, t) = 0$. Such nonintegrable equations of constraint are said to be "nonholonomic."

2. *Virtual work* (δW): Work done by specified forces on virtual displacements
3. *Active forces:* Those forces which produce nonzero net virtual work
4. *Ideal mechanical systems:* Systems in which constraints are maintained by forces which do no virtual work (e.g., frictionless contacts, rigid links)

5. *Generalized forces:* Terms which multiply the virtual generalized displacements (δq_r) in the expression for virtual work of the form $\delta W = \Sigma_{r=1}^{E} Q_r\, \delta q_r$
6. *Equilibrium:* State in which the resultant force on each particle of the system vanishes
7. *Generalized equilibrium:* State in which all the generalized forces vanish

12.3.3 The Principle of Virtual Work

The principle of virtual work may be viewed as a basic *postulate* from which all the well-known laws of vector statics may be deduced, or it may be proved directly from the laws of vector statics. The following two-part statement of principle is proved the latter way in Sec. 11.30 of Ref. 66:

1. If an ideal mechanical system is in equilibrium, the net virtual work of all the active forces vanishes for every set of virtual displacements.
2. If the net virtual work of all the active forces vanishes, for every set of virtual displacements, an ideal mechanical system is in a state of generalized equilibrium.

It is pointed out in Ref. 66 that no distinction need be made between generalized equilibrium and equilibrium for systems which are at rest (relative to an inertial frame) before the active forces are applied.

For a system which consists of n rigid bodies, we may express the resultant external force acting on a typical body by $\mathbf{X}^p$ acting through a reference point fixed in the body which momentarily occupies the point $\mathbf{x}^p$ measured in the global coordinate frame. Similarly the active torque $\mathbf{M}^p$ is assumed to act on the same body. The virtual displacements of the reference points are denoted by $\delta\mathbf{x}^p$, and the corresponding rotations of each rigid body may be expressed by the infinitesimal rotation vector

$$\delta\boldsymbol{\theta}^p = (\delta\theta_1^p, \delta\theta_2^p, \delta\theta_3^p) \tag{12.75}$$

where $\delta\theta_i^p$ is the rotation component about an axis, through the point $\mathbf{x}^p$, parallel to the global x_i axis. The virtual work of all the active forces is therefore given by

$$\delta W = \sum_{p=1}^{n} \sum_{i=1}^{3} (X_i^p\, \delta x_i^p + M_i^p\, \delta\theta_i^p) \tag{12.76}$$

From Eqs. (12.50), and (12.56) we know that

$$\delta x_i^p = \sum_{j=1}^{F} U_{ij}^p\, \delta q_j \tag{12.77}$$

$$\delta\theta_i^p = \sum_{j=1}^{F} W_{ij}^p\, \delta q_j \tag{12.78}$$

Upon substitution of Eqs. (12.77) and (12.78) into Eq. (12.76) we find that

$$\delta W = \sum_{j=1}^{F} Q_j\, \delta q_j \tag{12.79}$$

where

$$Q_j = \sum_{p=1}^{n} \sum_{i=1}^{3} (U_{ij}^p X_i^p + W_{ij}^p M_i^p) \qquad j = 1, \ldots, F \tag{12.80}$$

Equation (12.80) is an explicit formula for the calculation of generalized forces, although they can usually be calculated in more direct, less formal, ways as described in Sec. 11.22 of Ref. 66.

From part 1 of the statement of the principle of virtual work, it follows that if an ideal system is in equilibrium, then for all possible choices of δq_i

$$\delta W = \sum_{j=1}^{F} Q_j \, \delta q_j = 0 \tag{12.81}$$

In the absence of temporal constraints, the δq_j are all-independent, and it follows from Eq. (12.81) that for such cases

$$Q_j = 0 \qquad j = 1, 2, \ldots, F \tag{12.82}$$

Thus, the vanishing of all generalized forces is a necessary condition for equilibrium and a sufficient condition (by definition) for generalized equilibrium.

For examples of the method of virtual work applied to the statics of machine systems, see Sec. 11.30 of Ref. 66.

For those cases where the generalized coordinates q_r are not independent, but are related by constraint relations expressed in the form of Eq. (12.71) (arising either from temporal or nonholonomic constraints), we cannot reduce Eq. (12.81) to the simple form of Eq. (12.82). For such cases we can use the method of Lagrange multipliers.

12.3.4 Lagrange Multipliers

When the F generalized coordinates are not independent, but are related by equations of constraint such as Eq. (12.74), i.e.,

$$\sum_{j=1}^{F} \alpha_{ij} \, \delta q_j \qquad i = 1, \ldots, N_t \tag{12.83}$$

we say that the generalized coordinates form a "redundant," or "superfluous," set.

However, we may choose to think of all the q_i as independent coordinates if we conceptually remove the physical constraints (e.g., hinge pins or slider guides) and consider the reactions produced by the constraints as external forces. In other terms, we must supplement the generalized forces Q_i of the active loads by a set of forces Q'_i, associated with the reactions produced by the (now deleted) constraints. Since the reactions in an ideal system, by definition, produce no work on any displacement compatible with the constraints, it is necessary that the supplemental forces Q'_i satisfy the condition

$$\sum_{j=1}^{F} Q'_j \, \delta q_j = 0 \tag{12.84}$$

where the displacements δq_j satisfy the constraint equation represented by Eq. (12.83).

It may be shown (Ref. 100, p. 562) that in order for Eqs. (12.83) and (12.84) to be satisfied, it is necessary that

$$Q'_j = \sum_{i=1}^{N_t} \lambda_i \alpha_{ij} \tag{12.85}$$

where λ_i are unknown functions of time called "lagrangian multipliers." Accordingly, Eq. (12.85) provides the desired supplements to the generalized forces, and the equations of equilibrium assume the more general form

$$Q_j + Q_j' = Q_j + \sum_{i=1}^{N_t} \lambda_i \alpha_{ij} = 0 \qquad j = 1, \ldots, F \tag{12.86}$$

If we consider that Q_j and α_{ij} are known functions of the q_r, then Eqs. (12.86) and (12.69) provide for (holonomic constraints) $F + N_t$ equations for determining the unknowns $q_1, \ldots, q_F$ and $\lambda_1, \ldots, \lambda_N t$. However, in practice, it is likely that Q_j and α_j are expressed as functions of $\psi_1, \ldots, \psi_M$. Then it is necessary to utilize the M additional equations consisting of Eqs. (12.24) and (12.31). For nonholonomic constraints, equations of the form of Eq. (12.69) are not available, and one must utilize instead differential equations of the form of Eq. (12.71). Such problems must usually be solved by a process of numerical integration. Numerical integration will be discussed in Sec. 12.5.

12.3.5 Conservative Systems

If the work done on an ideal mechanical system by a set of forces is a function $W(q_1,\ldots, q_F)$ of the generalized coordinates only, the system is said to be conservative. Such forces could arise from elastic springs, gravity, and other field effects.

The increment in W due to a virtual displacement δq_i is

$$\delta \overline{W} (\partial \overline{W}/\partial q_1)\partial q_1 + \cdots + (\partial \overline{W}/\partial q_F)\partial q_F \tag{12.87}$$

By tradition, one defines the "potential energy" or "potential" of the system as

$$V(q) \equiv - \overline{W}(q) \tag{12.88}$$

Therefore

$$\delta \overline{W} = - (\partial V/\partial q_1)\partial q_1 - \cdots - (\partial V/\partial q_F)\delta q_F$$

If, in addition, a set of active forces (not related to $\overline{W}$) acts on the system and produces the generalized forces Q_r*, the net work done during a virtual displacement is

$$\delta W = (Q_1^* - \partial V/\partial q_1)\partial q_1 - \cdots - (Q_F^* - \partial V/\partial q_F)\partial q_F = 0 \tag{12.89}$$

where the right-hand zero is a consequence of the principle of virtual work. From the definition of Q_r (Sec. 12.3.2) it follows that

$$Q_r = Q_r^* - \partial V/\partial q_r \tag{12.90}$$

If all the δq_r are independent (i.e., no temporal or nonholonomic constraints exist) then Eq. (12.89) predicts that

$$Q_r \equiv Q_r^* - \partial V/\partial q_r = 0 \qquad r = 1, \ldots, F \tag{12.91}$$

If the δq_r are related, as in Eq. (12.83),

$$\sum_{j=1}^{F} \alpha_{ij}\, \delta q_j = 0 \qquad i = 1, 2, \ldots, N_t \tag{12.92}$$

it follows from the discussion leading to Eq. (12.86) that

$$Q_j^* - \frac{\partial V}{\partial q_j} + \sum_{i=1}^{N_t} \lambda_i \alpha_{ij} = 0 \qquad j = 1, \ldots, F \tag{12.93}$$

It is only necessary to define V within an additive constant, since only the derivative of V enters into the equilibrium equations.

Some examples of the form of V for some common cases follow:

1. A linear spring with free length s_f, current length s, and stiffness K has potential

$$V(s) = \frac{1}{2} K(s - s_e)^2 \tag{12.94}$$

2. A system of mass particles close to the surface of the earth (where the acceleration of gravity g is essentially constant) has potential

$$V(\bar{z}) = W\bar{z} \tag{12.95}$$

where $\bar{z}$ is the elevation of center of gravity of the system vertically above an arbitrary datum plane, and W is the total weight of all particles.

3. A mass particle is located at a distance r from the center of the earth. If its weight is W_e when the particle is at a distance r_e from the earth's center (e.g., at the surface of the earth), its potential in general is

$$V(r) = -W_e r_e^2 / r \tag{12.96}$$

In all these examples, V has been expressed in terms of a lagrangian variable. More generally, V is of the form

$$V = V(\psi_1, \ldots, \psi_M) \tag{12.97}$$

The required derivatives of V are of the form

$$\frac{\partial V(q)}{\partial q_r} = \sum_{j=1}^{M} \frac{\partial V(\psi)}{\partial \psi_j} \frac{\partial \psi_j}{\partial q_r} = \sum_{j=1}^{M} \frac{\partial V(\psi)}{\partial \psi_r} C_{jr} \tag{12.98}$$

and C_{jr} are components of the matrix **C** defined by Eq. (12.39). Illustrations of the use of potential functions in problems of statics of machines are given in Sec. 11.42 of Ref. 66.

12.3.6 Reactions at Joints

By Vector Statics. Perhaps the most straightforward way to compute reactions is to establish the three equations of equilibrium for each link, as exemplified by Eqs. (12.67). The equations are linear in the desired reactions and are readily solved by the same linear equation subroutine that is needed for other purposes in a numerical analysis. This method has been described in depth for planar mechanisms in Secs. 11.50 and 11.51 of Ref. 66, and has been used in the author's computer programs STATMAC[70] and DYMAC.[68] It was also used in Ref. 72 and in the computer programs MEDUSA[18] and VECNET.[2]

By Principle of Virtual Work. In order for the internal reactions at a specific joint to appear in the expression for virtual work, it is necessary to imagine that the joint is "broken" and then to introduce, as active forces, the unknown joint reactions needed to maintain closure. This technique is briefly described in Ref. 64, and in more detail by Denavit et al.[16] and by Uicker,[97] who combines the method of virtual work with Lagrange's equations of second type (see Sec. 12.4.2).

By Lagrange Multipliers. As shown by Chace et al.[10] and Smith,[85] it is possible to find certain crucial joint reactions, simultaneously with the solution of the general dynamics problem, by the method of Lagrange multipliers. A brief description of this method is also given in Ref. 64.

12.3.7 Effects of Friction in Joints

Up to this point, it has been assumed that all internal reactions are workless. If the friction at the joints is velocity-dependent, we can place a damper (not necessarily linear) at the joints; the associated pairs of velocity-dependent force or torque reactions are then included in the active forces $\mathbf{X}^p$ and $\mathbf{M}^p$ as defined in Sec. 12.3.3. These forces, which appear linearly in the generalized forces Q [see Eq. (12.80)], result in velocity-dependent generalized forces.

However, when Coulomb friction exists at the joints, the associated active forces become functions of the unknown joint reactions. These forces complicate the solution of the equations, but they can be accounted for in a number of approximate ways. A detailed treatment of frictional effects in statics of planar mechanisms is given in Secs. 11.60 to 11.63 of Ref. 66. Frictional effects in the dynamics of planar machines are discussed in Sec. 12.70 of Ref. 66. Greenwood[30] (p. 271) utilizes Lagrange multipliers to solve a problem of frictional effects between three bodies moving in a plane with translational motion only.

12.4 KINETICS OF MACHINE SYSTEMS

12.4.1 Lagrange's Form of d'Alembert's Principle

We now assume that active forces $\mathbf{X}^i \equiv (X_1, X_2, X_3)^i$ are applied to the mass center $\mathbf{x}^i \equiv (x_1, x_2, x_3)^i$ of link i and an active torque $\mathbf{M}^i \equiv (M_1, M_2, M_3)^i$ is exerted about the mass center. Any set of active forces which does not pass through the mass center may be replaced by a statically equipollent set of forces which does. In general, $\mathbf{X}^i$ and $\mathbf{M}^i$ may depend explicitly upon position, velocity, and time. Massless springs may be modeled by pairs of equal, but opposite, collinear forces which act on the links at the terminal points of the springs. The spring forces may be any specified linear or nonlinear function of the distance between the spring's terminals. Dampers may be modeled by similar pairs of equal and opposite collinear forces which are dependent on the rate of relative extension of the line joining the damper terminals. Examples of such springs and dampers are given in Sec. 14.43 of Ref. 66.

A typical link is subjected to inertia forces $-m\ddot{\mathbf{x}}$ as well as the real forces $\mathbf{X}$. Thus the virtual work due to *both* these "forces" is given by

$$\delta W_F = \delta \mathbf{x}^T(\mathbf{X} - m\ddot{\mathbf{x}}) \tag{12.99}$$

From Eqs. (12.48) and (12.50) we observe that

$$\ddot{\mathbf{x}} = \mathbf{U}\ddot{\mathbf{q}} + \mathbf{v} \tag{12.100a}$$

$$\delta\mathbf{x} = \mathbf{U}\,\delta\mathbf{q} \qquad \therefore\ \delta\mathbf{x}^T = \delta\mathbf{q}^T\mathbf{U}^T \tag{12.100b}$$

where superscript T denotes a matrix transpose. Therefore the virtual work due to "forces" on the rigid body is

$$\delta W_F = \delta\mathbf{q}^T[\mathbf{U}^T\mathbf{X} - m\mathbf{U}^T\mathbf{v} - m\mathbf{U}^T\mathbf{U}\ddot{q}] \tag{12.101}$$

To this must be added the virtual work due to the real and inertial torques which act about the mass center. To help calculate these, we introduce the following notation:

G = The mass center of a rigid body
(ξ_1, ξ_2, ξ_3) = Local orthogonal reference axes through G aligned with the principal axes of inertia of the body
$\bar{J}_1, \bar{J}_2, \bar{J}_3$ = Principal moments of inertia of the body, referred to axes ξ_1, ξ_2, ξ_3
$\bar{M}_1, \bar{M}_2, \bar{M}_3$ = Components of external torque in directions ξ_1, ξ_2, ξ_3
$\bar{\omega}_1, \bar{\omega}_2, \bar{\omega}_3$ = Components of angular velocity in directions ξ_1, ξ_2, ξ_3
$\delta\bar{\theta}_1, \delta\bar{\theta}_2, \delta\bar{\theta}_3$ = Infinitesimal rotation components about axes ξ_1, ξ_2, ξ_3
D_{ij} = Direction cosine between (local) axis ξ_i and (global) axis x_j

Note that an overbar is used on vector components referred to the local (ξ) axes. Corresponding symbols without overbars refer to global (x) axes. The two systems are related by matrix relationships of the type

$$\bar{\mathbf{M}} = \mathbf{DM} \tag{12.102}$$

$$\bar{\boldsymbol{\omega}} = \mathbf{D}\boldsymbol{\omega} \tag{12.103}$$

$$\delta\bar{\boldsymbol{\theta}} = \mathbf{D}\,\delta\boldsymbol{\theta} \tag{12.104}$$

It may be shown[29,53] that the direction cosines are related to the Euler angles as follows:

$$\mathbf{D} \equiv [D_{ij}] = \begin{bmatrix} C\Psi C\Phi - C\Theta S\Phi S\Psi & C\Psi S\Phi + C\Theta C\Phi S\Psi & S\Psi S\Theta \\ -S\Psi C\Psi - C\Theta S\Phi C\Psi & -S\Psi S\Phi + C\Theta C\Phi C\Psi & C\Psi S\Theta \\ S\Theta S\Phi & -S\Theta C\Phi & C\Theta \end{bmatrix} \tag{12.105}$$

where

$$C\Psi = \cos\Phi \qquad C\Phi = \cos\Phi \qquad C\Theta = \cos\Theta$$
$$S\Psi = \sin\Psi \qquad S\Phi = \sin\Phi \qquad S\Theta = \sin\Theta \tag{12.106}$$

and **D** is an orthogonal matrix, i.e.,

$$\mathbf{D}^T = \mathbf{D}^{-1} \tag{12.107}$$

We now take note of Euler's equation of motion for rigid bodies expressed in the form[53,94]

$$\overline{M}_i + \overline{T}_i = 0 \qquad i = 1, 2, 3 \tag{12.108}$$

where the *inertia torque* components $\overline{T}_i$ are given by

$$\overline{T}_1 = -\overline{J}_1\dot{\overline{\omega}}_1 + (\overline{J}_2 - \overline{J}_3)\overline{\omega}_2\overline{\omega}_3 \tag{12.109a}$$

$$\overline{T}_2 = -\overline{J}_2\dot{\overline{\omega}}_2 + (\overline{J}_3 - \overline{J}_1)\overline{\omega}_3\overline{\omega}_1 \tag{12.109b}$$

$$\overline{T}_3 = -\overline{J}_3\dot{\overline{\omega}}_3 + (\overline{J}_1 - \overline{J}_2)\overline{\omega}_1\overline{\omega}_2 \tag{12.109c}$$

In matrix form, we can write

$$\overline{\mathbf{T}} = -\overline{\mathbf{J}}\dot{\overline{\boldsymbol{\omega}}} + \boldsymbol{\tau} \tag{12.110}$$

where

$$\overline{\mathbf{J}} = \begin{bmatrix} \overline{J}_1 & 0 & 0 \\ 0 & \overline{J}_2 & 0 \\ 0 & 0 & \overline{J}_3 \end{bmatrix} \tag{12.111}$$

$$\boldsymbol{\tau} = \begin{Bmatrix} (\overline{J}_2 - \overline{J}_3)\overline{\omega}_2\overline{\omega}_3 \\ (\overline{J}_3 - \overline{J}_1)\overline{\omega}_3\overline{\omega}_1 \\ (\overline{J}_1 - \overline{J}_2)\overline{\omega}_1\overline{\omega}_2 \end{Bmatrix} \tag{12.112}$$

From Eq. (12.54) we recall that

$$\boldsymbol{\omega} = \mathbf{W}\dot{\mathbf{q}} \tag{12.113}$$

Hence, the local velocity components are given by Eq. (12.103) in the form

$$\overline{\boldsymbol{\omega}} = \mathbf{D}\boldsymbol{\omega} = \mathbf{D}\mathbf{W}\dot{\mathbf{q}} \tag{12.114}$$

The corresponding acceleration components are therefore

$$\dot{\overline{\boldsymbol{\omega}}} = \overline{\mathbf{W}}\ddot{\mathbf{q}} + \overline{\mathbf{V}} \tag{12.115}$$

where

$$\overline{\mathbf{W}} = \mathbf{D}\mathbf{W} \tag{12.116}$$

$$\overline{\mathbf{V}} = (\mathbf{D}\dot{\mathbf{W}} + \dot{\mathbf{D}}\mathbf{W})\dot{\mathbf{q}} \tag{12.117}$$

Thus the inertia torque given by Eq. (12.110) can be expressed as

$$\overline{\mathbf{T}} = -\overline{\mathbf{J}}\overline{\mathbf{W}}\ddot{\mathbf{q}} - \overline{\mathbf{J}}\overline{\mathbf{V}} + \boldsymbol{\tau} \tag{12.118}$$

In accordance with d'Alembert's principle, we may therefore express the virtual work of the active and inertia torques in the form

$$\delta W_M = \sum_{i=1}^{3} \delta\overline{\theta}_i(\overline{M}_i + \overline{T}_i) = \delta\overline{\boldsymbol{\theta}}^T(\overline{\mathbf{M}} + \overline{\mathbf{T}}) \tag{12.119}$$

Upon utilization of Eqs. (12.104) and (12.118) we find

$$\delta W_M = (\mathbf{D}\delta\boldsymbol{\theta})^T(\overline{\mathbf{M}} + \boldsymbol{\tau} - \overline{\mathbf{J}}\,\overline{\mathbf{V}} - \overline{\mathbf{J}}\,\overline{\mathbf{W}}\ddot{\mathbf{q}}) \tag{12.120}$$

Since $\delta\boldsymbol{\theta} = \mathbf{W}\,\delta\mathbf{q}$ and $\overline{\mathbf{W}} = \mathbf{DW}$, we can write

$$(\mathbf{D}\delta\boldsymbol{\theta})^T = (\mathbf{DW}\delta\mathbf{q})^T = (\overline{\mathbf{W}}\delta\mathbf{q})^T = \delta\mathbf{q}^T\overline{\mathbf{W}}^T \tag{12.121}$$

Now we can write Eq. (12.120) in the form

$$\delta W_M = \delta\mathbf{q}^T[\overline{\mathbf{W}}^T(\overline{\mathbf{M}} + \boldsymbol{\tau} - \overline{\mathbf{J}}\,\overline{\mathbf{V}}) - \overline{\mathbf{W}}^T\overline{\mathbf{J}}\,\overline{\mathbf{W}}\ddot{\mathbf{q}}] \tag{12.122}$$

Adding δW_M to δW_F given by Eq. (12.101), the virtual work of all the real and inertial forces and torques is seen to take the form

$$\delta W_M + \delta W_F = \delta\mathbf{q}^T[(\mathbf{U}^T\mathbf{X} + \overline{\mathbf{W}}^T\overline{\mathbf{M}}) - m\mathbf{U}^T\mathbf{v} + \overline{\mathbf{W}}^T(\boldsymbol{\tau} - \overline{\mathbf{J}}\,\overline{\mathbf{V}}) - (m\mathbf{U}^T\mathbf{U} + \overline{\mathbf{W}}^T\overline{\mathbf{J}}\,\overline{\mathbf{W}})\ddot{\mathbf{q}}] \tag{12.123}$$

If we use a superscript k to denote the virtual work done on the kth rigid body in the system, we can express the virtual work of all the real and inertial forces of the complete system by

$$\delta W = \sum_{k=1}^{B} (\delta W_M + \delta W_F)^k = \delta\mathbf{q}^T(\mathbf{Q} + \mathbf{Q}^\dagger - \mathbf{I}\ddot{\mathbf{q}}) = 0 \tag{12.124}$$

where we have utilized the following definitions:

$$\mathbf{I} \equiv \sum_{k=1}^{B} (m\mathbf{U}^T\mathbf{U} + \overline{\mathbf{W}}^T\overline{\mathbf{J}}\,\overline{\mathbf{W}})^k \tag{12.125}$$

$$\mathbf{Q}^\dagger = \sum_{k=1}^{B} [-m\mathbf{U}^T\mathbf{v} + \overline{\mathbf{W}}^T(\boldsymbol{\tau} - \overline{\mathbf{J}}\,\overline{\mathbf{V}})]^k \tag{12.126}$$

$$\mathbf{Q} = \sum_{k=1}^{B} (\mathbf{U}^T\mathbf{X} + \overline{\mathbf{W}}^T\overline{\mathbf{M}})^k = \sum_{k=1}^{B} (\mathbf{U}^T\mathbf{X} + \mathbf{W}^T\mathbf{M})^k \tag{12.127}$$

B is the number of rigid bodies present. It is worth noting that the definition of $\mathbf{I}$ in Eq. (12.125) implies that $\mathbf{I}$ is symmetric, and that the rightmost expression in Eq. (12.127) follows from Eqs. (12.102), (12.107), and (12.116).

We have set $\delta W = 0$ in Eq. (12.124) in accordance with the generalized version of the principle of virtual work which includes the inertial forces. This form of the principle is referred to as Lagrange's form of d'Alembert's principle.[30,64]

When there are no constraints relating the generalized coordinates q_i, the rightmost factor in Eq. (12.124) must vanish, and we have the equations of motion in their final form:

$$\mathbf{I}\ddot{\mathbf{q}} = \mathbf{Q} + \mathbf{Q}^\dagger \tag{12.128a}$$

or

$$\sum_{j=1}^{F} I_{ij}\ddot{q}_j = Q_i + Q_i^\dagger \qquad i = 1, \ldots, F \tag{12.128b}$$

Now consider the case where the generalized velocities are related by N_t constraint equations of the form

$$\sum_{j=1}^{F} \alpha_{ij}\dot{q}_j + \beta_i = 0 \qquad i = 1, 2, \ldots, N_t \tag{12.129}$$

where α_{ij} and β_i may all be functions of the lagrangian coordinates ψ_r and of the time t. These relationships may be nonintegrable (nonholonomic) or derivable from holonomic constraints of the form

$$g_i(\psi_1, \ldots, \psi_M, t) = 0 \tag{12.130}$$

In either case, the corresponding virtual displacements must satisfy (by definition) the equation

$$\sum_{j=1}^{F} \alpha_{ij}\delta q_j = 0 \tag{12.131}$$

Using the same argument which led to Eq. (12.85), we conclude that the generalized forces must be supplemented by the "constraint" forces

$$Q_j' = \sum_{i=1}^{N_t} \lambda_i \alpha_{ij} \qquad j = 1, 2, \ldots, F \tag{12.132a}$$

or

$$\mathbf{Q}' = \boldsymbol{\alpha}^T\boldsymbol{\lambda} \tag{12.132b}$$

where the Lagrange multipliers $\lambda_1, \ldots, \lambda_{Nt}$ are unknown functions of time to be determined. Therefore the equations of motion, Eqs. (12.128), may be written in the more general form

$$\mathbf{I}\ddot{\mathbf{q}} = \mathbf{Q} + \mathbf{Q}^\dagger + \boldsymbol{\alpha}^T\boldsymbol{\lambda} \tag{12.133a}$$

or

$$\sum_{j=1}^{F} I_{ij}\ddot{q}_j = Q_i + Q_i^\dagger + \sum_{j=1}^{N_t} \alpha_{ji}\lambda_j \qquad i = 1, \ldots, F \tag{12.133b}$$

These F equations together with the N_t constraint equations in the form of Eq. (12.129) suffice to find the unknowns $q_1, \ldots, q_F$ and $\lambda_1, \ldots, \lambda_{Nt}$.

From previous definitions, it may be verified that the terms $Q_i^\dagger$ are all quadratic forms in the generalized velocities $\dot{q}_r$, and therefore reflect the effects of the so-called centrifugal and Coriolis (inertia) forces. However, no use is made of this fact in the numerical methods discussed below.

Equations (12.133) form the basis of the author's general-purpose computer program DYMAC (*dy*namics of *mac*hinery).[1,66,68]

12.4.2 Equations of Motion from Lagrange's Equations

The kinetic energy T of a system of B rigid bodies is given by

$$T = \frac{1}{2}\sum_{k=1}^{B} (m^k\dot{\mathbf{x}}^T\dot{\mathbf{x}} + \overline{\boldsymbol{\omega}}^T\overline{J}\overline{\boldsymbol{\omega}}) \tag{12.134}$$

where all terms have been defined above. Upon recalling that $\dot{\mathbf{x}} \equiv \mathbf{U}\dot{\mathbf{q}}$ and $\overline{\boldsymbol{\omega}} = \overline{W}\dot{\mathbf{q}}$, it follows that

$$T = \tfrac{1}{2}\,\dot{\mathbf{q}}^T\mathbf{I}^k\dot{\mathbf{q}} = \frac{1}{2}\sum_{i,j=1}^{F} I_{ij}\dot{q}_i\dot{q}_j \tag{12.135}$$

where the generalized inertia coefficients I_{ij} are precisely those defined by Eq. (12.125). Since the I_{ij} are functions of position only, they may be considered implicit functions of the generalized coordinates q, i.e.,

$$I_{ij}[\psi(q)] \tag{12.136}$$

Until further notice we will assume that I_{ij} is known explicitly in terms of the generalized coordinates [i.e., $I_{ij} = I_{ij}(q)$] and that the q_r are all independent.

Now we may make use of the well-known Lagrange equations (of second type) for independent generalized coordinates*:

$$\frac{d}{dt}\frac{\partial T}{\partial \dot{q}^r} - \frac{\partial T}{\partial q^r} = Q_r = Q_r^* - \frac{\partial V}{\partial q_r} \tag{12.137}$$

where $\mathbf{Q}$ is the generalized force corresponding to $\mathbf{q}$, and $\mathbf{Q}^*$ and $\mathbf{V}$ are as defined in Sec. 12.3.5.

Making use of Eq. (12.135) and the fact that $I_{ij} = I_{ji}$, we find that

$$\frac{\partial T}{\partial \dot{q}_r} = \sum_{j=1}^{F} I_{rj}\dot{q}_j \tag{12.138}$$

and

$$\frac{\partial T}{\partial q_r} = \sum_{i,j=1}^{F} \tfrac{1}{2} I_r^{ij}\dot{q}_i\dot{q}_j \tag{12.139}$$

where

$$I_r^{ij} \equiv \frac{\partial I_{ij}(q)}{\partial q_r} \tag{12.140}$$

Equation (12.138) leads to the further result

$$\frac{d}{dt}\left(\frac{\partial T}{\partial \dot{q}_r}\right) = \sum_{j=1}^{F} I_{rj}\ddot{q}_j + \sum_{j=1}^{F} \dot{I}_{rj}\dot{q}_j \tag{12.141}$$

We may represent $\dot{I}_{rj}$ in the form

$$\dot{I}_{rj} = \sum_{i=1}^{F} \frac{\partial I_{rj}(q)}{\partial q_i}\dot{q}_i = \sum_{i=1}^{F} I_i^{rj}\dot{q}_i \tag{12.142}$$

Upon substitution of Eqs. (12.138) to (12.142) into Lagrange's equation (12.137), we find

$$\sum_{j=1}^{F} I_{rj}\ddot{q}_j + \sum_{i=1}^{F}\sum_{j=1}^{F} G_r^{ij}\dot{q}_i\dot{q}_j = Q_r \qquad r = 1, 2, \ldots, F \tag{12.143}$$

where

$$G_r^{ij} \equiv I_i^{rj} - \tfrac{1}{2} I_r^{ij} \tag{12.144}$$

Equation (12.143) represents a set of second-order differential equations of motion in the generalized coordinates. However, it can be cast in a somewhat more symmetric

*These equations may be found in any text on analytical mechanics or variational mechanics, e.g., Refs. 29, 30, 40, 46, 61, and 102.

form by a rearrangement of the coefficients G_r^{ij}. Toward this end, we note that an interchange of the dummy subscripts i and j on the left-hand side of the equation

$$\sum_{i,j=1}^{F} I_i^{rj}\dot{q}_i\dot{q}_j \equiv \sum_{i=1}^{F}\sum_{j=1}^{F} I_j^{ri}\dot{q}_j\dot{q}_i \tag{12.145}$$

results identically in the right-hand side. Hence the double sum in Eq. (12.143) can be expressed in the form

$$\sum_{i=1}^{F}\sum_{j=1}^{F} G_r^{ij}\dot{q}_i\dot{q}_j = \sum_{i=1}^{F}\sum_{j=1}^{F}\begin{bmatrix} ij \\ r \end{bmatrix}\dot{q}_i\dot{q}_j \tag{12.146}$$

where

$$\begin{bmatrix} ij \\ r \end{bmatrix} \equiv \tfrac{1}{2}\,(I_i^{rj} + I_j^{ri} - I_r^{ij}) \tag{12.147}$$

is the "Christoffel symbol of the first kind" associated with the matrix of inertia coefficients I_{ij}.

Upon substitution of Eq. (12.146) into Eq. (12.143) we find

$$\sum_{j=1}^{F} I_{rj}\ddot{q}_j + \sum_{i=1}^{F}\sum_{j=1}^{F}\begin{bmatrix} ij \\ r \end{bmatrix}\dot{q}_i\dot{q}_j = Q_r \tag{12.148}$$

These equations are called the "explicit form of Lagrange's equations" by Whittaker.[102]

Although they are a particularly good starting point for numerical integration of the equations of motion, these "explicit" equations have not been widely used in the engineering literature. They are mentioned by Beyer,[6] who found it necessary to evaluate the coefficients I_{rj} semigraphically and to evaluate coefficients of the type I_r^{ij} by an approximate finite difference formula.

Upon comparing Eqs. (12.128) and (12.148), we see that the terms $Q_r\dagger$ defined by Eq. (12.126) play precisely the same role, in the equations of motion, as the double sum in Eq. (12.148). The great advantage of the form involving $Q_r^\dagger$ is that these terms can be found without knowing explicit expressions for I_{ij} in terms of q_i, as would be required for the calculation of the Christoffel symbols via Eq. (12.140). In order to integrate Eq. (12.148) numerically, it is necessary to solve it explicitly for the acceleration $\ddot{q}_i$. This may be done by any elimination procedure or by inversion of the matrix $[I_{rs}]$.

If we denote by J_{pr} the elements of the matrix $[I_{pr}]^{-1}$, we have

$$\sum_{r=1}^{F} J_{pr}I_{rj} \equiv \delta_{pj} \tag{12.149}$$

where $\delta_{pj} = 1$ if $p = j$ and $\delta_{pj} = 0$ otherwise. Upon multiplying Eq. (12.148) through by J_{pr} and summing over r, we find[64]

$$\ddot{q}_p = \sum_{r=1}^{F} J_{pr}Q_r - \sum_{i=1}^{F}\sum_{j=1}^{F}\begin{Bmatrix} p \\ ij \end{Bmatrix}\dot{q}_i\dot{q}_j \tag{12.150}$$

where

$$\begin{Bmatrix} p \\ ij \end{Bmatrix} \equiv \sum_{r=1}^{F} J_{pr}\begin{bmatrix} ij \\ r \end{bmatrix} \tag{12.151}$$

is the Christoffel symbol of second kind.

When the generalized velocities are related by constraints of the form of Eq. (12.129), Lagrange multipliers may be introduced to generate the supplemental forces of Eqs. (12.132). Then it is only necessary to add the term

$$Q'_r = \sum_{i=1}^{N_t} \lambda_i \alpha_{ir} \tag{12.152}$$

to the right-hand side of Eq. (12.148).

Lagrange's equations (with multipliers) are used in several general-purpose computer programs, such as DRAM,[12,85] ADAMS,[58,60] DADS,[32,101] AAPD,[47] IMP,[81,82] and IMPUM.[81] Lagrange multipliers are not necessary for treating open-loop systems such as robots[34,73,98] or other systems with treelike topology[105] as covered by the program UCIN.[37,38]

12.4.3 Kinetics of Single-Freedom Systems

For systems with one DOF there is a single generalized coordinate, and we may write, for brevity,

$$q_1 \equiv q \qquad Q_1 \equiv Q \qquad Q_1^\dagger \equiv Q^\dagger \qquad I_{11} \equiv I \tag{12.153}$$

Then the equations of motion, Eqs. (12.128), reduce to the single differential equation

$$I\ddot{q} = Q + Q^\dagger \tag{12.154}$$

This result also follows directly from the principle of power balance, which states that the power of the active forces ($Q\dot{q}$) equals the rate of increase of the kinetic energy (T). From Eq. (12.135) we see that

$$T \equiv \tfrac{1}{2} I \dot{q}^2 \tag{12.155}$$

Therefore, the power balance principle is

$$dT/dt = (d/dt)\tfrac{1}{2} I\dot{q}^2 = Q\dot{q} \tag{12.156}$$

$$I\dot{q}\ddot{q} + \tfrac{1}{2}(dI/dq)(dq/dt)\dot{q}^2 = Q\dot{q} \tag{12.157}$$

Hence

$$I\ddot{q} = Q - C\dot{q}^2 \tag{12.158}$$

where

$$C \equiv \tfrac{1}{2} dI/dq \tag{12.159}$$

Equation (12.158) is clearly of the form of Eq. (12.154) with

$$Q^\dagger \equiv -C\dot{q}^2 \tag{12.160}$$

Using the simplified notation of Sec. 12.2.6, we can express I, via Eq. (12.125), in the form

$$I = \sum_{k=1}^{B} \sum_{i=1}^{3} (mu_i^2 + \bar{J}_1 \bar{w}_i^2)^k \tag{12.161}$$

Therefore

$$C = \frac{1}{2}\frac{dI}{dq} = \sum_{k=1}^{B} \sum_{i=1}^{3} (mu_i u_i' + \bar{J}_i \bar{w}_i \bar{w}_i')^k \tag{12.162}$$

where
$$u_i' = \sum_{j=1}^{M} \frac{\partial u_i(\psi)}{\partial \psi_j} \frac{\partial \psi_j}{\partial q} = \sum_{j=1}^{M} \frac{\partial u_i(\psi)}{\partial \psi_j} c_j \tag{12.163}$$

$$\overline{w}_i' = \sum_{j=1}^{M} \frac{\partial \overline{w}_i(\psi)}{\partial \psi_j} c_j \tag{12.164}$$

Similarly, the generalized force given by Eq. (12.127) can be expressed in the form

$$Q = \sum_{k=1}^{B} \sum_{i=1}^{3} (u_i X_i + w_i M_i)^k \tag{12.165}$$

For *planar motions,* in which all particles move in paths parallel to the global (x, y) axes, two of the Euler angles shown in Fig. 12.2 vanish (say $\Theta = \Phi = 0$). It then follows from Eq. (12.105) that **D** is a unit matrix, and from Eq. (12.116) that there is no distinction between the angular velocity coefficients w_i and $\overline{w}_i$.

Numerous examples of the application of these equations to planar single-freedom systems are given in Chap. 12 of Ref. 66. For related approaches to the use of the generalized equation of motion, (12.158), see Refs. 21 and 84.

For *conservative systems* (see Sec. 12.3.5) with one DOF, where all generalized forces are derived from a potential function $V(q)$, the generalized forces can be expressed in the form $Q = -dV/dq$. Therefore, the power balance theorem can be expressed in the form

$$\int_{T_0}^{T} dT = \int_{q_0}^{q} Q\,dq = -\int_{V_0}^{V} dV \tag{12.166}$$

where q_0 is some reference state where the potential energy is V_0 and the kinetic energy is T_0.

The integrated form of this equation is

$$T - T_0 = -(V - V_0) \tag{12.167}$$

or

$$T + V = T_0 + V_0 \equiv E \tag{12.168}$$

where E (a constant) is the total energy of the system. Equation (12.168) is called the "energy integral" or the "first integral" of the equation of motion.

Recalling Eq. (12.155), we can write Eq. (12.168) in the form

$$\frac{dq}{dt} = \dot{q} = \pm (2T/I)^{1/2} = \pm \{2[E - V(q)]/I\}^{1/2} \tag{12.169}$$

Finally, a relation can be found between t and q by means of a simple quadrature expression (e.g., by Simpson's rule) of Eq. (12.169) written in the form $dt = dq/\dot{q}$; i.e.,

$$t - t_0 = \int_{q_0}^{q} \frac{dq}{\dot{q}} = \pm \int_{q_0}^{q} \left\{ \frac{I(q)}{2[E - V(q)]} \right\}^{1/2} dq \tag{12.170}$$

where t_0 is the time at which $q = q_0$. The correct choice of + or − in Eqs. (12.169) and (12.170) is what makes $\dot{q}$ consistent with the given "initial" velocity at $t = t_0$; thereafter, a reversal of sign occurs whenever $\dot{q}$ becomes zero. Examples of the use of these equations are given in Sec. 12.50 of Ref. 66, where it is shown, among other results, that a stable oscillatory motion can occur in the neighborhood of equilibrium points (where $dV/dq = 0$) only if V is a local minimum at this point.

Many slightly different versions of the method of power balance have been published.[3,21,28,33,74] A brief review of the differences among these versions is given in Ref. 64.

12.4.4 Kinetostatics and Internal Reactions

If all the external forces acting on the links of a machine are given, the consequent motion can be found by integrating the generalized differential equation of motion, Eq. (12.133), or any of its equivalent forms. This is the *general problem of kinetics.*

However, one is frequently required to find those forces which must act on a system or subsystem (e.g., a single link) in order to produce a *given motion.* This problem can be solved, as we shall see, by using the methods of statics and is therefore called a problem of "kinetostatics."

Suppose, for example, that the general problem of kinetics has already been solved for a given machine (typically by a numerical solution of the differential equations as explained in Sec. 12.5). This means that the positions, velocities, and accelerations of all points are known at a given instant of time, and the corresponding inertia force $-m\ddot{\mathbf{x}}$ is known. Similarly, one may now calculate the inertia torque $\overline{\mathbf{T}}$, whose components, referred to principal central axes, are given by Eq. (12.109) or (12.118) for each link. We may then use the matrix **D** [defined by Eq. (12.105)] to express the components of inertia torque T_i along the global (x_i) axes as follows:

$$\mathbf{T} = \mathbf{D}^T\overline{\mathbf{T}} \tag{12.171}$$

Furthermore, let $\mathbf{X}'$ and $\mathbf{M}'$ be the net force and moment (about the mass center) exerted on the link by all the other links in contact with the subject link. For example, if the subject link k has its mass center at $\mathbf{x}^k$ and is in contact with link n which exerts the fore $\mathbf{X}'^{kn}$ and moment $\mathbf{M}'^{kn}$ at point $\mathbf{x}^{kn}$, we can write

$$\mathbf{X}'^k = \sum_n \mathbf{X}'^{kn} \tag{12.172}$$

$$\mathbf{M}'^k = \sum_n [\mathbf{M}'^{kn} + (\mathbf{x}^n - \mathbf{x}^k) \times \mathbf{X}'^{kn}] \tag{12.173}$$

where the summation is made over all links which contact link k. Using this notation, we may write, for each link k, the following vector equations of "dynamic equilibrium" for forces and moments respectively:

$$\mathbf{X}^k - m^k\ddot{\mathbf{x}}^k + \sum_n \mathbf{X}'^{kn} = \mathbf{0} \tag{12.174a}$$

$$\mathbf{M}^k + \mathbf{T}^k + \sum_n [\mathbf{M}'^{kn} + (\mathbf{x}^n - \mathbf{x}^k) \times \mathbf{X}'^{kn}] = \mathbf{0} \tag{12.174b}$$

Two such *vector equations* of kinetostatics can be written for each body in the system ($k = 1, 2, \ldots, B$), leading to a set of $6B$ scalar equations which are linear in the unknown reaction components $X_i'^{kn}$ and $M_i'^{kn}$. If one expresses Newton's law of action and reaction in the form

$$X_i'^{kn} = -X_i'^{kn} \qquad M_i'^{kn} = -M_i'^{kn}$$

there will be a sufficient number of linear equations to solve for all the unknown reaction force and moment components.[1,64]

When temporal constraints are present, the applied forces $\mathbf{X}^k$ and moments $\mathbf{M}^k$ are not all known a priori, and some of the scalar quantities X_i^k and M_i^k (or some linear combination of them) must be treated as unknowns in Eqs. (12.174*a*) and (12.174*b*). For example, if a motor exerts an unknown torque on link 7 about the local axis ξ_2^7 in order to maintain a constant angular speed $\overline{\omega}_2^7$ about that axis, the required *control torque* $\overline{T}_2^7$ is unknown a priori, and the following terms will appear in Eq. (12.174*b*) as unknown contributions to $\mathbf{M}^7$:

$$\begin{bmatrix} \Delta M_1^7 \\ \Delta M_2^7 \\ \Delta M_3^7 \end{bmatrix} = \begin{bmatrix} D_{11}^7 & D_{21}^7 & D_{31}^7 \\ D_{12}^7 & D_{22}^7 & D_{32}^7 \\ D_{13}^7 & D_{23}^7 & D_{33}^7 \end{bmatrix} \begin{bmatrix} 0 \\ \overline{T}_2^7 \\ 0 \end{bmatrix} = \begin{bmatrix} D_{21}\overline{T}_2^7 \\ D_{22}\overline{T}_2^7 \\ D_{23}\overline{T}_2^7 \end{bmatrix}$$

A systematic procedure for the automatic generation and solution of the appropriate linear equations, when such *control forces* must be accounted for, is given in Ref. 1.

12.4.5 Balancing of Machinery

From elementary mechanics, it is known that the net force acting on a system of particles with fixed total mass equals the total mass times the acceleration of the center of mass. If this force is provided by the foundation of a machine system, the equal but opposite force felt by the foundation is called the "shaking force." For a system of B rigid bodies, the shaking force is precisely the resultant inertia force $\mathbf{F} = -\Sigma m^k\ddot{\mathbf{x}}^k$.

Similarly, a net "shaking couple" $\mathbf{T}$ is transmitted to the foundation where $\mathbf{T}$ is the resultant of all the inertia torques defined by Eqs. (12.109) for an individual link.

The problem of *machine balancing* is usually posed as one of *kinetostatics,* wherein the acceleration of every mass particle in the system is given (usually in the steady-state condition of a machine), and one seeks to determine the size and location of *balancing weights* which will reduce $|\mathbf{F}|$ and $|\mathbf{T}|$ to acceptable levels. The subject is usually discussed under the two categories of *rotor balancing* and *linkage balancing.* Because of the large body of literature on these subjects, we shall only present a brief qualitative discussion of each, which follows closely that in Ref. 67.

Rotor Balancing.[8,44,45,48,54,66,67,104] A rotating shaft supported by coaxial bearings, together with any attached mass (e.g., a turbine disk or motor armature), is called a "rotor." If the center of mass (CM) of a rotor is not located exactly on the bearing axis, a net centrifugal force will be transmitted, via the bearings, to the foundation. The horizontal and vertical components of this periodic shaking force can travel through the foundation to create serious vibration problems in neighboring components. The magnitude of the shaking force is $F = me\omega^2$, where m = rotor mass, ω = angular speed, e = distance (called "eccentricity") of the CM from the rotation axis. The product me, called the "unbalance," depends only on the mass distribution and is usually nonzero because of unavoidable manufacturing tolerances, thermal distortion, etc. When the CM lies exactly on the axis of rotation ($e = 0$), no net shaking force occurs and the rotor is said to be in "static balance." Static balance is achieved in practice by adding or subtracting balancing weights at any convenient radius until the rotor shows no tendency to turn about its axis, starting from any given initial orientation. Static balancing is adequate for relatively thin disks or short rotors (e.g., automobile wheels). However, a long rotor (e.g., in a turbogenerator) may be in static balance and still exert considerable forces upon individual bearing supports. For example, suppose that a long shaft is supported on bearings which are coaxial with the ζ axis, where ξ, η, ζ

are body-fixed axes through the CM of a rotor, and ζ coincides with the fixed (global) z axis. If the body rotates with constant angular speed $|\omega| \equiv \omega_3$ about the ζ axis, the Euler equations (see Ref. 94 or Sec. 13.10 of Ref. 66) show that the inertial torques about the rotating ξ and η axes are

$$\bar{T}_1 = \bar{I}_{\zeta\eta}\omega^2$$

$$\bar{T}_2 = -\bar{I}_{\zeta\xi}\omega^2$$

$$\bar{T}_3 = 0$$

where $\bar{I}_{\zeta\eta}$ and $\bar{I}_{\zeta\xi}$ are the (constant) products of inertia referred to the (ξ, η, ζ) axes. From these equations it is clear that the resultant inertia torque rotates with the shaft and produces harmonically fluctuating shaking forces on the foundation. However, if $\bar{I}_{\zeta\eta}$ and $\bar{I}_{\zeta\xi}$ can be made to vanish, the shaking couple will likewise vanish. It is shown in the references cited that any *rigid* shaft may be *dynamically balanced* (i.e., the net shaking force *and* the shaking couple can be simultaneously eliminated) by adding or subtracting a definite amount of mass at any convenient radius in each of two arbitrary transverse cross sections of the rotor. The "balancing planes" selected for this purpose are usually located near the ends of the rotor where suitable shoulders or "balancing rings" have been machined to permit the convenient addition of mass (e.g., lead weights or calibrated bolts) or the removal of mass (e.g., by drilling or grinding). Long rotors, running at high speeds, may undergo appreciable deformations. For such *flexible rotors* it is necessary to utilize more than two balancing planes.

Balancing Machines and Procedures. The most common types of rotor *balancing machines* consist of bearings held in pedestals which support the rotor, which is spun at constant speed (e.g., by a belt drive, or compressed airstream). Electromechanical transducers sense the unbalanced forces (or associated vibrations) transmitted to the pedestals and electric circuits automatically perform the calculations necessary to predict the location and amount of balancing weight to be added or subtracted in preselected balancing planes.

For very large rotors, or for rotors which drive several auxiliary devices, commercial balancing machines may not be convenient. The *field balancing* procedures for such installations may involve the use of accelerometers on the bearing housings, along with vibration meters and phase discriminators (possibly utilizing stroboscopy) to determine the proper location and amount of balance weight. For an example used in field balancing of flexible shafts, see Ref. 26.

Committees of the International Standards Organization (ISO) and the American National Standards Institute (ANSI) have formulated recommendations for the allowable quality grade $G = e\omega$ for various classes of machines.[54]

Balancing of Linkage Machinery. The literature in this field pertaining to the slider-crank mechanism is huge, because of its relevance to the ubiquitous internal combustion engine.[8,14,17,35,39,43,66,79,91] For this relatively simple mechanism (even with multiple cranks attached to the same crankshaft), the theory is well understood and readily applied in practice. However, for more general linkages, the theory and reduction to practice has not yet been as thoroughly developed. The state of the field up to the 1960s has been reviewed periodically by Lowen, Tepper, and Berkof,[4,49,50] and practical techniques for balancing the shaking forces in four-bar and six-bar mechanisms have been described.[5,66,88,95,96,99]

It should be noted that the addition of the balancing weights needed to achieve net *force* balance will tend to increase the shaking moment, bearing forces, and required input torque. The size of the balancing weights can be reduced if one is willing to accept a *partial force balance.*[92,102] Because of the many variables involved in the linkage balancing problem, a number of investigations of the problem have been made from the point of view of optimization theory.[13,96,103]

12.4.6 Effects of Flexibility and Joint Clearance

In the above discussion, it has been assumed that all the links are perfectly rigid and that no clearance (backlash) exists at the joints. A great deal of literature has appeared on the influence of both these departures from the commonly assumed ideal conditions. In view of the extent of this literature, we will point specifically only to surveys of the topics and to recent representative papers which contain additional references.

Publications on link flexibility (up to 1971) have been reviewed by Erdman and Sandor,[23] and more recent work in this area has been described in several papers[7,55,56,90] dealing with the finite element method of analyzing link flexibility. A straightforward, practical, and accurate approach to this problem has been described by Amin,[1] who showed how to model flexible links so that they may be properly included in a dynamic situation by the general-purpose computer program DYMAC.[68]

A number of papers dealing with the effects of looseness and compliance of joints have been published by Dubowsky et al.,[19] and the entire field has been surveyed by Haines.[31] A later discussion of the topic was given by Ehle and Haug.[20]

12.5 NUMERICAL METHODS FOR SOLVING THE DIFFERENTIAL EQUATIONS OF MOTION

A system of n ordinary differential equations is said to be in "standard form" if the derivatives of the unknown functions $w_1, w_2, \ldots, w_n$, with respect to the independent variable t are given in the form

$$dw_i/dt = f_i(w_1, w_2, \ldots, w_n, t) \qquad i = 1, \ldots, n \tag{12.175}$$

If the initial values $w_1(0), w_2(0), \ldots,$ are all given we are faced with an *initial value problem,* in standard form, of order n. Fortunately, a number of well-tested and efficient computer subroutines are available for the solution of the problem posed. One major class of such programs is based on the so-called one-step or direct methods such as the Runge-Kutta methods. These methods have the advantage of being self-starting, are free of iterative processes, and readily permit arbitrary spacing Δt between successive times where results are needed.

A second major class of numerical integration routines belong to the so-called predictor-corrector or multistep methods. These are iterative procedures whose main advantage over the one-step procedures is that they permit more direct control over the size of the numerical errors developed. Most predictor-corrector methods suffer, however, from the drawback of starting difficulties and an inability to change the step size t once started. However, "almost automatic" starting procedures for multistep methods have been devised.[27] These use variable-order integration rules for starting or changing interval size. For a review of the theory, and a FORTRAN listing, of the Runge-Kutta algorithm, see Appendix G of Ref. 66. For examples of the use of other types of differential equation solvers see Refs. 10 and 101.

12.5.1 Determining Initial Values of Lagrangian Coordinates

We will assume that we are dealing with a system described by M lagrangian coordinates $\psi_1, \ldots, \psi_M$ which are related by spatial and temporal constraint equations of the form

$$f_i(\psi_1, \ldots, \psi_M) = 0 \qquad i = 1, 2, \ldots, N_s \tag{12.176}$$

$$g_k(\psi_1, \ldots, \psi_M, t) = 0 \qquad k = 1, 2, \ldots, N_t \tag{12.177}$$

In order for this system of $N_s + N_t$ equations in M unknowns to have a solution at time $t = 0$, it is necessary that F_R of the ψ_i be given, where

$$F_R \equiv M - N_s - N_t$$

is the reduced degree of freedom discussed in Sec. 12.2.2. Assuming that F_R of the ψ_i are given at $t = 0$, we may solve the $N_s + N_t$ equations for the unknown initial coordinates, using a numerical method, such as the Newton-Raphson algorithm (see Chap. 9 and Appendix F of Ref. 66). At this point, all values of ψ_i will be known.

12.5.2 Initial Values of Lagrangian Velocities

When there are no temporal constraints ($N_t = 0$), we must specify the F independent initial velocities $\dot{q}_1, \ldots, \dot{q}_F$ at time $t = 0$. Then all the lagrangian velocities at $t = 0$ follow from Eq. (12.38) in the form

$$\dot{\psi}_j = \sum_{k=1}^{F} C_{jk}\dot{q}_k \qquad j = 1, \ldots, M \tag{12.178}$$

When there exist N_t temporal relations expressed in the form of Eq. (12.177), we can express them in their rate form [see Eq. (12.71)]

$$\sum_{k=1}^{F} \alpha_{jk}\dot{q}_k + \beta_j = 0 \qquad j = 1, \ldots, N_t \tag{12.179}$$

These temporal constraints permit us to specify only F_R ($= F - N_t$) generalized velocities, rather than F of them. With N values of $\dot{q}_k$ left unspecified, we may solve Eqs. (12.178) and (12.179) for all the $M + N_t$ unknown values of $\dot{\psi}_i$ and $\dot{q}_j$. Thus, all values of both ψ_i and $\dot{\psi}_i$ may be found at time $t = 0$, and the numerical integration can proceed.

12.5.3 Numerical Integration

Let us introduce the notation

$$\{\psi_1, \psi_2, \ldots, \psi_M, \dot{q}_1, \ldots, \dot{q}_F\} \equiv \{w_1, w_2, \ldots, w_M, w_{M+1}, \ldots, w_{M+F}\} \tag{12.180}$$

Now we may rewrite the governing equations (12.178), (12.133), and (12.179), respectively, as

$$\dot{w}_j = \sum_{k=1}^{F} C_{jk} w_{M+K} \qquad j = 1, 2, \ldots, M \qquad (12.181)$$

$$\sum_{k=1}^{F} I_{jk} \dot{w}_{M+K} = Q_j + Q_j^{\dagger} + \sum_{m=1}^{N_t} \alpha_{mj} \lambda_m \qquad j = 1, \ldots, F \qquad (12.182)$$

$$\sum_{k=1}^{F} \alpha_{jk} \dot{w}_{M+k} = -\beta_j \qquad j = 1, \ldots, N_t \qquad (12.183)$$

We now note that Q_j, $Q_j\dagger$, and β_j are all known functions of $w_1, \ldots, w_{M+F}$ whenever the state $(\boldsymbol{\psi}, \dot{\boldsymbol{\psi}})$ is known (e.g., at $t = 0$). Then it is clear that we may solve the $F + N_t$ governing equations [Eqs. (12.182) and (12.183)] for the $F + N_t$ unknowns $\dot{w}_{M+1}, \ldots, \dot{w}_{M+F}$ and $\lambda_1, \ldots, \lambda_{Nt}$. Then we may solve Eq. (12.181) for the M remaining velocities $\dot{w}_1, \dot{w}_2, \ldots, \dot{w}_M$. In short, Eqs. (12.181) and (12.183) enable us to calculate all the $M + F$ derivatives $\dot{w}_i$ from a knowledge of the w_i. That is, they constitute a system of $M + F$ first-order differential equations in standard form. Any of the well-known subroutines for such initial-value problems may be utilized to march out the solution for all the ψ_i and $\dot{\psi}_i$, at user-specified time steps. The lagrangian accelerations may be computed at any time step from Eq. (12.44), and the displacements, velocities, and accelerations for any point of interest may be found from Eqs. (12.2), (12.46), and (12.48).

12.6 SURVEY OF MACHINE DYNAMICS LITERATURE AND COMPUTER PROGRAMS

In this section, we shall give a brief review of the large body of literature dealing with our subject that has appeared since the publication of the Second Edition of the *Mechanical Design and Systems Handbook* (1985). Much of the essential historical background on the development of multibody dynamics, including the areas of aerospace and biomechanics, is given by Roberson and Schwertassek[A29] and Huston.[A15] A compact summary of the field's literature is given in Paul[A25] under the headings of "Open-Loop Mechanisms" ("Satellite Controls," "Human Body Models"), "Closed-Loop Mechanisms," "Elastodynamics," and "Machine Balancing."

12.6.1 Computer Programs

A detailed discussion, by their respective authors, of 20 general-purpose multibody dynamics computer programs will be found in Schiehlen,[A33] from which the list in Table 12.1 has been taken.

12.6.2 Survey Papers

Like most rapidly growing scientific fields, that of multibody dynamics has been surveyed on several occasions. One of the earliest such surveys (in the area of theory of machines) was given by Paul.[64] An excellent review of the state of the art in the early 1980s is provided by the proceedings of a NATO sponsored symposium (Haug[A12]). A more recent review was given by Schwertassek and Roberson[A34] who present an in-depth

TABLE 12.1 General-Purpose Multibody Dynamics Computer Programs

Program	Principal contributors	Country
ADAMS	M. A. Chace, N. Orlandea, R. R. Ryan	United States
AUTODYN	P. Maes, J. C. Samin, P. Y. Willems	Belgium
AUTOLEV	D. A. Levinson, T. R. Kane	United States
CAMS	L. Lilov, B. Bekjarov, M. Lorer	Bulgaria
COMPAMM	J. M. Jimenez, A. Avello, A. Garcia-Alonso, J. Garcia de Jalon	Spain
DADS	E. J. Haug, R. L. Smith	United States
DISCOS	H. Frisch	United States
DYMAC	B. Paul, A. Amin, G. Hud	United States
DYNOCOMBS	R. L. Huston, T. P. King, J. W. Kamman	United States
DYSPAM	B. Paul, R. Schaffa	United States
MEDYNA	O. Wallrapp, C. Führer	Germany
MESA VERDE	J. Wittenburg, U. Wolz, A. Schmidt	Germany
NBOD	H. Frisch	United States
NEWEUL	E. Kreuzer, W. Schiehlen	Germany
NUBEMM	E. Pankiewicz	Germany
PLEXUS	A. Barraco, E. Cuny	France
ROBOTRAN	(see entry under AUTODYN)	Belgium
SIMPACK	W. Rulka	Germany
SPACAR	J. B. Jonker, J. P. Meijaard	Netherlands
SYM	M. Vukobratovic, N. Kirkanski, A. Timcenko, M. Kirkanski	Yugoslavia

comparison of the various dynamic formulations used. Many of the books mentioned below also contain extensive literature surveys.

12.6.3 Books on Multibody Dynamics

Suh and Radcliffe[89] broke the mold of the textbook literature by introducing the use of computers for *kinematics* of mechanisms. Their book deals primarily with spatial *kinematics,* although it does contain a relatively brief discussion of *dynamic* analyses for particular mechanisms, utilizing the Newton-Euler formulation. The first book to contain a systematic treatment of the *dynamics* of machinery, formulated in a manner best suited to solution by computer, was Paul.[66] Although this book was focussed on planar problems, the fundamental tools of analytical dynamics (especially the d'Alembert-Lagrange principle) which it expounds are applicable to spatial problems as well. Erdman and Sandor[A6] and Sandor and Erdman[A31] each contain a relatively brief discussion of planar dynamics, but from the more traditional point of view, except for the inclusion of some problems of elastokinetics (contributed by A. Midha). More recently there has appeared a spate of books which treat multibody dynamics from a modern, computer-oriented viewpoint, including Roberson and Schwertassek,[A29] Nikravesh,[A22] Shabana,[A35] Haug,[A13] Schiehlen,[A33] and Huston.[A15] Reference A29 contains a particularly comprehensive list of references, and useful discussions of the historical record.

12.6.4 Books on Dynamics of Robots

There now exists a substantial number of publications which deal with the dynamics of robots, and manipulators. Perhaps the first book to focus on three-dimensional problems of the kinematics and dynamics of robots was that of R. P. Paul.[73] A good idea of the state of the art at that time may be obtained from the collection edited by Brady et al.[A2] This book contains reprints of significant papers, as well as tutorial articles on dynamics, feedback control, trajectory planning, compliance, and task planning. Among the huge number of books on robotics that have since been published, the following are representative books that have significant material on the kinematics, statics, and dynamics of robots: Asada and Slotine,[A2] Fu, Gonzalez, and Lee,[A9] Wolovich,[A40] and Yoshikawa.[A41]

REFERENCES

1. Amin, A.: "Automatic Formulation and Solution Techniques in Dynamics of Machinery," Ph.D. Dissertation, University of Pennsylvania, Philadelphia, 1979.
2. Andrews, G. C., and H. K. Kesavan: "The Vector Network Model: A New Approach to Vector Dynamics," *Mechanism and Machine Theory*, vol. 10, pp. 57–75, 1975.
3. Benedict, C. E., and D. Tesar: "Dynamic Response Analysis of Quasi-rigid Mechanical Systems Using Kinematic Influence Coefficients," *J. Mechanisms*, vol. 6, pp. 383–403, 1971.
4. Berkof, R. S., G. Lowen, and F. R. Tepper: Balancing of Linkages, *Shock & Vibration Dig.*, vol. 9, no. 6, pp. 3–10, 1977.
5. Berkof, R. S., and G. Lowen: A New Method for Completely Force Balancing Simple Linkages, *J. Eng. Ind., Trans. ASME*, ser. B, vol. 91, pp. 21–26, 1969.
6. Beyer, R.: "Kinematisch-getriebedynamisches Praktikum," Springer, Berlin, 1960.
7. Bhagat, B. M., and K. D. Wilmert: "Finite Element Vibration Analysis of Planar Mechanisms," *Mechanism and Machine Theory*, vol. 11, pp. 47–71, 1976.
8. Biezeno, C. B., and R. Grammel: "Engineering Dynamics, Vol. IV—Internal-Combustion Engines," (transl. by M. P. White), Blackie & Son, Ltd., Glasgow, 1954.
9. Bocher, M.: "Introduction to Higher Algebra," The Macmillan Company, New York, pp. 46, 47, 1907.
10. Chace, M. A., and Y. O. Bayazitoglu: "Development and Application of a Generalized d'Alembert Force for Multi-Freedom Mechanical Systems," *J. Eng. Ind., Trans. ASME*, ser. B, vol. 93, pp. 317–327, 1971.
11. Chace, M. A., and P. N. Sheth: "Adaptation of Computer Techniques to the Design of Mechanical Dynamic Machinery," ASME Paper 73-DET-58, ASME, New York, 1973.
12. Chace, M. A., and D. A. Smith: "DAMN—Digital Computer Program for the Dynamic Analysis of Generalized Mechanical Systems," *SAE Trans.*, vol. 80, pp. 969–983, 1971.
13. Conte, F. L., G. R. George, R. W. Mayne, and J. P. Sadler: "Optimum Mechanism Design Combining Kinematic and Dynamic-Force Considerations," *J. Eng. Ind., Trans. ASME*, ser. B, vol. 97, pp. 662–670, 1975.
14. Crandall, S. H.: "Rotating and Reciprocating Machines," in "Handbook of Engineering Mechanics," W. Flugge, ed., McGraw-Hill Book Company, Inc., New York, chap. 58, 1962.
15. Crandall, S. H., D. C. Karnopp, E. F. Kurtz, Jr., and D. C. Pridmore-Brown: "Dynamics of Mechanical and Electromechanical Systems," McGraw-Hill Book Company, Inc., New York, 1968.
16. Denavit, J., R. S. Hartenberg, R. Razi, and J. J. Uicker, Jr.: "Velocity, Acceleration, and Static-Force Analysis of Spatial Linkages," *J. Appl. Mech.*, vol. 32, *Trans. ASME*, ser. E, vol. 87, pp. 903–910, 1965.
17. Den Hartog, J. P.: "Mechanical Vibrations," 4th ed., McGraw-Hill Book Company, Inc., New York, 1956.
18. Dix, R. C., and T. J. Lehman: "Simulation of the Dynamics of Machinery," *J. Eng. Ind. Trans. ASME*, ser. B, vol. 94, pp. 433–438, 1972.

19. Dubowsky, S., and T. N. Gardner: "Design and Analysis of Multilink Flexible Mechanisms with Multiple Clearance Connections," *J. Eng. Ind. Trans. ASME,* ser. B, vol. 99, pp. 88–96, 1977.

20. Ehle, P. E., and E. J. Haug: "A Logical Function Method for Dynamic Design Sensitivity Analysis of Mechanical Systems with Intermittent Motion," *J. Mechanical Design, Trans. ASME,* vol. 104, pp. 90–100, 1982.

21. Eksergian, R.: "Dynamical Analysis of Machines" (a series of 15 installments), appearing in *J. Franklin Inst.,* vols. 209, 210, 211, 1930–1931.

22. Erdman, A. G., and J. E. Gustavson: "LINCAGES: Linkage Interaction Computer Analysis and Graphically Enhanced Synthesis Package," ASME Paper 77-DET-5, 1977.

23. Erdman, A. G., and G. N. Sandor: "Kineto-Elastodynamics—A Review of the Art and Trends," *Mechanism and Machine Theory,* vol. 7, pp. 19–33, 1972.

24. Freudenstein, F.: "On the Variety of Motions Generated by Mechanisms," *J. Eng. Ind., Trans. ASME,* ser. B, vol. 84, pp. 156–160, 1962.

25. Freudenstein, F., and G. N. Sandor: "Kinematics of Mechanisms," sec. 5 of "Mechanical Design and Systems Handbook," 2d ed., H. A. Rothbart, ed., McGraw-Hill Book Company, Inc., New York, 1985.

26. Fujisaw, F., et al.: "Experimental Investigation of Multi-Span Rotor Balancing Using Least Squares Method," *J. Mechanical Design, Trans. ASME,* vol. 102, pp. 589–596, 1980.

27. Gear, C. W.: "Numerical Initial Value Problems in Ordinary Differential Equations," Prentice-Hall, Inc., Englewood Cliffs, N.J., 1971.

28. Givens, E. J., and J. C. Wolford: "Dynamic Characteristics of Spatial Mechanisms," *J. Eng. Ind., Trans. ASME,* ser. B, vol. 91, pp. 228–234, 1969.

29. Goldstein, H.: "Classical Mechanics," 2d ed., Addison-Wesley Publishing Company, Inc., Reading Mass., 1980.

30. Greenwood, D. T.: "Classical Dynamics," Prentice-Hall, Inc., Englewood Cliffs, N.J., 1977.

31. Haines, R. S.: "Survey: 2-Dimensional Motion and Impact at Revolute Joints," *Mechanism and Machine Theory,* vol. 15, pp. 361–370, 1980.

32. Haug, E. J., R. A. Wehage, and N. C. Barman: "Dynamic Analysis and Design of Constrained Mechanical Systems," *J. of Mech. Design, Trans. ASME,* vol. 104, pp. 778–784, 1982.

33. Hirschhorn, J.: "Kinematics and Dynamics of Plane Mechanisms," McGraw-Hill Book Company, Inc., New York, 1962.

34. Hollerback, J. M.: "A Recursive Lagrangian Formulation of Manipulator Dynamics and a Comparative Study of Dynamics Formulation," *IEEE Trans. on Systems, Man and Cybernetics,* SMC-10, 11, pp. 730–736, November 1980.

35. Holowenko, A. R.: "Dynamics of Machinery," John Wiley & Sons, Inc., New York, 1955.

36. Hud, G. C.: "Dynamics of Inertia Variant Machinery," Ph.D. Dissertation, University of Pennsylvania, Philadelphia, 1976.

37. Huston, R. L.: "Multibody Dynamics Including the Effects of Flexibility and Compliance," *Computers and Structures,* vol. 14, pp. 443–451, 1981.

38. Huston, R. L., C. E. Passerello, and M. W. Harlow: Dynamics of Multirigid-Body Systems, *J. Appl. Mech.,* vol. 45, pp. 889–894, 1978.

39. Judge, A. W.: "Automobile and Aircraft Engines," vol. 1, "The Mechanics of Petrol and Diesel Engines," Pitman Publishing Corporation, New York, 1947.

40. Kane, T. R.: "Dynamics," Holt, Rinehart and Winston, New York, 1968.

41. Kane, T. R., and C. F. Wang: "On the Derivation of Equations of Motion," *J. Society for Industrial and Applied Math,* vol. 13, pp. 487–492, 1965.

42. Kaufman, R. E.: "Kinematic and Dynamic Design of Mechanisms," in "Shock and Vibration Computer Programs Reviews and Summaries," W. & B. Pilkey, ed., Shock and Vibration Information Center, U.S. Naval Research Laboratory, Code 8404, Washington, D.C., 1975.

43. Kozesnik, J.: "Dynamics of Machines," Ervin P. Noordhoff, Ltd., Groningen, Netherlands, 1962.

44. Kroon, R. P.: "Balancing of Rotating Apparatus—I," *J. Appl. Mech.*, vol. 10, *Trans. ASME*, vol. 65, pp. A225–A228, 1943.

45. Kroon, R. P.: "Balancing of Rotating Apparatus—II," *J. Appl. Mech.*, vol. 11, *Trans. ASME*, vol. 66, A47–A50, 1944.

46. Lanczos, C.: "The Variational Principles of Mechanics," 3d ed., University of Toronto Press, Toronto, 1966.

47. Langrana, N. A., and D. L. Bartel: "An Automated Method for Dynamic Analysis of Spatial Linkages for Biomechanical Applications," *J. Eng. Ind., Trans. ASME*, ser. B, vol. 97, pp. 556–574, 1975.

48. Loewy, R. G., and V. J. Piarulli: "Dynamics of Rotating Shafts," Shock and Vibration Information Center, NRL, Washington, D.C., 1969.

49. Lowen, G. G., and R. S. Berkof: "Survey of Investigations into the Balancing of Linkages," *J. Mechanisms*, vol. 3, pp. 221–231, 1968.

50. Lowen, G. G., F. R. Tepper, and R. S. Berkof: "Balancing of Linkages—An Update," *Mechanisms and Machine Theory*, vol. 18, pp. 213–220, 1983.

51. Mabie, H. H., and F. W. Ocvirk: "Mechanisms and Dynamics of Machinery," John Wiley & Sons, Inc., New York, 1975.

52. Martin, G. H.: "Kinematics and Dynamics of Machines," 2d ed., McGraw-Hill Book Company, Inc., New York, 1982.

53. Mitchell, T.: "Mechanics," sec. 4, "Mechanical Design and Systems Handbook," 2d ed., H. A. Rothbart, ed., McGraw-Hill Book Company, Inc., New York, 1985.

54. Muster, D., and D. G. Stadelbauer: "Balancing of Rotating Machinery," in "Shock and Vibration Handbook," 2d ed., C. M. Harris and C. E. Crede, eds., McGraw-Hill Book Company, Inc., New York, 1976, chap. 39.

55. Nath, P. K., and A. Ghosh: "Kineto-Elastodynamic Analysis of Mechanisms by Finite Element Method," *Mechanism and Machine Theory*, vol. 15, pp. 179–197, 1980.

56. Nath, P. K., and A. Ghosh: "Steady State Response of Mechanisms with Elastic Links by Finite Element Methods," *Mechanisms and Machine Theory*, vol. 15, pp. 179–197, 1980.

57. Nikravesh, P. E., and I. S. Chung: "Application of Euler Parameters to the Dynamic Analysis of Three-Dimensional Constrained Systems," *J. Mechanical Design, Trans. ASME*, vol. 104, 1982.

58. Orlandea, N.: "Node-Analogous, Sparsity-Oriented Methods for Simulation of Mechanical Systems," Ph.D. Dissertation, University of Michigan, Ann Arbor, 1973.

59. Orlandea, N., and M. A. Chace: "Simulation of a Vehicle Suspension with the ADAMS Computer Program," SAE Paper No. 770053, Society of Automotive Engineers, Warrendale, Pa., 1977.

60. Orlandea, N., M. A. Chace, and D. A. Calahan: "A Sparsity-Oriented Approach to the Dynamic Analysis and Design of Mechanical Systems—Parts I and II," *J. Eng. Ind., Trans. ASME*, ser. B, vol. 99, pp. 733–779, 780–784, 1977.

61. Pars, L. A.: "A Treatise on Analytical Dynamics," John Wiley & Sons, Inc., New York, 1968.

62. Paul, B.: A Unified Criterion for the Degree of Constraint of Plane Kinematic Chains," *J. Appl. Mech.*, vol. 27, *Trans. ASME* vol. 82, ser. E, pp. 196–200, 1960. See discussion, same volume, pp. 751–753.

63. Paul, B.: "On the Composition of Finite Rotations," *Amer. Math. Monthly*, vol. 70, no. 8, pp. 859–862, 1963.

64. Paul, B.: "Analytical Dynamics of Mechanisms—A Computer-Oriented Overview," *Mechanism and Machine Theory*, vol. 10, pp. 481–507, 1975.

65. Paul, B.: "Dynamic Analysis of Machinery Via Program DYMAC." SAE Paper 770049, Society of Automotive Engineers, Warrendale, Pa., 1977.

66. Paul, B.: "Kinematics and Dynamics of Planar Machinery," Prentice-Hall, Inc., Englewood Cliffs, N.J., 1979.

67. Paul, B.: "Shaft Balancing," in "Encyclopedia of Science and Technology," McGraw-Hill Book Company, Inc., New York, 1993.

68. Paul, B., a nd A. Amin: "User's Manual for Program DYMAC-G4 (DYnamics of MAChinery—General Version 4)," available from B. Paul, Philadelphia, Pa., 1979.

69. Paul, B., and A. Amin: "User's Manual for Program KINMAC (KINematics of MAChinery)," available from B. Paul, Philadelphia, Pa., 1979.

70. Paul, B., and A. Amin: "User's Manual for Program STATMAC (STATics of MAChinery)," available from B. Paul, Philadelphia, Pa., 1979.

71. Paul, B., and D. Krajcinovic: "Computer Analysis of Machines with Planar Motion—Part 1: Kinematics," *J. Appl. Mech.,* vol. 37, *Trans. ASME,* ser. E, vol. 92, pp. 697–702, 1970a.

72. Paul, B., and D. Krajcinovic: "Computer Analysis of Machines with Planar Motion—Part 2: Dynamics," *J. Appl. Mech.,* vol. 37, *Trans. ASME,* ser. E, vol. 92, pp. 703–712, 1970b.

73. Paul, R. P.: "Robot Manipulators: Mathematics, Programming, and Control," The MIT Press, Cambridge, Mass., 1981.

74. Quinn, B. E.: "Energy Method for Determining Dynamic Characteristics of Mechanisms," *J. Appl. Mech.,* vol. 16, *Trans. ASME,* vol. 71, pp. 283–288, 1949.

75. Reuleaux, F.: "The Kinematics of Machinery" (transl. and annotated by A. B. W. Kennedy), reprinted by Dover Publications, Inc., New York, 1963, original date 1876.

76. Roger, R. J., and G. C. Andrews: "Simulating Planar Systems Using a Simplified Vector-Network Method," *Mechanism and Machine Theory,* vol. 10, pp. 509–519, 1975.

77. Rogers, R. J., and G. C. Andrews: "Dynamic Simulation of Planar Mechanical Systems with Lubricated Bearing Clearances Using Vector Network Methods," *J. Eng. Ind., Trans. ASME,* ser. B, vol. 99, pp. 131–137, 1977.

78. Rongved, L., and H. J. Fletcher: "Rotational Coordinates," *J. Franklin Institute,* vol. 277, pp. 414–421, 1964.

79. Root, R. E.: "Dynamics of Engine and Shaft," John Wiley & Sons, Inc., New York, 1932.

80. Rubel, A. J., and R. E. Kaufman: "KINSYN III: A New Human Engineered System for Interactive Computer-Aided Design of Planar Linkages," ASME Paper No. 76-DET-48, 1976.

81. Sheth, P. N.: "A Digital Computer Based Simulation Procedure for Multiple Degree of Freedom Mechanical Systems with Geometric Constraints," Ph.D. thesis, University of Wisconsin, Madison, 1972.

82. Sheth, P. N., and J. J. Uicker, Jr.: "IMP (Integrated Mechanisms Program), A Computer-Aided Design Analysis System for Mechanisms and Linkage," *J. Eng. Ind., Trans. ASME,* ser. B, vol. 94, pp. 454–464, 1972.

83. Shigley, J. E., and J. J. Uicker, Jr.: "Theory of Machines and Mechanisms," McGraw-Hill Book Company, Inc., New York, 1980.

84. Skreiner, M.: "Dynamic Analysis Used to Complete the Design of a Mechanism," *J. Mechanisms,* vol. 5, pp. 105–109, 1970.

85. Smith, D. A.: "Reaction Forces and Impact in Generalized Two-Dimensional Mechanical Dynamic Systems," Ph.D. dissertation, Mech. Eng., University of Michigan, Ann Arbor, 1971.

86. Smith, D. A.: "Reaction Force Analysis in Generalized Machine Systems," *J. Eng. Ind., Trans. ASME,* ser. B, vol. 95, pp. 617–623, 1973.

87. Smith, D. A., M. A. Chace, and A. C. Rubens: "The Automatic Generation of a Mathematical Model for Machinery Systems," ASME Paper 72-Mech-31, ASME, New York, 1972.

88. Stevensen, E. N., Jr.: "Balancing of Machines," *Trans. ASME, of J. Eng. Ind.* vol. 95, ser. B, pp. 650–656, 1973.

89. Suh, C. H., and C. W. Radcliffe: "Kinematics and Mechanisms Design," John Wiley & Sons, Inc., New York, 1978.

90. Sunoda, W. H., and S. Dubowski: "The Application of Finite Element Methods to the Dynamic Analysis of Flexible Spatial and Co-planar Linkage Systems," *J. Mech. Des., Trans. ASME,* vol. 103, pp. 643–651, 1981.

91. Taylor, C. F.: "The Internal Combustion Engine in Theory and Practice," The MIT Press, Cambridge, Mass., vol. I, 1966; vol. II, 1968.

92. Tepper, F. R., and G. G. Lowen: "Shaking Force Optimization of Four-Bar Linkages with Adjustable Constraints on Ground Bearing Forces," *J. Eng. Ind., Trans. ASME,* ser. B, vol. 97, pp. 643–651, 1975.

93. Thomson, W. (Lord Kelvin), and P. G. Tait: "Treatise on Natural Philosophy, Part I," Cambridge University Press, Cambridge, 1879, p. vi.

94. Timoshenko, S. P., and D. H. Young: "Advanced Dynamics," McGraw-Hill Book Company, Inc., New York, 1948.

95. Tricamo, S. J., and G. G. Lowen: "A New Concept for Force Balancing Machines for Planar Linkages, Part I: Theory," *J. Mechanical Design, Trans. ASME,* vol. 103, pp. 637–642, 1981.

96. Tricamo, S. J., and G. G. Lowen: "A New Concept for Force Balancing Machines for Planar Linkages, Part 2: Application to Four-Bar Linkages and Experiment," *J. Mechanical Design, Trans. ASME,* vol. 103, pp. 784–792, 1981.

97. Uicker, J. J. Jr.: "Dynamic Force Analysis of Spatial Linkages," *J. Appl. Mech.,* vol. 34, *Trans. ASME,* ser. E, vol. 89, pp. 418–424, 1967.

98. Vukobratovic, M., and V. Potkonjak: "Dynamics of Manipulation Robots," Springer Verlag, New York, 1982.

99. Walker, M. J., and K. Oldham: "A General Theory of Force Balancing Using Counterweights, Mechanism and Machine Theory," vol. 13, pp. 175–185, 1978.

100. Webster, A. G.: "Dynamics of Particles and of Rigid Elastic and Fluid Bodies," Dover Publication reprint, New York, 1956.

101. Wehage, R. A., and E. J. Haug: "Generalized Coordinate Partitioning for Dimension Reduction in Analysis of Constrained Dynamic System," *J. Mechanical Design, Trans. ASME,* vol. 104, 1982, pp. 247–255.

102. Whittaker, E. T.: "A Treatise on the Analytical Dynamics of Particles and Rigid Bodies," 4th ed., Dover Publications, Inc., New York, 1944.

103. Wiederrich, J. L., and B. Roth: "Momentum Balancing of Four-Bar Linkages," *J. Eng. Ind., Trans. ASME,* ser. B, vol. 98, pp. 1289–1295, 1976.

104. Wilcox, J. B.: "Dynamic Balancing of Rotating Machinery," Sir Isaac Pitman & Sons, Ltd., London, 1967.

105. Wittenburg, J.: "Dynamics of Systems of Rigid Bodies," B. G. Teubner, Stuttgart, 1977.

SUPPLEMENTARY REFERENCES

A1. Asada, H., and J. E. Slotine: "Robot Analysis and Control," John Wiley & Sons, Inc., New York, 1986.

A2. Brady, M., J. M. Hollerback, T. J. Johnson, Lozano-Pérez, and M. T. Mason (eds.): "Robot Motion: Planning and Control," The MIT Press, Cambridge, Mass., 1982.

A3. Denavit, J., and R. S. Hartenberg: "A Kinematic Notation for Lower-Pair Mechanisms Based on Matrices," *Trans. ASME, J. of Appl. Mech.,* vol. 22, pp. 215–221, 1955.

A4. Desloge, E. A.: "Relationship between Kane's Equations and the Gibbs-Appell Equations," *J. Guidance, Control and Dynamics,* vol. 1, pp. 120–122, 1987.

A5. Erdman, A. G. (ed.): "Modern Kinematics: Developments in the Last Forty Years," John Wiley & Sons, Inc., New York, 1993.

A6. Erdman, A. G., and G. N. Sando: "Mechanism Design Analysis and Synthesis," 2d ed., vol. I, Prentice-Hall, Inc., Englewood Cliffs, N.J., 1991.

A7. Fletcher, H., L. Rongved, and E. Yu: "Dynamics Analysis of a Two-Body Gravitationally Oriented Satellite," *Bell System Tech. J.*, vol. 42 (no. 5, September), pp. 2239–2266, 1963.

A8. Freudenstein, F., J. P. Macey, and E. R. Maki: "Optimum Balancing of Combined Pitching and Yawing Movement in High Speed Machinery," *Trans. ASME, J. Mechanical Design*, vol. 103, pp. 571–577, 1981.

A9. Fu, K., R. Gonzalez, and C. Lee: "Robotics: Control, Sensing, Vision, and Intelligence," McGraw-Hill Book Company, Inc., New York, 1987.

A10. Fujisaw, F. E. A.: "Experimental Investigation of Multi-Span Rotor Balancing Using Least Squares Method," *Trans. ASME, J. Mechanical Design*, vol. 102, pp. 589–596, 1980.

A11. Gabriele, G. A. (ed.), et al.: "Optimization in Mechanisms," chap. 11, Ref. A5, 1992.

A12. Haug, E. J. (ed.): "Computer Aided Analysis and Optimization of Mechanical System Dynamics," NATO ASI Series F, vol. 9, Springer-Verlag, Berlin, 1984.

A13. Haug, E. J.: "Computer Aided Kinematics and Dynamics of Mechanical Systems," vol. I: "Basic Methods," Allyn and Bacon, Boston, 1989.

A14. Hollerback, J. M.: "A Recursive Lagrangian Formulation of Manipulator and a Comparative Study of Dynamics Formulation," *IEEE Transactions on Systems, Man and Cybernetics*, SMC-10,11, 1980.

A15. Huston, R.: "Multibody Dynamics," Butterworth-Heinemann, Stoneham, Mass., 1990.

A16. Huston, R. L., and C. Passerello: "On the Dynamics of a Human Body Model," *J. Biomechanics*, vol. 4, pp. 369–378, 1976.

A17. Huston, R. L., and C. Passerello: "On Multi-Rigid-Body System Dynamics," *Computers and Structures*, vol. 10, pp. 439–446, 1979.

A18. Kane, T. R., and D. A. Levinson: "Dynamics: Theory and Application," McGraw-Hill Book Company, Inc., New York, 1985.

A19. Lo, C-K.: "A General Purpose Comprehensive Dynamics Simulation of Serial Robot Manipulators," Ph.D. dissertation, University of Pennsylvania, Philadelphia, 1990.

A20. Luh, J. Y. S., M. W. Walker, and R. P. C. Paul: "On-Line Computational Scheme for Mechanical Manipulators," *Trans. ASME, J. Dynamic Systems, Measurement and Control*, vol. 102, pp. 69–76, 1980.

A21. Midha, A. M. (ed.), et al.: "Elastic Mechanisms," chap. 9, Ref. A5.

A22. Nikravesh, P. E.: "Computer-Aided Analysis of Mechanical Systems," Prentice-Hall, Inc., Englewood Cliffs, N.J., 1988.

A23. Norton, R. L. (ed.), et al.: "Cams and Cam Followers," chap. 7, Ref. A5.

A24. Paul, B.: "DYMAC (DYnamics of MAChinery)," pp. 305–322, Ref. A33.

A25. Paul, B.: "Dynamics of Multibody Mechanical Systems, A Forty Year History," chap. 8, Ref. A5.

A26. Paul, B., and R. Schaffa: "DYSPAM (DYnamics of SPatial Mechanisms)," pp. 323–340, Ref. A33, 1990. (*Note:* The name of coauthor R. Schaffa was omitted from the text of Ref. A33 due to an editorial error.)

A27. Paul, B., J. W. West, and E. Y. Yu: "A Passive Gravitational Attitude Control System for Satellites," *Bell System Tech. J.*, vol. 42, pp. 2195–2238, 1963.

A28. Pisano, A., and F. Freudenstein: "An Experimental and Analytical Investigation of the Dynamic Response of High-Speed Cam-Follower Systems—Parts I, II," *Trans. ASME, J. Mechanisms, Transmissions and Automation in Design*, vol. 105, pp. 692–698, 699–704, 1983.

A29. Roberson, R. E., and R. Schwertassek: "Dynamics of Multibody Systems" Springer-Verlag, Berlin, New York, 1988.

A30. Roberson, R., and J. Wittenburg: "A Dynamical Formalism for an Arbitrary Number of Interconnected Rigid Bodies, with Reference to the Problem of Satellite Attitude Control," vol. 1, book 3, paper 46D, in *Proc. 3d Congr. Int. Fed. Autom. Control*, Butterworth, 1967, London, 1966.

A31. Sandor, G. N., and A. G. Erdman: "Advanced Mechanism Design, Analysis and Synthesis," vol. II, Prentice-Hall, Inc., Englewood Cliffs, N.J., 1984.

A32. Schaffa, R.: "Dynamics of Spatial Mechanisms and Robots," Ph.D. dissertation, University of Pennsylvania, Philadelphia, 1984.

A33. Schiehlen, W. (ed.): "Multibody Systems Handbook," Springer-Verlag, Berlin, New York, 1990.

A34. Schwertassek, R., and R. E. Roberson: "A perspective on computer oriented multibody dynamic formalisms and their implementations," pp. 261–273 in G. Bianchi and W. Schiehlen, eds., *Proc. IUTAM/IFToMM Symposium on Dynamics of Multibody Systems,* (September 1985, Udine, Italy), Springer-Verlag, Berlin, 1986.

A35. Shabana, A. A.: "Dynamics of Multibody Systems," John Wiley & Sons, Inc., New York, 1989.

A36. Sommerfeld, A.: "Mechanics," Academic Press, Inc., New York, 1964, p. 187.

A37. Vance, J. M.: "Rotordynamics of Turbo Machinery," John Wiley & Sons, Inc., New York, 1988.

A38. Walker, M. W., and D. E. Orin: "Efficient Dynamic Computer Simulation of Robot Mechanisms," *Trans. ASME, J. Dynamic Systems, Measurement, Control,* vol. 104, pp. 205–211, 1982.

A39. Wittenburg, J.: "Dynamics of Systems of Rigid Bodies," B. G. Teubner, Stuttgart, 1977.

A40. Wolovich, W.: "Robotics: Basic Analysis and Design," Holt, Rinehart and Winston, New York, 1987.

A41. Yoshikawa, T.: "Foundations of Robotics," The MIT Press, Cambridge, Mass., 1990.

CHAPTER 13
SYSTEM RELIABILITY

Kailash C. Kapur, Ph.D., P.E.
Professor and Director of Industrial Engineering
University of Washington
Seattle, Wash.

Leonard R. Lamberson, Ph.D., P.E.
Professor and Dean
College of Engineering and Applied Sciences
Western Michigan University
Kalamazoo, Mich.

13.1 INTRODUCTION 13.1
13.2 RELIABILITY MEASURES 13.2
13.2.1 Reliability 13.2
13.2.2 Maintainability 13.4
13.2.3 Availability 13.6
13.2.4 Durability 13.6
13.3 LIFE DISTRIBUTIONS IN RELIABILITY 13.6
13.4 RELIABILITY ACTIVITIES DURING THE PRODUCT DESIGN CYCLE 13.8
13.4.1 Design Review 13.8
13.4.2 Failure Mode, Effects, and Criticality Analysis 13.8
13.4.3 Fault-Tree Analysis 13.9
13.4.4 Reliability Design Guidelines 13.9
13.4.5 Probabilistic Approach to Design 13.10
13.5 SYSTEM RELIABILITY ANALYSIS 13.12
13.5.1 Reliability Block Diagram 13.12
13.5.2 Static Reliability Models 13.12
13.5.3 Dynamic Reliability Models 13.12
13.6 RELIABILITY TESTING 13.14
13.6.1 Exponential Distribution 13.14
13.6.2 Weibull Distribution 13.16
13.6.3 Success-Failure Testing 13.18
13.7 SOURCES OF FAILURE DATA 13.19
13.7.1 MIL-HDBK 217E 13.20
13.7.2 Nonelectronic Parts Reliability Data 13.20
13.7.3 The Government/Industry Data Exchange Program 13.20
13.8 ELEMENTS OF A RELIABILITY PROGRAM 13.20
13.8.1 Reliability Data-Collection Systems 13.21

13.1 INTRODUCTION

"Reliability" is concerned with failure-free performance of a system and the duration of time over which this failure-free performance is maintained. Reliability is established at the design stage, and subsequent testing and production will not raise the reliability without basic design changes. Reliability is difficult to conceptualize as part of the usual design calculations; yet it is largely determined by these calculations.

The term often used to describe the overall capability of a system to perform its intended function is "system effectiveness." System effectiveness is defined as the probability that the system can successfully meet an operational demand within a given time when operating under specified conditions. For consumer products the term also reflects customer satisfaction or overall total quality and the degree to which it can be achieved. Effectiveness is influenced by the way the system is designed, manufactured, used, and maintained, and thus is a function of all life-cycle activities.

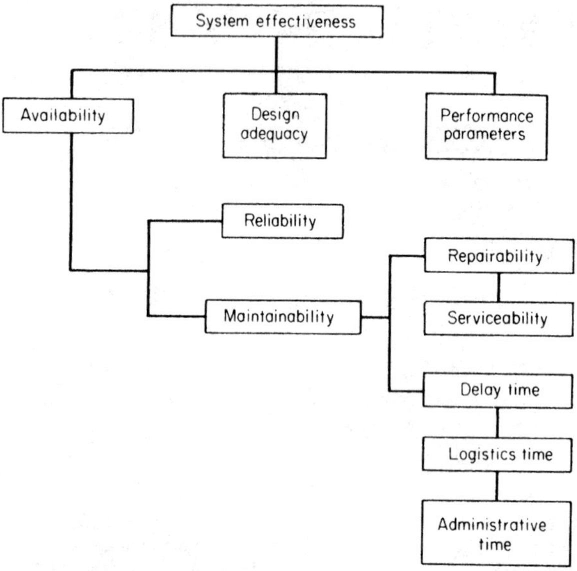

FIG. 13.1 Components of system effectiveness.

The effectiveness of a system is a function of several attributes, such as design adequacy, performance measures, safety, reliability, quality, and maintainability. Availability, reliability, and maintainability are time-oriented qualities of a system. Figure 13.1 gives the relationships among some of the components of system effectiveness.

Recently, the term "assurance sciences"[2,4] has been used to address the overall effectiveness of any system. The assurance sciences are (engineering) disciplines which help us with the attainment of product integrity (that is, the product does what it says it is supposed to do). The term "product assurance" is also used by some companies. Product assurance activities are performed throughout the life cycle of a product.

13.2 RELIABILITY MEASURES

Various components of system effectiveness are defined in this section.

13.2.1 Reliability

Reliability is defined as the probability that a given system will perform its intended function satisfactorily for a specified period of time under specified operating conditions. Thus, reliability is related to the probability of successful performance for any system. It is clear that we must define what successful performance for any system is or what we mean by the failure of the system; otherwise, it is not possible to predict when a system will fail to perform its intended function. The time to failure or "life" of a system cannot be deterministically defined and, hence, it is a random variable. Thus, we must quantify reliability by assigning a probability function to the time-to-failure random variable.

Let T be the time-to-failure random variable (rv). Then reliability at time t, $R(t)$, is the probability that the system will not fail by time t, or

$$\begin{aligned} R(t) &= P(T > t) \\ &= 1 - P(T \leq t) \\ &= 1 - F(t) \\ &= 1 - \int_0^t f(\tau)\, d\tau \end{aligned} \tag{13.1}$$

where $f(t)$ and $F(t)$ are the probability density function (pdf) and cumulative distribution function (cdf), respectively, for the rv T. For example, if T is exponentially distributed, then

$$f(t) = \lambda e^{-\lambda t} \qquad t \geq 0$$

$$\lambda > 0$$

$$F(t) = \int_0^t \lambda e^{-\lambda \tau}\, d\tau = 1 - e^{-\lambda t} \qquad t \geq 0$$

$$R(t) = e^{-\lambda t} \qquad t \geq 0$$

Another measure that is frequently used as an indirect indicator of system reliability is called the "mean time to failure" (MTTF), which is the expected or mean value of the time-to-failure random variable. Thus, the MTTF is theoretically defined as

$$\text{MTTF} = E(T) = \int_0^\infty t f(t)\, dt = \int_0^\infty R(t)\, dt \tag{13.2}$$

Sometimes the term "mean time between failures" (MTBF) is also used to denote $E(T)$. The problem with only using the MTTF as an indicator of system reliability is that it only uniquely determines reliability if the underlying time-to-failure distribution is exponential. If the failure distribution is other than exponential, the MTTF does not give us the whole picture and can produce erroneous comparisons.

If we have a large population of the items whose reliability we are interested in studying, then for replacement and maintenance purposes we are interested in the rate at which the items in the population, which have survived at any point in time, will fail. This is called the "failure rate" or "hazard rate" and is given by the following relationship:

$$h(t) = f(t)/R(t) \tag{13.3}$$

The failure rate for most components follows the curve shown in Fig. 13.2, which is called the "life-characteristic curve" or "bathtub curve." Shown in Fig. 13.2 are three types of failures, namely quality failures, stress-related failures, and wear-out failures. The sum total of these failures gives the overall failure rate.

The failure-rate curve in Fig. 13.2 has three distinct periods. The initial decreasing failure rate is termed "infant mortality" and is due to the early failure of substandard products. Latent material defects, poor assembly methods, and poor quality control can contribute to an initially high failure rate. A short period of in-plant product testing, termed "burn-in," is used by manufacturers to eliminate these early failures from the consumer market. The flat, middle portion of the failure-rate curve represents the design failure rate for the specific product as used by the consumer market. During the useful-life portion, the failure rate is relatively constant. It might be decreased by redesign or restricting usage. Finally, as products age they reach a wear-out phase characterized by an increasing failure rate. See Refs. 9, 16, and 17 for further details.

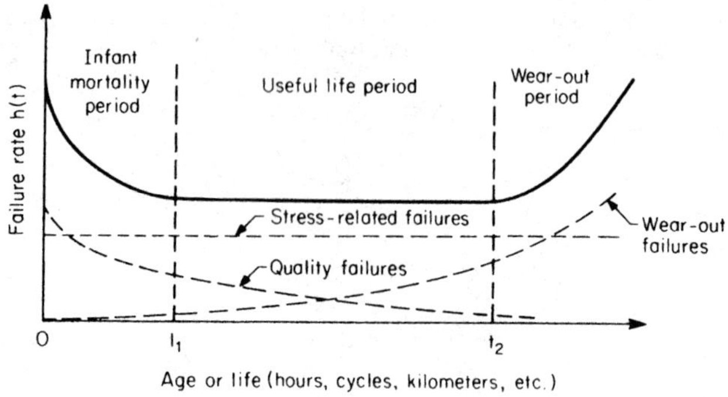

FIG. 13.2 Failure rate vs. life-characteristic curve.

To help the reader understand the notion of failure rate or hazard rate, basic mathematical relations are given below.

The "hazard rate" is defined as the limit of the instantaneous failure rate given no failure up to time t and is given by[8,11]

$$h(t) = \lim_{\Delta t \to 0} \frac{P(t < T \leq t + \Delta t | T > t)}{\Delta t} = \lim_{\Delta t \to 0} \frac{R(t) - R(t + \Delta t)}{\Delta t \, R(t)} \tag{13.4}$$

$$= \frac{1}{R(t)} \left[- \frac{d}{dt} R(t) \right] = \frac{f(t)}{R(t)}$$

Also,

$$f(t) = h(t) \exp \left[- \int_0^t h(\tau) \, d\tau \right] \tag{13.5}$$

and thus

$$R(t) = \exp \left[- \int_0^t h(\tau) \, d\tau \right] \tag{13.6}$$

The term "mission reliability" refers to the reliability of a product for a specified time period t^*, i.e.,

$$\text{Mission reliability} = R(t^*) = P(T > t^*) \tag{13.7}$$

Another term that is used to characterize the reliability of a system is the "B life," which is defined as

$$F(B_\alpha) = \frac{\alpha}{100}$$

Thus B_{10} is the 10th percentile for the life of a system.

13.2.2 Maintainability

"Maintainability" is a system design parameter which has a great impact on system effectiveness. Failures will occur even in highly reliable systems, and the ability of a system to recover is often as important to system effectiveness as is its reliability. Maintainability is defined as the probability that a failed system will be restored to a satisfactory operating condition within a specified interval of downtime. The ease of fault detection, isolation, and repair are all influenced by system design and are the main factors contributing to maintainability. Also contributing are the supply of spare parts, the supporting repair organization, and the preventive maintenance practices. Good maintainability may somewhat offset low reliability. Hence, if the desired reliability cannot be met because of performance constraints, improvements in the system maintainability should be considered.

Let T be the time-to-repair random variable. Then the maintainability function $M(t)$ is given by

$$M(t) = P(T \leq t) \tag{13.8}$$

If the repair time T follows the exponential distribution with mean time to repair (MTTR) of $1/\mu$, where μ is the repair rate, then

$$M(t) = 1 - \exp(-t/\text{MTTR}) \tag{13.9}$$

In addition to the maintainability function, we also use median (50th percentile) and M_{max} (90th or 95th percentile) for the time-to-repair rv as measures of maintainability. The maintenance ratio (MR), which is defined as the total maintenance labor-hours per system operating hour, is also used as a measure of maintainability.

MTTR, which is the mean of the distribution of system repair time, can be evaluated by

$$\text{MTTR} = \frac{\sum_{i}^{n} = 1\, \lambda_i t_i}{\sum_{i}^{n} = 1\, \lambda_i} \tag{13.10}$$

where n = number of components in the system
λ_i = failure rate of the ith repairable component
t_i = time required to repair the system when the ith component fails

In addition, other quantities, such as mean active corrective maintenance time and mean active preventive maintenance time, are used to measure maintainability.

Serviceability affects maintainability, as shown in Fig. 13.1, and is very important from the customer viewpoint. Serviceability is a characteristic of design, just as is reliability, and must be considered at the design stage. Serviceability is defined as the degree of ease (or difficulty) with which a system can be repaired. This measure specifically considers fault detection, isolation, and repair. Repairability considers only the actual repair time and is defined as the probability that a failed system will be restored to operation within a specified interval of active repair time. Access covers, plug-in modules, or other features to allow easy removal and replacement of failed components improve the repairability and serviceability.

Maintainability and serviceability can be improved by consideration of some of the following factors during system design:

Accessibility: Provide sufficient working space around subsystems to allow ease of entry for maintenance personnel and tools in order to diagnose, troubleshoot, and complete maintenance activities.

Built-in test/diagnostics: Diagnostic devices indicating the status of equipment should be built-in to support maintainability processes. These devices should indicate the specific failure component for replacement.

Captive hardware: Attach/detach hardware should provide for quick and easy replacement of components, panels, brackets, and chassis. The hardware should be attached so that it is not lost during maintenance activities.

Color coding: Common color coding can greatly speed up maintenance activities. Color coding includes such things as lubrication information, orientation, timing marks, size, etc.

Modularity: Designs should be divided into physically and functionally distinct units for rapid removal and replacement to minimize on-line downtime.

Standardization: Design equipment with subsystems/components that are commercial standards, readily available, and common to the largest extent possible.

Standardized maintenance tools: Equipment should be maintainable with commonly available tools. This decreases tooling expense and training of maintenance personnel.

13.2.3 Availability

Availability is a function of both reliability and maintainability and answers the question, "Is a system available to perform its intended function when it is needed?" Availability measures are time-related, and some of the time elements are (1) storage, free, or off-time; (2) operating time; (3) standby time; and (4) downtime consisting of corrective and preventive maintenance as well as logistics and administrative delay time. "Inherent availability" A_I considers only operating time and corrective maintenance time and is given by

$$A_I = \mathrm{MTBF}/(\mathrm{MTBF} + \mathrm{MTTR}) \tag{13.11}$$

where MTBF = mean time between failures
MTTR = mean time to repair

Inherent availability is used to measure the design of a system. Achieved availability A_a is used for development and initial production of a system and is given by

$$A_a = \mathrm{OT}/(\mathrm{OT} + \mathrm{TPM} + \mathrm{TCM}) \tag{13.12}$$

where OT = operating time
TPM = total preventive maintenance time
TCM = total corrective maintenance time

Operational availability (A_O) covers all segments of time and is given by

$$A_O = (\mathrm{OT} + \mathrm{ST})/(\mathrm{OT} + \mathrm{ST} + \mathrm{TPM} + \mathrm{TCM} + \mathrm{TALDT}) \tag{13.13}$$

where ST = standby time
TALDT = total administrative and logistics delay time

13.2.4 Durability

Durability is a special case of reliability. It is defined as the probability that an item will successfully survive its projected life, overhaul point, or rebuild point (whichever is the more appropriate durability measure for the item) without a durability failure.

13.3 LIFE DISTRIBUTIONS IN RELIABILITY

In this section we summarize the pdf, cdf, reliability function, hazard-rate function, and MTBF for some of the well-known distributions that are used in reliability and maintainability.

Exponential Distribution

$$f(t) = \lambda e^{-\lambda t} \qquad t \geq 0 \tag{13.14}$$

$$F(t) = 1 - e^{-\lambda t} \qquad t \geq 0 \tag{13.15}$$

$$R(t) = e^{-\lambda t} \qquad t \geq 0 \tag{13.16}$$

$$h(t) = \lambda \tag{13.17}$$

$$\text{MTBF} = \theta = 1/\lambda \tag{13.18}$$

Thus, the failure rate for the exponential distribution is always constant.

Normal Distribution

$$f(t) = (1/\sigma\sqrt{2\pi}) \exp\{-\tfrac{1}{2}[(t-\mu)/\sigma]^2\} \qquad -\infty < t < \infty \tag{13.19}$$

$$F(t) = \Phi[(t-\mu)/\sigma] \tag{13.20}$$

$$R(t) = 1 - \Phi[(t-\mu)/\sigma] \tag{13.21}$$

$$h(t) = \frac{\phi[(t-\mu)/\sigma]}{\sigma R(t)} \tag{13.22}$$

$$\text{MTBF} = \mu \tag{13.23}$$

$\Phi(Z)$ is the cumulative distribution function and $\phi(Z)$ is the probability density function, respectively, for the standard normal variate Z. The failure rate for the normal distribution is a monotonically increasing function.

Log-Normal Distribution

$$f(t) = (1/\sigma t\sqrt{2\pi}) \exp[-\tfrac{1}{2}[(\ln t-\mu)/\sigma]^2] \qquad t \geq 0 \tag{13.24}$$

$$F(t) = \Phi[(\ln t-\mu)/\sigma] \tag{13.25}$$

$$R(t) = 1 - \Phi[(\ln t-\mu)/\sigma] \tag{13.26}$$

$$h(t) = \frac{\phi[(\ln t-\mu)/\sigma]}{t\sigma R(t)} \tag{13.27}$$

$$\text{MTBF} = \exp(\mu + \sigma^2/2) \tag{13.28}$$

The failure rate for the log-normal distribution is neither always increasing nor always decreasing but is increasing on the average.[1] It takes different shapes depending on the parameters μ and σ.

Weibull Distribution

$$f(t) = [\beta(t-\delta)^{\beta-1}/(\theta-\delta)^{\beta}] \exp\{-[(t-\delta)/(\theta-\delta)]^{\beta}\} \qquad t \geq \delta \geq 0 \tag{13.29}$$

$$F(t) = 1 - \exp\{-[(t-\delta)/(\theta-\delta)]^{\beta}\} \tag{13.30}$$

$$R(t) = \exp\{-[(t-\delta)/(\theta-\delta)]^{\beta}\} \tag{13.31}$$

$$h(t) = \beta(t - \delta)^{\beta - 1}/(\theta - \delta)^{\beta} \tag{13.32}$$

$$\text{MTBF} = \theta\Gamma(1 + 1/\beta) \tag{13.33}$$

The failure rate for the Weibull distribution is decreasing when $\beta < 1$, is constant when $\beta = 1$ (same as the exponential distribution), and is increasing when $\beta > 1$.

Gamma Distribution

$$f(t) = [\lambda^{\eta}/\Gamma(\eta)]t^{\eta-1}e^{-\lambda t} \qquad t \geq 0 \tag{13.34}$$

$$F(t) = \sum_{k=n}^{\infty} \frac{(\lambda t)^k \exp(-\lambda t)}{k!} \qquad \text{when } \eta \text{ is integer} \tag{13.35}$$

$$R(t) = \sum_{k=0}^{\eta-1} \frac{(\lambda t)^k \exp(-\lambda t)}{k!} \qquad \text{when } \eta \text{ is integer} \tag{13.36}$$

$$h(t) = f(t)/R(t) \qquad \text{[using Eqs. (13.34) and (13.36)]} \tag{13.37}$$

$$\text{MTBF} = \eta/\lambda \tag{13.38}$$

The failure rate for the gamma distribution is decreasing when $\eta < 1$, is constant when $\eta = 1$, and is increasing when $\eta > 1$.

13.4 RELIABILITY ACTIVITIES DURING THE PRODUCT DESIGN CYCLE

The reliability of a product is determined by the design, so the most economical and effective reliability activities are those that occur early in the design cycle. Such activities must be based largely on experience, since no hard data are available for a quantitative reliability assessment. Some of the procedures used to stimulate reliability improvements throughout the design cycle will now be covered.

13.4.1 Design Review

A design review is a formalized, documented, and systematic review of the design by senior company personnel. The review from a reliability, maintainability, and serviceability standpoint must be a technical review aimed at ferreting out weak points in the design and concerned with design improvements that will ensure early product maturity and improve reliability. Reference 7 gives more on this subject.

13.4.2 Failure Mode, Effects, and Criticality Analysis

For failure prevention during system design, two techniques are presented: failure mode, effects, and criticality analysis and fault-tree analysis. Failure mode, effects, and criticality analysis (FMECA) represents a "bottom-up" approach while fault-tree analysis (FTA) is a "top-down" approach. Both represent qualitative as well as quantitative approaches for assessing the consequences of failure and determining means for prevention or mitigation of the failure effect.

Failure mode and effects analysis (FMEA) is a procedure by which potential failure modes in a system are identified and analyzed. Since design is usually approached from an optimistic viewpoint, the FMEA procedure assumes a pessimistic viewpoint to identify potential product weaknesses. The terms FMEA; design FMA (DFMA); and failure mode, effects, and criticality analysis (FMECA) are also used. The main purpose of the technique is to identify and eliminate failure modes early in the design cycle where they are most economically dealt with. A documented FMECA procedure can be found in MIL-STD-1629A (1980).[14] The essential steps will be included here.

FMEA starts with the selection of a subsystem or component and then identifies and documents all potential failure modes. The effect of each failure mode is traced to other higher-level systems. A documented worksheet is used to record the following:

Function: A concise definition of the functions that the component must perform.

Failure mode: A particular way in which the component can fail to perform a function.

Failure mechanism: A physical or chemical process or hardware deficiency causing the failure mode.

Failure cause: The agent activating the failure mechanism; e.g., saltwater seepage owing to an inadequate seal might produce corrosion as a failure mechanism, though the failure cause was inadequate seal.

Identification of effects of higher-level systems: This determines whether the failure mode is localized, causes higher-level damage, or creates an unsafe condition.

Criticality rating: A measure of severity, probability of failure occurrence, and detectability used to assign priority design actions.

13.4.3 Fault-Tree Analysis

Fault-tree analysis is a technique used for systems safety and reliability analysis.[6] The analysis proceeds from a designated "top event" to basic failure causes called "primary events." A fault tree is a model that graphically portrays the combination of events leading up to the undesirable top event.

13.4.4 Reliability Design Guidelines

We will now discuss some basic principles of reliability in design that are useful for the designer. Each concept is briefly discussed in terms of its role in the design of reliable systems.

Simplicity. Simplification of system configuration contributes to reliability improvement by reducing the number of failure modes. A common approach is called component integration, which is the use of a single part to perform multiple functions.

Use of Proven Components and Preferred Designs. (1) If working within time and cost constraints, use proven components because this minimizes analysis and testing to verify reliability. (2) Mechanical and fluid system design concepts can be categorized and proven configurations given first preference.

Stress-Strength Design.[8] The designer should use various sources of data on strength of materials and strength degradation with time related to fatigue. The traditional and common uses of safety factors do not address reliability, and new techniques such as the probabilistic design approach should be used. The designer would use derating factors, proper reliability or safety margins, and develop stress-strength testing to determine stress and strength distributions. The probabilistic design approach is explained in the following section.

Redundancy. Redundancy sometimes may be the only cost-effective way to design a reliable system from less reliable components.

Local Environmental Control. A severe local environment sometimes prevents the achievement of required component reliability. In that case, the environment should be modified to achieve high reliability. Some typical environmental problems are (1) shock and vibration, (2) heat, and (3) corrosion.

Identification and Elimination of Critical Failure Modes. This is accomplished through FMECA and also by fault-tree analysis.

Self-Healing. A design approach which has possibilities for future development is the use of self-healing devices. Automatic sensing and switching devices represent a form of self-healing.

Detection of Impending Failures. Achieved reliability in the field can be improved by the introduction of methods and/or devices for detecting impending failures. Some of the examples are (1) screening of parts and components, (2) periodic maintenance schedules, and (3) monitoring of operations.

Preventive Maintenance. Preventive-maintenance procedures can enhance the achieved reliability, but the procedures are sometimes difficult to implement. Hence, effective preventive-maintenance procedures must be considered at the design stage.

Tolerance Evaluation. In a complex system, it is necessary to consider the expected range of manufacturing process tolerances, operational environmental, and all stresses, as well as the effect of time. Some of the tolerance evaluation methods are worst-case tolerance analysis, statistical tolerance analysis, and marginal checking.

Human Engineering. Human activities and limitations may be very important to system reliability. The design engineer must consider factors which directly refer to human aspects, such as (1) human factors, (2) human-machine interface, (3) evaluation of the person in the system, and (4) human reliability.

13.4.5 Probabilistic Approach to Design

Reliability is basically a design parameter and must be incorporated in the system at the design stage. One way to quantify reliability during design and to design for reliability is the probabilistic approach.[10] The design variables and parameters are random variables and, hence, the design methodology must consider them as random variables. The reliability of any system is a function of the reliabilities of its components. In order to analyze the reliability of the system, we have to first understand how to compute the reliabilities of the components. The basic idea in reliability analysis from the probabilistic design methodology viewpoint is that a given component has certain strengths which, if exceeded, will result in the failure of the component. The factors which determine the strengths of the component are random variables as well as the factors which determine the stresses or loading acting on the component. "Stress" is used to indicate any agency that tends to induce failures, while "strength" indicates any agency resisting failure. "Failure" itself is taken to mean failure to function as intended; it is said to have occurred when the actual stress exceeds the actual strength for the first time.

Let $f(x)$ and $g(y)$ be the pdf's for the stress random variable X and the strength random variable Y, respectively, for a certain mode of failure. Also, let $F(x)$ and $G(y)$ be the cumulative distribution functions for the random variables X and Y, respectively. Then the reliability R of the component for the failure mode under consideration, with the assumption that the stress and the strength are independent random variables, is given by

$$R = P(Y > X) \tag{13.39}$$

$$= \int_{-\infty}^{\infty} g(y)\left[\int_{-\infty}^{y} f(x)\,dx\right]dy \tag{13.40}$$

$$= \int_{-\infty}^{\infty} g(y)F(y)dy \tag{13.41}$$

$$= \int_{-\infty}^{\infty} f(x)\left[\int_{x}^{\infty} g(y)\,dy\right]dx \tag{13.42}$$

$$= \int_{-\infty}^{\infty} f(x)[1 - G(x)]dx \tag{13.43}$$

For example, suppose the stress rv X is normally distributed with a mean value of μ_X and standard deviation of σ_X, and the strength random variable is also normally distributed with parameters μ_Y and σ_Y. Then the reliability R is given by

$$R = \Phi[(\mu_Y - \mu_X)/\sqrt{\sigma_Y^2 + \sigma_X^2}] \tag{13.44}$$

where $\Phi(\,)$ is the cumulative distribution function for the standard normal variable.

EXAMPLE

$$\mu_Y = 40{,}000 \qquad \mu_X = 30{,}000 \qquad \sigma_Y = 4000 \qquad \sigma_X = 3000$$

Then, factor of safety = 40,000/30,000 = 1.33 and

$$R = \Phi[(40{,}000 - 30{,}000)/\sqrt{(4000)^2 + (3000)^2}] = \Phi(2) = 0.97725$$

If we change μ_X to 20,000 by increasing the factor of safety to 2, we have

$$R = \Phi[(40{,}000 - 20{,}000)/\sqrt{(4000)^2 + (3000)^2}] = \Phi(4) = 0.99997$$

There are four basic ways in which the designer can increase reliability:

1. Increase mean strength: this can be achieved by increasing size, weight, using stronger material, etc.
2. Decrease average stress: this can be done by controlling loads or using higher dimensions.
3. Decrease stress variations: this variation is harder to control but can be effectively truncated by putting limitations on use conditions.
4. Decrease strength variation: the inherent part-to-part variation can be reduced by improving the basic process, controlling the process, and utilizing tests to eliminate the less desirable parts.

13.5 SYSTEM RELIABILITY ANALYSIS

A system reliability model is used as the basis for reliability analysis and apportionment. The analysis is usually based on a block diagram that represents system success according to the definition of system reliability. This section will consider the reliability block diagram and its analysis using both the static and dynamic approaches. Reliability allocation will also be considered.

13.5.1 Reliability Block Diagram

The reliability block diagram (RBD) is constructed from an engineering analysis of the system considering modes of failure and understanding the grouping of similar parts for analysis purposes.

13.5.2 Static Reliability Models

System reliability with static models uses the RBD with a component reliability R_i assigned to each block. The system reliability R_s is then calculated. In the following formulations, n is the number of components.

Series Configuration. In a series configuration, all subsystems must survive for the system to operate successfully. For the series configuration as shown in Fig. 13.3*a,* the system reliability is

$$R_s = \prod_{i=1}^{n} R_i \tag{13.45}$$

This is sometimes called the product rule for system reliability.

Parallel Configuration. In a parallel configuration (Fig. 13.3*b*) all subsystems must fail before the system is considered to be in a failed state. The system reliability is calculated by

$$R_s = 1 - \prod_{i=1}^{n} (1 - R_i) \tag{13.46}$$

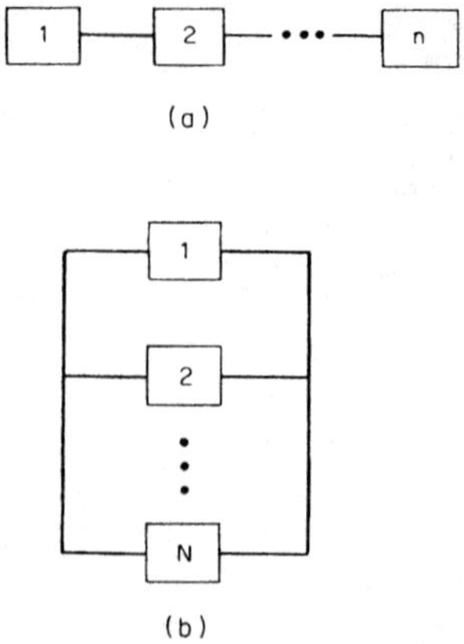

FIG. 13.3 System reliability block diagrams. (*a*) Series system. (*b*) Parallel system.

A system model may consist of both series and parallel subsystems. The system reliability can be calculated by re-peated applications of the above series and parallel equations.

13.5.3 Dynamic Reliability Models

Dynamic reliability models are an extension of the static models where time-dependent reliability functions are used for each subsystem. The following notation is defined:

$R_i(t)$ = the reliability function for the ith subsystem
$R_s(t)$ = the system reliability function

These will be used to reformulate series and parallel systems.

Series Configuration. The system reliability is given by

$$R_s(t) = \prod_{i=1}^{n} R_i(t) \tag{13.47}$$

where n = the number of subsystems. The failure rate $h_s(t)$ for the series system is given by

$$h_s(t) = \sum_{i=1}^{n} h_i(t) \tag{13.48}$$

where $h_i(t)$ is the failure rate for the ith subsystem.

If all subsystems have an exponentially distributed time to failure, then

$$h_s(t) = \sum_{i=1}^{n} \lambda_i \tag{13.49}$$

where λ_i is the failure rate for the ith subsystem.

Parallel Configuration. The system reliability for a parallel configuration where all parallel subsystems are activated when the system is turned on is given by

$$R_s(t) = 1 - \prod_{i=1}^{n} [1 - R_i(t)] \tag{13.50}$$

A standby parallel redundant system is depicted in Fig. 13.4. In this system the standby unit is activated by the switching mechanism S. For a system with two units (i.e., one standby unit), the system reliability is

$$R_s^2(t) = R_1(t) + \int_0^t f_1(t_1)R_2(t - t_1)dt_1 \tag{13.51}$$

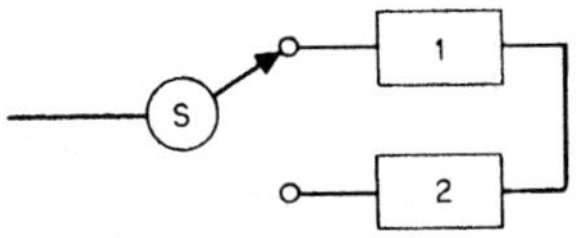

FIG. 13.4 Standby redundant configuration.

where it is assumed that the switch cannot fail. Here $f_1(t)$ is the pdf for device 1 and $R_2(t)$ is the reliability function for device 2.

For the special case where each subsystem is identical with an exponential time to failure, the system reliability is

$$R_s^2(t) = e^{-\lambda t}(1+\lambda t) \qquad t \geq 0 \tag{13.52}$$

If the switch has a reliability of P_s, then the dynamic model for the two-unit system is

$$R_s^2(t)' = R_1(t) + P_s\int_0^t f_1(t_1)R_2(t - t_1)dt_1 \tag{13.53}$$

For identical, exponential time-to-failure subsystems

$$R_s^2(t)' = e^{-\lambda t}[1 + P_s \lambda t] \qquad t \geq 0 \qquad (13.54)$$

Many other dynamic model situations are possible. (See Ref. 8.)

13.6 RELIABILITY TESTING

Reliability can be assessed through product testing. The reliability numbers that one obtains from testing are dependent on test conditions, so obviously if one wants to assess product performance in the field by testing, then one must take great care to duplicate field conditions and usage.

It must be remembered that reliability numbers are always relative to the practical implementation of the definition of reliability with respect to what constitutes a failure and the specified environmental and usage conditions. Products that deteriorate over time rather than fail catastrophically are more difficult to quantify. What is usually done is that some degree of loss of function (or degradation) is set as failure.

The data resulting from testing can be used to (1) estimate the reliability (point estimation), (2) set a confidence limit on reliability, and (3) test the hypothesis that a reliability goal has been met. The following will cover the various approaches using the exponential and Weibull distributions as time-to-failure models. The binomial distribution will also be covered for success-failure testing. This coverage should be sufficient for most situations that one might encounter in practice.

13.6.1 Exponential Distribution

The exponential distribution is a very popular and easy-to-use model to represent time to failure. Selection of the exponential distribution as an appropriate model implies that the failure rate is constant over the range of predictions. For certain failure situations and over certain portions of product life, the assumption of a constant failure rate may be appropriate.

Statistical Properties. The pdf for an exponentially distributed time-to-failure random variable T is given by

$$f(t,\lambda) = e^{-\lambda t} \qquad t \geq 0 \qquad (13.55)$$

where the parameter λ is the failure rate. The reciprocal of the failure rate ($\theta = 1/\lambda$) is the mean or expected life. For products that are repairable, the parameter θ is referred to as the mean time between failures (MTBF) and for nonrepairable products θ is called the mean time to failure (MTTF). The parameter λ must be known (or estimated) for any specific application situation.

The reliability function is given by

$$R(t) = e^{-\lambda t} \qquad t \geq 0 \qquad (13.56)$$

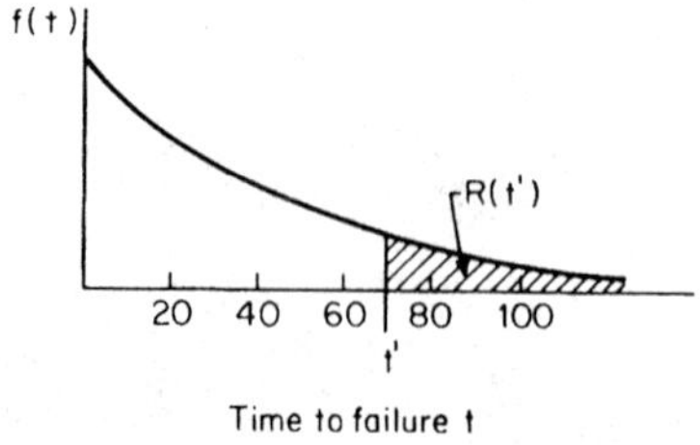

FIG. 13.5 The exponential distribution.

The relationship between $f(t)$ and $R(t)$ is illustrated in Fig. 13.5. For any value of t the quantity $R(t)$ provides the chance of survival beyond time t. If the reliability function is evaluated at the MTBF, it should be noted that $R(\theta) = 0.368$, or there is only a 36.8 percent chance of surviving the mean life.

Point Estimation. The estimator for the mean life parameter θ is given by

$$\hat{\theta} = T/r \tag{13.57}$$

where T = the total accumulated test time considering both failed and unfailed items and r = the total number of failures. Then the estimator for λ is $\hat{\lambda} = 1/\hat{\theta}$.

The reliability is estimated by

$$\hat{R}(t) = e^{-\hat{\lambda}t} \qquad t \geq 0 \tag{13.58}$$

and if one wants to estimate a time for a given reliability level R, this is obtained by

$$\hat{t} = \hat{\theta} \ln (1/R) \tag{13.59}$$

Confidence-Interval Estimates. The confidence intervals for the mean life or for reliability depend on the testing situation. A life-testing situation where n devices are placed on test and it is agreed to terminate the test at the time of the rth failure ($r < n$) is called failure censored and is termed type II life testing. A time-censored life test is one where the total accumulated time (T) is specified. This is termed type I life testing. In type I testing T is specified and r occurs as a result of testing, whereas in a type II situation r is specified and T results from testing. The confidence limits are slightly different for each situation.

Failure-Censored Life Tests. In this situation the number of failures r at which the test will be terminated is specified with n items placed on test. The $100(1 - \alpha)$ percent two-sided confidence interval for θ on the mean life is given by

$$2T/\chi^2_{\alpha/2,2r} \leq \theta \leq 2T/\chi^2_{1-\alpha/2,2r} \tag{13.60}$$

The quantities $\chi^2_{\alpha\nu}$ are the $1 - \alpha$ percentiles of a chi-square distribution with ν degrees of freedom.

The one-sided lower $100(1 - \alpha)$ percent confidence limit is

$$2T/\chi^2_{\alpha,2r} \leq \theta \tag{13.61}$$

The confidence limits on reliability for any specified time can be found by substituting the above limits on θ into the reliability function.

Time-Censored Life Tests. In this situation the accumulated test time T is specified for the test and the test produces r failures. The $100(1 - \alpha)$ percent two-sided confidence interval on the mean life is given by

$$2T/\chi^2_{\alpha/2,2(r+1)} \leq \theta \leq 2T/\chi^2_{1-\alpha/2,2r} \tag{13.62}$$

If only the one-sided $100(1 - \alpha)$ percent lower confidence interval is desired, it is given by

$$2T/\chi^2_{\alpha,2(r+1)} \leq \theta \tag{13.63}$$

Hypothesis Testing. Hypothesis testing is another approach to statistical decision making. Whereas confidence limits offer some degree of protection against meager statistical knowledge, the hypothesis-testing approach is easily misused. To use this approach one should be very familiar with the inherent errors involved in applying the hypothesis-testing procedure.

Consider a type II testing situation where n items are placed on test and the test is terminated at the time of the rth failure. An MTBF goal θ_g is to be verified. The hypotheses for this situation are

$$H_0: \theta \leq \theta_g$$
$$H_1: \theta > \theta_g \tag{13.64}$$

If H_0 is rejected, the conclusion is that the goal has been met.

The statistic calculated from test data is

$$\chi_c^2 = 2T/\theta_g \tag{13.65}$$

and the decision criteria are to reject H_0 and assume that the goal has been met if $\chi_c^2 > \chi_{\alpha,2r}^2$. Here the level of significance is taken as α.

The probability of accepting a design, that is, concluding that the goal has been met where the design has a true MTBF of θ_1, is

$$P_1 = P[\chi_{2r}^2 \geq (\theta_g/\theta_1)\chi_{\alpha,2r}^2] \tag{13.66}$$

and can be looked up in χ^2 tables.

13.6.2 Weibull Distribution

The Weibull distribution is considerably more versatile than the exponential distribution and can be expected to fit many different failure patterns. However, when applying a distribution the failure pattern should be carefully studied and the mixture of failure modes noted. The selection and application of a distribution should then be based on this study and on any knowledge of the underlying physical failure phenomena.

Graphical procedures for the Weibull distribution are attractive in that they provide practitioners with a visual representation of the situation. Although the graphical approach will be covered in the following, it should be recognized that there are better statistical estimation procedures; however, these procedures require the use of a computer, so the availability of computer programs will also be covered.

Statistical Properties. The reliability function for the three-parameter Weibull distribution is given by

$$R(t) = \exp\{-[(t-\delta)/(\theta-\delta)]^\beta\} \qquad t \geq \delta \tag{13.67}$$

where δ = the minimum life ($\delta \geq 0$)
θ = the characteristic life ($\theta > \delta$)
β = the Weibull slope or shape parameter ($\beta > 0$)

The two-parameter Weibull distribution has a minimum life of zero, and the reliability function is

$$R(t) = \exp[-(t/\theta)^\beta] \qquad t \geq 0 \tag{13.68}$$

where θ and β are as previously defined ($\theta > 0$). The term characteristic life resulted from the fact that $R(\theta) = 0.368$; or, there is a 36.8 percent chance of surviving the characteristic life for any Weibull distribution.

The hazard function for the Weibull distribution is given by

$$h(t) = (\beta/\theta^{\beta})t^{\beta-1} \qquad t \geq 0 \tag{13.69}$$

It can be seen that the hazard function will decrease for $\beta < 1$, increase for $\beta > 1$, and remain constant for $\beta = 1$.

The expected or mean life for the two-parameter Weibull distribution is given by

$$\mu = \theta\Gamma(1 + 1/\beta) \tag{13.70}$$

where $\Gamma(\,)$ is a gamma function as found in gamma tables. The standard deviation for the Weibull distribution is

$$\sigma = \theta\sqrt{\Gamma(1 + 2/\beta) - \Gamma^2(1 + 1/\beta)} \tag{13.71}$$

Graphical Estimation. The Weibull distribution is very amenable to graphical estimation. This procedure will now be illustrated.

The cumulative distribution for the two-parameter Weibull distribution is given by

$$F(t) = 1 - \exp[-(t/\theta)^{\beta}] \tag{13.72}$$

then by rearranging and taking logarithms one can obtain

$$\ln(\ln\{1/[1 - F(t)]\}) = \beta \ln t - \beta \ln \theta \tag{13.73}$$

Weibull paper is scaled such that t_j and $P_j = F(t_j)$ can be plotted directly and a straight line fitted to the data. A convenient way to assign values of P_j for plotting is to calculate

$$P_j = (j - 0.3)/(n + 0.4) \tag{13.74}$$

where j is the order of magnitude of the observation and n is the sample size.[8]

EXAMPLE The design for an aluminum flexible drive hub on computer disk packs is under study. The failure mode of interest is fatigue. Data from 12 hubs placed on an accelerated life test follow:

j	Cycles to failure	p_j
1	93,000	5.6
2	147,000	14
3	192,000	22
4	214,000	30
5	260,000	38
6	278,000	46
7	297,000	54
8	319,000	62
9	349,000	70
10	388,000	78
11	460,000	86
12	510,000	94.4

The p_j values were calculated using Eq. (13.74).

The plotted data with a visually fitted line are shown on Weibull paper in Fig. 13.6. The slope of this line provides an estimate of β, which in this case is about 2.3. Most commercially available Weibull papers will have a special scale for estimating β.

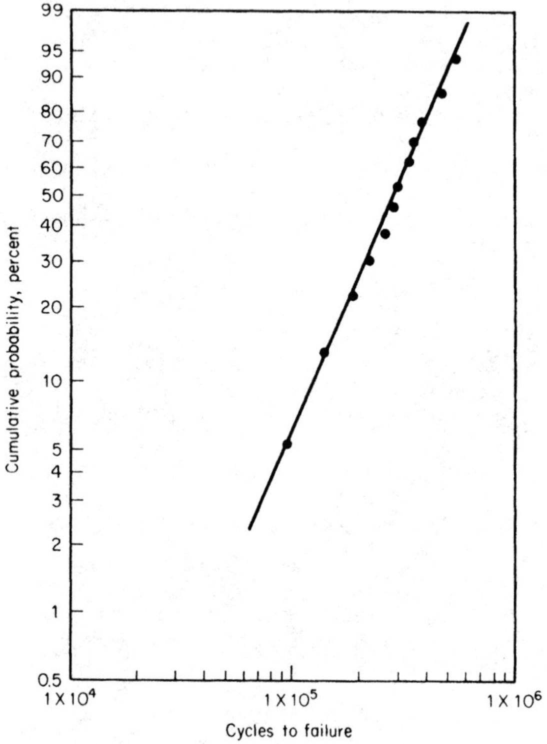

FIG. 13.6 Weibull probability paper.

The characteristic life can be estimated by recalling that $R(\theta) = 0.368$; or, then $F(\theta) = 0.632$. So, one can locate 63.2 percent on the cumulative probability scale, project across to the plotted line, and then project down to the time-to-failure axis. In Fig. 13.6, the characteristic life is about 330,000 cycles.

Weibull paper offers a quick and convenient method for analyzing a failure situation. The population line plotted on the paper can be used to estimate either percent failure at a given time or the time at which a given percentage will fail. Also, a concave plot is indicative of a nonzero minimum life.

13.6.3 Success-Failure Testing

Success-failure testing describes a situation where a product (component, subsystem, etc.) is subjected to a test for a specified length of time T (or cycles, stress reversals, miles, etc.). The product either survives to time T (i.e., survives the test) or fails prior to time T.

Testing of this type can frequently be found in engineering laboratories where a test "bogy" has been established and new designs are tested against this bogy. The bogy will specify a set number of cycles in a certain test environment and at predetermined stress levels.

The probability model for this testing situation is the binomial distribution given by

$$p(y) = \binom{n}{y} R^y(1 - R)^{n-y} \qquad y = 0, 1, 2, \ldots, n \tag{13.75}$$

where R = the probability of surviving the test
n = the number of items placed on test
y = the number of survivors

The value R is the reliability which is the probability of surviving the test. Also,

$$\binom{n}{y} = \frac{n!}{y!(n-y)!} \tag{13.76}$$

Procedures for estimating product reliability R based on this testing situation will now be covered.

Point Estimate. The point estimate of reliability is simply calculated as

$$\hat{R} = y/n \tag{13.77}$$

Confidence Limit Estimate. The 100(1 − α) percent lower confidence limit on the reliability R is calculated by

$$R_L = \frac{y}{y + (n - y + 1)F_{\alpha,2(n-y+1),2y}} \tag{13.78}$$

where $F_{\alpha,2(n-y+1),2y}$ is obtained from F tables. Here again n = the number of items placed on test and y = the number of survivors.

Success Testing. In receiving inspection and sometimes in engineering test labs, one encounters a situation where a no-failure ($r = 0$) test is specified. The concern is usually on ensuring that a reliability level has been achieved at a specified confidence level. A special adaptation of the confidence-limit formula can be derived for this situation.

For the special case where $r = 0$ (i.e., no failures), the lower 100(1 − α) percent confidence limit on the reliability is

$$R_L = \alpha^{1/n} \tag{13.79}$$

where α = the level of significance and n = the sample size (i.e., number placed on test). Then with 100(1 − α) percent confidence, we can say that

$$R_L \leq R$$

where R is the true reliability.

If we let $C = 1 - \alpha$ be the desired confidence level (i.e., 0.80, 0.90, etc.), then the necessary sample size to demonstrate a desired reliability level R is

$$n = \ln(1 - C)/\ln R \tag{13.80}$$

For example, if $R = 0.80$ is to be demonstrated with 90 percent confidence, we have

$$n = \ln(0.10)/\ln(0.80) = 11$$

Thus, we place 11 items on test and allow no failures. This is frequently referred to as success testing.

13.7 SOURCES OF FAILURE DATA

Relatively few sources for failure-rate data are available. There are some data on electronic components, particularly in military applications. However, on mechanical components practically no good data are commercially available. The use of any existing data bank to obtain failure-rate data on any particular design should be done with great caution and skepticism. The applicability of any past history of failure rate to a current design depends on the degrees of similarity in the design, the environment, and the definition of failure. With these words of caution three sources of failure-rate information are covered in the following sections.

13.7.1 MIL-HDBK 217E[12]

This handbook is concerned with reliability prediction for electronic systems. The handbook contains two methods of reliability prediction: the part-stress-analysis and the parts-count methods. These methods vary in complexity and in the degree of information needed to apply them. The part-stress-analysis method requires the greatest amount of information and is applicable during the later design stages when actual hardware and circuits are being designed. To apply this method, a detailed parts list including part stresses must be available.

The parts-count method requires less information and is relatively easy to apply. The information needed to apply this method is (1) generic part type (resistor, capacitor, etc.) and quantity, (2) part quality levels, and (3) equipment environment. Tables are provided in the handbook to determine the factors in a failure-rate model that can be used to predict the overall failure rate. The parts-count method is obviously easier to apply and, one would assume, less accurate than the stress-analysis method.

13.7.2 Nonelectronic Parts Reliability Data[15]

The *Nonelectronic Parts Reliability Data Handbook* (1981) was prepared under the supervision of the Reliability Analysis Center at Griffiss AFB. It is intended to complement MIL-HDBK-217C in that it has information on some mechanical components. Specifically, the handbook provides failure-rate and failure-mode information for mechanical, electrical, pneumatic, hydraulic, and rotating parts. Again, little is known about the environment, design specifics, etc., that produced the failure-rate data in the handbook.

13.7.3 The Government/Industry Data Exchange Program[13]

The Government/Industry Data Exchange Program (GIDEP, 1974) is a cooperative venture between government and industry participants that provides a means to exchange certain types of technical data. Participants in GIDEP are provided with access to various data banks. The Failure Experience Data Bank contains failure information generated when significant problems are identified on systems in the field. These data are reported to a central data bank by the participants.

The Reliability-Maintainability Data Bank contains failure-rate and failure-mode data on parts, components, and systems based on field operations. The data are reported by the participants. Here again the problem is one of identifying the exact operating conditions that caused failure.

13.8 ELEMENTS OF A RELIABILITY PROGRAM

Management and control of system reliability must be based on a recognition of the system's life cycle beginning at concept, extending through design and production, continuing with the usage of the system, and ending with the removal of the system from the inventory. We wish to achieve a high level of operational reliability in the field, and in order to achieve this objective, various reliability tasks have to be completed throughout the life cycle of the system, starting with the concept stage. Some of the tasks are given below:

1. *Tasks during design and development:*
   a. Initiate reliability activities during conceptual phase, such as reliability and cost trade-off studies, reliability models, and programs.
   b. Develop safety margins from reliability viewpoint.
   c. Predict component reliability from the data bank on the failure rates.
   d. Prepare parts program.
   e. Compute system reliability from component reliability.
   f. Perform worst-case and parts-tolerance analyses.
   g. Determine amount of redundancy needed to achieve a reliability goal.
   h. Determine reliability allocation.
   i. Provide input to human engineering.
   j. Interact with value engineering.
   k. Evaluate design changes.
   l. Perform trade-off analysis.
   m. Compare two or more designs.
   n. Prepare design specifications.
   o. Prepare guidelines for design review.
   p. Perform failure-mode analysis.
   q. Work with cost-reduction programs.

2. *Tasks during development and testing:*
   a. Establish reliability growth curves for the development-and-testing phase.
   b. Develop guidelines for the amount of testing.
   c. Develop bathtub curve based on the failure-rate data and the test data.
   d. Participate in the development of failure definition/scoring criteria document.
   e. Participate in scoring of the test failure.

3. *Tasks during manufacturing:*
   a. Establish guidelines for manufacturing processes.
   b. Provide input to quality control.
   c. Develop input to guidelines to evaluate suppliers and vendors.
   d. Develop product burn-in or debugging time.

4. *Tasks during field usage and maintenance:*
   a. Establish warranty cost and help reduce it.
   b. Optimize the length of warranty.
   c. Reduce inventory costs.
   d. Develop maintenance procedures, both corrective and preventive.
   e. Provide input to the spare parts allocation models.
   f. Participate in the collection and analysis of the field data.
   g. Participate in the feedback process to report and correct the field failures.

13.8.1 Reliability Data-Collection Systems

The ultimate testing of a product occurs in the field under varied customer usages and environments. In order to effectively and knowledgeably guide future design efforts, field information must be accurately collected, summarized, and disseminated to the design function. This information can guide future design approaches, material selection, component or vendor selection, and other activities in the evolution of future products.

Rather than attempt to obtain data from an inaccurate warranty cost-reporting system, it is better to effectively and accurately track a small, representative sample of products. Usually some form of stratified random sampling[3] is needed to include different environmental extremes and different customer usage modes. Each failure should be analyzed and categorized by a qualified technician. For consistency in reporting, a failure-modes dictionary should be developed. This dictionary should have descriptions of the various modes of failure to be used in the reporting procedures. The purpose of the dictionary is to promote consistent reporting among the technicians who analyze the failures.

A computerized data-base-management system (DBMS) must be set up to manage the data. Routine report-generating formats with easy to understand descriptors of reliability should be preestablished. Ultimately, a design engineer should be able to query the DBMS from a terminal to review past product performance.

REFERENCES

1. Barlow, R. E., and F. Proschan: "Statistical Theory of Reliability and Life Testing," Holt, Rinehart and Winston, New York, 1975.
2. Carrubba, E. R., R. D. Gordon, and A. C. Spann: "Assuring Product Integrity," Lexington Books, Lexington, Mass., 1975.
3. Cocharn, W. G.: "Sampling Techniques," 2d ed., John Wiley & Sons, Inc., New York, 1963.
4. Halpern, S.: "The Assurance Sciences," Prentice-Hall, Inc., Englewood Cliffs, N.J., 1978.
5. Haugen, E. G.: "Probabilistic Approach to Design," John Wiley & Sons, Inc., New York, 1968.
6. Henley, E. J., and H. Kumamoto: "Reliability Engineering and Risk Assessment," Prentice-Hall, Inc., Englewood Cliffs, N.J., 1981.
7. Juran, J. M., and F. M. Gryna, Jr.: "Quality Planning and Analysis," 2d ed., McGraw-Hill Book Company, Inc., New York, 1980.
8. Kapur, K. C., and L. R. Lamberson: "Reliability in Engineering Design," John Wiley & Sons, Inc., New York, 1977.
9. Kapur, K. C.: "Reliability and Maintainability," in "Industrial Engineering Handbook," G. Salvendi, ed., John Wiley & Sons, Inc., New York, 1982.
10. Kececioglu, D., and D. Cormier: "Designing a Specified Reliability Directly into a Component," *Proc. 3rd Annu. Aerospace Reliability and Maintainability Conference*, pp. 520–530, 1968.
11. Mann, N. R., R. E. Schafer, and N. D. Singpurwalla: "Methods for Statistical Analysis of Reliability and Life Data," John Wiley & Sons, Inc., New York, 1974.
12. MIL-HDBK-217E: U.S. Department of Defense, "Military Standardization Handbook," Reliability Prediction of Electronic Equipment, April 9, 1979.
13. MIL-STD-1556A: U.S. Department of Defense, Government-Industry Data Exchange Program (GIDEP), U.S. Air Force, February 29, 1976.
14. MIL-STD-1629A: U.S. Department of Defense, Military Standard, "Procedures for Performing a Failure Mode, Effects and Criticality Analysis," November 24, 1980.
15. NPRD-Z: "Nonelectronic Parts Reliability Data," Reliability Analysis Center, Rome Air Development Center, Griffiss Air Force Base, N.Y., 1981.
16. "Quality Assurance-Reliability Handbook," AMC Pamphlet No. 702-3, U.S. Army Materiel Command, Va., October 1968.
17. RDG-376: "Reliability Design Handbook," Reliability Analysis Center, Rome Air Development Center, Griffiss Air Force Base, N.Y., March 1976.

18. Weibull, W.: "A Statistical Distribution Function of Wide Applicability," *J. Appl. Mech.*, pp. 293–296, 1951.

19. Wingo, D. R.: "Solution of the Three-Parameter Weibull Equations by Constrained Modified Quasilinearization (Progressively Censored Samples)," *IEEE Trans. Rel.*, vol. R-22, pp. 96–102, 1973.

P • A • R • T • 3

MECHANICAL SUBSYSTEM COMPONENTS

CHAPTER 14
CAM MECHANISMS

Harold A. Rothbart, D.Eng.

Dean Emeritus
College of Science and Engineering
Fairleigh Dickinson University
Teaneck, N.J.

NOTATIONS AND DEFINITIONS 14.3
14.1 INTRODUCTION 14.4
14.2 PRESSURE ANGLE 14.5
14.3 CURVATURE 14.6
14.4 COMPLEX NOTATION 14.6
14.5 BASIC DWELL-RISE-DWELL CURVES 14.7
14.6 CAM LAYOUT 14.8
14.7 ADVANCED CURVES 14.8
14.7.1 Polynomial and Trigonometric 14.12
14.7.2 Finite Differences 14.12
14.7.3 Blending Portions of Curves 14.13
14.8 CAM DYNAMICS 14.15
14.8.1 Introduction 14.15
14.8.2 Lumped-Model Rigid Camshaft System 14.15
14.8.3 Lumped-Model Flexible Camshaft System: Spring Windup 14.18
14.9 CAM PROFILE ACCURACY 14.19
14.10 CAM LOADS 14.20

NOTATIONS AND DEFINITIONS

A = cross-sectional area, in^2
A = acceleration, in/s^2
a, b = distance, in
B = follower bearing length, in
C = constant
c = damping constant, lb·s/in
E = modulus of elasticity, lb/in^2
f = external forces, lb
f = friction force, lb
F = force normal to cam profile, lb
F_{st} = static friction
g = gravitational constant
h = maximum displacement of follower, in
I = moment of inertia, $lb/(in \cdot s^2)$
k = spring constant, lb/in
K = constant
l = length of link, in
L = length of contact between cam and roller followers, in
m = equivalent mass of follower, $lb \cdot s^2/in$
n = integer
N_1, N_2 = forces normal to translating follower stem, lb
r = a number
r = radius from cam center to center of curvature, in
r_c = radius from cam center to center of roller follower, in
R_a = radius of prime circle (smallest circle to the roller center), in
R_p = radius of pitch circle (circle drawn to roller center having maximum pressure angle), in
R_r = radius of follower roller, in
t = time for cam to rotate angle θ, s
T = time for cam to rotate angle θ_0, s

T = torques lb in
T_p = period
u = follower overhang, in
x = displacement, in
$\dot{x}$ = velocity, in/s
$\ddot{x}$ = acceleration in/s^2
W = safe endurance load, lb
y = cam function (follower) displacement, in
$\dot{y}$ = cam function (follower) velocity, in/s
$\ddot{y}$ = cam function (follower) acceleration, in/s^2
α, β, γ = angles, rad
ψ = pressure angle, rad
ψ_n = maximum pressure angle, rad
$\theta_0 = \omega_t$ − cam angle rotation for rise h, rad
$\theta = \omega_t$ − cam angle rotation for displacement y, rad
μ = coefficient of friction
ρ = radius of curvature of pitch curve (path of roller center), in
ρ_c = radius of curvature of cam surface at point of maximum stress (positive for concave and negative for convex), in
τ = angle between radius of curvature and follower motion, rad
ω = cam angular velocity, rad/s
ω_n = system fundamental frequency, rad/s

14.1 INTRODUCTION[28,29,45,48]

A "cam" is a mechanical member for transmitting a desired motion to a follower by direct contact. The driver is called a cam and the driven member is called the "follower." Either may remain stationary, translate, oscillate, or rotate. The motion is given by $y = f(\theta)$.

Kinematically speaking, in its general form the plane cam mechanism (Fig. 14.1) consists of two shaped members A and B connected by a fixed third body C. Either body A or body B may be the driver with the other the follower. We may at each instant replace these shaped bodies by an equivalent mechanism. These are pin-jointed at the instantaneous centers of curvature, 1 and 2, of the contacting surfaces. In general, at any other instant the points 1 and 2 are shifted and the links of the equivalent mechanism have different lengths. Figure 14.2 shows the two most popular cams.

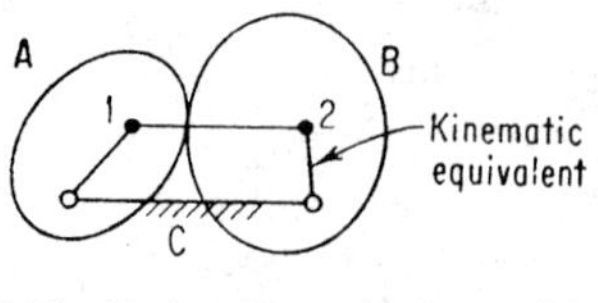

FIG. 14.1 Basic cam mechanism and its kinematic equivalent (points 1 and 2 are centers of curvature) of the contact point. (*Courtesy of John Wiley & Sons, Inc.*[1])

Cam synthesis may be accomplished by (1) shaping the cam body to some known curve, such as involute, spirals, parabolas, or circular arcs; (2) mathematically controlling and establishing the follower motion and forming the cam by tabulated data of this action; (3) establishing the cam contour in parametric form; and (4) layout of cam profile by eye. This last method is acceptable only for low speeds in which a smooth "bumpless" curve may fulfill the requirements. But as either loads, mass, speed, or elasticity of the members increases, a detailed study must be made of both the dynamic aspects of the cam curve and the accuracy of cam fabrication. References 2–5, 16, 27, 31, 32, 40, 41, 45, 47, 48, and 54 discuss the exact synthesis of cam-follower design kinematics. Reference 20 shows cam-computing mechanisms of all sorts.

The roller follower is most frequently employed to distribute and reduce wear between the cam and follower. Obviously the cam and follower must be constrained

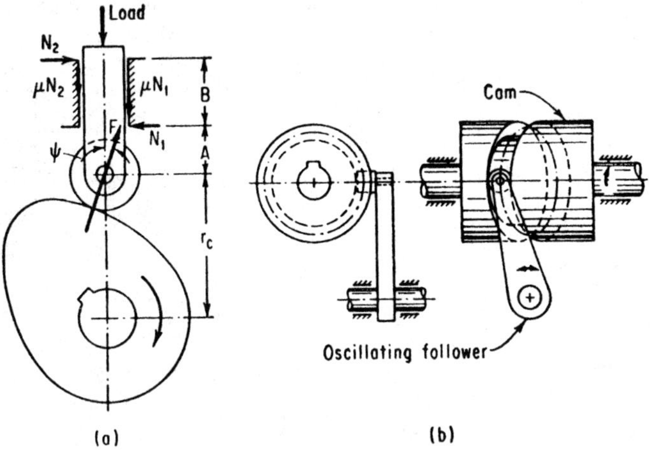

FIG. 14.2 Popular cams. (*a*) Radial cam—translating roller follower (open cam). (*b*) Cylindrical cam—oscillating roller follower (closed cam). (*Courtesy of John Wiley & Sons, Inc.*[1])

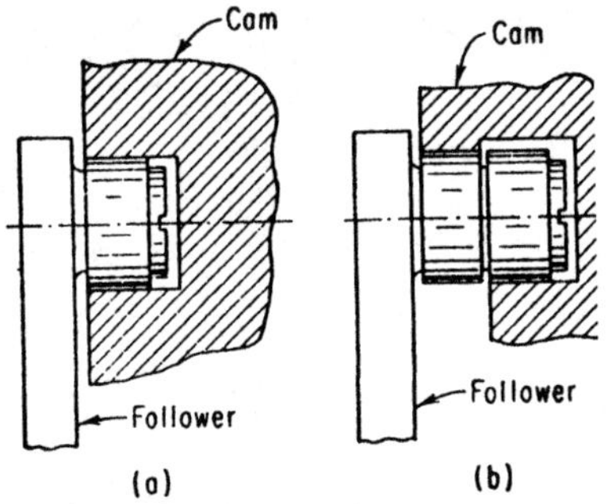

FIG. 14.3 Some roller and groove designs. (*a*) Single roller. (*b*) Double roller, undercut groove. (*Courtesy of John Wiley & Sons, Inc.*[1])

at all speeds. A preloaded compression spring (with an open cam) or a positive drive is used. Positive-drive action is accomplished by either a cam groove or a conjugate follower or followers in contact with opposite sides of a single or double cam. Figure 14.3*a* shows a single roller in a groove. Backlash and sliding become serious at high speed and may necessitate the double roller in a relieved groove (Fig. 14.3*b*). Tapered rollers have been used.

14.2 PRESSURE ANGLE[1,6]

The pressure angle is the angle between the normal to the cam profile and the instantaneous direction of the follower. It is related to the follower force distribution. With flat-faced followers, the pressure-angle force distribution rarely is of concern. With oscillating roller followers, the distance from the cam center to the follower pivot center and the length of the follower arm are pertinent to this study. A radial translating roller follower having rigid members has a theoretical limiting pressure angle based on the forces shown in Fig. 14.2*a*.

$$\psi_m < \arctan\,[B/\mu(2A + B)] \qquad (14.1)$$

To be practical, the pressure angle is generally limited to 30° for the translating roller follower because of the elasticity of the follower stem and the clearance between the stem in its guide bearing and the friction. We can show that the pressure angle $\tan \psi = \dot{y}/r_c\omega$. Symmetrical cam curves have the point of maximum pressure angle approximately at the midpoint of the rise

$$\tan \psi_m = \dot{y}_m/R_p\omega \tag{14.2}$$

14.3 CURVATURE[1,7]

The cam curvature is pertinent in establishing the cam-follower surface stresses, life, and undercutting. Undercutting is a phenomenon in which inadequate cam curvature yields incorrect follower movement. For a translating follower the curvature of pitch curve (path of roller center) at any point

$$\rho = \frac{[(R_a + y)^2 + (\dot{y}/\omega)^2]^{3/2}}{(R_a + y)^2 + 2(\dot{y}/\omega)^2 - (R_a + y)(\ddot{y}/\omega^2)} \tag{14.3}$$

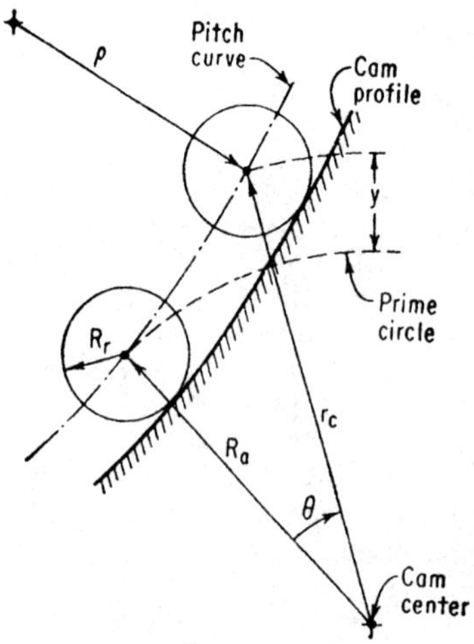

FIG. 14.4 Curvature cam profile terminology.

If ρ is positive, we have a convex cam profile, and if ρ is negative, we have a concave cam profile (Fig. 14.4). For a convex curved cam to prevent undercutting, which generally occurs in the point of maximum negative acceleration, $\rho > R_r$. In all cams the minimum radius of curvature should be determined to establish that the cam-follower system does not have excessive stresses or undercutting.

14.4 COMPLEX NOTATION[17,19]

Complex numbers can be employed to obtain the curvature, profile, cutter location, etc., for any kind of cam and follower. Use of a high-speed computer will aid in its use. Let us take the disk cam with an oscillating roller follower as an example (Fig. 14.5). The angular displacement of the follower from initial position β_0

$$\beta = \beta_0 + f(\theta)$$

The relations to the roller or cutter center

$$\mathbf{r}_c = r_c e^{i\theta_c} = re^{i\alpha} + \rho e^{i\varphi} = a + ib + le^{i\beta} \tag{14.4}$$

The equation of the cam contour measured from the initial position

$$r_c e^{i(\theta+\theta_c)} = (a + ib + le^{i\beta})e^{i\theta} \tag{14.5}$$

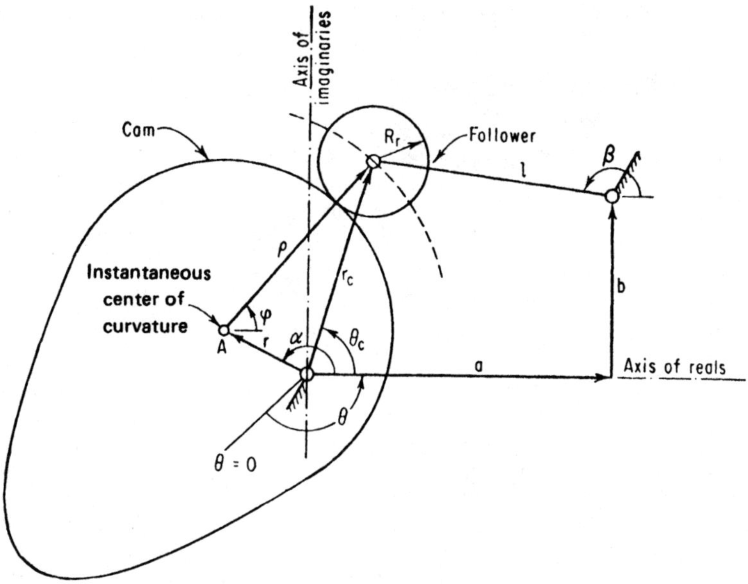

FIG. 14.5 Complex notation for oscillating roller follower.

Solving for r_c and θ_c yields the information necessary to construct the cam profile. Separating Eq. (14.4) into real and imaginary parts

$$r \cos \alpha + \rho \cos \varphi = a + l \cos \beta \tag{14.6}$$

$$r \sin \alpha + \rho \sin \varphi = b + l \sin \beta \tag{14.7}$$

Differentiating, we obtain the radius of curvature

$$\rho = \frac{(E^2 + D^2)^{3/2}}{(E^2 + D^2)[1 + f'(\theta)] - (aE + bD)f'(\theta) + (a \sin \beta - b \cos \beta)lf''(\theta)} \tag{14.8}$$

where

$$E = [1 + f'(\theta)]l \cos \beta + a \tag{14.9}$$

$$D = [1 + f'(\theta)]l \sin \beta + b \tag{14.10}$$

to avoid undercutting $\rho > R_r$.

14.5 BASIC DWELL-RISE-DWELL CURVES

The most popular method in establishing the cam shape is by first choosing a simple curve which is easy to construct and analyze. Conventionally such curves are of two families: the simple polynomial, i.e., parabolic and cubic, and the trigonometric, i.e., simple harmonic, cycloidal, etc. The polynomial family

$$y = C_0 + C_1\theta + C_2\theta^2 + C_3\theta^3 + \cdots + C_n\theta^n \tag{14.11}$$

of which the basic curve is generally with a low integer

$$y = C_n\theta^n \tag{14.12}$$

The trigonometric family

$$y = C_0 + C_1 \sin \omega t + C_2 \cos \omega t + C_3 \sin^2 \omega t + \cdots \tag{14.13}$$

is reduced to the basic curves using only the lowest-order terms. The trigonometric are better since they yield smaller cams, lower follower side thrust, cheaper manufacturing costs, and easier layout and duplication. A summary of equations and curves is shown in Fig. 14.6 and Tables 14.1 and 14.2.

The parabolic curve is constructed (Fig. 14.6*a*) by dividing any line *OB* into odd incremental spacings, i.e., 1, 3, 5, 5, 3, 1, the same number as the abscissa spacing. This line is projected to rise *h* and then to spacings. The simple harmonic curve is constructed by dividing the *h* diameter into equal parts and projecting.

14.6 CAM LAYOUT

The fundamental basis for cam layout is one of inversion in which the cam profile is developed by fixing the cam moving the follower to its respective relative positions.

EXAMPLE A radial cam rotating at 180 rpm is driving a ¾-in-diameter translating roller follower. Construct the cam profile with the pressure angle limited to 20° on the rise. The motion is as follows: (1) rise of 1 in with simple harmonic motion in 150° of cam rotation, (2) dwell for 60°, (3) fall of 1 in with simple harmonic motion in 120° of cam rotation, (4) dwell for the remaining 30°.

solution The maximum pressure angle for the rise action occurs at the transition point where $\theta = 75°$. From Eq. (14.8) we find $R_p = 1.64$ in which 1¾ in is chosen.

The simple harmonic displacement diagram is constructed in Fig. 14.7, and the cam angle of 150° is divided into equal parts, i.e., 6 parts at 25° on either side of the line of action *A*3″. This gives radial lines *A*0″, *A*1″, *A*2″, etc. From the cam center *A*, arcs are swung from points 0′, 1′, 2′, 3′, etc., intersecting respective radial lines, locating points 0, 1, 2, 3, etc. Next we draw the smooth pitch curve, roller circles, and cam profile tangent to the rollers. We continue in the same manner to complete the cam. Computer-aided design of cam profile is shown in Refs. 38 and 50.

14.7 ADVANCED CURVES[15]

Special curve shapes are obtained by (1) employing polynomial or trigonometric relations previously mentioned, (2) numerical procedure by the method of finite differences,

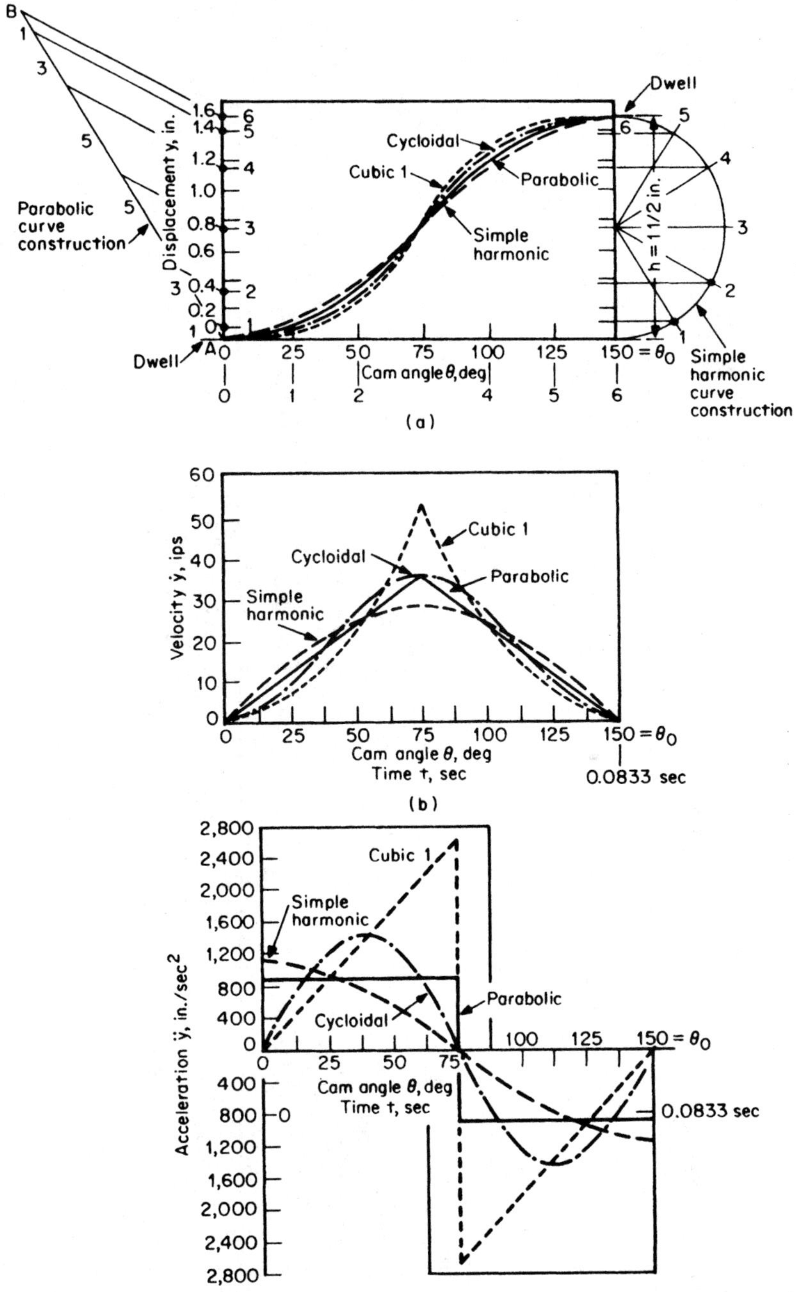

FIG. 14.6 Comparison of basic curves—dwell-rise-dwell cam. (Follower has 1½-in rise in 150° of cam rotation at 300 r/min.) (*a*) Displacement. (*b*) Velocity. (*c*) Acceleration. (*Courtesy of John Wiley & Sons, Inc.*[1])

TABLE 14.1 Characteristic Equations of Basic Curves*

Curves		Displacement y, in.	Velocity $\dot{y}$, ips	Acceleration $\ddot{y}$, in./sec²
Straight line		$h\theta/\theta_0$	$\omega h/\theta_0$	0
Simple harmonic motion (SHM)		$h/2[1 - \cos(\pi\theta/\theta_0)]$	$(h\pi\omega/2\theta_0)\sin(\pi\theta/\theta_0)$	$(h/2)(\pi\omega/\theta_0)^2\cos(\pi\theta/\theta_0)$
Cycloidal		$h/\pi[\pi\theta/\theta_0 - \frac{1}{2}\sin(2\pi\theta/\theta_0)]$	$h\omega/\theta_0[1 - \cos(2\pi\theta/\theta_0)]$	$2h\pi\omega^2/\theta_0^2\sin(2\pi\theta/\theta_0)$
Parabolic or constant acceleration	$\theta/\theta_0 \lesseqgtr 0.5$	$2h(\theta/\theta_0)^2$	$4h\omega\theta/\theta_0^2$	$4h(\omega^2/\theta_0^2)$
	$\theta/\theta_0 \gtreqless 0.5$	$h[1 - 2(1 - \theta/\theta_0)^2]$	$(4h\omega/\theta_0)(1 - \theta/\theta_0)$	$-4h(\omega^2/\theta_0^2)$
Cubic 1 or constant pulse 1	$\theta/\theta_0 \lesseqgtr 0.5$	$4h(\theta/\theta_0)^3$	$(12h\omega/\theta_0)(\theta/\theta_0)^2$	$(24h\omega^2/\theta_0^2)(\theta/\theta_0)$
	$\theta/\theta_0 \gtreqless 0.5$	$h[1 - 4(1 - \theta/\theta_0)^3]$	$(12h\omega/\theta_0)(1 - \theta/\theta_0)^2$	$-(24h\omega^2/\theta_0^2)(1 - \theta/\theta_0)$

where h = maximum rise of follower, in
θ_0 = cam angle of rotation to give rise h, rad
ω = cam angular velocity, rad/s
θ = cam-angle rotation for follower displacement y, rad

*Courtesy of John Wiley & Sons, Inc.[1]

TABLE 14.2 Comparison of Symmetrical Curves*

		Displacement y	Velocity	Acceleration	Pulse ÿ	Comments
Polynomial	Parabolic		2, $\theta_0=1$	4, 4, $\theta_0=1$	∞, ∞, $\theta_0=1$	Backlash serious
	Cubic No. 1		3	12, $\theta_0=1$, 12	∞, 24	Not suggested
	Cubic No. 2		1½	6.4, 6.4	∞, ∞, 12	Not suggested for high speeds
	2-3 polynomial		1½	6, 6	∞, ∞, 8	
	3-4-5 polynomial	Dwell, Rise, h=1 total rise, Dwell, ⊢$\theta_0=1$⊣, Total angle	1.9	5.8, 5.8	60, 30	Similar to the cycloidal
	4-5-6-7 polynomial		2.4	7.3, 7.3	42, 52	Doubtful high-speed improvement over the 3-4-5 polynomial
Trigonometric	Simple harmonic motion		1.6	4.9, 4.9	∞, ∞, 15.5	
	Cycloidal		2	6.3, 6.3	40, 40	Excellent for high speeds
	Double harmonic		2	5.5, 9.9	∞, 20, 40	Best as a dwell-rise-return-dwell cam
Combination	Trapezoidal acceleration		2	5.3, 5.3, $\beta/8$	44, 44	Excellent for high speeds; may have machining difficulty
	Modified trapezoidal acceleration		2	4.9, 4.9, $\beta/8$	61, 61	Excellent vibratory and easier fabrication may prove better than cycloidal

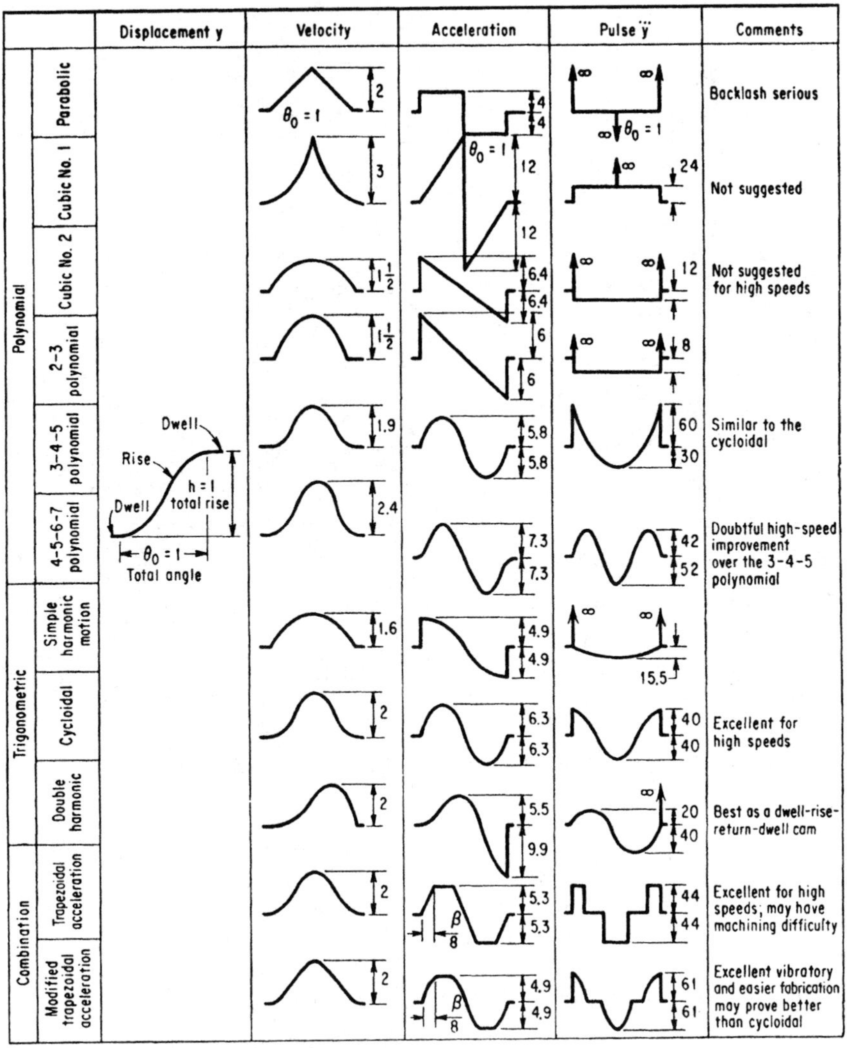

*Courtesy of John Wiley & Sons, Inc.[1]

(3) blending portions of curves, and (4) trial and error employment of a high-speed computer starting with the fourth-derivative (d^4y/dt^4) curve.

Rise or fall portions of any curve are each composed of a positive acceleration period and a negative acceleration period. For any smooth curve, we know that $\int_0^{\theta/\omega} \ddot{y}\, dt = 0$. Therefore, for any rise or fall curve, the area under the positive-acceleration portion equals the area under the negative-acceleration portion for each curve. Reference 37 shows a method for controlling the harmonics of a cam profile based on a minimum squared error performance index over a prescribed speed range.

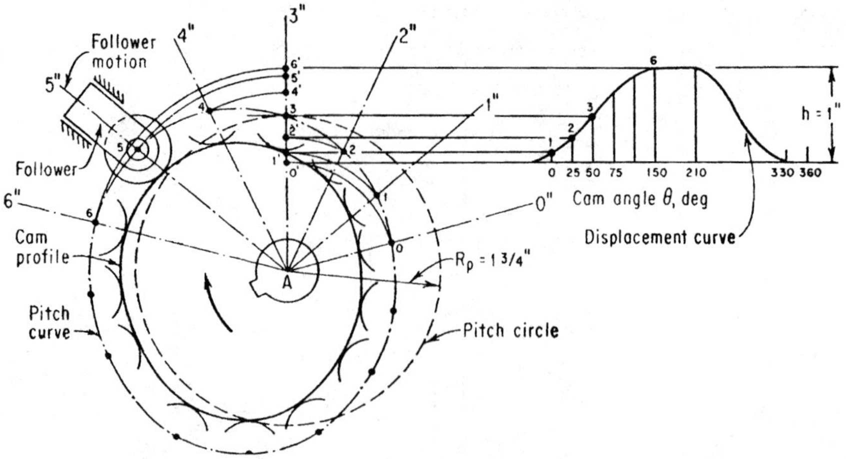

FIG. 14.7 Radial cam—translating roller. Follower example (scale ⅝ in = 1 in). (*Courtesy of John Wiley & Sons, Inc.*[1])

14.7.1 Polynomial and Trigonometric

Equations (14.11) and (14.13) show the general expressions. The symmetrical dwell-rise-dwell cam curve can be developed from a Fourier series with only odd multiples of the fundamental frequency[14]

$$\ddot{y} = \sum_{k\ odd}^{n} C_k \sin \omega_k t = C_1 \sin \omega_1 t + C_3 \sin \omega_3 t + C_5 \sin \omega_5 t + \cdots \tag{14.14}$$

where $\omega_k = 2\pi k/T$ and C_r's are coefficients.

If $k = 1$, it yields the well-known cycloidal curve; if $k = \infty$, we obtain the parabolic curve.

For $n > 1$ we can progressively flatten curves for any shape desired.

14.7.2 Finite Differences[1,8,38,39]

In the application of finite differences the displacement y is known for pivotal points equally spaced by Δt. The derivation of formulas by Taylor-series expansion of $y(t + \Delta t)$ about t is

$$y(t + \Delta t) = y(t) + \Delta t\, \dot{y}(t) + [(\Delta t)^2/2!]\ddot{y}(t) + [(\Delta t)^3/3!]\dddot{y}(t) + \cdots \tag{14.15}$$

For simplicity we take a three-point approximation which consists of passing a parabola through the three pivotal points. At any interval i

$$y_{i+1} \simeq y_i + \Delta t\, \dot{y}_i + (\Delta t)^2/2!\ddot{y}_i$$

$$y_{i-1} \simeq y_i - \Delta t\, \dot{y}_i + (\Delta t)^2/2!\ddot{y}_i$$

Solving for the velocity and acceleration, respectively,

$$\dot{y}_i \simeq (y_{i+1} - y_{i-1})/2\Delta t \tag{14.16}$$

$$\ddot{y}_i \simeq (y_{i+1} + y_{i-1} - 2y_i)/(\Delta t)^2 \tag{14.17}$$

Equations (14.16) and (14.17) can be used to determine the numerical relationships between y, $\dot{y}$, $\ddot{y}$ characteristics curves for incremental steps to include the cam accuracy if desired.

14.7.3 Blending Portions of Curves[1,33,34]

The primary condition for these combination curves is that (1) there must be no discontinuities in the acceleration curve and (2) the acceleration curve has a smooth shape. Therefore, weaker harmonics result. Table 14.2 gives a comparison of all the basic curves and some combination curves.

As an illustration of an advanced curve choice, automobile and aircraft designers are employing an unsymmetrical cam acceleration curve in which the maximum positive acceleration is much greater than the maximum negative acceleration (Fig. 14.8). This is a high-speed cam giving minimum spring size, maximum cam radius of curvature, and longer surface life.

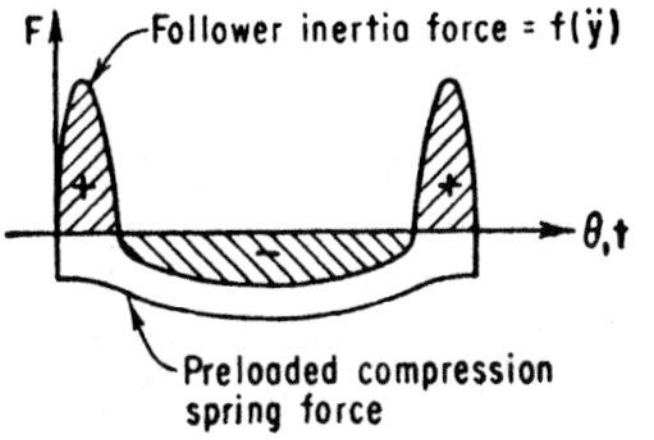

FIG. 14.8 Typical high-speed cam curve used for automotive or aircraft engine valve system.

EXAMPLE Derive the relationship for the excellent dwell-rise-fall-dwell cam curve shown in Fig. 14.9 having equal maximum acceleration values. Portions I and III are harmonic; portions II and IV are horizontal straight lines.

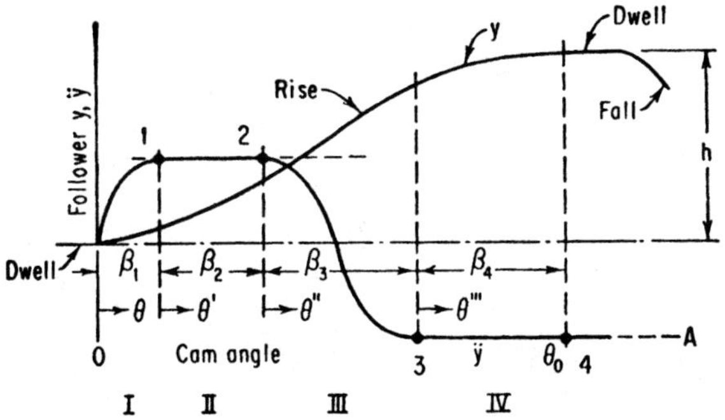

FIG. 14.9 Example of symmetrical dwell-rise-fall-dwell cam curve.

solution Let the θ's and β's be the angles for each portion shown. Note for velocity and acceleration one should multiply values by ω and ω^2, respectively, and the boundary conditions are $y(0) = 0$, $\dot{y}(0) = 0$, and $\dot{y}(\beta_4) = 0$ and $y(\beta_4) =$ total rise h.

Portion I

$$\ddot{y} = A \sin(\pi\theta/2\beta_1)$$

$$\dot{y} = (2\beta_1 A/\pi)[1 - \cos(\pi\theta/2\beta_1)]$$

$$y = -A(2\beta_1/\pi)^2 \sin(\pi\theta/2\beta_1) + (2A\beta_1/\pi)\theta$$

$$\dot{y}_1 = 2A\beta_1/\pi$$

$$y_1 = (2A\beta_1^2/\pi)(1 - 2/\pi)$$

Portion II

$$\ddot{y} = \dot{A}$$

$$\dot{y} = A\theta' + \dot{y}_1$$

$$y = A(\theta')^2/2 + \dot{y}_1\theta' + y_1$$

$$\dot{y}_2 = A\beta_2 + 2A\beta_1/\pi$$

$$y_2 = A\beta_2^2/2 + 2A\beta_1\beta_2/\pi + 2A\beta_1^2/\pi(1 - 2/\pi)$$

Portion III

$$\ddot{y} = A \cos(\pi\theta''/\beta_3)$$

$$\dot{y} = A\beta_3/\pi \sin(\pi\theta''/\beta_3) + \dot{y}_2$$

$$y = -A(\beta_3/\pi)^2[1 - \cos(\pi\theta''/\beta_3)] + \dot{y}_2\theta'' + y_2$$

$$\dot{y}_3 = \dot{y}_2$$

$$y_3 = 2A(\beta_3/\pi)^2 + A\beta_2\beta_3 + 2A\beta_1\beta_3/\pi + A\beta_2{}^2/2 + 2A\beta_1\beta_2/\pi + (2A\beta_1{}^2/\pi)(1 - 2/\pi)$$

Portion IV

$$\ddot{y} = -A$$

$$\dot{y} = -A\theta''' + \dot{y}_3$$

$$y = -[A(\theta''')^2/2] + \dot{y}_3\theta''' + y_3$$

$$\dot{y}_4 = -A\beta_4 + \dot{y}_3 = -A\beta_4 + A\beta_2 + 2A\beta_1/\pi = 0$$

$$\beta_4 = \beta_2 + 2\beta_1/\pi$$

Total Rise

$$h = y_4 = -(A\beta_4^2/2) + (A\beta_2 + 2A\beta_1/\pi)\beta_4 + y_3$$

Substituting

$$A = \frac{h}{[-\beta_4^2/2 + \beta_2\beta_4 + 2\beta_1\beta_4/\pi + (\beta_3/\pi)^2 + \beta_2\beta_3 + 2\beta_1\beta_3/\pi + \beta_2^2/2 + 2\beta_1\beta_2/\pi + (2\beta_1^2/\pi)(1 - 2/\pi)]}$$

For a given total rise h for angle θ_0 and any two of the β angles given one can solve for all angles and all values of the derivative curves.

14.8 CAM DYNAMICS

14.8.1 Introduction

Classical cam design and analysis, which treats the cam mechanism as a single-degree-of-freedom rigid body, is in general inadequate to describe the dynamics of cam mechanisms which are inherently compliant. A measure of system compliance is the extent of the deviation between the dynamic response and the intended kinematic response, which is the vibration level.

This deviation tends to increase (1) as any of the input harmonics of cam motion approaches the fundamental frequency of the mechanism, (2) with the increase of the maximum value of the cam function third-time derivative (jerk), and (3) as the cam surface irregularities become more severe. The deleterious effects of vibration are well known. Vibration causes motion perturbations, excessive component stresses, noise, chatter, wear, and cam surface erosion. This wear and cam surface erosion feeds back to further exacerbate the problem. References 1, 9 to 14, 18, 26, 29, 30, 49, and 51 to 54 give more on the subject.

Models. Modeling[21–23] transforms the system into a set of tenable mathematical equations which describe the system in sufficient detail for the accuracy required. Various tools of analysis are at the disposal of the designer-analyst, ranging in complexity from classical linear analysis for single-degree-of-freedom systems to complex multiple-degree-of-freedom nonlinear computer programs.

14.8.2 Lumped-Model Rigid Camshaft System

The cam mechanism consists of a camshaft, cam, follower train including one or more connecting links, and springs terminating in a load mass or force. For lumped systems, links or springs are divided into two or more ideal mass points interconnected with weightless springs and dampers. Ball bearings are modeled as point masses and/or springs due to radial or axial compliance of the balls and axial spring preloading.

Springs. The spring force generally follows the law

$$F = kx \pm \sum_{n=1} a_n x^{2n+1}$$

where terms under the Σ sign are nonlinear. Rarely does one go beyond the cubic term a_1x^3 when defining a nonlinear spring.

Dampers. Damping take the general form

$$C(\dot{x}) = C\dot{x}|\dot{x}|^{r-1}$$

$$\text{For } r = \begin{cases} 0 & C(\dot{x}) \text{ is coulomb damping} \\ 1 & C(\dot{x}) \text{ is viscous damping} \\ 2 & C(\dot{x}) \text{ is quadratic damping} \end{cases}$$

"Stiction" is accounted for as the breakaway force, $\mathcal{F}$, in one element i sliding along another element j.

$$\begin{aligned} F_{st} &= 0 && \text{for } \dot{x}_i \neq \dot{x}_j \\ &\leq \mathcal{F} && \text{for } \dot{x}_i = \dot{x}_j \\ &= \mathcal{F} && \text{at breakaway} \end{aligned}$$

Damping in cam-follower systems ranges from 5 to 15 percent of critical damping.

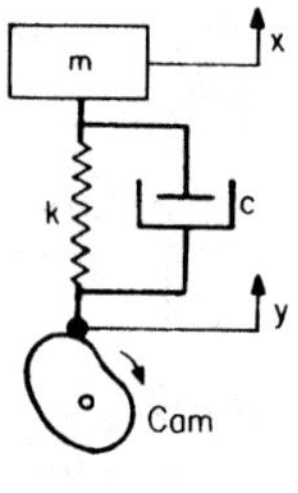

FIG. 14.10 Single-degree-of-freedom system.

Single-Degree-of-Freedom System. In its simplest form, the model is the one-degree-of-freedom system which lumps the follower train and mass load into a single equivalent mass with equivalent springs and dampers connected to the cam, as shown in Fig. 14.10.

$$m\ddot{x} + c(\dot{x} - \dot{y}) + k(x - y) = 0 \tag{14.18}$$

The vibration form of Eq. (14.18) is

$$m(\ddot{x} - \ddot{y}) + c(\dot{x} - \dot{y}) + k(x - y) = -m\ddot{y} \tag{14.18a}$$

Substituting $z = x - y$ yields

$$m\ddot{z} + c\dot{z} + kz = m\ddot{y} \tag{14.19}$$

which places in evidence the prominent role of cam function acceleration y on the vibrating system, similar to a system response to a shock input at its foundation.

The general solution to Eq. (14.19) consists of the complementary or transient solution plus the particular solution. The complementary solution is the solution to the homogeneous equation

$$m\ddot{z} + c\dot{z} + kz = 0$$

$$z + 2\zeta\omega_n z + \omega_n^2 = 0$$

$$\text{where } \omega_n = (k/m)^{1/2}$$

$$\zeta = \tfrac{1}{2} c(1/km)^{1/2}$$

and is

$$z_c = Ae^{-\zeta\omega_n t} \sin[(1 - \zeta^2)^{1/2}\omega_n]t + Be^{-\zeta\omega_n t} \cos[(1 - \zeta^2)^{1/2}\omega_n]t$$

where A and B are determined in conjunction with the particular solution and the initial conditions. One form of the particular solution is derivable from the impulse response.

$$z_p = \frac{1}{(1 - \zeta^2)^{1/2}\omega_n} \int_0^t \ddot{y}\, e^{-\zeta\omega_0(t - \tau)} \sin[(1 - \zeta^2)^{1/2}\omega_n](t - \tau)d\tau$$

For finite $\ddot{y}$, the case here, $z_p(0) = \dot{z}_p(0) = 0$, so that the general solution to Eq. (14.19) is

$$z = z_p + \left\{ \left[\frac{\dot{z}(0) + z(0)\zeta\omega_n}{(1 - \zeta^2)^{1/2}\omega_n} \right] \sin [(1 - \zeta^2)^{1/2}\omega_n]t + z(0) \cos [(1 - \zeta^2)^{1/2}\omega_n]t \right\} e^{-\zeta\omega_n t} \qquad (14.20)$$

where the initial conditions $z(0)$ and $\dot{z}(0)$ are subsumed in z_c. Accordingly, for a system initially at rest, the z_p solution becomes the complete solution. However, the difficulty with this solution is that $\ddot{y}$ is periodic and integration would have to be performed over many cycles depending on the amount of damping present. Solutions would be analyzed first to determine the input "transient" peak response. Finally the "steady-state" response would be reached when

$$z(nt) = z(n + 1)T \qquad \text{and} \qquad \dot{z}(nt) = \dot{z}(n + 1)T$$

Typical transient-response curves for some one-degree-of-freedom systems are shown in Fig. 14.11, where it is noted that discontinuities in the acceleration cause

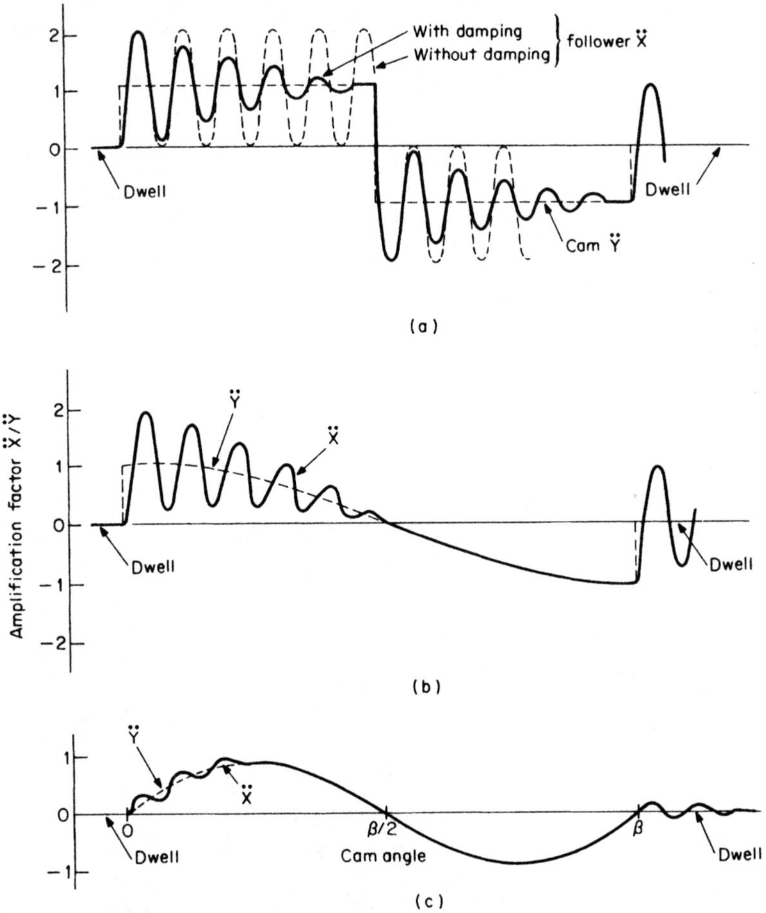

FIG. 14.11 Single-degree-of-freedom response of basic dwell-rise-dwell cam acceleration curves (*a*) Parabolic. (*b*) Simple harmonic. (*c*) Cycloidal.[1]

large amplification factors. For improved accuracy, the model may be developed in more detail, leading to the more general multiple-degree-of-freedom system. Reference 25 discusses matrix methods in modeling.

14.8.3 Lumped-Model Flexible Camshaft System: Spring Windup[24]

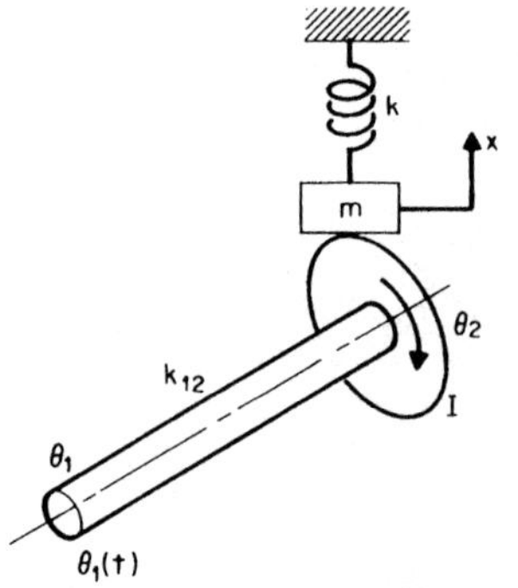

FIG. 14.12 Flexible camshaft system.

The system takes account of camshaft flexibility and leads to a set of nonlinear equations owing to the fact that the driving-cam function is no longer a known function of time.

As an example, consider the case shown in Fig. 14.12, where $\theta_1(t)$, the input shaft rotating at a known angular rate, is not equal to the cam angle θ_2, or $\theta_1 \neq \theta_2$. This is due to camshaft flexibility, k_{12} is the camshaft spring rate and the cam-follower rise is x. A convenient method of formulation is derivable from Lagrange's equations (see Chap. 4) as follows:

$$T = \tfrac{1}{2} M\dot{x}^2 + \tfrac{1}{2} I\dot{\theta}_2^2$$

$$U = \tfrac{1}{2} k_{12}(\theta_2 - \theta_1)^2 + \tfrac{1}{2} Kx^2$$

$$\mathcal{L} = T - V$$

The follower is constrained by the cam function.

$$x = x(\theta_2)$$

where $\theta_1 = \theta_1(t)$, by the Lagrange multiplier method

$$\lambda_1\, dx - \lambda_1 x'\, d\theta_2 = 0;\ x' = dx(\theta_2)/d\theta_2$$

$$\lambda_2\, d\theta_1 - \lambda_2 \theta_1'\, dt = 0;\ \theta' = d\theta_1/dt$$

$$\left(\frac{d}{dt}\frac{\partial L}{\partial x} - \frac{\partial L}{\partial \dot{x}}\right) dx + \lambda_1 dx = 0 \rightarrow m\ddot{x} + kx + \lambda_1 = 0 \tag{14.24}$$

$$\left(\frac{d}{dt}\frac{\partial L}{\partial \dot{\theta}_2} - \frac{\partial L}{\partial \theta_2}\right) d\theta_2 - \lambda_1 x' d\theta_2 = 0 \rightarrow I\ddot{\theta}_2 + k_{12}(\theta_2 - \theta_1) - \lambda_1 x' = 0 \tag{14.25}$$

$$\left(\frac{d}{dt}\frac{\partial L}{\partial \dot{\theta}_1} - \frac{\partial L}{\partial \theta_1}\right) d\theta_1 + \lambda_2 = 0 \rightarrow k_{12}(\theta_1 - \theta_2) + \lambda_2 = 0 \tag{14.26}$$

where λ_1 = contact force on cam
$\lambda_1 x'$ = torque on cam due to contact force
λ_2 = torque on shaft

Elimination of λ_1 between Eqs. (14.24) and (14.25) and taking two time derivatives of $x(\theta_2)$ yields

$$m\ddot{x}x' + kxx' + I\ddot{\theta}_2 + k_{12}(\theta_2 - \theta_1) = 0$$

$$\ddot{x} = x''\dot{\theta}_2^2 + \ddot{\theta}_2 x'$$

whence

$$(I + mx'^2)\ddot{\theta}_2 + mx'x''\dot{\theta}_2^2 + kxx' + k_{12}\theta_2 = k_{12}\theta_1(t) \qquad (14.27)$$

characterized as the spring-windup equation with one degree of freedom.

14.9 CAM PROFILE ACCURACY[1,8,43]

A surface may appear smooth to the eye and yet have poor dynamic properties, since the determination of whether a cam is satisfactory is based on the acceleration curve. The acceleration being the second derivative of lift, the large magnification of error will appear on it. Two methods for establishing the accuracy of the cam fabrication and its dynamic effect are (1) employing the mathematical numerical method of finite differences, and (2) utilizing electronic instrumentation of an accelerometer pickup (located on the cam follower), and cathode-ray oscilloscope.

Reference 42 contains a review of cam manufacturing methods, including a treatment of accuracy. Reference 44 applies stochastic and probabilistic techniques to the study of error in cams. Reference 36 uses a FORTRAN program to compute values to minimize errors in cam fabrication.

Many types of errors were investigated on measured shapes. All cams employed the basic cycloidal curve. In Fig. 14.13 we see a cam investigated at 300 r/min. The

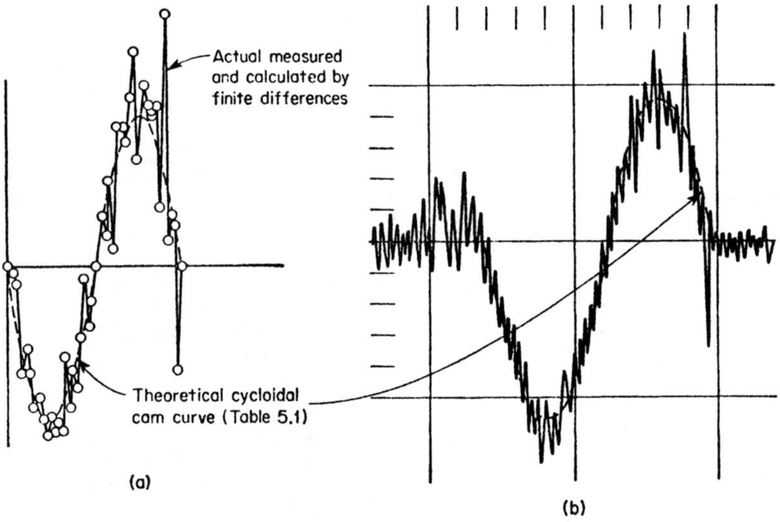

FIG. 14.13 Accuracy investigation of cam turning 300 r/min; in 45°, rise = ¾ in. (*a*) Finite-difference calculation. (*b*) Experimental with accelerometer and oscilloscope.

greatest error in fabrication exists at about the 40° cam angle Correlation is observed between the finite-difference and the experimental results. The theoretical maximum acceleration was calculated from the cycloidal formula to be 19.5*g* and the maximum acceleration of cam with errors is about 34*g*. On a broad average the maximum acceleration difference for the error only is 12*g* per 0.001-in error at a speed of 300 r/min. Figure 14.14 shows the dimension of a high-speed cam for a textile loom.

14.10 CAM LOADS[1,35]

Pertinent aspects of cam-follower design are the load and life of the cam and follower contacting surfaces having the parameters of surface roughness, waviness and stresses, prior history of machining, moduli of elasticity, friction (the amount of rolling and sliding), choice of materials, elastohydrodynamic lubrication, corrosion, heat treatment, materials in combination, the area and radii of contacting surfaces, the number of performance cycles, and the contaminants in the environment that may collect on the surfaces.

When the roller follower in contact with the cam is subjected to significant changes in rolling speed, skidding occurs. This skidding action may produce scuffing wear on the concave side of the cam surface at the point of maximum acceleration. Note that proper lubrication is necessary to prevent excessive wear, welding, and scoring of the materials, i.e., generally the cam surface.

Surface fatigue is the predominant cause of failure in cams. In Eq. (14.28) the usual choice of a hardened-steel roller follower is shown in combination with different cam materials with a percent sliding and 10^8 cycles. The test data based on hertzian equations and surface fatigue give the surface endurance load

$$w = \frac{KL}{1/R_r + 1/\rho_c} \tag{14.28}$$

where L is contact length, in inches.

Table 14.3 shows the values of K for various suggested cam materials in contact with a cam roller made of tool steel hardened to Rockwell C 60 to 62. The data shown should be modified depending on experience, the amount of sliding, and the number of cycles during performance. Steels in combination are usually chosen for heavier loads. Although class 20 gray irons have a low endurance limit, classes 30, 45, and others show advantages over free-machining steel. Austempering of Meehanite, or class 30, may be a good choice because of the five-grain structure, excessive dispersion of graphite, and conversion of released austenite. Sometimes bronze or phenolics are selected. Failure of the phenolics and other polymers was by material flows and surface cracking rather than by fatigue.

Alignment between cam and follower is critical. Dynamic deflection of the roller support system can seriously shorten the life of the materials in combination. This is caused by a smaller area of surfaces in contact showing wear on the edges of the cam and follower. Also harmful effects of stress concentrations should be avoided by eliminating sharp cam or follower edges by using generous fillet radii. Reference 46 demonstrates the use of sequential random search techniques for determining cam design parameters in order to minimize contact stress.

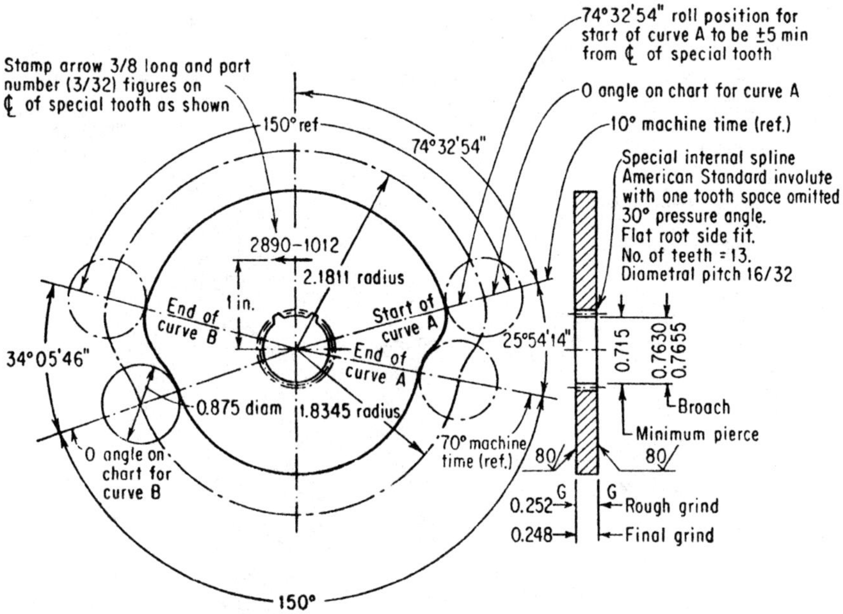

One required, No. A4140 steel 5/16 in. thick. Blank and pierce allow 3/64 in. on contour for finishing cam to be flat within ±0.010 after blanking. Heat-treat before machining and after blanking to 30 to 34 Rockwell C.

Induction-harden cam periphery 50 to 55 Rockwell C. Tolerance on cam curve must not vary in excess of ±0.001 from true curve, profile surface 32 μin.

Cam profile (roller center)			
Points for curve *A*		Points for curve *B*	
Angle	Radius	Angle	Radius
0° 0′ 0″	2.1811	0° 0′ 0″	1.8345
0° 30′ 0″	2.1809	0° 22′ 30″	1.8345
1° 0′ 0″	2.1801	0° 45′ 2″	1.8345
1° 30′ 0″	2.1788	1° 07′ 38″	1.8346
2° 0′ 0″	2.1769	1° 30′ 18″	1.8349
2° 30′ 0″	2.1746	1° 53′ 5″	1.8353
3° 0′ 0″	2.1707	2° 16′ 1″	1.8359
Etc.	Etc.	Etc.	Etc.
23° 34′ 14″	1.8388	32° 35′ 28″	2.1807
24° 24′ 14″	1.8369	32° 58′ 8″	2.1810
24° 54′ 14″	1.8355	33° 20′ 44″	2.1810
25° 24′ 14″	1.8347	33° 43′ 15″	2.1811
25° 54′ 14″	1.8345	34° 35′ 46″	2.1811
150° from end of curve *A* to start of curve *B* at 1.8345 rad		150° from end of curve *B* to start of curve *A* at 2.1811 rad	

FIG. 14.14 Cam for loom. (*Courtesy of Warner & Swasey Co., Cleveland, Ohio*)

TABLE 14.3 Cam Load Constant K

Material run against tool steel roll-hardened to Rockwell C 60–62

Material	K
Steel	
1020, 130–150 Bhn	1,700
1020, carburized, 0.045-in minimum depth, Rockwell C 50–60	10,000
4150, heat-treated, 220–300 Bhn, phosphate-coated	7,000
4340, heat-treated, 270–300 Bhn	5,000
4340, induction-hardened, 0.045-in minimum depth, Rockwell C 50–58	9,000
6150, 270–300 Bhn	2,000
Gray iron	
Class 20, 140–160 Bhn	700
Class 30, heat-treated (austempered), 255–300 Bhn, phosphate-coated (Meehanite)	2,500
Class 35, 225–245 Bhn	2,000
Class 45, 220–240 Bhn	1,000
Nodular iron	
Grade 80-60-03, 170–241 Bhn	2,000
Grade 100-70-03, heat-treated, 240–260 Bhn	3,600
Bronze (phosphorous), sand casting, 65–75 Bhn	400
Zinc, die casting, 70 Bhn	200
Phenolic-cotton base	900
Phenolic-linen base, laminate	700
Resin-acetal	600
Polyethylene	400

REFERENCES

1. Rothbart, H. A.: "Cams," John Wiley & Sons, Inc., New York, 1956.
2. Beyer, R.: "Kinematische Getriebesynthese," Springer-Verlag OHG, Berlin, 1953.
3. Artobolevskii, I. I.: "Theory of Mechanisms and Machines" (in Russian), Chief State Publishing House for Technical Theoretical Literature, Moscow, 1953.
4. Chen, F. Y.: "Kinematic Synthesis of Cam Profiles for Prescribed Acceleration by a Finite Integration Method," *Trans. ASME, J. Eng. Ind.*, vol. 95B, p. 769, 1973.
5. Jansen, B.: "Dynamik der Kurvengetriebe," VDI-Berichte 127, 1969.
6. Kloomok, M., and R. V. Muffley: "Determination of Pressure Angles for Swinging-Follower Cam Systems," *Trans. ASME*, vol. 70, p. 473, 1948.
7. Kloomok, M., and R. V. Muffley: "Determination of Radius of Curvature for Radial and Swinging Follower Cam Systems," ASME Paper 55-SA-89, 1955.
8. Johnson, R. C.: "Method of Finite Differences for Cam Design," *Mach. Des.*, vol. 27, p. 195, November 1955.
9. Jehle, F., and W. R. Spiller: "Idiosyncrasies of Valve Mechanisms and Their Causes," *Trans. SAE*, vol. 24, p. 197, 1929.
10. Turkish, M. C.: "Valve Gear Design," Eaton Mfg. Co., Detroit, Mich., 1946.
11. Dudley, W. M.: "New Methods in Valve Cam Design," *Trans. SAE*, vol. 2, p. 19, January 1948.
12. Mitchell, D. B.: "Tests on Dynamic Response of Cam-Follower Systems," *Mech. Eng.*, vol. 72, p. 467, June 1950.
13. Hrones, J. A.: "Analysis of Dynamic Forces in a Cam-Driven System," *Trans. ASME*, vol. 70, p. 473, 1948.
14. Freudenstein, F.: "On the Dynamics of High-Speed Cam Profiles," *Int. J. Mech. Sci.*, vol. 1, pp. 342–349, 1960.

15. Rothbart, H. A.: "Cam Dynamics," *Trans. 1st Int. Conf. Mechanisms,* Yale University, New Haven, Conn., March 27, 1961.
16. Beggs, J. S.: "Mechanism," McGraw-Hill Book Company, Inc., New York, pp. 139–141, 1955.
17. Raven, F. H.: "Analytical Design of Disk Cams and Three-Dimensional Cams by Independent Position Equations," *Trans. ASME,* vol. 26, series E, no. 1, pp. 18–24, March 1959.
18. Rothbart, H. A.: "Limitations on Cam Pressure Angles," *Prod. Eng.,* vol. 29, no. 19, p. 193, 1957.
19. Hinkle, R.: "Kinematics of Machines," Prentice-Hall, Inc., Englewood Cliffs, N.J., pp. 148–161, 1960.
20. Rothbart, H.: "Mechanical Control Cams," *Control Eng.,* vol. 7, no. 6, pp. 118–122, June 1960; vol. 8, no. 11, pp. 97–101, November 1961.
21. Chen, F. Y.: "A Survey of the State of the Art of Cam System Dynamics," *Mech. Mach. Theory,* vol. 12, p. 201, 1977.
22. Chen, F. Y., and N. Polanich: "Dynamics of High Speed Cam Driven Mechanisms," *Trans. ASME, J. Eng. Ind.,* p. 769, August 1975.
23. Mathew, G. K., and O. Tesar: "The Design of Modeled Cam Systems," *Trans. ASME, J. Eng. Ind.,* p. 1175, November 1975.
24. Seckallas, L. E., and M. Savage: "The Characterization of Cam Drive System Windup," *Trans. ASME, J. Mech. Des.,* vol. 102, p. 278, April 1980.
25. Winfrey, R. C.: "Elastic Link Mechanisms Dynamics," *Trans. ASME, J. Eng. Ind.,* vol. 93, no. 1, p. 268, February 1971.
26. Neklutin, C. N.: "Vibration Analysis of Cams," *Trans. 2d Conf. Mechanisms,* Penton Publishing Co., Cleveland, Ohio, p. 6, 1954.
27. Tesar, D.: "The Dynamic Synthesis, Analysis and Design of Modeled Cam Systems," Lexington Books, Lexington, Mass., 1976.
28. Chakraborty, J.: "Kinematics and Geometry of Planar and Spatial Cam Mechanisms," John Wiley & Sons, Inc., New York, 1977.
29. Barkan, P., and McGarrity, R. V.: "A Spring-Actuated Cam-Follower System," *Trans. ASME, J. Eng. Ind.,* vol. 67, ser. B, no. 1, p. 279, 1965.
30. Winfrey, R. C., R. V. Anderson, and C. W. Gnikla: "Analysis of Elastic Machinery with Clearances," *Trans. ASME, J. Eng. Ind.,* vol. 95, no. 2, p. 695, 1973.
31. Huey, C. O., Jr., and M. W. Dixon: "The Cam-Link Mechanism for Structural Error-Free Path and Friction Generation," *Mech. Mach. Theory,* vol. 9, p. 367, 1974.
32. Duca, C. D., and I. Simionescu: "The Exact Synthesis of Single-Disk Cams with Two Oscillating Rigidly Connected Roller Followers," *Mech. Mach. Theory,* vol. 15, p. 213, 1980.
33. Pagel, P. A.: "Custom Cams from Building Blocks," *Mach. Des.,* vol. 50, p. 86, May 11, 1978.
34. Weber, T., Jr.: "Simplifying Complex Cam Design," *Mach. Des.,* vol. 51, p. 115, March 22, 1979.
35. Morrison, R. A." "Load/Life Curve for Gear and Cam Materials," *Mach. Des.,* vol. 40, p. 102, August 1, 1968.
36. Giordana, F., V. Rognoni, and G. Ruggieri: "On the Influence of Measurement Errors in the Kinematic Analysis of Cams," *Mech. Mach. Theory,* vol. 14, p. 337, 1979.
37. Weiderrich, J. L.: "Dynamic Synthesis of Cams Using Finite Trigonometric Series," ASME Paper No. 74-Det-2, 1974.
38. Miske, C. R.: "An Introduction to Computer Aided Design," Prentice-Hall, Inc., Englewood Cliffs, N.J., 1968.
39. Shipley, J. E.: "Kinematic Analysis of Mechanisms," 2d ed., McGraw-Hill Book Company, Inc., New York, 1969.

40. Jackowski, C. S., and J. F. Dubil: "Single Disk Cams with Positively Controlled Oscillating Followers," *Trans. ASME, J. Appl. Mech.*, vol. 34, no. 2, p. 157, 1967.

41. Wunderlick, W.: "Contributions to the Geometry of Cam Mechanisms with Oscillating Followers," *Trans. ASME, J. Appl. Mech.*, vol. 38, no. 1, p. 1, 1971.

42. Grant, B., and A. H. Soni: "A Survey of Cam Manufacture Methods," *Trans. ASME, J. Mech. Des.*, vol. 21, p. 445, July 1979.

43. Kim, H. R., and W. R. Newcombe: "Stochastic Error Analysis in Cam Mechanisms," *Mech. Mach. Theory*, vol. 13, p. 631, 1978.

44. Tuttle, S. B.: "Error Analysis of Mechanisms," *Mach. Des.*, vol. 38, p. 152, June 1966.

45. Neklutin, C. N.: "Mechanisms and Cams for Automatic Machines," Elsevier North Holland, Inc., New York, 1967.

46. Chen, F. Y., and A. H. Soni: "On Harrisberger's Adjustable Trapezoidal Motion Program for Cam Design," *Proc. Appl. Mech. Conf.*, Paper No. 41, Oklahoma State University, Oklahoma City, 1971.

47. Sandler, B. Z.: "Suboptimal Mechanism Synthesis," *Israel J. Tech.*, vol. 12, p. 160, 1974.

48. Levitskii, N. I.: "Cam Mechanisms" (in Russian), *Mach. Tech. Publ.*, Moscow, 1964.

49. Koster, M. P.: "Vibration of Cam Mechanisms," Macmillan Publishers, Ltd., London, 1974.

50. Lafuente, J. M.: "Interactive Graphics in Data Processing Cam Design on Graphic Console," *IBM Syst. J.*, vol. 304, p. 335, 1968.

51. Hart, F. D., and C. F. Forowski: "Coupled Effects of Preloads and Damping on Dynamic Cam-Follower Separation," ASME Paper No. 64-Mech. 18, 1964.

52. Chen, F. Y.: "Analysis and Design of Cam-Driven Mechanisms with Nonlinearities," *Trans. ASME, J. Eng. Ind.*, vol. 78, no. 2, p. 685, August 1976.

53. Baumgarten, J. R.: "Preload Force Necessary to Prevent Separation of Follower from Cam," *Trans. 7th Conf. Mechanisms*, Purdue University, West Lafayette, Ind., 1962.

54. Chen, F. Y., "Mechanics and Design of Cam Mechanisms," Pergamon Press, New York, 1982.

55. Kim, H. R., and W. R. Newcombe: "The Effect of Cam Profile Errors and System Flexibility on Cam Mechanism Output," *Trans. ASME, Mechanisms and Machine Theory*, vol. 17, no. 1, p. 57, 1982.

56. Soni, A. H., and B. Grant: "Cam Design Survey," *Trans. ASME, Design Technology Transfer*, p. 177, 1974.

CHAPTER 15
ROLLING-ELEMENT BEARINGS

Bernard J. Hamrock, Ph.D.
Professor of Mechanical Engineering
Ohio State University
Columbus, Ohio

William J. Anderson, M.S.M.E.
Vice President
NASTEC Inc.
Cleveland, Ohio

SYMBOLS 15.1
15.1 INTRODUCTION 15.3
15.2 BEARING TYPES 15.4
15.2.1 Ball Bearings 15.7
15.2.2 Roller Bearings 15.9
15.3 KINEMATICS 15.13
15.4 MATERIALS AND MANUFACTURING PROCESSES 15.18
15.4.1 Ferrous Alloys 15.19
15.4.2 Ceramics 15.21
15.5 SEPARATORS 15.21
15.6 CONTACT STRESSES AND DEFORMATIONS 15.22
15.6.1 Elliptical Contacts 15.22
15.6.2 Rectangular Contacts 15.25
15.7 STATIC LOAD DISTRIBUTION 15.26
15.7.1 Load-Deflection Relationships 15.27
15.7.2 Radially Loaded Ball and Roller Bearings 15.28
15.7.3 Thrust-Loaded Ball Bearings 15.30
15.7.4 Preloading 15.33
15.8 ROLLING FRICTION AND FRICTION IN BEARINGS 15.34
15.8.1 Rolling Friction 15.34
15.8.2 Friction Losses in Rolling-Element Bearings 15.36
15.9 LUBRICANTS AND LUBRICATION SYSTEMS 15.37
15.9.1 Solid Lubrication 15.38
15.9.2 Liquid Lubrication 15.38
15.10 ELASTOHYDRODYNAMIC LUBRICATION 15.40
15.10.1 Relevant Equations 15.41
15.10.2 Dimensionless Grouping 15.42
15.10.3 Minimum-Film-Thickness Formula 15.43
15.10.4 Pressure and Film-Thickness Plots 15.43
15.11 BEARING LIFE 15.45
15.11.1 Lundberg-Palmgren Theory 15.45
15.11.2 ANSI/AFBMA Standards 15.47
15.12 DYNAMIC ANALYSES AND COMPUTER CODES 15.50
15.12.1 Quasi-Static Analyses 15.50
15.12.2 Dynamic Analyses 15.52
15.13 APPLICATION 15.52

SYMBOLS

a_1, a_2, a_3 = life factors
B = total conformity of bearing
b = semiminor axis of roller contact, m
C = dynamic load capacity, N
$c_1, \ldots, c_4$ = constants
D = distance between race curvature centers, m
$\tilde{D}$ = material factor
D_x = diameter of contact ellipse along x axis, m
D_y = diameter of contact ellipse along y axis, m
d = rolling-element diameter, m
d_a = overall diameter of bearing (Fig. 15.5), m

d_b = bore diameter, mm
d_e = pitch diameter, m
d_i = inner-race diameter, m
d_o = outer-race diameter, m
E = modulus of elasticity, N/m^2
E' = effective elastic modulus, $2/[(1 - \nu_a^2)/E_a + (1 - \nu_b^2)/E_b]$, N/m^2
$\tilde{E}$ = metallurgical processing factor
$\mathcal{E}$ = elliptic integral of second kind
$\overline{\mathcal{E}}$ = approximate elliptic integral of second kind
e = percentage of error
$\tilde{F}$ = lubrication factor
F = applied load, N
F' = load per unit length, N/m
F_e = bearing equivalent load, N
F_r = applied radial load, N
F_t = applied thrust load, N
$\mathcal{F}$ = elliptic integral of first kind
$\overline{\mathcal{F}}$ = approximate elliptic integral of first kind
f = race conformity ratio
f_a = rms surface finish of rolling element
f_b = rms surface finish of race
f_c = coefficient dependent on materials and bearing type (Table 15.14)
G = dimensionless materials parameter, $\xi E'$
$\tilde{G}$ = speed-effect factor
H = dimensionless film thickness, h/R_x
$\tilde{H}$ = misalignment factor
H_{min} = dimensionless minimum film thickness
h = film thickness, m
J = number of stress cycles
K = load-deflection constant
K_1 = load-deflection constant for roller bearing
$K_{1.5}$ = load-deflection constant for ball bearing
k = ellipticity parameter, D_y/D_x
$\bar{k}$ = approximate ellipticity parameter
L = fatigue life
L_A = adjusted fatigue life
L_{10} = fatigue life where 90 percent of bearing population will endure
L_{50} = fatigue life where 50 percent of bearing population will endure
l = bearing length, m
l_r = roller effective length, m
l_t = roller length, m
l_v = length dimension in stress volume, m
M = probability of failure
m = number of rows of rolling elements
N = rotational speed, rpm
n = number of rolling elements
P = dimensionless pressure, p/E'
P_d = diametral clearance, m
P_e = free endplay, m
p = pressure, N/m^2
q = constant, $\pi/2-1$
R = curvature sum, m
R = reliability factor, percent
R_x = effective radius in x direction, m
R_y = effective radius in y direction, m
r = race curvature radius, m
r_a = ball radius, m
r_c = roller corner radius, m
r_y = radius of roller in y direction, m
S = probability of survival
s = shoulder height, m
T = tangential force, N
U = dimensionless speed parameter, $u\eta_0/E'R_x$
u = mean surface velocity in direction of motion, $(u_a + u_b)/2$, m/s
V = stressed volume, m^3
v = elementary volume, m^3
W = dimensionless load parameter, $F/E'R_x^2$
X, Y = factors for calculation of equivalent load

x, y, z = coordinate system
Z = constant defined by Eq. (15.28)
Z_0 = depth of maximum shear stress, m
α = radius ratio, R_y/R_x
β = contact angle, degrees
β' = iterated value of contact angle, degrees
β_f = free or initial contact angle, degrees
Γ = curvature difference
δ = total elastic deformation, m
$\bar{\delta}$ = approximate total elastic deformation, m
ξ = pressure-viscosity coefficient of lubrication, m^2/N
η = absolute viscosity at gage pressure, $N \cdot s/m^2$
η_0 = viscosity at atmospheric pressure, $N \cdot s/m^2$
θ = angle used to define shoulder height
Λ = film parameter (ratio of film thickness to composite surface roughness)
μ = coefficient of rolling friction
ν = Poisson's ratio
ρ = lubricant density, $N \cdot s^2/m^4$
ρ_0 = density at atmospheric pressure, $N \cdot s^2/m^4$
σ_{max} = maximum hertzian stress, N/m^2
γ_0 = maximum shear stress, N/m^2
ψ = angular location
ψ_l = limiting value of ψ
ω = angular velocity, rad/s
ω_B = angular velocity of rolling-element–race contact, rad/s
ω_b = angular velocity of rolling element about its own center, rad/s
ω_c = angular velocity of rolling element about shaft center, rad/s

Subscripts

a = solid a
b = solid b
i = inner race
o = outer race
x, y, z = coordinate system

Superscript

$(\bar{\ })$ = approximate

15.1 INTRODUCTION

Ball bearings are used in many kinds of machines and devices with rotating parts. The designer is often confronted with decisions on whether a rolling-element or hydrodynamic bearing should be used in a particular application. The following characteristics make ball bearings *more desirable* than hydrodynamic bearings in many situations: (1) low starting and good operating friction, (2) the ability to support combined radial and thrust loads, (3) less sensitivity to interruptions in lubrication, (4) no self-excited instabilities, and (5) good low-temperature starting. Within reasonable limits, changes in load, speed, and operating temperature have but little effect on the satisfactory performance of ball bearings.

The following characteristics make ball bearings *less desirable* than hydrodynamic bearings: (1) finite fatigue life subject to wide fluctuations, (2) larger space required in the radial direction, (3) low damping capacity, (4) higher noise level, (5) more severe alignment requirements, and (6) higher cost. Each type of bearing has its particular strong points, and care should be taken in choosing the most appropriate type of bearing for a given application.

Useful guidance on the important issue of bearing selection has been presented by the Engineering Science Data Unit.[17,18] These Engineering Science Data Unit documents provide an excellent guide to the selection of the type of journal or thrust bearing most likely to give the required performance when considering the load, speed, and geometry of the bearing. The following types of bearings were considered:

1. Rubbing bearings, where the two bearing surfaces rub together [e.g., unlubricated bushings made from materials based on nylon, polytetrafluoroethylene (PTFE), and carbon]
2. Oil-impregnated porous metal bearings, where a porous metal bushing is impregnated with lubricant and thus gives a self-lubricating effect (as in sintered-iron and sintered-bronze bearings)
3. Rolling-element bearings, where relative motion is facilitated by interposing rolling elements between stationary and moving components (as in ball, roller, and needle bearings)
4. Hydrodynamic film bearings, where the surfaces in relative motion are kept apart by pressures generated hydrodynamically in the lubricant film

Figure 15.1 gives a guide to the typical load that can be carried at various speeds, for a nominal life of 10,000 h at room temperature, by journal bearings of various types of shafts of the diameters quoted. The heavy curves indicate the preferred type of journal bearing for a particular load, speed, and diameter and thus divide the graph into distinct regions. From Fig. 15.1 it is observed that rolling-element bearings are preferred at lower speeds and hydrodynamic oil-film bearings are preferred at higher speeds. Rubbing bearings and oil-impregnated porous metal bearings are not preferred for any of the speeds, loads, or shaft diameters considered. Also, as the shaft diameter is increased, the transitional point at which hydrodynamic bearings are preferred over rolling-element bearings moves to the left.

The applied load and speed are usually known, and this enables a preliminary assessment to be made of the type of journal bearing most likely to be suitable for a particular application. In many cases, the shaft diameter will already have been determined by other considerations, and Fig. 15.1 can be used to find the type of journal bearing that will give adequate load capacity at the required speed. These curves are based on good engineering practice and commercially available parts. Higher loads and speeds or smaller shaft diameters are possible with exceptionally high engineering standards or specially produced materials. Except for rolling-element bearings, the curves are drawn for bearings with a width equal to the diameter. A medium-viscosity mineral-oil lubricant is assumed for the hydrodynamic bearings.

Similarly, Fig. 15.2 gives a guide to the typical maximum load that can be carried at various speeds for a nominal life of 10,000 h at room temperature by thrust bearings of various diameters. The heavy curves again indicate the preferred type of bearing for a particular load, speed, and diameter and thus divide the graph into major regions. As with the journal bearing results (Fig. 15.1), the hydrodynamic bearing is preferred at higher speeds and the rolling-element bearing is preferred at lower speeds. A difference between Figs. 15.1 and 15.2 is that at very low speeds there is a portion of the latter figure in which the rubbing bearing is preferred. Also, as the shaft diameter is increased, the transitional point at which hydrodynamic bearings are preferred over rolling-element bearings moves to the left. Note also from this figure that oil-impregnated porous metal bearings are not preferred for any of the speeds, loads, or shaft diameters considered.

15.2 BEARING TYPES

A great variety of both design and size ranges of ball and roller bearings is available to the designer. It is the intent of this section to not duplicate the complete descriptions

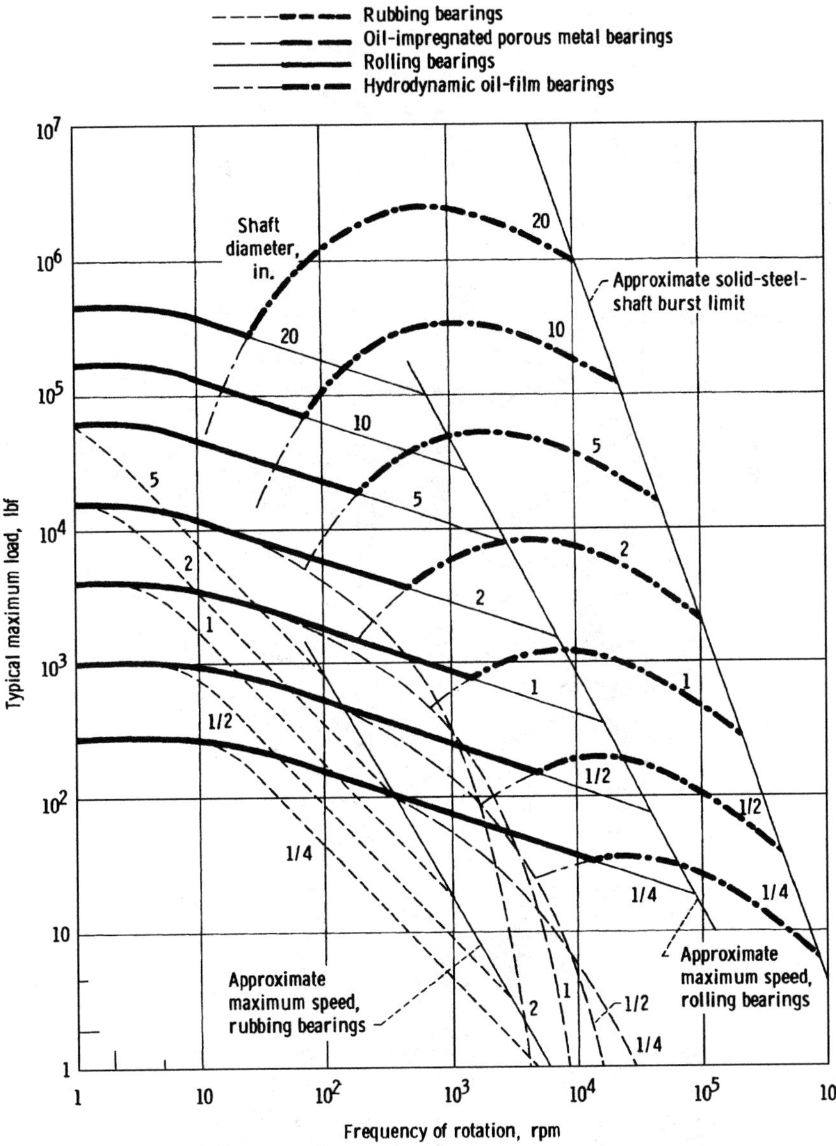

FIG. 15.1 General guide to journal bearing type. (Except for roller bearings, curves are drawn for bearings with width equal to diameter. A medium-viscosity mineral-oil lubricant is assumed for hydrodynamic bearings.) (*From Engineering Science Data Unit.*[17])

given in manufacturers' catalogs, but rather to present a guide to representative bearing types along with the approximate range of sizes available. Tables 15.1 to 15.9 illustrate some of the more widely used bearing types. In addition, there are numerous types of specialty bearings available; space does not permit a complete cataloging of all available bearings. Size ranges are given in metric units. Traditionally, most rolling-element bearings have been manufactured to metric dimensions, predating the efforts toward a metric standard.

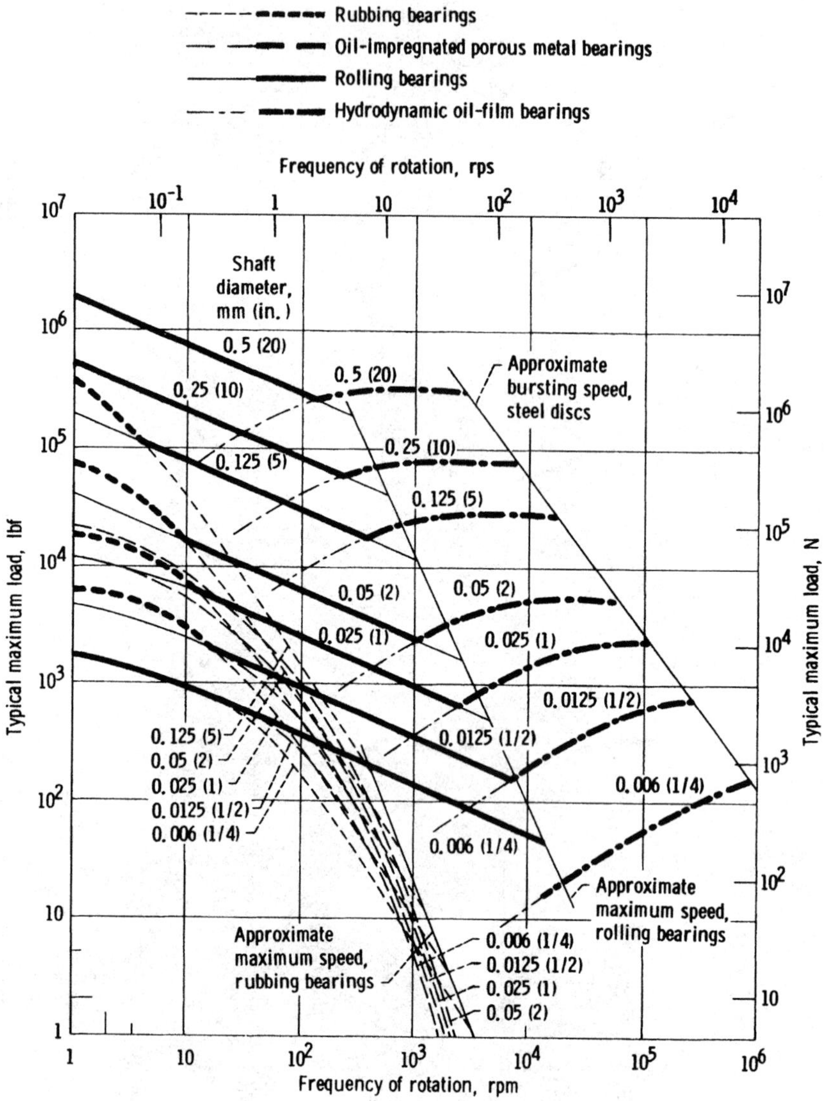

FIG. 15.2 General guide to thrust bearing type. Except for roller bearings, curves are drawn for typical ratios of inside to outside diameter. A medium-viscosity mineral-oil lubricant is assumed for hydrodynamic bearings. (*From Engineering Science Data Unit.*[18])

In addition to available bearing types and approximate size ranges, Tables 15.1 to 15.9 also list approximate relative load-carrying capabilities, both radial and thrust, and, where relevant, approximate tolerances to misalignment.

Rolling bearings are an assembly of several parts—an inner race, an outer race, a set of balls or rollers, and a cage or separator. The cage or separator maintains even spacing of the rolling elements. A cageless bearing, in which the annulus is packed with the maximum rolling-element complement, is called a "full-complement bearing."

Full-complement bearings have high load capacity but lower speed limits than bearings equipped with cages. Tapered-roller bearings are an assembly of a cup, a cone, a set of tapered rollers, and a cage.

15.2.1 Ball Bearings

Ball bearings are used in greater quantity than any other type of rolling bearing. For an application where the load is primarily radial with some thrust load present, one of the types in Table 15.1 can be chosen. A Conrad, or deep-groove, bearing has a ball complement limited by the number of balls that can be packed into the annulus between the inner and outer races with the inner race resting against the inside diameter of the outer race. A stamped and riveted two-piece cage, piloted on the ball set, or a machined two-piece cage, ball piloted or race piloted, is almost always used in a Conrad bearing. The only exception is a one-piece cage with open-sided pockets that

TABLE 15.1 Representative Radial Ball Bearings

Type	Approx. range of bore sizes, mm		Relative capacity		Limiting speed factor	Tolerance to misalignment
	Min.	Max.	Radial	Thrust		
Conrad or deep groove 	3	1060	1.00	0.7 (2 directions)	1.0	±0°15′
Maximum capacity or filling notch 	10	130	1.2–1.4	0.2 (2 directions)	1.0	±0°3′
Magneto or counterbored outer 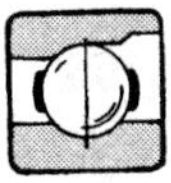	3	200	0.9–1.3	0.5–0.9 (1 direction)	1.0	±0°5′
Airframe or aircraft control 	4.826	31.75	High static capacity	0.5 (2 directions)	0.2	0°
Self-aligning, internal 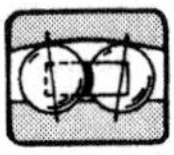	5	120	0.7	0.2 (1 direction)	1.0	±2°30′
Self-aligning, external 	—	—	1.0	0.7 (2 directions)	1.0	High
Double row, maximum 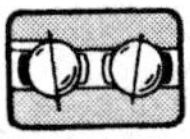	6	110	1.5	0.2 (2 directions)	1.0	±0°3′
Double row, deep groove 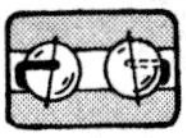	6	110	1.5	1.4 (2 directions)	1.0	0°

TABLE 15.2 Representative Angular-Contact Ball Bearings

Type		Approx. range of bore sizes, mm		Relative capacity		Limiting speed factor	Tolerance to misalignment
		Min.	Max.	Radial	Thrust		
One-directional thrust		10	320	1.00–1.15*	1.5–2.3* (1 direction)	1.1–3.0*	±0°2′
Duplex, back-to-back	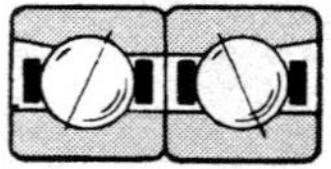	10	320	1.85	1.5 (2 directions)	3.0	0°
Duplex, face-to-face	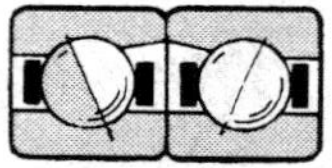	10	320	1.85	1.5 (2 directions)	3.0	0°
Duplex, tandem	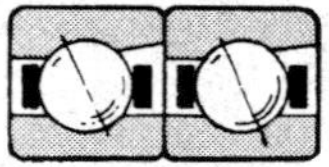	10	320	1.85	2.4 (1 direction)	3.0	0°
Two-directional or split ring		10	110	1.15	1.5 (2 directions)	3.0	±0°2′
Double row	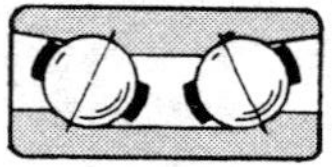	10	140	1.5	1.85 (2 directions)	0.8	0°
Double row, maximum	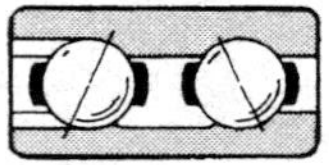	10	110	1.65	0.5 (in 1 direction) 1.5 (in other direction)	0.7	0°

*Depends on contact angle.

is snapped into place. A filling-notch bearing has both inner and outer races notched so that a ball complement limited only by the annular space between the races can be used. It has low thrust capacity because of the filling notch.

The self-aligning internal bearing shown in Table 15.1 has an outer-race ball path ground in a spherical shape so that it can accept high levels of misalignment. The self-aligning external bearing has a multipiece outer race with a spherical interface. It too can accept high misalignment and has higher capacity than the self-aligning internal bearing. However, the external self-aligning bearing is somewhat less self-aligning than its internal counterpart because of friction in the multipiece outer race.

Representative angular-contact ball bearings are illustrated in Table 15.2. An angular-contact ball bearing has a two-shouldered ball groove in one race and a single-shouldered ball groove in the other race. Thus it is capable of supporting only a unidirectional thrust load. The cutaway shoulder allows assembly of the bearing by snapping over the ball set after it is positioned in the cage and outer race. This also permits use of a one-piece, machined, race-piloted cage that can be balanced for high-speed operation. Typical contact angles vary from 15 to 25°.

Angular-contact ball bearings are used in duplex pairs mounted either back to back or face to face as shown in Table 15.2. Duplex bearing pairs are manufactured so that they "preload" each other when clamped together in the housing and on the shaft. The use of preloading provides stiffer shaft support and helps prevent bearing skidding at light loads. Proper levels of preload can be obtained from the manufacturer. A duplex pair can support bidirectional thrust load. The back-to-back arrangement offers more resistance to moment or overturning loads than does the face-to-face arrangement.

Where thrust loads exceed the capability of a simple bearing, two bearings can be used in tandem, with both bearings supporting part of the thrust load. Three or more bearings are occasionally used in tandem, but this is discouraged because of the difficulty in achieving good load sharing. Even slight differences in operating temperature will cause a maldistribution of load sharing.

The split-ring bearing shown in Table 15.2 offers several advantages. The split ring (usually the inner) has its ball groove ground as a circular arc with a shim between the ring halves. The shim is then removed when the bearing is assembled so that the split-ring ball groove has the shape of a gothic arch. This reduces the axial play for a given radial play and results in more accurate axial positioning of the shaft. The bearing can support bidirectional thrust loads but must not be operated for prolonged periods of time at predominantly radial loads. This results in three-point ball-race contact and relatively high frictional losses. As with the conventional angular-contact bearing, a one-piece precision-machined cage is used.

Ball thrust bearings (Table 15.3), which have a 90° contact angle, are used almost exclusively for machinery with vertically oriented shafts. The flat-race bearing allows eccentricity of the fixed and rotating members. An additional bearing must be used for radial positioning. It has low load capacity because of the very small ball-race contacts and consequent high hertzian stress. Grooved-race bearings have higher load capacities and are capable of supporting low-magnitude radial loads. All of the pure thrust ball bearings have modest speed capability because of the 90° contact angle and the consequent high level of ball spinning and frictional losses.

15.2.2 Roller Bearings

Cylindrical-roller bearings (Table 15.4) provide purely radial load support in most applications. An N- or U-type bearing will allow free axial movement of the shaft relative to the housing to accommodate differences in thermal growth. An F- or J-type bearing will support a light thrust load in one direction; and a T-type bearing, a light bidirectional thrust load.

TABLE 15.3 Representative Thrust Ball Bearings

Type	Approx. range of bore sizes, mm		Relative capacity		Limiting speed factor	Tolerance to misalignment
	Min.	Max.	Radial	Thrust		
One directional, flat race 	6.45	88.9	0	0.7 (1 direction)	0.10	0° (accepts eccentricity)
One directional, grooved race 	6.45	1180	0	1.5 (1 direction)	0.30	0°
Two directional, grooved race 	15	220	0	1.5 (2 directions)	0.30	0°

Cylindrical-roller bearings have moderately high radial load capacity as well as high speed capability. Their speed capability exceeds that of either spherical or tapered-roller bearings. A commonly used bearing combination for support of a high-speed rotor is an angular-contact ball bearing or duplex pair and a cylindrical-roller bearing.

As explained in Sec. 15.3 on bearing geometry, the rollers in cylindrical-roller bearings are seldom pure cylinders. They are crowned or made slightly barrel shaped to relieve stress concentrations of the roller ends when any misalignment of the shaft and housing is present.

Cylindrical-roller bearings may be equipped with one- or two-piece cages, usually race piloted. For greater load capacity, full-complement bearings can be used, but at a significant sacrifice in speed capability.

Spherical-roller bearings (Tables 15.5 to 15.7) are made as either single- or double-row bearings. The more popular bearing design uses barrel-shaped rollers. An alternative design employs hourglass-shaped rollers. Spherical-roller bearings combine very high radial load capacity with modest thrust load capacity (with the exception of the thrust type) and excellent tolerance to misalignment. They find widespread use in heavy-duty rolling mill and industrial gear drives, where all of these bearing characteristics are requisite.

Tapered-roller bearings (Table 15.8) are also made as single- or double-row bearings with combinations of one- or two-piece cups and cones. A four-row bearing assembly with two- or three-piece cups and cones is also available. Bearings are made with either a standard angle for applications in which moderate thrust loads are present or with a steep angle for high thrust capacity. Standard and special cages are available to suit the application requirements.

Single-row tapered-roller bearings must be used in pairs because a radially loaded bearing generates a thrust reaction that must be taken by a second bearing. Tapered-roller bearings are normally set up with spacers designed so that they operate with some internal play. Manufacturers' engineering journals should be consulted for proper setup procedures.

Needle-roller bearings (Table 15.9) are characterized by compactness in the radial direction and are frequently used without an inner race. In the latter case, the shaft is hardened and ground to serve as the inner race. Drawn cups, both open and closed end, are frequently used for grease retention. Drawn cups are thin walled and require substantial support from the housing. Heavy-duty roller bearings have relatively rigid

TABLE 15.4 Representative Cylindrical Roller Bearings

Type	Approx. range of bore sizes, mm		Relative capacity		Limiting speed factor	Tolerance to misalignment
	Min.	Max.	Radial	Thrust		
Separable outer ring, nonlocating (RN, RIN) 	10	320	1.55	0	1.20	±0°5′
Separable inner ring, nonlocating (RU, RIU) 	12	500	1.55	0	1.20	±0°5′
Separable outer ring, one-direction locating (RF, RIF) 	40	177.8	1.55	Locating (1 direction)	1.15	±0°5′
Separable inner ring, one-direction locating (RJ, RIJ) 	12	320	1.55	Locating (1 direction)	1.15	±0°5′
Self-contained, two-direction locating 	12	100	1.35	Locating (2 directions)	1.15	±0°5′
Separable inner ring, two-direction locating (RT, RIT) 	20	320	1.55	Locating (2 directions)	1.15	±0°5′
Nonlocating, full complement (RK, RIK) 	17	75	2.10	0	0.20	±0°5′
Double row, separable outer ring, nonlocating (RD) 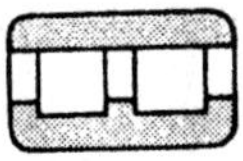	30	1060	1.85	0	1.00	0°
Double row, separable inner ring, nonlocating 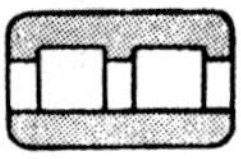	70	1060	1.85	0	1.00	0°

TABLE 15.5 Representative Spherical Roller Bearings

Type		Approx. range of bore sizes, mm		Relative capacity		Limiting speed factor	Tolerance to misalignment
		Min.	Max.	Radial	Thrust		
Single row, barrel or convex	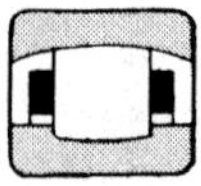	20	320	2.10	0.20	0.50	± 2°
Double row, barrel or convex	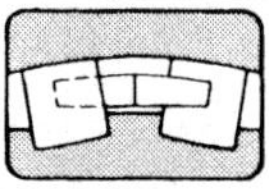	25	1250	2.40	0.70	0.50	± 1°30′
Thrust		85	360	0.10* 0.10†	1.80* 2.40†	0.35–0.50	± 3°
Double row, concave	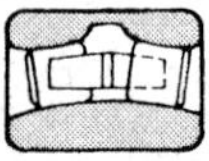	50	130	2.40	0.70	0.50	± 1°30′

*Symmetric rollers.
†Asymmetric rollers.

TABLE 15.6 Standardized Double-Row Spherical Roller Bearings

Type		Roller design	Retainer design	Roller guidance	Roller-race contact
SLB	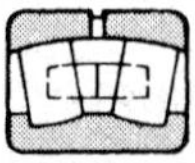	Symmetric	Machined, roller piloted	Retainer pockets	Modified line, both races
SC	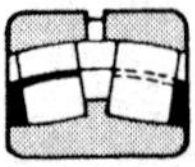	Symmetric	Stamped, race piloted	Floating guide ring	Modified line, both races
SD	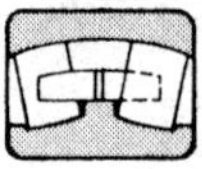	Asymmetric	Machined, race piloted	Inner-ring center rib	Line contact, outer; point contact, inner

races and are more akin to cylindrical roller bearings with high length-to-diameter ratio rollers.

Needle-roller bearings are more speed limited than cylindrical roller bearings because of roller skewing at high speeds. A high percentage of needle-roller bearings are full-complement bearings. Relative to a caged needle bearing, these have higher load capacity but lower speed capability.

There are many types of specialty bearings available other than those discussed here. Aircraft bearings for control systems, thin-section bearings, and fractured-ring bearings are some of the more widely used bearings among the many types manufactured. A complete coverage of all bearing types is beyond the scope of this chapter.

TABLE 15.7 Characteristics of Spherical Roller Bearings

Series	Types	Approx. range of bore sizes, mm		Approx. relative capacity*		Limiting speed factor
		Min.	Max.	Radial	Thrust	
202	Single-row barrel	20	320	1.0	0.11	0.5
203	Single-row barrel	20	240	1.7	0.18	0.5
204	Single-row barrel	25	110	2.1	0.22	0.4
212	SLB	35	75	1.0	0.26	0.6
213	SLB	30	70	1.7	0.53	0.6
22, 22K	SLB, SC, SD	30	320	1.7	0.46	0.6
23, 23K	SLB, SC, SD	40	280	2.7	1.0	0.6
30, 30K	SLB, SC, SD	120	1250	1.2	0.29	0.7
31, 31K	SLB, SC, SD	110	1250	1.7	0.54	0.6
32, 32K	SLB, SC, SD	100	850	2.1	0.78	0.6
39, 39K	SD	120	1250	0.7	0.18	0.7
40, 40K	SD	180	250	1.5	—	0.7

*Load capacities are comparative within the various series of spherical roller bearings only. For a given envelope size, a spherical roller bearing has a radial capacity approximately equal to that of a cylindrical roller bearing.

Angular-contact ball bearings and cylindrical-roller bearings are generally considered to have the highest speed capabilities. Speed limits of roller bearings are discussed in conjunction with lubrication methods. The lubrication system employed has as great an influence on bearing limiting speed as does the bearing design.

15.3 KINEMATICS

The relative motions of the separator, the balls or rollers, and the races of rolling-element bearings are important to understanding their performance. The relative velocities in a ball bearing are somewhat more complex than those in roller bearings, the latter being analogous to the specialized case of a zero- or fixed-value-contact-angle ball bearing. For that reason, the ball bearing is used as an example here to develop approximate expressions for relative velocities. These are useful for rapid but reasonably accurate calculation of elastohydrodynamic film thickness, which can be used with surface roughnesses to calculate the lubrication life factor.

The precise calculation of relative velocities in a ball bearing in which speed or centrifugal force effects, contact deformations, and elastohydrodynamic traction effects are considered requires a large computer to numerically solve the relevant equations. The reader is referred to the growing body of computer codes discussed in Sec. 15.12 for precise calculations of bearing performance. Such a treatment is beyond the scope of this section. However, approximate expressions that yield answers with accuracies satisfactory for many situations are available.

When a ball bearing operates at high speeds, the centrifugal force acting on the ball creates a divergency of the inner- and outer-race contact angles, as shown in Fig. 15.3, in order to maintain force equilibrium on the ball. For the most general case of rolling and spinning at both inner- and outer-race contacts, the rolling and spinning velocities of the ball are as shown in Fig. 15.4.

The equations for ball and separator angular velocity for all combinations of inner- and outer-race rotation were developed in Ref. 40. Without introducing additional relationships to describe the elastohydrodynamic conditions at both ball-race contacts,

TABLE 15.8 Principal Types of Tapered Roller Bearings

Type	Subtype	Approx. range of bore sizes, mm
Single row		8–1690
TS 	TST—tapered bore	24–430
	TSS—steep angle	16–1270
	TS—pin cage	—
	TSE, TSK—keyway cones	12–380
	TSF, TSSF—flanged cup	8–1070
	TSG—steering gear (without cone)	—
Two-row, double cone, single cups		30–1200
TDI 	TDIK, TDIT, TDITP—tapered bore	30–860
	TDIE, TDIKE—slotted double cone	24–690
	TDIS—steep angle	55–520
Two-row, double cup, single cones, adjustable	TDO	8–1830
TDO 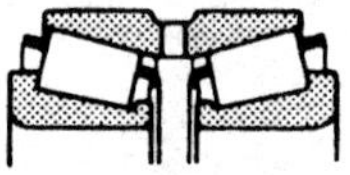	TDOS—steep angle	20–1430
Two-row, double cup, single cones, nonadjustable	TNA	20–60
TNA 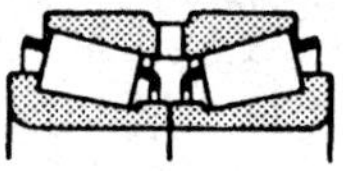	TNASW—slotted cones	30–260
	TNASWE—extended cone rib	20–305
	TNASWH—slotted cones, sealed	8–70
	TNADA, TNHDADX—self-aligning cup AD	—
Four-row, cup adjusted		70–1500
TQO 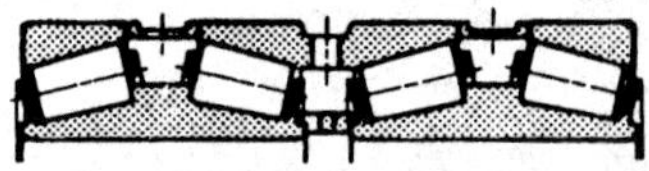	TQO, TOOT—tapered bore	250–1500
Four-row, cup adjusted		
TQI 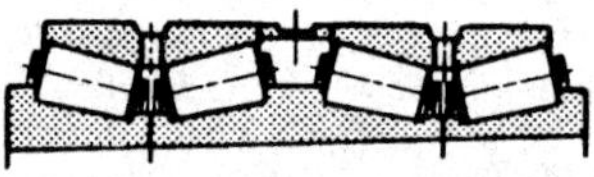	TQIT—tapered bore	—

however, the ball spin-axis orientation angle θ cannot be obtained. As mentioned, this requires a lengthy numerical solution except for the two extreme cases of outer- or inner-race control. These are illustrated in Fig. 15.5.

Race control assumes that pure rolling occurs at the controlling race, with all of the ball spin occurring at the outer-race contact. The orientation of the ball rotational axis is then easily determinable from bearing geometry. Race control probably occurs only in dry bearings or dry-film-lubricated bearings where Coulomb friction conditions exist in the ball-race contact ellipses. The moment-resisting spin will always be greater at one of the race contacts. Pure rolling will occur at the race contact with the higher magnitude of moment-resisting spin. This is usually the inner race at low speeds and the outer race at high speeds.

In oil-lubricated bearings in which elastohydrodynamic films exist in both ball-race contacts, rolling with spin occurs at both contacts. Therefore, precise ball motions can

TABLE 15.9 Representative Needle Roller Bearings

Type		Bore sizes, mm		Relative load capacity		Limiting speed factor	Misalignment tolerance
		Min.	Max.	Dynamic	Static		
Drawn cup, needle	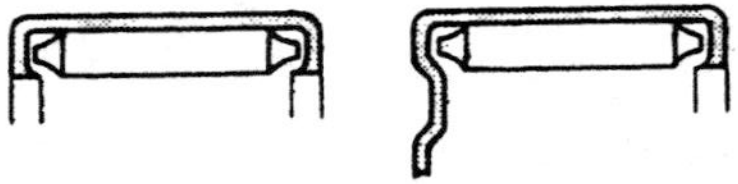	3	185	High	Moderate	0.3	Low
Drawn cup, needle, grease retained		4	25	High	Moderate	0.3	Low
Drawn cup, roller		5	70	Moderate	Moderate	0.9	Moderate
Heavy-duty roller		16	235	Very high	Moderate	1.0	Moderate
Caged roller		12	100	Very high	High	1.0	Moderate
Cam follower		12	150	Moderate to high	Moderate to high	0.3–0.9	Low
Needle thrust		6	105	Very high	Very high	0.7	Low

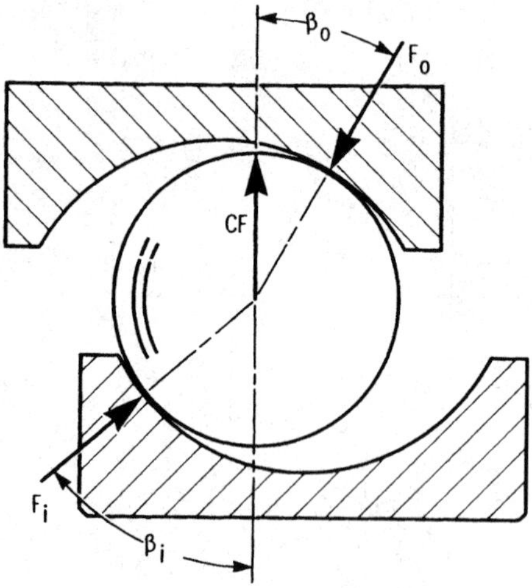

FIG. 15.3 Contact angles in a ball bearing at appreciable speeds.

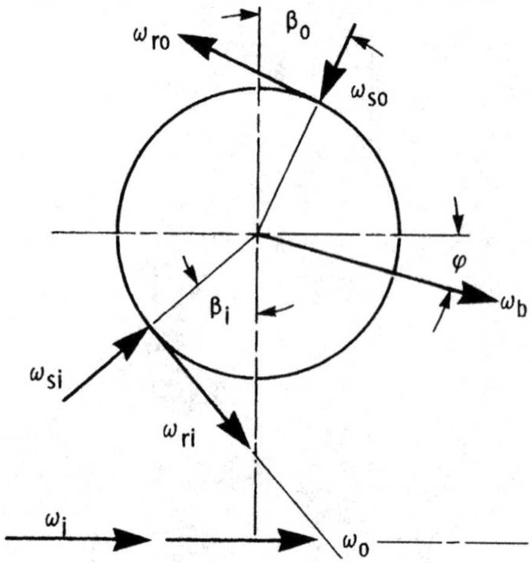

FIG. 15.4 Angular velocities of a ball.

only be determined through use of a computer analysis. We can approximate the situation with a reasonable degree of accuracy, however, by assuming that the ball rolling axis is normal to the line drawn through the centers of the two ball-race contacts.

The angular velocity of the separator or ball set ω_c about the shaft axis can be shown to be (Ref. 4),

$$\omega_c = \frac{(v_i + v_o)/2}{d_e/2} = \frac{1}{2}\left[\omega_i\left(1 - \frac{d \cos \beta}{d_e}\right) + \omega_o\left(1 + \frac{d \cos \beta}{d_e}\right)\right] \tag{15.1}$$

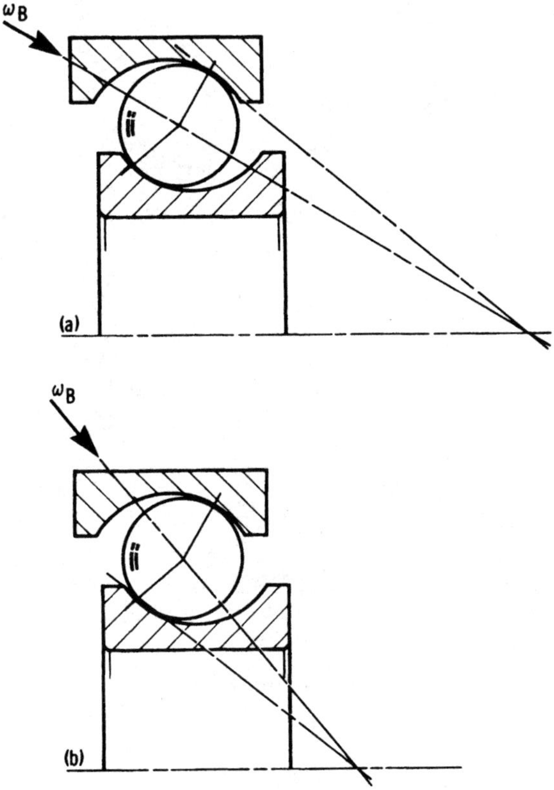

FIG. 15.5 Ball spin-axis orientations for outer- and inner-race control. (*a*) Outer-race control. (*b*) Inner-race control.

where v_i and v_o are the linear velocities of the inner and outer contacts. The angular velocity of a ball about its own axis ω_b is

$$\omega_b = \frac{v_i - v_o}{d_e/2} = \frac{d_e}{2d}\left[\omega_i\left(1 - \frac{d\cos\beta}{d_e}\right) - \omega_o\left(1 + \frac{d\cos\beta}{d_e}\right)\right] \tag{15.2}$$

To calculate the velocities of the ball-race contacts, which are required for calculating elastohydrodynamic film thicknesses, it is convenient to use a coordinate system that rotates at ω_c. This fixes the ball-race contacts relative to the observer. In the rotating coordinate system, the angular velocities of the inner and outer races become

$$\omega_{ir} = \omega_i - \omega_c = [(\omega_i - \omega_o)/2]\,[1 + (d\cos\beta)/d_e]$$

$$\omega_{or} = \omega_o - \omega_c = [(\omega_o - \omega_i)/2]\,[1 - (d\cos\beta)/d_e]$$

The surface velocities entering the *ball–inner-race contact* for pure rolling are

$$u_{ai} = u_{bi} = [(d_e - d\cos\beta)/2]\omega_{ir} \tag{15.3}$$

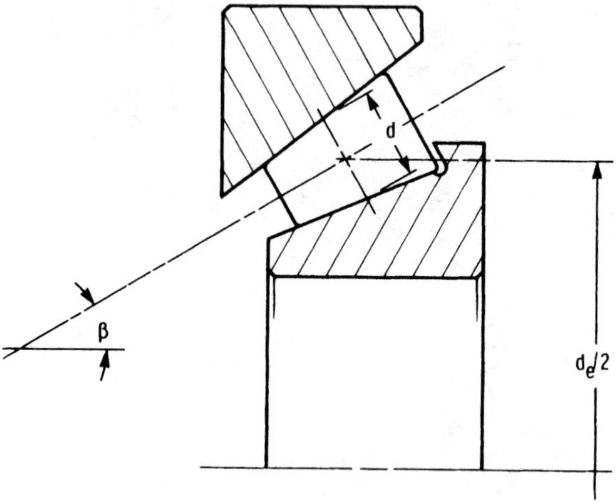

FIG. 15.6 Simplified geometry for tapered-roller bearing.

or

$$u_{ai} = u_{bi} = [d_e(\omega_i - \omega_o)/4][1 - (d^2 \cos^2 \beta)/d_e^2] \quad (15.4)$$

and those at the *ball–outer-race contact* are

$$u_{ao} = u_{bo} = [(d_e + d \cos \beta)/2]\omega_{or}$$

or

$$u_{ao} = u_{bo} = [d_e(\omega_o - \omega_i)/4][(1 - d^2 \cos^2 \beta)/d_e^2] \quad (15.5)$$

For a cylindrical roller bearing, $\beta = 0°$ and Eqs. (15.1), (15.2), (15.4), and (15.5) become, if d is roller diameter,

$$\omega_c = \tfrac{1}{2}[\omega_i(1 - d/d_e) + \omega_o(1 + d/d_e)]$$

$$\omega_R = (d_e/2d)[\omega_i(1 - d/d_e) + \omega_o(1 + d/d_e)]$$

$$u_{ai} = u_{bi} = [d_e(\omega_i - \omega_o)/4](1 - d^2/d_e^2) \quad (15.6)$$

$$u_{ao} = u_{bo} = [d_e(\omega_o - \omega_i)/4](1 - d^2/d_e^2)$$

For a tapered-roller bearing, equations directly analogous to those for a ball bearing can be used if d is the average diameter of the tapered roller, d_e is the diameter at which the geometric center of the rollers is located, and β is the angle as shown in Fig. 15.6.

15.4 MATERIALS AND MANUFACTURING PROCESSES

Until about 1955, rolling-element bearing materials technology did not receive much attention from materials scientists. Bearing materials were restricted to SAE 52100 and some carburizing grades, such as AISI 4320 and AISI 9310, which seemed to be adequate for most bearing applications, despite the limitation in temperature of about 176°C (350°F) for 52100 steel. A minimum acceptable hardness of Rockwell C 58 should be specified. Experiments indicate that fatigue life increases with increasing hardness.

The advent of the aircraft gas turbine engine, with its need for advanced rolling-element bearings, provided the major impetus for advancements in rolling-element bearing materials technology. Increased temperatures, higher speeds and loads, and the need for greater durability and reliability all served as incentives for the development and evaluation of a broad range of new materials and processing methods. The combined research efforts of bearing manufacturers, engine manufacturers, and government agencies over the past three decades have resulted in startling advances in rolling-element bearing life and reliability and in performance. The discussion here is brief in scope. For a comprehensive treatment of the research status of current bearing technology and current bearing designs, refer to Ref. 10.

15.4.1 Ferrous Alloys

The need for higher temperature capability led to the evaluation of a number of available molybdenum and tungsten alloy tool steels as bearing materials (Table 15.10). These alloys have excellent high-temperature hardness retention. Such alloys melted and cast in an air environment, however, were generally deficient in fatigue resistance because of the presence of nonmetallic inclusions. Vacuum processing techniques can reduce or eliminate these inclusions. Techniques used include vacuum induction melting (VIM) and vacuum arc remelting (VAR). These have been extensively explored, not only with the tool steels now used as bearing materials, but with SAE 52100 and some of the carburizing steels as well. Table 15.10 lists a fairly complete array of ferrous alloys, both fully developed and experimental, from which present-day bearings are fabricated. AISI M-50, usually VIM-VAR or consumable electrode vacuum melted (CEVM) processed, has become a very widely used quality bearing material. It is usable at temperatures to 315°C (600°F), and it is usually assigned a materials life factor

TABLE 15.10 Typical Chemical Compositions of Selected Bearing Steels[10]

Alloying element percent by weight

Designation	C	P (max)	S (max)	Mn	Si	Cr	V	W	Mo	Co	Cb	Ni
SAE 52100	1.00	0.025	0.025	0.35	0.30	1.45						
MHT*	1.03	0.025	0.025	0.35	0.35	1.50						
AISI M-1	0.80	0.030	0.030	0.30	0.30	4.00	1.00	1.50	8.00			
AISI M-2	0.83	↓	↓	0.30	0.30	3.85	1.90	6.15	5.00			
AISI M-10	0.85	↓	↓	0.25	0.30	4.00	2.00	—	8.00			
AISI M-50	0.80	↓	↓	0.30	0.25	4.00	1.00	—	4.25			
T-1 (18-4-1)	0.70	↓	↓	0.30	0.25	4.00	1.00	18.0				
T15	1.52	0.010	0.004	0.26	0.25	4.70	4.90	12.5	0.20	5.10		
440C	1.03	0.018	0.014	0.48	0.41	17.30	0.14	—	0.50			
AMS 5749	1.15	0.012	0.004	0.50	0.30	14.50	1.20	—	4.00			
Vasco matrix II	0.53	0.014	0.013	0.12	0.21	4.13	1.08	1.40	4.80	7.81	—	0.10
CRB-7	1.10	0.016	0.003	0.43	0.31	14.00	1.03	—	2.02	—	0.32	
AISI 9310†	0.10	0.006	0.001	0.54	0.28	1.18	—	—	0.11	—	—	3.15
CBS 600†	0.19	0.007	0.014	0.61	1.05	1.50	—	—	0.94	—	—	0.18
CBS 1000M†	0.14	0.018	0.019	0.48	0.43	1.12	—	—	4.77	—	—	2.94
Vasco X-2†	0.14	0.011	0.011	0.24	0.94	4.76	0.45	1.40	1.40	0.03	—	0.10

*Also contains 1.36 percent Al.
†Carburizing grades.

of 3 to 5 (Chap. 6). T-1 tool steel has also come into fairly wide use, mostly in Europe, in bearings. Its hot hardness retention is slightly superior to that of M-50 and approximately equal to that of M-1 and M-2. These alloys retain adequate hardness to about 400°C (750°F).

Surface-hardened or carburized steels are used in many bearings where, because of shock loads or cyclic bending stresses, the fracture toughness of the through-hardened steels is inadequate. Some of the newer materials being developed, such as CBS 1000 and Vasco X-2 have hot hardness retention comparable to that of the tool steels (Fig. 15.7). They too are available as ultraclean, vacuum-processed materials and should offer adequate resistance to fatigue. Carburized steels may become of increasing importance in ultrahigh-speed applications. Bearings with through-hardened steel races are currently limited to $d_bN \approx 2.5 \times 10^6$ (where d_b is bore diameter in millimeters and N is rotational speed in revolutions per minute) because, at higher d_bN values, fatigue cracks propagate through the rotating race as a result of the excessive hoop stress present.[9]

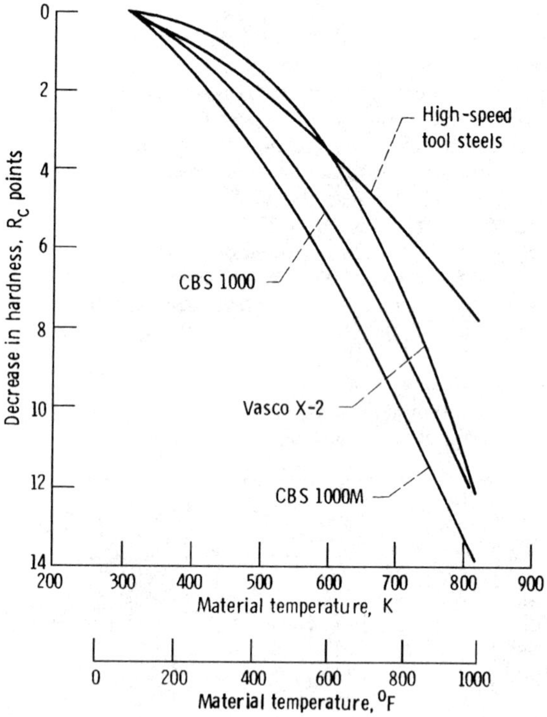

FIG. 15.7 Hot hardness of CBS 1000, CBS 1000M, Vasco X-2, and high-speed tool steels. (*From Anderson and Zaretsky.*[3])

In applications where the bearings are not lubricated with conventional oils and protected from corrosion at all times, a corrosion-resistant alloy should be used. Dry-film-lubricated bearings, bearings cooled by liquefied cryogenic gases, and bearings exposed to corrosive environments such as very high humidity and salt water are applications where corrosion-resistant alloys should be considered. Of the alloys listed in Table 15.10, both 440C and AMS 5749 are readily available in vacuum-melted heats.

In addition to improved melting practice, forging and forming methods that result in improved resistance to fatigue have been developed. Experiments indicate that fiber or grain flow parallel to the stressed surface is superior to fiber flow that intersects the stressed surface.[7,63] Forming methods that result in more parallel grain flow are now being used in the manufacture of many bearings, especially those used in high-load applications.

15.4.2 Ceramics

Experimental bearings have been made from a variety of ceramics including alumina, silicon carbide, titanium carbide, and silicon nitride. The use of ceramics as bearing materials for specialized applications will probably continue to grow for several reasons. These include:

1. *High-temperature capability:* Because ceramics can exhibit elastic behavior to temperatures beyond 1000°C (1821°F), they are an obvious choice for extreme-temperature applications.
2. *Corrosion resistance:* Ceramics are essentially chemically inert and able to function in many environments hostile to ferrous alloys.
3. *Low density:* This can be translated into improved bearing capacity at high speeds, where centrifugal effects predominate.
4. *Low coefficient of thermal expansion:* Under conditions of severe thermal gradients, ceramic bearings exhibit less drastic changes in geometry and internal play than do ferrous alloy bearings.

At the present time, silicon nitride is being actively developed as a bearing material.[15,53] Silicon nitride bearings have exhibited fatigue lives comparable to and, in some instances, superior to that of high-quality vacuum-melted M-50. Two problems remain: (1) quality control and precise nondestructive inspection techniques to determine acceptability, and (2) cost. Improved hot isostatic compaction, metrology, and finishing techniques are all being actively pursued. However, silicon nitride bearings must still be considered experimental.

15.5 SEPARATORS

Ball- and roller-bearing separators, sometimes called "cages" or "retainers," are bearing components that, although never carrying load, are capable of exerting a vital influence on the efficiency of the bearing. In a bearing without a separator, the rolling elements contact each other during operation and in so doing experience severe sliding and friction. The primary function of a separator is to maintain the proper distance between the rolling element and to ensure proper load distribution and balance within the bearing. Another function of the separator is to maintain control of the rolling elements in such a manner as to produce the least possible friction through sliding contact. Furthermore, a separator is necessary for several types of bearings to prevent the rolling elements from falling out of the bearing during handling. Most separator troubles occur from improper mounting, misaligned bearings, or improper (inadequate or excessive) clearance in the rolling-element pocket.

The materials used for separators vary according to the type of bearing and the application. In ball bearings and some sizes of roller bearings, the most common type of separator is made from two strips of carbon steel that are pressed and riveted together. Called ribbon separators, they are the least expensive to manufacture and are entirely suitable for many applications. They are also lightweight and usually require little space.

The design and construction of angular-contact ball bearings allow the use of a one-piece separator. The simplicity and inherent strength of one-piece separators permit their fabrication from many desirable materials. Reinforced phenolic and bronze are the two most commonly used materials. Bronze separators offer strength and low-friction characteristics and can be operated at temperatures to 230°C (450°F). Machined, silver-plated ferrous alloy separators are used in many demanding applications. Because reinforced cotton-base phenolic separators combine the advantages of low weight, strength, and nongalling properties, they are used for such high-speed applications as gyro bearings. In high-speed bearings, lightness and strength are particularly desirable, since the stresses increase with speed but may be greatly minimized by reduction of separator weight. A limitation of phenolic separators, however, is that they have an allowable maximum temperature of about 135°C (275°F).

15.6 CONTACT STRESSES AND DEFORMATIONS

The loads carried by rolling-element bearings are transmitted through the rolling element from one race to the other. The magnitude of the load carried by an individual rolling element depends on the internal geometry of the bearing and the location of the rolling element at any instant. A load-deflection relationship for the rolling-element–race contact is developed in this section. The deformation within the contact is a function of, among other things, the ellipticity parameter and the elliptic integrals of the first and second kinds. Simplified expressions that allow quick calculations of the stresses and deformations to be made easily from a knowledge of the applied load, the material properties, and the geometry of the contacting elements are presented in this section.

15.6.1 Elliptical Contacts

When two elastic solids are brought together under a load, a contact area develops, the shape and size of which depend on the applied load, the elastic properties of the materials, and the curvatures of the surfaces. The shape of the contact area may be elliptical, with D_y being the diameter in the y direction (transverse direction) and D_x being the diameter in the x direction (direction of motion). For the special case where $r_{ax} = r_{ay}$ and $r_{bx} = r_{by}$, the resulting contact is a circle rather than an ellipse. Where r_{ay} and r_{by} are both infinity, the initial line contact develops into a rectangle when load is applied.

The contact ellipses obtained with either a radial or a thrust load for the ball–inner-race and ball–outer-race contacts in a ball bearing are shown in Fig. 15.8. Inasmuch as the size and shape of these contact areas are highly significant in the successful operation of rolling elements, it is important to understand their characteristics.

The ellipticity parameter k is defined as the elliptical-contact diameter in the y direction (transverse direction) divided by the elliptical-contact diameter in the x direction (direction of motion), or $k = D_y/D_x$. If the equation $1/r_{ax} + 1/r_{bx} > 1/r_{ay} + 1/r_{by}$ is satisfied and $\alpha \geq 1$, the contact ellipse will be oriented so that its major diameter will be transverse to the direction of motion, and consequently $k \geq 1$. Otherwise, the major diameter would lie along the direction of motion with both $\alpha \leq 1$ and $k \leq 1$. Figure 15.9 shows the ellipticity parameter and the elliptic integrals of the first and second kinds for a range of the curvature ratio ($\alpha = R_y/R_x$) usually encountered in

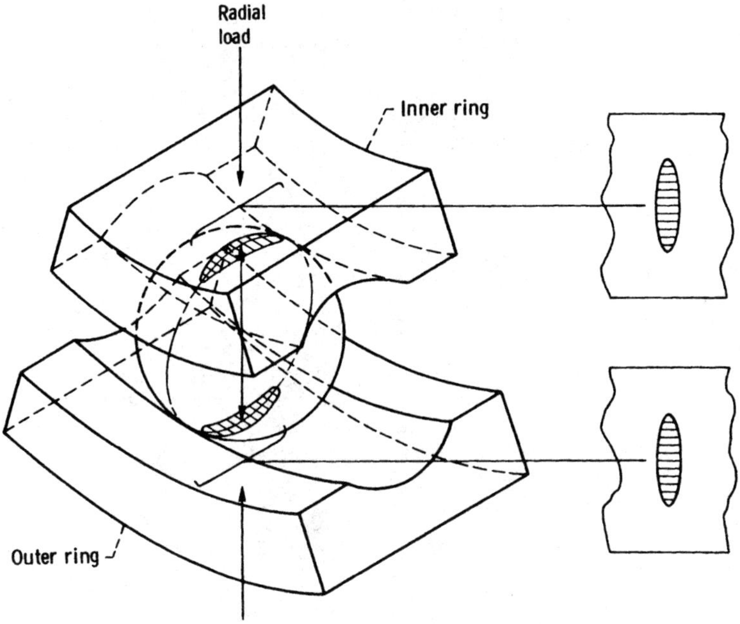

FIG. 15.8 Contact areas in a ball bearing.

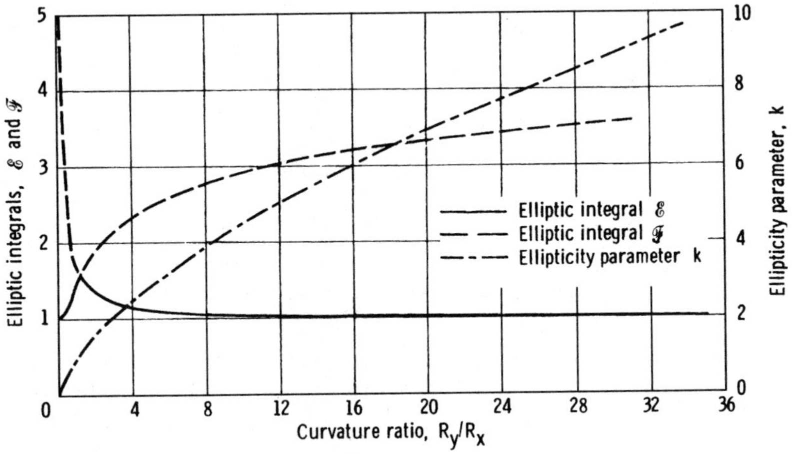

FIG. 15.9 Ellipticity parameter and elliptic integrals of the first and second kinds as a function of the curvature ratio.

concentrated contacts.

Simplified Solutions for $\boldsymbol{\alpha \geq 1}$**.** The classical hertzian solution requires the calculation of the ellipticity parameter k and the complete elliptic integrals of the first and second kinds $\mathcal{F}$ and $\mathcal{E}$. This entails finding a solution to a transcendental equation relating k, $\mathcal{F}$, and $\mathcal{E}$ to the geometry of the contacting solids. This is usually accomplished by some iterative numerical procedure, as described in Ref. 25, or with the aid of charts, as shown in Ref. 37. Reference 26 provides a shortcut to the classical hertzian solution for the local stress and deformation of two elastic bodies in contact. The shortcut is accomplished by using simplified forms of the ellipticity parameter and the complete elliptic integrals, expressing them as functions of the geometry. The results are summarized here.

Table 15.11 shows various values of radius-of-curvature ratios and corresponding values of k, $\mathcal{F}$, and $\mathcal{E}$ obtained from the numerical procedure given in Ref. 25. For the set of pairs of data $[(k_i, \alpha_i), i = 1, 2, \ldots, 26]$, a power fit using linear regression by the method of least squares resulted in the following equation:

$$\bar{k} = \alpha^{2/\pi} \qquad \text{for } \alpha \geq 1 \tag{15.7}$$

The asymptotic behavior of $\mathcal{E}$ and $\mathcal{F}$ ($\alpha \to 1$ implies $\mathcal{E} \to \mathcal{F} \to \pi/2$, and $\alpha \to \infty$ implies $\mathcal{F} \to \infty$ and $\mathcal{E} \to 1$) was suggestive of the type of functional dependence that $\mathcal{E}$ and $\mathcal{F}$ might follow. As a result, an inverse and a logarithmic curve fit were tried for $\mathcal{E}$ and $\mathcal{F}$, respectively. The following expressions provided excellent curve fits:

TABLE 15.11 Comparison of Numerically Determined Values with Curve-Fit Values for Geometrically Dependent Variables

(R_x = 1.0 cm)

Radius-of-curvature ratio α	Elliptical parameter k	Ellipticity		Complete elliptic integral of first kind			Complete elliptic integral of second kind		
		$\bar{k}$	Percent error e	$\mathcal{F}$	$\bar{\mathcal{F}}$	Percent error e	$\mathcal{E}$	$\bar{\mathcal{E}}$	Percent error e
1.00	1.0000	1.0000	0	1.5708	1.5708	0	1.5708	1.5708	0
1.25	1.1604	1.1526	0.66	1.6897	1.6892	−0.50	1.4643	1.4566	0.52
1.50	1.3101	1.2945	1.19	1.7898	1.8022	−0.70	1.3911	1.3805	0.76
1.75	1.4514	1.4280	1.61	1.8761	1.8902	−0.75	1.3378	1.3262	0.87
2.00	1.5858	1.5547	1.96	1.9521	1.9664	−0.73	1.2972	1.2854	0.91
3.00	2.0720	2.0125	2.87	2.1883	2.1979	−0.44	1.2002	1.1903	0.83
4.00	2.5007	2.4170	3.35	2.3595	2.3621	−0.11	1.1506	1.1427	0.69
5.00	2.8902	2.7860	3.61	2.4937	2.4895	0.17	1.1205	1.1142	0.57
6.00	3.2505	3.1289	3.74	2.6040	2.5935	0.40	1.1004	1.0951	0.48
7.00	3.5878	3.4515	3.80	2.6975	2.6815	0.59	1.0859	1.0815	0.40
8.00	3.9065	3.7577	3.81	2.7786	2.7577	0.75	1.0751	1.0713	0.35
9.00	4.2096	4.0503	3.78	2.8502	2.8250	0.88	1.0666	1.0634	0.30
10.00	4.4994	4.3313	3.74	2.9142	2.8851	1.00	1.0599	1.0571	0.26
15.00	5.7996	5.6069	3.32	3.1603	3.1165	1.38	1.0397	1.0381	0.15
20.00	6.9287	6.7338	2.81	3.3342	3.2807	1.60	1.0296	1.0285	0.10
25.00	7.9440	7.7617	2.29	3.4685	3.4081	1.74	1.0236	1.0228	0.07
30.00	8.8762	8.7170	1.79	3.5779	3.5122	1.84	1.0196	1.0190	0.05
35.00	9.7442	9.6158	1.32	3.6700	3.6002	1.90	1.0167	1.0163	0.04
40.00	10.5605	10.4689	0.87	3.7496	3.6764	1.95	1.0146	1.0143	0.03
45.00	11.3340	11.2841	0.44	3.8196	3.7436	1.99	1.0129	1.0127	0.02
50.00	12.0711	12.0670	0.03	3.8821	3.8038	2.02	1.0116	1.0114	0.02
60.00	13.4557	13.5521	−0.72	3.9898	3.9078	2.06	1.0096	1.0095	0.01
70.00	14.7430	14.9495	−1.40	4.0806	3.9958	2.08	1.0082	1.0082	0.01
80.00	15.9522	16.2759	−2.03	4.1590	4.0720	2.09	1.0072	1.0071	0.01
90.00	17.0969	17.5432	−2.61	4.2280	4.1393	2.10	1.0064	1.0063	0
100.00	18.1871	18.7603	−3.15	4.2895	4.1994	2.10	1.0057	1.0057	0

$$\bar{\mathscr{E}} = 1 + q/\alpha \qquad \text{for } \alpha \geq 1 \tag{15.8}$$

$$\bar{\mathscr{F}} = \pi/2 + q \ln \alpha \qquad \text{for } \alpha \geq 1 \tag{15.9}$$

where $q = \pi/2 - 1$

Values of $\bar{k}$, $\bar{\mathscr{E}}$, and $\bar{\mathscr{F}}$ are presented in Table 15.11 and compared with the numerically determined values of k, $\mathscr{E}$, and $\mathscr{F}$. Table 15.11 also gives the percentage of error determined as

$$e = (\bar{z} - z)100/z \qquad z = \{k, \mathscr{E}, \mathscr{F}\} \qquad \bar{z} = \{\bar{k}, \bar{\mathscr{E}}, \bar{\mathscr{F}}\}$$

When the ellipticity parameter k [Eq. (15.7)], the elliptic integrals of the first and second kinds [Eqs. (15.9) and (15.8)], the normal applied load F, Poisson's ratio ν, and the modulus of elasticity E of the contacting solids are known, we can write the major and minor axes of the contact ellipse and the maximum deformation at the center of the contact, from the analysis of Ref. 34, as

$$D_y = 2(6\bar{k}^2\bar{\mathscr{E}}FR/\pi E')^{1/3} \tag{15.10}$$

$$D_x = 2(6\bar{\mathscr{E}}FR/\pi\bar{k}E')^{1/3} \tag{15.11}$$

$$\delta = \bar{\mathscr{F}}[(9/2\bar{\mathscr{E}}R)(F/\pi\bar{k}E')^2]^{1/3} \tag{15.12}$$

$$E' = \frac{2}{(1 - \nu_a^2)/E_a + (1 - \nu_b^2)/E_b} \tag{15.13}$$

In these equations, D_y and D_x are proportional to $F^{1/3}$ and δ is proportional to $F^{2/3}$.

The maximum hertzian stress at the center of the contact can also be determined by using Eqs. (15.10) and (15.11).

$$\sigma_{max} = 6F/\pi D_x D_y$$

***Simplified Solutions for* $\alpha \leq 1$.** Table 15.12 gives the simplified equations for $\alpha < 1$ as well as for $\alpha \geq 1$. It is important to make the proper evaluation of α since it has a great significance in the outcome of the simplified equations. Figure 15.10 shows three diverse situations in which the simplified equations can be usefully applied. The locomotive wheel on a rail, Fig. 15.10*a,* illustrates an example in which the ellipticity parameter k and the radius ratio α are less than 1. The ball rolling against a flat plate, Fig. 15.10*b,* provides pure circular contact (i.e., $\alpha = k = 1.0$). Figure 15.10*c* shows how the contact ellipse is formed in the ball–outer-ring contact of a ball bearing. Here the semimajor axis is normal to the direction of rolling and consequently α and k are greater than 1. Table 15.13 uses this figure to show how the degree of conformity affects the contact parameters.

15.6.2 Rectangular Contacts

For this situation, the contact ellipse discussed in the preceding section is of infinite length in the transverse direction ($D_y \to \infty$). This type of contact is exemplified by a cylinder loaded against a plate, a groove, or another, parallel, cylinder or by a roller loaded against an inner or outer ring. In these situations, the contact semiwidth is given by

$$b = R_x \sqrt{8W/\pi}$$

TABLE 15.12 Simplified Equations

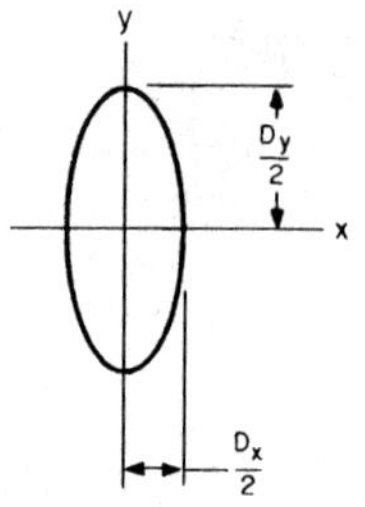	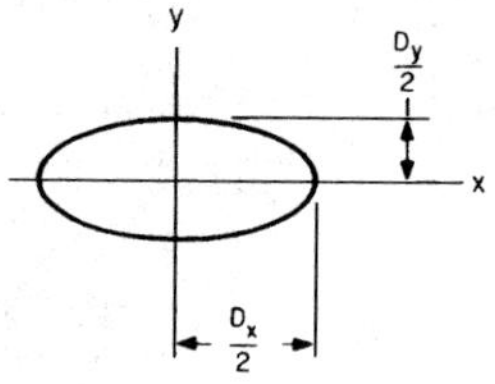
$\alpha \geq 1$	$\alpha < 1$
$\bar{k} = \alpha^{2/\pi}$	$\bar{k} = \alpha^{2/\pi}$
$\bar{\mathcal{F}} = \frac{\pi}{2} + q \ln \alpha$	$\bar{\mathcal{F}} = \frac{\pi}{2} + q \ln \alpha$
where $q = \frac{\pi}{2} - 1$	where $q = \frac{\pi}{2} - 1$
$\bar{\mathcal{E}} = 1 + \frac{q}{\alpha}$	$\bar{\mathcal{E}} = 1 + q\alpha$
$D_y = 2\left(\frac{6\bar{k}^2\bar{\mathcal{E}}FR}{\pi\mathcal{E}'}\right)^{1/3}$	$D_y = 2\left(\frac{6\bar{k}\bar{\mathcal{E}}FR}{\pi E'}\right)^{1/3}$
where $R^{-1} = R_x^{-1} + R_y^{-1}$	where $R^{-1} = R_x^{-1} + R_y^{-1}$
$D_x = 2\left(\frac{6\bar{\mathcal{E}}FR}{\pi\bar{k}E'}\right)^{1/3}$	$D_x = 2\left(\frac{6\bar{\mathcal{E}}FR}{\pi E'\bar{k}^2}\right)^{1/3}$
$\delta = \bar{\mathcal{F}}\left[\left(\frac{4.5}{\bar{\mathcal{E}}R}\right)\left(\frac{F}{\pi\bar{k}E'}\right)^2\right]^{1/3}$	$\delta = \bar{\mathcal{F}}\left[\left(\frac{4.5}{\bar{\mathcal{E}}R}\right)\left(\frac{F\bar{k}}{\pi E'}\right)^2\right]^{1/3}$

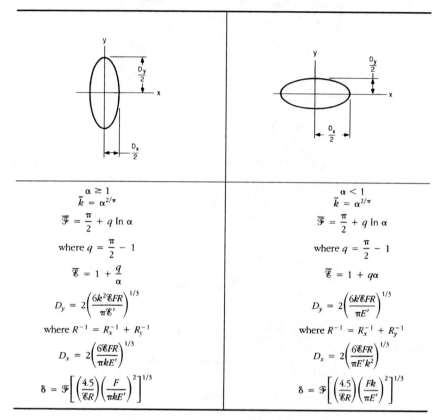

where the dimensionless load

$$W = F'/E'R_x$$

and F' is the load per unit length along the contact. The maximum deformation for a rectangular contact can be written as (Ref. 19)

$$\delta = (2WR_x/\pi)[2/3 + \ln(4r_{ax}/b) + \ln(4r_{bx}/b)] \qquad (15.14)$$

The maximum hertzian stress in a rectangular contact can be written as

$$\sigma_{max} = E'\sqrt{W/2\pi}$$

15.7 STATIC LOAD DISTRIBUTION

With the simple analytical expression for deformation in terms of load defined in Sec. 15.6, it is possible to consider how the bearing load is distributed among the rolling elements. Most rolling-element bearing applications involve steady-state rotation of either the inner or outer race or both; however, the speeds of rotation are usually not so great as to cause ball or roller centrifugal forces or gyroscopic moments of significant magnitudes. In analyzing the loading distribution on the rolling elements, it is usually satisfactory to ignore these effects in most applications. In this section the load-deflection relationships for ball

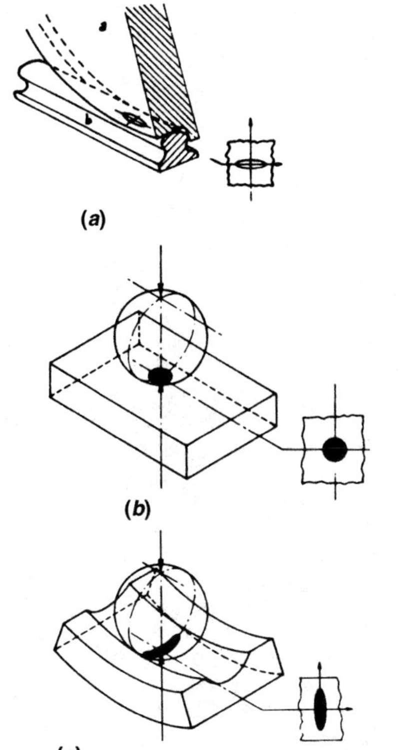

FIG. 15.10 Three degrees of conformity. (*a*) Wheel on rail. (*b*) Ball on plane. (*c*)Ball–outer-race contact.

and roller bearings are given, along with radial and thrust load distributions of statically loaded rolling elements.

15.7.1 Load-Deflection Relationships

For an elliptical contact, the load-deflection relationship given in Eq. (15.12) can be written as

$$F = K_{1.5}\delta^{3/2} \tag{15.15}$$

where $K_{1.5} = \pi k E' \sqrt{2\mathscr{E}R/9\mathscr{F}^3}$ (15.16)

Similarly, for a rectangular contact, Eq. (15.14) gives

$$f = K_1\delta$$

$$\text{where } K_1 = \frac{\pi l E'/2}{\tfrac{2}{3} + \ln(4r_{ax}/b) + \ln(4r_{bx}/b)} \tag{15.17}$$

In general then,

$$F = K_j\delta^j \tag{15.18}$$

TABLE 15.13 Practical Applications for Differing Conformities

($E' = 2.197 \times 10^7$ N/cm²)

Contact parameters	Wheel on rail	Ball on plane	Ball–outer-race contact
F, N	1.00×10^5	222.4111	222.4111
r_{ax}, cm	50.1900	0.6350	0.6350
r_{ay}, cm	∞	0.6350	0.6350
r_{bx}, cm	∞	∞	−3.8900
r_{by}, cm	30.0000	∞	−0.6600
α	0.5977	1.0000	22.0905
k	0.7099	1.0000	7.3649
$\bar{k}$	0.7206	1.0000	7.1738
$\mathscr{E}$	1.3526	1.5708	1.0267
$\bar{\mathscr{E}}$	1.3412	1.5708	1.0258
$\mathscr{F}$	1.8508	1.5708	3.3941
$\bar{\mathscr{F}}$	1.8645	1.5708	3.3375
D_y, cm	1.0783	0.0426	0.1842
$\bar{D}_y$, cm	1.0807	0.0426	0.1810
D_x, cm	1.5190	0.0426	0.0250
$\bar{D}_x$, cm	1.4997	0.0426	0.0252
δ, cm	0.0106	7.13×10^{-4}	3.56×10^{-4}
$\bar{\delta}$, cm	0.0108	7.13×10^{-4}	3.57×10^{-4}
σ_{max}, N/cm²	1.66×10^5	2.34×10^5	9.22×10^4
$\bar{\sigma}_{max}$, N/cm²	1.1784×10^5	2.34×10^5	9.30×10^4

in which $j = 1.5$ for ball bearings and 1.0 for roller bearings. The total normal approach between two races separated by a rolling element is the sum of the deformations under load between the rolling element and both races. Therefore,

$$\delta = \delta_o + \delta_i \tag{15.19}$$

where

$$\delta_o = [F/(K_j)_0]^{1/j} \tag{15.20}$$

$$\delta_i = [F/(K_j)_i]^{1/j} \tag{15.21}$$

Substituting Eqs. (15.19) through (15.21) into Eq. (15.18) gives

$$K_j = \{[1/(K_j)_o]^{1/j} + [1/(K_j)_i]^{1/j}\}^{-j}$$

Recall that $(K_j)_o$ and $(K_j)_i$ are defined by Eq. (15.16) or (15.17) for an elliptical or rectangular contact, respectively. From these equations, we observe that $(K_j)_o$ and $(K_j)_i$ are functions of only the geometry of the contact and the material properties. The radial and thrust load analyses are presented in the following two sections and are directly applicable for radially loaded ball and roller bearings and thrust-loaded ball bearings.

15.7.2 Radially Loaded Ball and Roller Bearings

A radially loaded rolling element with radial clearance P_d is shown in Fig. 15.11. In the concentric position shown in Fig. 15.11*a*, a uniform radial clearance between the rolling element and the races of $P_d/2$ is evident. The application of a small radial load to the shaft causes the inner ring to move a distance $P_d/2$ before contact is made between a rolling element located on the load line and the inner and outer races. At any angle, there will still be a radial clearance c that, if P_d is small compared with the radius of the tracks, can be expressed with adequate accuracy by $c = (1 - \cos \psi)P_d/2$. On the load line where $\psi = 0$, the clearance is zero, but where $\psi = 90°$, the clearance retains its initial value of $P_d/2$.

The application of further load will cause elastic deformation of the balls and the elimination of clearance around an arc $2\psi_c$. If the interference or total elastic compression on the load line is δ_{max}, the corresponding elastic compression of the ball δ_ψ along a radius at angle ψ to the load line will be given by

$$\delta_\psi = (\delta_{max} \cos \psi - c) = (\delta_{max} + P_d/2) \cos \psi - P_d/2$$

This assumes that the races are rigid. Now, it is clear from Fig. 15.11*c* that $(\delta_{max} + P_d/2)$ represents the total relative radial displacement of the inner and outer races. Hence,

$$\delta_\psi = \delta \cos \psi_d - P/2 \tag{15.22}$$

The relationship between load and elastic compression along the radius at angle ψ to the load vector is given by Eq. (15.18) as

$$F_\psi = K_j \delta_\psi^{\,j}$$

Substituting Eq. (15.22) into this equation gives $F_\psi = K_j(\delta \cos \psi - P_d/2)^j$.

For static equilibrium, the applied load must equal the sum of the components of the rolling-element loads parallel to the direction of the applied load. $F_r = \Sigma F_\psi \cos \psi$. Therefore,

$$F_r = K_j \Sigma(\delta \cos \psi - P_d/2)^j \cos \psi \tag{15.23}$$

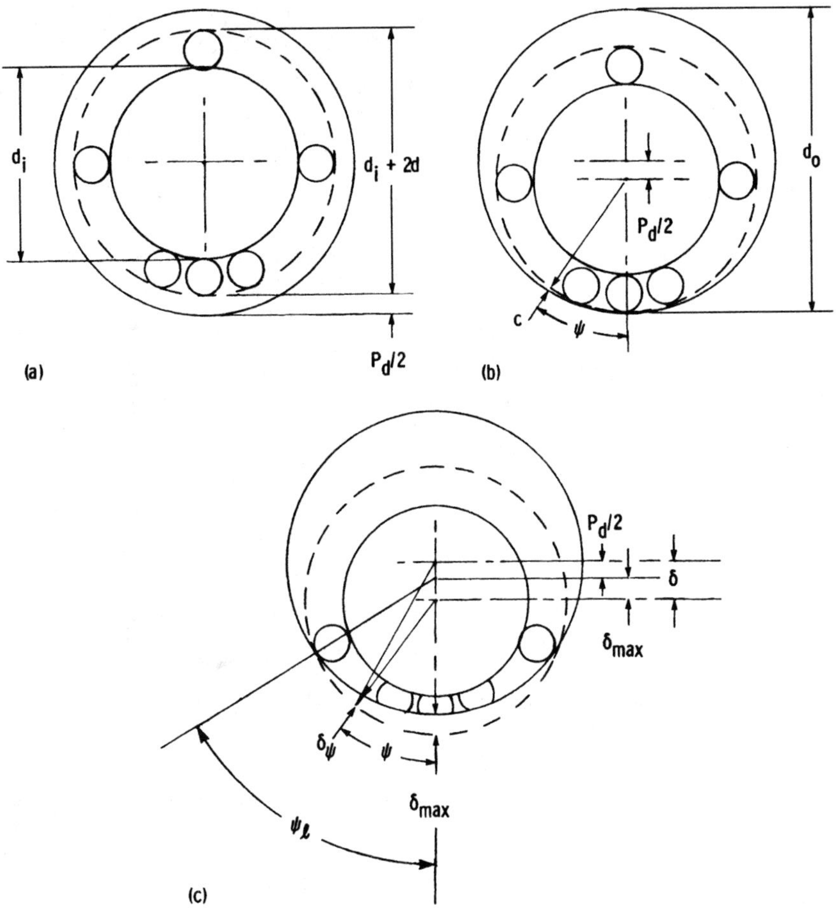

FIG. 15.11 Radially loaded rolling-element bearing. (*a*) Concentric arrangement. (*b*) Initial contact. (*c*) Interference.

The angular extent of the bearing arc $2\psi_p$ in which the rolling elements are loaded, is obtained by setting the root expression in Eq. (15.23) equal to zero and solving for ψ:

$$\psi_l = \arccos(P_d/2\delta)$$

The summation in Eq. (15.23) applies only to the angular extent of the loaded region. This equation can be written for a *roller bearing* as

$$F_r = [\psi_l - (P_d/2\delta)\sin\psi_l](nK_1\delta/2\pi) \tag{15.24}$$

and similarly in integral form for a *ball bearing* as

$$F_r = \frac{n}{\pi} k_{1.5}\delta^{3/2} \int_0^{\psi l} \left(\cos\psi - \frac{P_d}{2\delta}\right)^{3/2} \cos\psi \, d\psi$$

The integral in the equation can be reduced to a standard elliptic integral by the hypergeometric series and the beta function. If the integral is numerically evaluated directly, the following approximate expression is derived:

$$\int_0^{\psi_l} \left(\cos\psi - \frac{P_d}{2\delta}\right)^{3/2} \cos\psi \, d\psi = 2.491 \left\{ \left[1 + \left(\frac{P_d/2\delta - 1}{1.23} \right)^2 \right]^{1/2} - 1 \right\}$$

This approximate expression fits the exact numerical solution to within ± 2 percent for a complete range of $P_d/2\delta$.

The load carried by the most heavily loaded ball is obtained by substituting $\psi = 0°$ in Eq. (15.23) and dropping the summation sign

$$F_{max} = K_j \delta^j (1 - P_d/2\delta)^j \qquad (15.25)$$

Dividing the maximum ball load, Eq. (15.25), by the total radial load for a *roller bearing*, Eq. (15.24), gives

$$F_r = \frac{[\psi_l - (P_d/2\delta)\sin\psi_l](nF_{max}/2\pi)}{1 - P_d/2\delta} \qquad (15.26)$$

and similarly for a *ball bearing*

$$F_r = nF_{max}/Z \qquad (15.27)$$

where

$$Z = \frac{\pi(1 - P_d/2\delta)^{3/2}}{2.491\{[1 + ((1 - P_d/2\delta)/1.23))^2]^{1/2} - 1\}} \qquad (15.28)$$

For *roller bearings* when the diametral clearance P_d is zero, Eq. (15.26) gives

$$F_r = nF_{max}/4 \qquad (15.29)$$

For *ball bearings* when the diametral clearance P_d is zero, the value of Z in Eq. (15.27) becomes 4.37. This is the value derived for ball bearings of zero diametral clearance.[57] The approach used was to evaluate the finite summation for various numbers of balls. The celebrated Stribeck equation for static load-carrying capacity was then derived by writing the more conservative value of 5 for the theoretical value of 4.37:

$$F_r = nF_{max}/5 \qquad (15.30)$$

In using Eq. (15.30), it should be remembered that Z was considered to be a constant and that the effects of clearance and applied load on load distribution were not taken into account. However, these effects were considered in obtaining Eq. (15.27).

15.7.3 Thrust-Loaded Ball Bearings

The "static-thrust-load capacity" of a ball bearing may be defined as the maximum thrust load that the bearing can endure before the contact ellipse approaches a race shoulder, as shown in Fig. 15.12, or the load at which the allowable mean compressive stress is reached, whichever is smaller. Both the limiting shoulder height and the mean compressive stress must be calculated to find the static-thrust-load capacity.

The contact ellipse in a bearing race under a load is shown in Fig. 15.12. Each ball is subjected to an identical thrust component F_t/n, where F_t is the total thrust load. The initial contact angle before the application of a thrust load is denoted by β_f. Under load, the normal ball thrust load F acts at the contact angle β and is written as

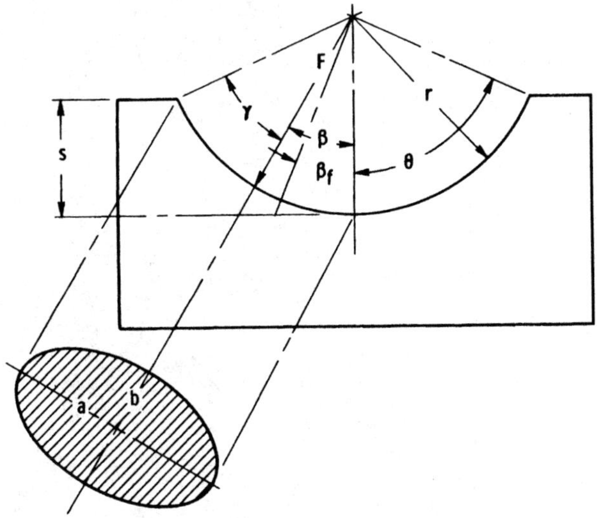

FIG. 15.12 Contact ellipse in bearing race.

$$F = F_t/(n \sin \beta) \tag{15.31}$$

A cross section through an angular-contact bearing under a thrust load F_t is shown in Fig. 15.13. From this figure, the contact angle after the thrust load has been applied can be written as

$$\beta = \arccos\left(\frac{D - P_d/2}{D + \delta}\right) \tag{15.32}$$

The initial contact angle is given in the equation $\cos \beta_f = (D - P_d/2)/D$. Using this equation and rearranging terms in Eq. (15.32) gives, solely from geometry (Fig. 15.13).

$$\delta = D(\cos \beta_f/\cos \beta - 1)$$

$$\delta = \delta_o + \delta_i$$

$$\delta = [F/(K_j)_o]^{1/j} + [F/(K_j)_i]^{1/j}$$

$$K_j = \{[1/(K_j)_o]^{1/j} + [1/(K_j)_i]^{1/j}\}^{-j}$$

$$= \{[4.5\overline{\mathscr{F}}_0^{\,3}/\pi \bar{k}_o E_o'(R_o\overline{\mathscr{E}}_o)^{1/2}]^{2/3} + [4.5\overline{\mathscr{F}}_i^{\,3}/\pi \bar{k}_i \bar{E}_i(R_i\overline{\mathscr{E}}_i)^{1/2}]^{2/3}\}^{-1} \tag{15.33}$$

$$F = K_j D^{3/2}(\cos \beta_f/\cos \beta - 1)^{3/2} \tag{15.34}$$

where $$K_{1.5} = \pi \bar{k} E'(R\overline{\mathscr{E}}/4.5\overline{\mathscr{F}}^3)^{1/2} \tag{15.35}$$

and k, $\overline{\mathscr{E}}$, and $\overline{\mathscr{F}}$ are given by Eqs. (15.7), (15.8), and (15.9), respectively.

From Eqs. (15.31) and (15.34), we can write

$$F_t/(n \sin \beta) = F$$

$$F_t/nK_jD^{3/2} = (\sin \beta)(\cos \beta_f/\cos \beta - 1)^{3/2} \tag{15.36}$$

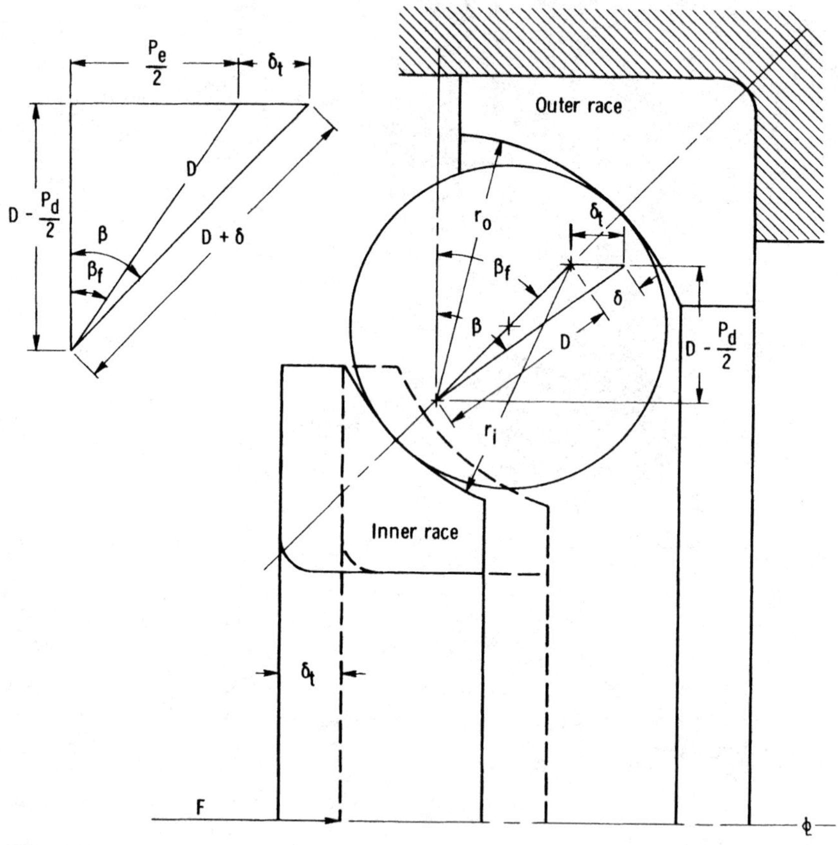

FIG. 15.13 Angular-contact ball bearing under thrust load.

This equation can be solved numerically by the Newton-Raphson method. The iterative equation to be satisfied is

$$\beta' - \beta = \frac{F_t/nK_{1.5}D^{3/2} - (\sin\beta)(\cos\beta_f/\cos\beta - 1)^{3/2}}{(\cos\beta)(\cos\beta_f/\cos\beta - 1)^{3/2} + \frac{3}{2}(\cos\beta_f)(\tan^2\beta)(\cos\beta_f/\cos\beta - 1)^{1/2}} \tag{15.37}$$

which converges and is satisfied when $\beta' - \beta$ becomes essentially zero.

When a thrust load is applied, the shoulder height is limited to the distance by which the pressure-contact ellipse can approach the shoulder. As long as the following inequality is satisfied, the pressure-contact ellipse will not exceed the shoulder-height limit $\theta > \beta + \arcsin(D_v/fd)$. The angle used to define the shoulder height θ may be written as $\theta = \arccos(1 - s/fd)$. The axial deflection δ_t corresponding to a thrust load may be written as

$$\delta_t = (D + \delta)\sin\beta - D\sin\beta_f \tag{15.38}$$

Substituting Eq. (15.33) into Eq. (15.38) gives

$$\delta_t = [D\sin(\beta - \beta_f)]/(\cos\beta)$$

Having determined β from Eq. (15.37) and β_f from Eq. (15.19), we can easily evaluate the relationship for δ_r.

15.7.4 Preloading

In Sec. 15.2.1, the use of angular-contact bearings as duplex pairs preloaded against each other is discussed. As shown in Table 15.2, duplex bearing pairs are used in either back-to-back or face-to-face arrangements. Such bearings are usually preloaded against each other by providing what is called "stickout" in the manufacture of the baring. This is illustrated in Fig. 15.14 for a bearing pair used in a back-to-back arrangement. The magnitude of the stickout and the bearing design determine the level of preload on each bearing when the bearings are clamped together as in Fig. 15.14. The magnitude of preload and the load-deflection characteristics for a given bearing pair can be calculated by using Eqs. (15.15), (15.31), (15.33), (15.34), (15.35), and (15.36).

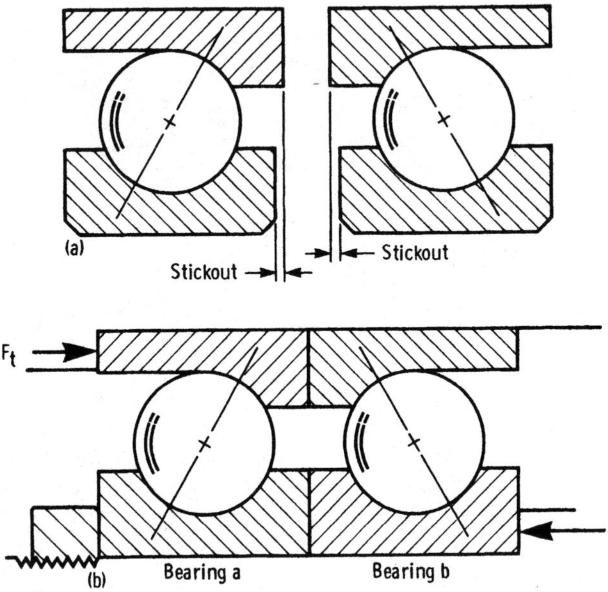

FIG. 15.14 Angular-contact bearings in back-to-back arrangement, shown individually as manufactured and as mounted with preload. (*a*) Separated. (*b*) Mounted and preloaded.

The relationship of initial preload, system load, and final load for bearings *a* and *b* is shown in Fig. 15.15. The load-deflection curve follows the relationship $\delta = KF^{2/3}$. When a system thrust load F_t is imposed on the bearing pairs, the magnitude of load on bearing *b* increases while that on bearing *a* decreases until the difference equals the system load. The physical situation demands that the *change* in each bearing deflection be the same (Δa and Δb in Fig. 15.15). The increments in bearing load, however, are *not* the same. This is important because it always requires a system thrust load far greater than twice the preload before one bearing becomes unloaded. Prevention of bearing unloading, which can result in skidding and early failure, is an objective of preloading.

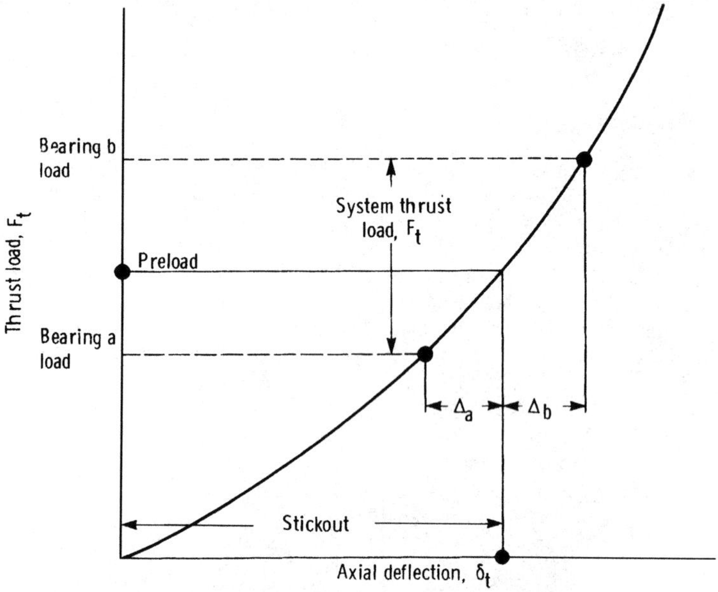

FIG. 15.15 Thrust load–axial deflection curve for a typical ball bearing.

15.8 ROLLING FRICTION AND FRICTION IN BEARING

15.8.1 Rolling Friction

The concepts of rolling friction are important, generally, in understanding the behavior of machine elements in rolling contact and particularly because rolling friction influences the overall behavior of rolling-element bearings. The theories of Refs. 33 and 50 attempted to explain rolling friction in terms of the energy required to overcome the interfacial slip that occurs because of the curved shape of the contact area. As shown in Fig. 15.16, the ball rolls about the *y* axis and makes contact with the groove from *a* to *b*. If the groove is fixed, then for zero slip over the contact area, no point within the area should have a velocity in the direction of rolling. The surface of the contact area is curved, however, so that points *a* and *b* are at different radii from the *y* axis than are points *c* and *d*. For a rigid ball, points *a* and *b* must have different velocities with respect to the *y* axis than do points *c* and *d*, because the velocity of any point on the ball relative to the *y* axis equals the angular velocity times the radius of the *y* axis. Slip must occur at various points over the contact area unless the body is so elastic that yielding can take place in the contact area to prevent this interfacial slip. The theories assumed that this interfacial slip took place and that the forces required to make a ball roll were those required to overcome the friction due to this interfacial slip. In the contact area, rolling without slip will occur at a specific radius from the *y* axis. Where the radius is greater than this radius to the rolling point, slip will occur in one direction; where it is less than this radius to the rolling point, slip will occur in the other direction. In Fig. 15.16, the lines to points *c* and *d* represent the approximate location of the rolling bands, and the arrows shown in the three portions of the contact area represent the directions of interfacial slip when the ball is rolling into the paper.

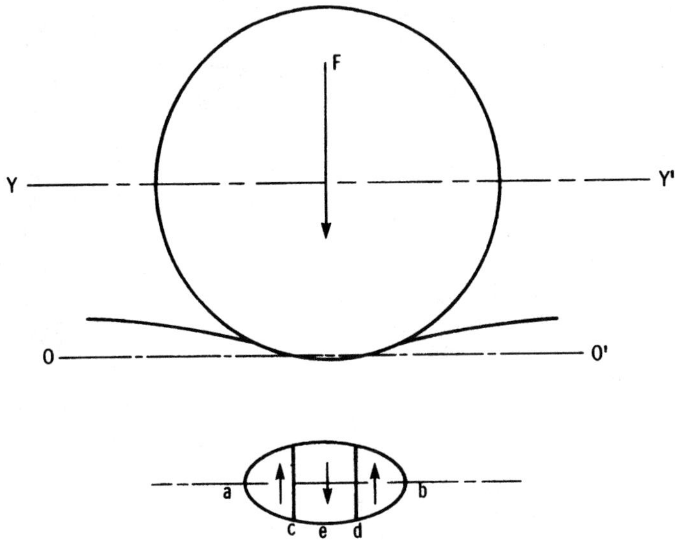

FIG. 15.16 Differential slip due to curvature of contact ellipse.

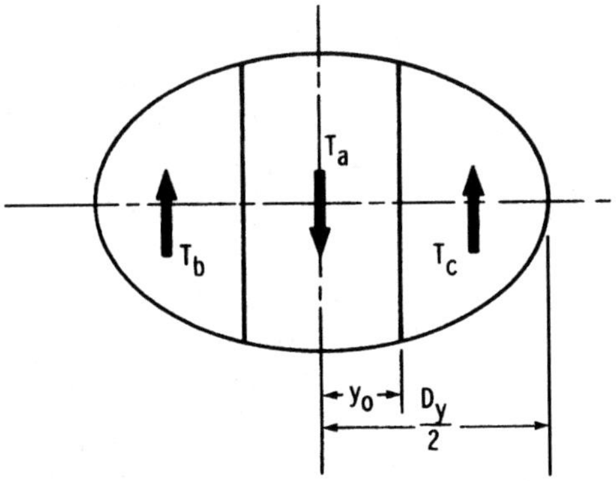

FIG. 15.17 Friction forces in contact ellipse.

The location of the two rolling bands relative to the axes of the contact ellipse can be obtained by summing the forces acting on the ball in the direction of rolling. In Fig. 15.17, these are $2T_b - T_a = \mu F$, where μ is the coefficient of rolling friction. If a hertzian ellipsoidal pressure distribution is assumed, the location of the rolling bands can be determined (Ref. 11). For bearing steels $y_o/D_y \simeq 0.174$, where D_y is the diameter of the contact ellipse along the y axis [Eq. (15.10)].

In actuality, all materials are elastic so that areas of no slip as well as areas of microslip exist within the contact, as pointed out in Ref. 36. Differential strains in the materials in contact will cause slip unless it is prevented by friction. In high-conformity contacts, slip is likely to occur over most of the contact region, as shown in Fig. 15.18. The limiting value of friction force T is $T = 0.2\mu D_y^2 F/r_{ax}^2$.

A portion of the elastic energy of compression in rolling is always lost because of hysteresis. The effect of hysteresis losses on rolling resistance has been studied in Ref. 58, in which the following expression was developed:

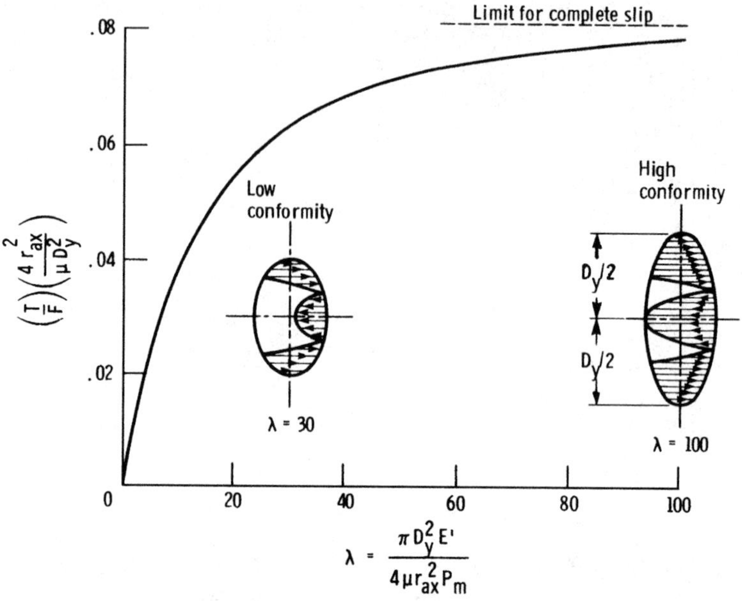

FIG. 15.18 Frictional resistance of ball in conforming groove.

$$T = c_4 \lambda F D_y / r_{ax}$$

where $\lambda = \pi D_y^2 E' / 4\mu r_{ax}^2 p_m$
p_m = mean pressure within the contact
c_4 = ⅓ for rectangular contacts and ³⁄₃₂ for elliptical contacts

Two hard-steel surfaces show a value λ of about 1 percent.

Of far greater importance in contributing to frictional losses in rolling-element bearings, especially at high speeds, is ball spinning. Spinning as well as rolling takes place in one or both of the ball–race contacts of a ball bearing. The situation is shown schematically in Fig. 15.4.

High speeds cause a divergency of the contact angles β_i and β_o. In Fig. 15.5*b* the ball is rolling with inner-race control, so that approximately pure rolling takes place at the inner-race contact. The rolling and spinning vectors ω_r and ω_s at the outer-race contact are shown in Fig. 15.4. The higher the ratio ω_s/ω_r, the higher the friction losses.

15.8.2 Friction Losses in Rolling-Element Bearings

Some of the factors that affect the magnitude of friction losses in rolling bearings are (1) bearing size, (2) bearing type, (3) bearing design, (4) load (magnitude and type, either thrust or radial), (5) speed, (6) oil viscosity, and (7) oil flow. In a specific bearing, friction losses consist of the following:

1. Sliding friction losses in the contacts between the rolling elements and the races. These losses include differential slip and slip due to ball spinning. They are complicated by

the presence of elastohydrodynamic lubrication films. The shearing of elastohydrodynamic lubrication films, which are extremely thin and contain oil whose viscosity is increased by orders of magnitude above its atmospheric pressure value, accounts for a significant fraction of the friction losses in rolling-element bearings.

2. Hysteresis losses due to the damping capacity of the race and ball materials.
3. Sliding friction losses between the separator and its locating race surface and between the separator pockets and the rolling elements.
4. Shearing of oil films between the bearing parts and oil churning losses caused by excess lubricant within the bearing.
5. Flinging of oil off the rotating parts of the bearing.

Because of the dominant role played in bearing frictional losses by the method of lubrication, there are no quick and easy formulas for computing rolling-element bearing power loss. Coefficients of friction for a particular bearing can vary by a factor of 5, depending on lubrication. A flood-lubricated bearing may consume five times the power of one that is merely wetted by oil–air mist lubrication.

As a rough but useful guide, we can use the friction coefficients given in Ref. 46. These values were computed at a bearing load that will give a life of 1×10^9 revolutions for the respective bearings. The friction coefficients for several different bearings are shown below.

Bearing type	Friction coefficient
Self-aligning ball	0.0010
Cylindrical roller, with flange-guided short rollers	0.0011
Thrust ball	0.0013
Single-row, deep-groove ball	0.0015
Tapered and spherical roller, with flange-guided rollers	0.0018
Needle roller	0.0045

All of the friction coefficients are referenced to the bearing bore.

More accurate estimates of bearing power loss and temperature rise can be obtained by using one or more of the available computer codes that represent the basis for current design methodology for rolling-element bearings. These methodologies are discussed in Refs. 5 and 49, and are briefly reviewed later in this chapter.

15.9 LUBRICANTS AND LUBRICATION SYSTEMS

A liquid lubricant has several functions in a rolling-element bearing. It provides the medium for the establishment of separating films between the bearing parts (elastohydrodynamic between the races and rolling elements and hydrodynamic between the cage or separator and its locating surface). It serves as a coolant if circulated through the bearing either to an external heat exchanger or simply brought into contact with the bearing housing and machine casing. A circulating lubricant also serves to flush out wear debris and carry it to a filter where it can be removed from the system. Finally, it provides corrosion protection. The different methods of providing liquid lubricant to a bearing are discussed here individually.

15.9.1 Solid Lubrication

An increasing number of rolling-element bearings are lubricated with solid-film lubricants, usually in applications such as extreme temperature or the vacuum of space where conventional liquid lubricants are not suitable.

Success in cryogenic applications where the bearing is cooled by the cryogenic fluid (liquid oxygen or hydrogen) has been achieved with transfer films of polytetrafluoroethylene (Ref. 52). Bonded films of soft metals, such as silver, gold, and lead, applied by ion plating as very thin films (0.2 to 0.3 μm) have also been used (Ref. 60). Silver, and lead in particular, have found use in bearings used to support the rotating anode in x-ray tubes. Very thin films are required in rolling-element bearings in order not to significantly alter the bearing internal geometry and to retain the basic mechanical properties of the substrate materials in the hertzian contacts.

15.9.2 Liquid Lubrication

The great majority of rolling-element bearings are lubricated by liquids. The liquid in greases can be used or liquid can be supplied to the bearing from either noncirculating or circulating systems.

Greases. The most common and probably least expensive mode of lubrication is grease lubrication. In the strictest sense, a grease is not a liquid, but the liquid or fluid constituent in the grease is the lubricant. Greases consist of a fluid phase of either a petroleum oil or a synthetic oil and a thickener. The most common thickeners are sodium-, calcium-, or lithium-based soaps, although thickeners of inorganic materials such as bentonite clay have been used in synthetic greases. Some discussion of the characteristics and temperature limits of greases is given in Refs. 11 and 44.

Greases are usually retained within the bearing by shields or seals that are an integral part of the assembled bearing. Since there is no recirculating fluid, grease-lubricated bearings must reject their heat by conduction and convection. They are therefore limited in operating speeds to maximum d_bN values of 0.25×10^6 to 0.4×10^6.

The proper grease for a particular application depends on the temperature, speed, and ambient pressure environment to which the bearing is exposed. Reference 44 presents a comprehensive discussion useful in the selection of a grease; the bearing manufacturer can also recommend the most suitable grease and bearing type.

Nonrecirculating Liquid Lubrication Systems. At low to moderate speeds, where the use of grease lubrication is not suitable, other methods of supplying lubricant to the bearing can be used. These include splash or bath lubrication, wick, oil-ring, and oil–air mist lubrication. Felt wicks can be used to transport oil by capillary action from a nearby reservoir. Oil rings, which are driven by frictional contact with the rotating shaft, run partially immersed in an oil reservoir and feed oil mechanically to the shaft, which is adjacent to the bearing. The bearing may itself be partially immersed in an oil reservoir to splash-lubricate itself. All of these methods require very modest ambient temperatures and thermal conditions as well as speed conditions equivalent to a maximum d_bN of about 0.5×10^6. The machinery must also remain in a fixed-gravity orientation.

Oil–air mist lubrication supplies atomized oil in an airstream to the bearing, where a reclassifier increases the droplet size, allowing it to condense on the bearing surfaces. Feed rates are low and a portion of the oil flow escapes with the feed air to the atmosphere. Commercial oil–air mist generators are available for systems ranging from a single bearing to hundreds of bearings. Bearing friction losses and heat generation with mist lubrication are low, but ambient temperatures and cooling requirements must be moderate because oil–air mist systems provide minimal cooling. Many bearings, especially small-bore bearings, are successfully operated at high speeds (d_bN values to greater than 10^6) with oil–air mist lubrication.

Jet Lubrication. In applications where speed or heat-rejection requirements are too high, jet lubrication is frequently used to lubricate and control bearing temperatures. A number of variables are critical to achieving not only satisfactory but near-optimal performance and bearing operation. These include the placement of the nozzles, the number of nozzles, the jet velocity, the lubricant flow rate, and the scavenging of the lubricant from the bearing and immediate vicinity. The importance of proper jet lubricating system design is shown in Ref. 43 and is summarized in Fig. 15.19.

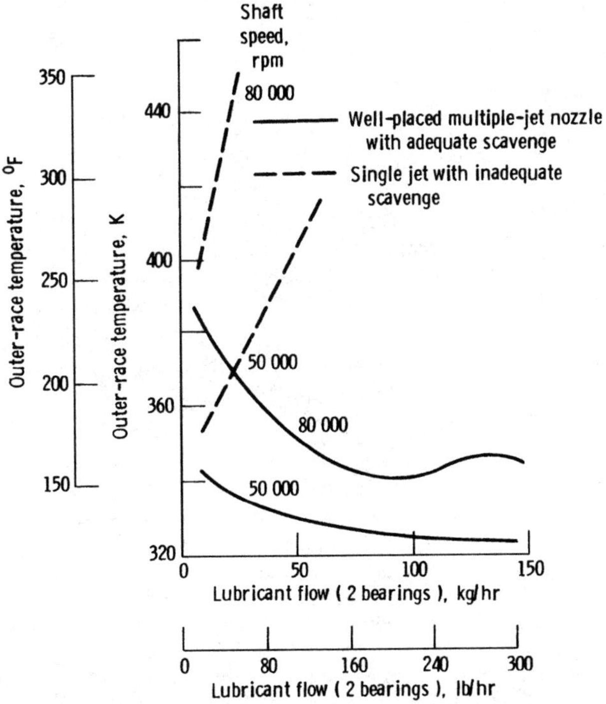

FIG. 15.19 Effectiveness of proper jet lubrication. Test bearings, 20-mm-bore angular-contact ball bearings; thrust load, 222 N (50 lbf). (*From Matt and Gianotti.*[43])

Proper placement of the jets should take advantage of any natural pumping ability of the bearings. Figure 15.20 (Ref. 47) illustrates jet lubrication of ball and tapered-roller bearings. Centrifugal forces aid in moving the oil through the bearing to cool and lubricate the elements. Directing jets at the radial gaps between the cage and the races achieves maximum penetration of oil into the interior of the bearing. References 6, 45, 48, and 64 present useful data on the influence of jet placement and velocity on the lubrication of several types of bearings.

Under-Race Lubrication. As bearing speeds increase, centrifugal effects become more predominant, making it increasingly difficult to effectively lubricate and cool a

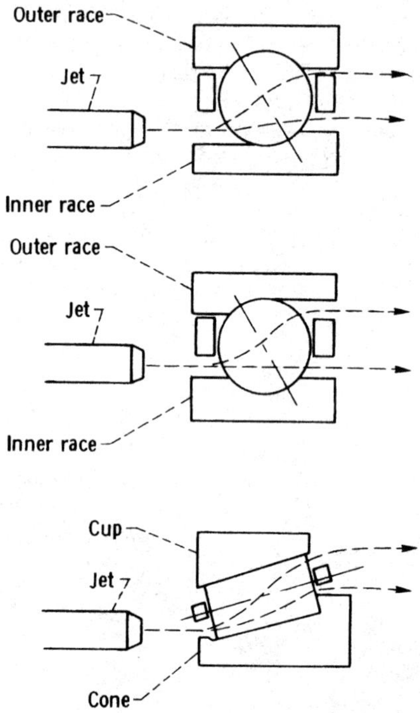

FIG. 15.20 Placement of jets for ball bearings with relieved rings and tapered-roller bearings.

bearing. The jetted oil is thrown off the sides of the bearing rather than penetrating to the interior. At extremely high d_bN values (2.4×10^6 and higher), jet lubrication becomes ineffective. Increasing the flow rate only adds to heat generation through increased churning losses. Reference 12 describes an under-race oiling system used in a turbofan engine for both ball and cylindrical roller bearings. Figure 15.21 illustrates the technique. The lubricant is directed radially under the bearings. Centrifugal effects assist in pumping oil out through the bearings through suitable slots and holes, which are made a part of both the shaft and mounting system and the bearings themselves. Holes and slots are provided within the bearing to feed oil directly to the ball-race and cage-race contacts.

This lubrication technique has been thoroughly tested for large-bore ball and roller bearings up to $d_bN = 3 \times 10^6$. Pertinent data are reported in Refs. 13, 51, 54, and 55.

15.10 ELASTOHYDRODYNAMIC LUBRICATION

Elastohydrodynamic lubrication is a form of fluid-film lubrication where elastic deformation of the bearing surfaces becomes significant. It is usually associated with highly stressed machine components such as rolling-element bearings. Historically, elastohydrodynamic lubrication may be viewed as one of the major developments in the field of lubrication in the twentieth century. It not only revealed the existence of a previously unsuspected regime of lubrication in highly stressed nonconforming machine elements,

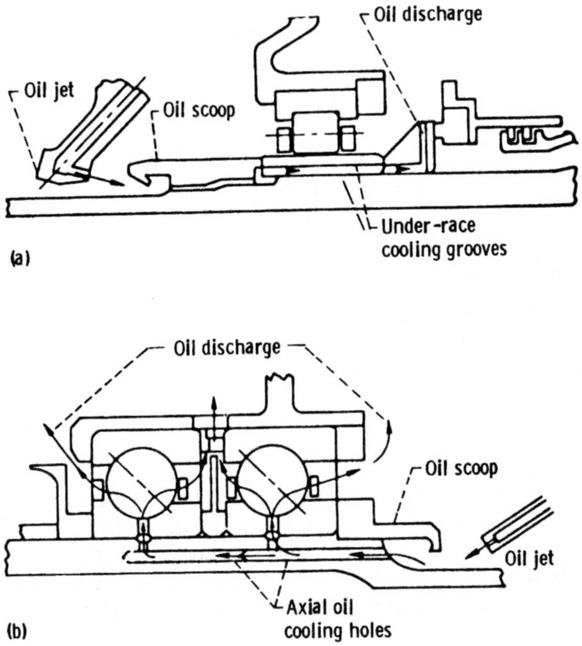

FIG. 15.21 Under-race oiling system for main shaft bearings on turbofan engine. (*a*) Cylindrical-roller bearing. (*b*) Ball thrust bearing. (*From Brown, 1970.*[12])

but it also brought order to the complete spectrum of lubrication regimes, ranging from boundary to hydrodynamic.

15.10.1 Relevant Equations

The relevant equations used in elastohydrodynamic lubrication are given here.

Lubrication Equation (Reynolds Equation)

$$\frac{\partial}{\partial x}\left(\frac{\rho h^3}{\eta}\frac{\partial p}{\partial x}\right) + \frac{\partial}{\partial y}\left(\frac{\rho h^3}{\eta}\frac{\partial p}{\partial y}\right) = 12u\frac{\partial}{\partial x}(\rho h)$$

where $u = (u_a + u_b)/2$.

Viscosity Variation

$$\eta = \eta_0 e^{\xi p}$$

where η_0 = coefficient of absolute or dynamic viscosity at atmospheric pressure and ξ = pressure-viscosity coefficient of the fluid.

Density Variation (for Mineral Oils)

$$\rho = \rho_0\left(1 + \frac{0.6p}{1 + 1.17p}\right)$$

where ρ_0 = density of atmospheric conditions.

Elasticity Equation

$$\delta = \frac{2}{E'} \iint \frac{p(x,y)\,dx\,dy}{\sqrt{(x - x_1)^2 + (y - y_1)^2}}$$

where $$E' = \frac{2}{(1 - \nu_a^2)/E_a + (1 - \nu_b^2)/E_b}$$

Film Thickness Equation

$$h = h_0 + x^2/2R_x + y^2/2R_y + \delta(x,y)$$

where $$1/R_x = 1/r_{ax} + 1/r_{bx} \tag{15.39}$$

$$1/R_y = 1/r_{ay} + 1/r_{by} \tag{15.40}$$

The elastohydrodynamic lubrication solution therefore requires the calculation of the pressure distribution within the conjunction, at the same time allowing for the effects that this pressure will have on the properties of the fluid and on the geometry of the elastic solids. The solution will also provide the shape of the lubricant film, particularly the minimum clearance between the solids. A detailed description of the elasticity model one could use is given in Ref. 16, and the complete elastohydrodynamic lubrication theory is given in Ref. 29.

15.10.2 Dimensionless Grouping

The variables resulting from the elastohydrodynamic lubrication theory are

E' = effective elastic modulus, N/m^2

F = normal applied load, N

h = film thickness, m

R_x = effective radius in x (motion) direction, m

R_y = effective radius in y (transverse) direction, m

u = mean surface velocity in x direction, m/s

ξ = pressure-viscosity coefficient of fluid, m^2/N

η_0 = atmospheric viscosity, N·s/m^2

From these variables the following five dimensionless groupings can be established:

Dimensionless Film Thickness

$$H = h/R_x$$

Ellipticity Parameter

$$k = D_y/D_x = (R_y/R_x)^{2/\pi}$$

Dimensionless Load Parameter

$$W = F/E'R_x^2 \tag{15.41}$$

Dimensionless Speed Parameter

$$U = \eta_0 u / E' R_x \tag{15.42}$$

Dimensionless Materials Parameter

$$G = \xi E' \tag{15.43}$$

The dimensionless minimum film thickness can thus be written as a function of the other four parameters:

$$H = f(k, U, W, G)$$

The most important practical aspect of elastohydrodynamic lubrication theory is the determination of the minimum film thickness within a conjunction. That is, maintaining a fluid-film thickness of adequate magnitude is extremely important to the operation of such machine elements as rolling-element bearings. Specifically, elastohydrodynamic film thickness influences fatigue life, as discussed in Sec. 15.11.

15.10.3 Minimum-Film-Thickness Formula

By using the numerical procedures outlined in Ref. 27, the influence of the ellipticity parameter and the dimensionless speed, load, and materials parameters on minimum film thickness has been investigated in Ref. 28. The ellipticity parameter k was varied from 1 (a ball-on-plate configuration) to 8 (a configuration approaching a rectangular contact). The dimensionless speed parameter U was varied over a range of nearly two orders of magnitude, and the dimensionless load parameter W over a range of one order of magnitude. Situations equivalent to using solid materials of bronze, steel, and silicon nitride and lubricants of paraffinic and naphthenic oils were considered in the investigation of the role of the dimensionless materials parameter G. Thirty-four cases were used in generating the minimum-film-thickness formula given here:

$$H_{\text{min}} = 3.63\, U^{0.68} G^{0.49} W^{-0.073} (1 - e^{-0.68k}) \tag{15.44}$$

In this equation, the most dominant exponent occurs on the speed parameter, and the exponent on the load parameter is very small and negative. The materials parameter also carries a significant exponent, although the range of this variable in engineering situations is limited.

15.10.4 Pressure and Film-Thickness Plots

A representative contour plot of dimensionless pressure is shown in Fig. 15.22 for $k = 1.25$, $U = 0.168 \times 10^{-11}$, $W = 0.111 \times 10^{-6}$, and $G = 4522$. In this figure and in Fig. 15.23, the + symbol indicates the center of the hertzian contact zone. The dimensionless representation of the x and y coordinates causes the actual hertzian contact ellipse to be a circle, regardless of the value of the ellipticity parameter. The hertzian contact circle is shown by asterisks. There is a key in this figure that shows the contour labels and each corresponding value of dimensionless pressure. The inlet region is to the left and the exit region is to the right. The pressure gradient at the exit end of the conjunction is much larger than that in the inlet region. In Fig. 15.22 a pressure spike is visible at the exit of the contact.

Contour plots of the film thickness are shown in Fig. 15.23 for $k = 1.25$, $U = 0.168 \times 10^{-11}$, $W = 0.111 \times 10^{-6}$, and $G = 4522$. In this figure two minimum

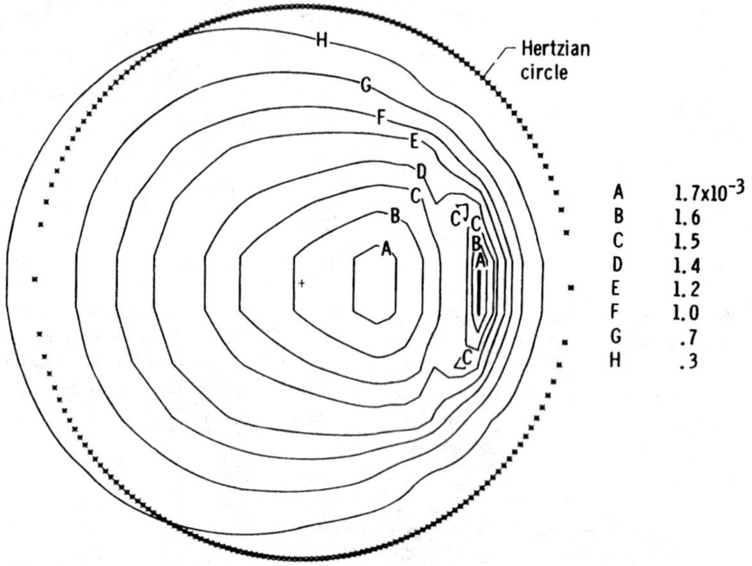

FIG. 15.22 Contour plot of dimensionless pressure. $k = 1.25$; $U = 0.168 \times 10^{-11}$; $W = 0.111 \times 10^{-6}$; $G = 4522$.

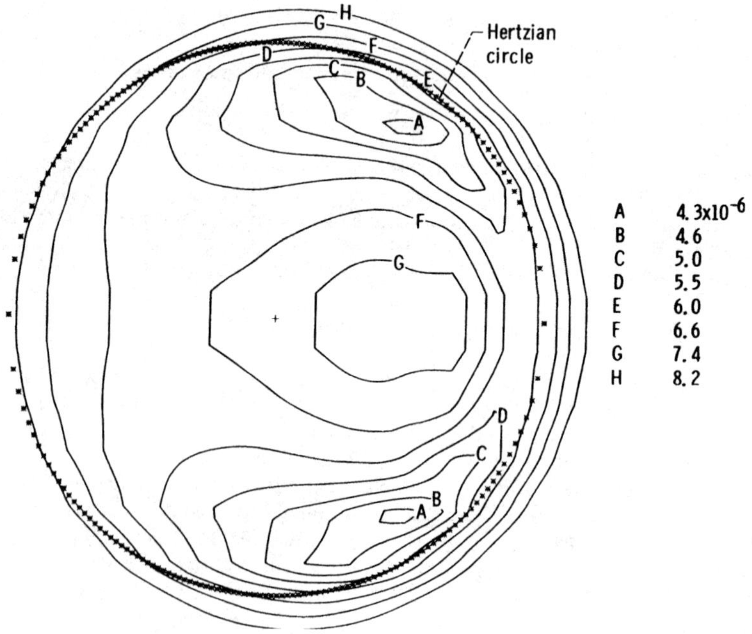

FIG. 15.23 Contour plot of dimensionless film thickness. $k = 1.25$; $U = 0.168 \times 10^{-11}$; $W = 0.111 \times 10^{-6}$; $G = 4522$.

regions occur in well-defined side lobes that follow, and are close to, the edge of the hertzian contact circle. These results produce all the essential features of previously reported experimental observations based on optical interferometry.[14]

15.11 BEARING LIFE

15.11.1 Lundberg-Palmgren Theory

Calculations of bearing life are based upon the assumption that a bearing will fail from rolling-element fatigue.

Rolling fatigue is a material failure caused by the application of repeated stresses to a small volume of material. It is a unique failure type. It is essentially a process by which the first failure occurs at the weakest point. A typical spall is shown in Fig. 15.24. We can surmise that on a microscale there will be a wide dispersion in material strength or resistance to fatigue because of inhomogeneities in the material. Because bearing materials are complex alloys, we would not expect them to be homogeneous or equally resistant to failure at all points. Therefore, the fatigue process can be expected to be one in which a group of supposedly identical specimens exhibit wide variations in failure time when stressed in the same way. For this reason, it is necessary to treat the fatigue process statistically.

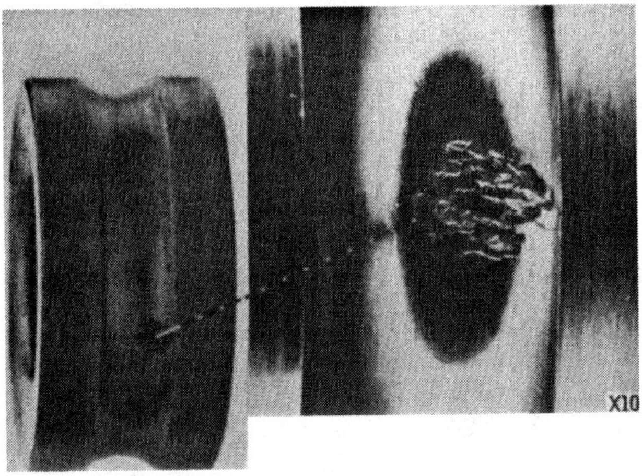

FIG. 15.24 Typical fatigue spall.

To be able to predict how long a particular bearing will run under a specific load, we must have the following two essential pieces of information: (1) an accurate, quantitative estimate of the life dispersion or scatter, and (2) the life at a given survival rate or reliability level. This translates into an expression for the "load capacity," or the ability of the bearing to endure a given load for a stipulated number of stress cycles or revolutions. If a group of supposedly identical bearings is tested at a specific load and speed, there will be a wide scatter in bearing lives, as shown in Fig. 15.25. This distribution is exponential and is typical for most rolling-element bearings. Using Weibull analysis (Ref. 62), the distribution shown in Fig. 15.25 can be plotted linearly on a

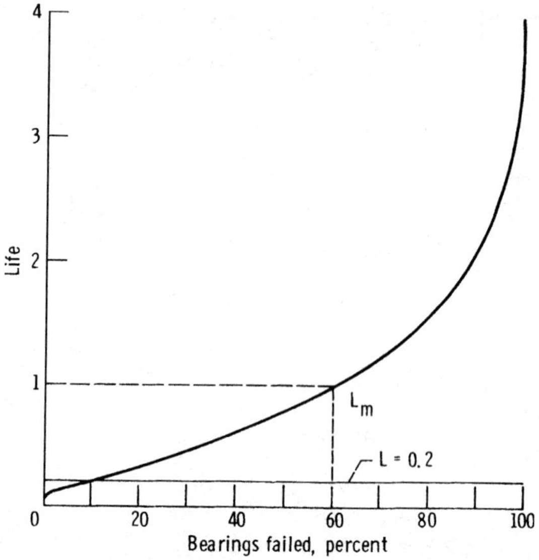

FIG. 15.25 Distribution of bearing fatigue failures.

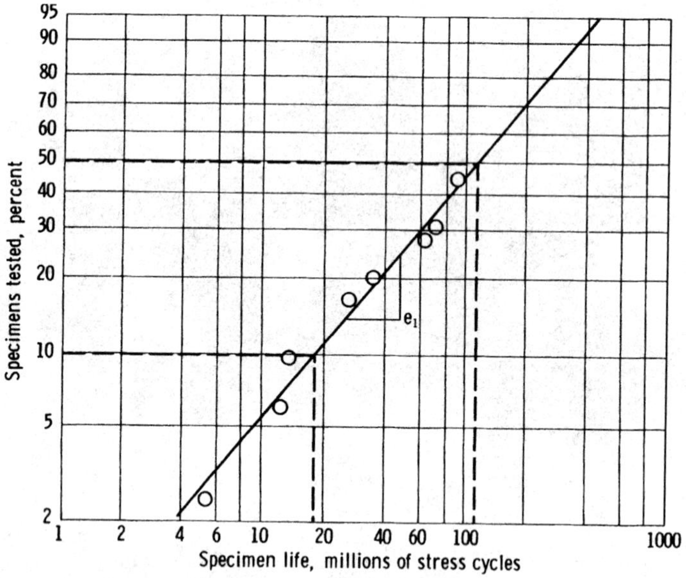

FIG. 15.26 Typical Weibull plot of bearing fatigue failures.

Weibull plot shown in Fig. 15.26. The formula for this distribution can be expressed as a Weibull function,

$$\ln \ln (1/S) = e_1 \ln (L/A) \tag{15.45}$$

where S is the probability of survival expressed as a fraction, e is the Weibull slope, L is the life or time at a probability of survival S, and A is the characteristic life or time at a probability of survival of 0.368. The slope e_1 for most bearings ranges from 1 to 2 and can be assumed for purposes of most analysis to be 1.1.

Lundberg and Palmgren (Refs. 41 and 42) using Weibull analysis developed the Lundberg-Palmgren theory on which bearing life and ratings are based, where

$$\ln (1/S) = \frac{\tau_0^{c_1} L^{e_1}}{Z_0^{c_3}} V \tag{15.46}$$

and

$$V = D_y Z_0 l_v \tag{15.47}$$

Then Eq. (15.46) becomes

$$\ln (1/S) = \frac{\tau_0^{c_1} L^{e_1}}{Z_0^{c_3 - 1}} D_y l_v \tag{15.48}$$

The concept of a bearing dynamic-load capacity C is introduced by Lundberg and Palmgren. This is the theoretical load placed on a bearing which will produce 1 million inner-race revolutions at a probability of survival of 90 percent, or $S = 0.9$. This S of 0.9 is also referred to as the 10 percent or L_{10} life where 10 percent of a group of bearings can be expected to fail before reaching a given time. By substituting the appropriate values in Eq. (15.48) this value of C can be calculated.

Factors on which dynamic load capacity and bearing life depend are:

1. Size of rolling element
2. Number of rolling elements per row
3. Number of rows of rolling elements
4. Conformity between rolling elements and races
5. Contact angle under load
6. Material properties

For different size and types of bearings, values of C can be obtained using AFBMA and ISO ratings (Refs. 1 and 2) and bearing manufacturers' catalogs. Knowing C, the L_{10} life of any other load F_e can be calculated using the following equation from Lundberg-Palmgren:

$$L = (C/F_e)^m \tag{15.49}$$

where C = basic dynamic capacity or load rating
F_e = equivalent bearing load
m = 3 for ball bearings and 10/3 for roller bearings
L = life, millions of race revolutions

The equivalent bearing load F_e can be calculated from the following equation:

$$F_e = XF_r + YF_1 \tag{15.50}$$

Factors X and Y are given in bearing manufacturers' catalogs for specific bearings.

15.11.2 ANSI/AFBMA Standards

The ANSI/AFBMA standards (Refs. 1 and 2) became effective in 1950 for ball bearings and in 1953 for roller bearings. By the early 1960s many bearing manufacturers began to incorporate life factors into their bearing catalog ratings. These life factors are not recognizable to the casual bearing user. Therefore, life estimates obtained from different bearing manufacturers' catalogs can be significantly different.

The Lundberg-Palmgren life theory incorporated into its equations a material factor that reflected bearing technology and life existing prior to 1950. This material factor was based on groups of bearings that were laboratory life-tested over several decades. When these equations were incorporated into the ANSI/AFBMA standards, the material factor used by Lundberg and Palmgren was also incorporated into the standards.

Research in steel metallurgy and processing, lubrication and lubricants, and bearing-manufacturing process has resulted in significant improvement in bearing life over that obtained in the early 1950s. To account for these technology advancements, the ANSI/AFBMA and ISO standards combine three factors into the following formula to adjust for life:

$$L_{na} = a_1 a_2 a_3 L_{10} = a_1 a_2 a_3 \left(\frac{C}{F_e} \right)^m \tag{15.51}$$

where L_{na} = adjusted life, millions of revolutions
C = dynamic load capacity, lb
e = equivalent bearing load, lb
a_1 = adjustment factor for reliability
a_2 = adjustment factor for materials and processing
a_3 = adjustment factor for operating conditions

Zaretsky, in "STLE Life Factors for Rolling Bearings,"[65] lists the variables affecting bearing life and gives specific values as they may apply to specific applications. These variables are listed in Table 15.14. (Reference 65 can be obtained from The Society of Tribologist and Lubrication Engineers, Park Ridge, Ill.)

The value for reliability is obtained from the Weibull equation (15.45) and is in the ANSI/AFBMA standards:

$$a_1 = 4.48 \left(\ln \frac{100}{R} \right)^{2/3} \tag{15.52}$$

where R = reliability factor, percent. Research has shown that at values of S equal 0.999, a_1 = 0.053. At values of S above 0.999, no failure due to fatigue would be expected. At values of S equal 0.9, a_1 = 1.[65]

Material and processing variables affect life and are reflected in life factor a_2. For over a century, AISI 52100 steel has been the predominant material for rolling-element bearings. In fact, the basic dynamic capacity as defined by AFBMA in 1949 is based on an air-melted 52100 steel, hardened to at least Rockwell C 58. Since that time, as discussed in Sec. 15.4.1, better control of air-melting processes and the introduction of vacuum remelting processes have resulted in more homogeneous steels with fewer impurities. Such steels have extended rolling-element bearing fatigue lives to several times the AFBMA or catalog life. Life improvements of 3 to 8 times are not uncommon. Other steel compositions, such as AISI M-1 and AISI M-50, chosen for their higher temperature capabilities and resistance to corrosion, also have shown greater resistance to fatigue pitting when vacuum melting techniques are employed. Case-hardened materials, such as AISI 4620, AISI 4118, and AISI 8620, used primarily for roller bearings, have the advantage of a tough, ductile steel core with a hard, fatigue-resistant surface.

Life factor a_3 reflects the effect of lubricant on bearing life. Until approximately 1960, the role of the lubricant between surfaces in rolling contact was not fully appreciated.

TABLE 15.14 Summary of Life Factors (LF) for Rolling-Element Bearings (From Zaretsky, Ref. 65)

a_1 (reliability)	a_2 (materials and processing)	a_3 (operating conditions)
Probability of failure	Bearing steel* Melting process† Metalworking Material hardness Residual stress Carbide factor‡ Nonferrous materials Ceramic hybrid bearings Bearing refurbishment and restoration Coatings and surface treatments	Load: Heavy load and ring distortion Roller bearing in housing with rigid legs Dish or sag of horizontally mounted large-diameter ball bearing Misalignment Housing clearance Axially loaded cylindrical bearings Rotordynamics Hoop stresses Speed Temperature: Steel Ceramic Lubrication: Lubricant film thickness Starvation Surface finish Grease Additives Traction Water Filtration

*To be used with melting-process LF.
†To be used with bearing-steel LF.
‡To be used in place of bearing-steel LF but with melting-process LF.

Metal-to-metal contact was presumed to occur in all applications, with attendant required boundary lubrication. The development of elastohydrodynamic lubrication theory showed that lubricant films of thicknesses of the order of microinches and tens of microinches occur in rolling contact. Since surface finishes are of the same order of magnitude as the lubricant film thicknesses, the significance of rolling-element bearing surface roughnesses to bearing performance became apparent. Reference 59 first reported on the importance on bearing life of the ratio of elastohydrodynamic lubrication film thickness to surface roughness. Figure 15.27 shows life as a percentage of calculated L_{10} as a function of Λ, where

$$\Lambda = h_{\min}/\sqrt{f_a^2 + f_b^2} \qquad (15.53)$$

Figure 15.28, from Ref. 8, presents a curve of the recommended $\tilde{F}$ factor as a function of the Λ parameter. A mean of the curves presented in Ref. 59 for ball bearings and in Ref. 56 for roller bearings is recommended for use. A formula for calculating the minimum film thickness $h_{\min}$ is given in Eq. (15.44).

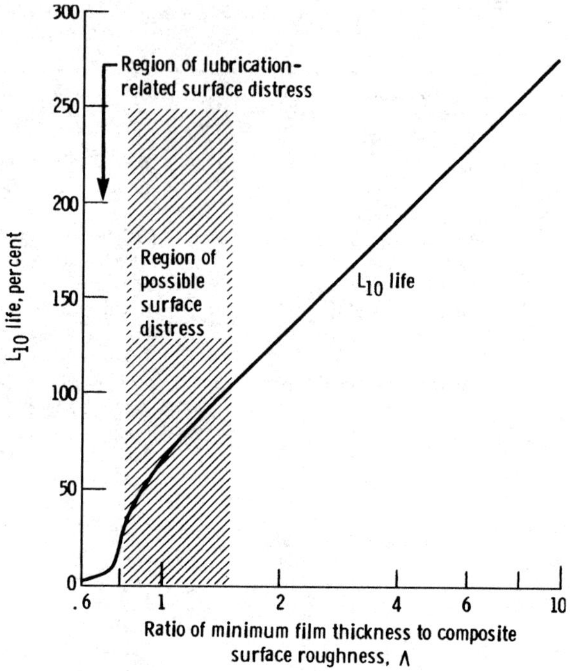

FIG. 15.27 Group fatigue life L_{10} as function of Λ. (*From Tallian, 1967.*[54])

15.12 DYNAMIC ANALYSES AND COMPUTER CODES

As has been stated, a precise analysis of rolling-bearing kinematics, stresses, deflections, and life can be accomplished only by using a large-scale computer code. Presented here is a very brief discussion of some of the significant analyses that have led to the formulation of modern bearing analytical computer codes. The reader is referred to COSMIC, University of Georgia, Athens, Ga., for the public availability of program listings.

15.12.1 Quasi-Static Analyses

Rolling-element bearing analysis began with the work of Jones on ball bearings.[38,39] This work was done before there was a general awareness of elastohydrodynamic lubrication and assumed Coulomb friction in the race contacts. This led to the commonly known "race control" theory, which assumes that pure rolling (except for Heathcote interfacial slip) can occur at one of the ball-race contacts. All of the spinning required for dynamic equilibrium of the balls would then take place at the other, or "noncontrolling," race contact. Jones' analysis proved to be quite effective for predicting fatigue life but less useful for predicting cage slip, which usually occurs at

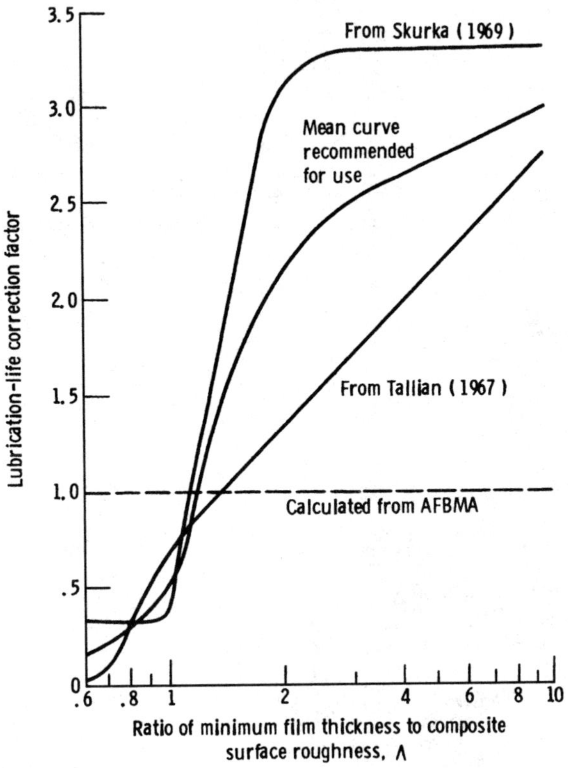

FIG. 15.28 Lubrication-life correction factor as function of Λ. (*From Bamberger.*[8])

high speeds and light loads. Reference 32 extended the analysis, retaining the assumption of Coulomb friction, but allowing a frictional resistance to gyroscopic moments at the noncontrolling as well as the controlling race contact. Reference 32 is adequate for predicting bearing performance under conditions of dry-film lubrication or whenever there is a complete absence of any elastohydrodynamic film.

Reference 31 first incorporated elastohydrodynamic relationships into a ball-bearing analysis. A revised version of Harris' computer program,[32] called SHABERTH, was developed that incorporated actual traction data from a disk machine. The program SHABERTH has been expanded until today it encompasses ball, cylindrical roller, and tapered-roller bearings.

Reference 30 first introduced elastohydrodynamics into a cylindrical roller-bearing analysis. The initial analysis has been augmented with more precise viscosity-pressure and temperature relationships and traction data for the lubricant. This has evolved into the program CYBEAN, and more recently has been incorporated into SHABERTH. Parallel efforts by Harris' associates have resulted in SPHERBEAN, a program that can be used to predict the performance of spherical roller bearings. These analyses can range from relatively simple force balance and life analyses through a complete thermal analysis of a shaft bearing system in several steps of varying complexity.

15.12.2 Dynamic Analyses

The work of Jones and Harris[32,38,39] is categorized as quasi-static because it applies only when steady-state conditions prevail. Under highly transient conditions, such as accelerations or decelerations, only a true dynamic analysis will suffice. Reference 61 made the first attempt at a dynamic analysis to explain cage dynamics in gyro-spin-axis ball bearings. Reference 20 solved the generalized differential equations of motion of the ball in an angular-contact ball bearing. The work was continued for both cylindrical roller bearings[21,22] and for ball bearings[23,24] and is available in the program DREB.

15.13 APPLICATION

Consider a single-row, radial, deep-groove ball bearing with the following dimensions:

Dimensions	Minimum elastohydrodynamic film thickness
Inner-race diameter d_i, m	0.052291
Outer-race diameter d_o, m	0.077706
Ball diameter d, m	0.012700
Number of balls in complete bearing, n	9
Inner-groove radius r_i, m	0.006604
Outer-groove radius r_o, m	0.006604
Contact angle β, degrees	0
RMS surface finish of balls, f_b, μm	0.0625
RMS surface finish of races, f_a, μm	0.0175

A bearing of this kind might well experience the following operation conditions:

Operating condition	Result
Radial load F_r, N	8900
Inner-race angular velocity ω_i, rad/s	400
Outer-race angular velocity ω_o, rad/s	0
Lubricant viscosity at atmospheric pressure and effective operating temperature of bearing, η_0, N·s/m^2	0.04
Viscosity-pressure coefficient ξ, m^2/N	2.3×10^{-8}
Modulus of elasticity for both balls and races, E, N/m^2	2×10^{11}
Poisson's ratio for both balls and races, ν	0.3

The essential features of the geometry of the inner and outer conjunctions can be ascertained as follows:

Pitch Diameter

$$d_e = 0.5(d_o + d_i) = 0.065 \text{ m}$$

Diametral Clearance

$$P_d = d_o - d_i - 2d = 1.5 \times 10^{-5} \text{ m}$$

Race Conformity

$$f_i = f_o = r/d = 0.52$$

Equivalent Radius

$$R_{x,i} = d(d_e - d)/2d_e = 0.00511 \text{ m}$$

$$R_{x,o} = d(d_e + d)/2d_e = 0.00759 \text{ m}$$

$$R_{y,i} = f_i d/(2f_i - 1) = 0.165 \text{ m}$$

$$R_{y,o} = f_o d/(2f_o - 1) = 0.165 \text{ m}$$

The curvature sum

$$1/R_i = 1/R_{x,i} + 1/R_{y,i} = 201.76$$

gives $R_i = 4.956 \times 10^{-3}$ m, and the curvature sum

$$1/R_o = 1/R_{x,o} + 1/R_{y,o} = 137.81$$

gives $R_o = 7.256 \times 10^{-3}$ m. Also, $\alpha_i = R_{y,i}/R_{x,i} = 32.35$ and $\alpha_o = R_{y,o}/R_{x,o} = 21.74$.
The nature of the hertzian contact conditions can now be assessed.

Ellipticity Parameters. From Eq. (15.7),

$$\bar{k}_i = \alpha_i^{2/\pi} = 9.42 \qquad \bar{k}_o = \alpha_o^{2/\pi} = 7.09$$

Elliptic Integrals

$$q = \pi/2 - 1$$

From Eq. (15.8),

$$\bar{\mathcal{E}}_i = 1 + q/\alpha_i = 1.0188 \qquad \bar{\mathcal{E}}_o = 1 + q/\alpha_o = 1.0278$$

From Eq. (15.9),

$$\bar{\mathcal{F}}_i = \pi/2 + q \ln \alpha_i = 3.6205 \qquad \bar{\mathcal{F}}_o = \pi/2 + q \ln \alpha_o = 3.3823$$

The effective elastic modulus E' is given by

$$E' = \frac{2}{(1 - \nu_a^2)/E_a + 1 - \nu_b^2/E_b} = 2.198 \times 10^{11} \text{ N/m}^2$$

To determine the load carried by the most heavily loaded ball in the bearing, it is necessary to adopt an iterative procedure based on the calculation of local static compression and the analysis presented in Sec. 15.7. Reference 57 found that the value of Z was about 4.37 by using the expression [Eq. (15.27)]

$$F_{max} = ZF_r/n$$

where F_{max} = load on most heavily loaded ball
F_r = radial load on bearing
n = number of balls

However, it is customary to adopt a value of $Z = 5$ in simple calculations in order to produce a conservative design, and this value will be used to begin the iterative procedure.

Stage 1. Assume $Z = 5$. Then

$$F_{max} = 5F_r/9 = \tfrac{5}{9}(8900) = 4944 \text{ N}$$

From Eq. (15.12) the maximum local elastic compression is

$$\delta_i = \bar{\mathcal{F}}_i[(9/2\bar{\mathcal{E}}_i R_i)(F_{max}/\pi \bar{k}_i E')^2]^{1/3} = 2.902 \times 10^{-5} \text{ m}$$

$$\delta_o = \bar{\mathcal{F}}_o[(9/2\bar{\mathcal{E}}_o R_o)(F_{max}/\pi \bar{k}_o E')^2]^{1/3} = 2.877 \times 10^{-5} \text{ m}$$

The sum of the local compressions on the inner and outer races is

$$\delta = \delta_i + \delta_o = 5.779 \times 10^{-5} \text{ m}$$

A better value for Z can now be obtained from

$$Z = \frac{\pi(1 - P_d/2\delta)^{3/2}}{2.491(\{1 + [(1 - P_d/2\delta)/1.23]^2\}^{1/2} - 1)}$$

since $P_d/2\delta = (1.5 \times 10^{-5})/(5.779 \times 10^{-5}) = 0.1298$. Thus,

$$Z = 4.551$$

Stage 2

$$Z = 4.551$$

$$F_{max} = (4.551)(8900)/9 = 4500 \text{ N}$$

$$\delta_i = 2.725 \times 10^{-5} \text{ m}$$

$$\delta_o = 2.702 \times 10^{-5} \text{ m}$$

$$\delta = 5.427 \times 10^{-5} \text{ m}$$

$$P_d/2\delta = 0.1382$$

Thus, $Z = 4.565$.

Stage 3

$$Z = 4.565$$

$$F_{max} = (4.565)(8900)/9 = 4514 \text{ N}$$

$$\delta_i = 2.731 \times 10^{-5} \text{ m}$$

$$\delta_o = 2.708 \times 10^{-5} \text{ m}$$

$$\delta = 5.439 \times 10^{-5} \text{ m}$$

$$P_d/2\delta = 0.1379$$

and hence $Z = 4.564$.

This value is very close to the previous value of 4.565, from stage 2, and a further iteration confirms its accuracy.

Stage 4

$$Z = 4.564$$

$$F_{max} = (4.564)(8900)/9 = 4513 \text{ N}$$

$$\delta_i = 2.731 \times 10^{-5} \text{ m}$$

$$\delta_o = 2.707 \times 10^{-5} \text{ m}$$

$$\delta = 5.438 \times 10^{-5} \text{ m}$$

$$P_d/2\delta = 0.1379$$

and hence $Z = 4.564$.

The load on the most heavily loaded ball is thus 4513 N.

Elastohydrodynamic Minimum Film Thickness. For pure rolling, from Eq. (15.3),

$$u = |\omega_o - \omega_i|(d_e^2 - d^2)/4d_e = 6.252 \text{ m/s}$$

The dimensionless load, speed, and materials parameters for the inner- and outer-race conjunctions thus become [from Eqs. (15.41) to (15.43), respectively]

$$W_i = F/E'(R_{x,i})^2 = 4513/(2.198 \times 10^{11})(5.11)^2 \times 10^{-6} = 7.863 \times 10^{-4}$$

$$U_i = \eta_0 u/E'R_{x,i} = (0.04)(6.252)/(2.198 \times 10^{11})(5.11 \times 10^{-3}) = 2.227 \times 10^{-10}$$

$$G_i = \xi E' = (2.3 \times 10^{-8})(2.198 \times 10^{11}) = 5055$$

$$W_o = F/E'(R_{x,o})^2 = 4513/(2.198 \times 10^{11})(7.59)^2(10^{-6}) = 3.564 \times 10^{-4}$$

$$U_o = \eta_0 u/E'R_{x,o} = (0.04)(6.252)/(2.198 \times 10^{11}) \times (7.59 \times 10^{-3}) = 1.499 \times 10^{-10}$$

$$G_o = \xi E' = (2.3 \times 10^{-8})(2.198 \times 10^{11}) = 5055$$

The dimensionless minimum elastohydrodynamic film thickness in a fully flooded elliptical contact is given by Eq. (15.44):

$$H_{min} = h_{min}/R_x = 3.63U^{0.68}G^{0.49}W^{-0.073}(1 - e^{-0.68k})$$

Ball–Inner-Race Conjunction. From Eq. (15.44),

$$(h_{min})_i/R_{x,i} = (3.63)(2.732 \times 10^{-7})(65.29)(1.685)(0.9983) = 1.09 \times 10^{-4}$$

Thus, $$(h_{min})_i = (1.09 \times 10^{-4})R_{x,i} = 0.557 \ \mu\text{m}$$

The lubrication factor Λ was found to play a significant role in determining the fatigue life of rolling-element bearings. In this case, from Eq. (15.53),

$$\Lambda_i = (h_{min})_i/\sqrt{f_a^2 + f_b^2} = (0.557 \times 10^{-6})/\{[(0.175)^2 + (0.0625)^2]^{1/2} \times 10^{-6}\} = 3.00 \quad (15.53)$$

Ball–Outer-Race Conjunction. From Eq. (15.44),

$$(H_{min})_o = (h_{min})_o/R_{x,o} = 3.63U_o^{0.68}G^{0.49}W^{-0.073}(1 - e^{-0.68ko})$$

$$= (3.63)(2.087 \times 10^{-7})(65.29)(1.785)(0.9919) = 0.876 \times 10^{-4}$$

Thus, $$(h_{\min})_o = (0.876 \times 10^{-4})R_{x,o} = 0.665\ \mu\text{m}$$

In this case, the lubrication factor Λ is given by Eq. (15.53):

$$\Lambda = (0.665 \times 10^{-6})/\{[(0.175)^2 + (0.0625)^2]^{1/2} \times 10^{-6}\} = 3.58$$

Once again, it is evident that the smaller minimum film thickness occurs between the most heavily loaded ball and the inner race. However, in this case the minimum elastohydrodynamic film thickness is about 3 times the composite surface roughness, and the bearing lubrication can be deemed to be entirely satisfactory. Indeed, it is clear from Fig. 15.28 that very little improvement in the lubrication factor $\tilde{F}$ and thus in the fatigue life of the bearing could be achieved by further improving the minimum film thickness and hence Λ.

REFERENCES

1. "Load Ratings and Fatigue Life for Ball Bearings," ANSI/AFBMA 9-1990, The Anti-Friction Bearing Manufactures Association, Washington, D.C., 1990.
2. "Load Ratings and Fatigue Life for Roller Bearings," ANSI/AFBMA 11-1990, The Anti-Friction Bearing Manufacturers Association, Washington, D.C., 1990.
3. Anderson, N. E., and E. V. Zaretsky: "Short-Term Hot Hardness Characteristics of Five Case-Hardened Steels," NASA TN D-8031, National Aeronautics and Space Administration, 1975.
4. Anderson, W. J.: "Elastohydrodynamic Lubrication Theory as a Design Parameter for Rolling Element Bearings," ASME Paper 70-DE-19, American Society of Mechanical Engineers, New York, 1970.
5. Anderson, W. J.: "Practical Impact of Elastohydrodynamic Lubrication," in Proc 5th Leeds-Lyon Symposium in Tribology on Elastohydrodynamics and Related Topics," D. Dowson, C. M. Taylor, M. Godet, and D. Berthe, eds., Mechanical Engineering Publication, Bury St. Edmunds, Suffolk, England, p. 217, 1979.
6. Anderson, W. J., E. J. Macks, and Z. N. Nemeth: "Comparison of Performance of Experimental and Conventional Cage Designs and Materials for 75-Millimeter-Bore Cylindrical Roller Bearings at High Speeds," NACA TR-1177, National Advisory Committee for Aeronautics, 1954.
7. Bamberger, E. N.: "Effect of Materials—Metallurgy Viewpoint," in "Interdisciplinary Approach to the Lubrication of Concentrated Contacts," P. M. Ku, ed., NASA SP-237, National Aeronautics and Space Administration, 1970.
8. Bamberger, E. N.: "Life Adjustment Factors for Ball and Roller Bearings—An Engineering Design Guide," American Society for Mechanical Engineers, New York, 1971.
9. Bamberger, E. N., E. V. Zaretsky, and H. Signer: "Endurance and Failure Characteristics of Main-Shaft Jet Engine Bearings at 3×10^6 *DN*," *ASME Trans., J. Lubr. Technol.*, ser. F, vol. 98, no. 4, p. 510, 1976.
10. Bamberger, E. N., et al.: "Materials for Rolling Element Bearings," in "Bearing Design—Historical Aspects, Present Technology and Future Problems," W. J. Anderson, ed., American Society of Mechanical Engineers, New York, p. 1, 1980.
11. Bisson, E. E., and Anderson, W. J.: "Advanced Bearing Technology," NASA SP-38, National Aeronautics and Space Administration, 1964.
12. Brown, P. F.: "Bearing and Dampers for Advanced Jet Engines," SAE Paper 700318, Society of Automotive Engineers, 1970.
13. Brown, P. F., L. C. Dobek, F. C. Hsing, and J. R. Miner: "Mainshaft High Speed Cylindrical Roller Bearings for Gas Engines," PWA-FR-8615, Pratt & Whitney Group, West Palm Beach, Fla., 1977.

14. Cameron, A., and R. Gohar: "Theoretical and Experimental Studies of the Oil Film in Lubricated Point Contact," *Proc. Roy. Soc. London,* ser. A, vol. 291, p. 520, 1976.
15. Cundill, R. T., and F. Giordano: "Lightweight Materials for Rolling Elements in Aircraft Bearings," in "Problems in Bearing and Lubrication," AGARD Conf. Preprint No. 323, Advisory Group for Aeronautical Research and Development (NATO), Paris, p. 6–1, 1982.
16. Dowson, D., and B. J. Hamrock: "Numerical Evaluation of the Surface Deformation of Elastic Solids Subjected to a Hertzian Contact Stress," *ASLE Trans.,* vol. 19, no. 4, p. 279, 1976.
17. "General Guide to the Choice of Journal Bearing Type," Engineering Science Data Unit, item 65007, Institution of Mechanical Engineers, London, 1965.
18. "General Guide to the Choice of Thrust Bearing Type," Engineering Science Data Unit, item 67033, Institution of Mechanical Engineers, London, 1967.
19. "Contact Stresses," Engineering Science Data Unit, item 78035, Institution of Mechanical Engineers, London, 1978.
20. Gupta, P. K.: "Transient Ball Motion and Skid in Ball Bearings," *Trans. ASME, J. Lubr. Technol.,* vol. 97, no. 2, p. 261, 1975.
21. Gupta, P. K.: "Dynamics of Rolling Element Bearings—Part I, Cylindrical Roller Bearings Analysis," *Trans. ASME, J. Lubr. Technol.,* vol. 101, no. 3, p. 293, 1979.
22. Gupta, P. K.: "Dynamics of Rolling Element Bearings—Part II, Cylindrical Roller Bearing Results," *Trans. ASME, J. Lubr. Technol.,* vol. 101, no. 3, p. 305, 1979.
23. Gupta, P. K.: "Dynamics of Rolling Element Bearings—Part III, Ball Bearing Analysis," *Trans. ASME, J. Lubr. Technol.,* vol. 101, no. 3, p. 312, 1979.
24. Gupta, P. K.: "Dynamics of Rolling Element Bearings—Part IV, Ball Bearing Results," *Trans. ASME, J. Lubr. Technol.,* vol. 101, no. 3, p. 319, 1979.
25. Hamrock, B. J., and W. J. Anderson: "Analysis of an Arched Outer-Race Ball Bearing Considering Centrifugal Forces," *Trans. ASME, J. Lubr. Technol.,* vol. 95, no. 3, p. 265, 1973.
26. Hamrock, B. J., and D. Brewe: "Simplified Solution for Stresses and Deformations," *Trans. ASME, J. Lubr. Technol.,* vol. 105, p. 171, 1983.
27. Hamrock, B. J., and D. Dowson: "Isothermal Elastohydrodynamic Lubrication of Point Contacts—Part I, Theoretical Formulation," *ASME J. Lubr. Technol.,* vol. 98, no. 22, p. 223, 1976.
28. Hamrock, B. J., and D. Dowson: "Isothermal Elastohydrodynamic Lubrication of Point Contacts—Part III, Fully Flooded Results," *ASME J. Lubr. Technol.,* vol. 99, no. 2, p. 264, 1977.
29. Hamrock, B. J., and D. Dowson: "Ball Bearing Lubrication—The Elastohydrodynamics of Elliptical Contacts," John Wiley & Sons, Inc., New York, 1981.
30. Harris, T. A.: "An Analytical Method to Predict Skidding in High Speed Roller Bearings," *ASLE Trans.,* vol. 9, p. 229, 1966.
31. Harris, T. A.: "An Analytical Method to Predict Skidding in Thrust-Loaded, Angular-Contact Ball Bearings," *Trans. ASME, J. Lubr. Technol.,* vol. 93, no. 1, p. 17, 1971.
32. Harris, T. A.: "Ball Motion in Thrust-Loaded, Angular Contact Bearings with Coulomb Friction," *Trans. ASME, J. Lubr. Technol.,* vol. 93, no. 1, p. 32, 1971.
33. Heathcote, H. L.: "The Ball Bearing in the Making, Under Test and in Service," *Proc. Inst. Automot. Engrs.,* vol. 15, p. 569, London, 1921.
34. Hertz, H.: "The Contact of Elastic Solids," *J. Reine Angew, Math.,* vol. 92, p. 156, 1881.
35. "Rolling Bearings, Dynamic Load Ratings and Rating Life," ISO/TC4/JC8, rev. of ISOR281, International Organization for Standardization, Technical Committee ISO/TC4, 1976.
36. Johnson, K. L., and D. Tabor: "Rolling Friction," *Proc. Inst. Mech. Engs.,* vol. 182, part 3A, p. 168, 1967/1968.
37. Jones, A. B.: "Analysis of Stresses and Deflections," New Departure Engineering Data, General Motors Corp., Bristol, Conn., 1946.

38. Jones, A. B.: "Ball Motion and Sliding Friction in Ball Bearings," *Trans. ASME, J. Basic Eng.*, vol. 81, no. 1, p. 1, 1959.

39. Jones, A. B.: "A General Theory for Elastically Constrained Ball and Radial Roller Bearings Under Arbitrary Load and Speed Conditions," *Trans. ASME, J. Basic Eng.*, vol. 82, no. 2, p. 309, 1960.

40. Jones, A. B.: "The Mathematical Theory of Rolling Element Bearings," in "Mechanical Design and Systems Handbook," 1st ed., H. A. Rothbart, ed., McGraw-Hill Book Company, Inc., New York, p. 13-1, 1964.

41. Lundberg, G., and A. Palmgren: "Dynamic Capacity of Rolling Bearings," *Acta Polytech.* (Mechanical Engineering Series), vol. I, no. 3, 1947.

42. Lundberg, G., and A. Palmgren: "Dynamic Capacity of Rolling Bearings," *Acta Polytech.* (Mechanical Engineering Series), vol. II, no. 4, 1952.

43. Matt, R. J., and R. J. Gianotti: "Performance of High Speed Ball Bearings with Jet Oil Lubrication," *Lub. Eng.*, vol. 22, no. 8, p. 316, 1966.

44. McCarthy, P. R.: "Greases," in "Interdisciplinary Approach to Liquid Lubricant Technology," P. M. Ku, ed., NASA SP-318, p. 137, National Aeronautics and Space Administration, 1973.

45. Miyakawa, Y., K. Seki, and M. Yokoyama: "Study on the Performance of Ball Bearings at High *DN* Values," NAL-TR-284, National Aerospace Lab., Tokyo, (NASA TTF-15017, 1973), 1972.

46. Palmgren, A.: "Ball and Roller Bearing Engineering," 3d ed., SKF Industries, Inc., Philadelphia, Pa., 1959.

47. Parker, R. J.: "Lubrication of Rolling Element Bearings," in "Bearing Design Historical Aspects, Present Technology and Future Problems," W. J. Anderson, ed., American Society of Mechanical Engineers, New York, p. 87, 1980.

48. Parker, R. J., and H. R. Signer: "Lubrication of High-Speed, Large Bore Tapered-Roller Bearings," *Trans. ASME, J. Lubr. Technol.*, vol. 100, no. 1, p. 31, 1978.

49. Pirvics, J.: "Numerical Analysis Techniques and Design Methodology for Rolling Element Bearing Load Support Systems," in "Bearing Design Historical Aspects, Present Technology and Future Problems," W. J. Anderson, ed., American Society of Mechanical Engineers, New York, p. 47, 1980.

50. Reynolds, O.: "On Rolling Friction," *Phil. Trans. Roy. Soc. London,* part 1, vol. 166, p. 155, 1875.

51. Schuller, F. T.: "Operating Characteristics of a Large-Bore Roller Bearing to Speed of 3×10^6 *DN*," NASA TP-1413, 1979.

52. Scibbe, H. W.: "Bearing and Seal for Cryogenic Fluids," SAE Paper 680550, 1968.

53. Sibley, L. W.: "Silicon Nitride Bearing Elements for High Speed High Temperature Applications," in "Problems in Bearing Lubrication," AGARD Conf., no. 323, Advisory Group for Aeronautical Research and Development (NATO), Paris, 1982.

54. Signer, H. R., E. N. Bamberger, and E. V. Zaretsky: "Parametric Study of the Lubrication of Thrust Loaded 120-mm Bore Ball Bearings to 3 Million *DN*," *Trans. ASME, J. Lubr. Technol.*, vol. 96, no. 3, p. 515, 1974.

55. Signer, H. R., and F. T. Schuller: "Lubrication of 35-Millimeter-Bore Ball Bearings of Several Designs at Speeds to 2.5 Million *DN*," in "Problems in Bearing and Lubrication," AGARD Conf., no. 323, Advisory Group for Research and Development (NATO), Paris, France, p. 8-1, 1982.

56. Skurka, J. C.: "Elastohydrodynamic Lubrication of Roller Bearings," *Trans. ASME, J. Lubr. Technol.*, vol. 92, no. 2, p. 281, 1970.

57. Stribeck, R.: "Kugellager fur beliebige Belastungen," *Z. Ver. dt. Ing.*, vol. 45, no. 3, p. 73, 1901.

58. Tabor, D.: "The Mechanism of Rolling Friction. II. The Elastic Range," *Proc. Roy. Soc. London,* part A, vol. 229, p. 198, 1955.

59. Tallian, T. E.: "On Competing Failure Modes in Rolling Contact," *Trans. ASLE,* vol. 10, p. 418, 1967.

60. Todd, M. J., and R. H. Bentall: "Lead Film Lubrication in Vacuum," *Proc. 2d ASLE Int Conf. Solid Lubrication,* American Society of Lubrication Engineers, p. 148, 1978.

61. Walters, C. T.: "The Dynamics of Ball Bearings," *Trans. ASME, J. Lubr. Technol.,* vol. 93, no. 1, p. 1, 1975.

62. Weibull, W.: "A Statistical Representation of Fatigue Failures in Solids," *Trans. Roy. Inst. Technol.,* Stockholm, vol. 27, 1949.

63. Zaretsky, E. V., and W. J. Anderson: "Material Properties and Processing Variables and Their Effect on Rolling-Element Fatigue," NASA TM X-52226, National Aeronautics and Space Administration, 1966.

64. Zaretsky, E. V., H. Signer, and E. N. Bamberger: "Operating Limitations of High-Speed Jet-Lubricated Ball Bearings," *Trans. ASME, J. Lubr. Technol.,* vol. 8, no. 1, p. 32, 1976.

65. Zaretsky, E. V.: "STLE Life Factors for Rolling Bearings," Society of Tribologist and Lubrication Engineers, Park Ridge, Ill., 1992.

CHAPTER 16
POWER SCREWS

John E. Johnson
Manager, Mechanical Model Shops
T.R.W. Corp.
Redondo Beach, Calif.

16.1 THREADS 16.1
16.1.1 V Threads 16.1
16.1.2 Square Threads 16.1
16.1.3 Acme Threads 16.2
16.1.4 Buttress Threads 16.2
16.1.5 Multiple Threads 16.2
16.2 FORCES 16.2
16.3 FRICTION 16.4
16.4 DIFFERENTIAL AND COMPOUND SCREWS 16.5
16.4.1 Differential Screw 16.5
16.4.2 Compound Screw 16.6
16.5 EFFICIENCY 16.7
16.6 DESIGN CONSIDERATIONS 16.8
16.6.1 Thread Bearing Pressure 16.8
16.6.2 Tensile and Compressive Stresses 16.9
16.6.3 Shear Stresses 16.9
16.6.4 Deflection 16.10
16.6.5 Collar Friction 16.10
16.6.6 Efficiency 16.10
16.7 ROLLING-ELEMENT BEARING POWER SCREWS 16.10

Power screws are used to translate rotary motion into uniform longitudinal motion. Common applications include jacks, valves, machine tools, and presses. Screws also permit very accurate position adjustment because they provide a high reduction ratio from rotational to longitudinal displacement.[8–10]

16.1 THREADS

16.1.1 V Threads

The efficiency of a power screw depends upon the profile angle of the thread: the larger the angle, the lower the efficiency. V threads are therefore not well suited for transmission of great loads. They are generally used only where accuracy of adjustment and low production cost are required and the power demands are quite small.

16.1.2 Square Threads

The square thread (Fig. 16.1) has the greatest efficiency (zero profile angle). However, it is costly to manufacture, because it cannot be cut with dies, and it is difficult to engage with a moving split nut as is sometimes required.

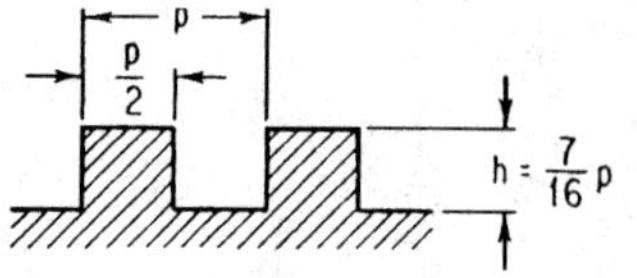

FIG. 16.1 Square thread.

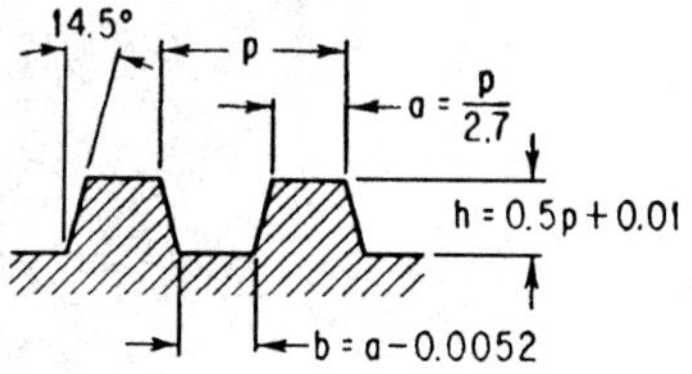

FIG. 16.2 Acme thread, profile angle β = 14.5°.

16.1.3 Acme Threads

The Acme thread (Fig. 16.2) is often used in order to overcome the difficulties associated with the square thread.

While its efficiency is lower than that of a square thread, it has the advantage that lost motion resulting from manufacturing tolerances or wear can be taken out by using a split nut.

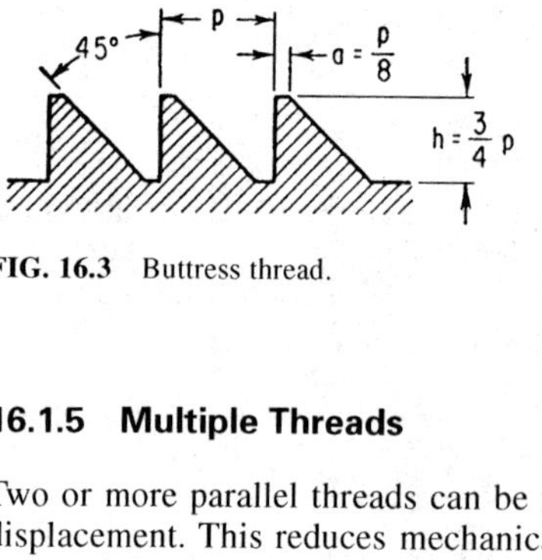

FIG. 16.3 Buttress thread.

16.1.4 Buttress Threads

Where unidirectional power transmission is required and the nut returns with little or no load, the buttress thread (Fig. 16.3) can be used. Because its square face is used for power transmission, it has the square-thread efficiency but slightly lower manufacturing cost.

16.1.5 Multiple Threads

Two or more parallel threads can be used to reduce the ratio of screw rotation to nut displacement. This reduces mechanical advantage but increases efficiency because of increased helix angle.

16.2 FORCES

The pitch of a thread p is the distance from a point on one thread to the corresponding point on an adjacent thread regardless of whether the screw has a single or multiple thread. The displacement d_s of the nut or screw resulting from one full turn of either is the lead l. Thus for a multiple thread of m threads,

$$l = mp \tag{16.1}$$

and the displacement for n revolutions of either screw or nut

$$d_s = nl = nmp \tag{16.2}$$

The angle of the thread at its mean diameter with respect to a normal to the screw axis is the helix angle α. To determine the force P required to overcome a certain load Q it

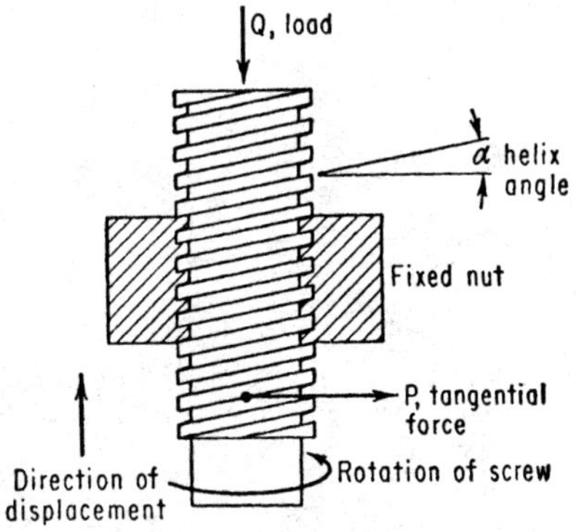

FIG. 16.4 Force P required to overcome load Q.

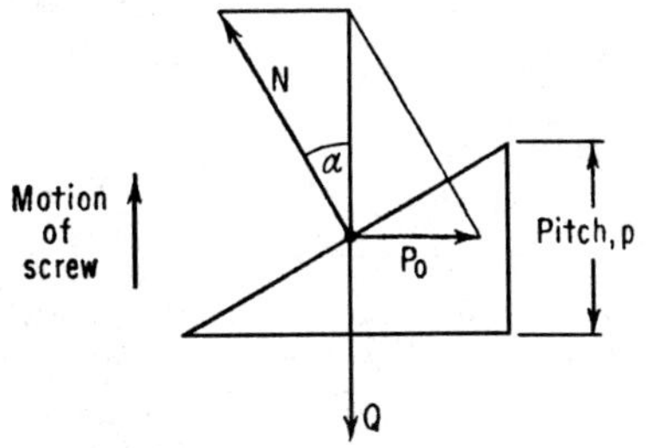

FIG. 16.5 Vector solution for force P_0 required to overcome load Q, neglecting friction.

is necessary to observe the relation of the load direction with respect to the displacement direction. If the load opposes the direction of motion (Fig. 16.4) the force required to overcome it, neglecting friction, is given by

$$P_0 = Q \tan \alpha \qquad \text{(see Fig. 16.5)} \tag{16.3}$$

However, friction displaces the normal N (Fig. 16.6) by angle ϕ, and P_1 is given by

$$P_1 = Q \tan(\alpha + \phi) = Q(\tan \alpha + \tan \phi)/(1 - \tan \alpha \tan \phi) \tag{16.4}$$

Replacing $\tan \phi$ with μ, the coefficient of friction,

$$P_1 = Q(\tan \alpha + \mu)/(1 - \mu \tan \alpha) \tag{16.5}$$

Because of the thread profile angle β, the resultant R should be replaced by $R/\cos \beta$. This affects only the friction terms since friction gave rise to R to begin with; hence these must be divided by $\cos \beta$. Thus

$$P_1 = Q(\mu \sec \beta + \tan \alpha)/(1 - \mu \sec \beta \tan \alpha) \tag{16.6}$$

When a load is applied to the screw in Fig. 16.4 but P is removed, friction alone keeps the screw from turning and moving in the direction of Q. Hence, when motion in the direction of the load is required, the applied force need only be large enough to overcome the friction (Fig. 16.7).

$$P_2 = Q \tan(\phi - \alpha) \tag{16.7}$$

For a profile angle β

$$P_2 = Q(\mu \sec \beta - \tan \alpha)/(1 + \mu \sec \beta \tan \alpha) \tag{16.8}$$

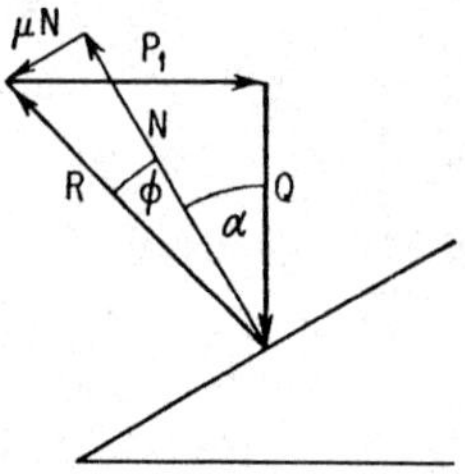

FIG. 16.6 Vector solution for force P_1 required to overcome load Q, including frictional force μN.

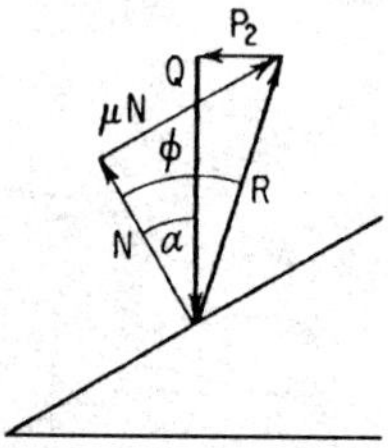

FIG. 16.7 Vector solution for force P_2 required to overcome the frictional force when motion is in the direction of the load.

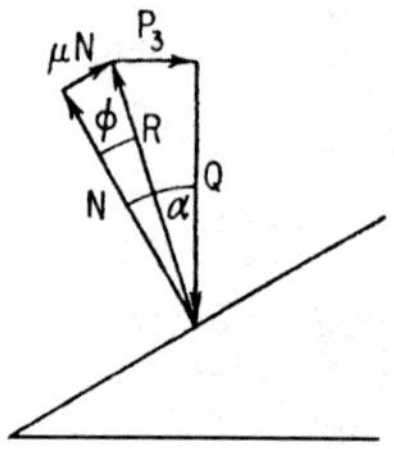

FIG. 16.8 Force P_3 required to prevent motion of screw or nut due to an applied load.

When motion of the screw or nut may be caused by the load applied to either, the screw and nut are not self-locking and a tangential force will be required to prevent motion (Fig. 16.8). This force is given by

$$P_3 = Q(\tan\alpha - \mu \sec\beta)/(1 + \mu \sec\beta \tan\alpha) \quad (16.9)$$

When $P_3 = 0$ the friction force will just cancel the tangential force produced by the load, and the screw will be self-locking. Setting $P_3 = 0$, the helix angle α at which the screw is self-locking is given by

$$\tan\alpha = \mu \sec\beta \quad (16.10)$$

Note that, for the helix angle at the equilibrium point when the screw is just self-locking, force P_1, to move against the load, will be

$$P_1 = Q \tan 2\alpha \quad (16.11)$$

The torque required to overcome the load Q on the screw is given by

$$T_1 = \tfrac{1}{2} D_s P_1 \qquad \text{for motion against the load} \quad (16.12)$$

$$T_2 = \tfrac{1}{2} D_s P_2 \qquad \text{for motion with the load} \quad (16.13)$$

where D_s is the mean diameter of the screw.

16.3 FRICTION

From Eq. (16.10), it can be seen that the helix angle at which a screw will be self-locking depends upon the coefficient of friction as well as the thread profile. For an average coefficient of friction $\mu = 0.150$, the helix angle must be at least 9° for square

TABLE 16.1 Coefficient of Friction μ

Screw	Nut			
	Steel	Brass	Bronze	Cast iron
Steel, dry	0.15–0.25	0.15–0.23	0.15–0.19	0.15–0.25
Steel, machine oil	0.11–0.17	0.10–0.16	0.103–0.15	0.11–0.17
Bronze	0.08–0.12	0.04–0.06	....	0.06–0.09

and buttress and 10° for Acme threads. These values allow for a small (½°) margin of safety so that the screw will not keep turning under load for kinetic coefficient of friction slightly less than 0.150 if the applied force P is removed. Coefficients of friction are given in Table 16.1.[2]

The effect of friction in bearings and thrust collars, which must always be used either on the nut or the screw depending upon application, was not included in the preceding considerations. The force necessary to overcome these frictional forces must be determined separately and added to Eqs. (16.6), (16.7), and (16.8).

Where two surfaces are in sliding contact, good design practice requires that nut and screw be made of different materials in order to reduce both wear and friction. Because the nut is usually smaller and easier to replace than the screw, it is made of the softer material, usually high-grade bronze or brass where loads are light.

Workmanship also has an important effect upon friction. A 30- to 50-μin finish will result in a coefficient of friction about one-third lower than a finish of 100 to 125 μin.[3]

16.4 DIFFERENTIAL AND COMPOUND SCREWS

The displacement of the nut or screw depends upon the pitch. From Eqs. (16.1) and (16.2),

$$d_s = nmp \tag{16.2}$$

where n is the number of turns of the screw and m the number of threads. If small displacement per turn is required, m must equal 1 and $d = p$. However, very small displacements require a very small pitch, and this results in a weak thread. This difficulty can be overcome to some extent by using a differential screw.

When a fast motion is required, a multiple thread of $m = 2, 3$, or more can be used. However, machining is expensive because each thread must be separately cut. Also, since the helix angle on multiple-thread screws is quite large, the screw will not be self-locking. Here, the remedy may be the use of a compound screw.

16.4.1 Differential Screw

The differential screw has two threads in series. Both are of the same hand but different pitch. Every revolution of the screw will move the two nuts toward or away from each other by an amount equal to the difference in pitch. For the arrangement in Fig. 16.9 the nuts C and F will separate, and with the coarser pitch at the fixed nut

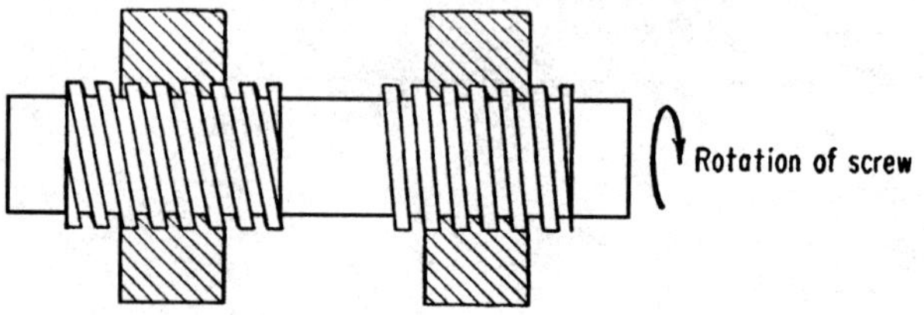

FIG. 16.9 Differential screw.

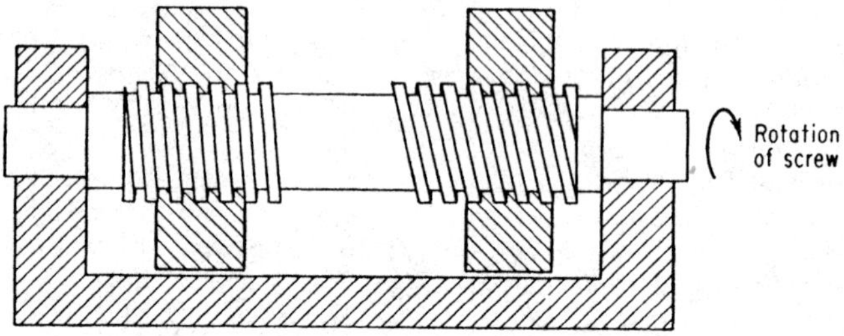

FIG. 16.10 Differential screw.

$$d_s = p_C - p_F \tag{16.14}$$

If the fixed nut has the finer pitch, then, for the same sense of rotation for the screw,

$$-d_s = p_F - p_C \tag{16.15}$$

that is, the nuts will move toward each other. Another arrangement (Fig. 16.10) shows the nuts F and C approaching each other. Their relative displacement

$$d_R = p_C - p_F \tag{16.16}$$

If the rotation of the screw is reversed, the nuts will separate:

$$-d_R = p_F - p_C \tag{16.17}$$

16.4.2 Compound Screw

If as in the arrangements in Fig. 16.11 the threads are of opposing hands, the result will be a compound screw. The displacement will then be the sum of the two pitches.

$$d_s = p_C + p_F \tag{16.18}$$

Note that for the arrangement in Fig. 16.10 interchange of coarse and fine pitch between the nuts will not result in a reversal of the displacement direction as it would

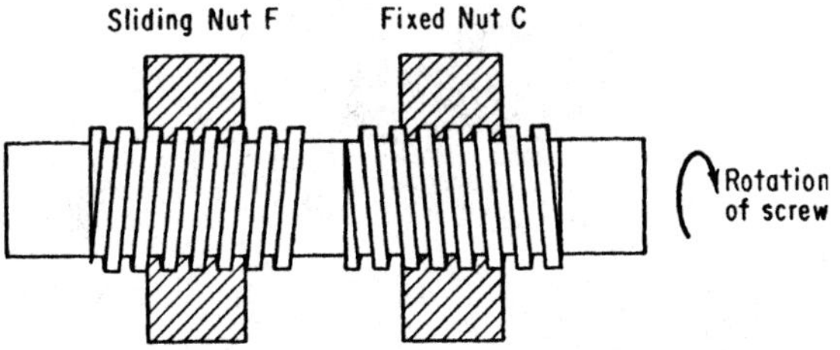

FIG. 16.11 Compound screw.

for a differential screw. Also note that a differential screw must always have threads of different pitch, whereas a compound screw may have both threads of the same pitch, but they must always be of opposite hand.

16.5 EFFICIENCY

The efficiency of a screw is the ratio of the force required for motion against load without friction to that required when friction is present:

$$e_1 = P_0/P_1 = (\tan \alpha)(1 - \mu \sec \beta \tan \alpha)/(\tan \alpha + \mu \sec \beta) \tag{16.19}$$

$$e_1 = (\cos \beta - \mu \tan \alpha)/(\cos \beta + \mu \cot \alpha) \tag{16.20}$$

Thus the efficiency of a square thread ($\beta = 0$),

$$e_{\text{sq. thread}} = (1 - \mu \tan \alpha)/(1 + \mu \cot \alpha) \tag{16.21}$$

From Eq. (16.20) it is seen that max e occurs when $\alpha = 45°$. The maximum attainable efficiency

$$e_{\max} = (\cos \beta - \mu)/(\cos \beta + \mu) \tag{16.22}$$

To obtain the greatest efficiency, μ should be as small as possible.

When the motion occurs in the direction of the load, it is meaningless to speak of efficiency unless it is proposed to convert an axial load into a tangential force with a non-self-locking screw, a most unlikely design application. When the screw is self-locking, efficiency must be considered negative because a tangential force is required to do what the axial load cannot accomplish by itself. Zero efficiency then represents the equilibrium condition given by Eqs. (16.9) and (16.10). However, the efficiencies are required to consider differential and compound screws, and mathematical expression can be derived provided the foregoing is kept in mind. When the screw is not self-locking the efficiency is given by

$$e_3 = P_3/P_0 = (\cos \beta - \mu \cot \alpha)/(\cos \beta + \mu \tan \alpha) \tag{16.23}$$

The 100 percent efficient case occurs when P_3 equals P_0, to keep the nut or screw from overrunning. Zero efficiency represents the equilibrium case when screw and nut are just self-locking. For the latter, setting Eq. (16.23) equal to zero,

$$\cos\beta = \mu\cot\alpha \tag{16.24}$$

or

$$\tan\alpha = \mu\sec\beta \tag{16.25}$$

For a self-locking nut, the negative efficiency is given by

$$e_2 = 1 - \frac{P_0}{P_0 + P_2} = 1 - \frac{\cos\beta + \mu\tan\alpha}{\mu(\tan\alpha + \cot\alpha)} \tag{16.26}$$

For the equilibrium case of self-locking, when $e = 0$,

$$\tan\alpha = \mu\sec\beta \tag{16.27}$$

For a differential screw, note that, regardless of rotational direction, one thread will always cause displacement against the load, while the other causes displacement with the load. For the screw in Fig. 16.9, nut F may move with the load with respect to the screw thread, and the nut at C against the load. Reversal of rotation reverses the motions as well. Therefore, the efficiency for a differential screw must be checked for both directions of rotation. The efficiency is given by

$$e_4 = e_1 e_2 \tag{16.28}$$

when both threads are self-locking. Care must be taken to substitute the proper values of α, μ, and β for each thread into the respective efficiency relationships. When only one thread is self-locking,

$$e_5 = e_1 e_3 \tag{16.29}$$

for one direction of motion, and e_4 from Eq. (16.28) for the other. When neither thread is self-locking, Eq. (16.29) applies for either direction.

While the efficiency for a differential screw is quite low for self-locking threads, it can be improved by making one or both threads overrunning. With proper design the tangential force P_3 resulting from a load Q on one thread is not large enough to cause motion against the load on the other thread; so that, even if either thread is separately not self-locking, the combined result in a differential screw produces the desired self-locking feature.

Since the threads are of opposite hand in a compound screw, the displacement with respect to the load is the same for both threads. The efficiency is therefore the product of the efficiency of each thread.

$$e_5 = e_{1,\text{thread1}} e_{1,\text{thread2}} \tag{16.30}$$

16.6 DESIGN CONSIDERATIONS

16.6.1 Thread Bearing Pressure

Design of power screws, as with all other screws, is based upon the assumption that the load is uniformly distributed over all threads. This assumption is not true. When the screw is in tension and the nut in compression, only the first threads will carry the load. The other threads merely maintain contact with one another. This is especially true when the threads are new and not worn. After the threads have undergone some plastic and elastic deformation, some of the load will be carried by the other threads too, but the load distribution nevertheless is far from uniform.

When both screw and nut are in tension or compression, the load is distributed among all threads in engagement, but the distribution again is not uniform. It depends upon the total number of threads in engagement. For three threads, the distribution for a screw in tension is $\frac{2}{3}Q$ from the first to the second thread and $\frac{1}{3}Q$ from the second to the third thread. The distribution in the nut is the same in reverse order.[4] This condition can be alleviated by using a nut of variable cross section. The outside of the nut is parabolic; uniform pressures produced in such nuts greatly reduce wear.[5] To simplify design, however, the assumption of uniformly distributed load is usually made, allowance being made by selecting low values of bearing pressure. The required number of threads is then given by

$$Q = 0.785(D^2 - d^2)p_bN \tag{16.31}$$

where D is the outside diameter of the screw, d the inside diameter of the nut (root diameter of the screw), and N the number of threads required. Values for p_b may be taken from Table 16.2)[4,6]

TABLE 16.2 Safe Bearing Pressures

Screw	Nut	Safe bearing pressure p_b, psi	Speed range
Steel..........	Bronze	2,500–3,500	Low speed
Steel..........	Bronze	1,600–2,500	Not over 10 fpm
	Cast iron	1,800–2,500	Not over 8 fpm
Steel..........	Bronze	800–1,400	20–40 fpm
	Cast iron	600–1,000	20–40 fpm
Steel..........	Bronze	150–240	50 fpm and over

16.6.2 Tensile and Compressive Stresses

The screw tensile stress is based upon the cross-sectional area at the root diameter. A factor of 4 to 5 is recommended for stress concentrations in the absence of fillets at the root of the thread.

Compressive stresses also cause stress concentrations, but these are not so dangerous and may be neglected.

If the screw is longer than 6 times its root diameter, it should be treated as a short column.

16.6.3 Shear Stresses

The shear produced at the thread or root in a direction parallel to the screw axis is generally not dangerous.

The number of threads in the nut, where shear occurs at the major diameter, is usually controlled by wear considerations.

16.6.4 Deflection

An applied turning moment produces torsional shear, the effect of which should be considered. It must also be borne in mind that torsional, bending, and axial deflections are not negligible. A torsional deflection of 1° causes a pitch variation of 0.00005 in. For a precision screw, where the pitch tolerance may be ±0.0002, this represents a 25 percent deviation. When the screw is a beam, the bending deflection will cause the threads on the compression side to move together and those on the tension side to separate. When the effect of all three deflections is added, the resulting deflection may be quite large (0.008 to 0.015 in is not unusual on heavily loaded screws) and must not be neglected lest undue wear result.[3]

16.6.5 Collar Friction

The frictional forces associated with thrust collars must be determined separately and added to the frictional forces arising from thread loads in order to determine the torque required to turn the screw or nut. Since power screws almost always move with higher than negligible speed, the equations used to determine the static-friction forces at the head of a screw or nut fastening should not be used. A useful relationship is $pv = 20{,}000$, where p is the allowable pressure, lb/in², and v the velocity at the mean diameter of the collar, ft/min. The bearing pressure thus determined should not exceed the value associated with load Q over the thrust-collar area. Coefficients of collar friction are found in Table 16.3.[2]

TABLE 16.3 Coefficients of Friction for Thrust Collars

Material	Running	Starting
Soft steel on cast iron	0.12	0.17
Hard steel on cast iron	0.09	0.15
Soft steel on bronze	0.08	0.10
Hard steel on bronze	0.06	0.08

16.6.6 Efficiency

Advantageous use of Eq. (16.10) should be made in order to obtain the most efficient thread for a self-locking screw. The helix angle so derived will always be the most efficient. When the screw can be overrunning, the most efficient thread will have a helix angle as close to 45° as possible.

16.7 ROLLING-ELEMENT BEARING POWER SCREWS[11,13–15]

To reduce friction, special threads using a ball or roller between the screw and nut are employed. The efficiency of these devices is about 90 percent. The most popular kinds are *ball screws* and *roller screws.* In addition there are *reversing screws, linear ball mechanisms, ball splines, ball plates,* and *ball bushings.* All replace sliding action

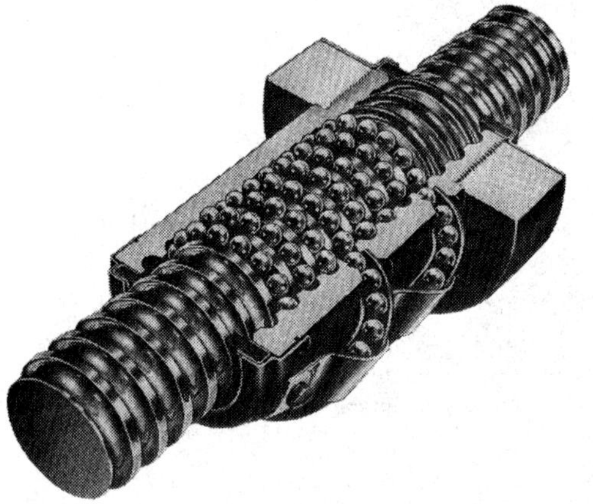

FIG. 16.12 Ball screw. (*Courtesy Warner Electric/Dana, South Beloit, Ill.*)

with lower rolling-element friction. *Ball screws* (Fig. 16.12) have semicircular grooves and the nut has rows of bearing balls. The nut has grooves so that the ball at the end of the nut returns to the starting thread in the nut. *Roller screws* are similar to ball screws except that they utilize threaded rollers and have an efficiency slightly less than ball screws. *Reversing screws* are ball screws that provide a steady reciprocating motion with a constant rotation input. The lead angle of these screws is between 15 and 30°, used in fishing reels. Another application of reversing screws is the rotating-output screwdriver with a lead angle between 50 and 80°. *Linear ball mechanisms*[12] are not power screws. They are mechanisms that permit axial movement by use of rolling elements.

It is suggested to utilize manufacturers' catalogs for details of size and dynamic performance. Proper lubrication is essential: oil bath or oil mist is desired as lubricant. A second choice is to use sodium- or lithium-based greases. The life in distance traveled of the bearing is $L = (C/F_e)^3$ where F_e is the equivalent load on the bearing and C is the dynamic constant given in the manufacturer catalog. Rolling-element bearing power screws do not handle shock loads very well. For more on rolling-element bearings see Chap. 15. References 10 and 11 show design techniques to approach the problems of thermal expansion and internal friction.

REFERENCES

1. *Engineering Forum,* vol. 21, July 1960, Eaton Manufacturing Co., Cleveland, Ohio.
2. Ham, C. W., and D. G. Ryan: *Univ. Illinois Bull.,* vol. 29, no. 81, June 1932.
3. Hieber, G. E.: "Power Screws," *Mach. Des.,* November 1953.
4. Maleev, V. L., and J. B. Hartman: "Machine Design," 3d ed., International Textbook Co., Scranton, Pa., p. 385, 1954.
5. Timoshenko, S., and J. M. Lessels: "Applied Elasticity," Westinghouse Press, East Pittsburgh, Pa., 1925.
6. Vallance, A., and V. L. Doughtie: "Design of Machine Members," 3d ed., McGraw-Hill Book Company, Inc., New York, p. 169, 1951.

7. Faires, V. M.: "Designs of Machine Elements," 3d ed., The Macmillan Co., New York, 1955.
8. Niemann, G.: "Maschinen Elemente," Springer-Verlag OHG, Berlin, p. 161, 1960.
9. Spotts, M. F.: "Design of Machine Elements," 2d ed., Prentice-Hall, Inc., Englewood Cliffs, N.J., 1953.
10. Ninomiya, M.: "Maintaining Ball Screw Precision," *Mach. Des.,* vol. 51, no. 8, p. 105, Apr. 12, 1979.
11. "Ball Screws," ANSI B5.48, American Society of Mechanical Engineers, 1977.
12. Linear Motion Systems, Catalog No. 100-IDE, THK Co., Ltd., Tokyo, 1993.
13. Ball Bearing Screws Master Catalog, P-978, Warner Electric Corp., Walterboro, S.C., 1993.
14. Advanced Linear Motion Systems, Form T1-001B, Thomson Industries Inc., Port Washington, N.Y., 1992.
15. "Linear Actuator Technology Guide," TS9105248, Thomson Saginaw, Port Washington, N.Y., 1993.

CHAPTER 17
FRICTION CLUTCHES

Thomas A. Dow, Ph.D.
Professor of Mechanical and Aerospace Engineering
North Carolina State University
Raleigh, N.C.

17.1 INTRODUCTION 17.1
17.2 TYPES OF FRICTION CLUTCHES 17.1
17.2.1 Spring-Loaded Dry Clutch 17.1
17.2.2 Over-Center Dry Clutch 17.3
17.2.3 Over-Center Power Takeoff Dry Clutch 17.3
17.2.4 Electromagnetic Single-Plate Dry Clutch 17.3
17.2.5 Mechanically Actuated Wet Clutches 17.4
17.2.6 Hydraulic Multiple-Disk Clutch 17.4
17.2.7 Electromagnetic Multiple-Disk Clutches 17.4
17.2.8 Centrifugal-Force Clutch 17.4
17.2.9 Wrapped-Spring Clutch 17.5
17.3 CLUTCH MATERIALS 17.5
17.3.1 Dry-Clutch Lining Materials 17.5
17.3.2 Wet-Clutch Lining Materials 17.5
17.4 TORQUE EQUATIONS 17.8
17.5 THERMAL PROBLEMS IN CLUTCH DESIGN 17.11
17.5.1 Derivation of Temperature Equation 17.13
17.5.2 Application 17.15
17.5.3 Variation of Maximum Surface Temperature with Plate Thickness 17.18
17.5.4 Optimum Radial Dimensions of a Clutch Plate 17.19
17.5.5 Flash Temperatures 17.19
17.5.6 Bulk Temperature of Clutch Assembly 17.21
17.6 DYNAMICS OF SYSTEMS EMPLOYING FRICTION CLUTCHES 17.26
17.6.1 Transient Analysis of Mechanical Power Transmission Containing a Friction Clutch 17.29
17.6.2 Simplified Approach to Dynamic Problems Involving Friction Clutches 17.38
17.7 ELECTROMAGNETIC FRICTION CLUTCHES 17.41

17.1 INTRODUCTION

The operation of a friction clutch is dependent upon the phenomenon of friction. A more detailed discussion of this subject can be found in Chap. 7 on friction, lubrication, and wear, and in Chap. 18 on friction brakes. The key parameters of a successful clutch design are its operating temperature and its dynamic engagement characteristics. These two parameters are related, since the temperature has an effect on the static/dynamic friction coefficients which influence the engagement and the smoothness of operation.

The following sections describe the common types of friction clutches, the important properties of the friction materials, methods of determining operating temperatures, and the dynamics of friction clutch systems.

17.2 TYPES OF FRICTION CLUTCHES[1,2]

17.2.1 Spring-Loaded Dry Clutch

The spring-loaded dry clutch is widely used in automomiles for spring-applied and manually released applications. The output shaft is connected to a hub assembly faced

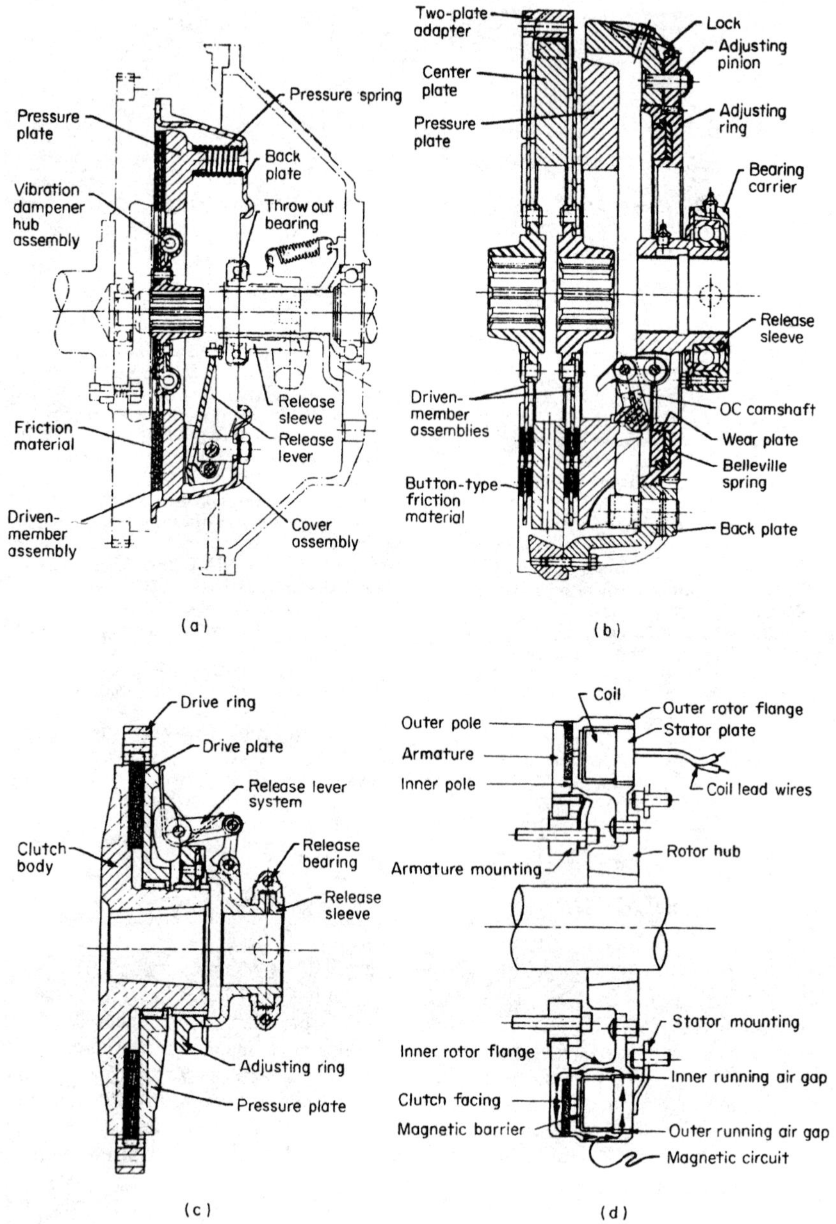

FIG. 17.1 Types of friction clutches.[1] (*a*) Single-plate clutch. (*b*) Twin-plate clutch. (*c*) Power takeoff over-center clutch. (*d*) Single-plate electromagnetic clutch.

with a friction material, usually an asbestos-based material. This output shaft is then connected to the input shaft by clamping the hub assembly between two rotating plates: the flywheel of the prime mover and the pressure plate of the clutch assembly. The clamping force is obtained by coil or Belleville springs which hold the clutch in the applied state. The clutch is released by levers on the rotating plate which compress the springs. A throw-out bearing is necessary to connect the external clutch release mechanism to this rotating assembly.

The parts of such a clutch are shown in Fig. 17.1. The pressure plate must be massive enough to prevent distortion and to provide thermal mass to absorb the heat generated during engagement without overheating the clutch facing. The springs must be designed to provide sufficient clamping pressure over the range of motion caused by clutch-facing wear, and not be excessively influenced by clutch temperature. In some designs, the friction material is isolated from the output shaft by torsional dampers. The purpose of these dampers is to prevent the torsional vibrations of the engine from being transmitted through the clutch at frequencies which would create objectionable gear and driveline noises.

17.2.2 Over-Center Dry Clutch

The over-center dry clutch differs from the spring-loaded design in the method by which clamping force is generated. The lever system incorporates an over-center toggle lever action or cam to develop the clamping force, as shown in Fig. 17.1*b*. The over-center clutch, therefore, will remain in either the released or the engaged position with no external force being applied. This clutch is usually activated by a hand lever and is utilized in many farm and off-road applications.

17.2.3 Over-Center Power Takeoff Dry Clutch

The over-center power takeoff dry clutch is usually used in conjunction with a power takeoff on an engine to attach auxiliary or stationary equipment. This clutch does not use the engine flywheel as part of the friction surface, but contains both clamping faces. The drive is typically through a spline or gear, as shown in Fig. 17.1*c*. Some applications of such clutch systems[3,4] also use the clutch to protect the driveline by slipping when a critical torque is exceeded. The operation of the clutch under these conditions is somewhat different than normal engagement because the overload condition occurs at maximum clamping load.

17.2.4 Electromagnetic Single-Plate Dry Clutch

The electromagnetic single-plate dry clutch utilizes an electromagnet to provide the clamping force, and the single-plate design has only a single friction surface, as compared to the two surfaces of a mechanical single-plate clutch (see Fig. 17.1*d*). Nevertheless, such a clutch can provide fairly high torque capacity for its size. Electromagnetic clutches are attractive for automatically actuated operations such as for air-conditioning compressor drives. The major limitation is the heat capacity, since the clutch operating temperature is limited by the temperature rating of the coil insulation (see Sec. 17.7).

17.2.5 Mechanically Actuated Wet Clutches

The design of mechanically actuated wet clutches is similar to dry clutches, but requires oiltight cover assemblies, proper oil circulation, and adequate release mechanisms. The wet clutch[5] is a much better heat exchanger and relies on the oil to carry much of the heat away during engagement, making it less sensitive than the dry clutch to total heat generation. Since the coefficient of friction of materials operating in oil is lower than under dry conditions, it is necessary to provide higher clamping force, more friction faces, or larger clutch diameter to have the same torque capacity. Clutch assemblies with three or more plates are commonly used to develop sufficient torque in a minimum size. The unit pressure for a wet clutch is typically higher than for a dry clutch, ranging from a high of 200 lb/in^2 in a mechanically actuated wet clutch to seldom greater than 50 lb/in^2 in a dry-clutch design.

17.2.6 Hydraulic Multiple-Disk Clutch

The hydraulic multiple-disk clutch is finding its way into many modern applications. With proper valving it can be controlled either manually or automatically through auxiliary controls. It is widely used in heavy-duty power-shift transmissions for earth moving, construction equipment, farm tractors, and motor trucks. It has also been applied to cooling-fan drives.[6]

The basic designs used for the other clutches apply for this case, but the actuation utilizes hydraulic pressure. The effect of the rotation of the clutch on the hydraulic pressure must be considered when the disengagement mechanisms of such clutches are designed, and balance pistons or return springs are built into most designs.

17.2.7 Electromagnetic Multiple-Disk Clutch

The electromagnetic multiple-disk clutch is an electrically operated oil-bath clutch, using a friction disk pack like the hydraulic clutch. It is used in machine-tool transmissions which require power shifting for automatic sequence operation. However, as with other electromagnetic designs, this clutch has a limited torque capacity when compared to a hydraulically actuated clutch of the same size. The additional requirement of a separate electrical power supply as well as a hydraulic system to circulate the fluid makes this design less attractive for heavy-duty applications. However, the adaptability to automatic control of the electrical clamping force makes this clutch an attractive choice for some applications.

17.2.8 Centrifugal-Force Clutch

Centrifugal force is often utilized to generate the clamping force when automatic clutch engagement and release dependent upon prime-mover speed are desired. This permits load-free start-up of the prime mover and can prevent stalling. The rotating weights of such designs often work against spring load, so that at a predetermined speed the centrifugal-force effects balance the spring preload. Any further increase in speed will result in a gradual clamp force buildup and a soft start of the driven member. Such a gradual start is desirable, but high torque at low speed can cause excessive slip and thermal destruction of the clutch plates and facing.

17.2.9 Wrapped-Spring Clutch

The clamping force for this clutch is produced by twisting a coil spring so that its inside diameter is reduced and it clamps to a rotating central shaft. Such clutches are typically used in low-power applications[7] where they provide a compact, efficient drive system.

17.3 CLUTCH MATERIALS

17.3.1 Dry-Clutch Lining Materials[8]

Friction materials used for clutch facings are similar to those employed for brakes and, at the present time, three different types of materials are used: organic, sintered iron, and ceramic.

By far, the most widely used is the organic material which consists mainly of asbestos yarn, woven with copper or brass wire. The maximum operating pressure is 30 lb/in^2 for these materials, and the friction coefficient ranges from 0.2 to 0.45. A good design estimate for friction coefficient is 0.25. The torque requirement of the clutch will determine the size and the number of plates. These materials have changed very little over the past 25 years, and still perform satisfactorily at low cost in many applications.

Sintered iron and ceramic materials are capable of withstanding operating pressures and temperatures far in excess of organic facings. Organic facing materials are limited to a maximum operating temperature of approximately 450°F, and rapid wear will result from operation over this temperature (see discussion in Chap. 18.1 on brake wear). Ceramic and sintered iron will operate at twice this temperature, but their cost is substantially higher. The ceramic material also has another advantage—light weight—and clutches using such facings have a low moment of inertia. Such clutch plates are made by riveting small buttons of the ceramic facing material to the disk. The number of buttons will depend upon the type of application and the size of the clutch. The result is an assembly which is 10 percent lighter than a similar disk faced with organic material, and 45 percent lighter than one faced with sintered iron. This puts less load on synchronizer rings, and more rapid shifts are possible. The button design also allows more area for heat transfer from the clutch disk which reduces its operating temperature. However, the trade-off of increased life must be compared to the increased initial cost of iron and ceramic materials.

17.3.2 Wet-Clutch Lining Materials

The main application of wet friction clutches began with the GM prototype automatic transmission in 1938. Dry-type friction materials—asbestos-based sintered metal and woven products—were used in this application. Although these materials performed satisfactorily, the desire for lower cost materials with improved performance led to new friction products. The next development was the use of semimetallic linings after World War II. These materials gave higher static and dynamic friction coefficients, but were less durable. These early friction materials for wet clutches were hard and dense because such properties were desirable for dry applications. However, resilient materials are more desirable for wet-clutch designs (natural cork has this property and a high wet friction coefficient in oil of 0.13). Cork was used in automatic transmissions into the 1950s. The next development was Bakelite-type materials with a phenolic binder. These friction materials had enough physical strength that they could also be used as structural elements and thus eliminate the steel carrier. However, higher horsepower applications soon made these materials inadequate and a new type of wet friction material was developed in the late 1950s. This class of friction materials is known as "paper" type because it is manufactured on a paper-making machine. The process uses a water slurry of materials (asbestos, cellulose, and fillers) which are dried on a screen into a continuous sheet of material. This paper is blanked into the appropriate shape, saturated with oil or liquid resins, cured, and bonded to a steel core. "Paper" materials are noted for their low cost, high dynamic friction coefficient, and extremely low static/dynamic coefficient ratios.

Figure 17.2 shows the engagement characteristics (in oil) of the older wet friction materials and the "paper" type. The sintered metal and the semimetallic friction material have low dynamic friction coefficients and sharp friction peaks as the clutch approaches static contact. The static/dynamic friction ratio for these materials is high. On the other hand, the "paper" material exhibits uniform friction and a smooth transition to static contact. These characteristics make the "paper" material among the smoothest and quietest wet friction materials available.

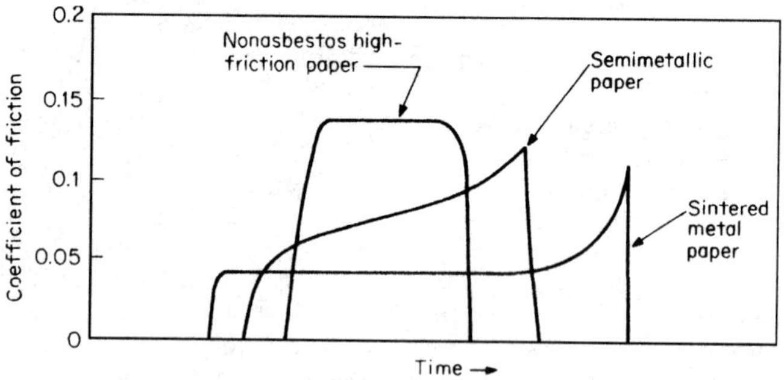

FIG. 17.2 Dynamic engagement characteristics (wet friction trace) of wet friction material.

In the 1960s, "paper" materials were improved with the addition of graphite. This change made the paper more tolerant to surface finish of the mating steel plates, and also allowed higher horsepower transmission because of the high specific heat and conductivity of the graphite. These benefits led to the development of a new family of friction materials known as "graphitics." These materials are superior for high-energy transmission applications.

Another class of materials known as "polymerics" is the extension of rubber compounding technology into the friction materials industry. These materials are a blend of elastomer, fibers (ceramic and organic), fillers, lubricants, friction particles, and a curing system. The material is formed to shape, cured, and bonded to a steel core. The advantages of polymerics are high resilience and durability, high friction coefficient, and good engagement properties. These materials have the highest power-absorbing capabilities of any current clutch material.

Table 17.1 compares the characteristics of different clutch materials operating in oil. The key operating characteristics for smooth, rapid engagement are high dynamic friction and medium-to-high static/dynamic friction ratio. However, these factors must be weighed against the cost and durability of the friction material for a specific application.

TABLE 17.1 Characteristics of Various Types of Friction Materials Currently Available[9]

Characteristic	Molded semi-metallic	Sintered metal	Woven	"Paper" Low static	"Paper" High static	"Paper" Graphitic	Graphitic	Polymeric
Dynamic friction	Low	Low to medium	Medium	High	High	High	Medium	Medium to high
Static/dynamic ratio	High	High	Medium	Low	Very high	Medium	Medium	Medium
Resistance to heat	Good	Excellent	Good	Good	Fair	Very good	Excellent	Good
Durability	Good	Excellent	Very good	Good	Fairly good	Very good	Excellent	Excellent
Resistance to mating plate distortion	Fair	Fair	Good	Very good	Very good	Very good	Very good	Very good
Compressive strength	High	High	Medium	Low	Low	Medium	Medium	High
Facing thickness, in	0.030–0.075	0.025–0.060	0.125–0.300	0.020–0.045	0.020–0.045	0.020–0.040	0.020–0.060	0.020–0.100
Dynamic friction	0.06–0.08	0.05–0.07	0.09–0.11	0.13–0.15	0.12–0.14	0.10–0.12	0.09–0.11	0.11–0.13
Static friction:								
Dexron oil	0.12–0.15	0.11–0.14	0.13–0.16	0.13–0.15	0.18–0.21	0.13–0.15	0.12–0.15	0.14–0.17
Engine oil	0.18–0.20	0.17–0.20	0.19–0.21	0.18–0.20	0.25–0.30	0.18–0.20	0.18–0.20	0.17–0.19
Mating surface requirement	Rough to medium	Rough	Medium	Very smooth	Very smooth	Smooth	Medium	Rough to smooth
Optimum grooving configuration	Spiral or spiral with radial	Sunburst or spiral with radial	Multiple parallel large land	Full waffle or multiple parallel	Full waffle or multiple parallel	Full waffle or multiple parallel	Sunburst or multiple parallel	Spiral/radial sunburst full waffle

Table 17.1 also shows the friction coefficient for the different materials as a function of facing thickness, and the roughness requirements[10] of the mating surfaces from the point of view of retaining reasonable clutch facing wear.

The grooving pattern on the plates of a lubricated clutch will have a significant effect on the friction coefficient. The optimum grooving pattern is shown in Table 17.1 for the different friction materials.

The oil used for clutch applications is a highly compounded hydrocarbon lubricant. The additives include viscosity-index improvers, antifoam agents, antioxidants, and extreme-pressure lubricants. These additives modify the frictional characteristics of the clutch materials, and have an important effect on the life of the clutch.[11,12] As shown in Table 17.1, the choice of oil can have a significant impact on the friction coefficient; for example, the static friction of a semi-metallic material is increased by 50 percent when the lubricant is changed from Dexron to engine oil. Therefore, it is essential that both the friction material and the fluid be considered in a new clutch design.

17.4 TORQUE EQUATIONS

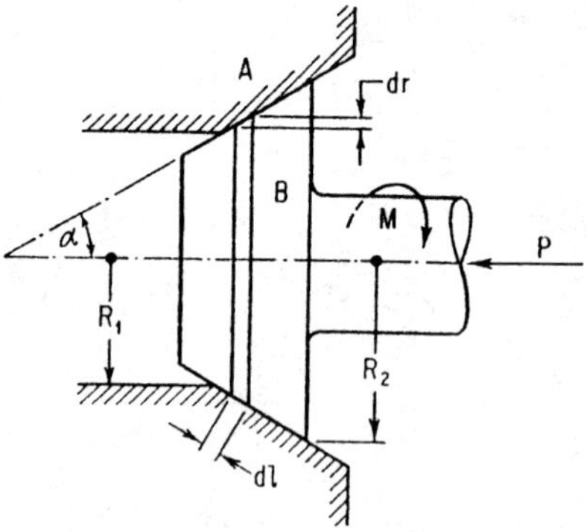

FIG. 17.3 Friction clutch.

Friction clutches are used in power-transmission systems (1) to transmit the desired torque from one point in the mechanism to the other and simultaneously to limit this torque to a desired value, and (2) to couple the prime mover to the load in a manner which allows the acceleration of the load to be controlled.

Consider the arrangement shown in Fig. 17.3, in which it is desired to determine the torque M due to frictional forces existing at the interface of members A and B, and due to the axial force P.

The normal load acting on the differential strip of width dl is $2\pi rp\, dl$, where p is the pressure. Since $dr = dl \sin \alpha$, the axial load supported by the strip

$$dP = 2\pi rp\, dr \tag{17.1}$$

so that

$$P = 2\pi \int_{R_1}^{R_2} pr\, dr \tag{17.2}$$

The frictional force acting on the strip dl

$$\mu\, dP = 2\,\pi\mu pr\, dl \tag{17.3}$$

where μ = coefficient of friction.

The associated moment

$$dM = 2\,\pi\mu pr^2 \csc \alpha\, dr \tag{17.4}$$

and the total frictional moment

$$M = 2\pi \csc \alpha \int_{R_1}^{R_2} \mu p r^2 \, dr \tag{17.5}$$

Before Eq. (17.2) can be integrated, the dependence of μ and p on r and the relationship connecting μ and p must be known.

Variation of μ with r. If friction is independent of sliding velocity, μ is independent of r. If, as is commonly the case in practice, friction is a decreasing function of linear sliding velocity v, the relationship can be written

$$\mu = f(v) = f(\omega_r r) \tag{17.6}$$

where ω_r is the relative angular velocity between the members A and B of the clutch in Fig. 17.3.

In case μ varies linearly with v, Eq. (17.6) can be written

$$\mu = \mu_0 - mv = \mu_0 - m\omega_r r \tag{17.7}$$

where μ_0 = static coefficient of friction and m = slope of friction speed characteristic of the interface (depends on material combination).

Variation of p and r. There is, in any practical situation, considerable uncertainty as to the distribution of axial load P over the contact area. Because of the difficulty of obtaining perfect geometrical fit (especially in the case of cone clutches) between the two surfaces, and particularly with relatively hard materials, such as sintered metals and ceramics, contact between the surfaces is far from uniform and pressure distribution highly irregular. When dealing with resilient friction materials such as granulated cork or paper, the pressure is reasonably independent of r, and the assumption of uniform contact is safe.

Relationship between μ and p. As previously indicated, μ may be a function of p. If p is invariant with r, the expression for the clutch frictional moment will not be affected by the variation of μ with p. However, the design of the clutch (i.e., the number of plates required, etc.) will be different for different values of p.

If p varies with r and μ is a function of p as in

$$\mu = \mu_p - ap \tag{17.8}$$

and if we assume that the wear rate is constant, i.e., the pressure times the surface speed (proportional to radius) is a constant C, one obtains[13]

$$\mu = \mu_p - aC/r \tag{17.9}$$

Examination of the characteristics of those materials in which μ varies with p shows that Eq. (17.8) and hence Eq. (17.9) can be expected to hold only over a limited range of values of p. A nonlinear representation of μ as a function of p will probably be necessary in most cases.

In subsequent discussion the variation of friction with sliding velocity is of importance. If the variation is linear, Eq. (17.7) may be used, and depending on which of the two assumptions, that of the uniform rate of wear or of the constancy of p over the frictional area, is taken as the basis upon which Eqs. (17.2) and (17.5) are integrated, one obtains two expressions for the frictional moment of the clutch:

1. p invariant with r, $\mu = \mu_0 - mr\omega_r$:

$$M = \frac{2nP\mu_0}{3\sin\alpha}\frac{R_2^3 - R_1^3}{R_2^2 - R_1^2} - \frac{nmP}{2\sin\alpha}(R_2^2 + R_1^2)\omega_r \tag{17.10}$$

2. Uniform rate of wear, $pr = C_3$, $\mu = \mu_0 - mr\omega_r$:

$$M = \frac{Pn\mu_0}{2\sin\alpha}(R_1 + R_2) - \frac{nmP}{3\sin\alpha}\frac{R_2^3 - R_1^3}{R_2 - R_1}\omega_r \tag{17.11}$$

where n = number of pairs of frictional surfaces in contact.

It is seen that Eqs. (17.10) and (17.11) have the form

$$M = M_0 - \alpha\omega_r \tag{17.12}$$

where M_0 is recognized as the static torque capacity of a friction clutch

$$M_0 = (2nP\mu_0/3\sin\alpha)/[(R_2^3 - R_1^3)/(R_2^2 - R_1^2)]$$

or

$$M_0 = (nP\mu_0/2\sin\alpha)(R_1 + R_2) \tag{17.13}$$

depending on the assumption made in integrating Eqs. (17.2) and (17.5).

Simple calculation will show that for clutches so designed that R_2 is not very much larger than R_1, both equations (17.10) and (17.11) give values of M which, for all practical purposes, are identical. In general, M obtained from Eq. (17.11) has a slightly lower value than that given by Eq. (17.10).

Cone Clutches. Equations (17.10) and (17.11) can be used as they stand for computation of M.

Disk Clutches. In this case, $\alpha = \pi/2$, $\sin\alpha = 1$, and Eqs. (17.10) and (17.11) become, respectively,

$$M = \tfrac{2}{3}nP\mu_0[(R_2^3 - R_1^3)/(R_2^2 - R_1^2)] - \tfrac{1}{2}nmP(R_2^2 + R_1^2)\omega_r \tag{17.14}$$

$$M = \tfrac{1}{2}nP\mu_0(R_1 + R_2) - \tfrac{1}{3}nmP[(R_2^3 - R_1^3)/(R_2 - R_1)]\omega_r \tag{17.15}$$

Multiple-Disk Clutches. In the case of multiple-disk clutches Eqs. (17.10) through (17.13) require correction in that not all the disks in the assembly are subject to the same externally applied axial clamping force P_0.

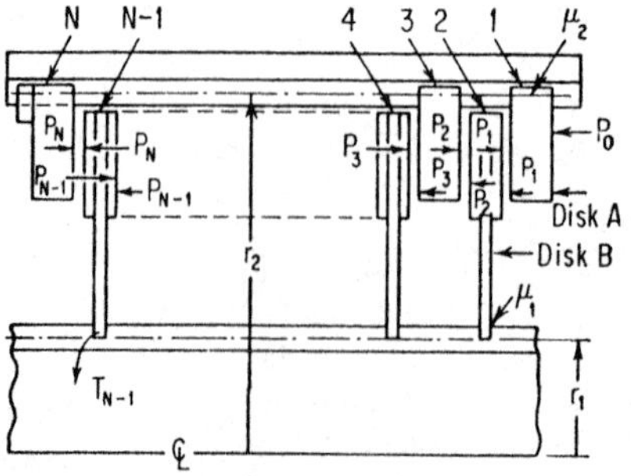

FIG. 17.4 Multiple-disk clutch.

Referring to Fig. 17.4, individual disks are in equilibrium under the action of three axial forces P_N, P_{N-1}, and the force originating at the hub spline because of friction between the disk and the hub as a consequence of torque transmitted by that disk, T_{N-1}/r_1.

Considering disk 1, the following relationship holds between forces acting on this disk when the clutch transmits full rated torque:

$$P_1 = P_0 - T_1\mu_2/r_2 \tag{17.16}$$

where T_1 = frictional moment transmitted by disk 1
μ_2 = coefficient of friction between disk 1 and its housing
r_2 = radius, as indicated in Fig. 17.4

But
$$T_1 = kP_1 \tag{17.17}$$

where, for simplicity, k may be taken from Eq. (17.13).

$$k = \tfrac{1}{2}\,\mu_0(R_1 + R_2)$$

for $\alpha = \pi/2$ and $n = 1$.

Thus,

$$P_1 = \frac{1}{1 + k\mu_2/r_2} \tag{17.18}$$

Considering now the conditions under which disk 2 remains in balance one obtains

$$P_2 = P_1 - (\mu_1/r_1)(T_1 + T_2)$$

which becomes

$$P_2 = \frac{1 - k\mu_1/r_1}{1 + k\mu_1/r_1} \qquad P_1 = \frac{1}{1 + k\mu_2/r_2}\,\frac{1 - k\mu_1/r_1}{1 + k\mu_1/r_1}\,P_0 \tag{17.19}$$

where μ_1 = coefficient of friction between disk 2 and the hub
r_1 = radius, as indicated in Fig. 17.4

Extending this procedure to include all disks in the assembly and expressing axial forces in terms of P_0, one can write the equivalent load P

$$\frac{P}{P_0} = \frac{1}{(N-1)(1+a)}\left(1 + \frac{1-b}{1+b}\right)\sum_{n=1}^{j}\left[\frac{(1-b)(1-a)}{(1+b)(1+a)}\right]^{n-1} \tag{17.20}$$

where $a = k\mu_2/r_2$
$b = k\mu_1/r_1$
$j = (N-1)/2$
N = total number of disks (both sets)

With P_0 known, P is computed through Eq. (17.20) and then inserted into Eqs. (17.10) through (17.13) to compute the value of the total frictional moment of the clutch.

17.5 THERMAL PROBLEMS IN CLUTCH DESIGN

When one solid body slides over another, most of the work done against frictional forces opposing the motion will be liberated as heat at the interface. Consequently the temperature of the rubbing surfaces will increase and may reach values high enough to destroy the clutch.

Even under moderate loads and slow speeds, high instantaneous values of surface temperature are reached at points of actual contact of the two surfaces. Thermoelectric methods employed in the measurement of these transient local temperatures indicate values as high as 1000°C, and their duration has been observed to be of the order of 10^{-4} s. The results are similar when the surfaces are lubricated with mineral oil except that peak temperatures are reduced. Also, it has been observed that, when large quantities of heat are liberated at the rubbing interface during a short time interval, high surface temperatures are generated while the bulk of the two bodies in contact remains relatively cool.

Most premature clutch failures can be attributed to excessive surface temperatures generated during slipping under axial load. In the case of metallic clutch plates, high temperatures existing at the rubbing interface may cause the individual plates to be welded together. When nonmetallic or semimetallic plates are used with steel or cast-iron separating plates, the high temperatures can cause excessive wear. Other effects of high surface temperatures are distortion of the shape of the plates, surface cracks in solid metallic plates caused by thermal stresses, appreciable fluctuation in the value of the friction coefficient ("fading"), and in the case of lubricated clutches, oxidation of oil resulting in the formation of deposits on the working surfaces and in grooves.

In order to design a clutch that is satisfactory from the thermal point of view, it is necessary to know or estimate the value of surface temperature considered safe, and the maximum value of surface temperature likely to occur in a given assembly operating under known conditions of loading.

Because the required information is not readily obtained, it is probably best to perform a carefully executed series of tests. Direct measurement of surface temperatures in a clutch assembly presents great difficulties, however, especially when the friction material is a poor electrical conductor. Furthermore, experimentation with different clutch designs working under different conditions is rather expensive and time-consuming.

If a model is created, based on some simplifying assumptions, surface temperature of the clutch plates sliding relative to each other may be easily computed. The calculated values may then be compared with the safe temperature values usually supplied by friction-material manufacturers. Complete correlation cannot be expected if this procedure is followed, but in most situations a good criterion for establishing the size of the total frictional area of the clutch is thus obtained.

The imperfection with which the model represents the actual system should be borne in mind. The results obtained analytically should be regarded as an approximation. Even though the calculated values of the surface temperature of a clutch plate may differ significantly from the actual transient local values, the equations express the manner in which temperature will vary with system parameters and thus will show which of them should be altered to obtain desired change in temperature.

One widely used method of evaluating the thermal capacity of a friction clutch consists of calculating the average rate of energy dissipation per unit area of the clutch frictional surface during the slip period. The answer, usually stated in terms of hp/in^2 (often of the order of 0.3 to 0.5 hp/in^2), is compared with a similar figure computed for some other successfully operating design and is adjusted accordingly by variation of plate area and number of plates in the assembly. This method affords a quick and rough estimation of the extent of thermal loading to which the active surfaces are subjected and can serve as a very approximate basis for comparing two similar assemblies operating under identical conditions. This method completely neglects such factors of utmost importance in heat transfer as conductivity, specific heat, and plate thickness which have large effects on the surface temperature.

The temperature of the clutch plates will be evaluated by solving the Fourier conduction equation, and the necessary assumptions discussed below.

1. The heat flux generated at the rubbing interface is uniformly distributed over the total frictional area of the clutch; i.e., the rate of energy dissipation is a function of time only. This assumption is based upon the premise that the rate of wear of the rubbing surfaces and not the intensity of pressure is uniformly distributed over the area of the plate. This assumption is justified as follows: First, in a properly designed system, the rate of heat dissipation is a function of time—it decreases from an initial maximum value at the beginning of the slip period to zero at the end, i.e., at the time when the two shafts are effectively coupled by the clutch. Secondly, most clutches are so designed that the ratio of the outside diameter to the inside diameter of the plate is not very much greater than unity (about 1.3) and, therefore, the effect of the heat-flux variation with radius (if any) on plate temperature will be small. Finally, the manner in which the thermal parameters affect the surface temperature is not greatly influenced by the flux distribution.

2. Heat transfer from the clutch plates to the surroundings during slipping is negligible.

This assumption is well justified when dealing with slip periods of short duration, of the order of several seconds or less. Also, in most clutches the edge area of the plate is much smaller than the area receiving heat; thus the heat flow in the radial direction is small and can be neglected in comparison with axial direction.

3. The effect of the lubricating oil is neglected. The effect of the lubricant is difficult to account for in the analysis, mainly because, during the time interval in which the plates move closer together under the action of the axial force, lubrication changes from hydrodynamic to boundary while the oil is squeezed out of the space between the plates. The presence of oil generally lowers surface temperature considerably, depending on the type of lubricant. Under boundary lubrication the temperature can be expected to be between 30 to 60 percent lower than in the case of dry friction.

17.5.1 Derivation of Temperature Equation

First to be considered is the form of the expression for the rate of heat dissipation to be used in subsequent analysis. Consider a basic power-transmission system shown in Fig. 17.5.

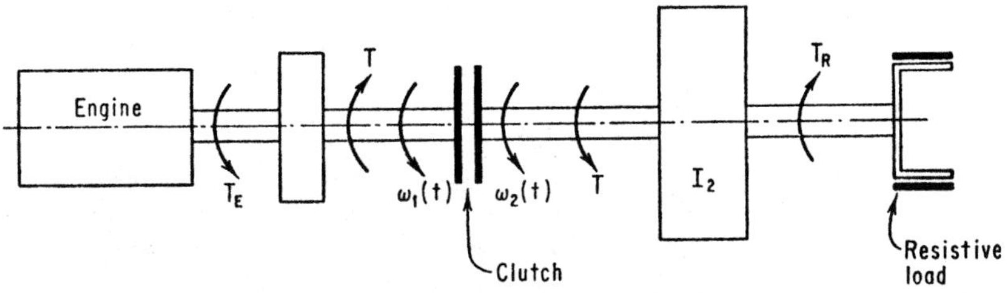

FIG. 17.5 Basic power-transmission system.

T_E is the engine output torque, T the clutch frictional torque, T_R the resistive load torque, and $\omega_1(t)$ and $\omega_2(t)$ are angular velocities of engine and load sides of clutch, respectively. All torques acting on the system will, for simplicity, be assumed constant.

It is easily shown that, when the compliance of the system is neglected,

$$\omega_1(t) = \frac{T_E - T}{I_1} t + \Omega_1 \qquad \text{rad/s} \tag{17.21}$$

$$\omega_2(t) = \frac{T - T_R}{I_2}\,t + \Omega_2 \qquad \text{rad/s} \tag{17.22}$$

where Ω_1 and Ω_2 are initial velocities of the two shafts.

The rate at which energy is dissipated in the clutch during slipping is given by

$$q_0(t) = T[\omega_1(t) - \omega_2(t)] \qquad \text{lb·ft/s} \tag{17.23}$$

$$q_0(t) = T(\{[I_2T_E + I_1T_R - T(I_1 + I_2)]/I_1I_2\}t + \Omega_1 - \Omega_2) \tag{17.24}$$

The duration of the slip period is given by

$$t_0 = I_1I_2(\Omega_1 - \Omega_2)/[T(I_1 + I_2) - (I_2T_E + I_1T_R)] \qquad \text{s} \tag{17.25}$$

The total amount of energy dissipated during slipping (in one cycle) is obtained by integrating Eq. (17.24) between the limits $t = 0$ and $t = t_0$. Thus

$$Q = T(\Omega_1 - \Omega_2)^2 I_1I_2/[2T(I_1 + I_2) - (I_2T_E + I_1T_R)] \qquad \text{lb·ft} \tag{17.26}$$

In the special case, when $T_R = T_E = 0$, i.e, when the system consists of two flywheels I_1 and I_2 rotating at different speeds in the absence of external torques, Eqs. (17.24), (17.25), and (17.26) reduce to

$$q_0(t) = \{\Omega_1 - \Omega_2 - T[(I_1 + I_2)/I_1I_2]t\}\,T \qquad \text{lb·ft/s} \tag{17.27}$$

$$t_0 = I_1I_2(\Omega_1 - \Omega_2)/T(I_1 + I_2) \qquad \text{s} \tag{17.28}$$

$$Q = I_1I_2(\Omega_1 - \Omega_2)^2/2(I_1 + I_2) \qquad \text{lb·ft} \tag{17.29}$$

There is considerable experimental evidence to suggest that the rate of heat dissipation varies approximately linearly with time and thus has the general form

$$-At + B \tag{17.30}$$

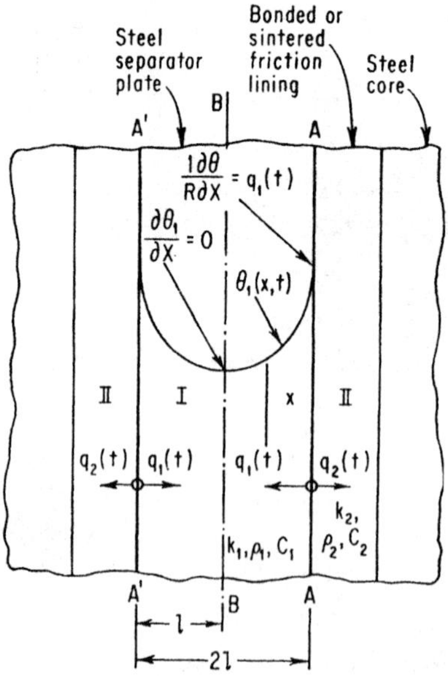

FIG. 17.6 Two clutch plates in contact under the action of the axial clamping force.

which is identical with Eqs. (17.24) and (17.27).

Figure 17.6 shows the two clutch plates in contact under the action of the axial clamping force. Surface *A-A*, where rubbing occurs, can be thought of as a plane source of heat producing $q(t)$, Btu/s·ft². At the rubbing interface the surface temperature of both plates I and II must be the same. The rates of heat flow into I and II will not, in general, be equal but will depend on the thermal properties of the two materials in contact.

While the manner of division of the total flux is yet to be determined, the mathematical form of solution of the heat-conduction equation can be written

$$q_s t = m(-at + b)$$

$$= -\alpha_s t + \beta_s \tag{17.31}$$

$$q_f(t) = (1 - m)(-at + b)$$

$$= -\alpha_f t + \beta_f \tag{17.32}$$

where m is as yet undetermined and $q_s(t)$ = heat flux entering steel plate, $q_f(t)$ = heat flux entering friction lining. The heat-diffusion equation written for the steel plate is[14]

$$\partial^2\theta_s/\partial x^2 = C_s R_s(\partial\theta_s/\partial t) \tag{17.33}$$

where $C_s = c_s\rho_s$
$R_s = 1/k_s$
θ_s = temperature of the steel plate over ambient temperature
c_s = specific heat of steel
k_s = thermal conductivity of steel
ρ_s = density of steel

Using Eq. (17.31) together with the boundary conditions,

$$\begin{aligned} -(1/R_s)(\partial\theta_s/\partial x) &= q_s(t) \qquad \text{at } x = 0 \\ \partial\theta_s/\partial x &= 0 \qquad \text{at } x = l \end{aligned} \tag{17.34}$$

From the solution of Eq. (17.33), the surface temperature of a clutch plate as a function of time is determined.

$$\begin{aligned}\theta_s(0,t) = {} & \zeta\left(\frac{t}{l\sqrt{R_sC_s}} + \frac{l^3\sqrt{R_sC_s}}{3} - \frac{2l\sqrt{R_sC_s}}{\pi^2}\sum_{n=1}^{\infty}\frac{1}{n^2}\exp - \frac{n^2\pi^2}{l^2R_sC_s}t\right) \\ & - \nu\left[\frac{t^2}{2l\sqrt{R_sC_s}} + \frac{l\sqrt{R_sC_s}}{3} - \frac{l^3(C_sR_s)^{3/2}}{45} + \frac{2l^3(R_sC_s)^{3/2}}{\pi^4}\sum_{n=1}^{\infty}\frac{1}{n^4}\exp - \frac{n^2\pi^2}{l^2R_sC_s}t\right]\end{aligned} \tag{17.35}$$

where $\zeta = \beta_s\sqrt{R_s/C_s}$ and $\nu = \alpha_s\sqrt{R_s/C_s}$.

For very large values of $l\sqrt{R_sC_s}$ or very small values of t, Eq. (17.35) is not convenient for computation of the surface temperature. It can be shown that, for very large plate thickness, such that reflection from the opposite side can be neglected,

$$\theta_s(0,t)_{l\to\infty} = 2\zeta\sqrt{t/\pi} - \tfrac{4}{3}(\nu t_{3/2}/\sqrt{\pi}) \tag{17.36}$$

In this case, the maximum temperature occurs at the time $t_m = \tfrac{1}{2}t_0$ and its value is $\sqrt{2}$ times that at $t = t_0$, i.e., the temperature at the end of engagement period.

The value of m in Eqs. (17.31) and (17.32) is determined by considering that the temperature at the rubbing interface ($x = 0$) must be the same for both the steel plate and the friction lining, i.e.,

$$\theta_s(0,t) = \theta_f(0,t) \tag{17.37}$$

The result, obtained by using any one of the temperature expressions, is

$$m = \frac{\sqrt{R_f/C_f}}{\sqrt{R_s/C_s} + \sqrt{R_f/C_f}} \tag{17.38}$$

It can be seen from Eq. (17.38) that the thermal properties and the densities of both mating materials are important in determining the value of the surface temperature.

17.5.2 Application

Equation (17.38) rewritten in more familiar notation

$$m = \sqrt{c_f \rho_f k_f} / (\sqrt{c_f \rho_f k_f} + \sqrt{c_s \rho_s k_s}) \qquad (17.39)$$

shows that the total heat flux divides equally between the two contacting plates when the products of their thermal properties and densities are equal.

From Eqs. (17.31), (17.32), (17.35), and (17.39) it can be shown that the temperature at the interface is proportional to $(\sqrt{c_f \rho_f k_f} + \sqrt{c_s \rho_s k_s})^{-1}$. High values of thermal conductivity, specific heat, and high density will decrease the temperature at the rubbing interface. These properties, however, are not the only criteria to be used in the selection of materials when designing a clutch.

Other factors, such as coefficient of friction, fading, crushing strength, wear characteristics, and conformability, must also be considered when designing a friction clutch. Other criteria are discussed in the section on temperature flashes.

In practice, many clutches, especially those of multiple-disk type, are so designed that the thickness of the plate $2l$ does not exceed 0.10 in. When l (0.05 in) is substituted in Eq. (17.35) the constant in the exponent is

$$n^2\pi^2/l^2 R_s C_s = 71.3n^2 \qquad \text{for steel}$$

Thus, even with $t = 0.1$ and $n = 1$, $e^{-7.13} = 9 \times 10^{-4}$ and for all practical purposes the contribution of the infinite series in n can be neglected. Thus

$$\theta_s(0, t) = -\frac{vt^2}{2l\sqrt{R_s C_s}} + \left(\frac{\zeta}{l\sqrt{R_s C_s}} - \frac{vl\sqrt{R_s C_s}}{3}\right)t + \left[\frac{\zeta l\sqrt{R_s C_s}}{3} + \frac{vl^3(R_s C_s)^{3/2}}{45}\right] \qquad (17.40)$$

In order to find the maximum surface temperature of the plates, we differentiate Eq. (17.40) and set $\partial\theta_s/\partial t = 0$. The time of incidence of the maximum surface temperature is thus given by

$$t_m = \zeta/v - l^2 R_s C_s/3 \qquad \text{s} \qquad (17.41)$$

or, since $\zeta/v = t_0$,

$$t_m = t_0 - l^2 R_s C_s/3 \qquad \text{s} \qquad (17.42)$$

Substituting the value of t_m from Eq. (17.41) into Eq. (17.40), the maximum surface temperature is found to be (for small l)

$$\theta_{s,\max} = \tfrac{1}{2}\zeta^2/vl\sqrt{R_s C_s} + \tfrac{7}{90}\, vl^3(R_s C_s)^{3/2} \qquad (17.43)$$

A detailed examination of Eq. (17.43) will yield interesting and useful results depending on the nature of the system in which the clutch is used. Thus, if the system consists of two flywheels rotating at different speeds in the absence of external torques, and the purpose of the clutch is to transfer energy from one flywheel to the other (such systems form the basis of various forms of inertia starters), Eq. (17.29) shows that the total amount of energy dissipated in the clutch during the process of coupling is independent of the clutch frictional torque T. On the other hand, the duration of the slip period t_0, according to Eq. (17.28), is inversely proportional to T. The question may be raised, how does the change in t_0, caused by varying T, affect the maximum surface temperature of the plates? It is to be understood that the variation of clutch frictional torque is effected by varying the axial clamping force F only, and that clutch dimensions, number of plates, and their frictional and thermal properties are held constant.

This problem may arise when it is desired to adapt a clutch of certain thermal and mechanical capacity to a system demanding a higher-capacity clutch, and the deficiency is to be compensated for by increased pressure on the frictional surfaces.

It can be shown that the amount of energy absorbed by the steel plate per unit surface area is given by

$$Q_s = \bar{\mu}\, \delta(Q/A) \tag{17.44}$$

where δ = conversion factor from lb·ft to Btu = 1.285×10^{-3}
A = total frictional area of clutch, ft^2

and $\bar{\mu} = \dfrac{\sqrt{R_f/C_f}}{\sqrt{R_f/C_f} + \sqrt{R_s/C_s}}$

Q was defined in Eq. (17.29).

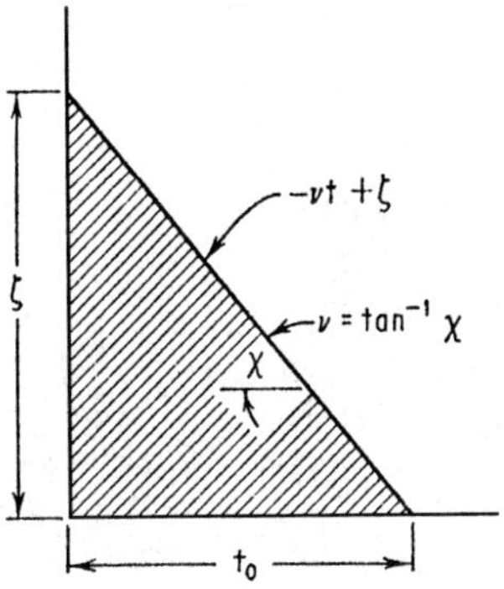

FIG. 17.7 Area number $-\nu t + \zeta$ plot which is proportional to Q_s.

Also, it is not difficult to verify that Q_s will be proportional to the shaded area in Fig. 17.7, and since Q_s is independent of either T or t_0, we may write

$$\zeta t_0 = K_1 = \text{const} \tag{17.45}$$

where $K_1 = 2Q_s\sqrt{R_s/C_s}$

Now
$$\zeta/t_0 = \nu \tag{17.46}$$

Then, from Eqs. (17.45) and (17.46),

$$\zeta = K_1/t_0 \tag{17.47}$$

$$\nu = K_1/t_0^2 \tag{17.48}$$

$$\zeta^2/\nu = K_1 \tag{17.49}$$

but $t_0 = K_2/T$ where $K_2 = I_1I_2(\Omega_1 - \Omega_2)/(I_1 + I_2)$. Substituting for t_0 into Eq. (17.48),

$$\nu = (K_1/K_2)T^2 \tag{17.50}$$

Substitution of ζ^2/ν from Eq. (17.49) and ν from Eq. (17.50) into Eq. (17.43) results in

$$\theta_{s,\max} = \frac{Q_s}{lC_s} + \frac{7}{45}\,\frac{Q_sT^2l^3R_s^2(I_1 + I_2)^2}{I_1^2I_2^2(\Omega_1 - \Omega_2)^2} \tag{17.51}$$

Although the contribution of the second term in Eq. (17.51) to the value of θ_s max becomes appreciable only at very high values of T, it will be observed that surface temperature is proportional to the square of the clutch frictional torque. Thus, in the case when the clutch is used to start or stop pure inertia loads, too small and *not* too large values of t_0 will tend to damage the clutch.

Another much more frequently used system is one in which external driving and resistive torques, as well as the clutch frictional torques, act on the rotating inertias. In such systems, Eqs. (17.24), (17.25), and (17.26) would be used to calculate the rate of energy dissipation, duration of slip period, and the total amount of energy dissipated per cycle. It will be observed that, in the present case, Q is no longer independent of T and that both $t_0 \to \infty$ and $Q \to \infty$ when $T = (I_2T_E + I_1T_R)/(I_1 + I_2)$.

We again wish to examine the effect of varying clutch frictional torque on maximum surface temperature when the change in T is brought about by varying axial force on the clutch plates only.

From Eqs. (17.24) and (17.26), it follows that

$$\alpha_s = \frac{\bar{\mu}\delta}{A}\,\frac{T^2(I_1 + I_2) - T(I_2T_E + I_1T_r)}{I_1I_2} \qquad \text{Btu/s}^2\cdot\text{ft}^2 \tag{17.52}$$

$$\beta_s = (\bar{\mu}\,\delta/A)T(\Omega_1 - \Omega_2) \qquad \text{Btu/s}^2\cdot\text{ft}^2 \tag{17.53}$$

then

$$\zeta_s = (\bar{\mu}\,\delta/A)T(\Omega_1 - \Omega_2)\sqrt{R_s/C_s} \tag{17.54}$$

$$\nu_s = (\bar{\mu}\,\delta/A)[(T^2 - MT)/P]\sqrt{R_s/C_s} \tag{17.55}$$

where

$$M = (I_1T_E + I_1T_R)/(I_1 + I_2)$$

$$P = I_1I_2/(I_2 + I_2)$$

Substituting for ζ_s and ν_s from Eqs. (17.54) and (17.55) into Eq. (17.43), one obtains

$$\theta_{s,\max} = \frac{\delta\bar{\mu}T(\Omega_1 - \Omega_2)^2P}{2Al(T - M)C_s} + \frac{7\bar{\mu}\,\delta T(T - M)l^3R_s^2C_s}{90PA} \tag{17.56}$$

Equation (17.56) expresses analytically what may, perhaps, have been arrived at by some qualitative speculation, namely, that when driving and resistive torques are present in the system shown in Fig. 17.5, maximum surface temperature will reach very high values when the frictional clutch torque approaches the value of the driving torque. The same equation also shows that, when $T >> (I_2T_E + I_1T_R)/(I_1 + I_2)$(i.e., when t_0 is very small), the contribution of the second term to the value of θ_s max may become appreciable; so that, in spite of decreased slip period, the surface temperature of clutch plates may begin to increase.

17.5.3 Variation of Maximum Surface Temperature with Plate Thickness

If in the previously discussed system it is desired to lower the value of the maximum surface temperature by properly designing the friction elements, two methods are generally available: (1) increase in the total clutch frictional area A by increasing the number of plates or their size, and (2) increase in the thickness of the steel separator plate.

In applying the first of these methods, Eq. (17.56) shows that the maximum surface temperature is inversely proportional to the total frictional area A. Thus it follows that virtually any desired value of maximum surface temperature may be obtained by assigning proper value to A. The addition of one or more sets of plates, however, will generally be more expensive than the increase in thickness of the steel separator plates in an already designed clutch.

A general criterion for the maximum useful plate thickness when heated on both sides is based upon plotting the nondimensional Fourier number $\mathcal{F} = t_0/l^2CR$ as a function of plate thickness. It can be shown that any increase beyond the value corresponding approximately to $1/\mathcal{F} = 1.75$ ceases to be significant:

$$l_{max} = \sqrt{1.75t_0/C_sR_s} = \sqrt{1.75t_0k_s/c_s\rho_s} \tag{17.57}$$

where $l_{\max} = \frac{1}{2}$ maximum useful plate thickness when heated on both sides.

In the case when the plate is heated on one side only, Eq. (17.57) gives the actual maximum useful plate thickness.

17.5.4 Optimum Radial Dimensions of a Clutch Plate

The optimum dimensions of a clutch plate are dependent on the criterion used to define the meaning of the word "optimum." If the criterion is formulated on the basis that the ratio of frictional torque to the clutch-plate surface temperature should be a maximum one can obtain the relationship between the inside and outside radii as follows:

$$T \propto R_o + R_i \tag{17.58}$$

where T = torque
R_o = outside radius of plate
R_i = inside radius of plate

The surface temperature is inversely proportional to the area of the plate

$$\theta_s \propto [1/(R_o^2 - R_i^2)] \tag{17.59}$$

$$T/\theta_s = (1 - k^2)(1 + k) \tag{17.60}$$

where $k = R_i/R_o$, and from which T/θ_s is a maximum when $k = ⅓$ or

$$R_i = ⅓\, R_o \tag{17.61}$$

17.5.5 Flash Temperatures

Very high values of transient surface temperature occur at the interface between two sliding bodies, these temperature flashes occurring at points of intimate contact between the two surfaces.[15–17]

The effective or microscopic contact between the two surfaces is confined to surface asperities, and it can be shown that the relationship between the effective and the nominal or macroscopic area of contact is

$$A'/A = p/p_m \tag{17.62}$$

where p_m = mean yield pressure of the softer material
p = nominal pressure (average pressure)
A' = effective frictional area over which contact actually occurs

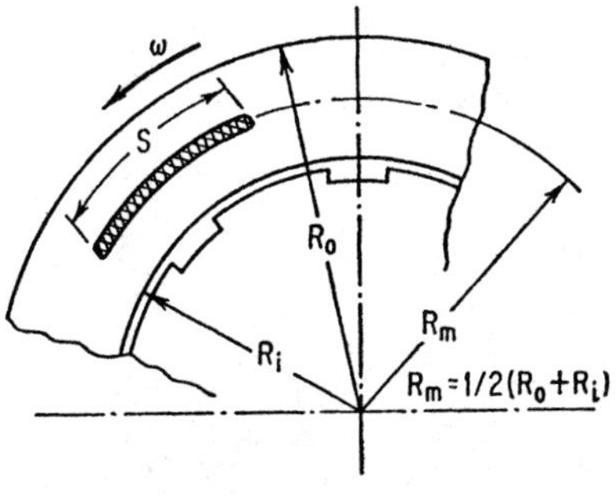

FIG. 17.8 A portion of a clutch plate showing the distance covered by an isolated hot spot.

If contact occurs at n spots, and if these spots are assumed to be circular in shape and equal in size

$$A' = \pi n r^2 \tag{17.63}$$

where r = radius of the spot.

When relative motion exists between the surfaces, it has been observed that a number of luminous points appear at the rubbing interface and that the position of these spots changes rapidly as points of intimate contact wear away and new asperities come into contact. Referring to Fig. 17.8, which represents a portion of a clutch plate, let S be the distance covered by such an isolated hot spot in the small time interval t_s. Then

$$t_s = S/\omega R_m \tag{17.64}$$

where ω = angular velocity and R_m = mean radius of the plate.

Also,

$$S \,{}^{\prime}\, r \tag{17.65}$$

If a clutch with only one frictional surface is considered, the rate at which heat enters each spot is given by

$$q_r = \delta T\omega/\pi n r^2 \tag{17.66}$$

Because t_s is very small, q_r will remain essentially constant. Also, the effect of heat reflection from the other side of the plate can be neglected. This allows the use of the simplest expression for the surface temperature, which is derived for the case of a semi-infinite solid receiving heat at a constant rate:

$$\theta_f(t) \propto \sqrt{R_s/C_s} q_r \sqrt{t_c} \tag{17.67}$$

From $T = \mu pAR_m$, $A \propto R2m$, and Eqs. (17.62) through (17.67), the following expression is obtained:

$$\theta_f(t) \propto \sqrt{R_s/C_s}(\mu p_m{}^{3/4} R_m p^{1/4} \omega^{1/2}/n^{1/4}) \tag{17.68}$$

Equation (17.68) indicates that, in order to decrease the value of flash temperature occurring at the areas of intimate contact, the following points should be considered when selecting friction material and clutch-plate size:

1. Soft friction materials having good surface conformability are preferable.
2. The coefficient of friction should be low. Since, in order to minimize clutch size, high values of μ are usually preferred, a material having sharp fade characteristics at high temperatures should be used.
3. Contact between the mating plates should be as uniform as possible; i.e., the number of contact spots n should be as large as possible.
4. For a clutch of given nominal frictional area A, the mean radius of the plate should be kept small. Since, for the plate of a given outside diameter, the frictional torque of the clutch is proportional to R_m, a compromise decision regarding the value of R_m should be made which takes into account both flash temperatures and the value of the frictional torque desired.

The foregoing analysis to a certain extent explains why substitution of hard sintered-bronze plates for those coated with such materials as cork or wood flour and asbestos mixtures does not always result in solving the high-temperature problems in a poorly designed clutch. Even though, according to Eq. (17.35), the use of sintered bronze should, because of its higher thermal conductivity, result in lower surface temperature than that obtained with cork lining, this advantage is offset by high values of surface-temperature flashes if the contact between the rubbing plates is not uniform. Therefore, when using metallic friction materials, great care should be taken in the preparation and finish of working surfaces.

17.5.6 Bulk Temperature of Clutch Assembly

When in a given power-transmission system the clutch is cycled continuously, it is necessary to have some means of computing the value of the average temperature at which the assembly will operate. This temperature is based upon radiative and convective heat transfer.

Problems encountered in analysis include the following:

1. Although most clutch assemblies are similar in outside appearance, i.e., cylindrical, the details of their design vary widely and may, in many cases, influence the process of heat transfer considerably (see Sec. 17.2).
2. Heat-transfer data, which could be directly applied to conditions under which the clutch is required to perform, are not available (e.g., shape, large diameter/length ratio, oil-mist atmosphere, end effects).
3. When formulas and methods applicable to simple and idealized cases are used to describe complicated systems errors result, mainly because many of the so-called "constants" rarely remain truly constant, and also some of the effects which can be considered negligible in idealized cases may become quite prominent in an actual system.

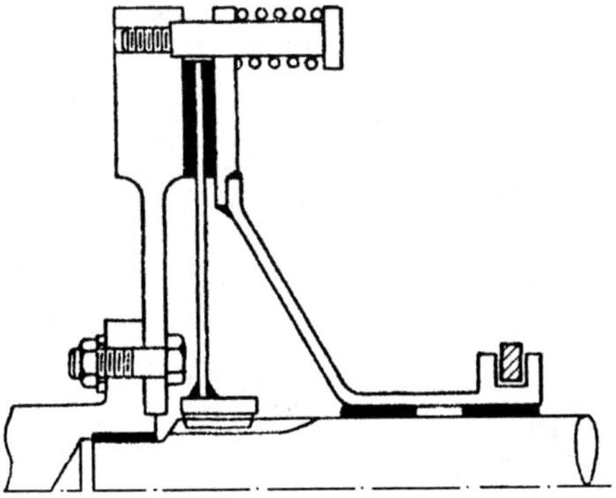

FIG. 17.9 Design where Newton's "law of cooling" may be used.

Two approaches are possible toward the estimation of the bulk temperature of the clutch, depending on its construction. When the design is such as shown in Fig. 17.9, i.e., when the heat generated at the rubbing surfaces is transmitted directly by conduction through a homogeneous material to the outer surfaces and from there transferred to the surroundings by the mechanism of forced convection, Newton's "law of cooling" may be used. This method is strictly correct only when the temperature everywhere within the body, including the surface, is the same and the surface coefficient of heat transfer is a constant independent of temperature.

In practice, the above conditions are approximated if the thermal conductivity of the material in question is high, when temperature does not vary too rapidly, and when the relative size of the body is small.

When a solid body of weight W, specific heat c, having an exposed surface area A cools slowly from an initial temperature θ'', the temperature at any subsequent time t is given by

$$\theta - \theta_a = (\theta'' - \theta_a) \exp(-\alpha t) \qquad (17.69)$$

where θ_a = ambient temperature
$\alpha = Ah/Wc$
h = surface coefficient of heat transfer

The temperature rise per cycle is

$$\theta = \theta'' - \theta' = \delta Q/Wc \qquad {}^\circ\text{F} \qquad (17.70)$$

where θ' = temperature at the beginning of the slip period (or cycle) and θ'' = temperature at the end of the slip period.

It is assumed that the clutch attains the temperature θ'' instantaneously, which is approximately correct when the time during which it is allowed to cool is large in comparison with the time required for the heat to diffuse throughout the bulk of the plates.

If t_0 is the equal time interval between successive applications of the clutch, the temperatures θ' and θ'' are given by the following equations. First cycle,

$$\theta' = \theta_a \qquad \theta'' = \theta_a + \Delta\theta$$

Second cycle,

$$\theta_2' = \theta_a + \Delta\theta \exp(-\alpha t_c) \qquad \theta''_2 = \theta_a + \Delta\theta[1 + \exp(-\alpha t_c)]$$

*n*th cycle,

$$\theta_n' = \theta_a + \Delta\theta[\exp(-\alpha)t_c + \exp(-2\alpha)t_c + \cdots + \exp(-n\alpha t_c)]$$

$$= \theta_a + \Delta\theta \frac{\exp[-(n-1)\alpha t_c] - 1}{\exp(-\alpha t_c) - 1} \exp(-\alpha t_c)$$

$$\theta_n'' = \theta_a + \Delta\theta[1 + \exp(-\alpha t_c) + \exp(-2\alpha t_c) + \cdots + \exp(-n\alpha t_c)]$$

$$= \theta_a + \Delta\theta \frac{\exp(-n\alpha t_c) - 1}{\exp(-\alpha t_c) - 1}$$

The steady-state values of θ_n' and θ_n'' are obtained by making $n \to \infty$. Thus

$$\theta_{ss}' = \theta_a + \Delta\theta \exp(-\alpha t_c)/[1 - \exp(-\alpha t_c)] \tag{17.71}$$

$$\theta_{ss}'' = \theta_a + \Delta\theta[1 - \exp(-\alpha t_c)]^{-1} \tag{17.72}$$

Before Eqs. (17.71) and (17.72) can be used, the value of h must be determined. This can be approximately obtained by the use of the charts given in Figs. 17.10 and 17.11, taken from Refs. 18 and 19. Strictly speaking, the data contained in Fig. 17.10 pertain to rotating cylinders in which the length/diameter ratio is large, so that end effects are negligible, which is not quite the case when dealing with clutches. Figure 17.11 represents the results of experiments designed to establish the heat transfer in rotating electrical machines. Figure 17.12 gives those properties of air which are needed to evaluate Re and Nu.

The other method of evaluating the average temperature of the clutch assembly may be applied in cases where the preceding exponential development cannot be used. Consider that the clutch shown in Fig. 17.13 (disregard, for this purpose, the presence of cooling oil) is intended to be cycled continuously and at a regular time interval of 60-s duration. The clutch cylinder rotates at a constant speed of 4000 r/min and is completely enclosed in a cylindrical housing made of good heat-conducting material. The ambient temperature outside the housing is 100°F. Assume that the average rate of energy dissipation at the rubbing surfaces during the slip period is 0.5 hp/in^2 per cycle and that t_0 = 0.8 s. With A = 117 in^2, the total amount of energy liberated per cycle in the form of heat is 33.1 Btu. Thus in the steady state the average rate of heat rejection from the clutch assembly is q = 33.1/60 = 0.55 Btu/s.

Because of the high thermal conductivity of the housing material and its small wall thickness, the temperature drop across it will be neglected in the foregoing.

Referring to Fig. 17.14, the heat-dissipating areas of the housing and of the clutch cylinder are not equal:

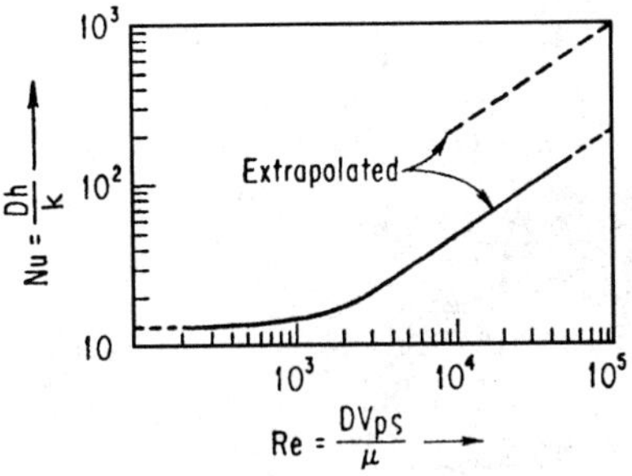

FIG. 17.10 Correlation of Nu (Nusselt number) vs. Re (Reynolds number) for a rotating cylinder without cross flow in air. D = diameter; h = coefficient of heat transfer; k = thermal conductivity of air; V_p = peripheral velocity; s = specific weight of air; μ = viscosity of air.

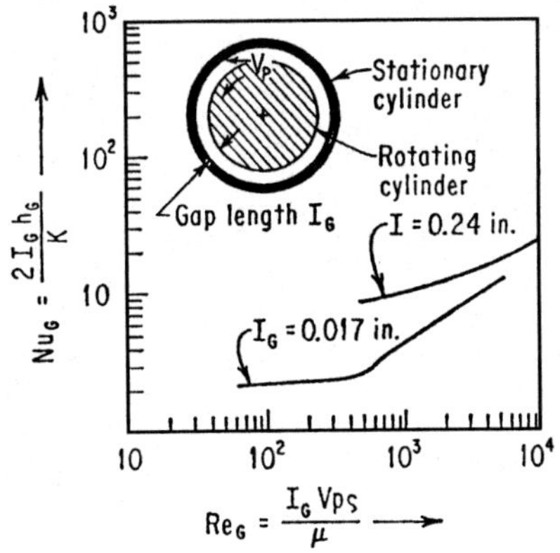

FIG. 17.11 Correlation of Nu_G (gap Nusselt number) vs. Re_G (gap Reynolds number) for a pair of coaxial cylinders. Inner cylinder rotating in air without axial flow (heat transfer from rotor surface to stator surface). Note curves show heat transfer across gap from rotating to stationary cylinder.

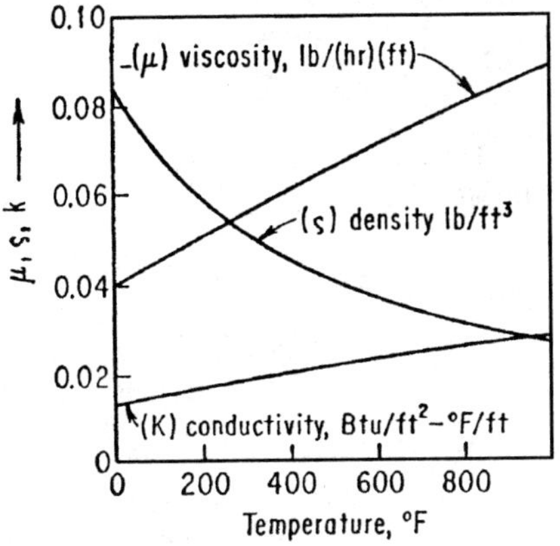

FIG. 17.12 Properties of air.

$$q_{12} = q/A_{\text{I}} = 0.55/0.44 = 1.25 \text{ Btu/(s·ft}^2)$$

$$q_{23} = q/A_{\text{II}} = 0.55/0.52 = 1.06 \text{ Btu/(s·ft}^2)$$

where q_{12} = rate of heat flow per square foot from I to II
q_{23} = rate of heat flow per square foot from II to III
A_{I} and A_{II} = heat-dissipating areas of clutch cylinder and housing, respectively

Now
$$q_{12} = h_g(\theta_{\text{I}} - \theta_{\text{II}}) + (E_{\text{I}} - E_{\text{II}})\epsilon_c \tag{17.73}$$

where h_g = gap heat-transfer coefficient, Btu/(ft²·s²·°F)
θ_{I} = surface temperature (°F) of the clutch cylinder
θ_{II} = temperature of the housing
E_{I} = rate at which energy is radiated from the surface of I at a temperature θ_{I} (°F) (black-body emission)
E_{II} = rate at which energy is reflected from II back to I at a temperature θ_{II} (°F) (black-body reflection)
ϵ_c = combined emissivity of I and II (associated with geometry of system)

$$\epsilon_c = \frac{\epsilon_{\text{I}}\epsilon_{\text{II}}}{\epsilon_{\text{I}}(A_{\text{I}}/A'_{\text{II}}) + \epsilon_{\text{II}} - \epsilon_{\text{I}}\epsilon_{\text{II}}(A_{\text{I}}/A'_{\text{II}})}$$

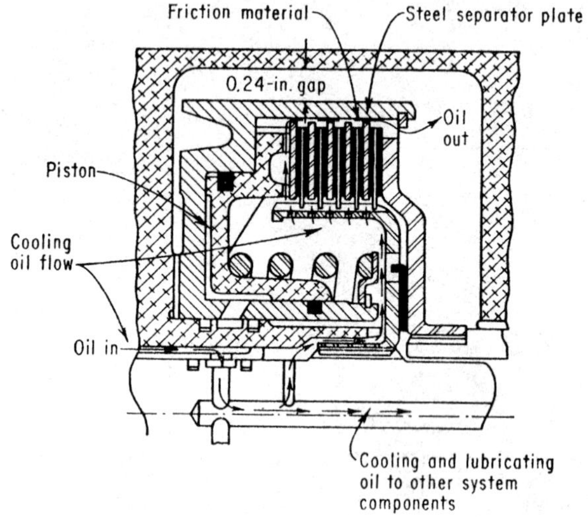

FIG. 17.13 Design where Newton's "law of cooling" cannot be used.

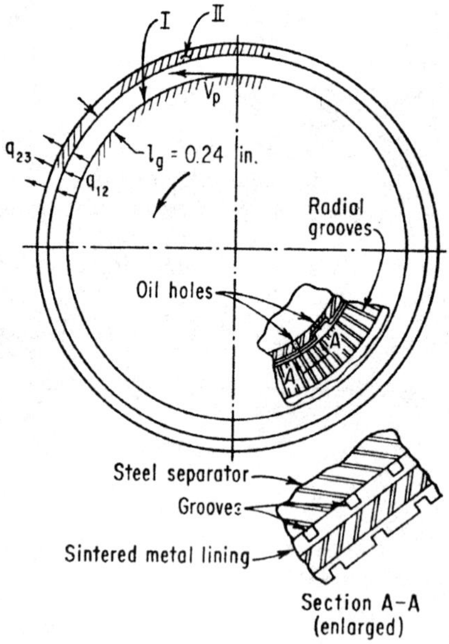

FIG. 17.14 Heat dissipation from outer surface of clutch cylinder for clutch shown in Fig. 22.13. I = clutch cylinder outer surface; II = housing; III = surrounding air.

where ϵ_{I} = emissivity of I

ϵ_{II} = emissivity of II

A_{I} = outside area of I

A'_{II} = inside area of II

The rate of heat flow from II to III is given by

$$q_{23} = h_c(\theta_{II} - \theta_{III}) + (E_{II} - E_{III})\epsilon_{II} \quad (17.74)$$

where h_c = coefficient of heat transfer due to natural (or forced) convection, Btu/(ft²·s·°F)

θ_{III} = ambient temperature, °F

E_{III} = rate of radiant-heat reflection from the surroundings

The value of the coefficient h_c may be obtained from Fig. 17.15 computed from the correlation of Nu vs. Pr × Gr for horizontal cylinders in air. The coefficient h_g is obtained from Fig. 17.11. The values of black-body radiant-heat transfer are plotted in Fig. 17.16 in a manner designed to facilitate computation. Values of ϵ_I and ϵ_{II} may be obtained from Ref. 14.

All five quantities, h_g, h_c, E_I, E_{II}, and E_{III}, are, for any physical configuration, functions of temperatures θ_I, θ_{II}, and θ_{III}. The simplest method of solution is outlined below:

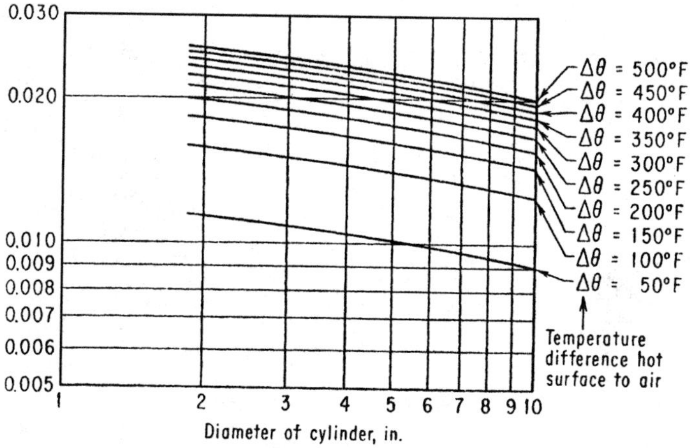

FIG. 17.15 Surface coefficient of heat transfer h_c, Btu/(ft²·s·°F).

1. With θ_{III} known, assume a series of several values of θ_{II} and calculate corresponding values of q_{23}. Plot these values on a temperature scale as a function of θ_{II}. The resulting graph is shown in Fig. 17.17, curve *A*. The intersection of this curve (point *X*) with the horizontal line q_{23} = 1.06 Btu/(ft²·s) determines the value of θ_{II}.
2. The same procedure is applied in order to determine the surface temperature of the clutch cylinder θ_I. The point of intersection of curve *B* with the horizontal line q_{12} = 1.25 Btu/(ft²·s) determines the surface temperature of the rotating clutch cylinder, which in this case is 537°F.

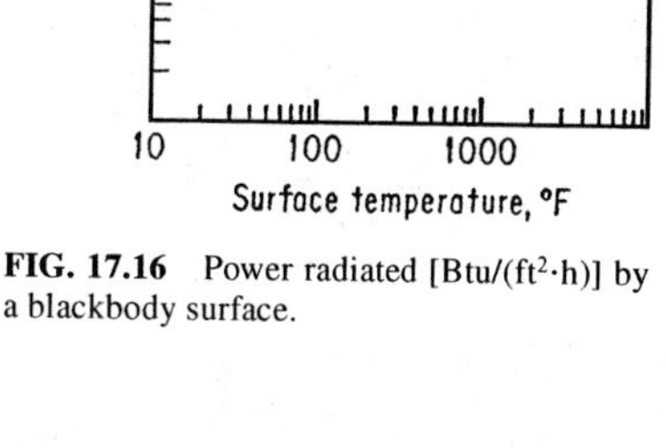

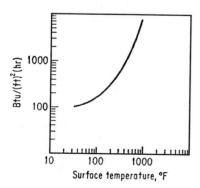

FIG. 17.16 Power radiated [Btu/(ft²·h)] by a blackbody surface.

With this value of the surface temperature of the clutch cylinder, the temperature existing at the rubbing surfaces of the plates during the slip period would be several hundred degrees higher, and thus steps must be taken to provide added cooling of the clutch assembly. In the case of lubricated clutches, such as shown in Fig. 17.13, this is most effectively accomplished by providing sufficient amounts of cooling oil with passages arranged so that the flow is directed at the surfaces where heat is generated. The flow of oil should be continuous, and the clutch so designed that the oil is in direct contact with the plates prior to, during, and after the slip period. In a heavy-duty clutch, sintered-metal friction plates would probably be used and, in this case, radial grooves, such as shown in Fig. 17.14, can be most easily and inexpensively provided.

Heat transfer from the clutch assembly by conduction to other system components (shafts, gears, etc.) has been neglected. This would probably amount to some 10 to 15 percent of the average rate of heat generation; so that the calculated value of θ_I = 537°F may be too high.

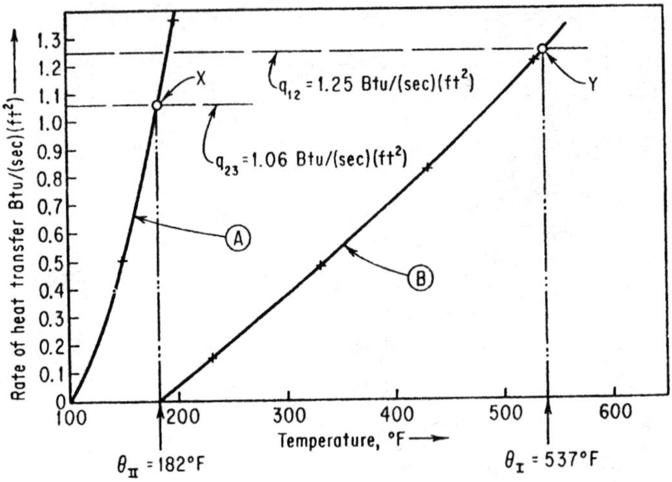

FIG. 17.17 Plot of values of q from assumed values of θ in order to find θ_I and θ_{II}.

17.6 DYNAMICS OF SYSTEMS EMPLOYING FRICTION CLUTCHES

A rigorous dynamic analysis of a mechanical power-transmission system containing a friction clutch presents a very difficult problem in derivation and solution of the differential equations describing the motion of the system.[20–22] This is because these systems are usually characterized by several degrees of freedom and are generally nonlinear. Because simple and general analytical techniques for dealing with nonlinear differential equations of higher order than the second are not available, there will generally be some sacrifice of mathematical rigor and elegance in the analysis.

The problem is to determine the behavior of a system following application of the clutch, given a friction clutch of certain specified characteristics, and given a set of initial conditions.

As a first step in arriving at a mathematical representation of a friction clutch it can be stated that a device whose operation depends upon friction exerts a frictional force or torque on the system only when relative motion exists between contacting members.[23]

This frictional force or torque is, in general, a function of both the magnitude and the direction of the relative velocity, and it may also be made time-dependent by introducing external control.

Thus a friction clutch may be characterized

$$M_c = f(\Delta\dot{\theta}, t) \tag{17.75}$$

where M_c = frictional torque
$\Delta\dot{\theta}$ = relative angular velocity
t = time

The variation in the friction coefficient with variables such as temperature, humidity, duty cycle, and loading history is slow in comparison with the frequencies encountered in vibrating mechanical systems.

For a clutch in which the torque is not externally controlled,

$$M_c = f(\dot{\theta}_1 - \dot{\theta}_2) \tag{17.76}$$

where $\dot{\theta}_1$ and $\dot{\theta}_2$ are the velocities of input and output plates, respectively. This relationship could be represented graphically as shown in Fig. 17.18.

It should be recognized that Fig. 17.18, to a different scale, provides the variation of frictional coefficient of the materials used in the construction of the clutch. This information will normally be available in graphical form as a result of testing. Equation (17.76) may be represented analytically:

$$M_c = \frac{\dot{\theta}_1 - \dot{\theta}_2}{|(\dot{\theta}_1 - \dot{\theta}_2)|} M_c(0) + \sum_{n=1}^{\infty} \frac{1}{n!} (\dot{\theta}_1 - \dot{\theta}_2)^n M_c^{(n)}(0) \tag{17.77}$$

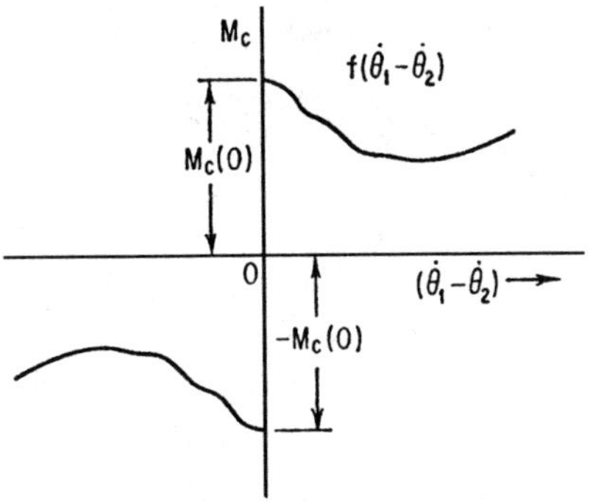

FIG. 17.18 Curve of $M_c = f(\theta_1 - \theta_2)$ for a clutch in which the torque is not externally controlled.

where $M_c(0)$ is the value of the clutch torque when $(\dot{\theta}_1 - \dot{\theta}_2) = 0$ (i.e., "static" friction torque) and primes denote derivatives with respect to $(\dot{\theta}_1 - \dot{\theta}_2)$. In practice, as many terms of the series (17.77) are used as are necessary to obtain an adequate "fitting" to the curve in Fig. 17.18. It should be noted that Eq. (17.77) is nonlinear not only by virtue of the terms $(\dot{\theta}_1 - \dot{\theta}_2)^2$, $(\dot{\theta}_1 - \dot{\theta}_2)^3$, ... $(\dot{\theta}_1 - \dot{\theta}_2)^n$, but also because the term $M_c(0)$ is a function of the relative velocity $\dot{\theta}_1 - \dot{\theta}_2$.

Since the torque-speed curve of a clutch may be characterized over the whole or part of the range by negative slope, it is of interest to see what effect this may have on the behavior of the dynamical system of which the clutch is a part.

Consider a simple mechanical oscillator consisting of mass, m, spring of constant k, and a damper of constant α. The equation of motion is

$$m\ddot{x} + \alpha\dot{x} + kx = 0 \tag{17.78}$$

Multiply Eq. (17.78) by x and integrating from 0 to τ

$$\left[\frac{m(\dot{x})^2}{2}\right]_0^\tau + \left[\frac{kx^2}{2}\right]_0^\tau = -\alpha \int_0^\tau (\dot{x})^2\,dt \tag{17.79}$$

The sum of the left-hand side of Eq. (17.79) represents the variation of the total energy, kinetic plus potential, during this time. It can be seen that this variation may be either positive or negative depending upon the sign of α. If α is positive, the variation in the total energy is negative, i.e., the system loses energy as $\tau \to \infty$. If α is negative the system gains energy as $\tau \to \infty$. In a system without a source of energy this is not physically realizable. α represents the slope of the friction-speed characteristic of the damper.

A power-transmission system to be considered here consists of three essential elements: (1) a source of energy (in the form of some type of prime mover on the input to the clutch), (2) a clutch controlling the flow of power, and (3) the load on the output side of the clutch.

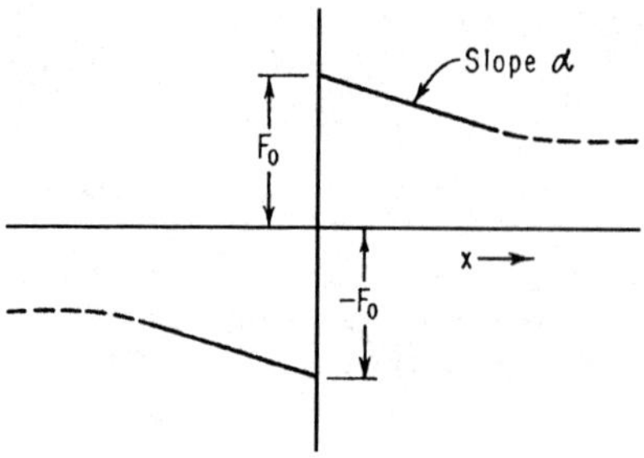

FIG. 17.19 Characteristics of damper of mechanical oscillator.

Since the presence of inertia and compliance is inevitable in mechanical systems, the entire system may become oscillatory following the application of the clutch. If the clutch characteristic has anywhere a portion with negative slope, these oscillations will tend to build up as suggested by Eq. (17.78) until limited by some system nonlinearity. In this case, the nonlinearity which eventually limits the growth of the oscillation and causes the amplitude to decrease to zero in a properly designed system is the term $[(\dot{\theta}_1 - \dot{\theta}_2)/|(\dot{\theta}_1 - \dot{\theta}_2)|]\,M_c(0)$ in Eq. (17.77).

Consider now a simple mechanical oscillator, letting the damper have a characteristic shown in Fig. 17.19.

This characteristic has been constructed by retaining only the first and the second terms of Eq. (17.77), i.e.,

$$F = (\dot{x}/|\dot{x}|)F_0 - \alpha\dot{x} \tag{17.80}$$

Equation (17.78) is now written

$$\ddot{x} + (1/m)[(\dot{x}/|\dot{x}|)F_0 - \alpha\dot{x}] + \omega^2 x = 0 \tag{17.81}$$

where $\omega = \sqrt{k/m}$ = natural frequency of the oscillator.

If oscillations are admitted they will be of the form $x = A(t)\sin(\omega t + \Phi)$. If the rate of variation of $A(t)$ is small in comparison with ω, we find

$$\frac{d[A(t)]}{dt} = -\frac{1}{2\pi\omega m}\left[\int_0^{\pi/2}(F_0 - \alpha A\omega\cos\psi\cos\psi\,d\psi - \int_{\pi/2}^{3\pi/2}(-F_0 - \alpha A\omega\cos\psi)\cos\psi\,d\psi - \int_{3\pi/2}^{2\pi}(F_0 - \alpha A\omega\cos\psi)\cos\psi\,d\psi\right] \tag{17.82}$$

Integration yields

$$d[A(t)]/dt = \alpha A/2m - 2F_0/\pi\omega m \tag{17.83}$$

Equation (17.83), a linear differential equation of the first order, is solved to obtain

$$A(t) = F_0/\pi\alpha\omega + (A_0 - F_0/\alpha\pi\omega)\exp(\alpha/2m) \tag{17.84}$$

where A_0 = initial amplitude [value of $A(t)$ at $t = 0$].

The damper characteristic given by Eq. (17.80) is a mathematical function used to obtain Eq. (17.84). A more realistic case would be represented by the dashed extension of the curves in Fig. 17.19. Thus Eq. (17.84) should be used with the restriction that $|A_0| < |4F_0/\alpha\pi\omega|$, which implies that the case of $|\alpha\,\dot{x}| > |F_0|$ is inadmissible.

Examining Eq. (17.84) in the light of restrictions just imposed, it is seen that the amplitude of the oscillation decreases exponentially despite the fact that the damper is characterized by negative slope.

When $F_0 = 0$, the amplitude increases exponentially, e.g.,

$$A(t)|_{F_0 = 0} = A_0\exp[(\alpha/2m)t] \tag{17.85}$$

which shows that it is the nonlinearity of the fractional characteristics of the damper which has the effect of positive damping.

17.6.1 Transient Analysis of Mechanical Power Transmission Containing a Friction Clutch

A mechanical power-transmission system to be discussed in this section consists of a prime mover, a friction clutch, and a load. It will also be characterized by the presence of a multiplicity of energy-storage elements in the form of inertias, compliances, and dissipative elements. Included will be various nonlinear and linear coupling elements.

Backlash and hysteresis in various coupling elements will not be considered, but otherwise no restriction is made as to the linearity of the components present in the system, with the exception that inertia is assumed to remain invariant.

The following method of analysis is general enough to permit extension to systems employing any number of energy-storage elements and any number of dissipative elements.

Consider a system shown in Fig. 17.20.

I_1 = moment of inertia of prime mover
I_2 = moment of inertia of input plate of clutch
I_3 = moment of inertia of output plate of clutch
I_4, I_5 = moment of inertia of load side of clutch
M_1 = driving torque
M_c = clutch frictional torque
M_2 = load torque
$f_1(\theta_1 - \theta_2), f_2(\theta_2 - \theta_4), f_3(\theta_4 - \theta_5)$ = equivalent characteristics of various shafts; torque as a function of the angle of twist
C_1, C_2, C_3, C_4, C_5 = coefficients of viscous resistance corresponding to $\dot{\theta}_1, \dot{\theta}_2, \dot{\theta}_3, \dot{\theta}_4$, and $\dot{\theta}_5$, respectively.
$g(\dot{\theta}_4$ and $\dot{\theta}_5)$ = characteristic of a damper placed in the system as shown in Fig. 17.20, torque as a function of relative velocity $\dot{\theta}_4 - \dot{\theta}_5$

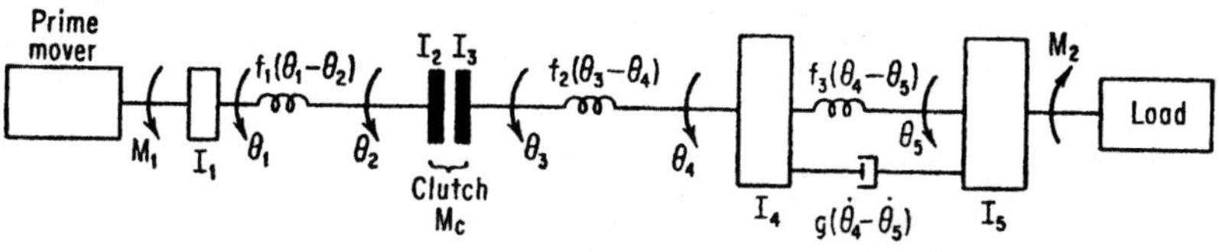

FIG. 17.20 Mechanical power-transmission system with a friction clutch.

Having described the lumped parameters of the transmission system, consider next the characteristics of the prime mover and the load. In the general case, torques M_1 and M_2 can be considered functions of time, angular velocity, and angular displacement.

In practice, the torque pulsations of multicylinder engines are small in comparison with the average value of torque and can be neglected in the present analysis.

If the torque M_1 is considered a function of speed only, it can be represented analytically by an infinite series.

$$M_1(\dot{\theta}_1) = M_1(0) + \sum_{n=1}^{\infty} \frac{1}{n!} (\dot{\theta}_1)^n M_1^{(n)}(0) \tag{17.86}$$

where $M_1(0)$ is the value of $M_1(\dot{\theta}_1)$ at $\dot{\theta}_1 = 0$.

Generally, the load torque M_2 can be considered a function of both speed $\dot{\theta}_5$ and the angular position of the shaft θ_5 and can be represented by the sum of two infinite series of the same form

$$M_2(\dot{\theta}_5, \theta_5) = M_2(0) + \sum_{n=1}^{\infty} \frac{1}{n!} (\dot{\theta}_5)^{(n)} M_2^{(n)}(0) + \overline{M}_2(0) + \sum_{r=1}^{\infty} \frac{1}{r!} (\theta_5)^r \overline{M}_2^{(r)}(0) \qquad (17.87)$$

In certain situations it may be more convenient to replace the second term on the right-hand side of Eq. (17.87) with a Fourier series, e.g., if $M_2(\dot{\theta}_5, \theta_5)$ is periodic in θ_5.

In Eq. (17.87) $M_2(0)$ is the value of the load torque at $\dot{\theta}_5 = 0$ and $M_2(0)$ is the value of the load torque at $\theta_5 = 0$.

For a mechanical system with k degrees of freedom the form of Lagrange's equations will be suited for deriving the equations of motion if the system of Fig. 17.20 is written

$$(d/dt)(\partial T/\partial \dot{q}_k) - \partial T/\partial q_k + \partial V/\partial q_k + \partial F/\partial \dot{q}_k = Q_k \qquad (17.88)$$

where T = total kinetic energy of the system
V = total potential energy of the system
F = dissipation function
Q_k = generalized force
q_k = generalized position coordinate
$\dot{q}_k$ = generalized velocity coordinate
k = 1,2,3, ... = number of independent coordinates

The computation of T is normally quite straightforward and needs no special explanation. In the computation of V should be included all elastic forces whether internal or external that are functions of displacement and all constant forces. All external and internal forces and moments that are functions of velocities must be included in the computation of F. Q_k includes all external forces and moments which are functions of time.

The coordinates used below follow:

Velocities	Positions	k
$\dot{q}_1, \dot{\theta}_1$	q_1, θ_1	1
$\dot{q}_2, \dot{\theta}_2$	q_2, θ_2	2
$\dot{q}_3, \dot{\theta}_3$	q_3, θ_3	3
$\dot{q}_4, \dot{\theta}_4$	q_4, θ_4	4
$\dot{q}_5, \dot{\theta}_5$	q_5, θ_5	5

In the case considered presently $Q_k = 0$, since it was assumed that no time-dependent torques act on the system. The potential energy stored in a single elastic element whose torque-deflection characteristic is given by $f(\theta_1 - \theta_2)$ can be expressed as

$$V = \int_0^{\theta_1 - \theta_2} f(\theta_1 - \theta_2)\, d(\theta_1 - \theta_2) \qquad (17.89)$$

Similarly the dissipation function or the energy dissipated in a damper whose torque-velocity characteristic is $g(\dot{\theta}_1 - \dot{\theta}_2)$ is given by

$$F = \int_0^{\dot{\theta}_1 - \dot{\theta}_2} g(\dot{\theta}_1 - \dot{\theta}_2)\, d(\dot{\theta}_1 - \dot{\theta}_2) \qquad (17.90)$$

T, V, and F may now be established:

$$T = \tfrac{1}{2}\,[I_1(\dot\theta_1)^2 + I_2(\dot\theta_2)^2 + I_3(\dot\theta_3)^2 + I_4(\dot\theta_4)^2 + I_5(\dot\theta_5)^2] \tag{17.91}$$

$$V = \int_0^{\theta_1-\theta_2} f_1(\theta_1-\theta_2)\,d(\theta_1-\theta_2) + \int_0^{\theta_3-\theta_4} f_2(\theta_3-\theta_4)\,d(\theta_3-\theta_4)$$
$$+ \int_0^{\theta_4-\theta_5} f_3(\theta_4-\theta_5)\,d(\theta_4-\theta_5) + \int_0^{\theta_5}\left[M_2(0) + \sum_{r=1}^{\infty}\frac{1}{r!}(\theta_5)^r\overline{M}_2^{(r)}(0)\right]d\theta_5 \tag{17.92}$$

Note that the part of the load torque M_2 which is a function of position θ_5 has been included in the computation of V.

In cases dealing with linear springs the first three terms of Eq. (17.92) become $\tfrac{1}{2}k_1(\theta_1 - \theta_2)^2 + \tfrac{1}{2}k_2(\theta_3 - \theta_4)^2 + \tfrac{1}{2}k_3(\theta_4 - \theta_5)^2$, where k_1, k_2, and k_3 are spring constants.

The dissipation function for this system is given by

$$F = \tfrac{1}{2}\,[C_1(\dot\theta_1)^2 + C_2(\dot\theta_2)^2 + C_3(\dot\theta_3)^2 + C_4(\dot\theta_4)^2 + C_5(\dot\theta_5)]$$
$$+ \int_0^{\dot\theta_2-\dot\theta_3} M_c(\dot\theta_2-\dot\theta_3)\,d(\dot\theta_2-\dot\theta_3) + \int_0^{\dot\theta_4-\dot\theta_5} g(\dot\theta_4-\dot\theta_5)\,d(\dot\theta_4-\dot\theta_5)$$
$$- \int_0^{\dot\theta_1} M_1(\dot\theta_1)\,d\dot\theta_1 + \int_0^{\dot\theta_5} M_2(\dot\theta_5, \theta_5)\,d\dot\theta_5 \tag{17.93}$$

In computing F the driving torque $M_1(\dot\theta_1)$ supplies energy to the system. Therefore, the associated dissipation function is negative.

It should be noted that dissipation due to load torque $M_2(\dot\theta, \theta_5, \theta_5)$ is obtained by integration with respect to $\dot\theta_5$ only.

In order to obtain the dynamical equations of motion for the system, Eqs. (17.91), (17.92), and (17.93) are substituted into Eq. (17.75) and indicated partial differentiation performed in order for $k = 1, 2$, etc.

The result is

$$I_1\ddot\theta_1 + \frac{\partial}{\partial\theta_1}\int_0^{\theta_1-\theta_2} f_1(\theta_1-\theta_2)\,d(\theta_1-\theta_2) + C_1\dot\theta_1 - M_1(\dot\theta_1) = 0$$

$$I_2\ddot\theta_2 + \frac{\partial}{\partial\theta_2}\int_0^{\theta_1-\theta_2} f_1(\theta_1-\theta_2)\,d(\theta_1-\theta_2)$$
$$+ \frac{\partial}{\partial\dot\theta_2}\int_0^{\dot\theta_2-\dot\theta_3} M_c(\dot\theta_2-\dot\theta_3)\,d(\dot\theta_2-\dot\theta_3) + C_2\dot\theta_2 = 0$$

$$I_3\ddot\theta_3 + \frac{\partial}{\partial\theta_3}\int_0^{\theta_3-\theta_4} f_2(\theta_3-\theta_4)\,d(\theta_3-\theta_4) \tag{17.94}$$
$$+ \frac{\partial}{\partial\dot\theta_3}\int_0^{\dot\theta_2-\dot\theta_3} M_c(\dot\theta_2-\dot\theta_3)\,d(\dot\theta_2-\dot\theta_3) + C_3\dot\theta_3 = 0$$

$$I_4\ddot\theta_4 + \frac{\partial}{2\theta_4}\int_0^{\theta_3-\theta_4} f_2(\theta_3-\theta_4)\,d(\theta_3-\theta_4) + \frac{\partial}{\partial\theta_4}\int_0^{\theta_4-\theta_5} f_3(\theta_4-\theta_5)\,d(\theta_4-\theta_5)$$
$$+ \frac{\partial}{\partial\dot\theta_4}\int_0^{\dot\theta_4-\dot\theta_5} g(\dot\theta_4-\dot\theta_5)\,d(\dot\theta_4-\dot\theta_5) + C_4\dot\theta_4 = 0$$

$$I_5\ddot{\theta}_5 + \frac{\partial}{\partial \theta_5}\int_0^{\theta_4-\theta_5} f_3(\theta_4 - \theta_5)\, d(\theta_4 - \theta_5)$$

$$+ \frac{\partial}{\partial \dot{\theta}_5}\int_0^{\dot{\theta}_4-\dot{\theta}_5} g(\dot{\theta}_4 - \dot{\theta}_5)\, d(\dot{\theta}_4 - \dot{\theta}_5) + C_5\dot{\theta}_5 + M_2(\dot{\theta}_5, \theta_5) = 0 \quad (17.94)$$

Equations (17.94) apply only during the transient, i.e., only when $(\dot{\theta}_2 - \dot{\theta}_3) \neq 0$. When $(\dot{\theta}_2 - \dot{\theta}_3) = 0$ and the clutch is engaged permanently a different set of equations must be derived to represent the system.

In deriving the equations of motion, the coefficient of the highest derivative, no matter how small, must not be neglected or the resulting equations will not properly represent the system.

Equations (17.94) represent a starting point in the analysis of power transmissions utilizing friction clutches when the various functions describing system parameters are known. If it is desired to obtain explicit solutions for the various speeds, shaft deflections, and torques existing in the system and their variation in time, because of the complexity of the mathematics recourse must be made to analog or digital computers.

In order to gain some approximate information regarding the behavior of the system, Eqs. (17.94) can be "linearized" and some useful techniques normally applicable to sets of linear differential equations can be applied with the understanding that only a limited amount of information can be obtained in this way.

If the system does not contain any highly nonlinear springs (or couplings), or if these can be represented by means of piecewise linear functions, the functions $f_1(\theta_1 - \theta_2)$, etc., can be considered linear of the form $k_1(\theta_1 - \theta_2)$. The same pertains to various dissipation functions, which can also be made representable by linear functions or combinations of linear functions.

It is often useful to determine, even approximately, the natural frequencies of the system. A method of calculating these is given below and is based upon the assumption that Eqs. (17.94) can be linearized.

Equations (17.94) are rewritten, omitting all dissipation terms and all external torques acting on the system and considering all springs to be linear.

$$
\begin{aligned}
I_1\ddot{\theta}_1 + k_1(\theta_1 - \theta_2) &= 0 \\
I_2\ddot{\theta}_2 - k_1(\theta_1 - \theta_2) &= 0 \\
I_3\ddot{\theta}_3 + k_2(\theta_3 - \theta_4) &= 0 \\
I_4\ddot{\theta}_4 - k_2(\theta_3 - \theta_4) + k_3(\theta_4 - \theta_5) &= 0 \\
I_5\ddot{\theta}_5 - k_3(\theta_4 - \theta_5) &= 0
\end{aligned}
\quad (17.95)
$$

When the expression for the clutch torque is omitted, as it is in Eqs. (17.95), there is no coupling between the input and output sides of the clutch and the transmission system consists of two dynamically independent subsystems. Since the presence of the dissipation term which acts as a coupling term between the two subsystems will not greatly influence their natural frequencies, they remain essentially independent during the transient. The situation changes after the engagement of the clutch is completed since then $\dot{\theta}_2 = \dot{\theta}_3$ and I_2 and I_3 form one flywheel.

In order to obtain the natural frequencies of the system, solutions of the form $\theta_k = A_k \sin(\omega t + \phi_k)$ are sought, where A_k is the amplitude and ϕ_k is the phase angle of the kth harmonic, with $\omega = 2\pi f$ the natural angular frequency of the system. Substituting θ_k for θ_1 and θ_2 in Eqs. (17.95), a set of equations is obtained

$$(k_1 - \lambda I_1)A_1 - k_1A_2 = 0$$

$$-k_1A_1 + (k_1 - \lambda I_2)A_2 = 0 \tag{17.96}$$

where $\lambda = \omega^2$, which results in a quadratic equation in λ which, when solved, yields the roots

$$\lambda_1 = \omega_1^2 = 0$$

$$\lambda_2 = \omega_2^2 = k_1(I_1 + I_2)/I_1I_2$$

The natural frequency of the input side of the clutch is given by

$$f_2 = (1/2\pi)\sqrt{k_1(I_1 + I_2)/I_1I_2} \tag{17.97}$$

Similarly,

$$f_3 = (1/2\pi)[(a + \sqrt{a^2 - 4b})/2]^{1/2} \tag{17.98}$$

$$f_4 = (1/2\pi)[(a - \sqrt{a^2 - 4b})/2]^{1/2} \tag{17.99}$$

where

$$a = [k_2(I_3I_5 + I_4I_5) + k_3(I_3I_5 + I_3I_4)]/I_3I_4I_5$$

$$b = k_2k_3(I_3 + I_4 + I_5)/I_3I_4I_5$$

Equations (17.94) can be useful in another way to gain more information about the dynamic behavior of the system during the transient following the application of the clutch.

Earlier, a simple mechanical oscillator was analyzed from the standpoint of stability, and in this connection, a question may well be asked how will a more complicated system behave under similar conditions.

For the purpose of the present discussion it is more convenient to take as a model

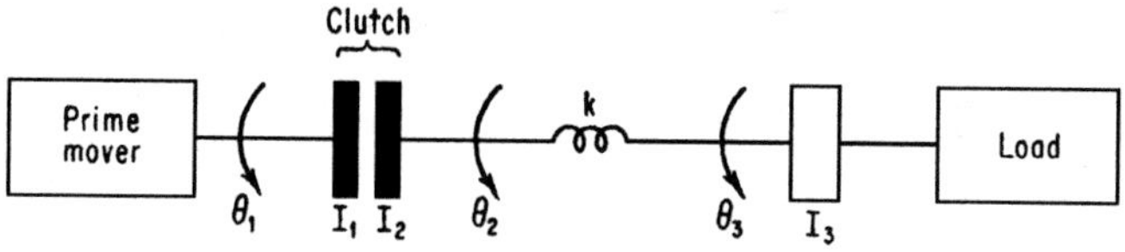

FIG. 17.21 Simplified power-transmission system.

the simpler system shown in Fig. 17.21.

The stiffness of the shaft connecting the prime mover to the input member of the clutch is made very large in comparison with the shaft connecting the clutch and the load. Such a system approximately represents a power transmission consisting of an engine, a clutch, flexible coupling of stiffness k, a gearbox, and a load. It will be assumed that the prime mover, the clutch, and the load have torque-speed characteristics as shown in Fig. 17.22.

Equations (17.77), (17.86), and (17.87) yield the following, respectively: clutch torque, driving torque, load torque

$$M_c = M_c(0)(\dot{\theta}_1 - \dot{\theta}_2)/|(\dot{\theta}_1 - \dot{\theta}_2)| + \alpha(\dot{\theta}_1 - \dot{\theta}_2) \tag{17.100}$$

$$M_1 = M_1(0) + a\dot{\theta}_1 \tag{17.101}$$

$$M_2 = M_2(0) + b\dot{\theta}_3 \tag{17.102}$$

where there is no restriction placed on the algebraic sign of α, a, and b.

Furthermore, if the discussion is limited to apply only for values of $0 < (\dot{\theta}_1 - \dot{\theta}_2) < \Omega$

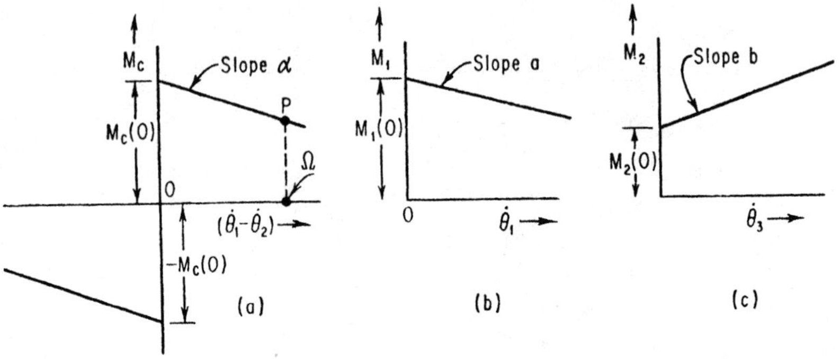

FIG. 17.22 Torque-speed characteristics for clutch, prime mover, and load shown in Fig. 17.21. (*a*) Clutch characteristics. (*b*) Prime-mover characteristics. (*c*) Load characteristics.

where Ω is the value of the relative velocity of clutch plates corresponding to some point P on the torque-speed characteristic of the clutch in Fig. 17.22*a,* set of equations of motion for the system can be obtained which will be linear in the interval specified. With these restrictions, the expression for the clutch torque becomes

$$M_c = M_c(0) + \alpha(\dot{\theta}_1 - \dot{\theta}_2) \tag{17.103}$$

and the equations of motion are written

$$\begin{aligned} I_1\ddot{\theta}_1 + c_1\dot{\theta}_1 - [M_1(0) + a\dot{\theta}_1] + [M_c(0) + \alpha(\dot{\theta}_1 - \dot{\theta}_2)] &= 0 \\ I_2\ddot{\theta}_2 + c_2\dot{\theta}_2 + k(\theta_2 - \theta_3) - [M_c(0) + \alpha(\dot{\theta}_1 - \dot{\theta}_2)] &= 0 \\ I_3\ddot{\theta}_3 + c_3\dot{\theta}_3 - k(\theta_2 - \theta_3) + [M_2(0) + b\dot{\theta}_3] &= 0 \end{aligned} \tag{17.104}$$

Expansion of the determinant resulting from Eq. (17.104) leads to

$$\begin{aligned} p_2\{I_1I_2I_3p^4 &+ [I_1(I_2R_3 + I_3R_2) + R_1I_2I_3]p^3 \\ &+ [I_1(I_2k + R_2R_3 + I_3k) + R_1(I_2R_3 + I_3R_2) - \alpha^2I_3]p^2 \\ &+ [I_1k(R_2 + R_3) + (I_2k + R_2R_3 + I_3k) + \alpha R_3]p \\ &+ k[R_1(R_2 + R_3) + \alpha]\} = 0 \end{aligned} \tag{17.105}$$

where

$$p = d/dt$$

$$R_1 = c_1 - a + \alpha \tag{17.106}$$

$$R_2 = c_2 + \alpha \tag{17.107}$$

$$R_3 = c_3 + b \tag{17.108}$$

If the system represented by Eqs. (17.104) is to be stable, the roots of Eq. (17.105) must have all their real parts negative. The presence of two equal roots $p^2 = 0$ merely signifies that the system is capable of rotation in space without executing any oscillations, and is of no consequence in this discussion.

In general, in order to discover whether or not an algebraic equation of the form

$$a_0p^n + a_1p^{n-1} + a_2p^{n-2} + \cdots + a_n = 0 \tag{17.109}$$

possesses roots having positive real parts, an array of coefficients is formed. The coefficients are first written in two rows:

Row 1: $a_0 \quad a_2 \quad a_4 \quad \cdots$

Row 2: $a_1 \quad a_3 \quad a_5 \quad \cdots$

By cross multiplication of the first column with each succeeding column in turn a further $n-1$ rows are formed:

Row 3: $b_1 = (a_1a_2 - a_0a_3) \qquad b_2 = (a_1a_4 - a_0a_5) \qquad \cdots$

Row 4: $c_1 = (b_1a_3 - a_1b_2) \qquad c_2 = (b_1a_5 - a_1b_3) \qquad \cdots$

Equation (17.109) contains no roots with positive real parts if, and only if, all terms in the first column of the array (i.e., a_0, a_2, b_2, c_1, etc.) are positive for positive a_0, or negative for negative a_0. The number of times a change in sign takes place in going from one term to the next in the first column gives the number of roots with positive real parts. This is known as "Routh's stability criterion."

The first conclusion that can be made is that a necessary but not sufficient condition for stability is that all coefficients of p in Eq. (17.109) must be present and that they must all have the same algebraic sign. Since inertia and stiffness in a mechanical system will generally be positive, this statement imposes certain restrictions on the values of R_1, R_2, and R_3 in Eq. (17.105), and consequently on the values of α, a, and b in Eqs. (17.100), (17.101), and (17.102). Normally the values of the coefficients of viscous resistance c_1, c_2, and c_3 will be made as small as possible in an efficient power-transmission system, so that the three other dissipation terms a, α, and b which characterize the prime mover, the clutch, and the load will play an important part in determining the dynamic behavior of the system during the transient following the application of the clutch.

Since it is desirable from the viewpoint of stability that R_1, R_2, and R_3 in Eq. (17.105) be positive the following statements can be made:

1. The output torque of the prime mover should decrease with speed.
2. The torque-speed characteristic of the clutch should be a curve having positive slope.
3. The torque-speed characteristic of the load should be a curve having positive slope, but the value of the load torque at any particular speed within the range of operation should not exceed that of the static clutch torque, $M_c(0)$.

In most cases of practical importance statements 1 and 3 above are satisfied, but the case of a friction clutch exhibiting a torque-speed characteristic with positive slope is rare. In a friction clutch, the choice of friction materials, the preparation of surfaces, lubrication, temperature, and pressure on the frictional surfaces will all influence the torque-speed characteristic, which in some cases can be made, if not increasing with speed, at least of zero slope over most of the operating range.

If the transmission system is such that the vibrations of the drive line are particularly disturbing during the transient caused by the application of the clutch, some form of tuned and damped vibration absorber is necessary if the change in the torque-speed characteristic of the clutch is inadmissible. Drive-line vibration frequency caused by the action of a friction clutch usually falls within the lower end of the audio spectrum and may also have values low enough to be quite unpleasant to a human being. Examples of this are clutch "chatter"[24] in various forms of motor vehicles when starting from rest and the noise occurring during the shift from one geared ratio to the next in automatic transmissions.

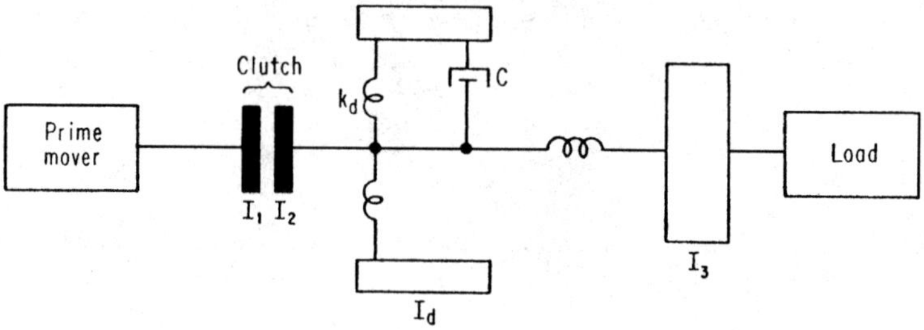

FIG. 17.23 Transmission system shown in Fig. 17.21 with damped vibration absorber added.

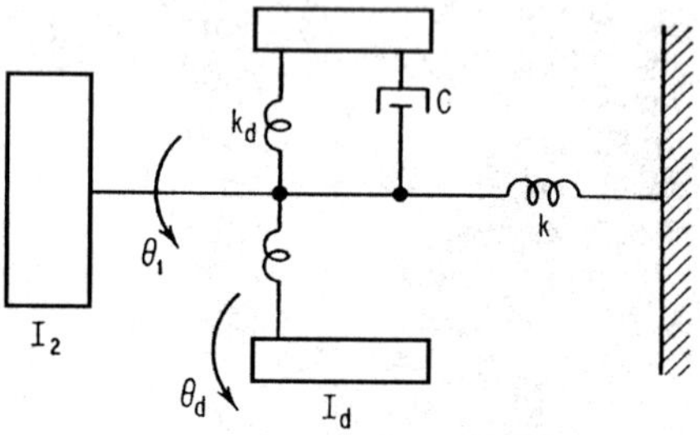

FIG. 17.24 Output end of clutch shown in Fig. 17.23.

In order to develop criteria for the design of a damped vibration absorber to be used with a friction clutch, consider a system shown in Fig. 17.23, which is essentially the same as that shown in Fig. 17.21, the difference being that another oscillating system consisting of a small flywheel has been flexibly attached to the clutch plate and a viscous damper connected between the clutch plate and the absorber inertia.

If, as is usually the case with mechanical power transmission, $I_3 >> I_2$, the natural frequency of the system on the output side of the clutch is very closely given by $\sqrt{k/I_2}$. The output end of the clutch can thus be represented as in Fig. 17.24.

I_2 = moment of inertia of drive clutch plate(s)

c = viscous damping coefficient of the absorber

k_1 = spring constant of the main drive shaft

α = slope of the torque-speed characteristic of the clutch, equivalent dynamically to viscous damping coefficient

k_d = spring constant of the absorber

The total kinetic energy of the system is given by

$$T = \tfrac{1}{2} I_2(\dot{\theta}_2)^2 + \tfrac{1}{2} I_d(\dot{\theta}_d)^2 \tag{17.110}$$

The total potential energy of the system is given by

$$V = \tfrac{1}{2} k_1(\theta_2)^2 + \tfrac{1}{2}k_d(\theta_2 - \theta_d)^2 \tag{17.111}$$

The dissipation function is

$$F = \tfrac{1}{2}\,\alpha(\dot{\theta}_2)^2 + \tfrac{1}{2}\,c(\dot{\theta}_2 - \dot{\theta}_d)^2 \tag{17.112}$$

Substituting Eqs. (17.110) through (17.112) into Eq. (17.88) yields, in the absence of any external forces,

$$\begin{aligned} I_2\ddot{\theta}_1 + k_1\theta_1 + k_d(\theta_1 - \theta_d) + \alpha\dot{\theta}_1 + c(\dot{\theta}_1 - \dot{\theta}_d) &= 0 \\ I_d\ddot{\theta}_d - k_d(\theta_1 - \theta_d) - c(\dot{\theta}_1 - \dot{\theta}_d) &= 0 \end{aligned} \tag{17.113}$$

Equations (17.113) describe the motion of the system of Fig. 17.24.

Making α negative in Eqs. (17.113) and expanding the determinant associated with Eqs. (17.113), an equation in p is obtained:

$$I_1I_dp^4 + [c(I_1 + I_d) - \alpha I_d]p^3 + [k_1I_d + k_d(I_1 + I_d) - c\alpha]p^2 + (k_1c - k_d\alpha)p + k_1k_d = 0 \tag{17.114}$$

Applying Routh's criterion to Eq. (17.114) yields the following array of coefficients:

Row 1: I_2I_d

Row 2: $c(I_2 + I_d) - I_d\alpha$

Row 3: $[c(I_2 + I_d) - I_d\alpha][k_2I_2 + k_2(I_2 + I_d) - c\alpha] - I_2I_d(k_1c - k_2\alpha)$

Row 4: $[c(I_2 + I_d) - I_d\alpha][k_1I_d + k_2(I_2 + I_d) - c\alpha](k_1c - k_2\alpha)$
$- [c(I_2 + I_d) - I_d\alpha]^2k_1k_2 - I_2I_d(k_1c - k_2\alpha)^2$

Since $I_2I_d > 0$, all the subsequent rows must be positive if the system is to be dynamically stable. This affords a means of obtaining a range of values for the viscous damping coefficient of the vibration absorber c which will result in stable operation.

In order for all the terms in the array to be positive, all the coefficients of all powers of p in Eq. (17.114) must be positive. This gives a very rough indication of the range of acceptable values of c.

$$\begin{aligned} &c > I_d\alpha/(I_2 + I_d) \\ &c > (k_d/k_1)\alpha \\ &c < [k_1I_d + k_d(I_2 + I_d)]\alpha \end{aligned} \tag{17.115}$$

The vibration absorber must also be proportioned so that the natural frequency of the inertia I_d (the damper flywheel) is the same as that of I_2. This imposes a second condition

$$k_1/I_2 = k_d/I_d \tag{17.116}$$

In cases where I_3 in Fig. 17.23 is not very large in comparison with I_2, for the natural frequencies to be the same,

$$k_d/I_d = k_1\, I_2I_3/(I_2 + I_3) \tag{17.117}$$

Substituting Eq. (17.116) into Eq. (17.115) yields

$$c > I_d\alpha/(I_2 + I_d) \qquad c > (k_d/k_1)\alpha \qquad c < (2I_2 + I_d)k_d/\alpha \tag{17.118}$$

17.6.2 Simplified Approach to Dynamic Problems Involving Friction Clutches

In designing power-transmission systems containing friction clutches situations frequently arise in which it is required to calculate energy dissipation in the clutch during the engagement period or simply to compute the inertial torques acting on the system during a transition from one ratio to the other. These problems can be greatly simplified if the compliance is neglected.

This procedure is justified in cases where the natural frequencies of the system are relatively high and the resulting amplitude of oscillation small in comparison with the changes in velocities which take place as a consequence of the application of the clutch.

Consider a simple two-inertia system shown in Fig. 17.25. Such systems frequently form the basis of various forms of inertia starters.

It is convenient here to use velocities as coordinates rather than angular displacements, because the compliance is neglected.

The coefficient of friction will be assumed to be a constant, independent of speed. Assume also that no external torques act on the system.

In Fig. 17.25, the two flywheels initially rotate at two different angular velocities Ω_1 and Ω_2. Let the clutch be instantaneously applied at time $t = 0$ and let the torque exerted by it on the two flywheels be M_c = constant. This torque will act on the system only as long as there exists a difference in speeds between the two shafts, as shown in Fig. 17.26.

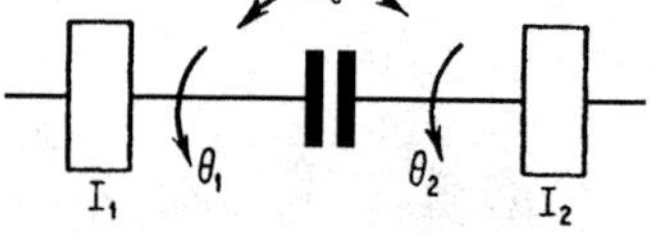

FIG. 17.25 Simple two-inertia system.

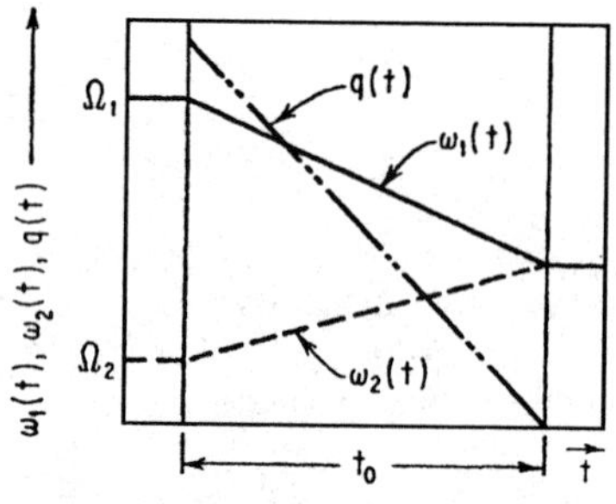

FIG. 17.26 Plot showing time t_0, which torque will be exerted, and the rate of energy dissipation.

Equations for speed, rate of heat dissipation, etc., are therefore valid for time $0 < t < t_0$, where t_0 is the time required to couple the two shafts.

The equations of motion for the two sides of the clutch are

$$I_1(d\omega_1/dt) = -M_c \tag{17.119}$$

$$I_2(d\omega_2/dt) = M_c \tag{17.120}$$

where $\omega_1 = \dot{\theta}_1$ and $\omega_2 = \dot{\theta}_2$, the angular velocities of I_1 and I_2, respectively. Integrating Eqs. (17.119) and (17.120) and applying the conditions at $t = 0$, $\omega_1(t) = \Omega_1$, and $\omega_2(t) = \Omega_2$ yields

$$\omega_1(t) = -(M_c/I_1)t + \Omega_1 \tag{17.121}$$

$$\omega_2(t) = (M_c/I_2)t + \Omega_2 \tag{17.122}$$

The relative velocity of I_1 with respect to I_2 is given by

$$\omega_1(t) = \omega_2(t) = \omega_r(t) = -M_c[(I_1 + I_2)/I_1 I_2]t + \Omega_1 - \Omega_2 \tag{17.123}$$

The rate at which energy is dissipated in the clutch during the engagement period is given by $q(t) = T\omega_r(t)$.

$$q(t) = M_c\{\Omega_1 - \Omega_2 - M_c[(I_1 + I_2)/I_1 I_2]t\} \qquad \text{lb·in/s} \tag{17.124}$$

The duration of the engagement period, found by considering that when $t = t_0$, $\omega_1(t) = \omega_2(t)$, is given by

$$t_0 = I_1 I_2(\Omega_2 - \Omega_2)/M_c(I_1 + I_2) \qquad \text{s} \tag{17.125}$$

The total energy dissipated is obtained by integration of Eq. (17.124):

$$Q = \int_0^{t_0} q(t)\, dt = \frac{I_1 I_2(\Omega_1 - \Omega_2)^2}{2(I_1 + I_2)} \qquad \text{lb·in} \tag{17.126}$$

The efficiency with which energy is transferred from I_1 to I_2 during the clutching operation is

$$\eta = 1 - Q/E_0 \tag{17.127}$$

where E_0 = kinetic energy of the system at a time $t = 0$. Thus

$$\eta = (I_1\Omega_2 + I_2\Omega_2)^2/(I_1 + I_2)(I_1\Omega_1^2 + I_2\Omega_2^2) \tag{17.128}$$

If $\Omega_2 = 0$, this expression reduces to

$$\eta = I_1/(I_1 + I_2)$$

and if $I_1 = I_2 = I = \Omega_2 \neq 0$

$$\eta = (\Omega_1 + \Omega_2)^2/2(\Omega_1^2 + \Omega_2^2)$$

The case where the frictional clutch torque is a function of relative velocity of clutch plates is analyzed next, assuming negligible compliance and the dependence of friction on velocity represented by a piecewise linear function, shown in Fig. 17.27.

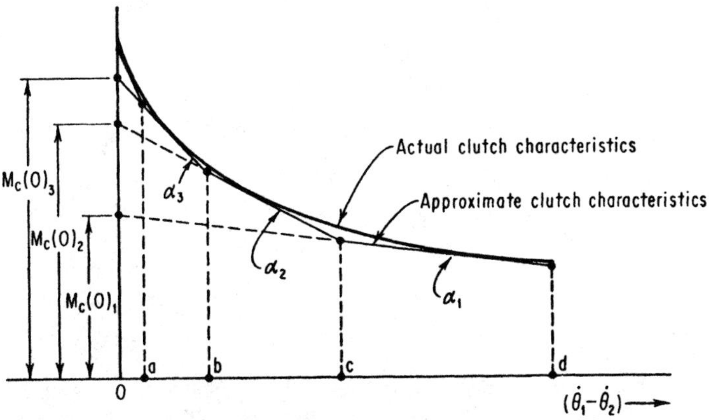

FIG. 17.27 Actual and approximate clutch characteristics.

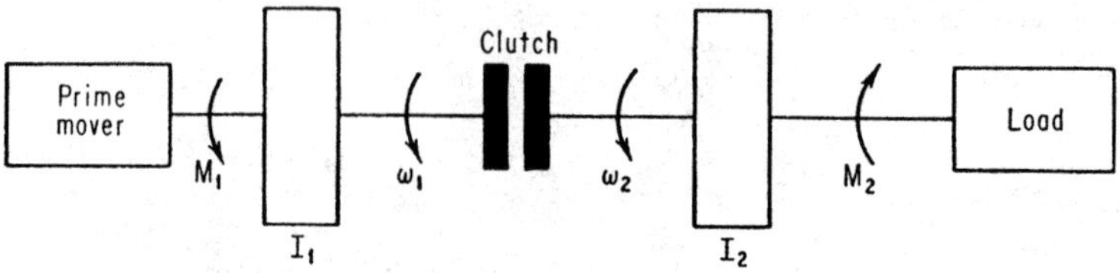

FIG. 17.28 Two-inertia transmission system.

For the system shown in Fig. 17.28, the equations of motion are

$$I_1\dot{\omega}_1 - \alpha_1\omega_1 + \alpha_1\omega_2 = M_1 - M_c(0)_1 \tag{17.129}$$

$$I_2\dot{\omega}_2 - \alpha_1\omega_2 + \alpha_1\omega_1 = M_c(0)_1 - M_2 \tag{17.130}$$

In Eqs. (17.129) and (17.130), α is the slope of the clutch frictional characteristic corresponding to the interval $c < (\omega_1 - \omega_2) < d$, and $M_c(0)$ is the point on the M_c axis obtained by intersection of the line of slope α_1 with the axis.

Solution of Eqs. (17.129) and (17.130) yields

$$\omega_1(t) = [(A_1\beta^2 + B_1\beta + C_1)\exp(\beta t) - \beta(C_1 t + B_1) - C_1]/I_1I_2\beta^2 \tag{17.131}$$

$$\omega_2(t) = [(A_2\beta^2 + B_2\beta + C_1)\exp(\beta t) - \beta(C_1 t + B_2) - C_1]/I_1I_2\beta^2 \tag{17.132}$$

where

$$\begin{aligned} A_1 &= I_1I_2\omega_1(0) \\ B_1 &= I_2[M_1 - M_c(0)_1] - \alpha_1[I_1\omega_1(0) + I_2\omega_2(0)] \\ C_1 &= M_2 - M_1 \\ \beta &= \alpha_1(I_1 + I_2)/I_1I_2 \\ A_2 &= I_1I_2\omega_2(0) \\ B_2 &= I_1[M_c(0)_1 - M_2] - \alpha_1[I_1\omega_1(0) + I_2\omega_2(0)] \end{aligned} \tag{17.133}$$

In Eqs. (17.133) $\omega_1(0)$ and $\omega_2(0)$ are the values of the angular velocities $\omega_1(t)$ and $\omega_2(t)$ at $t = 0$, i.e., at point (*d*) in Fig. 17.27.

Equations (17.131) and (17.132) are used to calculate the time which must elapse for the two velocities to reach the value $\omega_1(t) - \omega_2(t) = \omega_r(c)$ (where ω_r denotes relative velocity), the point on the $\omega_1(t) - \omega_2(t)$ axis in Fig. 17.27 at which the slope of the clutch characteristic changes discontinuously. This value of time is given by

$$t_c = \frac{1}{\beta}\ln\frac{\omega_r(c)I_1I_2\beta + (B_1 + B_2)}{(A_1 - A_2)\beta + (B_1 - B_2)} \tag{17.134}$$

The computation is carried out along the line of slope α_2 with α_2 substituted for α_1, $\omega_1(t_c)$ and $\omega_2(t_c)$ for $\omega_1(0)$ and $\omega_2(0)$, respectively, and $M_c(0)_2$ for $M_c(0)_1$ in Eqs. (17.133). The process is repeated at points (*b*), (*a*), and (0), each time using appropriate constants in Eqs. (17.133) and finding the time the system takes to arrive at the appropriate point on the $\omega = \omega_1 - \omega_2$ axis in Fig. 17.27.

The rate at which the energy is dissipated in the clutch when operating along the linear segment of its characteristic is given by

$$q_0(t) = \{M_c(0) - \alpha[\omega_1(t) - \omega_2(t)]\}\,[\omega_1(t) - \omega_2(t)] \tag{17.135}$$

Since the coefficient of friction is in this case a variable depending on the relative speed between clutch plates, it is of interest to examine the conditions under which engagement of the clutch can occur.

To this end, Eqs. (17.131) and (17.132) are differentiated, and by setting $t = 0$, expressions for accelerations of the two

$$\dot{\omega}_1(0) = \{M_1 - [M_2 - \alpha(\Omega_1 - \Omega_2)]\}/I_1 \tag{17.136}$$

$$\dot{\omega}_2(0) = (M_1 - M_2)/I_2 - \{M_1 - [M_2 - \alpha(\Omega_1 - \Omega_2)]\}/I_2 \tag{17.137}$$

Examining Eqs. (17.136) and (17.137), it can be noted that:

1. The most desirable condition for fast and positive engagement occurs when the following inequality is satisfied:

$$M_1 < \{M_c(0)_1 - \alpha_1[\omega_1(0) - \omega_2(0)]\} \tag{17.138}$$

2. When

$$M_1 = \{M_c(0)_1 - \alpha_1[\omega_1(0) - \omega_2(0)]\} \tag{17.139}$$

engagement will be slower but will be accomplished in a finite length of time, depending on the value of $M_1 - M_2$. $M_1 - M_2$ must be positive in order that engagement may be accomplished.

The rate at which $\omega_2(t)$ will begin to increase immediately after the clutch is applied is determined by the quantity $(M_1 - M_2)/I_2$.

The increase of speed of the output side of the clutch decreases the relative velocity of clutch plates, thus increasing clutch torque. Any increase in clutch torque above the value given by Eq. (17.139) will cause the deceleration of input shaft. Thus the effect of the initial increase in speed of the load side of clutch is cumulative and leads to positive engagement.

3. When the prime-mover output torque is greater than the clutch torque at the instant of clutch application, i.e., when

$$M_1 > \{M_c(0)_1 - \alpha_1[\omega_1(0) - \omega_2(0)]\} \tag{17.140}$$

engagement is possible only when the following inequality is satisfied:

$$(M_1 - \{M_c(0)_1 - \alpha_1[\omega_1(0) - \omega_2(0)]\})/I_1$$
$$< (\{M_c(0)_1 - \alpha_1[\omega_1(0) - \omega_2(0)]\} - M_2)/I_2 \tag{17.141}$$

Nevertheless, the analysis provides a basis for continued study of friction clutches and associated control components pertinent to any specific design through the use of modern computing equipment which makes it possible to include in the analysis the dependence of friction on speed, temperature, etc., while simultaneously greatly reducing the time required for computation.

17.7 ELECTROMAGNETIC FRICTION CLUTCHES

A method widely used for controlling friction clutches depends on the electromagnetic force of attraction between two magnetized iron parts. The axial clamping force on the friction disks is thus a function of exciting current, and current may be made time-dependent, thus controlling the torque transmitted through the clutch. Electromagnetic clutches are frequently used in applications requiring close control of acceleration of rotating masses because the exciting current is easily and accurately controlled.

Electromagnetic clutches can be operated dry or in oil and can be of single- and multiple-disk design. The clutch may be so designed that the magnetic flux either (1) passes through the disk stack or (2) does not pass through the stack. The latter case provides more flexibility in design, since the disk stack can then be formed and friction materials selected without regard to the geometry of the magnetic circuit and the magnetic properties of the disk materials. Figure 17.29 shows one example of such a clutch.

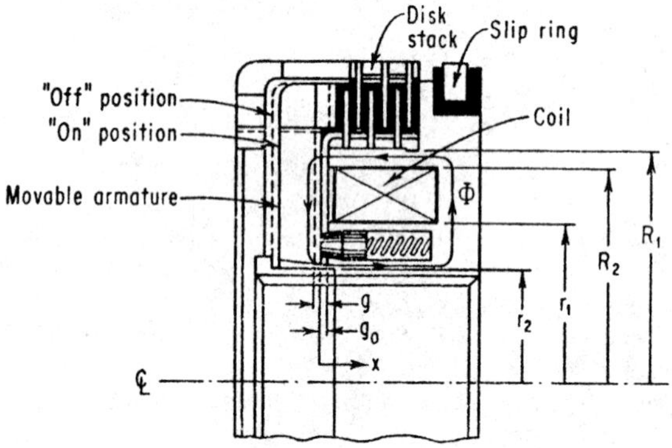

FIG. 17.29 Electromagnetic clutch.

In the following analysis, it will be assumed that effects of saturation, leakage flux, flux fringing, eddy currents, and hysteresis are all negligible, and that the ampere-turns required to maintain the flux in the iron parts of the magnetic circuit are very small compared with the ampere-turns required to maintain the field across the air gap. Let

α = mechanical friction coefficient (armature parts)

d = value of x at which spring force equals zero

E = voltage applied to the coil

N = number of turns in the coil

i = coil current, A

μ = magnetic permeability of air (rationalized mks)

Φ = magnetic flux, Wb

λ = flux linkages, Wb·turn

L = inductance of the coil, H

R = resistance of the coil, Ω

x = displacement of armature, m

g = maximum air-gap length, m

g_0 = minimum air-gap length, m

m = mass of the armature, kg

k = return spring elastance, N/m

R_1, R_2, r_1, r_2 are defined in Fig. 17.29.

Under the assumptions made above (using rationalized mks units),

$$\Phi = \mu A_1 A_2 Ni/(g - x)(A_1 + A_2) \qquad \text{Wb} \tag{17.142}$$

where $A_1 = \pi(R_1^2 - R_2^2)$ and $A_2 = \pi(r_1^2 - r_2^2)$. Also

$$\lambda = N\Phi \tag{17.143}$$

The self-inductance of the coil can be defined as

$$L = L_0/(g - x) \qquad \text{H} \tag{17.144}$$

where

$$L_0 = \mu A_1 A_2 N^2/(A_1 + A_2) \tag{17.145}$$

The magnetic energy stored in the field existing in the air gap is given by

$$W_m = \tfrac{1}{2} Li^2 \qquad \text{J} \tag{17.146}$$

From Eq. (17.146) the electromagnetic force acting on the armature is

$$F_m = \partial W_m/\partial x = -[L_0 i^2/2(g - x)^2] \qquad \text{N} \tag{17.147}$$

Equation (17.147) shows that the force, and also the torque transmitted by the clutch, are very sensitive to change in the coil current and to the difference $(g - x)$. Whereas in the case of the hydraulically actuated clutch the position of the actuating piston relative to the main body of the clutch is immaterial when computing torque transmitted, the minimum distance of the armature from the body of the electromagnetic clutch is an important parameter.

The applied coil voltage is given by

$$E = Ri + d\lambda/dt \tag{17.148}$$

However, λ is a function of x and i, and both x and i are functions of time. Therefore,

$$E = R_i + (dL/dx)(dx/dt) + L(di/dt) \tag{17.149}$$

Summing all forces acting on the armature and performing the differentiation indicated by Eq. (17.149) yields

$$M(d^2x/dt^2) + \alpha(dx/dt) - k(d + x) + L_0 i^2/2(g - x)^2 = 0 \tag{17.150}$$

$$-[L_0 i/(g - x)^2](dx/dt) + [L_0/(g - x)](di/dt) + R_i = E \tag{17.151}$$

Equations (17.150) and (17.151) describing dynamic behavior of the clutch system are nonlinear and cannot be solved by employing elementary methods. However, they provide a reasonable basis for studying the response of the system to the control voltage E using an analog computer, in which case the nonlinearity of the magnetic circuit (saturation and hysteresis) can also be taken into account when formulating system equations.

It is difficult to calculate complete response of the system from the time the voltage E is applied across the coil, but when it is realized that the clutch begins to transmit torque only after the armature has been attracted and the clamping force applied to the disk stack, Eqs. (17.150) and (17.151) can be simplified considerably and can be used to examine the factors affecting the torque capacity of the clutch.

After the motion of the armature has ceased $d^2x/dt^2 = dx/dt = 0$ and $g - x = g_0$ and Eqs. (17.150) and (17.151) become, respectively,

$$-k(d + g - g_0) + L_0 i^2/2(g_0)^2 = F_c \tag{17.152}$$

$$L(di/dt) + Ri = E \tag{17.153}$$

where $L = L_0/g_0$ and F_c = net axial clamping force.

Solution of Eq. (17.153) yields

$$i(t) = E/R - (E/R - I_0) \exp[-(R/L)t] \tag{17.154}$$

where I_0 is the value of i at the time $t = 0$, i.e., at the time when $x = g - g_0$ and armature motion ceased.

Substituting for $i(t)$ from Eq. (17.154) into (17.152) the response of the axial clamping force to the sudden application of control voltage E is obtained:

$$F_c = [L_0/2(g_0)^2]\{(E/R)^2 - 2(E/R)[(E/R) - I_0] \exp[-(R/L)t] + [(E/R) - I_0]^2 \exp[(2R/L)t]\} - K(d + g - g_0) \quad (17.155)$$

For sufficiently large values of t, the steady-state value of the clamping force on the disk stack is given by

$$F_c = [L_0/2(g_0)^2](E/R) - K(d + g - g_0) \quad (17.156)$$

To disengage the clutch, the control voltage E is reduced to zero, and a similar procedure leads to the expression for F_c

$$F_c = \frac{L_0}{2(g_0)^2}\left(\frac{E}{R + r}\right)^2 \exp\left[-\frac{2(R + r)}{L}t\right] - K(d + g - g_0) \quad (17.157)$$

Equation (17.157) shows how the clamping force F_c, acting on the disk stack, decreases with time following disappearance of the control voltage E at a time $t = 0$. In this equation, r is any external resistance through which the coil discharges and which may be added to decrease the disengagement time by decreasing the time constant of the system.

If the coefficient of friction of the disk surfaces is a reasonably invariant quantity, Eqs. (17.155), (17.156), and (17.157) also show how the clutch torque varies with time. In practice, these equations may represent clutch torque variation with time to a very good approximation in the case of dry clutches, but when dealing with oil-lubricated clutches additional factors must be considered which pertain to the mechanical-design details of friction disks.

Frictional torque characteristics of magnetic clutches have been studied experimentally and the results of these investigations are discussed in Ref. 25.

The value of frictional torque developed by the clutch and its variation with time depends not only on the magnetizing current but also on the kind of disks and disk material used. Similarly, the decrease of the clutch torque besides being a function of time such as indicated by Eq. (17.157) also depends on adhesion between the disks, which in turn is a function of disk design and material.

REFERENCES

1. Harting, G. R.: "Design and Application of Heavy-Duty Clutches," *SAE Trans.*, vol. 72, pp. 405–432, 1964.
2. Steinhagen, H. G.: "The Plate Clutch," SAE Paper 800978, 1980.
3. Crolla, D. A.: "Friction Materials for Overload Clutches," *Tribology Int.*, vol. 12, no. 4, pp. 155–160, Aug. 1979.
4. Chestney, A. A. W., and D. A. Crolla: "The Effect of Grooves on the Performance of Frictional Materials for Overload Clutches," *Wear*, vol. 53, no. 1, pp. 143–163, 1979.
5. Harting, G. R.: "The Why, Status, and Future of Wet Clutches for Trucks," SAE Paper 700875, 1970.
6. Cummins, G. F.: "A New Wet Clutch Fan Drive System," ASME Paper 74-DE-12, 1974.

7. Mitchell, R. K., T. R. Kosmatka, and O. Braunschweig: "Clutch for Lawn Mower Engine Auxiliary Shaft," SAE Paper 800039, 1980.

8. Hermanns, M. J.: "Heavy-Duty, Long-Life, Dry Clutches," SAE Paper 700874, 1970.

9. Lloyd, F. A., and M. A. Dipino: "Advances in Wet Friction Materials—75 Years of Progress," SAE Paper 800977, 1980.

10. Fish, R. L., and F. A. Lloyd: "Surface Finish Requirements of Spacer Plates for Paper Friction Applications," SAE Preprint 730840, 1973.

11. Fish, R. L.: "Wet Friction Material—Some Modes of Failure and Methods of Correction," SAE Paper 760664, 1976.

12. Anderson, A. E.: "Friction and Wear of Paper Type Wet Friction Elements," SAE Paper 720521, 1972.

13. Morse, I. E., and R. T. Hinkle: "Taking Guesswork out of Disk Clutch Design," *Mach. Des.*, vol. 47, no. 28, pp. 64–65, Nov. 27, 1965.

14. Jakob, M.: "Heat Transfer," vols. 1 and 2, John Wiley & Sons, Inc., New York, 1949, 1957.

15. Blok, H.: "Measurement of Temperatures Flashes on Gear Teeth under Extreme Pressure Conditions," *Proc. Gen. Disc. Lubrication, Inst. Mech. Eng.* (*London*), vol. 2, pp. 14–20, 1937.

16. Barber, J. R.: "Thermoelastic Instabilities in the Sliding of Conforming Solids," *Proc. Roy. Soc. London,* ser. A, vol. 312, pp. 381–394, 1969.

17. Dow, T. A.: "Thermoelastic Effects in a Thin Sliding Seal—A Review," *Wear,* vol. 59, pp. 31–52, 1980.

18. Gazley, C.: "Heat Transfer Characteristics of the Rotational and Axial Flow between Coaxial Cylinders," ASME Paper 56-A-128, 1958.

19. Kays, W. M., and I. S. Bjorklund: "Heat Transfer from a Rotating Cylinder with and without Crossflow," ASME Paper 56-A-71, 1957.

20. Andronow, A. A., and C. E. Chaikin: "Theory of Oscillations," Princeton University Press, Princeton, N.J., 1949.

21. MacLachlan, N. W.: "Ordinary Nonlinear Differential Equations," 2d ed., Oxford University Press, New York, 1956.

22. Timoshenko, S.: "Vibration Problems in Engineering," 3d ed., D. Van Nostrand Co., Inc., Princeton, N.J., 1955.

23. Bunda, T., et al.: "Friction Behavior of Clutch-Facing Materials: Friction Characteristics in the Low-Velocity Slippage," SAE Paper 720522, 1972.

24. Jarvis, R. P., and R. M. Oldershaw: "Clutch Chatter in Automobile Drivelines," *Proc. Inst. Mech. Eng.* (*London*), vol. 187, no. 27, pp. 369–379, 1973.

25. Nitsche, C.: "Electromagnetic Multi-Disk Clutches," ASME Paper 58-5A-61, 1957.

CHAPTER 18
FRICTION BRAKES

Thomas A. Dow, Ph.D.
Professor of Mechanical and Aerospace Engineering
North Carolina State University
Raleigh, N.C.

18.1 FRICTION MATERIALS 18.1
18.1.1 Types and Physical Properties 18.1
18.1.2 Brake Friction 18.2
18.1.3 Brake Wear 18.5
18.2 VEHICLE BRAKING 18.6
18.3 BRAKE TEMPERATURE 18.7
18.3.1 Heat Input 18.8
18.3.2 Average Brake Temperature 18.8
18.3.3 Brake Surface Temperature 18.9
18.4 SIMPLE BRAKES 18.13
18.4.1 Block Brakes 18.13
18.4.2 Band Brakes 18.16
18.5 ANALYSIS OF SELF-ACTUATING DRUM BRAKES 18.17
18.5.1 The Internal Drum Brake 18.18
18.5.2 Theory of Resultant Forces on Shoe 18.18
18.5.3 Effect of Moderate Distortion of Drum and Shoe: Uniform Pressure 18.20
18.5.4 The Actuation Equation: Effectiveness 18.20
18.5.5 Drum and Shoe Distortion 18.25
18.5.6 Cam Brakes: Truck Brakes 18.26
18.6 DISK BRAKES 18.26
18.6.1 Types of Disk Brakes 18.28
18.6.2 Actuation 18.29
18.6.3 Design 18.29
18.7 BRAKE SQUEAL 18.29
18.7.1 Sprag-Slip Theory 18.30

18.1 FRICTION MATERIALS

18.1.1 Types and Physical Properties

The logical selection of a brake material for a specific application requires a knowledge of friction material formulation. The "formulation of a brake lining" is defined as the specific mixture of materials and the sequence of production steps which together determine the characteristics of the brake lining. There are three basic types of linings[1,2] currently in use.

Organic Linings. Organic linings are generally compounded of six basic ingredients.

1. *Asbestos:* for heat resistance and high coefficient of friction
2. *Friction modifiers:*[3] for example, the oil of cashew nut shells to give desired friction qualities
3. *Fillers:* for example, rubber chips to control noise
4. *Curing agents:* to produce the desired chemical reactions during manufacture
5. *Other materials:* for example, powdered lead, brass chips, and aluminum powders for improved overall braking performance
6. *Binders:* phenolic resins for holding the ingredients together

Asbestos has characteristics that make it well suited for friction applications: thermal stability, adequate wear resistance, and strength. For these reasons it has found universal acceptance as the basic ingredient in brake linings. However, because occupational exposure to asbestos during lining manufacture has been shown to be a health hazard, the lining industry is looking for replacements. One possibility[4] is Kevlar aramid fiber which has been shown to boost the performance of nonasbestos friction materials to that currently available for asbestos materials.

Semimetallic Linings.[5] These linings substitute iron, steel, and graphite for part of the asbestos and organic components of an organic lining. The reasons for this substitution are:

1. Improved frictional stability and high-temperature performing (fade resistance)
2. Excellent compatibility with rotor and high-temperature wear resistance at temperatures greater than 450°F (230°C)
3. High performance with minimal noise

Such linings were first released for police cars and taxi cabs and their use has spread to some original-equipment passenger cars.

Metallic Linings. These linings have received attention for special application involving large heat dissipation and high temperatures. Sintered metallic–ceramic friction materials[6] have been successfully applied to jet aircraft brakes,[7] heavy-duty clutch facings, and race and police car brakes.

Two methods are in use for making brake linings—weaving and molding. Both types are basically asbestos with binders to hold the asbestos fibers together. For the woven-type material, asbestos thread and fine copper wire are woven together with the other components, cured, and cut to the appropriate shape. For the organic molded type, asbestos fibers and other compounds are ground and pressed into the final shape. Molded linings are more commonly used.

Molded organic linings may be formed from a dry mix or a wet mix. The dry mix consists of asbestos compounds, filler materials, and powdered resins. These ingredients are mixed, preformed to shape, and heated under pressure to form a hard slatelike board. This board is then cut and drilled for specific pad shapes. The wet mix, on the other hand, consists of asbestos compounds, organic fillers, and liquid resins. High-pressure extruding, screw extruding, calendering, and other processes are used to form the brake pad from this mix. In general, the softer the friction material, the greater the friction coefficient and the shorter the life of the lining. The hard ceramic and metallic linings have comparatively low friction coefficients but have high wear resistance and can survive in adverse operating conditions, for example, at high temperature. Because many linings are riveted to the backing plate, the material must have sufficient strength[8] to remain firmly attached to the backing plate under panic stop conditions. Properties[9,10] of different friction materials are shown in Table 18.1.

Composition brake blocks have a number of ingredients, and quality control[11] of the final product is a continuing problem. Quality control typically consists of 100 percent inspection of the pads for electrical conductivity (shows whether the binder is fully cured), metal content (checks for the presence of metal impurities), and visual imperfections (determines the overall integrity of the pads). Partial inspection is also carried out for specific gravity and hardness.

18.1.2 Brake Friction

The general concepts of friction have been developed over many years.[12–14] As applied to a composition brake lining on a steel or cast-iron brake surface, the main sources of friction are:[15,16]

TABLE 18.1 Properties of Friction Materials

Type	Woven lining (flexible)	Molded lining (sheeter molded)	Molded block (rigid molded)
Compressive strength, lb/in^2	10,000–15,000	10,000–18,000	10,000–15,000
Shear strength, lb/in^2	3500–5000	8000–10,000	4500–5500
Tensile strength, lb/in^2	2500–3000	4000–5000	3000–4000
Modulus of rupture, lb/in^2	4000–5000	5000–7000	5000–7000
Hardness, Brinell	*	*	20–28
Specific gravity	2.0–2.5†	1.7–2.0	2.1–2.5
Thermal expansion, in/(in.°F) (mean to 350°F)	*	1.5×10^{-5}	1.3×10^{-5}
Thermal conductivity, Btu·in/(h·ft^2·°F) (125°F mean)	5.5	3	3.5–4
Principal type of service	Industrial	Automotive (passenger)	Automotive (truck and bus) and railway
Specific heat			
Maximum continuous operating temperature, °F	400–500	500	750
Maximum rubbing speed, ft/min	7500	5000	7500
Maximum continuous operating pressure, lb/in^2	50–100	100	150
Average friction coefficient (350°F, 50 lb/in^2, 600 ft/min)	0.45	0.47	0.40–0.45

*Nonhomogeneous.
†High wire-mesh content.

1. *Adhesion:*[17] As the lining material moves over the drum or rotor surface, the metallic constituents in the lining weld to the rotor and drum material. Shearing of these junctions produces a frictional force.
2. *Shear deformation:* The coefficient of friction increasing with increasing temperature suggests that deformation effects play an important role since the resin softens with increasing temperature. It is believed that the deformation effect is due to the formation of wavelike surface deformation[18] rather than elastic hysteresis loss.
3. *Plowing:* In the process of tangential motion of the surfaces, the roughness protrusions on the disk-drum interlock with particles of ingredients and displace them. When the ultimate strength is exceeded, the polymer structure is disrupted and particles of ingredients are torn away. Long asbestos or steel fibers supply the mechanical strength to prevent excessive loss of material during plowing.
4. *Hysteresis:* The energy losses involved with elastically stressing and unstressing the brake linings produce a small source of brake friction.
5. *Surface films:* Contamination of the surface from decomposed lining materials or fluid-borne contaminants will greatly affect the friction coefficient by reducing adhesion and shear deformation.

The importance of each component of friction discussed[19] above will vary over the life of the lining. The initial operation of the system may involve high plowing losses due to the high original surface roughness. As the roughness is reduced due to wear, the positive effect of increased adhesive bonds will become more important as will the adverse effects of surface-contamination films.

The friction coefficient[20] for a brake material rubbing against cast iron is a function of load, speed, and bulk temperature. A general expression for the friction force F can be written as

$$F = K(T)\, P^{a(T)}\, V^{b(T)} \tag{18.1}$$

where $K(T)$ = constant, dependent upon temperature
P = normal load
$a(T)$ = exponent of load dependent upon temperature (varies between +0.8 and +1.25)
V = sliding speed
$b(T)$ = exponent of speed dependent upon temperature (varies between −0.25 and +0.25)

The influence of load, speed, and temperature for one asbestos friction material is shown in Figs. 18.1 and 18.2. Figure 18.1 shows the influence of bulk temperature and normal load on the friction coefficient. Increasing the load causes a slight decrease in friction coefficient (except at low loads and 600°F). As a function of temperature, this friction material shows a peak in friction coefficient at a temperature between 400 and 500°F. At a 100-lb normal load (200 lb/in^2), the friction coefficient varies from 0.25 to 0.4, depending on the lining temperature. An increase in sliding speed results in a decrease in friction coefficient, as shown in Fig. 18.2.

Because of the strong influence of surface temperature on the friction coefficient, considerable care must be exercised in looking at independent influences of speed and load on friction. For the experiments described above, a small specimen (0.7 in × 0.7 in) of brake lining was loaded against the inside of a 11-in-diameter cast-iron drum (SAE standard J661). The temperature was measured by a thermocouple embedded in the drum slightly below its surface. The experimental conditions were such that heat was added to the system to achieve the test temperatures, and the thermocouple was used to control the heating system. This situation was selected to minimize the influence of frictionally generated heat on the surface temperature. However, such experimental conditions may not guarantee identical surface temperatures for different speeds and loads. Since the frictional heat is generated at the surface, even though it is a small part of the total heat input, the frictional heating (due to increased speed or load) may produce a different temperature profile on the lining and be responsible for part of the decrease in the friction coefficient measured in Figs. 18.1 and 18.2.

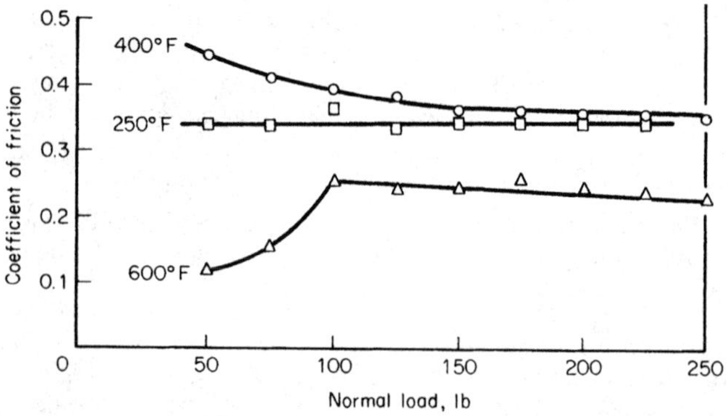

FIG. 18.1 Friction coefficient as function of normal load for asbestos friction material, 300 r/min speed.[20]

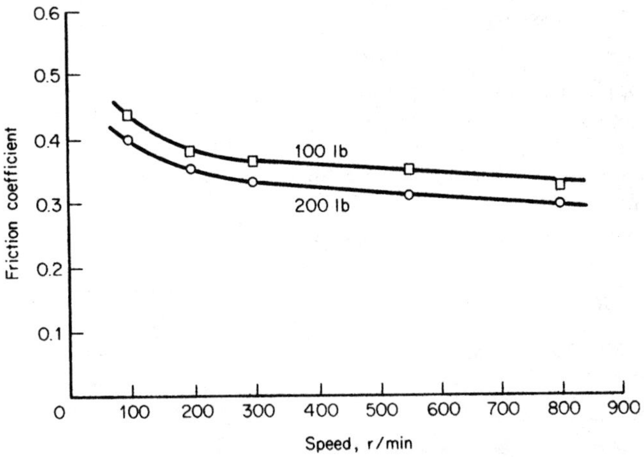

FIG. 18.2 Friction coefficient as function of speed for asbestos friction material, 350°F drum temperature.[20]

Because of these complexities, the determination of the coefficients of Eq. (18.1) must be corroborated by experimental measurements.

18.1.3 Brake Wear

The wear[21–24] of composition brake materials when pressed against a steel or cast-iron rotating surface is a complex mixture of adhesive and oxidative wear and surface fatigue. At low temperatures (below 450°F) the controlling factor for wear is adhesive wear and surface fatigue. However, above 450°F, the dominant wear mechanisms are thermal degradation of the resin binder[25] (phenolic) and mechanical and thermal changes in the asbestos reinforcing fibers.[26]

One effect which makes wear prediction difficult is that during braking, the friction material rubs against a reaction product[27] considerably different from that of the virgin cast iron. The reaction product is a layer on the rotor surface from 50 to 250 μin thick which influences both the friction and wear behavior of the brake. Wear data have been compared for stainless-steel and quartz rotors, with each of two different surface finishes, and the results indicated that surface finish and conductivity of the rotor were more important than the material in determining the transfer film and therefore the friction and wear behavior.

Wear models[28–31] have been developed as a result of extensive dynomometer and field experiments. The results of these experiments show that the wear of organic and semimetallic friction materials can be modeled by a modified Archard wear law, separated into two temperature regimes. At temperatures less than 450°F,

$$\text{Wear} = P^{\alpha} V^{\beta} t \tag{18.2}$$

At temperatures greater than 450°F,

$$\text{Wear} = P^{\alpha} V^{\beta} t e^{E/RT} \tag{18.3}$$

where W = weight loss from lining material
P = load
α, β = material constants
V = sliding speed
t = time
E = activation energy, Btu/mol
R = universal gas constant = 1.986 Btu/(mol·°R)
T = absolute temperature, °R = °F + 460

Values of α, β, and E for selected brake materials[29] are shown in the following table:

Lining material	α	β	E, Btu/mol
Commercial lining	0.42		38.1
Secondary lining*	0.63	1.45	30.6
Disk pad*	0.28		15.9

*High-temperature materials.

The reason for the two ranges of temperature has to do with the modes of wear which take place under these conditions. At low temperatures (below 450°F) the wear is a result of plowing, cutting, three-body abrasive wear, and adhesive wear. For the high-temperature region, the wear is controlled by thermal processes. As a result the wear of friction materials increases exponentially above some critical temperature. For most of the current composition materials held together with phenolic resins, the critical temperature is approximately 450°F.

The exponential term in Eq. (18.3) includes the influence of pyrolysis by using the activation energy of the brake material and the absolute temperature of the brake surface. Using Eqs. (18.2) and (18.3), the wear of different operating conditions can be compared. For example, changing the operating temperature of secondary lining shown in Table 18.1 from 450 to 600°F will increase the wear by a factor of 3. However, the absolute value of the wear rate cannot be accurately predicted without experimental corroboration.

18.2 VEHICLE BRAKING

A friction brake turns the kinetic energy of the vehicle into heat. But because of the design of most vehicles, this heat is not equally distributed to the wheels. The heat dissipated at each brake will be a function of the static and dynamic weight distribution on the wheels and the design of the brake system. The dynamic loading will depend upon the design of the vehicle (the static weight distribution, the height of the center of gravity, and the wheelbase) and the deceleration. The sum of the forces during braking shows that the deceleration of the vehicle in percentages of the acceleration of gravity g must be less than or equal to the friction coefficient between tire and road. This friction coefficient[32] will depend upon the tire size and construction, the road surface, and the relative slip between tire and road.

If the weight is uniformly distributed right to left, the front and rear wheel loading (L_F and L_R) can be written as

$$\begin{aligned} L_F &= W(F + \mu h/d) \\ L_R &= W(R - \mu h/d) \end{aligned} \tag{18.4}$$

where F = static front wheel loading = d_R/d
R = static rear wheel loading = d_F/d
d = wheel base
d_F = distance from center of gravity to front wheel
d_R = distance from center of gravity to rear wheel
μ = friction coefficient, equivalent to deceleration expressed as a multiple of g
h = vertical distance from road to center of gravity

This expression can be used to estimate the change in wheel loading due to friction forces at the road during a stop. For short, high vehicles significant weight transfer will occur, while for long, low vehicles, smaller percentages of weight will be transferred. For example, a typical front-wheel-drive sedan of 2500 lb will have a static weight distribution of 60/40 front to rear. During a $0.8g$ stop, this distribution will change to about 80/20.

The balance[33] of braking between front and rear is an important design consideration. The brake system could be designed so that the front brakes produced 4 times the torque of the rear; and for the $0.8g$ stop described above, maximum tire friction at both ends of the car could be utilized. However, if under wet conditions, the friction coefficient between tire and road were reduced to 0.2, the dynamic weight distribution would be 65/35. Under these conditions, the brake system balanced 80 percent front to 20 percent rear would cause the front wheels to slide. If the brake system were balanced for this lower-deceleration dynamic weight distribution, the rear wheels would slide first during maximum deceleration for dry conditions.

To make a decision as to the design of the brake balance, the influence of wheel slide on vehicle control must be considered. The control[34,35] of a vehicle is related to wheel slide in the following way: Locking only the rear wheels produces a condition which results in partial or complete loss of control of the vehicle. Depending upon the vehicle's characteristics, this situation could lead to a spin. Locking only the front wheel produces a stable condition where the vehicle travels in a straight line but nearly all steering control is lost. The result is that for most vehicles it is best to balance the brake system so that the front wheels lock first.

To improve vehicle control during braking, antiskid braking systems have been developed. These systems measure the relative wheel and vehicle speed and modulate the brake pressure to keep each wheel (or each set of wheels) at the limit of adhesion[36,37] without sliding. The maximum friction coefficient for the tire on the road occurs at a low percentage of slip, which is nearly rolling conditions, rather than gross sliding which would occur if the brake were locked. Thus an antiskid system can be designed to produce maximum brake torque, while still retaining full steering control, a situation which would otherwise only exist for the most experienced driver.

18.3 BRAKE TEMPERATURE

The actual temperature at the lining-rotor interface plays a key role in the friction and wear associated with a brake material.[40–42] It is at this interface that the frictional heat is generated and the highest temperatures occur. The surface temperature of the lining material determines (1) the mode of wear and (2) the film present on the surface which influences the friction coefficient (see Sec. 18.1). The equilibrium temperature of the brake will be influenced by the heat input (proportional to the vehicle weight, initial speed, and stopping frequency) and the magnitude of the heat dissipation from the brake. The heat is lost through conduction to the brake assembly as well as convection and radiation to the surroundings.

18.3.1 Heat Input

The instantaneous heat input to the brakes q is equal to the change in kinetic energy of the vehicle:

$$q = \Delta KE = \frac{\partial}{\partial t} KE = \frac{\partial}{\partial t}\left(\frac{1}{2}mv^2\right) \tag{18.5}$$

where q = rate of heat input to brakes, Btu/s
KE = kinetic energy of vehicle, Btu
m = mass of vehicle = weight ÷ 32.2 ft/s^2
v = instantaneous speed of vehicle, ft/s

If we assume that the deceleration of the vehicle is uniform, the change in speed will be a linear function of time, i.e.,

$$v(t) = v(0)(1 - t/t_s) \tag{18.6}$$

where t_s = stopping time = $\sqrt{2D/ag}$
D = stopping distance, ft
a = deceleration rate of vehicle, percent of g
g = acceleration of gravity, ft/s^2
$v(0)$ = initial speed of vehicle, ft/s

The heat generation rate can then be written as

$$q = mv(t)\, dv(t)/dt = magv(t) \tag{18.7}$$

$$= magv(0)(1 - t/t_s) \tag{18.8}$$

The total heat generated over the stop will be the sum of the instantaneous heat rates or the total change in kinetic energy of the vehicle, which for a stop to rest will be its initial kinetic energy

$$Q = \tfrac{1}{2}\, mv(0)^2 \qquad \text{Btu} \tag{18.9}$$

For the vehicle discussed, the instantaneous heat input will be a maximum at the beginning of the stop and will have a value of $q(0) = 212$ Btu/s and will linearly decrease over the stopping time. The total heat generated over the stop will be $Q =$ 390 Btu.

The design of the brake system will determine the percentage of the total heat generated which will be dissipated at each wheel. If the system is balanced 60/40 front to rear, then the heat input to one front brake will be 30 percent of the total, or 117 Btu.

18.3.2 Average Brake Temperature

The change in the average brake temperature for a single stop can be calculated from

$$\Delta T = Q/M_B c \qquad °\text{F} \tag{18.10}$$

where Q = heat input to brake, Btu
M_B = mass of brake, lb
c = specific heat of brake rotor = 0.1 Btu/(lb·°F) for cast iron

For a single stop from ambient conditions, the bulk of the heat will be absorbed by the rotor or drum, since it has much higher conductivity than the lining material. For a 20-lb cast-iron front disk brake, the average temperature rise for the single stop described previously would be about 60°F. The temperature of the brake after repeated stops will depend upon how much of the heat generated is lost due to conduction, convection, and radiation. Another significant factor will be the residual brake torque. While in itself this torque will not generate high temperatures, it will reduce the heat losses from the brake and greatly change the equilibrium temperature after multiple stops.

18.3.3 Brake Surface Temperature

The surface temperature of the disk or drum and the lining will be significantly higher than the average brake temperature. Peak surface temperatures on disk brakes have been measured in the range of 1200 to 1400°F. It is this temperature which influences the wear of the lining material since the wear particles are generated at the surface. Other effects of high brake surface temperatures are distortion of the drum or disk or surface cracking in these members as a result of plastic deformation from high thermal stresses. However, surface temperature is difficult to predict because of the nature of the contact. The rotor surface and the lining actually make contact at small localized regions (asperities) and the peak temperatures for these contacts are quite high. Some operating conditions result in larger localized "hot spots" or "fire-banding"[43,44] which also produce high localized surface temperatures. Therefore, the problem of predicting the surface temperature becomes one of determining the size and duration of the contacts.

For a linearly decreasing heat input into the surface of a semi-infinite body, the surface temperature rise can be estimated from the expression[45]

$$T_s(t) = [2\,q(0)\sqrt{t}/A\sqrt{\pi K\rho c}](1 - 2t/3t_s) \tag{18.11}$$

where $q(0)$ = initial heat input, Btu/s
t = time, s
t_s = total stopping time, s
K = thermal conductivity of body, Btu/(in·s·°F)
ρ = density of body, lb/in^3
c = specific heat of body, Btu/(lb·°F)
A = surface area, in^2

The maximum value of the temperature occurs after half of the stopping time and can be written as

$$T_{max} = [0.53q(0)/A]\sqrt{t_s/K\rho c} \tag{18.12}$$

The difficulty in applying the above expressions comes in deciding upon an appropriate value of the area over which it acts. The percentage of the heat that flows into the disk or into the pad varies with the operating conditions. Therefore, the decision cannot be made arbitrarily, but requires a complete analysis of the problem.[46–48] However, several approximations can be made which illustrate the range of surface temperatures possible.

Average Surface Temperature. One estimate of the surface temperature can be made by assuming that the heat uniformly enters the entire swept area of the disk. For the 2500-lb vehicle previously discussed, with ventilated cast-iron disks, the values for Eq. (18.11) will be

$q(0)$ = heat input rate per brake surface
= 59 Btu/s
t_s = 3.5 s
$K = 6.9 \times 10^{-4}$ Btu/(in·s°F)

ρ = 0.26 lb/in^3
c = 0.1 Btu/(lb·°F)
A = 34 in^2

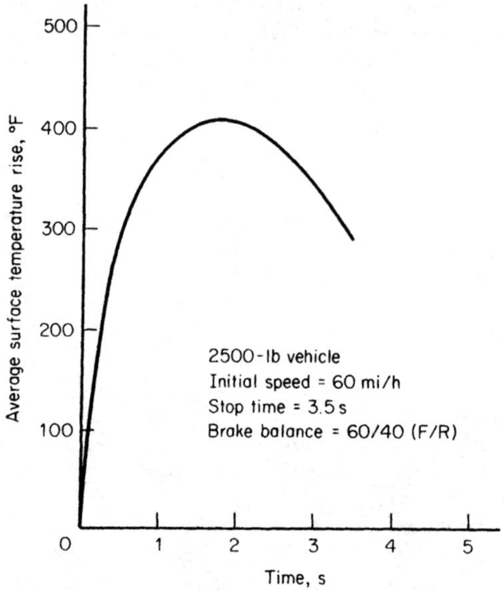

FIG. 18.3 Disk surface-temperature rise for a single stop.

The surface temperature as a function of time is plotted in Fig. 18.3, and the maximum temperature from Eq. (18.12) is 407°F. This temperature is significantly hotter than the bulk temperature rise of the disk calculated from Eq. (18.10) for the same stop.

Peak Surface Temperature. Real surfaces contact at small isolated regions defined by plastic deformation of the softer material. For brakes, the softer material is the brake lining and its hardness is a function of temperature. Studies[30] of debris generated during braking experiments showed it to be generated at near 900°F. The Meyer's hardness associated with this temperature is 5700 lb/in^2. Therefore, real area of contact, A_R, between lining material and disk at this temperature can be written as

$$A_R = L/H_M \tag{18.13}$$

where L = load on brake pad and H_M = Meyer's hardness.

Typical brake loading for a current 2500-lb automobile will be 5000 lb over a pad area of 12 in^2. From Eq. (18.13), the real area of contact will be 0.88 in^2 or about 7 percent of the total pad area. This real area will be divided into small contact regions and part of the heat will enter the disk through each of these regions. If we assume that all of these regions are collected into a circular region of area A_R at the center of the pad, the real area of the brake surface, A_{SR}, swept by the contact in one revolution can be written as

$$A_{SR} = d_A(4\pi A_R)^{1/2} \tag{18.14}$$

where d_A = average diameter of disk or drum surface.

When the real swept area from Eq. (18.14) is substituted to find the maximum surface temperature from Eq. (18.12), the result is a surface rise temperature of 760°F. If this temperature rise is added to the ambient conditions (100°F), the temperature will be similar to that expected by the structure of the wear debris.

Equilibrium Temperature. The heat dissipated during a stop is transferred to the disk and then lost to the surroundings by conduction, convection, and radiation. The heat loss is a function of disk temperature so that there will be an equilibrium temperature for intermittent braking where the energy gained during a single brake application will be lost before the next application. This equilibrium temperature will be influenced by the weight of the vehicle, the deceleration, the brake design and balance, and the time between stops. The loss of heat from a disk brake due to convection is significantly improved by the use of a ventilated rotor. Figure 18.4 shows the results[49] of a series of brake applications at a constant disk speed. The ventilated rotor reached an equilibrium temperature of 700°F after eight cycles, whereas the solid disk of the same diameter and mass reached a significantly higher equilibrium temperature (1100°F) after 20 cycles. In the case of the solid rotor, more of the heat energy was retained during each cooling cycle, and subsequent brake applications produced a higher rotor equilibrium temperature. Figure 18.5 shows the definite advantage in cooling rate of the ventilated disk. This advantage is due to the increased surface and of the ventilated disk as well as the air-pumping properties of that design. The convection heat transfer coefficient[46] for a ventilated rotor is shown in Fig. 18.6. These values can be used to estimate the heat lost to convection from the rotor. Radiation and conduction losses must be taken into account.

Brake Fade. Brake fade has been the main contributor to the nearly total replacement of drum brakes with disk brakes on the front wheels of new cars. It is caused by two factors. The first cause is the drop in the coefficient of friction at high temperatures. The friction coefficient for an asbestos-based friction material is shown in

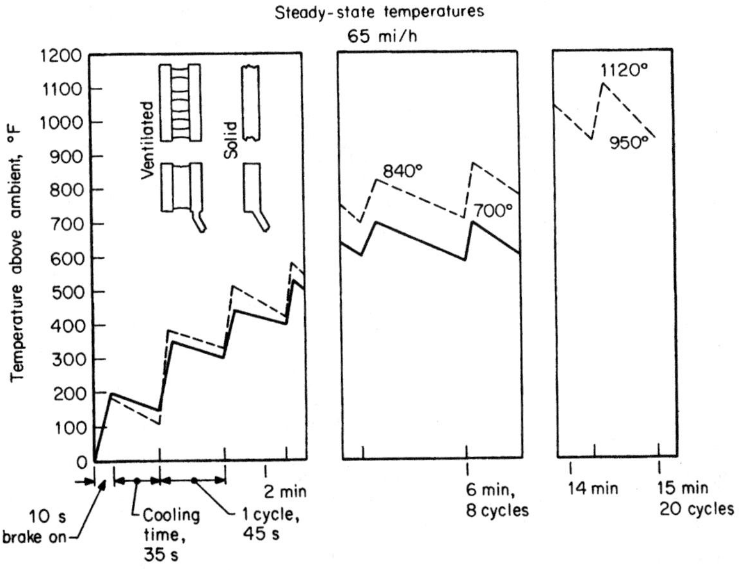

FIG. 18.4 Stabilization temperature curves for solid and ventilated rotors of equal mass and diameter.[49]

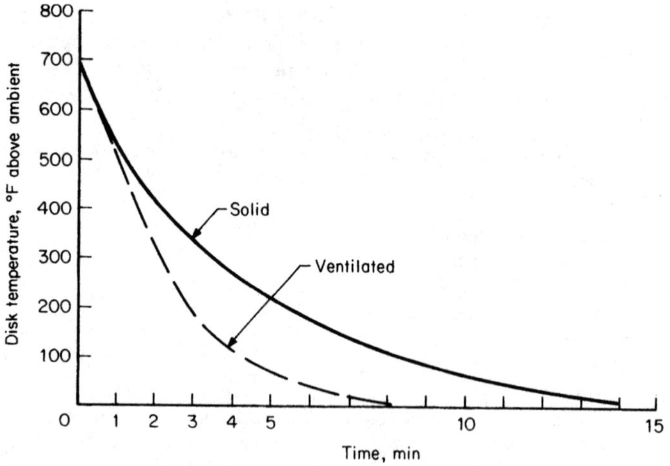

FIG. 18.5 Cooling temperature curves for solid and ventilated rotors of equal mass and diameter.[49] Disks have common initial temperatures. Speed is 65 mi/h.

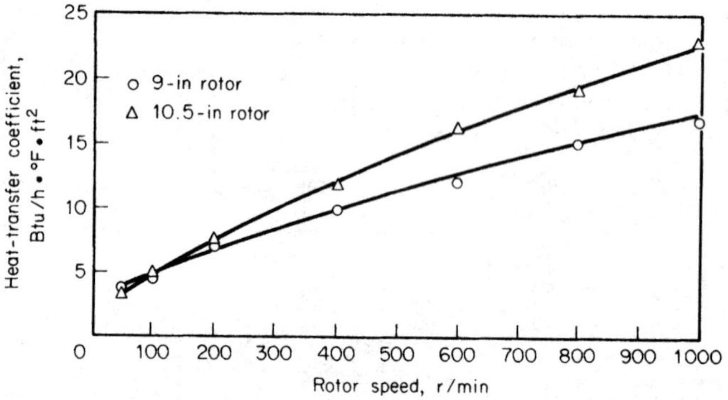

FIG. 18.6 Vent heat-transfer coefficient versus rotor speed in r/min for cast-iron rotors with radial vanes.[46]

Fig. 18.7 for several unit loads. For all cases, the friction coefficient drops dramatically above 500°F. At this temperature, the phenolic binder and other lining constituents decompose, oxidize, or melt. The result is an increase in wear and a reduction in friction coefficient known as "fade." Fade behavior has been attributed to the formation of a liquid or low-friction solid phase at the interface as a result of the high surface temperature. This is purely a thermal effect which depends upon the lining and will occur for any brake design. The second cause of fade is mechanical loss of contact which occurs in drum brakes. During severe braking, the effective drum radius will grow

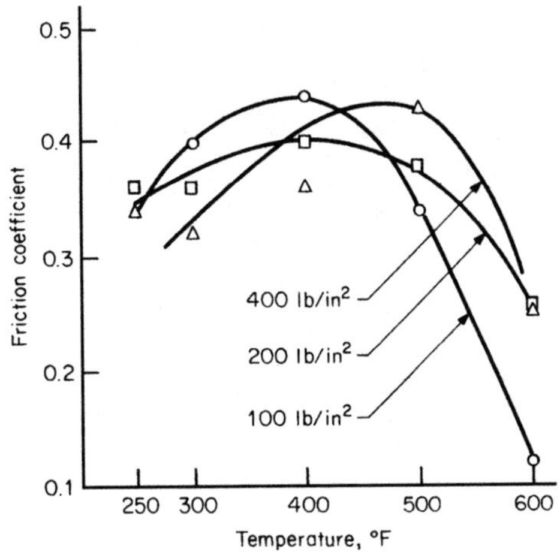

FIG. 18.7 Variation of friction coefficients with temperature for different unit loads.[20]

more than the brake shoe radius due to the larger thermal conductivity of the brake drum.[50] The result will be a change in pressure distribution over the brake lining and a reduction in brake torque of 20 percent for the same actuating force.

The performance of a drum brake suffers more under severe operating conditions than the disk brake. Since a drum brake is self-energizing, a reduction in friction coefficient with high temperature will cause a relatively larger reduction in brake torque at constant operating pressures than for the non-self-energizing disk brake. Then the additional mechanical change in the contact geometry of the drum and shoe lower the brake torque even more. Attempts to combat fade on drum brakes by using hard linings that retain friction coefficient at high temperatures were not too successful. The reason was that the fade caused by changes in geometry was more pronounced when unyielding linings were used, even though the fade due to the friction reduction was somewhat alleviated.

The disk brake does not experience mechanical fade so that disk linings can be hard with high friction coefficient at high temperatures. Semimetallic and ceramic linings have these qualities and have been used for disk brakes operating at severe conditions. The disk brake still experiences some fade[51,52] due to friction reduction with temperature, but a modern automobile braking system with ventilated disks will show no fade (negligible increase in pedal effort) for six consecutive 0.5g stops from 60 mi/h.

18.4 SIMPLE BRAKES

18.4.1 Block Brakes

The block brake is the simplest form of brake. Single-lever block brakes are not widely used because the normal braking force results directly in shaft bearing pressure and shaft deflection. Shaft bending is prevented by arranging two brake blocks opposite one another in a double block brake as shown in Fig. 18.8.

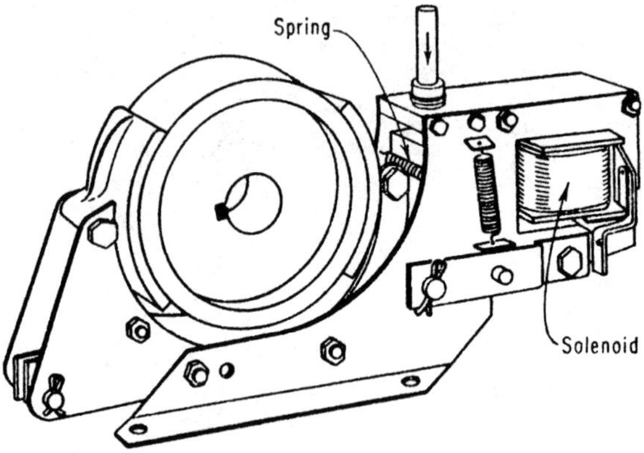

FIG. 18.8 Double block brake.

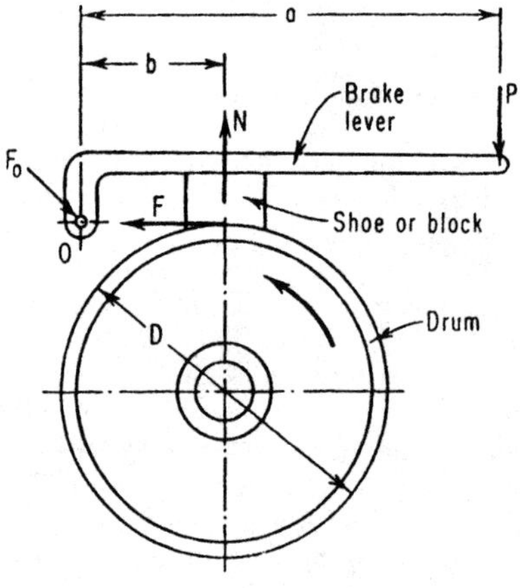

FIG. 18.9 Single-lever block brake (case I).

Single-Lever Block Brake (Case I). Shoe subtends small angle on brake drum. Line of action F passes through fulcrum O as shown in Fig. 18.9. Summing moments about O and equating to zero yields

$$\Sigma M_0 = Pa - Nb = Pa - (F/\mu)b = 0$$

or

$$F = \mu(a/b)P \tag{18.15}$$

The braking torque

$$T = FD/2 = \mu a\, DP/2b \tag{18.16}$$

where N = normal reaction between drum and shoe
P = applied load
F = friction force = μN
μ = coefficient of friction

The above equations apply to drum rotation in either direction.

Single-Lever Block Brake (Case II). Shoe subtends small angle on brake drum. Line of action F passes a distance e below fulcrum O. For counterclockwise drum rotation,

$$\Sigma M_0 = Pa - (F/\mu)b + Fe = 0$$

or

$$F = \frac{Pa}{b/\mu - e} \tag{18.17}$$

$$T = \frac{Pa}{b/\mu - e}\,\frac{D}{2} \tag{18.18}$$

The frictional force in this case helps to apply the brake. The brake is therefore "self-energizing."

If $e > b/\mu$, the brake is self-locking, and some force P is necessary to disengage the brake. For clockwise drum rotation,

$$F = \frac{Pa}{b/\mu + e} \tag{18.19}$$

Single-Lever Block Brake (Case III). Shoe subtends small angle on brake drum. Line of action F passes a distance e above fulcrum O. For counterclockwise drum rotation,

$$\Sigma M_0 = Pa - (F/\mu)b - Fe = 0 \tag{18.20}$$

$$F = \frac{Pa}{b/\mu + e}$$

For clockwise drum rotation,

$$F = \frac{Pa}{b/\mu - e} \tag{18.21}$$

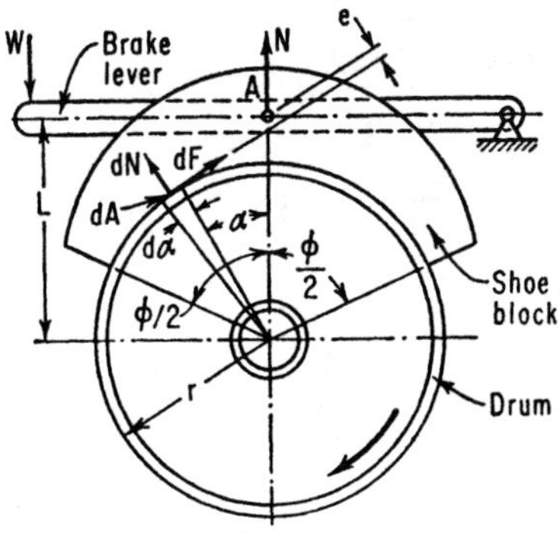

FIG. 18.10 Single-lever block brake with long pivoted shoe (case IV).

Single-Lever Block Brake, Long Pivoted Shoe (Case IV). (See Fig. 18.10.) Assume the distribution of normal pressure p on the shoe to be

$$p = P \cos \alpha$$

where P is a constant. For a face width w, the differential area of the shoe

$$dA = wr\, d\alpha$$

and the normal and frictional forces are, respectively,

$$dN = pwr\, d\alpha = Pwr \cos \alpha\, d\alpha \tag{18.22}$$

$$dF = \mu Pwr \cos \alpha\, d\alpha \tag{18.23}$$

The moment of dF about A (not necessarily the center of the pin) is

$$dM_A = e\, dF = \mu Pwr(L \cos^2 \alpha - r \cos \alpha)\, d\alpha$$

$$M_A = \int dM_A = \mu Pwr(L\alpha/2 + L \sin 2\alpha/4 - r \sin \alpha)_{-\phi/2}^{\phi/2} = 0 \tag{18.24}$$

and

$$L = \frac{4r \sin \phi/2}{\phi + \sin\phi} \tag{18.25}$$

where L is the distance from the drum center to the line of action of F. Point A is the center of pressure. If the pivot pin is located at A, the shoe will not tip. The normal force N is given by

$$N = \int dN = Pwr \int_{-\phi/2}^{\phi/2} \cos \alpha \, d\alpha = 2\, Pwr \sin \phi/2 \tag{18.26}$$

and the braking torque is approximately

$$T = \int r\, dF = \int \mu pwr^2\, d\alpha = \mu Pwr^2 \int \cos \alpha \, d\alpha = \mu Nr \tag{18.27}$$

18.4.2 Band Brakes

In a band brake, braking action is obtained by pulling a band tightly against the drum. The braking force F_t is defined as the difference between the tensions F_1 and F_2 at the two ends of the band. Thus

$$F_t = F_1 - F_2 \tag{18.28}$$

Referring to Fig. 18.11, summation of the horizontal and vertical components of the forces acting on a differential element of band length yields, respectively,

$$\mu F_N - dF \cos (d\theta/2) = 0$$

and

$$F_N - (2F + dF) \sin (d\theta/2) = 0$$

where F_N = force between band and drum
F = band tension
μF_N = frictional force
θ = angle of contact

Making the small-angle approximations $\sin (d\theta/2) = d\theta/2$ and $\cos (d\theta/2) = 1$ and eliminating F_N from the equilibrium equations yields

$$\mu F\, d\theta - dF = 0$$

Integrating,

$$\int_{F_2}^{F_1} \frac{dF}{F} = \mu \int_0^{\theta} d\theta$$

Thus

$$F_1/F_2 = e^{\mu\theta} \qquad \text{or} \qquad F_1 = F_t e^{\mu\theta}/(e^{\mu\theta} - 1) \tag{18.29}$$

Figure 18.12 shows a differential band brake. Equating to zero the moments acting on the lever about point A,

$$M_A = Pa + F_1 b_1 - F_2 b_2 = 0$$

and substituting Eq. (18.29), the solution for P is

$$P = F_t(b_2 - e^{\mu\theta} b_1)/a(e^{\mu\theta} - 1) \tag{18.30}$$

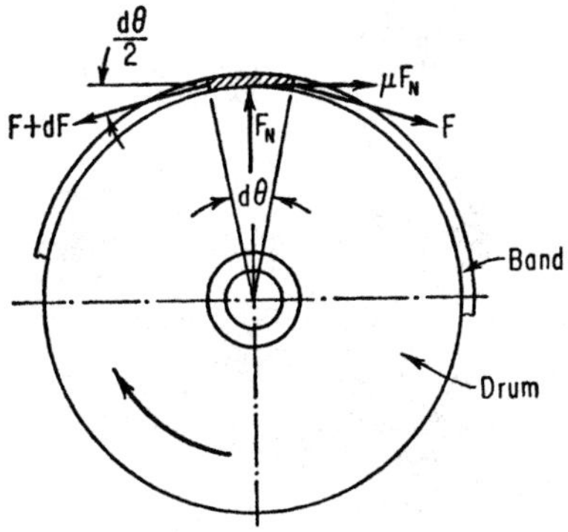

FIG. 18.11 Band tension.

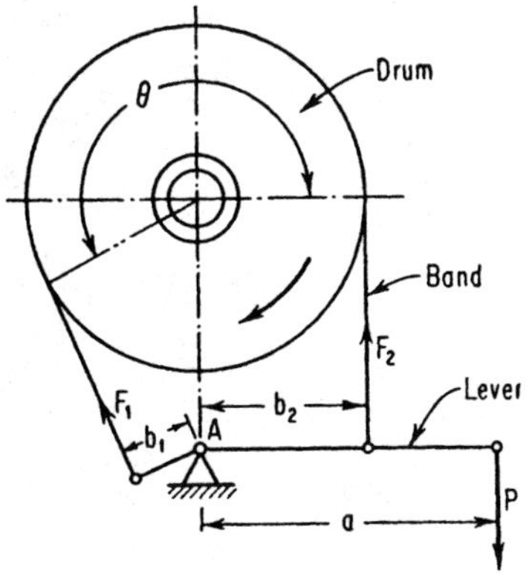

FIG. 18.12 Differential band brake.

Normal operation of the arrangement shown in Fig. 18.12 requires that $b_2 > b_1 e^{\mu\theta}$. If $b_2 = b_1 e^{\mu\theta}$, the brake is self-locking. If $b_2 < b_1 e^{\mu\theta}$, a force must be applied in the opposite direction in order to permit drum rotation. Similar analyses are used for other combinations of direction of P, fulcrum location, and direction of drum rotation.

For $b_1 = 0$ and clockwise drum rotation

$$P = F_t b_2 / a(e^{\mu\theta} - 1) \tag{18.31}$$

and the arrangement is termed a simple band brake.

18.5 ANALYSIS OF SELF-ACTUATING DRUM BRAKES

Automotive brakes commonly used in the United States are self-actuating or "self-energizing" brakes. They utilize wedging action to cause the normal force exerted by the lining on the drum to increase nonlinearly with the coefficient of friction. The "caliper" disk brake is not self-actuating. The normal force is independent of the coefficient of friction and depends only on the mechanical design.

Self-actuating brakes are rated by their effectiveness, the ratio of friction force to applied force.* The effectiveness of the drum brake is complicated by its geometry. Since the friction force developed by self-actuating brakes is not proportional to the coefficient of friction, measurement of the coefficient on self-actuating brakes is impractical. The drum brake is particularly unsuited for this purpose, because its geometry and hence its effectiveness for a given coefficient of friction may alter with applied load because of drum and shoe distortion.

*The effectiveness is often defined as the ratio of vehicle deceleration to hydraulic pressure. This quantity is directly proportional to the effectiveness as defined above, the proportionality factor varying with type of brake and weight of vehicle.

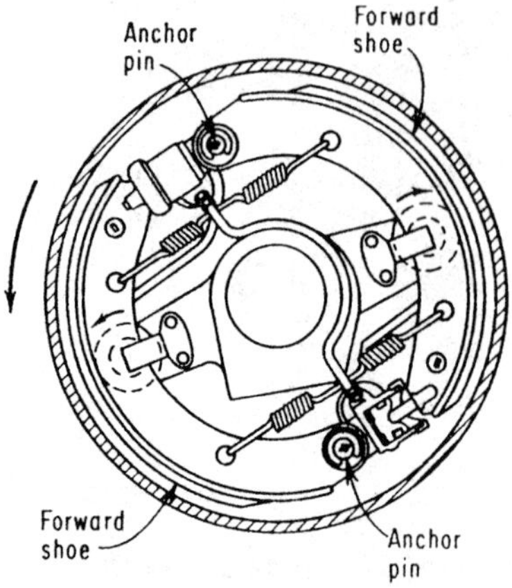

FIG. 18.13 Fixed-anchor internal drum brake.

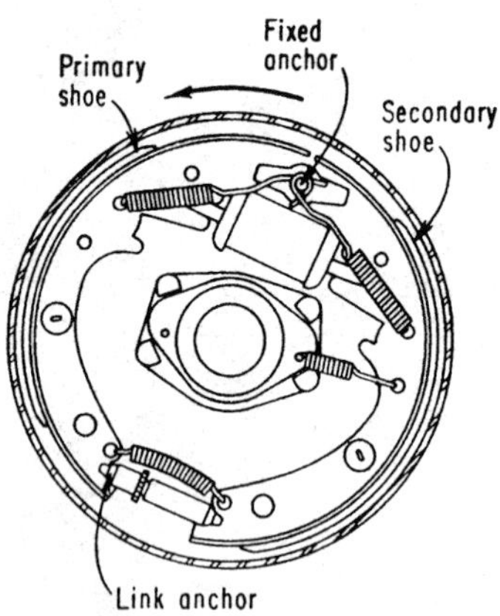

FIG. 18.14 Movable- or link-anchor internal drum brake (Duo-Serve).

18.5.1 The Internal Drum Brake

Self-actuating drum brakes are of two types: (1) fixed anchor (Fig. 18.13) and (2) movable or link anchor (Fig. 18.14). Each brake contains two shoes. Both shoes may be self-actuating, called "forward shoes," or one shoe may be anti-self-actuating, called the "drag shoe." In passenger car brakes, the load is applied through hydraulic wheel cylinders and is given by the product of the fluid pressure and wheel-cylinder area. In many truck brakes, the load is applied through cams and must be measured.

Several methods are in use for calculating brake effectiveness. They usually assume negligible drum and shoe distortion, an assumption valid for light loads only. They also assume a uniform coefficient of friction over the surface of the lining, valid for most practical purposes. The equations derived herein are based upon a graphical method described by Fazekas.[53]

Accurate methods involve integration of the friction and normal forces on an element of lining. Some methods require the calculation of a definite integral for each type of brake and for each change of dimensions. The present method applies to all cylinder-loaded brakes and avoids special integration. The actuation equation gives the effectiveness, in most cases more quickly and accurately than graphical analysis. Structural and finite-element analyses[54–56] have also been applied to drum brakes to study the role of contact geometry and distortion on the torque characteristics.

18.5.2 Theory of Resultant Forces on Shoe

Three forces act on the shoe (Figs. 18.18 and 18.19), the applied load L, the reaction R of the anchor, and the vector resultant reaction V of the friction and normal forces exerted by the drum through the lining. The force V acts through the so-called "drag point" or "center of pressure" (CP). In most practical cases, the CP lies outside the drum of any internal curved-shoe brake. With negligible distortion, the pressure distribution is harmonic. With moderate distortion, it is practically uniform over the lining.

Moment of Friction Forces about Center of Drum. Figure 18.15 is a schematic diagram of the circular surface of the lining of angular arc 2ϕ. The origin of coordinates is the center of the brake. Let p_m be the normal (radial) pressure at θ_m, locating

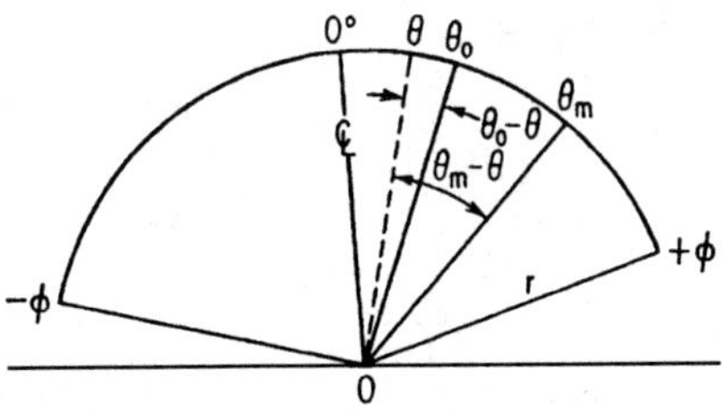

FIG. 18.15 Schematic diagram of the circular surface of the lining of angular arc 2ϕ.

the line of maximum pressure (LMP), θ_m being measured from the center line of the lining. The pressure at an angle θ from the center line is

$$p = p_m \cos(\theta_m - \theta) \tag{18.32}$$

The normal force on an element of area of lining is $pwr\, d\theta$, where w is the width of lining and r is the drum radius. The corresponding friction force in the direction of rotation of the drum is $\mu pwr\, d\theta$ and the moment of this force about the center is $\mu pwr^2\, d\theta$, where μ is the coefficient of friction.

Integrating these moments over the lining gives the total friction moment

$$Fr = \int_{-\phi}^{+\phi} \mu p_m wr^2 \cos(\theta_m - \theta)\, d\theta$$

$$Fr = 2\mu p_m wr^2 \cos\theta_m \sin\phi \tag{18.33}$$

Resultant Normal Force. The normal forces $pwr\, d\theta$ acting together produce a vector resultant normal force N along a drum radius at some angle θ_0 from the center line of the lining. The component of the pressure p in the direction θ_0 is $p\cos(\theta_0 - \theta)$, so that the component of the normal force in this direction is $pwr\cos(\theta_0 - \theta)\, d\theta$.

Integrating these forces over the lining gives the resultant normal force

$$N = p_m wr \int_{-\phi}^{+\phi} \cos(\theta_m - \theta)\cos(\theta_0 - \theta)\, d\theta$$

$$N = p_m wr[\cos\theta_m \cos\theta_0(\phi + \sin\phi\cos\phi) + \sin\theta_m \sin\theta_0(\phi - \sin\phi\cos\phi)] \tag{18.34}$$

Relation of θ_0 to θ_m. By definition, the resultant normal force perpendicular to the direction θ_0 is zero. The component of the normal force perpendicular to θ_0 is $pwr \sin(\theta_0 - \theta)\, d\theta$. Integrating these forces over the lining gives

$$\int_{-\phi}^{+\phi} \cos(\theta_m - \theta)\sin(\theta_0 - \theta)\, d\theta = 0$$

$$\cos\theta_m \sin\theta_0(\phi + \sin\phi\cos\phi) - \sin\theta_m \cos\theta_0(\phi - \sin\phi\cos\phi) = 0$$

Transposing and dividing gives

$$\tan\theta_m = \tan\theta_0[(\phi + \sin\phi\cos\phi)/(\phi - \sin\phi\cos\phi)] \tag{18.35}$$

Center-of Pressure Circle: CP Locus. The resultant friction force μN acts through the CP at right angles to N. The moment of μN about the center of the drum $\mu Nc = Fr$, where c is the distance of the CP from the center of the drum. Using the values for Fr and N from Eqs. (18.33) and (18.34) together with Eq. (18.35),

$$c = Fr/\mu N = (2r\sin\phi\cos\theta)/(\phi + \sin\phi\cos\phi) \tag{18.36}$$

This is the polar equation of a circle with the origin on its circumference, its center at point $d/2$, 0, and of diameter

$$d = (2r \sin \phi)/(\phi + \sin \phi \cos \phi) \tag{18.37}$$

The relationship between d/r and ϕ is shown in Fig. 18.16, curve A.

Relation of Center of Pressure to Line of Maximum Pressure. This relationship is given in Eq. (18.35) which, employing Eq. (18.37), may be written as

$$\tan \theta_0 = \tan \theta_m[1 - (d/r) \cos \phi] \tag{18.38}$$

The relationship between $\tan \theta_m/\tan \theta_0$ and ϕ is shown in Fig. 18.17.

18.5.3 Effect of Moderate Distortion of Drum and Shoe: Uniform Pressure

Assume the CP is initially at the center line of the lining, i.e., $\theta_0 = 0$, for the usual coefficient of friction. This gives the minimum variation of pressure over the lining and consequently the most uniform heating and wear, and maximum effectiveness. In well-adjusted brakes, the wear is often uniform, i.e., the pressure is uniform over the lining. This means that the CP remains at the center line after moderate distortion. This will not necessarily be the case if the coefficient changes considerably, since, in general, the CP is at the center line for only one coefficient of friction. With uniform pressure p,

$$Fr = \int_{-\phi}^{+\phi} \mu p w r^2 \, d\theta = 2\mu p w r^2 \phi$$

The component of p in the direction θ_0 is $p \cos \theta$, since $\theta_0 = 0$. Therefore,

$$N = \int_{-\phi}^{+\phi} p w r \cos \theta \, d\theta = 2pwr \sin \phi$$

The distance of the CP from the center of the drum

$$\begin{aligned} c &= Fr/\mu N \\ &= 2\mu p w r^2 \phi / 2\mu p w r \sin \phi \\ &= r\phi/\sin \phi \end{aligned} \tag{18.39}$$

This is a circle of radius c with its center at the drum center. The function c/r versus 2ϕ is shown plotted in Fig. 18.16, curve B.

18.5.4 The Actuation Equation: Effectiveness

The above relations, together with the known geometry of a given brake, suffice to obtain the actuation equation and hence the effectiveness, usually without graphical construction. Either the LMP (θ_m) or the CP (θ_0) can always be found from the geometry of the brake, and the other found from Fig. 18.17.

Fixed-anchor brakes are the most stable. Each shoe has a separate anchor (Fig. 18.13) which allows rotation about a fixed axis. The actuation equation for the drag shoe, anchored at the leading end, is obtained by reversing the sign of the actuation factor. The anchored end is herein called the "heel" and the loaded end the "toe."

Figure 18.18 is a schematic diagram showing the lining of a forward shoe of an

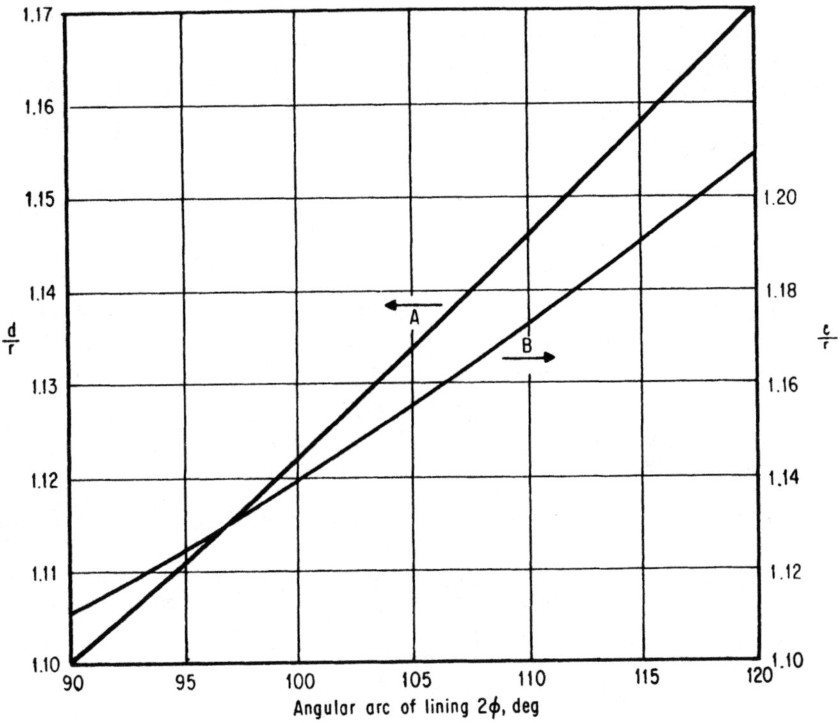

FIG. 18.16 Relationship between *d/r*, *c/r*, and ϕ.

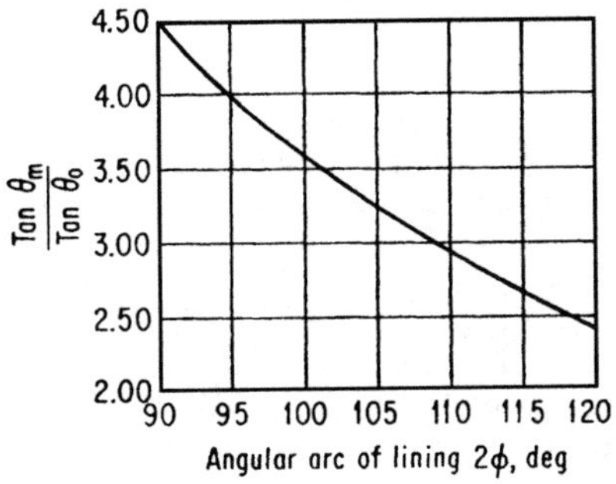

FIG. 18.17 Relationship between tan θ_m/tan θ_0, and ϕ.

11-in-diameter brake and the angles and dimensions required to obtain the actuation equation. The drum reaction *V* is drawn through the CP, making the friction angle 17° = arctan 0.3 with the normal force *N*. The anchor reaction *R* is drawn through the anchor and the intersection of *V* with the load *L* (off the diagram). The magnitudes of *R* and *V* may be obtained by constructing a force triangle on *L*.

The angles θ_m and θ_0 are measured from the lining center line. All other lining angles are measured from the anchor line, a line from the center of the brake to the anchor. From Fig. 18.16 for 2ϕ = 117°, the CP-circle diameter *d* = 6.4 in for no drum distortion. The CP circle is drawn as defined by Eq. (18.37), with its center on the center line of the lining at 85½°.

In a fixed-anchor brake, the LMP necessarily lies on a line at right angles to the anchor line.[53] Therefore, θ_m = 90 − 85½° = 4½°. From Fig. 18.17 θ_0 = 1°46′.

The CP lies on the CP circle at its intersection with the drum radius inclined at 1°46′ to the center line of the lining. The pressure distribution on this lining is thus very nearly symmetrical.

Since the LMP is fixed in a fixed-anchor brake, the location of the CP is independent of the coefficient of friction and depends only on the length and position of the lining. This explains its greater stability. In link- and sliding-anchor brakes, the CP and the LMP both move as the coefficient of friction varies.

It is convenient to write the actuation equation in terms of the ratio of the braking torque to the applied torque about the anchor:

$$\frac{Fr}{Lr_2} = \frac{\mu Nc}{Na - \mu Nb} = \frac{\mu(c/a)}{1 - \mu(b/a)} \tag{18.40}$$

where μ = coefficient of friction
a = moment arm of the normal force N about the anchor
b = moment arm of the friction force μN about the anchor
c = moment arm of the friction force μN about the center of the brake
r = radius of the brake drum
r_2 = moment arm of the load L about the anchor

The ratio b/a is the actuation constant Q. The factor $\mu(b/a)$ is the actuation factor A. The shoe locks when the drum reaction V passes through the anchor, i.e., when $\mu = a/b = \tan \angle (V - N) = \tan 34\frac{1}{2}° = 0.687$, in this case.

In a drag shoe, the algebraic sign of the actuation factor is changed. It cannot lock and its effectiveness is less than if non-self-actuating.

Figure 18.18 illustrates the graphical method. The quantities needed to obtain the actuation equation may be measured on a full-scale drawing, or they may be obtained analytically. From Fig. 18.18,

$$\begin{aligned} a &= r_1 \cos(\theta_m - \theta_0) \\ b &= c - r_1 \sin(\theta_m - \theta_0) \\ c &= d \cos \theta_0 \end{aligned} \tag{18.41}$$

where r_1 is the length of the anchor line. These quantities can all be obtained from Figs. 18.16 and 18.17 and the geometry of the brake and lining, without graphical construction.

The effectiveness of two forward shoes of the 11-in brake, referred to a single cylinder load L, is plotted in Fig. 18.20, curve B. For any coefficient, the effectiveness is a maximum when the CP lies on the center line of the lining. The magnitude and distribution of pressure on the lining can be calculated from Eqs. (18.22), (18.33), (18.40), and (18.41).

The link-anchor brake, exemplified by the Bendix Duo-Servo brake[57] (Fig. 18.14), has one movable-anchor shoe, called the primary, and one fixed-anchor shoe, called the secondary. The load is applied near the toe of the primary and the heel of the secondary through hydraulic cylinders placed back to back. The heel of the primary is connected to the toe of the secondary through a rigid, hinged link. The end of the link attached to the primary is the link anchor which can move toward or away from the drum as the primary rotates about it.

Figure 18.19 is a schematic diagram of the primary in the 12-in Bendix brake. The forces shown are those exerted on the primary shoe and lining. The drum rotation moves the primary out of contact with the fixed anchor. The load L acts at a right angle to the center line of the brake. The reaction R through the link is, for all practical purposes, parallel to L. Therefore, $V = R + L$. The moments about the link anchor $Vr_3 = Lr_2$. These two equations give the magnitude of V and R. The pressure distribution is harmonic, but the LMP is not known a priori as for a fixed anchor.

In the graphical method, the CP is found by drawing a line parallel to L through a point P located on the CP circle at an angular distance $2\theta_\mu$ from the lining center line.

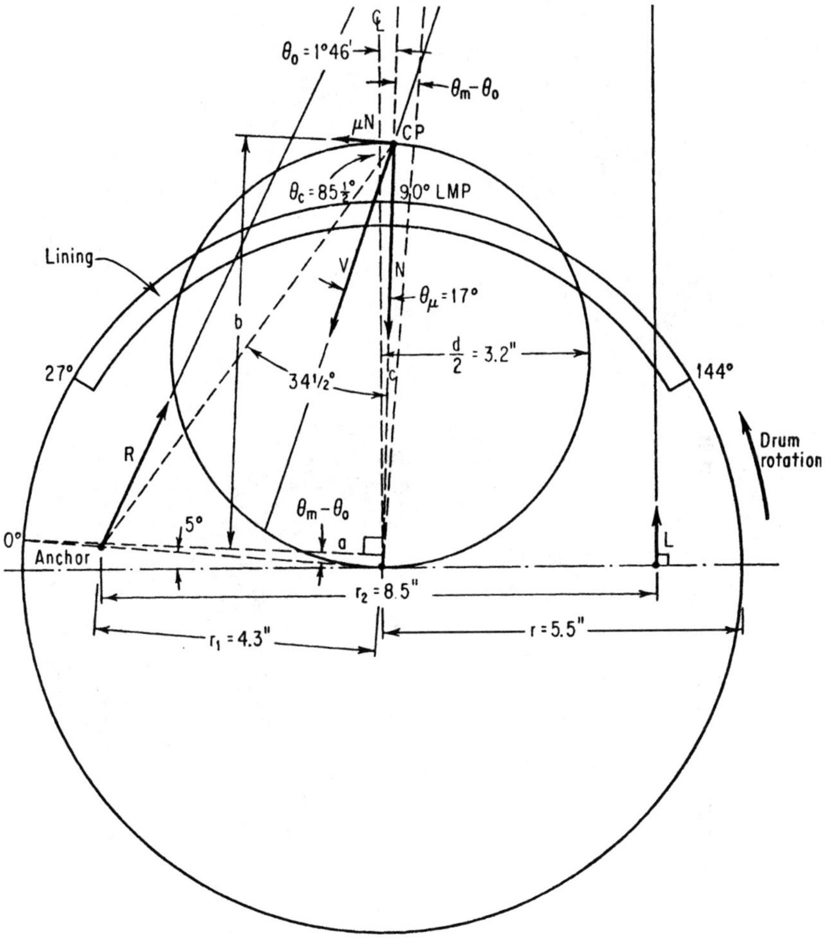

FIG. 18.18 Forces on the fixed-anchor brake.

This line also intersects the CP circle at the CP. Since an inscribed angle is measured by half the intercepted arc, this line makes the angle θ_μ with the drum radius through the CP, thus defining the position and direction of *V* and *N*.

Analytically, from Fig. 18.19,

$$\theta_a = 74 - \theta_\mu = 52° \tag{18.42}$$

$$\theta_0 = \theta_c - \theta_a = \theta_c + \theta_\mu - 74° = 18° \tag{18.43}$$

where θ_a = angular position of the CP measured from the anchor line
θ_μ = friction angle, arctan μ
θ_c = angular position of the lining

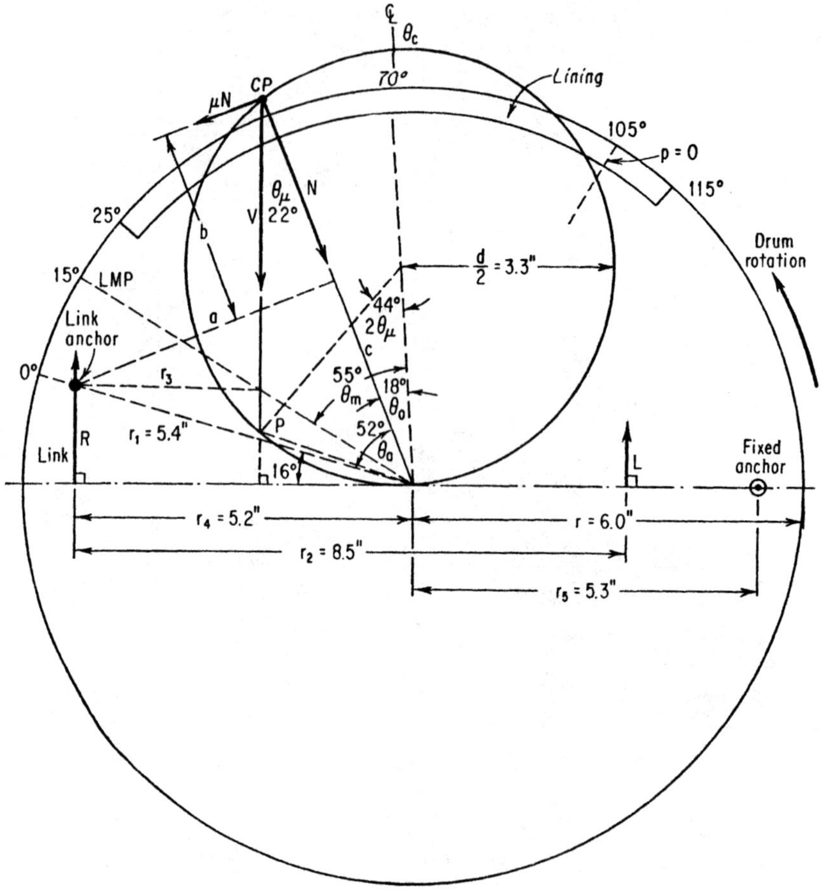

FIG. 18.19 Forces on the primary shoe of link-anchor brake.

The LMP (θ_m) is found from Eq. (18.43) and Figs. 18.15 and 18.16 to be 15° from the anchor line, i.e., 10° beyond the heel of the lining. At 90° from this, or 10° in from the toe, the pressure is zero. The remainder of the toe is lifted off the drum.

The pressure distribution is extremely asymmetrical with the lining in this position. For $\mu = 0.4$, the pressure distribution would be symmetrical if the lining were moved 18° toward the link anchor. In the symmetrical-pressure position, the effectiveness is 10 percent higher than in the position of Fig. 18.17.

The actuation equation is, as before,

$$\frac{Fr}{Lr_2} = \frac{\mu(c/a)}{1 - \mu(b/a)} \tag{18.44}$$

From Fig. 18.19,

$$\begin{aligned} a &= r_1 \sin \theta_a \\ b &= c - r_1 \cos \theta_a \\ c &= d \cos \theta_0 \end{aligned} \tag{18.45}$$

Since θ_0 and θ_a are both functions of μ, the effectiveness is a complicated function of μ. In general, both the numerator and denominator of the right-hand side of Eq. (18.44) are quadratic functions of μ. This shoe cannot lock, since the drum reaction V cannot pass through the anchor; the quadratic denominator of the right-hand side of Eq. (18.44) has no real roots. The secondary shoe, however, can lock.

In this case, the friction force in the primary can be found by a simpler method without using the actuation equation. Since V is normal to the brake center line, the moment of V about the center of the brake is $Vc \sin \theta_\mu = (Lr_2/r_3)c \sin \theta_\mu$, which, in turn, is equal to Fr.

The secondary is analyzed as is any other fixed-anchor shoe. The drum rotation forces the secondary against the fixed anchor. Two loads are exerted on the secondary, the reaction R of the primary and the applied load L, both opposite in direction to that shown in Fig. 18.19. The total torque exerted by these two loads about the fixed anchor is equal to that exerted by V.

The total torque T applied to the secondary about the fixed anchor is therefore

$$T = L(r_2/r_3)(r_5 + c \sin \theta_\mu) \tag{18.46}$$

where r_2 = moment arm of L about the link anchor
r_3 = moment arm of V about the link anchor = $r_4 - c \sin \theta_\mu$
r_4 = distance from the link anchor to the center of the brake
r_5 = distance from the fixed anchor to the center of the brake

The actuation equation is

$$\frac{Fr}{T} = \frac{\mu(c/a)}{1 - \mu(b/a)} \tag{18.47}$$

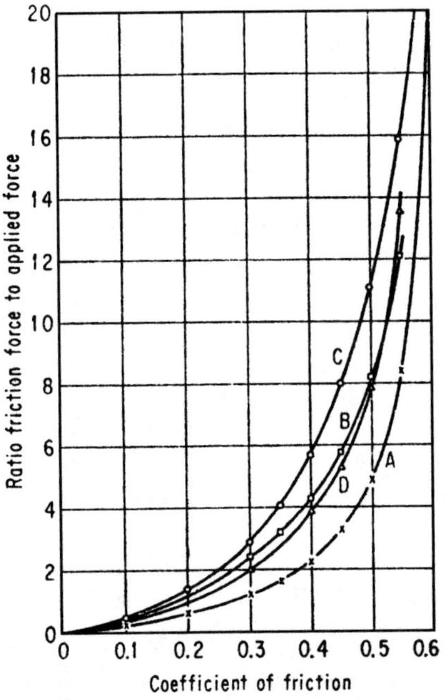

FIG. 18.20 Ratio of friction force to applied force vs. coefficient of friction.

where a, b, and c have the values for a fixed-anchor brake [Eq. (18.41)].

For the values given in Fig. 18.19, $T = 22.8L$, which may be compared with $Lr_2 = 8.5L$ for the primary. The secondary lining commonly has a lower coefficient but greater arc length than the primary. Since this brake is symmetrical, except for the linings, its behavior is identical in reverse, the primary and secondary becoming interchanged.

The nearly threefold greater braking torque exerted by the secondary results in an asymmetrically loaded and distorted drum. The pressure on the secondary is three to four times that on the primary and the rate of wear correspondingly greater. The combined effectiveness of the two shoes of the 12-in Duo-Servo brake is shown in curve C of Fig. 18.20.

18.5.5 Drum and Shoe Distortion

Distortion, which can greatly alter the magnitude and distribution of pressure on the lining and, to some extent, the effectiveness, depends upon the applied load, the rigidity of the drum and shoe, and the compressibility of the lining.

Moderate distortion produces the desirable condition of uniform pressure, and its effect can be calculated (see Sec. 18.5.3). Severe distortion, caused by the high loads required by a low lining coefficient, markedly increases the pressure on the toe and heel and decreases the pressure on the center of the lining. The changes in pressure are minimized by a flexible shoe and a soft lining which tend to conform to the drum when it distorts. The effect on brake performance must be measured.

Computation of lining pressure for a distorted drum required measurement of lining compressibility and numerical integration of the moment equations. The computed effectiveness for the distorted drum was about 20 percent greater than when undistorted, for all coefficients of friction.

The effect of distortion on the Bendix brake is reported by Winge.[50] He finds that the effectiveness is increased more for lower coefficients than for higher coefficients. The wide variation of effectiveness with coefficient is thus somewhat less than shown in Fig. 18.20, curve *C*.

18.5.6 Cam Brakes: Truck Brakes

Cam-operated drum brakes have one forward and one drag shoe loaded by a double cam[58] mounted on a single pivot and actuated through a lever arm. The load and hence the effectiveness of cam brakes are difficult to calculate, because the rotation of the drum pushes the drag shoe against the cam, thereby increasing the load, and tends to pull the forward shoe away from the cam, thereby decreasing the load. The amount of this effect varies with the coefficients of friction between lining and drum and between cam and shoe, the clearance in the cam pivot, the compressibility of the lining, and the distortion of the drum and shoes. The usual result is that the effectiveness of the two shoes is about equal.

Figure 18.21 shows the forces and lining pressure distribution in a Timken 14½-in-diameter cam-and-roller-type brake. A similar brake is the cam-and-plate type.

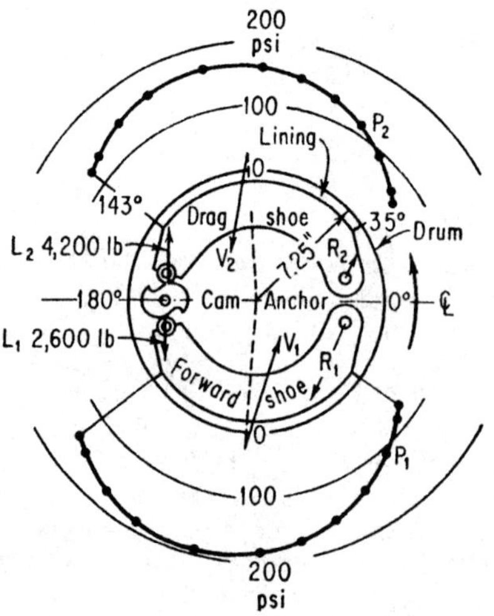

FIG. 18.21 Forces and lining pressure distribution in a Timken brake.

18.6 DISK BRAKES

The disk-brake concept is among the oldest brake designs,[59] the first patent having been issued in 1902. However, because of their lack of self-energization, disk brakes were applied only to aircraft before 1940. After World War II, the development of disk brakes was accelerated because the increases in vehicle weight and speeds demanded a brake with high heat-dissipation capabilities. The disk brake provided a solution to this need. The Crosley Hot Shot of the early 1950s was the first American production car to come equipped with disk brakes. They were applied to heavier passenger cars in 1965 and are standard front brakes for most automobiles at the present time. Rear disk brakes[60,61] have also been applied to many automobiles, but the problems of increased cost, parking brake complications, and reduced heat-dissipation requirements at the rear wheels have limited their application.

Disk brakes have become popular because they overcome some basic disadvantages of the drum brakes they replace. Ironically, the basic mechanical advantage of drum brakes (self-energization), which spurred their development is the source of their main problem. This problem is related to the influence of friction coefficient on the ratio of friction force to applied force, which is defined as the effectiveness of the brake. The drum brake is extremely sensitive to changes in friction coefficient, as shown in Fig. 18.22. Small changes in friction due to moisture, high temperature, or speed require a large change in pedal pressure for the same braking torque. A change in friction coefficient from 0.3 to 0.4 (which is not unlikely, as shown in Sec. 18.1) will require twice the normal force on the brake shoe for the same braking torque. This behavior is not desirable for automobile brakes, where predictable performance is a requirement. The disk brake overcomes this problem because its torque output is linearly proportional to friction coefficient; thus, a change in friction from 0.3 to 0.4 requires only a 33 percent increase in pedal effort. However, because no self-energization occurs, the disk brake requires power assist to provide reasonable pedal effort on vehicles over 3500 lb.

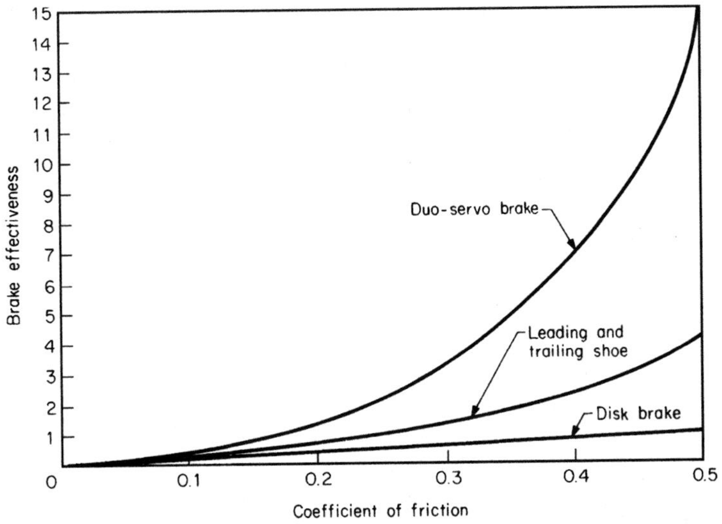

FIG. 18.22 Influence of friction coefficient[59] on brake torque for drum and disk brakes.

Basically, the advantages[59] of the disk brake over the drum brake stem from three factors:

1. Linear relationship between torque output and friction coefficient
2. Higher heat-transfer rates from radiation and convection (ventilated rotors)
3. Lack of mechanical fade (due to distortion of drum brakes)

The disadvantages of disk brakes are higher cost due to the number of castings required and the necessity of power assistance for many applications.

18.6.1 Types of Disk Brakes

Fixed Caliper. Early disk brakes had a caliper rigidly mounted to the vehicle suspension straddling the disk (Fig. 18.23). A pair of hydraulic pistons mounted in the caliper applied the clamping force to the disk from each side. To apply this design to heavier American vehicles in a limited space, a four-piston caliper was developed, but has largely been superseded by the floating-caliper brake.

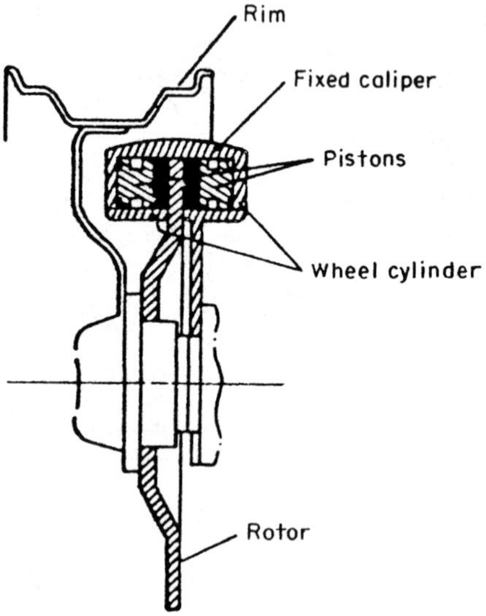

FIG. 18.23 Opposing piston—fixed-caliper disk brake.

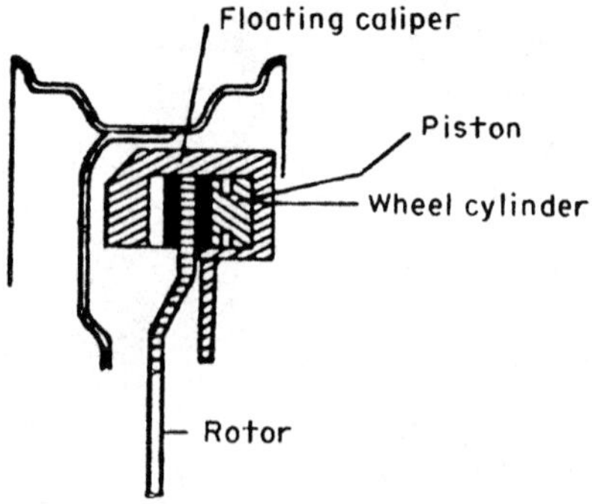

FIG. 18.24 Single piston—floating-caliper disk brake.

Floating Caliper. Most disk brakes now use a floating caliper (Fig. 18.24) which can move perpendicular to the disk surface. This caliper supports the inner and outer linings and contains a single piston, usually inboard of the disk. Upon application of pressure to the piston, a force is exerted on one shoe and the reaction on the caliper forces it to slide and apply the same force to the other shoe. Thus a single cylinder clamps both linings against the disk. The floating caliper is necessary to accommodate the wear of the linings.

18.6.2 Actuation

Because no self-actuation occurs in most disk brakes, a significant mechanical advantage is required to apply the clamping force. Typically for automotive applications, vacuum-boosted hydraulic systems are utilized; however, straight hydraulic actuation is used on lighter vehicles. Disk brakes are also being applied to railroad vehicles and large trucks. These applications typically utilize an air-mechanical or air-over-hydraulic system for actuation.

18.6.3 Design[62,63]

Disk. The disk size is based upon the space available and the energy-dissipation requirements of the vehicle. The brake duty cycle can be divided into two states: the transient state, as found in rapid deceleration, and the steady state, as experienced in prolonged applications. In the transient state, the heat enters the disk more rapidly than it can be dissipated and the disk becomes a heat reservoir. The mass of the disk must be such that the temperature will not be excessively high. Under steady-state conditions, or repeated transient states, the heat will be lost to the surroundings by radiation and convection. Ventilated disk brakes with properly designed and ducted rotors, can appreciably increase the cooling rate of the disk (see Sec. 18.3).

The disk material must have a high conductivity so that the heat generated at the disk-lining interface is quickly conducted to the disk interior. For the initial brake application, the temperature of the disk will be a function of the vehicle speed, brake balance, and disk material. For subsequent stops, the equilibrium temperature will depend upon the heat losses from the disk and the zero-pressure (residual) brake torque. This latter quantity, while it does not generate large quantities of heat, influences the loss of heat from the disk between stops. Most modern disk brakes use the hydraulic cylinder seals[59] to back the pads away from the disk when the pressure is released. Thus the residual brake torque for a well-maintained system will be quite low. However, corrosion or wear may influence the motion of the pads and increase the residual torque.

Lining. The face area of the lining material and its wrap on the disk face are important considerations for the design of disk brakes. The face area of the lining is usually specified in the form of maximum horsepower dissipation per square inch of each pad surface. For heavy-duty applications[49] this can be as much as 8 hp/in^2, but lower values are more commonly used (2 hp/in^2) for automotive applications. The wrap of the lining should not exceed one-third of the swept area in order to retain the benefits of heat loss from the disk face outside the lining, which accounts for about 15 percent[64] of the total heat loss.

18.7 BRAKE SQUEAL

Brake squeal has been a problem since the early days of motoring. However, it has been an elusive problem, and the occurrence of squeal, even for the same vehicle, will depend[65] upon the driver and the history of brake operation. The elimination of squeal has generally been tackled empirically on a case-by-case basis, and a number of ways of reducing squeal on existing brake systems have been utilized. Techniques used to reduce squeal include (1) applying boundary lubricants on the lining surface or backing plate to reduce friction coefficient, (2) inserting antisqueal shims between piston and back plates on disk brakes, (3) adding weight to shoes or calipers at strategic locations, or (4) changing the stiffness of the pad. Other solutions have included changes in the disk or drum material, such as the use of pearlitic-gray iron because of its high internal damping.

Theoretical explanations of the problem have been varied, including stick-slip theory,[66–68] variable dynamic friction coefficient,[69] sprag-slip theory,[70] and the theory of geometric coupling.[71] The current understanding of squeal relates its occurrence to a coupling of vibrations within the brake assembly. The most important parameters are the coefficient of friction of the lining material (higher coefficients leading to higher probability of squeal), the geometry of the brakes, and, in particular, the location of the areas of contact between the lining and the disk or drum. (A leading angle of contact, that is, the contact a specific distance ahead of the pivot location, can lead to brake squeal.) The mechanism for squeal is an increase in normal load (and therefore friction) and the release of that load due to geometric changes or deflections. Measurements of accelerations[65,72] of brake components have shown that during squeal the friction material is deflected elastically along the pad surface by the frictional forces. That deflection causes a second deflection with a component vertical to the surface of the pad. This vertical component reduces the friction force; thus, the stored energy returns to the first configuration. The cycle is then repeated. The phase difference between the coupled vibrations necessary to maintain squeal arises from the inertia of the components in the system.

Holographic measurements[65] of a disk brake have shown the modes of vibration present and the main sources of noise. Under squealing conditions, the disk can have from 6 to 10 circumferential modes. However, the magnitude of the disk vibrations is not the main source of vibration. The friction pads and the calipers have higher vibration amplitudes and these components transmit the main part of the noise energy. The pad vibration occurs perpendicular to the disk surface and has several mode shapes, but, in general, the leading edge of the pad has the greatest amplitude of vibration (frequently more than 0.0001 in). This result reinforces the idea of sprag-slip motion as being the key explanation of the squeal behavior.

18.7.1 Sprag-Slip Theory

The prevailing understanding of squeal is based, in part, on a model[70] first published in 1961. As shown in Fig. 18.25, the model consists of a block at A in contact with a plane sliding to the right with speed V. The arm AO' is rigid and connected to a pivot which is mounted in a flexible support. A second pivot exists at O'', also flexibly mounted. An increase in friction force will, because of the geometry, produce an increase in normal load on the contact. The increase in load due to friction force, L_f, can be written from the sum of the moments about O' as

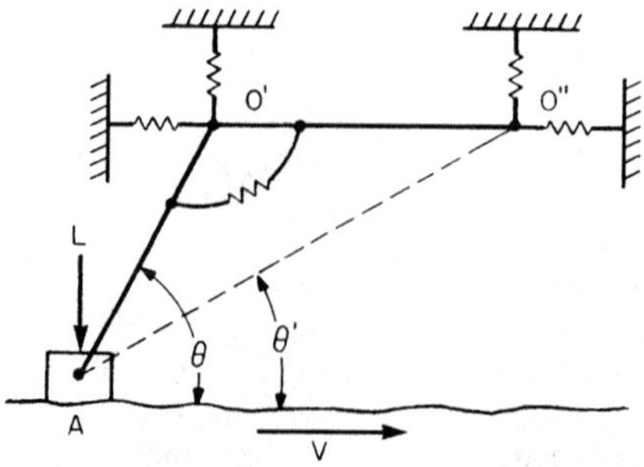

FIG. 18.25 Geometry for sprag-slip model of brake contact.

$$L_f = F_F \tan \theta \tag{18.48}$$

where F_F = friction force and θ = leading angle for front strut. Then the sum of the horizontal forces at the contact yields

$$F_F = L\mu/(1 - \mu \tan \theta) \tag{18.49}$$

where L = initial load on block and μ = friction coefficient

The value of friction goes unbounded as the friction coefficient approaches cot θ, which is called the "spragging angle". When this condition occurs, there is no slip between the block and the moving surface; A is therefore displaced elastically to the right and θ increases slightly. This motion reduces the pivot angle at O' and the elastic moment at O' becomes large enough such that $AO'O''$ becomes like a rigid triangle pivoting ATO'', which can be replaced by AO''. For this case, the angle θ' is less than the spragging angle and the normal, and friction force is reduced as in Eq. (18.49). This action releases the elastic energy at O' and slip occurs. This "sprag-slip" behavior leads to a variation in the normal load on the brake pad, setting up vibrations of the pad and its support which account for the squeal behavior.

This contact behavior can occur for internal and external drum brakes and disk brakes. For disks, the vibration mode consists of motion of the pad perpendicular to the disk and vibration of the pad and the caliper. For the drum brakes, the vibration involves the shoe and the backing plate motion.

Solutions to squeal problems are as varied as their sources. Since the contact of the pad and disk or drum varies with (1) the life of the pad, (2) the current operating temperature, (3) the pad stiffness (dependent upon pressure), and (4) the thermal distortion of the system, no universal fix for squealing problems is available. However, observations as to the sources of vibration and therefore noise generation have been published. These results were generated using holographic[65] pictures of squealing disk brakes, and the conclusions included the following:

1. Increasing stiffness of brake pad is likely to reduce squeal.
2. Brake pads can be supported at antinodes (points of minimum vibration) to reduce transmitted vibration.
3. The natural frequencies of pads and calipers should be made significantly different.
4. Reduction in friction coefficient will reduce tendency to squeal.

REFERENCES

1. Joshi, M. N.: "Disk Linings—Analytical Study and Selection Criteria," SAE Paper 800782, 1980.
2. Newcomb, T. P., and R. T. Spurr: "Friction Materials for Brakes," *Tribology (London)*, vol. 4, no. 2, pp. 75–81, May 1971.
3. Schiefer, H. M., and G. V. Kubczak: "Controlled Friction Additives for Brake Pads and Clutches," SAE Paper 790717, 1979.
4. Token, H. Y.: "Asbestos Free Brakes and Dry Clutches Reinforced with Kevlar Aramid Fiber," SAE Paper 800667, 1980.
5. Aldrich, F. W.: "Semi-Metallics: A New Type of Friction Material," SAE Paper 710591, 1971.
6. Rhee, S. K.: "Ceramics in Automotive Brake Materials," *Am. Ceram. Soc. Bull.*, vol. 55, no. 6, pp. 585–588, June 1976.
7. Ho, T. L., F. E. Kennedy, and M. B. Peterson: "Evaluation of Aircraft Brake Materials," *ASLE Trans.*, vol. 22, no. 1, pp. 71–78, Jan. 1979.
8. Knapp, R. A., and A. E. Anderson: "Friction Material Riveting Studies: Load-Deflection Characteristics and Finite Element Modeling," SAE Paper 790716, 1979.

9. Unpublished Reports. As manufactured and tested by Johns Manville.
10. Anderson, A. E., and R. A. Knapp: "Brake Lining Mechanical Properties, Laboratory Specimen Studies," SAE Paper 790715 for Passenger Car Meeting, June 1979.
11. Russell, G. R.: "The Manufacture of Disk Brake Linings," SAE Paper 750228, 1975.
12. Bowden, F. P., and D. Tabor: "Friction and Lubrication of Solids," Part 1, 1954 and Part 2, 1964, Oxford University Press, New York.
13. Rabinowicz, E.: "Friction and Wear of Materials," John Wiley & Sons, Inc., New York, 1965.
14. Kragelskii, I. V.: "Friction and Wear," Butterworth, Washington, D.C., p. 301, 1965.
15. Tanaka, K., S. Veda, and N. Noguchi: "Fundamental Studies on the Brake Friction of Resin-Based Friction Materials," *Wear,* vol. 23, no. 2, pp. 349–365, March 1973.
16. Scieszka, S. F.: "Tribological Phenomena in Steel Composite Brake Material Friction Pairs," *Wear,* vol. 64, no. 2, pp. 367–378, November 1980.
17. Bros, J., and S. F. Scieszka: "The Investigation of Factors Influencing Dry Friction in Brakes," *Wear,* vol. 44, no. 2, pp. 271–286, 1977.
18. Tanaka, K.: "Friction and Deformation of Polymers," *J. Phys. Soc. Japan,* vol. 16, p. 2003, 1961.
19. Rhee, S. K., and P. A. Thesier: "Effects of Surface Roughness of Brake Drums on Coefficient of Friction and Lining Wear," SAE Paper 720449, 1972.
20. Rhee, S. K.: "Friction Coefficient of Automotive Friction Materials—Its Sensitivity to Load, Speed, and Temperature," *SAE Trans.,* Paper 740415, pp. 1575–1580, 1974.
21. Libsch, T. A., and S. K. Rhee: "Microstructural Changes in Semi-Metallic Disk Brake Pads Created by Low Temperature Dynomometer Testing," *Wear,* vol. 46, no. 2, pp. 203–212, 1978.
22. Rhee, S. K.: "Wear Mechanisms for Asbestos-Reinforced Automotive Friction Materials," *Wear,* vol. 29, no. 3, pp. 391–393, 1974.
23. Pogosian, A. K., and N. A. Lambarian: "Wear and Thermal Processes in Asbestos-Reinforced Friction Materials," *Trans. ASME, J. Lubr. Technol.,* vol. 101, no. 4, pp. 451–485, October 1979.
24. Weintraub, M. H., A. E. Anderson, and R. L. Gealer: "Wear of Resin Asbestos Friction Materials," *Polym. Sci. Technol.,* vol. 5B, pp. 623–649, 1974.
25. Liu, T., and S. K. Rhee: "High Temperature Wear of Semi-Metallic Disk Brake Pads," *Wear,* vol. 46, no. 2, pp. 213–218, 1978.
26. Jacko, M. G.: "Physical and Chemical Changes of Organic Disk Pads in Service," *Wear,* vol. 46, no. 1, pp. 163–175, 1978.
27. Liu, T., and S. K. Rhee: "A Study of Wear Rates and Transfer Films of Friction Materials," *Wear,* vol. 60, no. 1, pp. 1–12, 1980.
28. Rhee, S. K.: "Wear Mechanism at Low-Temperature for Metal-Reinforced Phenolic Resins," *Wear,* vol. 23, no. 2, pp. 261–263, 1973.
29. Liu, T., and S. K. Rhee: "High Temperature Wear of Asbestos-Reinforced Friction Materials," *Wear,* vol. 37, no. 2, pp. 291–297, 1976.
30. Bush, H. D., D. M. Rowson, and S. E. Warren: "The Application of Neutron-Activation Analysis to the Measurement of the Wear of a Frictional Material," *Wear,* vol. 20, no. 2, pp. 211–225, June 1972.
31. Pavelescu, D., and M. Musat: "Some Relations for Determining the Wear of Composite Brake Materials," *Wear,* vol. 20, no. 1, pp. 91–97, 1974.
32. Kant, S., et al.: "Prediction of the Coefficient of Friction for Pneumatic Tires on Hard Pavement," *Proc. Inst. Mech. Eng. (London),* vol. 189, no. 34, pp. 259–266, 1975.
33. Limpert, R., and C. Y. Warner: "Proportional Braking of Solid-Frame Vehicles," SAE Paper 710047, 1971.
34. Campbell, C.: "The Sports Car," Robert Bentley, Inc., Cambridge, Mass., p. 244, 1970.

35. Yim, B., et al.: "Highway Vehicle Stability in Braking Maneuvers," SAE Paper 700515, 1970.

36. Hartley, J.: "Anti-Lock Braking," *Automot. Des. Eng.,* vol. 13, pp. 15–19, June 1974.

37. Clemett, H. R., and J. W. Moules: "Brakes and Skid Resistance," *Highway Res. Rec.,* no. 477, pp. 27–33, 1973.

38. Harned, J. L., L. E. Johnson, and G. Scharpf: "Measurement of Tire Brake Force Characteristics as Related to Anti-Lock Control System Design," SAE Paper 690214, 1969.

39. Bernard, J. E., L. Segel, and R. E. Wild: "Tire Shear Force Generation During Combined Steering and Brake Maneuvers," SAE Paper 770852, 1977.

40. Jacko, M. G., W. M. Spurgeon, and S. B. Catalano: "Thermal Stability and Fade Characteristics of Friction Materials," SAE Paper 680417, 1968.

41. Ho, T., M. B. Peterson, and F. F. Ling: "Effect of Frictional Heating on Brake Materials," *Wear,* vol. 30, no. 1, pp. 73–91, October 1974.

42. Bark, L. S., D. Moran, and S. J. Percival: "Polymer Changes During Frictional Material Performance," *Wear,* vol. 41, no. 2, pp. 309–314, February 1977.

43. Wetenkamp, H. R., and R. M. Kipp: "Hot Spot Heating by Composition Shoes," *Trans. ASME,* ser. B, vol. 98, no. 2, pp. 453–458, May 1976.

44. Santini, J. J., and F. E. Kennedy: "Experimental Investigation of Surface Temperature and Wear in Disk Brakes," *Lubr. Eng.,* vol. 31, no. 8, pp. 402–404 and 413–417, August, 1975.

45. Rowson, D. M.: "The Interfacial Surface Temperature of a Disk Brake," *Wear,* vol. 47, pp. 323–328, 1978.

46. Sisson, A. E.: "Thermal Analysis of Vented Brake Rotors," SAE Paper 780352, pp. 1685–1694, 1978.

47. Limpert, R.: "Thermal Performance of Automotive Disk Brakes," SAE Paper 750873, pp. 2355–2368, 1975.

48. Morgan, S., and R. W. Dennis: "A Theoretical Prediction of Disk Brake Temperatures and a Comparison with Experimental Data," SAE Paper 720090, 1972.

49. Soltis, P. J.: "Straight Air Disk Brakes," SAE Paper 791041, 1979.

50. Winge, J. L.: "Instrumentation and Methods for the Evaluation of Variables in Passenger Car Brakes," SAE Paper 361 B, 1961.

51. Secrist, D. A., and R. W. Hornbeck: "Analysis of Heat Transfer and Fade in Disk Brakes," ASME Paper 75-DE-A, 1975.

52. Newcomb, T. P.: "Stopping Revolutions: Developments in the Braking of Cars from the Earliest Days," *Proc. Inst. Mech. Eng.,* vol. 195, p. 139, 1981.

53. Fazekas, G. A. G.: "Graphical Shoe-Brake Analysis," *Trans. ASME,* vol. 79, p. 1322, August 1957, also "Some Basic Properties of Shoe Brakes," *J. Appl. Mech.,* vol. 25, p. 7, March 1958.

54. Millner, N., and B. Parsons: "Effect of Contact Geometry and Elastic Deformations on the Torque Characteristics of a Drum Brake," *Proc. Inst. Mech. Eng. (London),* vol. 187, no. 26, pp. 317–331, 1973.

55. Petrof, R. C.: "Structural Analysis of Automotive Brake Drums," SAE Paper 760788, 1976.

56. Day, A. J., P. R. Harding, and T. P. Newcomb: "Finite Element Approach to Drum Brake Analysis," *Automot. Eng. (London),* vol. 5, no. 5, pp. 21–22, October/November, 1980.

57. Lupton, C. R.: "Drum Brakes," *Mach. Des.,* vol. 22, p. 146, July 1950.

58. Frazee, I., and E. L. Bedell: "Automotive Brakes and Power Transmission Systems," Chap. 1, American Technical Society, Chicago, Ill., 1956.

59. Lueck, F. E.: "Why Disk Brakes," SAE Paper 1004B, 1965.

60. Ballard, C. E., R. W. Emmons, F. L. Janoski, and K. Goering: "Rear Disk Brake for American Passenger Cars," SAE Paper 741064, 1974.

61. Klein, H. C., and Rowell, J. M.: "Development of 4-wheel Disk Brake Systems in Europe," SAE Paper 741065, 1974.

62. Dike, G.: "On Optimum Design of Disk Brakes," *J. Eng. Inc., Trans. ASME,* ser. B, vol. 96, no. 3, pp. 863–870, August 1974.

63. Cords, F. W., and J. B. Dale: "Designing the Brake System Step-by-Step," SAE Paper 760637, 1976.

64. Airheart, F. B.: "Design Approaches to Truck Disk Brakes," SAE Paper 740604, 1974.

65. Felske, A., G. Hoppe, and H. Matthai: "Oscillation in Squealing Disk Brakes—Analysis of Vibration Modes by Holographic Interferometry," SAE Paper 786333, p. 1577, 1978.

66. Blok, H. "Fundamental Mechanical Aspects of Boundary Lubrication," *Automot. Eng.,* vol. 46, p. 54, 1940.

67. Sinclair, D.: "Frictional Vibrations," *J. Appl. Mech.,* pp. 207–214, 1955.

68. Rabinowicz, E.: "A Study of Stick-Slip Process," *Proc. Symp. Friction and Wear,* Detroit, Mich., 1957.

69. Rabinowicz, E.: *Proc. Symp. Friction and Wear,* Elsevier Publishing Co., Amsterdam, The Netherlands, p. 149, 1959.

70. Spurr, R. T. "A Theory of Brake Squeal," *Proc. Inst. Mech. Eng.,* no. 1, pp. 33–40, 1961/1962.

71. Jarvis, R. P., and B. Mills: "Vibration Induced by Dry Friction," *Proc. Inst. Mech. Eng.,* vol. 178, Part 1, p. 847, 1963/1964.

72. Millner, N.: "An Analysis of Disk Brake Squeal," SAE Paper 780332, p. 1565, 1978.

CHAPTER 19
BELTS

William H. Baier, Ph.D.

Director of Engineering
The Fitzpatrick Co.
Elmhurst, Ill.

19.1 INTRODUCTION 19.1
19.2 FLAT BELTS 19.2
19.2.1 Forces 19.2
19.2.2 Action of Belt on Pulley 19.3
19.2.3 Belt Stresses 19.4
19.2.4 Belt Geometry 19.4
19.2.5 Coefficient of Friction 19.6
19.2.6 Creep and Slip 19.6
19.2.7 Angle Drives 19.7
19.2.8 Flat-Belt Fasteners 19.8
19.3 FLAT-BELT MATERIALS 19.8
19.3.1 Leather Belting 19.8
19.3.2 Nylon-Core Belting 19.9
19.3.3 Rubber Flat Belts 19.10
19.3.4 Cotton and Canvas Belting 19.10
19.3.5 Balata Belting 19.11
19.3.6 Steel Belting 19.11
19.3.7 V-Ribbed Belting (Grooved, Poly-V) 19.12
19.4 FLAT-BELT WORKING STRESSES 19.12
19.4.1 Design Stresses and Loads 19.12
19.4.2 Horsepower Ratings of Flat Belts 19.12
19.5 FLAT-BELT DESIGN 19.13
19.6 FLAT-BELT PULLEYS AND IDLERS 19.16
19.6.1 Pulleys 19.16
19.6.2 Idlers 19.17
19.7 V BELTS 19.17
19.7.1 Introduction 19.17
19.7.2 Forces 19.18
19.7.3 Geometry 19.18
19.7.4 Sheaves 19.19
19.7.5 V-Flat Drives 19.19
19.7.6 Quarter-Turn V-Belt Drives 19.20
19.8 V-BELT TYPES 19.20
19.8.1 Heavy Duty Narrow (Industrial Narrow) 19.20
19.8.2 Heavy Duty Standard (Classical/Conventional) 19.20
19.8.3 Heavy Duty Super (Heavy Duty Classical) 19.20
19.8.4 Light Duty (Standard Light, Single-V) 19.20
19.8.5 Agricultural Belts 19.21
19.8.6 Automotive Belts 19.21
19.8.7 Open End V Belts 19.21
19.8.8 Link V Belts 19.21
19.8.9 Joined V Belts 19.21
19.8.10 Double-Angle V Belts 19.21
19.8.11 Low-Stretch Cable (Steel Cable) Belts 19.22
19.8.12 Wide Angle V Belts 19.22
19.8.13 Flexible Sidewall V Belts 19.22
19.9 V-BELT DESIGN 19.22
19.10 SYNCHRONOUS BELTS (TIMING, POSITIVE) 19.23
19.11 VARIABLE-SPEED BELTS 19.24
19.12 ENDLESS BELTS 19.25
19.12.1 Film Belts 19.25
19.12.2 Woven Belts 19.25
19.12.3 Round Belts 19.26
19.12.4 Elastomeric Round Belts 19.26
19.12.5 Cast Polyurethane Belts 19.27
19.13 POSITIVE-DRIVE BELTS 19.27
19.13.1 Miniature Sprocket Chains and Pulley Belts 19.27
19.13.2 Spur-Gear Cable-Drive Chains 19.28
19.13.3 Spur-Gear Drive Belts (Pinned Drive) 19.28
19.13.4 Metal Belts 19.28
19.13.5 Bead-Chain Belts 19.29
19.14 PIVOTED MOTOR BASES 19.29
19.15 BELT DYNAMICS 19.30
19.15.1 Transverse Vibration 19.30
19.15.2 Torsional Rigidity of Belt Drives 19.31

19.1 INTRODUCTION

Belt drives, employing flat, V, or other belt cross sections, are used in the transmission of power between shafts which may be located some distance apart. There are basically two types of belting: friction and positive drive.

Friction belting transmits power by means of friction between belt and pulley or sheave; hence, because of slippage, the velocity ratio between shafts is not constant. Positive drives have projections which mesh with toothed or grooved pulleys.

A number of belt materials are currently in use. Among these are leather, natural and synthetic rubber, cotton, canvas, nylon, and metal. Shaft centerline position may be fixed or may vary.

Belting is generally operated at speeds less than 6500 ft/min. Flat woven or synthetic endless belts are used at speeds up to 18,000 ft/min, film belts to 47,000 ft/min.

19.2 FLAT BELTS

19.2.1 Forces

In transmitting power from one shaft to another by means of a flat belt and pulleys, the belt must have an initial tension T_0. When power is being transmitted, the tension T_1 in the tight side exceeds the tension T_2 in the slack side (Fig. 19.1).

The tension in a belt transmitting power is determined by considering belt deformation under load. For leather belts the stress-strain relationship is not linear. If it is assumed that the total stretch of the belt is the same when it is transmitting power as it is when it is at rest, T_1, T_2, and T_0 are related by the following equations. For vertical (and short horizontal) belts,

$$T_1^{1/2} + T_2^{1/2} = 2T_0^{1/2} \tag{19.1}$$

and for horizontal belts,

$$T_1^{1/2} + T_2^{1/2} = 2T_0^{1/2} + \frac{w^2 E_1 C^2 A^{1/2}}{24}\left(\frac{1}{T_1^2} + \frac{1}{T_2^2} - \frac{2}{T_0^2}\right) \tag{19.2}$$

where w = weight of belt, lb/ft (Sec. 19.3)

E = elastic constant for leather, determined[1] to be in the range 860 to 900 $(\text{lb/in}^2)^{1/2}$

A = the cross-sectional area, in^2

C = center distance, in

The last term in Eq. (19.2) is introduced by the catenary effect resulting from belt sag. For a given T_1 the resulting T_2 is higher for a horizontal belt.

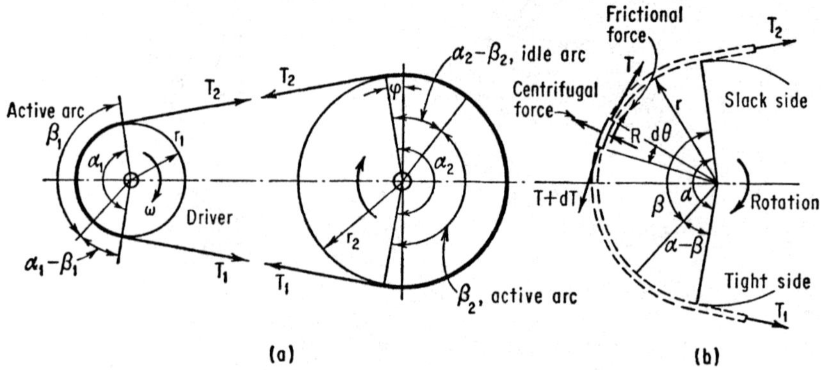

FIG. 19.1 Action of belts on pulleys.

For belts having a linear stress-strain relation the above equations become, for vertical (and short horizontal) belts,

$$T_1 + T_2 = 2T_0 \tag{19.3}$$

and for horizontal belts,

$$T_1 + T_2 = 2T_0 + \frac{w^2EC^2A}{24}\left(\frac{1}{T_1^2} + \frac{1}{T_2^2} - \frac{2}{T_0^2}\right) \tag{19.4}$$

where E is the modulus of elasticity of the belt material, lb/in².

The horsepower transmitted by the belt is given by

$$\text{hp} = (T_1 - T_2)V/33{,}000 \tag{19.5}$$

where V = belt velocity, ft/min.

19.2.2 Action of Belt on Pulley

Because the tension T_1 in the tight side of the belt is greater than that in the slack side T_2, the belt material undergoes a change in strain as it passes around the belt. The belt has a small relative motion, called "creep," with respect to the pulley to compensate for these different strains. Except when the drive delivers maximum power, this creep occurs only on the so-called active arc β of the pulley (see Fig. 19.1). The active arc increases as the effective tension ($T_1 - T_2$) increases until the active arc equals the entire arc of contact.

The forces acting on the belt are shown in Fig. 19.1*b*. Summing forces in the tangential and normal directions and substituting and integrating the tangential forces over the active arc, 0 to β, yields

$$(T_1 - wv^2/g)/(T_2 - wv^2/g) = e^{\mu\beta} \tag{19.6}$$

where wv^2/g accounts for centrifugal effects, v = belt velocity, ft/s, and μ = coefficient of friction (Table 19.1). The output of the belt is shown to decrease with increasing belt velocities as a result of centrifugal effects. Equation (19.6) indicates that, at some particular speed, the output drops to zero. Effects other than those considered allow the belt to transmit power even at these speeds and the equation therefore gives conservative results, especially in the case of heavy belts.

Flat belts in line-shaft and machine-belting service are generally operated in the 1000 to 4500 ft/min range. While operation in the 3500 to 4500 ft/min range results in the highest horsepower capacity per dollar expended for belts and pulleys, drive efficiency is less than that obtained when operating at lower speed. Total annual cost of lower-speed (1000 to 3000 ft/min) drives is in most cases lower.

To utilize the weight of the belt in maintaining tension (by increasing the arc of contact) the driving member should rotate so that the tension side of the belt is at the bottom in a horizontal drive. When the drive is vertical, steeply inclined, or horizontal with short centerline distance, no such advantage can be gained from the belt weight.

A number of methods are in use to increase contact pressure or angle of contact to enable the belt to transmit a greater effective tension. One such method utilizes an idler pulley between driver and driven pulleys. While effective tension is increased, belt life is reduced because of the reversed flexing introduced by the idler.

TABLE 24.1 Coefficients of Friction μ for Belts and Pulleys*

Belt material	Pulley material					
	Iron, steel	Wood	Paper	Wet iron	Greasy iron	Oily iron
Oak-tanned leather	0.25	0.30	0.35	0.20	0.15	0.12
Mineral-tanned leather	0.40	0.45	0.50	0.35	0.25	0.20
Canvas stitched	0.20	0.23	0.25	0.15	0.12	0.10
Balata	0.32	0.35	0.40	0.20		
Cotton woven	0.22	0.25	0.28	0.15	0.12	0.10
Camel hair	0.35	0.40	0.45	0.25	0.20	0.15
Rubber, friction	0.30	0.32	0.35	0.18		
Rubber, covered	0.32	0.35	0.38	0.15		
Rubber on fabric	0.35	0.38	0.40	0.20		

*Belt dressings or cork inserts tend to raise these values.

In a vertical drive, the belt tension at the top of the belt on each side of the pulley is greater than the tension at the bottom, by an amount equal to the weight of the freely hanging belt. It is therefore preferable to have the larger of the two pulleys at the bottom, so that the loss of tension may be offset by an increased contact arc.

19.2.3 Belt Stresses

Figure 19.2 shows the tensile stress acting in an operating belt. Note the location of the maximum stress.

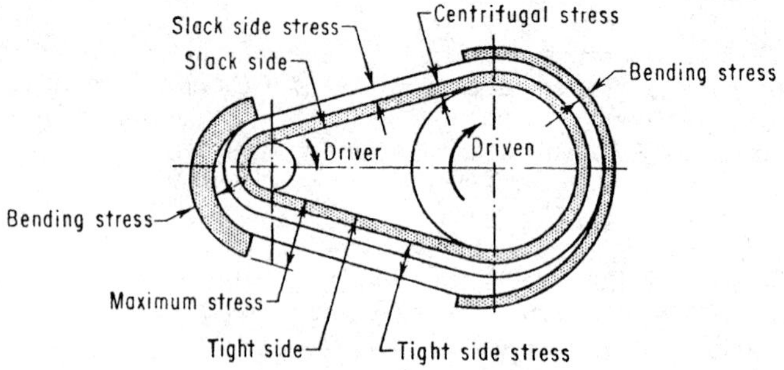

FIG. 19.2 Belt stresses.

19.2.4 Belt Geometry

Open and Crossed Belts. The length of belt L required for an open drive (Fig. 19.3) is given by the series

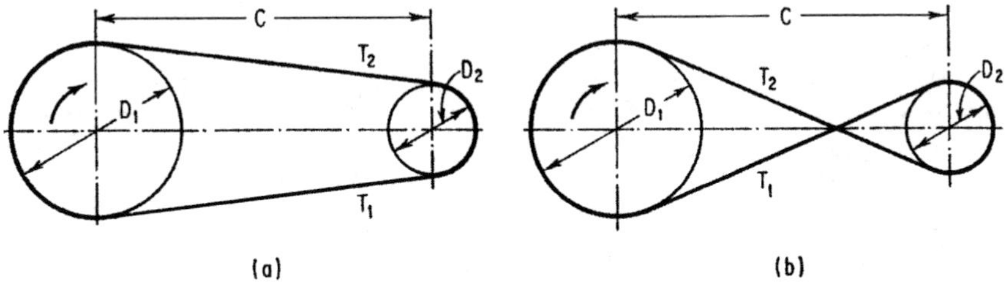

FIG. 19.3 Belt drives.

$$L = 2C + (\pi/2)(D_1 + D_2) + (D_1 - D_2)^2/4C + \cdots \tag{19.7}$$

where C is the center distance, and D_1, D_2 are the diameters of the large and small pulleys, respectively. The length of a crossed belt is given by the series

$$L = 2C + (\pi/2)(D_1 + D_2) + (D_1 + D_2)^2/4C + \cdots \tag{19.8}$$

For rough calculations, the first two terms provide sufficient accuracy. For increased accuracy, D_1 and D_2 should be taken as the pulley diameter plus one belt thickness. For large center distances, it is evident that the same belt may be used for either an open or a crossed drive. Center distance for an open belt is given by

$$4C = b + [b^2 - 2(D_1 - D_2)^2]^{1/2} \tag{19.9}$$

where $b = L - (\pi/2)(D_1 + D_2)$. The above discussion neglects the effects of belt sag.

With drives employing a short center distance and a high speed ratio, the angle of contact on the small pulley is decreased and therefore the capacity of the drive is reduced.

To provide reasonable values of effective tension in drives having angles of contact less than 165° on the small pulley requires high belt tensions. Contact angles less than 155° are never recommended.

Cone Pulleys. Cone or stepped pulleys (Fig. 19.4) may be used to obtain a variable velocity ratio. Generally, the velocity ratios are chosen so that the driven velocities increase in a geometric ratio from step to step. That is, if the angular velocity of the first pulley is a, that of the second is ka, of the third k^2a, etc. In practice, k is in the range 1.25 to 1.75. The value[4] for machine tools is about 1.2.

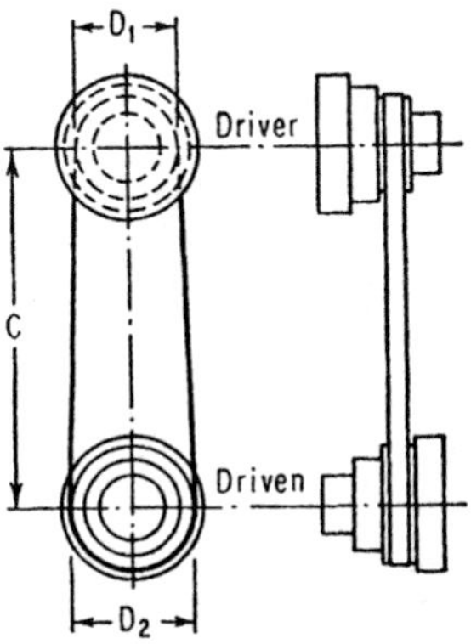

FIG. 19.4 Cone or stepped pulleys.

For crossed belts, for a constant belt length, $D_{11} + D_{21} = D_{1i} + D_{2i}$, where D_{11}, D_{21} are the diameters of the first set of pulleys and D_{1i}, D_{2i} are the diameters of any succeeding pair. In addition, for the velocity ratio n_{11}/n_{2i}, where n_{11} is the angular velocity of the driver shaft and n_{2i} is the angular velocity of the ith driven pulley, the diameters D_{1i} and D_{2i} must satisfy the ratio $D_{2i}/D_{1i} = n_{11}/n_{2i}$. These two relations must be solved simultaneously to determine D_{1i}, D_{2i}.

A more complex relation exists for open belts. If the required diameter ratio is substituted in the open-belt-length equation, the result is

$$L = 2C + (\pi/2)(1 + n_{11}/n_{2i})D_{1i} + (1 - n_{11}/n_{2i})^2(D_{1i}^2/4C) + \cdots \qquad (19.10)$$

This equation is solved for D_{1i} and D_{2i} obtained from the required ratio of diameters. Graphical methods for solution of this equation are described in books on mechanisms.

An alternative approximate method for open belts, accurate to four decimal places, follows. Let $m = 1.58114C - D_0$, where D_0 is the diameter of either pulley when $D_2/D_1 = 1$. Then $(D_1 + m)^2 + (k^nD_1 + m)^2 = 5L^2$, where k is the geometric ratio between steps. This equation is solved for D_1 after D_0, C, and k have been chosen. C is the center distance, L the belt length.

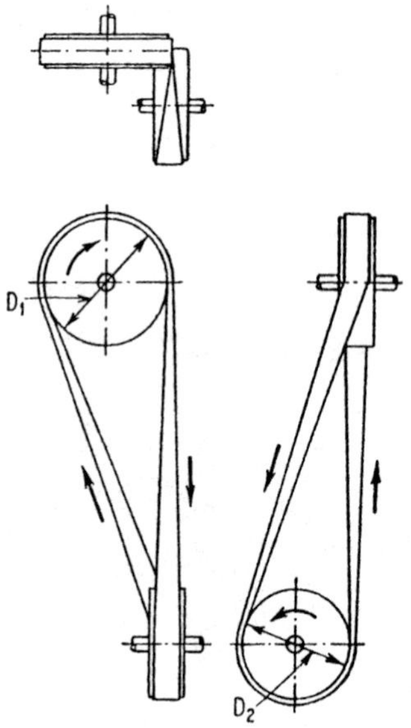

FIG. 19.5 Quarter-turn drive.

Quarter-Turn Drive. The length of belt L for a quarter-turn drive (Fig. 19.5) is given by

$$L = (\pi/2)(D_1 + D_2) + (C^2 + D_1^2)^{1/2} + (C^2 + D_2^2)^{1/2} \qquad (19.11)$$

Arc of Contact. Arc of contact for an open drive is

$$\theta = \pi \pm \sin^{-1}(D_1 - D_2)/C \qquad (19.12)$$

where the plus sign corresponds to the larger pulley and the minus sign to the smaller pulley. Arc of contact for both pulleys in a crossed drive is

$$\theta = \pi + \sin^{-1}(D_1 + D_2)/C \qquad (19.13)$$

19.2.5 Coefficient of Friction

The coefficient of friction between belt and pulley depends upon the materials involved and the condition of the surfaces. Coefficients of friction, for design purposes, for a number of belt and pulley combinations are given in Table 19.1. Barth has shown that the coefficient varies with velocity. Experimental data for leather belts on iron pulleys gave the equation

$$\mu = 0.54 - 140/(500 + V) \qquad (19.14)$$

The coefficient of friction may also be influenced by the cleanliness of the surrounding air, its temperature and humidity, the area of surface contact, the belt-to-pulley-thickness-diameter ratio, and slip. The coefficient rises rapidly after installation of a new belt; changes from 0.25 to 0.65 in a short time have been observed. The variation of the coefficient with belt tension has not been determined. For design purposes, it is ordinarily assumed that the coefficient remains constant.

19.2.6 Creep and Slip

The tension in a belt as it passes around the driver pulley increases from slack-side tension to the tight-side tension. At the driven pulley, the reverse is true. Consequently, the belt elongates as it passes around the driver and shortens at the driven. A greater length of belt leaves the driver than is delivered by the driven pulley; hence the belt must move relative to the surface of both pulleys. This motion, caused by elongation, is called "creep."

At low values of effective tension, creep occurs over only a small arc of the pulley. This arc, called the "active arc of contact," increases with increasing effective tension until the belt creeps over the entire pulley. If the effective tension is further increased, the belt begins to slide on the pulley. The latter motion is called "slip."

A creep velocity of 2 percent of belt velocity is considered reasonable. Slip usually begins at a creep velocity of about 1 to 3 percent of belt velocity. The equation for creep velocity in a leather belt is

$$v = (V_1/2)(\sigma_1^{1/2} - \sigma_2^{1/2})/(830 + \sigma_2^{1/2}) \tag{19.15}$$

where σ_1 and σ_2 are the unit stresses (lb/in^2) in the tight and slack sides, respectively, and V_1 is the tight-side velocity.

19.2.7 Angle Drives

Open or crossed-belt drives are used to greatest advantage where center distances are large. In many instances, however, space is at a premium and power must be transmitted between shafts whose center lines are not parallel. Shafts which lie at an angle to each other may lie in the same plane, in which case they would intersect if extended, or they may not lie in the same plane. The most common arrangements are parallel and right-angle or quarter-turn shafting.

Parallel shafting on close centers can be connected by means of the nonreversible drive shown in Fig. 19.6. The effect is similar to a crossed drive, the shafts rotating in opposite directions. The drive of Fig. 19.6 is nonreversible because pulley *B* does not deliver the belt in the plane of guide pulley *C*. A reversible drive can be made, however, but the guide pulleys will no longer have a common shaft.

One of the most common angular drives is the quarter turn, in which the driven pulley is mounted at right angles to the driver. As the belt leaves one pulley, it must be turned so as to run onto the other pulley correctly. When rotation is in one direction, guides are usually not required. Guide pulleys are required for reversible operation.

To overcome the belt distortion introduced in quarter-turn drives, the belt is turned through a 180° angle so that, at the joint, the grain side is adjacent to the flesh side. With this construction, wear is equalized on the faces and edges of the belt. An alternative method of combating belt distortion is to make one edge longer than the other. For heavy belts, this method is preferred.

A reversible one-pulley quarter-turn drive is shown in Fig. 19.7. A two-guide-pulley arrangement, not shown, has an advantage in that the belt leaves the pulley in a straighter condition, reducing side stretch. The possibility of the belt's rubbing on itself is eliminated, and tension adjustment is possible if one guide pulley is made movable.

If the shafts intersect in a horizontal plane, the guide pulleys may have to be carried on a vertical axis. In this case, they are called "mule pulleys" and must be heavily crowned to hold the belt. For long belts, crowning will not hold the belt on the pulley and flanges must be provided. The flanges must be made with a reentrant fillet to reduce the tendency for the belt to ride upon the flanges. A guide plate mounted on the shaft adjacent to the pulley may be substituted for the flange. The flange or plate should be regarded as a safety measure; the crowning should be sufficient to keep the belt on the pulley under ordinary circumstances.

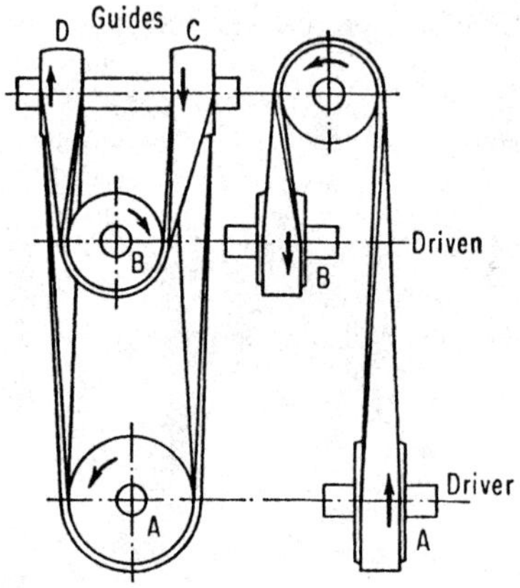

FIG. 19.6 Reversible drive between close parallel shafts.

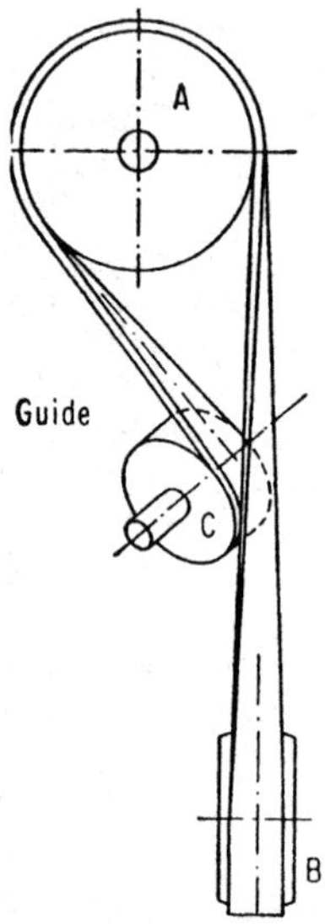

FIG. 19.7 Reversible quarter-turn drive, one guide pulley.

19.2.8 Flat-Belt Fasteners

Belts may be made endless by splicing and cementing, or the ends may be joined by means of belt fasteners. The strength of a properly made splice is equal to that of the belt, while the use of a fastener decreases belt capacity from 10 to 50 percent dependent upon the fastener used. Fasteners offer a more convenient method of joining belt ends, however, when it becomes necessary to take up stretch or repair a damaged section.

Fasteners are made in four general classes: laces, hooks, plates, and the pressure type in which the belt's ends are held together. Laces are made of rawhide or wire. Hooks are formed from wire and plates are of solid or hinged construction.

Metallic fasteners are fabricated from ordinary steel for normal applications. Stainless steel and monel are used for highly corrosive or abrasive conditions; Everdur (copper, silicon, and manganese alloy) where electrical static protection is required.

19.3 FLAT-BELT MATERIALS

19.3.1 Leather Belting

Leather belting is vegetable-tanned, chrome-tanned, or combination-tanned rawhide, or semi-rawhide. Leather belting comes in the sizes shown in Table 19.2. The weight of leather is about 0.035 lb/in^3. Oak-tanned belts are preferred in dry applications, chrome-tanned and rawhide in damp areas. Waterproof belts can be used in damp or oily applications. Leather belts can be joined by cementing. Such joints, when properly made, have a strength equal to that of the belt. They can also be joined by leather lacing or riveted joints which have a strength one-third to two-thirds of the belt

TABLE 19.2 Horsepower per Inch of Width, Oak-Tanned Leather Belting†

Belt speed, fpm		Single ply		Double ply			Triple ply	
		11/64 medium	13/64 heavy	18/64 light	20/64 medium	23/64 heavy	30/64 medium	34/64 heavy
1,000		1.8	2.1	2.6	3.1	3.6	4.1	4.5
2,000		3.5	4.1	4.9	6.0	6.9	8.1	8.9
3,000		5.2	5.9	7.2	8.7	10.0	11.6	12.8
4,000		6.4	7.4	9.0	10.9	12.6	14.5	16.0
5,000		7.4	8.4	10.3	12.5	14.3	16.5	18.2
6,000		7.8	8.9	10.9	13.2	15.2	17.6	19.3
	rpm							
Min pulley diam,* in.	0–2,500	2½	3	4	5	8	16	20
	2,500–4,500	3	3½	4½	6	9	18	22
	4,000–6,000	3½	4	5	7	10	20	24

*Belts to 8 in wide. For wider belts add 2 in to minimum pulley diameter for single- and double-ply belts. Add 4 in for triple-ply.

†From National Industrial Leather Association.

strength, and wire-laced joints which have a strength of 85 to 90 percent of the belt strength. Oak-tanned leather can be used up to 110°F, chrome or retanned leather to a somewhat higher temperature. Efficiency is about 98 percent. Belts in which the surface is ribbed or treated to increase friction are available in mineral-tanned or oak-tanned forms. They are not recommended for installations where the belt can slip momentarily. Leather belts are also available as V belts, laminated V belts, block V belts, and round, oval, solid, or plied belting. On step-cone pulleys the narrowest, thickest belt that is applicable should be chosen. Pulley faces should be 0.5 to 2 in wider than the belt.

19.3.2 Nylon-Core Belting

These belts have a nylon-core load-carrying member. Leather or rubberized friction material is used in both surfaces or on the pulley side with nitrile rubber, nylon cloth, or nylon on the other side to provide protection. Lightweight belts are available without facing. Drives range from those which transmit a few inch-ounces of torque to over 1000 hp. Belt speeds of 60,000 ft/min are possible. The high tensile strengths of the core, 40,000 lb/in^2, results in a thin belt in which the energy required to flex the belt and centrifugal forces are low.

Dehydration and moist heat will result in premature aging. Normal operating temperatures are −20 to 212°F while some plastic-covered belts can stand brief exposures to 300°F. Available friction materials have resistance to oil, gasoline, and static. Nylon core thicknesses of 0.020 to 0.217 in, which corresponds to belt thickness of 0.125 to 0.3125 in, lengths to 75 in, and widths to 36 in are available. Nylon belts can be supplied endless or with ends prepared for hot or cold cementing in the field. Some nylon belts are suitable for joining with fasteners.

Proper initial tension is attained when the belt is elongated as shown in the following table:

Condition	Elongation, %
Normal	1.5–2
Shock loads, heavy starting	2–3
High pulley ratios, dusty pulleys, cone pulleys, variable humidity and temperature	3–4

When the center distance is adjustable, initial tension should be just great enough to transmit the peak load without slipping. Belt pitch length is the measured steel-tape length plus 1 in for 0.020-in belts and 1.75 in for 0.217-in belts.

The thickest suitable belt should be chosen, except where flexibility is important, as in very short drives where the pulleys are almost touching or in multiple-pulley drives. For additional design information, see Refs. 8 to 10 and 14.

19.3.3 Rubber Flat Belts

Such belts are made up of plies of fabric or cord load-carrying members impregnated or laminated with protective or friction-improving materials. Load-carrying materials include prestretched cotton, rayon, nylon, polyester, or steel cables. Because of their distinctive properties, nylon-core belts are discussed in Sec. 19.3.2. Belts in which the load-carrying member consists of plies of rubberized fabric permit the use of metal belt fasteners. When this member consists of cords it must be supplied in endless form or an oil-field-type clamp used. Some belts are spliced in place in the field.

Laminating materials include neoprene, polyvinyl chloride (PVC), rubber, chrome leather, hypalon, and polyurethane. Belts without laminating materials are used where high strength and flexibility are of primary importance. Neoprene or hypalon are used in the place of rubber where oil or grease is present. Neoprene belts can operate at 250°F. Static-resistant belts and dust-jacketed belts, which reduce wear resistance on the edges and provide chemical resistance, are available. A disadvantage of rubberized belts is the decrease in friction that occurs when the rubber lamination wears away.

Initial tension should be in the range of 15- to 25-lb/ply per inch of width, or such that the belt elongates 1 percent for normal conditions and 1.5 percent for severe load. These percentages should be reduced 50 percent under humid conditions. Ultimate tensile strength ranges from 280 to 600 lb/ply or more per inch of width. A take-up of 2 to 4 percent of belt length should be provided for stretch in service or for manufacturing variations. Depending on construction, widths to 60 in are available. Belt speeds in the range 2500 to 3000 ft/min are considered low speed; 3000 to 4000 ft/min, medium speed; and 4000 to 5500 ft/min, high speed. Such belts are limited to 6000 ft/min. Maximum power is about 500 hp. Belts with cord load-carrying members are well suited to serpentine drives.

19.3.4 Cotton and Canvas Belting

Fabric belts of this type are made up of layers of woven material held together by gums, rubbers, or synthetic resins which also serve to protect the fabric. Stitched canvas belts are inexpensive belts which need little protection against oil or moisture. Friction coefficient is low (0.15 to 0.22 against a steel pulley); hence pulley bearing pressure will be high. Resistance to high temperature is good. Bituminous-coated belts resist moisture and are recommended for rough usage. Combination leather and cotton belts consist of a single or double ply of oak or chrome leather cemented to a woven cotton backing.

Initial tension should be 20 to 25 lb/ply per inch of width.

Because of the lower coefficient of friction, the largest possible pulleys which do not produce excessive belt speeds should be used to reduce bearing loads. Cotton and canvas belts are limited to belt velocities of 5000 to 6000 ft/min.

19.3.5 Balata Belting

Balata belts are made from closely woven duck of high tensile strength impregnated with balata gum obtained from South American trees. Balata is tough, stretches very little, is waterproof, resistant to aging, and affected by mineral oil but not by animal oils or humidity. Maximum allowable ambient temperature is 100 to 120°F. Initial tension is ordinarily in the range 22 to 25 lb/ply/per inch of width. Balata belting can be laced by metal fasteners or made endless. Its horsepower capacity is similar to that of canvas stitched or rubber belts. The coefficient of friction remains high even under damp conditions.

The specific weight of balata and other belts in lb/in^3 is as follows: leather ranges from 0.035 to 0.045, while canvas is 0.044, rubber 0.041, balata 0.040, single-woven cotton belt, 0.042, and double-woven cotton belt 0.045.

19.3.6 Steel Belting

Thin steel belts are generally used in installations where belt speed is high. Initial tension is generally high because of the lower coefficient of friction. Stretch is low because of the high elastic modulus of steel. Steel belts provide very high power transmission for a given belt cross section. Belt material is similar to that used for clock springs. Carbon content is high; the material is drawn and rolled, and ground to size. It is available in widths from ½ to 3 in in 0.01-in thickness. Tensile strength is in excess of 300,000 lb/in^2. Belt material has rounded edges to reduce the hazard while running.

Joints may be riveted or silver-soldered. Rivets can be phosphor bronze or cold-drawn iron wire. Joints are about 80 percent efficient, failure occurring by shear. Silver-soldered butt joints, with 60° beveled edges, which can be used with small pulleys, are about 65 percent efficient, rupture occurring in the adjoining belting, which is softened by the joining process. Cover-plate joints are bent to conform to the smallest pulley.

Hampton,[3] using 0.01- by 0.75-in steel belting on cork pulleys, found that horsepower transmitted increased in proportion to slip velocity until the tension in the slack side approached that due to centrifugal effects, at which point slip velocity increased rapidly without a corresponding increase in horsepower transmitted. This breakaway point occurs at higher horsepowers for higher belt speeds because effective belt pull for a given horsepower is lower at higher speeds. The power in the range before slip occurred is given by

$$\text{hp} = 0.17T_1^{0.65}V_s \tag{19.16}$$

where V_s = slip velocity and T_1 = tight-side tension.

Friction coefficient f rises with slip velocity, but drops with increasing belt speed. For the belts of the previous reference,

$$f = V_s^{1.37} \times 3.2 \times 10^5/(p^{0.5} \times V_d^{1.5}) \tag{19.17}$$

where p = unit pressure on the driver pulley face, lb/in^2, and V_d = belt velocity, ft/min. Drive efficiency is over 98 percent. Slip velocities to approximately 3 ft/min are considered reasonable. Young[4] showed that the horsepower which can be transmitted by a steel belt running on wood pulleys is

$$\text{hp} = 1.51p^{0.81}/1000V_d^{0.78} \tag{19.18}$$

19.3.7 V-Ribbed Belting (Grooved, Poly-V)

These belts are flat belts with a ribbed underside. The ribs, which run in grooves in the sheaves, provide some wedging action. V-ribbed belts require less tension than flat belts, but more than V belts. They are usually more efficient than a flat belt and sometimes a V belt. The thin belt construction, which minimizes flexing, permits the use of small pulleys. They may be used on vertical and turned drives and, with back-bend idlers, on serpentine drives. They are not used in multibelt drives. If used in cross drives, a steel plate, located at the point of belt intersection, must be provided to stop the belt from rubbing on itself. Standard designations for these belts are J, L, and M. Designation H is used on belts which can run on small pulleys and K for automotive applications. Because the ribs bottom in the groove, the drive ratio does not change as the belt wears, as is the case with V belts.

19.4 FLAT-BELT WORKING STRESSES

19.4.1 Design Stresses and Loads

Allowable tension for flat-belt materials is customarily expressed in terms of lb/in width or lb/ply/in width rather than in terms of allowable stress, lb/in^2. Ultimate strength for oak-tanned leather belts is in the range 3000 to 4500 lb/in^3; for chrome-tanned leather, 4000 to 5500 lb/in^2; for rubber, balata, cotton, or canvas belts, 900 to 1500 lb/in^2; and for nylon-core belts, 35,000 lb/in^2.

As belt thickness and number of plies increase, recommended maximum allowable tension, lb/in width, increases. For leather belting particularly, however, recommended maximum allowable stress, lb/in^2, and maximum allowable tension per ply, lb/ply/in width, decrease as a result of increased bending stresses at the pulleys.

Recommended maximum values of allowable tension (lb/in width) for oak-tanned leather belts range from 108 for single-ply medium to 275 for triple-ply heavy; for rubber, canvas, and balata, 32-oz duck, they range from 75 for 3-ply to 300 for 12-ply. Recommended maximum stress for oak-tanned leather belts ranges from 620 lb/in^2 for single-ply medium to 520 lb/in^2 for triple-ply heavy. For rubber, balata, or canvas belts the corresponding value is 500 lb/in^2.

19.4.2 Horsepower Ratings of Flat Belts

The Goodyear Company has developed the following empirical relation relating service life I to d, the small pulley diameter; l, the belt length; S, the belt speed; N, the belt thickness; and T, the tight-side tension (Ref. 2):

$$I = kd^{5.35}l/S^{0.5}N^{6.27}T^{4.12} \tag{19.19}$$

where k is a constant dependent upon the particular belt characteristics. From Eq. (19.19) it is seen that a reduction in small pulley diameter, an increase in belt thickness, or an increase in tight-side tension have the greatest effect in decreasing belt life. A 50 percent decrease in pulley diameter will decrease drive life to $\frac{1}{32}$ its original value; a 10 percent increase in tension will decrease belt life 60 percent.

The difference in horsepower recommendations among several manufacturers for a given type of belt may result from a difference in choice of service life. In addition some manufacturers base their recommended horsepower for smaller belts on a shorter service life because ordinarily the user does not expect so long a life from such drives.

19.5 FLAT-BELT DESIGN

Design techniques for all types of flat belts are basically similar; however, variations are encountered as outlined in this section. Current design is generally based on empirical data concerning the load-carrying capacity of a given belt. These recommended horsepowers are chosen to provide reasonable life under most circumstances. As may be deduced from Eq. (19.19), belt horsepower capacity is a function of desired service life, pulley diameters, belt length, belt speed, belt thickness, and tight-side tension. In addition, such factors as service conditions and contact angle must be included.

Existing practice is such that the effect of some factors such as belt length is ignored. In the case of leather belts, that of contact angle is neglected, while for nylon belts, no account is taken of service conditions. These omissions are permissible because of their comparatively minor effects or because recommended horsepower values are somewhat conservative.

In the design of a flat-belt drive, the diameters of the driver and driven pulleys, center distance, horsepower, and belt speed are first determined. Then, in accordance with the recommendations of Table 19.2 for leather belting; Table 19.3 for rubber, balata, woven cotton, or canvas stitched belting; and Table 19.4 for nylon belting, a belt of proper thickness is chosen. From these tables, the rated horsepower per inch of width (hp_R) is also obtained.

To account for variations from the normal conditions assumed in Tables 19.2, 19.3, and 19.4 for (hp_R), these values must be modified by service factors to obtain the belt design horsepower per inch of width (hp_D). Recommendations regarding appropriate service factors vary for different types of belting as listed below.

1. Leather:

   F, special operating conditions (Table 19.5)

   M, motor and starting method (Table 19.5)

   P, small pulley diameter (Table 19.5)

   Accordingly, for leather belts,

$$hp_D = hp_R(P/FM) \tag{19.20}$$

2. Rubber, cotton, canvas, balata:

   K_1, arc of contact (Table 19.6)

   F, special operating conditions (Table 19.5)

   M, motor and starting method (Table 19.5)

TABLE 19.3 Horsepower per Inch of Width: Rubber, Balata, Cotton Duck, and Canvas Stitched Belts

35-oz silver hard duck, ply	Small pulley diam, in.	Belt speed, fpm 1,000	2,000	4,000	6,000
4	4	1.07	1.87	2.75	2.31
	8	1.98	3.52	5.72	6.49
	12	2.20	4.18	7.04	8.47
	14+	2.20	4.29	7.59	9.24
6	8	1.87	3.41	5.17	4.84
	12	2.86	5.06	8.36	9.46
	16	3.30	6.05	10.01	11.77
	28+	3.30	6.49	11.33	13.86
8	14	2.86	5.17	8.14	8.36
	20	4.07	7.37	12.21	13.97
	26	4.40	8.58	14.52	17.49
	36+	4.40	8.58	15.18	18.37
10	24	4.18	7.48	11.88	12.76
	30	5.39	9.68	17.16	18.81
	36	5.50	10.67	17.82	21.45
	48+	5.50	10.78	18.92	22.99
32-oz duck:					
4	4	0.9	1.2	2.2	1.8
	8	1.5	2.8	4.5	5.1
	12	1.8	3.5	5.8	7.0
	18+	1.8	3.6	6.3	7.7
6	8	1.4	2.5	3.9	3.5
	12	2.3	4.1	6.6	7.5
	16	2.7	4.9	8.1	9.6
	26+	2.7	5.3	9.4	11.5
8	16	2.5	4.4	7.0	7.4
	22	3.5	6.3	10.3	12.1
	28	3.6	6.9	11.5	13.9
	36+	3.6	7.1	12.5	15.3
10	28	3.7	6.7	10.9	12.2
	34	4.5	8.6	14.4	17.3
	40	4.5	8.9	15.2	18.4
	44+	4.5	8.9	15.7	19.2

Accordingly, for such belts,

$$\text{hp}_D = \text{hp}_R(K_1/FM) \qquad (19.21)$$

3. Nylon core:

K_1, arc of contact (Table 19.6)

Accordingly, for nylon-core belts,

$$\text{hp}_D = \text{hp}_R(K_1) \qquad (19.22)$$

When the allowable design horsepower per inch of width (hp_D) is established, the necessary belt width t is obtained from

$$t = \text{hp}_T/\text{hp}_D \qquad (19.23)$$

where hp_T is the horsepower being transmitted by the belt. For more information on flat-belt design, see Refs. 5–7.

TABLE 19.4 Horsepower Capacity per Inch of Width, Nylon-Core Belts

Nylon-core thickness, in., approx	Small pulley diam, in.	Belt speed, fpm									
		1,000	2,000	3,000	4,000	5,000	6,000	7,000	8,000	9,000	10,000
0.020	½	0.15	0.20	0.30	0.40	0.50					
	1	0.20	0.30	0.50	0.62	0.81					
	2	0.30	0.55	0.85	1.12	1.45					
0.030	1	0.55	1.0	1.62	2.02	2.62	3.25	3.37	3.5	3.62	3.7
	4	1.12	2.25	3.12	4.0	5.25	6.0	6.5	6.7	6.8	6.9
	8	1.3	2.37	3.62	4.7	5.7	6.7	7.5	8.0	8.1	8.2
0.060	2	1.5	2.75	4.37	5.02	6.7	8.0	9.2	10.0	10.5	10.6
	4	1.87	4.6	6.0	6.5	10.0	11.7	13.2	13.7	15.2	15.4
	12	2.75	5.0	8.0	10.0	12.5	15.0	17.5	18.7	20.0	20.2
0.090	4	3.75	5.62	9.0	11.7	13.7	17.0	18.7	21.2	22.5	23.0
	8	4.12	6.87	10.5	13.7	17.5	20.0	22.5	25.0	27.6	28.0
	20	4.5	8.12	13.0	16.0	21.2	25.0	27.5	31.2	33.7	35.0
0.120	8	5.5	9.5	14.5	18.7	22.7	27.7	32.5	35.5	40.0	42.0
	15	6.35	11.87	18.7	23.8	29.0	35.0	37.0	40.0	44.0	46.0
	40	7.0	13.0	20.0	25.5	32.5	37.5	42.5	45.0	48.0	50.0

Note: Published data vary. Other recommendations range from 75 to 100 percent of these values.

TABLE 19.5 Correction Factors for Belt Drives

Factor	Condition	Factor *F*
F, special operating conditions	Oily, wet, dusty	1.35
	Vertical drive	1.2
	Jerky loads	1.2
	Shock and reversing	1.4
	Method	**Factor *M***
M, motor and starting method	Squirrel cage, compensator start	1.5
	Squirrel cage, line start	2.0
	Slip ring and high start torque	2.5
	Diam, in.	**Factor *P***
P, small pulley diameter	0–4	0.5
	4.5–8	0.6
	9–12	0.7
	13–16	0.8
	17–30	0.9
	30+	1.0

TABLE 19.6 Arc Correction Factor K_1

Arc of contact, deg*	*Factor*
180	1.00
174	0.99
168	0.97
162	0.96
156	0.94
150	0.92
144	0.90
138	0.88
132	0.87
126	0.85
120	0.83
114	0.80
108	0.78
102	0.75
96	0.72
90	0.69

*Small sheave.

19.6 FLAT-BELT PULLEYS AND IDLERS

19.6.1 Pulleys

As the belt passes around the pulley, it flexes, and bending stresses are induced. As a result of the reduced radius of curvature, bending stresses are higher for small pulleys. Because of the reduced fatigue stresses, a large pulley results in an increased belt life. Pulleys should not be so large, however, as to cause excessive belt speeds.

Table 19.2 lists minimum pulley sizes for leather belting. If pulley diameters less than those shown are used, belt life will be greatly reduced as a result of the higher bending stresses and greater elongation of the outer belt fiber.

In order to aid in keeping the belt on the pulley face, it is recommended that the face be crowned as shown in Fig. 19.8, which also shows approximate pulley proportions. An alternate form of crowning consists of making the central half of the pulley face cylindrical and incorporating a slight straight or curved taper in the face at either side of the cylindrical portion.

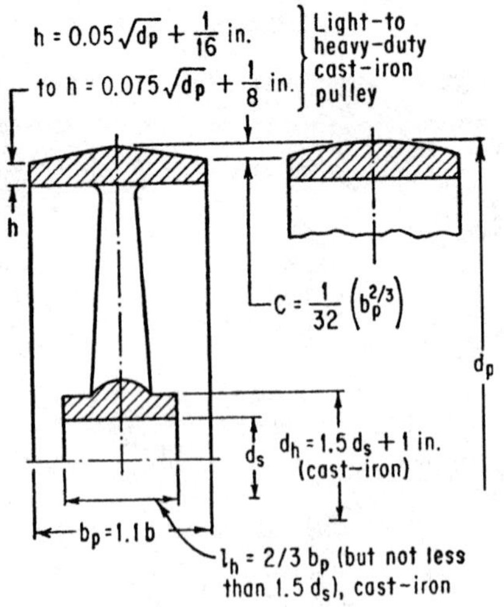

FIG. 19.8 Crowned pulley proportions.

19.6.2 Idlers

Idlers are always flat-faced and located on the slack side of the belt next to the small pulley, whether the small pulley is driving or driven.

Only where the drive is steady and the small pulley is driven may the idler be fixed.

Spring-loaded idlers used as gravity idlers have the disadvantage of relieving the spring tension on the belt when the slack increases upon application of load.

The clearance between the small pulley and the idler should not exceed 2 in, and the idler should be positioned so that the belt wrap around the small pulley is 225 to 245°.

Idlers should be well balanced.

19.7 V BELTS

19.7.1 Introduction

The cross section of a V belt (Fig. 19.9) is such that belt tension wedges the belt into the sheave groove, resulting in increased frictional forces. When compared with a flat-belt drive, a V-belt drive can operate at a reduced angle of contact, with lower initial tension, is usually more compact, and has less severe pulley alignment requirements. The load-carrying member consists of one or more layers of cord composed of nylon, rayon, cotton, glass, or steel. The cords are often encased in a soft rubber matrix, the cushion section. This section is encased in harder synthetic or natural rubber. The entire belt is enclosed in an abrasion-resistant rubberized canvas cover.

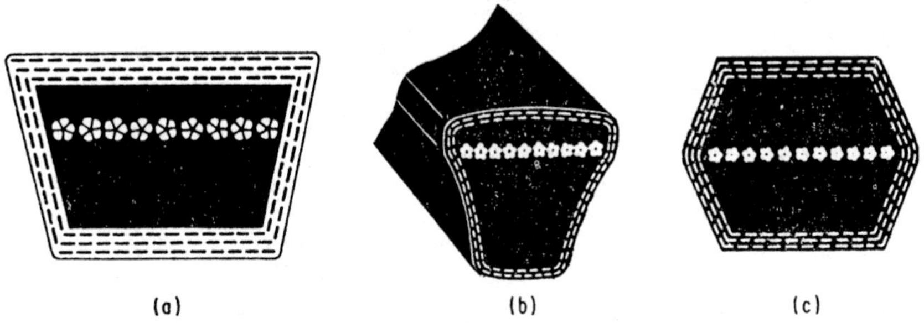

FIG. 19.9 V-belt construction.

V belts may be used for small center-distance drives without idlers, for speed ratios as high as 10:1, and for belt speeds to 15,000 ft/min. Belts operate most efficiently at belt speeds of 1500 to 6000 ft/min, and at temperatures in the range −30 to 140°F. Belts are available for operation at −60 and 185°F. Because they are subject to creep, they should not be used on synchronous drives.

If the belt squeals in service, belt tension should be increased. Belt dressing, oil, or gasoline should not be applied. Soap and water should be used for cleaning. Grooved idlers should be used on the inside surface of the belt. Flat idlers, whose diameter is 1.33 times that of the smallest pulley, should be used on the back surface. A tight-to-slack-side tension ratio of 5:1 is used on 180° arc of contact drives. V belts are made

TABLE 19.7 Properties of V Belts

Belt section	Width a, in	Thickness b, in
Heavy-duty standard and super		
A	0.50	0.31
B	0.66	0.41
C	0.88	0.53
D	1.25	0.75
E	1.50	0.91
Heavy-duty narrow		
3V	0.38	0.32
5V	0.62	0.54
8V	1.0	0.88
Light-duty		
2L	0.25	0.16
3L	0.38	0.22
4L	0.50	0.31
5L	0.66	0.38

Note: The included angle of the sheave is less than that of the belt, so that wedging will increase the effective friction. The difference should not be too great or the force required to pull the belt free will be excessive.

in the standard letter sizes of Table 19.7 and numerous other sizes. For information regarding V belt drives, see Refs. 11 to 13.

19.7.2 Forces

Forces normal to the sides of the grooves act on a V belt as shown in Fig. 19.10. Referring to Fig. 19.10, the normal force on the groove face

$$P_n = P/2 \sin \beta \tag{19.24}$$

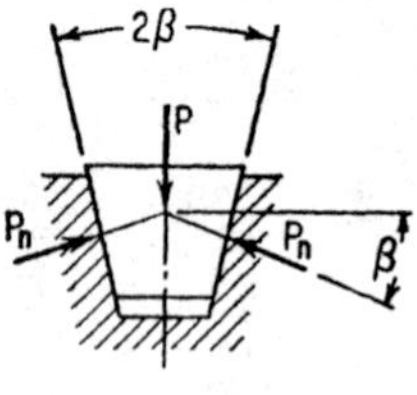

FIG. 24.10 V-belt forces.

The tractive force

$$F = 2\mu P_n = 2\mu P/2 \sin \beta = \mu_e P \tag{19.25}$$

where P = the sum of the tight- and slack-side tensions = $T_1 + T_2$
μ = coefficient of friction
μ_c = effective coefficient of friction = $\mu/\sin \beta$

The relation between tight and slack tensions [Eq. (19.6)] is applicable to V belts when μ_e replaces μ in that equation.

19.7.3 Geometry

Center Distance. Drive center distance is generally larger than the larger sheave diameter, but less than the sum of the larger and smaller sheave diameters. Because the life of the V belt is substantially shortened by slack-side shock and vibration, long center distances, in excess of 2½ to 3 times the large sheave diameter, are usually not recommended for V belts.

Provision should be made to permit a small increase or decrease in the center distance to accommodate the nearest standard size of belt.

Larger center distances may be used with link-type V belts which, because of better balance, are subject to less vibration.

The drive should make provision for the adjustment of center distance above and below the nominal value for belt take-up after stretching and installation, respectively. Take-up allowance should be at least 2½ percent of the belt length for belts up to 200 in long, slightly less for longer belts. The application allowance should be at least equal to the belt thickness, and up to three times this long for heavier sections and longer belts.

In multiple-belt drives, all belts should elongate equally in order to assure an equal distribution of load. When one belt breaks, all the belts should be replaced in order that all the belts will have nearly equal properties.

The adjustable sheave in a multiple-belt drive should permit an increase in center distance of at least 5 percent above the nominal value and should also permit a decrease in this distance by an amount at least equal to the belt thickness to permit installation.

19.7.4 Sheaves

Sheaves are generally made of wood, stamped sheet steel, or cast iron. For higher speeds, cast-steel sheaves are used.

A minimum clearance of ⅛ in is required between the inner surface or base of the belt and the bottom of the sheave groove. This clearance prevents the belt from bottoming as it becomes narrower from wear.

It is important that sheaves be statically balanced for applications up to 6500 ft/min and dynamically balanced for higher speeds.

Sheaves are available which permit the groove width to be adjusted, thus varying the effective pitch diameter of the sheave and permitting moderate changes in the speed ratio.

The diameter of the small sheave should not be less than the values given in Table 19.7.

Sheaves should be as large as possible; however, if possible, the belt speed should not exceed 6500 ft/min.

19.7.5 V-Flat Drives

Where the speed ratio is 3:1 or more and the center distance is short (equal to or slightly less than the diameter of the large sheave), it is possible to develop the required tractive force and maintain efficiency while omitting the grooving from the face of the large sheave. In this case, the belt contact is at its inner surface and the cost of cutting the grooves of the large sheave is eliminated. The maximum tractive power of the belts is obtained when the contact angle on the larger sheave is in the range 240 to 250° and that on the smaller sheave is 110 to 120°.

When power must be transmitted to grooved sheaves from both the top and bottom of the belt, double V belts, made to fit standard V grooves, are used (see Fig. 19.9*c*).

19.7.6 Quarter-Turn V-Belt Drives

Quarter-turn drives may be used with V belts. The horsepower rating of a V belt used in this arrangement is 75 percent of its rating in a straight drive under the same load conditions.

If the sheaves are relatively small, and take-up can be provided on one of the shafts, an idler may be omitted. The minimum recommended center distance in this case is $6(D + b)$, where D is the large sheave diameter and b is the belt width.

An idler is necessary to keep the belt on the sheaves when a larger speed ratio is used. The center distance between the idler and the small sheave should not be less than $8b$.

19.8 V-BELT TYPES

19.8.1 Heavy Duty Narrow (Industrial Narrow)

This line of three belts, 3V, 5V, and 8V, whose dimensions are given in Table 19.7, serves the same application area as classical V belts. While their constructions are generally similar, the narrower wedge shape of the narrow belt results in a greater power-transmitting capability. A narrow belt drive can transmit up to three times the horsepower of a classical belt of the same size. They have high resistance to oil, temperature, and environmental conditions, and can run at speeds to 6500 ft/min without dynamically balanced pulleys.

19.8.2 Heavy Duty Standard (Classical, Conventional)

This line of five standard cross sections A, B, C, D, and E, whose dimensions are given in Table 19.7, is normally used in multiple belt installations. They are made with one or more layers of load-carrying cords encased in gum rubber. Single-layer cord belts are used in high-speed, short-center-distance drives with small sheaves. Temperature range is from −30 to 145°F. Standard belts are ordinarily run at speeds of less than 6000 ft/min.

19.8.3 Heavy Duty Super (Heavy Duty Classical)

This line of five belts has a construction similar to conventional belts and the same code designation, as given in Table 19.7. They provide about 30 percent more power transmission capability than a conventional belt of the same size. Some are notched to permit running over small pulleys.

19.8.4 Light Duty (Standard Light, Single-V)

The dimensions of this line of four belts, 2L, 3L, 4L, and 5L, whose construction is similar to and slightly thinner than conventional belts, are given in Table 19.7. They are intended for fractional horsepower applications where one belt is required. Because they have only one load-carrying layer encased in rubber and are usually enclosed in a fabric layer, they can flex around small pulleys. Service is usually intermittent and overall operating life shorter than in industrial applications.

19.8.5 Agricultural Belts

Agricultural V belts are available in nine cross sections. Five sections, HA, HB, HC, HD, and HE, have the same cross sections as classical V belts. Four sections, HAA, HBB, HCC, and HDD, are available in a double-V or hexagonal cross section. To meet agricultural application requirements, construction differs from that of the classical belt. These requirements include intermittent shock loads and the ability to bend in reverse over small-diameter sheaves. As a result, agricultural belts incorporate an undercord which acts to prevent the belt from wedging into the sheave when subjected to shock loads, and a flexible jacket. Many agricultural drives have more than one driven shaft and, in addition, have idlers. Double-V belts are intended for such drives. Agricultural belts are available in the joined configuration for multiple-belt drives. These belts operate in classical V-belt drive sheaves.

19.8.6 Automotive Belts

These belts, intended for automotive accessory applications, where space is limited, have a narrow cross section. They must transmit high horsepower over small-diameter pulleys. They are available in six sizes, having top widths of 0.38 to 1 in. The two smallest sizes, SAE sizes 0.380 and 0.500, where the designation equals the top width in inches, are most widely used. Because several accessories are often driven from a single belt, design problems are complex and similar to those for agricultural belts.

19.8.7 Open-End V Belts

These belts, which have cross sections similar to those of conventional belts, have prepunched holes. They are cut to length and the open ends connected by a fastener. The load-carrying members consists of layers of fabric which hold the fasteners securely. They are limited to low-power applications and to speeds under 4000 ft/min.

19.8.8 Link V Belts

These belts consist of a series of links, having a V cross section, which are fastened together with metallic fasteners. They may be joined together in the field. Such belts run without vibration. They tend to elongate until all links are run in and seated. The higher-power types are designated by a T symbol (TD and TE, for example).

19.8.9 Joined V Belts

These belts consist of two or more heavy duty conventional or narrow V belts bonded together at their upper surfaces. They tend to reduce lateral vibrations and the possibility of the belt turning over.

19.8.10 Double-Angle V Belts

These belts are essentially two heavy duty conventional belts molded back to back with the load-carrying layer at the junction. They may be used on serpentine drives containing reverse bends or where power is to be transmitted from the top and bottom of the belt. The five belts in this series are designated AA, BB, CC, DD, and EE. The symbol DG, used together with the belt symbol, indicates that the belt runs in a deep-groove pulley.

19.8.11 Low-Stretch Cable (Steel Cable) Belts

These belts are similar in construction to standard belts. They have high-strength steel or glass-fiber cords to provide for high power-transmission capability. They are notched for greater flexibility and can run efficiently at speeds to 10,000 ft/min. Their construction provides little shock-absorbing capability and is intolerant of misalignment.

19.8.12 Wide Angle V Belts

These belts have a large V angle that results in reduced wedging action and more dependence on the capacity of the load-carrying members. The high friction coefficient of the polyurethane encasing material compensates for the reduced wedging action. Applications include fractional horsepower office equipment, automotive, and light industrial equipment. They run over pulleys as small as 0.67 inch in diameter and at belt speeds over 10,000 ft/min. There are four belts in this metric series: 3M, 5M, 7M, and 11M.

19.8.13 Flexible Sidewall V Belts

The load-carrying member of these belts is a prestretched aircraft cable around which is molded a polyurethane, nylon, or other plastic V-shaped member. The cross section of the V-shaped member is such that the sidewalls are free to flex to permit the belt to conform to sheaves of different included groove angles. The belt consists of a series of individual V-blocks spaced along the load-carrying member. The individual blocks are free to move relative to each other to permit flexing around smaller pulleys than conventional V belts. The belts are available in four sizes, 2V, 3V, 4V, and 5V, having the same cross section as the corresponding L-series light-duty belt. Published data indicate the belts may be run at speeds to 6500 ft/min and transmit up to about 15 hp.[19]

19.9 V-BELT DESIGN

Current design techniques for V belts are based on empirical data concerning the load-carrying capacity of a given belt. Drives designed in accordance with these recommendations will have reasonable life under most circumstances. These techniques are described in detail in Refs. 11, 20, and in many V belt manufacturers' catalogs. A summary is given in the next paragraph.

In the design of a drive, the pitch diameters of the driving and driven pulley, center distance, given horsepower, belt speed, and speed ratio are first determined. If the driver is an electric motor, the diameter of the smallest pulley must exceed those recommended by manufacturers of V belts and motors. The proper belt section, drive service factor, and degree-of-maintenance service factor are determined from Ref. 20 or manufacturers' catalogs. The horsepower for which the drive is designed is called the transmitted horsepower hp_T. It is determined by multiplying the given horsepower by the appropriate service factors.

The belt length is determined from Eqs. (19.8) or (19.9), where D and d are the resulting pitch diameters. The nearest standard belt length is chosen and the resulting center distance calculated. The arc of contact for the smaller sheave is calculated from Eq. (19.12) and the arc of contact factor K_1 determined from Table 19.6. A length correction factor K_2, the additional horsepower for the design speed ratio, and the nominal horsepower per belt hp_B for a drive having an arc of contact of 180° are determined from Ref. 20 or manufacturers' catalogs. The additional horsepower for the design speed ratio is added to the nominal horsepower belt to obtain the rated horsepower per belt hp_R. This sum is multiplied by K_1 and K_2 to obtain the design horsepower per belt hp_D.

Accordingly, design horsepower for V belts is

$$hp_D = hp_R(K_1K_2) \tag{19.26}$$

where the rated horsepower per belt is based on the rating diameter.

The number of belts required, n, is determined from

$$n = hp_T/hp_D \tag{19.27}$$

where hp_T is the horsepower transmitted by the drive.

19.10 SYNCHRONOUS BELTS (TIMING, POSITIVE)

In these belts, one surface is formed into teeth which mesh with similar teeth in the mating sheave. They do not depend on friction to transmit power; speed is uniform, there being no chordal rise and fall of pitch line. The synchronous belt, shown in Fig. 19.11, has load-carrying members which can be carbon steel, rayon, glass, or stainless steel. Encasing material is usually neoprene. The teeth, molded integrally with the encasing material, can be a moderately hard rubber or neoprene which may be nylon-jacketed. Power may also be transmitted by frictional means from the back of the belt; however, such transmission will not be synchronous. Stock belts can transmit up to 275 hp, special belts to 800 hp. Speeds to 16,000 ft/min are possible, although the larger pitch belts, 0.875- and 1.25-in pitch, should not be run over 8000 ft/min. Stock neoprene belts can be run at temperatures from −30 to over 185°F. Special encasing materials permit operation from −65 to 260°F, and may be static-conductive or resistant to oil and chemicals.

There are two series of synchronous belts, the conventional trapezoidal tooth series and the newer high-torque drive (HTD) curvilinear tooth series, which was developed

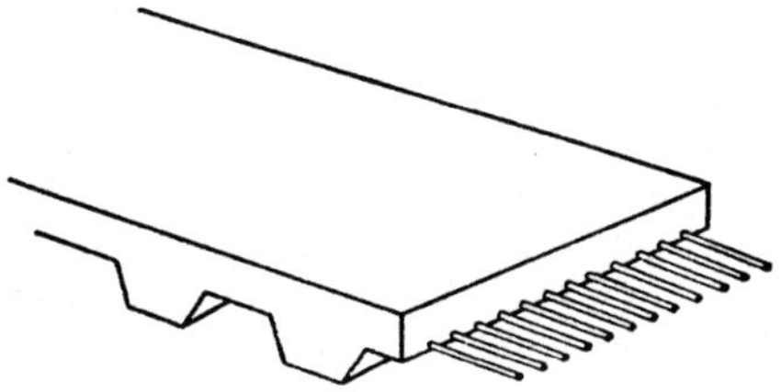

FIG. 19.11 Synchronous belt construction.

TABLE 19.8 Synchronous-Belt Sizes

Conventional belts		HTD belts
Designation	Pitch, in	Pitch, mm
MXL	0.080	3
XL	$\frac{1}{5}$	5
L	$\frac{3}{8}$	8
H	$\frac{1}{2}$	14
XH	$\frac{7}{8}$	
XXH	$1\text{-}\frac{1}{4}$	

for applications where chains can be used. Conventional belts are available in six pitches; HTD belts, in four pitches, as shown in Table 19.8.

Synchronous belts do not require an initial tension. The two largest conventional pitch belts can be installed slightly slack if shock loads are high. Belts must be taut enough to avoid contact between tight and slack side teeth. Belt width should be approximately equal to the diameter of the smaller pulley. If the belt is too wide, the inherent side thrust of these belts may be excessive. For short drives, with center distance less than 8 times the small pulley diameter, and with both shafts horizontal, one pulley should be flanged. For all drives in which the shafts are vertical and for long horizontal-shaft drives, both pulleys should be flanged. When the drive has more than two pulleys, every other pulley should be flanged. Idlers may contact either face to adjust belt length, for power takeoff, to increase arc of contact, and to prevent belt whip. In the last case, the idler is placed on the slack side on either face. If the idler contacting the inside face of the belt is at least 6 in in diameter, it may be a noncrowned flat pulley. All flat belt idlers should be noncrowned.

To reduce slippage on a high-ratio flat-belt application, a synchronous belt running over a synchronous-belt small pulley and an uncrowned flat-belt large pulley may be used. Synchronous belts are not recommended for quarter-turn, mule, and crossed drives. If they are used, center distance should be large and the belts themselves as narrow as possible. Quarter-turn drives above 2.5:1 ratio should be replaced by a two-step drive consisting of a 1:1 quarter-turn drive, followed by a conventional two-pulley reducing drive. In high-speed drives, the smallest pitch belt that will transmit the power should be used to keep centrifugal forces low.

Conventional belts can be ordered in other pitches, widths, and lengths. Widths to 14 in are available. Belts over 180 in long are available in the two largest pitches; however, they are factory-spliced and have reduced power-transmitting capability. HTD 14-mm belts are available in lengths up to 270 in. Synchronous belts, with teeth on both sides, are available in both systems for use in serpentine drives.

Belts are available in 40DP, 0.0816-in pitch with polyester load-carrying materials and molded polyurethane encasing material. A design recently introduced may be spliced to any length in the field. No pulley side flanges are required because the load member, which is encased in polyurethane, runs in a groove in the pulleys.

19.11 VARIABLE-SPEED BELTS

V belts may be used in variable-speed drives whose speed variation is affected by moving one sheave side wall with respect to the other. The speed is infinitely variable within the specified range. In simple drives, the drive must be stopped to change speed. Drives are available in which speed is adjusted manually or automatically in response to speed or torque requirements without stopping the drive.

Five types of V belt are used in variable-speed drives. Where the range of speed variation is limited, classical V belts are often used. Usually only the motor pulley has variable pitch. Under these conditions, speed variation is limited to 1:1.4. If both pulleys have variable pitch, variation can be 1:1.96. Sections A, B, and C are used in these drives.

Industrial variable-speed belts have thin, wide sections which permit speed variations to 8:1 and horsepowers per belt of 32 or more. There are many cross sections having top widths from 0.875 to 3.0 in. Some are standardized including P, Z, R, T, and W.

Agricultural variable-speed belts are thicker than industrial variable-speed belts to handle the higher power requirements of such applications. There are five standard cross sections having top widths from 1.0 to 2.0 in. If both pulleys have variable pitch, the maximum speed variation is 1:3.67.

Recreational-vehicle belts are used in lightweight, highly responsive variable-speed transmissions. They are subject to high, rapidly changing loads with sidewall temperatures reaching 400°F. Belts are customized using special rubber compounds and tensile reinforcement.

Automotive variable-speed belts have been in limited use for main traction and accessory drives for many years. Development is being accelerated because of the high efficiency of such a drive when compared to existing automatic transmission. Improved materials and configurations are aiding the development of these custom belts.

Permissible speed variation increases as belt thickness decreases, width increases, and as groove angle increases. Because a wide, thin belt required to produce a large speed variation tends to collapse under heavy loading, a compromise between speed variation and load-carrying capacities is necessary.

19.12 ENDLESS BELTS

Endless belts are made in a variety of constructions, shapes, and materials. Load-carrying members are both natural and synthetic yarns. Coating materials include butyl, neoprene, acrylonitriles, and fluorocarbon rubbers; vinyl and acrylic resins; and polyurethanes and polyethylenes whose durometer measurements range from 50 to 90 on the Shore A scale. They can be made in widths to 72 in and lengths to 900 in. While many are used in instrumentation, computers, and fractional-horsepower drives, they can be used to transmit a significant amount of power.

19.12.1 Film Belts

Film belts are thin (0.5 to 14 mils), flat belts available in widths of 0.125 to 2 in and lengths of 4 to 50 in. They are made of polyesters, Mylar or Kapton, or polyimides. Drive velocity may be as high as 47,000 ft/min and pulley diameters as small as 0.050 in. Application temperature may reach 225°F.

The largest permissible pulleys should be used to reduce belt tension. The design should be based on steady-state torque because the belt will slip under starting or shock loading. The length-to-width ratio should exceed 30:1 to promote good tracking. The diameter of the smallest pulley should be at least 100 times the belt thickness. The thinnest belt possible should be used.

Applications for film belts include lightly loaded, high-speed drives in business machines and instruments.

19.12.2 Woven Belts

In woven endless belts, the carcass or load-carrying member is woven as an endless belt in that it has yarns running in both the warp and fill directions. These belts are smooth-operating and are stable in both the lateral and longitudinal directions.

Multi-ply belts are available with the plies either stitched or cemented together. Natural cotton and synthetic yarns such as nylon are used as carcass materials. Typical applications include business machines and portable power tools. Single-ply belts 0.025 in thick, made of Dacron or Kevlar nylon, are run at belt speeds of 20,000 ft/min. Thinner single-ply belts, 0.0010 to 0.0015 in thick, are used in the same low-power, precision drives as film belts. Single-ply elastic and semielastic belts, 0.010 to 0.040 in thick, are used in fixed center distance applications and where their elasticity dampens vibrations induced by the driving motor.

Some woven belts are furnished with one surface rough and the other smooth. The rough side contacts the pulley in oily applications, the smooth side when the application is dry. Carcass materials are impregnated with a number of materials, including black, white, and red neoprene and synthetic resins. Widths to 60 in and lengths to 900 in are available. Tight-to-slack-side tension ratio should be in the range 2 to 4. For a ratio over 4, efficiency decreases.

19.12.3 Round Belts

Round belts are available as all-yarn, yarn-and-elastomer, and all-elastomer types. They may be purchased endless or made endless in the field by splicing. Round belts can operate on a drive in which the pulleys are in several planes. They are available in diameters from 0.03125 to 0.75 in, and lengths are unlimited. Pulley pitch diameter may be as small as 0.5 in.

Various natural and synthetic yarns are used for the carcass. In wound construction, yarn or braid is wound in a continuous helix, with or without a cover. If the winding is in one direction, stretch is about 1 percent; if alternate windings are in opposite directions, about 3 to 5 percent. Some round belts can run in V-belt as well as V-grooved pulleys. A tight-side-to-slack-side ratio of about 3.5 is most efficient. Elastic round belts have helically wound carcasses that can stretch. They are used in two-stage drives and applications similar to those for O-ring belts.

Two types of round belts, elastomeric and cast polyethylene, are treated in Secs. 19.12.4 and 19.12.5.

19.12.4 Elastomeric Round Belts

Elastomeric round belts, called "O-ring belts," are usually used in sound reproduction applications. They should be able to withstand oil, ozone, and weather, and be stretchable. Compounds used are neoprene, urethane, natural rubber, ethylene-propylene-terpolymer (EPT), and ethylene-propylene-diene-monomer (EPDM). Neoprene is a general-purpose material. EPT and EPDM are low-cost materials. Urethane is expensive and has high strength and abrasion resistance. Natural rubber can be run over small pulleys at speeds to 2000 ft/min. Compounding improves its poor aging properties.

In designing belt drives several factors should be considered. The minimum pulley diameter should be 6 times the cross-sectional diameters of the belt. Belt life is about inversely proportional to the fifth power of the pulley diameter. Pulley grooves should be semicircular with a radius of $0.45d$, where d is the O-ring diameter. Belt stretch should be 16 percent to optimize belt life and efficiency. Overall efficiency is about 80 percent. O-ring diameters of less than 0.070 in should not be used. O-ring belts will not run well in V grooves.

19.12.5 Cast Polyurethane Belts

Polyurethane belts are made in flat, round, V, and special cross sections. They are available in solid urethane and endless woven fabric load-carrying members. The urethane can be solid or foam. These belts have high resistance to abrasion, dirt, oil, water, most chemicals, shock, and vibration. Normal operating range is −20 to 150°F, with a maximum of 190°F. The coefficient of friction is high, 0.7 on steel. They can be used in limited space, reversed bends, and serpentine drives. The surface can be ground to ± 0.0005 in. Solid urethane belts may be purchased endless or spliced, hot or cold, to length, in the field.

Because of their high coefficient of friction, urethane belts should not be used on applications which require the belt to slip on the pulley. On cross drives the belt strands must not touch. A steel plate must be placed between the strands at the point of intersection. Belt-tightening drives, such as belt tighteners, must not be used. Urethane V belts can be run in V belt pulleys. Round belts can be run in V belt pulleys or round-belting pulleys.

Pulleys must be carefully aligned. Because of the high coefficient of friction of urethane, the belt will turn over if this is not the case. If pulleys must be misaligned, round belts should be used. A small amount of petroleum jelly on the belt will assist in preventing the belt from climbing out of the pulley grooves.

Flat belts should be operated on crowned pulleys. The crown should be 2° except for a level section of 0.2 of the belt width at the center of the pulley. The pulley width should be 0.125 in more than the belt width. Flat belting cannot be used for converging or diverging applications. They can be used on twist belt drives if both pulley shafts are in parallel planes and the pulley faces have a common centerline. For installations involving multiple flat belts on common shaft, maximum shaft deflection must not exceed 0.2 percent of the shaft length between bearings.

Because of the high hysteresis of urethane, the allowable minimum pulley diameter is a function of belt speed, length, and number of pulley diameters. Urethane belts should be installed with an initial stretch of 5 to 10 percent, with 12 percent as a maximum. Proper belt tension is that which just eliminates slack in the belt adjacent to the drive pulley on the slack side.

19.13 POSITIVE-DRIVE BELTS

Positive-drive belts maintain a positive rotational relationship between shafts by means of projections on the belt which fit into depressions in the pulley. One form of positive drive belt, the synchronous belt, is examined in Sec. 19.10.

19.13.1 Miniature Sprocket Chains and Pulley Belts

Two closely related means of power transmission, miniature sprocket chains and pulley belts, are used in instrumentation, servo, and other low-power installations. They have 0.03125-in-diameter stainless-steel load-carrying members jacketed in polyurethane. The chains, which are similar in appearance to roller chains, are available in single and double configurations, having two and three load-carrying members, respectively, joined by crosspins. The belts have a single load-carrying member and either a single side cog extending to either side of this member or two sets of side cogs at right angles to each other. The latter, referred to as a three-dimensional belt, provides a positive drive in the applications where a round belt is ordinarily used. Because chains are flexible laterally and belts can twist, chain sprockets and belt pulleys can be at an angle to each other. Two pitches are available, 20 pitch (0.15708 in) and 32 pitch (0.0982 in), a standard gear pitch. Sprockets and grooved gears are available in pitch diameters from 0.5 to 4.0 in.

Plastic chains and belts produce low noise even at high speeds and they dampen shock. Wear rate is low and they do not require lubrication. In most drives, the small pulley rotates at 6000 r/min or less; speeds of 15,000 r/min have been attained. Maximum recommended horsepower capacity for pulley belts is 1.5; for single-strand sprocket chains, 2.5; and for double chains, 7.5. Ratios are limited only by available sprockets. No pretensioning is necessary. The belts may be produced in endless form or the ends may be joined in the field, at a loss of strength of about 50 percent, using a soft sintered cast bushing.[17]

19.13.2 Spur-Gear Cable-Drive Chains

This positive drive has two prestretched aircraft standard cables as load-carrying members which are connected by crosspins. Encasing material is polyurethane, nylon, or fire-resistant plastic. It is available in three pitches, 12 (0.2168 CP), 16 (0.1965 CP), and 24 (0.1309 CP), where CP = circular pitch in inches, and mates with standard 14.5 and 20° spur gears of these pitches. Maximum horsepower at 1800 r/min is 12 pitch, 10.2 hp; 16 pitch, 48 hp; and 12 pitch, 123 hp. No lubrication is required.[17]

19.13.3 Spur-Gear Drive Belts (Pinned Drive)

This positive drive has features similar to those of the spur-gear cable-drive chain, but has only one load-carrying member. It is available in two pitches: 0.1 CP and 32 (0.0982 CP). Both are rated at a maximum of 5.10 hp at 1800 r/min. To mate with these belts, standard pulleys with 0.1875-in-wide faces are modified. The modification consists of a groove, 0.0937 in wide by 0.077 in deep, cut in the face for the load-carrying member to ride in.[17]

19.13.4 Metal Belts

These very thin belts have high strength-to-weight ratios and stretch very little. They are made of stainless steel. They are available in thicknesses of 0.002 to 0.030 in, widths to 36 in, and circumferences down to 5 in. They can be obtained plain or in various hole configurations. Standard perforations for 16- and 35-mm tape drives and others specified by the user are available. Holes can be produced mechanically or chemically. Such belts are intended for precision electromechanical drive systems, encoder drives, and XY plotters.[15]

NASA has patented two steel positive drive belts, originally developed for spacecraft instrument drives. In a chain-type version, the load-carrying members are wires wrapped around metal rods which run at right angles to them. These rods serve as link pins. In another design, a steel strip is formed into a series of U-shaped teeth. The steel is flexible enough to flex around a sprocket but resists stretching. Both belts are coated with plastic to reduce noise and wear.[16]

19.13.5 Bead-Chain Belts

The load member of these belts consists of a stainless-steel or Aramid cable covered with polyurethane. At specified intervals, this coating is formed into beads which seat into conical recesses in the pulley. Because the belt is flexible, it can be used with misaligned or skewed pulleys, with guide tubes for the belt, and to provide rotary or oscillatory motion. Because the beads may come out of the sprockets under heavy loading, sprocket chains and pulley belts were developed to deal with this problem.[16]

These belts are used in electronically controlled, fast film transport and positioning. A soft polyurethane web 0.180 by 0.020 in, reinforced by 0.004-in stainless-steel wire, supports the teeth. One tooth system, on an outer pitch line, engages the film, while the other, on an inner pitch line, engages the drive and driven sprockets. The belts are produced in circular form.[18]

19.14 PIVOTED MOTOR BASES

With small-diameter pulleys necessary with high-speed motors and short center-distance drives, it is impractical to maintain belt tension by belt sag, so pivoted motor bases and idler pulley drives are employed. Pivoted motor bases, which automatically maintain correct belt tension, are of two types, gravity and reaction-torque.

The gravity type (Fig. 19.12) utilizes part or all of the weight of the driving motor to maintain belt tension. In horizontal drives, with the driven pulley above the driver, motor weight keeps the belt in tension as it stretches because of the torque transmitted. Vertical drives are also used.

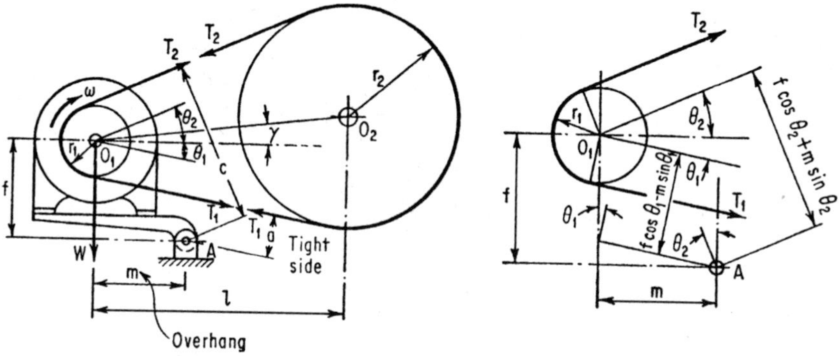

FIG. 19.12 Gravity-type pivoted motor base.

Reaction-torque motor bases are of all-steel construction. The standard base (Fig. 19.13) is designed to function satisfactorily only with the tight side nearest the pivot axis. Special bases are available for reverse-direction drives. Reaction torque on the motor stator tends to rotate the stator about the pivot axis in the opposite direction to the belt pull, thus tightening the belt.

Vertical drives with the motor mounted above the driven pulley are called "down drives." A counterweight is utilized to balance the motor weight.

The spring-type motor base maintains belt tensions by means of springs built into the motor base.

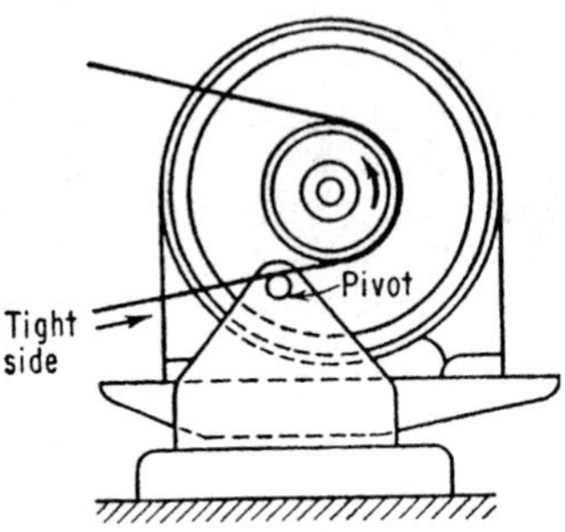

FIG. 19.13 Reaction-torque pivoted motor base.

Belt tension in a gravity-type pivoted motor base drive (Fig. 19.12) is found by taking moments about the pivot A. The resulting moment equation,

$$Wm = T_1 a + T_2 c$$

is solved simultaneously with the equation for tension ratio, $T_1/T_2 = e^{\mu\beta}$, to find the belt tensions. Maximum power, which can be transmitted when $T_2 = 0$, is

$$\text{hp} = mWNd_1/126{,}000a$$

where N is the motor r/min and all dimensions are in inches. During operation, only about 75 percent of this horsepower can be transmitted by the drive. For ordinary oak-tanned-leather belt drives values of $T_1/T_2 = 5$ at 180° arc of contact and $T_1/T_2 = 3$ at 120° are recommended. For high-capacity belts, these values may be increased to 6.5 and 3.5, respectively. Centrifugal effects in the belt are usually neglected because the motor weight maintains belt pulley contact. From Fig. 19.12, the moment arms a and c are given by $a = f\cos\theta_1 - m\sin\theta_1 - r_1$, and $c = f\cos\theta_2 + m\sin\theta_2 + r_1$.

Belts should be thin and wide to reduce flexing around the small pulley. Pulleys should be as large as possible consistent with reasonable belt speeds, and little, if any, crown should be provided for wider belts. V belts can also be used with pivoted motor bases.

19.15 BELT DYNAMICS

19.15.1 Transverse Vibration

When the belt is excited by forces which have a frequency corresponding to its transverse natural frequency, it may undergo serious flapping vibrations. It is accordingly desirable to determine the natural frequency of the belt to avoid such operation. In the treatment below, it is assumed that the flexural rigidity of the belt is negligible.

The velocity of sound u, which is equal to the velocity of propagation of a disturbance along the belt, is given by

$$u = (Pg/\rho A)^{1/2} \qquad \text{in/s} \tag{19.28}$$

where P is the tension in the belt, lb; g the gravitational constant, in/s^2; ρ the density of the belt, lb/in^3; and A the cross-sectional area, in^2.

The velocity v of the belt is

$$v = 2\pi(RN/60) \qquad \text{in/s} \tag{19.29}$$

where R and N are the radius and r/min of the sheave. The time for a disturbance to travel up and down the belt is $t = 2L_S u/(u^2 - v^2)$ s, where L_S is the distance between points of tangency of the belt, so that the frequency of the fundamental mode is

$$F_1 = 60/t = 60(u^2 - v^2)2L_S u \qquad \text{cycles/min} \tag{19.30}$$

In addition to the fundamental mode, the belt may execute higher modes of vibration whose frequency is given by

$$F_i = mF_1 \tag{19.31}$$

where $m = 1, 2, 3$, etc. It should be noted that the natural frequency of the slack side differs from that of the tight side and that this frequency differs with belt speed.

To account for its flexural rigidity, the belt is assumed to be a simply supported beam in lateral vibration whose natural frequency

$$F_2 = 30(m^2/L_S^2)(EIg/A)^{1/2} \qquad \text{cycles/min} \tag{19.32}$$

where $m = 1, 2, 3$, etc. For a steel flat belt this reduces to

$$F_2 = 5.5 \times 10^6 bm^2/L_S^2 \tag{19.33}$$

where b is the belt thickness.

The approximate natural frequency F of a belt having flexural rigidity is given by

$$F^2 = F_1^2 + F_2^2 \tag{19.34}$$

The effect of flexural rigidity, even for a steel belt, is generally negligible.

Excitation of lateral vibrations in a belt drive can arise as a joint, debris, or thickened area of the belt passes over a pulley. The excitation frequency in this case is obtained by dividing the linear velocity of the belt by the total length of the belt, or

$$F_e = 2\pi RN_R/L_t \qquad \text{cycles/min} \tag{19.35}$$

19.15.2 Torsional Rigidity of Belt Drives

The torsional rigidity of a belt drive is the ratio of the moment applied to one pulley to the angular deflection undergone by the pulley. The torsional rigidity for the open drive may be written

$$C_R = \frac{R^2\beta}{e[L_S + (R+r)/3]} \qquad \text{in·lb/rad} \tag{19.36}$$

$$C_r = \frac{r^2\beta}{e[L_S + (R+r)/3]} \qquad \text{in·lb/rad} \tag{19.37}$$

where C_R, C_r are the torsional rigidities referred to the larger and smaller pulleys, respectively, $e = 1/AE$ where A is the cross-section area, E the modulus of elasticity of the belt, L_S the distance between points of tangency of the belt

$$L_S = [C^2 - (R-r)^2]^{1/2}$$

and R and r are the radii of the large and small pulley, respectively. The empirical factor $(R + r)/3$ is included to account for stretching effects as the belt passes around the pulley and may be omitted in approximate calculations. β is a constant whose value depends on the relation between steady-state and vibratory torques imposed. If the magnitude of the vibratory force is less than the magnitude of the steady-state force in the slack side, $\beta = 2$. If the vibratory torque is large enough that the force in the slack side is negligible throughout the cycle, $\beta = 1$. If the force in the slack side is zero for ½ cycle, $\beta = 1.5$ (approximately). For steady-state conditions, with no vibratory torque, $\beta = 2$.

The belt may be converted to an equivalent length of either drive or driven shafting for frequency calculations. This value is, when referred to the larger pulley,

$$L_E = (\pi/32)(GD^4/\beta EAR^2)L_S \tag{19.38}$$

or, when referred to the smaller pulley,

$$L_e = (\pi/32)(Gd^4/\beta EAr^2)L_S \qquad (19.39)$$

where G is the torsional rigidity of the shaft and D and d are the diameters of large and small pulleys, respectively.

EXAMPLE Calculate the torsional rigidity of a steel belt 1.0 in wide and 0.004 in thick; diameter of both pulleys is 6 in and center distance is 7 in. With these values, $e = 1/(1.0 \times 0.004) \times 30 \times 10^6 = 250/30 \times 10^6$.

$$C_r = C_r = \frac{2 \times 9 \times 30 \times 10^6}{250 \times 7 + (3 + 3)/3} = 240{,}000 \text{ in}\cdot\text{lb/rad}$$

The value of E for leather is usually taken as 25,000 lb/in².

When calculating the fundamental frequency of a multimass system in which a belt drive is included, it is usually assumed that this mode of vibration has a node in the belt, whose stiffness is low compared with that of other members of the system.

Excitations at this frequency can be caused by sheave unbalance or a bent or misaligned shaft. Excitations at twice this frequency are produced by out-of-round sheaves or by shaft misalignment. Excitations at ½, 1, and 1½ times this frequency arise when the belt is driven by four-stroke-cycle internal-combustion engines.

REFERENCES

1. Barth, C. G.: "The Transmission of Power by Leather Belting," *Trans. ASME,* vol. 31, p. 29, 1909.
2. "Handbook of Power Transmission," "Flat Belting," p. 11, Publication S-5119, The Goodyear Tire and Rubber Co., Akron, Ohio, 1954.
3. Hampton, F. G., C. F. Leh, and W. E. Helmick: "An Experimental Investigation of Steel Belting," *Mech. Eng.*, vol. 42, p. 369, July 1920.
4. Young, G. L., and G. V. D. Marx: "Performance Tests of Steel Belts with Compressed Spruce Pulleys," *Mech. Eng.*, vol. 45, p. 246, 1923.
5. "1981–1982 Power Transmission Design Handbook," Penton IPC, p. C/108.
6. Erickson, W.: "Straight Talk about Belt Drives," *Mach. Des.*, p. 199, April 21, 1977.
7. "Flat Belts," *Mach. Des.*, (Mechanical Drives Reference Issue), p. 22, June 3, 1976.
8. Page-Lon Catalog, Page Belting Company, Concord, N.H.
9. Nycor-M Catalog, L. H. Shingle Co., Worcester, Mass.
10. Foulds Vitalastic Horsepower Tables, I. Foulds & Sons, Inc., Hudson, Mass.
11. Dodge Engineering Catalog, D78, Dodge Division, Reliance Electric Co., Mishawaka, Ind., 1978.
12. Erickson, W.: "New Standards for Power Transmission Belts," *Mach. Des.*, p. 83, January 22, 1981.
13. Hitchcox, A.: "V-belts—Designed to Deliver," *Power Transm.* p. 22, November 1981.
14. Tantastic Nylon Core Belting, Catalog 3M, J. E. Rhoads & Sons, Inc., Wilmington, Del., March 1977.
15. "Drive Bands Data," Metal Belts Inc. Agawam, Mass., 1981.
16. "Durable, Nonslip, Stainless Steel Drive Belts," *Mech. Eng.*, p. 48, July 1979.
17. "Getting in Step With Hybrid Belts," *Des. Eng.*, p. 63, April 1981.

18. Stefanides, E. J.: "Polyurethane Belt Safely Links Servo Drive," *Des. News,* p. 44, December 12, 1978.

19. Catalog Supplement, W. B. Berg, Inc., East Rockaway, N.Y., 1976.

20. "Narrow Multiple V-Belts" (3V, 5V, and 8V Cross Sections), IP-22, Rubber Manufacturers Association, Washington, D.C., 1977.

21. Sun, D. C.: "Performance Analysis of Variable Speed-Ratio Metal V-Belt Drive," *Trans. ASME, J. Mechanisms, Transmissions and Automation in Design,* vol. 110, p. 472, 1988.

CHAPTER 20
CHAINS

George V. Tordion, Ing.P.
Professor of Mechanical Engineering
Université Laval
Quebec, Canada

20.1 INTRODUCTION 20.1
20.2 NOMENCLATURE 20.1
20.3 DESIGN OF ROLLER CHAINS 20.2
20.3.1 Chain Length 20.3
20.3.2 Ratings 20.4
20.3.3 Roller Impact Velocity 20.6
20.3.4 Centrifugal Force 20.6
20.3.5 Force Distribution on the Sprocket Teeth 20.6
20.4 SYSTEM ANALYSIS 20.7
20.4.1 Chordal Action 20.7
20.4.2 Transverse Vibrations of Chains 20.8
20.4.3 Longitudinal Vibrations of the Tight Strand 20.9
20.4.4 Equivalent Stiffness of Chain for Torsional-Vibration Calculations 20.10
20.4.5 Noise 0.10
20.5 SILENT OR INVERTED-TOOTH CHAIN 20.10

20.1 INTRODUCTION (Refs. 4, 6, 8, 9, 12, 13, 15–18, 23–27, 29)

Power-transmission chains are primarily of two kinds, roller and silent. They transmit power in a positive manner through sprockets rotating in the same plane. This fairly old transmission element (sketches of its design have been found in notebooks of Leonardo da Vinci) has been improved in the course of time to a very high standard of precision and quality.

Chain transmission is positive and there is no slip present as in belt drives. Large center distances can be dealt with more easily, with fewer elements and in less space than with gears. Chain drives have high efficiency. No initial tension is necessary and shaft loads are therefore smaller. The only maintenance required, after a careful alignment of elements, is lubrication. Chains as well as sprockets are thoroughly standardized in ASA standard B.29.1. The primary specifications of the standard single-strand roller chain are given in Table 20.1.

20.2 NOMENCLATURE

p = pitch, in

T = chain tension, lb

T_c = chain tension due to centrifugal forces, lb

w = chain weight per unit length, lb/in

$m = w/g$ = chain mass per unit length, $\text{lb}\cdot\text{s}^2/\text{in}^2$

D_1, D_2 = sprocket diameters, in

N_1, N_2 = sprocket-teeth numbers

n_1, n_2 = revolutions per minute of the sprockets

$\rho = n_2/n_1$ = speed ratio

V = chain speed, in/s in equations, ft/min in specifications

TABLE 20.1 American Standard Roller Chain

Pitch, in	ASA No.	Weight, lb/ft	Avg ultimate strength, lb	Roller width, in	Roller diam, in	Side plate	
						Thickness, in	Height, in
$\frac{1}{4}$	25	0.085	875	$\frac{1}{8}$	0.130	0.03	0.23
$\frac{3}{8}$	35	0.22	2,100	$\frac{3}{16}$	0.200	0.050	0.36
$\frac{1}{2}$	41	0.28	2,000	$\frac{1}{4}$	0.306	0.050	0.39
$\frac{1}{2}$	40	0.41	3,700	$\frac{5}{16}$	$\frac{5}{16}$	0.060	0.46
$\frac{5}{8}$	50	0.68	6,100	$\frac{3}{8}$	0.400	0.080	0.59
$\frac{3}{4}$	60	0.96	8,500	$\frac{1}{2}$	$\frac{15}{32}$	0.094	0.68
1	80	1.70	14,500	$\frac{5}{8}$	$\frac{5}{8}$	0.125	$\frac{7}{8}$
$1\frac{1}{4}$	100	2.70	24,000	$\frac{3}{4}$	$\frac{3}{4}$	0.156	$1\frac{5}{32}$
$1\frac{1}{2}$	120	4.00	34,000	1	$\frac{7}{8}$	0.187	$1\frac{13}{32}$
$1\frac{3}{4}$	140	5.20	46,000	1	1	0.218	$1\frac{5}{8}$
2	160	6.80	58,000	$1\frac{1}{4}$	$1\frac{1}{8}$	0.250	$1\frac{7}{8}$
$2\frac{1}{4}$	180	9.10	76,000	$1\frac{13}{32}$	$1\frac{13}{32}$	0.281	$2\frac{1}{8}$
$2\frac{1}{2}$	200	10.80	95,000	$1\frac{1}{2}$	$1\frac{9}{16}$	0.312	$2\frac{5}{16}$
3	240	16.5	130,000	$1\frac{7}{8}$	1.875	0.375	2.8

$V_0 = \sqrt{T/m}$ = velocity of propagation of transverse waves, in/s

$L = L'p$ = chain length, in

$C = C'p$ = center distance of the sprockets, in

a = strand length between seated rollers, in

h = chord length of the strand between seated rollers, in

s = strand maximum sag, in

γ = pressure angle

$\alpha_1 = 180°/N_1$

$\alpha_2 = 180°/N_2$

ω_n = natural circular frequency of vibrations, rad/s

E = Young's modulus, lb/in^2

f = roller width, in

20.3 DESIGN OF ROLLER CHAINS (Fig. 20.1)

The speed ratio is given by a $\rho = n_2/n_1 = N_1/N_2$. For one-step transmission, it is recommended that $\rho < 7$. Values between 7 and 10 may be used at low speeds (< 650 ft/min). The minimum wrap angle of the chain on the smaller sprocket is 120°. A smaller angle may be used on idler sprockets used for adjustment of the chain slack,

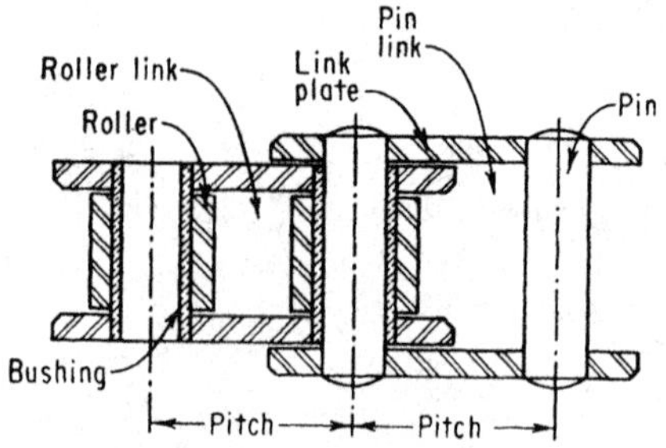

FIG. 20.1 Roller chain.

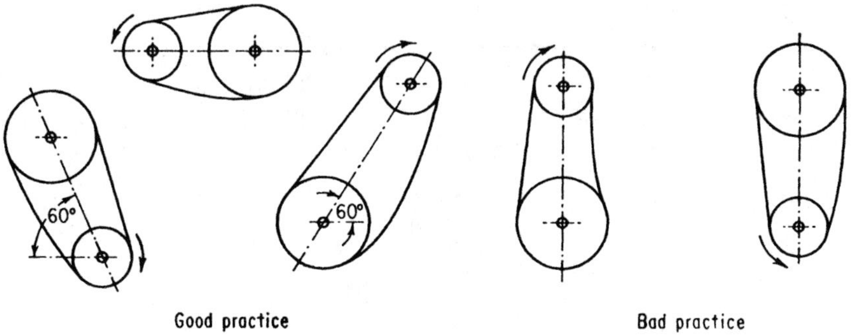

FIG. 20.2 Chain-drive arrangements.

where the center distance is not adjustable. Horizontal drive is recommended, in which case the power should be transmitted by the upper strand. Preferred inclined-drive arrangements are shown in Fig. 20.2. Vertical drives should be used with idlers to prevent the chain from sagging and to avoid disengagement from the lower sprocket. When running outside the chain, idlers should be located near the smaller sprocket.

20.3.1 Chain Length

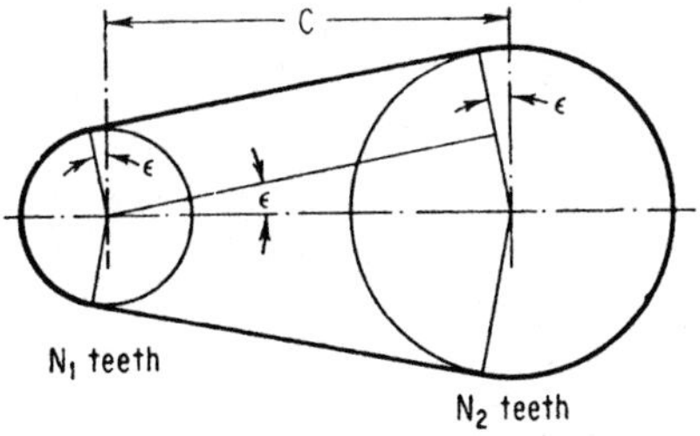

FIG. 20.3 Angle ϵ used in chain-length calculation.

The chain length L' in pitch numbers, for a given distance C' in pitch numbers, is exactly (Fig. 20.3)

$$L' = 2C' \cos \epsilon + (N_1 + N_2)/2 + (N_2 - N_1)\epsilon/\pi \tag{20.1}$$

where

$$\epsilon = \arcsin \frac{1/\sin \alpha_2 - 1/\sin \alpha_1}{2C'} \tag{20.2}$$

Since ϵ is generally small, it is sufficient in most cases to use the approximation $\epsilon = (N_2 - N_1)/(2C')$. Then

$$L' = 2C' + (N_1 + N_2)/2 + S/C' \tag{20.3}$$

where $S = [(N_2 - N_1)/2\pi]^2$. The values of S are tabulated in Table 20.2. The actual chain length is $L = L'p$, and the actual sprocket center distance $C = C'p$. It is recommended that C' lie between 30 and 50 pitches. If the center distance is not given, the designer is free to fix C' and to calculate L' from Eq. (20.3). The nearest larger (preferably even) integer L' should be chosen. In this case, an offset link is avoided. With L' an integer, the center distance becomes exactly

$$C' = e/4 + \sqrt{(e/4)^2 - S/2} \tag{20.4}$$

with $e = L' - (N_1 + N_2)/2$. C' should be decreased by about 1 percent to provide slack in the nondriving chain strand. For horizontal drive, this will result in a chain sag of some 2 percent of the strand length.

The chain speed is $V = \pi n_1 D_1/12$ (ft/min), or approximately $V = n_1 p N_1/12$ (ft/min). The sprocket diameters are $D_i = p/\sin \alpha_i$ (in), $i = 1$ or 2.

TABLE 20.2 Values of S for Different $N_2 - N_1$

$N_2 - N_1$	S	$N_2 - N_1$	S	$N_2 - N_1$	S	$N_2 - N_1$	S
1	0.02533	16	6.4846	31	24.342	46	53.599
2	0.10132	17	7.3205	32	25.938	47	55.955
3	0.22797	18	8.2070	33	27.585	48	58.361
4	0.40528	19	9.1442	34	29.282	49	60.818
5	0.63326	20	10.132	35	31.030	50	63.326
6	0.91189	21	11.171	36	32.828	51	65.884
7	1.2412	22	12.260	37	34.677	52	68.493
8	1.6211	23	13.400	38	36.577	53	71.153
9	2.0518	24	14.590	39	38.527	54	73.863
10	2.5330	25	15.831	40	40.528	55	76.624
11	3.0650	26	17.123	41	42.580	56	79.436
12	3.6476	27	18.466	42	44.683	57	82.298
13	4.2808	28	19.859	43	46.836	58	85.211
14	4.9647	29	21.303	44	49.039	59	88.175
15	5.6993	30	22.797	45	51.294	60	91.189

20.3.2 Ratings

The transmitted horsepower is related to the required horsepower rating by the following:

$$\text{Required hp rating} = \frac{\text{hp transmitted} \times \text{service factor}}{\text{multiple-strand factor}} \tag{20.5}$$

where the service factors and the strand factors are given in Tables 20.3 and 20.4. The ratings are given in extensive tables in Ref. 2. They are partially reproduced in graphical form in Fig. 20.4. For every chain pitch and number of teeth in the small sprocket, the horsepower ratings are given as a function of speed in revolutions per minute. The service life expectancy is approximately 15,000 hr. The lubrication conditions are: type I, manual lubrication; type II, drip lubrication from a lubricator; type III, bath lubrication or disk lubrication; type IV, forced-circulation lubrication.

The rating graphs have the peculiar "tent form." At lower speeds chain fatigue is responsible for the chain failures. From a certain speed the roller impact resistance becomes responsible for roller breakage, and the rated horsepower falls rapidly until, at the ultimate speed, failure is caused by joint galling.[3] Even with the new reliable

TABLE 20.3 Service Factor[2]

Type of driven load	Type of input power		
	Internal-combustion engine with hydraulic drive	Electric motor or turbine	Internal-combustion engine with mechanical drive
Smooth	1.0	1.0	1.2
Moderate shock	1.2	1.3	1.4
Heavy shock	1.4	1.5	1.7

TABLE 20.4 Strand Factor[2]

No. of strands	*Multiple-strand factor*
2	1.7
3	2.5
4	3.3

ratings, careful consideration should be given the application. The design is a compromise between life and cost. A large pitch provides more bearing area than a small one for the same load, but requires fewer teeth on the sprocket. This leads to strong chordal action and thus to large dynamic effects, which, in turn, results in premature wear. A multiple strand with smaller pitch is preferable in such cases. Reference 23 gives more information.

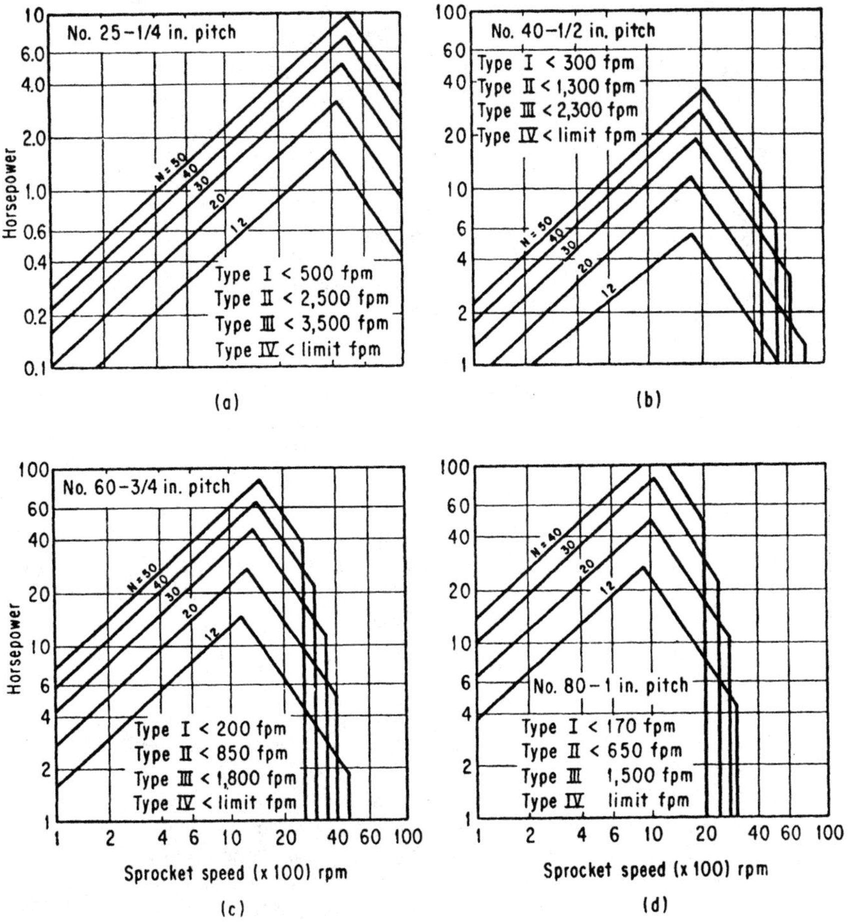

FIG. 20.4 Horsepower ratings vs. sprocket speed.

20.3.3 Roller Impact Velocity

The roller impact velocity is given by $\omega p \sin(2\alpha + \gamma)$. How much of the chain mass is involved in the impact is unknown. It is sometimes assumed[1,5] that two links of mass mp are involved. In this case, the kinetic energy is $mp\omega^2 p^2 \sin^2(2\alpha + \gamma)$. The strain energy is $\frac{1}{2}P\Delta$ (P = impact force, Δ = local deformation). Using the approximation[5] $\Delta = 3P/fE$, the strain energy becomes $3P^2/2fE$ where f is the roller width. If the kinetic energy is completely transformed into strain energy, the impact force

$$P = (2\pi V/N_1)\sqrt{2wpfE/3g}\,\sin(2\alpha+\gamma) \qquad (20.6)$$

A sprocket having a small number of teeth, running at high speeds, leads to roller breakage. Lubricant viscosity also plays an important part in roller fatigue because the rollers must squeeze the lubricant before they are completely seated, and squeezing provides good damping.

20.3.4 Centrifugal Force

The chain loop is subjected to a uniformly distributed additional tension $T_c = mV^2$ caused by centrifugal force. T_c is generally low and does not affect sprocket tooth pressure but it must be added to the useful tension because it affects chain-joint pressure.

20.3.5 Force Distribution on the Sprocket Teeth

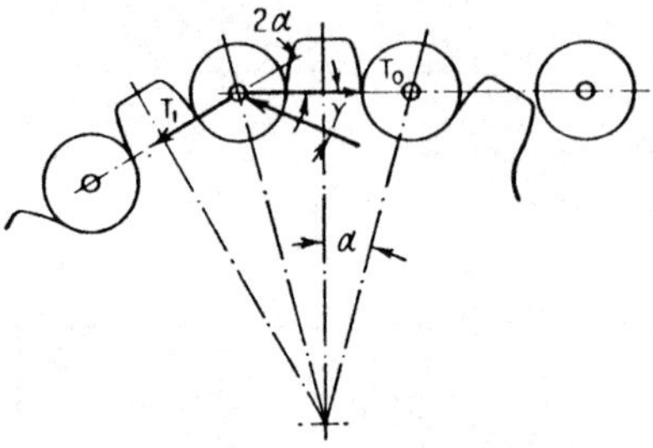

FIG. 20.5 Forces on sprocket teeth.

Referring to Fig. 20.5 and providing that the pressure angle is constant for all seated rollers,

$$T_n = T_0\{(\sin\gamma)/[\sin(2\alpha + \gamma)]\}^n \qquad (20.7)$$

where n is the number of seated rollers. For example, for $N = 18$, $n = 6$, and $\gamma = 25°$, $T_6 = 0.045T_0$. Therefore, T_6 is only 4.5 percent of T_0. The tension in the chain decreases very rapidly but never becomes zero unless the pressure angle is zero. For a new chain the pressure angle is given by ASA standard:

$$\gamma = 35° - 120°/N$$

As the chain wears, the pitch of each link is increased, but because wear has a random nature, it is difficult to predict the pressure angles. No actual measurements of γ under running conditions have been made. For elongated chains, the roller engages the top curve of the teeth farther and farther away from the working curve, leading to a dangerous situation. An elongation limit of $\Delta L = 80/N_2$ percent has been proposed, but a maximum of 3 percent is admitted in the new ratings for 15,000 h of service. Increased elongation indicates that the hardened layers of pins and bushings are worn away and the soft cores reached. A very rapid wear increase follows at this stage.

20.4 SYSTEM ANALYSIS

20.4.1 Chordal Action

Chordal action, or polygonal effect, is the variation of the velocity of the driven sprocket which results because the actual instantaneous velocity ratio is not constant but a function of the sprocket angular position $\rho = d\varphi_2/d\varphi_1 = f(\varphi_1)$. Therefore, we have the angular-displacement equation $\dot{\varphi}_2 \int f(\varphi_1)d\dot{\varphi}_1$ + constant, the velocity equation $\dot{\varphi}_2 = \rho(\varphi_1)\dot{\varphi}_1$, and the acceleration equation $\ddot{\varphi}_2 = \rho(\varphi_1)\ddot{\varphi}_1 + (d\rho/d\varphi_1)\dot{\varphi}_1^2$. From the last equation, it is seen that, even in the case of constant driving speed $\dot{\varphi}_1$, the angular acceleration of the driven sprocket is not zero. This produces dynamic loads, vibrations, and premature chain wear. The analysis of chordal action is made under the assumption that the chain strand acts like a connecting rod in a four-bar linkage. Experiments at low speeds[7] yield excellent agreement with the theoretical results. At higher speeds the chain vibrations are sufficient to alter the static behavior of the chordal action. Nevertheless this action still remains a main source of excitation. Analysis shows[1,7] that the speed variation is maximum for a strand length equal to an *odd multiple of half pitches*. For this case,

$$(\Delta\omega/\omega)_{\max} = (N_2/N_1)(\tan\alpha_2/\sin\alpha_1 - \sin\alpha_2/\tan\alpha_1) \tag{20.8}$$

If the strand length is equal to an *even number of half pitches*, the speed variation is close to a minimum:

$$(\Delta\omega/\omega)_{\min} = (N_2/N_1)(\sin\alpha_2/\sin\alpha_1 - \tan\alpha_2/\tan\alpha_1) \tag{20.9}$$

For $N_1 = N_2$, and an even number of half pitches, the speed variation is zero. Generally, for a fractional number of half pitches the speed variation fluctuates between maximum and minimum (Fig. 20.6). For large N_2/N_1 the values of $\Delta\omega/\omega$ in both cases approach the same limit (Fig. 20.7). There is a considerable advantage in having as many teeth as possible on the small sprocket. Chordal action, in general, may be ignored if the minimum number of sprocket teeth is 15.

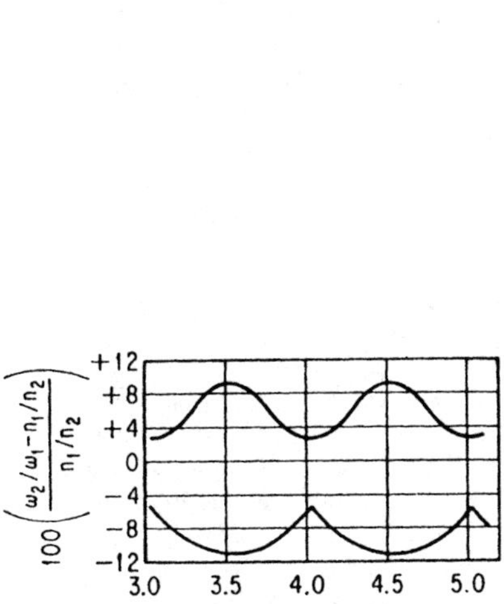

FIG. 20.6 Speed variation vs. center distance in pitches.

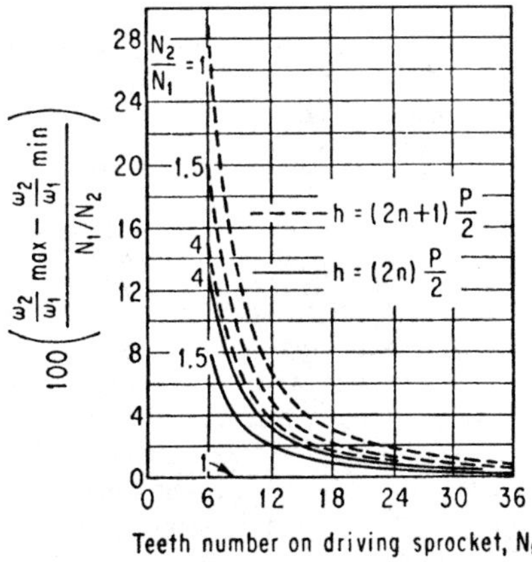

FIG. 20.7 Speed fluctuation vs. number of teeth on driving sprocket.

20.4.2 Transverse Vibrations of Chains

The power-transmitting strand may be considered as a tight string with a uniformly distributed mass. The small, transverse, undamped free vibrations $y(x, t)$ are described by the differential equation[10]

$$(V^2 - V_0^2)(\partial^2 y/\partial x^2) + 2V(\partial^2 y/\partial x\, \partial t) + \partial^2 y/\partial t^2 = 0 \qquad (20.10)$$

which has the solution

$$y = C_1 \cos [\omega t + \omega x/(V_0 - V) + \phi_1] + C_2 \cos [\omega t - \omega x/(V_0 + V) + \phi_2] \qquad (20.11)$$

The natural frequency, for the boundary conditions $y = 0$ at $x = 0$ and $x = a$, is

$$\omega_n = (n\pi V_0/a)[1 - (V/V_0)^2] \qquad n = 1, 2, 3,\ldots \qquad (20.12)$$

To each ω_n corresponds a mode of vibration. For strings with continuously distributed mass, the number of modes is infinite. For a chain, there is a limit to this number because the strand is composed of a finite number of rollers s. The string approximation is therefore unsuitable for n close to s, and for small s, in general. Considering the chain as a tight string of zero mass with point masses m_0 at every pitch distance, the natural frequencies at zero speed

$$\omega_n = (n\pi V_0/a)(2s/n\pi) \sin [n\pi/(2s + 2)] \qquad n = 1, 2,\ldots, s \qquad (20.13)$$

For $s \to \infty$ Eq. (20.13) becomes identical with Eq. (20.12) with $V = 0$. For $s = 19$ and $s = 44$ the errors are 5 percent and 2.2 percent, respectively. One important source of vibration excitation is the chordal action discussed above. The fundamental frequency of this action is the tooth-engagement circular frequency $2\pi V/p$; harmonics are also possible. Neglecting harmonics, the critical chain speeds

$$V_{crit} = (aV_0/np)[\sqrt{1 + (pn/a)^2} - 1] \cong nV_0 p/2a \qquad n = 1, 2, 3,\ldots \qquad (20.14)$$

Another important source of vibration excitation is the runout of the shaft or sprocket. Its frequency is $2\pi V/Np$. The critical speed attributable to eccentricity is therefore

$$V_{crit} = (aV_0/npN)[\sqrt{1 + (Npn/a)^2} - 1] \qquad n = 1, 2, 3,\ldots \qquad (20.15)$$

The additional tension $T_c = mV^2$ changes the natural frequency of the strand, since $V_0^2 = T/m + mV^2$, and therefore

$$\omega_n = \frac{n\pi}{a} \frac{T/m}{(T/m + V^2)^{1/2}} \qquad (20.16)$$

but the critical speeds [Eq. (20.14)] remain unchanged. All critical speeds must be avoided, even when a resonance does not result in infinite vibration amplitudes; damping is always present and reduces resonant amplitudes. For the computation of transverse amplitudes as a function of $V \neq V_{crit}$, see Ref. 10. Resonance may be remedied by varying tight-strand tension and length, and the possible use of guides or idlers.

The traveling curve of the slack strand under the action of gravity is the catenary of a heavy chain in equilibrium. The catenary is disturbed by the chordal action at the ends. The stability criteria are formulated in Ref. 14, where the problem is reduced, after linearization, to a Mathieu differential equation of the type

$$\ddot{y} + (b - 2q \cos 2t)y = 0$$

The dimensionless parameters

$$b = [(\omega_n/V)^2 - (7.5/a)^2](p/\pi)^2$$

and

$$q = 2(\omega_n^2/g)(p^2/D_1) \tag{20.17}$$

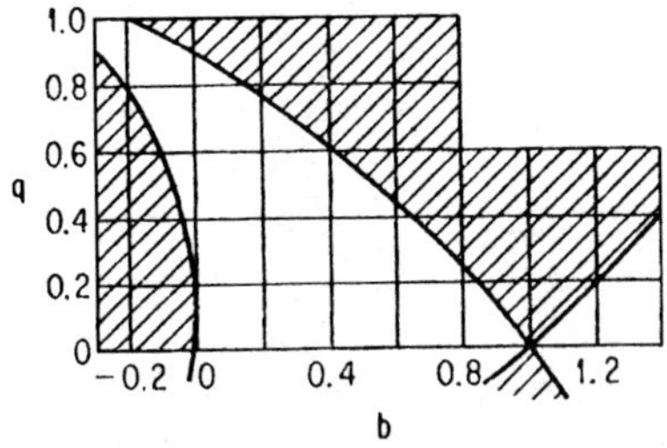

FIG. 20.8 Plot of the dimensionless parameters b and q of the Mathieu differential equation.

are shown in Fig. 20.8. The shaded regions represent unstable running conditions and must be avoided if possible. The slack strand ω_n can be determined either experimentally or from Table 20.5. The given values are valid for a horizontal arrangement of the suspension points (seated rollers) and small s/h, when the catenary of the heavy chain can be approximated by a parabola. The values were experimentally verified to within 5.7 percent.

TABLE 20.5 Values of ω_n

h/a	0.95	0.96	0.97	0.98	0.99	0.998
s/h	0.140	0.125	0.110	0.088	0.060	0.030
$\omega_n^2 a/g$	10.3	14.2	18	20.4	29	91.5

20.4.3 Longitudinal Vibrations of the Tight Strand

The chain is elastic in tension also, and the cushioning of shocks between the driving and driven shafts is one of the advantages of chain transmissions. The elongation of a chain under 50 percent of ultimate load has been measured with great care.[11] A statistical average for all pitches is 0.081 in/ft. Not more than 5 percent of chains of a given pitch will vary from this average by more than 0.014 in/ft. The values of the elastic constant given in Table 20.6 are based upon an assumed linear relationship between load and elongation.

TABLE 20.6 Values of the Elastic Constant K

p, in.	3/8	1/2	5/8	3/4	1	1 1/4	1 1/2	1 3/4	2
$K/10^6$ lb-in./in.	0.163	0.269	0.420	0.604	1.075	1.68	2.41	3.3	4.3

The natural frequency at zero speed is

$$\omega_n = (\pi n/a)\sqrt{K/m} \qquad n = 1, 2, 3, \ldots \tag{20.18}$$

while the critical speeds due to the tooth-engagement frequency become

$$V_{crit} = (np/2a)\sqrt{K/m} \qquad n = 1, 2, 3, \ldots \tag{20.19}$$

20.4.4 Equivalent Stiffness of Chain for Torsional-Vibration Calculations

The arrangement to be considered is that of a chain drive with two sprockets. The torsional stiffness depends upon whether the shaft of the first or second sprocket is taken as a reference value. With reference to the first shaft, the torsional stiffness of the tight strand is $K_1 = (K/a)R_1^2$(lb·in/rad). With reference to the second shaft, it is $K_2 = (K/a)R_2^2 = K_1(R_2/R_1)^2$(lb·in/rad). In both equations K is the tensile elastic constant of the chain. The equivalent shaft length in terms of a reference diameter d_0 and shear modulus G (lb/in²) is

$$L_{1,\text{eq}} = \pi d_0^4 G/32K_1 \quad \text{in}$$

or

$$L_{2,\text{eq}} = \pi d_0^4 G/32K_2 \quad \text{in} \tag{20.20}$$

Further development follows the standard torsional-vibration procedure. For sufficiently high torsional amplitudes, however, the tight strand can become slack during a half cycle of vibration, introducing a nonlinear element into the computations. A theory taking this phenomenon into account has not yet been developed.

20.4.5 Noise

The transverse, longitudinal, and torsional vibrations radiate noise. The most important sources of their excitation are the roller impacts on the sprocket (high-frequency bursts at meshing-frequency rate) and the polygonal action with its very many harmonics, even if their kinematic effect is negligible. The lubrication is an excellent damper of noise. Some qualitative relations are given in Ref. 28.

20.5 SILENT OR INVERTED-TOOTH CHAIN

Silent chains consist of inverted-toothed links alternately assembled on articulating-joint parts. There are two types of chain, (1) flank contact and (2) heavy-duty chordal-action-compensating.

The *flank-contact-type chains*[19,20] are shown in Fig. 20.9*a*. The chain width may be as great as 16 times the pitch, and speed-reduction ratios as high as 8 are employed. The speed limit is about 6000 r/min for ⅜-in-pitch chain and 900 r/min for 2-in-pitch chain. The minimum and maximum number of sprocket teeth suggested is 17 and 150, respectively. The sprockets have a pitch diameter approximately equal to the outside diameter.

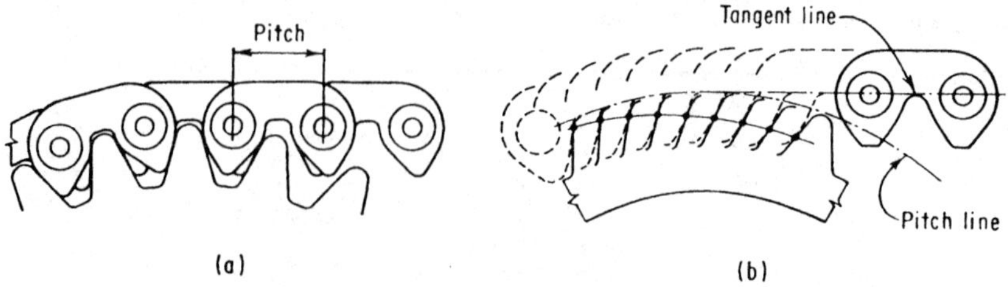

FIG. 20.9 Types of silent chain. (*a*) Flank-contact. (*b*) Chordal-action-compensating.

These chains are subject to the same speed variations and chordal rise and fall due to the polygonal chordal action as roller chains. In general, they are suggested for slow to moderate speed and 20,000 or more hours of life expectancy. They have an approximate ultimate tensile strength of 12,500 × pitch × width.

Heavy-duty chordal-action-compensating-type chains are designed to obtain the optimum in link strength and to eliminate the effects of chordal action. Figure 20.9*b* shows how the chain pitch line is maintained in a constant tangential relation to the sprocket pitch circle. This characteristic is obtained by generating the link profile for conjugate action with the sprocket tooth and/or by special articulating-joint design. These chains are used for a life requirement of less than 20,000 h and/or widely varying load conditions. They have an approximate ultimate tensile strength of 20,000 × pitch × width.

For further information on silent-tooth chains see Refs. 19 to 22.

REFERENCES

1. Binder, R. C.: "Mechanics of the Roller Chain Drive," Prentice-Hall, Inc, Englewood Cliffs, N.J., 1956.
2. "New Horsepower Ratings of American Standard Roller Chains," proposed by ARSCM, Diamond Chain Company, Inc., Indianapolis, Ind., 1960.
3. Frank, J. F., and C. O. Sundberg: "New Roller Chain Horsepower Ratings," *Mach. Des.*, July 6, 1961.
4. Kuntzmann, P.: "Roller Chain Drives" (French), Dunod, Paris, 1961.
5. Niemann, G.: "Machine Design" (German), vol. 2. Springer-Verlag OHG, Berlin, 1960.
6. Jackson and Moreland: "Design Manual for Roller and Silent Chain Drives," prepared for ARSCM, 1955.
7. Bouillon, G., and G. V. Tordion: "On Polygonal Action in Roller Chain Drives," *Trans. ASME, J. Eng. Ind.*, p. 243, May 1965.
8. Morrison, R.: "Polygonal Action in Chain Drives," *Mach. Des.*, vol. 24, no. 9, September 1952.
9. Mahalingham, S.: "Polygonal Action in Chain Drives," *J. Franklin Inst.*, vol. 265, no. 1, January 1958.
10. Mahalingham, S.: "Transverse Vibrations of Power Transmission Chains," *Brit. J. Appl. Phys.*, vol. 8, April 1957.
11. Whitney, L. H., and P. M. MacDonald: "Elastic Elongation of Chains," *Prod. Eng.*, February 1952.
12. Germond, H. S.: "Wear Limits for Roller and Silent Chain Drives," ASME Paper 60-WA-8, 1960.
13. Schakel, R. A., and C. O. Sundberg: "Proven Concepts in Oil Field Roller Chain Drive Selection," ASME Paper 57-PET-24, 1957.
14. Ignatenko, V. V.: "Variable Forces in the Strand of a Chain Drive" (Russian), *Bull. Inst. Higher Educ.*, no 4, 1961.
15. "Roller Chains and Sprockets," Catalog 8 ACME Chain Corp., Holyoke, Mass.
16. "Stock Power Transmission and Conveyor Products," Catalog 760, Diamond Chain Co., Inc., Indianapolis, Ind.
17. Radzimovsky, E. I.: "Eliminating Pulsations in Chain Drives," *Prod. Eng.*, July 1955.
18. Hofmeister, W. F., and H. Klaucke: "Dynamic Check Point Way to Longer Chain Life," *Iron Age*, August 1956.

19. American Standards Association: ASA B29.1, "Transmission Roller Chains and Sprocket Teeth" (SAE SP-69).
20. American Standards Association: ASA B29.3, "Double Pitch Power Transmission Chains and Sprockets" (SAE SP-69).
21. American Standards Association: ASA B29.2, "Inverted Tooth (Silent) Chains and Sprocket Teeth" (SAE SP-68).
22. American Standards Association: ASA B29.9, "Small Pitch Silent Chains and Sprocket Tooth Form (Less Than ⅜ Inch Pitch)" (SAE TR-96).
23. Rudolph, R. O., and P. J. Imse: "Designing Sprocket Teeth," *Mach. Des.*, pp. 102–107, Feb. 1, 1962.
24. Turnbull, S. R., and J. N. Fawcett: "An Approximate Kinematic Analysis of the Roller Chain Drive," *Proc. Inst. Mech. Eng. 4th World Cong. Theory Mach. Mech.*, p. 907, 1975.
25. Shimizu, H., and A. Sueoka: "Nonlinear Free Vibration of Roller Chain Stretched Vertically," *Bull. Japan Soc. Mech. Eng.*, vol. 19, no. 127, p. 22, January 1976.
26. Marshek, K. M.: "On the Analysis of Sprocket Load Distribution," *Mech. Mach. Theory*, vol. 14, pp. 135–139, 1979.
27. Fawcett, J. N., and S. W. Nicol: "A Theoretical Investigation of the Vibration of Roller Chain Drives," *ASME Proc. 5th World Cong. Theory Mach. Mech.*, vol. 2, p. 1482, 1979.
28. Uehara, K., and T. Nakajima: "On the Noise of Roller Chain Drives," *ASME Proc. 5th World Cong. Theory Mach. Mech.*, p. 906, 1979.
29. Fawcett, J. N., and S. W. Nicol: "Vibration of a Roller Chain Drive Operating at Constant Speed and Load," *Proc. Inst. Mech. Eng.*, vol. 194, p. 97, 1980.
30. Lee, T. W.: "Automated Dynamic Analysis of Chain-Driven Mechanical Systems," *Trans. ASME J. Mechanisms, Transmissions and Automation in Design*, vol. 105, p. 362, 1983.

CHAPTER 21
GEARING

John J. Coy, M.S.M.E., Ph.D., P.E.
Chief of Mechanical System Technology Branch
NASA Lewis Research Center
Cleveland, Ohio

Dennis P. Townsend, B.S.M.E.
Senior Research Engineer
NASA Lewis Research Center
Cleveland, Ohio

Erwin V. Zaretsky, B.S.M.E., J.D., P.E.
Chief Engineer of Structures
NASA Lewis Research Center
Cleveland, Ohio

21.1 SYMBOLS 21.2
21.2 INTRODUCTION 21.4
21.2.1 Early History of Gearing 21.4
21.2.2 Beginning of Modern Gear Technology 21.5
21.3 TYPES AND GEOMETRY 21.6
21.3.1 Parallel Shaft Gear Types 21.6
21.3.2 Intersecting Shaft Gear Types 21.17
21.3.3 Nonparallel, Nonintersecting Shafts 21.22
21.4 PROCESSING AND MANUFACTURE 21.27
21.4.1 Materials 21.27
21.4.2 Metallurgical Processing Variables 21.34
21.4.3 Manufacturing 21.38
21.5 STRESSES AND DEFLECTIONS 21.39
21.5.1 Lewis Equation Approach for Bending Stress Number 21.40
21.5.2 Allowable Bending Stress Number 21.45
21.5.3 Other Methods 21.50
21.6 GEAR LIFE PREDICTIONS 21.53
21.6.1 Theory of Gear Tooth Life 21.54
21.6.2 Life for the Gear 21.57
21.6.3 Gear System Life 21.57
21.6.4 Helical Gear Life 21.58
21.6.5 Bevel and Hypoid Gear Life 21.59
21.7 LUBRICATION 21.62
21.7.1 Lubricant Selection 21.63
21.7.2 Elastohydrodynamic Film Thickness 21.63
21.7.3 Boundary Lubrication 21.67
21.7.4 Lubricant Additive Selection 21.68
21.7.5 Jet Lubrication 21.69
21.7.6 Gear Tooth Temperature 21.72
21.8 POWER-LOSS PREDICTIONS 21.74
21.8.1 Sliding Loss 21.75
21.8.2 Rolling Loss 21.76
21.8.3 Windage Loss 21.77
21.8.4 Other Losses 21.77
21.8.5 Optimizing Efficiency 21.78
21.9 OPTIMAL DESIGN OF SPUR GEAR MESH 21.79
21.9.1 Minimum Size 21.80
21.9.2 Specific Torque Capacity 21.82
21.10 GEAR TRANSMISSION CONCEPTS 21.83
21.10.1 Series Trains 21.83
21.10.2 Multispeed Trains 21.84
21.10.3 Epicyclic Gearing 21.85
21.10.4 Split-Torque Transmissions 21.88
21.10.5 Differential Gearing 21.89
21.10.6 Closed-Loop Trains 21.90

21.1 SYMBOLS

A = Cone distance, m (in)
a = Semimajor axis of hertzian contact ellipse, m (in)
B = Backlash, m (in)
B_{10} = Bearing life at 90 percent reliability level
b = Semiminor axis of hertzian contact ellipse, m (in)
C = Center distance, m (in)
C_1 to C_7 = Special constants (see Table 21.19)
c = Distance from neutral axis to outermost fiber of beam, m (in)
c = Exponent
D = Diameter of pitch circle, m (in)
D_G = Pitch diameter of gear wheel, m (in)
D_W = Pitch diameter of worm gear, m (in)
d = Impingement depth, m (in)
E = Young's modulus of elasticity, N/m² (lb/in²)
E' = Effective elastic modulus, N/m² (lb/in²)
e = Weibull slope, dimensionless
F = Face width of tooth, m (in)
F_e = Effective face width, m (in)
f = Coefficient of friction
G = Dimensionless materials parameter
G_{10} = Life of gear at 90 percent reliability level, in millions of revolutions or hours
H = Dimensionless film thickness
H_B = Brinell hardness
h = Film thickness, m (in)
h = Exponent
I = Area moment of inertia, m⁴ (in⁴)
J = Geometry factor
k = Ellipticity ratio, dimensionless
K = Load intensity factor, N/m² (lb/in²)
K_a = Application factor
K_L = Life factor
K_m = Load distribution factor
K_R = Reliability factor
K_s = Size factor
K_T = Temperature factor
K_v = Dynamic factor
K_x = Cutter radius factor
L = Life, hours or millions of revolutions or stress cycles
L_W = Lead of worm gear, m (in)
l = Length, m (in)
M = Bending moment, N·m (lb·in)
m_c = Contact ratio
m_g = Gear ratio, N_2/N_1
N = Number of teeth
N_e = Equivalent number of teeth
n = Angular velocity, r/min
O = Centerpoint location
P = Diametral pitch, m^{-1} (in^{-1})
P_n = Normal diametral pitch, m^{-1} (in^{-1})
P_t = Transverse diametral pitch, m^{-1} (in^{-1})
p = Circular pitch, m (in)
p_a = Axial pitch, m (in)
p_b = Base pitch, m (in)
p_n = Normal circular pitch, m (in)
p_t = Transverse circular pitch, m (in)
Q = Power loss, kW (hp)
q = Number of planets
R_c = Rockwell C scale hardness
r = Radius, m (in)
r_a = Addendum circle radius, m (in)
r_b = Base circle radius, m (in)
r_c = Cutter radius, m (in)
r_p = Pitch circle radius, m (in)
r_{av} = Mean pitch cone radius, m (in)
S = Probability of survival
S = Stress, N/m² (lb/in²)

S_{at} = Allowable tensile stress, N/m^2 (lb/in^2)
S_{ay} = Allowable yield stress, N/m^2 (lb/in^2)
S_c = Maximum hertzian compressive stress, N/m^2 (lb/in^2)
S_r = Residual shear stress, N/m^2 (lb/in^2)
S_T = Probability of survival of complete transmission including bearings and gears
s = Distance measured from the pitch point along line of action, m (in)
T = Torque, N·m (lb·in)
T_{10} = Tooth life at 90 percent reliability, in millions of stress cycles or hours
t = Tooth thickness, m (in)
U = Dimensionless speed parameter
u = Integer number
V = Stressed volume, m^3 (in^3)
$\mathbf{v}$ = Velocity vector, m/s (ft/s)
v_s = Sliding velocity, m/s (ft/s)
v_j = Jet velocity, m/s (ft/s)
v_r = Rolling velocity, m/s (ft/s)
W = Dimensionless load parameter
W_N = Load normal to surface, N (lb)
W_r = Radial component of load, N (lb)
W_t = Transverse or tangential component of load, N (lb)
W_x = Axial component of load, N (lb)
w = Semiwidth of path of contact, m (in)
X_1 = Distance from line of centers to impingement point on pinion, m (in)
X_2 = Same as X_1, but for gear
Y = Lewis form factor, dimensionless
Z = Length of line of action, m (in)
z = Depth to critical shear stress, m (in)
α = Pressure-viscosity exponent, m^2/N (in^2/lb)
Γ = Pitch angle, degrees
γ = Angle, degrees
Δc = Change in center distance, m (in)
$\Delta \rho$ = Curvature difference, dimensionless
Δp = Differential pressure between oil and ambient, N/m^2 (lb/in^2)
δ = Tooth deflection, m (in)
ϵ = Involute roll angle, degrees (rad)
ϵ_H = Increment of roll angle for which a single pair of teeth is in contact, degrees (rad)
ϵ_L = Increment of roll angle during one of the periods of two pairs of teeth in contact, degrees (rad)
ϵ_c = Involute roll angle to lowest point of double tooth pair contact
η = Life, millions of stress cycles
θ = Angle of rotation, rad
ξ = Efficiency
Λ = Film thickness parameter, ratio of film thickness to rms surface roughness, dimensionless
λ_w = Lead angle of worm gear, degrees
μ = Lubricant absolute viscosity, $N \cdot s/m^2$ ($lb \cdot s/in^2$)
μ_o = Absolute viscosity at standard atmospheric pressure, $N \cdot s/m^2$ ($lb \cdot s/in^2$)
ν = Poisson's ratio, dimensionless
ρ = Radius of curvature, m (in)
Σ = Shaft angle, degrees
$\Sigma \rho$ = Curvature sum, m^{-1} (in^{-1})
σ = Surface rms roughness, m (in)
τ = Shear stress, N/m^2 (lb/in^2)
τ_{max} = Maximum shear stress, N/m^2 (lb/in^2)
τ_o = Maximum subsurface orthogonal reversing shear stress, N/m^2 (lb/in^2)
Υ = Parameter defined in Table 21.15
ϕ = Pressure angle, degrees
ϕ_n = Normal pressure angle, degrees
ϕ_t = Transverse pressure angle, degrees
ϕ_w' = Angle between load vector and tooth centerline (Fig. 21.36), degrees
ψ = Helix angle, degrees
ψ = Spiral angle for spiral bevel gear, degrees
ω = Angular velocity, rad/s

Subscripts

1 = Pinion
2 = Gear
a = At addendum, at addendum circle
B = Bearing
G = Gear wheel
j = Jet
m = Mesh
p = Planet
r = Rolling, ring
s = Sliding, sun
t = Threads
T = Total
w = Windage
W = Worm gear
x, y = Mutually perpendicular planes in which the minimum and maximum (principal) radii of surface curvature lie

21.2 INTRODUCTION

Gears are the means by which power is transferred from source to application. Gearing and geared transmissions drive the machines of modern industry. Gears move the wheels and propellers that transport us through the sea, land, and air. A very sizable section of industry and commerce in today's world depends on gearing for its economy, production, and likelihood.

The art and science of gearing has its roots before the common era. Yet many engineers and researchers continue to delve into the regions where improvements are necessary, seeking to quantify, establish, and codify the methods to make gears meet the ever-widening needs of advancing technology.[1] References 135–142 are recommended for further reading on gearing.

21.2.1 Early History of Gearing

The earliest written descriptions of gears are said to have been made by Aristotle in the fourth century B.C.[2] It has been pointed out[3,4] that the passage attributed to Aristotle by some[2] was actually from the writings of his school, in "Mechanical Problems of Aristotle" (ca. 280 B.C.). In the passage in question, there was no mention of gear teeth on the parallel wheels, and they may just as well have been smooth wheels in frictional contact. Therefore, the attribution of gearing to Aristotle is, most likely, incorrect.

The real beginning of gearing was probably with Archimedes who about 250 B.C. invented the endless screw turning a toothed wheel, which was used in engines of war. Archimedes also used gears to simulate astronomical ratios. The Archimedian spiral was continued in the hodometer and dioptra, which were early forms of wagon mileage indicators (odometer) and surveying instruments. These devices were probably "thought" experiments of Heron of Alexandria (ca A.D. 60), who wrote on the subjects of theoretical mechanics and the basic elements of mechanism. The oldest surviving relic containing gears is the Antikythera mechanism, so named because of the Greek island of that name near which the mechanism was discovered in a sunken ship in 1900. Professor Price[3] of Yale University has written an authoritative account of this mechanism. The mechanism is not only the earliest relic of gearing, but it also is an extremely complex arrangement of epicyclic differential gearing. The mechanism is identified as a calendrical computing mechanism for the sun and moon, and has been dated to about 87 B.C.

The art of gearing was carried through the European dark ages after the fall of Rome, appearing in Islamic instruments such as the geared astrolabes which were used to calculate the positions of the celestial bodies. Perhaps the art was relearned by the clock- and instrument-making artisans of fourteenth-century Europe, or perhaps some crystallizing ideas and mechanisms were imported from the East after the crusades of the eleventh through the thirteenth centuries.

It appears that the English abbot of St. Alban's monastery, born Richard of Wallingford, in A.D. 1330, reinvented the epicyclic gearing concept. He applied it to an astronomical clock, which he began to build at that time and which was completed after his death.

A mechanical clock of a slightly later period was conceived by Giovanni de Dondi (1348–1364). Diagrams of this clock, which did not use differential gearing,[3] appear in the sketchbooks of Leonardo da Vinci, who designed geared mechanisms himself.[3,5] In 1967 two of da Vinci's manuscripts, lost in the National Library in Madrid since 1830,[6] were rediscovered. One of the manuscripts, written between 1493 and 1497 and known as "Codex Madrid I,"[7] contains 382 pages with some 1600 sketches. Included among this display of Leonardo's artistic skill and engineering ability are his studies of gearing. Among these are tooth profile designs and gearing arrangements that were centuries ahead of their "invention."

21.2.2 Beginning of Modern Gear Technology

In the period 1450 to 1750, the mathematics of gear-tooth profiles and theories of geared mechanisms became established, Albrecht Dürer is credited with discovering the epicycloidal shape (ca. 1525). Philip de la Hire is said to have worked out the analysis of epicycloids and recommended the involute curve for gear teeth (ca. 1694). Leonard Euler worked out the law of conjugate action (ca. 1754).[5] Gears designed according to this law have a steady speed ratio.

Since the industrial revolution in the mid-nineteenth century, the art of gearing blossomed, and gear designs steadily became based on more scientific principles. In 1893 Wilfred Lewis published a formula for computing stress in gear teeth.[8] This formula is in wide use today in gear design. In 1899 George B. Grant, the founder of five gear manufacturing companies, published "A Treatise on Gear Wheels."[9] New inventions led to new applications for gearing. For example, in the early part of this century (1910), parallel shaft gears were introduced to reduce the speed of the newly developed reaction steam turbine enough to turn the driving screws of ocean-going vessels. This application achieved an overall increase in efficiency of 25 percent in sea travel.[2]

The need for more accurate and quiet-running gears became obvious with the advent of the automobile. Although the hypoid gear was within our manufacturing capabilities by 1916, it was not used practically until 1926, when it was used in the Packard automobile. The hypoid gear made it possible to lower the drive shaft and gain more usable floor space. By 1937 almost all cars used hypoid-geared rear axles. Special lubricant antiwear additives were formulated in the 1920s which made it practical to use hypoid gearing. In 1931 Earle Buckingham, chairman of an American Society of Mechanical Engineers (ASME) research committee on gearing, published a milestone report on gear-tooth dynamic loading.[10] This led to a better understanding of why faster-running gears sometimes could not carry as much load as slower-running gears.

High-strength alloy steels for gearing were developed during the 1920s and 1930s. Nitriding and case-hardening techniques to increase the surface strength of gearing were introduced in the 1930s. Induction hardening was introduced in 1950. Extremely clean steels produced by vacuum melting processes introduced in 1960 have proved effective in prolonging gear life.

Since the early 1960s there has been increased use of industrial gas turbines for electric power generation. In the range of 1000 to 14,000 hp, epicyclic gear systems have been used successfully. Pitch-line velocities are from 50 to 100 m/s (10,000 to 20,000 ft/min). These gear sets must work reliably for 10,000 to 30,000 h between overhauls.[1]

In 1976 bevel gears produced to drive a compressor test stand ran successfully for 235 h at 2984 kW (4000 hp) and 200 m/s (40,000 ft/min).[11] From all indications these gears could be used in an industrial application if needed. A reasonable maximum pitch-line velocity for commercial spiral-bevel gears with curved teeth is 60 m/s (12,000 ft/min).[12]

Gear system development methods have been advanced in which lightweight, high-speed, highly loaded gears are used in aircraft applications. The problems of strength and dynamic loads, as well as resonant frequencies for such gearing, are now treatable with techniques such as finite-element analysis, siren and impulse testing for mode shapes, and application of damping treatments where required.[13]

The material contained in this chapter will assist in the design, selection, applications, and evaluation of gear drives. Sizing criteria, lubricating consideration, material selection, and methods to estimate service life and power loss are presented.

21.3 TYPES AND GEOMETRY

References 14 and 15 outline the various gear types including information on proper gear selection. These references classify single-mesh gears according to the arrangement of their shaft axes in a single-mesh gear set. These arrangements are *parallel shafts; intersecting shafts*; and *nonparallel, nonintersecting shafts.* Table 21.1 lists and compares these gear sets for a single mesh.

21.3.1 Parallel Shaft Gear Types

Spur Gears. Spur gears are the most common type of gears. A representative pair of external spur gears is shown in Fig. 21.1. The basic geometry of spur gears is shown in Fig. 21.2. Teeth are straight and parallel to the shaft axis. This is consequently the simplest possible type of gear. The smaller of two gears in mesh is called the "pinion." The larger is customarily designated as the gear. In most applications (for speed reduction) the pinion is the driving element whereas the gear is the driven element.

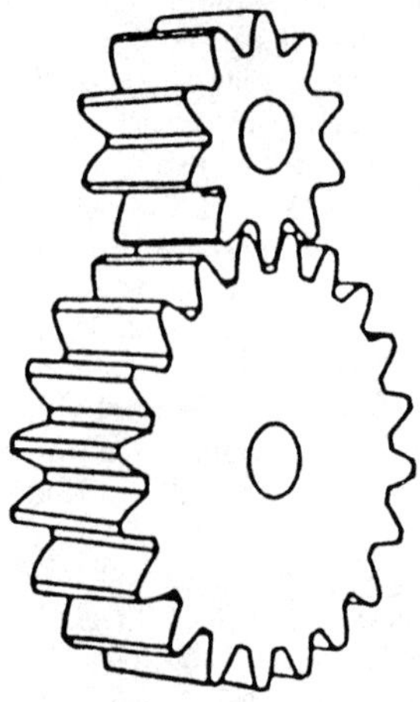

FIG. 21.1 Spur gears.

Most spur gear tooth profiles are cut to conform to an involute curve which ensures conjugate action. "Conjugate action" is defined as a constant angular velocity ratio between two meshing gears. Conjugate action may be obtained with any tooth profile shape for the pinion, provided the mating gear is made with a tooth shape that is conjugate to the pinion tooth shape. The conjugate tooth profiles are such that the common normal at the point of contact between the two teeth will always pass through a fixed point on the line of centers. The fixed point is called the "pitch point."

TABLE 21.1 Comparison of Single-Mesh Gears

Type of gear	Gear ratio range (nominal)	Maximum pitch line velocity, ft/min: Aircraft, high precision	Maximum pitch line velocity, ft/min: Commercial	Nominal efficiency at rated power, %	Remarks
Parallel shafts					
Spur	1:1–5:1	20,000	4,000	97–99.5	
Helical	1:1–10:1	40,000	4,000	97–99.5	15–35° helix angle
Double helical "herringbone"	1:1–20:1	40,000	4,000	97–99.5	
Conformal	1:1–12:1	—	—	97–99.5	
Intersecting shafts					
Straight bevel	1:1–8:1	10,000	1,000	97–99.5	
Spiral bevel	1:1–8:1	25,000	12,000	97–99.5	35° spiral angle most common
Zerol®* bevel	1:1–8:1	10,000	1,000	97–99.5	
Nonintersecting and nonparallel shafts					
Worm	3:1–100:1	10,000	5,000	50–90	
Double enveloping worm	3:1–100:1	10,000	4,000	50–98	
Crossed helical	1:1–100:1	10,000	4,000	50–95	
Hypoid	1:1–50:1	10,000	4,000	50–98	Higher ratios have lower efficiency

*Zerol is a registered trademark of Gleason Works, Rochester, N.Y.

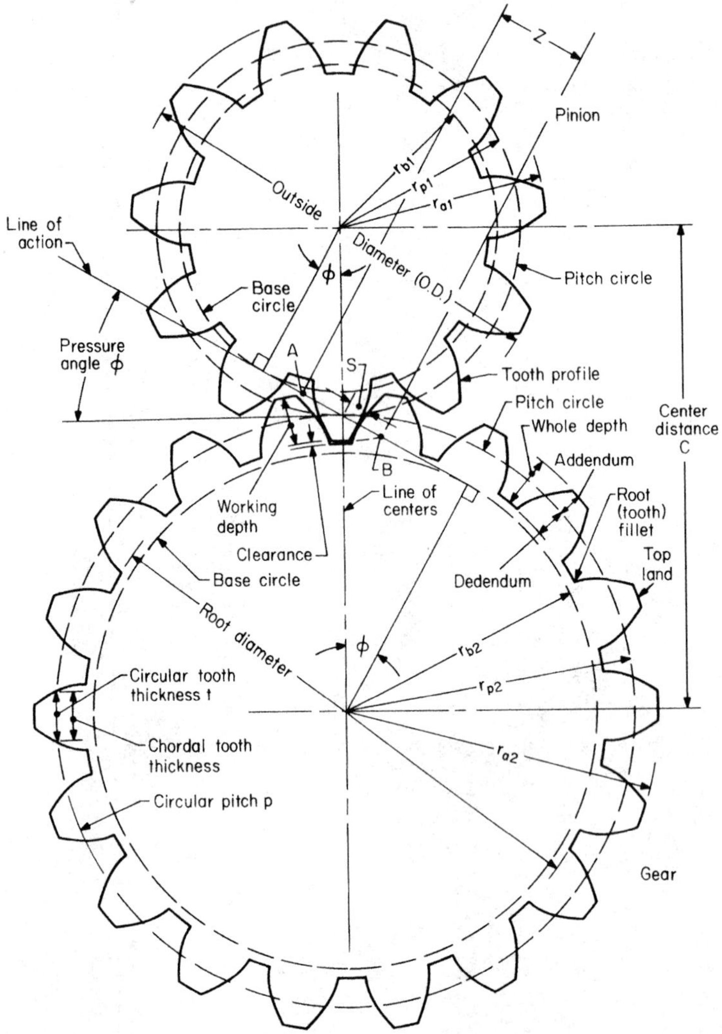

FIG. 21.2 External spur gear geometry.

Referring to Fig. 21.2, the line of action or pressure line is shown as the line which is tangent to both base circles. This is the line along which all points of contact between the two teeth will lie. The pressure angle ϕ is defined as the acute angle between the line of action and a line perpendicular to the line of centers. While the pressure angle can vary for different applications, most spur gears are cut to operate at pressure angles of 20 or 25°.

The circular pitch p of a spur gear is defined as the distance, on the pitch circle, from a point on a tooth to the corresponding point of the adjacent tooth. The circular pitch

$$p = \frac{2\pi r_p}{N} \tag{21.1}$$

where r_p is the pitch circle radius and N is the number of teeth. Similarly, the base pitch p_b is the distance on the base circle from a point on one tooth to the corresponding point on an adjacent tooth. Hence, base pitch

$$p_b = \frac{2\pi r_b}{N} \tag{21.2}$$

where r_b is the base circle radius. Also, from geometry,

$$p_b = p \cos \phi \tag{21.3}$$

The diametral pitch P is defined as the number of teeth on the gear divided by the pitch circle diameter in inches (determines the relative sizes of gear teeth):

$$P = \frac{N}{D} \tag{21.4}$$

where D is the pitch circle diameter and

$$P \cdot p = \pi \tag{21.5}$$

For a given size gear diameter, the larger the value for P, the greater the number of gear teeth and the smaller their size. Module is the reciprocal of diametral pitch in concept; but whereas diametral pitch is expressed as number of teeth per inch, the module is generally expressed in millimeters per tooth. Metric gears (in which tooth size is expressed in module) and American standard-inch-diametral pitch gears are not interchangeable.

Referring to Fig. 21.2, the distance between the centers of two gears is

$$C = \frac{D_1 + D_2}{2} \tag{21.6}$$

The space between the teeth must be larger than the mating gear tooth thickness in order to prevent jamming of the gears.

The difference between the tooth thickness and tooth space as measured along the pitch circle is called *backlash.* Backlash may be created by cutting the gear teeth slightly thinner than the space between teeth or by setting the center distance slightly greater between the two gears. In the second case, the operating pressure angle of the gear pair is increased accordingly. The backlash for a gear pair must be sufficient to permit free action under the most severe combination of manufacturing tolerances and operating temperature variations. Backlash should be very small in positioning control systems, and it should be quite generous for single-direction power gearing. Table 21.2 gives recommended backlash amounts. If the center distance between mating external

TABLE 21.2 Recommended Backlash for Assembled Gears

P	Backlash, in	P	Backlash, in
1	0.025–0.040	5	0.006–0.009
$1\frac{1}{2}$	0.018–0.027	6	0.005–0.008
2	0.014–0.020	7	0.004–0.007
$2\frac{1}{2}$	0.011–0.016	8–9	0.004–0.006
3	0 009–0.014	10–13	0.003–0.005
4	0.007–0.011	14–32	0.002–0.004

gears is increased by an amount ΔC, the resulting increase in backlash ΔB is given by the following equation:

$$\Delta B = \Delta C \times 2 \tan \phi \tag{21.7}$$

For the mesh of internal gears, Eq. 21.7 gives the decrease in backlash for an increase in center distance.

Referring to Fig. 21.2, the distance between points A and B is the length of contact along the line of action, which is defined as the distance between where the gear outside circumference intersects the line of action and where the pinion outside circumference intersects the line of action. This is the total length along the line of action for which there is tooth contact. This distance can be determined by the various radii and the pressure angle as follows:

$$Z = [r_{a1}^2 - (r_{p1} \cos \phi)^2]^{1/2} + [r_{a2}^2 - (r_{p2} \cos \phi)^2]^{1/2} - C \sin \phi \tag{21.8}$$

The radii of curvature of the teeth when contact is at a distance s along the line of action from the pitch point can be represented for the pinion as

$$\rho_1 = r_{p1} \sin \phi \pm s \tag{21.9}$$

and for the gear

$$\rho_2 = r_{p2} \sin \phi \pm s \tag{21.10}$$

Where s is positive for Eq. (21.9), it must be negative for Eq. (21.10) and vice versa.

In order to determine how many teeth are in contact, the contact ratio of the gear mesh must be determined. The contact ratio m_c is defined as

$$m_c = \frac{Z}{p_b} \tag{21.11}$$

Gears are generally designed with contact ratios between 1.2 and 1.6. A contact ratio of 1.6, for example, means that for 40 percent of the time, there will be one pair of teeth in contact, and for 60 percent of the time, two pairs of teeth will be in contact. A contact ratio of 1.2 means that 80 percent of the time one pair of teeth will be in contact and two pairs of teeth will be in contact 20 percent of the time.

Gears with contact ratios greater than 2 are referred to as *high-contact ratio gears.* For high-contact ratio gears there are never fewer than two pairs of teeth in contact. A contact ratio of 2.2 means that for 80 percent of the time there will be two pairs of teeth in contact and for 20 percent of the time there will be three pairs of teeth in contact. High-contact ratio gears are generally used in select applications where long life is required. Figure 21.3*a* shows a high-contact-ratio (2.25) modified tooth, and Fig. 21.3*b* shows a normal-contact-ratio (1.3) involute tooth. Analysis should be cautious in designing high-contact ratio gearing because of higher bending stresses which may occur in the tooth dedendum region. Also, higher sliding in the tooth contact may contribute to distress of the tooth surfaces. In addition, higher dynamic loading may occur with high-contact ratio gearing.[16]

Interference of the gear teeth is an important consideration. The portion of the spur gear below the base circle is sometimes cut as a straight radial line, but never as an involute curve (Fig. 21.4). Hence, if contact should occur below the base circle, nonconjugate action (interference) will occur. The maximum addendum radius of the gear without interference is the value r_a' calculated from the following equation:

$$r_a' = (r_b^2 + c^2 \sin^2 \phi)^{1/2} \tag{21.12}$$

FIG. 21.3 Comparison of gear contact ratio. (*a*) Modified tooth profile; contact ratio 2.25. (*b*) Conventional involute tooth; contact ratio 1.3.

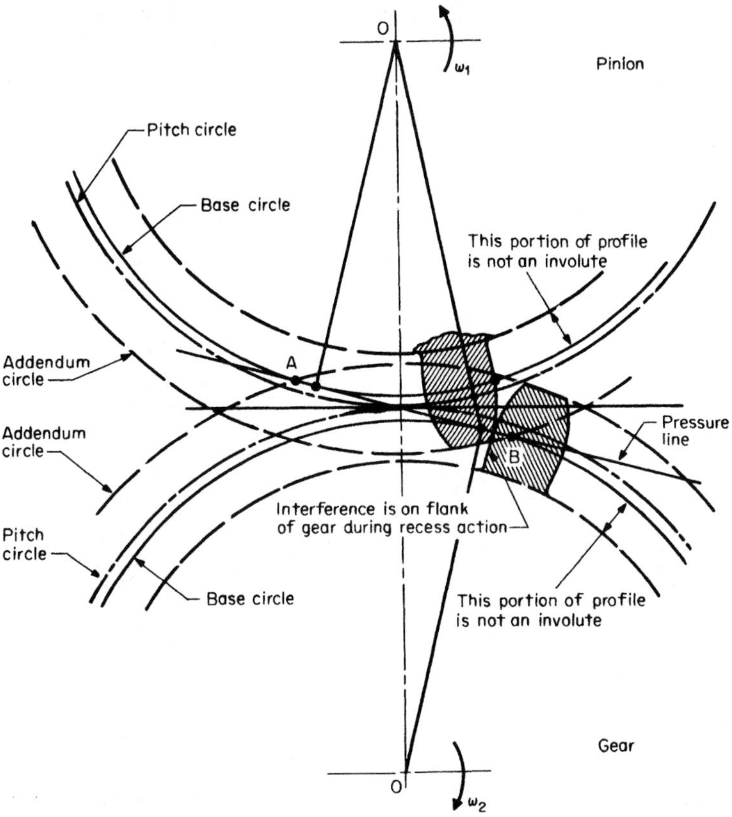

FIG. 21.4 Interference in the action of gear teeth. (Interference is on flank of pinion during approach action and flank of gear during recess action.)

where r'_a is the radial distance from the gear center to the point of tangency of the mating gear's base circle with the line of action. Should interference be indicated, there are several methods to eliminate it. The center distance may be increased, which will also increase the pressure angle and change the contact ratio. Also, the addendum may be shortened, with a corresponding increase in the addendum of the mating member. The method used depends upon the application and experience.

The geometry for *internal gearing* where teeth are cut on the inside of the rim is shown in Fig. 21.5. The pitch relationships previously discussed for external gears apply also to internal gears [Eqs. (21.1) through (21.5)]. A secondary type of involute interference called "tip fouling" may also occur. The geometry should be carefully checked in the region labeled F in Fig. 21.5 to see if this occurs.

Helical Gears. Figure 21.6 shows a helical gear set. These are cylindrical gears with the teeth cut in the form of a helix; one gear has a right-hand helix and the mating gear a left-hand helix. Helical gears have a greater load-carrying capacity than equivalent-size spur gears. Although these gears produce less vibration than spur gears because of overlapping of the teeth, a high thrust load is produced along the axis of rotation. This results in high rolling-element bearing loads which may reduce the life and reliability of a transmission system. This end thrust increases with helix angle.

In order to overcome the axial-thrust load, double-helical gear sets (herringbone gears) are used (Fig. 21.7). The thrust loads are counterbalanced, so that no resultant axial load is transmitted to the bearings. Space is sometimes provided between the two

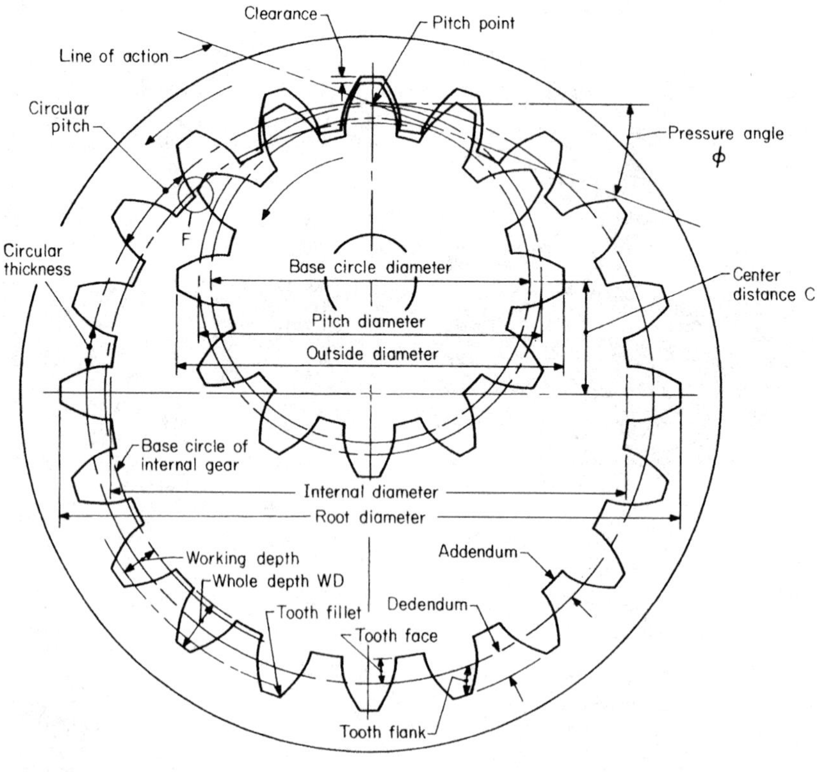

FIG. 21.5 Internal spur gear geometry.

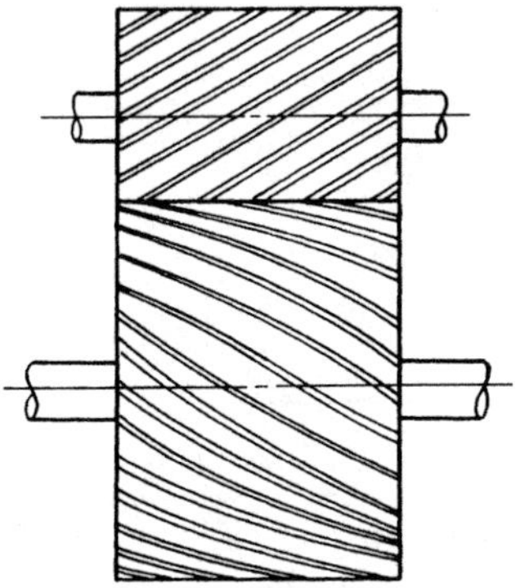

FIG. 21.6 Single-helical gear set.

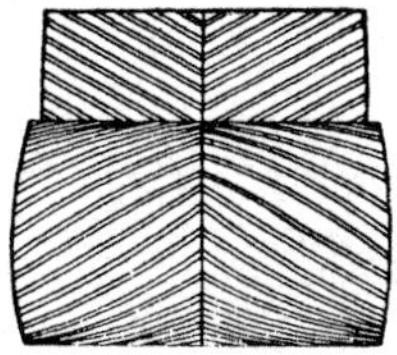

FIG. 21.7 Double-helical, or herringbone, gear set.

sets of teeth to allow for runout of the tooth-cutting tool, or the gear is assembled in two halves. The space should be small because it reduces the active face width. Double-helical gears are used for the transmission of high torques at high speeds in continuous service.

The geometry of a helical gear is shown in Fig. 21.8. While spur gears have only one diametral and circular pitch, helical gears have two additional pitches. The normal circular pitch p_n is the distance between corresponding points of adjacent teeth as measured in plane *BB* of Fig. 21.8, which is perpendicular to the helix.

$$p_n = p_t \cos \psi \tag{21.13}$$

where p_t is the transverse circular pitch and ψ is the helix angle.

The transverse circular pitch p_t is measured in plane *AA* of Fig. 21.8 perpendicular to the shaft axis. The axial pitch p_a is a similar distance measured in a plane parallel to the shaft axis:

$$p_a = p_t \cot \psi \tag{21.14}$$

The diametral pitch p_t in the transverse plane is

$$p_t = \frac{N}{D} \tag{21.15}$$

where N is the number of teeth of the gear and D is the diameter of the pitch circle.

The normal diametral pitch is

$$p_n = \frac{p_t}{\cos \phi} \tag{21.16}$$

To maintain contact across the entire tooth face, the minimum face width F must be greater than the axial pitch p_a. However, to assure a smooth transfer of load during tooth engagement the face width F should be at least 1.2 to 2 times the axial pitch p_a.

The two pressure angles associated with the helical gear are the transverse pressure angle ϕ_t measured in plane *AA* (Fig. 21.8) and the normal pressure angle ϕ_n measured in plane *BB* can be related as follows:

$$\tan \varphi_n = \tan \phi_t \cos \psi \tag{21.17}$$

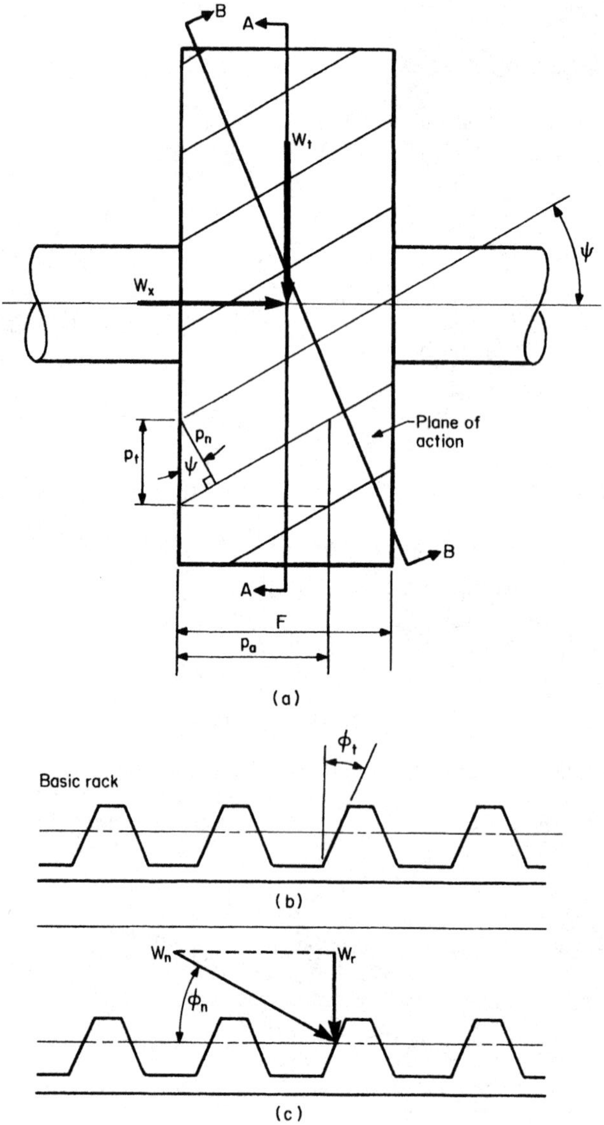

FIG. 21.8 Helical gear geometry. (*a*) Gear. (*b*) Section *AA* (transverse plane). (*c*) Section *BB* (normal plane).

The three components of normal load W_N acting on a helical gear tooth can be written as follows:

Tangential: $$W_t = W_N \cos \phi_n \cos \psi \tag{21.18}$$

Radial: $$W_r = W_N \sin \varphi_n \tag{21.19}$$

Axial: $$W_x = W_N \cos \varphi_n \sin \psi \tag{21.20}$$

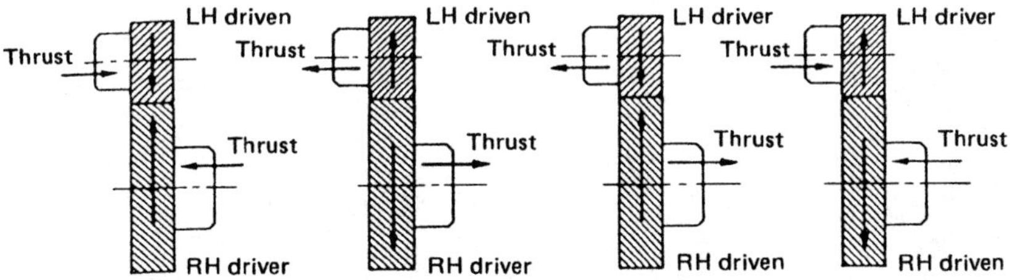

FIG. 21.9 Direction of the axial thrust load for helical gears on parallel shafts. LH indicates left-hand and RH indicates right-hand.

where W_t is the tangential load acting at the pitch circle perpendicular to the axis of rotation. Figure 21.9 shows the direction of the axial-thrust loads for pairs of helical gears, considering whether the driver has a right-hand or left-hand helix and the direction of rotation. These loads must be considered when selecting and sizing the rolling-element bearings to support the shafts and gears.

The plane normal to the gear teeth, *BB* in Fig. 21.8, intersects the pitch cylinder. The gear tooth profile generated in this plane would be a spur gear, having the same properties as the actual helical gear. The number of teeth of the equivalent spur gear in the normal plane is known as either the "virtual" or "equivalent" number of teeth. The equivalent number of teeth is

$$N_e = \frac{N}{\cos^3 \psi} \tag{21.21}$$

Conformal (Wildhaber-Novikov) Gears. In 1923, E. Wildhaber filed a U.S. patent application[17] for helical gearing of a circular arc form. In 1926 he filed a patent application for a method of grinding of this form of gearing.[18] Until Wildhaber, helical gears were made only with screw involute surfaces. However, Wildhaber-type gearing failed to receive the support necessary to develop it into a working system.

In 1956, M. L. Novikov in the U.S.S.R. was granted a patent for a similar form of gearing.[19] Conformal gearing then advanced in the U.S.S.R., but the concept found only limited application. Conformal gearing was also the subject of research in Japan and China but did not find widespread industrial application.

The first and probably only application of conformal gearing to aerospace technology was by Westland Helicopter in Great Britain.[20,21] Westland modified the geometry of the conformal gears as proposed by Novikov and Wildhaber and successfully applied the modified geometry to the transmission of the Lynx helicopter. The major advantage demonstrated on this transmission system is increased gear load capacity without gear tooth fracture. It would be anticipated that the gear system would exhibit improved power-to-weight ratio and/or increased life and reliability. However, these assumptions have neither been analytically shown nor experimentally proved.

The tooth geometry of a conformal gear pair is shown in Fig. 21.10. To achieve the maximum contact area, the radii of two mating surfaces, r_1 and r_2, should be identical. However, this is not practical. Inaccuracies in fabrication will occur on the curvatures and in other dimensions such as the center distance. Such inaccuracies can cause stress concentrations on the edges and tips of the gear teeth. To prevent this edge loading, the radii are made slightly unequal and are struck from different centers. Although this slight mismatch decreases the stress concentrations, the contact surface area is also reduced, which amplifies the hertzian contact stress. The resultant contact stresses are dependent on the helix angle.

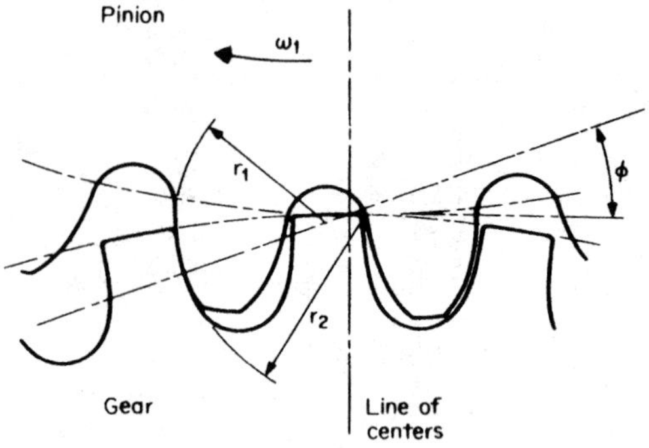

FIG. 21.10 Conformal tooth geometry.

As a historical note, in early designs thermal expansion in gearcases and shafts was sufficient to affect the gear pair center distances; the contact stresses were thus increased, leading to early failure.[22]

As was discussed for involute gearing, a constant velocity ratio exists for spur gear tooth pairs at every angular position because the teeth "roll" over each other. However, with conformal gearing, there is only one angular position where a tooth is in contact with its opposite number. Immediately before and after that point there is no contact. To achieve a constant velocity ratio, the teeth must run in a helical form across the gear and an overlap must be achieved from one tooth pair to the next as the contact area shifts across the gear, from one side to the other, during meshing.[21] The sweep velocity of the contact area across the face width is often thought of as being a pure

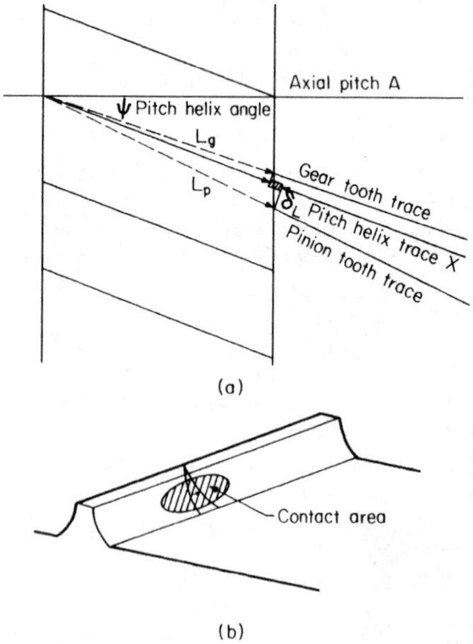

FIG. 21.11 Conformal gearing tooth contact trace diagram. (*a*) Sliding components in rolling action. (*b*) Relative sliding velocities in contact area.

rolling action, because no physical translation of the metal occurs in this direction. However, there is a small sliding component acting along the tooth.

If one considers the pitch surfaces of the mating gears, the length of the meshing helices would be the same on both the pinion and the gear. But the conformal system uses an all-addendum pinion and all-dedendum wheel, so when the contact area traverses one axial pitch, it must sweep over a longer distance on the pinion and a shorter distance on the gear. Thus, a small sliding component exists as shown in Fig. 21.11*a*. The magnitude of the sliding can vary with tooth design. However, since the sliding velocity is dependent on the displacement from the pitch surface, the percentage sliding will vary across the contact area, as shown in Fig. 21.11*b*. The main sliding velocity which occurs at the tooth contact acts up and down the tooth height. The magnitude of this sliding component can vary up to approximately 15 percent of the pitch helix sweep velocity. Thus, in comparison with involute gears, the slide-roll ratios are significantly lower, although the two motions being compared are in different directions. This lower slide-roll ratio seems to indicate why slightly lower power losses can be attained using conformal gears.[21]

21.3.2 Intersecting Shaft Gear Types

There are several types of gear systems which can be used for power transfer between intersecting shafts. The most common of these are straight-bevel gears and spiral-bevel gears. In addition, there are special bevel gears which accomplish the same or similar results and are made with special geometrical characteristics for economy. Among these are Zerol* gears, Conifex* gears, Formate* gears, Revacycle* gears, and face gears. For the most part, the geometry of spiral-bevel gears is extremely complex and does not lend itself to simplified formulas or analysis.

The gear tooth geometry is dictated by the machine tool being used to generate the gear teeth and the machine tool settings. The following are descriptions of these gears:

Straight-Bevel Gears. Bevel gear arrangements are shown in Fig. 21.12. Figure 21.13 is a drawing of a straight-bevel gear showing the terminology and important physical

*Registered trademarks of Gleason Works, Rochester, N.Y.

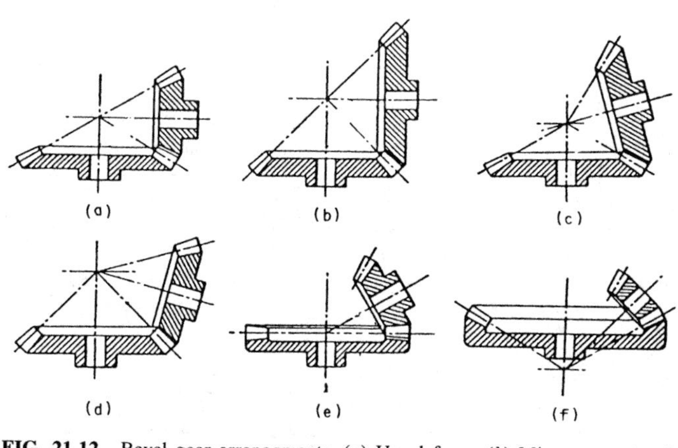

FIG. 21.12 Bevel gear arrangements. (*a*) Usual form. (*b*) Miter gears. (*c, d*) Other forms. (*e*) Crown gear. (*f*) Internal bevel.

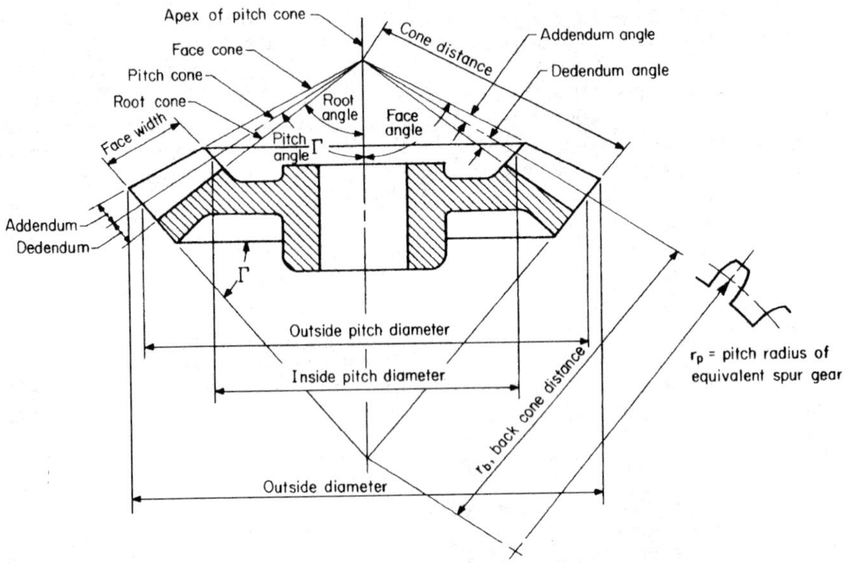

FIG. 21.13 Straight-bevel gear terminology.

dimensions. Straight-bevel gears are used generally for relatively low speed applications with pitch line velocities up to 1000 ft/min and where vibration and noise are not important criteria. However, with carefully machined and ground straight-bevel gears it may be possible to achieve speeds up to 15,000 ft/min.

Bevel gears are mounted on intersecting shafts at any desired angle, although 90° is most common. They are designed and manufactured in pairs and as a result are not always interchangeable. One gear is mounted on the cantilevered or outboard end of the shaft. Because of the outboard mounting, the deflection of the shaft where the gear is attached can be rather large. This would result in the teeth at the small end moving out of mesh. The load would thus be unequally distributed, with the larger ends of the teeth taking most of the load. To reduce this effect, the tooth face width is usually made no greater than one-third the cone distance. Bevel gears are usually classified according to their pitch angle. A bevel gear having a pitch angle of 90° and a plane for its pitch surface is known as a *crown gear.*

When the pitch angle of a bevel gear exceeds 90°, it is called an *internal bevel gear.* Internal bevel gears cannot have a pitch angle very much greater than 90° because of problems incurred in manufacturing such gears. These manufacturing difficulties are the main reason why internal bevel gears are rarely used.

Bevel gears with pitch angles less than 90° are the type most commonly used. When two meshing bevel gears have a shaft angle of 90° and have the same number of teeth, they are called *miter gears.* In other words, miter gears have a speed ratio of 1. Each of the two gears has a 45° pitch angle.

The relationship (Tregold's approximation) between the actual number of teeth for the bevel gear, the pitch angle, and the virtual or equivalent number of teeth is

$$N_e = \frac{N}{\cos \Gamma} \tag{21.22}$$

where Γ is the pitch angle. The back cone distance becomes the pitch radius for the equivalent spur gear.

Face Gears. These gears have teeth cut on the flat face of the blank. The face gear meshes at right angles with a spur or helical pinion. When the shafts intersect, it is known as an "on-center" face gear. These gears may also be offset to provide a right-angle nonintersecting shaft drive.

Coniflex Gears. These are straight-bevel gears whose teeth are crowned in a lengthwise direction to accommodate small shaft misalignments.

Formate Gears. These have the gear member of the pair with nongenerated teeth, usually with straight tooth profiles. The pinion member of the pair has generated teeth that are conjugate to the rotating gear.

Revacycle Gears. These are straight-bevel gears generated by a special process, with a special tooth form.

Spiral-Bevel Gears. A spiral-bevel gear pair is shown in Fig. 21.14*a*. The teeth of spiral-bevel gears are curved and oblique. These gears are suitable for pitch-line velocities up to 12,000 ft/min. Ground teeth extend this limit to 25,000 ft/min.

There are different types of spiral-bevel gears depending on the method used to generate the gear tooth surfaces. Among these are spiral-bevel gears made by the Gleason method, the Klingelnberg system, and the Oerlikon system. These companies have developed specified and detailed directions for the design of spiral-bevel gears which are related to the respective method of manufacture. However, there are some general considerations which are common for the generation of all types of spiral-bevel gears (Fig. 21.14*b*) such as the concept of pitch cones, the generating gear, and the conditions of force transmission. The actual tooth gear geometry is dictated by the machine tool settings supplied by the respective manufacture.

The standard spiral-bevel gear has a pressure angle of 20°, although 14½ and 16° angles are used. The usual spiral angle is 35°. Because spiral gears are much stronger than similar-sized straight or Zerol gears, they can be used for large speed reduction ratio applications at a reduced overall installation size. The hand of the spiral should be selected so as to cause the gears to separate from each other during rotation and not to force them together, which might cause jamming.

Thrust loads produced in operation with spiral-bevel gears are greater than those produced with straight-bevel gears. An analogy for the relationship that exists between straight- and spiral-bevel gears is that existing between spur and helical gears. Localized tooth contact is also achieved. This means that some mounting and load deflections can occur without a resultant load concentration at the end of the teeth. The total force W_N acting normal to the pinion tooth and assumed concentrated at the average radius of the pitch cone may be divided into three perpendicular components. These are the transmitted, or tangential, load W_t; the axial, or thrust, component W_x; and the separating, or radial, component W_r. The force W_t may, of course, be computed from the following:

$$W_t = \frac{T}{r_{\text{av}}} \tag{21.23}$$

where T is the input torque and r_{av} is the radius of the pitch cone measured at the midpoint of the tooth. The forces W_x and W_r depend upon the hand of the spiral and the direction of rotation. Thus there are four possible cases to consider. For a right-hand pinion spiral with clockwise pinion rotation and for a left-hand spiral with counterclockwise rotation. For gears with 90° between shaft axes, the equations are

(a)

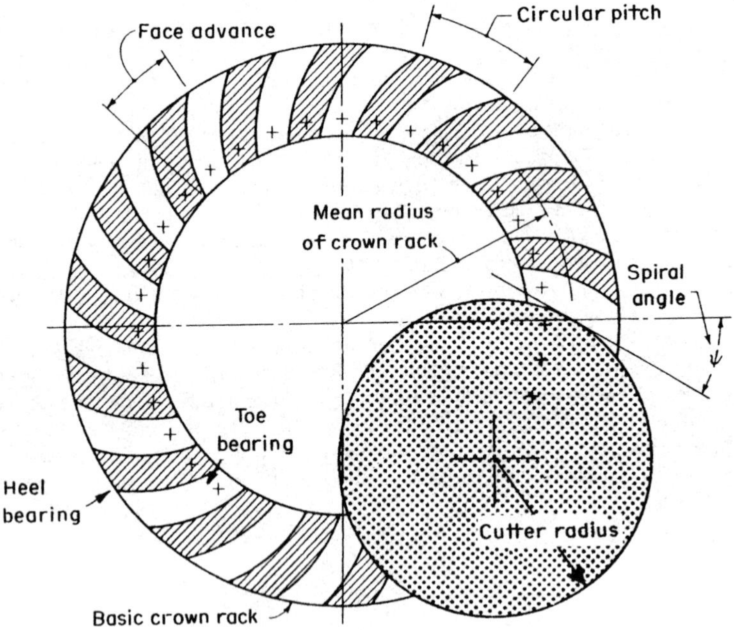

(b)

FIG. 21.14 Spiral-bevel gears. (*a*) Spiral-bevel gear pair. (*b*) Cutting spiral gear teeth on basic crown rack.

$$W_x = \frac{W_t}{\cos\psi}(\tan\phi_n \sin\Gamma - \sin\psi\cos\Gamma) \tag{21.24}$$

$$W_r = \frac{W_t}{\cos\psi}(\tan\phi_n \cos\Gamma + \sin\psi\sin\Gamma) \tag{21.25}$$

The other two cases are a left-hand spiral with clockwise rotation and a right-hand spiral with counterclockwise rotation. For these two cases, the equations are

$$W_x = \frac{W_t}{\cos\psi}(\tan\phi_n \sin\Gamma + \sin\psi\cos\Gamma) \tag{21.26}$$

$$W_r = \frac{W_t}{\cos\psi}(\tan\phi_n \cos\Gamma - \sin\psi\sin\Gamma) \tag{21.27}$$

where ψ = spiral angle
Γ = pinion pitch angle
ϕ_n = normal pressure angle

and the rotation is observed from the input end of the pinion shaft. Equations (21.24) to (21.27) give the forces exerted by the gear on the pinion. A positive sign for either W_x or W_r indicates that it is directed away from the cone center. The forces exerted by the pinion on the gear are equal and opposite. Of course, the opposite of an axial pinion load is a radial gear load, and the opposite of a radial pinion is an axial gear load.

The applications in which Zerol bevel gears are used are very much the same as those for straight-bevel gears (Fig. 21.15). The minimum suggested number of teeth is 14, one or more teeth should always be in contact, and the basic pressure angle is 20°, although angles of 22½ or 25° are sometimes used to eliminate undercutting.

FIG. 21.15 Zerol gears. (*Courtesy of Gleason Works, Rochester, N.Y.*)

21.3.3 Nonparallel, Nonintersecting Shafts

Nonparallel, nonintersecting shafts lie in parallel planes and may be skewed at any angle between 0 and 90°. Among the gear systems used for the purpose of power-transfer between nonparallel, nonintersecting shafts are hypoid gears, crossed-helical gears, worm gears, Cone-Drive* worm gears, face gears, Spiroid† gears, Planoid† gears, Helicon† gears, and Beveloid‡ gears. The following are descriptions of these gears.

Hypoid Gears. A hypoid gear pair is shown in Fig. 21.16. Hypoid gears are very similar to spiral-bevel gears. The main difference is that their pitch surfaces are hyperboloids rather than cones. As a result, their pitch axes do not intersect, the pinion axis being above or below the gear axis. In general, hypoid gears are most desirable for those applications involving large speed reduction ratios, those having nonintersecting shafts, and those requiring great smoothness and quietness of operation. Hypoid gears are almost universally used for automotive applications, allowing the drive shaft to be located underneath the passenger compartment floor. They operate more smoothly and quietly than spiral-bevel gears and are stronger for a given ratio. Because the two supporting shafts do not intersect, bearings can be mounted on both sides of the gear to provide extra rigidity. The pressure angles usually range between 19 and 22½°. The minimum number of teeth suggested is 8 for speed ratios greater than 6:1, and 6 for smaller ratios. High-reduction hypoids permit ratios between 10:1 and 120:1 and even as high as 360:1 in fine pitches.

Crossed-Helical Gears. A pair of crossed-helical gears, also known as spiral gears, are shown in Fig. 21.17. They are in fact ordinary helical gears used in nonparallel shaft

*Registered trademark, Michigan Tool Company, Detroit, Michigan.
†Registered trademark, Illinois Tool Works, Chicago, Illinois.
‡Registered trademark, Vinco Corporation, Detroit, Michigan.

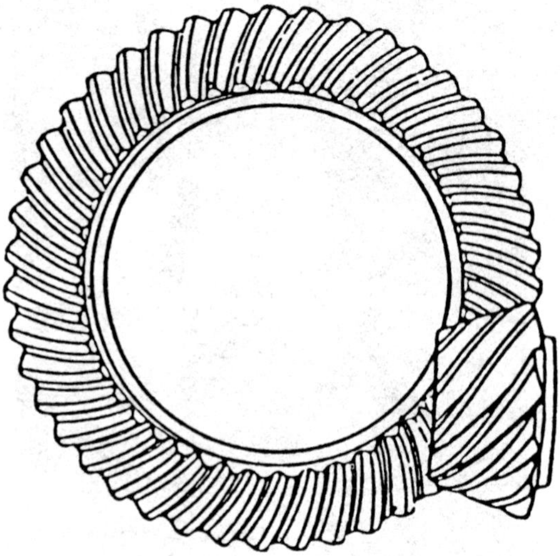

FIG. 21.16 Hypoid gears.

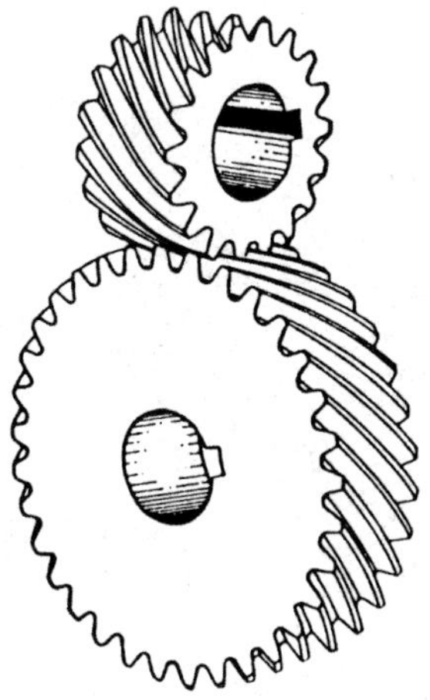

FIG. 21.17 Crossed-helical gears.

applications. They transmit relatively small amounts of power because of sliding action and limited tooth contact area. However, they permit a wide range of speed ratios without change of center distance or gear size. They can be used at angles other than 90°. In order for two helical gears to operate as crossed-helical gears, they must have the same normal diametral pitch P_n and normal pressure angle ϕ_n. The gears do not need to have the same helix angle or be opposite of hand. In most crossed gear applications, the gears have the same hand.

The relationship between the helix angles of the gears and the angle between the shafts on which the gears are mounted is shown in Fig. 21.17 and given by

$$\Sigma = \psi_1 + \psi_2 \tag{21.28}$$

for gears having the same hand and

$$\Sigma = \psi_1 - \psi_2 \tag{21.29}$$

for gears having opposite hands, where Σ is the shaft angle. For crossed-helical gears where the shaft angle is 90°, the helical gears must be of the same hand.

The center distance between the gears is given by the following relation:

$$C = \frac{1}{2P_n}\left(\frac{N_1}{\cos\psi_1} + \frac{N_2}{\cos\psi_2}\right) \tag{21.30}$$

The sliding velocity v_s acts at the pitch point tangentially along the tooth surface (Fig. 21.18), where by the law of cosines

$$v_s = [v_1^2 + v_2^2 - 2v_1v_2\cos\Sigma]^{1/2} \tag{21.31}$$

For shaft angles of 90°, the sliding velocity is given by

$$v_s = \frac{v_1}{\cos\psi_2} = \frac{v_2}{\cos\psi_1} \tag{21.32}$$

Referring to Fig. 21.19, the direction of the thrust load for crossed-helical gears may be determined. The forces on the crossed-helical gears are

Transmitted force: $$W_t = W_N\cos\phi_n\cos\psi \tag{21.33}$$

Axial thrust load: $$W_x = W_N\cos\phi_n\sin\psi \tag{21.34}$$

Radial or separating force: $$W_r = W_N\sin\phi_n \tag{21.35}$$

where ϕ_n is the normal pressure angle.

The pitch diameter is

$$D = \frac{N}{P_n\cos\psi} \tag{21.36}$$

Since the pitch diameters are not directly related to the tooth numbers, they cannot be used to obtain the angular velocity ratio. This ratio must be obtained from the ratio of the tooth numbers.

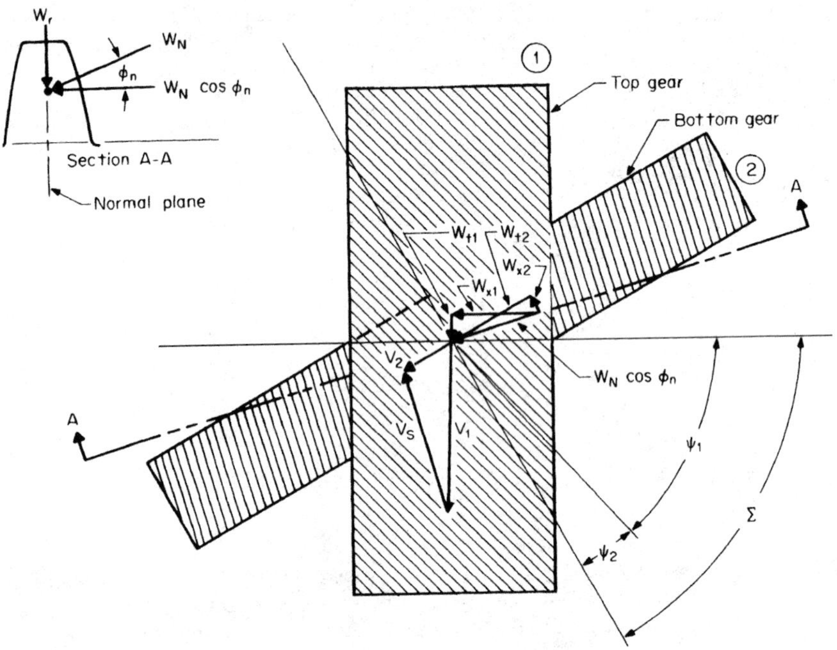

FIG. 21.18 Meshing crossed-helical gears showing the relationship between the helix angles, shaft angle, and load and velocity vectors.

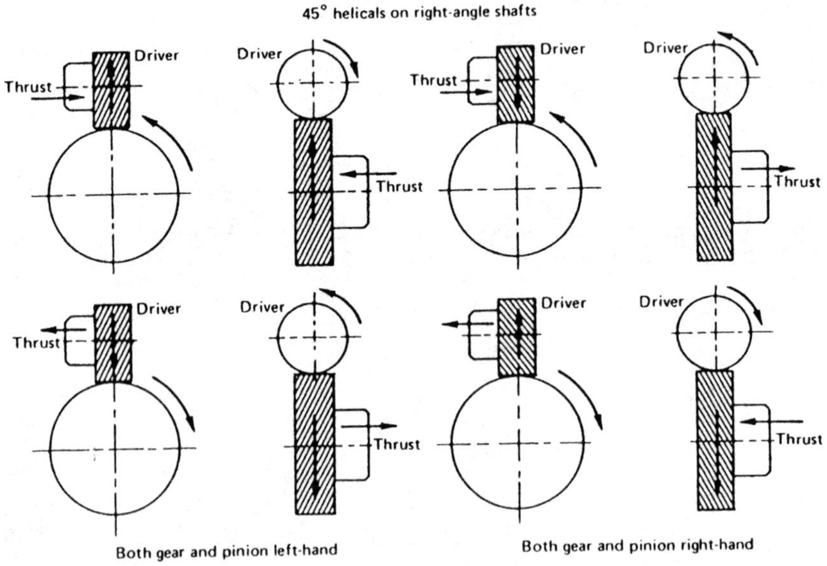

FIG. 21.19 Direction of thrust load for two meshing crossed-helical gears mounted on shafts that are oriented 90° to each other.

Worm Gears. A worm gear set is depicted in Fig. 21.20 showing forces which are discussed later. Referring to Fig. 21.21, the worm gear set is comprised of a worm which resembles a screw and a worm wheel which is a helical gear. The shafts on which the worm and wheel are mounted are usually 90° apart. Because of the screw action of worm gear drives, they are quiet, vibration-free, and produce a smooth output. On a given center distance, much higher ratios can be obtained through a worm gear set than with other conventional types of gearing. The contact (hertz) stresses are lower with worm gears than with crossed-helical gears. Although there is a large component of sliding motion in a worm gear set (as in crossed-helical gears), a worm gear set will have a higher load capacity than a crossed-helical gear pair.

Worm gear sets may be either single- or double-enveloping. In a single-enveloping set, Fig. 21.20, the worm wheel has its width cut into a concave surface, thus partially enclosing the worm when in mesh. The double-enveloping set, in addition to having the helical gear width cut concavely, has the worm length cut concavely. The result is that both the worm and the gear partially enclose each other. A double-enveloping set will have more teeth in contact and will have area rather than line contact, thus permitting greater load transmission.

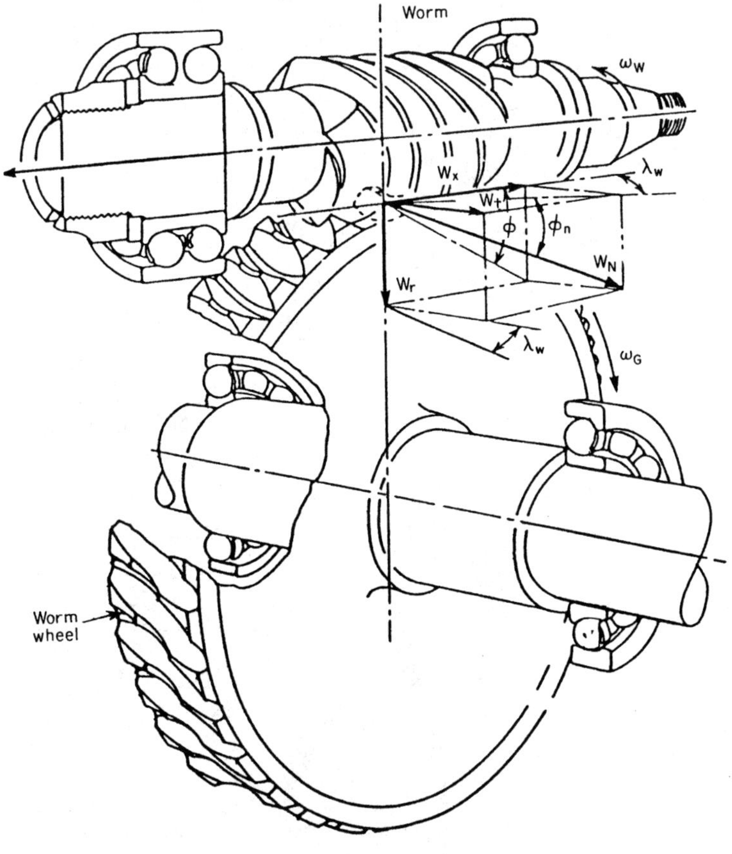

FIG. 21.20 Worm gear set showing load vectors.

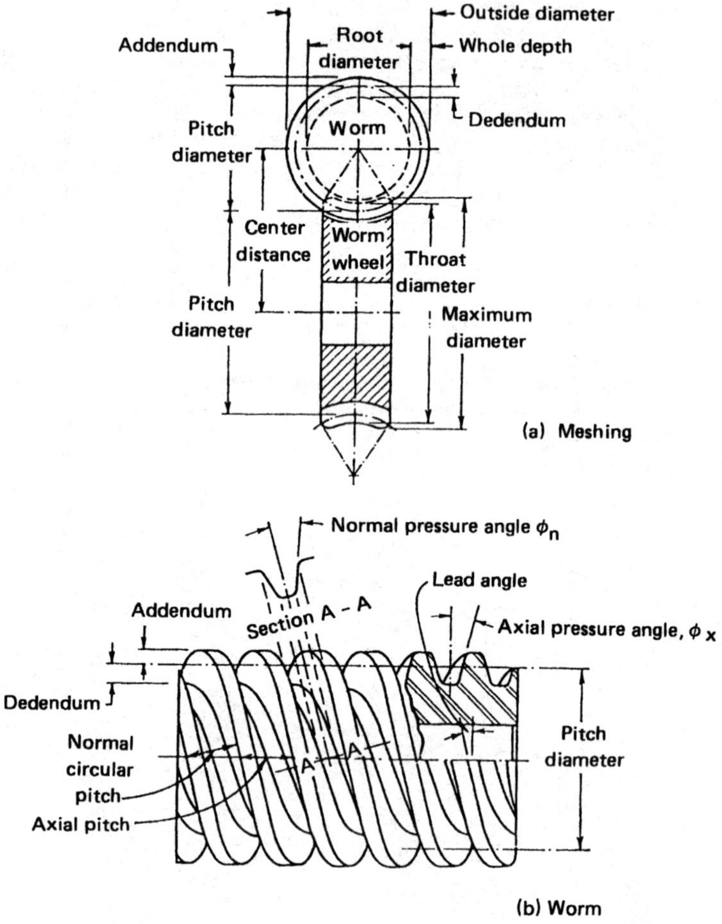

FIG. 21.21 Worm gear set nomenclature. (*a*) Meshing. (*b*) Worm.

All worm gear sets must be carefully mounted in order to assure proper operation. The double-enveloping or cone type is much more difficult to mount than the single-enveloping.

Referring to Figs. 21.20 and 21.21, the relationships to define the worm gear set are as follows.

The lead angle of the worm is

$$\lambda_w \arctan \frac{L_W}{\pi D_W} \tag{21.37}$$

This is the angle between a tangent to the pitch helix and the plane of rotation. If the lead angle is less than 5°, a rotation worm gear set cannot be driven backwards or, in other words, is self-locking. The pitch diameter is D_W. The lead is the number of teeth or threads multiplied by the axial pitch of the worm, or

$$L_W = N_W \times p_a \tag{21.38}$$

The lead and helix angles of the worm and worm wheel are complementary:

$$\lambda_W + \psi_W = 90° \tag{21.39}$$

where ψ_W is the helix angle of the worm. Also

$$\lambda_G + \psi_G = 90° \tag{21.40}$$

and for 90° shaft angles,

$$\lambda_W = \psi_G \tag{21.41}$$

The center distance is given by

$$C = \frac{D_W + D_G}{2} \tag{21.42}$$

where D_G is the pitch diameter of the wheel.

The AGMA recommends the following equation be used to check the magnitude of D_W:

$$D_W \geq \frac{C^{0.875}}{2.2} \approx 3P_G \tag{21.43}$$

where P_G is the circular pitch of the worm wheel in inches. The gear ratio is given by

$$m_g = \frac{\omega_W}{\omega_G} = \frac{N_G}{N_W} = \frac{\pi D_G}{L_W} \tag{21.44}$$

Referring again to Fig. 21.20, the forces acting on the tooth surface are as follows if friction is neglected:

W_t = transmitted force which is tangential to worm and axial to the worm wheel

$$= W_N \cos \phi_n \sin \lambda_w \tag{21.45}$$

W_x = axial thrust load on worm and tangential or transmitted force on worm wheel

$$= W_N \cos \phi_n \cos \lambda_w \tag{21.46}$$

$$W_r = \text{the radial or separating force} = W_N \sin \phi_n \tag{21.47}$$

The directions of the thrust loads for worm gear sets are given in Fig. 21.22.

21.4 PROCESSING AND MANUFACTURE

21.4.1 Materials

A wide variety of gear materials are available today for the gear designer. Depending on the application, the designer may choose from materials such as wood, plastics, aluminum, magnesium, titanium, bronze, brass, gray cast iron, nodular and malleable iron, and a range of low-, medium-, and high-alloy steels. In addition, there are many different ways of modifying or processing the materials to improve their properties or to reduce the cost of manufacture. These include reinforcements for plastics, coating and processing for aluminum and titanium, and hardening and forging for many of the iron-based (or ferrous) gear materials.

When selecting a gear material for an application, the gear designer will first determine the actual requirements for the gears being considered. The design requirements for a gear in a given application will depend on such things as accuracy, load, speed, material, and noise limitations. The more stringent these requirements are, the more costly the gear will become. For instance, a gear requiring high accuracy because of

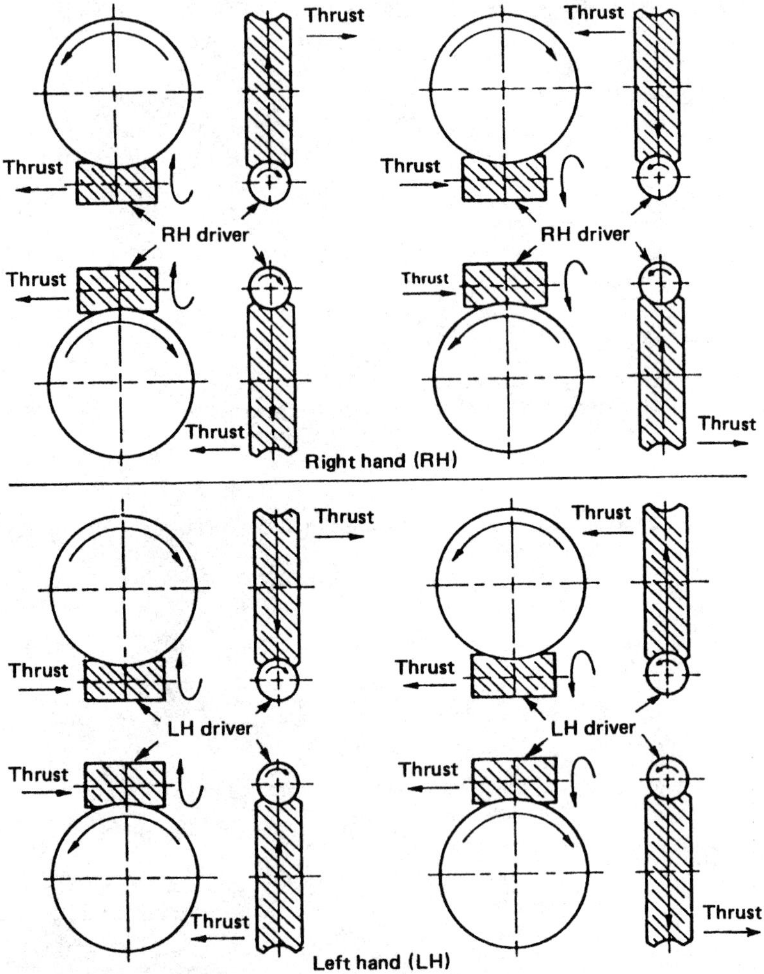

FIG. 21.22 Direction of thrust load for a worm gear set.

speed or noise limitations may require several processing operations in its manufacture. Each operation will add to the cost. Machined gears, which are the most accurate, can be made from materials with good strength characteristics. However, these gears are very expensive. The cost is further increased if hardening and grinding are required, as in applications where noise limitation is a critical requirement. Because of cost, machined gears should be a last choice for a high-production gear application. Figure 21.23 shows the relative costs for high-volume gearing.

Some of the considerations in the choice of a material include allowable bending and hertzian stress, wear resistance, impact strength, water and corrosion resistance, manufacturing cost, size, weight, reliability, and lubrication requirements. Steel, under proper heat treatment, meets most of these qualifications. Where wear is relatively severe, as with worm gearing, a high-quality, chill-cast nickel bronze may be used for rim material. The smaller worm gears may be entirely of nickel bronze.

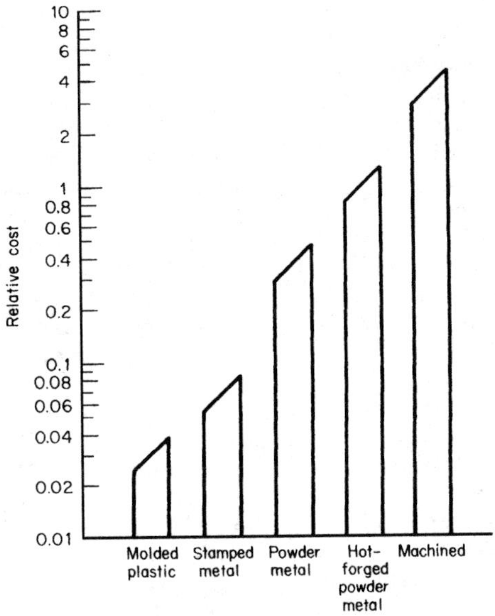

FIG. 21.23 Relative costs of gear materials.

In aircraft applications, such as helicopter, vertical or short take-off and landing (V/STOL), and turboprop aircraft, the dominant factors are reliability and weight. Cost is of secondary importance. Off-the-road vehicles, tanks, and some actuators may be required to operate at extremely high loads with a corresponding reduction in life. These loads may produce bending stresses in excess of 150,000 lb/in^2 and hertzian stresses in excess of 400,000 lb/in^2 for portions of the duty cycle. This may be acceptable because of the relatively short life of the vehicle and a deemphasis on reliability. (As a contrast, aircraft gearing typically operates at maximum bending stresses of 65,000 lb/in^2 and maximum hertzian stresses of 180,000 lb/in^2.)

Plastics. There has always been a need for lightweight, low-cost gear material for light-duty applications. In the past, gears were made from wood or phenolic-resin-impregnated cloth. However, in recent years with the development of many new polymers, many gears are made of various "plastic" materials. Table 21.3 lists plastic

TABLE 21.3 Properties of Plastic Gear Materials

Property	ASTM	Acetate	Nylon	Polyimide
Yield strength, lb/in^2	D 638	10,000	11,800	10,500
Shear strength, lb/in^2	D 732	9510	9600	11,900
Impact strength (Izod)	D 256	1.4	0.9	0.9
Elongation at yield, percent	D 638	15	5	6.5
Modulus of elasticity, lb/in^2	D 790	410 000	410 000	460 000
Coefficient of linear thermal expansion, in/in·°F	D 696	4.5×10^{-5}	4.5×10^{-5}	2.8×10^{-5}
Water absorption (24 h), percent	D 570	0.25	1.5	0.32
Specific gravity	D 792	1.425	1.14	1.43
Temperature limit, °F	—	250	250	600

materials used for molded gears. The most common molded plastic gears are the acetate and nylon resins. These materials are limited in strength, temperature resistance, and accuracy. The nylon and acetate resins have a room-temperature yield strength of approximately 10,000 lb/in^2. This is reduced to approximately 4000 lb/in^2 at their upper temperature limit of 250°F. Nylon resin is subject to considerable moisture absorption, which reduces its strength and causes considerable expansion. Larger gears are made with a steel hub that has a plastic tire for better dimensional control. Plastic gears can operate for long periods in adverse environments, such as dirt, water, and corrosive fluids, where other materials would tend to wear excessively. They can also operate without lubrication or can be lubricated by the processed material as in the food industry. The cost of plastic gears can be as low as a few cents per gear for a simple gear on a high-volume production basis. This is probably the most economical gear available. Often a plastic gear is run in combination with a metal gear to give better dimensional control, low wear, and quiet operation.

Polyimide is a more expensive plastic material than nylon or acetate resin, but it has an operating temperature limit of approximately 600°F. This makes the polyimides suitable for many adverse applications that might otherwise require metal gears. Polyimides can be used very effectively in combination with a metal gear without lubrication because of polyimide's good sliding properties.[23–25] However, polyimide gears are more expensive than other plastic gears because they are difficult to mold and the material is more expensive.

Nonferrous Metals. Several grades of wrought and cast aluminum alloys are used for gearing. The wrought alloy 7075-T6 is stronger than 2024-T4 but is also more expensive.

Aluminum does not have good sliding and wear properties. However, it can be anodized with a thin, hard surface layer that will give it better operating characteristics. Anodizing gives aluminum good corrosion protection in salt water applications. But the coating is thin and brittle and may crack under excessive load.

Magnesium is not considered a good gear material because of its low elastic modulus and other poor mechanical properties. Titanium has excellent mechanical properties, approaching those of some heat-treated steels with a density nearly half that of steel. However, because of its very poor sliding properties, producing high friction and wear, it is not generally used as a gear material. Several attempts have been made to develop a wear-resistant coating, such as chromium plating, iron coating, and nitriding for titanium.[26–29] Titanium would be an excellent gear material if a satisfactory coating or alloy could be developed to provide improved sliding and wear properties.

Several alloys of zinc, aluminum, magnesium, brass, and bronze are used as die-cast materials. Many prior die-cast applications now use less expensive plastic gears. The die-cast materials have higher strength properties, are not affected by water, and will operate at higher temperatures than the plastics. As a result, they still have a place in moderate-cost, high-volume applications. Die-cast gears are made from copper or lower-cost zinc or aluminum alloys. The main advantage of die casting is that the requirement for machining is either completely eliminated or drastically reduced. The high fixed cost of the dies makes low production runs uneconomical. Some of the die-cast alloys used for gearing are listed in Refs. 30 and 31.

Copper Alloys. Several copper alloys are used in gearing. Most are the bronze alloys containing varying amounts of tin, zinc, manganese, aluminum, phosphorus, silicon, lead, nickel, and iron. The brass alloys contain primarily copper and zinc with small amounts of aluminum, manganese, tin, and iron. The copper alloys are most often used in combination with an iron or steel gear to give good wear and load capacity, especially in worm gear applications where there is a high sliding component. In these cases, the steel worm drives a bronze gear. Bronze gears are also used where corrosion

TABLE 21.4 Properties of Copper Alloy Gear Materials

Material	Modulus of elasticity, 10^6 lb/in^2	Yield strength, lb/in^2	Ultimate strength, lb/in^2
Yellow brass	15	50,000	60,000
Naval brass	15	45,000	70,000
Phosphor bronze	15	40,000	75,000
Aluminum bronze	19	50,000	100,000
Manganese bronze	16	45,000	80,000
Silicon bronze	15	30,000	60,000
Nickel-tin bronze	15	25,000	50,000

and water are a problem. Several copper alloys are listed in Table 21.4. The bronze alloys are either aluminum bronze, manganese bronze, silicon bronze, or phosphorus bronze. These bronze alloys have yield strengths ranging from 20,000 to 60,000 lb/in^2 and all have good machinability.

Cast Iron. Cast iron is used extensively in gearing because of its low cost, good machinability, and moderate mechanical properties. Many gear applications can use gears made from cast iron because of its good sliding and wear properties, which are in part a result of the free graphite and porosity. There are three basic cast irons distinguished by the structure of graphite in the matrix of ferrite. These are (1) gray cast iron, where the graphite is in flake form; (2) malleable cast iron, where the graphite consists of uniformly dispersed fine, free-carbon particles or nodules; and (3) ductile iron, where the graphite is in the form of tiny balls or spherules. The malleable iron and ductile iron have more shock and impact resistance. The cast irons can be heat-treated to give improved mechanical properties. The bending strength of cast iron[32] ranges from 5000 to 25,000 lb/in^2, and the surface fatigue strength[33] ranges from 50,000 to 115,000 lb/in^2. In many worm gear drives a cast iron gear can be used to replace the bronze gear at a lower cost because of the sliding properties of the cast iron.

Sintered Powder Metals. Sintering of powder metals is one of the more common methods used in high-volume, low-cost production of medium-strength gears with fair dimensional tolerance.[34] In this method a fine metal powder of iron or other material is placed in a high-pressure die and formed into the desired shape and density under very high pressure. The unsintered (green) part has no strength as it comes from the press. It is then sintered in a furnace under a controlled atmosphere to fuse the powder together for increased strength and toughness. Usually, an additive (such as copper in iron) is used in the powder for added strength. The sintering temperature is then set at the melting temperature of the copper to fuse the iron powder together for a stronger bond than would be obtained with the iron powder alone. The parts must be properly sintered to give the desired strength.

There are several materials available for sintered powder-metal gears that give a wide range of properties. Table 21.5 lists properties of some of the more commonly used gear materials, although other materials are available. The cost for volume production of sintered powder-metal gears is an order of magnitude lower than that for machined gears.

A process that has been more recently developed is the hot-forming powder-metal process.[35,36] In this process a powder-metal preform is made and sintered. The sintered powder-metal preformed part is heated to forging temperature and finish-forged. The hot-formed parts have strengths and mechanical properties approaching the ultimate

TABLE 21.5 Properties of Sintered Powder-Metal Alloy Gear Materials

Composition	Ultimate tensile strength, psi	Apparent hardness, Rockwell	Comment
1 to 5Cu-0.6C-94Fe	60,000	B 60	Good impact strength
7Cu-93Fe	32,000	B 35	Good lubricant impregnation
15Cu-0.6C-84Fe	85,000	B 80	Good fatigue strength
0.15C-98Fe	52,000	A 60	Good impact strength
0.5C-96Fe	50,000	B 75	Good impact strength
2.5Mo-0.3C-1.7N-95Fe	130,000	C 35	High strength, good wear
4Ni-1Cu-0.25C-94Fe	120,000	C 40	Carburized and hardened
5Sn-95Cu	20,000	H 52	Bronze alloy
10Sn-87Cu-0.4P	30,000	H 75	Phosphorous-bronze alloy
1.5Be-0.25Co-98Cu	80,000	B 85	Beryllium alloy

mechanical properties of the wrought materials. A wide choice of materials is available for the hot-forming powder-metal process. Since this is a fairly new process, there will undoubtedly be improvements in the materials made from this process and reductions in the cost. Several promising high-temperature, cobalt-base alloy materials are being developed.

Because there are additional processes involved, hot-formed powder-metal parts are more expensive than those formed by the sintered powder-metal process. However, either process is more economical than machining or conventional forging and produces the desired mechanical properties. This makes the hot-forming powder-metal process attractive for high-production parts where high strength is needed, as in the automotive industry.

Accuracy of the powder-metal and hot-formed processes is generally in the AGMA class 8 range. Better accuracy can be obtained with special precautions if die wear is limited. This tends to increase the cost. Figure 21.23 shows the relative costs of some of the materials or processes for high-volume, low-cost gearing.

Hardened Steels. A large variety of iron or steel alloys are used for gearing. The choice of which material to use is based on a combination of operating conditions such as load, speed, lubrication system, and temperature plus the cost of producing the gears. When operating conditions are moderate, such as medium loads with ambient temperatures, a low-alloy steel can be used without the added cost of heat treatment and additional processing. The low-alloy material in the non-heat-treated condition can be used for bending stresses in the 20,000 lb/in^2 range and surface durability hertzian stresses of approximately 85,000 lb/in^2. As the operating conditions become more severe, it becomes necessary to harden the gear teeth for improved strength and to case-harden the gear tooth surface by case-carburizing or case-nitriding for longer pitting fatigue life, better scoring resistance, and better wear resistance. There are several medium-alloy steels that can be hardened to give good load-carrying capacity with bending stresses of 50,000 to 60,000 lb/in^2 and contact stresses of 160,000 to 180,000 lb/in^2. The higher alloy steels are much stronger and must be used in heavy-duty applications. AISI 9310, AISI 8620, Nitralloy N, and Super-Nitralloy are good materials for these applications and can operate with bending stresses of 70,000 lb/in^2 and maximum contact (hertzian) stresses of 200,000 lb/in^2. These high-alloy steels should be case-carburized (AISI 8620 and 9310) or case-nitrided (Nitralloy) for a very hard wear-resistant surface. Gears that are case-carburized will usually require grinding after the hardening operation because of distortion during the heat-treating process. The nitrided materials offer the advantage of much less distortion during nitriding and therefore can be used in the as-nitrided condition without additional finishing. This is very helpful for large gears with small cross sections where distortion can be a problem. Since case depth for nitriding is limited to approximately 0.020 in, case crushing can occur if the load is too high. Some of the steel alloys used in the gearing industry are listed in Table 21.6.

Gear surface fatigue strength and bending strength can be improved by shot peening.[37,38] The 10 percent surface fatigue life of the shot-peened gears is 1.6 times that of the standard ground gears.[37]

The low- and medium-alloy steels have a limited operating temperature above which they begin to lose their hardness and strength, usually around 300°F. Above this temperature, the material is tempered, and early bending failures, surface pitting failures, or scoring will occur. To avoid this condition, a material is needed that has a higher tempering temperature and that maintains its hardness at high temperatures. The generally accepted minimum hardness required at operating temperature is Rockwell C 58. In recent years several materials have been developed that maintain a Rockwell C 58 hardness at temperatures from 450 to 600°F.[39] Several materials have shown promise of improved life at normal operating temperature. The hot hardness data indicate that they will also provide good fatigue life at higher operating temperatures.

AISI M-50 has been used successfully for several years as a rolling-element bearing material for temperatures to 600°F.[40–44] It has also been used for lightly loaded accessory gears for aircraft applications at high temperatures. However, the standard AISI M-50 material is generally considered too brittle for more heavily loaded gears. AISI M-50 is considerably better as a gear material when forged with integral teeth. The grain flow from the forging process improves the bending strength and impact resistance of the AISI M-50 considerably.[45]

The M-50 material can also be ausforged with gear teeth to give good bending strength and better pitting life.[46,47] However, around 1400°F, the ausforging temperature is so low that forging gear teeth is very difficult and expensive. As a result, ausforging

TABLE 21.6 Properties of Steel Alloy Gear Materials

Material	Tensile strength, lb/in²	Yield strength, lb/in²	Elongation in 2 in, percent
Cast iron	45,000	—	—
Ductile iron	80,000	60,000	3
1020	80,000	70,000	20
1040	100,000	60,000	27
1066	120,000	90,000	19
4146	140,000	128,000	18
4340	135,000	120,000	16
8620	170,000	140,000	14
8645	210,000	180,000	13
9310	185,000	160,000	15
440C	110,000	65,000	14
416	160,000	140,000	19
304	110,000	75,000	35
Nitralloy 135M	135,000	100,000	16
Super-Nitralloy	210,000	190,000	15
CBS 600	222,000	180,000	15
CBS 1000M	212,000	174,000	16
Vasco X-2	248,000	225,000	6.8
EX-53	171,000	141,000	16
EX-14	169,000	115,000	19

for gears has considerably limited application.[46] Test results show that the forged and ausforged gear can give lifetimes approximately 3 times those of the standard AISI 9310 gears.[46]

Nitralloy N is a low-alloy nitriding steel that has been used for several years as a gear material. It can be used for applications requiring temperatures of 400 to 450°F. A modified Nitralloy N called Super-Nitralloy, or 5Ni-2A1 Nitralloy, was used in the U.S. supersonic aircraft program for gears. It can be used for gear applications requiring temperatures to 500°F. Surface fatigue data for the Super-Nitralloy gears and other steels are shown in Table 21.7.

Two materials that were developed for case-carburized tapered roller bearings but also show promise as high-temperature gear materials are CBS 1000M and CBS 600.[48,49] These materials are low- to medium-alloy steels that can be carburized and hardened to give a hard case of Rockwell C 60 with a core of Rockwell C 38. Surface fatigue test results for CBS 600 and AISI 9310 are shown in Table 21.7. The CBS 600 has a medium fracture toughness that could cause fracture failures after a surface fatigue spall has occurred.

Two other materials that have recently been developed as advanced gear materials are EX-53 and EX-14. Reference 50 reports that the fracture toughness of EX-53 is excellent at room temperature and improves considerably as temperature increases. The EX-53 surface fatigue results show a 10 percent life that is twice that of the AISI 9310. Vasco X-2 is a high-temperature gear material[51] that is used in advanced CH-47 helicopter transmissions. This material has an operating temperature limit of 600°F and has been shown to have good gear load-carrying capacity when properly heat treated. The material has a high chromium content (4.9 percent) that oxidizes on the surface and can cause soft spots when the material is carburized and hardened. A special process has been developed that eliminates these soft spots when the process is closely followed.[52] Several groups of Vasco X-2 with different heat treatments were surface fatigue-tested in the NASA gear test facility. All groups except that with the special processing gave poor results.[53] Vasco X-2 has a lower fracture toughness than AISI 9310 and is subject to tooth fractures after a fatigue spall.

21.4.2 Metallurgical Processing Variables

Research reported in the literature on gear metallurgical processing variables is not as extensive as that for rolling-element bearings. However, an element of material in a hertzian stress field does not recognize whether it is in a bearing or a gear. It only recognizes the resultant shearing stress acting on it. Consequently, the behavior of the

TABLE 21.7 Relative Surface Pitting Fatigue Life Compared to VAR AISI 9310 Steel for Aircraft-Quality Gear Steels (Rockwell C 59–62)

Steel	10 percent relative life
VAR AISI 9310	1
VAR AISI 9310 (shot-peened)	1.6
VAR Carpenter EX-53	2.1
CVM CBS 600	1.4
VAR CBS 1000	2.1
CVM VASCO X-2	1.0
CVM Super-Nitralloy (5Ni-2A1)	1.3
VIM-VAR AISI M-50 (forged)	3.2
VIM-VAR AISI M-50 (ausforged)	2.4

material in a gear will be very similar to that in a rolling-element bearing. The metallurgical processing variables to be considered are

1. Melting practice, such as air, vacuum induction, consumable electrode vacuum melt (CVM), vacuum degassing, electroslag (electroflux) remelt, and vacuum induction melting–vacuum arc remelting (VIM-VAR)
2. Heat treatment which gives hardness, differential hardness, and residual stress
3. Metalworking consisting of thermomechanical working and fiber orientation

The above items can significantly affect gear performance. It should be mentioned that some factors that can also significantly affect gear fatigue life and that have some meaningful documentation are not included. These are trace elements, retained austenite, gas content, inclusion type, and content. While any of these can exercise some effect on gear fatigue life, they are, from a practical standpoint, too difficult to measure or control by normal quality control procedures.

Heat-treatment procedures and cycles, per se, can also affect gear performance. However, at present no controls, as such, are being exercised over heat treatment. The exact thermal cycle is left to the individual producer with the supposition that a certain grain size and hardness range must be met. Hardness is discussed in some detail in reference to gear life. In the case of grain size, lack of any definitive data precludes any meaningful discussion.

Melting Practice. Sufficient data and practical experience exist to indicate the use of vacuum-melted materials and, specifically, CVM can increase gear surface pitting fatigue life beyond that obtainable by air-melted materials.[54–59] Since in essentially all critical applications such as gears for helicopter transmissions and turboprop aircraft, vacuum-melt material is specified, a multiplication factor can be introduced into the life estimation equation.

Life improvements up to 13 times by CVM processing[41–44] and up to 100 times by VIM-VAR processing[59] over air-melted steels are indicated in the literature. However, it is recommended that conservative estimates for life improvements be considered such as a factor of 3 for CVM processing and a factor of 10 for VIM-VAR processing. While these levels may be somewhat conservative, the confidence factor for achieving this level of improvement is high. Consequently, the extra margin of reliability desired in critical gear applications will be inherent in the life calculations.

Data available on other melting techniques such as vacuum induction, vacuum degassing, and electroslag remelting would indicate that the life improvement approaches or equals that of the CVM process. However, it is also important to differentiate between CVM and CVD (carbon vacuum degassing). The CVM process yields cleaner and more homogeneous steels than CVD.

Heat Treatment. Reference 14 discusses the various methods of heat treatment as applied to gears. Heat treatment is accomplished by either furnace hardening, carburizing, induction hardening, or flame hardening. Gears are either case-hardened, through-hardened, nitrided, or precipitation-hardened for the proper combination of toughness and tooth hardness.

The use of high-hardness, heat-treated steels permits smaller gears for a given load. Also, hardening can significantly increase service life without increasing size or weight. But the gear must have at least the accuracies associated with softer gears and, for maximum service life, even greater precision.

Heat-treatment distortion must be minimized if the gear is to have increased service life. Several hardening techniques have proved useful. For moderate service-life increases, the gears are hardened but kept within the range of machinability so that distortion produced by heat treatment can be machined away.

Where maximum durability is required, surface hardening is necessary. Carburizing, nitriding, and induction hardening are generally used. However, precision gearing (quality 10 or better) can only be assured by finishing after hardening. Full-contour induction hardening is an economical and effective method for surface-hardening gears. The extremely high but localized heat allows small sections to come to hardening temperatures while the balance of the gear dissipates heat. Thus, major distortions are eliminated.

While conventional methods such as flame hardening reduce wear by hardening the tooth flank, gear strength is not necessarily improved. In fact, stresses built up at the juncture of the hard and soft material may actually weaken the tooth. Induction hardening provides a hardened tooth contour with a heat-treated core to increase both surface durability and tooth strength. The uniformly hardened tooth surface extends from the flank, around the tooth, to the flank. No stress concentrations are developed to impair gear life.

Nitriding is a satisfactory method of hardening small- and medium-size gears. Distortion is minimal because furnace temperatures are comparatively low. Hardening pattern is uniform but depth of hardness is limited. Best results are achieved when special materials, suited to nitriding, are specified.

Most gear manufacturing specifications do not designate heat treatment, but rather call for material characteristics, i.e., hardness and grain size, which are controlled by the heat-treatment cycle. Hardness is the most influential heat-treatment-induced variable.[60–62] It is recommended that Rockwell C 58 be considered the minimum hardness required for critical gear applications.

A relationship has been proposed in Ref. 42 which approximates the effect of hardness on surface fatigue life:

$$\frac{L_2}{L_1} = e^{0.1(RC_2 - RC_1)} \tag{21.48}$$

where L_1 and L_2 are 10 percent fatigue lives at gear hardness R_{C1} and R_{C2}, respectively. Although this relationship was obtained for AISI 52100, it may be extended to other steels. The life-vs.-hardness curve in Fig. 21.24 represents Eq. (21.48) with the relative life at Rockwell C 60 equal to 1.0. The hardness-vs.-temperature curves in Fig. 21.24 indicate the decrease in hardness with increased temperature for various initial room-temperature material hardnesses.

To use the nomograph (Fig. 21.24), first determine the gear operating temperature. Then follow a horizontal line to the appropriate room-temperature hardness curve. Then move vertically to the life-hardness curve and read the relative life at that point. See the dashed line example for a 300°F operating temperature and Rockwell C 64 at room temperature hardness. Relative life for this example is approximately 1.3.

Another concept to be considered is the effect of differences in hardness between the pinion and the gear.[63,64] Evidence exists to show that hardness differences between the mating components can positively affect system life by inducing compressive residual stresses during operation.[64] Differential hardness (ΔH) is defined as the hardness of the larger of two mating gears minus the hardness of the smaller of the two. It appears that a differential hardness (ΔH) of two points Rockwell C may be an optimum for maximum life. For critical applications, as a practical matter, it would be advisable to match the hardness of the mating gears to assure a ΔH of zero and at the same time assure that the case hardness of the gear teeth at room temperature is the maximum attainable. This will allow for maximum elevated operating temperature and maximum life. While the ΔH effect has been verified experimentally with rolling-element bearings, there is no similar published work with gears.

As previously discussed, residual stresses can be induced by the heat-treatment process, differential hardness, and/or shot peening. There is no analytical method to predict the amount of residual stress in the subsurface region of gear tooth contact.

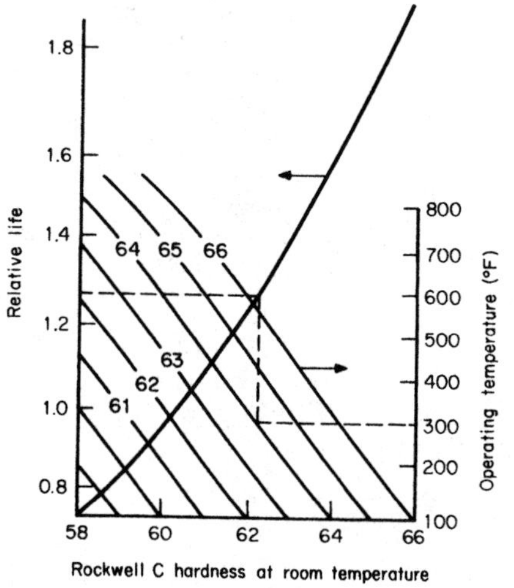

FIG. 21.24 Nomograph for determining relative life at operating temperature as a function of room-temperature hardness.

However, these residual stresses can be measured in test samples by x-ray diffraction methods. The effect of these residual stresses on gear pitting fatigue life can be determined by the following equation[37]:

$$\text{Life} \propto \left(\frac{1}{\tau_{\max} - S_{ry}/2}\right)^9 \tag{21.49}$$

where $\tau_{\max}$ is the maximum shearing stress (45° plane) and S_{ry} is the measured compressive residual stress below the surface at the location of $\tau_{\max}$.

Metalworking. Proper grain flow or fiber orientation can significantly affect pitting fatigue life[65,66] and may improve bending strength of gear teeth. Proper fiber orientation can be defined as grain flow parallel to the gear tooth shape. Standard forging of gears with integral gear teeth as opposed to machining teeth in a forged disk is one way of obtaining proper fiber orientation.[46] A controlled-energy-flow forming (CEFF) technique can be used for this purpose. This is a high-velocity metalworking procedure that has been a production process for several years.

AISI M-50 steel is a through-hardened material often used for rolling-element bearings in critical applications. Test gears forged from M-50 steel yielded approximately 3 times the fatigue life of machined vacuum-arc-remelted (VAR) AISI 9310 gears.[46] Despite the excellent fatigue life, M-50 is not recommended for gears because its low fracture toughness makes gears prone to sudden catastrophic tooth fracture after a surface fatigue spall has begun rather than the gradual failure and noisy operation typical of surface pitting. It is expected that forged AISI 9310 (VAR) gears would achieve similar life improvement while retaining the greater reliability of the tougher material.

Ausforging, a thermomechanical fabrication method, has potential for improving strength and life of gear teeth. Rolling-element tests with AISI M-50 steel (not recommended for gears—see above paragraph) show that 75 to 80 percent work (reduction of area) produces maximum benefit.[67–69] The suitability of candidate steels to the ausforging process must be individually evaluated. AISI 9310 is not suitable because of its austenite-to-martensite transformation characteristics. Tests reported in Ref. 46 found no statistically significant difference in lives of ausforged and standard-forged AISI M-50 gears. The lack of improvement of the ausforged gears is attributed to the final machining required of the ausforged gears which removed some material with preferential grain flow. Reference 46 also reported a slightly greater tendency toward tooth fracture. This tendency is attributed to better grain flow obtained in the standard forged gears. The energy required limits the ausforging process to gears no larger than 90 mm (3½ in) in diameter.

21.4.3 Manufacturing

Gears can be formed by various processes that can be classified under the general head of milling, generating, and molding.

Milling. Almost any tooth form can be milled. However, only spur, helical, and straight-bevel gears are usually milled. Surface finish can be held to 125 μin rms.

Generating. In the generating process, teeth are formed in a series of passes by a generating tool shaped somewhat like a mating gear. Either hobs or shapers can be used. Hobs resemble worms that have cutting edges ground into their teeth. Hobbing can produce almost any external tooth form except bevel gear teeth, which are generated with face-mill cutters, side-mill cutters, and reciprocating tools. Hobbing closely controls tooth spacing, lead, and profile. Surface finishes as fine as 63 μin rms can be obtained. Shapers are reciprocating pinion or rack-shaped tools. They can produce external and internal spur, helical, herringbone, and face gears. Shaping is limited in the length of cut it can produce. Surface finishes as fine as 63 μin rms are possible.

Molding. Large-volume production of gears can often be achieved by molding. Injection molding produces light gears of thermoplastic material. Die casting is a similar process using molten metal. Zinc, brass, aluminum, and magnesium gears are made by this process.

Sintering is used in small, heavy-duty gears for instruments and pumps. Iron and brass are the materials most used. Investment casting and shell molding produce medium-duty iron and steel gears for rough applications.

Gear Finishing. To improve accuracy and finish, gears may be shaved. Shaving removes only a small amount of surface metal. A very hard mating gear with many small cutting edges is run with the gear to be shaved. Surface finish can be as fine as 32 μin rms.

Lapping corrects minute heat-treatment distortion errors in hardened gears. The gear is run in mesh with a gear-shaped lapping tool, or another mating gear. An abrasive lapping compound is used between them. Lapping improves tooth contact but does not increase accuracy of the gear. Finish is on the order of 32 μin rms.

Grinding is the most accurate tooth-finishing process. Profiles can be controlled or altered to improve tooth contact. For example, barreling or crowning the flanks of teeth promotes good center contact where the tooth is strong, and minimizes edge and corner contact where the tooth is unsupported. Surface finishes as fine as 16 μin rms or better can be obtained. However, at surface finishes better than 16 μin rms, tooth errors can be induced which may affect gear tooth dynamic loading and, as a result, nullify the advantage of the improved surface finish.

21.5 STRESSES AND DEFLECTIONS

There are several approaches to determining the stresses and deflections in gears. The one used most commonly for determining the bending stress and deflection is a modified form of the Lewis equation.[8] This approach considers the gear tooth to be a cantilevered beam or plate. Modifying factors based on geometry or type of application are then used to amend the calculated stress and allowed design strength. It is now known that the calculated stress may not accurately represent the true stress. But since a large amount of experience has accrued across the gearing industry, compatible "allowed stress" has been compiled, and the modified Lewis equation is used successfully to design gears. The Lewis method should be thought of as a means of comparing a proposed new gear design for a given application with successful operation in the past of similar designs. If the computed stress number is less than the computed allowable stress number, the design will be satisfactory. The American Gear Manufacturer's Association has published a standard (AGMA 218.01)[70] for calculating the bending strength for spur and helical gears, as well as many other standards for gear design and applications. A second method of calculation of stresses and deflections comes from classical theory on elasticity. Several examples of such methods are found in the literature.[71–73] The methods involve the use of complex variable theory to map the shape of the gear tooth onto a semi-infinite space in which the stress equations are solved. The methods have the advantage of computing accurate stresses in the regions of stress concentration. The methods do, however, seem limited to plane stress problems. Therefore, they will not work for bevel or helical gears where the effect of the wide teeth and distributed tooth load is important.

The most powerful method for determining accurate stress and deflection information is the finite-element method. However, the method is too expensive and cumbersome to use in an everyday design situation. While at first the method would seem to answer all problems of computational accuracy, it does not. Research is continuing in this promising area.[74–79]

The versatility of the finite-element method introduces other questions of how best to use the method. These questions include (1) how many elements there are and the most efficient arrangement of elements in the regions of stress concentration, (2) how to represent the boundary support conditions, and (3) how to choose the aspect ratios of the solid brick element in three-dimensional stress problems. Pre- and postprocessors specifically for gears have been developed.[79] Various experiments have been conducted around specific questions such as the effect of rim thickness on critical stress in lightweight gears.[80,81] It is expected that methods such as the finite element method and the classical theory of elasticity will continue to be used in a research type mode, with research results being used to supply application factors and stress modifying factors for the modified Lewis (AGMA) method. One such approach has been suggested for the effect of the ratio of rim thickness to gear tooth height.[81]

Experimental methods and testing of all proposed gear applications cannot be overemphasized. The foregoing discussion has centered around the bending fatigue strength of the gear teeth, where the stress causing failure is at the gear tooth fillet between the tooth profile and root of the gear. Surface pitting fatigue and scoring types of failure are discussed in Sec. 21.6 and are also likely modes of failure. Experimental testing of gear prototype designs will reveal any weakness in the design which could not be anticipated by conventional design equations, such as where special geometry factors will cause the failure-causing stress to appear at locations other than the tooth fillet.* In critical applications such as aircraft, full-scale testing of every

*In Ref. 82 it is shown that for thin-rimmed internal gears, the maximum tensile stress occurs several teeth away from the loaded tooth. In Ref. 79 it is shown that for thin-rimmed external gears the bending stress in the root can be a higher stress than the fillet stress. These observations emphasize the need for care in applying conventional design equations with a cavalier confidence in their universal applicability.

prototype gearbox is done in ground-based test rigs, and every piece of a flight hardware gearbox assembly is tested in a "green run" transmission test rig as a distinct step in the production process for the completed assembly. An experimental verification of a proposed gear design is essential to a complete design process.

21.5.1 Lewis Equation Approach for Bending Stress Number (Modified by the AGMA)

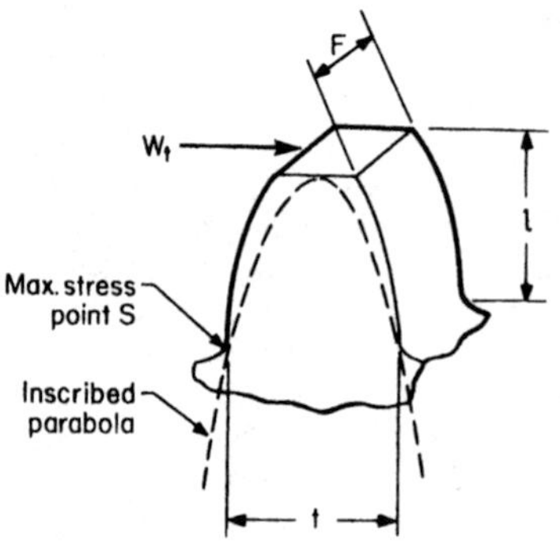

FIG. 21.25 Gear tooth load.

Figure 21.25 shows a gear tooth loaded by the transmitted tangential load W_t, which is assumed to act at the tip of the tooth. The Lewis formula for the stress at the fillet follows the simple strength of materials formula for bending stress in a beam:

$$S = \frac{Mc}{I} = \frac{(W_t l) \times t/2}{\frac{1}{12}Ft^3} \tag{21.50}$$

The values for the tooth thickness at the critical section and the beam length are obtained from inscribing a parabola inside the tooth, tangent to the fillet, with the apex at the point of load application. The procedure is based on the fact that a parabolic-shaped beam is a beam of constant bending stress. For a family of geometrically similar tooth forms, a dimensionless tooth form factor Y is defined as follows:

$$Y = \frac{t^2 P}{6l} \tag{21.51}$$

where P is the diametral pitch.

Using the form factor, the tooth bending stress is calculated by

$$S = \frac{W_t}{F}\frac{P}{Y} \tag{21.52}$$

where W_t is transmitted tangential load and F is face width. The assumptions from which Eq. (21.52) was derived are (1) the radially directed load was neglected; (2) only one tooth carried the full transmitted load; (3) there was uniform line contact between teeth, causing a uniform load distribution across the face width; and (4) the effect of stress concentration at the fillet was neglected.

A more comprehensive equation (defined by the AGMA) for bending stress number is

$$S = \frac{W_t}{F}\frac{P}{J}\frac{K_a K_s K_m}{K_v} \tag{21.53}$$

where S = the bending stress number, lb/in² (MPa)
K_a = application factor
K_s = size factor
K_m = load distribution factor
K_v = dynamic factor or velocity factor
J = geometry factor

The *application factor* K_a is intended to modify the calculated stress number, as a result of the type of service the gear will see. Some of the pertinent application influences include type of load, type of prime mover, acceleration rates, vibration, shock, and momentary overloads. Application factors are established after considerable field experience is gained with a particular type application. It is suggested that designers establish their own values of application factors based on past experience with actual designs that have run successfully. If they are unable to do this, suggested factors from past experience are available in Table 21.8.

The *size factor* K_s reflects the influence of nonhomogeneous materials. On the basis of the weakest link theory, a larger section of material may be expected to be weaker than a small section because of the greater probability of the presence of a "weak" link. The size factor K_s corrects the stress calculation to account for the known fact that larger parts are more prone to fail. At present, a size factor $K_s = 1$ is recommended for spur and helical gears by the AGMA, and for bevel gears the Gleason Works recommends

$$K_s = P^{-0.25} \qquad P < 16$$
$$= 0.5 \qquad P > 16 \tag{21.54}$$

The *load distribution factor* K_m is to account for the effects of possible misalignments in the gear which will cause uneven loading and a magnification of the stress above the uniform case. If possible, keep the ratio of face width to pitch diameter small (less than 2) for less sensitivity to misalignment and uneven load distribution due to load-sensitive deflections. A second rule of thumb is to keep the face width between 3 and 4 times the circumferential pitch. Often gear teeth are crowned or edge-relieved to diminish the effect of misaligned axes on tooth stresses. Normally, the range of this factor can be approximately 1 to 3. Use a factor larger than 2 for gears mounted with less than full face contact. When more precision is desired and enough detail of the proposed design is in hand to make the necessary calculations, the AGMA standards should be consulted. For most preliminary design calculations the data in Table 21.9 for spur and helical gears and Table 21.10 and Fig. 21.26 for bevel gears should be used.

The *dynamic factor* K_v is intended to correct for the effects of the speed of rotation and the degree of precision in the gear accuracy. As a first approximation K_v may be taken from the chart in Fig. 21.27. This chart is intended only to account for the effect of tooth inaccuracies. It should not be used for lightly loaded, lightly damped gears, or resonant conditions. If the gear system approaches torsional resonance, or if the gear blank is near resonance, then computerized numerical methods must be used such as

TABLE 21.8 Suggested Application Factors K_a (for Speed Increasing and Decreasing Drives)*

	Load on driven machine		
Power source	Uniform	Moderate shock	Heavy shock
Uniform	1.00	1.25	1.75 or higher
Light shock	1.25	1.50	2.00 or higher
Medium shock	1.50	1.75	2.25 or higher

*For speed increasing drives of spur and bevel gears (but not helical and herringbone gears), add $0.01(N_2/N_1)^2$ to the factors where N_1 = number of teeth in pinion, N_2 = number of teeth in gear.

Source: AGMA

TABLE 21.9 Spur and Helical Gear Load Distribution Factor K_m

	Face width							
	2-in face and under		6-in face		9-in face		16-in face and over	
Condition of support	Spur	Helical	Spur	Helical	Spur	Helical	Spur	Helical
Accurate mounting, low bearing clearances, minimum elastic deflection, precision gears	1.3	1.2	1.4	1.3	1.5	1.4	1.8	1.7
Less rigid mountings, less accurate gears, contact across full face	1.6	1.5	1.7	1.6	1.8	1.7	2.0	2.0
Accuracy and mounting such that less than full face contact exists			Over 2.0					

Source: AGMA

TABLE 21.10 Bevel Gears Load Distribution Factor K_m

Application	Both members straddle-mounted	One member straddle-mounted	Neither member straddle-mounted
General industrial	1.00 to 1.10	1.10 to 1.25	1.25 to 1.40
Automotive	1.00 to 1.10	1.10 to 1.25	—
Aircraft	1.00 to 1.25	1.10 to 1.40	1.25 to 1.50

Source: Gleason Works, Rochester, N.Y.

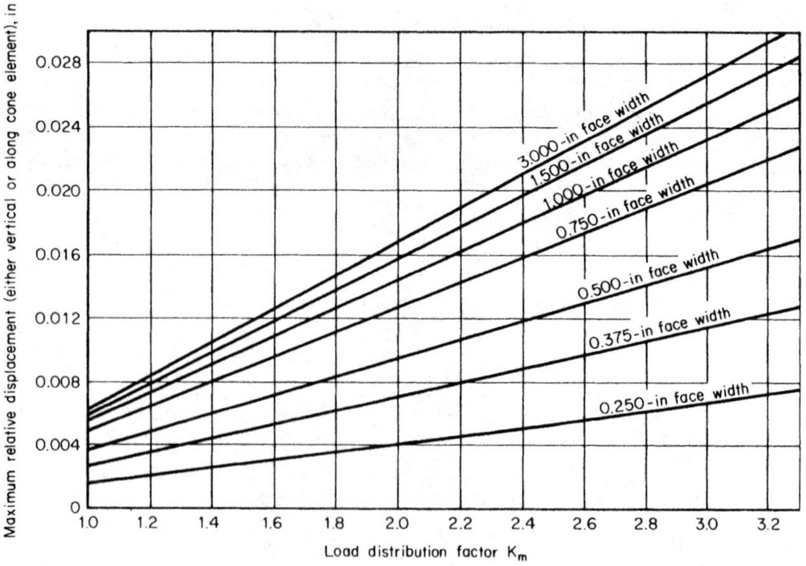

FIG. 21.26 Bevel gear load distribution factor K_m. *(Courtesy of Gleason Works, Rochester, N.Y.)*

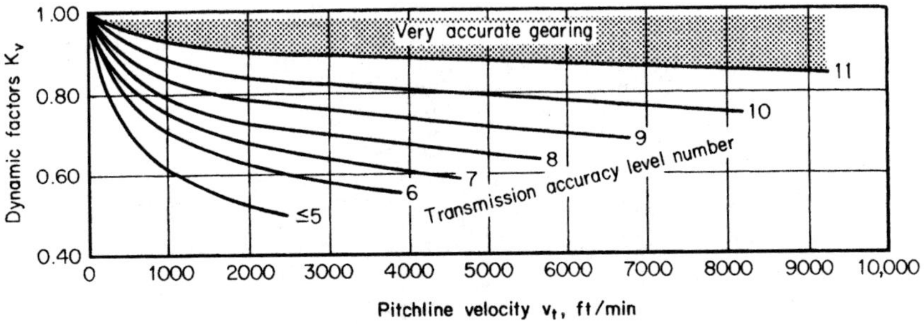

FIG. 21.27 Gear dynamic factor K_v. Transmission accuracy level number specifies the AGMA class for accuracy of manufacture. (*Courtesy of AGMA.*)

presented in Refs. 83 and 84. The chart in Fig. 21.28 shows the effect of damping on the dynamic factor for a certain specific design. No comprehensive data for a broad range of parameters is yet available. However, the computer program TELSGE (*t*hermal *e*lastohydrodynamic *l*ubrication of *s*pur *ge*ars) and GRDYNSING (*gear dyn*amic analysis for *sing*le mesh) are available from the NASA computer program library.*

The *geometry factor J* is really just a modification of the form factor *Y* to account for three more influences. They are stress concentration, load sharing between the teeth, and changing the load application point to the highest point of single tooth contact. The form factor *Y* can be found from a geometrical layout of the Lewis parabola within the gear tooth outline (Fig. 21.25). The load-sharing ratio may be worked out using a statically indeterminate analysis of the pairs of teeth in contact, considering the flexibilities of the teeth in contact, and stress concentration factors from the work of Dolan and Broghammer[85] may be applied to get *J*. The *J* factors have already been

*Available from the Computer Software Management and Information Center (COSMIC) Computing and Information Services, University of Georgia, Athens.

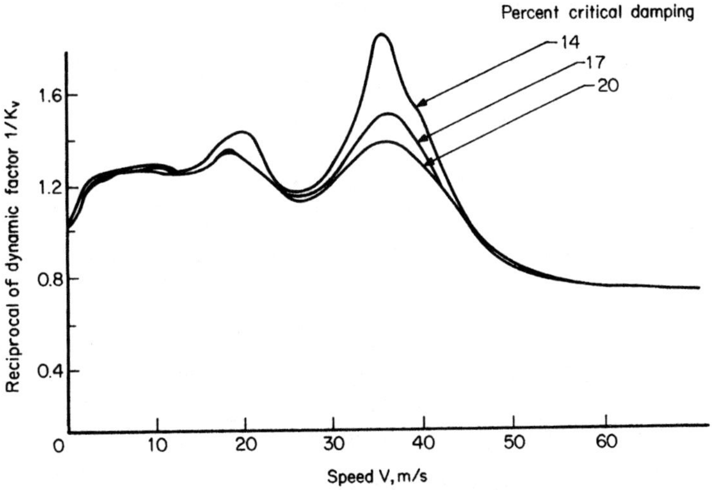

FIG. 21.28 Effect of damping ratio on dynamic load factor K_v (m_g = 1, N = 28, P = 8).

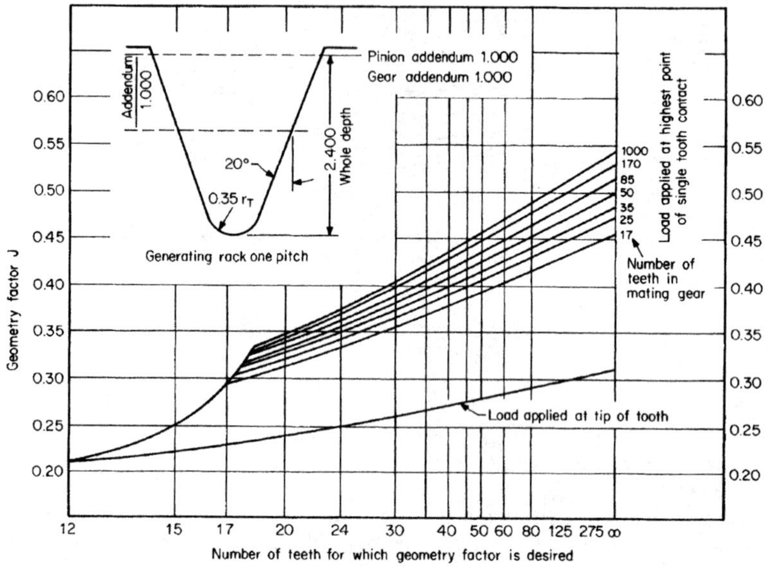

FIG. 21.29 Geometry factor J for 20° spur gear with standard addendum. (*From AGMA 218.01, December 1982.*)

computed for a wide range of standard gears and are available in the AGMA standards. For gear applications, the charts presented in Figs. 21.29 through 21.34 for spur, helical, and bevel gears may be used.

The *transmitted tooth load* W_t is equal to the torque divided by the pitch radius for spur and helical gears. For bevel gears the calculation should use the large-end pitch cone radius.

A *rim thickness factor* K_b which multiplies the stress computed by the AGMA formula [Eq. (21.53)] has been proposed in Ref. 81. The results in Fig. 21.35 show that for thin-rimmed gears, the calculation stress should be adjusted by a factor K_b if the backup ratio is less than 2.

For spiral-bevel gears, *a cutter radius factor* K_x (a stress correction factor) has been given by Gleason to account for the effects of cutter radius curvature, which is a measure of the tooth lengthwise curvature:

$$K_x = 0.2111\left(\frac{r_c}{A}\right)^{0.2788/\log \sin \psi} + 0.7889 \qquad (21.55)$$

where A = gear mean cone distance
ψ = gear mean spiral angle
r_c = cutter radius

Normally (r_c/A) ranges from 0.2 to 1 and ψ from 10 to 50°, giving a range of K_x from 1 to 1.14. Recommended practice is to select the next larger nominal cutter that has a radius given by

$$r_c = 1.1A \sin \psi \qquad (21.56)$$

The cutter radius factor should be entered in the denominator of the stress equation [Eq. (21.53)].

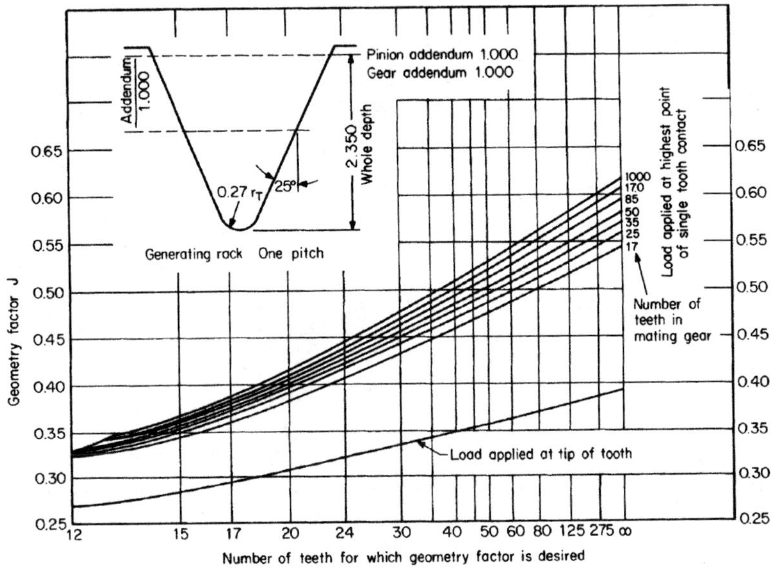

FIG. 21.30 Geometry factor J for 25° spur gear with standard addendum. (*From AGMA 218.01, December 1982.*)

21.5.2 Allowable Bending Stress Number

The previous section has presented a consistent method for calculating a bending stress index number. The stresses calculated by Eq. (21.53) may be much less than stresses measured by strain gauges or calculated by other methods such as finite-element methods. However, there is a large body of allowable stress data available in the AGMA standards which is consistent with the calculation procedure of Eq. (21.53). If the calculated stress number S is less than the allowed stress number S_{at}, the design

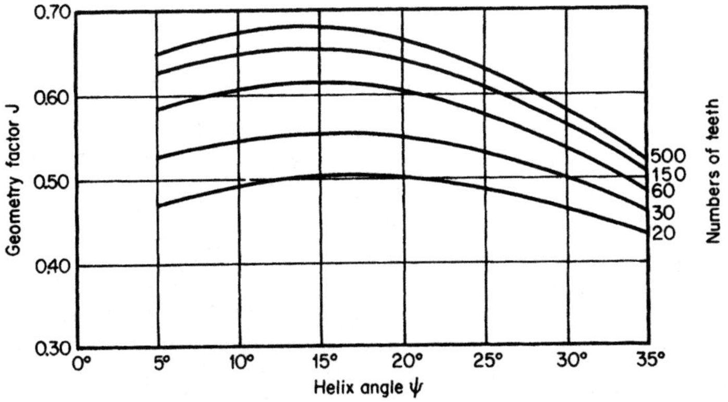

FIG. 21.31 Helical gear geometry factor J for 20° normal pressure angle, standard addendum, full fillet hole. (*From AGMA 218.01, December 1982.*)

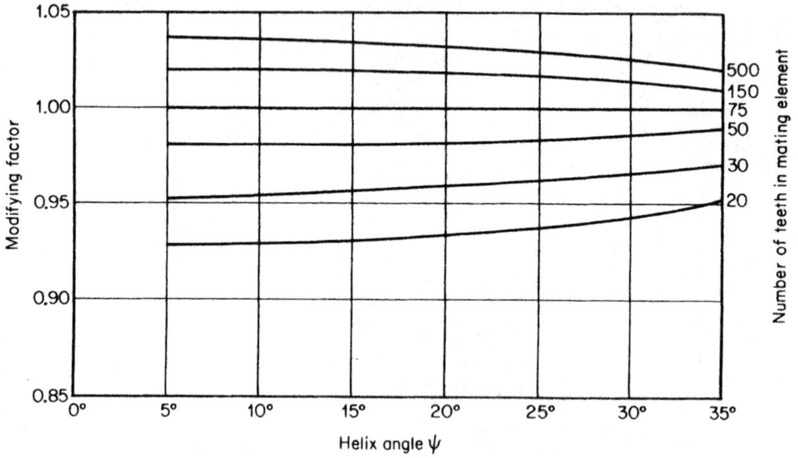

FIG. 21.32 Helical gear J factor multipliers, 20° normal pressure angle. (*From AGMA 218.01, December 1982.*)

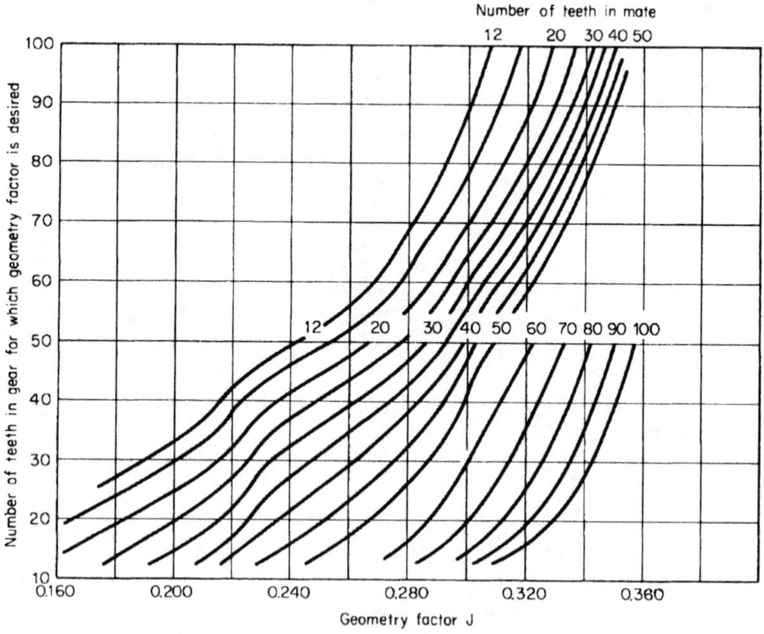

FIG. 21.33 Spiral bevel gear geometry factor for 20° pressure angle, 35° spiral angle, and 90° shaft angle. (*From Gleason publication SD 3103, April 1981.*)

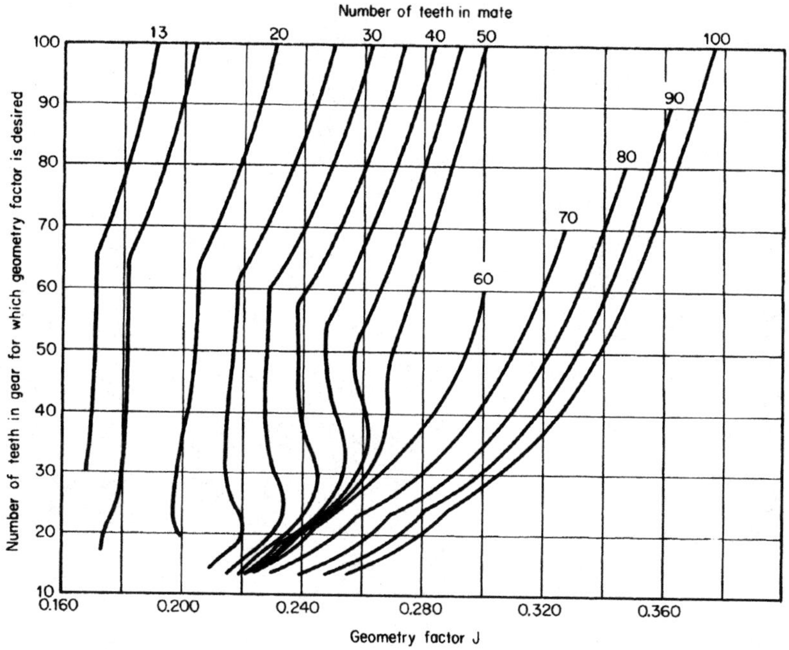

FIG. 21.34 Straight-bevel gear geometry factor for 20° pressure angle and 90° shaft angle. (*From Gleason publication SD 3103E, April 1981.*)

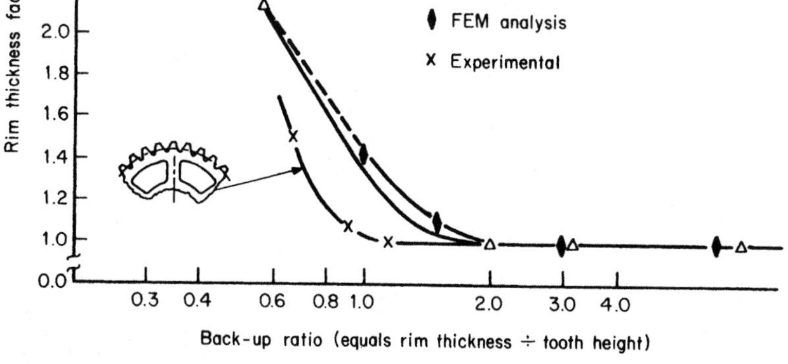

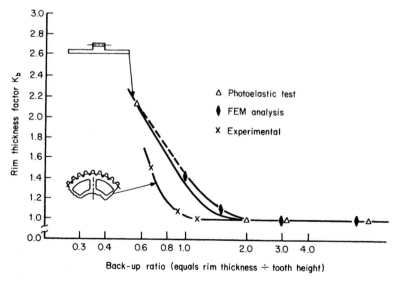

FIG. 21.35 Gear rim thickness factor K_b.

TABLE 21.11 Allowable Bending Stress Number, S_{at} for 10^7 Stress Cycles

Material	AGMA class	Commercial designation	Heat treatment	Minimum hardness surface	Cores	S_{at}, lb/in² Spur and helical	Bevel
Steel	A-1 through A-5	—	Through-hardened and tempered	180 Brinell	—	25–33,000	14,000
				240 Brinell	—	31–41,000	
				300 Brinell	—	36–47,000	19,000
				360 Brinell	—	40–52,000	
				400 Brinell	—	42–56,000	
			Flame or induction-hardened section	50–54 Rockwell C	—	45–55,000	—
				55 Rockwell C min		—	30,000
			Flame or induction-hardened section		—	22,000	13,500
			Carburized and case-hardened	55 Rockwell C	—	55–65,000	27,500
				60 Rockwell C	—	55–70,000	30,000

		AISI 4140	Nitrided	48 Rockwell C	300 Brinell	34–45,000	20,000
		AISI 4340	Nitrided	46 Rockwell C	300 Brinell	36–47,000	
		Nitralloy 135M	Nitrided	60 Rockwell C	300 Brinell	38–48,000	
		$2\frac{1}{2}$% chrome	Nitrided	54–60 Rockwell C	350 Brinell	55–65,000	
Cast Iron	20		As cast	—	—	5000	2,700
	30		As cast	175 Brinell	—	8500	4,600
	40		As cast	200 Brinell	—	13,000	7,000
Nodular (ductile) iron	A-7-a	ASTM 60-40-18	Annealed	140 Brinell	—	15,000	8,000
	A-7-c	ASTM 80-55-06	Quenched and tempered	180 Brinell	—	20,000	11,000
	A-7-d	ASTM 100-70-03	Quenched and tempered	230 Brinell	—	26,000	14,000
	A-7-e	ASTM 120-90-02	Quenched and temepred	270 Brinell	—	30,000	18,500
Malleable iron (pearlitic)	A-8-c	45007	—	165 Brinell	—	10,000	—
	A-8-e	50005	—	180 Brinell	—	13,000	—
	A-8-f	53007	—	195 Brinell	—	16,000	—
	A-8-i	80002	—	240 Brinell	—	21,000	—
Bronze	Bronze 2	AGMA 2C	Sand cast Sand cast	Tensile strength min. 40,000 lb/in^2 (275 MPa)		5,700	3,000
	Al/Br 3	ASTM B-148-52 Alloy 9C	Heat-treated	Tensile strength min. 90,000 lb/in^2 (620 MPa)		23,600	12,000

Source: AGMA and Gleason Works, Rochester, N.Y.

TABLE 21.12 Gear Life Factor K_L

	Spur, helical, and herringbone				Bevel gears
Number of cycles	160 Brinell hardness	250 Brinell hardness	450 Brinell hardness	Case-carburized*	Case-carburized*
Up to 1000	1.6	2.4	3.4	2.7	4.6
10,000	1.4	1.9	2.4	2.0	3.1
100,000	1.2	1.4	1.7	1.5	2.1
1 million	1.1	1.1	1.2	1.1	1.4
10 million	1.0	1.0	1.0	1.0	1.0
100 million and over	1.0–0.8	1.0–0.8	1.0–0.8	1.0–0.8	1.0

*Rockwell C 55 min.
Source: AGMA.

should be satisfactory. The equation is

$$S \leq S_{at}\left(\frac{K_L}{K_T K_R}\right) \tag{21.57}$$

where the K factors are to modify the allowable stress to account for the effects that alter the basic allowed stress in bending. Table 21.11 shows the basic allowable stress number S_{at} in bending fatigue, which is for 10^7 stress cycles for single-direction bending. Use 70 percent of these values for reversed bending such as in an idler gear. If there are momentary overloads, the design should be checked against the possibility of exceeding the yield strength and causing permanent deformation of the gear teeth. For through-hardened steel the allowable yield strength S_{av} is dependent on Brinell hardness H_B, according to the following empirical equation:

$$S_{ay} = (482H_B - 32{,}800) \text{ psi} \tag{21.58}$$

The *life factor* K_L may be selected from Table 21.12 and is used to adjust the allowed stress number for the effect of designing to lives other than 10^7 stress cycles.

The *temperature factor* K_T derates the strength to account for loss of basic material strength at elevated temperature. For temperatures up to 250°F the factor is unity. For higher temperatures there is loss of strength due to the tempering effect, and in some materials over 350°F the value of K_T must be greater than unity. One alternative method that is used widely is to form the K_T factor as follows:

$$K_T = \frac{460 + {}^\circ\text{F}}{620} \tag{21.59}$$

where the temperature range is between 160 and 300°F. Operating at temperatures above 300°F is normally not recommended.

The *reliability factor* K_R is used to adjust for desired reliability levels either less or greater than 99 percent, which is the level for the allowed bending strength data in Table 21.11. If actual statistical data on the strength distribution of the gear material are in hand, a suitable K_R factor may be selected. In lieu of this, use the values from Table 21.13.

21.5.3 Other Methods

There is no available methodology that applies to all types of gears for stress and deflection analysis. However, some selected methods that have a limited application are practical for design use and for gaining some insight into the behavior of gears under load.

TABLE 21.13 Reliability Factor K_R

Requirements of application	K_R	Reliability
Fewer than one failure in 10,000	1.50	0.9999
Fewer than one failure in 1000	1.25	0.999
Fewer than one failure in 100	1.00	0.99
Fewer than one failure in 10	0.85*	0.9

*At this value plastic flow might occur rather than pitting.
Source: AGMA.

A method for determining the stress in bending has been successfully applied to spur gears.[86] The stress is calculated by the following equations:

$$S = \frac{W_n \cos \phi_W'}{F}\left(1 + 0.26\frac{t_s^{0.7}}{2r}\right)\left[\frac{6l_S'}{t_s^2} + \left(\frac{0.72}{t_s l_s}\right)^{1/2}\left(1 - \frac{t_W}{t_s}\nu \tan \phi_W'\right) - \frac{\tan \phi_W'}{t_s}\right] \quad (21.60)$$

where S = root fillet tensile stress at location indicated in Fig. 21.36
F = tooth face width measured parallel to gear axis
W_N = instantaneous force normal to tooth surface transmitted by tooth
$\nu \approx \frac{1}{4}$

The remaining quantities in Eq. (21.60) are defined in Fig. 21.36. Angle γ in Fig. 21.36 defines the point where the root fillet tensile stress is calculated by Eq. (21.60).

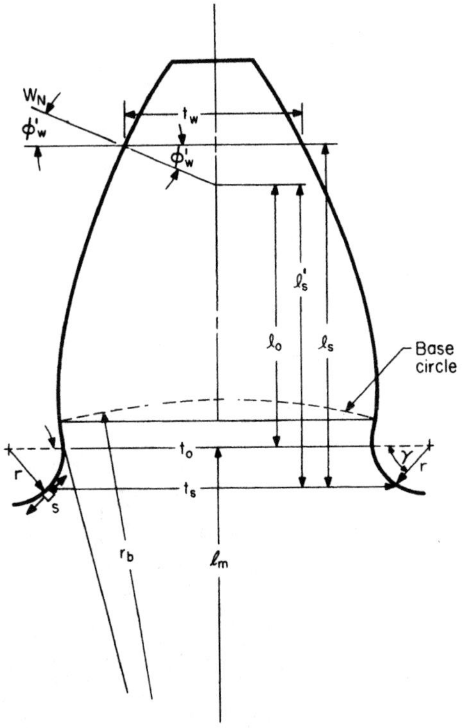

FIG. 21.36 Tooth geometry for evaluation of root bending stress formula parameters.

Reference 86 provides an equation for the value of γ that locates the position of the maximum root stress:

$$\tan \gamma_{i+1} = \frac{(1 + 0.16A_i^{0.7})A_i}{B_i(4 + 0.416A_i^{0.7}) - (\tfrac{1}{3} + 0.016A_i^{0.7})A_i \tan \phi_W'} \tag{21.61}$$

where
$$A_i = \frac{t_o}{r} + 2(1 - \cos \gamma_i) \tag{21.62}$$

and
$$B_i = \frac{l_o}{r} + \sin \gamma_i \tag{21.63}$$

where t_o and l_o are defined in Fig. 21.36, and subscripts i and $i + 1$ denote that the transcendental equation [Eq. (21.61)] can be solved iteratively for γ with i counting the steps in the iteration procedure.

Once the angle γ is determined, the dimension t_s shown in Fig. 21.36 also is determined by the formula

$$t_s = t_o + 2r(1 - \cos \gamma) \tag{21.64}$$

Deflections of gear teeth at the load contact point are important to determining the correct amount of tip relief necessary to minimize dynamic loads. A finite-element study has been done for a gear with a fully backed-up tooth (thick rim).[84] Figure 21.37 shows the results. The ordinate is the dimensionless deflection $\delta EF/W_N$, where δ is

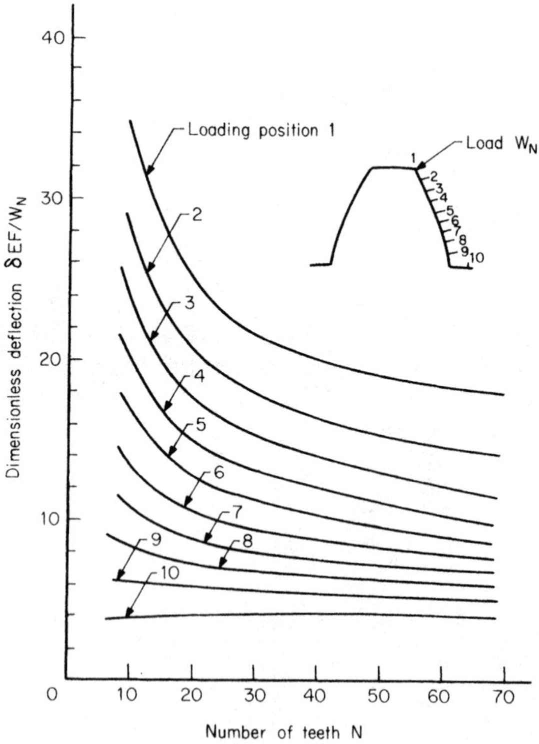

FIG. 21.37 Gear tooth dimensionless deflection as a function of number of teeth and loading position.

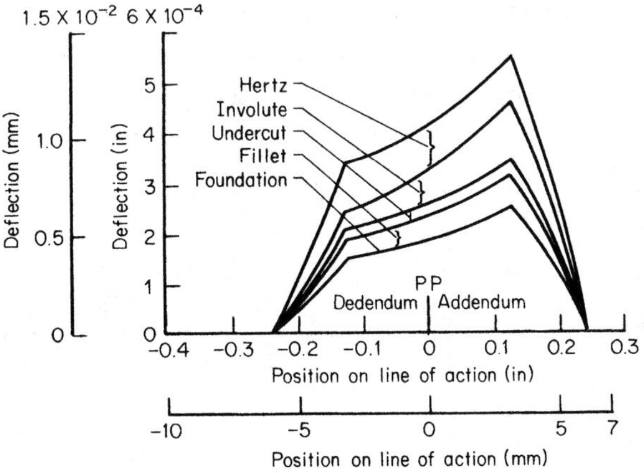

FIG. 21.38 Tooth deflection calculated by the method of Cornell.[87]

deflection, W_N/F is the load per unit of face width, and E is Young's modulus. The dimensionless deflection depends only on the load position and the tooth-gear blank configuration and not on the size of the gear. The load positions are numbered 1 to 10 with the pitch point being load position 5.5. These deflection values are also useful for the construction of a mathematical model of torsional vibration in the gear train.

The methods presented in Ref. 86 allow study of the separate effects contributing to gear tooth deflection. The method is based on beam deflection theory with corrections included to account for hertz deflection, shear deformation, and the flexibility effects of the gear blank itself. The results of the method applied to a 3.5-in pitch diameter gear with 28 teeth, 8 diametral pitch, 0.25-in face width, and solid gear blank are shown in Fig. 21.38. This figure gives an idea of the relative contributions of individual flexibility effects to the total deflection. The hertzian deflection is approximately 25 percent of the total. The gear blank deflection is approximately 40 percent. The beam bending effect of the tooth is approximately 35 percent of the total. The beam bending effect occurs in three parts: the part in the fillet region; the zone of blending between fillet and involute profile (called "undercut" in Fig. 21.38); and the tooth proper (labeled "involute" in the figure). Figure 21.38 shows deflection as a function of load position on the tooth. The discontinuities in slope are the points where the load sharing changes from single tooth pair contact to double pair contact.

21.6 GEAR LIFE PREDICTIONS

Gears in operation are subjected to repeated cycles of contact stress. Even if the gears are properly designed to have acceptable bending stresses to avoid bending fatigue failure, there remains the likelihood of failure by surface contact fatigue. Contact fatigue is unlike bending fatigue in that there is no endurance limit for contact fatigue failure. Eventually a pit will form when the gear has been stressed by enough repeated load cycles. This section shows how to use the method that was developed at NASA for prediction of surface fatigue life for gears.[87–91] The method is based on the assumption that pitting fatigue failures for gears and bearings are similar. A short explanation of the theoretical basis of the life prediction formulas presented in this section follows.

21.6.1 Theory of Gear Tooth Life

The fatigue-life model proposed in 1946 by Lundberg and Palmgren[92] is the commonly accepted theory to determine the fatigue life of rolling-element bearings. The probability of survival as a function of stress cycles is expressed as

$$\log \frac{1}{S} \alpha \frac{\tau_o^c \eta^e V}{z^h} \tag{21.65}$$

Hence, if the probability of survival S is specified, the life η in millions of stress cycles for the required reliability S can be considered a function of the stressed volume V, the maximum critical shearing stress τ_o, and the depth to the critical shearing stress z. As a result, the proportionality can be written as

$$\eta \propto \frac{z^{h/e}}{\tau_o^{c/e} V^{1/e}} \tag{21.66}$$

Equation (21.66) reflects the idea that greater stress shortens life. The depth below the surface z at which the critical stress occurs is also a factor. A microcrack beginning at a point below the surface requires time to work its way to the surface. Therefore, we expect that life varies by power of depth to the critically stressed zone.

The stressed volume V is also an important factor. Pitting initiation occurs near any small stress-raising imperfection in the material. The larger the stressed volume, the greater the likelihood of fatigue failure. By the very nature of the fatigue-failure phenomenon, it is the repetitive application of stress that causes cumulative damage to the material. The greater the number of stress cycles, the greater the probability of failure. By experience it has been found that the failure distribution follows the Weibull* model, with $e = 2.5$ for gears. Experiments have determined that exponent $c = 23.25$ and exponent $h = 2.75$. Based on direct application of statistical fundamentals and using the hertzian stress equations, the following formulas enable the straightforward calculation of fatigue life for spur gears. The expression for reliability as a function of stress cycles on a single gear tooth is

$$\log \frac{1}{S} = \left(\frac{\eta}{T_{10}}\right)^e \log\left(\frac{1}{0.9}\right) \tag{21.67}$$

where S is the probability of survival and T_{10} is the life corresponding to a 90 percent reliability. The dispersion or scatter parameter e (Weibull exponent) based on NASA fatigue tests is equal to 2.5. The single tooth life T_{10} is determined from

$$T_{10} = 9.18 \times 10^{18} W_N^{-4.3} F_e^{3.9} \Sigma \rho^{-5} l^{-0.4} \tag{21.68}$$

where T_{10} = millions of stress cycles
W_N = normal load, lb
F_e = loaded or effective face width, in
l = loaded profile length, in
$\Sigma \rho$ = curvature sum

*The Weibull distribution is defined as $S = \exp\,[(n/n_c)^e]$ where e is the Weibull exponent and n_c is the characteristic life.

The curvature sum is given by

$$\Sigma \rho = \left(\frac{1}{r_{p1}} + \frac{1}{r_{p2}}\right)\frac{1}{\sin \phi} \tag{21.69}$$

where r_{p1} and r_{p2} are the pitch radii in inches and ϕ is the pressure angle. The factor 9.18×10^{18} is an experimentally obtained constant based on NASA gear fatigue tests using aircraft-quality gears of vacuum-arc-remelted (VAR) AISI 9310 steel. If another material with a known fatigue life expectancy relative to the AISI 9310 material is selected, then the life calculated by Eq. (21.68) for T_{10} should be multiplied by an appropriate factor. Table 21.7 lists relative life for several steels based on NASA gear fatigue data. These values can be used as life adjustment factors. A similar treatment of life adjustment factors was made for rolling-element bearings for effects of material, processing, and lubrication variables.[93] As always, the designer should base the life calculations on experimental life data whenever possible.

An alternate procedure for estimating life is given by the American Gear Manufacturers' (AGMA) Standard 218.01.[70] The method is based on allowable stress which may be corrected for certain application and geometry effects. Because of the wide availability of Standard 218.01 the method will not be repeated here. Unlike the Lundberg and Palmgren[92] approach presented herein, the AGMA method does not include stressed volume directly and depth to critical stress sensitivities.

EXAMPLE A gear is to be made of an experimental steel with a known life of three times that of AISI 9310 (based on experimental surface fatigue results with a group of specimens). The gear tooth geometry has a 20° pressure angle, 8 pitch, 28 teeth, face width equals 0.375 in, standard addendum, and meshes with an identical gear. Determine the expected life at the 90 percent reliability level and the 99 percent reliability level for a single tooth, if the mesh is to carry 200 hp at 10,000 r/min.

solution First the necessary geometry must be worked out. The pitch radius is

$$r_{p1} = r_{p2} = \frac{28 \text{ teeth}}{8 \text{ teeth/in of pitch dia.}} \times \frac{1}{2} \times 1.75 \text{ in}$$

The load normal to the gear tooth surface is determined from the power. The torque is

$$T = 63{,}000 \times \frac{\text{hp}}{\text{r/min}} = 63{,}000 \times \frac{200}{10{,}000} = 1260 \text{ lb}\cdot\text{in}$$

The normal load is

$$W_N = \frac{T}{r_{p1} \cos \phi} = \frac{1260}{1.75 \cos 20°} = 766 \text{ lb}$$

The curvature sum is

$$\Sigma \rho = \left(\frac{1}{1.75} + \frac{1}{1.75}\right)\frac{1}{\sin 20°} = 3.34 \text{ in}^{-1}$$

The contact ratio is defined as the average number of tooth pairs in contact. It can be calculated by dividing the contact path length by the base pitch. The standard addendum gear tooth has an addendum radius $r_a = r_p + 1/P$. Therefore, the addendum radius is 1.88 in. The length of contact path along the line of action is

$$Z = [r_{a1}^2 - (r_{p1} \cos \phi)^2]^{1/2} + [r_{a2}^2 - (r_{p2} \cos \phi)^2]^{1/2} - (r_{p1} + r_{p2})\sin \phi$$

$$= 0.604 \text{ in}$$

The base pitch is

$$p_b = 2\pi r_{p1} \frac{\cos \varphi}{N}$$

$$= 2\pi \frac{1.75}{28} \cos 20^\circ$$

$$= 0.369 \text{ in}$$

The contact ratio is 1.64, which means that the gear mesh alternates between having one pair of teeth and two pairs of teeth in contact at any instant. This means that the load is shared for a portion of the time. Since the load is most severe for the single pair of teeth in contact, the length of loaded profile length corresponds to the profile length for single-tooth pair contact. If the contact ratio is above 2.0, then two separate zones of double-tooth pair contact contribute to the loaded profile length. For the current example, the pinion roll angle increment for the single-pair contact (heavy load zone) is

$$\epsilon_H = \frac{2p_b - Z}{r_{p1} \cos \phi}$$

$$= \frac{(2 \times 0.369) - 0.604}{1.75 \cos 20^\circ}$$

$$= 0.081 \text{ rad}$$

The pinion roll angle increment for the double-tooth pair contact (light load zone) is

$$\epsilon_L = \frac{Z - p_b}{r_{p1} \cos \phi}$$

$$= \frac{0.604 - 0.369}{1.75 \cos 20^\circ}$$

$$= 0.143 \text{ rad}$$

The pinion roll angle from the base of the involute to where the load begins (lowest point of double-tooth contact) is

$$\epsilon_c = \frac{(r_{p1} + r_{p2}) \sin \varphi - [r_{a2}^2 - (r_{p2} \cos \phi)^2]^{1/2}}{r_{p1} \cos \phi}$$

$$= \frac{(1.75 + 1.75) \sin 20^\circ - [1.88^2 - (1.75 \cos 20^\circ)^2]^{1/2}}{1.75 \cos 20^\circ}$$

$$= 0.174 \text{ rad}$$

The length of involute corresponding to the single-tooth pair contact is

$$l_1 = r_{p1} \cos \phi \, \epsilon_H (\epsilon_c + \epsilon_L + \tfrac{1}{2}\epsilon_H)$$

$$= 1.75 \cos 20^\circ (0.081)(0.174 + 0.143 + 0.081/2)$$

$$= 0.048 \text{ in}$$

Finally the life is obtained as follows:

$$T_{10} = 9.18 \times 10^{18} (766)^{-4.3} (0.375)^{3.9} (3.34)^{-5} (0.048)^{-0.4} = 643 \text{ million stress cycles}$$

In hours, the life is

$$T_{10} = \frac{643 \times 10^6 \text{ r}}{10{,}000 \text{ r/min} \times 60 \text{ min/h}} = 1072 \text{ h}$$

Conversion to the basis of 99 percent reliability gives

$$T_{10} = \left(\frac{\log 1/0.99}{\log 1/0.9}\right)^{1/2.5} = 419 \text{ h}$$

21.6.2 Life for the Gear

From basic probability notions, the life of the gear is based on the survival probabilities of the individual teeth. For a simple gear having N teeth meshing with another gear, the gear life corresponding to the tooth life at the same reliability level is given by the following relation:

$$G_{10} = T_{10} N^{-1/e} \qquad \text{simple gear} \tag{21.70}$$

where e = Weibull slope (2·5)
G_{10} = 10 percent life for gear
T_{10} = 10 percent life for single tooth as determined by previous equations
N = number of teeth.

For the case of an idler gear, where the power is transferred from an input gear to an output gear through the idler gear, both sides of the tooth are loaded once during each revolution of the idler gear. This is equivalent to a simple gear with twice the number of teeth. Therefore, the life of an idler gear with N teeth is

$$G_{10} = T_{10}(2N)^{-1/e} \qquad \text{idler gear} \tag{21.71}$$

when it meshes with gears of equal face width.

The life of a power-collecting (bull) gear, where there are u number of identical pinions transmitting equal power to the bull gear which has N teeth, is

$$G_{10} = \frac{T_{10}}{u}(N)^{-1/e} \qquad \text{bull gear} \tag{21.72}$$

21.6.3 Gear System Life

From probability theory, it can be shown that the life of two gears in mesh is given by the following:

$$L = (L_1^{-e} + L_2^{-e})^{-1/e} \tag{21.73}$$

where L_1 and L_2 are the gear lives.

Equation (21.73) may be generalized in n gears, in which case

$$L = \left(\sum_{i=1}^{n} L_i^{-e}\right)^{-1/e} \tag{21.74}$$

If the power transmission system has rolling-element bearings, then the system life equations must be solved with different values for the Weibull scatter parameter. For cylindrical or tapered bearings $e = 3/2$ and for ball bearings $e = 10/9$. The probability distribution for the total power transmission is obtained from the following equation[94]:

$$\log\left(\frac{1}{S_T}\right) = \log\left(\frac{1}{0.9}\right)\left[\left(\frac{L}{L_1}\right)^{e_1} + \left(\frac{L}{L_2}\right)^{e_2} + \left(\frac{L}{L_3}\right)^{e_3} + \cdots\right] \tag{21.75}$$

where L_1, L_2, L_3, etc., are the component lives corresponding to 90 percent reliability. Information on life calculation for bearings may be found in the chapters dealing with bearings in this handbook.

EXAMPLE A gearbox has been designed and the component lives have been determined as follows. Find the system life at the 95 percent reliability level.

Bearing	e	B_{10} life	Gear	G_{10} life
No. 1 ball	(10/9)	1200 h	No. 1 pinion	(2.5) 1000 h
No. 2 ball	(10/9)	1500 h		
No. 3 roller	(3/2)	1300 h	No. 2 gear	(2.5) 1900 h
No. 4 roller	(3/2)	1700 h		

solution

$$\log\left(\frac{1}{0.95}\right) = \left(\log\frac{1}{0.9}\right)\left[\left(\frac{L}{1200}\right)^{10/9} + \left(\frac{L}{1500}\right)^{10/9} + \left(\frac{L}{1300}\right)^{3/2} + \left(\frac{L}{1700}\right)^{3/2} + \left(\frac{L}{1000}\right)^{2.5} + \left(\frac{L}{1900}\right)^{2.5}\right]$$

Using a programmable calculator, iteration yields the following result for the life at the 95 percent reliability level:

$$L = 249 \text{ h}$$

This result is surprisingly low and emphasizes the need for high-reliability components if the required system reliability and life is to be met.

21.6.4 Helical Gear Life

Life calculation of helical gear teeth is presented in this section. The approach is to think of the helical gear as a modified spur gear, with the tooth profile in the plane normal to the gear axis being preserved but with the spur tooth being sliced like a loaf of bread and wrapped around the base circle along a helix of angle ψ. The helical tooth life may be obtained as a "corrected" spur tooth life where adjustment factors for each of the factors which influence life in the previous equations are given as follows:[88]

Load adjustment:
$$W_{\text{helical}} = \frac{W_{\text{spur}}\, p_b}{0.95Z\cos\psi} \tag{21.76}$$

Curvature sum adjustment:
$$\Sigma\rho_{\text{helical}} = (\Sigma\rho_{\text{spur}})\cos\psi \tag{21.77}$$

Face width in contact adjustment:
$$F_{\text{helical}} = F_{\text{spur}}\sec\psi \tag{21.78}$$

No adjustment factor for involute length l is recommended. The above adjustment factors are recommended for helical gears with axial contact ratios above 2. For lesser axial contact ratio, the life should be calculated as if the gear were a spur gear.

EXAMPLE A gear has the following properties: standard addendum, transverse pressure angle, 20°; diametral pitch, 8; number of teeth, 28; face width, 1 in; helix angle, 45°; and material, AISI 9310. Calculate the tooth life at 90 percent reliability if 766 lb is transmitted along the pitch line. (The properties in the transverse plane are taken from the example problem for spur gear life.)

solution From the preceding example problem, the transverse base pitch is 0.369 in and the length of the contact path is 0.604 in. The axial pitch p_a is calculated as follows:

$$p_a = \frac{p_b}{\tan\psi} = \frac{0.369}{\tan 45°} = 0.369 \text{ in}$$

The axial contact ratio is 2.71, which allows use of the defined correction factors:

$$\begin{aligned} W_{\text{helical}} &= \frac{W_{\text{spur}} p_b}{0.95 Z \cos\psi} \\ &= (766 \times 0.369)/(0.95 \times 0.604 \times \cos 45°) \\ &= 697 \text{ lb} \\ \Sigma\,\rho_{\text{helical}} &= \Sigma\,\rho_{\text{spur}} \cos\psi \\ &= 3.34 \cos 45° \\ &= 2.362 \text{ in}^{-1} \\ F_{\text{helical}} &= F_{\text{spur}} \sec\psi \\ &= 1/\cos 45° \\ &= 1.414 \text{ in} \end{aligned}$$

The above terms are now substituted into the tooth life equation as follows:

$$\begin{aligned} T_{10} &= 9.18 \times 10^{18} W_{\text{helical}}^{-4.3} F_{\text{helical}}^{3.9} \Sigma\,\rho_{\text{helical}}^{-5} l^{-0.4} \\ &= 9.18 \times 10^{18} (697)^{-4.3} (1.414)^{3.9} (2.362)^{-5} (0.048)^{-0.4} \\ &= 9.66 \times 10^{5} \text{ million stress cycles} \end{aligned}$$

21.6.5 Bevel and Hypoid Gear Life

For bevel and hypoid gears, the contact pattern may be determined from a tooth contact analysis (TCA). The TCA is done with a computer program developed by the Gleason Works.[95] The mathematics is very complicated,[96,97] but the needed parameters for a life analysis are simply the surface curvatures and path of contact (from which the hertzian stress may be calculated). Assuming that the result from a TCA-type analysis is available, the bevel or hypoid tooth life is calculated by the following formulas, which are adapted from Ref. 98. Reference 99 shows an alternate method of assessing surface durability:

$$T_{10} = \left(\frac{3.58 \times 10^{56} z^{7/3}}{\tau^{31/3} V} \right)^{1/2.5} \qquad \text{where } V = wzl \tag{21.79}$$

TABLE 21.14 Values of Contract Stress Parameters

Δe	a^*	b^*	ν
0	1	1	1.2808
0.1075	1.0760	0.9318	1.2302
0.3204	1.2623	0.8114	1.1483
0.4795	1.4556	0.7278	1.0993
0.5916	1.6440	0.6687	1.0701
0.6716	1.8258	0.6245	1.0517
0.7332	2.011	0.5881	1.0389
0.7948	2.265	0.5480	1.0274
0.83495	2.494	0.5186	1.0206
0.87366	2.800	0.4863	1.0146
0.90999	3.233	0.4499	1.00946
0.93657	3.738	0.4166	1.00612
0.95738	4.395	0.3830	1.00376
0.97290	5.267	0.3490	1.00218
0.983797	6.448	0.3150	1.00119
0.990902	8.062	0.2814	1.000608
0.995112	10.222	0.2497	1.000298
0.997300	12.789	0.2232	1.000152
0.9981847	14.839	0.2070	1.000097
0.9989156	17.974	0.18822	1.000055
0.9994785	23.55	0.16442	1.000024
0.9998527	37.38	0.13050	1.000006
1	∞	0	1.000000

$$\Sigma \rho = \frac{1}{\rho_{1x}} + \frac{1}{\rho_{2x}} + \frac{1}{\rho_{1y}} + \frac{1}{\rho_{2y}} \quad \text{curvature sum} \tag{21.80}$$

$$\Delta \rho = (1/\rho_{1x} + 1/\rho_{2x}) - (1/\rho_{1y} + 1/\rho_{2y}) \quad \text{curvature difference} \tag{21.81}$$

From Table 21.14 we get a^*, b^*, ν:

$$a = a^*\left[\frac{3W_N}{2\Sigma\rho}\frac{(1-\nu_1^2)}{E_1} + \frac{(1-\nu_2^2)}{E_2}\right]^{1/3} \tag{21.82}$$

$$b = b^*\left[\frac{3W_N}{2\Sigma\rho}\frac{(1-\nu_1{}^2)}{E_1} + \frac{(1-\nu_2^2)}{E_2}\right]^{1/3} \tag{21.83}$$

$$S_c = \frac{3W_N}{2\pi ab} \tag{21.84}$$

$$\tau_o = \frac{(2\nu - 1)^{1/2}S_c}{2\nu(Y+1)} \tag{21.85}$$

$$z = \frac{b}{(\nu+1)(2\nu-1)^{1/2}} \tag{21.86}$$

EXAMPLE From the results of a TCA study, it is found that the contact path length on the bevel pinion is approximately 0.3 in long. At the central point the principal radii of curvature are as follows:

Pinion	Gear
$\rho_{1x} = 2$ in	$\rho_{2x} = \infty$
$\rho_{1y} = 4$ in	$\rho_{2y} = -5$ in

The load normal to the tooth surface is 700 lb. Estimate the pinion tooth life.

solution

1. Calculate the curvature sum.

$$\Sigma\,\rho = \frac{1}{\rho_{1x}} + \frac{1}{\rho_{2x}} + \frac{1}{\rho_{1y}} + \frac{1}{\rho_{2y}}$$

$$\Sigma\,\rho = \frac{1}{2} + \frac{1}{\infty} + \frac{1}{4} - \frac{1}{5}$$

$$= 0.550 \text{ in}^{-1}$$

2. Calculate the curvature difference.

$$\Delta\rho = \frac{\left(\dfrac{1}{\rho_{1x}} + \dfrac{1}{\rho_{2x}}\right) - \left(\dfrac{1}{\rho_{1y}} + \dfrac{1}{\rho_{2y}}\right)}{\Sigma\,\rho}$$

$$= \frac{\left(\dfrac{1}{2} + \dfrac{1}{\infty}\right) - \left(\dfrac{1}{4} - \dfrac{1}{5}\right)}{0.550} = 0.818$$

3. From Table 21.14, by interpolation,

$$a^* = 2.397$$

$$b^* = .531$$

$$\nu = 1.023$$

4. Calculate the dimensions of the hertz ellipse, assuming the properties of steel, $E = 30 \times 10^6$ lb/in^2, $\nu = 0.25$.

$$a = a^*\left[\frac{3W_N}{2\Sigma\rho}\left(\frac{(1-\nu_1^2)}{E_1} + \frac{(1-\nu_2^2)}{E_2}\right)\right]^{1/3}$$

$$= 2.397\left[\frac{3 \times 700}{2 \times 0.550}\left(\frac{1-0.25^2}{30 \times 10^6}\right) \times 2\right]^{1/3} = 0.118 \text{ in}$$

$$b = b^*\left[\frac{3W_N}{2\Sigma\rho}\left(\frac{(1-\nu_1^2)}{E_1} + \frac{(1-\nu_2^2)}{E_2}\right)\right]^{1/3}$$

$$= 0.531\left[\frac{3 \times 700}{2 \times 0.550}\left(\frac{1-0.25^2}{30 \times 10^6}\right) \times 2\right]^{1/3} = 0.026 \text{ in}$$

5. Calculate the maximum hertzian contact stress.

$$S_c = \frac{3W_N}{2\pi ab} = \frac{3 \times 700}{2\,\pi \times 0.118 \times 0.026} = 108{,}000 \text{ lb/in}^2$$

6. Calculate the maximum reversing orthogonal shear stress and depth under the surface at which it occurs.

$$\tau_o = \frac{(2\nu - 1)^{1/2}}{2Y(\nu + 1)} S_c$$

$$= \frac{(2 \times 1.023 - 1)^{1/2}}{2 \times 1.023(1.023 + 1)} \times 108{,}000 = 26{,}700 \text{ psi}$$

$$z = \frac{1}{(\nu + 1)(2\nu - 1)^{1/2}} b$$

$$= \frac{1}{(1.023 + 1)[2(1.023) - 1]^{1/2}} \times 0.026 = 0.013 \text{ in}$$

7. Assume that the semiwidth of the hertzian contact ellipse is representative of the width of the contact path across the tooth, and calculate the stressed volume as follows:

$$V = wzl \approx azl$$

$$= 0.118 \times 0.013 \times 0.3$$

$$= 460 \times 10^{-6} \text{ in}^3$$

8. Finally the tooth life at 90 percent reliability is calculated as follows:

$$T_{10} = \left(\frac{3.58 \times 10^{56} z^{7/3}}{\tau_o^{31/3} V}\right)^{1/2.5}$$

$$= \left[\frac{3.58 \times 10^{56}(0.013)^{7/3}}{(26{,}700)^{31/3} 460 \times 10^{-6}}\right]^{1/2.5}$$

$$= 7.94 \times 10^{9} \text{ stress cycles}$$

21.7 LUBRICATION

In gear applications, fluid lubrication serves four functions:

1. Provides a separating film between rolling and sliding contacting surfaces, thus preventing wear
2. Acts as a coolant to maintain proper gear temperature
3. Protects the gear from dirt and other contaminants
4. Prevents corrosion of gear surfaces

The first two lubrication functions are not necessarily separable. The degree of surface separation affects friction mode and the magnitude of friction force. This in turn influences frictional heat generation and gear temperatures.

Until the last two decades the role of lubrication between surfaces in rolling and sliding contact was not fully appreciated. Metal-to-metal contact was presumed to occur in all applications with attendant required boundary lubrication. An excessive quantity of lubricant sometimes generated excessive gear tooth temperatures causing thermal failure. The development of the elastohydrodynamic lubrication theory showed that lubricant films of thicknesses of the order of microinches and tens of microinches occur in rolling contact. Since surface finishes are of the same order of magnitude as the lubricant film thickness, the significance of gear tooth finish to gear performance became apparent.

21.7.1 Lubricant Selection

The useful bulk-temperature limits of several classes of fluid lubricants in an oxidative environment are given in Table 21.15.[100] Grease lubricants are listed in Table 21.16.[100] The most commonly used lubricant is mineral oil, both in liquid and grease form (Table 21.17).[100] As a liquid, mineral oil usually has an antiwear or extreme pressure (EP) additive, an antifoam agent, and an oxidation inhibitor. In grease, the antifoam agent is not required.

Synthetic lubricants have been developed to overcome some of the harmful effects of lubricant oxidation. However, synthetic lubricants should not be selected over readily available and invariably less-expensive mineral oils if operating conditions do not require them. Incorporation of synthetic lubricants in a new design is usually more easily accomplished than conversion of an existing machine to their use.

The heat-transfer requirements of gears dictate whether a grease lubricant can be used. Grease lubrication permits the use of simplified housing and seals.

21.7.2 Elastohydrodynamic Film Thickness

Elastohydrodynamics is discussed in detail in Chap. 15 for rolling-element bearings. Rolling elements, which gear teeth can be regarded as, are separated by a highly compressed, extremely thin lubricant film. Because of the presence of high pressures in the contact zone, the lubrication process is accompanied by some elastic deformation

TABLE 21.15 Lubricants

Lubricant type and specification	Bulk temperature limit in air, °F
Mineral oil MIL-L-6081	200+
Diester MIL-L-7808	300+
Type II ester MIL-L-23699	425
Triester MIL-L-9236B	425
Super-refined and synthetic mineral oils	425
Fluorocarbon*	550
Polyphenyl ether*	600

*Not recommended for gears.

TABLE 21.16 Greases

Type	Specification	Recommended temp. range, °F	General composition
Aircraft: High-speed, ball and roller bearing	MIL-G-38220	− 40 to +400	Thickening agent and fluorocarbon
Aircraft: Synthetic, extreme pressure	MIL-G-23827	−100 to +250	Thickening agent, low-temperature synthetic oils, or mixtures EP additive
Aircraft: Synthetic, molybdenum disulfide	MIL-G-21164	−100 to +250	Similar to MIL-G-23827 plus MoS_2
Aircraft: General purpose, wide temp. range	MIL-G-81322	− 65 to +350	Thickening agent and syntehtic hydrocarbon
Helicopter: Oscillating, bearing	MIL-G-25537	− 65 to +160	Thickening agent and mineral oil
Plug valve: Gasoline- and oil-resistant	MIL-G-6032	+ 32 to +200	Thickening agent, vegetable oils, glycerols and/or polyesters
Aircraft Fuel- and oil-resistant	MIL-G-27617	− 30 to +400	Thickening agent and fluorocarbon or fluorosilicone
Ball and roller bearing: Extreme high temperature	MIL-G-25013	−100 to +450	Thickening agent and silicone fluid

of the contact surface. Accordingly, this process is referred to as elastohydrodynamic (EHD) lubrication. Ertel[101] and later Grubin[102] were among the first to identify this phenomenon. Hamrock and Dowson[103] have derived a simplified approach to calculating the EHD film thickness as follows:

$$H_{\min} = 3.63U^{0.68}G^{0.49}W^{-0.073}(1 - e^{-0.68k}) \tag{21.87}$$

In this equation the most dominant exponent occurs on the speed parameter U, and the exponent on the load parameter W is very small and negative. The materials parameter G also carries a significant exponent, although the range of this variable in engineering situations is limited.

The variables related in Eq. (21.87) consist of five dimensionless groupings. These are

Dimensionless film thickness:

$$H = \frac{h}{\rho_x} \tag{21.88}$$

Ellipticity parameter:

$$k = \left(\frac{\rho_y}{\rho_x}\right)^{2/\pi} \tag{21.89}$$

Dimensionless load parameter:

$$W = \frac{W_N}{E'\rho_x^2} \tag{21.90}$$

TABLE 21.17 Mineral Oil Classification and Comparative Viscosities

Category	SAE number	ASTM grade*	AGMA gear oil	Approx. viscosity (CS) 100°F	210°F
	—	32	—	2	—
	—	40	—	3–5.5	—
	—	60	—	8.5–12	—
	—	75	—	12–16	—
	—	105	—	19–24	—
	—	150	—	29–35	—
Extra light	—	215	—	42–51	—
	10W	—	1	45	4.2
	—	315	—	61–75	—
Light	20W	—	2	69	5.7
	—	465	—	90–110	—
Medium	30W	—	—	—	10
	—	—	3	128	—
	—	700	—	130–166	—
Medium heavy	40	—	—	—	13
	—	—	4	183	—
Heavy	50	—	—	—	17
	—	1000	—	194–237	18
	—	—	5	—	—
	—	1500	—	291–356	—
	—	2150	—	417–525	—
	—	—	6	—	24
	140	—	—	—	25
	—	3150	—	630–780	—
Super heavy	—	—	7†	—	29
	—	4650	—	910–1120	—
	—	—	8†	—	36
	250	—	—	—	43
	—	—	8A†	—	47
	—	7000	—	1370–1670	—
	—	—	9	—	97
	—	—	10	—	227
	—	—	11	—	464

*Grade number is equivalent to average Saybolt universal viscosity (SUS) at 100°F.
†Compounded with fatty oil.

Dimensionless speed parameter:

$$U = \frac{\mu_o v}{E' \rho_x} \tag{21.91}$$

Dimensionless materials parameter:

$$G = \alpha E' \tag{21.92}$$

where E' = effective elastic modulus, N/m^2 (lb/in^2)
W_N = normal applied load, N (lb)
h = film thickness, m (in)
ρ_x = effective radius of curvature in x (motion) direction, m in
ρ_y = effective radius of curvature in y (transverse) direction, m in
v = mean surface velocity in x direction, m/s (in/s)
α = pressure-viscosity coefficient of fluid, m^2/N (in^2/lb)
μ_o = atmospheric viscosity, $N\cdot s/m^2$ ($lb\cdot s/in^2$)

$$E' = \frac{2}{(1 - \nu_1^2)/E_1 + (1 - \nu_2^2)/E_2} \tag{21.93}$$

$$\frac{1}{\rho_x} = \frac{1}{\rho_{1x}} + \frac{1}{\rho_{2x}} \qquad \frac{1}{\rho_y} = \frac{1}{\rho_{1y}} + \frac{1}{\rho_{2y}} \tag{21.94}$$

where E_1, E_2 = elastic modulus of bodies 1 and 2
ν_1, ν_2 = Poisson's ratios of bodies 1 and 2
ρ_{1x}, ρ_{2x} = radius in x (motion) direction of bodies 1 and 2
ρ_{1y}, ρ_{2y} = radius in y (transverse) direction of bodies 1 and 2

For typical steels, E' is 33×10^6 lb/in^2, and, for mineral oils, a typical value of α is 1.5×10^{-4} in^2/lb. Thus, for mineral-oil-lubricated rolling elements, $G = 5000$.

For bodies in sliding and rolling contact with parallel axes of rotation, the tangential surface velocities are

$$v_1 = \omega_1 R_{1x} \qquad v_2 = \omega_2 R_{2x} \tag{21.95}$$

where R_{1x} and R_{2x} are the radii from the centers of rotation to the contact point.

The geometry of an involute gear contact at distance s along the line of action from the pitch point can be represented by two cylinders rotating at the angular velocity of the respective gears. The distance s is positive when measured toward the pinion (member 1). Equivalent radius, from Eq. (21.94), is

$$\rho_x = \frac{(r_{p1} \sin \phi - s)(r_{p2} \sin \phi + s)}{(r_{p1} + r_{p2}) \sin \phi} \tag{21.96}$$

Contact speeds from Eq. (21.95) are

$$v_1 = \omega_1 (r_{p1} \sin \phi - s) \qquad v_2 = \omega_2 (r_{p2} \sin \phi + s) \tag{21.97}$$

Surface topography is important to the EHD lubrication process. EHD theory is based on the assumption of perfectly smooth surfaces, that is, no interaction of surface asperities. Actually, of course, this is not the case. An EHD film of several millionths of an inch can be considered adequate for highly loaded rolling elements in a high-temperature environment. However, the calculated film might be less than the combined surface roughness of the contacting elements. If this condition exists, surface asperity contact, surface distress (in the form of surface glazing and pitting), and surface smearing or deformation can occur. Extended operation under these conditions can result in high wear, excessive vibration, and seizure of mating components. A surface-roughness criterion for determining the extent of asperity contact is based upon the ratio of film thickness to a composite surface roughness. The film parameter Λ is

$$\Lambda = \frac{h}{\sigma} \tag{21.98}$$

where composite roughness σ is

$$\sigma = \left(\sigma_1^2 + \sigma_2^2\right)^{1/2} \tag{21.99}$$

and σ_1 and σ_2 are the rms roughnesses of the two surfaces in contact. Figure 21.39 is a plot of relative life as a function of the film parameter Λ. Although the curve of Fig. 21.39 has been prepared using the rms estimate, an arithmetic average of *AA* might be used.

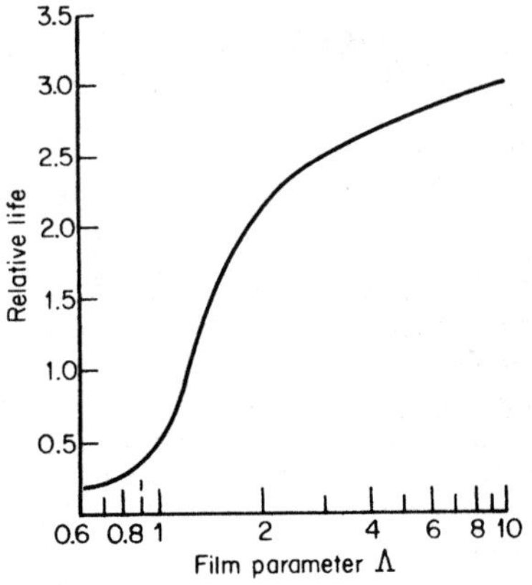

FIG. 21.39 Gear relative life as a function of film parameter Λ.

In addition to life, the Λ ratio can be used as an indicator of gear performance. At values of Λ less than 1, surface scoring, smearing, or deformation, accompanied by wear, will occur on the gear surface. When Λ is between 1 and 1.5, surface distress can occur accompanied by superficial surface pitting. For values between 1.5 and 3 some surface glazing can occur with eventual gear failure by classical subsurface-originated rolling-element (pitting) fatigue. At values of 3 or greater, minimal wear can be expected with extremely long life. Failure will eventually be by classical subsurface-originated rolling-element fatigue.[93]

21.7.3 Boundary Lubrication

Extreme-pressure lubricants can significantly increase the load-carrying capacity of gears. The extreme-pressure additives in the lubricating fluid form a film on the surfaces by a chemical reaction, adsorption, and/or chemisorption. These boundary films can be less than 1 μin to several microinches thick.[104] These films may be from the chemical reaction of sulfur[105] or from the chemisorption of iron stearate. Figure 21.40 shows the range of film thicknesses for various types of films.[105] The films can provide separation of the metal surfaces when the lubricant becomes thin enough for the asperities to interact. The boundary film probably provides lubrication by microasperity-elastohydrodynamic lubrication,[104] where the asperities deform under the load. The boundary film prevents contact of the asperities and at the same time

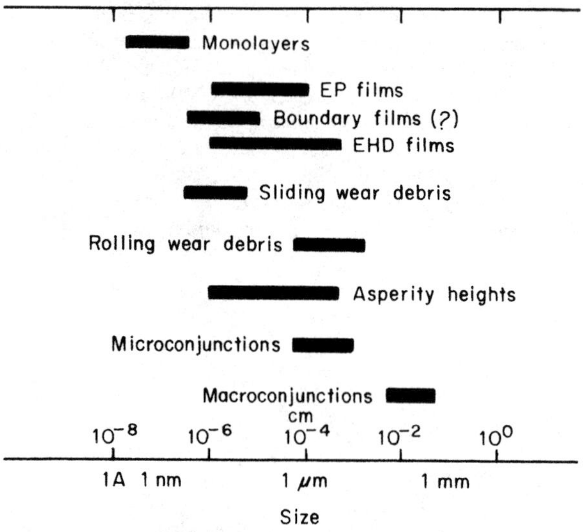

FIG. 21.40 Range of film thickness for various types of lubricant films, with relation to gear surface roughness and wear.

provides low-shear-strength properties that prevent shearing of the metal and reduce the friction coefficient over that of the base metal. These boundary films provide lubrication at different temperature conditions, depending on the materials used. For example, some boundary films will melt at a lower temperature than others[105] and will then fail to provide protection of the surfaces. The "failure temperature" is the temperature at which the lubricant film fails. In extreme-pressure lubrication this failure temperature is the temperature at which the boundary film melts. The melting point or thermal stability of surface films appears to be one unifying physical property governing failure temperature for a wide range of materials.[105] It is based on the observation[106] that only a film which is solid can properly interfere with potential asperity contacts. For this reason, many extreme-pressure lubricants contain more than one chemical for protection over a wide temperature range. For instance, Borsoff[107,108] found that phosphorus compounds were superior to chlorine and sulfur at slow speeds, while sulfur was superior at high speeds. He explains this as a result of the increased surface temperature at the higher speeds. It should be remembered that most extreme-pressure additives are chemically reactive and increase their chemical activity as temperature is increased. Horlick[109] found that some metals such as zinc and copper had to be removed from their systems when using certain extreme-pressure additives.

21.7.4 Lubricant Additive Selection

Some of the extreme-pressure additives commonly used for gear oils are those containing one or more compounds of chlorine, phosphorus, sulfur, or lead soaps.[110] Many chlorine-containing compounds have been suggested for extreme-pressure additives, but few have actually been used. Some lubricants are made with chlorine containing molecules where the Cl_3—C linkage is used. For example, either tri(trichloroethyl) or tri(trichlortert butyl) phosphate additives have shown high load-carrying capacity. Other chlorine-containing additives are chlorinated paraffin or petroleum waxes and hexachloroethene.

The phosphorus-containing compounds are perhaps the most commonly used additives for gear oils. Some aircraft lubricants have 3 to 5 percent tricresyl phosphate or tributyl phosphite as either an extreme-pressure or antiwear agent. Other phosphorus extreme-pressure agents used in percentages of 0.1 to 2.0 percent could be dodecyl dihydrogen phosphate; diethyl-, dibutyl-, or dicresyl-phenyl trichloroethyl phosphite; and a phosphate ester containing a pentachlorophenyl radical. Most of the phosphorus compounds in gear oils also have other active elements. The sulfur-containing extreme-pressure additives are believed to form iron sulfide films that prevent wear up to very high loads and speeds. However, they give higher friction coefficients and are therefore usually supplemented by other boundary film forming ingredients that reduce friction. The sulfur compounds should have controlled chemical activity (such as oils containing dibenzyl disulfide of 0.1 or more percent). Other sulfur-containing extreme-pressure additives are diamyl disulfide, dilauryl disulfide, sulfurized oleic acid and sperm oil mixtures, and dibutylxanthic acid disulfide.

Lead soaps have been used in lubricants for many years. They resist the wiping and sliding action in gears, and they help prevent corrosion of steel in the presence of water. Some of the lead soaps used in lubricants are lead oleate, lead fishate, lead-12-hydroxystearate, and lead naphthenate. The lead soap additive most often used is lead naphthenate because of its solubility. Lead soaps are used in concentrations of from 5 to 30 percent.

There are other additive compounds that contain combinations of these elements and most extreme-pressure lubricants contain more than one extreme-pressure additive. Needless to say, the selection of a proper extreme-pressure additive is a complicated process. The word susceptibility is frequently used with reference to additives in oils to indicate the ability of the oil to accept the additive without deleterious effects. Such things as solubility, volatility, stability, compatibility, load-carrying capacity, cost, etc., must be considered.

Many gear oil compounds depend upon the use of proprietary or packaged extreme-pressure additives. As a result, the lubricant manufacturer does not evaluate the additives' effectiveness. Because of this, any selection of extreme-pressure additives should be supported by an evaluation program to determine their effectiveness for a given application. However, a few firms have considerable background in the manufacture and use of extreme-pressure additives for gear lubrication, and their recommendations are usually accepted without question by users of gear oils.

21.7.5 Jet Lubrication

For most noncritical gear applications where cooling the gear teeth is not an important criterion, splash or mist lubrication is more than adequate to provide acceptable lubrication. However, in critical applications such as helicopter transmission systems and turboprop aircraft gear boxes, heat rejection becomes important. Hence, jet lubrication is used. It is important to have the oil penetrate into the dedendum region of the gear teeth in order to provide for cooling and lubrication. Failure to do so can result in premature gear failure primarily by scoring and wear. Methods of oil jet lubrication were analytically determined and experimentally verified by Akin and Townsend.[111–117]

Out-of-Mesh. An analysis along with experimental results was given in Refs. 111 and 114 for out-of-mesh jet lubrication of spur gears. The analysis determines the impingement depth for both gear and pinion when the jet is pointed at the out-of-mesh location and directed at the pitch point of the gear and pinion. Figure 21.41 shows the oil jet as it clears the gear tooth and impinges on the pinion tooth and also gives the nomenclature for the equations. The time of flight for the oil jet to clear the gear tooth and impinge on the pinion tooth must equal the time of rotation of the gears during

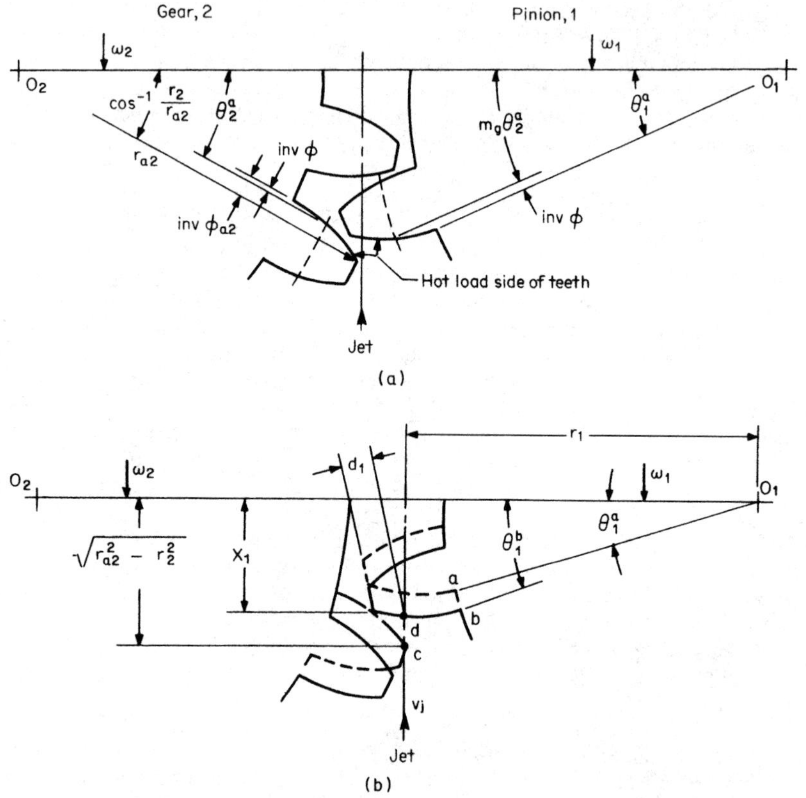

FIG. 21.41 Model for jet lubrication of out-of-mesh gears. (*a*) Gear tooth is at position where oil jet just clears the gear tooth and begins flight toward the pinion tooth. Pinion is shown in position *a*. (*b*) As jet moves from point *c* on the gear to point *d* on the pinion, the pinion rotates from position *a* to *b*.

this time and is given by the following equation:

$$\text{Time of rotation} = \text{time of flight}$$

$$\{\tan^{-1} x_1/r_1 + \text{inv}\{\cos^{-1}[r_{b1}/(r_1{}^2 + x_1{}^2)^{1/2}]\} - m_g(\cos^{-1} r_2/r_{a2} - \text{inv}\ \phi_{a2} + \text{inv}\ \phi) - \text{inv}\ \phi\}/w_1 = [(r_{a2}^2 - r_2^2)^{1/2} - x_1]/v_j \qquad (21.100)$$

where inv $\phi = \tan\phi - \phi$.

Equation (21.100) must be solved iteratively for x_1 and substituted into the following equation to determine the pinion impingement depth:

$$d_1 = r_{a1} - (r_1^2 + x_1^2)^{1/2} \qquad (21.101)$$

A similar equation is used to determine the time of flight and rotation of the gears for the impingement on the unloaded side of the gear tooth:

$$[\tan^{-1}(x_2/r_2) + \text{inv}\{\cos^{-1}[r_{b2}/(r_2^2 + x_2^2)^{1/2}\} - \{\cos^{-1}(r_1/r_{a1}) - \text{inv}\ \phi_{a1} + \text{inv}\ \phi\}/m_g - \text{inv}\ \phi]/\omega_2 = [(r_{a1}^2 - r_1^2)^{1/2} - x_2]v_f \qquad (21.102)$$

Solving for x_2 iteratively and substituting in the following equation gives the impingement depth for the gear:

$$d_2 = r_{a2} - (r_2^2 + x_2^2)^{1/2} \tag{21.103}$$

At high gear ratios, it is possible for the gear to shield the pinion entirely so that the pinion would receive no cooling unless the jet direction and/or position is modified. The following three tests may be used to determine the condition for zero impingement depth on the pinion, when

$$\tan^{-1}\frac{x_1}{r_1} < \cos^{-1}\frac{r_1}{r_{a1}} \qquad \text{or} \qquad x_1 < (r_{a1}^2 - r_1^2)^{1/2}$$

and when $d_1 \leq 0$.

The pinion tooth profile is not impinged on directly for speed reducers with ratios above about 1.25, depending upon the jet velocity ratio. Gear impingement depth is also limited by leading-tooth blocking at ratios above about 1.25 to 1.5.

Into-Mesh. In Refs. 116 and 117 it was shown for into-mesh lubrication there is an optimum oil jet velocity to obtain best impingement depth and therefore best lubrication and cooling for the gear and pinion. This optimum oil jet velocity is equal to the gear and pinion pitch line velocity. When the oil jet velocity is greater or less than pitch line velocity, the impingement depth will be less than optimum. Also the oil jet should be pointed at the intersections of the gear and pinion outside diameters and should intersect the pitch diameters for best results. A complete analytical treatment for into-mesh lubrication is given in Refs. 116 and 117. Into-mesh oil jet lubrication will give much better impingement depth than out-of-mesh jet lubrication. However, in many applications, there will be considerable power loss from oil being trapped in the gear mesh. Radial oil jet lubrication at the out-of-mesh location will give the best efficiency, lubrication, and cooling.[112]

Radial Oil Jet Lubrication. The vectorial model for impingement depth using a radially directed oil jet from Ref. 113 is shown in Fig. 21.42. The impingement depth d can be calculated to be

$$d = \frac{1.57 + 2\tan\phi + BP/2}{P\{nD/[2977(\Delta p)^{1/2}] + \tan\phi\}} \tag{21.104}$$

where the oil jet velocity $V_j = 13\sqrt{\Delta p}$ ft/s with Δp in pounds per square inch gauge. If the required d is known, then the ΔP that is required to obtain that impingement depth is

$$\Delta p = \left\{\frac{ndPD}{2977[1.57 + BP/2 + (2 - dP)\tan\phi]}\right\}^2 \tag{21.105}$$

High-speed motion pictures were used to determine the impingement depth experimentally for various oil pressures and gear revolutions per minute. Figure 21.43 is a plot of calculated and experimental impingement depth vs. jet nozzle pressure for gear speeds of 2560 and 4920 r/min. At both speeds there is good agreement between the calculated and experimental impingement depths at higher pressures. However, at the lower pressures, there is considerable difference between the calculated and experimental impingement depths. Most of this difference in impingement depth is due to viscous losses in the nozzle with the very viscous oil used.

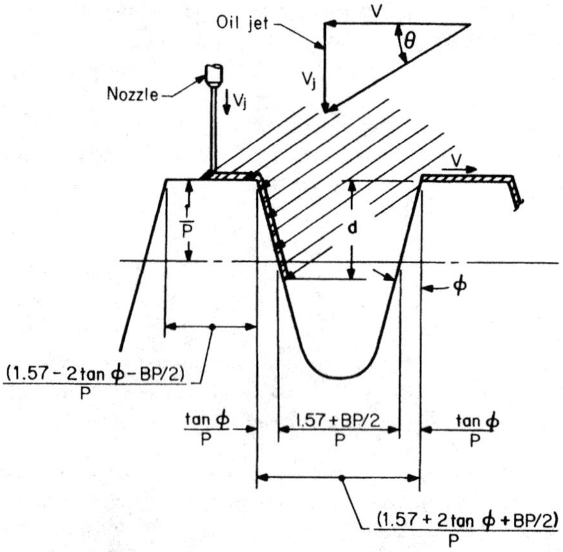

FIG. 21.42 Vectorial model for oil jet penetration depth.

21.7.6 Gear Tooth Temperature

A computer program was developed using a finite element analysis to predict gear tooth temperatures.[83,84,118] However, this program does not include the effects of oil jet cooling and oil jet impingement depth. It uses an average surface heat-transfer coefficient for surface temperature calculation based on the best information available at that time.

In order to have a method for predicting gear tooth temperature more accurately, it is necessary to have an analysis that allows for the use of a heat transfer coefficient for oil jet cooling coupled with a coefficient for air–oil mist cooling for that part of the

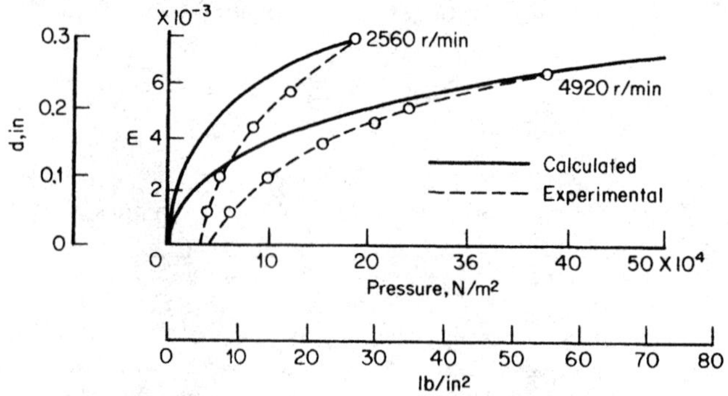

FIG. 21.43 Calculated and experimental impingement depth vs. oil jet pressure at 4920 and 2560 r/min.

time that each condition exists. Once the analysis can make use of these different coefficients, it can be combined with a method that determines the oil jet impingement depth to give a more complete gear temperature analysis program.[115] However, both the oil jet and air–oil mist heat transfer coefficients are unknowns and must be determined experimentally.

Once the heat generation and the oil jet impingement depth have been calculated, the heat transfer coefficients are either calculated or estimated. Then, a finite element analysis can be used to calculate the temperature profile of the gear teeth as shown in Fig. 21.44. The isotherms in this figure are the temperature differences between the calculated gear tooth temperatures and the inlet cooling oil.

Infrared temperature measurements of the gear tooth during operation verified the accuracy of the calculated values for the gear tooth surface.[112,115]

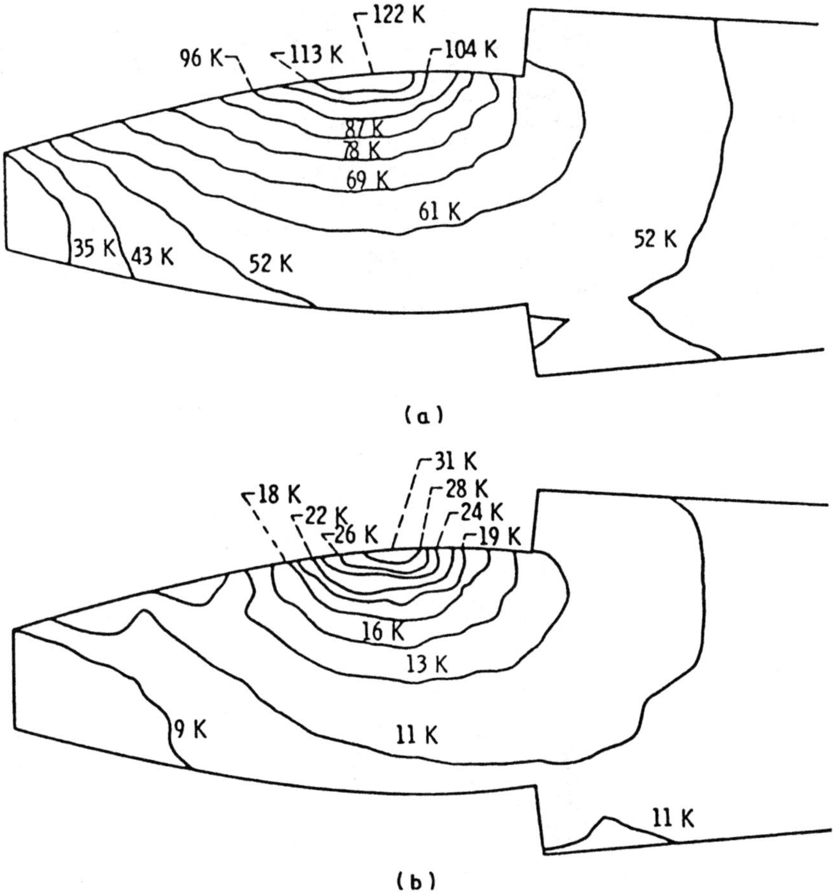

FIG. 21.44 Calculated gear tooth temperature increase above oil-in temperature at a speed of 10,000 r/min and load of 5903 N/cm (3373 lb·in) with 3½-in pitch diameter gear. (*a*) Zero impingement depth. (*b*) 87.5 percent impingement depth.

21.8 POWER-LOSS PREDICTIONS

Figure 21.45 is an experimental plot of efficiency as a function of input torque for a 317-hp, 17:1 ratio helicopter transmission system comprising a spiral-bevel gear input and a three-planet system. The nominal maximum efficiency of this transmission is 98.4 percent. The power loss of 1.6 percent of 317 hp is highly dependent on load and almost independent of speed. Table 21.1 gives maximum efficiency values for various types of gear systems. By multiplying the efficiency for each gear stage, the overall maximum efficiency of the gear system may be estimated. Unfortunately, most transmission systems do not operate at maximum rated torque. As a result, the percent power-loss prediction at part load will generally be high. In addition, gear geometry can significantly influence power losses and, in turn, the efficiency of gear sets.

As an example, consider a set of spur gears with a 4-in (10-cm) pinion. A reduction in gear diametral pitch from 32 to 4 can degrade peak efficiency from 99.8 to 99.4 percent under certain operating conditions. Although this reduced efficiency may at first glance appear to be of little significance, the change does represent a 200 percent increase in power loss.

Although some analytical methods have been developed for predicting gear power losses, most of them do not conveniently account for gear-mesh geometry or operating conditions. Generally, these predictive techniques base power-loss estimates on the friction coefficient at the mesh, among other parameters. Experience shows that the accuracy of these analytical tools often leaves much to be desired, especially for estimating part-load efficiency.

Figure 21.46 shows a typical power-loss breakdown for a gear system. An evaluation of all the power-loss components as a function of torque and speed shows that an unloaded gear set rotating at moderate to high speeds can comprise more than half of the total power losses at full load. Although the sliding loss is dominant at low operating speeds, it becomes only a moderate portion of the total gear-system losses at higher speeds. Both the rolling loss and support ball bearing losses increase with operating speed.

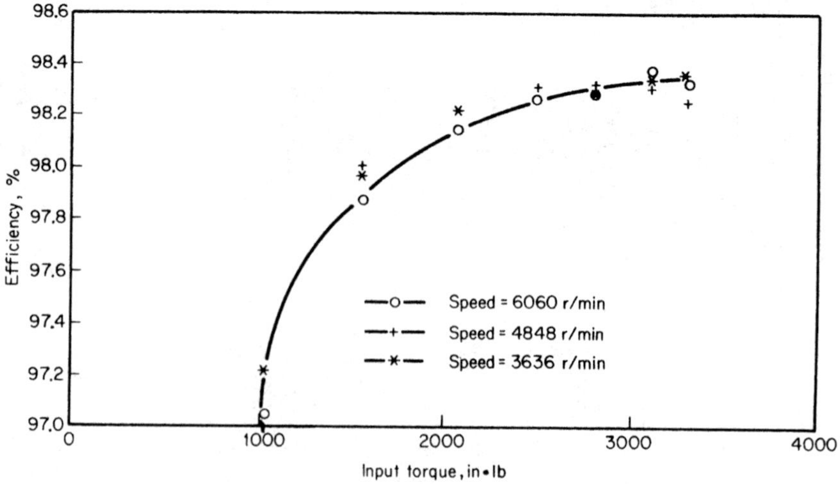

FIG. 21.45 Experimental input efficiency of a three-planet 317-hp helicopter transmission with a spiral-bevel gear input and an input-output ratio of 17:1.

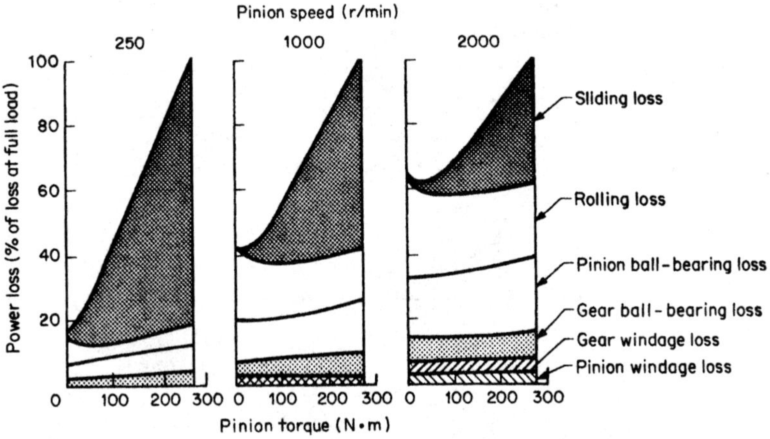

FIG. 21.46 Typical power-loss breakdown for a gear system. Gear set parameters include pitch diameter, 15.2 cm; gear ratio, 1.66; diametral pitch, 8; pinion face width-to-diameter ratio, 0.5; lubricant viscosity, 30 cP.

Good estimates of these speed-dependent losses are vital for accurately determining the power consumption of gearboxes. Anderson and Loewenthal[119–122] have developed an analytical method to predict power losses accurately in spur gear sets which have been correlated with experimental data. Their method applies to spur gears which have standard tooth proportions and are jet- or splash-lubricated. Moreover, it considers the individual contributions of sliding, rolling, and windage to power loss at the gear mesh. Some of the power-loss computations required by the method involve the use of mathematical expressions based on average operating conditions where the friction coefficient f, average sliding velocity v_s, average total rolling velocity v_r, and average EHD film thickness h are all evaluated at the point halfway between the pitch point and the start of engagement along the line of action. The mesh losses are determined simply by summing the sliding, rolling, and windage power loss components. Thus,

$$Q_m = Q_s + Q_r + Q_{w1} + Q_{w2} \tag{21.106}$$

It should be noted that the churning loss of gears running submerged in oil is not accounted for in this analysis.

21.8.1 Sliding Loss

The sliding loss Q_s at the mesh results from frictional forces that develop as the teeth slide across one another. This loss can be estimated as

$$Q_s = C_1 f \overline{W}_N \bar{v}_s \tag{21.107}$$

Values of C_1 through C_7 are given in Table 21.18. Friction coefficient f, average normal load $\overline{W}_N$, and average sliding velocity $\bar{v}_s$ are

$$f = 0.0127 \log\left(\frac{C_6 \overline{W}_N}{F\mu\, \bar{v}_s \bar{v}_r^2}\right) \tag{21.108}$$

TABLE 21.18 Constants for Computing Gear-Mesh Power Losses

Constant	Value for SI metric units	Value for U.S. customary units
C_1	2×10^{-3}	3.03×10^{-4}
C_2	9×10^{4}	1.970
C_3	1.16×10^{-8}	1.67×10^{-14}
C_4	0.019	2.86×10^{-9}
C_5	39.37	1.0
C_6	29.66	45.94
C_7	2.05×10^{-7}	4.34×10^{-3}

$$\overline{W}_N = \frac{T_1}{D_1 \cos \phi} \tag{21.109}$$

$$\bar{v}_s = |\mathbf{v}_2 - \mathbf{v}_1| \quad \text{(in the vector sense)} \tag{21.110}$$

$$= 0.0262 n_1 \left(\frac{1 + m_g}{m_g} \right) Z \text{ ft/s}$$

It should be noted that the expression for f is based upon test data applied to gear sliding-loss computations involving the elastohydrodynamic (EHD) lubrication regime, where some asperity contact occurs. Such a lubrication regime generally can be considered as the common case for gearset operation. Parameter f should be limited to a range from 0.01 to 0.02.

For computations of $\bar{v}_s$ and f, the contact-length path Z and average total rolling velocity $\bar{v}_r$ are

$$Z = 0.5 \left\{ \left[\left(D_1 + \frac{2}{C_5 P} \right)^2 - (D_1 \cos \phi)^2 \right]^{1/2} + \left[\left(D_2 + \frac{2}{C_5 P} \right)^2 - (D_2 \cos \phi)^2 \right]^{1/2} - (D_1 + D_2) \sin \phi \right\} \tag{21.111}$$

$$\bar{v}_r = 0.1047 n_1 \left[D_1 \sin \phi - \frac{Z}{4} \left(\frac{m_g - 1}{m_g} \right) \right] \tag{21.112}$$

21.8.2 Rolling Loss

Rolling loss occurs during the formation of an EHD film when oil is squeezed between the gear teeth, then subsequently pressurized. The rolling loss Q_r is a function of the average EHD film thickness h and the contact ratio m_c, which is the average number of teeth in contact. Therefore,

$$Q_r = C_2 \bar{h} \bar{v}_r F m_c \tag{21.113}$$

where the "central" EHD film thickness $\bar{h}$ for a typical mineral oil lubricating steel gear is

$$\bar{h} = C_7 (\bar{v}_r \mu)^{0.67} \overline{W}_N^{-0.067} \rho_{\text{eq}}^{0.464} \tag{21.114}$$

and the contact ratio m_c is

$$m_c = \frac{C_5 Z P}{\pi \cos \phi} \qquad (21.115)$$

The expression for $\bar{h}$ in Eq. (21.114) does not consider the thermal and starvation effects occurring at high operating speeds, generally above 40 m/s (8000 ft/min) pitch line velocity. These effects typically tend to constrain the film thickness to some limiting value. For computing h, the equivalent-contact rolling radius is

$$\rho_{eq} = \frac{\left[D_1(\sin\phi) + \frac{Z}{2}\right]\left[D_2(\sin\phi) - \frac{Z}{2}\right]}{2(D_1 + D_2)\sin\phi} \qquad (21.116)$$

21.8.3 Windage Loss

In addition to sliding and rolling losses, power losses due to pinion and gear windage occur. Such losses can be estimated from expressions based on turbine-disk drag test data. Thus, the pinion and gear windage losses are approximated as

$$Q_{w1} = C_3\left(1 + 4.6\frac{F}{D_1}\right)n_1^{2.8}D_1^{4.6}(0.028\mu + C_4)^{0.2} \qquad (21.117)$$

$$Q_{w2} = C_3\left(1 + 4.6\frac{F}{D_2}\right)\left(\frac{n_2}{m_g}\right)^{2.8}D_2^{4.6}(0.028\mu + C_4)^{0.2} \qquad (21.118)$$

These windage loss expressions apply for an air-oil atmosphere, commonly found in typical gearboxes. The expressions assume a oil specific gravity of 0.9 and constant values for air density and viscosity at 339 K (150°F). To further account for the oil-rich gearbox atmosphere, both density and viscosity of the atmosphere are corrected to reflect a 34.25:1 air-oil combination, which often is reported for helicopter transmissions lubricated with oil jets. Most of the preceding expressions involve special constants, which are labeled as C_1 through C_7. The appropriate values for these constants can be obtained from Table 21.18.

21.8.4 Other Losses

To determine the total gear-system loss, power losses due to the support bearings also should be considered. Both hydrodynamic and rolling-element bearing losses often can equal or exceed the gear-mesh losses. Bearing losses can be quantified either theoretically or by testing.

Total gear-system loss Q_T is

$$Q_T = Q_m + Q_h \qquad (21.119)$$

where Q_m is gear-mesh losses and P_B is total bearing power losses. Given Q_T and input power Q_{in}, gear-system efficiency is predicted to be

$$\zeta = \frac{Q_{in} - Q_T}{Q_{in}} \times 100 \qquad (21.120)$$

21.8.5 Optimizing Efficiency

The method described for predicting spur-gear power losses also can be employed in a repetitive manner for a parametric study of the various geometric and operating variables for gears. Typical results of such parametric studies are summarized by the two efficiency maps shown in Figs. 21.47 and 21.48, commonly called *carpet plots*, for light and moderate gear tooth loads. These carpet plots describe the simultaneous effects on gear-mesh efficiency when the diametral pitch, pinion pitch diameter, and pitchline velocity are varied. The computed gear-mesh efficiencies do not include support-bearing losses.

On these plots, the three key variables are represented along orthogonal intersecting planes for three different values of diametral pitch, pinion pitch diameter, and pitchline velocity. The gear-mesh efficiency accompanying any combination of these three gear parameters is represented by an intersecting point for the three planes. Efficiencies at intermediate values can be readily found by interpolation among the planes.

The light and moderate spur gear loads are specified in terms of a contact load factor, usually called a *K* factor. The *K* factor is a widely used parameter that normalizes the degree of loading on gears of different size and ratio. Essentially, the *K* factor describes the load intensity on a gear tooth. It is evaluated as

$$K = \frac{\overline{W}_t(m_g + 1)}{FD_1 m_g} \tag{21.121}$$

The two carpet plots of Figs. 21.47 and 21.48 were generated for light and moderate loads where $K = 10$ and 300 lb/in^2, respectively. Allowable *K* factors for spur gears generally range from about 100 lb/in^2 for low-hardness steel gears with generated teeth to 1000 lb/in^2 for aircraft-quality, high-speed gears that are case-hardened and ground. A nominal *K* factor for a general-purpose industrial drive with 300 Bhn steel gears, carrying a uniform load at a pitchline velocity of 15 m/s (3000 ft/min) or less, typically ranges from 275 to 325 lb/in^2.

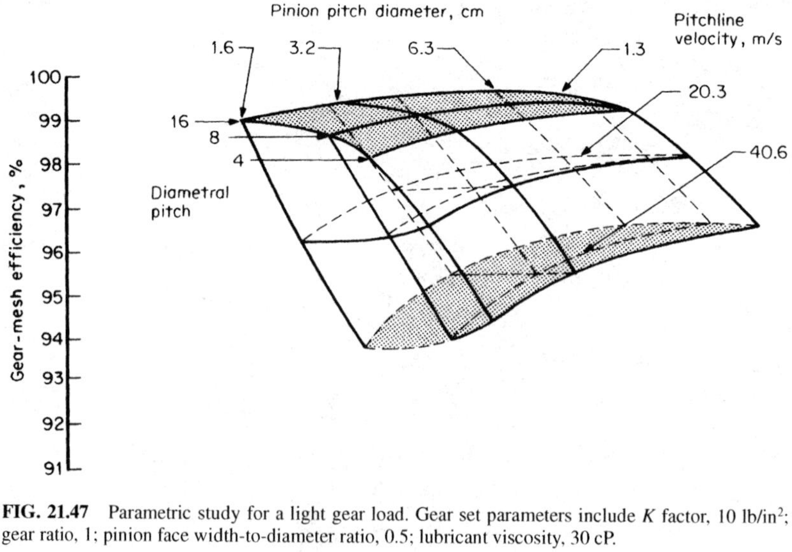

FIG. 21.47 Parametric study for a light gear load. Gear set parameters include *K* factor, 10 lb/in^2; gear ratio, 1; pinion face width-to-diameter ratio, 0.5; lubricant viscosity, 30 cP.

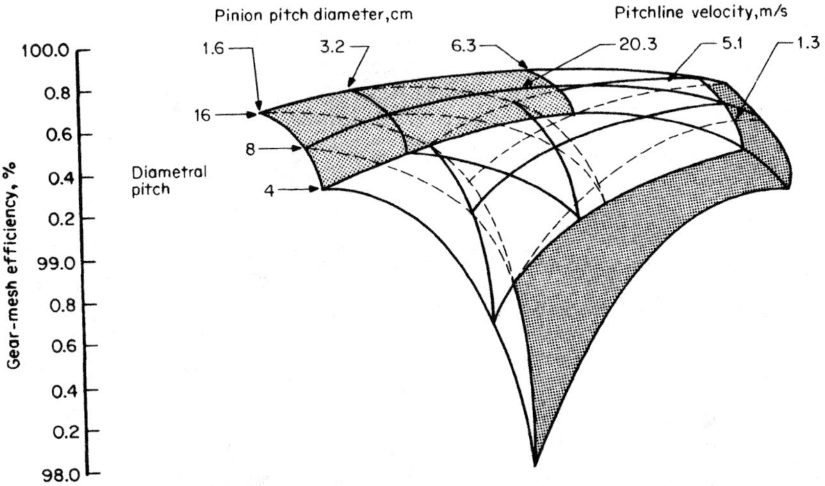

FIG. 21.48 Parametric study for a moderate gear load. Gear set parameters include *K* factor, 300 lb/in^2; gear ratio, 1; pinion face width-to-diameter ratio, 0.5; lubricant viscosity, 30 cP.

The plot for a light load indicates that increasing diametral pitch (resulting in smaller gear teeth) and decreasing pitchline velocity tends to improve gear-mesh efficiency. For a moderate to heavy load, increasing both diametral pitch and pitchline velocity improves efficiency. The reason for this reversal in efficiency characteristics is that speed-dependent losses at light loads dominate, whereas sliding loss is more pronounced at higher loads. Also, at higher speed, the sliding loss is reduced because the sliding-friction coefficient is lower.

Essentially, these two parametric studies indicate that large-diameter, fine-pitch gears tend to be most efficient when operated under appreciable load and high pitchline velocity. Consequently, to attain best efficiency under these operating conditions, the gear geometry should be selected as noted.

21.9 OPTIMAL DESIGN OF SPUR GEAR MESH

When designing gear systems, it is cost-effective to make the design efficient and long-lived with a minimum weight-to-power ratio from some standard point of view. This introduces the concept of optimization in the design, selecting as a point of reference either minimum size (which relates also to minimum weight) or maximum strength. While the designer is seeking to optimize some criterion index, all the basic design requirements (constraints) such as speed ratio, reliability, producibility, acceptable cost, and no failures due to wear, pitting, tooth breakage, or scoring must also be met. The design process must carefully assess all of the possible modes of failure and make adjustments in the allowable design parameters in order to achieve the optimum design. Figure 21.49 presents a qualitative diagram of expected failure modes for operation in various regimes of load and speed. The job of the designer is to anticipate the type of load and speed conditions the gears will see in service and choose a design that will not fail and is optimum in some sense.

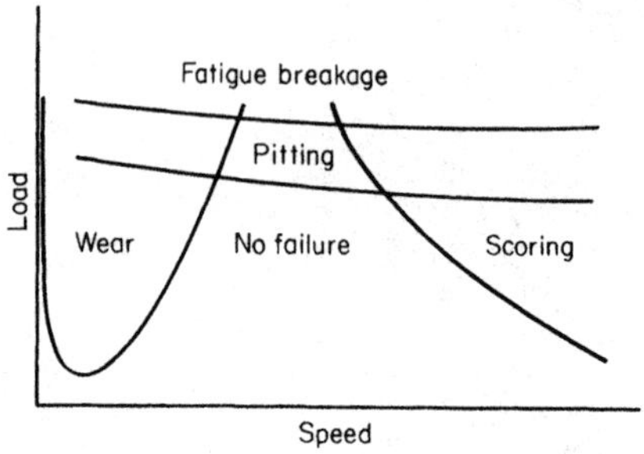

FIG. 21.49 Various failure regimes encountered in gear teeth.

21.9.1 Minimum Size

The index for optimization is often minimum size. A measure of size is center distance. A procedure for optimization where minimum center distance was the objective has been developed.[123,124] In the procedure, the following items were the constraints: (1) allowed maximum hertzian (normal) stress S_{max} equals 200,000 lb/in^2, which affects scoring and pitting; (2) the allowed bending stress, S_b = 60,000 lb/in^2, which affects tooth breakage due to fatigue; (3) the geometric constraint for no involute interference, which is the contact between the teeth under the base circle, or tip fouling in the case of internal gears; and (4) to help in keeping the misalignment factor to a minimum, a ratio of 0.25 of tooth width to pinion pitch diameter was selected. The results of the optimization study for a gear ratio of 1 are shown in Fig. 21.50. After consideration of all the design parameters it was found that for a given set of allowed stresses, gear ratio, and pressure angle, the design space could be shown on a plane with the vertical axis as the number of teeth on the pinion and the horizontal axis as the diametral pitch. In Fig. 21.50, the region for designs that meet acceptable conditions is shown. But not all the designs resulting from those combinations of number of teeth and pitch are best in approaching a minimum center-distance design.

The combinations of number of teeth and diametral pitch that fall on any straight line emanating from the origin will give a constant-sized design. The slope of the line is directly related to the center distance. The minimum slope line that falls into the region of acceptable designs gives the optimum choice of pitch and tooth number for an optimally small (light) design.

The data in Fig. 21.51 summarize results from a large number of plots similar to those in Fig. 21.50. Figure 21.51 may be used to select trial values of designs that can lead to optimum final designs. The value of these charts is that initial designs may be quickly selected, and then a more detailed study of the design should be made, using the more detailed methods for calculating bending fatigue stress, life, scoring, and efficiency previously described. In Fig. 21.50 the point labeled *A* is the point nearest optimum on the minimum center distance criterion since that is the point which lies on the minimum slope straight line through the origin and yet contains points in the regions of acceptable designs. The point *B* is the point that would be chosen in a selection based only on bending strength criteria and kinematic conditions of no involute interference.

In the consideration of the gear and pinion, the hertzian stress is common to both gear and pinion tooth, while the bending strength criterion was applied only to the pinion since it is the weaker element of the pinion and gear teeth.

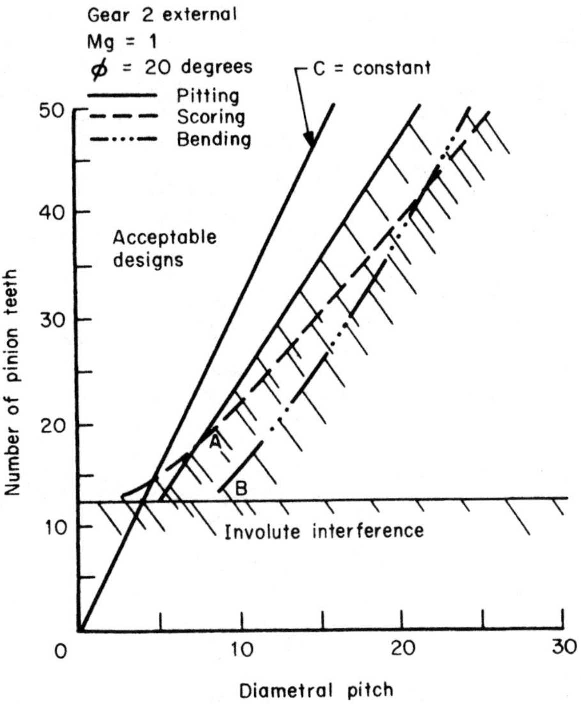

FIG. 21.50 External gear mesh design space ($F/D = 0.25$, $T_1 = 113$ n·m (1000 lb·in), $S_{max} = 1.38$ GPa (200 ksi), $S_b = 414$ MPa (60 ksi), $E = 205$ GPa (30×10^6 lb/in²), and $\nu = 0.25$).

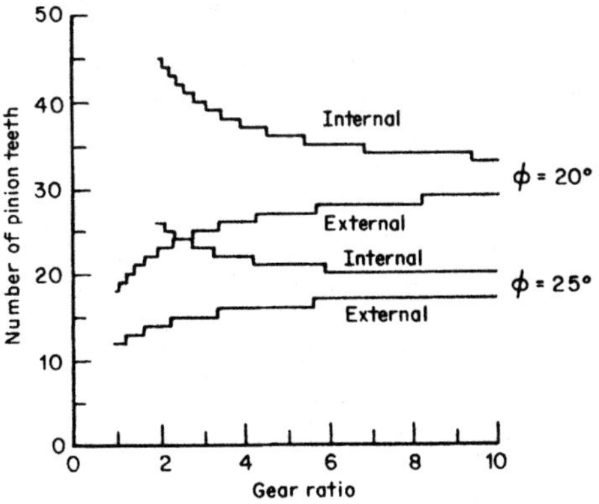

FIG. 21.51 Optimal number of pinion gear teeth for pressure angle ϕ of 20 and 25°.

Other approaches to optimization have included the balanced strength approach, where the pinion was made thicker and with shorter addendum while the gear was made with a longer addendum. The idea was to balance the strength of the pinion tooth with the gear tooth. In Ref. 125 this idea of balanced design was applied to scoring temperature index balance as well as bending strength balance. In Ref. 126 long and short addenda gear sets were examined for their optimum tooth numbers. In Ref. 127 a direct-search computer algorithm was applied to the minimum center distance design problem.

21.9.2 Specific Torque Capacity

Criteria for comparing two designs have been described in the literature as *specific torque capacity* (*STC*).[21] The STC rating of a single gear is defined as the torque transmitted by that gear divided by the superficial volume of its pitch cylinder. That is, torque per unit volume is equal to

$$\frac{T}{V} = \frac{4T}{\pi D^2 F} \tag{21.122}$$

where D is pitch circle diameter, and F is face width.

For a pair of gears with a 1:1 ratio, the same STC value can be attributed to the pair as that for each gear individually. Thus, for such a gear pair, the average STC value is

$$\frac{T_1/V_1 + T_2/V_2}{2} = \frac{1}{2}\left(\frac{4T_1}{\pi D_1^2 F_1} + \frac{4T_2}{\pi D_2^2 F_2}\right) \tag{21.123}$$

But if T_1 and T_2 are the torque loadings of a pinion and a wheel, then $T_2 = m_g T_1$, where m_g is the gear ratio. Similarly, $D_2 = m_g D_1$, but $F_2 = F_1$. Thus,

$$\text{STC}_{\text{av}} = \frac{2}{\pi}\left(\frac{T_1}{D_1^2 F_1} + \frac{m_g T_1}{m_g^2 D_1^2 F_1}\right)$$

or

$$\text{STC}_{\text{av}} = \frac{2T_1}{\pi D_1^2 F_1}\,\frac{1 + m_g}{m_g} \tag{21.124}$$

But $T = W_t R/2$, where W_t is total tangential load acting at the pitch surface. Thus

$$\text{STC}_{\text{av}} = \frac{W_t}{F_1 D_1 \pi}\,\frac{1 + m_g}{m_g} \tag{21.125}$$

This expression bears a close resemblance to the AGMA K factor for contact stress, where

$$K = \frac{W_t}{F_1 D_1}\,\frac{1 + m_g}{m_g} \tag{21.126}$$

Hence, $\text{STC}_{\text{av}} = K/\pi$

Thus, although K has been considered as a surface-loading criterion for a gear pair, it also has a direct relationship to the basic torque capacity of a given volume of gears. It can, therefore, be used as a general comparator of the relative load capacity of gear pairs, without any specific association with tooth surface or root stress conditions. The STC value thus becomes a very useful quantity for comparing the performance of gears based on completely different tooth systems, even where the factors that limit performance are not consistent.

21.10 GEAR TRANSMISSION CONCEPTS

21.10.1 Series Trains

Speed Reducers. The overall ratio of any reduction gear train is the input shaft speed divided by the output speed. It is also the product of the individual ratios at each mesh, except in planetary arrangements. The ratio is most easily determined by dividing the product of the numbers of teeth of all driven gears by the product of the numbers of teeth of all driving gears. By manipulating numbers, any desired ratio can be obtained, either exactly or with an extremely close approximation.

In multiple-mesh series trains the forces transmitted through the gear teeth are higher at the low-speed end of the train. Therefore, the pitches and face widths of the gears are usually not the same throughout the train. In instrument gears, which transmit negligible power, this variation may not be necessary.[14] Table 21.19 gives several possible speed ratios with multiple meshes.

Speed Increasers. Speed-increasing gear trains require greater care in design, especially at high ratios. Because most gear sets and gear trains are intended for speed reduction, standards and published data in general apply to such drives. It is not safe to assume that these data can be applied without modification to a speed-increasing drive. Efficiency is sometimes lower in an increasing drive, requiring substantial input torques to overcome output load; in extreme instances, self-locking may occur.[14] Traction and hybrid drives should be considered as reasonable alternatives to gear drives for this purpose.

Reverted Trains. When two sets of parallel-shaft gears are so arranged that the output shaft is concentric with the input, the drive is called a "reverted train." The requirement of equal center distance for the two trains complicates determination of how many teeth should be in each gear to satisfy ratio requirements with standard

TABLE 21.19 Gear Speed Ratios with Multiple Meshes

Type of gear	Ratio
Helical, double-helical, herringbone gears:	
Double reduction	10:1–75:1
Triple reduction	75:1–350:1
Spiral bevel (first reduction) with helical gear:	
Double reduction	9:1–50:1
Triple reduction	50:1–350:1
Worm gears:	
Double reduction	100:1–8000:1
Helicals with worm gears:	
Double reduction	50:1–270:1
Planetary gear sets:	
Simple	4:1–10:1
Compound	10:1–125:1

pitches. Helical gears provide greater design freedom through possible variation of helix angle.[14]

21.10.2 Multispeed Trains

Sliding Gears. Speed adjustment is effected by sliding gears on one or more intermediate parallel shafts (Fig. 21.52). Shifting is generally accomplished by disengaging the input shaft. Sliding-gear transmissions are usually manually shifted by means of a lever or a handwheel. A variety of shaft arrangements and mountings are available.[128]

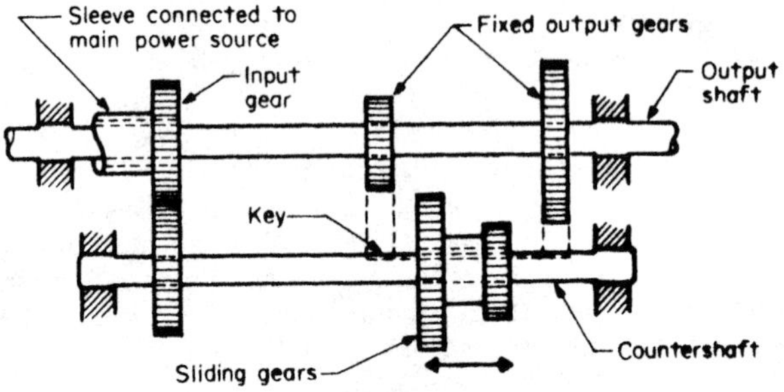

FIG. 21.52 Sliding gear assembly.

Constant Mesh. Several gears of different sizes mounted rigidly to one shaft mesh with mating gears free to rotate on the other shaft (Fig. 21.53). Speed adjustment is obtained by locking different gears individually on the second shaft by means of splined clutches or sliding couplings.

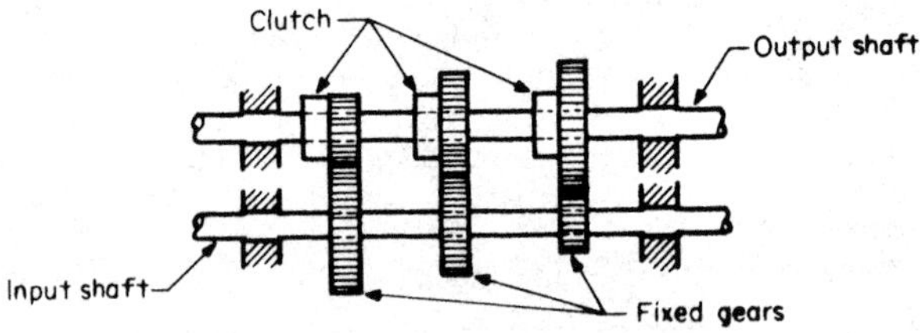

FIG. 21.53 Constant-mesh gear assembly.

Constant-mesh gears are used in numerous applications, among them heavy-duty industrial transmissions. This arrangement can use virtually any type of gearing—for example, spur, helical, herringbone, and bevel gears.

Manually shifted automotive transmissions combine two arrangements. The forward speeds are constant-mesh helical gears, whereas reverse uses sliding spur gears.[128]

Idler Gears. An adjustable speed drive comprising an idler gear consists of one shaft which carries several different-size gears rigidly mounted as shown in Fig. 21.54.

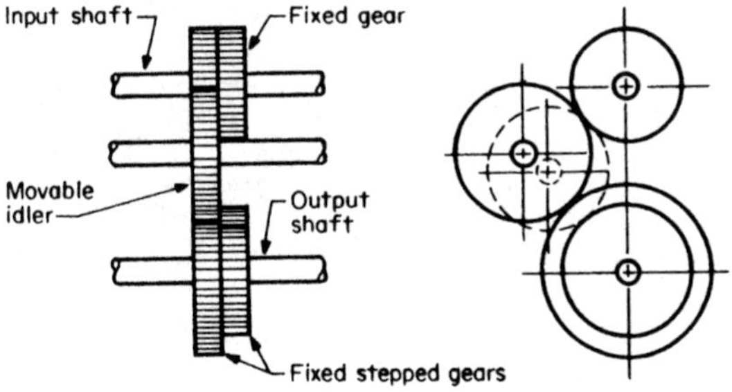

FIG. 21.54 Idler gear assembly.

Speed adjustment is through an adjustable arm which carries the idler gear to connect with fixed or sliding gears on the other shaft.

This arrangement is used to provide stepped speeds in small increments and is frequently found in machine tools. A transmission of this type can be connected to another multispeed train to provide an extremely wide range of speeds with constant-speed input.

Shifting is usually manual. The transmission is disengaged and allowed to come to a stop before the sliding idler gear is moved to the desired setting.[128]

21.10.3 Epicyclic Gearing

An epicyclic gear train such as shown in Fig. 21.55 is a reverted-gear arrangement in which one or more of the gears (planets) moves around the circumference of coaxial gears, which may be fixed or rotating with respect to their own axes. The planet gears have a motion consisting of rotation about their own axes and rotation about the axes of the coaxial gears.[14] The arrangement shown in Fig. 21.55 consists of a central sun gear with external teeth, a ring gear with internal teeth, revolving planet pinions which engage the sun gear and the internal ring gear, and a planet carrier in which the planet pinions are supported.[14] Epicyclic trains may incorporate spur or helical gears, external or internal, or bevel gears arranged in numerous ways either for purposes of fixed ratio or for multispeed applications.

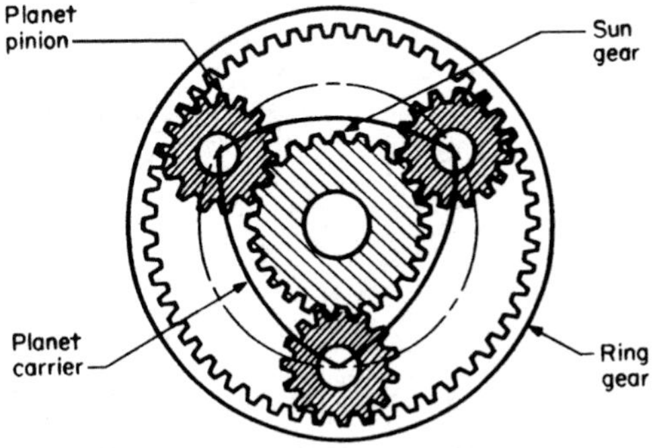

FIG. 21.55 Planetary gears.

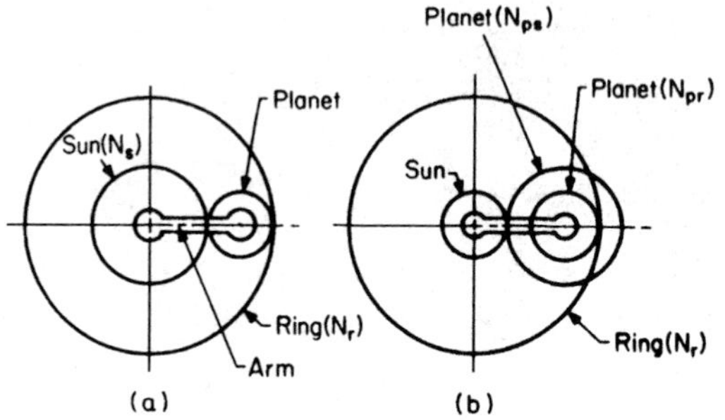

FIG. 21.56 Epicyclic gear trains. (*a*) Simple train. (*b*) Compound train (planet gears keyed to same shaft).

Single Epicyclic Trains. A simple single epicyclic train is shown in Fig. 21.56*a*. The coaxial sun and ring gears are connected by a single intermediate planet gear carried by a planet carrier arm. A compound epicyclic train is shown in Fig. 21.56*b*. The intermediate planet pinions are compound gears. Table 21.19 gives possible speed ratios which can be attained. For single planetary arrangements, to make assembly possible, $(N_r + N_s)/q$ must be a whole number, where N_r and N_s are the numbers of teeth in the ring and sun gears, and q is the number of planets equally spaced around the sun.[14] In order to make assembly possible for a compound planetary arrangement $(N_r N_{ps} - N_s N_{pr})/q$ must equal a whole number,[14] where N_{ps} and N_{pr} are the numbers of teeth in the planet gears in contact with the sun and ring, respectively.

Coupled Epicyclic Trains. Coupled epicyclic trains consist of two or more single epicyclic trains arranged so that two members in one train are common to the adjacent train.[14]

Multispeed Epicyclic Trains. Multispeed epicyclic trains are the most versatile and compact gear arrangement for a given ratio range and torque capacity. Tables 21.20 and 21.21 show the speed ratios possible with simple and compound epicyclic trains when one member is fixed and another is driving. With a suitable arrangement of clutches and brakes, an epicyclic train can be the basis of a change-speed transmission. With the gears locked to each other, an epicyclic train has a 1:1 speed ratio.[14]

TABLE 21.20 Simple Epicyclic Train Ratios (See Fig. 21.56*a*)

	Revolution of		
Condition	Sun	Arm	Ring
Sun fixed	0	1	$1 + \frac{N_s}{N_r}$
Arm fixed	1	0	$-\frac{N_s}{N_r}$
Ring fixed	$1 + \frac{N_r}{N_s}$	1	0

Note: In the equations, N_s and N_r denote either pitch diameter or number of teeth in the respective gears.

TABLE 21.21 Compound Epicylic Train Ratios (See Fig. 21.56*b*)

Condition	Revolution of Sun	Arm	Ring
Sun fixed	0	1	$1 + \frac{N_s N_{pr}}{N_{ps} N_r}$
Arm fixed	1	0	$-\frac{N_s N_{pr}}{N_{ps} N_r}$
Ring fixed	$1 + \frac{N_{ps} N_r}{N_s N_{pr}}$	1	0

Note: In the equations, N_r, N_s, N_{pr} and N_{ps} denote either pitch diameter or number of teeth in the respective gears.

Most planetary-gear transmissions are of the "automatic" type, in which speed changes are carried out automatically at a selected speed or torque level. At the same time it is also the most expensive, because of the clutching and braking elements necessary to control operation of the unit. In addition, practical ratios available with planetary sets are limited.[128]

Bearingless Planetary Trains. The self-aligning bearingless planetary transmission (Fig. 21.57) covers a variety of planetary-gear configurations, which share the common characteristic that the planet carrier, or spider, is eliminated, as are conventional planet-mounted bearings. The bearings are eliminated by load-balancing the gears, which are separated in the axial direction. All forces and reactions are transmitted through the gear meshes and contained by simple rolling rings. The concept was first demonstrated by Curtis Wright Corporation under sponsorship of the U.S. Army Aviation Research and Development Command.[129,130]

Hybrid Transmission. Geared planetary transmissions have a limitation on the speed ratio that can be obtained in a single stage without the planets interfering with each

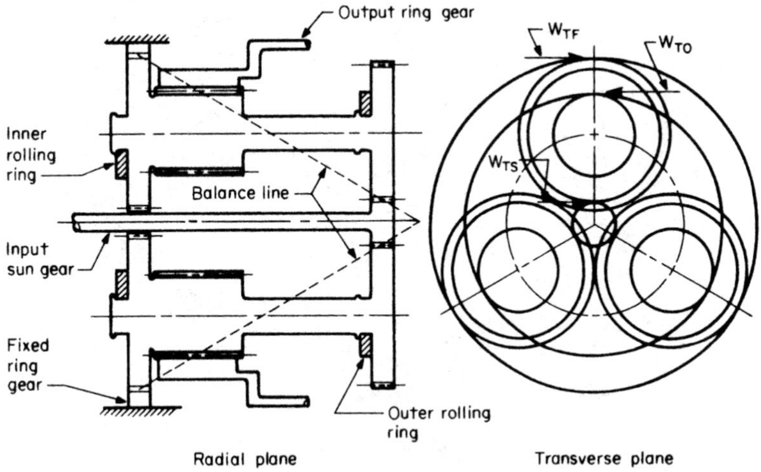

FIG. 21.57 A 500-hp bearingless planetary transmission.

other. For example, a four-planet drive would have a maximum speed ratio of 6.8 before the planets interfered with each other. A five-planet drive would be limited to a ratio of 4.8, and so on. A remedy to the speed-ratio and planet number limitations of simple, single-row planetary systems was devised by A. L. Nasvytis.[131] His drive system used the sun and ring roller of the simple planetary traction drive, but he replaced the single row of equal-diameter planet rollers with two or more rows of stepped, or dual diameter, planets. With this new, multiroller arrangement, practical speed ratios of 250:1 could be obtained in a single stage with three planet rows. Furthermore, the number of planets carrying the load in parallel could be greatly increased for a given ratio. This resulted in a significant reduction in individual roller contact loading with a corresponding improvement in torque capacity and fatigue life.

To further reduce the size and the weight of the drive for helicopter transmission applications, pinion gears in contact with a ring gear are incorporated with the second row of rollers (Fig. 21.58). The ring gear is connected through a spider to the output rotor shaft. The number of planet-roller rows and the relative diameter ratios at each contact are variables to be optimized according to the overall speed ratio and the uniformity of contact forces. The traction-gear combination is referred to as the hybrid transmission. The potential advantages of the hybrid transmission over a conventional planetary are higher speed ratios in a single stage, higher power-to-weight ratios, lower noise by replacing gear contacts with traction rollers, and longer life because of load sharing by multiple power paths.

21.10.4 Split-Torque Transmissions

A means to decrease the weight-to-power ratio of a transmission or to decrease the unit stress of gear teeth is by load sharing through multiple power paths. This concept is referred to as the *split-torque transmission.*[132–134] The concept is illustrated in Fig. 21.59

FIG. 21.58 A 500-hp hybrid helicopter transmission.

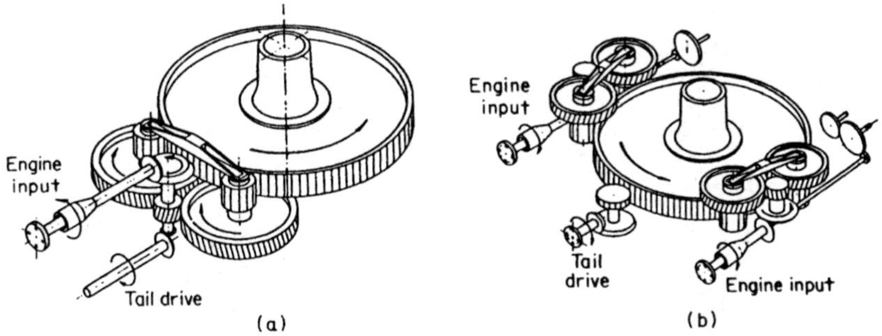

FIG. 21.59 Conceptual sketch of a split-torque transmission. (*a*) Single input. (*b*) Dual input.

for single-input and double-input variants. Instead of a planetary-gear arrangement, the input power is split into two or more power paths and recombined in a bull gear to the output power shaft. This concept appears to offer weight advantages over conventional planetary concepts without high-contact-ratio gearing. The effect of incorporating high-contact-ratio gearing into the split-torque concept is expected to further reduce transmission weight.

21.10.5 Differential Gearing

Differential gearing is an arrangement in which the normal ratio of the unit can be changed by driving into the unit with a second drive. This arrangement, or one having two outputs and one input, is used to vary ratio. It is called the *free type.* Simple differentials may use bevel gears (Fig. 21.60*a*) or spur gears (Fig. 21.60*b*). The bevel-gear type is used in automotive rear-end drives. The input-output speed relationship is

$$2\omega = \omega_1 - \omega_2 \tag{21.127}$$

where ω is the speed of the arm, and ω_1, ω_2 are the two shaft speeds.

Another type of differential (called the *fixed type*) has a large, fixed ratio. Such a drive is an evolution of the compound epicyclic train in Fig. 21.56*b*. If the ring gear is replaced with a sun gear which meshes with planet N_{pr}, the equations in Table 21.21 apply, except that the plus signs change to minus. If the sun and planet gears are made

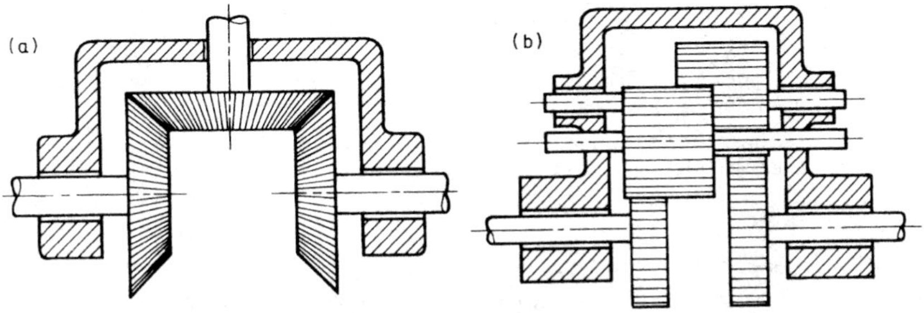

FIG. 21.60 Simple differentials. (*a*) Bevel gears. (*b*) Spur gears.

almost but not exactly equal, the output speed is the small difference between two terms that are almost equal. Such a drive is good for ratios from 10:1 to 3000:1.

Using two ring gears instead of a ring and a sun as in Fig. 21.56*b* results in a fixed-differential drive suitable for ratios of 15:1 to 100:1.

Any epicyclic gear train can be designed for differential operation. For instance, instead of one of the elements, such as a sun or ring gear, being fixed, two of the elements may be driven independently. Output speed is then the net result of the two inputs.

Compound epicyclic trains can produce several input or output speeds by the addition of extra sun or ring gears meshing with the planet pinions. Thus, the compound epicyclic train in Fig. 21.56*b* may have another sun gear meshing with N_{Pr} and another ring gear meshing with N_{Ps}. Any one of the four (two suns and two rings) could be fixed, and the others available for input, output, or free.[14]

21.10.6 Closed-Loop Trains

Epicyclic gear trains with more than one planet pinion meshing with the same sun and ring gears have parallel paths through which power can flow. Other gear trains, too, sometimes use multiple-tooth contact to increase the capacity within a given space. Such gearing is sometimes called "locked-train gearing."

Two considerations arise with multiple-tooth contact which are not present in open trains. One is the proper selection of tooth numbers and spacing to ensure assembly. The other is gear accuracy and adjustment to ensure equal distribution of the load to each mesh. Multiple-mesh or locked-train gearing requires careful attention to tooth accuracy and support of the gears. Backlash and backlash tolerances should be as nearly equal as possible at each mesh to ensure equal load distribution.[14]

REFERENCES

1. Coy, J. J.: "Geared Powered Transmission Technology," *Advanced Power Transmission Technology*, G. K. Fischer, ed., NASA CP-2210, pp. 49–77, 1983.
2. "The New Encyclopedia Britannica," 15th ed., Encyclopedia Britannica, Inc., 1974.
3. de Solla Price, Derek: "Gears from the Greeks," Science History Publications, 1975.
4. Drachmann, A. G.: "The Mechanical Technology of Greek and Roman Antiquity," Munksgeard Publisher, Copenhagen, 1963.
5. Dudley, D. W.: "The Evolution of the Gear Art," American Gear Manufacturers Association, 1969.
6. Reti, Ladislao: "Leonardo on Bearings and Gears," *Sci. Am.*, vol. 224, no. 2, pp. 100–110, February 1971.
7. da Vinci, Leonardo (L. Reti, trans.): "The Madrid Codices of Leonardo da Vinci," McGraw-Hill Book Company, Inc., New York, 1974.
8. Lewis, Wilfred: "Investigations of the Strength of Gear Teeth," *Proc. of the Engineers Club of Philadelphia*, Philadelphia, pp. 16–23, 1893.
9. Grant, George B.: "A Treatise on Gear Wheels," 21st ed., Philadelphia Gear Works, Inc., 1980.
10. Buckingham, Earle: "Dynamic Loads on Gear Teeth," *Am. Soc. Mech. Eng.*, 1931.
11. Tucker, A. I.: "Bevel Gears at 203 Meters per Second (40,000 FPM)," ASME paper 77-DET-178, 1977.

12. Haas, L. L.: "Latest Developments in the Manufacture of Large Spiral Bevel and Hypoid Gears," ASME paper 80C2/DET-119, August 1980.
13. Drago, R. J., and F. W. Brown: "The Analytical and Experimental Evolution of Resonant Response in High-Speed, Light-Weight, Highly Loaded Gearing," *J. Mech. Des.*, vol. 103, no. 2, pp. 346–356, April 1981.
14. Crawshaw, S. L., and Kron, H. O.: "Mechanical Drives," *Machine Design,* vol. 39, pp. 18–23, September 21, 1967.
15. "Gears and Gear Drives," *Machine Design,* vol. 55, no. 15, pp. 16–32, June 30, 1983.
16. Kasuba, R.: "A Method for Static and Dynamic Load Analysis of Standard and Modified Spur Gears," *Advanced Power Transmission Technology,* G. K. Fischer, ed., NASA CP-2210, 1983, pp. 403–419.
17. Wildhaber, E.: "Helical Gearing," U.S. Patent 1,601,750, 1926.
18. Wildhaber, E.: "Method of Grinding Gears," U.S. Patent 1,858, 568, 1932.
19. Novikov, M. L.: U.S.S.R. Patent 109,750, 1956.
20. Shotter, B. A.: "The Lynx Transmission and Conformal Gearing," SAE paper 7810141, 1979.
21. Shotter, B. A.: "Experiences with Conformal/W-N Gearing," *Machinery,* pp. 322–326, October 5, 1977.
22. Costomiris, G. H., D. P. Daley, and W. Grube: "Heat Generated in High Power Reduction Gearing," Pratt & Whitney Aircraft Division, PWA-3718, p. 43, June 1969.
23. Fusaro, R. L.: "Effects of Atmosphere and Temperature on Wear, Friction and Transfer of Polyimide Films," *ASLE Trans.,* vol. 21, no. 2, pp. 125–133, April 1978.
24. Fusaro, R. L.: "Tribological Properties at 25°C of Seven Polyimide Films Bonded to 440C High-Temperature Stainless Steel," NASA TP-1944, 1982.
25. Fusaro, R. L.: "Polyimides Tribological Properties and Their Use as Lubricants," NASA TM-82959, 1982.
26. Johansen, K. M.: "Investigation of the Feasibility of Fabricating Bimetallic Coextruded Gears," AFAPL-TR-73-112, (AD-776795), AiResearch Mfg. Co., December 1973.
27. Hirsch, R. A.: "Lightweight Gearbox Development for Propeller Gearbox System Applications Potential Coatings for Titanium Alloy Gears," AFAPL-TR-72-90, (AD-753417), General Motors Corp., December 1972.
28. Delgrosso, E. J., et al.: "Lightweight Gearbox Development for Propeller Gearbox Systems Applications," AFAPL-TR-71-41-PH-1, (AD-729839), Hamilton Standard, August 1971.
29. Manty, B. A., and H. R. Liss: "Wear Resistant Coatings for Titanium Alloys," FR-8400, (AD-A042443), Pratt & Whitney Aircraft, March 1977.
30. Dudley, D. W.: "Gear Handbook, The Design, Manufacture and Applications of Gears," McGraw-Hill Book Company, Inc., New York, 1962.
31. Michalec, G. W.: "Precision Gearing Theory and Practice," John Wiley & Sons, Inc., New York, 1966.
32. "AGMA Standard for Rating the Strength of Spur Gear Teeth," AGMA 220.02, American Gear Manufacturers Association, August 1966.
33. "AGMA Standard for Surface Durability (Pitting) of Spur Gear Teeth," AGMA 210.02, American Gear Manufacturers Association, January 1965.
34. Smith, W. E.: "Ferrous-Based Powder Metallurgy Gears," *Gear Manufacture and Performance,* P. J. Guichelaar, B. S. Levy, and N. M. Parikh, eds., American Society for Metals, pp. 257–269, 1974.
35. Antes, H. W.: "P/M Hot Formed Gears," *Gear Manufacture and Performance,* P. J. Guichelaar, B. S. Levy, and N. M. Parikh, eds., American Society for Metals, pp. 271–292, 1974.
36. Ferguson, B. L., and D. T. Ostberg: "Forging of Powder Metallurgy Gears," TARADCOM-TR-12517, TRW-ER-8037-F, TRW, Inc., May 1980 (AD-A095556).

37. Townsend, D. P., and E. V. Zaretsky: "Effect of Shot Peening on Surface Fatigue Life of Carburized and Hardened AISI 9310 Spur Gears," NASA TP-2047, 1982.
38. Straub, J. C.: "Shot Peening in Gear Design," AGMA Paper 109.13, June 1964.
39. Anderson, N. E., and E. V. Zaretsky: "Short-Term, Hot-Hardness Characteristics of Five Case-Hardened Steels," NASA TN D-8031, 1975.
40. Hingley, C. G., H. E. Southerling, and L. B. Sibley: "Supersonic Transport Lubrication System Investigation" (AL65T038 SAR-1, SKF Industries, Inc.; NASA Contract NAS3-6267), NASA CR-54311, May 1965.
41. Bamberger, E. N., E. V. Zaretsky, and W. J. Anderson: "Fatigue Life of 120-mm-Bore Ball Bearings at 600°F and Fluorocarbon, Polyphenyl Ether and Synthetic Paraffinic Base Lubricants," NASA TN D-4850, 1968.
42. Zaretsky, E. V., W. J. Anderson, and E. N. Bamberger: "Rolling-Element Bearing Life from 400° to 600°F," NASA TN D-5002, 1969.
43. Bamberger, E. N., E. V. Zaretsky, and W. J. Anderson: "Effect of Three Advanced Lubricants on High-Temperature Bearing Life," *J. Lubr. Technol., Trans. ASME,* vol. 92, no. 1, pp. 23–33, January 1970.
44. Bamberger, E. N., and E. V. Zaretsky: "Fatigue Lives at 600°F of 120-Millimeter-Bore Ball Bearings of AISI M-50, AISI M-1 and WB-49 Steels," NASA TN D-6156, 1971.
45. Bamberger, E. N.: "The Development and Production of Thermo-Mechanically Forged Tool Steel Spur Gears," (R73AEG284, General Electric Co.; NASA Contract NAS3-15338), NASA CR-121227, July 1973.
46. Townsend, D. P., E. N. Bamberger, and E. V. Zaretsky: "A Life Study of Ausforged, Standard Forged, and Standard Machined AISI M-50 Spur Gears," *J. Lubr. Technol.,* vol. 98, no. 3, pp. 418–425, July 1976.
47. Bamberger, E. N.: "The Effect of Ausforming on the Rolling Contact Fatigue Life of a Typical Bearing Steel," *J. Lubr. Technol.,* vol. 89, no. 1, pp. 63–75, January 1967.
48. Jatczak, C. F.: "Specialty Carburizing Steels for Elevated Temperature Service," *Met. Prog.,* vol. 113, no. 4, pp. 70–78, April 1978.
49. Townsend, D. P., R. J. Parker, and E. V. Zaretsky: "Evaluation of CBS 600 Carburized Steel as a Gear Material," NASA TP-1390, 1979.
50. Culler, R. A., E. C. Goodman, R. R. Hendrickson, and W. C. Leslie: "Elevated Temperature Fracture Toughness and Fatigue Testing of Steels for Geothermal Applications," TERRATEK Report No. Tr 81-97, Terra Tek, Inc., October 1981.
51. Roberts, G. A., and J. C. Hamaker: "Strong, Low Carbon Hardenable Alloy Steels," U.S. Patent No. 3,036,912, May 29, 1962.
52. Cunningham, R. J., and W. N. J. Lieberman: "Process for Carburizing High Alloy Steels," U.S. Patent No. 3,885,995, May 27, 1975.
53. Townsend, D. P., and E. V. Zaretsky: "Endurance and Failure Characteristics of Modified Vasco X-2, CBS 600 and AISI 9310 Spur Gears," *J. Mech. Des., Trans. ASME,* vol. 103, no. 2, pp. 506–515, April 1981.
54. Walp, H. O., R. P. Remorenko, and J. V. Porter: "Endurance Tests of Rolling-Contact Bearings of Conventional and High Temperature Steels under Conditions Simulating Aircraft Gas Turbine Applications," (AD-212904) TR 58-392, Wright Air Development Center, Dayton, Ohio, 1959.
55. Baile, G. H., O. G. Gustafsson, H. Mahncke, and L. B. Sibley: "A Synopsis of Active SKF Research and Development Programs of Interest to the Aerospace Industry," AL-63M003, SKF Industries, Inc., King of Prussia, Pa., July 1963.
56. Scott, D.: "Comparative Rolling Contact Fatigue Tests on En 31 Ball Bearing Steels of Recent Manufacture," NFL 69T96, National Engineering Laboratory, Glasgow, Scotland, 1969.
57. "1968 Aero Space Research Report," AL 68Q013, SKF Industries, Inc., King of Prussia, Pa., 1968.

58. Scott, D., and J. Blackwell, J.: "Steel Refining as an Aid to Improved Rolling Bearing Life," NEL Report No. 354 National Eng. Lab.

59. Bamberger, E. N., E. V. Zaretsky, and H. Signer: "Endurance and Failure Characteristic of Main-Shaft Jet Engine Bearing at 3×10^6 DN," *J. Lubr. Technol., Trans. ASME,* vol. 98, no. 4, pp. 580–585, October 1976.

60. Carter, T. L., E. V. Zaretsky, and W. J. Anderson: "Effect of Hardness and Other Mechanical Properties on Rolling-Contact Fatigue Life of Four High-Temperature Bearing Steels," NASA TN D-270, 1960.

61. Jackson, E. G.: "Rolling-Contact Fatigue Evaluations of Bearing Materials and Lubricants," *ASLE Transactions,* vol. 2, no. 1, pp. 121–128, 1959.

62. Baughman, R. A.: "Effect of Hardness, Surface Finish, and Grain Size on Rolling-Contact Fatigue Life of M-50 Bearing Steel," *J. of Basic Engineering,* vol. 82, no. 2, pp. 287–294, June 1960.

63. Zaretsky, E. V., R. J. Parker, W. J. Anderson, and D. W. Reichard: "Bearing Life and Failure Distribution as Affected by Actual Component Differential Hardness," NASA TN D-3101, 1965.

64. Zaretsky, E. V., R. J. Parker, and W. J. Anderson: "Component Hardness Differences and Their Effect on Bearing Fatigue," *J. Lubr. Technol.,* vol. 89, no. 1, January 1967.

65. Anderson, W. J., and T. L. Carter: "Effect of Fiber Orientation, Temperature and Dry Powder Lubricants on Rolling-Contact Fatigue," *ASLE Transactions,* vol. 2, no. 1, pp. 108–120, 1959.

66. Carter, T. L.: "A Study of Some Factors Affecting Rolling Contact Fatigue Life," NASA TR R-60, 1960.

67. Bamberger, E. N.: "The Effect of Ausforming on the Rolling Contact Fatigue Life of a Typical Bearing Steel," *J. Lubr. Technol.,* vol. 89, no. 1, pp. 63–75, January 1967.

68. Bamberger, E. N.: "The Production, Testing, and Evaluation of Ausformed Ball Bearings," Final Engineering Report, AD-637576, June 1966, General Electric Co., Cincinnati, Ohio.

69. Parker, R. J., and E. V. Zaretsky: "Rolling Element Fatigue Life of Ausformed M-50 Steel Balls," NASA TN D-4954, 1968.

70. "AGMA Standard for Rating the Pitting Resistance and Bending Strength of Spur and Helical Involute Gear Teeth," AGMA 218.01, American Gear Manufacturers Association, December 1982.

71. Cardou, A., and G. V. Tordion: "Numerical Implementation of Complex Potentials for Gear Tooth Stress Analysis," *ASME J. Mech. Des.,* vol. 103, no. 2, pp. 460–466, April 1981.

72. Rubenchik, V.: "Boundary-Integral Equation Method Applied to Gear Strength Rating," *ASME Journal of Mechanisms, Transmissions, and Automation in Design,* vol. 105, no. 1, pp. 129–131, March 1983.

73. Alemanni, M., S. Bertoglio, and P. Strona: "B.E.M. in Gear Teeth Stress Analysis: Comparison with Classical Methods," *Proc. of the International Symposium on Gearing and Power Transmissions, Japan Soc. of Mech. Engrs., Tokyo,* vol. 2, pp. 177–182, 1981.

74. Wilcox, L., and W. Coleman: "Application of Finite Elements to the Analysis of Gear Tooth Stresses," *ASME Journal of Engineering for Industry,* vol. 95, no. 4, pp. 1139–1148, November 1973.

75. Wilcox, L. E.: "An Exact Analytical Method for Calculating Stresses in Bevel and Hypoid Gear Teeth," *Proc. of the International Symposium on Gearing and Power Transmissions, Japan Soc. of Mech. Engrs., Tokyo,* vol. 2, pp. 115–121, 1981.

76. Frater, J. L., and R. Kasuba: "Extended Load-Stress Analysis of Spur Gearing: Bending Strength Considerations," *Proc. of the International Symposium on Gearing and Power Transmissions, Japan Soc. of Mech. Engrs., Tokyo,* vol. 2, pp. 147–152, 1981.

77. Coy, J. J., and C. H. Chao: "A Method of Selecting Grid Size to Account for Hertz Deformation in Finite Element Analysis of Spur Gears," *J. Mech. Des., ASME Trans.,* vol. 104, no. 4, pp. 759–766, 1982.

78. Drago, R. J.: "Results of an Experimental Program Utilized to Verify a New Gear Tooth Strength Analysis," AGMA technical paper P229.27, October 1983.

79. Drago, R. J., and R. V. Lutthans: "An Experimental Investigation of the Combined Effects of Rim Thickness and Pitch Diameter on Spur Gear Tooth Root and Fillet Stresses," AGMA technical paper P229.22, October 1981.

80. Chang, S. H., R. L. Huston, and J. J. Coy: "A Finite Element Stress Analysis of Spur Gears Including Fillet Radii and Rim Thickness Effects," *ASME Trans., Journal of Mechanisms, Transmissions, and Automation in Design*, vol. 105, no. 3, pp. 327–330, 1983.

81. Drago, R. J.: "An Improvement in the Conventional Analysis of Gear Tooth Bending Fatigue Strength," AGMA technical paper P229.24, October 1982.

82. Chong, T. H., T. Aida, and H. Fujio: "Bending Stresses of Internal Spur Gear," *Bulletin of the JSME*, vol. 25, no. 202, pp. 679–686, April 1982.

83. Wang, K. L., and H. S. Cheng: "A Numerical Solution to the Dynamic Load, Film Thickness, and Surface Temperatures in Spur Gears, Part I—Analysis," *J. Mech. Des.*, vol. 103, pp. 177–187, January 1981.

84. Wang, K. L., and H. S. Cheng: "A Numerical Solution to the Dynamic Load, Film Thickness, and Surface Temperatures in Spur Gears, Part II—Results," *J. Mech. Des.*, vol. 103, pp. 188–194, January 1981.

85. Dolan, T. J., and E. L. Broghamer: "A Photoelastic Study of the Stresses in Gear Tooth Fillets," University of Illinois Engineering Experiment Station, Bulletin No. 335, 1942.

86. Cornell, R. W.: "Compliance and Stress Sensitivity of Spur Gear Teeth," *J. Mech. Des.*, vol. 103, no. 2, pp. 447–459, April 1981.

87. Coy, J. J., D. P. Townsend, and E. V. Zaretsky: "Analysis of Dynamic Capacity of Low-Contact-Ratio Spur Gears Using Lundberg-Palmgren Theory," NASA TN D-8029, 1975.

88. Coy, J. J., and E. V. Zaretsky: "Life Analysis of Helical Gear Sets Using Lundberg-Palmgren Theory," NASA TN D-8045, 1975.

89. Coy, J. J., D. P. Townsend, and E. V. Zaretsky: "Dynamic Capacity and Surface Fatigue Life for Spur and Helical Gears," *ASME Trans., J. Lubr. Technol.*, vol. 98, no. 2, pp. 267–276, April 1976.

90. Townsend, D. P., J. J. Coy, and E. V. Zaretsky: "Experimental and Analytical Load-Life Relation for AISI 9310 Steel Spur Gears," *J. Mech. Des.*, vol. 100, no. 1, pp. 54–60, January 1978. (Also NASA TM X-73590, 1977.)

91. Coy, J. J., D. P. Townsend, and E. V. Zaretsky: "An Update on the Life Analysis of Spur Gears," *Advanced Power Transmission Technology*, NASA CP 2210, AVRADCOM TR 82-C-16, pp. 421–433, 1983.

92. Lundberg, G., and A. Palmgren: "Dynamic Capacity of Rolling Bearings," *Acta Polytech., Mech. Eng. Ser.*, vol. 1, no. 3, 1947.

93. Bamberger, E. N., et al.: "Life Adjustment Factors for Ball and Roller Bearings," *An Engineering Design Guide*, American Society of Mechanical Engineers, 1971.

94. Savage, M., C. A. Paridon, and J. J. Coy: "Reliability Model for Planetary Gear Trains," *ASME Trans., J. Mechanisms, Transmissions, and Automation in Design*, vol. 105, no. 5, pp. 291–297, 1983.

95. Krenzer, T. J., and R. Knebel: "Computer-Aided Inspection of Bevel and Hypoid Gears," SAE Paper 831266, 1983.

96. Litvin, F. L., and Y. Gutman: "Methods of Synthesis and Analysis for Hypoid Gear-Drives of 'Formate' and 'Helixform,'" *ASME Trans., J. Mech. Des.*, vol. 103, no. 1, pp. 83–113, January 1981.

97. Litvin, F. L., and Y. Gutman: "A Method of Local Synthesis of Gears Grounded on the Connections Between the Principal and Geodetic Curvatures of Surfaces," *ASME Trans., J. Mech. Des.*, vol. 106, no. 1, pp. 114–125, January 1981.

98. Coy, J. J., D. A. Rohn, and S. H. Loewenthal: "Constrained Fatigue Life Optimization for a Nasvytis Multiroller Traction Drive," *ASME Trans., J. Mech. Des.*, vol. 103, no. 2, pp. 423–429, April 1981.

99. Coleman, W.: Bevel and Hypoid Gear Surface Durability: Pitting and Scuffing, "Source Book on Gear Design, Technology and Performance," M. A. H. Howes, ed., Am. Soc., for Metals, Metals Park, Ohio 44073, pp. 243–258, 1980.

100. Anderson, W. J., and E. V. Zaretsky: "Rolling-Element Bearings," *Machine Design,* vol. 42, no. 15, pp. 21–37, June 18, 1970.

101. Ertel, A. M.: "Hydrodynamic Lubrication Based on New Principles," *Prikl. Mat. Melch.,* vol. 3, no. 2, 1939 (in Russian).

102. Grubin, A. N.: "Fundamentals of the Hydrodynamic Theory of Lubrication of Heavily Loaded Cylindrical Surfaces," "Investigation of the Contact of Machine Components," Kh. F. Ketova, ed., translation of Russian Book No. 30, Central Scientific Institute for Technology and Mechanical Engineering, Moscow, 1949, chap. 2. (Available from Dept. of Scientific and Industrial Research, Great Britain, Transl. CTS-235 and Special Libraries Assoc., Transl. R-3554.)

103. Hamrock, B. J., and D. Dowson: "Minimum Film Thickness in Elliptical Contacts for Different Regimes of Fluid-Film Lubrication," NASA TP 1342, October 1978.

104. Fein, R. S.: "Chemistry in Concentrated-Conjunction Lubrication," *Interdisciplinary Approach to the Lubrication of Concentrated Contacts,* SP-237, NASA, Washington, D.C., pp. 489–527, 1970.

105. Godfrey, D.: "Boundary Lubrication," *Interdisciplinary Approach to Friction and Wear,* SP-181, NASA, Washington, D.C., pp. 335–384, 1968.

106. Bowden, F. P., and Tabor, D: "The Friction and Lubrication of Solids," vol. 2, Clarendon Press, Oxford, 1964.

107. Borsoff, V. A.: "Fundamentals of Gear Lubrication," Annual Report, Shell Development Co., Emeryville, Calif. [Work under Contract NOa(s) 53-356-c], June 1955.

108. Borsoff, V. N., and R. Lulwack: "Fundamentals of Gear Lubrication," Final Report, Shell Development Co., Emeryville, Calif. [Work under Contract NOa(s) 53-356-c], June 1957.

109. Horlick, E. J., and D. E. O'D. Thomas: "Recent Experiences in the Lubrication of Naval Gearing," *Gear Lubrication Symposium,* Institute of Petroleum, 1966.

110. Boner, C. J.: "Gear and Transmission Lubricant," Reinhold Publishing Company, New York, 1964.

111. Townsend, D. P., and L. S. Akin: "Study of Lubricant Jet Flow Phenomena in Spur Gears—Out of Mesh Condition," Advanced Power Transmission Technology, NASA CP-2210, G. K. Fischer, ed., pp. 461–476, 1983.

112. Townsend, D. P., and L. S. Akin: "Gear Lubrication and Cooling Experiment and Analysis," Advanced Power Transmission Technology, NASA CP-2210, G. K. Fischer, ed., pp. 477–490, 1983.

113. Akin, L. S., J. J. Mross, and D. P. Townsend: "Study of Lubricant Jet Flow Phenomena in Spur Gears," *ASME Trans., J. Lubr. Technol.,* vol. 97, no. 2, pp. 283–288, 295, April 1975.

114. Townsend, D. P., and L. S. Akin: "Study of Lubricant Jet Flow Phenomena in Spur Gears—Out of Mesh Condition," *ASME Trans., J. Mech. Des.,* vol. 100, no. 1, pp. 61–68, January 1978.

115. Townsend, D. P., and L. S. Akin: "Analytical and Experimental Spur Gear Tooth Temperature as Affected by Operating Variables," *ASME Trans., J. Mech. Des.,* vol. 103, no. 1, pp. 219–226, January 1981.

116. Akin, L. S., and D. P. Townsend: "Into Mesh Lubrication of Spur Gears with Arbitrary Offset Oil Jet. Part 1: For Jet Velocity Less Than or Equal to Gear Velocity," *ASME Trans., J. Mech. Trans. and Automation in Design,* vol. 105, no. 4, pp. 713–718, December 1983.

117. Akin, L. S., and D. P. Townsend: "Into Mesh Lubrication of Spur Gears with Arbitrary Offset Oil Jet. Part 2: For Jet Velocities Equal to or Greater Than Gear Velocity," *ASME Trans., J. Mech. Trans. and Automation in Design,* vol. 105, no. 4, pp. 719–724, December 1983.

118. Patir, N., and H. S. Cheng: "Prediction of Bulk Temperature in Spur Gears Based on Finite Element Temperature Analysis," *ASLE Trans.,* vol. 22, no. 1, pp. 25–36, January 1979.

119. Anderson, N. E., and S. H. Loewenthal: "Spur-Gear-System Efficiency at Part and Full Load," NASA TP-1622, 1980.

120. Anderson, N. E., and S. H. Loewenthal: "Effect of Geometry and Operating Conditions on Spur Gear System Power Loss," *ASME Trans., J. Mech. Des.*, vol. 103, no. 1, pp. 151–159, January 1981.

121. Anderson, N. E., and S. H. Loewenthal: "Design of Spur Gears for Improved Efficiency," *ASME Trans., J. Mech. Des.*, vol. 104, no. 3, pp. 283–292, October 1982.

122. Anderson, N. E., and S. H. Loewenthal: "Selecting Spur-Gear Geometry for Greater Efficiency," *Machine Design*, vol. 55, no. 10, pp. 101–105, May 12, 1983.

123. Savage, M., J. J. Coy, and D. P. Townsend: "Optimal Tooth Numbers for Compact Standard Spur Gear Sets," *ASME Trans., J. Mech. Des.*, vol. 104, no. 4, pp. 749–758, October 1982.

124. Savage, M., J. J. Coy, and D. P. Townsend: "The Optimal Design of Standard Gearsets," Advanced Power Transmission Technology, NASA CP-2210, G. K. Fischer, ed., pp. 435–469, 1983.

125. Lynwander, P.: "Gear Tooth Scoring Design Considerations," AGMA paper P219.10, October 1981.

126. Savage, M., J. J. Coy, and D. P. Townsend: "The Optimal Design of Involute Gear Teeth with Unequal Addenda," NASA TM-82866, AVRADCOM TR 82-C-7, 1982.

127. Carroll, R. K., and G. E. Johnson: "Optimal Design of Compact Spur Gear Sets," ASME paper 83-DET-33, 1983.

128. Wadlington, R. P.: "Gear Drives," *Machine Design*, vol. 39, pp. 24–26, September 21, 1967.

129. DeBruyne, Neil A.: "Design and Development Testing of Free Planet Transmission Concept," USAAMRDL-TR-74-27, Army Air Mobility Research and Development Laboratory, 1974 (AD-782857).

130. Folenta, D. J.: "Design Study of Self-Aligning Bearingless Planetary Gear (SABP)," *Advanced Power Transmission Technology*, G. K. Fischer, ed., NASA CP-2210, 1983, pp. 161–172.

131. Nasvytis, A. L.: "Multiroller Planetary Friction Drives," SAE Paper 660763, October 1966.

132. White, G.: "New Family of High-Ratio Reduction Gears with Multiple Drive Paths," *Proc. Inst. Mech. Eng. (London)*, vol. 188, no. 23/74, pp. 281–288, 1974.

133. Lastine, J. L., and G. White: "Advanced Technology VTOL Drive Train Configuration Study," USAAVLABS-TR-69-69, Army Aviation Material Laboratories, 1970 (AD-867905).

134. White, G.: "Helicopter Transmission Arrangements with Split-Torque Gear Trans," *Advanced Power Transmission Technology*, G. K. Fischer, ed., NASA CP-2210, pp. 141–150, 1983.

135. Buckingham, E.: "Analytical Mechanics of Gears," Dover Publications, New York, 1963.

136. Boner, C. J.: "Gear and Transmission Lubricants," Reinhold Publishing, New York, 1964.

137. Deutschman, A., W. J. Michels, and C. E. Wilson: "Machine Design," The Macmillan Company, Inc., New York, 1975, pp. 519–659.

138. Dudley, D. W.: "Practical Gear Design," McGraw-Hill Book Company, Inc., New York, 1954.

139. Dudley, D. W.: "Gear Handbook," McGraw-Hill Book Company, Inc., New York, 1962.

140. Khiralla, T. W.: "On the Geometry of External Involute Spur Gears," T. W. Khiralla, Studio City, Calif., 1976.

141. Michalec, G. W.: "Precision Gearing: Theory and Practice," John Wiley & Sons, Inc., New York, 1966.

142. "Mechanical Drives Reference Issue," *Machine Design*, 3d ed., vol. 39, no. 22, September 21, 1967.

CHAPTER 22
SPRINGS*

Abilio A. Relvas, B.A., B.S.M.E.
Manager—Technical Assistance
Associated Spring
Barnes Group Inc.
Bristol, Conn.

22.1 INTRODUCTION

A spring is a device which stores energy or provides a force over a distance by elastic deflection. Energy may be stored in a compressed gas or liquid, or in a solid that is bent, twisted, stretched, or compressed. The energy is recoverable by the elastic return of the distorted material. Spring structures are characterized by their ability to withstand relatively large deflections elastically. The energy stored is proportional to the volume of elastically distorted material, and in the case of metal springs, the volume of distorted material is limited by the spring configuration and the stress-carrying capacity (elastic limit) of the most highly strained portion. For this reason, most commercial spring materials come from the group of high-strength materials, including high-carbon steel, cold-rolled and precipitation-hardening stainless and nonferrous alloys, and a few specialized nonmetallics such as laminated fiberglass. Thus spring design is basically an analysis of machine elements which undergo relatively large elastic deflections and are made of materials capable of withstanding high stress levels without yielding.

Symbols. Following are the letter symbols for equations and their corresponding units. Where a symbol can have more than one interpretation its meaning is made explicit in each section where it is used.

b = breadth or width, in
C = spring index, dimensionless
d = wire or bar diameter, in
D = diameter, in
E = modulus of elasticity, lb/in^2
f = frequency of vibration, Hz (cycles per second)
g = gravitational constant, 386.4 in/s^2
G = modulus of elasticity in shear, lb/in^2
h = height, in
k = rate, lb/in or in·lb/rad
K = factor, used variously
ln = natural logarithm
L = length, in
M = moment of force, in·lb
N = number of active coils in helically wound springs, dimensionless
p = pitch, in
P = force, lb
r = radius, in
R = radius, in
t = thickness, in
T = number of turns, dimensionless
U = energy, in·lb
v = velocity, in/s
V = volume, in^3
δ = linear deflection, in
θ = angular deflection, rad
μ = Poisson's ratio, dimensionless
γ = density, lb/in^3
σ = normal stress, lb/in^2
τ = shear stress, lb/in^2
ϕ = angle, °

The following terms are peculiar to spring engineering. Definitions of general terms are not given.

Rate (see Sec. 22.2): Also referred to as scale, gradient, and load factor.

Spring index: Applies to helical-type springs. For springs wound with circular wire, spring index $C = D/d$; for springs wound with rectangular wire $C = D/t$ if t is in the radial direction, or $C = D/b$ if b is in the radial direction. See Figs. 22.7 and 22.11*b*.

Active coils: Coils free to deflect under load.

Inactive coils: Coils not free to deflect under load.

Total coils: The sum of active and inactive coils.

Free-length or height: A length or height measurement of a spring in the free (unloaded) condition.

Solid length or height: A length or height measurement of a compression spring when no further deflection is possible.

Solid stress: The stress corresponding to deflection to solid length or height.

Set: A permanent distortion resulting when a spring is stressed beyond the elastic limit.

Initial tension: The force holding coils closed in the free or unloaded condition of helical extension or torsion springs.

Sizes: For economy and availability, preferred sizes for spring steel wire and strip should be used. Sizes are usually specified in thousandths of an inch.

22.2 SPRING CHARACTERISTICS

The load-deflection curve is the most fundamental of all spring properties. For a load-deflection curve of the form $P = f(\delta)$ the force P will cause a linear deflection δ in the direction of P; when the load is a moment of force M, the load-deflection curve has the form $M = f(\theta)$ and M is in the direction of θ. Four types of load-deflection curve are represented in Fig. 22.1*a* to *d*.

22.2.1 Rate

By definition, rate is the slope of the load-deflection curve

$$k = dP/d\delta \qquad \text{or} \qquad k = dM/d\theta \tag{22.1}$$

When the load-deflection curve is linear, rate is constant so that

$$k = \Delta P/\Delta\delta \qquad \text{or} \qquad k = \Delta M/\Delta\theta \tag{22.2}$$

For a linear load-deflection curve with no initial load, Eq. (22.2) reduces to

$$k = P/\delta \qquad \text{or} \qquad k = M/\theta \tag{22.3}$$

22.2.2 Operating Range

Many springs operate over only a part of their useful load-deflection curve. The operating range is defined by the limits of load and deflection as shown in Fig. 22.1*e*.

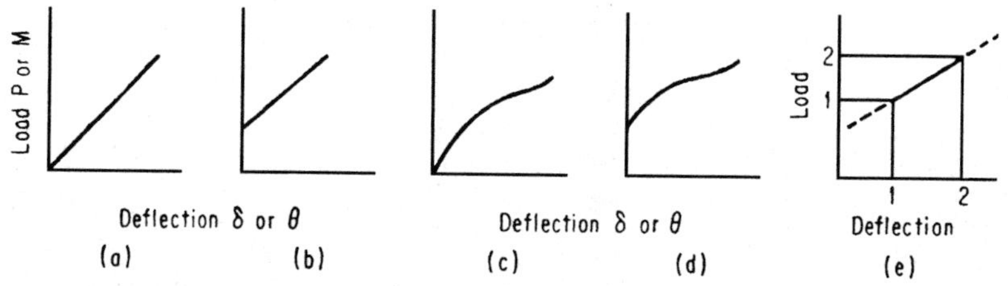

FIG. 22.1 Types of load-deflection curve (*a*) Linear, no initial load. (*b*) Linear, initial load. (*c*) Nonlinear, no initial load. (*d*) Nonlinear, initial load. (*e*) Operating range.

22.2.3 Energy Storage

The energy stored by deflecting a spring over its operating range,

$$U = \int_{\delta_1}^{\delta_2} P\, d\delta \qquad \text{or} \qquad U = \int_{\theta_1}^{\theta_2} M\, d\theta \tag{22.4}$$

For any linear load-deflection curve,

$$U = (P_2 + P_1)(\delta_2 - \delta_1)/2 \qquad \text{or} \qquad U = (M_2 + M_1)(\theta_2 - \theta_1)/2 \tag{22.5}$$

If in addition to the load-deflection curve being linear, the initial load is zero, Eq. (22.5) reduces to

$$U = P\delta/2 \qquad \text{or} \qquad U = M\theta/2 \tag{22.6}$$

22.2.4 Stress

Values of maximum allowable load are determined by equations relating stress and load. Calculated stresses differ from true stresses because of the approximate nature of the stress formulas, because these formulas do not take into account residual stresses induced in manufacturing, and because of the difficulty in evaluating stress concentrations. For these reasons, limits for calculated stresses are not always based merely on the permissible stress values for spring materials as determined by standard tests. Rather, the amount by which calculated stresses can approach permissible values for spring materials often depends on factors devised from tests.

Residual stresses exist. They may be either beneficial or detrimental, depending upon the direction of the applied service stresses. Favorable residual stresses may be deliberately added to the system already present from the cold-forming operations.

As a general rule, plastic flow caused by forming in the same direction as in service will result in a favorable residual-stress system, one in which the pertinent residual stresses are opposite in sign to the service stresses. These residual stresses may be both tension and shear, and calculations must be made on a combined-stress basis. For example, a compression spring will be able to support a higher load elastically if it has been previously set, removed, or overloaded into the plastic region by a compression load. If, however, it has been stretched out previously (to meet a blueprint dimension), the residual-stress system is unfavorable, and the elastic limit in compression will be below expectations unless the spring is again stress-relieved. Similarly, if a flat spring is formed by bending pretempered material, the residual stresses will be favorable to loads tending to close up the bend and unfavorable to loads opening the bend. A cold-wound torsion spring is best stressed when its coils tend to wind down onto an arbor and unfavorably stressed when the coils unwind.

For best spring design it is thus necessary to predict the type and direction of the residual stresses and add them to the service stresses. In extreme cases of unfavorable residual stresses, hardening the spring after forming may be necessary. However, it is always best to attempt to design so that residual stresses can be utilized effectively; if not, the spring manufacturer must have the necessary design information so that he can stress-relieve the springs at the highest possible temperature.

The simple spring formulas do not take into account the residual-stress system which is always present in cold-formed springs. Most stress calculations are therefore directed toward determining relative rather than absolute stress figures. Charts of allowable design stresses take into account residual stresses.

Conditions of stress concentrations introduced by notches, holes, and curvature not otherwise included in stress equations should be included in stress calculations.

22.2.5 Energy Storage Capacity

The energy storage capacity (ESC)[2] of a spring is the amount of energy that can be stored per unit volume of active spring material when the spring is loaded so that stress is the maximum allowable. For springs with load-deflection curves of the type shown in Fig. 22.1*a* and subject to a static load, the resilience

$$U/V = K(\max \sigma)^2/E \qquad \text{or} \qquad U/V = K(\max \tau)^2/G \tag{22.7}$$

V is the volume of active spring material. Values of K for common spring types are given in Table 22.1; K is an index of volume efficiency.

TABLE 22.1 Values of K

Spring type	K
1. Helical extension and compression, round wire:	
C-3	1/5.4
C-6	1/4.7
C-9	1/4.5
C-12	1/4.3
2. Helical torsion, round wire	$\frac{1}{8}$
3. Helical torsion, rectangular wire	$\frac{1}{6}$
4. Cantilever leaf, uniform rectangular section	$\frac{1}{18}$
5. Triangular cantilever leaf, rectangular section	$\frac{1}{6}$
6. Spiral spring, rectangular section	$\frac{1}{8}$
7. Torsion bar, circular section	$\frac{1}{4}$
8. Belleville spring	$\frac{1}{40}$ to $\frac{1}{10}$

22.3 SPRINGS IN COMBINATION

Springs are often used in combination because of space limitations, because a combination may be more efficient than a single equivalent spring or because a combination will give a load-deflection curve or dynamic characteristics not possible with a single spring of standard design.

For simplicity, the examples shown are of combinations having the fewest number of springs; these examples can be easily generalized to apply to larger combinations.

22.3.1 Series and Parallel Combinations

Figure 22.2 shows schematically how springs are combined in parallel and series. For the parallel combination,

$$P_{12} = P_1 + P_2 \qquad \delta_{12} = \delta_1 = \delta_2 \qquad k_{12} = k_1 + k_2 \tag{22.8}$$

For the series combination,

$$P_{12} = P_1 = P_2 \qquad \delta_{12} = \delta_1 + \delta_2 \qquad 1/k_{12} = 1/k_1 + 1/k_2 \tag{22.9}$$

A parallel-series combination is shown in Fig. 22.2*c*.

When the operating range of a combination is within the operating range of each component spring, the resulting load-deflection curve will not have any abrupt changes. However, for springs combined as in Fig. 22.3, the load-deflection curve will

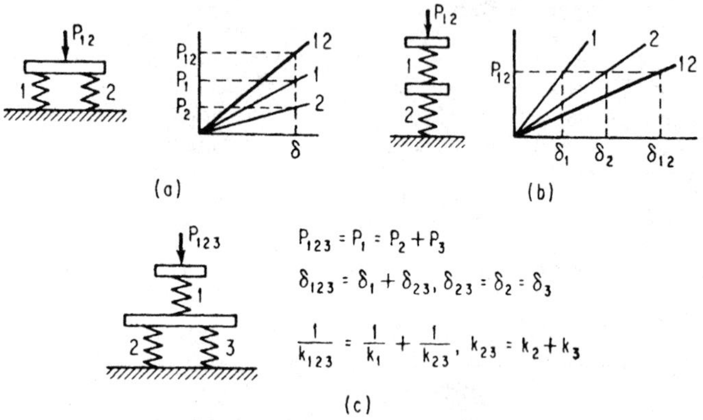
P_{12}
P_{12}
P_1
P_2
12
1
2
δ
(a)
P_{12}
1
2
12
δ_1 δ_2 δ_{12}
(b)
P_{123}
1
2
3
$P_{123} = P_1 = P_2 + P_3$
$\delta_{123} = \delta_1 + \delta_{23},\ \delta_{23} = \delta_2 = \delta_3$
$\frac{1}{k_{123}} = \frac{1}{k_1} + \frac{1}{k_{23}},\ k_{23} = k_2 + k_3$
(c)

FIG. 22.2 Spring combination. (*a*) Parallel springs, linear load deflection. $P = 0$ where $\delta = 0$. (*b*) Series springs, linear load deflection. $P = 0$ where $\delta = 0$. (*c*) Springs in series-parallel combinations.

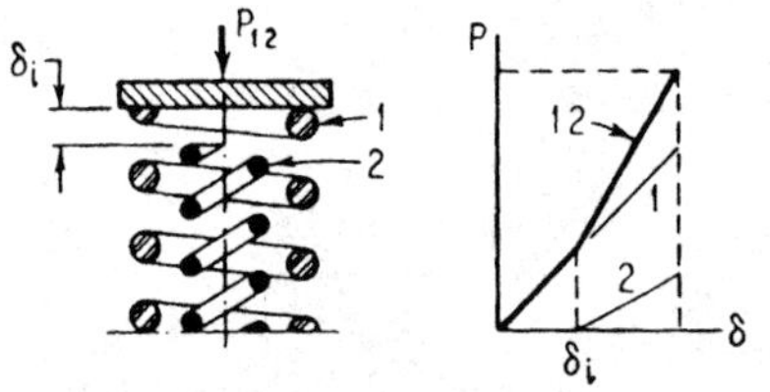
δ_i
P_{12}
1
2
P
12
1
2
δ
δ_i

FIG. 22.3 Nest of helical springs. Inner spring shorter than outer results in abrupt change in load-deflection curve. Note that springs are wound opposite in hand to prevent tangling.

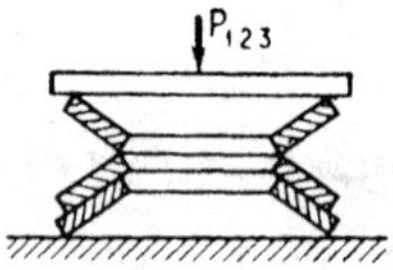
P_{123}

FIG. 22.4 Nest of Belleville springs.

change abruptly when the inner spring begins to deflect. Similarly, the load-deflection curve for the combination of Belleville springs shown in Fig. 22.4 will change abruptly when the single spring has deflected to the flat condition.

22.3.2 Energy Storage

The energy stored within the operating range of a combination of springs may be determined from the load-deflection characteristics of the combination or by adding together the energy stored by the separate springs.

22.4 SPRING MATERIALS AND OPERATING CONDITIONS

In this section properties of some common spring materials are given. Data applicable only to specific types of springs are given in the sections devoted to these springs. For further data consult Ref. 5.

22.4.1 Common Spring Materials

Designations for and properties of typical common spring materials are given in Table 22.2.

22.4.2 Strength of Spring Wire and Strip

Figure 22.5 shows the dependence of tensile strength on wire diameter. Table 22.3 gives recommended working stresses for flat-spring materials in static applications. Data apply to springs subject to static and low-cycle duty at normal temperatures. Recommendations as to the use of the data are given under specific spring types.

22.4.3 Cyclic Loading

Fatigue failures usually start at a surface defect or point of stress concentration. Metals hardened by heat treatment show typical fatigue fractures. Some hard-drawn wires develop long longitudinal cracks, which are often mistaken for metallurgical seams; hence tests of springs made from hard-drawn wire can be misleading if not properly monitored, as a crack can grow over several hundred thousand cycles without external sign of failure.

It is difficult to apply to springs much of the data obtained from the usual reverse-bending and torsional-fatigue tests. There are numerous reasons for this difficulty. The chief reasons are: residual stresses are inherent in springs but not found in test specimens; the hardness at which metals are used in springs makes them more sensitive to stress concentrations than the same metals fatigue-tested at lower hardnesses; most springs are not subject to complete stress reversals as is the case with test specimens.

The difference between stresses at the extremes of the operating range is called the *stress range.* Allowable stress ranges must be evaluated by using Goodman or Soderberg diagrams, *SN* curves, etc. Because of the variables involved in cyclic loading, the number of cycles to failure between similar springs may differ by a factor of 5 or more.

Springs in cyclic service may also fail by setting, particularly when highly stressed, as small amounts of plastic flow undetectable in a single cycle can accumulate over many operations.

22.4.4 Selection of Materials

When fatigue resistance is important and wire sizes are small, use:

1. Music wire where space is limited and stresses are high
2. Stainless steel when corrosive conditions exist or temperatures are as high as 600°F
3. Oil-tempered or hard-drawn wire when space is available and costs must be kept low
4. Chrome-vanadium steel when temperature approaches 400°F
5. Phosphor bronze when electrical conductivity is important

When fatigue resistance is important and wire sizes are above $\frac{1}{8}$ in, use:

TABLE 22.2 Typical Properties of Common Spring Materials

Common name, specification	E, 10^6 lb/in^2	G, 10^6 lb/in^2	Density γ, lb/in^3	Max service temp., °F	Electrical conductivity*	Principal characteristics
			High-carbon steels			
1. Music wire, ASTM A228	30	11.5	0.283	250	7	High strength, excellent fatigue life
2. Hard drawn, ASTM A227	30	11.5	0.283	250	7	General-purpose use, poor fatigue life
3. Valve spring, ASTM A230	30	11.5	0.283	300	7	For excellent fatigue life at average stress range
4. Oil tempered, ASTM A229	30	11.5	0.283	300	7	For moderate stress; low impact or shock
			Alloy steels			
5. Chrome vanadium, ASTM A232	30	11.5	0.283	425	7	High strength, resists shock, good fatigue life. ASTM A59 used as less expensive substitute for ASTM A232
6. Chrome silicon, ASTM A401	30	11.5	0.283	475	5	
7. Silicon manganese, ASTM A59	30	11.5	0.283	450	4.5	

Stainless steels						
8. Martensitic, AISI 410, 420	29	11	0.280	500	2.5	Unsatisfactory for subzero applications
9. Austenitic, AISI 301, 302	28	10	0.282	600	2	Good strength at moderate temperatures, low stress relaxation
10. Precipitation hardening, 17-7PH	29.5	11	0.286	600	2	Long life under extreme service conditions
11. High temperature, A286	29	10.4	0.290	950	2	For elevated-temperature applications
Copper-base alloys						
12. Spring brass, ASTM B134	16	6	0.308	200	17	Low cost, high conductivity; poor mechanical properties
13. Phosphor bronze, ASTM B159	15	6.3	0.320	200	18	Able to withstand repeated flexures. Popular alloy
14. Silicon bronze, ASTM B99	15	6.4	0.308	200	7	Used as less expensive substitute for phosphor bronze
15. Beryllium copper, ASTM B197	19	6.5	0.297	400	21	High elastic and fatigue strength; hardenable
Nickel-base alloys						
16. Inconel 600	31	11	0.307	600	1.5	Good strength, high corrosion resistance
17. Inconel X-750	31	11	0.298	1100	1	Precipitation hardening; for high temperatures
18. Ni-Span C	27	9.6	0.294	200	1.6	Constant modulus over a wide temperature range

*Electrical conductivity given as a percent of the conductivity of the International Annealed Copper Standard.

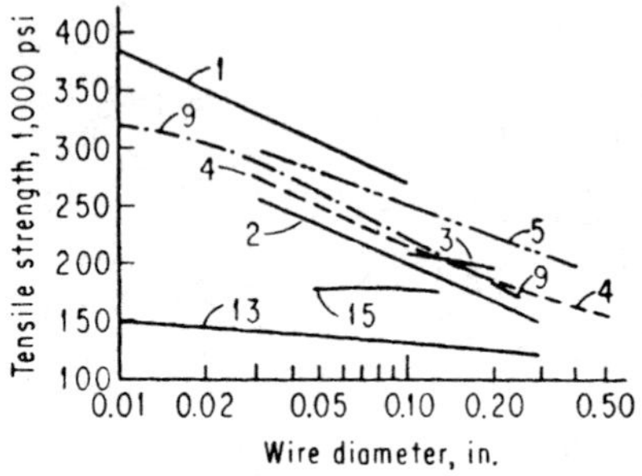

FIG. 22.5 Tensile strength of spring wires vs. diameter. Numbers correspond to listing in Table 22.2. (1) Music wire ASTM A222. (2) Hard-drawn ASTM A227. (3) Valve spring ASTM A230. (4) Oil-tempered ASTM A229. (5) Chrome vanadium ASTM A232. (9) Stainless steel AISI 302. (13) Phosphor bronze ASTM B159. (15) Beryllium copper ASTM B197.

TABLE 22.3 Recommended Working Stresses for Flat-Spring Materials in Static Applications[3]

Material, specifications	1000 lb/in² σ
Stainless steel, AISI 302, hard	130
Stainless steel 17-7 PH, hardened	160
Phosphor bronze (521),* spring temper	70
Beryllium copper (172),* ½ hard	90
Beryllium copper (172),* hard	110
Inconel 600 spring temper	90
Inconel X-750 at 500°F	100
At 700°F	90
At 900°F	75

*Numbers are of the Copper Development Association Inc. standard alloys.

1. Valve-spring wire for critical applications
2. Oil-tempered or hard-drawn wire when space is available and costs must be kept low
3. Chrome-vanadium steel when temperatures approach 400°F
4. Chrome-silicon steel when temperatures are as high as 500°F
5. Stainless steel when corrosive conditions exist
6. Phosphor bronze when electrical conductivity is important

When resistance to set in static load is important, use:

1. Music wire where space is limited and wire size small
2. Oil-tempered wire, lower stress than music wire
3. Hard-drawn wire where low cost is essential
4. Chrome-vanadium steel when temperatures are as high as 400°F
5. Chrome-silicon steel when temperatures are as high as 500°F
6. Stainless steel for corrosion resistance and temperatures approaching 600°F
7. Inconel, A286 for temperatures 500 to 800°F
8. Inconel X for temperatures above 800°F
9. Beryllium copper when electrical conductivity is important and stresses are high

When fatigue resistance or high resistance to setting is important, use (flat springs):

1. AISI 1095 for high-duty applications
2. AISI 1075 for general application
3. Stainless steel for corrosive conditions
4. Phosphor bronze for electrical conductivity
5. Beryllium copper for electrical conductivity and shapes that demand forming while soft followed by hardening

Formability requirements may dictate the need to use annealed materials and hardening and tempering after forming. It would not apply to 300 stainless steel and phosphor bronze. When low cost is essential, use:

1. AISI 1050 for low-cost clips where volume is substantial
2. AISI 1065
3. Brass for decorative purposes

22.4.5 Tolerances

Allowable variations in commercial spring wire sizes are given in Table 22.4. Standard commercial tolerances for coil diameter, free length, and load are given in Tables 22.5 through 22.7. Tolerances on squareness is 3°.

22.4.6 Springs at Reduced and Elevated Temperatures

The mechanical properties of spring metals vary with temperature. For most spring metals E and G decrease almost linearly with temperature within their useful ranges of application. One exception is the family of nickel-base alloys which includes Elinvar and Ni-Span C whose composition is specifically designed to confer a near zero temperature coefficient of spring rate over a limited temperature range around room temperature. This anomalous behavior is caused by a gradual and reversible loss of ferromagnetic effect as the material is raised in temperature toward its Curie point. Once the Curie point is neared (at temperatures above 200°F), the temperature coefficient of rate is comparable with that of other spring alloys.

Reduced Temperature. The body-centered-cubic materials such as spring steel are normally considered to be brittle at subzero temperatures, and the plain-carbon and low-alloy steels are not generally recommended for such service. However, a spring structure is designed to be resilient, and thus the spring configuration may protect the spring material from brittle failure. In special cases of low-temperature application where a large quantity of springs is required and the cost must be low, ordinary spring materials should not be disregarded without a service test. Otherwise, the face-centered-cubic materials, such as copper-base, nickel-base, or austenitic iron-base alloys, should be specified.

Elevated Temperature. Springs used at temperatures above room temperature must be designed with allowance for the rate change associated with modulus decrease, if accuracy of rate is essential.

TABLE 22.4 Allowable Variations in Commercial Spring-Wire Sizes

Material	ASTM Specification No.	Wire diam, in.	Permissible variation, in.
Hard-drawn spring wire	A227	0.028–0.072	±0.001
Chrome-vanadium spring wire	A231	0.073–0.375	±0.002
Oil-tempered wire	A229	0.376 and over	±0.003
Music wire	A228	0.026 and under	±0.0003
		0.027–0.063	±0.0005
		0.064 and over	±0.001
Carbon-steel valve spring wire	A230	0.093–0.148	±0.001
		0.149–0.177	±0.0015
Chrome-vanadium valve spring wire	A232	0.178–0.250	±0.002

TABLE 22.5 Coil-Diameter Tolerances, Helical Compression and Extension Springs, ± in

Wire diameter, in	Spring index D/d						
	4	6	8	10	12	14	16
0.015	0.002	0.002	0.003	0.004	0.005	0.006	0.007
0.023	0.002	0.003	0.004	0.006	0.007	0.008	0.010
0.035	0.002	0.004	0.006	0.007	0.009	0.011	0.013
0.051	0.003	0.005	0.007	0.010	0.012	0.015	0.017
0.076	0.004	0.007	0.010	0.013	0.016	0.019	0.022
0.114	0.006	0.009	0.013	0.018	0.021	0.025	0.029
0.171	0.008	0.012	0.017	0.023	0.028	0.033	0.038
0.250	0.011	0.015	0.021	0.028	0.035	0.042	0.049
0.375	0.016	0.020	0.026	0.037	0.046	0.054	0.064
0.500	0.021	0.030	0.040	0.062	0.080	0.100	0.125

TABLE 22.6 Free-Length Tolerances, Closed-and-Ground Helical Compression Springs, ± in/in of Free Length

Number of active coils per inch	Spring index D/d						
	4	6	8	10	12	14	16
0.5	0.010	0.011	0.012	0.013	0.015	0.016	0.016
1	0.011	0.013	0.015	0.016	0.017	0.018	0.019
2	0.013	0.015	0.017	0.019	0.020	0.022	0.023
4	0.016	0.018	0.021	0.023	0.024	0.026	0.027
8	0.019	0.022	0.024	0.026	0.028	0.030	0.032
12	0.021	0.024	0.027	0.030	0.032	0.034	0.036
16	0.022	0.026	0.029	0.032	0.034	0.036	0.038
20	0.023	0.027	0.031	0.034	0.036	0.038	0.040

For springs less than ½ in long, use the tolerances for ½ in.
For closed ends not ground, multiply above values by 1.7.

TABLE 22.7 Load Tolerances, Helical Compression Springs, ± Percent of Load
(Start with Tolerances from Table 22.6 Multiplied by Free Length)

Length tolerance,	Deflection from free length to load, in														
± in	0.050	0.100	0.150	0.200	0.250	0.300	0.400	0.500	0.750	1.000	1.500	2.000	3.000	4.000	6.000
0.005	12	7	6	5											
0.010		12	8.5	7	6.5	5.5	5								
0.020		22	15.5	12	10	8.5	7	6	5						
0.030			22	17	14	12	9.5	8	6	5					
0.040				22	18	15.5	12	10	7.5	6	5				
0.050					22	19	14.5	12	9	7	5.5				
0.060					25	22	17	14	10	8	6	5			
0.070						25	19.5	16	11	9	6.5	5.5			
0.080							22	18	12.5	10	7.5	6	5		
0.090							25	20	14	11	8	6	5		
0.100								22	15.5	12	8.5	7	5.5		
0.200										22	15.5	12	8.5	7	5.5
0.300											22	17	12	9.5	7
0.400												21	15	12	8.5
0.500												25	18.5	14.5	10.5

First load test at not less than 15% of available deflection. Final load test at not more than 85% of available deflection.

The type of failure characteristic of springs at elevated temperature is setting or loss of load. The setting of a spring is the result of two types of flow—plastic and anelastic. Plastic flow of the spring material is simply the result of exceeding the elastic limit or yield point at the service temperature. This level limits the allowable stress at any given temperature, regardless of length of time under load. The loss of load caused by plastic flow is not recoverable.

In addition, a spring under load for a period of time will lose more load or set more, by a process called *anelastic flow*. This flow is time- and temperature-dependent and may be recoverable when the load is removed. The net effect in spring service is that a spring will lose load or relax in service at a rate which is stress-, time-, and temperature-dependent. If the spring is unloaded at temperature, the load loss may be partially recoverable. The complete relaxation curve for a spring loaded to a constant deflection at elevated temperature consists of a very rapid first stage characterized by plastic flow caused by the lower yield point at elevated temperature, and a second stage of anelastic flow at a rate that is dependent on the log of the time under load. There has been no evidence of a third stage or that relaxation ever stops, although the rate is decreasing. If the spring is loaded under constant-load (creep) conditions, instead of constant-deflection (relaxation) conditions, the rate of load loss is far higher, particularly at longer test times.[4]

The accurate and efficient design of a high-temperature spring is a difficult undertaking. Assuming that operating temperatures and spring-loading conditions are well known (and they seldom are), the designer must know what relaxation he can accept within his specifications of load loss and time of service, for zero change is close to unobtainable. The more stable the spring must be and the longer the time of service, the lower the design stress must be. On the other hand, if the service is intermittent with periods of no load at temperature, the spring may recover a portion of its load loss in the "rest" periods and relax far less than might be anticipated. Because of these variables, data plots can only indicate the worst service condition, that of constant loading for the times and temperatures indicated.

If the spring requirements for minimum relaxation are severe, heat setting may be used. This is a process for inducing favorable residual stresses during manufacture by preloading the spring at higher than service temperatures. Such a process can contribute markedly to the apparent relaxation resistance of a spring, but since the heat setting is actually an anelastic-flow process, its effects are recoverable. In other words, a heat-set spring will grow and lose its stability if it is unloaded at the service temperature. Furthermore, if the heat-setting operation is improperly carried out, its effects may be transient, and the spring will behave unpredictably.

22.4.7 Constancy of Rate

Some springs are such that their rate is constant over the entire usable range of deflection. For example, a helical extension spring has an essentially constant rate after the initial tension has been exceeded and before the yield point is reached. A torsion spring exhibits a constant rate within the elastic range, because the decrease of diameter as the spring winds down on an arbor is compensated for by the increase in number of coils. In both examples cited, minor deviations from a straight-line rate may be caused by end or loop deflection.

A helical compression spring will show a constant rate only in the central portion of its load-deflection plot. Initially, nonuniform compression of the end coils appears to produce a lower rate than calculated, and as the spring approaches solid, the rate increases. A constant-rate design should utilize only the center 70 percent of the total deflection. If the compression spring must have a constant rate over a major portion of its deflection range, certain special techniques can achieve this at added cost. The closed-end coils may be soldered or brazed together to reduce the effect of early end closure. A more exact solution is to braze an insert into the dead coils.

Some springs are deliberately designed with a nonconstant rate, e.g., variable-pitch compression springs, conical springs, and volute springs. These are useful in shock-absorbing operations because their rate continues to increase with deflection. No helical compression spring can be designed with a decreasing rate.

A few specialized spring designs may have a rate close to zero, a feature desirable in counterbalance springs and springs used in seals. Belleville springs of special configurations, the constant-force spring and a buckling column, fall in this category. Belleville springs can also be designed with a decreasing rate or even a negative rate over a given deflection range, so that snap-through action will occur. Buckling flat springs, which will snap through, are also a common spring application of negative rate.

The torque output of a motor spring decreases with each turn, and the rate of change depends upon the spring design and manufacturing technique. A constant-force spring can be made to have a torque output that is nearly constant over the useful range of the spring.

22.5 HELICAL COMPRESSION SPRINGS

22.5.1 Principal Characteristics

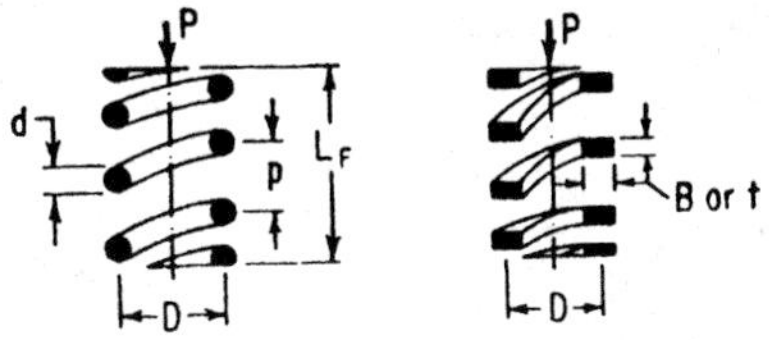

FIG. 22.6 Helical compression springs. L_F is the free length. For rectangular wire $b > t$; $b = t$ if the wire has a square section.

The load-deflection curve is of the type shown in Fig. 22.1*a*. For springs wound from circular wire (see Fig. 22.6).

$$P = \delta G d^4 / 8ND^3 \qquad (22.10)$$

$$\tau = 8K_A PD/\pi d^3 \qquad (22.11)$$

For springs wound from rectangular wire

$$P = K_1 \delta G b t^3 / ND^3 \qquad (22.12)$$

$$\tau = K_A K_2 PD/bt^2 \qquad (22.13)$$

Values for stress factor K_A are given in Table 22.8. Values for the shape factors K_1 and K_2 are given in Table 22.9.[6]

TABLE 22.8 Stress Factors for Helical Springs

C	3	4	5	6	8	10	12	16
K_A	1.58	1.40	1.31	1.25	1.18	1.15	1.12	1.09
K_B	1.33	1.23	1.17	1.14	1.10	1.08	1.07	1.05
K_C	1.29	1.20	1.15	1.12	1.09	1.07	1.05	1.04

TABLE 22.9 Shape Factors K_1 and K_2 for Helical Compression Springs Wound from Rectangular Wire

b/t	1.00	1.50	1.75	2.00	2.50	3.00	4.00	6.00
K_1	0.180	0.250	0.272	0.292	0.317	0.335	0.358	0.381
K_2	2.41	2.16	2.09	2.04	1.94	1.87	1.77	1.67

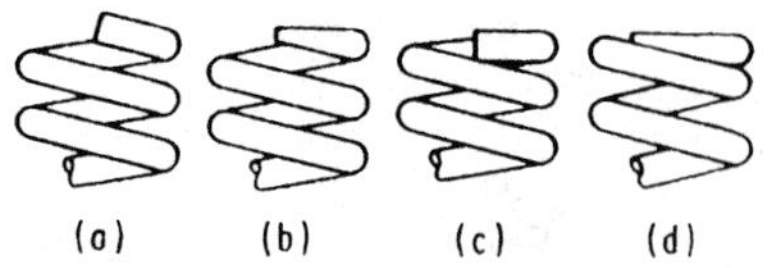

FIG. 22.7 Types of ends for helical compression springs. (*a*) Open, not ground. (*b*) Open and ground. (*c*) Closed, not ground. (*d*) Closed and ground.

Types of ends for helical compression springs are shown in Fig. 22.7. *Open ends not ground* are satisfactory only if accuracy of load is not important. Springs of this type tangle easily when loose packed for shipment or storage. *Open ends ground* also tangle easily and are generally specified only where it is necessary to obtain as many active coils as possible in a limited space. *Closed ends* not ground are preferred for wire sizes less than 0.020 in diameter or thickness, or if $C > 12$. Closed ends ground are preferred for wire sizes greater than 0.020 in diameter or thickness and if $C < 12$. Springs of this type are usually ground to a bearing surface of 270 to 330. Relationships between types of ends and spring measurements are given in Table 22.10.

TABLE 22.10 Helical Compression Spring Measurements for Common Types of Ends

Type of ends	L_F	N_T	L_S
Open, not ground	$pN + d$	N	$dN + d$
Open and ground	$pN + p$	$N + 1$	$dN + d$
Closed, not ground	$pN + 3d$	$N + 2$	$dN + 3d$
Closed and ground	$pN + 2d$	$N + 2$	$dN + 2d$

L_F = free length, N_T = total number of coils, L_S = solid height.

Table applies to springs wound from rectangular wire if b or t as appropriate is substituted for d. However, rectangular wire keystones in coiling and the axial dimension increase must be allowed for in the solid height calculation, according to

$$t_1 = t(C + 0.5)/C \quad \text{or} \quad b_1 = b(C + 0.5)/C$$

where t_1 or b_1 is the thickness of the inner edge of the material after coiling and t or b the thickness before coiling.

22.5.2 Lateral Loading

Occasionally an application requires lateral as well as axial loading of a helical compression spring. Although the axial rate of a spring is constant for all loads, the lateral rate varies with compression. An adjustable spring-rate system can thus be designed to take advantage of this fact.

The combined stress of a laterally loaded spring

$$\tau = \tau_A(1 + F_L/D + P_L L/PD) \tag{22.14}$$

where τ_A is the axial stress, P the axial load, L the length of the spring under load, and F_L and P_L the lateral deflection and lateral load, respectively.

22.5.3 Buckling and Squareness in Helical Springs

If the free length of a compression spring is more than four times the spring diameter, its stability under load may become critical, and the spring may buckle as a column. The lateral stability of a spring depends on the ratio of its free length to mean diameter L_F/D. A slender spring will buckle under load when the ratio of deflection to free length δ/L_F exceeds a critical value. Figure 22.8 can be used to evaluate lateral stability of closed and ground springs. If possible, the spring should be so designed that it will not buckle under service conditions; otherwise the spring must be guided within a tube or on a rod. The friction between spring and guides will interfere somewhat with the accuracy of the load-deflection curve and reduce the spring endurance limit (see Table 22.11).

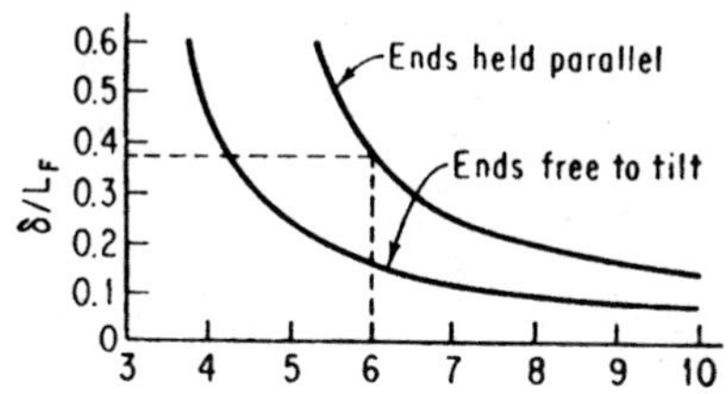

FIG. 22.8 Criterion for stability. Example: A spring having the proportions $L_F/D > 6$ and ends held parallel will buckle if $\delta/L_F > 3.7$.

Torsion springs, if long and slender, may also buckle. This may sometimes be prevented by properly clamping the ends or by using initial tension between coils; if the spring still buckles, guides or arbors must be used.

A coiled compression spring does not exert force directly along its axis; the loading is usually eccentric. If the spring must be square under load, as where uniform pressure must be exerted on a valve in guides, special design considerations are used. It is difficult to design a spring which will be square when unloaded and square throughout its loading cycle. Squareness under a specific load is often achieved at the expense of squareness at other deflections. Practical experience in spring design is far ahead of theory in this area at present.

The effect of eccentric loading is to cause an increase in torsional stress on one side of a compression spring and a decrease in stress on the other side.

An approximate relationship for the load eccentricity

$$e = 1.12R(0.504/N + 0.121/N^2 + 2.06/N^3) \tag{22.15}$$

TABLE 22.11 Factors for Determining Limiting Stress Values for Helical Compression Springs

Material	Factor (before set removed)	Factor (after set removed)
Music wire	0.45	0.65–0.75 for all materials
Valve spring	0.50	
Chrome vanadium	0.50	
Oil-tempered	0.50	
Hard-drawn	0.45	
Phosphor bronze	0.35	
Beryllium copper	0.35	
Stainless steel	0.35	

Equation (22.15) indicates that the eccentricity is especially important when there are few active coils since

$$\frac{\text{Stress in eccentrically loaded spring}}{\text{Stress in spring under pure axial load}} = 1 + \frac{e}{R} \tag{22.16}$$

where e is a measure of load displacement from the spring axis and R is the spring mean radius. However, this stress correction is usually ignored unless the number of active turns is two or less. Since the eccentricity varies with load, these calculations are only approximate.

22.5.4 Change in Diameter during Deflection

A helical spring will change in diameter during deflection, which may be significant if the spring is functioning under severe space restrictions. If the spring ends are not allowed to unwind during compression the maximum increase in diameter of a compression spring deflected to solid height is given by

$$\Delta D_{max} = 0.05(p^2 - d^2)/D \tag{22.17}$$

where pitch p and mean diameter D are those of the spring in the free condition.

22.5.5 Surge and Vibration

When a spring is impact loaded, maximum stress is limited not by total deflection but by the impact velocity of loading v

$$\tau \cong v\sqrt{2\gamma G/g} \tag{22.18}$$

For a steel spring

$$\tau = 131v \text{ (Ref. 7)} \tag{22.19}$$

Surge waves can travel through a spring at a frequency characteristic of the spring. The fundamental frequency of a helical compression spring wound from circular wire

$$f = (Kd/9ND^2)\sqrt{gG/\gamma} \tag{22.20}$$

For springs with one end fixed and one free, $K = \frac{1}{2}$ for springs with both ends fixed, $K = 1$. For a steel spring

$$f = 13{,}900Kd/ND^2 \tag{22.21}$$

To avoid vibration, the fundamental frequency of a spring should be at least 13 times the frequency of the operating cycle.

22.5.6 Limiting-Stress Values for Static and Low-Cycle Duty

The value of τ calculated by Eq. (22.11) or (22.13) must be less than the permissible limiting-stress value of the spring materials by a suitable margin of safety. Values of permissible limiting stress are determined by multiplying the tensile strength obtained from Fig. 22.5 by the factors given in Table 22.11. Preferred practice is to design for deflection to the solid height.

Set removal increases the load carrying ability of springs in static applications by inducing favorable residual stresses. This allows the use of a higher factor as indicated in Table 22.11. For springs with set removed, the stress correction factor K_A in Eq. (22.11) should be replaced by $1 + 0.5/C$.

22.5.7 Limiting-Stress Values for Cyclic Loading

The value of τ calculated by Eq. (22.11) should not exceed the permissible limiting stress determined by multiplying the tensile strength by the factor given in Table 22.12. Values in Table 22.12 are guidelines and are based on a stress ratio (minimum stress to maximum stress) of zero, no surging, room temperature, and noncorrosive environment. To estimate the life at any other stress ratio, a combined *S-N* curve and modified Goodman diagram can be used. See Refs. 5 and 6.

Shot peening improves fatigue life, as shown in Table 22.12.

TABLE 22.12 Factors for Determining Limiting-Stress Values for Round Wire Helical Compression Springs in Cyclic Loading

Fatigue life, cycles	ASTM A228, austenitic stainless steel, and nonferrous		ASTM A230 and A232	
	Not shot-peened	Shot-peened	Not shot-peened	Shot-peened
10^5	0.36	0.42	0.42	0.49
10^6	0.33	0.39	0.40	0.47
10^7	0.30	0.36	0.38	0.46

22.6 HELICAL EXTENSION SPRINGS (Fig. 22.9)

22.6.1 Principal Characteristics

For close-wound springs the load-deflection curve is of the type shown in Fig. 22.1*b* and

$$P = P_i + \delta G d^4/8ND^3 \qquad \tau = 8K_A PD/\pi d^3 \tag{22.22}$$

P_i is the initial tension wound into the spring and τ_i the corresponding torsional stress,

$$P_i = \pi\tau_i d^3/8D \tag{22.23}$$

Recommended values of τ_i depend on the spring index (see Fig. 22.10).

For open-wound springs τ_i and P_i are zero and the load-deflection curve is of the type shown in Fig. 22.1*a*.

22.6.2 Stresses in Hooks

Critical stresses may occur at sections *A* and *B* as shown in Fig. 22.9. At section *A* the stress is due to bending, and at section *B* the stress is due to torsion:

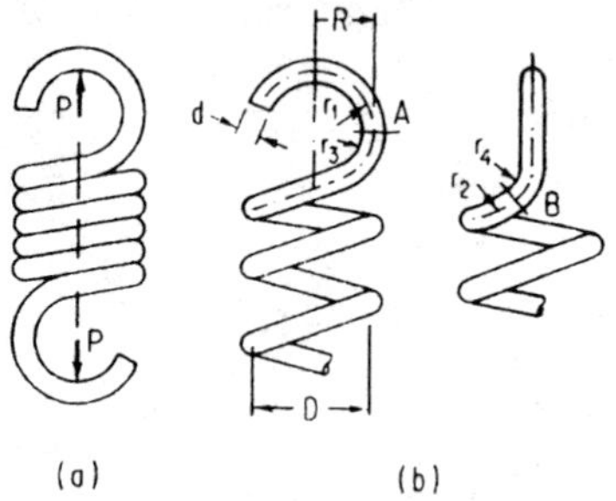

FIG. 22.9 Helical extension springs. (*a*) Close-wound. (*b*) Open-wound.

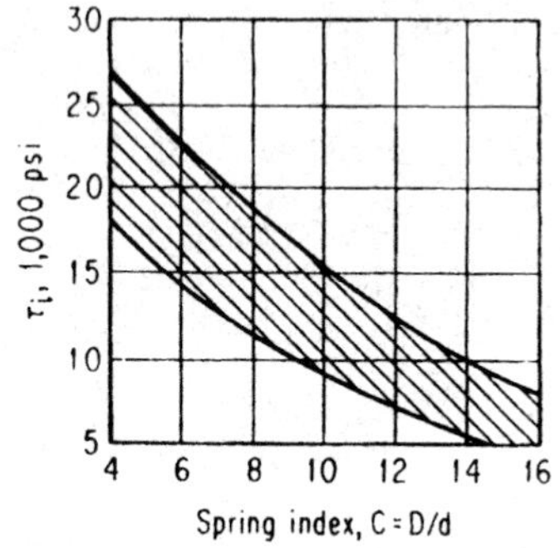

FIG. 22.10 Ranges of recommended values for stress due to initial tension in close-wound helical extension springs.[9]

$$\sigma = (32PR/\pi d^3)(r_1/r_3) \qquad \tau = (16PR/\pi d^3)(r_2/r_4) \tag{22.24}$$

Recommended practice is that $r_4 > 2d$. By winding the last few coils with a smaller diameter than the main part of the spring so that R is reduced, the stresses in hooks can be minimized.

22.6.3 Limiting-Stress Values for Static and Low-Cycle Duty

Values τ and σ calculated by Eqs. (22.22) and (22.24) must be less than the permissible limiting stress of the spring by a suitable margin of safety. Values of permissible limiting stress are determined by multiplying the value of tensile strength obtained from Fig. 22.5 by the multiplying factor given in Table 22.13. Recommended values in Table 22.13 are based on set not removed and low-temperature heat treatment.

TABLE 22.13 Factors for Determining Limiting-Stress Values for Helical Extension Springs

Material	Factor for τ		Factor for σ
	Body	Loop	(loop)
Music wire	0.45–0.50	0.40	0.75
Valve spring	0.45–0.50	0.40	0.75
Chrome vanadium	0.45–0.50	0.40	0.75
Oil-tempered	0.45–0.50	0.40	0.75
Hard-drawn	0.45–0.50	0.40	0.75
Phosphor bronze	0.35	0.30	0.55
Beryllium copper	0.35	0.30	0.55
Stainless steel	0.35	0.30	0.55

22.6.4 Limiting-Stress Values for Cyclic Loading

Calculated stresses should not exceed the recommended permissible limiting stresses determined by multiplying the tensile strength by the factors given in Table 22.14. Values in Table 22.14 are based on a stress ratio of 0, no surging, no shot peening, ambient environment, and low-temperature heat treatment.

TABLE 22.14 Factors for Determining Limiting-Stress Values for ASTM A 228 and Type 302 Stainless Steel Helical Extension Springs in Cyclic Loading

Fatigue life, cycles	Factor for τ Body	Factor for τ Loop	Factor for (loop)
10^5	0.36	0.34	0.51
10^6	0.33	0.30	0.47
10^7	0.30	0.28	0.45

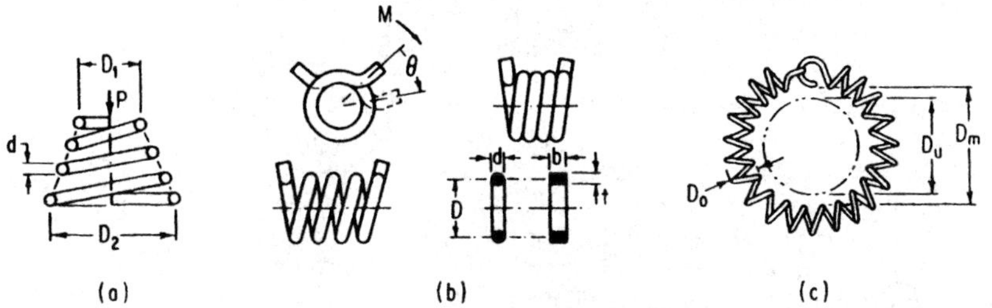

FIG. 22.11 Wound springs. (*a*) Conical spring. (*b*) Helical torsion spring. (*c*) Garter spring.

22.7 CONICAL SPRINGS (Fig. 22.11*a*)

22.7.1 Principal Characteristics

The load deflection curve shown in Fig. 22.1*a* applies only until the bottom coil closes:

$$P = \delta G D^4 / 8 N D^3 \qquad \tau = 8 K_A D P / \pi d^3 \tag{22.25}$$

A conical spring of uniformly changing diameter can be calculated using the average diameter as the mean diameter, or

$$D = (D_1 + D_2)/2$$

provided the bottom coil does not close. After the bottom coil closes, the conical spring should be calculated on a turn-by-turn basis as if the spring were a series of separate coils. The highest stress at a given load should be calculated by using the mean diameter of the largest active coil at load.

The conical spring can be designed so that each coil nests wholly or partially in the adjacent larger coil so that the solid height of the spring can be as small as one wire diameter. Conical springs can be designed to have a uniform rate by varying the pitch.

22.7.2 Limiting-Stress Values

For static and cyclic loading, limiting-stress values may be determined on the same basis as for helical compression springs.

22.8 HELICAL TORSION SPRINGS (Fig. 22.11*b*)

22.8.1 Principal Characteristics

For open-wound springs the load-deflection curve is of the type shown in Fig. 22.1*a*. For springs wound from circular wire

$$M = \theta E d^4 / 67.8ND \qquad \sigma = 32K_B M / \pi d^3 \tag{22.26}$$

The factor 67.8 is greater than the theoretical factor of 64 and has been found satisfactory in practice.

For springs wound from rectangular wire,

$$M = \theta E b t^3 / 13.2\pi ND \qquad \sigma = 6K_C M / bt^2 \tag{22.27}$$

The factor 13.2 is greater than the theoretical factor of 12 and has been found satisfactory in practice. Deflection is given in radians to facilitate the calculation of energy storage. However, it is often convenient to express deflection in number of turns $T = \theta/(2\pi)$. Stress factors K_B and K_C at the spring ID are given in Table 22.8. For stress factors at the spring OD see Ref. 5.

For close-wound springs there may be an initial turning moment due to friction between the coils. Friction between coils will affect rate and cause considerable hysteresis in a complete deflection cycle.

22.8.2 Changes of Dimension with Deflection

As a helical torsion spring winds up, mean diameter decreases and the number of coils increases. After T turns the mean diameter $D_T = DN/(T + N)$ and the number of coils $N_T = T + N$; D and N are the mean diameter and number of coils before windup.

22.8.3 Limiting-Stress Values for Static and Low-Cycle Duty

The value of σ calculated by Eq. (22.26) or (22.27) can be as high as 100 percent of the tensile strength if the residual stress is favorable. If it is unfavorable, use 60 to 80 percent of the value for stress-relieved springs.

22.8.4 Limiting-Stress Values for Cyclic Loading

Calculated stresses should not exceed the recommended permissible limiting stresses determined by multiplying the tensile strength by the factors given in Table 22.15.

TABLE 22.15 Factors for Determining Limiting-Stress Values for Helical Torsion Springs in Cyclic Loading

Fatigue life, cycles	ASTM A228 and type 302 stainless steel	ASTM A230 and A232
10^5	0.53	0.55
10^6	0.50	0.53

Values in Table 22.15 are based on no shot peening, no surging, and springs in the "as stress-relieved" condition.

22.9 GARTER SPRINGS (Fig. 22.11*c*)

22.9.1 Principal Characteristics

Garter springs are special forms of helical extension springs with ends fastened together to form garterlike rings. The inside diameter D_u of an unmounted spring is made to assemble over a larger mounting diameter D_m so that the circumferential load P_c lb exerted by the spring per inch of circumference,

$$P_c = 2\pi K - 2\pi D_u K/D_m + 2P_i/D_m \tag{22.28}$$

22.9.2 Application

Garter springs are used in mechanical seals for shafting and to hold circular segments in place.

22.10 LEAF SPRINGS

The term "leaf spring" as it is used here applies to springs whose load-deflection and load-stress formulas are based on beam equations. Load-deflection curves are of the type shown in Fig. 22.1*a*.

22.10.1 Single-Leaf Cantilever Springs (Fig. 22.12)

For springs of either rectangular or tapered section

$$P = \delta Ebt^3/4L^3K \qquad \sigma = 6PL/bt^2 \tag{22.29}$$

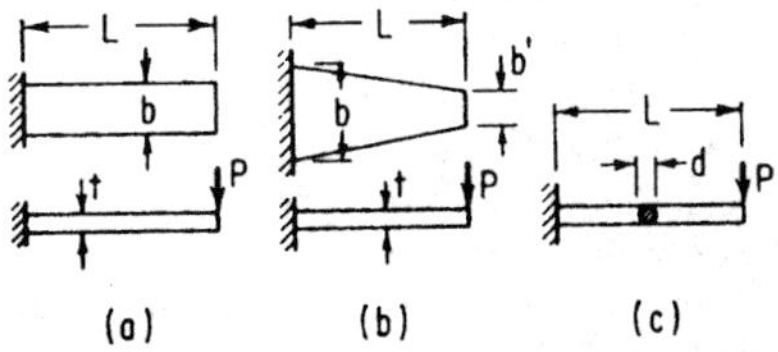

FIG. 22.12 Single-leaf cantilever spring. (*a*) Uniform cantilever section. (*b*) Tapering uniform rectangular section. (*c*) Uniform circular section.

If the rectangular section is uniform as shown in Fig. 22.12*a*, $K = 1$. A more economical use is made of material where a cantilever leaf spring is tapered as shown in Fig. 22.12*b*. For tapered springs, values of K are given in Table 22.16.

For cantilever springs made from circular wire

$$P = 3\pi\delta Ed^4/64L^3 \qquad \sigma = 32PL/\pi d^3 \tag{22.30}$$

TABLE 22.16 Values of K for Tapered Leaf Springs[6]

b'/b	0	0.1	0.2	0.3	0.4	0.5	0.6	0.7	0.8	0.9	1.0
K	1.50	1.39	1.32	1.25	1.20	1.16	1.12	1.09	1.05	1.03	1.00

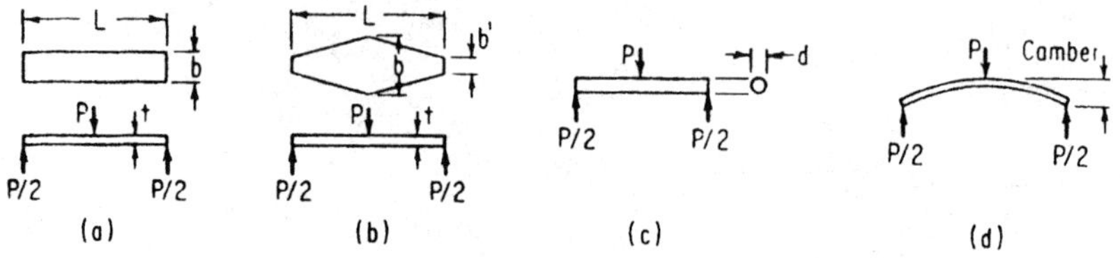

FIG. 22.13 Single-leaf end-supported springs. (*a*) Uniform rectangular section. (*b*) Tapering rectangular section. (*c*) Uniform circular section. (*d*) Elliptical type.

22.10.2 Single-Leaf End-Supported Springs (Fig. 22.13)

For springs of rectangular or tapered section

$$P = 4\delta Ebt^3/L^3K \qquad \sigma = 3PL/2bt^2 \tag{22.31}$$

If the rectangular spring is uniform as shown in Fig. 22.13*a*, $K = 1$. A more economical use is made of material when an end-supported leaf spring is tapered as shown in Fig. 22.13*b*. For tapered springs, values of K are given in Table 22.16.

For end-supported springs made from round wire

$$P = 3\pi\delta Ed^4/4L^3 \qquad \sigma = 8PL/\pi d^3 \tag{22.32}$$

Springs initially bowed as shown in Fig. 22.13*d* are called *elliptical springs*. Camber at no load is often made so the spring will be flat under load.

22.10.3 Multiple-Leaf Springs (Fig. 22.14)

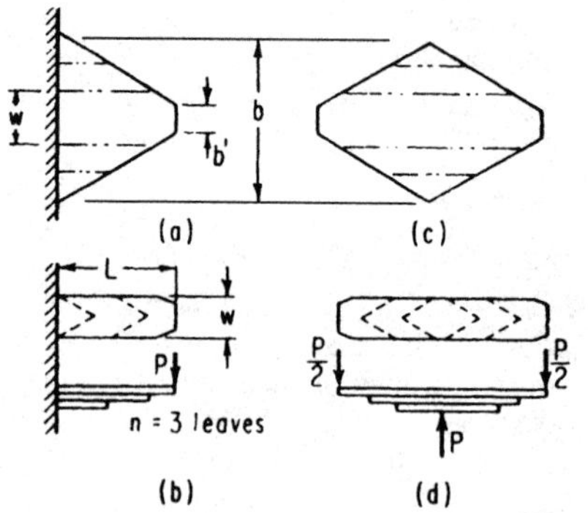

FIG. 22.14 Multiple-leaf springs and their equivalent single-leaf springs.

The formulas for multiple-leaf springs are based on those for equivalent single-leaf springs. For example, the single-leaf cantilever spring shown in Fig. 22.14*a* is equivalent to a multiple-leaf cantilever spring for which $b = nw$, n being the number of leaves.

In using equations for equivalent single-leaf springs, allowance must be made for interleaf friction, which may increase the load-carrying capacity by from 2 to 12 percent depending on the number of leaves and conditions of lubrication.

22.10.4 Limiting-Stress Values for Static and Low-Cycle Duty

Values of calculated stress can be as high as 100 percent of tensile strength of ferrous material and as high as 80 percent of tensile strength for nonferrous material provided residual stress is favorable.

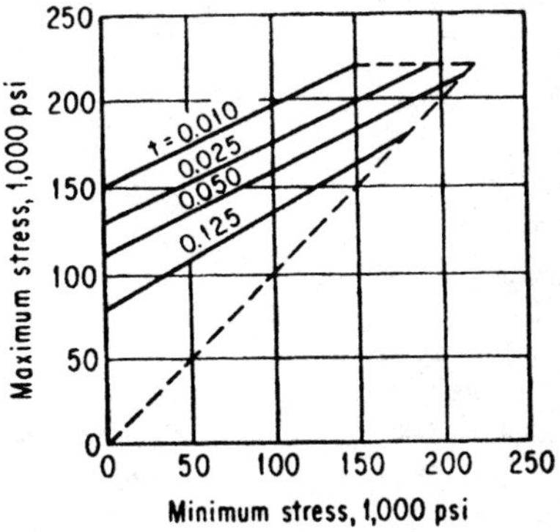

FIG. 22.15 Design chart for cyclically loaded flat steel springs, applies to 0.65 and 0.80 carbon steel at 45 to 48 Rockwell C hardness.[5] Data are for 10^6 cycles.

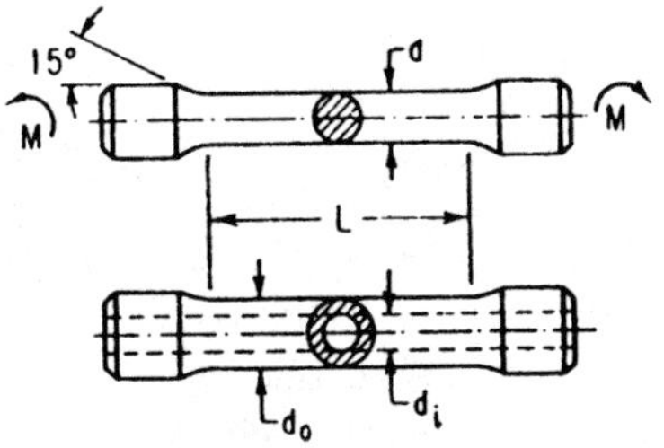

FIG. 22.16 Torsion-bar springs.

22.10.5 Cyclic Loading

Maximum and minimum values of stress for cyclic loading must fall within the permissible limits defined by a Goodman or similar diagram. The design chart shown in Fig. 22.15 applies to carbon-steel springs.

22.11 TORSION-BAR SPRINGS (Fig. 22.16)

22.11.1 Principal Characteristics

The load-deflection curve is of the type shown in Fig. 22.1*a*. For torsion bars with hollow circular sections,

$$M = \pi\theta G(d_0^4 - d_i^4)/32L \qquad \tau = 16Md_0/\pi(d_0^4 - d_i^4) \tag{22.33}$$

For torsion bars with solid circular sections, $d_i = 0$ and $d_0 = d$. Torsion-bar ends should be stronger than the body to prevent failure at the ends.

22.11.2 Material

For highly stressed applications, torsion bars are generally made of silicomanganese. Fatigue resistance is increased by shot peening. Presetting increases strength in the direction of preset but should not be used if bars are loaded in both directions.

22.11.3 Application

Torsion-bar springs are used where highly efficient energy-storage devices are required. The most familiar applications are in automotive equipment.

22.12 SPIRAL TORSION SPRINGS (Fig. 22.17)

22.12.1 Principal Characteristics

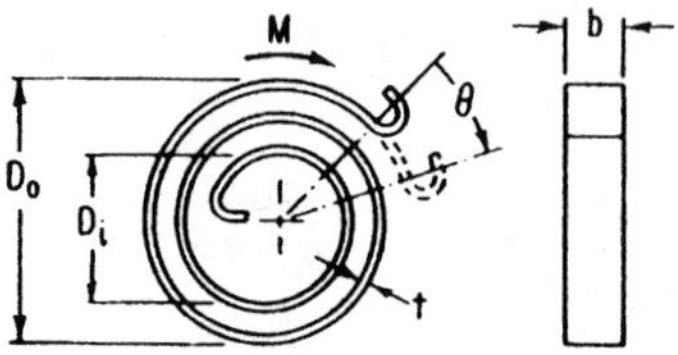

FIG. 22.17 Spiral torsion spring.

The load-deflection characteristic is of the type shown in Fig. 22.1*a* and

$$M = \theta Ebt^3/12L \qquad \sigma = 6M/bt^2 \qquad (22.34)$$

The length of active material $L \approx \pi n(D_0 + D_i)/2$; n is the number of coils. Deflection in Eq. (22.34) is expressed in radians to facilitate the calculation of energy storage. The corresponding number of rotations $T = \theta/(2\pi)$.

Stresses at the spring ends may exceed the values of stress calculated by Eq. (22.34) depending on the design of the end and its treatment in manufacture.

22.12.2 Limiting-Stress Values for Static and Low-Cycle Duty

Calculated stress should be less than 80 percent of the tensile strength.

22.12.3 Application

Spiral torsion springs are widely used for locks in automobiles and to maintain pressure on carbon brushes of electric motors and generators.

22.13 POWER SPRINGS (Fig. 22.18*a*)

22.13.1 Principal Characteristics

The power or clock spring is a special form of the spiral torsion spring. However, Eq. (22.34) applies only to maximum torque when the spring is fully wound. Figure 22.18*b*

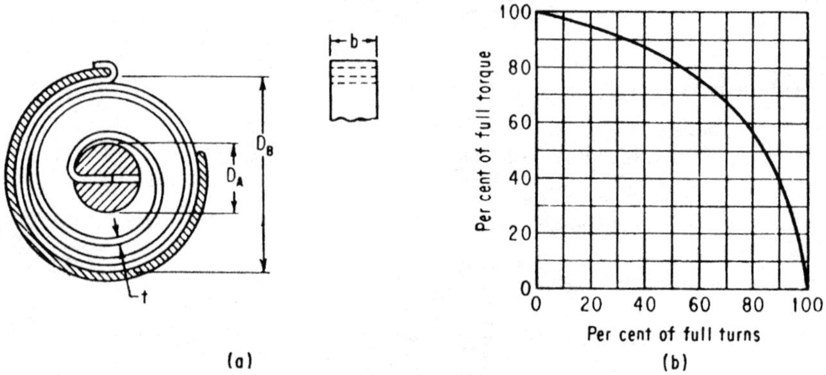

FIG. 22.18 Power spring. (*a*) Orientation. (*b*) Load-deflection curve.

indicates the form of a typical load-deflection curve. To deliver the maximum number of turns, the space occupied by the spring coil should be half the space available in the drum so that

$$Lt = (\pi/8)(D_B^2 - D_A^2) \tag{22.35}$$

in which case the number of initial coils n_A wound tight on the arbor, and the number of rotations T that can be made by the arbor before the spring is unwound solid against the drum,

$$n_A = \frac{-D_A + \sqrt{1/2(D_A^2 + D_B^2)}}{2t} \qquad T = \frac{-D_A - D_B + \sqrt{2(D_A^2 + D_B^2)}}{2.55t} \tag{22.36}$$

The arbor diameter D_A should be from $15t$ to $25t$. The ratio of length to thickness L/t is generally between 3000 and 4000. If an extremely large number of operating cycles are desired, L/t should be less than 1000.

22.13.2 Materials and Recommended Working Stresses

Equation (22.34) is used for calculation of maximum stress when the spring is tightly wound on its arbor so that M is a maximum. Power springs are usually made from cold-rolled tempered spring steel AISI 1074 or cold-rolled tempered spring steel AISI 1095. Recommended maximum values of working stress for these materials depend on desired life. They may be as high as 80 percent of tensile strength for conventional power springs and as high as 100 percent of tensile strength for prestressed power springs.

22.14 CONSTANT-FORCE SPRINGS

Constant-force springs are made from flat stock that has been given a constant natural curvature R by prestressing. A force is required to increase R_n during extension. The principle of the constant-force spring is applied to extension and motor types.

22.14.1 Extension Type (Fig. 22.19*a, b*)

In its relaxed condition the spring will form a tightly wound spiral. An extension spring is mounted on a spool of radius $R_2 > R_n$. The force P is virtually constant; it will vary somewhat as the coil unwinds and the radius R decreases:

$$P = (Ebt^3/26.4)(2/R_1R_n - 1/R_1) \qquad R_1 = R_2 + nt \tag{22.37}$$

where n is the number of coils still wound on the roller. R_2 should be about $1.2R_n$. The length of prestressed extension springs should be

$$L = \delta + 10R_2 \tag{22.38}$$

A stress factor K relates spring material, proportions, load, and number of operations[1]

$$P = Kbt$$

Values of K for materials commonly used for constant-force springs are given in Fig. 22.19*c*. Recommended practice is that $50 < b/t < 200$, $b/t = 100$ being a good median value.

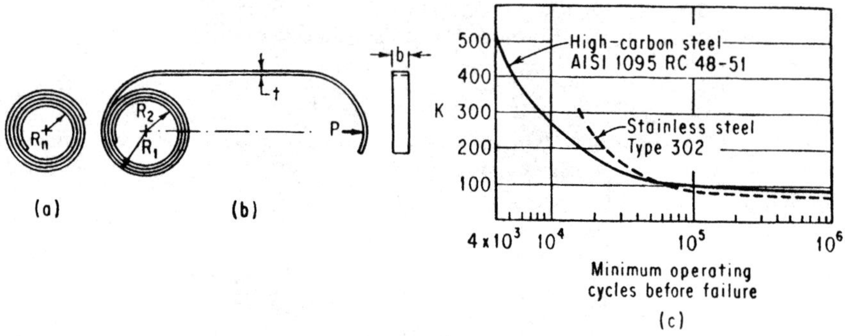

FIG. 22.19 Constant-force spring. (*a*) Unmounted coil. (*b*) Extension type. (*c*) Stress factors.[1]

Constant-force extension springs are used where large deflections and a zero rate are required as in counterbalancing devices.

22.14.2 Motor Type (Fig. 22.20)

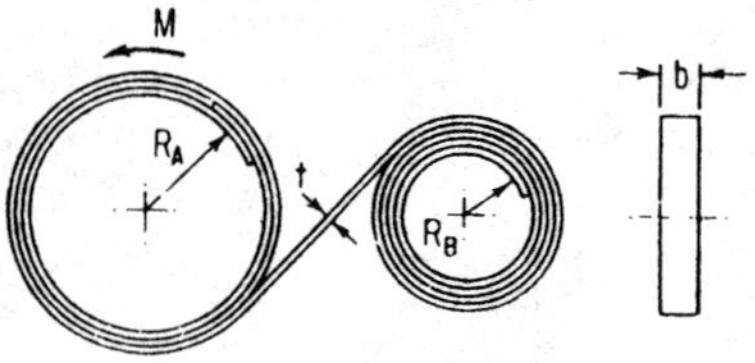

FIG. 22.20 Motor-type constant-force spring.

A constant-force spring reverse-wound on the larger of two spools will wind up on the smaller spool, driving the larger spool with a torque approximated by the equation

$$M = ER_A bt^3/24(1/R_A + 1/R_B)^2 \quad (22.39)$$

Spring load and stress are related by the equation

$$\sigma = 2Et(1/R_n + 1/R_A) \quad (22.40)$$

Recommended practice is that $50 < b/t < 200$, $b/t = 100$ being a good median value.

22.15 BELLEVILLE WASHERS (Fig. 22.21*a*)

22.15.1 Principal Characteristics

The load-deflection curve is of the type shown in Fig. 22.1*c*. Belleville washers give relatively high loads for small deflections. By a suitable choice of spring proportions a wide variation of load-deflection curves can be obtained as shown in Fig. 22.21*b*. Rate is almost constant when $h/t = 0.4$ and close to zero for a large part of the deflection range when $h/t = 1.4$. When $h/t > 2.8$, a Belleville washer becomes unstable and will snap over. The general equations for Belleville washers are[9]

$$P = \frac{4E\delta}{(1 - \mu^2)K_1 D_0^2}\left[(h - \delta)\left(h - \frac{\delta}{2}\right)t + t^3\right] \quad (22.41)$$

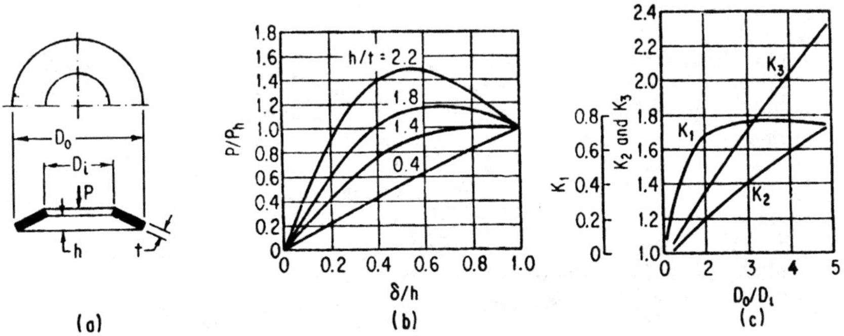

FIG. 22.21 Belleville washer. (*a*) Orientation. (*b*) Load-deflection curve P/P_h is the ratio of any P to the load P_h when deflected to the flat condition. (*c*) Factors K_1, K_2, and K_3 for spring equation.

$$\sigma = \frac{4E\delta}{(1-\mu^2)K_1D_0^2}\left[K_2\left(h-\frac{\delta}{2}\right)+K_3t\right] \tag{22.42}$$

The factors K_1, K_2, and K_3 are all functions of D_0/D_i as shown in Fig. 22.21*c*.

Simpler equations are possible if washers are designed on the basis of selected proportions. The ratio h/t as shown in Fig. 22.21*b* determines load-deflection characteristics. The ratio of diameters D_0/D_i is usually restricted by assembly requirements; one of the ratios 1.2, 1.8, 2.4, or 3.0 will ordinarily be found suitable. If the most efficient use of spring material is a factor, then D_0/D_i should be 1.8. For washers designed to deflect to the flat positions so that σ is a solid stress

$$P = \frac{\sigma^2(1-\mu^2)K_4D_0^2}{4E} \qquad t = \left[\frac{\sigma(1-\mu^2)}{4E}\right]^{1/2} K_5D_0 \tag{22.43}$$

For steel springs $\mu = 0.3$ and $E = 30 \times 10^6$ lb/in^2; hence

$$P = 75.8 \times 10^{-10}K_4\sigma^2D_0^2 \qquad t = 87.0 \times 10^{-6}K_5\sigma^{1/2}D_0 \tag{22.44}$$

Factors K_4 and K_5 are given in Table 22.17.

TABLE 22.17 Factors K_4 and K_5 for Simplified Belleville-Washer Equations

D_o/D_i	K	$h/t = 0.4$	$h/t = 1.4$	$h/t = 1.8$	$h/t = 2.2$
1.2	K_4	0.475	0.068	0.043	0.029
	K_5	0.770	0.346	0.290	0.248
1.8	K_4	0.681	0.103	0.064	0.043
	K_5	1.024	0.465	0.388	0.336
2.4	K_4	0.598	0.089	0.056	0.038
	K_5	1.039	0.467	0.391	0.338
3.0	K_4	0.480	0.075	0.048	0.033
	K_5	0.990	0.452	0.381	0.328

22.15.2 Materials and Recommended Working Stresses

The value of stress calculated by Eqs. (22.42) and (22.43) is a compressive stress at the upper edge of the inside diameter. This stress usually determines the load-carrying ability of the washer and can be as high as 120 percent of tensile strength when set is not removed. In cyclic service cracks may start at the upper edge of the inside diameter but will usually not progress. Actual fatigue failures start as cracks at the outer diameter in a zone stressed in tension. To determine the stress at these locations see Ref. 5. Stresses in static applications for carbon or alloy steel can be as high as 120 percent of the tensile strength when set is not removed and 275 percent when set is removed. For nonferrous metals and austenitic stainless steel, stresses can be as high as 95 percent of the tensile strength when set is not removed and 160 percent when set is removed.

22.15.3 Application

Belleville washers are used where a high load with a relatively small deflection is required or where their unique load-deflection characteristics can be used to advantage. Belleville washers are used for preloading bearings in spindles, as pressure disks for power brakes, and in buffer assemblies for absorbing impact loads.

22.16 WAVE-WASHER SPRINGS (Fig. 22.22*a*)

22.16.1 Principal Characteristics

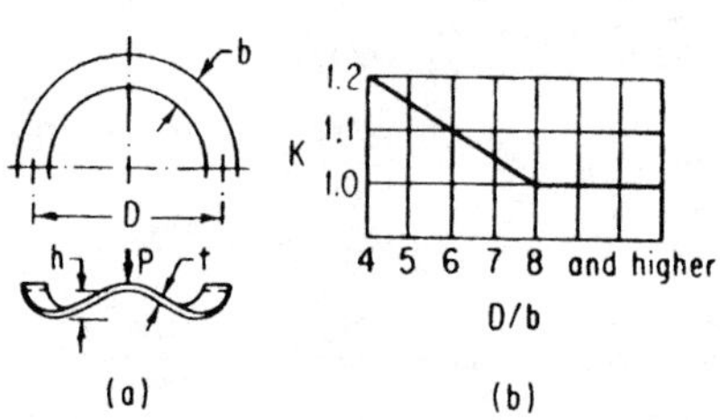

FIG. 22.22 Wave-washer spring. (*a*) Orientation. (*b*) Factor *K* for wave-washer spring equation.

The load-deflection curve is of the type shown in Fig. 22.1*a* and

$$P = 13K\delta Ebt^3n^4/\pi^3D^3$$
$$\sigma = 3\pi PD/4bt^2n^2 \qquad (22.45)$$

The factor K is plotted in Fig. 22.22*b*. The number of waves n can be three or more. Usual practice is to have three, four, or six waves.

Allowance must be made in the assembly of wave washers for the slight decrease in diameters from the loaded to the unloaded condition.

22.16.2 Limiting-Stress Values for Static and Low-Cycle Duty

Wave washers are similar to leaf springs; hence the recommendations in Sec. 22.10.4 apply. Maximum stress should be based on deflection to the solid height, in which case

$$\sigma = 39KhEtn^2/4\pi^2D^2 \qquad (22.46)$$

22.16.3 Application

Wave-washer springs are used because of their compact form where a static load or small deflection range is required. A common use for wave washers is to assemble them between parts mounted on a shaft in order to compensate for variations in assembly clearances.

22.17 NONMETALLIC SYSTEMS

22.17.1 Elastomer Springs

An elastomer is rubber or a rubberlike material which can be stretched to at least twice its free length at normal temperatures and will quickly return to its original length when the load is released. The ability to undergo very large deformations makes elastomers ideally suited to applications where energy absorption is important. Because of their high damping properties elastomers are used for resilient mountings in applications requiring vibration isolation. The stress-strain characteristics of elastomers are not linear; hence load-deflection curves are of the type shown in Fig. 22.1*c*.

Elastomers can be applied to a variety of spring designs. Only the simplest types of elastomer springs are considered here. The load-deflection curves for these simple types are nearly linear for limited deflections. For more complete information on elastomer springs, see Refs. 6, 10, and 11.

Compression Cylinders (Fig. 22.23*a*). Principal formulas are

$$P = \pi \delta E D^2 / 4h \qquad \sigma = 4P/\pi D^2 \tag{22.47}$$

The value of E is not constant but depends on the relative area available for bulging as measured by the form factor.[10]

$$K = [(\pi/4)D^2]/\pi D h = D/4h$$

The relationship between E and K is given by Fig. 22.24*a*. Recommended practice is for $\delta < 0.2h$ to avoid creep.

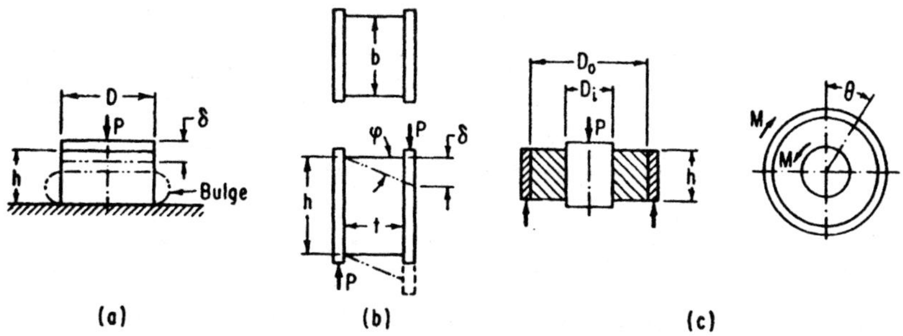

FIG. 22.23 Elastomer springs. (*a*) Compression cylinder. (*b*) Cylindrical bushing—force applied for direct shear loading—applied torque. (*c*) Shear sandwich.

Shear Sandwiches (Fig. 22.23*c*). For $\phi < 20°$ the load-deflection curve is nearly linear. Principal formulas are

$$P = \delta bhG/t \qquad \tau = P/bh \tag{22.48}$$

Recommended practice is that $t < h/4$ or $t < b/4$, whichever is the least value.

Cylindrical Bushings (Fig. 22.23*b*). Cylindrical bushings can be loaded in either direct shear or torsion. For direct-shear loading

$$\delta = P \ln(D_0/D_i)/2\pi hG \qquad \tau = P/\pi hD \tag{22.49}$$

Equation (22.49) is sufficiently accurate for practical purposes if $P/(\pi hGD_0) < 0.4$. Shear stress calculated by Eq. (22.49) refers to the bond between elastomers and the surface at D_i. For torsion loading

$$\theta = (M/\pi hG)(1/D_i^2 - 1/D_0^2) \qquad \tau = 2M/\pi hD_i^2 \tag{22.50}$$

Equation (22.50) is valid for $\theta < 0.7$ rad (40°). Shear stress calculated by Eq. (22.50) is for the bond between the elastomer and the surface at D_i.

Mechanical Properties and Recommended Working Stresses. Both E and G depend on hardeners; values for static loading are shown in Fig. 22.24*b* to have different values for dynamic loading.

For compression cylinders subject to static or low-cycle loading, maximum recommended compressive stress is 700 lb/in^2. Lower values should be used for cyclic duty.

For shear sandwiches and cylindrical bushings loaded in direct shear, recommended maximum design stress based on the bond strength is 35 lb/in^2; for cylindrical bushings loaded in torsion, recommended design stress based on the bond strength is 50 lb/in^2. Durometer hardness of 30 to 60 is usual for elastomer springs subject to shear loading.

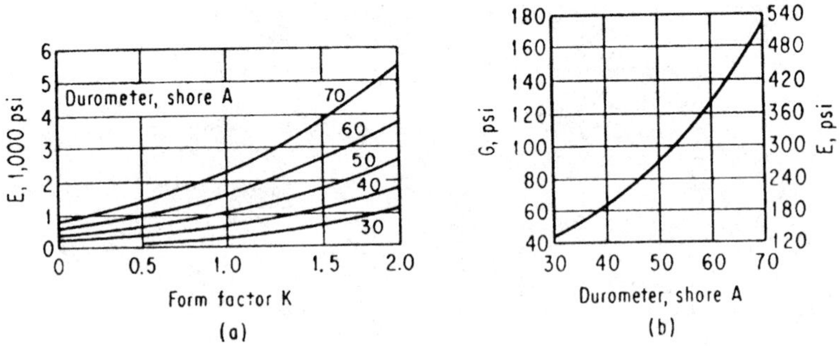

FIG. 22.24 Natural-rubber charts. (*a*) Modulus of elasticity vs. form factor *K*. (*b*) *G* and *E* vs. durometer.

22.17.2 Compressible Liquids

Special spring devices utilize the compressibility of silicone-base liquids, which can be compressed to 90 percent of their volume at pressures maintainable in production units. These compact springs will sustain high loads in a much smaller space than possible with other spring systems, because of their high spring rate. They are costly because of the necessity of excellent pressure seals. They are also sensitive to temperature variations. These springs are applicable where space is at a premium, as in aircraft landing gear, presses, and dies. Operating frequency generally should be kept below 1 (Hz).

Compressible-liquid springs may also be used as shock absorbers with excellent damping qualities attained by utilizing dashpot action.

22.17.3 Compressible Gases

The air springs in use in many shock-absorbing applications utilize gas compressibility.[12] Many applications take advantage of the increasing rate of such a spring as the deflection continues. The spring rate can be changed at will by intentionally altering the gas pressure, which also affords a method of adjusting the loaded height of the spring system. There are problems associated with the design of a leakproof flexible gas container and with increasing temperatures. Generally spring rates are low, deflections high, and considerable space is needed.

REFERENCES

1. Mechanical Drawing Requirements for Springs, Military Standard, MIL-STD-29A, March 1962.
2. Maier, Karl W.: "Springs That Store Energy Best," *Prod. Eng.,* Nov. 10, 1958, pp. 71–75.
3. Carson, Robert W.: "Flat Spring Materials," *Prod. Eng.,* June 11, 1962, pp. 68–80.
4. Crooks, R. D., and W. R. Johnson: "The Performance of Springs at Temperatures above 900°F," *Trans. SAE,* vol. 69, pp. 325–330, 1961.
5. "Design Handbook," Associated Spring: Barnes Group Inc., Bristol, Conn., 1981.
6. Wahl, A. M.: "Mechanical Springs," 2d ed., McGraw-Hill Book Company, Inc., New York, 1963.
7. Maier, Karl W.: "Dynamic Loading of Compression Springs," part II, *Prod. Eng.,* March 1955, pp. 162–174.
8. Chironis, Nicholas P. (ed).: "Spring Design and Application," McGraw-Hill Book Company, Inc., New York, 1961.
9. Almen, J. O., and A. Laszlo: "The Uniform Section Disc Spring," *Trans. ASME,* vol. 58, no. 4, pp. 305–314, May 1936.
10. Gobel, E. F.: "Berechnung and Gestaltung von Gummifedern," Springer-Verlag OHG, Berlin, 1955.
11. McPherson, A. T., and A. Klemmin (eds.): "Engineering Uses of Rubber," Reinhold Publishing Corporation, New York, 1956.
12. Deist: "Air Springs Cushion the Ride," *Mech. Eng.,* June 1958, p. 61.

APPENDIX A
ANALYTICAL METHODS FOR ENGINEERS

Saul K. Fenster, Ph.D.
President Emeritus
New Jersey Institute of Technology
Newark, N.J.

Herbert H. Gould, Mech.E.
Chief, Vehicle Technology Branch
Transportation Systems Center
U.S. Department of Transportation
Cambridge, Mass.

A.1 MATHEMATICAL NOTATION

$+$ = plus
$-$ = minus
$\times, \cdot$ = multiplied by
$\div$ = divided by
$=$ = equals
$\neq$ = is not equal to
$\cong, \approx$ = approximately equals
$\equiv$ = is identical to
$>$ = is greater than
$<$ = is less than
$\geqq$ = equals or is greater than
$\leqq$ = equals or is less than
$\propto$ = varies directly as
$:$ = is to (proportion), as in $x : y = u : v$ or $x/y = u/v$
x^n = x raised to the nth power
$x^{\frac{1}{n}}$ = the nth root of $x^{1/n}$
$\sqrt[n]{x}$ = the nth root of x

$\sqrt{x}$ = the square root of $x (n = 2)$

$\sqrt[3]{x}$ = the cube root of $x (n = 3)$

$\sqrt{-1}, i, j$ = imaginary unit

$\log, \log_{10}$ = logarithm to the base 10, common logarithm

$\log_e, \ln$ = logarithm to the base e, natural logarithm, napierian logarithm, hyperbolic logarithm

$\log^{-1} x$ = number whose logarithm is x

e = base of natural logarithms
$e = 2.7182818285$

π = pi, $\pi = 3.1415926536$

$n!, n$ = n factorial, $1 \cdot 2 \cdot 3 \cdot 4 \cdot \cdots \cdot (n - 1) \cdot n$

$P(n, r), nPr, P^n r$ = permutations of n different things taken r at a time

$C(n, r), nCr, \binom{n}{r}$ = combinations of n different things taken r at a time

D_r^n = combinations with repetition of n things taken r at a time

$\angle$ = angle

$\llcorner$ = right angle

$\triangle$ = triangle

$\odot$ = circle

$\square$ = parallelogram

$\perp$ = perpendicular to

$\parallel$ = parallel to

rad = radian, 57.29578 degrees, 2π rad = 360 degrees

$^\circ$, deg = degree

$'$, min, ft = minute, feet

$''$, s, in = second, inches

$u = f(x, y, z)$ = the variable u is a function of x, y, and z

dy = differential of y

$dy/dx, y'$ = derivative of y taken with respect to x; $y = f(x)$

$d^2y/dx^2, y''$ = second derivative of y taken with respect to x

$d^ny/dx^n, y^{(n)}$ = nth derivative of y taken with respect to x

$\partial y/\partial x, f_x$ = partial derivative of y taken with respect to x; $y = f(x, z, \ldots)$

Δ = delta, increment of

δ = delta, variation of

D = D operator, $Dy = dy/dx$

$\rightarrow$ = approaches as a limit

lim = limit

Σ = summation

$\sum_{i=a}^{b}$ = summation over i from a to b

$\prod_{b}^{a} A_i$ = product taken in order, $A_b \cdot A_{b-1} \cdot A_{b-2} \cdots A_u$.

$\int$ = integral of

$\int_a^b$ = definite integral between limits a and b

$\iint$ = double integral

$\mathbf{a}$ = indicates vector quantity
$\mathbf{a} \cdot \mathbf{b}$ = scalar product
$\mathbf{a} \times \mathbf{b}$ = vector product
( ) = parentheses
[ ] = brackets
{ } = braces
∞ = infinity
$|a|$ = absolute value of a
$f(s)$, $\mathcal{L}\{F(t)\}$ = Laplace transformation of $F(t)$
$fs(n)$, $S\{F(x)\}$ = finite sine transform of $F(x)$
$fc(n)$, $C\{F(x)\}$ = finite cosine transform of $F(x)$

A.2 MATHEMATICAL TABLES

TABLE A.1 Binomial Coefficients

n	$\binom{n}{0}$	$\binom{n}{1}$	$\binom{n}{2}$	$\binom{n}{3}$	$\binom{n}{4}$	$\binom{n}{5}$	$\binom{n}{6}$	$\binom{n}{7}$	$\binom{n}{8}$	$\binom{n}{9}$	$\binom{n}{10}$
0	1										
1	1	1									
2	1	2	1								
3	1	3	3	1							
4	1	4	6	4	1						
5	1	5	10	10	5	1					
6	1	6	15	20	15	6	1				
7	1	7	21	35	35	21	7	1			
8	1	8	28	56	70	56	28	8	1		
9	1	9	36	84	126	126	84	36	9	1	
10	1	10	45	120	210	252	210	120	45	10	1
11	1	11	55	165	330	462	462	330	165	55	11
12	1	12	66	220	495	792	924	792	495	220	66
13	1	13	78	286	715	1287	1716	1716	1287	715	286
14	1	14	91	364	1001	2002	3003	3432	3003	2002	1001
15	1	15	105	455	1365	3003	5005	6435	6435	5005	3003
16	1	16	120	560	1820	4368	8008	11440	12870	11440	8008
17	1	17	136	680	2380	6188	12376	19448	24310	24310	19448
18	1	18	153	816	3060	8568	18564	31824	43758	48620	43758
19	1	19	171	969	3876	11628	27132	50388	75582	92378	92378
20	1	20	190	1140	4845	15504	38760	77520	125970	167960	184756

TABLE A.2 Complete Elliptic Integrals, K and E, for Different Values of the Modulus k*

$\sin^{-1}k$	K	E	$\sin^{-1}k$	K	E	$\sin^{-1}k$	K	E
0°	1.5708	1.5708	50°	1.9356	1.3055	81°.0	3.2553	1.0338
1	1.5709	1.5707	51	1.9539	1.2963	81.2	3.2771	1.0326
2	1.5713	1.5703	52	1.9729	1.2870	81.4	3.2995	1.0314
3	1.5719	1.5697	53	1.9927	1.2776	81.6	3.3223	1.0302
4	1.5727	1.5689	54	2.0133	1.2681	81.8	3.3458	1.0290
5	1.5738	1.5678	55	2.0347	1.2587	82.0	3.3699	1.0278
6	1.5751	1.5665	56	2.0571	1.2492	82.2	3.3946	1.0267
7	1.5767	1.5649	57	2.0804	1.2397	82.4	3.4199	1.0256
8	1.5785	1.5632	58	2.1047	1.2301	82.6	3.4460	1.0245
9	1.5805	1.5611	59	2.1300	1.2206	82.8	3.4728	1.0234
10	1.5828	1.5589	60	2.1565	1.2111	83.0	3.5004	1.0223
11	1.5854	1.5564	61	2.1842	1.2015	83.2	3.5288	1.0213
12	1.5882	1.5537	62	2.2132	1.1920	83.4	3.5581	1.0202
13	1.5913	1.5507	63	2.2435	1.1826	83.6	3.5884	1.0192
14	1.5946	1.5476	64	2.2754	1.1732	83.8	3.6196	1.0182
15	1.5981	1.5442	65	2.3088	1.1638	84.0	3.6519	1.0172
16	1.6020	1.5405	65.5	2.3261	1.1592	84.2	3.6852	1.0163
17	1.6061	1.5367	66.0	2.3439	1.1545	84.4	3.7198	1.0153
18	1.6105	1.5326	66.5	2.3622	1.1499	84.6	3.7557	1.0144
19	1.6151	1.5283	67.0	2.3809	1.1453	84.8	3.7930	1.0135
20	1.6200	1.5238	67.5	2.4001	1.1408	85.0	3.8317	1.0127
21	1.6252	1.5191	68.0	2.4198	1.1362	85.2	3.8721	1.0118
22	1.6307	1.5141	68.5	2.4401	1.1317	85.4	3.9142	1.0110
23	1.6365	1.5090	69.0	2.4610	1.1272	85.6	3.9583	1.0102
24	1.6426	1.5037	69.5	2.4825	1.1228	85.8	4.0044	1.0094
25	1.6490	1.4981	70.0	2.5046	1.1184	86.0	4.0528	1.0086
26	1.6557	1.4924	70.5	2.5273	1.1140	86.2	4.1037	1.0079
27	1.6627	1.4864	71.0	2.5507	1.1096	86.4	4.1574	1.0072
28	1.6701	1.4803	71.5	2.5749	1.1053	86.6	4.2142	1.0065
29	1.6777	1.4740	72.0	2.5998	1.1011	86.8	4.2744	1.0059
30	1.6858	1.4675	72.5	2.6256	1.0968	87.0	4.3387	1.0053
31	1.6941	1.4608	73.0	2.6521	1.0927	87.2	4.4073	1.0047
32	1.7028	1.4539	73.5	2.6796	1.0885	87.4	4.4811	1.0041
33	1.7119	1.4469	74.0	2.7081	1.0844	87.6	4.5609	1.0036
34	1.7214	1.4397	74.5	2.7375	1.0804	87.8	4.6477	1.0031
35	1.7312	1.4323	75.0	2.7681	1.0764	88.0	4.7427	1.0026
36	1.7415	1.4248	75.5	2.7998	1.0725	88.2	4.8478	1.0021
37	1.7522	1.4171	76.0	2.8327	1.0686	88.4	4.9654	1.0017
38	1.7633	1.4092	76.5	2.8669	1.0648	88.6	5.0988	1.0014
39	1.7748	1.4013	77.0	2.9026	1.0611	88.8	5.2527	1.0010
40	1.7868	1.3931	77.5	2.9397	1.0574	89.0	5.4349	1.0008
41	1.7992	1.3849	78.0	2.9786	1.0538	89.1	5.5402	1.0006
42	1.8122	1.3765	78.5	3.0192	1.0502	89.2	5.6579	1.0005
43	1.8256	1.3680	79.0	3.0617	1.0468	89.3	5.7914	1.0004
44	1.8396	1.3594	79.5	3.1064	1.0434	89.4	5.9455	1.0003
45	1.8541	1.3506	80.0	3.1534	1.0401	89.5	6.1278	1.0002
46	1.8691	1.3418	80.2	3.1729	1.0388	89.6	6.3509	1.0001
47	1.8848	1.3329	80.4	3.1928	1.0375	89.7	6.6385	1.0001
48	1.9011	1.3238	80.6	3.2132	1.0363	89.8	7.0440	1.0000
49	1.9180	1.3147	80.8	3.2340	1.0350	89.9	7.7371	1.0000

Note:

$$K = \int_0^{\pi/2} \frac{dx}{\sqrt{1 - k^2 \sin^2 x}} \qquad E = \int_0^{\pi/2} \sqrt{1 - k^2 \sin^2 x}\, dx$$

*From R. S. Burington, "Handbook of Mathematical Tables and Formulas," 5th ed., McGraw-Hill Book Company, Inc., New York, 1972.

TABLE A.3 Error Function or Probability Integral,* $\operatorname{erf} x = \frac{2}{\sqrt{\pi}} \int_0^x e^{-n^2} dn$

x	0	1	2	3	4	5	6	7	8	9
0.0		**.01128**	**.02256**	**.03384**	**.04511**	**.05637**	**.06762**	**.07886**	**.09008**	**.10128**
0.1	.11246	.12362	.13476	.14587	.15695	.16800	.17901	.18999	.20094	.21184
0.2	.22270	.23352	.24430	.25502	.26570	.27633	.28690	.29742	.30788	.31828
0.3	.32863	.33891	.34913	.35928	.36936	.37938	.38933	.39921	.40901	.41874
0.4	.42839	.43797	.44747	.45689	.46623	.47548	.48466	.49375	.50275	.51167
0.5	.52050	.52924	.53790	.54646	.55494	.56332	.57162	.57982	.58792	.59594
0.6	.60386	.61168	.61941	.62705	.63459	.64203	.64938	.65663	.66378	.67084
0.7	.67780	.68467	.69143	.69810	.70468	.71116	.71754	.72382	.73001	.73610
0.8	.74210	.74800	.75381	.75952	.76514	.77067	.77610	.78144	.78669	.79184
0.9	.79691	.80188	.80677	.81156	.81627	.82089	.82542	.82987	.83423	.83851
1.0	**.84270**	**.84681**	**.85084**	**.85478**	**.85865**	**.86244**	**.86614**	**.86977**	**.87333**	**.87680**
1.1	.88021	.88353	.88679	.88997	.89308	.89612	.89910	.90200	.90484	.90761
1.2	.91031	.91296	.91553	.91805	.92051	.92290	.92524	.92751	.92973	.93190
1.3	.93401	.93606	.93807	.94002	.94191	.94376	.94556	.94731	.94902	.95067
1.4	.95229	.95385	.95538	.95686	.95830	.95970	.96105	.96237	.96365	.96490
1.5	.96611	.96728	.96841	.96952	.97059	.97162	.97263	.97360	.97455	.97546
1.6	.97635	.97721	.97804	.97884	.97962	.98038	.98110	.98181	.98249	.98315
1.7	.98379	.98441	.98500	.98558	.98613	.98667	.98719	.98769	.98817	.98864
1.8	.98909	.98952	.98994	.99035	.99074	.99111	.99147	.99182	.99216	.99248
1.9	.99279	.99309	.99338	.99366	.99392	.99418	.99443	.99466	.99489	.99511
2.0	**.99532**	**.99552**	**.99572**	**.99591**	**.99609**	**.99626**	**.99642**	**.99658**	**.99673**	**.99688**
2.1	.99702	.99715	.99728	.99741	.99753	.99764	.99775	.99785	.99795	.99805
2.2	.99814	.99822	.99831	.99839	.99846	.99854	.99861	.99867	.99874	.99880
2.3	.99886	.99891	.99897	.99902	.99906	.99911	.99915	.99920	.99924	.99928
2.4	.99931	.99935	.99938	.99941	.99944	.99947	.99950	.99952	.99955	.99957
2.5	.99959	.99961	.99963	.99965	.99967	.99969	.99971	.99972	.99974	.99975
2.6	.99976	.99978	.99979	.99980	.99981	.99982	.99983	.99984	.99985	.99986
2.7	.99987	.99987	.99988	.99989	.99989	.99990	.99991	.99991	.99992	.99992
2.8	.99992	.99993	.99993	.99994	.99994	.99994	.99995	.99995	.99995	.99996
2.9	.99996	.99996	.99996	.99997	.99997	.99997	.99997	.99997	.99997	.99998
3.0	**.99998**									

*From H. B. Dwight, "Mathematical Tables of Elementary and Some Higher Mathematical Functions," 3d ed., Dover Publications, Inc., New York, 1961.

A.3 PHYSICAL TABLES

Force, mass, and acceleration are related by Newton's law, which states that the force acting on an object is proportional to the product of its mass and its acceleration, $F \propto ma$. This relation can be made an equality by the introduction of $1/g_c$, a constant of proportionality, i.e., $F = ma/g_c$. By substituting definitions of certain standard masses and forces into the above equation the value of g_c can be determined.

For example, the standard pound mass (lbm), a quantity of matter defined in terms of the standard kilogram (1 lbm = 0.4536 kg), in a standard gravitational field of 32.1740 ft/s^2 (980.665 cm/s^2) is acted upon by a force of 1 pound (lbf). The value of g_c is determined as follows:

$$F = ma/g_c \qquad 1 \text{ lbf} = 1 \text{ lbm} \times \frac{32.1740 \text{ ft/s}^2}{g_c}$$

Therefore, $g_c = 32.1740 \text{ lbm ft/lbf} \cdot \text{s}^2$

When a pound force acts upon a mass of one slug, the resulting acceleration will be 1 ft/s^2. Thus,

$$1 \text{ lbf} = \frac{1 \text{ slug} \times 1 \text{ ft/s}^2}{g_c}$$

Therefore, $g_c = 1 \text{ slug} \cdot \text{ft/lbf} \cdot \text{s}^2$

When a mass of one pound is accelerated at the rate of 1 ft/s^2, the force acting is one poundal (1 pdl). Thus,

$$1 \text{ pdl} = \frac{1 \text{ lbm} \times 1 \text{ft/s}^2}{g_c}$$

Therefore, $g_c = 1 \text{ lbm} \cdot \text{ft/pdl} \cdot \text{s}^2$

In the metric system, a dyne is defined as the force required to accelerate a mass of 1 g at the rate of 1 cm/s^2. Thus,

$$1 \text{ dyne} = 1 \frac{\text{g} \cdot \text{cm/s}^2}{g_c}$$

and, therefore,

$$g_c = \frac{1 \text{ g} \cdot \text{cm/s}^2}{\text{dyne}}.$$

It is important to note that g_c is a constant that depends upon the specific units employed and not upon the magnitude of the local gravitational acceleration g. Table A.4 summarizes various systems of units. Conversion factors are given in Table A.5.

TABLE A.4 Systems of Units

Quantity	International System (SI) of Units	Engineering English	Technical English	Absolute English	Absolute metric
Mass	Kilogram	Pound mass	Slug	Pound mass	Gram
Force	Newton	Pound force	Pound force	Poundal	Dyne
Length	Meter	Foot	Foot	Foot	Centimeter
Time	Second	Second	Second	Second	Second
Energy	Joule	lbf · ft	lbf · ft	Foot-poundal	Erg
g_c	1 (kg · m/N · s^2)	32.1740 (lbm · ft/lbf · s^2)	1 (slug · ft/lbf · s^2)	1 (lbm · ft/pdl · s^2)	1 (g · cm/dyne · s^2)

TABLE A.5 Conversion Factors

A To convert A into B	B	Multiply A by	C For the converse operation, multiply B by C
Acres	Square feet	4.356×10^{4}	2.296×10^{-5}
Atmospheres	Dynes per cm^2	1.0132×10^{6}	0.98697×10^{-6}
Atmospheres	Feet of water at 4°C	33.90	2.950×10^{-2}
Atmospheres	Inches of mercury at 0°C	29.92	3.342×10^{-2}
Atmospheres	Kilograms per square meter	1.0332×10^{4}	9.678×10^{-5}
Atmospheres	Millimeters of mercury at 0°C	760	1.316×10^{-3}
Atmospheres	Newtons per square meter	1.0133×10^{5}	9.869×10^{-6}
Atmospheres	Pounds per in.^{2}	14.696	6.804×10^{-2}
British thermal units (Btu)	Foot-pounds	778.26	1.2849×10^{-3}
British thermal units (Btu)	Horsepower-hours	3.929×10^{-4}	2,545
British thermal units (Btu)	Joules	1,054.8	9.480×10^{-4}
British thermal units (Btu)	Kilogram-calories	0.2520	3.969
British thermal units (Btu)	Kilowatthours	2.930×10^{-4}	3,413
Btu per minute	Horsepower	0.02358	42.40
Btu per minute	Watts	17.58	0.05688
Btu per square foot per minute	Watts per in.^{2}	0.1221	8.190
Bushels	Cubic feet	1.2445	0.8036
Centimeters	Feet	3.281×10^{-2}	30.48
Centimeters	Inches	0.3937	2.540
Centimeters	Meters	0.010	100.00
Centimeters	Millimeters	10.00	0.10
Centimeters	Mils (10^{-3} in.)	393.7	2.540×10^{-3}
Centimeters per second	Feet per minute	1.969	0.5080
Centimeters per second per second	Feet per second per second	3.281×10^{-2}	30.48
Circular mils	Square centimeters	5.067×10^{-6}	1.973×10^{5}
Circular mils	Square inches	7.854×10^{-7}	1.273×10^{6}
Circular mils	Square mils	0.7854	1.273
Cubic inches	Cubic centimeters	16.39	6.102×10^{-2}
Cubic inches	Cubic feet	5.787×10^{-4}	1,728
Cubic inches	Cubic meters	1.639×10^{-5}	6.1023×10^{4}
Cubic inches	Cubic yards	2.143×10^{-5}	4.6656×10^{4}
Cubic inches	Gallons	4.329×10^{-3}	231
Cubic inches	Liters	1.639×10^{-2}	61.02
Cubic inches	Quarts (liquid)	0.01732	57.75
Degrees (angular)	Minutes	60	1.667×10^{-2}
Degrees (angular)	Radians	1.745×10^{-2}	57.30
Degrees (angular)	Seconds	3600	2.778×10^{-4}
Degrees Fahrenheit	Degrees centigrade	°C = $\frac{5}{9}$(°F − 32)	°F = $\frac{9}{5}$°C + 32
Dynes	Gram cm per sec^2	1	1
Dynes	Grams at standard gravity	1.020×10^{-3}	980.7
Dynes	Pounds	2.248×10^{-6}	4.448×10^{5}

TABLE A.5 Conversion Factors *(Continued)*

A	B	Multiply A by	C
To convert A into B			For the converse operation, multiply B by C
Dynes	Poundals	7.233×10^{-5}	1.3826×10^{4}
Dyne-centimeters	Pound-feet	7.376×10^{-8}	1.356×10^{7}
Ergs	Btu	9.480×10^{-11}	1.054×10^{10}
Ergs	Dyne-centimeters	1	1
Ergs	Foot-pounds	7.376×10^{-8}	1.356×10^{7}
Ergs	Joules	10^{-7}	10^{7}
Feet	Inches	12	8.333×10^{-2}
Feet	Miles	1.894×10^{-4}	5,280
Feet	Yards	0.333	3
Foot-pounds	Horsepower-hours	5.050×10^{-7}	1.98×10^{6}
Foot-pounds	Kilowatthours	3.766×10^{-7}	2.655×10^{6}
Grams	Kilograms	10^{-3}	10^{3}
Grams	Milligrams	10^{3}	10^{-3}
Grams	Ounces (avoir.)	3.527×10^{-2}	28.35
Grams	Ounces (Troy)	3.215×10^{-2}	31.10
Grams per cm^3	Pounds per ft^3	62.43	1.602×10^{-2}
Grams per cm^3	Pounds per $in.^3$	3.613×10^{-2}	27.68
Gram-cm^2	Pound-ft^2	2.37285×10^{-6}	4.21434×10^{5}
Gram-cm^2	Pound-$in.^2$	3.4169×10^{-4}	2.9266×10^{3}
Gram-cm^2	Slug-ft^2	7.37507×10^{-8}	1.3559×10^{7}
Horsepower	Foot-pounds per minute	33,000	3.030×10^{-5}
Horsepower	Foot-pounds per second	550	1.818×10^{-3}
Horsepower	Horsepower (metric)	1.014	0.9863
Horsepower	Kilowatts	0.7457	1.341
Hours	Day	4.167×10^{-2}	24
Hours	Minutes	60	1.667×10^{-2}
Hours	Seconds	3,600	2.778×10^{-4}
Hours	Weeks	5.952×10^{-3}	168
Inches	Centimeters	2.540	0.3937
Inches	Mils	10^{3}	10^{-3}
Joules (int.)	Foot-pounds	0.7376	1.356
Joules (int.)	Watthours	2.778×10^{-4}	3,600
Kilograms	Pounds	2.205	0.4536
Kilograms	Tons (short, 2,000 lb avoir.)	1.102×10^{-3}	907.2
Kilograms	Tons (long, 2,240 lb avoir.)	9.482×10^{-4}	1,016
Kilogram-calories	Foot-pounds	3,088	3.238×10^{-4}
Kilogram-calories	Horsepower-hours	1.560×10^{-3}	641.1
Kilogram-calories	Joules	4,186	2.389×10^{-4}
Kilogram-calories	Kilowatthours	860	1.163×10^{-3}
Kilometers	Feet	3,281	3.048×10^{-4}
Kilometers	Meters	10^{3}	10^{-3}
Kilowatts	Watts	10^{3}	10^{-3}
Knots (speed)	Feet per second	1.688	0.5925
Knots (speed)	Kilometers per hour	1.853	0.5396
Knots (speed)	Miles per hour	1.1508	0.869
Liters	Cubic centimeters	10^{3}	10^{-3}
Liters	Cubic inches	61.02	1.639×10^{-2}
Liters	Gallons (U.S. liquid)	0.2642	3.785
Meters	Yards	1.094	0.9144

TABLE A.5 Conversion Factors *(Continued)*

A	B	Multiply A by	C
To convert A into B			For the converse operation, multiply B by C
Meters per second	Feet per minute	196.8	5.080×10^{-3}
Miles (nautical)	Feet	6,076.1	1.646×10^{-4}
Miles (statute)	Feet	5,280	1.894×10^{-4}
Miles per hour	Feet per minute	88	1.136×10^{-2}
Millimeters	Meters	10^{-3}	10^{3}
Minutes (angular)	Radians	2.909×10^{-4}	3,438
Minutes (angular)	Seconds	60	1.667×10^{-2}
Newtons	Dynes	10^{5}	10^{-5}
Newton meters	Dyne-centimeters	10^{7}	10^{-7}
Newton meters	Pound-feet	0.7376	1.356
Ounces (avoir.)	Grains	437.5	2.285×10^{-3}
Ounces (avoir.)	Grams	28.35	0.03527
Ounces (avoir.)	Pounds	0.0625	16
Ounces (fluid)	Cubic inches	1.805	0.5540
Ounces (fluid)	Liters	0.02957	33.81
Ounces (fluid)	Quarts	3.125×10^{-2}	32
Ounces (Troy)	Grams	31.10	0.03215
Ounces (Troy)	Pounds (Troy)	0.08333	12
Pints (liquid)	Gallons	0.125	8
Pints (liquid)	Liters	0.4732	2.113
Pints (liquid)	Quarts	0.5000	2
Poundals	Dynes	1.3826×10^{4}	7.233×10^{-5}
Poundals	Pounds	3.108×10^{-2}	32.17
Pounds (avoir.)	Grams	453.6	2.205×10^{-3}
Pounds (avoir.)	Ounces (avoir.)	16	0.0625
Pounds per ft²	Feet of water at 4°C	1.602×10^{-2}	62.43
Pounds per ft²	Inches of mercury	1.414×10^{-2}	70.73
Pounds per ft²	Pounds per in.²	6.944×10^{-3}	144
Radians per second	Revolutions per minute	9.549	0.1047
Radians per second	Revolutions per second	0.1592	6.283
Revolutions per minute per minute	Radians per second per second	1.745×10^{-3}	573.0
Slugs	Pounds (avoir.)	32.174	3.108×10^{-2}
Square centimeters	Square feet	1.076×10^{-3}	929.0
Square centimeters	Square inches	0.1550	6.452
Square centimeters	Square meters	10^{-4}	10^{4}
Square centimeters	Square millimeters	10^{2}	10^{-2}
Square feet	Square inches	144	6.944×10^{-3}
Square feet	Square meters	0.09290	10.76
Square inches	Square mils	10^{6}	10^{-6}
Square millimeters	Circular mils	1.973×10^{3}	5.067×10^{-4}
Watts	Ergs per second	10^{7}	10^{-7}
Watts	Foot-pounds per minute	44.26	2.260×10^{-2}
Watts	Horsepower	1.341×10^{-3}	745.7

A.4 MATHEMATICS

A.4.1 Algebra

Basic

Fundamental Laws

Commutative law: $a + b = b + a \qquad ab = ba$

Associative law: $a + (b + c) = (a + b) + c \qquad a(bc) = (ab)c$

Distributive law: $a(b + c) = ab + ac$

Sums of Numbers. The sum of the first n numbers:

$$\sum_{1}^{n}(n) = \frac{n(n+1)}{2}$$

The sum of the squares of the first n numbers:

$$\sum_{1}^{n}(n^2) = \frac{n(n+1)(2n+1)}{6}$$

The sum of the cubes of the first n numbers:

$$\sum_{1}^{n}(n^3) = \frac{n^2(n+1)^2}{4}$$

Progressions

Arithmetic Progression

$$a, a + d, a + 2d, a + 3d, \ldots$$

where a = first term
d = common difference
n = number of terms
S = sum of n terms
l = last term
$l = a + (n - 1)d$
$S = (n/2)(a + l)$

Arithmetic mean of a and $b = (a + b)/2$

Geometric Progression

$$a, ar, ar^2, ar^3, \ldots$$

where a = first term
r = common ratio
n = number of terms
S = sum of n terms
l = last term
$l = ar^{n-1}$
$S = a\dfrac{r^n - 1}{r - 1} = \dfrac{rl - a}{r - 1}$; for $r^2 < 1$ and $n = \infty$, $S = \dfrac{a}{1 - r}$

Geometric mean of a and $b = \sqrt{ab}$

Powers and Roots

$$a^x a^y = a^{x+y}$$
$$a^x/a^y = a^{x-y}$$
$$(ab)^x = a^x b^x$$
$$(a^x)^y = a^{xy}$$
$$a^0 = 1 \text{ if } a \neq 0$$
$$a^{-x} = 1/a^x$$
$$a^{x/y} = \sqrt[y]{a^x}$$
$$a^{1/y} = \sqrt[y]{a}$$
$$\sqrt[x]{ab} = \sqrt[x]{a}\sqrt[x]{b}$$
$$\sqrt[x]{a/b} = \sqrt[x]{a}/\sqrt[x]{b}$$

Binomial Theorem

$$(a \pm b)^n = a^n \pm na^{n-1}b + \frac{n(n-1)}{2!}a^{n-2}b^2 \pm \frac{n(n-1)(n-2)}{3!}a^{n-3}b^3 + \cdots$$
$$+ (\pm 1)^m \frac{n(n-1)\cdots(n-m+1)}{m!}a^{n-m}b^m + \cdots$$

where $m! = \underline{|m} = 1 \cdot 2 \cdot 3 \cdots (m-1)m$

The series is finite if n is a positive integer. If n is negative or fractional, the series is infinite and will converge for $|b| < |a|$ only.

Absolute Values. The numerical or absolute value of a number n is denoted by $|n|$ and represents the magnitude of the number without regard to algebraic sign. For example, $|-3| = |+3| = 3$

Logarithms. Definition of a logarithm: If $N = b^x$, the exponent x is the logarithm of N to the base b and is written $x = \log_b N$. The number b must be positive, finite, and different from unity. The base of common, or briggsian, logarithms is 10. The base of natural, napierian, or hyperbolic logarithms is 2.7182818 . . ., denoted by e.

Laws of Logarithms

$$\log_b MN = \log_b M + \log_b N \qquad \log_b 1 = 0$$
$$\log_b M/N = \log_b M - \log_b N \qquad \log_b b = 1$$
$$\log_b N^m = m \log_b N \qquad \log_b 0 = +\infty,\ 0 < b < 1$$
$$\log_b \sqrt[r]{N^m} = m/r \log_b N \qquad \log_b 0 = -\infty,\ 1 < b < \infty$$
$$\log_b N = \log_a N/\log_a b$$

Important Constants

$$\log_{10} e = 0.4342944819$$
$$\log_{10} x = 0.4343 \log_e x = 0.4343 \ln x$$
$$\ln 10 = \log_e 10 = 2.3025850930$$
$$\ln x = \log_e x = 2.3026 \log_{10} x$$

Synthetic Division. "Synthetic division" is a process by which a polynomial $f(x)$ is divided by a binomial $x - a$. (It is a simplified form of long division.)

The following procedure can be followed:

1. Arrange the coefficients $c_0, c_1, c_2, \ldots, c_n$, including the zeros, of the polynomial $f(x) = c_0x^n + c_x^{n-1} + \cdots + c_n$ on the first line.
2. Write a at the right and c_0 in the first place on the third line.
3. Multiply c_0 by a and place the product (c_0a) on the second line under c_1.
4. Write the sum of c_1 and (c_0a) in the second place on the third line.
5. Multiply this sum by a and add to c_2.
6. Write this sum in the third place on the third line, and so forth.
7. The *remainder* is the last number in the third line. The other numbers on the third line are the coefficients of the powers of x, in descending order, in the quotient.
8. The binomial which is the divisor must be of the form $x - a$. If the binomial has the form $kx - b = k(x - b/k)$, then divide synthetically by $x - b/k$ and then divide the resulting quotient by k.

EXAMPLE Divide $3x^4 - 4x^2 + x$ by $x + 2$.

$$\begin{array}{r} 3 + 0 - 4 + 1 + 0 \,\lfloor\!\underline{-2} \\ - 6 + 12 - 16 + 30 \\ \hline 3 - 6 + 8 - 15 \quad 30 \end{array}$$

$$\text{Quotient} = 3x^3 - 6x^2 + 8x - 15$$

$$\text{Remainder} = 30$$

Partial Fractions. A "rational fraction" is the ratio of two polynomials. If the numerator of the algebraic fraction is of lower degree than the denominator, it is termed a "proper fraction" and can be resolved into "partial fractions." If the numerator is of higher degree than the denominator, division of numerator by denominator will result in the sum of a proper fraction and a polynomial, termed a "mixed fraction."

A number of commonly found proper fractions are given below together with the form of the partial fractions into which they can be resolved.

1. If the denominator is factorable into real linear factors $F_1, F_2, F_3, \ldots$ $(F = ax + b)$, each different, then corresponding to each factor there exists a partial fraction $A/(ax + b)$. Thus

$$\frac{\text{Numerator}}{F_1 \cdot F_2 \cdot F_3 \cdots} = \frac{\text{numerator}}{(a_1x + b_1)(a_2x + b_2)(a_3x + b_3) \cdots}$$
$$= \frac{A}{a_1x + b_1} + \frac{B}{a_2x + b_2} + \frac{C}{a_3x + b_3} + \cdots$$

where $A, B, C, \ldots$ are constants which can be obtained by clearing fractions and equating coefficients of like powers of x.

2. If the denominator is factorable into real linear factors, one or more of which is repeated, then corresponding to each repeated factor there exist the partial fractions

$$\frac{A}{a_1x + b_1} + \frac{B}{(a_1x + b_1)^2} + \cdots + \frac{C}{(a_1x + b_1)^n}$$

where n equals the number of times the factor appears in the denominator.

3. If the denominator is factorable into quadratic factors $ax^2 + bx + c$, none of which is repeated or factorable into real linear factors, then corresponding to each quadratic factor there exists a partial fraction

$$\frac{Ax + B}{ax^2 + bx + c}$$

4. If the denominator is factorable into quadratic factors, one or more of which is repeated and none of which is factorable into real linear factors, then corresponding to each repeated factor there exist the partial fractions

$$\frac{Ax + B}{a_1x^2 + b_1x + c} + \frac{Cx + D}{(a_1x^2 + b_1x + c)^2} + \cdots + \frac{Ex + F}{(a_1x^2 + b_1x + c_1)^n}$$

where n equals the number of times the factor appears in the denominator.

Quadratic Equations. The general form of a quadratic equation is

$$f(x) = ax^2 + bx + c = 0$$

This equation has two roots, x_1 and x_2:

$$x_1, x_2 = (-b \pm \sqrt{b^2 - 4ac}\,)/2a$$

where $b^2 - 4ac$ is called the "discriminant."
If $b^2 - 4ac > 0$, the roots are real and unequal.
If $b^2 - 4ac = 0$, the roots are real and equal.
If $b^2 - 4ac < 0$, the roots are imaginary.

Cubic Equations. The general form of a cubic equation is

$$f(x) = a_0x^3 + a_1x^2 + a_2x + a_3 = 0$$

It may also be written in the form $f(x) = ax^3 + 3bx^2 + 3cx + d = 0$. Let

$$\alpha = ac - b^2$$
$$\beta = \tfrac{1}{2}(3abc - a^2d) - b^3$$
$$\gamma_1 = (\beta + \sqrt{\alpha^3 + \beta^2})^{1/3}$$
$$\gamma_2 = (\beta - \sqrt{\alpha^3 + \beta^2})^{1/3}$$

The roots are

$$x_1 = (\gamma_1 + \gamma_2 - b)/a$$
$$x_2 = (1/a)\left[-\tfrac{1}{2}(\gamma_1 + \gamma_2) + (\sqrt{-3}/2)(\gamma_1 - \gamma_2) - b\right]$$
$$x_3 = (1/a)\left[-\tfrac{1}{2}(\gamma_1 + \gamma_2) - (\sqrt{-3}/2)(\gamma_1 - \gamma_2) - b\right]$$

The above solution for the roots x_1, x_2, x_3 is termed the "algebraic solution." If, however, $\alpha^3 + \beta^2 < 0$, the above formulas require the finding of the cube roots of complex numbers. In this case, and if desired, for any case, the "trigonometric solution" given below may be employed.

Write the roots as

$$x_1 = (u_1 - b)/a \qquad x_2 = (u_2 - b)/a \qquad x_3 = (u_3 - b)/a$$

and substitute the appropriate value for u_1, u_2, u_3 from Table A.6. If $\alpha^3 + \beta^2 = 0$, the three roots will be real and at least two will be equal. If $\alpha^3 + \beta^2 > 0$, one root is real and two are complex. If $\alpha^3 + \beta^2 < 0$, the three roots are real.

Combinations and Permutations

Permutation. A "permutation" is an arrangement with regard to order, of all or a part of a set of things. The number of permutations of n different things taken r at a time is

$$P(n, r) = {}_nP_r = P^n{}_r = n!/(n - r)!$$

Combination. A "combination" is an arrangement without regard to order of all or a part of a set of things. The number of combinations of n different things taken r at a time is

$$C(n, r) = {}_nC_r = \binom{n}{r} = \frac{P(n, r)}{r!} = \frac{n!}{r!(n - r)!}$$

The number of combinations, with repetition of n things taken r at a time, is

$$D^n{}_r = \frac{(n + r - 1)!}{r!(n - 1)!}$$

Matrices and Determinants

Definitions. A rectangular array of mn elements arranged in m rows and n columns is termed a "matrix" of order $m \times n$. A matrix may be represented in any of the following ways:

$$A = [a_{ij}] = \begin{bmatrix} a_{11} & a_{12} & \cdots & a_{1n} \\ a_{21} & a_{22} & \cdots & a_{2n} \\ \cdots & \cdots & \cdots & \cdots \\ a_{m1} & a_{m2} & \cdots & a_{mn} \end{bmatrix} = \begin{pmatrix} a_{11} & a_{12} & \cdots & a_{1n} \\ a_{21} & a_{22} & \cdots & a_{2n} \\ \cdots & \cdots & \cdots & \cdots \\ a_{m1} & a_{m2} & \cdots & a_{mn} \end{pmatrix}$$

$$= \left\| \begin{matrix} a_{11} & a_{12} & \cdots & a_{1n} \\ a_{21} & a_{22} & \cdots & a_{2n} \\ \cdots & \cdots & \cdots & \cdots \\ a_{m1} & a_{m2} & \cdots & a_{mn} \end{matrix} \right\|$$

where the first subscript of each element designates the row and the second subscript the column.

A "column matrix" consists of elements arranged in a single column, as

$$\begin{bmatrix} a_{11} \\ a_{21} \\ \cdot \\ \cdot \\ \cdot \\ a_{m1} \end{bmatrix}$$

A "row matrix" consists of elements arranged in a single row, as

$$[a_{11} \quad a_{12} \quad \cdots \quad a_{1n}]$$

TABLE A.6 For Trigonometric Solution of Cubic Equations

	For negative α		For positive α
	$\alpha^3 + \beta^2 \leq 0$	$\alpha^3 + \beta^2 \geq 0$	
u_1	$\pm 2\sqrt{-\alpha}\cos\left(\frac{1}{3}\cos^{-1}\frac{\pm\beta}{\sqrt{-\alpha^3}}\right)$	$\pm 2\sqrt{-\alpha}\cosh\left(\frac{1}{3}\cosh^{-1}\frac{\pm\beta}{\sqrt{-\alpha^3}}\right)$	$\pm 2\sqrt{\alpha}\sinh\left(\frac{1}{3}\sinh^{-1}\frac{\pm\beta}{\sqrt{\alpha^3}}\right)$
u_2	$\pm 2\sqrt{-\alpha}\cos\left(\frac{1}{3}\cos^{-1}\frac{\pm\beta}{\sqrt{-\alpha^3}} + \frac{2\pi}{3}\right)$	$\mp \sqrt{-\alpha}\cosh\left(\frac{1}{3}\cosh^{-1}\frac{\pm\beta}{\sqrt{-\alpha^3}}\right)$ $+ i\sqrt{-3\alpha}\sinh\left(\frac{1}{3}\cosh^{-1}\frac{\pm\beta}{\sqrt{-\alpha^3}}\right)$	$\mp \sqrt{\alpha}\sinh\left(\frac{1}{3}\sinh^{-1}\frac{\pm\beta}{\sqrt{\alpha^3}}\right)$ $+ i\sqrt{3\alpha}\cosh\left(\frac{1}{3}\sinh^{-1}\frac{\pm\beta}{\sqrt{\alpha^3}}\right)$
u_3	$\pm 2\sqrt{-\alpha}\cos\left(\frac{1}{3}\cos^{-1}\frac{\pm\beta}{\sqrt{-\alpha^3}} + \frac{4\pi}{3}\right)$	$\mp \sqrt{-\alpha}\cosh\left(\frac{1}{3}\cosh^{-1}\frac{\pm\beta}{\sqrt{-\alpha^3}}\right)$ $- i\sqrt{-3\alpha}\sinh\left(\frac{1}{3}\cosh^{-1}\frac{\pm\beta}{\sqrt{-\alpha^3}}\right)$	$\mp \sqrt{\alpha}\sinh\left(\frac{1}{3}\sinh^{-1}\frac{\pm\beta}{\sqrt{\alpha^3}}\right)$ $- i\sqrt{3\alpha}\cosh\left(\frac{1}{3}\sinh^{-1}\frac{\pm\beta}{\sqrt{\alpha^3}}\right)$

Note: The upper of the alternative algebraic signs above is used when β is positive and the lower when β is negative.

If $m = n$, the array is a "square matrix" of order m. A "diagonal matrix" is a square matrix in which the elements not on the principal diagonal are zero; i.e.,

$$\begin{bmatrix} a_{11} & 0 & \cdots & 0 \\ 0 & a_{22} & \cdots & 0 \\ \cdots & \cdots & \cdots & \cdots \\ 0 & 0 & \cdots & a_{nn} \end{bmatrix}$$

A "scalar matrix" is a diagonal matrix whose elements are equal. A "unit matrix," denoted by I, is a diagonal matrix whose nonzero elements are unity. A "symmetric matrix" is a square matrix in which $a_{ij} = a_{ji}$. A "skew-symmetric" or "antisymmetric matrix" is one in which $a_{ij} = -a_{ji}$ and $a_{ii} = 0$. If $a_{ii} \neq 0$, the array is a "skew matrix." A "zero" or "null matrix" is a matrix in which all the elements are zero.

Two matrices are equal only if they have the same number of rows and columns, and if corresponding elements are equal. If a matrix is obtained by interchanging the rows and columns of another matrix, the two matrices are "transposes" or "conjugates" of one another and are represented by the symbols A^T, A', or A.

The "determinant" of a square matrix of order n, denoted by the symbols $|A|$, $|a_{ij}|$, or

$$\begin{vmatrix} a_{11} & a_{12} & \cdots & a_{1n} \\ a_{21} & a_{22} & \cdots & a_{2n} \\ \cdots & \cdots & \cdots & \cdots \\ a_{n1} & a_{n2} & \cdots & a_{nn} \end{vmatrix}$$

is the sum of all possible $n!$ products $(-1)^k a_{i_1 1} a_{i_2 2} \cdots a_{i_n n}$ obtained by selecting one and only one element from each row and from each column. The algebraic sign of each product is determined by the integer k, which is the number of times the elements a_{ij} in $(-1)^k a_{i_1 1} a_{i_2 2} \cdots a_{i_n n}$ are interchanged in order to arrange the first subscripts (i_1, i_2, . . ., i_n) in the normal order 1, 2, 3, . . ., n. (The interchanges need not be of adjacent elements.) If the determinant $|a_{ij}|$ of a matrix $[a_{ij}]$ is zero, the matrix is "singular." A matrix is of "rank r" if it contains at least one r-rowed determinant that is not zero, while all its determinants of order higher than r are zero. The "minor" D_{ij} of the element a_{ij} in the square matrix $[a_{ij}]$ is the determinant which remains when the row and the column which contain the element a_{ij} are omitted from $[a_{ij}]$. A minor obtained by omitting the same rows as columns is termed a "principal minor." The "cofactor" A_{ij} of the element a_{ij} is the minor of a_{ij} multipled by $(-1)^{i+j}$.

$$A_{ij} = (-1)^{i+j} D_{ij}$$

The "adjoint" of the matrix $[a_{ij}]$ is

$$\text{adj } [a_{ij}] = \begin{bmatrix} A_{11} & \cdots & A_{n1} \\ A_{12} & \cdots & A_{n2} \\ \cdots & \cdots & \cdots \\ A_{1n} & \cdots & A_{nn} \end{bmatrix}$$

The "inverse" or "reciprocal" $[a_{ij}]^{-1}$ of the nonsingular square matrix $[a_{ij}]$ is

$$[a_{ij}]^{-1} = \frac{\text{adj } [a_{ij}]}{|a_{ij}|} = \begin{bmatrix} \dfrac{A_{11}}{|a_{ij}|} & \cdots & \dfrac{A_{n1}}{|a_{ij}|} \\ \cdots & \cdots & \cdots \\ \dfrac{A_{1n}}{|a_{ij}|} & \cdots & \dfrac{A_{nn}}{|a_{ij}|} \end{bmatrix}$$

If the elements of a determinant M, of order k, are common to any k columns and any k rows in a determinant $|a_{ij}|$ of order n, then M is a "kth minor" of $|a_{ij}|$. The elements of $|a_{ij}|$ which are not included in M form a determinant of order $n - k$ that is termed the "complementary minor" of M. The complementary minor of M multiplied by $(-1)^{i_1+i_2+\cdots+i_k+j_1+j_2+\cdots+j_k}$ where the $i_1, \ldots, i_k$ and $j_1, \ldots, j_k$ denote the rows and columns of $|a_{ij}|$ that contain M, is termed the "algebraic complement" of the minor M.

Operations with Determinants. The expansion of a second-order determinant is obtained by taking diagonal products with algebraic signs as indicated below:

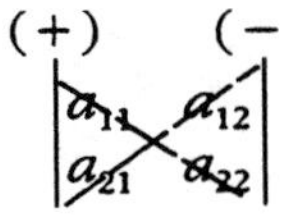

$$\begin{vmatrix} a_{11} & a_{12} \\ a_{21} & a_{22} \end{vmatrix} = a_{11}a_{22} - a_{12}a_{21}$$

The expansion of a third-order determinant is obtained by taking diagonal products, with the algebraic signs as indicated below (note the use of the repeated columns):

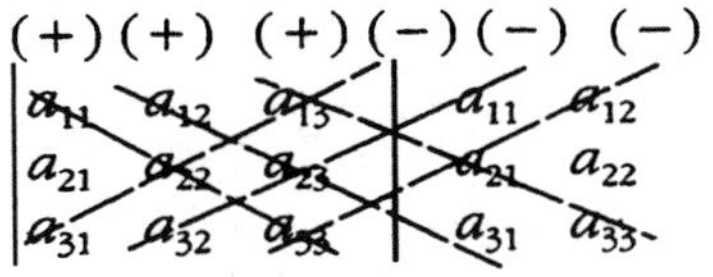

Thus

$$\begin{vmatrix} a_{11} & a_{12} & a_{13} \\ a_{21} & a_{22} & a_{23} \\ a_{31} & a_{32} & a_{33} \end{vmatrix} = a_{11}a_{22}a_{33} + a_{12}a_{23}a_{31} + a_{13}a_{21}a_{32} - a_{13}a_{22}a_{31} - a_{11}a_{23}a_{32} - a_{12}a_{21}a_{33}$$

The method of taking diagonal products as shown above in the expansion of second- and third-order determinants does not apply to the expansion of higher-order determinants. The following rules and methods can be employed in the expansion of higher-order determinants:

1. A determinant is identically zero if all the elements in any row or any column are zero. For example

$$\begin{vmatrix} 6 & 5 \\ 0 & 0 \end{vmatrix} = 0$$

2. A determinant is identically zero if corresponding elements of two rows or two columns are proportional or equal. For example,

$$\begin{vmatrix} 4 & 6 \\ 2 & 3 \end{vmatrix} = 0 \quad \text{(row 1 is twice row 2)}$$

3. The value of a determinant remains unchanged if the rows and columns are interchanged in the same order.

4. The value of a determinant remains unchanged if, to each element of a row or column, a constant multiple of the corresponding elements of another row or column is added.

For example,

$$\begin{vmatrix} 3 & 1 \\ 4 & 2 \end{vmatrix} = \begin{vmatrix} 3 + (2)(1) & 1 \\ 4 + (2)(2) & 2 \end{vmatrix} = \begin{vmatrix} 5 & 1 \\ 8 & 2 \end{vmatrix} = 2$$

5. The sign of a determinant is changed if any two rows or any two columns are interchanged.

6. The determinant is multiplied by k if all the elements in one row or one column are multiplied by k.

7. A determinant containing a row or column of binomials can be written as the sum of two determinants as follows:

$$\begin{vmatrix} a_{11} & \cdots & a_{1i} + b_{1i} & \cdots & a_{1n} \\ a_{21} & \cdots & a_{2i} + b_{2i} & \cdots & a_{2n} \\ a_{31} & \cdots & a_{3i} + b_{3i} & \cdots & a_{3n} \\ \cdots & \cdots & \cdots & \cdots & \cdots \\ a_{11} & \cdots & a_{ni} + b_{ni} & \cdots & a_{nn} \end{vmatrix}$$

$$= \begin{vmatrix} a_{11} & \cdots & a_{1i} & \cdots & a_{1n} \\ a_{21} & \cdots & a_{2i} & \cdots & a_{2n} \\ a_{31} & \cdots & a_{3i} & \cdots & a_{3n} \\ \cdots & \cdots & \cdots & \cdots & \cdots \\ a_{n1} & \cdots & a_{ni} & \cdots & a_{nn} \end{vmatrix} + \begin{vmatrix} a_{11} & \cdots & b_{1i} & \cdots & a_{1n} \\ a_{21} & \cdots & b_{2i} & \cdots & a_{2n} \\ a_{31} & \cdots & b_{3i} & \cdots & a_{3n} \\ \cdots & \cdots & \cdots & \cdots & \cdots \\ a_{n1} & \cdots & b_{ni} & \cdots & a_{nn} \end{vmatrix}$$

8. A determinant equals the sum of the products of the elements of any row or column multiplied by the corresponding cofactors. For example,

$$\begin{vmatrix} 1 & 2 & 3 & 4 \\ 3 & 1 & 1 & 1 \\ 2 & 4 & 5 & 7 \\ 0 & 0 & 2 & 5 \end{vmatrix} = 0 + 0 + 2(-1)^{4+3} \begin{vmatrix} 1 & 2 & 4 \\ 3 & 1 & 1 \\ 2 & 4 & 7 \end{vmatrix} + 5(-1)^{4+4} \begin{vmatrix} 1 & 2 & 3 \\ 3 & 1 & 1 \\ 2 & 4 & 5 \end{vmatrix}$$

by expanding the last row.

9. A determinant can be expanded by choosing any k rows or columns and by summing the products of all the kth minors of $|a_{ij}|$ contained in the k rows or columns multiplied by their algebraic complements. For example, expanding by minors of the first two rows:

$$\begin{vmatrix} 1 & 6 & 1 & 1 \\ 5 & 7 & 1 & 2 \\ 4 & 5 & 0 & 2 \\ 3 & 2 & 5 & 1 \end{vmatrix} = (-1)^{1+2+1+2} \begin{vmatrix} 1 & 6 \\ 5 & 7 \end{vmatrix} \begin{vmatrix} 0 & 2 \\ 5 & 1 \end{vmatrix} + (-1)^{1+2+1+3} \begin{vmatrix} 1 & 1 \\ 5 & 1 \end{vmatrix} \begin{vmatrix} 5 & 2 \\ 2 & 1 \end{vmatrix}$$

$$+ (-1)^{1+2+1+4} \begin{vmatrix} 1 & 1 \\ 5 & 2 \end{vmatrix} \begin{vmatrix} 5 & 0 \\ 2 & 5 \end{vmatrix} + (-1)^{1+2+2+3} \begin{vmatrix} 6 & 1 \\ 7 & 1 \end{vmatrix} \begin{vmatrix} 4 & 2 \\ 3 & 1 \end{vmatrix}$$

$$+ (-1)^{1+2+2+4} \begin{vmatrix} 6 & 1 \\ 7 & 2 \end{vmatrix} \begin{vmatrix} 4 & 0 \\ 3 & 5 \end{vmatrix} + (-1)^{1+2+3+4} \begin{vmatrix} 1 & 1 \\ 1 & 2 \end{vmatrix} \begin{vmatrix} 4 & 5 \\ 3 & 2 \end{vmatrix}$$

$$= \begin{vmatrix} 1 & 6 \\ 5 & 7 \end{vmatrix}\begin{vmatrix} 0 & 2 \\ 5 & 1 \end{vmatrix} - \begin{vmatrix} 1 & 1 \\ 5 & 1 \end{vmatrix} + \begin{vmatrix} 1 & 1 \\ 5 & 2 \end{vmatrix}\begin{vmatrix} 5 & 0 \\ 2 & 5 \end{vmatrix}$$

$$+ \begin{vmatrix} 6 & 1 \\ 7 & 2 \end{vmatrix}\begin{vmatrix} 4 & 2 \\ 3 & 1 \end{vmatrix} - \begin{vmatrix} 6 & 1 \\ 7 & 2 \end{vmatrix}\begin{vmatrix} 4 & 0 \\ 3 & 5 \end{vmatrix} + \begin{vmatrix} 1 & 1 \\ 1 & 2 \end{vmatrix}\begin{vmatrix} 4 & 5 \\ 3 & 2 \end{vmatrix}$$

(The above method is called Laplace's development.)

10. A determinant $|a_{ij}|$ can also be expanded by the following numerical pivotal-element method:

a. Reduce some element a_{ij}, if necessary, to unity by dividing either the ith row or jth column by a_{ij} and then multiplying the determinant by the same factor. This element is the *pivotal element*.

b. Strike out the row and column containing the pivotal element.

c. Subtract from each remaining element in the determinant the product of the two crossed-out elements which are in line, horizontally and vertically, with this remaining element.

d. Multiply the new determinant, which is of order one less than the original determinant by $(-1)^{i+j}$ where i and j are the row and column numbers of the pivotal element. For example,

$$\begin{vmatrix} 4 & 4 & 3 & 8 \\ 3 & 3 & 2 & 9 \\ 2 & 2 & 6 & 2 \\ 0 & 4 & 5 & 2 \end{vmatrix} = 2\begin{vmatrix} 4 & 4 & 3 & 8 \\ 3 & 3 & 2 & 9 \\ 1 & 1 & 3 & 1 \\ 0 & 4 & 5 & 2 \end{vmatrix}$$

$$= 2(-1)^{3+2}\begin{vmatrix} 4-(4)(1) & 3-(4)(3) & 8-(4)(1) \\ 3-(3)(1) & 2-(3)(3) & 9-(3)(1) \\ 0-(4)(1) & 5-(4)(3) & 2-(4)(1) \end{vmatrix}$$

$$= -2\begin{vmatrix} 0 & -9 & 2 \\ 0 & -4 & 6 \\ -4 & -7 & -2 \end{vmatrix} = (-2)(-4)(-1)^{3+1}\begin{vmatrix} -9 & 2 \\ -4 & 6 \end{vmatrix}$$

$$= 8(-54+8) = -368$$

Matrix Operations

1. The *sum* (or *difference*) of two $m \times n$ matrices is a matrix, the elements of which are the sums (or differences) of the corresponding elements of the two matrices. Thus,

$$[a_{ij}] \pm [b_{ij}] = [a_{ij} \pm b_{ij}]$$

For example,

$$\begin{bmatrix} 6 & 3 \\ 1 & 0 \end{bmatrix} + \begin{bmatrix} 1 & 5 \\ 2 & 3 \end{bmatrix} = \begin{bmatrix} 7 & 8 \\ 3 & 3 \end{bmatrix}$$

2. If the number of columns of matrix A equals the number of rows of matrix B, the two matrices are *conformable* and can be multiplied in the order $A \times B$. The *product* $A \times B$ is a matrix P in which the ijth element is obtained by selecting the ith row of

A and the jth column of B and then summing the products of their corresponding elements:

$$P_{ik} = \sum_{j=1}^{n} a_{ij} b_{jk}$$

For example,

$$\begin{bmatrix} 1 & 2 & 3 \\ 4 & 5 & 6 \end{bmatrix} \begin{bmatrix} 1 & 2 & 4 & 6 \\ 8 & 1 & 1 & 0 \\ 1 & 2 & 1 & 4 \end{bmatrix} = \begin{bmatrix} (1)(1) + (2)(8) + (3)(1) & (1)(2) + (2)(1) + (3)(2) & \cdots \\ (4)(1) + (5)(8) + (6)(1) & (4)(2) + (5)(1) + (6)(2) & \cdots \end{bmatrix}$$

Multiplication of matrices is not in general commutative, even when the two matrices are conformable in either order, that is, $AB \neq BA$. Multiplication of matrices is associative. That is, $(AB)C = A(BC)$ if the matrices are conformable. Multiplication of matrices is distributive as long as the matrices being added have the same number of rows and columns, and the matrices being multiplied are conformable. That is, $AB + AC = A(B + C)$.

3. The product of a matrix $|a_{ij}|$ and a scalar S is the matrix $[Sa_{ij}]$. For example,

$$5\begin{bmatrix} 4 & 3 \\ 6 & 3 \end{bmatrix} = \begin{bmatrix} (5)(4) & (5)(3) \\ (5)(6) & (5)(3) \end{bmatrix}$$

4. If the product of two or more matrices vanishes, it does not imply that one of the matrices is zero. The factors are termed "divisors of zero." The product of a column and a row results in a matrix with proportional rows and proportional columns. The product of a matrix and a column results in a column. The product of a matrix and a row is a row. The product of a row and a column is a scalar. The product of a matrix and its inverse is a unit matrix:

$$AA^{-1} = A^{-1}A = I$$

Solution of n Simultaneous Linear Equations in n Unknowns. A system of linear simultaneous equations is said to be "consistent" if at least one common solution exists. The solution of n unknowns requires n independent equations which are derived from n independent conditions.

In the case of systems of two equations in two unknowns, a graphical solution is possible. The two equations are plotted on a common coordinate system and the coordinates of any point of intersection of the graphs are solutions of the system. (If the graphs intersect at one point, the equations of the system are "independent" and "consistent." If the graphs coincide, the equations of the system are "dependent" and "consistent." If the graphs are parallel, the equations are "inconsistent.")

Given the following system of homogeneous equations:

$$\begin{aligned} a_{11}x_1 + a_{12}x_2 + \cdots + a_{1n}x_n &= 0 \\ a_{21}x_1 + a_{22}x_2 + \cdots + a_{2n}x_n &= 0 \\ \cdots\cdots\cdots\cdots\cdots\cdots\cdots\cdots \\ a_{n1}x_1 + a_{n2}x_2 + \cdots + a_{nn}x_n &= 0 \end{aligned}$$

the system will have a solution other than the trivial solution $x_1 = x_2 = x_n = 0$ if the determinant of the coefficient matrix $|a_{ij}|$ is zero.

The following system of nonhomogenous equations, where at least one b is nonzero, has a single solution provided that the determinant of the coefficient matrix $|a_{ij}|$

is not zero.

$$\begin{aligned} a_{11}x_1 + \cdots + a_{1n}x_n &= b_1 \\ a_{21}x_1 + \cdots + a_{2n}x_n &= b_2 \\ \cdots\cdots\cdots\cdots&\cdots\cdots \\ a_{n1}x_1 + \cdots + a_{nn}x_n &= b_n \end{aligned}$$

Methods of Solution

1. Cramer's rule. Given the system of equations

$$[a_{ij}]\{x\} = \{b\} \qquad |a_{ij}| \neq 0$$

then

$$x_i = |B|/|a_{ij}|$$

where $|B|$ is the same determinant as $|a_{ij}|$ except that the column of the coefficients of x_i in $|a_{ij}|$ is replaced by the constants of column $\{b\}$. For example,

$$\begin{bmatrix} 3 & 1 \\ 2 & 1 \end{bmatrix} \begin{Bmatrix} x_1 \\ x_2 \end{Bmatrix} = \begin{Bmatrix} 3 \\ 1 \end{Bmatrix}$$

$$x_1 = \frac{\begin{vmatrix} 3 & 1 \\ 1 & 1 \end{vmatrix}}{\begin{vmatrix} 3 & 1 \\ 2 & 1 \end{vmatrix}} = \frac{3 - 1}{3 - 2} = 2$$

2. Coefficient matrix inversion method. Given

$$[a_{ij}]\{x\} = \{b\}$$

then

$$\{x\} = [a_{ij}]^{-1}\{b\}$$

where

$[a_{ij}]^{-1}$ is the inverse of $[a_{ij}]$.

3. Elimination method. Given, for example,

$$\begin{aligned} 2x_1 + 4x_2 + 8x_3 &= 6 \\ x_1 + 3x_2 + 7x_3 &= 10 \\ 3x_1 + 12x_2 + 6x_3 &= 12 \end{aligned}$$

The following solution illustrates the method. Rewrite the array of coefficients and constants as follows:

$$\begin{array}{ccc:c} 2 & 4 & 8 & 6 \\ 1 & 3 & 7 & 10 \\ 3 & 12 & 6 & 12 \end{array}$$

Divide each equation by its leading coefficient. The array thus becomes

$$\begin{array}{ccc:c} 1 & 2 & 4 & 3 \\ 1 & 3 & 7 & 10 \\ 1 & 4 & 2 & 4 \end{array}$$

Subtract the first equation from each of the remaining equations

$$\begin{array}{rrr|r} 1 & 2 & 4 & 3 \\ 0 & 1 & 3 & 7 \\ 0 & 2 & -2 & 1 \end{array}$$

The last two rows constitute a system of equations of one order lower than the original system. Apply the same technique to these (and so on in higher-order systems) to obtain the value of x_3 (or x_n). Thus,

$$\begin{array}{rrr|r} 1 & 2 & 4 & 3 \\ 0 & 1 & 3 & 7 \\ 0 & 1 & -1 & 0.5 \\ \hline 1 & 2 & 4 & 3 \\ 0 & 1 & 3 & 7 \\ 0 & 0 & -4 & -6.5 \\ \hline 1 & 2 & 4 & 3 \\ 0 & 1 & 3 & 7 \\ 0 & 0 & 1 & -6.5/-4 \end{array}$$

Substitute the value of x_3 in the second equation and solve for x_2, and then substitute the values of x_3 and x_2 in the first equation. The solution is obtained.

A.4.2 Mensuration

In the following formulas P = perimeter and A = area.

General Triangle (Fig. A.1)

$$\text{Length of line to side } c \text{ bisecting angle } \lambda = \frac{\sqrt{ab[(a+b)^2 - c^2]}}{a+b}$$

$$\text{Length of median line to side } c = \tfrac{1}{2}\sqrt{2(a^2+b^2) - c^2}$$

$$A = \frac{1}{2}bh = \frac{1}{2}ba\sin\gamma = \frac{1}{2}\frac{a^2\sin\gamma\sin\beta}{\sin\alpha} = \sqrt{\frac{P}{2}\left(\frac{P}{2}-a\right)\left(\frac{P}{2}-b\right)\left(\frac{P}{2}-c\right)}$$

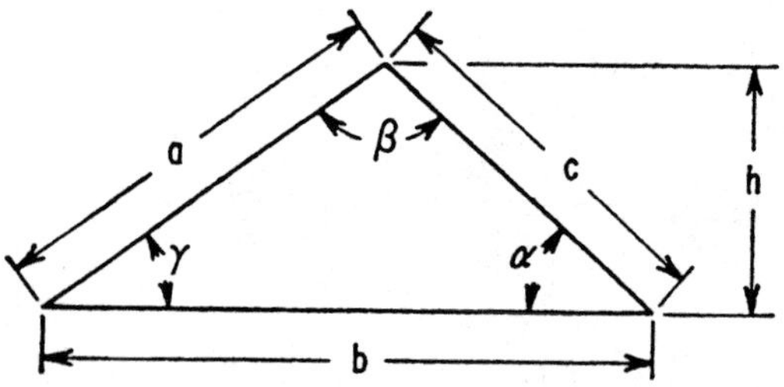

FIG. A.1 General triangle.

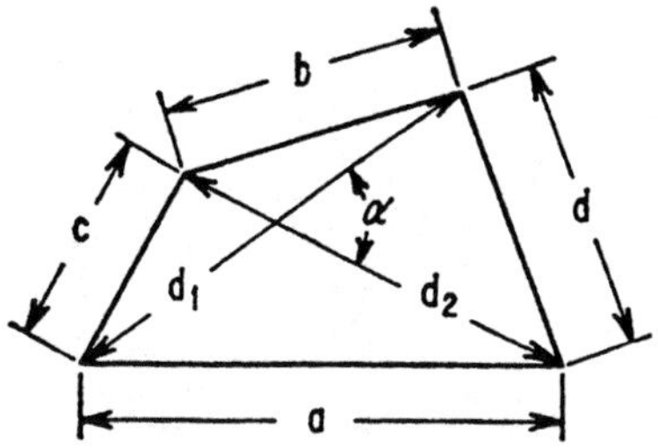

FIG. A.2 General quadrilateral or trapezium.

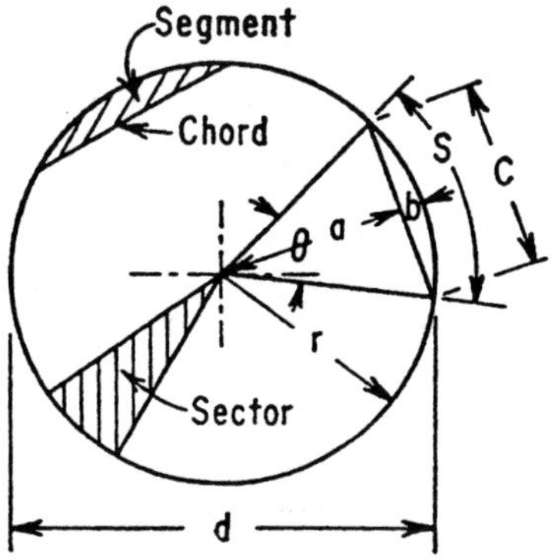

FIG. A.3 Circle.

where $P = a + b + c$.
For an *isosceles triangle*, $a = c$ $\gamma = \alpha$
For an *equilateral triangle*, $a = b = c$ $\alpha = \beta = \gamma$

General Quadrilateral or Trapezium (Fig. A.2)

$$P = a + b + c + d$$

$$A = \frac{1}{2} d_1 d_2 \sin\alpha$$

(or the total area may be divided into several simple geometric figures, such as two triangles).

Circle (Fig. A.3)

$C = 2r \sin(\theta/2)$ θ = central angle, rad

$S = r\theta$

$P = 2\pi r = \pi d$

$b = r \mp \sqrt{r^2 - C^2/4}$ $-$for $\theta \leqq 180°$ (π rad), + for $\theta \geqq 180°$ (π rad)

$b = r[1 - \cos(\theta/2)] = r \text{ versin } (\theta/2)$

Area of sector $= \frac{1}{2}rS = \frac{1}{2}r^2\theta$

Area of segment $= \frac{1}{2}r^2(\theta - \sin\theta)$

Area of circle $= \frac{1}{4}\pi d^2 = \pi r^2$

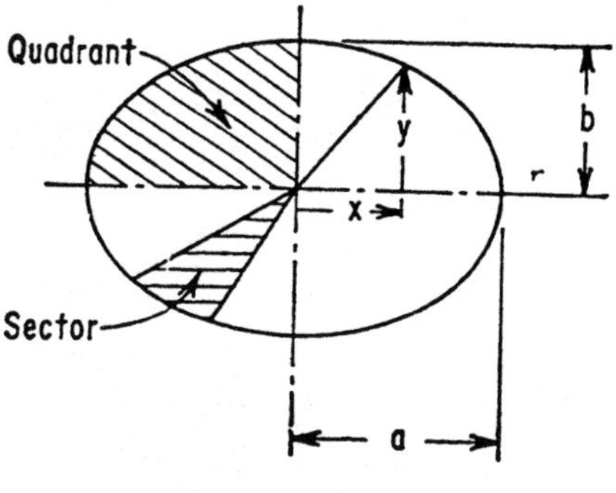

FIG. A.4 Ellipse.

Ellipse (Fig. A.4)

$$P = (4a)E$$

where

$$E = \int_0^{2\pi} \sqrt{1 - k^2 \sin^2 x\, dx}$$

a complete elliptic integral (see Table A.5), and $k = 1 - b^2/a^2$.

$$P = \pi(a + b)(1 + \gamma^2/4 + \gamma^4/64 + \cdots)$$

where $\gamma = (a - b)/(a + b)$.

$$P = \pi[(a + b)/4][3(1 + \lambda) + 1/(1 - \lambda)]$$

where $\lambda = [(a - b)/2(a - b)]^2$.

$$\text{Area of ellipse} = \pi ab$$

Regular Polygon—Equal sides, equal angles (Fig. A.5)

$$\alpha = 360/n \quad \text{degrees}$$
$$= 2\pi/n \quad \text{rad}$$
$$\beta = [(n - 2)/n] \times 180 \quad \text{degrees}$$
$$r = \frac{1}{2} a \cot(180°/n) = \text{radius of inscribed circle}$$
$$R = \frac{1}{2} a \csc(180°/n) = \text{radius of circumscribed circle}$$
$$a = 2r \tan(\alpha/2) = 2R \sin(\alpha/2)$$
$$A = \frac{1}{4} na^2 \cot 180°/n = nr^2 \tan(\alpha/2) = \frac{1}{2} nR^2 \sin\alpha$$
$$n = \text{number of sides}$$

Catenary (Fig. A.6)

$$S = \text{arc } ABC \cong l\left[1 + \tfrac{2}{3}(2h/l)^2\right] \qquad \text{if } h \ll l$$

Parabola (Fig. A.7)

$$S = \tfrac{1}{2}\sqrt{16H^2 + L^2} + \frac{L^2}{8H} \ln\left(\frac{4H + \sqrt{16H^2 + L^2}}{L}\right)$$

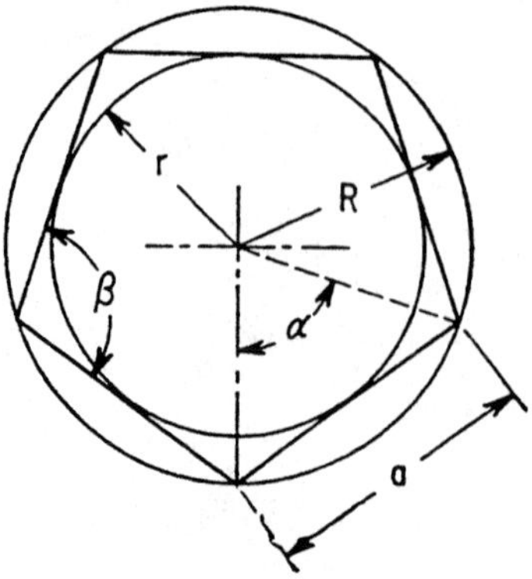

FIG. A.5 Regular polygon.

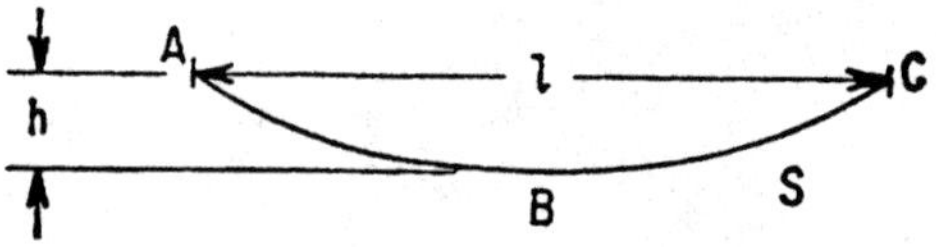

FIG. A.6 Catenary.

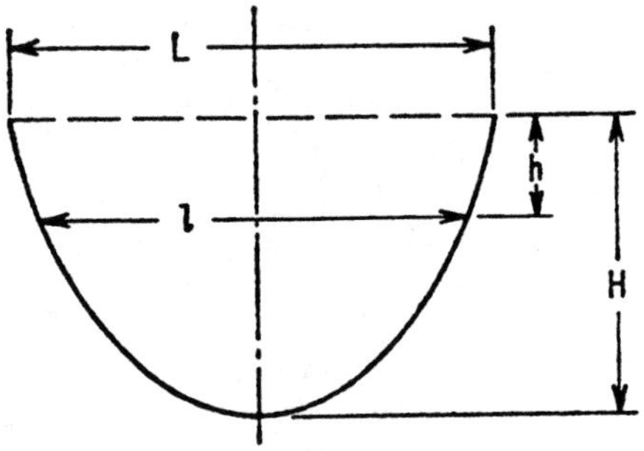

FIG. A.7 Parabola.

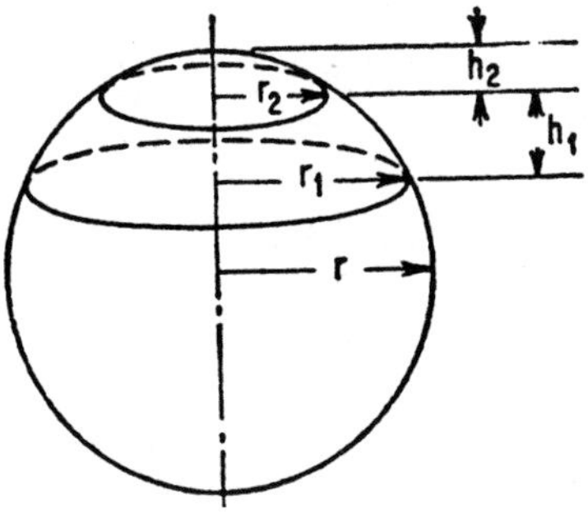

FIG. A.8 Sphere.

$$= L\left[1 + \frac{2}{3}\left(\frac{2H}{L}\right)^2 - \frac{2}{5}\left(\frac{2H}{L}\right)^4 + \cdots\right]$$

$$h = (H/L^2)(L^2 - l^2)$$

$$A = \tfrac{2}{3}HL$$

Sphere (Fig. A.8)

$$\text{Area of sphere} = 4\pi r^2 = \pi d^2$$

$$\text{Volume of spherical segment (one base)} = \tfrac{1}{6}\pi h_2\left(3r_2^2 + h_2^2 = \tfrac{1}{3}\pi h_2^2(3r - h_2)\right.$$

$$\text{Volume of spherical segment (two bases)} = \tfrac{1}{6}\pi h_1\left(3r_2^2 + 3r_1^2 + h_1^2\right)$$

$$\text{Volume of sphere} = \tfrac{4}{3}\pi r^3 = \tfrac{1}{6}\pi d^3$$

Ellipsoid (Fig. A.9)

$$V = \tfrac{4}{3}\pi abc$$

Paraboloid of Revolution (Fig. A.10)

$$A = \frac{2\pi d_1}{3(h_1 + h_2)^2}\left\{\left[\frac{d_1^2}{16} + (h_1 + h_2)^2\right]^{3/2} - \left(\frac{d_1}{4}\right)^3\right\}$$

$$\text{Volume of paraboloid of revolution (segment of one base)} = \tfrac{1}{2}\pi r_2^2 h_2$$

$$\text{Volume of paraboloid of revolution (segment of two bases)} = \tfrac{1}{2}\pi h_1\left(r_1^2 + r_2^2\right)$$

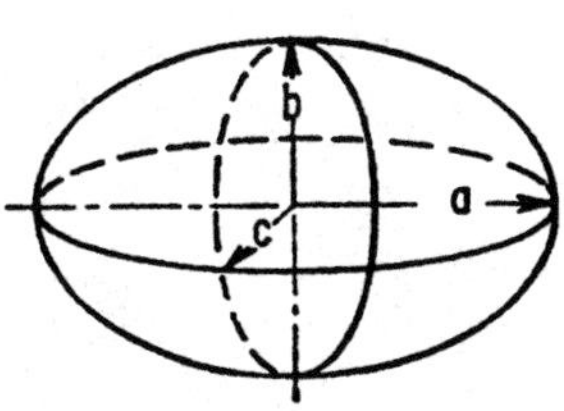

FIG. A.9 Ellipsoid.

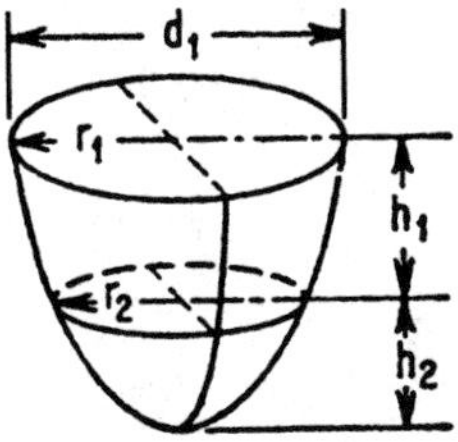

FIG. A.10 Paraboloid of revolution.

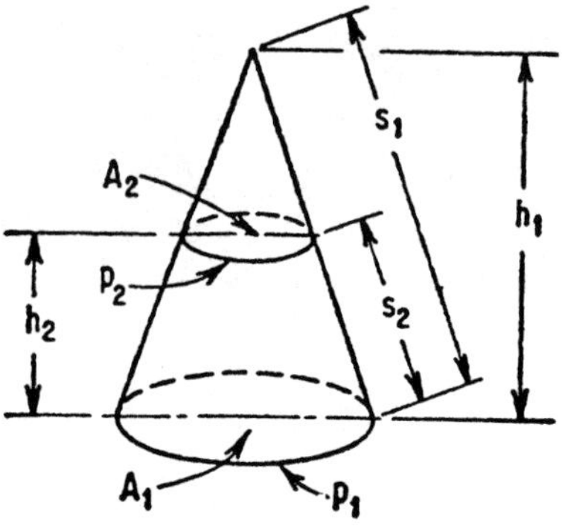

FIG. A.11 Cone.

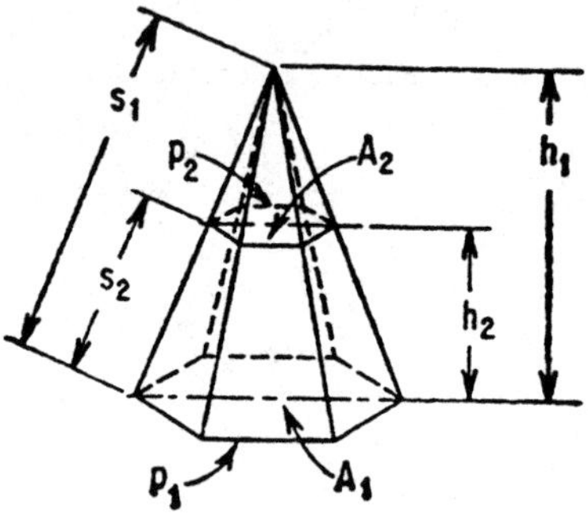

FIG. A.12 Pyramid.

Cone and Pyramid (Figs. A.11 and A.12)

$$\text{Lateral area of regular frustum of cone or pyramid} = \frac{1}{2}(p_1 + p_2)s_2$$

$$\text{Lateral area of regular figure} = \frac{1}{2}\text{ slant height} \times \text{perimeter of base} = \frac{1}{2}s_1 p_1$$

$$\text{Volume of frustum of cone or pyramid} = \frac{1}{3}(A_1 + A_2 + \sqrt{A_1A_2})h_2$$

$$\text{Volume of cone or pyramid} = \frac{1}{3}\text{ altitude} \times \text{area of base}$$

$$= \frac{1}{3}h_1A_1$$

Cylinder and Prism (Figs A.13 and A.14)

$$\text{Lateral area} = \text{lateral edge} \times \text{perimeter of right section}$$

$$V = \text{altitude} \times \text{area of base}$$

Surface of Revolution (Fig. A.15)

$$\text{Area of surface} = 2\pi RS$$

where R = distance from axis of rotation to the center of gravity of arc of length S. The axis of rotation is in the plane of the arc and does not cross it.

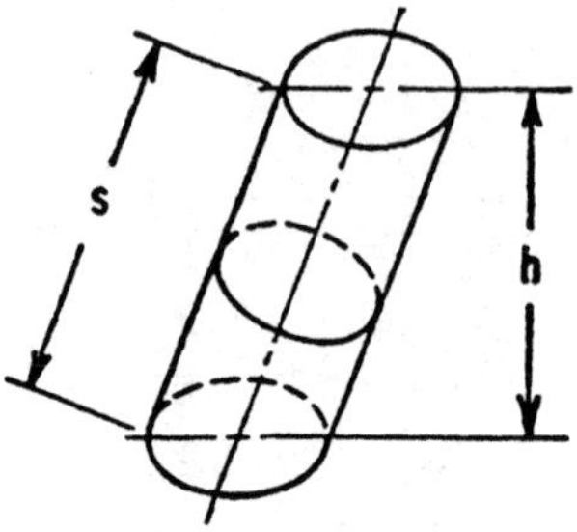

FIG. A.13 Cylinder.

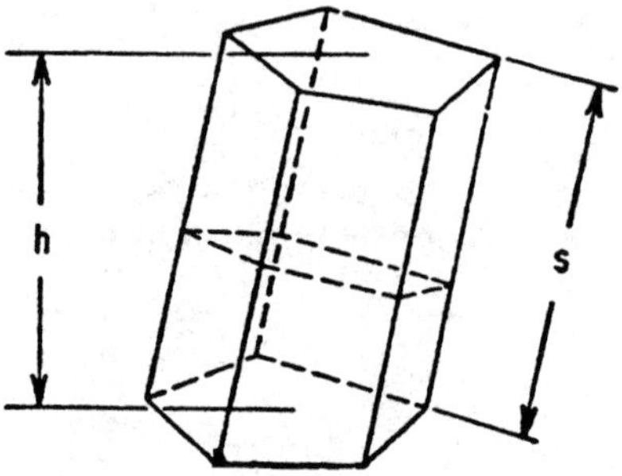

FIG. A.14 Prism.

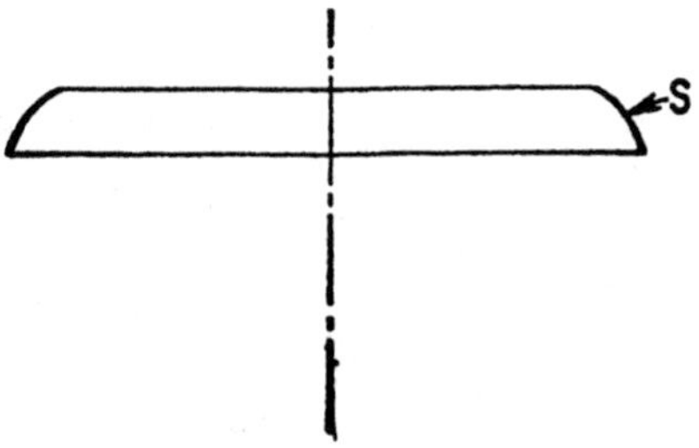

FIG. A.15 Surface of revolution.

Volume of Revolution (Fig. A.16)

$$V = 2\pi RA$$

where R = distance from the axis of rotation to the center of gravity of the area A. The axis of rotation is in the plane of area A and does not cross it.

Torus (Fig. A.17)

$$\text{Area of surface} = 4\pi^2 Rr$$

$$V = 2\pi^2 Rr^2$$

A.4.3 Properties of Plane Sections (see Table A.7)

The following symbols are used:

A = area

x_c, y_c = coordinates of centroid of section in xy coordinate system

I_{x_c}, I_{y_c} = moment of inertia about an axis through the centroid parallel to the x and y axes

r_{x_c}, r_{y_c} = radius of gyration of the section with respect to the centroidal axes parallel to the x and y axes

$I_{x_c y_c}$ = product of inertia with respect to the centroidal axes parallel to the x and y axes

I_x, I_y = moment of inertia with respect to the x and y axes shown

r_x, r_y = radius of gyration of the section with respect to the x and y axes shown

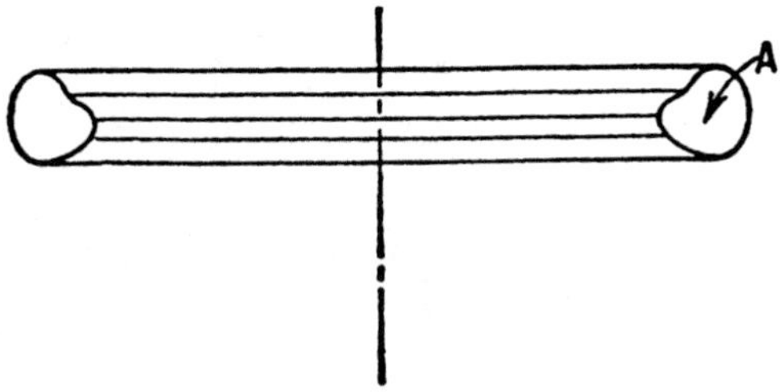

FIG. A.16 Volume of revolution.

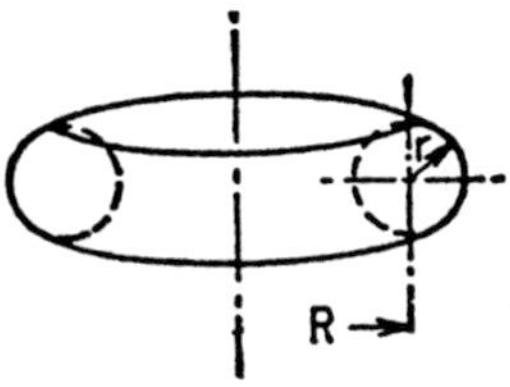

FIG. A.17 Torus.

TABLE A.7 Properties of Plane Sections*

Figure	Area and centroid	Moment of inertia	r^2	Product of inertia
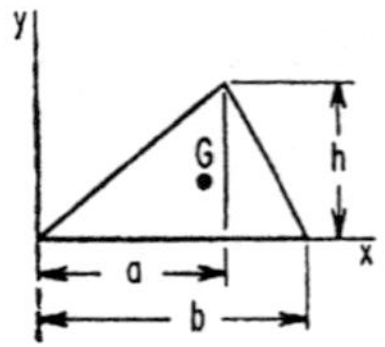 Triangle	$A = \frac{1}{2}bh$ $x_c = \frac{1}{3}(a + b)$ $y_c = \frac{1}{3}h$	$I_{xc} = \frac{bh^3}{36}$ $I_{yc} = \frac{bh}{36}(b^2 - ab + a^2)$ $I_x = \frac{bh^3}{12}$ $I_y = \frac{bh}{12}(b^2 + ab + a^2)$	$r_{xc}^2 = \frac{1}{18}h^2$ $r_{yc}^2 = \frac{1}{18}(b^2 - ab + a^2)$ $r_x^2 = \frac{1}{6}h^2$ $r_y^2 = \frac{1}{6}(b^2 + ab + a^2)$	$I_{xcyc} = \frac{Ah}{36}(2a - b) = \frac{bh^2}{72}(2a - b)$ $I_{xy} = \frac{Ah}{12}(2a + b) = \frac{bh^2}{24}(2a + b)$
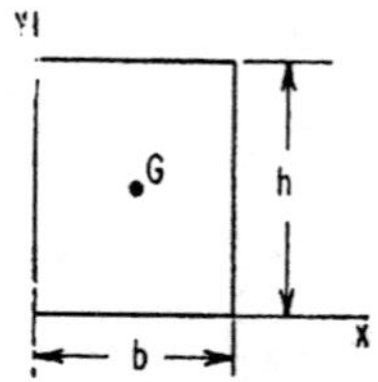 Rectangle	$A = bh$ $x_c = \frac{1}{2}b$ $y_c = \frac{1}{2}h$	$I_{xc} = \frac{bh^3}{12}$ $I_{yc} = \frac{b^3h}{12}$ $I_x = \frac{bh^3}{3}$ $I_y = \frac{b^3h}{3}$ $I_P = \frac{bh}{12}(b^2 + h^2)$	$r_{xc}^2 = \frac{1}{12}h^2$ $r_{yc}^2 = \frac{1}{12}b^2$ $r_x^2 = \frac{1}{3}h^2$ $r_y^2 = \frac{1}{3}b^2$ $r_P^2 = \frac{1}{12}(b^2 + h^2)$	$I_{xcyc} = 0$ $I_{xy} = \frac{A}{4}bh = \frac{b^2h^2}{4}$
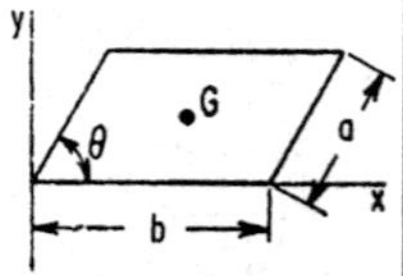 Parallelogram	$A = ab \sin\theta$ $x_c = \frac{1}{2}(b + a\cos\theta)$ $y_c = \frac{1}{2}(a \sin\theta)$	$I_{xc} = \frac{a^3b}{12}\sin^3\theta$ $I_{yc} = \frac{ab}{12}\sin\theta(b^2 + a^2\cos^2\theta)$ $I_x = \frac{a^3b}{3}\sin^3\theta$ $I_y = \frac{ab}{3}\sin\theta\,(b + a\cos\theta)^2 - \frac{a^2b^2}{6}\sin\theta\cos\theta$	$r_{xc}^2 = \frac{1}{12}(a\sin\theta)^2$ $r_{yc}^2 = \frac{1}{12}(b^2 + a^2\cos^2\theta)$ $r_x^2 = \frac{1}{3}(a\sin\theta)^2$ $r_y^2 = \frac{1}{3}(b + a\cos\theta)^2 - \frac{1}{6}(ab\cos\theta)$	$I_{xcyc} = \frac{a^2b}{12}\sin^2\theta\cos\theta$

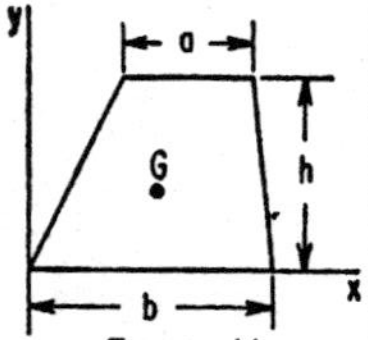 Trapezoid	$A = \frac{1}{2}h(a+b)$ $y_c = \frac{1}{3}h\,\frac{2a+b}{a+b}$	$I_{x_c} = \frac{h^3(a^2+4ab+b^2)}{36(a+b)}$ $I_x = \frac{h^3(3a+b)}{12}$	$r_{x_c}^2 = \frac{h^2(a^2+4ab+b^2)}{18(a+b)^2}$ $r_x^2 = \frac{h^2(3a+b)}{6(a+b)}$	
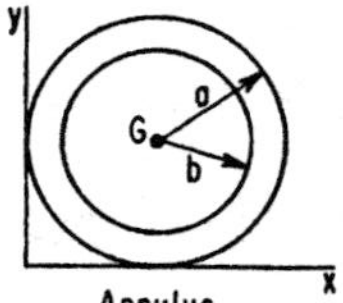 Annulus	$A = \pi(a^2-b^2)$ $x_c = a$ $y_c = a$	$I_{x_c} = I_{y_c} = \frac{\pi}{4}(a^4-b^4)$ $I_x = I_y = \frac{5}{4}\pi a^4 - \pi a^2b^2 - \frac{\pi}{4}b^4$ $I_P = \frac{\pi}{2}(a^4-b^4)$	$r_{x_c}^2 = r_{y_c}^2 = \frac{1}{4}(a^2+b^2)$ $r_x^2 = r_y^2 = \frac{1}{4}(5a^2+b^2)$ $r_P^2 = \frac{1}{2}(a^2+b^2)$	$I_{x_cy_c} = 0$ $I_{xy} = Aa^2 = \pi a^2(a^2-b^2)$
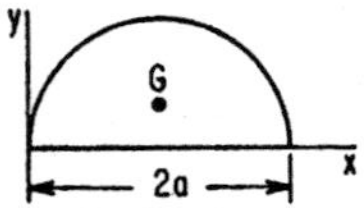 Semicircle	$A = \frac{1}{2}\pi a^2$ $x_c = a$ $y_c = \frac{4a}{3\pi}$	$I_{x_c} = \frac{a^4(9\pi^2-64)}{72\pi}$ $I_{y_c} = \frac{1}{8}\pi a^4$ $I_x = \frac{1}{8}\pi a^4$ $I_y = \frac{5}{8}\pi a^4$	$r_{x_c}^2 = \frac{a^2(9\pi^2-64)}{36\pi^2}$ $r_{y_c}^2 = \frac{1}{4}a^2$ $r_x^2 = \frac{1}{4}a^2$ $r_y^2 = \frac{5}{4}a^2$	$I_{x_cy_c} = 0$ $I_{xy} = \frac{2}{3}a^4$
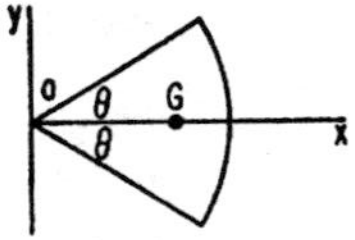 Circular sector	$A = a^2\theta$ $x_c = \frac{2a}{3}\,\frac{\sin\theta}{\theta}$ $y_c = 0$	$I_x = \frac{1}{4}a^4(\theta - \sin\theta\cos\theta)$ $I_y = \frac{1}{4}a^4(\theta + \sin\theta\cos\theta)$	$r_x^2 = \frac{1}{4}a^2\,\frac{\theta - \sin\theta\cos\theta}{\theta}$ $r_y^2 = \frac{1}{4}a^2\,\frac{\theta + \sin\theta\cos\theta}{\theta}$	$I_{x_cy_c} = 0$ $I_{xy} = 0$

TABLE A.7 Properties of Plane Sections *(Continued)*

Figure	Area and centroid	Moment of inertia	r^2	Product of inertia
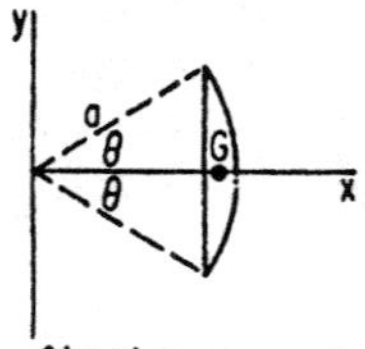 Circular segment	$A = a^2(\theta - \frac{1}{2}\sin 2\theta)$ $x_c = \frac{2a}{3}\frac{\sin^3\theta}{\theta - \sin\theta\cos\theta}$ $y_c = 0$	$I_x = \frac{Aa^2}{4}\left[1 - \frac{2\sin^3\theta\cos\theta}{3(\theta - \sin\theta\cos\theta)}\right]$ $I_y = \frac{Aa^2}{4}\left(1 + \frac{2\sin^3\theta\cos\theta}{\theta - \sin\theta\cos\theta}\right)$	$r_x^2 = \frac{a^2}{4}\left[1 - \frac{2\sin^3\theta\cos\theta}{3(\theta - \sin\theta\cos\theta)}\right]$ $r_y^2 = \frac{a^2}{4}\left(1 + \frac{2\sin^3\theta\cos\theta}{\theta - \sin\theta\cos\theta}\right)$	$I_{x_cy_c} = 0$ $I_{xy} = 0$
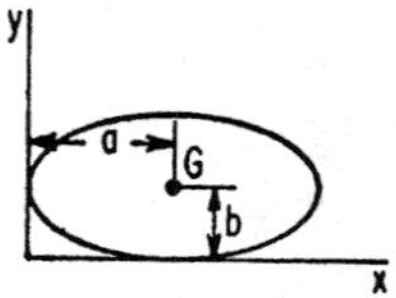 Ellipse	$A = \pi ab$ $x_c = a$ $y_c = b$	$I_{x_c} = \frac{\pi}{4}ab^3$ $I_{y_c} = \frac{\pi}{4}a^3b$ $I_x = \frac{5}{4}\pi ab^3$ $I_y = \frac{5}{4}\pi a^3b$ $I_P = \frac{\pi ab}{4}(a^2 + b^2)$	$r_{x_c}^2 = \frac{1}{4}b^2$ $r_{y_c}^2 = \frac{1}{4}a^2$ $r_x^2 = \frac{5}{4}b^2$ $r_y^2 = \frac{5}{4}a^2$ $r_P^2 = \frac{1}{4}(a^2 + b^2)$	$I_{x_cy_c} = 0$ $I_{xy} = Aab = \pi a^2b^2$
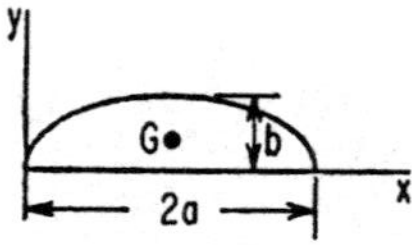 Semiellipse	$A = \frac{1}{2}\pi ab$ $x_c = a$ $y_c = \frac{4b}{3\pi}$	$I_{x_c} = \frac{ab^3}{72\pi}(9\pi^2 - 64)$ $I_{y_c} = \frac{\pi}{8}a^3b$ $I_x = \frac{\pi}{8}ab^3$ $I_y = \frac{5}{8}\pi a^3b$	$r_{x_c}^2 = \frac{b^2}{36\pi^2}(9\pi^2 - 64)$ $r_{y_c}^2 = \frac{1}{4}a^2$ $r_x^2 = \frac{1}{4}b^2$ $r_y^2 = \frac{5}{4}a^2$	$I_{x_cy_c} = 0$ $I_{xy} = \frac{2}{3}a^2b^2$

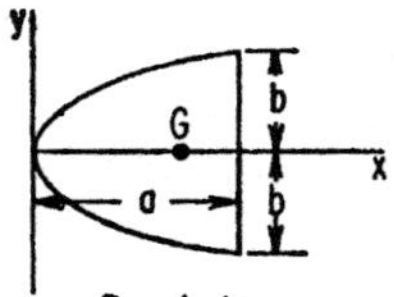 Parabola	$A = \frac{4}{3}ab$ $x_c = \frac{3}{5}a$ $y_c = 0$	$I_{xc} = I_x = \frac{4}{15}ab^3$ $I_{yc} = \frac{16}{175}a^3b$ $I_y = \frac{4}{7}a^3b$	$r_{xc}^2 = r_x^2 = \frac{1}{5}b^2$ $r_{yc}^2 = \frac{12}{175}a^2$ $r_y^2 = \frac{3}{7}a^2$	$I_{xcyc} = 0$ $I_{xy} = 0$
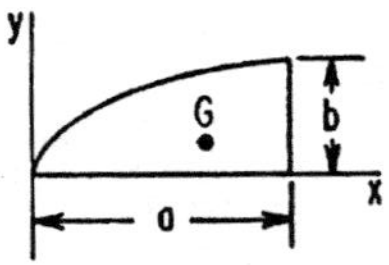 Semiparabola	$A = \frac{2}{3}ab$ $x_c = \frac{3}{5}a$ $y_c = \frac{3}{8}b$	$I_x = \frac{2}{15}ab^3$ $I_y = \frac{2}{7}a^3b$	$r_x^2 = \frac{1}{5}b^2$ $r_y^2 = \frac{3}{7}a^2$	$I_{xy} = \frac{A}{4}ab = \frac{1}{6}a^2b^2$

*From G. W. Housner and D. E. Hudson, "Applied Mechanics," vol. II, "Dynamics," D. Van Nostrand Co., Inc., Princeton, N.J., 1959.

I_{xy} = product of inertia with respect to the x and y axes shown

I_p = polar moment of inertia about an axis passing through the centroid

r_p = radius of gyration of the section about the polar axis passing through the centroid

G marks the centroid.

A.4.4 Properties of Homogeneous Bodies (see Table A.8)

The following symbols are used:

ρ = mass density

M = mass

x_c, y_c, z_c = coordinates of centroid in xyz coordinate system

$I_{x_c}, I_{y_c}, I_{z_c}$ = moment of inertia about an axis through the centroid parallel to the x, y, and z axes shown

$r_{x_c}, r_{y_c}, r_{x_c}$ = radius of gyration of the body with respect to the centroidal axes parallel to the x, y, and z axes shown

$I_{x_c y_c}, I_{x_c z_c}$, etc. = product of inertia with respect to the centroidal axes parallel to the x, y, and z axes shown

I_x, I_y, I_z = moment of inertia with respect to the x, y, and z axes shown

r_x, r_y, r_z = radius of gyration of the body with respect to the x, y, and z axes shown

I_{xy}, I_{xz}, etc. = product of inertia with respect to the x, y, and z axes shown

I_{AA}, r_{AA} = moments of inertia and radii of gyration with respect to special axes shown

G marks the centroid.

A.4.5 Analytic Geometry

Rectangular-Coordinate System. The rectangular-coordinate system in space is defined by three mutually perpendicular coordinate axes which intersect at the origin O as shown in Fig. A.18. The position of a point $P(x, y, z)$ is given by the distances x, y, z from the coordinate planes ZOY, XOZ, and XOY, respectively.

Cylindrical-Coordinate System. The position of any point $P(r, \theta, z)$ is given by the polar coordinates r and θ, the projection of P on the XY plane, and by z, the distance from the XY plane to the point (Fig. A.19).

Spherical-Coordinate System. The position of any point $P(r, \phi, \theta)$ (Fig. A.20) is given by the distance $r(=\overline{OP})$, the angle ϕ which is formed by the intersection of the X coordinate and the projection of OP on the XY plane, and the angle θ which is formed by $\overline{OP}$ and the coordinate z.

Relations between Coordinate Systems. Rectangular and cylindrical:

$$x = r\cos\theta \qquad y = r\sin\theta \qquad z = z \qquad \overline{OP} = \sqrt{r^2 + z^2}$$

Rectangular and spherical:

$$x = r\sin\theta\cos\phi \qquad y = r\sin\theta\sin\phi \qquad z = r\cos\theta$$

TABLE A.8 Properties of Homogeneous Bodies*

Body	Mass and centroid	Moment of inertia	r^2	Product of inertia
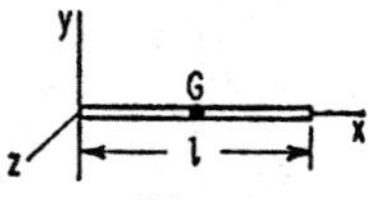 Thin rod	$M = \rho l$ $x_c = \frac{1}{2}l$ $y_c = 0$ $z_c = 0$	$I_x = I_{x_c} = 0$ $I_{y_c} = I_{z_c} = \frac{M}{12} l^2$ $I_y = I_z = \frac{M}{3} l^2$	$r_x^2 = r_{x_c}^2 = 0$ $r_{y_c}^2 = r_{z_c}^2 = \frac{1}{12}l^2$ $r_y^2 = r_z^2 = \frac{1}{3}l^2$	$I_{x_c y_c}$, etc. $= 0$ I_{xy}, etc. $= 0$
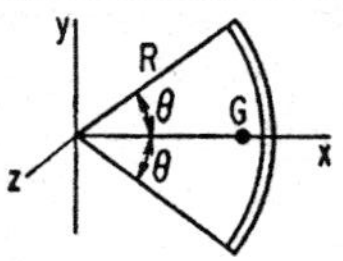 Thin circular rod	$M = 2\rho R\theta$ $x_c = \frac{R \sin\theta}{\theta}$ $y_c = 0$ $z_c = 0$	$I_x = I_{x_c}$ $= \frac{MR^2(\theta - \sin\theta\cos\theta)}{2\theta}$ $I_y = \frac{MR^2(\theta + \sin\theta\cos\theta)}{2\theta}$ $I_z = MR^2$	$r_x^2 = r_{x_c}^2 = \frac{R^2(\theta - \sin\theta\cos\theta)}{2\theta}$ $r_y^2 = \frac{R^2(\theta + \sin\theta\cos\theta)}{2\theta}$ $r_z^2 = R^2$	$I_{x_c y_c}$, etc. $= 0$ I_{xy}, etc. $= 0$
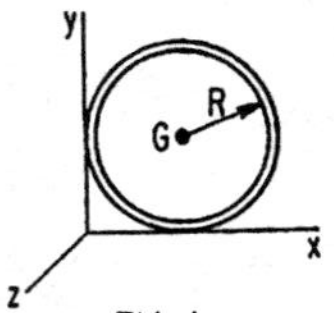 Thin hoop	$M = 2\pi\rho R$ $x_c = R$ $y_c = R$ $z_c = 0$	$I_{x_c} = I_{y_c} = \frac{M}{2} R^2$ $I_{z_c} = MR^2$ $I_x = I_y = \frac{3}{2}MR^2$ $I_z = 3MR^2$	$r_{x_c}^2 = r_{y_c}^2 = \frac{1}{2}R^2$ $r_{z_c}^2 = R^2$ $r_x^2 = r_y^2 = \frac{3}{2}R^2$ $r_z = 3R^2$	$I_{x_c y_c}$, etc. $= 0$ $I_{xy} = MR^2$ $I_{xz} = I_{yz} = 0$
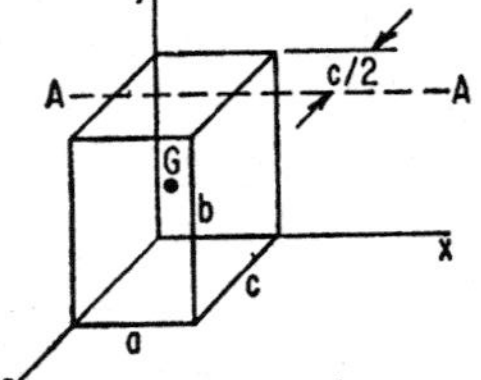 Rectangular prism	$M = \rho abc$ $x_c = \frac{1}{2}a$ $y_c = \frac{1}{2}b$ $z_c = \frac{1}{2}c$	$I_{x_c} = \frac{1}{12}M(b^2 + c_2)$ $I_x = \frac{1}{3}M(b^2 + c^2)$ $I_{AA} = \frac{1}{12}M(4b^2 + c^2)$	$r_{x_c}^2 = \frac{1}{12}(b^2 + c^2)$ $r_x^2 = \frac{1}{3}(b^2 + c^2)$ $r_{AA}^2 = \frac{1}{12}(4b^2 + c^2)$	$I_{x_c y_c}$, etc. $= 0$ $I_{xy} = \frac{1}{4}Mab$ $I_{xz} = \frac{1}{4}Mac$ $I_{yz} = \frac{1}{4}Mbc$

TABLE A.8 Properties of Homogeneous Bodies (*Continued*)

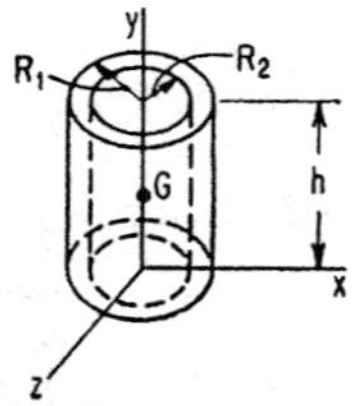 Hollow right circular cylinder	$M = \pi\rho h(R_1^2 - R_2^2)$ $x_c = 0$ $y_c = \frac{1}{2}h$ $z_c = 0$	$I_{x_c} = I_{z_c}$ $= \frac{1}{12}M(3R_1^2 + 3R_2^2 + h^2)$ $I_{y_c} = I_y = \frac{1}{2}M(R_1^2 + R_2^2)$ $I_x = I_z$ $= \frac{1}{12}M(3R_1^2 + 3R_2^2 + 4h^2)$	$r_{x_c}^2 = r_{z_c}^2 = \frac{1}{12}(3R_1^2 + 3R_2^2 + h_2)$ $r_{y_c}^2 = r_y^2 = \frac{1}{2}(R_1^2 + R_2^2)$ $r_x^2 = r_z^2 = \frac{1}{12}(3R_1^2 + 3R_2^2 + 4h^2)$	$I_{x_c y_c}$, etc. $= 0$ I_{xy}, etc. $= 0$
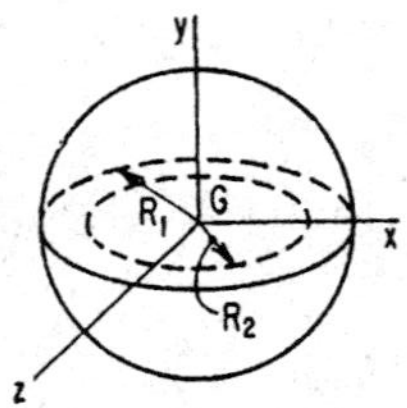 Hollow sphere	$M = \frac{4}{3}\pi\rho(R_1^3 - R_2^3)$ $x_c = 0$ $y_c = 0$ $z_c = 0$	$I_x = I_y = I_z = \frac{2}{5}M\frac{R_1^5 - R_2^5}{R_1^3 - R_2^3}$	$r_x^2 = r_y^2 = r_z^2 = \frac{2}{5}\frac{R_1^5 - R_2^5}{R_1^3 - R_2^3}$	I_{xy}, etc. $= 0$
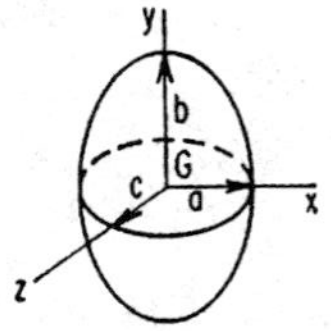 Ellipsoid	$M = \frac{4}{3}\pi\rho abc$ $x_c = 0$ $y_c = 0$ $z_c = 0$	$I_x = \frac{1}{5}M(b^2 + c^2)$ $I_y = \frac{1}{5}M(a^2 + c^2)$ $I_z = \frac{1}{5}M(a^2 + b^2)$	$r_x^2 = \frac{1}{5}(b^2 + c^2)$ $r_y^2 = \frac{1}{5}(a^2 + c^2)$ $r_z^2 = \frac{1}{5}(a^2 + b^2)$	I_{xy}, etc. $= 0$

*From G. W. Housner and D. E. Hudson, "Applied Mechanics," vol. II, "Dynamics," D. Van Nostrand Co., Inc., Princeton, N.J., 1959.

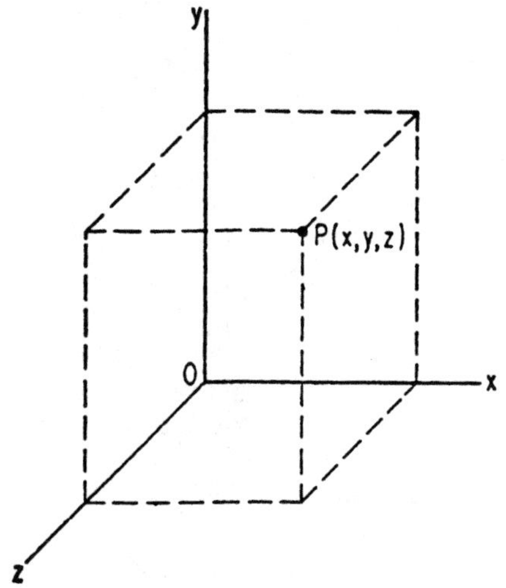

FIG. A.18 Rectangular-coordinate system.

A.4.6 Differential Calculus

Definitions. The derivative of a function $y = f(x)$ of a single variable x is

$$\frac{dy}{dx} = \lim_{\Delta x \to 0} \frac{\Delta y}{\Delta x} = \lim_{\Delta x \to 0} \frac{f(x + \Delta x) - f(x)}{\Delta x} = f'(x)$$

i.e., the derivative of the function y is the limit of the ratio of the increment of the function Δy to the increment of the independent variable Δx as the increment of x-varies and approaches zero as a limit. The derivative at a point can also be shown to equal the slope of the tangent line to the curve at the same point; i.e., $dy/dx = \tan \theta$, as shown in Fig. A.21. The derivative of $f(x)$ is a function of x and may also be differentiated with respect to x. The first differentiation of the first derivative yields the second derivative of the function d^2y/dx^2 or $f''(x)$. Similarly, the third derivative d^3y/dx^3 or $f'''(x)$ of the function is the first derivative of d^2y/dx^2, and so on.

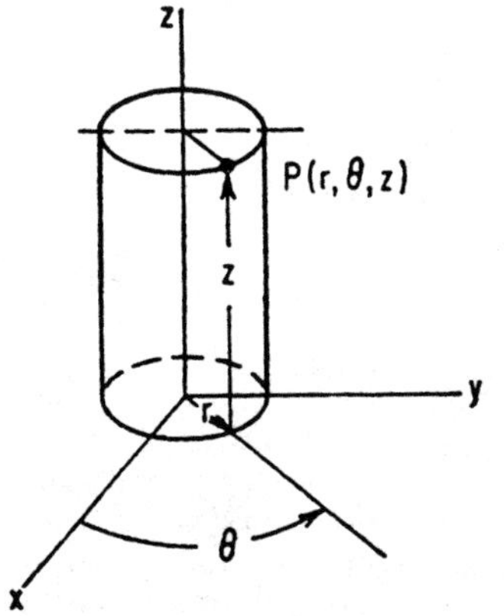

FIG. A.19 Cylindrical-coordinate system.

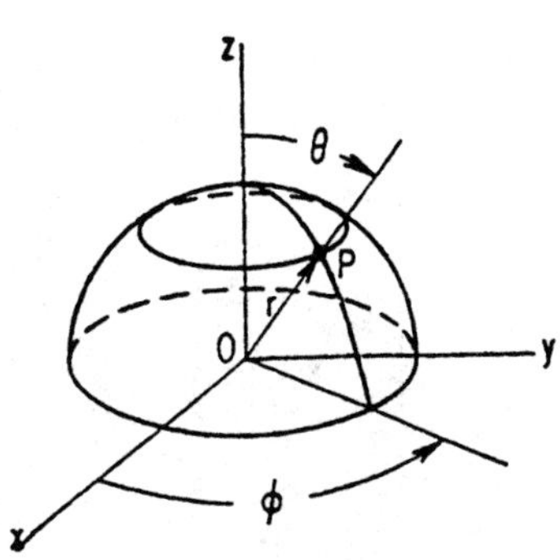

FIG. A.20 Spherical-coordinate system.

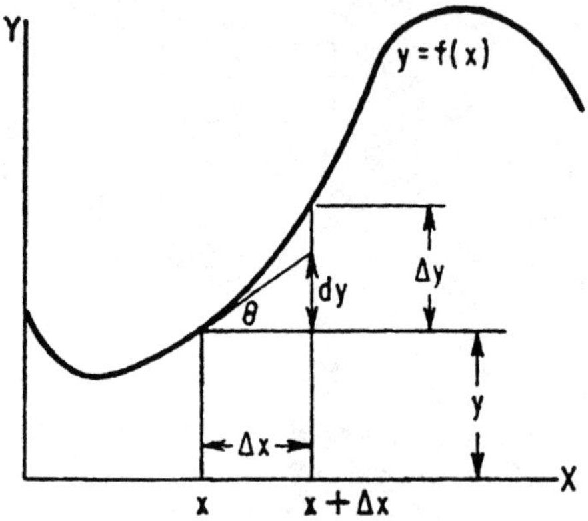

FIG. A.21 Derivative representation.

From Fig. A.21 it is seen that the function $f(x)$ possesses a maximum value where the derivative is zero and the concavity is downward, and the function possesses a minimum value where the slope is zero and the curve has an upward concavity. If the function $f(x)$ is concave upward, the second derivative will have a positive value; if negative, the curve will be concave downward. If the second derivative equals zero at a point, that point is a point of inflection. More particularly, where the nature of the curve is not well known, Table A.9 may be used to adjudge the significance of the derivatives.

The "partial derivative" of a function $u = u(x, y)$ of two variables, taken with respect to the variable x, is defined by

$$\frac{\partial u}{\partial x} = \lim_{\Delta x \to 0} \frac{u(x + \Delta x, y) - u(x, y)}{\Delta x} = u_x = D_x u$$

The partial derivatives are taken by differentiating with respect to one of the variables only, regarding the remaining variables as momentarily constant. Thus the partial derivative of u with respect to x of the function $u = 2xy^2$ is equal to $2y^2$; similarly $\partial u/\partial y = 4xy$. Higher derivatives are similarly formed. Thus

$$\partial^2 u/\partial y^2 = u_{yy} = 4x \qquad \partial^2 u/\partial x^2 = u_{xx} = 0$$

$$\partial^2 u/\partial x\, \partial y = u_{xy} = 4y \qquad \text{and} \qquad \partial^2 u/\partial y\, \partial x = 4y$$

TABLE A.9 Significance of the Derivative

$f'(x)\Big]_{x_0}$	$f''(x)\Big]_{x_0}$	$f'''(x)\Big]_{x_0}$	$f''''(x)\Big]_{x_0}$	Comment
0	<0			x_0 a maximum point
0	>0			x_0 a minimum point
0	0	≠0		x_0 a point of inflection
0	0	0	<0	x_0 a maximum point
0	0	0	>0	x_0 a minimum point

TABLE A.10 Conditions for Maxima, Minima, and Saddle Points

$\left.\frac{\partial u}{\partial x}\right]_{x_0,y_0}$	$\left.\frac{\partial u}{\partial y}\right]_{x_0,y_0}$	$\left.\frac{\partial u^2}{\partial x^2}\right]_{x_0,y_0}$	$\left.\frac{\partial^2 u}{\partial y^2}\right]_{x_0,y_0}$	Comments
0	0	<0	<0	$u(x,y)$ maximum at x_0, y_0 if $(u_{xx})(u_{yy}) - (u_{xy})^2 > 0$
0	0	>0	>0	$u(x,y)$ minimum at x_0, y_0 if $(u_{xx})(u_{yy}) - (u_{xy})^2 > 0$
0	0			Saddle point if u_{xx} and u_{yy} are of different sign

The order of differentiation in obtaining the mixed derivatives in immaterial if the derivatives are continuous.

Table A.10 gives the conditions required to determine maxima, minima, and saddle points for $u = u(x, y)$

Implicit Functions. If y is an implicit function of x, as, for example,

$$xy - 5x^3y^2 = 3$$

and if it is difficult to solve the equation for y (or x), differentiate the terms as given, treating y as a function of x and solving for dy/dx. Thus, taking the derivatives of the above expression,

$$(d/dx)(xy) - (d/dx)(5x^3y^2) = (d/dx)(3)$$

$$x(dy/dx) + y - 5x^3 2y(dy/dx) - y^2 15x^2 = 0$$

$$dy/dx = (15x^2y^2 - y)/(x - 10x^3y)$$

Another approach utilizes the relationship $dy/dx = -f_x/f_y$. Thus, in the foregoing example,

$$f_x = y - y^2 15x^2 \qquad f_y = x - 10x^3y$$

Curvature. The curvature K at a point P of a curve $y = f(x)$ (Fig. A.22)

$$K \equiv \left| \lim_{\Delta s \to 0} \frac{\Delta \gamma}{\Delta s} \right| = \frac{d\gamma}{ds}$$

A working expression for the curvature is

$$K = y''/[1 + (y')^2]^{3/2}$$

where the derivatives are evaluated at the point P. For a curve described in polar coordinates, the corresponding expression for the curvature is

$$K = (\rho^2 + 2\rho'^2 - \rho\rho'')/(\rho^2 + \rho'^2)^{3/2}$$

where ρ' and ρ'' represent the first and second derivatives of ρ with respect to θ.

The circle of curvature tangent, on its concave side, the curve $y = f(x)$ at P. The circle of curvature and the given curve had equal curvature at P. The circle of curvature is described by a radius of curvature R located at the center of curvature.

$$R = 1/K = (1 + y'^2)^{3/2}/y''$$

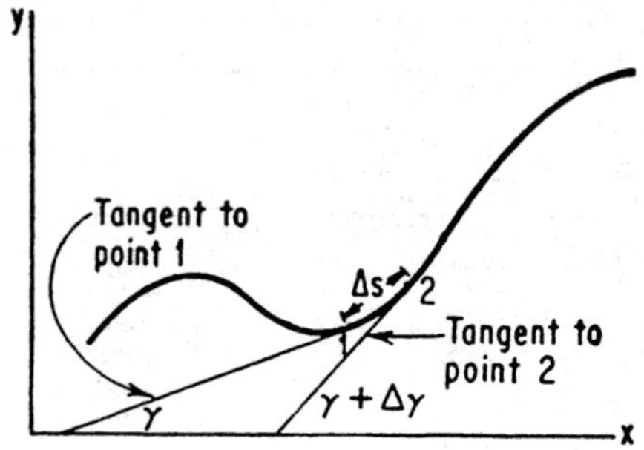

FIG. A.22 Curvature.

The center of curvature is located at (α, β) given by the following expressions:

$$\alpha = x - y'(1 + y'^2)/y'' \qquad \beta = y + (1 + y'^2)/y''$$

where the expressions are evaluated at the point P.

Table A.11 shows derivatives of functions.

Differentials. The differential of a function is equal to the derivative of the function multiplied by the differential of the independent variable. Thus,

$$dy = dy/dx\, dx = f'(x)\, dx$$

The total differential dz of a function of two variables $z = z(x, y)$ is

$$dz = (\partial z/\partial x)\, dx + (\partial z/\partial y)\, dy$$

$$dz/dt = (\partial z/\partial x)(dx/dt) + (\partial z/\partial y)(dy/dt)$$

For a function of three variables $u = u(x, y, z)$,

$$du = (\partial u/\partial x)\, dx + (\partial u/\partial y)\, dy + (\partial u/\partial z)\, dz$$

$$du/dt = (\partial u/\partial x)(dx/dt) + (\partial u/\partial y)(dy/dt) + (\partial u/\partial z)(dz/dt)$$

x, y, z being functions of the independent variable t.

In the following relationships for differential of arc in rectangular coordinates, ds represents the differential of arc and τ is the angle of the tangent drawn at the point in question; i.e., $\tan \tau$ = slope

$$ds^2 = dx^2 + dy^2$$

$$ds = [1 + (dy/dx)^2]^{1/2}\, dx = [1 + (dx/dy)^2]^{1/2}\, dy$$

$$dx/ds = \cos \tau = 1/(1 + y'^2)^{1/2} \qquad dy/ds = \sin \tau = y'/(1 + y'^2)^{1/2}$$

In the following relationships for differential of arc in polar coordinates for the function $\rho = \rho(\theta)$, ds represents the differential of arc:

$$ds = \sqrt{d\rho^2 + \rho^2\, d\theta^2} \qquad ds = [\rho^2 + (d\rho/d\theta)^2]^{1/2}\, d\theta$$

Indeterminate Forms. The function $f(x) = u(x)/v(x)$ has an indeterminate form 0/0 at $x = a$ if $u(x)$ and $v(x)$ each approach zero as x approaches a through values greater

TABLE A.11 Derivatives u, v, and w are functions of x

$$\frac{d}{dx}(a) = 0 \qquad a = \text{constant}$$

$$\frac{d}{dx}(x) = 1$$

$$\frac{dy}{dx} = \frac{dy}{dv}\frac{dv}{dx} \qquad y = y(v)$$

$$\frac{d}{dx}(au) = a\frac{du}{dx}$$

$$\frac{dy}{dx} = \frac{1}{dx/dy} \qquad \text{if } \frac{dx}{dy} \neq 0$$

$$\frac{d}{dx}(\pm u \pm v \pm \cdots) = \pm\frac{du}{dx} \pm \frac{dv}{dx} \pm \cdots$$

$$\frac{d}{dx}(u^n) = nu^{n-1}\frac{du}{dx}$$

$$\frac{d}{dx}(uv) = u\frac{dv}{dx} + v\frac{du}{dx}$$

$$\frac{d}{dx}\frac{u}{v} = \frac{v\,du/dx - u\,dv/dx}{v^2}$$

$$\frac{d}{dx}(u^v) = vu^{v-1}\frac{du}{dx} + u^v \ln u \frac{dv}{dx}$$

$$\frac{d}{dx}(a^u) = a^u \ln a \frac{du}{dx}$$

$$\frac{d}{dx}(e^u) = e^u\frac{du}{dx}$$

$$\frac{d}{dx}(\ln u) = \frac{1}{u}\frac{du}{dx}$$

$$\frac{d}{dx}(\log_a u) = \frac{\log_a e}{u}\frac{du}{dx}$$

$$\frac{d}{dx}(\sin u) = \cos u\frac{du}{dx}$$

$$\frac{d}{dx}(\cos u) = -\sin u\frac{du}{dx}$$

$$\frac{d}{dx}(\tan u) = \sec^2 u\frac{du}{dx}$$

$$\frac{d}{dx}(\csc u) = -\csc u \cot u\frac{du}{dx}$$

$$\frac{d}{dx}(\sec u) = \sec u \tan u\frac{du}{dx}$$

$$\frac{d}{dx}(\cot u) = -\csc^2 u\frac{du}{dx}$$

$$\frac{d}{dx}(\text{vers } u) = \sin u\frac{du}{dx}$$

$$\frac{d}{dx}\sin^{-1} u = \frac{1}{\sqrt{1-u^2}}\frac{du}{dx} \qquad \frac{-\pi}{2} \leqq \sin^{-1} u \leqq \frac{\pi}{2}$$

$$\frac{d}{dx}\cos^{-1} u = -\frac{1}{\sqrt{1-u^2}}\frac{du}{dx} \qquad 0 \leqq \cos^{-1} u \leqq \pi$$

$$\frac{d}{dx}\tan^{-1} u = \frac{1}{1+u^2}\frac{du}{dx}$$

TABLE A.11 Derivatives *u*, *v*, and *w* are functions of *x* (*Continued*)

$\frac{d}{dx}\sinh^{-1} u = \frac{1}{\sqrt{u^2+1}}\frac{du}{dx}$	
$\frac{d}{dx}\cosh^{-1} u = \frac{1}{\sqrt{u^2-1}}\frac{du}{dx}$	$u > 1$
$\frac{d}{dx}\tanh^{-1} u = \frac{1}{1-u^2}\frac{du}{dx}$	
$\frac{d}{dx}\operatorname{csch}^{-1} u = -\frac{1}{u\sqrt{u^2+1}}\frac{du}{dx}$	
$\frac{d}{dx}\operatorname{sech}^{-1} u = -\frac{1}{u\sqrt{1-u^2}}\frac{du}{dx}$	$u > 0$
$\frac{d}{dx}\coth^{-1} u = \frac{1}{1-u^2}\frac{du}{dx}$	
$\frac{d}{dx}\csc^{-1} u = -\frac{1}{u\sqrt{u^2-1}}\frac{du}{dx}$	$-\pi < \csc^{-1} u \leqq -\frac{\pi}{2}, 0 < \csc^{-1} u \leqq \frac{\pi}{2}$
$\frac{d}{dx}\sec^{-1} u = \frac{1}{u\sqrt{u^2-1}}\frac{du}{dx}$	$-\pi \leqq \sec^{-1} u < -\frac{\pi}{2}, 0 \leqq \sec^{-1} u < \frac{\pi}{2}$
$\frac{d}{dx}\cot^{-1} u = \frac{-1}{1+u^2}\frac{du}{dx}$	
$\frac{d}{dx}\operatorname{vers}^{-1} u = \frac{1}{\sqrt{2u-u^2}}\frac{du}{dx}$	$0 \leqq \operatorname{vers}^{-1} u \leqq \pi$
$\frac{d}{dx}\sinh u = \cosh u \frac{du}{dx}$	
$\frac{d}{dx}\cosh u = \sinh u \frac{du}{dx}$	
$\frac{d}{dx}\tanh u = \operatorname{sech}^2 u \frac{du}{dx}$	
$\frac{d}{dx}\operatorname{csch} u = -\operatorname{csch} u \coth u \frac{du}{dx}$	
$\frac{d}{dx}\operatorname{sech} u = -\operatorname{sech} u \tanh u \frac{du}{dx}$	
$\frac{d}{dx}\coth u = -\operatorname{csch}^2 u \frac{du}{dx}$	

than a ($x \to a+$). The function $f(x)$ is not defined at $x = a$, and therefore, it is often useful to assign a value to $f(a)$. L'Hôpital's rule is readily applied to indeterminacies of the form 0/0:

$$\lim_{x\to a+} u(x)/v(x) = \lim_{x\to a+} u'(x)/v'(x)$$

L'Hôpital's rule may be reapplied as often as necessary, but it is important to remember to differentiate numerator and denominator separately. The above discussion is equally valid if $x \to a-$.

Other indeterminate forms such as ∞/∞, $0 \cdot \infty$, $\infty - \infty$, 0^0, ∞^0, and 1^∞ may also be evaluated by L'Hôpital's rule by changing their forms. For example, in order to evaluate the indeterminate form $0 \cdot \infty$, the function $u(x)v(x)$ may be written $u(x)/[1/v(x)]$ and the same technique employed as before.

A.4.7 Integral Calculus

Indefinite Integrals. The inverse operation of differentiation is integration; i.e., given a differential of a function, the process of integration yields the original function. This process is denoted by the integral sign in front of the differential of the function. Thus

$$\int f(x)dx = F(x) + C$$

if

$$f(x)\,dx = dF(x)$$

or

$$dF(x)/dx = f(x)$$

C is an arbitrary constant of integration, and the expression $F(x) + C$ is the indefinite integral of $f(x)$ [or $F'(x)$]. The constant of integration is required because any constant term in the function $F(x)$ does not appear in $f(x)$. Table A.12 shows Table of Integrals.

Definite Integrals. The difference between the indefinite integral evaluated at b, i.e., the variable equated to b, and the same integral evaluated at a is a definite integral with upper limit b and lower limit a. This is termed "integration between limits." Since the constant of integration cancels in subtraction, it need not be introduced. Thus,

$$\int_a^b f(x)\,dx = F(x)\Big|_a^b = F(b) - F(a)$$

which is always a definite number, can be interpreted geometrically as the area bounded by the x axis, the lines $x = a$ and $x = b$, and the curve $y = f(x)$ provided a and b are finite, $f(x)$ does not cross the x axis, and $f(x)$ is not infinite between a and b. The following fundamental theorems apply to definite integrals:

$$\int_a^b f(x)\,dx = -\int_b^a f(x)\,dx$$

$$\int_a^b f(x)\,dx = \int_a^\phi f(x)\,dx + \int_\phi^b f(x)\,dx$$

$$\int_a^b Kf(x)\,dx = K\int_a^b f(x)\,dx$$

where K = a constant

$$\int_a^b [f_1(x) + f_2(x) + \cdots + f_n(x)]\,dx = \int_a^b f_1(x)\,dx + \int_a^b f_2(x)\,dx + \cdots + \int_a^b f_n(x)\,dx$$

$$\frac{d}{dx}\int_a^x f(\xi)\,d\xi = f(x)$$

Improper Integrals

$$\int_a^{+\infty} f(x)\,dx = \lim_{b\to+\infty}\int_a^b f(x)\,dx$$

if the limit exists

$$\int_{-\infty}^b f(x)\,dx = \lim_{a\to-\infty}\int_a^b f(x)\,dx$$

TABLE A.12 Table of Integrals

Fundamental Indefinite Integrals

1. $\int a\,dx = ax$
2. $\int (u + v + w + \cdots)\,dx = \int u\,dx + \int v\,dx + \int w\,dx + \cdots$
3. $\displaystyle\int x^n\,dx = \frac{x^{n+1}}{n+1} \qquad (n \neq -1)$
4. $\displaystyle\int \frac{dx}{x} = \log_e x + c = \log_e c_1 x \qquad [\log_e x = \log_e -x) + (2k+1)\pi i]$
5. $\displaystyle\int e^{ax}\,dx = \frac{1}{a} e^{ax}$
6. $\displaystyle\int a^x\,dx = \frac{a^x}{\log_e a}$
7. $\int a^x \log_e a\,dx = a^x$
8. $\displaystyle\int \sin ax\,dx = -\frac{1}{a}\cos ax$
9. $\displaystyle\int \sin^2 ax\,dx = \frac{1}{2}x - \frac{1}{2a}\sin ax \cos ax$
10. $\displaystyle\int \sin^{-1} ax\,dx = x\sin^{-1} ax + \frac{1}{a}\sqrt{1 - a^2x^2}$
11. $\displaystyle\int \cos ax\,dx = \frac{1}{a}\sin ax$
12. $\displaystyle\int \cos^2 ax\,dx = \frac{1}{2}x + \frac{1}{2a}\sin ax \cos ax$
13. $\int \cos^{-1} ax\,dx = x\cos^{-1} ax - 1/a\sqrt{1 - a^2x^2}$
14. $\int \tan ax\,dx = -1/a \log_e \cos ax$
15. $\int \tan^2 ax\,dx = 1/a \tan ax - x$
16. $\int \tan^{-1} ax\,dx = x\tan^{-1} ax - 1/2a \log_e (1 + a^2x^2)$
17. $\int \cot ax\,dx = 1/a \log_e \sin ax$
18. $\int \cot^2 ax\,dx = -1/a \cot ax - x$
19. $\int \cot^{-1} ax\,dx = x\cot^{-1} ax + 1/2a \log_e (1 + a^2x^2)$
20. $\int \sec ax\,dx = 1/a \log_e (\sec ax + \tan ax)$
21. $\int \sec^2 ax\,dx = 1/a \tan ax$
22. $\int \sec^{-1} ax\,dx = x\sec^{-1} ax - 1/a \log_e (ax + \sqrt{a^2x^2 - 1})$
23. $\displaystyle\int \csc ax\,dx = \frac{1}{a}\log_e (\csc ax - \cot ax)$
24. $\int \csc^2 ax\,dx = -1/a \cot ax$
25. $\int \csc^{-1} ax\,dx = x\csc^{-1} ax + 1/a \log_e (ax + \sqrt{a^2x^2 - 1})$
26. $\displaystyle\int \sinh ax\,dx = \frac{1}{a}\cosh ax$
27. $\displaystyle\int \cosh ax\,dx = \frac{1}{a}\sinh ax$
28. $\displaystyle\int \tanh ax\,dx = \frac{1}{a}\log_e (\cosh ax)$
29. $\displaystyle\int \coth ax\,dx = \frac{1}{a}\log_e (\sinh ax)$
30. $\displaystyle\int \operatorname{sech} ax\,dx = \frac{1}{a}\sin^{-1} (\tanh ax)$
31. $\displaystyle\int \operatorname{csch} ax\,dx = \frac{1}{a}\log_e \left(\tanh \frac{ax}{2}\right)$

TABLE A.12 Table of Integrals (*Continued*)

Fundamental Indefinite Integrals (Continued)

32. $\int \frac{dx}{a^2 + x^2} = \frac{1}{a} \tan^{-1} \frac{x}{a}$

33. $\int \frac{dx}{\sqrt{a^2 - x^2}} = \sin^{-1} \frac{x}{a}$

34. $\int \frac{dx}{\sqrt{2x - x^2}} = \text{versin}^{-1} x, \text{ or } -\text{coversin}^{-1} x$

35. $\int \frac{dx}{x\sqrt{x^2 - 1}} = \sec^{-1} x, \text{ or } -\csc^{-1} x$

36. $\int u \frac{dv}{dx} dx = uv - \int v \frac{du}{dx} dx$

37. $\int u\, dv = uv - \int v\, du$

Integrals Involving $\sin^n ax$ *and* $\sin^{-1} x$

38. $\int \sin ax \sin bx\, dx = \frac{\sin (a - b)x}{2(a - b)} - \frac{\sin (a + b)x}{2(a + b)} \quad (a^2 \neq b^2)$

39. $\int \sin nx \sin^m x\, dx = \frac{1}{m + n}\left[-\cos nx \sin^m x + m \int \cos (n - 1)x \sin^{m-1} x\, dx\right]$

40. $\int \sin ax \sin bx \sin cx\, dx = -\frac{1}{4}\left[\frac{\cos (a - b + c)x}{a - b + c} + \frac{\cos (b + c - a)x}{b + c - a} + \frac{\cos (a + b - c)x}{a + b - c} - \frac{\cos (a + b + c)x}{a + b + c}\right]$

41. $\int (\sin^{-1} x)^2\, dx = x(\sin^{-1} x)^2 - 2x + 2\sqrt{1 - x^2} \sin^{-1} x$

42. $\int x \sin^{-1} x\, dx = \frac{1}{4}[(2x^2 - 1) \sin^{-1} x + x\sqrt{1 - x^2}]$

43. $\int x^n \sin^{-1} x\, dx = \frac{1}{n + 1}\left(x^{n+1} \sin^{-1} x - \int \frac{x^{n+1}\, dx}{\sqrt{1 - x^2}}\right)$

44. $\int \sin^3 ax\, dx = -\frac{1}{a} \cos ax + \frac{1}{3a} \cos^3 ax$

45. $\int \sin^4 ax\, dx = \frac{3}{8} x - \frac{1}{4a} \sin 2ax + \frac{1}{32a} \sin 4ax$

46. $\int \sin^n ax\, dx = -\frac{\sin^{n-1} ax \cos ax}{na} + \frac{n - 1}{n} \int \sin^{n-2} ax\, dx \quad (n = \text{positive integer})$

47. $\int x \sin ax\, dx = \frac{\sin ax}{a^2} - \frac{x \cos ax}{a}$

48. $\int x^2 \sin ax\, dx = \frac{2x}{a^2} \sin ax - \left(\frac{x^2}{a} - \frac{2}{a^3}\right) \cos ax$

49. $\int x^3 \sin ax\, dx = \left(\frac{3x^2}{a^2} - \frac{6}{a^4}\right) \sin ax - \left(\frac{x^3}{a} - \frac{6x}{a^3}\right) \cos ax$

50. $\int x^n \sin ax\, dx = -\frac{x^n}{a} \cos ax + \frac{n}{a} \int x^{n-1} \cos ax\, dx \quad (n > 0)$

51. $\int e^{ax} \sin bx\, dx = \frac{e^{ax}}{a^2 + b^2} (a \sin bx - b \cos bx)$

52. $\int xe^{ax} \sin bx\, dx = \frac{xe^{ax}}{a^2 + b^2} (a \sin bx - b \cos bx) - \frac{e^{ax}}{(a^2 + b^2)^2} [(a^2 - b^2) \sin bx - 2ab \cos bx]$

53. $\int \frac{dx}{\sin^n ax} = -\frac{1}{a(n - 1)} \frac{\cos ax}{\sin^{n-1} ax} + \frac{n - 2}{n - 1} \int \frac{dx}{\sin^{n-2} ax} \quad (n \text{ integer} > 1)$

TABLE A.12 Table of Integrals (*Continued*)

Integrals Involving $\sin^n ax$ *and* $\sin^{-1} x$ *(Continued)*

54. $\int \frac{x\,dx}{\sin^2 ax} = -\frac{x}{a}\cot ax + \frac{1}{a^2}\log_e \sin ax$

55. $\int \frac{dx}{1+\sin ax} = -\frac{1}{a}\tan\left(\frac{\pi}{4}-\frac{ax}{2}\right)$

56. $\int \frac{dx}{1-\sin ax} = \frac{1}{a}\cot\left(\frac{\pi}{4}-\frac{ax}{2}\right)$

57. $\int \frac{x\,dx}{1+\sin ax} = -\frac{x}{a}\tan\left(\frac{\pi}{4}-\frac{ax}{2}\right) + \frac{2}{a^2}\log_e \cos\left(\frac{\pi}{4}-\frac{ax}{2}\right)$

58. $\int \frac{x\,dx}{1-\sin ax} = \frac{x}{a}\cot\left(\frac{\pi}{4}-\frac{ax}{2}\right) + \frac{2}{a^2}\log_e \sin\left(\frac{\pi}{4}-\frac{ax}{2}\right)$

59. $\int \frac{dx}{b+d\sin ax} = \frac{-2}{a\sqrt{b^2-d^2}}\tan^{-1}\left[\sqrt{\frac{b-d}{b+d}}\tan\left(\frac{\pi}{4}-\frac{ax}{2}\right)\right] \quad (b^2 > d^2)$

60. $\int \frac{dx}{b+d\sin ax} = \frac{-1}{a\sqrt{d^2-b^2}}\log_e \frac{d+b\sin ax+\sqrt{d^2-b^2}\cos ax}{b+d\sin ax} \quad (d^2 > b^2)$

61. $\int \frac{\sin ax}{x^n}\,dx = -\frac{1}{n-1}\frac{\sin ax}{x^{n-1}} + \frac{a}{n-1}\int \frac{\cos ax}{x^{n-1}}\,dx$

62. $\int \frac{\sin^n x\,dx}{x^m} = \frac{1}{(m-1)(m-2)}\left[-\frac{\sin^{n-1} x((m-2)\sin x + nx\cos x)}{x^{m-1}} - n^2\int \frac{\sin^n x\,dx}{x^{m-2}} + n(n-1)\int \frac{\sin^{n-2} x\,dx}{x^{m-2}}\right]$

63. $\int \frac{\sin nx\,dx}{\sin^m x} = 2\int \frac{\cos(n-1)x\,dx}{\sin^{m-1} x} + \int \frac{\sin(n-2)x\,dx}{\sin^m x}$

64. $\int \frac{\sin x}{x}\,dx = x - \frac{x^3}{3\times 3!} + \frac{x^5}{5\times 5!} - \frac{x^7}{7\times 7!} + \frac{x^9}{9\times 9!}\cdots$

65. $\int \frac{\sin x}{x^m}\,dx = -\frac{1}{m-1}\frac{\sin x}{x^{m-1}} + \frac{1}{m-1}\int \frac{\cos x}{x^{m-1}}\,dx$

66. $\int \frac{x\,dx}{\sin x} = x + \frac{x^3}{3\times 3!} + \frac{7x^5}{3\times 5\times 5!} + \frac{31x^7}{3\times 7\times 7!} + \frac{127x^9}{3\times 5\times 9!} + \cdots$

Integrals Involving $\cos^n ax$ *and* $\cos^{-1} x$

67. $\int \cos ax\cos bx\,dx = \frac{\sin(a-b)x}{2(a-b)} + \frac{\sin(a+b)x}{2(a+b)} \quad (a^2 \neq b^2)$

68. $\int \cos nx\cos^m x\,dx = \frac{1}{m+n}\left[\sin nx\cos^m x + m\int \cos(n-1)x\cos^{m-1} x\,dx\right]$

69. $\int \cos ax\cos bx\cos cx\,dx = \frac{1}{4}\left[\frac{\sin(a+b+c)x}{a+b+c} + \frac{\sin(b+c-a)x}{b+c-a} + \frac{\sin(a-b+c)x}{a-b+c} + \frac{\sin(a+b-c)x}{a+b-c}\right]$

70. $\int(\cos^{-1} x)^2\,dx = x(\cos^{-1} x)^2 - 2x - 2\sqrt{1-x^2}\cos^{-1} x$

71. $\int x\cos^{-1} x\,dx = \frac{1}{4}[(2x^2-1)\cos^{-1} x - x\sqrt{1-x^2}]$

72. $\int x^n\cos^{-1} x\,dx = \frac{1}{n+1}\left(x^{n+1}\cos^{-1} x + \int \frac{x^{n+1}\,dx}{\sqrt{1-x^2}}\right)$

73. $\int \cos^3 ax\,dx = \frac{1}{a}\sin ax - \frac{1}{3a}\sin^3 ax$

74. $\int \cos^4 ax\,dx = \frac{3}{8}x + \frac{1}{4a}\sin 2ax + \frac{1}{32a}\sin 4ax$

TABLE A.12 Table of Integrals (*Continued*)

Integrals Involving $\cos^n ax$ *and* $\cos^{-1} x$ *(Continued)*

75. $$\int \cos^n ax\,dx = \frac{\cos^{n-1} ax \sin ax}{na} + \frac{n-1}{n}\int \cos^{n-2} ax\,dx \quad (n = \text{positive integer})$$

76. $$\int x \cos ax\,dx = \frac{\cos ax}{a^2} + \frac{x \sin ax}{a}$$

77. $$\int x^2 \cos ax\,dx = \frac{2x}{a^2}\cos ax + \left(\frac{x^2}{a} - \frac{2}{a^3}\right)\sin ax$$

78. $$\int x^3 \cos ax\,dx = \left(\frac{3x^2}{a^2} - \frac{6}{a^4}\right)\cos ax + \left(\frac{x^3}{a} - \frac{6x}{a^3}\right)\sin ax$$

79. $$\int x^n \cos ax\,dx = \frac{x^n \sin ax}{a} - \frac{n}{a}\int x^{n-1}\sin ax\,dx \quad (n > 0)$$

80. $$\int e^{ax}\cos bx\,dx = \frac{e^{ax}}{a^2+b^2}(a\cos bx + b\sin bx)$$

81. $$\int xe^{ax}\cos bx\,dx = \frac{xe^{ax}}{a^2+b^2}(a\cos bx + b\sin bx) - \frac{e^{ax}}{(a^2+b^2)^2}[(a^2-b^2)\cos bx + 2ab\sin bx]$$

82. $$\int \frac{\cos ax}{x^n}\,dx = -\frac{1}{n-1}\frac{\cos ax}{x^{n-1}} - \frac{a}{n-1}\int \frac{\sin ax}{x^{n-1}}\,dx$$

83. $$\int \frac{dx}{\cos^n ax} = \frac{1}{a(n-1)}\frac{\sin ax}{\cos^{n-1} ax} + \frac{n-2}{n-1}\int \frac{dx}{\cos^{n-2} ax} \quad (n \text{ integer} > 1)$$

84. $$\int \frac{x\,dx}{\cos^2 ax} = \frac{x}{a}\tan ax + \frac{1}{a^2}\log_e \cos ax$$

85. $$\int \frac{dx}{1+\cos ax} = \frac{1}{a}\tan\frac{ax}{2}$$

86. $$\int \frac{dx}{1-\cos ax} = -\frac{1}{a}\cot\frac{ax}{2}$$

87. $$\int \frac{x\,dx}{1+\cos ax} = \frac{x}{a}\tan\frac{ax}{2} + \frac{2}{a^2}\log_e\cos\frac{ax}{2}$$

88. $$\int \frac{x\,dx}{1-\cos ax} = -\frac{x}{a}\cot\frac{ax}{2} + \frac{2}{a^2}\log_e\sin\frac{ax}{2}$$

89. $$\int \frac{dx}{b+d\cos ax} = \frac{2}{a\sqrt{b^2-d^2}}\tan^{-1}\left(\sqrt{\frac{b-d}{b+d}}\tan\frac{ax}{2}\right) \quad (b^2 > d^2)$$

90. $$\int \frac{dx}{b+d\cos ax} = \frac{1}{a\sqrt{d^2-b^2}}\log_e\frac{d+b\cos ax+\sqrt{d^2-b^2}\sin ax}{b+d\cos ax} \quad (d^2 > b^2)$$

91. $$\int \frac{\cos^n x\,dx}{x^m} = \frac{1}{(m-1)(m-2)}\left[\frac{\cos^{n-1} x(nx\sin x - (m-2)\cos x)}{x^{m-1}} - n^2\int \frac{\cos^n x\,dx}{x^{m-2}} + n(n-1)\int \frac{\cos^{n-2} x\,dx}{x^{m-2}}\right]$$

92. $$\int \frac{\sin nx\,dx}{\cos^m x} = 2\int \frac{\sin (n-1)x\,dx}{\cos^{m-1} x} - \int \frac{\sin (n-2)x\,dx}{\cos^m x}$$

93. $$\int \frac{\cos x}{x}\,dx = \log x - \frac{x^2}{2\times 2!} + \frac{x^4}{4\times 4!} - \frac{x^6}{6\times 6!} + \frac{x^8}{8\times 8!}\cdots$$

94. $$\int \frac{\cos x}{x^m}\,dx = -\frac{1}{m-1}\frac{\cos x}{x^{m-1}} - \frac{1}{m-1}\int \frac{\sin x}{x^{m-1}}\,dx$$

95. $$\int \frac{x\,dx}{\cos x} = \frac{x^2}{2} + \frac{x^4}{4\times 2!} + \frac{5x^6}{6\times 4!} + \frac{61x^8}{8\times 6!} + \frac{1{,}385\,x^{10}}{10\times 8!} + \cdots$$

TABLE A.12 Table of Integrals (*Continued*)

Integrals Involving $\sin^n ax$ *and* $\cos^n ax$

96. $$\int \sin ax \cos bx\,dx = -\frac{1}{2}\left[\frac{\cos(a-b)x}{a-b} + \frac{\cos(a+b)x}{a+b}\right] \quad (a^2 \neq b^2)$$

97. $$\int \sin^n ax \cos ax\,dx = \frac{1}{a(n+1)} \sin^{n+1} ax \quad (n \neq -1)$$

98. $$\int \sin ax \cos^n ax\,dx = -\frac{1}{a(n+1)} \cos^{n+1} ax \quad (n \neq -1)$$

99. $$\int \sin^2 ax \cos^2 ax\,dx = \frac{x}{8} - \frac{\sin 4ax}{32a}$$

100. $$\int \sin^n ax \cos^m ax\,dx = -\frac{\sin^{n-1} ax \cos^{m+1} ax}{a(n+m)} + \frac{n-1}{n+m}\int \sin^{n-2} ax \cos^m ax\,dx$$ $(m, n \text{ positive})$

101. $$\int \frac{dx}{\sin ax \cos ax} = \frac{1}{a}\log_e \tan ax$$

102. $$\int \frac{dx}{b \sin ax + d \cos ax} = \frac{1}{a\sqrt{b^2+d^2}} \log_e \tan \tfrac{1}{2}\left(ax + \tan^{-1}\frac{d}{b}\right)$$

103. $$\int \frac{\sin ax}{b + d \cos ax}\,dx = -\frac{1}{ad}\log_e (b + d \cos ax)$$

104. $$\int \frac{\cos ax}{b + d \sin ax}\,dx = \frac{1}{ad}\log_e (b + d \sin ax)$$

105. $$\int \frac{\sin^n ax}{\cos^m ax}\,dx = \frac{\sin^{n+1} ax}{a(m-1)\cos^{m-1} ax} - \frac{n-m+2}{m-1}\int \frac{\sin^n ax}{\cos^{m-2} ax}\,dx$$ $(m, n \text{ positive}, m \neq 1)$

106. $$\int \frac{\cos^m ax}{\sin^n ax}\,dx = \frac{-\cos^{m+1} ax}{a(n-1)\sin^{n-1} ax} + \frac{n-m-2}{(n-1)}\int \frac{\cos^m ax}{\sin^{n-2} ax}\,dx$$ $(m, n \text{ positive}, n \neq 1)$

107. $\int \sin x \cos x\,dx = \tfrac{1}{2}\sin^2 x$

108. $$\int \frac{dx}{\sin^m x \cos^n x} = \frac{1}{n-1}\,\frac{1}{\sin^{m-1} x \cos^{n-1} x} + \frac{m+n-2}{n-1}\int \frac{dx}{\sin^m x \cos^{n-2} x}$$
$$= -\frac{1}{m-1}\,\frac{1}{\sin^{m-1} x \cos^{n-1} x} + \frac{m+n-2}{m-1}\int \frac{dx}{\sin^{m-2} x \cos^n x}$$

109. $$\int \sin ax \cos bx \cos cx\,dx = -\frac{1}{4}\left[\frac{\cos(a+b+c)x}{a+b+c} - \frac{\cos(b+c-a)x}{b+c-a} + \frac{\cos(a+b-c)x}{a+b-c} + \frac{\cos(a+c-b)x}{a+c-b}\right]$$

110. $$\int \cos ax \sin bx \sin cx\,dx = \frac{1}{4}\left[\frac{\sin(a+b-c)x}{a+b-c} + \frac{\sin(a-b+c)x}{a-b+c} - \frac{\sin(a+b+c)x}{a+b+c} - \frac{\sin(b+c-a)x}{b+c-a}\right]$$

111. $$\int e^{ax} \sin^m x \cos^n x\,dx = \frac{1}{(m+n)^2 + a^2}$$
$$\left[e^{ax} \sin^m x \cos^{n-1} x[a \cos x + (m+n)\sin x] - ma\int e^{ax} \sin^{m-1} x \cos^{n-1} x\,dx + (n-1)(m+n)\int e^{ax} \sin^m x \cos^{n-2} x\,dx\right]$$

Integrals Involving $\tan^n ax$, $\sec^n ax$, $\cot^n ax$, *and* $\csc^n ax$

112. $$\int \tan^n ax\,dx = \frac{1}{a(n-1)}\tan^{n-1} ax - \int \tan^{n-2} ax\,dx \quad (n \text{ integer} > 1)$$

TABLE A.12 Table of Integrals (*Continued*)

Integrals Involving $\tan^n ax$, $\sec^n ax$, $\cot^n ax$, *and* $\csc^n ax$ (*Continued*)

113. $\int \tan ax \sec ax \, dx = \frac{1}{a} \sec ax$

114. $\int \tan^n ax \sec^2 ax \, dx = \frac{1}{a(n+1)} \tan^{n+1} ax \quad (n \neq -1)$

115. $\int \frac{\sec^2 ax \, dx}{\tan ax} = \frac{1}{a} \log_e \tan ax$

116. $\int \cot^n ax \, dx = -\frac{1}{a(n-1)} \cot^{n-1} ax - \int \cot^{n-2} ax \, dx \quad (n \text{ integer} > 1)$

117. $\int \cot ax \csc ax \, dx = -\frac{1}{a} \csc ax$

118. $\int \cot^n ax \csc^2 ax \, dx = -\frac{1}{a(n+1)} \cot^{n+1} ax \quad (n \neq -1)$

119. $\int \frac{\csc^2 ax}{\cot ax} dx = -\frac{1}{a} \log_e \cot ax$

120. $\int \sec^n ax \, dx = \frac{1}{a(n-1)} \frac{\sin ax}{\cos^{n-1} ax} + \frac{n-2}{n-1} \int \sec^{n-2} ax \, dx \quad (n \text{ integer} > 1)$

121. $\int \csc^n ax \, dx = -\frac{1}{a(n-1)} \frac{\cos ax}{\sin^{n-1} ax} + \frac{n-2}{n-1} \int \csc^{n-2} ax \, dx \quad (n \text{ integer} > 1)$

122. $\int \frac{dx}{b + d \tan ax} = \frac{1}{b^2 + d^2} \left[bx + \frac{d}{a} \log_e (b \cos ax + d \sin ax) \right]$

123. $\int \frac{dx}{\sqrt{b + d \tan^2 ax}} = \frac{1}{a\sqrt{b-d}} \sin^{-1} \left(\sqrt{\frac{b-d}{b}} \sin ax \right) \quad (b \text{ positive}, b^2 > d^2)$

124. $\int x \tan^{-1} x \, dx = \frac{1}{2}[(x^2+1) \tan^{-1} x - x]$

125. $\int x \cot^{-1} x \, dx = \frac{1}{2}[(x^2+1) \cot^{-1} x + x]$

126. $\int x \sec^{-1} x \, dx = \frac{1}{2}(x^2 \sec^{-1} x - \sqrt{x^2 - 1})$

127. $\int x \csc^{-1} x \, dx = \frac{1}{2}(x^2 \csc^{-1} x + \sqrt{x^2 - 1})$

Integrals Involving $\sinh^n x$, $\cosh^n x$, *and* $\tanh^n x$

128. $\int \sinh^n x \, dx = \frac{1}{n} \sinh^{n-1} x \cosh x - \frac{n-1}{n} \int \sinh^{n-2} x \, dx$
$= \frac{1}{n+1} \sinh^{n+1} x \cosh x - \frac{n+2}{n+1} \int \sinh^{n+2} x \, dx$

129. $\int x \sinh x \, dx = x \cosh x - \sinh x$

130. $\int x^2 \sinh x \, dx = (x^2 + 2) \cosh x - 2x \sinh x$

131. $\int x^n \sinh x \, dx = x^n \cosh x - nx^{n-1} \sinh x + n(n-1) \int x^{n-2} \sinh x \, dx$

132. $\int \cosh^n x \, dx = \frac{1}{n} \sinh x \cosh^{n-1} x + \frac{n-1}{n} \int \cosh^{n-2} x \, dx$
$= -\frac{1}{n+1} \sinh x \cosh^{n+1} x + \frac{n+2}{n+1} \int \cosh^{n+2} x \, dx$

133. $\int x \cosh x \, dx = x \sinh x - \cosh x$

134. $\int \sinh x \cosh x \, dx = \frac{1}{4} \cosh (2x)$

135. $\int \sinh (mx) \sinh (nx) \, dx = \frac{1}{m^2 - n^2} [m \sinh (nx) \cosh (mx) - n \cosh (nx) \sinh (mx)]$

136. $\int \cosh (mx) \cosh (nx) \, dx = \frac{1}{m^2 - n^2} [m \sinh (mx) \cosh (nx) - n \sinh (nx) \cosh (mx)]$

137. $\int \cosh (mx) \sinh (nx) \, dx = \frac{1}{m^2 - n^2} [m \sinh (nx) \sinh (mx) - n \cosh (nx) \cosh (mx)]$

TABLE A.12 Table of Integrals (*Continued*)

Integrals Involving $\sinh^n x$, $\cosh^n x$, *and* $\tanh^n x$ *(Continued)*

138. $\int \sinh x \cos x \, dx = \frac{1}{2}(\cosh x \cos x + \sinh x \sin x)$
139. $\int \sinh x \sin x \, dx = \frac{1}{2}(\cosh x \sin x - \sinh x \cos x)$
140. $\int \cosh x \cos x \, dx = \frac{1}{2}(\sinh x \cos x + \cosh x \sin x)$
141. $\int \cosh x \sin x \, dx = \frac{1}{2}(\sinh x \sin x - \cosh x \cos x)$
142. $\int \sinh^{-1} x \, dx = x \sinh^{-1} x - \sqrt{1 + x^2}$
143. $\int \cosh^{-1} x \, dx = x \cosh^{-1} x - \sqrt{x^2 - 1}$
144. $\int \tanh^2 x \, dx = x - \tanh x$

Integrals Involving Algebraic Functions

145. $$\int (ax + b)^n \, dx = \frac{1}{a(n+1)} (ax + b)^{n+1} \quad (n \neq -1)$$

146. $$\int x(ax + b)^n \, dx = \frac{1}{a^2(n+2)} (ax + b)^{n+2} - \frac{b}{a^2(n+1)} (ax + b)^{n+1} \quad (n \neq -1, -2)$$

147. $$\int \frac{dx}{ax + b} = \frac{1}{a} \log_e (ax + b)$$

148. $$\int \frac{x \, dx}{ax + b} = \frac{x}{a} - \frac{b}{a^2} \log_e (ax + b)$$

149. $$\int \frac{x \, dx}{(ax + b)^2} = \frac{b}{a^2(ax + b)} + \frac{1}{a^2} \log_e (ax + b)$$

150. $$\int \frac{x^2 \, dx}{(ax + b)^2} = \frac{1}{a^3} \left[(ax + b) - 2b \log_e (ax + b) - \frac{b^2}{ax + b} \right]$$

151. $$\int \frac{x^2 \, dx}{ax + b} = \frac{1}{a^3} \left[\frac{1}{2} (ax + b)^2 - 2b(ax + b) + b^2 \log_e (ax + b) \right]$$

152. $$\int \frac{dx}{x(ax + b)} = \frac{1}{b} \log_e \frac{x}{ax + b}$$

153. $$\int \frac{dx}{x(ax + b)^2} = \frac{1}{b(ax + b)} - \frac{1}{b^2} \log_e \frac{ax + b}{x}$$

154. $$\int (ax^2 + b)^n \, x \, dx = \frac{1}{2a} \frac{(ax^2 + b)^{n+1}}{n + 1} \quad (n \neq -1)$$

155. $$\int \frac{dx}{ax^2 + b} = \frac{1}{\sqrt{ab}} \tan^{-1} \left(x \sqrt{\frac{a}{b}} \right) \quad (a \text{ and } b \text{ positive})$$

156. $$\int \frac{dx}{ax^2 + b} = \frac{1}{2\sqrt{-ab}} \log_e \frac{x\sqrt{a} - \sqrt{-b}}{x\sqrt{a} + \sqrt{-b}} \quad (a \text{ positive}, b \text{ negative})$$

$$= \frac{1}{2\sqrt{-ab}} \log_e \frac{\sqrt{b} + x\sqrt{-a}}{\sqrt{b} - x\sqrt{-a}} \quad (a \text{ negative}, b \text{ positive})$$

157. $$\int \frac{dx}{(ax^2 + b)^n} = \frac{1}{2(n-1)b} \frac{x}{(ax^2 + b)^{n-1}} + \frac{2n - 3}{2(n-1)b} \int \frac{dx}{(ax^2 + b)^{n-1}} \quad (n \text{ integer} > 1)$$

158. $$\int \frac{x^2 \, dx}{ax^2 + b} = \frac{x}{a} - \frac{b}{a} \int \frac{dx}{ax^2 + b}$$

159. $$\int \frac{x^2 \, dx}{(ax^2 + b)^n} = -\frac{1}{2(n-1)a} \frac{x}{(ax^2 + b)^{n-1}} + \frac{1}{2(n-1)a} \int \frac{dx}{(ax^2 + b)^{n-1}}$$

$(n \text{ integer} > 1)$

160. $$\int \frac{dx}{ax^2 + bx + d} = \frac{1}{\sqrt{b^2 - 4ad}} \log_e \frac{2ax + b - \sqrt{b^2 - 4ad}}{2ax + b + \sqrt{b^2 - 4ad}} \quad (b^2 > 4ad)$$

161. $$\int \frac{dx}{ax^2 + bx + d} = \frac{2}{\sqrt{4ad - b^2}} \tan^{-1} \frac{2ax + b}{\sqrt{4ad - b^2}} \quad (b^2 < 4ad)$$

TABLE A.12 Table of Integrals (*Continued*)

Integrals Involving Algebraic Functions (Continued)

162. $\displaystyle\int \frac{dx}{ax^2 + bx + d} = -\frac{2}{2ax + b} \quad (b^2 = 4ad)$

Integrals Involving Irrational Algebraic Functions

163. $\displaystyle\int \sqrt{ax + b}\, dx = \frac{2}{3a}\sqrt{(ax + b)^3}$

164. $\displaystyle\int x\sqrt{ax + b}\, dx = -\frac{2(2b - 3ax)\sqrt{(ax + b)^3}}{15a^2}$

165. $\displaystyle\int x^2\sqrt{ax + b}\, dx = \frac{2(8b^2 - 12abx + 15a^2x^2)\sqrt{(ax + b)^3}}{105a^3}$

166. $\displaystyle\int \frac{\sqrt{ax + b}}{x}\, dx = 2\sqrt{ax + b} + \sqrt{b}\log_e \frac{\sqrt{ax + b} - \sqrt{b}}{\sqrt{ax + b} + \sqrt{b}} \quad (b \text{ positive})$

167. $\displaystyle\int \frac{\sqrt{ax + b}}{x}\, dx = 2\sqrt{ax + b} - 2\sqrt{-b}\tan^{-1}\sqrt{\frac{ax + b}{-b}} \quad (b \text{ negative})$

168. $\displaystyle\int \frac{dx}{x\sqrt{ax + b}} = \frac{1}{\sqrt{b}}\log_e \frac{\sqrt{ax + b} - \sqrt{b}}{\sqrt{ax + b} + \sqrt{b}} \quad (b \text{ positive})$

169. $\displaystyle\int \frac{dx}{x\sqrt{ax + b}} = \frac{2}{\sqrt{-b}}\tan^{-1}\sqrt{\frac{ax + b}{-b}} \quad (b \text{ negative})$

170. $\displaystyle\int \frac{dx}{x^2\sqrt{ax + b}} = -\frac{\sqrt{ax + b}}{bx} - \frac{a}{2b\sqrt{b}}\log_e \frac{\sqrt{ax + b} - \sqrt{b}}{\sqrt{ax + b} + \sqrt{b}} \quad (b \text{ positive})$

171. $\displaystyle\int \frac{dx}{x^2\sqrt{ax + b}} = -\frac{\sqrt{ax + b}}{bx} - \frac{a}{b\sqrt{-b}}\tan^{-1}\sqrt{\frac{ax + b}{-b}} \quad (b \text{ negative})$

172. $\displaystyle\int \sqrt{ax^2 + b}\, dx = \frac{x}{2}\sqrt{ax^2 + b} + \frac{b}{2\sqrt{a}}\log_e \frac{x\sqrt{a} + \sqrt{ax^2 + b}}{\sqrt{b}} \quad (a \text{ positive})$

173. $\displaystyle\int \sqrt{ax^2 + b}\, dx = \frac{x}{2}\sqrt{ax^2 + b} + \frac{b}{2\sqrt{-a}}\sin^{-1}\left(x\sqrt{-\frac{a}{b}}\right) \quad (a \text{ negative})$

174. $\displaystyle\int \frac{dx}{\sqrt{ax^2 + b}} = \frac{1}{\sqrt{a}}\log_e\left(x\sqrt{a} + \sqrt{ax^2 + b}\right) \quad (a \text{ positive})$

175. $\displaystyle\int \frac{dx}{\sqrt{ax^2 + b}} = \frac{1}{\sqrt{-a}}\sin^{-1}\left(x\sqrt{-\frac{a}{b}}\right) \quad (a \text{ negative})$

176. $\displaystyle\int \frac{\sqrt{ax^2 + b}}{x}\, dx = \sqrt{ax^2 + b} + \sqrt{b}\log_e \frac{\sqrt{ax^2 + b} - \sqrt{b}}{x} \quad (b \text{ positive})$

177. $\displaystyle\int \frac{\sqrt{ax^2 + b}}{x}\, dx = \sqrt{ax^2 + b} - \sqrt{-b}\tan^{-1}\frac{\sqrt{ax^2 + b}}{\sqrt{-b}} \quad (b \text{ negative})$

178. $\displaystyle\int \frac{x\, dx}{\sqrt{ax^2 + b}} = \frac{1}{a}\sqrt{ax^2 + b}$

179. $\displaystyle\int x\sqrt{ax^2 + b}\, dx = \frac{1}{3a}(ax^2 + b)^{3/2}$

180. $\displaystyle\int \frac{dx}{x\sqrt{ax^n + b}} = \frac{1}{n\sqrt{b}}\log_e \frac{\sqrt{ax^n + b} - \sqrt{b}}{\sqrt{ax^n + b} + \sqrt{b}} \quad (b \text{ positive})$

181. $\displaystyle\int \frac{dx}{x\sqrt{ax^n + b}} = \frac{2}{n\sqrt{-b}}\sec^{-1}\sqrt{-\frac{ax^n}{b}} \quad (b \text{ negative})$

TABLE A.12 Table of Integrals (*Continued*)

Integrals Involving Irrational Algebraic Functions (Continued)

182. $\int \sqrt{ax^2 + bx + d}\, dx = \frac{2ax + b}{4a}\sqrt{ax^2 + bx + d} + \frac{4ad - b^2}{8a}\int \frac{dx}{\sqrt{ax^2 + bx + d}}$

183. $\int x\sqrt{ax^2 + bx + d}\, dx = \frac{(ax^2 + bx + d)^{3/2}}{3a} - \frac{b}{2a}\int \sqrt{ax^2 + bx + d}\, dx$

184. $\int \frac{dx}{\sqrt{ax^2 + bx + d}} = \frac{1}{\sqrt{a}}\log_e (2ax + b + 2\sqrt{a(ax^2 + bx + d)})$ (*a* positive)

185. $\int \frac{dx}{\sqrt{ax^2 + bx + d}} = \frac{1}{\sqrt{-a}}\sin^{-1}\frac{-2ax - b}{\sqrt{b^2 - 4ad}}$ (*a* negative)

186. $\int \frac{dx}{x\sqrt{ax^2 + bx + d}} = -\frac{1}{\sqrt{d}}\log_e\left(\frac{\sqrt{ax^2 + bx + d} + \sqrt{d}}{x} + \frac{b}{2\sqrt{d}}\right)$ (*d* positive)

187. $\int \frac{dx}{x\sqrt{ax^2 + bx + d}} = \frac{1}{\sqrt{-d}}\sin^{-1}\frac{bx + 2d}{x\sqrt{b^2 - 4ad}}$ (*d* negative)

188. $\int \frac{x\, dx}{ax^2 + bx + d} = \frac{1}{2a}\log_e (ax^2 + bx + d) - \frac{b}{2a}\int \frac{dx}{ax^2 + bx + d}$

189. $\int \frac{x\, dx}{\sqrt{ax^2 + bx + d}} = \frac{\sqrt{ax^2 + bx + d}}{a} - \frac{b}{2a}\int \frac{dx}{\sqrt{ax^2 + bx + d}}$

Integrals Involving e, x^n, a^x, ln x

190. $\int xe^{ax}\, dx = \frac{e^{ax}}{a^2}(ax - 1)$

191. $\int x^m e^{ax}\, dx = \frac{x^m e^{ax}}{a} - \frac{m}{a}\int x^{m-1}e^{ax}\, dx$

192. $\int \frac{e^{ax}}{x^m}\, dx = \frac{1}{m - 1}\left(-\frac{e^{ax}}{x^{m-1}} + a\int \frac{e^{ax}\, dx}{x^{m-1}}\right)$

193. $\int x^n a^x\, dx = \frac{a^x x^n}{\log a} - \frac{na^x x^{n-1}}{(\log a)^2} + \frac{n(n - 1)a^x x^{n-2}}{(\log a)^3} \cdots \pm \frac{n(n - 1)(n - 2) \cdots 2 \cdot 1 a^x}{(\log a)^{n+1}}$

194. $\int \frac{a^x\, dx}{x} = \log x + x \log a + \frac{(x \log a)^2}{2 \times 2!} + \frac{(x \log a)^3}{3 \times 3!} + \cdots$

195. $\int \frac{dx}{1 + e^x} = \log \frac{e^x}{1 + e^x}$

196. $\int \frac{dx}{a + be^{mx}} = \frac{1}{am}[mx - \log (a + be^{mx})]$

197. $\int x^m \log x\, dx = x^{m+1}\left[\frac{\log x}{m + 1} - \frac{1}{(m + 1)^2}\right]$

198. $\int (\log x)^n\, dx = x(\log x)^n - n\int (\log x)^{n-1}\, dx$

199. $\int x^m(\log x)^n\, dx = \frac{x^{m+1}(\log x)^n}{m + 1} - \frac{n}{m + 1}\int x^m(\log x)^{n-1}\, dx$

200. $\int \frac{(\log x)^n\, dx}{x} = \frac{(\log x)^{n+1}}{n + 1}$

201. $\int x^m \log (a + bx)\, dx = \frac{1}{m + 1}\left[x^{m+1}\log (a + bx) - b\int \frac{x^{m+1}\, dx}{a + bx}\right]$

TABLE A.12 Table of Integrals (*Continued*)

Definite Integrals

202. $\int_0^{\frac{\pi}{2}} \sin^n x\,dx = \int_0^{\frac{\pi}{2}} \cos^n x\,dx = \frac{1 \times 3 \times 5 \cdots (n-1)}{2 \times 4 \times 6 \cdots (n)} \frac{\pi}{2}$ if n is an even integer

$= \frac{2 \times 4 \times 6 \cdots (n-1)}{1 \times 3 \times 5 \times 7 \cdots n}$ if n is an odd integer

$= \frac{1}{2}\sqrt{\pi}\,\frac{\Gamma\left(\frac{n+1}{2}\right)}{\Gamma\left(\frac{n}{2}+1\right)}$ for any value of n greater than -1

203. $\int_0^\infty \frac{\sin mx\,dx}{x} = \frac{\pi}{2}$, if $m > 0$; 0, if $m = 0$; $-\frac{\pi}{2}$, if $m < 0$

204. $\int_0^\infty \frac{\sin x \cos mx\,dx}{x} = 0$, if $m < -1$ or $m > 1$; $\frac{\pi}{4}$, if $m = -1$ or $m = 1$; $\frac{\pi}{2}$, if $-1 < m < 1$

205. $\int_0^\infty \frac{\sin^2 x\,dx}{x^2} = \frac{\pi}{2}$

206. $\int_0^\pi \sin^2 mx\,dx = \int_0^\pi \cos^2 mx\,dx = \frac{\pi}{2}$

207. $\int_0^\infty \frac{\cos x\,dx}{\sqrt{x}} = \int_0^\infty \frac{\sin x\,dx}{\sqrt{x}} = \sqrt{\frac{\pi}{2}}$

208. $\int_0^\infty e^{-ax} \sin mx\,dx = \frac{m}{a^2+m^2}$, if $a > 0$

209. $\int_0^\infty e^{-ax} \cos mx\,dx = \frac{a}{a^2+m^2}$, if $a > 0$

210. $\int_0^{\frac{\pi}{2}} \log \sin x\,dx = \int_0^{\frac{\pi}{2}} \log \cos x\,dx = -\frac{\pi}{2} \log 2$

211. $\int_0^\pi x \log \sin x\,dx = -\frac{\pi^2}{2} \log 2$

212. $\int_0^\pi \sin kx \sin mx\,dx = \int_0^\pi \cos kx \cos mx\,dx = 0$, if k is different from m

213. $\int_0^\infty x^n e^{-ax}\,dx = \frac{\Gamma(n+1)}{a^{n+1}} = \frac{n!}{a^{n+1}}$ $\quad n > -1,\ a > 0$

214. $\int_0^\infty x^{2n} e^{-ax^2}\,dx = \frac{1 \times 3 \times 5 \cdots (2n-1)}{2^{n+1} a^n} \sqrt{\frac{\pi}{a}}$

215. $\int_0^\infty e^{-a^2x^2}\,dx = \frac{1}{2a}\sqrt{\pi} = \frac{1}{2a}\Gamma(\frac{1}{2})$ $\quad a > 0$

216. $\int_0^\infty \frac{e^{-nx}}{\sqrt{x}}\,dx = \sqrt{\frac{\pi}{n}}$

217. $\int_0^1 \frac{\log x}{1+x}\,dx = -\frac{\pi^2}{12}$

218. $\int_0^1 \frac{\log x}{1-x}\,dx = -\frac{\pi^2}{6}$

219. $\int_0^1 \frac{\log x}{1-x^2}\,dx = -\frac{\pi^2}{8}$

if the limit exists

$$\int_a^b f(x)\,dx = \lim_{\epsilon\to 0}\int_{a+\epsilon}^b f(x)\,dx$$

$f(x)$ discontinuous at $x = a$, $a < b$, ϵ positive, and limit exists

$$\int_a^b f(x)\,dx = \lim_{\epsilon\to 0}\int_a^{b-\epsilon} f(x)\,dx$$

$f(x)$ discontinuous at $x = b$, $a < b$, ϵ positive, and limit exists

Elliptic Integrals (See Table A.2)

Elliptic Integral of the First Kind

$$F(\phi, k) = \int_0^\phi \frac{d\theta}{\sqrt{1 - k^2\sin^2\theta}} = \int_0^x \frac{dt}{\sqrt{(1 - t^2)(1 - k^2t^2)}} \qquad k^2 < 1$$

$$x = \sin\phi$$

Elliptic Integral of the Second Kind

$$E(\phi, k) = \int_0^\phi \sqrt{1 - k^2\sin^2\theta}\,d\theta = \int_0^x \frac{\sqrt{1 - k^2t^2}}{\sqrt{1 - t^2}}\,dt \qquad k^2 < 1$$

$$x = \sin\phi$$

Elliptic Integral of the Third Kind

$$\Pi(\phi, n, k) = \int_0^\infty \frac{d\theta}{(1 + n\sin^2\theta)\sqrt{1 - k^2\sin^2\theta}}$$

$$= \int_0^x \frac{dt}{(1 + nt^2)\sqrt{(1 - t^2)(1 - k^2t^2)}} \qquad k^2 < 1$$

$$x = \sin\phi$$

The complete elliptic integrals are

$$K = F\left(\frac{\pi}{2}, k\right)$$

$$= \frac{\pi}{2}\left[1 + \left(\frac{1}{2}\right)^2 k^2 + \left(\frac{1\times 3}{2\times 4}\right)^2 k^4 + \left(\frac{1\times 3\times 5}{2\times 4\times 6}\right)^2 k^6 + \cdots\right] \qquad k^2 < 1$$

$$E = E\left(\frac{\pi}{2}, k\right)$$

$$= \frac{\pi}{2}\left[1 - \left(\frac{1}{2}\right)^2 k^2 - \left(\frac{1\times 3}{2\times 4}\right)^2 \frac{k^4}{3} - \left(\frac{1\times 3\times 5}{2\times 4\times 6}\right)^2 \frac{k^6}{5} - \cdots\right] \qquad k^2 < 1$$

Approximate Numerical Value of an Elliptic Integral of the First and Second Kind (Landen's Method). Elliptic integral of the first kind:

$$F(\phi, k) = \int_0^\phi \frac{d\theta}{\sqrt{1 - k^2\sin^2\theta}} \qquad (k^2 < 1)$$

$$= (1 + k_1)(1 + k_2)(1 + k_3)\cdots(1 + k_n)\frac{F(k_{n_1}\phi_n)}{2^n}$$

where $$k_1 = \frac{1 - \sqrt{1 - k^2}}{1 + \sqrt{1 - k^2}}$$

and in general

$$k_m = \frac{1 - \sqrt{1 - k_{m-1}}}{1 + \sqrt{1 - k_{m-1}}}$$

$$\tan(\phi_1 - \phi) = \sqrt{1 - k^2}\tan\phi$$

and in general

$$\tan(\phi_m - \phi_{m-1}) = \sqrt{1 - k_{m-1}^2}\tan\phi_{m-1}$$

Also, since $k_n \to 0$ and $F(k_n, \phi_n) \to \phi_n$ as $n \to \infty$, the integral can be approximated by

$$F(\phi, k) \cong (1 + k_1)(1 + k_2)(1 + k_3)\cdots(1 + k_n)\phi_n/2^n$$

Elliptic integral of the second kind:

$$\begin{aligned} E(\phi, k) &= \int_0^\phi \sqrt{1 - k^2\sin^2\theta}\,d\theta \\ &= F(\phi, k)\left[1 - \frac{k^2}{2}\left(1 + \frac{k_1}{2} + \frac{k_1k_2}{2^2} + \frac{k_1k_2k_3}{2^3} + \cdots\right)\right] \\ &\quad + k\left[\left(\frac{\sqrt{k_1}}{2}\right)\sin\phi_1 + \left(\frac{\sqrt{k_1k_2}}{2^2}\right)\sin\phi_2 + \left(\frac{\sqrt{k_1k_2k_3}}{2^3}\right)\sin\phi_3 + \cdots\right] \end{aligned}$$

where the recursion formulas from above apply.

The Gamma Function. The "gamma" or "generalized factorial" function is defined by

$$\Gamma(n) = \int_0^\infty x_{n-1}e^{-x}\,dx \qquad \text{for } n > 0$$

and by means of the recursion formula

$$\Gamma(n) = \Gamma(n + 1)/n$$

for negative values of n.

The gamma function is infinite when n assumes the value of a negative integer or zero (Fig. A.23). When n is a positive integer, the relation

$$\Gamma(n + 1) = n! \qquad n = 1, 2, 3, \ldots$$

holds. It also follows that

$$\begin{aligned} 0! = \Gamma(1) &= 1 \\ \Gamma(2) &= 1 \\ \Gamma(3) &= 2 \\ \Gamma(4) &= 6 \\ &\cdots\cdots \end{aligned}$$

The Beta Function. The beta function is defined by

$$\beta(m, n) = \int_0^1 x^{m-1}(1 - x)^{n-1}dx \qquad m, n > 0$$

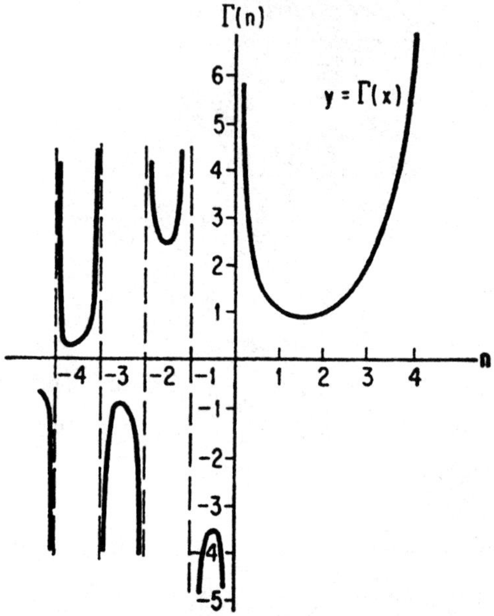

FIG. A.23 The gamma function.

or by letting

$$x = \sin^2 \phi$$

$$\beta(m, n) = 2 \int_0^{\pi/2} (\sin \phi)^{2m-1} (\cos \phi)^{2n-1}\, d\phi$$

The beta function and gamma function are connected by the expression

$$\beta(m, n) = \Gamma(m)\Gamma(n)/\Gamma(m + n)$$

The Error Function. The error function is defined by

$$\text{erf}\, x = \frac{2}{\sqrt{\pi}} \int_0^x e^{-n^2}\, dn \qquad \text{(see Table A.3)}$$

If follows from the definition that

$$\text{erf}\, 0 = 0$$

$$\text{erf}(-x) = -\text{erf}\, x$$

$$\text{erf}\, \infty = 1$$

$$\text{erf}\, iz = \frac{2i}{\sqrt{\pi}} \int_0^z e^{n^2}\, dn \qquad \text{where } i = \sqrt{-1}$$

Integration by Parts

$$\int u\,dv = uv - \int v\,du \qquad \text{where } u = u(x) \quad \text{and} \quad v = v(x)$$

For example, to evaluate $\int x \cos x \, dx$, let $u = x$ and $dv = \cos x \, dx$. Then

$$du = dx \quad and \quad v = \int \cos x \, dx = \sin x$$

Thus $$\int x \cos x \, dx = x \sin x - \int \sin x \, dx = x \sin x + \cos x + C$$

Also $$\int_a^b u \, dv = uv \Big|_a^b - \int_a^b v \, du$$

Multiple Integrals. Multiple integrals are of the form

$$\iint f(x, y) \qquad \iiint f(x, y, z) \qquad \text{etc.}$$

Two successive integrations, for example, an integration with respect to y holding x constant, and an integration with respect to x between constant limits, will yield the value for the double integral

$$\int_a^b \int_{y_1(x)}^{y_2(x)} f(x, y) \, dy \, dx$$

Similarly a triple integral is evaluated by three successive single integrations. The order of integration can be reversed if the function $f(x, y, \ldots)$ is continuous.

A.4.8 Differential Equations

Introduction. A differential equation describes some fundamental relationship between independent variables $x, y, z, \ldots$ and dependent variables $u, v, w, \ldots$ and some of the derivatives of $u, v, w, \ldots$ with respect to $x, y, z, \ldots$. An ordinary differential equation is one which contains an independent variable, a dependent variable, and various orders of derivatives of the dependent variable. The order of a differential equation is the order of the highest derivative which it contains. A general solution of an ordinary differential equation of nth order, $F(x, y, y', \ldots, y^{(n)}) = 0$, is a family of curves $G(x, y, C_1, \ldots, C_n) = 0$, each curve of which is a solution of the differential equation.

A partial differential equation contains a function of more than one independent variable and its partial derivatives of various order.

The degree of a differential equation is the power to which the derivative of highest order is raised.

Methods of Solution of Ordinary Differential Equations of First Order

Exact Equations. An exact differential equation is of the form

$$(dy/dx)N(x, y) + M(x, y) = 0 \quad \text{or} \quad M(x, y) \, dx + N(x, y) \, dy = 0$$

for which $$\partial M(x, y)/\partial y = \partial N(x, y)/\partial x$$

The solution of the exact equation is

$$\int M(x, y) dx + \int [N(x, y) - (\partial/\partial y) \int M(x, y) \, dx] \, dy = C$$

A differential equation, though not initially exact, may be rendered exact by multiplication through by a suitable function $u(x, y)$ called an integrating factor. For example, a differential equation containing the form $(x\,dx + y\,dy)$ may often be made exact by using the integrating factor $x^2 + y^2$ or a function of $x^2 + y^2$. Likewise, corresponding to the form $x\,dy + y\,dx$, try xy or a function of xy, and for $x\,dy - y\,dx$, the integrating factors $1/x^2$, $1/y^2$, $1/xy$, $1/(x^2 + y^2)$, or $1/x^2 f(y/x)$ may be tried.

Differential Equations with Separable Variables. If the differential equation is of the form $M(x)\,dx + N(y)\,dy = 0$, it is not only exact but may be integrated directly because the variables are separable. The solution is

$$\int M(x)\,dx + \int N(y)\,dy = 0 \quad \text{or} \quad F(x) = G(y) = C$$

Linear Equations. The differential equation of the form $dy/dx + P(x)y = Q(x)$ is a linear equation of first order. If $Q(x) = 0$, the equation is linear homogeneous.
The solution of the general case is

$$y = e^{-\int P\,dx}\left(\int Q e^{\int P\,dx}\,dx + c\right)$$

EXAMPLE Solve the equation

$$dy/dx + (1/x)y = 2x$$
$$y = e^{-\int dx/x}\left(\int 2xe^{\int dx/x}\,dx + c\right)$$
$$y = \tfrac{2}{3}x^2 + C/x$$

since $\qquad e^{\ln x} = x$

Bernoulli's equation is of the form

$$dy/dx + P(x)y = Q(x)y^n$$

By substituting $v = y^{1-n}$, the equation is reducible to the following linear differential equation in v and x:

$$dy/dx + (1 - n)P(x)v = (1 - n)Q(x)$$

Clairaut's equation is of the form $y = xp + f(p)$, where $p = dy/dx$. The solution is $y = Cx + f(C)$, which is the equation of a family of straight lines obtained when an arbitrary constant C is substituted for P, and the curve obtained by eliminating p from the two simultaneous equations $x + df(p)/dx = 0$ and $y = xp + f(p)$.

Ordinary Differential Equations of higher Than First Order. The equation

$$a_n(x)d^ny/dx^n + a_{n-1}(x)d^{n-1}y/dx^{n-1} + \cdots + a_1(x)dy/dx + a_0(x)y = f(x)$$

is the general representation of a linear differential equation of order n. When $f(x) = 0$, the equation is homogeneous; otherwise it is termed a nonhomogeneous equation. The complete solution of the general differential equation may be written

$$y = y_c + y_p$$

where y_c is the solution of the homogeneous equation corresponding to the general equation, and y_p is a particular solution of the nonhomogeneous equation.

The D-Operator Method. This method is frequently employed to solve linear homogeneous differential equations with constant coefficients,

$$a_n d^n y/dx^n + a_{n-1} d^{n-1}y/dx^{n-1} + \cdots + a_1 dy/dx + a_0 y = 0$$

The operator D is defined by the equation $Dy \equiv dy/dx$ or $D = d/dx$. Rewriting the homogeneous equation in operator form, we obtain

$$(a_n D^n + a_{n-1} D^{n-1} + \cdots + a_1 D + a_0)y = 0$$

Let $\alpha_n, \alpha_{n-1}, \alpha_{n-2}, \ldots, \alpha_1$ be the roots of the auxiliary algebraic equation

$$a_n D^n + a_{n-1} D^{n-1} + \cdots + a_1 D + a_0 = 0$$

The solution of the differential equation if all the roots are real and distinct (none repeated) is

$$y = C_1 e^{\alpha_1 x} + C_2 e^{\alpha_2 x} + \cdots C_n e^{\alpha_n x}$$

If q roots are equal, the remaining being real and distinct, the solution is

$$y = e^{\alpha_1 x} (C_1 + C_2 x + C_3 x^2 + \cdots + C_q x^{q-1}) + \cdots + C_{n-1} e^{\alpha_{n-1} x} + C_n e^{\alpha_n x}$$

If two roots are conjugate imaginary,

$$\alpha_1 = \beta + \gamma i \qquad \alpha_2 = \beta - \gamma i$$

the solution is $y = e^{\beta x}(C_1 \cos \gamma x + C_2 \sin \gamma x) + C_3 e^{\alpha_3 x} + \cdots + C_n e^{\alpha_n x}$. If there are two conjugate imaginary double roots,

$$\alpha_1 = \alpha_2 = \beta + \gamma i \qquad \alpha_3 = \alpha_4 = \beta - \gamma i$$

the solution is $y = e^{\beta x}[(C_1 + C_2 x) \cos \gamma x + (C_3 + C_4 x) \sin \gamma x]$.

Method of Undetermined Coefficients. This approach to the solution of linear nonhomogeneous differential equations with constant coefficients may be employed where it is possible to guess at the form of the particular solution. The following example will serve to illustrate the method of solution.

Solve the equation

$$d^2y/dx^2 - y = \sin x$$

The solution to the homogeneous equation $y'' - y = 0$ is obtained by the D-operator method. The auxiliary equation being $D^2 - 1 = 0$, $D = \pm 1$; therefore,

$$y_c = C_1 e^x + C_2 e^{-x}$$

To obtain the particular solution, assume the form of solution to be

$$y_p = A \sin x + B \cos x$$

Differentiating y_p and substituting into the original equation, we obtain

$$-2A \sin x - 2B \cos x = \sin x$$

from which $$A = -\frac{1}{2} \quad \text{and} \quad B = 0 \quad y_p = -\frac{1}{2} \sin x$$

and the complete solution is

$$y = C_1 e^x + C_2 e^{-x} - \frac{1}{2} \sin x$$

In the event the trial particular solution has a component which solves the homogeneous equation, the particular solution should be multiplied by x. If this function also has a component which solves the homogeneous equation, the original trial y_p should be multiplied by x^2, and so on.

If the right-hand side of the differential equation contains terms of more than one type, the trial particular integral will be the sum of trial terms for each type. The trial terms in Table A.13 are suggested for the type listed.

Taylor Series Solution. If the solution $y(x)$ of a linear differential equation can be expressed as a Taylor series expansion valid in some neighborhood of the initial point $x = x_0$, then the first n coefficients $y(x_0), y'(x_0)/1!, \ldots, y^{n-1}(x_0)/(n-1)!$ of the series $y(x) = y(x_0) + [y'(x_0)/1!](x - x_0) + [y''(x_0)/2!](x - x_0)^2 + \cdots$ are obtained from the n initial conditions and additional coefficients are obtained by differentiating the given differential equation $(n - 1)$ times and solving these simultaneously with the original differential equation. For example, given the equation

$$x\, d^2y/dx^2 + x^2\, dy/dx - y = 0$$

and the initial conditions

$$y(1) = 0 \quad y'(1) = 1$$

the first four terms of the expansion are required.

The expansion of $y(x)$ is

$$y(x) = y(1) + \frac{y'(1)}{1!}(x - 1) + \frac{y''(1)}{2!}(x - 1)^2 + \frac{y'''(1)}{3!}(x - 1)^3 + \cdots$$

The coefficients $y(1)$ and $y'(1)$ are given by the initial conditions as zero and one, respectively. The remaining two coefficients are obtained from differentiating the

TABLE A.13 Trial Terms

Type of term in f(x)	*Form to be included in y_p*
c (constant)	C (constant)
$c \sin x$ or $c \cos x$	$C_1 \sin x + C_2 \cos x$
cx^n	$C_0x^n + C_1x^{n-1} + C_2x^{n-2} + \cdots + C_{n-1}x + C_n$
ce^{kx}	Ce^{kx}
cx^ne^{kx}	$e^{kx}(C_0x^n + C_1x^{n-1} + \cdots + C_n)$
$e^{kx} \sin \alpha x$ or $e^{kx} \cos \alpha x$	$e^{kx}(C_1 \sin \alpha x + C_2 \cos \alpha x)$
$cx^ne^{kx} \sin \alpha x$ or $cx^ne^{kx} \cos \alpha x$	$(C_0x^n + \cdots + C_{n-1}x + C_n)e^{kx} \sin \alpha x + (C'_0x^n + \cdots$ $C'_{n-1}x + C'_n)e^{kx} \cos \alpha x$

differential equation. Thus the equations

$$xy'' + x^2y' - y = 0$$

$$xy''' + (x^2 + 1)y'' + (2x - 1)y' = 0$$

yield after substituting $x = 1$, $y = 0$, and $y' = 1$

$$y'' = -1 \qquad y''' = 1$$

The first four terms of the expansion are therefore

$$y = 0 + (x - 1) - \tfrac{1}{2}(x - 1)^2 + \tfrac{1}{6}(x - 1)^3 + \cdots$$

Variations of Parameters. This method of solution yields the particular integral of a differential equation of which the coefficients need not be constant. It is required that the complementary solution be known before this approach may be used.

EXAMPLE Given the differential equation

$$x^2y'' + xy' - y = 1 + x$$

which has a complementary solution $y = C_1x + C_2/x$ of the corresponding homogeneous equation $x^2y'' + xy' - y = 0$, find the particular integral.

solution The arbitrary constants C_1 and C_2 are replaced by functions of x, $A(x)$, and $B(x)$ such that $y = A(x)x + B(x)/x$ is a solution of the given nonhomogeneous, equation. Differentiate to obtain

$$y' = A'x + A + B'/x - B/x^2$$

Since there are two functions $A(x)$ and $B(x)$, whose identity must be established, two sets of conditions related to these functions must be written. The first condition is the nonhomogeneous equation itself. The remaining condition (there will be $n - 1$ conditions remaining where there are n functions, in general) may be assigned arbitrarily. Thus we select

$$A'x + B'/x = 0$$

Now differentiate $y = A(x)x + B(x)/x$ to obtain

$$y' = A'x + A + B'/x - B/x^2$$

which together with the above yields

$$y' = A - B/x^2$$

Differentiating, we obtain

$$y'' = A' - B'/x^2 + 2B/x^3$$

which is substituted in the original differential equation

$$x^2(A' - B'/x^2 + 2B/x^3) + x(A - B/x^2) - Ax - B/x = 1 + x$$

or

$$A'x^2 - B' = 1 + x$$

which together with $A'x + B'/x = 0$, yields

$$A' = \tfrac{1}{2}(1/x^2 + 1/x) \qquad A = \tfrac{1}{2}(-1/x + \ln x) + C_3$$

$$B' = -\tfrac{1}{2}(1 + x) \qquad B = -x/2(1 + x/2) + C_4$$

so that

$$y(x) = C_1 x + C_2/x + Ax + B/x$$

Systems of Linear Differential Equations with Constant Coefficients. One method of solution of a system of n linear differential equations with constant coefficients is the *symbolic algebraic method.* The system has the form

$$P_1(D)x + Q_1(D)y + \cdots = F_1(t)$$

$$P_2(D)x + Q_2(D)y + \cdots = F_2(t)$$

$$\cdots\cdots\cdots\cdots\cdots\cdots\cdots\cdots$$

where $D = d/dt$.

The solution may be found by treating these equations as algebraic equations. If the elimination of $x(t)$ from the first two equations is desired, one may operate on the first equation by $P_2(D)$ and on the second equation by $P_1(D)$. Then by subtraction $x(t)$ can be eliminated. For example, given the two simultaneous equations

$$(D + 1)x + Dy = 0$$

$$(D - 1)x + (D + 1)y = 0$$

we obtain, after operating on the first equation with $(D - 1)$ and on the second equation with $(D + 1)$,

$$(D - 1)(D + 1)x + (D - 1)(D)y = 0$$

$$(D + 1)(D - 1)x + (D + 1)^2 y = 0$$

By subtracting the second equation from the first we obtain

$$(3D + 1)y = 0$$

which has only one dependent variable. Solving for $y(t)$ in this equation and substituting back into one of the original equations, $x(t)$ can be found.

Solutions to Important Differential Equations in Engineering. *Legendre's Equation*

$$(1 - x^2)\, d^2y/dx^2 - 2x\, dy/dx + n(n + 1)y = 0$$

where n is a positive integer. The solution is $y(x) = C_1 P_n(x) + C_2 Q_n(x)$. $P_n(x)$ is a Legendre polynomial of degree n (Legendre function of the first kind). $Q_n(x)$ is the Legendre function of the second kind.

Bessel Differential Equation

$$x^2\, d^2y/dx^2 + x\, dy/dx + (x^2 - n^2)y = 0$$

n may take on fractional, integral, positive, negative, or complex values. The general solution of the Bessel equation is

$$y = C_1 y_I + C_2 y_{II}$$

where y_I and y_{II} are the Bessel functions of the first and second kind, respectively, and the order of the Bessel functions will depend upon the value of n.

Consider the zero-order Bessel equation obtained by setting $n = 0$,

$$d^2y/dx^2 + (1/x)(dy/dx) + y = 0$$

The solution is

$$y = C_1 J_0(x) \quad \text{and} \quad C_2 Y_0(x)$$

where $J_0(x)$ and $Y_0(x)$ are the zero-order Bessel functions of the first and second kinds, respectively.

$$J_0(x) = 1 - \frac{x^2}{2^2} + \frac{x^4}{2^2 \times 4^2} - \frac{x^6}{2^2 \times 4^2 \times 6^2} + \cdots$$

$$Y_0(x) = J_0(x) \ln x + \frac{x^2}{2^2} - \frac{x^4}{2^2 \times 4^2}\left(1 + \frac{1}{2}\right) + \frac{x^6}{2^2 \times 4^2 \times 6^2}\left(1 + \frac{1}{2} + \frac{1}{2}\right) - \cdots$$

The general solution of an nth-order Bessel equation is

$$y = C_1 J_n(x) + C_2 Y_n(x) \quad n = 0, 1, 2, 3, \ldots$$

The general equation for the Bessel functions of the first kind for n equal to zero or a positive integer is

$$J_n(x) = \frac{(x/2)^n}{n!}\left[1 - \frac{x^2}{2(2n + 2)} + \frac{x^4}{2 \times 4(2n + 2)(2n + 4)} + \cdots\right] n = 0, 1, 2, 3, \ldots$$

For the Bessel function of the second kind we obtain

$$Y_n(x) = J_n(x) \ln x - \frac{1}{2}\sum_{p=0}^{n-1} \frac{(n - p - 1)!}{p!}\left(\frac{x}{2}\right)^{2p-n}$$
$$- \frac{1}{2}\sum_{p=0}^{\infty} \frac{(-1)^n}{p!(n + p)!}\left(\frac{x}{2}\right)^{2p+n} [f(P) + f(n + p)]$$

where

$$[f(P) + f(n + p)] = {}^1 + \frac{1}{2} + \cdots + 1/p + {}^1 + \cdots + 1/(n + p) n = 0, 1, 2, 3, \ldots$$

Recursion Formulas

$$d/dx\, J_0(x) = -J_1(x) \qquad J_1(-x) = -J_1(x)$$

$$d/dx\, J_n(x) = n/x\, J_n(x) - J_{n+1}(x) \qquad J_{-n}(x) = (-1)^n J_n(x) \quad n = 0, 1, 2, \ldots$$

$$d/dx\, Y_0(x) = -Y_1(x) \qquad Y_{-n}(x) = (-1)^n Y_n(x) \quad n = 0, 1, 2, \ldots$$

$$J_0(-x) = J_0(x)$$

The Modified Bessel Equation

$$x^2\, d^2y/dx^2 + x\, dy/dx + m^2x^2y = 0 \qquad m = \text{const}$$

Making the substitution $\xi = mx$, we obtain

$$\xi^2\, d^2y/d\xi^2 + \xi\, dy/d\xi + \xi^2 y = 0$$

whose solution, already discussed, is

$$y = C_1J_0(\xi) + C_2Y_0(\xi) = C_1J_0(mx) + C_2Y_0(mx)$$

The equation

$$x^2\, d^2y/dx^2 + x\, dy/dx - (x^2 + n^2)y = 0$$

can be transformed to the form

$$x^2\, d^2y/dx^2 + x\, dy/dx + (x^2 - n^2)y = 0$$

by making the substitution $x = i\xi$:

$$\xi^2\, d^2y/d\xi^2 + \xi\, dy/d\xi + (\xi^2 - n^2)y = 0$$

For integral n,

$$I_n(x) = i^{-n}J_n(ix)$$

$$K_n(x) = i^{n+1}\pi/2[J_n(ix) + iY_n(ix)]$$

The solution for the modified Bessel equation is therefore

$$y = C_1I_n(x) + C_2K_n(x) \qquad n = 0, 1, 2, \ldots$$

Two useful relationships are

$$I_{-n}(x) = I_n(x) \qquad n = 0,1,2, \ldots$$

$$K_{-n}(x) = K_n(x) \qquad n = 0,1,2, \ldots$$

Solution of Partial Differential Equations. The only means of solution that will be discussed is the separation-of-variables method. For the following equation,

$$f_1(x)\, \partial^2z/\partial x^2 + f_2(x)\, \partial z/\partial x + f_3(x)z + g_1(y)\, \partial^2z/\partial y^2 + g_2(y)\, \partial z/\partial y + g_3(y)z = 0$$

the solution of $z = z(x, y)$ will be assumed to have the form of a product of two functions $X(x)$ and $Y(y)$, which are functions of x and y only, respectively,

$$z = z(x, y) = X(x)Y(y)$$

Substituting $z = XY$ into the original differential equation, we obtain

$$-\frac{1}{X}\left[f_1(x)\frac{\partial^2X}{\partial x^2} + f_2(x)\frac{\partial X}{\partial x} + f_3(x)X\right] = \frac{1}{Y}\left[g_1(y)\frac{\partial^2Y}{\partial y^2} + g_2(y)\frac{\partial Y}{\partial y} + g_3(y)Y\right]$$

Note that the left-hand side contains the function of x only and that the right-hand side contains functions of y only. Since the right- and left-hand sides are independent of x and y, respectively, they must be equal to a common constant, called a separation constant α; thus,

$$f_1(x)\, d^2X/dx^2 + f_2(x)\, dX/dx + [f_3(x) + \alpha]X = 0$$

and

$$g_1(y)\, d^2Y/dy^2 + g_2(y)\, dY/dy + [g_3(y) - \alpha]Y = 0$$

Once the solutions to the above have been obtained, the product solution XY is obtained. The method outlined may be extended to additional variables. A differential equation may be separable, but where the product solution will not satisfy initial and boundary conditions, the foregoing method will not yield a solution.

A.4.9 Operational Mathematics

The Laplace Transform. The Laplace transform $f(s)$ of a function $F(t)$ defined for all positive t is

$$\mathcal{L}\{F(t)\} = f(s) = \int_0^\infty e^{-st}F(t)\, dt$$

In order for a Laplace transform of the function $F(t)$ to exist, the function must be sectionally continuous in every finite interval in $t \geqq 0$ and must be of exponential order. A function $F(t)$ is of exponential order if these exists a constant α such that $e^{-\alpha t}|F(t)|$ is bounded for all t greater than some finite number T.

Given the function $f(s)$, the original function $F(t)$ is obtained from the inversion integral

$$F(t) = \mathcal{L}^{-1}\{f(s)\} = \frac{1}{2\pi i}\int_{\gamma - i\infty}^{\gamma + i\infty} e^{st}f(s)\, ds$$

An extremely useful theorem for the solution of differential equations is

$$\mathcal{L}\{F^{(n)}(t)\} = s^n f(s) - s^{n-1}F(+0) - s^{n-2}F'(+0) - s^{n-3}F''(+0) - \cdots - F^{(n-1)}(+0)$$

In the foregoing equation, the function $F(t)$ has a continuous derivative of order $n - 1$, $F^{(n-1)}(t)$, and a sectionally continuous derivative $F^{(n)}(t)$ in every finite interval $0 \leqq t \leqq T$. In addition $F(t)$, $F'(t) \cdots F^{(n-1)}(t)$ are of exponential order (of order $e^{\alpha t}$) and $s > \alpha$.

The foregoing theorem in conjunction with Tables A.14 and A.15 can be used to solve a variety of differential equations encountered in engineering problems. For example, solve the differential equation

$$d^2X/dt^2 + K/mX = 0$$

given the boundary conditions $X(0) = C_1$ and $X'(0) = C_2$. The differential equation may also be written

$$X''(t) + K^2X(t) = 0$$

TABLE A.14 Laplace Transforms—Operations*

	$f(s) \quad (\text{Re}\, s > \alpha)$	$F(t) \quad (t > 0)$
1	$\int_0^\infty e^{-st} F(t)\, dt = L\{F\}$	$F(t)$
2	$f(s)$	$\frac{1}{2\pi i} \lim_{\beta \to \infty} \int_{\gamma - i\beta}^{\gamma + i\beta} e^{tz} f(z)\, dz$
3	$sf(s) - F(0)$	$F'(t)$
4	$s^n f(s) - s^{n-1}F(0) - s^{n-2}F'(0) - \cdots - F^{(n-1)}(0)$	$F^{(n)}(t)$
5	$\frac{1}{s} f(s)$	$\int_0^t F(\tau)\, d\tau$
6	$e^{-bs} f(s)\ (b > 0)$	$\begin{cases} F(t - b) & \text{if } t > b \\ 0 & \text{if } t < b \end{cases}$
7	$f(s)g(s)$	$\int_0^t F(\tau) G(t - \tau)\, d\tau$
8	$f'(s)$	$-tF(t)$
9	$f^{(n)}(s)$	$(-1)^n t^n F(t)$
10	$\int_s^\infty f(x)\, dx$	$\frac{1}{t} F(t)$
11	$f(s - a)$	$e^{at} F(t)$
12	$f(cs)\ (c > 0)$	$\frac{1}{c} F\left(\frac{t}{c}\right)$
13	$\frac{\int_0^a e^{-st} F(t)\, dt}{1 - e^{-as}}\ (a > 0)$	$F(t)$ if $F(t + a) = F(t)$
14	$\frac{\int_0^a e^{-st} F(t)\, dt}{1 + e^{-as}}\ (a > 0)$	$F(t)$ if $F(t + a) = -F(t)$
15	$\sum_{n=0}^{\infty} \frac{a_n}{s^{n+k}}\ (k > 0)$	$\sum_{n=0}^{\infty} \frac{a_n}{\Gamma(n + k)} t^{n+k-1}$

*From R. V. Churchill, "Operational Mathematics," 3d ed., McGraw-Hill Book Company, Inc., New York, 1972.

where $k^2 = K/m$. Applying the foregoing theorem we obtain

$$\mathcal{L}\{X''(t) + k^2 X(t)\} = \mathcal{L}\{X''(t)\} + k^2 \mathcal{L}\{X(t)\} = 0$$
$$\mathcal{L}\{X''(t)\} = s^2 x(s) - sX(+0) - X'(+0)$$
$$k^2 \mathcal{L}\{X(t)\} = k^2 x(s)$$

Substituting, we obtain

$$s^2 x(s) - sX(+0) - X'(+0) + k^2 x(s) = 0$$
$$s^2 x(s) - sC_1 - C_2 + k^2 x(s) = 0$$
$$x(s) = sC_1/(s^2 + k^2) + C_2/(s^2 + k^2)$$

TABLE A.15 Laplace Transforms‡

	$f(s)$	$F(t)$
1	$\frac{1}{s}$	1
2	$\frac{1}{s^2}$	t
3	$\frac{1}{s^n}\ (n = 1, 2, \ldots)$	$\frac{t^{n-1}}{(n-1)!}$
4	$\frac{1}{\sqrt{s}}$	$\frac{1}{\sqrt{\pi t}}$
5	$s^{-\frac{3}{2}}$	$2\sqrt{\frac{t}{\pi}}$
6	$s^{-(n+\frac{1}{2})}\ (n = 1, 2, \ldots)$	$\frac{2^n t^{n-\frac{1}{2}}}{1 \times 3 \times 5 \cdots (2n-1)\sqrt{\pi}}$
7	$\frac{\Gamma(k)}{s^k}\ (k > 0)$	t^{k-1}
8	$\frac{1}{s-a}$	e^{at}
9	$\frac{1}{(s-a)^2}$	te^{at}
10	$\frac{1}{(s-a)^n}\ (n = 1, 2, \ldots)$	$\frac{1}{(n-1)!}t^{n-1}e^{at}$
11	$\frac{\Gamma(k)}{(s-a)^k}\ (k > 0)$	$t^{k-1}e^{at}$
12*	$\frac{1}{(s-a)(s-b)}$	$\frac{1}{a-b}(e^{at} - e^{bt})$
13*	$\frac{s}{(s-a)(s-b)}$	$\frac{1}{a-b}(ae^{at} - be^{bt})$
14*	$\frac{1}{(s-a)(s-b)(s-c)}$	$-\frac{(b-c)e^{at} + (c-a)e^{bt} + (a-b)e^{ct}}{(a-b)(b-c)(c-a)}$
15	$\frac{1}{s^2+a^2}$	$\frac{1}{a}\sin at$
16	$\frac{s}{s^2+a^2}$	$\cos at$

*Here a, b, and (in entry 14) c represent distinct constants.

TABLE A.15 Laplace Transforms (*Continued*)

	$f(s)$	$F(t)$
17	$\frac{1}{s^2 - a^2}$	$\frac{1}{a} \sinh at$
18	$\frac{s}{s^2 - a^2}$	$\cosh at$
19	$\frac{1}{s(s^2 + a^2)}$	$\frac{1}{a^2}(1 - \cos at)$
20	$\frac{1}{s^2(s^2 + a^2)}$	$\frac{1}{a^3}(at - \sin at)$
21	$\frac{1}{(s^2 + a^2)^2}$	$\frac{1}{2a^3}(\sin at - at \cos at)$
22	$\frac{s}{(s^2 + a^2)^2}$	$\frac{t}{2a} \sin at$
23	$\frac{s^2}{(s^2 + a^2)^2}$	$\frac{1}{2a}(\sin at + at \cos at)$
24	$\frac{s^2 - a^2}{(s^2 + a^2)^2}$	$t \cos at$
25	$\frac{s}{(s^2 + a^2)(s^2 + b^2)} \quad (a^2 \neq b^2)$	$\frac{\cos at - \cos bt}{b^2 - a^2}$
26	$\frac{1}{(s - a)^2 + b^2}$	$\frac{1}{b} e^{at} \sin bt$
27	$\frac{s - a}{(s - a)^2 + b^2}$	$e^{at} \cos bt$
28	$\frac{3a^2}{s^3 + a^3}$	$e^{-at} - e^{at/2}\left(\cos \frac{at\sqrt{3}}{2} - \sqrt{3} \sin \frac{at\sqrt{3}}{2}\right)$
29	$\frac{4a^3}{s^4 + 4a^4}$	$\sin at \cosh at - \cos at \sinh at$
30	$\frac{s}{s^4 + 4a^4}$	$\frac{1}{2a^2} \sin at \sinh at$
31	$\frac{1}{s^4 - a^4}$	$\frac{1}{2a^3}(\sinh at - \sin at)$
32	$\frac{s}{s^4 - a^4}$	$\frac{1}{2a^2}(\cosh at - \cos at)$
33	$\frac{8a^3 s^2}{(s^2 + a^2)^3}$	$(1 + a^2 t^2) \sin at - at \cos at$
34*	$\frac{1}{s}\left(\frac{s - 1}{s}\right)^n$	$L_n(t) = \frac{e^t}{n!} \frac{d^n}{dt^n}(t^n e^{-t})$
35	$\frac{s}{(s - a)^{3/2}}$	$\frac{1}{\sqrt{\pi t}} e^{at}(1 + 2at)$
36	$\sqrt{s - a} - \sqrt{s - b}$	$\frac{1}{2\sqrt{\pi t^3}}(e^{bt} - e^{at})$

*$L_n(t)$ is the Laguerre polynomial of degree n.

TABLE A.15 Laplace Transforms (*Continued*)

	$f(s)$	$F(t)$
37	$\dfrac{1}{\sqrt{s}+a}$	$\dfrac{1}{\sqrt{\pi t}} - ae^{a^2t}\operatorname{erfc}(a\sqrt{t})$
38	$\dfrac{\sqrt{s}}{s-a^2}$	$\dfrac{1}{\sqrt{\pi t}} + ae^{a^2t}\operatorname{erf}(a\sqrt{t})$
39	$\dfrac{\sqrt{s}}{s+a^2}$	$\dfrac{1}{\sqrt{\pi t}} - \dfrac{2a}{\sqrt{\pi}}e^{-a^2t}\displaystyle\int_0^{a\sqrt{t}} e^{\lambda^2}\,d\lambda$
40	$\dfrac{1}{\sqrt{s}\,(s-a^2)}$	$\dfrac{1}{a}e^{a^2t}\operatorname{erf}(a\sqrt{t})$
41	$\dfrac{1}{\sqrt{s}\,(s+a^2)}$	$\dfrac{2}{a\sqrt{\pi}}e^{-a^2t}\displaystyle\int_0^{a\sqrt{t}} e^{\lambda^2}\,d\lambda$
42	$\dfrac{b^2-a^2}{(s-a^2)(b+\sqrt{s})}$	$e^{a^2t}[b - a\operatorname{erf}(a\sqrt{t})] - be^{b^2t}\operatorname{erfc}(b\sqrt{t})$
43	$\dfrac{1}{\sqrt{s}\,(\sqrt{s}+a)}$	$e^{a^2t}\operatorname{erfc}(a\sqrt{t})$
44	$\dfrac{1}{(s+a)\sqrt{s+b}}$	$\dfrac{1}{\sqrt{b-a}}e^{-at}\operatorname{erf}(\sqrt{b-a}\sqrt{t})$
45	$\dfrac{b^2-a^2}{\sqrt{s}\,(s-a^2)(\sqrt{s}+b)}$	$e^{a^2t}\left[\dfrac{b}{a}\operatorname{erf}(a\sqrt{t}) - 1\right] + e^{b^2t}\operatorname{erfc}(b\sqrt{t})$
46*	$\dfrac{(1-s)^n}{s^{n+\frac{1}{2}}}$	$\dfrac{n!}{(2n)!\sqrt{\pi t}}H_{2n}(\sqrt{t})$
47	$\dfrac{(1-s)^n}{s^{n+\frac{3}{2}}}$	$-\dfrac{n!}{\sqrt{\pi}\,(2n+1)!}H_{2n+1}(\sqrt{t})$
48†	$\dfrac{\sqrt{s+2a}}{\sqrt{s}} - 1$	$ae^{-at}[I_1(at) + I_0(at)]$
49	$\dfrac{1}{\sqrt{s+a}\sqrt{s+b}}$	$e^{-\frac{1}{2}(a+b)t}I_0\left(\dfrac{a-b}{2}t\right)$
50	$\dfrac{\Gamma(k)}{(s+a)^k(s+b)^k}\ (k>0)$	$\sqrt{\pi}\left(\dfrac{t}{a-b}\right)^{k-\frac{1}{2}}e^{-\frac{1}{2}(a+b)t} \times I_{k-\frac{1}{2}}\left(\dfrac{a-b}{2}t\right)$
51	$\dfrac{1}{(s+a)^{\frac{1}{2}}(s+b)^{\frac{3}{2}}}$	$te^{-\frac{1}{2}(a+b)t}\left[I_0\left(\dfrac{a-b}{2}t\right) + I_1\left(\dfrac{a-b}{2}t\right)\right]$
52	$\dfrac{\sqrt{s+2a}-\sqrt{s}}{\sqrt{s+2a}+\sqrt{s}}$	$\dfrac{1}{t}e^{-at}I_1(at)$

*$H_n(x)$ is the Hermite polynomial, $H_n(x) = e^{x^2}(d^n/dx^n)(e^{-x^2})$.
†$I_n(x) = i^{-n}J_n(ix)$, where J_n is Bessel's function of the first kind.

TABLE A.15 Laplace Transforms (*Continued*)

	$f(s)$	$F(t)$
53	$\dfrac{(a-b)^k}{(\sqrt{s+a}+\sqrt{s+b})^{2k}}\ (k>0)$	$\dfrac{k}{t}\,e^{-\frac{1}{2}(a+b)t}I_k\left(\dfrac{a-b}{2}\,t\right)$
54	$\dfrac{(\sqrt{s+a}+\sqrt{s})^{-2\nu}}{\sqrt{s}\,\sqrt{s+a}}\ (\nu>-1)$	$\dfrac{1}{a^\nu}\,e^{-\frac{1}{2}at}I_\nu\left(\dfrac{1}{2}\,at\right)$
55	$\dfrac{1}{\sqrt{s^2+a^2}}$	$J_0(at)$
56	$\dfrac{(\sqrt{s^2+a^2}-s)^\nu}{\sqrt{s^2+a^2}}\ (\nu>-1)$	$a^\nu J_\nu(at)$
57	$\dfrac{1}{(s^2+a^2)^k}\ (k>0)$	$\dfrac{\sqrt{\pi}}{\Gamma(k)}\left(\dfrac{t}{2a}\right)^{k-\frac{1}{2}}J_{k-\frac{1}{2}}(at)$
58	$(\sqrt{s^2+a^2}-s)^k\ (k>0)$	$\dfrac{ka^k}{t}\,J_k(at)$
59	$\dfrac{(s-\sqrt{s^2-a^2})^\nu}{\sqrt{s^2-a^2}}\ (\nu>-1)$	$a^\nu I_\nu(at)$
60	$\dfrac{1}{(s^2-a^2)^k}\ (k>0)$	$\dfrac{\sqrt{\pi}}{\Gamma(k)}\left(\dfrac{t}{2a}\right)^{k-\frac{1}{2}}I_{k-\frac{1}{2}}(at)$
61	$\dfrac{e^{-ks}}{s}$	$S_k(t)=\begin{cases}0 \text{ when } 0<t<k\\ 1 \text{ when } t>k\end{cases}$
62	$\dfrac{e^{-ks}}{s^2}$	$\begin{cases}0 & \text{when } 0<t<k\\ t-k & \text{when } t>k\end{cases}$
63	$\dfrac{e^{-ks}}{s^\mu}\ (\mu>0)$	$\begin{cases}0 & \text{when } 0<t<k\\ \dfrac{(t-k)^{\mu-1}}{\Gamma(\mu)} & \text{when } t>k\end{cases}$
64	$\dfrac{1-e^{-ks}}{s}$	$\begin{cases}1 \text{ when } 0<t<k\\ 0 \text{ when } t>k\end{cases}$
65	$\dfrac{1}{s(1-e^{-ks})}=\dfrac{1+\coth\frac{1}{2}ks}{2s}$	$1+[t/k]=n$ when $(n-1)\,k<t<nk$ $(n=1,2,\ldots)$
66	$\dfrac{1}{s(e^{ks}-a)}$	$\begin{cases}0 \text{ when } 0<t<k\\ 1+a+a^2+\cdots+a^{n-1}\\ \quad\text{when } nk<t<(n+1)k\\ \quad(n=1,2,\ldots)\end{cases}$
67	$\dfrac{1}{s}\tanh ks$	$M(2k,t)=(-1)^{n-1}$ when $2k(n-1)<t<2kn$ $(n=1,2,\ldots)$
68	$\dfrac{1}{s(1+e^{-ks})}$	$\dfrac{1}{2}M(k,t)+\dfrac{1}{2}=\dfrac{1-(-1)^n}{2}$ when $(n-1)k<t<nk$
69	$\dfrac{1}{s^2}\tanh ks$	$H(2k,t)$

TABLE A.15 Laplace Transforms (*Continued*)

	$f(s)$	$F(t)$
70	$\dfrac{1}{s \sinh ks}$	$F(t) = 2(n-1)$ when $(2n-3)k < t < (2n-1)k$ $(t > 0)$
71	$\dfrac{1}{s \cosh ks}$	$M(2k, t+3k) + 1 = 1 + (-1)^n$ when $(2n-3)k < t < (2n-1)k$ $(t > 0)$
72	$\dfrac{1}{s} \coth ks$	$F(t) = 2n - 1$ when $2k(n-1) < t < 2kn$
73	$\dfrac{k}{s^2 + k^2} \coth \dfrac{\pi s}{2k}$	$\lvert \sin kt \rvert$
74	$\dfrac{1}{(s^2+1)(1-e^{-\pi s})}$	$\begin{cases} \sin t \text{ when } (2n-2)\pi < t < (2n-1)\pi \\ 0 \quad \text{when } (2n-1)\pi < t < 2n\pi \end{cases}$
75	$\dfrac{1}{s} e^{-(k/s)}$	$J_0(2\sqrt{kt})$
76	$\dfrac{1}{\sqrt{s}} e^{-(k/s)}$	$\dfrac{1}{\sqrt{\pi t}} \cos 2\sqrt{kt}$
77	$\dfrac{1}{\sqrt{s}} e^{k/s}$	$\dfrac{1}{\sqrt{\pi t}} \cosh 2\sqrt{kt}$
78	$\dfrac{1}{s^{3/2}} e^{-(k/s)}$	$\dfrac{1}{\sqrt{\pi k}} \sin 2\sqrt{kt}$
79	$\dfrac{1}{s^{3/2}} e^{k/s}$	$\dfrac{1}{\sqrt{\pi k}} \sinh 2\sqrt{kt}$
80	$\dfrac{1}{s^{\mu}} e^{-(k/s)} \; (\mu > 0)$	$\left(\dfrac{t}{k}\right)^{(\mu-1)/2} J_{\mu-1}(2\sqrt{kt})$
81	$\dfrac{1}{s^{\mu}} e^{k/s} \; (\mu > 0)$	$\left(\dfrac{t}{k}\right)^{(\mu-1)/2} I_{\mu-1}(2\sqrt{kt})$
82	$e^{-k\sqrt{s}} \; (k > 0)$	$\dfrac{k}{2\sqrt{\pi t^3}} \exp\left(-\dfrac{k^2}{4t}\right)$
83	$\dfrac{1}{s} e^{-k\sqrt{s}} \; (k \geqq 0)$	$\operatorname{erfc}\left(\dfrac{k}{2\sqrt{t}}\right)$
84	$\dfrac{1}{\sqrt{s}} e^{-k\sqrt{s}} \; (k \geqq 0)$	$\dfrac{1}{\sqrt{\pi t}} \exp\left(-\dfrac{k^2}{4t}\right)$
85	$s^{-3/2} e^{-k\sqrt{s}} \; (k \geqq 0)$	$2\sqrt{\dfrac{t}{\pi}} \exp\left(-\dfrac{k^2}{4t}\right)$ $- k \operatorname{erfc}\left(\dfrac{k}{2\sqrt{t}}\right)$
86	$\dfrac{a e^{-k\sqrt{s}}}{s(a+\sqrt{s})} \; (k \geqq 0)$	$-e^{ak} e^{a^2 t} \operatorname{erfc}\left(a\sqrt{t} + \dfrac{k}{2\sqrt{t}}\right)$ $+ \operatorname{erfc}\left(\dfrac{k}{2\sqrt{t}}\right)$

TABLE A.15 Laplace Transforms (*Continued*)

	$f(s)$	$F(t)$
87	$\dfrac{e^{-k\sqrt{s}}}{\sqrt{s}\,(a+\sqrt{s})}\quad (k \geqq 0)$	$e^{ak}e^{a^2t}\,\text{erfc}\left(a\sqrt{t}+\dfrac{k}{2\sqrt{t}}\right)$
88	$\dfrac{e^{-k\sqrt{s(s+a)}}}{\sqrt{s(s+a)}}$	$\begin{cases} 0 & \text{when } 0 < t < k \\ e^{-\frac{1}{2}at}I_0(\frac{1}{2}a\sqrt{t^2-k^2}) & \text{when } t > k \end{cases}$
89	$\dfrac{e^{-k\sqrt{s^2+a^2}}}{\sqrt{s^2+a^2}}$	$\begin{cases} 0 & \text{when } 0 < t < k \\ J_0(a\sqrt{t^2-k^2}) & \text{when } t > k \end{cases}$
90	$\dfrac{e^{-k\sqrt{s^2-a^2}}}{\sqrt{s^2-a^2}}$	$\begin{cases} 0 & \text{when } 0 < t < k \\ I_0(a\sqrt{t^2-k^2}) & \text{when } t > k \end{cases}$
91	$\dfrac{e^{-k(\sqrt{s^2+a^2}-s)}}{\sqrt{s^2+a^2}}\quad (k \geqq 0)$	$J_0(a\sqrt{t^2+2kt})$
92	$e^{-ks}-e^{-k\sqrt{s^2+a^2}}$	$\begin{cases} 0 & \text{when } 0 < t < k \\ \dfrac{ak}{\sqrt{t^2-k^2}}J_1(a\sqrt{t^2-k^2}) & \text{when } t > k \end{cases}$
93	$e^{-k\sqrt{s^2-a^2}}-e^{-ks}$	$\begin{cases} 0 & \text{when } 0 < t < k \\ \dfrac{ak}{\sqrt{t^2-k^2}}I_1(a\sqrt{t^2-k^2}) & \text{when } t > k \end{cases}$
94	$\dfrac{a^\nu e^{-k\sqrt{s^2+a^2}}}{\sqrt{s^2+a^2}\,(\sqrt{s^2+a^2}+s)^\nu}\quad (\nu > -1)$	$\begin{cases} 0 & \text{when } 0 < t < k \\ \left(\dfrac{t-k}{t+k}\right)^{\frac{1}{2}\nu} J_\nu(a\sqrt{t^2-k^2}) & \text{when } t > k \end{cases}$
95	$\dfrac{1}{s}\log s$	$\Gamma'(1)-\log t\quad [\Gamma'(1) = -0.5772]$
96	$\dfrac{1}{s^k}\log s\ (k > 0)$	$t^{k-1}\left\{\dfrac{\Gamma'(k)}{[\Gamma(k)]^2}-\dfrac{\log t}{\Gamma(k)}\right\}$
97*	$\dfrac{\log s}{s-a}\quad (a > 0)$	$e^{at}[\log a - \text{Ei}\,(-at)]$
98†	$\dfrac{\log s}{s^2+1}$	$\cos t\ \text{Si}\ t - \sin t\ \text{Ci}\ t$
99	$\dfrac{s\log s}{s^2+1}$	$-\sin t\ \text{Si}\ t - \cos t\ \text{Ci}\ t$
100	$\dfrac{1}{s}\log(1+ks)\ (k > 0)$	$-\text{Ei}\left(-\dfrac{t}{k}\right)$

*The exponential-integral function Ei $(-t)$. For tables of this function and other integral functions see Jahnke, Eugene, and Fritz Emde, "Tables of Functions," 4th ed., Dover Publications, Inc., New York, 1945.

†The cosine-integral and sine-integral functions Ci t and Si t.

TABLE A.15 Laplace Transforms (*Continued*)

	$f(s)$	$F(t)$
101	$\log \frac{s-a}{s-b}$	$\frac{1}{t}(e^{bt} - e^{at})$
102	$\frac{1}{s} \log (1 + k^2 s^2)$	$-2 \operatorname{Ci}\left(\frac{t}{k}\right)$
103	$\frac{1}{s} \log (s^2 + a^2)\ (a > 0)$	$2 \log a - 2 \operatorname{Ci}(at)$
104	$\frac{1}{s^2} \log (s^2 + a^2)\ (a > 0)$	$\frac{2}{a} [at \log a + \sin at - at \operatorname{Ci}(at)]$
105	$\log \frac{s^2 + a^2}{s^2}$	$\frac{2}{t}(1 - \cos at)$
106	$\log \frac{s^2 - a^2}{s^2}$	$\frac{2}{t}(1 - \cosh at)$
107	$\arctan \frac{k}{s}$	$\frac{1}{t} \sin kt$
108	$\frac{1}{s} \arctan \frac{k}{s}$	$\operatorname{Si}(kt)$
109	$e^{k^2 s^2} \operatorname{erfc}(ks)\ (k > 0)$	$\frac{1}{k\sqrt{\pi}} \exp\left(-\frac{t^2}{4k^2}\right)$
110	$\frac{1}{s} e^{k^2 s^2} \operatorname{erfc}(ks)\ (k > 0)$	$\operatorname{erf}\left(\frac{t}{2k}\right)$
111	$e^{ks} \operatorname{erfc} \sqrt{ks}\ \ (k > 0)$	$\frac{\sqrt{k}}{\pi \sqrt{t}(t + k)}$
112	$\frac{1}{\sqrt{s}} \operatorname{erfc}(\sqrt{ks})$	$\begin{cases} 0 & \text{when } 0 < t < k \\ (\pi t)^{-\frac{1}{2}} & \text{when } t > k \end{cases}$
113	$\frac{1}{\sqrt{s}} e^{ks} \operatorname{erfc}(\sqrt{ks})\ (k > 0)$	$\frac{1}{\sqrt{\pi(t + k)}}$
114	$\operatorname{erf}\left(\frac{k}{\sqrt{s}}\right)$	$\frac{1}{\pi t} \sin(2k\sqrt{t})$
115	$\frac{1}{\sqrt{s}} e^{k^2/s} \operatorname{erfc}\left(\frac{k}{\sqrt{s}}\right)$	$\frac{1}{\sqrt{\pi t}} e^{-2k\sqrt{t}}$
116*	$K_0(ks)$	$\begin{cases} 0 & \text{when } 0 < t < k \\ (t^2 - k^2)^{-\frac{1}{2}} & \text{when } t > k \end{cases}$
117	$K_0(k\sqrt{s})$	$\frac{1}{2t} \exp\left(-\frac{k^2}{4t}\right)$
118	$\frac{1}{s} e^{ks} K_1(ks)$	$\frac{1}{k}\sqrt{t(t + 2k)}$
119	$\frac{1}{\sqrt{s}} K_1(k\sqrt{s})$	$\frac{1}{k} \exp\left(-\frac{k^2}{4t}\right)$
120	$\frac{1}{\sqrt{s}} e^{k/s} K_0\left(\frac{k}{s}\right)$	$\frac{2}{\sqrt{\pi t}} K_0(2\sqrt{2kt})$

TABLE A.15 Laplace Transforms (*Continued*)

	$f(s)$	$F(t)$
121	$\pi e^{-ks} I_0(ks)$	$\begin{cases} [t(2k-t)]^{-\frac{1}{2}} & \text{when } 0 < t < 2k \\ 0 & \text{when } t > 2k \end{cases}$
122	$e^{-ks} I_1(ks)$	$\begin{cases} \dfrac{k-t}{\pi k\sqrt{t(2k-t)}} & \text{when } 0 < t < 2k \\ 0 & \text{when } t > 2k \end{cases}$
123	$-e^{as}\,\mathrm{Ei}\,(-as)$	$\dfrac{1}{t+a}\ (a > 0)$
124	$\dfrac{1}{a} + se^{as}\,\mathrm{Ei}\,(-as)$	$\dfrac{1}{(t+a)^2}\ (a > 0)$
125	$\left(\dfrac{\pi}{2} - \mathrm{Si}\, s\right) \cos s + \mathrm{Ci}\, s \sin s$	$\dfrac{1}{t^2+1}$

‡From R. V. Churchill, "Operational Mathematics," 3d ed., McGraw-Hill Book Co. Inc., New York, 1972.

From Table A.15 we find that

$$\mathcal{L}^{-1}\left\{\frac{s}{s^2+k^2}\right\} = \cos kt \quad \text{and} \quad \mathcal{L}^{-1}\left\{\frac{1}{s^2+k^2}\right\} = \frac{1}{k}\sin kt$$

The solution is therefore

$$X(t) = C_1 \cos kt + (C_2/k) \sin kt$$

In the case of a nonhomogeneous equation, the Laplace transform of both sides of the equation is obtained, and the procedure is identical with that indicated above.

Often the function $f(s)$ is such that it is adaptable to the inversions of Table A.15 by use of partial fractions. Thus,

$$f(s) = \frac{s+4}{s^2+2s} = \frac{s+4}{s(s+2)} = \frac{A}{s} + \frac{B}{s+2}$$

Multiplying through by $(s^2 + 2s)$, we obtain $s + 4 = (A + B)s + 2A$. Equating coefficients of like powers of s, the constants are evaluated: $A = 2$, $B = -1$. Therefore, $f(s) = 2/s - 1/(s + 2)$. From Table A.15 we find that $F(t) = 2 - e^{-2t}$. For a more complete discussion of partial fractions, see Sec. A.4.1.

The *convolution* of two functions, $F_1(t)$ and $F_2(t)$ is denoted $F_1 * F_2$ and is defined as follows:

$$F_1 * F_2 = \int_0^t F_1(t-\tau)F_2(\tau)\,dr$$

The correspondence between the transforms of two functions $f_1(s) \cdot f_2(s)$ and the convolution of the functions $F_1(t)$ and $F_2(t)$ is given by

$$f_1(s) \cdot f_2(s) = \mathcal{L}\{F_1(t) * F_2(t)\}$$

Thus,

$$\mathcal{L}^{-1}\{f_1(s) \cdot f_2(s)\} = F_1(t) * F_2(t) = \int_0^t F_1(t-\tau)\,F_2(\tau)\,dr$$

Finite Fourier Sine Transforms (See Table A.16). The finite Fourier sine transform of a sectionally continuous function $F(x)$ is defined by the following relationship:

$$f_s(n) = S\{F(x)\} = \int_0^{\pi} F(x) \sin nx\, dx \qquad \begin{matrix} n = 1, 2, \ldots \\ 0 < x < \pi \end{matrix}$$

If $F'(x)$ is sectionally continuous and the function $F(x)$ is defined at each point of discontinuity by

$$F(x_0) = \tfrac{1}{2}\left[F(x_0 + 0) + F(x_0 - 0)\right] \qquad 0 < x_0 < \pi$$

then the inversion relationship for the finite sine transform is

$$S^{-1}\{f_s(n)\} = F(x) = \frac{2}{\pi}\sum_{1}^{\infty} f_s(n) \sin nx \qquad 0 < x < \pi$$

If $F(x)$ for $0 < x < \pi$ has a sectionally continuous derivative of order $2v$ ($v = 1, 2, \ldots$) and a continuous derivative of order $2v - 1$, then $S\{F^{(2v)}(x)\}$ is given by

$$\begin{aligned} S\{F^{(2v)}(x)\} &= (-n^2)^v f_s(n) - (-1)^v n^{2v-1}[F(0) - (-1)^n F(\pi)] \\ &\quad - (-1)^{v-1} n^{2v-3}[F''(0) - (-1)^n F''(\pi)] \\ &\quad - \cdots + n[F^{(2v-2)}(0) - (-1)^n F^{(2v-2)}(\pi)] \end{aligned}$$

Finite Fourier Cosine Transforms (See Table A.17). The finite Fourier cosine transform of a sectionally continuous function $F(x)$ is defined by the following relationship:

$$f_c(n) = C\{F(c)\} = \int_0^{\pi} F(x) \cos nx\, dx \qquad \begin{matrix} n = 1, 2, \ldots \\ 0 < x < \pi \end{matrix}$$

If $F(x)$ and $F'(x)$ are sectionally continuous, then the inversion relationship for the finite cosine transform is

$$C^{-1}\{f_c(n)\} = F(x) = \frac{1}{\pi} f_c(0) + \frac{2}{\pi}\sum_{1}^{\infty} f_c(n) \cos nx$$

If $F(x)$ for $0 < x < \pi$ has a sectionally continuous derivative or order $2v$ ($v = 1, 2, \ldots$) and a continuous derivative of order $2v - 1$, then $C\{F^{(2v)}(x)\}$ is given by

$$\begin{aligned} C\{F^{(2v)}(x)\} &= (-n^2)^v f_c(n) - (-1)^{v-1} n^{2v-2}[F'(0) - (-1)^n F'(\pi)] \\ &\quad - (-1)^{v-2} n^{v-4}[F''(0) - (-1)^n F'''(\pi)] \\ &\quad - \cdots - [F^{(2v-1)}(0) - (-1)^n F^{(2v-1)}(\pi)] \end{aligned}$$

A.4.10 Complex Variables

Complex Numbers. A complex number z consists of a real part x and an imaginary part y and is represented as

$$z = x + iy$$

where $\quad i = \sqrt{-1} \quad (i^2 = -1)$

TABLE A.16 Finite Sine Transforms*

	$f_s(n) \quad (n = 1, 2, \ldots)$	$F(x) \quad (0 < x < \pi)$
1	$\int_0^{\pi} F(x) \sin nx\, dx = S_n\{F\}$	$F(x)$
2	$f_s(n)$	$\frac{2}{\pi} \sum_{n=1}^{\infty} f_s(n) \sin nx$
3	$-n^2 f_s(n) + n[F(0) - (-1)^n F(\pi)]$	$F''(x)$
4	$\frac{1}{n^2} f_s(n)$	$\frac{x}{\pi} \int_0^{\pi} (\pi - r)F(r)dr - \int_0^{x} (x - r)F(r)\, dr$
5	$(-1)^{n+1} f_s(n)$	$F(\pi - x)$
6	$2f_s(n) \cos nc$	$F_1(x + c) + F_1(x - c)$
7	$\frac{2}{n} f_s(n) h_s(n)$	$\int_0^{\pi} H(y) \int_{x-y}^{x+y} F_1(r)dr\, dy$
8	$\frac{\pi}{n}$	$\pi - x$
9	$\frac{1}{n}(-1)^{n+1}$	$\frac{x}{n}$
10	$\frac{1}{n}[1 - (-1)^n]$	1
11	$\frac{\pi}{n} \cos nc \quad (0 < c < \pi)$	$\begin{cases} -x & \text{if } x < c \\ \pi - x & \text{if } x > c \end{cases}$
12	$\frac{\pi}{n^2} \sin nc \quad (0 < c < \pi)$	$\begin{cases} (\pi - c)x & \text{if } x \leqq c \\ (\pi - x)c & \text{if } x \geqq c \end{cases}$
13	$\frac{6\pi}{n^3}$	$x(\pi - x)(2\pi - x)$
14	$\frac{2}{n^3}[1 - (-1)^n]$	$x(\pi - x)$
15	$\frac{\pi^2}{n}(-1)^{n+1} - \frac{2}{n^3}[1 - (-1)^n]$	x^2
16	$\pi(-1)^n\left(\frac{6}{n^3} - \frac{\pi^2}{n}\right)$	x^3
17	$\frac{n}{n^2 + c^2}[1 - (-1)^n e^{c\pi}]$	e^{cx}
18	$\frac{n}{n^2 + c^2} \quad (c \neq 0)$	$\frac{\sinh c(\pi - x)}{\sinh c\pi}$
19	$\frac{n}{n^2 + c^2}[1 - (-1)^n \cosh c\pi]$	$\cosh cx$
20	$\frac{n}{n^2 - k^2} \quad (k \neq 0, +1, +2, \ldots)$	$\frac{\sin k(\pi - x)}{\sin k\pi}$
21	0 if $n \neq m$; $f_s(m) = \frac{\pi}{2}$	$\sin mx \quad (m = 1, 2, \ldots)$
22	$\frac{n}{n^2 - k^2}[1 - (-1)^n \cos k\pi]$	$\cos kx \quad (k \neq \pm 1, \pm 2, \ldots)$
23	$n\frac{1 - (-1)^{m+n}}{n^2 - m^2}$ if $n \neq m$; $f_s(m) = 0$	$\cos mx \quad (m = 1, 2, \ldots)$

*From R. V. Churchill, "Operational Mathematics," 3d ed., McGraw-Hill Book Company, Inc., New York, 1972.

TABLE A.16 Finite Sine Transforms (*Continued*)

	$f_s(n)$ $(n = 1, 2 \ldots)$	$F(x)$ $(0 < x < \pi)$
24	$\frac{2kn}{(n^2 - k^2)^2}$ $(k \neq 0, \pm 1, \pm 2, \ldots)$	$\frac{\partial}{\partial k}\left[\frac{\sin k(\pi - x)}{\sin k\pi}\right]$
25	$\frac{2cn}{(n^2 + c^2)^2}$ $(c \neq 0)$	$\frac{\partial}{\partial c}\left[\frac{\sinh c(x - \pi)}{\sinh c\pi}\right]$
26	$\frac{4c^2(\sin^2 c\pi + \sinh^2 c\pi)n}{n^4 + 4c^4}$ $(c \neq 0)$	$\sin cx \sinh c(2\pi - x) - \sin c(2\pi - x) \sinh cx$
27	$\frac{2k^2 n(-1)^n}{n^4 - k^4}$ $(k \neq 0, \pm 1, \pm 2, \ldots)$	$\frac{\sinh kx}{\sinh k\pi} - \frac{\sin kx}{\sin k\pi}$
28	$\frac{c^2(-1)^n}{n(n^2 + c^2)}$ $(c \neq 0)$	$\frac{\sinh cx}{\sinh c\pi} - \frac{x}{\pi}$
29	$\frac{k^2(-1)^n}{n(n^2 - k^2)}$ $(k \neq 0, \pm 1, \pm 2, \ldots)$	$\frac{x}{\pi} - \frac{\sin kx}{\sin k\pi}$
30	b^n $(-1 < b < 1)$	$\frac{2}{\pi} \frac{b \sin x}{1 + b^2 - 2b \cos x}$
31	e^{-ny} $(y > 0)$	$\frac{1}{\pi} \frac{\sin x}{\cosh y - \cos x}$
32	$\frac{b^n}{n}$ $(-1 < b < 1)$	$\frac{2}{\pi} \arctan \frac{b \sin x}{1 - b \cos x}$
33	$\frac{1}{n} e^{-ny}$ $(y > 0)$	$\frac{2}{\pi} \arctan \frac{\sin x}{e^y - \cos x}$
34	$\frac{1 - (-1)^n}{n} b^n$ $(-1 < b < 1)$	$\frac{2}{\pi} \arctan \frac{2b \sin x}{1 - b^2}$
35	$\frac{1 - (-1)^n}{n} e^{-ny}$ $(y > 0)$	$\frac{2}{\pi} \arctan \frac{\sin x}{\sinh y}$

The conjugate $\bar{z}$ of a complex number is defined as

$$\bar{z} = x - iy$$

Two complex numbers are equal only if their real parts are equal and their imaginary parts are equal; i.e.,

$$x_1 + iy_1 = x_2 + iy_2$$

only if $x_1 = x_2$ and $y_1 = y_2$

Also $x + iy = 0$

only if $x = 0$ and $y = 0$

Complex numbers satisfy the distributive, associative, and commutative laws of algebra. Complex numbers may be graphically represented on the $z(x - y)$ plane or in polar (r, θ) coordinates. The polar coordinates of a complex number are

$$r = \sqrt{x^2 + y^2} = |z| = \text{mod } z \qquad r \geqq 0$$

where mod = modulus, and

$$\theta = \tan^{-1}(y/x) = \arg z = \text{amp } z$$

TABLE A.17 Finite Cosine Transforms*

	$f_c(n) \quad (n = 0, 1, 2, \ldots)$	$F(x) \quad (0 < x < x)$
1	$\int_0^x F(x)\cos nx\,dx = C_n\{F\}$	$F(x)$
2	$f_c(n)$	$\frac{f_c(0)}{\pi} + \frac{2}{\pi}\sum_{n=1}^{\infty} f_c(n)\cos nx$
3	$-n^2 f_c(n) - F'(0) + (-1)^n F'(\pi)$	$F''(x)$
4	$\frac{1}{n^2} f_x(n)$ if $n = 1, 2, \ldots$	$\int_x^{\pi}(x - r)F(r)\,dr + \frac{f_c(0)}{2\pi}(x - \pi)^2 + A$
5	$(-1)^n f_c(n)$	$F(\pi - x)$
6	$2f_c(n)\cos nc$	$F_2(x + c) + F_2(x - c)$
7	$2S_n\{F\}\sin nc$	$F_1(x + c) - F_1(x - c)$
8	$\frac{1}{n}S_n\{F\}$ if $n = 1, 2, \ldots$	$-\int_0^x F(r)\,dr + A$
9	$2f_c(n)h_c(n)$	$\int_{-x}^{x} F_2(x - r)H_2(r)\,dr$
10	$f_c(n)$ if $n \neq 0$; $f_c(0) + A\pi$ if $n = 0$	$F(x) + A$
11	0 if $n = 1, 2, \ldots$; $f_c(0) = \pi$	1
12	$\frac{1}{n}$ if $n = 1, 2, \ldots$; $f_c(0) = 0$	$-\frac{2}{\pi}\log\left(2\sin\frac{x}{2}\right)$
13	$\frac{2}{n}\sin nc$ if $n \neq 0$; $f_c(0) = 2c - \pi$	$\begin{cases} 1 & \text{if } 0 < x < c \\ -1 & \text{if } c < x < x \end{cases}$
14	$\frac{(-1)^n - 1}{n^2}$ if $n \neq 0$; $f_c(0) = \frac{\pi^2}{2}$	x
15	$\frac{2\pi}{n^2}(-1)^n$ if $n \neq 0$; $f_c(0) = \frac{\pi^3}{3}$	x^2
16	$\frac{1}{n^2}$ if $n \neq 0$; $f_c(0) = 0$	$\frac{(\pi - x)^2}{2\pi} - \frac{\pi}{6}$
17	$-\frac{24\pi}{n^4}(-1)^n$ if $n \neq 0$; $f_c(0) = 0$	$x^4 - 2\pi^2x^2 + \frac{7}{15}\pi^4$
18	$\frac{1}{n^2 + c^2} \quad (c \neq 0)$	$\frac{\cosh c(\pi - x)}{c\sinh c\pi}$
19	$\frac{(-1)^n e^{cx} - 1}{n^2 + c^2} \quad (c \neq 0)$	$\frac{1}{c}e^{cx}$
20	$\frac{1}{n^2 - k^2} \quad (k \neq 0, \pm 1, \pm 2, \ldots)$	$-\frac{\cos k(\pi - x)}{k\sin k\pi}$
21	$\frac{(-1)^n\cos k\pi - 1}{n^2 - k^2} \quad (k \neq 0, \pm 1, \pm 2, \ldots)$	$\frac{\sin kx}{k}$
22	$\frac{(-1)^{m+n} - 1}{n^2 - m^2}$ if $n \neq m$; $f_c(m) = 0$	$\frac{\sin mx}{m} \quad (m = 1, 2, \ldots)$
23	$\frac{(-1)^{n+1}}{n^2 - 1}$ if $n \neq 1$, $f_c(1) = -\frac{1}{4}$	$\frac{1}{\pi}x\sin x$
24	0 if $n \neq m$; $f_c(m) = \frac{\pi}{2}$	$\cos mx \quad (m = 1, 2, \ldots)$
25	$\frac{2c(-1)^{n+1}}{(n^2 + c^2)^2} \quad (c \neq 0)$	$\frac{\partial}{\partial c}\left(\frac{\cosh cx}{c\sinh c\pi}\right)$

TABLE A.17 Finite Cosine Transforms (*Continued*)

	$f_c(n)$ $(n = 0, 1, 2, \ldots)$	$F(x)$ $(0 < x < \pi)$
26	b^n $(-1 < b < 1, b \neq 0)$	$\frac{1}{\pi}\frac{1 - b^2}{1 + b^2 - 2b\cos x}$
27	e^{-ny} $(y > 0)$	$\frac{1}{\pi}\frac{\sinh y}{\cosh y - \cos x}$
28	$\frac{b^n}{n}$ if $n \neq 0$; $f_c(0) = 0$ $(-1 < b < 1)$	$-\frac{1}{\pi}\log(1 + b^2 - 2b\cos x)$
29	$\frac{\pi}{n}e^{-ny}$ if $n \neq 0$; $f_c(0) = 0$ $(y > 0)$	$y - \log(2\cosh y - 2\cos x)$
30	ne^{-ny} $(y > 0)$	$\frac{1}{\pi}\frac{\partial}{\partial y}\left(\frac{\sinh y}{\cos x - \cosh y}\right)$
31	$\frac{\pi}{2}\frac{b^n}{n!}$ if $n \neq 0$; $f_c(0) = \pi$	$\exp(b\cos x)\cos(b\sin x)$

*From R. V. Churchill, "Operational Mathematics," 3d ed., McGraw-Hill Book Company, Inc., New York, 1972.

where arg = argument and amp = amplitude. arg z is multiple-valued, but for an angular interval of range 2π there is only one value of θ for a given z.

BIBLIOGRAPHY

Mathematical Tables

Jahnke, Eugene, and Fritz Emde: 'Tables of Functions with Formulae and Curves," 4th ed., Dover Publications, Inc., New York, 1945.

Algebra

Frazer, R. A., W. J. Duncan, and A. R. Collar: "Elementary Matrics, Cambridge University Press, New York, 1952.

Peterson, Thurman S., and Charles R. Hobby: "College Algebra," 3d ed., Happer & Row, Publishers, Inc., New York, 1978.

Trigonometry

Kells, Lyman M., Willis F. Kern, and James R. Bland: "Plane and Spherical Trigonometry." 3d ed., McGraw-Hill Book Company, Inc., New York, 1951.

Analytic Geometry

Hill, M. A., and J. B. Linker: "Brief Course in Analytics," 3d ed., Holt, Rinehart and Winston, Inc., New York. 1960.

Wilson, W. A. and J. I. Tracey: "Analytic Geometry," alternate ed., D. C. Heath and Company, Boston, 1937.

Differential and Integral Calculus

Byerly, W. E: "Elements of the Integral Calculus," 2d ed. (1888), reprinted by G. E. Stechert & Company, New York, 1941.

Franklin, P.: "Methods of Advanced Calculus," McGraw-Hill Book Company, Inc., New York, 1964.

Granville, W. A., P. F. Smith, and W. R. Longley: "Elements of the Differential and Integral Calculus," rev. ed., Ginn and Company, Boston, 1962.

Peirce, B. O.: "A Short Table for Integrals," 4th rev. ed., Ginn and Company, Boston, 1956.

Differential Equations

Agnew, Ralph Palmer: "Differential Equations," 2d ed., McGraw-Hill Book Company, Inc., New York, 1960.

Bowman, Frank: "Introduction to Bessel Functions," Dover Publications, Inc., New York, 1958.

Forsyth, Andrew Russell: "Theory of Differential Equations," vols. 1–6, Dover Publications, Inc., New York, 1959.

Golomb, Michael, and Merrill Shanks: "Elements of Ordinary Differential Equations," 2d ed., McGraw-Hill Book Company, Inc., New York, 1955.

Miller, K. S.: "Partial Differential Equations in Engineering Problems," Prentice-Hall, Inc., Englewood Cliffs, N.J., 1953.

Philips, H. B.: "Differential Equations," 4th ed., John Wiley and Sons, Inc., New York; also Chapman and Hall, Ltd., London, 1951.

Salvadori, M. G., R. J. Schwarz: "Differential Equations in Engineering Problems," Prentice-Hall, Inc., Englewood Cliffs, N.J., 1954.

Sommerfeld, Arnold: "Partial Differential Equations in Physics," Academic Press, Inc., New York, 1949.

Webster, A.G.: "Partial Differential Equations of Mathematical Physics," 2d ed., Dover Publications, Inc., New York, 1955.

Operational Mathematics

Churchill, R. V.: "Operational Mathematics," 3d ed., McGraw-Hill Book Co. Inc., New York, 1972.

Complex variables

Churchill, R. V., and J. W. Brown: "Complex Variables and Applications," 3d ed., McGraw-Hill Book Co. Inc., New York, 1974.

APPENDIX B
NUMERICAL METHODS FOR ENGINEERS

C. C. Wang, Ph.D., P.E.
Senior Staff Engineer
Central Engineering Laboratories
FMC Corporation
Santa Clara, Calif.

B.1 NUMERICAL TECHNIQUES

A computation may not have practical engineering value unless it is cost-effective, relatively reliable, and obtainable in a timely fashion. Meeting these requirements depends heavily on the art and science of numerical techniques. Since analytical engineers have overall responsibility for the engineering computations they make, they cannot blame computational inefficiency or errors on their computer or on their software suppliers; consequently, it becomes very important for engineers who deal with machine computations to have a fundamental understanding of the basis of their computations.

A few selective topics in numerical techniques are provided below for handy reference. More extensive references in numerical methods are given at the end of this section.

B.1.1 Evaluation of Polynomials by Horner's Rule

The use of indexed constants and variable names, as in matrix operations, is one of the most efficient types for human-machine communication. However, thinking of alternative ways of doing the same thing may lead to further computational shortcuts.

For evaluating

$$F(x) = a_1x^n + a_2x^{n-1} + \cdots + a_{n+1} \tag{B.1}$$

the FORTRAN code can be written as

```
      F = A(N + 1)
      DO 15 I = 1, N
      F = F + A(I)*X**(N - I + 1)
   15 CONTINUE
```

This usually requires $n(n + 1)/2$ multiplications and n additions. The same task can be done using Horner's rule, whereby $F(x)$ is rewritten

$$F(x) = a_{n+1} + x(\dot{a}_n + x(\ldots(a_2 + a_1x)\ldots)) \tag{B.2}$$

This can be programmed in FORTRAN as

```
      F = A(1)
      DO 15 I = 1, N
      F = F *X + A(I + 1)
   15 CONTINUE
```

The program requires only n multiplications and n floating-point additions to achieve the same results.

B.1.2 Polynominal Interpolation

Let $x_0, x_1, \ldots, x_n$ be $n + 1$ distinct points on the real axis, and $F(x)$ be a real-valued function defined in some interval $R = [a, b]$ containing these points. It is then desired to construct a polynomial $P(x)$ of degree $\leq n$ that interpolates $F(x)$ at the points $X_0, \ldots, x_n$, such that

$$P(x_i) = F(x_i) \qquad i = 0, \ldots, n \tag{B.3}$$

and we may write

$$P(x) = \sum_{i=0}^{n} F(x_i)L_i(x) \tag{B.4}$$

where $$L_i(x) = \prod_{\substack{j=0 \\ j \neq i}}^{n} \frac{x - x_j}{x_i - x_j} \qquad i = 0, 1, \ldots, n$$

For example, if we consider the case $n = 1$, namely, we are given $F(x)$ and two distinct points x_0, x_1, then

$$L_0(x) = (x - x_1)/(x_0 - x_1) \qquad L_1(x) = (x - x_0)/(x_1 - x_0)$$

and $$P(x) = F(x_0)L_0(x) + F(x_1)L_1(x)$$

$$= F(x_0) + [F(x_1) - F(x_0)](x - x_0)/(x_1 - x_0)$$

This is the familiar form of a linear interpolation.

B.1.3 Newton–Raphson's Method for Finding Roots for an Algebraic or Transcendental Equation

To locate the roots for the equation

$$F(x) = 0 \tag{B.5}$$

start with a trial value that is sufficiently close to the root to be evaluated. In order to bring the trial value close enough for the iteration to become convergent, it is sometimes necessary to find a usable starting value by another method, such as the bisection method.

The iteration formula has the form:

$$x_{n+1} = x_n - F(x_n)/F'(x_n) \tag{B.6}$$

If the root is known to be p-multiple, Eq. B.6 can be modified to improve the convergence speed:

$$x_{n+1} = x_n - pF(x_n)/F'(x_n) \tag{B.7}$$

If multiplicity is not certain, we can instead search for a zero of the function $F(x)/F'(x)$. If $F'(x)$ is finite everywhere, this function will have the same roots as $F(x)$, the only difference being that all the roots are now simple. This leads to the formula

$$x_{n+1} = x_n - F(x_n)F'(x_n)/\{[F'(x_n)]^2 - F(x_n)F''(x_n)\} \tag{B.8}$$

B.1.4 The Bisection Method

This is a slow iteration procedure for root finding. However, it is a good alternative should other methods fail.

GIVEN $F(x)$ is continuous in a region $R[a, b]$, and either $F(a)F(b) < 0$ or $s(a) \geq k$ and $s(b) < k$, where s is the Sturm sequence property, the number of agreements in polarity count (for details, see Sec. B.1.9).

FIND A root of $F(x)$ located within the region R between a and b. To iterate, start with points a and b and compute the midpoint $c = (a + b)/2$, compute $F(c)$ or $s(c)$, and, if $F(c)F(a) < 0$ or $s(c) < k$, replace the value b by c; or, if $F(c)F(b) < 0$ or $s(c) \geq k$, replace the value a by c. Repeat the iteration until $F(c)$ is equal to zero, or the latest interval $[a, b]$ is sufficiently small. The width of the region after p iterations is $(b_0 - a_0)/2^p$, where $(b_0 - a_0)$ is the width of initial region.

B.1.5 The Runge-Kutta Algorithm

First Order

GIVEN A first-order differential equation

$$y' = F(x, y) \tag{B.9}$$

together with a value of y' at point (x_n, y_n)

$$y'_n = F(x_n, y_n) = F_n \tag{B.10}$$

FIND The value y at $x_{n+1} = x_n = h$, where h is a small increment of x. For an algorithm up to order $2(h^2)$, the formula is

$$y_{n+1} = y_n + 0.5(k_1 + k_2) \tag{B.11}$$

where $\quad k_1 = hF_n \qquad k_2 = hF(x_n + h, y_n + k_1).$

For an algorithm up to order $4(h^4)$, the formula is

$$y_{n+1} = y_n + (k_1 + 2k_2 + 2k_3 + k_4)/6 \tag{B.12}$$

where

$$k_1 = hF(x_n, y_n)$$

$$k_2 = hF(x_n + 0.5h, y_n + 0.5k_1)$$

$$k_3 = hF(x_n + 0.5h, y_n + 0.5k_2)$$

$$k_4 = hF(x_n + h, y_n + k_3)$$

Higher Order. For higher-order algorithms, first consider the simultaneous differential equations of first order, then extend this to higher-order cases.

GIVEN The simultaneous differential equations

$$y' = F(x, y, z)$$

and

$$z' = G(x, y, z) \tag{B.13}$$

where $y' = dy/dx$, $z' = dz/dx$, and the value set x, y, z at point n are known and denoted as x_n, y_n, z_n.

FIND The values of y and z at

$$x_{n+1} = x_n + h$$

The algorithms for this are

$$y_{n+1} = y_n + k_1$$

and

$$z_{n+1} = z_n + k_2, \tag{B.14}$$

where

$$k_1 = [k_{11} + 2k_{21} + 2k_{31} + k_{41}]/6$$

$$k_2 = [k_{12} + 2k_{22} + 2k_{32} + k_{42}]/6,$$

in which

$$k_{11} = hF(x_n, y_n, z_n)$$

$$k_{12} = hG(x_n, y_n, z_n)$$

$$k_{21} = hF(x_n + 0.5h, y_n + 0.5k_{11}, z_n + 0.5k_{12})$$

$$k_{22} = hG(x_n + 0.5h, y_n + 0.5k_{11}, z_n + 0.5k_{12})$$

$$k_{31} = hF(x_n + 0.5h, y_n + 0.5k_{21}, z_n + 0.5k_{22})$$

$$k_{32} = hG(x_n + 0.5h, y_n + 0.5k_{21}, z_n + 0.5k_{22})$$

$$k_{41} = hF(x_n + h, y_n + k_{31}, z_n + k_{32})$$

$$k_{42} = hG(x_n + h, y_n + k_{31}, z_n + k_{32})$$

To extend these results to higher-order cases, consider the following: Given a second-order differential equation of the form

$$y'' + E_1(x, y)y' + E_0(x, y)y = E(x, y) \tag{B.15}$$

let us define

$$z = y' = F(x, y, z) \tag{B.16}$$

and

$$w = z' = G(x, y, z) = E(x, y) - E_0(x, y)y - E_1(x, y)z \tag{B.17}$$

The second-order differential equation then can be treated as two simultaneous equations, and the results obtained for the latter can be applied to the former. The extension to higher-order differential equations can be achieved in a similar manner. For example, for

$$y''' + E_2(x, y)y'' + E_1(x, y)y' + E_0(x, y)y = E(x, y) \tag{B.18}$$

we may start with

$$z = y' = F(x, y, z, w) \tag{B.19}$$

$$w = z' = G(x, y, z, w) \tag{B.20}$$

and

$$v = w' = H(x, y, z, w) = E(x, y) - E_0(x, y)y - E_1(x, y)z - E_2(x, y)w \tag{B.21}$$

The previous expressions will be valid with minor modifications, such as

$$k_{21} = hF(x_n + 0.5h, y_n + 0.5k_{11}, z_n + 0.5k_{12}, w_n + 0.5k_{13})$$

$$k_{22} = hG(x_n + 0.5h, y_n + 0.5k_{11}, z_n + 0.5k_{12}, w_n + 0.5k_{13})$$

$$k_{23} = hH(x_n + 0.5h, y_n + 0.5k_{11}, z_n + 0.5k_{12}, w_n + 0.5k_{13})$$

Others are done similarly.

B.1.6 The Singular-Value Decomposition of a Matrix

A "matrix" is defined as an ordered rectangular array of numbers. We denote the number of rows by m and the number of columns by n. The whole matrix is usually written as

$$\mathbf{A} = \begin{bmatrix} a_{11} & a_{12} & \cdots & a_{1n} \\ a_{21} & a_{22} & \cdots & a_{2n} \\ \cdots & \cdots & \cdots & \cdots \\ a_{m1} & a_{m2} & \cdots & a_{mn} \end{bmatrix} \tag{B.22}$$

Sometimes it is written in the more compact form $\mathbf{A} = (a_{ik})$. If $m = n$, the matrix is quadratic; when $m = 1$, we have a row vector; and for $n = 1$, we have a column vector. In a quadratic matrix, the elements a_{ii} form a diagonal when all diagonal elements assume value 1; when all nondiagonal elements assume value 0, the matrix is called a "unit matrix" or "identity matrix" **I**. The sum of the diagonal elements is called the "trace of the matrix":

$$\mathrm{Tr}\,\mathbf{A} = \sum_{i=1}^{n} a_{ii} \tag{B.23}$$

The transpose of **A**, denoted by $\mathbf{A}^{\mathrm{T}}$, means the matrix is obtained from the exchange of rows and columns from matrix **A**. Two vectors, **u** and **v**, are orthogonal if their inner product is zero; that is, if

$$\mathbf{u}^T\mathbf{v} = 0 \tag{B.24}$$

The length of vector **u** is 1 if $\mathbf{u}^T\mathbf{u} = 1$.

A square (quadratic) matrix is called orthogonal if its columns are mutually orthogonal vectors, each of length 1. Thus, a matrix **U** is orthogonal if

$$\mathbf{U}^T\mathbf{U} = \mathbf{I} \tag{B.25}$$

where **I** is an identity matrix.

The determinant of a matrix **A** is a scalar number denoted by

$$\det \mathbf{A} = \sum_{i=1}^{n} a_{ik}\mathbf{A}_{ik} = \sum_{k=1}^{n} a_{ik}\mathbf{A}_{ik} \tag{B.26}$$

where $\mathbf{A}_{ik}$ is the algebraic complement of an element a_{ik}.

If $\det \mathbf{A} = 0$, the matrix **A** is said to be singular. The orthogonal matrix is automatically nonsingular, since $\mathbf{U}^{-1} = \mathbf{U}^T$. The simplest examples of orthogonal matrices are planar rotations of the form:

$$\mathbf{U} = \begin{bmatrix} \cos\theta & \sin\theta \\ -\sin\theta & \cos\theta \end{bmatrix} \tag{B.27}$$

It is easily proved that $\det \mathbf{U} = 1$. Therefore, **U** is also called a "unitary matrix." If **v** is a vector in two-dimensional space, then **Uv** is the same vector rotated through an angle θ. The parameter θ then can be used to control the results of matrix transformation.

Multiplication by orthogonal matrices does not alter the important geometric quantities, such as vector length, or the angle between two vectors; furthermore, this type of multiplication can be done without magnifying errors due to computation. For any matrix **A** and any two orthogonal matrices **U** and **V**, let us define a matrix defined by

$$\mathbf{Z} = \mathbf{U}^T\mathbf{A}\mathbf{V} \tag{B.28}$$

If $\mathbf{u}_j$ and $\mathbf{v}_j$ are the columns of **U** and **V**, respectively, then the individual components of **Z** are

$$\mathbf{s}_{ij} = \mathbf{u}_i^T\mathbf{A}\mathbf{v}_j$$

Such manipulation is important because by proper selection of **U** and **V**, it is possible to reduce most of $\mathbf{s}_{ij}$ to zero. In this way, transformation of matrices can be done while completely preserving the properties of the matrices. This type of manipulation may lead to many important applications.

A "singular-value decomposition" (SVD) of an $m \times n$ matrix **A** is defined as any factorization of the form

$$\mathbf{A} = \mathbf{U}\mathbf{Z}\mathbf{V}^T \tag{B.29}$$

where **U** is an $m \times m$ orthogonal matrix, **V** is an $n \times n$ orthogonal matrix, and **Z** is an $m \times n$ diagonal matrix with $\mathbf{s}_{ij} = 0$, if $i \neq j$ and $\mathbf{s}_{ii} = \mathbf{s}_i \geq 0$. The quantities $\mathbf{s}_i$ are called the singular values of **A**, and the columns of **U** and **V** are called the left and right singular vectors.

To explain the usefulness of SVD, examine the general linear equation

$$\mathbf{A}\mathbf{x} = \mathbf{b} \tag{B.30}$$

where **A** is an $m \times n$ matrix (assuming $m \geq n$), **b** is a given set of m vectors, and **x** is a set of n unknown vectors. Note that the important case where **A** is square and possibly singular is included. The questions that arise are: Are the equations consistent? Does any solution exist? Is the solution unique? Does $\mathbf{A}x = 0$ have nontrivial solutions? And finally, can a general solution be formulated to answer all these questions for all the equations? Although theoretically there are many different algorithms that can answer these questions, when the imperfections of the computational reality such as inexact data and imprecise arithmetic are considered, SVD is essentially the only known method that is reliable. Using Eqs. (B.29) and (B.30) lead to

$$\mathbf{UZV}^T\mathbf{x} = \mathbf{b}$$

or

$$\mathbf{ZV}^T\mathbf{x} = \mathbf{U}^T\mathbf{b} \tag{B.31}$$

and hence

$$\mathbf{Zy} = \mathbf{d} \tag{B.32}$$

where $\mathbf{y} = \mathbf{V}^T\mathbf{x}$ and $\mathbf{d} = \mathbf{U}^T\mathbf{b}$. Since **Z** is diagonal, the left-hand side of Eq. (B.32) is diagonal and, hence, can be easily studied.

B.1.7 Remarks on Eigenvalues and Eigenvectors

Definition of Eigenvalues and Eigenvectors. Given a linear transformation (or equation set) $\mathbf{u} = \mathbf{Ax}$, a nonzero vector **x** is called an "eigenvector" of **A** if

$$\mathbf{Ax} = \lambda\mathbf{x} \tag{B.33}$$

where λ is a real number. The number λ is called an "eigenvalue" of the transformation **A**. Physically, Eq. (B.33) stands for a characteristic equation for a system with n degrees of freedom. If **A** is an $n \times n$ matrix, then the equation has n roots of λ, and each root is called an "eigenvalue" of the system. For each λ the matrix equation (B.33) can be solved for vector **x** for the relative values of n components, which constitute one set of eigenvectors.

Up to now there has been no single (or pair of) eigenvalue-extraction method(s) that has been found to be satisfactory with respect to efficiency, reliability and generality of application for all situations. Prior to the computer age it could be painstaking to solve an eigenvalue problem with more than four degrees of freedom. Now, because of the improvement in methods and computers, we are able to solve reasonable-size problems rather efficiently and economically. On the one hand, engineers, with the help of machine computation, are now capable of doing much more than could be done in the past. But, on the other hand, even with the help of the most powerful computers and computer programs, engineers seem to rely on their engineering judgment as much today as they did years ago. The depth of engineering problems has changed; the nature of engineering as an art has not. The eigenvalue-extraction method is one of the most extensive and important numerical procedures in engineering computation. Engineers who encounter eigenvalue problems most frequently are those who deal with structural and machine dynamics, buckling and elastic stability, etc. It is important for engineers to acquire adequate numerical insight so they can make appropriate judgments about when and why a computer result becomes unacceptable.

Reasons Why the Direct Method is Undesirable. Equation (B.33), when fully expanded, is a polynomial of nth power. Unless n is less than 4, trying to solve characteristic polynomial equations directly has long been regarded as an overwhelming task. Even with the advent of digital computers, this remains true. Floating-point digits work only within a finite magnitude and with limited digits of precision. Raising the power for large numbers (of positive or negative sign) will quickly make the computation exceed the computer's capability, and searching for the location of the roots compounds the difficulty. Other prevailing reasons are: (1) the roots of a polynomial can be very sensitive to the values of the coefficients of the equation, and the accurate values of these coefficients are difficult to obtain, and (2) it is less efficient to evaluate the roots directly from the characteristic equation. Although these statements are basically true, ironically they are not universally correct. The exception for (1) is that when **A** is in tridiagonal form, the coefficients of the polynomial can be evaluated very accurately. Because the algorithms have been found, the coefficient of each polynomial term can be reduced to the summation of the values that have the same sign. The exception for (2) is that when the characteristic equation can be written in Tuplin's form, it may be more favorable to solve the equation directly through an algebraic maneuver. Consequently, the absolutely necessary rules for governing good computational practices can be reduced to one: live with the computer's limitations according to the resources available, and try to use them in a most efficient way. The prevailing beliefs of the past should be noted, but not necessarily followed.

B.1.8 The Power Method

The power method is probably the simplest method to explain. It is more suitable to a situation where only the first few eigenvalues (or a small number of eigenvalues out of all eigenvalues that exist in the dynamic system under consideration) are desired. We assume that the eigenvalues of **A** are $\lambda_1, \lambda_2, \ldots, \lambda_n$, where $|\lambda_1| > |\lambda_2| \geq \cdots \geq |\lambda_n|$. Now we let **A** repeatedly operate on a vector **v**, which we express as a linear combination of the eigenvectors

$$\mathbf{v} = c_1\mathbf{v}_1 + c_2\mathbf{v}_2 + \cdots + c_n\mathbf{v}_n \tag{B.34}$$

Then we have

$$\mathbf{A}\mathbf{v} = c_1\mathbf{A}\mathbf{v}_1 + c_2\mathbf{A}\mathbf{v}_2 + \cdots + c_n\mathbf{A}\mathbf{v}_n = \lambda_1\left(c_1\mathbf{v}_1 + c_2\frac{\lambda_2}{\lambda_1}\mathbf{v}_2 + \cdots + c_n\frac{\lambda_n}{\lambda_1}\mathbf{v}_n\right)$$

Iterating this p times, we obtain

$$\mathbf{A}^p\mathbf{v} = \lambda^p\left[c_1\mathbf{v}_1 + c_2\left(\frac{\lambda_2}{\lambda_1}\right)^p\mathbf{v}_2 + \cdots + c_n\left(\frac{\lambda_n}{\lambda_1}\right)^p\mathbf{v}_n\right] \tag{B.35}$$

When p is large enough and if $(\lambda_2/\lambda_1) < 1$, the vector

$$c_1\mathbf{v}_1 + c_2\left(\frac{\lambda_2}{\lambda_1}\right)^p\mathbf{v}_2 + \cdots + c_n\left(\frac{\lambda_n}{\lambda_1}\right)^p\mathbf{v}_n$$

will converge toward, $c_1\mathbf{v}_1$, which is the eigenvector of λ_1. The eigenvalue is determined by

$$\lambda_1 = \lim_{p\to\infty}\frac{(\mathbf{A}^{p+1}\mathbf{v})_j}{(\mathbf{A}^p\mathbf{v})_j} \qquad j = 1, 2, 3, \ldots, n \tag{B.36}$$

where the index number j denotes the jth component in the corresponding vector. This iteration method will attain the largest eigenvalue first. Normally, when the smallest eigenvalue is of interest, the inverse power method is used. One would then operate with $\mathbf{B} = \mathbf{A}^{-1}$ instead of **A**. In other words, in a typical matrix equation

$$(\mathbf{K} - \lambda\mathbf{M})\mathbf{v} = 0$$

the term $\mathbf{A} = \mathbf{M}^{-1}\mathbf{K}$, leads to $\mathbf{Av} = \lambda\mathbf{v}$, and $\mathbf{B} = \mathbf{A}^{-1} = \mathbf{MK}^{-1}$, leads to $\mathbf{Bv} = (1/\lambda)\mathbf{v} = \lambda'\mathbf{v}$.

To prevent the vector $\mathbf{v}$ from becoming very large or very small because of repeated multiplications, it should be normalized (divided by the component with the largest absolute value) after each operation. The algorithm for this can be formulated as follows:

$$\begin{aligned} \mathbf{y}_{k+1} &= \mathbf{Ax}_k \\ m &= \max_j |(y_{k+1})_j| \\ \mathbf{x}_{k+1} &= \mathbf{y}_{k+1}/m \end{aligned} \tag{B.37}$$

This method, as it is, has very little practical value for the following reasons: (1) the speed of convergence is slow for closely spaced eigenvalues, (2) accuracy diminishes as the process continues toward finding more eigenvalues, and (3) the procedure is awkward for zero eigenvalues (rigid-body structural modes).

To counteract all these disadvantages, a modification with a shift of origin should be used. The simplest polynomials in $\mathbf{A}$ are those of the form $(\mathbf{A} - s\mathbf{I})$. If λ_i is an eigenvalue of $\mathbf{A}$, then obviously $(\lambda_i - s)$ is the corresponding eigenvalue in $(\mathbf{A} - s\mathbf{I})$. Therefore, if s is chosen appropriately, the convergence to an eigenvector may be accelerated. Since the speed of convergence depends on the ratio of the dominant eigenvalue to the next closest eigenvalue (which is worst when the value approaches unity), a shift of the origin can change this ratio. The inverse power method with shifts is particularly effective for problems that are formulated by the displacement approach, when only a small fraction of all of the eigenvalues are of interest.

B.1.9 The Sturm Sequence Property for Characteristic Equations

To apply the Sturm sequence property in a matrix, note that each leading principal minor is a factor of the matrix. The sequence of leading principal minors of a matrix $\mathbf{A} = (a_{ij})$ is defined by the determinants

$$D_0 = 1 \quad D_1 = a_{11} \quad D_2 = \begin{bmatrix} a_{11} & a_{12} \\ a_{21} & a_{22} \end{bmatrix} \quad D_3 = \begin{bmatrix} a_{11} & a_{12} & a_{13} \\ a_{21} & a_{22} & a_{23} \\ a_{31} & a_{32} & a_{33} \end{bmatrix} \cdots D_n = [\mathbf{A}] \tag{B.38}$$

If we now define $\mathbf{A} = \mathbf{K} - \lambda\mathbf{M}$, where λ is an arbitrarily chosen test eigenvalue, then the Sturm sequence formed from this polynomial equation of variable λ has the property that the number of agreements in polarity (the sign of the result) of consecutive elements of the sequence is equal to the number of eigenvalues greater than the chosen test value. By repeated use of this procedure, one can pinpoint the location of the roots of interest. The algorithm applies the gaussian elimination method to reduce each element D to a triangular form. The value for each element does not have to be determined, but the following are of interest: (1) seeing whether any diagonal element is zero, in which case the test value is a root of the equation; (2) monitoring the sign of each diagonal element and from this process determining the sign of each determinant D; and (3) counting the number of agreements in polarity (sign) of consecutive D's.

The above mentioned operation may be combined with the bisection method to form a general matrix eigenvalue-extraction routine, but it is more efficient to use it just as a method for checking missing roots or for searching for a preliminary root.

B.1.10 The Jacobi Method

The Jacobi method is one of the efficient methods for eigenvalue extraction for real, symmetric matrices: $\mathbf{A} = \mathbf{A}^T$. This method preserves the eigenvalues during matrix transformation by using plane rotation (see Sec. B.1.6). For every matrix **A**, another matrix **X** exists such that

$$\mathbf{X}\mathbf{X}^T = \mathbf{I} \qquad \text{or} \qquad \mathbf{X}^T = \mathbf{X}^{-1} \tag{B.39}$$

and

$$\mathbf{X}^{-1}\mathbf{A}\mathbf{X} = \mathbf{D} \tag{B.40}$$

To prove that **D** and **A** have the same eigenvalues, let $\mathbf{Y} = \mathbf{X}^{-1}\mathbf{A}\mathbf{X}$. Then

$$\mathbf{X}\mathbf{Y} = (\mathbf{X}\mathbf{X}^{-1})\mathbf{A}\mathbf{X} = \mathbf{A}\mathbf{X}$$

and

$$\mathbf{X}\mathbf{Y}\mathbf{X}^{-1} = \mathbf{A}(\mathbf{X}\mathbf{X}^{-1}) = \mathbf{A}$$

Therefore,

$$\begin{aligned}\mathbf{A} - \lambda\mathbf{I} &= \mathbf{X}\mathbf{Y}\mathbf{X}^{-1} - \lambda\mathbf{I} = \mathbf{X}\mathbf{Y}\mathbf{X}^{-1} - \lambda\mathbf{X}\mathbf{X}^{-1} = \mathbf{X}(\mathbf{Y}\mathbf{X}^{-1} - \lambda\mathbf{X}^{-1})\\ &= \mathbf{X}(\mathbf{Y} - \lambda\mathbf{I})\mathbf{X}^{-1}\end{aligned}$$

so that

$$\begin{aligned}|\mathbf{A} - \lambda\mathbf{I}| &= |\mathbf{X}(\mathbf{Y} - \lambda\mathbf{I})\mathbf{X}^{-1}| = |\mathbf{X}\mathbf{X}^{-1}|\,|\mathbf{Y} - \lambda\mathbf{I}|\\ &= |\mathbf{I}|\,|\mathbf{Y} - \lambda\mathbf{I}| = |\mathbf{Y} - \lambda\mathbf{I}|\end{aligned} \tag{B.41}$$

Since the determinants are equal, they lead to the same characteristic equation. Thus **D** and **A** have the same eigenvalues. The rest of the task is to describe in practical terms execution of the concept defined by Eqs. (B.39) and (B.40). Jacobi suggested that an identity matrix can be chosen for **X**, in which any four elements at the vertices of a rectangle are replaced by $\cos\theta$, $-\sin\theta$, $\sin\theta$, $\cos\theta$ in positions *pp*, *pq*, *qq*, and *qp*. For example,

$$\mathbf{X} = \begin{bmatrix} 1 & 0 & 0 & 0 & 0\\ 0 & \cos\theta & 0 & -\sin\theta & 0\\ 0 & 0 & 1 & 0 & 0\\ 0 & \sin\theta & 0 & \cos\theta & 0\\ 0 & 0 & 0 & 0 & 1 \end{bmatrix} \tag{B.42}$$

The θ is used as a parameter of the undetermined coefficient, the value of which is to be determined such that the pre- and postmultiplication made by $\mathbf{X}^{-1}$ and **X**, with respect to a symmetrical matrix of the same order, will cause the terms *pq* and *qp* of the resulting matrix to be reduced to zero. Geometrically, this is equivalent to performing a two-dimensional plane rotation; thus, Eq. (B.42) is also called a rotation matrix. To illustrate this operation with a 3×3 matrix, let $c = \cos\theta$ and $s = \sin\theta$. We then have

$$\begin{bmatrix} c & s & 0\\ -s & c & 0\\ 0 & 0 & 1 \end{bmatrix} \begin{bmatrix} a_{11} & a_{12} & a_{13}\\ a_{21} & a_{22} & a_{23}\\ a_{31} & a_{32} & a_{33} \end{bmatrix} \begin{bmatrix} c & -s & 0\\ s & c & 0\\ 0 & 0 & 1 \end{bmatrix} \tag{B.43}$$

Elements at (1, 2) and (2, 1) of the resulting matrix become

$$(a_{22} - a_{11})\sin\theta\cos\theta + a_{12}(\cos^2\theta - \sin^2\theta) \tag{B.44}$$

Equating Eq. (B.44) to zero leads to

$$\tan 2\theta = \frac{2a_{12}}{a_{11} - a_{22}} \qquad \text{or} \qquad \theta = 0.5\tan^{-1}\left(\frac{2a_{12}}{a_{11} - a_{22}}\right) \tag{B.45}$$

For the general case, the above expression remains correct if the subscripts are replaced by pq, pp, qq for 12, 11, and 22. Therefore, if in Eq. (B.43) the angle θ is chosen to satisfy the condition Eq. (B.45), the elements in the resulting matrix at locations (p, q) and (q, p) are annihilated. Knowing this, we can assign parameters c and s in Eq. (B.43) the necessary values in reference to Eq. (B.45).

When this procedure is applied repeatedly to the off-diagonal elements, the resulting matrix can be made as close to the diagonal matrix D as desired. Thus, by using the Jacobi method we can obtain the eigenvalues all at once. The advantages of this procedure are simplicity, stability, and the ability to deal with zero eigenvalues.

B.1.11 The Givens Method

The drawback of Jacobi's method is that although annihilation takes place at the corners of the rectangle, every other element of the matrix is also altered. Therefore, every new step of annihilation may spoil some portion of the old effort. This prolongs the computational procedure. The Givens method avoids this drawback by not seeking diagonalization. Instead, Givens proposed an alternative scheme: instead of the elements in positions (p, q) and (q, p), elements in positions $(p - 1, q)$ and $(q, p - 1)$ are annihilated. The following example shows that $\mathbf{A}$ is a 5×5 real, symmetric matrix. For $p = 2$, and $q = 3$,

$$\begin{bmatrix} 1 & 0 & 0 & 0 & 0 \\ 0 & c & s & 0 & 0 \\ 0 & -s & c & 0 & 0 \\ 0 & 0 & 0 & 1 & 0 \\ 0 & 0 & 0 & 0 & 1 \end{bmatrix} \begin{bmatrix} a_{11} & a_{12} & a_{13} & a_{14} & a_{15} \\ a_{21} & a_{22} & a_{23} & a_{24} & a_{25} \\ a_{31} & a_{32} & a_{33} & a_{34} & a_{35} \\ a_{41} & a_{42} & a_{43} & a_{44} & a_{45} \\ a_{51} & a_{52} & a_{53} & a_{54} & a_{55} \end{bmatrix} \begin{bmatrix} 1 & 0 & 0 & 0 & 0 \\ 0 & c & -s & 0 & 0 \\ 0 & s & c & 0 & 0 \\ 0 & 0 & 0 & 1 & 0 \\ 0 & 0 & 0 & 0 & 1 \end{bmatrix} \tag{B.46}$$

The result is

$$\begin{bmatrix} a_{11} & ca_{12} + sa_{13} & ca_{13} - sa_{12} & a_{14} & a_{15} \\ ca_{12} + sa_{31} & c^2a_{22} + 2sca_{23} + s^2a_{33} & (c^2 - s^2)a_{23} + sc(a_{33} - a_{22}) & ca_{24} + sa_{34} & ca_{25} + sa_{35} \\ ca_{31} - sa_{21} & (c^2 - s^2)a_{23} + sc(a_{33} - a_{23}) & c^2a_{33} - 2sca_{23} + s^2a_{22} & ca_{34} - sa_{24} & ca_{35} - sa_{25} \\ a_{41} & ca_{42} + sa_{43} & ca_{43} - sa_{42} & a_{44} & a_{45} \\ a_{51} & ca_{52} + sa_{53} & ca_{53} - sa_{52} & a_{54} & a_{55} \end{bmatrix}$$

We now annihilate the elements in positions (1, 3) and (3, 1) by letting

$$ca_{13} - sa_{12} = ca_{31} - sa_{21} = 0$$

or

$$a_{13}\cos\theta - a_{12}\sin\theta = 0$$

This leads to

$$\tan\theta = a_{13}/a_{12} \qquad \text{or} \qquad \theta = \tan^{-1}(a_{13}/a_{12})$$

Varying the index number p and q for the rotation matrix and repeating the same procedure, one can reduce matrix **A** to a tridiagonal form in an orderly fashion without the problem of new annihilation spoiling a portion of the effort of the old one. The rotation matrix setup for various steps can be summarized as follows:

To eliminate the off-tridiagonal elements for column i and row i of an $n \times n$ symmetric matrix at the qth element		Angle of rotation
Set	$p = i + 1$ $q = p + 1$ to n, step 1	$\theta = \tan^{-1}(a_{iq}/a_{ip})$; all a's are based on the current value

Finally, the eigenvalues are extracted one by one from the tridiagonal matrix **S**, using one of the root's solvers. Givens further suggested that the successive principal minors $F_i(\lambda)$ of $\lambda\mathbf{I} - \mathbf{S}$, and the characteristic equation of $F_n(\lambda) = 0$ (which is the same as that of **A**), can be obtained through the recursion formula

$$F_i(\lambda) = (\lambda - d_i)F_{i-1}(\lambda) - (e_{i-1})^2 F_{i-2}(\lambda) \tag{B.47}$$

where $F_{-1} = 0$
$F_0 = 1$
$i = 1, 2, 3, \ldots, n$
d = diagonal element
e = off-diagonal element

From this point on, the Sturm sequence property may guide root finding.

B.1.12 The Householder Method

The Householder method uses a somewhat involved unitary, orthogonal, real, symmetric matrix to reduce matrix **A** to a tridiagonal form identical to that of Givens. The major differences in comparison with Givens' method are: (1) the Givens method annihilates one pair of off-tridiagonal elements at a time, whereas Householder reduces one row and one column of off-tridiagonal elements in one step, and (2) Givens uses one adjustable parameter θ to determine the values of the elements for a unitary matrix, whereas Householder uses as many undetermined parameters as needed, so they can provide the algorithms for forming the required unitary matrix that can annihilate the target elements. Here is the basic procedure with a 4×4 matrix **A**:

$$\begin{bmatrix} a_{11} & a_{12} & a_{13} & a_{14} \\ a_{21} & a_{22} & a_{23} & a_{24} \\ a_{31} & a_{32} & a_{33} & a_{34} \\ a_{41} & a_{42} & a_{43} & a_{44} \end{bmatrix}$$

Let us define a unitary matrix **U**,

$$\mathbf{U} = \mathbf{I} - 2\mathbf{w}\mathbf{w}^T \tag{B.48}$$

where $\mathbf{w} = \begin{bmatrix} w_1 \\ w_2 \\ w_3 \\ w_4 \end{bmatrix}$ and $\mathbf{w}^T = (w_1, w_2, w_3, w_4)$

We now impose the condition $\mathbf{w}^T\mathbf{w} = 1$, so that

$$\mathbf{UU}^T = (\mathbf{I} - 2\mathbf{ww}^T)(\mathbf{I} - 2\mathbf{ww}^T) = \mathbf{I} - 4\mathbf{ww}^T + 4\mathbf{w}(\mathbf{w}^T\mathbf{w})\mathbf{w}^T \qquad \text{(B.49)}$$

$$= \mathbf{I} - 4\mathbf{ww}^T + 4\mathbf{ww}^T = \mathbf{I}$$

Equation (B. 48), when written out, is

$$\mathbf{U} = \begin{bmatrix} 1 - 2w_1^2 & -2w_1w_2 & -2w_1w_3 & -2w_1w_4 \\ -2w_1w_2 & 1 - 2w_2^2 & -2w_2w_3 & -2w_2w_4 \\ -2w_1w_3 & -2w_2w_3 & 1 - 2w_3^2 & -2w_3w_4 \\ -2w_1w_4 & -2w_2w_4 & -2w_3w_4 & 1 - 2w_4^2 \end{bmatrix} \qquad \text{(B.50)}$$

Obviously, if we let $\mathbf{w}_1 = 0$, Eq. (B.50) reduces to

$$\mathbf{U} = \begin{bmatrix} 1 & 0 & 0 & 0 \\ 0 & 1 - 2w_2^2 & -2w_2w_3 & -2w_2w_4 \\ 0 & -2w_2w_3 & 1 - 2w_3^2 & -2w_3w_4 \\ 0 & -2w_2w_4 & -2w_3w_4 & 1 - 2w_4^2 \end{bmatrix} \qquad \text{(B.51)}$$

where the subscript 1 for **U** indicates the unitary matrix for the first reduction step. (For the second step, one may assign both w_1 and w_2 to zero, and so on.) Therefore,

$$\mathbf{A}_1 = \mathbf{U}_1\mathbf{A}\mathbf{U}_1 \qquad \text{(B.52)}$$

The first row of the resulting matrix will be a'_{11}, a'_{12}, a'_{13} and a'_{14}, which is equal to

$$a_{11} \quad a_{12} - 2vw_2 \qquad a_{13} - 2vw_3 \qquad a_{14} - 2vw_4 \qquad \text{(B.53)}$$

where $v = a_{12}w_2 + a_{13}w_3 + a_{14}w_4$.

Squaring and adding the last 3 elements of Eq. (B.53), and using the fact that

$$\mathbf{w}^T\mathbf{w} = w_2^2 + w_3^2 + w_4^2 = 1$$

we conclude that the sum of the squares of the new elements is equal to the sum of the squares of the old elements,

$$\pm\, S = a'^2_{12} + a'^2_{13} + a'^2_{14} = a_{12}^2 + a_{13}^2 + a_{14}^2 = S^2 \qquad \text{(B.54)}$$

Our intention is to make the last two elements vanish. This requirement leads to

$$a'_{12} = \sqrt{a_{12}^2 + a_{13}^2 + a_{14}^2} = a_{12} = 2vw_2 \qquad \text{(B.55)}$$

and

$$a_{13} - 2vw_3 = 0 \qquad \text{(B.56)}$$

$$a_{14} - 2vw_4 = 0 \qquad \text{(B.57}a\text{)}$$

Multiplying Eq. (B.55) by w_2, Eq. (B.56) by w_3, and Eq. (B.57) by w_4 and summing gives

$$\pm\, Sw_2 = a_{12}w_2 + a_{13}w_3 + a_{14}w_4 - 2v(w_1^2 + w_2^2 + w_3^2) = -v \qquad \text{(B.57}b\text{)}$$

Thus, the four unknowns w_2, w_3, w_4, and v can be solved using the four equations above. For elimination of the off-tridiagonal elements of the second row and column, the unitary matrix assumes the form

$$\mathbf{U}_2 = \begin{bmatrix} 1 & 0 & 0 & 0 \\ 0 & 1 & 0 & 0 \\ 0 & 0 & 1 - 2w_3^2 & -2w_3w_4 \\ 0 & 0 & -2w_3w_4 & 1 - 2w_4^2 \end{bmatrix} \tag{B.58}$$

To determine the values of w_3 and w_4, a similar procedure applies. In general,

$$S_i = (\sum a_{ij}^2)^{1/2} \qquad j = i + 1, \ldots, n \tag{B.59}$$

$$w_i^2 = 0.5(1 + a_{ij}/S_i) \qquad j = i + 1 \tag{B.60}$$

$$w_k = a_{ik}/2w_iS_i \qquad k = i + 2, \ldots, n \tag{B.61}$$

where the subscript i is the step of reduction, n is the order of matrix **A**, and the a's refer to the element values of the last step. After **A** reduces to a tridiagonal form, the eigenvalue extraction can be carried out as usual.

B.1.13 The *QR* Iteration

The basic step in the QR iteration is the decomposition of matrix **A** into the form

$$\mathbf{A} = \mathbf{QR} \tag{B.62}$$

where **Q** is an orthogonal matrix and **R** is an upper triangular matrix. Premultiplying Eq. (B.62) by $\mathbf{Q}^T$, and postmultiplying by **Q**, we obtain

$$\mathbf{Q}^T\mathbf{A}\mathbf{Q} = \mathbf{Q}^T\mathbf{Q}\mathbf{R}\mathbf{Q} = \mathbf{R}\mathbf{Q} \tag{B.63}$$

Therefore, by computing **RQ**, we, in fact, perform the transformation of Eq. (B.28)

The factorization in Eq. (B.62) could be done in various ways, for example, by systematically premultiplying the rotary matrix **X** in the form similar to Eq. (B.42), or the unitary matrix in the form similar to Eq. (B.51). This is the same as carrying out the triangularization by means of Jacobi's, Givens', or Householder's reduction. In this way, we evaluate

$$\mathbf{P}_{n,n-1}^T \cdots \mathbf{P}_{j,i}^T \cdots \mathbf{P}_{2,1}^T\mathbf{A} = \mathbf{R} \tag{B.64}$$

where the matrix $\mathbf{P}_{j,i}^T$ is selected to annihilate the element (j, i), as in Givens' reduction. Comparing Eq. (B.62) with Eq. (B.64) leads to

$$\mathbf{Q} = \mathbf{P}_{2,1} \cdots \mathbf{P}_{j,i} \cdots \mathbf{P}_{n,n-1} \tag{B.65}$$

Using the subscript for the matrices to represent the steps of the iterations, we denote

$$\mathbf{A}_k = \mathbf{Q}_k\mathbf{R}_k$$

and

$$\mathbf{A}_{k+1} = \mathbf{R}_k\,\mathbf{Q}_k \tag{B.66}$$

As this procedure continues, $\mathbf{A}_{k+1}$ will be diagonalized and represent eigenvalues, and $\mathbf{Q}_1 \cdots \mathbf{Q}_{k-1}\mathbf{Q}_k$ will become eigenvectors, as $(k + 1) \to \infty$.

REFERENCES

1. Davis, P. J.: "Interpolation and Approximation," Dover Publication, Inc., New York, 1975.
2. Forsythe, G. E., M. A. Malcolm, and C. B. Moler: "Computer Methods for Mathematical Computations," Prentice-Hall, Inc., Englewood Cliffs, N.J., 1977.

3. Bathe, K.-J., and E. L. Wilson: "Numerical Methods in Finite Element Analysis," Prentice-Hall, Inc., Englewood Cliffs, N.J., 1976.
4. Wilkinson, J. H.: "The Algebraic Eigenvalue Problem," Clarendon Press, Oxford, England, 1965.
5. Parlett, B. N.: "The Symmetric Eigenvalue Problem," Prentice-Hall, Inc., Englewood Cliffs, N.J., 1980.
6. Hageman, L. A., and D. M. Young: "Applied Iterative Methods," Academic Press, Inc., New York, 1981.
7. Dodes, I. A: "Numerical Analysis for Computer Science," North-Holland Publishing Company, New York, 1978.
8. Conte, S. D., and C. de Boor: "Elementary Numerical Analysis—An Algorithmic Approach," 3d ed., McGraw-Hill Book Company, New York, 1980.
9. Wendroff, B.: "Theoretical Numerical Analysis," Academic Press, Inc., New York, 1966.
10. Rust, B. W., and W. R. Burrus: "Mathematical Programming and the Numerical Solution of Linear Equations," American Elsevier Publishing Company, Inc., New York, 1972.
11. Householder, A. S.: "The Theory of Matrices in Numerical Analysis," Dover Publications, Inc., New York, 1964.
12. Lanczos, C.: "Applied Analysis," Prentice-Hall, Inc., Englewood Cliffs, N.J., 1961.
13. Bodewig, E.: "Matrix Calculus," North-Holland Publishing Company, Amsterdam, The Netherlands, 1959.
14. LaFara, R. L.: "Computer Methods for Science and Engineering," Hayden Book Company, Inc., Rochelle Park, N.J., 1973.
15. Froberg, C.-E.: "Introduction to Numerical Analysis," 2d ed., Addison-Wesley Publishing Company, Inc., Reading, Mass., 1969.
16. Vemuri, V., and W. J. Karplus: "Digital Computer Treatment of Partial Differential Equations," Prentice-Hall, Inc., Englewood Cliffs, N.J., 1981.
17. Wilkinson, J. H., and C. H. Reinsch: "Handbook for Automatic Computation," vol. 2. "Linear Algebra," Springer-Verlag, New York, 1971.
18. Golub, G. H., and C. Reinsch: "Singular Value Decomposition and Least Squares Solutions," in J. H. Wilkinson and C. Reinsch, eds., "Handbook for Automatic Computation," vol. 2, "Linear Algebra," Springer-Verlag, New York, 1971.
19. Givens, W.: "Numerical Computation of the Characteristic Values of a Real Symmetric Matrix," Oak Ridge National Laboratory report no. ORNL 1574, Oak Ridge, Tenn., 1954.
20. Householder, A. S., and F. L. Bauer: "On Certain Methods for Expanding the Characteristic Polynomial," *Numer. Math.*, vol. 1, 1959, p. 29.
21. Givins, W.: "The Linear Equation Problem," technical report no. 3, Applied Mathematics and Statistics Laboratory, Stanford University, December 1959.
22. Jacobs, D., ed.: "The State of the Art in Numerical Analysis," Academic Press, Inc., London, 1977.
23. Wang, C. C.: "Calculating Natural Frequencies with Extended Tuplin's Method," *Trans. ASME, J. Mech. Design,* vol. 103, 1981, pp. 379–386.
24. Jacobi, C. G. J., "Concerning an Easy Process for Solving Equations Occurring in the Theory of Secular Disturbances," *J. Reine Angew. Math.*, vol. 30, 1846, p. 51.

INDEX

ABOUT THE EDITORS

Harold A. Rothbart is a noted consulting engineer and lecturer. He is the author or editor of numerous books, including the *Mechanical Design and Systems Handbook*. Dr. Rothbart was formerly Dean of the College of Science and Engineering at Fairleigh Dickinson University.

Thomas H. Brown, Jr., is a distinguished engineer who specializes in stress analysis and machine design. He is the co-editor of the Third Edition of the *Standard Handbook of Machine Design* and the author of *Marks' Calculations for Machine Design*.